Integrative and Functional Medical Nutrition Therapy

Diana Noland
Jeanne A. Drisko
Leigh Wagner
Editors

Integrative and Functional Medical Nutrition Therapy

Principles and Practices

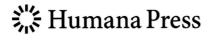

Editors
Diana Noland
Noland Nutrition
Burbank, CA, USA

Leigh Wagner
Department of Dietetics & Nutrition
University of Kansas Medical Center
Kansas City, KS, USA

Jeanne A. Drisko
Professor Emeritus
School of Medicine
University of Kansas Health System
Kansas City, KS, USA

ISBN 978-3-030-30729-5   ISBN 978-3-030-30730-1 (eBook)
https://doi.org/10.1007/978-3-030-30730-1

© Springer Nature Switzerland AG 2020

This work is subject to copyright. All rights are reserved by the Publisher, whether the whole or part of the material is concerned, specifically the rights of translation, reprinting, reuse of illustrations, recitation, broadcasting, reproduction on microfilms or in any other physical way, and transmission or information storage and retrieval, electronic adaptation, computer software, or by similar or dissimilar methodology now known or hereafter developed.
The use of general descriptive names, registered names, trademarks, service marks, etc. in this publication does not imply, even in the absence of a specific statement, that such names are exempt from the relevant protective laws and regulations and therefore free for general use.
The publisher, the authors, and the editors are safe to assume that the advice and information in this book are believed to be true and accurate at the date of publication. Neither the publisher nor the authors or the editors give a warranty, express or implied, with respect to the material contained herein or for any errors or omissions that may have been made. The publisher remains neutral with regard to jurisdictional claims in published maps and institutional affiliations.

This Humana imprint is published by the registered company Springer Nature Switzerland AG
The registered company address is: Gewerbestrasse 11, 6330 Cham, Switzerland

I wish to dedicate this book to the enlightened mentors in the field of integrative and functional medicine with whom I have had the honour to interact: Doctors Jeff Bland, Jeanne Drisko, Sam Queen, Roger Newton, Sidney MacDonald-Baker, Douglas Hunt and Robert P Heaney. Through their gifts of knowledge, they have provided the foundation for the powerful impact nutrition is experiencing in influencing human health. This textbook is a tribute to their genius.

–Diana Noland

This textbook is dedicated to all the integrative practitioners who taught me how to care for patients as people, those practitioners who are living and those who have slipped away and especially my students who taught me to keep up and stay on my toes. But most of all, what is important for practitioners who are finding their way, "Listen with your heart".

–Jeanne A. Drisko

This book is dedicated to my family, friends and close colleagues who've provided unconditional love and support throughout this and all my work; it is for my teachers, professors and students who've been my patient and unwavering mentors. Finally, this is for my patients and clients who have taught me more than I imagined possible about what it means to be a healthcare provider.

–Leigh Wagner

Foreword

The incidence of almost every chronic disease has increased relentlessly in every age group for the past 50 years. Diabetes which was once rare (<0.7% of the population) is now projected to affect 35% of the people in their lifetime. For the first time in American history, life expectancy has gone down several years in a row. Healthcare costs (or, more accurately, disease treatment costs) are now 20% of the GDP, increasing inevitably and bankrupting the country. Why is the healthcare system—which obviously has several areas of great success—failing dramatically everywhere else?

Because the common causes of disease are not being addressed.

Conventional medicine has been incredibly effective in many areas, such as injuries, overwhelming infections, congenital malformations and some cancers. However, this model does not work for everyday health and disease prevention. Unfortunately, the burden of chronic disease has become so widespread and severe that most medical treatment is now primarily for symptom control and prevention of even worse sequelae. Very little of medicine as actually practiced today addresses the real reasons why people are suffering such a huge and progressively increasing disease burden.

What are those unaddressed causes? To quote an inspirational functional medicine expert Sid Baker, MD: "Most of medicine is quite simple, get into people what they uniquely need and get out what they uniquely do not need". In other words, nutrition and detoxification—*according to patient's unique biochemical needs*.

Nutritional deficiencies are rampant in the population and are getting worse. Not only is the general public choosing food with lower nutrient density but also hybridization and synthetic fertilizers have decreased the trace nutrient content of the food. Almost the entire population is deficient in at least one nutrient and over half has multiple nutritional deficiencies according to conventional standards. But even worse, the environment has become increasingly contaminated by toxic metals and chemicals that displace nutrients, poison enzymes, damage DNA, disrupt cellular communications, increase inflammation and dramatically increase oxidative stress in all systems of the body.

With the genomics revolution, we now know that there is a huge variation between nutritional needs and toxin susceptibility. For example, some people's single nucleotide polymorphisms (SNPs) result in the lower functioning of their vitamin D receptor sites and need ten times the recommended daily allowance (RDA) of vitamin D to maintain their bones. Another example is the huge 1000-fold variation in the activity of liver Phase I detoxifying enzyme CYP2D6. This enzyme detoxifies 25% of prescription drugs and a number of environmental toxins. Those with a poorly functioning version of this detoxification enzyme are not only much more likely to have an adverse drug reaction to the standard dose of a prescription but also more likely to suffer a number of diseases, such as Parkinson's disease, if they are exposed to neurotoxins. With over 2 million SNPs, there are a huge number of examples of SNP variants requiring specialized attention.

This is why *Integrative and Functional Medicine Nutrition Therapy: Principles and Practices* is such an important textbook. Every clinician who wants to care for their patients in a curative way must recognize and understand the clinical presentations of *functional* nutritional deficiencies and toxin overload. Relying on blood (or other tissue) levels of nutrients or environmental toxins is an ineffective diagnostic strategy for many reasons. For example, measuring blood levels of B-vitamins is very misleading. Using the conventional range standards, only a small percentage of the public is deficient. But measuring levels of toxic metabolites, such as homocysteine and methylmalonic acid, which build up when a person has a functional deficiency of B-vitamins, elevated levels indicate increased risk of tissue damage. This means increased risk for heart disease, Alzheimer's disease, osteoporosis and other chronic degenerative diseases.

We see the same pattern with environmental toxins. For example, the "safe" levels of most toxins are established according to population norms. But the "normal" population suffers a huge disease burden! And as mentioned above, using absolute levels of toxin misses the huge variation in detoxification function found in the general population. Some people can smoke their entire life without apparent problems while a non-smoking spouse gets lung cancer. There are numerous examples of common chronic diseases being primarily caused by the combination of nutritional deficiencies and toxin load in the context of biochemical susceptibility. Drugs do not reverse deficient nutrients or help increase toxin excretion. In fact, almost all drugs are yet another load on the already overloaded detoxification systems.

The only effective clinical strategy is that which thoughtfully *integrates* multiple diverse interventions that not only address the actual causes of disease but also utilizes safe therapies that optimize each person's unique physiological function.

This is the premise and promise of IFMNT.

Joseph Pizzorno, ND
Co-Author, Textbook of Natural Medicine
Chair of Board, Institute for Functional Medicine
Seattle, WA, USA

Preface

Over the past 50–60 years, the world has seen an unparalleled and unprecedented growth in non-communicable (chronic) disease. Chronic disease has replaced acute infectious disease and trauma as the dominant influence on healthcare delivery. Our current healthcare model is no longer sustainable, i.e. the captain of the ship making a brilliant diagnosis and delivering the silver bullet cure or surgically repairing the problem. No silver bullet exists that satisfactorily addresses the chronic, complex disorders that currently clog our healthcare systems. The health of our nation and world has changed rapidly while healthcare delivery has remained stagnant. But change, it must!

This urgently needed change does not appear to be coming from within large healthcare systems. However, healthcare consumers, especially those with chronic and life-threatening disorders, are demanding a different delivery of care. As one patient with chronic disease relayed, her primary care physician told her that she had so many problems he wouldn't be in practice long enough to solve them all. This is a message devoid of hope and care. In this setting, "care" is a misnomer as it has been stripped from healthcare by the abbreviated allotted visit times, and with practitioners no longer allowed to interact with those entrusted to their charge and prevented from knowing their patients as people.

The unifying theme of this textbook is the understanding of the nutritional status of patients. Without a strong nutritional foundation, healing of chronic disease is impossible. Unlike conventional medical care that narrowly defines nutrition, dismisses its importance and marginalizes its delivery, we propose that nutrition become a fundamental part of healthcare providers' broad scope. Without this foundation, our ability to successfully deliver care is limited. You will find within these pages authors with all types of backgrounds who have found success using integrative and functional medicine nutrition therapies when faced with chronic health conditions.

We wrote this textbook to help practitioners, health system leaders and policy makers to understand that there is a different and novel approach to address complex chronic disorders. Human physiology has not changed. What has changed are those factors acting on human physiology in the form of poor nutrition, environmental toxicant exposures, fractured human interactions, infrequent physical movement, stress, poor sleep and other lifestyle problems.

As editors with a collective 65 years of practice in integrative and functional medicine nutrition therapies, we have witnessed (and practiced) a new healthcare delivery system that incorporates important advances of conventional medical care that, combined with the original roots of healthcare, returns the patient back to the centre of the process. Partnering with patients fosters better understanding of the chronic disorders that plague them. We have witnessed how teams of practitioners with diverse healthcare backgrounds work together to find solutions to our looming healthcare crisis.

Although our backgrounds are in integrative and functional medicine, the model we present transcends and defies all labels and simply should be called "good medicine". In this textbook, we provide principles to illuminate understanding of how healthcare can be delivered to get to the root of chronic disorders. After the principles are laid out, the practice of "good healthcare" is described in detail. The reader will see that there is again hope for finding satisfaction in healthcare careers when burnout rates are soaring.

Within, you will find the underlying principles and practices laid out that give you, the reader, an understanding of the science of integrative and functional medicine nutrition therapies with real-world examples. The authors were selected because of their notable experience in integrative and functional medical nutrition therapies. This textbook will augment your journey back to the heart and soul of the practice of good healthcare.

The 10-year journey to writing *Integrative and Functional Medicine Nutrition Therapy: Principles and Practices* was the brainchild of editor Diana Noland. Diana collaborated with colleagues Jeanne Drisko and Leigh Wagner to successfully launch a master-level certificate program in integrative nutrition at the University of Kansas Health System in conjunction with the Department of Dietetics and Nutrition and the University of Kansas Integrative Medicine. We crafted materials to be used in the education of dietetic students but also used the materials in training all practitioners and students under our guidance. Their care for patients has thrived with this underpinning. The development of this textbook provides a roadmap for care.

The editors and authors of this textbook have written this book in the hope that the knowledge it provides will help to achieve the day when all medical healthcare practices appreciate and include nutrition and lifestyle assessment and intervention for each unique individual.

We are thankful that this journey was guided by Coco and Roger Newton, who shared our vision for Integrative and Functional Medicine Nutrition Therapy: Principles and Practices. We also acknowledge our families who, for the past 3 years, have stood by us, supported our efforts and nurtured us while we wrote and edited, even during nights, weekends and family vacations. Special thanks to our spouses, Steve Noland, Robert Drisko II and Rob Bauer, who gave us insights and advice along this journey.

Many thanks to our faithful and accomplished copy editors, Matt Erickson and Jeffrey Field, who polished every chapter over the 2-year project.

Thanks to Michael D. Sova, our Springer developmental editor, who conducted the conversion from a vision to a reality. He brought together the work of individual authors to a cohesive collection of contributions toward providing the science and the practical aspects of integrative and functional medical nutrition therapy. Michael guided us in the preparation for a finished product and became an integral member of our team.

And finally, we would like to thank our friend, Dr. Joe Pizzorno, for writing the Foreword of the textbook as he joins us to express his common vision of the importance of nutrition therapy in the quest to quell the epidemics of chronic diseases in the twenty-first century.

Diana Noland
Burbank, CA, USA

Jeanne A. Drisko
Kansas City, KS, USA

Leigh Wagner
Kansas City, KS, USA

Contents

I Global Healthcare Challenge of the Twenty-First Century and the Future of Chronic Disease

1 The History and Evolution of Medicine 3
Jeanne A. Drisko

2 Influences of the Nutrition Transition on Chronic Disease 17
Sudha Raj

3 Nutritional and Metabolic Wellness 31
Diana Noland

4 Nutritional Ecology and Human Health 39
David Raubenheimer and Stephen J. Simpson

5 The Radial: Integrative and Functional MNT 57
Kathie M. Swift, Elizabeth Redmond, and Diana Noland

6 The Power of Listening and the Patient's Voice: "Please Hear Me" 73
Carolyn L. Larkin

II Metabolic Characteristics and Mechanisms of Chronic Disease

7 Metabolic Correction Therapy: A Biochemical–Physiological Mechanistic Explanation of Functional Medicine 89
Michael J. Gonzalez

8 The Nutrition Assessment of Metabolic and Nutritional Balance 99
Margaret Gasta

9 IFMNT NIBLETS Nutrition Assessment Differential 123
Robyn Johnson and Lauren Hand

10 Nutritional Role of Fatty Acids 135
Vishwanath M. Sardesai

11 Lipidomics: Clinical Application 151
Diana Noland

12 Structure: From Organelle and Cell Membrane to Tissue 173
David Musnick, Larissa Severson, and Sarah Brennan

13 Protective Mechanisms and Susceptibility to Xenobiotic Exposure and Load 191
Robert H. Verkerk

14 Detoxification and Biotransformation 205
Janet L. Black

15 Drug–Nutrient Interactions 213
Mary Demarest Litchford

16	The Enterohepatic Circulation...	221
	Robert C. Barton Jr.	
17	A Nutritional Genomics Approach to Epigenetic Influences on Chronic Disease...............	235
	Christy B. Williamson and Jessica M. Pizano	
18	Nutritional Influences on Methylation..	269
	Jessica M. Pizano and Christy B. Williamson	
19	The Immune System: Our Body's Homeland Security Against Disease	285
	Aristo Vojdani, Elroy Vojdani, and Charlene Vojdani	
20	Nutritional Influences on Immunity and Infection	303
	Joel Noland and Diana Noland	
21	Body Composition...	323
	Sue Ward and Diana Noland	
22	The Therapeutic Ketogenic Diet: Harnessing Glucose, Insulin, and Ketone Metabolism	335
	Miriam Kalamian	
23	The GUT-Immune System...	367
	Elizabeth Lipski	
24	Centrality of the GI Tract to Overall Health and Functional Medicine Strategies for GERD, IBS, and IBD..	379
	Ronald L. Hoffman	
25	The Microbiome and Brain Health..	391
	Sharon L. Norling	
26	The Role of Nutrition in Integrative Oncology...	407
	Cynthia Henrich	
27	The Microenvironment of Chronic Disease	437
	Steven Gomberg	
28	Chronic Pain	447
	Jena Savadsky Griffith	
29	Nutrition and Behavioral Health/Mental Health/Neurological Health.....................	473
	Ruth Leyse Wallace	
30	Neurodevelopmental Disorders in Children...	493
	Mary Anne Morelli Haskell	
31	Nutritional Influences on Hormonal Health	517
	Filomena Trindade	
32	Nutritional Influences on Reproduction: A Functional Approach	533
	Brandon Horn and Wendy Yu	
33	Lifestyle Patterns of Chronic Disease..	563
	Sarah Harding Laidlaw	

| 34 | Circadian Rhythm: Light-Dark Cycles | 577 |

Corey B. Schuler and Kate M. Hope

| 35 | Nutrition with Movement for Better Energy and Health | 595 |

Peter Wilhelmsson

| 36 | Mental, Emotional, and Spiritual Imbalances | 613 |

Muffit L. Jensen

III IFMNT Nutrition Care Process

| 37 | The IFMNT Practitioner | 621 |

Robin L. Foroutan

| 38 | The Patient Story and Relationship-Centered Care | 633 |

Leigh Wagner

| 39 | The Nutrition-Focused Physical Exam | 637 |

Mary R. Fry

| 40 | Modifiable Lifestyle Factors: Exercise, Sleep, Stress, and Relationships | 695 |

Margaret Christensen

| 41 | Developing Interventions to Address Priorities: Food, Dietary Supplements, Lifestyle, and Referrals | 715 |

Aarti Batavia

| 42 | Therapeutic Diets | 743 |

Tracey Long and Leigh Wagner

| 43 | Dietary Supplements: Understanding the Complexity of Use and Applications to Health | 755 |

Eric R. Secor

| 44 | Clinical Approaches to Monitoring and Evaluation of the Chronic Disease Client | 769 |

Cynthia Bartok and Kelly Morrow

| 45 | Ayurvedic Approach in Chronic Disease Management | 783 |

Sangeeta Shrivastava, Pushpa Soundararajan, and Anjula Agrawal

IV Cases & Grand Rounds

| 46 | Cardiometabolic Syndrome | 801 |

Anup K. Kanodia and Diana Noland

| 47 | Revolutionary New Concepts in the Prevention and Treatment of Cardiovascular Disease | 823 |

Mark C. Houston

| 48 | Immune System Under Fire: The Rise of Food Immune Reaction and Autoimmunity | 843 |

Aristo Vojdani, Elroy Vojdani, and Charlene Vojdani

49	Amyotrophic Lateral Sclerosis (ALS): The Application of Integrative and Functional Medical Nutrition Therapy (IFMNT)	863
	Coco Newton	
50	Gastroenterology	913
	Jason Bosley-Smith	
51	Respiratory	927
	Julie L. Starkel, Christina Stapke, Abigail Stanley-O'Malley, and Diana Noland	
52	The Skin, Selected Dermatologic Conditions, and Medical Nutrition Therapy	969
	P. Michael Stone	
53	Movement Issues with Chronically Ill or Chronic Pain Patients	1003
	Judy Hensley, Julie Buttell, and Kristie Meyer	

V Practitioner Practice Resources

54	Systems Biology Resources	1015
	Jeanne A. Drisko, Diana Noland, and Leigh Wagner	
55	Initial Nutrition Assessment Checklist	1019
	Leigh Wagner, Diana Noland, and Jeanne A. Drisko	
56	Nutritional Diagnosis Resources	1043
	Leigh Wagner, Diana Noland, and Jeanne A. Drisko	
57	Specialized Diets	1045
	Leigh Wagner, Diana Noland, and Jeanne A. Drisko	
58	Motivational Interviewing	1051
	Leigh Wagner, Diana Noland, and Jeanne A. Drisko	
59	Authorization for the Release of Information	1055
	Leigh Wagner, Diana Noland, and Jeanne A. Drisko	
60	Patient Handouts	1057
	Leigh Wagner, Diana Noland, and Jeanne A. Drisko	

Supplementary Information

Glossary	1074
References	1076
Index	1077

Contributors

Anjula Agrawal, BDS, RAP
Private Practice
Rancho Mission Viejo, CA, USA
anjulaag@gmail.com

Cynthia Bartok, PhD, RDN
Department of Nutrition and Exercise Science
Bastyr University
Kenmore, WA, USA
cbartok@bastyr.edu

Robert C. Barton Jr., MD
Family Medicine
Northridge, CA, USA
rcbmd@bartonshealth.com

Aarti Batavia, MS, RDN, CLT, CFSP, IFMCP
Nutrition and Wellness Consulting
Novi, MI, USA
aartibatavia@gmail.com

Janet L. Black, MSN, MPH, FNP certification, BSN
LLC, Candler, NC, USA
healthfromjanet@gmail.com

Jason Bosley-Smith, MS, LDN, CNS, FDN
Center for Integrative Medicine, University of Maryland
Baltimore, MD, USA
jbosley-smith@som.umaryland.edu

Sarah Brennan
Portland State University
Portland, OR, USA
sar26@pdx.edu

Julie Buttell
Department of Physical Therapy
Redefine Wellness and Physical Therapy
Overland Park, KS, USA
Julie1buttell@yahoo.com

Margaret Christensen, MD
Carpathia Collaborative
Dallas, TX, USA
drmargaret@carpathiacollaborative.com

Jeanne A. Drisko, MD, CNS, FACN
Professor Emeritus, School of Medicine,
University of Kansas Health System
Kansas City, KS, USA
jdrisko@kumc.edu

Robin L. Foroutan, MS, RDN
New York, NY, USA
rforoutan@mac.com

Mary R. Fry, BSC, ND
Department of Nutrition & Integrative Health
Maryland University of Integrative Health (MUIH)
Laurel, MD, USA
mfry@muih.edu

Margaret Gasta, RDN, DCN
Nutrition By Margo, LLC
Boulder, CO, USA
Margo@nutritionbymargo.com

Steven Gomberg, LAc, MTOM
Lotus Center for Integrative Medicine
Los Angeles, CA, USA
taohongjing@yahoo.com

Michael J. Gonzalez, DSc, NMD, PhD
Human Development, Nutrition Program
University of Puerto Rico, Medical Sciences Campus,
School of Public Health
San Juan, PR, Puerto Rico
michael.gonzalez5@upr.edu

Jena Savadsky Griffith, RDN, IHC
Aroda, VA, USA
jenas_mailbox@yahoo.com

Lauren Hand, MS, RDN, LD
Exercise, Diet, Genitourinary, & Endocrinology (EDGE)
Research Lab, Department of Dietetics and Nutrition
University of Kansas Medical Center
Kansas City, KS, USA
lhand2@kumc.edu

Mary Anne Morelli Haskell, DO
Osteopathic Center of Coronado
Coronado, CA, USA
www.drmaryanne.com

Cynthia Henrich, BA
Certified Integrative Nutritionist
Woodcliff Lake, NJ, USA
cynthiafhenrich@gmail.com

Judy Hensley, PT, DPT, MHS, OCS, MTC
Transformative Physical Therapy
Lenexa, KS, USA
judyahensley@gmail.com

Ronald L. Hoffman, MD, PLLC
Private Practice, New York, NY, USA
drrhoffman@aol.com

Kate M. Hope, MS, CNS
Department of Nutrition
Atlantic Medicine and Wellness
Wall, NJ, USA
khope@amwwall.com

Brandon Horn, PhD, JD, Lac
Department of Obstetrics and Gynaecology
Lotus Center for Integrative Medicine
Los Angeles, CA, USA

Department of Anesthesiology and Critical Care Medicine
Children's Hospital of Los Angeles
Los Angeles, CA, USA
bhorn@lotuscenter.com

Mark C. Houston, MD, MS, MSc, FACP, FAHA, FASH
Department of Medicine
Vanderbilt Medical School and Hypertension Institute,
Saint Thomas Hospital
Franklin, TN, USA
boohouston@comcast.net

Muffit L. Jensen, DC
Integrated Wellness Solutions
Burbank, CA, USA
info@drmuffitjensen.com

Robyn Johnson, MS, RDN, LD
Nutrition by Robyn, LLC
Bend, Oregon, USA
robynjohnson.rdn@gmail.com

Miriam Kalamian, EdM, MS, CNS
Dietary Therapies LLC
Hamilton, MO, USA
info@dietarytherapies.com

Anup K. Kanodia, MD, MPH
Department of Family Medicine
The Ohio State University Wexner Medical Center
Columbus, OH, USA
doctor@kanodiamd.com

Sarah Harding Laidlaw, MS, RDN, CDE
Peak Nutrition Consultants
Montrose, CO, USA
peaknut70@gmail.com

Carolyn L. Larkin
Human Perspectives Int'l
Laguna Niguel, CA, USA
carolyn.larkin@hpiintl.com

Ruth Leyse Wallace, PhD, RDN
Chandler, AZ, USA
ruthwallace78@gmail.com

Elizabeth Lipski, PhD, CNS, BCHN, IFMCP
Department of Nutrition
Maryland University of Integrative Health
Portland, OR, USA
llipski@muih.edu
LL@innovativehealing.com

Mary Demarest Litchford, PhD, MS, RDN, LDN
CASE Software & Books
Greensboro, NC, USA
mdlphd@casesoftware.com

Tracey Long, MPH, RDN
Big Picture Health LLC
Hendersonville, NC, USA
tlong@bigpicturehealth.com

Kristie Meyer, PT, MPT, COMT
Department of Physical Therapy
Elite Sports Medicine & Physical Therapy
Kansas City, MO, USA
meyer_kristie@yahoo.com

Kelly Morrow, MS, RDN
Department of Nutrition and Exercise Science
Bastyr University
Kenmore, WA, USA
kmorrow@bastyr.edu

David Musnick, MD
Peak Medicine
Bellevue, WA, USA
office@peakmedicine.com

Coco Newton, MPH, RD, CNS
Lifetime Nutrition, LLC
Maple City, MI, USA
coco@coconewton.com

Contributors

Diana Noland, MPH, RDN, CCN, IFMCP, LD
Noland Nutrition
Burbank, CA, USA
diana@diananoland.com

Joel Noland, ND, LAc
Sequoia Family Medicine, LLC
Burbank, CA, USA
drjoel@sequoiamedicine.com

Sharon L. Norling, MD, MBA, ABIHM, DABMA
Mind Body Spirit Center
Hendersonville, NC, USA
dr_norling@msn.com

Jessica M. Pizano, MS, CNS
Mast Cell Advanced Diagnostics/
Zebra Diagnostics/SNPed
Chesterfield, VA, USA
jessicapizano@comcast.net

Joseph Pizzorno, ND
Co-Author, Textbook of Natural Medicine
Chair of Board, Institute for Functional Medicine
Seattle, WA, USA
www.drpizzorno.com

Sudha Raj, PhD, RD, FAND
Department of Nutrition and Food Studies
Syracuse University
Syracuse, NY, USA
sraj@syr.edu

David Raubenheimer, D Phil
Charles Perkins Centre and School of Life and Environmental Sciences, The University of Sydney
Camperdown, NSW, Australia
david.raubenheimer@sydney.edu.au

Elizabeth Redmond, PhD, MMSc, RDN
Nutrition Provisions, LLC
Atlanta, GA, USA
bredmond@nutritionprovisions.com

Vishwanath M. Sardesai, PhD
Department of Surgery
Wayne State University School of Medicine
Detroit, MI, USA
vsardesa@med.wayne.edu

Corey B. Schuler, RN, MS, CNS, LN, DC
Department of Clinical Affairs
Integrative Therapeutics, Green Bay, WI, USA
corey.schuler@integrativepro.com

Eric R. Secor, ND, PhD, MPH, MS, LAc, NCCAOM
Department of Integrative Medicine
Hartford HealthCare Cancer Institute, Hartford Hospital
Avon, CT, USA
eric.secor@hhchealth.org

Larissa Severson, BS
Bastyr University
Seattle, WA, USA
larissa.severson@bastyr.edu

Sangeeta Shrivastava, PhD, RDN
California State Polytechnic University, Pomona
Irvine, CA, USA
a.sangeeta.aa@gmail.com

Stephen J. Simpson, PhD
Charles Perkins Centre and School of Life and Environmental Sciences, The University of Sydney
Camperdown, NSW, Australia
stephen.simpson@sydney.edu.au

Pushpa Soundararajan, MBA, RDN, LD, AHE, AFNC
Private Practice, Medical Nutrition Therapy
Willowbrook, IL, USA
vpknutrition@yahoo.com

Abigail Stanley-O'Malley, MS, RDN, LD, CLT
Revive Nutrition Solutions, LLC
St. Joseph, MO, USA
dietitianabby@gmail.com

Christina Stapke, RDN, CD
Christina Stapke, RDN , LLC
Seattle, WA, USA
christinastapke@gmail.com

Julie L. Starkel, MS, MBA, RDN, CD
Starkel Nutrition
Seattle, WA, USA
julie@starkelnutrition.com

P. Michael Stone, MD, MS
Ashland Comprehensive Family Medicine-Stone Medical
Ashland, OR, USA

Institute for Functional Medicine
Federal Way, WA, USA

Asante Ashland Community Hospital
Ashland, OR, USA
mstone@ashlandmd.com

Kathie M. Swift, MS
Integrative and Functional Nutrition Academy
Saratoga Springs, NY, USA
kathie@kathieswift.com

Filomena Trindade, MD, MPH, ABFM, FAARM, IFMCP
Saudade Wellness
Capitola, CA, USA
www.drtrindade.com
info@drtrindade.com

Robert H. Verkerk, PhD
Science Unit, Alliance for Natural Health International
Dorking, Surrey, UK
rob@anhinternational.org

Aristo Vojdani, PhD, MSc, CLS
Immunosciences Lab., Inc.
Los Angeles, CA, USA
drari@msn.com

Charlene Vojdani, PsyD
Immunosciences Lab., Inc.
Los Angeles, CA, USA
charlenevojdani@gmail.com

Elroy Vojdani, MD
Regenera Medical
Los Angeles, CA, USA
evojdani@gmail.com

Leigh Wagner, PhD, MS, RDN, LD
Department of Dietetics & Nutrition
University of Kansas Medical Center
Kansas City, KS, USA
lwagner@kumc.edu

Sue Ward, MS, CNS
Sanoviv Medical Institute
Rosarito, BC, Mexico
sward8@mac.com

Peter Wilhelmsson, MA, ND
IFM Klinik, Falun, Sweden
pow@alpha-plus.se

Christy B. Williamson, DCN, MS, CNS
Nutritional Genomics Institute, SNPed, OmicsDx
Chesterfield, VA, USA
DrChrissie@NGIva.com
www.nutritionalgenomicsinstitute.com
www.omicsdna.com
www.snppros.com
smsilva10@gmail.com

Wendy Yu, MS, LAc
Department of OBGYN
Lotus Center for Integrative Medicine
Los Angeles, CA, USA
wyu@lotuscenter.com

Abbreviations

5-HIAA	5-Hydroxyindoleacetic acid	DHF	Dihydrofolate
5-MTHF	5-Methylenetetrahydrofolate	DHFR	Dihydrofolate reductase
5-OH-TRP	5-Hydroxytryptophan	DHGL/DGLA	Dihomo-gamma-linolenic acid
		DMG	Dimethylglycine
A1AT	Alpha-1 antitrypsin	DMT2	Diabetes mellitus type 2
AA	Arachidonic acid	DXA/DEXA	Dual-energy X-ray absorptiometry
AADC	Aromatic L-amino acid decarboxylase		
ABPM	Allergic bronchopulmonary mycosis	ECW	Extracellular water
ACD	Anemia of chronic disease	ED	Erectile dysfunction
ACE	American Council on Exercise	EEI	Equilibrium Energy Intake
ACSM	American College of Sports Medicine	EFA	Essential Fatty Acids
AERD	Aspirin-exacerbated respiratory disease	eNCPT	Electronic Nutrition Care Process Terminology
ALA	Alpha-lipoic acid		
ALA	α-linolenic acid	eNOS	Endothelial nitric oxide synthase
ALS	Amyotrophic lateral sclerosis	EPA	Eicosapentaenoic acid
AMDR	Approximate Macronutrient Distribution Ranges	ET	Exercise training
ARDS	Acute/Adult Respiratory Distress Syndrome	FAD	Flavin Adenine Dinucleotide
ATP	Adenosine triphosphate	FDA	US Food and Drug Administration
		FeNO	Exhaled fractional nitric oxide
BCM	Body cell mass	FEV	Forced expiratory volume
BCMO1	β-carotene 15,15′-monooxygenase	FIGLU	Formiminoglutamate
BCNH	Bastyr Center for Natural Health	FIP	Familial interstitial pneumonia
BH4	Tetrahydrobiopterin	FPF	Familial pulmonary fibrosis
BHMT	Betaine homocysteine methyltransferase	FUT2	Fucosyltransferase 2
BIA	Bioelectrical impedance analysis		
BIS	Bioimpedance spectroscopy	GBS	Guillain Barre syndrome
BMI	Body mass index (body weight (kg)/height (m)^2)	GERD	Gastroesophageal reflux disease
		GFN	Geometric Framework for Nutrition
BMR	Basal metabolic rate	GGT	*Gamma-glutamyl transpeptidase*
BRCA1	Breast cancer type 1 susceptibility protein	GLA	Gamma-linolenic acid
BRCA2	Breast cancer type 2 susceptibility protein	GMP	Guanosine monophosphate
		GPx	*Glutathione peroxidase*
C	Capacitance	GSH	Reduced glutathione
CBS	Cystathione beta synthase	GSSG	Oxidized glutathione
CF	Cystic fibrosis	GTP	Guanosine triphosphate
CGL	Cystathionine gamma-lyase		
cGMP	Cyclic guanosine monophosphate	HAP	Hospital-acquired pneumonia
CH3	Methyl group	HDL	High-density lipoprotein
CHD	Coronary heart disease	henceforth NPE	Non-protein energy
CLA	Conjugated linoleic acid	HFE	Hemochromatosis gene
CLIA	Clinical Laboratory Improvement Amendments	HGBA$_1$C	Hemoglobin A$_1$c
COMT	Catechol-O-Methyltransgerase	HIIT	High-intensity interval training
COOH	Carboxyl group	HTN	Hypertension
COPD	Chronic obstructive pulmonary disease	HVA	Homovanillate
CT	Computed tomography		
CTH	Cystathionine gamma-lyase	ICW	Intracellular water
CVD	Cardiovascular disease	IFMNT	Integrative and Functional Medical Nutrition Therapy
DASH	Dietary Approaches to Stop Hypertension	IFN	Integrative and functional nutrition
DHA	Docosahexaenoic acid	ILD	Interstitial lung disease

IP3	Inositol triphosphate	P	Protein
IPF	Idiopathic pulmonary fibrosis	P5P	Pyridoxal-5-phosphate
IR	Insulin resistance	PA	Phase angle
IRDS	Infant Respiratory Distress Syndrome	PAH	Phenylalanine hydroxylase
		PAH	Pulmonary arterial hypertension
ISMET	Integrated standing, movement, and exercise training	PDE5	Phosphodiesterase type 5
		PDXK	Pyridoxal kinase
		PES statement	Problem-etiology-signs and symptoms
LA	Linoleic acid	PG	Prostaglandins
LC	Lung cancer	PGE	Prostaglandin E_2
LDL	Low-density lipoprotein	PGG	Prostaglandin G_2
LT	Leukotrienes	PGH	Prostaglandin H_2
LTB	Leukotriene B_4 (LTB_4)	PLH	Protein leverage hypothesis
LTD	Leukotriene D_4 (LTD_4)	PN	Parenteral nutrition
LTE	Leukotriene E_4 (LTE_4)	PNMT	Phenylethenolamine N-methyltransferase
		PUFA	Polyunsaturated fatty acids
MAO-A	Monoamine Oxidase A		
MAO-B	Monoamine oxidase-B	RA	Rheumatoid arthritis
MCT	Medium-chain triglycerides	RBC	Red blood cell
MetSyn	Metabolic syndrome	RMT	Right-angled mixture triangle
MMA	Methylmalonic acid		
MNT	Medical nutrition therapy	SAH	S-Adenosylhomocysteine
MRI	Magnetic resonance imaging	SAM	S-Adenosylmethionine
MSQ	Medical symptoms questionnaire	SCFA	Short-chain fatty acids
MTHFR	Methylenetetrahydrofolate reductase	SCLC	Small-cell lung cancer
		SIT	Sprint intensity training
MTR	Methionine synthase	SMA	α-smooth muscle actin
MTRR	Methionine synthase reductase	SMART goal	Specific, measurable, accountable, reachable/realistic, and timely
MUFA	Monounsaturated fatty acids		
		SNP	Single-nucleotide polymorphism
NAC	N-acetyl cysteine	SOD	Superoxide dismutase
NADPH, NADP+, iNOS	Inducible nitric oxide synthase	sTFR	Soluble transferrin receptor
NAFLD	Nonalcoholic fatty liver disease	SUOX	Sulfite oxidase
NASH	Nonalcoholic steatohepatitis		
NCP	Nutrition Care Process	TBW	Total body water
NIBLETS	Nutrition, inflammation, biochemical individuality, lifestyle, energy and metabolism, toxic load, and stress	TETR	Therapeutic ET for Rehabilitation
		TG	Triglycerides
		TH	Tyrosine hydroxylase
NIH	National Institute of Health	THF	Tetrahydrofolate
nNOS	Nitric oxide synthase 1	TIBC	Total iron binding capacity
NO	Nitric oxide	TPH	Tryptophan hydroxylase
NOS	Nitric oxide synthase	TX	Thromboxanes
NOS1	Nitric oxide synthase 1	TXA	Thromboxane A
NOS2	Nitric oxide synthase 2		
NOS3	Nitric oxide synthase 3	VAP	Ventilator-associated pneumonia
NPE	Nutrition physical exam		
NSAID	Nonsteroidal anti-inflammatory drugs	W_3	Omega-3
		W_6	Omega-6
NSCLC	Non-small-cell lung cancer	WHR	Waist-to-hip ratio

Global Healthcare Challenge of the Twenty-First Century and the Future of Chronic Disease

Contents

Chapter 1   **The History and Evolution of Medicine – 3**
Jeanne A. Drisko

Chapter 2   **Influences of the Nutrition Transition on Chronic Disease – 17**
Sudha Raj

Chapter 3   **Nutritional and Metabolic Wellness – 31**
Diana Noland

Chapter 4   **Nutritional Ecology and Human Health – 39**
David Raubenheimer and Stephen J. Simpson

Chapter 5   **The Radial: Integrative and Functional MNT – 57**
Kathie M. Swift, Elizabeth Redmond, and Diana Noland

Chapter 6   **The Power of Listening and the Patient's Voice: "Please Hear Me" – 73**
Carolyn L. Larkin

The History and Evolution of Medicine

Jeanne A. Drisko

1.1 Introduction: The Problem – 4

1.2 Brief History of Medicine – 6

1.3 Changes Needed: Nutrition Education Across All Medical Education Systems, Interprofessional Teams, and New Systems of Care – 8
1.3.1 Nutrition Education Across All Medical Education Systems: The Foundation – 8
1.3.2 Interprofessional Teams: The Way Forward – 11
1.3.3 New Systems of Care: The Way Forward – 12

1.4 Conclusion – 14

References – 14

1.1 Introduction: The Problem

At an interim meeting of the American Medical Association in November 2018, AMA President Barbara McAneny, MD, called the current US health system "dysfunctional" [1]. As Dr. McAneny stated, "the current system is working to improve hospital and insurance-payer bottom lines, but not working as intended for patients or the doctors who care for them." She went on to describe how the system, with its ever-growing appetite for consolidating businesses and health systems, is getting in the way of actual healthcare. As insurance payers place egregious restrictions on healthcare providers, the health plan determines the course of care, regardless of the patient's needs or the provider's preferred interventions. Dr. McAneny stated, "We are witness to greater concentration of wealth and power in the hands of ever-larger corporations with more and more middlemen pulling down large salaries while our patients go broke and physician practices struggle to survive."

An example of a recent merger is the pharmacy giant, CVS, joining insurance company, Aetna [2]. CVS CEO Larry Merlo said he envisions "CVS Pharmacy evolving from not just a store that happens to have a pharmacy and products" into "more of a healthcare destination." Merlo added that CVS will continue to offer retail products, though its retail sales grew only 1.5% in the first 9 months of 2018. CVS already has MinuteClinic locations at many of its stores, but the company plans to branch out into other comprehensive, community-based healthcare services in the wake of the Aetna deal.

With more than 500 hospital mergers between 2010 and 2015, nearly half of the hospital markets in the United States are now highly concentrated, with many areas of the country dominated by one or two large health systems [3]. Large systems are less nimble and have difficulty responding to market shifts. As healthcare markets concentrate, CEOs focus more on growth than cost-cutting. In addition, hospital mergers have been leading to ever-increasing prices for healthcare consumers. Healthcare systems are also facing challenges to their business models because new types of healthcare venues will shake up the market. These new venues include retail clinics, outpatient urgent care centers, laboratory service-based providers in shopping centers, upcoming projects from tech giants like Amazon and Apple, telehealth services, and the potential for additional mergers between payers and retail pharmacies [3].

The insurance payer market is also consolidating, with many large metropolitan areas dominated by companies that make up at least 50% of the market share [3, 4]. As reimbursement rates to healthcare consumers decline, many now pay more out-of-pocket expenses and are willing to switch healthcare providers to get faster appointments, more convenient locations, better services, and greater cost transparency. This chaos will certainly tax the large conglomerates to provide models of healthcare delivery and payment that will further stress their infrastructure. Currently, the dominant payer system keeps healthcare providers dependent on its strategies instead of emphasizing quality and value.

Electronic health record (EHR) technology, another growing concern for medical practices, grew out of billing software and works for payers whether they are private or federal [1, 3]. The EHR industry remains fragmented across the United States with large vendors dominating but not controlling the market [1, 3]. A single EHR seems unlikely and, as a result, chronic care providers will continue to struggle to capture, share, and use data from the wide number of disparate EHR systems. Most current EHR system users aren't satisfied and many patients question who will have access to their personal data. We need technology that works for patient care, not proprietary software that blocks information gathering between different healthcare providers. The archaic paper documentation, data entry, and faxing must cease [1]. A RAND Corporation study of physician payment methods found an overemphasis on EHR data entry payment models rather than documentation of the healthcare provider-patient relationship that is essential for the success of patient care [5].

The gold standard for clinical decision-making, evidence-based medicine (EBM), has been called into question with its narrowly designed definition and increasing conflicts of interest. As early as 2005, concerns about falsification in published research were raised [6]. The author went on to claim there was proof that most research findings are false. Subsequently, Horton's 2015 Lancet editorial made a straight-forward case against EBM in which he said, "much of the scientific literature, perhaps half, may simply be untrue" [7]. Specific concerns that included invalid exploratory analysis, flagrant conflicts of interest, and an obsession for pursuing fashionable trends in science of dubious importance led him to write, "[S]cience has taken a turn towards darkness." The editorial goes on to highlight the fact that research universities are in a struggle for money and talent and these and other reductive metrics incentivize bad practices. One of the most significant concerns about bias in EBM is financial conflicts of interest [8, 9]. As the growing cost of science and research has outstripped the ability of federal funding, industry and the private sector have stepped into the gap. Unfortunately, those in industry who give the research dollars and support the infrastructure also have a hand in research design, conduct, and publication of findings with significant concerns regarding bias. Financial conflicts of interest with ties to industry are placing undue influence on researchers. A recent prominent example is José Baselga, MD, who was forced to resign as chief medical officer of Memorial Sloan Kettering after it came to light that he failed to disclose millions of dollars in payments from pharmaceutical companies in dozens of published research articles [10]. Although Dr. Baselga claimed the oversight was inadvertent, *The New England Journal of Medicine* editors described his failure as a "breach of trust" [11]. This and other breaches of trust have confirmed that EBM has been hijacked, placing limitations on use of the published evidence that degrade the doctor-patient relationship and flagrantly disregard patient values [9]. As top researchers, federal funders, and most journals call for action, the bad news is no one seems ready to act.

The above changes and challenges have resulted in greater frustrations for practitioners in the trenches caring for patients, leading to burnout and increased suicides [1, 12, 13]. Healthcare providers feel a loss of control over delivery of care with more patient encounters shoehorned into the workday. For every patient they see, they spend two or three times as much time on EHR documentation, with approximately 25% of it spent outside of the clinic. Healthcare providers are not feeling good about the entire structure at the end of the day. It is this loss of autonomy and control over the practice of medicine that is leading to the crisis. And, as Dr. McAneny says, "Yoga at lunch won't fix this" [1].

Healthcare costs at the current growth rate are unsustainable at $3 trillion, and chronic noncommunicable disease accounts for the vast majority of every healthcare dollar expenditure [14–17]. It is time to reimagine medical education, chronic disease care, and EHR data liquidity and to eliminate dysfunctional healthcare. The traditional healthcare model focuses on acute care with a singular focus on hospital-centered care built around ever-expanding large health systems [4]. This current system seeks to continue the status quo and is attempting to improve the current processes within an unsustainable environment. To correct this destructive path, there is a call to change medical education with focus on chronic disease inside this current model. This will include short-term supplementation of the current conventional care system models with alternative care models. However, there are no processes in place or systems built in to help patients navigate and take charge of their healthcare and financial responsibilities. A futuristic view of medicine demands a consumer-centric healthcare delivery system, and the question remains if the current system can provide it [4].

The long-term goal for care delivery is change in the model and a predominant focus on preventive care [3, 4]. The patient will be a consumer of the healthcare system and will demand transparency in billing and access to medical information (see ◘ Fig. 1.1). This will require engaged medical personnel who are transparent and willing to embrace necessary change, especially in technology with interactive platforms and computer software that interfaces with medical records, physician offices, and all members of the health team. This will include integrated devices that transmit patient health data to providers. New ways to manage chronic disease effectively and efficiently while keeping patients at home will be required. Consumer input will be expected, making clear what they need and want from their healthcare providers. Guidance will come from leading consumer-centered practices from industries outside of healthcare. In the end, there will be transparency of pricing, quality data, and technology that engages patients, as well as personalized care delivery. When patients and their healthcare providers are engaged in this process, it will become a powerful disruptive force to the current model. The mandate is on healthcare systems to transform or become obsolete. It is important for change agents and disruptors to take risks, partner with innovators, be creative, and engage healthcare team members in the process.

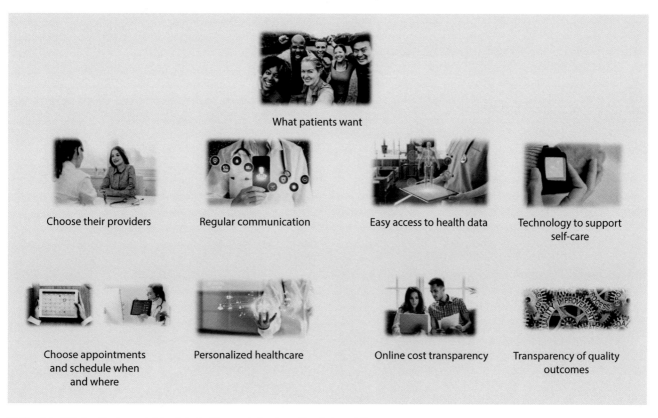

◘ Fig. 1.1 What patients want. (Reprinted with permission from iStockphoto LP)

Who will lead the charge for change? It will not be Big Pharma or healthcare insurance plans or health systems. It is the burden of medical professionals and consumers to become the agents for change and disruptors of this dysfunctional system. Healthcare providers need training to intervene early in the continuum of chronic disease and be able to teach patients to stay healthy, exercise, and choose nutritious foods. The healthcare workforce needs to become partners and educators of patients and speak up for patient equity and ethical treatment and demand that patients have a voice. We have been called to a mission to be dedicated medical professionals and, as such, willing to design a system that values health over money, power, and politics.

1.2 Brief History of Medicine

A brief history of medicine is in order to understand its evolution and how we have found ourselves at this crossroad. Medicine arose from a superstitious approach of ancient civilizations with the healer as a philosopher, physician, and priest [18]. This mindset continued until the Renaissance and blossoming of scientific approaches during the Enlightenment. Medicine likely developed out of the primal sympathy of man for fellow man and out of the desire to help those in sorrow, need, and sickness. This basis of compassionate medicine continues and is rooted in sympathy and the desire to help others.

For many millennia, priests and physicians were co-identified in the practice of medicine and medicine was never fully dissociated from religion. Because of evil spirits and angry gods, man needed emetics, purgatives, enemas, diuretics, and bleeding to rid these demons from the ailing body [18]. A rich pharmacopeia was available during this time as has been identified from recovered papyri. The community used animal body parts and secretions and other elements derived from nature such as opium, hemlock, copper, salts, and castor oil. Crude surgical procedures were employed with historical records of some of these interventions surviving to modern times.

Medicine passed from the rich Egyptian culture to the Assyrians and Babylonians with the ascendancy of these later cultures and the decline of Egypt. Medicine developed to a higher stage but continued to ascribe to evil spirits and demons as the cause of illness and use practices to predict the future. One practice was for physicians to examine animal organs, particularly the liver, which was believed to be an important seat of the soul, vitality, mind, and emotions. This led to anatomical models and the understanding and observation of anatomy. These were related to the priest/healer's yearning to peer into the minds of the gods.

Greek medical schools arose as early as the fifth century BCE where the appreciation between anatomy and physiology was recognized [18]. It was at this time that the doctrine of humors arose. Healers strove to keep equilibrium of the humors – blood, phlegm, yellow bile, and black bile – because health maintenance and disturbances were equated with disease. It was also during the time of the Greeks that shrines and temples arose where climate, diet, and hygiene became important in healing. These locations were often associated with mineral springs and incorporated exercise, baths, and the use of oils and ointments on the body. Prescribed diets, along with prayers and sacrifice, were administered. Libraries with records of treatments, therapies, operations, and all types of medicines were established with some of these records surviving to modern times.

Hippocrates is the father of modern medicine and he richly influenced the evolution of its science and art. He noted that the healer provides service to the community and must aim to be useful and make men better. Hippocrates directed the healer to correct lives and ethically conduct his own life. He must make the beautiful soul harmonize with the beautiful body, as it is the duty of the healer to respect and care for the body [18]. Hippocrates instructed healers to employ consultations when necessary and pushing for fees was discouraged. The healer was to choose the least sensational method and not pretend to be infallible. Above all, cleanliness was to be valued. Because of the high order of mandates established by Hippocrates, bylaws were instituted regulating the personal behavior of physicians. By setting aside the roots of mysticism, superstition, and religious ritual, Greeks began looking to nature and the science of man to base accurate observations. There is no mention of divination, incantations, or charms from the Greeks. They recognized that disease is part of nature, not divine or sacred, and the old beliefs were ascribed to ignorance.

The Greeks expanded the understanding of anatomy as it pertained to the heart, circulation, and pulmonary system, although accurate circulation was not described until much later. The Alexandrian school under the Ptolemies opened a university with the study of literature, science, arts, and medicine, and early understanding of the structure of the human body was defined by limited dissection. Physical exam, to include pulses and other observations, was developed to look for the cause of disease. Galen's program began teaching by example, visiting the ill at the bedside, and this method of ward rounds continues to this day.

Medieval medicine saw the light of learning burn low and flicker almost to extinction. The gifts of knowledge from Athens and Alexandria were deliberately cast aside after the barbarians invaded Rome and civilization was reduced to a wilderness [18]. During this time, Christianity wrought changes in the approach to medicine by reviling the body and all evil on earth. This had profound negative effects by stagnating the evolution of medicine. Science was disregarded as unnecessary. The light of scientific medicine from the ancients was almost lost in the morass of the Middle Ages, but was revived in the Renaissance. With the reintroduction of universities in the thirteenth century, guilds of students were attached to a famous teacher, and the study of anatomy was revived [18]. The rising of the sun of science dawned, and of many, one notable, Roger Bacon, advocated experimental science. Treatments were prescribed accordingly and some of these doctrines reach into modern days.

As the Greek writers of medicine were rediscovered in the fifteenth century, physicians became naturalists and botanists and interested in humanities. The sixteenth and seventeenth centuries saw the foundation laid for accurate knowledge of the structure and function of the human body and how it should be studied. Chemistry was elevated with many important discoveries advanced. It was during this time that Vesalius, a Belgian by birth but a teacher in Padua, began the study of anatomy through dissection of the human body with one of the greatest books of the world published with his anatomical dissections. However, Vesalius was reviled for his work because it was so revolutionary and unfavorable false reports about his character were spread widely [18].

William Harvey was trained by the Paduan Italian anatomists and returned to England, eventually teaching at the College of Physicians in 1604. He was the first to thoroughly describe structure and function of the organs of the thorax with accurate understanding of the workings of the heart and circulation of the blood. It has been stated what Vesalius did for anatomy, Harvey did for physiology [18]. Harvey had a special gift to prove experimentally what he hypothesized and developed the spirit of modern sciences. It was through luminaries like Harvey, Bacon, and Descartes that experimental research became the mode, with superstition and religion finally removed. The beginnings of modern medicine began in the 1680s by incorporating anatomy and physiology into education, and the use of postmortem dissection facilitated the association of illness with death. Systematic description and experiments ruled with truth overcoming dogma.

Herman Boerhaave's Dutch school was the epicenter of European medical learning in the early 1700s, with training in medicine, scientific method, and patient observations. Students were attracted to the Dutch school from all over the civilized world during this period. The Dutch method of instruction was brought back to Edinburgh by John Rutherford in 1747, augmenting the rise of the age of the Scottish Enlightenment. Rutherford taught at the Royal infirmary and was followed by William Cullen and Robert Whytt. It is important to note that Benjamin Rush, a young American physician and one of the signers of the Declaration of Independence, studied under Cullen in Edinburgh before returning to Philadelphia and establishing American medicine [19]. John Hunter organized pathological processes and determined that they were governed by laws of nature and subsequently accumulated, tabulated, and systematized this field of thought. Hunter reestablished the close union between medicine and the natural sciences and laid the foundation for collections and museums connected to medical schools. The systematic teaching of clinical medicine by ward rounds followed by amphitheater lectures became the standard mode of medical education.

By the end of the 1700s and through the 1800s, many great medical advances were made with only a few noted here. This included the birth of smallpox eradication by Edward Jenner through inoculation with cowpox. Jenner contributed to the beginning understanding of acquired immunity. Rene Laennec described the use of the stethoscope and began the practice of rigorous physical diagnosis. Richard Bright described renal disorders; Rudolph Virchow became the father of cellular pathology; Humphrey Davy described the use of nitrous oxide; and William Morton used ether, beginning the advent of surgical anesthesia. Louis Pasteur developed the understanding of plagues, fevers, and pestilence and described the epidemic spread of infection, which destroyed the spontaneous generation theory. Robert Koch showed the life cycle of infectious disease and developed Koch's postulate, while Joseph Lister developed the germ theory and destroyed the belief in miasmas or bad air causing disease. Great strides were taken at the close of the 1800s in the area of tuberculosis, malaria, sleeping sickness, and syphilis, and the ancient doctrine of Celsus was proved true: to determine the cause of the disease often leads to the remedy [18]. Thomas Addison, a physician at Guy's Hospital, described the adrenal glands and their function, which was followed closely by others with descriptions of the pancreas and its relation to diabetes, the thyroid, and the pituitary. Out of this grew chemistry and laboratory medicine with the foundation of the physiology and composition of expired air, blood, food components, excrement, urine, and metabolism.

With the understanding of the germ theory and attention to sanitation, preventive medicine was born. The attempts to fight poverty, filth, poor living conditions, and other public health measures helped stamp out typhus and reduce infections of malaria and yellow fever. Sanitation and improved living conditions reduced the burden of tuberculosis from about one million in the eighteenth century to half that number by 1911. The dawn of modern medicine through the first half of the twentieth century saw the reduction in acute illness or communicable disease and the ascendancy of chronic disease or noncommunicable disease that now plagues mankind worldwide [14].

The history of twentieth-century medicine would not be complete without discussion of the Flexner report. Published in 1910 after the Carnegie Foundation commissioned the study of US medical education [20], the project's overarching goal was to weed out medical schools that had no prerequisite educational requirements and no preparation in the basic sciences, those not associated with a university, and those with no systematic teaching of patient care by hospital rounds. Abraham Flexner felt his mission was to eliminate those who called themselves physicians but had received very little scientific education before or during medical school. Certainly, there was the need to elevate medical teaching and training with expansion of the medical sciences.

In Chapter X of the groundbreaking report, he detailed "illegitimate nonscientific" approaches that he called "Medical Sects" [20]. These groups included naturopaths or eclectic practitioners, herbalists, psychologists, homeopaths, chiropractors, and osteopaths, and these groups very quickly found their education and practice driven underground by the polemics of Flexner's disdain [20, 21]. The report also marginalized practices he felt to be unscientific, like social work and other disciplines like nutrition [20, 22].

Flexner advocated for the state board system that continues to license medical practitioners to this day [20, 23]. He noted that the state boards could reject applicants as an indirect method of discrediting the school which had given them their medical degree and, in his words, "[T]he board should summarily refuse to entertain the applicant's petition because his medical education rests upon no proper preliminary training. The full weight of its refusal would fall with crushing effect upon the school which sent him forth." In 1912, a group of the state licensing boards joined together to create the Federation of State Medical Boards [23]. This is neither a federal agency nor an independent governing body as the group voluntarily agreed to base its accreditation policies on academic standards determined by the American Medical Association's Council on Medical Education policies. Consequently, these decisions became the law of the medical land. A great deal of political power was placed in the hands of the American Medical Association that continued to suppress competing forms of medical practice well into the late twentieth century.

Although many excellent advances occurred because of improvements in the medical education system, there was a chilling effect of contracting medicine to reductionist strategies with established hierarchies. The job of medical care morphed in the early twentieth century after the development of drugs and advances in surgery. This led to drugs and surgery as the purview of physicians in dealing with acute care until the shift to chronic diseases made these approaches less successful. What suffered with the Flexner report was patient-centered humanistically oriented community medicine [21, 22]. Medical schools to this day continue to struggle to overcome the effects of standardization of American medical education [23]. All accredited American medical schools apply Flexner's "uniformly arduous and expensive" goals in medical education [23]. Unfortunately, with the rise in cost of chronic healthcare, many schools have been forced to make curricular changes and form corporate alliances because of the need to balance academic ideals with economic and social realities.

It is clear our current education and treatment systems are administered on the old foundations built during the era of acute communicable disease with heavy reliance on hospital rounds. But acute communicable disease has given way to dominance of chronic noncommunicable disease. Medical education must now include a curriculum that addresses the major cause of mortality and morbidity in the population that the health workforce serves and in the systems where they practice [22]. Will we be able to rebuild within the current system or find the need to construct a new model that will give home to the treatment of chronic noncommunicable disease? It is without question that medical education needs to undergo radical change and this poor foundation in education and treatment of chronic disease is largely to blame for the dysfunctional medical system, burnout of its caregivers, dissatisfied patients, and skyrocketing costs. Even Flexner believed that medical education must be accountable to the society that it serves [22]. Just as the Greeks and those physicians in the Renaissance and Enlightenment let go of the ancient ways while retaining the good parts, so should we let go of the acute care model that is collapsing and join forces with all types of practitioners to build a new model, with patient-centered humanistically oriented community medicine focusing on subduing chronic disease and supporting health-oriented citizens.

1.3 Changes Needed: Nutrition Education Across All Medical Education Systems, Interprofessional Teams, and New Systems of Care

1.3.1 Nutrition Education Across All Medical Education Systems: The Foundation

Worldwide, chronic disease will cause $17.3 trillion of accumulated economic losses over the next 20 years [16, 17]. These economic burdens are related to healthcare expenditures, reduced productivity, and lost capital and are the leading priorities of our time. The predominant risk factor for chronic disability and death worldwide is related to suboptimal diet and poor lifestyle choices [14, 24]. Diet-related coronary heart disease, stroke, type II diabetes, and obesity produce ever-growing global health burdens [14]. It is now evident that poor dietary habits contribute to obesity, low-density lipoprotein cholesterol increases, increased blood pressure, glucose-insulin irregularities, oxidative stress, inflammation, poor endothelial health, fatty liver, irregular adipocyte metabolism, abnormalities in hormone pathways of weight regulation, visceral adiposity, and destruction of the gut microbiome [17]. Unfortunately, the focus for decades has been a preoccupation on dietary recommendations that limit fat and cholesterol. It is now acknowledged the full health impact of fat-limited diets extend far beyond cholesterol pathways.

Dietary research and interventions evaluating overall metabolic risks have up to the present focused on single isolated nutrients rather than broad dietary patterns. Rather than isolated nutrients, a food-based approach better serves the general populations and minimizes industry manipulation [17]. Pushed by burgeoning obesity rates, nutrition research is beginning to focus on the quality and types of foods consumed that influence diverse pathways such as glucose-insulin response, hepatic lipogenesis, fat storage, and the microbiome. Armed with this new information, it is up to the healthcare team to work with individual patients and meet them in their sociocultural communities to create lasting changes in chronic disease burdens. However, new paradigms in nutrition are not being translated into clinical practice because of limits in medical education.

Medical education has a mandate to prepare students to face the types of clinical problems they will see in their everyday practices [25]. Medical education has predominantly occurred in a hospital setting with didactic lectures in amphitheaters and ward rounds consisting of education at a patient's bedside. There is a failure of medical education, and as a result the healthcare delivered, to adapt to transformations that have occurred in the past 50 years, with chronic disease replacing acute disease as the dominant problem. While acute disease is episodic, commonly associated with hospitalization and a return to baseline health, chronic disease is constant with the prevailing belief there is rarely a cure. It can be argued that in order to find relief and possibly cures, chronic disease requires a change in delivery of care and its location with a broader emphasis on lifestyle and diet. Diet and nutrition are the foundation of health. Currently, medical education does not value nutrition education or understand its contributing role in chronic disease.

Poor nutrition as a leading cause of morbidity and mortality is competing with and perhaps outpacing the deleterious effects of tobacco and physical inactivity, but is largely unrecognized by conventional primary care providers related to limited nutrition education opportunities. As early as 1963, there were reports of deficiencies in nutrition education in conventional medical schools [26]. The authors acknowledged, "a close relationship between nutrition and the individual's response to pathologic stress, such as severe infections or trauma." They also acknowledged the relationship between nutrition and degenerative disease, atherosclerosis, hypertension, and neoplasia. At the time, there was a call to improve medical education and medical practice to keep "abreast of the tremendous advances in nutritional knowledge" [26]. Yet this call was not heeded.

Over the ensuing decades, the increasing burden of chronic, diet-related disease, including the obesity epidemic, was not matched with nutrition education in conventional medical education [27–29]. The neglect in teaching nutrition is a growing problem as the workforce is not armed with tools to address patients' lifestyle issues. The average amount of required nutrition education, at the time of one report, was noted to be 19.6 hours, while the number of schools offering a dedicated nutrition course fell 25% [28]. Rather than measuring hours of nutrition education and coursework, the focus should be on competencies. In fact, most medical education across disciplines does not even teach diet and nutrition beyond a few metabolic pathways. Barriers to the inclusion of nutrition in medical education include such factors as the focus on treating disease with drugs and surgery and the erroneous belief system that nutrition is insufficiently evidenced-based [28].

The research evidence base for nutrition is only in its infancy, having begun in earnest in the beginning of the twentieth century. It wasn't until the 1980s that US dietary guidelines focused on the nutritional basis of chronic disease [24, 27–29]. However, because of the historical emphasis in nutrition research on deficiency disease, the single-nutrient paradigm continued to dominate research approaches. As a result, the literature oversimplified the impact of dietary factors in chronic disease [17]. In addition, nutrition epidemiology is in need of "radical reform" [6]. As an example of the absurdity of focusing on the previous research paradigm, the author states " [E]ating 12 hazelnuts daily (1 oz) would prolong life by 12 years (i.e., one year per hazelnut), drinking 3 cups of coffee daily would achieve a similar gain of 12 extra years, and eating a single mandarin orange daily (80 g) would add 5 years of life. Conversely, consuming 1 egg daily would reduce life expectancy by 6 years, and eating 2 slices of bacon (30 g) daily would shorten life by a decade, an effect worse than smoking. Could these results possibly be true?" New paradigms of nutrition research are desperately needed.

As nutrition science research matures, there is agreement that diet, nutritional status, and lifestyle choices predispose or protect against many chronic diseases including heart disease, diabetes, and cardiovascular disease. Yet, uncertainty remains in evaluating dietary evidence for federal policy recommendations and broad public health statements. Unlike the pharmaceutical industry, where evidence-based medicine (EBM) evaluates drug effects and has done so for many years, this type of research and evidence has proved challenging when attempting to adapt it to nutrition and diet research. Notwithstanding the concerns regarding traditional EBM [6–9], researchers have not established clear guidelines to point the way to study the effects of nutrients in complex human systems combined with a lack of consensus on how to evaluate findings and establish policy and dietary guidelines.

It is imperative to remember that pharmaceutical drugs are often studied in a therapeutic intervention context, that is, to treat, cure, or mitigate disease. However, nutrients are studied with a focus on health promotion or disease risk reduction, again supporting the idea that there needs to be fundamentally different approaches in researching the two models. What is needed is evidence-based nutrition not built on principles of EBM or reliance on randomized controlled trials [30]. When comparing food and nutrients to pharmaceuticals, it is best to remember that drugs usually have single and targeted effects, are not homeostatically controlled by the body, and can be contrasted with true placebo. By way of comparison, nutrients work in complex systems, within an epigenetic framework, and in concert with other nutrients and simultaneously affect multiple cells, tissues, and organs because they are homeostatically controlled. An added consideration is how the research participant's baseline nutrient status will affect the response to nutrient intervention. Finally, in the EBM research model, single isolated nutrients are required to be evaluated; however, this approach could never be effective in the complex biochemical mammalian environment. Nutrients function in a vastly complex biochemical network subject to interplay of genetics, diet, lifestyle, activity, stress, and environmental toxins. Single-nutrient studies can never answer questions in this complex system, and their effects in long-latency chronic disease cannot be adequately assessed. Until a new paradigm is developed, it is likely that healthcare providers in the trenches caring for those with chronic disease will be reduced to technicians delivering evidence-based algorithms and guidelines. It is up to leaders in the field of nutrition to establish evidence-based nutrition research guidelines.

An example of this misguided research approach occurred in the 1980s and 1990s. Policymakers and healthcare practitioners believed saturated fats and cholesterol were the cause of coronary heart disease and obesity [17]. Over the past 30 years, the substitution of fats with simple carbohydrates and sugar in processed food led to the rapid rise in the obesity epidemic and skyrocketing type II diabetes rates. Based on evidence, the 2015 Dietary Guidelines Advisory Committee concluded that low-fat diets had no effect on cardiovascular disease and recommended focus on healthful food-based dietary recommendations rich in healthful fats [31]. Fortunately, more sophisticated approaches are being discussed with a rapid transition away from the single-nutrient theories and surrogate outcomes.

We see a transition in a more modern understanding of diet in its relationship to chronic disease. The science of obesity in more educated circles is shifting away from simplistic ideas of energy balance and calorie counting with new research focused on the effects of food and diet patterns on complex physiologic pathways. Healthful food-based patterns of intake have been shown to be the most important for fighting obesity [31]. Poor dietary quality is associated with an increase in processed foods, resulting in loss of nutrients and fiber, including polyphenols, minerals, essential fatty acids, vitamins, and other bioactive food ingredients. Poor diet quality is a driver of excess dietary intake, which in turn influences metabolic risk and the increase in abdominal adiposity. Independent of calories, poor dietary quality strongly influences metabolic dysfunction and drives the increased risk of diabetes. With increases in processed carbohydrates and simple sugars, these diets contribute to insulin dysregulation and resistance over time [17].

Evidence shows that an educated patient achieves better outcomes and a collaborative medical environment further enhances this [32]. Patients need education in healthful dietary choices. Sadly, the public considers physicians to be one of the most trusted sources of nutrition-related information [25, 27–29]. Yet physicians and other healthcare providers lack core competencies to properly counsel and educate patients about nutrition and diet-related disease. Additional barriers beyond education in instituting these goals in the conventional medical encounter include shrinking medical encounter time and increased EHR burdens, leaving little room for discussion of diet and lifestyle interventions. There is a need for creativity and innovation for approaches in chronic disease intervention, beginning with medical education and extending into the care setting. Imaginative ways of interacting with patients besides a traditional physical encounter should be employed such as telemedicine, telephone, and email [4]. This model needs to be incorporated in conventional medical education across all disciplines, and these shared learning experiences need to be coupled with new understanding and new behaviors [25, 32]. This in no way conflicts with medical science or the care of acute disease in the hospital setting but rather allows the best of medical science and technology to care for all patients.

The Institute of Medicine report, *Crossing the Quality Chasm* [32] reported, "[T]he fundamental approach to medical education has not changed since 1910." There is now a call for medical care teams to be educated and to work together in a manner that shares management responsibilities and decisions for these new challenges. With this shared responsibility, patients with chronic disease will be better cared for. Educating patients and their families in self-management methods will make them a valuable member of this team as well.

Successful and sustainable improvements in patients with chronic disease can occur but will require close collaboration among multiple stakeholders, including interprofessional healthcare teams and the health systems they work in, policymakers, researchers, community organizations, schools, farmers, retailers, restaurants, and food manufacturers. In the clinic setting, evidence-based dietary interventions are needed for individual behavior changes. These approaches include goal-setting, in collaboration with the patient, that includes a personalized diet plan; self-monitoring that could be achieved with electronic diaries through web-based apps; regularly scheduled follow-ups with feedback on progress through telehealth, telephone, or electronic feedback; the use of interventions that include motivational interviewing; and consistent use of family and peer support. See ◘ Table 1.1 [17]. This will necessitate training interprofessional teams on

Table 1.1 Evidence-based approaches for individual behavior change in the clinic setting

Intervention	
1	*Specific, proximal, shared goals.* Set specific, proximal goals in collaboration with the patient, including a personalized plan to achieve the goals (e.g., increase fruits by 1 serving/day over the next 3 months)
2	*Self-monitoring.* Establish a strategy for self-monitoring, such as a dietary or physical activity diary or web-based or mobile phone application
3	*Scheduled follow-up.* Schedule regular follow-up (in-person, telephone, written, or electronic), with clear frequency and duration of contacts, to assess success, reinforce progress, and set new goals as necessary
4	*Regular feedback.* Provide feedback on progress toward goals, including using in-person, telephone, or electronic feedback
5	*Self-efficacy.* Increase the patient's perception that they can successfully change their behavior
6	*Motivational interviewing.* Use motivational interviewing when patients are resistant or ambivalent about behavior change
7	*Family and peer support.* Arrange long-term support from family, friends, or peers for behavior change, such as in workplace, school, or community programs
8	*Multicomponent approaches.* Combine two or more of the above strategies into the behavior change effort

Adapted from Mozaffarian [17]. With permission from Wolters Kluwer Health

evidence-based behavior change strategies and adopting new techniques such as telehealth and electronics systems to help assess, track, and report what specific dietary behaviors are occurring.

1.3.2 Interprofessional Teams: The Way Forward

Lifestyle and behavioral risks resulting in complex chronic disease along with shortages in the health workforce are taxing the medical system. It has been argued that even though the 1910 Flexner report sparked groundbreaking reforms in medical education, medical education has not kept pace with the current challenges that have produced ill-equipped graduates [33]. Because of the structure of medical education post-Flexner report, there was a tendency of various medical professionals to act in isolation from or even in competition with one another. There has been a call to redesign professional health education to capitalize on the interdependence of health professionals in this new age of complex chronic disease. The idea of a team of medical professionals joining forces to deliver patient-centered care for complex medical disorders in the community setting has been discussed for several decades [32, 34–38].

Chronic disease evolves over time with waxing and waning severity. Compared to acute disease in a hospital setting where the patient submits themselves to treatment, complex chronic disease demands involving the patient as an active partner in their care; and because of the complexity of chronic illness, it is essential that a coordinated team of healthcare providers unite in its management. The team provides continuity and organized integration of care that are essential elements and helps create a healing relationship with the patient. Because the patient is now an active partner, education of the patient as well as family members or other social support groups is paramount. It is important for the professional group of providers, the interprofessional team, to understand that their role has changed from the time when acute disease treatment was delivered to the patient to our current era of chronic disease where the patient is part of the decision-making team. The interprofessional team members are now professional guides and advisors, teaching the patient healthcare skills, and there is shared decision-making between the patient and the healthcare team demanding a treatment program tailored to the patient's specific needs and wishes. In addition, the location of treatment intervention changes from the hospital ward to the ambulatory setting, which can take on a variety of appearances and locations such as traditional outpatient clinics, health clubs, home settings, community centers, telemedicine, email, or group educational programs [4, 5, 39]. The location is of less importance than the partnership and relationship of the patient and the interprofessional team, which is central to the healing encounter.

In 2009, six organizations representing schools of health education came together to form the Interprofessional Education Collaborative (IPEC) with the goal of working together to promote and encourage interprofessional learning experiences [38]. This vision grew out of a desire to form an interprofessional collaborative practice to deliver safe, high-quality, accessible, patient-centered care. The intent was not to harmonize the educational experience but to build on each profession's expected competencies to enhance the interprofessional experience in patient care. To enable the vision to be fulfilled, it was acknowledged that students needed to interact with one another in a learning environment so that they would effectively become members of a clinical team. This was to help prepare future health professions for enhanced team-based care of patients and improve population health outcomes particularly with the growth of chronic disease. A report published in 2011 detailed the core competencies for interprofessional collaborative practice [38]. This report and its goal to change were widely accepted and adopted with dissemination of the competencies across multiple colleges of medical learning with increased incorporation as part of the required curriculum.

Between 2011 and 2016, the interprofessional collaborative team grew from 6 core members to 15, along with input from 60 other health professions [40], with expanded goals to create a shared taxonomy among all the health professions and to streamline and synergize educational activities, related assessments, and evaluation efforts. The updated 2016 report highlighted the Triple Aim, which includes improve the patient experience of care, improve the health of populations, and reduce the per capita cost of healthcare [40]. The group articulated shared ideals that translated to competencies and include professionalism and respect, using knowledge for the benefit of patients and populations, improving communication to promote and maintain health and prevent disease, and using teams to plan, deliver, and evaluate patient-centered care that is safe, timely, efficient, effective, and equitable [40]. But can the current conventional health system with shrinking office visit times and increasing EHR demands successfully deliver interprofessional team medicine?

It is the belief of this author and the authors of this textbook that an effective model of interprofessional teams has been in existence for several decades. An integrative and functional healthcare team has brought together a partnership of multiple types of providers with the focus on the patient [39, 41]. With the patient at the center, attention is on the roots of chronic disease with diet as the foundation, requiring a team of professionals with a wide variety of expertise. This type of practice is bringing enjoyment back into the practice of medicine. Integrative and functional medicine is using paradigms to address chronic disease, while conventional medicine remains stuck in its acute care model. As spiraling rates of burnout and suicides occur, conventional providers are finding their way into new models of practice, and as a result, the dominance of the acute care model will wane.

1.3.3 New Systems of Care: The Way Forward

In the Introduction, ◘ Fig 1.1 highlighted what patients want from their healthcare providers. Patients want to be part of the healthcare team with their preferences valued. Information sharing is the answer and educating patients is the key. The future of medicine depends on the partnership of the interprofessional team with patients while avoiding methods that limit patient access to self-care information. A startling example of the attempt to reduce patient self-care is seen in the recent report stating patients with type II diabetes are testing their blood sugar by glucometer more often than they need to [42]. Three medical societies, the American Academy of Family Physicians, the Society of General Internal Medicine, and the Endocrine Society, released a statement that type II diabetics do not benefit from tracking daily blood sugars. The information was collected from insurance data with a look at expense. This flies in the face of good medicine and reduces the partnership with patients. Patients are no longer to glean information from self-testing to learn what effects food and lifestyle factors have on blood sugar levels. As conventional frontline medical practitioners are no longer encouraged to partner with their patients, there has been the growth of other types of healthcare delivery that look at root cause with more effective models of intervention. An example of one of these models is the Virta Health system that was developed to intervene in type II diabetes with diet, lifestyle, and patient education [43]. When patients understand their disease and can discuss information with their healthcare providers, remarkable changes can occur even in such significant debilitating chronic illnesses as type II diabetes. Drugs and surgery are held in reserve and used only when necessary. These alternate models of healthcare delivery treat the root cause of chronic disease rather than immediately intervening with drugs and surgery, long the prevailing paradigm in conventional medicine.

As the Centers for Disease Control and Prevention (CDC) warns, obesity, cardiovascular disease, and diabetes are the leading causes of preventable mortality [24]; personalized lifestyle medicine is a growing paradigm that delivers a different approach to conventional acute care medicine. Personalized medicine is specifically tailored to the individual patient's lifestyle, environment, genetics, and preferences to develop strategies for improving health outcomes and managing chronic disease [44–46]. To address the crisis, personalized lifestyle medicine requires a new set of healthcare practitioner skills and competencies to address multiple risky behaviors and improve patient self-management. This can lead to designing patient-specific prescriptions for diet, exercise, stress reduction, and avoiding environmental toxicity by taking advantage of identification of genetic variants and functional laboratory biomarkers. This comprehensive personalized approach promises to promote the safety of conventional therapeutics while reducing the long-term cost of chronic disease care. Information gathered from this approach empowers patients to have more control over their health with partnership at the center in the therapeutic setting. However, a barrier to implementing this style of medicine is modest research underpinnings and tepid recommendations from policymakers.

Another barrier to instituting a patient-centered diet and lifestyle approach includes the challenge in dealing with the complexity and the often-insidious transition from health to chronic disease with late onset of symptoms. This requires a workforce willing and trained to reconsider reactionary treatments and interventions as first-line treatments and to shift toward diet and lifestyle interventions first, with an eye on long-term prevention. A framework for this type of medicine has been proposed and is labeled "P4" for predictive, preventive, personalized, and participatory [46]. The authors state this system was developed to move away from a reactive medical approach with expensive episodic acute care interventions delivered in conventional settings that results in minimal interactions between specialists and general practitioners, fragmented healthcare approaches with multiple prescription medications, scattered follow-up, and a suboptimal cost-effectiveness ratio of care delivered [46]. Chronic disease care under these new models can be preventive with drug and surgical interventions delayed until significantly later in life. This is proposed by adoption of a healthy lifestyle throughout the lifespan resulting in an extended healthspan with the duration of one's life spent in a state of wellness. The P4 framework also believes in reversing chronic disease once it has begun by lifestyle and dietary interventions, but must include the patient as a partner. It is also acknowledged that at some point in the continuum of chronic disease, it may be necessary to intervene with traditional healthcare with pharmacotherapy, surgery, and other interventions as management tools.

The P4 model and other models such as integrative medicine, naturopathy, functional medicine, nutritional medicine, and other systems capture this belief and put it into practice. Moving toward a modern, proactive healthcare system, different levels of intervention must be clearly defined, and the list of stakeholders invested in the implementation of this model must be expanded. Training programs have sprung up to fulfill the need for retooling the interprofessional workforce, but effective change demands an orchestrated language to help researchers, healthcare professionals, and stakeholders across a multitude of sectors to collaborate as efficiently as possible [44–46]. Businesses are being developed to help practitioners become financially successful in establishing these types of clinics but will certainly expand over the next decade [3, 4, 39].

A "Blueprint" for health system sustainability was recently released as a community-based model to help reduce the burden of preventable disease on the financially strapped National Health Service in the United Kingdom, but could also serve as a blueprint for the world as a whole [47]. The Blueprint identifies 12 domains of human health thrown out of balance by contemporary lifestyles (see ◘ Fig. 1.2) [47]. The most effective way of treating lifestyle disorders is with appropriate lifestyle changes that are tailored to individuals, their needs, and their circumstances. Such approaches,

Fig. 1.2 The ecological terrain of an individual including 12 inter-connected domains of health and resilience

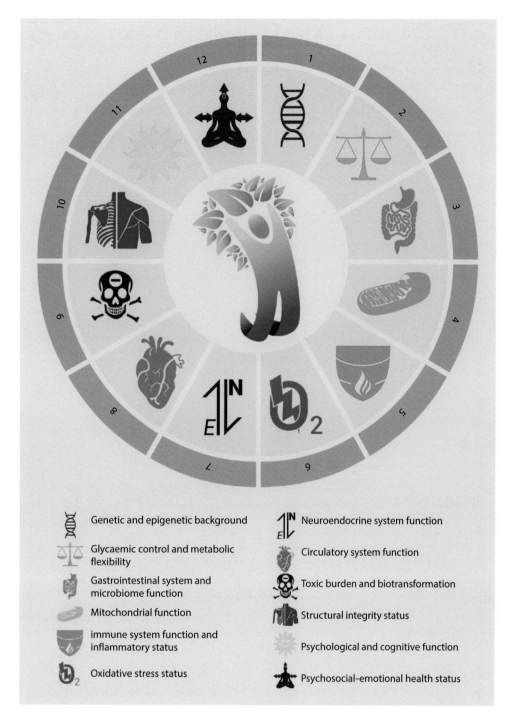

guided by the trained interprofessional team, tend to be far more economical and more sustainable as a means of maintaining or restoring people's health. At the core of a sustainable health system is one with individuals more responsible for maintaining their own health and where more effort is invested earlier in an individual's life prior to the manifestation of downstream chronic, degenerative, and preventable diseases. The report outlines hallmarks of sustainable health systems that are centered on the needs of individuals while also being focused on health creation or regeneration. Hallmarks include significantly reduced pharmaceutical use as first-line treatment for dietary and lifestyle-mediated disorders, financial and social frameworks that encourage the use of non-drug, lifestyle "prescriptions," and much greater engagement and autonomy by the individual.

The Blueprint acknowledges that implementing such seismic changes in the way healthcare is managed will be met with opposition [47], notably from those with interests in maintaining the status quo; however, identifying and addressing the barriers could transform some into opportunities. The most substantial and complex impediment will likely arise from the scientific community with the associated and limiting model of evidence-based medicine (EBM) as outlined above. New approaches to building evidence to guide clinical decisions are being called for by federal agencies [48, 49]. It is imperative that we place the patient and practitioner back to the center of the decision-making process so that clinical choices can be based on evidence, personal experience of the doctor, and expectations of the patient [9]. Just as the promise of EBM has been the use of the best evidence, personalized precision medicine must also reintroduce the use of judgment. As one size does not fit all, it must be remembered that evidence-based medicine regards disease at the population level with minimal consideration of individuals. It is the call for personalized precision medicine approaches that need to be implemented allowing us to tailor interventions to our individual patients [9].

As patient-centered care in this current age requires a team, an effort to reconfigure and redesign health systems will require support from a team drawn from industry, politics, scientific and education communities, medical providers, and the consumers who ultimately will benefit. As the Blueprint [47] states, "To help develop a consensus approach that will allow progress towards more sustainable health systems…," transdisciplinary stakeholder working groups need to be formed that will develop assessment tools for outcome studies of health and resilience, economics of the proposed approaches, policy and regulatory measures, and approaches that minimize social inequality and maximize sustainability. It follows that the working groups will recommend pilot trials to establish proof of concept and the setting will include both clinical and nonclinical settings to evaluate effectiveness, value, and sustainability of the collaborative and participatory approaches. It is the belief of this author that subsequent larger-scale trials evaluating the proposed model of healthcare approaches will likely find reduced healthcare costs while improving health outcomes, value, and quality of life.

The final call is to forthrightly overcome biases and form consensus on what is required for the creation and management of necessary health systems. Well stated in the Blueprint [47], "We owe it to future generations to work urgently, earnestly and cooperatively together to develop and thoroughly evaluate new ways of managing and creating health in our society."

1.4 Conclusion

The time of crisis in healthcare is upon us, and difficult decisions need to be made to preserve good medicine for future generations. The acute care model in education and patient care will need to be moved aside to make room for solutions for chronic diseases that are swamping the system. Patients are demanding a place at the table to craft a way forward that engages them in the healing process. Disruptive elements in the technology world and in large corporations outside of medicine will make the current healthcare model obsolete. It will take brave actions and bold solutions. Will healthcare providers rise to the moment?

Seth Godin's Blog: Respect Difficult Problems [50]

» They're difficult because they resist simple solutions. Glib answers and over-simplification have been tried before, and failed.

» People have tried all of the obvious solutions. They haven't worked. That's why we've resorted to calling them difficult problems.

» Difficult problems require emotional labor, approaches that feel risky and methods that might not work. They reward patience, nuance and guts, and they will fight off brute force all day long.

References

1. Minemyer P. Practices: AMA outlines initiatives aimed at fixing 'dysfunctional' healthcare system at interim meeting. [Internet 11/12/2018] Accessed 11/27/2018: https://www.fiercehealthcare.com/practices/ama-outlines-initiatives-aimed-at-fixing-dysfunctional-healthcare-system-at-interim.
2. Vartorella L. What CVS stores will look like after the Aetna deal. [Internet 11/28/2018] Accessed 12/1/2018: https://www.beckershospitalreview.com/hospital-management-administration/what-cvs-stores-will-look-like-after-the-aetna-deal.html.
3. Philips Population Health Management. How to harness market chaos and lead: Controlling technology fragmentation, leveraging payer relationships and managing competition with actionable market insights. [Internet 2018] Accessed 11_27_2018: https://www.wellcentive.com/resource/harness-market-chaos-white-paper/.
4. Bates M, Deao C, Seabrook H. for the Huron/Studer Group. The future of healthcare leadership: creating the consumer-centric organization. [Internet 2018] Accessed 12/1/2018: https://www.studergroup.com/future-of-healthcare-leadership?utm_source=fierce&utm_medium=disp&utm_term=2018-hc-sg-ebook&utm_content=disp-1-hc-sg-future-healthcare-leadership&utm_campaign=7010B000001gfgQ.
5. Friedberg MW, Chen PG, White C, Jung O, Raaen L, Hirshman S, et al. Effects of health care payment models on physician practice in the United States. www.rand.org/t/rr869. Santa Monica: RAND corporation Publishing; 2015.
6. Ioannidis JPA. Why most published research findings are false. PLoS Med. 2005;2(8):e124.
7. Horton R. Offline: what is medicine's 5 sigma? Lancet. 2015;385(9976):P1380. https://doi.org/10.1016/S0140-6736(15)60696-1.
8. Fava GA. Evidence-based medicine was bound to fail: a report to Alvan Feinstein. J Clin Epidemiol. 2017;84:3–7. https://doi.org/10.1016/j.jclinepi.2017.01.012.
9. Tebala GD. The Emperor's new clothes: a critical appraisal of evidence-based medicine. Int J Med Sci. 2018;15(12):1397–405. https://doi.org/10.7150/ijms.25869.
10. Thomas K, Ornstein C. Top cancer doctor resigns as editor of medical journal [Internet December 18, 2018] Accessed 12/22/2018: https://www.nytimes.com/2018/12/19/health/baselga-cancer-conflict-disclosure.html.
11. Editorial Staff. Correction. NEJM. 2018;379(16):1585.
12. National Academy of Medicine. Action collaboration on clinician well-being and resilience. [Internet 2017] Accessed 11/27/2018: https://nam.edu/initiatives/clinician-resilience-and-well-being/.
13. Swensen S, Strongwater S, Mohta NS. Physician burnout and resilience. leadership survey: immunization against burnout. NEJM Catalyst. [Internet April 12, 2018] Accessed 11/27/2018: https://catalyst.nejm.org/survey-immunization-clinician-burnout/.

14. World Health Organization. Preventing chronic diseases: a vital investment. [Internet 2005] Accessed 11/27/2018: https://www.who.int/chp/chronic_disease_report/contents/part1.pdf?ua=1.
15. Norris J, Miller P, Hatch M, Gray J, Spantchak Y, Finkelstein S. The macroeconomic consequences of chronic diseases in emerging market economies: phase I report. Washington DC: The Center for Science in Public Policy; 2009. [Internet January 22, 2009] Accessed 11/27/2018: http://www.paho.org/hq/dmdocuments/2009/macroeconomic-report-novartis.pdf.
16. Martin AB, Hartman M, Washington B, Catlin A. National Health Care Spending in 2017: growth slows to post-great recession rates; Share of GDP Stabilizes. Health Aff. 2019;38(1):1–11. https://doi.org/10.1377/hlthaff.2018.05085.
17. Mozaffarian D. Dietary and policy priorities for cardiovascular disease, diabetes, and obesity a comprehensive review. Circulation. 2016;133:187–225. https://doi.org/10.1161/CIRCULATIONAHA.115.018585.
18. Keller C, Widger D editors: Sir William Osler. The evolution of modern medicine a series of lectures delivered at Yale University on the Silliman Foundation in April, 1913. [internet 2006] Accessed 08/2018: http://www.gutenberg.org/files/1566/1566-h/1566-h.htm. (Historical Note: Sir William Osler delivered a series of lectures at Yale University in 1913 on the history and evolution of modern medicine. The lectures were submitted to Yale University Press for publication and Dr. Osler corrected and revised the galleys until the beginning of the Great War. During the war years, he devoted himself to military service and public duties and as a result never completed the proofreading and intended revisions. Editors subsequently completed the revisions and the document is available online. The document is described as a sweeping panoramic survey of the whole vast field of medicine with an extraordinary variety of detail. Dr. Osler drew upon wide culture and the availability of the best literature on the subjects ranging from ancient medicine to the modern time in 1913).
19. Fried S. Rush: revolution, madness, and the visionary doctor who became a founding father. New York: Crown Publishing; 2018.
20. Flexner A. Medical Education in the United States and Canada: a report to the Carnegie Foundation for the advancement of teaching. [full report Internet 1910] Accessed 12_04_2018: http://archive.carnegiefoundation.org/pdfs/elibrary/Carnegie_Flexner_Report.pdf.
21. Stannisch FW, Verhoef M. The Flexner report of 1910 and its impact on complementary and alternative medicine and psychiatry in North America in the 20th century. Evid Based Complement Altern Med. 2012:Article ID 647896.,, 10 pages. https://doi.org/10.1155/2012/647896.
22. Maeshiro R, Johnson I, Koo D, Parboosingh J, Carney JK, Gesundheit N, Ho ET, et al. Medical education for a healthier population: reflections on the Flexner report from a public health perspective. Academ Med. 2010;85(2):211–9.
23. Beck AI. The Flexner report and the standardization of American Medical Education. JAMA. 2004;291(17):2139–40.
24. Centers for Disease Control and Prevention. Deaths and mortality. 4 Mar 2013. http://www.cdc.gov/nchs/fastats/deaths.htm. Accessed 3 Apr 2013.
25. Holman H. Chronic disease-the need for a new clinical education. J Am Med Assoc. 2004;292(9):1057–9.
26. Council on Foods and Nutrition. Nutrition teaching in medical schools. JAMA. 1963;183(1):955–7.
27. Weinsier RL, Boker JR, Brooks CM, et al. The priorities for nutrition content in a medical school curriculum: a national consensus of medical educators. Am J Clin Nutr. 1989;50(4):707–12.
28. Nestle M, Baron RB. Nutrition in medical education: from accounting hours to measuring competence. JAMA Intern Med. 2014;174(6):843–4.
29. Morris NP. The neglect of nutrition in medical education: a firsthand look. JAMA Intern Med. 2014;174(6):841–2.
30. Shao A, MacKay D. A Commentary on the Nutrient-Chronic Disease Relationship and the New Paradigm of Evidence-Based Nutrition: a discussion regarding the relationship between predisposition to many chronic diseases and the diet, nutritional status, and lifestyle of an individual. Nat Med J. [internet December 2010;2(12)] Accessed 11/27/2018: https://www.naturalmedicinejournal.com/journal/2010-12/commentary-nutrient-chronic-disease-relationship-and-new-paradigm-evidence-based.
31. Dietary Guidelines Advisory Committee. Scientific report of the 2015 Dietary Guidelines Advisory Committee 2015. http://www.health.gov/dietaryguidelines/2015-scientific-report/. Accessed 25 Mar 2015.
32. Institute of Medicine. Crossing the quality chasm: a new health system for the 21st-century. Washington, DC: National Academies Press; 2001.
33. Frenk J, Chen L, Bhutta ZA, Cohen J, Crisp N, Evans T, et al. Health professionals for a new century: transforming education to strengthen health systems in an interdependent world. Lancet. 2010;376(9756):1923–58.
34. Institute of Medicine. Educating for the health team. Washington, DC: National Academy of Sciences; 1972.
35. Institute of Medicine. To err is human: building a safer health system. Washington, DC: National Academy Press; 2000.
36. Institute of Medicine. Health professions education: a bridge to quality. Washington, DC: The National Academies Press; 2003.
37. Framework for Action on Interprofessional Education & Collaborative Practice: Health Professions Network Nursing and Midwifery Office within the Department of Human Resources for Health. Framework for Action on Interprofessional Education & Collaborative Practice. World Health Organization, Department of Human Resources for Health, CH-1211 Geneva 27, Switzerland. 2010. http://www.who.int/hrh/nursing_midwifery/en/.
38. Interprofessional Collaborative: American Association of Colleges of Nursing, American Association of Colleges of Osteopathic Medicine, American Association of Colleges of Pharmacy, American Dental Education Association, Association of American Medical Colleges, Associations of Schools of Public Health. Core competencies for interprofessional collaborative practice. [Internet February 2011]. Accessed 11/27/2018: https://nebula.wsimg.com/3ee8a4b5b5f7ab794c742b14601d5f23?AccessKeyId=DC06780E69ED19E2B3A5&disposition=0&alloworigin=1.
39. Maskell J. The evolution of medicine: join the movement to solve chronic disease and fall back in love with medicine. Bedford: Knew Books Publishing; 2016.
40. Interprofessional Collaborative: American Association of Colleges of Nursing, American Association of Colleges of Osteopathic Medicine, American Association of Colleges of Pharmacy, American Dental Education Association, Association of American Medical Colleges, Associations of Schools of Public Health, American Association of Colleges of Podiatric Medicine, American Council of Academic Physical Therapy, American Occupational Therapy Association, American Psychological Association, Association of American Veterinary Medical Colleges, Association of Schools and Colleges of Optometry, Association of Schools of Allied Health Professions, Council on Social Work Education, Physician Assistant Education Association. Core competencies for interprofessional collaborative practice: 2016 update. [Internet 2016]. Accessed 11/27/2018: https://nebula.wsimg.com/2f68a39520b03336b41038c370497473?AccessKeyId=DC06780E69ED19E2B3A5&disposition=0&alloworigin=1.
41. Wagner LE, Evans RG, Noland D, Barkley R, Sullivan DK, Drisko JA. The next generation of dietitians: implementing dietetics education and practice in integrative medicine. J Am Coll Nutr. 2015;34(5):430–5. https://doi.org/10.1080/07315724.2014.979514.
42. Platt KD, Thompson AN, Lin P, Basu T, Linden A, Fendrick AM. Assessment of self-monitoring of blood glucose in individuals with type 2 diabetes not using insulin. JAMA Int Med,

Published online December 10. 2018; https://doi.org/10.1001/jamainternmed.2018.5700. Accessed 20 Dec 2018.
43. Virta Health. [Internet December 2018] Accessed 20 Dec 2018: https://www.virtahealth.com/.
44. Minich DM, Bland JS. Personalized lifestyle medicine: relevance for nutrition and lifestyle recommendations. Sci World J. 2013;2013:Article ID 129841,14 pages. https://doi.org/10.1155/2013/129841.
45. Kushner RF, Sorensen KW. Lifestyle medicine: the future of chronic disease management. Curr Opin Endocrinol Diabetes Obes. 2013;20(5):389–95. https://doi.org/10.1097/01.med.0000433056.76699.5d.
46. Sagner M, McNeila A, Puska P, Auffray C, Price ND, Hoode L, Lavie CJ, Hang ZG, et al. The P4 Health Spectrum – a predictive, preventive, personalized and participatory continuum for promoting healthspan. Prog Prevent Med. 2017;2(1):e0002. https://doi.org/10.1097/pp9.0000000000000002.
47. Alliance for Natural Health – UK. A blueprint for health sustainability in the UK: towards a bottom-up solution to the impending health and care crisis. [Internet December 2018] Accessed 18 Dec 2018: https://www.anhinternational.org/campaigns/health-sustainability-blueprint/
48. National Academies of Sciences, Engineering, and Medicine. Examining the impact of real-world evidence on medical product development: II. Practical approaches: proceedings of a workshop in brief. Washington, DC: The National Academies Press; 2018. https://doi.org/10.17226/25176.
49. National Academies of Sciences, Engineering, and Medicine 2011. Learning what works infrastructure required for comparative effectiveness research workshop summary. Institute of Medicine (US) Roundtable on Value & Science-Driven Health Care. Washington (DC): National Academies Press (US); 2011. ISBN-13: 978-0-309-12068-5 ISBN-10: 0-309-12068-3.
50. Godin S. Respect difficult problems. [Internet December 13, 2018] Accessed 20 Dec 2018: https://seths.blog/2018/12/respect-difficult-problems/

Influences of the Nutrition Transition on Chronic Disease

Sudha Raj

2.1 Nutrition Transition: A Model of Changing Dietary Patterns – 18

2.2 The Changing Face of Food Over Time and Space – 19

2.3 The Globalization of Food – 19
2.3.1 Changing Trajectory of Populations' Lifestyles – 20
2.3.2 Is the Nutrition Transition Experience Different Between the Developed and Developing World? – 21
2.3.3 A Growing Double Burden of Disease – 21
2.3.4 The Overlap Between Undernutrition and Overnutrition – 22
2.3.5 Metabolic Programming and NCDs – 22

2.4 Pathophysiological Consequences of NT – 23

2.5 Anthropogens and Chronic Inflammation – 23
2.5.1 Behavior Change for Positive Vitality – 24
2.5.2 Initiatives Addressing the Nutrition Transition – 24

2.6 Conclusion – 26

References – 26

© Springer Nature Switzerland AG 2020
D. Noland et al. (eds.), *Integrative and Functional Medical Nutrition Therapy*,
https://doi.org/10.1007/978-3-030-30730-1_2

2.1 Nutrition Transition: A Model of Changing Dietary Patterns

Nutrition transition (NT) is a complex conceptual framework. It is rooted in the historic processes of demographic and epidemiological transition theories that describe trajectories in fertility, mortality, disease patterns, and causes of morbidity and mortality, respectively [1]. Nutrition transition is described by Barry Popkin as evolutionary stages or patterns (◘ Fig. 2.1) that occur globally in a nonlinear fashion over time and space. Stage 1 is referred to as the "hunter-gatherer"; Stages 2 and 3 are termed "famine" and "receding famine"; Stages 4 and 5 are termed as "degenerative disease" and "behavioral change," respectively [please refer to ◘ Table 2.1] [2]. A majority of the writings on the subject draw attention to Stages 4 and 5, where dramatic shifts in diets, activity patterns, and health outcomes have been observed in different parts of the world over the last three centuries [3, 4]. Each stage is viewed as a pattern of food use and corresponding nutrition-related disease that affects individuals in the short term and populations in the long term [3]. In this view, each stage is a roadmap that highlights historical developments that occur in global societies during different time periods. Each stage is marked by specific patterns of food acquisition, food use, and physical activity with the ensuing stature, body composition, and nutrition-related health outcomes that affect individuals and, consequently, populations [5]. Transitory changes in food acquisition exist along a continuum, beginning with subsistence agriculture, progressing through industrialized agriculture to a globalized food system. Concomitant changes related to nutritional status lie along a spectrum of malnutrition ranging from undernutrition and nutritional deficiency to overnutrition

◘ Table 2.1 Nutrition transition stages

Stage	Description
Stage 1	Hunter-gatherer
Stage 2	Famine
Stage 3	Receding famine
Stage 4	Degenerative disease
Stage 5	Behavioral change

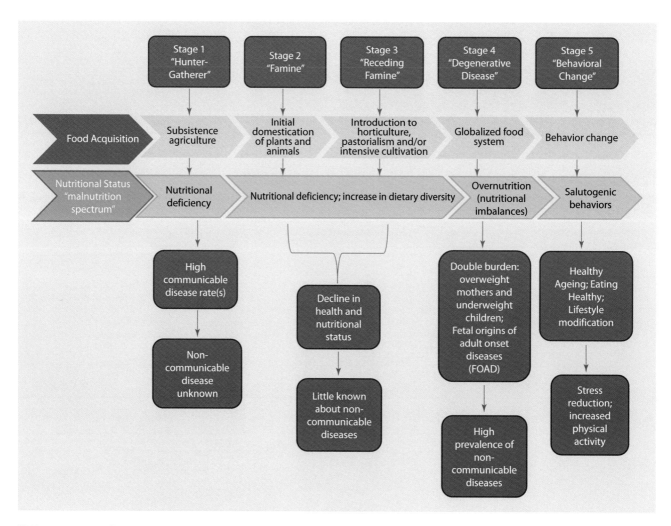

◘ Fig. 2.1 Stages of nutrition transition

and non-communicable diseases. The timing and degree to which these changes occur depends on the degree of development experienced by various global societies. Sociocultural and political factors, environmental conditions, degree of urbanization, industrialization, population demographics, migration dynamics, physical activity patterns, dietary acculturation, and the ability of the individual country's healthcare infrastructure to simultaneously handle infectious and chronic disease prevalence and incidence mediate these transitions and their health outcomes [6, 7].

2.2 The Changing Face of Food Over Time and Space

Stage 1 Stage 1 describes the myriad ways in which early global societies or "hunter-gatherers" engaged in food procurement and consumption practices. *Stages 2 and 3* mark the transition from a "hunting-gathering" society to a "food-producing" milieu. The latter was created with the advent of agriculture and the industrial revolution, along with the climatic and production fluctuations associated with these respective milestones. Regardless of location, dietary choices in *Stage 1* were limited to wild plant and animal foods indigenous to the geographic areas of inhabitance. For example, high-latitude populations such as the *Inuits* relied on seafood with almost no plant foods, while populations in the lower latitudes depended on fruits, nuts, roots, berries, and hunted or scavenged animal foods [8]. Stage 1 was noted for its (1) high degree of ecological availability, nutrient density, and need for dietary flexibility and (2) energy-intense food procurement activities. In general, humans experienced a shorter lifespan and non-communicable diseases were unknown [9]. However, at that time, there was a high probability of mortality from communicable diseases, inhospitable environmental conditions, and dangerous encounters [8].

Stage 2 The transition to *Stage 2* occurred with the initial domestication of plants and animals around 11,000 BCE, followed by the introduction of horticulture, pastoralism, and/or intensive cultivation.

Stage 3 In *Stage 3*, the domestication of plants and animals and the cultivation of one or more plant foods such as rice, wheat, or maize intensified. This shift is believed to have occurred in various locations around the globe at different times [8]. These activities provided populations with a plethora of locally harvested traditional or indigenous foods. Foods produced were native to the geographic area and not necessarily uniform across the globe; however, they were adequate in quantity and quality to meet the nutritional needs of the local population. For example, the *Inuits*' seafood-dominated diets were high in saturated fats with few carbohydrates, while the Balinese rice farmers [10], the Nigerian *Kofyars* [11], and the Papua New Guinea *Tsembaga* [12] relied on diets that were high in carbohydrates, including fiber, but low in fat, particularly saturated fats. Stability and abundance was ensured with the establishment and specialization of agriculture for the expanding populations. Cereals, legumes, and starchy foods became predominant dietary staples. Food was polysemic and served roles beyond sustenance. For instance, cultural anthropologists note that the first domesticated animals and plants were primarily bred and cultivated to be used as foods during feasts and for trading; they were symbols of power and status in addition to serving a hunger need [13]. Although the agriculture-oriented food system at that time ensured food security, it was limited in its resiliency to withstand shocks and vulnerabilities, such as droughts or floods that led to production swings and unpredictable availability [14]. *Stages 2 and 3* were marked by gradual decreases in diet diversity compared to *Stage 1*. Populations were subject to cycles of plenty and scarcity caused by migrations, political conflicts, overuse and degradation of natural resources, including soil fertility, arability, and unpredictable climatic conditions [8]. History is replete with instances of geographical regions subjected to such vulnerabilities resulting in famines, increased nutritional deficiencies, and a generalized weakened food supply with a growing dependence on foods from elsewhere [14]. Concomitantly, demographic changes marked by expanding populations and sociocultural and economic factors led to stratification by gender and social status, inequities, and unequal access to food and other basic necessities [15]. Historical records note the overall decline in health and nutritional status, citing maladies such as dental caries, anemia, and enamel defects, reduced stature and cortical bone thickness in various geographical locations [16, 17], yet through this period, little was known about the incidence and prevalence of non-communicable diseases.

2.3 The Globalization of Food

By the turn of the twentieth century, food procurement no longer remained an isolated activity that fostered self-reliance and kinship. Instead, the era's prevailing political, socioeconomic, and technological environments influenced food-related activities. For instance, the industrial revolution, rise of mechanization, new seed varieties, and the introduction of synthetic fertilizers caused massive shifts in the production, processing, distribution, and consumption of foods. The Green Revolution in South Asia in the 1960s and 1970s was credited as a success for its increased global yields; however, the following decades witnessed its consequences in the form of greater socioeconomic divides, poverty, catastrophic environmental degradation, and farmer suicides [18]. Parallel advancements in medicine; immunizations; antibiotics; sanitation and access to safe, good-quality water; the discovery of micronutrients; and, more recently, the expansion of healthcare technologies, e.g., telemedicine [19, 20], have since facilitated gradual improvements in the public's health.

Agriculture in the 1980s became a "global phenomenon" as its focus shifted from one of "traditional or indigenous foods" to a "cash crop" economy. Sugarcane, palm oil, wheat, soy, and animal source foods became agricultural trading commodities moving to the forefront to meet processing demands [21]. At present, the scope of the food system has broadened from one of ensuring adequate food supplies to that of a financial enterprise governed by consolidation, economies of scale, money, and markets [22–24]. Selective breeding techniques, biotechnology, genetic engineering, fortification, enrichment strategies, expansion of the functional foods and nutraceutical categories, novel food processing and packaging technologies, and the introduction of information and communication technologies (ICTs) have facilitated increased yields of agricultural commodities, product innovation, diversification, and volume production of value-added processed and ultra-processed foods [25]. Precision agriculture or site-specific farming that combines global positioning systems (GPS) and geographic information systems (GIS) to monitor climatic conditions and provide farmers site-specific information is one such innovation in the agriculture realm [26, 27]. Farmers use this information to efficiently utilize and monitor their agricultural environments for crop planning, soil sampling, pest control, application, and yield mapping, mitigating several climatic fluctuations of previous centuries.

Advances in retailing, marketing, advertising, and the introduction of formalized markets provide a platform to inform and sell consumers the cornucopia of processed foods with different qualitative and nutritional attributes. Improved transportation, the development of domestic and international markets fueled by the expansion of global trade, and the introduction of high speed connectivity ensure better global food systems management while guaranteeing food access to large populations worldwide [27]. Continued urbanization, expanding populations consequent to global migration, and natural increases provide the necessary consumer base that sustains demand for these products. Further, the liberalization of trade policies is a key facet of the twenty-first-century globalization phenomenon facilitating trade across national borders. Low- and middle-income countries (LMICs), such as China, India, and Brazil, recognize the incentives and economic advantages of market-oriented agricultural policies [22] and have gradually liberalized their agricultural markets both domestically and internationally. Imports, exports, consolidation and merger activity in the food system, foreign direct investments (FDI), and the establishment of transnational food and beverage companies [28] have boosted economic developments in their food sectors [22]. Transnational entities such as Coca-Cola, PepsiCo, Nestle, Kraft, Carrefour, Danone, and Walmart have consequently expanded their markets and gained a significant foothold in emerging economies such as China, India, Brazil, and Mexico, thereby changing the scope and nature of these countries' food systems [29, 30].

2.3.1 Changing Trajectory of Populations' Lifestyles

The evolution of the globalized food system has proven to be a double-edged sword. Depending on the degree, pace of industrialization and globalization, different parts of the globe have experienced a mixed bag of positive and negative consequences and health outcomes [31, 32]. On the one hand, enormous economic developments and benefits to health and nutritional status have accrued in industrialized societies and the LMICs transitioning to *Stage 4* of the NT process [3]. Incomes have risen for many, poverty has declined, standards of living have improved, healthcare is more accessible, and these societies have seen a wider variety of processed foods, energy-saving devices, and technologies. Easy access to a globalized food system has fostered increased and varied consumption patterns worldwide, changing the trajectory of cultural foodways, value systems, human behaviors, and lifestyles of global populations [33, 34].

On the other hand, myriad challenges signaled by nutritional imbalances consequent to dietary pattern shifts, occupation-related stress, sharp divides in socioeconomic status, inequities in healthcare access, and chronic disease trajectories have risen for selected population segments such as the poor, women, and children [35]. A major consequence is the increased reliance on and dietary convergence of food components such as fats and oils, sugars, animal products, and processed and prepared products, all at the expense of traditional, indigenous foods [36]. High-, middle-, and low-income countries experience concomitant dietary pattern changes, such as increased portion sizes, eating out, and snacking frequency at varying rates [37, 38]. These patterns, coupled with a high degree of dietary convergence, are collectively referred to as the "Westernized diet or dietary patterns" [21, 36]. For instance, in the United States between 1977 and 2006, the overall energy intake from sweetened beverages increased by 135% while that from milk decreased by 38% [39, 40]. Sugar-sweetened beverages contributed a substantial amount of energy to the diet of Australian children, with mean intakes ranging from 4% in children 2–3 years of age to 7.5% in adolescents [41]. Mexico had the largest per capita (163 liters) intake of soft drinks in 2011 [42]. A market data analysis between 2000 and 2013 for ultra-processed food consumption by Euromonitor International for four lower-middle-income, three upper-middle-income, and five high-income Asian countries raised concerns about the growth in the carbonated soft drinks sector in the LMICs while their sales were declining or stagnating in high-income countries such as the United States [34]. Specifically, soft drink sales and volume were reported to be high in Thailand and in the Philippines, while sales of oils and fats were high in Malaysia.

Since the mid-1950s, developed countries have used specialized technologies such as oilseed breeding and oil extraction techniques to increase seed oil content, contributing to the widespread availability of cheap vegetable oils such as soybean, palm, and canola oils in LMICs of Asia and Latin America [43]. Between 1985 and 2010, vegetable oil consumption increased three- to sixfold in LMICs such as India and China. In India, the consumption of edible vegetable oils in urban and rural areas rose from 24 g/day to 36 g/day and 36 g/day to 48 g/day, respectively [44]. Data from China indicate that between 1994 and 2004, edible oil production increased nearly twofold in China, and 83% of the population had cooking oil intakes over 28 g/day by 2010 [45]. Shifts in easy accessibility to vegetable oil [46], increasing market concentration in other food sectors, e.g., the beverage industry, coupled with growing fast food sales [21] have and continue to contribute to increased dietary fat intakes and global obesity rates.

2.3.2 Is the Nutrition Transition Experience Different Between the Developed and Developing World?

Despite varied and limited global epidemiological data on diet and physical activity, certain key differences exist between developed and developing countries. Developed countries, like the United States, carry a large burden of heart disease, cancer, diabetes, chronic pulmonary, and mental disease with a lower proportion of infectious diseases [47].

Stage 4 Since the 1980s, these countries have experienced *Stage 4* of the transition phenomenon with massive shifts in dietary and physical activity patterns resulting in body composition changes [32]. Adiposity, marked by a high BMI with the associated chronic inflammation and oxidative stress, is one of the well-documented risk factors at the nexus of the NCD concerns in this part of the world. Considerable scientific evidence reports that the high glycemic load, undesirable dietary fatty acid composition, altered macro- and micronutrient status, disturbed acid-base balance, unbalanced sodium and potassium ratios, and decreased dietary fiber content exacerbate the morbidities associated with NCDs [36].

Increasing dietary convergence to a more processed, westernized diet, a decline in traditional, indigenous food consumption, sedentary lifestyles, changes in average stature and body composition, and NCD morbidity and mortality are reflective of the LMICs transition phenomenon [21]. The complexity of these transitory changes and their contextual occurrence are noted by Corinna Hawkes as particularly pertinent to the nutrition-related non-communicable diseases or *Stage 4* in the nutrition transition spectrum. In a recent ecological analysis using multiple regression and cluster analysis on a sample of 98 countries across the globe, Oggioni et al. studied the association of obesity and diabetes with the agricultural, transitional, and Westernized dietary patterns. The Westernized dietary pattern had a direct dose-response association with diabetes prevalence, while the agricultural diet had the lowest prevalence of obesity and diabetes association [48]. Yet the NT experience of the LMICs is different and unique for several reasons.

First, the NT and disease pattern shifts in the LMICs of Asia, Africa, the Middle East, Latin America, and Oceania are occurring at an accelerated pace. Compared to the nearly five decades of transitory changes in the industrialized world, the LMIC transition has occurred over the last two decades. Further, the simultaneous population expansion consequent to a demographic transition from a high fertility and mortality pattern to one of low fertility and mortality has exacerbated the situation [49].

Second, according to Popkin [31, 32], there is considerable intra- and inter-country heterogeneity in transition patterns and health outcomes within and between different population segments. Ecological, internationalization, political, technological, and socioeconomic climates that prevail in the individual countries and/or regions during the transition phases are responsible for this heterogeneity. For instance, within Asia, China experienced great shifts in dietary and physical activity patterns between 1985 and 2000, while India's transition is still in the early stages and gaining momentum in urban areas [50, 51]. While the Chinese experience a high burden of hypertension followed by diabetes and cardiovascular disease, diabetes rates and susceptibility to cardiovascular disease in the South Asian region have skyrocketed in the last decade. India ranks highest in the world for diabetes incidence; Bangladesh accounts for 40% of all diabetes among the least developed countries and 10% of Pakistanis suffer from diabetes [52–54].

Third, obesity rates in many LMICs are two to five times greater than those of the industrialized countries with stark changes in specific subsets of the population – the poor, women, and children [55]. The UNICEF–World Bank–WHO group reports that the overweight prevalence in children is on the rise; between 1990 and 2014, the numbers have risen from 31 million to 41 million [56]. Since 2000, obesity rates for children under 5 years of age have risen by more than 50% in Africa and 40% in Asia [57]. Myriad reasons are fueling the obesity pandemic across the lifespan, increasing vulnerability and impacting health outcomes. These include unhealthy food choices, energy imbalances caused by sedentarism, chronic stress, and genetic and ethnic body composition predispositions that favors adiposity [58, 59]. Further exposure to environmental toxins, inequities in income and healthcare, healthcare systems that are unable to simultaneously handle chronic diseases caused by under- and overnutrition, and epigenetic fetal programming changes only make the situation worse [59].

2.3.3 A Growing Double Burden of Disease

The rising prevalence of obesity promotes a misconception that communicable and nutritional deficiency diseases are eradicated and supplanted by obesity and NCDs. However, this is far from reality [60]. Instead, global epidemiological data reflect the concurrent existence of undernutrition and overnutrition referred to as *the double burden of disease*. Compared to the developed world, the LMICs experience disproportionately large problems given the rapid globalization, economic growth, and population expansion these countries face [61, 62]. LMICs have always experienced a high incidence and prevalence of undernutrition, infectious and deficiency diseases driven by socioeconomic challenges, poverty, food insecurity, famine, and poor healthcare quality, access, and utilization. However, with nutrition and epidemiological transitions, LMICs face the additional challenge of a rapid increase in obesity and overweight status. Accompanied by pathophysiological metabolic risk mediators such as hs-CRP, IL-6, obesity increases vulnerability to NCDs across the lifespan [63]. According to the annual WHO report, nearly 40 million people succumb to NCDs, accounting for 70% of all global deaths. Approximately 17 million die before age 70 and 87% of these premature deaths occurred in LMICs [64]. The metabolic syndrome, in which abdominal obesity and insulin resistance play a central role, is associated with a doubling of cardiovascular disease risk [65].

In a recent pooled analysis of 1698 population-based studies consisting of over 19 million male and female participants in over 180 countries, the prevalence of overweight and underweight using body mass indexes (BMIs) was studied over the 1975–2014 period. BMIs increased from 21.3 (1975) to 24.2 (2014) in men and from 21.7 to 24.6 in women, respectively [66], with some regions experiencing more accelerated rates than others. For instance, regional BMIs were highest for men (21.4–29.2) and (21.8–32.8) women in South Asia and the Polynesian islands, respectively. The prevalence of being underweight globally decreased from 13.8% to 8.8% in men and 14.6% to 9.7% in women; however, South Asia had the highest prevalence of underweight individuals at 23.4% and 24% in men and women, respectively. In 2014, of the 667 million children under five in the world, 159 million were stunted and 50 million were wasted [56].

Short-term consequences of the double burden of disease include malnutrition-related stunting and premature child deaths and compromised immunity, physical development, and cognitive abilities; long-term consequences include obesity, NCD-associated morbidities, and mortality. Ultimately this results in a lower productivity and quality of life in adulthood and higher healthcare costs [67]. The impact of these changes is variable between urban and rural populations, escalating with increased migration and changing food environments. The healthcare systems in LMICs lack the capacity and governance to address the double burden efficiently and adequately and the costs of treating NCDs are rising, consuming larger proportions of health budgets in LMICs. Initiatives such as the Millennium Development Goals, which were originally designed to combat undernutrition, are underway and need to keep pace with the NCD trajectory [68].

2.3.4 The Overlap Between Undernutrition and Overnutrition

The causes and consequences of both undernutrition and overnutrition as distinct health status entities are well described [48]; however, recent WHO reports and research from Africa, Mexico, the Middle East, and South Asia [69] highlight overlap of these two conditions, sometimes within the same household or individual. While the poor and uneducated are disproportionately affected, of particular concern are the long-term consequences in women at the population level [70–72]. At the individual level, the common manifestation of the double burden is the dual occurrence of energy imbalance and micronutrient deficiency. For example, between 1990 and 2005, obesity rates in West Africa increased by 114%, with more women affected than men [73]. In urban Burkina Faso, 73 out of 310 apparently healthy adults had one marker of overnutrition concurrently with at least one nutritional deficiency [61]. Studies using nationally representative data from three developing countries, Mexico, Peru, and Egypt, showed that overweight women were deficient in iron and other micronutrients such as vitamin A [74]. Reports from Sub-Saharan Africa note that while obesity is more prevalent among the affluent, more than 20% of women have BMIs of less than 18.5, 57.1% are anemic, and 18.5% are deficient in vitamin A [75, 76]. At the household level, the most common occurrence is a stunted child coexisting with an overweight mother [77]. The WHO classifies a country as experiencing a double burden of disease when at least 30% of children under 5 years old are stunted with an age-adjusted overweight rate for females above 25% [69].

Researchers interested in the fetal origins of adult-onset disease view undernutrition and overnutrition as being linked at the level of developmental programming and metabolic adaptation in the fetus, with important health consequences as individuals age and their environments change [78, 79]. In this regard, the double burden of disease in LMICs poses a threat both as a metabolic programming consequence of maternal undernutrition and a cause for NCDs in adult life. Epidemiologists attribute the NCD incidence in LMICs to perinatal and postnatal influences and reduced birth weight [80–82]. While public health focus continues to center on the mitigation of undernutrition [83], there is consensus that undernutrition, obesity, and NCDs are linked through the processes of developmental programming and metabolic adaptation that require immediate attention and action [84].

2.3.5 Metabolic Programming and NCDs

Developmental programming and metabolic adaptation are the major principles underlying the fetal origins of adult-onset diseases (FOAD) hypothesis, also known as the developmental origins of adult disease proposed by Barker in 1986 [85]. According to Barker, "adverse influences early in development and particularly during intrauterine life can result in permanent changes in physiology and metabolism, leading to an increased disease risk in adulthood" [86]. Prenatal insults include acute and chronic nutritional deficiencies, environmental influences such as smoking, exposure to endocrine disruptors, limitations in maternal physical stature, exposure to steroid hormone excesses, and maternal physiological and psychosomatic stress marked by elevated glucocorticoid hormones. These adverse influences pose several constraints during critical periods of fetal development, consequently altering tissue and organ development, structure and function which are also referred to as developmental programming [75, 87]. For instance, lowered intra-uterine protein availability can modify pancreatic islet cell proliferation, causing future alterations in glucose homeostasis [88]. Undesirable fetal developmental programming in a compromised nutritional environment results in dysfunctional metabolic capacity and physiological function as future consequences. For instance, in utero stresses alter mitochondrial activity that influence the function of oxidative phosphorylation linked to skeletal muscle insulin resistance in type 2 diabetics [89]. The ensuing metabolic maladaptations in early life owing to inadequate maternal nutrition reflect the development of a thrifty phenotype with an altered metabolism that is beneficial to the survival of the fetus and anticipatory of a similar future postnatal environment. However, a mismatch occurs when there is a discrepancy in this expectation upon exposure to a food and a nutrient-rich environment, resulting in a greater susceptibility eventually culminating in obesity and metabolic disease [90]. While epidemiological and animal studies have garnered a large body of evidence, the mechanisms are yet to be elucidated. Suggested mechanisms include permanent alterations in cell numbers affecting structure and function of organs, inheritable epigenetic changes that affect DNA methylation patterns consequent to an altered maternal nutrient supply. These alterations potentially mediate the metabolic priming process very subtly through gene expression modulations [87, 91].

2.4 Pathophysiological Consequences of NT

Just as the germ theory provided a monocausal focus for infectious and communicable diseases [92], modern lifestyles are proposed as a major contributor to the NCD pandemic. Egger and Dixon [93], in a recent review on chronic disease determinants, describe the role of "anthropogens" in the etiology of meta-inflammation [94]. Anthropogens are man-made environments, their by-products and/or lifestyles encouraged by these, some of which may be detrimental to health [93]. Meta-inflammation is a low-grade, chronic systemic inflammation linked to a dysfunctional immune response. It differs from the classic, acute inflammatory response to infection and injury initiated by the body's innate immune system [95] in the following ways: first, chronic inflammation is low grade with systemic effects causing small rises in immune system markers such as proinflammatory cytokines, e.g., TNF alpha, hs-CRP, and interleukin 6; second, it is persistent, resulting in a chronic activation of the immune and neuroendocrine systems in an effort to bring about homeostasis; and third, it perpetuates a chronic, dysmetabolic state induced by anthropogens [93]. Although obesity is thought of as a prerequisite [96, 97], several behaviors linked to post-industrial, globalized lifestyles, e.g., poor diets, inactivity, stress, inadequate sleep, occupation, prescription and nonprescription drugs, smoking, toxic exposures, dysfunctional relationships, the social factors that encourage such lifestyle patterns over time, and the obesogenic environments in which they occur, e.g., technology, environment, occupation, air pollution [98], and endocrine-disrupting chemicals [99] are also pro-inflammatory. Viewed in the broader social, economic, and environmental contexts, anthropogens are extremely detrimental to human health. Anthropogens foster a chronic stress milieu, promoting a low-level, sustained chronic inflammation response. The innate immune system and the central stress axes are constantly activated without opportunity for the resolution and termination of the natural inflammatory response as seen in the acute stress response [95]. The proinflammatory cytokines either cause or are caused by dysmetabolic responses, leading to core imbalances in one or more of the body's physiological systems [100]. For example, unhealthy, calorie-dense, highly processed diets increase NCD risk, while feasting, cultural foodways, and changing food habits can act as moderators to enhance or decrease risk; chronic stress determined by an individual's coping capacity leads to elevated adrenocortical hormones and activation of the hypothalamic–pituitary–adrenal axis resulting in a host of vascular, metabolic, and inflammatory processes; obesogenic endocrine-disrupting chemicals (EDCs) and persistent organic pollutants (POPs) are documented to cause significant physiological and behavioral changes, e.g., increased appetite, ultimately contributing to obesity. Individual responses to the anthropogens are not uniform and dictated by genetic and epigenetic predispositions, as well as fetal programming [93, 101].

2.5 Anthropogens and Chronic Inflammation

Inflammation is the body's natural response to damage after injury or pathogen invasion. It is a self-limiting process consisting of an initiation, a resolution, and a termination phase. Its action is controlled by the synergistic roles of the innate immune system, the sympathetic nervous system, and the hypothalamus pituitary adrenal axis [95, 102]. The stress axes are involved in the response primarily through the action of norepinephrine and cortisol by modulating insulin and cortisol sensitivity. The efficiency of the response depends on the capability of the glucocorticoid and catecholamine receptors of the innate immune system. The inflammatory response is initiated by polymorphonuclear neutrophils that generate proinflammatory cytokines like leukotriene B4 and prostaglandins from arachidonic acid with the help of lipoxygenase 5 and cyclooxygenase 2 enzymes. While the leukotrienes exert their strong chemo-toxic response to the invading pathogen or stimulus, the prostaglandins regulate the switch to the second or resolution phase when their concentration equals that of the leukotrienes primarily by limiting the lipoxygenase enzyme activity. This phase is marked by high anti-inflammatory activity as seen in the production of specialized pro-resolving mediators (SPMs) such as *lipoxins, resolvins, protectins, and maresins* from eicosapentaenoic acid (EPA) and docosahexaenoic acid (DHA) [103, 104]. SPMs are described recently to be involved in myriad mechanisms that promote the resolution and termination of the inflammatory response, such as switching off the stress axes, enhancing microbial clearance through generation of noncytotoxic macrophages. Both omega-6 and omega-3 fatty acids play critical roles in their biosynthesis and are expected to be of potential therapeutic value in microbial defense, pain, organ protection and tissue regeneration, wound healing, cancer, reproduction, and cognition [105–107].

2.5.1 Behavior Change for Positive Vitality

Stage 5 A growing awareness of the negative consequences of NCDs and/or the diagnosis of a chronic disorder such as diabetes prompts motivated individuals and communities to try and adopt a healthy lifestyle through behavior modification strategies. Stage 5, referred to as the behavior change stage, encompasses the various initiatives at the individual and population level that prevent or delay degenerative diseases and prolong life through healthy aging. It includes the practice of salutogenic behaviors that promote robust lifestyles, such as eating healthy foods, stress reduction, and increased physical activity. Healthy aging is "the condition of being alive, while having a highly preserved functioning of the metabolic, hormonal and neuroendocrine control systems at the organ, tissue and molecular levels" [108, 109]. Maintaining functional organ reserves and biological resiliency or the ability to adapt and/or withstand environmental stressors are at the core of the healthy aging concept. Furthermore, it is well recognized that healthy aging can be successful, provided a holistic multidimensional approach that addresses environmental, physiological, and psychological factors and healthy lifestyles is taken. For instance, a multifactorial behavior modification approach that includes programmed exercise activities, stress-reduction techniques like yoga and meditation, eating healthy foods like fruits and vegetables, and purchasing organic foods due to environmental concerns [110] can independently and synergistically contribute toward building and sustaining biological resiliency over the course of a lifetime. Personalized lifestyle medicine is one such approach that builds on the healthy aging concept using a combination of functional medicine principles [111] and innovative emerging *omic* technologies such as genomics, epigenetics, and diagnostic assessment tools [100]. Identifying the root cause of the disease rather than a symptom resolution approach, assessing for antecedents, triggers, and mediators, and keeping the biological uniqueness of the individual in perspective are major principles of functional medicine [112, 113]. Specialized assessment techniques such as nutrition-focused physical assessments and other functional diagnostic assessments, such as organic acid testing for assessing mitochondrial health and stool tests for gut permeability, are then employed to analyze an individual's NCD health metrics. Genomic analysis and testing [114, 115], molecular diagnostics, and tailored biomarkers are examples of specialized functional diagnostic techniques that provide personalized assessments and advice so that treatments can be individualized. They have the potential to improve health outcomes by developing sustainable lifestyle medicine-oriented strategies, empowering individuals so that they can be in control of their health [116–118] as well as facilitate the identification of clinical imbalances at an earlier stage before they become pathological.

2.5.2 Initiatives Addressing the Nutrition Transition

Nutrition transition underscores the need for all nations to build healthcare capacity and infrastructure, regardless of their economic status, to address the dual burden of malnutrition. This requires immediate action which necessitates healthcare and nutrition policies in developing countries to offer a more comprehensive integrated approach to address both undernutrition and overnutrition simultaneously. In a recent review of nutrition policies of 36 low-income, 48 low- and middle-income, and 55 upper-middle-income countries, only 39.6% had nutrition policies that addressed the dual burden of disease, while a majority of the countries' nutrition policies continue to focus on the mitigation of undernutrition [119]. The study further pointed out that having a nutrition policy in place did not necessarily translate into positive health outcomes. This highlights two important issues.

The first is the importance of a strong governance to facilitate the translation of the nation's policies into sustainable action and initiatives that are context-specific to individual countries [119]. Strong governance entails (1) stewardship; (2) development of surveillance systems that facilitate early detection of NCDs; (3) collaborative partnerships between stakeholders such as government, industry, health sector, consumers, and policymakers; (4) learning from previous experiences in other countries; and (5) strengthening the evidence base to enhance and support the design and implementation of health-promoting interventions [120–123].

The second is the need to focus policies that take on a life course approach to a healthy diet and physical activity lifestyle using culturally appropriate methods and messages initiated early in life [124]. This is the focus of the World Cancer Research Fund International's ongoing NOURISHING framework [125] for obesity prevention in 11 high-, middle-, and low-income countries (◘ Fig. 2.2). The policy recommendations span across three critical factors that impact healthy eating and physical activity: food environment, food system, behavior change, and communication. This framework informs governments in these countries regarding appropriate interventions, such as restricting marketing of unhealthy foods to children, salt reduction strategies, using existing policies designed for combating micronutrient deficiency to simultaneously address the obesity problem by promoting local fruits and vegetables via local farm networks. Another major initiative is the Global Strategy on Diet, Physical activity, and Health action plan aimed at prevention and control of NCDs between 2013 and 2020. The action plan resulted from a multi-stakeholder consultation process consisting of the WHO member states, relevant UN agencies, funds and programs, international financial institutions,

Fig. 2.2 Nourishing framework. (Used with permission from World Cancer Research Fund International. Wcrforg. 2017. Available at: ► http://www.wcrf.org/int/policy/nourishing-framework)

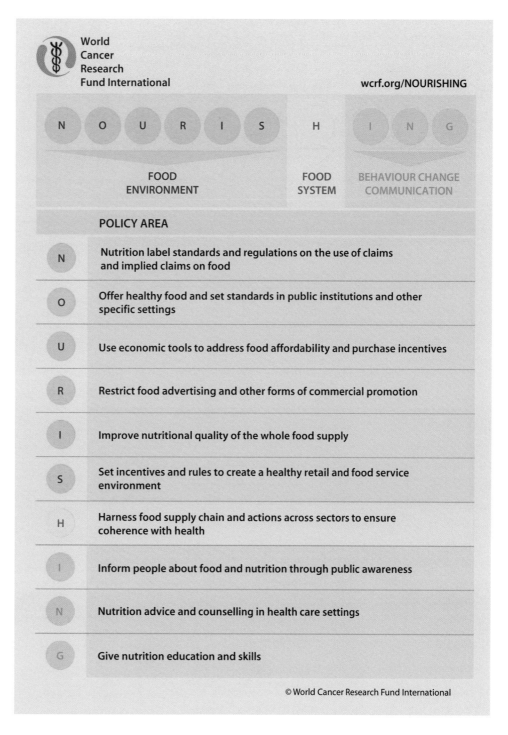

banks, NGOs, professionals, academicians, the civil society, and the private sector [126]. It operationalizes the tasks articulated in the Political Declaration of the General Assembly's initiative on the prevention and control of NCDs. The action plan focuses on four major chronic diseases, namely, cardiovascular disease, cancer, chronic obstructive pulmonary disease, and diabetes; several chronic conditions, e.g., mental illness; disabilities; and four shared behavioral risk factors – tobacco use, unhealthy diet, physical inactivity, and harmful use of alcohol [126–128]. While the WHO leads and coordinates the plan through engagement, international cooperation, and collaboration, the individual governments are responsible for action and monitoring. The overarching principles are as follows:

1. Highest quality and standard of health is a fundamental human right.
2. Social determinants of health should be addressed to create equitable, productive healthy societies.
3. Member governments and international agencies should create alliances and foster high-level multi-sectoral engagement.
4. A life course approach starting at preconception moving through healthy aging in later life will lead to sustainable health outcomes.

5. Communities should be engaged at all stages of planning, implementing, evaluating, and monitoring to ensure empowerment and motivation for sustained success.
6. Evidence-based and/or experiential-based best practices that are cost-effective, affordable, and culturally congruent should be implemented.
7. All segments of the society should have access to safe, affordable, effective, and quality-based promotive, preventive, curative, and rehabilitative services.

2.6 Conclusion

It is apparent that NCDs are the most pressing challenge at the present time. They pose a threat to healthy aging for global communities, irrespective of their stage of socioeconomic development. At the same time, it is important to not lose sight of the communicable and nutritional deficiency diseases that persist across large segments of the globe. It is important to recognize that malnutrition at either end of the spectrum is distinctive in etiology and physiological manifestations yet, in fact two sides of the same coin. Undernutrition and overnutrition share certain common risk factors and are multifactorial in their etiology, and the need for their control and prevention cannot be underscored.

References

1. Omran AR. The epidemiological transition: a theory of the epidemiology of population change. Milbank Q. 1971;49:509–38.
2. Popkin BM. An overview on the nutrition transition and its health implications: the Bellagio meeting. Public Health Nutr. 2002;5(1A):93–103.
3. Popkin BM. The nutrition transition in low income countries: an emerging crisis. Nutr Rev. 1994;52(9):285–98.
4. Popkin BM. Global changes in diet and activity patterns as drivers of the nutrition transition. Nestle Nutr Workshop Ser Pediatr Program. 2009;63:1–10;discussion 10–4, 259–68. https://doi.org/10.1159/000209967.
5. Popkin BM, Adair LS, Ng SW. Global nutrition transition and the pandemic of obesity in developing countries. Nutr Rev. 2012;70(1):3–21. https://doi.org/10.1111/j.1753-4887.2011.00456.x.
6. Popkin BM. Global dynamics: the world is shifting rapidly toward a diet linked with non-communicable diseases. Am J Clin Nutr. 2006;84(2):289–98.
7. Kennedy G, Nantel G, Shetty P. Globalization of food systems in developing countries: a synthesis of case studies. FAO Food Nutr Pap. 2004;83:1–24.
8. Bender RL, Dufour DL. Nutrition transitions: a view from anthropology. In: Dufour DL, Goodman AH, Pelto GH, editors. Nutritional anthropology biocultural perspectives on food and nutrition. 2nd ed. New York: Oxford University Press; 2012. p. 372–82.
9. Russel LB. Chronic health effects of dispossession and dietary change: lessons from North American hunter-gatherers. Med Anthropol. 1999;18(2):135–61.
10. Lansing S. Priests and programmers: technologies of power in the Engineered Landscape in Bali. Princeton: Princeton University Press; 1991. p. 183.
11. Netting RM, Stone MP, Stone GD. Kofyar cash cropping: choice and change in indigenous agricultural development. In: Bates DG, Lees SH, editors. Case studies in human ecology. New York: Plenum Press; 1996. p. 327–48.
12. Rappaport RA. Pigs for the ancestors: ritual in the ecology of a new Guinea people. 2nd ed. New Haven: Yale University Press; 1968. p. 501.
13. Hayden B. Were luxury foods the first domesticates? Ethnoarchaeological perspectives from Southeast Asia. World Archaeol. 2003;34:458–69.
14. Cox GW. The ecology of famine: an overview. Ecol Food Nutr. 1978;6(4):207–20.
15. Dietler M. Feasts and commensal politics in the political economy: food, power and status in pre-historic Europe. In: Weissner P, Schieffenhovel W, editors. Food and the status quest: an interdisciplinary perspective. Providence: Berghahn Books; 1996. p. 87–125.
16. Larsen CS. Biological changes in human populations with agriculture. Ann Rev Anthropol. 1995;24:185–213.
17. Gurven M, Kaplan H. Longevity among hunter-gatherers: a cross-cultural examination. Popul Dev Rev. 2007;33(2):321–65.
18. Sebby K. The green revolution of the 1960's and its impact on small farmers in India [undergraduate thesis]. Lincoln: University of Nebraska; 2010. p. 1–25.
19. Vimarlund V, Le Rouge C. Barriers and opportunities to the widespread adoption of telemedicine: a bi-country evaluation. Stud Health Technol Inform. 2013;192:933.
20. Centers for Disease Control and Prevention. Ten great public health achievements- United States- 2001-2010. MMWR. 2011;60(19):619–623.
21. Hawkes C. Uneven dietary development: linking the policies and processes of globalization with the nutrition transition, obesity and diet related chronic diseases. Global Health. 2006;2:4. https://doi.org/10.1186/1744-8603-2-4.
22. Hawkes C. The role of foreign direct investment in the nutrition transition. Public Health Nutr. 2005;8(4):357–65.
23. Mody A. Is FDI integrating the world economy? World Econ. 2004;27(8):1195–222.
24. Dube L, Pingali P, Webb P. Paths of convergence for agriculture, health and wealth. Proc Natl Acad Sci U S A. 2012;109(31):12294–301. https://doi.org/10.1073/pnas.0912951109.
25. Hamid El Bilali Allahyari MS. Transition towards sustainability in agriculture and food systems: role of information and communication technologies. Inf Process Agric. 2018; https://doi.org/10.1016/j.inpa.2018.06.006.
26. GPS.gov: Agricultural applications. http://www.gps.gov/applications/agriculture. Last accessed 21 Apr 2017.
27. Hammonds T. Use of GIS in Agriculture April 3, 2017. https://smallfarms.cornell.edu/2017/04/use-of-gis/.
28. Hawkes C. Marketing activities of global soft drink and fast food companies in emerging markets: a review. In: Globalization, diets and non-communicable diseases. Geneva: World Health Organization; 2002.
29. Monteiro CA, Cannon G. The impact of transnational "big food" companies on the south: a view from Brazil. PLoS Med. 2012;9(7):e1001252. https://doi.org/10.1371/journal.pmed.1001252.
30. Hawkes C, Chopra M, Friel S. Globalization, trade and the nutrition transition. In: Labonté R, Schrecker T, Packer C, Runnels V, editors. Globalization and health: pathways, evidence and policy. New York: Routledge; 2009. p. 235–62.
31. Popkin BM, Adair LS, Ng SW. Now and then: the global nutrition transition: the pandemic of obesity in developing countries. Nutr Rev. 2012;70(1):3–21. https://doi.org/10.1111/j.1753-4887.2011.00456x.
32. Popkin BM, Gordon-Larsen P. The nutrition transition: worldwide obesity dynamics and their determinants. Int J of Obes Relat Metab Disord. 2004;28:S2–9.
33. Drewnowski A, Popkin BM. The nutrition transition: new trends in the global diet. Nutr Rev. 1997;55(2):31–43.

34. Baker P, Friel S. Food systems transformations, ultra-processed food markets and the nutrition transition in Asia. Glob Health. 2016;12(1):80. https://doi.org/10.1186/s12992-016-0223-3.

35. Gersh BJ, Sliwa K, Mayosi BM, Yusuf S. *Novel therapeutic concepts*: the epidemic of cardiovascular disease in the developing world: global implications. Eur Heart J. 2010;31(6):642–8. https://doi.org/10.1093/eurheartj/ehq030.

36. Cordain L, Eaton SB, Sebastian A, Mann N, Lindeberg S, Watkins BA, et al. Origins and evolution of the Western diet: health implications for the 21st century. Am J Clin Nutr. 2005;81(2):341–54.

37. Popkin BM. The nutrition transition: an overview of world patterns of change. Nutr Rev. 2004;62(7 Pt 2):S140–3. https://doi.org/10.1301/nr.2004.jul.S140–S143.

38. Popkin BM, Doak CM. The obesity epidemic is a worldwide phenomenon. Nutr Rev. 1998;56:106–14.

39. Nielsen SJ, Popkin BM. Changes in beverage intake between 1977 and 2001. Am J Prev Med. 2004;27(3):205–10.

40. Duffey KJ, Popkin BM. Shifts in patterns and consumption of beverages between 1965 and 2002. Obesity (Silver Spring). 2007;15(11):2739–47.

41. Hafekost K, Mitrou F, Lawrence D, Zubrick SR. Sugar sweetened beverage consumption by Australian children: implications for public health strategy. BMC Public Health. 2011;11:950. https://doi.org/10.1186/1471-2458-11-950.

42. Colchero MA, Popkin BM, Rivera JA, Ng SW. Beverage purchases from stores in Mexico under the excise tax on sugar sweetened beverages: observational study. BMJ. 2016;352:h 6704. https://doi.org/10.1136/bmj.h6704.

43. Hawkes C. The influence of trade liberalization and global dietary change: the case of vegetable oils, meat and highly processed foods. In: Hawkes C, Blouin C, Henson S, Drager N, Dubé L, editors. Trade, food, diet and health: perspectives and policy options. Chichester: Wiley; 2010. p. 35–59.

44. Gayathri R, Ruchi V, Mohan V. Impact of nutrition transition and resulting morbidities on economic and human development. Curr Diabetes Rev. 2017;13(6) https://doi.org/10.2174/1573399812666160901095534.

45. Min Y, Jiang LX, Yan LF, Wang LH, Basu S, Wu YF, et al. Tackling China's noncommunicable diseases: shared origins, costly consequences and the need for action. Chin Med J. 2015;128(6):839–43. https://doi.org/10.4103/0366-6999.152690.

46. Beckman C. Vegetable oils: competition in a global market. Biweekly Bulletin: Agri Agri-Food Canada. 2005;18(11):1–6.

47. Sassi F. How U.S. obesity compares with other countries [Internet]. PBS Newshour. 2013; [cited 21 April 2017]. Available from: http://www.pbs.org/newshour/rundown/how-us-obesity-compares-with-other-countries/.

48. Oggioni C, Lara J, Wells JC, Soroka K, Siervo M. Shifts in population dietary patterns and physical inactivity as determinants of global trends in the prevalence of diabetes: an ecological analysis. Nutr Metab Cardiovasc Dis. 2014;24(10):1105–11. https://doi.org/10.1016/j.numecd.2014.05.005.

49. Popkin BM. The shift in stages of the nutrition transition in the developing world differs from past experiences. Part II. What is unique about the experience in lower- and middle income less-industrialized countries compared with the very-high income industrialized countries? Public Health Nutr. 2002;5(1A):205–14.

50. Popkin BM. Synthesis and implications: China's nutrition transition in the context of changes across other low- and middle-income countries. Obes Rev. 2014;15:60–7. https://doi.org/10.1111/obr.12120.

51. Bankman J. India and the Hidden Consequences of Nutrition Transition [Internet]. Civil Eats. 2013. [cited 17 April 2017]. Available from: http://civileats.com/2013/08/27/india-and-the-hidden-consequences-of-nutrition-transition/.

52. Bishwajit G. Nutrition transition in South Asia: the emergence of non-communicable chronic diseases. F1000Res. 2015;4:8. https://doi.org/10.12688/f1000research.5732.2.

53. Ramachandran A, Mary S, Yamuna A, Murugesan N, Snehalatha C. High prevalence of diabetes and cardiovascular risk factors associated with urbanization in India. Diabetes Care. 2008;31(5):893–8.

54. Saquib N, Saquib J, Ahmed T, Khanam MA, Cullen MR. Cardiovascular diseases and type 2 diabetes in Bangladesh: a systematic review and meta-analysis of studies between 1995 and 2010. BMC Public Health. 2012;12:434.

55. Ford ND, Patel SA, Narayan KM. Obesity in low- and middle-income countries: burden, drivers, and emerging challenges. Annu Rev Public Health. 2017;38:145–64. https://doi.org/10.1146/annurev-publhealth-031816-044604.

56. Levels and trends in child malnutrition: Key findings of the 2015 edition. http://www.who.int/nutgrowthdb/jme_brochure2015.pdf?ua=1. Last accessed 21 Apr 2017.

57. Levels and trends in child malnutrition: Key findings of the 2016 edition. http://www.who.int/nutgrowthdb/jme_brochure2016.pdf?ua=1. Last accessed 21 Apr 2017.

58. Swinburn BA, Sacks G, Hall KD, McPherson K, Finegood DT, Moodie ML, et al. The global obesity pandemic: shaped by global drivers and local environments. Lancet. 2011;378(9793):804–14. https://doi.org/10.1016/S0140-6736(11)60813-1.

59. Kim J. Are genes destiny? Exploring the role of intrauterine environment in moderating genetic influences on body mass. Am J Hum Biol. 2019;8:e23354. https://doi.org/10.1002/ajhb.23354. [Epub ahead of print].

60. Fall CH. Fetal programming and the risk of noncommunicable disease. Indian J Pediatr. 2013;80(1):S13–20. https://doi.org/10.1007/s12098-012-0834-5.

61. Zeba AN, Delisle HF, Renier G, Savadogo B, Baya B. The double burden of malnutrition and cardiometabolic risk widens the gender and socio-economic health gap: a study among adults in Burkina Faso (West Africa). Public Health Nutr. 2012;15(12):2210–9. https://doi.org/10.1017/S1368980012000729. Epub 2012 Mar 30.

62. Abegunde DO, Mathers CD, Adam T, Ortegon M, Strong K. The burden and costs of chronic diseases in low-income and middle-income countries. Lancet. 2007;370(9603):1929–38.

63. Miranda JJ, Kinra S, Casa JP, Davey Smith G, Ebrahim S. Non-communicable diseases in low and middle income countries: context, determinants and health policy. Tropical Med Int Health. 2008;13(10):1225–34. https://doi.org/10.1111/j.1365-3156.2008.02116.x.

64. Noncommunicable diseases. http://www.who.int/mediacentre/factsheets/fs355/en/. Last accessed 21 Apr 2017.

65. Nishida K, Otsu K. Inflammation and metabolic cardiomyopathy. Cardiovasc Res. 2017;113(4):389–98. https://doi.org/10.1093/cvr/cvx012.

66. NCD Risk Factor Collaboration (NCD-RisC). Trends in adult body-mass index in 200 countries from 1975 to 2014: a pooled analysis of 1698 population-based measurement studies with 19.2 million participants. Lancet. 2016;387(10026):1377–96.

67. Kankeu HT, Saksena P, Xu K, Evans DB. The financial burden from non-communicable diseases in low- and middle-income countries: a literature review. Health Res Policy Syst. 2013;11:31. https://doi.org/10.1186/1478-4505-11-31.

68. Lomazzi M, Borisch B, Laaser U. The millennium development goals: experiences, achievements and what's next. Global Health Action. 2014;7:23695. https://doi.org/10.3402/gha.v7.23695.

69. Haddad L, Cameron L, Barnett I. The double burden of malnutrition in SE Asia and the Pacific: priorities, policies and politics. Health Policy Plan. 2015;30(9):1193–206. https://doi.org/10.1093/heapol/czu110.

70. Shrimpton R, Rokx C. The double burden of malnutrition: a review of global evidence. In: Health, Nutrition and Population (HNP) discussion paper. Washington, DC: World Bank; 2012. p. 1–59. Available from: http://documents.worldbank.org/curated/en/905651468339879888/The-double-burden-of-malnutrition-a-review-of-global-evidence.

71. Black RE, Victora CG, Walker SP, Bhutta ZA, Christian P, de Onis M, et al. Maternal and child undernutrition and overweight in low-

72. The double burden of malnutrition. Case studies from six developing countries. FAO Food Nutr Pap. 2006;84:1–334.
73. Abubakari AR, Lauder W, Agyemang C, Jones M, Kirk A, Bhopal RS. Prevalence and time trends in obesity among adult West African populations: a meta-analysis. Obes Rev. 2008;9(4):297–311.
74. Eckhardt CL, Torheim LE, Monterrubio E, Barquera S, Ruel MT. The overlap of overweight and anaemia among women in three countries undergoing the nutrition transition. Eur J Clin Nutr. 2008;62(2):238–46.
75. Black RE, Allen LH, Bhutta ZA, Caulfield LE, de Onis M, Ezzati M, et al. Maternal and child undernutrition: global and regional exposures and health consequences. Lancet. 2008;371(9608):243–60. https://doi.org/10.1016/S0140-6736(07)61690-0.
76. West KP Jr. Extent of vitamin A deficiency among preschool children and women of reproductive age. J Nutr. 2002;132(9 Suppl):857S–2866S.
77. Barquera S, Peterson KE, Must A, Rogers BL, Flores M, Houser R. Coexistence of maternal central adiposity and child stunting in Mexico. Int J Obes. 2007;31(4):601–7.
78. Uauy R, Kain J, Corvalan C. How can the Developmental Origins of Health and Disease (DOHaD) hypothesis contribute to improving health in developing countries? Am J Clin Nutr. 2011;94(6 Suppl):1759s–64S. https://doi.org/10.3945/ajcn.110.000562.
79. Elshenawy S, Simmons R. Maternal obesity and prenatal programming. Mol Cell Endocrinol. 2016;435:2–6.
80. Zambrano E, Ibáñez C, Martinez-Samayoa PM, Lomas-Soria C, Durand-Carbajal M, Rodríquez-González GL. Maternal obesity: lifelong metabolic outcomes for offspring from poor developmental trajectories during the perinatal period. Arch Med Res. 2016;47(1):1–12. https://doi.org/10.1016/j.arcmed.2016.01.004.
81. Segovia SA, Vickers MH, Gray C, Reynolds CM. Maternal obesity, inflammation, and developmental programming. Biomed Res Int. 2014;2014:418975. https://doi.org/10.1155/2014/418975.
82. Nathanielsz PW, Ford SP, Long NM, Vega CC, Reyes-Castro LA, Zambrano E. Interventions designed to prevent adverse programming outcomes resulting from exposure to maternal obesity during development. Nutr Rev. 2013;71(01):S78–87. https://doi.org/10.1111/nure.12062.
83. Kapil U, Sachdev HP. Urgent need to orient public health response to rapid nutrition transition. Indian J Community Med. 2012;37(4):207–10. https://doi.org/10.4103/0970-0218.103465.
84. Bates I, Boyd A, Aslanyan G, Cole DC. Tackling the tensions in evaluating capacity strengthening for health research in low- and middle-income countries. Health Policy Plan. 2015;30(3):334–44. https://doi.org/10.1093/heapol/czu016.
85. De Boo HA, Harding JE. The developmental origins of adult disease (Barker) hypothesis. Aust N Z J Obstet Gynaecol. 2006;46(1):4–14.
86. Barker DJ. The origins of the developmental origins theory. J Intern Med. 2007;261(5):412–7.
87. Padmanabhan V, Cardoso RC, Puttabyatappa M. Developmental programming a pathway to disease. Endocrinology. 2016;157(4):1328–40. https://doi.org/10.1210/en.2016-1003.
88. Bruce KD, Hanson MA. The developmental origins, mechanisms, and implications of metabolic syndrome. J Nutr. 2010;140(3):648–52. https://doi.org/10.3945/jn.109.111179.
89. Befroy DE, Petersen KF, Dufour S, Mason GF, de Graaf RA, Rothman DL. Impaired mitochondrial substrate oxidation in muscle of insulin-resistant offspring of type 2 diabetic patients. Diabetes. 2007;56(5):1376–81. https://doi.org/10.2337/db06-0783.
90. Chmurzynska A. Fetal programming: link between early nutrition, DNA methylation and complex diseases. Nutr Rev. 2010;68(2):87–98. https://doi.org/10.1111/j.1753-4887.2009.00265.x.
91. Tutino GE, Tam WH, Yang X, Chan JC, Lao TT, Ma RC. Diabetes and pregnancy: perspectives from Asia. Diabet Med. 2014;31(3):302–18. https://doi.org/10.1111/dme.12396.
92. Egger G. In search of a germ theory equivalent for chronic disease. Prev Chronic Dis. 2012;9(11):1–7.
93. Egger G, Dixon J. Beyond obesity and lifestyle: a review of 21st century chronic disease determinants. Biomed Res Int. 2014;2014:731685. https://doi.org/10.1155/2014/731685.
94. Hotamisligil GS. Inflammation and metabolic disorders. Nature. 2006;444(7121):860–7.
95. Bosma-den Boer MM, van Wetten ML, Pruimboom L. Chronic inflammatory diseases are stimulated by current lifestyle: how diet, stress levels and medication prevent our body from recovering. Nutr Metab (Lond). 2012;9(1):32. https://doi.org/10.1186/1743-7075-9-32.
96. Barbaresko J, Koch M, Schulze MB, Nothlings U. Dietary pattern analysis and biomarkers of low grade inflammation: a systematic literature review. Nutr Rev. 2013;71(8):511–27. https://doi.org/10.1111/nure.12035.
97. Gregor MF, Hotamisligil GS. Inflammatory mechanisms in obesity. Annu Rev Immunol. 2011;29:415–45.
98. Laumbach RJ, Kipen HM. Acute effects of motor vehicle traffic-related air pollution exposures on measures of oxidative stress in human airways. Ann N Y Acad Sci. 2010;1203:107–12. https://doi.org/10.1111/j.1749-6632.2010.05604.x.
99. Kelley AS, Banker M, Goodrich JM, Dolinoy DC, Burant C, Domino SE, Padmanabhan V. Early pregnancy exposure to endocrine disrupting chemical mixtures are associated with inflammatory changes in maternal and neonatal circulation. Sci Rep. 2019;9(1):5422. https://doi.org/10.1038/s41598-019-41134-z.
100. Hyman MA. Functional diagnostics: redefining disease. Altern Ther Health Med. 2008;14(4):10–4.
101. Sebert S, Sharkey D, Budge H, Symonds ME. The early programming of metabolic health: is epigenetic setting the missing link? Am J Clin Nutr. 2011;94(6 Suppl):1953S–8S. https://doi.org/10.3945/ajcn.110.001040.
102. Cohen S, Janicki-Deverts D, Doyle WJ, Miller GE, Frank E, Rabin BS, et al. Chronic stress, glucocorticoid receptor resistance, inflammation, and disease risk. Proc Natl Acad Sci U S A. 2012;109(16):5995–9. https://doi.org/10.1073/pnas.1118355109.
103. Fredman G, Spite M. Specialized pro-resolving mediators in cardiovascular diseases. Mol Aspects Med. 2017. pii:S0098-2997(17)30017–1. https://doi.org/10.1016/j.mam.2017.02.003.
104. Serhan CN, Chiang N, Dalli J, Levy BD. Lipid mediators in the resolution of inflammation. Cold Spring Harb Perspect Biol. 2014;7(2):a016311. https://doi.org/10.1101/cshperspect.a016311.
105. Kuda O. Bioactive metabolites of docosahexaenoic acid. Biochimie. 2017;136:12–20. https://doi.org/10.1016/j.biochi.2017.01.002.
106. Sansbury BE, Spite M. Resolution of acute inflammation and the role of resolvins in immunity, thrombosis, and vascular biology. Circ Res. 2016;119(1):113–30. https://doi.org/10.1161/CIRCRESAHA.116.307308.
107. Bannenberg GL. Therapeutic applicability of anti-inflammatory and proresolving polyunsaturated fatty acid derived lipid mediators. Sci World J. 2010;10:676–712. https://doi.org/10.1100/tsw.2010.57.
108. Franco OH, Karnik K, Osborne G, Ordovas JM, Catt M, van der Ouderaa F. Changing course in ageing research: the healthy ageing phenotype. Maturitas. 2009;63(1):13–9. https://doi.org/10.1016/j.maturitas.2009.02.006.
109. Kiefete-de Jong JC, Mathers JC, Franco OH. Nutrition and healthy ageing: the key ingredients. Proc Nutr Soc. 2014;73(2):249–59. https://doi.org/10.1017/S0029665113003881.
110. Mozaffarian D. Dietary and policy priorities for cardiovascular disease, diabetes, and obesity – a comprehensive review. Circulation. 2016;133(2):187–225. https://doi.org/10.1161/CIRCULATIONAHA.115.018585.
111. Minich DM, Bland JS. Personalized lifestyle medicine: relevance for nutrition and lifestyle recommendations. ScientificWorldJournal. 2013;2013:129841. https://doi.org/10.1155/2013/129841.
112. Jones DS, Quinn S. AAPI's nutrition guide to optimal health: using principles of functional medicine and nutritional genomics [Inter-

net]. 2012. Chapter 1, Functional medicine; [cited 21 April 2017]; p. 6–12. Available from www.aapiusa.org.
113. Baker SM. The metaphor of an oceanic disease. Integr Med. 2008;7(1):40–5.
114. Shah T, Zabaneh D, Gaunt T, Swerdlow DI, Shah S, Talmud PJ, et al. Gene-centric analysis identifies variants associated with interleukin-6 levels and shared pathways with other inflammation markers. Circ Cardiovasc Genet. 2013;6(2):163–70. https://doi.org/10.1161/CIRCGENETICS.112.96425.
115. Rana S, Kumar S, Rathore N, Padwad Y, Bhushana S. Nutrigenomics and its impact on life style associated metabolic diseases. Curr Genomics. 2016;17(3):261–78. https://doi.org/10.2174/13892029 17666160202220422.
116. Leischow SJ, Best A, Trochim WM, Clark PI, Gallagher RS, Marcus SE, et al. Systems thinking to improve the public's health. Am J Prev Med. 2008;35(20):S196–203. https://doi.org/10.1016/j.amepre.2008.05.014.
117. Bloch P, Toft U, Reinbach HC, Clausen LT, Mikkelsen BE, Poulsen K, et al. Revitalizing the setting approach – supersettings for sustainable impact in community health promotion. Int J Behav Nutr Phys Act. 2014;11:118. https://doi.org/10.1186/s12966-014-0118-8.
118. Fardet A, Rock E. Toward a new philosophy of preventive nutrition: from a reductionist to a holistic paradigm to improve nutritional recommendations. Adv Nutr. 2014;5(4):430–46. https://doi.org/10.3945/an114.006122.
119. Sunguya BF, Ong KI, Dhakal S, Mlunde LB, Shibanuma A, Yasuoka J, et al. Strong nutrition governance is a key to addressing nutrition transition in low and middle-income countries: review of countries' nutrition policies. Nutr J. 2014;13:65. https://doi.org/10.1186/1475-2891-13-65.
120. Buckinx F. The public health challenge of ending malnutrition: the relevance of the World Health Organization's. Asia Pac J Public Health. 2018;30(7):624–8. https://doi.org/10.1177/1010539518800341. Epub 2018 Sep 15.
121. Hawkes C. Food policies for healthy populations and healthy economies. BMJ. 2012;344:e2801. https://doi.org/10.1136/bmj.e2801.
122. Sacks G, Swinburn B, Lawrence M. Obesity policy action framework and analysis grids for a comprehensive policy approach to reducing obesity. Obes Rev. 2009;10(1):76–86. https://doi.org/10.1111/j.1467-789X.2008.00524.x.
123. Capacci S, Mazzocchi M, Shankar B, Macias JB, Verbeke W, Pérez-Cueto FJ, et al. Policies to promote healthy eating in Europe: a structured review of policies and their effectiveness. Nutr Rev. 2012;70(3):188–200. https://doi.org/10.1111/j.1753-4887.2011.00442.x.
124. Gayathri R, Ruchi V, Mohan V. Impact of nutrition transition and resulting morbidities on economic and human development. Curr Diabetes Rev. 2017;13(6):00–0. https://doi.org/10.2174/15733998 12666160901095534.
125. Hawkes C, Jewell J, Allen K. A food policy package for healthy diets and the prevention of obesity and diet related non-communicable diseases: the NOURISHING framework. Obes Rev. 2013;14(2):159–68. https://doi.org/10.1111/obr.12098.
126. 2008–2013 Action Plan for the Global Strategy for the Prevention and Control of Noncommunicable Diseases. http://www.who.int/nmh/publications/ncd_action_plan_en.pdf. Last accessed 21 Apr 2017.
127. Mattei J, Malik V, Wedick N, Hu FB, Spiegelman D, Willett WC, et al. Reducing the global burden of type 2 diabetes by improving the quality of staple foods: the global nutrition and epidemiologic transition initiative. Glob Health. 2015;11:23. https://doi.org/10.1186/s12992-015-0109-9.
128. Doak C. Large scale interventions and programmes addressing nutrition related chronic diseases and obesity: examples from 14 countries. Public Health Nutr. 2002;5(1A):275–7.

Nutritional and Metabolic Wellness

Diana Noland

3.1 Introduction to Human Wellness – 32

3.2 What Is Wellness? – 32

3.3 Biomarkers of Wellness – 33

3.4 Biochemical Individuality/Health Standards Through the Lifespan – 33
3.4.1 In Utero – 34
3.4.2 Infant (From Birth to 6–24 Months) – 34
3.4.3 Young Adult (20–34 Years) – 35

3.5 Nutritional Wellness – 36

3.6 Community of Wellness – 36

3.7 Summary – 37

References – 37

Health is a state of complete harmony of the body, mind and spirit. When one is free from physical disabilities and mental distractions, the gates of the soul open.
–B.K.S. Iyengar, yoga master

Learning Objectives
- Current "normal" has transitioned away from genuine wellness.
- Genuine wellness must remain the plumb line for human wellness.
- Salutogenesis is an important foundation for wellness healthcare.

3.1 Introduction to Human Wellness

Have you noticed the trend to perceive type 2 diabetes as a "normal" condition after age 50? Have you noticed the trend to perceive a little arthritis pain in the joints as a "normal" condition after age 40? These conditions have become so prevalent that adults are trending to consider these conditions as the "normal" for states of health. The "normal" today in this time of epidemic chronic disease is promoting amnesia of true wellness.

3.2 What Is Wellness?

Wellness certainly means freedom from the debilitating, weakening effects of chronic disease. As a side product of this level of wellness, one feels dynamic, energetic, alive, vital, and vibrant. From this healthy state, we can respond effectively to environmental stress, toxins, or infections, quickly returning to our previous state of health and wellness [1].

Health and vitality refer to your life energy and the power to live, grow, know the purpose in life, and express your maximum potential as a human being. Individual lives are made up of many facets, all interconnected. Nutrition provides the energetic and physical foundation for achieving wellness as defined by Hershoff in all its dimensions.

Before exploring the principles surrounding restoration of wellness from chronic disease, it is important to recognize that we must set a clear vision of the health goals for each patient. Throughout human history, humanity has faced many challenges in the search for health and longevity, and people have culturally developed ways of life to function successfully. There is ongoing speculation by experts in human biology that the current state of global health is being challenged by the beginning of declining lifespan [2, 3] and fertility [4].

It is important that we identify the biomarkers that define and characterize the human wellness we are striving to restore in our patients. The "normal" of current populations has already deviated significantly from states of wellness. There is no better example than the current epidemic of obesity that is being predicted as the twenty-first century begins: Upwards of three-fourths or more of industrialized countries' populations are predicted to be obese (>30% ideal body weight) by 2030 [5]. In 1962, research statistics showed that 13% of America's population was obese. By 1980 it had risen to 15%, to 23% by 1994, and to an unprecedented 39.8% of adults by 2016. According to the Centers for Disease Control and Prevention, the percentage of overweight children ages 6–11 has nearly doubled since the early 1980s, while the percentage of overweight adolescents has almost tripled [6]. These overwhelming research statistics reveal an alarming obesity trend, the need for diagnosis, and a call to action [7, 8] (◘ Fig. 3.1).

CDC research statistics on American obesity tell us that 63% of adult Americans and 18.5% of children and adolescents have a body mass index (BMI) in excess of 25.0 and are therefore overweight; more than a quarter surpass 30.0, qualifying as obese [9]. Obesity is becoming the "norm," even

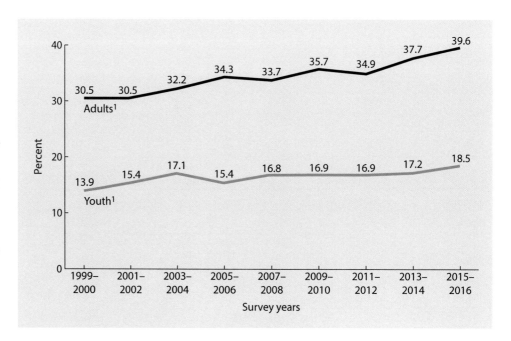

◘ Fig. 3.1 Trends in obesity prevalence among adults aged 20 and over (age-adjusted) and youth aged 2–19 years: United States, 1999–2000 through 2015–2016 (Hales 2017). ¹Significant increasing linear trend from 1999–2000 to 2015–2016. NOTES: All estimates for adults are age adjusted by the direct method to the 2000 US census population using the age groups 20–39, 40–59, and 60 and over. Access data table available at ▶ https://www.cdc.gov/nchs/data/databriefs/db288_table.pdf#5. (SOURCE: NCHS, National health and Nutrition Examination Survey, 1999–2016)

though it is not an example of wellness. That is a staggering thought when we realize the relatively short period of time during which this epidemic has occurred. Obesity is strongly related to the incidence of chronic disease. As the obesity trend increases, human models of wellness are becoming rarer. Humans tend to gravitate toward acceptance of what is prevalent as "normal" and may tend to forget the real wellness we are seeking to restore. Health practitioners must keep the definition of true wellness before them as a standard with which to compare their patients.

3.3 Biomarkers of Wellness

What are the biomarkers for wellness? This chapter presents suggestions for a framework for markers to assess wellness, for which further development is welcome. At least it is a good foundation developed from many studies in the last century that evaluated common characteristics of human populations who achieved excellent function, life fulfillment, and good reproduction capacity during their relative longevity. Currently, it is proposed within the longevity science community that a 115-year lifespan is a realistic goal [10].

Thanks to the efforts of various scientists in the past 70 years, there has been research stimulated by scientific curiosity regarding the populations who have stretched longevity and wellness to the upper limits. Humans confirmed to have attained the age of 110 years or more are referred to as *supercentenarians*. Identifying factors that help people remain healthy, vigorous, and disability-free at older ages is a major research priority of longevity scientists. The most recent study is that of Dan Buettner named *The Blue Zones* – those pockets of societies with the most healthy centenarians and generally healthy populations [11].

Common factors that have measurable biomarkers among the healthiest societies are repeatedly found to be:
1. Unprocessed, whole foods, plant-rich diet (diet history; nutrient status)
2. Caloric and nutrient intake so as to maintain a healthy Weight (anthropometrics)
3. Regulating insulin production [12] (blood glucose/insulin fasting; HgbA1C)
4. Moderate daily physical activity (minutes per day or week; handgrip strength)
5. Small amount of alcohol frequently (daily female </= 1 serving; men 1-2 servings daily)
6. Strong community and social connectivity (1–10; 10 highest)
7. Meditation and spiritual beliefs (time per week)
8. Feeling of purpose in life (1–10; 10 highest)

The diets of these longevity-rich societies vary according to cultural traditions, but all contain a high intake of whole vegetables and fruits and beneficial food oils. Interestingly, these foods are rich in the phytonutrients, the most recently identified nutrient group [11].

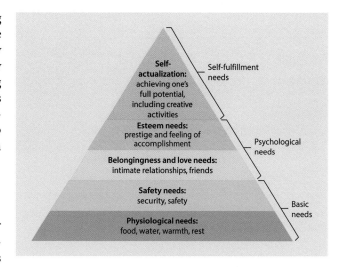

 Fig. 3.2 Hierarchy of needs

The way you think, the way you behave, the way you eat, can influence your life by 30 to 50 years. –Deepak Chopra

Throughout the community of integrative and functional medicine practitioners' concepts like Maslow's hierarchy of needs have developed [13] has agreement that the following factors are key influencers that summarize the findings of many studies on longevity and wellness:

"Basic needs": Biological, physiological, and safety needs
1. Foods (protein, fats, carbohydrates, fiber)
2. Vitamins, minerals, accessory, or conditionally essential nutrients
3. Light, water, and air
4. Movement rhythm
5. Circadian rhythm balance
6. "Mind-body needs": Love, belongingness, self-esteem, cognitive, aesthetic, and self-actualization needs
7. Meaning and purpose
8. Love, community, connection

All seven are inherently interrelated in the context of the human experience that affects wellness (Fig. 3.2).

3.4 Biochemical Individuality/Health Standards Through the Lifespan

As the professional rules of thumb for nutritional status assessment are developed in subsequent chapters, it will be seen that the environmental conditions necessary to produce the identified markers of wellness are biochemical, clinical, behavioral, and functional and change throughout developmental stages of the lifespan. The following are some of the primary stages of the lifespan with suggestions for considerations to assess and manage to promote wellness for each stage. These are of particular importance to increase the health of a person and their nutritional status within the area of chronic disease management:

3.4.1 In Utero

- Fetal nutrition: promote healthy musculoskeletal-organ-tissue fetal growth
- Maternal:
    - Hormonal health: adrenal, thyroid, insulin, and others
    - Blood glucose management
    - Emotional environment: calm and joyful
    - Rest: adequate rest, sleep, and exercise
    - Nutrition status: adequate vitamin D25-hydroxy, iodine, DHA, iron, folic acid, protein, minerals, beneficial oils, antioxidants, and phytonutrients, to promote good growth and protect tissue from free radical damage [14]
    - Avoid toxin exposure
- Grandmaternal nutrition status: same as mother, avoidance of toxins to avoid epigenetic transgenerational effects
- Paternal nutrition status: important 2 years prior to conception, avoid toxins, and especially 14 days prior to fertilization

3.4.2 Infant (From Birth to 6–24 Months)

- Breastfeeding infant nutrition: promote healthy musculoskeletal-organ-tissue fetal growth
- Maternal lactation and caretaker:
    - Emotional environment: calm and joyful
    - Rest: adequate rest, sleep, exercise
    - Hormonal health: adrenal, thyroid, insulin especially
    - Nutrition status: adequate vitamin D25-hydroxy, iodine, DHA, iron, protein, minerals, antioxidants, and phytonutrients to promote good growth and protect tissue from free radical damage and adequate fluid intake
- Home environment: stable, caring, joyful
- Infant 6 months + oral nutrition: balanced organic whole foods, hypoallergenic diet
- Infant:
    - Vaccination wisdom
        1. Never when sick
        2. Wait until 4+ months if possible
        3. Mercury- and formaldehyde-free
        4. Extra nutritional antioxidant support and folic acid prior to a vaccination
        5. Consider vaccination alternatives for at-risk infants with poor methylation family history or genetics
    - Safe sun exposure for vitamin D production (never burn)
    - Fresh air and exercise and play

3.4.2.1 Toddler (12–36 Months)
- Positive parenting.
- Seriously consider vaccination need/avoid unnecessary/mercury- and formaldehyde-free.
- Child safety environment.
- Nutrition: Regular meals of organic/low-toxin food containers and nutrient-dense, balanced whole foods.
    - Protein
    - Calcium- and mineral-rich foods
    - Nuts/seeds
    - Fish or fish oils/EFA balance
    - Whole grains
    - Sea salt/herbs/spices
    - Fruits and vegetables
    - Onions/garlic (sulfur/bioflavonoid)
    - Cruciferous vegetables (sulfur/indole-3-carbinol)
    - Leafy greens (folic acid, chlorophyll, magnesium, etc.)
    - Fermented or cultured foods (yogurt/kefir/sauerkraut/miso/tempeh)
    - Sea vegetables (iodine/minerals)
    - Vitamin A- and D-rich foods/supplements if indicated
- Sun exposure (promote Vitamin D) (or supplementation cod liver oil >40° latitude).
- Play and exercise.

3.4.2.2 Childhood (3–12 Years)
- Continue toddler recommendations.
- Educational programs.
- Self-esteem-building lifestyle.
- Nutrition: Ensure adequate protein/calcium/minerals for rapid musculoskeletal growth – possible increased needs during this growth period.
- Sun exposure almost daily – cod liver oil during winter if >40° latitude.
- Develop food preparation self-skills.
- Develop hygiene practices – especially oral and dental care habits.
- Sleep: 9–10 hours with bedtime routine/quiet and dark environment.
- Focus on oral/dental care: avoid mercury amalgams, orthodontics that include removal of teeth, as well as prevent periodontal diseases and decay by daily dental hygiene and regular dental cleanings.

3.4.2.3 Teenage (13–19 Years) (◘ Fig. 3.3)
- Continue childhood nutrition recommendations.
- Strong family base/good relationships with parents and teachers.
- Avoid "junk food" and "toxin exposure" (cigarettes, drugs, environmental).
- Possible need for extra sleep.
- Weight-bearing regular exercise (peak time for bone density building for adulthood).
- Develop advanced food preparation and self-care skills.
- Creative outlets for interests.
- Begin to develop purpose in life.
- Abstinence/avoid communicable diseases/avoid birth control pills (upsets hormonal balance, depletes folic acid).

Nutritional and Metabolic Wellness

Fig. 3.3 Teenage energy and vitality. (2005© Diana Noland)

- Focus on oral/dental care: avoid mercury amalgams, orthodontics that include removal of teeth, as well as prevent periodontal diseases and decay by daily dental hygiene and regular dental cleanings.

3.4.3 Young Adult (20–34 Years)

- Continue teenage nutrition recommendations.
- Whole foods; low-toxin organic foods, containers, and cookware; balanced, nutrient-rich diet.
- Moderate daily exercise.
- Avoid toxins.
- Seriously consider vaccination need/avoid unnecessary/ mercury- and formaldehyde-free.
- Prepare health for possible pregnancy or fathering – even if not planning. Caution about use of birth control pills (BCP) related to the nutrient-drug depletions of folic acid, magnesium, zinc, and other B complex vitamins to promote select malnutrition (i.e., those depleted in folic acid are at more risk for HPV infection which is related to increased sexual activity, more prevalent with use of BCP).

3.4.3.1 Middle-Age Adult (35–54 Years)

- Continue young adult recommendations with nutrition focus.
- Focus on nutrition to support healthy aging with phytonutrients, healthy fatty acid balance, minerals, methylation and sulfation nutrients, and hormonal and immune support nutrition like vitamin D.
- Focus on oral/dental care: avoid periodontal diseases/ decay/misalignment.
- Continue to focus on weight-bearing exercise to maintain good bone density throughout the lifespan.

3.4.3.2 Seniors (54–74 Years)

- Continue to focus on nutrition to support healthy aging with phytonutrients, healthy fatty acid balance, minerals, methylation and sulfation nutrients, and hormonal and immune support nutrition like vitamin D.
- Focus on gastrointestinal health as digestive function wanes naturally in elderly.
- Good oral hygiene and health.
- Maintain healthy percentage of body fat composition and weight.
- Weight-bearing exercise.
- Emphasize social connectedness, especially around food quality and meal time.
- Stress management.
- Adequate sleep.

3.4.3.3 Old-Old Seniors [15] (75–99 Year)

- Continue to focus on nutrition to support healthy aging with phytonutrients, healthy fatty acid balance, minerals, methylation and sulfation nutrients, and hormonal and immune support nutrition like vitamin D.
- Focus on gastrointestinal health as digestive function wanes naturally in elderly.
- Weight-bearing exercise.
- Safe environment to meet basic needs.
- Emphasize social connectedness, especially around food quality and meal time.
- Maintain healthy percentage of body fat composition and weight.
- Weight-bearing exercise.
- Continue to focus on nutrition to support healthy aging with phytonutrients, healthy fatty acid balance, minerals, methylation and sulfation nutrients, and hormonal and immune support nutrition like vitamin D.
- Safe environment to meet basic needs.
- Residential support system with family or caring friends (► Boxes 3.1 and 3.2).

> **Box 3.1 Key Markers of Childhood Wellness**
> - No cavities
> - Sturdy
> - Strong
> - Cheerful disposition
> - Not overweight
> - No allergies
> - Manages stress
> - Emotionally stable
> - Sleeps soundly
> - Straight teeth
> - Learns easily
> - Good concentration
> - Optimistic
> - Lots of energy
> - Rarely sick
> - Strong digestion

> **Box 3.2 Key Markers of Human Wellness**
> 1. HEALTHY BODY WEIGHT: +/− 10%; %BODY FAT: M < 18%/F < 25% with no observable central adiposity
> 2. HEALTHY SKIN: color, tone, texture, free of lesions, abnormal moles, itching, pain
> 3. HEALTHY DIET: balanced whole foods and low-toxin diet, water
> 4. HEALTHY EMOTIONS: caring, social connectedness, relationships, purpose in life, self-esteem, community/family involvement, appropriate emotions
> 5. HEALTHY ORGAN SYSTEMS FUNCTION: including blood glucose management resulting in stable insulin blood levels
> 6. HEALTHY IMMUNE SYSTEM: infrequent illness, illness only of mild nature with quick recovery, absence of chronic disease, absence of allergies
> 7. HEALTHY MOVEMENT: adequate energy and ability to exercise, run, jump, play, and lift appropriate weight, no pain
> 8. HEALTHY MUSCULOSKELETAL: Strong grip, strong muscles for walking and running, no aches and pain, no broken bones or sprains
> 9. HEALTHY REPRODUCTION/FERTILITY: Throughout lifespan normal gonadal development, fertile, no pain with menses (F), no STDs, sexual health, healthy birth delivery, symptom-free menopause
> 10. HEALTHY GASTROINTESTINAL SYSTEM: 2–3 BM/day, good digestion, no flatulence
> 11. LONGEVITY to >110 years old with high-quality function and life
> 12. LOW-TOXIN exposure lifestyle

3.4.3.4 Centenarians/Supercentenarians (100+ Years)

- Emphasize social connectedness, especially around food quality and meal time.
- Maintain healthy percentage of body fat composition and weight.
- Weight-bearing exercise.
- Continue to focus on nutrition to support healthy aging with phytonutrients, healthy fatty acid balance, minerals, methylation and sulfation nutrients, and hormonal and immune support nutrition like vitamin D.

- Safe environment to meet basic needs.
- Residential support system with family or caring friends.

3.5 Nutritional Wellness

Nutrients come from food, and food comes from the environment. While taken for granted at an intellectual level, this nutrient-food/food-environment relationship is not something citizens of industrialized nations usually consider in their everyday lives [16]. Consumer awareness of food-toxic harmful events over the past 30 years due to media reporting has increased consumer food safety concerns. The increasing frequency of recalls of contaminated food, findings of animals with "mad cow disease," chickens with avian flu viruses, the fish supply testing high in mercury and PCBs, endocrine disruptors, and others presents a growing concern for the quality of the food supply. The message is coming through loud and clear to society that environment plays an integral part in food safety.

The human community must ensure safe food and agricultural practices as a priority in this new era of toxicity, superbugs, and depleted soils, ensure to reverse trends, ensure environmental safety of the food supply, and enable humanity to have wellness within its grasp.

A nutritional environment for wellness results from an attitude of guardianship over the food supply. A wellness environment requires a global food supply of diversity in geography, climate, exposure to toxins, and many other variables. It involves ensuring the food is not depleted and does not contain dangerous level of toxins [17]. The food that produces wellness requires human care in policies and agriculture in every stage from soil to plate.

3.6 Community of Wellness

Throughout history, humans have been brought together around the table, a symbol of congregation, family, and community, where we celebrate loved ones. Food is a central part of a healthy community. It provides not only bodily nourishment but also nourishment to the soul from the care taken in raising, gathering, and preparing the food and in the relationships and connectedness forged and strengthened during the meal. In the larger community of a nation, the societal mores about food and the wellness of the population affect people's health. It is important to elevate the societal value of health and wellness within global communities so they guard the soil, water, food supply, food industry, healthcare systems, and governmental policies. The tremendous knowledge of human metabolism gained by each generation has the potential to positively influence populations to improve every individual's opportunity to be well and live a full life. There is an imminent need for an alignment of society's thinking with the newer discoveries of nutrition science to implement lifestyle basics and policies that are key to thriving in wellness.

Salutogenesis is a concept that literally means "that which gives birth to health." It is a term that has not been well known since its creation by Antonovsky in 1979 [18]. In traditional public health and community medicine approaches, the focus is the opposite: on disease or illness and its prevention or treatment. This latter focus most often dominates interventions. In the communities where longevity and quality of life are most prevalent, the culture of salutogenesis prevails. Adoption of a salutogenic perspective highlights the importance of starting from a consideration of how health is created and maintained. Contrary to industrialized societies where medical care and "healthcare" focus on "pathogenesis," salutogenesis is the inverse [19].

The salutogenic paradigm needs to suffuse all members of a community as it focuses on the promotion of global well-being rather than the origins of specific disease processes. The focus on wellness strategies and lifestyle choices empowers individuals to experience the full spectrum of the human experience. A community of interconnected individuals and families is necessary to activate the force of wellness and develop the ability to empower others to lead salutogenic lives and to transform our culture into one of worldwide wellness.

3.7 Summary

- True wellness achieves a life that is dynamic, energetic, alive, vital, and vibrant.
- Salutogenesis is a paradigm of a healthcare approach of how to create and maintain health.
- Twelve markers of human wellness.
- The healthiest societies on earth have a good diet and lifestyle in common.

References

1. Hershoff A. Homeopathic remedies: a quick and easy guide to common disorders and their homeopathic treatments. Avery Trade: New York; 2000. p. 10.
2. Olshansky SJ, Passaro J, Hershow RC, Layden J, Carnes BA, Brody J, Hayflick L, Butler RN, Allison DB, Ludwig D. A potential decline in life expectancy in the United States in the 21st century. N Engl J Med. 2005;352:1138–45.
3. Lederberg J, Shope RE, Oaks SC Jr, editors. Emerging infections, microbial threats to health in the United States. Institute of Medicine (US) Committee on Emerging Microbial Threats to Health. Washington, DC: National Academies Press (US); 1992. ISBN-10: 0-309-04741-2.
4. W.H.O. Infertility is a global public health issue. http://www.who.int/reproductivehealth/topics/infertility/perspective/en/ Accessed 30 Nov 2018.
5. de Martel C, Ferlay J, et al. Global burden of cancers attributable to infections in 2008: a review and synthetic analysis. Lancet Oncol. 2012;13(6):607–15. Epub 2012 May 9. -Associated Press posted Friday June 1st, 2012.
6. National Obesity Rates & Trends. https://stateofobesity.org/obesity-rates-trends-overview/ Accessed 10 Jan 2019.
7. Carmona RH statement: The Obesity Crisis in America: Testimony before Subcommittee on Education Reform Committee on Education and the Workforce US House of Representatives. July 16, 2003/ revised January 8, 2007. https://www.surgeongeneral.gov/news/testimony/obesity07162003.html. Accessed 10 Jan 2019.
8. Ogden CL, Fakhouri TH, Carroll MD, Hales CM, Fryar CD, Li X, Freedman DS. Prevalence of obesity among adults, by household income and education — United States, 2011–2014. MMWR Morb Mortal Wkly Rep. 2017;66(50):1369–73.
9. CDC Childhood Obesity Facts. https://www.cdc.gov/obesity/data/childhood.html Accessed 1 Jan 2019.
10. Ben-Haim MS, Kanfi Y, Mitchell SJ, et al. Breaking the ceiling of human maximal life span. J Gerontol A Biol Sci Med Sci. 2018;73(11):1465–71. https://doi.org/10.1093/gerona/glx219.
11. Buettner D. The blue zones: lessons for living longer from the people who've lived the longest. National Geographic; 2nd ed. (November 6, 2012).
12. Gerlini R, Berti L, Darr J, et al. Glucose tolerance and insulin sensitivity define adipocyte transcriptional programs in human obesity. Mol Metab. 2018;18:42–50.
13. Hoffman E. The right to be human: a biography of Abraham Maslow. Los Angeles: Jeremy P. Tarcher, Inc.; 1988.
14. Ames B. Prolonging healthy aging: longevity vitamins and proteins. Proc Natl Acad Sci. 2018;115(43):10836–44. Published by National Academy of Sciences.
15. N., Pam M.S., "OLD OLD" in PsychologyDictionary.org, April 7, 2013, https://psychologydictionary.org/old-old/. Accessed 9 May 2019.
16. Levin B. Environmental nutrition: understanding the link between environment, food quality, and disease. Vashon Island, Wash: HingePin Integrative Learning Materials; 1999.
17. Hartman Group. The Hartman report. Food and the environment: a consumer's perspective. Phase II. Bellevue: The Hartman Group; 1997. p. 53.
18. Antonovsky A. Health, stress, and coping: new perspectives on mental and physical Well-being. San Francisco: Jossey-Bass; 1979.
19. Kent DC, Christopher, Salutogenesis: Giving Birth To Health, The American Chiropractor, 2007;29(9).

Nutritional Ecology and Human Health

David Raubenheimer and Stephen J. Simpson

4.1   Introduction – 40

4.2   "Nutrition" and "Ecology" – 40
4.2.1 Human Nutrition – 40
4.2.2 Ecology and Anthropology – 41
4.2.3 Nutritional Ecology and Human Health – 41

4.3   The Importance of Appetite – 41
4.3.1 Multiple Appetites – 42
4.3.2 Appetites Interact – 42

4.4   From Concepts to Models: Introduction to the Geometric Framework for Nutrition – 43
4.4.1 Model Selection – 43
4.4.2 Selecting an Intake Target: Nutritionally Balanced and Complementary Foods – 44
4.4.3 Negotiating a Compromise: When the Intake Target Cannot Be Reached – 44

4.5   The Geometry of Nutrition in Humans: Protein Leverage – 45
4.5.1 Do Humans Select an Intake Target? – 45
4.5.2 What Is the P:NPE Rule of Compromise in Humans? – 46

4.6   Beyond Appetites – 47
4.6.1 The Geometry of Mixtures: Three Components in Two Dimensions – 47
4.6.2 A Hierarchy of Mixtures – 47
4.6.3 Dietary Macronutrient Balance – 48
4.6.4 Relationships Between Macronutrient Balance and Energy Intake – 49
4.6.5 Energy Balance – 49
4.6.6 Beyond Energy: Protein Intake – 50
4.6.7 Interactions of Appetite with the Food Environment – 51

4.7   Conclusions – 52

       References – 53

© Springer Nature Switzerland AG 2020
D. Noland et al. (eds.), *Integrative and Functional Medical Nutrition Therapy*,
https://doi.org/10.1007/978-3-030-30730-1_4

4.1 Introduction

Barely a year into the United Nations Decade of Action on Nutrition, the global community of researchers and practitioners in nutrition and associated fields has its work cut out. Malnutrition, which includes overnutrition, undernutrition, and imbalanced nutrition, affects at least a third of the earth's population and is by far the greatest contributor to the global burden of disease [1]. For the period 2014–2016, it is estimated that more than 794 million people (10.9% of the world population [2]) were undernourished and concurrently 2.1 billion were overweight or obese [3]. Both overweight and underweight are significantly associated with premature death [4], as are many other diet-related factors including excess sodium intake and micronutrient deficiencies [5]. Every year, more than 3 million children die of causes related to undernutrition [6], and in 2010 alone, overweight conditions and obesity are estimated to have resulted in 3.4 million deaths [3].

To appreciate the scale of these statistics, consider that the 3.4 million deaths attributed to overweight and obesity in 2010 is 1432 times the number of airline fatalities in the worst-ever year of commercial aviation disasters (1972) and 42 times the *total* number in 74 years of commercial aviation history [7]. It is equivalent to 6800 jumbo jets fully laden with 500 passengers each – 18 jets for every day of the year. Adding childhood deaths due to undernutrition, and we are staring in the face of what amounts to 12,800 fatal jumbo jet crashes a year, or 35 daily. Aviation-related losses would not be allowed to reach a small fraction of these numbers; in fact, since 1978 they have steadily decreased, notwithstanding the steep increase in the numbers of flights [7].

A question that is increasingly attracting attention is why have we not done a better job of preventing illness and deaths due to malnutrition? There are many views, but four themes repeatedly emerge. First, nutritional behavior and its causes and consequences are highly complex [8, 9], and the dominant models of nutrition research and practice do not pay sufficient respect to its complexity [10–12]. Second, nutrition is an intrinsically interdisciplinary problem requiring expertise from many areas – e.g., chemistry, physiology, and psychology to economics, law, and politics – but insufficient attention is paid to this fact in the training of nutrition scientists. Third, a systems perspective is needed to bring together the diverse parts of the complex multifaceted nutrition system and to identify how they interact to determine the important outcomes. Finally, we need to evolve a food and nutrition system that respects optimization goals beyond minimum cost and maximum pleasure and profit, including human health, equity, and planetary sustainability. Considerably less well-developed than the articulation of where we have gone wrong in nutrition are concrete suggestions for how we can implement a systems-based interdisciplinary agenda in nutrition science and practice to put the problem right.

In this chapter, we introduce a field from the natural sciences, called nutritional ecology, which we believe can contribute toward bridging the gap between the food and nutrition system that we currently have and the system we would prefer. We begin by clarifying how we use the term "nutritional ecology" to place it in the context of related approaches. We then discuss some biological insights from nutritional ecology that we think can make a significant contribution to nutrition research and practice and thereafter introduce a geometric framework for implementing this theory. We end with examples showing how the implementation of biological thinking via nutritional geometry can provide a concrete step toward a fresh, systems-based, view of human nutrition.

4.2 "Nutrition" and "Ecology"

An important first step is to define what is meant by the term "nutritional ecology" as used in the ecological sciences (and this chapter) and how it relates to the use of the same term and the closely related term "nutrition ecology" in human nutrition.

4.2.1 Human Nutrition

In human nutrition, the first published use of either term of which we are aware was in a collection of essays titled *The Feeding Web: Issues in Nutritional Ecology* [13], which examined the impacts on health and ecological sustainability of the industrialization of the US and global food systems. Elaborating on this approach, the term "nutrition ecology" was introduced in 1986, referring to an "interdisciplinary scientific discipline that incorporates the entire food chain as well as its interactions with health, the environment, society, and the economy" where the food chain encompasses "production, harvesting, preservation, storage, transport, processing, packaging, trade, distribution, preparation, composition, and consumption of foods, as well as disposal of waste materials along the food path" [14]. Nutrition ecology is considered a new nutrition science [15], which differs from conventional human nutrition science in two key respects. Firstly, it takes a broader view that extends beyond nutritional biology to encompass also societal and ecological issues. Secondly, it emphasizes interdisciplinary systems science as a framework for dealing with the broad scope of the subject [16].

Nutrition ecology is considered to also be distinct from the field of "nutritional ecology" that has developed in the ecological sciences and anthropology [16]. We are, however, unaware of any explicit discussions on the similarities and differences between the nutritional ecology frameworks in these fields. Since "nutritional ecology" sensu the ecological sciences has recently come to focus on humans [12, 17], as the closely related framework in anthropology has done for some time now (see below), it is worth briefly examining how this term is used in those two fields.

4.2.2 Ecology and Anthropology

An early use of the term "nutritional ecology" is Schneider (1967) [18]. While based in the clinical and public health-related field of nutritional immunology [19], this prescient paper advocates the use of experimental designs in epidemiology derived from "an analysis of the problem in ecological concepts that goes beyond classical concepts in the fields of nutrition and microbiology" [18]. As we discuss below, this adumbrated a central defining feature of the ecological field of nutritional ecology, namely, the importance of applying ecological and evolutionary theory to understand the nutritional biology of animals [20].

The earliest use of the word term "nutritional ecology" that we could locate in the ecological literature is a treatise on grasshopper ecology, published in a 1962 volume of the Memoirs of the Indian Museum [21]. Relevant subsequent uses include Von Goldschmidt-Rothschild and Lüps (1976) [22] and Stanley Price (1978) [23]. The subjects of these publications could not have been more different from the context in which Gussow (1978) [13] used the term in relation to industrialized human environments. The study of Von Goldschmidt-Rothschild and Lüps (1976), for example, examined the extent to which domestic cats in Switzerland prey on small game birds, by dissecting the gut contents of 257 cats that had been shot while hunting [22]. Stanley Price's study assessed the nutritional status of wild hartebeest [23]. As in Von Goldschmidt-Rothschild and Lüps' cat study, Price did this by dissecting out the gut contents of shot animals but went further by measuring the nutritional composition of the eaten foods. He also performed digestion trials that compared the nutrient content of food eaten by captive hartebeest and sheep with the nutrient content of the feces they produced. Given how disparate Stanley Price's study is from Gussow's book, which appeared in the same year, it is not surprising that these works did not refer to each other, and neither did Gussow refer to any of the earlier ecological studies nor Schneider's (1967) paper [13, 18, 23]. This suggests that the term "nutritional ecology" was probably derived independently in nutritional immunology, human nutrition, and ecology.

Over subsequent years, "nutritional ecology" became established as a subfield of ecology in its own right, being applied across animals in general, including insects [24], mammals [25], birds [26], and fish [27], as well as plants [28]. Nutritional ecology in this sense has been defined as "the science of relating an animal to its environment via nutritional interactions" [29]. These interactions involve behavioral and physiological aspects of food and nutrient acquisition and their relationship to the health, growth, and reproduction of animals [30]. A strong emphasis is on the ways that individuals adapt to nutrition-related aspects of the environment, which takes place across various timescales from short-term homeostatic responses to environmental changes, through developmental adaptation and Darwinian adaptation by natural selection [31].

The term "nutritional ecology" has been used in anthropology in much the same way that it developed in ecological sciences [32]. Jenike (2001) defined it as "…the interaction of diet, somatic maintenance, physical activity, and pathogenic agents as they relate to growth, body composition, development, and function in a constraining social, political and natural environment" [32]. This definition was modified by Hockett and Haws (2003), who defined the approach as "…the study of the relationship between essential nutrient intake and its effects on overall human health, including growth and maintenance in individuals and general demographic trends in populations" [33]. Both definitions share, in common with the ecological sciences, an emphasis on the individual within the context of the environment.

4.2.3 Nutritional Ecology and Human Health

The origins of nutrition ecology (in human nutrition) and nutritional ecology (in the ecological sciences and anthropology) thus differed, with the former being focused more on the broader human nutrition system and the latter on the individual within that system. In the terms of ecological science, this broadly reflects the difference between the sub-fields of community ecology (emphasis on community-level issues, such as food webs) and functional ecology (emphasis on the characteristics of individuals that adapt them to their environments) [34]. This difference is not, however, absolute, but more an issue of emphasis. Nutrition ecologists do not exclusively consider the social and environmental dimensions of nutrition, but also the biological dimension [15]. Likewise, nutritional ecologists are aware of the importance of the biological characteristics of individual organisms in shaping the ecological communities within which they exist [20, 35]. Issues of emphasis are, nonetheless, important because they influence the questions, theory, and methods that direct progress in a scientific field.

In the remainder of this chapter, we demonstrate using our own work how the biological theory and methods from nutritional ecology can provide new insight into human nutrition and its links to health [12]. The perspective that we bring derives from almost three decades of research into the nutritional ecology of non-human organisms, ranging from slime molds, yeast, and insects in laboratory studies to giant pandas, monkeys, gorillas, and orangutans in the wild [36]. We do not view this approach to be an alternative to nutrition ecology, but rather a means to expand human nutrition science to integrate the interests and expertise of nutrition ecology with the biological theory and the comparative perspective of nutritional ecology.

4.3 The Importance of Appetite

Among the most emphatic messages to emerge from our studies of non-human animals is that appetite is paramount for understanding foraging, feeding, and its impacts on the animal and, ultimately, on the community of which it is a part. This might seem obvious, except that biological theory combined with studies on the nutritional ecology of non-human animals has provided a more nuanced and powerful concept of appetite than currently exists in human nutrition science.

In this section, we summarize the biological background to the nutritional ecology view of appetite and, in the following section, introduce an approach for modeling and measuring appetite in this sense. Thereafter, we show how these models can help to integrate appetite within the broader human nutrition system to help bridge the gap between the biological theory and methods of nutritional ecologists and the integrative goals of nutrition ecology [15].

4.3.1 Multiple Appetites

To appreciate the subtleties of appetite, we need to step back and ask the question "what is appetite for?" This is a good stage to point out that the "what for?" question is a hallmark of the evolutionary framework on which nutritional ecology is based; it aims to understand biological traits partly through knowledge of what they evolved to achieve [37]. The power of this approach is that it provides an expectation of, firstly, how the biological trait should be designed to achieve its evolved function and, secondly, how it should respond in various circumstances. For example, it can help to understand how the human appetite works and how it might respond to the industrialized food environments of which Gussow (1978) wrote in the book that first used the term "nutritional ecology" in the context of human nutrition [13].

The simple answer to the question "what is appetite for" is that it evolved as a "control center" that directs the animal to meet its nutritional needs, in the same way that thirst directs it to drink and the sensation of cold causes shivering and heat-seeking behavior. To achieve its purpose, however, appetite should be considerably more complex than motivations such as thirst or temperature-related discomfort, for several reasons. First, thirst and temperature regulation have simple endpoints (sufficient hydration and optimal temperature, respectively), whereas an animal's nutrient needs are complex, involving many nutrients, each required at its own particular level. Second, the relative needs for different nutrients change over time and with the circumstances of the animal; for example, female birds have elevated protein and calcium requirements for producing eggs [38], whereas more fat and carbohydrates are needed to fuel the costs of long-distance migration [39]. Meeting multiple and changing nutritional needs in this way, in complex and changing environmental circumstances, is a substantial challenge.

An animal could not deal with this challenge if it had only a single appetite that, for example, caused a bird to eat the same diet regardless of whether reproducing or preparing for migration, any more than we would expect a motor car to have only a single warning light that did not distinguish between an oil and gasoline shortage. Theory therefore predicts that animals should have separate appetites for specific nutrients to help ensure that they eat the specific blend of nutrients that is appropriate for their particular needs at a given time. Such "nutrient-specific appetites" are known to exist for many animals, including humans (see ▶ Sect. 4.5). Although the details of which nutrients are regulated by specific appetites are expected to differ with the ecological and evolutionary circumstances of different species, the universal importance of the macronutrients protein, carbohydrates, and fat means that most animals have separate appetites to regulate either two or all three of these [36]. Some also have specific appetites for the minerals calcium [40] and sodium [41].

4.3.2 Appetites Interact

Important as they are, the possession of nutrient-specific appetites is not on its own sufficient to ensure that animals satisfy their nutrient needs [42]. An added complexity is that most nutrients are not available separately in the environment, but come packaged in mixtures called foods. If a food contains the same ratio of different nutrients as is needed by the animal – for example, high protein relative to carbohydrate and fat when reproducing – then the complex nature of foods is a benefit, because this enables the animal to satisfy its requirements for all nutrients from one source. It is, however, seldom the case that the composition of foods exactly matches the mix of nutrients needed, and this complicates nutrition considerably, both for animals and for attempts by nutritional ecologists to understand animal nutrition.

Where possible, animals deal with this challenge through "complementary feeding," in which specific combinations of nutritionally imbalanced foods are mixed in the right proportions to balance the diet overall. This requires that the appetites for the different nutrients cooperate to ensure that each is eaten at a level that meets, but does not exceed, the respective requirements. Many animals have been shown to balance their diets in this way, both in laboratory experiments and in the wild [36], and so too do humans (see ▶ Sect. 4.5.1).

In reality, however, ecology is not always so obliging as to provide combinations of complementary foods that can be mixed to obtain a balanced diet. Rather, animals often find themselves in situations where the only foods available restrict them to eating a diet that is imbalanced with respect to their nutrient needs. For example, many primates that need both fruits, which are high in sugars and fats, and protein-rich leaves to balance their diet, endure periods of fruit scarcity in which they are forced to eat a leaf-rich diet containing excess protein [43].

In this situation, the appetites for protein and non-protein energy enter into conflict, because the target intakes for both cannot be achieved simultaneously. Rather, the nutritional options available to the animal are to eat the target level of protein and suffer a shortage of non-protein energy, to eat the target level of non-protein energy while overeating protein, or to settle on an intermediate outcome in which it has both a moderate excess of protein and a moderate shortage of non-protein energy. Research has shown that different species resolve this competition between appetites in different ways, which likely reflect the relative costs of under- or overeating the two nutrients [36].

While such interactions between appetite systems are complex, they are also extremely important for understanding how diet influences the behavior, the physiology, and the health of animals. To help simplify the challenge, we have invented a modeling approach called the Geometric Framework for Nutrition (GFN). In the following section, we introduce the basic concepts of GFN and show how they have been applied to understand macronutrient regulation in humans.

4.4 From Concepts to Models: Introduction to the Geometric Framework for Nutrition

The value of the ecology- and evolution-inspired concepts from nutritional ecology is substantially enhanced if these concepts can be expressed in quantitative models. Quantitative models provide a framework within which the various relevant factors can be measured and their relationships explored to interpret the ways in which they interact to influence important outcomes such as energy intake and health [44]. We now demonstrate how the nutritional ecology ideas from the previous section can be modeled using the simple geometry of GFN (◘ Fig. 4.1). Thereafter, we show how these models have been applied to understand macronutrient regulation and its consequences for energy intake in humans.

4.4.1 Model Selection

The first step in constructing a GFN model is to select the nutrients that are relevant to the problem under investigation. In so doing, special care should be taken to heed the dictum attributed to Einstein that "things should be as simple as possible … but no simpler": we should include in the model only those nutrients that we suspect play an important role in the problem, but ensure that all of the nutrients that do so are included. A common problem in human nutrition science is the tendency to attribute outcomes such as obesity to individual nutrients, usually fats or carbohydrates, without regard for how they interact with other nutrients in exerting their influence [12, 45] (please see ▶ Chap. 8).

Since our example concerns obesity, we will focus on the energetic macronutrients protein, fat, and digestible carbo-

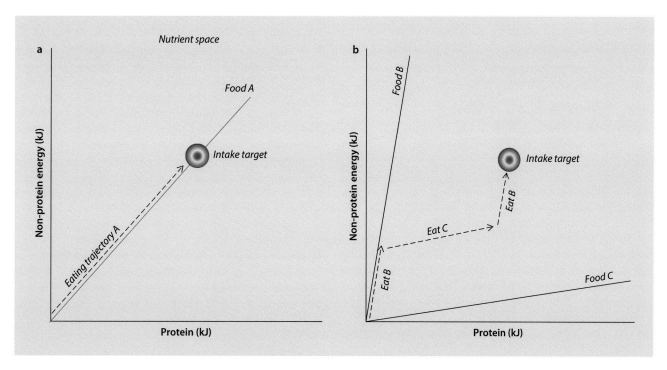

◘ Fig. 4.1 Basic concepts in the Geometric Framework for Nutrition. The "nutrient space" is the space formed by the two nutrient axes, in this hypothetical model protein (horizontal) and combined fats and digestible carbohydrates (non-protein energy) [vertical]. An "intake target" is plotted within the nutrient space, representing the amount and balance of nutrients (in this case protein and non-protein energy) that are targeted by the animal's regulatory systems. Foods are represented as lines, called "nutritional rails," which originate at the origin and project into the graph at angles that represent the ratio of the nutrients that each contains. As the animal eats, it ingests the nutrients in the same ratio as they are present in the food it is eating, and its nutritional state thus changes along a trajectory identical to the rail for that food. The animal can therefore reach its intake target either by selecting a food that has the same ratio of nutrients as the intake target (i.e., a nutritionally balanced food) (food A, in **a**) or by switching between foods that are individually imbalanced but together nutritionally complementary (foods B and C in **b**)

hydrates, although in different contexts we might be interested in other nutrients (e.g., [46]). We will also initially simplify the model by combining fats and carbohydrates into a single category to address the question of how the human appetite for protein (P) interacts with non-protein energy (henceforth NPE) to drive energy intake (Fig. 4.1). Our reasons for doing this are that, firstly, in some contexts, fats and carbohydrates can be regarded as interchangeable sources of metabolic energy for humans [47] and, secondly, as we show in the next section, when all three macronutrients are considered in their own right, distinguishing fats and carbohydrates in the data that we present as an example does not provide a better explanation for energy eaten than does considering them combined. We note, however, that in many other contexts – for example, appetite interactions in mice [48] and the causes of cardiovascular disease in humans [49] – it is important to distinguish fats and carbohydrates and even the different sub-categories of these nutrients. Geometric models can readily be applied in this context.

Having selected the nutrients of interest, the next step is to make a graph where each nutrient is represented on its own axis, as shown in Fig. 4.1. Here, too, there is an important decision to make, regarding the units in which to represent the different nutrients – for example, whether mass (grams) or energy (kilojoules). In our model, we have chosen energy units, because the aim is to understand how different blends of macronutrients influence energy intake.

We have now constructed a *nutrient space*, which provides the platform on which we model how the appetites for different macronutrients interact to influence energy intake (Fig. 4.1).

4.4.2 Selecting an Intake Target: Nutritionally Balanced and Complementary Foods

An important reference point in modeling appetite interactions is the *intake target*, or the point in nutrient space that shows the amounts and balance of the nutrients that the appetite systems will target under circumstances in which they are unconstrained by the quality or quantity of available food (Fig. 4.1). Theory predicts, and studies have shown, that in many cases the selected intake target provides a diet that optimally satisfies the animal's needs for the different nutrients [50–52]. The animal is able to reach its intake target through the appetites for the different nutrients working in harmony to meet the respective intakes that are best for the animal.

The way that the animal satisfies its nutrient needs is, of course, by eating foods, but the important question is *which* foods the animal can eat to reach its intake target. To examine this, we can plot in the model the composition of various foods, as lines called *nutritional rails*, which project from the origin into the nutrient space at an angle determined by the ratio of the nutrients that each contains. As the animal eats, it ingests the nutrients in the same ratio as is contained in the food it is eating. Its nutritional state can therefore be viewed as "moving" through the nutrient space at an angle equivalent to the rail for the food it is eating and over a distance determined by how much of the food it eats. Figure 4.1a shows that food A contains the same ratio of the nutrients as does the intake target (the rail passes through the intake target), and in the spatial metaphor of the model, the animal can therefore "navigate" to the target by eating this food (feeding trajectory A in the figure). This food is thus *nutritionally balanced* with respect to the animal's requirements for protein and non-protein energy.

Foods B and C, by contrast, do not pass through the intake target – they are *nutritionally imbalanced* with respect to P and NPE (Fig. 4.1b). However, since food C contains too much P relative to NPE (falls to the right of the target) and food B too much NPE relative to P (to the left of the target), the two foods are nutritionally *complementary* and can be combined in appropriate proportions to provide a balanced diet. Complementary feeding is shown in the nutrient space as a zigzag trajectory, where each leg represents an amount eaten of the respective foods (Fig. 4.1b).

4.4.3 Negotiating a Compromise: When the Intake Target Cannot Be Reached

Animals can therefore reach their intake target by selecting balanced foods (e.g., Fig. 4.1a) or mixing their intake from nutritionally complementary foods (Fig. 4.1b). What options does the animal have when confined to a single nutritionally imbalanced food or non-complementary combination of imbalanced foods? As discussed above, in this case the target point cannot be reached for both nutrients, and the appetite systems need to negotiate a compromise that minimizes the cost to the animal of its dietary predicament. Such strategies are known as *rules of compromise*.

A distinguishing feature of GFN is that it enables us to model the various rules of compromise and thereby to understand the strategies that animals have evolved to deal with dietary imbalance. Figure 4.2 shows three examples, representing extreme responses chosen from a wide range of possible strategies [36, 53]. In the blue strategy, protein wins over entirely – the animal eats to the point on the nutritional rail where its need for P is met, regardless of whether NPE is under-eaten (on high P:NPE diets) or overeaten (on low P:NPE diets). This pattern, known as *protein prioritization*, has been observed in wild spider monkeys [54] and orangutans [55]. The green strategy is the opposite, namely, prioritization of non-protein energy, where P is over- or under-eaten to meet the NPE target. This pattern is shown by wild mountain gorillas, which overeat protein to obtain the target level of fats and carbohydrates in periods when fruit shortage commits these animals to a high-protein diet [56]. It is has also been observed in several carnivores, including mink [57] and predatory beetles [52]. The red strategy represents the situation where neither nutrient group dominates over the other, but the appetite systems give equal weighting to

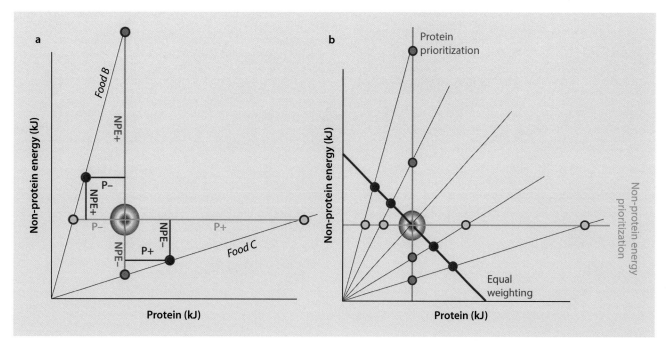

Fig. 4.2 **a** Schematic showing geometric representation of possible regulatory responses to nutrient imbalance. When confined to a single nutritionally imbalanced food (i.e., with a rail that doesn't intersect the intake target), the animal needs to resolve a trade-off between over-ingesting one nutrient and under-ingesting the other. By feeding to the green point, it meets its target for non-protein energy (NPE, comprising fats and carbohydrate combined) at the cost of suffering a shortage of protein of magnitude P- on low P:NPE diets (e.g., if restricted to food B) or a protein excess (P+) on high P:NPE diets (food C). The converse would be true if the animal ate to the blue points – it would meet its protein target, but to do so would have to ingest an excess or deficit of NPE (NPE+ and NPE-, respectively). By feeding to either of the red points, the animal would meet its target for neither nutrient, but would ingest a moderate excess of NPE and a deficit of P on food B or a moderate excess of P and a deficit of NPE on food C. **b** Testing different experimental groups, each on one of a range of foods varying in nutrient balance, provides a description of how the animal resolves the trade-off between over- and under-ingesting nutrients when confined to imbalanced foods, termed a rule of compromise (ROC). Three possibilities are illustrated: the blue symbols represent absolute prioritization of protein (i.e., feeding to the target coordinate for protein regardless of whether this involves over- or under-eating NPE), the green symbols represent NPE prioritization, and the red symbols represent an intermediate response in which the regulatory systems assign equal weighting to excesses and deficits of the two nutrients (P- = NPE+ on food B and P+ = NPE- on food C). Many other configurations are possible

each: it eats to the point on the imbalanced nutritional rail where the ingested excess of one nutrient exactly matches the deficit of the other. This pattern has been observed in wild rhesus macaque monkeys [58] and several species of generalist-feeding herbivorous insects [36].

4.5 The Geometry of Nutrition in Humans: Protein Leverage

How do humans regulate macronutrient intake, and can this help to understand obesity? These questions have been addressed in several independent studies, including randomized control trials [59–61], analyses of data compiled from the literature [62], and survey data of human populations [63–65]. Results of these studies have consistently highlighted the importance of appetite interactions in driving energy intake.

4.5.1 Do Humans Select an Intake Target?

A recent study has addressed this question, with striking results. Following protocols developed by Gosby et al. (2010), Campbell et al. (2016) presented 63 Jamaican volunteers with 3 menus from which to select a diet over 3 days within a residential experimental facility [61, 66]. All 3 menus contained the same 29 dishes, but the compositions of the dishes were manipulated using added protein and carbohydrate such that all options on one menu had 10% of energy from protein, a second had 15%, and a third had protein at 25%. Fat was held constant at 30% for all dishes and menus. Since the human diet seldom contains less than 10% protein or more than 25%, we reasoned that if an intake target exists, then it lies somewhere between these extremes and can be reached by composing a diet from the experimental menus through complementary feeding. The menu with 15% protein represented our best guess at what the composition of the target diet is, because this is close to the mean intake levels found in Western diets [67].

Results are plotted in Fig. 4.3. The three solid radial lines are nutritional rails representing the compositions of the experimental menus, and the blue-shading shows the area that could potentially be reached by subjects in this experiment. Despite the wide range of possibilities, it is striking that all subjects, whether male or female, clustered tightly around a line representing 14.7% protein, although males ate

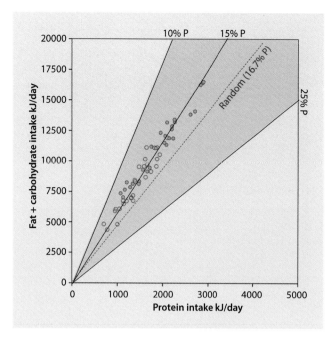
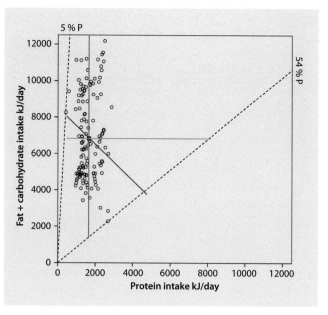

Fig. 4.3 Daily protein and non-protein energy intakes self-selected by 63 adult Jamaican volunteers, averaged for each subject over the 3-day experimental period. Red symbols are females, and blue symbols are males. The solid radial lines are nutritional rails representing the compositions of the three experimental menus from which the diets could be freely selected (10, 15, and 25% protein by energy). The pale blue region represents the range of diets that could theoretically be selected from these menus, and the dotted red line represents the expected outcome if the subjects mixed the diets randomly (16.7% P). The data show that subjects selected a diet of 14.7% protein, which was significantly different from the random outcome. (Adapted from Campbell et al. [61]. With permission from Creative Commons License)

Fig. 4.4 Interaction of human appetite systems with dietary macronutrient ratios. Data are protein (x-axis) and non-protein (fat + carbohydrate, y-axis) ad libitum energy intakes by subjects restricted to 1 of 138 experimental diets [62, 69]. The black dashed radials represent the nutritional rails for the diets with the highest (54%) and lowest (5%) proportional protein content. The area between these radials is the region of the nutrient space within which points for nutrient intakes are constrained to lie, with the pattern of actual intakes being determined by the ways that appetites for protein, fat, and carbohydrate interact. The blue, red, and green lines represent the protein prioritization, NPE prioritization, and equal weighting models from Fig. 4.2. The analysis shows that humans maintain absolute protein intake relatively tightly, with non-protein energy intake varying more passively with dietary macronutrient ratios. (Adapted from Raubenheimer and Simpson [12]. With permission from Annual Reviews)

more of the diet overall than females, which is not surprising given their larger body size.

One possible explanation for this pattern is that the subjects did not distinguish between the three diets based on their macronutrient content, but simply ate equal amounts of each. It can be calculated that if this were the case, the intake points would align along the dashed radial line, representing a diet with 16.7% protein. However, the actual intakes differed with a high degree of statistical certainty from this line, showing that they represent macronutrient regulation to a P:NPE intake target. An earlier experiment by Simpson et al. (2003) showed similar results [68].

4.5.2 What Is the P:NPE Rule of Compromise in Humans?

Several studies, including randomized control trials [59, 60, 68], analyses of experimental data compiled from the literature [62], and observational studies using diet surveys [63], have addressed the question of how humans respond to macronutrient-imbalanced diets. Since the results all tell the same story, we will present only the data of the literature compilation of Gosby et al. (2013), as expanded by Raubenheimer et al. (2015) [62, 69].

What all of these studies in the compilation have in common is that each experimental subject was restricted to one of a range of single diets, each with fixed P:NPE ratio, and allowed to eat as much of their respective diet as they wished. The experiments therefore test how the human appetite systems interact to determine protein, fat, and carbohydrate intake as dietary macronutrient balance varies – i.e., the rule of compromise for these nutrients. The result is shown in Fig. 4.4. Protein intake remained relatively stable over a wide range of diet compositions, while the intake of fats and carbohydrates increased with decreasing dietary P:NPE ratio. Humans thus show the protein prioritization pattern of macronutrient regulation (Fig. 4.2).

This analysis shows that information about the dynamics of human appetite systems is essential for understanding why we eat the amounts of nutrients and energy that we do. The main conclusion that it presents is by no means obvious: it suggests that humans will overeat fats and carbohydrates not because they have a particularly strong drive to eat these nutrients, but because of a strong appetite for *protein*. On the other hand, we should not interpret this to suggest that the human appetite is exclusively about protein. We know that this is not the case, because when allowed to select a diet from nutritionally complementary foods, humans regulate the intake of both P and NPE (Fig. 4.3). Rather, in circumstances where it is possible, the appetites for different macronutrients cooperate to select a balanced diet, but when limits on available foods prevent this, protein regulation overrides and fat and carbohydrate intakes follow more passively.

This phenomenon, where the appetite for protein influences the intake of other nutrients such as fats and carbohydrates, has been termed *protein leverage* [47].

4.6 Beyond Appetites

Our discussion to this point illustrates the logical progression of research within the nutritional ecology framework. We began by showing that evolutionary and ecological reasoning predicts that animals should have separate appetites for different nutrients and that these appetite systems should interact to determine nutrient intakes. We then introduced nutritional geometry as an approach for measuring such appetite interactions and demonstrated how it has been applied to humans. Results showed that the humans in our study selected an intake target of approximately 15% energy from protein, and when restricted to macronutrient-imbalanced diets that prevented them from achieving this target, they showed the protein prioritization rule of compromise, in which fats and carbohydrates are passively over- or under-eaten as the percentage of dietary protein varies.

This is, principally, an examination of organismal (in this case human) biology, which, as we commented above, is an important starting point for nutritional ecology research. The next step is to expand the model to understand how the trait, in this case appetite interactions, engages with broader aspects of the animal's nutritional biology, including specific nutrient requirements and the food environment that nutrition ecologists have emphasized in their writing.

One hypothesis that addresses this is the *protein leverage hypothesis* (PLH) [47]. PLH proposes that the protein prioritization pattern of macronutrient regulation has interacted with reductions in the P:NPE ratio of the human diet to drive fat and carbohydrate overconsumption and obesity. This hypothesis can potentially provide a powerful bridge between human biology, modern human environments, and health because, if true, it simplifies the search for the causes of the obesity epidemic. It does this by focusing attention on a very simple question about the role of environmental change in driving this epidemic: *what is the cause of protein dilution in the human diet that leads our appetite systems to overeat fats and carbohydrates?*

Simple questions are not, however, necessarily simple to answer, and this is no exception. Like human biology, modern industrialized human food environments are extremely complex, and the interactions between human biology and modern environments potentially all the more so. To help deal with this complexity, we have adopted from nutritional ecology a form of nutritional geometry called the right-angled mixture triangle (RMT); [70]) (Fig. 4.5). In the remainder of this chapter, we introduce the RMT and demonstrate in the context of the protein leverage hypothesis how it can provide an aid for understanding the biology-environment interactions that influence health in industrialized food environments.

4.6.1 The Geometry of Mixtures: Three Components in Two Dimensions

As its name implies, the RMT is an approach for modeling mixtures. The axes are therefore scaled as the proportional (or %) contribution of each nutrient to the overall mixture, rather than absolute amounts, as is the case in the GFN models of appetite interactions (e.g., kilojoules eaten per day, as in Figs. 4.3 and 4.4). The key difference between these two variants of nutritional geometry is thus that one is proportions-based and the other is amounts-based. Beyond that, they share much in common and are, in fact, complementary approaches for modeling nutrition.

As is the case for amounts-based geometric models, the first step in building an RMT model is to decide which nutrients are most relevant to the problem. Since we are extending the analysis of how macronutrient appetites interact to influence energy intake, we will include in our model the macronutrients protein, fats, and carbohydrates expressed in energy units. An important advantage of the RMT approach, however, is that it enables all three nutrients to be modeled in a simple two-dimensional graph. Amounts-based geometry can also cope with more than two nutrients, but generally only by simplifying the model (e.g., combining two nutrients into a single axis, as for fats and carbohydrates in Figs. 4.3 and 4.4) or else by plotting three two-dimensional graphs (protein vs. fat, protein vs. carbohydrates, and fat vs. carbohydrates).

To illustrate how three components are represented in a two-dimensional RMT, consider the black point in Fig. 4.5a, representing the macronutrient composition of a sample of rice. The percentage contribution of protein to total macronutrient energy in the rice is 5% (x-axis) and of fat is 10% (y-axis). Since % protein + % fat + % carbohydrate must add up to 100% of macronutrient energy, it is easy to see that % carbohydrate = 100 − (%protein + %fat) = 100 − (5 + 10) = 85. Geometrically, 85% carbohydrate is represented by a diagonal line that connects the same value on the x- and y-axes and intersects the point representing the sample of rice. Any mixture of macronutrients that comprises 85% of energy from carbohydrates will fall on this line, which is plotted in the figure as the black dotted diagonal labeled "85."

By the same logic, it can be seen at a glance that the peas in the plot comprise 25% of energy from protein, 5% from fat, and 70% from carbohydrate, and the steak comprises 40% protein, 60% fat, and 0% carbohydrate.

4.6.2 A Hierarchy of Mixtures

Just as the macronutrients combine in specific proportions in foods, so too do foods combine into meals, meals into diets,

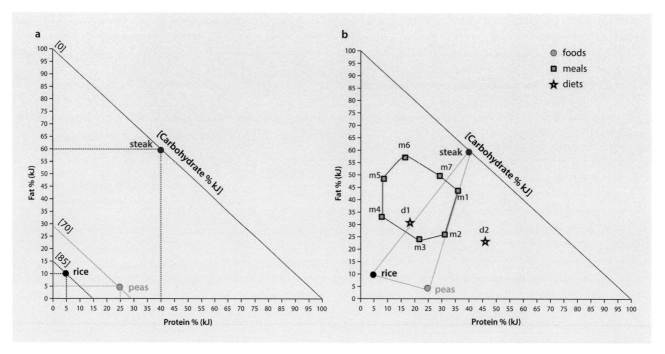

Fig. 4.5 **a** Right-angled mixture triangle [70] illustrating how components (in this case the macronutrients) combine into foods (rice, peas, and steak). Points represent the percentage contributed by each component (protein, fat, and carbohydrate) to the sum of the three. Thus, the macronutrient composition of rice is 5% protein and 10% fat, and since protein, fat, and carbohydrate sum to 100%, carbohydrate = 100 − (5 + 10) = 85%. This value is represented by the negative dashed diagonal joining 15% on the x- and y-axes, such that any mixture of macronutrients containing 85% of carbohydrate will fall on that line. Likewise, the peas contain 25% protein, 5% fat, and 70% carbohydrate, and the steak contains 40% protein, 60% fat, and 0% carbohydrate. **b** Foods combine into meals (m1–m7), and meals combine into diets (d1 and d2). A meal composed of two foods (e.g., peas and steak) is constrained to fall on the line connecting those foods (e.g., m1 and m2), with the exact position along the line being determined by the proportion of the foods in the meal. Adding a third food (e.g., rice) expands the set of possibilities to a triangle (meals m1, m2, and m3 can be composed from the three foods, but m4–m7 cannot). By extension, diets d1 can be composed from meals m1 to m7, but d2 cannot. (Adapted from Raubenheimer and Simpson [12]. With permission from Annual Reviews)

and diets into dietary patterns. Recent work in nutritional ecology has emphasized the value of considering all of these levels for understanding the nutritional strategies of animals [71]. This is critically important in human nutrition, because each level in the hierarchy engages in different ways with the complex organism-environment interface of humans in industrialized food environments [12].

For example, nutrients, the base level in the mixture hierarchy, interact with physiology, by engaging taste receptors, appetites systems, and numerous physiological pathways relevant to health (e.g., the insulin signaling system). It is not, however, nutrients that we buy, but principally foods, and to understand our shopping choices, we need to consider also this level in the hierarchy. Although some foods are eaten directly, the greatest portion of the human diet is eaten as mixtures of foods, called meals. Meals, therefore, are important levels of focus for understanding human eating choices. And yet neither foods nor meals are the primary link between nutrition and health; for that we need to consider the long-term cumulative intakes of foods and meals, namely, diets. To close the circle, diets impact health and disease principally via their primary components, the nutrients.

A powerful aspect of RMT plots is that they can model all levels in this hierarchy of mixtures, as illustrated in Fig. 4.5b. Consider, for example, a meal comprised of two foods, peas and steak. The macronutrient composition of this meal is constrained to fall on the line connecting these foods, with the exact position determined by the proportions of the two foods in the meals. Thus, meal compositions labeled m1 and m2 are attainable from peas and steak, but m3–m7 are not. If additional foods are included in the diet, then the set of possible meal compositions expands to a space. For example, a meal consisting of peas, steak, and rice can take on any macronutrient composition that falls within the triangle formed by these foods (e.g., m3), but no composition that falls outside of the triangle (m4–m7). By the same logic, meal m1–m7 could combine into a weekly diet that falls within the polygon formed by connecting these meals (e.g., d1), but not outside of it (d2).

4.6.3 Dietary Macronutrient Balance

Distinguishing and interrelating different levels of the dietary mixture hierarchy in this way provides important benefits for examining the ways that human nutritional biology engages with industrialized food environments. However, to realize the potential of this approach, we need to move beyond describing mixtures such as foods, meals, and diets and examine how they link to human biology and to the food environments with which our biology interacts. An important first step is to relate these compositional data to nutrient requirements.

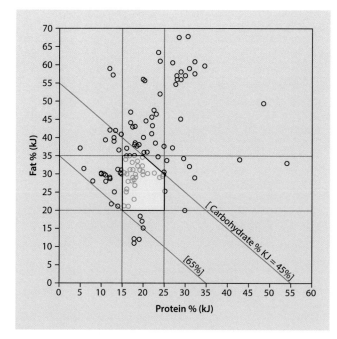

◘ Fig. 4.6 Macronutrient compositions of the 138 experimentally fixed diets plotted from ◘ Fig. 4.4. The yellow polygon is an integrated representation of Australian/New Zealand Acceptable Macronutrient Distribution Range (AMDR; %P = 15–25%, %F = 20–35%, %C = 45–65%), such that diet points falling within this polygon represent macronutrient-balanced diets and those falling outside are macronutrient-imbalanced. (Adapted from Raubenheimer and Simpson [12]. With permission from Annual Reviews)

To illustrate, ◘ Fig. 4.6 presents in RMT format the same data as plotted in ◘ Fig. 4.4, representing the compositions of experimental diets compiled from the literature. As the figure shows, the data spanned a wide range of protein–fat–carbohydrate mixtures. To examine how these mixtures relate to proportional macronutrient requirements, we used for reference the Acceptable Macronutrient Distribution Ranges (AMDR) for Australia and New Zealand [72]. According to these recommendations, 15–25% of energy intake should come from protein, 20–35% from fat, and 45–65% from carbohydrate.

Delineating these individual ranges in ◘ Fig. 4.6 enables us to identify the region representing diets that satisfy all three recommendations, plotted as yellow polygon. Any diet with a composition that falls within the yellow region is thus macronutrient-balanced with respect to the AMDR, and any diet that falls outside is macronutrient-imbalanced.

4.6.4 Relationships Between Macronutrient Balance and Energy Intake

Central to the protein leverage hypothesis is the question of how energy intakes relate to dietary macronutrient ratios. We already have shown that low dietary protein leverages the intake of excess fat and carbohydrate and now address the implications for total energy intake. To do this, in ◘ Fig. 4.7, we have constructed a response surface onto the data from ◘ Fig. 4.6 that relates the ad libitum energy intakes (which was a voluntary response of the subjects) to macronutrient ratios of the diet (the experimentally fixed variable) [69].

The analysis shows that total energy intakes increased (intake values grade from blue to red) as the percentage of energy contributed to the diet by protein decreased (movement from right to left on the protein axis). This result substantiates the protein leverage effect shown in ◘ Fig. 4.4, but takes it further. First, it shows that the leveraging by protein of fat and carbohydrate intake (◘ Fig. 4.4) translates into increased total energy consumption, as predicted by the protein leverage hypothesis. Second, energy intake changed as dietary protein varied (along the x-axis), but remained relatively constant along the fat axis. This suggests that, for these data at least, the main determinant of energy intake was the protein energy ratio (x-axis), with little effect of the relative proportion of fat:carbohydrate, and justifies our decision, discussed above, to combine fat and carbohydrate into a single axis (◘ Fig. 4.4). Finally, plotting the data in this way helps to integrate additional factors into the model, such as energy balance.

4.6.5 Energy Balance

An important reference point for predicting the effects of protein leverage on obesity is Equilibrium Energy Intake (EEI), or the point at which energy intake matches energy expenditure. Energy intakes that exceed EEI signify positive energy balance and risk of fat accumulation, while lower intakes signify negative energy balance.

To incorporate the concept of energy balance into the model relating macronutrient ratios to energy intake, we estimated the EEI (given body size, sex, and expected activity levels under the experimental conditions) for the subjects from the experiments represented in the analysis to be 8813 kJ/day [69]. This value can be incorporated into the model as an EEI contour, represented in ◘ Fig. 4.7a as the bold dashed line. Intakes to the left of this line represent positive energy balance, and intakes to the right represent negative energy balance.

We can now relate across the experimental population dietary macronutrient balance (position in relation to the AMDR polygon) to energy intakes and energy balance, to predict the compositions of diets that will cause human appetites to drive energy overconsumption. An interesting feature of the model is that the EEI contour passes through the AMDR region. Assuming the same applies more generally within the relevant populations, this suggests that following the official New Zealand and Australian recommendations for proportional macronutrient intakes would spontaneously lead to balanced energy intake.

For comparison, we have also plotted in ◘ Fig. 4.7a the US AMDR [73]. The recommendations for carbohydrate and fat are the same as the Australian and the New Zealand AMDR, but the range for protein is broader, spanning 10–35% of energy intake. The model suggests that following the low end of the protein range (10–15% of energy intake from protein) would lead to energy overconsumption and

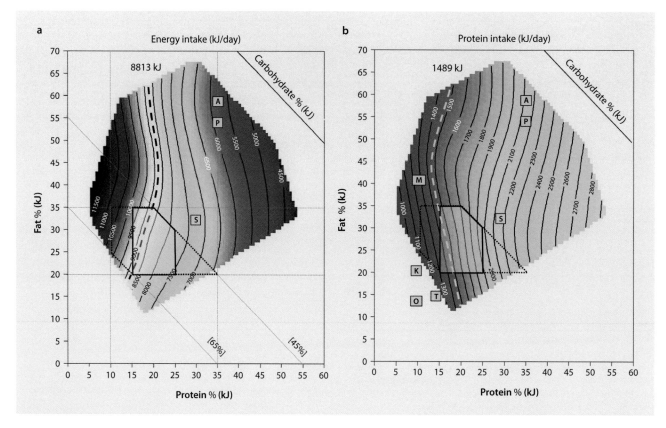

Fig. 4.7 **a** Response surface showing ad libitum daily energy intakes associated with the experimental diet compositions plotted in Figs. 4.4 and 4.6. The dashed contour represents estimated equilibrium energy requirements (8813 kJ) for sex and weight assuming a physical activity level (PAL) of 1.5, which is commensurate with activity levels in the experimental subjects. The data suggest that energy equilibrium was achieved on diets with 15–20% protein, with energy balance being negative and positive for diets with higher and lower % protein, respectively. The model is consistent with the association between weight loss and high-protein diets, such as the Atkins (A), Protein Power (P), and Sugar Busters (S) diets: their macronutrient compositions fall within the blue region of low ad libitum energy intakes. The dotted polygon represents the AMDR for the USA, which has the same ranges for fat and carbohydrate as the Australia/New Zealand AMDR, but a wider protein range (spanning 10–35%). **b** Response surface showing ad libitum protein intakes associated with the data in **a**. The dashed contour represents approximate average protein requirements for the study population (1489 kJ). The figure shows that protein intakes considerably higher than estimated requirements are associated with diets having macronutrient compositions equivalent to high-protein weight loss diets (A Atkins, P Protein Power, and S Sugar Busters). Conversely, low protein intakes are likely to be associated with the macronutrient composition of the diets associated with exceptionally healthy and long-lived human populations, the Mediterranean (M), Kitavan Islander (K), Tsimane (T), and traditional Okinawan (O) diets. (Adapted from Raubenheimer et al. [69]. With permission from The Obesity Society)

positive energy balance. In contrast, the high end of the US range for protein (25–35%) corresponds with low energy intakes and negative energy balance (the blue region). Indeed, the protein leverage effect can help to explain why many popular high-protein weight loss diets including the Atkins, Paleo, and Sugar Busters, which are shown in the figure, fall within this region.

4.6.6 Beyond Energy: Protein Intake

While the above analysis suggests a reason based on human appetite interactions (protein leverage) why high-protein diets are effective for weight loss (in the short term at least), we caution that low energy intake is not the only effect of consuming diets with high protein energy ratios. Another outcome is that compensation for the low levels of fats and carbohydrates results in protein overconsumption, albeit to a smaller extent than low protein leads to fat and carbohydrate overconsumption [12].

To illustrate, in Fig. 4.7b we have plotted the corresponding protein intake surface for the data shown in Figs. 4.6 and 4.7a. We have also calculated the Estimated Average Protein Requirements for the subjects in the experimental trials to be 1489 kJ and plotted this as a contour equivalent to the EEI contour in Fig. 4.7a. Viewing the data in this way clearly shows the increase in absolute protein intakes with increasing dietary % protein (left to right on the x-axis). It also shows that estimated dietary protein requirements are met for diets with approximately 15–20% protein energy, and at higher dietary protein densities, excess protein is ingested.

There is now strong evidence that excess protein intakes are associated with negative cardiometabolic profiles and accelerated aging, especially when coupled with low carbohydrate intakes [45, 74]. Consistent with this is the observation that the healthiest dietary patterns, including the Mediterranean, traditional Okinawan [75], Kitavan Islanders [76], and Tsimane [77] diets, are associated with low dietary protein densities and low protein intakes, as shown in ◘ Fig. 4.7b. This should caution against high-protein diets, such as the Atkins, high-protein Paleo, and Sugar Busters diets (◘ Fig. 4.7), except as therapeutic interventions for weight loss. It also raises questions about the high end of the protein range sanctioned by the US AMDR.

In comparing ◘ Fig. 4.7a, b, the alert reader might have noted an apparent inconsistency. The Mediterranean, Kitavan Islander, traditional Okinawan, and Tsimane diets all have low protein energy ratios (between 10% and 15%) (◘ Fig. 4.7b) and under the model presented in ◘ Fig. 4.7a should thus be associated with excess energy intake, and yet obesity is not a problem within these societies. The reason for this apparent inconsistency is that the parameters of such models, including the shape of the surface relating energy consumption-to-macronutrient ratios, are population-specific and might be influenced by differences in nutrient and other aspects of the respective food environments. For example, the low-protein dietary patterns in ◘ Fig. 4.7b are associated with high fiber content compared with Westernized diets to which the model in ◘ Fig. 4.7a applies. High fiber is likely to induce satiety [78] at lower levels of protein (and energy) intakes than are low fiber diets [69], as has been demonstrated in mice [45]. The combination of low protein and high fiber thus has the double health benefits of limiting protein intake while avoiding energy overconsumption. The extension to consider also fiber emphasizes the importance of matching the model to the context and also demonstrates how these models can be built incrementally to incorporate multiple food constituents. Equally important is their extension to include broader components of the food environment, beyond diet composition.

4.6.7 Interactions of Appetite with the Food Environment

To this point we have built a model that integrates human appetite interactions with a range of factors relevant to the relationships between diet and health, including the nutrient-food-meal-diet mixture hierarchy, dietary macronutrient ratios, intakes of energy and protein, recommended dietary macronutrient proportions and protein intakes, and energy balance. The model demonstrates one advantage of doing this: it helps to identify how these factors interrelate to explain the links between diet, health and disease.

Another advantage of building such a model is that it provides a context for identifying important aspects of our food environment that might influence the relationships within the model. For example, among the most salient and influential aspects of industrialized food environments is economics, giving rise to the question of whether the cost of foods might play a role in influencing the macronutrient composition of our diets. To address this, Brooks et al. (2010) calculated the relationship between the concentration (g/100 g) of protein, fats, and carbohydrates and the cost (in US dollars/100 g) of 106 supermarket foods [79]. Results showed that the cost of supermarket foods is positively related to their protein content (◘ Fig. 4.8). This suggests that economic considerations might be one factor that contributes toward diluting dietary protein content in industrialized food environments, an influence that our model has shown is transduced via the protein leverage effect into increased energy intake. In this way, protein leverage might help to explain the well-established association between lower socioeconomic status and obesity [80].

We might likewise address the question of why the USDA AMDR spans such a wide range of dietary protein densities, encompassing both low-protein diets (10–15% protein), which our model suggests are likely to be associated with excess energy intake (◘ Fig. 4.7a), and the high end (25–35%), associated with excess protein intake and premature aging (◘ Fig. 4.7b). One possibility is that this reflects influence on research and government policy by the food industry, rather than health considerations. For example, the sugar and affiliated industries selectively sponsor research that casts doubt on recommended upper limits to sugar intake, and the meat, dairy, and egg industries do the same for protein [81, 82]. These industries also exert influence on dietary guidelines through political lobbying [82].

Several other possible facets of industrialized food environments that might interact with human appetite systems to influence health have been identified. These include the influx of low-protein processed foods into the human food

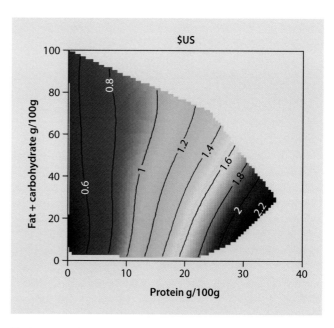

◘ **Fig. 4.8** Relationship between the concentration (g/100 g) of macronutrients (protein, fat, and non-structural carbohydrates) and the cost ($US/100 g) of 106 supermarket foods. Cost increases from dark blue to red. The graph suggests that the cost of food increases with protein density but is unaffected by fat and carbohydrates. (Reprinted with permission from Raubenheimer et al. [17])

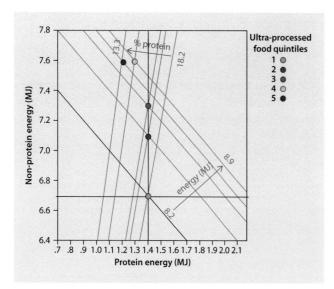

● Fig. 4.9 Relationship between ultra-processed food consumption and protein leverage. The symbols represent protein and non-protein energy intakes for the lowest (green) to highest (red) quintiles of ultra-processed food (UPF) consumption reported in the National Health and Nutrition Examination Survey 2009–2010. The negatively sloped diagonals represent daily total energy intakes (calculated as the sum of X + Y), and the positive radials represent the dietary protein:non-protein energy ratio (X/Y). The dark vertical, horizontal, and diagonal lines represent alternative models to explain the data, as in ● Figs. 4.2 and 4.4. The data show that increased inclusion of UPF in the diet corresponds with reduced dietary protein density (18.3–13.3%) and increased total energy intake (8.2–8.9 MJ), as predicted by the protein leverage hypothesis. (Reprinted with permission from Martinez Steele et al. [64])

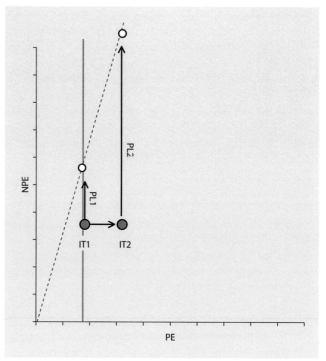

● Fig. 4.10 Schematic showing the effect on protein leverage of an increase in the protein coordinate of the intake target, as might come about through decreased protein efficiency. The x-axis represents protein energy (PE) and the y-axis energy from carbohydrates and fat (NPE). The dashed radial shows the macronutrient composition of a food that has a lower PE:NPE ratio than intake target IT1. The arrow labeled PL1 denotes the extent to which surplus intake of carbohydrate and fat is leveraged by the mismatch between the PE:NPE ratio of the food relative to target IT1. For the same food, protein leverage is greatly exacerbated (PL2) for a small change in the protein coordinate of the intake target (IT1 increases to IT2). (Reprinted with permission from Raubenheimer et al. [17])

supply (● Fig. 4.9) [64] and the reduction in plant protein that may be associated with rising atmospheric carbon dioxide [17], both of which could potentially have the effect of diluting the energy contribution of protein, thereby increasing energy intake.

In addition to such influences on the compositions of foods and diets, another interesting possibility is that environmental factors might interact directly with human biology to influence the parameters of protein leverage. For example, any factor that reduces protein efficiency could exacerbate protein leverage, by increasing the intake target of the protein appetite, thus requiring a stronger compensatory response to protein dilution (● Fig. 4.10) [47]. Such environment-induced variation in protein leverage might explain a number of poorly understood correlations between obesity and environmental factors. One example is the association of obesity with recent cultural transitions from traditional high-protein diets to Westernized diets rich in fats, oils, and simple carbohydrates [47]; another is and the vulnerability to later-life obesity of infants fed high-protein milk formulas [17]. In both cases, high-protein diets are hypothesized to developmentally program low-protein efficiency, thus exacerbating protein leverage (● Fig. 4.10).

We cite these examples to illustrate how an understanding of the dynamics of human appetite systems might help to illuminate the factors that have driven the epidemic of obesity and associated disease in recent decades. The broader message, however, is that a nutritional ecology perspective, which emphasizes the interaction between biological traits and food environments, can provide a structured research framework for generating and testing hypotheses regarding the causes of health and disease in our radically altered industrialized food environments.

4.7 Conclusions

Statistics such as those with which we opened this chapter leave no doubt that nutrition science ought to be doing a better job of preventing premature deaths that are associated with malnutrition and chronic disease. The challenge, however, is to understand how the diverse and complex set of interacting causes drives the problem and to identify the key control points that are amenable to intervention to improve nutrition-related health. There is now widespread recognition that success will require a systems-based approach, which recognizes that relevant causal components are distributed across and between domains representing diverse academic disciplines and societal sectors [8–12, 83]. The leadership, however, must come from nutrition science, which at present is ill-equipped for the task.

We have suggested in this chapter that a constructive branch for interdisciplinary engagement is with the basic biological sciences. Although nutrition science clearly has drawn heavily on chemistry, molecular biology, and physiology, its engagement with the core theory of biology – evolutionary and ecological theory – has been rudimentary. Such theory can provide a powerful framework for identifying pivotal systems components and interactions, and in this way direct and simplify the task, in much the same way that aerodynamics theory can help to direct aviation research. Ecology and evolution also provide a broad comparative perspective, which helps to identify patterns and generalities across a wide range of species and environments and, in this way, enrich the understanding of human-environment interactions. Nutritional ecology is the branch of the natural sciences that applies this approach in the context of nutrition.

As an example, we have shown how biological theory predicts that separate appetites would exist for particular nutrients and that these appetites would interact to broker beneficial outcomes across the range of varied food environments within which they evolved. We introduced nutritional geometry as an approach for investigating appetite interactions and examining how they are linked to broader aspects of human biology and industrialized food environments. Our analysis suggests, somewhat counter-intuitively, that the human propensity to overeat fats and carbohydrates is closely linked to our appetite for protein, via protein leverage. This, in turn, suggests a different focus for examining the causes of obesity, through drawing attention to the factors that influence dietary protein density. Our analysis also cautions against the common tendency to assume that if "too little is bad, a lot must be good," by highlighting the dangers both of diets with too low and too high protein energy density. It emphasizes the importance of dietary *balance*.

In closing, we emphasize that our main goal is not to suggest that we have solved the problem of energy overconsumption, obesity, and related diseases, but rather to introduce a biologically inspired approach that can help to structure nutrition research. Beyond the macronutrients and their different types and constituents, other dietary components such as fiber and micronutrients clearly are relevant to the problem, and likewise, many nutrient combinations are important for various other aspects of health. We suggest, however, that these relationships are best examined in a framework that is guided by biological theory and which examines the interactions among nutrients rather than considering them separately.

References

1. IFPRI. 2016 Global nutrition report. From promise to impact: ending malnutrition by 2030. 2016.
2. FAO. The State of Food Insecurity in the World 2015. 2015.
3. Ng M, Fleming T, Robinson M, Thomson B, Graetz N, Margono C, et al. Global, regional, and national prevalence of overweight and obesity in children and adults during 1980-2013: a systematic analysis for the Global Burden of Disease Study 2013. Lancet. 2014;384(9945):766–81.
4. Di Angelantonio E, Bhupathiraju SN, Wormser D, Gao P, Kaptoge S, de Gonzalez AB, et al. Body-mass index and all-cause mortality: individual-participant-data meta-analysis of 239 prospective studies in four continents. Lancet. 2016;388(10046):776–86.
5. Lim SS, Vos T, Flaxman AD, Danaei G, Shibuya K, Adair-Rohani H, et al. A comparative risk assessment of burden of disease and injury attributable to 67 risk factors and risk factor clusters in 21 regions, 1990–2010: a systematic analysis for the Global Burden of Disease Study 2010. Lancet. 2012;380(9859):2224–60.
6. Dangour AD, Mace G, Shankar B. Food systems, nutrition, health and the environment. Lancet Planet Health. 2017;1:e8.
7. Network AS. Food systems, nutrition, health and the environment. https://www.aviation-safety.net/statistics/, downloaded March 30 2017.
8. Hummel E, Hoffmann I. Complexity of nutritional behavior: capturing and depicting its interrelated factors in a cause-effect model. Ecol Food Nutr. 2016;55:241–57.
9. Bennett BJ, Hall KD, Hu FB, McCartney AL, Roberto C. Nutrition and the science of disease prevention: a systems approach to support metabolic health. Ann N Y Acad Sci. 2015;1352:1–12.
10. Allison DB, Bassaganya-Riera J, Burlingame B, Brown AW, le Coutre J, Dickson SL, et al. Goals in nutrition science 2015-2020. Front Nutr. 2015;2:26.
11. Tapsell LC, Neale EP, Satija A, Hu FB. Foods, nutrients, and dietary patterns: interconnections and implications for dietary guidelines. Adv Nutr. 2016;7(3):445–54.
12. Raubenheimer D, Simpson SJ. Nutritional ecology and human health. Annu Rev Nutr. 2016;36:603–26.
13. Gussow JD. The feeding web: issues in nutritional ecology. Palo Alto: Bull Publishing Company; 1978. p. 457.
14. Leitzmann C. Nutrition ecology: the contribution of vegetarian diets. Am J Clin Nutr. 2003;78(3):657S–9S.
15. Leitzmann C, Cannon G. Dimensions, domains and principles of the new nutrition science. Public Health Nutr. 2005;8(6A):787–94.
16. Schneider K, Hoffmann I. Nutrition ecology - a concept for systemic nutrition research and integrative problem solving. Ecol Food Nutr. 2011;50(1):1–17.
17. Raubenheimer D, Machovsky-Capuska GE, Gosby AK, Simpson S. Nutritional ecology of obesity: from humans to companion animals. Brit J Nutr. 2014;113:S26–39.
18. Schneider HA. Ecological ectocrines in experimental epidemiology. A new class, the "pacifarins", is delineated in the nutritional ecology of mouse salmonellosis. Science. 1967;158(3801):597–603.
19. Beisel WR. History of nutritional immunology - introduction and overview. J Nutr. 1992;122(3):591–6.
20. Raubenheimer D, Simpson SJ, Mayntz D. Nutrition, ecology and nutritional ecology: toward an integrated framework. Funct Ecol. 2009;23(1):4–16.
21. Misra SD. Nutritional ecology of the clear-winged grasshopper, *Camnula pellucida* (Scudder) (Orthoptera, Acrididae). Mem Indian Mus. 1962;14(3):87–172.
22. von Goldschmidt-Rothschild B, Lüps P. Nutritional ecology of "wild" domestic cats (*Felis silvestris f. catus* L.) in the Canton of Bern (Switzerland). Rev Suisse Zool. 1976;83(3):723–35.
23. Stanley Price MR. The nutritional ecology of coke's hartebeest (Alcelaphus buselaphus cokei) in Kenya. J Appl Ecol. 1978;154(1):33–49.
24. Scriber JM, Slansky F. The nutritional ecology of immature insects. Annu Rev Entomol. 1981;26:183–211.
25. van Soest PJ. Nutritional ecology of the ruminant. Ithaca/New York: Cornell University Press; 1994.
26. Carey C, editor. Avian energetics and nutritional ecology. New York: Springer; 1996.

27. Clements KD, Raubenheimer D, Choat JH. Nutritional ecology of marine herbivorous fishes: ten years on. Funct Ecol. 2009;23:79–92.
28. Yanbuaban M, Nuyim T, Matsubara T, Watanabe T, Osaki M. Nutritional ecology of plants grown in a tropical peat swamp. Tropics. 2007;16(1):31–9.
29. Parker KL. Advances in the nutritional ecology of cervids at different scales. Ecoscience. 2003;10(4):395–411.
30. McWilliams SR. Ecology of vertebrate nutrition. eLS: Macmillan Publishers; 2002.
31. Raubenheimer D, Simpson SJ, Tait AH. Match and mismatch: conservation physiology, nutritional ecology and the timescales of biological adaptation. Phil Trans R Soc B. 2012;367(1596):1628–46.
32. Jenike MR. Nutritional ecology: diet, physical activity and body size. In: Panter-Brick C, Layton RH, Rowley-Conwy P, editors. Hunter-gatherers: an interdisciplinary perspective. Cambridge: Cambridge University Press; 2001. p. 205–38.
33. Hockett B, Haws J. Nutritional ecology and diachronic trends in paleolithic diet and health. Evol Anthr. 2003;12(5):211–6.
34. Raubenheimer D, Boggs CL. Nutritional ecology, functional ecology and Functional Ecology. Funct Ecol. 2009;23(1):1–3.
35. Simpson SJ, Raubenheimer D, Charleston MA, Clissold FJ, ARC-NZ Vegetation Function Network Herbivory Working Group. Modelling nutritional interactions: from individuals to communities. Trends Ecol Evol. 2010;25(1):53–60.
36. Simpson SJ, Raubenheimer D. The nature of nutrition: a unifying framework from animal adaptation to human obesity. Princeton: Princeton University Press; 2012. p. 239.
37. Tinbergen N. On aims and methods in ethology. Zeitschrift fur Tierpsychologie. 1963;20(4):410–33.
38. Robbins CT. Estimation of the relative protein cost of reproduction in birds. Condor. 1981;83(2):177–9.
39. Bairlein F, Fritz J, Scope A, Schwendenwein I, Stanclova G, van Dijk G, et al. Energy expenditure and metabolic changes of free-flying migrating Northern bald ibis. PLoS One. 2015;10(9):e0134433.
40. Tordoff MG. Calcium: taste, intake, and appetite. Physiol Rev. 2001;81(4):1567–459.
41. Geerling JC, Loewy AD. Central regulation of sodium appetite. Exp Physiol. 2008;93(2):177–209.
42. Raubenheimer D, Simpson SJ. Hunger and Satiety: linking mechanisms, behaviour and evolution. In : Choe JC, editor. Encyclopedia of Animal Behaviour 2nd ed. Amsterdam: Elsevier; 2019.
43. Lambert JE, Rothman JM. Fallback foods, optimal diets, and nutritional targets: primate responses to varying food availability and quality. In: Brenneis D, Strier KB, eds. Annu Rev Anthr. 2015;44: 493–512.
44. Dunbar RIM. Modelling primate behavioral ecology. Int J Primatol. 2002;23(4):785–819.
45. Simpson SJ, Le Couteur DG, Raubenheimer D. Putting the balance back in diet. Cell. 2015;161(1):18–23.
46. Blumfield M, Hure A, MacDonald-Wicks L, Smith R, Simpson S, Raubenheimer D, et al. The association between the macronutrient content of maternal diet and the adequacy of micronutrients during pregnancy in the Women and Their Children's Health (WATCH) Study. Nutrients. 2012;4(12):1958–76.
47. Simpson SJ, Raubenheimer D. Obesity: the protein leverage hypothesis. Obes Rev. 2005;6(2):133–42.
48. Solon-Biet SM, McMahon AC, Ballard JW, Ruohonen K, Wu LE, Cogger VC, et al. The ratio of macronutrients, not caloric intake, dictates cardiometabolic health, aging, and longevity in ad libitum-fed mice. Cell Metab. 2014;19(3):418–30.
49. Wang DD, Hu FB. Dietary fat and risk of cardiovascular disease: recent controversies and advances. Annu Rev Nutr. 2017;37:423. https://doi.org/10.1146/annurev-nutr-071816-064614.
50. Simpson SJ, Sibly RM, Lee KP, Behmer ST, Raubenheimer D. Optimal foraging when regulating intake of multiple nutrients. Anim Behav. 2004;68(6):1299–311.
51. Lee KP, Simpson SJ, Clissold FJ, Brooks R, Ballard JWO, Taylor PW, et al. Lifespan and reproduction in drosophila: new insights from nutritional geometry. Proc Natl Acad Sci. 2008;105(7):2498–503.
52. Jensen K, Mayntz D, Toft S, Clissold FJ, Hunt J, Raubenheimer D, et al. Optimal foraging for specific nutrients in predatory beetles. Proc R Soc B. 2012;279(1736):2212–8.
53. Raubenheimer D, Simpson SJ. The geometry of compensatory feeding in the locust. Anim Behav. 1993;45(5):953–64.
54. Felton AM, Felton A, Raubenheimer D, Simpson SJ, Foley WJ, Wood JT, et al. Protein content of diets dictates the daily energy intake of a free-ranging primate. Behav Ecol. 2009;20(4):685–90.
55. Vogel ER, Rothman JM, Moldawer AM, Bransford TD, Emery-Thompson ME, Van Noordwijk MA, et al. Coping with a challenging environment: nutritional balancing, health, and energetics in wild Bornean orangutans. Am J Phys Anthr. 2015;156:314–5.
56. Rothman JM, Raubenheimer D, Chapman CA. Nutritional geometry: gorillas prioritize non-protein energy while consuming surplus protein. Biol Lett. 2011;7(6):847–9.
57. Mayntz D, Nielsen VH, Sorensen A, Toft S, Raubenheimer D, Hejlesen C, et al. Balancing of protein and lipid by a mammalian carnivore, the mink (Mustela vison). Anim Behav. 2009;77(2):349–55.
58. Cui ZW, Wang ZL, Shao Q, Raubenheimer D, Lu JQ. Macronutrient signature of dietary generalism in an ecologically diverse primate in the wild. Behav Ecol. 2018;29:804–13.
59. Gosby AK, Conigrave AD, Lau NS, Iglesias MA, Hall RM, Jebb SA, et al. Testing protein leverage in lean humans: a randomised controlled experimental study. PLoS One. 2011;6(10):e25929.
60. Martens EA, Lemmens SG, Westerterp-Plantenga MS. Protein leverage affects energy intake of high-protein diets in humans. Am J Clin Nutr. 2013;97(1):86–93.
61. Campbell CP, Raubenheimer D, Badaloo AV, Gluckman PD, Martinez C, Gosby A, et al. Developmental contributions to macronutrient selection: a randomized controlled trial in adult survivors of malnutrition. Evol Med Public Health. 2016;2016(1):158–69.
62. Gosby AK, Conigrave A, Raubenheimer D, Simpson SJ. Protein leverage and energy intake. Obes Rev. 2013;15(3):183–99.
63. Martinez-Cordero C, Kuzawa CW, Sloboda DM, Stewart J, Simpson SJ, Raubenheimer D. Testing the protein leverage hypothesis in a free-living human population. Appetite. 2012;59(2):312–5.
64. Martinez Steele E, Raubenheimer D, Simpson SJ, Galastri Baraldi L, Monteiro CA. Ultra-processed foods, protein leverage and energy intake in the US. Public Health Nutrition. 2018;21(1):114–24.
65. Bekelman TA, Santamaria-Ulloa C, Dufour DL, Marin-Arias L, Dengo AL. Using the protein leverage hypothesis to understand socioeconomic variation in obesity. Am J Human Biol. 2017;29(3):e22953.
66. Gosby AK, Soares-Wynter S, Campbell C, Badaloo A, Antonelli M, Hall RM, et al. Design and testing of foods differing in protein to energy ratios. Appetite. 2010;55(2):367–70.
67. Bilsborough S, Mann N. A review of issues of dietary protein intake in humans. Int J Sport Nutr Exercise Metabol. 2006;16:129–52.
68. Simpson SJ, Batley R, Raubenheimer D. Geometric analysis of macronutrient intake in humans: the power of protein? Appetite. 2003;41(2):123–40.
69. Raubenheimer D, Gosby AK, Simpson SJ. Integrating nutrients, foods, diets, and appetites with obesity and cardiometabolic health. Obesity. 2015;23(9):1741–2.
70. Raubenheimer D. Toward a quantitative nutritional ecology: the right-angled mixture triangle. Ecol Monogr. 2011;81(3):407–27.
71. Machovsky-Capuska GE, Senior AM, Simpson SJ, Raubenheimer D. The multidimensional nutritional niche. Trends Ecol Evol. 2016;31(5):355–65.
72. NHMRC. Nutrient reference values for Australia and New Zealand including recommended dietary intakes. National Health and Medical Research Council: Canberra; 2006.
73. Food and Nutrition Board. Dietary reference intakes for energy, carbohydrate, fiber, fat, fatty acids, cholesterol, protein, and amino acids. Washington D.C: The National Academies Press; 2005.

74. Le Couteur DG, Solon-Biet S, Cogger VC, Mitchell SJ, Senior A, de Cabo R, et al. The impact of low-protein high-carbohydrate diets on aging and lifespan. Cell Mol Life Sci. 2016;73(6):1237–52.
75. Le Couteur DG, Solon-Biet S, Wahl D, Cogger VC, Willcox BJ, Willcox DC, et al. New horizons: dietary protein, ageing and the Okinawan ratio. Age Ageing. 2016;45(4):443–7.
76. Lindeberg S, Berntorp E, Nilssonehle P, Terent A, Vessby B. Age relations of cardiovascular risk factors in a traditional melanesian society: the kitava study. Am J Clin Nutr. 1997;66(4):845–52.
77. Kaplan H, Thompson RC, Trumble BC, Wann LS, Allam AH, Beheim B, et al. Coronary atherosclerosis in indigenous South American Tsimane: a cross-sectional cohort study. Lancet. 2017;389(10080):1730–9.
78. Chambers L, McCrickerd K, Yeomans MR. Optimising foods for satiety. Trends Food Sci Technol. 2015;41(2):149–60.
79. Brooks RC, Simpson SJ, Raubenheimer D. The price of protein: combining evolutionary and economic analysis to understand excessive energy consumption. Obes Rev. 2010;11(12):887–94.
80. Krueger PM, Reither EN. Mind the gap: race/ethnic and socioeconomic disparities in obesity. Curr Diab Rep. 2015;15(11):95.
81. Mozaffarian D. Conflict of interest and the role of the food industry in nutrition research. J Am Med Assoc. 2017;317(17):1755–6.
82. Nestle M. Food politics: how the food industry influences nutrition and health. Revised and expanded 10th anniversary edition. Berkeley: University of California Press; 2013.
83. Hammond RA, Dube L. A systems science perspective and transdisciplinary models for food and nutrition security. Proc Natl Acad Sci U S A. 2012;109:12356–63.

The Radial: Integrative and Functional MNT

Kathie M. Swift, Elizabeth Redmond, and Diana Noland

5.1 Background – 58

5.2 Overview – 58

5.3 Core of the Radial: Personalized Nutrition Care – 58

5.4 Mind, Body, Spirit, Community, and Earth – 59

5.5 DNA Strands and Microbes – 59

5.6 Food, Environment, and Lifestyle – 59

5.7 Nutrition Physical Exam and Signs and Symptoms – 60

5.8 Biomarkers – 61

5.9 Laboratory Testing – 61

5.10 Metabolic Pathways and Networks – 64

5.11 Systems – 66

5.12 Gastrointestinal System – 67

5.13 Immune System: Defense and Repair – 67

5.14 Cardiovascular System: Cardiometabolic Comorbidities – 67

5.15 Endocrine System: Hormonal Health Influences – 67

5.16 Respiratory System: Lung and Sinus Illness – 67

5.17 Potential Triggers – 68

5.18 Stress – 69

5.19 Toxins and Toxicants – 69

5.20 Pathogens (See ▶ Chap. 21) – 69

5.21 Food Allergies and Intolerances – 70

5.22 Conclusion – 70

References – 70

© Springer Nature Switzerland AG 2020
D. Noland et al. (eds.), *Integrative and Functional Medical Nutrition Therapy*,
https://doi.org/10.1007/978-3-030-30730-1_5

5.1 Background

Radial (rā′dē-əl)
radiates from or converges to a common center.
American Heritage Medical Dictionary, 2007

As the field of integrative and functional nutrition took root, a concrete model to help practitioners gain a deep understanding of this approach was needed. The ability to connect, synthesize, and apply information coherently using an integrated, whole systems-based lens to nutrition care led to the development of the Integrative and Functional Medical Nutrition Therapy (IFMNT) Radial [1]. Recent updates to the IFMNT Radial (referred to as the Radial) were based on emerging data on nutritional science, systems biology medicine, omics technologies, and the microbiome.

5.2 Overview

The Radial is a conceptual framework for critical thinking and clinical investigation that graphically depicts the multidimensional facets of a systems-based nutrition assessment in delivering medical nutrition therapy (MNT). The term medical nutrition therapy is defined as "nutritional diagnostic, therapy, and counseling services for the purpose of disease management, which are furnished by a registered dietitian or nutrition professional" [2]. The counseling component of MNT has been described as a "supportive process to set priorities, establish goals and create individualized action plans which acknowledge and foster responsibility for self-care" [3]. The circular architecture of the Radial depicts a person-centered process that allows for the evaluation of complex interactions and interrelationships. The Radial purposefully integrates the evidence-based Nutrition Care Process (NCP) for clinicians to apply an integrative and functional methodology to provide high-quality nutrition care. The radial is the core, five sphere, and four potential triggers.

The five primary spheres of influence in the Radial are as follows:
- Food, lifestyle, and environment
- Nutrition physical, signs, and systems
- Biomarkers
- Metabolic pathways and networks
- Systems (◘ Fig. 5.1)

5.3 Core of the Radial: Personalized Nutrition Care

Personalized nutrition care highlighting the NCP is featured at the core of the Radial. It is centrally positioned and shows the four distinct, interrelated NCP steps including nutrition assessment, diagnosis, intervention, and monitoring/evaluation. Applying the NCP ensures that nutrition care is individualized and holistic using the highest quality information

◘ Fig. 5.1 The Radial. (Courtesy of Kathie Madonna Swift, Diana Noland, Elizabeth Redmond)

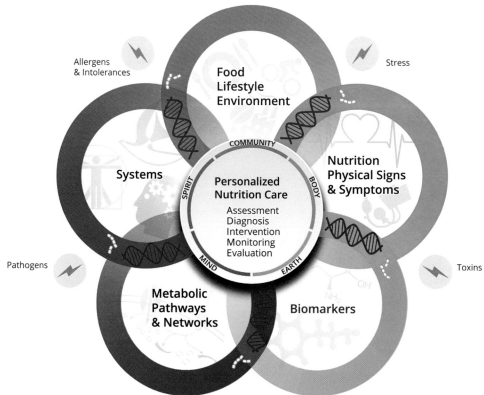

© 2010, 2018 Copyright. All Rights Reserved. KM Swift, D. Noland, E. Redmond

available to the clinician providing the service. A more detailed description of the NCP can be explored elsewhere [4].

A holistic approach to nutrition care embraces the assimilation of multiple components that influence one's well-being. Health is viewed as a dynamic quality where mind, body, and spirit are engaged in the fullness of life [5].

5.4 Mind, Body, Spirit, Community, and Earth

The Radial core illustrates the intermingling of mind, body, spirit, community, and earth and its association with personalized nutrition care, since all of these factors influence one's health and healing. Mind, body, and spirit are viewed as wholeness versus distinct and separate physiological, psychological, and spiritual units. The value of community and social networks as a component of health and wellness must be considered since social contexts influence biological systems [6]. An appreciation of our intimate connection to the earth and the healing power of nature to foster one's health is also valued as an integrative concept of food and sustainable nutrition.

5.5 DNA Strands and Microbes

The sequencing of the human genome and growing knowledge regarding genetic variation and biotechnological omics advances is contributing to the development of personalized nutrition. The Radial depicts genetic influence by the strands of DNA that surround the core. Personalized nutrition is an application of nutritional genomics and seeks to enhance health through the understanding of "the functional interaction between bioactive food components with the genome at the molecular, cellular, and systemic level in order to understand the role of nutrients in gene expression and… how diet can be used to prevent or treat disease" [7]. Diversity in the genetic profile among individuals impacts nutrient requirements, metabolism, and response to dietary, nutritional, and lifestyle interventions [8].

The microbiome revolution has established that microbial signatures vary between individuals far more than genetics [9]. Dietary habits and lifestyle affect the gut microbiota composition in dramatic ways. This vital influence must be taken into account and is represented by the microbes in the graphic that, like the DNA strands, radiate to all five spheres of the Radial. The composition of the microbiome and its diverse activities are involved in most, if not all, of the biological processes, and thus, it is a key player in health and disease at all stages of the life cycle [9].

The concept of "food as information" influencing genetic expression, manipulating the microbiome and, consequently, supporting or hijacking the patient's physiological systems is cardinal to the integrative and functional model. The Systems sphere radiates from the core of the Radial and underscores the imbalances that are created by poor diet, unhealthy lifestyle behaviors, and environmental exposures such as chronic inflammation, oxidative stress, digestive and detoxification disturbances, metabolic chaos, neuro-endocrine-immune disruption, and nutritional depletion.

5.6 Food, Environment, and Lifestyle

Food is the ideal starting point in an IFMNT assessment since it is a powerful determining factor in health and disease. Hippocrates is credited with the "food as medicine" philosophy, although scholars debate the origin of this theme since nutrition has been a central premise in ancient, traditional forms of medicine [9, 10]. A modern, integrated concept of food as medicine emphasizes scientific understanding of nutrition with lifestyle behaviors in relation to a person's ability to realize vital goals of healthy living. Food is valued as an instrument for vibrant health and an essential tool to kindle healing by restoring nutritional integrity.

Changes in the human diet and food systems have altered crucial nutritional attributes that impact biological systems including glycemic and insulinogenic load, carbohydrate quality, fiber content, fatty acid composition, macronutrient balance, micronutrient density, phytonutrient richness, sodium-potassium ratio, and acid-base balance [11]. A theme of healthful eating that considers these nutritional factors has been described with the strength of evidence supporting a plant-based diet with foods close to nature [12]. Other elements of diverse diets and healthy dietary patterns include limited refined starches, added sugars including fructose, avoidance of ultra-processed foods, and limited intake of certain fats.

Personalized nutrition extends beyond these general principles of healthy eating and targets the underlying root causes of system imbalances, is specific to the individual's unique nutrient requirements, and is designed with biological compatibility and lifestyle in mind [13]. It seeks to understand how diet and lifestyle-related modulators act together and co-conspire to create the perfect storm of complex, chronic disease. For example, personalized nutrition recognizes that following general nutrition guidelines may result in high glycemic responses in some individuals, accelerating metabolic decline and that there can be unique dietary responses because of biochemical uniqueness. Thus, consideration of personal variables including sex, age, genetics, biomarkers, lifestyle, and environmental factors must be taken into account in tailoring a personalized nutrition prescription [14]. Food intolerances such as gluten, fructose, FODMAPs, food chemicals, and histamine are other examples of the dietary discord that can impact the health of some individuals while not affecting others. The age-old saying, "one person's meat is another person's poison," is now viewed in light of the scientific revelation that the multi-tasking, metabolically active microbes play a significant role in the unique interpretation by the body of food as friend or foe [15].

To be effective, personalized nutrition care must integrate the scientific approach with cultural, emotional, ethical, spiritual, social, and sensual understandings of food [16]. The *science* of nutrition and the *art* of dietary behavior change are indispensable partners in personalized nutrition care that is relational in nature. The heart of nutrition assessment lies in a therapeutic relationship that values the patient's story (see ► Chap. 39). This means that the nutrition professional is able to listen reflectively to the patient's narrative, absorb what the person is sharing, and understand their concerns. It must go beyond root-cause analysis of medical history, anthropometrics, symptomatology, testing, and other metrics and focus on the patient's humanity and their desire to shape their wellbeing. This requires skills and competencies in addressing health behaviors and engaging the patient in self-care management of mind, body, and spirit. It implies attention of the patient on their personal health goals, increased self-awareness, and commitment on ways to achieve and maintain optimal health. Shared decision-making by both the clinician and patient is essential in the design of a nutrition intervention plan that is truly personalized, actionable, practical, sustainable, and, ultimately, successful.

Integrative health questionnaires can assist the clinician in the personalized nutrition care process. Supportive tools include food records, activity journals, circadian patterns, timeline of significant medical events, medication and dietary supplement lists, symptom appraisal forms, nutrition-focused physical findings, and biomarker results from both conventional and functional laboratory testing. Individual health risks of diet, physical activity, substance abuse, excess stress, sleep deprivation, environmental toxins, pathogens, food allergens, and intolerances must also be unearthed. Open-ended questions can reveal aspects of the patient's story that may not have been revealed in specific data collection tools and techniques. These questions can help the clinician and patient as co-investigators in uncovering unique and deeply personal themes that affect an individual's health:

- *Did something trigger a change in your health?*
- *What makes you feel better?*
- *What makes you feel worse?*
- *Have you made any changes in your eating habits or lifestyle because of your health?*
- *What do you think would make the most difference in your health and well-being?*
- *In your own words, tell me your story…*

The personalized nutrition care process, when delivered skillfully, engages the patient in self-care to attain their highest health potential. A thorough understanding of how the trilogy of food, environment, and lifestyle impact physiological systems must be realized for the clinician to apply this in practice and support the patient with a truly personalized care plan.

5.7 Nutrition Physical Exam and Signs and Symptoms

Most patient encounters begin by hearing the patient's story with appropriate history taking, identifying the signs and symptoms with which they are presenting, and the nutrition physical exam (NPE). Universally, basic anthropometrics are measured: blood pressure, height, weight, and calculating BMI. For the IFMNT approach, the NPE can be expanded with tools that add more detailed information for assessing the nutritional and body composition status. These tool options to consider collecting nutrition data are:

- Tape measure: measures waist circumference
- Bioelectric impedance analysis (BIA) (full body): measures several parameters described in ► Chap. 22 on Body Composition. One of the measurements most commonly appreciated is percentage of body fat/lean mass and BMI.
- Fingertip Pulse Oximeter Blood Oxygen Saturation Monitor: measures oxygen saturation and pulse rate
- Oral/axillary thermometer: measures body temperature
- Calipers: measures body fat

Signs and symptoms can be documented by patient questionnaires completed prior to the client visit, or upon arrival at the appointment. The medical symptoms questionnaire (MSQ) commonly used in integrative and functional medicine practice measures symptoms based on systems biology, according to body systems. The practitioner's assessment of the questionnaire can be done quickly to easily see where the system priorities are to consider for diagnosis and intervention for the client. The systems and related symptoms often included on the questionnaire are:

- *HEAD*: headache, faintness, dizziness, insomnia
- *EYES*: Watery or itchy eyes; swollen, reddened/sticky eyelids; bags, dark circles; blurred or tunnel vision (does not include near or far-sightedness)
- *EARS*: Itchy ears; earaches, ear infections; drainage from ear; ringing/hearing loss
- *NOSE*: Stuffy nose, sinus problems, hay fever, sneezing attacks, excessive mucus
- *MOUTH/THROAT*: Chronic coughing; gagging/throat clearing; sore throat, hoarseness; swollen/discolored tongue, gums, lips; canker sores
- *HEART*: Irregular/skipped beats, rapid/pounding beats, chest pain
- *SKIN*: Acne; hives, rashes, dry skin; hair loss; flushing, hot flashes; excessive sweating
- *LUNGS*: Chest congestion; asthma, bronchitis; shortness of breath; difficulty breathing
- *DIGESTIVE TRACT*: Nausea, vomiting; diarrhea; constipation; bloated feeling; belching, passing gas; heartburn; intestinal/stomach pain
- *JOINTS/MUSCLE*: Pain or aches in joints, arthritis, stiffness/limited movement, pain or aches in muscles, feeling of weakness or tiredness
- *WEIGHT*: Binge eating/drinking, craving certain foods, excessive weight, compulsive eating, water retention, underweight
- *ENERGY/ACTIVITY*: Fatigue/sluggishness; apathy, lethargy; hyperactivity; restless leg; jetlag
- *MIND*: Poor memory; confusion, poor comprehension; poor concentration; poor physical coordination;

difficulty making decisions; stuttering or stammering; slurred speech; learning disabilities
- *EMOTIONS*: Mood swings; anxiety, fear, nervousness; anger, irritability, aggressiveness; depression
- *OTHER*: Frequent illness; frequent or urgent urination; genital itch or discharge; bone pain
- *PAST SURGERIES AND DATES*: See ▶ Chap. 56 for a sample questionnaire.

5.8 Biomarkers

Nutrition-related biomarkers are biological indicators. While a nutrition assessment is essential to assess what nutrients the diet is providing, an evaluation of biomarkers may provide a more detailed picture of what the client is actually digesting, absorbing, and utilizing. Diet assessments may not always coordinate with laboratory values of individual nutrients. This may be due to many factors, such as an individual having a greater need for a nutrient than the RDA, poor-quality food or supplements, a genetic variation, impaired absorption, exposure to a specific toxin whose detoxification requires specific nutrients, etc.

5.9 Laboratory Testing

Most clinicians are familiar with conventional laboratory testing of standard markers. In addition to conventional lab testing, integrative and functional clinicians also often utilize specialty diagnostic laboratories, which are often early adopters of new testing. Functional medicine was an early adopter of the concept of utilizing organic acid measurements for assessment of biochemical pathways and metabolomics. Functional medicine clinicians also incorporate a systems-based interpretation of standard laboratory tests.

There are two primary types of laboratory tests: indirect (also referred to as functional) or direct.

A functional test is an evaluation of an analyte that is dependent on the marker. For example, levels of methylmalonic acid (MMA) measure functional need for vitamin B12. Vitamin B12 is a nutrient cofactor needed for the enzyme methylmalonyl-CoA mutase. It breaks down L-methylmalonyl-CoA to succinyl-CoA, which then goes into the Krebs cycle. If vitamin B12 is not available in adequate amounts, the pathway will stall and MMA will build up. Thus, elevated MMA levels in both urine and blood have been correlated to vitamin B12 deficiency [17].

Direct laboratory testing identifies levels that are actually there, such as serum 25-hydroxy vitamin D. Though there has been significant disagreement on the "ideal" vitamin D level, as the level needed for an individual may vary significantly, there is strong consensus on the specimen selection of serum 25-hydroxy vitamin D.

Specimen selection should generally be based on what literature has identified. For example, urine amino acids can identify what a person has eaten over the last few days, while serum amino acids can tell more about overall amino acid status. Additionally, whole blood lead may not be the best way to assess lead levels, but it is the way public health organization and researchers check lead levels, and thus if you are evaluating lead levels, it is best to check whole blood lead so comparisons can be made.

Beyond specimen selection, clinicians should be aware of reference ranges and how they are set. While some biomarkers have medical decision points identifying what each level means, such as levels of A1C, other markers do not have a standard accepted reference range or decision point. For many markers, the reference range is represented by what has been noted within a "normal" population. A patient is considered abnormal if they are more than 2 standard deviations (>95.4%) from the mean within the population utilized to set the reference range. Researchers may correlate values with disease status, but if comparisons are made to research studies, the laboratory technique and reference range population must be compared. For example, elevated alpha-hydroxybutyrate has been associated with insulin resistance, but the specific level is not established.

Integrative and functional clinicians look at several areas of laboratory assessments, including metabolism, inflammation and immune reactions, nutrition, nutrigenomics, digestion and absorption, biotransformation, etc.

Metabolism: Metabolism includes energy metabolism and metabolomics. Metabolomics accesses metabolite profiles to detect which biomarkers or biomarker patterns are associated with a disease in the hope of better defining a specific metabolic profile for each condition or disease. It has also been referred to as the specific metabolic "fingerprint" that is unique to each person and disease. In other words, metabolomics takes into account the impact of lifestyle factors like diet, movement, and the environment to illustrate overall health.

Inflammation: Evaluation of inflammation can identify systemic inflammation, such as hsCRP, or it can be site-specific, such as fecal calprotectin to evaluate gastrointestinal inflammation. Different specimen types can identify the type of inflammation, lipid peroxides identify fatty acid membrane oxidation, and 8-hydroxy-2-deoxyguanosine evaluates oxidative damage to DNA. Oxidative stress and inflammation are key markers of nutrient deficiencies [18]. Nutrients are the substrates for many immune reactions, such as fatty acids and immune cytokines. Nutrients can also modulate the immune system, for example, antioxidants tempering immune responses to inflammation. While classic inflammation is short term, chronic "metaflammation" is a low-level long-term inflammation that is associated with inflammation-related diseases. Chronic inflammation differs because there is only a small rise in immune markers (i.e., a four- to sixfold increase versus a several hundred-fold increase seen in acute reactions) which can lead to systemic effects and is associated with a reduced metabolic rate [19].

Testing can also identify the immune system's reaction to components in foods, such as proteins, toxins, and other bioactive components. Functional medicine evaluates a patient's

Table 5.1 Micronutrient laboratory assessment

Minerals	IOM assessment [22–24]	Possible functional or other testing options
Magnesium	A serum magnesium concentration of less than 0.75 mmol/liter (1.8 mg/dl) is thought to indicate magnesium depletion.	Serum/plasma magnesium concentration, red blood cell (RBC) magnesium concentration, and urinary magnesium excretion appear to be useful biomarkers of magnesium status in the general population [25].
Zinc	Factorial analysis was used to set the Estimated Average Requirement (EAR).	Physical growth response to zinc supplementation [24]. While both plasma and serum zinc concentrations are used as indicators of zinc status, plasma zinc concentration is preferable because of the lack of contamination of zinc from the erythrocyte [26].
Copper	The primary criterion used to estimate the EAR for copper is a combination of indicators, including plasma copper and ceruloplasmin concentrations, erythrocyte superoxide dismutase activity, and platelet copper concentration in controlled human depletion/repletion studies.	Low serum copper. Serum copper and ceruloplasmin levels may fall 30% in deficiency [27]. An elevated homovanillate/vanilmandelate has been reported to identify copper need since the conversion of dihydroxyphenylalanine (DOPA) to epinephrine requires copper, though research is limited. Homovanillate is the breakdown product of DOPA and vanilmandelate is the breakdown product of epinephrine and nor-epinephrine. The ratio has been used as a screening for Menkes [28].
Vitamins	**IOM assessment**	**Possible functional or other testing options**
Vitamin K	Due to insufficient laboratory values to estimate levels, an Adequate Intake (AI) was set based on representative dietary intake data from healthy individuals.	A direct measure of vitamin K is not of value as both serum and plasma phylloquinone reflect recent intakes (24 hours) and do not responds to changes in dietary intake [24]. Carboxylated Gla (Gla) or undercarboxylated osteocalcin (uc-OC) may be used. A deficiency of vitamin K upregulates the level of serum undercarboxylated osteocalcin (ucOC), and serum ucOC has been found to correlate with fracture risk [29].
Folate	The primary indicator used to estimate the RDA for folate is erythrocyte (RBC) folate in conjunction with plasma homocysteine and folate concentrations.	Serum folate, without a concurrent vitamin B12 deficiency, is a useful biomarker for folate deficiency [30]. Formiminoglutamic acid (FIGLU) is an intermediate metabolite in L-histidine catabolism in the conversion of L-histidine to L-glutamic acid. It may be an indicator of vitamin B12, folic acid deficiency, or liver disease [31]. Measurement of urinary FIGLU excretion after a histidine load has been used as a marker of folate in those with adequate B12 levels [32].
B3 Niacin	The most reliable and sensitive measures of niacin status, also used to set the RDA, was urinary excretion of the two major methylated metabolites, N1-methyl-nicotinamide and its 2-pyridone derivative (N1-methyl-2-pyridone-5-carboxamide).	An increase in urinary excretion of kynurenic acid and a decreased in quinolinic acid were found in pellagra that was corrected with niacin supplementation [33]. Urinary branched chain keto acids (organic acids) positively identified subjects with B vitamin-complex deficiency; those with the highest level of urine branched chain amino acids (BCAA) were more likely to be B vitamin deficiency [34].

reaction to foods or environment. By definition, IgE reactions are true allergic reactions, while other reactions are food intolerances or sensitivities, measured with immunoglobulins (IgG) or white cells, and have significantly less literature support. IgG testing (ELISA) includes IgG1, IgG2, IgG3, and IgG4. IgG testing is controversial, as limited evidence from peer-reviewed research has found correlations with migraines and IBS [20]. Some researchers believe IgG is identifying exposure, since "healthy" people can have elevated levels. Leukocyte testing includes both the mediator release and the leucocyte activation test. Though both evaluate changes in white blood cells, they are different processes. Both are also controversial with limited peer-reviewed literature, though, like IgG, many clinicians claim to find them helpful in clinical practice [21].

Nutrition: As noted, personalized nutrition is specific to the individual's unique nutrient requirements. Table 5.1 identifies examples of both laboratory markers used by the IOM in setting recommended levels, and functional or other testing options. Though there is often controversy with the specific tests selected by the IOM, it is important to know the technique and specimen used, and how other testing options compare to it.
- Micronutrients are common enzyme cofactors. Identifying inadequate levels of individual micronutrients may also identify possible blocks in biochemical pathways (Table 5.2).

Nutrigenomics: Nutrigenomics is the study of the interaction of nutrition and genes and has been shown to personalize

Table 5.2 Functional nutrition approach to macronutrients

Protein	Functional assessment goes beyond total protein evaluation and identifies individual amino acids and types of protein (plant or animal). Plasma assessment of individual amino acids helps to identify overall level of protein intake and processing. Plasma identifies longer term status and use. Urine identifies intake in the last 24 hours and some metabolic issues. Both plasma and urine assessments are best done in a stable dietary intake. Blood amino acids can be measured with a blood draw or from a finger stick.
Fat	Functional assessment goes beyond evaluating standard blood lipids such as TAG, LDL, and HDL and includes individual fatty acids, such as the polyunsaturated fats omega-3 and omega-6, as well as monounsaturated fats and saturated fats. Specimen types include whole blood, plasma, or RBC and are available in blood draw or finger stick. Fatty acids are measured as a percentage of total or as a concentration, and direct comparisons should not be made. Further, the different methods of expressing fatty acids can lead to dissimilar correlations between blood lipids and certain fatty acids [35]. Evaluation of fatty acids of varying carbon chain length and degree of saturation can help to identify dietary intake and have been noted as biomarkers of various conditions. A review of individual fatty acids helps to evaluate function of desaturase and elongase enzymes, which can impact treatment. For example, the conversion of alpha-linolenic acid (ALA) to eicosapentaenoic acid (EPA) is not always efficient and supplementation with flax, high in ALA, may not result in expected increases in EPA. Testing would aid in identifying those who are not making the conversion.
Carbohydrates	Conventional assessment of carbohydrates is generally the assessment of the body's blood sugar response and includes glucose and A1c. Functional medicine also evaluates additional early markers of biochemical disruption using metabolomics, such as identifying alpha-hydroxybutyrate or impaired plasma levels of branched-chain amino acids, which may identify risk of T2D and CVD [36]. The composition of the microbiome can also identify the impact of carbohydrate and fiber intake. Plant-focused diets high in fiber are associated with greater microbial diversity and higher levels of *Prevotella* over *Bacteroides*. Low levels of SCFAs (acetate, propionate, butyrate) in stool can identify poor carbohydrate (fiber) intake or inadequate gut bacterial function [37]. Additionally, high-protein diets can increase the level of bacterial products of protein breakdown, including the short-chain proteolytic fatty acids, valerate, iso-butyrate, and iso-valerate [38, 39].

For comparison, the IOM Recommended Dietary Intake (RDI) for macronutrients in adults is 45–65% of their calories from carbohydrates, 20–35% from fat, and 10–35% from protein [40]

treatment. Test profiles are often buccal swabs and packaged to evaluate a specific area such as weight loss, immune function, nutrient absorption, etc. Previously, the US Food and Drug Administration raised concerns about the validity of the information and the potential for inappropriate medical actions, highlighting the need for practitioners to be aware of the literature of individual single nucleotide polymorphisms (SNPs) [41].

Digestion and Absorption: Functional assessment of digestion and absorption includes tests related to the gastrointestinal tract and its function. Assessment of a patient's digestion and absorption can be assessed in a variety of tests. Two examples are pancreatic elastase 1 (PE1) and fecal fats. PE1 is a proteolytic enzyme that identifies the function of the exocrine pancreas (not to be confused with the endocrine pancreas). The exocrine pancreas produces 3 types of enzymes: amylase, protease, lipase. If it is impaired, digestion may be impaired. Exocrine pancreatic insufficiency is identified with an elastase level < 200 μg/g stool [42]. Fecal fats are used to evaluate fat malabsorption. The gold standard for fat malabsorption is a 3-day quantitative determination of fecal fat. It is cumbersome and takes significant dietary preparation, so a single fecal assessment is often done as an initial evaluation [43].

Assessment of gut bacteria or the microbiome can aid clinicians understanding of the impact of individual bacteria and overall diversity. Assessment is generally done with fecal samples, though breath tests can also identify levels of some gut bacteria. The microbiome is a primary player in the immune system and is heavily impacted by diet, lifestyle, and environment. There are many culture-independent techniques in use, such as fluorescent in situ hybridization (FISH), polymerase chain reaction (PCR), next- and third-generation sequencing, etc. While each technique is valid, the results are not comparable. Even laboratories that do the same stated process such as PCR assessment are generally not comparable due to a lack of standardization. Evaluation of microbiome levels with disease associations or conditions needs to be done using comparable testing.

Intestinal permeability is another concern that functional clinicians evaluate in a full assessment. The intestine is lined with a single layer of epithelial cells held together with tight junctions (TJ), which coordinate exchange between lumen and tissues. Disruption of the intestinal barrier impairs function and may increase the risk of specific disease. Zonulin is a physiologic modulator of intercellular intestinal tight junctions. An increase in zonulin has been proposed to identify a "leaky gut" and has been related to autoimmune disease, metabolic disorders, heart disease, and intestinal disease such as IBS, IBD, and non-celiac gluten [44, 45]. Unfortunately, zonulin has been difficult to test accurately in commercially available tests [46].

Detoxification (also known as biotransformation): Biotransformation refers to the process of transforming toxins, hormones, etc. through the phases of detoxification. Functional clinicians can support the process by ensuring adequate nutritional status of detoxification pathways and identifying toxin levels when needed. There is no single test to evaluate detoxification phases. Phase I is generally done with genetic SNP testing and phase II includes status of amino acid substrates and nutrient co-factors. Detoxification is heavily reliant on enzymes, such as cytochrome P450 oxidase enzymes (CYP450) and nutrient status. Diet can significantly impact the ability to detoxify because it provides both needed nutrients and is a leading contributor of body toxins. Testing for the presence of toxins is available from several laboratories. The ability to compare toxin levels to population data, such as NHANES, allows clinicians to compare client's levels to the general population.

5.10 Metabolic Pathways and Networks

Understanding biochemical pathways and how they function is essential in functional medicine. Biomarkers are not just levels of individual analytes; they include markers of status and pathway function. Individual nutrients often work as nutrient cofactors, and can impact the ability of a pathway to flow smoothly. In order to fully evaluate a patient's biochemical status, clinicians must know the key pathways and networks, such as methylation, conjugation, nutrient breakdown, urea cycle, etc., including substrates, products, enzymes, and cofactors. After a full evaluation of a patient's history, signs and symptoms, anthropometrics, diet assessment, and laboratory evaluations, clinicians can take the next step of identifying key pathways that could be impaired. Several examples are listed below, including vitamin B12 and MMA, and breakdown of fatty acids and BCAA.

Vitamin B12 and Methylmalonic Acid (MMA): When MMA levels are elevated, it may identify a need for vitamin B12 due to inadequate intake or increased need. If diet assessment identifies an adequate intake, there may be increased needs due to other factors, such as decreased stomach acid or increased methylation issues. As seen in the simplified diagram below, L-methylmalonyl-CoA (substrate) is converted to succinyl-CoA (product) via the enzyme methylmalonyl-CoA mutase. The enzyme requires a nutrient-cofactor, adenosylcobalamin (vitamin B12). If the enzyme function is decreased, the L-methylmalonyl-CoA will build up and be excreted as MMA. The enzyme may have decreased function due to a genetic SNP or decreased levels of its nutrient cofactor, vitamin B12. As with most pathways, it is important to know that there are additional related pathways that can also be impacted. For example, converting homocysteine to methionine (methylation) may also be increased with a need for B12 (◘ Fig. 5.2).

Fatty Acid Breakdown Eicosanoids are signaling molecules that result in a range of processes generally related to immune and inflammation. They are derived from omega-3 and omega-6 fatty acids, as seen in ◘ Fig. 5.2. Diet provides the fatty acids and nutrients support the breakdown pathways.

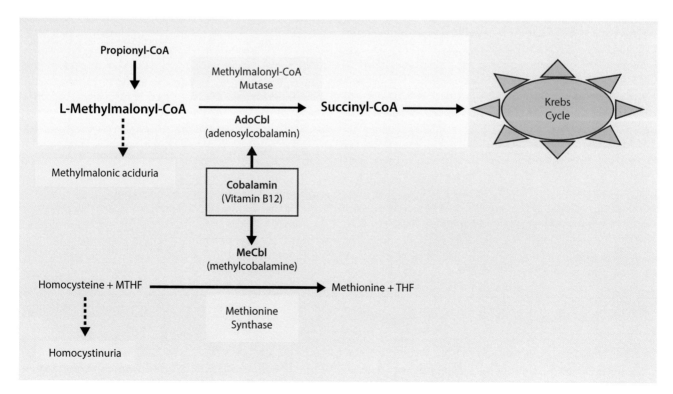

◘ Fig. 5.2 Methylmalonic acid (MMA) pathway. (Courtesy of E. Redmond, PhD, RDN)

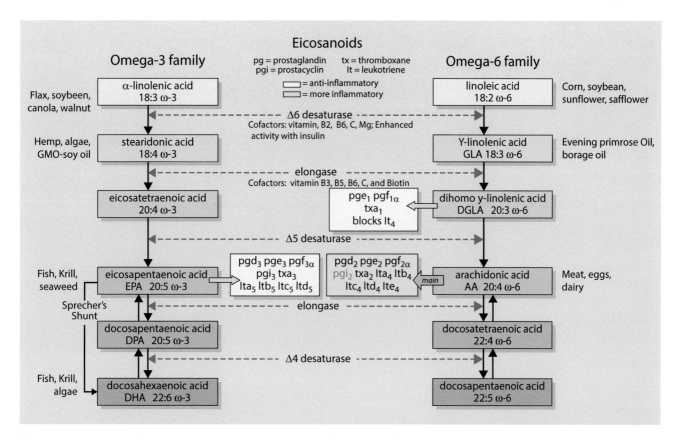

● Fig. 5.3 Eicosanoid pathway [47]. (Adapted with permission from: ► https://commons.wikimedia.org/wiki/File:EFA_to_Eicosanoids.svg)

Altered fatty acids have been associated with several metabolic diseases, including diabetes, metabolic syndrome, hypertension, alcohol, radiation exposure, and others. The eicosanoid pathway is heavily reliant on elongase and desaturase enzymes, which are dependent on adequate nutrient cofactors. Though a conclusive list has not been developed the following are thought to play a role in desaturase and elongase enzyme function, vitamins B2, B3, B5, B6, C, and biotin, and minerals zinc and magnesium. Additionally insulin is believed to impact function [48–50] (● Fig. 5.3).

Furthermore, evaluations of specific monounsaturated and saturated fatty acids can also identify functional impairments. The Δ9-desaturase enzyme or stearoyl-CoA desaturase (SCD) catalyzes the synthesis of monounsaturated fatty acids, primarily oleate (18:1) and palmitoleate (16:1), from saturated fatty acids, palmitate (16:0), and stearate (18:0). Stearic acid is correlated to cholesterol level and Δ9 may be impacted by insulin; thus, an evaluation may help to identify early biochemical function and possible increased risk [51, 52]. A decrease in insulin activity reduces the activity of desaturase enzymes that convert saturated to monounsaturated fats, which can be identified by increases in blood saturated fat levels.

Branched-chain Amino Acids (BCAA) Breakdown: Leucine, isoleucine, and valine are BCAAs. The BCAAs differ from other essential amino acids in that the liver lacks the enzymes to break them down or catabolize them. They are broken down to their keto-acids with transaminase enzymes and the nutrient cofactor vitamin B6. If vitamin B6 is not in adequate supplies, the BCAA amino acids may build up. The next step utilizes a dehydrogenase enzyme and its required nutrient cofactors, vitamins B1, B2, B3, B5, and lipoic acid (LA), which then flow into the Krebs cycle in the mitochondria. Identification of elevated keto-acids may signal a need for B-complex vitamins [34]. Thus, an impairment in the BCAA process has been proposed to impair basic mitochondrial function. Research has found that levels of BCAAs were higher in some individuals with obesity and have been associated with worse metabolic health and future insulin resistance or type 2 diabetes mellitus (T2DM). Insulin resistance may increase protein degradation since insulin normally suppresses it. In clients with metabolic issues, functional clinicians may evaluate plasma BCAAs as well as their keto acids to identify issues in the pathways (● Fig. 5.4 and ● Table 5.3).

The evaluation of pathways and metabolites is continuously being researched. Key resources include the following
— The Scripps Center for Metabolomics: METLIN metabolites database ► https://metlin.scripps.edu/index.php
— Expasy: omics scientific databases and tools ► https://www.expasy.org/
— Canada's Human Metabolome Database (HMDB) site for the human metabolome markers: ► http://www.hmdb.ca/

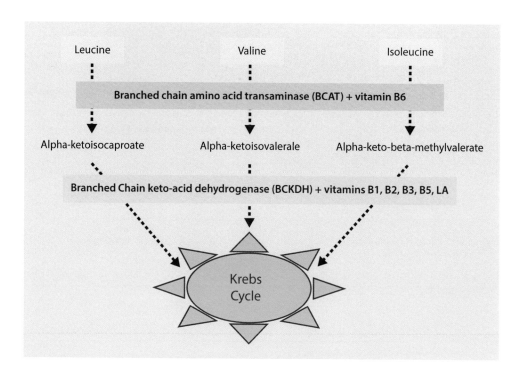

◻ Fig. 5.4 Krebs cycle. (Courtesy of E. Redmond, PhD, RDN)

◻ Table 5.3 Branched-chain amino acids (BCAA)

BCAA)	Enzyme plus cofactors	BCAA keto-acid	Enzyme plus cofactors
Valine	BCAA transaminase + vitamin B6	Alpha-ketoisovalerate	Branched-chain keto-acid dehydrogenase + B1, B2, B3, B5, LA
Leucine		Alpha-ketoisocaproate	
Isoleucine		Alpha-keto-beta-methylvalerate	

5.11 Systems

The foundation of integrative and functional medicine is systems biology [53]. There is a broad consensus as to the definition of systems biology. The biological systems described in IFMNT recognize that the whole human organism is comprised of systems working together contributing to total function. The systems included in assessment of the whole patient story are those where disease can be promoted with the individual's identified pathophysiological symptoms and imbalances. In systems biology, of three primary principles, the first is *diversity*, where there is the appreciation of the complex molecular and genetic uniqueness of each individual. The depth of grasping uniqueness has been enhanced with the discovery and identification of the human genome.

The second key characteristic of systems biology is the recognition of the *complexity* of biological systems and the perturbations of those systems that can promote disease. In contrast to the reductionist medical acute-care model of quickly finding the diagnosis and applying a standard of care, a systems biology-based medicine and medical nutrition therapy applies a whole-systems approach to identification of "root causes" that is a better fit to treatment of chronic diseases.

Thirdly, *simplicity* reflects the application of systems biology to manage the complexity of metabolism to create manageable healthcare procedures that can be applied clinically. Simplicity refers to prioritizing the systems that most impact the individual. Once systems of concern are ranked in importance, they enable a realistic analysis of "root" causes of a chronic disease condition. Prioritizing and simplifying systems that need modulation increases the possibility of successful outcomes for such patients. Prioritized systems require practitioner skills in modulating the systems' biochemical networks, especially in the use of nutrients as cofactors and structural components. This knowledge is foundational for simplifying the development of effective interventions addressing "root causes" with food, dietary supplementation, and lifestyle education. This knowledge depends on the principle that one disease can have multiple influences, with nutrient cofactors being primary when searching for "root causes."

Chronic diseases are diet- and lifestyle-related with the influence of an individual's unique genotype. Thirty years ago, the Institute for Functional Medicine (IFM) was the first to provide a framework, "The IFM Matrix," for how to harness the complexity of chronic diseases into a manageable clinical tool that simplifies primary priorities needing metabolic correction (see ▶ Chaps. 7 and 47). Later, IFMNT practitioners implemented the conceptual diagram, The Radial (see ◘ Fig. 5.1), to target the nutrition component in the practice of patient care. For a systems biology approach to be effective, the uniqueness of each individual becomes "patient-centered care" [54], using tools like the Radial to "extract simplicity out of chaos" [53].

Working knowledge of the following systems equips the IFMNT practitioner with advanced skills in "metabolic correction" once an imbalance is identified (see ▶ Chap. 7).

As we review the systems below, it is obvious that there is interaction among all the systems to produce a personalized phenotype, but at any time, there are usually two or three of the systems raising a flag indicating dysfunction and needing restoration back to wellness.

5.12 Gastrointestinal System

> All disease starts in the gut. –Hippocrates

The gastrointestinal tract provides the largest interaction of the human with the environment primarily through ingested food, water, and beneficial fermented bacteria to promote a healthy gut ecology. The gut houses more than 70% of the immune system in its lymphoid tissues. It is also the secretor of many neurotransmitters and other life-giving metabolites like the short-chain fatty acid butyric acid, the primary repair fuel for colonocytes (colon cells), which promotes increased energy and cell proliferation and may protect against colon cancer. Key GI tract interventions are based on a whole foods low-antigen diet, with adequate prebiotics of soluble fiber and fermented foods or supplemental probiotics. A common thoughtful functional medicine approach to gut restoration is the 5R program [55]. It begins with *removing* the offender – whatever is in excess or an antigen for an individual; then *replacing* digestive secretions/enzymatic activity and fiber if indicated, repletion of insufficient or deficient nutrients; next, *reinoculating* to restore the balance of good bacteria, pre- and probiotics, adequate hydration; *repairing* the gut barrier that may be compromised from long-standing insults like antibiotics, antigen exposure, toxins, and emotional conflicts; and finally *rebalancing*. This type of protocol provides a significant tool to provide treatment for an individual who is experiencing gastrointestinal distress [56].

5.13 Immune System: Defense and Repair

With chronic systemic inflammation being a common denominator with all chronic diseases, it is important to focus on the condition of a person's immune system. This system, when perturbed, responds with inflammation that can lead to recoverable illness like flu and autoimmunity, as well as life-threatening cancer. The IFMNT practitioner can assess if the immune system is a priority for an individual by using the tools of nutrition physical exam, laboratory biomarkers of inflammation like C-reactive-protein-high sensitivity, sedimentation rate, differential with hearing the patient's history, and diagnosis/symptoms. Dietary history can reveal a diet of pro-inflammatory character that can be modulated to improve the immune response for the individual. Recent recognition of genomic polymorphisms that may increase risk of inflammation includes people with HLA DQ2 and DQ8 who have increased predisposition to celiac disease-related inflammation (see ▶ Chap. 49 Autoimmune).

5.14 Cardiovascular System: Cardiometabolic Comorbidities

According to the World Health Organization, heart and cardiovascular diseases are the most common chronic diseases worldwide. The bigger picture is the pathophysiology of the cardiometabolic system that underlays the trend toward the high risk of cardio diseases for an individual.

Inflammation is foundational to the early stage development of atherosclerosis, along with contributing chronic infections, pro-inflammatory dietary choices, poor sleep, endocrine-disrupting chemical toxins, etc. Nutritional and lifestyle interventions provide powerful change agents to modulate the cardiovascular system.

5.15 Endocrine System: Hormonal Health Influences

The endocrine system's response to our environment to message cellular activity is important to consider when assessing a patient. Two glands in jeopardy today are the adrenal and pancreatic glands. The adrenal gland responds to stress and lack of quality sleep-producing cortisol (stimulatory) elevations. The concerns for the hormones produced by the pancreas are the influence of insulin which, when in excess, is involved in promoting sarcopenic obesity (loss of muscle, gain of body fat) and metabolic syndrome. The pancreas also has an important influence by secretion of the digestive enzymes: lipases, amylases, and proteases to digest the dietary intake of the macronutrients fat, carbohydrate, and proteins. Other important glandular systems are thyroid, parathyroid, hippocampus, and pineal. All of the glandular hormones dance together, interacting to guide the human body to allostasis – the best metabolic balance to survive (see ▶ Chap. 47).

5.16 Respiratory System: Lung and Sinus Illness

The oxygenation of the body is dependent on the respiratory system exchange of oxygen from the air in the atmosphere with the metabolic waste of carbon dioxide – a rhythm critical throughout life; going without for approximately 6 min or more can result in death. The health of the sinus areas as the

Table 5.4 Genomic considerations for restrictive respiratory diseases

Condition	Gene(s)	Key nutrient considerations [57]
Asthma	Alpha 1 antitrypsin [Smoking increases risk]	Vitamins C, A, D Zinc Biotin Essential fatty acids
Emphysema		
Chronic obstructive pulmonary disease (COPD)		
Cystic fibrosis	Cystic fibrosis transmembrane conductance regulator (CFTR)	Vitamins C, A, D Zinc Biotin Essential fatty acids
Lung cancer	[Smoking increases risk] ROS1 symptoms attributed to a respiratory infection	Zinc Vitamins D, A, E Essential fatty acids Immune support
Sarcoidosis	Small groups of inflammatory cells to grow in lungs	Vitamins D*, A, E Essential fatty acids (*Caution using vitamin D)
Idiopathic pulmonary fibrosis	Scarring in the lungs	Vitamins C, A, D Zinc Biotin Essential fatty acids Phospholipids
Granulomatosis with polyangiitis (GPA) ("Wegener's granulomatosis")	Chronic inflammation of lungs, kidney, and cells usually related to infection; a vasculitis; rare multisystem autoimmune disease	Vitamins C, A, D Zinc Biotin Essential fatty acids Phospholipids Immune support

first air entry point is important as hair filters in the nose protect from most particles being inhaled into the lungs. The lung has nutrient requirements of fats, oils, and proteins to maintain the structure of the lung barrier along with nutrient cofactors vitamin A, zinc, and vitamin C influencing lung metabolism. As the exchange occurs, the oxygen entering the bloodstream connects with the RBC heme to be carried and distributed to cells and tissues throughout the body. With many unique genomics that can affect the lung, knowledge of those increases the skill of the nutrition practitioner to intervene successfully for each individual (see ▶ Chap. 52) (◘ Table 5.4).

Genotype-related risk of metabolic perturbation pathway systems and genes:
- Methylation genomics: the genomic expression related to B vitamins, phospholipids, and vitamin D. Genes: *MTHFR 677C/1298C, MMTR, MTR*, etc.
- Detoxification genomics: Glutathione-S-transferase genes (*GSTM1, GSTP1, GST*), *CYP*, and *COMT*
- Vitamin D receptor: *VDR RXR*. Vitamin D is dependent on magnesium status for its function. Vitamin D is a hormone and immune modulator in addition to the other roles in mineral metabolism and inflammation.

Because of the recognition of the unique biochemical phenotype and genotype for each individual, the current trend in medicine is toward personalized medicine [58]. Often referred to as "P4 medicine" (predictive, preventive, personalized, and participatory), the personalized paradigm for chronic disease healthcare is gaining support from practitioners of diagnostic medicine.

Biological systems are groupings of molecules, tissues, hormones, and organs that are affected by lifestyle choices and that work together to perform a common function to express as one whole human organism. Nutrients are basic co-factors for all those systems that determine their function. Common systems, such as the circulatory, respiratory, and nervous systems, need to be appreciated in the context of their interactions. The concept of systems biology focuses on interacting systems, in contrast to conventional acute-care medicine that generally considers systems as isolated entities without appreciating their interrelationships.

5.17 Potential Triggers

The five spheres of the Radial can be impacted by genetics, epigenetics, biochemical uniqueness, and microbial interactions and interplay. Imbalances can be triggered by chronic stress, toxins, pathogens, food allergens and intolerances resulting in a disruption in cells, tissues, and organ systems. These potential biological triggers are illustrated in the external section of the Radial and can collude to harm health and fuel chronic diseases such as obesity, diabetes, cardiovascular, neurodegeneration, mental illness, autoimmune conditions, allergies, asthma, and more. Evaluating the five spheres of influence and identifying pathological insults is essential in order to reestablish systems balance, foster metabolic integrity, and promote nutritional resilience.

5.18 Stress

Chronic stress is an antecedent to health issues and is represented in the outside of the Radial. There have been many definitions of stress in the medical literature from Hans Selye's original description as "the nonspecific response of the body to any demand made upon it" to R.S. Lazarus describing stress as "a circumstance external to a person, which makes unusual or extraordinary demands on him, or threatens him in some way" [59]. The detrimental impact of chronic stress has been well documented and causes major disruptions in the neuro-endocrine-gastro-immune and musculoskeletal systems. Chronic adverse life experiences, especially psychosocial stress, have been shown to induce destructive changes to the microbiome [60].

The nutrition professional should assess the role stress plays in a patient's life by incorporating some questions in the health questionnaire that are specific to stress. This can be done with a stress rating scale where the patient scores their chronic stress level (e.g., 1 = no stress to 5 = major stress) in categories such as health, finances, relationships, and work. Referrals to clinical social workers, behavioral psychologists, integrative yoga therapists, or mind body skills groups should be made if the individual is unable to effectively manage chronic stress. In addition, some integrative nutrition professionals have advanced training in behavioral therapies and mind-body medicine and can apply those skills to a person-centered nutrition care process (see ▶ Chap. 47).

5.19 Toxins and Toxicants

It can be difficult to know when a level of a toxin will have a physiologic impact on a single person, as response to toxins is individual. Nutritional interventions have been proposed as a key prevention strategy. The impact of toxins is being identified beyond just those with work-related exposures, and evidence suggests a significant impact in seemingly everyday exposures [61]. Two primary references are the Centers for Disease Control and Prevention's (CDC) Agency for Toxic Substances and Disease Registry (ATSDR) ToxFacts sheets, which address where toxins come from and how to best test for them, and the National Health and Nutrition Examination Survey (NHANES) Fourth Report evaluates toxins assessed by NHANES from population studies, allowing the clinician to make comparisons against population values [62, 63].
- ◻ Table 5.1 lists a truncated list of the ▶ Box 5.1 ATSDR 2013 Substance Priority List.

> **Box 5.1 The ATSDR 2013 Substance Priority List [62]**
>
> **Arsenic**
> *Testing*: Urine arsenic is the most reliable test for recent exposure to arsenic (within a few days prior). Tests of hair and fingernails can measure exposure over the last 6–12 months. Most tests measure the total amount of arsenic in urine. This can sometimes be misleading because there are non-harmful forms of arsenic in fish and shellfish, which can give a high reading even if you have not been exposed to a toxic form of arsenic.
>
> **Lead**
> *Testing*: A blood test measures the amount and estimates recent exposure. Lead in blood is rapidly taken up by red blood cells and referred to as blood lead concentration (PbB) which is the most widely used and reliable biomarker for general clinical use, and reflects recent exposure, <30 days. Venous sampling of blood is preferable to finger prick sampling. Urinary lead excretion reflects, mainly, recent exposure.
>
> **Mercury**
> *Testing*: Blood or urine samples are used to test for exposure to metallic mercury and to inorganic forms of mercury. Mercury in whole blood or in hair from the scalp is measured to determine exposure to methylmercury.
>
> **Polychlorinated biphenyls (PCB)**
> *Testing*: There are more than 200 PCBs. Testing can identify if PCB levels are elevated, which would indicate past exposure to above-normal levels of PCBs but cannot determine time and length of exposure.
>
> **Benzene**
> *Testing*: Measuring benzene in blood is a common test. However, since benzene disappears rapidly from the blood, it is only useful for recent exposures. Benzene is converted to the metabolite S-phenylmercapturic acid in urine, which is a sensitive indicator of benzene exposure. Since benzene is lipophilic, it is preferentially distributed to lipid-rich tissues, so blood tests should be lipid adjusted.
>
> **Cadmium**
> *Testing*: Cadmium can be measured in blood, urine, hair, or nails. Blood shows your recent exposure to cadmium. Urine shows both recent and past exposure and can reflect the amount of cadmium in the body. However, urine test results can be impacted by kidney function.

This priority list is not a list of the most toxic substances, but rather a prioritization of substances based on a combination of their frequency, toxicity, and potential for human exposure. A biomarker of toxic exposure is a xenobiotic substance, its metabolite(s), or the product of an interaction between a xenobiotic agent.

5.20 Pathogens (See ▶ Chap. 21)

Pathogens, by definition, are a bacterium, virus, fungus, prion, or other microorganism that can cause disease. Parasites are generally larger organisms like worms, ticks, or insects that can also cause disease. The concern for parasites is often decided based on their level of virulence. Common bacterial pathogens are *Pseudomonas*, *Shigella*, and *Salmonella*. Viral pathogens include influenza, adenovirus, rubella, and others. Examples of pathogenic fungi include *Candida*, *Aspergillus*, and *Cryptococcus*. The best-known prionic pathogen is bovine spongiform encephalopathy, also known as "mad cow disease." Pathogens can be difficult to identify and treat. In 2016, the CDC debuted the MicrobeNet Pathogen database [64], which contains information on pathogens and treatment protocols. Undernutrition puts people at a greater risk of infection and increased severity. Adequate nutritional status helps to support the negative physiologic impact of a parasite, such as increased metabolic rate, loss of appetite, immune responses, and specific nutrient requirements [65].

5.21 Food Allergies and Intolerances

There has been a notable rise in food allergies and intolerances around the globe. A food allergy is defined as an adverse health effect arising from a specific IgE immune response that occurs reproducibly on exposure to a given food. Symptoms can range from urticaria to life-threatening anaphylaxis. A food intolerance is an adverse reaction to a food or food component that lacks an identified immunologic pathophysiology. It results from the body's inability to digest, absorb, or metabolize a food or component of the food. These non-immune mediated reactions are caused by metabolic, toxicological, pharmacological, microbial, and undefined mechanisms [21] (see ► Chap. 20).

5.22 Conclusion

The Radial assimilates all the elements of a systems biology approach to healthcare (also referred to as P4 medicine) [58]. It is a paradigm-shifting model for nutrition professionals to apply in practice to engage and empower patients in self-care to achieve their highest health potential and realize their vital goals. Poor diet coupled with harmful environmental exposures and unhealthy lifestyle behaviors are causative factors driving the global burden of chronic disease.

> » The truly competent physician (practitioner) is the one who sits down, senses the 'mystery' of another human being, and offers with an open hand the simple gifts of personal interest and understanding [66].
> –Harold S. Jenkins, MD

References

1. Noland D, Raj S. Academy of nutrition and dietetics: revised 2019 standards of practice and standards of professional performance for registered dietitian nutritionists (Competent, Proficient, and Expert) in nutrition in integrative and functional medicine. J Acad Nutr Diet. 119(6):1019–1036.e47.
2. U.S. Department of Health and Human Services: Final MNT regulations. CMS-1169-FC. Federal Register, 1 November 2001. 42 CFR Parts 405, 410, 411, 414, and 415.
3. Morris SF, Wylie-Rosett J. Medical nutrition therapy: a key to diabetes management and prevention. Clin Diab. 2010;28(1):12–18 referencing American Dietetic Association: Comparison of the American Dietetic Association (ADA) Nutrition Care Process for nutrition education services and the ADA Nutrition Care Process for medical nutrition therapy (MNT) services.
4. "Nutrition Care Process." Eatrightpro.Org, 2019, www.eatrightpro.org/practice/practice-resources/nutrition-care-process. Accessed 16 June 2019.
5. Bradley KL USA (Ret), Goetz T and Viswanathan S. Toward a contemporary definition of health. Mil Med. 2018;183(suppl_3):204–7.
6. McCowan B, Beisner B, Bliss-Moreau E, Vandeleest J, et al. Connections matter: social networks and lifespan health in primate translational models. Front Psychol. 2016;22(7):433. https://doi.org/10.3389/fpsyg.2016.00433. eCollection 2016.
7. Castle D, Cline C, Daar AS, Tsamis C, Singe PA. Science, society, and the supermarket. The opportunities and challenges of nutrigenomics. Hoboken: Wiley; 2007. p. 3.
8. Ferguson LR, DeCaterina R, Görman U, Allayee H, et al. Guide and position of the International Society of Nutrigenetics/Nutrigenomics on personalised nutrition: part 1-fields of precision nutrition. J Nutrigenet Nutrigenomics. 2016;9:12–27.
9. Witkamp RF, van Norren K. Let thy food be thy medicine....when possible. Eur J Pharm. 2018;836:102–14.
10. Nordström K, Coff C, Jönsson H, Nordenfelt L, Görman U. Food and health: individual, cultural, or scientific matters? Genes Nutr. 2013;8:357–63.
11. Cordain L, Eaton SB, Sebastian A, Mann N, et al. Origins and evolution of the Western diet: health implications for the 21st century. Am J Clin Nutr. 2005;81(2):341–54.
12. Katz DL, Meller S. Can we say what diet is best for health? Annu Rev Public Health. 2014;35:83–103.
13. Personal communication-biological compatibility, John Bagnulo, PhD, 2018.
14. Magni P, Bier DM, Pecorelli S, Agostoni C, et al. Perspective: improving nutritional guidelines for sustainable health policies: current status and perspectives. Adv Nutr. 2017;8:532–45.
15. Ash C. One person's meat is another's poison. Science. 2017;356(6344):1243–5.
16. Nordström K, Coff C, Jönsson H, Nordenfelt L, Görman U, Nordstram K, et al. Food and health: individual, cultural, or scientific matters? Genes Nutr. 2013;8:336.
17. Methylmalonic acidemia diagnosis by laboratory methods. Rep Biochem Mol Biol. 2016;5(1):1–14.
18. Penberthy WT, Tsunoda I. The importance of NAD in multiple sclerosis. Curr Pharm Des. 2009;15(1):64–99.
19. Egger G. In search of a germ theory equivalent for chronic disease. Prev Chronic Dis. 2012;9:E95.
20. Lee HS, Lee KJ. Alterations of food-specific serum IgG4 Titers to common food antigens in patients with irritable bowel syndrome. J Neurogastroenterol Motil. 2017;23(4):578–84.
21. Ali A, et al. Efficacy of individualised diets in patients with irritable bowel syndrome: a randomised controlled trial. BMJ Open Gastroenterol. 2017;4(1):e000164.
22. Institute of Medicine (US) Standing Committee on the Scientific Evaluation of Dietary Reference Intakes and its Panel on Folate, O.B.V., and Choline. Dietary Reference Intakes for Thiamin, Riboflavin, Niacin, Vitamin B6, Folate, Vitamin B12, Pantothenic Acid, Biotin, and Choline., Dietary Reference Intakes for Thiamin, Riboflavin, Niacin, Vitamin B6, Folate, Vitamin B12, Pantothenic Acid, Biotin, and Choline. National Academies Press (US), 1998.
23. Institute of Medicine (US) Standing Committee on the Scientific Evaluation of Dietary Reference Intakes for Calcium, Phosphorus, Magnesium, Vitamin D, and Fluoride. Washington (DC); 1997.

24. Institute of Medicine (US) Standing Committee on the Scientific Evaluation of Dietary Reference Intakes for Vitamin A, Vitamin K, Arsenic, Boron, Chromium, Copper, Iodine, Iron, Manganese, Molybdenum, Nickel, Silicon, Vanadium, and Zinc. National Academy Press; 2001. https://www.nap.edu/read/10026/chapter/1. Assessed 5.4.2018.
25. Witkowski M, Hubert J, Mazur A. Methods of assessment of magnesium status in humans: a systematic review. Magnes Res. 2011;24(4):163–8.
26. Hess SY, et al. Use of serum zinc concentration as an indicator of population zinc status. Food Nutr Bull. 2007;28(3 Suppl):S403–29.
27. Linus Pauling Institute, Micronutrient Information Center. http://lpi.oregonstate.edu/mic/ Accessed 5 April 2018.
28. Lee T, et al. Standard values for the urine HVA/VMA ratio in neonates as a screen for Menkes disease. Brain and Development. 2015;37(1):114–9.
29. Suzuki Y, et al. Level of serum undercarboxylated osteocalcin correlates with bone quality assessed by calcaneal quantitative ultrasound sonometry in young Japanese females. Exp Ther Med. 2017;13(5):1937–43.
30. Antony AC. Evidence for potential underestimation of clinical folate deficiency in resource-limited countries using blood tests. Nutr Rev. 2017;75(8):600–15. https://doi.org/10.1093/nutrit/nux032.
31. U.S. National Library of Medicine: FIGLU test MeSH Descriptor Data 2019. https://meshb.nlm.nih.gov/record/ui?name=FIGLU%20Test. Assessed 15 June 2019.
32. Cooperman JM, Lopez R. The role of histidine in the anemia of folate deficiency. Exp Biol Med (Maywood). 2002;227(11):998–1000.
33. Shibata K, Yamazaki M, Matsuyama Y. Urinary excretion ratio of xanthurenic acid/kynurenic acid as a functional biomarker of niacin nutritional status. Biosci Biotechnol Biochem. 2016:1–9.
34. Shibata K, Sakamoto M. Urinary branched-chain 2-oxo acids as a biomarker for function of B-group vitamins in humans. J Nutr Sci Vitaminol (Tokyo). 2016;62(4):220–8.
35. Sergeant S, Ruczinski I, Ivester P, Lee TC. Impact of methods used to express levels of circulating fatty acids on the degree and direction of associations with blood lipids in humans. Br J Nutr. 2016;115(2):251–61.
36. Tobias DK, et al. Circulating branched-chain amino acids and incident cardiovascular disease in a prospective cohort of US women. Circ Genom Precis Med. 2018;11(4):e002157.
37. Simpson HL, Campbell BJ. Review article: dietary fibre-microbiota interactions. Aliment Pharmacol Ther. 2015;42(2):158–79.
38. Relevance of protein fermentation to gut health. Mol Nutr Food Res. 2012;56:184–96; November 1, 2010 vol. 299 no. 5 G1030-G1037.
39. Yao CK, Muir JG, Gibson PR. Review article: insights into colonic protein fermentation, its modulation and potential health implications. Aliment Pharmacol Ther. 2016;43(2):181–96. Epub 2015 Nov 2.
40. Dietary Reference Intakes for Energy, Carbohydrate, Fiber, Fat, Fatty Acids, Cholesterol, Protein, and Amino Acids: Health and Medicine Division. Nationalacademies.Org, 2019, www.nationalacademies.org/hmd/Reports/2002/Dietary-Reference-Intakes-for-Energy-Carbohydrate-Fiber-Fat-Fatty-Acids-Cholesterol-Protein-and-Amino-Acids.aspx.
41. Ferguson LR, et al. Guide and position of the international society of nutrigenetics/nutrigenomics on personalised nutrition: part 1 - fields of precision nutrition. J Nutrigenet Nutrigenomics. 2016;9(1):12–27.
42. Parsons K, et al. Novel testing enhances irritable bowel syndrome medical management: the IMMINENT study. Glob Adv Health Med. 2014;3(3):25–32.
43. Pezzilli R, et al. Exocrine pancreatic insufficiency in adults: a shared position statement of the Italian Association for the study of the pancreas. World J Gastroenterol. 2013;19(44):7930–46.
44. Fasano A. Zonulin, regulation of tight junctions, and autoimmune diseases. Ann N Y Acad Sci. 2012;1258:25–33.
45. Sturgeon C, Fasano A. Zonulin, a regulator of epithelial and endothelial barrier functions, and its involvement in chronic inflammatory diseases. Tissue Barriers. 2016;4(4):e1251384.
46. Scheffler L, Crane A, Heyne H, Tönjes A, Schleinitz D, Ihling CH, et al. Widely used commercial ELISA does not detect precursor of haptoglobin2, but recognizes properdin as a potential second member of the Zonulin Family. Front Endocrinol (Lausanne). 2018;9:22.
47. Eicosanoid. Wikipedia, Wikimedia Foundation, 25 May 2019, en.wikipedia.org/wiki/Eicosanoid#/media/File:EFA_to_Eicosanoids.svg. Accessed 16 June 2019.
48. Das UN. A defect in Δ^6 and Δ^5 desaturases may be a factor in the initiation and progression of insulin resistance, the metabolic syndrome and ischemic heart disease in South Asians. Lipids Health Dis. 2010;9:130.
49. Yary T, Voutilainen S, Tuomainen TP, Ruusunen A, Nurmi T, Virtanen JK. Omega-6 polyunsaturated fatty acids, serum zinc, delta-5- and delta-6-desaturase activities and incident metabolic syndrome. J Hum Nutr Diet. 2017;30(4):506–14.
50. Tsoukalas D, et al. Application of metabolomics part II: Focus on fatty acids and their metabolites in healthy adults. Int J Mol Med. 2019;43(1):233–42.
51. Ntambi JM, Miyazaki M. Recent insights into stearoyl-CoA desaturase-1. Curr Opin Lipidol. 2003;14(3):255–61.
52. Cho JS, et al. Serum phospholipid monounsaturated fatty acid composition and Delta-9-desaturase activity are associated with early alteration of fasting glycemic status. Nutr Res. 2014;34(9):733–41.
53. Breitling R. What is systems biology? Front Physiol. 2010;1:9. Published 2010 May 21. https://doi.org/10.3389/fphys.2010.00009.
54. Chiauzzi E, Rodarte C, DasMahapatra P. Patient-centered activity monitoring in the self-management of chronic health conditions. BMC Med. 2015;13:77.
55. Bennet P, et al. Chapter 28 clinical approaches to gastrointestinal imbalances. In: Textbook of functional medicine. 2nd ed: Example Product Manufacturer; 2010.
56. The 5Rs. Lipski L. Digestive wellness: strengthen the immune system and prevent disease through healthy digestion, 5th ed. McGraw-Hill Education; 2019.
57. Escott-Stump S. Nutrition & diagnosis-related care. 8th ed. Philadelphia: Wolters Kluwer; 2015.
58. Sagner M, McNeil A, Puska P, Auffray C, Price ND, Hood L, et al. The P4 health spectrum - a predictive, preventive, personalized and participatory continuum for promoting healthspan. Prog Cardiovasc Dis. 2017;59(5):506–21. Epub 2016 Aug 18.
59. Lazarus RS. Patterns of adjustment and human effectiveness. New York: McGraw-Hill; 1969.
60. Langgartner D, Lowry CA, Reber SO. Old friends, immunoregulation and stress resilience. Pflugers Arch. 2019;471(2):237–69.
61. Hennig B, et al. Using nutrition for intervention and prevention against environmental chemical toxicity and associated diseases. Environ Health Perspect. 2007;115(4):493–5.
62. Substance Priority List | ATSDR. Cdc.Gov, 2017, www.atsdr.cdc.gov/spl/index.html. Accessed 16 June 2019.
63. ATSDR - Toxicological and Public Health Professionals Home Page. Cdc.Gov, 2011, www.atsdr.cdc.gov/substances/ToxHealthReferences.asp. Accessed 16 June 2019.
64. Pathogens and Protocols. 2019., www.cdc.gov/microbenet/pathogens-protocols.html. Accessed 16 June 2019.
65. Hall A, et al. The role of nutrition in integrated programs to control neglected tropical diseases. BMC Med. 2012;10(1):25. www.ncbi.nlm.nih.gov/pmc/articles/PMC3378428/, https://doi.org/10.1186/1741-7015-10-41. Accessed 16 June 2019.
66. Jenkins HS. A piece of my mind. The morning after. J Am Med Assoc. 2002;287(2):161–2.

The Power of Listening and the Patient's Voice: "Please Hear Me"

Carolyn L. Larkin

6.1 Introduction: Solutions to Help Build Trust with Patients – 74

6.2 The Need for Improved Practitioner-Patient Communication: Practitioner Know Thyself! – 74

6.3 Behavioral Competency Soft Skills Improve the Practitioner-Patient Experience – 75

6.4 Leading the Interdisciplinary Medical Team – 79

6.5 The Powerful Art of Listening: Learning to Adapt Practitioner Behaviors to Meet the Needs of the Patient – 79

6.6 Telling the Patient's Story – 79

6.7 Naturally Occurring Patient Tribes: The Power of Communities – 81

6.8 Conclusion – 82

References – 83

> **Therapeutic Relationship**
> This means that the doctor (or practitioner) can imagine the inner experience of the patient and practices clinically sound medicine in a way that respects that inner experience. In other words, patient-centeredness is a mindset expressed in professional behavior; and consulting skills are the translation tool between the mindset and the behavior [1].

6.1 Introduction: Solutions to Help Build Trust with Patients

While growing research is focused on how to build a therapeutic practitioner-patient relationship, the current medical paradigm has a less-than-complete understanding of how to do so [1]. Ideally, practitioners should treat each patient as a client and as such provide their clients superior customer service. Superior customer service requires individualization of patient care, which should be the focus for all practitioners. It also demands a high level of engagement with the patient, one that can lead to mutual trust. However, a high level of engagement is only possible when the practitioner is trained in the best way to interact with the patient as an individual [2, 3]. In the end, the practitioner's goal should be to foster excellent relationships with patients to deliver superior, customized personalized patient healthcare [3].

Teaching behavioral competencies, also referred to as *soft skills* (see ◘ Table 6.1 for list of soft skills), has been an important part of leadership development in the corporate world for many years [1, 2]. The development of leadership soft skills and the teaching of soft skills competencies is a missing element in healthcare training and education [1]. Providing this model of education to all types of medical practitioners would improve patient interaction, communication, and engagement while increasing effectiveness in delivering personalized patient care in a time-efficient manner [2]. Exploring the importance of understanding and using soft skills within medical practice will be our starting point. To understand soft skills, practitioners need to learn how to use multiple assessments, including those that measure personal behaviors, team behaviors, emotional intelligence, resilience, and empathy.

This chapter is intended to demonstrate useful techniques for creating stronger trust relationships with patients that will become an integral part of patient therapy. A well-prepared practitioner who understands interpersonal skills and who communicates sensitively and effectively with patients of all personality types will improve their practitioner-patient relationships. We encourage practitioners to be involved in a holistic healing process encompassing mind-body-soul to deliver personalized patient care and to encourage families to become part of the healing process. We cannot continue to confine the healing process inside of disconnected silos. Everything related to the patient healing process should be part of the same system. We need to define and improve connectivity of the totality of the health system.

> Trust impacts us 24/7, 365 days a year. It undergirds and affects the quality of every relationship, every communication, every work project, every business venture, every effort in which we are engaged. It changes the quality of every present moment and alters the trajectory and outcome of every future moment of our lives – both personally and professionally [4].

6.2 The Need for Improved Practitioner-Patient Communication: Practitioner Know Thyself!

Educating practitioners to understand their behavior styles and to help them to improve communication skills is a missing element of healthcare education curricula [2]. When practitioners understand their preferred work behaviors, medical teams can come together and adapt their behaviors to coworkers while meeting the needs of their patients [3]. This can help teams have a more cohesive, less stressful environment while interacting with all types of patients with all types of needs. If practitioners were to be taught new interpersonal skills or soft skills for effectively communicating with patients, the practice of medicine could significantly improve. ◘ Table 6.1 lists a set of useful soft skills, compared with more commonly understood hard skills that are usually acquired through education, training, and/or life experiences. In this chapter, we focus on communication and interpersonal skills.

> The most effective use of behavioral soft skills occurs when practitioners understand how to adapt their own behavioral style to meet the needs of the patient.

◘ **Table 6.1** Description of hard skills and soft skills competencies

Hard skills	Soft skills
A degree or certificate program	Communication skill
Technology experience	Interpersonal skill
Proficiency is a foreign language	Teamwork
Machine operation knowledge	Leadership
Coding ability	Problem solving

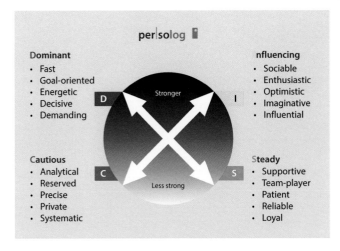

Fig. 6.1 Four behavioral styles originally described by William M. Marston in his book *Emotions of Normal People* [5] and illustrated by Persolog [6]. (Used with permission from Persolog.)

Starting in the mid-1920s, personality researchers began to develop descriptions of human personality types as they relate to personal interactions. This work was brought into focus by William M. Marston in his seminal book *The Emotions of Normal People* [5]. After being refined through further research, the description of behavioral styles has settled on terms denoted by the acronym "DISC," standing for dominant, influencing, cautious, and steady [6]. The characteristics of each behavioral style are shown in Fig. 6.1 [5, 6].

> Using skills related with human behavior in the health field is an effective way to create a stronger therapeutic relationship between practitioners and patients.

There are specific and identifiable styles of communication between practitioners and patients that can be characterized using the DISC paradigm. What follows are examples of each style of either practitioner or patient behavior. These four styles are commonly seen in healthcare settings and offer some clues toward understanding behaviors of both the patient and the practitioner:

Dominant Behavior (D) This style of patient communication is interested in receiving bottom-line information from the practitioner. These patients prefer a brief conversation and limited social interaction. They are not interested in a lot of data – just enough to hear the news and what they will have to do with the news. They may test the practitioner by asking direct questions while evaluating if the practitioner knows what they are doing. They have a need to control their environment and have a tendency to act as the authority figure. When unable to take control, they can be aggressive and demanding.

Influential Behavior (I) This type of patient seeks more communication than in the dominant type above and may want hand-holding, reassurance, and their practitioner's understanding and acceptance of their emotional state. They want to hear information in simple words and will ask a lot of questions to help them understand the diagnosis. They may want to talk at length with the practitioner and may be emotional. They don't respond well when facing a practitioner who is all business. They prefer a practitioner they consider a friend.

Steady Behavior (S) Patients with this communication style will listen and take time to digest information. They limit conversation, are cooperative, and will want to know their prognosis. Their focus when listening to the practitioner will be mainly on how their health status will impact their family. They will want to protect and remove any risk to their loved ones and want to put things in order, should their health be in danger. They do not like conflict and are congenial and thoughtful.

Conscientious Behavior (C) Patients with this style of communication focus on accuracy of the diagnosis and will review all available data shared by the practitioner. They will conduct independent research to validate the data given by the practitioner. They want the diagnostic data numbers to be very clear. If they have access to a computer during a hospitalization, they may check the information they receive from the practitioner against online sources.

For an example of two different behavioral styles and their interactions, consider a practitioner who is a dominant A-type personality, or DISC-style type D, working with a patient who has a social-interactive style type I. The practitioner will be direct and conclude the conversation in a short amount of time. The patient's expectation of how the practitioner should communicate might include asking the patient questions such as, "How are you feeling today? Have you talked to your family about your situation? Do you have any questions about what I said?" Many of these questions from the practitioner will satisfy, reduce fear, and build trust and confidence with the patient. Some practitioners naturally connect with patients. For others, informal conversation is not easy. Soft skills education will help practitioners understand numerous communication styles.

> The objective of learning to understand behavioral styles for better patient care is to increase engagement between caregivers and patients, to further build stronger relationships.

6.3 Behavioral Competency Soft Skills Improve the Practitioner-Patient Experience

Time spent in the therapeutic encounter continues to be reduced. Today, a major opportunity in healthcare presents itself through teaching behavioral competency techniques that will mitigate the patient's frustration level when the

Table 6.2 The most important behavioral competencies (soft skills)

Competencies	Some critical behaviors practitioners need to strengthen
Leadership	Remaining calm and effective in high-pressure situations Taking initiative – doing what needs to be done without being asked to do so Demonstrating high standards of ethical conduct Dealing fairly with all people Showing sensitivity and compassion for people
Teamwork	Treating coworkers with courtesy and respect Supporting decisions made by the team Putting team goals before individual goals Helping coworkers when they are having difficulty Asking coworkers about their projects and priorities
Communication	Expressing thoughts, feelings in words Explaining the reasoning behind own opinions Listening to others without interrupting Communicating in a clear, logical, and organized manner Exhibiting an open mind when hearing people's opinions
Focus on patient	Placing a high priority on improving patient service Recognizing and rewarding people who deliver excellent patient service Encouraging employees to contact and listen to patients Identifying and understanding the patient needs and service expectations Appropriately communicating with patients to keep them informed on a regular basis
Interpersonal skills	Establishing effective relationships with coworkers Showing knowledge and respect for people's responsibilities throughout the organization Establishing trust with people at all levels Taking time to establish relationships with people at all levels Demonstrating an appreciation of the value a diversity of people in the workforce

practitioner does not have enough time to listen. See ◘ Table 6.2. Using interpersonal skills adapted to human behavior is an effective way to create stronger relationships between practitioners and patients [7, 8].

Behavioral competencies are identified in terms of both hard skills and soft skills. Soft skills are frequently used in the business realm, but currently less so in medicine. Soft skills are important competencies for practitioners to have, understand, and use when caring for patients. Improving behavioral competencies (listed in ◘ Table 6.2) will improve the practitioner-patient relationship.

Using soft skills allows practitioners to consider how the patient is receiving a difficult diagnosis, for example, while simultaneously allowing the patient to understand what the medical expert is saying. If behavioral competency soft skills are not in place, the patient receiving the medical diagnosis may not be able to understand the important information the practitioner needs to deliver. Health practitioners frequently convey diagnoses to their patients using medical jargon or vocabulary that non-healthcare providers may be unfamiliar with, making it more difficult for patients to accept or process the information. If the medical practitioner does not consider how the patient is able to receive the information, then the patient can experience confusion, fear, worry, stress, and denial. The intent may be good on the part of the practitioner, but the message's impact on the patient will be ineffective.

> In healthcare, practitioners would be more effective if additional educational curriculum included competencies focused on personalized patient communication.

There are definite challenges and barriers between practitioners, team members, and the patient; these challenges and barriers are, in part, communication barriers. Strengthening soft skills and enhancing communication reduces conflict and initiates cooperation and collaboration. It turns disagreement into discussion, which leads to creative solutions. Communication skills are sharpened to humanize the delivery of information to the patient and to help the patient understand the practitioner more clearly. Besides words and body language, better communication also includes allowing the patient to be heard. This reduces patient fear, resistance, and shame in the clinical setting [9–12].

The responsibility for effective communication rests with the practitioner. Focusing on meeting the needs of the patient through effective communication can help improve patient care, satisfaction, and outcomes. Enabling practitioners to learn and develop specific and critical behavioral competencies may significantly improve the level of engagement in patient interaction. In some ways, the healing process begins when patients feel they have a voice in the conversation and an audience for their story, one that listens, rather than directs, how the conversation should go.

The patient who believes their practitioner is listening to them will begin to trust and be more willing to reveal information. Information such as the possible root cause of their ailment, a trauma that could have initiated the beginning of the disease, emotional upheaval, stress they live with, fears, denial, and/or concern for their family's reaction will be shared fully. A psychologist is not necessary to uncover such background. A trained, patient-focused team can do it when it works in synergy. A term, *situational communication*, has been developed to describe this type of teamwork [12]. Please refer to the Case Study.

Case: Effective Communication to Build Strong and Trusting Patient Relationships: A Way to Personalized Patient Care

Optimal health depends on a robust practitioner–patient relationship. This relationship involves the critical need for patients and their families to have their often-unheard voices heard. Information shared in an open and safe environment contributes to the formula that will effectively change the current model of the practitioner–patient relationship. Paramount in this process is the need to examine the importance of storytelling as a means to build trust and the significance of identifying support communities when striving toward optimal health. These communities have been identified as tribes and are a rich resource and support for the patient and their families. As a new type of practitioner, it is critical to be armed with the communication resources to facilitate these trusting, healing connections.

Below is a shared story; with changes in details and variations in setting, thousands of people with similar life-altering experiences have told this type of life narrative. Important in this story is the focus on the conversation and allowing the storyteller to speak uninterrupted so as to value their experience [14, 15].

The storyteller was first exposed to a family member's serious health challenge when she was a young girl. She had a younger brother 10 years her junior who was ill from birth, and she felt compelled to help her mother care for him. He had been born with a congenital urinary disorder that was not diagnosed until he was a toddler and was eventually found to have an obstruction of the urethra, resulting in reflux with injury to the bladder and kidneys. The obstruction was surgically removed, but recurred quickly. As a result, he underwent many surgical procedures over the following years. The mother became his nurse, changing catheters, while the sister was designated as her "little helper," sterilizing the equipment and assisting in other ways. The family members were not specifically trained, but knew these interventions were necessary and critical to save the child. They did the best they could to understand the complex conversations and directions for interventions the practitioners shared, including information regarding prognosis, what treatments he required, and what to perform at home. The storyteller recalled her mother "doing everything she could possibly do to keep him alive," in addition to answering the questions about her brother's condition in a way the family could understand.

What the storyteller remembers vividly is the frustration her mother expressed in this situation, the isolation she experienced, and feelings of being incapable of providing accurate information to family and friends. She describes these feelings as "palpable." The storyteller discusses being involved throughout the remainder of her childhood until she was a young adult and no longer living at home, when she describes her support changing from direct involvement to moral support. She describes her mother providing 27 years of effective care for her brother until he died.

As the only girl in the family of four children, she had an early education in "caretaking," but believes this early education allowed her to acquire many competencies that proved to be extremely useful later in life. As the storyteller describes, "These skill sets were ingrained in me and allowed me to face one of the most challenging health crisis I never expected to deal with again. A crisis that changed my life in ways I never expected, never thought I could endure, nor survive."

By the age of 27, the storyteller was the mother of four children, all born within six years. She describes herself as a child raising children. During that time and before the development of the rubella vaccine, a rubella epidemic swept the United States. Over that short period, there were 12.5 million cases of rubella. Around 20,000 children were affected with congenital rubella syndrome (CRS), which left more than 11,000 children deaf, 3,500 blind, and 1,800 with mental retardation. There were 2,100 neonatal deaths and more than 11,000 abortions, some spontaneous because of prenatal rubella infection and others performed surgically after women were informed of the serious risks of prenatal exposure. The storyteller states 1965 as the worst year of her life and goes on to describe:

» "My one son was two years old at the time he contracted the disease (rubella). Unbeknownst to me, I was 1 ½ months pregnant. During this time, primary doctors made house calls and the doctor came to my home to treat my son. A month later, again not knowing I was pregnant, I called my doctor to inform him that I wasn't feeling good. I had a fever and felt like I was getting the flu. He suggested I take some aspirin, a warm bath, and try to get some sleep. I did exactly that, however I noticed that I had small bumps on parts of my body, so I called once again to inform him and was told that I most likely contracted rubella. After the rash had cleared, I went into the office for a blood test. It was during that visit that I found out I was pregnant. Little did I know that this event would change my life dramatically. My baby girl was born with multiple medical problems; problems that I was not aware could happen because of the rubella infection."

» "At the time of her birth, she weighed in at 3 lbs. 11 oz. and was 21 inches in length." When she reached 8 pounds, she had her first open-heart surgery at Boston Children's Hospital. At that time, neonatal care was just emerging. It was not until 1965 that the first American (neonatal) intensive care unit (NICU) was opened in New Haven, Connecticut and in 1975 the American Board of Pediatrics established sub-board certification for neonatology."

» "I had an appointment with the heart specialist to review her prognosis and prepare for her surgery. With my baby in my arms, terrified and alone, the doctor entered the room and introduced himself."

» "The doctor was all business and spoke to me in terms I didn't understand. The questions I asked were uninformed, too vague for the doctor to understand, and I was not confident enough to continue to probe. I just listened… my anxiety was at an all-time high. After about 20 minutes of listening to the doctor's information, he stood up and said his intern would be meeting with me to give me some printed information regarding my child's surgery. The surgery was scheduled for the next day. The doctor left the room, while I sat crying, holding my baby in my arms, wishing my resilient mother were with me.

» "The next person to enter the room was the intern. He introduced himself and handed me instructions and directions for the next day. I didn't hear much of what he had to say, I was still in shock at the reality of what was happening. My heart was racing, tears were flowing; I felt like a baby trying to save my child. Then, the most frightening words I ever heard came from the intern. He said, 'I'm sorry there is nothing that we can offer you to help you with your child's additional health problem.' Already confused, I said, 'What additional health problem?" He answered, 'Your child is deaf. Didn't your pediatrician inform you about this?' My reaction was filled with rage and emotion, 'You have the wrong chart. My child is NOT deaf! What kind of hospital is this? My child has a heart problem… That's all. I need to speak to the doctor!'

> "The doctor came back into the room. He tried to explain that my daughter had both health challenges, while trying to reassure me that everything would be fine. At that point, it fell on *my* deaf ears. I went into total denial. I would not accept the fact that my daughter was deaf. Even though her heart surgery was successful, I lost trust and confidence with the doctor, the hospital, and the staff. I was filled with fear for my daughter's survival. I left the hospital and drove back home in a rainstorm, with my tears flowing faster than the windshield wipers could wipe the rain away. As soon as I got home from the meeting I tried to prove to myself that my daughter was not deaf. I spent most of the rest of the day slamming cabinet doors and hitting pots together intensely watching my daughter's reaction, trying to prove that she was not deaf. I was in shock, ultimately ignorant about raising a deaf child. I had zero exposure or experience with her health challenge. The worst nightmare had just begun, and I had no way out."

The storyteller goes on to describe that 50 years later, her daughter, now a grown woman with multiple prior invasive heart procedures behind her, was facing another open-heart surgery. Due to the difficult and complex surgery she was facing, most of the extended family members were present. When the surgeon spoke to the family before entering the operating room, he was asked by the storyteller to take good care of their "little girl." She reports he replied, "I'll do my best, but this will not be easy," words not welcomed or comforting. The presurgical ritual was all too familiar for this mother as she accompanied her daughter into the OR and translated through sign language what was being communicated. Words of reassurance were signed to the daughter and received with absolute trust in her mother. Everyone in the room was supportive and kind, and the storyteller reported feeling satisfied with how events were proceeding at this point.

Although the surgery was technically a success, the storyteller reported waiting for hours not knowing if there were complications or even if her daughter had survived. Nothing from the surgeon. Finally, the family was notified that the surgery was over, led to a conference room, and anxiously waited for the surgeon to enter. When he did arrive, the storyteller reported he looked at them, did not say a word, sat down on a chair, and reviewed his paperwork. They waited until he finally said, "Oh, the surgery went well." Aside from the communication skills lacking on the part of the surgeon, the family burst into tears of relief. With each medical procedure the patient endured, she insisted the storyteller be by her side not only to interpret by signing but also to interpret the nuances of what was being communicated – or poorly communicated.

By telling her story, this storyteller was able to impart the nuanced richness of her life and life experiences and better able to project whom she is based on these experiences. If not allowed the space to tell her story, an important element of care would be lost. The practice of medicine has improved and will continue to improve, and medical practitioners are dedicated to providing best-known solutions for their patients. Patients cannot expect medical practitioners to excel in patient care when medical curriculum does not include what is currently unknown to them. Unknown curriculum subjects such as storytelling and practitioner-know-thyself training accompanied by important applicable self-assessments should be mandated in medical curriculum in order for practitioners to have the tools to be successful while providing personalized patient care.

Trust needs to be established between the medical practitioner, the patient, and the family members. Trust is inherent when communication is valued. Just as this storyteller describes, examples of poor communication can isolate patients and family members. When they are not heard, healing is not facilitated. As highlighted by this story, ineffective communication practices existed throughout the medical establishment; unfortunately, they still exist. Education to teach practitioners to hear the stories is paramount. As a new breed of practitioners, we need more focus on listening with our hearts, using better ways to acknowledge the patient's voice, and finding solutions to capture the stories, in order to realistically achieve personalized patient care.

Situational communication is a clear, concise, and focused communication strategy that maximizes the ability of a minimum amount of time to achieve successful results and effective relationships [12]. Although it may be used in personal settings, situational communication is mostly reserved for business and professional interactions, especially those that are challenging, where the points of view, interests, or preferred solutions of the communicators are different or in conflict. Leaders must be capable of planning for and executing situational communication in order to be both successful and effective. These interactions differ greatly from our mostly effortless run-of-the-mill interpersonal interactions. Situational communication represents purposeful, thoughtful engagement. In the case of situational communication, there is a planned payoff. These types of interactions are more formal, structured, and well-planned, requiring significant energy and focus [13].

In most businesses, the organization strives to help its human talent develop necessary competencies in order to be successful. In the healthcare industry, practitioners would benefit by employing the same approach. This is accomplished by adding educational curriculum that includes behavioral competencies focused on personalized patient communication with a desired effect of improved patient outcomes. Developing soft skills competencies that focus on communication skills needed by the practitioners will improve the patient's experience and make the job of the practitioner less frustrating and more rewarding.

Experts in situational communication know how to engage in robust and fruitful conversations that are tailored to the individual patient. As previously noted (Table 6.2), There are other soft skills practitioners should develop that impact the practitioner-patient relationship in addition to communication skills. All of these important skills promote a stronger trusting therapeutic relationship and improve the biological function of the human body through mind-body interactions.

Behavioral competencies should be required as part of medical curricula across all medical professions, and integrative and functional practitioners have been leading the way. When the interaction is personalized, patients can express themselves in ways that lead practitioners to understand the patient's status in a clear and concise manner. Behavioral competencies and communication skills are mandatory in any industry; however, behavioral competencies and situational communication skills will enhance the development of the therapeutic relationships that leads to successful patient outcomes.

6.4 Leading the Interdisciplinary Medical Team

There are unique competencies that all members of the medical team bring to their specific roles. One important role is leadership. Many practitioners have not been schooled in leadership competencies. However, leadership competencies are highly relevant when it comes to patient care. Members of the entire medical team (such as dieticians, physical therapists, psychologists, health coaches, pharmacists, nurses, doctors), referred to as an interdisciplinary team, need to function together in relating to a patient. The interdisciplinary team has shared leadership responsibilities, starting with understanding the principles of behavioral competency soft skills in relationship to patients.

Research states that the parts of the roles practitioners are highly competent to fill are usually roles that are personally motivating for them [13]. Excellent leaders understand that some of their roles are not self-motivating and frequently ask for feedback from the interdisciplinary team in order to gauge their own effectiveness. They seek out feedback on their leadership style. They also self-evaluate their level of effectiveness. Ineffectiveness demands leadership changes. Without the willingness to change, everything becomes static [13]. Competencies associated with corporate business are transferable throughout many industries and can be adapted to healthcare to improve patient care. Improving patient care is the ultimate outcome healthcare should strive to accomplish.

6.5 The Powerful Art of Listening: Learning to Adapt Practitioner Behaviors to Meet the Needs of the Patient

The objective of learning to understand behavioral styles is to increase engagement between caregivers and patients, ultimately building stronger relationships. Effective communication occurs when both parties have a clear understanding of their own preferred communication style. To achieve that understanding, behavioral profiling tools can help and allow the practitioner and patient to become a cohesive team. Once practitioners are trained to understand their personal behavioral style, they can quickly assess the patient's basic style, enabling both parties to better work together. Understanding the behaviors of their patients allows practitioners to change their behaviors from expecting the patient to listen and understand them, to practitioners listening and understanding the patient. More importantly, behavior assessment tools can assist interdisciplinary medical teams to understand their own preferred work behavior, so they can also learn how to adapt their behaviors to meet the needs of the patients through stronger teamwork. Utilizing assessment tools like those described below can reduce stress for the interdisciplinary team when dealing with very ill patients and challenging situations.

All of the major psychometric assessment tools in worldwide use employ the basic behaviors established by Dr. John Geier, considered the leader in human assessment science [16]. A combination of psychometric behavioral assessments (Table 6.3) can provide important data needed to properly train and prepare medical teams. Practitioners and interdisciplinary teams gain insight into their behaviors and the effect on the patient, which allows the team to be of better service. Such assessments may include emotional intelligence, DISC profile, 360-degree assessments, and performance evaluations, if appropriate. Behavioral evaluation report results are directly available to each practitioner, which opens the door for the practitioner to share results with the team to facilitate improved cohesiveness. Patients may also be assessed, and this allows the practitioner to understand his/her own behavioral style, to recognize the patient's style, and how to adapt his/her communication behavior to best address issues with the patient [18]. If medical practitioners are educated in soft skills assessments, like the ones mentioned in Table 6.2, it would enable them to recognize how individual patients choose to hear information, resulting in improved communication between the practitioner and patient. The most effective use of behavioral soft skills occurs when practitioners understand how to adapt their own behavioral style to meet the needs of the patient as discussed.

Obtaining practitioner behavioral assessments for each member of the interdisciplinary medical team enables a team-building educational process. Tools, including training modules, can teach practitioner teams how to effectively improve their communication within their medical team while understanding and respecting everyone's behavioral style. Competencies within the team are strengthened and create more concise, positive outcomes in the management of the difficult patient. In addition, for complicated patients, the combined process of tools and training is not restricted to any particular disease or diagnosis. Any medical team can benefit from data-driven assessments. These, combined with educational methodologies, should be a necessary form of education for the entire health industry.

6.6 Telling the Patient's Story

For centuries, stories have been shared through the spoken word, interpersonal conversations, diaries, and the writing of history. Storytelling has the potential to reveal emotion and psychological trauma. People remember stories. They repeat them, often linking them to their own similar experiences. No matter what the story is, if it is told, millions of people in many different languages around the world will have gone through something similar. It is that sense of connection to community that says, "I've been there; I know what that is like." Sharing stories can be an experience of enlightenment as emphasized in the Case Study.

Table 6.3 Psychometric behavioral assessments commonly used worldwide

Assessments	What the tool evaluates	Description of tools
H R Tools, Inc. ▶ http://HRToolsINC.com	The following tools can be located on the ▶ HRToolsINC.com link	See the following tools
Emotional Intelligence	Ability to characterize emotions in oneself and in others, ability to use feelings constructively and to understand and analyze emotions and solve emotional problems, ability to take responsibility for one's emotions, attaining emotional growth and maturity	The person takes the assessment and identifies what are his/her strong skills and which are the ones he/she needs to develop. Then work with a coach or participate in a training program that teaches them how to develop each skill and work on that process
DISC Profile	The DISC profile describes human behavior on the basis of four behavioral styles, dominant (D), influencing (I), steady (S), and cautious (C), and aims at a better understanding of one's own needs and those of others	There are many different programs where participants take the assessment and then work in many different training programs (also in coaching works perfectly) like teamwork, leadership, communication, interpersonal skills, and many others. The model adapts well for all these methodologies because it is based on what is my behavioral style, how I can identify other styles, and how I can adapt to them [16]
360 Assessments	This methodology consists in receiving feedback from employee's managers, peers, and direct reports. The feedback is provided by them using an online tool that ensures the confidentiality of the feedback. The questionnaire of the feedback is based on competencies and behaviors that measure those competencies or soft skills	After the 360 assessment is completed, the person receives the feedback and a consultant or coach helps him to understand it and how to work in that, as well as how to create a development plan
E4 Method	The E4 method facilitates the connection through communication and identifying teaching materials	A teaching methodology tool for practitioners to develop a standard of patient communication principles. The E4 Method was developed by Butler and Keller [17] in 1999 and is a commonly used tool

In many ways, social media holds the potential to be a healing vehicle. It provides humans an environment wherein they can tell their stories. Postings can quickly go viral; as stories are shared, experiences connect us. In this era of advanced communication technology, we have the freedom to share information and stories that instantly reach every corner of the globe.

> When a patient shares their story, it helps the caregiver to understand the patient's journey and health challenges through the lens of the past and enlightens the present state of health.

As stated by Hall and colleagues, medicine is an art whose magic and creative ability have long been recognized as residing in the interpersonal aspects of patient-practitioner relationships [19]. In the healing encounter, sharing stories between practitioners and patients is one of the most important parts of the healing process. When a patient shares a story, it helps the caregiver to understand the patient's journey and health challenges through the lens of the past and enlightens the present state of health. It is critical that the practitioner understand the patient's stories so the two can build a strong connection. Much of the psychological healing process depends on this relationship. With shared experiences through storytelling, synergy and mutual trust are created. The practitioner needs to understand the patient's feelings of fear, grief, or loss, and these feelings may be mitigated if received with the type of communication the patient needs. The practitioner and patient may then connect on another level where the patient feels understood, and, in those moments of connection, the practitioner is truly serving the patient in a partnership leading to the best outcome.

To facilitate the connection through communication, teaching materials are needed. The E4 Method developed by Butler and Keller [17] is a commonly used tool for practitioners to develop a standard of patient communication principles. The E4 Method includes the following 4 Es:
- *Engagement* – skills that support development of rapport with patients
- *Empathy* – skills that help clinicians reflect concern for the patient's condition
- *Education* – skills needed for discovering and developing the patient's understanding of his/her condition
- *Enlistment* – skills that help in motivating and changing behavior

As the craft of storytelling expands, education in the field of storytelling is emerging and accessible [17]. Medical practitioners need to be educated and trained through storytelling curriculum, which teaches understanding of the process and the uncovering of hidden meanings. Materials like the E4 Method, developed through evidence-based research, are needed to build powerful relationships and would be a beneficial addition to the medical education curriculum.

As highlighted by the E4 Method, communication between the practitioner and patient is more effective when the practitioner allows the patient to construct a timeline of their story uninterrupted. A brief conversation with the patient is often not enough and may not be satisfactory for the patient in the moment. This storytelling process has power to reveal traumas and triggers and provides a window into the current state of the patient's physical and mental health. After the patient reveals significant psychological events and stressors by sharing life's timeline, the practitioner can pinpoint important information to help in the treatment of the patient's disease. The practitioner, indicating a degree of empathy for the situation, reflects the important events back to the patient.

As a cohesive partnership develops between the practitioner and the patient, family members can be enlisted as part of the team. This will serve to educate the patient and the family about the health condition in a safe and supportive environment. Enlisting the patient and family as a team will help with motivating changes that need to be made. The team will be in sync working together toward the accomplishment of the best possible outcome. It is time for medical practitioners to reap the rewards from the soft skill of listening…listening to the stories waiting to be told, shared, understood, and accepted.

> It is time for medical practitioners to reap the rewards from the soft skill of listening…listening to the stories waiting to be told, shared, understood, and accepted.

Barriers to implementing therapeutic storytelling include significant time constraints. The burden of the current healthcare paradigm with mandated abbreviated appointment times puts practitioners in a time-crunch that is not conducive to meeting the needs for personalized patient care. Practitioners expect patients to listen intently to what they have to say in this limited time, but the infrastructure of conventional care does not afford the practitioner similar time to listen to the patient. By the act of listening with intention to the patient's story and basing decisions on the shared conversation, some measure of relief of symptoms can be expected [20]. This can have the effect of reducing symptomatic complaints, shortening the length of the disease course, improving outcomes, and, in the long-term, reducing the burden of time with the patient.

Another solution to the problem of implementing and retaining the personalization in a therapeutic encounter is to depend on the contributions of the interdisciplinary healthcare team members. With common training of each team member in the art of storytelling and listening, the entire team can be called upon to effectively help when other members are unavailable. There can be building of trust, safety, and healing by wise use of team members' time.

Perceived barriers to training may also include the belief system that the practitioner is sacrificing time and energy when faced with coursework to develop communication skills. Training must emphasize the payoff of improved patient relationships, reduction in practitioner stress, improved effectiveness of time management, and healthier outcomes for patients and their caregivers. Training should be emphasized as a method to develop the skill to listen to and appreciate the patient's story which will pay dividends by providing the needed context to the patient's situation. Practitioners need to understand that this education would enable personalized patient care with effective interventions and enhancement of their skills.

We are overdue in providing the type of training necessary to change, improve, and close the gap between mediocre communication skills and high-performance communication skills. Connecting with the patient's story is critical. If practitioners are not trained in the necessary skills to connect with patients, those stories will never be heard. Training health practitioners to become proficient in storytelling and story listening is a viable potential solution to the current healthcare crisis.

6.7 Naturally Occurring Patient Tribes: The Power of Communities

Articles and methodologies are available that build on patient communities and patient support groups [21, 22]. Naturally occurring patient tribes are forming throughout the globe, and physicians, health practitioners, and health communities are not necessarily included in them. This model of patient support is becoming a powerful new way for patients to share their specific experiences and deepen their medical knowledge.

A naturally occurring tribe can be formed anywhere, anytime, and in any location. The beginning of a tribe often occurs when patients are gathered in clinical settings while waiting for treatment. In these situations, patient and caregiver tribes begin to form. What makes patient tribes different is the conversation shared between patients with shared medical challenges. Such conversations can be highly valuable for the patient and may even be more informative than a patient having a conversation with their health practitioner.

Naturally occurring patient tribes may be useful and effective for patients because individuals dealing with similar health challenges often find support in these informal connections. There may be an unspoken acceptance to engage in difficult conversations, yet facing the difficult conversation is necessary and consoling. There is a level of trust that the tribe environment creates. The trust is based on sharing information and like experiences, such as fears, doubts, and challenges with practitioners.

Patients may seek out guidance from each other, learn tips for managing their condition, and educate one another to understand the medical terminology associated with their diagnoses. People may not feel free to ask questions of their medical team, yet have no hesitancy in asking the patients or families of patients sitting nearby. There is power in this shared experience. The tribe members are on the same journey and perhaps have learned valuable things along the way. The tribe can help each other almost like a mentor helps a mentee understand specific situations and offers solutions that may be found to be extremely effective. There are many conversations about how relationships are built, how powerful they are, and how they naturally create tribes of like minds [14, 15]. This connection to the tribe is reinforced and strengthened as the relationship blossoms.

Tribes also connect and form in other ways. In today's world of technology, shared experiences go viral, and powerful connections are achieved on an unprecedented scale never before seen. Social media is an easy-to-access "share button" that connects to the world of human experiences. This global technological connectivity vehicle allows millions of people who do not know each other the ability to communicate, share stories, reveal information, and connect both intellectually and emotionally because of the thread of shared experiences.

> We want to encourage practitioners to be involved in a holistic healing process for personalized patient care and very importantly for families who should be part of the healing process.

This global phenomenon has informed many shared stories containing private information and diverse conversations in every language. This global engagement between people is powerful and has the ability to bond strangers together to become both a tribe of like minds and extended families. Evidence has emerged in recent years that reveals how powerful relationships are [21, 22]. The relationships you have and the tribe you spend time with and relate to are more powerful than even what you eat.

6.8 Conclusion

In conclusion, we cannot expect health practitioners to have skills that they have not been taught. Some health practitioners have a level of intuitive empathy and are able to fill the gap between formal education and understanding how to excel in patient communication, interaction, behavioral styles, and skills. However, all healthcare practitioners should be given the opportunity to be trained to understand the human dynamic and the diversity of each human experience. This training will be an asset to practitioners, patients, medical teams, and organizations that employ them. Communication skills are transferable and long-lasting. They may apply to other professional settings, as well as personal relationships.

When patients need information, the practitioner must give it in a clear and supportive manner. When patients need support, the practitioner must let them know they will be there for them [23]. In order to fulfill these needs and build a more effective practitioner-patient relationship, the practitioner can learn effective listening. When using effective listening, the practitioner maximizes valuable time over the long term. Effective listening allows the dialogue between practitioner and patient to be focused, informative, and positive.

In challenging settings where there may be time constraints or an undercurrent of psychological issues, the dialogue can evolve and become more productive when the practitioner attempts to understand the needs of the patient and puts the patient first. By being sensitive about word choice learned through training sessions, practitioners can make headway toward fruitful communication and let the patient know they understand where the patient is in the moment. It is important for many patients to hear and understand this. In addition, practitioners would do well to communicate with patients' families. Family members often deal with the shock of the illness of their family member and may put their feelings aside in order to cope and manage the medical situation. This situation can be acknowledged and validated. The practitioner needs to realize the family members have their own stories and can offer important insight into the history and status of the patient. Stories reveal important information indicative of their culture, family medical history, and genetic information. Families share stories. Stories have power.

The tendency to organically form support groups or tribes in medical office waiting rooms, social media, or other social venues is an important part of patient healing that healthcare teams could tap. If practitioners can develop ways to understand their patients' tribes and reinforce information shared, positive clinical outcomes will be fostered and enhance the practitioner's effectiveness and efficiency.

The time is overdue to identify and adjust the human factors that create barriers to build strong relationships. The differences in human behaviors are vast, unique, diverse, and can be complicated. However, as human beings we strive to be understood, respected, and included. It will take a concentrated effort to work to eliminate the barriers that separate us. The solution is to first understand our own behaviors and behavioral limitations, and work to improve them. "Knowing thyself" can be daunting and frightening. It will however be enlightening.

As practitioners, taking accountability for how we behave when we communicate to patients puts the onus of responsibility where it should be – on ourselves! Honing communication skills is worth the effort. These skills will improve our own lives, as well as the lives of others. When in the driver's seat, we hold the power to make or break relationships. The reward for taking responsibility and accountability for the effective way we communicate, listen, and respond to each other is a powerful reward for all of us humans. It will unite us in our effort to change and build a powerful and healing tribe of like minds.

References

1. Neighbour R. The inner consultation: how to develop and effective and intuitive consulting style. 2nd ed. Boca Raton: CRC Press, Taylor & Francis Group; 2005.
2. Jones D, Hoffman L, Quinn S. 21st century medicine: a new model for medical education and practice. Gig Harbor: Institute for Functional Medicine; 2009.
3. Janicik R, Kalet AL, Schwartz MD, Zabar S, Lipkin M. Using bedside rounds to teach communication skills in the internal medicine clerkship. Med Educ Online. 2007;12(1):4458. https://doi.org/10.3402/meo.v12i.4458.
4. Covey SMR, Merrill RR. The speed of trust: the one thing that changes everything, updated edition. New York: Free Press An Imprint of Simon and Schuster, Inc; 2018.
5. Marston WM. Emotions of normal people. London: Kegan Paul Trench Trubner and Company Publishing Co; 1928.
6. Persolog Personality Factor. [Internet]. www.persolog.com. Last accessed 23 Aug 2018.
7. Goold SD, Lipkin M. The doctor–patient relationship: challenges, opportunities, and strategies. J Gen Intern Med. 1999;14(Suppl 1):S26–33. www.ncbi.nlm.nih.gov/pmc/articles/PMC1496871/.
8. Ha JF, Longnecker N. Doctor-patient communication: a review. Ochsner J. 2010;10(1):38–43. https://www.ncbi.nlm.nih.gov/pmc/articles/PMC3096184/.
9. Choudhary A, Gupta V. Teaching communications skills to medical students: introducing the fine art of medical practice. Int J Appl Basic Med Res. 2015;5(Suppl 1):S41–4.
10. Ranjan P, Kumari A, Chakrawarty A. How can doctors improve their communication skills? J Clin Diagn Res. 2015;9(3):JE01–4.
11. Stewart K. Can you hear me now? Improving practitioner-patient. Today's practitioner [Internet June 9, 2017]. Integrative care practice, integrative medicine. Last accessed 23 Aug 2018 from: http://todayspractitioner.com/integrative-medicine/can-you-hear-me-now-improving-practitioner-patient-communication/.
12. Zalihić A, Obrdalj EC. "Fundamental communication skills in medical practice" as minor elective subject. Acta Medica Academica. 2014;43(1):87–91.
13. Riordan CM. How to juggle multiple roles. Harvard Business Review [Internet]. 2013. Last accessed 23 Aug 2018 from: https://hbr.org/2013/10/how-to-juggle-multiple-roles.
14. Patterson K, Grenny J, McMillan R, Switzler R. Crucial conversations: tools for talking when stakes are high. New York: McGraw-Hill Education; 2012.
15. Renter E. Why nice doctors are better doctors. US News. 2015, Apr 20 Health. [Internet]. Last accessed 24 Aug 2018 from: https://health.usnews.com/health-news/patient-advice/articles/2015/04/20/why-nice-doctors-are-better-doctors.
16. Geier JG. Geier learning international: DISC profiling system [Internet]. Accessed 11/29/19 https://www.persolog.de.
17. Hall JA, Roter DL, Rand CS. Communication of affect between patient and physician. J Health Soc Behav. 1981;22(1):18–30.
18. Butler J, Keller V. Words and attitudes: a better office visit for doctor and patient. Managed Care. May 1, 1999. [Internet]. Last accessed 24 Aug 2018 from: https://www.managedcaremag.com/archives/1999/5/better-office-visit-doctor-and-patient.
19. Chamorro/Premuzic T. Social skills and leadership in healthcare: the case for boosting doctors' EQ. Forbes. 2014 Oct 26. [Internet]. Last accessed 24 Aug 2018 from: https://www.forbes.com/sites/tomaspremuzic/2014/10/26/social-skills-leadership-in-healthcare-the-case-for-boosting-doctors-eq/#763207492240.
20. Durance B. Stories at work. Train Dev. 1997;51(2):25–9.
21. Van der Horst M, Coffé H. How friendship network characteristics influence subjective well-being. Soc Indic Res. 2012;107(3):509–29.
22. Berkman LF, Glass T, Brissette I, Seeman TE. From social integration to health: Durkheim in the new millennium. Soc Sci Med. 2000;51(6):843–57.
23. Covey SR. The 7 habits of highly effective people: powerful lessons in personal change. New York: Fireside Books; 1990.

Metabolic Characteristics and Mechanisms of Chronic Disease

Contents

Chapter 7   Metabolic Correction Therapy: A Biochemical–Physiological Mechanistic Explanation of Functional Medicine – 89
Michael J. Gonzalez

Chapter 8   The Nutrition Assessment of Metabolic and Nutritional Balance – 99
Margaret Gasta

Chapter 9   IFMNT NIBLETS Nutrition Assessment Differential – 123
Robyn Johnson and Lauren Hand

Chapter 10   Nutritional Role of Fatty Acids – 135
Vishwanath M. Sardesai

Chapter 11   Lipidomics: Clinical Application – 151
Diana Noland

Chapter 12   Structure: From Organelle and Cell Membrane to Tissue – 173
David Musnick, Larissa Severson, and Sarah Brennan

Chapter 13   Protective Mechanisms and Susceptibility to Xenobiotic Exposure and Load – 191
Robert H. Verkerk

Chapter 14   Detoxification and Biotransformation – 205
Janet L. Black

Chapter 15   Drug–Nutrient Interactions – 213
Mary Demarest Litchford

Chapter 16   The Enterohepatic Circulation – 221
Robert C. Barton Jr.

Chapter 17   **A Nutritional Genomics Approach to Epigenetic Influences on Chronic Disease – 235**
Christy B. Williamson and Jessica M. Pizano

Chapter 18   **Nutritional Influences on Methylation – 269**
Jessica M. Pizano and Christy B. Williamson

Chapter 19   **The Immune System: Our Body's Homeland Security Against Disease – 285**
Aristo Vojdani, Elroy Vojdani, and Charlene Vojdani

Chapter 20   **Nutritional Influences on Immunity and Infection – 303**
Joel Noland and Diana Noland

Chapter 21   **Body Composition – 323**
Sue Ward and Diana Noland

Chapter 22   **The Therapeutic Ketogenic Diet: Harnessing Glucose, Insulin, and Ketone Metabolism – 335**
Miriam Kalamian

Chapter 23   **The GUT-Immune System – 367**
Elizabeth Lipski

Chapter 24   **Centrality of the GI Tract to Overall Health and Functional Medicine Strategies for GERD, IBS, and IBD – 379**
Ronald L. Hoffman

Chapter 25   **The Microbiome and Brain Health – 391**
Sharon L. Norling

Chapter 26   **The Role of Nutrition in Integrative Oncology – 407**
Cynthia Henrich

Chapter 27   **The Microenvironment of Chronic Disease – 437**
Steven Gomberg

Chapter 28   **Chronic Pain – 447**
Jena Savadsky Griffith

Chapter 29   **Nutrition and Behavioral Health/Mental Health/Neurological Health – 473**
Ruth Leyse Wallace

Chapter 30 Neurodevelopmental Disorders in Children – 493
Mary Anne Morelli Haskell

Chapter 31 Nutritional Influences on Hormonal Health – 517
Filomena Trindade

Chapter 32 Nutritional Influences on Reproduction: A Functional Approach – 533
Brandon Horn and Wendy Yu

Chapter 33 Lifestyle Patterns of Chronic Disease – 563
Sarah Harding Laidlaw

Chapter 34 Circadian Rhythm: Light-Dark Cycles – 577
Corey B. Schuler and Kate M. Hope

Chapter 35 Nutrition with Movement for Better Energy and Health – 595
Peter Wilhelmsson

Chapter 36 Mental, Emotional, and Spiritual Imbalances – 613
Muffit L. Jensen

Metabolic Correction Therapy: A Biochemical–Physiological Mechanistic Explanation of Functional Medicine

Michael J. Gonzalez

7.1 Introducing Metabolic Correction – 90

7.2 Implications of Subclinical Nutrient Deficiencies (Nutritional Insufficiencies) – 90

7.3 Nutrient Requirements of Disease States – 90

7.4 Vitamins and Minerals and Hidden Hunger – 91

7.5 Why Metabolic Correction? – 92
7.5.1 Inferior Nutritional Value of Food and Low Availability of Nutrient-Dense Foods – 92
7.5.2 Medication-Induced Nutrient Depletion, Adverse Side Effects of Medication, and Iatrogenic Deaths – 92
7.5.3 To Compensate for the Increased Demand of Nutrients due to the Disease State – 93
7.5.4 Biochemical Mechanism of Metabolic Correction: Molecular Concentrations and Rate of Reactions – 93

7.6 B-Complex and Metabolism Briefing – 94

7.7 B-Complex and Metabolic Correction – 95

7.8 Ten Principles of the Concept of Metabolic Correction in Disease Therapy – 95

7.9 Beyond Metabolic Correction: Improving the Healthcare Management Model – 96

7.10 Conclusion – 96

References – 97

© Springer Nature Switzerland AG 2020
D. Noland et al. (eds.), *Integrative and Functional Medical Nutrition Therapy*,
https://doi.org/10.1007/978-3-030-30730-1_7

7.1 Introducing Metabolic Correction

Metabolic correction (MC) is the utilization of a synergistic combination of micronutrients and cofactors in their active forms and proper doses to maximize the function of metabolic enzymes [1–3]. MC helps improve or rectify biochemical disturbances that disrupt normal physiological processes and may originate degenerative diseases. MC is the fine-tuning of the cellular biochemistry by means of specific, targeted nutritional supplementation with the goal of improving cellular, tissue, organ, system, and organism function [1–3]. Impaired or incomplete cellular biochemical reactions are amended with MC.

MC is a functional biochemical/physiological concept that explains how improvements in cellular biochemistry help the body achieve metabolic or physiologic optimization [1–3]. The MC concept provides the biochemical elucidation of the utilization of nutrients for preventive and therapeutic purposes against disease. The MC concept becomes important since our food is decreasing in nutritional value [4]; diseases increase the demand for nutrients, and medications can deplete nutrients [1]. These nutrient insufficiencies are causing enormous costs due to increased morbidity and mortality. In summary, MC increases enzymatic function that enhances biological functions contributing to better health and well-being.

> » Metabolic correction is a functional biochemical/physiological concept that explains how improvements in cellular biochemistry help the body achieve metabolic or physiologic optimization.
>
> Michael J Gonzalez, DSc, NMD, PhD, FACN
> University of Puerto Rico Medical Sciences Campus,
> Schools of Public Health[1], Department of Human Development, Nutrition Program
> San Juan PR

a unique genetic or acquired dependency on supra-dietary levels of one or more nutrients [1].

Clinical assessment of nutritional status has long been exclusively focused on detection of absolute nutrient deficiencies by relying on clinical evidence of signs and symptoms of classic nutritional deficiency states (e.g., bleeding gums with vitamin C deficiency or mouth redness and sores with b-vitamin deficiencies). Yet, relative nutrient deficiencies or insufficiencies due to disease are equally important, as is the detection of nutrient imbalances before clinical evidence of deficiency is present. At the point where altered cellular function has evolved into clinical manifestations of nutrient deficiency, cellular activity will have been compromised for some time, and the compensatory responses that might have allowed a temporary adjustment to the deficiency would no longer be effective. Detection of subclinical changes in cell processes early in the course of a nutrient deficiency, when cell damage is minor and more reversible, can have a considerable impact on the prevention and treatment of disease. If subclinical deficiency is not corrected by providing the lacking nutrient, then prolonged marginalization of cellular activity may not only increase vulnerability to disease, but it may also exacerbate progression of existing disease and interfere with effectiveness of treatment, since all drugs require some level of metabolic support to achieve their desired therapeutic effects.

> » Detection of subclinical changes in cell processes early in the course of a nutrient deficiency, when cell damage is minor and more reversible, can have a considerable impact on the prevention and treatment of disease.
>
> Michael J Gonzalez, DSc, NMD, PhD, FACN
> University of Puerto Rico Medical Sciences Campus,
> Schools of Public Health[1], Department of Human Development, Nutrition Program
> San Juan PR

7.2 Implications of Subclinical Nutrient Deficiencies (Nutritional Insufficiencies)

A nutritional deficiency is defined as an inadequate supply of essential nutrients (vitamins and minerals) in the diet, resulting in malnutrition or disease. Tissue levels are low enough in one or more nutrients that pathology results primarily from the nutritional deficiency and from secondary complications. Nutrient deficiency is described in medical pathology textbooks; unfortunately, this is the start and end to nutritional education for most physicians. A *nutritional insufficiency* is a subtle deficiency of a nutrient sufficient to affect health but not severe enough to cause classic index deficiency symptoms. Problems here are more of a functional/biochemical nature rather than pathologic, but they may lead to the disease state. A *nutrient dependency* refers to

7.3 Nutrient Requirements of Disease States

Diet and nutrition are important factors in the promotion and maintenance of good health throughout the entire life course. A change in metabolism occurs following injury, stress, or infection, characterized by the need for increased macronutrients and enzymatic cofactors (micronutrients). The involvement of nutrients in cellular defense and repair systems suggests that nutrient requirements are modified by pathology. In disease, nutrient-dependent activities are expanded to include effects that enhance cellular responsiveness to treatment and accommodate accelerated rates of metabolic activity. Although specific nutrient requirements for different diseases have not been established, they may be imputed from current knowledge of the chemical, physical, and biological properties of each nutrient, the nature of the disease, the physiology of tissues involved, the type of treatment indicated, and the cellular activities targeted by the treatment.

7.4 Vitamins and Minerals and Hidden Hunger

Vitamins fall into two general categories: water-soluble and fat-soluble. Water-soluble vitamins are found mainly in watery or starchy foods such as grains, fruits, and vegetables, while fat-soluble vitamins are found mainly in fatty foods such as butter, nuts, olives, seafood, and organ meats. Only water-soluble vitamins function as coenzymes, while cofactors can also be minerals and other micronutrients (◘ Fig. 7.1).

Vitamins, minerals, and other micronutrients play critical roles in a wide variety of highly complex and integrated cellular processes in human biochemistry. The rate and extent of the enzymatic activity that determines these processes depend on the bioavailability of these micronutrients. The healthy state where optimal (maximum) functioning, health, and well-being are achieved may be attained by *metabolic optimization therapy*. Metabolic optimization is achieved when we provide the metabolism the necessary components (form and dose) to reach full velocity and completion of biochemical reactions to attain an optimal metabolic equilibrium. Thus, metabolic optimization therapy uses a combination of genetic makeup, diet, trauma, diseases, infections, toxins, and environmental stressors, among others, which will often elevate the demand of nutrients in order to achieve this optimal metabolic equilibrium. It can be argued that the true importance of vitamins in human biochemistry is far from fully elucidated, simply because of the high complexity of cellular processes. What is commonly ignored and not fully appreciated is the essential role that various vitamins and minerals play in functional human biochemistry. Critical enzymes require vitamins and minerals as an integral part of their molecular structure and/or functional mechanism of action. Enzymes play a critical role in regulating and orchestrating the rates of the multitude of biochemical reactions that take place in living organisms.

Vitamin deficiencies or coenzyme deficiency can lead to serious health disorders because important biological processes break down when a lack of coenzymes prevents enzymes from catalyzing essential chemical reactions [5]. Two well-known coenzyme vitamins are thiamin and niacin. Thiamin compounds serve as coenzymes for a variety of reactions involving cellular energy production, protein synthesis, and brain function. Thiamin deficiency causes a disorder known as beriberi, with symptoms such as irritability, weakness, and even heart failure. Niacin is needed for numerous reactions related to energy production and fatty-acid synthesis. Deficiency causes pellagra, which leads to dementia, skin problems, weight loss, and eventually death.

Inadequate or insufficient dietary intakes of vitamins and minerals are widespread [4], most likely due to excessive consumption of calorie-rich, nutrient-poor, refined food. Suboptimal intake of micronutrients often accompanies caloric excess (hidden hunger). *Hidden hunger* (or occult hunger) is a form of undernutrition in which a chronic lack of vitamins and minerals has no visible warning signs. Hidden hunger refers to subclinical deficiencies or nutrient insufficiencies from which our population suffers because of our excessive consumption of calorie-rich, nutrient-poor, refined food. This may be the basis of metabolic disruptions that underlie many pathological/disease states. Since we are not eating enough of the proper nutrient-dense foods, what we end up eating contains insufficient vital nutrients and excessive calories. The Standard American Diet lacks essential nutrients. This state of nutritional insufficiency is a possible reason why millions walk around with headaches, body aches, digestive upset, skin problems, sinus problems, frequent colds, and other signs and symptoms that may quickly disappear when you start taking necessary vitamins and minerals. Nutrition is enhanced through supplementation.

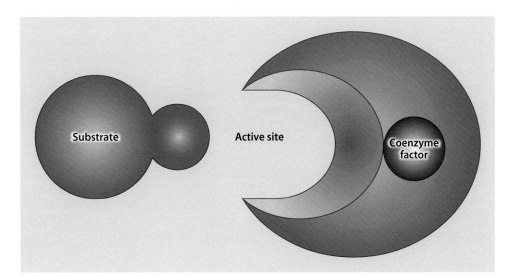

◘ Fig. 7.1 Water-soluble vitamins function as coenzymes

Hidden hunger can lead to mental impairment, poor health and productivity, or even death.

These inadequate intakes may result in metabolic disruptions [6]. Episodic shortages of micronutrients were common during evolution. Natural selection favors short-term survival at the expense of long-term health [6]. Short-term survival was achieved by allocating scarce micronutrients by nutritional *triage* [6]. As micronutrients become scarce, a triage mechanism for allocating scarce micronutrients is activated. This triage (mechanism) means prioritization of relatively scarce nutrients to the most fundamental life-preserving functions. In metabolic reactions, enzymes involved in adenosine triphosphate (ATP) synthesis would be favored over deoxyribonucleic acid (DNA) repair enzymes, as well as over the production of immune system components and neurological chemicals [5]. When there is a lack of synergistic components of the metabolic network, an array of negative metabolic repercussions arise, eventually leading to loss of healthy physiological equilibrium and the development and progression of degenerative diseases.

> » This triage (mechanism) means prioritization of relatively scarce nutrients to the most fundamental life-preserving functions.
>
> Michael J Gonzalez, DSc, NMD, PhD, FACN
> University of Puerto Rico Medical Sciences Campus,
> Schools of Public Health[1], Department of Human Development, Nutrition program
> San Juan PR

The triage theory of optimal nutrition states that the human body prioritizes the use of vitamins and minerals when it is getting an insufficient amount of them to be able to keep functioning [6]. Triage means deciding which patient to treat when faced with limited resources. When nutritional resources are limited, physiology (biological intelligence) must decide which biological functions to prioritize to give the total organism and the species the best chance to survive and reproduce. While short-term deficiencies or insufficiencies are common, they are often not taken seriously by mainstream medicine. Under such a limited scenario, the body will always direct nutrients toward short-term health and survival capability and away from regulation and repair of cellular DNA and proteins, which ultimately optimize health and increase longevity. Dr. Ames's research shows how bodily insults accumulate over time as a result of vitamin and mineral insufficiencies and can lead directly to age-related diseases. The Triage hypothesis states that the risk of degenerative diseases (those associated with aging, including cancer, cognitive decline, and immune dysfunction) can be decreased by ensuring adequate intake of micronutrients [5–9].

7.5 Why Metabolic Correction?

Metabolic correction therapy is indispensable because of these specific issues:

7.5.1 Inferior Nutritional Value of Food and Low Availability of Nutrient-Dense Foods

We need to eat a wide variety of food to obtain the nutrients we need. A big problem we face is that the nutritional values of foods that people eat may be inferior to the listed values given in food tables. Foods today have less nutrient content than foods 50 years ago. A study that assessed this issue showed declines in protein (−6%), calcium (−16%), phosphorus (−9%), iron (−15%), riboflavin (−38%), and vitamin C (−20%) [4]. There is a dilution effect, in which yield-enhancing methods such as fertilization and irrigation may decrease nutrient concentrations. Recent evidence suggests that genetically based yield increases may have the same result, a genetic dilution effect. Modern crops that grow larger and faster are not necessarily able to acquire all the necessary nutrients, whether by synthesis or from the water and soil. Today's foods are not as nutritious as those eaten in the past. A report from the US and UK Government statistics shows a decline in trace minerals of up to 76% in fruit and vegetables over the period from 1940 to 1991 [10, 11]. The nutritional decline findings alone give reason to eat organic fruits and vegetables. In fact, for nearly all nutrients, organic fruits and vegetables remain the most nutrient-dense foods [12]. This information makes the updated nutritional recommendations not so much current as reflective of the need for an increase in fruits and vegetables in order to get the same nutritional benefits as in the past. It is likely that Americans do not come close to the recommendations to limit added sugars, refined carbohydrates, and added fats and oils.

7.5.2 Medication-Induced Nutrient Depletion, Adverse Side Effects of Medication, and Iatrogenic Deaths

There are more than 100,000 deaths annually due to medication properly prescribed and taken as directed by the health provider [13], resulting in extremely high incidence of serious and fatal adverse side effects in US hospitals – more frequently than is generally recognized. Fatal adverse side effects appear to be the fourth leading cause of death in the USA [13]. If medication is necessary, providing metabolic correction principles may reduce medication requirements, reduce adverse side effects, and improve treatment outcomes [14].

7.5.3 To Compensate for the Increased Demand of Nutrients due to the Disease State

The body's nutritional demands increase in acute and chronic illness [15]. These imbalances are mainly caused by either deficiencies or insufficiencies of one or more nutrients. Nutrients function in disease by mechanisms that differ substantially from those of pharmacologic agents. Nutrients will modify nutrient fluxes and metabolic activities that are part of cellular processes needed for physiological repair, correction, and/or balance, whereas drugs will bind to membrane receptors and modify their activity to alter cell responsiveness. Nutrient requirements in the presence of disease are considerably higher than those that have been established to prevent the symptoms of the classic deficiency diseases. These requirements can increase incrementally by as much as 10 to more than 100 times the usual amounts (orthomolecular medicine) [16]. At these levels of intake (some at pharmacological doses), the roles for most nutrients are expanded to include functions that are not typically observed at normal intakes. The higher requirements for nutrients in disease are needed to support the accelerated rate of metabolic activity that cellular systems demand in order to reduce the potential for permanent damage from the pathophysiological processes associated with the disease.

Nutrient imbalances impose a metabolic burden on all organ systems, with the greatest burden on those systems responsible for achieving and maintaining metabolic equilibrium. Long-term disruption of metabolic equilibrium will most often adversely impact the cardiovascular, pulmonary, renal, gastrointestinal, neurological, and/or musculoskeletal systems. In the absence of an adequate supply of nutrients to satisfy normal physiological requirements or adjust to increased metabolic demand, compensatory mechanisms involving one or more of these systems must be initiated to re-establish homeostasis. As with metabolic adjustments to address short-term nutrient deficiencies, these compensatory responses are important for correction of temporary imbalances, but if sustained over the long term, they may become maladaptive and contribute to the degenerative changes responsible for development or worsening of chronic diseases. For example, burns lead to loss of protein and essential nutrients [17]. Surgery increases the need for zinc, vitamin C, and other nutrients involved in cellular tissue repair [18]. Broken bones need calcium, magnesium, and vitamin C for healing [19]. Infections challenge the immune system and place high demands on nutritional resources such as zinc, B-complex vitamins, and vitamin C [20]. The same nutritional demand is present when exposed to chemical, physical, and emotional stress.

People taking medications for chronic diseases are at higher risk of interactions between these drugs and nutrients. There are thousands of conceivable genetic defects (inborn or acquired), so it is likely that many people have higher genetic requirements for many micronutrients. This discussion seeks to provide a needed better understanding of the interrelationship between nutritional biochemistry and the disease-pathological state.

7.5.4 Biochemical Mechanism of Metabolic Correction: Molecular Concentrations and Rate of Reactions

The majority of the chemical reactions in living organisms are catalyzed by enzymes. The mechanisms of enzyme-catalyzed reactions involve [1] the formation of a complex between the enzyme and a substrate and [2] the breakdown of this complex to form the products of the reaction. The rate-determining step is usually the breakdown of the complex to form the products. Under conditions such that the concentration of the complex corresponds to equilibrium with the enzyme and the substrate, the rate of the reaction is given by the Michaelis–Menten equation [21].

A certain amount of energy, known as activation energy, is needed to initiate any chemical reaction. The fundamental purpose of enzymes is to facilitate reactions by lowering the activation energy. Enzymes accomplish this by binding to reactant molecules and allowing them to interact in a more energy-efficient manner. Reactant molecules bind to enzymes at an intricately structured location known as an active site, and the molecule involved in this binding process is called the substrate. Coenzymes, some of which are vitamins and some of which are synthesized directly from vitamins, activate enzymes by helping the enzyme to bind to its substrate.

Cofactors (vitamins, minerals, and other compounds) work as enzyme assistants. Coenzymes activate enzymes primarily by assisting the transfer of specific particles or compounds involved in the chemical reaction. For example, some coenzymes facilitate enzymatic reactions by carrying electrons and hydrogen ions from one atom to another, while others are involved in transporting the entire atoms or larger molecules. Explained another way, an enzyme might not be a perfect fit for the intended substrate unless the active site is modified by the addition of a coenzyme (◘ Fig. 7.2).

The rate of an enzyme-catalyzed reaction is approximately proportional to the concentration of the reactant, until concentrations that largely saturate the enzyme are reached. The saturating concentration is larger for a defective enzyme with decreased combining power for the substrate than for the normal enzyme. For such a defective enzyme, the catalyzed reaction could be made to take place at or near its normal rate by an increase in the substrate concentration. This mechanism of action of gene mutation is only one of several that lead to disadvantageous manifestations that could be overcome by an increase in the concentration of enzymatic cofactors. These binding problems may result in metabolic inefficiency with the accumulation of metabolic

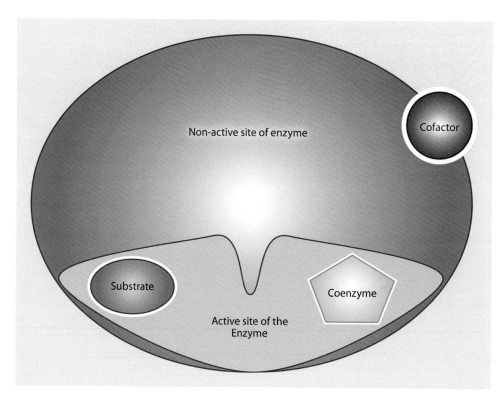

□ Fig. 7.2 Coenzymes bind to the reactant active site substrate of the enzyme, with cofactors as "helper molecules" that assist in biochemical transformations

byproducts. In general, this is the *Law of Mass Action*: As the vitamin and mineral concentration increases, enzyme efficiency increases. These considerations suggest a rationale for metabolic correction to provide the needed cofactors in the amount needed to improve function. This increased enzyme efficiency may allow a genetic defect to be overcome. This biochemical activity follows the chemical *Principle of Le Chatelier*, which states that when stress is applied in an equilibrium situation, it will move to the direction to minimize stress. In this case, there is an unfavorable equilibrium of active enzyme that, with the addition of the necessary nutrients, will be moved toward a more physiologically favorable metabolic state [22].

Many human genetic diseases due to defective enzymes can be remedied or ameliorated by the administration of high doses of the vitamin component of the corresponding coenzyme (metabolic correction), which can partially restore the enzymatic activity [5]. For several single-nucleotide polymorphisms in which the variant amino acid reduces coenzyme binding, enzymatic activity can be remedied by raising cellular concentrations of the cofactor through high-dose nutrient therapy (orthomolecular therapy).

Inadequate intakes of vitamins and minerals from food can lead to DNA damage, mitochondrial decay, and other pathologies [3]. Ames suggests that evolutionary allocation of scarce micronutrients by enzyme triage is an explanation of why DNA damage is commonly found along with micronutrient deficiency [3]. This particular situation favors the development and progression of diseases. Moreover, Motulsky has argued that many of the common degenerative diseases are the result of the imbalance of nutritional intake and genetically determined needs [23, 24].

Chromosome breaks lead to mutations that precede tissue damage and disease. Many types of physiological impairments due to inadequacy of vitamins and minerals can lead to suboptimal organ system function, including poor drug metabolism, insufficient neurotransmitter production, and impaired immune defenses. Chronic vitamin–mineral undernutrition reduces immune competency and central nervous system efficiency; it may increase morbidity, which may lead to degenerative diseases. This approach to optimize health by improving enzyme efficiency, and thereby metabolism and physiology, is the basis of metabolic correction as a therapeutic approach.

7.6 B-Complex and Metabolism Briefing

The basic B-complex consists of a group of eight water-soluble compounds: thiamin (vitamin B1), riboflavin (vitamin B2), niacin (vitamin B3), pantothenic acid (vitamin B5), pyridoxine (vitamin B6), cobalamin (vitamin B12), folate (vitamin B9), and biotin (vitamin B7). Although each B vitamin is chemically distinct, they often work together in various biochemical functions throughout the body, from cellular energy production to healthy red blood cell formation to healthy neurological function. Most B vitamins, with the exception of vitamin B12, are not stored in the body. They must be acquired daily from the diet in order to maintain optimal health.

Supplementing with bioactive B vitamins (e.g., methyl folate) is important to everyone, especially individuals who may not be able to convert inactive or synthetic vitamins to their active forms in the liver because of compromised liver function, poorly functioning enzymes, digestive disturbances, genetics, or age. Dietary supplements containing these active cofactors have enhanced bioavailability, ensuring the body gets the nutrients it needs.

Every process that goes on inside our bodies requires energy, specifically metabolic energy. When the body does not have enough energy to function properly, different components of the body may malfunction in their own ways. For example, if the brain does not have enough energy, cognitive processes, such as memory and focus, may become impaired. The body converts fats, sugars, and proteins into ATP that is then used for energy. However, there are other factors involved that can affect how well our body can make this conversion into the ATP molecules. The thyroid hormones are essential in maintaining and regulating the body's metabolism. The thyroid gland makes the hormone T4 (thyroxine). T4 converts to T3 (triiodothyronine) and RT3 (reverse T3, rT3). T3 turns on the ATP (energy), while RT3 is a way to get rid of any unneeded T4, thus reducing energy output. This happens because the adrenal glands are too weak to handle the stress of the body's normal metabolic energy and force a downregulation of energy production.

Triiodothyronine (T3) is the most active thyroid hormone. Approximately 85% of circulating T3 is produced by mono-deiodination of thyroxine (T4) in tissues such as the liver, muscle, and kidney. Production of these thyroid hormones is controlled by TSH (thyroid-stimulating hormone), which is released by the pituitary gland in the brain. The pituitary takes its orders from the hypothalamus. The adrenal glands, located on top of each kidney, help the body deal with stress. If the metabolic activity is excessive, the adrenals perceive this as a stress. In response to this stress, the hypothalamus will signal the pituitary to produce less TSH, thus producing decreased T4 and thyroid activity. This metabolic control activity utilizes various enzymes whose main cofactors are B-complex vitamins.

Lack of B vitamins may lead to hypothyroidism because the thyroid gland cannot make enough T4 [25]. Hypothyroidism is a condition in which an underactive thyroid gland produces less than optimal amounts of thyroid hormone. A lack of the thyroid hormones can lead to fatigue, constipation, hoarse voice, puffy face, unexplained weight gain, pain and stiffness in the joints, muscle weakness, sensitivity to cold, elevated cholesterol levels, brittle nails, and depression. Hypothyroidism has also been linked to other health issues, such as heart disease, infertility, autoimmunity, and obesity.

7.7 B-Complex and Metabolic Correction

B-complex vitamins are essential for optimum metabolism. Michaelis developed a set of mathematical expressions to calculate enzyme activity in terms of reaction speed. The Michaelis constant Km is defined as the substrate concentration at half the maximum velocity. For practical purposes, Km is the concentration of substrate that permits the enzyme to achieve half of its maximum activity.

An example of metabolic correction is that high-dose B vitamins can counteract a poor Km. As many as one-third of mutations in a gene result in the corresponding enzyme having an increased Km (decreased binding affinity) for a coenzyme, causing a lower rate of reaction [5, 7]. About 50 different human genetic diseases are due to a poorer-binding affinity of the mutant enzyme for its coenzyme. This can be corrected by feeding high-dose B vitamins, which raise levels of the corresponding coenzyme; many polymorphisms also result in a lowered affinity of enzyme for coenzyme [5, 7] and thus may be, in part, remediable.

To summarize, metabolic correction has three important biological actions:
1. Optimizes cellular function by improving enzymatic efficiency
2. Produces a pharmacological effect to correct abnormal cell function due to biochemical disarray occasioned by the disease process
3. Provides a compensatory response to nutrient imbalances in order to re-establish physiological homeostasis

An optimum intake of micronutrients and metabolites – which varies with age, environmental factors, and genetics – should tune up metabolism and markedly increase health at a modest cost, particularly for the poor, obese, and elderly [7].

7.8 Ten Principles of the Concept of Metabolic Correction in Disease Therapy

1. *Metabolic correctors*, along with proper nutrition, should come first in medical treatment. Knowledge of the safe and effective use of nutrient combination, enzymes, hormones, and other naturally occurring molecules in their bioactive forms is essential to ensure an effective outcome. However, some patients may need more acute treatment for their particular condition, for which pharmacological therapy is recommended.
2. *Metabolic correctors have a low risk of toxicity*. Pharmacological drugs often carry a risk of negative side effects and, in a chronic condition, should be the second choice if there is a metabolic correction alternative available.

3. *Detect nutrient deficiencies*: Some laboratory tests might be useful in identifying the nutritional needs of some patients. These tests could be of special importance to patients who present genetotrophic diseases or genetic polymorphism associated with specific conditions. However, some laboratory tests do not necessarily reflect nutrient and enzyme levels within specific organs or tissues, particularly in the nervous system. The need for laboratory testing for nutrients varies among individuals. For many patients, therapeutic trial and dose titration is often the most practical therapy approach, especially when utilizing synergistic metabolic correction formulations.
4. *Biochemical individuality* is a central precept of metabolic correction. Hence, the search for optimal nutrient combination doses is a practical issue. Doses of nutrients and their combinations above the recommended daily allowances are often effective. Many patients tolerate optimal doses and respond well; however, dose titration is indicated in otherwise unresponsive cases.
5. *Recommended daily allowances (RDA) for diseased individuals.* RDAs for nutrients are intended for normal, healthy people. By definition, diseased patients are not normal or healthy and not likely to be adequately served by obtaining just the recommended daily allowances. Practically every person is deficient or insufficient in a nutrient at some level due to an insufficient diet among other limiting factors (genetics, medication, toxins, etc.).
6. *Environmental pollution* of air, water, and food is an increasing problem and more common than is generally recognized, posing a very important risk factor for mitochondrial damage and related diseases such as cancer and neurometabolic disorders [26, 27]. Diagnostic search for toxic pollutants and treatment is necessary to identify these factors and design a proper treatment approach.
7. *Monitor and update metabolic correction over time.* Optimal health is a lifetime challenge. Biochemical needs change, and our metabolic correction prescriptions need to change based upon follow-up, repeated testing, and therapeutic trials to permit fine-tuning of each prescription and to provide the best possible health outcome.
8. *Nutrient-related disorders are always treatable*, and deficiencies and insufficiencies are curable. Most diseases encounter some nutrient-related disruption. To ignore their existence is malpractice.
9. *Nutrigenomics and pharmacogenomics.* Genetic and hereditary disorders are often responsive to metabolic correction because it takes advantage of nutrigenomics and pharmacogenomics [28].
10. *Inspire active role-taking responsibility for your health.* Inspire patients to understand that health is not merely the absence of disease, but the positive attainment of optimal function and well-being. This requires an individual to take an active role in necessary lifestyle changes, and it requires a commitment to continuous education along with a responsible attitude about health.

7.9 Beyond Metabolic Correction: Improving the Healthcare Management Model

Metabolic correction should be a fundamental consideration of a complete healthcare plan, and the fundamental root causes of the prevalent conditions must be identified and addressed to achieve optimal therapeutic results. Unfortunately, medical guidelines often do not assess these root causes of disease in each individual case. Therefore, comprehensive laboratory testing is mandatory to identify etiological factors and monitor progress toward the desired outcomes. This approach allows for correlating objective laboratory tests with functional evaluations and patient signs and symptoms. We know that a certain combination of nutrients can improve physiological function. However, giving the necessary nutrients will have limited results if environmental contaminants such as pesticides, heavy metals, medications, or an undetected chronic infection are affecting the patient. Therefore, to achieve metabolic correction and the best physiological function possible, it is necessary to correct the underlying cause of the problem. This involves providing the necessary nutrients for repair, correction, and improved enzymatic activity, which must include removing the contaminants present [29].

> » Metabolic correction should be a fundamental consideration of a complete healthcare plan, and the fundamental root causes of the prevalent conditions must be identified and addressed in order to achieve optimal therapeutic results.
>
> Michael J. Gonzalez, DSc, NMD, PhD, FACN
> University of Puerto Rico Medical Sciences Campus,
> Schools of Public Health[1], Department of human Development, Nutrition Program
> San Juan PR

To improve outcomes beyond current standards, it is necessary to improve our health models to include detection of all significant underlying causes of disease. In addition to a comprehensive medical history and a complete physical examination, there should be comprehensive laboratory testing that includes whole-genome sequencing, inflammation, immune testing, heavy metals, xenobiotics, nutrient, food intolerance/sensitivities, and metabolic panels for a complete patient assessment.

7.10 Conclusion

To encourage the most efficient metabolism, we need basic macronutrients required for fuel: fat, protein, and carbohydrate. But we also need about 15 vitamins that are coenzymes and about 15 minerals that are required for enzyme function.

We also need two essential fatty acids (omega-3 and omega-6) and seven or eight essential amino acids. In addition, other important nutrients – such as coenzyme Q10, acetyl-L-carnitine, and lipoic acid – must also be considered in our quest for physiological optimization. Virtually every metabolic pathway requires micronutrients.

What determines the optimal concentration of a nutrient is its physiological functionality. While most people function below 100% efficiency, they nevertheless do not present any detectable disease or significant symptoms. Yet, we can improve their functionality if we supply them with the needed micronutrient substances in optimum concentrations. Certain individuals have a greater need than that supplied by the diet (even if a good dietary regime is followed). This could be caused by an array of variables (digestive problems, malabsorption, food sensitivities, metabolic dysfunction, low levels of neurotransmitter precursors, etc.). This lack of needed micronutrient cofactors manifests insidiously and is difficult to identify. Some vague symptoms may be present, such as lethargy, irritability, insomnia, and difficulty in concentrating. This also affects the body's ability to resist disease and infection, its ability to recover from exercise, surgery, or disease, and the ability of the brain to function at an optimal level. Detecting and treating disease at its earliest stages of cellular biochemical abnormality, rather than waiting for clear clinical symptoms, is cost-effective and of benefit to the patient.

Nutrient deficiency diseases are the end product of a long and complex series of nutrient depletion reactions. We need to abandon outdated paradigms framing nutrient intake as needed merely to prevent deficiencies and expand them to include preventing chronic degenerative diseases and achieving optimal health. Metabolic correction provides the biochemical–physiological explanation of how to achieve this state of metabolic harmony and health. Metabolic correction therapy used as a comprehensive model can provide a systematic, rational, evidence-based, proactive, individualized, and integrated decision-making process. That process should be evaluated by a trained health professional understanding metabolic correction for its potential to elevate our standards for cost-effectiveness and outcome improvement.

References

1. González MJ, Miranda-Massari JR. Metabolic correction: a functional explanation of orthomolecular medicine. J Orthomolec Med. 2012;27(1):13–20.
2. Gonzalez MJ, Miranda-Massari JR, Duconge J, Allende-Vigo MZ, Jiménez-Ramírez FJ, Cintrón K, Rodríguez-Gómez JR, Rosario G, Ricart C, Santiago-Cornier JA, Zaragoza-Urdaz R, Vázquez A, Hickey S, Jabbar-Berdiel M, Riordan N, Ichim T, Santiago O, Alvarado G, Vora P. Metabolic correction: a functional biochemical mechanism against disease–part 1: concept and historical background. P R Health Sci J. 2015;34(1):3–8.
3. Miranda-Massari JR, Gonzalez MJ, Duconge J, Allende-Vigo MZ, Jiménez-Ramírez FJ, Hickey S, Vázquez A, Berdiel MJ, Cintrón K, Rodríguez-Gómez JR. Metabolic correction: a functional biochemical mechanism against disease–part 2: mechanisms and benefits. P R Health Sci J. 2015;34(1):9–13.
4. Davis DR, Epp MD, Riordan HD. Changes in USDA food composition data for 43 garden crops, 1950 to 1999. J Am Coll Nutr. 2004;23:669–82.
5. Ames BN, Elson-Schwab I, Silver EA. High-dose vitamin therapy stimulates variant enzymes with decreased coenzyme-binding affinity (increased Km): relevance to genetic disease and polymorphisms. Am J Clin Nutr. 2002;75:616–58.
6. Ames BN. Low micronutrient intake may accelerate the degenerative diseases of aging through allocation of scarce micronutrients by triage. Proc Natl Acad Sci U S A. 2006;103:17589–94.
7. Ames BN. The metabolic tune-up: metabolic harmony and disease prevention. J Nutr. 2003;133(5 Suppl 1):1544S–8S.
8. Ames BN. A role for supplements in optimizing health: the metabolic tune-up. Arch Biochem Biophys. 2004;423:227–34.
9. Ames BN, Suh JH, Liu J. Enzymes lose binding affinity for coenzymes and substrates with age: a strategy for remediation. In: Rodriguez JKR, Kaput J, editors. Nutrigenomics: discovering the path to personalized nutrition. Hoboken: Wiley; 2006. p. 277–93.
10. Worthington V. Nutritional quality of organic versus conventional fruits, vegetables, and grains. J Altern Compliment Med. 2001;7:161–73.
11. Leape LL. Institute of medicine medical error figures are not exaggerated. JAMA. 2000;284:95–7.
12. The Organic Center. State of science review: nutritional superiority of organic foods. 2008. Retrieved: https://www.organic-center.org/reportfiles/NutrientContentReport.pdf. Accessed 29 Jul 2017.
13. Lazarou J, Pomeranz BH, Corey PN. Incidence of adverse drug reactions in hospitalized patients: a meta-analysis of prospective studies. JAMA. 1998;279:1200–5.
14. Miranda-Massari JR, Gonzalez MJ, Jimenez FJ, et al. Metabolic correction in the management of diabetic peripheral neuropathy: improving clinical results beyond symptom control. Curr Clin Pharmacol. 2011;6:260–73.
15. Richardson RA, Davidson HIM. Nutritional demands in acute and chronic illness. Proc Nutr Soc. 2003;62:777–81.
16. Williams RJ, Kalita DK, editors. A physician's handbook on orthomolecular medicine. Oxford: Pergamon Press; 1978.
17. Prins A. Nutritional management of the burn patient. S Afr J Clin Nutr. 2009;22:9–15.
18. Rahm DH, Labovitz JM. Perioperative nutrition and the use of nutritional supplements. Clin Podiatr Med Surg. 2007;24:245–59.
19. Kakar S, Einhorn T. Importance of nutrition in fracture healing. In: Holick M, Dawson-Hughes B, editors. Nutrition and bone health. Totowa: Humana Press, Inc; 2004. p. 85–103.
20. Bendich A. Antioxidant vitamins and human immune responses. Vit Horm. 1996;52:35–62.
21. Pauling L. Orthomolecular psychiatry. Varying the concentrations of substances normally present in the human body may control mental disease. Science. 1968;160:265–71.
22. Bland J. The justification for vitamin supplementation. J Holistic Med. 1981;3:12–22.
23. Motulsky A. Human genetic variation and nutrition. Am J Clin Nutr. 1987;45:1108–13.
24. Motulsky A. Nutrition and genetic susceptibility to common diseases. Am J Clin Nutr. 1992;55:1244S–5S.
25. Jabbar A, Yawar A, Waseem S, et al. Vitamin B12 deficiency common in primary hypothyroidism. J Pak Med Assoc. 2008;58(5):258–61.
26. SH Reuben for the President's Cancer Panel, US Department of Health and Human Services, National Institutes of Health, National Cancer Institute. Reducing environmental cancer risk. What we can do now. 2010. http://deainfo.nci.nih.gov/advisory/pcp/annualReports/pcp08-09rpt/PCP_Report_08-09_508.pdf.
27. Sly PD, Carpenter DO, Van den Berg M, et al. Health consequences of environmental exposures: causal thinking in global environmental epidemiology. Ann Glob Health. 2016;82(1):3–9.
28. González MJ, Miranda-Massari JR, Duconge J, Rodríguez JW, Cintrón K, Berdiel MJ, Rodríguez JR. Nutrigenomics, metabolic correction and disease: the restoration of metabolism as a regenerative medicine perspective. J Restor Med. 2015;4:74–82.
29. Miranda-Massari JR, Gonzalez MJ, Duconge J, Berdiel M, Alfaro I, Grace K. Metabolic correction therapy: changing the healthcare management paradigm. Integr Food Nutr Metab. 2016;3(2):281–2.

The Nutrition Assessment of Metabolic and Nutritional Balance

Margaret Gasta

8.1 Introduction – 101

8.2 The Microbiome – 101

8.3 Fiber – 102

8.4 Iodine – 103

8.5 B Vitamins – 104
8.5.1 Thiamine (Vitamin B1) – 107
8.5.2 Riboflavin (Vitamin B2) – 107
8.5.3 Niacin (Vitamin B3) – 107
8.5.4 Pantothenic Acid (B5) – 107
8.5.5 Vitamin B6 – 107
8.5.6 Folate (Vitamin B9) – 108
8.5.7 Vitamin B12 – 108

8.6 Fat-Soluble Vitamins – 109
8.6.1 Vitamin D and Vitamin K – 109
8.6.2 Vitamin A and Vitamin D – 109
8.6.3 Tocopherols and Tocotrienols – 110

8.7 Minerals – 110
8.7.1 Calcium and Magnesium – 110
8.7.2 Sodium-to-Potassium Ratio and Hypertension – 112
8.7.3 Zinc and Copper – 112
8.7.4 Assessing Zinc Status – 114
8.7.5 Copper-to-Zinc Ratio in Cancer – 114

© Springer Nature Switzerland AG 2020
D. Noland et al. (eds.), *Integrative and Functional Medical Nutrition Therapy*,
https://doi.org/10.1007/978-3-030-30730-1_8

8.8	**Fatty Acids and Phospholipids – 114**
8.8.1	Fatty Acid and Phospholipid Balance – 114
8.8.2	Gamma-Linolenic Acid – 116
8.8.3	Conjugated Linoleic Acid – 116
8.8.4	Phospholipids – 117
8.8.5	Short-Chain Fatty Acids – 117
8.8.6	Increasing Beneficial Fatty Acids – 118
8.8.7	Assessing Erythrocyte Fatty Acid Profiles – 118
8.8.8	Hydrogenated Oils – 119
8.8.9	Overall Diet and Macronutrient Distribution – 119
8.9	**Summary – 119**
	References – 120

8.1 Introduction

The field of integrative and functional nutrition is expected, by its very nature, to be progressive and on the cutting edge of new concepts in nutrition. Patients in desperate need of healing seek out reliable, valid, and progressive solutions from their nutritionist. The constant barrage of changing nutritional recommendations should teach integrative and functional nutrition practitioners to hold the principle of nutritional balance at the core while new findings continue to come forth.

Fad diets and supplements will come and go through the decades, and it is important to always question the long-term consequences of every proposed new nutritional concept. Concurrently, nutritionists must keep an open mind for something that may help a patient with a puzzling condition. What may seem extreme and/or inappropriate for one person may be lifesaving for another. Many human genetic diseases, due to defective enzymes, may be ameliorated by high doses of specific nutrients required as cofactors for the enzyme to partially restore activity of that enzyme [1]. Over the lifespan, metabolic and biochemical needs change and nutritional balance should be periodically reassessed and rebalanced [1]. Furthermore, detecting and addressing disease at its earliest stage of biochemical imbalance, rather than waiting for the condition to progress to a diagnosed disease, is a cost effective and physiologically beneficial approach for the patient.

"Biochemical individuality," first proposed by Roger J. Williams in 1947, refers to differing metabolic needs for optimal physiological function and infers that nutritional balance for one person may require different amounts of nutrients than for another person [1]. One of the foundational principles of integrative and functional nutrition is to recognize each person's unique nutritional needs. If practitioners rely on nutritional studies as a guide to establishing optimal nutrient intake for individuals, as opposed to using individualized nutrient assessment, this may pose challenges. For instance, many nutrition studies fail to show a benefit of supplementation due to flawed study design, as investigators often fail to establish baseline nutrient status of study participants to accurately measure outcomes [2]. Testing and retesting may be one of the most helpful tools to keep clients' nutritional status in balance and provide metabolic correction. Integrative and functional nutritionists have a vast array of tools to choose from in this regard.

To assess nutritional imbalance, a clinician can use a nutritional physical exam, conventional and integrative laboratory testing, as well as signs and symptoms. When imbalance is found, such as a zinc deficiency, calling for diet change and supplement intervention, one must then pay attention to the balance of other nutrients that may be negatively affected, such as copper status. Another example of this is the critical role that magnesium plays in the synthesis and metabolism of vitamin D and parathyroid hormone [3].

> The key considerations for nutritional balance are as follows:
> - The microbiome: use of probiotics and foods in different gastrointestinal conditions
> - Fiber: appropriate needs for different diagnoses
> - Iodine: finding the right balance and removing antagonistic toxic halogens
> - B vitamins: appropriate amounts and forms based on biochemical individuality
> - Mineral balance: sodium, potassium, zinc, copper, magnesium, calcium iron
> - Vitamin D status: associated requirement for magnesium
> - Fat-soluble vitamins: individual requirements
> - Omega-6 and Omega-3 fatty acids: ratios, adequate gamma-linolenic acid (GLA), and specialized proresolving mediators (SPM)
> - Micronutrient and macronutrient ratios in different disease processes

8.2 The Microbiome

One of the core principles of Integrative and Functional Medicine Nutrition Therapy (IFMNT) is biochemical individuality, and recommendations for the feeding and care of individual microbiome are no exception. Although we are far from fully understanding the microbiome, certain research-guided patterns have begun to emerge. For instance, the conditions of obesity, type II diabetes, and nonalcoholic liver disease are associated with a higher abundance of the phylum Firmicutes compared with *Bacteroides* [4]. In certain disease states, we might find that overgrowth or subclinical bowel infections of normally occurring commensal bacteria, such as the *Klebsiella* microbes, may be associated with autoimmune diseases such as Sjogren's syndrome, ankylosing spondylitis, rheumatoid arthritis, or Crohn's disease [5].

Research suggests that changing the diet to one that is plant-based and high in fiber is a rapid, effective way to cause a beneficial change in our intestinal microbiome [6]. Conversely, for those with irritable bowel syndrome (IBS), small intestinal bacterial overgrowth (SIBO), or inflammatory bowel disease (IBD), this approach may not be appropriate. Foods high in certain types of fiber such as highly fermentable oligosaccharides, monosaccharides, disaccharides, and polyols (FODMAPs) may result in rapid gas production and discomfort for those with IBS and possibly those with SIBO or IBD [7]. Additionally, therapeutic diets for autoimmune disease may consist of grain-free, nut-free, dairy-free, egg-free, and legume-free diets which would make it almost impossible to successfully implement a solely plant-based diet. However, many plant foods can be incorporated into an autoimmune diet to promote a healthy microbiome. Diets need to be individualized in every circumstance.

Stool cultures may show that certain strains of beneficial bacteria, such as *Bifidobacterium*, are low, and a specific supplemental probiotic strain might be indicated. Recurrent infections such as *H. pylori* or *Strep pyogenes* may also indicate specific strains of probiotics to be used.

8.3 Fiber

Fiber intake is critical for gut ecology because of its role as fuel for the microbiome [8]. Many food sources of fiber contain a mix of both soluble and insoluble fiber. Soluble fiber becomes gel-like and is fermented by friendly bacteria in the colon to make short-chain fatty acids (SCFA). Insoluble fiber is not only indigestible but also important for stool bulking and better gut motility. Eating a wide variety of plant foods ensures a wide variety of the different types of fibers including pectin, gum, mucilage, cellulose, hemicellulose, lignin, and soluble fiber. The peels of fruits and vegetables are high in fiber, although all peels may not be edible or palatable.

Beneficial bacteria ferment soluble fiber, which produces short-chain fatty acids (SCFA) that, in turn, serve as a fuel source to colonocytes [8]. The fermentation of soluble dietary fiber by the microbiota produces SCFA that promote increased gut cellular proliferation and differentiation, lower intestinal pH which makes the gut more resistant to pathogens, and regulates inflammation. Fiber improves the health of the gut barrier by promoting tight junctions and increasing the production of mucins that play a role in forming mucus to protect the gut barrier [8].

Butyrate, a SCFA produced by fermentation of fiber, may decrease the risk of colon cancer by inducing apoptosis in tumor cells [9]. The 2011 Colorectal Cancer Report found evidence that for every 10 grams per day increase in fiber intake, there was a 10% decrease in colorectal cancer risk [9]. Fiber aids healthy elimination of toxins through the enterohepatic circulatory system with higher elevation of SCFAs while adding bulk to the stool to assist motility through the gut [8]. In general, consuming a variety of dietary fiber helps promote diversity of gut microbiota, which supports a more robust state of health [8]. A 2017 report of the gut microbiome of the Hadza hunter–gatherers of Tanzania who average 100–150 gm of fiber/day showed much greater microbiota diversity when compared with Americans who consume only about 15 grams of fiber per day [10]. Fiber intake from legumes, fruits, vegetables, and nuts had a protective effect against CVD events, and cereal fiber was associated with a lower risk of stroke and ischemic stroke [11]. Fiber intake is inversely associated with cardiovascular disease, body weight, type II diabetes, some cancers, and chronic diseases [12].

The recommended amount of fiber in the UK is 30 grams per day, while it is 14 grams for every 1000 calories in the United States [12]. The Institute of Medicine recommends 38 grams per day for men aged 50 and younger and 30 grams for males over 50. For women, the recommendation is 25 grams daily for those 50 and younger and 21 grams daily for those over 50 [9]. The amount of fiber recommended for children is the child's age plus 5 grams per day; an 8-year-old child, for instance, is recommended to consume 13 grams of fiber per day. However, Americans average 15 grams of fiber per day, which is very low, and only 3% of the population meets the recommendations for fiber intake [9, 12].

Most plant foods are a mix of soluble and insoluble fiber [13]. Some key sources of insoluble fiber include wheat bran, psyllium, quinoa, nuts–seeds, pine nuts, and flaxseed. Key sources of soluble fiber include beans, oats and oat bran, psyllium, barley, prunes, figs, pears, leafy greens, cauliflower, broccoli, flaxseed, acorn squash, and potatoes. Cozma-Petrut and colleagues suggest a new classification of fiber that includes solubility, fermentability, viscosity, and gel formation [13]. Examples of these categories include insoluble poorly fermented (wheat bran), soluble, nonviscous readily fermented (inulin), soluble, viscous gel-forming readily fermented (beta-glucans), soluble viscous–gel-forming low or non-fermentability (psyllium) [13].

Several studies suggest that the best tolerated and most effective fiber for IBS are those with low fermentability, such as psyllium [13]. For IBS, soluble fiber was found to be better tolerated and possibly helpful compared with insoluble fiber such as wheat bran, which may worsen symptoms [7]. For IBS, an intake of 20–30 grams per day of soluble viscous fibers with low rate of fermentation, such as psyllium and ground flax seeds, starting with a low dose and slowly increasing is recommended, at a rate of not more than 5 grams per week [7]. Psyllium husks and powder should always be mixed in plenty of fluid.

A novel concept regarding fiber intake for those with chronic idiopathic constipation is that these individuals may have worsening symptoms when consuming a high-fiber diet. Ho et al. [14] conducted a prospective longitudinal case study on subjects with chronic idiopathic constipation that showed marked improvement in constipation and bloating in subjects who consumed a low-fiber or no-fiber diet compared with those who stayed on a high-fiber diet. The authors point out that most individuals who visit medical practitioners for chronic constipation are already consuming a very high-fiber diet to self-treat their constipation [14]. Those who improve on a low-fiber diet may have bacteria that are fermenting the fiber which may contribute to bloating and constipation [14]. Fiber creates bulk in the stool that can make it more difficult to expel and can result in straining and fissures [14]. Fiber can slow peristalsis and result in a buildup of gas that becomes difficult to expel [14].

When adding fiber supplements to the diet, nutritionists need to use caution and start with low amounts and increase water intake so as not to cause constipation or a bezoar (intestinal blockage of a solid mass of indigestible material that accumulates in the digestive tract). In addition, if clients complain of severe constipation or of not having had a bowel movement for days or weeks, do not recommend fiber supplements until the constipation has resolved. It may be prudent to recommend they visit a physician to uncover the cause of the protracted constipation. In this author's clinical experience, prunes, magnesium, vitamin C, flaxseed oil, or the herbal blend triphala, and aloe vera juice can be helpful in these cases, although caution is recommended for the use of aloe vera juice. At times, a physician-prescribed laxative is needed to clear a blockage. The nutritionist can follow this intervention with constipation prevention techniques (e.g., diet, 5-HTP, magnesium, etc.). In addition, referral to an integrative or naturopathic practitioner may help to manage lifelong constipation issues with other interventions such as homeopathics, herbals, acupuncture, and the like.

Notwithstanding all the benefits of fiber, the nutritionist may find it challenging to help patients consume a wide variety of fiber when certain gastrointestinal conditions exist. Judicious use of different fibers will be required for different gastrointestinal disease states and symptomatology. Determining the cause of a person's distress may be helpful in determining which type of fibers will be tolerated. An example is someone with severe IBS. Certain fermentable fiber products may worsen the symptoms. Instead of improving the person's health, these products may cause increased distress and worsen their constipation or diarrhea. Those with celiac disease (CD) might be made worse with oat fiber and certainly with wheat bran. Standard high-fiber foods often recommended such as beans, oats, apples, pears, plums, cashews, and pistachios might worsen constipation in someone with IBS, SIBO, or aggravate cases of IBD. In addition, healthy high-fiber seeds such as chia seeds, hemp seeds, and flax seeds might be damaging to sensitive individuals. Being familiar with FODMAPs, SIBO diets, the specific carbohydrate diet, and a low-residue diet is essential in helping patients with certain GI conditions to recover. Combing the market for suitable fiber supplements can be an arduous task but highly recommended, as one size does not fit all.

In summary, dietary fiber is a critical component to maintaining balance and diversity in gut ecology. Fiber can help normalize gut transit time, promote mucus production, and strengthen the gut barrier. Promoting the proper intake of fiber and water is crucial to improving the health of our clients. For those with IBS or chronic constipation, instead of using highly fermentable fiber grains, flax seeds, chia seeds, or fermentable fiber supplements, clinicians may simply want to recommend a diet that uses FODMAP-compliant fruits, vegetables, and increasing water intake. For these individuals, adding *Bifidobacter* probiotic supplements with the fiber may also be essential. If psyllium is to be added, start with very low doses; this author recommends 1–2 grams/day to start and slowly build up depending on individual tolerance. The herbal blend called triphala, as well as supplemental magnesium citrate, can be helpful with chronic constipation.

8.4 Iodine

Iodine is an essential element needed for thyroid hormone synthesis and fetal neurodevelopment [15]. The World Health Organization (WHO) and the US Institutes of Medicine recommend 90–150 mcg/day of iodine for the general population, 220–250 mcg/day during pregnancy, and 250–290 mcg/day of iodine during lactation [15]. The WHO uses median spot urinary iodine concentrations of 100 mcg/L to represent an intake of 150 mcg/day and 100–190 mcg/L per day to define adequate intake for a nonpregnant population [15]. Excessive iodine intake is classified by the WHO as the median urinary iodine excretion >300 mcg/L for general population and >500 mcg/L for pregnant women [16].

To assess iodine status, a clinician may use clinical manifestations such as thyroid size, combined with detailed history focused on the dietary iodine intake and a history of frequent or current infection, and laboratory testing for serum thyroglobulin, thyroid-stimulating hormone (TSH), and urinary iodine [15]. Urinary iodine reflects recent iodine intake within days, whereas thyroglobulin represents iodine intake over a period of months, and thyroid size assessment represents iodine intake for a period of years [15]. Either spot urine or 24-hour urine collection iodine testing may be used. Spot urinary iodine is more useful for assessing populations rather than the individual patient, as anything less than 10 urine samples in an individual is considered misleading due to diurnal variations [16]. Urine iodine levels also vary greatly in and between individuals based on the amount of iodine consumed in a day and the level of fluids ingested by the individual [15]. Expressing the ratio of urinary iodine to creatinine may be especially useful when estimating 24-hour urinary iodine [15].

Those at risk with iodine-deficient diets include vegans and people who avoid dairy or iodized salt, as well as athletes who experience excessive sweating [15]. Alternatively, people who consume kelp may have excessive iodine intakes [15]. Goiter is thought to be an adaptation to chronic iodine deficiency as low iodine intake leads to reduced thyroid hormone production, which stimulates TSH production from the pituitary gland. Goiter results as increased TSH increases iodine uptake, which stimulates thyroid growth [15]. In adults, the thyroid may develop nodules as it enlarges [15]. For those with overt hypothyroidism due to iodine deficiency, Niwattisaiwong et al. recommend initiating levothyroxine treatment along with iodine supplementation by using iodized salt or a multivitamin containing approximately 150 mcg of iodine [15]. When urine iodine has normalized and goiter has decreased, consider decreasing the thyroid treatment, but reassessing thyroid status in 4–6 weeks after stopping levothyroxine [15]. While mild gestational iodine deficiency does not result in cretinism (severe mental retardation and other neurologic or physical defects), children born to mothers with mild gestational iodine deficiency were found to have reductions in spelling, grammar, and English literacy performance [15].

In the recent past, high-dose supplemental iodine between 12 mg and 50 mg (12,000 mcg– 50,000 mcg) was a common recommendation in the integrative and functional medicine community for preventing breast cancer and helping with subclinical or overt hypothyroidism. While adequate amounts of iodine are required for healthy thyroid function, a need for caution exists if using supplemental iodine. The American

Thyroid Association (ATA) recently released a statement advising against the use of >500 mcg of iodine daily through dietary supplements [16]. Excessive amounts of iodine can be an environmental factor linked to the development of autoimmune thyroiditis and can cause hypothyroidism, hyperthyroidism, cancers, and autoimmune thyroid disease [17]. This can especially be a concern for autoimmune-prone individuals who are at risk of developing thyroid autoantibodies [17]. Iodine causes cytokine and chemokine-mediated lymphocyte infiltration in autoimmune-prone individuals, which is a key element in the production of thyroid autoantibodies [17]. Excess iodine is also thought to produce oxidative stress-related injury to thyrocytes, and active lipid peroxidation can occur after high-dose iodine administration [17].

Excessive iodine, an example of poor nutritional balance, can easily occur and is becoming a more common concern due to high levels of salt iodization, iodine in supplements, and regular consumption of iodine-rich foods [17]. Iodine toxicity has been reported due to overconsumption of seaweed in Asian countries [17]. Drinking water can also be a source of iodine excess in places such as Somalia, Saharawi, and Europe and can occur from the use of water purification systems that contain iodine [17]. In Western countries, dairy can be a source of excessive iodine from the animal feed and equipment-cleaning products used in the dairy industry [17]. Iodized salt may contain excessive or deficient amounts of iodine as diligent monitoring of iodine levels in salt does not always occur [17]. Multivitamin supplements testing in the United States had ranges for iodine between 11 and 610 mcg, and 15 brands had higher iodine levels than what was listed on the label [17]. Pharmaceutical products are another source of excessive iodine as Amiodarone contains 37% iodine; one tablet can contain several hundred times the recommended dose [17]. Contrast agents used for diagnostic radiology can contain hundreds of thousands of times the recommended daily amounts for iodine in one single dose. Transdermal antiseptic cleaners can also be a source of excess iodine for patients and healthcare workers [17]. It can take more than 1 month for the iodine levels in the body to normalize following exposure [17] (See Table 8.1).

To protect thyroid health, one lifestyle recommendation integrative and functional clinicians make is to limit exposure to the halogens consisting of fluoride, chlorine, and bromine. Iodine is a halogen as well, and excessive exposure to fluoride, chlorine, and bromine may result in the thyroid absorbing and storing these halogens, and thus displacing iodine. Halogens have the potential to interfere with the production of thyroid hormone, iodine metabolism, and may contribute to hypothyroidism or thyroid hormone derangement. For example, excessive ingestion of fluoride, even in the presence of adequate dietary iodine intake, may induce thyroid disturbances by interfering with enzymes such as deiodinases that are required for metabolizing thyroxine into its derivative forms [19]. Fluoride may also interfere with iodide transport and displace iodide resulting in accumulation in the thyroid [19]. Halogens are found in flame retardants, dioxins, pesticides, polychlorinated biphenyls, and fluoride. Chlorine is present in swimming pools and chlorinated drinking water [20].

Current recommendations point to iodine supplementation between 300 and 500 mcg per day with diet and lifestyle changes to support healthy thyroid function. If the patient has autoimmune thyroiditis, it is best to limit iodine to iodized salt and dietary intake and steer clear of supplemental iodine. Some autoimmune patients seem to do better eliminating iodized salt, but every patient is different, and individual needs should be assessed.

Table 8.1 Tolerable upper intake levels for chronic iodine ingestion

Age	Male	Female	Pregnancy	Lactation
Birth to 1 year	200 mcg	200 mcg		
1–3 years	300 mcg	300 mcg		
9–13	600 mcg	600 mcg		
14–18	900 mcg	900 mcg	900 mcg	900 mcg
≥19	1100 mcg	1100 mcg	1100 mcg	1100 mcg

Based on data from Institute of Medicine (US) Panel on Micronutrients [18]
Note: The American Thyroid Association recently advised against consuming >500 mcg of iodine daily in supplements [16]

8.5 B Vitamins

The concept of nutritional balance is important with dosing of B vitamin supplements. The B vitamins consist of eight water-soluble vitamins interacting together as coenzymes for a variety of catabolic and anabolic enzyme reactions [21]. Collectively, B vitamins have effects in proper brain function, energy production, DNA and RNA synthesis and repair, genomic and non-genomic methylation, and the synthesis of numerous neurochemicals and signaling molecules [21]. Although water-soluble and generally regarded as nontoxic, over-supplementation of a complex of B vitamins or isolated B vitamins such as B6 or folate is a concern. Conversely, underdosing by adhering strictly to RDA levels for certain B vitamins may leave some portions of the population at risk for insufficiency [21] (See Table 8.2).

Homocysteine, a potentially toxic amino acid, is thought to accumulate when vitamins B12, folate, B6, and/or trimethylglycine (TMG) are insufficient. Elevated homocysteine is theorized to increase oxidative stress, inhibit methylation reactions, increase damage to DNA and dysregulation of its repair, promote atherosclerosis [28], and direct and indirect neurotoxicity, leading to cell death and apoptosis [21]. These processes are thought to lead to the detrimental health conditions seen with elevated homocysteine including accumula-

Table 8.2 Tolerable upper limits of B vitamins

B Vitamin	RDA		Upper limit	Additional comments
B1 thiamine	*RDA for adults* [17]		Not established as no adverse effects have been found with daily doses of 50 mg [22]	Common dose of 50–100 mg for adults unless under special medical conditions exist such as alcoholism, malabsorption, the elderly, HIV/AIDS, or after bariatric surgery [22].
	Male	1.2 mg		
	Females	1.1 mg		
	Pregnancy and lactation	1.4 mg		
B2 riboflavin	Adults 19–50 years [23]		Studies have not shown adverse effects from high riboflavin intakes of 400 mg/day and there is no established UL [23]	Essential component of the two coenzymes FMN and FAD that play major roles in energy production, cellular function, growth and development, and metabolism of fats, drugs, and steroids [23]. FAD is required for the conversion of tryptophan to niacin [23]. FMN is required for the conversion of vitamin B6 to coenzyme pyridoxal 5′-phosphate [23]. Also used in the metabolism of homocysteine. Clinically it may be helpful in higher doses with prevention of migraine headaches [24]. People at risk for deficiency include vegetarians, athletes, pregnant and lactating women and their infants, vegan and dairy-free diet followers, infantile Brown–Vialetto–Van Laere syndrome [23].
	Males	1.3 mg		
	Females	1.1 mg		
	Pregnancy	1.4 mg		
	Lactation	1.6 mg		
B3 niacin	RDA for adults [25]: 16 mg males 14 mg females 18 mg pregnancy 17 mg lactation		35 mg [25]	Temporary flushing of the skin may occur at doses of 100 mg [21]. Nausea, vomiting, diarrhea, and very rarely liver damage have occurred at extended doses of 1 gram of more [21]. Doses of up to 250 mg have been used in Parkinson's disease [21].
B5 pantothenic acid	Adequate intake for adults [25]: 5 mg 6 mg pregnant 7 mg lactation		Not established [25]	Essential coenzyme in the mitochondria for formation of adenosine triphosphate (ATP) along with thiamine, riboflavin and niacin [21].
B6 pyridoxine (pyridoxal-5′-phosphate)	RDA for adults [25]: 1.3 mg 1.7 mg 51+ years 1.9 mg pregnancy 2.0 mg lactation		100 mg [25]	50–100 mg [16]. Essential in the metabolism of homocysteine. Doses over 100 mg have been known to cause neurosensory issues [21].
B9 Folate	RDA for adults [25]: 400 mcg 600 mcg pregnancy 500 mcg lactation		1000 mcg [25]	1000 mcg is the daily upper limit; however, 200–400 mcg is often recommended as research is unclear about supplemental folic acid and cancer [21]. Exceptions include the use of certain medications and higher levels than the upper limit may be indicated in some cases of high homocysteine [21]. High doses of folate can exacerbate the effects of B12 deficiency [21]. Folate is essential in the metabolism of homocysteine [24].
B12	RDA for adults [25]: 2.4 mcg 2.6 mcg pregnancy 2.8 mcg lactation		Not established [25]	Dependent on the individual physiological needs and conditions involving absorption [24]. Essential in the metabolism of homocysteine.
Inositol	No RDA set		Not established	Inositol can be synthesized in the body from glucose. Found in the germ and bran of grains, beans, nuts, seeds, and citrus fruit [26].
Biotin	Adequate intake for adults [25]: 30 mcg 35 mcg lactation		Not established [25]	Biotin, thiamine, and B12 interrelate in the citric acid cycle [16]. Biotin can be deficient genetically susceptible individuals or with dysbiosis. Biotin is formed by gut microbiota [27].

tion of beta-amyloid, hyper-phosphorylation of tau, brain tissue atrophy, compromised cerebrovascular circulation, cardiovascular disease, and compromised cognitive function and dementia [21]. Elevated plasma homocysteine levels can be due to renal insufficiency, deficiencies of folate, vitamin B12, vitamin B6 and vitamin B2 [29]. Plasma homocysteine levels were categorized by the 2009 US National Academy of Clinical Biochemistry Laboratory Medicine Practice Guidelines on Emerging Biomarkers of Cardiovascular Disease and Stroke as (umol/L): desirable <10, intermediate

>10 to <15, high >15, very high >30 [29]. For biomarkers for B vitamin assessment, see ◘ Table 8.3.

Elevated homocysteine and low levels of vitamins B12, B6, and folate have also been associated with bone loss and structural deterioration of bone tissue [35]. Deficiency of vitamin B12 is associated with lower blood levels of osteocalcin and alkaline phosphatase and may point to the activity of osteoblasts and bone metabolism being affected by vitamin B12 status [35].

While it is pertinent to address elevated homocysteine levels, the homocysteine hypothesis has resulted in limited B-vitamin research in other conditions with nutritional supplementation research focused mainly on folic acid and vitamin B12, only occasionally including vitamin B6 [3]. This has resulted in virtually ignoring the importance and interconnectivity of the role of the entire spectrum of B vitamins as essential in maintaining optimal physiological and neurological function [21].

Deficiency of even one B vitamin will negatively affect the ability to generate energy in the cell [21]. For example, the status of folate, B6, and B12 is dependent on levels of flavoproteins derived from riboflavin. Riboflavin is also essential to homocysteine metabolism [21]. Niacin, vitamin B3, serves as a necessary cofactor in the folate–tetrahydrobiopterin and methionine cycles [21]. Prolonged elevated folate status is associated with protected cognitive function only in those with normal B12 status, whereas high folate status exacerbated the detrimental effects of low B12 status [21]. Supplementation with folic acid also contributed to riboflavin deficiency in participants in one study [21].

The active forms of thiamine, riboflavin, niacin, and pantothenic acid are essential coenzymes in the mitochondria to

◘ **Table 8.3** Biomarkers to assess B vitamin status

B1	B2	B3	B6	Folate	B12
Whole blood thiamine pyrophosphate (TPP) available at labcorp. NL 66.5–200 nmol/L. Retrieved from ▶ https://www.labcorp.com/test-menu/36661/vitamin-bsub1-sub-whole-blood# on 2/17/2019. Urinary amino acids test or organic acids test can also be helpful	Riboflavin status is not routinely measured. Please refer to Riboflavin Fact Sheet for Health Professionals for more information [23]. Urinary organic acids test can be helpful	Most reliable: urinary excretion of niacin's two major methylated metabolites, N1-methyl-nicotinamide and N1-methyl-2-pyridone-5-carboxamide [30]. Levels in adults: Normal niacin status is >17.5 micromol/day of these two metabolites [30]. Low niacin status: Excretion rates between 5.8 and 17.5 micromol/day [30]. Deficient niacin status: urinary-excretion rates are less than 5.8 micromol/day [30]	Direct measurement: plasma pyridoxal 5′phosphate (P5P) >20 nmol/L [31]. This is affected by inflammation markers and albumin concentration [32]. Functional biomarkers may be better assessed by integrative testing	RBC folate is considered to be the most robust marker for long-term folate status. RBC folate for populations should be around 1000–1300 nmol/L. Woolf et al. used >140 ng/mL [31]. WHO recommends <906 nmol/L (>400 μg/L) in women of reproductive age to prevent neural tube defects [29]. Evaluate homocysteine and MMA. If homocysteine is high and MMA is normal, this phenomenon may reflect a folate deficiency [33]. Urinary FIGLU normal: not detected. Elevated levels can represent folate deficiency [34]. This case can be assessed through organic acids testing	Serum B 12 marker. Subclinical deficiency: <200 pmol/L. Low is 150–249 pmol/L. Acute deficiency is <149 pmol/L. This marker has limited diagnostic ability. Serum B12 does not represent cellular B12 levels [33]. Severe B12 deficiency has been documented with normal or high serum levels of B12 [33]. Plasma homocysteine is high at >13 μmol/L, can be used along with the most specific test for functional B12 deficiency which is elevated serum methylmalonic acid at >260–350 nmol/L, (118 pmol/L), or 0.80 μmol/L [33]. Reduced kidney function may affect the clearance of both homocysteine and MMA, thus resulting in higher levels. Another marker to consider in these cases is holo-transcobalamin (holo-TC) [33] Normal holo-TC level is 20–125 pmol/L. Below 20 pmol/L would be correlate with B12 deficiency [33]

Note: Iron deficiency should always be taken into account when assessing folate and B12 status. Iron deficiency causes microcytosis, whereas anemia of folate and B12 can cause macrocytosis [33]. Iron-deficiency anemia can mask macrocytosis and megaloblastic anemia from B12 [33].

make adenosine triphosphate (ATP), the cell's energy currency. Acetyl CoA provides the main substrate for this cycle and relies on pantothenic acid [21]. Thiamine, biotin, and B12 also interrelate in the citric acid cycle and electron transport chain. B vitamins have a fundamental impact on brain function and are actively transported across the blood–brain barrier with the concentration of methyltetrahydrofolate, biotin, and pantothenic acid found in the brain at levels much higher than plasma [21]. B vitamins can generally be consumed in amounts much higher than the RDA and may be necessary for optimal health. To date, only vitamins B6, B12, and folate have an established upper daily limit.

8.5.1 Thiamine (Vitamin B1)

Thiamine plays a role in the synthesis of fatty acids, steroids, nucleic acids, and aromatic acid precursors and in the synthesis of neurotransmitters and bioactive compounds essential for brain function [21]. Thiamine also plays a neuromodulatory role in the acetylcholine neurotransmitter system and can relieve fatigue associated with hypothyroidism [36]. Thiamine can be deficient in grain-free diets and depleted with high intake of alcohol [37]. In a study enrolling young women with adequate thiamine stores but given 50 mg of thiamine or placebo for 2 months, the thiamine-supplemented group reported improved mood as assessed by the Profile of Mood States. The thiamine-treated group also demonstrated improved attention evidenced by faster decision-making on reaction time tasks [21]. Thiamine plays a role in glucose metabolism. Between 17% and 79% of obese patients examined for bariatric surgery were found to be deficient in thiamine, and this may suggest a connection between thiamine status, blood sugar regulation, and obesity. In conditions of fatigue, neurological conditions, or hypothyroidism, higher supplemental doses of thiamine (50–100 mg) may be beneficial [19, 25]. Alcoholism and bariatric surgery will likely require far higher doses delivered under medical supervision.

8.5.2 Riboflavin (Vitamin B2)

Riboflavin is required for the synthesis of two flavoprotein coenzymes, FMN and FAD, which are rate-limiting factors in most cellular enzymatic processes [21]. The flavoproteins are required for the synthesis, conversion, and recycling of niacin, folate, B6, the synthesis of heme proteins, nitric oxide synthesis, P450 enzymes, and proteins involved in electron transfer and oxygen transport and storage. The flavoproteins are also involved in fatty acid metabolism in brain lipids, the absorption and utilization of iron, and the regulation of thyroid hormones. For many reasons, riboflavin deficiency would negatively impact brain function. Clinically, higher doses of riboflavin at 400 mg are helpful with preventing migraine headaches [24]. Recent randomized controlled trials have demonstrated that riboflavin may play a novel role as a modulator of blood pressure, specifically in individuals with the MTHFR 677TT genotype, and can reduce systolic BP by 5–13 mmHg in these genetically at-risk adults [38].

8.5.3 Niacin (Vitamin B3)

Niacin-derived nucleotides, such as nicotinamide adenine dinucleotide (NAD) and NAD phosphate (NADP), are critical for enzymes involved in every aspect of peripheral and brain cell function [21]. Niacin receptors are found in immune cells and adipose tissue as well. Niacin plays a role in neuromodulation of inflammatory cascades and antiatherogenic lipolysis in adipose tissue. People with Parkinson's disease are often low in niacin. One study found this population to benefit from 250 mg niacin supplementation, which resulted in attenuation of the disturbed sleep architecture associated with Parkinson's disease [21].

8.5.4 Pantothenic Acid (B5)

Pantothenic acid is required for the synthesis of coenzyme A (CoA). CoA plays a role in oxidative metabolism and contributes to the structure and function of the brain via its role in the synthesis of cholesterol, amino acids, phospholipids (PLs), and fatty acids. Through the action of CoA, pantothenic acid also helps with the synthesis of neurotransmitters and steroid hormones [21].

8.5.5 Vitamin B6

Vitamin B6 plays an essential role in the folate cycle and amino acid metabolism and is a rate-limiting cofactor in the synthesis of neurotransmitters including dopamine, serotonin, GABA, noradrenaline, and the hormone melatonin [21]. Neurotransmitter synthesis is especially sensitive to B6 deficiency. Even a mild deficiency can result in downregulation of GABA and serotonin synthesis. When GABA is unable to participate in its inhibitory role on neural activity, disordered sleep, behavior changes, cardiovascular function, and loss of hypothalamus–pituitary control of hormone secretion can result.

Low B6 status has been found in oral contraceptive users, smokers, and people with celiac disease, alcoholism, and diabetes [32]. B6 deficiency usually does not occur in isolation, but rather, with other B-vitamin deficiencies. Assessing B6 status can be difficult. Pyridoxal phosphate (PLP), the active form of B6, is often used to assess vitamin B6 by laboratory testing. PLP testing appears to be accurate in healthy individuals and has been found to reflect the vitamin B6 content in the liver and correlate with vitamin B6 dietary intake [32]. However, plasma PLP levels can be affected by albumin concentration, alkaline phosphatase activity, inflammation, and alcohol consumption [32]. Also of concern, low plasma PLP has been found with inflammation such as rheumatoid arthritis, inflammatory bowel disease, cardiovascular disease, diabetes, deep vein thrombosis, and cancer [32]. Plasma PLP shows an inverse relationship with markers of inflammation such as C-reactive protein and other markers of acute phase reactants [32].

The activated form of B6, pyridoxal-5′-phosphate is downregulated during times of inflammation and therefore may contribute to dementia and cognitive decline due to the essential role it plays in brain glucose regulation, immune function, and gene transcription–expression [21]. Vitamin B6 can be over-supplemented and cause sensory neuropathy that is usually reversible. Evidence from case studies suggests that this can happen with as little as 100 mg per day [19]. Keeping the total amount of B6 to 50–100 mg per day is recommended [21]. If higher doses need to be used, consider this as a short-term intervention of 2–3 months, and a lower dose of around 50 mg should be recommended after this time period. There are, however, clinical trials that have used B6 doses as high as 750 mg over several years with no reports of neuropathy [21]. Individuals with certain single nucleotide polymorphisms in the CBS genes and MTHFR will have different requirements for vitamin B6. Please refer to ▶ Chap. 18 for this discussion.

8.5.6 Folate (Vitamin B9)

Folate and B12 are linked due to their complementary roles in the folate and methionine cycles [21]. If B12 is deficient, folate can become trapped as methyltetrahydrofolate, resulting in a functional folate deficiency [21]. Requirements for folate may also differ based on genetics. Either an actual or functional deficiency of folate may hamper DNA stability and repair as well as gene expression and transcription. This, in turn, affects neuronal differentiation and repair, which promotes hippocampal atrophy and demyelination. Ultimately, this compromises the integrity of membrane phospholipids resulting in the impaired action potential of the neuron [21].

A lack of folate may result in decreased synthesis of proteins and nucleotides required for DNA and RNA synthesis. This can have negative ramifications for rapidly dividing tissues, such as fetal development, and can lead to megaloblastic anemia and neuronal dysfunction [21]. Additionally, the folate cycle needs to function effectively to synthesize and recycle tetrahydrobiopterin, which is an essential cofactor for enzymes that convert amino acids to monoamine neurotransmitters such as serotonin, melatonin, dopamine, noradrenaline, adrenaline, and nitric oxide [21].

The upper daily limit for folic acid is set at 1000 mcg/day. This upper level is related to folate's ability to mask vitamin B12 deficiency, resulting in irreversible damage [21]. Detrimental effects of high doses of folic acid are related to high levels of unmetabolized folic acid on normal folate metabolism and immune function [21].

The use of folate comes with caveats. Individuals with MTHFR single nucleotide polymorphisms often require higher doses of folate [39]. Caution is advised in using high doses of folic acid in combination with antifolate medications, such as methotrexate, prescribed for conditions such as rheumatoid arthritis, psoriasis, cancer, bacterial infections, and malaria [21]. In addition, supplemental folate may confer protection against cancer at lower doses but may cause increased risk of cancer at higher doses, but there is no consensus on blood levels of folate that may cause harm.

As many patients turn to vitamins and supplements to enhance energy, relieve fatigue, or generally feel better, it is important to understand the connection between the B vitamins and psychiatric symptomatology. Vitamins B6, B8, and B12 have been shown not only to reduce psychiatric symptoms but also shorten the duration of illness [39]. However, when patients lack a specific genetic enzyme which converts folate–folic acid to its most usable form, L-methylfolate, the neuroprotective, and neuropsychiatric benefits are lost. L-methylfolate allows for the synthesis of the three major neurochemicals—serotonin, norepinephrine, and dopamine—across the blood–brain barrier [39]. Exploring the conversion of folate–folic acid into L-methylfolate and the various polymorphisms of the MTHFR gene while examining the B vitamins associated with the treatment of psychiatric symptoms allows integrative and functional practitioners to treat patients with the appropriate B vitamins [39]. For an in-depth analysis of how to assess folate status, please see ◘ Table 8.2 for biomarker assessment. For a detailed discussion, see Nutragenomics in ▶ Chap. 17.

8.5.7 Vitamin B12

Vitamin B12 is protective against neurological deterioration, and deficiency of B12 is associated with peripheral neuropathy, cognitive impairment, and neurodegenerative disease [33]. Causes of B12 deficiency are largely related to absorption in the GI tract, lack of intrinsic factor, or dietary deficiency such as with vegan diets [33]. Autoimmune pernicious anemia, intestinal surgery such as bariatric surgery, and chronic gastritis from *H. pylori* infections all decrease the release of intrinsic factor which can result in B12 malabsorption. Undiagnosed celiac disease can also result in malabsorption of B12 [33]. Some medications interfere with absorption and metabolism of B12, including metformin. Metformin decreases serum B12, but in some studies, it was shown to decrease plasma methylmalonic acid (MMA) and increase intracellular B12, although there are conflicting reports [33]. Metformin may alter B12 homeostasis and tissue distribution, but the clinical consequences remain to be determined. Proton pump inhibitors and other medications that reduce the production of hydrochloric acid are also associated with B12 deficiency [33]. B12 deficiency can be masked by folate supplementation and can cause severe neuropathy and neurodegeneration. An effective test for B12 sufficiency is MMA in blood or urine. Refer to ◘ Table 8.2 for assessment of B12 levels.

Doses of vitamin B12 that far exceed the RDA are commonly used in the integrative and functional nutrition realm. Anecdotally, nutrition practitioners have found that genetically susceptible individuals with catechol-O-methyltransferase (COMT) single nucleotide polymorphisms have a decreased ability to metabolize catecholamines, and supplementation with the methylcobalamin form of vitamin B12 may aggravate symptoms of anxiety and insomnia. Some of these individuals do better with hydroxocobalamin or adenosylcobalamin. To counter untoward effects when beginning B12 supplementation, start with a lower dose, such as 500 mcg, and slowly titrate dose upward.

The take-home message with B vitamins is to supplement all the B vitamins (not just one) because they work synergistically, even if the exact mechanisms has not been fully elucidated in research.

8.6 Fat-Soluble Vitamins

Fat-soluble vitamins are at risk of deficiency with certain states of malabsorption and malnutrition. When used in supplemental form, these vitamins need to stay in balance with each other, notably vitamin K1–K2 and vitamin D and vitamin D and vitamin A. Vitamin E is best provided in a natural full-spectrum combination as delta, beta, gamma, d-alpha tocopherols and the four tocotrienols, the way it is found in food.

8.6.1 Vitamin D and Vitamin K

One of the risks of using supplemental vitamin D is the increased gastrointestinal absorption of calcium that results in high serum levels of calcium and increased vascular calcification [40]. Theoretically, if vitamin D supplementation is balanced with supplementation of vitamin K1 and K2 in the forms of menaquinone-4 (MK 4) and menaquinone-7 (MK 7), this will decrease the likelihood of soft tissue calcification and of bone fracture [40]. In addition, a few vitamin K-dependent small proteins act to inhibit soft tissue calcification and include osteocalcin (bone Gla protein), matrix Gla protein (MGP), and possibly Gla-rich protein (GRP) [41]. The role of these proteins has been elucidated in studies monitoring the effects of treatment with oral anticoagulants that are known vitamin K antagonists (VKA).

With regard to VKA, all vitamin K-dependent calcification inhibitors will remain uncarboxylated and inactive [41]. Compared with controls, subjects taking anticoagulants had significantly higher calcification of arteries and the aortic valve [41]. Despite previous publications that reported vitamin K1 had no effect on bone mineral density, in a recent study, vitamin K1 deficiency was the strongest predictor of vertebral fractures and vascular calcification in chronic kidney disease. In postmenopausal women, 5 mg of vitamin K1 protected postmenopausal women from bone fractures, despite having no positive effect on bone mineral density [42]. Long-term, three-year supplementation with menaquinone (K2) of 180 mcg in 120 healthy postmenopausal women resulted in decreased arterial stiffness [40].

Vitamin K2 includes several different vitamers, of which MK7 is the most well-known. One study found that plasma concentrations of albumin and vitamin K1 were significant predictors of hip fracture in the general population, whereas MK7 was not [42]. To promote a decreased risk of bone fracture while protecting from vascular calcification, it is likely prudent to include both vitamin K1 and vitamin K2, specifically MK 4 and MK 7, when recommending calcium and vitamin D [42].

8.6.2 Vitamin A and Vitamin D

Vitamins A and D are important synergistic partners due to their shared binding of the nuclear retinoid X receptors (RXR), resulting in synergistic effects of one vitamin with the other. In nature, vitamins A and D are found together in balance such as in cod liver oil, egg yolk, and organ meats. Vitamins A and D function in many systems throughout the body beyond the eyes and bones. Furthermore, vitamins A and D are important as immune and hormone modulators as well as affecting structural forms such as bones, cell membranes, tissues, etc.

The effects of vitamins A and D supplementation on bone mineral density have been controversial. Observational studies have reported an association between higher serum retinol concentration and risk of bone fracture; another study reported that high retinol concentration in the presence of vitamin D deficiency can increase the risk of osteoporosis in menopausal women [42]; another recent study found that higher intake of vitamin A, when combined with sufficient intake of vitamin D, can promote increased bone density [43]. In a Korean population, a higher dietary vitamin A intake does not appear to negatively affect bone mineral density when 25 (OH) D levels are moderate at 50–75 nmol/L [43].

When assessing an individual's vitamins A and D status, it is beneficial to get a baseline of blood vitamin A retinol and vitamin D 25–OH, vitamin D1, 25-OH, and PTH, at minimum, to develop more targeted interventions and improve outcomes. If genomic information is available for vitamin D receptor and BCMO1 genes, genomic differences that influence the ability to establish balance between vitamin A and vitamin D might come to light. This highlights the fact that there are unique requirements for individuals to maintain nutrient balance.

Serum retinol and serum 25(OH)D can both be tested to ensure safe and adequate levels are maintained. Normal levels of serum vitamin A range from 50 to 200 mcg/dL or 1.75–6.98 micromol/L [6]. Interestingly, in nature, vitamins A and D are found together such as in egg yolk, liver, and butter. Overall, keeping the vitamin D level adequate appears to be protective to bones when supplementing vitamin A.

8.6.3 Tocopherols and Tocotrienols

Vitamin E is a family of fat-soluble antioxidants that mainly refers to alpha tocopherol, but naturally includes several tocopherols and tocotrienols [40]. Vegetable oils contain higher amounts of tocopherols, while tocotrienols are found in palm oil [44]. Both tocopherols and tocotrienols have four homologues consisting of alpha, beta, gamma, and delta [44]. Gamma-tocopherol is known mainly for its beneficial function in maintaining cardiovascular health, whereas the tocotrienols have shown more diverse application and protection against cancer, cardiovascular disease, neurodegeneration, oxidative stress, fertility, and immune regulation [44].

Tocotrienols exert different effects. Delta- and gamma-tocotrienols were more potent in cancer studies, while alpha-tocotrienols are more effective with neuroprotection. In a trial, subjects supplemented with gamma- and delta-tocotrienols showed a significant reduction in triglycerides and very low-density lipoprotein with no change in total cholesterol, LDL and HDL cholesterol, whereas the other homologues, alpha and gamma, did not show any effect on lipid profiles [44].

In preclinical cancer studies, tocotrienols have been shown to have antiproliferative, antiangiogenic, proapoptotic, and immune-enhancing properties [44]. The first clinical trial of tocotrienols involved a five-year, placebo-controlled, double-blinded study. Participants showed decreased number of deaths and incidence of recurrence in the tocotrienol group. Gamma-tocopherol showed increased apoptosis in a study on pancreatic ductal neoplasia.

Tocopherols and tocotrienols have neuroprotective properties and were shown to reduce glutamate toxicity, and subsequent damage to neurons and astrocytes and the use of vitamin E for the management of Alzheimer's disease is a topic of interest [45]. In a two-year study, daily supplementation of 400 mg of tocotrienol-rich fraction (TRF) resulted in statistically significant reduction in white matter lesions, compared with placebo group [44].

At this writing, there is no general consensus on recommendations for vitamin E intake as it is dependent on age, lifestyle, fat malabsorption, individual differences in vitamin E metabolism, and interaction with pharmaceuticals, such as blood thinners and statins [40]. Vitamin E may become a prooxidant when administered at high levels, but this effect may be mitigated by the coadministration of other antioxidants, such as vitamin C [40]. There is recent speculation that supplemental tocotrienol (T3) exerts more benefit than supplemental tocopherol and tocopherols should be solely obtained through food sources.

8.7 Minerals

8.7.1 Calcium and Magnesium

Calcium and magnesium are important nutrient partners that require balance, as elevation of one can cause suppression of the other. Integrative and functional medicine clinicians recommend supplementing calcium–magnesium at a ratio of 2:1 to 1:1. Supplemental calcium continues to be controversial as it may be associated with an increased risk for myocardial infarction, stroke, and cardiovascular mortality and an increased risk for kidney stones and soft tissue calcification [46] and more recently continues to be controversial as it may be associated with an increased risk for myocardial infarction, stroke, and cardiovascular mortality and an increased risk for kidney stones and soft tissue calcification [46]. There is growing evidence for the need for calcium supplementation to be in tandem and balanced with magnesium. Integrative and functional nutritionists often see clients with malabsorption, and part of the dietary intervention may include eliminating dairy products, often without recommending supplemental calcium or calcium-rich foods. This may have unintended harmful consequences (See Tables 8.4 and 8.5).

Using celiac disease (CD) as an example, calcium malabsorption and insufficient intakes of calcium may result in hypocalcemia, which produces a compensatory increase in serum levels of parathyroid hormone. This, in turn, leads to increased bone turnover and ultimately bone loss, with bone resorption being faster than new bone formation [47]. It is possible that the malabsorption of calcium is the main contributing factor to secondary hyperparathyroidism seen in CD [48, 49]. In addition, whenever parathyroid hormone is elevated, there is rapid conversion of 25 OH vitamin D to 1,

Table 8.4 High calcium foods to recommend to patients on dairy-free diets

1 cup okra, cooked	150 mg
1 cup spinach, cooked	350 mg
1 cup kale, cooked	94 mg
1 cup broccoli, raw	42 mg
1 cup collard greens, cooked	150 mg
1 cup turnip greens, boiled	190 mg
1 cup bok choy, raw	74 mg
3 oz canned sardines with bones	325 mg
3 oz canned salmon with bones	190 mg
½ cup firm tofu made with calcium sulfate	253 mg
1 cup calcium-fortified milk alternative	300 mg
Calcium-fortified juice	261 mg
½ cup soy beans and white beans	80–90 mg
Luna bar	350 mg
Power bar	300 mg
1 cup calcium-fortified cereal	250–300 mg
2 tablespoons of Tahini or sesame seeds	120–180 mg

Based on data from National Institutes for Health Calcium Fact Sheet for Health Professionals retrieved on 8/14/2017 from
► https://ods.od.nih.gov/factsheets/Calcium-HealthProfessional/

Table 8.5 Recommended daily allowances (RDAs) for calcium intake

Age	Male	Female	Pregnant	Lactating
0–6 months[a]	200 mg	200 mg		
7–12 months[a]	600 mg	600 mg		
1–3 years	700 mg	700 mg		
4–8 years	1000 mg	1000 mg		
9–13 years	1300 mg	1300 mg		
14–18 years	1300 mg	1300 mg	1300 mg	1300 mg
19–50 years	1000 mg	1000 mg	1000 mg	1000 mg
51–70 years	1200 g	1200 mg		

[a]AI (adequate intake) National Institutes for Health Calcium Fact Sheet for Health Professionals retrieved on 8/14/2017 from ► https://ods.od.nih.gov/factsheets/Calcium-HealthProfessional/

25 OH vitamin D (the active form of vitamin D) to enhance calcium absorption [48]. Clinically, you may see a low 25 OH and an elevated 1, 25 OH. Regarding CD, 75% of untreated adult CD patients with overt malabsorption, and about half of those with subclinical CD presenting with minimal symptoms or asymptomatic CD patients will have bone loss. In 2000, the British Society of Gastroenterology published guidelines for treating osteoporosis in CD that includes a daily calcium intake of 1500 mg and vitamin D supplementation [47]. Malabsorption of other micronutrients may also contribute to altered bone metabolism and increased pro-inflammatory cytokines that enhance osteoclastogenesis and bone resorption [49].

In cases where malabsorption is suspected, it may be prudent to request the patient have a parathyroid hormone test. It may be possible that secondary hyperparathyroidism may occur with low, normal, or high serum calcium levels. High serum calcium levels may be the result of long-term calcium malabsorption resulting in secondary hyperparathyroidism. Have clients work with a physician to see if they are a candidate for calcium and vitamin D supplementation. Clinically, it may be safe to assume that individuals at risk for calcium and vitamin D malabsorption include those with CD, inflammatory bowel disease, bariatric surgery (gastric bypass surgery) and other GI surgeries, eating disorders, and the presence of gastrointestinal pathogens [49]. Seeing a pattern of low normal serum protein, low normal albumin, low ferritin, or other markers of iron status, elevated MCV (suggesting deficiencies of folate and vitamin B12) and low normal serum calcium might suggest a pattern of malabsorption. Various medications, such as steroids, proton pump inhibitors, and loop diuretics, also increase the need for calcium supplementation [49]. The overarching recommendation is to dose calcium based on individual needs rather than blanket recommendations for the general population.

Magnesium is often left undiscussed in treating bone health and cardiovascular disease. Low magnesium status may lead to a greater risk of metabolic syndrome, type 2 diabetes, cardiovascular disease, skeletal disorders, chronic obstructive pulmonary disease, depression, decreased cognition, and vitamin D deficiency [3]. Magnesium plays a critical synergistic role in the synthesis and metabolism of parathyroid hormone (PTH), vitamin D-binding protein (VDBP), and three major enzymes that determine 25-OH vitamin D concentrations [3]. Magnesium deficiency leads to decreased levels of 1, 25 $(OH)_2$ D, the active form of vitamin D, and subsequent impaired PTH response [3]. Additionally, a specific form of vitamin D-resistant rickets that is magnesium-dependent exists and only responds to vitamin D therapy when magnesium is concurrently administered [3]. Readers will note other nutrients besides calcium, magnesium, and vitamin D that play a role in bone density include protein, vitamin K, zinc, copper, and boron.

Two small clinical studies on magnesium-deficient patients show that magnesium administered alone is not enough to raise vitamin D levels and that administration of vitamin D3 combined with magnesium infusion resulted in a substantial increase in both 25(OH)D and 1, 25 $(OH)_2$ compared with the infusion of magnesium alone. A recent NHANES study found that a high intake of magnesium was independently and significantly associated with a reduced risk of vitamin D deficiency [3]. Although more studies will elucidate the relationship between magnesium and vitamin D status, it may be prudent for clinicians to correct both vitamin D status and magnesium status concurrently. If clinicians experience vitamin D-resistant patients, the patient may be deficient in magnesium as well. Supplementing with vitamin D and calcium in the presence of suboptimal magnesium intake may be affecting health through interactions between the three nutrients that have not yet been fully explained [3].

Since 1977, the United States has increased both dietary calcium and magnesium intake, but dietary calcium intake has increased at a rate of 2–2.5 times that of dietary magnesium [3]. Some US studies have shown that since 2000, the calcium-to-magnesium ratio has increased ≥3 times and coincides with increasing rates of type 2 diabetes and colorectal cancer [3]. High dietary calcium-to-magnesium ratio has been proposed as an explanation for the striking variation in incidence of postmenopausal breast cancer associated with geographic location [50], the lowest calcium intakes being in Asia where the incidence of breast cancer is much lower compared to North America and northern Europe [50]. An imbalance of the calcium–magnesium intake may lead to irregularities in DNA repair, cell proliferation, differentiation, and carcinogenesis [50].

Reduced magnesium intake may be due to increased intake of refined and processed foods, softening of hard water, chronic alcohol ingestion, and gastrointestinal disorders causing malabsorption and certain medications. In addition, calcium supplementation may accentuate the problem of reduced magnesium intake [50]. Subclinical dietary magnesium deficiency was shown to increase calcium retention, and once calcium is high, magnesium absorption can be significantly depressed. Moderate alcohol consumption is associated with breast cancer and alcohol exacerbates magnesium deficiency [50].

Higher self-reported dietary and supplemental magnesium intakes were associated with lower levels of coronary artery calcification, which is a sensitive, discriminating measure of subclinical cardiovascular disease [51]. Huang et al. found that consumption of moderate amount of calcium and an adequate amount of magnesium with maintenance of calcium-to-magnesium ratio of 2.0–2.5 are important for reducing cardiovascular risks in older patients with diabetes [52]. Individuals who consume 1000–1200 mg of calcium per day need to increase their magnesium intake to maintain the calcium-to-magnesium ratio of 2.0–2.5 [52].

Traditional advice is to maintain calcium-to-magnesium ratios for optimal health [3], and the authors of a 2007 colorectal neoplasia study suggested the optimal dietary calcium-to-magnesium ratio be <2.8. It is thought that calcium-to-magnesium ratios >2.6–2.8 can have detrimental health effects, and it has also been questioned if a calcium-to-magnesium ratio of <2.0 may also be detrimental [3]. More research is needed to establish a beneficial ratio.

8.7.2 Sodium-to-Potassium Ratio and Hypertension

The effect of sodium-to-potassium (Na–K) ratio on hypertension is perhaps more important than looking separately at the effect of individual sodium or potassium intake on hypertension. Inconsistencies have been found in observational studies around the world showing a relationship between high sodium intake and hypertension, as well as high Na–K ratios and hypertension [53]. The mechanisms of sodium and potassium on blood pressure are multiple. In salt-sensitive individuals, sodium intake results in sodium and water retention and extracellular volume expansion. The extracellular volume expansion results in release of substances that increase heart and blood vessel contraction and affect the renin–aldosterone system [53]. Potassium increases urinary sodium excretion, which lowers serum sodium levels, and is thought to induce vascular smooth muscle relaxation or widening of the blood vessels to lower blood pressure [53].

Potassium is an electrolyte needed for normal cellular function and is easily excreted by healthy functioning kidneys rather than stored in the body [54]. Therefore, humans need a constant supply of potassium through the diet. Adequate intake of fruits and vegetables is a major source of potassium [54]. The average potassium consumption is at 54% of the US recommended intake [54]. Low potassium intake is also correlated with central obesity and metabolic syndrome [54]. A caveat is high potassium intake which is contraindicated in renal disease because of poor excretion and potential for elevated levels.

Potassium supplementation has consistently been shown to lower blood pressure, and low dietary potassium is associated with an increased risk of developing hypertension [55].

The joint effect of high sodium and low potassium intakes may have a greater effect on hypertension than elevated sodium intakes or low potassium intakes alone [53]. In a Korean study by Park et al., dietary intake of both sodium and potassium was evaluated, and the authors showed that an increased ratio of sodium to potassium was correlated with increased prevalence of hypertension [53].

Study subjects who had a lower prevalence of hypertension had a Na–K ratio of 1.21:1, and with increased prevalence of hypertension, there was a Na–K ratio of 2.56:1 [56]. Additionally, blood pressure results had almost a linear dose response to increasing ratio of Na–K [53]. The blood pressure lowering effects of potassium were greatest in those with the highest intake of sodium, and the ratio of Na–K had a greater effect on the risk of cardiovascular disease than sodium or potassium alone [53]. Recommending a diet high in fresh fruits and vegetables, including low-fat dairy products, and consuming a whole-food diet, is likely the most effective approach for lowering the Na–K ratio [53]. Caution is advised when looking at sodium-to-potassium ratios as great variability in sodium sensitivity exists between different ethnic groups [53].

8.7.3 Zinc and Copper

Zinc and copper are another example of nutrient partners that need to be metabolically balanced. Excessive elevation of either one can suppress the other, much like the relationships described above.

Zinc is used clinically in supplement form to facilitate wound healing; decrease skin inflammation; support immune function, tissue growth, and maintenance of thyroid function; promote GI tract healing; protect against such ocular diseases as macular degeneration; and promote testosterone balance. Zinc supplementation can induce copper deficiency that may result in serious neurological conditions, hematological abnormalities, and possibly thyroid abnormalities; therefore, zinc needs to be balanced with copper [57, 58]. High tissue–copper levels may be associated with inflammation and certain disease states, such as cancer. "The RDA recommendations for zinc and copper intake are in a ratio of 9:1 [59]". When working with medical conditions, it is important to measure a baseline blood test for both nutrients to determine if supplementation of either is needed. (See ◘ Table 8.6 Assessing zinc and copper status.)

Measuring the serum copper to zinc ratio is a helpful parameter in states of disease and inflammation. The normal plasma zinc to serum copper ratio in children and adults is 1:1 [64]. The increment of this ratio of the opposite (copper-zinc) above 2.0 in the elderly usually reflects an inflammatory response or decreased zinc status [65]. High serum copper to zinc ratio has been associated with cardiovascular death, malignancy, and all-cause mortality in the elderly [65]. During the acute phase of various diseases, systemic mechanisms decrease zinc and increase copper. Using the ratio of serum copper to zinc may be more predictive of inflamma-

Table 8.6 Summary of copper and zinc assessment

Copper reference ranges	Zinc reference ranges
Serum free copper: 1.6–2.4 µmol/L or 10–15 µg/dL [60] *Total copper*: 10–22 µmol/L or 63.7–140.12 µg/dL [60] *Serum ceruloplasmin*: 2.83–5.50 µmol/L or 18–35 µg/dL [60] <20 mg/dL ceruloplasmin signifies Wilson's disease along with Kayser–Fleischer rings [61]. Wilson's disease is a genetic defect in copper excretion. *24-hour urine copper*: 0.3–0.8 µmol or 20–50 µg [60] <40 mcg/day [61] >100 mcg/day occurs in Wilson's disease [61] *Liver copper*: 0.3–0.8 µmol/g of tissue or 20–50 µg/g of tissue [60]	*Normal serum zinc*: 60–120 µg/dl [62] *Plasma zinc is deficient if below*: 60 µg/dL [63] *Alkaline phosphatase (ALP) normal range*: 45–115 units/liter (U/L) [62] <45 U/L ALP indicates zinc and/or magnesium deficiency [62] * For reference in assessing low ALP, normal serum magnesium: 1.3–2.5 mEq/L [62]

tion than using other inflammatory biomarkers such as CRP and ESR [65]. Nutritional factors such as increased copper intake and decreased zinc intake are not thought to be the cause of an elevated copper-to-zinc ratio during states of inflammation, but rather, other systemic mechanisms may be the cause. Zinc is carried on albumin, and copper is carried on ceruloplasmin (Cp). Synthesis of these proteins and the regulation of serum copper and zinc occurs mainly at the hepatic level.

During oxidative stress, the albumin-bound zinc in the plasma decreases while labile zinc relocates to peripheral tissues [65]. This labile zinc induces the antioxidant metallothionein (MT). Infections and inflammatory conditions which induce oxidative stress cause copper–ceruloplasmin levels to rise. During inflammation and aging, the cytokines interleukin (IL) 6, IL-1 beta, tumor necrosis factor-alpha, and interferon gamma (IFN-gamma) are known to suppress the synthesis of albumin (i.e., bound to zinc) and increase the synthesis of ceruloplasmin. This mechanism may be protective in response to oxidative stress, infection, and low-grade inflammation. Impaired insulin action also decreases the synthesis of albumin, which may, in turn, decrease plasma levels of zinc.

Following inflammatory stimuli, zinc is redistributed from the plasma to the liver, thymus, and marrow. This may be a mechanism to restrict zinc from being used by invading pathogens, protect the liver and tissues from oxidative stress by producing MT, and manufacture lymphocytes to protect against invading pathogens [65]. Increased levels of copper in the serum may help the antimicrobial function of macrophages. Very low or undetectable levels of zinc are found in infected tissue, whereas copper accumulates at sites of infection where macrophages are present in high amounts [65].

Copper deficiency impairs oxidative phosphorylation, cellular antioxidant defense, collagen and elastin biosynthesis, production of several metalloenzymes such as copper–zinc superoxide dismutase, and affects levels of selenium-dependent glutathione peroxidase [64]. Copper deficiency impairs blood coagulation, has adverse effects on blood pressure and heart function, affects cholesterol and glucose regulation, contributes to neurodegeneration, causes anemia, affects cross-linking of connective tissue, and causes mineralization of the bone, ultimately leading to osteoporosis [66]. Copper deficiency results in optic neuropathy [67]. Anemia that is refractory to iron supplementation is a classic, well-documented sign of chronic copper deficiency [68]. One study in frail elderly men found decreased hematocrit and serum iron levels in subjects who had an elevated serum copper to zinc ratio [68]. Additional symptoms of copper deficiency include leukopenia, neutropenia, decreased superoxide dismutase, decreased ceruloplasmin, increased plasma cholesterol and LDL–HDL ratio, and abnormal cardiac function [64].

Copper intoxication rarely occurs in humans, as it is excreted in the bile [64]. However, elevated copper levels are associated with infections, inflammation, Wilson's disease, trauma, systemic lupus erythematosus (SLE), and autism [64]. Both zinc and molybdenum deficiency may be a risk factor for copper toxicity. Due to its role as a cofactor, copper toxicity has been associated with such neurological diseases as amyotrophic lateral sclerosis, Alzheimer's disease, and Creutzfeldt–Jakob disease [64]. Physiological factors independent of copper intake can affect copper levels. Plasma copper levels are higher in women than men due to estrogen [59]. Infection, inflammation, and estrogen levels increase plasma copper. Corticosteroid and adrenocorticotropic hormone lower copper concentrations.

Copper is bound to the protein ceruloplasmin that is regulated by homeostatic mechanisms [59]. High serum copper concentration with elevated copper-to-ceruloplasmin ratio is indicative of copper excess [59]. In Wilson's disease, a condition of copper excess, free copper is elevated, and ceruloplasmin is lowered [59]. Ceruloplasmin is an acute phase reactant and may be raised in response to hepatic inflammation, pregnancy, estrogen use, or infection [61]. Falsely low ceruloplasmin levels may occur with any protein deficiency state including nephrotic syndrome, malabsorption, protein-losing enteropathy, and malnutritions [61].

A review of methods of assessment of copper status in humans suggests that serum copper seems to be the most accurate biomarker to assess copper status in humans, reflecting changes in status in both depleted and replete individuals [69]. Measuring copper levels may be elusive, but low plasma or serum levels of copper indicate depletion. Hair and urinary copper are not useful indicators of copper status [59]. Total ceruloplasmin protein level is related to copper status, but deficiency reflects changes in highly depleted individuals only [69]. One of the first signs of copper deficiency is a drop in ceruloplasmin. Ceruloplasmin accounts for 90% of total plasma copper [70]. Ceruloplasmin and serum copper are acute phase reactants and can rise in response to inflammation, even in states of copper deficiency [66].

The dosing range for copper in adults is 2–10 mg/day with monitoring of zinc status during supplementation [59]. In states of copper deficiency, typically 2 mg/day of copper will reverse the hematological abnormalities in the early stages [58].

8.7.4 Assessing Zinc Status

Symptoms of severe zinc deficiency include hypogonadism, dwarfism, growth-retarded infants and children, dermatitis, diarrhea, alopecia, and loss of appetite [64]. More moderate zinc deficiency can result in decreased immune function, increased mortality due to infections, and brain damage in a fetus when the pregnant mother is zinc deficient [64]. Zinc-repletion dosing ranges anywhere from 5 mg/day to 50 mg/day with monitoring of copper status during supplementation [59].

There is currently no specific sensitive biochemical or functional indicator of zinc status [71]. Accurate measures of plasma zinc are complicated by the body's homeostatic control of zinc levels and factors affecting zinc status that are unrelated to nutritional status [71]. Plasma zinc is bound to its carrier protein albumin; therefore, anything that alters albumin levels will alter plasma zinc levels [59]. There is no functional reserve of zinc in the body, and the body's zinc levels are maintained by conservation and tissue redistribution [71].

Cessation of growth in children is an example of severe zinc deficiency [71]. When dietary zinc intake is low, fecal zinc excretion may be reduced by 60%, coupled with increased intestinal zinc absorption. In mild zinc deficiency, plasma zinc can be maintained at the expense of zinc from other tissues [71]. Therefore, plasma zinc is not a reliable measurement of zinc intake or whole-body zinc status [71]. Plasma zinc levels in mildly zinc-deficient growth-retarded children who responded to zinc supplementation were not significantly different from normally developed children, before or after zinc supplementation [71].

Decreased plasma zinc levels may result from stress, infection, inflammation, use of estrogen, oral contraceptives, and corticosteroids. Plasma zinc levels are known to fall 15–20% after a meal [71]. Increased zinc levels may occur from fasting or red blood cell hemolysis [59]. A better alternative to plasma zinc levels might be red blood cell zinc levels, as RBCs contain zinc-dependent proteins. RBCs turn over every 120 days; hence, RBC zinc status would reflect whole-body zinc status over a longer period of time than plasma zinc and is not prone to recent changes in whole-body zinc status [59].

Zinc-containing enzymes are another possible marker for zinc status, and alkaline phosphatase has been studied as a biomarker for zinc status. In a study in Guatemalan children, low serum zinc was associated with low serum albumin and low serum alkaline phosphatase [72]. When assessing ALP, it is important to differentiate if low alkaline phosphatase is due to zinc or magnesium deficiency [62]. When zinc was repleted, the serum zinc increased, as did alkaline phosphatase. Other studies in Guatemalan children have shown an association between serum zinc and alkaline phosphatase [72]. However, it was noted in children and adolescents depending on age, serum alkaline phosphatase levels have a wide variation and therefore should be interpreted cautiously.

8.7.5 Copper-to-Zinc Ratio in Cancer

Functional medicine experience-based practice recommendation in serum copper-to-zinc ratio is 1:1 for optimal health, although this author could not find research to corroborate at this time. When serum copper-to-zinc ratio is being assessed during cancer treatment, serum zinc is used. Numerous studies have found an association between elevated copper-to-zinc ratio and disease severity. Inducing copper deficiency to prevent angiogenesis in cancer has been proposed as an adjunctive treatment for cancer [70]. Currently, it is uncertain if enough research exists to put this into clinical practice. A phase II trial of advanced kidney cancer looked at inducing copper deficiency by using TM, a novel antiangiogenic agent, to chelate copper, with the goal of lowering ceruloplasmin level to 5–15 mg/dl for 90 days [70]. This resulted in disease stabilization rather than a reduction in disease burden. The authors proposed that this may only be useful for a patient with minimal disease.

Diez et al. reported on the ratio of serum copper to zinc in diagnosis of lung cancer and found ratios of 2.34 ± 0.78 in the lung cancer cohort, 1.62 ± 0.23 in patients with benign pulmonary lesions, 1.43 ± 0.29 in patients with benign lung diseases, and 0.188 (not significant) in healthy subjects [73]. A cutoff value of Cu–Zn ratio of 1.72 resulted in sensitivity of 89%, specificity of 84%, positive predictive value of 7%, and negative predictive value of 92% between lung cancer patients and healthy subjects [73]. Lung cancer patients with pretreatment Cu–Zn concentrations equal to or above 2.25 had 24-month survival rates of less than 5%. By contrast, pretreatment Cu–Zn levels below 1.72 were associated with 24-month survival rate of 70% [73].

Functional medicine doctors working in oncology typically strive for a serum Cu–Zn ratio of 1.0, although this author could not find adequate studies supporting this ratio.

Serum Cu–Zn ratio should be interpreted cautiously in cases of suspected malignancy that are accompanied by infections, liver diseases, use of oral contraceptives, or other conditions known to affect copper.

8.8 Fatty Acids and Phospholipids

8.8.1 Fatty Acid and Phospholipid Balance

A balanced ratio of omega-6 to omega-3 fatty acids is important for overall health related to its impact on inflammation, immune balance, obesity, depression, and cardiovascular disease. See also ▶ Chaps. 10 and 11. Omega-6 and omega-3 fatty acids cannot interconvert into one another; they are metabolically and functionally distinct and have important, opposing physiological effects [74]. Alpha-Linolenic acid (omega-3) and linoleic acid (omega-6) are the two essential fatty acids that must be obtained through the diet; they cannot be endogenously formed by humans and other mammals due to a lack of enzymes for desaturation [74, 75].

The balance of omega-6 compared with omega-3 fatty acids influences the metabolic pathway of each of these fatty acids [76]. Omega-6 and omega-3 fatty acids produce different eicosanoid products that result in different effects on inflammation. The need for balance between the eicosanoid families is critical for metabolism involving inflammation, resolving inflammation, structural composition, hormones, and immune functions. The nutritionist should be aware of the balance of the eicosanoids, their downstream metabolites, and the functions of such prostaglandins as PG1, PG2, PG3, and specialized proresolving mediators (SPM).

In the past three decades, because of public health prescriptions for low-fat diets, the intake of overall fat in the Western diet has decreased. However, proinflammatory omega-6 fatty acid consumption has increased, while omega-3 fatty acid consumption has decreased [74]. The ratio of omega-6–omega-3, historically, was 1:1–4:1, and today it is reported to be around 20:1. This change in consumption correlates with the rising epidemic of obesity and inflammatory disorders [74]. Arachidonic acid (AA), an omega-6 fatty acid, is a critical fatty acid for cell membrane structure and resolution of inflammation as it promotes specialized SPM. However, when it is out of balance, excessive AA can become pro-inflammatory. As with all molecules and metabolites throughout metabolism, altered metabolic balance can promote dysfunction.

AA is used as a substrate to make eicosanoids such as prostaglandin E2 and leukotriene B4 [74]. These cytokines, when in excess, are pro-inflammatory and more potent mediators of thrombosis and inflammation than cytokines derived from omega-3 PUFAs. Eicosanoids from AA are biologically active in small amounts, and when present in large amounts, they contribute to the formation of thrombus and atheromas, allergic and inflammatory disorders, and proliferation of cells [74].

EPA and DHA, both omega-3 essential fats, suppress the production of proinflammatory cytokines IL-1B, Il-6, and TNF alpha [75]. EPA inhibits AA synthesis from linoleic acid and competes with AA for enzymatic conversion, resulting in reducing AA conversion to pro-inflammatory molecules [75]. The ingestion of fish or fish oil results in the EPA and DHA partially replacing omega-6 in the cell membranes of platelets, erythrocytes, neutrophils, monocytes, and liver cells [74]. In humans, the cerebral cortex, retina, testis, and sperm are particularly rich in DHA, and DHA is abundant in the brain's structural lipids [74].

During evolution, omega-3 fatty acids were found in almost all foods consumed including meat, fish, wild plants, nuts, and berries [74]. The dietary and agricultural changes over the last 100–150 years have resulted in less omega-3 available in foods and an increased risk for changing the physiological state of the body to a more inflammatory, atherogenic state. The balance of omega-6 to omega-3 can affect gene expression, prostaglandin, and leukotriene metabolism and interleukin-1 production [74]. Modern agriculture and aquaculture have changed the omega-3 fatty acid content for the worse in many foods by replacing them with omega-6 fats.

Depression is associated with low plasma phospholipid and erythrocyte levels of EPA and DHA [75]. It is possible that omega-3 fatty acids aid in mitigating depression by decreasing inflammation systemically. Evidence points to utilizing supplemental omega-3 fatty acids for therapeutic intervention of depression as a monotherapy or in addition to antidepressant medications, although results have been mixed [75]. It has yet to be determined what blood levels of EPA and DHA might be associated with improvement in depression. Therapeutic levels of omega-3 depend on diet, genetic variation in omega-3 fatty acid metabolism, and the rate of absorption and incorporation into biologic systems [75]. In both short- and long-term studies, a fixed daily dose of omega-3 results in a large variation of individual omega-3 blood levels, possibly due to genetic polymorphisms and dietary intake. In a study on supplemental omega-3 fatty acids and depression, researchers found that participants whose depression remitted had higher baseline omega-3 levels than those whose depression did not go into remission [75].

RBC levels are less sensitive to recent intake of omega-3 compared with plasma levels and may provide a more reliable estimate of omega-3 levels over time [75]. A level of 5–6% of omega-3 in RBC may be a reasonable target to provide a therapeutic level. However, those who experienced remission of depression had levels of omega-3 in RBC at 8% with the highest probability of remission at 9–10% [75]. This finding is similar to results of a review on omega-3s to improve cardiovascular outcomes [75]. Overall, looking at individual RBC levels to obtain therapeutic efficacy might be a better way to dose supplemental omega-3 than providing a fixed dose for everyone.

Omega-3 fatty acid-enriched diets improve inflammatory status of metabolic dysfunction and reduction in adipose tissue if kept in balance with other essential fats [76]. Supplementation with omega-3 fatty acids and certain doses of omega −6 fatty acids was shown to help with decreasing body fat mass and hip circumference loss [76]. Dietary interventions optimizing the ratio of omega-6 to omega-3 have shown significant reduction in low-density cholesterol as well. Omega-6 versus omega-3 essential fats have different influences on body fat mass through mechanisms of adipogenesis, lipid homeostasis, brain–gut–adipose tissue axis, and systemic inflammation [74]. In summary, lowering omega-6 and increasing omega-3 may decrease the risk for weight gain and associated inflammation.

Testing RBC fatty acid profiles is helpful to clinicians when assessing fatty acid balance and is important because AA and DHA both have critical but different biological functions. There may be instances where omega-3 fatty acid levels are too high compared with levels of arachidonic acid (AA), which may have untoward consequences. The eicosanoids produced from AA are important for immunity and immune response and serve as mediators and regulators of inflammation [77]. AA is critical for infant growth, brain development, and health. Too much DHA may suppress benefits provided by AA [77]. DHA controls signaling membranes in the photoreceptor, brain, and nervous system, whereas AA has a role in the vasculature and certain aspects of immunity [77]. Animal studies have provided evidence that both preformed DHA and AA are required for optimal cognitive function [77]. More about omega-3 dominance is found in the section on gamma-linolenic acid.

8.8.2 Gamma-Linolenic Acid

Gamma-linolenic acid (GLA) and its downstream metabolite di-homo-gamma-linolenic acid (DGLA) are two omega-6 fatty acids important to any discussion of fatty acids (please see ▶ Chaps. 10 and 11). They are often grouped into the "too much omega-6 statements" without addressing their critical importance in nutritional therapy. GLA is not present in the human diet and is formed from the essential fatty acid cis-linoleic acid [78]. It is often supplemented in the form of borage oil, evening primrose oil (EPO), or black currant seed oil. As a metabolite of the essential fatty acid linoleic acid (LA), GLA, in turn, produces DGLA. DGLA is available as a substrate to produce AA or be directed to form the prostaglandin 1 series (PG1) metabolites. The balance needed between the prostaglandin groups is another critical consideration when assessing an individual's fatty acid balance. For example, autoimmune and skin conditions are dependent on adequate PG1 molecules. Most of the PGs, TXs, and LTs are proinflammatory. AA products generally enhance inflammation as the precursor of 2-series PGs and thromboxanes (TXs) and 4-series leukotrienes (LTs) [79]. Therefore, the balance of AA to DGLA in the body may be a critical factor in regulating inflammatory processes [78].

GLA metabolism in humans is complex as all the cellular compartments metabolize GLA differently [79]. This difference is due to differential expression of PUFA-metabolizing enzymes. GLA supplementation can lead to increased DGLA levels in certain inflammatory cells but can increase both DGLA and AA in circulating lipids [79]. The impact of AA accumulation in the body remains controversial. When GLA is given with omega-3 fatty acids, the conversion of DGLA (from GLA) to AA is inhibited, and the accumulation of serum AA appears to be prevented [79].

Inflammation is the immune system's response to infection and injury and facilitates the removal of the cause of the inflammation (the offending agent) and the restoration of tissue structures and physiological function [80]. Inflammatory prostaglandins derived from AA play a key role in necessary inflammation. Therefore, our goal is not to eliminate inflammation but, rather, keep it in check. Persistent and dysfunctional inflammation may promote cancer, cardiovascular disease, autoimmune disease, and the loss of organ function [80]. The nutritional balance between GLA, DGLA, and AA and their metabolites may be key to maintaining a healthy balance of inflammation in the body.

Deficiency of fatty acids is often due to dietary insufficiency, but excessive supplementation of flax or fish oils leads to omega-3 fatty acid dominance [81]. Potential laboratory findings will reveal high omega-3 fatty acids with low AA fatty acids and elevated EPA–DGLA ratio as well as depressed AA–EPA ratios. This imbalance will produce a lowering of class 2 eicosanoid signals, which may lead to blunting of immune responses [81]. Supplementing omega-3 fatty acids in an individual with a deficiency of n-6 fatty acids can further exacerbate clinical outcomes by competing for desaturase enzymes. Adding GLA-rich evening primrose oil may improve the clinical outcome in these individuals [81]. A diet low in LA can be corrected by supplementing with LA-rich oils and/or correcting for malabsorption.

Another risk of omega-3-dominant conditions is lipid peroxidation of cell membranes and increased risk of oxidative damage in the body that may give way to serious health conditions, including heart and neurological diseases. In general, it is recommended to supplement with antioxidant nutrients when using omega-3 fatty acids to protect against lipid peroxidation and monitor fatty acid profiles and antioxidant nutrients [81].

To prevent an imbalance in fatty acids, consider supplementing the full range. When GLA, DGLA, EPA, and DHA are given together, the overproduction of pro-inflammatory cytokines IL-6 and TNFα is suppressed [78]. The combination of GLA, EPA, and DHA in supplemental form would likely induce a powerful combination of anti-inflammatory and necessary inflammatory metabolites [79]. Common genetic and epigenetic variations affect the rate of conversion of long-chain PUFAs and their metabolites and are strongly related to ethnicity, suggesting that "one size fits all" dietary and supplement recommendations for fatty acids may not be appropriate, with laboratory testing critical [79].

8.8.3 Conjugated Linoleic Acid

Conjugated linoleic acid (CLA) is another fatty acid that may have anti-inflammatory and anti-cancer benefits. CLAs are a series of isomers of LA and chemically do not belong to the omega-6 family; however, they can originate from the endogenous biohydrogenation of LA by gastrointestinal tract bacteria [82]. CLAs have demonstrated a benefit in inflammatory bowel disease (IBD) and colon cancer and may serve as an agonist for peroxisome proliferator-activated receptor gamma (PPARγ) [83]. PPARs are nuclear receptors for endogenous lipid molecules such as prostaglandins or hydroxyl-containing PUFAs [83]. Their main biological function is to sense intracellular nutrient concentrations and regulate gene expression involved in maintaining metabolic and tissue homeostasis. When CLA is used in conjunction with omega-3 fatty acids, it may prevent or ameliorate IBD in animal models [83]. In addition to activating PPAR γ, CLA can also modulate the production of AA metabolites, which may reduce the production of inflammatory lipid mediators. Several CLA isomers are found naturally in milk, cheese, and ruminant products [83]. Grass-fed beef and fatty milk products from grass-fed beef are good dietary sources of CLA.

8.8.4 Phospholipids

Any discussion on fatty acids must include a discussion on the importance of phospholipids. Phospholipids (PLs) are amphiphilic lipids (containing both hydrophilic and hydrophobic properties) found in all animal and plant cell membranes [84]. PLs are arranged as lipid bilayers with the hydrophilic regions reaching toward the outer surface of the cell membrane and the hydrophobic properties reaching toward the inner membrane compartment. PLs can be glycerophospholipids (GPLs) found in most cell membranes or sphingophospholipids (i.e., sphingomyelin or SPM), commonly found in high quantities in the brain and neural tissues [84]. Cell membranes are also composed of glycolipids and cholesterol and proteins.

GPLs can be extracted from soybeans, egg yolk, milk, or marine organisms such as fish, roe, or krill [84]. GPLs can have different fatty acid compositions. Soybean PL consists of GPLs with a high content of unsaturated FA such as LA (n-6 FA); egg yolks have mainly phosphatidylcholine (PC) as their source of PL and contain unsaturated FA, mainly oleic acid; milk GPLs have both PC and phosphatidylethanolamine (PE) as their main PL classes in addition to SPM, and the FA content consists of both unsaturated and saturated; marine GPL is mainly composed of PC and binds the unsaturated eicosapentaenoic acid (EPA) and docosahexaenoic acid (DHA) [84]. Consuming dietary GPLs with a specific FA composition may alter the FA composition of membrane PLs and affect cellular function such as signaling, transport, and activity of membrane enzymes. The possibility for modulating cell function from specific PLs could contribute to health benefits [84].

A number of possible clinical applications of using phospholipids exists. They involve reducing inflammation; improving lipid profile; increasing production of cytoprotective mucosal PGE2 in the GI tract, which may help with ulcerative colitis or the side effects of using NSAIDs; and improving cognitive function, visual function, and memory [84].

Phosphatidylserine (PS), the best studied GPL for brain performance, may be especially protective against age-related cognitive decline. PS has been shown to revitalize memory, learning, concentration, and vocabulary skills and assist with managing stress hormones and depression [84]. PS also stimulates acetylcholine synthesis, which triggers the release of neurotransmitters. Changes in the composition of cell membranes with age include an increased cholesterol–PL ratio, which may reduce the immunological function of lymphocytes. In a study on rats, lymphocyte number and macrophage phagocytic capacity were significantly improved with soy PC supplementation. PLs (purified extract of PPC from soybeans) have shown a beneficial effect with viral hepatitis and alcohol-induced liver damage. PLs have also been demonstrated to reduce total liver lipid, liver TG, and total cholesterol [84].

Chronic alcohol ingestion depletes PL in brain cell membranes and depletes antioxidant systems from the cell membranes which, in turn, promotes lipid peroxidation in the brain and other cells such as enterocytes [84].

8.8.5 Short-Chain Fatty Acids

Short-chain fatty acids (SCFA) are the major metabolic products formed by anaerobic bacterial fermentation of soluble fiber in the gastrointestinal tract and may be the link between microbiota and host tissues [85]. SCFA include acetate (C_2), propionate (C_3), and butyrate (C_4) [85]. Acetate is utilized for lipogenesis in the liver and as a fuel source once it enters the peripheral circulation; propionate is also used in the liver as a substrate for hepatic gluconeogenesis; butyrate is primarily used as a fuel source for the colonocytes [86]. Besides serving as a fuel source for the intestinal epithelial cells, SCFA also modulate electrolyte and water absorption in the GI tract and regulate the inflammatory process in the GI tract [85]. The concentration of SCFA in the GI tract and blood may predispose to or prevent illnesses such as inflammatory bowel disease (IBD), cancer, and diabetes [85]. SCFA modulate different processes such as cell proliferation and differentiation, hormone secretion of leptin and peptide YY, and immune and inflammatory responses and serve as an energy source for colonocytes, the liver, and muscle [85]. SCFA levels have been found to be lower with increasing age in humans, dogs, and mice [86].

Butyrate is the most widely studied SCFA and is predominantly produced by *Faecalibacterium prausnitzii* (*F. prausnitzii*). Butyrate serves as an energy source for colonocytes and exerting an anti-inflammatory effect in the colon [87]. Studies reported benefit when butyrate was administered for IBD-related lesions and symptoms [87]. A study by Zhang et al. in rats found that oral administration of butyrate resulted in increased percentage of butyric acid, fecal concentration of butyric acid, and overall total SCFA [87]. Studies in mice have suggested that the obese microbiota, which has a higher ratio of *Firmicutes* to *Bacteroidetes* compared with lean microbiota, produce a higher amount of SCFA compared with leaner counterparts, potentially contributing to obesity [86].

The reasons for higher SCFA in obese individuals may be due in part to lower colonic absorption of SCFA, reduced colonic transit time, or increased SCFA production due to differences in dietary intake [86]. The ratio of *Firmicutes* to *Bacteroidetes* in humans is not always higher in obese individuals; thus, the higher SCFA theory in obese individuals is not a consistent finding [86]. A study by Rahat-Rozenbloom et al. comparing obese versus lean subjects suggested that

SCFA production was higher in obese individuals, but this was not due to decreased absorption of SCFA or differences in dietary intake. Rather, this may be due to the differences in the microbiome with the obese microbiome being higher in *Firmicutes* than the lean microbiome [86]. Further research will be critical in finding the important relationships.

Increasing the production of SCFA is largely accomplished by eating a whole-foods diet high in plant foods and fiber. Butter is a natural food source of butyrate. Butter from grass-fed cows would provide the additional benefit of the anti-inflammatory effects of the fatty acid CLA. In cases such as IBD, a high-fiber diet may not be tolerated during an exacerbation of the illness, and this is where the oral or enema administration of butyrate might be of benefit. Working with an experienced integrative and functional medicine practitioner in the use of butyrate is recommended.

8.8.6 Increasing Beneficial Fatty Acids

Increasing omega-3 fatty acid intake might include eating grass-fed beef (also high in CLA), wild-caught fish, eggs from free-range chickens with omega-3 fatty acids in their feed, and supplementing with omega-3 fatty acids. Simopoulos et al. recommend lowering omega-6 by changing vegetable oils from corn, sunflower, safflower, cottonseed, and soybean oils to oils that are high in omega-3, such as flax, perilla, chia, rapeseed, and oils that are high in monounsaturated fats such as olive oil, macadamia nut oil, hazelnut oil, and increasing fish intake to two to three times per week while decreasing meat intake [74]. Processed foods are notoriously high in omega-6; therefore, consuming a whole-food, unprocessed diet is also an important factor in balancing the ratio of omega-6–omega-3 fatty acids. Increasing GLA involves supplementing with a good source of GLA, such as evening primrose oil, black currant seed oil, or borage oil. Increasing CLA can be done through supplemental sources that include a variety of isomers or eating grass-fed beef and whole-fat dairy products from grass-fed beef, such as butter, yogurt, cheese, or milk, and enhancing the microbiome through diet to support the endogenous biohydration of LA. Increasing phospholipids such as choline can occur through eating sunflower seeds, egg yolks, fish, and milk. Sunflower or soy lecithin are both sources of phospholipids (sunflower lecithin is typically recommended over soy lecithin) and dietary supplementation of phosphatidylserine and phosphatidylcholine.

8.8.7 Assessing Erythrocyte Fatty Acid Profiles

Because a "one-size-fits-all" approach to fatty acid supplementation is not optimal for balancing fatty acid nutritional status, testing erythrocyte fatty acid profiles is a helpful tool as a part of nutritional assessment. Assessing fatty acid profiles in packed erythrocytes is the most common procedure and is representative of longer-term dietary intake rather than recent dietary intake as is represented in plasma profiles of fatty acids [81]. Fatty acid analysis can also help pinpoint certain key vitamin and mineral needs. Laboratory profiles of fatty acids include more than 40 analytes that may be evaluated using patterns within families rather than assessing each individual fatty acid [81]. Individual variability of intake, digestion, absorption, and degradation can produce different fatty acid profiles and support the notion of not using a "one-size-fits-all" approach to recommending supplemental fatty acids [81]. The assessments of patterns include general fatty acid deficiency, omega-3 deficiency or excess, omega-6 deficiency or excess, hydrogenated oil toxicity, micronutrient deficiency, metabolic and genetic disorders, and fatty acid ratios and indices [81].

RBC fatty acid profiles can indicate certain micronutrient deficiencies. For example, when biotin is deficient, the ratio of vaccenic acid to palmitoleic acid is significantly lower. Vaccenic acid has large effects on membrane fluidity and inhibition of tumor growth in cell culture [81]. B12 deficiency is associated with abnormal fatty acid synthesis that results in odd-chain fatty acids building up in the lipids of the nervous system. This results in altered myelin integrity and demyelination, leading to impaired nervous system functioning [81]. Moreover, this may in part explain the neuropathy associated with cobalamin deficiency [81].

To further elucidate this metabolic process, the production of fatty acids with odd numbers of carbon atoms is initiated by propionic acid [81]. Propionic acid requires B12 to be converted into succinate; therefore, a deficiency of vitamin B12 results in the accumulation of propionate and a build-up of odd-numbered fatty acids [81]. In animal studies of biotin deficiency, abnormalities from the accumulation of propionate have been shown to occur. These abnormalities result in the buildup of odd-chain fatty acids in the plasma phospholipids, plasma, and liver in experimental animals [81]. Gut bacteria in ruminants also produce high amounts of propionate. Consequently, eating animal and dairy products may result in accumulation of odd-numbered fatty acids [81].

Zinc may be the best-known nutritional deficiency that can show up in a RBC fatty acid analysis. The desaturase enzymes needed for the conversion of fatty acid substrates into products rely on zinc. The delta-6-desaturase enzyme that converts LA to DGLA will be under-functioning in zinc deficiency; thus, an elevated ration of LA–DGLA is a sensitive marker for zinc deficiency [81]. In cases where LA-rich foods are restricted and extra flaxseed oil is used, the ALA–zinc ratio may also be elevated as that pathway requires zinc [81].

Essential fatty acid deficiency will show up as an elevation in mead acid [81]. The production of mead acid rises as EFA intake falls. Mead acid is an omega-9 PUFA produced by repeated desaturation of nonessential fatty acids. Mead acid cannot participate in eicosanoid formation but mimics membrane fluidity action of PUFAs derived from EFA.

When supplemental fatty acids are used, they must be kept from air and light to keep oxidation low. In addition, these supplements often contain vitamin E to help prevent oxidative damage to the fatty acids [81]. Erythrocyte fatty acid follow-up testing is recommended to be done after 90 days of starting interventions [81].

8.8.8 Hydrogenated Oils

Hydrogenated oils are harmful fats also known as trans-fatty acids. Hopefully, trans-fatty acids will be less of a concern in the United States due to the phasing out of these fats in the food manufacturing industry. Trans-fatty acids are harmful because, although listed as unsaturated fats, they behave like saturated fats and lead to higher cholesterol levels, and they mimic unsaturated fats by binding to desaturase enzymes and interfering with the normal production of necessary substances [81]. Trans-fatty acids have been shown to contribute to the risk of heart disease and cancer [81]. Trans-fatty acids are found in margarine, hydrogenated peanut butter, baked goods, desserts, snack foods, and crackers [81]. The main trans-fatty acid type found in hydrogenated foods is C-18 trans-fatty acids including elaidic acid, trans-vaccenic, and trans-petroselinic, followed by palmitelaidic acid.

Milk products and beef can contain some naturally occurring trans-fatty acids (elaidic acid) formed from the bacteria in the gastrointestinal tract of ruminant animals [81]. Naturally occurring trans-fatty acids are not a health concern [81].

8.8.9 Overall Diet and Macronutrient Distribution

Diet is the core treatment for many diseases, including obesity, hypertension, hyperglycemia, and dyslipidemia. Many interventions are beneficial, including replacing harmful fats with health-promoting fats; increasing fiber; increasing phytonutrient content such as flavonoids, polyphenols, and antioxidants from plant-based foods; reducing salt; and restricting calories [88]. The Mediterranean diet is one of the most studied diets and recommends high consumption of extra virgin olive oil, fruits, vegetables, nuts, seeds, legumes, cereals, moderate fish, poultry, dairy, and red wine and lower consumption of eggs, red meat, processed meat, and processed foods. The Mediterranean diet improves blood pressure, lipid profiles, insulin sensitivity, CRP, oxidative stress, atherosclerotic disease, and cognitive function [88]. The Mediterranean, pescetarian, vegetarian, or vegan diets also offer environmental benefits with decreased greenhouse gas emissions [88]. People who eat an organic, Mediterranean diet may contribute to a reduction in global environmental impact [88]. In terms of lowering of HbA1C, a systematic review found that a low carbohydrate Mediterranean diet and low-fat vegan diet were possibly the most effective [89].

To reemphasize, a one-size-fits-all diet or macronutrient recommendation is not plausible. The percentage of calories from carbohydrate, fat, and protein needs to be determined on an individual basis. Specific conditions such as chronic kidney disease will benefit from a controlled protein intake, and epilepsy can benefit from a very low-carbohydrate, high-fat ketogenic diet. For years, the field of nutrition has been bombarded with claims of various adjustments in macronutrient ratios having benefit. Nutritionists need to take into account individual disease states and give the patient the most sustainable, realistic plan to improve health.

While the ketogenic diet (low-carbohydrate, high-fat) is showing promise for various conditions including epilepsy, cancer, weight loss, type 2 diabetes, metabolic syndrome, dementia, and neurological diseases [90], the ketogenic diet can be difficult for the average person to sustain. The guidance of a ketogenic-trained nutritionist can assist in compliance. This dietary regimen can be a critical intervention to produce a successful outcome. It is important that a ketogenic or modified ketogenic diet regimen be developed for the individual and their medical condition. In this author's opinion, intermittent fasting (IF) and time-restricted feeding (TRF) may prove to have as important cardiometabolic and neurological benefits as the ketogenic diet and may be appropriate for those who cannot sustain a ketogenic diet.

Time-restricted feeding (TRF) may be a feasible alternative to calorie restriction, ketogenic diet, and intermittent fasting when compliance with more restrictive eating regimes is not likely or not recommended. TRF is an eating pattern based around circadian rhythms that occurs within a limited time span (usually 8–12 hours), with a span of 4 hours in between meals with no attention paid to calorie intake [91]. In one study, eight overweight participants consumed their entire caloric intake within a 10–11-hour window. These participants consumed 20% less calories (due to eliminating alcohol and late-night snacks) and lost 4% body weight in 16 weeks that they sustained for 1 year. The participants reported improved sleep and elevated alertness during the day [91]. TRF can impart other benefits as well. Prolonged overnight fasting (>13 hours) correlated with reduced risk of breast cancer [91]. TRF is still being investigated, and more studies are needed to validate its effectiveness.

> *Correcting nutritional imbalances by designing individualized food and dietary supplementation recommendations are the cornerstone of functional nutrition.*

8.9 Summary

Nutrient intake is a contributing factor to living with or without disease. Correcting nutritional imbalances by designing individualized food and dietary supplementation recommendations are the cornerstone of integrative and functional nutrition. When using supplements, always keep balance in mind. Retest lab parameters to ensure clients are staying in balance. Adjust dietary macronutrient ratios according to individual needs instead of using a one-size-fits-all approach (See ▶ Box 8.1).

Box 8.1 Recommendations for Assessment-Based Interventions

- Using the ingestion–digestion–utilization (IDU) worksheet can help to pinpoint which additional tests need to be ordered to better assess nutritional balance.
- Dietary intake may be assessed to see what nutrients are low in the overall dietary intake.
- A micronutrient test can look at individual nutrients in the red blood cell or plasma and RBC fatty acid status.
- Plasma or urinary amino acid testing can be used to assess amino acid status as it relates to behavioral and neurotransmitter function and overall adequacy of nutritional intake.
- Waist circumference and BMI can be used to determine what macronutrient distribution, and calorie level might be best suited for an individual.
- Nutrition physical exam and conventional lab tests such as a CBC can give an indication of protein status, electrolytes, iron, and vitamin B12 status:
  - HBA1C and fasting insulin can be a guide toward adjusting carbohydrate in the diet.
  - Lipid profiles and RBC fatty acids can be combined together to determine the type and quantity of fat and carbohydrate that needs to be included in the diet.
- Comprehensive stool and digestive analysis can give an indication as to how the person is digesting and absorbing their food and the need for digestive or pancreatic enzymes:
  - This test can also identify any pathogens that might be contributing to malabsorption.
  - The need for hydrochloric can also be determined by assessing symptoms such as gas and bloating right after meals, nail strength, easy hair pluckability, serum protein and albumin as well as looking for any protein fragments in the stool analysis.
  - Fat malabsorption may be assessed by looking at fat in the stool and serum levels of cholesterol and triglycerides that are often low normal in conditions of fat malabsorption.
- Antioxidant status can be assessed by integrative nutritional testing that looks at lipid peroxidation and glutathione status and levels of CoQ10, vitamin E, selenium, and zinc.
  - Simple questioning that can assess the quantity of fruits and vegetables, nuts, and legume servings can also give an indication of the level of antioxidants and phytonutrients the person is consuming.
- Vitamins A and D may be assessed by testing serum retinol levels and vitamin D3 (25, OH).
- Ferritin levels may give a more accurate assessment of iron status than hemoglobin or hematocrit.
- Dietary and laboratory assessment and a nutrition physical exam are key tools for identifying clinical and subclinical nutrient deficiencies that could be derailing an individual's ability to achieve an optimal and more functional state of health:
  - Optimizing nutritional status can be pivotal for helping people turn the corner with their health challenges.

References

1. Gonzalez MJ, Miranda Massari JR. Metabolic correction: a functional explanation of orthomolecular medicine. J Orthomole Med. 2012;27(1):13–20.
2. Heaney RP. Guidelines for optimizing design and analysis of clinical studies of nutrient effects. Nutr Rev. 2014;72(1):48–54.
3. Rosanoff A, Dai Q, Shapses SA. Essential nutrient interactions: does low or suboptimal magnesium status interact with vitamin D and/or calcium status? Adv Nutr. 2016;7(1):25–43.
4. Arslan N. Obesity, fatty liver disease and intestinal microbiota. World J Gastroenterol. 2014;20(44):16452–63. https://doi.org/10.3748/wjg.v20.i44.16452.
5. Rashid T, Ebringer A. Autoimmunity in rheumatic diseases is induced by microbial infections via crossreactivity or molecular mimicry. Autoimmun Dis. 2012;2012:539282.
6. David LA, Maurice CF, Carmody RN, Gootenberg DB, Button JE, Wolfe BE, et al. Diet rapidly and reproducibly alters the human gut microbiome. Nature. 2014;505(7484):559–63.
7. El-Salhy M, Ystad SO, Mazzawi T, Gundersen D. Dietary fiber in irritable bowel syndrome (review). Int J Mol Med. 2017;40(3):607–13.
8. Kieffer DA, Martin RJ, Adams SH. Impact of dietary fibers on nutrient management and detoxification organs: gut, liver, and kidneys. Adv Nutr. 2016;7(6):1111–21.
9. Navarro SL, Neuhouser ML, Cheng TD, Tinker LF, Shikany JM, Snetselaar L, et al. The interaction between dietary fiber and fat and risk of colorectal cancer in the women's health initiative. Nutrients. 2016;8(12)
10. Smits SA, Leach J, Sonnenburg ED, Gonzalez CG, Lichtman JS, Reid G, et al. Seasonal cycling in the gut microbiome of the Hadza hunter-gatherers of Tanzania. Science (New York, NY). 2017;357(6353):802–6.
11. Mirmiran P, Bahadoran Z, Khalili Moghadam S, Zadeh Vakili A, Azizi F. A prospective study of different types of dietary fiber and risk of cardiovascular disease: Tehran Lipid and Glucose Study. Nutrients. 2016;8(11)
12. Kranz S, Dodd KW, Juan WY, Johnson LK, Jahns L. Whole grains contribute only a small proportion of dietary fiber to the U.S. diet. Nutrients. 2017;9(2)
13. Cozma-Petrut A, Loghin F, Miere D, Dumitrascu DL. Diet in irritable bowel syndrome: what to recommend, not what to forbid to patients! World J Gastroenterol. 2017;23(21):3771–83.
14. Ho KS, Tan CY, Mohd Daud MA, Seow-Choen F. Stopping or reducing dietary fiber intake reduces constipation and its associated symptoms. World J Gastroenterol. 2012;18(33):4593–6.
15. Niwattisaiwong S, Burman KD, Li-Ng M. Iodine deficiency: clinical implications. Cleve Clin J Med. 2017;84(3):236–44.
16. Prete A, Paragliola RM, Corsello SM. Iodine supplementation: usage "with a grain of salt". Int J Endocrinol. 2015;2015:312305.
17. Luo Y, Kawashima A, Ishido Y, Yoshihara A, Oda K, Hiroi N, et al. Iodine excess as an environmental risk factor for autoimmune thyroid disease. Int J Mol Sci. 2014;15(7):12895–912.
18. Institute of Medicine (US) Panel on Micronutrients. Dietary reference intakes for vitamin A, vitamin K, arsenic, boron, chromium, copper, iodine, iron, manganese, molybdenum, nickel, silicon, vanadium, and zinc. Washington, DC: National Academy Press; 2001.

19. Hosur MB, Puranik RS, Vanaki S, Puranik SR. Study of thyroid hormones free triiodothyronine (FT3), free thyroxine (FT4) and thyroid stimulating hormone (TSH) in subjects with dental fluorosis. Eur J Dent. 2012;6(2):184–90.
20. Butt CM, Stapleton HM. Inhibition of thyroid hormone sulfotransferase activity by brominated flame retardants and halogenated phenolics. Chem Res Toxicol. 2013;26(11):1692–702.
21. Kennedy DO. B vitamins and the brain: mechanisms, dose and efficacyDOUBLEHYPHENa review. Nutrients. 2016;8(2):68.
22. National Institutes of Health OoDS. Thiamin Fact Sheet for Health Professionals [cited 2017 9/22/2017]. Available from: https://ods.nih.gov/factsheets/Thiamin-HealthProfessional.
23. National Institutes of Health OoDS. Riboflavin fact sheet for health professionals: National Institutes of Health; 2018 [cited 2017 9/22/2017]. Available from: https://ods.nih.gov/factsheets/Riboflavin-HealthProessional.
24. Boehnke C, Reuter U, Flach U, Schuh-Hofer S, Einhaupl KM, Arnold G. High-dose riboflavin treatment is efficacious in migraine prophylaxis: an open study in a tertiary care centre. Eur J Neurol. 2004;11(7):475–7.
25. Lab C. Recommended daily intakes and upper limits for nutrients [cited 2017 9/22/2017]. Available from: https://www.consumerlab.com/RDAs.
26. Clements RS Jr, Darnell B. Myo-inositol content of common foods: development of a high-myo-inositol diet. Am J Clin Nutr. 1980;33(9):1954–67.
27. Lord RB, Biotin JA. Laboratory evaluations for integrative and functional medicine. 2nd ed. Duluth: Metametrix Institute; 2008. p. 39.
28. McCully KS. Homocysteine and the pathogenesis of atherosclerosis. Expert Rev Clin Pharmacol. 2015;2(Mar 8):211–9.
29. Bailey LB, Stover PJ, McNulty H, Fenech MF, Gregory JF 3rd, Mills JL, et al. Biomarkers of nutrition for development-folate review. J Nutr. 2015;145(7):1636S–80S.
30. National Institutes of Health OoDS. Niacin Fact Sheet for Health Professionals. 2019.
31. Woolf K, Hahn NL, Christensen MM, Carlson-Phillips A, Hansen CM. Nutrition assessment of B-vitamins in highly active and sedentary women. Nutrients. 2017;9(4):329.
32. Ueland PM, Ulvik A, Rios-Avila L, Midttun Ø, Gregory JF. Direct and functional biomarkers of vitamin B6 status. Annu Rev Nutr. 2015;35:33–70.
33. Hannibal L, Lysne V, Bjørke-Monsen A-L, Behringer S, Grünert SC, Spiekerkoetter U, et al. Biomarkers and algorithms for the diagnosis of Vitamin B12 deficiency. Front Mol Biosci. 2016;3:27.
34. Majumdar R, Yori A, Rush PW, Raymond K, Gavrilov D, Tortorelli S, et al. Allelic spectrum of formiminotransferase-cyclodeaminase gene variants in individuals with formiminoglutamic aciduria. Mol Genet Genomic Med. 2017;5(6):795–9.
35. Fratoni V, Brandi ML. B vitamins, homocysteine and bone health. Nutrients. 2015;7(4):2176–92.
36. Costantini A, Pala MI. Thiamine and Hashimoto's thyroiditis: a report of three cases. J Altern Complementary Med. 2014;20(3):208–11.
37. Sachdeva A, Chandra M, Choudhary M, Dayal P, Anand KS. Alcohol-related dementia and neurocognitive impairment: a review study. Int J High Risk Behav Addict. 2016;5(3):e27976.
38. McAuley E, McNulty H, Hughes C, Strain JJ, Ward M. Riboflavin status, MTHFR genotype and blood pressure: current evidence and implications for personalised nutrition. Proc Nutr Soc. 2016;75(3):405–14.
39. Leahy L. Vitamin B supplementation: what's the right choice for your patients? J Psychosoc Nurs Ment Health Serv. 2017;55(7):7–11.
40. Mozos I, Stoian D, Luca CT. Crosstalk between vitamins A, B12, D, K, C, and E status and arterial stiffness. Dis Markers. 2017;2017:8784971.
41. Theuwissen E, Smit E, Vermeer C. The role of vitamin K in soft-tissue calcification. Adv Nutr. 2012;3(2):166–73.
42. Fusaro M, Noale M, Viola V, Galli F, Tripepi G, Vajente N, et al. Vitamin K, vertebral fractures, vascular calcifications, and mortality: Vitamin K Italian (VIKI) dialysis study. J Bone Miner Res. 2012;27(11):2271–8.
43. Joo NS, Yang SW, Song BC, Yeum KJ. Vitamin A intake, serum vitamin D and bone mineral density: analysis of the Korea National Health and nutrition examination survey (KNHANES, 2008-2011). Nutrients. 2015;7(3):1716–27.
44. Meganathan P, Fu JY. Biological properties of tocotrienols: evidence in human studies. Int J Mol Sci. 2016;17(11)
45. Selvaraju TR, Khaza'ai H, Vidyadaran S, Abd Mutalib MS, Vasudevan R. The neuroprotective effects of tocotrienol rich fraction and alpha tocopherol against glutamate injury in astrocytes. Bosn J Basic Med Sci. 2014;14(4):195–204.
46. Tankeu AT, Ndip Agbor V, Noubiap JJ. Calcium supplementation and cardiovascular risk: a rising concern. J Clin Hypertens (Greenwich). 2017;19:640.
47. Di Stefano M, Mengoli C, Bergonzi M, Corazza GR. Bone mass and mineral metabolism alterations in adult celiac disease: pathophysiology and clinical approach. Nutrients. 2013;5(11):4786–99.
48. Selby PL, Davies M, Adams JE, Mawer EB. Bone loss in celiac disease is related to secondary hyperparathyroidism. J Bone Miner Res. 1999;14(4):652–7.
49. Mirza F, Canalis E. Management of endocrine disease: secondary osteoporosis: pathophysiology and management. Eur J Endocrinol. 2015;173(3):R131–51.
50. Sahmoun AE, Singh BB. Does a higher ratio of serum calcium to magnesium increase the risk for postmenopausal breast cancer? Med Hypotheses. 2010;75(3):315–8.
51. Hruby A, O'Donnell CJ, Jacques PF, Meigs JB, Hoffmann U, McKeown NM. Magnesium intake is inversely associated with coronary artery calcification: the Framingham Heart Study. JACC Cardiovasc Imaging. 2014;7(1):59–69.
52. Huang JH, Tsai LC, Chang YC, Cheng FC. High or low calcium intake increases cardiovascular disease risks in older patients with type 2 diabetes. Cardiovasc Diabetol. 2014;13:120.
53. Park J, Kwock CK, Yang YJ. The effect of the sodium to potassium ratio on hypertension prevalence: a propensity score matching approach. Nutrients. 2016;8(8)
54. Cai X, Li X, Fan W, Yu W, Wang S, Li Z, et al. Potassium and obesity/metabolic syndrome: a systematic review and meta-analysis of the epidemiological evidence. Nutrients. 2016;8(4):183.
55. Kieneker LM, Gansevoort RT, de Boer RA, Brouwers FP, Feskens EJ, Geleijnse JM, et al. Urinary potassium excretion and risk of cardiovascular events. Am J Clin Nutr. 2016;103(5):1204–12.
56. Jenkins DJ, Jones PJ, Frohlich J, Lamarche B, Ireland C, Nishi SK, et al. The effect of a dietary portfolio compared to a DASH-type diet on blood pressure. Nutr Metab Cardiovasc Dis. 2015;25(12):1132–9.
57. Plum LM, Rink L, Haase H. The essential toxin: impact of zinc on human health. Int J Environ Res Public Health. 2010;7(4):1342–65.
58. Rowin J, Lewis SL. Copper deficiency myeloneuropathy and pancytopenia secondary to overuse of zinc supplementation. J Neurol Neurosurg Psychiatry. 2005;76(5):750–1.

59. Lord RB, Bralley JA. Laboratory evaluations for integrative and functional medicine. Duluth: Metametrix Institute; 2008. p. 94–101.
60. Sloan J, Feyssa E. Copper. Medscape. Feb. 16th 2017. Retrieved from: https://emedicine.medscape.com/article/2087780-overview.
61. Gilroy RK. Wilson disease workup: Medscape; 2016 [updated October 18, 2016; cited 2017 September 9, 2017]. Available from: https://emedicine.medscape.com/article/183456-workup.
62. Ray CS, Singh B, Jena I, Behera S, Ray S. Low alkaline phosphatase (ALP) in adult population an indicator of zinc (Zn) and magnesium (mg) deficiency. Curr Res Nutr Food Sci. 2017;5(3):347–52.
63. Kogan S, Sood A, Granick M. Zinc and wound healing. Wounds. 2017;29(4):102–6.
64. Bjorklund G. The role of zinc and copper in autism spectrum disorders. Acta Neurobiol Exp. 2013;73(2):225–36.
65. Malavolta M, Piacenza F, Basso A, Giacconi R, Costarelli L, Mocchegiani E. Serum copper to zinc ratio: relationship with aging and health status. Mech Ageing Dev. 2015;151:93–100.
66. Collins JF, Klevay LM. Copper. Adv Nutr. 2011;2(6):520–2.
67. Moss HE. Bariatric surgery and the neuro-ophthalmologist. J Neuro Ophthalmol. 2016;36(1):78–84.
68. Gaier ED, Kleppinger A, Ralle M, Mains RE, Kenny AM, Eipper BA. High serum Cu and Cu/Zn ratios correlate with impairments in bone density, physical performance and overall health in a population of elderly men with frailty characteristics. Exp Gerontol. 2012;47(7):491–6.
69. Harvey L, Ashton K, Hooper L, Casgrain A, Fairweather-Tait S. Methods of assessment of copper status in humans: a systematic review. Am J Clin Nutr. 2009;89:2009S–20024S.
70. Goodman VL, Brewer GJ, Merajver SD. Copper deficiency as an anticancer strategy. Endocr Relat Cancer. 2004;11(2):255–63.
71. Lee R, Neiman D. Nutritional assessment. 6th ed. New York: McGraw-Hill; 2013.
72. Bui VQ, Marcinkevage J, Ramakrishnan U, Flores-Ayala RC, Ramirez-Zea M, Villalpando S, et al. Associations among dietary zinc intakes and biomarkers of zinc status before and after a zinc supplementation program in Guatemalan schoolchildren. Food Nutr Bull. 2013;34(2):143–50.
73. Diez M, Cerdan F, Arroyo M, Balibrea L. Use of the copper/zinc ratio in the diagnosis of lung cancer. Cancer. 1989;63(4):726–30.
74. Simopoulos AP. An increase in the Omega-6/Omega-3 fatty acid ratio increases the risk for obesity. Nutrients. 2016;8(3):128.
75. Carney RM, Steinmeyer BC, Freedland KE, Rubin EH, Rich MW, Harris WS. Baseline blood levels of omega-3 and depression remission: a secondary analysis of data from a placebo-controlled trial of omega-3 supplements. J Clin Psychiatry. 2016;77(2):e138–43.
76. Cahyaningrum F, Permadhi I, Ansari MR, Prafiantini E, Rachman PH, Agustina R. Dietary optimisation with omega-3 and omega-6 fatty acids for 12-23-month-old overweight and obese children in urban Jakarta. Asia Pac J Clin Nutr. 2016;25(Suppl 1):S62–s74.
77. Hadley KB, Ryan AS, Forsyth S, Gautier S, Salem N Jr. The essentiality of arachidonic acid in infant development. Nutrients. 2016;8(4):216.
78. Das UN. n-3 fatty acids, gamma-linolenic acid, and antioxidants in sepsis. Crit Care. 2013;17(2):312.
79. Sergeant S, Rahbar E, Chilton FH. Gamma-linolenic acid, Dihommo-gamma linolenic, eicosanoids and inflammatory processes. Eur J Pharmacol. 2016;785:77–86.
80. Ricciotti E, FitzGerald GA. Prostaglandins and inflammation. Arterioscler Thromb Vasc Biol. 2011;31(5):986–1000.
81. Lord RB, Bralley JA. Fatty acids. In: Institute M, editor. Laboratory evaluations for integrative and functional medicine. 2nd ed. Duluth: Metametrix Institute; 2008. p. 269–317.
82. Xu Y, Qian SY. Anti-cancer activities of omega-6 polyunsaturated fatty acids. Biom J. 2014;37(3):112–9.
83. Bassaganya-Riera J, Hontecillas R. Dietary conjugated linoleic acid and n-3 polyunsaturated fatty acids in inflammatory bowel disease. Curr Opin Clin Nutr Metab Care. 2010;13(5):569–73.
84. Kullenberg D, Taylor LA, Schneider M, Massing U. Health effects of dietary phospholipids. Lipids Health Dis. 2012;11:3.
85. Vinolo MA, Rodrigues HG, Nachbar RT, Curi R. Regulation of inflammation by short chain fatty acids. Nutrients. 2011;3(10):858–76.
86. Rahat-Rozenbloom S, Fernandes J, Gloor GB, Wolever TM. Evidence for greater production of colonic short-chain fatty acids in overweight than lean humans. Int J Obes. 2014;38(12):1525–31.
87. Zhang M, Zhou Q, Dorfman RG, Huang X, Fan T, Zhang H, et al. Butyrate inhibits interleukin-17 and generates Tregs to ameliorate colorectal colitis in rats. BMC Gastroenterol. 2016;16(1):84.
88. Wade AT, Davis CR, Dyer KA, Hodgson JM, Woodman RJ, Keage HA, et al. A mediterranean diet to improve cardiovascular and cognitive health: protocol for a randomised controlled intervention study. Nutrients. 2017;9(2)
89. Dussaillant C, Echeverria G, Urquiaga I, Velasco N, Rigotti A. Current evidence on health benefits of the mediterranean diet. Revista medica de Chile. 2016;144(8):1044–52.
90. Boison D. New insights into the mechanisms of the ketogenic diet. Curr Opin Neurol. 2017;30(2):187–92.
91. Longo VD, Panda S. Fasting, circadian rhythms, and time-restricted feeding in healthy lifespan. Cell Metab. 2016;23(6):1048–59.

IFMNT NIBLETS Nutrition Assessment Differential

Robyn Johnson and Lauren Hand

9.1 Introduction – 124

9.2 N: Nutrient Deficiencies/Insufficiencies – 124

9.3 I: Inflammation/Immunity – 125

9.4 B: Biochemical Individuality and Genetic/Epigenetic Influences on Chronic Disease – 127

9.5 L: Lifestyle Factors – 127

9.6 E: Energy – 128

9.7 T: Toxic Load – 129

9.8 S: Stress and Sleep – 129

9.9 Sleep – 130

9.10 Conclusion – 131

References – 131

© Springer Nature Switzerland AG 2020
D. Noland et al. (eds.), *Integrative and Functional Medical Nutrition Therapy*,
https://doi.org/10.1007/978-3-030-30730-1_9

9.1 Introduction

An integrative assessment and diagnosis of each individual is fundamental to identifying root causes for an appropriate intervention. Because of the complexity of human metabolism and the diverse influences, the NIBLETS assessment was developed as a way to assess factors that are affecting nutrient metabolism. The NIBLETS model is named for its seven components:

- *Nutrient deficiencies/insufficiencies*
- *Inflammation/immunity*
- *Biochemical individuality (genetic/epigenetic influences on chronic disease)*
- *Lifestyle factors*
- *Energy*
- *Toxic load*
- *Stress and sleep*

Historically, nutrition assessment focused on evidence-based research, nutrient requirements of populations, and dietary intake. IFMNT expands this population-focused nutrition assessment by assessing the individual and their unique nutrition status at the time of the assessment. The individual's nutrition status becomes the foundation on which interventions are based, within the safety of evidence-based medicine and clinical expertise. Many of the tools that IFMNT practitioners have found useful in performing a NIBLETS assessment are included in this textbook, with further description of each of the NIBLETS categories.

One of the most beneficial tools used by the IFMNT practitioner when assessing nutrient status is laboratory testing of various components of the body: blood, saliva, urine, breath, stool, and others. Integrative laboratory testing can reveal the deeper, unseen, and otherwise unidentified issues that linger underneath the surface. However, a practitioner can begin a differential nutrition assessment through the NIBLETS lens prior to receiving any test results, immediately after hearing a patient's story. Whereas the patient's story will cover their medical history, this part of the assessment considers the current modifiable factors and clinical imbalances that influence an individual's health. To treat a person rather than a condition requires a thorough assessment of external factors negatively impacting that person's health. Regardless of the advances in pharmaceutical and nutraceutical support the medical community is able to make, patients will find themselves in a similar position once again if interventions are not made to change the patterns that helped cause the problem in the first place. In order to inform these changes, a thorough assessment and diagnosis must first be made. In this section we will outline the seven components of the NIBLETS model and discuss the benefit of using the integrative NIBLETS assessment for determining an individual's nutrition and metabolic status.

9.2 N: Nutrient Deficiencies/Insufficiencies

Medical history/current condition/diagnosis When assessing an individual patient, the patient's medical history and current condition or diagnosis can guide initial considerations, based on evidenced nutrient foundations for that condition or diagnosis. For example, for a patient presenting with a diagnosis of heart disease, nutritional-metabolic concerns would focus on lipid metabolism, vitamin D, and inflammation. If a patient presented with type 2 diabetes mellitus, one would consider carbohydrate metabolism, vitamin D, and nutrients required for glucose/insulin metabolism. Each health condition will have nutrient-related considerations that have shown particularly strong associations or causal relationships. The assessment investigation can begin with these considerations.

Dietary intake The foundation of any person's nutritional sufficiency is his/her dietary intake to meet individual biochemical nutrient needs, which are influenced by genetics and environmental stressors. The Standard American Diet, consisting of mostly nutrient-poor processed and fast foods, has left the majority of the public overfed and undernourished. While over 70% of US adults are considered clinically overweight or obese, theoretically consuming more calories than needed, they are still not consuming adequate amounts of essential nutrients. Three-quarters of Americans consume less than the recommended intake of vegetables and fruits, and that is when fried potatoes and refined juices are included in these counts [1]. The majority of Americans are not even reaching the estimated average requirement or adequate intake of potassium, fiber, choline, magnesium, calcium, and vitamins A, D, E, and C [1]. Meanwhile, these standards for nutrient consumption are currently based on average needs for our population to remain in a disease-free state. For many individuals, nutrient needs may actually be much higher for optimal function, driving the chasm between need and intake even wider.

Reliance on convenience foods and the dissipation of the family dinner table is certainly a lifestyle factor affecting our nutrient status. Within the span of less than 30 years starting in the late 1970s, Americans' consumption of food prepared outside the home, as a percentage of total calories consumed, nearly doubled [2]. This has a direct correlation with decline in diet quality [3]. However, even within the home, the standard American way of eating is heavy in refined grains and added sugars and increasingly lacking in whole, nutrient-dense foods. The current hurried lifestyle pattern in the USA leaves very little time for the food preparation involved in cooking tasty meals containing nutrient-dense foods such as quality proteins and vegetables. In 2014, the average American spent roughly half an hour each day on preparing food and cleaning up. However, between those who used any restaurant foods on a regular basis and those who did not, there was a 30-minute discrepancy in daily time spent on food preparation [4]. Considering our currently overworked, under-rested, hurried lifestyles, it is no wonder why these convenience foods are highly sought after. For this reason, intervening to improve diet quality will undoubtedly require directives on addressing these time and budget constraints (see ▶ Chap. 2).

In addition to the typical dietary patterns of the Standard American Dieter, those who follow restrictive diets such as vegetarianism and veganism are at risk for further nutrient deficiencies. While many factors may lead a person to this dietary pattern including moral, religious, or even health-based motivations, it is still worth noting the increased risk. In addition to the aforementioned nutritional deficiencies for those with a Western-style diet, those who avoid animal products are at further risk of inadequate intakes of protein, omega-3 fatty acids, vitamin B12, iron, zinc, vitamin K2, and iodine. While a nutritionally adequate diet can still be achieved, special attention should be given to those adopting this dietary pattern to address the higher risk.

Laboratory testing Lab testing is a valuable component of a nutrition assessment, as it can provide insight and information that is hard to identify through symptoms alone (i.e., infections, nutrient deficiencies, hormonal imbalances, etc.). While there are many different forms of testing – such as micronutrients, stool, urine, and saliva – not all are equal, nor do they test the same things. For example, a urine organic acid test is valuable to assess a patient's methylation function, while a comprehensive stool analysis is valuable to assess a patient's gut health and microbiome status. Both tests are valuable, but both might not be necessary for every patient. During a nutrition assessment, the practitioner can gather information and use clinical judgment to determine which laboratory testing would provide additional data necessary for that patient.

There is no "perfect" method of testing for nutrients. Many forms of testing are currently available, including serum, red blood cell (RBC), lymphocyte, stool, urine, and hair. Labs are a tool to be used in conjunction with the rest of the nutrition assessment. If laboratory testing is ordered, even if it isn't direct nutrient testing, the results can provide guidance to the practitioner on what dietary, lifestyle, or supplement changes need to be made.

IFMNT practitioners look beyond population guidelines. Instead, they look at the individual and work with each individual to optimize the body. Identification of an individual's current nutrient status via laboratory testing can provide answers and guidance for a proper intervention for the patient.

9.3   I: Inflammation/Immunity

The inflammatory and immune responses to injury and infection are essential for host defense and survival. However, these responses carry a significant metabolic burden, driving the need for certain nutrients. Furthermore, both micro- and macronutrient deficiencies can cause immune dysfunction, increasing the likelihood of inappropriate inflammation or infection. Therefore, since the relationship between nutrient status and immunity is bidirectional, assessing a patient for both is crucial to determining an appropriate and effective intervention.

Acute immune response requires many different nutrients for both initiation and resolution. Adequate nutrition is needed for both adaptive and innate immune responses. Undernutrition is recognized worldwide as a major contributor to poor survival in developing countries. However, even in populations of adequate food access, hospitals are screening for malnutrition because of its association with increased complications and mortality [5]. In the case of starvation, immune function can be impaired quickly because of the lack of nutritional reserve within immune cells. Undernutrition is known to alter both innate and adaptive immune responses, including T-cell proliferation and function [6]. However, micronutrient status has been one vastly overlooked area, particularly in acute settings, though two-thirds of patients hospitalized in an infectious disease clinic were found to be deficient in at least one nutrient [7]. Vitamin A, vitamin D, vitamin E, zinc, omega-3 fatty acids, and omega-6 fatty acids are among the most-studied nutrients active in the immune response, though many other nutrients play integral parts in the optimal function of the immune system. Therefore, a complete nutritional assessment would include looking for macro- and micronutrient deficiencies.

However, the relationship between inflammation and nutrition is rather complex and can lead to a cyclical issue. While malnutrition certainly increases susceptibility to illness, inflammation can in turn promote nutrient deficiencies. Many immune or inflammatory processes drive consumption of certain nutrients, such as in the case of conditionally essential amino acids. Other inflammatory conditions, such as in the case of infection, can also indirectly cause nutrient deficiencies by increasing intestinal permeability or anorexia. Therefore, using an integrative assessment can help pinpoint the root cause of the issue so it can most effectively be addressed.

Prolonged inflammation contributes to the progression of many chronic diseases, such as types 1 and 2 diabetes mellitus, cardiovascular disease, Alzheimer's, psoriasis, lupus, and more. Modifiable lifestyle factors including diet, smoking, and exercise have profound impacts on these levels of inflammation. While acute inflammation is necessary for healing, chronic, low levels of inflammation appear to be major players in chronic disease etiology. Immune dysregulation and oxidative stress are closely related to inflammation in the body, with common etiology and positive feedback loops, compounding the detrimental impacts. For many, this inflammation may go undetected for years, leading it to be termed the "silent killer" by scientific and media reports. Regardless of a symptom manifestation, chronic inflammation poses a damaging impact, increasing risk of chronic disease and mortality. Western lifestyle behaviors such as poor diet, inactivity, and smoking are often underlying root causes of this inflammation.

Diet is a major player in this underlying inflammation. Red meat consumption, for instance, has shown to stimulate the production of the prooxidant trimethylamine-N-oxide (TMAO) 3.7 times more than chicken in rat models [8]. Excess consumption alone can be a stimulant for inflammation. Hyperinsulinemia caused by excess carbohydrates and calories has a pro-inflammatory effect. High-carbohydrate diets have been associated with a greater likelihood of having an elevated hs-CRP in postmenopausal women [9]. Furthermore, excess adiposity, particularly in regard to white and/or visceral adipose tissue, contributes to the circulation of pro-inflammatory adipokines such as leptin and is associated with increased levels of interleukin-6 and tumor necrosis factor alpha [10].

Immune responses to both acute and chronic inflammatory responses involve complex molecular signaling. One major pathway at play involves the eicosanoids, which are derived from polyunsaturated fatty acids. The relative production of anti-inflammatory prostaglandin eicosanoid (PGE)-1 or PGE-3 versus pro-inflammatory PGE-2 directly modulates the inflammatory state. Because PGE-3 is derived from omega-6 fatty acids and PGE-2 from omega-3 fatty acids, balancing fatty acid intake is crucial to creating an optimal environment. These two fatty acids and their subsequent metabolites compete for space within the cellular membrane, cleavage from the cell by phospholipase A2, and further metabolism by delta-5 and -6 desaturases. While current dietary guidelines for Americans recommend increasing intake of all polyunsaturated fatty acids, evidence suggests that increasing omega-6s without increasing omega-3s increases the risk of CVD events, further confirming how critical the appropriate ratio is for balancing inflammation. In fact, having a higher omega-6/omega-3 ratio in the diet has shown association with a higher level of CRP [11]. Though determining an optimal ratio for PUFAs is difficult, experts suggest a range between 1:1 and 4:1 in omega-6 to omega-3 FAs [12]. Because of the omega-6 fatty acids in cheap oils such as corn, soybean, and cottonseed used in processed and restaurant foods, the Western diet has an estimated 20:1 ratio, leaving its consumers with cells primed for propagating inflammation.

Conversely, diet can be used therapeutically to reduce chronic inflammation. Those who followed a Mediterranean or Paleolithic pattern of dieting have been found to have lower levels of hs-CRP [13]. Similarly, those who had higher scores on the Alternative Healthy Eating Index or were able to improve their scores over a 6-year period had lower levels of IL-6 [14]. This is likely related to the high antioxidant and fiber intake found within these dietary patterns. Antioxidants work to quench free radicals, halting the propagation of lipid peroxidation. Having a higher intake of foods rich in antioxidants relative to those devoid of antioxidants alone is associated with a lower C-reactive protein level, as well as a decrease in systolic and diastolic blood pressure [15]. As noted previously, omega-3 fatty acids have a large anti-inflammatory effect because of their eicosanoid metabolites and downstream resolvins and protectins, which have an inflammation-resolving impact. Furthermore, negative attributes of the Western lifestyle may act synergistically to either counteract or compound the inflammation. Mice exposed to cigarette smoke have shown increased oxidation levels when fed a high-refined-carbohydrate diet [16] and protection from the oxidative damage when given omega-3 fatty acids [17].

Beyond the direct influence on inflammation, lifestyle factors also influence our immune system through various mechanisms. Generally speaking, obesity preferences Th1 and Th17 dominance over Th2, promoting inflammation. Visceral adiposity, in particular, appears to contribute to this immune and inflammatory dysregulation because of the macrophage infiltration of the adipocytes. These macrophages promote the Th1 dominance through the release of IL-6 and IL-1 and TNF-alpha [18]. Furthermore, the health of the gastrointestinal system is a major factor in determining the health of our immune system. It is known as our first line of defense against pathogens. Unfortunately, the Western style of living with stress, poor diet, and toxin exposure has led to the rise in intestinal permeability, impacting the integrity of the gut lining. This leads to a greater susceptibility of pathogens to escape defenses of the gut and enter circulation. Diet, by promotion of a healthy microbiome, affects adaptive immunity. A healthy diet rich in fiber leads to a rise in beneficial bacteria that produce metabolites acting as ligands on aryl hydrocarbon receptors (AHR) found in the mucosa of the intestinal lining. The activation of AHR promotes the differentiation of T cells into T-regs. Nutrients such as vitamins A and D also have direct impact on the adaptive immune system. Vitamin A is needed for stability of Th1 response, in such a way that a deficiency has been shown to stimulate Th1 in allergic diseases. Synergistically, vitamin D when binding to a vitamin D receptor heterodimerizes with vitamin A receptor RXR, promoting T-cell differentiation to T-regs and inhibiting IgE synthesis [18].

Considering the complexities of the relationship between inflammation and nutrition, a deeper, integrative assessment of an individual is necessary for determining nutrient status. First of all, the discovery of chronic inflammation in most cases should indicate a greater nutrient need, due to its impact on malabsorption in the gut. However, determining adequate status can also prove difficult. For instance, several nutrient laboratory values are known to be altered in inflammatory conditions, mostly because of changes in acute-phase reactants. Vitamin A is a one example of this. Serum levels of vitamin A do not accurately reflect liver stores during an inflammatory response. Therefore, assessment of vitamins without knowledge of inflammatory status could lead to inappropriate supplementation. Conversely, ferritin, the storage form of iron, is elevated during times of inflammation, potentially masking a deficiency. For this reason, an integrative assessment demands looking at the inflammatory and immune status in conjunction with nutrient status to determine the patient need (Table 9.1).

Table 9.1 Assessment for inflammatory/immune status

Assessment category	Markers of inflammation/immune status
Biomarkers	Hs-CRP, fibrinogen, myeloperoxidase, WBC,
Physical exam	Large waist circumference, excess adiposity, edema
Symptoms	Chronic pain/aches, bloating, swelling, weight gain

9.4 B: Biochemical Individuality and Genetic/Epigenetic Influences on Chronic Disease

IFMNT skills start with the medical condition or symptoms, and then an assessment is completed using the NIBLETS differential. The evidence gathered is the baseline to consider corrective interventions where nutritional injuries/altered nutrition status have been identified. With technological and scientific discoveries of the past few decades, the tools of assessment have expanded tremendously. Biochemist Roger Williams was the first to gain recognition for the term "biochemical individuality" after he published his book *Biochemical Individuality* in 1965. He led a movement toward understanding the origin of disease through the concept of biochemical individuality, which he defines as "the anatomical and physiological variations among people and how they relate to their individual responses to the environment." He then related this individuality to differing nutritional needs for optimal wellness. Another breakthrough in healthcare came with the recognition that nutritional status and needs can be influenced by genetics. This is another important component in biochemical individuality and has received a significant amount of attention and research over the past several decades. It is well recognized that genetics affect disease risk and are influenced by environment, lifestyle, and nutritional factors.

Before nutrigenomic data was available, inherited or disease tendencies within a family were gleaned assembling a "family history," which would hint at metabolic tendencies that should be considered for an individual. With the discovery of the Human Genome in 2008, the research on nutrigenomics has grown, and so has the availability of genetic testing and analysis (see ▶ Chaps. 17 and 18). Genetic testing can be extremely valuable information for individualizing or personalizing assessment, as it gives the provider an understanding of molecular-level interactions between nutrients and the patient's genome. If a patient is not responding to lifestyle and diet changes, genetic data can provide more explanations as to why or provide more information on the next intervention to make. Genetic and epigenetic information is an extremely valuable component of the patient's story.

Though genetics are not modifiable, lifestyle habits can modify gene expression through epigenetics. As with any other body system, practitioners must not solely treat the genes but must also consider the many additional factors that influence gene expression. Environment, food, water, stress, gut health, and toxin load can act on various genes both directly and indirectly and alter gene expression. The degree to which any lifestyle factors influence disease states may depend on an individual's genetic makeup.

Genetic information provides additional answers to the question of "why?" Gene expression can be altered by SNPs, nutrient insufficiency or toxicity, toxin overload, and many other factors. Aberrant genes do not, in and of themselves, cause disease. It is now recognized that human genotypes are transformed into human phenotypes as a consequence of environment and lifestyle factors that determine health or disease patterns. The topic of epigenetics and genomics is covered in this textbook in ▶ Chaps. 17 and 18.

The combination of certain genetic polymorphisms and epigenetic factors may require nutrient support beyond the recommended daily allowance (RDA) if genetic polymorphisms are present and the nutrition assessment or lab results indicate need for nutrient therapy. In these cases, modifications in diet and supplementation can be made. However, just because an individual has certain genetic SNPs does not necessarily mean any intervention is necessary. As with the other components of IFMNT, genetics is one tool in the toolbox for the health provider to utilize but should be considered with the rest of the information gathered in an assessment.

9.5 L: Lifestyle Factors

One key component of an integrative assessment is taking the time to hear a patient's story. In order to truly assess root causes of any health needs, including nutritional deficiencies, a practitioner must also consider influencing lifestyle factors. Sleep and stress are two major lifestyle factors influencing nutrient balance covered later in this chapter; however, other factors such as relationships, activity, and even substance abuse should also be evaluated for their nutritional impact.

Relationships are a major part of an individual's health and well-being. Often, a community or intimate relationships can have a direct influence on diet choices. On a very practical level, positive relationships can allow for easier implementation of nutritional interventions, particularly in the case of caregiver support. Spouses committed to health changes can promote reciprocal behavior in their partners [19]. While they can provide a great support particularly in a battle against chronic disease, stressed relationships can also prevent adequate nourishment and be a burden to one's health. One study showed that men who became divorced or widowed consumed fewer vegetables [20]. Even beyond diet choices, induced stress, potentially from a relational cause, can alter digestive, lipid, glycemic, and inflammatory responses to meals [21]. Considering these relational influences on a patient's dietary choices and metabolism allows for more practical and effective interventions for improving nutritional status.

Physical activity is an integral part of an individual's nutritional status. Physical activity has a long-established role in body composition and nutrient metabolism. However, physical activity is less considered in the context of its influence on nutrient status. Prolonged exercise, because of its metabolic demand, can lend to greater nutrient needs, particularly in the case of electrolytes. For instance, plasma magnesium is depleted following intensive exercise, yet generally rebounds within hours [22]. Therefore, assessment of magnesium status and supplementation interventions should also consider exercise timing. Evidence shows intensive exercise can promote intestinal permeability and, therefore, could influence nutrient status indirectly as well, potentially warranting gut-healing protocols [23, 24]. Furthermore, both excessive exercise and sedentary lifestyles can induce oxidative stress, potentially driving the need for antioxidant nutrients.

Substance abuse is another lifestyle factor likely to influence nutritional status. Alcoholism gives way to susceptibility to multiple nutrient deficiencies via primary or secondary pathways of malnutrition. While many alcoholics simply have inadequate diets by the preference of alcohol over other food and drinks, alcohol can also interfere with the absorption and metabolism of nutrients. The most widely recognized nutrient deficiency common in alcoholics is thiamine. Alcohol inhibits absorption of thiamine, and severe deficiencies can lead to Wernicke-Korsakoff's syndrome [25]. Though less recognized, many other deficiencies are common in alcoholics. Folate deficiency has been found in 80% of hospitalized alcoholics and may be of particular interest due to its role in the hepatic methylation cycle and the observed influence of folate deficiency on progression of liver disease in animal models [26]. Because of the impact of liver disease on production of bile and vitamin carrier proteins (see ▶ Chap. 16), alcoholics are also at risk for depletion of fat-soluble vitamins; however, current recommendations suggest using caution with replacing any of them without initially testing an individual's blood levels [25]. In regards to vitamin A, fatty liver from alcoholism prevents the conversion of beta-carotene to retinol and promotes the excretion of vitamin A from the liver. Therefore, alcoholism can promote functional deficiencies of vitamin A while maintaining serum levels. Understandably, then, correcting the deficiency proves difficult and is cautioned for fear of toxicity [27]. This highlights the need for monitoring blood levels of the fat-soluble vitamins periodically when intervening to replete. Deficiencies in vitamin C and B12, magnesium, and zinc have also been reported in alcoholics, but their direct relationship with alcohol consumption is less understood [25]. Additionally, cigarette smoking appears to directly impact antioxidant levels independent of dietary intake. The evidence is so compelling in this regard that smokers have higher requirements for vitamin C [28]. Additionally, levels of provitamin A carotenoids may also be compromised in smokers [29]. Identifying these modifiable lifestyle factors should be a first-line approach to addressing the nutritional deficiencies at the root cause.

9.6 E: Energy

When assessing the nutrition status of an individual, the energy systems of the human metabolism must be considered. This starts at the most basic level of energy production, within the mitochondria, and expands to a larger assessment of the entire organism's energy balance, which determines an individual's body composition. Nutrient sufficiency is needed at every level of energy production, greatly influencing metabolic pathways and determining phenotypes.

Starting at the cellular level, an integrative practitioner considers the nutrient cofactors and structural components of the mitochondria and cell membrane, where the molecular production of the energy unit of ATP occurs. The practitioner needs to have working knowledge of the primary biochemical systems involved in energy production, such as the citric acid or Kreb's cycle, gluconeogenesis, management of glucose and insulin, and energy fuel sources found in fats. Emerging evidence supports supplementing key nutrients to improve outcomes in conditions of mitochondrial dysfunction such as Alzheimer's and Parkinson's diseases [30]. In fact, a recent trial suggests using serum folate and red cell hemoglobin content levels as a predictive marker of amyloid plaque burden [31]. One of the most-studied nutrients involved in mitochondrial function is coenzyme Q10 (CoQ10). CoQ10 is required for ATP production via the electron transport chain and also has a powerful antioxidant function [32]. Deficiencies in CoQ10 caused by metabolic disorders or statin medications can result in mitochondrial dysfunction and reactive oxygen species [32]. Assessment and repletion of the status of this antioxidant can reduce inflammatory markers [33] and even reduce the atherogenic lipoprotein (a) [34].

Beyond the molecular level, further assessment of a patient's phenotype indicates how those energy systems contribute to chronic disease when energy metabolism is altered. Increased body fat, particularly within visceral adipose tissue (VAT), and the body shape that ensues give clinical evidence that energy is altered in an inflammatory and unhealthful direction. Though very basic in their assessment, anthropometrics can provide a global scale view of energy balance. These assessments include height, weight, body mass index (BMI), and circumferences including arm, waist, and hip. In addition to general energy balance, these measurements have also been indicative of cardiometabolic risk factors, including insulin resistance, inflammation, and triglyceride level in certain populations [35]. Body composition analysis, obtained through a number of techniques such as skin folds, bioelectrical impedance analysis (BIA), and dual-energy X-ray absorptiometry (DXA), can provide a more complete look at body structure beyond a height-to-weight comparison, including distribution of weight among fat and lean tissues and distribution of water in relation to the cell. Because body weight can be deceiving in regard to health status, a body composition analysis providing breakdown of mass will assess excess adiposity or inadequate muscle mass even in an otherwise "healthy"-weight individual. In fact, BMI alone has shown inaccuracy in predicting cardiometabolic risk in healthy-weight, overweight, and obese individuals [36]. Depending on the body composition test taken, more data may also be available suggestive of a global nutritional balance.

While the assessments at this level tend to lack specificity, they lay the foundation to help direct and prioritize the subsequent steps of the integrative assessment, even in investigating nutrient sufficiency. For instance, obesity alone warrants a deeper look into key nutrients such as vitamin D, whereas wasting or low muscle mass would trigger an investigation of the slew of nutrients associated with undernutrition. In body composition assessments that evaluate hydration status, such as BIA, an imbalance in intracellular vs. extracellular water could suggest evaluating electrolytes. Additionally, the BIA also includes a phase angle, a marker of cell membrane integrity indicative of status of minerals such as magnesium [37, 38], and has also been shown to be predictive of severity of certain conditions such as psoriasis [39] and chronic obstructive pulmonary disease [40]. By including this deeper assessment of metabolism and considering the underlying nutrients driving their functions, an integrative approach expands on the conventional assessment of energy balance for a more complete analysis.

9.7 T: Toxic Load

Just as many nutrients function as cofactors and substrates for gene function, many toxins act as inhibitors or inducers as they impact function and metabolism detrimentally. Both past and present environmental exposures can have an impact on gene expression [41–43] and disease risk as well as a buildup of endogenous toxins overloading an individual's metabolism.

Environmental/exogenous toxins These toxins affect people in different ways, due to differences in body chemistry and effects from the food they eat, genetics, and lifestyle choices. These toxins include pesticides, air pollutants, heavy metals, excessive medications, and various toxins in our food and water supply, as well as the air we breathe. We've recently become aware that the "toxic" effects of electromagnetic frequencies (EMF) on the metabolic pathways may add to a person's total toxin load. Excessive amounts of toxins can impact gene expression and affect many biochemical pathways. If an individual has ineffective and inefficient detoxification processes for reasons such as genetic SNPs, poor nutrient status, or organ dysfunction or removal, or the toxin load is too high for the body to combat, disease may arise. Laboratory testing can help identify levels of some specific toxins, such as heavy metals, and can aid in intervention development.

Though past environmental exposures cannot be prevented, current xenobiotic/exogenous toxins in an individual's environment can be identified and further exposure eliminated. Many of the toxins can be identified through various available testing modalities that can quantify and reveal the impact of residual tissue toxins from past exposures on a person's health.

Endogenous toxins Optimizing management and detoxification of endogenous toxins such as hormonal metabolites, excess stress toxins, and elimination organ processes is directly related to nutrients, food, and lifestyle.

All sources of toxins contribute to the total toxin load, which can benefit by interventions with food, dietary supplements, and other naturopathic interventions that enhance the body's detoxification pathways (see ▶ Chap. 13). Dietary interventions utilizing genetic information, individual nutrition requirements, and other lifestyle factors can be used to prevent, mitigate, and improve chronic disease affected by toxin exposure.

When it comes to total toxin load for an individual, many factors will contribute. Those factors will include some of the more obvious toxin exposures such as cigarette smoke, alcohol, and various drugs, but it will also include car exhaust, heavy metals, paint fumes, Teflon, aluminum cookware, mold, dry-cleaned clothes, pet dander, pesticides, nail polish, hair dyes, perfumes, fertilizers, plastics, etc.

Exposure to polycyclic aromatic hydrocarbons (PAHs) has been shown to impair the immune system through epigenetic modifications [44]. Toxin exposure affects methylation and enzymatic activity, which then influences compound and nutrient production [45]. During an assessment and/or with laboratory testing, the practitioner can get information on the patient's total toxin load and begin to recognize its implications. If the patient has a high toxin load combined with a poor nutrition quality of life and possibly even genetic polymorphisms, then their system will not be working optimally, increasing their risk for disease.

9.8 S: Stress and Sleep

We use the word "stress" in everyday life, yet the meaning is ambiguous. Stress can be traumatic, or it can be eustress (good stress), simply reflecting the daily grind of life. The role of stress in health is well established, yet all too often overlooked by practitioners. Stress has many different triggers with many different manifestations. Therefore, identifying presence of points of stress can be difficult.

Psychoneuroimmunology is a term used to describe the impact of mental attitudes on the body's resistance to disease, especially with respect to the links among and between the mind, the brain, and the immune system. The central nervous system and the immune system have constant communication [46]. Research on the brain-immune system connection has emerged over the last decade [46–48], giving practitioners more evidence to give appropriate attention to the beliefs and feelings of the patient. Cognitive states such as perceived control of a situation, of one's health, view of self, and views of the future have been linked with immune responses. During an assessment, the conversation about psychological stressors can be initiated through initial paperwork or the flow of the discussion, but for the provider, it's a component that cannot be overlooked.

While short-term/acute stress can be healthy and important for our health (e.g., exercise), chronic/long-term stress is another factor increasing risk for development of chronic disease. After an acute stress response, the main stress hormone, cortisol, lowers again, and the body can return to a balanced state. However, in times of chronic stress, cortisol remains elevated, which has many negative effects on the body [49]. During times of stress, the adrenal glands are working harder to produce cortisol. In this case, they also need additional nutrient support. Nutrients such as magnesium, potassium, and sodium are in higher demand when the body is under stress.

The brain first perceives a stressor and determines what is threatening. It then determines the behavioral and physiological responses to the stressor. Physiologically, the sympathetic and parasympathetic systems, hypothalamic-pituitary-adrenal (HPA) axis, immune system, metabolic hormones, and molecular processes within all organs adjust to combat the response [50]. That is, these organs activate to achieve stability. These adjustments are manageable in the short term, but when they are overused or imbalanced for too long, the body can become overloaded. This overload or imbalance can affect systemic physiology via neuroendocrine, autonomic, immune, and metabolic mediators. Nutritionally, this overload requires more nutrient support [51]. Studies have shown nutrient therapy such as with B vitamins has a beneficial effect on perceived stress, mild psychiatric symptoms, and aspects of everyday mood to support those under chronic stress [52].

9.9 Sleep

Having a properly functioning circadian rhythm is critical to healthy hormonal balance, metabolism function, memory and mental performance, optimal cellular energy production, and the immune system. In fact, one study showed people with chronic insomnia have a three times greater risk of dying from any cause [53].

The modern world sends mixed messages to the brain about whether it's day or night, and the line between sleep and awake can easily be blurred. When the body receives mixed messages, insomnia and fatigue are inevitable results. In contrast, when the right messages are sent, healthy sleep patterns result.

Sleep is essential to health and even if nutrition is optimal but sleep is poor, one's health status will quickly decline. In our modern world and fast-paced society, there are many expectations to do more, which for some people means trading sleep for productivity. This is common for college students, parents of newborns, and many people with demanding careers. Sleep hygiene affects nutrition intake and status, but nutrition intake and status also affect the ability to establish proper sleep hygiene. For some individuals, it's difficult to know which is the root cause of the problem and which should be addressed first or most aggressively. A 1999 study conducted at the University of Chicago concluded that restricting sleep to just 4 hours per night for 7 days led to increased insulin sensitivity and characteristics of diabetes [54]. Additionally, research shows sleep deprivation increases various inflammatory markers including IL-6 and CRP-hs [55]. Research by Van Cauter shows that individuals who are sleep-deprived have an increased appetite, which aligns with other findings that have identified the relationship between short sleep duration and increased intake of unhealthy food with more sedentary habits [56, 57]. Leptin has been proposed as one possible mechanism linked to the increasing rates of obesity. Various studies have indicated a rise in leptin levels after sleep deprivation [58, 59]. This is promising as we know the cravings for unhealthy food are strong and in an environment dominated by processed food, it's difficult to practice willpower, especially when sleep-deprived or stressed. The current research on the neurological connection between sleep, food consumption, and obesity is promising and will, like many other factors in reaching health, require lifestyle changes. A review by Dashti et al. [60] concluded that short sleep duration is associated with higher total caloric intake, higher total fat intake, protein intake, lower intake of fruits and vegetables, and diets of lower quality. Dashti also concluded that behavior and meal timing may also be impacted by sleep. Even changing the timing of meals and/or altering the ratios of macronutrients in those meal times can be a starting place for some people and start the process of reaching better alignment with their natural circadian rhythm. Evidence points to the fact that eating behaviors deviate from the traditional three meals per day to more frequent energy-dense and highly palatable snacks throughout the day and at night. The mechanisms behind these findings linking sleep and dietary intake are likely multifactorial but are important to consider during a nutrition assessment.

When it comes to sleep and nutrition, some nutrients have been studied more than others. Caffeine, for example, has plenty of literature showing an inverse relationship [61]. Not surprisingly, the function of caffeine is evidenced to disrupt sleep quality and quantity. Conversely, vitamin D has also been shown to promote circadian rhythm at a cellular level [62].

Light exposure is the primary regulator of circadian rhythm. Too much light after sunset affects melatonin production [63]. The blue light spectrum from many artificial light sources such as TVs, cell phones, and computers enters the eye and feeds a signal to the brain that "it's daytime." The combination of poor sunlight during the day and excessive blue light at night has major implications on circadian rhythm. Many other factors also contribute to circadian rhythm, including activity, stress, micronutrient status, and transition activities before bed, but each client will have different levels of readiness on where to start.

Bringing awareness to the importance of sleep is step one and can begin during an assessment when collecting information on their habits.

Gut motility is mediated by specific genes expressed in intestinal epithelial cells and in the enteric neurons, which help control circadian rhythm [64]. Colonic motility is faster in the morning and slower at night [65]. Disruption of the circadian cycle can provoke changes in gut motility. Circadian rhythm shifts that occur with traveling, night shifts, or other sleep disruptions can lead to gut symptoms, including bloating, abdominal pain, diarrhea, or constipation [64] (see ► Chap. 35).

9.10 Conclusion

A complete and thorough assessment is the crux of an effective practitioner. In order to intervene at the root cause of any condition, the status of the building blocks must first be thoroughly evaluated. Nutrients are the foundation for the body's metabolic processes that ultimately determine one's health. While a conventional nutritional assessment holds merit, it rarely moves beyond intake and expenditure to truly evaluate nutrient utilization and metabolism. Assessing nutritional status through an integrative look at nutrient sufficiency, immunity, biochemical individuality, lifestyle, energy balance, sleep, and stress provides a more complete picture of what interventions need to be made to meet nutritional demands and truly push a patient toward health.

References

1. U.S. Department of Health and Human Services and U.S. Department of Agriculture. 2015–2020 dietary guidelines for Americans. 8th ed; 2015.
2. United States Department of Agriculture (USDA). Food-away-from-home. Food Consumption & Demand. 2016. https://www.ers.usda.gov/topics/food-choices-health/food-consumption-demand/food-away-from-home.aspx.
3. Todd JE, Mancino L, Lin BH. The impact of food away from home on adult diet quality. In: Agriculture USDo, editor. Economic Research Service; 2010.
4. Hamrick KS, McClelland K. Americans' eating patterns and time spent on food: the 2014 eating & health module data. In: Agriculture USDo, editor. Economic Research Service; 2016.
5. Sorensen J, Kondrup J, Prokopowicz J, Schiesser M, Krahenbuhl L, Meier R, Liberda M. Euro Oops study group. EuroOOPS: an international, multicentre study to implement nutritional risk screening and evaluate clinical outcome. Clin Nutr. 2008;27(3):340–9. https://doi.org/10.1016/j.clnu.2008.03.012.
6. Cohen S, Danzaki K, MacIver NJ. Nutritional effects on T-cell immunometabolism. Eur J Immunol. 2017;47(2):225–35. https://doi.org/10.1002/eji.201646423.
7. Dizdar OS, Baspinar O, Kocer D, Dursun ZB, Avci D, Karakukcu C, Celik I, Gundogan K. Nutritional risk, micronutrient status and clinical outcomes: a prospective observational study in an infectious disease clinic. Nutrients. 2016;8(3):124. https://doi.org/10.3390/nu8030124.
8. Van Hecke T, Jakobsen LM, Vossen E, Gueraud F, De Vos F, Pierre F, Bertram HC, De Smet S. Short-term beef consumption promotes systemic oxidative stress, TMAO formation and inflammation in rats, and dietary fat content modulates these effects. Food Funct. 2016;7(9):3760–71. https://doi.org/10.1039/c6fo00462h.
9. Alves BC, Silva TR, Spritzer PM. Sedentary lifestyle and high-carbohydrate intake are associated with low-grade chronic inflammation in post-menopause: a cross-sectional study. Rev Bras Ginecol Obstet. 2016;38(7):317–24. https://doi.org/10.1055/s-0036-1584582.
10. Asghar A, Sheikh N. Role of immune cells in obesity induced low grade inflammation and insulin resistance. Cell Immunol. 2017;315:18–26. https://doi.org/10.1016/j.cellimm.2017.03.001.
11. Julia C, Meunier N, Touvier M, Ahluwalia N, Sapin V, Papet I, Cano N, Hercberg S, Galan P, Kesse-Guyot E. Dietary patterns and risk of elevated C-reactive protein concentrations 12 years later. Br J Nutr. 2013;110(4):747–54. https://doi.org/10.1017/S0007114512005636.
12. Yehuda S, Rabinovitz S, Mostofsky DI. Modulation of learning and neuronal membrane composition in the rat by essential fatty acid preparation: time-course analysis. Neurochem Res. 1998;23(5):627–34.
13. Whalen KA, McCullough ML, Flanders WD, Hartman TJ, Judd S, Bostick RM. Paleolithic and mediterranean diet pattern scores are inversely associated with biomarkers of inflammation and oxidative balance in adults. J Nutr. 2016;146(6):1217–26. https://doi.org/10.3945/jn.115.224048.
14. Akbaraly TN, Shipley MJ, Ferrie JE, Virtanen M, Lowe G, Hamer M, Kivimaki M. Long-term adherence to healthy dietary guidelines and chronic inflammation in the prospective Whitehall II study. Am J Med. 2015;128(2):152–60.e4. https://doi.org/10.1016/j.amjmed.2014.10.002.
15. Pounis G, Costanzo S, di Giuseppe R, de Lucia F, Santimone I, Sciarretta A, Barisciano P, Persichillo M, de Curtis A, Zito F, et al. Consumption of healthy foods at different content of antioxidant vitamins and phytochemicals and metabolic risk factors for cardiovascular disease in men and women of the Moli-sani study. Eur J Clin Nutr. 2013;67(2):207–13. https://doi.org/10.1038/ejcn.2012.201.
16. Pena KB, Ramos CO, Soares NP, da Silva PF, Bandeira AC, Costa GP, Cangussu SD, Talvani A, Bezerra FS. The administration of a high refined carbohydrate diet promoted an increase in pulmonary inflammation and oxidative stress in mice exposed to cigarette smoke. Int J Chron Obstruct Pulmon Dis. 2016;11:3207–17. https://doi.org/10.2147/COPD.S119485.
17. Wiest EF, Walsh-Wilcox MT, Walker MK. Omega-3 polyunsaturated fatty acids protect against cigarette smoke-induced oxidative stress and vascular dysfunction. Toxicol Sci. 2017;156(1):300–10. https://doi.org/10.1093/toxsci/kfw255.
18. De Rosa V, Galgani M, Santopaolo M, Colamatteo A, Laccetti R, Matarese G. Nutritional control of immunity: balancing the metabolic requirements with an appropriate immune function. Semin Immunol. 2015;27(5):300–9. https://doi.org/10.1016/j.smim.2015.10.001.
19. Perry B, Ciciurkaite G, Brady CF, Garcia J. Partner influence in diet and exercise behaviors: testing behavior modeling, social control, and normative body size. PLoS One. 2016;11(12):e0169193. https://doi.org/10.1371/journal.pone.0169193.
20. Eng PM, Kawachi I, Fitzmaurice G, Rimm EB. Effects of marital transitions on changes in dietary and other health behaviours in US male health professionals. J Epidemiol Community Health. 2005;59(1):56–62. https://doi.org/10.1136/jech.2004.020073.
21. Kiecolt-Glaser JK. Stress, food, and inflammation: psychoneuroimmunology and nutrition at the cutting edge. Psychosom Med. 2010;72(4):365–9. https://doi.org/10.1097/PSY.0b013e3181dbf489.
22. Terink R, Balvers MGJ, Hopman MT, Witkamp RF, Mensink M, Jmtk G. Decrease in ionized and total magnesium blood concentrations in endurance athletes following an exercise bout restores within hours-potential consequences for monitoring and supplementation. Int J Sport Nutr Exerc Metab. 2017;27(3):164–70. https://doi.org/10.1123/ijsnem.2016-0284.
23. Pals KL, Chang RT, Ryan AJ, Gisolfi CV. Effect of running intensity on intestinal permeability. J Appl Physiol (1985). 1997;82(2):571–6.

24. Roberts JD, Suckling CA, Peedle GY, Murphy JA, Dawkins TG, Roberts MG. An exploratory investigation of endotoxin levels in novice long distance triathletes, and the effects of a multi-strain probiotic/prebiotic, antioxidant intervention. Nutrients. 2016;8(11):E733. https://doi.org/10.3390/nu8110733.
25. Rossi RE, Conte D, Massironi S. Diagnosis and treatment of nutritional deficiencies in alcoholic liver disease: overview of available evidence and open issues. Dig Liver Dis. 2015;47(10):819–25. https://doi.org/10.1016/j.dld.2015.05.021.
26. Halsted CH, Villanueva JA, Devlin AM, Niemela O, Parkkila S, Garrow TA, Wallock LM, Shigenaga MK, Melnyk S, James SJ. Folate deficiency disturbs hepatic methionine metabolism and promotes liver injury in the ethanol-fed micropig. Proc Natl Acad Sci U S A. 2002;99(15):10072–7. https://doi.org/10.1073/pnas.112336399.
27. Lieber CS. Relationships between nutrition, alcohol use, and liver disease. Alcohol Res Health. 2003;27(3):220–31. Version current Internet: https://pubs.niaaa.nih.gov/publications/arh27-3/220-231.htm. Aaccessed 2017.
28. German Nutrition Society. New reference values for Vitamin C intake. Ann Nutr Metab. 2015;67(1):13–20. https://doi.org/10.1159/000434757.
29. Alberg A. The influence of cigarette smoking on circulating concentrations of antioxidant micronutrients. Toxicology. 2002;180(2):121–37.
30. Liu J, Ames BN. Reducing mitochondrial decay with mitochondrial nutrients to delay and treat cognitive dysfunction, Alzheimer's disease, and Parkinson's disease. Nutr Neurosci. 2005;8(2):67–89. https://doi.org/10.1080/10284150500047161.
31. Yoshinaga T, Nishimata H, Kajiya Y, Yokoyama S. Combined assessment of serum folate and hemoglobin as biomarkers of brain amyloid beta accumulation. PLoS One. 2017;12(4):e0175854. https://doi.org/10.1371/journal.pone.0175854.
32. Acosta MJ, Vazquez Fonseca L, Desbats MA, Cerqua C, Zordan R, Trevisson E, Salviati L. Coenzyme Q biosynthesis in health and disease. Biochim Biophys Acta. 2016;1857(8):1079–85. https://doi.org/10.1016/j.bbabio.2016.03.036.
33. Fan L, Feng Y, Chen GC, Qin LQ, Fu CL, Chen LH. Effects of coenzyme Q10 supplementation on inflammatory markers: a systematic review and meta-analysis of randomized controlled trials. Pharmacol Res. 2017;119:128–36. https://doi.org/10.1016/j.phrs.2017.01.032.
34. Sahebkar A, Simental-Mendia LE, Stefanutti C, Pirro M. Supplementation with coenzyme Q10 reduces plasma lipoprotein(a) concentrations but not other lipid indices: a systematic review and meta-analysis. Pharmacol Res. 2016;105:198–209. https://doi.org/10.1016/j.phrs.2016.01.030.
35. Araujo AJ, Santos AC, Prado WL. Body composition of obese adolescents: association between adiposity indicators and cardiometabolic risk factors. J Hum Nutr Diet. 2016;30(2):193–202. https://doi.org/10.1111/jhn.12414.
36. Tomiyama AJ, Hunger JM, Nguyen-Cuu J, Wells C. Misclassification of cardiometabolic health when using body mass index categories in NHANES 2005-2012. Int J Obes. 2016;40(5):883–6. https://doi.org/10.1038/ijo.2016.17.
37. Matias CN, Monteiro CP, Santos DA, Martins F, Silva AM, Laires MJ, Sardinha LB. Magnesium and phase angle: a prognostic tool for monitoring cellular integrity in judo athletes. Magnes Res. 2015;28(3):92–8. https://doi.org/10.1684/mrh.2015.0389.
38. Fein P, Suda V, Borawsky C, Kapupara H, Butikis A, Matza B, Chattopadhyay J, Avra MM. Relationship of serum magnesium to body composition and inflammation in peritoneal dialysis patients. Adv Perit Dial. 2010;26:112–5.
39. Barrea L, Macchia PE, Di Somma C, Napolitano M, Balato A, Falco A, Savanelli MC, Balato N, Colao A, Savastano S. Bioelectrical phase angle and psoriasis: a novel association with psoriasis severity, quality of life and metabolic syndrome. J Transl Med. 2016;14(1):130. https://doi.org/10.1186/s12967-016-0889-6.
40. Maddocks M, Kon SS, Jones SE, Canavan JL, Nolan CM, Higginson IJ, Gao W, Polkey MI, Man WD. Bioelectrical impedance phase angle relates to function, disease severity and prognosis in stable chronic obstructive pulmonary disease. Clin Nutr. 2015;34(6):1245–50. https://doi.org/10.1016/j.clnu.2014.12.020.
41. Cardenas A, Koestler DC, Houseman EA, Jackson BP, Kile ML, Karagas MR, Marsit CJ. Differential DNA methylation in umbilical cord blood of infants exposed to mercury and arsenic in utero. Epigenetics. 2015;10(6):508–15. https://doi.org/10.1080/15592294.2015.1046026.
42. Shen W, Zhang B, Liu S, Wu H, Gu X, Qin L, Tian P, Zeng Y, Ye L, Ni Z, et al. Association of blood lead levels with methylenetetrahydrofolate reductase polymorphisms among Chinese pregnant women in Wuhan city. PLoS One. 2015;10(2):e0117366. https://doi.org/10.1371/journal.pone.0117366.
43. Austin DW, Spolding B, Gondalia S, Shandley K, Palombo EA, Knowles S, Walder K. Genetic variation associated with hypersensitivity to mercury. Toxicol Int. 2014;21(3):236–41. https://doi.org/10.4103/0971-6580.155327.
44. Hew KM, Walker AI, Kohli A, Garcia M, Syed A, McDonald-Hyman C, Noth EM, Mann JK, Pratt B, Balmes J, et al. Childhood exposure to ambient polycyclic aromatic hydrocarbons is linked to epigenetic modifications and impaired systemic immunity in T cells. Clin Exp Allergy. 2015;45(1):238–48. https://doi.org/10.1111/cea.12377.
45. Varela-Rey M, Woodhoo A, Martinez-Chantar ML, Mato JM, Alcohol LSC. DNA methylation, and cancer. Alcohol Res. 2013;35(1):25–35.
46. Ziemssen T, Kern S. Psychoneuroimmunology--cross-talk between the immune and nervous systems. J Neurol. 2007;254(Suppl 2):II8–11. https://doi.org/10.1007/s00415-007-2003-8.
47. Kemeny ME, Gruenewald TL. Psychoneuroimmunology update. Semin Gastrointest Dis. 1999;10(1):20–9.
48. Dinan TG, Cryan JF. Microbes, immunity, and behavior: psychoneuroimmunology meets the microbiome. Neuropsychopharmacology. 2017;42(1):178–92. https://doi.org/10.1038/npp.2016.103.
49. Fraser R. Disorders of the adrenal cortex: their effects on electrolyte metabolism. Clin Endocrinol Metab. 1984;13(2):413–30.
50. Bruce ME. Neurobiological and systemic effects of chronic stress. Chronic Stress. 2017;1:1–11. https://doi.org/10.1177/2470547017692328.
51. Stough C, Scholey A, Lloyd J, Spong J, Myers S, Downey LA. The effect of 90 day administration of a high dose vitamin B-complex on work stress. Hum Psychopharmacol. 2011;26(7):470–6. https://doi.org/10.1002/hup.1229.
52. Long SJ, Benton D. Effects of vitamin and mineral supplementation on stress, mild psychiatric symptoms, and mood in nonclinical samples: a meta-analysis. Psychosom Med. 2013;75(2):144–53. https://doi.org/10.1097/PSY.0b013e31827d5fbd.
53. Wehrens SM, Hampton SM, Finn RE, Skene DJ. Effect of total sleep deprivation on postprandial metabolic and insulin responses in shift workers and non-shift workers. J Endocrinol. 2010;206(2):205–15. https://doi.org/10.1677/JOE-10-0077.
54. Knutson KL, Spiegel K, Penev P, Van Cauter E. The metabolic consequences of sleep deprivation. Sleep Med Rev. 2007;11(3):163–78. https://doi.org/10.1016/j.smrv.2007.01.002.
55. Miller MA, Kandala NB, Kivimaki M, Kumari M, Brunner EJ, Lowe GD, Marmot MG, Cappuccio FP. Gender differences in the cross-sectional relationships between sleep duration and markers of inflammation: Whitehall II study. Sleep. 2009;32(7):857–64.
56. Garaulet M, Ortega FB, Ruiz JR, Rey-Lopez JP, Beghin L, Manios Y, Cuenca-Garcia M, Plada M, Diethelm K, Kafatos A, et al. Short sleep duration is associated with increased obesity markers in European adolescents: effect of physical activity and dietary habits. The HELENA study. Int J Obes. 2011;35(10):1308–17. https://doi.org/10.1038/ijo.2011.149.
57. Grandner MA, Jackson N, Gerstner JR, Knutson KL. Dietary nutrients associated with short and long sleep duration. Data from a nation-

ally representative sample. Appetite. 2013;64:71–80. https://doi.org/10.1016/j.appet.2013.01.004.
58. Van Cauter E, Knutson KL. Sleep and the epidemic of obesity in children and adults. Eur J Endocrinol. 2008;159(Suppl 1):S59–66. https://doi.org/10.1530/EJE-08-0298.
59. Simpson NS, Banks S, Dinges DF. Sleep restriction is associated with increased morning plasma leptin concentrations, especially in women. Biol Res Nurs. 2010;12(1):47–53. https://doi.org/10.1177/1099800410366301.
60. Dashti HS, Scheer FA, Jacques PF, Lamon-Fava S, Ordovas JM. Short sleep duration and dietary intake: epidemiologic evidence, mechanisms, and health implications. Adv Nutr. 2015;6(6):648–59. https://doi.org/10.3945/an.115.008623.
61. Golem DL, Martin-Biggers JT, Koenings MM, Davis KF, Byrd-Bredbenner C. An integrative review of sleep for nutrition professionals. Adv Nutr. 2014;5(6):742–59. https://doi.org/10.3945/an.114.006809.
62. Gutierrez-Monreal MA, Cuevas-Diaz Duran R, Moreno-Cuevas JE, Scott SP. A role for 1alpha,25-dihydroxyvitamin d3 in the expression of circadian genes. J Biol Rhythm. 2014;29(5):384–8. https://doi.org/10.1177/0748730414549239.
63. Figueiro MG, Wood B, Plitnick B, Rea MS. The impact of light from computer monitors on melatonin levels in college students. Neuro Endocrinol Lett. 2011;32(2):158–63.
64. Hoogerwerf WA. Role of clock genes in gastrointestinal motility. Am J Physiol Gastrointest Liver Physiol. 2010;299(3):G549–55. https://doi.org/10.1152/ajpgi.00147.2010.
65. Kumar D, Wingate D, Ruckebusch Y. Circadian variation in the propagation velocity of the migrating motor complex. Gastroenterology. 1986;91(4):926–30.

Other Suggestions: These Might Lead to Some Other Resources

http://www.dairyherd.com/dairy-news/nine-out-of-10-americans-fall-short-of-key-nutrients-they-need-new-study-concludes-114565944.html
http://campaignforessentialnutrients.com/
http://agriculture.columbia.edu/projects/nutrition/nutrient-gap-analysis/

Nutritional Role of Fatty Acids

Vishwanath M. Sardesai

10.1 Introduction – 136

10.2 Saturated Fatty Acids – 137
10.2.1 Short-Chain Saturated Fatty Acids (SCFA): C_1–C_6 [6, 7] – 137
10.2.2 Medium Chain Triglycerides (MCT): C6–C12 with an Aliphatic Tail [12] – 137
10.2.3 Long-Chain C_{13}–C_{21} and Very-Long-Chain Fatty Acids C_{22} or More – 138

10.3 Monounsaturated Fatty Acids – 139

10.4 Polyunsaturated Fatty Acids – 139
10.4.1 Biosynthesis – 139
10.4.2 Functions – 140
10.4.3 Deficiency – 143
10.4.4 Requirements – 144
10.4.5 W_3 Fatty Acids and Health – 144
10.4.6 Food Sources – 146

10.5 *Trans*-Fatty Acids – 146

10.6 Conjugated Fatty Acids – 147

10.7 Conclusions – 147

References – 148

© Springer Nature Switzerland AG 2020
D. Noland et al. (eds.), *Integrative and Functional Medical Nutrition Therapy*,
https://doi.org/10.1007/978-3-030-30730-1_10

10.1 Introduction

Fatty acids are key components of lipids. They are important dietary sources of energy and have roles in modulating acute and prolonged inflammation [1] and as key components of cell membrane structure [2]. They are called acids because of the organic acid group (COOH) that they contain. They are chains of covalently linked carbon atoms bearing hydrogen atoms. The naturally occurring fatty acids are, for the most part, unbranched and acyclic, but complex structures with branched or cyclic chains also occasionally occur. They have the basic formula $CH_3(CH_2)n \cdot COOH$ where n can be any number from 2 to 24. The bulk of the fatty acids in the human body have 16, 18, or 20 carbon atoms, but there are some with longer chains that occur principally in the lipids of the nervous system. One method of fatty acid classification is according to their chain length. Those containing 2–4 carbon atoms are called short-chain fatty acids, while those with 6–10 and 12–24 carbon atoms are called medium-chain and long-chain fatty acids, respectively. In most cases, naturally occurring fatty acids contain an even number of carbon atoms. Fatty acids are also classified as saturated with no double bonds, monounsaturated with one double bond, and polyunsaturated fatty acids (PUFA) with two or more double bonds. Depending on the number of double bonds from 2, 3, 4, 5, and 6, they are called dienoic, trienoic, tetraenoic, pentanoic, and hexaenoic, respectively. The carbon atoms of fatty acids may be numbered either with the carboxyl carbon as number 1 (Δ numbering system) or from the terminal methyl (CH_3) group carbon as number 1 (W or n numbering system).

The double bond locations are designated by the number of carbon atoms from the carboxyl side or from the methyl side.

Fatty acids are abbreviated in the Δ system by listing the total carbon number and the position of the double bonds. Palmitic acid (which contains no double bond) is abbreviated as C_{16}:0 or C_{16}:Δ_0, and palmitoleic acid which contains one double bond is C_{16}:1 or C_{16}:1 Δ_9. The number after the Δ in this classification system signifies the position of double bond relative to the carboxyl end. In palmitoleic acid, the double bond is in the ninth carbon atom from the carboxyl group. It is between carbon atoms 9 and 10 from the carboxyl carbon as number one. In the W numbering system, palmitoleic acid is abbreviated as C_{16}:1, W_7. This signifies that the fatty acid has 16 carbons and one double bond which is located 7 carbon atoms away from the W carbon counting the W carbon as number 1. In other words, the double bond is between carbon 7 and 8 counting from the W end as shown here:

$$CH_3 - CH_2 - CH_2 - CH_2 - CH_2 - CH = CH - (CH_2)_7 COOH$$
Palmitoleic acid

Table 10.1 lists unsaturated fatty acids of importance in human nutrition using the W numbering system.

Other relevant information is the stereochemical configuration of each double bond. Fatty acids that contain double bonds can exist in two geometric isomeric forms. These forms are known as *cis* and *trans* isomerism. When a double bond occurs along the length of a carbon chain, the pair of the hydrogen atoms on the participating carbon atom may lie on the same side of the double bond; in this case, they are described as *cis* or they may be on opposite sides of the double bond, in which case they are described as *trans*. Double bonds in naturally occurring fatty acids are of the *cis* configuration. Thus, oleic acid is the *cis* form and elaidic acid is its *trans* form [3]:

Δ numbering system (carboxyl side)

```
           ↘
16             4    3    2    1
CH3 — (CH2)11 — CH2 — CH2 — CH2 — COOH
1              13   14   15   16
↑
```

Table 10.1 Unsaturated fatty acids of importance in human nutrition

Common name	Systematic name	Structural formula
Palmitoleic	Hexadecenoic (W_7)	$CH_3(CH_2)_5CH = CH(CH_2)_7COOH$
Oleic	Octadecenoic (W_8)	$CH_3(CH_2)_7CH = CH(CH_2)_7COOH$
Linoleic	Octadecadienoic ($W_{6,9}$)	$CH_3(CH_2)_4CH = CHCH_2CH = CH(CH_2)_7COOH$
Linolenic	Octadecatrienoic ($W_{3,6,9}$)	$CH_3CH_2CH = CHCH_2CH = CHCH_2CH = CH(CH_2)_7COOH$
γ-Linolenic	Octadecatrienoic ($W_{6,9,12}$)	$CH_3(CH_2)_4CH = CHCH_2CH = CHCH_2CH = CH(CH_2)_4COOH$
Dihomo-γ-linolenic	Eicosatrienoic ($W_{3,6,9}$)	$CH_3CH_2CH = CHCH_2CH = CHCH_2CH = CH(CH_2)_9COOH$
Arachidonic	Eicosatetraenoic ($W_{6,9,12,15}$)	$CH_3(CH_2)_4CH = CHCH_2CH = CHCH_2CH = CHCH_2CH = CH(CH_2)_3COOH$
	Eicosapentaenoic ($W_{3,6,9,12,15}$)	$CH_3CH_2(CH = CHCH_2)_5CH_2CH_2COOH$
	Docosahexaenoic ($W_{3,6,9,12,15,18}$)	$CH_3(CH_2CH = CH)_6CH_2CH_2COOH$
Vaccenic	Octadecenoic (W_7)	$CH_3(CH_2)_5CH = CH(CH_2)_9COOH$
Nervonic	Tetracosenoic (W_9)	$CH_3(CH_2)_7CH = CH(CH_2)_{13}COOH$

Nutritional Role of Fatty Acids

```
H — C — (CH₂)₇ — CH₃        CH₃ — (CH₂)₇ — C — H
    ‖                                       ‖
H — C — (CH₂)₇COOH          H — C — (CH₂)₇COOH
        cis                         trans
     oleic acid                  elaidic acid
```

When unsaturated and PUFA are hydrogenated to make margarine and shortening, anywhere from 5% to 70% of the double bonds occur in *trans* form. The *trans*-fatty acids [4] account for 5–8% of the fat in the American diet.

When double bonds of the unsaturated fatty acids are conjugated, that is, the double bonds occur without an intervening carbon atom not part of a double bond, they are called conjugated fatty acids. Conjugated linoleic acid (CLA) occurs naturally in meat and milk food products derived from ruminant sources (e.g., beef, lamb, dairy) because of the process of bacterial isomerization of linoleic acid (present in food stuff for these animals) in the rumen. CLA is higher in grass-fed ruminant sources compared with grain-fed [5].

Fatty acids occur primarily as esters of glycerol when they are being used for energy storage and utilization. Esters of one, two, or three fatty acid residues are called monoglycerides, diglycerides, and triglycerides, respectively. Mono- and diglycerides occur only in minor amounts and largely as metabolic intermediates in the biosynthesis and degradation of glycerol-containing lipids. The bulk of the fatty acids in the human body exist as triglycerides in which all three hydroxyl groups on the glycerol are esterified with a fatty acid. Most triglyceride molecules contain two or more different fatty acid moieties (i.e., the fatty acids within triglycerides are not often repeated) and have the general composition as follows:

```
CH₂ — O — COF₁
 |
CH  — O — COF₂
 |
CH₂ — O — COF₃
```

where F1COOH, F2COOH, and F3COOH are fatty acid chains that may or may not all be the same.

Triglycerides account for about 90% of the fat in our food and more than 90% of the fat in the body.

There are five categories of fatty acids based on the degree of saturation and types of double bonds. They are saturated, monounsaturated, polyunsaturated, *trans*, and conjugated fatty acids.

10.2 Saturated Fatty Acids

When discussing saturated fats and their nutritional influences on human metabolism, it is important to differentiate between the three main saturated fat categories: short chain, medium chain, and long chain.

10.2.1 Short-Chain Saturated Fatty Acids (SCFA): C_1–C_6 [6, 7]

Understanding the role of SCFAs is important for nutritional modulation of the gut ecology. The short-chain fatty acid substrates result from fermentation between resistant starches and soluble fibers, and beneficial species and amounts of microflora present in the colon. The SCFAs are produced, primarily acetate, propionate, and butyrate, as end products. The SCFAs produced are the primary energy sources for colonocytes and are particularly important for colon health. These SCFAs have anticarcinogenic properties, inhibit the growth and proliferation of tumor cell lines in vitro, induce differentiation of tumor cells, and possess anti-inflammatory properties [8]. Maintaining a life-long colon of healthy cells depend on continual production of SCFAs [8, 9] (◘ Fig. 10.1).

- *Acetic acid* (C_2) does not occur in natural fats and oils, but vinegar contains approximately 5% of acetic acid.
- *Propionic Acid* (C_3)
- *Butyric acid* (C_4) is also uncommon in natural food fats except for milk, butterfat, and breastmilk. A primary endogenous source of butyric acid for its important role as a primary fuel for colonocyte repair is from the healthy colonic microbiome when dietary fiber is fermented in the colon [11].

10.2.2 Medium Chain Triglycerides (MCT): C6–C12 with an Aliphatic Tail [12]

Straight-chain MCFAs are as follows:
- Caproic acid C_6
- Caprylic acid C_8 (n-octanoic acid or fatty acid C8, goat milk, coconut oil, palm oil)
- Capric acid C_{10}
- Lauric acid C_{12}

MCTs do not require bile salts for digestion. The MCTs are absorbed directly into the portal vein and then transported rapidly to cells and the liver for β-oxidation. Thus, thermogenesis is increased. Because of the ability to be absorbed without requiring emulsification from bile, MCT oils are commonly used in enteral feedings, liver disease diets, weight loss, and others. Patients who have malnutrition, malabsorption, or particular fatty-acid metabolism disorders benefit from inclusion of MCTs in their diet as they do not require energy for absorption, use, or storage. More recently, the MCT oils are universally used in ketogenic diets of all types as they evidence beneficial contributions to brain/nerve metabolism [12]. Substantial amounts of saturated fatty acids are found in certain vegetable products such as coconut and palm oil.

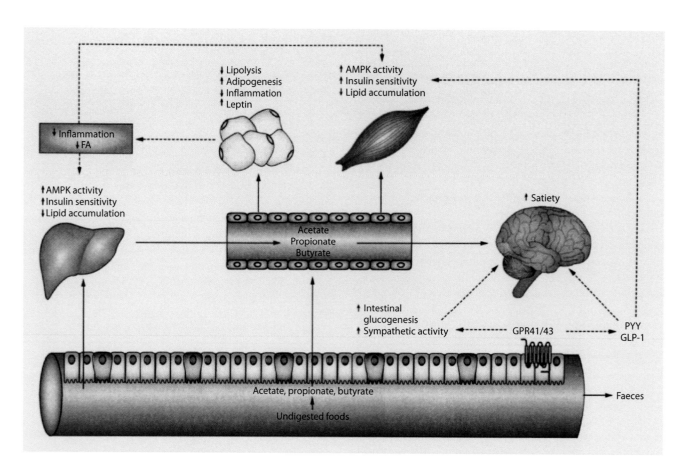

Fig. 10.1 SCFA and interorgan crosstalk [10]. SCFA and interorgan crosstalk fermentation of indigestible foods in the distal intestine results in the production of SCFA. The ratio of acetate to propionate to butyrate in the ileum, caecum, and colon is ~3:1:1. Butyrate and propionate are generally metabolized in the colon and liver and, therefore, mainly affect local gut and liver function. In the distal gut, SCFA bind to GPR41 and GPR43, which leads to the production of the gut hormones PYY and GLP-1 and affect satiety and glucose homeostasis. Furthermore, propionate and butyrate might induce intestinal gluconeogenesis and sympathetic activity, thereby improving glucose and energy homeostasis. Small amounts of propionate and butyrate and high amounts of acetate reach the circulation and can also directly affect the peripheral adipose tissue, liver, and muscle substrate metabolism and function. In addition, circulating acetate might be taken up by the brain and regulate satiety via a central homeostatic mechanism. Whether metabolic effects are mainly explained by direct effects of SCFA or indirectly via gut-derived signaling molecules remain unclear. Solid lines indicate direct SCFA effects, and dashed lines indicate indirect SCFA effects. *Abbreviations*: AMPK adenosine monophosphate-activated protein kinase, FA fatty acid, GLP-1 glucagon-like peptide-1, GPR G-protein coupled receptor, PYY peptide YY, SCFA short-chain fatty acid. (Reprinted from Canfora et al. [6]. With permission from Springer Nature)

10.2.3 Long-Chain C_{13}–C_{21} and Very-Long-Chain Fatty Acids C_{22} or More

These fatty acids are the most abundant in animal and human fats, accounting for 30–40% of the fatty acids in adipose tissue. Substantial amounts of longer-chain saturated fatty acids are found in meat and certain heat- and oxygen-processed oils that are used in hydrogenated shortenings and margarines (Table 10.2).

The saturated fatty acids are not only a source of body fuel but also are structural components of cell membranes. Various saturated fatty acids are also associated with proteins and are necessary for their normal function [14]. They are synthesized as needed by the body to provide an adequate level required for their physical and structural functions.

Saturated fatty acids vary in the degree of cholesterol influence. For example, stearic acid (C_{18}) has little or no effect

Table 10.2 Examples of long-chain and very-long-chain saturated fatty acids

Common name	Chemical structure	C:D
Myristic acid	$CH_3(CH_2)_{12}COOH$	14:0
Palmitic acid	$CH_3(CH_2)_{14}COOH$	16:0
Stearic acid	$CH_3(CH_2)_{16}COOH$	18:0
Arachidic acid	$CH_3(CH_2)_{18}COOH$	20:0
Behenic acid	$CH_3(CH_2)_{20}COOH$	22:0
Lignoceric acid	$CH_3(CH_2)_{22}COOH$	24:0
Cerotic acid	$CH_3(CH_2)_{24}COOH$	26:0

Based on data from Ref. [13]

on serum cholesterol, while lauric (C_{12}) and palmitic (C_{16}) dramatically increase cholesterol concentrations. Because palmitic acid is more abundant than lauric and myristic acids, it may be the primary fatty acid affecting cholesterol concentrations. Saturated fatty acids containing 10 carbons or less have not been shown to have any effect on cholesterol levels.

There is no dietary requirement for saturated fatty acids because they can be synthesized endogenously. The average American diet provides approximately 15% of calories in the form of saturated fat. The saturated fatty acid intake ranges from 21 to 34 g per day in men and 15 to 21 g per day in women.

10.3 Monounsaturated Fatty Acids

The liver is the main organ responsible for the endogenous synthesis of monounsaturated fatty acids (MUFA). Monounsaturation at the Δ_9 position is the rule, and in humans the double bond cannot be introduced between carbon 1 and 6 starting from the W carbon of the fatty acids. Therefore, there are only two saturated fatty acids available for desaturation: Palmitic to palmitoleic acid and stearic to oleic acid. On the one hand, MUFA with double bonds occurring before the Δ_9 position are not synthesized to a significant extent by animals and humans because the required desaturase is absent and trace amounts found in animal tissue lipids probably arise from the diet. Plants, on the other hand, can introduce double bonds between Δ_9 and the terminal methyl group.

Plant sources rich in MUFA include olive oil (about 75%). High oleic acid variety sunflower oil contains as much as 80–85% MUFA. Canola oil has about 58% MUFA. It is also found in red meat, whole milk products, olives, and avocados. Oleic acid accounts for about 90% of dietary MUFA. The MUFA, including oleic and nervonic acid (C_{24}:1, Δ_9), are important in membrane structural lipids, particularly nervous tissue myelin. Other MUFA, such as palmitoleic acid, are present in minor amounts in the diet. MUFA neither elevate nor lower the level of serum cholesterol; thus, a high-fat diet is not necessarily associated with a high level of serum cholesterol in the population. Inhabitants in Mediterranean countries consume large amounts of olive oil but tend to have low levels of serum cholesterol. The average American consumes MUFA which makes up about 50% of total fat and provides 20% of the calories in the diet.

10.4 Polyunsaturated Fatty Acids

Humans are not able to introduce double bonds closer to the methyl group than the W_9 position. Thus, two broad classes of biologically active polyunsaturated fatty acids (PUFA) – the W_6 linoleic acid (LA) C_{18}:2, W_6, and the W_3 linolenic acid also called α-linolenic acid (ALA) C_{18}:3, W_3 – cannot be synthesized by humans. These are required for normal growth and function of all tissues. These fatty acids must be supplied by a dietary source [15, 16] and are called essential fatty acids (EFA).

10.4.1 Biosynthesis

The liver is the site of most of the PUFA metabolism that transforms dietary 18-carbon EFA into long-chain PUFA with 20 or 22 carbons. The mammalian tissues contain four series of PUFA. The precursors of the two of these groups are the MUFA palmitoleic and oleic acids, which can be synthesized from saturated fatty acids. The precursors for the other two are necessarily derived from dietary source: LA and ALA. A complex series of desaturation and elongation reactions acting in concert transform the precursors to their higher polyunsaturated derivatives. These four precursors are alternatively desaturated (two hydrogens are removed to create a double bond) and chain elongated (addition of two carbons). The desaturations are catalyzed by Δ6, Δ5, and Δ4 desaturases, and the two carbon additions are catalyzed by elongases to form the principal PUFA found in tissues (◘ Fig. 10.2).

The same enzymes catalyze the equivalent steps in the W_7, W_9, W_6, and W_3 fatty acid pathways, and there is competition among substrates for these enzymes. The critical enzyme Δ6 desaturase displays highest affinity for the most highly unsaturated C_{18} substrate. The order of preference is ALA>LA > oleic acid>palmitoleic acid (provided the substrate concentrations are equal). In the presence of either of the two EFA, little desaturation of oleate occurs. ALA effectively inhibits the desaturation of LA (at equal concentrations). In the absence of the members of the W_6 and W_3 families, however, oleate is desaturated and members of the W_9 family, particularly C_{20}:3, W_9, appear in the tissues [17].

In the desaturation process, additional double bonds are inserted between the preexisting double bonds and the carboxyl group, and the chain elongation always proceeds by the addition of two carbon units to the carboxyl terminus of the fatty acid chain. Therefore, the position of the double bond counting from the methyl (W) end of the precursor fatty acid remains unaltered through all transformations. All products of oleic acid possess the W_9 configuration of oleate itself and are recognized as members of the W_9 family. Nervonic acid C_{18}:1, W_9, a component of nerve tissue lipid, is derived by chain elongation of oleate. Vaccenic acid C_{18}:1 W_7 that occurs in small amounts in tissue lipids is formed by chain elongation of palmitoleic acid. Similarly, LA and ALA give rise to W_6 and W_3 families of PUFA. No interconversion between these families occurs. In addition to desaturases, there is competition of the substrates for chain elongase enzymes and for the acetyl transferases involved in the formation of phospholipids (which require PUFA). Lower members of the family may also be able to compete with some of the products for enzyme sites and the expression of its family. Long-chain highly unsaturated fatty acids can be shortened by two car-

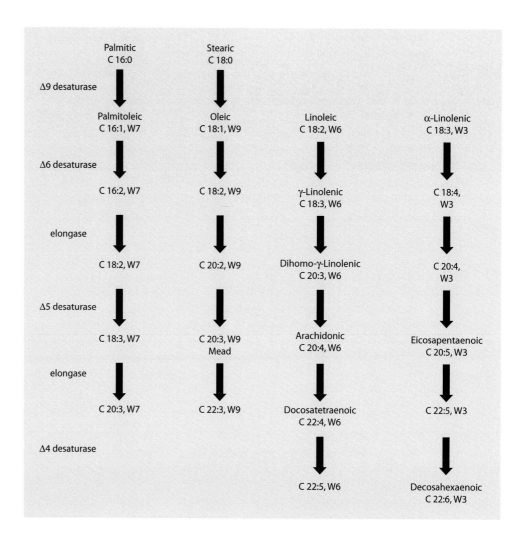

Fig. 10.2 Biosynthesis of polyunsaturated fatty acids. (Courtesy of Vishwanath M. Sardesai, PhD)

bons, a phenomenon called retroconversion. Because of the competition, retroconversion, etc., each family has characteristic end products that accumulate in tissue lipids, while intermediates are usually found in much smaller, often trace amounts. Thus, for oleate and palmitoleate, the major PUFA are the trienes C_{20}:3, W_7, and C_{20}:3, W_9, respectively. For linoleate, the major PUFA are arachidonic acid (AA) with four double bonds (tetraene) and some dihomo-gamma-linolenic acid (DHGL/DGLA). For linolenate, eicosapentaenoic acid (EPA) and docosahexaenoic acid (DHA) are the main end products. The 22-carbon hexanoic acid is the most unsaturated fatty acid commonly found in the lipids of higher animals.

Δ6 desaturase is considered one of the rate-limiting enzymes in the metabolic pathway. A typical western diet tends to be much higher in W_6 than in W_3 fatty acids. Although ALA is the preferred substrate for Δ6 desaturase, the excess of dietary LA compared with ALA results in greater formation of AA than EPA. The capacity for conversion of ALA to DHA is higher in women than men. Studies of ALA metabolism in healthy young men indicated that approximately 8% of ALA is converted to EPA and 0.4% is converted to DHA. In healthy young women, approximately 21% of ALA is converted to EPA and 9% is converted to DHA [18]. This better conversion efficiency in young women compared with men appears to be related to the effects of estrogen.

The data from experimental animals suggest that Δ6 desaturase is sensitive to several factors. Those that tend to increase the enzyme activity include a high-protein diet, insulin, and EFA deficiency, whereas fasting, zinc deficiency, aging, glucagon, glucocorticoids, and diabetes are known to decrease its activity. Reduction in the activity of Δ6 desaturase can limit the availability of DHGL and AA, which are required for normal physiologic functions. Although a good deal of AA is acquired from foods of animal origin such as meat and dairy products, little or no DHGL/DGLA is found in a normal diet.

10.4.2 Functions

After ingestion, LA and ALA are distributed between adipose tissue triglycerides, other tissue stores, and tissue structural lipids. A proportion of LA and ALA provides energy, and these are oxidized more rapidly than the saturated and MUFA. In contrast, long-chain PUFA derived from EFA are less rapidly oxidized. These acids when present preformed in the diet are incorporated into structural lipids about 20 times more efficiently than after synthesis from dietary LA and ALA. PUFA are the major components of structural lipids of membranes of cells, mitochondria, and nuclei, and they play a major and vital role in the properties of most biomembranes. In the phospholipids, the saturated fatty acids preferentially occupy the carbon 1 position and PUFA occupy carbon 2 of the glycerol moiety, although MFUA can also take the carbon 2 position.

Structure of phospholipids:

$$\begin{array}{l} {}^1CH_2 - O - COR \\ \quad | \\ R - CO - O - {}^2CH \\ \quad | \\ {}^3CH_2 - O - P - O - \text{nitrogenous compound} \\ \qquad\qquad\quad | \\ \qquad\qquad\quad OH \end{array}$$

↑ Essential Fatty Acid

RCOOH is the fatty acid

The physical properties (such as fluidity) of phospholipids are in large part determined by the chain length and the degree of unsaturation of their component fatty acids. The physical properties, in turn, affect the phospholipid's ability to perform the structural function, for example, the maintenance of normal activities of membrane-bound enzymes such as adenylyl cyclase and Na/K ATPase. Several cellular functions which include secretion, signal transmission, and susceptibility to microorganism invasion depend on membrane fluidity. The highly unsaturated fatty acids EPA and DHA are particularly concentrated where there is a requirement for rapid activity at the cellular level, such as may be required in transport mechanism in the brain, its synaptic junctions, and the retina, where only the long-chain PUFA derived from EFA are found. Other biological functions of EFA include stimulation of growth, maintenance of skin and hair growth, regulation of cholesterol metabolism, lipotropic activity, maintenance of reproductive performance, and other physiologic and pharmacologic effects. A few specific functions are given below.

10.4.2.1 Skin

One essential function of LA is to maintain the integrity of the epidermal water barrier in the skin. The physical structure of water barrier is ascribed to sheets of stacked bilayers that fill intercellular spaces of the uppermost layer of the epidermis. These lipid bilayers contain large amounts of sphingolipids which contain LA-rich ceramides. In EFA deficiency, LA is replaced by oleic acid which results in severe water loss from the skin.

10.4.2.2 Immunity and Infection

There is significant reduction in immune response in EFA deficiency. The data from animal studies have shown that these PUFA have powerful anti-inflammatory and immunomodulatory activities in a wide array of diseases (e.g., autoimmunity, arthritis, and infection) [19]. PUFA may cause change in the membrane fatty acid composition of lymphoid cells. This may cause a change in membrane fluidity leading to alteration of activity of enzymes, receptor expression, and intercellular signaling, which in turn can influence lymphocyte responsiveness. Patients with acquired immune deficiency syndrome (AIDS) exhibit significant reductions of 20 and 22 carbon fatty acids derived from EFA. EPA and DHA are thought to be anti-inflammatory nutrients with protective effects in inflammatory diseases including asthma and allergies [20].

Infection is a common clinical problem in patients undergoing fat-free hyperalimentation. Certain PUFA are known to be effective in killing those viruses that have a lipid component in the envelope. Recent study has shown that W_3 fatty acids from fish oil can prevent wound infection and can improve early wound healing. Combinations of W_3 fatty acids with arginine are much more effective in preventing infections than either one alone [21]. The combinations can reduce the incidence of adult respiratory distress syndrome and shorten the hospital stay in surgical patients. Other studies have shown that W_3 fatty acids could help the body fight lung infections common to inflammatory diseases such as chronic obstructive pulmonary disease.

10.4.2.3 Gene Expression

Cell culture and animal studies suggest that PUFA can modulate the expression of a number of genes including those involved in fatty acid metabolism and inflammation [22]. These PUFA may regulate gene expression by acting like steroid hormones [23].

10.4.2.4 Cholesterol

PUFA of both the W_6 (LA) and W_3 (ALA) family tend to lower plasma cholesterol levels, including the low-density lipoprotein (LDL) cholesterol fraction, possibly by increasing the activity of LDL receptors. Many epidemiologic and controlled interventional studies have shown the antiatherogenic effect of both ALA and its long-chain fatty acids. W_3 acids have unique triglyceride-lowering properties not shared by W_6 fatty acids. PUFA, although effective in lowering both total and LDL cholesterol, have a tendency to lower high-density lipoprotein (HDL) cholesterol, which is protective against coronary heart disease (CHD).

10.4.2.5 Eicosanoids

DHGL/DGLA, AA, and EPA are the precursors of biologically potent metabolites. These include prostaglandins, thromboxanes, prostacyclins, lipoxins, and other related compounds. They participate in many physiological and pathological processes. Eicosanoids are compounds with diverse physiological and pathological properties. They are derived from 20 carbon PUFA (e.g., DHGL/DGLA, AA, and EPA). The precursor fatty acids are normally not found in the free state to an appreciable extent in the cell, but occur as esterified components of phospholipids, preferentially located in the carbon-2 position of glycerol (of the membrane phospholipid). In humans, the tissue phospholipid content of AA is much higher than that of DHGL/DGLA and

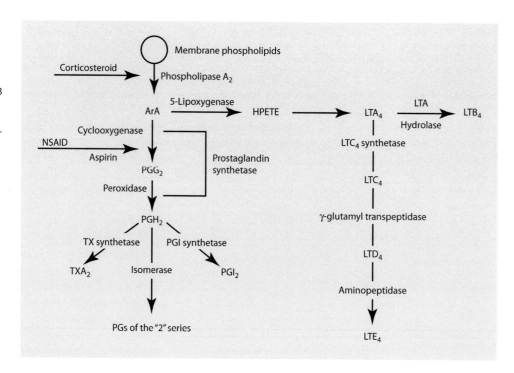

Fig. 10.3 Formation of eicosanoids. The pathway is for arachidonic acid (ArA in the figure). With DHGL/DGLA as the precursor, the PGs, TXs, and PGIs will be of 1 series and LTs of the 3 series; with EPA as the precursor, the PGs, TXs, and PGIs will be of 3 series and LTs of 5 series. (Courtesy of Vishwanath M. Sardesai, PhD)

EPA. Therefore, eicosanoids derived from AA dominate over those from other 2 precursors [24]. All cells except erythrocytes are able to produce one or more types of eicosanoids. They are not stored within cells but are synthesized as required and rapidly inactivated. Classic eicosanoids include prostaglandins (PG), thromboxanes (TX), prostacyclins (PGI), and leukotrienes (LT).

Perturbation of the cell membrane by mechanical, trauma, clinical, or other stimuli activates phospholipase A_2, a membrane-bound enzyme that is found in virtually every cell type and organ in the body, and liberates the precursor fatty acid from phospholipids. This step appears to be rate-limiting because the free fatty acid is readily converted to PGs and other eicosanoids. The synthesis is carried out by a membrane-bound PG synthetase complex which has two components: the cyclooxygenase component converts fatty acid to PGG, and the peroxidase component quickly converts PGG to PGH (Fig. 10.3).

The enzyme cyclooxygenase converts AA to PGH_2. During this process, two double bonds of the precursor fatty acid are lost, and the product has two double bonds (if the precursor is AA). The subscripts 1, 2, and 3 refer to the double bonds of the product. If the precursors were DHGL/DGLA and EPA, the products would be PGH_1 and PGH_3, respectively. Once formed, PGH is acted upon by a series of enzymes to produce PGs, TXs, and PGIs. PGH is converted to several PGs by individual isomerases which are found in all tissues, but some cells are highly selective in the metabolism of PGH.

PGs play important roles in several aspects of human physiology. They regulate a wide variety of processes including stomach secretion, uterus contraction, reproduction, blood pressure control, central nervous system, and inflammation.

In platelets, the principal product is TXA_2 which is formed by the action of TX synthetase on PGH_2 [25]. TXA_2 stimulates platelet aggregation and is also a potent vasoconstrictor. The product formed from PGH_1 is TXA_1, but is not of much significance because PGH_1 is a poor substrate for the enzyme. TXA_3 formed from PGH_3 is less potent as a vasoconstrictor than TXA_2 and has little platelet aggregating ability.

In endothelial cells, PGH_2 is converted to PGI_2 by the action of prostacyclin synthetase. It has an action dramatically opposite of TXA_2. It inhibits platelet aggregation and is a powerful vasodilator. PGH_1 does not have the Δ_5 double bond and therefore cannot be converted to PGI_1. PGI_3 formed from PGH_3 has actions similar to those of PGI_2.

Arachidonic acid can be converted to LTB_4, and cysteine-containing LTC_4, LTD_4, and LTE_4 by the action of 5-lipoxygenase followed by other enzymes. LTB_4 is a potent chemotactic agent attracting neutrophils and macrophages and causing aggregation at sites of infection or injury. LTB_3 produced from DHGL/DGLA has action similar to LTB_4 and is equally potent in its biological activity. The LTE_5 formed from EPA is 10- to 100-fold less active than LTB_4. The cysteinyl LTs are potent vasoconstrictors and bronchoconstrictors. They increase permeability in postcapillary venules and stimulate mucus secretion.

In addition to these classical eicosanoids, there are others produced by different pathways. The products include lipoxins and others with different properties.

Sometimes it becomes necessary to decrease or eliminate the formation of inflammatory eicosanoids. A number of compounds are known to interact at one or more of the many steps involved in the formation of eicosanoids. The steroidal anti-inflammatory agents, corticosteroids, inhibit phospholipase A_2 and block the release of PUFA from membrane phospholipids. This prevents the formation of all eicosanoids. Several synthetic corticosteroids have been developed and are used clinically. Drugs blocking prostanoid synthesis are usually preferred over the more powerful steroids because of their undesirable systemic effects. The nonsteroidal anti-inflammatory drugs (NSAIDs) such as aspirin and ibuprofen are potent inhibitors of cyclooxygenase. These drugs block the synthesis of PGG and hence of PGs, TXs, and PGIs. The NSAIDs have been used to reduce inflammation, pain, and fever as well as platelet stickiness in an effort to reduce the risk of CHD. Aspirin (low doses) is one of the most important and cost-effective drugs for arthritis and for the secondary prevention of CHD. It reduces the rate of arterial thrombotic events in high risk patients by at least 25%. Recent available research data of aspirin's use for primary prevention for high-risk CVD patients is showing a modest benefit. Recommendations are that future aspirin studies consider the extent of benefit–risk for healthy individuals that possibly can be partially counterbalanced by the risks [26, 27].

Dietary intake determines to a great extent the fatty acid composition of phospholipids in the plasma and cell membranes. The eicosanoids derived from AA are, in general, pro-inflammatory, while those derived from DHGL/DGLA and EPA are less inflammatory or anti-inflammatory. EPA can be increased by consuming fish regularly or taking fish oil supplements. DHGL/DGLA can be increased by taking borage oil, which is rich in gamma linolenic acid. It increases DHGL/DGLA which forms PGE_1, a potent antagonist to inflammatory PGE_2. There are several dietary constituents that can modulate the accumulation of PUFA in tissues and/or control the eicosanoid production.

10.4.2.6 Specific Role of W_3 Fatty Acids

ALA is not known to have any specific functions other than to serve as a precursor of EPA and DHA. Biological structures involved in fast movement or signal transmission appear to have a requirement for the highly unsaturated fatty acid DHA. This measure is important for visual and neurological development. In several species, including humans, the retinal rod outer segment disk membrane in which rhodopsin rests, the major phospholipid contains 40–60% of the total fatty acid as DHA. The retina conserves and recycles DHA even when fatty acid intake is low. It is required for normal development and function of the retina. The phospholipids of the brain gray matter contain high proportion of DHA, suggesting it may be important in central nervous system function. Brain DHA content may be particularly important because animal studies have shown that depletion of DHA in the brain can result in learning deficits.

10.4.3 Deficiency

Animal experiments have shown that the young are more susceptible to EFA deficiency than adults. Deficiency in humans is rare because most diets contain adequate EFA to meet the daily requirement. Because adipose tissue lipids in free-living healthy adults contain about 10% of total fatty acids as LA, biochemical and clinical signs of deficiency do not appear during dietary fat restriction or malabsorption when they are accompanied by an energy deficit. In this situation, release of LA and small amounts of AA from adipose tissue reserves may prevent development of EFA deficiency. Young, especially low-birth weight infants have limited body stores of EPA and are more susceptible to deficiency. Cow's milk has only 25% of the LA contained in human milk, but when ingested in normal amounts, it has enough LA to prevent deficiency. Babies fed formula low in LA, such as a skim milk formula, can develop deficiency [28].

In human infants, the deficiency signs include dermatitis, rough and scaly skin, unsatisfactory growth, and impaired water balance. The skin symptoms and water loss are both examples of a general derangement of membrane structure and consequent function [29]. Erythrocytes become more fragile and susceptible to osmotic hemolysis. There is elevated triene/tetraene ratio above 0.2. The most common cause of EFA deficiency in humans of all ages is the long-term intake of fat-free parenteral nutrition (PN). It is commonly administered as a continuous infusion of a glucose-containing solution, which results in a constant elevation of serum insulin. This depresses the release of fat from adipose stores, which in normal adults have a little more than 1000 g of EFA, enough to otherwise sustain dietary needs for more than 6 months. Because of the block in the release of EFA, continuous fat-free PN seems to provide optimal conditions for the development of deficiency [30]. In these patients, plasma-free fatty acids are derived from glucose and these do not include EFA. Glucose-free PN containing only amino acids and fat does not produce EFA deficiency.

Studies in infants maintained on long-term fat-free PN showed development of clinical signs together with biochemical evidence of deficiency (high triene/tetraene ratio). Some premature infants developed very rapid clinical signs and biochemical changes in plasma starting on the second and third day of life. Limited stores of EFA characteristic of the premature state and high caloric expenditure might have been responsible for the early development of deficiency. In the neonate, signs of deficiency may become apparent in 5–10 days on fat-free PN, whereas in adults, the biochemical evidence of deficiency may be seen by the end of 2 weeks after initiation of PN and by the end of 7 weeks; all patients exhibit clinical signs of deficiency. Manifestations of deficiency in these patients include alopecia, brittle nails, dermatitis, increased capillary fragility, indolent wound healing, and increased susceptibility to infection. Administration of diets containing LA reverses both clinical and biochemical abnormalities. To correct or prevent the deficiency when oral intake is denied, LA must be provided intravenously. The minimum LA dose to prevent deficiency state is about 5% of the total calories for adults and 2% for pediatric patients. Deficiency is accelerated by the increased metabolic demands associated with growth and hypermetabolism following injury, sepsis, or stress. These patients should receive 500 ml of 10% lipid emulsion 2–3 times per week.

The first case involving ALA deficiency was described in 1982. A young girl was maintained on PN-containing safflower oil emulsion rich in LA and poor in ALA. After 3 months, she developed visual problems and sensory neuropathy [31]. Serum analysis revealed very low levels of ALA and PUFA derived from it. PN was then changed to include soybean oil, which contains both LA and ALA. Within 3 months, all the symptoms of deficiency were resolved. Since then several other cases involving ALA deficiency have been reported [32, 33].

W_3 PUFA are low in children with attention-deficit hyperactivity disorder (ADHD) compared with other children. Patients with Zellweger syndrome have lower levels of DHA in the brain, retina, and erythrocytes. Recent studies show a correlation between low levels of DHA and certain behavioral and neurological conditions associated with aging, such as dementia, depression, memory loss, and visual problems.

10.4.4 Requirements

The exact requirement of EFA in humans is not clearly defined [34]. Clinical signs of EFA deficiency are generally found only in patients on PN and without a source of PUFA. The plasma triene/tetraene ratio is below 0.2 when dietary EFA are adequate and is increased above 0.2 in relation to the degree of deficiency. The optimum dietary LA required to give a ratio of less than 0.2 and to prevent symptoms of EFA deficiency is 1–2% of total calories. It has been suggested that those functions dependent on eicosanoid formation may show an optimum response with dietary LA at higher levels perhaps as high as 6–10% of total calories. An absolute amount required is not yet known, but because no ill effects have been reported up to this level, the tendency is to consider it as optimal. The average daily intake of LA by adults in most industrialized western countries is about 10 g. The recommended adequate intake is 17 g for young men and 12 g for young women. The quantitative requirement for ALA in humans is not known. Biochemical changes of W_3 fatty acid deficiency include a decrease in plasma and tissue DHA concentrations. There are no accepted cut-off concentrations of plasma or tissue DHA below which functions attributed to W_3 fatty acids, such as visual or neural functions, are impaired. However, on the basis of its derived products found in the blood of patients maintained on PN, it has been suggested that ALA requirement should be between 0.2% and 0.3% of total calories. The recommended adequate intake is 1.6 and 1.2 g per day for men and women, respectively.

In pregnancy, the accumulation of EFA is estimated to be about 620 g, which includes the demand for uterine, placenta, mammary gland, and fetal growth and the increased maternal blood volume. To meet these needs, 4.3% of the accepted calorie intake in the form of EFA is recommended during pregnancy. Approximately 4–5% of total energy in human milk is present as LA and ALA and 1% as long-chain PUFA derived from these acids amounting to about 6% of total energy as EFA and its metabolites. The fat stored during normal pregnancy is utilized during lactation at the rate of about 300 calories per day. Between 3 and 5 g of EFA are secreted in milk per day [35]. An additional 1% and 2% of energy in the form of EFA is recommended during the first 3 months of lactation, and an additional 2% and 4% of the energy above the basic requirements is recommended thereafter. The adequate intake for LA is set at 13 g per day and for ALA 1.3 g per day. For infants, the adequate level for LA is 4.4 g per day, and ALA is 0.5 g per day based on the average amount of LA provided by human milk. Human milk contains long-chain fatty acids including AA and DHA. Normal growth of infants depends on an adequate supply of EFA. The formula milk should have EFA, including AA and DHA similar to those found in human milk.

10.4.5 W_3 Fatty Acids and Health

The tissues of the body require W_3 PUFA for their proper functioning just as they need W_6 PUFA. The importance of W_3 PUFA is their likely role in the prevention of many chronic diseases, now rampant in our society [36]. These especially include CHD and even cancer. EPA and DHA have been assumed to reduce the risk of CHD and stroke by a multitude of mechanisms by preventing arrhythmias, reducing atherosclerosis, decreasing platelet aggregation, lowering plasma triglyceride concentrations, decreasing proinflammatory eicosanoids, and decreasing blood pressure in hypertensive individuals.

Many epidemiologic studies have used fish and fish oil intake as surrogates for W_3 fatty acids intake because of their high content of EPA and DHA found in fish. People who follow a Mediterranean diet are less likely to develop heart disease. The diet emphasizes foods that are rich in W_3 fatty acids. Alaska natives who get high amounts of W_3 fatty acids from eating daily fish also tend to have lower incidence of heart disease. The ratio of LA/ALA is important in the diet because LA and ALA compete for the same desaturase enzyme [37]. Thus, high ratio can inhibit conversion of ALA to DHA, while a low ratio inhibits conversion of LA to AA.

Clinical and epidemiological studies have addressed the W_6/W_3 fatty acid ratio with respect to the beneficial effects on the risk of certain diseases associated with W_3 fatty acids, including EPA and DHA [38]. Low rates of heart disease in Japan compared with the United States have been attributed, in part, to a total W_6/W_3 fatty acid ratio of 4:1 with about 5% energy as LA, 0.6% energy from ALA, and 2% energy from EPA + DHA in Japan, compared with intakes of 6% energy from LA, 0.7% energy from ALA, and less than 0.1% from EPA plus DHA in the United States. Similarly, an inverse association between the dietary total W_6/W_3 fatty acid ratio and cardiovascular disease, cancer, and all-cause mortality as well as between fish intake and CHD mortality has been reported. Based on studies in animals, children, and adults, a reasonable LA/ALA ratio of 5:1 to 10:1 has been recommended for adults.

Two recent studies show that people who eat substantial amounts of fish are greatly protected from sudden unexpected death caused by severely abnormal heart rhythms [39]. In the Physicians' Health Study, 22,000 men were divided in four groups based on the concentrations of W_3 fatty acids in blood. The men in the highest quartile had 81% lower risk of sudden death than those in the lowest quartile during a 17-year period of observation [40]. In the Nurses' Health Study, investigators used dietary information gathered in five interviews between 1980 and 1994 to estimate the fish intake of 85,000 female nurses. Those who ate fish once a week had a 30% lower risk of heart attack or death than those who never ate fish, and those who ate fish 5 times a week had a 34% lower risk. The data also indicated that similar to the mortality from CHD, all-cause mortality was lowest among groups of women who ate the most fish [41].

Results of these studies indicate that even in people with no history of CHD, high blood W_3 fatty acid levels can lower their risk of CHD death. The link to arrhythmias is the most substantiated [42], but fish oils also have other heart-friendly effects, such as lowering the levels of triglycerides in blood, reducing inflammation, slowing coronary artery thickening, and reducing the tendency of blood to clot.

It has been proposed that omega-3 index be used as a biomarker for cardiovascular disease risk [43]. The index is defined as the amount of EPA + DHA in red blood cell (RBC) membranes expressed as the percent of total RBC membrane fatty acids. The EPA + DHA content of RBCs correlates with that of cardiac muscle cell, and several observational studies indicate that a lower omega-3 index is associated with increased risk of CHD mortality [44]. The proposed zones being used are high risk <4%, intermediate risk 4–8%, and low risk >8%. Supplementation with fish oil capsules for a few months increases the omega-3 index.

The American Heart Association recommends that healthy adults, especially those at higher risk for heart disease, eat a variety of fish, preferably oily fish such as salmon, mackerel, and sardines, at least twice a week [45]. The association also recommends increasing the intake of ALA-rich foods such as walnuts, flax seeds, and canola and soybean oil. In September 2004, the US Food and Drug Administration (FDA) allowed the following "qualified" health claim for certain foods containing fish oils: "Supportive but not conclusive research shows that consumption of EPA and DHA W_3 fatty acids may reduce the risk of coronary heart disease" [46]. The FDA action marks the second time the agency has allowed a qualified health claim for a conventional food product: early the same year, the agency granted a claim linking walnuts and certain other nuts to a reduction in heart disease risk.

Oily fish and fish oil consumption have been associated with protection against the development of some types of cancer, such as colorectal, mammary, and prostate cancer [47]. A study has shown that consumption of fish constituting at least a moderate part of the diet significantly lowers the risk of prostate cancer. EPA and DHA have been shown to suppress neoplastic transformation, inhibit cell growth and proliferation, induce apoptosis, and inhibit angiogenesis by suppressing omega-3-derived eicosanoid products.

Several studies have reported a negative relationship between PUFA intake and risk of diabetes. Fish intake has specifically been reported to have a negative association. Rheumatoid arthritis (RA) is an autosomal disease that causes inflammation of joints. Fish oils have been found to reduce symptoms of RA, including joint pain and morning stiffness. People with RA who take ALA may be able to reduce their dose of anti-inflammatory drugs. Fish oil also may help people with osteoarthritis. One study in rats has suggested that diets containing omega-3 fatty acids lead to lower levels of fat accumulation compared with diets containing other fatty acids.

Population studies in Chicago have reported that people 65 and older who ate fish at least once weekly were 60% less likely to develop Alzheimer's disease than those who never or rarely ate fish [48]. Depression is associated with lower levels of W_3 fatty acids in RBC membranes. Countries with the highest rates of depression ate the least amount of fish while those with the lowest rates of depression ate the most fish [49, 50]. W_3 fatty acids may also protect the eyes. In one study, it has been reported that those who ate fish twice a week had a 36% lower risk of macular degeneration, the leading cause of blindness in old age [51].

There are also data suggesting W_3 fatty acids may be helpful for a raft of other ills such as hypertension, aggression, attention deficit disorder, autoimmune diseases, and several other diseases [52, 53]. Fish and fish oil thus qualify as functional foods.

Fish such as salmon and tuna are considered as optimal dietary sources of omega-3 fatty acids. The acceptable daily intake of these fatty acids is 1.6 g for men and 1.1 g for women, although more may be recommended for certain conditions such as heart disease and arthritis. Fish contain mercury and other unsafe pollutants which may limit the amount of fish that should be consumed, particularly for children and women who are pregnant or breastfeeding. Mercury is a known toxin for the human nervous system and is linked to learning and behavioral problems in children [54]. In adults, mercury can cause memory loss and other health problems. For those who do not eat fish, a fish oil supplement may be considered. The best fish oil supplements contain 1 g per capsule, which provides 440 mg of W_3 fatty acids. The most common fish oil capsules in the USA provide 180 mg of EPA and 120 mg of DHA. The recommended dosage is 1–2 capsules a day for adults. Large doses may exert a dose-related effect on lowering bleeding time.

With the increasing popularity of the vegetable diet and mounting focus on mercury and other contaminants in seafood, flaxseed oil which contains ALA has become a safer choice. We can make EPA and DHA from ALA, but the process is slow. Researchers at Dow AgroSciences inserted the algae genes required to make DHA into canola seeds. The enzyme that these genes produce allows the canola plant to synthesize the DHA from ALA [55]. Such plants produce canola oil that is enriched with DHA. Monsanto genetically engineered a soybean plant to produce W_3-derived products in soybean oil.

10.4.6 Food Sources

The most prevalent source of W_6 fatty acids is LA. It is present in almost all vegetable oils such as safflower oil, corn oil, sunflower oil, and soybean oil. Common fatty acids found in most land plants are generally not elongated above the 18-carbon level. Borage seed oil is the richest source of γ-linolenic acid (GLA) among vegetable oils. It contains 24% of GLA. Black currant is the next with 17% and evening primrose contains 8% of GLA. This fatty acid is an intermediate in the conversion of LA to AA. Dietary meat, poultry, and eggs provide small amounts of AA. The W_6 PUFA intake in the USA ranges from 12 to 17 g per day in men and 9–11 g per day in women. It accounts for 5–7% of total energy intake in diets of adults. Most W_6 PUFA (80–90%) are consumed in the form of LA. Others such as AA and GLA are in small amounts in the diets.

The major sources of W_3 fatty acids as ALA include vegetable oils such as soybean, canola, and flaxseed oil. Flaxseed is the rich source containing 50–60% ALA. ◘ Table 10.3 shows the linolenic acid content of some foods.

Leafy vegetables, such as spinach, kale, and romaine lettuce, contain small amounts. Fish oils provide a mixture of EPA and DHA, and fatty fish are the major dietary sources of EPA and DHA. ◘ Table 10.4 shows the content of EPA and DHA in some types of fish.

The intake of total omega-3 fatty acids in the United States is about 1.6 g per day (0.7% of energy intake), and of this, ALA accounts for 1.4 g/day and only 0.1–0.2 g/day of EPA and DHA.

◘ **Table 10.3** Selected food sources of ALA

Source	Serving size	Content in grams
Canola oil	1 tablespoon	1.30
Walnut oil	1 tablespoon	1.40
Flaxseed oil	1 tablespoon	7.48
Soybean oil	1 tablespoon	0.90
Flaxseed ground	1 tablespoon	7.48
Walnuts	1 ounce	2.60
Chia seeds	1 ounce	4.90
Soybeans	1 ounce	0.47
Hickory nuts	1 ounce	0.29

◘ **Table 10.4** Selected food sources of EPA and DHA content in grams

Source	Serving size (ounce)	EPA (g)	DHA (g)
Herring	3	1.06	0.75
Salmon	3	0.86	0.62
Sardines	3	0.45	0.74
Crab	3	0.24	0.10
Oysters	3	0.75	0.93
Tuna	3	0.40	0.44
Trout	3	0.45	0.74
Mackerel	3	0.43	0.59
Shrimp	3	0.07	0.10
Flounder	3	0.24	0.25

10.5 *Trans*-Fatty Acids

Due to fermentation in ruminant animals such as cows, small amounts of *trans*-fatty acids exist naturally in meat and dairy products. This was the source of *trans*-fatty acids in diets of humans until the beginning of the last century.

After the introduction of Crisco in 1911, partially hydrogenated fat was used to make margarine and numerous other products, including crackers, cookies, and other baked goods and fried snacks such as potato chips. Food manufacturers started using partially hydrogenated fats because of their favorable properties, such as longer shelf life, stability during frying, and palatability. Consumers preferred these products (usually considered to be "cholesterol-free" and high in PUFA) to lower cholesterol intake, theoretically reducing the risk of CHD. The consumption of these products continued to increase steadily [56].

Trans-fatty acids account for 3–8% of the fat in the American diet or 8.3–13.5 g per day. However, the intake may be much higher for persons consuming large amounts of commercially baked products and fried foods. A medium-size helping of French fries contains 5–6 g, a doughnut contains 2 g, and an ounce of crackers 2 g of *trans*-fatty acids.

Prior to 1980, there was generally no concern about the trend toward increased consumption of hydrogenated fat in the US diet, especially because it displaced fat relatively high in saturated fat. In the 1980s, studies showed hypercholesterolemic effects of *trans*-fatty acids in rabbits. Several studies in humans reported that a diet enriched with *trans*-fatty acids caused blood LDL cholesterol to increase and HDL cholesterol to decrease, resulting in less favorable total cholesterol/HDL ratio to increase [57]. Two studies also showed a rise in plasma lipoprotein(a) with relatively high consumption of *trans*-fatty acids. Similar to LDL, the concentration of lipoprotein(a) in plasma is directly associated with increased risk for CHD.

On a per calorie basis, *trans*-fatty acids currently appear to increase the risk of CHD more than any other macronutrient at low levels of consumption (1–3% of energy intake). The major evidence came from the Nurses' Health Study (NHS) – a cohort study that has been following approximately 120,000 female nurses since its inception in 1976. Researchers analyzed data from 900 coronary events from the NHS population during 14 years of follow-up and determined that a nurse's CHD risk roughly doubled for each 2% increase in *trans*-fat calories consumed (instead of carbohydrate calories). By contrast, it took a more than 15% increase in saturated fat calories (instead of carbohydrate calories) to produce a similar increase in risk. *Trans*-fatty acids behaved similar to saturated fatty acids by increasing plasma LDL cholesterol, but unlike saturated fat, it had the additional effect of lowering plasma HDL cholesterol levels [58]. Other adverse effects reported were on growth and development, inhibition of conversion of LA and ALA to AA and DHA, respectively, possible association with Alzheimer's disease, and increase in weight gain [59].

In recent years, a series of studies has provided unequivocal evidence that *trans*-fat adversely affects health [60]. In 2003, Denmark became the first country to ban *trans*-fat. Many other countries followed suit. In the United States, New York City passed such a ban in restaurants in 2006 and California did the same in 2008.

The US Food and Drug Administration addressed this issue by requiring disclosure of the *trans*-fat content in food in Nutrition Labels, beginning in 2006. On March 16, 2015, the FDA took further action that will significantly reduce the use of partially hydrogenated oils, the major source of artificial *trans*-fat in the food supply [61]. The FDA is providing the companies 3 years to either reformulate products without partially hydrogenated oils and/or petition FDA to permit specific uses. Food companies have already been working to remove partially hydrogenated oils from processed foods, and FDA anticipates that many may eliminate them ahead of the three-year compliance date. This action is expected to reduce cardiovascular disease risk. The Centers for Disease Control and Prevention estimates that this action could prevent as many as 20,000 coronary events and 7000 deaths from coronary causes each year in the USA.

10.6 Conjugated Fatty Acids

Conjugated linoleic acid (CLA) represents a collective term for a group of geometric and positional isomers of LA that contain a conjugated double bond system instead of the isolated double bonds. It has attracted a fair amount of attention and has been reported to have different biological effects in health-related disorders [62].

CLA is beneficial in lowering body fat while preserving muscle tissue. Individuals who took 3.2 g CLA per day had a drop in fat mass of about 0.2 lb. each week (i.e., about 1 lb. a month) compared with those given a placebo. Some studies indicate CLA has potent beneficial effects as antitumor, antiobesity, antiatherogenic, and antidiabetic activities. The molecules have been shown to modulate immune function mainly in animals and in in vitro studies [63].

Small amounts of CLA are present in all diets. It is usually found in beef and dairy products. The best sources of CLA are from grass-fed cattle and its products. The average intake in the USA is in the range of 151 to 212 mg/day. CLA is marketed in dietary supplement form for its anticancer benefit, for which there is some evidence, but no known mechanism, and as a body building aid. Despite the many claims made, good evidence of human health benefits remains scarce.

10.7 Conclusions

Fatty acids are integral components of the lipid macronutrients that have four major physiological roles. First, fatty acids are fuel molecules stored as triacylglycerols, or triglycerides, that have a neutral charge mainly stored in adipose tissue in adipocyte cells. The triacylglycerols are oxidized and able to provide energy needs of the cells or tissues. Second, fatty acids function as the base unit of phospholipids and glycolipids, colluding with cholesterol responsible for modulating cell membrane fluidity (see ▶ Chap. 22). These amphipathic molecules are important components of all the biological bilayer membranes that allow for highly selective permeability barriers. Membrane processes such as transport and cell sensing depend of the fluidity of the membrane modulated by cholesterol (see ▶ Chap. 12). Dietary intake of fats and cholesterol are primary determinants of the structure and composition of the membranes affecting their function. Third, fatty acids by a covalent attachment can modify proteins to position the location of the protein on the membrane structure. Specific proteins involved with mediating membrane functions such as energy transduction, transport, and cell-sensing communication. The fourth role of fatty acids is their synthesis to derivatives that serve as hormones and intracellular messengers. Fatty acids and their role in human metabolism are presented in this chapter to provide the science and principles for which the nutrition practitioner can guide disease management for assessment of an individual's fatty acid status and develop therapeutic interventions toward restoring optimal function (see ▶ Chap. 11).

Acknowledgments I wish to gratefully acknowledge the secretarial assistance of Cindy Luiz in the preparation of this manuscript. I also wish to thank Amie Dozier, Associate Director, Medical Education Support Group of the School, for preparing the illustrations and Cynthia Washell, Practice Administrator, Department of Surgery, for providing support.

References

1. Fritsche K. The science of fatty acids and inflammation. Adv Nutr. 2015;6(3):293S–301S. Published 7 May 2015. https://doi.org/10.3945/an.114.006940.
2. Alberts B, Johnson A, Lewis J, et al. Molecular biology of the cell. 4th ed. New York: Garland Science; 2002. The lipid bilayer. Available from: https://www.ncbi.nlm.nih.gov/books/NBK26871
3. Berg J, Tymoczko J, et al., editors. Biochemistry 9th ed. W.H. Freeman & Co.; New York: 2019. p. 860. Plus index. p. 4. ISBN: 978-1-319-11467-1.
4. Brody T. Nutritional biochemistry 2nd ed. Academic Press; 1999. p. 320. ISBN: 0121348369. Retrieved 21 Dec 2012.
5. Daley C, Abbott A, Doyle P, Nader G, Larson S. A review of fatty acid profiles and antioxidant content in grass-fed and grain-fed beef. Nutr J. 2010;9:10. Published 10 Mar 2010. https://doi.org/10.1186/1475-2891-9-10.
6. Canfora E, Jocken J, Blaak E. Short-chain fatty acids in control of body weight and insulin sensitivity. Nat Rev Endocrinol. 2015;11:577–91.
7. Greer JB, O'Keefe SJ. Microbial induction of immunity, inflammation, and cancer. Front Physiol. 2011;1:168.
8. Scheppach W. Effects of short chain fatty acids on gut morphology and function. Gut. 1994;35(1 Suppl):S35–8.
9. Andoh A, Tsujikawa T, Fujiyama Y. Role of dietary fiber and short-chain fatty acids in the colon. Curr Pharm Des. 2003;9(4):347–58.
10. Canfora E, Jocken J, Blaak E. Short-chain fatty acids in control of body weight and insulin sensitivity. Nat Rev Endocrinol. 2015;11:577–91. Figure 2. SCFA and interorgan crosstalk.
11. Wong JM, De Souza R, Kendall CW, Emam A, Jenkins DJ. Colonic health: fermentation and short chain fatty acids. J Clin Gastroenterol. 2006;40(3):235–43.
12. St-Onge MP, Bosarge A, Goree LL, Darnell B. Medium chain triglyceride oil consumption as part of a weight loss diet does not lead to an adverse metabolic profile when compared to olive oil. J Am Coll Nutr. 2008;27:547–52.
13. Altar T. More than you wanted to know about fats/oils. Sundance Natural Foods. Retrieved 31 Aug 2006.
14. Vallim T, Salter AM. Regulation of hepatic gene expression by saturated fatty acids. Prostaglandins Leukot Essent Fatty Acids. 2010;82(4–6):211–8.
15. Sardesai VM. The essential fatty acids. Nutr Clin Pract. 1992;7(4):179–86.
16. Holman RT. George O. Burr and the discovery of essential fatty acids. J Nutr. 1988;118(5):535–40.
17. Sardesai VM. Nutritional role of polyunsaturated fatty-acids. J Nutr Biochem. 1992;3(4):154–66.
18. Burdge G. Alpha-linolenic acid metabolism in men and women: nutritional and biological implications. Curr Opin Clin Nutr Metab Care. 2004;7(2):137–44.
19. Das UN. Infection, inflammation, and polyunsaturated fatty acids. Nutrition. 2011;27(10):1080–4.
20. Fritsche K. Fatty acids as modulators of the immune response. Annu Rev Nutr. 2006;26:45–73.
21. Alexander JW, Supp DM. Role of arginine and omega-3 fatty acids in wound healing and infection. Adv Wound Care (New Rochelle). 2014;3(11):682–90.
22. Jump DB, Tripathy S, Depner CM. Fatty acid-regulated transcription factors in the liver. Annu Rev Nutr. 2013;33:249–69.
23. Georgiadi A, Kersten S. Mechanisms of gene regulation by fatty acids. Adv Nutr. 2012;3(2):127–34.
24. Sardesai VM. Eicosanoids. In: Introduction to clinical nutrition, vol. xxix. 3rd ed. Boca Raton: Taylor & Francis/CRC Press; 2012. p. 674.
25. Sardesai VM. Biochemical and nutritional aspects of eicosanoids. J Nutr Biochem. 1992;3(11):562–79.
26. Brotons C, Benamouzig R, Filipiak KJ, Limmroth V, Borghi C. A systematic review of aspirin in primary prevention: is it time for a new approach? Am J Cardiovasc Drugs. 2014;15(2):113–33. https://doi.org/10.1007/s40256-014-0100-5.
27. McNeil JJ, Wolfe R, Woods RL, et al. Effect of aspirin on cardiovascular events and bleeding in the healthy elderly. N Engl J Med. 2018;379(16):1509–18. https://doi.org/10.1056/NEJMoa1805819.
28. Hansen AE, Wiese HF, Boelsche AN, Haggard ME, Adam DJD, Davis H. Role of linoleic acid in infant nutrition: clinical and chemical study of 428 infants fed on milk mixtures varying in kind and amount of fat. Pediatrics. 1963;31:171–92.
29. Holman RT, Johnson SB. Essential fatty acid deficiencies in man. In: Perkins EG, Visek WJ, editors. Dietary fats and health. Champaign: American Oil Chemists' Society; 1983. p. 247–66.
30. Hirono H, Suzuki H, Igarashi Y, Konno T. Essential fatty acid deficiency induced by total parenteral nutrition and by medium-chain triglyceride feeding. Am J Clin Nutr. 1977;30(10):1670–6.
31. Holman RT, Johnson SB, Hatch TF. A case of human linolenic acid deficiency involving neurological abnormalities. Am J Clin Nutr. 1982;35(3):617–23.
32. Bjerve KS. Alpha-linolenic acid deficiency in adult women. Nutr Rev. 1987;45(1):15–9.
33. Bjerve KS. n-3 fatty acid deficiency in man. J Intern Med Suppl. 1989;731:171–5.
34. Flock MR, Harris WS, Kris-Etherton PM. Long-chain omega-3 fatty acids: time to establish a dietary reference intake. Nutr Rev. 2013;71(10):692–707.
35. Hansen AE, Haggard ME, Boelsche AN, Adam DJ, Wiese HF. Essential fatty acids in infant nutrition. III. Clinical manifestations of linoleic acid deficiency. J Nutr. 1958;66(4):565–76.
36. Simopoulos AP. Essential fatty acids in health and chronic disease. Am J Clin Nutr. 1999;70(3 Suppl):560S–9S.
37. Lands WE, Hamazaki T, Yamazaki K, Okuyama H, Sakai K, Goto Y, Hubbard VS. Changing dietary patterns. Am J Clin Nutr. 1990;51(6):991–3.
38. Simopoulos AP. The importance of the omega-6/omega-3 fatty acid ratio in cardiovascular disease and other chronic diseases. Exp Biol Med (Maywood). 2008;233(6):674–88.
39. Albert CM, Campos H, Stampfer MJ, Ridker PM, Manson JE, Willett WC, Ma J. Blood levels of long-chain n-3 fatty acids and the risk of sudden death. N Engl J Med. 2002;346(15):1113–8.
40. Wilk JB, Tsai MY, Hanson NQ, Gaziano JM, Djousse L. Plasma and dietary omega-3 fatty acids, fish intake, and heart failure risk in the Physicians' Health Study. Am J Clin Nutr. 2012;96(4):882–8.
41. Hu FB, Bronner L, Willett WC, Stampfer MJ, Rexrode KM, Albert CM, Hunter D, Manson JE. Fish and omega-3 fatty acid intake and risk of coronary heart disease in women. JAMA. 2002;287(14):1815–21.
42. Yashodhara BM, Umakanth S, Pappachan JM, Bhat SK, Kamath R, Choo BH. Omega-3 fatty acids: a comprehensive review of their role in health and disease. Postgrad Med J. 2009;85(1000):84–90.
43. Harris WS. The omega-3 index as a risk factor for coronary heart disease. Am J Clin Nutr. 2008;87(6):1997S–2002S.
44. Harris WS. Are n-3 fatty acids still cardioprotective? Curr Opin Clin Nutr Metab Care. 2013;16(2):141–9.
45. American Heart Association. Fish and omega-3 fatty acids. Available from: https://www.heart.org/en/healthy-living/healthy-eating/eat-smart/fats/fish-and-omega-3-fatty-acids.
46. HIH/ODS. Omega-e fatty acids. Available from: https://ods.od.nih.gov/factsheets/Omega3FattyAcids-HealthProfessional/.
47. Gerber M. Omega-3 fatty acids and cancers: a systematic update review of epidemiological studies. Br J Nutr. 2012;107(Suppl 2):S228–39.
48. Morris MC, Evans DA, Bienias JL, Tangney CC, Bennett DA, Wilson RS, Aggarwal N, Schneider J. Consumption of fish and n-3 fatty acids and risk of incident Alzheimer disease. Arch Neurol. 2003;60(7):940–6.
49. Bruinsma KA, Taren DL. Dieting, essential fatty acid intake, and depression. Nutr Rev. 2000;58(4):98–108.

50. Sarris J, Schoendorfer N, Kavanagh DJ. Major depressive disorder and nutritional medicine: a review of monotherapies and adjuvant treatments. Nutr Rev. 2009;67(3):125–31.
51. Hodge WG, Barnes D, Schachter HM, et al. Evidence for the effect of omega-3 fatty acids on progression of age-related macular degeneration: a systematic review. Retina. 2007;27(2):216–21.
52. Riediger ND, Othman RA, Suh M, Moghadasian MH. A systemic review of the roles of n-3 fatty acids in health and disease. J Am Diet Assoc. 2009;109(4):668–79.
53. Miyata J, Arita M. Role of omega-3 fatty acids and their metabolites in asthma and allergic diseases. Allergol Int. 2015;64(1):27–34.
54. Mozaffarian D, Rimm EB. Fish intake, contaminants, and human health: evaluating the risks and the benefits. JAMA. 2006;296(15):1885–99.
55. Arnold C. Fish out of water. Chem Eng News. 2008;86(32):39–41.
56. Mozaffarian D, Katan MB, Ascherio A, Stampfer MJ, Willett WC. Trans fatty acids and cardiovascular disease. N Engl J Med. 2006;354(15):1601–13.
57. Willett WC, Stampfer MJ, Manson JE, Colditz GA, Speizer FE, Rosner BA, Sampson LA, Hennekens CH. Intake of trans-fatty-acids and risk of coronary heart-disease among women. Lancet. 1993;341(8845):581–5.
58. Oh K, Hu FB, Manson JE, Stampfer MJ, Willett WC. Dietary fat intake and risk of coronary heart disease in women: 20 years of follow-up of the nurses' health study. Am J Epidemiol. 2005;161(7):672–9.
59. Brownell KD, Pomeranz JL. The trans-fat ban–food regulation and long-term health. N Engl J Med. 2014;370(19):1773–5.
60. Dietz WH, Scanlon KS. Eliminating the use of partially hydrogenated oil in food production and preparation. JAMA. 2012;308(2):143–4.
61. Dennis B. FDA moves to ban trans fat from U.S. food supply. The Washington Post. 16 June 2015.
62. Belury MA. Dietary conjugated linoleic acid in health: physiological effects and mechanisms of action. Annu Rev Nutr. 2002;22:505–31.
63. Nagao K, Yanagita T. Conjugated fatty acids in food and their health benefits. J Biosci Bioeng. 2005;100(2):152–7.

Resources

Armstrong, D, editor. Lipidomics volume 1: methods and protocols. New York: Humana Press; 2012. p. 549. ISBN: 978-1-60761-322-0 https://doi.org/10.1007/978-1-60761-322-0.

Berg J, Tymoczko J, et al., editors. Biochemistry 9th ed. W.H. Freeman & Co.; 2019. p. 860. plus index. ISBN: 978-1-319-11467-1.

Chow C, editor. Fatty acids in foods and their health implication, 3rd ed. Boca Raton: CRC Press; 2008. p. 1281. ISBN: 978-0-8493-7261-3.

Horrobin D, editor. Omega-6 essential fatty acids. New York: Wiley-Liss; 1990. ISBN-10: 0-471-56693-4.

Mouritsen O, Bagatolli L. Life as a matter of fat, 2nd ed. Berlin: Springer; 2016. p. 298. ISBN: 978-3-319-22614-9.

Quinn P, Wang X, editors. Lipids in health and disease. New York: Springer; 2008. p. 597. ISBN: 978-1-4020-8830-8.

Sardesai V. Introduction to clinical nutrition, 3rd ed. Boca Raton: Taylor & Francis Group LLC/CRC Press; 2012. ISBN-10: 1-439-81818-5.

Lipidomics: Clinical Application

Diana Noland

11.1	Introduction – 153	
11.2	History of Dietary Fat – 153	
11.2.1	Human History – 153	
11.2.2	The Nutrition Transition of Oils and Fats in the Early 20th Century – 154	
11.3	The Lipidome and Clinical Application – 154	
11.3.1	Clinical Imbalances – 155	
11.4	Nutritional Influences on Body Composition and Function – 156	
11.4.1	Structure and Functions of the Cell Membrane – 156	
11.5	The Eicosanoid Cascade: Acute and Chronic Tissue Inflammation Management – 158	
11.5.1	Fatty Acid Elongation (See ◘ Fig. 11.4) – 159	
11.5.2	Fatty Acid Desaturation (See ◘ Fig. 11.4) – 159	
11.6	Metabolic Stressors – 160	
11.7	Tools of the Trade for Lipid Therapy – 160	
11.7.1	Laboratory Principles – 163	
11.7.2	Structural Integrity – 163	
11.7.3	Defense and Repair – 163	
11.8	Key Nutrient Cofactors and Foods Influencing the Eicosanoid Metabolism – 163	
11.8.1	Lipids – 163	
11.8.2	Sterols – 164	
11.8.3	Minerals – 164	
11.8.4	Methyl Nutrients – 164	
11.8.5	Phytonutrients: Protective Support for Lipid Structures – 165	
11.9	Key Lifestyle Factors Influencing the Risk of Lipid Damage – 165	
11.9.1	Sleep [74] (See ▶ Chap. 35) – 165	
11.9.2	Stress (See ▶ Chap. 47) – 165	
11.9.3	Movement (See ▶ Chaps. 36 and 54) – 165	

© Springer Nature Switzerland AG 2020
D. Noland et al. (eds.), *Integrative and Functional Medical Nutrition Therapy*,
https://doi.org/10.1007/978-3-030-30730-1_11

11.10	**Chronic Disease and Impaired Lipid Metabolism (See Tables ◘ 11.3 and 11.4) – 166**
11.10.1	Heart Disease/Cardiovascular Association with the Lipidome – 166
11.10.2	Oncology – 166
11.10.3	Neurological – 166
11.10.4	Respiratory – 167
11.10.5	Autoimmune – 167

11.11 **Case Reviews – 168**

11.12 **Conclusion – 170**

References – 170

Learning Objectives
1. The Lipidome and Clinical Application
2. Nutritional Influences on Composition and Function
3. Tools of the Trade for Lipid Therapy
4. Case Reviews

11.1 Introduction

The main global healthcare concern of the twenty-first century is chronic disease. Public health leaders around the world are working to quell the unsustainable widespread changing phenotype of populations. In the United States, more than two-thirds of the citizens are currently overweight or obese, and the trend is increasing. In 2012, there began an unprecedented decline in U.S. lifespan. Much of the trend toward chronic disease is due to dysfunction in the underlying mechanisms of cell signaling and the messages that operate at the organelle and cell membrane sites. This is where the DNA is expressed and put into action as body systems read the environment and determine how to respond to survive.

IFMNT is a person-centered approach with each individual assessed, and interventions developed, as personalized therapy based on the nutrition and medical data that is gathered during a comprehensive evaluation. For population of persons with the same diagnosis, there will be an equal number of uniquely designed protocols. Each disease has many causes, thus this chapter does not give a recipe for diagnosis and intervention rather provides principles and tools to draw from to form the best protocol to restore wellness.

In this chapter, we first examine the structure of the cell membrane, which is comprised of different regions of *microdomains* referred to as *membrane rafts* with varying percent lipids (e.g., fats, phospholipids, steroids), which are mainly formed from the food fats and oils we eat [1]. The structure of membranes determines the function of the "control tower" of our systems where communication and cell signaling occurs. Second, we assess function and how well the body can manage all its systems to enable them to produce a healthy organism. The evidence that the lipids we eat create the composition of our lipid structures and control our function is mounting in recognition of the metabolic plasticity of human metabolism [2, 3] (see ▶ Chap. 10). At the time of this publication, there is public and scientific interest in the comparison of the effects of high carbohydrate, low fat diets to those with low carbohydrates and high fat. The main thrust of this latter diet protocol is exchange of carbohydrates for fats. Nutrition professionals need to be aware of the metabolic implications of this dietary change and the science behind what is often referred to as paleo or ketogenic (keto) diets (see ▶ Chap. 23).

This chapter addresses emerging evidence about fatty acid metabolism and outdated misinformation that is still accepted dogma regarding dietary fat. Understandably, the public and even nutrition professionals are confused about dietary fats. We are on the brink of a new understanding of the science of lipids, food processing, and how one's unique personal genetics determine the type and amount of fats that may benefit an individual.

The goal of this chapter is to help the nutrition-trained professional understand the molecular structure of an individual's cellular membranes and how those structures affect the efficiency and proper functioning of metabolism. With that knowledge, one has the clinical skills in medical nutrition therapy for assessment and intervention. The practitioner must have working knowledge of foods and supplements, which support the body's *metabolic plasticity*. This approach is viewed through the lens of *systems biology*, which is the science of all systems interacting and influencing each other to produce the phenotype of an individual; it is often referred to as integrative and/or functional medicine (see ◘ Table 11.1).

◘ Table 11.1 Integrative and functional medicine principles
Focuses on patient/client-centered approach
Acknowledges biochemical uniqueness of each individual
Patient practitioner partnership
Appropriately uses conventional and complementary treatments
Considers genetic, environmental, and lifestyle factors
Seeks balance between mind, body, and spirit
Acknowledges all body systems are interconnected by physiological and metabolic factors
Identifies health as a positive vitality
Uses natural, effective, less invasive interventions, when possible

11.2 History of Dietary Fat

It can be enlightening to think back on historical diets and events that led to the current recommendations about fat. Around the 1950s, recommendations to Americans concerning fat were promoted by publications from Gofman [3] and Keys (1953) [4], in which the dogma was propagated that "saturated fat and cholesterol were bad." A connection was made with respect to the rise of heart disease and epidemiologic correlations between dietary saturated fat, cholesterol and heart disease. Not much has changed in the last 70 years in recommendations to reduce total fat, saturated fat, and cholesterol-rich foods. Until the 1940s in the United States, heart disease and cancer were minor contributors to mortality statistics, but they have continued to grow to the number-one and -two causes of death in industrialized countries.

11.2.1 Human History

Fats used for dietary ingestion were "all natural" (not processed) prior to ~1900. Lipids, oils, and fats commonly used were:

- Fat or oil-rich foods
  - Meat
  - Poultry and skin
  - Fish, shellfish, roe
  - Egg yolk
  - Olives, olive oil
  - Other plants: nuts, seeds, avocado
  - Essential and/or medicinal oils
- Cooking oils
  - Butter, ghee
  - Lard
  - Tallow
  - Coconut oil
  - Olive oil, foot-pressed

The nutrition transition to the standard American diet (SAD) [5] at the beginning of the twentieth century paralleled the increase in the onset and progression of chronic human diseases as the food supply changed in quality and quantity, giving rise to obesity and a mismatch of gene–diet interactions. Cooking oils intolerant to heat, processed foods containing trans fats, and damaged oils have become commonplace. There is currently a debate regarding the impact of the increase in omega-6 on human health [6] and newer processing methods in the food oil industry that are allowing "damaged" oils and "new-to-nature" forms of oils into the food supply (e.g., hydrogenated vegetable oils, high heat processing of vegetable oils, changing proportion of linoleic acid, and monounsaturated fats to "high oleic" vegetable oils). Issues that need to be addressed include the influence of ingested food oils on the structure and function of cellular health and how these food oil changes have contributed to the epidemic of chronic disease?

11.2.2 The Nutrition Transition of Oils and Fats in the Early 20th Century

Processed oils:
- Heat-processed vegetable oil
- High heated commercial oil used for deep frying
- Hydrogenation of vegetable oils
- Charred red meats or high-temp-oil deep frying produce trans fats or acrylamides

Specialty foods available after 1950:
- *Natural*: nitrogen-packed seed/nut oils, refrigerated
  - Sunflower, safflower, flax (refrigerated), avocado oil, flax–safflower oil 4:1, macadamia oil
- Processed:
  - Deep frying with vegetable oils
  - Hydrogenated or partially hydrogenated vegetable oils

The ratio between the types of fats has changed dramatically during the past 70–100 years with a three-fold increase in dietary levels of the omega-6 (n-6) 18 carbon (C18), polyunsaturated fatty acids (PUFA), linoleic acid (LA; 18:2n-6) [5].

Horrobin [6] proposed an ideal ω6:ω3 ratio of 4:1, and most fatty-acid scientists agree optimum ratio somewhere with the range of 1:1 to 5:1 ratios [7, 8].

Chronic diseases are characterized as long-latency lifestyle and diet-related. Chronic diseases all have inflammatory pathophysiology promoted by injury, infection, or biological stressors (e.g., chronic inflammation, high visceral adiposity, emotional stress). For example, an acute infection may either be resolved by a healthy immune system or survive and continue as a subclinical infection that is often not recognized but continues to wear on the immune system. Lipids that participate in eicosanoid metabolism are largely responsible for control of inflammation and the ability of the metabolic resolution of an inflammatory event.

The healthy human body is equipped with defense features from conception throughout life to interact with the environment to protect it from infection and injury. All of the barriers' defensive functions are a reflection of their lipid-dominant structure. Much of the defense starts with the lipid-rich skin barrier and microbiome at all body orifices to protect from pathogens entering and infecting or causing injury.

11.3 The Lipidome and Clinical Application

The application of lipidomics in clinical practice is an important new tool for the healthcare practitioner due to expanding knowledge of the structural and functional properties of fat. Dietary fat is a key topic in government food policy, research, public media, and in the homes of families preparing food during a period of dramatic change in thinking about how we incorporate fat into diets. Many people feel confused about the amount and types of fat to eat. Is saturated fat bad? Is a low-fat diet good? Dietitians and healthcare professionals are beginning to learn about the emerging science of lipidomics and how that applies to the science of lipid metabolism and the use of dietary and food supplement to maintain health.

Lipids are highly diverse molecules that are as important for life as proteins and genes [2] with critical roles in membrane structure, cell signaling, energy storage, inflammation regulation, and as base units for constructing messenger hormones. Maintaining lipid balance and homeostasis is within a practitioner's capability when they are skilled at recognizing lipid imbalances and their relationship to disease pathology. One of the profound roles of lipid metabolism involves metabolic regulation of inflammation. It is essential to understand how to modulate lipid metabolism considering that the global epidemic of chronic disease in healthcare reflects prolonged and unresolved inflammation. All the "ingredients" of lipids (fatty acids, phospholipids, sterols, sphingolipids) and the associated enzymes and nutrient cofactors affect the control and resolution of inflammation. Knowledge of lipid modulation by dietary and lifestyle changes is a powerful addition to the toolbox of the IFMNT practitioner.

The two functions of the lipids that can be modulated by nutrition therapy are membrane structure and inflammation control.

- Membrane structure: the membrane is at least 50–75% lipids with embedded protein structures forming receptors, channels, and other structures [6] "You are what you eat." What fats and oils and sterols you eat become the structural composition of your membranes and influence their function of cell signaling, communication, and transport.
- Inflammation control: The lipid eicosanoid molecules play a key role in our survival. They are the primary metabolites teaming with the immune system to manage the immune response and control inflammation. Dysregulated lipid metabolism and nutrient status are thought to play a major role in the pathophysiology of every chronic disease. Since chronic prolonged inflammation is present in every chronic disease, the eicosanoids become priority in supporting their metabolic function. And the most effective way of modulating the eicosanoid cascade is nutrition lipid therapy guiding dietary intake of fats and oils and nutrient cofactors.

Each individual is unique. IFMNT is a person-centered approach with each individual assessed, and interventions develop as personalized therapy based on the nutrition and medical data that are gathered in the initial interview. For a population with the same diagnosis, there will be that many different protocols recommended. For one disease, there are many causes. That is why this chapter does not include specific interventions for a diagnosis but provides principles to consider and draw from in forming the best regulation of structure and inflammation control to restore wellness.

11.3.1 Clinical Imbalances

The conceptual diagram below (see ◘ Fig. 11.1) provides a guide to hearing the patient's whole story, so that an assessment can be as comprehensive as possible to narrow down root causes of the patient's condition. In seeking root causes of the etiology of a disease condition, one must investigate the health and *structural integrity* of the cell membrane. All cells in the body share the basic structure and function of a

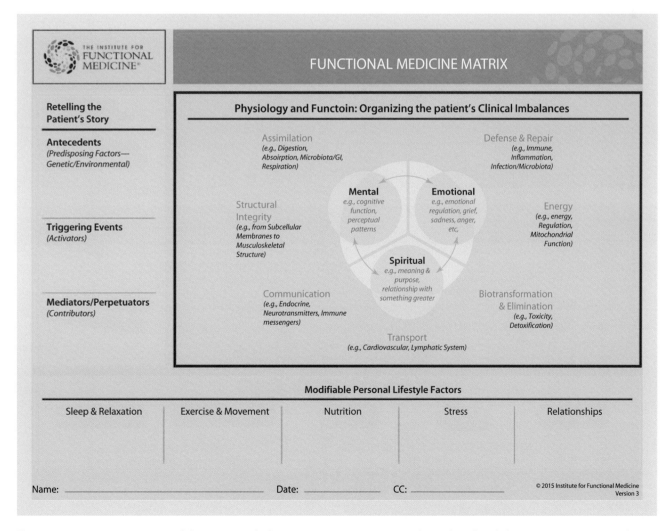

◘ Fig. 11.1 IFM Matrix™ Conceptual diagram to guide the practitioner assessment procedure to hear the whole patient's story preparing for improved diagnosis of root causes of health issues. (Used with permission from The Institute for Functional Medicine ©2015)

bilayer membrane made of 50–75% phospholipids, fatty acids, cholesterol with embedded proteins forming channels, and receptors to facilitate ports of entry and exit to and from the intracellular to extracellular compartments. There are some carbohydrate molecules functioning as "antennas" extending from the surface of the cell for messaging and cell signaling [9, 10].

The other core physiological system to be investigated when assessing an individual's fatty acid status is *defense and repair*. Since lipids and fatty acids of the eicosanoid molecules are the primary influencers and regulators of inflammation, treatment of the defense and repair systems begins to reveal the etiology of the inflammatory aspect of an individual's condition. Once identified, utilizing the skill set of modulating lipid and fatty acid status will help target the intervention needed to restore structure, function, and wellness. The cell membrane is a primary determinant of the quality of an individual's health (see ► Chaps. 12 and 19) (◘ Fig. 11.1).

11.4 Nutritional Influences on Body Composition and Function

11.4.1 Structure and Functions of the Cell Membrane

The prevailing concept of cell membrane structure is the phospholipid bilayer that is impermeable to most water-soluble molecules, often referred to as the fluid mosaic model [9]. Most of the phospholipids in the membrane are present as a biomolecular sheet, with the fatty acid chains in the interior and exterior of the bilayer. Membrane proteins are located either on the internal or external faces of the membrane or projecting from one side to the other. An important feature of the membrane is "membrane permeability," allowing flexibility for molecules to move around [2, 11] (◘ Fig. 11.2).

The structure and function of cells depend on membranes, which not only separate the interior of the cell from its environment but also define the internal compartments of eukaryotic cells, including the nucleus and cytoplasmic organelles. The formation of biological membranes is based on the properties of lipids, and all cell membranes share a common structural organization: bilayers of phospholipids with associated proteins. These membrane proteins are responsible for many specialized functions: some act as receptors that allow the cell to respond to external signals, some are responsible for the selective transport of molecules across the membrane, and others participate in electron transport and oxidative phosphorylation. In addition, membrane proteins control the interactions between cells of multicellular organisms. The common structural organization of membranes thus underlies a variety of biological processes and specialized membrane functions, which will be discussed in detail in later chapters [9].

The Cellular, Organelle, and Nuclear Bilayer Membranes of Each Cell [10]
- Protect and hold together each compartment.
- Protect cell compartments from their surrounding environment.
- Control movement of substances transported in and out of the cell.
- Manage immune responses regarding inflammation (eicosanoid metabolites).
- Maintain cell and mitochondrial membrane integrity—key to cell survival.
- Foundational to the core physiological clinical imbalances (see ◘ Fig. 11.1).
- Provide structural integrity

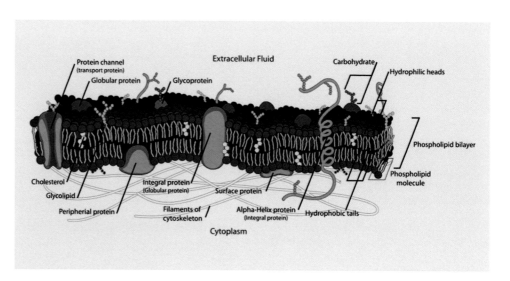

◘ Fig. 11.2 Fluid mosaic model of membrane structure. (Reprinted from OpenStax CNX [88]. With permission from Creative Commons License 4.0: ► https://creativecommons.org/licenses/by/4.0/)

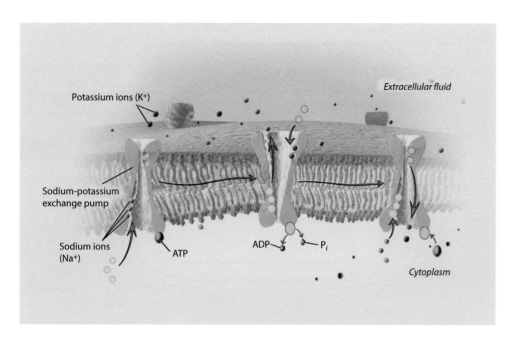

◘ **Fig. 11.3** Sodium-Potassium Pump. A transporter on the membrane that maintains high potassium and low sodium intracellular concentrations relative to the extracellular electrolyte concentrations. This pump functions at the expense of ATP energy and is influenced by the dietary intake of the minerals magnesium, potassium, and sodium. (Reprinted from ► Blausen.com staff [89]. With permission from Creative Commons License 3.0: ► https://creativecommons.org/licenses/by/3.0/deed.en.)

- Provide defense and repair.
- When compromised or damaged, allow healthy molecules to leave the cell and unwanted materials to enter. This can be referred to as "leaky cell membranes.
- Phospholipids, fatty acids, cholesterol, and proteins when in balance facilitate repair and maintenance of cell membranes.
- Sodium-potassium pump (see ◘ Fig. 11.3).
- Intracellular cytosol 97% potassium controlled by the sodium-potassium pump (◘ Fig. 11.3).

11.4.1.1 Cellular Hydration

Among the many properties of the cells and organelles is hydration. Measurable aspects of cell hydration include total body water, intracellular water, and extracellular water [2]. These measurements are clinically available using bioelectric impedance analysis (BIA) (see ► Chap. 22). All metabolic characteristics apply to the balance of water between intracellular and extracellular fluids [2, 9]. In healthy cells, there is opposite composition of the intracellular potassium concentration versus the extracellular sodium managed by the sodium-potassium pump (Na-K pump) embedded in the cell membrane [9]. ATP production is the energy driving the activity of the pump [2, 10].

- Intracellular matrix (cytosol) is 97% potassium-controlled by the Na-K pump [9]
- Extracellular matrix is 97% sodium-controlled by the Na-K pump [9]
- Magnesium rate-limiting nutrient for the Na-K pump [10]

A deficit of any of the three minerals will affect the function of the Na-K Pump and cellular hydration.

11.4.1.2 The Membrane Barriers: Organelle, Cell, Tissue, and Organs

All membrane barriers contain the basic bimolecular bilayer membrane structure. That basic structure is found in every type of cell, organelle, tissue, or organ function, with only slight variation. Examples are that neuron cells have a higher percentage of phospholipids than liver cells and heart and muscle cells have a greater percentage than brown fat cells (see ► Chap. 12).

The most important barriers to be considered when assessing body systems are:
- Skin (see ► Chap. 54)
- Gastrointestinal barrier, with small intestine housing about 70% of immune cells in the lymphoid tissues (see ► Chap. 24)
- Blood–brain barrier (see ► Chap. 12)
- Respiratory-lung barrier, comprised of bronchi, bronchioles, and alveoli cells which together are responsible for exchanging oxygen intake and carbon dioxide waste exhalation (see ► Chap. 52)

Significant differences in the membrane structures of various cells, besides phospholipid composition, are in the presence and amount of cholesterol. Mitochondria have almost no cholesterol embedded in their inner and outer membranes. In all other eukaryotic cells (complex cells with organelles), cholesterol is present and is important to the stability of the cell, the sorting of protein structures [2], and guarding from toxic substances entering the cell.

During the twentieth century, cholesterol developed a bad reputation driven by the Ancel Keys' research concerning its relationship to cardiovascular disease. More recently, the scientific community has challenged Keys' research on cholesterol and saturated fat [4].

It is important to consider the beneficial role of cholesterol in the structure of the membrane. Cholesterol is a "lipid of its own" [2]. Cholesterol is a sterol and is different from phospholipids. Cholesterol has a steroid ring structure and polar head group (-OH). As an amphipathic molecule possessing both lipophilic and hydrophilic properties, cholesterol easily incorporates into lipid bilayers. Cholesterol has a "bulky and stiff tail and small head," contributing a stabilizing order to the cell membrane structure, making the membranes "stiffer" but allowing the membrane permeability to function. Membrane permeability is an important issue of chronic diseases and that the membrane signaling, therefore cell metabolism and viability, depends on the phospholipid, cholesterol, and fatty acid composition [1, 12]. The permeability of the organelle and cell membranes requires a balance that is not too rigid and not excess (leaky membrane). Membrane permeability is affected by several factors like age, dietary history, activity level, and hydration. A practical assessment of cell membrane permeability is the measurement of the *phase angle* (PA), a quantitative measurement available using the bioelectric impedance analysis (BIA) technology (see ▶ Chap. 22). The PA is used as a marker of cell membrane integrity and permeability. Although the biological significance of the PA is not fully understood, many studies have recognized that low PA values are associated with a poor prognosis and as a prognostic indicator for some cancers [11, 13, 14]. The BIA instruments have been used in research and clinical practice for over 30 years and are easy to operate in a clinical setting and relatively inexpensive. This BIA data when added to the information from a red blood cell (RBC) fatty acid analysis, blood lipid panel, and disease condition gives some indication of cell membrane permeability.

Cholesterol is essential to life by providing the base unit for production of hormones, neurological cells (neurons, myelin, brain tissue, etc.), bile, and others. As with all natural components of the chemical body, each cholesterol molecule has multiple functions. Each function depends on the balance of the amount of cholesterol deposited in the cell membranes. This balance is foundational to optimized cell function and may be related to compromised metabolism when cholesterol is too low (studies suggest hypocholesterolemia is total cholesterol <120–150 mg/dL). On the low-end of the spectrum, hypocholesterolemia is associated with increased incidence of mood disorders like depression, as well as cancer, and sepsis [15]. During a nutritional assessment, biomarkers for cholesterol status are important to quantify and should consider any medications that influence cholesterol synthesis. The IFMNT practitioner should be skilled at restoring cholesterol balance and managing hypercholesterolemia along with helping manage clinical symptoms.

11.5 The Eicosanoid Cascade: Acute and Chronic Tissue Inflammation Management

The eicosanoid cascade is comprised of a complex group of organic molecules with multiple metabolic functions. This section will focus on eicosanoid functions and their influence on the immune response to initiate and resolve inflammation [1, 16]. The 20-carbon eicosanoids are derived from the 18-carbon fatty acids omega-6 linolenic acid (LA) and omega-3 alpha-linolenic acid (ALA) by catalytic action of two enzyme groups, desaturases and elongases. The dietary amounts of omega-6 and omega-3 fatty acids affect this process of eicosanoid production. The eicosanoid metabolites are signaling molecules that determine the function of many metabolic pathways. The families of eicosanoids include prostaglandins, prostacyclins, thromboxanes, and leukotrienes [10]. Within each family, there are many metabolites, including recently discovered *specialized pro-resolving mediators* (SPM), involved in resolving inflammation [1, 6, 17]. Eicosanoid families may either produce or reduce inflammation depending on which molecules are produced by the immune response signaling. Eicosanoids also help regulate blood pressure, modulate the immune system, and affect blood clotting [7].

The IFMNT practitioner applies their knowledge of the eicosanoid cascade biochemistry and rate-limiting nutrient cofactors and lifestyle habits that affect elongation and desaturase enzyme functions. With skill in managing chronic inflammation, one can target nutritional interventions to restore optimum balance of the eicosanoid metabolites derived from the essential fatty acids, *linoleic acid* (LA C18:2 ω 6), and *alpha-linolenic acid* (ALA C18:3 ω3) (see ◘ Fig. 11.4). To assess an individual's eicosanoid and metabolite status, obtain functional lab testing and a diet history to assess the patient's fatty acid status and metabolites affecting inflammation. The initial assessment best includes an RBC fatty acid profile, blood lipid panel, and dietary history of fat and oil rich foods (see ▶ Chap. 58 on Fats and Oils Survey). From this data, one can assess fatty acid status and recommend changes to support inflammation control.

The two arms of the eicosanoid cascade share the desaturase and elongation enzymes and compete for their use, with preference toward omega-3 metabolites [18]. Each of the omega-3 and omega-6 metabolite families influence each other and cannot be converted from one family to the other due to lack of the required enzymes in the human metabolism. Even though the LA and ALA are the two essential fatty acids, their important metabolites like γ-linolenic acid (GLA), di-homo-γ-linolenic acid (DGLA), arachidonic acid (AA) in the omega-6 family, and EPA and DHA in the omega-3 family can be obtained from some foods rich in the converted forms. For example, for a person who did not eat

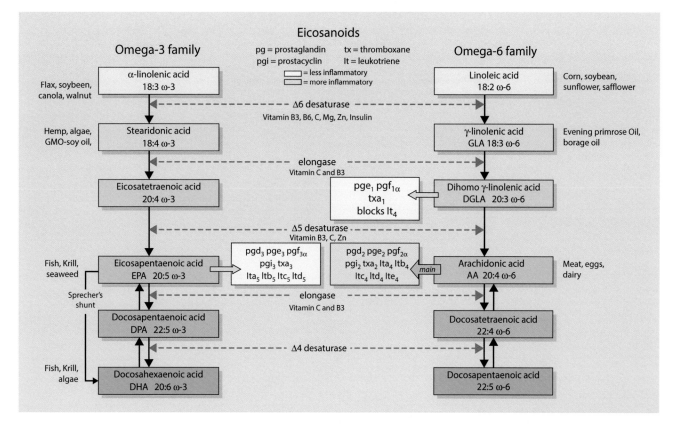

● **Fig. 11.4** The Eicosanoid Cascade of essential fatty acids, their metabolites, the nutrient cofactors, and foods rich in those cofactors. (Adapted with permission from: ► https://commons.wikimedia.org/wiki/File:EFA_to_Eicosanoids.svg)

any ALA sources like walnuts, flax oil, or vegetables, adequate EPA, and DHA could be provided by eating fish (EPA, DHA) or algae (DHA) sources. Another direct source of EPA, DHA, and LA is grass-fed or pasture-fed beef. If a person is depleted in the nutrient cofactors for the elongation and desaturase conversion biochemical steps, it would limit the efficiency of those enzymes. Repletion of those nutrient co-factors can increase the efficiency of the conversion enzymes.

11.5.1 Fatty Acid Elongation (See ● Fig. 11.4)

Elongation chemistry is dependent on the rate-limiting nutrients vitamin C and vitamin B3 (niacin). Any insufficiency or deficiency of those nutrients can significantly hamper achieving the balance of the eicosanoid metabolites. This balance is foundational to optimized cell function [19].

11.5.1.1 Elongase

The elongation of long chain fatty acids is derived from the omega-3 and omega-6 essential fats into very long chain fatty acids (VLCFA). The rate-limiting nutrient cofactors are zinc, magnesium, and vitamin B6. The elongation of very long (ELOVL) fatty acid by elongase enzymes is influenced by insulin and high carbohydrate diets [10]. It is important to assess glucose-insulin management when assessing of lipid status.

11.5.2 Fatty Acid Desaturation (See ● Fig. 11.4)

Desaturase enzymes delta-6-desaturase (D6D) and delta-5-desaturase (D5D) control the conversion of the essential fatty acids LA and ALA and eicosanoid metabolites. The key nutrient cofactors for the D6D and D5D are vitamins B3 (niacin), B6 (pyridoxyl-5-phosphate), C (ascorbate), and the minerals magnesium and zinc.

11.5.2.1 Delta-6-desaturase (D6D)

Interactions between dietary LA, D6D, and insulin resistance from the modern Western diet are associated with increased activity of the D6D enzymes, which result in the increased conversion of LA to excess pro-inflammatory AA [20, 21]. Alcohol and smoking can both suppress D6D activity, and those individuals may have poor conversion of LA to DGLA and AA. The GLA conversion to DGLA is a pivotal occurrence because the DGLA has two options for continued production of either the prostaglandin 1 series of metabolites or AA and the proinflammatory prostaglandin 2 series. The D6D and D5D enzymes determine how much the omega-6 metabolites proceed in either direction. If D6D is inhibited, not as much DGLA is produced. If DGLA is produced and D5D is increased in activity, DGLA will increase production of AA and lessen the ability to form the anti-inflammatory PG1 metabolites. For those who smoke and/or drink significant alcohol, the use of evening primrose or other GLA-rich oil can support increasing DGLA conversion toward anti-inflammatory prostaglandin 1 metabolites and bypass the poor activity of D6D [6]. The D6D is encoded by the fatty acid desaturase (FADS2) gene and its function is rate-limiting in polyunsaturated fatty acid biosynthesis [22].

11.5.2.2 Delta 5-desaturase (D5D)

D5D activity is responsible for the conversion of omega-6 DGLA to pro-inflammatory excess AA. Elevated glucose, insulin metabolism, and obesity promote D5D activity, increasing AA formation and pro-inflammatory conditions. Dyslipidemia and the use of prescription medications (statins) are usually involved with increased D5D activity [3, 23]. Because of the epidemic of sarcopenic obesity and obesity overall, awareness of interventions such as supplemental GLA-rich sources can direct the omega-6 DGLA to its alternative anti-inflammatory pathway to prostaglandin 1 series metabolites and contribute to decreasing insulin resistance [23]. D5D is also involved in gonadal hormone metabolism. An example is when estrogen-dominant conditions develop, the D5D is more activated. This may increase conversion of DGLA to AA instead of a balanced conversion to the PGE1 anti-inflammatory molecules. Weight reduction to ideal body weight and weight maintenance can improve the healthy function of D5D [24].

It is important to remember that the omega-6 and omega-3 fatty acids cascades compete for the desaturase and elongase enzyme activities, so inhibiting the omega-6 DGLA to AA conversion will also affect the D5D activity that converts omega-3 ALA to EPA and then DHA. When inhibiting D5D, there may be an increased need for dietary intake of EPA and DHA by eating fish and/or fish oil supplementation. D5D enzyme activity is a target of research for diabetes management. D5D inhibitors are currently a target of the pharmaceutical industry [24]. The potential of nutrition therapy to optimize fatty acid status, reduce simple carbohydrate intake, and achieve weight management can be powerful modulators of these enzymes.

More comprehensive lipid blood tests have recently become clinically available that can improve assessment of a patient's status. The arachidonic/di-homo-gamma-linolenic acid ratio (AA/DGLA ratio) is a meaningful biomarker for assessing balance between AA and DGLA concentrations that affect the inflammation process and the function of D5D [25] (See ◘ Fig. 11.5).

11.6 Metabolic Stressors

Systems biology emphasizes the interactions of all systems to influence the phenotype of an organism. New stressors have arisen in the past century from environmental toxicants, decreased physical activity, highly refined foods, increased carbohydrate intake, and increased occurrence of metabolic syndrome to enable dramatic changes in phenotype due to chronic disease. Some of these stressors are known to influence the activity of the elongases and delta-5 and delta-6-desaturases, resulting in altered eicosanoid metabolism. Obesity is one metabolic condition that is associated with disturbed lipid metabolism and low-grade inflammation in

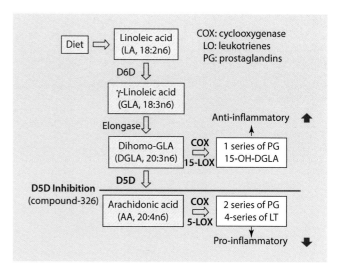

◘ Fig. 11.5 Biosynthesis pathway of n-6 polyunsaturated fatty acids. D5D is a key enzyme affected by glucose, insulin, and stress status of metabolism and can be mediated by diet and lifestyle choices. (Reprinted from Yashiro et al. [24]. With permission from Creative Commons License 4.0: ► https://creativecommons.org/licenses/by/4.0/)

tissues, and can vary based on the nutrigenomic profile of an individual for the FADS1 and FADS2 genes (See ► Chap. 17).

> The functional role of eicosanoids in the inflammatory etiology of diseases of metabolic syndrome (MetS) has been extensively studied in relation to immune cell recruitment and cytokine, chemokine production and their activation of inflammatory pathways in cancer, diabetes, and cardiovascular disease (CVD) [11, 26].

11.7 Tools of the Trade for Lipid Therapy

The toolbox for the IFMNT practitioner considering lipid therapy assessments and interventions includes all clinical and measureable parameters and modalities that influence lipid metabolism or are influenced by food intake, lifestyle, and/or environment. The following checklists and principles of gathering lipid data suggest laboratory testing and other clinical information to contribute to a comprehensive assessment, diagnosis, and intervention to promote the best outcome for an individual.

Toolbox to identify fatty acid/nutrition status
- Nutrition Physical Exam (see ► Chap. 40)
- Medical history: diagnoses, medical event history, residential location
- Signs and symptoms: Medical Symptoms Questionnaire (MSQ)
- Laboratory testing: basic nutrition, lipids, disease-specific and sometimes patient-specific markers
- Bioelectrical Impedance Analysis (BIA), if available

The nutrients in ◘ Table 11.2 are the most studied regarding lipid metabolism and should be considered as part of a nutritional assessment.

Table 11.2 Key lipid nutrient insufficiencies/deficiencies associated with subclinical or acute disease; key immune nutrients foundational for healthy lipid status

Nutrient	Clinical or biochemical implication of deficiency	Reference	Symptoms	Testing to consider
Immune modulators				
Omega-6 GLA	Eczema, Dermatitis, GERD, Viral infections	[6]	Broken skin lesions, Rash, diabetes, Long-term effects of chronic disease	RBC fatty acid, Gamma linolenic (GLA), di-homo-gamma linolenic acid
Eicosapentaenoic acid (EPA, EPA, DHA, LA, ALA)	Altered mood, Skin health, Cardiovascular, Cancer	[27]		RBC fatty acid
Malonaldehyde	Cancer, Altered fatty acid Metabolism	[6]		RBC fatty acid
Choline phosphatidylcholine phosphatidylethanolamine phosphatidylinositol	Poor membrane permeability, Impaired fat metabolism, Fatty liver, Mitochondrial dysfunction, Released alanine aminotransferase (ALT) from liver cells	[28, 29]	Neurological, Fat malabsorption, Muscle damage	Homocysteine, Creatine kinase, Liver enzyme ALT, Lipid panel, Folate, RBC folate
Vitamin D	Immune modulator	[30–32]	Mood disorders, Bone loss, Joint pain, Compromised immunity	Vitamin D25OH, Vitamin D 1,25OH, Parathyroid hormone (PTH), Ionized calcium
Vitamin A retinol	Mucosal immunity	[33, 34]	Mucosal bacterial/viral infections	Vitamin A, retinol, B-carotene
Eicosanoids	Anti-inflammatory, Pro-inflammatory, Immune regulation, Resolution biology	[35–37]	Pain, Swelling, Cancer	RBC fatty acid
Rate-limiting co-factors				
Folate	NK cell activity, Cytotoxic cellular immunity, Modulate T-cell responses	[38]	Fatigue, Infection, Mood disorder	Folate, serum, RBC folate, FIGLU
Zinc	Regulates intracellular signaling pathways in innate and adaptive immune cells	[38–40]	Skin conditions, Poor smell, Poor taste, Frequent illness, Nail spots	Zinc, serum, RBC zinc
Magnesium	>300 enzymes cofactor, Urine excretion under stress, Muscle tension, Camps	[10]	NHANES ~80% US population less than RDA magnesium	Magnesium, serum, RBC magnesium
Antioxidants				
Vitamin C	Regulates cellular humoral immune function, Increases macrophage, Antihistamine, Requirements increase during infection, Anti-viral	[41]	Connective tissue impairment, Skin conditions, Poor wound healing	Vitamin C, blood CPT 82180, Urine, functional need for vitamin C test: urine *p-hydroxyphenyllactate* (HPLA)

(continued)

Table 11.2 (continued)

Nutrient	Clinical or biochemical implication of deficiency	Reference	Symptoms	Testing to consider
Vitamin E	4 tocopherols + 4 tocotrienols protective from oxidative stress/lipid peroxides	[30, 42]	Oxidative stress Premature wrinkles Cysts Leg cramps	Serum: Alpha-tocopherol gamma-tocopherol
Selenium	Thyroid peroxidase metabolism with vitamin E Selenoproteins special effects on cellular immunity Resistance to viral infections Central to glutathione peroxidase structure	[38]	Hypothyroid Low glutathione blood levels Impaired detoxification	Selenium, serum RBC selenium
GUT secretions				
Short-chain fatty acids (SCFA)	Intestinal cells Anti-inflammatory Microbiome-gut- brain axis via immune system/vagal nerve Mucosal immunity	[43–48]	Inflammation Perturbed uric acid cycle Colon disease	Acetate Propionate (blood or fecal) Butyrate (blood or fecal)
Bile acids	Fat malabsorption Metabolic liver diseases Glyco and Taurochenodeoxycholic Acid	[49, 50]	Perturbed fat digestion	Bile acids US imaging gallbladder
Lipotoxic Conditions				
Dysfunctional regulation of lipid metabolism and homeostasis causes cellular lipotoxicity, impairs cellular processes and contributes to the pathogenesis of disorders such as obesity, atherosclerosis, and neurodegeneration	Neurological Cancer Autoimmune Developmental Cardiovascular Psychological	[51, 52]	Diagnosis specific	Diagnosis specific
Anti-Nutrients				
Endocrine-disruptors	Circadian rhythm disruption that can modulate immune function	[53, 54]	Hormonal imbalance Insomnia Cancer	GPL-TOX, Great Plains Lab toxic organic chemical profile
Damaged food components/Western Diet & food preparation Trans fat Oxidized fat Toxic metals Toxic chemicals "new to nature molecules" (cookware, pest control, pollutants, processed, etc.)	Most damaged or toxic substance *Lipophilic* and embed into the fatty membrane and other tissues Damaged high-heat foods Chemicals Toxic metals ingested in foods and from food utensils during food preparation	[55]	Chronic disease Specific toxic symptoms Weakened immune integrity	Diet History RBC fatty acid analysis
Environmental Toxins				
Mold/mycotoxins Natural gas carcinogenic Chemical vapors Pesticides	Neuropsychiatric Immune disruption Vulnerable to poor cell signaling Insulin resistance	[56, 57] [58] [59]	Toxicities Chronic kidney failure Infertility Cancer Developmental Frequent infections	*Mold, pathogenic bacteria:* Multiple Antibiotic Resistant Coagulase Negative Staphylococcus (MARCoNS) nasal swab GPL-TOX, Great Plains Lab toxic organic chemical profile (pesticides, toxic chemicals)

11.7.1 Laboratory Principles

With laboratory data available, a diagnostic profile can clarify the priorities of core physiological imbalances and provide clues regarding nutrition and lifestyle therapy to restore structure and dynamic function of the membrane. From the IFMNT nutrition assessment, consideration of the two most likely physiological clinical imbalances is *structural integrity* and *defense & repair* (see ◘ Fig. 11.3).

11.7.2 Structural Integrity

The membrane's structural integrity affects the transport and communication of the cell membrane and receptors. If there is a history of head, neck, dentition, or back injury where there may be a possible structural misalignment in the cervical vertebrae, the brainstem and vagal nerve may be impaired. Someone with this history should be referred to a cranial specialist for evaluation and have their lipid status reviewed.

11.7.2.1 Assessment Checklist for Structural Integrity
- Cell membrane permeability and integrity (BIA phase angle, fatty acid status) [11]
- Dental periodontitis: infection of the tissues that surround the teeth-inflammation
- Structural-spinal alignment: neuronal membrane [1, 60]
  - Cervical C1-C7 – brainstem & vagal assessment – check vagal tone
  - Thoracic T1-T5 – stenosis or injury increased pain resulting in exaggerated immune inflammatory response
  - Lumbar L1-L5 – stenosis or injury increased pain resulting in exaggerated immune inflammatory response

11.7.3 Defense and Repair

Defense and repair are managed by a dynamic immune system that responds to endogenous and exogenous infection, injury, malnutrition, altered gut microbiome, stress, and other potential inflammatory triggers.

11.7.3.1 Assessment Checklist for Defense and Repair
Blood Markers
Vitamin D 25-Hydroxy
Vitamin D has many functions. It is a powerful immune modulator that plays a role in defense and repair.

> Vitamin 25OH serum levels have been associated with overseeing the dynamics of the cell membrane. Vitamin D seems to stimulate anti-inflammatory processes [31].

Vitamin A (Retinol) [61]
Vitamin A is a nutrient cofactor for the eicosanoid desaturase enzymes. It is increasingly recognized in experimental and human studies to enable suppression of inflammatory reactions and plays a significant role in normal mucosal immunity, regulation of T cell-dependent responses, antiviral activity, and cell communication.

Adequate vitamin A status, whether from intake of preformed retinol (e.g., animal sources: egg yolk, organ meats, fish, shellfish, and roe) or from b-carotene (e.g., yellow and green vegetables) is important for preventing excessive or prolonged inflammatory reactions and supporting the eicosanoid cascade [61].

Gut microbiome [55]
Gut commensal bacteria making up the microbiota are critical for the health of the immune system by defending and repairing the intestinal barrier where >70% of the immune cells reside. Stool testing provides biomarkers to assess GI ecology. From this data one can develop an intervention plan to correct and repair imbalances in the gut microbiome (see ▶ Chap. 24).

Inflammatory Load Assessment
- High-sensitivity C-reactive protein (hs-CRP): acute-phase-reactant and marker of systemic inflammation. It is often elevated due to bacterial infection, central adiposity, fatty liver, neoplastic activity, or traumatic injury. All clinical laboratory references include normal hs-CRP as ≤1.0. Its elevation implies potential bacterial infection or physical trauma. If no dental/oral infection is identified, begin further investigation of the root cause. It is important to rule out a recent traumatic injury that may be related, and hs-CRP should be retested in a month or two to observe if injury has affected the hs-CRP.
- CBC with differential
- Complete metabolic panel (CMP)
- Lipid panel
- Erythrocyte sedimentation rate (sed rate)
- TSH
- Bacterial/viral evaluation if indicated; saliva or blood
- Genomic testing (saliva), if available

11.8 Key Nutrient Cofactors and Foods Influencing the Eicosanoid Metabolism

11.8.1 Lipids

Lipids play critical roles in membrane structure, cell signaling, energy storage, control of inflammation, and as base units for constructing messenger hormones [2]. Each member of the family of lipids has specialized functions. Some provide cell signaling, cellular component transporting and regulate immune response of inflammation, hormonal modulation, and other undiscovered functions. When there is poor structure, dysfunction occurs [62]. The following key nutrients below can be obtained from foods and/or dietary supplements:

11.8.1.1 Phospholipids (PL)

Phosphatidylcholine (PC), phosphatidylethanolamine (PE), phosphatidylinositol (PI), phosphatidylserine (PS). These phospholipids are ubiquitous in all membrane structures and especially important for the functions of neurological and mitochondrial membranes [6]. The body can synthesize them, and they can also be found in foods. Foods rich in phospholipids are animal meats and organs, egg yolk, and legumes.

11.8.1.2 Fatty Acids:

- *Omega 9* Monounsaturated fatty acids (MUFAs): These oils are stabilizing components of structures and also have an anti-inflammatory effect in metabolism. Foods rich in MUFAs are olive oil, avocado/avocado oil, almonds, sesame, and peanuts. They are mildly heat sensitive and best raw or used with low heat.
- *Omega 6* Linoleic Acid (LA) (essential) and the eicosanoid metabolites GLA, DGLA and AA. LA is rich in seeds, some nuts, greens, grains and grasses.
- *Omega 3* Alpha Linolenic Acid (ALA) (essential) and the eicosanoid metabolites EPA, DPA, DHA recognized for their anti-inflammatory effects on metabolism. The balance between omega 6 and omega 3 fatty acids is important with most fatty acid scientists proposing an optimum ω6:ω3 ratio range of 1:1 to 5:1.
- *Saturated fatty acids* have been blamed as being detrimental for humans. But better understanding that there are beneficial saturated fatty acids when maintained at about 10% of dietary fat [63]. Natural and beneficial forms of the saturated fats are:
    - *Short chain fatty acids* (SCFA) with many critical roles of anti-inflammatory and fuel for colonocytes. Many health benefits are recognized reducing risk of colon cancer, autoimmunity, and gastrointestinal disease.
        1. Acetate (C2) and Proprionate (C3) SCFAs are formed in a healthy gut.
        2. Butyric Acid: butyrate: C4: butyrate-rich sources are butter, mother's milk, healthy gut microbiome consuming resistant and soluble vegetable and fruit fibers to produce SCFAs, and sodium or calcium-magnesium butyrate dietary supplements. If there are infectious or antibiotic insults to the gut microbiome, the production of butyrate may be notably reduced [64].
    - *Medium chain triglycerides* (MCTs) are composed of a glycerol backbone with three medium-chain fatty acids (MCFAs) (C6-C12). The MCTs are beneficial as part of dietary intake. They are heat resistant and recommended as cooking oil. Rich sources are coconut oil, palm kernel oil, and ruminant animal milk (cow, sheep, goat, horse). Therapeutic use of MCTs for liver failure and other gastrointestinal conditions is common due to their rapid absorption and not requiring bile salts for digestion and can be an easily metabolized source of energy.
- *Saturated Fatty Acids (SFAs)(C13-16)* are made up of carbon chains with only single bonds. The body can synthesize SFAs and they are also found in SFA-rich foods such as animal meat fats like beef tallow, pork lard, poultry fat, and also cocoa butter.
- *"New-to-nature" Fats:* with high heat and hydrogenation processing of vegetable oils, aberrant fatty acids and lipid structures can be formed that have been recognized as unhealthy for human metabolism, with some identified as carcinogenic. Examples of these compounds are trans fats and acrylamides.

11.8.2 Sterols

Sterols are naturally occurring unsaturated steroid alcohols, waxy lipids. The primary sterol for human metabolism is cholesterol, which is the base unit for all hormone production and vitamin D. Cholesterol is also an important component of cell membrane structure [2]. Most endogenous cholesterol is synthesized by the liver, but dietary cholesterol can influence total cholesterol levels. Foods rich in cholesterol are of animal origin: fats from animal milk, meat, egg yolk, poultry, seafood, and organ meats.

11.8.3 Minerals

11.8.3.1 Zinc

Zinc is a nutrient cofactor for the desaturase and elongase enzymes and a nutrient partner with copper. Zinc and copper should always be balanced in body fluids (see ▶ Chap. 8). When copper is elevated, it promotes increased fat deposition in some organs like the liver. Increased zinc intake can modulate the copper to a healthy level and reduce the fat deposition [65].

11.8.3.2 Magnesium

Magnesium functions as a nutrient cofactor for both the desaturase and elongase enzymes for eicosanoid conversions. When it is insufficient or deficient, restrictions in function can occur. In the case of magnesium as a cofactor in the sodium-potassium pump, the rate of conversion is decreased during a deficiency of magnesium. Magnesium is the cofactor catalyzing the enzymes driving the sodium-potassium pump, moving substances in and out of the cell through the lipid membrane [9, 66, 67] (see ▶ Chap. 17).

11.8.4 Methyl Nutrients

Vitamins B6, B12 (-cobalamins), B9 (Folate), B2 (riboflavin), B3 (niacin), choline (betaine, phosphatidylcholine/phosphatidylethanolamine), SAMe, and related Vitamin C, B1, B5 [68, 69].

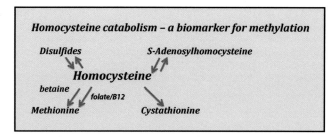

Fig. 11.6 Homocysteine major metabolic pathways in humans. (Adapted from Dudman et al. [70]. With permission from Oxford University Press)

Methyl nutrients support the process of methylation, having many roles within human metabolism with DNA methylation being the underlying mechanism, and they currently appear to be primary messengers of epigenetic expression (see ▶ Chap. 18) identified in the etiology of developing cardiovascular, cancer, and neurological disease conditions. Methyl nutrients are involved in the conversion of desaturase and elongase enzymes. Methylation involves biochemical pathways where the B vitamins and other cofactors like amino acids are rate-limiting cofactors [70] (◘ Fig. 11.6).

11.8.4.1 B12 and Folate Metabolism

B12 in the natural bioactive forms (methyl-, hydroxy-, or adenosylcobalamin) is a nutrient cofactor that inhibits the excessive formation of arachidonic acid. B vitamins team together, and folate is an especially important teammate with B12. Folate is critical to many metabolic pathways like nucleic acid precursors, several amino acids, and erythropoiesis, which is the process in which new erythrocytes are produced. A biomarker that can suggest folate deficiency is an elevated mean corpuscular volume (MCV) on a complete blood count quantitatively measuring size of RBCs. Folate deficiency can be part of the etiology of enlarged RBC or megaloblastic anemia, from ineffective erythropoiesis. Additionally, Vitamins B6 and B12 are cofactors involved in erythropoiesis [71].

11.8.4.2 Niacin, Vitamin A, Vitamin C, and Zinc

– Rate-limiting nutrients for DGLA conversion to anti-inflammatory Prostaglandin 1 (PG1) series
– PG1 anti-inflammatory action primarily involved in mediating conditions of allergy, viral, autoimmune

11.8.4.3 Vitamin D, A [72]

These fat-soluble vitamins have many metabolic roles. Their influence on structural integrity and defense and repair (e.g., inflammation and immune response) modulates the lipid environment and metabolic dynamics. The fat-soluble vitamins function synergistically, with the vitamin D and A receptors sharing their nuclear receptor, influencing each other. Vitamin D2/3 and A are found in their food-rich sources together (e.g., liver, caviar, /roe, egg yolk). Vitamin A retinol is one of the key nutrient cofactors for the desaturase enzyme activity.

11.8.5 Phytonutrients: Protective Support for Lipid Structures

– Inflammation: Phytonutrients are antioxidants that protect the lipid membrane from oxidative stress through their powerful polyphenols found in a variety of pigment-rich fruits, vegetables, grains, nuts, teas, herbal spices, and legumes that have anti-inflammatory properties. The mechanisms of the plant chemicals include antioxidants, antibacterial, and antiviral. Even though phytonutrients are not considered essential nutrients, the evidence is mounting for their critical role in health maintenance and anti-aging.
– Biomarkers of poor phytonutrient status: Poor dietary intake of high polyphenol foods. Significant biomarkers for inflammation can be related to poor vegetable and fruit intake and lack of (or imbalance in) dietary intake of healthy fats and oils (see ▶ Chap. 58 Fats & Oils Survey)(◘ Table 11.2)
– Resource: The Rainbow Diet. Color Can Heal Your Life [73].

11.9 Key Lifestyle Factors Influencing the Risk of Lipid Damage

11.9.1 Sleep [74] (See ▶ Chap. 35)

Sleep and circadian rhythm have a great influence on the integrity of the immune system. Much evidence has accumulated over the past decades associating poor sleep quantity and quality with weakening of the immune system, increasing vulnerability to the poor function of cellular structures.

11.9.2 Stress (See ▶ Chap. 47)

Chronic stress impacts every biological and psychological system. When the chemical microenvironment is under long-term stress, it pushes the immune system response into chronic inflammation and increased acidity. The vicious cycle continues until the threshold of resilience and adaptation is exceeded, leading to vulnerability to many chronic diseases including damage to lipid structures and influencing the eicosanoid metabolism.

11.9.3 Movement (See ▶ Chaps. 36 and 54)

As with all biological systems, there must be movement of structures like muscles, heart, lungs (breathing), as well as the fluids in the body to maintain health. Without movement, there is congestion that does not support healthy metabolism. The health of the immune system is dependent on the lymphatic system, which is supported by movement. The lymphatic circulatory system does not have a pump as compared with cardiovascular circulation. The lymphatic vessels are "pumped" by physical activity with arm and leg movement, abdominal breathing, laughing, etc.

11.10 Chronic Disease and Impaired Lipid Metabolism (See Tables 11.3 and 11.4)

11.10.1 Heart Disease/Cardiovascular Association with the Lipidome

Lipids manage and resolve inflammation via the eicosanoid cascade. Chronic inflammation is a hallmark of cardiovascular disease, so if it is well controlled and resolved, the risk of a cardiovascular event is decreased. The current recommendation for eating fish or fish oil supplementation to benefit the cardiovascular system is primarily as a modulator of the eicosanoids to lower the pro-inflammatory excess omega 6 arachidonic acid [78].

11.10.2 Oncology

Several of the eicosanoid metabolites are evidenced to be tumor suppressive [79]. Eicosanoids, including prostaglandins and leukotrienes, are biologically active lipids that have been associated with many of the pathologies of chronic disease such as inflammatory cancer. Eicosanoid metabolites and their function of inflammation control give the IFMNT practitioner the ability to develop a targeted intervention for cancer by assessing the patient fatty acid status and analyzing the eicosanoid metabolism status [80].

Prostaglandins and leukotrienes can modulate tumor epithelial cell proliferation and apoptosis.

When the prostaglandin 1, 2, and 3 series are in balance, they can provide a change in the microenvironment toward wellness [81].

11.10.3 Neurological

11.10.3.1 Mitochondrial Dysfunction

Most neurological conditions involve underlying mitochondrial dysfunction. The mitochondrial inner and outer membranes are sensitive to influences from dietary fat. The inner membrane especially requires the phospholipid derived from choline, phosphatidylethanolamine (PE), and gamma-linolenic acid (GLA). Lipid nutrition therapy can use seed oils like evening primrose, black currant, sea buckthorn, or borage to provide adequate GLA in interventions to support mitochondrial repair. Evening primrose oil has also been shown to induce apoptosis and tumor suppression for cancer patients [82].

11.10.3.2 Alzheimer's Disease

The brain is approximately 70% fat and phospholipids. Restoring optimum lipid balance can benefit the structure and function of the nervous system. Comprehensive medical nutrition therapy supports high-fat diets in Alzheimer's and other neurological conditions. The brain can use glucose or ketones for fuel. Research on the ketogenic diet, where

Table 11.3 Laboratory: recommended biomarkers for lipid assessment

Blood tissue testing [75]

Lipid panel

   Total cholesterol, HDL, LDL, triglycerides, lipoprotein particles [1]

RBC fatty acid analysis

   Blood biopsy

   Dietary fat intake reflected in RBC fatty acid membrane composition [76, 77]

Bioelectric impedance analysis (BIA) [11]

   Phase angle: cell membrane permeability of the lipid bilayer

   Capacitance: ionic potential, membrane surface biomarker

   Intracellular and total body water

Organic acids [75]

   Urine: first morning collection, fasting

   View of all major systems (conventional and functional tests as indicated)

Inflammatory load

   Clinical observation of any inflammation of the face, skin, pain, biomarkers?

Laboratory

   CBC with differential

   Sed rate

   C-reactive protein-high sensitivity

   Acute phase reactants per diagnosis or signs & symptoms

Nutrition physical exam: Barriers; rule out potential symptoms or history

   Skin: tone, color, texture, lesions, skin tags, abnormal pigmentation

   Lung capacity/O_2 Sat: optimum 98–100%; history lung disease or surgery

   Gastrointestinal health: Comprehensive Digestive Stool Analysis (CDSA) abnormalities

   Oral cavity: periodontal disease, swollen glands, tonsillectomy history

   Esophagus: symptoms of esophageal pain/irritation, dysphagia, *Heliobacter pylori*

   Stomach: pain, digestive upset, surgical history, vagal tone

   Duodenum: structural changes, Small Intestinal Bacterial Overgrowth (SIBO)

   Jejunum: structural changes

   Ileum: structural changes, abnormal BM/bile circulation reentry

   Colon: abnormal BM

   Rectum: abnormal BM, hx colonoscopy

   Pain: location, barrier involved

Fatigue

   Time of day, meal timing, sleep quality and quantity, sleep apnea

Table 11.4 Lipid-supportive foods, herbs, and dietary supplements

Foods

Whole-foods, pesticide-free, vegetables and fruits in a variety of colors, adequate protein, healthy fats & oils, herbs, fluids; Minimize or avoid processed and high-sugar foods and beverages; avoidance of antigenic foods.

Oil-/fat-rich foods

Plant-based: avocado, raw seeds, olives, hearts of palm, nuts: macadamia, pine nuts, almonds, Brazil nuts, coconut oil

Animal source: organ meats, meat, poultry, fish, shellfish, roe, krill

Herbs

Turmeric/curcumin

Proteolytic enzymes: Bromelain, papain, trypsin, etc. (contraindicated for Alpha-1-Antitrypsin deficiency+ genetics)

Resveratrol

Boswellia

Diet

Macronutrient distribution

Insulin-glucose management

Meal timing

Intermittent fasting

Calorie restriction

Dietary supplements

Lipids

Phospholipids: phosphatidylcholine/phosphatidylethanolamine

GLA: evening primrose oil; black currant, sea buckthorn, and borage

Arachidonic acid: grass-fed meats, poultry, egg yolk.

EPA/DHA: various ratios are available; DHA vegan/algae; fish (ideally, lower on food chain)

ALA: cold nitrogen processed seed and nut oils flax oil and/or ALA-rich raw/soaked nuts, seeds, vegetables; sensitive to heat, light and oxygen.

LA: cold nitrogen processed seed and nut oils: refrigerated; sensitive to heat, light and oxygen.

Butyrate: short chain fatty acid (SCFA); sodium-potassium butyrate, calcium-magnesium butyrate, sodium butyrate, Rx: glycerol phenylbutyrate (Ravicti®), sodium phenylbutyrate (Buphenyl®)

MCT Oil: medium chain triglyceride; sources: coconut, palm and breast milk fats; MCT: Caproic acid (C6:0), caprylic acid (C8:0), capric acid (C10:0), lauric acid (C12:0)

Table 11.4 (continued)

Co-nutrients

Vitamin C (contraindicated for hemochromatosis mutation genetics)

Adequate methyl nutrients: B6, folate, B12, choline (betaine) and related B1, B2, B3, B5, Biotin, Vitamin D3, vitamin A (if indicated per personalized assessment)

Zinc: rate limiting nutrient to D5D, D6D, and elongase enzyme

Magnesium: rate limiting nutrient to D5D, D6D, and elongase enzymes

Personalize recommendations based on patient assessment

ketones are produced from high-fat low-carb diets, inducing a "nutritional ketosis", has shown therapeutic benefit in neurological conditions (see ▶ Chap. 23) [83–85].

11.10.3.3 Developmental Plasticity

Fetal and early childhood metabolic plasticity includes rapid growth of cells requiring lipids. The rapidly maturing brain and neurological system are high-fat cells that need lipids for membrane structure, cell signaling, and development of the immune system. Maternal health and nutrition status are important to secretion of fat-rich mother's milk, which lays the foundation for fetal growth [86].

11.10.4 Respiratory

The lung has a major protective barrier function in the body, and the lipid composition of the membrane is integral to modulating inflammation and promoting optimal function. Assessing and prioritizing the structural integrity and inflammation load using diet history, nutrition physical exam, and testing for nutritional status provide the foundation for developing a targeted intervention.

11.10.5 Autoimmune

Autoimmunity is the loss of immune recognition of self and non-self with resulting self-damage. Ongoing research has identified genetic relationships that have susceptibility to the development of autoimmune conditions (see ▶ Chap. 49). Chronic inflammation is present in all autoimmune conditions with resulting oxidative stress and reactive oxygen species that can damage lipid structures. Nutritional therapy considerations to reduce the chronic inflammation can be reducing or eliminating identified antigenic foods and replete nutrients that are insufficient or deficient.

11.11 Case Reviews

The following cases give examples of clinical application of lipid nutrition therapy. Each patient case presenting to an IFMNT practitioner needs to be approached as unique. Every diagnosis or symptom can have a multitude of root causes. The patient's story begins with an investigation to identify metabolic priorities and the focus of interventions using food, dietary supplements, and lifestyle to restore health. It is important to monitor interventions to assess effectiveness and determine if adjustments should be made.

Case Review: Simple Childhood Asthma

Diagnosis: Severe asthma
Male: age 8
Medical History:
- Asthma medications: Singular, albuterol inhaler, glucocorticoid inhaler- Budesonide and formoterol (Symbicort), salmeterol (Serevent). Fluticasone and salmeterol (Advair Diskus)
- Asthma emergency hospital ER event monthly

Laboratory Significant: CBC WBC 3.8, MCV 99 HI, Vitamin D 12 ng/mL; +grass and + casein IGE allergens
  Nutrition Physical: depressed affect; depapillation pale tongue (low B6, B12, Folate, B2 and iron), corner lips cracked skin (low B2), pale skin, matty hair
  Diet and Supplement history: Eighty percent fast food (hamburgers, hot dogs, cold cereal, cow milk casein, soda, yogurt, ice cream, toast, candy). No dietary supplementation.
  Lifestyle: difficulty breathing, grass pollen allergy, no outside play (reduced sun exposure); poor sleep related to breathing difficulty.
  Metabolic Priorities and Intervention Plan
- Remove antigen: cow's milk
- Remove empty calories: sugar, processed and refined foods, damaged oils
- Gut ecology: diet associated with poor gut pre and probiotics
- Get nutrients:

- EPA 350 mg/DHA 250 mg/GLA 130 mg supplement daily
- Vit D3 4000 IU emulsified D3 daily
- B Complex: bioactive forms – chewable
- Probiotic 450 billion powder in 6 oz. coconut yogurt daily
- Whole grain gluten-free bread for sandwiches, toast with organic butter
- Beverages: water, stevia-sweetened sodas, almond or coconut milk

Monitor: 6-Week Follow-up
- Happy affect, started 2 weeks able to take gym classes outside
- Nutrition: decreased probiotic; continued diet and remaining supplements

Follow-up plan: 3 Months
- Doctor removed four of the medications
- No hospitalization for asthma event

Outcome
- High school soccer team player, no asthma medications, continued casein-milk free diet and avoid high sugar foods, resumed gluten foods
- Annual physical: recommend Vitamin D25OH and maintain 40–60 ng/mL using vitamin D3 supplements and safe sun exposure.

Case Review – Early-Onset Alzheimer's

Patient Story: A 64-year-old male who presented with memory problems and diagnosis of early onset Alzheimer's disease by neurologist. Two months previous he showed no interest in activities, could not work. He was a building contractor for 35 years.
  Medical History: Thyroid cancer, total thyroidectomy, and Hemachromatosis recessive.
  Medical Data: Anthropometrics: BMI 30; height 76.2"; weight 250 lbs,
  Dietary/Alimentation: High intake of soda (~2 quarts/day) and processed foods including gluten containing grains and starches; some candy, doughnuts, fried foods, fast food, and canned fruit; commercial lunch meats. Low intake of water, essential fatty acids, vegetables, and whole foods.
  Nutrition Physical Exam: Memory deficits reported by wife and observed during exam, wrinkles beyond age appropriate (implies high oxidative stress), talks very slow and only when asked a question, poor dentition and dental hygiene, bleeding gums, very coated tongue with central crack.
  Medications: Thyroid, Finasteride, Terazosin for prostate enlargement, and a baby aspirin daily.
  Genotypic Risks: Family history of cancer (7 of 8 siblings and father); brother and uncle with Alzheimer's; son Hemachromatosis (BB polymorphisms).
  Biochemical Lab: Mildly reduced GFR, high BUN, low TSH (0.03)/standard of care post thyroidectomy, PSA 1.19, high Homocysteine 18. Low albumin 3.9, Total Protein 6.3, Globulin 2.3.

Nutrition Assessment
- *Intake:* High sugar, processed foods and cured meats
- *Digestion & Assimilation & Elimination:* RZ reports a 1/week bowel movement
- *Utilization:* Cellular & Molecular Function:
- *Minerals:* Low magnesium, low-end blood electrolytes, BIA Capacitance 730 (goal ~ 1300)
- *Antioxidants:* Water Soluble: Vitamin C, Phytonutrients: severe skin wrinkles-poor Vit C, vegetable/fruit intake
- *Protein:* Low-end protein status; albumin 3.9
- *Vitamin D & Fat soluble vitamins:* Low vitamin D (25 ng/mL)
- *Oils/Fatty Acids:* Low omega-3 and omega-6:GLA intake foods; cholesterol panel is OK
- *Methylation*: elevated Homocysteine implying poor methylation; low folate, B12, MCV 101 HI, MCH 33

Plan: (Continued Early-Onset Alzheimer's)
1. *Elimination Diet*: Avoid sugars, soda-replace with tea or fresh lemonade; avoid cured meats, dairy, processed foods; eat more whole foods, sea salt, salads and no nitrate, eat high-quality meats; increase water intake to 2 qts/day (previous intake none). GOAL: reduce sugar intake and antigenic foods; increase phytonutrients, nutrient-dense-high-fiber food intake. Good family support.
2. *Improve Methylation and Oils/Fatty Acids*: Focused on neurological support;

Supplement with multivitamin, magnesium, EPA/DHA 1:1, GLA 370 mg QD, folate (5-MTHF 800 µg) QD, B12 (Methyl B12 1000 µg QD), D3 4000 IU QD; herbal laxative support until BM QD.
3. *Adjunctive Physician-Supported Nutrition Support Using IV Lipid Therapy:* [phosphatidylcholine (PC) × 2 months, Glutathione, phenylbutyrate, assessed methylation support] twice a week for 8 weeks then start oral protocol of PC, Glutathione and Butyrate, methylating nutrients (5-MTHF, B12) as needed)

Outcome: After implementation of herbal colon cleanse, magnesium and increased water intake, at 2 weeks bowel movements increased to 1/day; Memory improved ~70% within 4 weeks. Great family support. Client returned to work and driving at 6 months, showing significant improvement. Continue diet, oral supplements. Follow-up monthly for 14 months then every 6 months while continuing oral phospholipid, methyl nutrients and very low sugar nutrition program. Continues cognitive function as long as on supplements after 7 years. Blood homocysteine 9, MCV 92, MCV 30.5. Neurologist reversed Alzheimer diagnosis after 14 months.

Case Review: Complex Neurological Condition

Female AR, age 20
Weight 70 lb, 46 inches, BMI 15.7 Underweight
Diagnosis
- Cardiofaciocutaneous (CFC) syndrome (Q87.1 ICD10); Genes are: *BRAF* (~75%), *MAP 2K1* and *MAP 2K2* (~25%), and *KRAS* (<2%). [Congenital malformation syndromes predominantly associated with short stature]; 23andme.com nutrigenomics: homozygous MTHF +/+, heterozygous VDR –/+

The *cardiofaciocutaneous (CFC) syndrome* is a condition of sporadic occurrence, with patients showing multiple congenital anomalies and mental retardation. It is characterized by failure to thrive, relative macrocephaly, a distinctive face with prominent forehead, bitemporal constriction, absence of eyebrows, hypertelorism, downward-slanting palpebral fissures often with epicanthic folds, depressed nasal root and a bulbous tip of the nose. The cutaneous involvement consists of dry, hyperkeratotic, scaly skin, sparse and curly hair, and cavernous hemangioma. Most patients have a congenital heart defect, most commonly pulmonic stenosis and hypertrophic cardiomyopathy. ▶ https://www.cfcsyndrome.org/fact-sheet
- Epilepsy and recurrent seizures (G40 ICD10): ~200 seizures per month w/ anti-seizure Rx; three severe seizure hospitalizations per month.
- Severe mental impairment (*F79 ICD10*)
- Osteopenia (M85.80 ICD10)

Medical History
- Normal 9-month vaginal birth; 5 months developmental delay, 9 months diagnosed "failure to thrive"; started tube feeding; residential care at home parents
- Diet History: Tube feeding Infant Formula until 3 years old; 3–20 years old received Boost™ tube feeding; weight 78–80 lb. bedridden, Birth lab significant for hypercalcemia 10.9 HI (<10). No PTH was ordered. Hypothyroid Day 1 screen WNL.

Signs and Symptoms
- No eye contact
- Appears pain when touching arms and hands during bathing
- Tube feeding tolerated

Lab History: Significant Findings
- Methylmalonic acid 861 very HI (B12 functional marker)
- B12, serum >2000 pg/mL
- Lipid Panel: Total Cholesterol 128 mg/dL; LDL 55 mg/dL: Low-end
- Ferritin 24 – Low End
- Vitamin D >120
- GGT 88 HI (ideal <10) [can be suggestive of toxcity and/or low glutathione status]
- Differential: WBC 6.04; Neutrophils 40.2 low end/Lymphocytes 44.7 HI
- Specialty Lab: RBC Fatty acid analysis: Very low arachidonic acid, GLA, DGLA, Hi EPA, Hi DHA
- Specialty Lab: lymphocyte micronutrient analysis: LOW: vitamin D, B12, A, carnitine, zinc, magnesium, choline

PLAN: Complex Neurological Condition
Three-Month Plan
- Retest previous abnormal tests
- Monitor symptoms /behavior – report TF tolerance and any changes weekly
- Mother requested investigating possibility that AM and mother were hypothyroid - mother exploring possibility AM was hypothyroid and infant screening was done day of birth negative, but protocol is to be tested day 3 after birth. Early screening may have missed hypothyroid condition.

Three-Month Follow-up
- Review new lab. WNL: B12, serum, methylmalonic acid ABNORMAL:, +ANA Titer, +++ EBV, Total + IgG/+IgE, carnitine panel: low,
- Seizures reduced 50–60% no hospital trips, reduced anti-seizure Rx
- Eye contact; no pain appearance touching during bathing.

Nutrition Assessment Metabolic Priorities
- B12 deficiency; specific for adenosylcobalamin B12 2000 µg/day with L-5-MTHF (bioactive folate)
- Request doctor rule out subclinical infection with further viral/bacterial panel
- Depleted vitamin/minerals: D, A, B12, -5-MTHF, carnitine, phosphatidylcholine/phosphatidylethanolamine, Linoleic Acid,
- Home tube feeding Lipid Liquid Meal recipe 2 quarts
- 2 scoop vegan or beef protein powder
- 48 oz unsweetened almond or coconut milk; add water to achieve TF consistency
- 2 tsp Phosphotidylcholine/Phosphotidylethanolamine (mitochondrial support)
- 2 tsp evening primrose oil (GLA)
- Multivitamin/mineral powder 2 scoop (bioactive B vitamin forms)
- Active B12 w (adenosylcobalamin B12) w/ L-5-MTHF 1000 µg 1 tab in TF
- 1–500 mg L-carnitine – open capsule into TF
- 1 blanched egg yolk
- Blend and administer within 24 hours, refrigerate

Outcome 14 Months and beyond:
- Five months free of hospitalization due to severe seizures
- Eye contact daily
- Reach for objects
- Adult school; at 14 months mouthed quietly to Mom "I love you".
- Continues Lipid Shake daily, sometimes adds healthy variety of foods.

11.12 Conclusion

Lipids are the molecular components that comprise the *lipidome*—the complete lipid profile within the membrane, cell, tissue, or organism. The lipidome is in the dynamic metabolism of life expressing the genetic information in the DNA book of life. The membrane houses the receptors that receive information and direct the biochemistry to survive. The two functions of the lipids that can be modulated by nutrition therapy are membrane structure and inflammation control.

This chapter has described the association between a person's nutritional status and their vulnerability for impaired fatty acid status. Nutritional status of an individual is a determinant of how well their body can respond to and resolve an infection returning it to a state of wellness.

It is important to consider an individual's fatty acid status for several chronic disease conditions. Pathophysiology of chronic disease reveals its contrast to acute disease by its long-term development, often asymptomatic and not recognized. Subclinical chronic disease progression can "smolder" unrecognized for many years, even beginning in-utero and childhood. This biological stress often goes undiagnosed or with the patient being unaware of the significance. This chapter provides knowledge of lipid metabolism, as it relates to membrane function, the assessment of lipid imbalances, and specific membrane and anti-inflammatory nutrients that can contribute to targeted intervention and optimal health [87].

References

1. Watson RR, De Meester F, editors. Handbook of lipids in human function: fatty acids. Oxford: AOCS Press (by Elsevier); 2016.
2. Mouristen OG. Lipids in charge. In: Life-as a matter of fat. Heidelberg: Springer; 2005. p. 159–72.
3. Gofman JW, Jones HB, Lindgren FT, Lyon TP, Elliott HA, Strisower B. Blood lipids and human atherosclerosis. Circulation. 1950;2:161.
4. Keys A. Prediction and possible prevention of coronary disease. Am J Pub Health. 1953;43:1399–407.
5. Chilton FH, Murphy RC, Wilson BA, et al. Diet-gene interactions and PUFA metabolism: a potential contributor to health disparities and human diseases. Nutrients. 2014;6(5):1993–2022. Published 2014 May 21. https://doi.org/10.3390/nu6051993.
6. Horrobin DE. Omega-6 essential fatty acids: pathophysiology and roles in clinical medicine. Wiley-Liss: New York; 1990.
7. DiNicolantonio JJ, OKeefe J. Importance of maintaining a low omega-6/omega-3 ratio for reducing platelet aggregation, coagulation and thrombosis. Open Heart. 2019;6(1):e001011. Published 2019 May 2. https://doi.org/10.1136/openhrt-2019-001011.
8. Simopoulos AP. The importance of the omega-6/omega-3 fatty acid ratio in cardiovascular disease and other chronic diseases. Exp Biol Med (Maywood). 2008;233(6):674–88. https://doi.org/10.3181/0711-MR-311. Epub 11 Apr 2008.
9. Cooper GM, Hausman RE. The cell: a molecular approach. 6th ed. Sunderland: Sinauer Associates; 2013, Chapter 13.
10. Berg JM, Tmoczko JL, Gatto GJ, Stryer L. Biochemistry. 9th ed. W.H. Freeman: New York; 2019. p. 374–382.
11. Gonzalez MC, Barbosa-Silva TG, Bielemann RM, Gallagher D, Heymsfield SB. Phase angle and its determinants in healthy subjects: influence of body composition. Am J Clin Nutr. 2016;103(3):712–6. https://doi.org/10.3945/ajcn.115.116772.
12. Jaureguiberry MS, Tricerri MA, Sanchez SA, et al. Role of plasma membrane lipid composition on cellular homeostasis: learning from cell line models expressing fatty acid desaturases. Acta Biochim Biophys Sin Shanghai. 2014;46(4):273–82.
13. Barbosa-Silva MC, Barros AJ. Bioelectrical impedance analysis in clinical practice: a new perspective on its use beyond body composition equations. Curr Opin Clin Nutr Metab Care. 2005;8:311–7.
14. Lukaski H, Kyle U, Kondrup J. Assessment of adult malnutrition and prognosis with bioelectrical impedance analysis: phase angle and impedance ratio. Curr Opin Clin Nutr Metab Care. 2017;20(5):330–9. https://doi.org/10.1097/MCO.0000000000000387.
15. Elmehdawi R. Hypolipidemia: a word of caution. Libyan J Med. 2008;3(2):84–90. Published 1 Jun 2008. https://doi.org/10.4176/071221.
16. Levy BD. Resolvins and protectins: natural pharmacophores for resolution biology. Prostaglandins Leukot Essent Fatty Acids. 2010;82(4–6):327–32. https://doi.org/10.1016/j.plefa.2010.02.003.
17. Bannenberg GL. Resolvins: current understanding and future potential in the control of inflammation. Curr Opin Drug Discov Devel. 2010;13(1):136.
18. Ge C, Chen H, Mei T, et al. Application of a ω-3 desaturase with an arachidonic acid preference to eicosapentaenoic acid production in mortierella alpina. Front Bioeng Biotechnol. 2018;5:89. Published 22 Jan 2018. https://doi.org/10.3389/fbioe.2017.00089.
19. Leonard AE, Pereira SL, Sprecher H, Huang YS. Elongation of long-chain fatty acids. Prog Lipid Res. 2006;45(3):237–49. https://doi.org/10.1016/S0163-7827(03)00040-7.
20. Das UN. A defect in delta 6 and delta5 desaturases may be a factor in the initiation and progression of insulin resistance, the metabolic syndrome and ischemic heart disease in South Asians. Lipids Health Dis. 2010;9:130.
21. Mulligan C. Interaction of delta-6 desaturase activity and dietary fatty acids in determining cardiometabolic risk. Project end date: 14 Aug 2014. Colorado State University (n/a), Fort Collins, Food Science and Human Nutrition.
22. Vaittinen M, Walle P, Kuosmanen E, Männistö V, Käkelä P, Ågren J, et al. FADS2 genotype regulates delta-6 desaturase activity and inflammation in human adipose tissue. Lipid Res. 2016;57(1):56–65. First Published on 25 Nov 2015. https://doi.org/10.1194/jlr.M059113.
23. Tosi F, Sartori F, Guarini P, Olivieri O, Martinelli N. Delta-5 and delta-6 desaturases: crucial enzymes in polyunsaturated fatty acid-related pathways with pleiotropic influences in health and disease. In: Camps J, editor. Oxidative stress and inflammation in non-communicable diseases - molecular mechanisms and perspectives in therapeutics. Advances in experimental medicine and biology, vol. 824. Cham: Springer; 2014.
24. Yashiro H, Takagahara S, Tamura YO, et al. A novel selective inhibitor of delta-5 desaturase lowers insulin resistance and reduces body weight in diet-induced obese C57BL/6J mice. PLoS One. 2016;11(11):e0166198. Published 10 Nov 2016. https://doi.org/10.1371/journal.pone.0166198.
25. Sergeant S, Rahbar E, Chilton FH. Gamma-linolenic acid, dihomo-gamma linolenic, eicosanoids and inflammatory processes. Eur J Pharmacol. 2016;785:77–86. https://doi.org/10.1016/j.ejphar.2016.04.020.
26. Hardwick JP, Eckman K, Lee YK, et al. Eicosanoids in metabolic syndrome. Adv Pharmacol. 2013;66:157–266. https://doi.org/10.1016/B978-0-12-404717-4.00005-6.
27. Simopoulos A, Meester F. A balanced omega-6/omega-3 fatty acid ratio, cholesterol and coronary heart disease. Basel: Karger; 2009.
28. Spencer MD, Hamp TJ, Reid RW, Fischer LM, Zeisel SH, Fodor AA. Association between composition of the human gastrointestinal microbiome and development of fatty liver with choline deficiency. Gastroenterology. 2011;140(3):976–86. https://doi.org/10.1053/j.gastro.2010.11.049.

29. Kohlmeier M, da Costa KA, Fischer LM, Zeisel SH. Genetic variation of folate-mediated one-carbon transfer pathway predicts susceptibility to choline deficiency in humans. Proc Natl Acad Sci U S A. 2005;102(44):16025–30. https://doi.org/10.1073/pnas.0504285102.
30. Gough ME, Graviss EA, Chen T, Obasi EM, May EE. Compounding effect of vitamin D3 diet, supplementation, and alcohol exposure on macrophage response to mycobacterium infection. Tuberculosis. 2019;116S:S42–58. Available online 30 Apr 2019.
31. Palau EE, Martínez FS, Freud HK, Colomés JLL, Pérez AD. Tuberculosis: plasma levels of vitamin D and its relation with infection and disease. Med Clín (English Edition). 2015;144(3):111–4.
32. Keflie TS, Nölle N, Lambert C, Biesalski HK. Vitamin D deficiencies among tuberculosis patients in Africa: a systematic review author links open overlay panel. Nutrition. 2015;31(10):1204–12.
33. Aibana O, Franke MF, Huang CC, et al. Impact of vitamin a and carotenoids on the risk of tuberculosis progression. Clin Infect Dis. 2017;65(6):900–9. https://doi.org/10.1093/cid/cix476.
34. Stephensen CB, Vitamin A. Infection, and immune function. Annu Rev Nutr. 2001;21:167–92.
35. Lankinen M, Uusitupa M, Schwab U. Genes and dietary fatty acids in regulation of fatty acid composition of plasma and erythrocyte membranes. Nutrients. 2018;10(11):1785. https://doi.org/10.3390/nu10111785.
36. Leonard AE, Pereira SL, Sprecher H, Huang YS. Elongation of long-chain fatty acids. Prog Lipid Res. 2004;43(1):36–54.
37. Sergeanta S, Rahbarb E, Chilton FH. Gamma-linolenic acid, Dihomo-gamma linolenic, eicosanoids and inflammatory processes. Eur J Pharmacol. 2016;785:77–86. https://doi.org/10.1016/j.ejphar.2016.04.020.
38. Elmadfa I, Meyer AL. The role of the status of selected micronutrients in shaping the immune function. Endocr Metab Immune Disord Drug Targets. 2019;19:1. https://doi.org/10.2174/1871530319666190529101816.
39. Wu D, Lewis ED, Pae M, Meydani SN. Nutritional modulation of immune function: analysis of evidence, mechanisms, and clinical relevance. Front Immunol. 2019;9:3160. Published 15 Jan 2019. https://doi.org/10.3389/fimmu.2018.03160
40. Maywald M, Wessels I, Rink L. Zinc signals and immunity. Int J Mol Sci. 2017;18:E2222. https://doi.org/10.3390/ijms18102222.
41. Fowler AA 3rd, Syed AA, Knowlson S, et al. Phase I safety trial of intravenous ascorbic acid in patients with severe sepsis. J Transl Med. 2014;12:32. Published 31 Jan 2014. doi:10.1186/1479-5876-12-32
42. Chan JM, Darke AK, Penney KL, et al. Selenium- or vitamin E-related gene variants, interaction with supplementation, and risk of high-grade prostate cancer in SELECT. Cancer Epidemiol Biomark Prev. 2016;25(7):1050–8. https://doi.org/10.1158/1055-9965.EPI-16-0104.
43. Ohira H, Tsutsui W, Fujioka Y. Are short chain fatty acids in gut microbiota defensive players for inflammation and atherosclerosis? J Atheroscler Thromb. 2017;24(7):660–72. https://doi.org/10.5551/jat.RV17006.
44. Ratajczak W, Ryl A, Mizerski A, Walczakiewicz K, Sipak O, Laszczyńska M. Immunomodulatory potential of gut microbiome-derived short-chain fatty acids (SCFAs). Acta Biochim Pol. 2019;66(1):1–12. https://doi.org/10.18388/abp.2018_2648.
45. Stilling RM, de Wouw M, Clarke G, Stanton C, Dinan TG, Cryan JF. The neuropharmacology of butyrate: the bread and butter of the microbiota-gut-brain axis? Neurochem Int. 2016;99:110–32. https://doi.org/10.1016/j.neuint.2016.06.011.
46. Bourassa MW, Alim I, Bultman SJ, Ratan RR. Butyrate, neuroepigenetics and the gut microbiome: can a high fiber diet improve brain health? Neurosci Lett. 2016;625:56–63. https://doi.org/10.1016/j.neulet.2016.02.009.
47. Szentirmai É, Millican NS, Massie AR, Kapás L. Butyrate, a metabolite of intestinal bacteria, enhances sleep. Sci Rep. 2019;9(1):7035. Published 7 May 2019. doi:10.1038/s41598-019-43502-1.
48. Liu H, Wang J, He T, et al. Butyrate: a double-edged sword for health? Adv Nutr. 2018;9(1):21–9. https://doi.org/10.1093/advances/nmx009.
49. Biagioli M, Carino A. Signaling from intestine to the host: how bile acids regulate intestinal and liver immunity. Handb Exp Pharmacol. 2019;256:95–108. https://doi.org/10.1007/164_2019_225. [Epub ahead of print].
50. Chiang JYL, Ferrell JM. Bile acids as metabolic regulators and nutrient sensors. Annu Rev Nutr. 2019;39:175–200.
51. Wymann MP, Schneiter R. Lipid signaling in disease. Nat Rev Mol Cell Biol. 2008;9:162–76.
52. Schaffer JE. Lipotoxicity: many roads to cell dysfunction and cell death. J Lipid Res. 2016;57(8):1327–8.
53. Bansal A, Henao-Mejia J, Simmons RA. Immune system: an emerging player in mediating effects of endocrine disruptors on metabolic health. Endocrinology. 2018;159(1):32–45. https://doi.org/10.1210/en.2017-00882.
54. Papalou O, Kandaraki EA, Papadakis G, Diamanti-Kandarakis E. Endocrine disrupting chemicals: an occult mediator of metabolic disease. Front Endocrinol (Lausanne). 2019;10:112. Published 1 Mar 2019. https://doi.org/10.3389/fendo.2019.00112.
55. Statovci D, Aguilera M, MacSharry J, Melgar S. The impact of western diet and nutrients on the microbiota and immune response at mucosal interfaces. Front Immunol. 2017;8:838. Published 28 Jul 2017. https://doi.org/10.3389/fimmu.2017.00838.
56. Ratnaseelan AM, et al. Effects of mycotoxins on neuropsychiatric symptoms and immune processes. Clin Ther. 2018;40(6):903–17.
57. Akbari P, Braber S, Varasteh S, Alizadeh A, Garssen J, Fink-Gremmels J. The intestinal barrier as an emerging target in the toxicological assessment of mycotoxins. Arch Toxicol. 2017;91(3):1007–29. https://doi.org/10.1007/s00204-016-1794-8.
58. Elliott EG, Trinh P, Ma X, Leaderer BP, Ward MH, Deziel NC. Unconventional oil and gas development and risk of childhood leukemia: assessing the evidence. Sci Total Environ. 2017;576:138–47. https://doi.org/10.1016/j.scitotenv.2016.10.072.
59. Rull RP, Ritz B, Shaw GM. Neural tube defects and maternal residential proximity to agricultural pesticide applications. Am J Epidemiol. 2006;163(8):743–53. https://doi.org/10.1093/aje/kwj101.
60. Rosenberg S, Porges SW, Shield B. Accessing the healing power of the vagus nerve: self-help exercises for anxiety, depression, trauma, and autism: North Atlantic Books: Berkeley, CA; 2017.
61. Ross AC. Vitamin A and retinoic acid in T cell–related immunity. Am J Clin Nutr. 2012;96(Suppl):1166S–72S. https://doi.org/10.3945/ajcn.112.034637.
62. Pakiet A, Kobiela J, Stepnowski P, Sledzinski T, Mika A. Changes in lipids composition and metabolism in colorectal cancer: a review. Lipids Health Dis. 2019;18(1):29. Published 26 Jan 2019. https://doi.org/10.1186/s12944-019-0977-8.
63. USDA Dietary Guidelines for Americans 2015–2020. https://www.choosemyplate.gov/2015-2020-dietary-guidelines-answers-your-questions. Accessed 26 June 2019.
64. Liu H, et al. Butyrate: a double-edged sword for health? Adv Nutr. 2018;9:21–9.
65. Jenkins KJ, Kramer JKG. Influence of excess dietary copper on lipid composition of calf tissues. J Dairy Sci. 1989;72(10):2582–91.
66. Mahfouz MM, Kummerow FA. Effect of magnesium deficiency on $\Delta 6$ desaturase activity and fatty acid composition of rat liver microsomes. Lipids. 1989;24(8):727–32. https://doi.org/10.1007/BF02535212.
67. Halsted CH, Medici V. Vitamin-dependent methionine metabolism and alcoholic liver disease. Adv Nutr. 2011;2(5):421–7. https://doi.org/10.3945/an.111.000661.
68. Liu U, Bin P, Wang T, Ren W, Zhong J, Liang J, Hu CAA, Zeng Z, Yin Y. DNA methylation and the potential role of methyl-containing nutrients in cardiovascular diseases. Oxidative Med Cell Longev. 2017;2017:1670815.

69. Neidhart N. Methyl donors. In: DNA methylation and complex human disease; 2016. p. 429–39. https://doi.org/10.1016/B978-0-12-420194-1.00027-0.
70. Dudman NPB, Guo XW, Gordon RB, Dawson PA, Wilcken DEL. Human homocysteine catabolism: three major pathways and their relevance to development of arterial occlusive disease. J Nutr. 1996;126(Suppl_4):1295S–300S. https://doi.org/10.1093/jn/126.suppl_4.1295S.
71. Koury MJ, Ponka P. New insights into erythropoiesis: the roles of folate, vitamin B12, and iron. Annu Rev Nutr. 2004;24:105–31. (Volume publication date 14 Jul 2004) First published online as a Review in Advance on 10 Mar 2004. https://doi.org/10.1146/annurev.nutr.24.012003.132306.
72. Cheng T, Goodman G, Thornquist M, Barnett M, Beresford S, LaCroix A, Zheng Y, Neuhouser M. Estimated intake of vitamin D and its interaction with vitamin A on lung cancer risk among smokers. Int J Cancer. 2014;135(9):2135–45.
73. Minich D. The rainbow diet. Color can heal your life! https://www.deannaminich.com. Accessed 2 June 2019.
74. Scheiermann C, Kunisaki Y, Frenette PS. Circadian control of the immune system. Nat Rev Immunol. 2013;13(3):190–8. https://doi.org/10.1038/nri3386.
75. Lord B. Laboratory evaluations for integrative and functional medicine. Duluth, Georgia: Metametrix Institute; 2012.
76. Visentin S, Vicentin D, Magrini G, Santandreu F, Disalvoa L, Salaa M, et al. Red blood cell membrane fatty acid composition in infants fed formulas with different lipid profiles. Early Hum Dev. 2016;100:11–5. https://doi.org/10.1016/j.earlhumdev.2016.05.018.
77. Revskij D, Haubold S, Viergutz T, Kröger-Koch C, Tuchscherer A, Kienberger H, et al. Dietary fatty acids affect red blood cell membrane composition and red blood cell ATP release in dairy cows. Int J Mol Sci. 2019;20(11):2769. https://doi.org/10.3390/ijms20112769.. (29 April 2019/Revised: 29 May 2019/Accepted: 4 June 2019/Published: 5 June 2019).
78. Weitz D, Weintraub H, Fisher E, Schwartzbard AZ. Fish oil for the treatment of cardiovascular disease. Cardiol Rev. 2010;18(5):258–63. https://doi.org/10.1097/CRD.0b013e3181ea0de0.
79. Greene ER, Huang S, Serhan CN, Panigrahy D. Regulation of inflammation in cancer by eicosanoids. Prostaglandins Other Lipid Mediat. 2011;96(1–4):27–36. https://doi.org/10.1016/j.prostaglandins.2011.08.004.
80. Pidgeon GP, et al. Lipoxygenase metabolism: roles in tumor progression and survival. Cancer Metastasis Rev. 2007;26:503–24.
81. Wang D, DuBois RN. Eicosanoids and cancer. Nat Rev Cancer. 2010;10:181–93. https://doi.org/10.1038/nrc2809.
82. Lewandowska U, Owczarek K, Szewczyk K, Podsędek A, Koziołkiewicz M, Hrabec E. Influence of polyphenol extract from evening primrose (Oenothera paradoxa) seeds on human prostate and breast cancer cell lines. Postepy Hig Med Dosw (Online). 2014;68:110–118e. ISSN 1732-2693.
83. Zhu TB, Zhang Z, Luo P, Wang SS, Chen NH. Lipid metabolism in Alzheimer's disease. Brain Res Bull. 2019;144:68–74.
84. Peña-Bautista C, Baquero M, Vento M, Cháfer-Pericás C. Free radicals in Alzheimer's disease: lipid peroxidation biomarkers. Clin Chim Acta. 2019;491:85–90.
85. Fluegge K. Theoretical article-A model of lipid dysregulation and altered nutrient status in Alzheimer's disease. Alzheimer's Dementia Trans Res Clin Interventions. 2019;5:139–45.
86. Hanson MA, Gluckman PD. Early developmental conditioning of later health and disease: physiology or pathophysiology? Physiol Rev. 2014;94(4):1027–76. https://doi.org/10.1152/physrev.00029.2013.
87. Shao A, Drewnowski A, Willcox DC, et al. Optimal nutrition and the ever-changing dietary landscape: a conference report. Eur J Nutr. 2017;56(Suppl 1):1–21. https://doi.org/10.1007/s00394-017-1460-9.
88. OpenStax CNX. Chapter 5.1: Components and structure. In: OpenStax, Biology. Rice University; 2019. http://cnx.org/contents/185cbf87-c72e-48f5-b51e-f14f21b5eabd@11.10.
89. Blausen.com staff. Medical gallery of Blausen Medical 2014. WikiJournal Med. 2014;1(2). https://doi.org/10.15347/wjm/2014.010. ISSN 2002-4436. Retrieved from: https://commons.wikimedia.org/wiki/File:Blausen_0818_Sodium-PotassiumPump.png.

Resources

Armstrong D, editor. Methods in molecular biology. In: Lipidomics volume 1: methods and protocols. Humana Press: Totowa, NJ; 2009.
Calviello G, Serini S, editors. Diet and cancer 1: dietary omega-3 polyunsaturated fatty acids and cancer: Springer: Dordrecht; 2010.
Chow CK. Fatty acids in foods and their health implications. 3rd ed: CRC Press: Boca Raton London New York; 2008.
Das UN. Molecular basis of health and disease: Springer Science & Business Media: Springer Netherlands; 2011.
Horrobin DF, editor. Omega-6 essential fatty acids. New York: Wiley-Liss; 1990.
Mahan LK, Raymond J. Krause's: food & the nutrition care process. 14th ed: Elsevier: St. Louis, MO; 2017.
Mouritsen OG. The frontier collection: life—as a matter of fat: the emerging science of lipidomics: Springer: Germany (Springer Germany); 2005.
Quinn PJ, Wang X, editors. Lipids in health and disease. Basel: Springer Nature; 2018.
Simopoulos AP, De Meester F, editors. World review of nutrition and dietetics: a balanced omega-6/omega-3 fatty acid ratio, cholesterol and coronary heart disease: Karger AG: Basel (Switzerland); 2009.
Watson RR, De Meester F, editors. Handbook of lipids in human function: fatty acids. Oxford: AOCS Press (by Elsevier); 2016. p. 809.

Structure: From Organelle and Cell Membrane to Tissue

David Musnick, Larissa Severson, and Sarah Brennan

12.1 Introduction – 174

12.2 Part I: Membrane Structure – 174
12.2.1 Introduction – 174
12.2.2 Biological Structure – 174

12.3 Part II: Dietary and Lifestyle Influences – 177
12.3.1 Introduction – 177
12.3.2 Pathophysiology – 177
12.3.3 Evaluation/Assessment – 179
12.3.4 Prevention/Treatment – 179

12.4 Part III: Organ Structure and Function – 180
12.4.1 Introduction – 180
12.4.2 Eye – 180
12.4.3 Skin – 181
12.4.4 Brain – 182
12.4.5 Barriers – 185

12.5 Part V: Musculoskeletal Structure and Influences – 186
12.5.1 Introduction – 186
12.5.2 Evaluation/Assessment – 187
12.5.3 Prevention/Treatment – 187
12.5.4 Pathophysiology – 187
12.5.5 Evaluation/Assessment – 188
12.5.6 Treatment/Prevention – 188

12.6 Conclusion – 188

References – 188

© Springer Nature Switzerland AG 2020
D. Noland et al. (eds.), *Integrative and Functional Medical Nutrition Therapy*,
https://doi.org/10.1007/978-3-030-30730-1_12

12.1 Introduction

Cellular structure is the foundation on which biochemical and energetic functions occur. The vast number of processes that our tissues and organs are capable of accomplishing derives from the basic molecular structures of which they are composed. In other words, unless the individual cells, their organelles, and their membranes are supported by the proper nutrients and suitable environment, the body systems they comprise will be unable to function optimally.

Damage to the cellular structure can result in a multitude of dysfunctions. Dietary and lifestyle choices can influence the overall health of our bodies down to the structure of our cell membranes. Maintaining a diet full of nutrients such as essential omega-3s and omega-6s, and even cholesterol, can help structurally stabilize our cell membranes as well as reduce the susceptibility to leakage of beneficial nutrients out of the cell.

In any patient presenting with a constellation of symptoms, a clinician should ask if there is a structure, or structures, that should be evaluated and supported. When developing a plan to support a particular structure, both the clinician and the patient must be aware that this modification may take months to years. Short- and long-term goals should be established, and monitoring the structural health and integrity should be done periodically.

There are key questions to ask when evaluating for structure, including:
- What structures are compromised in my patient in relationship to their symptoms or aging process?
- Is there a cell type, organelle, membrane, tissue, organ, or other structure that is related to my patient's symptoms?
- From the cellular to the tissue/organ/skeletal level, what is the possible structural compromise/pathology?
- What is the level of function and dysfunction of these structures?
- If there is degeneration in the structure, what is the extent of degeneration?
- Is regeneration of this structure possible?
- What support could I give these structures?
- Can function be augmented without improving structure?
- What structural barriers (intestinal wall, blood–brain barrier, lung barrier, etc.) are compromised in my patient??
- Is there research to show that the membrane, tissue, or organ can be supported and improved in its structure and function?
- How much of a change in this structure and function can my intervention make?
- How long will it take to make changes in this structure?
- What are the most important interventions to improve the health of the cell, tissue, or system, and what is an estimated cost of these interventions?
- Can I evaluate the integrity of these structures with a lab test, imaging study, or assessment of nutrition status?

In this chapter, we will discuss structure, starting at the cell membrane and moving to musculoskeletal and other tissues, while examining key structural questions.

12.2 Part I: Membrane Structure

12.2.1 Introduction

The membranes of our cells are important in that their integrity is critical for cellular function. For this purpose, cell membranes are comprised of a plethora of components necessary to not only hold all of the vital cellular structures in but also have the ability to be selectively permeable in regard to what solutes can pass through the membrane, as well as provide a location for necessary biological reactions to occur. The central component that is arguably the most vital for the cell membrane to fulfill its specific demands are lipids.

12.2.2 Biological Structure

Our cell membranes are made up of two stacked sheets of lipids, known as the lipid bilayer, with proteins intermittently embedded within, at a ratio of about 1:1. Carbohydrates are also incorporated throughout, but they only constitute about 10% of the membrane [1]. Phospholipids are amphipathic molecules, meaning they have a nonpolar head that is hydrophobic attached by a phosphate ester bond to two polar tails that are hydrophilic. The lipids arrange themselves in such a way that hides their hydrophobic end from the surrounding water in our tissues and presents their hydrophilic end, forming a curved bilayer structure appearing as a circle. Their properties make them insoluble in water, proving them to be an effective candidate for protecting the fragile internal components of our cells from our surrounding fluids. Besides acting as a protective barrier, the membrane also provides an anchor for the internal cytoskeleton to grasp onto, creating a place for cell adhesion, allowing groups of cells to bind together to form tissues [2]. Another important function of the membrane is to act as a surface for important biological reactions to occur, including oxidative phosphorylation on the inner mitochondrial membrane [3]. The membrane also is home to numerous receptors.

Membranes thereby become the most abundant cellular structure in all living matter. They can be considered as nature's preferred mode of microencapsulation technology, developed as means of compartmentalizing living matter and protecting the genetic material. The biological membrane is the essential capsule of life. Many important biological processes in the cell either take place at membranes or are mediated by membranes, such as transport, growth, neural function, immunological response, signaling, and enzymatic activity. An important function of the lipid bilayer is to act as a passive permeability barrier to ions and other molecular substances and leave the transmembrane transport to active carriers and channels [1] (◘ Fig. 12.1).

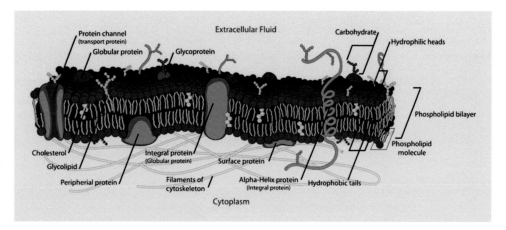

Fig. 12.1 Schematic model of the plasma membrane of a eukaryotic cell which highlights the membrane as a composite of a central lipid bilayer sandwiched between the carbohydrate glycocalyx (which consist of polysaccharides) on the outside and the rubber-like cytoskeleton (which is a polymeric protein network) on the inside. Intercalated in the lipid bilayer are shown various integral proteins and polypeptides. The membrane is subject to undulations and the lipid bilayer displays lateral heterogeneity, lipid domain formation, and thickness variations close to the integral proteins. Whereas the lipid molecules in this representation are given with some structural details, the membrane-associated proteins remain fairly featureless. In order to capture many different features in the same illustration, the different membrane components are not drawn to scale. (Reprinted from Benton et al. [78]. With permission from Wolters Kluwer Health, Inc.)

In order for a cell membrane, or a lipid bilayer, to function optimally, it requires a set of conditions that ultimately make the completion of the membrane's tasks easier and more likely to occur properly. Lipid bilayers are soft and pliable while retaining a structure in order to bend without breaking down [1]. In addition, they exist in specific structural compositions so that they are stable enough to hold their shape. The stability of the membranes is largely due to adequate cholesterol molecules embedded throughout the membrane as an important structure builder in all cells, making the membrane thicker and less "leaky." In addition, cholesterol is a basic unit of some vitamins and hormones. All animals contain cholesterol in the plasma membranes in amounts from 30% to 50% of total lipids. The organelle membranes contain very little: mitochondrial membranes less than 5%, Golgi membranes about 8%, and ER membranes around 10% [4]. Finally, membranes are particularly selective in the way that the individual lipids and transport proteins are organized within the bilayer so that they can control their permeability efficiently.

The length and degree of saturation of fatty acid chains in the phospholipids affects membrane properties, specifically fluidity. The breakdown of the different types of fats found in the membrane is about one-third cholesterol, one-third unsaturated fatty acids, and one-third saturated fatty acids, but these ratios fluctuate depending on other variables [5]. Unsaturation creates a kink in the fatty acid chain, which prevents the fatty acids from packing together as tightly, therefore increasing the fluidity of the membrane. The presence of cholesterol in the membrane also affects fluidity, which will be discussed later in this section. Membrane fluidity is carefully regulated since enzymatic activity as well as transport processes can stop working if the viscosity of the membrane gets too high. A common tool used in IFMNT practice is the bioelectric impedance analysis equipment measuring the phase angle, often described as a marker of cell membrane fluidity and nutritional status.

The lipid composition of the two monolayers of the lipid bilayer are characteristically different from one another. The phospholipids phosphatidylcholine (PC) and sphingomyelin, as well as cholesterol and glycolipids, are predominantly found on the outer monolayer. Phosphatidylethanolamine (PE) and phosphatidylserine (PS) are primarily found on the inner monolayer, facing the inside of the cell. The presence of PS on the inner monolayer creates a difference in charges between the two monolayers [2]. This is especially important in electrically excitable cells, like neurons and muscle cells, where changes in membrane potential are used to generate signals within the cell or to the neighboring cells.

12.2.2.1 Phosphatidylcholine

Phosphatidylcholine (PC) is the most common phospholipid contained in the bilayer cell membrane, and accounts for over half of the total membrane phospholipids primarily in the outer leaflet [6]. It is an essential nutrient that is consumed regularly in our diet, from eggs, raw dairy, and some cruciferous vegetables, as well as available as a dietary supplement or medical intravaneous infusion. PC is used to help maintain cell structure and aid in signaling, particularly in neurons [4, 6] (◘ Fig. 12.2). It organizes itself spontaneously into the bilayer, providing the membrane with the bulk of its structure. Besides playing an essential role in the cell membrane, phosphatidylcholine is also necessary for the secretion of important plasma lipoproteins VLDL and HDL, since both of these molecules are surrounded by a phospholipid monolayer [6]. Phosphatidylcholine is also secreted into bile, where it assists with the digestion of dietary fat [6].

Fig. 12.2 Lipid polar head of phospholipid groups. (Reprinted from Mouritsen et al. [4]. With permission from Springer Nature)

Substituent	Chemical formula[a]	Polar head group name	Ab[b]
Hydrogen	–H	Phosphatidic acid	PA
Choline	$-CH_2CH_2N(CH_3)_3^+$	Phosphatidylcholine	PC
Ethanolamine	$-CH_2CH_2NH_3^+$	Phosphatidylethanolamine	PE
Serine	$-CH_2CH(NH_3)COO^-$	Phosphatidylserine	PS
Glycerol	$-CH_2CH(OH)CH_2OH$	Phosphatidylglycerol	PG
myo-inositol	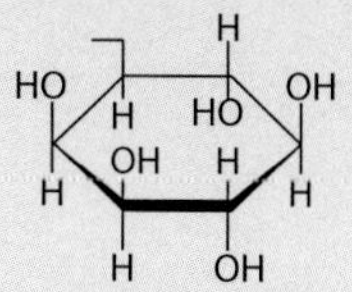 (inositol ring structure)	Phosphatidylinositol	PI

[a]Chemical formula for the substituent linked to the phosphate group at position 3 of the glycerol moiety
[b]Abbreviation for the polar head group nomenclature

12.2.2.2 Phosphatidylethanolamine

Phosphatidylethanolamine (PE) comprises about 25% of the phospholipids in most membranes, with about 37% PE in the inner mitochondrial membrane [7]. This percentage can be up to 45% in nervous tissue, such as neural tissue, nerves, white matter in the brain, and cells of the spinal cord [7]. It is involved in cell fusion and division as well as regulating membrane curvature [8]. PE contains a polar head group, which in turn lowers the melting point of the phospholipid, contributing to the viscosity and fluidity of the lipid membrane. Beyond the membrane, phosphatidylethanolamine is also involved in blood clotting and the secretion of lipoproteins in the liver [9].

12.2.2.3 Phosphatidylserine

Constituting 3–10% of the total lipids within the cell, phosphatidylserine (PS) plays an important role in cell-to-cell signaling and is involved in the proper localization and activation of many intracellular proteins [10]. The location of PS within the membrane directly affects the health of the cell. PS is located on the inner monolayer, and therefore faces the inside of the cell [10]. During the immune system response, apoptotic lymphocytes bring PS to the surface of the outer monolayer, where it becomes a signal for phagocytosis by activated macrophages with the process of autophagy and the cell is cleared from circulation [11]. This makes the differences between the two monolayers critical to the health and survival of the cell. Proper biological function depends on the elimination of unwanted, useless cells often referred to as "cellular debris"; therefore, the mechanisms of apoptotic cell recognition using PS and the subsequent removal by phagolysosomes are important in order to prevent excess inflammation and uncontrolled cell lysis [11]. The decision to use PS as a dietary supplement should consider the individual's membrane status and adequacy of other important phospholipids like PC and PE.

12.2.2.4 Cholesterol

Cholesterol is a vital component of all biological membranes. In fact, cholesterol is the most abundant type of lipid in our membranes, accounting for 30–50% of the lipid molecules [1]. Cholesterol is a lipid that is structurally very different from the phospholipids that are found in the membrane. It has a steroid ring structure as well as a hydroxyl group as its polar head. This structural difference makes cholesterol bulkier than its surrounding lipids, promoting stability in membrane shape. The presence of cholesterol in the membrane also makes the membrane less pliable, less susceptible to compression, and leads to a thickening of the membrane. This rigidity allows the membrane to maintain its integrity and shape, as well as making the membrane tighter and therefore less permeable.

In addition to providing stability, cholesterol also plays a vital role in protecting the cell membrane from nutrients that we consume that could be harmful to the cell. Many pharmaceutical drugs are comprised of amphiphilic molecules, meaning that their target sites within the body are places that are composed of similar molecules, that is, lipid bilayers. Such drugs that are intended to target membranes include, but are not limited to, antipsychotics, antitumoral drugs, antidepressants, tranquilizers, antihistamines, antifungals, and analgesics. As these drugs interact with the membrane, they cause it to lose stability, making it susceptible to compression and leakage. In addition to amphiphilic drugs, ethanol-based alcohols can also have a detrimental effect on the lipid bilayers of our membranes. Ethanol congregates on the glycerol backbone of the lipid bilayer, creating a buildup of pressure on the membrane. However, cholesterol can help protect our membranes from damage due to medication and alcohol consumption. The presence of cholesterol can help redistribute the pressure that ethanol places on the glycerol backbone of the lipids in the bilayer. In addition, due to its bulkiness in size and structure, it can help squeeze ethanol and other amphiphilic drugs out of the membrane.

12.2.2.5 Proteins

Embedded throughout the lipid bilayer are peripheral (temporary) and integral (permanent) proteins. Peripheral membrane proteins are almost exclusively found on the cytosolic-facing monolayer. Proteins have many metabolic functions in the cell, including providing structure, transferring molecules across the membrane in and out of the cell, acting as enzymes in biological processes, and making up receptors for molecules outside the cell, such as hormones.

12.2.2.6 Carbohydrates

Carbohydrates are also included in the structure and function of the membrane, but to a much smaller degree than fats and proteins. They are only present on the outer monolayer, facing the extracellular environment. Their function is to act as signaling molecules, allowing cells to be recognized by other cells. They do not exist in the membrane as sole molecules, but instead are attached to other proteins or lipids, creating glycoproteins and glycolipids.

All of these membrane components ultimately allow the cell to function optimally, which in turn allows the tissues and organs that these cells comprise to work effectively. With a properly working cell membrane, essential nutrients are able to be adequately absorbed into tissues, while harmful waste is removed and excreted. Cells are able to effectively communicate with one another, a necessity for cells to work together to form tissues. Hormone sensitivity and utilization increases when the proteins that form hormone receptors on the membrane are intact.

12.3 Part II: Dietary and Lifestyle Influences

12.3.1 Introduction

The overall health of our cell membranes, in addition to their ability to function at maximum capacity, is largely dependent on one's diet and lifestyle habits. However, many are unaware that even our most subtle nutritional choices have the power to influence our bodies down to the molecular level, particularly in regard to the state of our cell membranes. Knowing how vital the role of a cell membrane is in the grand scheme of the human body makes knowing what causes harm to these membranes equally as important.

12.3.2 Pathophysiology

Cell membrane damage occurs in a variety of different mechanisms, all of which can be detrimental to the body at the cellular and macromolecular levels. Considering cell membranes consist of a variety of different lipids and proteins, it is not difficult for one element to damage a vast majority of the phospholipid bilayer membrane once it has been introduced into just a portion of the membrane.

Since cell membranes are composed of over 50% lipids, consuming a diet that is low in fat is damaging to the membrane. A diet rich in monounsaturated fatty acids and essential fatty acids, especially a balance of the essential omega-3 and omega-6s producing the eicosanoid metabolites, while limiting long-chain saturated fat and avoiding ultra-processed vegetable oils with trans-fat and acrylamides supports healthy cell membranes [12]. The ratio of omega-3s to omega-6s of 5:1 to 1:1 should be properly maintained in balance, since a higher ratio of omega-3s in the body leads to the production of anti-inflammatory molecules, whereas an abundance of omega-6s leads to the production of inflammatory molecules [12]. Chronic stressors can also cause inflammation in the body, which can lead to programmed cell death, so it is important to manage stress.

In addition to lifestyle and dietary habits, certain inevitable biological processes have the ability to naturally damage cell membranes as well. Aging unfortunately has proven to have a negative influence on cell membrane health. As the human body ages, the amount of phosphatidylcholine that comprises our cell membranes decreases as the amount of cholesterol and sphingomyelin increases. Phosphatidylcholine is one of several phospholipids in the membrane that contribute to its fluidity and ability to react to differing internal conditions. When there is a low amount of phosphatidylcholine, cell membranes are more rigid, which ultimately diminishes their permeability and movement properties.

Many clinical conditions can also be attributed to membrane phospholipid oxidative damage. Damage to the membrane is destructive to the health of the overall cell and therefore harmful to the health of tissues and organs that these cells comprise. The body regulates each system and reaction under tight control to minimize any unintended consequences, but even under normal biological conditions, there is some level of uncontrolled reactions occurring, resulting in inadvertent outcomes. The production of reactive oxygen species is a prime example of unintended consequences formed by normal physiological processes.

12.3.2.1 Reactive Oxygen Species

Reactive oxygen species (ROS) are occasionally formed during the metabolism of molecular oxygen [13]. This is particularly present in the mitochondria during oxidative phosphorylation. The damage that reactive oxygen species cause can be detrimental to the cell's ability to function properly. It is important to limit the production of ROS, although some will always be generated as a natural byproduct of normal cell processes.

The properties of phospholipids as well as the chemical reactivity of fatty acids in the bilayer make the cell membrane an easy target for oxidation [14]. Oxygen and free radicals are more soluble in the fluid lipid bilayer than in the aqueous environment inside and out of the cell [14]. This attracts oxygen molecules to the bilayer, increasing the chances for oxidative damage [14]. Polyunsaturated fatty acids are extremely sensitive to oxygen, and the number of double bonds in their structure increases their susceptibility to attack by reactive oxygen [14]. Unsaturated fatty acids comprise about a third of the total fat content of the cell membrane, and this creates a large target for oxidation.

12.3.2.2 Mitochondria

In order to properly illustrate the impact that cell membrane health has on greater biological processes, we will explore the mitochondria as an example of an organelle whose energy production relies entirely on the proper function of its membrane.

Mitochondria are known as the powerhouse of the cell. Though other cellular processes may produce small amounts of adenosine triphosphate (ATP), or the cellular form of energy, the majority of it comes from the mitochondria via oxidative phosphorylation. The process of producing ATP occurs specifically within the inner mitochondrial membrane, not the cytoplasm [15]. Healthy and high-functioning mitochondria are imperative for the overall function of the cell and ultimately the greater biological structure because of its energy-producing properties.

Mitochondria contain two membranes, the inner and the outer mitochondrial membranes. This distinct structure provides plenty of space for mitochondria to function optimally. The outer mitochondrial membrane contains an abundance of lipids, while the inner mitochondrial membrane contains only about 20% lipids and instead has an abundance of proteins and phospholipids [16]. Oxidative phosphorylation occurs within the inner membrane, a process laden with enzymes (proteins) and other complexes. The membranes of mitochondria contain the same phospholipids as the cell membrane, with phosphatidylcholine dominating the outer membrane, and phosphatidylethanolamine dominating the inner mitochondrial membrane with lessor amounts of phosphatidylinositol, phosphotidic acid and phosphatidylserine, while also containing a mitochondria-specific phospholipid, cardiolipin, with a primary role in the mechanism of apoptosis [17]. The mitochondrial membranes contain low levels of cholesterol and sphingomyelin compared to the cell membrane [15].

There are many factors that can cause immense damage to the mitochondria. These include, but are not limited to, statin drugs, antibiotics, chronic stress, age, cigarettes, hyperglycemia, excessive arachidonic acid, excessive exercise, persistent organic pollutants (POPs), and heavy metals [18–22]. These all contribute to the formation of reactive oxygen species, but the greatest generator of ROS is mitochondrial oxidative phosphorylation, with about 1–5% of the oxygen consumed by the mitochondria converted to ROS. Oxidative phosphorylation includes the conversion of nutrient-derived substrates into ATP. This takes place on the inner mitochondrial membrane, through the action of five respiratory enzyme complexes. The main sites for ROS production are on complexes I and III of the mitochondrial electron transport chain, where electrons are being transferred from different oxygen molecules [23]. The desired outcome from this electron transfer is the generation of ATP to provide energy for cells. However, in some cases, the transfer can inadvertently generate superoxide anion, a dangerous ROS. Superoxide anion is produced from the one-electron reduction of oxygen and is a precursor for most reactive oxygen species.

The presence of cardiolipin, a phospholipid found in the inner mitochondrial membrane, increases the mitochondria's susceptibility to attack by reactive oxygen species. Cardiolipin is required for the electron transport chain (ETC) by connecting the four ETC complexes as well as supplying protons for ATP synthase [24]. In the ETC, hydrogen ions are pumped from the matrix of the mitochondria across the complexes located on the inner mitochondrial membrane (IMM) into the intermembrane space, creating a higher concentration of hydrogen ions within the intermembrane space than in the matrix. This gives the intermembrane space a positive charge and the matrix a negative charge, creating an electrochemical gradient. The IMM is not permeable to ions, preventing the abundance of hydrogen ions within the intermembrane space from crossing the IMM. With the help of ATP synthase, hydrogen ions are transported into the matrix of the mitochondria. The energy that is generated from this flow of ions against the electrochemical gradient phosphorylates adenosine diphosphate (ADP) into adenosine triphosphate, resulting in cellular energy.

While cardiolipin in the inner mitochondrial membrane is crucial for the generation of ATP, its proximity to the ROS that are generated within the mitochondria also makes it an easy target for oxidative damage. When cardiolipin and other membrane phospholipids are damaged by oxidation, they no longer form a tight ionic barrier from surrounding protons and ions, allowing these charged molecules to freely move across the membrane. This weakens the electrochemical gradient and therefore results in a loss of electron transport and therefore diminished cellular energy production.

It is imperative to protect mitochondria from damage, as harm to mitochondrial health can lead to impairment in their function. The more damage to the mitochondria, the less efficient they are at generating ATP, and therefore the more ROS they produce. Consequently, the more ROS that is produced, the more damage occurs, resulting in a vicious cycle of deterioration. Endogenous and supplemental CoQ10 are known protectors of the delicate mitochondria from excess ROS exposure.

There are many diseases and illnesses that can be traced back to membrane damage or dysfunction. Fatigue is often the first indication of cell damage, due to membrane malfunction. When the membrane is damaged, cells may not be able to signal efficiently and hormone receptors may not work as well, resulting in either the overstimulation of hormone targets or the lack of hormone activity.

Mitochondrial dysfunction is characterized by the loss of efficiency of the electron transport chain and therefore the reduction in synthesis of ATP [25]. This decrease in energy production can lead to symptoms of fatigue. While this seems like an obvious, expected outcome, there are numerous other illnesses that can result from dysfunctional mitochondria. Among these are neurodegenerative disorders, including Amyotrophic Lateral Sclerosis (ALS) (▶ Chap. 50), Alzheimer's, Parkinson's, Huntington's, bipolar, and autism spectrum disorders (▶ Chap. 31), as well as cardiovascular disorders (▶ Chap. 47), such as diabetes, metabolic syndrome, atherosclerosis, and obesity [25]. Mitochondrial dysfunction can also lead to gastrointestinal and musculoskeletal disorders, including fibromyalgia and muscular atrophy, as well as chronic infections and even cancer [25].

12.3.3 Evaluation/Assessment

There are many lab tests that can be used to determine cell and mitochondrial function. Some tests are used to determine how well certain cell components are working, including blood assessments of lactate and pyruvate levels, which can reflect how well mitochondria are oxidizing pyruvate [12]. An increase in lactate compared to pyruvate indicates a problem with pyruvate metabolism by the mitochondria. Other tests look for damaged cell components, which can be an indicator of oxidative damage. Two useful assessments for detecting oxidative damage include urinary analysis of 8-oxo-7, 8-dihydro-2′-deoxyguanosine, an oxidized nucleotide of DNA, which can be used to detect damaged nuclear or mitochondrial DNA [26]. Another test includes the serum level of the enzyme gamma-glutamyl transferase (GGT) [27]. Assessing antioxidant levels, including superoxide dismutase and catalase, can also be beneficial in determining the effects of ROS in the body. The organic acid test is another useful tool that evaluates for cell function by detecting blockages in the citric acid cycle.

12.3.4 Prevention/Treatment

There are a multitude of nutritional components that can enhance cellular membrane health. Since the cell membrane is made up primarily of lipids, consuming adequate amounts of fat in the diet is essential to maintaining proper structure and therefore optimal function of the cell. The current Acceptable Macronutrient Distribution Range (AMDR) for adults suggests that 25–35% of total calories should come from fat. Good sources include fatty fish, nuts, seeds, avocados, and oils.

12.3.4.1 Antioxidants

Considering reactive oxidants are a major detriment to the membrane, serious damage can be prevented by consuming adequate amounts of antioxidants, including vitamins E and C. The lipophilic properties of vitamin E enable it to fit comfortably in the membrane, protecting the phospholipids' unsaturated hydrophilic ends from free radical attachment and damage. These nutrients are critical in preventing initial oxidation, subsequent peroxidation, and the resulting oxidative chain reactions. The current Recommended Dietary Allowance (RDA) for vitamin E, found in nuts, seeds, and safflower oil, is 15 mg/day (or 22.5 IU/day) for adult men and women. It is not unreasonable to take 200–400 IU of mixed tocopherols with tocotrienols daily to protect cell membranes. The current RDA for vitamin C is 90 mg/day for men and 75 mg/day for women and the highest dietary sources include bell peppers, oranges, kiwifruit, and strawberries. Since vitamin C has a reasonably short half-life in the body, it would be reasonable to dose vitamin C two to three times a day. The RDA is recommended for the prevention of scurvy, not for the protection of cell membranes, therefore a good recommendation can be approximately 250 mg of vitamin C two to three times a day. Along with antioxidants, the cholesterol content of the membrane affects how phospholipids are packed in the membrane and therefore can affect the efficiency of free radical propagation through the lipid bilayer, so it is important to not shy away from it. There is no current RDA for cholesterol since our bodies synthesize it naturally. Cholesterol in the diet comes from animal sources, particularly meat, poultry, eggs, and dairy products.

12.3.4.2 CoQ10

Coenzyme Q10 (CoQ10) is a vital cofactor in the electron transport chain, facilitating the transfer of electrons between complexes I and II by acting as an electron donor or acceptor. In its oxidized form, ubiquinone, it accepts an electron from a molecule in the electron transport chain and is reduced to ubiquinol. When the reduced ubiquinol donates an electron to a neighboring molecule, it returns to its oxidized form of ubiquinone. The reduction and oxidation reactions that CoQ10 constantly undergoes make it a valuable antioxidant, quenching the harmful reactive oxygen species that are spontaneously generated during oxidative phosphorylation. CoQ10 also transfers protons across the inner mitochondrial membrane, creating a proton gradient, aiding in the generation of ATP. CoQ10 is the only fat-soluble antioxidant that can be generated by the body, and therefore may not need to be obtained from the diet or as a supplement. There is no RDA for CoQ10, however supplementation is recommended in order to treat a number of mitochondria-related conditions, including Parkinson's disease, migraines, and muscular dystrophy. The recommended supplemented intake of CoQ10 ranges from 100 mg a day to upwards of 300 mg a day, depending on the condition [28]. There are many foods that are high in this nutrient, particularly chlorophyll-rich vegetables, such as dark, leafy greens, fresh herbs, broccoli, peas, and asparagus, however it may be difficult to get adequate CoQ10 from diet alone when dealing with neurodegenerative or other disorders.

12.3.4.3 Lipoic Acid

Lipoic acid is another vital cofactor in mitochondria, aiding in the conversion of food into useful energy, reacting with reactive oxygen species, and protecting membranes by reacting with vitamin C and glutathione to recycle vitamin E [29, 30]. In these ways, it acts as an antioxidant, protecting mitochondria from ROS damage. Lipoic acid is synthesized in our bodies and studies have found that only 30–40% of supplemented lipoic acid is absorbed, meaning there is no recommended dietary allowance for lipoic acid [31]. Dietary sources include protein-rich foods, such as meat and some vegetables, including broccoli and spinach. Alpha lipoic acid may be more effective in reducing oxidative damage if taken it its R form, as R-ALA. Consider supplementing R-ALA in chronic neurological conditions or in other conditions in which oxidative stress is thought to be a significant issue.

12.3.4.4 Phosphotidylethanolamine

Primary mitochondrial inner membrane component influencing production of cardiolipin. PE is the second most abundant phospholipid in metabolism and a precursor to production of the important primary phospholipid PC. PE has many other critical functional roles in membrane metabolism including protein integration into membranes, autophagy, cell division, mitochondrial stability, as well as determing membrane conformational changes. Nutritional status of PE is dependent on the intake of the essential nutrient, choline.

12.3.4.5 L-carnitine

L-carnitine helps transport fatty acids into the mitochondria where they are burned as fuel during beta oxidation. The body sufficiently synthesizes and reabsorbs L-carnitine at a rate of about 95%, so supplementing this nutrient is not usually needed, unless severe deficiencies occur. Some dietary sources include meat, poultry, fish, and dairy products in high amounts, as well as fruits, vegetables, and grains in relatively small amounts. Research suggests that a therapeutic dose of L-carnitine to support mitochondria is 1000 mg three times a day, along with the supplementation of 400 mg of CoQ10 a day to prevent oxidative damage to mitochondria during oxidative phosphorylation [32].

12.3.4.6 Mitochondrial Biogenesis

Besides slowing the loss of mitochondria, it is also important to increase the number of functioning mitochondria and their size in various tissues. Type 1 and type 2 diabetic patients have fewer mitochondria as well as less efficient energy production in their mitochondria. This population should be given suggestions to augment mitochondrial mass and function. As people age, they accumulate higher amounts of mitochondrial DNA damage and this, along with blunted mitochondrial biogenesis, leads to a decline in the numbers of functioning mitochondria, especially in muscle cells. Type 1 muscle fibers, known as endurance or slow twitch, have a higher content of mitochondria than fast twitch fibers. As people age, the density of mitochondria decreases in their muscles as well as in other tissues. Aerobic exercise completed regularly can lead to increased density of mitochondria in muscle. Another method of mitochondrial biogenesis includes intermittent fasting; therefore, patients should be advised to not eat for 12–13 hours from the time of their evening meal to breakfast.

Resveratrol is a sirtuin activator and can be used as a supplement to augment mitochondrial mass. Researchers have determined that in order for sirtuin activators, like resveratrol and exercise, to adequately support healthy mitochondrial function, adequate NAD+ levels are necessary, and nicotinamide riboside has been shown to effectively increase NAD+ levels in humans [33]. Nicotinamide riboside can be supplemented in doses of 125–250 mg twice a day, along with 125 mg of resveratrol [34]. Pyrroloquinoline quinone (PQQ) is an antioxidant that can decrease mitochondrial oxidative damage, as well as augment mitochondrial biogenesis. Doses of 20 mg have been used in studies [35]. It has also been shown to have synergistic effects when combined with 300 mg of CoQ10 [35].

PQQ may be particularly relevant when treating neurological diseases, such as Parkinson's and Alzheimer's. PQQ can also be used to enhance cardiac function as well as decrease mitochondrial damage after an MI or when a patient is in heart failure.

Improving or maintaining mitochondrial health is key to overall wellbeing, particularly in treating chronic fatigue and many degenerative neurological conditions. The mitochondrial membrane can be supported with omega-3 fatty acids, and it may be worthwhile to use supplemental phosphatidylserine as well as phosphatidylcholine. Reducing exposure to damage caused by free radicals can be accomplished by consuming adequate antioxidants, while minimizing exposure to mitochondrial toxins, such as statin drugs and persistent organic pollutants, is an important first step in prevention, and can be helped by purchasing most, if not all, foods as organic. Promoting mitochondrial biogenesis will increase the mass of mitochondria in tissues.

The keys to focus on when it comes to maintaining mitochondrial health include reducing exposure to damaging factors, consuming nutrients that are beneficial to mitochondrial function, and preventing oxidative damage in the first place. Avoiding mitochondrial toxins, such as statin drugs, POPs, and chronic stress, while also consuming nutrients that are beneficial to mitochondrial function, such as CoQ10 and antioxidants, means that oxidative damage to the mitochondria can be reduced and therefore the diseases associated with mitochondrial dysfunction can be prevented.

12.4 Part III: Organ Structure and Function

12.4.1 Introduction

Disturbances to our major tissues and organs, especially those that inhibit their inability to function optimally, are often a result of dietary and lifestyle influences that impact cellular function. The eyes, the brain, and the gastrointestinal tract are three crucial components of the body in which any change to structure causes noticeable dysfunction. Since the ability of a tissue or organ to function at its full capacity relies on the basic cellular structures of which it is comprised, it is imperative that we analyze the components that have the means of diminishing or enhancing individual cells within these major systems.

12.4.2 Eye

12.4.2.1 Biological Structure

The eye is a prime example of how understanding structure allows us to understand the pathophysiology associated with this organ. The main disease affecting the eye is age-related macular degeneration (AMD), which is characterized by the gradual loss of central vision, particularly in those over the age of 55. The macula is a small, 3–5.5 mm area within the retina, containing the fovea at its center. It is a multilayered structure that is made up of millions of light-sensing cells that provide central vision. The macula and fovea are rich in photoreceptive cone cells, which are responsible for light sensing. The retina also contains the retinal pigment epithelium, which, among other functions, helps transport nutrients and water, is involved in the visual cycle, as well as protects the eye by absorbing light and preventing photo-oxidation, via the action of tight junctions [36].

12.4.2.2 Pathophysiology
There are many possible physiologic changes to the different structure in the eye that can lead to the development of age-related macular degeneration.

Dry Form
As the eye ages, the retina's ability to receive proper nourishment is hindered, leading to the accumulation of waste, and resulting amorphous deposits, or drusen, in the retina [37]. This causes the retinal pigment epithelium cells to degenerate and atrophy, resulting in the loss of central vision. This is classified as the dry, or atrophic, form of AMD. This form slowly progresses and can lead to total blindness in a span of 5–10 years [37].

Wet Form
The integrity of the retina can be attributed to the Bruch's membrane, which is the basal lamina of the retinal pigment epithelium, separating the retina and the choroid [38]. When the Bruch's membrane is broken, this can allow abnormal blood vessels to grow underneath the macula and retina [38]. When these blood vessels bleed, the macula bulges out, causing distorted vision. Resulting vision loss is rapid. This severe form of AMD is called wet, or exudative, and only makes up about 10–15% of AMD cases [39].

Reactive Oxygen Species
The constant exposure to light that the eyes are faced with leads to a high chance of photo-oxidation. The constant exposure to oxygen leads to a high rate of generation of reactive oxygen species. The role of antioxidants, both enzymatic (i.e., superoxide dismutase, catalase) and non-enzymatic (i.e., carotenoids lutein and zeaxanthin), are vital to prevent ROS damage.

12.4.2.3 Evaluation/Assessment
There are many risk factors for age-related macular degeneration, and fortunately many of them are modifiable and therefore can prevent or help delay the onset of the disease. These include smoking, elevated homocysteine levels, altered RBC fatty acid balance, obesity, and other cardiovascular-related risks, including hypertension, elevated total cholesterol, diabetes, stroke, and coronary artery disease [37, 40]. Another risk factor includes light exposure, particularly to blue light, as this generates free radicals and oxidatively damages the eye [41].

All major risk factors for developing AMD are associated with macular pigment as levels are significantly lower in patients with the disease [42]. Assessing visual acuity during a full eye examination is usually the first step toward diagnosis of AMD, along with retinal and choroidal angiography using green dyes [37].

12.4.2.4 Treatment/Prevention
While there is currently limited success with treating advanced age-related macular degeneration, there are many things that can help to prevent or slow the early progression of the disease.

Lutein and zeaxanthin are yellow-pigmented carotenoids that are found in high concentrations in the macula. In the fovea, they are present in amounts 1000 times higher than in other tissues in the body. These xanthophylls help to protect macular photoreceptors and improve vision [43]. Zeaxanthin absorbs damaging high-energy blue light and protects the macula from free radicals, acting as an antioxidant and subsequently protecting the central vision [43]. Carotenoids are not synthesized by the human body, so they must be obtained from the diet. Good dietary sources of lutein and zeaxanthin include leafy greens, such as kale, spinach, and lettuce, as well as broccoli, corn, peas, carrots, oranges, and eggs. There is currently no recommended dietary allowance for lutein or zeaxanthin; however, research has shown a decreased risk for AMD with intakes of only about 6 mg a day and a greater decrease in risk by consuming 10 mg a day of lutein and 2 mg a day of zeaxanthin [44].

Omega-3 fatty acids, including DHA and EPA, and carotenoids lutein and zeaxanthin have a protective effect on the eye, including anti-inflammatory as well as antioxidant [45]. DHA is found in large amounts in the eye, playing a role in phototransduction, as well as photoreceptor function [46]. Like cell membranes, photoreceptor membranes function optimally when they are fluid and selectively permeable—properties that are highly determined by their fat content. There is no RDA for omega-3 intake, but research has shown stabilization of AMD with supplementation of 120 mg DHA and 180 mg EPA a day in patients with AMD [47]. Good sources of omega-3s include fatty fish, such as salmon, and oils, including vegetable oils.

Zinc, cysteine, and alpha lipoic acid have been shown to prevent the progression of AMD and oxidative damage in the eye by acting as antioxidants, protecting tissues from free radical damage [42]. Cysteine can be obtained from consuming cysteine-rich foods, such as protein-rich foods, including meat, poultry, and eggs, or supplementing with N-acetylcysteine. Alpha lipoic acid also plays a role in regenerating other antioxidants, including glutathione, vitamin C, and vitamin E. Taurine is a cysteine-derived antioxidant that is found in high concentrations in the macula [42]. Taurine deficiency leads to retinal photoreceptor degeneration and impaired visual acuity and can be supplemented as 1000 mg twice a day [42].

12.4.3 Skin

12.4.3.1 Biological Structure
The largest of the membrane tissues in the body is the skin barrier which, when intact, provides primary protection from pathogens and undesirables from entering the body. When the skin barrier becomes compromised and loses its integrity, a variety of insults can appear.

12.4.3.2 Insults to Skin Barrier Integrity

Among the most concerning insults to the skin seen in healthcare settings are infection, skin cancers, allergies/sensitivities, and autoimmune diseases. For infections, underlying the vulnerability to protract these infectious insults can be an injury-wound opening access to the inner body. With skin cancers, allergies/sensitivities, and autoimmune conditions, the microenvironment terrain becomes the challenge to assess and know how to treat with considerations of potential antigenic dietary intake, environmental exposures, inflammation, pathogenic triggers of subclinical chronic infections, insufficient or deficient nutritional status, emotional and biological stresses, and genomic propensities.

- Wounds/wound chronic infection/post-op surgical wounds
- Skin cancers
  - *Basal cell carcinoma*
  - *Squamous cell carcinoma*
  - *Melanoma*
- Atopic dermatitis/Eczema
- Psoriasis
- Cracked lesions on or around lips, mouth
- Dermatitis Herpetiforme (extremities and/or truncal)
- Acne/Cystic acne
- Blistering diseases
- Pemphigus vulgaris
- Pemphigus
- Vitiligo

12.4.3.3 Evaluation/Assessment

Inflammation is a common denominator with the above skin conditions. The primary mechanism of healthy control of inflammation is from a balance of fatty acids, especially the omega-6 and omega-3 essential fatty acids and their metabolites that form the prostaglandins PG1, PG2 and PG3, and phospholipids. Obtaining an RBC fatty acid analysis is important to assess the fatty acid composition in order to develop a targeted intervention of food oils and lifestyle choices.

Phytonutrient phenol foods provide anti-inflammatory support to the role of the prostaglandins in inflammation control found in the rainbow of color in fruits and vegetables. The co-nutrients of minerals and vitamins are required for the above control systems to heal and keep inflammation at a healthy maintenance level.

12.4.3.4 Initial Biomarkers for Assessment of Skin Health Status

- Nutrition physical exam of skin and signs of nutritional deficits (face, tongue, skin)
- Skin scraping/biopsy pathology
- Blood
- RBC fatty acid analysis
- Lipid panel (TC, LDL, HDL, TG)
- hsCRP
- Sed rate
- IGG total
- IGE total
- Vitamin D25OH
- Vitamin A retinol
- DHEA-s (AM)
- Tissue transglutaminase IgG, IgA, IgM
- Celiac panel
- Dietary and supplement history
- Medication history

12.4.3.5 Treatment

- Vitamin C
- Biotin
- Fats and oils indicated from individual assessment
- Linoleic acid
- Gamma linolenic acid
- Arachidonic acid
- Saturated fats
- Short-chain fatty acids (SCFA)
- Medium-chain fatty acids (MCT)
- CoQ10
- Vitamin E, full spectrum tocopherols/tocotrienols
- Vitamin D3 (dose per blood test vitamin D25OH/VDR genomic)
- Vitamin A retinyl palmitate (if indicated by testing *vitamin A retinol*)
- Protein
- Amino acids
- Collagen
- Carnosine
- Honey, medical
- Aloe vera

Avoid: trans fats, acrylamide, rancid oils, hydrogenated fats, heat processed vegetable oils, high oleic-vegetable oils, antibiotics, and hormone - containing animal fats.

12.4.4 Brain

12.4.4.1 Biological Structure

The brain is composed of mostly fat, about 60%. Brain cells, otherwise known as neurons, rely entirely on the function of their membrane to serve their purpose of transmitting electrical signals. In order to communicate with one another, neurons generate an influx and/or outflux of specific ions across their membrane to produce an electrical signal to send to a neighboring neuron.

The key to proper function of the brain comes down to the integrity of the blood–brain barrier (BBB), as well as the integrity of axons, neurons, and mitochondrial membranes. The blood–brain barrier has the important function of letting nutrients and certain small molecules into the brain tissue while allowing waste to be eliminated. In addition, the composition of the cerebral spinal fluid (CSF) is important to consider along with its relationship with the glymphatic circulatory system (GCS) function of bathing the brain tissue with CSF four times daily. The glymphatic circulatory system has been more recently understood as separate from the body's lymphatic system (◘ Fig. 12.3).

12.4.4.2 Pathophysiology

The BBB can be damaged during concussions and other conditions and is also susceptible to damage from antibodies. If the BBB is disrupted, there can be an increase in autoimmunity to brain tissue, neural inflammation, and oxidative stress to cell and mitochondrial membranes.

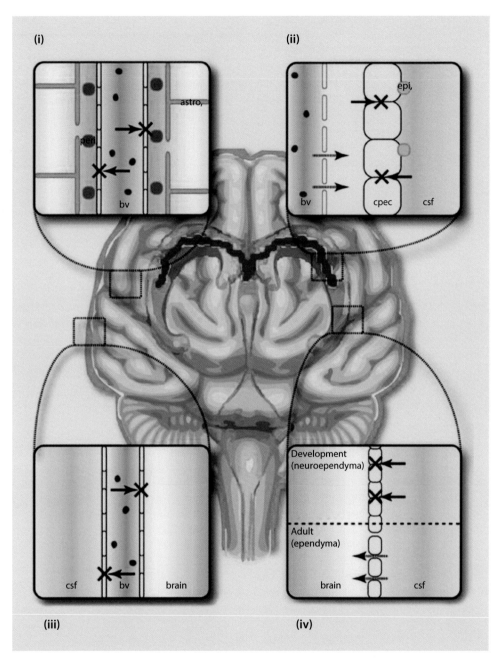

◘ **Fig. 12.3** Blood–brain barrier. Protective barriers of the brain. The collective term "blood–brain barrier" is used to describe four main interfaces between the central nervous system and the periphery. (i) The blood–brain barrier proper formed by tight junctions between the endothelial cells of the cerebral vasculature. It is thought that pericytes (peri.) are sufficient to induce some barrier characteristics in endothelial cells, while astrocytes (astro.) are able to maintain the integrity of the blood–brain barrier postnatally. (ii) The blood–CSF barrier formed by tight junctions between epithelial cells of the choroid plexus epithelial cells (note the plexus vasculature is fenestrated). Resident epiplexus (epi.) immune cells are present on the CSF surface of the plexus epithelium. (iii) The outer CSF–brain barrier and the level of the pia arachnoid, formed by tight junctions between endothelial cells of the arachnoid vessels. (iv) The inner CSF–brain barrier, present only in early development, formed by strap junctions between the neuroependymal cells lining the ventricular surfaces. In the adult, this barrier is no longer present. Both the blood–brain and CSF–brain barriers extend down the spinal cord. The CSF-filled ventricular system is depicted in blue, while CNS brain tissue is in brown. The lateral ventricular choroid plexuses are shown in red. Abbreviations: astro, astrocyte; bv, blood vessel; cpec, choroid plexus epithelial cell; csf, cerebrospinal fluid; peri, pericytes. (Reprinted from Stolp et al. [79]. With permission from Creative Commons License 3.0: ► https://creativecommons.org/licenses/by/3.0/)

If damage to the membranes of these neurons occurs to any extent, the entire cell will be absolutely compromised in its ability to complete its purpose in electrical signaling. There is a plethora of clinical conditions, many quite severe, which are associated with the degeneration or malfunction in the membranes of brain cells.

Neurodegenerative diseases are characterized by the progressive loss of neuronal structure or function, leading to neuronal dysfunction [16]. Proper neuronal function relies on all components to be present and intact. Amyotrophic lateral sclerosis, Alzheimer's, Parkinson's, and Huntington's diseases can all be traced back to misfolding and aggregation of certain proteins, resulting in the formation and deposition of fibrils, tangles, and plaques [16]. The accumulation of these harmful proteins is coupled by impairment of mitochondrial integrity, mutations in mitochondrial DNA, reduction of ATP synthesis, oxidative damage, and even cell death [16].

Structural concerns regarding brain function can be the vagus nerve, especially if cervical vertebrae (C1–C2 and C3–C7) are misaligned, impacting the brain stem and vagal nerve. The vagus nerve is the largest nerve network in the body impacting multiple systems including immune, gastrointestinal function, and others. Assessing and correcting the structural components is important as well as treating the function.

12.4.4.3 Evaluation/Assessment

BBB antibodies can be measured as antibodies to protein S100B. If this test is positive, the BBB should be treated to restore proper functions. Similar testing should be done for antibodies to zonulin and occludin, indicating a possible breach in small intestine barrier function. Increased permeability of the BBB is synergistically worsened by increased small intestinal permeability for a number of reasons. Excessive permeability in the small intestine can be a sign of leakage of bacterial endotoxin, as well as food allergens, into the brain, leading to more damage to brain tissue.

Patients with any brain-related matter should be initially tested for gluten antibodies with a complete gluten antibody panel. Patients should also be tested for food sensitivities and immune reactions to the foods that they consume.

It would be reasonable to test any patient with a significant neurological disease for antibodies to brain proteins. Assessing the health of the brain can be accomplished in various ways and is indirect. It can be useful to test for antibodies to the BBB as well as to various brain tissues. If antibodies are found, it is important to treat the cause as well as any symptoms. Quantitative MRI can be performed to assess for gross loss of brain tissue, and QEEG and SPECT scans can be done to look at how the different parts of the brain are functioning.

12.4.4.4 Treatment

The brain is 60% fat, so it makes sense that consuming adequate fat intake in the diet helps nourish the cells of the brain. It is also important to restore the BBB in order to prevent brain autoimmunity, brain damage from free radicals, neuro-inflammatory molecules, neurotoxins, and electromagnetic fields.

Nutritional support for the BBB includes optimizing fatty acids and phospholipid dietary and supplemental intake, increasing lipid protectors like fat soluble vitamins and CoQ10, and increasing polyphenols found in blueberries, especially wild blueberries, in order to decrease oxidative damage. Resveratrol can be used to regulate MMP9 and protect against OxLDL damage of the BBB. Curcumin, given 400 mg twice a day, should also be used in the Longvida form to inhibit microglial activation and MMP9, as well as to prevent damage to the tight junctions in the blood–brain barrier [48, 49]. R-lipoic acid taken in 150–200 mg doses twice a day can decrease oxidative stress. Sulforaphane inhibits MMP9 and activates NRF2, which can activate the brain's own anti-inflammatory and antioxidant systems [50]. Sulforaphane supplements may not be very effective. Glucoraphanin is the stored form of sulforaphane in cruciferous vegetables. In order to yield usable sulforaphane, glucoraphanin requires the enzyme myrosinase for it to be converted into sulforaphane. Cooking cruciferous vegetables destroys the myrosinase enzyme, thus very little sulforaphane can be obtained from steamed or fried cruciferous vegetables. It is important when eating raw cruciferous vegetables to have a healthy gut microbiome in order to enable some myrosinase activity. The best sources of glucoraphanin include broccoli sprouts and raw broccoli.

An important goal for supporting the brain should be to improve the structure and function of mitochondria and neuronal membranes. Omega-3 fatty acids with a higher percentage of DHA to EPA, and CoQ10 should be used, as well as providing a substrate for neuronal membrane phospholipids phosphotidylcholine (PC), phosphatidylethanolamine (PE), phosphotidylinositol (PI) and phosphatidylserine (PS). These phospholipid nutrients may also play a role in preventing synaptic degradation and reducing excitotoxic damage. PE, PI and PS are anionic phospholipids in the inner leaflet of the plasma membrane of neural tissues. They are made from phosphatidylcholine (PC) and phosphatidylethanolamine in the endoplasmic reticulum in the brain. DHA and PC can promote phosphatidylserine synthesis. Adequate doses of DHA may be in the range of 2–4 g/day, and doses of PC with PE may be in the range of 1–2 g/day [51]. PS can also be taken directly to improve neuronal membranes. After a head injury, PS doses of 300–600 mg twice a day may be used [51]. When used in the absence of a brain injury, lower doses, such as 100–200 mg twice a day, may be used [51]. The patient should also be encouraged to consume dietary sources of choline, which include egg yolks, poultry, collard greens, Brussels sprouts, broccoli, Swiss chard, cauliflower, and asparagus.

When considering maintenance or improvement in neuroplasticity, one must consider increasing the levels of various trophic factors, such as nerve growth factor, neurotrophin, and brain-derived neurotrophic factor (BDNF). Increasing levels of BDNF may lead to axonal and dendritic sprouting, nerve stem cell differentiation, and may enhance synaptogenesis. Aerobic exercise done in the training heart rate zone (220-age) 70–80% for 30–45 minutes can increase BDNF as well as protect loss of genetic telomere length [52]. Aerobic exercise can also activate NRF2 gene responses [52]. These gene responses can decrease brain inflammation and oxidative stress, which can improve neuronal and mitochondrial membranes.

Stimulation of neural stem cells and synaptogenesis in the brain, especially in the hippocampus, may be encouraged by the use of exercise and certain supplements [53]. These include melatonin 1.5–2 mg taken at night [54, 55], adequate vitamin D dosing to achieve levels of 50–70 ng/mL [56, 57], which would usually be about 5000 IU, and low dose lithium of 10–20 mg [58]. Low dose lithium aids repair of neurons as well as induces neurogenesis of hippocampal neurons and may lower toxicity of amyloid protein [58]. EGCG green tea extract is also beneficial, either by drinking green tea or by taking 200–400 mg/day [59].

Other nutritional interventions to increase BDNF include consuming blueberries with high polyphenol content. Wild blueberries contain the most of these beneficial nutrients. Encouraging patients to make smoothies with frozen wild blueberries is the most practical way to ensure patients are consuming polyphenols. High percent cacao and zinc may also promote BDNF, along with adequate sleep [60].

Taurine is an amino acid that can improve brain structure through a number of mechanisms. It can protect the brain against osmotic changes, has a neurotrophic effect, can activate nerve stem cells, and can enhance neurite (axon or dendrite) growth. Taurine can be safely dosed at 1000 mg twice a day.

Enhancing brain structure can also involve eliminating certain components from the diet. If there is a head injury, the patient should be taken off of gluten and all dairy products, because of cross-reactivity between these foods and neural tissue with subsequent autoimmune or inflammatory reactions in the brain. These reactions may be mediated by TH1, TH17, or microglial cells in the brain. Gliadin from gluten can cross-react with asialoganglioside GM1, myelin basic protein, synapsins, and cerebellar tissue. Milk butyrophilin can cross-react with cerebellar tissue and dairy casein can cross-react with synuclein and oligodendrocytes [61, 62].

Other dietary changes recommended after a RBC fatty acid analysis or fats and oils dietary survey are the intake of beneficial fat-rich foods and supplements when indicated for the individual.

12.4.5 Barriers

12.4.5.1 Biological Structure

The lining of the small intestine is a major barrier tissue whose function is directly supported by the integrity of its structure. The gastrointestinal mucosa forms a barrier between the body and the lumen of the intestines. The role of this barrier is to selectively allow the passage of nutrients and other small molecules across the epithelium of the intestine, while blocking larger, harmful molecules. Disruptions to this barrier can lead to possible autoimmune disease, food allergies, mood disorders, and other conditions.

The alimentary canal consists of the oral cavity, esophagus, stomach, and intestines and is lined with sheets of epithelial cells, forming the intestinal mucosa. The epithelial cells are held together by tight junctions, sealing off spaces between adjacent cells, forming an enclosed barrier between the lumen of the intestine and the surrounding tissues [63]. The epithelium is the innermost layer of the gastrointestinal tract and is responsible for most of the digestion, absorption, and secretion within the GI tract [64]. Maintaining proper integrity of the intestines comes down to the tight junctions holding the epithelial cells together while also regulating the permeability of water, ions, and nutrients.

Tight junctions consist of protein complexes located at the apical ends of the lateral membranes of the intestinal epithelial cells, forming the selectively permeable seal of the GI tract [63]. There are four specific complexes, including the occludin, claudin, junctional adhesion molecule, and tricellulin [63]. These proteins interact with the actin in the cytoskeleton, allowing a barrier to form [63].

12.4.5.2 Pathophysiology

Disruption to the intestinal epithelial barrier, namely increased permeability, leads to overall gut dysfunction and subsequent GI-related diseases, including irritable bowel syndrome, celiac disease, and leaky gut syndrome [12]. Factors that affect gut permeability include infection, inflammation, immune dysfunction, environmental toxins, medications, and the composition of the gut microbiota [12, 63]. Lipid and zinc deficiencies have also been shown to disrupt tight junctions, alter membrane permeability, and cause intestinal ulcers [12].

12.4.5.3 Evaluation/Assessment

There are many factors that provide some predictability for intestinal permeability, including the use of NSAIDs, bacterial infections, chronic stress, and hypochlorhydria, with an increased likelihood in those with IBS, type 1 diabetes, migraines, celiac disease, and food allergies [65]. Taking cyclooxygenase inhibitors and being deficient in essential fatty acids can also contribute to intestinal permeability [66].

If a patient displays any of these predictability factors, further lab tests can determine the presence of increased intestinal permeability. The lactulose/mannitol test consists of the patient drinking a specific amount of the disaccharide, lactulose, and the monosaccharide, mannitol [67]. The amount of these sugars that are excreted in the urine indicates how well each of these has been absorbed and therefore the degree of permeability in the intestine [12]. Disaccharides are absorbed through the paracellular junction complex within the intestine, indicating the permeability of larger molecules. Increased excretion of lactulose indicates increased absorption of larger sugars and therefore an increased permeability to molecules that are normally impermeable in the intestine [67].

12.4.5.4 Prevention/Treatment

While damage to the intestinal mucosa can be treated, it is beneficial to prevent disruption in the first place. The probiotic species *Bifidobacterium longum* has been shown to prevent damage to intestinal cells as well as increase the production of tight junction cell proteins, improving intestinal integrity [68]. It has also been shown to successfully treat increased permeability in patients with Crohn's disease and ulcerative colitis [68]. Antioxidants have also been shown to help prevent oxidative damage to the intestine. These include vitamin C, vitamin E, beta-carotene, grape seed extract, milk thistle, and quercetin [69]. Foods that can damage the gut include gluten, dairy, and sugar. Even in non-celiacs, gluten consumption may damage zonulin production, which is a protein that is crucial in tight junctions, increasing permeability to unwanted molecules [70]. Dairy can be inflammatory for a lot of people, so it is best to avoid it when trying to prevent damage to the gut. Too much sugar in the diet feeds the bad bacteria in the gut, causing an overgrowth of bacteria and degradation of gut permeability.

Nutrients that can provide some treatment to the intestine once it is compromised include L-glutamine, which is the primary amino acid source for intestinal cells and regulates intercellular junction integrity, and N-acetylglucosamine (NAG), which is a substrate for the glycosaminoglycans that are normally broken down in a leaky gut. A leaky gut diet should consist of bone broth, steamed vegetables, fermented foods, and healthy fats. Bone broth contains NAG, as well as collagen and glutamine, which are both elements that make up the gut, while steamed vegetables and healthy fats help provide essential nutrients, like L-glutamine, that keep the gut working properly. Fermented foods contain necessary probiotics, which keeps the microbiota in the gut healthy.

Preventing disease in the gastrointestinal tract boils down to the proper maintenance and regulation of the tight junctions that hold everything together. This is a clear example of how optimal function relies on proper, healthy structure.

12.5 Part V: Musculoskeletal Structure and Influences

12.5.1 Introduction

The musculoskeletal system, which is composed of bones, muscles, tendons, joints, and all other connective tissues, is responsible for providing structure and support in addition to functional movement of the body. Though aging has an inevitable effect on musculoskeletal health and function, there are several strategies that can be used to combat this degeneration. Consuming a balanced diet and maintaining a regular exercise routine can aid in healthy and efficient regeneration of musculoskeletal cells as well as maintain current bone density, joint flexibility, and muscle strength. However, there is a plethora of ways in which the musculoskeletal system can begin to develop structural dysfunction. Some of the most common structural changes in the musculoskeletal system are loss of muscle (sarcopenia), tendon degeneration (tendinosis), degradation of joint structure (osteoarthritis), and decreased bone density (osteopenia and osteoporosis). Of those large areas, all of them are modifiable with nutritional interventions, except for tendinosis.

12.5.1.1 Sarcopenia

Sarcopenia is a loss of muscle tissue and is most associated with aging. Muscle tissue may be replaced with fat and connective tissue. Aging also decreases mitochondrial density in muscle tissue. The loss of muscle structure can lead to injuries, falling, and decreased enjoyment of life and participation in sports and activities. Loss of muscle tissue can be related to a number of contributing factors, including hormones, inflammation, nutrition, toxins, and lack of exercise stimulus. The decline of key hormones, such as testosterone, growth hormone, and DHEA, can increase the rate of sarcopenia as well as ongoing systemic inflammation. Resistance training of large muscle groups (biceps, triceps, chest, upper back, thighs, and abdominal wall) to fatigue with sets of 8–12 repetitions should be done two to three times a week. Also, consuming 25–30 g of protein containing at least 2.5 g of leucine (a branched chain amino acid) can augment muscle building from resistance training and can slow age-related sarcopenia. Supplementing with 20 mg of PQQ one to two times per day can increase mitochondrial biogenesis so as to improve energy production from muscle [35]. It is also important for an individual to consume high quality protein on a daily basis of approximately 1 g per kilogram of lean body mass. Muscle content can be monitored with body composition equipment. One of the simplest ways is with a BIA device in which non-fat mass is evaluated (► Chap. 22).

12.5.1.2 Osteoarthritis

Osteoarthritis (OA) is the loss of chondrocytes and matrix from a joint leading to joint space narrowing, sclerosis, and deposits of calcium (osteophytes). Osteoarthritis can be isolated to one joint or be found in multiple weight-bearing joints of the body, including the knee, hips, and spine. Factors that contribute include inflammation, obesity, trauma, infection, lack of adequate precursor nutrients, and possible food allergies. Although the most common approach is to simply work on inflammation with an NSAID, this is an oversimplified and ineffective approach to OA of a joint. As it turns out, the pathophysiology of OA is more complex than simply an excess of prostaglandins. The following processes appear to be involved: inflammation (an extensive set of inflammatory biochemical pathways), lack of adequate precursor nutrients, enzyme breakdown of the joint, and dysfunctional synovial fluid.

12.5.2 Evaluation/Assessment

Evaluation of osteoarthritis is done using a combination of physical examination, since OA joints show warmth, swelling, deformity, or restricted range of motion, as well as imaging studies. X-rays are an excellent, low cost method and may reveal more details than an MRI. However, MRI would be important when evaluating details about labral cartilage or meniscus cartilage.

12.5.3 Prevention/Treatment

Any osteoarthritis intervention can take 8 weeks or more to show a decrease in pain and an improvement in function. For basic nutritional intervention, consuming sulfur-containing vegetables, such as from the cruciferous and onion families, on a daily basis is suggested. Providing sulfur to the chondrocytes in the joint can supply a building block for the joint matrix. It is reasonable to do a trial off of nightshade plants, such as tomatoes, potatoes, peppers, and eggplants, for about 1 month to determine if pain decreases or if function significantly improves. If the pain does improve on this diet, the patient should be kept off of nightshades long term.

There may be an effect from food sensitivities and food allergies. This would be related to an immune complex mediated inflammatory process. In a patient with OA in more than one joint or in a person with hand OA or rapidly progressive hip or knee OA, it is important to try an elimination diet. It is also important to do a food allergy test as well as gluten antibody test. The individual should be taken off of all highly reactive foods for about 8 weeks to see if this decreases pain and improves function.

It is noteworthy that weight loss in overweight or obese people can improve OA by more than simply mechanical factors. As it turns out, obesity can lead to dyslipidemia and oxidized LDL, which can trigger inflammatory pathways in the joints. There is also an alteration in adipokines, which can lead to joint degradation. Any successful approach to weight loss and achieving closer to ideal body weight can help OA of various joints.

12.5.3.1 Omega-3 Fatty Acids

Omega-3 fatty acids can often improve OA by decreasing inflammation. Omega-3 fatty acids can suppress key inflammatory mediators including IL1, TNF-alpha, PGE-2, 5-lox, and cox 2 [71]. They can also inhibit certain metalloproteinases that break down joints, including MMP-3 and 13, as well as aggrecanase. Doses of 3–4 g in two divided doses should be given [71]. Consideration should be given to digestive enzyme (lipase) support in the individual that has difficulty with fat digestion.

12.5.3.2 Vitamin C

A key vitamin for joint health is vitamin C. Adequate levels are necessary for collagen and proteoglycan synthesis as well as stabilization of the collagen fibril. Vitamin C should be dosed two to three times per day at doses of 250–500 mg [72].

12.5.3.3 Strontium

Strontium is a nutrient that may be used to increase OPG and IGF-1 and should be thought of in any individual with both OA and osteoporosis. The dose would be 680 mg/day.

12.5.3.4 Glucosamine Sulfate

Stabilized glucosamine sulfate is an essential supplement that should be used in any case of OA. It functions as a building block in the synthesis of structural cartilage matrix substrates, such as glycoproteins, glycolipids, GAGs, hyaluronate, and proteoglycans, and is required to manufacture joint lubricants and protective agents such as mucin and mucus secretions [73]. Glucosamine sulfate has been found to inhibit NF-kB and PGE2 and thus can have anti-inflammatory effects [74]. Crystalline stabilized glucosamine sulfate has been shown to inhibit IL-1beta-induced gene expression of matrix degradation factors MMP-3 (stromelysin-1) and ADAMTS5 (aggrecanase 2), thus having an effect on decreasing enzyme-mediated chondrocyte degradation [75]. Although glucosamine is not generally found in the human diet, it is made from the exoskeletons of shrimp, crabs, and lobsters for use in medical applications. It is also available in a vegan form that is made from corn. The clinician should recommend the sulfate form instead of the HCL form, as well as to make sure that the product is stabilized. It should be dosed one time per day at 1500 mg, unless there is more than one large joint involved, in which case it can be dosed 1500 mg twice a day [76, 77].

12.5.3.5 Osteoporosis and Osteopenia

Osteopenia is defined as less than 2.5 standard deviations below normal of bone density and osteoporosis is defined as greater than −2.5 standard deviations of decreased bone density. This condition can lead to compression fractures of the spine as well as fracture of the hip and forearm. Of interest is that the bone structure is an area that may store metal toxins and certain metals, including lead and cadmium, which may be released during periods of bone loss, such as in menopause and andropause. Conventional treatment programs will use only bisphosphonates; however, these drugs can lead to dysfunctional bones as well as side effects affecting the jawbone. Consideration of the use of dietary supplementation with a series of menaquinones (MK) (vitamin K2) in the form of MK4 and MK7 that promote healthy bone metabolsim.

12.5.4 Pathophysiology

There are many contributing factors related to loss of bone density, including declining hormone levels (estradiol and testosterone), inflammation (including an inflammatory diet), lack of adequate minerals and vitamins, malabsorption of nutrients, such as vitamin A and D, and inadequate loading of bones with exercise.

12.5.5 Evaluation/Assessment

The most appropriate assessment tools are DEXA scans, which will yield the T scores and urinary measures of bone turnover, such as DPD and CTX. Other useful measures are the serum vitamin D25OH test as well as calcium, RBC magnesium, vitamin D 1,25OH, PTH, calcitonin, osteocalcin, estradiol, and testosterone levels.

12.5.6 Treatment/Prevention

Therapeutic dietary measures can include using an anti-inflammatory diet, a more alkaline diet and a diet rich in certain minerals such as calcium, magnesium, vitamin C, and vitamin K. Food sources of calcium appear safer than supplements in regard to improving bone density without raising serum calcium too much (a risk for coronary arteries). One should have at least 500–1000 mg of food-based calcium. The most reasonable source would be dairy unless the individual is sensitive to dairy, in which case other calcium-rich foods can be eaten, such as collard and kale greens, broccoli, sardines with bones, or almonds. If supplements are used, calcium hydroxyapatite appears to have more efficacy than calcium alone in improving bone density. Vitamin D25OH serum levels should be measured and should be brought well above the minimum level. When treating with vitamin D3, an individual's vitamin A status should be assessed with a blood vitamin A retinol test. If depleted or low, vitamin A supplementation should be considered. It is not unreasonable in a patient with osteoporosis to obtain a whole genetic panel of tests as variations in vitamins A (related to the gene BCMO1) and D (related to the gene VDR) metabolism do occur.

Weight-bearing exercise should always augment any dietary approach. Every individual should be doing a walking program as well as resistance training to load the biceps and triceps.

Strong consideration should be given to assessment of estradiol and testosterone. Bio-identical hormone replacement in any woman or man with osteoporosis should be seriously considered.

12.6 Conclusion

Consideration of structural integrity is as important as assessment of function, from cell membranes all of the way up to the structure of skin, eyes, brain, muscles and joints. Improvement of structural integrity is lengthy, and the patient must be educated that it will take time to replace or improve the affected structure to a more healthy status.

References

1. Mouritsen OG, Bagatolli LA. Life from molecules. In: Life – As a matter of fat. The frontiers collection. Cham: Springer; 2016. p. 3–17.
2. Alberts B, Johnson A, Lewis J, et al. Molecular biology of the cell. 4th ed. New York: Garland Science; 2002.
3. Levine S. Repair the membrane, restore the body: a breakthrough discovery comes of age. Clin Ed [Internet]. 2012 [cited 15 Nov 2017]. Available from: https://www.clinicaleducation.org/news/repair-the-membrane-restore-the-body-a-breakthrough-discovery-comes-of-age/.
4. Mouritsen OG, Bagatolli LA. Head and tail. In: Life – As a matter of fat. The frontiers collection. Cham: Springer; 2016. p. 19–30.
5. Functional Forum. Journey to 100 session 3: food – Dr. Rupy Aujla [video on the Internet]. 4 Jul 2017. Available from: https://www.youtube.com/watch?v=B7EgR_auWSA.
6. Christie WW. Phosphatidylcholine and related lipids [Internet]. The Lipid Web. 2018 [cited 9 Nov 2017]. Available from: http://www.lipidhome.co.uk/lipids/complex/pc/index.htm.
7. Vance JE, Tasseva G. Formation and function of phosphatidylserine and phosphatidylethanolamine in mammalian cells. Biochim Biophys Acta [Internet]. 2012 [cited 9 Nov 2017];1831(3):543–54. Available from: http://www.sciencedirect.com/science/article/pii/S1388198112001874.
8. Emoto K, Kobayashi T, Yamaji A, Aizawa H, Yahara I, Inoue K, et al. Redistribution of phosphatidylethanolamine at the cleavage furrow of dividing cells during cytokinesis. Proc Nat Acad Sci USA [Internet]. 1996 [cited 9 Nov 2017];93(23):12867–72. Available from: https://www.ncbi.nlm.nih.gov/pmc/articles/PMC24012/.
9. Majumder R, Liang X, Quinn-Allen MA, Kane WH, Lentz BR. Modulation of prothrombinase assembly and activity by phosphatidylethanolamine. J Biol Chem [Internet]. 2011 [cited 15 Nov 2017];286(41):35535–42.
10. Kay JG, Koivusalo M, Ma X, Wohland T, Grinstein S. Phosphatidylserine dynamics in cellular membranes. Mol Biol Cell [Internet]. 2012 [cited 9 Nov 2017];23(11):2198–212. Available from: https://www.ncbi.nlm.nih.gov/pmc/articles/PMC3364182/.
11. Schlegel RA, Williamson P. Phosphatidylserine, a death knell. Cell Death and Diff [Internet]. 2001 [cited 9 Nov 2017];8(6):551–63. Available from: https://www.nature.com/articles/4400817.
12. Pizzorno JE, Katzinger J. Clinical pathophysiology: a functional perspective: a systems approach to understanding and reversing disease processes. Coquitlam: Mind; 2012.
13. Halliwell B, Gutteridge JMC. Free radicals in biology and medicine. 5th ed. Oxford: Oxford University Press; 1999.
14. Pamplona R. Membrane phospholipids, lipoxidative damage and molecular integrity: a causal role in aging and longevity. Biochim Biophys Acta [Internet]. 2008 [cited 9 Nov 2017];1777(10):1249–62. Available from: http://www.sciencedirect.com/science/article/pii/S0005272808006427.
15. Schenkel LC, Bakovic M. Formation and regulation of mitochondrial membranes. Int J Cell Biol [Internet]. 2014 [cited 15 Nov 2017]. Available from: https://www.ncbi.nlm.nih.gov/pmc/articles/PMC3918842/.
16. Aufschnaiter A, Kohler V, Diessl J, et al. Mitochondrial lipids in neurodegeneration. Cell Tissue Res [Internet]. 2017 [cited 15 Nov 2017];367(1):125–40. Available from: https://www.ncbi.nlm.nih.gov/pmc/articles/PMC5203858/. doi: https://doi.org/10.1007/s00441-016-2463-1.
17. Kim J, Minkler PE, Salomon RG, Anderson VE, Hoppel CL. Cardiolipin: characterization of distinct oxidized molecular species. J Lipid Res [Internet]. 2011 [cited 9 Nov 2017];52(1):125–35. Available from: https://www.ncbi.nlm.nih.gov/pmc/articles/PMC2999925/.
18. Byun HM, Panni T, Motta V, Hou L, Nordio F, Apostoli P, et al. Effects of airborne pollutants on mitochondrial DNA methylation. Part Fibre Toxicol [Internet]. 2013 [cited 9 Nov 2017];10:18. Available from: https://www.ncbi.nlm.nih.gov/pmc/articles/PMC3660297/.
19. Garrecht M, Austin DW. The plausibility of a role for mercury in the etiology of autism: a cellular perspective. Toxicol Environ Chem [Internet]. 2011 [cited 9 Nov 2017];93(5–6):1251–73. Available from: https://www.ncbi.nlm.nih.gov/pmc/articles/PMC3173748/.
20. Lee HK. Mitochondrial dysfunction and insulin resistance: the contribution of dioxin-like substances. Diabetes Metab [Internet]. 2011

21. La Guardia PG, Alberici LC, Ravagnani FG, Catharino RR, Vercesi AE. Mitochondrial dysfunction and insulin resistance: the contribution of dioxin-like substances. Diabetes Metab [Internet]. 2011 [cited 9 Nov 2017];35(3):207–15. Available from: https://www.ncbi.nlm.nih.gov/pmc/articles/PMC3138092/.
22. Kalghatgi S, Spina CS, Costello JC, Liesa M, Morones-Ramirez JR, Slomovic S, et al. Bactericidal antibiotics induce mitochondrial dysfunction and oxidative damage in mammalian cells. Sci Transl Med [Internet]. 2013 [cited 9 Nov 2017];5(192):192ra85. Available from: https://www.ncbi.nlm.nih.gov/pmc/articles/PMC3760005/.
23. Brand MD, Affourtit C, Esteves TC, Green K, Lambert AJ, Miwa S, et al. Mitochondrial superoxide: production, biological effects, and activation of uncoupling proteins. Free Radic Biol Med [Internet]. 2004 [cited 9 Nov 2017];37(6):755–67. Available from: http://www.sciencedirect.com/science/article/pii/S0891584904004538.
24. Paradies G, Paradies V, De Benedictis V, Ruggiero FM, Petrosillo G. Functional role of cardiolipin in mitochondrial bioenergetics. Biochim Biophys Acta [Internet]. 2014 [cited 9 Nov 2017];1837(4):408–17. Available from: http://www.sciencedirect.com/science/article/pii/S000527281300176X.
25. Nicolson GL. Mitochondrial dysfunction and chronic disease: treatment with natural supplements. Integr Med [Internet]. 2014 [cited 9 Nov 2017];13(4):35–43. Available from: https://www.ncbi.nlm.nih.gov/pmc/articles/PMC4566449/.
26. Valavanidis A, Vlachogianni T, Fiotakis C. 8-hydroxy-2′-deoxyguanosine (8-OHdG): a critical biomarker of oxidative stress and carcinogenesis. J Environ Sci Health C Environ Carcinog Ecotoxicol Rev [Internet]. 2009 [cited 9 Nov 2017];27(2):120–39. Available from: https://www.ncbi.nlm.nih.gov/pubmed/19412858.
27. Smith GS, Walker RM. Handbook of toxicologic pathway. 3rd ed. London: Academic Press; 2013.
28. Matthews RT, Yang L, Browne S, Baik M, Beal MF. Coenzyme Q10 administration increases brain mitochondrial concentrations and exerts neuroprotective effects. Proc Natl Acad Sci USA [Internet]. 1998 [cited 9 Nov 2017];95(15):8892–7. Available from: https://www.ncbi.nlm.nih.gov/pmc/articles/PMC21173/.
29. Salinthone S, Yadav V, Bourdette DN, Carr DW. Lipoic acid: a novel therapeutic approach for multiple sclerosis and other chronic inflammatory diseases of the CNS. Endocr Metab Immune Disord Drug Targets [Internet]. 2008 [cited 9 Nov 2017];8(2):132–42. Available from: http://www.eurekaselect.com/82738/article.
30. Packer L, Witt EH, Tritschler HJ. Alpha-lipoic acid as a biological antioxidant. Free Radic Biol Med [Internet]. 1995 [cited 9 Nov 2017];19(2):227–50. Available from: http://www.sciencedirect.com/science/article/pii/089158499500017R?via%3Dihub.
31. Peterson-Shay K, Moreau RF, Smith EJ, Smith AR, Hagen TM. Alpha-lipoic acid as a dietary supplements: molecular mechanisms and therapeutic potential. Biochim Biophys Acta [Internet]. 2009 [cited 9 Nov 2017];1790(10):1149–60. Available from: https://www.ncbi.nlm.nih.gov/pmc/articles/PMC2756298/.
32. DiMauro S, Mancuso M. Mitochondrial diseases: therapeutic approaches. Biosci Rep [Internet]. 2007 [cited 9 Nov 2017];27(1–3):125–37. Available from: http://www.bioscirep.org/content/27/1-3/125.long.
33. Belenky P, Racette F, Bogan K, MoClure LM, Smith JS, Brenner C. Nicotinamide riboside promotes Sir2 silencing and extends lifespan via Nrk and Urh1/Pnp1/Meu1 pathways to NAD+. Cell [Internet]. 2007 [cited 9 Nov 2017];129:473–84. Available from: http://www.cell.com/cell/fulltext/S0092-8674(07)00390-X.
34. Bogan K, Brenner C. Nicotinic acid, nicotinamide, and nicotinamide riboside: a molecular evaluation of NAD+ precursor vitamins in human nutrition. Annu Rev Nutr [Internet]. 2008 [cited 9 Nov 2017];28:115–30. Available from: http://www.annualreviews.org/doi/full/10.1146/annurev.nutr.28.061807.155443.
35. Chowanadisai W, Bauerly K, Tchaparian E, Rucker R. Pyrroloquinoline quinone (PQQ) stimulates mitochondrial biogenesis. FASEB J [Internet]. 2007 [cited 9 Nov 2017];21(6). Available from: https://www.ncbi.nlm.nih.gov/pmc/articles/PMC2804159/.
36. Simó R, Villarroel M, Corraliza L, Hernández C, Garcia-Ramírez M. The retinal pigment epithelium: something more than a constituent of the blood-retinal barrier—implications for the pathogenesis of diabetic retinopathy. J Biomed Biotechnol [Internet]. 2010 [cited 9 Nov 2017];2010:190724. Available from: https://www.ncbi.nlm.nih.gov/pmc/articles/PMC2825554/.
37. Chopdar A, Chakravarthy U, Verma D. Age related macular degeneration. BMJ [Internet]. 2003 [cited 9 Nov 2017];326(7387):485–8. Available from: http://www.bmj.com/content/326/7387/485.
38. Booij JC, Baas DC, Beisekeeva J, Gorgels TG, Bergen AA. The dynamic nature of Bruch's membrane. Prog Retin Eye Res [Internet]. 2010 [cited 9 Nov 2017];29:1–18. Available from: http://www.sciencedirect.com/science/article/pii/S1350946209000597.
39. Hageman GS, Gehrs K, Johnson LV, Anderson D. Age-related macular degeneration (AMD). Salt Lake City: University of Utah Health Sciences Center; 1995.
40. Thornton J, Edwards R, Mitchell P, Harrison RA, Buchan I, Kelly SP. Smoking and age-related macular degeneration: a review of association. Eye [Internet]. 2005 [cited 9 Nov 2017];19(9):935–44. Available from: https://www.nature.com/articles/6701978.
41. Sparrow JR, Nakanishi K, Parish CA. The lipofuscin fluorophore A2E mediates blue light induced damage to retinal pigmented epithelial cells. Invest Ophthalmol Vis Sci [Internet]. 2000 [cited 9 Nov 2017];41(7):1981–9. Available from: http://iovs.arvojournals.org/article.aspx?articleid=2123554.
42. Kohlstadt I. Advancing medicine with food and nutrients. Boca Raton: CRC Press, Taylor & Francis Group; 2013.
43. Youssef PN, Sheibani N, Albert DM. Retinal light toxicity. Eye (Lond) [Internet]. 2011 [cited 9 Nov 2017];25(1):1–14. Available from: https://www.ncbi.nlm.nih.gov/pmc/articles/PMC3144654/.
44. Scripsema N, Hu DN, Rosen RB. Lutein, zeaxanthin, and meso-zeaxanthin in the clinical management of eye disease. J Ophthalmol [Internet]. 2015 [cited 9 Nov 2017];2015:865179. Available from: https://www.ncbi.nlm.nih.gov/pmc/articles/PMC4706936/.
45. Querques G, Forte R, Souied E. Retina and omega-3. J Nutr Metab [Internet]. 2011 [cited 9 Nov 2017];2011:748361. Available from: https://www.ncbi.nlm.nih.gov/pmc/articles/PMC3206354/.
46. Shindou H, Koso H, Sasaki J, et al. Docosahexaenoic acid preserves visual function by maintaining correct disc morphology in retinal photoreceptor cells. J Biol Chem. 2017;292(29):12054–64.
47. Cangemi FE. TOZAL Study: an open case control study of an oral antioxidant and omega-3 supplement for dry AMD. BMC Ophthalmol [Internet]. 2007 [cited 9 Nov 2017];7:3. Available from: https://www.ncbi.nlm.nih.gov/pmc/articles/PMC1831760/.
48. Ding F, Li F, Li Y, Hou X, Ma Y, Zhang N, et al. HSP60 mediates the neuroprotective effects of curcumin by suppressing microglial activation. Exp Ther Med [Internet]. 2016 [cited 9 Nov 2017];12(2):823–8. Available from: https://www.ncbi.nlm.nih.gov/pmc/articles/PMC4950749/.
49. Begum AN, Jones MR, Lim GP, Morihara T, Kim P, Heath DD, et al. Curcumin structure-function, bioavailability, and efficacy in models of neuroinflammation and Alzheimer's disease. J Pharmacol Exp Ther [Internet]. 2008 [cited 9 Nov 2017];326(1):196–208. Available from: https://www.ncbi.nlm.nih.gov/pmc/articles/PMC2527621/.
50. Annabi B, Rojas-Sutterlin S, Laroche M, Lachambre MP, Moumdjian R, Beliveau R. The diet-derived sulforaphane inhibits matrix metalloproteinase-9-activated human brain microvascular endothelial cell migration and tubulogenesis. Mol Nutr Food Res [Internet]. 2008 [cited 9 Nov 2017];52(6):692–700. Available from: http://onlinelibrary.wiley.com/doi/10.1002/mnfr.200700434/abstract.
51. Amen DG, Wu JC, Taylor D, Willeumier K. Reversing brain damage in former NFL players: implications for traumatic brain injury and substance abuse rehabilitation. J Psychoactive Drugs [Internet]. 2011 [cited 9 Nov 2017];43(1):1–5. Available from: https://www.ncbi.nlm.nih.gov/pubmed/21615001.

52. Seifert T, Brassard P, Wissenberg M, Rasmussen P, Nordby P, Stallknecht B, et al. Endurance training enhances BDNF release from the human brain. Am J Physiol Regul Integr Comp Physiol [Internet]. 2010 [cited 9 Nov 2017];298(2):R372–7. Available from: http://ajpregu.physiology.org/content/298/2/R372.long.

53. Nokia MS, Lensu S, Ahtiainen JP, Johansson PP, Koch LG, Britton SL, et al. Physical exercise increases adult hippocampal neurogenesis in male rats provided it is aerobic and sustained. J Physiol [Internet]. 2016 [cited 9 Nov 2017];594(7):1855–73. Available from: https://www.ncbi.nlm.nih.gov/pmc/articles/PMC4818598/.

54. Shu T, Wu T, Pang M, Liu C, Wang X, Wang J, et al. Effects and mechanisms of melatonin on neural differentiation of induced pluripotent stem cells. Biochem Biophys Res Commun [Internet]. 2016 [cited 9 Nov 2017];474(3):566–71. Available from: http://www.sciencedirect.com/science/article/pii/S0006291X16306131.

55. Yu X, Li Z, Zheng H, Ho J, Chan MT, Wu WK. Protective roles of melatonin in central nervous system diseases by regulation of neural stem cells. Cell Prolif [Internet]. 2017 [cited 9 Nov 2017];50(2):e12323. Available from: http://onlinelibrary.wiley.com/doi/10.1111/cpr.12323/abstract.

56. Schlögl M, Holick M. Vitamin D and neurocognitive function. Clin Interv Aging [Internet]. 2014 [cited 9 Nov 2017];9:559–68. Available from: https://www.ncbi.nlm.nih.gov/pmc/articles/PMC3979692/.

57. Soni M, Kos K, Lang IA, Jones K, Melzer D, Llewellyn DJ. Vitamin D and cognitive function. Scand J Clin Lab Invest Suppl [Internet]. 2012 [cited 9 Nov 2017];243:79–82. Available from: https://www.ncbi.nlm.nih.gov/pubmed/22536767.

58. Leeds PR, Yu F, Wang Z, Chiu CT, Zhang Y, Leng Y, et al. A new avenue for lithium: intervention in traumatic brain injury. ACS Chem Neurosci [Internet]. 2014 [cited 9 Nov 2017];5(6):422–33. Available from: https://www.ncbi.nlm.nih.gov/pmc/articles/PMC4063503/.

59. Mancini E, Beglinger C, Drewe J, Zanchi D, Lang U, Borgwardt S. Green tea effects on cognition, mood and human brain function: a systematic review. Phytomed [Internet]. 2017 [cited 9 Nov 2017];34:26–37. Available from: http://www.sciencedirect.com/science/article/pii/S0944711317300867.

60. Yoneda M, Sugimoto N, Katakura M, et al. Theobromine up-regulates cerebral brain-derived neurotrophic factor and facilitates motor learning in mice. J Nutr Biochem. 2017;39:110–6. https://doi.org/10.1016/j.jnutbio.2016.10.002.

61. Vojdani A, O'Bryan T, Green JA, Mccandless J, Woeller KN, Vojdani E, et al. Immune response to dietary proteins, gliadin and cerebellar peptides in children with autism. Nutr Neurosci [Internet]. 2004 [cited 9 Nov 2017];7:151–61. Available from: http://www.tandfonline.com/doi/abs/10.1080/10284150400004155.

62. Vojdani A, Kharrazian D, Mukherjee PS. The prevalence of antibodies against wheat and milk proteins in blood donors and their contribution to neuroimmune reactivities. Nutrients [Internet]. 2014 [cited 9 Nov 2017];6(1):15–36. Available from: https://www.ncbi.nlm.nih.gov/pmc/articles/PMC3916846/.

63. Lee SH. Intestinal permeability regulation by tight junction: implication on inflammatory bowel diseases. Intest Res [Internet]. 2015 [cited 9 Nov 2017];13(1):11–8. Available from: https://www.ncbi.nlm.nih.gov/pmc/articles/PMC4316216/.

64. Groschwitz KR, Hogan SP. Intestinal barrier function: molecular regulation and disease pathogenesis. Curr Opin Allergy Cl [Internet]. 2009 [cited 2017 Nov];124(1):3–22. Available from: https://www.ncbi.nlm.nih.gov/pmc/articles/PMC4266989; https://doi.org/10.1016/j.jaci.2009.05.038.

65. Bjarnason I, Takeuchi K. Intestinal permeability in the pathogenesis of NSAID-induced enteropathy. J Gastroenterol [Internet]. 2009 [cited 9 Nov 2017];44(19):23–9. Available from: https://link.springer.com/article/10.1007%2Fs00535-008-2266-6.

66. Costantini L, Molinari R, Farinon B, Merendino N. Impact of omega-3 fatty acids on the gut microbiota. Int J Mol Sci. 2017;18(12):2645. Published 7 Dec 2017. https://doi.org/10.3390/ijms18122645.

67. Johnston SD, Smye M, Watson RG, McMillan SA, Trimble ER, Love AH. Lactulose-mannitol intestinal permeability test: a useful screening test for adult coeliac disease. Ann Clin Biochem [Internet]. 2000 [cited 9 Nov 2017];37(4):512–9. Available from: http://journals.sagepub.com/doi/abs/10.1177/000456320003700413.

68. Takeda Y, Nakase H, Namba K, Inoue S, Ueno S, Uza N, et al. Upregulation of T-bet and tight junction molecules by Bifidobacterium longum improves colonic inflammation of ulcerative colitis. Inflamm Bowel Dis [Internet]. 2009 [cited 9 Nov 2017];15(11):1617–8. Available from: http://journals.lww.com/ibdjournal/Citation/2009/11000/Upregulation_of_T_bet_and_tight_junction_molecules.4.aspx.

69. Amashesh M, Schlichter S, Amashesh S, Mankertz J, Zeitz M, Fromm M, et al. Quercetin enhances epithelial barrier function and increases claudin-4 expression in Caco-2 cells. J Nutr [Internet]. 2008 [cited 9 Nov 2017];138(6):1067–73. Available from: http://jn.nutrition.org/content/138/6/1067.long.

70. Fasano A. Surprises from celiac disease. Sci Amer [Internet]. 2009 [cited 9 Nov 2017];301(2):54–61. Available from: https://www.nature.com/scientificamerican/journal/v301/n2/full/scientificamerican0809-54.html.

71. Curtis CL, Rees SG, Little CB, Flannery CR, Hughes CE, Wilson C, et al. Pathologic indicators of degradation and inflammation in human osteoarthritic cartilage are abrogated by exposure to n-3 fatty acids. Arthritis Rheum [Internet]. 2002 [cited 9 Nov 2017];46(6):1544–53. Available from: http://onlinelibrary.wiley.com/doi/10.1002/art.10305/abstract;jsessionid=9F940A2A4DA672943202791559D4E90E.f04t01.

72. McAlindon TE, Jacques P, Zhang Y, Hannan MT, Aliabadi P, Weissman B, et al. Do antioxidant micronutrients protect against the development and progression of knee osteoarthritis? Arthritis Rheum [Internet]. 1996 [cited 9 Nov 2017];39(4):648–56. Available from: http://onlinelibrary.wiley.com/doi/10.1002/art.1780390417/abstract.

73. Matheson AJ and Perry CM. Glucosamine: a review of its use in the management of osteoarthritis. Drugs Aging [Internet]. 2003 [cited 9 Nov 2017];20(14):1041–60. Available from: https://www.ncbi.nlm.nih.gov/pubmed/14651444.

74. Largo R, Alvarez-Soria MA, Diez-Ortego I, Calvo E, Sanchez-Pernaute O, Egido J, et al. Glucosamine inhibits IL-1beta-induced NF kappaB activation in human osteoarthritic chondrocytes. Osteoarthritis Cartilage [Internet]. 2003 [cited 9 Nov 2017];11(4):290–8. Available from: http://www.oarsijournal.com/article/S1063-4584(03)00028-1/fulltext.

75. Chiusaroli R, Piepoli T, Zanelli T, Ballanti P, Lanza M, Rovati LC, et al. Experimental pharmacology of glucosamine sulfate. Int J Rheumatol [Internet]. 2011 [cited 9 Nov 2017], 939265. Available from: https://www.hindawi.com/journals/ijr/2011/939265/.

76. Pavelka K, Gatterova J, Olejarova M, Machacek S, Giacovelli G, Rovati L. Glucosamine sulfate use and delay of progression of knee osteoarthritis: a 3-year, randomized, placebo-controlled, double-blind study. Arch Intern Med [Internet]. 2002 [cited 9 Nov 2017];162(18):2113–23. Available from: https://jamanetwork.com/journals/jamainternalmedicine/fullarticle/213562.

77. Towheed TE, Anastassiades T. Glucosamine therapy for osteoarthritis: an update. J Rheumatol [Internet]. 2007 [cited 9 Nov 2017];34(9):1787–90. Available from: http://www.jrheum.org/content/34/9/1787.long.

78. Benton MJ, Whyte MD, Dyal BW. Sarcopenic obesity: strategies for management. Am J Nurs. 2011;111(12):38–44; quiz 45-6.

79. Stolp HB, Liddelow SA, Sá-Pereira I, Dziegielewska KM, Saunders NR. Immune responses at brain barriers and implications for brain development and neurological function in later life. Front Integr Neurosci. 2013;7:61. https://doi.org/10.3389/fnint.2013.00061.

Protective Mechanisms and Susceptibility to Xenobiotic Exposure and Load

Robert H. Verkerk

13.1 Introduction – 192

13.2 Biotransformation – 193

13.3 Pathophysiology – 194
13.3.1 Mechanisms – 194
13.3.2 Chronic Diseases Related to Xenobiotic Exposure – 195

13.4 Clinical Considerations – 198
13.4.1 Assessment of Xenobiotic Exposure, Historically and Presently – 198
13.4.2 Assessment of Genetic Susceptibility – 200
13.4.3 Assessment of Diet and Lifestyle – 200

13.5 Clinical Strategies – 200
13.5.1 Reducing or Avoiding Exposure to Xenobiotics – 200
13.5.2 Supporting the Body's Detoxification Capacity – 201

13.6 Conclusions – 202

References – 202

© Springer Nature Switzerland AG 2020
D. Noland et al. (eds.), *Integrative and Functional Medical Nutrition Therapy*,
https://doi.org/10.1007/978-3-030-30730-1_13

13.1 Introduction

Human exposure to exogenous toxin sources (xenobiotics) has increased dramatically over the last few decades as a result of industrialization and globalization. This results in exposures that may be greater, more frequent, and qualitatively different, especially with regard to exposure to new-to-nature substances, compared with exposures that have typified the greater part of our species' evolution prior to the Industrial Revolution.

More than 100 million substances (organic and inorganic chemicals) have been added to the Chemical Abstracts Service (CAS) registry system since its inception in 1965. About 75% of those were added in the last decade, exemplifying the exponential increase in registrations [1]. While the number of chemicals manufactured in high volumes and released into the environment represents a minor fraction of these, it is estimated that there are between 100,000 and 200,000 industrial chemicals in common circulation [2]. The toxicology of the vast majority of these isolated chemicals is either unknown or poorly understood. Even less is known about the effects of complex mixtures of compounds to which humans in industrial societies are routinely exposed.

By definition, *xenobiotics* are substances that are foreign to an organism, the term stemming from the Greek word *xenos* meaning foreigner and *bios*, life. In relation to human health, the term xenobiotic is typically used to refer to artificial substances, which did not exist in nature before their synthesis by humans (e.g. polychlorinated biphenyls, dioxins, pesticides). Alternatively, the term may be used to describe other exogenous toxin sources that are present in much higher concentrations than might be expected naturally (e.g. following consumption of cadmium or mercury-contaminated fish) or ones that would not be expected to be found within a human (e.g. bacterial toxins, mycotoxins).

Exposures are typically regarded as being either acute or chronic. In the case of the former, the toxicity usually manifests after a single, major exposure, and symptoms of toxicity in one or more organs (e.g. liver, kidney, brain, nervous system) are usually evident clinically within a short period (<24 h) following exposure. An example of an acute exposure includes an overdose of non-steroidal anti-inflammatory drugs (NSAIDs) associated with attempted suicide. Chronic toxicity, by contrast, is the result of repeated, lower-dose exposures over longer periods of time. Again, using NSAIDs as an example, long-term usage of this category of drugs can result in long-term damage to the liver [3] and gastrointestinal tract [4], especially the small intestine [5].

Exposure to some xenobiotics may lead concurrently to beneficial effects and adverse effects (e.g. pharmaceuticals). Exposure to xenobiotics may also yield no evident adverse or beneficial effect, owing to a low (i.e. sub-acute) exposure concentration or insufficient duration or frequency of exposure. Adverse effects, such as carcinogenicity, may arise from either acute or chronic exposure and may be delayed, taking years or decades to manifest clinically. Other categories of delayed adverse effect include mutagenicity (potential to cause mutations, as measured, for example, by the Ames test) [6], genotoxicity (potential to cause damage to a cell's DNA or RNA), reprotoxicity (potential to cause adverse effects on sexual function and fertility in males and females, developmental toxicity in the offspring, and effects through or via lactation) and teratogenicity (potential to cause birth defects, typically evaluated in laboratory animals).

The Globally Harmonised System of classification and labelling of chemicals (GHS) (revision 6, 2015) identifies 10 categories of health hazard, namely, acute toxicity, skin corrosion/irritation, serious eye damage/eye irritation, respiratory or skin sensitization, germ cell mutagenicity, carcinogenicity, toxic to reproduction (reprotoxicity), specific target organ toxicity/single exposure, specific target organ toxicity/repeated exposure, and aspiration hazard [7].

Organs and body systems that have specific sensitivities to xenobiotics include the liver, kidney, nervous system/brain, mitochondria, endocrine system, immune system, eyes, and skin. Substances that adversely affect one particular system are referred to accordingly, for example, hepatotoxins (liver), nephrotoxins (kidney), neurotoxins (nerves/brain), mitochondrial toxins, endocrine disruptors, immunotoxins, etc.

Xenobiotic exposure in a given individual may exceed the body's innate biotransformation capacities and contribute to a wide range of different pathologies. Some xenobiotics may affect quality of life, increase the risk of cancer, or impact reproductive potential. While the human body has been gifted with a multitude of different mechanisms and pathways to reduce body burdens of xenobiotics, these have evolved to cater for both the types and exposures of xenobiotic substances associated with the majority of our evolutionary history. Mammals such as humans are less likely to be able to adapt quickly to synthetic xenobiotics as compared with natural ones to which humans have been exposed during the majority of our species' evolution. Long generation times coupled with low selection pressure will limit or slow the rate of evolutionary adaptation to xenobiotics. Hence, herbivorous insect 'pests' that are pre-adapted to a multitude of host plant secondary metabolites (phytochemicals) have the capacity to rapidly develop *insecticide resistance*, a process aided by high selection pressure, rapid generational turnover rate, and prior adaptation of an array of detoxification enzymes [8]. Honeybees, by comparison, that have not needed to adapt to a high phytochemical load, have a much lesser array of protein coding genes, thus creating a marked reduction in the diversity of cytochrome P450 enzymes, glutathione-S-transferases (GSTs), and carboxyl/cholinesterases (CCEs) compared with herbivorous insects. This, in turn, likely accounts for the honeybee's extreme sensitivity to insecticides [9].

In human evolutionary terms, the time scale during which most adaptations evolved represents a period of some tens of thousands of years, excluding the most recent 250 years or so since the Industrial Revolution. The past 70 years has seen the rapid development of industries reliant on organic chemistry (e.g. industrial chemicals, food technology, plastics, agrochemicals, pharmaceuticals, personal hygiene, cosmetics) and biotechnology (e.g. nanomaterials, vaccines) that now represent important sources of xenobiotic exposure of humans. In addition, the growth, intensification and globalization of large-scale industry, continued reliance on fossil fuels as the primary energy source, increased human dependence on technology and the continuing expansion of polluting transportation systems (road, sea, and air) are associated with significantly increased indoor and outdoor pollution burdens compared with those that occurred over the majority of human evolutionary history.

Possible routes of exposure to xenobiotics are shown in ▶ Box 13.1.

> **Box 13.1 Routes of Exposure to Xenobiotics**
> Exposure to xenobiotics occurs via one or more of the following routes:
> 1. Prenatal [10]: Relevant for xenobiotics capable of placental barrier (e.g. tobacco smoke, mercury, lead, SSRI drugs)
> 2. Oral: Exposure via breast milk, food, water, beverages, drugs, supplements
> 3. Inhalation: Relating to both outdoor and indoor pollution
> 4. Dermal: Especially in relation to cosmetics, toiletries, washing water, medications exposed via the skin, eyes, vaginal, and other mucous membranes
> 5. Intramuscular: Vaccines and their associated adjuvants may represent important xenobiotic exposure that bypasses both the dermal and gastrointestinal barriers.

13.2 Biotransformation

A healthy human body, uncompromised by polymorphisms affecting critical enzymatic biotransformation (detoxification) pathways, is highly adapted to handling a diverse range of xenobiotic substances below dosage or exposure thresholds that might yield adverse effects. In fact, the body is gifted with an array of xenobiotic-sensing receptors, such as the pregnane X receptor (PXR) that has evolved to regulate genes involved in the metabolism and transport of xenobiotics absorbed from food or the environment and protect the body from their harmful effects [11].

The biotransformation process essentially involves two main phases, referred to as phase 1 and phase 2, respectively. In the former, non-polar, lipophilic xenobiotics are most commonly enzymatically converted to polar metabolites via a diverse family of cytochrome P450 enzymes (CYP), especially in the liver, and also in the kidney, lung, brain, adrenal gland, and gut. In some cases, the polar metabolites may be more cytotoxic than the original xenobiotic, for example, the biotransformation of the insecticide DDT to the metabolite DDE [12], or in the activation of polyaromatic hydrocarbons and nitrosamines in the diet to form carcinogens [13].

Other phase 1 enzymes include flavin-containing monooxygenase (FMO), hydrolyses, epoxide hydrolyses, aldehyde dehydrogenase, monoamine oxidases and xanthine oxidase [14]. In general, these phase 1 metabolites become substrates

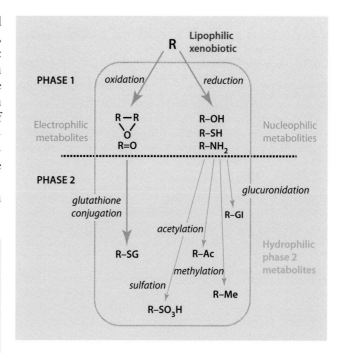

□ **Fig. 13.1** Summary of typical biotransformation of a lipophilic xenobiotic

for phase 2 conjugase enzymes, and following sulfation, amino acid conjugation, glutathione conjugation, glucuronidation, methylation, or acetylation are rendered both less toxic and more water soluble, thereby contributing to urinary or fecal (via the biliary route) excretion (□ Fig. 13.1) [15]. Chemically modified (more polar) xenobiotics may also be excreted via sweat, as volatile substance by lungs or in human milk [14].

There is increasing recognition of the existence of a complex active transporter (pump) system that is capable of acting on specific xenobiotics (most research having been carried out in relation to pharmaceutical drugs). These are sometimes classified into two discrete, additional biotransformation processes, referred to, respectively, as phase 0 and phase 3 [16, 17].

Both phase 1 and 2 enzymes are highly polymorphic [18]. Accordingly, genetic polymorphisms may contribute to significant inter-individual differences in xenobiotic clearance and responses [19]. A range of other factors also influence inter-individual variations in metabolism of, and response to, xenobiotics, including age, disease status, hormonal changes in the body, ingestion of medications, net exposure to environmental chemicals, and changes in lifestyle, including factors such as cigarette smoking, alcohol consumption, and diet [20, 21].

Given the continued unravelling of the science on biotransformation mechanisms and the growing body of evidence demonstrating the influence of diet and lifestyle on phase 1 and 2 biotransformation, more attention is being placed on dietary and lifestyle modifications that not only reduce the xenobiotic load (i.e. behavioural adaptation to xenobiotics) but also ones that enhance xenobiotic clearance via different and multiple biotransformation pathways.

Dietary composition and individual bioactive constituents can have particularly profound effects on the metabolism of xenobiotics. Animal studies have demonstrated that diets rich in specific saturated and polyunsaturated fats may alter CYP expression, notably of CYP2E1 [22, 23].

Inter-individual responses vary not only according to the potency of the xenobiotic agent(s) and the frequency of cumulative exposure, but also as to the individual's capacity to biotransform and eliminate the agent(s) at a given time. This capacity is dependent on numerous factors, including age, health (including inflammatory) status [24], body size/weight, nutrition, lifestyle, epigenetic background, and polymorphisms affecting biotransformation enzymes.

The clinical phenomenon of *multiple chemical sensitivity* is increasingly well recognized and was usefully defined at a workshop of experts, conducted at the request of the U.S. Environmental Protection Agency (EPA) in 1988, 'as an adverse reaction to ambient doses of toxic chemicals in our air, food, and water at levels which are generally accepted as subtoxic' [25]. The expert workshop concluded that adverse reactions manifest in susceptible individuals depending on a variety of factors, including:

1. The tissue or organ involved
2. The chemical and pharmacologic nature of the toxin
3. The individual susceptibility of the exposed person (genetic makeup, nutritional state, and total load at the time of exposure)
4. The length of time of the exposure
5. The amount and variety of other body stressors (total load) and synergism at the time of reaction
6. The derangement of metabolism that may occur from the initial insults [25]

Intra-individual variation in susceptibility to xenobiotics may also occur temporally, with some patients developing increasing tolerance, or, conversely, increased susceptibility, following continued or repeat exposure to particular xenobiotics.

13.3 Pathophysiology

13.3.1 Mechanisms

Given the huge array of xenobiotics to which humans are now exposed [26] and the general acceptance of their key importance in the pathogenesis of chronic diseases, such as certain types of cancer, it is perhaps surprising that so little, rather than so much, is known about the specific mechanisms by which their effects are mediated. Among the challenges to our improved understanding of the real-world interactions between xenobiotics and humans are the sheer number of xenobiotics to which humans are exposed (and the lack of toxicological knowledge about most of these); the quantitative and qualitative differences in chemical load over time; the challenges facing the study of the effects of exposure to complex mixtures as compared with isolated xenobiotics; [26] the complexity of multigene-environment and epigenetic interactions; the confounding effect of dietary and lifestyle choices; and profound inter-individual variations in susceptibility and tolerance [27].

Dysfunction in homeostatic processes often involve disturbances to the function of interrelated 'super-systems' (e.g. inflammatory, immune, endocrine, neurological) or they may be linked to specific organs or tissues (e.g. liver, kidney, mitochondria, motor neurons).

While there are very large gaps in our knowledge of the mechanisms by which xenobiotics induce adverse effects, three of the most well-researched mechanisms are as follows:

1. *Interference with critical biotransformation steps.* A number of xenobiotics are known to block critical steps in the production of biotransformation enzymes. For example, mercury (e.g. as a contaminant in food) or nitrous oxide (as a gaseous anaesthetic or airborne pollutant) act as potent inhibitors of cobalamin-dependent methionine synthase [28, 29], a critical intermediary in the methionine cycle that is required to synthesize endogenous glutathione, which has the capacity to detoxify both xenobiotics.

2. *Induction of supra-physiological oxidative stress.* Normal metabolic processes, exposure to xenobiotics in our food and environment generate both reactive oxygen species (ROS) and reactive nitrogen species (RNS) [30]. Radical ROS species, characterised by the presence of one or more unpaired electrons, are highly reactive, short-lived molecules, reacting especially with DNA, proteins, and lipids, causing an alteration in their function. While ROS are vital to numerous processes, including signalling cell growth and differentiation, regulating enzyme activity, vasodilation and protecting the host from pathogens and foreign particles, excessive oxidative stress may give rise to DNA, cellular or tissue damage, or to alterations to enzyme function or intracellular signalling pathways. This may, in turn, trigger a wide range of chronic diseases, including heart disease [31] or cancer [32].

3. *Dysregulation of xenobiotic nuclear receptors.* A variety of nuclear receptors, ligand-specific transcription factors, have evolved to sense the presence of toxic metabolites of endogenous metabolism as well as exogenous xenobiotics to which humans are exposed, most notably in the diet. They play a crucial role in biological development, differentiation, metabolic homeostasis, and protection against xenobiotic-induced stresses [33]. Depending on the ligand and the presence of specific cofactors, these nuclear receptors regulate transcription factors that, when functioning properly, control biological functions. However, when expression of these nuclear receptors is dysregulated, they are associated with a wide range of chronic diseases, including asthma, type 2 diabetes, obesity, atherosclerosis, osteoporosis, and cancer [34, 35].

In humans, nuclear receptors can be divided into two main groups according to their ligand-binding specificity [36]:
1. *Orphan receptors*, e.g. constitutive androstane receptor (CAR, NR1I3), pregnane X receptor (PXR, NR1I2), aryl hydrocarbon receptor (AhR), and peroxisome proliferator-activated receptors (PPAR), expressed particularly in the liver and intestines and also in a wide range of other tissues.

    These receptors express a broad range of biotransformation enzymes including CYP1A, CYP1B, CYP2B, CYP3A, CYP2Cs, CYP2A, GSTA1, ALDH1A, MRP3, and MDR1 [32], as well as phase-2 enzymes such as Uridine diphospho-glucuronosyltransferases (UDPGT), glutathione S-transferases (GSTs), and sulfotransferases (SULTs) [37].

    While it has been established that phenobarbital is a major ligand, these receptors have been found to be promiscuous, engaged in 'cross-talk' by stimulating expression of multiple genes, and their function may be promoted (agonist) or repressed (antagonist) by a very broad range of environmental, occupational, and natural products, including many pesticides, pharmaceuticals, dietary chemicals, herbal remedies, and industrial chemicals, typically at micromolar concentrations [36].

    Presently, more than 11,000 ligands have been added to the Orphan Nuclear Receptor Ligand Binding Database (ONRLDB) [▶ www.onrldb.org], with more than 6500 of these being unique. Orphan receptors for which endogenous ligands are later discovered are referred to as 'adopted orphan' receptors.

2. *Steroid receptors*, e.g. androgen receptor, estrogen receptor (ER), glucocorticoid receptor (GR), and vitamin D receptor (VDR). These receptors are responsive to steroid hormones and exposure to nanomolar concentrations of endocrine-disrupting chemicals (EDCs) such as xenoestrogens which may disrupt normal estrogen signalingsignalling and lead to disease (e.g. estrogen-related cancers) [38].

    Disruption of the function of these receptors and their cross-talk with a broad range of signalling pathways means that xenobiotics affecting steroid receptors may contribute to a daunting range of endocrine-related diseases including metabolic diseases such as cardiovascular, type 2 diabetes and obesity [36], and thyroid diseases [39].

13.3.2 Chronic Diseases Related to Xenobiotic Exposure

Chronic diseases are multifactorial and manifest following highly complex multi-gene/multi-environment interactions, usually over many decades. With limited exceptions (e.g. asbestos- or smoking-related cancers), given the plethora of possible causations, it is often difficult to identify with a high degree of certainty specific causes for particular chronic diseases, given that real-world interactions over multiple decades are likely to give rise to what has been referred to as *symphonic causation* [40].

Given also the vast array of environmental chemicals to which humans are now exposed, it is usually not possible to determine accurately the contribution of environmental chemicals to chronic disease. Notwithstanding this dilemma, exposure to some xenobiotics has been strongly related to specific chronic diseases.

One of the most comprehensive efforts to associate xenobiotic agents with genetic mediators of disease has been through the open-source Comparative Toxicogenomics Database (CTD) [▶ www.ctdbase.org], an NC State University initiative. The database divides chemicals for which an inferred relationship has been made with human diseases and specific genes into 13 groups and provides an inference score (high score = high inference), with links to the relevant peer-reviewed references. ◘ Table 13.1 provides examples of proven or inferred associations.

The great investment in cancer research over recent decades, the increasing recognition of the importance of environment factors as key triggers in carcinogenesis (as well as in the pathology of other inflammatory and metabolic diseases), along with the emergence of cancer as the leading cause of death in most industrialized, and increasingly in less-industrialized, countries, has stimulated increased interest in establishing scientific consensus over the carcinogenic status of xenobiotics. This role is largely fulfilled by the International Agency for Research on Cancer (IARC), an intergovernmental agency of the World Health Organization (WHO), which publishes comprehensive monographs of the present state of knowledge on carcinogens or potential carcinogens. ◘ Table 13.2 provides a summary of current classifications (including monograph 118) into the five IARC groups.

While the IARC has had a long history of criticism from independent quarters for making 'soft-touch' decisions that avoid negative impacts on the chemical or tobacco industry, it has committed to be more objective [41]. The 2015 decision to include processed meats in Group 1 and the world's top-selling herbicide, glyphosate, in Group 2A, are likely examples of this shift.

While the body of evidence linking a wide range of environmental chemicals to a variety of cancers is indisputable [42], the evidence for an association between environmental chemicals and metabolic diseases like obesity and cancer, as well as processes such as inflammation (refer to ◘ Table 13.1), a key mediator of most, if not all, chronic diseases [43], continues to grow.

Increasing evidence suggests that xenobiotics may interact adversely with the gastro-intestinal (GI) mucosa and microbiome, adversely affecting signalling in the immune, endocrine, and neurological super-system, as well as affecting nutrient assimilation and increasing the risk of a broad range of chronic diseases, including obesity, type 2 diabetes, non-alcoholic fatty liver disease (NAFLD), cardiovascular disease, cancer, and mental diseases [44].

Table 13.1 Chemicals for which associations with human diseases and specific genes have been inferred

CTD chemical category	Top interacting genes	Examples of strongly inferred chemical/human chronic disease relationships [no. genes associated]
Amino acids, peptides, and proteins	CASP3, TNF, GSTP1, IL6, CXCL8, IL1B, MAPK3, ABCB1, MAPK1, HMOX1	Glutathione/prostatic neoplasms [74 genes] Bleomycin/pulmonary fibrosis [35 genes] Cyclosporine/obesity [96]
Biological factors	TNF, IL6, IL1B, NOS2, PTGS2, IFNG, HMOX1, RELA, CXCL8, MAPK3	Lipopolysaccharides/inflammation [79 genes] Mycotoxins/inflammation [15 genes] Aflatoxins/liver neoplasms [2 genes]
Carbohydrates	TNF, NOS2, IL1B, IL6, PTGS2, INS, RELA, IFNG, CASP3, NFKBIA	Lipopolysaccharides/liver cirrhosis [117 genes] Fructose/diabetes mellitus [46 genes] Glucose/carcinoma [59 genes]
Chemical actions and uses	MGEA5, CYP19A1, TNF, IL1B, AR, CASP3, IL6, MAPK1, ACHE, ESR1	Estrogens/carcinoma (hepatocellular) [36 genes] Air pollutants/breast neoplasms [58 genes] Water-pollutant chemicals/breast neoplasms [51 genes] Pesticides/prostatic neoplasms [51 genes] Adjuvants (immunologic)/inflammation [12 genes]
Complex mixtures	TNF, IL6, CXCL8, IL1B, NFE2L2, PTGS2, CYP1A1, HMOX1, NOS2, CAT	Tobacco smoke pollution/stomach neoplasms [102 genes] Smoke/breast neoplasms [101 genes] Particulate matter [lung neoplasms] [79 genes] Chinese herbal drugs/carcinoma (hepatocellular) [55 genes] Vehicle emissions/breast neoplasms [250 genes] Petroleum/prostatic neoplasms [26 genes] Particulate matter/autoimmune diseases [18 genes]
Enzymes and coenzymes	POR, SLC5A6, AKR1B8, CAT, PPARA, CASP3, GAPDH, CYP3A4, NQO1, NQO2	NAD/obesity [8 genes] Thioctic acid/hypertension [41 genes] Leucovorin/heart diseases [2 genes]
Heterocyclic compounds	NOG, AHR, PPARA, CYP1A1, TNF, CASP3, MAPK1, MAPK3, HMOX1, CYP3A4	Tetrachlorodibenzodioxin (TCDD)/liver cirrhosis [763 genes] 2-amino-1-methyl-6-phenylimidazo(4,5-b)pyridine (heterocyclic amine)/carcinoma (multiple) [109+ genes] Nicotine/stomach neoplasms [65 genes]
Hormones, hormone substitutes, and hormone antagonists	ESR1, AR, ESR2, PGR, FSHB, EGF, MAPK1, MAPK3, LHB, TNF	Dihydrotestosterone/prostatic neoplasms [77 genes] Testosterone/breast neoplasms [173 genes] Estradiol/mammary neoplasms [112 genes] Estrogens/carcinoma (hepatocellular) [36 genes]
Inorganic chemicals	APP, CASP3, TNF, HMOX1, CAT, MAPK1, MAPK3, HIF1A, NOG, IL1B	Cadmium/prostatic neoplasms [166 genes] Asbestos/malignant mesothelioma [36 genes] Sodium chloride (dietary)/hypertension [52 genes] Sodium arsenite/carcinoma (hepatocellular) [147 genes] Arsenic/prostatic neoplasms [168 genes] Hexavalent chromium/lung neoplasms [42 genes]
Lipids	TNF, NOG, IL6, NOS2, IL1B, PTGS2, IFNG, RELA, PPARA, MAPK3	Dietary fats/prostatic neoplasms [222 genes] Arachidonic acid/inflammation [30 genes] Palmitic acid/insulin resistance [18 genes]
Nucleic acids, nucleotides, and nucleosides	CASP3, TP53, TNF, STAR, IL4, IFNA1, CDKN1A, IL6, IL1B, MAPK1	Decitabine (demethylation chemotherapy drug)/carcinoma (hepatocellular) [94 genes] Azathioprine (immunosuppressive drug)/colonic neoplasms [14 genes]
Organic chemicals	NOG, TNF, CASP3, MAPK1, PPARA, MAPK3, CYP1A1, AHR, PTGS2, ACHE	Benzo(a)pyrene/prostatic neoplasms [382 genes] Bisphenol A/prostatic neoplasms [462 genes] Diethylhexyl phthalate/breast neoplasms [112 genes] DDT/carcinoma (hepatocellular) [38 genes] Polychlorinated biphenyls/breast neoplasms [75 genes] Benzene/lung neoplasms [54 genes] Dieldrin Acrylamide/breast neoplasms [35 genes] 1,2,5,6-dibenzanthracene (polyaromatic hydrocarbon)/carcinoma [24 genes] Glyphosate/colonic neoplasms [13 genes]

Table 13.1 (continued)

CTD chemical category	Top interacting genes	Examples of strongly inferred chemical/human chronic disease relationships [no. genes associated]
Polycyclic compounds	AHR, ESR1, CYP1A1, TNF, CASP3, AR, HMOX1, MAPK1, MAPK3, ESR2	Benzo(a)pyrene/prostatic neoplasms [382 genes] Polycyclic hydrocarbons (aromatic)/breast neoplasms [28 genes] Simvastatin/liver cirrhosis [31 genes] Naphthalene/lung neoplasms [34 genes]

Based on data from: Comparative Toxicogenomics Database [CTD] [▶ www.ctdbase.org]

Table 13.2 International Agency for Research on Cancer (IARC) categorisation of carcinogens and examples

IARC category	Scientific basis of IARC classification	Number of entries (2017)	Examples
Group 1	Carcinogenic to humans	120	Alcoholic beverages, aflatoxins, aristolochic acid, arsenic, asbestos, benzene, benz(a)anthracene, cadmium, benzo(a)pyrene, coal, coal tar, chromium (VI) compounds, diesel exhaust, dioxin, ethanol in alcoholic beverages, lindane, Epstein-Barr virus, Estrogen-progestogen menopausal therapy (combined), Estrogen-progestogen oral contraceptives (combined), ethylene oxide, formaldehyde, *Helicobacter pylori* infection, Hepatitis B and C virus (chronic infection), human papillomavirus (HPV) types 16, 18, 31, 33, 35, 39, 45, 51, 52, 56, 58, 59, ionizing radiation, leather dust, untreated mineral oils, naphthylamine, nickel compounds, paints (occupational exposure of painters), polychlorinated biphenyls, outdoor air pollution, processed meat (consumption of), radionuclides, various forms of radium and their decay products, rubber manufacturing industry, salted fish (Chinese style), shale oils, crystalline silica dust, solar radiation, soot, Tamoxifen, tobacco (smoking, second-hand smoke, smokeless, chewing), trichloroethylene, ultraviolet-emitting tanning devices, vinyl chloride, wood dust, X- and Gamma-radiation
Group 2A	Probably carcinogenic to humans	81	Acrylamide, anabolic steroids, adriamycin, wood (and other biomass) fuels, bitumens, Captafol, chlorinated toluenes, chlorozotocin, Cisplatin, creosotes, cyclopentalpyrene, dibenzacridine, dibenzopyrene, dimethylhydrazene, dimethyl sulphate, ethyl carbamate (urethane), ethylene dibromide, emissions from high temperature frying, occupational exposure as hairdresser or barber, glyphosate, inorganic lead compounds, infection by *Plasmodium falciparum* (that causes malaria), mate (hot), Merkel cell polyomavirus, 5-methoxypsoralen, methyl methanesulfonate, N-methyl-N'-nitro-N-nitrosoguanidine (MNNG), ingested nitrates or nitrites (under conditions that result in endogenous nitrosation), nitrogen mustard, 1-nitropyrene, N-nitrosodimethylamine, N-nitrodimethylamine, 2-nitrotoluene, application of non-arsenical insecticides (occupational exposure), petroleum refining, polychlorinated biphenyls (PCBs), red meat (consumption of), shift work involving disruption of circadian rhythms, styrene-7-8-oxide, tetrachloroethylene (perchloroethylene), 1,2,3-trichloropropane, vinyl bromide, vinyl fluoride
Group 2B	Possibly carcinogenic to humans	294	Aflatoxin M1, acetaldehyde, acetamide, para-aminoazobenzene, anthraquinone, benzofuran, benzophenone, benzyl violet 4B, bitumens, occupational exposure to straight-run bitumens and their emissions during road paving, caffeic acid, carbon black, carbon tetrachloride, chloroform, cobalt and cobalt compounds, cobalt metal without tungsten carbide, coconut oil diethanolamine condensate, para-dichlorobenzene, diethanolamine, ethylbenzene, gasoline, human immunodeficiency virus type 2 (infection with), human papillomavirus types 26, 53, 66, 67, 70, 73, 82, lead, magnetic fields (extremely low-frequency), methylmercury compounds, metronidazole, mitoxantrone, naphthalene, nickel (metallic and alloys), nitrobenzene, ochratoxin A, pickled vegetables (traditional in Asia), phenobarbital, styrene, talc-based body powder (perineal use)

(continued)

Table 13.2 (continued)

IARC category	Scientific basis of IARC classification	Number of entries (2017)	Examples
Group 3	Not classifiable as to carcinogenicity in humans	505	Aciclovir, actinomycin D, amaranth, para-aminobenzoic acid, ampicillin, anaesthetics (volatile), arsenobetaine and other organic arsenic compounds that are not metabolized in humans, atrazine, benzoyl peroxide, bisphenol A, diglycidyl ether (Araldite), bisulfites, caffeine, carrageenan (native), chlorinated drinking water, chloroquine, cholesterol, chromium (metallic), coal dust, coumarin, crude oil, cyclamates (sodium cyclamate), diazepam, electric fields (extremely low-frequency), electric fields (static), ethylene, fluorides (inorganic, used in drinking-water), haematite, human papillomavirus genus beta (except types 5 and 8) and genus gamma, lead compounds, organic (NB: Organic lead compounds are metabolized at least in part, to ionic lead both in humans and animals. To the extent that ionic lead, generated from organic lead, is present in the body, it will be expected to exert the toxicities associated with inorganic lead), magnetic fields (static), mineral oils (highly refined), acetaminophen (paracetamol), polyethylene, polypropylene, polystyrene, saccharin and its salts, tea, temazepam, vitamin K substances
Group 4	Probably not carcinogenic to humans	1	Caprolactam

Emerging evidence suggests that xenobiotics may cause significant alteration to the microbiota in the human gut. Antibiotics and other oral pharmaceuticals may create significant short- or long-term changes in GI microbiome stability as well as changes to the relative abundance of particular bacterial taxa. Older patients on long-term prescriptions and polypharmacy may suffer reduced microbiota stability and diversity [45]. A study on the effect of the antibiotic cephalosporin on wild gorillas showed that the drug had a statistically significant tendency to increase *Firmicutes* (Gram positive) and decrease *Bacteroidetes* (Gram negative) colony numbers and species diversity, [46] a pattern that is associated with obesity in humans [47].

Further evidence implicates certain groups of pesticides, persistent organic pollutants (e.g. polychlorinated biphenyls), heavy metals (e.g. cadmium, mercury), food additives, and nanomaterials in further disturbances to the GI microbiota [48].

Xenobiotics, most notably excitotoxins and neurotoxins capable of passing the blood–brain barrier such as those transported by P-glycoprotein, are increasingly implicated in neurological diseases such as Parkinson's disease [49, 50], especially among those who are genetically more susceptible (e.g. organophosphate insecticide-exposed individuals homozygous for the paraoxonase 1 (PON1–55) gene) [51].

Other organs, tissues, and organelles that may be particularly vulnerable to xenobiotics are those directly involved in biotransformation (liver, kidney) [52], excretion (colon, bladder, urethra) [53], and energy production (mitochondria) [54].

13.4 Clinical Considerations

Where xenobiotics are thought to have been a trigger or mediator of a particular disease or condition, an integrative and functional medicine approach necessitates three main areas of investigation prior to the development of a treatment plan:

13.4.1 Assessment of Xenobiotic Exposure, Historically and Presently

This assessment, likely based on patient interview, should take into account known prenatal, childhood, occupational, and other lifetime exposures.

Xenobiotics may be categorized according to the CTD (□ Table 13.1); given the extreme sensitivity to xenoestrogens, consideration should be given to even very low levels of exposure to xenobiotic hormones or hormone analogues that act as endocrine disrupting chemicals (EDCs) even at nanomolar concentrations, close to the limit of analytical detection.

All five routes of potential exposure (▶ Box 13.1) should be considered, taking into account indoor pollutants (e.g. flame retardants, mycotoxins from moulds), xenobiotics in foods (e.g. preserved meats, polyaromatic hydrocarbons/heterocyclic amines on charred/high temperature cooked foods, food additives, sugar, pesticide contamination), outdoor pollutants, chlorinated/fluoridated drinking water, cosmetics, toiletries, etc.

Risk is determined by both the exposure (including dose and frequency) and an individual's susceptibility, the latter being heavily predicated genetically (□ Table 13.3).

Table 13.3 Important single nucleotide polymorphisms (SNPs) affecting biotransformation of xenobiotics

Phase	Gene	Gene variant	Risk allele	Example of impact	Reference
Phase 1	CYP1A1*1 (M1)	Msp1T>C	C	Metabolic of estrogens and polyaromatic hydrocarbons into carcinogenic reactive metabolites	Moorthy et al. (2015) [55], Sharma et al. (2014) [56]
	CYP1A1*2 (M2)	Ile462ValA>G	G		
	CYP1A2*1C	3858G>A	A	Metabolic activation of heterocyclic amines (HCAs) and aromatic amines (AAs) (e.g. in high temperature cooked meats)	Wang et al. (2012) [57]
	CYP1A2*1F	164A>C	C	CC (homozygote) individuals are 'slow' metabolizers of caffeine	Cornelis et al. (2006) [58]
	CYP2E1	96-bp insertion	N/a	Bioactivation of N-nitroso compounds derived from processed meats containing nitrite preservatives	Jiang et al. (2013) [59], Cross and Sinha (2014) [60]
Phase 2	COMT	Val158M	A	Slow COMT expression may lead to reduced methylation and increased DNA damage	Tahara et al. (2009) [61]
	MTFHR	C677T	T	Homozygote (and to a lesser extent heterozygote) individuals of each polymorphism have impaired methylation, increased risk of neurotransmitter disturbances and cardiovascular disease, and are slow (~70% reduced) metabolizers of folic acid to bioactive 5'-methyl-tetrahydrofolate	Stover (2011) [62], Alizadeh et al. (2016) [63]
	MTHFR	A1298C	T		
	N-acetyltransferase (NAT)2	Multiple, incl. 590A, 341C, 481T, 803G and 282T	Various	Holders of non-wild type alleles may have various combinations of alleles making them slow acetylators, affecting the metabolism of many drugs	Sabbagh et al. (2011) [64]
	Glutathione S-transferase (GST)M1	Deletion	Null	Deletions of these members of the GST gene (mu and theta 1 positions) superfamily are associated with reduced glutathione conjugation and elevated risk of some cancers	Bolt and Their (2006) [65]
	GSTT1	Deletion	Null		
	Aldehyde dehydrogenase (ALDH2)	rs671 G>A	A	Significantly reduced capacity to convert aldehydes (including from alcohol consumption) to acetate	Way et al. (2017) [66]
	PON1–55	55 L>M	M	MM homozygotes are more susceptible to adverse effects following exposure to organophosphate insecticides; associations with increased risk of Parkinson's disease	O'Leary et al. (2005) [67], Manthripragada et al. (2010) [68]
Phase 1/ Phase 2	SULT2B1	Multiple, including SULT2B1b and SULT2B1a	Various	Key member of the steroid metabolizing sulfotransferase (SULT) gene superfamily; imbalanced metabolism of hydroxysterois hormones and cholesterol	Ji et al. (2007) [69]

In some cases, it may be necessary to determine the presence of specific chemicals using relevant tests, e.g. lipid-soluble chemicals following fat biopsy, water-soluble chemicals via urine or sweat, or neurologically active pesticides using acetylcholinesterase assay.

13.4.2 Assessment of Genetic Susceptibility

An increasing array of genetic tests is commercially available to evaluate specific single nucleotide polymorphisms (SNPs) that increase (or decrease) susceptibility to xenobiotic agents (◘ Table 13.3).

Special consideration should be given to individuals expressing multiple high-impact polymorphisms relating to compromised biotransformation.

13.4.3 Assessment of Diet and Lifestyle

Of key importance are elements of diet and lifestyle that affect exposure to, or enhance, biotransformation of xenobiotics.

Diets including regular consumption of highly processed or ready-made foods, high-temperature cooked foods, and ones low in a diversity of vegetables and fruit generally contain larger amounts of synthetic additives, contaminants or other xenobiotic compounds as well as fewer disease protective compounds. Food and lifestyle diaries are a useful means of gaining information about a patient's habits and potential exposures.

13.5 Clinical Strategies

Key strategies may be divided into those that reduce total xenobiotic load.

13.5.1 Reducing or Avoiding Exposure to Xenobiotics

The most important way of modifying risk to environmental toxins is to avoid, or at least reduce, exposure to them. The following section draws on strategies proposed by renowned functional medicine doctor, Mark Hyman, MD [70].

Reduction or avoidance strategies include:
- Avoid processed foods; consume whole foods, home-prepared for freshness and to avoid nutrient loss where possible
- Consume organically certified or guaranteed pesticide-free produce. This is especially important when consuming fatty foods (e.g. dairy produce, vegetable oils, fatty meats) that tend to accumulate pesticides, veterinary drugs, and POPs
- Reduce or eliminate personal care products that contain harmful ingredients (e.g. phthalates, parabens, PEGs, propylene glycol)
- Eliminate or avoid excess exposure to petrochemicals, agrochemicals, and other sources of environmental toxin, for example, garden chemicals, dry cleaning, car exhaust, second-hand smoke
- Reduce or eliminate the use of toxic household cleaners (use low toxicity, environmentally friendly versions, wear gloves to avoid skin contact)
- Avoid unfiltered, municipal tap water. A reverse osmosis or distillation system are the only two systems that remove xenoestrogens, although it is advised to re-mineralize water (to at least pH 7.5) with a suitable mineral source prior to drinking
- HEPA/ULPA filters and ionizers can be helpful in reducing dust, moulds, volatile organic compounds, and other sources of indoor air pollution
- Avoid high-temperature cooking, such as frying and deep frying
- Avoid using PTFE-coated non-stick-treated pans (that may release fluorine gas during high-temperature cooking)
- Do not drink water or drinks from plastic bottles, unless they are guaranteed BPA-free (use glass bottles)
- Avoid storing food in plastic containers, or covering food in plastic wrapping, especially where food contact occurs, unless it is guaranteed to be phthalate-free (use glass or earthenware for food storage)
- Clean and monitor heating systems for release of carbon monoxide
- Include houseplants throughout house (including bedrooms) to help filter the air and increase oxygen concentration
- Air dry-cleaned clothes in well-ventilated space before wearing or storing
- Use solvent-free (water-based) paints if decorating interior spaces
- Avoid inhaling heavy traffic fumes, especially when exercising heavily (e.g. running, cycling). A respirator containing both particulate and carbon filters will reduce the level of harmful exposure, but filters should be changed regularly
- Understand all sources of possible workplace exposure and take action to avoid or minimise. In some cases, it may be helpful to engage the relevant trade union for assistance
- Use a carbon filter on baths or showers (and replace regularly according to manufacturer specifications) or reduce their duration
- Avoid chlorinated swimming pools; preferably, swim in sea water or other natural, open water or use seawater or ozone-treated pools
- Prospective mothers should ensure they have minimized exposure to environmental toxins 6–12 months before planning to get pregnant and should minimise exposure to xenobiotics throughout breastfeeding
- Avoid taking antacids, paracetamol, or other common over-the-counter medications and seek support for natural/non-drug alternatives
- Remove allergens and dust in living areas as much as possible

- Minimize exposure to electromagnetic radiation (EMR) from cellular or cordless phones by ensuring time spent with handset close to head or body is kept to a minimum. Do not carry phones in pockets or close to the body unless turned off. Do not sleep with phone near bedside if left on. Use 'air tubes' or speakers to reduce proximity of phone to head/body when talking. Use radiation protection cases or sleeves on mobile devices
- If working on computer, ensure screens and main computer are at least 30 cm from body. Use separate wired keyboard and low-radiation screen if laptop is main computer
- Do not use cordless telephones as most base stations emit EMR equivalent to transmission mast 250 m from house. Use corded phones for landlines
- Avoid excessive time (more than 1–2 h/day) watching television or using screens and sit more than 3 m away from television when watching
- Avoid use of microwave ovens
- Avoid excessive exposure to sun (avoid burning)
- Avoid any exposure to X-rays other than those regarded medically essential
- Reduce heavy metal exposure (predatory and river fish, some municipal drinking waters, lead paint, thimerosal-containing products, etc.)

13.5.2 Supporting the Body's Detoxification Capacity

There is a large body of research, as well as decades of clinical experience, supporting nutritional approaches to enhancing biotransformation (detoxification) processes in the body [71, 72].

13.5.2.1 Improve Elimination of Toxins

- Try to ensure 1–2 bowel movements a day
- Drink 6–8 glasses of clean drinking water a day
- Sweat regularly (use exercise, steam baths, and/or saunas to encourage sweating)
- Regular physical activity and exercise, yoga, or lymphatic massage can improve lymph flow and assist elimination of toxic metabolites
- Consume adequate soluble and insoluble fibre: approx. 30g/day
- Consume legumes (generally cooked to reduce/eliminate lectins), whole grains (preferably gluten-free), vegetables, fruits, nuts, and seeds
- Consume fermented foods as natural probiotic sources

13.5.2.2 Foods that Support Biotransformation

- Cruciferous vegetables (cabbage, broccoli, collards, kale, Brussels sprouts) containing indole-3-carbinol, sulforaphane, etc., at least 1–2 cups daily
- Garlic cloves (several daily) or garlic (preferably kyolic aged) supplement
- Decaffeinated green tea; preferably morning
- Freshly made vegetable juices, e.g. kale, celery, cilantro, beets, parsley, ginger, and carrot (the latter should be limited because of its high sugar content)
- Herbal detoxification teas, e.g. burdock root, dandelion root, ginger root, liquorice root, sarsaparilla root, cardamom seed, cinnamon (not cassia) bark, etc.
- High-quality, sulfur-containing proteins; eggs, plant protein (not soya) isolates, as well as garlic and onions
- Citrus peels, caraway and dill oil (limonene sources)
- Bioflavonoid/polyphenol-rich berries, grapes, citrus, and other fruits
- Dandelion greens may help in liver detoxification, improve the flow of bile and increase urine flow
- Celery may increase urine flow
- Fresh cilantro may help eliminate 'heavy metals'
- Rosemary, as fresh herb or extract, promotes expression of biotransformation enzyme genes, chelates heavy metals, antioxidant, anti-inflammatory
- Turmeric/curcuminoids (in fresh and dried turmeric and curry powders): exhibit multi-target functions including detoxification, antioxidant, and anti-inflammatory effects
- Chlorophyll in dark green leafy vegetables, wheat grass, etc.

13.5.2.3 Dietary/Food Supplements that May Support Enhanced Biotransformation

- Full-spectrum, high-quality multivitamin and mineral formula including bioavailable nutrient forms
- Buffered vitamin C (with mineral ascorbates): 1000–4000 mg a day in divided doses (to avoid loose stools) in powder, capsule, or tablet forms during periods of increased detoxification. If dosage causes loose stools, lower dose
- Milk thistle (*Silybum marianum*): 200–600 mg silymarin/day
- Rosemary (*Rosmarinus officinalis*) extract: 200–500 mg dry extract/day (standardized to 6–10% rosmarinic acid)
- Turmeric curcuminoids with bioavailability enhancer (e.g. turmeric essential oils, cyclodextrin, piperine): 200–600 mg curcuminoids/day, in divided doses
- Astaxanthin (from *Haematococcus pluvialis*): 5–20 mg/day
- Vitamin B6 (as pyridoxal 5′-phosphate): 10–25 mg/day
- Vitamin B12 (as methylcobalamin): 500–10,000 µg/day
- Folate as (6S)-5-methyltetrahydrofolate (glucosamine salt), calcium methylfolate, or food-form folates [73]: 1500 µg/day
- Omega-3 fatty acids (as EPA and DHA): 2000–5000 mg/day
- Liposomal glutathione: 400–800 mg/day

Additional supplements (for use under medical supervision):
- N-acetylcysteine: 500–1000 mg a day
- Amino acids: taurine 500 mg twice/day, glycine 500 mg twice/day
- Alpha-lipoic acid: 100–600 mg a day
- L-carnitine: 1000–2000 mg a day in divided doses
- Bioflavonoids (citrus, pine bark, grape seed, green tea): 50 mg/day

13.6 Conclusions

There is growing evidence that xenobiotics are playing an increasing role in a wide variety of chronic diseases and multi-morbidities that present the primary burdens on healthcare system [74]. The specific manifestation of disease in a given individual is dependent on extraordinarily complex and generally poorly understood gene-environment interactions, mediated by disrupted nuclear transcription factor trafficking and signalling pathways.

The huge, variable, and unpredictable array of xenobiotics to which individuals are exposed presently, coupled with the genetic and epigenetic variability, make it almost impossible to assess the net effect of xenobiotic load on an individual. This dilemma is compounded further by the absence of adequate toxicological and toxicogenomic data on environmental chemicals, acting both singly or, even more relevant to real-world situations, as mixtures.

Toxicogenomics offers a new lens through which to understand more about the effects of xenobiotic exposure mediated by effects on specific genes and signalling pathways. The clinical practice of integrative and functional medicine is unique in its emphasis on trying to establish causes, triggers, and mediators of chronic disease, often much earlier in the disease cycle than with conventional medical approaches.

Rapidly emerging omics sciences, including nutrigenomics and metabolomics, as well as cost-effective testing of SNPs for gene variants associated with compromised biotransformation, are further able to assist clinicians in their development of personalised protocols for their patients.

Despite these complexities, a number of robust strategies apply to most, if not all, cases: every effort should be made to help patients minimise total xenobiotic exposure and body load, while dietary and lifestyle patterns that promote effective biotransformation and elimination of metabolites should be strongly encouraged.

References

1. American Chemical Society media release: CAS Assigns the 100 Millionth CAS Registry Number® to a Substance Designed to Treat Acute Myeloid Leukemia. June 29th, 2015. [http://www.cas.org/news/media-releases/100-millionth-substance; Last accessed 04/17/17].
2. Nielsen KA, Elling B, Gigueroa M, Jelsøe E, editors. A new agenda for sustainability. Farnham: Ashgate Publishing Ltd; 2010. p. 303.
3. Hinson JA, Roberts DW, James LP. Mechanisms of acetaminophen-induced liver necrosis. Handb Exp Pharmacol. 2010;196:369–405.
4. García Rodríguez LA, Barreales Tolosa L. Risk of upper gastrointestinal complications among users of traditional NSAIDs and COXIBs in the general population. Gastroenterology. 2007;132(2):498–506.
5. Boelsterli UA, Redinbo MR, Saitta KS. Multiple NSAID-induced hits injure the small intestine: underlying mechanisms and novel strategies. Toxicol Sci. 2013;131(2):654–67.
6. Gee P, Maron DM, Ames BN. Detection and classification of mutagens: a set of base-specific Salmonella tester strains. Proc Natl Acad Sci U S A. 1994;91(24):11606–10.
7. United Nations Economic Commission for Europe (2015). Globally Harmonized System of classification and labelling of chemicals (GHS) (Revision 6). United Nations, New York & Geneva. 521 pp.
8. Schuler MA, Berenbaum MR. Structure and function of cytochrome P450S in insect adaptation to natural and synthetic toxins: insights gained from molecular modeling. J Chem Ecol. 2013;39(9):1232–45.
9. Claudianos C, Ranson H, Johnson RM, Biswas S, Schuler MA, Berenbaum MR, Feyereisen R, Oakeshott JG. A deficit of detoxification enzymes: pesticide sensitivity and environmental response in the honeybee. Insect Mol Biol. 2006;15:615–36.
10. Lee WC, Fisher M, Davis K, Arbuckle TE, Sinha SK. Identification of chemical mixtures to which Canadian pregnant women are exposed: the MIREC Study. Environ Int. 2017;99:321–30.
11. Kodama S, Negishi M. PXR cross-talks with internal and external signals in physiological and pathophysiological responses. Drug Metab Rev. 2013;45(3):300–10.
12. Lund BO, Bergman A, Brandt I. Metabolic activation and toxicity of a DDT-metabolite, 3-methylsulfonyl-DDE, in the adrenal zona fasciculata in mice. Chem Biol Interact. 1988;65(1):25–40.
13. Gonzalez FJ, Gelboin HV. Role of human cytochromes P450 in the metabolic activation of chemical carcinogens and toxins. Drug Metab Rev. 1994;26(1–2):165–83. Review.
14. Hodgson E, editor. Toxicology and human environments. New York: Academic Press; 2012. p. 450.
15. Caira MR, Ionescu C, editors. Drug metabolism: current concepts. Netherlands: Springer; 2005. p. 422.
16. Petzinger E, Burckhardt G, Tampé R. The multi-faceted world of transporters. Naunyn Schmied Arch Pharmacol. 2006;372:383–4.
17. Döring B, Petzinger E. Phase 0 and phase III transport in various organs: combined concept of phases in xenobiotic transport and metabolism. Drug Metab Rev. 2014;46(3):261–82.
18. Jancova P, Anzenbacher P, Anzenbacherova E. Phase II drug metabolizing enzymes. Biomed Pap Med Fac Univ Palacky Olomouc Czech Repub. 2010;154(2):103–16. Review.
19. Zhou SF, Liu JP, Chowbay B. Polymorphism of human cytochrome P450 enzymes and its clinical impact. Drug Metab Rev. 2009;41(2):89–295.
20. Conney AH, Kappas A. Interindividual differences in the metabolism of xenobiotics. Carcinog Compr Surv. 1985;10:147–66.
21. Yang CS, Brady JF, Hong JY. Dietary effects on cytochromes P450, xenobiotic metabolism, and toxicity. FASEB J. 1992;6(2):737–44.
22. Gonzalez FJ, Ueno T, Umeno M, Song BJ, Veech RL, Gelboin HV. Microsomal ethanol oxidizing system: transcriptional and post-transcriptional regulation of cytochrome P450, CYP2E1. Alcohol Alcohol Suppl. 1991;1:97–101.
23. Kraner JC, Lasker JM, Corcoran GB, Ray SD, Raucy JL. Induction of P4502E1 by acetone in isolated rabbit hepatocytes. Role of increased protein and mRNA synthesis. Biochem Pharmacol. 1993;45(7):1483–92.
24. Ganey PE, Roth RA. Concurrent inflammation as a determinant of susceptibility to toxicity from xenobiotic agents. Toxicology. 2001;169(3):195–208. Review.
25. National Research Council. Multiple chemical sensitivities: addendum to biologic markers in immunotoxicology. Washington DC: National Academy Press; 1992. p. 200.
26. Evans RM, Martin OV, Faust M, Kortenkamp A. Should the scope of human mixture risk assessment span legislative/regulatory silos for chemicals? Sci Total Environ. 2016;543(Pt A):757–64.
27. Alam G, Jones BC. Toxicogenetics: in search of host susceptibility to environmental toxicants. Front Genet. 2014;5:327.
28. Smith JR, Smith JG. Effects of methylmercury in vitro on methionine synthase activity in various rat tissues. Bull Environ Contam Toxicol. 1990;45(5):649–54.
29. Rowland AS, Baird DD, Weinberg CR, Shore DL, Shy CM, Wilcox AJ. Reduced fertility among women employed as dental assistants exposed to high levels of nitrous oxide. N Engl J Med. 1992;327(14):993–7.
30. Nimse SB, Pal D. Free radicals, natural antioxidants, and their reaction mechanisms. RSC Adv. 2015;5:27986–8006.
31. Chen Z, Shentu TP, Wen L, Johnson DA, Shyy JY. Regulation of SIRT1 by oxidative stress-responsive miRNAs and a systematic approach

to identify its role in the endothelium. Antioxid Redox Signal. 2013;19(13):1522–38.
32. Henkler F, Brinkmann J, Luch A. The role of oxidative stress in carcinogenesis induced by metals and xenobiotics. Cancers (Basel). 2010;2(2):376–96.
33. Li H, Wang H. Activation of xenobiotic receptors: driving into the nucleus. Expert Opin Drug Metab Toxicol. 2010;6(4):409–26.
34. Banerjee M, Robbins D, Chen T. Targeting xenobiotic receptors PXR and CAR in human diseases. Drug Disc Today. 2015;20(5):618–28.
35. Gao J, Xie W. Targeting xenobiotic receptors PXR and CAR for metabolic diseases. Trends Pharmacol Sci. 2012;33(10):552–8.
36. Sonoda J, Pei L, Evans RM. Nuclear receptors: decoding metabolic disease. FEBS Lett. 2008;582(1):2–9.
37. Hernandez JP, Mota LC, Baldwin WS. Activation of CAR and PXR by dietary, environmental and occupational chemicals alters drug metabolism, intermediary metabolism, and cell proliferation. Curr Pharmacogenomics Person Med. 2009;7(2):81–105.
38. Shanle EK, Xu W. Endocrine disrupting chemicals targeting estrogen receptor signaling: identification and mechanisms of action. Chem Res Toxicol. 2011;24(1):6–19.
39. Brent GA. Mechanisms of thyroid hormone action. J Clin Investig. 2012;122(9):3035–43.
40. Boyce WT. Biology and context: symphonic causation and the distribution of childhood morbidities (Chapter 5). In: Keating DP, editor. Nature and nurture in early child development: Cambridge University Press; 2011. p. 114–44.
41. Ferber D. Lashed by critics, WHO's cancer agency begins a new regime. Science. 2003;301(5629):36–7.
42. NTP (National Toxicology Program). 2016. Report on Carcinogens, Fourteenth Edition.; Research Triangle Park: U.S. Department of Health and Human Services, Public Health Service. [http://ntp.niehs.nih.gov/go/roc14].
43. Prasad S, Sung B, Aggarwal BB. Age-associated chronic diseases require age-old medicine: role of chronic inflammation. Prev Med. 2012;54(Suppl):S29–37.
44. Lu K, Mahbub R, Fox JG. Xenobiotics: interaction with the intestinal microflora. ILAR J. 2015;56(2):218–27.
45. Voreades N, Kozil A, Weir TL. Diet and the development of the human intestinal microbiome. Front Microbiol. 2014;5:494.
46. Vlčková K, Gomez A, Petrželková KJ, Whittier CA, Todd AF, Yeoman CJ, Nelson KE, Wilson BA, Stumpf RM, Modrý D, White BA, Leigh SR. Effect of antibiotic treatment on the gastrointestinal microbiome of free-ranging Western Lowland Gorillas (Gorilla g. gorilla). Microb Ecol. 2016;72(4):943–54.
47. Bervoets L, Van Hoorenbeeck K, Kortleven I, Van Noten C, Hens N, Vael C, Goossens H, Desager KN, Vankerckhoven V. Differences in gut microbiota composition between obese and lean children: a cross-sectional study. Gut Pathog. 2013;5:10.
48. Jin Y, Wu S, Zeng Z, Fu Z. Effects of environmental pollutants on gut microbiota. Environ Pollut. 2017;222:1–9.
49. Drożdzik M, Białecka M, Myśliwiec K, Honczarenko K, Stankiewicz J, Sych Z. Polymorphism in the P-glycoprotein drug transporter MDR1 gene: a possible link between environmental and genetic factors in Parkinson's disease. Pharmacogenetics. 2003;13(5):259–63.
50. O'Donoghue JL. Neurologic manifestations of organic chemicals. In: Johnston MV, Adams HP, Fatemi SA, editors. Neurobiology of disease. 2nd ed: Oxford University Press; 2016.
51. Manthripragada AD, Costello S, Cockburn MG, Bronstein JM, Ritz B. Paraoxonase 1, agricultural organophosphate exposure, and Parkinson disease. Epidemiology. 2010;21(1):87–94.
52. Gu X, Manautou JE. Molecular mechanisms underlying chemical liver injury. Expert Rev Mol Med. 2012;14:e4.
53. Clapp RW, Jacobs MM, Loechler EL. Environmental and occupational causes of cancer: new evidence 2005-2007. Rev Environ Health. 2008;23(1):1–37. Review.
54. Pereira CV, Moreira AC, Pereira SP, Machado NG, Carvalho FS, Sardão VA, Oliveira PJ. Investigating drug-induced mitochondrial toxicity: a biosensor to increase drug safety? Curr Drug Saf. 2009;4(1):34–54. Review.
55. Moorthy B, Chu C, Carlin DJ. Polycyclic aromatic hydrocarbons: from metabolism to lung cancer. Toxicol Sci. 2015;145(1):5–15.
56. Sharma KL, Agarwal A, Misra S, Kumar A, Kumar V, Mittal B. Association of genetic variants of xenobiotic and estrogen metabolism pathway (CYP1A1 and CYP1B1) with gallbladder cancer susceptibility. Tumour Biol. 2014;35(6):5431–9.
57. Wang J, Joshi AD, Corral R, Siegmund KD, Marchand LL, Martinez ME, Haile RW, Ahnen DJ, Sandler RS, Lance P, Stern MC. Carcinogen metabolism genes, red meat and poultry intake, and colorectal cancer risk. Int J Cancer. 2012;130(8):1898–907.
58. Cornelis MC, El-Sohemy A, Kabagambe EK, Campos H. Coffee, CYP1A2 genotype, and risk of myocardial infarction. JAMA. 2006;295(10):1135–41.
59. Jiang O, Zhou R, Wu D, Liu Y, Wu W, Cheng N. CYP2E1 polymorphisms and colorectal cancer risk: a HuGE systematic review and meta-analysis. Tumour Biol. 2013;34(2):1215–24.
60. Cross AJ, Sinha R. Meat-related mutagens/carcinogens in the etiology of colorectal cancer. Environ Mol Mutagen. 2004;44(1):44–55.
61. Tahara T, Shibata T, Arisawa T, Nakamura M, Yamashita H, Yoshioka D, Okubo M, Maruyama N, Kamano T, Kamiya Y, Fujita H, Nagasaka M, Iwata M, Takahama K, Watanabe M, Hirata I. Impact of catechol-O-methyltransferase (COMT) gene polymorphism on promoter methylation status in gastric mucosa. Anticancer Res. 2009;29(7):2857–61.
62. Stover PJ. Polymorphisms in 1-carbon metabolism, epigenetics and folate-related pathologies. J Nutrigenet Nutrigenomics. 2011;4(5):293–305.
63. Alizadeh S, Djafarian K, Moradi S, Shab-Bidar S. C667T and A1298C polymorphisms of methylenetetrahydrofolate reductase gene and susceptibility to myocardial infarction: a systematic review and meta-analysis. Int J Cardiol. 2016;217:99–108.
64. Sabbagh A, Darlu P, Crouau-Roy B, Poloni ES. Arylamine N-Acetyltransferase 2 (*NAT2*) genetic diversity and traditional subsistence: a worldwide population survey. PLoS One. 2011;6(4):e18507.
65. Bolt HM, Thier R. Relevance of the deletion polymorphisms of the glutathione S-transferases GSTT1 and GSTM1 in pharmacology and toxicology. Curr Drug Metab. 2006;7(6):613–28.
66. Way MJ, Ali MA, McQuillin A, Morgan MY. Genetic variants in ALDH1B1 and alcohol dependence risk in a British and Irish population: a bioinformatic and genetic study. PLoS One. 2017;12(6):e0177009.
67. O'Leary KA, Edwards RJ, Town MM, Boobis AR. Genetic and other sources of variation in the activity of serum paraoxonase/diazoxonase in humans: consequences for risk from exposure to diazinon. Pharmacogenet Genomics. 2005;15(1):51–60.
68. Manthripragada AD, Costello S, Cockburn MG, Bronstein JM, Ritz B. Paraoxonase 1 (PON1), agricultural organophosphate exposure, and Parkinson disease. Epidemiology. 2010;21(1):87–94.
69. Ji Y, Moon I, Zlatkovic J, Salavaggione OE, Thomae BA, Eckloff BW, Wieben ED, Schaid DJ, Weinshilboum RM. Human hydroxysteroid sulfotransferase SULT2B1 pharmacogenomics: gene sequence variation and functional genomics. J Pharmacol Exp Ther. 2007;322(2):529–40.
70. Hyman M. Systems biology, toxins, obesity, and functional medicine. Altern Ther Health Med. 2007;13(2):S134–9.
71. Liska D, Lyon M, Jones DS. Detoxification and biotransformational imbalances. Explore (NY). 2006;2(2):122–40.
72. Hodges RE, Minich DM. Modulation of metabolic detoxification pathways using foods and food-derived components: a scientific review with clinical application. J Nutr Metab. 2015;2015:760689.
73. Folate intake may be estimated using USDA Food Composition Database as a guide: [http://ndb.nal.usda.gov]. (Last accessed 18 April 2017).
74. Tinetti ME, Fried TR, Boyd CM. Designing health care for the most common chronic condition--multimorbidity. JAMA. 2012;307(23):2493–4.

Detoxification and Biotransformation

Janet L. Black

14.1 Introduction – 206

14.2 Primary Approach – Avoidance – 206

14.3 Secondary Approach – Caring for Organs of Detoxification – 207
14.3.1 Liver Considerations – 207
14.3.2 Kidney Considerations – 208

14.4 Tertiary Approach – Detoxification Support – 208
14.4.1 Start with Food – 208
14.4.2 Supplementation Use – 209

References – 210

14.1 Introduction

Detoxification is the process by which the body breaks down and removes substances that are undesirable and have potential for harm. These can be endogenous waste products or such exogenous substances as medications, environmental toxins or any substances that we ingest, inhale or absorb. Detoxification is a natural process of the body carried out continuously by the liver, the kidneys, the skin, and the lungs [1].

"Detox" is a popular topic and there is a lot of disagreement about what it is and how to do it. While many companies sell detoxification products, the claims that these products are essential to cleanse the body are often exaggerated. The body is designed to effectively eliminate toxins in the majority of cases. Using a radical detoxification product and then returning to the same poor dietary and lifestyle habits is like going on an extreme diet for weight loss and then returning to the prior way of eating. The approach of using an aggressive detoxification product is not necessarily effective to help the body function properly and may even result in harm. Better approaches to detoxification are discussed below.

The critical role of detoxification is handled primarily by the liver [2]. Rich blood flow is seen throughout the liver, bathing its columns of orderly arranged cells separated by spaces. As the blood is transported through the spaces between the liver cell units, it is filtered and chemicals, dead cells, drugs, microorganisms, and other debris are removed from the bloodstream. Important in this process, Kupffer cells are specialized macrophages found in the liver sinusoids and make up more than 80% of macrophages found in the entire body. Kupffer cells are especially important for elimination of infectious agents circulating in the blood filtered through the liver, but also important for other debris and toxin degradation [3]. A significant elaborate system of breakdown of the removed debris is handled by several step processes, beginning with the inherited cytochrome P450 enzymes that require important cofactors for function in the form of nutrients such as B vitamins [2]. (Please see ► Chap. 13). This critical first phase of the detoxification addresses fat-soluble toxins that are chemically converted into intermediate metabolites that become more water-soluble. The second phase completes the chemical transformation into water-soluble compounds on their way to excretion into bile and subsequently into the gastrointestinal tract for removal from the body. Amino acids, vitamins, and other cofactors are important for this second phase. More in-depth understanding can be gained from ► Chap. 13.

The kidney's filtering system is comprised of glomeruli that remove waste products and toxins from the blood with subsequent excretion into the urine [4]. Urinary excretion is one of the primary ways that toxins are eliminated. Kidneys not only eliminate wastes but also maintain levels of water and minerals, produce renin to manage blood pressure, synthesize erythropoietin to make red blood cells, and synthesize the active form of Vitamin D for strong bones and a variety of other functions. We want to do everything we can to keep these important organs healthy.

Encountering a patient immediately following an alcoholic binge, one notes a distinctive odor emanating from the skin; this is the body's way to eliminate alcohol metabolites. We often don't think of the skin as an organ of elimination, but waste products and toxins, such as heavy metals, phthalates, and other environmental toxins do diffuse into the sweat glands for elimination in the sweat [5, 6]. Saunas are often promoted for their detoxification effects due to the benefits of sweating. Vigorous exercise with sweating has similar outcomes.

The lungs are primarily designed to remove carbon dioxide in exhaled breath, but the lungs are also able to filter and remove other waste products [7]. A well-known example is a diabetic patient in ketosis exhaling ketones in the breath, with its characteristic odor. While this is often described as a fruity odor, the actual chemical odor being eliminated is acetone. Other metabolites are also eliminated by the lungs when in abundance, which brings up images of a breathalyzer in the hands of a police officer. However, exhaled breath has the potential to surpass blood and urine for toxicity testing because breath gives unlimited supply, does not require trained medical personnel, is noninvasive, and does not produce potentially infectious waste such as needles.

14.2 Primary Approach – Avoidance

The primary approach to detoxification is avoidance of the exposure when possible. Unfortunately, the majority of populations around the world are exposed to chemicals and foreign substances that were not present until the early twentieth century. Estimates suggest there are almost 100,000 chemicals in current use with only a fraction of them rigorously tested for safety in humans [8]. A further concern is that tested chemicals are evaluated based on a single exposure without interaction with other chemicals, unlike the daily experience where humans are exposed to a chemical soup from personal care products, food contaminants, drugs, pesticides, etc. Since exposure to multiple chemicals occurs daily, the risk of toxicity is actually much higher and taxes the detoxification system. This makes the concept of avoidance even more critical, and patients need to be educated about the potential sources of exposures. The more patients are aware of exposures, the better equipped they will be to reduce them. Ultimately this leads to reduced burden on the normal biological processes [9].

What we now call "organic food" was simply called "food" 100 years ago. People had relatively small plots of land where they usually had a combination of animals and crops. The manure from the animals and other organic waste were composted to provide fertilizer for the garden and field plantings. Currently, the majority of crops are farmed large in monoculture plots using chemical fertilizers, pesticides, herbicides, and fungicides. Monoculture farming and use of chemicals on the land have led to nutrient depletion of the soil [10, 11]. Some crops are "Roundup

Ready," meaning that they are genetically modified to withstand large amounts of herbicide (glyphosate). Besides being a risk for cancer, kidney disease and other health problems, glyphosate can bind with minerals in the soil including copper, iron, magnesium, manganese, nickel cobalt and zinc, making them less available in our food [12]. Roundup is also sprayed on other crops, such as wheat, as a desiccant prior to harvest [13, 14].

It is not only plant food that is modified with chemicals. Animals are factory farmed in crowded warehouses or pens, making them more susceptible to distress and disease [15]. They are fed food containing genetically modified (GMO) corn and soy and are given antibiotics and hormones to hasten weight gain. When cattle are crowded into pens full of urine and feces, is it any surprise that *E.coli*, a fecal-borne bacteria, can end up in the meat [16]? It is not a healthy situation for the animals or those who consume them.

Cattle are intended to eat grass and similar plant materials as they have four stomachs designed to break down the cellulose. Feeding cattle grain is not a healthful alternative. Nor is crowding the animals in pens without exercise options or giving them chemicals and antibiotics to produce rapid weight gain. These procedures produce a marbling appearance in meat that many people desire and create a milder flavor; but these practices do not contribute to animal health and well-being or health benefits for those who consume the meat produced in this manner.

When large amounts of urine and feces are produced in small contained areas by animal factory farming, there is often contamination of nearby groundwater [17]. When cattle, poultry or hogs are rotated on pastures in traditional farming techniques, their wastes fertilize the ground without overwhelming the area. Just as with human detoxification, nature has the means to handle the waste products in small quantities.

Our water supply can be contaminated by other sources of toxins. Herbicides such as Roundup, pesticides, lawn chemicals, factory farm waste, heavy metals, and prescription drugs can all be found in tap water [9]. If the water is from a metropolitan water supply (as opposed to a private well), it may also contain fluoride (a neurotoxin) and will have chlorine or chlorinated by-products that have been used to kill microorganisms. The list of possible contaminants is too long to include here, but the recommendation is to use filtered tap water [9]. Reverse osmosis systems are the most effective at removing these, but other filters are also helpful. Bottled water may be no different than tap water, plus there is the risk of leaching endocrine-disrupting contaminants like bisphenol-A (BPA) from the plastic bottle, especially if exposed to sunlight [18].

Common everyday sources of toxic chemicals are in personal care and cleaning products. We have become obsessed with cleanliness and with killing germs to a degree that is unhealthy. Our bodies are hosts to around 100 trillion microorganisms that live in the gastrointestinal tract and on the skin, and these are necessary for health [19]. Anything that disturbs the balance of these microorganisms can create problems. Therefore, any chemical-containing products, particularly antibacterial-containing products for everyday use, should be discouraged [9].

The Food and Drug Administration does not approve cosmetics or personal care product ingredients but is, by law, charged with regulating these ingredients [20]. For example, the FDA has banned the use of triclosan in soaps, but triclosan may still be found in other products such as toothpaste [21]. Otherwise, unless there is evidence of obvious harm, ingredients are not premarket-approved. Because we apply these products directly to the skin, any toxins contained in them can be absorbed. The European Union has banned more than 1100 chemicals that are allowed in the United States [22]. One way to avoid the most toxic products is to use the Environmental Working Group website (▶ ewg.org), which lists products and rates them according to ingredients.

Cleaning products are even more likely to contain toxic ingredients when compared to cosmetics and personal care products. Most cleaning products do not list their ingredients on the label, and many consumers are unaware of what their exposure may be. There are green products that do list ingredients and are less toxic, and can be found easily locally. Once again, the Environmental Working Group is a source of information that can guide buying decisions. Another option is to make products from safe ingredients and recipes are found on the internet for this purpose.

Are you aware that the "new car smell," the smell of fresh paint, the smell of new carpet and similar smells include chemical fragrances? These are volatile organic compounds (VOCs) and should be avoided as they can be toxic [23]. Air fresheners, fabric softeners, oven cleaners, flame retardants, and plastic food containers should also be avoided for similar reasons [9]. Unscented products and low VOC products are preferred. If fragrance is desired, essential oils provide a better alternative than artificial fragrances which contain multiple chemicals and phthalates (hormone-disrupting chemicals).

Now that we have discussed the first step, avoidance of toxins, let us look at ways that we can support the detoxification process by care of organs of detoxification and aiding excretion through liver and kidneys.

14.3 Secondary Approach – Caring for Organs of Detoxification

14.3.1 Liver Considerations

Because of poor dietary intake high in sugars and damaged fats, people may have impaired liver function. One very common problem is what is commonly called "fatty liver" or hepatic steatosis. Fatty liver may occur in those with excessive alcohol intake, but a more common problem is non-alcoholic fatty liver (NAFLD). This can be either isolated fatty liver, nonalcoholic steatohepatitis (NASH), or

NAFLD, which is more severe and associated with inflammation, obesity, and metabolic syndrome [24]. Patients often have no symptoms but will probably have abdominal obesity. It can be found via imaging (ultrasound, CT or MRI), or if more severe, via elevated liver enzymes. NASH and NAFLD can progress to cirrhosis as scarring accumulates in the liver tissue. The treatment for fatty liver is primarily lifestyle changes, including abstinence from alcohol, weight loss, and controlling blood sugar. High fructose corn syrup is especially detrimental to the liver and should be eliminated from the diet [25]. Coffee appears to reduce the build-up of scarring in NASH, so it is beneficial [26]. The first-line treatment for non-diabetic NASH is Vitamin E [27]. Other things that may be beneficial include omega-3 fatty acids, N-acetylcysteine (NAC), glycyrrhizin (from licorice root) and whey protein [28–33]. Both NAC and whey protein have high levels of cysteine which can increase glutathione, an antioxidant that is liver-protective. Whey protein may also aid in weight loss.

Other liver diseases include viral hepatitis, chemical hepatitis (such as results from acetaminophen overdose), cirrhosis, and cancer. The presence of any of these can affect the liver's ability to break down toxins by deficiencies in Phase I and Phase II nutrients and by generally overwhelming the detoxification pathways [2].

14.3.2 Kidney Considerations

Chronic kidney impairment can result from diabetes and/or hypertension [34]. The kidneys may also be damaged by direct injury, such as a severe blow or something that decreases blood flow such as an acute hemorrhagic injury that results in significant blood loss. A backup of urine may also damage the kidneys, such as with prostate enlargement, with certain medications or with kidney stones that block the urine outflow. Infections may start in the bladder and ascend into the kidneys which, if not treated promptly, can result in problems such as scarring. Infections from hepatitis, HIV, and strep also may affect the kidneys. Autoimmune diseases, like lupus, that attack body tissues can also damage kidneys. Long-term use of NSAIDS and lead poisoning are other risks.

Kidney impairment can be diagnosed by increased microalbumin in the urine and elevated creatinine in the blood [35]. The estimated glomerular filtration rate (eGFR) is calculated from the creatinine level and considers age, gender, and race. These should be monitored regularly in patients with diabetes and/or hypertension and other chronic medical conditions; keeping blood pressure and blood glucose under good control is paramount. Angiotensin-converting enzyme (ACE) inhibitors and angiotensin II receptor blockers (ARBs) are commonly used to control blood pressure and reduce proteinuria and are often prescribed for normotensive diabetics for their kidney protective effects.

14.4 Tertiary Approach – Detoxification Support

14.4.1 Start with Food

After protecting the organs of detoxification, the next step is aiding in support of the body's inherent detoxification pathways [36–38]. The baseline issues for patients to enhance detoxification are a clean, good diet and adequate hydration, with clean water providing the best source for fluid intake [36, 37]. Liquids containing caffeine with a diuretic action, such as coffee, soda and, to a lesser extent, tea, result in the body losing more water than is ingested. Caffeinated drinks should not be the primary source of dietary liquid intake. Alcohol also has a diuretic function and inhibits organs of detoxification. Fruit juices are generally high in sugars and not likely healthy choices for detoxification purposes. If patients prefer juice and do not want to eat adequate amounts of vegetables and whole fruit, a blender can be used to make a smoothie with predominantly vegetables and minimal fruit. A juicer that extracts the juice and leaves the fiber behind is not recommended. Fiber is a very important part of the diet in maintaining gastrointestinal health and slowing the absorption of sugars. Vegetables and fruit provide antioxidants and phytonutrients essential for good health [39, 40]. Vegetables should be the mainstay of the diet, especially colorful vegetables such as dark leafy greens, dark yellow, orange, and purple varieties. Fruit also provides important nutrients and fiber, but if used predominantly may provide too much sugar.

Sugar has an inflammatory effect and should be avoided as much as possible [41]. Artificial sweeteners, such as aspartame, have controversial yet concerning effects on human health and should be avoided during detoxification [42]. If added sugar is used, it should be organic cane sugar. Concerns are also raised regarding genetically modified foods (GMO), although this area also remains controversial [43, 44]. We just don't know enough about the safety of GMO foods, but there is emerging research to show there are differences that may be detrimental to health [44]. Sugar beets are a common GMO crop and most of the commercially available sugar uses GMO sugar beets. Corn sweeteners, such as high-fructose corn syrup, are also produced using GMO corn. Because many healthcare practitioners now advise against high-fructose corn syrup and many patients seek to avoid it, manufacturers may refer to it by a different name on food labels [45].

Stevia and monk fruit are safe sweeteners derived from plants [29]. Stevia comes in liquid and powder forms and is a safe sweetener but may have an aftertaste if too much is used. Other sweeteners considered low-glycemic are the GMO or Non-GMO birch or corn derived polyols (examples are xylitol, erythritol, malitol). They contain about 40% fewer calories than sugar and do not cause a rapid increase of blood sugar. They are partially absorbed in the gastrointestinal tract which can produce a laxative or loosening of the stool for some people.

In the 1950s, the role of sugar as an etiology of heart disease began to come to light. However, in 1965, the Sugar Research Foundation paid scientists to do a literature review that downplayed the role of sugar and laid the blame on fat and cholesterol [41]. This led to decades of poor nutritional advice and patients still follow low-fat, low-cholesterol diets to this day. Recommendations to follow a low-fat diet resulted in higher carbohydrate consumption, obesity, metabolic dysfunction, and NAFLD [24]. Rather than low-fat diets, food sources should be rich in beneficial fats. Healthy fats are an important part of the diet (See ► Chaps. 10 and 11). Healthy fat ratios should be higher in omega-3 fatty acids rather than omega-6 fatty acids and other fats, like medium-chain fats and omega-9 fats, are to be recommended. Unfortunately, healthy omega-6 fats are not consumed as often as the highly processed damaged vegetable oils which are high in inflammatory omega-6 fatty acids. Processed foods tend to contain fats that are the damaged omega-6 fatty acids and may even contain trans-fats and should be avoided, especially during detoxification. It is important to strike a balance between omega-3 fatty acids and unprocessed omega-6 fatty acids obtained from natural sources. We also now know that saturated fat and cholesterol are not the demons they have been made out to be, but have roles in healthy diets and normal metabolism. Recommended dietary sources of fats include, but are not limited to, coconut oil, olive oil, avocados, grass-fed butter, and fatty fish (wild-caught).

Wheat and gluten have been a subject of much interest in recent years with books like *Wheat Belly* and *Grain Brain* becoming best sellers [46, 47]. Gluten-free products are now found ubiquitously because of the popularity of gluten-free diets. Certainly, the wheat we are using today has been bred to contain more gluten, which is advantageous for bakers wanting to create light, fluffy bakery products. Unfortunately, the gluten, gliadin, and agglutinin found in wheat, especially the modern hybrids, are irritating to the digestive tract and can lead to intestinal permeability, AKA "leaky gut" [46]. The use of glyphosate as a desiccating agent prior to harvest has also been implicated in the rise of gluten intolerance [13, 14]. Intestinal permeability associated with gluten intolerance has been linked to the rise in autoimmune conditions, obesity, and other chronic health problems. Therefore, many detoxification programs recommend eliminating the use of grains, especially wheat.

While some people find that they must completely eliminate grains from their diet, others can tolerate limited amounts without any problems. One of the best ways to determine sensitivity is to try an elimination diet for a period of several weeks. When wheat or gluten is added back to the diet as a challenge, the subsequent reaction – or lack of one – can be evaluated and assessed (See ► Chap. 24 for guide to elimination diet). This author has found that, in many patients, eliminating sugars and grains result in a remission of some symptoms, such as gastrointestinal problems, allergic symptoms, and a reduction in chronic pain.

A recent study on long-term gluten-free diets and heart disease indicated that a gluten-free diet increases the risk of coronary heart disease [48]. They concluded that gluten-free diets should not be recommended for people without celiac disease. The answer to this dilemma may reside in the types of food used in substitution of wheat/gluten. Many gluten-free products available to consumers are processed and high in carbohydrates and simple sugars, but not high in fiber and nutrients. This dietary substitution is not particularly conducive to good health. The fiber content of whole grain is a big reason why it is routinely recommended, but it is possible to get high-quality fiber by eating plenty of high-fiber vegetables and fruit. Otherwise, the addition of such high-fiber substances as psyllium husk is an alternative dietary fiber source.

Dairy consumption is another concern during detoxification. In the case of lactose intolerance, dairy sensitivity may be easy to recognize. Casein, the protein in dairy, may also cause intolerance but is more difficult to assess. Regardless, most detoxification programs recommend eliminating dairy. Again, an elimination diet with subsequent challenge is a good way for the patient to see if dairy is contributing to their health problems. If dairy is used, it should be organic.

The diet should contain adequate protein with sources from meat, poultry, fish, eggs, dairy, nuts and seeds, and legumes [49]. The recommended daily allowance (RDA) for protein is 0.80 g/kg per day. This is a minimum amount and a higher intake of protein may be indicated in active individuals to maintain and promote muscle growth. However, most Americans eating a standard diet are easily exceeding the RDA for their protein requirement. Most can obtain adequate good quality protein while reducing the quantity of meat by choosing a predominantly plant-based diet [49]. Protein is essential to support enzymatic detoxification processes because of the amino acids supplied. Patients who might be at risk for not getting enough good quality protein include vegans or those who are severely restricting calories.

14.4.2 Supplementation Use

It is imperative that nutritional counseling be sought to ensure those preparing for detoxification are getting all the necessary nutrients in their diets to support the detoxification process. Because our food may not provide an optimal level of nutrients, a good multivitamin/mineral supplement provides a back-up for what is not received in the diet. Of importance for detoxification are the B vitamins, which are co-factors in many enzymatic reactions. One concern for restricted diets is vitamin B12. Since Vitamin B12 is only found in animal products, those who do not eat animal products will need to use a supplement. As people age, their absorption of Vitamin B12 is lower so older people may also need to take this as a supplement. Magnesium, selenium and zinc are minerals that are frequently deficient and critical for enzymatic conversion of endogenous and exogenous toxins.

Other critical yet deficient nutrients include vitamins D and C. Since a multivitamin/mineral supplement may not provide enough of these critical nutrients, additional supplementation may be needed and can be guided by diet history assessment and laboratory testing.

There are many supplements that are often recommended for detoxification. One common medicinal plant recommended is milk thistle, *Silybum marianum* [29, 50]. This herb has been used for detoxification because of its hepatoprotective properties. It acts as an antioxidant and inhibits the binding of toxins to the hepatic cell membranes and reduces fibrosis. It has been used to treat viral hepatitis as well as to reduce the injury caused by acetaminophen, alcohol, iron overload, and carbon tetrachloride [50].

Another popular medicinal plant, dandelion extract, has also been used for many years for its antioxidant properties and its usefulness in treating liver disorders. Research in rats has confirmed its ability to prevent fibrosis from chemical damage and to promote regeneration [51, 52]. This, in part, is related to its effects on Phase I detoxification pathways, particularly cytochrome P450 2E1. Ginger (*Zingiber officinale*) is another substance that is helpful to prevent fibrosis through its antioxidant capabilities [53–55].

Garlic and vitamin C are useful for many detoxification purposes, but also for reducing lead levels [56–61]. This is especially important for children whose neurodevelopment is damaged by lead toxicity [59]. Garlic has been found to be more effective than d-penicillamine in treating lead poisoning [60].

Glutathione (glutathione S-transferase or GSH) functions as an antioxidant and chelator of metals, including aluminum and mercury, so substances that increase glutathione help with detoxification [62–65]. Spices such as black pepper, curcumin or turmeric, rosemary, and ginger have many beneficial effects, including the potential to increase glutathione levels. Other natural products that may increase glutathione include alpha lipoic acid, chlorella, dark roast coffee, methylsulfonylmethane (MSM), N-acetyl cysteine (NAC), phosphotidylcholine and phosphotidylethanolamine, probiotics with soluble fiber, selenium, sulforaphane (found in cruciferous vegetables), whey protein, and zinc [66–77].

In summary, the approach to detoxification involves three steps: the primary approach – avoidance; the secondary approach – caring for organs of detoxification; and the tertiary approach – detoxification or metabolic biotransformation support. All of these steps promote the health of the detoxification organs and encourage the breakdown and excretion of toxins that will do much to transform the health of your patients and prevent chronic disease.

References

1. Liska D, Lyon M, Jones DS. Detoxification and biotransformational imbalances. Explore. 2006;2(2):122–40. https://doi.org/10.1016/j.explore.2005.12.009.
2. Meyer UA. Overview of enzymes of drug metabolism. J Pharmacokinet Biopharm. 1996;24(5):449–59.
3. Bilzer M, Roggel F, Gerbes AL. Role of Kupffer cells in host defense and liver disease. Liver Int. 2006;26(10):1175–86.
4. Pollak MR, Quaggin SE, Hoenig MP, Dworkin LD. The glomerulus: the sphere of influence. CJASN. 2014;9(8):1461–9.
5. Genuis S, Beesoon S, Lobo RB. Human elimination of phthalate compounds: blood, urine, and sweat (BUS) study. ScientificWorldJournal. 2012;2012:615068.
6. Genuis S, Birkholz D, Rodushkin I, Beesoon S. Blood, urine, and sweat (BUS) study: monitoring and elimination of bioaccumulated toxic elements. Arch Environ Contam Toxicol. 2011;61:344–57.
7. Pleil JD. Breath biomarkers in toxicology. Arch Toxicol. 2016;90:2669. https://doi.org/10.1007/s00204-016-1817-5.
8. Chemical Inspection & Regulation Service (CRS). US TSCA inventory. 2012. Retrieved from: http://www.cirs-reach.com/Inventory/US_TSCA_Inventory.html.
9. Black JL. Living in our toxic world: protect yourself and your family from environmental toxins and reduce your risk of cancer and chronic disease. Candler: Peaceful Heart Press; 2016.
10. Altieri MA, Rosset P. Ten reasons why biotechnology will not ensure food security, protect the environment and reduce poverty in the developing world. AgBioforum. 1999;2(3 & 4):155–62.
11. Altieri MA. The ecological impacts of transgenic crops on agroecosystem health. Ecosyst Health. 2000;6(1):13–23. http://www.ask-force.org/web/Biotech-Biodiv/Altieri-Ecological-Impacts-Transgenic-2000.pdf. Accessed Mar 2019.
12. Sebiomo A, Ogundero V, Bankole S. The impact of four herbicides on soil minerals. Res J Environ Earth Sci. 2012;4(6):617–24.
13. Bott S, Lebender U, Yoon D-J, et al. UC Davis: the proceedings of the international plant nutrition colloquium XVI. 20 May 2009. https://cloudfront.escholarship.org/dist/prd/content/qt25v599pr/qt25v599pr.pdf?t=kro6b6. Accessed Mar 2019.
14. Samsel A, Seneff S. Glyphosate's suppression of cytochrome P450 enzymes and amino acid biosynthesis by the gut microbiome: pathways to modern diseases. Entropy. 2013;15:1416–63. https://doi.org/10.3390/e15041416.
15. Pluhar EB. Meat and morality: alternatives to factory farming. J Agric Environ Ethics. 2010;23:455. https://doi.org/10.1007/s10806-009-9226-x.
16. Khaitsa ML, Smith DR, Stoner JA, Parkhurst AM, Hinkley S, Klopfenstein TJ, Moxley RA. Incidence, duration, and prevalence of Escherichia coli O157:H7 fecal shedding by feedlot cattle during the finishing period. J Food Prot. 2003;66(11):1972–7. https://doi.org/10.4315/0362-028X-66.11.1972.
17. Sampat P. editor, Uncovering groundwater pollution. In: State of the world 2001: a Worldwatch Institute report on progress toward a sustainable society, Lester R. Brown, Christopher Flavin, Hilary French. New York: W.W. Norton, 2001. 21–42.
18. Vom Saal FS, Nagel SC, Coe BL, Angle BM, Taylor A. The estrogenic endocrine disrupting chemical bisphenol A (BPA) and obesity. Mol Cell Endocrinol. 2012;354(1–2):74–84. https://doi.org/10.1016/j.mce.2012.01.001.
19. Zhu B, Wang X, Li L. Human gut microbiome: the second genome of human body. Protein Cell. 2010;1:718. https://doi.org/10.1007/s13238-010-0093-z.
20. FDA. FDA authority over cosmetics: how cosmetics are not FDA-approved, but are FDA-Regulated. 3 Mar 2005; updated 3 Aug 2013. https://www.fda.gov/cosmetics/guidanceregulation/lawsregulations/ucm074162.htm. Accessed Feb 2019.
21. FDA. 5 things to know about triclosan. 9 Dec 2017. https://www.fda.gov/ForConsumers/ConsumerUpdates/ucm205999.htm. Accessed Feb 2019.
22. IFLScience. Banned in Europe, safe in the U.S. 9 June 2014. https://www.iflscience.com/health-and-medicine/banned-europe-safe-us/. Accessed Feb 2019.
23. Ott WR, Roberts JW. Everyday exposure to toxic pollutants. Sci Am. 1998;278(2):86–91. Accessed Feb 2019: JSTOR, www.jstor.org/stable/26057669.

24. Marchesini G, Bugianesi E, Forlani G, Cerrelli F, Lenzi M, Manini R, Natale S, Vanni E, Villanova N, Melchionda N, Rizzetto M. Nonalcoholic fatty liver, steatohepatitis, and the metabolic syndrome. Hepatology. 2003;37:917–23. https://doi.org/10.1053/jhep.2003.50161.
25. Basaranoglu M, Basaranoglu G, Sabuncu T, Senturk H. Fructose as a key player in the development of fatty liver disease. World J Gastroenterol. 2013;19(8):1166–72.
26. Cavin C, Marin-Kuan M, Sangouet S, Bezencon C, Guignard G, Verguet C, Piguet D, Holzhäuser D, Cornaz R, Schilter B. Induction of Nrf2-mediated cellular defenses and alteration of phase I activities as mechanisms of chemoprotective effects of coffee in the liver. Food Chem Toxicol. 2008;46(4):1239–48.
27. Moyad M. The supplement handbook. New York: Rodale; 2014.
28. Allison Sarubin-Fragakis, The health professional's guide to popular dietary supplements. 3rd ed. Chicago: American Dietetic Association; 2006.
29. Murray MT, Pizzorno J. The encyclopedia of natural medicine third edition. New York: Atria Books (Simon and Schuster); 2012.
30. Hajiaghamohammadi A, Ziaee A, Samimi R. The efficacy of licorice root extract in decreasing transaminase activities in non-alcoholic fatty liver disease: a randomized controlled clinical trial. Phytother Res. 2012;26(9):1381–4.
31. Grey V, Mohammed S, Smountas A, Bahlool R, Lands L. Improved glutathione status in young adult patients with cystic fibrosis supplemented with whey protein. J Cyst Fibros. 2003;2(4):195–8.
32. Zavorsky G, Kubow S, Grey V, Riverin V, Lands L. An open-label dose-response study of lymphocyte glutathione levels in healthy men and women receiving pressurized whey protein isolate supplements. Int J Food Sci Nutr. 2007;58(6):429–36.
33. Chung R. Detoxification effects of phytonutrients against environmental toxicants and sharing of clinical experience on practical applications. Environ Sci Pollut Res Int. 2017;24(10):8946–56.
34. Levey AS, Atkins R, Coresh J, Cohen EP, Collins AJ, Eckardt KU, Nahas ME, Jaber BL, Jadoul M, Levin A, Powe NR, Rossert J, Wheeler DC, Lameire N, Eknoyan G. Chronic kidney disease as a global public health problem: approaches and initiatives – a position statement from Kidney Disease Improving Global Outcomes. Kidney Int. 2007;72(3):247–59. https://doi.org/10.1038/sj.ki.5002343.
35. Jha V, Garcia-Garcia G, Iseki K, Li Z, Naicker S, Plattner B, Saran R, Wang AY, Yang CW. Chronic kidney disease: global dimension and perspectives. Lancet. 2013;382(9888):260–72. https://doi.org/10.1016/S0140-6736(13)60687-X.
36. Sears M, Genuis S. Environmental determinants of chronic disease and medical approaches: recognition, avoidance, supportive therapy, and detoxification. Environ Public Health. 2012;2012:356798.
37. Klein A, Kiat H. Detox diets for toxin elimination and weight management: a critical review of the evidence. J Hum Nutr Diet. 2015;28(6):675–86.
38. Kim J, Kim J, Kang S. Effects of dietary detoxification program on serum glutamyltransferase, anthropometric data and metabolic biomarkers in adults. J Lifestyle Med. 2016;6(2):49–57.
39. Lee S, Kim C, Lee HJ, Yun J, Nho C. Induction of the phase II detoxification enzyme NQO1 in hepatocarcinoma cells by lignans from the fruit of Schisandra chinensis through nuclear accumulation of Nrf2. Planta Med. 2009;75(12):1314–8.
40. Rose P, Ong C, Whiteman M. Protective effects of Asian green vegetables against oxidant induced cytotoxicity. World J Gastroenterol. 2005;11(48):7607–14.
41. Kearns CE, Schmidt SA, Glantz SA. Sugar industry and coronary heart disease research: a historical analysis of internal industry documents. JAMA Intern Med. 2016;176(11):1680–5.
42. Tandel KR. Sugar substitutes: health controversy over perceived benefits. J Pharmacol Pharmacother. 2011;2(4):236–43. https://doi.org/10.4103/0976-500X.85936.
43. Hicks DJ. Epistemological depth in a GM crops controversy. Stud Hist Phil Biol Biomed Sci. 2015;50:1–12. https://doi.org/10.1016/j.shpsc.2015.02.002.
44. Druker SM. Altered genes, twisted truth. Salt Lake City: Clear River Press; 2015.
45. Goran MI, Ulijaszek SJ, Ventura EE. High fructose corn syrup and diabetes prevalence: a global perspective. Glob Public Health. 2013;8(1):55–64. https://doi.org/10.1080/17441692.2012.736257.
46. Davis W. Wheat belly. New York: Rodale; 2011.
47. Perlmutter D. Grain brain. New York: Little, Brown & Co.; 2013.
48. Lebwohl B, Cao Y, Zong G, Hu F, Green P, Neugut A, Rimm EB, Sampson L, Dougherty LW, Giovannucci EL, Willett WC, Qi S, Chan A. Long term gluten consumption in adults without celiac disease and risk of coronary heart disease: prospective cohort study. BMJ. 2017;357:j1892.
49. Rodriguez N, Miller S. Effective translation of current dietary guidance: understanding and communication the concepts of minimal and optimal levels of dietary protein. Am J Clin Nutr. 2015;101(6):1353–8S.
50. Abenavoli L, Capasso R, Milic N, Capasso F. Milk thistle in liver diseases: past, present, future. Phytother Res. 2010;24(10):1423–32.
51. Park CM, Cha YS, Youn HJ, Cho CW, Song YS. Amelioration of oxidative stress by dandelion extract through CYP2E1 suppression against acute liver injury induced by carbon tetrachloride in Sprague-Dawley rats. Phytother Res. 2010;24(9):1347–53. https://doi.org/10.1002/ptr.3121.
52. Domitrović R, Jakovac H, Romić Z, Rahelić D, Tadić Z. Antifibrotic activity of Taraxacum officinale root in carbon tetrachloride-induced liver damage in mice. J Ethnopharmacol. 2010;130(3):569–77. Epub 2 Jun 2010. PMID: 20561925.
53. Stoilova I, Krastanova A, Stoyanova A, Denev P, Gargova S. Antioxidant activity of a ginger extract (Zingiber officinale). Food Chem. 2007;102(3):764–70. https://doi.org/10.1016/j.foodchem.2006.06.023.
54. Motawi TK, Hamed MA, Shabana MH, Hashem RM, Aboul Naser AF. Zingiber officinale acts as a nutraceutical agent against liver fibrosis. Nutr Metab (Lond). 2011;8:40.
55. Sharma P, Singh R. Dichlorvos and lindane induced oxidative stress in rat brain: protective effects of ginger. Pharm Res. 2012;4(1):27–32.
56. Butt M, Sultan M, Butt M, Iqbal J. Garlic: nature's protection against physiological threats. Crit Rev Food Sci Nutr. 2009;49(6):538–51.
57. Chang H, Ko M, Ishizuka M, Fujita S, Yabuki A, Hossain M, Yamato O. Sodium 2-propenyl thiosulfate derived from garlic induces phase II detoxification enzymes in rat hepatoma H4IIE cells. Nutr Res. 2010;30(6):435–40.
58. El-Barbary M. Detoxification and antioxidant effects of garlic and curcumin in Oreochromis niloticus injected with aflatoxin B1 with reference to gene expression of glutathione peroxidase (GPx) by RT-PCR. Fish Physiol Biochem. 2016;42:617–29.
59. Ghasemi S, Hosseini M, Feizpour A, Alipour F, Sadeghi A, Vafaee F, Mohammadpour T, Soukhtanloo M, Ebrahimzadeh Bideskan A, Beheshti F. Beneficial effects of garlic on learning and memory deficits and brain tissue damages induced by lead exposure during juvenile rat growth is comparable to the effect of ascorbic acid. Drug Chem Toxicol. 2016;40:1–9.
60. Kianoush S, Balali-Mood M, Mosavi S, Moradi V, Sadeghi M, Dadpour B, Rajabi O, Shakeri M. Comparison of therapeutic effects of garlic and d-Penicillamine in patients with chronic occupational lead poisoning. Basic Clin Pharmacol Toxicol. 2012;110:476–81.
61. Najar-Nezhad V, Aslani M, Balali-Mood M. Evaluation of allicin for the treatment of experimentally induced subacute lead poisoning in sheep. Bio Trace Elem Res. 2008;126:141–7.
62. Kaur P, Aschner M, Syversen T. Glutathione modulation influences methyl mercury induced neurotoxicity in primary cell cultures of neurons and astrocytes. Neurotoxicology. 2006;27:492–500.

63. Lee M, Cho S, Roh K, Chae J, Park J, Park J, Lee MA, Kim J, Auh CK, Yeom CH, Lee S. Glutathione alleviated peripheral neuropathy in oxaliplatin-treated mice by removing aluminum from dorsal root ganglia. Am J Transl Res. 2017;9:926–39.
64. Lutgendorff F, Trulsson L, van Minnen L, Rijkers G, Timmerman H, Franzen L, Gooszen HG, Akkermans LM, Söderholm JD, Sandstrom P. Probiotics enhance pancreatic glutathione biosynthesis and reduce oxidative stress in experimental acute pancreatitis. Am J Physiol Gastrointest Liver Physiol. 2008;295:1111–21.
65. Zhang W, Joseph E, Hitchcock C, DiSilvestro R. Selenium glycinate supplementation increase blood glutathione peroxidase activities and decreases prostate-specific antigen readings in middle-aged US men. Nutr Res. 2011;31:165–8.
66. Agarwal R, Goel S, Behari J. Detoxification and antioxidant effects of curcumin in rats experimentally exposed to mercury. J Appl Toxicol. 2010;30(5):457–68.
67. Iqbal M, Sharma S, Okazaki Y, Fujisawa M, Okada S. Dietary supplementation of curcumin enhances antioxidant and phase II metabolizing enzymes in ddY male mice: possible role in protection against chemical carcinogenesis and toxicity. Pharmacol Toxicol. 2003;92:33–8.
68. Kotyczka C, Boettler U, Lang R, Stiebitz H, Bytof G, Lantz I, Hofmann T, Marko D, Somoza V. Dark roast coffee is more effective than light roast coffee in reducing body weight, and in restoring red blood cell vitamin E and glutathione concentrations in healthy volunteers. Mol Nutr Food Res. 2011;55(10):1582–6.
69. Singh A, Rao A. Evaluation of the modulatory influence of black pepper (Piper nigrum, L.) on the hepatic detoxification system. Cancer Lett. 1992;72:5–9.
70. Soyal D, Jindal A, Singh IG. Modulation of radiation-induced biochemical alterations in mice by rosemary (Rosemarinus officinalis) extract. Phytomedicine. 2007;14:701–5.
71. Abel E, Boulware S, Fields T, McIvor E, Powell K, Digiovanni J, Vasquez KM, Macleod MC. Sulforaphane induces phase II detoxification enzymes in mouse skin and prevents mutagenesis induced by a mustard gas analog. Toxicol Appl Pharmacol. 2013;266(3): 439–42.
72. Ibrahim K, Saleh Z, Garrag A, Shaban E. Protective effects of zinc and selenium against benzene toxicity in rats. Toxicol Ind Health. 2011;27(6):537–45.
73. Lee I, Tran M, Evans-Nguyen T, Stickle D, Kim S, Han J, Park JY, Yang M. Detoxification of chlorella supplement on heterocyclic amines in Korean young adults. Environ Toxicol Pharmacol. 2015;39:441–6.
74. Li L, Li W, Kim Y, Lee YW. Chlorella vulgaris extract ameliorates carbon tetrachloride-induced acute hepatic injury in mice. Exp Toxicol Pathol. 2013;65(1–2):73–80.
75. Lii C, Lui K, Cheng Y, Lin A, Chen H, Tsai C. Sulforaphane and alpha-lipoic acid upregulate the expression of the pi class of glutathione S-transferase through c-jun and Nrf2 activation. J Nutr Health Aging. 2010;140:219–23.
76. Sears M, Kerr K, Bray R. Arsenic, cadmium, lead, and mercury in sweat: a systematic review. J Environ Public Health. 2012;2012: 184745.
77. Yoshida K, Ushida Y, Ishijima T, Suganuma H, Inakuma T, Yajima N, Abe K, Nakai Y. Broccoli sprout extract induces detoxification-related gene expression and attenuates acute liver injury. World J Gastroenterol. 2015;21:10091–103.

Drug–Nutrient Interactions

Mary Demarest Litchford

15.1 Effect of Food and Nutrients on Drug Kinetics and Efficacy – 214

15.2 Effect of Drugs on Food and Nutrient Kinetics and Nutrition Status – 215

15.3 Role of the Nutrition Professional – 215

References – 218

Prescription and non-prescription drugs have improved the health and well-being of humans. Yet, the potential for alterations in drug performance and/or changes in nutritional status exist. A drug interaction is a situation in which a substance affects the kinetics of a drug or produces a new side effect. Bioactive components in some food have the potential to interact with drugs and either reduce or enhance pharmaceutical effects. Dietary supplements, herbals, and botanicals may contribute additional interactions with drugs. Typically, drug–nutrient interactions are considered adverse side effects [1–4].

> The International Dietetics & Nutrition Terminology Reference Manual [5] defines a food-medication interaction as an 'undesirable or harmful interaction(s) between food and over-the-counter medications, prescribed medications, herbals, botanicals and/or dietary supplements that diminishes, enhances, or alters the effect of nutrients and/or medications.'

Identification of potential risk for drug–nutrient interactions is an essential component of the comprehensive nutrition assessment performed during the nutrition care process. The likelihood of interactions may be increased when the patient is malnourished, has an underlying illness, takes botanical and herbal supplements, consumes alcohol daily, has food allergies or food intolerances, follows a restrictive therapeutic diet, has health beliefs that limit food choices, takes more than two medications, does not follow medication instructions, and/or is a growing child or older adult. Individuals with acute and chronic inflammatory conditions are at risk for sub-optimal serum albumin levels. Albumin is the most important drug-binding protein in the body. Hypoalbuminemia diminishes the number of drug-binding receptor sites and may result in reduced drug bioavailability [3].

It is in the patient's best interest to minimize drug–nutrient interactions. Patients who avoid these interactions are more likely to experience the drug's intended effect and less prone to discontinue taking the drug earlier than recommended. Avoiding drug-induced nutrient deficiencies helps to maintain nutritional status, avoid falls, and injuries that may be caused in part by nutrient imbalances [6, 7].

Not all patients have optimal nutritional status when a new drug is recommended. The undernourished individual may have nutrient insufficiencies that evolve into frank deficiencies due to a drug-induced adverse effect. Malnutrition with loss of lean muscle mass is of concern because of alterations in protein-binding, drug distribution and drug elimination. Drug distribution is the movement of an active drug from the bloodstream to the site of effect. It is affected by a number of factors, including lipophilicity and plasma protein binding. Drug elimination includes metabolism and excretion of the drug.

Malnourished individuals experience loss of fat mass, skeletal muscle mass and visceral muscle mass. The altered body composition has the potential to reduce transport proteins and regulatory hormones involved in drug distribution. The loss of visceral muscle mass contributes to the changes in cardiac output, reduced blood flow to the liver and reduced glomerular filtration rate that may alter drug elimination [8–10].

Individuals taking drugs for a long duration who experience insidious weight loss may be taking drug dosages based on a higher body weight. These individuals are at higher risk for drug–nutrient interactions. In obese patients, there is a risk for accumulation of fat-soluble drugs or a prolonged clearance of drugs resulting in increased risk for drug toxicity.

15.1 Effect of Food and Nutrients on Drug Kinetics and Efficacy

Food and dietary supplements may alter drug kinetics and bioavailability. The bioavailability of a drug is the amount of the drug that reaches systemic circulation. Drugs taken orally have a lower bioavailability than drugs administered intravenously.

The presence of food and nutrients in the stomach and small intestine may increase, decrease or have no effect on the bioavailability of the drug. For example, immediate-release bisphosphonates, such as alendronate sodium, taken with food, significantly reduce drug absorption [11]. However, delayed-release bisphosphonates, such as risedronate delayed-release, may be taken before or after a meal without significantly reducing drug absorption [12].

Furanocoumarins found in grapefruit segments, grapefruit juice, Seville oranges, tangelos, minneolas, and other exotic oranges inhibit the actions of cytochrome P450 enzymes required for oxidative metabolism of numerous drugs. This interaction is of greatest concern for oral drugs with low bioavailability. Moreover, the effects of grapefruit segments and grapefruit juice on the actions of cytochrome P450 enzymes can last up to 72 hours [3, 4].

The presence of food and nutrients in the gut may enhance the drug bioavailability. For example, the absorption of cefuroxime axetil (antibiotic) is increased when taken with a meal versus a fasting state [13]. The bioavailability of iron sulfate supplements is enhanced if taken with food or with ascorbic acid. However, certain food components and nutrients may inhibit iron absorption, including high phytate foods, bran, fiber supplements, coffee, tea, dairy products, and calcium supplements [14, 15].

Drug bioavailability may be altered when achlorhydria or hypochlorhydria persists either due to the action of another drug or because of a medical condition. Drugs used to treat chronic acid suppression raise the pH of the stomach. The higher pH prevents drugs such as ketoconazole (antifungal) from reaching its optimal effect [3, 4]. Table 15.1 summarizes common effects of food and nutrients on drug kinetics.

Table 15.1 Common effect of food and nutrients on drug kinetics

Drug	Food, macronutrient or micronutrient	Potential food–drug interaction
Antibiotics	Milk	Calcium and magnesium in milk may complex with drug and reduce bioavailability [16–18]
Anticonvulsants	Grapefruit juice, grapefruit segments, Seville oranges, tangelos, minneolas, and other exotic oranges	Reduce bioavailability by inhibiting the actions of the cytochrome P450 3A enzymes [19]
Antihypertensives	Licorice	Licorice may cause hypermineralocorticoidism with sodium retention, increased potassium loss, edema, increased blood pressure and depression of the renin-angiotensin-aldosterone system [20]
Calcium channel drugs with Calcium Channel drugs HMG-CoA Reductase Inhibitors	Grapefruit juice, grapefruit segments, Seville oranges, tangelos, minneolas, and other exotic oranges	Reduce bioavailability by inhibiting the actions of the cytochrome P450 3A enzymes required for oxidative metabolism of numerous drugs [3, 4, 19]
Celiprolol (beta-blocker)	Orange juice	Hesperidin, present in orange juice, is responsible for the decreased absorption [21]
Monoamine oxidase inhibitors	Tyramine-containing foods	Consuming foods containing tyramine with MAOI may trigger a hypertensive crisis [22]
Psychotropics	Grapefruit juice	Components in grapefruit juice interfere with the intestinal efflux transporter P-glycoprotein (P-gp) [23, 24]
Warfarin	Foods rich in vitamin K	Inconsistent intakes of vitamin K rich foods may alter the effectiveness and safety of warfarin [25, 26]
	Cranberry juice	Consumption of cranberry juice is reported to alter the effectiveness and safety of warfarin in some individuals [27, 28]

15.2 Effect of Drugs on Food and Nutrient Kinetics and Nutrition Status

Drugs have the potential to alter food and nutrient intake and kinetics. Nutrients are essential for metabolic processes, and micronutrients reserves, or pools, are quickly depleted when the metabolic rate is increased, absorption and utilization of key nutrients are reduced, or excretion of nutrients is increased.

Drugs have the potential to impact nutrition status in many ways. Many prescription and non-prescription drugs reduce appetite, which reduces total nutrient intake. Other drugs increase the appetite for all food or specific categories of foods, e.g., refined carbohydrates, resulting in excessive energy and refined sugar intake. Moreover, drugs can reduce the absorption of key nutrients in the gastrointestinal tract in a variety of ways, including altering the stomach pH, binding the nutrient into an unusable form, and damaging the absorptive surfaces.

Drugs may increase the metabolism of nutrients, thereby increasing requirements and depleting nutrient reserves. Moreover, drugs may block the conversion of a pre-vitamin to its active form. Key nutrients may be lost in urine and feces. Drugs may increase or decrease urinary excretion. An increase in urinary excretion is typically due to a reduction in reabsorption of the nutrient. Drugs that decrease normal nutrient excretion of sodium may result in water retention.

Drugs that cause damage to the absorptive surfaces have the greatest potential to affect nutrient absorption. Common offenders include chemotherapeutic agents, nonsteroidal anti-inflammatory drugs, and prolonged antibiotic therapy. Table 15.2 summarizes potential drug-induced nutrient deficiencies.

15.3 Role of the Nutrition Professional

Malnutrition and nutrient deficiencies are often viewed as problems unique to developing countries and regions of the world affected by environmental disasters, famine, or political unrest. However, malnutrition and nutrient deficiencies are seen globally. Malnutrition diagnoses may be overlooked because the medical team is not mindful of the potential for nutrient losses to occur. It is essential to recognize that some drug-induced nutrient insufficiencies and deficiencies are insidious and others develop quickly. Drug-induced nutrient deficiencies are compounded by malnutrition. The early signs and symptoms of nutrient insufficiencies and deficiencies are often nonspecific and may be overlooked or misdiagnosed. Laboratory assessments used concurrently with

Table 15.2 Common effects of drugs on food and nutrient kinetics

Drug category	Macronutrient or micronutrient loss	Potential consequences of food–drug interaction
Antacids, magnesium and aluminum, calcium carbonate, proton pump inhibitors (PPI)	Calcium, magnesium, phosphorus, folic acid, copper, iron, vitamin B12	Increased stomach pH and reduced absorption of key nutrients that are best absorbed in the duodenum with a low pH including folic acid [29], calcium, phosphorus, copper, and iron [30, 31]
		It is unclear how PPI's promote hypomagnesemia [30–32]
		Aluminum can bind the phosphate in small intestine, thus lowering serum levels. The body responds by releasing calcium and phosphorus stores from the bones. Calcium levels are tightly controlled in the blood. Excess calcium is excreted in the urine [30]
		Increased pH impairs the body's ability to cleave vitamin B12 from its protein carrier in order to be transported via intrinsic factor (IF). IF is synthesized by the parietal cells in the stomach in the presence of a low pH. An increased pH reduces the synthesis of IF, which will result in reduced absorption of vitamin B12 [33]
Antiarrhythmic: digoxin	Magnesium	Digoxin promotes increased renal excretion of magnesium [34]
Antibiotics, sulfonamide combination drugs	Folic acid	May interfere with folic acid metabolism if taken for a prolonged period of time [35]
Antibiotics, fluoroquinolones	Magnesium, iron, zinc, calcium	Drug binds to iron, magnesium, zinc, and calcium creating insoluble complexes [36, 37]
Antibiotics, tetracyclines	Magnesium, iron, zinc, calcium, vitamin K, B complex vitamins	Drugs binds to iron, magnesium, zinc, and calcium creating insoluble complexes. May reduce bacterial synthesis of vitamin K2, menaquinone, in the colon. Long-term use may result in depletion of B vitamin stores [38]
Anticonvulsants	Vitamin B6, vitamin B12, folate	May interfere with vitamin B6, vitamin B12, and folate absorption, resulting in lower serum levels [39, 40]
Anticonvulsants	Biotin	May accelerate catabolism of biotin resulting in lower serum levels [41]
Anticonvulsants	Vitamin D	Lower serum levels reported possibly related to low bone density and osteomalacia [42]
Anticonvulsants	Calcium	Reduced absorption possibly related to vitamin D deficiency [43]
Anticonvulsants	Vitamin K	Drugs may decrease half-life of vitamin K and impair its key functions [44, 45].
Antihyperglycemic metformin	Vitamin B12	Metformin appears to inhibit the absorption of vitamin B12 [46]
Antihypertensive: ACEI angiotensin-converting enzyme inhibitor; ARB, angiotensin receptor blocker	Zinc	ACEI and ARB therapy has been shown to increase urinary excretion of zinc [47]
Antihypertensive: ACEI angiotensin-converting enzyme inhibitor; ARB, angiotensin receptor blocker	Potassium	ACEI and ARBs are associated with increased serum potassium, which may or may not be offset by the reduction of potassium due to loop diuretics [48, 49]
Antihypertensive: hydralazine	Vitamin B6, copper	Hydralazine may interfere with vitamin B6 metabolism. It may promote increased excretion of copper [50, 51]
Antihypertensive: RAAS renin-angiotensin-aldosterone system	Potassium	RAAS have the potential to cause hyperkalemia by interfering with the production and secretion of aldosterone [52–56]
Antimanic: lithium	Sodium	Lithium may increase sodium excretion [57]

Drug–Nutrient Interactions

Table 15.2 (continued)

Drug category	Macronutrient or micronutrient loss	Potential consequences of food–drug interaction
Antineoplastic: methotrexate	Folic acid	Methotrexate is a folic acid antagonist that interferes with nutrient utilization [58, 59]
Antiplatelet agents	Iron, folic acid, sodium, potassium, vitamin B12	Long-term use associated with reduced levels of iron, folic acid, sodium, potassium, vitamin B12 [60, 61]
Antipsychotics, phenothiazine class, tricyclic antidepressants	Riboflavin	Drug increases the excretion of riboflavin that may lead to deficiency in individuals with insufficient riboflavin intakes [62]
Antitubular: isoniazid	Vitamin B6, niacin (B3), vitamin D, calcium, phosphate	Drug may deplete vitamin B6 and niacin stores resulting in peripheral neuropathy and pellagra [63, 64]
		May impair vitamin D metabolism and consequently reducing calcium and phosphate absorption [65]
Beta-adrenergic blockers (beta-blockers)	Potassium	Beta-blockers have the potential to cause hyperkalemia by causing redistribution of potassium from the intracellular space into the serum [66, 67]
Beta-2 agonists	Magnesium, potassium	Reduced serum levels of magnesium and potassium reported. The degree of deficiency is exacerbated when beta-2 agonist is taken with theophylline [68, 69]
Bile acid sequestrants	Vitamins A, D, E, K, beta-carotene, iron	Bile acid sequestrants bind fat soluble vitamins, beta-carotene, and iron [70]
Bile acid sequestrants	Magnesium, iron, calcium, zinc and folic acid	Alterations in calcium, magnesium, and zinc metabolism may be explained by inadequate vitamin D absorption in the duodenum followed by an increased secretion of parathyroid hormone [71]
Bronchodilator: theophylline	Vitamin B6, potassium, magnesium	Reduced levels of pyridoxal phosphate may be related to altered tryptophan metabolism or impaired vitamin B6 utilization. Reduced levels of potassium and magnesium have been reported, possibly related to increase urinary excretion [72–75]
Colchicine (antigout)	Vitamin B12	In animals, colchicine may reduce vitamin B12 absorption and efficiency of ileal receptor sites leading to a vitamin B12 insufficiency or deficiency [76, 77]
Diuretics: loop	Sodium	Loop diuretics reduce sodium reabsorption in the proximal tubule. Patients who are prescribed a sodium-restricted diet as part of medical management of hypertension are at greater risk of hyponatremia [57]
Diuretics: loop	Potassium	Loop diuretics reduce potassium reabsorption at the site of action and enhance potassium secretion in the distal tubules of the nephron. In addition, aldosterone can also contribute to hypokalemia after administration of loop diuretics [78]
Diuretics: loop	Magnesium	Loop diuretics reduce magnesium reabsorption in the loop of Henle and proximal tubule. It is also dependent on sodium and chloride concentrations. Magnesium depletion promotes the efflux of potassium from cells and subsequent urinary excretion [79–81]
Diuretics: loop	Thiamine	Long-term use is associated with reduced levels of thiamine. Loop diuretics promote thiamine losses up to twice baseline loss. Increased loss is associated with an increase in urine flow rate, but it is not related to sodium excretion. Up to 1/3 of CHF patients were found to be thiamine deficient [82–88]
Diuretics: loop	Zinc	Long-term use of loop diuretics reduce zinc reabsorption in the proximal tubule [89]
Diuretics: loop	Calcium	Loop diuretics reduce calcium reabsorption in the proximal tubule. It is also dependent on sodium and chloride concentrations [90]
Diuretics: thiazide	Calcium	Thiazide diuretics reduce calcium reabsorption in the proximal tubule. It is also dependent on sodium and chloride concentrations [90]

(continued)

Table 15.2 (continued)

Drug category	Macronutrient or micronutrient loss	Potential consequences of food–drug interaction
Diuretics: thiazide	Sodium	Thiazide diuretics reduce sodium reabsorption in the proximal tubule. Patients who are prescribed a sodium-restricted diet as part of medical management of hypertension are at greater risk of hyponatremia [57]
Diuretics: thiazide	Potassium	Thiazide diuretics reduce potassium reabsorption at the site of action and enhance potassium secretion in the distal tubules of the nephron. In addition, aldosterone can also contribute to hypokalemia after administration of loop diuretics [91]
Diuretics: thiazide	Magnesium	Thiazide diuretics reduce magnesium reabsorption in the loop of Henle and proximal tubule. It is also dependent on sodium and chloride concentrations. Magnesium depletion promotes the efflux of potassium from cells and subsequent urinary excretion [79–81]
Diuretics: thiazide	Zinc	Long-term use of thiazide diuretics reduce zinc reabsorption in the proximal tubule [92]
Glucocorticoids	Zinc	Glucocorticoids may promote development of zinc deficiency in some patients [89, 90]
Glucocorticoids	Calcium, Chromium	Glucocorticoids increase urinary losses of chromium and calcium [93, 94]
Immunosuppressant: cyclosporine	Magnesium	Cyclosporine promotes renal magnesium wasting [95]
Immunosuppressant: Hydroxychloroquine	Vitamin D	Hydroxychloroquine may inhibit the conversion of 25-hydroxyvitamin D to the active form, i.e., 1,25 dihydroxyvitamin D [96]
Laxatives, cathartics	Calcium, potassium	Laxatives reduce transit time in the gut leading to diarrhea and increased fecal loss of calcium and potassium [97]
Mineral oil laxatives	Vitamins A, D, E, K	Mineral oil-based laxatives may reduce the absorption of fat soluble vitamins [98]
Peripheral vasodilator	Vitamin B6	Peripheral vasodilators interfere with vitamin B6 metabolism and may result in lower levels [50, 99]

nutrition-focused physical exams are essential tools to detect drug-induced nutrient insufficiencies and deficiencies.

The nutrition professional's approach to detect drug-induced nutrient insufficiencies and deficiencies is determined by the patient's health history. For example, patients who are starting on a new drug are looked at prospectively. The nutrition professional uses clinical judgment to predict the potential for drug-induced nutrient insufficiencies and deficiencies by identifying specific strategies and interventions to prevent or compensate for nutrient losses. Moreover, foods or food intake patterns associated with reduced drug absorption are identified and discussed as part of patient education. The nutrition professional will monitor the readiness of the patient to incorporate specific strategies and interventions as well as the health outcomes. Adjustments in the interventions are often required.

Patients who have been on specific drugs for an extended period of time are assessed retrospectively. The nutrition professional uses clinical judgment to detect signs and symptoms of drug-induced nutrient insufficiencies and deficiencies using historical data. Trends in laboratory results may indicate suspected nutrient insufficiencies that are confirmed with nutrition-focused physical exam. Specific interventions are recommended by the nutrition professional to compensate for nutrient losses. Monitoring and evaluation of changes in nutrition status are essential to determine the efficacy of interventions.

The nutrition professional does not work in a vacuum. As a member of an integrative healthcare team, the nutrition professional provides valuable insight and findings to improve the health and well-being of patients.

References

1. Bushra R, Aslam N, Khan AY. Food-drug interactions. Oman Med J. 2011;26(2):77–83.
2. Pelton R, LaValle J. The nutritional cost of drugs. Englewood: Morton Publishing; 2004.
3. Pronsky Z, Elbe D, Ayoob K. Food medication interactions. 18th ed. Burchrunville: Author; 2015.
4. McCabe BJ, Frankel EH, Wolfe JJ. Monitoring nutritional status in drug regimens. In: BJ MC, Frankel EH, Wolfe JJ, editors. Handbook of food-drug interactions. Boca Raton: CRC Press; 2003. p. 73–108.
5. Academy of Nutrition & Dietetics. International dietetics & nutrition terminology reference manual. 4th ed. Chicago: Academy of Nutrition & Dietetics; 2013.
6. Dharmarajanm TS, et al. Anemia increases risk for falls in hospitalized older adults: an evaluation of falls in 362 hospitalized, ambulatory, long-term care, and community patients. JAMDA. 2007;8(3 Supplement 2):e9–e15.
7. Machon M, et al. Dietary patterns and their relationship with frailty in functionally independent older adults. Nutrients. 2018;10(4):406.

8. Boullata J. Drug disposition in obesity and protein–energy malnutrition. Proc Nutr Soc. 2010;69(4):543–50.
9. Santos CA, Boullata JI. An approach to evaluating drug–nutrient interactions. Pharmacotherapy. 2005;25:1789–800.
10. Mehta S. Malnutrition and drugs: clinical implications. Dev Pharmacol Ther. 1990;15:159–65.
11. Mitchell DY, Barr WH, Eusebio RA, Stevens KA, Duke FP, Russell DA, Nesbitt JD, Powell JH, Thompson GA. Risedronate pharmacokinetics and intra-and inter-subject variability upon single-dose intravenous and oral administration. Pharm Res. 2001;18(2):166–70.
12. Pazianas M, Abrahamsen B, Ferrari S, Russell RG. Eliminating the need for fasting with oral administration of bisphosphonates. Ther Clin Risk Manag. 2013;9:395.
13. Finn A, Straughn A, Meyer M, Chubb J. Effect of dose and food on the bioavailability of cefuroxime axetil. Biopharm Drug Dispos. 1987;8(6):519–26.
14. Hallberg L, Brune M, Erlandsson M, Sandberg AS, Rossander-Hulten L. Calcium: effect of different amounts on nonheme-and heme-iron absorption in humans. Am J Clin Nutr. 1991;53(1):112–9.
15. Hallberg L, Rossander-Hulthèn L, Brune M, Gleerup A. Inhibition of haem-iron absorption in man by calcium. Br J Nutr. 1993;69(02):533–40.
16. Papai K, Budai M, Ludányi K, Antal I, Klebovich I. In vitro food–drug interaction study: which milk component has a decreasing effect on the bioavailability of ciprofloxacin? J Pharm Biomed Anal. 2010;52(1):37–42.
17. Gurley BJ, Hagan DW. Herbal and dietary supplement interactions with drugs. In: BJ MC, Frankel EH, Wolfe JJ, editors. Handbook of food-drug interactions. Boca Raton: CRC Press; 2003. p. 259–93.
18. Cardona PD. Drug-food interactions. Nutr Hosp. 1999;(Suppl 2):129S–40S.
19. Li P, Callery PS, Gan LS, Balani SK. Esterase inhibition by grapefruit juice flavonoids leading to a new drug interaction. Drug Metab Dispos. 2007;35(7):1203–8.
20. Størmer FC, Reistad R, Alexander J. Glycyrrhizic acid in liquorice—evaluation of health hazard. Food Chem Toxicol. 1993;31(4):303–12.
21. Uesawa Y, Mohri K. Hesperidin in orange juice reduces the absorption of celiprolol in rats. Biopharm Drug Dispos. 2008;29(3):185–8.
22. Walker SE, Shulman KI, Tailor SA, Gardner D. Tyramine content of previously restricted foods in monoamine oxidase inhibitor diets. J Clin Psychopharmacol. 1996;16(5):383–8.
23. Genser D. Food and drug interaction: consequences for the nutrition/health status. Ann Nutr Metab. 2008;52(Suppl. 1):29–32.
24. Sharif SI, Ali BH. Effect of grapefruit juice on drug metabolism in rats. Food Chem Toxicol. 1994;32(12):1169–71.
25. Hornsby LB, Hester EK, Donaldson AR. Potential interaction between warfarin and high dietary protein intake. Pharmacotherapy. 2008;28(4):536–9.
26. Holt GA. Food & drug interactions, vol. 1998. Chicago: Precept Press; 1998. p. 293.
27. Zikria J, Goldman R, Ansell J. Cranberry juice and warfarin: when bad publicity trumps science. Am J Med. 2010;123(5):384–92.
28. Ansell J, McDonough M, Zhao Y, Harmatz JS, Greenblatt DJ. The absence of an interaction between warfarin and cranberry juice: a randomized, double-blind trial. J Clin Pharmacol. 2009;49(7):824–30.
29. Russell RM, Golner BB, Krasinski SD, Sadowski JA, Suter PM, Braun CL. Effect of antacid and H2 receptor antagonists on the intestinal absorption of folic acid. J Lab Clin Med. 1988;112(4):458–63.
30. Ito T, Jensen RT. Association of long-term proton pump inhibitor therapy with bone fractures and effects on absorption of calcium, vitamin B12, iron, and magnesium. Curr Gastroenterol Rep. 2010;12(6):448–57.
31. McColl KE. Effect of proton pump inhibitors on vitamins and iron. Am J Gastroenterol. 2009;104:S5–9.
32. Famularo G, Gasbarrone L, Minisola G. Hypomagnesemia and proton-pump inhibitors. Expert Opin Drug Saf. 2013;12(5):709–16.
33. Dharmarajan TS, Kanagala MR, Murakonda P, Lebelt AS, Norkus EP. Do acid-lowering agents affect vitamin B12 status in older adults? J Am Med Dir Assoc. 2008;9(3):162–7.
34. Flink EB. Magnesium deficiency in human subjects—a personal historical perspective. J Am Coll Nutr. 1985;4(1):17–31.
35. Zimmerman J, Selhub J, Rosenberg IH. Competitive inhibition of folate absorption by dihydrofolate reductase inhibitors, trimethoprim and pyrimethamine. Am J Clin Nutr. 1987;46(3):518–22.
36. Pletz MW, Petzold P, Allen A, Burkhardt O, Lode H. Effect of calcium carbonate on bioavailability of orally administered gemifloxacin. Antimicrob Agents Chemother. 2003;47(7):2158–60.
37. Neuhofel AL, Wilton JH, Victory JM, Hejmanowski LG, Amsden GW. Lack of bioequivalence of ciprofloxacin when administered with calcium-fortified orange juice: a new twist on an old interaction. J Clin Pharmacol. 2002;42(4):461–6.
38. Mapp RK, McCarthy TJ. The effect of zinc sulphate and of bicitropeptide on tetracycline absorption. South African Med J = Suid-Afrikaanse tydskrif vir geneeskunde. 1976;50(45):1829–30.
39. Hansson O, Sillanpaa M. Pyridoxine and serum concentration of phenytoin and phenobarbitone. Lancet. 1976;307(7953):256.
40. Tamura T, Aiso K, Johnston KE, Black L, Faught E. Homocysteine, folate, vitamin B-12 and vitamin B-6 in patients receiving antiepileptic drug monotherapy. Epilepsy Res. 2000;40(1):7–15.
41. Mock DM, Dyken ME. Biotin catabolism is accelerated in adults receiving long-term therapy with anticonvulsants. Neurology. 1997;49(5):1444–7.
42. Mikati MA, Dib L, Yamout B, Sawaya R, Rahi AC, Fuleihan GE. Two randomized vitamin D trials in ambulatory patients on anticonvulsants impact on bone. Neurology. 2006;67(11):2005–14.
43. Wahl TO, Gobuty AH, Lukert BP. Long-term anticonvulsant therapy and intestinal calcium absorption. Clin Pharmacol Ther. 1981;30(4):506–12.
44. Wilson AC, Park BK. The effect of phenobarbitone pre-treatment on vitamin K1 disposition in the rat and rabbit. Biochem Pharmacol. 1984;33(1):141–6.
45. Park BK, Wilson AC, Kaatz G, Ohnhaus EE. Enzyme induction by phenobarbitone and vitamin K1 disposition in man. Br J Clin Pharmacol. 1984;18(1):94–7.
46. Wulffele MG, Kooy A, Lehert P, Bets D, Ogterop JC, Burg B, Donker AJ, Stehouwer CD. Effects of short-term treatment with metformin on serum concentrations of homocysteine, folate and vitamin B12 in type 2 diabetes mellitus: a randomized, placebo-controlled trial. J Intern Med. 2003;254(5):455–63.
47. Koren-Michowitz M, Dishy V, Zaidenstein R, Yona O, Berman S, Weissgarten J, Golik A. The effect of losartan and losartan/hydrochlorothiazide fixed-combination on magnesium, zinc, and nitric oxide metabolism in hypertensive patients: a prospective open-label study. Am J Hypertens. 2005;18(3):358–63.
48. D'costa DF, Basu SK, Gunasekera NP. ACE inhibitors and diuretics causing hypokalaemia. Br J Clin Pract. 1990;44(1):26–7.
49. Griffing GT, Melby JC. Reversal of diuretic-induced secondary hyperaldosteronism and hypokalemia by trilostane, an inhibitor of adrenal steroidogenesis. Metabolism. 1989;38(4):353–6.
50. Raskin NH, Fishman RA. Pyridoxine-deficiency neuropathy due to hydralazine. N Engl J Med. 1965;273(22):1182–5.
51. Wester PO. The urinary excretion of trace elements before and during treatment with hydralazine. J Intern Med. 1975;197(1–6):307–9.
52. Konstam MA, Kronenberg MW, Rousseau MF, Udelson JE, Melin J, Stewart D, Dolan N, Edens TR, Ahn S, Kinan D. Effects of the angiotensin converting enzyme inhibitor enalapril on the long-term progression of left ventricular dilatation in patients with asymptomatic systolic dysfunction. SOLVD (Studies of Left Ventricular Dysfunction) investigators. Circulation. 1993;88(5):2277–83.
53. Cohn JN, Johnson G, Ziesche S, Cobb F, Francis G, Tristani F, Smith R, Dunkman WB, Loeb H, Wong M, Bhat G. A comparison of enalapril with hydralazine–isosorbide dinitrate in the treatment of chronic congestive heart failure. N Engl J Med. 1991;325(5):303–10.
54. Pitt B, Zannad F, Remme WJ, Cody R, Castaigne A, Perez A, Palensky J, Wittes J. The effect of spironolactone on morbidity and mortality in patients with severe heart failure. N Engl J Med. 1999;341(10):709–17.

55. Pitt B, Williams G, Remme W, Martinez F, Lopez-Sendon J, Zannad F, Neaton J, Roniker B, Hurley S, Burns D, Bittman R. The EPHESUS trial: eplerenone in patients with heart failure due to systolic dysfunction complicating acute myocardial infarction. Cardiovasc Drugs Ther. 2001;15(1):79–87.
56. Pfeffer MA, Swedberg K, Granger CB, Held P, McMurray JJ, Michelson EL, Olofsson B, Östergren J, Yusuf S. CHARM Investigators and Committees. Effects of candesartan on mortality and morbidity in patients with chronic heart failure: the CHARM-overall programme. Lancet. 2003;362(9386):759–66.
57. Liamis G, Milionis H, Elisaf M. A review of drug-induced hyponatremia. Am J Kidney Dis. 2008;52(1):144–53.
58. Morgan SL, Baggott JE, Vaughn WH, Austin JS, Veitch TA, Lee JY, Koopman WJ, Krumdieck CL, Alarcon GS. Supplementation with folic acid during methotrexate therapy for rheumatoid arthritis: a double-blind, placebo-controlled trial. Ann Intern Med. 1994;121(11):833–41.
59. Van Ede AE, Laan RF, Rood MJ, Huizinga TW, Van De Laar MA, Denderen CJ, Westgeest TA, Romme TC, De Rooij DJ, Jacobs MJ, De Boo TM. Effect of folic or folinic acid supplementation on the toxicity and efficacy of methotrexate in rheumatoid arthritis: a forty eight-week, multicenter, randomized, double-blind, placebo-controlled study. Arthritis Rheum. 2001;44(7):1515–24.
60. Ambanelli U. Changes in serum and urinary zinc induced by ASA and indomethacin. Scand J Rheumatol. 1982;11(1):63–4.
61. van Oijen MG, Laheij RJ, Peters WH, Jansen JB, Verheugt FW. Association of aspirin use with vitamin B 12 deficiency (results of the BACH study). Am J Cardiol. 2004;94(7):975–7.
62. Pinto JT, Rivlin RS. Drugs that promote renal excretion of riboflavin. Drug Nutr Interact. 1987;5(3):143–51.
63. Biehl JP, Vilter RW. Effect of isoniazid on pyridoxine metabolism. JAMA. 1954;156(17):1549–52.
64. Biehl JP, Vilter RW. Effect of isoniazid on vitamin B6 metabolism; its possible significance in producing isoniazid neuritis. Nutr Rev. 1982;40(6):183–6.
65. Brodie MJ, Boobis AR, Hillyard CJ, Abeyasekera G, MacIntyre I, Park BK. Effect of isoniazid on vitamin D metabolism and hepatic monooxygenase activity. Clin Pharmacol Ther. 1981;30(3):363–7.
66. Bakris GL, Siomos M, Richardson D, et al. ACE inhibition or angiotensin receptor blockade: impact on potassium in renal failure. VAL-K Study Group. Kidney Int. 2000;58(5):2084–92.
67. Clausen T, Flatman JA. Beta 2-adrenoceptors mediate the stimulating effect of adrenaline on active electrogenic Na-K-transport in rat soleus muscle. Br J Pharmacol. 1980;68(4):749–55.
68. Phillips PJ, Vedig AE, Jones PL, Chapman MG, Collins M, Edwards JB, Smeaton TC, Duncan BM. Metabolic and cardiovascular side effects of the beta 2-adrenoceptor agonists salbutamol and rimiterol. Br J Clin Pharmacol. 1980;9(5):483–91.
69. Smith SR, Gove I, Kendall MJ, Berkin KE, Mcinnes GT, Thomson NC, Ball SG, Cayton R. Beta agonists and potassium. Lancet. 1985;325(8442):1394–5.
70. West RJ, Lloyd JK. The effect of cholestyramine on intestinal absorption. Gut. 1975;16(2):93–8.
71. Watkins DW, Khalafi R, Cassidy MM, Vahouny GV. Alterations in calcium, magnesium, iron, and zinc metabolism by dietary cholestyramine. Dig Dis Sci. 1985;30(5):477–82.
72. Ubbink JB, Vermaak WH, Delport R, Serfontein WJ, Bartel P. The relationship between vitamin B6 metabolism, asthma, and theophylline therapy. Ann N Y Acad Sci. 1990;585(1):285–94.
73. Bartel PR, Ubbink JB, Delport R, Lotz BP, Becker PJ. Vitamin B6 supplementation and theophylline-related effects in humans. Am J Clin Nutr. 1994;60(1):93–9.
74. Weir MR, Keniston RC, Enriquez JI, McNamee GA. Depression of vitamin B6 levels due to theophylline. Ann Allergy. 1990;65(1):59–62.
75. Flack JM, Ryder KW, Strickland D, Whang R. Metabolic correlates of theophylline therapy: a concentration-related phenomenon. Ann Pharmacother. 1994;28(2):175–9.
76. Stopa EG, O'Brien R, Katz M. Effect of colchicine on Guinea pig intrinsic factor-vitamin B~2 receptor. Gastroenterology. 1979;76:309–14.
77. Gemici AI, Sevindik ÖG, Akar S, Tunca M. Vitamin B12 levels in familial Mediterranean fever patients treated with colchicine. Clin Exp Rheumatol. 2013;31(77):S57–9.
78. Henning R, Lundvall O. Evaluation in man of bumetanide, a new diuretic agent. Eur J Clin Pharmacol. 1973;6(4):224–7.
79. Ceremużyński L, Gębalska J, Wołk R, Makowska E. Hypomagnesemia in heart failure with ventricular arrhythmias. Beneficial effects of magnesium supplementation. J Intern Med. 2000;247(1):78–86.
80. Oladapo OO, Falase AO. Congestive heart failure and ventricular arrhythmias in relation to serum magnesium. Afr J Med Med Sci. 2000;29(3–4):265–8.
81. Swaminathan R. Magnesium metabolism and its disorders. Clin Biochem Rev. 2003;24(2):47.
82. Suter PM, Haller J, Hany A, Vetter W. Diuretic use: a risk for subclinical thiamine deficiency in elderly patients. J Nutr Health Aging. 1999;4(2):69–71.
83. Zenuk C, Healey J, Donnelly J, Vaillancourt R, Almalki Y, Smith S. Thiamine deficiency in congestive heart failure patients receiving long term furosemide therapy. The Canadian journal of clinical pharmacology=. J Can Pharmacol Clin. 2002;10(4):184–8.
84. Rieck J, Halkin H, Almog S, Seligman H, Lubetsky A, Olchovsky D, Ezra D. Urinary loss of thiamine is increased by low doses of furosemide in healthy volunteers. J Lab Clin Med. 1999;134(3):238–43.
85. Seligmann H, Halkin H, Rauchfleisch S, Kaufmann N, Tal R, Motro M, Vered ZV, Ezra D. Thiamine deficiency in patients with congestive heart failure receiving long-term furosemide therapy: a pilot study. Am J Med. 1991;91(2):151–5.
86. Sica DA. Loop diuretic therapy, thiamine balance, and heart failure. Congest Heart Fail. 2007;13(4):244–7.
87. Hanninen SA, Darling PB, Sole MJ, Barr A, Keith ME. The prevalence of thiamin deficiency in hospitalized patients with congestive heart failure. J Am Coll Cardiol. 2006;47(2):354–61.
88. Lubetsky A, Winaver J, Seligmann H, Olchovsky D, Almog S, Halkin H, Ezra D. Urinary thiamine excretion in the rat: effects of furosemide, other diuretics, and volume load. J Lab Clin Med. 1999;134(3):232–7.
89. Wester PO. Urinary zinc excretion during treatment with different diuretics. J Intern Med. 1980;208(1–6):209–12.
90. Friedman PA, Bushinsky DA. Diuretic effects on calcium metabolism. Semin Nephrol. 1999;19(6):551.
91. Hollifield JW, Slaton PE. Thiazide diuretics, hypokalemia and cardiac arrhythmias. J Intern Med. 1981;209(S647):67–73.
92. Goldey DH, Mansmann HC Jr, Rasmussen AI. Zinc status of asthmatic, prednisone-treated asthmatic, and non-asthmatic children. J Am Diet Assoc. 1984;84(2):157–63.
93. Ravina A, Slezak L, Mirsky N, Bryden NA, Anderson RA. Reversal of corticosteroid-induced diabetes mellitus with supplemental chromium. Diabet Med. 1999;16(2):164–7.
94. Reid IR, Ibbertson HK. Evidence for decreased tubular reabsorption of calcium in glucocorticoid-treated asthmatics. Horm Res Paediatr. 1987;27(4):200–4.
95. Allen RD, Hunnisett AG, Morris PJ. Cyclosporine and magnesium. Lancet. 1985;325(8440):1283–4.
96. Huisman AM, White KP, Algra A, Harth MA, Vieth RE, Jacobs JW, Bijlsma JW, Bell DA. Vitamin D levels in women with systemic lupus erythematosus and fibromyalgia. J Rheumatol. 2001;28(11):2535–9.
97. Coghill NF, McAllen PM, Edwards F. Electrolyte losses associated with the taking of purges investigated with aid of sodium and potassium radioisotopes. Br Med J. 1959;1(5113):14.
98. Haley's M. Drug and nutrient interactions. Am Fam Physician. 1991;44:1651–8.
99. Afifi AK, Sabra FA. Treatment of toxic and drug induced neuropathies. Mod Treat. 1968;5(6):1236–48.

The Enterohepatic Circulation

Robert C. Barton Jr.

16.1 Enterohepatic Circulation (EHC) - Definition – 222

16.2 Introduction – 222

16.3 Overview – 222

16.4 Details of Microanatomy, Physiology, and Biochemistry – 222
16.4.1 Portal Vein – 222
16.4.2 Portal Tracts – 222
16.4.3 Liver/Hepatic Lobule – 223
16.4.4 Hepatic Sinusoids and Hepatocytes – 223
16.4.5 Interaction of Sinusoidal Blood with the Hepatocyte Basolateral Cell Membranes – 223
16.4.6 Metabolic Zonation/Organization of Hepatocytes by their Function – 224
16.4.7 Functions of the Liver – 225
16.4.8 Xenobiotics – 225
16.4.9 Where Are Xenobiotics Found? – 226
16.4.10 A Few Examples of Problems Caused by Xenobiotics – 227
16.4.11 Biotransformation – 227
16.4.12 Cytochrome P450 Genes and Enzymes – 228
16.4.13 Single-Nucleotide Polymorphisms – 228
16.4.14 Bile and the Biliary System – 228
16.4.15 Small Intestine – 229
16.4.16 Microbiome/Metabolome – 230
16.4.17 Enteric Nervous System – 230
16.4.18 Gut-Associated Lymphoid Tissue – 231
16.4.19 Enteric Endocrine Cells – 231
16.4.20 Terminal Ileum – 231
16.4.21 Large Intestine – 231

16.5 Conclusion – 231

References – 232

© Springer Nature Switzerland AG 2020
D. Noland et al. (eds.), *Integrative and Functional Medical Nutrition Therapy*,
https://doi.org/10.1007/978-3-030-30730-1_16

> **Derivation of the name**
> Enteron - Intestine (Greek)
> Hepar - Liver (Greek)

16.1 Enterohepatic Circulation (EHC) - Definition

The circulation of bile acids, bilirubin, drugs, or other substances from the liver to the bile, followed by entry into the small intestine, absorption by the enterocyte and transport back to the liver. Recycling through liver by excretion in bile, reabsorption from intestines into portal circulation, passage back into liver, and re-excretion in bile [1].

16.2 Introduction

The purpose of this chapter is to provide insight into the functioning of the enterohepatic circulation (EHC) as it relates to the practice of clinical nutrition. The focus will not be on disease management; rather, it will be on ways in which improving the health and function of the tissues and organs of the EHC might be approached.

Health professionals, of any level of training or experience, are the intended audience. To be inclusive, the content may be helpful for those who aspire to become health professionals, those who want to understand human physiology in greater depth and anyone who is interested in learning how to create a better state of health. This chapter should help those with training in conventional Western medicine who employ a "mainstream" approach to healthcare, but who feel that there is more to patient care than they are equipped to deliver. It should help open channels of communication and new thought processes to investigate ways to deliver patient-focused healthcare and guide the practitioner to find a new group of like-minded professionals.

The foundation of what we do is basic science. Yet science is always growing. What we know now about the EHC is vastly different from what was known in 1923, the date of one of the earliest publications on the subject [2]. The role of the EHC in the recirculation and conservation of bile salts is foundational and was known from the beginning. But the awareness that the EHC also handles hormones, drugs, and other substances, both endogenous and exogenous, emerged over the next 40 years [3–8]. In the subsequent 50 years, these topics have been extensively studied and documented. The role of the EHC in the metabolism of xenobiotics (see below) began to emerge later. The first use of the term "xenobiotic" in the Medline database was in 1965 [9].

The term gut "microbiome" first appeared in 2002, and it became clear that colonization with healthy bacteria is critical for the normal structural and functional development, and the optimal function of the mucosal immune system [10]. Other areas of important research include the relationships between the microbiome, the gut-associated lymphoid tissue, the enteric nervous system, and the enteric endocrine cells in regard to their systemic effects [11, 12]. The metabolic effects of bile acids are also garnering increased attention [13, 14]. Thus, understanding of the functions of the EHC has increased dramatically over nearly 100 years, and we can only expect this to continue.

16.3 Overview

To refresh our understanding of the structure of the EHC, the following drawing and comments will provide an overview. The purpose of this section is to provide a basic, common understanding of the EHC as a system (Fig. 16.1).

16.4 Details of Microanatomy, Physiology, and Biochemistry

The following sections will present a much more detailed understanding of some of the individual components of the EHC, how they relate together, and how the EHC interacts with the rest of the body. This will help provide a framework for further study, and to provide a foundation for understanding why nutrition needs to be incorporated into the root of medical education. In one sense, nothing said here is new, but will aid in expanding understanding of nutritional therapies [15].

16.4.1 Portal Vein

The portal vein and its tributaries drain the entire intestinal tract from the gastroesophageal (GE) junction at the top of the stomach, all the way to the anus. This is a distance of 33 feet (11 yards) in the average adult. Through these veins, all the blood flowing from the stomach, small, and large intestines is carried directly to the liver. Veins, in general, carry blood toward the heart. In the case of the portal vein and its tributaries, the blood ultimately reaches the heart, but after being processed in the liver. This portal venous system is separate from and parallel to the systemic veins, which carry blood from all the other organs and tissues in the body directly to the heart.

16.4.2 Portal Tracts

In the liver, the portal vein branches like a large tree. The tiniest branches are called portal venules, which travel through the liver in a network of fibrous channels called portal tracts. In addition to the portal venules, the portal tracts carry tiny arterioles from the hepatic artery, which carry highly oxygenated blood directly from the heart. About 70% of the blood supply to the liver comes from the portal venules and so from the intestines, the remaining 30% from the hepatic arterioles (Fig. 16.2).

Also coursing through the portal tracts are bile ductules, which carry bile out of the liver; lymphatic channels, which

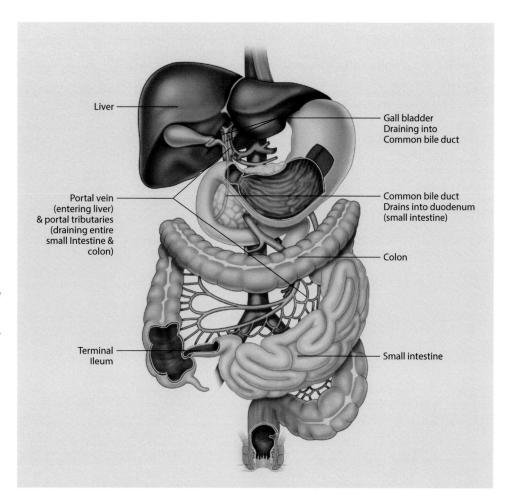

◻ Fig. 16.1 (1) *Portal vein & portal tributaries*. Carries blood from the intestinal tract into liver; (2) *Liver*. Functions: nutrients synthesized into metabolic compounds transported and utilized by organs; detoxification processes; stabilization of blood glucose; synthesization of ATP; bile production; (3) *Bile ducts and gallbladder*. Bile, the main secretory product of the liver, is stored and processed in the gallbladder; (4) *Small intestine*. Duodenum: first part of the small intestine receiving bile and pancreatic enzymes; jejunum and ileum: second and third portions of the small intestine where digestion occurs; (5) *Terminal ileum*. Final 6 inches of the small intestine where 95% of the bile is absorbed and recirculated to the liver, historically the origin of the term enterohepatic circulation; (6) *Large intestine or colon*. The final five feet of the intestine where stool is concentrated, other metabolic processes occur, and the resulting feces are discharged through the anus; further absorption of bile also occurs, entering the portal circulation. (Adapted from [49]. With permission from Wolters Kluwer Health)

carry lymphatic drainage away from the liver; and branches of the autonomic nervous system, evidence of the input of the autonomic nervous system into functioning of the liver on a cellular level.

16.4.3 Liver/Hepatic Lobule

Viewed under light microscopy, the liver parenchyma consists primarily of a single cell type, the hepatocyte or liver cell. Hepatocytes occur in extensive sheets of cells with an essentially uniform appearance, which are interwoven with the hepatic sinusoids. (See section below) The classic unit of hepatocyte organization is called the hepatic lobule. This structure is composed of six portal triads and a central vein, and can be seen under the microscope. However, because the lobule is a three dimensional structure in vivo , the classic hexagonal lobular structure is seldom seen in photomicrographs of a 2-dimensional cut section.

Each lobule consists of an average of six portal tracts forming the corners of a hexagon made up of hepatocytes, with a central hepatic venule (CV) or terminal hepatic venule (THV) located approximately in the center. Blood enters through the portal tracts at the periphery of the lobule, flows into the sinusoids adjacent to the hepatocytes, is processed while traversing the sinusoid, and then empties into the central hepatic venule.

The central hepatic venule is the smallest branch of the hepatic vein, which carries blood back to the heart.

16.4.4 Hepatic Sinusoids and Hepatocytes

Exiting the portal tracts, blood from the portal venules merges with blood from the hepatic arterioles and flows into a vast vascular web called the hepatic sinusoids, which extend throughout the liver (◻ Fig. 16.3). Blood flow into the sinusoids is regulated by specialized tissues in the portal venules and hepatic arterioles as they emerge from the portal tracts. Hormones, autonomic nervous input, and other factors influence this sinusoidal blood flow. Part of the cell membrane of each hepatocyte (the basolateral membrane) is adjacent to a sinusoid. It is through this basolateral membrane that transport of molecules into the hepatocytes occurs for processing.

16.4.5 Interaction of Sinusoidal Blood with the Hepatocyte Basolateral Cell Membranes

In the sinusoids, the liquid part of the blood, the serum, bathes the basolateral membranes of the hepatocytes. These cell membranes are highly specialized to carry out specific

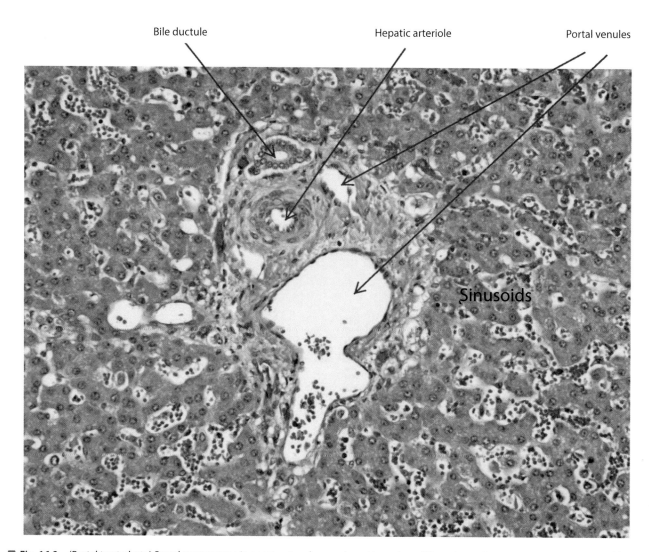

Fig. 16.2 (Portal tract photo) Portal tracts course in a vast network through the liver with the terminal branches of the portal venules carrying blood from the intestinal tract and hepatic arterioles carrying highly oxygenated blood from the heart. Portal blood mixes with hepatic arteriolar blood and flows into the sinusoids. Lymphatic channels and branches of the autonomic nervous system are not visible in this photomicrograph. The cells around the portal tract and adjacent to the sinusoids are periportal hepatocytes in which many synthetic and energy-producing steps occur. (Courtesy of Beverly B. Rogers MD, Emory University School of Medicine)

functions, and these functions vary as the blood flows through the sinusoids. As with cell membranes throughout the body, the basic hepatocyte cell membrane structure consists of molecules called phospholipids. Cholesterol is also a vital membrane component. Membranes are highly structured, yet fluid and flexible. Studding the membranes are hundreds of different types of transport channels, usually constructed from proteins, which allow molecules to move in and out of the hepatocytes in coordinated fashion.

Interactions of serum molecules with the basolateral cell membrane transport channels are not haphazard or random. As with cell membranes anywhere in the body, enzymes which comprise the transport channels seem to have a specific affinity for molecules they transport based on many factors, including pH, electrostatic forces, molecular size and folding (S Finnegan, Research on lactase and carbohydrase enzymes (Lactaid and Beano), Quality Control Director, Body Bio Corporation, personal communication, 2016; R Silva, Emeritus Professor of Chemistry, California State University Northridge, personal communication). In other words, interactions between serum molecules with the basolateral membrane of hepatocytes are not simply random collisions. There are dynamic and specific interactions which direct individual molecules to their corresponding transport channels so the processes of hepatic metabolism can proceed in orderly fashion. Other aspects of the sinusoidal anatomy (fenestrated endothelial cells, space of Disse, Kupffer cells, stellate cells, and others), although fundamental and vital, will not be discussed here, and the reader is referred to texts to expand knowledge.

16.4.6 Metabolic Zonation/Organization of Hepatocytes by their Function

The first article on metabolic zonation was published by Katz and Jungermann in 1976 [16]. Techniques have since been developed whereby the hepatocytes adjacent to the portal tract (periportal hepatocytes) and those adjacent to the cen-

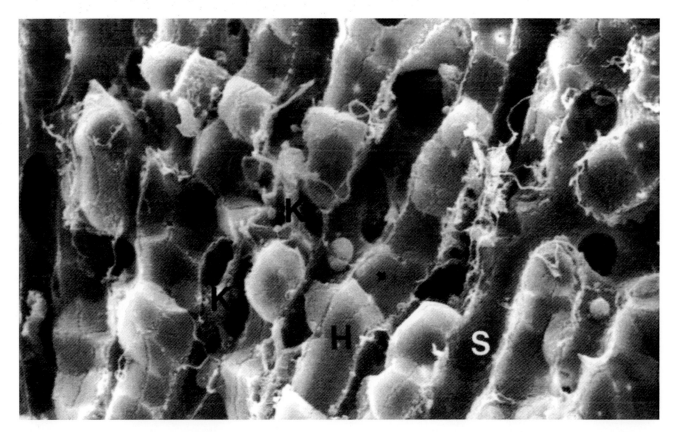

Fig. 16.3 (Sinusoid photo) Scanning electron micrograph showing rows of hepatocytes (H) adjacent to hepatic sinusoids (S) with bile canaliculi visible as small, winding grooves visible on the surface of some of the hepatocytes. (The cells designated K are kupffer cells, and are not discussed here.) (Reprinted from Schiff et al. [50]. With permission from Wiley)

tral vein (perivenular hepatocytes) can be isolated from one another, and their properties studied [17]. The hepatocytes in a liver lobule vary in their function, depending on their position along the path from the portal tract to the central venule. The generally uniform appearance of the liver under light microscopy does not accurately depict the highly specialized and tightly regulated functions of different hepatocyte populations (Fig. 16.4).

While the hepatic lobule describes structure, the concept of the liver acinus, first put forth by Rappaport [18], is related to hepatocyte function. Periportal hepatocytes have a higher oxygen content and perform different metabolic functions compared to perivenular hepatocytes. Rappaport called the periportal area, Zone 1, the perivenular area, Zone 3, with Zone 2 representing the intermediate area between Zones 1 and 3 [17, 18].

In their study on the microvasculature of the liver, Matsumoto and Kawakami found the organization of the vascular flow through the sinusoids is reflected in the precise detail seen on the microscopic level [19]. What might otherwise seem like a random flow of blood from the portal venules and hepatic arterioles into the liver parenchyma via the sinusoids is, in fact, a highly structured hemodynamic process which feeds the blood flow through the sinusoids in an organized fashion, with the resulting effluent flowing into the central hepatic venules [19, 20]. The results of their findings are congruent with the acinus concept, with periportal and perivenular hepatocytes performing different functions.

Zone 1 (Z1) is more highly oxygenated than Zone 3 (Z3) [17]. Many details of metabolism in Z1 and Z3 have been identified (Fig. 16.5), but there is much yet to be discovered. The details of this metabolic zonation can vary from time to time. Some of the driving forces have been identified, including oxygenation, pH, hormones, nutrients, metabolites, cytokines, and molecules called morphogens, the most well-known of which is named Wnt (beta) catenin, [17].

In summary, at the microscopic level of function, the complex processes of hepatic metabolism are proceeding and constantly adjusting to both what is delivered from the portal circulation and to inputs from the rest of the body, including hormonal signals and those from the autonomic nervous system.

16.4.7 Functions of the Liver

The liver is the most potent synthetic factory in the body and a significant organ for waste processing and disposal. See Table 16.1 for a summary of some of the important functions.

16.4.8 Xenobiotics

The term "xenobiotic" is derived from two Greek words, *xenos* meaning foreign or strange, and *bios* meaning life.

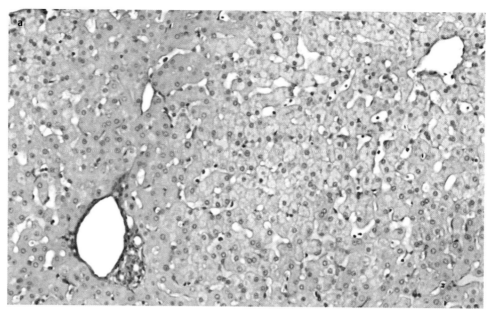

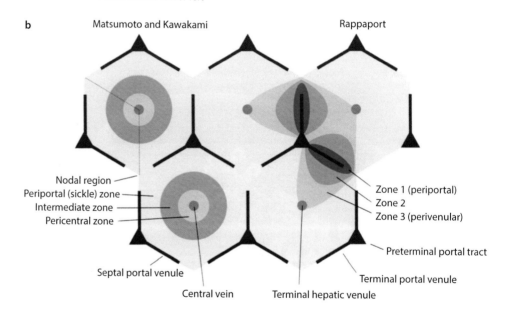

Fig. 16.4 [Hepatic lobule photomicrograph (**a**) and drawing (**b**)]. (**a**) Photomicrograph: Portal tract (lower left) and central venule (upper right) with sheets of hepatocytes interspersed with the intervening vascular sinusoids. Blood entering the liver through the portal tract flows into the sinusoids where it is processed and flows out of the liver through the central venule. Hepatic lobules typically consist of six portal tracts. (**b**) Diagram: Depiction of five classic hepatic lobules. Triangles correspond to portal tracts, from which blood flows to centrally located central venules; detailed microvascular work of Matsumoto and Kawakami contribute to concept of the hepatic acinus proposed by Rappaport; metabolic functions occur in these hepatic lobular structures. (**a**) (Courtesy of Beverly B. Rogers MD, Emory University School of Medicine). (**b**) (Reprinted from Schiff et al. [50]. With permission from Wiley)

20X original magnification
PAS with diastase stain of liver showing central vein (upper right) blending into sinusoid.
Portal tract is lower left

Xenobiotic can be defined as a chemical compound foreign to a given biological system. With respect to animals and humans, xenobiotics include drugs, drug metabolites, and environmental compounds such as pollutants that are not produced by the body. In the environment, xenobiotics include synthetic pesticides, herbicides, and industrial pollutants that would not be found in nature [21]. For further information see ▶ Chap. 13.

An important functional distinction is that molecules absorbed from the small intestine and transported to the liver can be viewed as either "nutrients" or "xenotiobitcs". A nutrient molecule, once it reaches the liver, does not require further breakdown or processing to be used metabolically. Examples of nutrients include amino acids from proteins, monosaccharides from complex carbohydrates, and cholesterol. In the hepatocyte, nutrients can be used in synthesis of macromolecules (building blocks for the body), metabolized through the electron transport chain to produce ATP, and for other purposes.

A xenobiotic, on the other hand, must undergo a process called biotransformation. The products of biotransformation are excreted from the liver in the bile, or in a minority of cases, sent into the systemic circulation, where some accumulate in other organs including the brain, liver, lung, fat, kidney, and/or bone (See ◘ Table 16.2).

16.4.9 Where Are Xenobiotics Found?

We are exposed daily to many harmful environmental chemical toxicants. The Centers for Disease Control and Prevention

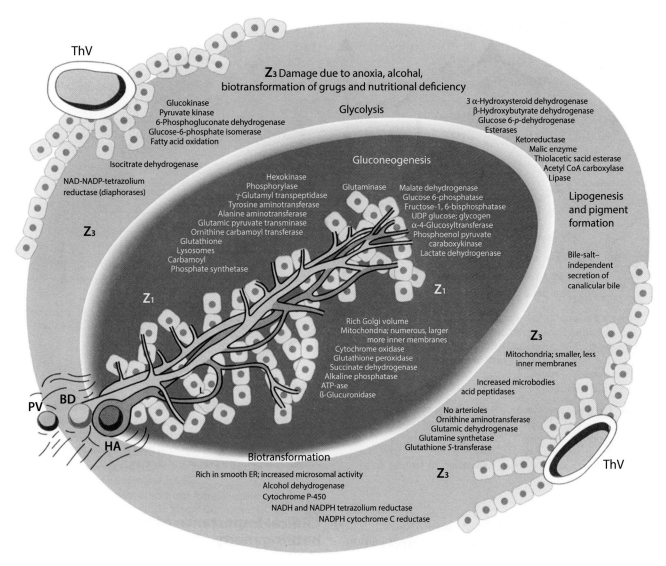

Fig. 16.5 (Schiff Z1 and Z3 characteristics) Diagram depicting some of the known metabolic processes occurring in Zone 1 (periportal) and Zone 3 (perivenular) of hepatic lobules. Zone 1, biosynthetic and energy producing processes; Zone 3, biotransformation enzymes which metabolize xenobiotics. Blood flowing from the portal tracts is first exposed to the hepatocytes in Zone 1, periportal areas, ultimately reaching Zone 3, the perivenular hepatocytes. (Reprinted from Schiff et al. [50]. With permission from Wiley)

(CDC) lists 265 environmental chemicals, many of which were shown to be present not just in the environment, but in the tissues of individuals studied [22]. It would be profitable for the reader to access the CDC report and scan through it to get a sense of the vast number of xenobiotics modern humankind is exposed to and understand this report gives only a partial listing. The report details xenobiotic exposures that may be present from multiple sources, including air, water, food, skincare products, and cleaning chemicals, and from a vast number of industrial sources that find their way into homes and the environment.

diethylstilbestrol which showed the exposed individual's subsequent risk in the development of vaginal cancer [23]. Subsequently, a consensus statement from the Wingspread Conference in 1991 began with this statement, "We are certain of the following: A large number of manmade chemicals that have been released into the environment, as well as a few natural ones, have the potential to disrupt the endocrine system of animals, including humans" [24]. One important approach to mitigating the problem of harmful xenobiotic exposures is addressing the liver's biotransformation enzyme systems, through which they are processed.

16.4.10 A Few Examples of Problems Caused by Xenobiotics

One of the first papers linking the harmful effects of environmental chemical exposures with fetal development was

16.4.11 Biotransformation

In the liver, xenobiotics, that is any molecules not treated as a nutrient, are processed through the biotransformation enzymes [25]. These biotransformation enzymes are located

Table 16.1 Important functions of the enterohepatic circulation and the liver

Proteins (including albumin), lipids (fats, including cholesterol) and complex carbohydrates are synthesized and transported out of the liver
Production of ATP, the principle energy storage molecule in the body, via the electron transport chain
Bile acids formed and secreted into the biliary system; also perform metabolic functions within the hepatocytes
Glucose homeostasis occurs in the balance between gluconeogenesis (formation of the storage molecule glycogen) and glycolysis (release of glucose from glycogen)
Foreign substances (xenobiotics) are metabolized, mostly into nontoxic substances, which can be secreted into the bile or sent to long-term storage depots elsewhere in the body
Important communication functions carried out by hormones and the autonomic nervous system

Table 16.2 Classes of xenobiotics

Food additives
Pharmaceuticals
Environmental pollutants, including many categories of synthetic molecules
Heavy metals
Genetically modified foods; unless the GMO modifications produce nutrients, which are native to the human body, these foods fall into the category of xenobiotics

almost exclusively in the perivenular hepatocytes, Z3, in the liver acinus. Most xenobiotics are fat-soluble, serve no useful physiological purpose, and must be eliminated from the body. In most situations, this occurs in a two-step process.

Phase I biotransformation (detoxification) occurs in the smooth endoplasmic reticulum of the perivenular hepatocytes. Most commonly, once a xenobiotic molecule enters the hepatocyte, it encounters the cytochrome P450 enzymes, which convert the molecule into a highly energetic intermediate. These intermediates are then metabolized by Phase II biotransformation enzyme pathways. The names of these pathways are sulfation, glucuronidation, acetylation, methylation, amino acid conjugation, and glutathione conjugation [25].

The majority of the products of Phase II enzymatic reactions are secreted into the bile. A small percentage of the excretory products from the hepatocyte do not go into the bile; they are secreted back through the basolateral membrane into the sinusoid, from which they enter the systemic circulation and are carried to other organs and tissues. This most likely represents the pathway by which heavy metals (lead, mercury, cadmium, etc.) reach and are ultimately stored in other organs, such as the brain, kidney, lung, bone, and the liver itself. Please see ▶ Chap. 13 for more in-depth discussion.

16.4.12 Cytochrome P450 Genes and Enzymes

The cytochrome P450 enzymes (CYP 450) are numbered to indicate a specific group within the gene family, a letter indicating the gene's subfamily, and a number assigned to the specific gene within the subfamily. A specific CYP gene codes for the corresponding CYP enzyme. As of 2012, there were 57 human CYP enzymes, each with its specific propensity to metabolize certain categories of xenobiotics [25–27]. One example would be CYP 450 2D6, which is one of the cytochromes responsible for metabolizing hypertensive drugs. Another is CYP 450 2E1, which plays a role in the hepatotoxicity of acetaminophen. There has been extensive research into the CYP 450 genes and enzymes, and there are extensive references regarding the CYP 450 enzymes and xenobiotic metabolism [24, 26, 28, 29].

16.4.13 Single-Nucleotide Polymorphisms

Where there are genes, there are gene mutations. These mutations code for altered proteins with varying structures and activities. A common form of gene mutation is the single nucleotide polymorphism (SNP), in which a single nucleotide in a gene is changed, which may result in altered enzyme function. There are extensive publications on the effects of these polymorphisms on disease patterns and susceptibility and drug metabolism [30, 31]. There are more than 6000 review articles in the PubMed database detailing polymorphisms and their clinical effects.

In summary, if a healthy biologic system depends on the normal function of its enzyme systems, mutated genes and altered enzymes may have adverse effects on health. These consequences may vary from inconsequential to devastating. Perhaps nowhere else in the human body is the problem with mutated genes and altered enzymes more prevalent than in the liver's biotransformation enzyme systems, which critically impact xenobiotic transformation. Also important to understand is that while the biotransformation systems in the liver are vast, they are not infinite. If our purpose is to create health, the focus should be on feeding the system substances that are nutrients, not xenobiotics, and allowing the pathways in the liver to function as efficiently as possible. This requires optimizing nutrient intake and minimizing exposure to unnecessary xenobiotics. Improving the function of Phase II biotransformation in an individual patient should serve to eliminate, or to ameliorate the adverse health effects of many harmful xenobiotics.

16.4.14 Bile and the Biliary System

Bile is secreted from each hepatocyte into the bile canaliculus through a secretory process requiring energy [13]. ▯ Figure 16.5 shows a bile canaliculus as it travels along adjacent hepatocytes on its way to the bile ductule in the portal tract. The bile ductules merge into the bile duct system, which transports bile out of the liver. Bile then flows down the common bile duct and through the sphincter of Oddi into the second part of the duodenum, where it mixes with intestinal content and begins its multifaceted function.

When there has not been a recent meal, the sphincter of Oddi is tightly closed, and the bile is diverted into the gallbladder. However, the gallbladder is more than a simple storage compartment. It concentrates bile and performs other important functions. When food is eaten, hormones, including cholecystokinin (CCK), are released, which cause gallbladder contraction, relaxation of the sphincter of Oddi, and a bolus of bile is released into the duodenum.

Bile is a micellar liquid. This means that the lipids/fats (cholesterol, phospholipids) and proteins secreted by the hepatocytes, when they reach a certain concentration in the bile, spontaneously form into round or cylindrical transport structures, which are carried in the ionic transport solution of the bile into the duodenum. Some molecules are carried in the micellar component of bile, some in the aqueous or liquid component. Bile contains bile acids which are essential to normal fat digestion, and have other metabolic properties. Bile is a potent antioxidant, and carries products from hepatic biotransformation out of the liver [32].

16.4.15 Small Intestine

The small and large intestines can be seen as the main digestive organs of the body (● Fig. 16.6). The small intestine includes the duodenum, jejunum and ileum for a combined length of about 25 feet. The large intestine or colon is about 5 feet long. In the small intestine, proteins, carbohydrates, fats, vitamins, minerals and trace elements are processed, and the products of digestion are absorbed through the intestinal wall.

The inner lining of the small intestine contains circular folds called plicae, the surface of which is studded with leaf- or-finger like projections called villi. Enterocytes, the absorptive cells in the small intestine, make up 90% of the cells lining the villi [33]. The apical surface of each enterocyte faces the intestinal lumen and is itself further folded into microscopic tubular structures called microvilli. Because of the plicae, villi and microvilli, the absorptive surface of the small intestine which is exposed to food and other ingested material is vastly increased - about the size of a tennis court in an average adult.

Considering the large number and varied types of ingested substances presented to the lining of the small intestine, and the large surface area involved, the small intestine performs a monumental task of identifying toxins and pathogens, digesting nutrients, maintaining the barrier function of

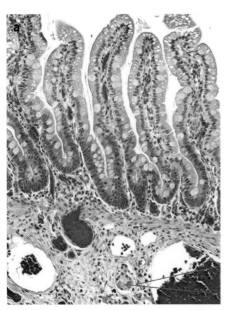

20X original magnification H&E

Ganglion cells

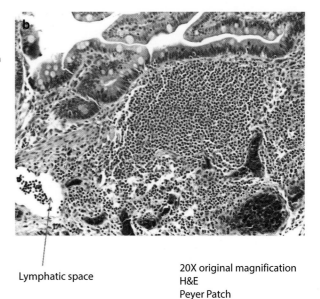

Lymphatic space

20X original magnification H&E
Peyer Patch

◘ **Fig. 16.6** Aspects of small intestine lining. (**a**) Photomicrograph of villi lining the mucosal surface of the small intestine. The predominant cell type is the enterocyte, through which nutrients and xenobiotics pass. Approximate surface area of the small intestinal lining in an adult is that of a tennis court, indicating the immense task of the enterocytes in distinguishing what should and should not be allowed to enter the submucosal space, and hence the portal circulation. Everything swallowed passes through the stomach into the small intestinal lumen, goes through processes of digestion, interacts with microbiome, and is sorted and evaluated, to determine if it will be allowed into the body or not. (**b**) Photomicrograph of submucosal lymphoid tissue known as a Peyer's patch, part of the GALT (gut-associated lymphoid tissue). Note lymphatic channels, and the capillaries which lead into the portal venous drainage. Fibers of the extensive enteric nervous system (the largest concentration of nervous tissue outside the central nervous system, and so called the "second brain" by some) are not visible in this exposure. Enteric endocrine cells (EEC), part of the intestinal lining, are likewise not specifically distinguished. (Courtesy of Beverly B. Rogers MD, Emory University School of Medicine)

the intestinal wall, and choosing which substances to allow to pass into the portal circulation.

16.4.16 Microbiome/Metabolome

After partially digested food from the stomach combines with bile and pancreatic secretions in the second part of the duodenum, it then passes into the jejunum and encounters the microbiome. In an average adult, the microbiome is four to five pounds of microorganisms which live in the lumen of the small and large intestines [34–37]. The microbiome begins to form immediately after birth. An infant's first stools are meconium, which is essentially digested amniotic fluid; cultures taken of meconium in healthy infants are sterile. Within a few days after delivery, microorganisms can be grown in culture. These microorganisms originate from the breast, skin, and secretions from the birth canal. Breast feeding promotes growth of *Bifidobacteria*, which have been associated with the healthy nature of stool flora in infants [38].

Bifidobacteria are well-known to be one of the more common probiotic species in the microbiome [38]. In adults, if the microbiome consists of healthy probiotic bacteria, such as *Lactobacillus*, *Bifidobacteria*, and others, this contributes significantly to healthy intestinal function and thus to the overall health of the individual. The microbiome assists in nutrient metabolism, protection against pathogens, processing of antigens and presenting them to the individual's immune system, processing of xenobiotics, and maintenance of a healthy barrier function of the intestinal wall [35]. However, if the constituents of the microbiome are pathogenic, the normal function of the microbiome is altered. The composition of the microbiome is affected by diet, antibiotics, stress, and other factors.

The term "Metabolome" refers to the multiple metabolic processes associated with the microbiome. There are more than 3000 articles referencing the metabolome, the first of which appeared in 2000 [39].

16.4.17 Enteric Nervous System

An important part of overall intestinal function is nerve cells in the wall of the small intestine forming a separate and intricate enteric nervous system (◘ Fig. 16.7). There are multiple neuron types in multiple locations. These neurons communicate from one part of the intestine to another and back and forth with the central and autonomic nervous systems. There is bidirectional information flow between the ENS and the CNS and between the ENS and the sympathetic prevertebral ganglia [11, 12].

Sensory distress from the enteric nervous system can lead to symptoms such as nausea, bloating and pain. But most of the sensory signals from the intestine are not consciously perceived. The enteric nervous system has been called the second brain [40] because of its many functions and its vast numbers of neurons. There are more neurons in the enteric nervous system than anywhere else in the body except the central ner-

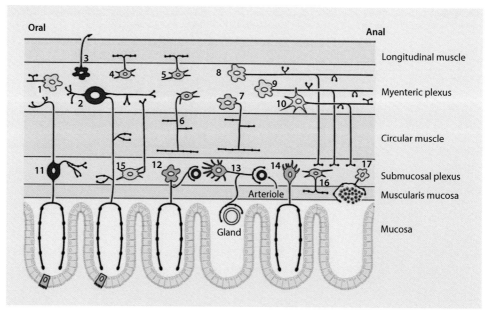

◘ Fig. 16.7 Neurons found in small intestine defined by function, cell body morphology, chemistry, neurotransmitters, and projections to targets. Neuron types: (1) ascending interneurons; (2) myenteric intrinsic primary afferent neurons (IPANs); (3) intestinofugal neurons; (4) excitatory longitudinal muscle motor neurons; (5) inhibitory longitudinal muscle motor neurons; (6) excitatory circular muscle motor neurons; (7) inhibitory circular muscle motor neurons; (8) descending interneurons (local reflex); (9) descending interneurons (secretomotor and motility reflex); (10) descending interneurons (migrating myoelectric complex); (11) submucosal IPANs; (12) non-cholinergic secretomotor/vasodilator neurons; (13) cholinergic secretomotor/vasodilator neuron; (14) cholinergic secretomotor (non-vasodilator) neurons; (15) uni-axonal neurons projecting to the myenteric plexus; (16) motor neuron to the muscularis mucosa; (17) innervation of Peyer's patches not illustrated, motor neurons to enteroendocrine cells. (Adapted from Furness [51]. With permission from Wiley)

vous system. A number of commonly used phrases reflect this plentiful innervation of the GI tract, and its function in interacting with the central and autonomic nervous systems: "I knew it in my gut," "I had a gut instinct," or "the bowels of compassion."

16.4.18 Gut-Associated Lymphoid Tissue

The small intestine has the job of sorting out whether molecules presented to its lining cells are friend or foe, as everything ingested is not an ideal nutrient. Some substances can trigger allergic reactions while some are frankly toxic. Some molecules may be a known nutrient, but trigger the immune system to develop an autoimmune disease, the most well-known being celiac disease and the reaction to gluten/gliadin proteins [41].

The lining of the small intestine serves as a barrier between the outside world (ingested substances) and the body. This is called the intestinal barrier function, or the intestinal firewall [42]. Sometimes the function of the lining of the small intestine in forming a barrier to the outside world is compromised in various ways. Molecules may translocate or leak through the mucosal membrane by alternate pathways and can end up in the intercellular space below the mucosal lining enterocytes. This leads to increased intestinal permeability (referred to by some as "leaky gut") [43].

When non-nutrient molecules pass through the intestinal lining, the gut-associated lymphoid (GALT) system is stimulated. In the small intestine, GALT resides throughout the intestinal wall in vast numbers of microscopic collections of lymphoid tissue. In the distal small intestine and colon, these cell clusters occur in larger structures called Peyer's patches.

16.4.19 Enteric Endocrine Cells

There are more hormones produced in specialized cells of the small intestine than in any other organ in the body [44, 45]. Hormones are messenger molecules that convey various messages between cells, between different parts of the small intestine, and between the small intestine and other organs.

16.4.20 Terminal Ileum

The last 6 inches of the small intestine are structurally different from the rest. This segment is called the terminal ileum. Vitamin B12 is absorbed at this location [46]. In this segment, about 95% of the bile, which has remained inside the intestinal lumen to perform its functions, is reabsorbed and returned to the liver through the portal circulation. This absorption and recirculation of bile was the origin of the name "entero-hepatic circulation".

16.4.21 Large Intestine

The large intestine, or colon, begins in the right lower abdomen in a bulging pouch called the cecum. The terminal ileum empties into the cecum through the ileocecal valve (IC). The structure of the IC is such that it prevents reflux of contents from the cecum back into the ileum. The colon is approximately 5 feet long, and digestion and nutrient absorption are essentially complete by the time the intestinal contents enter the cecum. There is a vibrant microbiome in the colon [47, 48], and no doubt there are many metabolic functions yet to be clarified. One main role of the colon is reabsorption of water and electrolytes, preparing the stool to be passed from the body through the anus.

16.5 Conclusion

Based on the structure and function of the EHC, the ingestion of proper nutrients, possessing a healthy microbiome, maintaining an intact intestinal barrier function, and keeping exposure to xenobiotics to a minimum, the whole EHC system will function as it was designed. However, if the microbiome is unhealthy, if the intestinal barrier function allows toxic molecules to be taken up, or if there are insufficient nutrients or an overload of xenobiotics, the system will not function well. This is a reasonable place to begin in patient care. Using this paradigm and approach, patient care becomes conceptually simple: restore a healthy microbiome, restore a healthy intestinal barrier function, ensure delivery of proper and optimal amounts of nutrients, take into account biochemical individuality, and minimize harmful xenobiotics. The goal becomes identifying ways in which function deviates from this norm and taking steps to restore normal function. This is a vastly different approach from waiting until an identifiable disease develops and is diagnosed, then treating that disease with medication and/or surgery. In many cases the approach discussed above, based on correcting altered function, deals with the problem long before it forms.

Integrative nutritional care use is the closest approach to returning to the roots of human physiology and biochemistry and allowing these principles, along with patients' needs, to guide practice. In this approach, nutritional principles are applied first, along with other principles of a healthy body. Other healthcare disciplines are included in the treatment process as they are found to be useful; none are excluded. The conversations may include naturopaths, chiropractors, herbalists, nutritionists, homeopaths, acupuncturists – anyone who seems to have something to offer. At times, pharmaceuticals are needed. From the author's observation, integrative nutritional therapy is not just a new buzz-word or catch-phrase. It is not a new discipline, to be replaced at some point by another. It represents the closest paradigm found leading to understanding the principles of human physiology and allowing the use of nutrition to emerge from that awareness. Integrative and functional nutrition properly applied leaves room for any approach, any discipline that has something to offer. With knowledgeable guidance and patient motivation, an immense amount can be accomplished.

References

1. Malik MY, Jaiswal S, Sharma A, Shukla M, Lal J. Role of enterohepatic recirculation in drug disposition: cooperation and complications. Drug Metab Rev. 2016;48(2):281–327. https://doi.org/10.3109/03602532.2016.1157600.
2. Broun GO, McMaster PD, Rous P. Studies on the total bile: IV. The enterohepatic circulation of bile pigment. J Exp Med. 1923;37(5):699–710. PMID: 19868754.
3. Manwaring WH. Enterohepatic circulation of estrogens. Cal West Med. 1943;59(5):257. PMID: 18746637.
4. Lorber SH, Shay H. Enterohepatic circulation of bromsulphalein. J Clin Invest. 1950;29(6):831. PMID: 15436781.
5. Siperstein MD, Hernandez HH, Chaikoff IL. Enterohepatic circulation of carbon 4 of cholesterol. Am J Phys. 1952;171(2):297–301. https://doi.org/10.1152/ajplegacy.1952.171.2.297.
6. Myant NB. Enterohepatic circulation of thyroxine in humans. Clin Sci. 1956;15(4):551–5. PMID: 13374940.
7. Acocella G, Muschio R. Demonstration of the existence, in man of an entero-hepatic circulation of rifamycin SV. Riforma Med. 1961;75:1490–3. PMID: 13859193.
8. Krebs M. The enterohepatic circulation of endogenous and exogenous substances. Med Welt. 1963;39:1967–72. PMID: 14087819.
9. Mason HS, North JC, Vanneste M. Microsomal mixed-function oxidases: the metabolism of xenobiotics. Fed Proc. 1965;24(5):1172–80. PMID:4378722.
10. Shanahan F. The host-microbe interface within the gut. Best Pract Res Clin Gastroenterol. 2002;16(6):915–31. PMID: 12473298.
11. Furness JB, Callaghan BP, Rivera LR, Cho HJ. The enteric nervous system and gastrointestinal innervation: integrated local and central control. In: Lyte M, Cryan JF, editors. Microbial endocrinology: the microbiota-gut-brain axis in health and disease. Advances in experimental medicine and biology, vol. 817. New York: Springer; 2014. p. 39–71.
12. Furness JB. Integrated neural and endocrine control of gastrointestinal function. In: Brierley S, Costa M, editors. The enteric nervous system, advances in experimental medicine and biology 891. Cham: Springer; 2016. p. 159–73.
13. Pavlovic N, Stanimirov B, Mikov M. Bile acids as novel pharmacological agents: the interplay between gene polymorphisms, epigenetic factors and drug response. Curr Pharm Des. 2017;23(1):187–215. https://doi.org/10.2174/1381612822666161006161409.
14. Zhu C, Fuchs CD, Halibasic E, Trauner M. Bile acids in regulation of inflammation and immunity: friend or foe? Clin Exp Rheumatol. 2016;34(4 Suppl 98):25–31. PMID: 27586800.
15. Feldman M, Friedman LS, Brandt LJ, editors. Sleisenger and Fordtran's gastrointestinal and liver disease: pathophysiology, diagnosis, management. 10th ed. Saunders; 2016.
16. Katz N, Jungermann K. Autoregulatory shift from fructolysis to lactate gluconeogenesis in rat hepatocyte suspensions. The problem of metabolic zonation of liver parenchyma. Hoppe Seylers Z Physiol Chem. 1976;357(3):359–75.
17. Gebhardt R, Matz-Soja M. Liver zonation: novel aspects of its regulation and its impact on homeostasis. World J Gastroenterol. 2014;20(26):8491–504.
18. Rappaport AM. The structural and functional unit in the human liver (liver acinus). Anat Rec. 1958;130(4):673–89.
19. Matsumoto T, Kawakami M. The unit-concept of hepatic parenchyma – a re-examination based on angioarchitectural studies. Acta Pathol Jpn. 1982;32(Suppl 2):285–314. PMID: 7187579.
20. Matsumoto T, Komori R, Magara T, Ui T, Kawakami M, Tokuda T, Takasaki S, Hayashi H, Jo K, Hano H, Fujino H, Tanaka H. A study on the normal structure of the human liver, with special reference to its angioarchitecture. Jikeikai Med. 1979;26:1–40.
21. Miller-Keane OM. Miller-Keane encyclopedia and dictionary of medicine, nursing, and allied health. 7th ed. Saunders; 2003.
22. Centers for Disease Control and Prevention (CDC). Fourth national report on exposure to environmental chemicals. February 2015. https://www.cdc.gov/biomonitoring/pdf/FourthReport_UpdatedTables_Feb2015.pdf.
23. Herbst AL, Ulfelder H, Poskanzer DC. Adenocarcinoma of the vagina. Association of maternal stilbestrol therapy with tumor appearance in young women. N Engl J Med. 1971;284(15):878–81. https://doi.org/10.1056/NEJM197104222841604.
24. Wingspread Conference 1991. Our stolen future: consensus statements. Wingspread Conference Center, Racine, Wisconsin. July 1991. Accessed: http://www.ourstolenfuture.com/consensus/wingspread1.htm.
25. Anzenbacher P, Zanger UM, editors. Metabolism of drugs and other xenobiotics. Weinheim: Wiley-VCH Verlag & Co. KGaA; 2012.
26. Genetics Home Reference. Your guide to understanding genetic conditions. U.S. National Library of Medicine. Accessed: https://ghr.nlm.nih.gov/
27. Sim SC, Ingelman-Sundberg M. The human cytochrome P450 (CYP) allele nomenclature website: a peer-reviewed database of CYP variants and their associated effects. Hum Genomics. 2010;4(4):278–81.
28. Rodriguez-Arcas MJ, Garcia-Jimenez E, Martinez-Martinez F, Conesa-Zamora P. Role of CYP450 in pharmacokinetics and pharmacogenetics of antihypertensive drugs. Farm Hosp. 2011;35(2):84–92.
29. Lee SST, Buters JTM, Pineau T, Fernandez-Salguero P, Gonzalez FJ. Role of CYP2E1 in the hepatotoxicity of acetaminophen. J Biol Chem. 1996;271:12063–7.
30. Iida A, Saito S, Harigae S, Osawa S, Mishima C, Kondo K, Kitamura Y, Nakamura Y. Catalog of 46 single-nucleotide polymorphisms (SNPs) in the microsomal glutathione S-transferase 1 (MGST1) gene. J Hum Genet. 2001;46(10):590–4.
31. Hollman AL, Tchounwou PB, Huang HC. The association between gene-environment interactions and diseases involving the human GST superfamily with SNP variants. Int J Environ Res Public Health. 2016;13(4):379.
32. Stocker R. Antioxidant activities of bile pigments. Antioxid Redox Signal. 2004;6(5):841–9.
33. Streutker CJ, Huizinga JD, Driman DK, Riddell RH. Interstitial cells of Cajal in health and disease. Part I: normal ICC structure and function with associated motility disorders. Histopathology. 2007;50(2):176–89. https://doi.org/10.1111/j.1365-2559.2006.02493.x.
34. Harmsen HJ, de Goffau MC. The human gut microbiota. Adv Exp Med Biol. 2016;902:95–108.
35. Jandhyala SM, Talukdar R, Subramanyam C, Vuyyuru H, Sasikala M. Role of the normal gut microbiota. World J Gastroenterol. 2015;21(29):8787–803.
36. Konkel L, Danska J, Mazmanian S, Chadwick L. The environment within: exploring the role of the gut microbiome in health and disease. Environ Health Perspect. 2013;121(9):a276–1281.
37. Krautkramer K, et al. Diet-microbiota interactions mediate global epigenetic programming in multiple host tissues. Mol Cell. 2016;64(5):982–92.
38. Langhendries JP, Paquay T, Hannon M, Darimont J. Intestinal flora in the neonate: impact on morbidity and therapeutic perspectives. Arch Pediatr. 1998;5(6):644–53. PMID: 9759211.
39. Kell DB, King RD. On the optimization of classes for the assignment of unidentified reading frames in functional genomics programmes: the need for machine learning. Trends Biotechnol. 2000;18(3):93–8. PMID: 10675895.
40. Gerson M. The second brain. New York: Harper Collins; 1998.
41. Parzanese I, Qehajaj D, Patrinicola F, Aralica M, Chiriva-Internati M, Stifter S, Elli L, Grizzi F. Celiac disease: from pathophysiology

to treatment. World J Gastrointest Pathophysiol. 2017;8(2):27–38. https://doi.org/10.4291/wjgp.v8.i2.27.
42. Bland J. The gut mucosal firewall and functional medicine. Integr Med (Encinitas). 2016;15(4):19–22.
43. Damms-Machado A, Louis S, Schnitzer A, Volynets V, Rings A, Basral M, Bischoff SC. Gut permeability is related to body weight, fatty liver disease, and insulin resistance in obese individuals undergoing weight reduction. Am J Clin Nutr. 2017;105(1):127–35.
44. Latorre R, Sternini C, De Giorgio R, Greenwood-Van Meerveld B. Enteroendocrine cells: a review of their role in brain-gut communication. Neurogastroenterol Motil. 2016;28(5):620–30. https://doi.org/10.1111/nmo.12754.
45. Campana A, editor. Steimer Th. Steroid hormone metabolism. Geneva Foundation for Medical Education and Research; 2016.
46. Cuvelier C, Demetter P, Mielants H, Veys EM, De Vos M. Interpretation of ileal biopsies: morphological features in normal and diseased mucosa. Histopathology. 2001;38(1):1–12.
47. Cho I, Blaser MJ. The human microbiome: at the interface of health and disease. Nat Rev Genet. 2012;13(4):260–70. https://doi.org/10.1038/nrg3182.
48. Cummings JG. The colon: absorptive secretory and metabolic functions. Digestion. 1975;13(4):232–40.
49. The digestive system anatomical chart. ISBN: 9781587790072.
50. Schiff ER, Maddrey WC, Sorrell MF, editors. Schiff's diseases of the liver. 11th ed. Wiley-Blackwell; 2012.
51. Furness JB. The enteric nervous system. Oxford: Wiley-Blackwell; 2006.

A Nutritional Genomics Approach to Epigenetic Influences on Chronic Disease

Christy B. Williamson and Jessica M. Pizano

17.1 What Is Epigenetics? – 236
17.1.1 The Epigenetics of Cardiometabolic Disease – 238
17.1.2 The Epigenetics of Psychiatric and Neurodegenerative Diseases – 240

17.2 The Epigenetics of Irritable Bowel Disease and Dysbiosis – 244
17.2.1 The Epigenetics of Nutrient-Associated Diseases – 259

17.3 Epigenetics of Cancer – 261
17.3.1 Epigenetics of Mitochondrial Insufficiency – 262

17.4 Introduction to Pharmacogenetics – 263
17.4.1 Final Thoughts – 264

References – 264

© Springer Nature Switzerland AG 2020
D. Noland et al. (eds.), *Integrative and Functional Medical Nutrition Therapy*,
https://doi.org/10.1007/978-3-030-30730-1_17

> "Nutritional genomics refers to the application of "omics" technologies, together with systems biology and bioinformatics tools, to understand how nutrients interact with the flow of genetic information to impact various health outcomes" [1]. Nutritional genomics is an umbrella term that encompasses two distinct but related fields: nutrigenomics and nutrigenetics. Nutrigenomics may be defined as the measurable effect of nutrients on the genome, proteome, microbiome and metabolome. Thus, the use of laboratory measures such as organic acids, amino acids, homocysteine, etc. may serve as an indicator of whether a gene is functioning or impaired due to a SNP, and to what degree. This clinical data may then be used to develop a personalized nutrition plan to influence the biochemical pathways in which the SNP interacts. Nutrigenetics is similar as it also explores the measurable interactions of nutrients on the genome, however, nutrigenetics is focused on the measurable interactions between diet and *disease risk* and which dietary interventions influence intervention outcomes. This may help predict how a patient may respond to a dietary intervention in terms of controlling or exacerbating disease risk. It is therefore instrumental in the design of personalized nutrition plans to ameliorate symptoms or in the prevention of the development of various disease types.

Nutrigenetics is focused on genetic variation and disease prediction or the way that diet influences the risk of developing a disease. Nutrigenetics is most often practiced by geneticists and genetic counselors who are skilled in computational statistics. Remember, this is *disease prediction*. Nutrigenomics is focused on how nutrients affect the genome and the biochemistry we can measure related to those lifestyle choices. Measurement in this case is typically through metabolomics and microbiome analyses. Due to the reliance on metabolomics for validation, advanced nutrition professionals are the primary nutrigenomics practitioners. These are two distinct fields under one large umbrella. Epigenetics intermingles easily through both fields, making the scope a bit more challenging to decipher.

In these next sections, we will discover the salient genetics associated with several conditions. Some conditions and interventions will be strictly nutrigenetic in nature while others are more nutrigenomic. There will be some overlap as that is the nature of the field. We will dive deeply into the nutritional genomics and epigenetics of cardiometabolic disease; neurodegenerative diseases like Alzheimer's disease and Parkinson's disease; common psychiatric conditions such as depression, anxiety, schizophrenia and bipolar disorder; understand the genomics of irritable bowel disease, mitochondrial insufficiency, cancer and nutrient-related autoimmune diseases followed by an introduction to pharmacogenomics. We will elucidate the effect that stress has on methylation and the epigenome and discuss how common toxic exposures can influence methylation. Remember, hypermethylation silences gene expression while hypomethylation activates it, thus in essence, hypomethylating means to turn on the gene, while hypermethylation will typically turn it off. The goal is to have proper methylation, not over or undermethylation. This chapter is an overview of the topics and is not an exhaustive review.

17.1 What Is Epigenetics?

The term epigenetics was originally coined by Waddington in 1942. It was derived from the Greek word "epigenesis," which described the influence of genetic processes on development [2, 3]. Modern day epigenetics is the study of gene expression through histone or methyl modulation along with RNA silencing rather than the alteration of the genetic code itself. Epigenes are additional instruction layered on top of our inherited genetic code (A, T, C, G, U); this additional instruction is known as the epigenome. If we think of the genetic code as the manuscript, the epigenetic information could be viewed as the highlighted or tagged sections. These "tags" or markers indicate to the instructional processors that something is either important or that it can be ignored. This is the science that turns "on" or "off" the genes, which can result in both positive and negative outcomes.

There are a multitude of "tags" that guide the genomic processors. Some are methyl groups, others are histone modifiers, and some even modify RNA, one of the processors! These tags work together to either increase or restrict access to the genome, therefore controlling the expression of the proteins found within the genome. The really interesting part about these tags is that they are not corrected like our genetic code. These tags or markers can change based on our lifestyle, experiences, diet, exercise habits, and exposure to chemicals. This fluidity allows the epigenome to "learn" and adapt to the current environmental circumstances. These epigenetic markers can even survive what is called global DNA demethylation, or the wiping clean of epigenetic markers from the gametes when a zygote is formed. Interestingly, there is a process called imprinting whereby one of the parent's genes is "dominant" and silences the other allele. These imprinted genes keep their epigenetic tags, even through global demethylation. These are the tags that survive from generation to generation and influence genetic expression. As the new embryo forms, other epigenetic tags physically record each cell's experiences on the genome. In this recording or writing stage, the epigenome ultimately stabilizes gene expression allowing for proper embryonic development [4].

Maternal diet, smoking status, mental state, and social environment can do one of two things. This environmental information can hypomethylate CpG islands, which in essence turns on the gene, or it can result in RNA silencing, thus turning off the gene [4]. This means that the trauma that your great grandmother experienced at the turn of the twentieth century could potentially increase your risk of developing a mental illness, such as depression. Fathers also have epigenetic transfer as well, thus, it also means that your "health-nut" father could also confer more favorable cancer and cardiovascular epigenetics [5].

There are three major systems that control epigenetics including DNA methylation, histone modification, and non-coding RNA (ncRNA)-associated gene silencing [4]. There are also less well-studied processes such as acetylation and ubiquitination. In addition to these processes, the "systems" are also modifiable. For example, histone proteins are responsible for packaging DNA via chromatin complexes into dense chromosomes. Histone structures may be modified by acetylation of lysine residues, methylation of lysine and arginine residues, phosphorylation of serine and threonine residues, ubiquitination of lysine residues on histone tails, sumoylation, and ADP ribosylation. Histone acetyltransferases (HATs) add acetyl groups to lysine residues on histone tails, whereas histone deacetylases (HDACs) can remove the acetyl groups [4, 6, 7].

DNA methylation typically occurs at CpG islands, which are most commonly found in gene promoter regions [6]. It is at this CpG island that DNA methyltransferases (DNMTs) coordinate the addition of a methyl group to the 5-carbon position of the cytosine residues where it regulates transcription. DNMT1 exclusively maintains normal methylation by exact duplication of DNA between cell generations. DNMT2, on the other hand, is responsible for methylation within embryonic stem cells. Methylation has also recently been discovered to occur in non-CpG cytosines within undifferentiated stem cells [4, 7–9]. Methylation is important as not all genes are expressed at all times. *In fact, most are repressed.* Methylation is an epigenetic means of keeping genes suppressed until they are needed. Remember, methylation, especially hypermethylation, silences gene expression. Pause and consider for a moment the use of methylated vitamins and what implications this may have on DNA expression. Further, methylation is important for determining chromosomal replication timing. During hypomethylation there is late-replication, which may be demethylated slowly during cell division. Hypermethylation, on the other hand, leads to much earlier replication during S phase. Those hypermethylated genes are typically repackaged with acetylated histones. Hypomethylation will cause the building of nucleosomes that contain deacetylated histones [9]. In this way, methylation may control the "turning off" of genes.

Lastly, small non-coding RNA molecules (MicroRNAs) are a range of molecules that also help to control the expression and function of genes. They are non-coding RNAs that regulate gene expression post-transcriptionally or after the protein has been processed by the ribosome. Typically, they repress protein production by altering the capabilities of messenger RNA and by a process called translational silencing [10]. In this way, they are able to repress gene function and expression. MicroRNAs are accountable for targeting approximately 30% of genes and can influence tumor suppression, apoptosis, cellular proliferation, and cell movement [6].

Agouti mice illustrate how nutrition may modify phenotypic expression, epigenetic expression, and disease risk out-

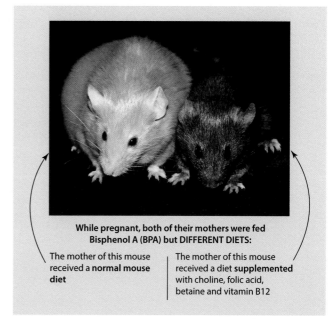

● Fig. 17.1 These two mice are genetically identical and of the same age. The larger yellow obese Agouti mouse on the left received a diet without methyl nutrients. The smaller brown healthy Agouti mouse on the right received a diet with methyl nutrients (choline, folic acid, betaine and B12). Both mice were fed BPA in each diet. In the laboratory, BPA appears to reduce methylation of the agouti gene. With the mother fed the regular (BPA-modified) diet, the pups were more likely to be yellow and obese and more prone to develop cancer and diabetes. When the mothers were fed the BPA-modified diet plus methyl nutrients, the pups were more likely to be born healthy and brown, and they were of ideal weight. The methyl nutrient supplementation counteracted the negative effects of the exposure. (Adapted from Jirtle R, Dolinoy D. Agouti Mice. Retrieved from: ▶ https://commons.wikimedia.org/wiki/File:Agouti_Mice.jpg. With permission from Creative Commons License 3.0: ▶ https://creativecommons.org/licenses/by/3.0/deed.en.)

come. Maternal dietary differences in methyl (CH_3) intake dictates the presence of a yellow hair color either as a band within primarily brown hair in wild-type mice, or as fully yellow hair in those with mothers who had decreased methylation. This is related to the agouti viable yellow (Avy) gene. This phenotype is accompanied by obesity and is a result of paracrine signaling issues resulting in hyperphagia in mice pups. The consequence of this epigene is that the mice expressing the aberrant agouti gene are more likely to develop diabetes and cancer later in life. The difference between pups expressing the wild-type genetics versus the polymorphism is the overmethylation status allowing for suppression of the *agouti* gene. To further influence the expression of this gene, supplementation with choline, vitamin B12, and folate both prior to and during pregnancy repress *agouti* expression [6, 11, 12] (● Fig. 17.1).

Another classic example of the influence of nutrition on epigenetic expression is the *Dutch Hunger Winter Study*. This study looked at a cohort of people conceived during a famine that occurred over 6 months during the last winter of World War II (1944–1945). During this time, the caloric intake decreased from approximately 1400 kilocalories in October

1944 to less than 1000 kilocalories towards the end of November 1944. Caloric intake further declined at the peak of the famine to 400–800 kilocalories from December 1944 to April 1945. Despite the plummeting amounts of kilocalories during this time, the proportion of fats, carbohydrates, and proteins remained the same [13–15]. The study found a positive correlation between famine, calorie restriction and obesity in the offspring of women pregnant during this time. This association was further correlated to obesity-related diseases in adulthood such as atherosclerosis, hyperlipidemia, coronary artery diseases, and increased risk of cardiovascular mortality [13–15]. As a segue into obesity and obesity-related disorders, we will explore specific epigenetic influences and single nucleotide polymorphisms (SNPs) associated with cardiometabolic disease.

17.1.1 The Epigenetics of Cardiometabolic Disease

Cardiometabolic disease encompasses a cluster of disease characteristics including atherosclerosis, dyslipidemia, diabetes mellitus type 2 (DM2), hypertension (HTN), increased waist circumference and obesity [16]. All of the aforementioned conditions have genetic etiologies that can increase the risk or prevalence of the disease in affected individuals.

To begin, let us look closely at genetics that increase the risk for cardiovascular disease (CVD). Apolipoprotein E (APOε) is a cholesterol carrier that assists in lipid transport [17]. It does so by merging with endogenous lipids to form lipoproteins that are responsible for restructuring other lipids and cholesterol. These lipoproteins also circulate these lipids in the bloodstream. The three isoforms of APOε are APOε2, APOε3, and APOε4. The APOε3 isoform is considered to be the neutral genotype and confers no additional CVD risk. However, the ε2/ε2 variation is associated with familial hypercholesterolemia, or genetically related high cholesterol [18]. Dysbetalipoproteinemia is a rare familial dyslipidemia characterized by approximately equally elevated serum cholesterol and triglyceride levels due to accumulated remnant lipoproteins in apolipoprotein ε2/ε2 homozygotes [18]. This genotype is very rare with the frequency in European ethnicity at 0.3%, African American ethnicity at 0.4%, and Asian ethnicity at 0.2% as compared to the ε3/ε3 genotype with 57.4%, 41.7% and 74.6%, respectively [19, 20]. This type of predictability is nutrigenetics and is not likely to be a candidate for epigenetic alteration. However, Yang, et al. (2007) found that for those with a single variant of APOε2, there was a 2.2-fold increased risk of myocardial infarction if there was a concurrent high saturated fat intake. They also found that for those that carry a single variation of the APOε4 gene, there was a 1.6-fold increased risk of myocardial infarction when paired with a high saturated fat diet. Compared to non-carriers, the APOε2 carriers who consumed a high saturated fat diet resulted in consistently elevated LDL cholesterol levels by (+ 17%) and carriers of the APOε4 variant had an increase of (+ 14%) [21]. This dietary intervention potentially activated these genes and increased the risk of CVD.

It has been suggested that those with the ε2/ε2 variation restrict their saturated fat intake to less than 12 g/day to prevent dysbetalipoproteinemia and myocardial infarctions [21]. However, there are further complications with suggested saturated fat intake recommendations. The apolipoprotein A2 (APOA2) gene determines the amount of saturated fat required to prevent dyslipidemia in those with DM2. Those with the CC genotype at rs5082 have an increased risk for dyslipidemia and should decrease saturated fat consumption to less than 10% [22]. The current nutritional United States Department of Agriculture (USDA) guideline already recommends the consumption of less than 10% of calories per day from saturated fats (20 grams) [23]. While this may seem auto-confirmatory, recently there is an extreme trend towards ketogenic, or very high fat diets. For those with either the ε2/ε2, ε4/ε4 or APOa2 genotypes, a high fat diet would clearly be contraindicated as it would significantly increase their risk of heart attack and cardiometabolic disease. The frequency of APO ε4/ε4 is much higher than APOε2. For European ethnicity, the frequency is 2.9%, African American ethnicity 3.6%, and Asian ethnicity is 1.0% while carriers of APOε4 are 24%, 34.1% and 15.4%, respectively [19, 20]. The Yang study reviewed carriers, thus the results are associated with having only one deleterious allele.

Hypertension is a hallmark diagnostic of cardiometabolic disease. For patients with hypertension and diabetes, treatment should be initiated when blood pressure is 140/90 mm Hg or higher, regardless of age [24]. Genomically, hypertension is related to the angiotensin-converting enzyme (ACE) and the angiotensinogen gene (AGT). Angiotensinogen, which is formed in the liver and controlled by AGT, is broken down by renin to form angiotensin I. Angiotensin II is formed from angiotensin I by ACE, which is then acted upon by the angiotensin receptor (AT1R) and ultimately converts to aldosterone. Angiotensin II regulates vasodilation and constriction, the sympathetic nervous system, antidiuretic hormone, and hormones in the adrenal glands. Aldosterone is the primary adrenal hormone for this system, as it regulates sodium retention, potassium excretion and ultimately fluid balance [25]. This is the underlying mechanism for hypertension drugs called ACE inhibitors, blocking angiotensin II production and thereby lowering aldosterone production and the above-described regulatory mechanisms. However, when there is a deletion in the ACE gene, there will be an increase in aldosterone and angiotensin II because the regulator has been removed [26]. This can result in anxiety, increased cortisol, memory problems, hypertension, and autoimmunity [27]. This imbalance in aldosterone will also alter the electrolyte balance, specifically, a decrease in sodium excretion, thus sodium retention and/or swelling and an increased excretion of potassium potentially resulting in hypokalemia [25]. Alterations in the AGT gene exacerbate the ACE deletion genotype while also carrying an inherent risk for pre-eclampsia, hypertension and insulin resistance. AGT directly increases angiotensinogen levels, thereby altering the renin-angiotensin-aldosterone system (RAAS) axis. Caproli, et al. found that almost 50% of all hypertensive cases were what they termed "salt-sensitive." They found that hypertensives carrying both the ACE and AGT polymorphisms responded favorably to a reduced sodium diet (less than 1500 mg per day). Those individuals who did not carry one or both of these polymorphisms are considered non-salt-sensitive and did not have any measurable effect when restricting sodium. This study elucidates that salt restriction is not always beneficial for all hypertensive patients [25].

> **Diagnostic and Testing Guidelines for Diabetes Mellitus Type 2**
> The diagnostic criteria for pre-DM2 consist of fasting plasma glucose (F/PG) 100 mg/dL (5.6 mmol/L) to 125 mg/dL (6.9 mmol/L) (impaired fasting glycaemia) OR 2-hour PG in the 75-g oral glucose tolerance test (OGTT) 140 mg/dL (7.8 mmol/L) to 199 mg/dL (11.0 mmol/L) (impaired glucose tolerance) OR hemoglobin A1C (A1C) 5.7–6.4%, while the diagnostics for DM2 are A1C $\geq$6.5% OR FPG $\geq$126 mg/dL (7.0 mmol/L).
> Fasting is defined as no caloric intake for at least 8 hour.∗OR 2-hour PG $\geq$200 mg/dL (11.1 mmol/L) during an OGTT. The test should be performed as described by the World Health Organization (WHO), using a glucose load containing the equivalent of 75 g anhydrous glucose dissolved in water OR in a patient with classic symptoms of hyperglycemia or hyperglycemic crisis, a random plasma glucose $\geq$200 mg/dL (11.1 mmol/L) [28].

Diabetes or dysregulated blood glucose is the other pillar of cardiometabolic disease. Type 2 diabetes (DM2) is a disease of insulin resistance, most often caused by central obesity and lack of exercise. This is considered a disease of lifestyle, thus epigenetically influenced. The idea that there may be an underlying genetic component to DM2 is appealing in many ways. Currently, the research is only suggestive of correlations rather than causations. One such example is the Transcription Factor 7 Like 2 (TCF7L2) locus. Those with the TT genotype at rs12255372 have a 67% increased risk of developing DM2. There is also preliminary evidence that this SNP could possibly increase the risk of breast cancer and aggressive prostate cancer [29]. Those with the genotype should include low glycemic index/glycemic load foods, reduce sugar consumption and limit processed grains to mediate this risk [29]. Interestingly, the global risk was not mediated in all study subjects, suggesting that the epigenome and environmental conditions were influencing the disease risk outcomes. This suggests an association between family history, genetics and disease outcomes.

Further, when considering diabetes risk associated with dyslipidemia, it is important to consider the gene Paraoxonase 1 (PON1). PON1 has both esterase and lactonase activity. The esterase enzymes assist in the catabolism of pesticides and certain pharmaceutical drugs while also protecting high and low-density lipoproteins from oxidation. Alterations in this gene could result in abnormally high levels of lipid peroxides, inflammation and compromised detoxification. The lactonase activity is specific to the catabolism of homocysteine thiolactones. When there is an increase in homocysteine, the result is an elevation of oxidizing thiolactones. When there are alterations in this gene, there is the potential for elevated levels of homocysteine and HDL-specific protein damage [30]. A specific SNP in PON1, rs662 (A) confers a higher risk of coronary heart disease and diabetes because it encodes for lower amounts of PON1 activity, thereby increasing oxidative stress and CVD-related disorders [31]. Specifically, when there are errors in PON1, HDL is oxidized, conferring a higher risk of cardiovascular disease. It has been shown that both Vitamin E supplementation and a diet higher in monounsaturated fats can help decrease both oxidized HDL and the activity of altered PON1. It should be noted that a diet high in saturated fat alone can inhibit the activity of PON1 even in the absence of genetic alteration, thus creating a functional enzyme deficiency. In this way, a diet high in saturated fats can increase the risk of both cardiometabolic disease and markers such as homocysteine [32, 33].

When presented with cardiometabolic disease, one of the most researched and used supplements is omega-3-fatty acid, often as fish oil. Typically, when we think of endothelial nitric oxide synthase (eNOS/NOS3), we are associating it with vasodilation, vascular smooth muscle relaxation via cGMP-mediated signal transduction pathway, vascular endothelial growth factor (VEGF)-induced angiogenesis in coronary vessels, and its ability to promote blood clotting via platelet activation. One intriguing hypothesis is that perhaps the mediating effects in CVD from NOS modulation depend on omega 3-fatty acid status. There is a SNP in NOS3 that determines whether increasing omega 3-fatty acids will decrease triglyceride levels in those with dyslipidemia. In rs1799983, it was found that subjects with the risk allele T (TT or GT) who had low levels of omega 3-fatty acids had 25% increases in serum triglyceride levels when compared to those with the GG genotype. This study suggested that those carrying the T allele increase their omega-3-fatty acid intake by 1.24 grams/day. Once subjects repleted with omega 3-fatty acids, their triglyceride levels normalized [34]. This may be the mechanism behind the plethora of research that indicates that a Mediterranean diet [35] or a diet high in fish like the Okinawan diet [36] confer such benefits in regards to CVD and cardiometabolic disease. It also may suggest a potential genetic component to omega-3 therapy in CVD and hypertriglyceridemia and should be a consideration in prescription/supplemental omega-3 therapy. It is also important to consider the corollary for this SNP, meaning those who do not have the risk allele and who may not be fish oil responders.

This brings us to the final pillar of cardiometabolic disease – obesity. It would be helpful for individuals and practitioners if there simply were a "fat" gene to determine whether dieting would be effective. Unfortunately, we have yet to find the genetic holy grail of obesity, but we have made some progress. Enter the fat mass and obesity gene (FTO). There has been quite an evolution of this gene and our understanding of its implications. To start, it was once a five-membered haplotype, meaning that there were five causally associated SNPs within the FTO locus that "predicted" the impact of

FTO [37]. Later, this group was revised to a three-membered haplotype, and then finally in 2015, a so-called causal variant was discovered [38]. SNP rs1421085 (C, C) confers a 1.7× increased obesity risk while the heterozygous form confers 1.3× increased obesity risk [39]. These are pretty low odds risks, and the magnitude does not exceed 4, meaning that this gene may play a minor role in obesity-related risks. This SNP disrupts the pathway for adipocyte thermogenesis involving ARID5B, IRX3, and IRX5, all regulatory genes for adipogenesis. Evidence suggests that because IRX3 binds so strongly to FTO, that other obesity-linked SNPs may be associated with IRX3 but not necessarily with FTO expression [40]. As you can see, we are merely scratching the surface. Unfortunately, we are unlikely to find a single answer to our complex obesity epidemic; instead, understanding and adjusting the epigenetics of obesity and cardiometabolic disease are far more promising. This epigenetic adjustment may include changes to diet and lifestyle, as well as optimizing methylation status.

Two conditions associated with cardiometabolic disease include polycystic ovarian syndrome (PCOS) and hemochromatosis (HFE). PCOS presents with infertility, hirsutism, polycystic ovaries, and insulin resistance, three out of five of these having considerable overlap with cardiometabolic syndrome [41]. The genomics of PCOS are heterogenous and have minimal GWAS (genome-wide associations). The SNPs that are associated with PCOS have been found only in non-obese PCOS patients, which is not the typical presentation of the disease [42]. There is, however, a sulfation gene, sulfotransferase 2A1 (SULT2A1), which may have a role in adrenal androgen excess in women with PCOS. Sulfotransferases are important for the metabolism of drugs and endogenous compounds. They convert them into more hydrophilic, water-soluble sulfate conjugates that then may be safely excreted. SULT2A1 specifically catalyzes the sulfation of steroids like DHEA and bile acids from the liver and adrenal glands. Alterations in these genes may increase the risk of PCOS by increasing endogenous levels of DHEA and/or be associated with hirsutism and insulin resistance [43].

Next is a clearly nutrigenetic condition, hemochromatosis (HFE). HFE C282Y accounts for 85% of all hemochromatosis cases. Another minor variant, HFE H63D, is responsible for the remaining 15% of cases, often with a milder presentation [44]. Hemochromatosis is an iron-overload syndrome that results in cirrhosis of the liver, diabetes, hypermelanotic pigmentation of the skin, heart disease, liver cancer, depression, and fatigue [45]. This condition can be exacerbated by hepatitis infections and it is associated with hypogonadism in males. When these genes are present, it is best to avoid excessive dosages of vitamin C as to not increase the absorption of non-heme iron. When presented with cases of liver disease, diabetes and cardiovascular disease, when the genomics does not necessarily make sense, it is important to rule out genetically inherited conditions such as HFE.

17.1.2 The Epigenetics of Psychiatric and Neurodegenerative Diseases

The gold standard of diagnostics for psychiatric diseases is the Diagnostic and Statistical Manual of Mental Disorders, 5th Edition, DSM-V. Dementia is characterized in the DSM-V; however, Alzheimer's disease (ALZ) is not specifically categorized [46]. This is a conundrum as more than half of dementia cases have ALZ as an etiology [47]. DSM-V characterizes dementia as having "multiple cognitive deficits, which include memory impairment and at least one of the following: aphasia, apraxia, agnosia or disturbance in executive functioning. Social or occupational function is also impaired. A diagnosis of dementia should not be made during the course of a delirium (a dementia and a delirium may both be diagnosed if the dementia is present at times when the delirium is not present) [46]." Interestingly, the same apolipoprotein (APOε) that increases the risk for cardiovascular disease (CVD) is also associated with ALZ. This raises the epigenetic question of true disease etiology: Is ALZ a disease of environmental exposures and lifestyle choices rather than Mendelian genetics? Many would argue yes, and some have even gone so far as to rename ALZ type 3 diabetes [48]. When we consider inheritable diseases, we should also remember that families often experience identical environmental influences. Therefore, it is challenging to decipher true genetic etiology and risk when the epigenome greatly confounds such concepts.

Let us consider the nutrigenetics, or Mendelian genetics, that are associated with ALZ. As with cardiovascular disease (CVD), the e3 variant in APOε is considered to be neutral with an odds ratio of 1 [49]. The ε2/ε2 genotype is actually protective of ALZ with approximately a 0.5 (0.22–1.1) odds ratio [49]. While the frequency of this ε2/ε2 genotype is quite rare and protective for ALZ, it also increases the risk for beta dyslipidemia. This is a case where the phenotype would present with CVD and not ALZ. However, the more common genotype, ε4, connects the two conditions clearly. Carrying one variant of ε4 is associated with approximately 3.4 (3–3.8) times increased odds of developing ALZ. Carrying two copies of the ε4 variant exponentially increases the risk resulting in an approximately 12.9 (10.2–16.2) times increased odds risk of developing ALZ in populations of European ancestry [49]. Thus, for those that carry even one of the ε4 variants, there is an increased risk for both CVD and ALZ. Further associating CVD with ALZ, Corneveaux et al. also found a consistent relationship with ALZ disease risk predictability to polymorphisms in angiotensin-converting enzyme (ACE), one of the hypertension-related SNPs [49]. There are many other genetic factors that play into the overall risk of developing ALZ that should also be considered outside of APOe, ACE, and CVD genetics. Studies have identified certain GWAS SNPs that are associated with both early and late onset ALZ, as well as many other regulating factors such as amyloid precursor protein (APP), Presenilin 1(PSEN1) and Presenilin 2 (PSEN2) [50]. The jury is still out on the etiology of ALZ and CVD; however, dietary and lifestyle choices clearly play an epigenetic role in the development of both diseases [49]

The DSM-V diagnostic criteria for generalized anxiety disorder (GAD) include the following: "The presence of excessive anxiety and worry about a variety of topics, events, or activities; the worry is experienced as very challenging to control; the anxiety and worry are associated with at least three of the following physical or cognitive symptoms (in children, only one symptom is necessary for a diagnosis of GAD): edginess or restlessness, tiring easily; more fatigued than usual; impaired concentration or feeling as though the mind goes blank; irritability (which may or may not be observable to others); increased muscle aches or soreness; difficulty sleeping (due to trouble falling asleep or staying asleep, restlessness at night, or unsatisfying sleep) [46]".

Generalized anxiety disorder (GAD) affects approximately 22% of the population, and more often in females than males [51]. With GAD affecting approximately one quarter of the population, there are epigenetic lifestyle factors and genomic alterations that need to be addressed.

First, our modern lifestyle prizes working excessive hours, typically away from home, while also trying to manage the basics of life, family, and household. Many work so many hours that they compromise their sleeping habits (impaired cortisol regulation), not to mention their diet (obesity, nutrient deficiency) and activities outdoors (vitamin D deficiency). Each of these factors are individually enough to alter the epigenetic activation of disease. In ▶ Chap. 18, the complex system of methylation is detailed. In brief, methylation is the process of moving methyl groups from one bio-molecule to the next, ultimately functioning in the regulatory capacity for everything from DNA synthesis, expression and modulation to other complex systems such as detoxification. Many of our modern diseases can be connected to alterations in this complex system due to the implications of regulating gene expression [52]. GAD, panic disorder and milder presentations of anxiety are not exceptions.

There is an enzyme called phenylethanolamine N-methyltransferase (PNMT) (notice the word, methyl-transferase; it transfers a methyl group from one molecule to another). In this case, PNMT specifically converts norepinephrine to epinephrine [53]. In more common terms, these bio-molecules are called noradrenaline and adrenaline. Chronic activation of this enzyme increases adrenaline and results in anxiety, insomnia, and ultimately adrenal fatigue [53]. The cofactor for this particular enzyme is the universal methyl donor, s-adenosyl-methyltransferase (SAM) [54]. When SAM releases its methyl group, it is first converted to s-adenosyl-homocysteine and ultimately into homocysteine, an inflammatory amino acid metabolite associated with neuroinflammation and CVD. Therefore, chronic stress can result in anxiety, elevated levels of homocysteine, CVD and a depletion of beneficial methyl donors, thereby increasing overall disease risk. This enzyme provides the biological basis for the connection between chronic stress and disease. Mind-body techniques such as yoga, tai-chi, meditation, gentle exercise, sound therapy and prayer are helpful strategies to reduce daily stress [55].

Delving a bit deeper into the genetics and epigenetics of anxiety, there are two major enzymes (genes) that can control either catecholamine regulation (dopamine, epinephrine and norepinephrine) or other excitatory neurotransmitters like glutamate. For more detailed information on catecholamine synthesis, please refer to ▶ Chap. 18. The regulatory enzyme for the catecholamines is called catechol-o-methyltransferase (COMT) (another methyl-transferase). COMT transfers methyl groups from SAM to the catecholamines dopamine, epinephrine and norepinephrine while assisting in their degradation (◘ Fig. 17.1) [56]. COMT also transfers other methyl groups, like those found in foods like green tea, citrus (quercetin), and potatoes, along with the hormone catechol estrogen [57]. The issue with this being a methyltransferase is that it is dependent on the flow of methyl groups generated in one-carbon metabolism, and recycled or processed in methylation pathways. Having both too little SAM and too much s-adenosyl homocysteine (SAH) can create inhibition or downregulation of enzymatic activity resulting in an increase of catechols. This can be independent of genetic alteration creating a functional inhibition of this enzyme [58]. This enzyme has broader applications and connects neurotransmitter regulation, estrogen metabolism and diet. Like most nutrigenomic enzymes, this one is dependent on a nutrient cofactor, or "catalyst" that is required for the enzyme to function properly (note: this is not a true catalyst in chemical terms as the catalyst is not always consumed, rather in these cases, we are describing a circumstance where the substrate is being converted to a product and this reaction requires a cofactor). The cofactor for COMT is the incredibly important mineral, magnesium [59]. In addition to serving as a cofactor for COMT, magnesium is also a cofactor for 300 other enzymatic processes in the body [60, 61]. The phenotype for this type of anxiety typically excludes panic attacks, but there is a ruminating presentation of anxiety and an increased risk for insomnia [62] and palpitations [63]. Anecdotally, those with COMT inhibition (functional or SNPs) typically have "type A" personalities due to increased dopamine and epinephrine levels, are often successful and may or may not have disordered sleep patterns.

Magnesium deficiency can be the result of many things, including downregulation of the COMT enzymes. If you consider basic sciences and relate that all enzymes have a substrate, a catalyst and a product, when there is an upregulation of this pathway, there will be a depletion of the substrate and catalyst with an increase in the product. If there is a downregulation, there is either a catalyst deficiency (nutrient deficiency) or a SNP. In the case of upregulation, COMT depletes methyl donors by moving them too swiftly away from the catecholamines, thus ultimately decreasing these levels while simultaneously decreasing its catalyst cofactor, magnesium. This may present with fatigue and potentially dopamine depression. To validate this on an organic acids test, the markers vanilmandelate (VMA) and homovanillate (HVA) would be decreased showing excessive breakdown. In

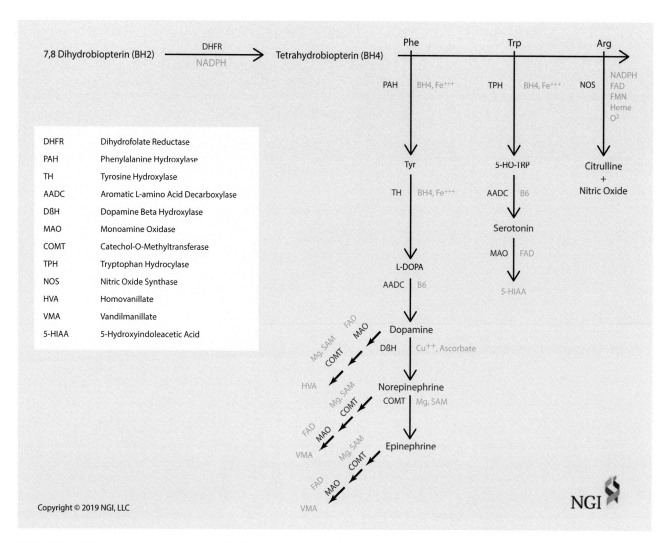

Fig. 17.2 BH4 and neurotransmitter synthesis. (Courtesy of Nutritional Genomics Institute, LLC)

downregulation, there is a "block" in the pathway, resulting in the inhibition of catecholamine breakdown, thereby resulting in increased levels of these neurotransmitters and the presentation of anxiety. The HVA/VMA pattern will be reversed in this case. If the presentation is consistent and there is a known alteration in COMT, magnesium threonate may ameliorate symptoms and anxiety. This particular form of magnesium has the potential to cross the blood-brain barrier and is considered to be a superior form for neurological conditions [64]. Interestingly, clinically, this pathway strives for homeostasis. Many patients will experience a refractory response or increased anxiety or insomnia when given magnesium. In these cases, it is important to investigate the other enzyme that also degrades these neurotransmitters, monoamine oxidase.

According to the DSM-V, panic disorder includes the following: "panic attacks must be associated with longer than 1 month of subsequent persistent worry about: (1) having another attack or consequences of the attack, or (2) significant maladaptive behavioral changes related to the attack. To make the diagnosis of panic disorder, panic attacks cannot directly or physiologically result from substance use (intoxication or withdrawal), medical conditions, or another psychiatric disorder. Other symptoms or signs may include headache, cold hands, diarrhea, insomnia, fatigue, intrusive thoughts, and ruminations." [46] (Fig. 17.2).

There is another presentation of anxiety as alluded to above. While this phenotype of anxiety can also include symptoms such as insomnia, it *includes* panic attacks associated with panic disorder. Panic attacks are associated with an enzyme that converts the excitatory neurotransmitter glutamate into the inhibitory neurotransmitter, γ-Aminobutyric acid (GABA). Not to be confused with generalized anxiety disorder, however, the name of this enzyme is glutamic acid decarboxylase 1 or GAD1. When there are alterations in these enzymes, there is a "block" in the conversion, resulting in an increase in glutamate and a decrease in GABA. This block is typically related to the limitation of the cofactor, vitamin B6 in the active form of pyridoxal-5-phosphate (P5P). To "validate" the decreased activity of this SNP, there is an organic acids marker called xanthurenate. When elevated, this is confirmation of a P5P deficiency [65].

Glutamate is found in a variety of foods, especially processed foods. Its most notorious conformation is monosodium glutamate or MSG. MSG can increase glutamate levels in the brain and increase both neuronal and gastrointestinal inflammation via nuclear factor kappa-beta (NF-κB) [66]. Glutamate is also a modulator of the kynurenine pathway, which is limited by tryptophan and modulated by the nutrients pyridoxal-5-phosphate (vitamin B6), niacin (vitamin B3), iron, and magnesium. This association is consistent with the presentation of anxiety with concurrent depression as tryptophan is the precursor to serotonin. Alterations in this pathway may result in an increase in quinolinate, a neuroinflammatory metabolite responsible for modulating N-methyl-D-aspartate (NMDA) receptors in the brain. Excessive stimulation of the NMDA receptors results in neuronal inflammation and is associated with conditions like autism spectrum disorder and ALZ [67]. These GAD1 isoforms provide potential etiologies for panic disorders being genetically inherited [68]. For those having alterations in GAD1, supplementation with P5P, magnesium, niacin, and potentially iron (if deficient) can help to modulate the panic attack phenotype. If there is concurrent depression, consider also supplementing with tryptophan or 5-hydroxytryptophan.

The GAD enzyme has another interesting association. In cases where there is a homozygous alteration in the GAD enzyme, there is an increased risk for glucose dysregulation. This particular alteration results in a peculiar presentation, whereas there are moderately elevated levels of plasma blood glucose with low levels of fasting insulin. This presentation typically has the phenotype of reactive hypoglycemia, thin habitus and an erroneous diagnosis of DM2 [69]. The connection between these two clinical presentations lies in the physiology of GAD being expressed in both the brain and the beta islet cells of the pancreas [70]. To correct the discrepancy, moderate to high dosages of P5P provide an intervention to stabilize the glucose dysregulation. Interestingly, ACE inhibitors also modulate this particular mechanism, contributing to the complexity of a clinical diagnosis [71]. An ACE deletion and homozygous GAD isoform would create the phenotype of cardiometabolic disease. Clinically, this would be difficult to distinguish, perhaps other than the thin habitus. The lesson is to not make assumptions about the etiology of disease based solely on clinical presentation. The reality is often complex.Depression (major depressive disorder, MMD) is defined by the DSM-V as: "depressed mood or a loss of interest or pleasure in daily activities for more than 2 weeks; mood represents a change from the person's baseline; impaired function: social, occupational, educational; specific symptoms, at least five of these nine, present nearly every day:

1. Depressed mood or irritable most of the day, nearly every day, as indicated by either subjective report (e.g., feels sad or empty) or observation made by others (e.g., appears tearful)
2. Decreased interest or pleasure in most activities, most of each day
3. Significant weight change (5%) or change in appetite
4. Change in sleep: Insomnia or hypersomnia
5. Change in activity: Psychomotor agitation or retardation
6. Fatigue or loss of energy
7. Guilt/worthlessness: Feelings of worthlessness or excessive or inappropriate guilt
8. Concentration: Diminished ability to think or concentrate, or more indecisiveness
9. Suicidality: Thoughts of death or suicide, or has suicide plan

DSM-V proposed (not yet adopted) anxiety symptoms that may indicate depression: irrational worry, preoccupation with unpleasant worries, trouble relaxing, feeling tense, fear that something awful might happen" [46].

Lastly, we will discuss depression. Please refer to ◘ Fig. 17.2 for additional information regarding the synthesis and catabolism of dopamine and serotonin along with specific variant information. Clinically, there are two types of depression [71]. The first and most commonly known is related to serotonin deficiency. This mechanism has spawned an entire class of pharmaceutical drugs called selective serotonin reuptake inhibitors (SSRIs). While success is found in a portion of MDD patients with SSRIs, there is a considerable portion of non-responders [72]. Clinically, serotonin deficiency-related depression encompasses feelings of worthlessness, a tendency to withdraw from social activities, lack of motivation and excessive sleeping. To validate serotonin depression, there is an organic acids marker called 5-hydroxyindoleacetic acid (5-HIAA), a serotonin catabolism metabolite. This marker will be low if there is a restriction in serotonin metabolism with concurrent suppression of the amino acid tryptophan [67]. There may also be variations in the enzymes that create serotonin, such as tryptophan hydroxylase. The cofactors for this enzyme are (tetrahydrobiopterin) BH4, P5P, riboflavin, and copper. Addressing any abnormalities found in these nutrients could potentially relieve the strain on the enzyme responsible for forming serotonin [67]. In cases such as these, consider supplementing with tryptophan, 5-hydroxytryptophan or St. John's wort. It should also be noted that in addition to the expected downregulation phenotype, there are also upregulations in the genes that regulate this pathway. Monoamine oxidase A (MAO-A) specifically regulates serotonin catabolism and is the enzyme that stimulated the class of medications known as MAOIs or monoamine oxidase inhibitors. In these cases, you may find that 5-HIAA is actually elevated along with the validation marker for riboflavin, glutaric acid [73].

The second, less frequently recognized form of depression involves dopamine deficiency. Clinically, these patients present with a decreased interest or pleasure in most activities. Often in these cases, there is an upregulation in dopamine beta hydroxylase (DβH), which converts dopamine to norepinephrine. DβH requires the cofactors of vitamin C and copper, elucidating the characteristic low copper phenotype in dopamine depression [74]. A blockage in this enzyme is associated with schizophrenia treatment outcomes as it can theoretically increase dopamine levels [75]. Dopamine can also be restricted by one of four dopamine receptors. There may be functional blocks in these receptors due to autoantibodies or polymorphism. Acute neuropsychiatric conditions like pediatric autoimmune neuropsychiatric disorders associated with streptococcal infections (PANDAS) and pediatric acute-onset neuropsychiatric syndrome (PANS) are also associated with autoantibodies to dopamine receptors. In the majority of these cases, supplementing with mucuna pruriens, also known as the velvet bean, supplies a direct precursor to dopamine, L-DOPA and resolves the symptoms of dopamine depression [76].

Sometimes, there will be abnormalities in both dopamine and serotonin. If there is evidence to support this conclusion, then there may be an enzyme deficiency in aromatic L-amino acid decarboxylase (AADC). This enzyme is P5P-dependent and is responsible for both the conversion of L-DOPA to dopamine and 5-hydroxytryptophan to tryptophan. Supplemental P5P is again the correct intervention for depression. It is also possible that there are genetic abnormalities or functional blocks in both the serotonin and dopamine pathways, thus this consideration should not be disregarded. For overall support, supplementation with P5P, magnesium and riboflavin are complementary for the resolution of anxiety and depression.

17.2 The Epigenetics of Irritable Bowel Disease and Dysbiosis

For every client that has a neurological condition, there are five more that have a variation of dysbiosis, leaky gut or irritable bowel disease (IBD). In this section, we will briefly discuss dysbiosis, environmental factors such as dietary influences and probiotic use associated with dysbiosis, and the genetics of celiac disease (CD). This is not an exhaustive review and is designed as an introduction. It is important to remember that humans have a symbiotic relationship with our microflora. While it is popularly believed that our microbes outnumber us in a ratio of 10:1, these numbers have recently been recalculated to show that while there are still more microbial species than there are human, the ratio is closer to 1:1 [77]. Regardless of these ratios, the concept that we equally share the control panel with the microbiome has considerable implications on our health, disease, and longevity.

Celiac disease affects only a small percentage of the population. It is estimated that between Europeans and Americans, there is an approximate 0.5–1.26% frequency of CD [78]. CD presents with damaged epithelial cells of the intestine, malnutrition, diarrhea, anemia, osteoporosis, dermatitis herpetiformis, dental enamel hypoplasia, an increased risk of cancer and neurological symptoms such as headache, brain fog, and paresthesia [78, 79]. The etiology of this disease is an autoimmune reaction to the gluten protein found in cereal grains such as wheat, barley and rye [79]. The current guidelines for the diagnosis of CD includes: Signs and symptoms suggestive of CD, anti-transglutaminase type 2 antibody (anti-TG2) with levels typically more than 10 times the upper limit of normal and a positive confirmation test of anti-endomysium-IgA antibodies (EMA) [79]. As we begin to explore the genetic susceptibility of CD, we will encounter a condition that has recently been coined to explain the sudden rise in gluten sensitivity in the absence of CD. This new condition has been named non-celiac gluten sensitivity (NCGS), and it is associated with heterozygous alterations in the human leukocyte antigen (HLA) genes that determine CD. HLAs are a highly polymorphic gene complex that encodes for the major histocompatibility complex (MHC) proteins. These MHC proteins are primarily responsible for immune regulation and are the genes that require matching for organ transplants [80]. Alterations in these genes increase the risk of several autoimmune diseases.

About 90–95% of CD patients carry a certain genotype called HLA-DQ 2.5. When there is a homozygous mutation for this genotype, there is a 50-fold increased risk of developing CD [81]. The other remaining 5–10% of CD patients have a milder alteration in HLA-DQ 8. The presence of a homozygous alteration in this genoset results in a 17-fold increased risk of CD development [81]. In some cases, especially in NCGS, there are heterozygous alterations in these SNPs or other related SNPs that result in a genotype called HLA-DQ 2.2. Please refer to the Epigenetic SNPs chart associated with this chapter for specific details, genes and SNPs regarding HLA-DQ 2.5, HLA-DQ 8, and HLA-DQ 2.2 (see Table 17.1).

Interestingly, 40% of the general population also has alterations in these DQ genes, yet they do not develop CD. NCGS can also be present in the absence of any known genomic alteration, which complicates the suspected etiologies [81]. Dr. Stephanie Seneff proposes that this dramatic rise in NCGS and CD is associated with the relatively recent advances in biotechnology. Glyphosate is commonly known as Roundup ® and is trademarked by Bayer AG [45]. This chemical is not only a ubiquitous herbicide, but the majority of our mega crops have been genetically altered to be "Roundup-ready" or have the ability for the plant organism to survive glyphosate application, while killing the other "weeds." These genetically modified organisms (GMOs) can potentially compromise the cytochrome p450 system, result in NCGS, irritable bowel syndrome and unfavorable alterations in the microbiome as well as negatively stimulate the immune system [82]. Unfortunately, GMO wheat, corn, and soy comprise a large portion of the American diet as recommended by the USDA [83]. Here we see can clearly see the advancements in crop productivity directly impacting the epigenome and risk for CD and NCGS. These environmental conditions (i.e., how much of these foods we consume) ultimately determine if these genes are to become epigenetically activated. Remember, the presence of the genetic alteration increases the risk of the disease, but it not solely diagnostic.

This balance of the microbiota is far more important than we imagined in the age of antibiotics. Further, there is a great deal of variability in the microbiomes of individuals. We have learned that these nuances relate to health status, age, diet, microbial interactions, and even host genotype.

Table 17.1 Epigenetic SNPs hart

SNPS	Co factors	Organic acids/clinical associations	Amino acids/other tests/interventions	rs Numbers
Epigenetics of cardiometabolic/neurodegenerative diseases				
APOε 2 *Apolipoprotein E 2* ε2/ε2 is protective of Alzheimer's disease but associated with familial hypercholesterolemia Dysbetalipoproteinemia is a rare familial dyslipidemia characterized by approximately equally elevated serum cholesterol and triglyceride levels due to accumulated remnant lipoproteins in apolipoprotein ε2/ε2 homozygotes	N/A	Dysbeta-lipoproteinemia ↑Cholesterol, ↑triglycerides	Lipid panel, CMP; those with this genotype should limit saturated fat intake to 12 g/day; advanced lipid panel/MPO	rs7412 (T) rs429358 (T)
APOε 3 *Apolipoprotein ε 3*	N/A	Considered to be the neutral genotype for both CVD and ALZ.	Lipid panel, CMP	rs7412 (C) rs429358 (T)
APOε 4 *Apolipoprotein ε 4* Carrying a single copy of the ε4 variant is associated with about 1.5–2 times increased odds of developing Alzheimer's disease (ALZ) Two copies of the ε4 variant is associated with approximately nine times increased odds risk of Alzheimer's disease populations of European ancestry	N/A	↑Cholesterol; a high saturated fat intake was associated with a 2.2-fold increased risk of MI among carriers of APO ε∗2 (OR = 3.17; 95% CI, 1.58–6.36) and with a 1.6-fold increase among carriers of the −491T and APOε ∗4 variants together (OR = 2.59; 95% CI, 1.38–4.87). Consistently, a high fat diet elicited a greater response in LDL cholesterol among carriers of APOε ∗2 (+17%) and APOε ∗4 (+14%) compared to noncarriers (+6%)	Lipid panel, CMP, investigate other ALZ-related genetics	rs7412 (C) rs429358 (C)
APOA2 *Apolipoprotein A2* In diabetics, this gene determines the amount of saturated fat required to prevent dyslipidemia	N/A	↑Cholesterol, ↑triglycerides Those with the CC genotype have an increased risk for dyslipidemia and should decrease saturated fat consumption to less than 10%	Lipid panel, CMP, advanced lipid panel/MPO	rs5082 (C)

(continued)

◨ Table 17.1 (continued)

SNPS	Co factors	Organic acids/clinical associations	Amino acids/other tests/interventions	rs Numbers
ACE del *Angiotensin converting enzyme* Deletion results in upregulation of activity Results in higher rate of conversion of angiotensin 1 to angiotensin 2 ↑aldosterone	N/A	↑Aldosterone can result in anxiety increased cortisol, learning, memory problems, hypertension and autoimmunity Electrolyte balance issues, specifically, ↓sodium excretion (so sodium retention/swelling) and ↑excretion of potassium excretion (low potassium); cardiovascular complications	Lipid panel, CMP, monitor blood pressure; modify stress	rs4343(A) is a proxy for ACE Del16; GA, AA genotypes should restrict sodium intake to less than 1500 mg/day
AGT *Angiotensinogen gene* Specifically increases angiotensinogen levels, exacerbating the ACE deletion; requires renin to degrade; works in the adrenals Associated with insulin resistance, especially when combined with ACE Associated with increased hypertension and its associated diseases like pre-eclampsia and heart disease	N/A	↑Insulin resistance, ↑hypertension, ↑angiotensinogen Electrolyte balance issues, specifically, ↓sodium excretion (so sodium retention/swelling) and ↑excretion of potassium excretion (low potassium); cardiovascular complications	Lipid panel, CMP, monitor blood pressure; modify stress	rs699, M235T (C) When combined with ACEdel16=salt sensitive hypertension— restrict sodium intake to less than 1500 mg/day
FTO *Fat mass and obesity* Causal variant that disrupts a pathway for adipocyte thermogenesis involving ARID5B, IRX3, and IRX5; mechanistic basis for the genetic association between FTO and obesity	Molecular oxygen, alpha-ketoglutarate and iron	↑Weight gain, ↑insulin resistance, ↑risk for DM2	Lipid panel, CMP, fasting Insulin	rs1421085(C)
TCF7L2 *Transcription factor 7 Like 2* Increased risk of diabetes mellites 2	N/A	TTs have a 67% increased risk of developing DM2. They should include low GI/GL foods, reduce sugar consumption and limit grains to reduce this risk	Lipid panel, CMP, HBA1C, fasting insulin	rs12255372(T); slightly increases (~1.5×) risk for type-2 diabetes and possibly breast cancer and aggressive prostate cancer
NOS1 *Nitric oxide synthase 1* Catalyzes the conversion of arginine to nitric oxide Also creates the byproduct citrulline	NADPH and NADP+	↑Orotate, ↑citrate, ↑isocitrate, ↑cis-aconitate	Potential abnormalities in arginine, citrulline, nitric oxide Lipid panel to screen for hypertension	rs7298903(C) rs3782206(T) rs2293054 [Val1353Val]

◘ Table 17.1 (continued)

SNPS	Co factors	Organic acids/clinical associations	Amino acids/other tests/interventions	rs Numbers
NOS2 *Nitric oxide synthase 2* Inducible nitric oxide synthase (iNOS) is considered a reactive free radical involved in neurotransmission along with antimicrobial and antitumoral activities Found in the immune and cardiovascular systems and is involved in immune defense against pathogens May be induced by lipopolysaccharides and cytokines Commonly found in inflammatory disorders	Manganese	↑Orotate, ↑citrate, ↑isocitrate, ↑cis-aconitate	Potential abnormalities in arginine, citrulline, nitric oxide	rs2297518(A/T) rs2248814(G/C) rs2274894(T)
SOD2 *Superoxide dismutase 2* Superoxide dismutase is an antioxidant enzyme required to catalyze the dismutation of two superoxide radicals into hydrogen peroxide and oxygen Protects tissues from oxidative stress	Manganese	Increased risk of ischemic heart disease in 9188 participants from the Copenhagen City Heart Study Later found to reduce the risk of chronic obstructive pulmonary disease and acute exacerbations of chronic obstructive pulmonary disease in three independent, large population studies.	RBC minerals, lipid panel; Do not practice high intensity exercise without guidance	rs1799895(G) R213G
PON1 *Paraoxonase* Esterase enzymes that break down pesticides and pharmaceutical drugs Involved in protecting both HDL and LDL from oxidation. Its lactonase activity is involved in the breakdown of homocysteine thiolactones	Displaces calcium	Those with alterations have an increased risk of pesticide poisoning, diabetes, atherosclerosis and heart disease High fat diet alone inhibits enzyme	Lipid panel, CMP, fasting insulin, Homocysteine; Increase MUFA and Vitamin E	rs662 (A) confers a higher risk of coronary heart disease
HFE-C282Y Accounts for 85% of all hemochromatosis cases Hemochromatosis results in cirrhosis of the liver, diabetes, hypermelanotic pigmentation of the skin, heart disease, liver cancer, depression and fatigue Made worse with hepatitis infections Associated with hypogonadism in males	Avoid excessive Vitamin C	↑Ferritin levels, sometimes accompanied with ↑transferrin saturation and ↑hepatic iron concentration (HIC)	Hepatic iron concentration in hemochromatosis- affected patients ranges from 5000 to 30,000 µg/g (normal values, 100–2200 µg/g) TIBC ranges from 200 to 300 µg/dL in hemochromatosis-affected patients (normal range, 250–400 µg/dL)	rs1800562 C282Y (AA)

(continued)

Table 17.1 (continued)

SNPS	Co factors	Organic acids/clinical associations	Amino acids/other tests/interventions	rs Numbers
HFE-H63D Accounts for 15% of all hemochromatosis cases. Hemochromatosis results in cirrhosis of the liver, diabetes, hypermelanotic pigmentation of the skin, heart disease, liver cancer, depression and fatigue Made worse with hepatitis infections Associated with hypogonadism in males Risk allele C can result in mild presentation, but was also found to be associated with iron deficiency in women	Avoid excessive Vitamin C	↑Ferritin levels, sometimes accompanied with ↑transferrin saturation and ↑hepatic iron concentration (HIC)	Hepatic iron concentration in hemochromatosis-affected patients ranges from 5000 to 30,000 µg/g (normal values, 100–2200 µg/g) TIBC ranges from 200 to 300 µg/dL in hemochromatosis-affected patients (normal range, 250–400 µg/dL)	rs1799945 – H63D (GG/CC) Risk allele C
SULT2A1 *Sulfotransferase 2A1* Sulfotransferases are important for the metabolism of drugs and endogenous compounds and convert them into more hydrophilic water-soluble sulfate conjugates that then may be excreted. SULT2A1 catalyzes the sulfation of steroids and bile acids in the liver and adrenal glands	N/A	May have a role in adrenal androgen excess in women with polycystic ovary syndrome (PCOS)	CMP, 24-hour urinary hormones test, FHS, LH, SHBG	rs11083907(A) rs11569679(T) rs2547231(C) rs296366 (T) rs4149449(T) rs4149452(T)
Epigenetics of psychiatric diseases				
GAD1 *Glutamic acid decarboxylase 1* Converts the excitatory neurotransmitter glutamate into the inhibitory neurotransmitter GABA GAD67 isoform encoded by the GAD1 gene is the rate-limiting enzyme in GABA synthesis from glutamate in the brain	P5P	↑Xanthurenate, mutations may result in anxiety/panic disorders, increased blood glucose levels and hemoglobin A1C, with decreased insulin levels	Fasting insulin, CMP, HBA1C	GAD1: rs2241165(C) rs2058725(C) rs3791850(A)
COMT *Catechol-O-Methyltransferase* Transfers methyl group from SAM to the catecholamines. Breaks down dopamine, epinephrine and norepinephrine. Transfers methyl group from SAM to catechols from foods, (green tea, potatoes), antioxidants (quercetin), and hormone catechols (estrogen)	Magnesium SAM	↓VMA, ↓HVA (downregulation; most common; suggests methyl trapping) ↑VMA, ↑HVA (Upregulation –suggests methyl dumping)	Consider RBC Magnesium	rs4680(A) rs4633(T)
PNMT *Phenylethanolamine N-methyltransferase* Converts norepinephrine to epinephrine	SAM/methyl donor	↑VMA, ↓SAM; Stress, insomnia, anxiety, methyl donor depletion The connection between stress and disease	Check COMT Status	rs149585781(A) rs3764351 rs773317399(A)

Table 17.1 (continued)

SNPS	Co factors	Organic acids/clinical associations	Amino acids/other tests/interventions	rs Numbers
DR1 *Dopamine Receptor 1* Most abundant dopamine receptor in the central nervous system Regulates neuronal growth and development Mediates some behavioral responses Modulates dopamine receptor D2-mediated events Reward and pleasure centers	Copper is the cofactor for DBH	↓HVA; Associated with bipolar disorders attention deficit hyperactivity disorder and alcoholism	RBC copper, ceruloplasmin, Iron panel with ferritin (iron and copper have the same transporter)	rs4867798(C) rs251937 (C/G) haplotype is: rs265981(C) rs4532(A) rs686(T)
DR2 *Dopamine receptor 2* DR2 is associated with adult walking behavior, associative learning, auditory behavior, behavioral responses to cocaine and ethanol, feeding and grooming behavior and myoclonic dystonia	Copper is the cofactor for DBH	↓HVA; Dopamine is required for fine motor control, cognition, mood and neurotransmitter balance. (High levels = schizophrenia or paranoia; low levels = Parkinson's)	RBC copper, ceruloplasmin, Iron panel with ferritin (iron and copper have the same transporter)	rs6277(C) rs1799732(−/−)
DR3 *Dopamine receptor 3* Localized to the limbic areas of the brain that are associated with cognitive, emotional and endocrine functions May be associated with susceptibility to hereditary essential tremor 1	Copper is the cofactor for DBH	↓HVA	RBC copper, ceruloplasmin, Iron panel with ferritin (iron and copper have the same transporter)	rs167771(G)
DR4 *Dopamine receptor 4* Responsible for neuronal signaling in the mesolimbic system of the brain Regulates emotion and complex behavior	Copper is the cofactor for DBH	↓HVA; Linked to many neurological and psychiatric conditions such as schizophrenia, bipolar disorder, addictive behaviors and eating disorders (i.e., anorexia nervosa) Target for drugs that treat schizophrenia and Parkinson's disease Involved in behavioral fear response; response to cocaine and ethanol; olfactory learning; response to amphetamines, histamine and steroid hormones; involved in short term memory and social behavior	RBC copper, ceruloplasmin, Iron panel with ferritin (iron and copper have the same transporter)	rs916457(T) rs752306(T) rs3758653(C) rs1800443(G) rs4331145(A)
Nutrient–specific disease associations				
VDR - TAQ *Vitamin D receptor* Its affinity for calcitriol is roughly 1000× greater than its affinity for calcidiol.	Vitamin D	VDR-TAQ SNPs are often linked to better tolerance of methyl donors (conflicting evidence)	Calcidiol	rs731236(A)

(continued)

Table 17.1 (continued)

SNPS	Co factors	Organic acids/clinical associations	Amino acids/other tests/interventions	rs Numbers
VDR-Bsm *Vitamin D receptor* Its affinity for calcitriol is roughly 1000× greater than its affinity for calcidiol	Vitamin D	Rapidly converts calcidiol to calcitriol Clinical picture includes consistently low calcidiol levels with elevated calcitriol levels Adverse effects are common when calcidiol is supplemented in higher levels (50,000 IU)	Calcidiol, calcitriol	rs1544410(A) Risk allele A: women have an increased risk of low bone mineral density. Conversely, A 26 study meta-analysis estimated a decreased risk of osteoporosis associated with the G;G genotype
VDR – GWAS *Vitamin D receptor – Genome-wide association*	Vitamin D	Homozygotes were associated with an ~2.5× higher risk of prostate cancer compared to homozygote carriers of the more common (C) allele in the 630 Caucasian patients studied	Calcidiol, calcitriol	rs2107301(T)
TCN1 *Transcobalamin I* Secreted by the salivary glands and is a B12 binding protein. TCN1 helps B12 survive the HCL in the stomach and the complex travels to the intestine	Vitamin B12; hydroxocobalamin bypasses this transporter	↑MMA; Transports up to 80% of vitamin B12. Thought to function as a circulating storage form and may prevent bacterial use of the vitamin	Serum B12; check FIGLU and serum folate to rule out folate trapping	rs526934(G)
TCN2 *Transcobalamin 2* Once inside the enterocytes of the ileum, B12 breaks apart from TCN1 to bind to TCN2, which then carries B12 to the liver	Vitamin B12; hydroxocobalamin bypasses this transporter	↑MMA; major part of the secondary granules found in neutrophils and facilitates the transport of cobalamin to the liver; 20–30% of cobalamin is transported on the TCN2	Serum B12; check FIGLU and serum folate to rule out folate trapping	rs1801198(G)
GIF *Gastric intrinsic factor* Is produced by the parietal cells and is also a B12 binding protein	Vitamin B12	↑MMA, pernicious anemia, familial pernicious anemia, acute lymphoblastic leukemia, high-altitude polycythemia In gastritis (non-*H. pylori* related), those with the FUT2 secretor variant had decreased amounts of GIF secretion. This was true even in heterozygous FUT2 mutations	Intrinsic factor, MMA, serum B12; check FIGLU and serum folate to rule out folate trapping Consider GI/stool testing	rs558660(G)
GSTT1Del *Glutathione S-transferase theta 1* Conjugates reduced glutathione Many exogenous and endogenous hydrophobic electrophiles	1200 mg Vitamin C/day	Potential ↑ in pyroglutamate, alpha-hydroxybutyrate and alpha-ketobutyrate	Potential abnormalities in glutamate, cysteine, glycine	rs79605217 (G/C) rs796052136(−/−)

Table 17.1 (continued)

SNPS	Co factors	Organic acids/clinical associations	Amino acids/other tests/interventions	rs Numbers
BCMO1 *Beta-carotene 15, 15´-monooxygenase 1* Converts beta-carotene into the active form of vitamin A, retinol (beta-carotene ->trans-retinol) Nomenclature: renamed BCO1 in early 2018	Iron	May increase plasma beta-carotene and decrease plasma retinol May cause between a 32% and 59% decrease in conversion of beta-carotene to trans-retinol	Signs and symptoms of retinol deficiency; check VDR and vitamin D status Plasma vitamin A	rs12934922(T) rs7501331(T) rs11645428 (A) rs6420424(A) rs6564851(G) confers higher beta carotene levels
Dysbiosis/GI disease SNPs				
FUT2 *Fucosyltransferase 2* Determines secretor status Secretor status allows for expression of the ABH and Lewis histo-blood group antigens in various secretions including the intestinal mucus	N/A	↓*Bifidobacteria*, ↑D-Lactate; ↑MMA, ↓serum B12, ↓microbiome diversity FUT2 non-secretors are at an increased risk of celiac disease FUT2 and MUC2 (mucin 2) polymorphism increases risk for colitis in those with Crohn's disease	Non-secretors are resistant to norovirus Serum B12 Consider stool testing	rs601338 (G>A) causes non-secretor status. Further linkage has been found with rs516246(T)
Celiac HLAs Those with these DQ genotypes have an increased risk for developing celiac disease	N/A	↑D-Lactate, ↑arabinitol, ↑benzoate, ↑phenylacetate, ↑p-hydroxybenzoate, ↑p-hydroxy-phenyl-acetate, ↑indican	Signs and symptoms suggestive of CD Anti-transglutaminase type 2 antibody (anti-TG2) levels more than 10 times the upper limit of normal Positive confirmation tests of anti-endomysium-IgA antibodies (EMA)	DQ 2.5 = rs2187668 (T) DQ7 = rs4639334 (A) DQ8 = rs7454108 (G) DQ 2.2 requires a test of three genes, two for inclusion, rs2395182 (T) and rs7775228 (C). DQ4 excludes a DQ2.2 as evidenced by rs4713586(C) The rs2040410(A) allele is associated with DRB1*0301, and the rs7454108(C) allele is associated with DQB1*0302 (and thus DQ8)
HLA-DQ2.5 90–95% of all celiac patients have this genotype	N/A	↑D-Lactate, ↑arabinitol, ↑benzoate, ↑phenylacetate, ↑p-hydroxybenzoate, ↑p-hydroxy-phenyl-acetate, ↑indican	Signs and symptoms suggestive of CD Anti-transglutaminase type 2 antibody (anti-TG2) levels more than 10 times the upper limit of normal Positive confirmation tests of anti-endomysium-IgA antibodies (EMA)	rs2187668 (T) (DQB1*0201) is linked to DQA1*0501; creating the DQ2.5 haplotype

(continued)

Table 17.1 (continued)

SNPS	Co factors	Organic acids/clinical associations	Amino acids/other tests/interventions	rs Numbers
HLA-DQ8 Half of the remaining 5–10% of celiac patients have DQ8	N/A	↑D-Lactate, ↑arabinitol, ↑benzoate, ↑phenylacetate, ↑p-hydroxy-benzoate, ↑p-hydroxy-phenyl-acetate, ↑indican	Signs and symptoms suggestive of CD Anti-transglutaminase type 2 antibody (anti-TG2) levels more than 10 times the upper limit of normal Positive confirmation tests of anti-endomysium-IgA antibodies (EMA)	rs7454108 (G) (DQB1∗0302)
HLA-DQ 2.2; GS221 Increased risk gluten intolerance and for autoimmune disorders such as celiac disease		↑D-Lactate, ↑arabinitol, ↑benzoate, ↑phenylacetate, ↑p-hydroxy-benzoate, ↑p-hydroxy-phenyl-acetate, ↑indican	Signs and symptoms suggestive of CD Anti-transglutaminase type 2 antibody (anti-TG2) levels more than 10 times the upper limit of normal Positive confirmation tests of anti-endomysium-IgA antibodies (EMA)	rs2395182(T), rs7775228(C), not rs4713586(C)
APB1 *AP2-like ethylene-responsive transcription factor* Responsible for extracellular histamine degradation (encodes the enzyme Diamine Oxidase)	Copper	↓ DAO, ↑Ulcerative colitis symptoms	RBS copper, cerruloplasmin	rs1049793(G/T)
Mitochondrial Disease				
NDUFs3 NADH: ubiquinone oxidoreductase, iron-sulfur *protein fraction 3; Complex 1* Cleavage of NDUFS3 is the first step in GZMA-induced (AKA blood coagulation factor IX/Christmas factor) cell death Leigh syndrome	N/A May be limited by NAD, electrons and ubiquinone Thiamine is the cofactor for several mitochondrial enzymes and stimulates energy conversion	Fatigue and exercise intolerance Homozygous mutations should respond well to treatment with riboflavin, thiamin, carnitine and higher doses of coQ10	Check CoQ10 status Deficiency is marked by elevated hydroxy-methyglutarate. Check fatty acid oxidation—may result in increases in adipate, suberate and ethylmalo-nate Elevated anion gap can indicate a need for thiamine	rs4147730(A)

Table 17.1 (continued)

SNPS	Co factors	Organic acids/clinical associations	Amino acids/other tests/interventions	rs Numbers
NDUFs7 *NADH: ubiquinone oxidoreductase, iron-sulfur protein fraction 7; Complex 1* Highest in heart and skeletal muscle Leigh Syndrome causes poor feeding, episodes of apnea and cyanosis, acute gastroenteritis, moderate hypertrophic obstructive cardiomyopathy, extensive white matter hypodensity, mild ventricular enlargement, and hypodense symmetric lesions in putamen and mesencephalon	N/A May be limited by NAD, electrons and ubiquinone Thiamine is the cofactor for several mitochondrial enzymes and stimulates energy conversion	Fatigue and exercise intolerance Homozygous mutations should respond well to treatment with riboflavin, thiamin, carnitine and higher doses of coQ10	Check CoQ10 status Deficiency is marked by elevated hydroxy-methyglutarate Check fatty acid oxidation, may result in increases in adipate, suberate and ethylmalonate Elevated anion gap can indicate a need for thiamine	rs1142530 (T) rs2332496(A)
NDUFs8 *NADH: ubiquinone oxidoreductase, iron-sulfur protein fraction 8; Complex 1* Required in the electron transfer process	N/A May be limited by NAD, electrons and ubiquinone Thiamine is the cofactor for several mitochondrial enzymes and stimulates energy conversion	Fatigue and exercise intolerance Homozygous mutations should respond well to treatment with riboflavin, thiamin, carnitine and higher doses of coQ10	Check CoQ10 status Deficiency is marked by elevated hydroxy-methyglutarate Check fatty acid oxidation, may result in increases in adipate, suberate and ethylmalonate Elevated anion gap can indicate a need for thiamine	rs1051806(T) rs2075626 (C)
COX5c *Cytochrome oxidase C subunit 5c* Cytochrome C (Complex 4) genes are the last step in the mitochondrial respiratory chain. This last step transfers all of that energy to oxygen, and having multiple SNPs in these genes creates free radicals	May be limited by ubiquinone Responds to riboflavin	COX5A is associated with sideroblastic anemia and cardio-encephalo-myopathy Glutaric Acid	CBC, Echocardiogram, EKG, 8-HO-2DG, Lipid peroxides	rs8042694(G)
COX6c *Cytochrome oxidase C subunit 6c* C Cytochrome C (Complex 4) genes are the last step in the mitochondrial respiratory chain. This last step transfers all of that energy to oxygen, and having multiple SNPs in these genes creates free radicals	May be limited by ubiquinone Responds to riboflavin	COX6C is associated with an increased risk for prostate cancer and kidney disease Glutaric Acid	CMP, 8-HO-2DG, Lipid peroxides	rs4626565(C)

(continued)

Table 17.1 (continued)

SNPS	Co factors	Organic acids/clinical associations	Amino acids/ other tests/ interventions	rs Numbers
ATP5c1 *ATP Synthase, H+ Transporting, Mitochondrial F1 Complex, Gamma Polypeptide* Encodes a subunit of mitochondrial ATP synthase Mainly expressed in heart Located in the mitochondrial matrix, complex V of oxidative phosphorylation	Hydrogen, phosphate group	Extreme fatigue, Fibromyalgia syndrome, decreased viability, decreased caspase 3/7 activity	Check all other mitochondrial subunits to ensure proper nutrient repletion	rs1244414(T)
ATP5g3 *ATP synthase, H+ transporting, mitochondrial Fo complex subunit C3 (Subunit 9)* Encodes a subunit of mitochondrial ATP synthase Mainly expressed in heart, located in the mitochondrial matrix, complex V of oxidative phosphorylation	Hydrogen, phosphate group	Extreme fatigue, fibromyalgia syndrome, decreased viability	Check all other mitochondrial subunits to ensure proper nutrient repletion	rs36089250(C)
Cancer–related SNPS				
TNFa *Tumor necrosis factor alpha* Pro-inflammatory cytokine that is associated with both anti and pro-cancer effects	N/A	AGONIST: Cannabidiol, echinacea, larch arabinogalactan ANTAGONIST: Astragalus, andrographis, resveratrol, alpha lipoic acid, Vitamin C, co-enzyme Q10, *Lactobacillus rhamnosus*, curcumin	Consider regular/ increased cancer screenings Consider stool testing	rs1800629(A) rs361525(A)
PTEN *Phosphatase and tensin homolog* Tumor suppressor gene that regulates apoptosis Diseases associated with PTEN: Bannayan-Riley-Ruvalcaba syndrome, Cowden disease, Cowden syndrome-like phenotype, Cowden syndrome, Endometrial carcinoma, Lhermitte-Duclos disease, Macrocephaly/autism syndrome, Malignant melanoma, Oligodendroglioma, PTEN hamartoma tumor syndrome with granular cell tumor, Prostate cancer, Proteus syndrome, Squamous cell carcinoma, head and neck, Vater association with hydrocephalus; In cancer, PTEN is frequently mutated or lost in human tumors	N/A	AGONIST: Astragalus, butyrate, honokiol, retinoic acid, Vitamin D; PTEN is increased by TNF-α (pro-inflammatory cytokine). ANTAGONIST: PTEN inhibition decreases nitric oxide (NO) production; high fat diet; resveratrol	Consider regular/ increased cancer screenings	rs701848(C) rs121909229 (A/C) rs121909233(A)

Table 17.1 (continued)

SNPS	Co factors	Organic acids/clinical associations	Amino acids/other tests/interventions	rs Numbers
NAT1/2 *N-acetyltransferase 1 protein* Involved in phase II xenobiotic metabolism and helps with the biotransformation of aromatic and heterocyclic amines Detoxifies hydrazine and acrylamine drugs	Acetylators like darkly colored berries and grapes (purple and red colored fruits and vegetables)	rs4986782 increases the risk for smoking-induced lung cancer, head and neck cancers Slow phenotype associated with increased risk of several cancers Fast phenotype associated with bladder cancer Remove charred meats and other xenobiotics	Check glutathione status and conjugation Check CYP enzymes to ensure proper detoxification	NAT2∗4 is wild type and considered to be a "rapid" metabolizer (rs1801279, rs1041983 and rs1801280) **Slow phenotype:** 1. rs1801289(C) + rs1799929(T) 2. rs1801280(C) + rs1799929(T) + rs1208(G) 3. rs1801280(C) + rs1208(G) 4. rs1801280(C) 5. rs1801280(C) + rs1799930(A) 6. rs1041983(T) + rs1801280(C) + rs1799929(T) + rs1208(G) 7. rs1041983(T) + rs1801280(C) + 1700020(T) + rs1208(G) 8. rs1041983(T) + rs1801280(C) + rs1799930(A) 9. rs1041983(T) + rs1799930(A) + rs1208(G) 10. rs1799929(T) + rs1799930(A)
Pharmacogenetics				
CYP1A1 Substrates: estrone, brassica vegetables containing diindolylmethane (DIM) or Indole-3-carbinole (I3C), polycyclic aromatic hydrocarbons, caffeine, aflatoxin B1, and acetaminophen	Heme	Fast metabolizer phenotype; associated with increased risk for breast cancer	24-hour urinary hormones testing	CYP1A1∗2C rs1048943(C) CYP1A1∗4 rs1799814(T)
CYP1A2 95% caffeine metabolism Other substrates: theophylline, phenacetin, acetaminophen, estrogen	Heme	GA: 26% increased risk for MI with two or more cups of coffee AA: 53% increased risk of MI with two or more cups of coffee Limit caffeine consumption to less than 200 mg/day High caffeine intake may result in increased estrogen levels and therefore hormone-related cancers (breast, ovarian, uterine, prostate, and testicular). See pharmacogenetics resources	Caffeine suppression test; 24-hour urinary hormones testing	rs2472300(A) rs762551 Risk Allele A increases activity (fast metabolizer) Risk Allele C decreases activity (slow metabolizer)

(continued)

◼ Table 17.1 (continued)

SNPS	Co factors	Organic acids/clinical associations	Amino acids/other tests/interventions	rs Numbers
CYP1B1 Substrates: hydroxylation of estrogens, oxidizes 17-beta-estradiol to the carcinogenic 4-hydroxy derivative, tamoxifen, polycyclincaromatic hydrocarbon, aflatoxins, theophylline, phenacetin, and warfarin	Heme	COMT SNPs (downregulation) may compound SNPs in CYP1B1 Increases risk for breast, ovarian and prostate cancers	RBC magnesium for COMT status and 24-hour urinary hormones testing	CYP1B1 L432V (rs1056836 (G) Is protective against prostate cancer) CYP1B1 N453S rs1800440(C) is associated with decreased levels of urinary 2-OHE and 16-alpha in premenopausal women and is associated with breast and endometrial cancers CYP1B1 R48G rs10012(C/G)
CYP2E1 Substrates: Acetaminophen, caffeine, alcohol, chlorzoxazone, enflurane, aniline, benzene, chlorzoxazon, dacarbazine, eszopiclone, halothane, isoflurane, isoniazid, methoxyflurane, paracetamol, sevoflurane, theophylline, trimethadione, zopiclone	Heme	Chronic detoxication challenges See pharmacogenetics resources	Check glutathione synthesis	CYP2W1∗1B C9896G (rs2070676): GG is induced by alcohol, coffee and smoking; CC is wild type with no up or downregulation CYP2E1∗4 A4768G (rs6413419 (A)) Works with other CYP genes to detoxify carcinogens.
CYP2D6 Substrates: antidepressants, antipsychotics, analgesics, antitussives, beta-blockers, antiarrythmics, and antiemetics	Heme	Very common in most populations; mixed regulation often results in adverse drug reactions See pharmacogenetics resources	N/A	CYP2D6 T100C (rs1065852): T risk allele is a non-functioning or partially functioning variant causing decreased activity CYP2D6 C2850T (rs16947) AA increased risk for upregulation CYP2D6 S486 S485T (rs1135840): G risk allele causes upregulation
CYP2C9 Substrates: warfarin, NSAIDs, amitriptyline, apixaban, azilsartan, clopidogrel, Benadryl, glibencalamide sulfonylurea glimepiride	Heme	SNPs indicate down-regulation See pharmacogenetics resources	N/A	CYP2C9∗2 C430T (rs1799853(T)) (~20% decrease) CYP2C9∗3 A1075C (rs1057910(C)) (~40% decrease)
CYP2A6 Substrates: nicotine, aflatoxin B1, cotinine, coumarin, dexmedtomidine, docetaxel, efavirenz, irinotecan, letrozole, methoxsalen, oxaliplatin, pilocarpine, tegafur, valproic acid, warfarin, methyl tert-buty (gasoline additive), halothane, methoxyflurane	Heme	Inhibited by grapefruit juice SNPs indicate down-regulation Chronic detoxification challenge phenotype See pharmacogenetics resources	Check VKORC1 status for warfarin sensitivity status	CYP2A6∗2 A1799T rs1801272(T) CYP2A6∗3 rs1057910(C)

Table 17.1 (continued)

SNPS	Co factors	Organic acids/clinical associations	Amino acids/other tests/interventions	rs Numbers
CYP2B6 Substrates: nicotine, alfentanil, bupropion, cyclophosphamide, ifosfamide, methadone, nevirapine, propfol, sertraline, sorafenib, tamoxifen, vaprioc acid, methoxetamine	Heme	SNPs indicate down-regulation Chronic detoxification challenge phenotype See pharmacogenetics resources	N/A	A136G rs3530384 C1132T rs34097093 C26470T rs8192719) G23280A rs2279344 G29435A rs7260329 I328T rs28399499 L262A rs2279343 Q172H rs3745274 R22C rs8192709 T1421C rs1042389 T20715C rs36079186 T23499C rs2279345
CYP2C19 Substrates: antidepressants, antiepileptics, proton pump inhibitors, clopidogrel, proguanil, propranolol, gliclazide, carisoprodol, chloramphenicol, cyclophosphamide, indomethacin, nelfinavir, nilutamide, progesterone, teniposide, warfarin	Heme	Mixed regulation may result in adverse drug reactions See pharmacogenetics resources	Urinary Hormones testing Check depression etiology (dopamine vs serotonin)	CYP2C19∗17 806C>T rs12248560 (T) Ultra rapid metabolizer CYP2C19∗2 rs4244285 (A) represents decreased activity or downregulation
CYP3A4 Substrates: Lidocaine, erythromycin, cyclosporine, ketoconazole, testosterone, estradiol, cortisone, alfentanil, alfuzosin, almotriptan, alprazolam, amiodarone, amitriptyline, amlodipine, anastrozole, aprepitant, aripiprazole, astemizole, astazanavir, atorvastatin, bepridil, bexarotene, etc.	Heme	Check CoQ10 markers such as hydroxy-methylglutarate Grapefruit causes decreased clearance of a substrate and higher plasma levels See pharmacogenetics resources	Urinary hormones Test; adrenal stress index test	CYP3A4∗1B 392G>A (rs2740574): GG involved in oxidative deactivation of testosterone GG/AG alleles confer 10X increased risk of aggressive prostate cancer in African American men (>54%) CYP3A4∗3 M455T rs4986910(C) involved in estrogen metabolism Significantly decreased risk of breast cancer with the minor allele T
CYP3A5 Substrates: Olanzapine, tacrolimus, nifedipine, cyclosporine, testosterone, progesterone, androstenedione	Heme	N/A See pharmacogenetics resources	Urinary hormones test	CYP3A5∗2 rs28365083(T) is a non-functioning allele CYP3A5∗3 rs776746(G) is associated with down regulation

Courtesy of Nutritional Genomics Institute, LLC

Fucosyltransferase 2 (FUT2) mutations, in particular, appear to be the cause of many abnormalities in the microbiome. FUT2 encodes the fucosyltransferase 2 enzyme, which determines secretor status. Secretor status allows for expression of the ABH and Lewis histo-blood group antigens in various secretions including the intestinal mucus. The presence of a particular SNP in FUT2, rs601338 (G>A) results in the non-secretor status. There is also another potential candidate SNP, rs516246 for which there has been considerable linkage to phenotypic expression. It appears that 20% of individuals who are of European descent have this non-secretor FUT2 polymorphism [84].

A 2011 study concluded that those with FUT2 polymorphisms (non-secretors) had only half of the *bifidobacterial* diversity and richness found in secretors. In particular, absence of bacterial denaturing gradient gel electrophoresis (DGGE) genotypes of species such as *Bifidobacterium adolescentis, Bifidobacterium catenulatum/pseudocatenulatum,* and *Bifidobacterium bifidum* were noted. In addition to the noted absence of beneficial bacteria, they also found that there are increased levels of potentially harmful bacteria such as *Blautia* et rel., *Dorea formicigenerans* et rel., *Ruminococcus gnavus* et rel., and *Clostridium sphenoides* et rel [84, 85]. Further, FUT2 non-secretor status also places individuals at higher risk for various diseases including Crohn's disease, ulcerative colitis, type 1 diabetes, vaginal candidiasis and urinary tract infections [84, 85]. Additionally, non-secretors have decreased carbohydrate availability in the intestine, which can cause increased risk of *Salmonella* and *C. difficile* following antibiotic treatment [84]. While more recent studies have called this relationship into question, supplementation with beneficial *bifidobacteria* can still be helpful in those with FUT2 polymorphisms.

A study that investigated Crohn's disease risk in those with FUT2 polymorphism found that mucin 2 (MUC2) might also play a role in this risk [86]. MUC2 is secreted in mucin in the colon and helps provide a barrier that excludes bacteria from the mucosal cell surface. This mechanism ultimately decreases the risk for colitis [86]. Aberrant glycosylation of MUC2 core proteins have been shown to cause spontaneous colitis in mice; thus, it appears that FUT2 negatively impacts microbial adhesion and/or the use of glycans that may lead to dysbiosis.

This association with dysbiosis was further related to the revelation that non-secretors were deficient in several pathways responsible for amino acid metabolism, but interestingly had higher metabolism rates for carbohydrates and lipids, cofactors, and vitamins and glycan biosynthesis [86]. These metabolic abnormalities can alter the integrity of the mucosal epithelium, which then alters the microbial composition. Interestingly, the decrease in the production of amino acids is often reported in those with inflammatory bowel disease (IBD). Alternatively, the metagenome shows an alternate pattern. It showed that while those with IBD had decreased amino acid biosynthesis and carbohydrate metabolism, these conditions resulted in an increased nutrient uptake. This could potentially suggest that the microbiota are compensating for the decreased availability of carbon sources. Further, a condition that results in decreased amino acids may cause a stress response in the individual and therefore the onset of autophagy of intestinal epithelial cells that may ultimately cause IBD [86]. It is important to note that these perturbations can result in sub-clinical intestinal inflammation even in apparently healthy individuals with FUT2 polymorphisms. Supplementation with zonulin tightening nutrients like zinc carnosine can be helpful in those with FUT2 associated intestinal permeability [87].

Lastly, to round out the genomic potential for IBD, specifically Crohn's disease, there is a gene called amiloride binding protein 1 (ABP1/AOC1). This gene is involved in histamine metabolism and poses an interesting intersection between diet and disease. A specific copper dependent SNP in this gene, rs1049793, is associated with an exacerbation in Crohn's disease symptoms. ABP1 is specifically responsible for the regulation of polarized epithelial cells, a mechanism that is dysregulated in IBD [88]. This effect can be compounded if there are also alterations in FUT2 and/or MUC2. It is important to assess copper and cobalamin status in these cases.

There are two mechanisms for histamine degradation, one working intracellularly and the other working via extracellular histamine degradation. Intracellular histamine metabolism is carried out by histamine methyltransferase (HNMT) while extracellular degradation occurs via another enzyme called diamine oxidase (DAO). ABP1 is one gene that encodes DAO, thus regulating extracellular histamine degradation [89]. Histamine is released in the body upon mast cell degranulation and there are four regulatory histamine receptors (H1R-H4R). H1 receptors are found throughout the body in smooth muscle, vascular endothelial cells, the heart and central nervous system. H2 receptors trigger the gastric secretion of histamine to regulate hydrochloric acid production. Less is known about H3 and H4 receptors; however, H1–H3 receptors are primarily found in the brain and H4 receptors are found in the periphery [90]. Histamine is responsible for a staggering number of biological effects, which vary based on receptor, cell location and target cell. Examples include gastric acid secretion, neurotransmitter release, smooth muscle constriction, vasodilation, tachycardia, arrhythmia, stimulation of nociceptive nerve fibers, and an increase in estrogen and endothelial permeability, which results in dysbiosis and an increase in mucus production [89].

The signs and symptoms associated with Crohn's disease include increases in *Bacteroides*, decreases in *Firmicutes* and anti-inflammatory *F. prausnitzii,* chronic diarrhea, increased smooth muscle contractions, excessive bowel mucus production, bleeding from the rectum, weight loss, and fever [91]. As evidenced, there is the potential for considerable overlap between histamine intolerance and Crohn's disease. Crohn's disease presentation and the histamine intolerance associated with the consumption of certain trigger foods appear to have a linear relationship, whereas alterations in ABP1 decrease DAO, which ultimately results in histamine intolerance and intestinal permeability [92]. There are several foods (i.e., epigenetic activators) that are thought to be high in histamine that may need to be avoided. Examples include fermented foods, alcohol, pickled foods, mature cheeses, smoked meats, shellfish, beans, nuts, and wheat. Some foods are also considered to be histamine liberators. These include most citrus fruits, strawberries, cocoa and chocolate, nuts, papaya, beans, tomatoes, wheat germ, and additives (benzoate, sulfites, nitrites, and glutamate). There are also foods that block diamine oxidase (DAO) which can be problematic, as a block in extracellular histamine degradation would ultimately result in excessive circulating histamine. Examples include alcohol, black tea, energy drinks, green tea, and mate tea. Most fresh meats, fruits, vegetables, eggs, grains, cooking oils, herbs, and non-citrus juices are low in histamine and are non-degranulating. Other known dietary triggers for histamine intolerance may include sulfur, gluten, oxalates, salicylates, and lectin, all of which may play a role in IBD [93].

The relationships between diet, genomics, microbiota, and disease are being discovered exponentially. Once we have mastered the other "omics," there is the potential for truly precise medicine, disease modulation and nutritional intervention.

17.2.1 The Epigenetics of Nutrient-Associated Diseases

To continue with the relationship between nutrients and diseases, we will discuss the relationship between vitamin B12, vitamin D, and autoimmune disease. These two vitamins are two of the most vital nutrients for not only proper one-carbon metabolism and methylation, but also for nutrient absorption and hormone regulation. The first of these is a vitamin that, if measured accurately, would be found to be deficient in such a large subset of the population that the governments would consider fortification. Enter the cobalamins, commonly known as vitamin B12.

There are technically four forms of cobalamin [94]. The most commonly supplemented is called cyanocobalamin. If we break apart the word into fundamental blocks, we have a cobalamin structure, which resembles hemoglobin, and this four-membered ring has a special center group, which in this case would be cyanide. This is the synthetic version of vitamin B12. That means that it is man-made, and is the cheapest supplement option. Biochemically, cyanocobalamin is the most stable form of B12. This molecular stability is the result of having the non-reactive cyanide molecule housed in the center [95]. While the amount of cyanide found in typical B12 supplementation is not inherently dangerous, it does require additional energy, such as ATP, for the body to safely excrete the cyanide from the body. This means that there is a potential net loss of ATP, or cellular energy, when this form of the vitamin is taken internally [96]. There are three better options: adenosylcobalamin, hydroxocobalamin and methylcobalamin.

To start, we need to understand vitamin B12 metabolism and deficiency. It can take a very long time to develop a vitamin B12 deficiency. This is not something that occurs overnight as vitamin B12 is stored long term in the liver, typically, anywhere from 3 to 5 years. Specifically, approximately 2–4 mg of vitamin B12 is stored in the body, of which 50% is found in the liver [95]. Of those 2–4 mg, adenosylcobalamin represents 70% of the vitamin B12 stored in the body, which is also mostly found in the liver. Humans cannot utilize a nutrient if it is only in the storage form, held captive by the liver. Thus, the non-storage form, methylcobalamin, is the main form of vitamin B12 found in the blood. Please take notice of the specific words here, this one is methylcobalamin; therefore, it has a methyl group that can be donated, located in the center of the cobalamin ring. This is important when we consider the vital role that vitamin B12 plays as the connection between one carbon metabolism and methylation. Methionine synthase and methionine synthase reductase (MTR and MTRR) recycle homocysteine back to methionine to produce SAM. This enzyme is the connection between the one-carbon cycle and methylation. MTR and MTRR must be able to convert adenosylcobalamin into methylcobalamin, gaining the methyl group from methylenetetrahydrofolate reductase (MTHFR). Lastly, there is the hydroxyl form of cobalamin, and as the name implies, it carries a hydroxyl group in the center of the cobalamin ring. A hydroxyl group is one molecule of hydrogen and one molecule of oxygen. If you remember from basic chemistry, water is made of two molecules of hydrogen bonded to one molecule of oxygen. Water is in the most stable electrochemical form; thus, a hydroxyl group will be searching for another hydrogen to stabilize its structure. This particular biochemistry allows this form of B12 to bypass several of the transport mechanisms for vitamin B12 [97]. Hydroxocobalamin and methylcobalamin are also stored, but to a lesser extent, in the muscles, bone, kidney, heart, brain, and spleen. When analyzing whole blood, methylcobalamin comprises 60–80% of vitamin B12 found in the blood and adenosylcobalamin comprises up to 20% of total plasma cobalamin [95]. It is important to remember this when measuring serum cobalamin as it is primarily a proxy for methylation status and not a true assay for cobalamin status or cellular levels.

Now that we understand the basics of cobalamin, we need to discuss how these molecules are transported and utilized in vivo. There are two major transporters for cobalamin. The first is transcobalamin I (TCN1 or haptocorrin), and it is a binding protein secreted by the salivary glands and assists in the complexes' survival from the acidity in the stomach. Once bonded with vitamin B12, and it has successfully survived digestion from hydrochloric acid (HCL), the complex travels to the intestine. Meanwhile, gastric intrinsic factor (GIF), a vitamin B12 binding protein which requires HCL for production, is produced by the parietal cells. After the TCN1/B12 complex has arrived in the less acidic environment of the intestine, the TCN1/B12 complex can then bind to the intrinsic factor formed from GIF and be absorbed in the ileum. Once inside the enterocytes of the ileum, vitamin B12 breaks apart from TCN1 to bind to TCN2, which then carries vitamin B12 to the liver [95].

TCN1 transports up to 80% of vitamin B12 and is thought to function as a circulating storage form. The remaining 20–30% of cobalamin is transported on TCN2. When cobalamin is bound to TCN1, it may prevent bacterial use of the vitamin, which has implications on and from the microbiome. In addition to being a B12 transporter, TCN2 is a major part of the secondary granules found in neutrophils connecting B12 status to the immune system [95]. If there are polymorphisms in either of these transporters, it may cause an increase in serum vitamin B12 (often even without supplementation) with an increase in the organic acid methylmalonic acid (MMA). This means that there will be high serum levels and low tissue concentration. Sometimes, if only TCN1 is present, serum levels may also be low. The use of hydroxocobalamin can bypass this transport issue and assist with repletion [95].

So, we have learned that TCN2 is involved in the function of neutrophils, but are there other, more overt autoimmune diseases associated with defects in any of these genes? The most notorious disease is pernicious anemia, which is the result of intrinsic factor deficiency or malfunction. Alterations in GIF are also associated with an increased prevalence of familial pernicious anemia, acute lymphoblastic leukemia, and high-altitude polycythemia. GIF deficiency is also responsible for many cases of non-*H. pylori* related gastritis [98]. This non-*H. pylori*-related gastritis is compounded by the presence of even a heterozygous FUT2 non-secretor variant, as FUT2 decreases GIF secretion. This circumstance would result in cobalamin deficiency [99]. On the contrary, the FUT2 secretor status is associated with elevated levels of plasma vitamin B12. This is because the non-secretor variants do not produce H-type antigens, the antigen which is present in blood group O. There is some association with decreased plasma vitamin B12 levels on non-type O blood [99].

Next, we will briefly discuss vitamin D, its receptors (VDR) and consequences associated with defects in these receptors. There are many additional resources on the associations between vitamin D and autoimmune disease beyond those discussed in this chapter. First, we will review vitamin D nomenclature. Calcidiol (25(OH)D) is the form typically measured in serum. This is a combination of both vitamin D2 and D3 levels from endogenous and exogenous sources. Next is calcitriol, or 1–25-OH(D), the active form of vitamin D. Calcidiol is converted into calcitriol and this mechanism is tightly controlled in the absence of VDR polymorphisms.

VDRs have many regulatory roles. To start, VDRs control genes and have hormone-binding and DNA-binding domains. Specifically looking at DNA, VDRs form a complex with the retinoid-X receptor (vitamin A receptor), and that heterodimer is what binds to DNA, either turning it on or turning it off (vitamin A and vitamin D are required cofactors) [100]. In most cases, VDRs activate transcription and gene expression, but VDRs have also been known to suppress transcription. The affinity VDRs have for calcitriol, the active form of Vitamin D, is roughly 1000×× greater than its affinity for calcidiol [100]. Thus, VDRs are specifically implicated in epigenetic control. Pollutants like smoke and lack of sunshine all contribute to the function of VDR and disease [100]. VDRs also control calcium homeostasis. Certain VDRs can theoretically elevate calcitriol while lowering calcidiol. Calcitriol increases the level of calcium in serum by increasing the uptake of calcium from the intestines, increasing the release of calcium from bones [100]. VDRs are also involved in tissue modulation. They can regulate apoptosis, cellular proliferation and differentiation via matrix metalloproteinases and plasminogen activators. They also modulate the immune system and have the potential to modulate B cells, T cells, dendritic cells, monocytes, and natural killer cells. Low levels of vitamin D, calcitriol specifically, are associated with autoimmune disease [100].

Here are four VDR SNPs that may be encountered. The first represents a causal link to cancer development. It was found that rs2107301 (T;T) homozygotes were associated with an ~2.5× higher risk of prostate cancer compared to homozygote carriers of the more common (C) allele in the 630 Caucasian patients studied [101]. Vitamin D status as an epigenetic influencer can ultimately increase one's risk for prostate cancer, elucidating the importance of vitamin D repletion. Interestingly, vitamin D status is also a regulator for melanoma, suggesting that those that spend more time indoors are ultimately at an increased risk for both forms of cancer [102].

There are two other *commonly reviewed* VDR SNPs; however, the evidence for and against them is inconclusive. These two SNPs are called VDR fok and VDR taq. The first has circumstantial evidence to suggest that polymorphisms can result in blood sugar regulatory issues. VDR-taq SNPs are also circumstantially linked to better tolerance of methyl donors and may cause a decrease in calcidiol levels [103, 104]. Unfortunately, there is insufficient evidence to support nutrigenomic interventions based on these SNPs.

Lastly is VDR-bsm. This VDR provides instructions for making nuclear vitamin D receptors and is involved in the binding of calcitriol and calcitriol receptor activity. When the risk allele A is found in rs1544410, women have an increased risk of low bone mineral density. Conversely, a 26 study meta-analysis estimated a decreased risk of osteoporosis associated with the G;G genotype [103, 104]. Clinically, this VDR often causes rapid conversion of vitamin D2/3 to the active form calcitriol. This over-conversion results in elevated calcitriol/1,25-dihydroxy vitamin D with normal to low 25(OH)D levels. This phenotype erroneously presents with an apparent vitamin D deficiency, because calcidiol is low. However, the gene activating, hormone modulating, immune system influencer calcitriol is too high. In these cases, the patient is often prescribed more supplemental vitamin D, which can result in adverse reactions like nausea, dizziness, syncope, tachycardia, and autoimmunity [105]. Calcitriol initiates the transcription of genes of your immune system (modulates B cells, T cells, dendritic cells, monocytes, and natural killer cells). If there is not enough calcitriol, these genes are not activated. However, if there is too much calcitriol from VDR bsm, these immune cells receptors are blocked. As you can see, at either end of the spectrum there is a connection between inflammation, autoimmunity and VDR bsm SNPs [106]. An interesting intersection to methylation is that excess calcitriol causes more copies of the cystathionine beta synthase (CBS) gene to be transcribed, thereby increasing hydrogen sulfide and may exacerbate transsulfuration/detoxification issues. Please see ▸ Chap. 18 for more information on CBS [107].

17.3 Epigenetics of Cancer

Cancer is a multifactorial disease state that, again, could have an entire textbook devoted to epigenetic regulation. In this section, we will bridge mitochondrial insufficiency and disease to key cancer pathway genomics. We will begin with an overview of glutathione synthesis. This begins in transsulfuration with an enzyme called CBS. This enzyme is responsible for the catabolism of a regulatory molecule homocysteine into our endogenous antioxidant, glutathione. When there is proper function of CBS, the results are taurine, sulfate and glutathione; all anti-inflammatory biomolecules, if within normal limits. In a normally functioning CBS, hydrogen sulfide enhances the production of reactive oxygen species (ROS) while simultaneously increasing glutathione, the cells' self-protection mechanism against ROS. This results in a net zero of ROS. Hydrogen sulfide in physiological dosages is beneficial to mitochondrial function as it protects it from cytotoxicity. In the case of ischemia, hydrogen sulfide actually blocks cytochrome oxidase, resulting in a decrease in mitochondrial damage. It also upregulates an important mitochondrial detoxification enzyme called superoxide dismutase (SOD) while concurrently decreasing ROS. Hydrogen sulfide is also a regulator of apoptosis, a precursor to sulfation and acts as a neuroprotectant by increasing glutathione and moving another important enzyme, cystathionine gamma lyase (CGL – also vitamin B6 dependent) to the mitochondria which results in an increase of cellular ATP [108].

However, there are variants for this CBS enzyme (◘ Fig. 17.3) that result in up or down regulation of this process. While somewhat controversial, clinically, CBS upregulation can result in sulfur sensitivity, meaning there is an increased sensitivity to sulfur-containing foods and drugs. In upregulation (CBS C699T), we end up with excess sulfate, glutamate, ammonia, and cortisol with decreased levels of glutathione, all of which are inflammatory. This results in a net increase of ROS and the excess hydrogen sulfide interferes with proper mitochondrial function [108]. In downregulation (CBS A360A), there is a functional block that results in elevated levels of homocysteine, a higher incidence of CVD, and decreased glutathione due to limitation [109]. Glutathione is also processed through the gamma-glutamyl cycle, which is a transport system for the amino acids that form glutathione, cysteine, glutamate, and glycine. If there is a decrease in glutathione in the absence of a CBS upregulation, investigating SNPs within this cycle could provide an etiology to the deficiency [110].

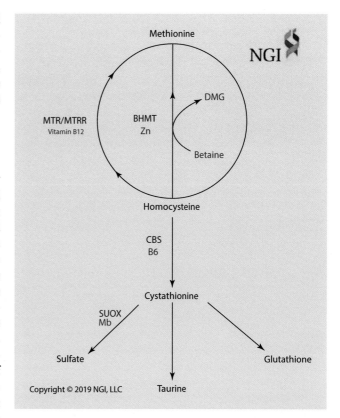

◘ Fig. 17.3 Transsulfuration and CBS upregulation. There are three regulatory pathways of homocysteine recycling to consider: MTR/MTRR, BHMT, and CBS; CBS is the only disposal route. CBS upregulation will result in decreased levels of homocysteine, whereas CBS downregulation will result in elevated levels of homocysteine. (Courtesy of Nutritional Genomics Institute, LLC)

Now that we know how glutathione is produced and have an introduction to a few of the regulatory mechanisms, what does glutathione have to do with cancer? There are two glutathione-S-transferase genes that are directly implicated in cancer prevalence. These enzymes are responsible for binding the powerhouse antioxidant, glutathione, in its special reduced form, to toxicants, which assists in the cytochrome P450 system's removal of them. These genes, rather the absence of them, have a defined role in human carcinogenesis.

Glutathione S-transferase theta 1 deletion (GSTT1) conjugates reduced glutathione along with many exogenous and reactive oxygen species (hydrophobic electrophiles). It is absent in 38% of the general population [111]. This enzyme is concentrated in the liver, heart, brain, skin, and blood and the deletion results in a higher prevalence of cancer development in these tissues [111]. Glutathione S-Transferase Mu 1 (GSTM1) also conjugates reduced glutathione and functions in the detoxification of electrophilic compounds, including carcinogens, therapeutic drugs, environmental toxins and products of oxidative stress [112]. Null mutations or deletions in this gene are associated with an increase in various cancers and compromised detoxification [112]. Both enzymes are associated with vitamin C deficiency [113]. Not having the coding genes for either of the enzymes places one at considerable risk for the development of several types of cancer, depending on the environmental toxin exposure. In addition to vitamin C deficiency status, both enzymes can be modulated by selenium and vitamin E [114].

In addition to these glutathione- and transsulfuration-related genes, there are also acetylation-related genomics. The N-acetyltransferase 1 and 2 (NAT1 and NAT2) proteins are involved in phase II xenobiotic metabolism and help with the biotransformation of aromatic and heterocyclic amines. It also detoxifies hydrazine and acrylamine drugs [115]. This is done via N-acetylation, which results in the detoxification of monocyclic aromatic amines. This detoxification process is ultimately responsible for the formation of DNA adducts, or a segment of DNA bound to a cancer-causing chemical that precipitates the onset of cancer development.

NAT2 is specifically responsible for the detoxification of smoke, caffeine, drugs, exhaust fumes and many other environmental toxins (heterocyclic aromatic amines). Drugs reported to be metabolized by NAT2 include isoniazid, sulfadimidine, hydralazine, dapsone, procaine amide, sulfapyridine, nitrazepam and some sulfa drugs [116]. These polymorphic conditions can result in fast and slow acetylators (N-acetyl transferase) phenotypes. The slow acetylators have a higher incidence of breast, lung, colon, head and neck, and bladder cancer, with the latter seeming to be the predominant cancer for this group [117]. The fast acetylator variants predominantly result in colon cancer [117]. The key to assisting these enzymes is to reduce the toxin exposure to reduce the stress on these enzymes. Avoidance of smoke, pesticides, insect sprays, charred meats, red meat, metal toxicity, chemicals, and solvents and the addition of acetylators like darkly colored fruits and vegetables high in anthocyanins balanced with adequate fiber will help mediate the cancer predisposition.

Next, let us look into regulatory pathways for cancer. The first is a tumor suppressor gene that regulates apoptosis, phosphatase and tensin homolog (PTEN). There are several diseases and cancers associated with a loss of PTEN function: Bannayan-Riley-Ruvalcaba syndrome, Cowden disease, Cowden syndrome-like phenotype, Cowden syndrome, endometrial carcinoma, Lhermitte-Duclos disease, macrocephaly/autism syndrome, malignant melanoma, oligodendroglioma, PTEN hamartoma tumor syndrome with granular cell tumor, prostate cancer, proteus syndrome, squamous cell carcinoma, head and neck and vertebral (V) abnormalities, anal (A) atresia, tracheoesophageal (T) fistula, esophageal (E) atresia, renal (R) abnormalities (VATER) associated with hydrocephalus [118]. There are several herbs and vitamins that can either upregulate or downregulate this enzyme. Substances that would upregulate this enzyme, or agonists, include astragalus, butyrate, honokiol, retinoic acid, and vitamin D [119, 120]. PTEN is also increased by TNF-α (pro-inflammatory cytokine), which we will discuss next. Substances that block PTEN, or antagonists, include a high fat diet and resveratrol [119, 120]. PTEN inhibition also decreases nitric oxide (NO) production, which could potentially result in hypertension and neurotransmitter imbalances.

Tumor necrosis factor alpha (TNF-α) is a pro-inflammatory cytokine that is associated with both anti- and pro-cancer effects. This means that in some cases, such as melanoma, we would not want to upregulate this enzyme [121]. However, the majority of the current research does suggest that for most cancers, it would be wise to upregulate this protein, as it is pro-inflammatory to the cancer itself. Like PTEN, there are several factors that can increase or decrease transcription. Agonists include cannabidiol, echinacea and larch arabinogalactan while antagonists include astragalus, andrographis, resveratrol, alpha lipoic acid, vitamin C, ubiquinol, curcumin, and *Lactobacillus rhamnosus* [122–127]. As evidenced, there are many common interventions that are actually epigenetic modulators that can ultimately increase or decrease one's overall risk of cancer development.

17.3.1 Epigenetics of Mitochondrial Insufficiency

Mitochondrial function was alluded to in regard to cystathionine gamma-lyase. Without a full review of mitochondrial function, we will focus on several key regulatory enzymes. As a refresher, mitochondria are the cells' powerhouse for ATP production. There are five major complexes, and the majority of these mitochondrial genes (mtDNA) are maternally inherited. Mitochondria convert energy into ATP through the process of oxidative phosphorylation via the electron transport chain, with the final conversion being carried out by an enzyme called ATP synthase. This process is dependent on hydrogen, ubiquinol, riboflavin, niacin, carnitine, thiamine, manganese, antioxidation, and succinate [128]. Any deficiency in these biomolecules can result in a functional mitochondrial deficiency.

The first enzyme we will discuss is part of complex 1. There are eight sulfur subunits in NADH ubiquinone oxidoreductase, iron-sulfur protein fraction; however, only three have actionable interventions (NADH-ubiquinone oxidoreductase 76 kDa subunits 3/7/8—NDUFS3/7/8). Children that have severe mutations in these genes have Leigh's syndrome, which is fatal in infancy. Non-ClinVar homozygous alterations in these subunits typically respond well to treatment with riboflavin, thiamin, niacin, carnitine, and higher doses of coenzyme Q10 (coQ10) [67]. Ultimately polymorphisms increase the production of free radicals (ROS) and addressing the subunit ahead in the chain (NDUF) will lessen stress on these enzymes. Broad-spectrum antioxidants are helpful in these cases, antioxidant vitamins such as C and E, and the mineral selenium help to quench the excess ROS. Minor alterations in these enzymes can result in fatigue and exercise intolerance and typically respond well to treatment with riboflavin, carnitine and higher doses of coQ10 [128–130].

SOD 2A16V is the antioxidant enzyme responsible for mitochondrial detoxification. It codes for the superoxide dismutase 2 enzyme and its role is to bind to the oxidant superoxide and convert it to less toxic by-products to be processed and removed. It has the cofactor of manganese and not having this functioning enzyme increases the risk of certain cancers and idiopathic cardiomyopathy. Unfortunately, intense exercise compromises this enzyme and, in terms of intense exercise, is related to a "dose-dependent" activity rate based on the number of substitutions [131]. Thus, those with alterations in this mitochondrial enzyme should take caution with high-intensity exercise and supplement with manganese to help assist with mitochondrial function.

The COX or cytochrome C oxidase genes code for the carrier protein between complex 3 and 4. This last transfer allows for all of the built-up energy in the form of hydrogen to bind with oxygen to make water. This mechanism is the cellular requirement for breathing. Like complex I insufficiency, having multiple SNPs in these genes creates free radicals. Specifically, alterations in COX6C are associated with prostate cancer and kidney disease and COX5A has associations with sideroblastic anemia and cardioencephalomyopathy. Both of these enzymes are concentrated in heart and skeletal muscle, elucidating the associated disease pathologies [132].

The last step is the conversion of this collected energy into ATP. ATP synthase is required for the final conversion of adenosine diphosphate (ADP) into the usable form of energy, ATP. There are SNPs in the ATP5C1 gene that code ATP synthase. Alterations in this gene result in a decreased output of ATP and increased ROS. Remember, ATP is our cellular currency, and a decrease in production will ultimately result in increased ROS. Alterations in ATP synthase have also been associated with accelerated aging, ALZ and an increased prevalence for cancer [133]. Overall, we want to upregulate mitochondrial function, starting with complex one and moving along through the complexes while assuring that there is adequate antioxidation. It is important to recognize that mitochondria are sequential chains. Therefore, nutrient cofactors should be added in order and not all at one time for optimal expression.

17.4 Introduction to Pharmacogenetics

Pharmacogenomics is one of the more defined areas of personalized medicine. It describes how genes may affect a person's response to a medication, combining pharmacology and genomics to tailor medical treatment on an individual basis. This field assists physicians on the selection of pharmaceuticals, length of treatment, and dosage. Studies in pharmacogenomics work to associate SNP biomarkers with pharmaceutical treatment outcomes. Challenges occur in that related pharmaceuticals such as statins have different degrees of heritability, meaning that each must be investigated separately [134]. The modern pharmaceutical industry will often develop drugs in a genetically guided manner. This has allowed for adoption of genetic testing to be used clinically prior to prescribing a medication [135].

Phase 1 detoxification is comprised of cytochrome P450, a superfamily of enzymes with many sub-divided families that are used in detoxification. All of the cytochrome enzymes are named using "CYP" for cytochrome P450 and are followed by an Arabic numeral (i.e., CYP1, CYP2, CYP3, etc.). These families are then further subdivided into subfamilies with the addition of a capital letter (i.e., CYP1A, CY1B, CYP1C, etc.). Individual members of each subfamily are then numbered in the order they were identified (i.e., CYP1A1, CYP1A2, CYP1A3, etc.). Phase 1 detoxification converts lipid-soluble molecules entering the liver into more water-soluble intermediary metabolites. These metabolites are often more toxic, not less. A range of substances including drugs, dietary components such as charcoal-broiled meats, steroid hormones, the vitamins niacin and riboflavin, as well as xenobiotics such as dioxin, exhaust and paint fumes, organophosphorus pesticides, and fragrances may induce P450 enzymes [136].

Phase 2 detoxification includes acetylation, glucuronidation, glutathione conjugation, peptide conjugation, methylation, and sulfation. Glutathione is our primary endogenous antioxidant that is essential for proper phase 2 detoxification. Glutathione is a tripeptide formed from the amino acids glutamate, cysteine, and glycine. It accounts for approximately half of our cysteine requirements. In phase 2 detoxification, glutathione is used to conjugate and excrete toxins and drugs, making them more water-soluble. Genetic SNPs in glutathione synthase (GSS) may impair glutathione production within the gamma-glutamyl cycle [137]. This enzyme aids in the production of glutathione and requires magnesium as its cofactor. When GSS levels diminish, it is associated with hemolytic anemia [138]. Further, mutations may lead to elevation of the urinary organic acid pyroglutamate [67]. Methylation connects to phase 2 detoxification via the adjoining pathway transsulfuration. Here we see the enzyme cystathionine beta-synthase (CBS) converts homocysteine to cystathionine, which ultimately leads to the production of taurine, sulfate, and glutathione (◘ Fig. 17.2) [139].

Acetylation is associated with the biotransformation of aromatic and heterocyclic amines. The enzymes N-acetyltransferase 1 and 2 (NAT1 and NAT2) determine whether an individual is considered to be a slow, intermediate, or rapid metabolizer. It requires two slow metabolizer alleles to result in the slow metabolizer phenotype [116, 117]. This means that the rapid metabolizer allele is dominant, and the slow metabolizer allele is recessive. Sulfation includes 12 phase 2 enzymes used in the biotransformation of drugs, hormones, and xenobiotics, as well as bioactivation of carcinogens. Sulfotransferases are important for the metabolism of drugs and endogenous compounds and convert them into more hydrophilic water-soluble sulfate conjugates that then may be excreted. Sulfotransferase 2A1 (SULT2A1) catalyzes sulfation of steroids and bile acids in the liver and adrenals [43].

Glucuronidation converts fat-soluble compounds to water-soluble compounds for excretion and conjugates drugs such as salicylates, morphine, acetaminophen, and benzodiazepines; xenobiotics such as phenols, polycyclic aromatic hydrocarbons, nitrosamines, aflatoxin, and heterocyclic amines; and dietary and endogenous substances such as bilirubin, melatonin, bile acid, steroid hormones, and fat-soluble vitamins [67].

Currently, pharmacogenomics is only used for a handful of current diseases and conditions such as depression, mood disorders, heart disease, cancer, asthma, and HIV/AIDS. For example, the breast cancer medication herceptin or trastuzumab, only works for women who have an overproduction of the herceptin 2 (HER2) protein [135]. Mercaptopurine is a medication used for acute lymphoblastic leukemia. It can only be used for those who do not have a genetic variant in NUDT15 c.515C>T, as it may interfere with the clearance of the drug. Improper dosing can result in severe side effects and increased risk for infection. Dosage may be adjusted to an individual's genetics. In the case of children with acute lymphoblastic leukemia, it may increase the risk of 6-mercaptopurine induced myelosuppression and may also cause liver function abnormalities [15, 135]. Lastly, the antiviral drug abacavir may only be used in HIV-infected patients without polymorphism to HLA-B*57:01, as this SNP can be associated with severe cutaneous adverse drug reactions (SCARs) which may be potentially life threatening [140].

Warfarin genetics are one of the few cases where there is more than one genetic variant related to understanding genetic risk. Both CYP2C9 and vitamin K epoxide reductase (VKORC1) variants influence metabolism, as well as risk for side effects such as hemorrhage [141]. It is important to understand that there are several variants within these classes of gene as well that help to predict the decrease in response. For example, CYP2C9*3 carriers have an average of 40% reduction in warfarin metabolism and CYP2C9*2 carriers have an average of 20% reduction in warfarin metabolism [142]. SNPs in CYP2D6, CYP1A2 and CYP3A4 can be particularly relevant in pharmacogenomics as these enzymes are important for the metabolism of nearly three quarters of medications [134, 143].

The quest to find ways of perfecting cholesterol levels has led to medications that target cholesteryl ester transfer protein (CETP) and proprotein convertase subtilisin/kexin type 9 (PCSK9). CETP is associated with high levels of high-density lipoprotein (HDL) and PCSK9 is associated with low levels of low-density lipoprotein (LDL). For this reason, medications have been created as inhibitors of CETP and PCSK9 in order to raise HDL and lower LDL [135]. To round out pharmacogenomics and cardiovascular disease, there is the antiplatelet drug clopidogrel. For patients who have suffered from acute coronary syndrome, stroke or percutaneous coronary intervention, clopidogrel is used as a blood thinner. CYP2C19 status relates to variability in drug metabolism as the CYP2C19*2 (rs4244285, c.681G>A) variant causes loss of function causing a decrease in the ability to metabolize clopidogrel and other medication inducers of CYP2C19. However, it is important to not view this enzyme–drug interaction in singularity. This one variant alone does not completely predict the drugs' efficacy. Other CYP2C19 variants may have differing effects, either inducing or inhibiting, while other genes such as PON1 and ATP binding cassette subfamily B member 1 (ABCB1) also influence the metabolism of this drug [143].

17.4.1 Final Thoughts

Epigenetics and the technology associated with the "omics" revolution have and will continue to shape the way medicine and precision nutrition is practiced. As with the FTO SNP revolution, this field requires constant review of (and contribution to) the literature and an ever-expansive open mind. It is imperative that practitioners view the complex interactions between diet, lifestyle, genomics, and medications to accurately practice precision medicine and nutritional genomics. In essence, one may not "treat" a SNP or combination of SNPs. Rather, practitioners of the future will be required to not only understand genetics, but also epigenetics, genomics and the complexity of influencers that determine gene expression.

References

1. Fenech M, El-Sohemy A, Cahill L, Ferguson LR, French T-A.C, Tai ES, Milner J, Koh W-P, Xie L, Zucker M, Buckley M, Cosgrove L, Lockett T, Fung K.Y.C, Head R. Nutrigenetics and Nutrigenomics: Viewpoints on the Current Status and Applications in Nutrition Research and Practice. Journal of Nutrigenetics and Nutrigenomics. 2011;4(2):69–89.
2. Waddington CH. The epigenotype. Endeavour. 1942;1:18–20.
3. Holliday R. Epigenetics: a historical overview. Epigenetics. 2006;1(2):76–80.
4. Kanherkar RR, Bhatia-Dey N, Csoka AB. Epigenetics across the human lifespan. Front Cell Dev Biol. 2014;2:49.
5. Dias BG, Ressler KJ. Parental olfactory experience influences behavior and neural structure in subsequent generations. Nat Neurosci. 2013;17(1):89–96.
6. Ruemmele FM, Garnier-Lengliné. Why are genetics important for nutrition? Lessons from epigenetic research. Ann Nutr Metab. 2012;60(suppl3):28–43.
7. Handy DE, Castro R, Loscalzo J. Epigenetic modifications: basic mechanisms and role in cardiovascular disease. Circulation. 2011;123(19):2145–56.
8. Devlin TM. Textbook of biochemistry with clinical correlations. 7th ed. Hoboken: John Wiley & Sons, Inc.; 2011.
9. Kubota T, Miyake K, Hirasawa T. The mechanisms of epigenetic modifications during DNA replication, the mechanisms of DNA replication, Stuart D, editor. InTech; 2013, https://doi.org/10.5772/51592.
10. Cannell IG, Kong YW, Bushell M. How do microRNAs regulate gene expression? Biochem Soc Trans. 2008;36(Pt 6):1224–31.
11. Ulrey CL, Liou L, Andrews LG, Tollefsbol TO. The impact of metabolism on DNA methylation. Hum Mol Genet. 2005;14(1):R139–47.
12. Weinhold B. Epigenetics: the science of change. Environ Health Perspect. 2006;114(3):A160–7.
13. van Abeelen AFM, Elias SG, Roseboom TJ, Bossuyt PMM, van der Schouw TYT, Grobbee DE, Uiterwaal CSPM. Postnatal acute famine and risk of overweight: the Dutch Hungerwinter Study. Int J Pediatrics. 2012;2012:936509. https://doi.org/10.1155/2012/936509.
14. Ekamper P, van Poppel F, Stein AD, Bijwaard GE, Lumey LH. Prenatal famine exposure and adult mortality from cancer, cardiovascu-

14. lar disease, and other causes through age 63 years. Am J Epidemiol. 2015;181(4):271–9.
15. Chiengthong K, Ittiwut C, Muensri S, et al. NUDT15 c.415C>T increases risk of 6-mercaptopurine induced myelosuppression during maintenance therapy in children with acute lymphoblastic leukemia. Haematologica. 2016;101(1):e24–6.
16. Ndisang JF, Rastogi S. Cardiometabolic diseases and related complications: current status and future perspective. Biomed Res Int. 2013;2013:467682.
17. Liu C-C, Kanekiyo T, Xu H, Bu G. Apolipoprotein E and Alzheimer disease: risk, mechanisms, and therapy. Nat Rev Neurol. 2013;9(2):106–18.
18. Kei A, Miltiadous G, Bairaktari E, Hadjivassiliou M, Cariolou M, Elisaf M. Dysbetalipoproteinemia: two cases report and a diagnostic algorithm. World J Clin Cases. 2015;3(4):371–6.
19. Logue MW, et al. A comprehensive genetic association study of Alzheimer disease in African Americans. Arch Neurol. 2011;68(12):1569–79.
20. Kuwano R, et al. Dynamin-binding protein gene on chromosome 10q is associated with late-onset Alzheimer's disease. Hum Mol Genet. 2006;15(13):2170–82.
21. Yang Y, Ruiz-Narvaez E, Kraft P, Campos H. Effect of apolipoprotein E genotype and saturated fat intake on plasma lipids and myocardial infarction in the Central Valley of Costa Rica. Hum Biol. 2007;79(6):637–47.
22. Curti MLR, Jacob P, Borges MC, Rogero MM, Ferreira SRG. Studies of gene variants related to inflammation, oxidative stress, dyslipidemia, and obesity: implications for a nutrigenetic approach. J Obes. 2011;2011:497401.
23. USDA Dietary Guidelines 2015-2020. Accessed from: https://www.choosemyplate.gov/2015-2020-dietary-guidelines-answers-your-questions.
24. James PA, Oparil S, Carter BL, et al. Evidence-based guideline for the management of high blood pressure in adults report from the panel members appointed to the Eighth Joint National Committee (JNC 8). JAMA. 2014;311(5):507–20.
25. Caprioli J, Mele C, Mossali C, Gallizioli LG, et al. Polymorphisms of EDNRB, ATG, and ACE genes in salt-sensitive hypertension. Can J Physiol Pharmacol. 2008;86:505–10.
26. Scharplatz M, Puhan MA, Steurer J, Bachmann LM. What is the impact of the ACE gene insertion/deletion (I/D) polymorphism on the clinical effectiveness and adverse events of ACE inhibitors?--protocol of a systematic review. BMC Med Genet. 2004;5:23. https://doi.org/10.1186/1471-2350-5-23.
27. Bernstein KE, Ong FS, Blackwell W-LB, Shah KH, Giani JF, et al. A modern understanding of the traditional and nontraditional biological functions of angiotensin-converting enzyme. Pharmacol Rev. 2013;65(1):1–46.
28. American Diabetes Association. (2) Classification and diagnosis of diabetes. Diabetes Care. 2015;38(Supplement 1):S8–S16. https://doi.org/10.2337/dc15-S005.
29. Cornelis MC, Qi L, Kraft P, Hu FB. TCF7L2, dietary carbohydrate, and risk of type 2 diabetes in US women. Am J Clin Nutr. 2009;89(4):1256–62. https://doi.org/10.3945/ajcn.2008.27058.
30. Jakubowski H. Homocysteine thiolactone: metabolic origin and protein homocysteinylation in humans. J Nutr. 2000;130(2S Suppl):377S–81S.
31. Serrato M, Marian AJ. A variant of human paraoxonase/arylesterase (HUMPONA) gene is a risk factor for coronary artery disease. J Clin Investig. 1995;96(6):3005–8.
32. Thomàs-Moyà E, Gianotti M, Proenza AM, Lladó I. Paraoxonase 1 response to a high-fat diet: gender differences in the factors involved. Mol Med. 2007;13(3-4):203–9.
33. Garner B, Witting PK, Waldeck AR, Christison JK, Raftery M, Stocker R. Oxidation of high density lipoproteins. I Formation of methionine sulfoxide in apolipoproteins AI and AII is an early event that accompanies lipid peroxidation and can be enhanced by alpha-tocopherol. J Biol Chem. 1998;273(11):6080–7.
34. Ferguson JF, et al. NOS3 gene polymorphisms are associated with risk markers of cardiovascular disease, and interact with omega-3 polyunsaturated fatty acids. Atherosclerosis. 2010;211(2):539–44.
35. José AA, et al. Adherence to the 'Mediterranean Diet' in Spain and its relationship with cardiovascular risk (DIMERICA Study). Nutrients. 2016;8(11):680.
36. Darwiche G, et al. An Okinawan-Based Nordic diet improves anthropometry, metabolic control, and health-related quality of life in Scandinavian patients with type 2 diabetes: a pilot trial. Food Nutr Res. 2016;60:32594.
37. Hinney A, Nguyen TT, Scherag A, Friedel S, Brönner G, et al. Genome Wide Association (GWA) Study for early onset extreme obesity supports the role of fat mass and obesity associated gene (FTO) variants. PLoS One. 2007;2(12):e1361.
38. Claussnitzer M, Dankel SN, Kim K-H, Quon G, Meuleman W, Haugen C, et al. FTO obesity variant circuitry and adipocyte browning in humans. N Eng J Med. 2015;373:895–907.
39. Dina C, Meyre D, Gallina S, Durand E, Körner A, et al. Variation in FTO contributes to childhood obesity and severe adult obesity. Nat Genet. 2007;39(6):724–6.
40. Landgraf K, et al. FTO obesity risk variants are linked to adipocyte *IRX3* expression and BMI of children - relevance of *FTO* variants to defend body weight in lean children? PLoS One. 2016;11(8):e0161739.
41. Sirmans SM, Pate KA. Epidemiology, diagnosis, and management of polycystic ovary syndrome. Clin Epidemiol. 2014;6:1–13.
42. Jones MR, et al. Systems genetics reveals the functional context of PCOS loci and identifies genetic and molecular mechanisms of disease heterogeneity. PLoS Genet. 2015;11(8):e1005455.
43. Thomae BA, Eckloff BW, Freimuth RR, Wieben ED, Weinshilboum RM. Human sulfotransferase SULT2A1 pharmacogenetics: genotype-to-phenotype studies. Pharmacogenomics J. 2002;2:48–56.
44. Bittencourt PL, Marin MLC, Couto CA, Cançado ELR, Carrilho FJ, Goldberg AC. Analysis of HFE and non-HFE gene mutations in Brazilian patients with hemochromatosis. Clinics (Sao Paulo). 2009;64(9):837–41.
45. Samsel A, Seneff S. Glyphosate, pathways to modern diseases II: celiac sprue and gluten intolerance. Interdiscip Toxicol. 2013;6(4):159–84.
46. DSM-V Criteria American Psychiatric Association. Diagnostic and statistical manual of mental disorders. 5th ed. Washington, D.C.: American Psychiatric Press; 2013.
47. American Academy of Neurology (AAN) Guidelines, American Academy of Neurology. Practice parameters for detection, diagnosis, and management of dementia (summary statements). Neurology. 2001;56:1133–1142, 1143–1153, 1154–1166.
48. la Monte D, Suzanne M, Wands JR. Alzheimer's disease is type 3 diabetes–evidence reviewed. J Diabetes Sci Technol. 2008;2(6):1101–13.
49. Corneveaux JJ, Myers AJ, Allen AN, Pruzin JJ, Ramirez M, Engel A, Nalls MA, Chen K, Lee W, Chewning K. Association of CR1, CLU and PICALM with Alzheimer's disease in a cohort of clinically characterized and neuropathologically verified individuals. Hum Mol Genet. 2010;19(16):3295–301. https://doi.org/10.1093/hmg/ddq221.
50. Scahill RI, Ridgway GR, Bartlett JW, et al. Genetic influences on atrophy patterns in familial Alzheimer's disease: a comparison of APP and PSEN1 mutations. J Alzheimers Dis. 2013;35(1):199–212. https://doi.org/10.3233/JAD-121255.2.
51. Wittchen HU. Generalized anxiety disorder: prevalence, burden, and cost to society. Depress Anxiety. 2002;16(4):162–71.
52. Conerly M, Grady WM. Insights into the role of DNA methylation in disease through the use of mouse models. Dis Model Mech. 2010;3(5–6):290–7.
53. Jirout ML, et al. Genetic regulation of catecholamine synthesis, storage and secretion in the spontaneously hypertensive rat. Hum Mol Genet. 2010;19(13):2567–80.

54. Mentch SJ, Locasale JW. One carbon metabolism and epigenetics: understanding the specificity. Ann N Y Acad Sci. 2016;1363(1):91–8.
55. Hewett ZL, et al. Effect of a 16-week Bikram yoga program on perceived stress, self-efficacy and health-related quality of life in stressed and sedentary adults: a randomised controlled trial. J Sci Med Sport. 2018;21(4):352–7.
56. Volavka J, Bilder R, Nolan K. Catecholamines and aggression: the role of COMT and MAO polymorphisms. Ann N Y Acad Sci. 2004;1036:393–8.
57. Shouman S, Wagih M, Kamel M. Leptin influences estrogen metabolism and increases DNA adduct formation in breast cancer cells. Cancer Biol Med. 2016;13(4):505–13.
58. Grossman MH, Emanuel BS, Budarf ML. Chromosomal mapping of the human catechol-O-methyltransferase gene to 22q11.1---q11.2. Genomics. 1992;12(4):822–5.
59. Segall SK, et al. Janus molecule I: dichotomous effects of COMT in neuropathic vs nociceptive pain modalities. CNS Neurol Disord Drug Targets. 2012;11(3):222–35.
60. Magnesium. In: Coates PM, Betz JM, Blackman MR, Cragg GM, Levine M, Moss J, White JD, editors. Encyclopedia of dietary supplements. 2nd ed. New York: Informa Healthcare; 2010. p. 527–537.
61. Magnesium. In: Ross AC, Caballero B, Cousins RJ, Tucker KL, Ziegler TR, editors. Modern nutrition in health and disease. 11th ed. Baltimore: Lippincott Williams & Wilkins; 2012. p. 159–175.
62. Staner L. Sleep and anxiety disorders. Dialogues Clin Neurosci. 2003;5(3):249–58.
63. Ershadifar T, Minaiee B, Gharooni M, Isfahani MM, Nikbakht Nasrabadi A, Nazem E, et al. Heart palpitation from traditional and modern medicine perspectives. Iranian Red Crescent Med J. 2014;16(2):e14301.
64. Li W, et al. Elevation of brain magnesium prevents synaptic loss and reverses cognitive deficits in Alzheimer's disease mouse model. Mol Brain. 2014;7:65.
65. Fuertig R, Azzinnari D, Bergamini G, Cathomas F, Sigrist H, Seifritz E, Vavassori S, Luippold A, Hengerer B, Ceci A, Pryce CR. Mouse chronic social stress increases blood and brain kynurenine pathway activity and fear behaviour: both effects are reversed by inhibition of indoleamine 2,3-dioxygenase. Brain Behav Immun. 2016;54:59–72.
66. Xu L, et al. Effect of glutamate on inflammatory responses of intestine and brain after focal cerebral ischemia. World J Gastroenterol: WJG. 2005;11(5):733–6.
67. Lord R, Bralley A. Laboratory evaluations for integrative and functional medicine. Duluth: Metametrix; 2012.
68. Weber H, Scholz CJ, Domschke K, Baumann C, Klauke B, et al. Gender differences in associations of glutamate decarboxylase 1 gene (GAD1) variants with panic disorder. PLoS One. 2012;7(5):1–7.
69. Miao D, Steck AK, Zhang L, Guyer KM, Jiang L, Armstrong T, et al. Electrochemiluminescence assays for insulin and glutamic acid decarboxylase autoantibodies improve prediction of type 1 diabetes risk. Diabetes Technol Ther. 2015;17(2):119–27. https://doi.org/10.1089/dia.2014.0186.
70. Yoon J-W, Yoon C-S, Lim H-W, Huang QQ, Kang Y, Pyun KH, Hirasawa K, Sherwin RS, Jun H-S. Control of autoimmune diabetes in NOD mice by GAD expression or suppression in beta cells. Science. 1999;284:1183–7.
71. Zahavi AY, et al. Serotonin and dopamine gene variation and theory of mind decoding accuracy in major depression: a preliminary investigation. PLoS One. 2016;11:e0150872.
72. Ji Y, et al. Pharmacogenomics of selective serotonin reuptake inhibitor treatment for major depressive disorder: genome-wide associations and functional genomics. Pharmacogenomics J. 2013;13(5):456–63.
73. Liang Y, Liu L, Wei H, Luo XP, Wang MT. Late-onset riboflavin-responsive multiple acyl-CoA dehydrogenase deficiency (glutaric aciduria type II). Zhonghua Er Ke Za Zhi. 2003;41(12):916–20.
74. Styczeń K, et al. Study of the serum copper levels in patients with major depressive disorder. Biol Trace Elem Res. 2016;174(2):287–93.
75. Yamamoto K, Cubells JF, Gelernter J, Benkelfat C, Lalonde P, Bloom D, Lal S, Labelle A, Turecki G, Rouleau GA, Joober R. Dopamine beta-hydroxylase (DBH) gene and schizophrenia phenotypic variability: a genetic association study. Am J Med Genet B Neuropsychiatr Genet. 2003;117B(1):33–8.
76. Pulikkalpura H, et al. Levodopa in *Mucuna Pruriens* and its degradation. Sci Rep. 2015;5:11078.
77. Sender R, Fuchs S, Milo R. Revised estimates for the number of human and bacteria cells in the body. PLoS Biol. 2016;14(8):e1002533.
78. Gujral N, Freeman HJ, Thomson ABR. Celiac disease: prevalence, diagnosis, pathogenesis and treatment. World J Gastroenterol: WJG. 2012;18(42):6036–59.
79. Lebwohl B, Ludvigsson JF, Green PHR. Celiac disease and non-celiac gluten sensitivity. BMJ. 2015;351:h4347.
80. Choo SY. The HLA system: genetics, immunology, clinical testing, and clinical implications. Yonsei Med J. 2007;48(1):11–23.
81. Selleski N, Almeida LM, Almeida FC, Gandolfi L, Pratesi R, Nobrega YKM. Simplifying celiac disease predisposing HLA-DQ alleles determination by the real-time PCR method. Arq Gastroenterol. 2015;52(2):143–6.
82. Samsel A, Seneff S. Glyphosate's suppression of cytochrome P450 enzymes and amino acid biosynthesis by the gut microbiome: pathways to modern diseases. Entropy. 2013;15(4):1416–63.
83. USDA Dietary guidelines 2015–2020. Accessed from: https://www.choosemyplate.gov/.
84. Wacklin P, Tuimala J, Nikkila J, et al. Faecal microbiota composition in adults is associated with the *FUT2* gene determining the secretor status. PLoS One. 2014;9(4):e94863.
85. Wacklin P, Makivuokko H, Alakulppi N, Nikkila J, et al. Secretor genotype (FUT2 gene) is strongly associated with the composition of Bifidobacteria in the human intestine. PLoS One. 2011;6(5):e20113.
86. Tong M, McHardy I, Ruegger P, Goudarzi M, Kashyap PC, Haritunians T, et al. Reprogramming of gut microbiome energy metabolism by the FUT2 Crohn's disease risk polymorphism. ISME J. 2014;8(11):2193–206.
87. Mahmood A, FitzGerald AJ, Marchbank T, Ntatsaki E, Murray D, Ghosh S, Playford RJ. Zinc carnosine, a health food supplement that stabilises small bowel integrity and stimulates gut repair processes. Gut. 2006;56(2):168–75.
88. Nakatsu F, Hase K, Ohno H. Review: the role of the clathrin adaptor AP-1: polarized sorting and beyond. Membranes. 2014;4:747–63.
89. Maintz L, Novak N. Histamine and histamine intolerance. Am J Clin Nutr. 2007;85:1185–96.
90. Lundius EG, et al. Histamine influences body temperature by acting at H1 and H3 receptors on distinct populations of preoptic neurons. J Neurosci. 2010;30(12):4369–81.
91. Wright EK, et al. Recent advances in characterizing the gastrointestinal microbiome in Crohn's disease: a systematic review. Inflamm Bowel Dis. 2015;21(6):1219–28. PMC.
92. Song WB, Lv YH, Zhang ZS, Li YN, et al. Soluble intercellular adhesion molecule-1, D-lactate and diamine oxidase in patients with inflammatory bowel disease. World J Gastroenterol: WJG. 2009;15(31):3916–9.
93. Zebra Diagnostics. Histamine and histamine intolerance. SNPed. Virginia: Richmond; 2016.
94. Adjalla C, Lambert D, Benhayoun S, Berthelsen JG, Nicolas JP, Guéant JL, Nexo E. Nutritional biochemistry forms of cobalamin and vitamin B12 analogs in maternal plasma, milk, and cord plasma. J Nutr Biochem. 1994;5(8):406–10.
95. Gropper S, Smith J. Advanced nutrition and human metabolism. 6th ed. Belmont: Wadsworth; 2013.
96. Mera PE, Escalante-Semerena JC. Multiple roles of ATP:cob(I)alamin adenosyltransferases in the conversion of B_{12} to coenzyme B_{12}. Appl Microbiol Biotechnol. 2010;88(1):41–8.

97. Froese DS, Gravel RA. Genetic disorders of vitamin B12 metabolism: eight complementation groups – eight genes. Expert Rev Mol Med. 2010;12:e37.
98. Desai HG, Gupte PA. Helicobacter pylori link to pernicious anaemia. J Assoc Physicians India. 2007;55:857–9.
99. Chery C, Hehn A, Mrabet N, Oussalah A, Jeannesson E, Besseau C, Alberto JM, Gross I, Josse T, Gérard P, Guéant-Rodriguez RM, Freund JN, Devignes J, Bourgaud F, Peyrin-Biroulet L, Feillet F, Guéant JL. Gastric intrinsic factor deficiency with combined GIF heterozygous mutations and FUT2 secretor variant. Biochimie. 2013;95(5):995–1001.
100. Sundar I, Rahman I. Vitamin D and susceptibility of chronic lung diseases: role of epigenetics. Front Pharmacol. 2011;2(50):1–10.
101. Holick CN, Stanford JL, Kwon EM, Ostrander EA, Nejentsev S, Peters U. Comprehensive association analysis of the vitamin D pathway genes, VDR, CYP27B1, and CYP24A1, in prostate cancer. Cancer Epidemiol Biomarkers Prev. 2007;16(10):1990–9.
102. Pandolfi F, Franza L, Mandolini C, Conti P. Immune modulation by vitamin D: special emphasis on its role in prevention and treatment of Cancer. Clin Ther. 2017;S0149-2918(17):30194–7.
103. Cooper GS, Umbach DM. The association between vitamin D receptor gene polymorphisms and bone mineral density at the spine, hip and whole-body in premenopausal women. Osteoporos Int. 1996;6(1):63–8.
104. Jia F, Sun RF, Li QH, Wang DX, Zhao F, Li JM, Pu Q, Zhang ZZ, Jin Y, Liu BL, Xiong Y. Vitamin D receptor BsmI polymorphism and osteoporosis risk: a meta-analysis from 26 studies. Genet Test Mol Biomarkers. 2013;17(1):30–4.
105. Zhang H, Zhuang XD, Meng FH, Chen L, Dong XB, Liu GH, Li JH, Dong Q, Xu JD, Yang CT. Calcitriol prevents peripheral RSC96 Schwann neural cells from high glucose & methylglyoxal-induced injury through restoration of CBS/H2S expression. Neurochem Int. 2016;92:49–57.
106. Aranow C. Vitamin D and the immune system. J Investig Med. 2011;59(6):881–6.
107. Kriebitzsch C, et al. 1,25-Dihydroxyvitamin D_3 influences cellular homocysteine levels in murine pre-osteoblastic MC3T3-E1 cells by direct regulation of cystathionine B-synthase. J Bone Miner Res Off J Am Soc Bone Miner Res. 2011;26(12):2991–3000.
108. Guo W, Kan J-t, Cheng Z-y, et al. Hydrogen sulfide as an endogenous modulator in mitochondria and mitochondria dysfunction. Oxidative Med Cell Longev. 2012;2012:1. https://doi.org/10.1155/2012/878052.
109. Kraus JP, Le K, Swaroop M, Ohura T, Tahara T, Rosenberg LE, Roper MD, Kožlch V. Human cystathionine β-synthase cDNA: sequence, alternative splicing and expression in cultured cells. Hum Mol Genet. 1993;2(10):1633–8.
110. Orlowski M, Meister A. The γ-Glutamyl cycle: a possible transport system for amino acids. Proc Natl Acad Sci U S A. 1970;67(3):1248–55.
111. NCBI (2017) GSTT1 glutathione S-transferase theta 1 [Homo sapiens (human)]. Gene ID: 2952, updated on 18-Nov-2019.
112. NCBI (2017) GSTM1 glutathione S-transferase mu 1 [Homo sapiens (human)].Gene ID: 2944, updated on 18-Nov-2019
113. Carr AC, Frei B. Toward a new recommended dietary allowance for vitamin C based on antioxidant and health effects in humans. Am J Clin Nutr. 1999;69(6):1086–107.
114. Mehlert A, Diplock AT. The glutathione S-transferases in selenium and vitamin E deficiency. Biochem J. 1985;227(3):823–31.
115. Morton LM, Schenk M, Hein DW, Davis S, Zahm SH, Cozen W, et al. Genetic variation in N-acetyltransferase 1 (NAT1) and 2 (NAT2) and risk of non-Hodgkin lymphoma. Pharmacogenet Genomics. 2006;16(8):537–45.
116. Harmer D, Evans DA, Eze LC, Jolly M, Whibley EJ. The relationship between the acetylator and the sparteine hydroxylation polymorphisms. J Med Genet. 1986;23(2):155–6.
117. Hein DW, Doll MA, Fretland AJ, Leff MA, Webb SJ, Xiao GH, Devanaboyina US, Nangju NA, Feng Y. Molecular genetics and epidemiology of the NAT1 and NAT2 acetylation polymorphisms. Cancer Epidemiol Biomark Prev. 2000;9(1):29–42.
118. OMIM (2017) Phosphatase and tensin homolog; PTEN.Access from: https://www.omim.org/entry/601728 on December 22, 2017.
119. Da Costa RM, Neves KB, Mestriner FL, Louzada-Junior P, Bruder-Nascimento T, Tostes RC. TNF-α induces vascular insulin resistance via positive modulation of PTEN and decreased Akt/eNOS/NO signaling in high fat diet-fed mice. Cardiovasc Diabetol. 2016;15(1):119.
120. McCormack D, McFadden D. A review of pterostilbene antioxidant activity and disease modification. Oxidative Med Cell Longev. 2013;2013:575482.
121. Waterson A, Bower M. TNF and cancer: the good and bad. Cancer Therapy. 2004;2:131–48.
122. Juknat A, Pietr M, Kozela E, Rimmerman N, Levy R, Gao F, Coppola G, Geschwind D, Vogel Z. Microarray and pathway analysis reveal distinct mechanisms underlying cannabinoid-mediated modulation of LPS-induced activation of BV-2 microglial cells. PLoS One. 2013;8(4):e61462.
123. Ye JF, Zhu H, Zhou ZF, Xiong RB, Wang XW, Su LX, Luo BD. Protective mechanism of andrographolide against carbon tetrachloride-induced acute liver injury in mice. Biol Pharm Bull. 2011;34(11):1666–70.
124. Hauer J, Anderer FA. Mechanism of stimulation of human natural killer cytotoxicity by arabinogalactan from Larix occidentalis. Cancer Immunol Immunother. 1993;36(4):237–44.
125. Bowie AG, O'Neill LA. Vitamin C inhibits NF-kappa B activation by TNF via the activation of p38 mitogen-activated protein kinase. J Immunol. 2000;165(12):7180–8.
126. Manna SK, Mukhopadhyay A, Aggarwal BB. Resveratrol suppresses TNF-induced activation of nuclear transcription factors NF-kappa B, activator protein-1, and apoptosis: potential role of reactive oxygen intermediates and lipid peroxidation. J Immunol. 2000;164(12):6509–19.
127. Jurenka JS. Anti-inflammatory properties of curcumin, a major constituent of Curcuma longa: a review of preclinical and clinical research. Altern Med Rev. 2009;14(2):141–53.
128. Taanman J-W, et al. The mitochondrial genome: structure, transcription, translation and replication. BBA-Bioenergetics. 1999;1410(2):103–23.
129. NCBI (2017) NDUFS3 NADH:ubiquinone oxidoreductase core subunit S3 [Homo sapiens (human)].Gene ID: 4722, updated on 3-Nov-2019
130. NCBI (2017) NDUFS7 NADH:ubiquinone oxidoreductase core subunit S7 [Homo sapiens (human)]. Gene ID: 374291, updated on 3-Nov-2019
131. Bresciani G, González-Gallego J, da Cruz IB, de Paz JA, Cuevas MJ. The Ala16Val MnSOD gene polymorphism modulates oxidative response to exercise. Clin Biochem. 2013;46(4–5):335–40.
132. Boczonadi V, Giunta M, Lane M, Tulinius M, Schara U, Horvath R. Investigating the role of the physiological isoform switch of cytochrome c oxidase subunits in reversible mitochondrial disease. Int J Biochem Cell Biol. 2015;63:32–40.
133. NCBI (2017) ATP5C1 ATP synthase, H+ transporting, mitochondrial F1 complex, gamma polypeptide 1 [Homo sapiens (human)]. Gene ID: 509, updated on 3-Nov-2019
134. Chhibber A, Kroetz DL, Tantisira KG, McGeachie M, Cheng C, et al. Genomic architecture of pharmacological efficacy and adverse events. Pharmacogenomics. 2014;15(16):2025–48.
135. Johnson JA. Pharmacogenetics in clinical practice: how far have we come and where are we going? Pharmacogenomics. 2013;14(7):835–43.
136. Ogu CC, Maxa JL. Drug interactions due to cytochrome P450. Proc (Bayl Univ Med Cent). 2000;13(4):421–3.
137. Lu SC. Regulation of glutathione synthesis. Mol Asp Med. 2009;30(1–2):42–59. https://doi.org/10.1016/j.mam.2008.05.005.

138. Beutler E, Gelbart T, Pegelow C. Erythrocyte glutathione synthetase deficiency leads not only to glutathione but also to glutathione-S-transferase deficiency. J Clin Investig. 1986;77(1):38–41.
139. De Stefano V, Dekou V, Nicaud V, Chasse JF, London J, Stansbie D, Humphries SE, Gudnason V. Linkage disequilibrium at the cystathionine beta synthase (CBS) locus and the association between genetic variation at the CBS locus and plasma levels of homocysteine. The Ears II Group. European Atherosclerosis Research Study. Ann Hum Genet. 1998;62(Pt 6):481–90.
140. Su SC, Hung SI, Fan WL, Dao RL, Chung WH. Severe cutaneous adverse reactions: the pharmacogenomics from research to clinical implementation. Int J Mol Sci. 2016;17(11):1890.
141. Dean L. Warfarin therapy and the genotypes CYP2C9 and VKORC1. 2012 Mar 8 [Updated 2016 Jun 8]. In: Pratt V, McLeod H, Dean L, et al., editors. Medical genetics summaries [Internet]. Bethesda: National Center for Biotechnology Information (US); 2012. Available from: https://www.ncbi.nlm.nih.gov/books/NBK84174/.
142. Scott SA, Edelmann L, Kornreich R, Desnick RJ. Warfarin pharmacogenetics: CYP2C9 and VKORC1 genotypes predict different sensitivity and resistance frequencies in the Ashkenazi and Sephardi Jewish populations. Am J Hum Genet. 2008;82(2):495–500.
143. Johnson JA, Cavallari LH. Pharmacogenetics and cardiovascular disease—implications for personalized medicine. Pharmacol Rev. 2013;65:987–1009.

Nutritional Influences on Methylation

Jessica M. Pizano and Christy B. Williamson

18.1 What Is Nutritional Genomics? – 270

18.2 How Are Diseases Inherited? – 270

18.3 Epigenetics and SNPs – 270

18.4 One-Carbon Metabolism Basics – 271

18.5 Introduction to Methylation – 273

18.6 Homocysteine Metabolism – 273

18.7 Connecting Neurotransmitters to Methylation – 275
18.7.1 Tetrahydrobiopterin (BH4) – 275
18.7.2 Catecholamines – 275
18.7.3 Serotonin – 276
18.7.4 Iron and Neurotransmitters – 276

18.8 Final Thoughts – 278

References – 283

© Springer Nature Switzerland AG 2020
D. Noland et al. (eds.), *Integrative and Functional Medical Nutrition Therapy*,
https://doi.org/10.1007/978-3-030-30730-1_18

18.1 What Is Nutritional Genomics?

Nutritional genomics is comprised of both nutrigenomics and nutrigenetics. Ordovas and Mooser define nutrigenomics as the exploration of the effect of nutrients on the genome, proteome, and metabolome, whereas nutrigenetics looks at the effect of genetics on the interaction between diet and disease [1]. Nutrigenomics falls clearly in the scope of practice of nutritionists, dietitians, and other health professionals that practice nutrition, whereas genetic counselors and geneticists more appropriately practice nutrigenetics as it involves diagnostics and disease prediction. Nutrigenomics is more about understanding the unique, *measurable* nutritional needs an individual might have based on single-nucleotide polymorphisms (SNPs). Simplistically, one can think of an SNP as a genetic spelling mistake where one letter is substituted for another. They may occur in coding regions, noncoding regions, or intergenic regions. In order to be classified as an SNP, two or more versions of a sequence must each be present in at least 1% of the general population [2]. An SNP is not altogether different from a point mutation in that both include a change in a base pair (adenine, guanine, cytosine, and thymine); however, an SNP is not always deleterious in nature [3]. As alluded to, SNPs can be variable in their effects. Some can cause positive changes while others may be more detrimental. Some SNPs increase enzyme activity while others decrease enzyme function, referred to respectively as up- or downregulation [3]. This change in enzyme activity can impact nutrition as each enzyme has one or more cofactors. Thomas Devlin in the *Textbook of Biochemistry with Clinical Correlations* defines a cofactor as an "ion or molecule that on binding to the catalytic site of an apoenzyme renders it active" [4]. Typically cofactors are vitamins and minerals; therefore, SNPs that have been epigenetically activated have the potential to cause micronutrient deficiencies. This is particularly true for upregulated variants as they will quickly use up body stores of the requisite nutrient. By increasing the intake of a required nutrient it is, therefore, possible to improve enzyme function and potentially compensate for a deleterious SNP.

18.2 How Are Diseases Inherited?

An autosomal gene is located on a numbered chromosome and typically affects both males and females in the same manner. For every gene there are two alleles. One allele is inherited from the mother and the other from the father. When there is an absence of genetic change in either allele, we refer to this as the wild type. When one contains a mutation and the other does not, this is called a heterozygous mutation. When both alleles contain an alteration, this is called a homozygous mutation. In autosomal inheritance, if both parents are heterozygotes, there is a 50% chance that the offspring will be heterozygous and, therefore, affected by the condition. In many diseases, this is considered to be the "carrier" status [5]. Conversely, there is also a 25% chance that the offspring will be homozygous recessive, and, therefore, be severely affected by the disease. Similarly, there is also a 25% chance that the offspring will not have any mutations at all. The probability of inheriting such a mutation is often plotted using a Punnett square (See **Table 18.1**). In 1912, Cambridge University professor of genetics Reginald Crundall Punnett created a square that illustrated Mendelian inheritance [6]. This is classical inheritance. The majority of SNPs analyzed in the field of nutrigenomics consist of epigenetically activated SNPs that may or may not follow Mendelian inheritance patterns; thus caution is advised in the interpretation of nutrigenomic SNPs in this classical fashion (**Table 18.1**).

Table 18.1 Example of a Punnett square with one parent's gene running horizontally and the other vertically

Father's genes	Mother's genes	
	a	a
A	Aa	Aa
a	aa	aa

The genetic possibilities of the offspring are created by taking a gene from each parent. In this example, there is a 50% chance that the offspring will be heterozygotes and a 50% chance they will be homozygotes

18.3 Epigenetics and SNPs

One may ask, is it altogether possible to be able to predict disease activity from the mere presence of an abnormality? The term "epigenetics," which was coined by Waddington in 1942, was derived from the Greek word "epigenesis" which originally described the influence of genetic processes on development [7]. Epigenetics is the study of the modification of gene expression rather than the alteration of the genetic code itself, and in this way can be considered "lifestyle" genetics. It is influenced by age, environment, diet, exercise habits, and disease state. This is the biochemistry that turns "on" or "off" the "good" or "bad" genes we all have.

If you will recall the basic variations of genomic alterations, an SNP and a point mutation both confer a change to a base pair; however, the point mutation is always deleterious [3]. The SNP may be beneficial, neutral, or deleterious [8]. Thus, herein lies the question of predictability. If we can make the assumption that most SNPs confer disease risk, then we would be likely to find a linear association to demonstrate this phenomenon. However, epigenetics, or the ability to turn on and off genetic expression via lifestyle alterations, explains why an individual might have an SNP and have no effect while another individual with the same SNP may experience a change. *In essence, there may be evidence that there is a deleterious SNP, but that does not make the SNP active.* For example, 80–85% of women who carry the BRCA1 or BRCA2 genes go on to develop breast cancer [8]. How then do the other 15–20% escape this fate? It is most often associated with their diet, lifestyle, and their parents' and grandparents' diets and lifestyles. The idea that your grandfather survived the potato famine and, therefore, he confers activated genetics to increase your risk of diabetes does seem a bit unfair [9], but not from the perspective of the genome. The alteration actually increased your grandfathers' survival during a time of stress, despite increasing your risk for diabetes in times of plenty [9].

Epigenetic influence can be measured by analyzing the metabolome, proteome, virome, microbiome, epigenome, and transcriptome [10]. Currently, in the realm of nutrition, there is a heavy focus on the microbiome and metabolome. In these analytes, we are able to pinpoint genomic expression, activation, and deactivation. The entirety of this practice is outside of the scope of this textbook; however, herein are a few examples to demonstrate the effectiveness of functional testing.

In the case of cystathionine beta synthase (CBS) upregulation, a practitioner would want to "validate" that a heterozygous or homozygous mutation was in fact epigenetically active. To do so, they would look at an organic acid marker called xanthurenate. This marker identifies functional pyroxidine (B6) deficiencies [11]. Pyroxidine (vitamin B6) in the active form is called pyridoxal-5-phosphate, and it is the cofactor for CBS. When the CBS enzyme is actively upregulated, there will be an elevation of xanthurenate which "confirms" that this cofactor is in fact being depleted. Further, in the case of CBS upregulation, one would also expect to find a decreased level of certain amino acids such as homocysteine, methionine, and potentially an increase in cystathionine [11]. CBS upregulation may also result in elevated levels of ammonia and/or glutamate, a potential outcome of excess cystathionine. Therefore, one would expect to find a supportive pattern in both organic acids and amino acids testing that points to ammonia excess. There would be potential amino acid abnormalities in glutamate, arginine, citrulline, aspartate, fumarate, and nitric oxide as well as elevations in the organic acid markers orotate, citrate, isocitrate, and cis-aconitate [11]. Why is this clinically important? It allows a practitioner to understand the importance of not formulating a nutrition plan simply by the genetics alone. Rather, it is essential to explore the confluence of the genome and the metabolome so that the nutrition plan reflects the current epigenetic expression and not simply the potential of what the genome could represent. One may have an upregulated CBS variant; however, unless the metabolomic scenario described above is present to confirm or validate the epigenetic activity of this SNP, there is no need to "treat" the SNP.

In the case of gastrointestinal dysbiosis, there is an organic acid marker called D-lactate. SNPs in fucosyltransferase 2 (FUT2) may result in an elevation of this marker [12]. A practitioner would want to "confirm" this type of genomic alteration by analyzing a stool sample, specifically looking for significantly decreased amount of *Bifidobacteria*. Decreased levels of *Bifidobacteria* inhibit bacterial competition, which results in *Lactobacillus* overgrowth [10]. In the case of elevated D-Lactate and decreased levels of *Bifidobacteria*, it would be contraindicated to recommend any probiotic strain or food item that contained *Lactobacilli* [11]. Signs and symptoms of D-lactic acidosis include altered mental status (drowsiness to coma), slurred speech, disorientation or impaired motor coordination, hostile, aggressive, abusive behavior, inability to concentrate, nystagmus, delirium, hallucinations, irritability, excessive hunger, headache, weak, droopy eyelids, hand tremor when the wrist is extended, and blurred vision [11]. As indicated, there are a wide variety of clinical signs and symptoms as well as confirmatory metabolomic and microbiome data that can indicate whether an SNP is "active" or deleterious. *Do not treat an SNP; first confirm that it is active.*

18.4 One-Carbon Metabolism Basics

One-carbon metabolism can be thought of as a folate conversion system. This system moves only one carbon at a time, creating oxidized or reduced forms of folate. This is beneficial because uncoupled carbons are highly volatile and result in oxidative stress. There are three molecules that participate in this process. The first is tetrahydrofolate (THF), the usable form of folate in the body. However, to generate THF, first dihydrofolate (DHF) or dietary folate must be converted into tetrahydrofolate (THF) by dihydrofolate reductase (DHFR) [13]. DHFR not only preferentially metabolizes dihydrofolate, but it is also subject to negative feedback inhibition via synthetic folic acid, and thus can be functionally blocked by excessive folic acid consumption [14]. If there is a polymorphism in DHFR, it is prudent to supplement with forms of folate that bypass this system, such as folinic acid. Folinic acid is available in a supplemental form as calcium folinate and as the prescription drug Leuvocorin [15]. The goal is to get to the least oxidized state; for one-carbon metabolism that is 5-methyltetrahydrofolate, the product of MTHFR. As a reminder, methylenetetrahydrofolate reductase (MTHFR) catalyzes the conversion of 5,10-methylenetetrahydrofolate (an intermediate level of oxidation) to 5-methyltetrahydrofolate (the lowest level of oxidation), a co-substrate for homocysteine remethylation to methionine [16]. The highest oxidative state for folates is the formino or formyl form. These metabolites are formed when there is an increase in cellular oxidation as a result of cellular folate deficiency, potentially caused from alterations in MTHFR and/or DHFR.

The next player on this carbon team is called S-adenosylmethionine or SAM. SAM is considered to be the universal methyl donor [17] as it is formed from methionine and then "donates" or releases the carried methyl donor for processes such as purine and neurotransmitter synthesis. Once SAM has released its methyl donor, it is converted into S-adenosylhomocysteine (SAH), the intermediary between SAM and homocysteine. At this point, homocysteine is either converted back to methionine via methionine synthase (MTR) or betaine homocysteine methyltransferase (BHMT) or it can enter the transsulfuration pathway to form other sulfur-

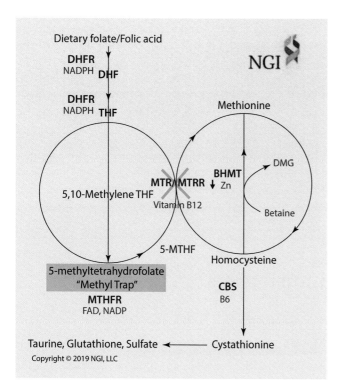

● Fig. 18.1 Methylation cycle: homocysteine sits at the bottom of the methylation pathway. It may be remethylated to methionine via MTR/MTRR or via BHMT. Alternatively, it may be converted to cystathionine via CBS and is the only true disposal route (transsulfuration)

containing amino acids such as cysteine and taurine [18]. This process of SAM moving methyl donors around to homocysteine is again called methylation (● Fig. 18.1). This is an oversimplification of this process as there are many additional fates of SAM; however, this is the basic process of recycling the methyl donor that has been created in one-carbon metabolism by MTHFR.

The last player is cobalamin or vitamin B12. MTR takes the methyl donor created by MTHFR (5-methyltetrahydrofolate) and converts the storage form of vitamin B12 known as adenosylcobalamin into a usable methyl recycler known as methylcobalamin. It is actually methylcobalamin that transfers the methyl donor from MTHFR to form methionine which in turn forms SAM. Methionine synthase and the methylcobalamin it forms are also partially responsible for homocysteine catabolism (● Fig. 18.2).

Within this system of one-carbon metabolism, there is a phenomenon referred to as "folate trapping." Referring back to the reaction of MTHFR, the conversion from 5,10-methylenetetrahydrofolate to 5-methyltetrahydrofolate is irreversible. This means that when 5-methyltetrahydrofolate tries to convert back into tetrahydrofolate it must be able to hand off the methyl group to successfully convert adenosylcobalamin into methylcobalamin. It must also convert back into the folate cycle substrate THF. If, for some reason, this conversion cannot occur, there could potentially be either a polymorphism in MTR or a vitamin B12 deficiency. In this circumstance, MTR cannot recycle vitamin B12 into the methylation cycle and the folate will get "trapped" in the

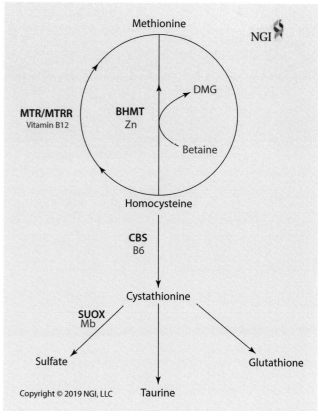

● Fig. 18.2 The pathway on the left is one-carbon metabolism. This pathway's ultimate goal is to use folate for formation of a methyl group via the creation of 5-methyltetrahydrofolate. This methyl group is then passed off to the pathway on the right, methylation. Here, the methyl group is accepted by vitamin B12, transforming it from the adenosylcobalamin form to the methylcobalamin form. Methylation then recycles this methyl donor to aid in DNA regulation (epigenetics). When there is insufficient vitamin B12 available to accept the methyl donor from folate, it becomes trapped in the 5-methyltetrahydrofolate form and is unable to be converted back to THF, the usable form of folate in the body. This is what is referred to as the folate trap. In summation, one-carbon metabolism is a methyl donor creation center, and methylation is a methyl donor recycling center with MTR/MTRR being the connection between the two pathways

methylated form. Please note the nomenclature; the only difference between *methyl*-tetrahydrofolate and tetrahydrofolate is the addition or subtraction of a *methyl* group [19]. Clinically this will result in an elevation of formiminoglutamate (FIGLU) (remember, the formino form is at the highest oxidative state), indicating a cellular folate deficiency along with oxidative stress [11]. Paradoxically, it will also increase serum folate levels. This occurs because serum folate measures all forms of folate, mostly methylfolate, and FIGLU is specifically measuring usable THF [11]. If you were to erroneously add in more folate, especially methylated folate, this will increase the amount of 5-methyltetrahydrofolate, thus exacerbating the discrepancy between cellular and serum folate levels. Other clinical associations include reduced methylation capacity as evidenced by a decrease in methionine, SAM, SAH, homocysteine, and ultimately, glutathione [11]. The correct intervention is not to add in more folate as indicated by the elevation in the urinary organic acid formiminoglutamate (FIGLU), rather it is to assist with the methyl transfer by correcting the cobalamin deficiency [11]. A cellular deficiency in cobalamin can be determined by an elevation in methylmalonic acid (MMA) [11]. In cases where there is a known polymorphism in DHFR, the importance of removing excess synthetic folate in the form of synthetic folic acid is incredibly important as a block in this enzyme will further restrict one-carbon metabolism, methylation, transsulfuration, and neurotransmitter synthesis [20].

18.5 Introduction to Methylation

Methylation is a biochemical pathway in the body that describes the adding or subtracting of a methyl group from different compounds. A methyl group is simply one carbon and three hydrogens (CH_3). At the center of methylation is homocysteine, an amino acid. The pathway works by either recycling homocysteine back to its precursor methionine or by disposing with homocysteine all together. When looking at the difference between homocysteine and its precursor methionine, it is important to understand that the only difference is the presence of an additional methyl group on methionine. In fact, methionine is methylated homocysteine. Methylation interacts with several additional pathways including transsulfuration, one-carbon metabolism, the tetrahydrobiopterin (BH4) cycle, and the urea cycle.

> Methionine = homocysteine + methyl group (CH_3)
> Methionine is synthesized from homocysteine and a methyl group (CH_3). This may occur by recycling homocysteine back to methionine via MTR/MTRR or via BHMT.

The National Institute of Health (NIH) defines methylation as "The attachment of methyl groups to DNA at cytosine bases; correlated with reduced transcription of the gene and thought to be the principal mechanism in X-chromosome inactivation and imprinting" [21]. Methylation is involved in gene regulation, biotransformation, neurotransmitter synthesis, estrogen catabolism, immune cell production (such as T cells and natural killer cells), DNA and histone synthesis, energy production, creation of the myelin sheath on nerve cells, and building and maintaining cell membranes via phosphatidylcholine. Nutritionally, methylation may be influenced by the presence or absence of sufficient nutrients that are required as either cofactors or substrates in the pathway. The genetics of the individual and presence of activated SNPs dictate the likelihood of requiring more or less of a particular nutrient. Pharmaceuticals such as antacids, methotrexate, and metformin may also influence methylation by decreasing the absorption of nutrients required as cofactors for the enzymes [22]. Use of high-dose niacin can also impede methylation as it absorbs S-adenosylmethionine (SAM), our universal methyl donor [23]. Exposure to environmental toxins such as heavy metals and chemicals may also inhibit methylation as these xenobiotics can block regulatory enzymes in the pathway as well as require additional glutathione production from an adjacent pathway (transsulfuration).

Stress can also have a negative impact on methylation. This relationship exists because stress influences the release of the neurotransmitters norepinephrine and epinephrine via cortisol [24]. Epinephrine, commonly known as adrenaline, is created from norepinephrine via an enzyme called phenylethanolamine N-methyltransferase (PNMT) [25]. This enzyme requires a methyl group and, when upregulated, can deplete overall methyl stores. Thus, an excessive need for conversion (stress) will place continuous demands on SAM and can negatively impact the other processes of methylation. Finally, having too much of a substrate (as in oversupplementation with methylated supplements) can cause negative feedback inhibition, which may also impede the methylation pathway by targeting certain SNPs and blocking their actions.

Remember, the presence of a genetic SNP alone is not sufficient to cause a decrease in enzyme function. An epigenetic influence must occur for this to happen.

18.6 Homocysteine Metabolism

Homocysteine is localized at the bottom of the methylation pathway as the segue to the transsulfuration pathway. It is produced from the amino acid methionine as it is methylated methionine. Homocysteine in excess or deficiency can have deleterious consequences. To address the most common issue with homocysteine (elevation), there are three routes of homocysteine metabolism. The two recycling routes are comprised of methionine synthase (MTR)/methionine synthase reductase (MTRR) and betaine homocysteine methyltransferase (BHMT). The only true disposal route is via cystathionine beta synthase (CBS) and provides a connection between the methylation and transsulfuration pathways.

MTR and MTRR recycle homocysteine back to methionine with the assistance of a methyl group produced by methylenetetrahydrofolate reductase (MTHFR). This accounts for approximately half of homocysteine catabolism. MTR is vitamin B12 dependent and assists in maintaining adequate intracellular methionine required for production of S-adenosyl methionine (SAM), the universal methyl donor [11]. It is also responsible for maintaining sufficient intracellular folate pools as it supplies a methyl donor to one-carbon metabolism [11]. In the folate cycle, MTHFR converts 5,10 methylenetetrahydrofolate into 5-methyl folate, a cosubstrate of homocysteine metabolism. Then, 5-methyl folate "hands off" its methyl donor to adenosylcobalamin, in effect pushing out the adenosyl group and replacing it with a methyl group to form methylcobalamin, the usable form of cobalamin required for methylation.

MTRR assists MTR in the maintenance of sufficient vitamin B12 to support appropriate homocysteine remethylation. If it cannot perform in this role, either via an SNP or functional B12 deficiency, the 5-methyl folate becomes "trapped" and is no longer able to participate in one carbon metabolism or methylation. There is a specific methionine synthase SNP that increases homocysteine levels and may also contribute to dyslipidemia in men, known as MTRR A66G [26]. This presentation is particularly true if there is a concurrent MTHFR C677T polymorphism as the result of this SNP also increases homocysteine.

BHMT provides a short cut through methylation. It supplies a methyl group that allows for the conversion of homocysteine back to methionine as well as converting betaine to dimethylglycine (DMG). While hypothetically BHMT polymorphisms could have an effect on homocysteine levels, there is conflicting evidence to support the idea that SNPs have any measurable effects. BHMT is thought to account for up to half of the homocysteine remethylation capacity, according to some studies [27], while it is thought to account for less than 1% of homocysteine recycling in others [28]. BHMT aids MTR in increasing the availability of methionine so that it may be used to produce SAM [29]. Zinc serves as a cofactor for BHMT and a polymorphism may increase an individual's zinc requirements. Betaine is also required for the production of DMG, thus it is a necessity for proper BHMT function. Betaine may be found in foods such as wheat, spinach, shellfish, and beets and may also be created from choline endogenously [28, 30]. This pathway also intersects with choline metabolism and more advanced methyltransferases, which ultimately regulate epigenetic expression (◘ Fig. 18.1).

CBS provides the only true disposal route for homocysteine and accounts for approximately half of all homocysteine catabolism. It catalyzes the conversion of homocysteine to cystathionine utilizing vitamin B6 and magnesium as its cofactors. This enzyme is important as it sits between the junction of methylation and transsulfuration, the pathway that ultimately leads to the generation of the endogenous antioxidant glutathione. In this way, it has the ability to influence detoxification via channeling sulfur down into the transsulfuration pathway and, ultimately, into the gamma glutamyl pathway, a mechanism for glutathione production. CBS generates both cystathionine and hydrogen sulfide (H2S) as its products. For CBS, there are both upregulated and downregulated variants. CBS C699T is considered to be upregulated and CBS A360A is downregulated [31] (◘ Fig. 18.3).

Upregulation of CBS will cause decreased levels of vitamin B6, magnesium, glutathione, and homocysteine [11]. Vitamin B6 deficiency should be confirmed via xanthurenate before supplementing in this sensitive population [11]. Upregulated CBS can compromise the ability to recycle

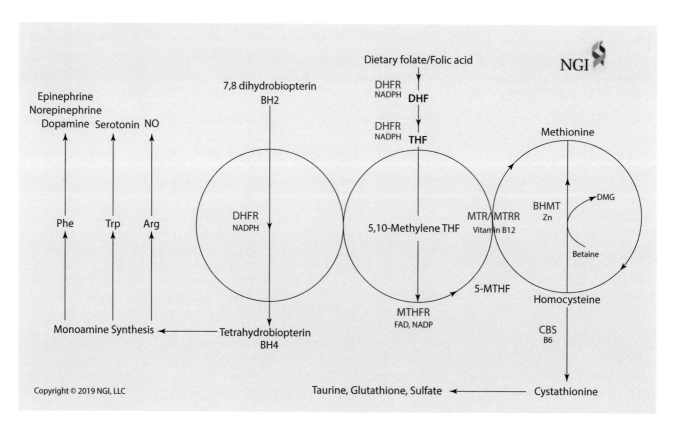

◘ Fig. 18.3 The intersection of major biochemical pathways including one-carbon metabolism, methylation, and transsulfuration are essential for the regulation of DNA expression. They also interact with additional pathways including the gamma glutamyl cycle, which produces glutathione to be used in glutathione conjugation (phase 2 detoxification), and the tetrahydrobiopterin pathway via DHFR, which is essential as a cofactor in the production of the monoamine neurotransmitters including dopamine, norepinephrine, epinephrine, serotonin, and nitric oxide

homocysteine back to methionine and to generate SAM. Further, there will be increased generation of hydrogen sulfide, which can increase the production of ammonia, cortisol, and glutamate resulting in excitotoxicity [32]. Further, ammonia depletes tetrahydrobiopterin (BH4) through increased production of nitric oxide [30]. In this way, polymorphisms can lead to insufficient production of serotonin, dopamine, norepinephrine, and epinephrine. The CBS C699T variant is often found in those with autism, Down syndrome, connective tissue disease, cancer, and homocystinuria [33]. A high organosulfur diet in this population will cause increased urinary sulfate levels but will typically not allow for the generation of glutathione. For this reason, it may be helpful to limit high organosulfur foods such as eggs, onion, garlic, and cruciferous vegetables to tolerable quantities for a short time to relieve stress on this enzyme.

The increased flux of sulfur through CBS forces sulfur too quickly downstream which taxes sulfate oxidase (SUOX), a lower capacity enzyme. This may cause increased requirements for the mineral molybdenum, SUOX's cofactor [34]. SUOX catalyzes the oxidation of sulfite to sulfate in the final reaction in the oxidative degradation of the sulfur amino acids cysteine and methionine [34]. SNPs in SUOX are rare and, when present, can result in neurological abnormalities that are often fatal at an early age.

CBS downregulation, on the other hand, limits the sulfur amino acids by slowing the conversion of homocysteine to cystathionine. The resultant elevation of homocysteine is, therefore, associated with hypertension and other cardiovascular disorders. Similar to upregulation, this will create an increased requirement for vitamin B6 (CBS's cofactor). Like those with upregulated variants, individuals who have CBS downregulation have limited availability to generate glutathione from sulfur amino acids such as cysteine [11]. To stimulate this enzyme to form glutathione, sulfate, and taurine, supplement with P5P and, potentially, cysteine in the form of N-acetyl cysteine.

Once cystathionine is generated through CBS, it then is converted to cysteine via cystathionine gamma-lyase (CGL) [11]. This enzyme also requires vitamin B6 as its cofactor and is known to cause significantly higher plasma homocysteine levels and cystathioninuria [13]. Further, polymorphisms are associated with hypertension as this enzyme regulates blood pressure via endogenous signaling of H2S [32]. Replenishing vitamin B6 as the active pyridoxal-5-phosphate (P5P) may help restore proper enzyme function and allow for proper sulfur influx into the gamma glutamyl cycle.

18.7 Connecting Neurotransmitters to Methylation

18.7.1 Tetrahydrobiopterin (BH4)

While DHFR is a key enzyme in one-carbon metabolism, it also is an important precursor to the tetrahydrobiopterin (BH4) cycle, one that determines cellular levels of several neurotransmitters. BH4 is a cofactor required for the hydroxylation of the aromatic amino acids (i.e., phenylalanine to tyrosine, tyrosine to DOPA, and tryptophan to serotonin). BH4 is also a necessary cofactor for the conversion of arginine into nitric oxide via nitric oxide synthase (NOS) [35]. In this way, BH4 serves an integral role in the production of all of the monoamines including dopamine, norepinephrine, epinephrine, and serotonin as well as nitric oxide. Lord and Bralley state, "Restricted BH4 cofactor availability has been suggested as an etiologic factor in neurological diseases, including DOPA-responsive dystonia, Alzheimer's disease, Parkinson's disease, autism and depression; as well as in other conditions such as insulin sensitivity and vascular disease" [11].

18.7.2 Catecholamines

The synthesis of all catecholamines starts with the essential amino acid phenylalanine, which is then converted to the amino acid tyrosine via phenylalanine hydroxylase (PAH). This reaction requires both BH4 and iron as cofactors. Tyrosine is then consequently converted to L-DOPA via tyrosine hydroxylase and is the rate-limited step also using the cofactors BH4 and iron [36]. Dopamine is then generated from L-DOPA via aromatic L-amino acid decarboxylase (AADC), a vitamin B6-dependent enzyme [37]. At this juncture, there are two paths for dopamine. It may be catabolized via monoamine oxidase B (MAO-B) using flavin adenine dinucleotide (FAD) as its cofactor along with catechol-O-methyltransferase (COMT) using magnesium and SAM as its cofactors to form the urinary organic acid homovanillate (HVA) [11, 36]. Alternatively, dopamine may also be used to generate norepinephrine via dopamine beta hydroxylase (DβH) [22]. This reaction requires copper and ascorbate as its cofactors. Norepinephrine may then be catabolized by COMT and MAO-B to the urinary organic acid vanilmandelate (VMA), or it may be converted via COMT to epinephrine, which ultimately is also catabolized by COMT and MAO-B to VMA [11, 22].

COMT is highly polymorphic and is a very common clinical finding. More often, heterozygosity is found rather than homozygosity. This enzyme is important not only for the transfer of methyl groups from SAM to the catecholamines but also from SAM to catechols from foods (e.g., green tea and potatoes), catechol antioxidants such as quercetin, and the hormone catechol estrogen [38]. Additionally, this enzyme is essential for the catabolism of dopamine, norepinephrine, and epinephrine [38]. Research has shown that the COMT Val allele is related to the presence of higher anxiety. This is likely due to the effect of COMT on the dopaminergic tone in the prefrontal cortex and the resultant effect on cognition [39]. There is some thought that gender may also be a factor in anxiety secondary to the COMT val158met polymorphism due to this enzyme's necessity in the catabolism of estrogen. Therefore, women are more likely to experience higher levels of anxiety than men [39].

Since methyl donors help to upregulate the production of monoamines, including the catecholamines, they may be problematic for those with epigenetically activated (downregulated) COMT SNPs. While methyl donors should not necessarily be avoided in the long run, it is important to insure proper COMT function prior to supplementation. When supplementing vitamin forms that contain methyl donors such as methylcobalamin, L-methylfolate, and trimethylglycine, many practitioners consider it important to dose only small amounts in the presence of COMT since hypothetically methyl donors may increase production of catecholamines, which could tax COMT. These supplements may be dosed several times per week rather than daily as another way of preventing increased anxiety.

Given that magnesium is the cofactor for COMT, the presence of a polymorphism supports investigating erythrocyte (RBC) magnesium levels and supplementing with a well-absorbed form of magnesium if needed. Some forms of magnesium, such as magnesium sulfate, oxide, and citrate work as osmotic laxatives and are not very well absorbed [40]. Magnesium glycinate is well absorbed and may provide an additional calming effect since it is chelated to the amino acid glycine, which also serves as an inhibitory neurotransmitter [41]. Another good option is magnesium threonate as it is the only form of magnesium that may cross the blood–brain barrier [42].

MAO-B catalyzes the oxidative deamination of biogenic and xenobiotic amines (i.e., histamine and tyramine), is a part of the metabolism of neuroactive and vasoactive amines in the central nervous system and peripheral tissues, and preferentially oxidizes dopamine (as opposed to MAO-A, which preferentially oxidizes serotonin and noradrenaline). Variants include rs1799836, rs10521432 (C112982T), and rs6651806 (T57758G). MAO-B inhibitors are a type of pharmaceutical that is often used for those with Parkinson's disease to slow the breakdown of dopamine.

18.7.3 Serotonin

In serotonin synthesis, the pathway begins with the amino acid tryptophan being converted to 5-hydroxy-tryptophan (5-OH-TRP) by tryptophan hydroxylase (TPH) using both BH4 and iron as its cofactors [11]. 5-OH-TRP is consequently converted to serotonin by AADC [11]. Monoamine oxidase A (MAO-A) is then used to break serotonin down into the urinary organic acid 5-hydroxyindoleacetic acid (5-HIAA) [11].

MAO-A SNP rs6323 is an upregulated variant causing increased enzyme activity leading to decreased amounts of amines. The G variant encodes for higher activity. A study by Slopien et al. in 2012 found that there was a particularly high rate of depression in postmenopausal women that had this variant and that there was MAO-A activity ranging more than 50-fold over controls [43]. This enzyme also has downregulated variants including rs5906883, rs5906957, rs909525, rs9593210, and rs2072743. These downregulated variants decrease enzyme activity leading to increased amounts of amines such as serotonin (Fig. 18.4).

Serotonin is considered to be the neurotransmitter most closely associated to depression. A common nutritional intervention for depression is to supplement folate. This intervention is effective because folate (5-MTHF) is responsible for regenerating oxidized BH4. When there is insufficient BH4, 5-MTHF may be substituted for BH4 at the enzymatic level. This is due to the similarity in chemical structures of BH4 and 5-MTHF, enough so that endothelial nitric oxide synthase (eNOS) can use 5-MTHF as a substitute cofactor [12]. This may account for folate's antidepressant effect.

18.7.4 Iron and Neurotransmitters

Iron is a cofactor, along with BH4, for PAH, TH, and TPH. This may account for why patients with iron-based anemia often struggle with mood. Without sufficient iron, the enzyme tryp-

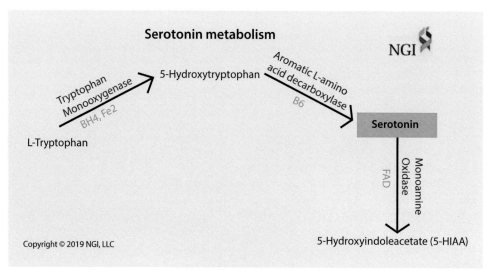

Fig. 18.4 Tryptophan is the precursor of serotonin and is metabolized by tryptophan monooxygenase with the cofactors of BH4 and iron to 5-HTP followed by AADC with the cofactor of P5P. Monoamine oxidase A preferentially catabolizes serotonin and uses the cofactor FAD to create the organic acid metabolite 5-HIAA

tophan dehydrogenase is unable to convert the amino acid tryptophan into the intermediate 5-hydroxy-tryptophan, which is ultimately converted to serotonin by aromatic L-amino acid decarboxylase (AADC) using its cofactor vitamin B6 [37]. This will cause a decreased capacity to synthesize serotonin which, when deficient, is the major neurotransmitter associated with depression. Some depression, however, may also be dopamine mediated. Iron deficiency anemia can also impact this type of depression as dopamine synthesis relies on the conversion of phenylalanine to tyrosine via the enzyme phenylalanine hydroxylase (PAH). This enzyme relies on both BH4 and iron as its cofactors [37]. Additionally, the conversion of tyrosine to L-DOPA is also iron and BH4 reliant as tyrosine hydroxylase requires these nutrients as cofactors. It is then L-DOPA that may be converted into dopamine via the enzyme AADC in both the central and peripheral nervous systems [22] (◘ Figs. 18.5 and 18.6).

Nitric oxide (NO) is synthesized from the amino acid arginine and is produced in the endothelial cells of blood vessels. It is a gaseous substance that diffuses readily through membranes and helps to regulate blood flow by dilation of muscles encircling blood vessels. Therefore, nutritionally, a relative arginine deficiency can occur in diseased coronary arteries. As such, replenishing arginine may allow for arterial dilation. NO can also help control glucose uptake by muscles [44]. This neurotransmitter is unstable and acts quickly, undergoing oxidation to nitrite or nitrate within seconds [44]. It plays an important role in neurotransmissions, blood clotting, and control of blood pressure (via endothelium-derived relaxing factor activity) [44]. Nitric oxide synthase (NOS's) ability to influence smooth muscle reaches beyond the cardiovascular system and allows for regulation of other smooth muscle actions such as peristalsis [44]. Further, NOS is associated with the neurotoxicity found in stroke and other

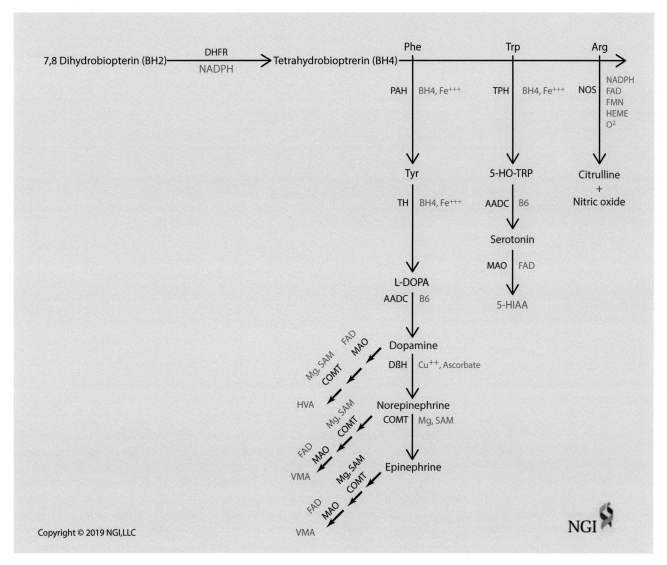

◘ Fig. 18.5 BH4 and neurotransmitter synthesis. DHFR dihydrofolate reductase, PAH phenylalanine hydroxylase, TH tyrosine hydroxylase, AADC aromatic L-amino acid decarboxylase, DβH dopamine beta hydroxylase, MAO monoamine oxidase; COMT catechol-O-methyltransferase, TPH Tryptophan hydroxylase, NOS Nitric oxide synthase, HVA homovanillate, VMA vanilmandelate, 5-HIAA 5-hydroxyindoleacetic acid

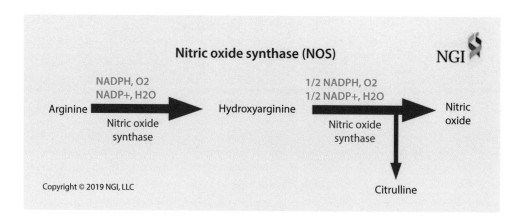

Fig. 18.6 Nitric oxide synthase pathway

neurodegenerative diseases [45]. In the immune system, it helps to stimulate immune cell activation and helps neurons regulate their function [4].

Low levels of NOS can impact male fertility by causing erectile dysfunction (ED) [46]. This is because NO is the primary biochemical way that an erection is stimulated via the corpus cavernosum [47]. Phosphodiesterase 5 (PDE5) inhibitor medications such as Viagra® capitalize on this mechanism, though they were not originally designed for this purpose. In fact, it was not until Viagra® was used in a phase I clinical trial for the treatment of angina that the medication's "side effect" of penile erection was discovered [15]. Viagra® allows for blood vessels within the erectile tissue to vasodilate in response to released NO as well as for the activation of guanylate cyclase (cGMP). Viagra® inhibits breakdown of cGMP causing increased and sustained levels of cGMP, thus causing vasodilation in erectile tissue [4].

There are three different forms of nitric oxide synthase (NOS 1–3). All three enzymes catalyze the conversion of arginine to nitric oxide using NADPH and NADP+ as cofactors [44]. This reaction also creates the byproduct citrulline [11]. It is possible to test NOS activity via urinary organic acid testing, specifically looking for increased levels of citrulline. NOS1 encodes neuronal nitric oxide synthase (nNOS) and is found in nervous tissue and skeletal muscle where it is involved in cell communication [48]. Key variants include rs7298903 (A57373G), rs3782206 (G59494A), and rs2293054 (T2202C). NOS2, or inducible nitric oxide synthase (iNOS), is considered a reactive free radical involved in neurotransmission along with antimicrobial and antitumor activities [48]. In this way, it may be induced by lipopolysaccharides and cytokines, allowing it to serve as an immune activator to help defend against pathogens. Found in both the immune and cardiovascular systems, elevations are commonly found in inflammatory disorders [49].

Via iNOS, NO may play a role in cancer, though this mechanism is not well understood. It may have the potential to stimulate tumor growth, development, and progression as well as angiogenesis [49]. Due to the precursory nature of arginine to NO, a low arginine diet has been suggested as a means of controlling cancer progression [50]. Key variants include rs2297518 (C1823T), rs2248814 (T32235C), and rs2274894 (T836165G). Endothelial nitric oxide synthase (eNOS) is encoded by NOS3 [51]. Found in the endothelium, it is involved in vasodilation, vascular smooth muscle relaxation via cGMP-mediated signal transduction, mediates vascular endothelial growth factor induced angiogenesis in coronary vessels, and promotes blood clotting via platelet activation [51]. Inside the muscle cell, NOS3 binds to guanylate cyclase, resulting in activation and, consequently, catalyzing the conversion of GTP to cyclic GMP. This process ultimately leads to the activation of cGMP-activated protein kinase and the subsequent attachment of a phosphate group to the inositol trisphosphate receptor (IP3), which then results in a decrease in calcium in the muscle cells' cytoplasm [52]. This is the mechanism by which NO results in arterial relaxation. Key variants include rs1800783 (A6251T), rs3918188 (C19635T), rs7830 (G10T), rs1800779 (G6797A), and rs2070744 (T786C) (Fig. 18.7).

Interestingly, manganese appears to be another nutrient that can be used to regulate NO production [53]. This is related to arginine's ability to be converted into either ornithine or citrulline to produce urea and nitric oxide, respectively [11]. If you inhibit production of ornithine via the enzyme arginase (converts arginine and water to ornithine and urea), it will increase nitric oxide production and decrease urea production. Since manganese is the cofactor for arginase, decreasing manganese intake may allow for increased nitric oxide production that could allow for vasorelaxation. This could have positive implications for childhood asthma [11].

18.8 Final Thoughts

Methylation has far-reaching implications and plays a central role in metabolism. Methylation contributes to epigenetic activation via turning "on" and "off" genes, synthesizes neurotransmitters, and is required for the production of glutathione required for detoxification. Understanding the nutritional implications of genetic SNPs while validating these alterations by correlating it to laboratory evaluations allows medical and nutrition professionals to directly influence methylation-related biological processes that may, in turn, improve the health of the patient. This may then help prevent and treat a host of chronic disease states as we are learning that all diseases are a confluence of nature and nurture or what we now call epigenetics Table 18.2.

■ Fig. 18.7 Guanosine triphosphate

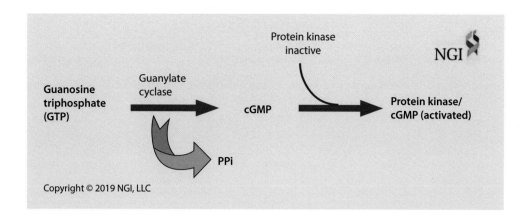

■ Table 18.2 Single-nucleotide polymorphisms (SNPs), cofactors, and functional markers

SNPs	Cofactors	Organic acids	Amino acids/other tests/interventions	rs Numbers
Methylation, transsulfuration, and one-carbon metabolism				▶ www.snppros.com
DHFR *Dihydrofolate reductase* Codes the enzyme dihydrofolate reductase used in conversion of dihydrofolate and folic acid into tetrahydrofolate	NADPH, BH4 re-coupling (BH4 requires magnesium)	↑FIGLU; ↓HVA, ↓VMA	Restricted neurotransmitters, ↓tyrosine, ↓tryptophan; restricted one-carbon metabolism, ↓methionine	rs1643649, rs1643659, rs1677693, rs1650697
MTHFR C677T *Methylene tetrahydrofolate reductase* Catalyzes the conversion of 5,10-methylenetetrahydrofolate to 5-methyltetrahydrofolate, a co-substrate for homocysteine remethylation to methionine; folate trapping may increase homocysteine levels and deplete cellular folate concentrations	Riboflavin, NADH, and ATP	↑FIGLU, ↑glutaric acid	↑Homocysteine; correct B12 status to avoid folate trapping	rs1801133
MTR/MTRR *Methionine synthase/ methionine synthase reductase* Responsible for B12 recycling and converting adenosyl-cobalamin into methyl-cobalamin as needed for methylation	B12, SAM, zinc, heme MTR cofactor: Zinc, blocked by lead, aluminum, and mercury MTRR cofactors: SAM,	↑MMA	↓Methionine, ↑↓homocysteine – Check other homocysteine catabolism SNPs like CBS and BHMT	rs1801394, rs1802059, rs162036

(continued)

Table 18.2 (continued)

SNPs	Cofactors	Organic acids	Amino acids/other tests/interventions	rs Numbers
	FAD, NAD, zinc, blocked by lead			
AHCY *Adenosylhomocysteinase* Regulates intracellular S-adenosylhomocysteine (SAH) concentration and assists with methionine catabolism	NAD, SAM	SAH regulator; SAH blocks COMT and is considered inflammatory	↑Methionine, ↑homocysteine	rs121918607
SHMT *Serine hydroxymethyl-transferase* Catalyzes the reversible, simultaneous conversions of L-serine to glycine and tetrahydrofolate to 5,10-methylenetetrahydrofolate (requires SAM); also catalyzes conversion of 5,10-methenyltetrahydrofolate to 10-formyltetrahydrofolate	P5P, SAM	↑FIGLU	Potential abnormalities in glycine and serine	rs9909104 rs1979277
CBS C699T *Cystathionine beta synthase* Converts homocysteine to cystathionine; sulfur sensitivity	P5P, magnesium, blocked by lead	↑Xanthurenate	↑Cystathionine, ↑cysteine, ↑cystine, ↓homocysteine, ↑glutamate, ↑ammonia (depletes BH4)	rs234706
CBS A360A *Cystathionine beta synthase* Converts homocysteine to cystathionine; cardiovascular complications	P5P, magnesium, blocked by lead	↑↓Xanthurenate	↑Homocysteine, ↓cystathionine, ↓cysteine, ↓cystine, ↑NO	rs1801181
CGL (CTH gene) *Cystathionine gamma lyase* Associated with cystathioninuria; converts cystathionine to cystine; limiting enzyme for glutathione synthesis	P5P, NADH	↓↑Xanthurenate, alpha-ketobutyrate, alpha-hydroxybutyrate	↑Homocysteine, ↑cystathionine, ↓cysteine, ↓cystine, ↑ammonia (depletes BH4)	Primary: rs1021737
PEMT *Phosphatidyl-ethanolamine N-methyltransferase* Required in choline synthesis, hepatocyte membrane structure, bile secretion, estrogen degradation and VLDL secretion	SAM, choline, folate	↓SAM: SAH; potential ↑ in hydroxyproline (membrane fluidity)	↓Phosphoethanolamine, potentially ↓ phosphatidyl-serine and ↓phosphatidyl-choline; pathway requires magnesium	rs12325817, rs7946, rs46464006, rs4244593
BHMT *Betaine homocysteine methyltransferase* Supplies a methyl group to convert homocysteine back to methionine and betaine to dimethylglycine (DMG)	Zinc, TMG; blocked by glucocorticoids	N/A Consider RBC zinc	↓Methionine, ↑homocysteine	rs3733890
SUOX *Sulfite oxidase* Is a mitochondrial enzyme; catalyzes the oxidation of sulfite to sulfate, the final reaction in the oxidative degradation of the sulfur amino acids cysteine and methionine	Molybdenum, hydroxocobalamin, vitamin E	↑sulfate, ↑pyroglutamate, potentially ↑alpha hydroxybutyrate	↓Homocysteine, ↓cystathionine, ↓cysteine, ↑taurine	rs705703

Table 18.2 (continued)

SNPs	Cofactors	Organic acids	Amino acids/other tests/interventions	rs Numbers
Neurotransmitters				
NOS1 *Nitric oxide synthase 1* Catalyzes the conversion of arginine to nitric oxide Also creates the byproduct citrulline	NADPH and NADP+	↑Orotate, ↑citrate, ↑isocitrate, ↑cis-aconitate	Potential abnormalities in arginine, citrulline, nitric oxide	rs7298903, rs3782206, rs2293054
NOS2 *Nitric oxide synthase 2* Inducible nitric oxide synthase (iNOS); considered a reactive free radical involved in neurotransmission along with antimicrobial and antitumoral activities	Manganese	↑Orotate, ↑citrate, ↑isocitrate, ↑cis-aconitate	Potential abnormalities in arginine, citrulline, nitric oxide; responds to antioxdiation	rs2297518, rs2248814, rs2274894
Found in the immune and cardiovascular systems; involved in immune defense against pathogens; maybe induced by lipopolysaccharides and cytokines; commonly found in inflammatory disorders				
NOS3 *Nitric oxide synthase 3* Endothelial nitric oxide synthase (eNOS); involved in vasodilation, vascular smooth muscle relaxation via cGMP-mediated signal transduction pathway; mediates vascular endothelial growth factor (VEGF)-induced angiogenesis in coronary vessels; promotes blood clotting via platelet activation	Manganese	↑Orotate, ↑citrate, ↑isocitrate, ↑cis-aconitate	Potential abnormalities in arginine, citrulline, nitric oxide	rs1800783, rs3918188, rs7830, rs1800779, rs2070744
PAH *Phenylalanine hydroxylase* Converts phenylalanine to tyrosine; associated with PKU	BH4, Fe^{+3}	↑Phenylacetic Acid, (potentially) ↓HVA, ↓VMA	↑Phenylalanine, ↓tyrosine	rs62507347, rs10860936, rs3817446, rs1722387, rs5030858, rs1522305, rs2037639, rs1718301, rs1522296, rs2245360, rs1801153, rs772897, rs1722392, rs1042503, rs5030849, rs1522307, rs11111419, rs10778209, rs1718312
TH *Tyrosine hydroxylase*	BH4, Fe^{+3}	Potentially ↓HVA, ↓VMA	↑Tyrosine, ↓DOPA	rs28934581, rs28934580,
Converts tyrosine to L-DOPA				rs2070762, rs7483056, rs6356
TPH *Tryptophan hydroxylase* Converts tryptophan to 5-HO-RP, a serotonin precursor	BH4, P5P, B2, copper	↑5-HIAA, ↑Glutaric acid	↓Tryptophan	rs623580
DBH *Dopamine beta hydroxylase* Converts dopamine to norepinephrine; may be associated with schizophrenia or depression	Copper, vitamin C	↓VMA, ↑HVA	↑↓Dopamine	rs1611115 rs1108580

(continued)

Table 18.2 (continued)

SNPs	Cofactors	Organic acids	Amino acids/other tests/interventions	rs Numbers
MOA-A *Monoamine oxidase A* Catalyzes the deamination of the amines dopamine, serotonin, epinephrine, norepinephrine Metabolizes the xenobiotic amines: Heterocyclic amines (meat), tyramines and histamine MAO-A preferentially oxidizes serotonin and noradrenaline	FAD	Upregulation: ↓5-HIAA, ↑glutaric acid Downregulation: ↑5-HIAA, ↑glutaric Acid	Potential ↓ in tryptophan, phenylalanine, and tyrosine	Upregulation: rs6323 Downregulation: rs5906883, rs5906957, rs909525, rs9593210, rs2072743
MOA-B *Monoamine oxidase B* Catalyzes the oxidative deamination of biogenic and xenobiotic amines (i.e., histamine and tyramine); part of the metabolism of neuroactive and vasoactive amines in the CNS and peripheral tissues; preferentially oxidizes dopamine; associated with depression	FAD	Upregulation: ↑HVA, ↑VMA, ↑glutaric acid	↓Tyrosine, ↓dopamine	rs1799836, rs10521432, rs6651806
COMT *Catechol-O-methyltransferase* Transfers methyl group from SAM to the catecholamines	Magnesium, SAM	Downregulation: ↓VMA, ↓HVA Upregulation: ↑HVA, ↑VMA	N/A Consider RBC magnesium	rs4680, rs769224, rs4633
Breaks down dopamine, epinephrine, and norepinephrine; transfers a methyl group from SAM to catechols from foods (green tea, potatoes), antioxidants (quercetin), and hormone catechols (estrogen)				
Glutathione				
GSS *Glutathione synthase* Found within the gamma-glutamyl cycle; aids in the production of glutathione	Magnesium	Potential ↑ in pyroglutamate, alpha-hydroxybutyrate and alpha-ketobutyrate	Potential abnormalities in glutamate, cysteine, glycine	rs22273684, rs6088659, rs2236270, rs28936396, rs6060124
GSTT1Del *Glutathione S-transferase theta 1* Conjugates reduced glutathione; has many exogenous and endogenous hydrophobic electrophiles	1200 mg vitamin C/day	Potential ↑ in pyroglutamate, alpha-hydroxybutyrate, and alpha-ketobutyrate	Potential abnormalities in glutamate, cysteine, glycine	rs796052137, rs796052136
GSTM1 *Glutathione-S-transferase Mu 1* Conjugates reduced glutathione; functions in the detoxification of electrophilic compounds including carcinogens, therapeutic drugs, environmental toxins, and products of oxidative stress	Vitamin C	Potential ↑ in pyroglutamate, alpha-hydroxybutyrate, and alpha-ketobutyrate	Potential abnormalities in glutamate, cysteine, glycine	rs2740574, rs4147565, rs4147567, rs4147568, rs1056806, rs12562055, rs2239892

References

1. Ordovas J, Mooser V. Nutrigenomics and nutrigenetics. Curr Opin Lipidol. 2004;15(2):10108.
2. Karki R, Pandya D, Elston RC, Ferlini C. Defining "mutation" and "polymorphism" in the era of personal genomics. BMC Med Genet. 2015;8:37. https://doi.org/10.1186/s12920-015-0115-z.
3. Albert PR. What is a functional genetic polymorphism? Defining classes of functionality. J Psychiatry Neurosci JPN. 2011;36(6):363–5.
4. Devlin TM. Textbook of biochemistry with clinical correlations. 7th ed. Hoboken, NJ: Wiley; 2011.
5. Li A, Meyre D. Jumping on the train of personalized medicine: a primer for non-geneticist clinicians: part 2. Fundamental concepts in genetic epidemiology. Curr Psychiatr Rev. 2014;10(2):101–17.
6. Edwards AWF. Reginald Crundall Punnett: first Arthur Balfour professor of genetics, Cambridge, 1912. Genetics. 2012;192:2–13.
7. Waddington CH. The epigenotype. Endeavour. 1942;1:18–20.
8. Howell A, Anderson AS, Clarke RB, Duffy SW, Evans DG, Garcia-Closas M, Gescher AJ, Key TJ, Saxton JM, Harvie MN. Risk determination and prevention of breast cancer. Breast Cancer Res BCR. 2014;16:446.
9. van Abeelen AFM, Elias SG, Roseboom TJ, Bossuyt PMM, van der Schouw YT, Grobbee DE, Uiterwaal CSPM. Postnatal acute famine and risk of overweight: the Dutch Hunger Winter study. Int J Pediatr. 2012;2012:936509.
10. Paul B, Barnes S, Demark-Wahnefried W, Morrow C, Salvador C, Skibola C, Tollefsbol TO. Influences of diet and the gut microbiome on epigenetic modulation in cancer and other diseases. Clin Epigenetics. 2015;7:112. https://doi.org/10.1186/s13148-015-0144-7.
11. Lord RS, Bralley JA. Laboratory evaluations for integrative and functional medicine. Revised. 2nd ed. Duluth: Metametrix Institute; 2012.
12. Lewis ZT, Totten SM, Smilowitz JT, Popovic M, Parker E, Lemay DG, et al. Maternal fucosyltransferase 2 status affects the gut bifidobacterial communities of breastfed infants. Microbiome. 2015;3:13.
13. Blom HJ, Smulders Y. Overview of homocysteine and folate metabolism. With special references to cardiovascular disease and neural tube defects. J Inherit Metab Dis. 2011;34(1):75–81.
14. Banerjee A, Ray S. Structural exploration and conformational transitions in MDM2 upon DHFR interaction from Homo sapiens: a computational outlook for malignancy via epigenetic disruption. Scientifica. 2016;2016:9420692.
15. Widemann BC, Balis FM, Kim A, Boron M, Jayaprakash N, Shalabi A, O'Brien M, Eby M, Cole DE, Murphy RF, Fox E, Ivy P, Adamson PC. Glucarpidase, leucovorin, and thymidine for high-dose methotrexate-induced renal dysfunction: clinical and pharmacologic factors affecting outcome. J Clin Oncol. 2010;28(25):3979–86.
16. Christensen KE, Mikael LG, Leung KY, Lévesque N, Deng L, Wu Q, Malysheva OV, Best A, Caudill MA, Rozen R. High folic acid consumption leads to pseudo-MTHFR deficiency, altered lipid metabolism, and liver injury in mice. Am J Clin Nutr. 2015;101(3):646–58.
17. King AL, Mantena SK, Andringa KK, Millender-Swain T, Dunham-Snary KJ, Oliva CR, Griguer CE, Bailey SM. The methyl donor S-adenosylmethionine prevents liver hypoxia and dysregulation of mitochondrial bioenergetic function in a rat model of alcohol-induced fatty liver disease. Redox Biol. 2016;9:188–97.
18. Stipanuk MH, Ueki I. Dealing with methionine/homocysteine sulfur: cysteine metabolism to taurine and inorganic sulfur. J Inherit Metab Dis. 2011;34(1):17–32.
19. Strickland KC, Krupenko NI, Krupenko SA. Molecular mechanisms underlying the potentially adverse effects of folate. Clin Chem Lab Med. 2013;51(3):607–16.
20. Banka S, Blom HJ, Walter J, Aziz M, Urquhart J, Clouthier CM, Rice GI, de Brouwer AP, Hilton E, Vassallo G, Will A, Smith DE, Smulders YM, Wevers RA, Steinfeld R, Heales S, Crow YJ, Pelletier JN, Jones S, Newman WG. Identification and characterization of an inborn error of metabolism caused by dihydrofolate reductase deficiency. Am J Hum Genet. 2011;88(2):216–25.
21. U.S. National Library of Medicine [Internet]. Genetics Home Reference; c2017. Available from: http://ghr.nlm.nih.gov/glossary=methylation.
22. Zeisel S. Choline, other methyl-donors and epigenetics. Nutrients. 2017;9(5):445.
23. Ranabir S, Reetu K. Stress and hormones. Indian J Endocrinol Metab. 2011;15(1):18–22. https://doi.org/10.4103/2230-8210.77573.
24. NCBI. PNMT phenylethanolamine N-methyltransferase [Homo sapiens (human)]. 2017. Retrieved from: https://www.ncbi.nlm.nih.gov/gene/5409.
25. Zhi X, Yang B, Fan S, Wang Y, Wei J, Zheng Q, Sun G. Gender-specific interactions of MTHFR C677T and MTRR A66G polymorphism with overweight/obesity on serum lipid levels in a Chinese Han populations. Lipids Health Dis. 2016;15:185.
26. Finkelstein JD, Martin JJ. Methionine metabolism in mammals. Distribution of homocysteine between competing pathways. J Biol Chem. 1984;259:9508–13.
27. Maron BA, Loscalzo J. The treatment of Hyperhomocysteinemia. Annu Rev Med. 2009;60:39–54.
28. Ganu RS, Ishida Y, Koutmos M, Kolokotronis S, Roca AL, Garrow TA, Schook LB. Evolutionary analyses and natural selection of betaine-homocysteine S-methyltransferase (BHMT) and BHMT2 genes. PLoS One. 2015;10(7):e0134084. https://doi.org/10.1371/journal.pone.0134084.
29. Zhou RF, Chen XL, Zhou ZG, Zhang YJ, Lan QY, Liao GC, Chen YM, Zhu HL. Higher dietary intakes of choline and betaine are associated with a lower risk of primary liver cancer: a case-control study. Sci Rep. 2017;7(1):679.
30. De Stefano V, Dekou V, Nicaud V, Chasse JF, London J, Stansbie D, Humphries SE, Gudnason V. Linkage disequilibrium at the cystathionine beta synthase (CBS) locus and the association between genetic variation at the CBS locus and plasma levels of homocysteine. The Ears II Group. European Atherosclerosis Research Study. Ann Hum Genet. 1998;62(Pt 6):481–90.
31. Wei G, Jun-tao K, Ze-yu C, et al. Hydrogen sulfide as an endogenous modulator in mitochondria and mitochondria dysfunction. Oxidative Med Cell Longev. 2012;2012:1. https://doi.org/10.1155/2012/878052.
32. Harding CO, Neff M, Wild K, Jones K, Elzaouk L, Thöny B, Milstien S. The fate of intravenously administered tetrahydrobiopterin and its implications for heterologous gene therapy of phenylketonuria. Mol Genet Metab. 2004;81(1):52–7.
33. NCBI. SUOX sulfite oxidase [Homo sapiens (human)]. 2017. Retrieved from: https://www.ncbi.nlm.nih.gov/gene/6821.
34. Gorren AC, Mayer B. Tetrahydrobiopterin in nitric oxide synthesis: a novel biological role for pteridines. Curr Drug Metab. 2002;3(2):133–57.
35. Daubner SC, Le T, Wang S. Tyrosine hydroxylase and regulation of dopamine synthesis. Arch Biochem Biophys. 2011;508(1):1–12. https://doi.org/10.1016/j.abb.2010.12.017.
36. Shih DF, Hsiao CD, Min MY, Lai WS, Yang CW, Lee WT, et al. Aromatic L-amino acid decarboxylase (AADC) is crucial for brain development and motor functions. PLoS One. 2013;8(8):e71741. https://doi.org/10.1371/journal.pone.0071741.
37. Rahman MK, Rahman F, Rahman T, Kato T. Dopamine-β-hydroxylase (DBH), its cofactors and other biochemical parameters in the serum of neurological patients in Bangladesh. Int J Biomed Sci. 2009;5(4):395–401.
38. Volavka J, Bilder R, Nolan K. Catecholamines and aggression: the role of COMT and MAO polymorphisms. Ann N Y Acad Sci. 2004;1036:393–8.
39. Lee LO, Prescott CA. Association of the Catechol-O-Methyltransferase (COMT) Val158Met polymorphism and anxiety-related traits: a meta-analysis. Psychiatr Genet. 2014;24(2):52–69.
40. Gropper SS, Smith JL. Advanced nutrition and human metabolism. 6th ed. Belmont: Wadsworth Cengage Learning; 2013.

41. Bear MF, Connors BW, Paradiso MA. Neuroscience exploring the brain. 3rd ed. Baltimore: Lippincott Williams & Wilkins; 2007.
42. Vink R. Magnesium in the CNS: recent advances and developments. Magnes Res. 2016;29(3):95–101.
43. Slopien R, Slopien A, Rozycka A, Warenik-Szymankiewicz A, Lianeri M, Jagodzinski PP. The c.1460C>T polymorphism of MAO-A is associated with the risk of depression in postmenopausal women. Sci World J. 2012;2012:194845.
44. Devlin TM. Textbook of biochemistry with clinical correlations. 7th ed. New York: Wiley; 2010.
45. Jin X, Yu ZF, Chen F, Lu GX, Ding XY, Xie LJ, Sun JT. Neuronal nitric oxide synthase in neural stem cells induces neuronal fate commitment via the inhibition of histone deacetylase 2. Front Cell Neurosci. 2017;11:66.
46. Elhwuegi AD. The wonders of Phosphodiesterase-5 inhibitors: a majestic history. Ann Med Health Sci Res. 2016;6(3):139–45.
47. Jung HC, Mun KH, Park TC, Lee YC, Park JM, Huh K, Seong DH, Suh JK. Role of nitric oxide in penile erection. Yonsei Med J. 1997;38(5):261–9.
48. Ramsey KH, Sigar IM, Rana SV, Gupta J, Holland SM, Byrne GI. Role for inducible nitric oxide synthase in protection from chronic chlamydia trachomatis urogenital disease in mice and its regulation by oxygen free radicals. Infect Immun. 2001;69(12):7374–9.
49. Palumbo P, Miconi G, Cinque B, Lombardi F, La Torre C, Dehcordi SR, Galzio R, Cimini A, Giordano A, Cifone MG. NOS2 expression in glioma cell lines and glioma primary cell cultures: correlation with neurosphere generation and SOX-2 expression. Oncotarget. 2017;8(15):25583–98.
50. Burrows N, Cane G, Robson M, Gaude E, Howat WJ, Szlosarek PW, Maxwell PH. Hypoxia-induced nitric oxide production and tumour perfusion is inhibited by pegylated arginine deiminase (ADI-PEG20). Sci Rep. 2016;6:622950.
51. Thameem F, Puppala S, Arar NH, Stern MP, Blangero J, Duggirala R, Abboud HE. Endothelial nitric oxide synthase (eNOS) gene polymorphisms and their association with type 2 diabetes related traits in Mexican Americans. Diab Vasc Dis Res. 2008;5(2):109–13. https://doi.org/10.3132/dvdr.2008.018.
52. Nelson DL, Cox MM. Lehninger principles of biochemistry. 4th ed. New York: W.H.Freeman; 2004.
53. Chang JY, Liu LZ. Manganese potentiates nitric oxide production by microglia. Brain Res Mol Brain Res. 1999;68(1–2):22–8.

The Immune System: Our Body's Homeland Security Against Disease

Aristo Vojdani, Elroy Vojdani, and Charlene Vojdani

19.1 Introduction – 286

19.2 The Mucosal Immune System: The First Line of Defense – 286
19.2.1 Intestinal Permeability to Large Macromolecules – 287
19.2.2 Diagnostic Features of the Mucosal Immune System – 287

19.3 An Innate Immune System – 288
19.3.1 Diagnostic Features of the Innate Immune System – 292

19.4 Acquired or Adaptive Immunity – 292
19.4.1 Primary and Secondary Immune Response – 293
19.4.2 IgE and Allergic Reactions – 294
19.4.3 Diagnostic Features of the Adaptive Immune System – 295

19.5 The Three Major Mechanisms of Protection against Autoimmunity – 295
19.5.1 Oral Tolerance – 295
19.5.2 Central Tolerance – 297
19.5.3 Peripheral Tolerance – 298

19.6 Dietary Intervention – 299

19.7 Conclusion – 300

References – 300

Learning Objectives

- The mucosal immune system as the first line of defense, our homeland security against invading pathogens and diseases.
- The importance of innate immunity in inflammation and autoinflammation.
- The role of adaptive or acquired immunity in health and disease.
- The three tolerance mechanisms that protect the body against immune disorders: The central, oral, and peripheral tolerance mechanisms provide an umbrella of protection against immune disorders.

19.1 Introduction

An immune system is a collection of organs, tissues, cells, and humoral factors that protects a living organism. The immune system shields the body against external harmful influences by developing several levels of protective weapons.

The three major protective layers of the immune system are:
1. The mucosal immune system or surface factors.
2. Innate immunity.
3. Adaptive or acquired immunity.

19.2 The Mucosal Immune System: The First Line of Defense

The mucosal immune system is our body's first line of defense. It allows the absorption of nutrients from our gastrointestinal tract while at the same time maintaining a constant alertness to a large number of harmful pathogens and antigens by using its antenna-like molecules or Toll-like receptors (TLRs) or antibodies such as immunoglobulin A (IgA) or immunoglobulin M (IgM). Using these and other molecules, the mucosal immune system is ready to launch an effective immune response against harmful environmental factors, including infections by viruses, bacteria, parasites, and fungi, as well as toxic chemicals and dietary proteins and peptides [1]. Mucosal immunity protects the epithelial barrier by eliciting pro-inflammatory responses against pathogens or mucosally induced oral tolerance toward food antigens and components of commensal microbiota. Secretory immunoglobulin A (SIgA), which is in all mucous secretions – including milk, saliva, and tears – as well as in GI secretions, the respiratory epithelium, and the genitourinary tract, plays a crucial role in the maintenance of immunological tolerance and protection against allergies and other inflammatory diseases [2]. This is done by inhibiting bacterial and fungal adhesion to the epithelial cells and neutralizing the viruses, antigens, enzymes, and toxins (see ◘ Fig. 19.1).

The role of the mucosal immune system in the body's health is so important that an entire society is dedicated to the promotion of its research, advancement, and literary and educational aspects. Part of the mission statement of the Society for Mucosal Immunology (SMI) states: "While general immunologists typically study immune responses in the spleen, lymph nodes or peripheral blood, mucosal immunologists focus on the sites at which most antigens enter – the mucosal surfaces of the gastrointestinal, respiratory, and urogenital tracts … Equally important is the study of the disease states that result when the mucosal immune system's ability to distinguish pathogens from innocuous antigens fails; examples include inflammatory bowel disease and food allergy."

IgA plays a critical role in mucosal immunity. Produced in the mucosal linings, more IgA is generated than all the other antibodies combined. About 3–5 g are secreted into the intestinal lumen daily [3, 4]. It is therefore the most abundant immunoglobulin in all mammals, secreted mostly via the mucous membranes, protecting us against the unceasing assault of pathogens and commensal microbes. The induc-

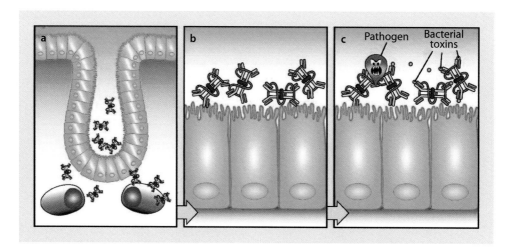

◘ **Fig. 19.1** The major antibody isotype present in the lumen of the gut is secretory polymeric IgA. **a** Secretory IgA is synthesized by plasma cells in the lamina propria and transported into the lumen of the gut through epithelial cells at the base of the crypts. **b** Polymeric IgA binds to the mucus layer overlying the gut epithelium and **c** acts as an antigen-specific barrier to the pathogens and toxins in the gut lumen

tion of IgA occurs via both T-dependent and T-independent mechanisms. After dendritic cells have sampled a tiny fraction of the intestinal antigens, IgA specifically is produced against the commensal microbes and food antigens.

In one study [5], IgA and IgG antigens specific to ovalbumin, β-lactoglobulin, or gliadin were measured in serum, saliva, colostrum, and breast milk from 40 healthy breastfeeding women, and specific IgA reactivity was found in all the samples analyzed. Levels were found to be highest in colostrum and milk, followed by saliva, and then in serum. Despite inclusion in the testing of three unrelated food antigens (gliadin, albumin, lactoglobulin), a correlation between the levels of specific IgA was found in the saliva, colostrum, and milk samples of all 40 subjects [5].

In individuals with IgA deficiency, a compensatory increase in secretory IgM has also been detected [6]. These findings together support the importance of measuring IgA and IgM antibodies against different components of gut microbiome and food antigens.

Furthermore, the increased production of IgA antibody in the blood, which is measured against food antigens and bacterial toxins, is the spillover from increased mucosal IgA production. In this process, naïve B cells generated in the bone marrow migrate to the inductive sites of mucosal immunity represented by the gut-associated lymphoid tissues (GALT or Peyer's patches; lymphoid follicles of the large bowel) or bronchus-associated lymphoid tissues, where they are stimulated by antigen-presenting cells (APCs) and cognate T-helper cells. Antigen-stimulated B and T cells migrate out through the draining lymph nodes and lymphatics and enter the bloodstream. During this process, a subset of these cells remains in the blood in the form of memory cells that recognize food antigens. The stimulated cells finally relocate or "home in" to the effector sites (including the lamina propria) of the gastrointestinal, respiratory, and genital tracts and various endocrine glands, where they secrete a significant amount of IgA antibody. This spillover from IgA production in saliva can be detected in the blood using different immunological assays.

Although production of IgA antibody in the secretion is considered protective, in the case of oral tolerance failure, the overproduction of IgA against food antigens and bacterial toxins may result in IgA immune complex formation. This may be followed by the production of pro-inflammatory cytokines, all of which can lead to increased intestinal permeability and increased antigen exposure. This will then lead to increased IgA and IgM production against dietary proteins and bacterial antigens in the blood [7, 8].

19.2.1 Intestinal Permeability to Large Macromolecules

A number of conditions and environmental triggers can result in damage to the intestinal barrier and increased permeability to large macromolecules, leading to the loss of mucosal immune tolerance and the passage of dietary proteins into the submucosa and regional lymph nodes and into the circulation. The ensuing immune response against these invading antigens results in the formation of polymeric IgA and immune complexes, which bind to the mucosal layer to protect against greater damage. However, these immune complexes, along with inflammatory cytokines, can lead to further increased gut permeability and to the entry of dietary proteins and peptides into the bloodstream and formation of antibodies against these antigens. Circulating memory B lymphocytes that have previously been exposed to these antigens upon repeated exposure to the same antigen(s) will cause the formation of plasma cells with J chains to produce antibodies in the blood in a much shorter period of time than the primary immune response. A J chain is a protein component that acts like a "glue" in IgA and IgM antibodies. When these antigens and peptides are presented to APCs, T cells, and B cells for the first time, the plasma cell formation can result in the production of both IgG and IgA against dietary proteins and peptides in the blood within 15 days. If these antigens share homology with different tissue antigens, they can initiate an autoimmune disease in different organs, such as the kidneys, liver, brain, skeletal muscles, and others [9, 10].

19.2.2 Diagnostic Features of the Mucosal Immune System

The importance of measuring the total and specific SIgA is well established in different disorders.

For example, testing for SIgA helps in determining:
- If the mucosal immune system is in a healthy and balanced state. Normal levels of SIgA may indicate a healthy mucosal immune system; very high or low SIgA may indicate inflammation, autoimmunity, and immunodeficiency.
- Elevated total SIgA with elevation of IgA or IgM antibodies against oral pathogens may indicate chronic infection in the oral cavity.
- Elevated total SIgA with elevation of IgA or IgM antibodies against bacterial cytotoxins and tight junction proteins may indicate intestinal barrier dysfunction to large molecules, irritable bowel syndrome, and small intestinal bacterial overgrowth (SIBO).
- Elevated total SIgA with elevation in IgA or IgM antibodies against a variety of food antigens may indicate breakdown in oral tolerance, enhanced intestinal permeability, food immune reaction, and possibly autoimmunities.
- Elevated total SIgA with elevation in IgA or IgM antibodies against gliadin and transglutaminase-2, actomyosin, *Saccharomyces cerevisiae,* and/or calprotectin may indicate the possibility of non-celiac gluten sensitivity (NCGS), celiac disease (CD), and its overlap with Crohn's disease and ulcerative colitis.
- Elevated total SIgA with elevation in IgA and IgM antibodies against different parasitic antigens may indicate parasitic infection [11–20].

This is why the textbook of Food Allergy and Intolerance [21] states that: "Mucosal production of IgA and IgM antibodies to dietary antigens, and particularly to gliadin, appears to be an early event in the gluten-sensitive enteropathy but does not initially cause a flat lesion."

19.3 An Innate Immune System

The mucosal immune system of higher mammals is covered by a single layer of epithelium, which secretes antimicrobial peptide and is topped by a thick layer of mucus. This system is fortified by both innate and adaptive immune responses to protect the host against the harsh environment of bacteria, toxins, antigens, and a variety of food components that varies from person to person. This interrelationship between the mucosal immune system and gut bacteria is bidirectional. For example, bacteria flourish on food components, such as carbohydrates, proteins, and vitamins, and in return reward the microbiota by developing oral tolerance and natural immunity through regulatory T cells (CD4+ cD25+) and regulatory cytokines (TGF-β and IL-10). Moreover, the host microbiota influences the development and maturation of lymphoid cells involved in mucosal immunity [22–24].

The innate immune system acts as the "first responders" against pathogens such as bacteria, viruses, fungi, and protozoans. This system is not unique to mammals; even invertebrates possess the ability to protect themselves against invaders. From an evolutionary point of view, this ancient mechanism was developed to protect multicellular organisms. However, starting with the fish, and going through amphibians and birds and all the way to mammals, the innate immune system provides this frontline barrier or defense. It very wisely stops the enemy to provide the required time for action by the adaptive immune system and the development of specific antibodies against the pathogens, which requires about 2 weeks [25]. Think of it as the frontline soldiers fighting a heroic holding action against the assaulting forces to give their reinforcements enough time to prepare the necessary weapons and bring them to the battle.

The innate immune system in the mucous membranes consists of several physiologic, anatomic, and cellular components. The anatomical aspects include mucosal surfaces of the skin, lungs, mammary, and gastrointestinal and urogenital tracts. Secretion of hydrochloric acid and digestive enzymes is a major physiologic function that – in collaboration with gut barrier structures – prevents the entry of food antigens, bacterial toxins, and whole pathogens into the body. Complement proteins and some cytokines are also considered components of innate immunity. Innate immunity as a whole is a nonselective, nonspecific immune system and does not modify itself based on the type of invaders. For example, most food antigens are digested by digestive enzymes and, other than *H. pylori*, most pathogens cannot withstand the low pH of the stomach.

However, the cellular component of innate immunity is more selective and constantly modifies itself to efficiently combat the enemy. The cellular component consists of mono-

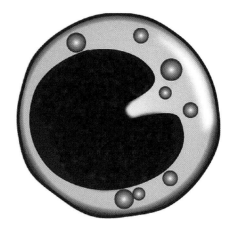

Fig. 19.2 Monocytes

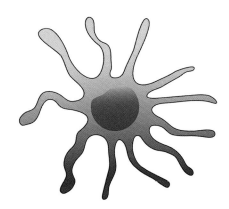

Fig. 19.3 Dendritic cell

cytes, dendritic cells, macrophages, innate lymphoid cells, natural killer cells, mast cells, basophils, neutrophils, eosinophils, epithelial cells, and paneth cells [26].

Monocytes are produced by the bone marrow (Fig. 19.2). The largest type of leukocyte, they are recognizable by their distinct kidney-shaped nucleus. They circulate in the bloodstream for 1–3 days before moving into the tissues throughout the body and differentiating into macrophages and myeloid lineage dendritic cells. Half of the body's monocytes remain stored in the spleen as a reserve. Monocytes and their macrophage and dendritic cell descendants have three main functions in the immune system: phagocytosis, antigen presentation, and cytokine production.

Dendritic cells (DCs) are antigen-presenting cells (APCs) that act as messengers between the innate and adaptive immune systems (Fig. 19.3). Their main function is to process antigenic material and present it on their cell surfaces to T cells. They can be found in the tissues that come in contact with the external environment, such as the skin, the inside of the nose, the lungs, the stomach, the intestines, and, in an immature state, in the blood. When activated, they migrate to the lymph nodes where they interact with B cells and T cells to craft the adaptive immune response. They grow the projections or dendrites that give them their name.

Macrophages are the junkyard dogs of the human body (Fig. 19.4). They are found in virtually all tissues, where

The Immune System: Our Body's Homeland Security Against Disease

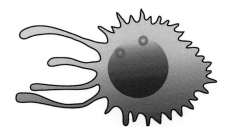

■ **Fig. 19.4** Macrophage

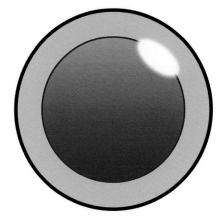

■ **Fig. 19.5** Innate lymphoid cell

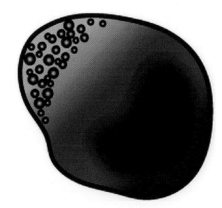

■ **Fig. 19.6** NK cell

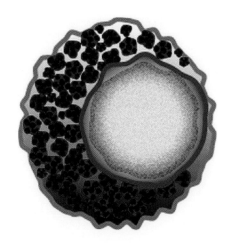

■ **Fig. 19.7** Mast cell

they patrol for potential pathogens such as cellular debris, foreign substances, microbes, cancer cells, and anything else that they don't identify as a healthy body cell. When they find an object that they identify as alien or harmful, they become true to their name (which means "big eater"), engulfing and digesting the offending antigen and presenting the digested bits to T cells. They have various forms and various names to go with them. M1 macrophages encourage inflammation, whereas M2 macrophages decrease inflammation and encourage tissue repair. Macrophages can also decrease immune reactions through the release of cytokines.

Innate lymphoid cells (ILCs) are a group of innate immune cells derived from a common lymphoid progenitor (CLP) (■ Fig. 19.5). They are identified by the absence of an antigen-specific B- or T-cell receptor because they lack the recombination-activating gene (RAG). They have various physiological functions and have an important role in protective immunity and the regulation of homeostasis and inflammation. Their dysregulation can lead to immune disorders and even autoimmune disease.

Natural killer cells (NK cells) are a type of cytotoxic lymphocyte (■ Fig. 19.6). Their main function is to kill tumor cells and virally infected cells. They are to the innate immune system what cytotoxic T cells are to the adaptive immune system. Normally, protective immune cells do their job by detecting the major histocompatibility complex (MHC) presented on infected cell surfaces, which cause them to trigger cytokine release and cause lysis or apoptosis. NK cells are unique in that they have the ability to recognize stressed cells even in the absence of antibodies or MHC. This gives them a much faster reaction time and makes them the only protective immune cell that can detect and destroy harmful cells that are missing MHC markers. They use perforin and granzyme B to kill their targets.

Mast cells (mastocytes, labrocytes) are granulocytes derived from the myeloid stem cell (■ Fig. 19.7). They contain many granules rich in histamine and heparin. Best known for their role in allergies, mast cells can bind to the fragment crystallizable (Fc) region of IgE, causing them to release histamine and other inflammatory mediators. However, they also play an important protective role and are critically involved in wound healing, angiogenesis, immune tolerance, anti-pathogen defense, and the blood-brain barrier function.

Basophils are the least common but largest of the granulocytes (■ Fig. 19.8). They are responsible for inflammatory reactions during immune response and for the formation of acute and chronic allergic diseases such as anaphylaxis, asthma, atopic dermatitis, and hay fever. On the protective side, they can perform phagocytosis and produce histamine and serotonin to reduce inflammation and heparin to prevent blood clotting. They were once thought to be mast cell precursors, but no longer. They have a distinct two-lobed nucleus surrounded by cytoplasmic granules.

Neutrophils or neutrocytes are the most abundant of the granulocytes and are an essential part of the innate immune

◘ **Fig. 19.8** Basophil

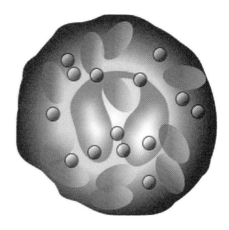

◘ **Fig. 19.10** Eosinophil

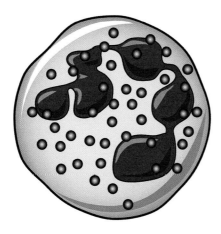

◘ **Fig. 19.9** Neutrophil

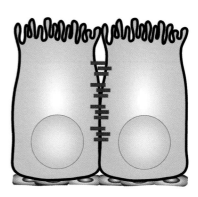

◘ **Fig. 19.11** Epithelial cells

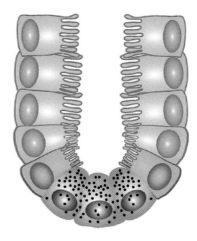

◘ **Fig. 19.12** Paneth cells

system (◘ Fig. 19.9). Formed from stem cells in the bone marrow, they are short-lived and highly motile and can enter parts of tissue that other cells or molecules can't. During the starting phase of inflammation, neutrophils are one of the first responders to migrate toward the trouble site, reaching it within minutes following the causative trauma. They are the predominant cells in pus and account for its whitish/yellowish appearance. They are part of the polymorphonuclear cell family along with the mast cells, basophils, and eosinophils.

Eosinophils are one of the immune system's components responsible for combating multicellular parasites and certain infections in vertebrates (◘ Fig. 19.10). They are granulocytes that develop during hematopoiesis in the bone marrow before migrating into blood. Like mast cells and basophils, they control mechanisms associated with asthma and allergy. Their granules contain many chemical mediators that are toxic to both parasite and host tissues. They are unique granulocytes in that they can survive for extended periods of time after maturation, persisting in the circulation for 8–12 hours and in tissue an additional 8–12 days without stimulation.

Epithelial cells (ECs) are one of the most important cells found in the human body. They are the first type of cells to encounter external stimuli (◘ Fig. 19.11). They line the cavities and surfaces of blood vessels and organs throughout the body. Epithelial cells come in different shapes, depending on where in the body they're found; the three basic shapes are called squamous, cuboidal, and columnar. Structurally, ECs rest on a basement membrane and are interconnected by cell junctions. The functions of ECs include secretion, selective absorption, protection, transcellular transport, sensing, and maintaining the integrity of the gut barriers.

Paneth cells are found at the bottom of the intestinal crypts in the small intestine (◘ Fig. 19.12). They are the key effectors of innate mucosal defense, producing large amounts

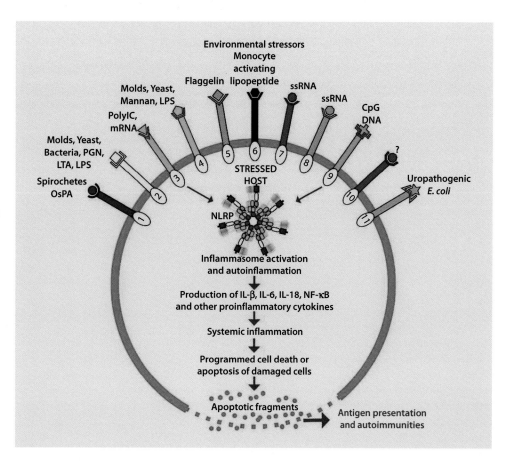

☐ **Fig. 19.13** How mammalian cells via the recognition of different patterns detect the presence of invading triggers. Environmental triggers bind to TLRs (1–11) present on the surface of macrophages and neutrophils. This initiates activation of inflammasomes and initiation of the autoinflammatory process. Inflammasomes activate the inflammatory cytokines, causing systemic inflammation and induction of programmed cell death of tissues that were affected by environmental stressors. PGN peptidoglycan, LTA lipoteichoic acid, LPS lipopolysaccharide, OsPA outer surface protein A

of α-defensins and other antimicrobial peptides, such as lysozymes and secretory phospholipase A2 (sPLA2). These substances are stored in secretory granules and released into the intestinal lumen upon stimulation with bacterial antigens such as lipopolysaccharide (LPS) and muramyl dipeptide (MDP).

This cellular component of innate immunity identifies foreign materials by the previously mentioned antenna-like structures known as TLRs. Because pathogens contain molecules such as lipoteichoic acid, lipopolysaccharides (LPS), and others which are not found on mammalian cells, the innate systems are able to recognize and detect the invading pathogens. These molecules are called pathogen-associated molecular patterns (PAMPs). The binding of PAMPs on the pathogens to TLRs on macrophages and neutrophils initiates the killing mechanism. An additional component of the innate system is an intracellular component known as the *inflammasome*. Inflammasomes are NOD-like receptors (NLRPs) which detect intracellular PAMPs after phagocytosis. This causes the release of proinflammatory cytokines from the activated cells and activation of the inflammasome.

The cytokines plus activated inflammasomes induce the apoptosis and death of infected cells. This way, cells that are heavily infected with pathogens are cleared from the system [27] (see ☐ Fig. 19.13). Overall, the inflammatory reaction is divided into exogenous and endogenous inflammation, both of which are associated with diseases:

The exogenous factors that induce inflammation-associated diseases are:
- Dietary components: gluten, casein, lectins, agglutinins, other food antigens, glucose, cholesterol.
- Toxic chemical exposure: drugs, smoking, pollution, plasticizer, adjuvants, pesticides, herbicides.
- Infections and their antigens: bacteria, viruses, yeasts, molds, parasites, spirochetes.

The endogenous inflammation includes autoinflammation and autoimmune inflammation. Inflammation underlies a wide variety of physiological and pathological conditions.
- Autoinflammatory diseases are diseases of innate immunity.
- Autoimmune diseases are diseases of both innate and adaptive immunity.

Inflammation is the immune system's natural response to infections, tissue damage, or metabolic disorders. The purpose of the inflammatory activity is to kill pathogenic microbes, repair injured tissue, and remove harmful metabolite deposits. Normally, tissue homeostasis is restored once the inflammation has done its work and run its course. However, the incomplete resolution of inflammatory responses along with chronic effects of immune stressors is known to be an important trigger of tissue pathology in numerous diseases, such as atherosclerosis, rheumatoid arthritis, psoriasis, and diabetes. When this happens in the brain as a result of astrocyte activation and astrocyte reaction to infectious agents (amyloid plaque formation), it is called neuroinflammation. Neuroinflammation could result in a decline of neurologic function. It accompanies a variety of neurodegenerative diseases, such as dementia, Alzheimer's, or Parkinson's; it has become increasingly evident that in many, or perhaps even all of these, neuroinflammation is not only a consequence but could also be a pathologic trigger. In many neurodegenerative pathologies, various triggers of inflammation are found and can actually be used as biomarkers for the particular disease. Therefore, rather than a late consequence, immune activation could be an early cause in neurodegenerative diseases; this suggests that anti-inflammatory therapies could be a promising treatment approach [28].

Both inflammatory and autoimmune diseases are very complex, and for their investigation, high-throughput methods are needed. For example, C-reactive protein (CRP) or sedimentation rate is not sensitive enough, but mediators such as IL-1β, TNF-α, IL-6, NF-κB, and other cytokines are very sensitive biomarkers. Furthermore, the study of the function of effectors such as T cells, B cells, NK cells, macrophages, neutrophils, epithelial cells, and endothelial cells helps in the investigation of the pathways involved in the physiological role of inflammatory pathophysiology [29]. The exact endogenous stimuli that induce autoinflammatory disorders is not known yet, but six different molecular mechanisms in the development of autoinflammatory diseases have been described [30].

1. *IL-1β activation disorders or inflammasomopathies.* The inflammasomes or NLRPs are the guardians of the body and are part of the inflammation engines. These engine-like materials are involved in the proteolytic activation of pro-inflammatory cytokines such as IL-1β and IL-18. One such disorder in which inflammasome activation is involved is gout, which could be initiated by dietary components (e.g., chicken), bacteria (such as *S. aureus*), viral DNA, toxic chemicals, skin irritants, UV, adjuvants, or food additives [27–30].
2. *NF-κB activation syndromes.* Loss-of-function mutations can cause dysfunctions in NF-κB's response to intracellular microbial products, leading to pro-inflammatory diseases.
3. *Protein misfolding disorder.* These mishaps in the cells of the innate immune system can have biological consequences. Generally speaking, misfolding disorders can trigger the wrong responses and cause inappropriate cytokine secretions. For instance, in TNF receptor-associated periodic syndrome (TRAPS), missense substitutions in the p55 TNF receptor lead to misfolding and ligand-independent activation of kinases and aberrant cytokine production.
4. *Complement regulatory diseases.* Complements are key components of innate immunity, and deficiencies in complement regulatory factors can lead to a classical autoimmune lupus-like picture, producing an autoinflammatory phenotype such as age-related macular degeneration and atypical hemolytic uremic syndrome.
5. *Disturbance in cytokine signaling.* Given the importance of cytokines in immunity, it is not surprising that erroneous cytokine signaling can lead to autoimmune disorders. One such example is cherubism, a relatively newly recognized autoinflammatory disorder of the bone. It is caused by mutations in an SH3-binding protein, which in animal models leads to heightened responsiveness to the cytokines M-CSF (macrophage colony-stimulating factor) and RANKL (receptor activator of NF-kappaB ligand), and increased osteoclastogenesis.
6. *Macrophage activation syndrome.* This syndrome is a common factor among many autoinflammatory disorders, such as Chediak-Higashi syndrome, familial hemophagocytic lymphohistiocytosis, and atherosclerosis. Among its causes are loss-of-function lesions in the adaptive immune system, the activation of effector cells in the innate immune system, and the elaboration of a pro-inflammatory cytokine milieu.

19.3.1 Diagnostic Features of the Innate Immune System

- Macrophage/monocyte function
- Neutrophil function (immune complexes), CD116/CD18
- Acute phase proteins and inflammation: CRP, serum amyloid A, fibrinogen, other clotting factors, vasoactive amines such as histamine and serotonin
- Complement hemolytic activity and complement components
- Pro-inflammatory cytokines and proteins: IL-1β, TNF-α, IL-6, NF-κB, MCP-1, IL-17A–IL-17F, INF-γ
- Natural killer cytotoxic activity

19.4 Acquired or Adaptive Immunity

The adaptive immune system is known as the second arm of immunity. It is characterized by cellular immune reaction against different antigens and production of specific antibodies against them. During this process, the pathogens, because of their large size, are phagocytosed and digested into very small pieces by macrophages and dendritic cells. These digested pieces of pathogens, which are named antigens, are then presented on the surfaces of antigen-presenting cells (APCs) to T-helper cells (Th cells) (see ◘ Fig. 19.14). The Th cells secrete different cytokines to communicate with B cells, stimulating them to go through blastogenesis, B-cell expansion, plasma cell formation, and the production of IgM followed by IgG antibodies.

Simultaneously, Th cells in response to interleukin-2 will develop into cytotoxic T lymphocytes, which express antibodies on their surface, guiding them to go after cells infected with pathogens, such as a virus. This way, the body gets rid of both pathogens and cells infected with pathogens [26].

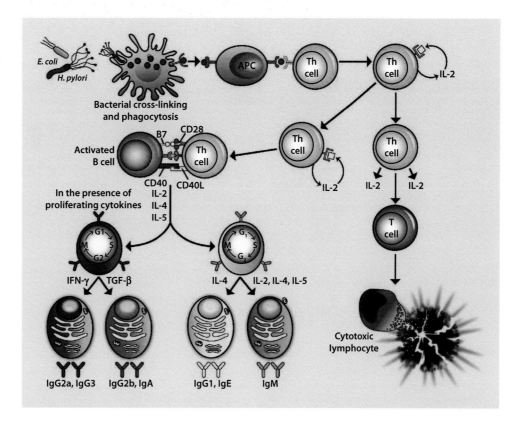

Fig. 19.14 The adaptive immune system. Invading pathogens are phagocytosed into smaller manageable pieces and presented by APCs to Th cells, causing them to secrete cytokines and undergo differentiation and other processes involved in immune response, including the production of different phenotypes of immunoglobulin antibodies

19.4.1 Primary and Secondary Immune Response

When an antigen comes into contact with the immune system for the first time, our body's response is to activate the *primary immune response*. At this time, the immune system still has to learn all it can about the invading pathogen so it can make the proper antibody against it. The lag time to form the response may take a while, from 4 to 7 days to sometimes even weeks or months, so that immunity takes longer to establish. The "first responders" to this antigenic attack are the *effector cells*, relatively short-lived cells designed for an initial response. Effector B cells or plasma cells secrete antibodies; effector T cells include cytotoxic T cells and helper T cells, which carry out cell-mediated responses. The first antibodies produced are mainly IgM, although small amounts of IgG also occur (see Fig. 19.15). The amount of antibody produced depends on the antigen but is typically low. The level of antibody peaks in 14 days and declines rapidly. The affinity of the antibody produced at this time for its antigen is lower. This primary antibody response appears mainly in the lymph nodes and spleen.

The *secondary immune response* is launched upon the immune system's second and subsequent exposure to the same previously encountered antigen. When the immune system first encounters the invading antigen, some of the responding naïve B cells and T cells become *memory cells*. Memory T cells and B cells are immune cells that are longer-lived than the other initial responder cells. They remain in the body after initial infection and retain a memory of the

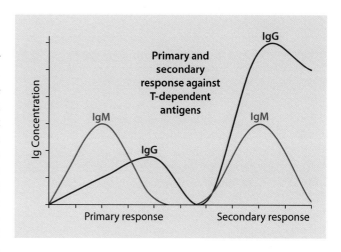

Fig. 19.15 IgM and IgG response. Th cells produce an immune response first in the form of IgM antibodies and then IgG antibodies

strategy used to defeat the pathogen the first time it was encountered. Once the immune system recognizes its old enemy, it fires up the memory cells for a fast and powerful secondary immune response. Thus, the response time is much shorter, from 1 to 4 days. This means that immunity takes a shorter time to establish. Primarily IgG antibody is produced, with some small amounts of IgM occasionally occurring as well (see Fig. 19.15). IgA and IgE are also produced. Usually 100–1000 times more antibodies are produced in the secondary response than in the first. The level of antibody peaks in 3–5 days and remains high for a longer period. The affinity of the antibody for its antigen is much

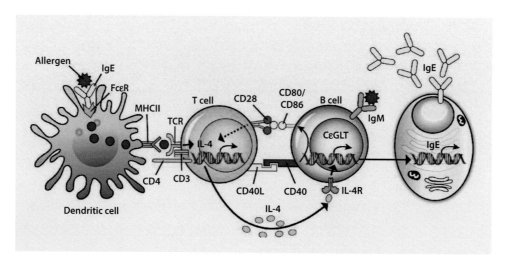

Fig. 19.16 Production of allergen-specific IgE by receptor upregulation in collaboration with dendritic cell, T cell, B cell, and plasma cell. Allergens are taken up by DCs and presented to T cells, stimulating the upregulation of CD40 ligand expression. This in turn stimulates the expression of CD80/CD86, which binds to CD28 and leads to the production of more IL-4. The IL-4 binds to the IL-4 receptor on B cells and activates the transcription factor, causing class-switch to IgE and the production of significant amounts of IgE, releasing the mediators involved in classical allergy

greater and therefore more effective. The secondary immune response appears mainly in the bone marrow, followed by the spleen and lymph nodes.

Antibodies or immunoglobulins (Ig) take a variety of forms or isotypes. The most common Ig isotypes are IgG, IgA, IgM, and IgE, or the easily remembered acronym GAME. IgM, known as acute antibodies, are the first to arrive. The IgM antibodies are relatively low-affinity, produced rapidly in order to control infection. IgG is the more specific isotype, consisting of IgG_1, IgG_2, IgG_3, and IgG_4 subclasses. For the production of IgG, additional time is required, and it appears in the blood when IgM levels are in the process of decline. IgG antibody levels peak at about 30 days and stay in the body for several months in order to protect it from future possible infections (◘ Fig. 19.15). The production of IgG, IgA, IgM, and IgE depends on the cytokine environment. For example, after collaboration between Th cells and B cells in the presence of proliferating cytokines, such as IL-2, IL-4, and IL-5, and of differentiation cytokines IFN-α and TGF-β, the plasma cells produce IgG_2, IgG_3, and IgA. With the help of IL-2, IL-4, and IL-5, the activated cells may produce IgM, and in the presence of Th2 cytokine IL-4, the plasma cells may produce IgG and IgE. The production of the IgE isotype of antibody, which is involved in allergy, depends on the cellular interactions that are important for IgE class switch recombination. During this process, the uptake of allergens such as peanut or house dust mites by dendritic cells allows antigenic presentation to the T cells. Stimulation of specific helper cells by these antigens or allergens leads to the production of Th2 cytokines (IL-4) and the upregulation of CD40 ligand expression by the T cells.

Binding of CD40L to CD40 on the B-cell membrane upregulates the expression of costimulatory molecules and CD80/CD86, which binds to CD28 receptor to stimulate further production of IL-4. This binding of CD 40 L to CD40 and CD28 to CD80/CD86, and the production of more IL-4, results in IL-4 binding to IL-4 receptor on B cells, signaling the activation of the transcription factor in B cells. Enhancement of transcription factor and activation-induced cytidine deaminase causes IgE class-switch recombination and the production of significant amounts of IgE (see ◘ Fig. 19.16). The binding of IgE to IgE receptor on mast cells and its bridging by the allergens result in the release of mediators that are involved in the classical allergy symptomologies [26].

19.4.2 IgE and Allergic Reactions

Distinctions should be made between sensitivities and allergies. True allergic reactions are triggered when allergens cross-link preformed IgE bound to the high-affinity receptor FcεRI on mast cells. Mast cells line the body surfaces and serve to alert the immune system to local infection. Once activated, they induce inflammatory reactions by secreting chemical mediators stored in preformed granules and by synthesizing leukotrienes and cytokines after activation occurs. In allergy, they provoke very unpleasant reactions to innocuous antigens that are not associated with invading pathogens that need to be expelled. The consequences of IgE-mediated mast-cell activation depend on the dose of antigen and its route of entry, with symptoms ranging from the irritating sniffles of hay fever when pollen is inhaled to the life-threatening circulatory collapse that occurs in systemic anaphylaxis. A more sustained inflammation known as the late-phase response follows the immediate allergic reaction caused by mast-cell degranulation. This late response involves the recruitment of other effector cells, notably Th2 lymphocytes, eosinophils, and basophils, which contribute significantly to the immunopathology of an allergic response.

Most antibodies are found in body fluids and engage effector cells, through receptors specific for the Fc constant regions, only after binding to a specific antigen through the antibody variable regions. IgE, however, is an exception, as it is captured by the high-affinity Fcε receptor in the absence of bound antigen. This means that IgE is mostly found fixed in the tissues on mast cells that bear the Fcε receptor, as well as on activated eosinophils and circulating basophils. The ligation of cell-bound IgE antibody by specific antigen triggers the activation of these cells at the site of antigen entry into the tissues. The release of inflammatory lipid mediators, cytokines, and chemokines at the sites of IgE-triggered reactions results in the recruitment of eosinophils and basophils to augment the type I response.

When the humoral and cellular components of adaptive immune response fail, the results could be allergies and autoimmunities.

19.4.3 Diagnostic Features of the Adaptive Immune System

- Immunoglobulins and immunoglobulin subclasses
- Immune response to vaccination or antigenic challenge
- Antibody titers, ANA, RF, ssDNA, and other tissue-specific antibodies against self-antigens (multiple autoimmune reactivity screen)
- Lymphocyte subpopulation including regulatory T cell
- T- and B-cell function (antigen/mitogen stimulation)
- Monitoring cell death (apoptosis)
- Cytokine production after antigen challenges

19.5 The Three Major Mechanisms of Protection against Autoimmunity

Oral tolerance, central tolerance, and peripheral tolerance are the three major mechanisms that protect the body against autoimmunities. Breakdown in any of these three mechanisms, especially failure in oral tolerance, can result in food immune reaction and associated autoimmunities [31].

19.5.1 Oral Tolerance

The gut mucosal immune system has to keep an intricate immune homeostasis by maintaining tolerance to harmless or even beneficial molecules in the gut while mounting an effective immune defense against pathogens [32]. The unresponsiveness to food antigens with subsequent downregulation of systemic immune response is what is characterized as oral tolerance. The failure of this system results in immune reactivities to the food we eat, sometimes with life-threatening consequences such as allergies and autoimmunities [32]. The revolutionary developments in the fields of mucosal immunology and microbiology of the gut in recent years are the best indication of the importance of commensal flora, gut barriers, and oral tolerance to human health and disease. The specific mechanisms of action that separate tolerance from effective immunity against various food and bacterial antigens have yet to be fully explored and are the subject of ongoing research.

When these different mechanisms of action fail to control ingested antigens, the first result can be a breakdown in tolerance to soluble antigens, which then triggers active secretory and systemic immune responses against food antigens. Indeed, individuals in whom the immune exclusion mechanism does not function may experience chronic hyperabsorption of macromolecules and the tendency to develop autoantibodies and even autoimmune disease [33, 34].

19.5.1.1 Exclusion of Various Antigens by Secretory IgA and IgM Antibodies to Modulate or Inhibit Colonization of Bacteria and Yeast and Dampen Penetration by Dangerous Soluble Luminal Agents

A child at birth has almost no protective immune system other than passive immunity and maternal transfer of IgG against various food antigens and infectious agents. Although born practically germ-free and with no microbiota in the gastrointestinal (GI) tract, the child's mucosae are immediately assaulted after birth by a motley horde of microorganisms originating, first, from the mother; secondly, from the air in the delivery room and the doctor and nurses present; thirdly, from breast milk or baby formula; and, lastly, from exposure to various food antigens upon the introduction of solid food. This is why the mucosal immune system has evolved into the two arms of defense, innate immunity and adaptive immunity, to handle these challenges [35].

Maternally acquired immunity is essential for the survival of newborns until endogenous immunity develops. These exogenous antibodies are acquired both prenatally through transplacental transfer and postnatally via breast-feeding and colostrum [36]. In fact, when breast-fed infants were compared with formula-fed babies, a more rapid increase in SIgA1, SIgA2, and total salivary IgA was observed during the first 6 months [37]. Furthermore, breast-fed infants also produce higher levels of SIgA in urine than formula-fed infants. Therefore, the importance of infant feeding practices cannot be underestimated, since there is significant association between feeding patterns, bacterial colonization, and immunological maturation – in particular with respect to IgA- and IgM-containing plasma cells in the gut lamina propria. This is why intravenous-fed fully developed infants lack these IgA- and IgM-producing plasma cells in their gut tissue, while the orally fed individuals have adult proportions of these immunocytes [38]. Furthermore, the initial bacterial colonization and subsequent antigenic challenge in the GI tract differ between breast-fed and formula-fed infants. In addition to providing secretory IgA and IgM, breast milk reinforces mucosal defenses by delivering antigens, immune complexes, regulatory cytokines, growth factors, and peribiotics such as oligosaccharides that promote the proliferation of friendly bacteria, which are part of the neonatal intestinal microbiota. This could be an explanation for the protective role breastfeeding plays when inflammatory bowel disease develops later in life. This emphasizes the impact of perinatal immune development, in particular IgA and IgM antibodies, on mucosal homeostasis and chronic inflammation [39–41].

Oral tolerance to dietary proteins is crucial to prevent the development of food immune reactivity. The mode of antigen uptake in the gut and different regulatory immune cells plays a role in its maintenance. In addition to the intestinal epithelial cells acting as nonprofessional APCS, DCs, CD8+ cells, and a variety of regulatory CD4+ cells – namely, TR1, Th3, or CD4+ CD25+ cells – play an important role in maintaining oral tolerance to low doses of antigen through suppression of immune responses. Other mechanisms are important in response to high antigen doses, including induction of lymphocyte anergy or deletion.

This induction of oral tolerance to soluble antigens is not limited only to the intestinal mucosa but can involve the entire body. The explanation for this is that through oral exposure, an antigen can gain access to the blood via the lymph. Indeed, food protein can be detected in the blood of mice and humans soon after eating [42]. This entry of undegraded food proteins into the circulation at low levels is a normal process, but in the presence of inactive enzymes or resistance of some dietary proteins to degradation, the level of dietary proteins in the blood is enhanced. Of course, this presence of food antigen in the blood does not go unnoticed by the immune system. If the antigen is taken up by the blood antigen-presenting cell, the result could be IgG or IgA antibody production.

19.5.1.2 Factors Involved in the Induction or Disturbance of Oral Tolerance

Several factors affect the induction of oral tolerance to a dietary antigen. Some are antigen-related, namely, the doses and nature of the antigen. Other factors are inherent to the host, including age, genetics, intestinal flora, diet, medication, and more. Epidemiological studies have revealed that there is one factor of particular importance in the development of oral tolerance [43, 44]. This is the period surrounding a child's birth. Early-life exposure to environmental triggers before birth (through the mother), during birth, and after birth acts as a priming period that shapes the future enteric microbiota and the innate and adaptive immune systems, which reach a lasting homeostatic equilibrium shortly after a child is born [43, 44]. This period could be called *the neonatal window of opportunity* (see ◘ Fig. 19.17).

19.5.1.3 Breach in Oral Tolerance and Its Association with Food Immune Reactivities

Oral tolerance is induced by multiple cellular and molecular processes to ensure lack of immune reactivity to harmless intestinally derived antigens both in the mucosa and in the systemic immune system [45]. Together, mucosal and

◘ Fig. 19.17 The neonatal window of opportunity. Early-life exposure to environmental triggers before birth, during birth, and after birth sets the foundations of the immune system to come

circulatory-induced tolerance appears to prevent intestinal disorders such as inflammatory bowel disease, food immune reactivity, and organ-specific and nonspecific autoimmunities. This process is carried out by a very special population of dendritic cells found in the microenvironment of mesenteric lymph nodes. The presence of antigen-specific T cells and nodes and cytokines such as TGF-β and IL-10 induces the generation and differentiation of these DCs into FOXP3+ regulatory T cells (Tregs). These committed Tregs home back to the intestinal lamina propria, where some of them may exit from the mucosa via the lymphatic system or bloodstream and disseminate throughout the immune system, promoting systemic oral tolerance [45]. The ability of oral tolerance to maintain an inhibitory environment by the Treg cells and the production of noninflammatory IgA against both dietary proteins and microbiota can prevent hyperimmune reactivities in the mucosa and in circulation [46, 47]. The perinatal period is therefore crucial for the establishment of oral tolerance and for the induction of food immune reactivities [48]. Food immune reactivities can result from many environmental factors that can disturb the homeostasis of the immune system, resulting in the penetration of dietary proteins and non-tolerogenic peptides to the submucosa. To avoid immune reactivity to food antigens, the body employs the inflammatory immune defenses, including secretory IgA (SIgA) antibodies and hyporesponsiveness to innocuous agents, particularly dietary antigens and the commensal gut microbiota [49, 50]. The induction of these homeostatic mechanisms depends on exogenous stimuli, and the neonatal period is particularly critical to this end. Both the intestinal surface barrier with its reinforcement by SIgA and the immunoregulatory network require adaptation.

In most cases, this adaptation is remarkably successful in view of the fact that a ton of food, perhaps including 100 kg

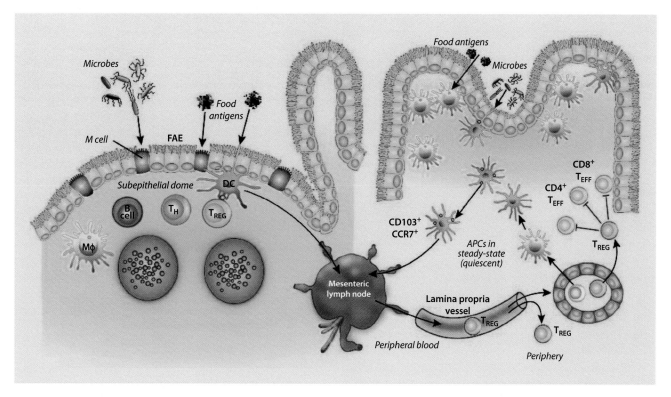

Fig. 19.18 The immunoregulatory network. Some antigen-presenting cells extend their dendrites between epithelial cells to sample luminal antigens. Such dendrites can also be seen in the follicle-associated epithelium (FAE) of gut-associated lymphoid tissue (GALT). Subepithelial APCs, mainly CD103 + CCR7+ dendritic cells (DCs) with captured antigen, migrate via draining lymph to mesenteric lymph nodes where they either mature to become active APCs that stimulate TEFF for productive immunity or become conditioned for tolerance via the generation and/or expansion of TREG cells. These inductile TREG cells migrate via efferent lymph to peripheral blood and then to the mucosa or the periphery where they exert anti-inflammatory control of CD4+ and CD8+ TEFF cells

of proteins, may pass through the gut of an adult human being every year without causing adverse reactions. Food immune reactivity reflects a lack of such homeostasis, either due to retarded immunological development with immaturity of the intestinal surface barrier or a persistently imbalanced immunoregulatory network. Both homeostatic deficiencies may be associated with immune reactivity, in particular IgA and IgG production against innocuous antigens, such as food proteins [51].

The mechanism of oral tolerance to food antigens and microbiota is shown in ◘ Fig. 19.18.

Understanding the mechanisms responsible for the induction of oral tolerance [52–54] is helpful in the design and introduction of vaccines for autoimmunities. For example, tolerance can be restored by sublingual immunotherapy [33, 55, 56], by a rush program of specific oral tolerance induction [57], and by introducing the triggering antigen nasally [58]. In a study on diabetes, intranasal administration of insulin, GAD-65, and even gliadin to 4-week-old, non-obese, diabetic (NOD) mice significantly reduced the diabetes incidence and lowered the insulitis. Likewise, another study [59] demonstrated that intranasal administration of glutamic acid decarboxylase 65 (GAD-65) could prevent murine, insulin-dependent diabetes in NOD mice.

It is interesting to note that gliadin has been shown to be cross-reactive to GAD-65 [60]. The information collected in this chapter shows that intranasal administration of either substance can be effective in reestablishing oral tolerance and either preventing or curing type 1 diabetes and other autoimmune diseases in which food antigens play a role.

19.5.2 Central Tolerance

Central tolerance refers to the tolerance established by certain events that occur in the early development stages of a lymphocyte. These events serve to focus these agents of the immune system onto pathogens (non-self) and away from innocent healthy tissue (self). This tolerance toward self is induced at the primary sites of lymphocyte development. For B cells, this is the bone marrow and lymphoid organs, while for T cells, it is the thymus [61, 62]. In these sites, maturing lymphocytes are exposed to self-antigens presented by medullary thymic epithelial cells and thymic dendritic cells, or bone marrow cells. Self-antigens are present due to endogenous expression, importation of antigen from peripheral sites via circulating blood, and –in the case of thymic stromal cells – expression of proteins of other non-thymic tissues by the action of the transcription factor named autoimmune regulator (AIRE). The lymphocytes will be unable to develop tolerance unless they encounter these antigens.

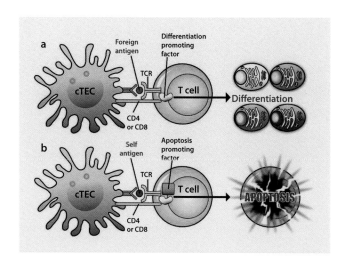

Fig. 19.19 How presentation of a foreign antigen or a self-antigen can lead to differentiation or cell death. **a** The cortical thymic epithelial cell (cTEC) presents a foreign peptide antigen to a T cell's TCR, sending signals that lead to differentiation. **b** Presenting a self-antigen peptide sends signals that lead to the T-cell lymphocyte's apoptosis or cell death

When a cortical thymic epithelial cell (cTEC) presents an antigen peptide, some lymphocytes have receptors that bind strongly to self-antigens, making them potentially strongly reactive to self-tissue. These autoreactive lymphocytes are removed before they develop into fully immunocompetent cells, for the most part by induction of apoptosis of the autoreactive cells (clonal deletion) and to a lesser extent by induction of anergy (a state of non-activity), or even receptor editing [61] (see Fig. 19.19).

The affinity of the peptide-MHC complex ligands also plays a role. If the peptide-MHC complex has low-affinity ligands, the T-cell receptor complex (TCR) will deliver signals that will promote the T cell's differentiation. However, if the peptide-MHC complex has high-affinity ligands, the TCR will deliver signals that promote the T cell's apoptosis (see Fig. 19.19).

The deletion threshold is much stricter for T cells than for B cells. This is because T cells alone can cause direct tissue damage, while B cells need costimulatory signals from T cells as well as a recognized antigen to proliferate and produce antibodies [63]. It is also more advantageous for the immune system to let its B cells recognize a wider variety of antigens so it can produce antibodies against a greater diversity of pathogens [64]. This process ensures that T and B cells that could initiate a potent immune response to the host's own tissues are eliminated while preserving the ability to recognize foreign antigens.

Weakly autoreactive B cells may also remain in a state of immunological ignorance, simply not responding to stimulation of their B-cell receptor. Alternatively, some weakly self-recognizing T cells are differentiated into natural regulatory T cells (*nTreg* cells), which act as sentinels in the periphery to calm down potential instances of T-cell autoreactivity.

19.5.3 Peripheral Tolerance

Peripheral tolerance describes the mechanisms that take place outside of primary lymphoid tissues to prevent lymphocytes from initiating potentially dangerous immune responses against the body's own tissues or against other harmless materials, such as food or commensal organisms. It takes place in the immune periphery after the T and B cells migrate from the primary lymphoid organs. Its main purpose is to ensure that self-reactive T and B cells that escaped the purging of central tolerance do not cause autoimmune disease [61, 65].

Peripheral tolerance is established by different but somewhat overlapping mechanisms that mostly involve T cells, in particular CD4+ helper T cells, which coordinate immune responses and give B cells the confirmatory signals they need in order to produce antibodies, as already mentioned above [33, 59]. T cells that have left the thymus are relatively but not completely safe; unwarranted immune response toward normal self-antigens that were not eliminated in central tolerance can still occur. Some of the mature T cells may have receptors (TCRs) sensitive to self-antigens present in such high concentration outside the thymus that they can bind to "weak" receptors. Alternatively, the T cell could react to a self-antigen it had not previously encountered in the thymus (tissue-specific molecules such as those from the islets of Langerhans, brain, or spinal cord not expressed by the AIRE transcription factor in the thymic tissues).

These self-reactive T cells that escaped clonal deletion in the thymus can inflict cell injury to self-tissues unless they are eliminated or effectively neutralized in the peripheral tissue, chiefly by the previously mentioned nTreg cells [64, 65].

Appropriate reactivity toward certain antigens can also be quieted by induction of tolerance after repeated exposure, or exposure in a certain context. In these cases, there is a differentiation of naïve CD4+ helper T cells into induced Treg cells (*iTreg* cells) in the peripheral tissue or nearby lymphoid tissue (lymph nodes, mucosal-associated lymphoid tissue, etc.). This differentiation is mediated by IL-2 produced upon T-cell activation and TGF-β from any of a variety of sources, including tolerizing dendritic cells (DCs), other antigen-presenting cells, or in certain conditions, surrounding tissue [66].

Adaptive Foxp3 + CD4+ regulatory T cells, otherwise known as induced or iTreg cells, develop outside the thymus under subimmunogenic antigen presentation, during chronic inflammation, and during normal homeostasis of the gut. They are essential in mucosal immune tolerance and in the control of severe chronic allergic inflammation [66].

The Foxp3+ iTreg cell repertoire is drawn from naive conventional CD4+ T cells, whereas natural Treg (nTreg) cells are selected by high-avidity interactions in the thymus [67].

There are other regulatory immune cells aside from Treg cells that mediate peripheral tolerance. These include T-cell subsets similar to but phenotypically distinct from Treg cells, such as TR1 cells that make IL-10 but do not express Foxp3, TGF-β-secreting TH3 cells, and other less well-characterized cells that help establish a local tolerogenic environment [68]. B cells also express CD22, a nonspecific inhibitor receptor that dampens B-cell receptor activation, and a subset of B regulatory cells makes IL-10 and TGF-β [69]. Some DCs can make indoleamine 2,3-dioxygenase (IDO), which depletes the amino acid tryptophan needed by T cells to proliferate, thereby reducing immune responsiveness. Additionally, DCs have the capacity to directly induce anergy in T cells that recognize antigens expressed at high levels and thus presented at steady state by DCs [70, 71].

19.6 Dietary Intervention

It is basic knowledge that eating bad food, the wrong food, can make us sick. Nowadays we can elaborate about sensitivities and immune reactivities and autoimmunities, but the easiest way to put it is: Bad food makes you sick. The converse is equally easy to understand; good foods, or the right foods, make you healthy. To elaborate on this in turn, the right foods can not only keep you healthy but can actually help to reestablish tolerance and fix or repair deficiencies or dysfunctions in your health and immune system.

Vitamins are pretty much taken for granted. Everyone knows they're generally good for you, but so what? People who consider themselves healthy don't think twice about not taking any vitamins at all. However, accumulated evidence shows that a healthy diet containing phytochemicals, fatty acids, and vitamins such as A, C, and D have a direct effect on the maintenance of our long-term health [72].

We have previously discussed innate lymphoid cells (ILCs). They are essential for orchestrating immune responses and discriminating between friendly bacteria, harmless food particles, and harmful food antigens. The binding of these cells to vitamins and their metabolites contributes significantly to immune homeostasis [73, 74].

Vitamin A or retinoic acid (RA) is an important requisite in the conversion of naïve CD4$^+$ cells into iTregs [75]. Metabolites of vitamin A, in particular all-trans RA, are an important determinant in mucosal immune homeostasis and the maintenance of oral tolerance. RA is key in the maintenance of mucosal immune homeostasis mediated by TGF-β1 and the IL-10 expression of T cells, which prevents immune reaction to normal food antigens and the development of gut inflammatory disorders [76].

Vitamin B comprises eight different members involved in various biochemical pathways in cell metabolism. Vitamin B6 influences cell growth and differentiation by enhancing the metabolism of amino acids, nucleic acids, and lipids. Vitamin B6 deficiency can lead to impairment of immune function, and, conversely, its supplementation can repair and even enhance weakened immune response. Vitamin B9, also known as folic acid, is essential for protein and nucleic acid synthesis. Low levels of this vitamin can alter the activity of NK cells and CD8$^+$ T cells. Vitamin B9 is essential for the activation of Bcl-2, the anti-apoptotic molecule that acts as a brake in cell survival; in the absence of vitamin B9, naïve T cells can differentiate into Treg cells but are unable to survive for a long period of time [77].

Vitamin C, or ascorbic acid, has been shown to be important in immunoregulation. Vitamin C deficiency can disrupt T- and B-cell function and NK cytotoxic activity [78–80]. A decrease in vitamin C may result in immune suppression [81]. Studies indicate therapeutic possibilities with vitamin C for the enhancement of NK-cell, T-cell, and B-cell function, as well as for immunoregulation and the induction of tolerance to autoantigens.

Vitamin D or 1,25-(OH)$_2$D$_3$ promotes the generation of tolerogenic or semi-mature DCs with a reduced ability to process and present the antigens of food and friendly bacteria to T cells. Vitamin D3 also promotes a general suppression of immune response and protects against various autoimmune disorders while enhancing the antimicrobial properties of monocytes and macrophages and increasing the body's defense against invading microorganisms. It also enhances the production of Toll-like receptors (TLRs) and triggering receptors expressed on myeloid cells (TREMs). Vitamin D has the potential to restore antigen-specific immune tolerance by inhibiting the maturation of dendritic cells or locking the cells in a semi-mature state so as to deprive them of their capacity to activate autoreactive T cells. Treatment with vitamin D is a promising tool for restoring the balance between immunogenicity and tolerogenicity in many autoimmune diseases [82, 83].

Aryl hydrocarbon receptor (AhR) is a ligand-activated transcription factor that is a crucial regulator in maintaining the number of intraepithelial lymphocyte (IEL) cells in the intestine. AhR is found in cells that are important in the defense against intestinal and extracellular pathogens and also helps in gene transcription and maintaining homeostasis in the immune system. It stands to reason that any food that helps AhR do its job is a food that's good to eat. Cruciferous vegetables such as broccoli, cabbage, cauliflower, kale, Brussels sprouts, and watercress, which contain *indole-3-carbinol*, are a major source of AhR ligands [84]. Plant-derived nutrients like indole-3-carbinol can work with AhR to shape the intestinal immune defenses and have an impact on the control of microbiota and overall immunity and health. However, if you're feeling dismay that your parents were right after all (they were) and you should be eating vegetables (you should), here's a bit of uplifting news.

Tryptophan is an essential amino acid that plays an important role in immune function. Yes, that tryptophan, the stuff in turkey that people blame when they feel sleepy after gorging themselves on Thanksgiving. Tryptophan in fact is found not just in turkey but in many plant and animal proteins, including chicken, beef, pork, fish, cheese, seeds, nuts, and beans [85, 86]. It is able to influence the immune system when it is converted into metabolites by the enzyme indoleamine 2,3-dioxygenase (IDO). This enzyme converts dietary tryptophan to kynurenine, hydroxykynurenine, and xanthurenic acid; these metabolites can regulate immune function, inducing Treg cells to produce IL-10, thereby preventing asthma and allergy.

Tryptophan metabolites and naturally occurring compounds found in fruits and vegetables, such as resveratrol in grapes and quercetin in apples, can act on the AhR on intraepithelial lymphocytes, innate lymphoid cells, Treg cells, and Th17 cells, thereby controlling their functions.

Probiotics are defined by the World Health Organization as live bacterial species that confer a health benefit when administered in adequate amounts [87]. These are commensal bacteria among the gut microbiome's bacterial population that have been identified to provide their hosts with many health benefits, including intestinal homeostasis, and blocking the harmful effects of other microbiota. They directly compete with pathogens for various nutrients, stimulate innate and adaptive immune responses, and promote tolerance.

Good bacteria can convert fiber-containing molecules such as starch pectins, fructan, and cellulose into three major fatty acid metabolites: acetate, butyrate, and propionate. These metabolites or short-chain fatty acids (SCFAs) act as ligands for the GP43 receptor (GP43R) on Treg cells. They activate the Treg cells into expanding and producing a significant amount of IL-10 and TGF-β, which do not only enhance the oral tolerance mechanisms but can also control inflammatory immune responses in the gut. Dietary intervention with probiotics could be used for the prevention and alleviation of food immune reactivity, intestinal barrier dysfunction, and even autoimmunities (see ◘ Fig. 19.20) [88].

19.7 Conclusion

The three major protective layers of the immune system (mucosal, innate, and adaptive) work together to protect an individual from the assault of environmental triggers such as infections, toxic chemicals, and immune-reactive foods. The mechanisms of the immune system comprise highly complex processes and myriad cells with a variety of functions, making it all too easy for an environmental trigger to cause a misstep or hiccup in the system.

As knowledge grows, the information gained is eventually put to use. The information we have presented here is gradually being applied in clinical practice. Such therapeutic interventions focused on the cells, molecules, and mechanisms of the immune system can have far-reaching benefits for a patient's immune system and health.

We now know that things as basic to life as foods are not to be taken for granted and that they can either help or hinder our immune system and, therefore, our health.

Understanding how the immune system works and how the mechanisms and components of the mucosal, innate, and adaptive immune systems all work together can help us to design modalities for the maintenance of immune homeostasis and the prevention of many immune disorders, including autoimmunities [89].

Acknowledgments We would like to acknowledge Joel Bautista for the preparation of this manuscript for publication and for the execution of the figures.

References

1. Macpherson AJ, Geuking MB, McCoy KD. Homeland security: IgA immunity at the frontiers of the body. Trends Immunol. 2012;33(4):160–7. https://doi.org/10.1016/j.it.2012.02.002. Epub 2012 Mar 10
2. Woof JM, Kerr MA. The function of immunoglobulin a in immunity. J Pathol. 2006;208(2):270–82.
3. Fagarasan S, Honjo T. Intestinal IgA synthesis: regulation of frontline body defenses. Nat Rev Immunol. 2003;3(1):63–72.
4. Brandtzaeg P, Pabst R. Let's go mucosal: communication on slippery ground. Trends Immunol. 2004;25(11):570–7.
5. Rumbo M, Chirdo FG, Añón MC, Fossati CA. Detection and characterization of antibodies specific to food antigens (gliadin, ovalbumin and β-lactoglobulin) in human serum, saliva, colostrum and milk. Clin Exp Immunol. 1998;112(3):453–8.
6. Cunningham-Rundles C. Physiology of IgA and IgA deficiency. J Clin Immunol. 2001;21(5):303–9.
7. Brandtzaeg P. Human secretory immunoglobulins V. Occurrence of secretory piece in human serum. J Immunol. 1971;106(2):318–23.
8. Waldman RH, Mach JP, Stella MM, Rowe DS. Secretory IgA in human serum. J Immunol. 1970;105(1):43–7.
9. Vojdani A. For the assessment of intestinal permeability, size matters. Altern Ther Health Med. 2013;19(1):12–24.
10. Vojdani A. A potential link between environmental triggers and autoimmunity. Autoimmune Diseases. 2014: 2014:437231, 18 pp. https://doi.org/10.1155/2014/437231.
11. Tlaskalová-Hogenováa H, Stepánkováa R, Hudcovica T, et al. Commensal bacteria (normal microflora), mucosal immunity and

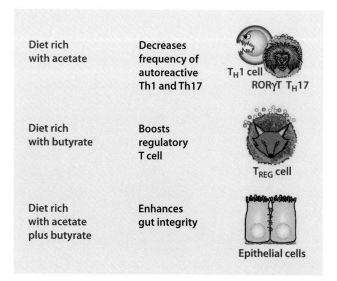

◘ Fig. 19.20 How dietary intervention can help the immune system. Eating the proper foods with the right chemicals/ligands/metabolites can help maintain or repair the immune system in various ways. Only a few of these are named in the figure

chronic inflammatory and autoimmune diseases. Immunol Lett. 2004;93(2–3):97–108.
12. Corthésy B. Role of secretory IgA in infection and maintenance of homeostasis. Autoimmun Rev. 2013;12(6):661–5.
13. Hakeem V, et al. Salivary IgA antigliadin antibody as a marker for coeliac disease. Arch Dis Child. 1992;67(6):724–7.
14. al-Bayaty HF, et al. Salivary and serum antibodies to gliadin in the diagnosis of celiac disease. J Oral Pathol Med. 1989;18(10):578–81.
15. Di Leo M, et al. Serum and salivary antiendomysium antibodies in the screening of coeliac disease. Panminerva Med. 1999;41(1):68–71.
16. Brandtzaeg P. Do salivary antibodies reliably reflect both mucosal and systemic immunity? Ann N Y Acad Sci. 2007;1098:288–311.
17. Bonamico M, et al. Radioimmunological detection of anti-transglutaminase autoantibodies in human saliva: a useful test to monitor coeliac disease follow-up. Aliment Pharmacol Ther. 2008;28(3):364–70.
18. Pastore L, et al. Orally based diagnosis of celiac disease: current perspectives. J Dent Res. 2008;87(12):1100–7.
19. Bonamico M, et al. Tissue transglutaminase autoantibody detection in human saliva: a powerful method for celiac disease screening. J Pediatr. 2004;144(5):632–6.
20. Bonamico M, et al. First salivary screening of celiac disease by detection of anti-transglutaminase autoantibody radioimmunoassay in 5000 Italian primary schoolchildren. J Pediatr Gastroenterol Nutr. 2011;52(1):17–2.
21. Brandtzaeg P. Food allergy and intolerance, 2nd edition by Brostoff J, Challacombe S. London: Elsevier Science; 2002. p. 479–95.
22. McGhee JR, Fujihashi K. Inside the mucosal immune system. PLoS Biol. 2012;10(9):e1001397.
23. Vojdani A. Oral tolerance and its relationship to food immunoreactivities. Altern Ther Health Med. 2015;21(Suppl 1):23–32.
24. Garrett WS, Gordon JI, Glimcher LH. Homeostasis and inflammation in the intestine. Cell. 2010;140(6):859–70.
25. Schröfelbauer B, Hoffmann A. How do pleiotropic kinase hubs mediate specific signaling by TNFR superfamily members? Immunol Rev. 2011;244(1):29–43.
26. Janeway CA Jr, Travers P, Walport MJ, Shlomchik M. Immunobiology: the immune system in health and disease. New York: Garland Publishing; 2005.
27. Guarda G, So A. Regulation of inflammasome activity. Immunology. 2010;130(3):329–36.
28. Heneka MT, Kummer MP, Latz E. Innate immune activation in neurodegenerative disease. Nat Rev Immunol. 2014;14:463–77.
29. Levy M, Shapiro H, Thaiss CA, Elinav E. NLRP6: a multifaceted innate immune sensor. Trends Immunol. 2017;38(4):248–60. https://doi.org/10.1016/j.it.2017.01.001. Epub 2017 Feb 15
30. Masters SL, Simon A, Aksentijevich I, Kastner DL. Horror autoinflammaticus: the molecular pathophysiology of autoinflammatory disease. Annu Rev Immunol. 2009;27:621–68.
31. Miranda PJ, Delgobo M, et al. The oral tolerance as a complex network phenomenon. PLoS One. 2015;10(6):e0130762.
32. Lim PL, Rowley D. The effect of antibody on the intestinal absorption of macromolecules and on intestinal permeability in adult mice. Int Arch Allergy Immunol. 1982;68:41–6.
33. Scurlock AM, Vickery BP, Hourihane JO, Burks AW. Pediatric food allergy and mucosal tolerance. Mucosal Immunol. 2010;3(4):345–54.
34. Verhasselt V. Oral tolerance in neonates: from basics to potential prevention of allergic disease. Mucosal Immunol. 2010;3(4):326–33.
35. Brandtzaeg PE. Current understanding of gastrointestinal immunoregulation and its relation to food allergy. Ann N Y Acad Sci. 2002;964:13–45.
36. Gleeson M, Cripps AW. Development of mucosal immunity in the first year of life and relationship to sudden infant death syndrome. FEMS Immunol Med Microbiol. 2004;42(1):21–33.
37. Fitzsimmons SP, Evans MK, Pearce CL, et al. Immunoglobulin A subclasses in infants' saliva and in saliva and milk from their mothers. J Pediatr. 1994;124(4):566–73.
38. Brandtzaeg P. Development and basic mechanisms of human gut immunity. Nutr Rev. 1998;56:S5–18.
39. Chehade M, Mayer L. Oral tolerance and its relation to food hypersensitivities. J Allergy Clin Immunol. 2005;115(1):3–12.
40. Chen Y, Inobe J, Marks R, Gonnella P, Kuchroo VK, Weiner HL. Peripheral deletion of antigen-reactive T cells in oral tolerance. Nature. 1995;376(6536):177–80.
41. Appleman LJ, Boussiotis VA. T cell anergy and costimulation. Immunol Rev. 2003;192:161–80.
42. Vojdani A, Erde J. Regulatory T cells, a potent immunoregulatory target for CAM researchers: modulating allergic and infectious disease pathology (II). ECAM. 2006;3(2):209–15.
43. Torow N, Hornef MW. The neonatal window of opportunity: setting the stage for life-long host-microbial interaction and immune homeostasis. J Immunol. 2017;198:557–63.
44. Grigg JB, Sonnenberg GF. Host-microbiota interactions shape local and systemic inflammatory diseases. J Immunol. 2017;198:564–71.
45. Zinselmeyer BH, Dempster J, Gurney AM, et al. In situ characterization of CD4+ T cell behavior in mucosal and systemic lymphoid tissues during the induction of oral priming and tolerance. J Exp Med. 2005;201(11):1815–23.
46. Goubier A, et al. Plasmacytoid dendritic cells mediate oral tolerance. Immunity. 2008;29(3):464–75.
47. Pabst O, Mowat AM. Oral tolerance to food protein. Mucosal Immunol. 2012;5:232–9.
48. Murai M, et al. Interleukin 10 acts on regulatory T cells to maintain expression of the transcription factor Foxp3 and suppressive function in mice with colitis. Nat Immunol. 2009;10(11):1178–84.
49. Föhse L, et al. High TCR diversity ensures optimal function and homeostasis of Foxp3+ regulatory T cells. Eur J Immunol. 2011;41(11):3101–13.
50. Lathrop SK, et al. Peripheral education of the immune system by colonic commensal microbiota. Nature. 2011;478(7368):250–4.
51. Turner JR. Intestinal mucosal barrier function in health and disease. Nat Rev Immunol. 2009;9(11):799–809.
52. Brandtzaeg P, Isolauri E, Prescott SL, editors. Microbial-host interaction: tolerance versus allergy. Nestlé Nutr Inst Workshop Ser Pediatr Program, vol 64. Basel: Nestec Ltd., Vevey/S. Karger AG © 2009. p. 23–43.
53. Brandtzaeg P. Mucosal immunity: induction, dissemination, and effector functions. Scand J Immunol. 2009 Dec;70(6):505–15.
54. Knutson TW, et al. Intestinal reactivity in allergic and nonallergic patients: an approach to determine the complexity of the mucosal reaction. J Allergy Clin Immunol. 1993 Feb;91(2):553–9.
55. Mempel M, Rakoski J, Ring J, Ollert M. Severe anaphylaxis to kiwi fruit: immunologic changes related to successful sublingual allergen immunotherapy. J Allergy Clin Immunol. 2003;111(6):1406–9.
56. Kerzl R, Simonowa A, Ring J, Ollert M, Mempel M. Life-threatening anaphylaxis to kiwi fruit: protective sublingual allergen immunotherapy effect persists even after discontinuation. J Allergy Clin Immunol. 2007;119(2):507–8.
57. Itoh N, Itagaki Y, Kurihara K. Rush specific oral tolerance induction in school-age children with severe egg allergy: one year follow up. Allergol Int. 2010;59(1):43–51.
58. Funda DP, Fundova P, Hansen AK, Buschard K. Prevention or early cure of type 1 diabetes by intranasal administration of gliadin in NOD mice. PLoS One. 2014;9(4):e94530.
59. Tian J, Atkinson MA, Clare-Salzler M, et al. Nasal administration of glutamate decarboxylase (GAD65) peptides induces Th2 responses and prevents murine insulin-dependent diabetes. J Exp Med. 1996;183(4):1561–7.
60. Vojdani A, Tarash I. Cross-reaction between gliadin and different food and tissue antigens. Food Nutr Sci. 2013;4(1):20–32.
61. Hogquist K, Baldwin T, Jameson S. Central tolerance: learning self-control in the thymus. Nat Rev Immunol. 2005;5(10):772–82.

62. Nemazee D. Mechanisms of central tolerance for B cells. Nat Rev Immunol. 2017;17:281–94.
63. Kindt TJ, Osborne BA, Goldsby RA. Kuby immunology. 6th ed. New York: W. H. Freeman; 2006.
64. Murphy K. Chapter 18. Janeway's immunobiology. 8th ed. London and New York: Garland Sciences; 2012. p. 275–334.
65. Murphy K. Chapter 15. Janeway's immunobiology. 8th ed. London and New York: Garland Sciences; 2012. p. 611–68.
66. Mueller DL. Mechanisms maintaining peripheral tolerance. Nat Immunol. 2010;11(1):21–7. https://doi.org/10.1038/ni.1817.
67. Curotto de Lafaille MA, Lafaille JJ. Natural and adaptive Foxp3+ regulatory T cells: more of the same or a division of labor? Immunity. 2009;30(6):626–35. https://doi.org/10.1016/j.immuni.2009.05.002.
68. Sakaguchi S, Miyara M, Costantino C, Hafler DA. Foxp3+ regulatory T cells in the human immune system. Nat Rev Immunol. 2010;10:490–500.
69. Vadasz Z, Haj T, Kessel A, Toubi E. B-regulatory cells in autoimmunity and immune mediated inflammation. FEBS Lett. 2013;587(13):2074–8.
70. Ganguly D, Haak S, Sisirak V, Reizis B. The role of dendritic cells in autoimmunity. Nat Rev Immunol. 2013;13:566–77.
71. Maher S, Toomey D, Condron C, Bouchier-Hayes D. Activation-induced cell death: the controversial role of Fas and Fas ligand in immune privilege and tumour counterattack. Immunol Cell Biol. 2002;80(2):131–7.
72. Veldhoen M, Brucklacher W. Dietary influences on intestinal immunity. Nat Rev Immunol. 2012;12:696–708.
73. Ebrel G, Littman DR. The role of the nuclear hormone receptor RORγt in the development of lymph nodes and Peyer's patches. Immunol Rev. 2003;195:81–90.
74. Wong SH, et al. Transcription factor RORα is critical for nuocyte development. Nat Immunol. 2012;13:229–36.
75. Strober W. Vitamin a rewrites the ABCs of oral tolerance. Mucosal Immunol. 2008;1:92–5.
76. Bakdash G, et al. Retinoic acid primes human dendritic cells to induce gut-homing, IL-10-producing regulatory T cells. Mucosal Immunol. 2015;8(2):265–78.
77. Kunisawa J, et al. A pivotal role of vitamin B9 in the maintenance of regulatory T cells in vitro and in vivo. PLoS One. 2012;7(2):e32094.
78. Heuser G, Vojdani A. Enhancement of natural killer cell activity and T and B cell function by buffered vitamin C in patients exposed to toxic chemicals: the role of protein kinase-C. Immunopharmacol Immunotoxicol. 1997;19:291–312.
79. Vojdani A, Namatalla G. Enhancement of human natural killer cytotoxic activity by vitamin C in pure and augmented formulations (ultra potent-C). J Nutrit Environ Med. 1997;7:187–95.
80. Vojdani A, Bazargan M, Wright J, Vojdani E. New evidence for antioxidant properties of vitamin C. Cancer Detect Prev. 2000;24:508–23.
81. Pavlovic V, et al. Ascorbic acid modulates spontaneous thymocyte apoptosis. Acta Medica Medianae. 2005;44:21–3.
82. Ferreira GB, et al. 1,25-dihydroxyvitamin D3 promotes tolerogenic dendritic cells with functional migratory properties in NOD mice. J Immunol. 2014;192:4210–20.
83. McMahon L, et al. Vitamin D-mediated induction of innate immunity in gingival epithelial cells. Infect Immun. 2011;79(6):2250–6.
84. Hooper LV. You AhR what you eat: linking diet and immunity. Cell. 2011;47:490–1.
85. Hayashi T, et al. Inhibition of experimental asthma by indoleamine 2,3-dioxygenase. J Clin Invest. 2004;114:270–9.
86. Matteoli G, et al. Gut CD103+ dendritic cells express indoleamine 2,3-dioxygenase which influences T regulatory/T effector cell balance and oral tolerance induction. Gut. 2010;59:595–604.
87. Food and Agriculture Organization/World Health Organization. Expert consultation on evaluation of health and nutritional properties of probiotics in food including milk powder with live lactic acid bacteria. http://www.who.int/entity/foodsafety/publications/fs_management/en/probiotics.pdf.
88. Kabat AM, et al. Modulation of immune development and function by intestinal microbiota. Trends Immunol. 2014;35(11):507–17.
89. David A. Horwitz, Tarek M. Fahmy, Ciriaco A. Piccirillo, Antonio La Cava. Rebalancing Immune Homeostasis to Treat Autoimmune Diseases. Trends in Immunology. 2019;40(10):888–908.

Nutritional Influences on Immunity and Infection

Joel Noland and Diana Noland

20.1 Introduction – 305

20.2 Impact of Infection on Health and Disease – 305

20.3 Key Metabolic Mechanisms for Defense and Repair – 306

20.4 Insults to Our Defense and Repair Systems – 306
20.4.1 Increased Toxin Load – 306

20.5 Antimicrobial Resistance (AMR) – 306

20.6 Gastrointestinal Dysbiosis: From Mouth to Anus – 306

20.7 Stressors – 306

20.8 Malnutrition, Inflammation, and the Infectious Processes – 307

20.9 Diagnosis of Nutrition Status and Infection-Related Diseases – 309

20.10 Differential for Nutritional Infection Risk – 310
20.10.1 Look for Evidence of Infections – 310

20.11 Chronic Diseases, Nutrition, Microbiome, and Infection – 311
20.11.1 Laboratory – 311
20.11.2 Assessment Laboratory and Clinical Tools – 313
20.11.3 Key Nutrients Influencing the Risk of Infectious Disease (◘ Fig. 20.8) – 315

20.12 Homocysteine Catabolism (◘ Fig. 20.9) – 315
20.12.1 Folate Metabolism – 315
20.12.2 Methionine Metabolism – 316

20.13 Key Lifestyle Factors Influencing the Risk of Infectious Disease – 316
20.13.1 Sleep (see ▶ Chap. 45) – 316
20.13.2 Stress (see ▶ Chap. 47) – 316

© Springer Nature Switzerland AG 2020
D. Noland et al. (eds.), *Integrative and Functional Medical Nutrition Therapy*,
https://doi.org/10.1007/978-3-030-30730-1_20

20.14	Movement (see ▸ Chap. 36)	– 316
20.14.1	Examples of Chronic Disease Connections with Infectious Disease	– 316
20.14.2	Vaccination	– 317
20.15	Conclusion	– 318
	References	– 319

Learning Objectives
- Impact of infection on health and disease
- Malnutrition, inflammation, and the infectious processes
- Chronic diseases, nutrition, microbiome, and infection

20.1 Introduction

The complex nutrition-immunity-microbiome-infection-genomic connection is presented in this chapter, bringing their relationship together in an integrated view with a focus on the role of nutrients in maintaining the integrity of the immune system. Infection has been a primary challenge to human health and disease throughout history. It is now recognized that an individual's vulnerability to infection is associated with nutritional status. Malnutrition increases the risk of infection and immune system compromise. The past century's renaissance of nutrition science has supplied the increasing evidence for the roles that essential nutrients, dietary intake, environmental exposures, and nutrigenomic influences play in the ability of the immune system to respond and resolve infectious insults. The recognition of the importance of nutrition within the scope of assessment and intervention, which is often overlooked in healthcare, is presented as having an overall impact on outcome. The current knowledge of the role of nutrients and their interrelationships with the immune system and chronic disease provides scientific support for the nutritionally trained practitioner. This chapter will describe some of the key mechanisms and nutrient influences on the immune response as well as dietary and lifestyle considerations for treatment.

Nutritional deficiencies and insufficiencies have known associations with increased susceptibility to infectious disease. Infection can also increase requirement for nutrients and produce further undernutrition, infection, and compromise of the immune system, setting up a vicious cycle between malnutrition and infection [1, 2]. Malnutrition is the primary global cause of immunodeficiency for all age groups, especially for infant mortality with poor nutrition promoting underweight, weak, and vulnerable children. Optimum nutrition status allows the metabolic function defense and repair mechanisms to increase immune integrity.

Optimal nutritional status contributes to health maintenance and the prevention of infection. The function of healthy cells is maintained by adequate nutrition, movement, and sleep routines. Primary and secondary malnutrition may occur when each individual lacks the available clean food, nutrients, and water that are required. The healthy immune system enables the body's ability to adapt, recover, and survive. When there is disruption in the nutrient intake, the malnutrition that ensues contributes to a cascade of adverse metabolic events leading to illness. The nutritionally trained practitioner assesses an individual to identify where malnutrition may be present and develops an intervention with early delivery of essential nutrients in an effective and comprehensive manner. Healthcare practitioners are often challenged to understand the importance of adequate nutritional support in the prevention and treatment of infection, multiple organ failure, and most life-threatening systemic sepsis [3].

The aim of this chapter is to provide the current science for and a description of the interaction between nutritional status of an individual and immunological susceptibility to infection, as well as integrative and functional approaches to interventions that may be considered to restore immune integrity and restore wellness. There will not be an in-depth review of immune function. Please refer to the excellent presentations of the immune system by Dr. Vodjani in ► Chaps. 19 and 49.

This chapter is divided into three sections.
1. Impact of infection on health and disease
2. Malnutrition, inflammation, and the infectious processes
3. Chronic diseases, nutrition, microbiome, and infection

20.2 Impact of Infection on Health and Disease

Until the beginning of the nineteenth century and the industrial revolution, illness was primarily impacted by physical injury and acute infections related to poor sanitation practices that allowed a higher prevalence of infection. Throughout thousands of years of history, the human body has survived through the strength of several defense mechanisms, including an alert immune system, the impermeable skin membrane, and gut barrier, and the more recently recognized microbiome that shields all body orifices. It has been well established, even by Hippocrates almost 3000 years ago, that these mechanisms are dependent on an optimal nutritional status in order to maintain health and to prevent infection.

> Illnesses do not come upon us out of the blue. They are developed from small daily sins against Nature. When enough sins have accumulated, illnesses will suddenly appear. –Hippocrates (c. 460 – c. 370 BC)

During the nineteenth century, outstanding scientific discoveries occurred and changes in philosophy took place regarding the biological functions within the human body, enabling greater appreciation of the fact that when the biochemical mechanisms of our defense system are disrupted, it increases vulnerability to infectious disease.

20.3 Key Metabolic Mechanisms for Defense and Repair

- Enterohepatic circulation (see ► Chap. 16)
- Phase I and phase II biotransformation/detoxification (see ► Chap. 14)
- Gastrointestinal, lung and skin/barrier integrity/membrane integrity (see ► Chap. 12)
- Mitochondrial function
- Methylation (see ► Chap. 18)
- Hormonal function (see ► Chap. 32)
- Autophagy (see ► Chap. 51.2.6)
- Happiness (see ► Chaps. 6 and 30)

20.4 Insults to Our Defense and Repair Systems

20.4.1 Increased Toxin Load

As the industrial revolution evolved, increased toxin exposure accumulated from the discovery and use of petroleum which released petrochemicals and mercury into the air, the discovery of mercury (e.g., used in making the British top hat), the common use in dentistry of mercury (comprising about 50% of dental amalgam material), and pollution from industrial practices into rivers and the resulting contaminated water supply. The trend for increased environmental toxin exposure has continued to grow, so that at the time of this publication there have been more than 80,000 chemicals created with relatively few tested for safety. A most pervasive and hazardous pesticide toxin, glyphosate, is now used nearly universally in agriculture, thereby ending up in our water and food supplies.

20.5 Antimicrobial Resistance (AMR)

There has been continuous improvement in sanitation practices, perhaps too robustly going beyond the abilities of human and animal life to balance with the microbial world to avoid development of "superbugs" that have become resistance to antibiotics. The challenge of acute infections is of great concern to public health because of antimicrobial resistance (AMR) and the rise of "superbugs" that are considered the biggest threats to modern healthcare [4]. The primary driver of AMR is thought to be the overuse of antibiotics in humans and agricultural animals and overuse of antibacterial hand soaps and gels.

20.6 Gastrointestinal Dysbiosis: From Mouth to Anus

The two largest defense barriers to infection are the skin and the gastrointestinal tract (gut). The mouth and oral cavity provide the beginning of defense and repair by mastication, saliva digestives, and endocrine immune glands (parotid, tonsils, adenoids) preparing food that enters the gut for digestion and absorption. The next step is the pH 1–3 acid bath the bolus of food passes through in the stomach with antimicrobial action suppressing any pathogens traveling in with your food. The small intestine is the next pass, bathing the bolus in bile, pancreatic digestive enzymes, and bicarbonate, ready for the serious work of digestion. This is the location where the gut houses more than 70% of our body's immune tissue (lymphoid tissues) and, when compromised by insults, we become more vulnerable to infection weakening our immune integrity. Those lymphoid immune cells also depend on a healthy microbiota's symbiotic relationship to optimally function. The gut is frequently referred to as "the second brain" [5] because of its generous secretion of neurotransmitters and direct connections to and from the brain, the enteric nervous system. When insults like antibiotics, emotional upsets, stress, infection, chemicals, or toxic metals enter the gut, the gut barrier breaks down and suffers intestinal permeability ("leaky gut") that allows for non-desirable molecules to be absorbed and enter the blood and lymphatic systems, triggering an immune response of loss of self-tolerance, or an antigenic response against one's own tissue [6]. Once someone experiences a "leaky gut," a pro-inflammatory cascade initiates, adding a burden to the immune system and increasing vulnerability to infection.

20.7 Stressors

Stress can insult our defense and repair in several forms: biological, emotional, and energetic stressors. As stressors increase, there is a resulting biological stress, along with a concurrent increase in nutritional requirements. If the food intake cannot keep up with meeting the nutritional needs under stress, malnutrition ensues. Unfortunately, with the increased consumption of the standard American diet (S.A.D.), the population eats more calorie-dense, nutrient-depleted, processed, and high-sugar foods. A majority of the US population does not have the available nutrients to meet the essentials of a generally more stressed society [7]. One of the most prevalent nutrient deficiencies and insufficiencies in the USA from NHANES studies are "40% deficient in Vitamin A, C, D & E, calcium or magnesium deficient and >90% do not get enough choline, fiber & potassium." [6] (◘ Fig. 20.1).

About 65% of the population does not even meet the recommended daily allowance (RDA) for magnesium [8, 9]. More than two-thirds of the US population are either overweight or obese. This population subgroup has a higher risk of several chronic diseases.

Early in the twentieth century, great trust and hope were generated by Alexander Fleming's discovery of penicillin and the world of antibiotics [10], with the anticipation of ending uncontrolled bacterial infection. Now in the twenty-first century, as science looks back on the use of antibiotics, recognition is increasing that the overuse of antibiotics is enabling the development of what are now termed "superbugs."

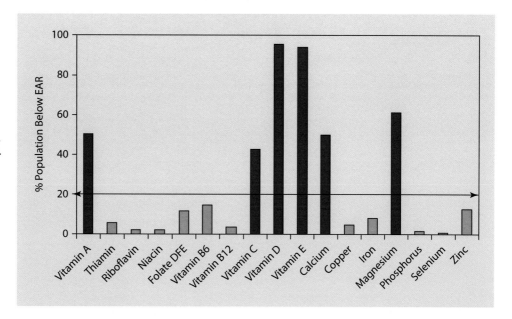

Fig. 20.1 NHANES 2001–2008 micronutrient deficits. Percentage of the adult population (aged 19 years) with vitamin and mineral intakes below the EAR for individuals (data from NHANES 2001–2008). Usual intakes from foods were estimated by using the National Cancer Institute (NCI) method [8]. (Reprinted from Agarwal et al. [8]. With permission from Taylor & Francis)

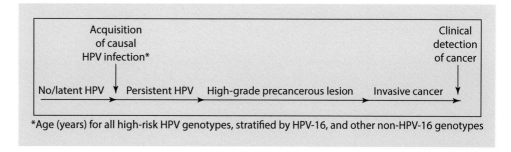

Fig. 20.2 HPV lifespan spectrum [12]. (Reprinted from Burger et al. [12]. With permission from Oxford University Press)

Superbugs are beginning to threaten the success of medicine and pharmacology. The antibiotics have weakened the defense mechanisms of the microbiome shield that humans have depended on throughout history. Public health concerns are increasing that the overuse of antibiotics in humans and animal husbandry, along with antimicrobial antiseptics, has become a health threat through the weakening of the microbiome protection of the population.

The discovery of subclinical unresolving infections that are associated with many of the chronic diseases is beginning to be appreciated by the global medical community. Infection increases infection-related morbidity and mortality [11]. Subclinical-level long-latency infections often go unnoticed while they alter and sometimes mutate tissue cells over time, leading to an acute disease. Examples of infection connections to chronic disease are:

- HPV virus causal of cervical cancer and neck/head cancers [12] (Fig. 20.2)
- *H. pylori* increased risk of gastric cancer [13] (Fig. 20.3)
- HSV-1 increasing risk of Alzheimer's disease [14]
- *Klebsiella pneumoniae* association with rheumatoid arthritis [15, 16]
- EBV and CMV combo-viral increasing risk of various cancers [17]
- *Chlamydia pneumonia* and atherosclerotic plaque formation [18]

With nutritional status being a major factor affecting host resistance to infection, this chapter focuses on how to assess a chronic disease sufferer's "infection load" or "infection status" using an integrative and functional lens, searching for the disease etiology and impaired resolution of inflammation (resolution biology) [19]. This assessment of infection status should become part of the nutritional and metabolic assessment differential. It rules out, or identifies, if infection is part of the etiology and pathophysiology of a disease condition. Once identified, targeted intervention proceeds to improve successful outcomes of restoring wellness (Fig. 20.4).

20.8 Malnutrition, Inflammation, and the Infectious Processes

Infection, as well as trauma or excessive visceral fat, appropriately perturbs the immune system into secretion of inflammatory molecules like cytokines, acute phase reactants, and others [11]. It is not possible to cover the scope of the physiology of the immune system and inflammation in this space; the focus of this chapter is to describe how malnutrition and insufficiencies of specific nutrient groups can perpetuate compromising the immune system and the microenvironment, so the natural immune mechanisms of defense and repair from infection are weakened (see ▶ Chaps.

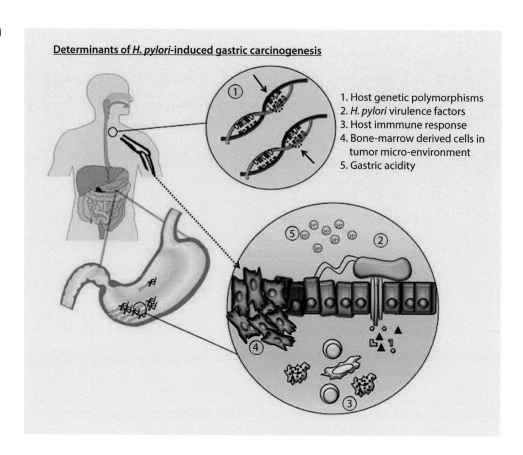

Fig. 20.3 *H. pylori* increased risk of gastric cancer [13]. (Reprinted from Kim et al. [13]. With permission from Elsevier)

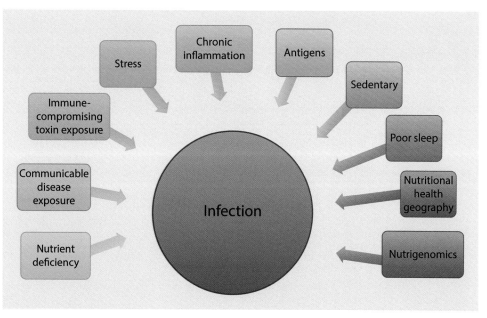

Fig. 20.4 Key nutritional, lifestyle, and environmental influences on infection

19 and 50). Inflammation almost always accompanies infection and, when prolonged, sets up susceptibility for all chronic diseases.

For acute infections, this inflammatory response is a critical part of tissue healing, with increased blood flow and heat. Increased heat can involve local tissue or produce systemic natural hyperthermia with fever. If infection continues to be unresolved, it can produce a prolonged state of inflammation with continuing subclinical infection(s) that over time can result in a loss of self-tolerance and perturb metabolic mechanisms toward diseased tissue that can lead to any of the chronic diseases. The inflammatory load of an individual should be assessed for each chronic disease and infection ruled out as a potential contributor to total body inflammation [20].

Tuberculosis (TB) is a leading cause of death worldwide, despite being preventable and often curable [21, 22]. It is prevalent in malnourished populations with poor sanitation [20].

Schwenk states that TB, also called "consumption," is predisposed by a state of macro- and micronutrient deficiencies.

There is a complex relationship between tuberculosis and malnutrition, in that TB can increase nutrient requirements and lead to a worsening of nutritional status. In high-TB rate countries, vitamin A, carotenoids, and vitamin D levels are found to be low and deficient. Nutritional correction of malnutrition factors is considered part of the best approach to the treatment of tuberculosis [20, 21].

20.9 Diagnosis of Nutrition Status and Infection-Related Diseases

There is much evidence for association or causal nutritional insufficiencies or deficiencies increasing the risk of someone becoming infected. Once infected, levels of tissue inflammation increase until clinically observable (see ◘ Fig. 20.2) and may linger long term unless resolved back to wellness. If the infection continues even at a subclinical level, it can continue to weaken the host integrity of the immune system [11].

Inflammation is the body's normal response and protects the body from infection from pathogens like bacteria or viruses, as well as injury. Inflammation can be acute, which should be short-lived as the body resolves an infection or injury, or chronic and long term and can be destructive, leading to chronic diseases. Examples of chronic diseases that can develop from a long-latency infection are periodontitis, asthma, inflammatory arthritis, and inflammatory bowel disease.

Conventional therapies for chronic inflammation use anti-inflammatories primarily from two categories of pharmaceuticals: steroids and nonsteroidal anti-inflammatory drugs (NSAIDs). NSAIDs counteract enzymes and prostaglandins. NSAID use of more than 10 days is not desirable, due to increased risk of stomach ulcers and gastrointestinal bleeding and sometimes adverse effects like worsening of asthma or kidney problems.

> **Cardinal signs and physiology of inflammation**
> *Rubor (redness):* increased blood flow
> *Tumor (swelling):* exudation of fluid
> *Calor (heat):* exudation of fluid, increased blood flow, release of inflammatory mediators
> *Dolor (pain):* chemical mediators; inflammatory exudates stretching pain receptors and nerves
> *Functio laesa (loss of function):* pain, fibroplasia, metaplasia, disruption of structure

For the IFMNT practitioner, changing the patient's dietary intake and use of dietary or herbal supplements to correct nutrient deficits or excesses can support the underlying systems promoting the inflammation. For instance, NSAIDs act on suppressing eicosanoid and prostaglandin metabolites. Assessment of blood RBC fatty acids (linoleic (LA) and alpha-linolenic acid (ALA) and their metabolites, gamma-linolenic acid (GLA), di-homo-gamma-linolenic (DGLA), eicosapentaenoic acid (EPA), and docosahexaenoic acid (DHA) reveals where there may be an underlying imbalance directly related to what they eat. Developing a diet plan to *rid* intake of inflammatory foods and *get* foods that are anti-inflammatory is a desirable goal. (see ▶ Chap. 43). Key foods include therapeutic use of food oils that can modulate a person's fatty acid status and provide ability to control inflammation. Phytonutrients rich in the variety of colorful fruits and vegetables have powerful anti-inflammatory action. Some herbs have excellent evidence of anti-inflammatory support with successful practice-based experience.

> **Anti-inflammatory foods, herbs, and dietary supplements targeted recommendations based on patient assessment**
> **Foods**
> Whole-foods, pesticide-free, vegetable, and fruit variety of color, adequate protein, healthy fats and oils, foods-rich in herbal components, and hydration; minimize or avoid processed and high-sugar foods and beverages; avoidance of identified antigenic foods. (see ▶ Chap. 43)
> **Herbs**
> - Turmeric/curcumin
> - Resveratrol
> - Boswellia
> - Artemisinin
> - Garlic
> - Quercetin (suppresses mast cells)
> - Proteolytic enzymes: bromelain, papain, trypsin, etc. (contraindicated for alpha-1-antitrypsin deficiency+ genetics)
>
> **Dietary supplements**
> - Vitamin C (contraindicated for hemochromatosis+ genetics)
> - Adequate methyl nutrients
> - Vitamin D3 (if indicated per personalized assessment)
> - Vitamin A (if indicated per personalized assessment)

Infectious processes are biological stressors that alter the requirement of an individual metabolism beyond the RDA/RDI recommendations. Recognition of this principle drove the emergence of the field of orthomolecular nutritional therapy early in the twentieth century, leading to provision of the right nutrients at a molecular and cellular level. Under high-stress conditions, micronutrient requirements are altered [23]. For example, vitamin C requirements as a primary cellular antioxidant vary for an individual depending on the stress load, diet, gastrointestinal function, and genomics [24] as stated by Dr. Tim Spector, "there is a lot of variability in the ways in which healthy people react to food (and nutrients)" [25]. With the recognition of how unique and diverse individual physiologic immune responses are to food and lifestyle, the inclusion of a per-

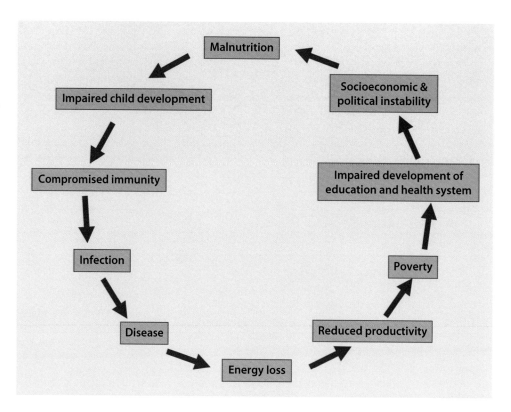

 Fig. 20.5 Protein energy malnutrition increases prevalence of infection, leading to energy loss for the individual [2]. (Reprinted from Schaible and Kaufmann [2]. With permission from Creative Commons License)

son's nutrition status as part of their differential exam for development of a comprehensive treatment plan is of utmost importance.

Over human history, many cultures have evolved food traditions to meet specific health needs such as pregnancy, or infection, or child development. Records from Hippocrates describe specific foods to treat various disease conditions. In the Middle Ages, eggs were soaked in vinegar to dissolve the eggshell, rich in calcium and minerals, to be given to a pregnant woman. Various herbal teas were given for various types of infections and many other conditions. Besides food being a source of nutrients, herbal and nutraceutical oral supplementation became prevalent in the eighteenth century. In 1831, the first intravenous (IV) technology was attempted for treatment of cholera by a Scottish doctor, Dr. Thomas Latta. It took another 100 years for further development of intravenous therapies and only became commonly available clinically by licensed practitioners in the 1960s. Once determined safe and clinically feasible, it was embraced by many medical specialties where patients were impacted by compromised gastrointestinal function and malnutrition [26]. Today, many hospital and clinic infusion centers administer IV vitamins, mineral and nutrient cocktails, and intramuscular (IM) nutrients to support nutritional status of individuals seeking prevention, or as prescribed due to compromised oral dietary intake. IV treatments can provide nutrients for individuals with increased nutrient requirements, especially with the biological stressors presented by acute or chronic infections.

One of the most life-threatening infectious conditions is severe sepsis [27] with no effective treatment options. In 2014, the results of one of the first clinical trials for IV ascorbic acid at Division of Pulmonary Disease and Critical Care Medicine, Department of Internal Medicine, School of Medicine, Virginia Commonwealth University were published.

The conclusion of the $n = 26$ human trials with severe sepsis and a variety of diagnoses of cancer, respiratory failure, and others was that infusion of intravenous ascorbic acid was safe and may positively influence patients when challenged with severe sepsis with multi-organ failure. The study showed improved lower biomarkers of inflammation, C-reactive protein (CRP), and procalcitonin (PCT) that correlate with the overall extent of infection. Higher levels of both have been biomarkers linked to higher incidences of organ injury and death in the critically ill. These two biomarkers proved accurate to assess effectiveness of the IV ascorbic acid. Thrombomodulin (TM) was the biomarker used to measure endothelial injury status and showed similar improvement Fig. 20.5.

20.10 Differential for Nutritional Infection Risk

20.10.1 Look for Evidence of Infections

Toolbox to identify infectious relationship with nutrition status:
- Nutrition physical exam (see ► Chap. 40)
- Medical history: diagnosis, medical event history, infectious history, residential location

- Signs and symptoms: medical symptoms questionnaire (MSQ)
- Laboratory and procedural testing (basic nutrition-related and, if applicable, for disease-specific and sometimes patient-specific markers)
- Bioelectrical impedance analysis (if available)
- Laboratory nutrition status and infection-related biomarkers (◘ Table 20.1 and ◘ Fig. 20.6).

The nutrients most studied regarding immunonutrition that should be considered in the assessment are listed in ◘ Table 20.2 and will be discussed below.

20.11 Chronic Diseases, Nutrition, Microbiome, and Infection

All chronic diseases have associations with acute or subclinical infections as potential etiologies for a chronic disease individual. Infections are caused by the exposure to such pathogens as virus, bacteria, parasites, fungi, or

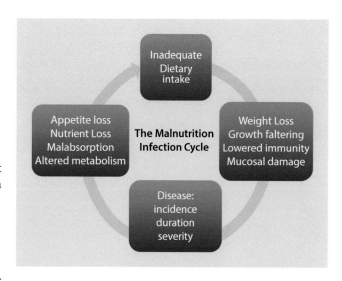

◘ **Fig. 20.6** The "vicious cycle" of malnutrition and infection. Spiral of malnutrition and infection [28]. (Adapted from Katona and Katona-Apte [28]. With permission from Oxford University Press)

prion. Many chronic diseases have known nutrient-microbiome-infection interactions [57]. When completing an initial assessment differential consideration of an infectious component of any patient presenting with a chronic disease, it should be part of an initial assessment differential to identify if infection could be part of the pathophysiology.

Malnutrition, altered microbiome, and infection interact to influence health and disease in the developed and developing world. Infectious morbidity is significant in the malnourished, whether nutrient insufficiencies or overnutrition. Infections significantly compromise utilization of oral nutrition and the immune lymphoid tissue in the gastrointestinal tract disturbing the microbiome. Malnutrition predisposes a person to infection, and restoring the injured nutritional status improves immune integrity. Improving nutritional status reduces risk of infection, and when one does contract infection, there may be a reduction in the severity of systems.

Nutritional assessment currently makes use of many new technological modalities. The Integrative and Functional Medical Nutrition Therapy assessment model identification of "root cause(s)" of a condition starts by hearing the patient's story (see ▶ Chap. 39). When was the last time they felt well? Family history? Signs and symptoms? Diet history? Medications? Supplements? Toxin exposure? Other issues? The model explores the question of how a person's metabolism evolved to the current disease condition.

20.11.1 Laboratory

With laboratory data to examine, a diagnostic profile begins to emerge to clarify the priorities of core physiological

◘ **Table 20.1** Immunonutrition assessment investigation tools

Test	Type	Practitioner options
Imaging	Diagnosis-related	Radiology
Functional testing	Digestive Hormonal Organic acids Structural	Functional medicine Endocrinology MD, DO, ND, LAc, RD, CNS, RN, NP, PA DC, DO, PT-classical
Mind-body	Psychology	MD, psychologist, family counselor therapist
Initial assessment and monitoring	Follow-up	Initial practitioner
	BIA	All; equipment required
	MSQ	All
	Monitoring abnormal test results	All
	Nutrition physical exam	All; more expanded with integrative and functional nutrition-trained practitioners
	Oxygen saturation monitor	All
	Thermometer	All; nutrition-trained practitioners will use temperature to consider thyroid functional status maintaining body temperature; rule out infection;

Table 20.2 Key immuno-nutrient insufficiencies/deficiencies associated with or causal for subclinical or acute infections; key immune-nutrients foundational for healthy immune endogenous defense

Nutrient	Infection: deficiency connection	Reference	Symptoms	Testing to consider
Immune modulators				
Vitamin D	Immune modulator	[29–31]	Mood disorders; bone loss; joint pain; compromised immunity	Vitamin D25OH
Vitamin A retinol	Mucosal immunity	[22, 24]	Mucosal bacterial/viral infections	Vitamin A, retinol; β-carotene
Phytonutrients polyphenols		[32, 33]		
Rate-limiting cofactors				
Folate	NK cell activity Cytotoxic cellular immunity Modulate T cell responses	[34]	Fatigue Infection Mood disorder	Folate, serum FIGLU
Zinc	Regulates intracellular signaling pathways in innate and adaptive immune cells	[34–36]	Skin conditions Poor smell Poor taste Frequent illness Nail spots	Zinc, serum Zinc, RBC
Magnesium	>300 enzymes cofactor Urine excretion under stress		One of the most deficient minerals in the USA Muscle tension Cramps	Magnesium, serum RBC magnesium
Iodine	Thyroid metabolism requirement	[37, 38] *UIC values* <20 μg/L (severe iodine deficiency) 20–49 μg/L (moderate iodine deficiency) 50–99 μg/L (mild iodine deficiency), 100–199 μg/L (adequate iodine intake) 200–299 μg/L (more than adequate iodine intake) >300 μg/L (excessive iodine intake)	Cysts Fibrocystic breasts Hypothyroid In utero developmental	Urinary iodine concentration (UIC) Iodine, random blood
Amino acids Arginine Lysine Glutamine	Nitric oxide (NO) Gut barrier support	Low grade evidence [39] Herpes Simplex [40]	Perturbed protein metabolism Poor healing Express herpes simplex skin rash Poor gut barrier repair	Plasma or urine amino acid profile
Antioxidants				
Vitamin C	Regulates cellular humoral immune function; increase macrophage Antihistamine requirements increase during infection Antiviral	[27]	Connective tissue impairment Skin conditions Poor wound healing	*Blood* Vitamin C CPT 82180

Table 20.2 (continued)

Nutrient	Infection: deficiency connection	Reference	Symptoms	Testing to consider
Vitamin E full-spectrum	4 tocopherols + 4 tocotrienols protective from oxidative stress/lipid peroxides	[29, 41]	Oxidative stress Premature wrinkles Cysts Leg cramps	Vitamin E (tocopherol) CPT 84446
Selenium	Thyroid peroxidase metabolism with vitamin E Selenoproteins special effects on cellular immunity Resistance to viral infections	[36, 41]	Hypothyroid Low glutathione blood levels Impaired detoxification	*Blood* Selenium, RBC Selenium CPT 84255
Gut secretions				
Short-chain fatty acids (SCFA)	Intestinal cells Anti-inflammatory Microbiome-gut-brain axis via immune system/vagal nerve Mucosal immunity	[42–47]	Inflammation Perturbed uric acid cycle Colon disease	Fecal collection
Bile acids	Fat emulsification duodenum Carrier of toxins/elimination	[48, 49]	Perturbed fat digestion Suppressed detoxification enterohepatic circulation	*Blood* Bile acids, total CPT 82239 or fecal collection
Anti-nutrients				
Endocrine disruptors	Circadian rhythm disruption that can modulate immune function	[50, 51]	Hormonal imbalance Insomnia Cancer	Blood or urine
Damaged food components/Western diet and food preparation Trans fat Oxidized fat Toxic metals Toxic chemicals "new-to-nature molecules" (cookware, pest control, pollutants, processed, etc.)	The damaged high-heat foods Chemicals Toxic metals ingested in foods and from food utensils during food preparation	[52]	Toxicity Symptoms vary	Blood or urine
Environmental toxins Mold/mycotoxins Natural gas carcinogenic Chemical vapors Pesticides	Neuropsychiatric immune disruption Vulnerable to infections increased risk for kidney failure, fertility problems, cancer, birth defects	[53–56]	Toxicities Chronic kidney failure Infertility Cancer Developmental Frequent infections	Specialty labs

imbalances for nutrition and lifestyle therapy. When focusing on assessing the existence of infectious activity, and the priority within the etiology of a disease condition, consideration of the three most likely physiological areas of immune imbalances are *defense and repair, assimilation, and structural integrity* (Fig. 20.7).

20.11.2 Assessment Laboratory and Clinical Tools

- Comprehensive Digestive Stool Analysis (CDSA) various labs provide CPT
- Ova and Parasitology (2–3 samples) CPT Code(s) 87177, 87209

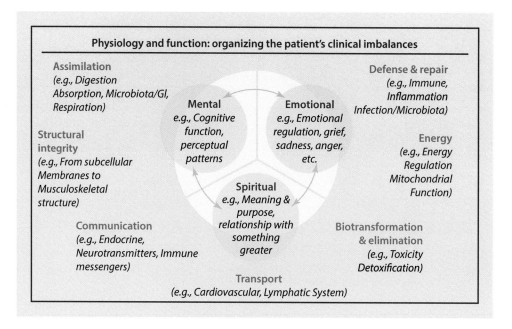

Fig. 20.7 IFM Matrix: physiology and function – organizing the patient's clinical imbalances [58]. (Used with permission from The Institute for Functional Medicine ©2015)

- Rule out other infections
- Calprotectin, fecal CPT 83993 [fecal calprotectin (FC)]
- *Clostridium difficile* toxin/GDH with reflex to PCR (if diarrhea)
- Lactoferrin, fecal CPT 83631: leukocyte marker; intestinal inflammation
- Differentiate IBD from irritable bowel syndrome (IBS)
- Monitor patients with IBD for treatment response and relapse
- Diagnose inflammatory bowel disease (IBD)
- Occult blood, fecal

Defense and Repair Immune, Inflammation, and Infection/Microbiota Assessment Biomarkers

Blood
- Vitamin D 25-hydroxy
  Vitamin 25-OH serum levels have been associated with several comorbidities, such as infectious, autoimmune and neurological diseases, as well as neuromuscular disorders, which can lead to increased pain sensitivity [59–61]. Regarding the mechanisms of pain sensitization, vitamin D seems to stimulate anti-inflammatory processes in some cases and thus to relieve the painful sensation of many diseases [62–64].

- Vitamin A retinol [65]
  Vitamin A retinol is increasingly recognized in experimental and human studies to suppress inflammatory reactions and plays a significant role in normal mucosal immunity, regulation of T cell-dependent responses, antiviral activity [66], and cell trafficking.
  Adequate vitamin A status, whether from intake of preformed retinol or β-carotene, is important for preventing excessive or prolonged inflammatory reactions and infectious events [66].

- high sensitivity CRP (hs-CRP)
  High-sensitivity C-reactive protein (hs-CRP) is an acute-phase-reactant marker of systemic inflammation most often promoted by bacterial infection, central adiposity, neoplastic activity, or traumatic injury. The ideal level of hs-CRP is ≤0.6. hs-CRP elevation implies potential bacterial infection. The most common infections with elevated hs-CRP are periodontitis or necrosis of the jawbone. If an elevated CRP >1.0 and clinical oral exam and report of bleeding upon flossing or brushing, appropriate referral to a biological dentist for evaluation is warranted. If no dental/oral infection is identified, further investigation as to the root cause of the elevated hs-CRP is warranted. It is important to rule out a recent traumatic injury that may be related, and hs-CRP should be retested in a month or two to observe if injury affected the hs-CRP.
- CBC with differential
- Complete metabolic panel (CMP)
- Lipid panel
- Sed rate
- TSH
- **Bacterial/viral evaluation**
  - CMV IgG Ab CPT 86644–0.5 ml red top serum
  - CMV IgM Ab CPT 86645–0.5 ml red top serum
  - Epstein-Barr virus (EBV) antibody panel ((IgM, VCA IgG, EBNA IgG) – CPT 86664, 86,665–1 ml red top serum
  - EBV early antigen D antibody (IgG) – CPT 86663–1 ml red top serum
  - *Chlamydophila pneumoniae* antibodies (IgG, IgA, IgM) – CPT 86631 86,632 1 ml red top serum
  - Mycoplasma IgG/IgM – CPT 8673–86,738 – 1 ml red top serum

- Herpesvirus 6 antibodies (IgG, IgM) – CPT 86790, 86,790 (×2)– 0.5 ml red top serum
- ASO CPT 86060–1 ml SST
- ANA W/RFX – CPT 86038–1 ml red top serum
- Immune function
  - Natural killer cells, functional – CPT 88184, 88,185– 10 ml (WB) green tube
- Toxin load
  - Heavy metals panel, blood – CPT 82175, 83,655, 83,825 – (WB) royal blue EDTA
    Includes: arsenic, lead, mercury
  - Cadmium, blood – CPT 82300 – (WB) royal blue EDTA trace element (REF)
- Fecal
  - Microbiology, fecal
  - Ova and parasitology
- Urine
  - Urinalysis (urine)
  - Organic acids (urine)
  - Complete hormone panel
- Saliva or blood Genomic testing
  - (saliva)

Structural integrity Membrane structure affects the function of transport and communication at the cell membrane site and receptors; review if history of head/neck/dentition/back injury may affect brainstem and vagal nerve immune-related function; structure, dysfunction occurs. Dietary intake of these lipid groups is reflected in their endogenous structure and function [67].

- Cell membrane fluidity (BIA phase angle, fatty acid status)
- Structural: spinal alignment
- Dental: periodontitis – infection of the tissues that surround the teeth
- Cervical C1–C7 – brainstem and vagal assessment – check vagal tone [68]
- Thoracic T1–T5 – stenosis or injury, increased pain resulting in exaggerated immune inflammatory response
- Lumbar L1–L5 – stenosis or injury increased pain resulting in exaggerated immune inflammatory response

20.11.3 Key Nutrients Influencing the Risk of Infectious Disease (◘ Fig. 20.8)

20.11.3.1 Vitamin D, A, E [69]
These fat-soluble vitamins have many metabolic roles, but for the focus on the immune system in this chapter, the role of immune modulation is discussed. The fat-soluble vitamins function synergistically; even the vitamin D and A receptors share the RXR nuclear receptor influencing each other. It is

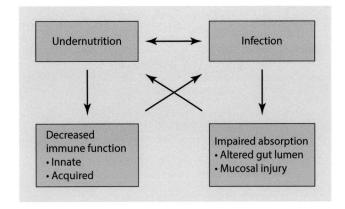

◘ Fig. 20.8 Interactions between malnutrition and infection [28]. (Adapted from Katona and Katona-Apte [28]. With permission from Oxford University Press)

worth noting in nature that vitamins D2/3 and A are found in their food-rich sources together (e.g., liver, caviar/roe, egg yolk, etc.).

Lipids Phospholipids, oils, and fat foods – RBC fatty acid, lipid panel tests. The phospholipids, sterols, and eicosanoid fatty acids and their metabolites give structural and functional influences on cell signaling and prostaglandin "hormone-like" regulation to transport of components in and out of the cell compartments to nourish and regulate immune response of inflammation, hormonal modulation, and other undiscovered functions. When there is poor structure, poor function follows (see ► Chap. 10).

Methyl nutrients Vitamins B6, folate, B12, riboflavin, betaine, biotin, choline, SAMe, and amino acids methionine, cysteine, serine, and glycine [70, 71].

Methyl nutrients support the process of methylation with many roles within human metabolism, with DNA methylation being the underlying mechanism, and currently appear to be the primary messengers of epigenetic expression (see ► Chap. 18) identified in the etiology of developing cardiovascular, cancer, and neurological disease conditions. Methyl nutrients include vitamins (folate, riboflavin, vitamin B12, vitamin B6, choline) and amino acids (methionine, cysteine, serine, glycine).

Methylation involves biochemical pathways where the B vitamins and other cofactors like amino acids are rate-limiting cofactors.

20.12 Homocysteine Catabolism (◘ Fig. 20.9)

20.12.1 Folate Metabolism

Folate is critical to many metabolic pathways like nucleic acid precursors, several amino acids, and erythropoiesis, the process in which new erythrocytes are produced. Elevated mean

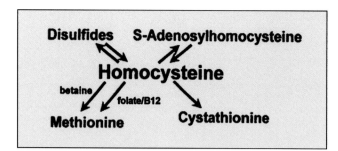

□ **Fig. 20.9** Homocysteine major metabolic pathways in humans [72]. (Reprinted from Dudman et al. [72]. With permission from Oxford University Press)

corpuscular volume (MCV) on a complete blood count can suggest folate deficiency. Folate deficiency can be part of the etiology of enlarged RBC, or megaloblastic anemia, from ineffective erythropoiesis. Vitamin B6 and B12 are cofactors also involved in erythropoiesis [73].

20.12.2 Methionine Metabolism

Methionine metabolism occurs predominantly in the liver tissue with two components: a transsulfuration pathway, involving homocysteine reduction to glutathione, and a transmethylation cycle with folate and methyl nutrients producing *S*-adenosylmethionine (SAMe). Thus, methionine metabolism is dependent on dietary intake of vitamins B12, B6, and folate. SAMe is key in regulating epigenetic expressions of multiple pathways and, when deficient, leads to ramifications of nutritional and immune injury [74] (see ► Chap. 17).

20.12.2.1 Phytonutrients
Inflammation: powerful pigment-rich polyphenols found in a variety of fruits, vegetables, grains, nuts, teas, herbal spices, and legumes have anti-inflammatory properties. Plant chemicals include antioxidants and antibacterial and antiviral mechanisms. Even though phytonutrients are not considered essential nutrients, the evidence is mounting of their critical role health maintenance and anti-aging.

Biomarkers of phytonutrient status: poor dietary intake of high polyphenol foods. Significant biomarkers for inflammation can be related to poor vegetable and fruit intake and lack of or imbalance in dietary intake of healthy fats and oils (see ► Chap. 57).

Resource: The Rainbow Diet. Color Can Heal Your Life [75]

20.12.2.2 Minerals: [76]
Zinc   critical role in the function of immune cells

Potassium   principal intracellular cytosol electrolyte

Iodine   thyroid hormone structure, brain development

Selenium   thyroid peroxidase metabolism with vitamin E, selenoproteins special effects on cellular immunity resistance to viral infections. Central to glutathione peroxidase structure

Iron   weaken cell-mediated immunity; decreases in neutrophil action

Magnesium   co-factor in >300 enzymes affecting all systems

20.13 Key Lifestyle Factors Influencing the Risk of Infectious Disease

20.13.1 Sleep (see ► Chap. 45)

Sleep and circadian rhythm have a great influence on the integrity of the immune system. Much evidence has accrued over the past decades associating poor sleep quantity and quality with weakening of the immune system, increasing vulnerability to infection.

20.13.2 Stress (see ► Chap. 47)

Chronic stress impacts every biological and psychological system. The chemical microenvironment under long-term stress pushes the immune system response into chronic inflammation. The vicious cycle continues until the threshold of resilience and adaptation is exceeded, leading to vulnerability to many chronic diseases including infection.

20.14 Movement (see ► Chap. 36)

20.14.1 Examples of Chronic Disease Connections with Infectious Disease

20.14.1.1 Heart Disease/Cardiovascular Association with Infectious Processes
- *Infectious risk*: *Chlamydia pneumonia*, group A *Streptococcus* [77].
- *Dental*: periodontitis, necrosis of the jaw.
- *Rheumatic heart disease*: [78].
- *Bacterial endocarditis* (cardiac inflammation and scarring triggered by an autoimmune reaction to infection with group A *Streptococci*).
- *Pancarditis* (involving inflammation of the myocardium, endocardium, and epicardium); follows pharyngitis infection without antibiotic treatment.
- *Rheumatic chronic disease*: mitral valve stenosis almost always originates from rheumatic conditions and expresses and perfusion insufficiency. Rheumatic conditions are highly associated with underlying infection [79]. A previous infective endocarditis should be ruled out if there are symptoms of unexplained fever, malaise, or weight loss [80].

20.14.1.2 Oncology
Etiology of cancer can be related to infectious disease.

Viral   EBV, CMV. Herpes simplex, herpes zoster, HIV, HPV

Bacterial   *H. pylori, Mycoplasma pneumoniae*

Fungal   Candida, mycotoxins

Mycoplasma   *M. pneumonia, M. genitalium, M. hominis, ureaplasma urealyticum, U. parvum*

Parasites   trichinosis [81], blastocystis hominis, tropical parasites

20.14.1.3 Neurological

Alzheimer's disease   Viruses of life-long carriage are typically asymptomatic, strongly associated immunologic, and virologic characteristics with Alzheimer disease neuropathology increasing amyloid plaque and neurofibrillary tangles (NFTs): herpes simplex virus-1 (HSV-1), long-term cytomegalovirus (CMV) [82].

Developmental plasticity   Due to fetal and early childhood metabolic plasticity influences by the environment, negative toxic exposures and infectious processes in utero or in childhood can influence the risk of later chronic adverse conditions, especially noncommunicable disease (NCD). This fact has driven the public health drive for vaccination programs during the first year of birth [83]. The importance of maternal health nutrition status and free from infectious disease lays the foundation for fetal growth. Developmental epigenetics studies provide insight into the importance of epigenetic marks occurring in utero and recognition of new biomarkers to provide interventions for prevention and treatment [83].

20.14.1.4 Respiratory (see ▶ Chap. 52)

Acute respiratory infection can derive from viral, bacterial, or mold/mycotoxin exposure. The risk for an individual can be related to exposure environment and genotype (e.g., cystic fibrosis, alpha-1-antitrypsin, Wegener's granulomatosis). More extensive discussion on respiratory conditions and infection is presented in ▶ Chap. 52.

20.14.1.5 Autoimmune (see ▶ Chap. 49)

Autoimmune conditions are inflammatory. Ongoing research has identified genetic relationships with susceptibility to developing autoimmune conditions, but coexisting infections can contribute to the etiology of an individual's disease. The human leukocyte antigen (HLA) group of genes is highly associated with risk of developing autoimmunity. HLA DQ2 and HLA DQ8 reside within the HLA gene group, and are known risks for developing celiac disease [84].

> **Comprehensive Human Leukocyte Antigen Panel**
> HLA DR1/3/4/5, DQ Intermediate Resolution CPT 81375
> LabCorp Specialty Labs

20.14.2 Vaccination

During the twentieth century, the discovery and use of antibiotics began the breakdown of natural microbiota throughout the gastrointestinal tract and other microbiome-containing orifices. It is recognized at the time of this publication that the current epidemic of "superbugs" has resulted from overuse and misuse of antibiotics and antimicrobials. After the discovery of antibiotics, more pharmaceuticals were discovered to be providing strong manipulation of body systems and perturbing nutrient functions, such as steroid medications weakening connective tissue resulting in the increased need for vitamin C, biotin, and zinc. In addition, after World War II, there was the introduction of many "new-to-nature" molecules to agricultural practices and as additives to the food supply [85]. All of these new introductions into the human environment have altered the immune system, contributing to weakening defense systems. The twenty-first century is introducing even more organic and inorganic molecules that need more extensive investigation about their safety.

Two examples of concern:

1. *Fecal microbiota for transplantation* (FMT) implanting foreign bacteria from a donor into the gastrointestinal tract of a patient. The US Food and Drug Administration (FDA) has not officially approved FMT procedures but supports the area of scientific discovery. In 2019, implants performed for two immune-compromised patients from a single donor contained drug-resistant organisms. One of the patients died. Clinical trials were suspended and the FDA has warned all fecal matter for FMT should be tested for drug-resistant bacteria [86, 87]. This event illustrates the potential risk of FMT as a source of life-threatening infections.

2. *Vaccinations* have strong pro and con public opinion, but all sides agree in vaccine safety monitored by the US FDA and the *Centers for Disease Control and Prevention* (CDC). The principle of vaccination is based on the body's healthy adaptive immune response when exposed to a pathogen to develop protective antibodies to the pathogen. The most prominent concerns about vaccine safety are the additives to the vaccine preparations for preservation and effectiveness, the age at which vaccine is administered, and the number of vaccines given simultaneously. The most commonly used preservative adjuvants are aluminum, mercury-containing thimerosal [88], and formaldehyde. These compounds have known adverse effects. In 2011, the *International Agency for Research on Cancer* (IARC) named formaldehyde "a known human carcinogen" [89]. The US FDA has a reporting site for *Vaccine Adverse Event Reporting System* (VAERS) and *Wide-ranging Online Data for Epidemiologic Research* (WONDER) to provide public health information [90].

Thimerosal contains mercury, of which the *World Health Organization* (WHO) says that exposure, even in small amounts, "may cause serious health problems and is a threat to the development of the child in utero and early in life." "Mercury is considered by WHO as one of the top ten chemicals or groups of chemicals of major public health concern" [91]. Aluminum is a known neurotoxin and can play a significant role in neurological diseases. Reported elevated levels of aluminum found in Alzheimer's patient brains increases public concern. There is no evidence of harm for aluminum content in single vaccines, but concern for the accumulation of total aluminum in multiple-dose vaccine vials is not known.

The use of vaccines has become a highly debated political issue with states overhauling fundamental changes in their vaccine laws toward mandatory vaccination. Integrative and functional medical practitioners tend to embrace "first do no harm" and at least recommend parents base their decisions with full knowledge of the pros and cons of vaccinations for their individual child. There is no clear evidence of why some children and adults have mild to life-threatening side effects after receiving a vaccine.

If there is a suspected reaction to a vaccine, the *National Vaccine Injury Compensation Program* (NVICP) exists to provide financial compensation to individuals who have documented injury [92].

General considerations for parents that may increase vaccine safety for a child:
- Is my child sick the day of a scheduled vaccine? If yes, best to reschedule.
- Has my child had a reaction to any previous vaccination?
- Do our family or I have a history of vaccine reactions, neurological disorders, or immune system conditions like Sjogren's, lupus, celiac, eczema, etc.? If yes, document your child's personal and family history [90].
- What are the vaccination laws in the state in which I live?

If agreeing to vaccination, recommend limiting vaccinations to one vaccine administration at a time instead of multiple vaccines. Single-use vials of vaccines, if available, can be considered to reduce exposure to preservatives like thimerasol (mercury-containing) and formaldehyde.

Genomic counselors can be sought to discuss currently identified *single-nucleotide polymorphisms* (SNPs) that may be associated with increased risk of vaccine reaction.

Ensure your child is in good nutrition status, eating a balanced whole-food, low-pesticide, fruit- and vegetable-rich, low-sugar diet with adequate intake of vitamins D, A, and C, essential fatty acids, and bioactive forms of B vitamins from food or supplements if needed.

The US FDA has a reporting site for *Vaccine Adverse Event Reporting System* (VAERS) and *Wide-ranging Online Data for Epidemiologic Research* (WONDER) to provide public health information [90].

> Several reports have shown that vitamin A deficiency results in a poor response to immunization, with generally low antibody responses to immunization with T cell-dependent antigens [93, 94].

20.15 Conclusion

This chapter reviews the importance of considering infection as a potential contributor to the etiology of any of the chronic diseases.

Infection can contribute to each type of chronic disease. Growing evidence is emerging that infections are most damaging to tissues and metabolism when they have continued as a prolonged burden on the immune system. Pathogens are generally attracted to specific tissue types and the disease that may develop for a unique individual may vary. Chronic diseases are characterized as long-latency, lifestyle, and diet-related diseases. An acute infection may either be resolved by a healthy immune system or survive and continue as a subclinical infection that is often not recognized but continues to wear on the immune system. Examples of commonly recognized subclinical infections that lead to increasing risk of a serious chronic disease: HPV risk of cervical or head/neck cancers, hepatitis C risk of liver cancer, Epstein-Barr virus risk of non-Hodgkin's lymphoma and some autoimmune conditions, and *C. pneumoniae* implicated in chronic illnesses, such as atherosclerosis, asthma, arthritis, multiple sclerosis, and many others. This chapter has described the association between a person's nutritional status and their vulnerability to infection. Individual nutritional status is a determinant of how well their body can respond to and resolve an infection, returning it to a state of wellness. The risk to get or not resolve an infection is greater from a compromised immune system when nutrient tissue levels are not optimized for an individual or able to provide adequate nutrient metabolic cofactors. Healthy nutritional status decreases risk of chronic diseases and vulnerability for long-latency prolonged infection [1, 3].

The healthy human body is equipped with defense features from conception throughout life to interact with the environment to protect from infection. Much of the defense starts with the skin barrier and microbiome at all body orifices to protect from pathogens entering and infecting. Another key is keeping the stomach acid-neutralizing pathogens or toxic organics from traveling further down the gastrointestinal tract where they could cause havoc and alter the powerful GI microbiome. These defenses guard pathogens and toxins from entering the blood and/or lymph circulation. Along with the internal milieu of the body, ethnic cultures have developed mores to support immune defense by traditions of toileting, diet, sleeping, sanitation (especially for food preparation and handwashing), and other complementary routines that minimize infection.

A growing recognition of the evidence of the pathophysiology of chronic disease reveals its contrast to acute disease by the long-term development often asymptomatic and not recognized. Subclinical infections can "smolder" unrecognized for many years, even beginning in utero and childhood, a stress on the immune system that often goes undiagnosed or there is unawareness of their significance. This chapter provides knowledge of assessment principles and specific immune-support nutrients and herbal components that can intervene if infections are identified. The increased clarity of scientific evidence supports "you and your genes are what you eat, digest, eliminate, sleep, move, avoid significant toxic exposure and live in a healthy relationship community" to live a long life that is vital and resilient and functional [95].

References

1. Calder PC, Kulkarni AD, editors. Nutrition, immunity and infection: Boca Raton: CRC Press; 2018.
2. Schaible UE, Kaufmann SHE. Malnutrition and infection: complex mechanisms and global impacts. PLoS Med. 2007;4(5):e115. https://doi.org/10.1371/journal.pmed.0040115.
3. Felblinger DM. Malnutrition, infection, and sepsis in acute and chronic illness. Crit Care Nurs Clin North Am. 2003;15(1):71–8.
4. Tackling AMR 2019-2024, the UK's five-year national plan. 24 Jan 2019. https://assets.publishing.service.gov.uk/government/uploads/system/uploads/attachment_data/file/773130/uk-amr-5-year-national-action-plan.pdf.
5. Gershon M. The second brain: a groundbreaking new understanding of nervous disorders of the stomach and intestine. Harper Perennial. 2019;
6. MEM O. Leaky Gut, Leaky Brain? Microorganisms. 2018;6(4):107. Published 2018 Oct 18. https://doi.org/10.3390/microorganisms6040107.
7. Grotto D, Zied E. The standard American diet and its relationship to the health status of Americans. Nutr Clin Pract. 2010;25(6)
8. Agarwal S, et al. Comparison of prevalence of inadequate nutrient intake based on body weight status of adults in the United States: an analysis of NHANES 2001–2008. J Am Coll Nutr. 2015;34(2):1–9.
9. Prevalent nutrient deficiencies in the US: more than 40% are Vitamin A, C, D & E, Calcium or Magnesium Deficient and >90% Don't Get Enough Choline, Fiber & Potassium. https://suppversity.blogspot.com/2015/01/prevalent-nutrient-deficiencies-in-us.html. Friday, January 16, 2015. Accessed June 10, 2019.
10. Tan SY, Tatsumura Y. Alexander Fleming (1881-1955): discoverer of penicillin. Singap Med J. 2015;56(7):366–7. https://doi.org/10.11622/smedj.2015105.
11. Bresnahan KA, Tanumihardjo SA. Undernutrition, the acute phase response to infection, and its effects on micronutrient status indicators. Adv Nutr. 2014;5(6):702–11. Published 2014 Nov 3. https://doi.org/10.3945/an.114.006361.
12. Burger EA Kim JJ, Sy S, Castle PE. Age of acquiring causal human papillomavirus (HPV) infections: leveraging simulation models to explore the natural history of HPV-induced cervical cancer. Clin Infect Dis. 2017;65(6):893–9. https://doi.org/10.1093/cid/cix475.
13. Kim SS, Ruiz VE, Carroll JD, Moss SF. *Helicobacter pylori* in the pathogenesis of gastric cancer and gastric lymphoma. Cancer Lett. 2011;305(2):228–38. https://doi.org/10.1016/j.canlet.2010.07.014.
14. Harris SA, Harris EA. Molecular mechanisms for herpes simplex virus type 1 pathogenesis in Alzheimer's disease. Front Aging Neurosci. 2018;10(48) https://doi.org/10.3389/fnagi.2018.00048.
15. Weber RG, Ansell BF. A report of a case of Klebsiella Pneumoniae arthritis and a review of extrapulmonary Klebsiella infections. Ann Intern Med. 1962;57(2_Part_1):281–9. https://doi.org/10.7326/0003-4819-57-2-281.
16. Zhang L, Zhang YJ, Chen J, Huang CX, et al. The association of HLA-B27 and *Klebsiella pneumoniae* in ankylosing spondylitis: a systematic review. Microb Pathog. 2018;117:49–54.
17. Castello JJ, Beltran BE, Miranda RN, Young KH, Chevez JC, Sotomayor EM. EBV-positive diffuse large B-cell lymphoma of the elderly: 2016 update on diagnosis, risk-stratification, and management. Hematology. 2016;91(5):529–37.
18. El Yazouli L, Criscuolo A, Hejaji H, Bouazza M, Elmdaghri N, Alami AA, Amraoui A, Dakka N, Radouani F. Molecular characterisation of *Chlamydia pneumoniae* associated to atherosclerosis. Pathog Dis. 2017;75(4):ftx039. https://doi.org/10.1093/femspd/ftx039.
19. Levy BD. Resolvins and protectins: natural pharmacophores for resolution biology. Prostaglandins Leukot Essent Fatty Acids. 2010;82(4–6):327–32. https://doi.org/10.1016/j.plefa.2010.02.003.
20. Schwenk A, Macallan DC. Tuberculosis, malnutrition and wasting. Curr Opin Clin Nutr Metab Care. 2000;3(4):285–91.
21. Global Health Policy 2019. http://events.r20.constantcontact.com/register/event?llr=8ogpai7ab&oeidk=a07efzxay49ae9e68e6. Accessed 1 June 2019.
22. Aibana O, Franke MF, Huang CC, et al. Impact of Vitamin A and carotenoids on the risk of tuberculosis progression. Clin Infect Dis. 2017;65(6):900–9. https://doi.org/10.1093/cid/cix476.
23. Orthomolecular Nutrition – Bing. Bing.Com, 2015., www.bing.com/search?q=orthomolecular%20nutrition&pc=cosp&ptag=C1N1234D010118A316A5D3C6E&form=CONBDF&conlogo=CT3210127&toHttps=1&redig=2A828873AF9D41E98AEE740BD5817EB6. Accessed 6.2.2019. Accessed 19 June 2019.
24. Stephensen CB, Vitamin A. Infection, and immune function. Annu Rev Nutr. 2001;21:167–92.
25. Spector T. Understand your body's unique responses to food. Predict2. Retrieved from: https://predict.study/. Accessed 2 June 2019.
26. Bistrian BR. Brief history of parenteral and enteral nutrition in the hospital in the USA. Clinical Nutrition, Beth Israel Deaconess Medical Center, Harvard Medical School, Boston, MA, USA.
27. Fowler AA 3rd, Syed AA, Knowlson S, et al. Phase I safety trial of intravenous ascorbic acid in patients with severe sepsis. J Transl Med. 2014;12:32. Published 2014 Jan 31. https://doi.org/10.1186/1479-5876-12-32.
28. Katona P, Katona-Apte J. The interaction between nutrition and infection. Clin Infect Dis. 2008;46(10):1582–8. https://doi.org/10.1086/587658.
29. Gough ME, Graviss EA, Chen T, Obasi EM, May EE. Compounding effect of vitamin D_3 diet, supplementation, and alcohol exposure on macrophage response to *Mycobacterium* infection. Tuberculosis. Available online 30 April 2019.
30. Esteve Palau E, Sánchez Martínez F, Knobel Freud H, López Colomés J-L, Diez Pérez A. Tuberculosis: plasma levels of vitamin D and its relation with infection and disease. Med Clín (English Edition). 2015;144(3):111–4.
31. Keflie TS, Nölle N, Lambert C, Nohr D, Biesalski HK. Vitamin D deficiencies among tuberculosis patients in Africa: a systematic review. Nutrition. 2015;31(10):1204–12.
32. Ghiringhelli F, Rebe C, Hichami A, Delmas D. Immunomodulation and anti-inflammatory roles of polyphenols as anticancer agents. Anti Cancer Agents Med Chem. 2012;12:852. https://doi.org/10.2174/187152012802650048.
33. Ding S, Jiang H, Fang J. Regulation of immune function by polyphenols. J Immunol Res. 2018;2018:1264074. Published 2018 Apr 12. https://doi.org/10.1155/2018/1264074.
34. Elmadfa I, Meyer AL. The role of the status of selected micronutrients in shaping the immune function. Endocr Metab Immune Disord Drug Targets. 2019;19(1) https://doi.org/10.2174/1871530319666190529101816.

35. Wu D, Lewis ED, Pae M, Meydani SN. Nutritional modulation of immune function: analysis of evidence, mechanisms, and clinical relevance. Front Immunol. 2019;9:3160. Published 2019 Jan 15. https://doi.org/10.3389/fimmu.2018.03160.
36. Maywald M, Wessels I, Rink L. Zinc signals and immunity. Int J Mol Sci. 2017;18:E2222. https://doi.org/10.3390/ijms18102222.
37. Mendieta I, Nuñez-Anita RE, Nava-Villalba M, et al. Molecular iodine exerts antineoplastic effects by diminishing proliferation and invasive potential and activating the immune response in mammary cancer xenografts. BMC Cancer. 2019;19(1):261. Published 2019 Mar 22. https://doi.org/10.1186/s12885-019-5437-3.
38. Xiu L, Zhong G, Ma X. Urinary iodine concentration (UIC) could be a promising biomarker for predicting goiter among school-age children: A systematic review and meta-analysis. [published correction appears in PLoS One. 2017 Jul 7;12 (7):e0181286]. PLoS One. 2017;12(3):e0174095. Published 2017 Mar 22. https://doi.org/10.1371/journal.pone.0174095.
39. Wijnands KA, Castermans TM, Hommen MP, Meesters DM, Poeze M. Arginine and citrulline and the immune response in sepsis. Nutrients. 2015;7(3):1426–63. Published 2015 Feb 18. https://doi.org/10.3390/nu7031426.
40. Cruzat V, Macedo Rogero M, Noel Keane K, Curi R, Newsholme P. Glutamine: metabolism and immune function, supplementation and clinical translation. Nutrients. 2018;10(11):1564. Published 2018 Oct 23. https://doi.org/10.3390/nu10111564.
41. Chan JM, Darke AK, Penney KL, et al. Selenium- or Vitamin E-related gene variants, interaction with supplementation, and risk of high-grade prostate cancer in SELECT. Cancer Epidemiol Biomark Prev. 2016;25(7):1050–8. https://doi.org/10.1158/1055-9965.EPI-16-0104.
42. Ohira H, Tsutsui W, Fujioka Y. Are short chain fatty acids in gut microbiota defensive players for inflammation and atherosclerosis? J Atheroscler Thromb. 2017;24(7):660–72. https://doi.org/10.5551/jat.RV17006.
43. Ratajczak W, Ryl A, Mizerski A, Walczakiewicz K, Sipak O, Laszczyńska MA. Immunomodulatory potential of gut microbiome-derived short-chain fatty acids (SCFAs). Acta Biochim Pol. 2019;66(1):1–12. https://doi.org/10.18388/abp.2018_2648.
44. Stilling RM, de Wouw M, Clarke G, Stanton C, Dinan TG, Cryan JF. The neuropharmacology of butyrate: the bread and butter of the microbiota-gut-brain axis? Neurochem Int. 2016;99:110–32. https://doi.org/10.1016/j.neuint.2016.06.011.
45. Bourassa MW, Alim I, Bultman SJ, Ratan RR. Butyrate, neuroepigenetics and the gut microbiome: Can a high fiber diet improve brain health? Neurosci Lett. 2016;625:56–63. https://doi.org/10.1016/j.neulet.2016.02.009.
46. Szentirmai É, Millican NS, Massie AR, Kapás L. Butyrate, a metabolite of intestinal bacteria, enhances sleep. Sci Rep. 2019;9(1):7035. Published 2019 May 7. https://doi.org/10.1038/s41598-019-43502-1.
47. Liu H, Wang J, He T, et al. Butyrate: a double-edged sword for health? Adv Nutr. 2018;9(1):21–9. https://doi.org/10.1093/advances/nmx009.
48. Biagioli M, Carino A. Signaling from intestine to the host: how bile acids regulate intestinal and liver immunity. Handb Exp Pharmacol. 2019; https://doi.org/10.1007/164_2019_225. [Epub ahead of print].
49. Chiang JYL, Ferrell JM. Bile acids as metabolic regulators and nutrient sensors. Annu Rev Nutr. 2019;39:1.
50. Bansal A, Henao-Mejia J, Simmons RA. Immune system: an emerging player in mediating effects of endocrine disruptors on metabolic health. Endocrinology. 159(1):32–45. https://doi.org/10.1210/en.2017-00882.
51. Papalou O, Kandaraki EA, Papadakis G, Diamanti-Kandarakis E. Endocrine disrupting chemicals: an occult mediator of metabolic disease. Front Endocrinol (Lausanne). 2019;10:112. Published 2019 Mar 1. https://doi.org/10.3389/fendo.2019.00112.
52. Statovci D, Aguilera M, MacSharry J, Melgar S. The impact of western diet and nutrients on the microbiota and immune response at mucosal interfaces. Front Immunol. 2017;8:838. Published 2017 Jul 28. https://doi.org/10.3389/fimmu.2017.00838.
53. Ratnaseelan AM, et al. Effects of mycotoxins on neuropsychiatric symptoms and immune processes. Clin Ther. 40(6):903–17.
54. Akbari P, Braber S, Varasteh S, Alizadeh A, Garssen J, Fink-Gremmels J. The intestinal barrier as an emerging target in the toxicological assessment of mycotoxins. Arch Toxicol. 2017;91(3):1007–29. https://doi.org/10.1007/s00204-016-1794-8.
55. Elliott EG, Trinh P, Ma X, Leaderer BP, Ward MH, Deziel NC. Unconventional oil and gas development and risk of childhood leukemia: assessing the evidence. Sci Total Environ. 2017;576:138–47. https://doi.org/10.1016/j.scitotenv.2016.10.072.
56. Rull RP, Ritz B, Shaw GM. Neural tube defects and maternal residential proximity to agricultural pesticide applications. Am J Epidemiol. 2006;163(8):743–53. https://doi.org/10.1093/aje/kwj101.
57. Hand TW, Vujkovic-Cvijin I, Ridaura VK, Belkaid Y. Linking the microbiota, chronic disease, and the immune system. Trends Endocrinol Metab. 2016;27(12):831–43. https://doi.org/10.1016/j.tem.2016.08.003.
58. IFM Matrix™ (2015) The Institute for Functional Medicine. www.functionalmedicine.org.
59. Dhesi JK, Bearne LM, Moniz C, Hurley MV, Jackson SH, Swift CG, Allain TJ. Neuromuscular and psychomotor function in elderly subjects who fall and the relationship with vitamin D status. J Bone Miner Res. 2002;17:891–7. https://doi.org/10.1359/jbmr.2002.17.5.891.
60. Orme RP, Bhangal MS, Fricker RA. Calcitriol imparts neuroprotection in vitro to midbrain dopaminergic neurons by upregulating GDNF in vitro to midbrain dopaminergic neurons by upregulating GDNF expression. PLoS ONE. 2013;8:e62040. https://doi.org/10.1371/journal.pone.0062040. Osunkwo I, Hodgman EI, Ch.
61. Lachmann R, Bevan MA, Kim S, Patel N, Hawrylowicz C, Vyakarnam A, Peters BS. A comparative phase 1 clinical trial to identify anti-infective mechanisms of vitamin D in people with HIV infection. AIDS. 2015;29:1127–35. https://doi.org/10.1097/QAD.0000000000000666.
62. Adorini L, Penna G. Control of autoimmune diseases by the vitamin D endocrine system. Nat Clin Pract Rheumatol. 2008;4:404–12. https://doi.org/10.1038/ncprheum0855.
63. Osunkwo I, Hodgman EI, Cherry K, Dampier C, Eckman J, Ziegler TR, Ofori-Acquah S, Tangpricha V. Vitamin D deficiency and chronic pain in sickle cell disease. Br J Haematol. 2011;153:538–40. https://doi.org/10.1111/j.1365-2141.2010.08458.x.
64. Le Goaziou MF, Kellou N, Flori M, Perdrix C, Dupraz C, Bodier E, Souweine G. Vitamin D supplementation for diffuse musculoskeletal pain: results of a before-and-after study. Eur J Gen Pract. 2014;20:3–9. https://doi.org/10.3109/13814788.2013.825769.
65. Ross AC. Vitamin A and retinoic acid in T cell-related immunity. Am J Clin Nutr. 2012;96(5):1166S–72S. https://doi.org/10.3945/ajcn.112.034637.
66. Elenius V, Palomares O, Waris M, Turunen R, Puhakka T, Rückert B, et al. The relationship of serum vitamins A, D, E and LL-37 levels with allergic status, tonsillar virus detection and immune response. PLoS One. 2017;12 https://doi.org/10.1371/journal.pone.0172350.
67. Pakiet A, Kobiela J, Stepnowski P, Sledzinski T, Mika A. Changes in lipids composition and metabolism in colorectal cancer: a review. Lipids Health Dis. 2019;18(1):29. Published 2019 Jan 26. https://doi.org/10.1186/s12944-019-0977-8.
68. Komegae EN, Farmer DGS, Brooks VL, McKinley MJ, McAllen RM. Vagal afferent activation suppresses systemic inflammation via the splanchnic anti-inflammatory pathway. Brain Behav Immun. 2018;73:441–9. https://doi.org/10.1016/j.bbi.2018.06.005.
69. de Oliveira DL, Hirotsu C, Tufik S, Andersen ML. The interfaces between vitamin D, sleep and pain. J Endocrinol. 2017;234(1):R23–36. https://doi.org/10.1530/JOE-16-0514. Epub 2017 May 23.
70. Liu G, Bin P, Wang T, Ren W, Zhong J, Liang J, Hu CAA, Zeng Z, Yin Y. DNA methylation and the potential role of methyl-containing nutrients in cardiovascular diseases. Oxidative Med Cell Longev. 2017;2017(1)

71. Neidhart M. Methyl donors, DNA methylation and complex human disease; 2016. p. 429–39. https://doi.org/10.1016/B978-0-12-420194-1.00027-0.
72. Dudman NPB, Guo XW, Gordon RB, Dawson PA, Wilcken DEL. Human homocysteine catabolism: three major pathways and their relevance to development of arterial occlusive disease. J Nutr. 1996;126(suppl_4):1295S–300S. https://doi.org/10.1093/jn/126.suppl_4.1295S.
73. New insights into erythropoiesis: The roles of folate, vitamin B_{12}, and iron. Annu Rev Nutr. 24:105–31. https://doi.org/10.1146/annurev.nutr.24.012003.132306.
74. Halsted CH, Medici V. Vitamin-dependent methionine metabolism and alcoholic liver disease. Adv Nutr. 2(5):421–7. https://doi.org/10.3945/an.111.000661.
75. Minich D. The rainbow diet. Color can heal your life! San Francisco: Conari Press; 2018.
76. Selected Micronutrients in Shaping the Immune Function. Endocr Metab Immune Disord Drug Targets. 2019; https://doi.org/10.2174/1871530319666190529101816.
77. McCarthy M. Superbugs The race to stop an epidemic. Avery; 1st ed; 2019. p. 304.
78. Allen Patrick Burke, MD. Pathology of rheumatic heart disease. TheHeart.org. Updated: Oct 15, 2015. Available from: https://emedicine.medscape.com/article/1962779-overview.
79. Boon NA, Bloomfield P. The medical management of valvar heart disease. Heart. 2002;87(4):395–400. https://doi.org/10.1136/heart.87.4.395.
80. Webb J, Arden C, Chambers JB. Heart valve disease in general practice: a clinical overview. Br J Gen Pract. 2015;65(632):e204–6. https://doi.org/10.3399/bjgp15X684217.
81. Velikyan I. Prospective of ^{68}Ga Radionuclide Contribution to the Development of Imaging Agents for Infection and Inflammation. Contrast Media Mol Imaging. 2018;2018:9713691. Published 2018 Jan 4. https://doi.org/10.1155/2018/9713691.
82. Lurain NS, Hanson BA, Martinson J, et al. Virological and immunological characteristics of human cytomegalovirus infection associated with Alzheimer disease. J Infect Dis. 2013;208(4):564–72. https://doi.org/10.1093/infdis/jit210.
83. Hanson MA, Gluckman PD. Early developmental conditioning of later health and disease: physiology or pathophysiology? Physiol Rev. 2014;94(4):1027–76. https://doi.org/10.1152/physrev.00029.2013.
84. Bottazzo GF, Hanafusa T, Pujol-Borrell R, Feldmann M. Role of aberrant hla-dr expression and antigen presentation in induction of endocrine autoimmunity. Lancet. 1983;322(8359):1115–9. https://doi.org/10.1016/S0140-6736(83)90629-3.
85. Bland J. Presentation PLMI annual conference 2014, Tucson, AZ.
86. Hou, Chia-Yi. FDA suspends clinical trials involving fecal transplants. The Scientist. The Scientist Magazine, 14 June 2019.
87. Safety Communication on Use of FMT and MDROs. U.S. Food and Drug Administration, 2019, www.fda.gov/vaccines-blood-biologics/safety-availability-biologics/important-safety-alert-regarding-use-fecal-microbiota-transplantation-and-risk-serious-adverse. Accessed 2 June 2019.
88. Provider Resources for Vaccine Conversations with Patients. Cdc.Gov, 2019, www.cdc.gov/vaccines/conversations. Accessed 19 June 2019.
89. Formaldehyde and Cancer Risk. National Cancer Institute, Cancer.gov, 2011, www.cancer.gov/about-cancer/causes-prevention/risk/substances/formaldehyde/formaldehyde-fact-sheet.
90. Vaccine Adverse Events. U.S. Food and Drug Administration, 2019, www.fda.gov/vaccines-blood-biologics/report-problem-center-biologics-evaluation-research/vaccine-adverse-events. Accessed 19 June 2019.
91. Mercury and health: key facts. https://www.who.int/en/news-room/fact-sheets/detail/mercury-and-health.
92. National Vaccine Injury Compensation Program. https://www.hrsa.gov/vaccine-compensation/index.html. Accessed 2 June 2019.
93. Pasatiempo AM, Kinoshita M, Taylor CE, Ross AC. Antibody production in vitamin A-depleted rats is impaired after immunization with bacterial polysaccharide or protein antigens. FASEB J. 1990;4:2518–27.
94. Sankaranarayanan S, Ma Y, Bryson MS, Li N-Q, Ross AC. Neonatal age treatment with vitamin a delays postweaning vitamin A deficiency and increases the antibody response to T-cell dependent antigens in young adult rats fed a vitamin A-deficient diet. J Nutr. 2007;137:1229–35.
95. Campos Ponce M, Polman K, Roos N, Wieringa FT, Berger J, Doak CM. What approaches are most effective at addressing micronutrient deficiency in children 0-5 years? A review of systematic reviews. Matern Child Health J. 2019;23(Suppl 1):4–17. https://doi.org/10.1007/s10995-018-2527-9.

Resources

Aggarwal BB, Heber D. Immunonutrition: Interactions of Diet, Genetics, and Inflammation 1st Ed, CRC Press (2014), Boca Raton, FL USA.

Bookchin D, Schumacher J. The virus and the vaccine: contaminated vaccine, deadly cancers, and government neglect. St. Martin's Griffin.New York 2004.

Calder PC, Field CJ, Gill HS. Frontiers in Nutritional Science, No. 1: Nutrition and Immune Function. CABI Publishing 2002, 2006.

Calder PC, Kulkarni AD. Nutrition, immunity and infection: CRC Press; 2018.

Daniel ES. Stealth germs in your body. New York/London: Union Square Press; 2008.

Das UN. Molecular basis of health and disease: Springer Science & Business Media; 2011.

Gershwin ME, Nestel P, Keen CL. Handbook of nutrition and immunity: Humana Press; 2004.

Pammi M, Vallejo JG, Abrams SA, editors. Nutrition-infection interactions and impacts on human health 1st edition. 1st ed: CRC Press; 2014.

Schmidt MA, Smith L, Sehnert KW. Beyond Antibiotics. 1993, 1994.

Body Composition

Sue Ward and Diana Noland

21.1 Introduction – 324

21.2 Body Composition Methods – 324

21.3 Field Methods – 324

21.4 Height, Weight, and Body Mass Index – 325

21.5 Girth Measurements – 325
21.5.1 Waist Circumference (WC) – 326
21.5.2 Waist-to-Hip Ratio – 326
21.5.3 Waist-to-Height Ratio (WHR) – 326
21.5.4 Upper Arm Circumference – 326
21.5.5 Using Circumference Measurements to Calculate Body Fat Percentage – 326
21.5.6 Skinfold Measurement – 327
21.5.7 Bioelectrical Impedance Analysis and Bioimpedance Spectroscopy – 327

21.6 Bioelectrical Impedance Analysis: Non-body Composition Parameters for Nutritional Assessment – 328

21.7 Intracellular Water (ICW)/Extracellular Water (ECW) Hydration – 329

21.8 Basal Metabolic Rate Analyzers – 330
21.8.1 Electrical Impedance Myography – 330
21.8.2 Consumer Apps for Smartphones – 330

21.9 Laboratory Methods – 331
21.9.1 Hydrostatic Weighing – 331
21.9.2 Air Displacement Plethysmography – 331
21.9.3 Dual-energy X-ray Absorptiometry (DXA/DEXA) – 331
21.9.4 Medical Imaging – Computed Tomography (CT) and Magnetic Resonance Imaging (MRI) – 332
21.9.5 Ultrasound – 332

21.10 Classification Based on Percent Fat – 332

21.11 Conclusion – 333

References – 333

© Springer Nature Switzerland AG 2020
D. Noland et al. (eds.), *Integrative and Functional Medical Nutrition Therapy*,
https://doi.org/10.1007/978-3-030-30730-1_21

21.1 Introduction

Anthropometric measurements, such as body size, weight, and proportions, are often used to evaluate and monitor the effects of nutritional interventions as well as reflect an individual's overall growth and development. Body composition is the proportion of fat, muscle, bone, and water that make up the human body and its measurement helps determine excesses or deficiencies of a component that is related to health risk. In the fitness and sports setting, body fatness is an indicator of physical fitness and may be assessed by personal trainers and athletic coaches since fat content can affect sports performance. Nutrition practitioners often assess body composition since certain proportions are associated with a variety of health problems. Obesity and the newer identification of sarcopenic obesity of increased body fat percentage and loss of muscle, independent of BMI, has been associated with a number of diseases such as type 2 diabetes, hypertension, heart disease, arthritis, autoimmunity, liver disease, cancer, and kidney disease.

At the other end of the spectrum, too little body fat is often seen in those with eating disorders, oligomenorrhea, exercise addiction, and certain diseases such as cystic fibrosis, Crohn's disease, and cancer. Since physiological dysfunction can occur with too much or too little body fat, sarcopenic obesity and its distribution in the body, assessment and monitoring of body composition has become widespread and important for health practitioners (◘ Fig. 21.1).

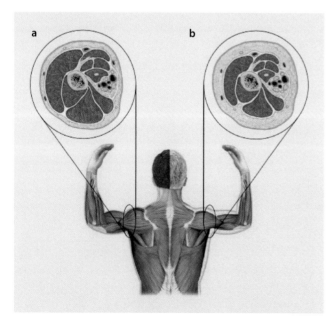

◘ **Fig. 21.1** Body Composition Changes with Sarcopenic Obesity. As people age they lose the lean muscle mass gained in young adulthood **a**, resulting in a higher proportion of fat mass **b**, even if the absolute amounts of body fat remain constant. This can lead to sarcopenic obesity – a loss of muscle and a concomitant increase in fat, often while the body weight remains stable or even decreases (Illustration by Anne Rains). (Reprinted from Benton et al. [1]. With permission from Wolters Kluwer Health)

Current methods for body composition assessment are characterized from simple to complex and all have some degree of measurement error. Some inherent problems with assessment techniques occur with the methodology, interpretation of data, and assumptions made with certain methods. The choice often depends on the intended purpose, cost, and available technology.

> Although decreases in muscle and increases in fat mass are considered part of the natural course of aging, to a great extent they're probably the results of inactivity and sedentary behaviors. – Melissa Benton [1]

21.2 Body Composition Methods

There are a variety of techniques used to estimate body composition as part of a nutritional assessment in the clinical setting. With all methods, strict adherence to the established protocols must be followed to obtain the most accurate results possible. Knowing the advantages and disadvantages of each method will help the practitioner decide wisely when choosing a method for body composition assessment.

One prevalent system of assessment is the two-compartment model, which assumes the body is made up of fat and fat-free compartments. The terms "fat-free mass" and "lean body mass" are often used incorrectly. Fat-free mass does not contain lipids, whereas lean mass includes approximately 2–3% for men and 5–8% for women. The lipids contained in lean mass include essential fat that serves as a structural component of cell membranes. Thus, the fat-free compartment is primarily made up of bone, muscle, other fat-free tissue, and body water. The water content of fat-free mass is about 72–74%. Except for cadaver studies, which provide a direct assessment of body composition, the methods discussed in this chapter are considered indirect measures and are appropriate for the clinical setting.

21.3 Field Methods

Field methods are commonly used for assessing body composition in both sport and health settings. Health practitioners use body composition in the assessment of nutritional and growth status, as well as in disease states and their treatment. Most measurements of body composition are made in the field or clinic, while more expensive laboratory methods are used primarily in research and for establishing the accuracy of the field methods [2]. Selecting a method is often difficult since practitioners want simple, inexpensive, and non-invasive options to measure body composition, but not at the expense of accuracy. Clinicians need to consider all factors and choose one or more methods that appropriately meet their needs.

21.4 Height, Weight, and Body Mass Index

Since loss of height may indicate certain health problems, the accuracy of the information on height is relevant for clinical practice. Measure stature, or height, using a wall-mounted stadiometer, or height rod, rather than asking the person his or her height. The wall-mounted measuring devices are preferred to the moveable rod on a platform scale, which lacks rigidity and can yield inaccuracy. Measurement of height can also be helpful in monitoring height loss in older individuals. In a recent study, the best predictors of a height loss of 3 cm or more were age, vertebral fractures, thoracic kyphosis, scoliosis, back pain, and osteoporosis [3]. Height loss may also be a useful tool in detecting vertebral fractures, low bone mineral density, and vitamin D deficiency [4].

Body weight should be measured using a standard physician's balance beam scale or a portable digital scale, which are accurate and consistent [5]. There is some belief that strain gauge scales may be more accurate than a conventional spring-type scale that simply has electronics to convert the result to a digital readout. Strain gauge scales use a strain gauge transducer to measure weight and display it digitally. This type of scale is inexpensive, accurate, reliable, and often portable. All scales should be placed on a flat, hard surface and checked for accuracy two times per year by using standard weights or the manufacturer recommended calibration methods.

From height and weight, body mass index (BMI), or Quetelet index, can be calculated and is a widely used clinical assessment to determine appropriateness of a person's body weight. It is currently the accepted method for interpreting the height-weight relationship and is commonly used in obesity research. BMI is calculated by dividing the weight in kilograms by height in meters squared. Although this is a quick and easy method for classifying weight, it does not assess actual body composition or distinguish between fat and muscle components. BMI tends to overestimate body fat levels in lean, muscular, or large-framed individuals, classifying them as overweight or obese [6]. A simple adjustment has been proposed that will allow the BMI of athletes to reflect the adiposity normally associated with non-athletic populations and provide the clinician or researcher with greater confidence when using BMI [7]. Body mass index also may not accurately reflect obesity and health risk in people with normal BMI values, but have a high percentage of body fat and visceral fat, especially among different ethnic groups [8]. Health practitioners should be aware of the limitations of using body mass index as the only assessment of body composition (◘ Table 21.1).

21.5 Girth Measurements

Several body and limb circumferences are often used to estimate body composition and describe body proportions. These methods provide quick and reliable information and are often used to track changes in body size during weight loss interventions. There are also consumer apps available for smartphones that often use circumferences to estimate body fat, but these give little direction about how to accurately measure. Some common circumference sites and measurement tips are as follows:

- Wrist – Measure the minimum circumference of the wrist (the tape may need to be moved up and down to identify the minimum circumference).
- Forearm – This girth is measured at the maximal circumference of the forearm, usually closer to the elbow.
- Upper arm – Measure at the midpoint between the acromial (bony part of the shoulder) and the olecranon (bony part of elbow) process.
- Waist – Measure at the midpoint between the lower margin of the last palpable rib and the top of the iliac crest.
- Abdomen – Measure the torso at the level of the umbilicus.
- Hips – Measure the maximal circumference of the buttocks above the gluteal fold (measure from the person's right or left side to better identify maximal circumference).
- Thigh – Measure the largest circumference of the right thigh just below the gluteal fold.
- Calf – Measure at the largest circumference of the calf.

How to take measurements:
1. The person to be measured should wear minimal or tight-fitting clothing.
2. Take all measurements on the right side of the body when measuring limb circumferences.
3. Obtain the measurement with the tape in contact with, but not compressing, the skin.

◘ Table 21.1 Body mass index

Body mass index (BMI) = body weight (kg)/height (m)^2
To determine body mass index, measure height to the nearest half inch and multiply by 0.0254 (to convert it to meters). Record weight to the nearest half pound and divide by 2.2 (to convert it to kilograms).

BMI	Classification
Under 18.5	Underweight
18.5–24.9	Normal
25.0–29.9	Overweight
30.0–34.9	Obesity – Class I
35.0–39.9	Obesity – Class II
≥40.0	Obesity – Class III (extreme obesity)

Based on data from Ref. [9]

4. Take waist measurements at the end of a normal expiration.
5. Take hip, thigh, and calf circumferences from the left or right side of the individual to ensure the tape measure is horizontal and parallel to the floor.
6. Record measurements to the closest line marking (i.e., fraction of an inch/cm).
7. Take each measurement is two times to ensure accuracy.

21.5.1 Waist Circumference (WC)

The measurement of waist circumference assesses an individual's pattern of fat distribution and visceral fat. In fact, obesity defined by waist circumference may be a better predictor of coronary artery calcification than BMI according to a recent study [10] that also showed those with BMIs in the overweight range and waist circumferences in the obese range had the second highest risk. Typically, waist circumference greater than 35 inches (88 cm) in women and greater than 40 inches (102 cm) in men significantly increases the risk for obesity-related diseases and may be a better predictor of the insulin-resistant phenotype and diabetes than BMI [11].

21.5.2 Waist-to-Hip Ratio

Waist-to-hip ratio (WHR) helps determine an individual's pattern of fat distribution and whether or not that pattern carries additional health risks. An excessive amount of visceral fat ("apple-shaped" bodies) has been associated with disorders of carbohydrate and fat metabolism and possible hypertension. Waist-to-hip ratio is also a marker of insulin resistance and hyperinsulinemia, which may assist the practitioner with recommendations for further laboratory testing. High insulin has been associated with increased risk of breast cancer and poorer outcomes after diagnosis. Elevated waist-to-hip ratio has been confirmed as a predictor of cancer mortality [12]. In one study, waist-to-hip ratio was a better predictor of obesity-related gastroesophageal reflux disease, also known as GERD [13]. Waist-to-hip ratio appears to be a better indicator for central body fat, regardless of waist circumference in patients with non-alcoholic fatty liver disease [14]. Those individuals with a healthy BMI or body fat percentage, but a high waist-to-hip ratio should be advised to further reduce overall body fat.

Calculate waist-to-hip ratio by dividing the waist girth by hip girth (waist ÷ hip). The World Health Organization (2008) has determined that a WHR of greater than 0.85 in women and greater than 0.90 in men shows an increased risk of metabolic complications.

21.5.3 Waist-to-Height Ratio (WHR)

Recent research has identified waist-to-height ratio (waist circumference more than half of your height) is considered as a simpler and more predictive indicator of early health risks associated with central obesity [15]. The use of waist-height ratio in public health screenings, with appropriate action, could help add years to life, according to a recent British study [16] and supports the simple message "keep your waist circumference to less than half your height." Waist-to-height ratio is a simple, quick, non-invasive method of assessment using measurements in centimeters or inches. The nutritionist should consider it as part of a standard assessment.

Waist-to-Height References (cm or inches) – ► omnicalculator.com
- Extremely slim – <0.34 both for men and women
- Healthy – 0.43–0.52 for men and 0.42–0.48 for women
- Overweight – 0.53–0.57 for men and 0.49–0.53 for women
- Very overweight – 0.58–0.62 for men and 0.54–0.57 for women
- Morbidly obese – >0.63 for men and > 0.58 for women

21.5.4 Upper Arm Circumference

Recently, some studies have shown the association of disease, biochemical changes, and nutritional status with upper-arm composition [17], which may provide a better assessment of muscularity and adiposity over other anthropometric measures, although it may not show short-term alterations in body composition [18]. One study found the combination of upper arm circumference and BMI good indicators of adiposity and showed arm circumference as a single measurement as a practical and good choice, correlating well with air displacement plethysmography [19]. Another study showed favorable results when predicting fat percentage using electrical resistivity of the upper arm [20]. Further research is needed to develop reliable normative data.

21.5.5 Using Circumference Measurements to Calculate Body Fat Percentage

One older formula (Katch, McArdle 1983) can be used to estimate body fat percentage if the clinician is looking for an easy way to assess fat along with monitoring changes in circumferences. The formulas below are based on age and gender and can be adjusted for athletes. Although the "athletic" adjustment was not defined for this formula, the practitioner can use the adjustment factor if the person regularly engages in moderate to vigorous physical activity or strenuous sports.
- *Younger Women (17–26 years)*
  % Body Fat = (abdominal circumference × 1.34) + (thigh circumference × 2.08) − (forearm circumference × 4.31) − 19.6
- *Older Women (over 26 years)*
  % Body Fat = (abdominal circumference × 1.19) + (thigh circumference × 1.24) − (calf circumference × 1.45) − 18.4

- *Younger Men (17–26 years)*
  |% Body Fat = (upper arm circumference × 3.70) + (abdominal circumference × 1.31) − (forearm circumference × 5.43) − 10.2
- *Older Men (over 26 years)*
  % Body Fat = (buttock circumference × 1.05) + (abdominal circumference × 0.90) − (forearm circumference × 3.00) − 15.0

Athletic Adjustment Factor: Subtract 4.0 for males and 3.0 for females from the percent body fat value.

21.5.6 Skinfold Measurement

Skinfold measurement, the thickness of a double fold of skin and compressed subcutaneous adipose tissue, is widely used in the clinical setting. The validity depends on the accuracy of the measurement and the skill of the technician. An improperly trained technician will introduce significant measurement error. Skinfold calipers are used and a variety of measurements are taken from different sites on the body. Common sites for measurement include the triceps, biceps, subscapular, suprailiac, abdomen, upper thigh, and calf. There are several different protocols utilizing these sites that yield an individual's estimate of body fat percent. Though skinfold measurement yields a percentage of body fat, the individual site measurements may also be a good way to monitor patient progress. It should be noted that this is a somewhat invasive measurement, since the fat must be pinched and pulled away from the muscle. This can cause some discomfort and embarrassment in some patients and is inappropriate for those who are severely obese. Individuals will have variations in skinfold thickness, compressibility of tissue, and hydration level, which is likely to affect the measurement.

There are many assumptions involved in using skinfold measurement to predict body fat [21]:
- Skinfold compressibility is constant.
- Skin thickness is constant (subscapular skin thickness was found to be greater than skin thickness at other sites).
- Subcutaneous adipose tissue patterning is constant (there is variability reinforcing the importance of using readings from multiple sites).
- Variability in fat content of adipose tissue is constant (estimated to be 20%).
- Internal and subcutaneous fat deposition is constant.

Biochemical individuality plays a major role in the accuracy and reliability of this method. While skinfold measurement is not the optimal method for body composition assessment when time and accuracy is a concern, practitioners should be aware of this technique, which has been widely used in a variety of health and fitness settings.

21.5.7 Bioelectrical Impedance Analysis and Bioimpedance Spectroscopy

Bioelectrical impedance analysis (BIA) is a non-invasive, useful tool for estimating body composition based on a two- or four-compartment model [22]. Numerous researchers have conducted studies on bioimpedance analysis and its applications in estimation of body composition and evaluation of clinical conditions [22]. This technique measures the resistance, called impedance, to a small electrical current as it travels through the body's water pool and is based on the principle that lean tissue has a higher electrical conductivity and lower impedance than fat tissue due to its electrolyte content relative to water. With this method, the instrument generates an alternating electrical current, which is passed through the body via electrodes that are either placed on the body or built into a scale or handheld instrument.

Bioimpedance spectroscopy (BIS) is known as full-body (hand-to-foot) multi-frequency BIA as it allows total body water to be differentiated further into intracellular and extracellular water compartments. This is useful when monitoring fluid shifts and hydration. BIS can also provide an estimate of body cell mass (BCM), an important indicator of nutritional status. BIS differs from the more widely recognized single-frequency BIA and does not require the use of statistically derived, population-specific prediction equations [23]. Some BIA instruments use the multi-segmental approach, which assumes the body is made up of a group of cylinders (left and right arms, left and right legs, and the total body) and are available in both single-frequency and multi-frequency systems.

The BIA is widely used in practice and body composition research. Its benefits of BIA include its portability, ease of use, low cost (depending on the instrument), and safety. It measures impedance with high precision, does not subject the person to ionizing radiation, and requires minimal subject involvement. It is useful in the clinical setting provided the individual follows important pre-test instructions:
- Avoid heavy exercise for 12 hours before testing.
- Refrain from consuming caffeine and alcohol the day before testing.
- Maintain good hydration (dehydration will overestimate fat mass).

Limitations of the BIA include poor accuracy in estimating fat-free mass and percent body fat, with the trunk being under-represented and the limbs over-represented [24]. Since BIA instruments rely on regression equations for their calculations, the method is good only if the appropriate equations are used. This is the reason why some instruments are better than others ranging from inexpensive, portable BIA scales to professional and expensive devices. These units can range in price from $50 to $30,000 (Figs. 21.2 and 21.3).

Bioelectrical impedance analysis has shown to be efficacious in assessing body composition in patients with eating

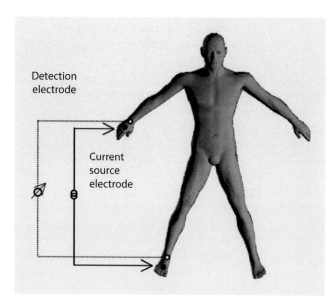

Fig. 21.2 Bioelectric Impedance Analysis standard placement of electrodes on hand-wrist-foot and ankle for tetrapolar single (SF-BIA) and multiple-frequency (MF-BIA) BIA. (Reprinted from Kyle et al. [25]. With permission from Elsevier)

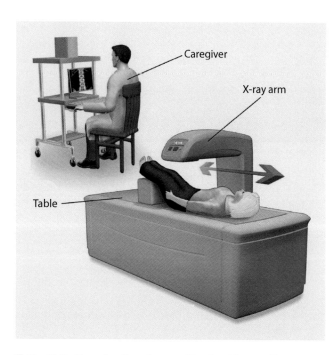

Fig. 21.3 Bone densitometry scan. (Reprinted from ▶ Blausen.com staff [26]. With permission from Creative Commons License 3.0: ▶ https://creativecommons.org/licenses/by/3.0/deed.en)

disorders during the treatment period [27], since only assessing weight changes does not reflect changes in various body compartments. BIA was studied as a tool to measure fluid changes in obese patients after bariatric surgery, but researchers found it impractical in the clinical setting for extremely obese patients due to individual variability [28].

BIA can also yield very beneficial information when used in wheelchair bound and bedridden [29] paraplegic and quadriplegic individuals and other special populations, especially when cognition or speech is impaired so a patient history cannot be reported for dietary and fluid intake.

21.6 Bioelectrical Impedance Analysis: Non-body Composition Parameters for Nutritional Assessment

In addition to body composition measurement, the bioelectrical impedance analysis (BIA) is also an important functional tool in evaluating the metabolic status of a patient. It provides data about body cell mass, cell membrane fluidity, and cell electrical potential.

Body cell mass (BCM) contains metabolically active components of the body such as muscle cells, organ cells, blood, and immune cells. BCM also includes intracellular water and the active portion of fat cells, but not the stored fat lipids. Since BCM is the functional mass of the body where most of the metabolic work is done, it is an indicator of overall nutrition status. In well-nourished individuals, the muscle tissue accounts for approximately 60% of the body cell mass, organ tissue for 20%, and the remaining 20% made up of red cells and tissue cells. When the body cell mass increases, it indicates an increase in muscle tissue. A body cell mass of about 40% is desirable.

Phase angle (PA) is a general indicator of overall health and is a measurement of cell membrane integrity and often referred to as a marker of cell membrane fluidity. "Phase angle (PA) is interpreted as an indicator of cell membrane integrity and a prognostic indicator in some clinical situations [30]." Without universal references available, and general recommendations developed from BIA research and clinical observation, a PA of 6–8 is optimal. As the membrane becomes rigid, receptor function and healthy exchange between intracellular and extracellular compartments decrease. Lower phase angles (<6.0) are consistent with an inability of cells to store energy and can indicate a breakdown in the permeability of cellular membranes. Since cell membranes have a high lipid content, phase angle provides an indication about lipid status. Low phase angles may reveal insulin resistance, malnutrition, infection, chronic disease, cancer, perturbed fatty acid status, reduced antioxidant status, and old age. Higher phase angles (>8.0) can be consistent with intact healthy cell membranes and body cell mass. Frequent clinical observation of higher phase angles is seen in those who exercise and consume a healthy diet, including antioxidant-rich fruits, vegetables, and healthy fats. Higher than 8 is excellent and 6 or lower can indicate a serious energy deficiency or chronic disease state.

Body Capacitance (C) is an indicator of the cell wall ionic potential, the ability of the non-conducting object to save electrical charges. All movement of energy and compounds require biochemical ionic charge. This value is primarily dependent on nutritional influences of mineral electrolytes and potential of hydrogen (pH) on establishing the capacitance (C), which is the ability of the non-conducting object to save electrical charges. When capacitance or electrical charges are low, it has clinically been beneficial to check the individual's mineral status (serum electrolytes, calcium, magnesium, and blood pressure). If the person's condition clinically supports low mineral status, consider increasing dietary sea salt and magnesium, and reduce calcium intake, if elevated. If capacitance is high, it is suggested to check serum calcium, electrolytes, blood pressure, and dietary intake of calcium. Oxidative stress, free radical damage, toxicity, and low antioxidant reserve also affect this result. Optimum numbers reflect a healthy cell. The body capacitance number can move up or down rapidly. Optimum capacitance for males is about 1300 and for females is about 900.

It is important to note regarding inverse relationships with calcium and other nutrient minerals with RBC and urine excretion testing results. Calcium is primarily an extracellular element because cells carefully regulate its entry into the cell. When BIA capacitance readings occasionally measure higher than optimum reference, it can be inferred that there may be a high calcium-to-magnesium and potassium ratio.

Total Body Water as a percentage of fat-free mass is a marker of hydration. If this result is <69% the patient should be put on hydration protocol and retested in 24–48 hours as values below this level indicate dehydration. This is an accurate indicator of hydration even when the person is significantly overweight.

Total Body Water (TBW/total weight) as a percentage of total weight can show dehydration but is also related to percentage of body fat. If a person's percent fat falls in the overweight or obese categories, then consider the inverse relationship between total body fat and TBW (total body water may show lower values). This is due to lean body mass being approximately 70% water and body fat being around 10% water. This value may decline with age.

Consumption of food or beverages in preparation is important to consider with the recommendations as food or fluid intake before BIA measurement affects TBW and ECW from studies in the 1990s.

» Significant alteration in body hydration, fluid distribution, and differences in the ratio of ECW to ICW caused by a medical condition will affect impedance measurements. Among those conditions, the most significant confounding variable is edema of the distal extremities, which is mainly caused by peripheral venous insufficiency. This insufficiency may result from congestive heart failure, cirrhosis, nephrotic syndrome, hypoalbuminemia, and lymphoedema [31]. Other medical conditions, which affect BIA validity, include cutaneous disease that may alter electrode-skin electrical transmission in patients with amputations, poliomyelitis, and muscular dystrophies. These conditions will have significant effects on the application of BIA in the clinical population [25].

21.7 Intracellular Water (ICW)/Extracellular Water (ECW) Hydration

The assessment of extra-, intracellular, and total body water (ECW, ICW, TBW) is important in many clinical situations [32]. For the nutritional perspective to be clinically useful, it is important to review the physiology of ICW and ECW. ICW is 97% potassium and ECW 97% sodium and calcium. The homeostasis between the two compartments is regulated by the action of the sodium-potassium pump (Na^+/K^+ pump) that was discovered in 1957 by Danish scientist Jens Christian Skou. The pump is an active process driven by the energy of ATP with magnesium being the primary driver of the pump. Thus, these three mineral-electrolytes become a focus when assessing nutritional balance of the cellular compartments, pumping potassium into cells while pumping sodium out of cells against their concentration gradients. The Na^+/K^+ pump supports maintenance of the resting potential affecting transport between the compartments while regulating cellular volume. Magnesium status of an individual is an important consideration in assessing a person's nutrition status since magnesium is important to the function of the pump. NHHANES studies have repeatedly identified that about 80% of the U.S. population consumes less than the RDA of magnesium (NHHANES).

The Na^+/K^+ pump has many functions with some of the most important the import of glucose, amino acids, and other nutrients into the cell using the energy of the sodium gradient.

Thus, the individual's data of ICW and ECW from a BIA measurement can be supportive of further investigation of mineral-electrolyte status with biomarkers in blood testing like: RBC minerals (magnesium, calcium, potassium), electrolyte panel and urinalysis-specific gravity, along with nutrition physical exam of skin assessing potential edema or dehydration issues (see ► Chap. 39). Assessing the dietary history for intake of magnesium and potassium-rich foods, and salt and high-sodium food intake can provide guidance, if needed, to improve better balance of these mineral-rich foods. Assessing intake of fatty acids and phospholipid-rich foods is also valuable to consider cell membrane fluidity influencing the activity of the Na^+/K^+ pump. During development of the nutrition intervention plan for an individual, improving the dietary intake (food, fluids, and potential supplements) based on the hydration measurement markers of the BIA ICW and ECW status can affect physiological function of all cells. Repeated BIA testing during monitoring and evaluation of a patient demonstrates if the nutrition intervention is effective in improving cellular hydration (◘ Fig. 21.4).

Basal metabolic rate (BMR) represents the estimated number of calories metabolized at rest during a 24-hour period. About 70% of a human's total daily energy expenditure is due to the basal life processes within the organs of the body. About 20% of one's energy expenditure comes from physical activity and another 10% from ► thermogenesis. BMR decreases with age and the loss of lean body mass and

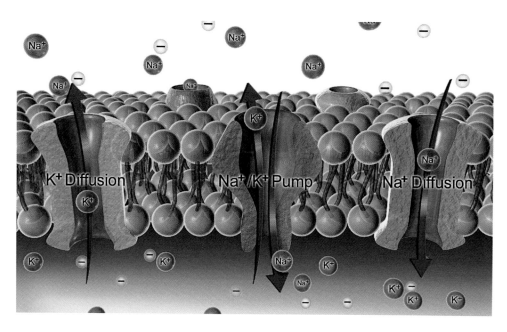

Fig. 21.4 Sodium-potassium pump. The Na+/K+-ATPase, as well as effects of diffusion of the involved ions maintain the resting potential across the membranes. (Reprinted from ▶ Blausen.com staff [26]. With permission from Creative Commons License 3.0: ▶ https://creativecommons.org/licenses/by/3.0/deed.en)

increases as lean body mass increases, since lean mass is metabolically active. The BMR value is important for providing patients with estimated daily calorie intake plus caloric need adjustments for physical activity or energetic requirements (like infection).

21.8 Basal Metabolic Rate Analyzers

- *Digital Calculator* ▶ calculator.net/bmr-calculator
- *Ergospirometry Indirect calorimetry*
- *PNOE cardio-metabolic analysis* (AKA VO2max, metabolic testing, Cardiopulmonary)
- *KORR Metabolic Test Equipment* ▶ korr.com
- *MedGem / BodyGem Indirect Calorimeter* ▶ measurermr.com
- *Metabolic Cart* – indirect calorimetry using a metabolic cart [33]

21.8.1 Electrical Impedance Myography

Electrical impedance myography (EIM) is a technique for the evaluation of neuromuscular diseases that may have future significance for the measurement of body composition. EIM technology is similar to bioelectrical impedance technology in that it sends an electrical current into the limb (usually arm or leg), but unlike BIA devices that send a current through lean mass, EIM measures how current flows through muscle and fat tissue. A new consumer product, Skulpt®, was recently introduced as a way to measure muscle quality along with body fat percent to help people monitor progress during exercise training. Muscle quality is defined as the force a muscle produces relative to its size and is a scientific metric for the fitness of muscles. This could have clinical significance not only for sports enthusiasts, but also for those with chronic diseases.

In addition to muscle quality, the device measures overall fat percentage and local fat percentage in 24 different areas of the body. The device was originally made to monitor muscular degeneration disorders, but using a complex algorithm that does not require age, gender, or weight, the device can estimate overall body fat percentage as well as in local areas. The company manufacturing the Skulpt® device was founded in 2009 by Dr. Seward Rutkove, a Harvard Medical School neurology professor, and Dr. Jose Bohorquez, an electrical engineering graduate from Massachusetts Institute of Technology (MIT). Most of the research involves EIM's use in neurodegenerative diseases such as amyotrophic lateral sclerosis (ALS) and muscular dystrophy. Internal research by the manufacturer on body composition shows a high correlation when compared with dual-energy X-ray absorptiometry (DEXA) measurements. Since this is the only device on the market that measures muscle quality, it may have important clinical significance for health practitioners. Although further research is needed, practitioners may consider using this low-cost device (about $100) to monitor changes resulting from weight loss and exercise interventions. The Skulpt® device is designed to be used with smartphones or tablets and is extremely portable.

21.8.2 Consumer Apps for Smartphones

There are a number of smartphone apps for consumers to assess body composition. Many use circumferences, skinfolds, bioelectrical impedance technology, or are an extension of Wi-Fi-connected bathroom scales. Most allow users to track body mass index (BMI). While many of these do not provide a high degree of accuracy, the health practitioner may find these useful for monitoring client progress. In addition, clients may find such apps motivational, which may improve compliance with weight loss interventions.

21.9 Laboratory Methods

Laboratory methods for assessing body composition are used often in research or when evaluating athletes, but vary in their accuracy and precision [24]. Some of these methods are used as criterion to validate field methods. Health practitioners may consider selecting a laboratory method depending on the size of the clinic and the demographics of patients. For example, a dual-energy X-ray absorptiometry (DXA/DEXA) scanner may be chosen if the practitioner wants to assess and monitor bone mineral density in addition to body composition. The main disadvantage of laboratory methods is the high cost of the equipment.

21.9.1 Hydrostatic Weighing

A technique for determining whole body density is hydrostatic, or underwater, weighing. This method is based on Archimedes' principle, which states that the volume of an object submerged in water equals the volume of water it displaces. This method yields a more accurate estimation of body fat percentage and is considered the criterion method, but is impractical in a clinical setting. Hydrostatic weighing is based on the two-compartment model of body composition and assumes that fat-free mass has a constant level of hydration and a constant proportion of bone mineral to muscle. Other studies have shown the density of fat-free mass is not constant [34] and the variation in bone mineral density affects the density of fat-free mass. This technique also requires cooperation from the person being measured since he or she must be submerged in water and remain still long enough for an accurate measurement to be taken. Hydrostatic weighing remains the standard laboratory technique to which most other methods for assessing body composition are often compared.

21.9.2 Air Displacement Plethysmography

The fundamental principle behind the air displacement plethysmography (ADP) method for measuring body composition uses the inverse relationship between pressure and volume to measure total body densitometry. This method is an alternative to underwater weighing and is better tolerated by individuals. Air displacement plethysmography is determined by a commercially available product called the BOD POD®, which measures body composition in adults and children. The PEA POD® has neonatal applications. One advantage of this method is that it is more convenient and comfortable than underwater weighing and more practical in the clinical setting. In general, the air-displacement plethysmography method shows agreement between two other criterion methods (hydrostatic weighing and dual-energy X-ray absorptiometry); however, some studies show significantly different estimates of body composition. According to a recent study, clinicians using this method should perform two trials and report the average result [35].

Both air displacement plethysmography and hydrostatic weighing correlate remarkably well with computed tomography (CT) algorithms for measurement of body composition [36]. Of all the methods, air displacement plethysmography appears to be the best instrument for tracking body composition from infancy to adulthood, when compared to bioelectrical impedance, dual-energy X-ray absorptiometry, hydrometry, and magnetic resonance imaging [37]. This is likely because it is the only practical clinical tool available for use in infants.

21.9.3 Dual-energy X-ray Absorptiometry (DXA/DEXA)

Dual-energy X-ray absorptiometry (DXA/DEXA) was originally developed to assess bone mineral density and is currently considered the gold standard technique for the diagnosis of osteopenia and osteoporosis. Recently, it has become a widely used approach for determining fat mass and fat-free mass. It uses a three-compartment model rather than two-compartment measuring bone mineral density, fat, and fat-free soft tissue. DXA/DEXA is non-invasive, quick, and reliable, though the equipment is relatively expensive.

The International Society for Clinical Densitometry (ISCD) recently developed official positions regarding the use of the DXA technique for body composition analysis. The report limited indications to three conditions: HIV patients treated with antiretroviral agents associated with a risk of lipoatrophy; obese patients undergoing treatment for high weight loss; and patients with sarcopenia or muscle weakness [38]. However, DXA/DEXA offers diagnostic and research possibilities for a variety of diseases and has the potential to monitor body composition changes with various nutritional interventions. The ongoing National Health and Nutrition Examination Survey (NHANES) uses this technique for bone density and body composition assessment.

As with most methods for assessment of body composition, DXA/DEXA has some disadvantages and limitations. The radiation exposure from a whole-body scan ranges from 0.04 to 0.86 millirem, an extremely low dose when compared to the exposure of a standard chest X-ray [39]. For this reason, repeated, frequent measurements are not recommended. Other limitations include weight and height limits on the scanning bed with some units and the field of view cannot accommodate large individuals. Differences in the hydration of lean tissue may affect the results as well as trunk or limb thickness. Despite its limitations, some studies show it is more accurate than densitometry methods [40], which has led to the use of DXA/DEXA as a reference laboratory method.

When DXA/DEXA was compared to air-displacement plethysmography (BOD POD®), researchers concluded that the BOD POD® was more accurate when measuring people closer to a healthy BMI, but it was less accurate than DXA/DEXA when it came to measuring lean individuals [41]. Those with low body fat percentages were more likely to get a higher percent fat reading from the BOD POD®, and people with higher body fat percentages were more likely to see a lower result. Both techniques had smaller discrepancies in normal weight and overweight individuals.

21.9.4 Medical Imaging – Computed Tomography (CT) and Magnetic Resonance Imaging (MRI)

Computed tomography (CT) scans are capable of estimating subcutaneous and intra-abdominal fat and compare closely with direct measurements in cadavers [42]. The disadvantages of using CT for assessing body composition include high levels of radiation exposure, cost, and limited availability of equipment. Magnetic resonance imaging (MRI) allows both imaging of the body and in vivo chemical analysis without ionizing radiation. It is a valuable tool for assessing regional fat distribution and muscle mass, but it is a costly method for health practitioners. Therefore, neither CT or MRI is practical for everyday body composition assessment.

21.9.5 Ultrasound

Ultrasound technology is a widely used imaging method and has been studied as a possible tool in nutritional assessment. It is a low cost, somewhat portable, alternative method for assessing body composition. In a recent study, when a specific regression equation was applied, ultrasound showed a strong correlation with DXA/DEXA [31]. Ultrasound can provide a similar degree of accuracy and reliability when compared with DXA/DEXA and air displacement plethysmography for tracking group-based body fat changes over time [43, 44]. One advantage of ultrasound is that it can provide information about fatty liver disease in obese patients and shows a high correlation with other markers such as waist circumference, trunk fat mass, and intra-abdominal adipose tissue [45]. In addition to assessment of body composition, ultrasound has also been suggested as a method for visceral fat mass evaluation. Estimates of total body fat based on ultrasound will result in higher accuracies when compared with skinfold measurements and bioelectrical impedance [24].

In the sport and athletic setting, ultrasound offers highly accurate and reliable field measurement essential for protection of the athlete's health and optimizing performance [46]. One study showed the validity of a portable ultrasound device questionable in female athletes, but due to its excellent reliability, practitioners and coaches should consider this method for assessing body composition changes [47].

21.10 Classification Based on Percent Fat

There are several different systems currently used for classifying body fat percent. The American College of Sports Medicine (ACSM) and American Council on Exercise (ACE) are among those most commonly used in fitness and are presented below. It should be noted that a certain amount of fat is essential for normal bodily functions. This is the fat in the bone marrow, heart, lungs, liver, muscle, and lipid-rich tissues throughout the body, including cell membrane. Essential fat in females is approximately 10–13% and in males is 2–5% (◘ Tables 21.2 and 21.3).

Some classifications have been presented that are adjusted for age, since body fat often increases and lean mass decreases as part of the aging process. ACSM also offers age-adjusted norms for body fat percent and are presented in ◘ Table 21.4.

Once a person's body fat has been determined, the clinician can use the following equation to predict a healthy weight based on body fat percent. This equation assumes that the person's lean mass remains the same.

$$\text{Desired Weight} = \text{Current Fat} - \text{free Mass in pounds} / (1 - \text{desired\%})$$

Example:
- Female with a current percent fat = 31% and body weight of 170 pounds.
- 170 × 0.31 = 52.7 pounds fat mass
- 170−52.7 = 117.3 pounds of fat-free mass
- Her desired percent fat is 25% − (1−0.25) = 0.75
- *Add to the equation:*
- 117.3 (current fat-free mass) divided by 0.75 = 156.4
- 156.4 would be her desired weight (25%) if her fat-free mass remains the same

◘ Table 21.2 ACSM body fat classifications

ACSM	Percent fat	
Body type	Females (%)	Males (%)
Athlete	<17	<10
Lean	17–22	10–15
Normal	22–25	15–18
Above average	25–29	18–20
Over-fat	29–35	20–25
Obese	>35	>25

◘ Table 21.3 ACE body fat classifications

ACE	Percent fat	
Body type	Females (%)	Males (%)
Essential fat	10–13	2–5
Athletes	14–20	6–13
Fitness	21–24	14–17
Acceptable	25–31	18–25
Obese	>31	>25

Table 21.4 ACSM age-adjusted body fat classifications for fitness

Females	Age				
Fitness category	20–29	30–39	40–49	50–59	60+
Essential fat	10–13	10–13	10–13	10–13	10–13
Excellent	14.5–17	15.5–17.9	18.5–21.2	21.6–24.9	21.1–25
Good	17.1–20.5	18–21.5	21.3–24.8	25–28.4	25.1–29.2
Average	20.6–23.6	21.6–24.8	24.9–28	28.5–31.5	29.3–32.4
Below average	23.7–27.6	24.9–29.2	28.1–32	31.6–35.5	32.5–36.5
Poor	>27.7	>29.3	>32.1	>35.6	>36.6
Males	Age				
Fitness category	20–29	30–39	40–49	50–59	60+
Essential fat	2–5	2–5	2–5	2–5	2–5
Excellent	7.1–9.3	11.3–13.8	13.6–16.2	15.3–17.8	15.3–18.3
Good	9.4–14	13.9–17.4	16.3–19.5	17.9–21.2	18.4–21.9
Average	14.1–17.5	17.5–20.4	19.6–22.4	21.3–24	22–25
Below average	17.6–22.5	20.5–24.1	22.5–26	24.1–27.4	25.1–28.4
Poor	>22.6	>24.2	>26.1	>27.5	>28.5

21.11 Conclusion

Body composition measurement helps determine excesses or deficiencies of components that relate to health risk. Significant changes in body composition are strongly correlated with the pathophysiology of chronic diseases. The importance of measuring body composition beyond height and weight anthropometrics is becoming increasingly appreciated in research and healthcare practices. The parameters of phase angle, capacitance, percentage of body fat, body cell mass, and distribution of water (ICW/ECW) provide beneficial data for assessment and developing intervention. Body fat percentage is one of the most appreciated body composition measurements, as evidenced by all chronic diseases exacerbated with the increase in body fat percentage and distribution of fat toward central adiposity. Whichever method is used to determine percent body fat, it is best for the practitioner to select that same method for repeat measurements. Some practitioners have experimented with the above techniques and measured themselves using a variety of methods, all within a 30-minute period. This type of experiment yields vastly different results, similar to what has been observed in research when comparing these methods to each other. Other body composition parameters also have valuable inferences about the health of the individual that can influence more effective interventions and monitoring of progress.

The field methods of body composition available are ideal for the nutrition practitioner whether using in-office full-body units or using the data from more expensive requiring equipment like *hydro-densitometry*, DXA/DEXA, Bod Pod, and Pea Pod that are considered for larger clinics or hospitals. The clinician should consider cost, ease of measurement, and ability to monitor progress in patients undergoing interventions.

References

1. Benton M, Whyte MD, Dyal BW. Sarcopenic obesity: strategies for management. AJN. 2011;111(12):38–44.
2. Norgan NG. Laboratory and field measurements of body composition. Public Health Nutr. 2005;8(7A):1108–22.
3. Briot K, Legrand E, Pouchain D, Monnier S, Roux C. Accuracy of patient-reported height loss and risk factors for height loss among postmenopausal women. CMAJ. 2010;182(6):558–62.
4. Mikula AL, Hetzel SJ, Binkley N, Anderson PA. Validity of height loss as a predictor for prevalent vertebral fractures, low bone mineral density, and vitamin D deficiency. Osteoporosis Int: a journal established as result of cooperation between the European Foundation for Osteoporosis and the National Osteoporosis Foundation of the USA. 2017;28:1659.
5. Yorkin M, Spaccarotella K, Martin-Biggers J, Quick V, Byrd-Bredbenner C. Accuracy and consistency of weights provided by home bathroom scales. BMC Public Health. 2013;13:1194.
6. Witt KA, Bush EA. College athletes with an elevated body mass index often have a high upper arm muscle area, but not elevated triceps and subscapular skinfolds. J Am Diet Assoc. 2005;105(4):599–602.
7. Nevill AM, Winter EM, Ingham S, Watts A, Metsios GS, Stewart AD. Adjusting athletes' body mass index to better reflect adiposity in epidemiological research. J Sports Sci. 2010;28(9):1009–16.
8. Kesavachandran CN, Bihari V, Mathur N. The normal range of body mass index with high body fat percentage among male residents of Lucknow city in north India. Indian J Med Res. 2012;135:72–7.
9. World Health Organization Global Database on Body Mass Index.

10. Park J, Lee ES, Lee DY, Kim J, Park SE, Park CY, et al. Waist circumference as a marker of obesity is more predictive of coronary artery calcification than body mass index in apparently healthy Korean adults: the Kangbuk Samsung Health Study. Endocrinol Metab (Seoul). 2016;31(4):559–66.
11. Seo DC, Choe S, Torabi MR. Is waist circumference >/=102/88cm better than body mass index >/=30 to predict hypertension and diabetes development regardless of gender, age group, and race/ethnicity? Meta-analysis. Prev Med. 2017;97:100–8.
12. Borugian MJ, Sheps SB, Kim-Sing C, Olivotto IA, Van Patten C, Dunn BP, et al. Waist-to-hip ratio and breast cancer mortality. Am J Epidemiol. 2003;158(10):963–8.
13. Ringhofer C, Lenglinger J, Riegler M, Kristo I, Kainz A, Schoppmann S. Waist to hip ratio is a better predictor of esophageal acid exposure than body mass index. Neurogastroenterol Motil. 2017;29(7).
14. Pimenta NM, Santa-Clara H, Melo X, Cortez-Pinto H, Silva-Nunes J, Sardinha LB. Waist-to-hip ratio is related to body fat content and distribution regardless of the waist circumference measurement protocol in nonalcoholic fatty liver disease patients. Int J Sport Nutr Exerc Metab. 2016;26(4):307–14.
15. Ashwell M, Gibson S. Waist-to-height ratio as an indicator of 'early health risk': simpler and more predictive than using a 'matrix' based on BMI and waist circumference. BMJ Open. 2016;6(3):e010159.
16. Ashwell M, Mayhew L, Richardson J, Rickayzen B. Waist-to-height ratio is more predictive of years of life lost than body mass index. PLoS One. 2014;9(9):e103483.
17. Jaswant S, Nitish M. Use of upper-arm anthropometry as measure of body-composition and nutritional assessment in children and adolescents (6-20 years) of Assam, Northeast India. Ethiop J Health Sci. 2014;24(3):243–52.
18. Friedl KE, Westphal KA, Marchitelli LJ, Patton JF, Chumlea WC, Guo SS. Evaluation of anthropometric equations to assess body-composition changes in young women. Am J Clin Nutr. 2001;73(2):268–75.
19. Varghese DS, Sreedhar S, Balakrishna N, Venkata Ramana Y. Evaluation of the relative accuracy of anthropometric indicators to assess body fatness as measured by air displacement plethysmography in Indian women. Am J Hum Biol. 2016;28(5):743–5.
20. Biggs J, Cha K, Horch K. Electrical resistivity of the upper arm and leg yields good estimates of whole body fat. Physiol Meas. 2001;22(2):365–76.
21. Martin AD, Ross WD, Drinkwater DT, Clarys JP. Prediction of body fat by skinfold caliper: assumptions and cadaver evidence. Int J Obes. 1985;9(Suppl 1):31–9.
22. Khalil SF, Mohktar MS, Ibrahim F. The theory and fundamentals of bioimpedance analysis in clinical status monitoring and diagnosis of diseases. Sensors (Basel). 2014;14(6):10895–928.
23. Earthman C, Traughber D, Dobratz J, Howell W. Bioimpedance spectroscopy for clinical assessment of fluid distribution and body cell mass. Nutr Clin Pract. 2007;22(4):389–405.
24. Ackland TR, Lohman TG, Sundgot-Borgen J, Maughan RJ, Meyer NL, Stewart AD, et al. Current status of body composition assessment in sport: review and position statement on behalf of the ad hoc research working group on body composition health and performance, under the auspices of the I. OC Med Comm Sports Med. 2012;42(3):227–49.
25. Kyle UG, Bosaeus I, De Lorenzo AD, Deurenberg P, Elia M, Gomez JM, Heitmann BL, Kent-Smith L, Melchior JC, Pirlich M, Scharfetter H, Schols AM, Pichard C. Bioelectrical impedance analysis – part I: review of principles and methods. Clin Nutr. 2004;23:1226–43.
26. Blausen.com staff (2014). Medical gallery of blausen medical 2014. Wikijournal Med 1 (2). Doi:https://doi.org/10.15347/wjm/2014.010. Issn 2002-4436. https://commons.wikimedia.org/w/index.php?curid=31574254.
27. Saladino CF. The efficacy of Bioelectrical Impedance Analysis (BIA) in monitoring body composition changes during treatment of restrictive eating disorder patients. J Eat Disord. 2014;2(1):34.
28. Mager JR, Sibley SD, Beckman TR, Kellogg TA, Earthman CP. Multifrequency bioelectrical impedance analysis and bioimpedance spectroscopy for monitoring fluid and body cell mass changes after gastric bypass surgery. Clin Nutr. 2008;27(6):832–41.
29. Mulasi U, Kuchnia AJ, Cole AJ, Earthman CP. Bioimpedance at the bedside: current applications, limitations, and opportunities. Body Comp/Phys Asses. 2015;30(2):180–93.
30. da Silva KT, Berbigier MC, de Almeida Rubin B, Moraes RB, Souza GC, Perry IDS. Phase angle as a prognostic marker in patients with critical illness. Nutr Clin Pract. 2015;30(2):261–5. Body Composition/Physical Assessment.
31. Ripka WL, Ulbricht L, Menghin L, Gewehr PM. Portable A-mode ultrasound for body composition assessment in adolescents. J Ultrasound Med. 2016;35(4):755–60.
32. Moissl UM, Wabel P, Chamney PW, Bosaeus I, Levin NW, Bosy-Westphal A, Korth O, Muller MJ, Ellegard L, Malmros V. Body fluid volume determination via body composition spectroscopy in health and disease. Physiol Measure. 2008;27(9):921.
33. Fujii T, Phillips B. Quick review: the metabolic cart. Int J Int Med. 2002;3(2):1–4.
34. Clark RR, Kuta JM, Sullivan JC. Prediction of percent body fat in adult males using dual energy X-ray absorptiometry, skinfolds, and hydrostatic weighing. Med Sci Sports Exerc. 1993;25(4):528–35.
35. Gibson AL, Roper JL, Mermier CM. Intraindividual variability in test-retest air displacement plethysmography measurements of body density for men and women. Int J Sport Nutr Exerc Metab. 2016;26(5):404–12.
36. Gibby JT, Njeru DK, Cvetko ST, Heiny EL, Creer AR, Gibby WA. Whole-body computed tomography-based body mass and body fat quantification: a comparison to hydrostatic weighing and air displacement plethysmography. J Comput Assist Tomogr. 2016;41(2):302–8.
37. Fields DA, Gunatilake R, Kalaitzoglou E. Air displacement plethysmography: cradle to grave. Nutr Clin Pract. 2015;30(2):219–26.
38. Messina C, Monaco CG, Ulivieri FM, Sardanelli F, Sconfienza LM. Dual-energy X-ray absorptiometry body composition in patients with secondary osteoporosis. Eur J Radiol. 2016;85(8):1493–8.
39. Lee SY, Gallagher D. Assessment methods in human body composition. Curr Opin Clin Nutr Metab Care. 2008;11(5):566–72.
40. Prior BM, Cureton KJ, Modlesky CM, Evans EM, Sloniger MA, Saunders M, et al. In vivo validation of whole body composition estimates from dual-energy X-ray absorptiometry. J Appl Physiol (1985). 1997;83(2):623–30.
41. Lowry DW, Tomiyama AJ. Air displacement plethysmography versus dual-energy x-ray absorptiometry in underweight, normal-weight, and overweight/obese individuals. PLoS One. 2015;10(1):1–8.
42. Rossner S, Bo WJ, Hiltbrandt E, Hinson W, Karstaedt N, Santago P, et al. Adipose tissue determinations in cadavers--a comparison between cross-sectional planimetry and computed tomography. Int J Obes. 1990;14(10):893–902.
43. Schoenfeld BJ, Aragon AA, Moon J, Krieger JW, Tiryaki-Sonmez G. Comparison of amplitude-mode ultrasound versus air displacement plethysmography for assessing body composition changes following participation in a structured weight-loss programme in women. Clin Physiol Funct Imaging. 2017;37(6):663–68.
44. Smith-Ryan AE, Fultz SN, Melvin MN, Wingfield HL, Woessner MN. Reproducibility and validity of A-mode ultrasound for body composition measurement and classification in overweight and obese men and women. PLoS One. 2014;9(3):1–8.
45. Monteiro PA, Antunes Bde M, Silveira LS, Christofaro DG, Fernandes RA, Freitas Junior IF. Body composition variables as predictors of NAFLD by ultrasound in obese children and adolescents. BMC Pediatr. 2014;14:25.
46. Muller W, Horn M, Furhapter-Rieger A, Kainz P, Kropfl JM, Ackland TR, et al. Body composition in sport: interobserver reliability of a novel ultrasound measure of subcutaneous fat tissue. Br J Sports Med. 2013;47(16):1036–43.
47. Wagner DR, Cain DL, Validity CNW. Reliability of A-mode ultrasound for body composition assessment of NCAA Division I Athletes. PLoS One. 2016;11(4):e0153146.

The Therapeutic Ketogenic Diet: Harnessing Glucose, Insulin, and Ketone Metabolism

Miriam Kalamian

22.1 The Current Paradigm – 338

22.2 Factors in Choosing an "Ideal" Diet – 338
22.2.1 Genetic and Epigenetic Influences – 339
22.2.2 The Inflammatory Response – 339
22.2.3 Insulin Resistance Adds Fuel to the Flames – 339
22.2.4 Anticipated Weight Loss – 339
22.2.5 Mitochondrial Health [13] – 340

22.3 Glucose and Insulin – 340

22.4 Ketogenesis – 340
22.4.1 Step One [20] – 341
22.4.2 Step Two [20] – 341
22.4.3 Step Three – 342
22.4.4 Step Four – 342
22.4.5 Ketogenic Metabolic Pathway – 343

22.5 History of Use in Epilepsy – 343

22.6 Diseases of Insulin Resistance – 344
22.6.1 Obesity – 344
22.6.2 Diabetes – 344
22.6.3 Polycystic Ovary Syndrome – 345

22.7 Cancer – 345
22.7.1 Cancer Cells Are Reliant on Fermentable Fuels – 345

22.8 Neurodegenerative Diseases – 347

22.9 Risk/Benefit Analysis – 348
22.9.1 Baseline Laboratory Evaluation – 348
22.9.2 Blood Panel – 348
22.9.3 Absolute Contraindications – 348
22.9.4 Relative Contraindications – 349

© Springer Nature Switzerland AG 2020
D. Noland et al. (eds.), *Integrative and Functional Medical Nutrition Therapy*,
https://doi.org/10.1007/978-3-030-30730-1_22

22.10 Macronutrient Calculations – 349
22.10.1 Protein Target – 349
22.10.2 Carbohydrate – 350
22.10.3 Fats – 350
22.10.4 The Question of Calorie Restriction – 350
22.10.5 Food Trackers and Meal Planners – 351

22.11 Food and Supplement Recommendations – 352
22.11.1 Eliminate the Following – 352
22.11.2 Allowed Foods – 352
22.11.3 Diet Supplementation – 353

22.12 Variations of the Ketogenic Diet – 353
22.12.1 Atkins Diet [44, 67] – 354
22.12.2 Modified Atkins Diet – 354

22.13 Diet Ratio – 354

22.14 Diet Macros – 355

22.15 Short-Term Versus Long-Term Maintenance – 355

22.16 Transitioning to a Ketogenic Diet – 356
22.16.1 Option #1: Fasting – 356
22.16.2 Option #2: A Rigorous Ketogenic Diet – 356
22.16.3 Option #3: A Slow Transition to Ketosis – 356

22.17 Potential Side Effects of the Transition – 357
22.17.1 Hunger and Cravings – 357
22.17.2 "Keto Flu" – 357
22.17.3 Hypoglycemia – 357
22.17.4 Acidosis – 357
22.17.5 Dizzy, Lightheaded, or Shaky – 358
22.17.6 Constipation – 358
22.17.7 Heart Rate or Rhythm Changes, Including Palpitations – 358
22.17.8 Change in Exercise Tolerance or Physical Performance – 358
22.17.9 "Keto Rash" – 358
22.17.10 Increased Risk of Kidney Stones and Gout – 358

22.18 The First Few Weeks: The "Make or Break" Period – 359
22.18.1 Testing Glucose and Ketones – 359
22.18.2 Testing Blood Glucose – 359
22.18.3 Ketone Testing Tools [73] – 360

22.19	Troubleshooting – 360	
22.19.1	Ongoing Flu-like Symptoms – 360	
22.19.2	Hunger between Meals – 360	
22.19.3	Food Cravings – 360	
22.19.4	Unsustainable or Rapid Weight Loss – 361	
22.19.5	Spikes in Glucose (a Rise of >25–30mg/dL) with a Meal or Snack – 361	
22.19.6	Steroid Medications (e.g., Prednisone, Dexamethasone, or Hydrocortisone) – 361	
22.19.7	Dropping out of Ketosis – 361	
22.19.8	Hidden Carbohydrates in Medicines, Supplements, or Hygiene Products – 361	
22.19.9	Nausea or Vomiting – 361	
22.19.10	Dissatisfaction with Food Choices – 362	
22.19.11	High Blood Glucose Levels – 362	
22.20	Conclusion – 362	
	References – 363	

Learning Objectives

1. Name at least two disease-promoting pathways that are downregulated through implementation of a therapeutic ketogenic diet.
2. Describe the steps that characterize the metabolic shift to ketosis.
3. Describe the rationale for use of a therapeutic ketogenic diet for expanded applications including (1) diseases of insulin resistance, (2) cancer, and (3) neurodegenerative disease.
4. Know what demographic and health history information must be included in order to begin a risk/benefit analysis.
5. Describe the calculations needed to develop an individualized macronutrient prescription for a therapeutic ketogenic diet.

22.1 The Current Paradigm

There is a pervasive belief that glucose is the optimal source of energy for the brain and central nervous system. However, the reliance on glucose to meet brain and central nervous system (CNS) energy requirements is only true if the intake of dietary carbohydrates is sufficiently high to keep homeostatic control of glucose dependent on the opposing actions of the hormones insulin and glucagon. When glucose availability is low—as it is during fasts or very-low-carbohydrate diets—cerebral energy metabolism shifts to the efficient utilization of ketone bodies. This makes evolutionary sense given the frequent disruption of the food supply that has marked most of human history prior to the transition from hunter-gatherer to agrarian societies.

The brain and CNS of most adults requires the *energy equivalent* of 130 g of carbohydrate per day. In a standard diet, it is assumed that this need will be met through the dietary intake of carbohydrates without considering the potential of ketone bodies as an energy source. This absolute requirement, designated as the Recommended Dietary Allowance (RDA), has been incorporated into a set of government guidelines known as Dietary Reference Intakes (DRIs) [1]. However, signs are appearing that government nutrition authorities are beginning to recognize that essential glucose energy is still available during ketosis:

» Diets contain a combination of carbohydrate, fat, and protein, and therefore available glucose is not limiting to the brain unless carbohydrate energy intake is insufficient to meet the glucose needs of the brain. Nevertheless, it should be recognized that the brain can still receive enough glucose from the metabolism of the glycerol component of fat and from the gluconeogenic amino acids in protein when a very low carbohydrate diet is consumed [1].

This statement acknowledges the *physiological fact* that very-low-carbohydrate diets will shift metabolism away from reliance on glucose obtained from exogenous sources. Subsequent to that shift, the majority of tissue with high energy needs, such as muscle, heart, and most brain tissue, will preferentially take up ketone bodies. During starvation or with adherence to a ketogenic diet, ketone bodies are produced in sufficient amounts to easily meet more than 60% of the energy needs of the brain and CNS. This amount was determined by a rigorous human fasting study conducted by Dr. George Cahill and colleagues, published in 1967 in the *Journal of Clinical Investigation* [2]. Unfortunately, most conventional nutrition texts either imply or implicitly state that the presence of more than trace amounts of ketone bodies reflects an invariably aberrant and potentially dangerous metabolic state. This fails to acknowledge the science supporting nearly a century of use of a therapeutic ketogenic diet as a medical nutritional therapy for epilepsy [3, 4].

The DRIs for glucose have been superseded by another set of government guidelines from the Food and Nutrition Board: the Acceptable Macronutrient Distribution Range (AMDR). The AMDR recommends that carbohydrate-containing foods make up 45–65% of total calories [5]: In a 2000 calorie-per-day plan, that percentage translates to an intake of 225–325 grams of carbohydrate, an amount far greater than the DRI. As previously noted, this recommendation is not due to the need for carbohydrate as an essential nutrient. Instead, it exists based on its role as a source of kilocalories to maintain body weight [6]. In fact, science makes clear that there is no requirement for dietary glucose even in populations with extremely high energy needs, such as ultra-athletes. A recent paper illustrates this point by stating: "less appreciated is the perspective that there is no essential requirement for dietary carbohydrate because humans possess a robust capacity to adapt to low carbohydrate availability" [7].

> **Key Point: Unlike Fat or Protein, Dietary Intake of Glucose Is Not Essential to Human Life**
> In the absence of sufficient dietary intake, glucose homeostasis is maintained at *normal physiological levels* primarily through gluconeogenesis: Glycerol backbones stripped from triglycerides as part of fatty acid oxidation become the major substrate with contributions from amino acid and the recycling of lactate through the Cori cycle.

22.2 Factors in Choosing an "Ideal" Diet

The ideal diet is one that ensures adequate nutrition, lowers inflammation, improves insulin sensitivity, and normalizes body weight. It must be sustainable long-term and flexible enough to accommodate food allergies, aversions, and intolerances. It must accommodate the fact that religious, cultural, and personally held beliefs, such as the philosophical decision to avoid all animal products, will influence food choices. Beyond the basics, though, there is no consensus as to what constitutes the ideal diet.

22.2.1 Genetic and Epigenetic Influences

Genetic variants can alter an individual's response to foods, even to those that are generally regarded as healthy. Traits such as lactose intolerance or glucose-6-phosphate dehydrogenase (G6PD) deficiency require the avoidance of certain foods or beverages. For example, the diet for an individual with lactose intolerance would limit milk and a diet for an individual with G6PD would exclude fava beans, sulfites, and quinine.

Changes in gene expression also arise from epigenetic influences. Interactions of genes with the environment may include but are not limited to those induced by environmental toxins (lead, smoke, radon, pesticides), lifestyle impacts (night shift work, poor sleep hygiene, food choices), and psychological stressors (chronic disease, unrelenting stress). Complexity increases when age, gender, nutritional insufficiencies, and comorbidities are factored into the situation. As a result, primary prevention might dictate one set of rules while tertiary care demands another.

Consuming certain foods can alter gene expression. The constituents in a food can turn on or turn off the signaling responsible for controlling vital activities at the cellular level. Research in the field of nutrigenomics is illuminating the impact of micronutrients and other food constituents while inquiry into the impact of dietary manipulation of macronutrients arises primarily from the study of biochemistry and physiology. One example is the abundance of research on the role of protein and constituent amino acids in regulating cellular pathways such as the mammalian (or mechanistic) target of rapamycin (mTOR) (see below). In muscle, the presence of abundant amino acids (particularly branched-chain amino acids) generally signals muscle tissue to increase anabolic activities that stimulate the accretion of muscle mass which is considered highly valuable. However, those same signals are also known to stimulate growth and proliferation in cancer cells characterized by upregulated mTOR activity. Changes in protein intake and meal timing are known to downregulate this activity but conventional cancer researchers remain focused on the development of drugs intended to inhibit this pathway. Unfortunately, these drug therapies, such as rapamycin, all have dire downstream effects.

It is understandable that the role of epigenetic influences in disease states is often overlooked as it involves the accumulation of numerous small changes occurring over years or even decades. Yet it is increasingly clear that these changes eventually tip the scales, causing an imbalance in the system that manifests as disease. This appears to be the case with insulin resistance, cancer, and neurological disease. As noted earlier, the interaction of genes with the environment is complex and multifactorial, which makes it exceedingly difficult to isolate which factors exert the most influence.

22.2.2 The Inflammatory Response

The production of inflammatory cytokines is an essential component of the immune response, but prolonged overproduction of these proteins may result from repeated exposure to antigens [8]. Inflammatory cytokines are also secreted by insulin-resistant adipose tissue. This is increasingly being shown to play a critical role in obesity-related insulin resistance, and it wreaks havoc in the form of chronic inflammation, a prime driver in many diseases. Chronic inflammation at any site damages cell membranes and DNA while simultaneously altering normal cell molecular signaling and repair mechanisms. The presence of ketone bodies lowers the levels of some of these cytokines. [9].

22.2.3 Insulin Resistance Adds Fuel to the Flames

In individuals consuming a standard diet (with carbohydrate intake of 45–65% of total calories) each meal or snack results in a significant rise in serum glucose, which in turn stimulates beta cells in the pancreas to secrete insulin. Over time, a persistent pattern of frequent spikes in glucose and insulin is likely to result in some degree of insulin resistance. Insulin receptors become less responsive to insulin signaling so high levels of glucose remain in circulation for a longer period of time. High fasting blood glucose is one of the markers used to diagnose diseases of insulin resistance, including prediabetes, type 2 diabetes, and polycystic ovary syndrome. Insulin resistance is also associated with obesity, which is a risk factor in certain cancers, including (but not limited to) cancers of the breast, prostate, and colon. More recently, insulin resistance has also been identified as a risk factor in Alzheimer's disease [10–12].

22.2.4 Anticipated Weight Loss

Weight loss is one of the most widely documented side effects of a ketogenic diet. Obviously, this is not always desirable, especially for individuals who are malnourished, already below a desirable weight, or have health issues (such as bowel disease or advanced cancer) that place them at high risk for sarcopenia or cachexia. Although weight stabilization is often possible, weight gain on a ketogenic diet may be difficult to accomplish unless the individual can promote muscle mass accretion through resistance training. There is also speculation that weight gain may interfere with the therapeutic impact of the diet in cancer. Accommodating medical conditions like these underscore the need for trained nutrition and health professionals that can individualize the plan and closely monitor the impact of any nutritional intervention.

22.2.5 Mitochondrial Health [13]

Structurally and functionally healthy mitochondria are essential to maintaining health throughout the lifespan. It is increasingly clear that mitochondrial health for most individuals is determined by more than the genetic code handed down at birth—instead, it is strongly influenced by epigenetic changes, including those related to the type and timing of food intake. Changes in the gut microbiome, in part altered by food choices, can also influence how genes interact with the environment. There is an urgent need to discover new ways to push back against the rising tide of metabolic diseases. Ketogenic metabolic therapy [14] offers great potential here as an adjunct to the current standard of care, even in treatment-resistant diseases, such as glioblastoma multiforme. Moving forward, it is essential that effective prevention strategies incorporate sound nutrition science.

22.3 Glucose and Insulin

Glucose is essential to human life and biology has developed a plethora of mechanisms to facilitate the movement of glucose from the bloodstream into cells. Once inside the cytoplasm of the cell, each glucose molecule is split into two molecules of pyruvate in a process known as glycolysis. In normal cells, most of the pyruvate then crosses the mitochondrial membrane where it is oxidized to generate energy within the Krebs cycle and electron transport chain. Glycolysis also converts nicotinamide adenine dinucleotide (NAD) to its reduced form, NADH, a coenzyme needed to facilitate mitochondrial activities.

When blood glucose levels rise, beta cells in the pancreas secrete insulin, which in turn facilitates the movement of glucose across cell membranes through membrane-embedded protein transporters. Interestingly, changes at the cellular level, including alterations in genetic expression, can alter the number of embedded glucose transporters as well as the amount of glucose that is oxidized relative to the amount that is fermented. Most of these transporters are insulin-independent, though some of their signaling may be activated through insulin-sensing pathways. Glucose transporter 1 (GLUT1), present in most cells and highly expressed in glucose-hungry erythrocytes and endothelial cells of the blood–brain barrier, allows insulin-independent facilitated diffusion across cell membranes based on the concentration gradient. Glucose transporter 2 (GLUT2) is highly expressed in hepatic tissue as well as intestine, kidney, and pancreatic islet beta cells, where it acts as a glucose sensor to stimulate insulin secretion. It is also expressed in neurons and astrocytes. Glucose transporter 3 (GLUT3), another insulin-independent transporter, is essential to the movement of glucose into neurons [15].

The presence of insulin, sensed through insulin receptors embedded in cell membranes, also stimulates the translocation of glucose transporter 4 (GLUT4) from cellular vesicles within the cytoplasm to the surface of the cell, which in turn facilitates the movement of glucose across the cell membrane. GLUT4 is highly concentrated in muscle, cardiac, and adipose tissue, allowing for rapid uptake of glucose. However, persistently high levels of circulating insulin are known to contribute to insulin resistance of these tissues. Exercise can improve insulin sensitivity and transport of glucose through cell membranes thus reducing the amount of insulin needed. Glucose transporter 5 (GLUT5) is specific to fructose rather than glucose and is active in skeletal muscle [16].

Insulin is essential to human life [17]. Individuals who experience acute destruction of pancreatic beta cells, through an autoimmune reaction or potentially through certain enteroviral infections, lose the ability to produce insulin and quickly progress to type 1 diabetes (formerly referred to as "juvenile-onset diabetes"). Before Frederick Banting's discovery of insulin in the 1920s, type 1 diabetes was inevitably fatal. In contrast, type 2 diabetes is most often characterized by prolonged high levels of circulating glucose and insulin due to insulin resistance, primarily in muscle, adipose, and liver tissue. This inability to move glucose into cells stimulates secretion of glucagon (insulin's counterregulatory hormone) from alpha cells in the pancreas. Glucagon has several actions, including mobilization of glycogen (glycogenolysis), further compounding the problems associated with high blood glucose levels. In type 2 diabetes, beta cell function can also deteriorate over time as a result of a disease process that allows amyloid plaques to accumulate in pancreatic islets.

Damage to cellular structure and metabolic function results in further dysregulation of regulatory and counterregulatory hormones that impact metabolism including (but not limited to) glucose homeostasis, fatty acid oxidation, and ketogenesis. Although the mechanisms for these degenerative changes are multifactorial, the expression of these genes can be altered by lifestyle changes that serve to improve insulin sensitivity, such as physical activity. It is now increasingly evident that the adoption of very-low-carbohydrate dietary patterns can also improve insulin sensitivity and glucose homeostasis while significantly reducing the need for drugs and insulin.

Insulin also plays a role in cellular signaling though this is complex and still poorly understood. That said, it is well-documented [18] that high carbohydrate intake and the associated rise in insulin initiates lipogenesis in liver and adipose tissue. These newly synthesized fatty acids are then converted into triglycerides for transport and storage. Upregulation of this activity is known to contribute to a number of metabolic diseases [19], including obesity, hepatic steatosis, diabetes, cardiovascular diseases, and the cluster of symptoms comprising metabolic syndrome.

22.4 Ketogenesis

Ketogenesis involves the biosynthesis of small energy molecules that can then be utilized alongside (or in place of) glucose, fatty acids, and amino acids. Ketogenesis is a normal and natural adaptation that ensures a continuous supply of metabolic fuels during periods of limited food availability or

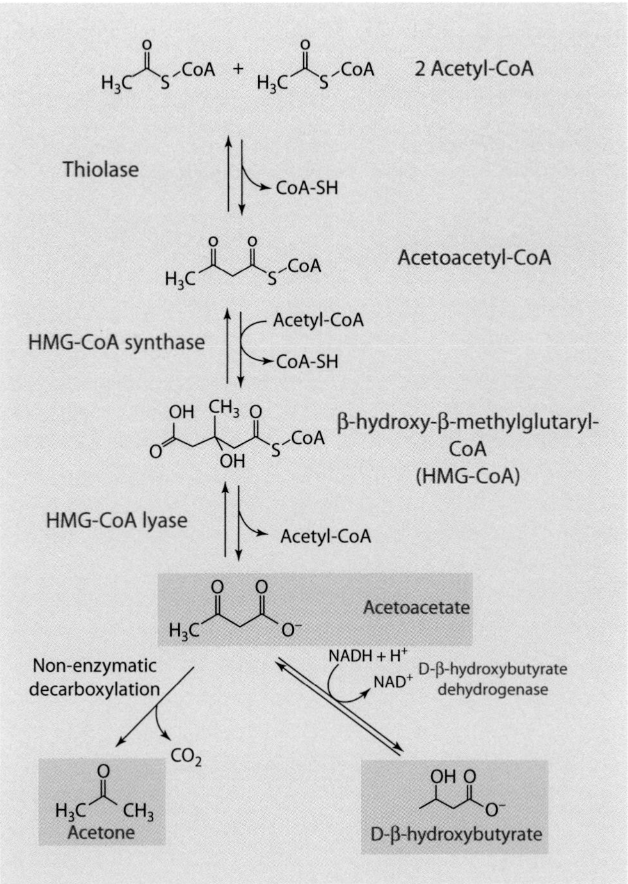

The Therapeutic Ketogenic Diet: Harnessing Glucose, Insulin, and Ketone Metabolism

Fig. 22.1 Ketogenesis pathway [21]. The three ketone bodies (acetoacetate, acetone, and beta-hydroxy-butyrate) are marked within an orange box. (Reprinted with permission from ▶ https://en.wikipedia.org/wiki/File:Ketogenesis.svg)

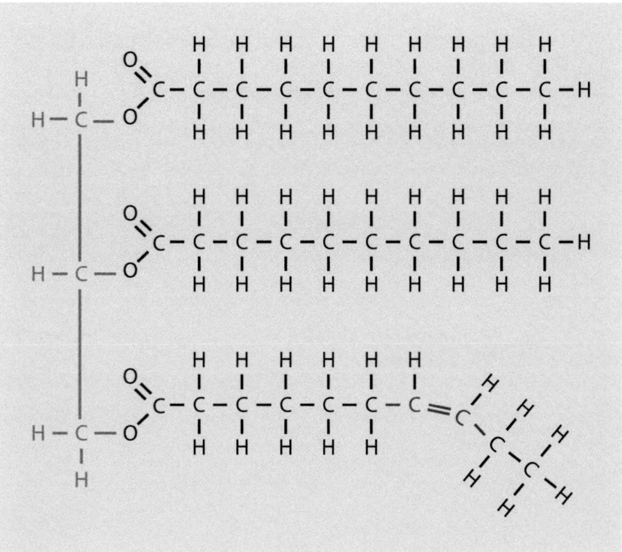

Fig. 22.2 Schematic of triglyceride ester. (Reprinted from Bayly [22]. With permission from Elsevier)

even frank starvation. However, the metabolic shift to ketosis is also possible in a well-fed state that limits carbohydrate intake. The shift from a standard diet begins with either a short fast or a rigorous restriction in carbohydrate intake. The change in energy substrates from "glucocentric" to "adipocentric" involves a series of predictable metabolic changes that accompany the adoption of a ketogenic dietary pattern [20] (◘ Fig. 22.1).

22.4.1 Step One [20]

Carbohydrate restriction results in a drop in blood glucose levels which in turn reduces insulin secretion. Low glucose levels also stimulate the secretion of the hormone, glucagon, which stimulates glycogenolysis; specifically, the breakdown of glucose stored in the liver as glycogen which in turn liberates approximately 100 g (400 calories) of stored glucose. Glucose is also stored as glycogen in muscle tissue. Muscle glycogen consists of an additional 400–500 g of glucose (1600–2000 calories) but it can only be utilized by the muscle tissue in which it is stored. In essence, muscle glycogen serves as reserve fuel for muscle which is crucial during activities that require a burst of energy, such as sprinting.

22.4.2 Step Two [20]

Depletion of liver glycogen stores, typically within 18–24 hours, upregulates gluconeogenic activity, primarily in hepatocytes but also in renal cortex. This endogenous production of glucose replaces dietary intake as the major contributor to blood glucose during periods of starvation or low carbohydrate intake. During the transition, low blood glucose levels also stimulate the production of steroid hormones (primarily adrenaline and norepinephrine) which stimulate an increase in gluconeogenic activity, thereby restoring glucose levels to physiological norms. Starvation and fasting are potent initiators but on a smaller scale, gluconeogenesis occurs many times throughout the day and during the overnight fast, effectively maintaining glucose homeostasis. Depending on availability at any given time, lactate, amino acids, and the glycerol backbones from triglycerides can all be used in the endogenous production of glucose.

22.4.2.1 Simple Schematic of a Triglyceride Ester (◘ Fig. 22.2)

A Closer Look at Gluconeogenesis
Most textbook descriptions of the shift to ketosis point to muscle mass degradation as the main source of the amino acids used in gluconeogenesis during fasting or starvation. The amount of degradation has been quantified and, in total starvation, an estimate can predict how long an individual can survive based solely on body fat stores. However, muscle mass degradation does not continue at that same unsustainable rate when dietary intake of fats and protein replace carbohydrates in a ketogenic fed state. In fact, the science suggests that ketone bodies spare both glucose and protein [23]. Fol-

> lowing a few days of carbohydrate restriction, gluconeogenesis shifts away from reliance on amino acids obtained from muscle and instead uses the abundant supply of glycerol backbones stripped from triglycerides (described in step 3 of this shift to ketosis). Limiting the discussion to fasting or starvation points to a poor understanding of the adaptations that occur in a very-low-carbohydrate (ketogenic) fed state.

22.4.3 Step Three

By itself, gluconeogenic activity cannot continue to meet the energy needs of the system as this would be dependent on the unsustainable breakdown of lean body tissue. With continued low carbohydrate intake, even in a fed state, fatty acid oxidation is upregulated, flipping the metabolic switch to reliance on energy stored as triglycerides in adipose tissue or provided through diet.

Lipases strip the backbone from triglycerides, releasing free fatty acids into the bloodstream where they are carried throughout most of the body. Exception: most long-chain fatty acids cannot cross the blood–brain barrier (BBB). In a resting state [24], heart and skeletal muscle prefer fatty acids over glucose: "Unlike skeletal muscle, heart muscle functions almost exclusively aerobically, as evidenced by the density of mitochondria in heart muscle. Moreover, the heart has virtually no glycogen reserves. Fatty acids are the heart's main source of fuel, although ketone bodies as well as lactate can serve as fuel for heart muscle. In fact, heart muscle consumes acetoacetate in preference to glucose." All but the most compromised individuals maintain fat stores that are adequate enough to survive for weeks or months during periods of total starvation [25, 26]. Glycerol released in this process is diverted to the liver and renal cortex where it is used primarily as a substrate for gluconeogenesis.

> **Analogy to a Hybrid Engine**
> It is easy to lose sight of the fact that humans have historically experienced frequent disruptions in food supply. In modern times and in affluent nations that shortfall has been reduced to a narrow window of time, usually limited to a few hours between meals or overnight.
> "Hybrid engine" is often used as a metaphor in describing the flexibility of evolutionary adaptations that ensure utilization of energy reserves. Normal cells and tissue function like tiny hybrid engines maintaining the metabolic flexibility that allows for a seamless integration of fuels from a variety of sources. In that context, it makes sense that a shortage of dietary carbohydrate would require both the endogenous production of new glucose and the breakdown of stored fats into free fatty acids and ketone bodies. (Accumulated mutations in cancer cells render them less metabolically flexible [27]; for example, many cancer cells are unable to upregulate production of key ketolytic enzymes, one of the characteristics that allow them to be targeted through a change in diet.)

22.4.4 Step Four

There are limitations in the utilization of longer chain fatty acids as a source of metabolic fuel for the brain and central nervous system: that is, they cannot cross the BBB, the semipermeable protective interface of highly specialized endothelial cells with uniquely tight junctions that line the blood vessels leading to the brain. Fatty acids also cannot be utilized by cells that lack mitochondria, such as mature red blood cells and lens tissue. For most other tissue, an evolutionary adaptation fills the energy gap through the conversion of fatty acids to ketone bodies [28]; small water-soluble energy molecules that are biosynthesized primarily in hepatocytes and, to a lesser degree, the renal cortex. Certain amino acids can also be converted to ketone bodies. These include L-lysine and L-leucine, which are exclusively ketogenic, along with five others that are conditionally ketogenic.

Ketone bodies are preferentially taken up by heart and muscle tissue but even more important, they readily cross the BBB, serving as the only metabolic fuel other than glucose that can meet the high energy needs of normal brain and central nervous system tissue [29]. This shift to adipocentric fuels plays a critical role in mitochondrial energy metabolism by reducing the demand for both glucose and protein during periods of fasting, starvation, or very low carbohydrate intake.

> **Is Ketosis Safe?**
> The level of ketosis achieved through adherence to a therapeutic ketogenic diet represents a normal and desirable adaptation to low availability of dietary carbohydrates. In contrast, dangerously high blood levels of ketone bodies seen in diabetic and alcoholic ketoacidosis are the result of a metabolic derangement that is not associated with the change in diet. Differential characteristics are described in the section on side effects.

Mitochondria in hepatocytes lack the enzyme needed to utilize ketone bodies [30]; therefore, virtually all of the ketones produced there are exported into the bloodstream for distribution throughout the body. As ketone bodies enter circulation, they are immediately taken up and utilized for energy by most normal cells that contain mitochondria. Most importantly, due to their size and solubility, ketone bodies readily cross the BBB, providing up to two-thirds [31] of the brain's ongoing energy needs. This represents roughly 20% of the body's total energy needs. In a ketogenic diet, endogenous ketone production relies on dietary intake of fats as well

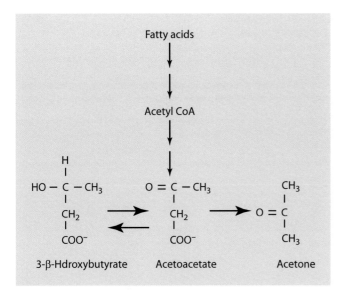

Fig. 22.3 Ketone bodies. (Reprinted from Laffel [31]. With permission from John Wiley & Sons, Inc.)

as robust lipolysis of triglycerides stored in adipose tissue. There are three types of ketone bodies [31]:

- *Acetoacetate* is preferentially taken up by muscle tissue and oxidized as energy. The brain also uses acetoacetate for energy [32, 33]. With keto-adaptation, most acetoacetate is converted (reduced) to D-beta-hydroxybutyrate, primarily by liver and muscle tissue, and returned to circulation.
- *D-beta-hydroxybutyrate* (D-βHB) is initially produced in almost equal proportions as acetoacetate. As the body adapts to ketosis, interconversion of acetoacetate favors the production of D-βHB, a preferred fuel for heart, muscle, and brain tissue.
- *Acetone* is usually produced from the spontaneous breakdown of acetoacetate. Only a small amount of acetone undergoes reactions that allow it to be used for energy. Most of it is simply eliminated from the body as fruity smelling "keto breath" (◘ Fig. 22.3).

Nutritional ketosis is defined as a metabolic state in which beta-hydroxybutyrate (βHB) is present in the blood at a level between 0.5 and 5.0 mmol/L [29, 34]. This represents the body's normal physiological response to a very-low-carbohydrate diet. If levels drop below this threshold, people are no longer in a ketogenic state. Levels above 5.0 mmol/L are not usually seen in adults on standard diets except during fasting, intense exercise, or with exogenous ketone supplementation.

22.4.5 Ketogenic Metabolic Pathway

Adaptation to starvation, fasting, or a diet very low in carbohydrates involves the upregulation of activity in the ketogenic metabolic pathway in order to meet the specific energy needs of tissues and organs. Conventional nutrition textbooks include information on ketone metabolism but it is most often presented as a temporary state with the sole purpose of providing energy during periods of starvation. Another limitation of such descriptions is that they typically blur the distinction between the *physiologically normal* levels of ketones that result from intentional dietary manipulation and the *pathologically high* levels of ketones seen in aberrant metabolic conditions such as alcoholic or diabetic ketoacidosis [7]. This does not acknowledge the unique nature of the ketogenic pathway: that is, as a stunningly sophisticated adaptation not only to periods of prolonged fasting or frank starvation but also to the voluntary restriction of carbohydrate intake that is adopted as part of a ketogenic diet or other very-low-carbohydrate pattern [7].

Utilization of newly synthesized ketone bodies begins as soon as they enter the bloodstream. Beta-hydroxybutyrate and acetoacetate pass from the bloodstream into the cytoplasm of cells via monocarboxylate transporters. These same transporters also allow passage of pyruvate and lactate (which are also monocarboxylates). From the cytoplasm, ketone bodies then pass through the mitochondrial membranes where they are oxidized for energy [29, 35], generating less reactive oxygen species (ROS) than either glucose or fatty acid metabolism. Next, a sustained rise in serum ketone levels drives the upregulation of ketolytic enzymes which serve to oxidize ketone bodies within the mitochondria of normally functioning cells.

In a standard diet, carbohydrate intake makes up more than half of total calories, much of that coming from grains, starches, and added sugars. In contrast, a classic ketogenic diet limits carbohydrate to an average of 6% of total calories [36, 37]. With protein kept low (but adequate in meeting the body's needs), the amount of fat in the diet can rise to over 80% of calories [20, 37].

22.5 History of Use in Epilepsy

The ketogenic diet has a century-long history of use as a dietary therapy for intractable epilepsy. By the start of the twenty-first century, it had passed through the clinic trial process, which led to greater acceptance as an evidence-based treatment [36–38]. There is general consensus on the initiation protocol for the classic ketogenic diet and nearly 100 hospitals in the United States [37, 39] alone are staffed with professionals, including neurologists, epileptologists, and registered dietitians that are trained in diet implementation and support. However, significant barriers to resources remain as most of the centers are concentrated in urban areas, leaving large regions of the country, including the states of Alaska and Hawaii, without access to clinical services.

Liberalized forms of the diet have emerged, including:
1. Modified Atkins (no tracking of protein or fats)

☐ **Table 22.1** Macronutrients as a percentage of total daily calories in a eucaloric ketogenic diet

	Fat (%)	Protein (%)	Carbohydrate (%)
Standard diet (AMDR)	20–35	10–35	45–65
Ketogenic diet (rigorous to moderate)	78–86	8–12	2–10

Based on data from Ref. [42]
These percentages represent a typical range—individual patterns may vary

2. Modified ketogenic diet (lower ratio of fat to carbohydrate/protein)
3. Low glycemic index diet (includes limited portions of carbohydrates, such as bread and pasta, with a GI index of <50)
4. Medium-chain triglyceride (MCT) oil diet (allows greater portions of vegetables, berries, and protein given that ketosis is enhanced by the inclusion of ketogenic MCT oil) [34, 40]

Although the level of ketosis is not as robust with liberalized versions as it is with the classic ketogenic diet, long-term adherence may be more realistic, especially in adolescents and adults. The ratio of fat grams to carbohydrate/protein grams is lower in modified versions, thus reducing the number of calories needed from fat [41]. Another benefit to liberalized plans is that diet initiation is usually more gradual, reducing the incidence and severity of adverse effects such as nausea, hypoglycemia, or mild metabolic acidosis that are commonly seen in patients who begin the diet with a fast. In fact, most adult patients who opt for less rigorous versions are not routinely hospitalized at initiation which in turn has greater appeal to professionals, families, and insurers (☐ Table 22.1).

As noted, the ketogenic diet has nearly a century of use as a medical dietary therapy. There has recently been a surge in research assessing the downstream metabolic effects of ketosis at both the system and cellular level, and a corresponding surge of interest in widespread implementation of the diet as an adjuvant or stand-alone treatment for a variety of seemingly unrelated diseases. Further progress in exploring the diet's potential will most likely arise from the demand for nutritional therapy from an informed and empowered public.

22.6 Diseases of Insulin Resistance

Recent results from both preclinical and clinical research provide evidence to support prior speculation that diseases of insulin resistance can be successfully treated using ketogenic diet therapy [43–46]. Despite mounting evidence, the majority of practitioners in the conventional medical and nutrition communities still view ketogenic diets as a fad, often labeling them as "too extreme" or "dangerous."

22.6.1 Obesity

The bias against ketogenic diet therapy in obesity medicine continues despite research showing a clear advantage for low-carbohydrate patterns in weight loss. In a 2007 paper published in the *Journal of the American Medical Association*, a randomized trial known as "The A to Z Weight Loss Study" tested four dietary and lifestyle interventions: Atkins ("induction" at 20 g carbohydrate per day for several months then an "ongoing weight loss" phase at 50 g/day), LEARN (standard guidelines plus lifestyle recommendations), Ornish (10% fat/low-protein/very-high-carbohydrate and primarily plant-based), and Zone (40% carbohydrate/30% protein/30% fat) [44]. At 2 and 6 months, weight loss was significantly greater, and, at the endpoint of 12 months, weight loss was still greatest in the Atkins group. The authors also assessed other biomarkers of health, including those assumed to be negatively impacted by a very-low-carbohydrate diet, and the findings did not confirm this bias:

> Many concerns have been expressed that low-carbohydrate weight-loss diets, high in total and saturated fat, will adversely affect blood lipid levels and cardiovascular risk. These concerns have not been substantiated in recent weight-loss diet trials. The recent trials, like the current study, have consistently reported that triglycerides, HDL-C, blood pressure, and measures of insulin resistance either were not significantly different or were more favorable for the very-low-carbohydrate groups. [44]

Given the rising tide of obesity and an increase in obesity-linked comorbidities, there is an urgent need to act on the evidence that supports a sea change in obesity medicine guidelines and practice. Well-formulated and nutritionally complete ketogenic diets, together with education and support provided by clinicians trained in proper implementation, are essential to moving ketogenic diets forward as the preferred nutritional approach to obesity treatment. This could include private models (i.e., corporate wellness programs) as well as traditional reimbursement schedules through private insurers.

22.6.2 Diabetes

Low carbohydrate and ketogenic diets have experienced a recent resurgence as the preferred dietary pattern for individuals with diabetes. In fact, the evidence from decades of research is so strong that it has prompted a group of obesity medicine researchers and clinicians to unequivocally support "the use of low-carbohydrate diets as the first approach to treating type 2 diabetes and as the most effective adjunct to pharmacology in type 1," even taking this a step further to state that "the burden of proof rests with those who are opposed" [45].

Results of a 10-week interventional trial aimed at reducing hemoglobin A1c level, medication use, and weight in individuals with type 2 diabetes [46]. In this study, researchers compared an onsite program with remote means using an online platform to deliver a comprehensive intervention that included a ketogenic diet, behavioral counseling, digital coaching, and physician-guided medication management. Results from both groups (totaling 238 participants) who completed the study included a 1% reduction from baseline in HbA1c, mean body mass reduction of 7.2%, and a reduction in the number and/or dosage of diabetes medications in the majority of participants. What is intriguing here is that these results support the development of medically supervised programs that utilize digital coaching to expand the implementation of ketogenic diet therapy beyond the geographical limitations imposed by reliance on traditional brick-and-mortar clinics.

22.6.3 Polycystic Ovary Syndrome

Potential therapeutic applications of ketogenic diets include polycystic ovary syndrome (PCOS). This is understandable given the high prevalence of insulin resistance in this population (65–70% of all women with PCOS, rising to 70–80% of obese women with this disease). The authors of a review in the *European Journal of Clinical Nutrition* stated that "although we have only preliminary evidence…there are clear mechanisms that are consistent with the physiological plausibility of such dietary therapy" [47]. The preliminary evidence referred to in this review included results of a pilot study of five obese or overweight women with PCOS who completed a 6-month trial of a low-carbohydrate (20 g/day) ketogenic diet. When compared to baseline, there was a 54% reduction in fasting insulin levels [48] with improvements in other markers of the disease as well. (Of note, two of the women became pregnant during this study despite a prior history of infertility.)

22.7 Cancer

There is mounting preclinical evidence to support the use of a ketogenic diet as an adjunct in cancer treatment [49–51]. Clinical trials are currently investigating the effects of combining diet therapy with standard of care. Unfortunately, most of these clinical studies are underpowered and poorly funded.

22.7.1 Cancer Cells Are Reliant on Fermentable Fuels

In 2005, biologist Dr. Thomas Seyfried and his colleague Dr. Purna Mukherjee were leaders in reexamining a characteristic of most cancer cells: That is, to produce energy, they preferentially ferment glucose in the cell's cytoplasm even when there is sufficient oxygen to support energy generation using the more efficient mitochondrial pathway. This particular characteristic of cancer is known as the "Warburg effect" [49], named after German biochemist and Nobel laureate Otto Warburg.

> **Dr. Otto Warburg: Nobel Laureate**
> Otto Warburg was a prominent German biochemist of the early twentieth century with a special interest in the metabolism of cancer cells. His famous observation, now known as the Warburg effect, was that cancer cells increase their rate of glycolysis even in the presence of oxygen. Dr. Seyfried sums it up this way:
> > Warburg's theory was based on his findings that all cancer cells have some defect in their ability to use oxygen to obtain energy through mitochondrial respiration. As a result, cancer cells relied more on fermentation than on respiration to obtain energy in order to compensate for their defective respiration. The reliance on fermentation even in the presence of sufficient oxygen was referred to as the Warburg effect. [50]

In a review published in the journal *Nutrition and Metabolism*, Seyfried and Mukherjee presented a strong case in support of Warburg's central theory that cancer is the downstream effect of damage to mitochondria [51]. They also describe the mechanism by which ROS from damaged respiration could produce mutations in the nucleus. Put simply, inefficiencies in cellular energy metabolism can be compensated for by an increase in glucose fermentation and, if left unchecked, ultimately transforms a normal cell into a cancer cell [51]. This line of reasoning suggests that cancer is primarily a mitochondrial metabolic disease that can be targeted with therapies, including ketogenic diet, that exploit this vulnerability. At the same time, Seyfried and Mukherjee also reiterated a fact well-known in cancer research and practice: namely, that the amino acid glutamine is another primary source of fuel that drives cancer progression.

Fermentation is a primitive process that meets bacteria's simple needs for energy. In humans, fermentation by itself usually contributes relatively little to overall energy production. That said, all cells ferment a small percentage of pyruvate, but the rate of glycolysis in cancer cells is typically 10–15 times the rate in a normal cell [52]. For this switch in the fate of glucose to occur, cells must also upregulate activities that allow more glucose to enter the cancer cell. This is accomplished through an increase in the number of glucose transporters and insulin receptors on the cell's surface. Fermentation produces lactic acid which, in large amounts, is toxic to cells, so there are mechanisms in place in all cells to quickly shunt lactic acid to the microenvironment. This increased acidity in the microenvironment of cancer cells increases inflammation, which in turn facilitates both local progression and metastatic spread of the disease. Fermentation also results in many more ROS molecules which inflict further damage to already dysfunctional mitochondria leading to even more cellular disruption and disease progression. This effect of ROS is not specific to cancer but instead is relevant in a wide array of other chronic and degenerative conditions.

Ketogenic diets also dampen spikes in glucose [53]. This not only reduces insulin levels but also lowers levels of an associated hormone, insulin-like growth factor 1 (IGF-1). IGF-1 and its cell receptor, IGF-1R, are upregulated in cancer and both can drive disease progression. In fact, activity in this pathway can be two to three times greater than in normal cells, making IGF-1 a target of many cancer drugs. Ketogenic diets that are calculated to meet but not exceed protein requirements may influence epigenetic changes that lead to the downregulation of activity in the mTOR pathway [54, 55]. This is significant given that upregulated mTOR in cancer is associated with disease progression (◘ Figs. 22.4 and 22.5).

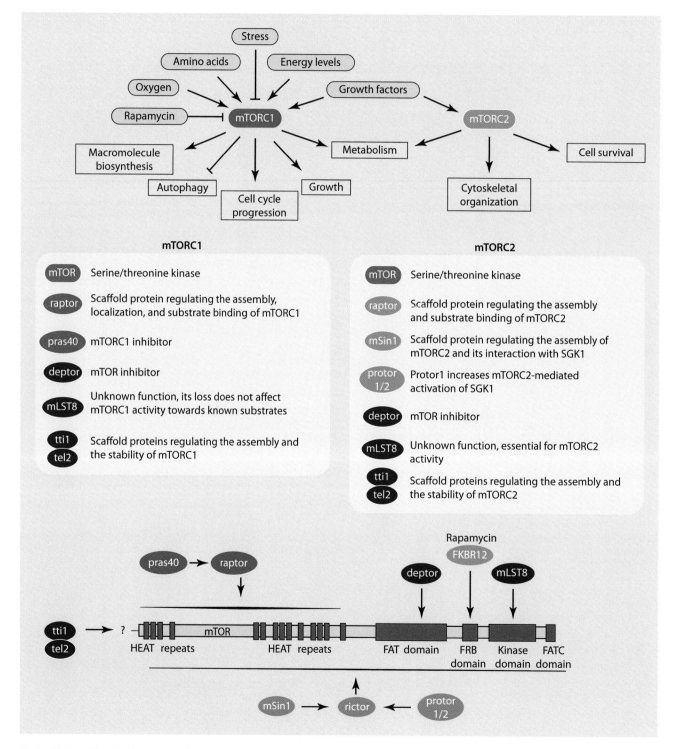

◘ Fig. 22.4 mTOR signaling in growth control and disease. (Reprinted from Laplante and Sabatini [55]. With permission from Elsevier)

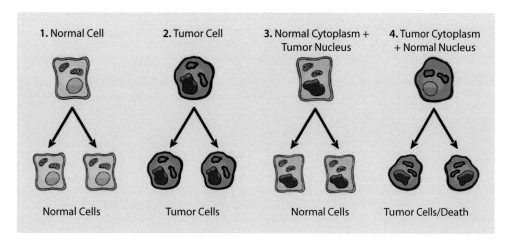

Fig. 22.5 Cybrid diagram [56]. (Courtesy of Jeffrey Ling and Thomas N. Seyfried)

Cancer as a Mitochondrial Metabolic Disease
Dr. Thomas Seyfried and colleagues describes some intriguing research involving nuclear cytoplasm transfer experiments in a paper entitled *Cancer as a mitochondrial metabolic disease* published in *Frontiers in Cell and Developmental Biology* and in *Cancer as a Metabolic Disease* [56, 57]. In nuclear transfer experiments, cytoplasmic hybrid cells (cybrids) were created by transferring *tumor* nuclei into cells with *normal* cytoplasm. The result was cells that replicated and survived, despite having a nucleus with mutated DNA. Other cybrids were created by transferring *normal* nuclei into tumor cells containing *cytoplasm that displayed the Warburg effect*. These cells also replicated but the majority did not survive. According to Dr. Seyfried: "The results suggest that nuclear genomic defects alone cannot account for the origin of tumors, and that normal mitochondria can suppress tumorigenesis. In essence, the evidence is consistent with the theory that cancer is primarily a mitochondrial metabolic disease." [56].

22.8 Neurodegenerative Diseases

Ketone bodies, particularly beta-hydroxybutyrate and acetoacetate, are a preferred energy metabolite for much of the brain and central nervous system, and a large body of evidence combined with a century-long history of use have consistently demonstrated neuroprotective and neurotherapeutic benefits of ketogenic diets in epilepsy [58]. Preliminary research indicates that these benefits may extend to applications in other neurological diseases and disorders, including autism spectrum disorder, traumatic brain injury, spinal cord injury, mood disorders, and even migraines. Several of these studies are included in a book edited by Susan Masino, PhD [59].

The Masino book also highlights and expands on earlier research pointing to therapeutic benefit of ketogenic therapies for neurodegenerative diseases, including amyotrophic lateral sclerosis [59, 60], Parkinson's disease [61], Alzheimer's disease [62], and mild cognitive impairment [62]. It is important to note that existing drug therapies offer only modest and temporary benefit in slowing disease progression and there is no curative therapy for any of these devastating diseases.

Recently, another mechanism underlying improved cognition has been elucidated by research, that is, the understanding that a ketogenic diet and/or ketone supplementation spares NAD+ (the oxidized form of NAD) [63, 64]. This is significant given that cell senescence and degeneration is associated with a decrease in the NAD+ pool, which in turn epigenetically impacts downstream cell signaling, particularly in activities related to SIRT1.

RECHARGE: Mild Cognitive Impairment and Alzheimer's Disease
Alzheimer's disease is characterized as a disorder of impaired cerebral glucose metabolism. Ironically and very tragically, the drug most commonly prescribed to treat this disease targets that deficiency by making *even more* glucose available to neurons. Instead, it makes sense to investigate ways to reduce reliance on glucose as the primary fuel by replacing glucose with ketone bodies. Ketone bodies are preferentially taken up and readily metabolized by most normal brain tissue and there is a large body of evidence that clearly demonstrates neuroprotective and neurotherapeutic properties. This could amount to a win/win/win in Alzheimer's disease [64, 65].

The MEND trial investigated the effects of changes in diet and lifestyle on symptoms of mild cognitive impairment and early Alzheimer's in a small group of older adults [65]. Interventions included (but were not limited to) lowering the intake of grains and refined starches, adding MCT and coconut oil, optimizing sleep, practicing daily intermittent fasting, and reducing stress through mind/body practices. Subjects could pick and choose from a smorgasbord of these options that as single therapies would have little total impact on disease but taken together improved outcomes for a majority of participants. This emphasis on individualizing protocols has recently been rebranded as the Bredesen Protocol™ [63, 65].

Combination therapies rather than miracle drug single agents have proven to be the solution for long-term management of other diseases, most notably HIV-AIDS. It is worthy to note that the development of that successful protocol was not made within the context of conventional medical model that tests one drug or therapy at a time in clinical trials. In fact, that model had hindered progress but was ultimately overshadowed by a grassroots movement by impassioned researchers, doctors, and the empowered lay people that combined scientific research with political activism.

22.9 Risk/Benefit Analysis

It is essential to conduct a risk/benefit analysis before initiating a ketogenic diet. This includes assessment through an intake that gathers information on the patient's health history, including demographics, present and past diagnoses, interventions, comorbidities, hospitalizations, clinical symptoms, lab results, and medications and supplements. A nutrition intake should include information about current dietary pattern and history of diets for weight loss or health improvement. It should identify conditions that may limit food choices, such as food allergies, intolerances, aversions, FODMAP issues, philosophy, and if applicable, religious dietary laws. Potential impacts of any physical or cognitive limitations, social environment, family responsibilities, financial limitations, cultural influences, and level of support need to be assessed as well.

22.9.1 Baseline Laboratory Evaluation

At a minimum, the following labs should be drawn prior to initiating a diet and the results evaluated alongside other information used in the risk/benefit analysis. Disease-specific biomarkers, such as carcinoembryonic antigen (CEA) in cancer or thyroid autoantibodies (anti-thyroid peroxidase and anti-thyroglobulin) in suspected autoimmune thyroid disease, should also be tested at baseline. Other labs may be ordered by the integrative or functional doctor to test for metabolic issues that may impact implementation or compliance. For example, an organic acids test may identify a pre-existing, yet previously undiagnosed, problem that could result in a debilitating symptom such as fatigue if the underlying disorder is not addressed prior to initiation of a ketogenic diet. Other tests include but are not limited to ferritin, uric acid, urine organic acids (Genova), comprehensive fatty acid profiles (such as the one offered through Mayo Medical Laboratories), and nutrient assessments including selenium, B12, methylmalonic acid, and folate [38].

22.9.2 Blood Panel

- Complete blood count (CBC)
- Comprehensive metabolic panel (CMP)
- Full thyroid panel (including reverse T3)
- 25-hydroxyvitamin D (optimal levels: 50–80 ng/dL)
- Red blood cell (RBC) magnesium
- Selenium
- Vitamin B12, serum
- Methylmalonic acid
- Folate
- Thiamin
- Riboflavin
- Niacin
- Pantothenate
- Biotin
- Glycated hemoglobin (HbA1c)
- Fasting insulin (as a baseline for comparison and to detect hyperinsulinemia)
- Carnitine panel (vegans are at risk of low levels due to the lack of dietary intake of carnitine; individuals experiencing catabolic events may also be low at baseline; high values for acyl/esterified are acceptable in those who are already in ketosis)
- Advanced lipid panel (e.g., NMR) including particle count and particle density
- High sensitivity C-reactive protein (hsCRP)
- Homocysteine (optimal level: low end of the normal range)

Note that the ranges typically used in certain labs, such as 25-hydroxyvitamin D and homocysteine, may not be optimal. Over time, levels of carnitine and insulin may drop below the optimal range, so retesting at intervals is essential to proper monitoring.

22.9.3 Absolute Contraindications

In 2008, a special report was published entitled "Optimal clinical management of children receiving the ketogenic diet: Recommendations of the International Ketogenic Diet Study Group" [38]. The 24 authors identified a set of absolute contraindications to the use of the diet. These should be ruled out during any risk/benefit analysis completed prior to initiating the diet. Absolute contraindications include:
- Carnitine deficiency (primary)
- Carnitine palmitoyltransferase (CPT) I or II deficiency

- Carnitine translocase deficiency
- β-oxidation defects
  - Medium-chain acyl dehydrogenase deficiency (MCAD)
  - Long-chain acyl dehydrogenase deficiency (LCAD)
  - Short-chain acyl dehydrogenase deficiency (SCAD)
  - Long-chain 3-hydroxyacyl-CoA deficiency
  - Medium-chain 3-hydroxyacyl-CoA deficiency.
- Pyruvate carboxylase deficiency
- Porphyria

> **How Common Are Contraindications?**
> With the exception of adult-onset porphyria, most of these limiting conditions are identified in childhood [36]. However, secondary carnitine deficiency can develop in individuals due to an increase in the use of carnitine predicated by a ketogenic diet as carnitine is needed to support the activity of the mitochondrial carnitine shuttle. It can also develop as a result of surgeries, serious injuries, or catabolic treatments, such as chemotherapy. Those who follow primarily plant-based diets and those with genetic variants that negatively impact the biosynthesis of carnitine are also at a higher risk of deficiency.

22.9.4 Relative Contraindications

In addition to absolute contraindications, there are a significant number of diseases, conditions, and situations that preclude or complicate initiation and compliance. The functional or integrative team should consider these in the planning process [36–41].
- Conditions that require medical and nutritional monitoring by a skilled integrative team
- Conditions that interfere with the ability to keep glucose low and steady
- Inability to comply with implementation guidelines

Potential side effects also need to be addressed prior to initiating the diet. These can be reduced in number and severity by proactive education and support during the transition. These side effects are discussed in detail in the section on transitioning to the ketogenic diet.

22.10 Macronutrient Calculations

The ketogenic diet specialist aids the integrative or functional medicine team in developing an initial macronutrient prescription. At best, this should be viewed as simply an estimate of where to start. Expect to periodically review and refine this prescription.

22.10.1 Protein Target

Most estimates of protein needs are very broad and inclusive. The Recommended Dietary Allowance for protein stands at 0.8 g of protein per kilogram of body weight, an intake that the Food and Nutrition Board of the Institute of Medicine has determined as adequate for over 97% of healthy Americans [66]. However, there are other considerations, including: (1) this amount was determined by studies of people following a standard diet, (2) this recommendation is based on *actual* body weight instead of *ideal* weight, and (3) protein needs are higher in subsets of the population including athletes and the elderly, as well as individuals who are undergoing cancer treatment or recovering from surgery. Women who are pregnant or lactating need to follow the medical and nutrition advice of their gynecologist. Aside from these exceptions, patients following ketogenic diets may need to adjust protein intake to support tissue repair and maintenance or to maintain muscle mass but there are simply too many variables here to come up with a single recommendation that can be generalized to everyone in ketosis.

> **Variables to Consider in Setting a Protein Target**
> - Older adults may benefit from more protein as greater intake is associated with reduced mortality, but this may simply reflect confounders that impact nutrient intake and digestion rather than an increased need for protein.
> - Those who engage in moderate or vigorous activity may benefit from a slight increase in protein intake, especially in the form of branched-chain amino acids before or during the activity. The needs of elite athletes should be calculated by a nutritionist that specializes in low carbohydrate nutrition for this special population.
> - Excess protein is converted to glucose via gluconeogenesis. This can interfere with ketosis.
> - Excess protein stimulates mTOR and IGF-1 activity which may be detrimental in individuals with cancer [54, 55].
> - Glutamine is fermented for energy in the mitochondria of many types of cancer cells.
> - A single high-protein meal can stimulate an insulin response sufficient to suppress ketogenesis. This is especially evident with dairy and egg protein.
> - The RDA for protein for a 5'9" individual weighing 220 lbs. (100 kg) is 80 g and the RDA for a 5'9" individual weighing an ideal 154 lbs. (70 kg) is 56 g. *Where is the evidence* that a heavier person needs an additional 24 g of protein to support what is basically the same lean body mass? Some may argue that more protein is needed to maintain the muscle tissue needed to support the extra weight, but realistically, the extra amount of protein—if needed at all—is likely to be very small, perhaps a few extra grams per day.

22.10.2 Carbohydrate

Determining carbohydrate intake does not entail a calculation as there is no requirement for dietary intake. Instead, the limit is based on an individual's response to carbohydrate intake [47]. Most people are able to maintain nutritional ketosis with carbohydrate intake between 20 and 50 net grams. Adopting a target at the lower end of the range may help to facilitate adaptation to the diet, particularly in the early days and weeks.

Patients following the ketogenic diet should be counseled to eliminate non-foods like coffee creamers that contain processed fats and to focus instead on consuming nutrient-dense non-starchy vegetables, nuts, and seeds that are high in fat and fiber. Consider these factors in setting carbohydrate limits:
- The most rigorous ketogenic diets (i.e.. for epilepsy or cancer) generally limit net carbohydrates (total carbohydrates minus fiber) to 12–16 g per day.
- Patients with thyroid disease or other hormone imbalances may be more successful and feel better during the transition if the carbohydrate limit is set at 20–25 g.
- Patients recovering from surgery, or those currently receiving chemotherapy or radiation, may benefit from a higher carbohydrate intake (20–40 g) to allow for a more liberal intake of nutrient dense foods.
- A more liberal intake of salad or sauté greens is not likely to interfere with ketosis, especially when consumed with oil-based sugar-free dressings.

22.10.3 Fats

While protein targets and carbohydrate limits are likely to remain stable over the early weeks and months of the diet, the amount of fat needed will increase or decrease in response to overall energy needs [38, 42, 47]. As stated earlier, the initial macronutrient prescription is only an estimate based on patient demographics, level of physical activity, and accommodations needed during active treatment of specific disease such as cancer or diabetes. The amount and/or type of fat initially prescribed may also reflect initial patient tolerance to an intake that may be roughly two-and-a-half times that of the baseline pattern. Variation in intake is amplified in those with high energy needs as they may require up to 300 g of fat per day to maintain current weight (if that is a goal). Comorbidities such as gallbladder disease may also impact digestion and tolerance and should be addressed proactively.

Patient education should include information on which fats and oils to include in the diet and which to eliminate. Patients may also need guidance on how to incorporate fats and oils into meals and snacks.

Dairy fats have traditionally played a large role in ketogenic diets for children. However, many functional and integrative practitioners may suggest a dairy-free diet for their dairy-sensitive patients, and this will require even more intensive support from the nutritionist. There is also some concern over estrogen metabolites in dairy, particularly for those with hormone-sensitive cancers.

> **Key Points**
> - Most protein- and fat-rich animal and plant foods contain a complex blend of fats and oils with varying amounts of saturated and unsaturated fatty acid chains
> - The ratio of omega-6 to omega-3 Intake appears to influence systemic inflammation: High ratios seen in most standard diets (i.e., 15:1) are assumed to be pro-inflammatory [57]
> - A subset of fatty acids referred to as essential fatty acids (EFAs) cannot be biosynthesized efficiently in humans and need to be included in the diet.

Table 22.2; Please see ▶ Chap. 10.

22.10.4 The Question of Calorie Restriction

The benefits of a calorie-restricted ketogenic diet include (1) intentional weight loss, (2) lower blood glucose and insulin, (3) more rapid transition to ketosis, (4) improved mitochondrial health, and (4) greater initial impact on disease-promoting pathways, such as angiogenesis, mTOR, and IGF-1 in cancer. That said, it is crucial to ensure that caloric restriction does not provoke or worsen malnutrition. All factors that impact GI functions must be evaluated and addressed prior to setting the diet prescription.

Weight loss is almost always present in the early days and weeks of the diet as the switch involves removing approximately half of total calories previously consumed as carbohydrate-containing food. Most patients adapt slowly to this change; in effect, they may inadvertently restrict calories. The integrative or functional team needs to assess baseline body fat percentage and lean body mass (via BIA, BodPod, DEXA/DXA, calipers) then carefully monitor weight loss to ensure that it is healthy and sustainable. Inadvertent weight loss is especially problematic in those who need to maintain or even gain weight during this transition so once again, proactive measures need to be in place before the transition to a ketogenic diet. Variations in approach to initiating the diet will be discussed in the following section.

It is important to note that many of the same benefits of caloric restriction can be achieved—often without weight loss—through the use of a fasting regimen. One of the simplest to initiate and maintain is daily intermittent fasting (referred to now as "time-restricted eating"). This is easily accomplished by avoiding all food for 3–4 hours prior to bedtime and refraining from carbohydrate- or protein-containing foods for the first few hours after awakening. Limited amounts of fats, such as MCT oil and/or ghee, are often consumed in a morning beverage to help extend the modified fast to 14–16 hours. This facilitates autophagy while also increasing morning ketone levels and simultaneously keeping glucose low and steady.

Table 22.2 Table of fats and oils

Sources	Choose most often	Limited amounts	Eliminate
Omega-3 (rich in EPA, DHA, ALA)			
Animal	Wild-caught cold-water fish including sardines, mackerel, salmon, and trout (rich in EPA/DHA)	Fish high up in the food chain (high concentrations of heavy metals and low in EPA/DHA) tuna, swordfish, halibut	Farmed fish with heavy contamination including tilapia, salmon, trout
Plant	Flaxseed, chia seed, hemp, walnut	Flaxseed oil (easily oxidized)	GMO flaxseed (due to crop handling practices)
Omega-6 (rich in LA, GLA)			
Animal	Fresh or frozen grass-fed pastured proteins/fats including beef, lamb, pork, poultry, dairy, properly sourced crustaceans	Processed animal products including deli meats (even from grass-fed and pastured sources)	Fresh/frozen/processed CAFO beef, pork, poultry, eggs or questionably sourced crustaceans
Plant	Small amounts of GLA-rich oils including borage, black currant, evening primrose	Oils including cold pressed non-GMO sunflower, sesame, canola	Soybean, corn, and all other GMO heat-damaged, solvent refined oils
Other mono- and polyunsaturated oils			
Animal	Animal fat from responsibly raised animals and fish oil from wild-caught (preferably) cold water fish and shellfish	n/a	Fats and oils from animals and fish/shellfish coming from CAFOs, fish farms, and contaminated fisheries
Plant	Nuts and seeds; oils including cold-pressed and minimally refined olive and avocado	Brazil nuts (concerns over radium and excess selenium)	Hydrogenated lard
Medium-chain triglyceride oils (uniquely ketogenic)			
Animal	Dairy fats from goats (when available)	n/a	n/a
Plant	Coconut and palm oil (including red palm oil)	n/a	Hydrogenated coconut and palm oils
Saturated fats (most often some combination of saturated and unsaturated fats)			
Animal	Animal and dairy fats from responsibly raised animals	Caveat for individuals with hormone-sensitive cancers: dairy products contain estrogen metabolites	Processed and packaged products containing fats, starches, and sugars
Plant	See medium-chain triglycerides		Same as above
Trans fats			
Animal	Small amounts naturally occurring in meat and dairy	n/a	Hydrogenated animal fats including commercial lard
Plant	n/a	n/a	Hydrogenated margarine and oils (mostly found in processed foods)

Based on data from Ref. [66]
EPA eicosapentaenoic acid (omega-3), *DHA* docosahexaenoic acid (omega-3), *ALA* alpha-linolenic acid (omega-3), *LA* linoleic acid (omega-6), *GLA* gamma-linolenic acid (omega-6), *AA* Arachidonic acid (omega-6), *CAFO* Concentrated animal feeding operation

22.10.5 Food Trackers and Meal Planners

Trackers and planners are important tools to both practitioners and clients/patients. When used properly, accountability for food choices is visible, which can help with compliance. There has been a rapid rise in the number and quality of food tracking apps and patients should be encouraged to use some method of recording intake (though some will still prefer to use written food logs). Used properly, trackers can provide direct feedback to the client/patient that they can then use to refine the plan. For example, Cronometer displays details on essential amino acids, comparing actual intake to the RDA. At a glance, clients/patients can see if they are meeting that requirement. Meal planning tools are not as readily available as trackers. Currently, the best and most accurate ketogenic meal planner is the KetoDietCalculator developed and maintained by The Charlie Foundation [39]. Although this is an important tool for dietitians and nutritionists who work with children, it is not used as frequently with adults given the additional steps needed to access it and the steep learning curve needed to work with this program. Accessing a patient's food records through one of these tools can help the practitioner spot areas of concern, such as inadvertent non-compliance or a lack of variety in food choices.

22.11 Food and Supplement Recommendations

22.11.1 Eliminate the Following

- *Sugars*
Read ingredient labels carefully. Eliminate agave nectar, honey, molasses, and evaporated cane juice. Many other sugars can be identified by their "-ose" endings (sucrose, dextrose, maltose).

- *Grains*
Eliminate wheat, corn, oats, rye, barley, spelt, triticale, quinoa, and bulgur.

- *Starchy Vegetables*
Eliminate potatoes, sweet potatoes, yams, winter squash, peas, and root vegetables.

- *Starchy or High-Glycemic Fruits*
Eliminate bananas, citrus, pears, grapes, pineapple, and watermelon. (Small amounts of most berries and a slice or two of some fruits can be keto-friendly if carbohydrate content is kept low.)

- *Legumes*
Eliminate peanuts, soy, garbanzos, beans, dried peas, and lentils. (Low glycemic index or vegetarian plans may include legumes.)

- *Milk* – It contains milk sugar (lactose).
- *Eliminate trans fats, heat-extracted and solvent-refined plant oils, and oxidized oils.*
- *Sugar Alcohols*
Eliminate polyol sugar alcohols in foods with ingredients that ends in "-ol" (sorbitol, mannitol, maltitol). These interfere with ketosis. *Exception*: Small amounts of erythritol or xylitol are allowed. (Note that xylitol is toxic to dogs.)

- *Alcohol (during the Transition)*
After keto-adaptation, a single shot of straight spirits or a glass of dry red wine may be well tolerated. Track its effect on ketones.

22.11.2 Allowed Foods

- *Vegetables* (not a complete list):
  - Asparagus
  - Broccoli
  - Brussels sprouts
  - Cabbage
  - Cauliflower
  - Celery
  - Cucumber
  - Kale
  - Mushrooms
  - Salad greens
  - Sauté greens (such as chard, beet greens, mustard greens)
  - Zucchini
- *After keto-adaptation, expand the options:*
  - Eggplant
  - Garlic
  - Onions
  - Peppers
  - Tomatoes
  - Turnips
  - Winter squash
- *Fruits:*
  - Apple (a few very thin slices)
  - Avocados (Hass)
  - Berries (keep it to <2 g of carb per meal)
  - Grapefruit (a few sections)
  - Olives (use more like a condiment)
- *Proteins:*
  - Beef
  - Eggs (preferably from free-range hens)
  - Lamb
  - Pork (including limited amounts of bacon and sausage)
  - Poultry
  - Seafood (wild-caught fish and shellfish)
  - Wild game meats
- *Dairy (limit in hormone-sensitive cancers):*
  - Cheese (hard cheeses, such as cheddar or Parmesan or soft, high-fat cheeses, such as Brie)
  - Full-fat "original" cream cheese
  - Heavy whipping cream
  - Sour cream (cultured, without added starches or fillers)
- *Nuts and seeds:*
  - Almonds (including almond butter, almond flour, almond milk)
  - Brazil nuts (rich in selenium—limit of two per day)
  - Chia seeds
  - Coconut (including unsweetened meat, milk, cream, or flour)
  - Flaxseed (rich in healthy omega-3s and fiber; may not be advisable in some cancers)
  - Hazelnuts
  - Hemp hearts/seeds
  - Macadamias (good choice: high in healthy fat; low in carbs and protein)
  - Pecans
  - Walnuts (good choice: fewer omega-6s than most nuts)

- *Fats and oils:*
  - Lard, tallow, or other saturated animal fats (such as duck fat)
  - Butter or ghee (if including dairy fats in the plan)
  - Buttery spreads (such as Earth Balance or Melt if substitute for dairy fat is needed)
  - Coconut oil and MCT oil
  - Avocado or extra virgin olive oil (for dressings or homemade mayonnaise)
  - Omega-3 fish oils (either as fresh fish—e.g., wild-caught salmon—or in purified supplements)
  - Salad dressings and mayonnaise (organic or homemade)
  - Small amounts of other oils based on personal preferences (e.g., almond, macadamia)
- *Natural sweeteners:*
  - Stevia (liquid drops)
  - Monk fruit (also known as luo han guo)

22.11.3 Diet Supplementation

Rigorous ketogenic diets (10–12 net grams of carbohydrate) are known to be deficient in some nutrients. More moderate ketogenic diets may also fail to provide all the vitamins and minerals considered essential to health. Nutritional sufficiency can be assessed with any number of tools used by the functional or integrative team. Each team will have their own preferred system for optimizing health through supplementation. What follows are some general guidelines:

22.11.3.1 Vitamins

Ensuring adequate intake of fat-soluble vitamins A (retinol), D_3, gamma tocopherol vitamin E, and K2 is important in any diet. Serum levels of these and other vitamins can be included in baseline assessments. Routine monitoring of serum levels of 25-hydroxyvitamin D to ensure optimal (as opposed to "sufficient") levels is already included in most functional and integrative protocols.

22.11.3.2 Minerals

Ketogenic diets have a diuretic effect and it is common, especially in the early weeks and months, to note some disruption in electrolyte balance. Calcium, magnesium, potassium, and sodium may all need to be supplemented, preferably as food (e.g., salted bone broth). However, food sources may not provide sufficient intake. Consider supplementation after screening an individual with blood tests:
- Electrolyte panel
- Calcium, serum
- Calcium, ionized
- RBC magnesium
- Parathyroid hormone (PTH)—If calcium is >10, consider PTH to rule out secondary or primary hyperparathyroidism.

22.11.3.3 Gut Health

As mentioned previously, comorbidities such as gallbladder disease, history of cholecystectomy or poor fat digestion should be addressed proactively. At this time, there are few guidelines specifically addressing the use of prebiotics or probiotics or constipation for people following ketogenic diets. Instead, each practitioner should address the needs of the individual in determining recommended foods and supplements. For example, clients/patients with insufficient diamine oxidase enzyme may not tolerate certain histamine-producing species of bacteria. Risk/benefit assessment here is also crucial for those in active cancer treatment or with impaired immune function.

22.11.3.4 Digestive Aids

Some individuals will need supplemental digestive enzymes to support fat digestion. For example, certain patients may benefit from a pancreatic enzyme that is high in lipase or from the inclusion of an emulsifier, such as non-GMO sunflower oil lecithin. Ox bile is another consideration for individuals post-cholecystectomy or with bile deficiency; however, this may not be well-tolerated and should be monitored for side effects. Other digestive aids are situation-specific: for example, some individuals may benefit from betaine HCl supplementation.

22.11.3.5 Herbs and Botanicals

Oversight by the functional or integrative team is essential particularly if patients are working with multiple teams and providers to address complex health issues. All contraindications (including drug/herb and herb/herb interactions) also need to be assessed and monitored at initiation and over time.

22.11.3.6 Antioxidants

The ketogenic diet has unique anti-inflammatory effects that should be considered when developing protocols using antioxidants. The ketogenic diet's decreased fruit and vegetable intake presents a common lack of intake of phytonutrients, fiber, and vitamin C that should be considered when developing an intervention. Also, antioxidant supplementation may be contraindicated in those undergoing pro-oxidant cancer treatments.

22.12 Variations of the Ketogenic Diet

There are some scenarios where adopting a variation of the classic ketogenic dietary may better fit the patient's needs, especially when fat is not well-tolerated or weight loss is not desirable [67]. Patients should still be counseled to avoid sugars and grains but may opt to include small portions of legumes, fruits, or root vegetables. Another approach to limiting weight loss is to allow a greater percentage of calories from protein (though this may not be recommended for patients with cancer). A liberalized plan may also be the best option for active or athletic individuals, especially if there is a substantially higher energy requirement. Here, a moderate plan might include more protein and carbohydrate as well as more total calories as fats and oils at each meal. Commercial ketone supplements are another option, especially for athletes who need a high-calorie snack before an event or workout.

22.12.1 Atkins Diet [44, 67]

The Atkins diet is a low-carbohydrate plan primarily aimed at those who want to lose weight. In Phase I (Induction), net carbohydrate intake is kept at or below 20 total grams and there is no requirement to divide that amount equally between meals. Atkins provides a list of acceptable induction foods. There are no specific guidelines for quantity or quality of fat or protein intake although more recent iterations of the diet do emphasize better quality choices. Despite detractors' claims that the Atkins plan is a kidney-damaging high-protein diet, years of research suggest that most people do not consume significantly more protein than they do on a standard diet. The combination of carbohydrate restriction along with a moderate increase in fat and protein intake results in a calorie deficit that most times leads to weight loss. Since the diet also has a diuretic effect, weight loss in the first few weeks can be exhilarating for people who have tried— and failed—other weight loss diets. It is important to note, though, that most individuals following an Atkins pattern do not achieve the low glucose or insulin levels seen with a more rigorous adherence to a ketogenic diet.

22.12.2 Modified Atkins Diet

The modified Atkins is a version of the Atkins plan that was developed and is now promoted by Dr. Eric Kossoff, an epileptologist at Johns Hopkins Hospital [68]. He was responding to the growing demand for a simpler ketogenic plan that did not require children to be hospitalized or fasted during initiation of the diet. His plan is also more user-friendly with adolescents and adults, few of whom were compliant to the restrictions used with younger children. In this version, carbohydrate intake is still low—initially 10–15 g of carbohydrate per day [68]. There are no specific guidelines for either fat or protein. This more relaxed approach to meal planning also makes it easier for families to dine out and travel. The modified Atkins for epilepsy is still viewed as a medical diet, and as such it is overseen by a team of specialists that include (at a minimum) a neurologist and a registered dietitian.

Over a decade of experience with the modified Atkins has now shown it to be a safe and efficacious alternative to the classic ketogenic diet therapy in epilepsy [67]. A recent 4-year study describes the challenges as well as the side effects of therapeutic ketogenic diets, including the modified Atkins. The authors concluded that these diets may be feasible, effective, and safe as long-term therapies in adults, which addresses some of the concerns of medical professionals who question the safety of keeping carbohydrate intake low over an extended period of time [67].

Published data also indicates that blood glucose levels range higher and ketosis is not as strong when the diet is liberalized to include more protein. This appears to improve compliance and has minimal impact on outcomes in adults choosing this pattern as a dietary therapy for epilepsy. The modified Atkins may also be a better choice than the classic ketogenic diet for those with compromised health status or complex medical conditions that prevent them from adhering to a more rigorous plan. Less frequently used variations include low-glycemic and medium-chain triglyceride versions of the diet (see ▶ Sect. 22.4).

> **The Cyclical Ketogenic Diet and Feast/Famine Cycling [7, 69, 70]**
> There is a great deal of interest in this option, which is promoted primarily through online forums and books directed at athletes and fitness enthusiasts. Simply put, the cyclical ketogenic diet is a technique to enhance muscle glycogen storage by maintaining a ketogenic diet during 5 days of intensive training then rebuilding stores with 2 days of carbohydrate-loading. Cycling may have merit for that application, but one possible downside is that keto-adaptation may take months, and cycling in carbohydrates may interfere with this process. Cycling may also lower resolve and commitment to a ketogenic plan.
>
> Feast/famine cycling has been proposed as an option for individuals with metabolic disorders other than cancer, such as diseases of insulin resistance. This plan allows a day or two every week of higher carbohydrate and protein intake followed by a return to a low-carbohydrate or ketogenic plan. There is little agreement as to which foods—and in what amounts—should be included on feast days. One of the downsides of cycling is that it may disrupt the metabolic reprogramming associated with keto-adaptation. It also has the potential to lower resolve and commitment to the ketogenic plan.

22.13 Diet Ratio

Ketogenic diets are calculated using either a macronutrient *ratio* or a macronutrient *distribution*. From the beginning of its use in epilepsy, diet *ratio* reigned. Children were most often started on a classic 3:1 or 4:1 ketogenic diet, meaning that every meal was carefully planned so that the amount of fat *grams* was 3 or 4 times the number of combined carbohydrate-plus-protein *grams*. [38].

Diet ratio can be confusing, primarily because the ratio is based on a system—gram weights of macronutrients—which is foreign to the public at large, which is more accustomed to thinking of diets in terms of calories, not grams. Diet ratio was developed as a tool for registered dietitians to use in calculating a ketogenic diet prescription for a child with epilepsy based on that child's individual protein and energy needs. Discussions of diet ratio migrated from the epilepsy world to scientific research where it is still sometimes used to describe diets used in both animal models and humans.

22.14 Diet Macros

More recently, another way of designating diet macronutrients has grown in popularity as use of the ketogenic diet has expanded to applications beyond epilepsy. In the fitness world, macronutrients (or "macros" as they are more commonly called) quickly became the preferred way to describe the distribution of macronutrients in low carbohydrate and ketogenic diets. In this model, macros are assigned a percentage based on *calories*, not *grams*. For example, in a 1600 calorie ketogenic diet, carbohydrates typically make up 5% of total calories (i.e., 80 calories, which represents 20 g). Obesity medicine specialist Dr. Eric Westman, along with researchers Eric Kossoff, Jeff Volek, and Stephen Phinney, refer to diet macronutrients in grams [20, 38, 48, 53].

Macronutrient distribution is most often viewed as more intuitive and user-friendly than diet ratio. Plus, the conventional nutrition community was already familiar with the Acceptable Macronutrient Distribution Ranges (AMDR) guidelines so it did not require further explanation to make it understandable to the public. Having a common language also made it much easier to compare the standard distribution with that of a ketogenic diet.

As illustrated above, each macronutrient is assigned a distribution range that represents its percentage of total calories in the diet. For example, 80% fat in a 1500-calorie plan amounts to 1200 calories (approximately 133 g of fat), while 80% fat in a 2500-calorie plan is 2000 calories (approximately 222 g of fat). Even though both of these diets are 80% fat, the increase in calories in the latter requires another 89 g of fat—an increase of more than 6 tablespoons of fat per day.

Protein as a percentage of the total calories may vary as well, but the absolute amount is in a narrower range as protein targets in a ketogenic diet are determined either by ideal body weight or lean body mass, not on total energy needs. Carbohydrates, as a percentage of the total calories, vary depending on two factors: total calories and diet rigor.

> **Ketogenic Diet Therapy for Children and Adolescents**
>
> There is a large body of work, including a consensus statement [38–40], that sets clear guidelines and protocols for initiation of the ketogenic diet for children and adolescents with intractable epilepsy. In the US, The Charlie Foundation is the major online resource for families and practitioners interested in working with this very special population. In the UK, the partner foundation is Matthew's Friends. Implementation of the diet for other applications requires cooperation and support of the child's medical team which should include an experienced keto dietitian or nutritionist. A specialized meal planning tool, the KetoDietCalculator (KDC), developed by Beth Zupec-Kania, consultant nutritionist for The Charlie Foundation, is available to registered clinics and dietitian/nutritionists. Using patient demographics, the KDC is used to calculate the initial diet prescription. Other features include a moderated helpline for questions from nutrition professionals, a vetted macronutrient database, lists of low- and no-carb dietary supplements, a selection of standard ketogenic meals, and flexibility to create new meals and meal plans based on a child's individual needs. The dietitian or nutritionist overseeing the child's diet can grant access to the KDC to allow families to create their own meals that adhere to the diet prescription. (Use of the KDC has expanded to include adults.)

22.15 Short-Term Versus Long-Term Maintenance

The clinical trials of the late twentieth century identified the potential side effects of the ketogenic diet when used for up to 2 years in children with epilepsy [38]. (It is important to note that a subset of children receive enteral nutrition with commercial formulas rather than whole foods.)

Reported side effects include:

- Minor metabolic abnormalities (hyperuricemia, hypocalcemia, hypomagnesemia, decreased amino acid levels, acidosis)
- Gastrointestinal symptoms (vomiting, constipation, diarrhea, abdominal pain)
- Carnitine deficiency
- Hypercholesterolemia
- Renal calculi

Other observations that may or may not be related to the diet include slowed linear growth in children, pancreatitis, and a few reported cases of cardiomyopathy and elongated QT interval (associated with selenium deficiency) [71, 72]. There are also reports in the literature of side effects, including osteoporosis, associated with long-term maintenance on the diet, but these are mostly associated with syndromes or inborn errors of metabolism, most of which limit physical activity or involve the use of medications that deplete or interfere with mineral absorption. Food allergies to dairy proteins would also limit the amount of calcium available through diet.

The growing body of research encompassing preclinical and anecdotal reports suggests that the diet is safe in the short-term but research to examine long-term effects lags behind. In fact, the lack of research identifying effects and outcomes from long-term adherence is one of the main criticisms of the growing number of studies that show a clear benefit to carbohydrate-restricted and/or ketogenic diets as treatment for type 2 diabetes and metabolic syndrome. Concerns about long-term side-effects of ketogenic diets have been raised and addressed in part by a long-term study conducted by physicians affiliated with the Johns Hopkins University School of Medicine and the Johns Hopkins Adult Epilepsy Diet Center. Their findings that ketogenic and modified Atkins diets were safe, feasible, and efficacious in the study population were published in the journal *Epilepsy Behavior* [68]. More questions will be answered and others will be framed within the context of studies similar to this one.

22.16 Transitioning to a Ketogenic Diet

Adaptations to ketosis fundamentally alter the manner in which the body processes nutrients. The impact of the transition must be considered in light of each individual's health history and current status. In order to ensure the best chance of success, the nutritionist should counsel the patient prior to initiating the diet and provide ready access to the team via email, phone calls, texts, or other predetermined means of communication.

22.16.1 Option #1: Fasting

Fasting has known therapeutic benefits that are beyond the scope of this chapter. One such benefit is the reduction or elimination of seizures in individuals with severe epilepsy. In fact, a short fast was part of the original Wilder/Peterman protocol [73] developed in the 1920s for initiating the diet as a treatment for epilepsy. Although most ketogenic centers for epilepsy no longer require a fast, it is still commonly used in other applications of the diet, particularly for weight loss and cancer. A fast quickly lowers serum levels of glucose and insulin while simultaneously upregulating fatty acid oxidation and ketogenesis. Fasts have a diuretic effect, which is motivating to individuals with a weight loss goal. A fast can also help cancer patients retain or regain a sense of control over a health crisis that is otherwise overwhelming. Engaging in an extended water-only fast may inhibit angiogenesis, which may in turn slow progression of the disease. Modified fasts are a less rigorous option and can include fluids such as salted broth with added fats or oils.

The benefits of fasting need to be weighed against the risks and potential downsides. The functional or integrative team needs to consider the patient's health and nutrition status, contraindications, comorbidities, and social support before recommending a fast. Before initiating a fast, the patient needs to be educated regarding what to expect and how to interpret the physical symptoms that they are likely to experience. They should be counseled to end the fast if continuing would endanger themselves or a person under their care, such as a child or disabled adult. They should also have some prior practice in the use of a home blood glucose meter. Ketone levels can also be assessed, either with blood ketone testing or urine test strips. And finally, the patient needs to have a plan in place to maintain ketosis once the fast has ended.

Despite the patient or practitioner's enthusiasm for a fast, there are medical reasons (such as gallbladder disease) why fasting may be a poor choice. For example, fasting is not a viable option if the patient is older, in frail health, recovering from surgery, or in need of medication that must be taken with food. Individuals should proceed cautiously if they are already below normal weight or have experienced a recent unintended weight loss. Consider also the risks for those with a history of heart arrhythmia or palpitations, or if prior drug therapies have caused an elongated QT interval, as fasting can cause some initial disruption in electrolyte balance and heart rhythm. Fasting should not be practiced by women who are pregnant or lactating. This is only a partial list—further investigation into health history is strongly recommended.

22.16.2 Option #2: A Rigorous Ketogenic Diet

Although fasting is the quickest way to reach ketosis, starting with a classic ketogenic diet will also facilitate a quick shift to ketosis [73]. The patient should be encouraged to eliminate all non-essential carbohydrates from the diet in order to keep net carbs at or below 16–25 g per day. Another factor in this decision relates to the glycemic index and glycemic load of the foods that the integrative or functional team is recommending. For example, most carbohydrates from leafy greens and a few cruciferous vegetables, such as cabbage, have a negligible impact on glucose levels. In contrast, blueberries will elicit more of a glycemic response.

The transition to the ketogenic diet will be smoother if the patient is educated as to which foods are included in a very-low-carbohydrate diet and which should be limited or eliminated entirely (see the food lists in ▶ Sect. 22.11). Those with prior experience in whole-food cooking will already have most of the skills needed to plan and prepare ketogenic meals. Many favorite recipes can be modified by substituting keto-friendly ingredients for the more traditional carbohydrate-loaded ones; for example, "riced" cauliflower can be used in place of traditional rice. Protein portions may need to be tapered down during this transition. Tracking food intake with an app or other diary is highly recommended as the feedback this provides is invaluable to both the patient and the practitioner. As with fasting, it is important to monitor blood glucose and track ketones.

22.16.3 Option #3: A Slow Transition to Ketosis

Patients are often nutritionally compromised by their diseases or treatments or otherwise unable to make all the changes needed in order to begin with a rigorous ketogenic diet [73]. A slower transition may also be more appropriate if the patient is moving to a ketogenic plan directly from a standard diet that is heavily reliant on packaged, processed, or pre-prepared foods. They will often need more time to properly stock their kitchen and pantry.

Patients can start by eliminating grains, starchy gluten-free flours, and added sugars of any kind. They should also reduce portions and servings of starchy vegetables, legumes, and fruits while simultaneously adding a tablespoon or two of a healthy fat or oil to each meal. Once they are comfortable with the added fats, they can begin to phase out other concentrated sources of carbohydrates. This is also a good time to introduce medium-chain triglyceride oil to the diet. Protein targets should be clearly prescribed. With ongoing support and monitoring from the nutrition specialist, the patient is less likely to experience side effects that often interfere with compliance.

22.17 Potential Side Effects of the Transition

The integrative or functional medicine team should educate each patient as to potential side effects of the transition. Addressing these proactively can ease the burden of symptoms and improve patient compliance to the program.

22.17.1 Hunger and Cravings

One of the most vexing problems for people following standard high-carbohydrate diets is the impact of hormonal signaling on appetite. This signaling is subjectively experienced as hunger and cravings. However, a ketogenic diet is associated with changes in hormone signaling that effectively dampen appetite, which greatly reduces these symptoms. If the patient is generally healthy and at (or above) a normal weight, this reduction in appetite is welcome. However, if weight loss is contraindicated, efforts should focus on ensuring an adequate intake of nutrients prior to Day One of the transition to the diet.

22.17.2 "Keto Flu"

In the first few days or weeks, many people experience symptoms that are collectively (and vaguely) referred to as "keto flu." This usually involves some combination of headaches, fatigue, lightheadedness, poor concentration, muscle cramps, body aches, insomnia, and constipation. At the system level, the decrease in insulin alters the manner in which the kidneys handle sodium, which in turn leads to a greater loss of fluids and electrolytes in urine. Patients should be counseled to replace fluids and replenish electrolytes, especially sodium [74]. At the cellular level, epigenetic changes prompted by the shift to ketosis result in an increase in the number of transporter proteins that allow for a greater influx of ketone bodies into the cytoplasm. Within days or weeks, mitochondria respond to this change by upregulating production of the enzyme needed to metabolize ketone bodies for energy which in turn helps to resolve these symptoms.

22.17.3 Hypoglycemia

Within hours of the last moderate- to high-carbohydrate meal, a drop in blood glucose levels will stimulate glycogenolysis [23]. Once liver glycogen is depleted, glucose homeostasis is maintained by the endogenous production of glucose, mostly through gluconeogenic activity in the liver. Some people experience a transient dizziness or a lightheaded feeling during the transition due to low blood glucose levels. Most often, glucose levels self-correct quickly. Patients should be provided with a blood glucose home testing kit (meter, strips, finger stick device, lancets) and instructed in best practices for testing. A low glucose reading (50–60 mg/dL) *accompanied by symptoms such as shaking, sweating, or lethargy* can be treated by dosing with two tablespoons of apple or orange juice. If low levels do not resolve in 30 minutes or if symptoms persist, a second dose of two tablespoons may correct the problem. If hypoglycemia persists, the patient should be seen by a physician to rule out acidosis or a previously undiagnosed metabolic disorder. (Readings of 50–60 mg/dL are virtually without symptoms once the brain begins utilizing ketone bodies.) Be aware that children deplete glycogen reserves more quickly and are more likely to be symptomatic. For this and other safety reasons, it is highly recommended that children starting therapeutic ketogenic diets do *not* start with a fast or a sharp reduction in carbohydrates unless they are under the watchful eye of a physician.

> **One Essential Caveat for Those with Diabetes**
> Individuals with type 1 diabetes or poorly controlled type 2 diabetes, with or without frequent hypoglycemic episodes, MUST NOT adopt a ketogenic diet unless they are under the supervision of an experienced ketogenic team. Some diabetics may also be at a higher risk of "hypoglycemic unawareness"—another reason to work with a specialist.

22.17.4 Acidosis

Mild and temporary acidosis may occur during the transition to a ketogenic diet, especially if it is initiated with a fast [7]. This is usually limited to a slight and easily reversible disruption in blood pH. This uncomplicated metabolic disorder is not the life-threatening condition known as diabetic ketoacidosis (DKA), a severe derangement that manifests as extremely high blood glucose (usually well over 200 mg/dL) along with extremely high levels of blood ketones (exceeding 14 mmol/L). DKA most often results from a shortage of insulin in people with type 1 diabetes or poorly controlled type 2 diabetes [74]. Another type of metabolic acidosis is associated with the use of a class of diabetes medications called SGLT2 inhibitors (such as Invokana) which should be discontinued—*under medical supervision*—prior to initiating any low-carbohydrate diet. In fact, practitioners managing a patient with diabetes need to be fully informed about any proposed change in diet.

Alcoholic ketoacidosis (AKA) is a form of metabolic acidosis characterized by an elevated anion gap, elevated serum ketone level, and normal or low blood glucose levels [74]. Practitioners should screen for prior history of alcohol abuse and/or AKA and advise against drinking alcohol during the transition.

Acidosis may also occur during the transition or at times of illness or other metabolic stress that results in a sharp increase in ketones (above 5 mmol/L), usually accompanied by dehydration. High ketone levels by themselves may not indicate acidosis but should be suspected when seen concurrent with symptoms that include extreme lethargy, nausea, vomiting, flushing, panting, and a rapid or irregular heart rate. Mild acidosis can be treated with two tablespoons of apple or orange juice as well as ½ teaspoon of baking soda in a glass of water. However, if symptoms persist, the patient will need to be seen by a physician that can assess blood chemistries, such as CO_2 levels and anion gap.

As is the case with hypoglycemia, children are more likely than adults to experience acidosis during the transition or as a result of loss of fluids (diarrhea or vomiting) during an illness [4, 37]. Parents should be educated about this and provided with a sick day protocol and an emergency phone contact.

22.17.5 Dizzy, Lightheaded, or Shaky

As noted earlier, these symptoms may indicate hypoglycemia, but they can also be due to a drop in blood pressure during the transition. Patients should be advised against engaging in activities that require balance or a sudden increase in effort, such as climbing ladders or working out with weights. If these symptoms persist for more than a few days, a healthcare practitioner should assess blood pressure [67, 72–74]. It is also crucial that medical professionals monitor all individuals who are taking medications to control high blood pressure. Another consideration: The diet has a diuretic effect which may overlap with the impact of diuretics used to manage high blood pressure.

22.17.6 Constipation

Most people have fewer bowel movements on a ketogenic diet than they may have had previously. This anticipated change in what is known as "regularity" is not the same health issue as true constipation (straining to pass hard stool). That said, true constipation may worsen with the initiation of a ketogenic diet and must be addressed proactively. Patients should be counseled to stay well-hydrated and to choose foods that are naturally high in fiber. (Use caution with fiber supplements.) Certain medications and nutritional supplements (such as calcium carbonate) can also contribute to constipation, while other supplements (such as certain forms of magnesium) can ease symptoms by drawing more water into the colon [75].

22.17.7 Heart Rate or Rhythm Changes, Including Palpitations

A change in blood volume brought on by dehydration and the loss of electrolytes in urine may cause the heart to beat faster and/or harder [76]. Salted broth and proper hydration can help to replenish fluids and electrolytes. Dehydration and an imbalance in electrolytes can trigger an episode in those with pre-existing heart disease or rhythm disorders, such as atrial fibrillation. Also, risk/benefit assessment should include a cardiac evaluation in individuals who have received medications known to prolong QT interval—this potentially fatal condition should be ruled out before transitioning a patient to a ketogenic diet [71, 72].

22.17.8 Change in Exercise Tolerance or Physical Performance

Patients should be advised to limit physical activities until transition symptoms have resolved. Instead of intense workouts or training, they can stay active with gentle walking, cycling, or lap swimming. However, some individuals will choose instead to engage in more vigorous activity. For them, the transition may be a source of frustration as their muscle tissue adapts to the increase in fatty acid oxidation and the availability of ketone bodies as a preferred metabolic fuel. Low carbohydrate and ketogenic diets appear to confer a metabolic edge in many sports, especially endurance events [7].

22.17.9 "Keto Rash"

Pruritis pigmentosa, a.k.a. "keto rash," is an uncommon and poorly understood potential side effect of the diet. There is some speculation as to cause—either acetone in sweat or a change in the skin microbiome. Reducing the level of ketosis or discontinuing the diet will resolve the symptoms [73, 74].

22.17.10 Increased Risk of Kidney Stones and Gout

There is an increased risk for certain types of kidney stones, especially for those with a personal or family history. This can be addressed prophylactically with a prescription medication containing citric acid and potassium citrate and dietary supplementation of magnesium citrate and vitamin K2. However, there are many contraindications to use especially for those with compromised renal function [76].

In the first few weeks or months of the diet, there may be a rise in uric acid levels due to competition between uric acid and ketone bodies for excretion in the urine [77]. Although this is often expressed as a concern by physicians, there is no evidence that it will progress to gout in individuals without a personal history of this disease [77].

22.18 The First Few Weeks: The "Make or Break" Period

The first few weeks of the ketogenic diet is often the "make or break" period so it is crucial to foster a strong commitment at the start. Along with commitment, the patient needs the following tools and strategies that enhance both compliance and accountability [73, 78]. Along with these items, an accessible nutrition coach can answer the questions that inevitably arise during this period.

> **Keep It Simple**
> The transition to a ketogenic diet usually involves a considerable amount of effort on the part of the patient. Changes can feel overwhelming at the start, so it is often best to offer simple recipes and meal preparation tips that allow the patient to put a meal on the table with a minimum amount of time and effort. Variety is not usually a priority in the first few weeks and many people will opt to recycle the ketogenic meals that they find most palatable and easiest to prepare. Ongoing education and support will help to move them toward more variety and better quality once they are comfortable with basic meal planning.
>
> Patients can start with one fully ketogenic meal per day and build from there. Most people find breakfast the easiest meal to adapt; for example, eggs cooked in butter garnished with grated cheese and perhaps a strip or two of bacon served with a steamed or sautéed vegetable and a drizzle of added oil.

Review the following actions:
- Have they completed a pantry sweep, removing or relocating non-keto foods?
- Did they buy new keto-friendly food staples?
- Have they obtained a blood glucose meter? If so, have they practiced with it yet?
- What is the plan for testing glucose and ketones in these first few weeks?
- Do they have a kitchen gram scale? If so, are they familiar with how to use it?
- Do they have access to meal planning tools and recipe resources?
- If the idea of structured meal planning is likely to be too daunting at the moment, have they been provided with a simple meal template customized to their macronutrient prescription?

22.18.1 Testing Glucose and Ketones

Testing blood glucose and ketones is strongly encouraged in the first few weeks of the transition as it offers an objective and quantifiable means of assessing whether or not the patient is achieving the goal of lowered blood glucose and elevated ketones. Glucose and ketone measures also help the patient and practitioner to troubleshoot issues that arise [73].

22.18.2 Testing Blood Glucose

There are currently two methods of testing blood glucose: glucometers and continuous glucose monitors. By far, the least expensive and most convenient testing method is with the use of a glucometer, also known as a home blood glucose meter. Specialized testing tools, such as watches and contact lenses, are currently in development and will reduce or eliminate the need for finger stick testing.

> **Patient Instructions for Testing Blood Glucose in the First Few Weeks of the Diet**
> Although blood testing of glucose and ketones is not essential in every application of a ketogenic diet, the practitioner may still recommend testing as it provides clear, objective data. What follows is one recommendation for a testing schedule:
> - Using the blood glucose meter, test and record fasting blood glucose (FBG)
>   - In these early weeks, expect fasting glucose measurement to be erratic
>   - Expect levels to become low and steady with keto-adaptation
> - Test and record blood glucose in the middle of the afternoon, either before a meal or least 2 hours after a meal
> - Test blood glucose before bedtime
> - If the glucose measurement is uncharacteristically high or low, immediately retest using a new drop of blood
>   - If the two measurements are close, take the average
>   - If the second measurement matches the general trend, record that one
>
> *If using a proprietary drink formula with oils, fats, and coffee to delay the first meal of the day:*
> - Test and record fasting blood glucose then test again just before the first meal. Most likely, the second measurement will be lower than FBG. If the pre-meal test is *higher* than FBG then the length of the fast may be too long. (Do not rely on a single day's numbers: test several days before making changes in meal timing.)

22.18.3 Ketone Testing Tools [73]

As described earlier, there are three types of ketone bodies: beta-hydroxybutyrate (detected in blood), acetoacetate (detected in urine), and acetone (detected in breath).

22.18.3.1 Blood Ketones

Testing blood levels of beta-hydroxybutyrate is the standard used in research and this method has gained widespread acceptance in therapeutic ketogenic diets. This requires the use of ketone test strips that are paired with a meter that tests both glucose and ketones.

> **Testing Blood Ketones**
> - Test and record fasting ketone levels
> - Also test ketones an additional two or three times a week—either in the middle of the afternoon (before a meal or at least 2 hours after a meal) or at bedtime
> - The threshold for nutritional ketosis is 0.5 mmol/L and the upper level is considered to be in the range of 3–5 mmol/l
> - Levels at 5–7 mmol/L are associated with fasting or starvation
> - Levels above 7 mmol/L are not common in adults unless they are supplementing with exogenous ketones.

22.18.3.2 Urine Testing

Urine test strips use nitroprusside to detect the presence of acetoacetate in the urine. Nitroprusside changes to a pink tone ranging from a blush of color to deep purple—the darker the shade of purple, the more acetoacetate is present. Urine test strips are inexpensive, non-invasive, and easy to use. While urine ketones are not as accurate a measure of ketosis as blood testing, the strips can still be used to check compliance and improve accountability. (Note that testing within a few hours of vigorous exercise or when hydration levels are either higher or lower than the norm may not yield accurate results.) Results using urine strips are typically more valid in the early weeks and months, before keto-adaptation is complete.

22.18.3.3 Breath Analyzer

Specialized breath analyzers can detect the presence of acetone, another of the ketone bodies. This technology uses a small handheld device which connects to a computer or other USB-supported device. Although these devices are convenient and reusable, they do not always correlate with blood levels of beta-hydroxybutyrate. The later iterations are more precise tools though they are more expensive to maintain. Some people, notably athletes, prefer this type of testing as part of their biohacking routine as it is a somewhat quantifiable marker of ketosis. However, some of the first-generation devices may not be good tools for individuals whose lung capacity is impaired by asthma, age, or disease.

22.19 Troubleshooting

Challenges that arise in the first few weeks or months of the diet need to be addressed on a case-by-case basis before they become obstacles to compliance. What follows is an overview of the most common transition symptoms along with suggested actions.

22.19.1 Ongoing Flu-like Symptoms

The decrease in insulin levels alters the way in which kidneys handle sodium, producing a diuretic effect and concurrent loss of electrolytes [74]. Sodium and chloride can be replenished by adding salt to homemade broth. Potassium supplements, including potassium citrate, may also be used along with the broth but may not be appropriate for patients taking potassium-sparing drugs, such as Lisinopril. Magnesium supplementation should be considered for all patients but may be contraindicated in those with certain kidney diseases [75, 76]. Also, ensure adequate intake of calcium, preferably through food.

22.19.2 Hunger between Meals

Hunger and craving are common in the first few weeks. If they continue, it may indicate that the patient is not yet in sustained ketosis. Increase the frequency of ketone testing to detect any drop in ketone levels below the threshold of 0.5 mmol/L. Hunger may also persist in those who are slower to adapt to the changes in hunger signaling associated with keto-adaptation. Another common reason for persistent hunger is inadvertent non-compliance to the diet. Review the food diary for sources of hidden carbohydrates especially in beverages and food-based supplements. Also confirm that protein intake is within the suggested limits. Caloric restriction adopted along with a therapeutic ketogenic diet may result in lingering hunger—ensure that adequate amounts of fat are included in the plan.

22.19.3 Food Cravings

Determine which foods the patient is craving. If it is a sweet treat, provide suggestions for keto-friendly substitutes, such as fat bombs or keto treats. If the craving is for salty or crunchy foods, suggest sour pickles, fried pork rinds, or keto-friendly crackers with a crunch. Some people crave protein: recheck macro calculations and food records to be sure that the protein target is accurate and met through the diet. Protein foods, especially protein in eggs and dairy, will also elicit an insulin response even in the absence of carbohydrates. In some individuals, this response may be exaggerated enough to suppress ketone production. (Cravings specific to cheese may be influenced by other factors, such as opiate-like receptors for beta-casomorphins in the brain.) [79].

22.19.4 Unsustainable or Rapid Weight Loss

Weight reduction is a goal for many people who adopt a ketogenic diet, but unsustainable or rapid weight loss is not the goal of a therapeutic ketogenic diet [73]. Weight loss that is greater than two pounds a week, or weight loss of any amount in someone who is already below ideal weight, must be slowed or stopped. This is especially important post-surgery or with metabolic dysregulation associated with sarcopenia and cancer cachexia.

Resistance training is anabolic in working muscle. Therefore, it is crucial to advise patients to consider some type of strength training to preserve or build muscle mass especially during periods of rapid weight loss. This may be as basic as the use of resistance bands or light weights ideally under the watchful eye of an experienced fitness trainer that can provide feedback as to proper form and ideal degree of resistance. Exercises may need to be modified to match them to the energy level and current fitness of each individual.

To slow weight loss (or encourage slight weight gain), consider adding a small serving (50–60 g) of a low glycemic index fruit or legume (GI less than 40) to one meal a day. The patient should assess their glucose response to that meal by testing blood glucose just before the meal and again 45 minutes to an hour after the meal. Expect to see a rise in glucose of about 25–30 mg/dL—this should be sufficient to briefly stimulate insulin. Most likely, this will temporarily reduce ketone levels, but this downside may be offset by the opportunity to preserve fat stores, restore glycogen, or regain lost weight. If glucose rises by more than 25–30 mg/dL, then reduce the portion slightly and test again on another day. Test only one meal a day. This will have less of a detrimental effect on ketosis.

Weight loss may accompany reliance on MCT oil as a primary source of dietary fats. MCT is readily converted to ketones but excess ketones are excreted in the urine which reduces the energy value. If weight loss is too rapid or otherwise undesirable, counsel patients to consider MCT oil as a supplement, or "bonus."

22.19.5 Spikes in Glucose (a Rise of >25–30mg/dL) with a Meal or Snack

Check the labels on packaged foods or low-carb snack bars to identify any ingredients that might be causing glucose spikes. Look for sugars or problematic fibers, such as isomaltooligosaccharides (these are often assumed to be indigestible sugars but are known to raise blood glucose in some individuals) [78, 80]. Sauces and dressings added to restaurant meals often contain sugars or flours. Also, be wary of foods promoted as "sugar-free" and marketed to people with diabetes, as these may not be keto-friendly.

22.19.6 Steroid Medications (e.g., Prednisone, Dexamethasone, or Hydrocortisone)

Prednisone and dexamethasone are commonly prescribed to control inflammation, and hydrocortisone may be prescribed as a replacement hormone if cortisol levels are low [81]. All steroid medications cause a rise in glucose levels and should only be used under the supervision of a medical doctor. Furthermore, any attempt to discontinue or wean from these drugs must be done under the direction of the medical team.

22.19.7 Dropping out of Ketosis

Use of an online tracking tool, such as Cronometer, can help patients with both compliance and accountability by identifying food choices or portion amounts that are not in keeping with their ketogenic plan. Check to be sure that macronutrient intake is divided fairly evenly between meals. Women may also drop out of ketosis during the menstrual cycle due to normal shifts in hormones that may be associated with transient insulin resistance [78].

22.19.8 Hidden Carbohydrates in Medicines, Supplements, or Hygiene Products

A pharmacist can provide information about the carbohydrate content of medications. Also look for sugars, starches, and sugar alcohols (aside from xylitol or erythritol) in the ingredients list on food and supplement labels. Visit the Charlie Foundation website [39] and download their list of low-carb and carb-free supplements and personal hygiene products.

22.19.9 Nausea or Vomiting

Gastrointestinal diseases and disorders can trigger the release of stress hormones that stimulate gluconeogenesis. If nausea or vomiting is due to cancer or cancer treatment, the doctor or oncology nurse is likely to prescribe antiemetics. However, if nausea appears to be diet-related, reduce the amount of fats and oils to a point that does not cause symptoms then gradually increase the amount, perhaps adding a high-lipase pancreatic enzyme, lecithin, and/or ox bile that will aid in digesting fat. Other strategies include spreading fat intake out over several small meals and snacks.

22.19.10 Dissatisfaction with Food Choices

Ketogenic diets are restrictive and patients often express boredom or dissatisfaction with the limited choices, especially at breakfast. Remind patients that they are not limited to traditional breakfast foods; any keto meal can take its place. Also, there are many keto cookbooks and online recipes that can spark new interest in those who are bored with current choices.

22.19.11 High Blood Glucose Levels

In the context of a ketogenic diet, fasting blood glucose levels over 85 mg/dL, or random blood glucose levels that remain over 90 mg/dL several hours after finishing a meal are considered high. Work through these questions:

- *Is it only fasting blood glucose that is high?* This can simply be due to the dawn effect, the normal circadian rise in cortisol that stimulates a release of glucose. Another possibility, but one that is poorly understood, may rest with what is referred to as *physiological insulin resistance* (more recently referred to as adaptive glucose sparing). Put simply, low fasting insulin concurrent with a rise in early-morning blood glucose may result in a higher than expected fasting number. High fasting numbers due to either the dawn effect or physiological insulin resistance typically drop within a few hours of waking, especially if the first meal of the day is delayed, as in time-restricted feeding. If this is the case, then no intervention is needed. However, high fasting glucose can also be due to consuming too much protein at the evening meal, particularly if that meal (or other food) is eaten within a few hours of bedtime.
- *Is the patient receiving chemotherapy or radiation, or recovering from surgery?* These are all systemic "injuries" and the body's response here is the same as for all other illnesses or injuries. Expect glucose levels to be higher than they would be in individuals who are not facing these additional challenges. Glucose levels will also be higher in individuals with extensive metastatic disease in cancer—lactic acid produced as metabolic waste from excessive fermentation of glucose is converted back into glucose by the liver. (High lactate also occurs in non-disease states; for example, as a byproduct of metabolizing muscle glycogen during intense anaerobic activities.)
- *Does the patient have a history of insulin resistance?* If so, circulating levels of glucose and insulin might remain higher than expected because cells are still less sensitive to insulin. These individuals may take longer to adapt to the diet. The chronic presence of insulin also suppresses ketone production, creating a vicious cycle. (Metformin, a prescription medication that helps lower blood glucose levels, also helps to restore insulin and is sometimes used concurrent with a therapeutic ketogenic diet. Note that individuals taking Metformin may see a drop in HDL cholesterol and serum levels of vitamin B_{12}.)
- *Is the patient consuming too few calories?* Rapid weight loss (greater than 5–7 pounds in the first 2 weeks) can be a red flag that calorie intake is too low. This may trigger hormones that upregulate glycogenolysis or gluconeogenesis. Reassess the diet prescription calculations and/or check the food diary to be sure that targets are being met.
- *Is the patient waiting too long between meals?* The ideal window for time-restricted feeding varies widely among individuals. Some people can routinely fast for 16–18 hours a day while others may be physically stressed by periods greater than 12–14 hours. Even in those who maintain a wider eating window, the length of time between meals may stimulate the production of hormones that contribute to a rise in glucose. To determine the ideal amount of fasting, the patient should be advised to take several blood glucose readings at various points in the day, preferably before meals, to detect any trends related to meal timing.
- *How well is the patient coping with stress?* External stressors directly and indirectly influence blood glucose levels. For example, fear activates the sympathetic nervous system which stimulates the production of epinephrine and release of glucose through glycogenolysis. Psychological stressors can raise cortisol levels which then upregulate gluconeogenic activity. Indirect stressors, such as poor sleep, can interfere with hormone regulation and insulin sensitivity. Physical activity, social connectivity, reduction in exposure to stressful environments, improved sleep, and meditation/prayer can all be effective ways to lower stress and improve blood glucose control and insulin sensitivity.

22.20 Conclusion

The development of the infrastructure needed to support widespread application of metabolic therapies, including the ketogenic diet, lags behind the science. Acceptance by a broader contingent of medical professionals remains elusive in large part due to the high cost of conducting human clinical trials that form the basis of most medical practice guidelines. As a result, insurance providers have been slow to provide reimbursement for healthcare professionals who offer these therapies. At the heart of this problem lies a pervasive belief, even among integrative practitioners, that ketogenic plans are inherently inferior to those that are viewed as a "balanced diet." This current nutrition paradigm should be re-evaluated within the context of the body of evidence pointing to an underlying mechanism common to most disease states: specifically, that epigenetic changes in gene expression induced by an overabundant supply of glucose contribute to a broad spectrum of conditions, including obesity, diabetes, cancer, and neurodegenerative disease.

Consumer demand not only in epilepsy, but in metabolic syndrome, neurodegeneration, cancer, and in high-performance athletes have pushed research further than previously and this is poised to continue.

References

1. Food and Nutrition Board. Dietary reference intakes for energy, carbohydrate, fiber, fat, fatty acids, cholesterol, protein, and amino acids. Washington, D.C.: National Academies Press; 2005. p. 289.
2. Owen OE, Morgan AP, Kemp HG, Sullivan JM, Herrera MG, Cahill GF. Brain metabolism during fasting∗. J Clin Invest. 1967;46(10):1589–95. https://doi.org/10.1172/jci105650.
3. Cahill GF. Fuel metabolism in starvation. Ann Rev Nutrition. 2006;26(1):1–22.
4. Freeman JM, Kossoff EH, Hartman AL. The ketogenic diet: one decade later. Pediatrics. 2007;119:535–43. https://doi.org/10.1542/peds.2006-2447.
5. Institute of Medicine (US) Committee to Review Dietary Reference Intakes for Vitamin D and Calcium; Ross AC, Taylor CL, Yaktine AL, et al., editors. Washington (DC): National Academies Press (US); 2011. https://www.ncbi.nlm.nih.gov/books/NBK56068/table/summarytables.t5/?report=objectonly.
6. USDA. Dietary reference intakes: macronutrients. Retrieved from: https://www.nal.usda.gov/sites/default/files/fnic_uploads//macronutrients.pdf.
7. Volek JS, Freidenreich DJ. Metabolic characteristics of keto-adapted ultra-endurance runners. Metab Clin Exp. 2016;65:100–10.
8. Xu H, Barnes G, Yang Q, Guo G, Yang D, Chou C, et al. Chronic inflammation in fat plays a crucial role in the development of obesity-related insulin resistance. J Clin Invest. 2003;112(12):1821–30. https://doi.org/10.1172/JCI19451. Available from: https://www.jci.org/articles/view/19451.
9. Youm HY, Nguyen KY, Grant RW, Goldberg EL, Bodogai M, Kim D, et al. The ketone metabolite β-hydroxybutyrate blocks NLRP3 inflammasome–mediated inflammatory disease. Nat Med. 2015;21:263–9. https://www.nature.com/articles/nm.3804.
10. Talbot K, Wang HY, Kazi H, Han LY, Bakshi KP, Stucky A, et al. Demonstrated brain insulin resistance in Alzheimer's disease patients is associated with IGF-1 resistance, IRS-1 dysregulation, and cognitive decline. J Clin Invest. 2012;122(4):1316–38. https://doi.org/10.1172/JCI59903.
11. Ho L, Qin W, Pompl PN, Xiang Z, Wang J, Zhao Z, et al. Diet-induced insulin resistance promotes amyloidosis in a transgenic mouse model of Alzheimer's disease. FASEB J. 2004;18(7):902–4.
12. Cunnane SC, Courchesne-Loyer A, Vandenberghe C, St-Pierre V, Fortier M, Hennebelle M, et al. Can ketones help rescue brain fuel supply in later life? Implications for cognitive health during aging and the treatment of Alzheimer's disease. Front Mol Neurosci. 2016;9(53). https://www.ncbi.nlm.nih.gov/pmc/articles/PMC4937039/.
13. Pizzorno J. Mitochondria-fundamental to life and health. Integr Med (Encinitas). 2014;13(2):8–15.
14. Winter SF, Loebel F, Dietrich J. Role of ketogenic metabolic therapy in malignant glioma: a systematic review. j.critrevonc. 2017;112:41–58. https://doi.org/10.1016/j.critrevonc.2017.02.016.
15. Navale AM, Paranjape AN. Glucose transporters: physiological and pathological roles. Biophys Rev. 2016;8(1):5–9. https://doi.org/10.1007/s12551-015-0186-2.
16. Ebert K, Ewers M, Bisha I, Sander S, Rasputniac T, Daniel H, et al. Identification of essential amino acids for glucose transporter 5 (GLUT5)-mediated fructose transport. J Biol Chem. 2018;293:2115–24.
17. Hormone Health Network. What is insulin?. 2019. Retrieved from: https://www.hormone.org/hormones-and-health/hormones/insulin?
18. Chascione C, Elwyn DH, Davila M, Gil KM, Askanazi J, Kinney JM. Effect of carbohydrate intake on de novo lipogenesis in human adipose tissue. ajpendo. 1987;253(6):E664–9. https://doi.org/10.1152/ajpendo.1987.253.6.E664.
19. Wong, Roger HF; Sul, Hei Sook. Insulin signaling in fatty acid and fat synthesis: a transcriptional perspective, Curr Opin Pharmacol. DeepDyve 2010. https://www.deepdyve.com/lp/elsevier/insulin-signaling-in-fatty-acid-and-fat-synthesis-a-transcriptional-ghlncULxNq. Accessed 16 Nov 2017.
20. Westman EC, Feinman RD, Mavropoulos JC, Vernon MC, Volek JS, Wortman JA. Low-carbohydrate nutrition and metabolism. Am J Clin Nutr. 2007;86(2):276–84.
21. Ketogensis. Wikipedia. Retrieved from: https://en.wikipedia.org/wiki/Ketogenesis.
22. Bayly GR. Lipids and disorders of lipoprotein metabolism. In: Marshall W, Lapsley M, Day A, Ayling R, editors. Clinical biochemistry: metabolic and clinical aspects. 3rd ed. London: Churchill Livingstone; 2014. p. 702–36.
23. Manninen A. Metabolic effects of the very-low-carbohydrate diets: misunderstood "villains" of human metabolism. J Int Soc Sports Nutr. 2004;1(7). https://jissn.biomedcentral.com/articles/10.1186/1550-2783-1-2-7.
24. Berg JM, Tymoczko JL, Stryer L. Each organ has a unique metabolic profile. In: Freeman WH, editor. Biochemistry. 5th ed; 2002. Section 30.2 https://www.ncbi.nlm.nih.gov/books/NBK22436/.
25. Cahill GF Jr. Starvation in man. New Engl J Med. 1970;282:668–75.
26. Ahmed S, Singh D, Khattab S, Babineau J, Kumbhare D. The effects of diet on the proportion of intramuscular fat in human muscle: a systematic review and meta-analysis. Front Nutr. 2018;5:7. https://doi.org/10.3389/fnut.2018.00007. eCollection 2018. Review.PMID: 29516003.
27. Chang HT, Olson L, Schwartz KA. Ketolytic and glycolytic enzymatic expression profiles in malignant gliomas: implication for ketogenic diet therapy. Nutr Metab. 2013;10(1):47. https://doi.org/10.1186/1743-7075-10-47.
28. Leiter LA, Marliss EB. Survival during fasting may depend on fat as well as protein stores. JAMA. 1982;248(18):2306–7. https://doi.org/10.1001/jama.1982.03330180066037.
29. Ward C. Ketone body metabolism [internet]. 2015; Diapedia 51040851169 rev. no. 29. Available from: https://doi.org/10.14496/dia.51040851169.29
30. Rui L. Energy metabolism in the liver. Compr Physiol. 2014;4(1):177–97. https://doi.org/10.1002/cphy.c130024.
31. Laffel L. Ketone bodies: a review of physiology, pathophysiology and application of monitoring to diabetes. Diabetes Metab Res Rev. 1999;15:412–26.
32. Noh HS, Hah Y-S, Nilufar R, et al. Acetoacetate protects neuronal cells from oxidative glutamate toxicity. J Neurosci Res. 2006;83(4):702–9. https://doi.org/10.1002/jnr.20736.
33. Rho JM. Substantia(ting) ketone body effects on neuronal excitability. Epilepsy Curr. 2007;7(5):142–4. https://doi.org/10.1111/j.1535-7511.2007.00206.x.
34. Volek J, Phinney SD, Kossoff E, Eberstein JA, Moore J.. The art and science of low carbohydrate living: an expert guide to making the life-saving benefits of carbohydrate restriction sustainable and enjoyable. Lexington: Beyond Obesity. 2011. Pg 5.
35. Cunnane SC, Courchesne-Loyer A, Vandenberghe C, et al. Can ketones help rescue brain fuel supply in later life? Implications for cognitive health during aging and the treatment of Alzheimer's disease. Front Mol Neurosci. 2016;9:53. https://doi.org/10.3389/fnmol.2016.00053.
36. Vining EPG, Freeman JM, Ballabin-Gil K, et al. A multicenter study of the efficacy of the ketogenic diet. Arch Neurol. 1998;55:1433–7.

37. Neal EG, Chaffe H, Schwartz RH, et al. The ketogenic diet for the treatment of childhood epilepsy: a randomised controlled trial. Lancet Neurol. 2008;7:500–6.
38. Kossoff EH, Zupec-Kania BA, Amark PE, Balaban-Gil KR, Bergqvist AGC, Blackford R, et al. Optimal clinical management of children receiving the ketogenic diet: recommendations of the International Ketogenic Diet Study Group. Epilepsia. 2009;50(2):304–17. https://doi.org/10.1111/j.1528-1167.2008.01765.x.
39. Charlie Foundation. Worldwide Keto Hospital Directory. Retrieved from: http://charliefoundation.org/resources-tools-home/resources-find-hospitals/united-states/united-states-list.
40. Zupec-Kania BA, Spellman E. An overview of the ketogenic diet for pediatric epilepsy. Nutr Clin Pract. 2008 Dec–2009;23(6):589–596. doi: https://doi.org/10.1177/0884533608326138.
41. Wirrell EC. Ketogenic ratio, calories, and fluids: do they matter? Epilepsia. 2008;49(Suppl 8):17–9. https://doi.org/10.1111/j.1528-1167.2008.01825.x.
42. Kalamian M. Keto for cancer: ketogenic metabolic therapy as a targeted nutritional therapy. White River Junction: Chelsea Green Publishing; 2017. p 149.
43. Kinzig KP, Honors MA, Hargrave SL. Insulin sensitivity and glucose tolerance are altered by maintenance on a ketogenic diet. Endocrinology. 2010;151(7):3105–14. https://doi.org/10.1210/en.2010-0175.
44. Gardner CD, Kiazand A, Alhassan S, Kim S, Stafford RS, Balise RR, et al. Comparison of the Atkins, Ornish, Zone, and LEARN diets for change in weight and related risk factors among overweight premenopausal women. JAMA. 2007;297(9):969–77. https://doi.org/10.1001/jama.297.9.969.
45. Feinman RD, Pogozelski WK, Astrup A, Bernstein RK, Fine EJ, Westman EC, Accurso A, Frassetto L, Gower BA, McFarlane SI, Nielsen JV, Krarup T, Saslow L, Roth KS, Vernon MC, Volek JS, Wilshire GB, Dahlqvist A, Sundberg R, Childers A, Morrison K, Manninen AH, Dashti HM, Wood RJ, Wortman J, Worm N. Dietary carbohydrate restriction as the first approach in diabetes management: critical review and evidence base. Nutrition. 2015;31(1):1–13. https://doi.org/10.1016/j.nut.2014.06.011.
46. McKenzie AL, Hallberg SJ, Creighton BC, Volk BM, Link TM, Abner MK, et al. A novel intervention including individualized nutritional recommendations reduces hemoglobin A1c level, medication use, and weight in type 2 diabetes. JMIR Diabetes. 2017;2(1):e5. https://doi.org/10.2196/diabetes.6981.
47. Paoli A, Rubini A, Volek JS, Grimaldi KA. Beyond weight loss: a review of the therapeutic uses of very-low-carbohydrate (ketogenic) diets. Eur J Clin Nutr. 2013;67(8):789–96. https://doi.org/10.1038/ejcn.2013.116.
48. Mavropoulos JC, Yancy WS, Hepburn J, Westman EC. The effects of a low-carbohydrate, ketogenic diet on the polycystic ovary syndrome: a pilot study. Nutrition & Metabolism. 2005;2:35. https://doi.org/10.1186/1743-7075-2-35.
49. Liberti MV, Locasale JW. The Warburg effect: how does it benefit cancer cells? Trends Biochem Sci. 2016;41(3):211–8. https://doi.org/10.1016/j.tibs.2015.12.001.
50. Thomas N. Seyfried, PhD, personal communication regarding Warburg in a professional review of the manuscript for Keto for Cancer (Chelsea Green Publishing, 2017) March 19, 2017.
51. Seyfried TN, Mukherjee P. Targeting energy metabolism in brain cancer: review and hypothesis. Nutrition & Metabolism. 2005;2:30. https://doi.org/10.1186/1743-7075-2-30.
52. Gatenby RA, Gillies RJ. Why do cancers have high aerobic glycolysis? Nat Rev Cancer. 2004;4:891–9.
53. Westman EC, Yancy WS Jr, Mavropoulos JC, Marquart M, McDuffie JR. The effect of a low-carbohydrate, ketogenic diet versus a low-glycemic index diet on glycemic control in type 2 diabetes mellitus. Nutrition & Metabolism. 2008;5(36) https://doi.org/10.1186/1743-7075-5-36.
54. McDaniel SS, Rensing NR, Thio LL, Yamada KA, Wong M. The ketogenic diet inhibits the mammalian target of rapamycin (mTOR) pathway. Epilepsia. 2011;52(3):e7–e11. https://doi.org/10.1111/j.1528-1167.2011.02981.x.
55. Laplante M, Sabatini DM. mTOR signaling in growth control and disease. Cell. 2012;149(2):274–93.
56. Seyfried T. Cancer as a mitochondrial metabolic disease. Front Cell Dev Biol. 2015;3(43) https://doi.org/10.3389/fcell.2015.00043.
57. Seyfried TN. Mitochondria: the ultimate tumor suppressor. In: Cancer as a metabolic disease: on the origin, management, and prevention of Cancer. Hoboken: Wiley; 2012d. p. 195–205.
58. Gasior M, Rogawski MA, Hartman AL. Neuroprotective and disease-modifying effects of the ketogenic diet.
59. Masino SA, editor. Ketogenic diet and metabolic therapies: expanding roles in health and disease, Oxford: Oxford University Press; 2017.
60. Zhao Z, Lange DJ, Voustianiouk A, MacGrogan D, Ho L, Suh J, et al. A ketogenic diet as a potential novel therapeutic intervention in amyotrophic lateral sclerosis. BMC Neurosci. 2006;7(29):29. https://doi.org/10.1186/1471-2202-7-29.
61. Vanitallie TB, Nonas C, Di Rocco A, Boyar K, Hyams K, Heymsfield SB. Treatment of Parkinson disease with diet-induced hyperketonemia: a feasibility study. Neurology. 2005;64(4):728–30. https://doi.org/10.1212/01.WNL.0000152046.11390.45.
62. Reger MA, Henderson ST, Hale C, Cholerton B, Baker LD, Watson GS, Hyde K, Chapman D, Craft S. Effects of beta-hydroxybutyrate on cognition in memory-impaired adults. Neurobiol Aging. 2004;25(3):311–4. https://doi.org/10.1016/S0197-4580(03)00087-3.
63. Bredesen DE. Reversal of cognitive decline: a novel therapeutic program. Aging (Albany NY). 2014;6(9):707–17. https://doi.org/10.18632/aging.100690.
64. Elamin M, Ruskin DN, Masino SA, Sacchetti P. Ketone-based metabolic therapy: is increased NAD^+ a primary mechanism? Front Mol Neurosci. 2017;10(377) https://doi.org/10.3389/fnmol.2017.00377.
65. Bredesen DE, Amos EC, Canick J, et al. Reversal of cognitive decline in Alzheimer's disease. Aging (Albany NY). 2016;8(6):1250–8. https://doi.org/10.18632/aging.100981.
66. USDA. Dietary reference intakes: Macronutrients (n.d.). National Agricultural Library. (n.d.). https://www.nal.usda.gov/sites/default/files/fnic_uploads//macronutrients.pdf Accessed 29 Jan 2017.
67. Kossoff EH, Dorward JL. The modified Atkins diet. Epilepsia. 2008;49(Suppl 8):37–41. https://doi.org/10.1111/j.1528-1167.2008.01831.x.
68. Cervenka MC, Henry BJ, Felton EA, Patton K, Kossoff EH. Establishing an adult epilepsy diet center: experience, efficacy and challenges. Epilepsy Behav. 2016;58:61–8. https://doi.org/10.1016/j.yebeh.2016.02.038.
69. Hartman AL, Rubenstein JE, Kossoff EH. Intermittent fasting: a "new" historical strategy for controlling seizures? Epilepsy Res. 2013;104(3):275–9. https://doi.org/10.1016/j.eplepsyres.2012.10.011.
70. Patterson RE, Sears DD. Metabolic effects of intermittent fasting. Annu Rev Nutr. 2017;37:371–93. https://doi.org/10.1146/annurev-nutr-071816-064634.
71. Bank IM, Shemie SD, Rosenblatt B, Bernard C, Mackie AS. Sudden cardiac death in association with the ketogenic diet. Pediatr Neurol. 2008;39(6):429–31. https://doi.org/10.1016/j.pediatrneurol.2008.08.013.
72. Best TH, Franz DN, Gilbert DL, Nelson DP, Epstein MR. Cardiac complications in pediatric patients on the ketogenic diet. Neurology. 2000;54(12):2328–30.
73. Neal E. Editor: Dietary treatment of epilepsy: practical implication of ketogenic diet. Oxford: Wiley-Blackwell for John Wiley and Sons; 2012.

74. Phinney S, Volek J. Sodium, nutritional ketosis, and adrenal function. Virta Health. 2017. Retrieved from: https://blog.virtahealth.com/sodium-nutritional-ketosis-keto-flu-adrenal-function/
75. Yancy WS Jr, Olsen MK, Guyton JR, Bakst RP, Westman EC. A low-carbohydrate, ketogenic diet versus a low-fat diet to treat obesity and hyperlipidemia: a randomized, controlled trial. Ann Intern Med. 2004;140(10):769–77.
76. Phinney S, Virta Team. How does a ketogenic diet affect kidney stones? Virta Health. 2017. Retrieved from: https://blog.virtahealth.com/ketogenic-diet-kidney-stones/
77. Goldberg EL, Asher JL, Molony RD, Shaw AC, Zeiss CJ, Wang C, et al. β-Hydroxybutyrate deactivates neutrophil NLRP3 inflammasome to relieve gout flares. Cell Rep. 2017;8(9):2077–87. https://doi.org/10.1016/j.celrep.2017.02.004.
78. Phinney S. Nutritional ketosis, treating type 2 diabetes and more—A Q&A with Dr. Stephen Phinney. Virta Health. 2018. Retrieved from: https://blog.virtahealth.com/ketogenic-nutrition-type-2-diabetes-dr-steve-phinney/
79. Brantl V, Teschemacher H, Bläsig J, Henschen A, Lottspeich F. Opioid activities of β-casomorphins. Life Sci. 1981;28(17):1903–9. https://doi.org/10.1016/0024-3205(81)90297-6.
80. Kohmoto K, Tsuji K, Kaneko T, Shiota M, Fukui F, Takaku H, et al. Metabolism of ^{13}C-Isomaltooligosaccharides in healthy men. Biosci Biotechnol Biochem. 1992;56(6):937–40. https://doi.org/10.1271/bbb.56.937.
81. Bruns CM, Kemnitz JW. Sex hormones, insulin sensitivity, and diabetes mellitus. ILAR J. 2004;45(2):160–9. https://doi.org/10.1093/ilar.45.2.160.

The GUT-Immune System

Elizabeth Lipski

23.1 Introduction – 368

23.2 The Digestive System – 368

23.3 The Digestive Process – 368

23.4 The Stomach and the Immune System – 369

23.5 Functional Laboratory Testing – 369
23.5.1 Key Dietary and Lifestyle Recommendations to Raise Gastric Acid – 369

23.6 The GALT and the MALT – 370
23.6.1 Key Dietary and Lifestyle Recommendations to Raise IgA – 370
23.6.2 Key Dietary and Lifestyle Recommendations to Break Down Immune Complexes – 370

23.7 Small Intestine – 371

23.8 The Gut Microbiota – 372

23.9 The Microbiome: Bacteria – 372

23.10 Biofilm Layer – 373

23.11 Healing the Gut: The 5R Protocol – 373

23.12 Role of Nutrition in Balancing the Gut-Immune System: Therapeutic Diets – 373

23.13 Prebiotics and Benefits – 375

23.14 Selected Therapeutic Foods – 375

23.15 Therapeutic Role of Probiotic Supplements – 375

23.16 Conclusion – 376

References – 376

© Springer Nature Switzerland AG 2020
D. Noland et al. (eds.), *Integrative and Functional Medical Nutrition Therapy*,
https://doi.org/10.1007/978-3-030-30730-1_23

The gut immune system has the challenge of responding to the pathogens while remaining unresponsive to food antigens and the commensal flora. In the developed world, this ability appears to be breaking down, with chronic inflammatory diseases of the gut commonplace in the apparent absence of overt infections. –Thomas T. MacDonald/ Giovanni Monteleone (2005). Immunity, inflammation, and allergy in the gut. *Science*, 307(5717), 1920–1925.

23.1 Introduction

The focus of this chapter is to explore the synergistic relationships, between the gastrointestinal system and the immune system, and their joint effects on systemic health and inflammation. Research in these areas is mushrooming. Current research indicates that 70% of the immune system is located in the gastrointestinal system (GIT) and in the cells in the gut lumen, which is the opening in the middle of the intestines [1]. This is called the gut-associated immune system (GALT).

Food plays an enormous role in our immune response. Matzinger's Danger model of immunity states that the function of the immune system is to distinguish between exposures that are dangerous and safe. Our tissues have two main responses to foreign exposures: to be tolerant or to send alarm signals. The immune system is tolerant to exposures nearly all of the time, yet when it determines that something is dangerous, it reacts [2]. Each day we eat pounds of food, which is a "foreign matter" to the body. Therefore, the food we eat each day is a main driver of immune response. The typical Western diet is inflammatory and begins the process of digestive imbalance.

Digestion is the river of life. When we cannot digest, absorb, and utilize nutrients from food, all tissues are compromised. Other players in the gut-immune system include the integrity of the gut barrier, the role of the gut microbiome, the cascade of immune responses that signal for healing and inflammation, gut motility, and the ability of the liver, gallbladder, and pancreas to perform. This review will provide a clinical introduction.

23.2 The Digestive System

The digestive tract is comprised of the mouth, esophagus, stomach, gallbladder, liver, pancreas, small intestine, large intestine, and anus. Its job is to take food we eat and to nourish each cell in the human body. This is a finely orchestrated process and, when working correctly, is something we pay little attention. When out of balance, we experience distress in the form of pain, gas, distension, inflammation, changes in bowel movements, and immune system activation.

23.3 The Digestive Process

Eating, also called ingestion or the cephalic phase, is voluntary when materials are put in the mouth. This is our portal for nearly all nutrients to enter the body and involves the mouth, teeth, tongue, parotid, and other salivary glands. Food choices are related to lifestyle, personal values, and cultural customs. This is our first immune contact: Does the food smell good, look good, and taste good? Is thorough mastication occurring?

Digestion occurs in the mouth, stomach, and small intestine and requires cooperation from the liver and pancreas. Mechanically, foods are chewed in the mouth and churned in the small and large intestine to break the foods into molecules that can be absorbed into the bloodstream and cells. The proper levels of hydrochloric acid, bile, enzymes, and intestinal bacteria are critical for full digestive capacity. Digestion can be supported with digestive enzymes, bitters, relaxation, and thorough chewing.

Secretion Every day, the walls of the digestive tract secrete about seven quarts of water, acid, buffers, and enzymes into the lumen (inside) of the digestive tract. Secretion occurs throughout the entire digestive tract. These secretions help maintain pH levels and send water into the gut to lubricate and keep things moving. They also provide enzymes to digest foods and facilitate the digestive process. Hydration is essential for this phase of digestion to work properly.

Motility Whatever we eat is squeezed through the digestive system by a rhythmic muscular contraction called peristalsis. Sets of smooth muscles throughout the digestive tract contract and relax, alternately pushing food through the esophagus to the stomach and through the intestines. Think of a snake swallowing a mouse. The process helps the food become acidified, liquefied, neutralized, and homogenized until it's broken down into usable particles. From the time you swallow, this process is involuntary and can occur even if you stand on your head. Irritable bowel syndrome, small intestinal bacterial overgrowth (SIBO), and gastroparesis are examples of altered motility disorders. Ginger, fiber, garlic, and an herbal product called Iberogast all help to stimulate motility.

Absorption occurs in the small intestine. Digested food molecules are taken through the epithelial cell lining of the small intestine into the bloodstream and through the portal vein to the liver, where it is filtered. From the bloodstream, it passes to the cells. Until food is absorbed, it is essentially outside the body – it is in a tube going through it. If the gut is inflamed, as with gluten intolerance or celiac disease, there can be malabsorption, which is typified by anemia, diarrhea, an inability to gain weight, and a lack of growth in children. Absorption is enhanced by healing intestinal permeability and utilizing a low-antigenic, low-inflammatory therapeutic diet until this healing has occurred.

Assimilation is the process by which fuel and nutrients enter the cells. The reason we eat food is to nourish each individual cell. Diabetes is a condition of poor assimilation. Assimilation is also enhanced by healing intestinal permeability and utilizing a low-antigenic, low-inflammatory therapeutic diet until this healing has occurred.

Elimination In digestion, we excrete wastes by having bowel movements (defecation). These wastes are comprised of indigestible food components, waste products, bacteria, cells from the mucosal lining being sloughed off, and food that has not been absorbed. Constipation and diarrhea are both examples of imbalanced elimination. Chronic diarrhea contributes to malnourishment. Chronic constipation is a sign of poor motility and can increase levels of beta-glucuronidase, which implies impaired phase II liver detoxification. Hydration, exercise, dietary fiber, probiotics, and prebiotics all enhance elimination.

When microbes enter the digestive system, they are confronted with several nonspecific and antigen-specific defense mechanisms (innate immune system) including peristalsis, bile secretion, hydrochloric acid, mucus, antibacterial peptides, and immunoglobulin A, also known as IgA. This stops most microbes and parasites from infecting the body.

23.4 The Stomach and the Immune System

The stomach contains two main immune protective substances: gastric acids – also called hydrochloric acid – and pepsin. After ingestion of food, this becomes the second line of immune defense.

Protein molecules are composed of chains of up to 200 amino acids strung together. Hydrochloric acid (HCl), produced by millions of parietal cells in the stomach lining, begins to break apart these protein chains. HCl also digests bacteria, fungi, and parasites that come in with our food, effectively sterilizing the food bolus. HCl is so strong that it would burn our skin and clothing, if spilled. However, a thick coating of mucus (mucopolysaccharides) protects the stomach and keeps the acid from burning through its lining. Prostaglandins, small chemical messengers, help keep the mucus layer active by sending messages to replace and repair the stomach lining and provide a protective coating. When this mucus layer breaks down, HCl burns a hole in the stomach lining, causing a gastric ulcer.

Stomach acid also provides our first defense against food poisoning, *Helicobacter pylori*, and parasitic and other infections. Low levels of stomach acid, hypochlorhydria, have been associated with common health issues, including food poisoning, small intestinal bacterial overgrowth (SIBO), and small intestinal fungal overgrowth (SIFO). Stomach acid is also necessary for absorption of many minerals, so mineral depletion may occur with hypochlorhydria.

A normal stomach acid level is a pH of 1.5–2.5. As we age, the parietal cells in the stomach lining produce less hydrochloric acid [3]. Use of acid-blocking medications increases stomach pH to 3.5 or higher, decreasing mineral absorption and opening us up to opportunistic infection.

The stomach also makes pepsin, a protein-splitting enzyme, that cuts the bonds between specific amino acids and breaks them down into short chains of just 4–12. The stomach also produces small amounts of lipase, enzymes that digest fat. Most foods are digested and absorbed farther down the gastrointestinal tract, but alcohol, water, and certain salts are absorbed directly from the stomach into the bloodstream.

23.5 Functional Laboratory Testing

- Heidelberg capsule test. This test is typically performed in an integrative medicine clinic.
- Smart pill: This is typically performed by a gastroenterologist.
- Empiric testing: Utilize betaine HCl supplementation to see if it resolves symptoms (see below).

23.5.1 Key Dietary and Lifestyle Recommendations to Raise Gastric Acid

Betaine HCl Begin with one capsule of betaine HCl with a protein-containing meal. Most people will respond with discomfort, burning, or warmth in their belly. Any sign of discomfort indicates that there is sufficient acid already. Stop the experiment by neutralizing the acid with 1 tsp of baking soda in water or milk. For people who experience no discomfort, titrate the dosage up slowly to a maximum of 2500 mg of betaine HCl per protein-containing meal. Any sign of discomfort indicates too much acid, so you cut the dosage. If symptoms resolve, this supplement can be continued. If symptoms do not resolve, stop. Betaine HCl is an acid and can create an ulcer, if used improperly.

Umeboshi plum Umeboshi plums/apricots are a traditional condiment used throughout Japan, Korea, and China for their health benefits, which include improvements in periodontal disease, improved motility, and inhibition of *H. pylori*. The ume plum is picked unripe, dried in the sun, and then pickled in a brine of sea salt and shiso leaves. The net result is a highly alkaline, naturally fermented pickle that is rich in enzymes and probiotics and has high antioxidant and antibiotic properties. It has been used traditionally for hangovers, liver support, detoxification, nausea, bad breath, dysentery, typhoid, paratyphoid, appetite stimulant, and skin diseases such as eczema [4–8].

Umeboshi plums can be eaten in many ways. They are used as a condiment on vegetables or rice. Plums are very salty but can be eaten whole from the jar. Umeboshi vinegars are also available for use in salad dressings or on rice or other grains. One can also take whole plums or umeboshi paste and drink as a tea. Just let it steep in boiled water for 5 minutes and then drink. This is a very restorative tonic.

Increase acidity with diluted vinegar Dilute 1 teaspoon of vinegar in 10 teaspoons of water and drink with each meal. Gradually increase the amount of vinegar, up to 10 teaspoons. If you experience burning, immediately neutralize the acid by drinking a glass of milk or taking a teaspoon of baking soda in water.

Swedish bitters Bitters are a long-standing remedy for poor digestion in Europe. Combinations of herbs such as gentian, ginger, cinnamon, cardamom, and others stimulate production of HCl and bile, act as antimicrobials, and increase motility. Take bitters either in tablet or liquid form as needed.

Change your eating habits and reduce stress Chew food thoroughly and eat small meals frequently. Small meals are easier to digest. Avoid drinking liquids with meals because fluids dilute stomach acid. Regularly practice stress management by decluttering your calendar, meditating, receiving acupuncture or massage, or doing tai chi or qigong.

23.6 The GALT and the MALT

Throughout our digestive systems, we have immune tissue called gut-associated lymphoid tissue (GALT) and, in the mucosal lining of the digestive lumen, the mucosal-associated lymphoid tissue (MALT). The MALT also encompasses your nose, bronchia, and, in women, the vulvovaginal areas. Altogether, 70% of your immune system lies within the GALT and digestive MALT, which protect us from antigens and other foreign invaders. The tonsils, appendix, and the Peyer's patches within the small intestine are examples of GALT.

Secretory IgA (sIgA) is the primary way that the MALT conveys the message of immune assault. Secretory IgA are antibodies – sentinels on constant alert for foreign substances present in the gut mucosa. When challenged by foreign molecules, sIgA forms immune complexes with allergens and microbes [4, 5, 9]. Immune complexes are clumps of IgA or IgG antibodies that signal the immune system to respond. If these immune complexes get deposited in organs, they can cause disease themselves. Some IgA-associated diseases are IgA nephropathy, vasculitis, lupus, rheumatoid arthritis, scleroderma, and Sjogren's syndrome. Immune complexes signal cytokines to begin the inflammatory process designed to rid our bodies of antigenic materials, a response of the adaptive immune system. Without sufficient sIgA, the MALT cannot work properly [5, 6, 10].

Secretory IgA response is the initial protective antibody which comprises 80% of those in the gut. It is a protective "scout" that signals trouble. Deficiency of sIgA is the most common immunodeficiency. IgA can be assessed in serum and stool. Low levels of sIgA make us more susceptible to infection and may be a fundamental cause of asthma, autoimmune diseases, candidiasis, celiac disease, chronic infections, food allergies, and more. In other words, if the sentry isn't standing at the gate, anyone can come in!

23.6.1 Key Dietary and Lifestyle Recommendations to Raise IgA

Rest and relaxation Ask your clients to rest for 2 hours during daylight every day. Typically, within a few weeks, they are beginning to feel amazingly better.

23.6.1.1 Dietary Supplementation Recommendations

These are suggestions to consider, but avoid if an individual is sensitive or allergic to a product.
- *Saccharomyces boulardii*. Typical dosage 250 mg three times daily [6–8, 11, 12].
- *Vitamin A*. Typical dosage 5000 IU daily. It is recommended to obtain a baseline *vitamin A retinol* blood level before beginning vitamin A retinyl palmitate supplementation to ensure it is indicated for an individual. Avoid recommending dosages of retinol if a person already has a high-end or elevated vitamin A retinol, because vitamin A supplementation above 10,000 IU daily can be toxic; monitor serum retinol (vitamin A) levels every 2–3 months when supplementing.
- *Bovine colostrum:* Typical dosage 500–2000 mg daily.
- *Transfer factor:* Dosage 12.5–75 mg.
- *Medium chain triglycerides (MCT oil)*: Typical dosage up to 2 tbsp. Daily. MCT oil is available as C8, C8–C10, oil, and powder; or eat coconut or drink coconut milk or fresh coconut juice [8, 9, 13].

23.6.2 Key Dietary and Lifestyle Recommendations to Break Down Immune Complexes

23.6.2.1 Dietary Supplementation Recommendations

These are suggestions to consider, but avoid if an individual is sensitive or allergic to a product.

Proteolytic enzymes are taken between meals to increase absorption to assist autophagy processes to break down the blood-clotting enzyme, fibrin, decreasing hypercoagulation, and cleaning up cell debris. Protease enzymes taken between meals can lessen food reactions and decrease systemic inflammation.
- *Porcine proteolytic enzymes*: Contain chymotrypsin and trypsin along with other combinations of proteolytic enzymes
- *Vegan proteolytic enzymes*: Most commonly contain vegan sources of bromelain (pineapple) and papain (papaya), along with other enzymes combinations
- *Nattokinase enzyme:* Speeds up biochemical reactions; extracted from fermented soybeans with a bacterium *Bacillus natto*. Used commonly for cardiovascular heart and circulatory system diseases at 540 FU (fibrinolytic units) dosage providing the ability to break down the blood-clotting enzyme fibrin
- *Boluoke and lumbrokinase* oral enzymes: Derived from earthworms; fibrin-dissolving enzyme used commonly for heart disease and hypercoagulation of cancer, as well as general inflammation

23.7 Small Intestine

The small intestine is hardly small. If this coiled-up garden hose were stretched out, it would average 15–20 feet long. If spread flat, it would cover a surface the size of a tennis court. In the small intestine, food is completely digested and nutrients are absorbed through cells called "enterocytes." The enterocytes are covered with hundreds of small fingerlike folds called villi, which in turn are covered by millions of microvilli. Think of them as small loops on a velvety towel that then have smaller threads projecting from them. The enterocyte layer is only one cell layer thick, but it performs multiple functions of producing digestive enzymes, absorbing nutrients, and blocking absorption of substances that are not useful to the body. The enterocytes are coated with a mucous layer called the "biofilm."

The enterocyte lining is semipermeable. The spaces between the cells are called "tight junctions" and are regulated by occludens and zonulin. Absorption of nutrients and substances into the bloodstream typically occurs through each cell; some occurs between the tight junctions. Zonulin is a molecule that opens the gates between the tight junctions. To date, there are two known activators of zonulin: gluten-containing grains in people who have celiac disease and lipopolysaccharides from gram-negative bacteria (LPS). High levels of zonulin have been reported in autoimmune disease, lung conditions, cancers, and neurological conditions such as multiple sclerosis and schizophrenia [14–16]. When you have high levels of LPS, it is called "endotoxemia." LPS particles are released from mostly gram-negative bacteria as they die and go into circulation, binding to TLR-4 receptors on muscles and organs. This triggers inflammatory molecules such as IL-6 and inflammatory cytokines, ultimately damaging tissue. Elevated LPS has also been associated with changes in tryptophan metabolism, increasing quinolinic acid, kynurenine, cortisol, and inflammation with simultaneous decreases in insulin sensitivity, thyroid function, melatonin, and glutathione.

The intestinal lining repairs and replaces itself every 3–5 days. The intestinal lining has a paradoxical function: it allows nutrients to pass into the bloodstream while blocking the absorption of foreign substances found in chemicals, bacterial products, and other large molecules found in food. If maintenance is delayed due to stress, injury, infection, toxic exposure, medications (such as antibiotics steroids, oral contraceptives), or illness, there will be increased intestinal permeability, also called "leaky gut." When this occurs, large molecules such as toxins, microbes, chemicals, and molecules from food enter the bloodstream, triggering an immune system reaction (Fig. 23.1).

Leaky gut syndrome puts an extra burden on the liver. All foods pass directly from the bloodstream through the liver for filtration. The liver "humanizes" the food and either lets it pass or changes it, breaking down or storing all toxic or foreign substances. Water-soluble toxins are easily excreted, but the breakdown of fat-soluble toxins is a two-stage process that requires energy. When the liver is bombarded by inflammatory irritants from incomplete digestion, it has less energy to neutralize chemical substances. When overwhelmed, the liver stores toxins in fat cells to deal with later. If the liver has time later, it can deal with the stored toxins, but most commonly it is busy processing what is newly coming in and never catches up. These toxins provide a continued source of inflammation to the body. Increased intestinal permeability is the driving factor in liver diseases, such as cirrhosis and nonalcoholic fatty liver disease (NAFLD) [29–32].

Stop the mediators that perpetuate the problem:
- *Excess stress.* Create the space in your life to relax and renew every day. Nap; meditate; do abdominal breathing exercises, tai chi, qigong, or a hobby; or watch birds in your yard. You choose, but do choose.
- *Poor sleep.* Your body needs 7–9 hours of sleep each night. When healing, often more. Go to bed early enough to achieve this. It is hard to heal when you are sleep-deprived.
- *Pain medications* that injure GI lining, such as nonsteroidal anti-inflammatories (NSAID), i.e., ibuprofen and aspirin.
- *Use of birth control pills, steroid medications.*
- *Environmental contaminants.*

Fig. 23.1 Increased intestinal permeability associated with many health conditions. (Based on data from Refs. [14, 17–28])

Increased Intestinal Permeability Associated with Many Health Conditions

Acute pancreatitis	Chronic Fatigue Syndrome	Kidney Stones
Asthma		Migraines
Autism	Eczema	Multi-Organ Failure
Autoimmune hepatitis	Food Intolerances	Multiple sclerosis
Autoimmune disease	Fibromyalgia	Non-Alcoholic Liver Disease
Bechet's disease	IgA nephropathy	Primary biliary cirrhosis
Burn patients	Inflammatory bowel diseases	Primary sclerosing cholangitis
Celiac Disease	Irritable Bowel Syndrome	Psoriasis

- *Overconsumption of alcohol.*
- *Poor food choices.*
- *Treat infection.*

Diet:
- Avoid foods that you are sensitive to.
- Eat gut-healing foods. Homemade chicken stock, beef stock, and fish stocks are inexpensive and contain nutrients, such as gelatin, glucosamine, and chondroitin, that help heal leaky gut better than anything else. It's a food, so consume to taste.

Supplements: Can be found individually or in combination products.
- L-glutamine (glutamine): Dosages vary 1–30 g daily. Begin with 1–3 g daily. Heat destroys its properties, so take with cold or cool beverage. Best on empty stomach. Many people find that glutamine enhances their muscle endurance, which is a lovely side benefit. Too much glutamine can be constipating, so use that as an indicator. Occasionally glutamine makes someone feel wired or anxious. If so, stop taking it; it is not for you. If you have kidney disease, be cautious in taking more than 1–3 g daily. Caution when recommending glutamine for oncology patients, especially later-stage progression as some cancers use glutamine as a fuel, along with glucose [33, 34].
- Zinc or zinc carnosine: Typical dose 75 mg of zinc carnosine twice daily (totals 34 mg of zinc) or 25–50 mg of other type of zinc.
- Aloe vera. Dosage varies. If utilizing fresh aloe, use one to two pieces the size of a thumb. Peel and utilize only the inside pulp.
- Colostrum: 1000–3000 mg daily.
- Probiotics. *Lactobacillus plantarum* is specifically soothing to the small intestine. Take 1 daily.
- Quercetin. Dosage: 500–3000 mg daily.
- Digestive enzymes with meals

Optional supplements:
- Gamma oryzanol. Dosage 100 mg three times daily
- Fish peptides. Brand: Seacure. Six capsules daily in divided doses. Keep in freezer to make it more palatable.
- Vitamin A. 5000 IU of preformed vitamin A: retinol. Do not take >5000 IU if you are pregnant or planning to become pregnant. Best to test vitamin A retinol status before determining need for supplementation.
- Increase antioxidants. Vitamin C: dosage 500–10,000 mg daily, and/or a full-spectrum antioxidant supplement.
- Deglycyrrhized licorice. Dosage two tablets between meals as needed, up to four times daily.
- Phosphatidylcholine. 2000–4000 mg daily.
- Herbal blends with combinations of marshmallow root, slippery elm, arabinogalactan, ashwagandha, ginger, MSM, etc. usually powdered or capsules.

23.8 The Gut Microbiota

One of the newest discoveries in the digestive system is the microbiome. Each of us has 2–7 pounds of microbes in our digestive system, about the size of the brain or liver. While these live throughout the entire digestive tract from the mouth to the anus, the great majority live in the colon and large intestine.

These microbes are comprised of bacteria, beneficial viruses known as bacteriophages or phage, fungi, archaea, and parasites that constitute more than 99% of the DNA in the human body [35]. This is a complex community that functions to stabilize overall health and keep us robust. We have as many bacteria in our microbiome as cells in our body [36]. Viruses equal or outnumber bacteria in the microbiome and regulate cell signaling of interferons and cytokines [37, 38]. The mycobiome are the fungi in our microbiome; they have not been well-studied as of this writing. *Saccharomyces boulardii* is the best-studied of the beneficial fungi. It is typically used as a prebiotic and to treat diarrhea from all causes [39–43]. Current research suggests that imbalances in this "organ" enhance nutrient absorption; synthesize vitamins; and drive obesity, type 2 diabetes, nonalcoholic fatty liver disease, mental health issues, drug and toxin metabolism, and intolerance to pain (Fig. 23.2).

23.9 The Microbiome: Bacteria

Of microbes, there are many families and phyla, with the phyla *Bacteroidetes* and *Firmicutes* predominant. These families are comprised of hundreds of subfamilies and types of bacteria. No matter their names, they fall into three categories: symbiotic bacteria, commensals, and pathogens. Symbiotic bacteria provide a benefit to the host, such as *Lactobacilli* and *Bifidobacterium* species. When used in supplemental form, we call these symbiotic microbes "probiotics." Commensals enhance digestion of fats, proteins, and carbohydrates and synthesis of SCFA and vitamins. They also aid detoxification and protect us from pathogenic microbes [44]. As more is learned about the specific roles of each commensal, many will be recategorized as symbiotic. Pathogens, which have gathered the most attention because of their devastating effects, comprise a smaller group of disease-causing microbes. In a healthy human gut, these live in balanced communities. When out of balance, we call this "dysbiosis," which is a general disordered microbiome rather than an infection with a specific pathogen.

In the digestive system alone, you have many different microbiomes. For example, your gingiva, tonsils, tongue, teeth, and saliva each have their own microbiomes. If you kiss someone, do you collect some of their microbiota? Research indicates you receive 80 million bacteria for every 10 second kiss [45]!

The microbiome begins to develop in the womb and matures during the first 3 years of life. Research indicates that this development is associated with how well our immune system and metabolism function. This process can be

Fig. 23.2 The microbiome in health and disease

The Microbiome in Health & Disease

IN HEALTH
- Digestive Health: motility, function, digestion
- Drug Metabolism
- Heavy Metal Detoxification
- Increased nutrient absorption: fats, protein, minerals
- Metabolism
- Modulate Immunity /Inhibit Pathogens
- Overall Health
- Production of Short-Chain-Fatty Acids (SCFA)
- Synthesis of vitamins: biotin, folate, B1, B2, B3, B6, B12, and K.

OUT OF BALANCE / DYSBIOSIS
- Atopic Illness
- Autoimmune Disease
- Cancers
- Cardiovascular disease
- Chronic kidney disease
- Dermatological issues
- Diabetes
- Digestive Conditions
- Heart Failure
- Liver diseases
- Mental Health: Depression, Schizophrenia, Anxiety, Autism, ADHD
- Obesity
- Pain Tolerance
- Sleep disruption

enhanced with vaginal delivery and breastfeeding, having a furry pet in the home [46], playing outdoors in dirt, and having older siblings. This development may be hampered by birth by C-section, bottle-feeding, antibiotics, and use of other medications [47–49] (Fig. 23.3).

23.10 Biofilm Layer

The gut microbes live in communities called biofilms. The biofilm layer contains antibacterial proteins and acts as a protective barrier for the microbial communities, protecting the gut lumen. Mucin proteins contain glycosylated carbohydrates that are used as fuel by commensal and symbiotic microbes, converting them into SCFA, which provide energy for the enterocytes.

When the microbial communities are out of balance, infection can set in the biofilm layer and also in the epithelium to which they adhere. Commensal and pathogenic microbes typically stay inside the gut lumen, yet this is an imperfect system and some continuously make their way into the gut-associated lymphatic tissue (GALT), triggering an immune and inflammatory response.

Although we have no real research on protection of the biofilm layer, gut-soothing foods and herbs have been used traditionally to enhance mucus production and soothe the gut. In herbal medicine, these are called "demulcents." These include licorice, slippery elm, marshmallow root (*Althaea officinalis*), fenugreek tea, figs, almonds, oats, and okra. Utilizing a gut-healing therapeutic dietary approach will also be of benefit. (See later section in this chapter.)

23.11 Healing the Gut: The 5R Protocol

The 5R protocol is widely utilized in integrative and functional medicine.

- *Remove*: Remove stressors such as infection, toxins, foods that provoke reactions, processed foods, refined sugars, nonessential medications, and molds.
- *Replace*: Enhance digestion by utilizing digestive enzymes, betaine hydrochloride, and bile salts when needed. Replace an inflammatory Western diet with a whole-food diet or gut-healing therapeutic dietary plan.
- *Reinoculate*: Eat probiotic- and prebiotic-rich foods. Take probiotic and/or prebiotic supplements containing immune-regulating species such as *L. acidophilus*, *L. reuteri*, *Bifidobacterium*, *Streptococcus thermophilus*, and *Bacilli* species.
- *Repair*: Eat gut-healing foods such as bone stock and vegetable broths. Eat at least nine servings (4.5 cups) daily of vegetables and fruit. Polyphenols and plant bioactives regulate inflammatory mediators. Increase dietary intake of gut-healing nutrients such as zinc, L-glutamine, vitamin A, and omega-3 fatty acids. Utilize nutritional supplementation as discussed for increased intestinal permeability: aloe vera, quercetin, colostrum, etc.
- *Rebalance*: When someone is ill and improves, they will come to a "new normal" which may be healthier or less robust than before becoming ill. This may affect positively or negatively what one can achieve in a day, week, or year and will require adjustment. Examine lifestyle issues, such as sleep, exercise, and stress management.

23.12 Role of Nutrition in Balancing the Gut-Immune System: Therapeutic Diets

Research also indicates that changing what we eat has the greatest impact on our microbiome [50, 51]. Food can have both inflammatory and healing effects. In your initial assessment, you will look for macronutrient, micronutrient, fiber, and hydration adequacy through a food diary and any available clinical assessments. If someone is fatty acid- or protein-insufficient, digestion and absorption will not be optimized.

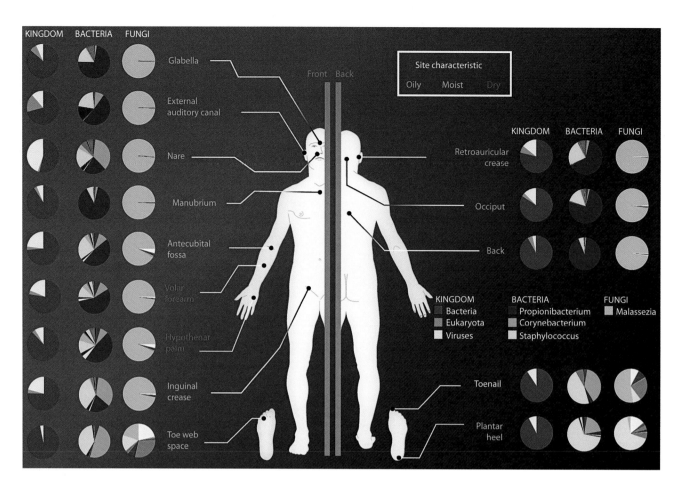

Fig. 23.3 Microbiome sites: the microbiota composition of each tissue is unique. (Reprinted from: ▶ https://commons.wikimedia.org/wiki/File:Microbiome_Sites_(27058471125).jpg. With permission from Creative Commons Attribution: ▶ https://creativecommons.org/licenses/by/2.0/)

Recommendation of an anti-inflammatory, low-antigenic food plan can decrease inflammation better than medications. The standard Western diet is inflammatory, so begin with a whole-food, low-processed elimination diet protocol. Utilizing a therapeutic elimination diet is simple: Avoid foods that potentially cause inflammation and eat only foods that have a low possibility of provoking a reaction. This is a therapeutic approach and used for a minimum of 2 weeks to a maximum of 3 months with a reintroduction phase to determine which foods provoke reactivation of symptoms. In some cases, some people may need to stay on a modified therapeutic dietary program for up to 2 years until the gut heals completely. Often the foods people react to are the ones that we least want to live without. If a patient says "I just couldn't live without dairy," dairy may be that patient's problem food. The food groups with the most common reactions include refined sugar of all types, gluten-containing grains, dairy products, soy-containing foods, and eggs.

There are many types of elimination diets. What they have in common is that they are based in whole foods and have the effect of enhancing digestion and absorption, reducing inflammation, helping to balance the microbiome, increasing intestinal permeability, and reducing toxic burden. Many also restrict carbohydrates. The best studies and most frequently utilized approaches include the comprehensive elimination diet, FODMAP diet, and the specific carbohydrate diet. Although prebiotic, probiotic, and high-fiber foods are health-promoting when we are healthy, these foods feed the microbial imbalance when we have dysbiosis. Therefore, avoiding them for a period of time helps starve out the problematic microbes. These are therapeutic diets and not meant to become a lifestyle. It's important to heal the other underlying factors, so that people can begin eating more normally again. One caution, is that these diets can feed into disordered eating patterns. So, good counseling about their temporary nature is important. Diet, along with herbal or pharmaceutical treatment for dysbiosis, and healing permeability, provides a synergistic approach for healing.

For people in whom inflammation, pain, and/or mental health issues dominate, try a comprehensive elimination diet. For people with dysbiosis or those who feel worse eating any carbohydrates, eating a low-carbohydrate, low-starch diet works best. Examples of this include the candida diet, body ecology diet, and the gut and psychology syndrome diet (GAPS), although currently we have no published research on the benefits.

While many foods may be unmasked during the elimination/provocation challenge, others may remain hidden. To enhance this process, you may also want to have a blood test for food sensitivities, allergies, and intolerances. Looking at antibodies is examining the adaptive/acquired immune system. IgG, IgG4, IgA, and IgM tests report on food sensitivities, although IgG or IgG4 are most common. IgE testing reports on true allergic reactions. Testing for lactose, fructose, and celiac disease reports on enzymatic insufficiencies. Tests such as the MRT-LEAP and ALCAT tests report on the adaptive immune process. There is no standardization of food sensitivity testing; yet, any of these tests can be clinically useful by lowering the threshold of activation.

Sensitivities are rated from normal to severe reactions. In addition to a detailed readout documenting individualized reactions, most laboratories also include a list of foods which contain hidden sources of the offending foods, a rotation menu, and other educational material to help patients in the healing process.

23.13 Prebiotics and Benefits

Prebiotics are food for the microbes of the microbiome. The average person today eats few prebiotic-rich foods, although historically we ate many. The definition of a prebiotic is: "A substrate that is selectively utilized by host microorganisms conferring a health benefit" [52]. In other words, prebiotics are food for the symbiotic and commensal microorganisms of the microbiome. Typically, they are "eaten" by the microbes which activates the bioactive food components. They contribute to colon health by synthesizing short-chain fatty acids: butyrate, valerate, and propionate. Prebiotics have been reported to have many benefits (◘ Fig. 23.4). Some diets, such as the low-FODMAP diet, temporarily restrict many prebiotic-rich foods.

Breast milk is a rich source of prebiotics and helps to establish the infant microbiome. Common prebiotic foods include kitchen herbs and spices, legumes, pulses, jicama, Jerusalem artichoke, onions, chicory, garlic, leeks, plantain, unripe bananas, fruit, burdock root, asparagus, radishes, dandelion root, peas, honey, green tea, black tea, cultured dairy products, artichoke, tomatoes, beets, and whole grains such as rye, barley, and sprouted wheat.

23.14 Selected Therapeutic Foods

Kefir: Supports immune health with a variety of prebiotics and probiotic microbes. Research reports that it supports gut-immune health, improves lactose digestion, has cancer-protective properties, and has been used to heal peptic and duodenal ulcers [55]. Kefir can be made from animal milk (cow, goat, sheep, buffalo, yak) or nondairy liquids such as water, nut milk, or seed milk.

Kimchi: Reported to help regulate bowel movements, improve metabolism and weight, reduce high-serum cholesterol levels, provide anticancer nutrients, promote immune and brain health, and reduce fibrinogen levels [56].

Ginger: Ginger has been reported in many papers to be effective for digestive conditions including colic, nausea, appetite stimulation, and digestion enhancement. It is used as a prokinetic to enhance motility in small intestinal bacterial and fungal overgrowth [57–59].

Cabbage juice: Research reports that cabbage juice is an effective immune modulator of peptic ulcers, diabetes, cirrhosis, arthritis, and cancers [60]. It has also been reported to enhance motility and gastric secretion. It contains the nutrients glutamine, methionine, and sulforaphanes. The dosage utilized was a quart of fresh cabbage juice daily for 7–10 days in divided disease. The taste is not pleasant, so diluting it with cherry or pomegranate or other juice may be beneficial for compliance. It is also effective as a dried powder in capsules [61–67].

23.15 Therapeutic Role of Probiotic Supplements

The definition of a probiotic is: "live microorganisms that, when administered in adequate amounts, confer a health benefit on the host" [68]. Probiotics from food and supplements modulate immune and inflammatory responses. Probiotics also improve lactose tolerance, enhance metabolism, regulate serum lipids, enhance digestive function,

◘ Fig. 23.4 Prebiotic benefits. (Based on data from Refs. [52–54])

Prebiotic Benefits	
• Improved bowel function • Promote growth of Bifidobacteria, • Lactobacilli and other beneficial microbes • Colon pH • Protect against negative effects of bile acids • Substrate for SCFA • Improves intestinal permeability • Improved Metabolism of microbiota • Adds sweetness to food • > Satiety • Stimulates neurochemical production in the gut	• > bone density (+ calcium) • > serum cholesterol and triglycerides • Cancer protective • Used in treatment of atherosclerosis • Immune function • Neural and cognitive function • Skin • > insulin sensitivity & glucose regulation (in all and Type 2 DM) • > mineral absorption • Small but sig. effects body weight

help control inflammatory bowel disease and irritable bowel syndrome, improve innate immunity, inhibit infection, and balance the microbiome [69]. Probiotic supplementation has demonstrated benefits in virtually all health conditions. They have been shown to be of benefit in digestive and immune conditions including celiac disease, parasitic infection, *H. pylori* infection, inflammatory bowel disease, bariatric surgery, irritable bowel syndrome, cardiometabolic syndrome, allergy, lupus erythematosus, rheumatoid arthritis, and lower serum cortisol levels. They can also help protect from a variety of cancers. What is less exactly known is what strains, dosages, and timing of specific probiotics to use in specific people at a specific moment in time. Dosages range from millions of colony-forming units (CFU) daily to trillions of CFU of supplemental probiotics daily.

23.16 Conclusion

The gut-immune system is an enormous topic; this chapter looks at some key points in a concise manner and provides clinical pearls to direct the nutrition care process. Research in the area of the gut-immune system, its relationship to systemic balance and disease, and the role of food, nutrition, and lifestyle is ballooning exponentially and will continue to evolve. Common themes for balancing the gut-immune system include stress management, increases in dietary polyphenols and plant bioactive components, use of digestive healing therapeutic dietary approaches, and inclusion of probiotic- and prebiotic-rich foods and supplements to modulate immune expression and promote health. Of critical importance is optimizing barrier function of the enterocytes to prevent and reduce immune and inflammatory responses. Diet and lifestyle play a major role in maintaining gut-immune health. Utilizing integrative and functional approaches promotes health maintenance and helps restore function throughout the body.

References

1. Vighi G, Marcucci F, Sensi L, Di Cara G, Frati F. Allergy and the gastrointestinal system. Clin Exp Immunol. 2008;153(Suppl 1):3–6.
2. Matzinger P. The evolution of the danger theory. Interview by Lauren Constable, Commissioning Editor. Expert Rev Clin Immunol. 2012;8(4):311–7.
3. Britton E, McLaughlin JT. Ageing and the gut. Proc Nutr Soc. 2013;72(1):173–7.
4. Kuleshnyk L. Umeboshi plum for a hang-over cure! macrobiotic guide [Internet]. [cited assessed 16 July 2008]. Available from: http://macrobiotics.co.uk/articles/umeboshi.htm.
5. Morimoto-Yamashita Y, Kawakami Y, Tatsuyama S, Miyashita K, Emoto M, Kikuchi K, et al. A natural therapeutic approach for the treatment of periodontitis by MK615. Med Hypotheses. 2015;85(5):618–21.
6. Maekita T, Kato J, Enomoto S, Yoshida T, Utsunomiya H, Hayashi H, et al. Japanese apricot improves symptoms of gastrointestinal dysmotility associated with gastroesophageal reflux disease. World J Gastroenterol. 2015;21(26):8170–7.
7. Tamura M, Ohnishi Y, Kotani T, Gato N. Effects of new dietary fiber from Japanese Apricot (Prunus mume Sieb. et Zucc.) on gut function and intestinal microflora in adult mice. Int J Mol Sci. 2011;12(4):2088–99.
8. Enomoto S, Yanaoka K, Utsunomiya H, Niwa T, Inada K, Deguchi H, et al. Inhibitory effects of Japanese apricot (Prunus mume Siebold et Zucc.; Ume) on Helicobacter pylori-related chronic gastritis. Eur J Clin Nutr. 2010;64(7):714–9.
9. Lord RBJ, editor. Laboratory evaluations for integrative and functional medicine. 2nd ed. Duluth: Metametrix Institute; 2009.
10. Grethlein SJ, Besa EC. Mucosa-associated lymphoid tissue. Emedicine [Internet]. 2008. [cited assessed 29 Jan 2010]. https://emedicine.medscape.com/article/207891-overview.
11. Jahn HU, Ullrich R, Schneider T, Liehr RM, Schieferdecker HL, Holst H, et al. Immunological and trophical effects of Saccharomyces boulardii on the small intestine in healthy human volunteers. Digestion. 1996;57(2):95–104.
12. Buts JP, Bernasconi P, Vaerman JP, Dive C. Stimulation of secretory IgA and secretory component of immunoglobulins in small intestine of rats treated with Saccharomyces boulardii. Dig Dis Sci. 1990;35(2):251–6.
13. Kono H, Fujii H, Asakawa M, Maki A, Amemiya H, Hirai Y, et al. Medium-chain triglycerides enhance secretory IgA expression in rat intestine after administration of endotoxin. Am J Physiol Gastrointest Liver Physiol. 2004;286(6):G1081–9.
14. Rittirsch D, Flierl MA, Nadeau BA, Day DE, Huber-Lang MS, Grailer JJ, et al. Zonulin as prehaptoglobin2 regulates lung permeability and activates the complement system. Am J Physiol Lung Cell Mol Physiol. 2013;304(12):L863–72.
15. Fasano A. Intestinal permeability and its regulation by zonulin: diagnostic and therapeutic implications. Clin Gastroenterol Hepatol. 2012;10(10):1096–100.
16. Fasano A. Zonulin and its regulation of intestinal barrier function: the biological door to inflammation, autoimmunity, and cancer. Physiol Rev. 2011;91(1):151–75.
17. Fasano A. Zonulin, regulation of tight junctions, and autoimmune diseases. Ann N Y Acad Sci. 2012;1258:25–33.
18. Hunt PW. Leaky gut, clotting, and vasculopathy in SIV. Blood. 2012;120(7):1350–1.
19. Lee SH. Intestinal permeability regulation by tight junction: implication on inflammatory bowel diseases. Intest Res. 2015;13(1):11–8.
20. Lerner A, Matthias T. Possible association between celiac disease and bacterial transglutaminase in food processing: a hypothesis. Nutr Rev. 2015;73(8):544–52.
21. Li X, Atkinson MA. The role for gut permeability in the pathogenesis of type 1 diabetes--a solid or leaky concept? Pediatr Diabetes. 2015;16(7):485–92.
22. Lin R, Zhou L, Zhang J, Wang B. Abnormal intestinal permeability and microbiota in patients with autoimmune hepatitis. Int J Clin Exp Pathol. 2015;8(5):5153–60.
23. Luther J, Garber JJ, Khalili H, Dave M, Bale SS, Jindal R, et al. Hepatic injury in nonalcoholic steatohepatitis contributes to altered intestinal permeability. Cell Mol Gastroenterol Hepatol. 2015;1(2):222–32.
24. Mulak A, Bonaz B. Brain-gut-microbiota axis in Parkinson's disease. World J Gastroenterol. 2015;21(37):10609–20.
25. Raftery T, Martineau AR, Greiller CL, Ghosh S, McNamara D, Bennett K, et al. Effects of vitamin D supplementation on intestinal permeability, cathelicidin and disease markers in Crohn's disease: results from a randomised double-blind placebo-controlled study. United European Gastroenterol J. 2015;3(3):294–302.
26. Nouri M, Bredberg A, Westrom B, Lavasani S. Intestinal barrier dysfunction develops at the onset of experimental autoimmune encephalomyelitis, and can be induced by adoptive transfer of auto-reactive T cells. PLoS One. 2014;9(9):e106335.
27. Yacyshyn B, Meddings J, Sadowski D, Bowen-Yacyshyn MB. Multiple sclerosis patients have peripheral blood CD45RO+ B cells and increased intestinal permeability. Dig Dis Sci. 1996;41(12):2493–8.

28. Maes M, Mihaylova I, Leunis JC. Increased serum IgA and IgM against LPS of enterobacteria in chronic fatigue syndrome (CFS): indication for the involvement of gram-negative enterobacteria in the etiology of CFS and for the presence of an increased gut-intestinal permeability. J Affect Disord. 2007;99(1–3):237–40.
29. Pacifico L, Bonci E, Marandola L, Romaggioli S, Bascetta S, Chiesa C. Increased circulating zonulin in children with biopsy-proven nonalcoholic fatty liver disease. World J Gastroenterol. 2014;20(45):17107–14.
30. Wong VW, Wong GL, Chan HY, Yeung DK, Chan RS, Chim AM, et al. Bacterial endotoxin and non-alcoholic fatty liver disease in the general population: a prospective cohort study. Aliment Pharmacol Ther. 2015;42(6):731–40.
31. Betrapally NS, Gillevet PM, Bajaj JS. Changes in the intestinal microbiome and alcoholic and nonalcoholic liver diseases: causes or effects? Gastroenterology. 2016;150(8):1745–55.e3.
32. Nier A, Engstler AJ, Maier IB, Bergheim I. Markers of intestinal permeability are already altered in early stages of non-alcoholic fatty liver disease: studies in children. PLoS One. 2017;12(9):e0183282.
33. Altman BJ, Stine ZE, Dang CV. From Krebs to clinic: glutamine metabolism to cancer therapy. Nat Rev Cancer. 2016;16(10):619–34.
34. DeBerardinis RJ, Lum JJ, Hatzivassiliou G, Thompson CB. The biology of cancer: metabolic reprogramming fuels cell growth and proliferation. Cell Metab. 2008;7(1):11–20.
35. Wexler HM. Bacteroides, the good, the bad, and the nitty-gritty. Clin Microbiol Rev. 2007;20:593–621. https://doi.org/10.1128/CMR.00008-07.
36. Sender R, Fuchs S, Milo R. Revised estimates for the number of human and bacteria cells in the body. PLoS Biol. 2016;14(8):e1002533.
37. Foca A, Liberto MC, Quirino A, Marascio N, Zicca E, Pavia G. Gut inflammation and immunity: what is the role of the human gut virome? Mediat Inflamm. 2015;2015:326032.
38. Cadwell K. The virome in host health and disease. Immunity. 2015;42(5):805–13.
39. Stier H, Bischoff SC. Saccharomyces boulardii CNCM I-745 influences the gut-associated immune system. MMW Fortschr Med. 2017;159(Suppl 5):1–6.
40. Zeber-Lubecka N, Kulecka M, Ambrozkiewicz F, Paziewska A, Lechowicz M, Konopka E, et al. Effect of saccharomyces boulardii and mode of delivery on the early development of the gut microbial community in preterm infants. PLoS One. 2016;11(2):e0150306.
41. Swidsinski A, Loening-Baucke V, Schulz S, Manowsky J, Verstraelen H, Swidsinski S. Functional anatomy of the colonic bioreactor: impact of antibiotics and Saccharomyces boulardii on bacterial composition in human fecal cylinders. Syst Appl Microbiol. 2016;39(1):67–75.
42. Consoli ML, da Silva RS, Nicoli JR, Bruna-Romero O, da Silva RG, de Vasconcelos Generoso S, et al. Randomized clinical trial: impact of oral administration of saccharomyces boulardii on gene expression of intestinal cytokines in patients undergoing colon resection. JPEN J Parenter Enteral Nutr. 2016;40(8):1114–21.
43. More MI, Swidsinski A. Saccharomyces boulardii CNCM I-745 supports regeneration of the intestinal microbiota after diarrheic dysbiosis - a review. Clin Exp Gastroenterol. 2015;11:237–55.
44. Sompayrac L, editor. How the immune system works. 5th ed. Chichester: Wiley; 2016.
45. Kort R, Caspers M, van de Graaf A, van Egmond W, Keijser B, Roeselers G. Shaping the oral microbiota through intimate kissing. Microbiome. 2014;2:41.
46. Tun HM, Konya T, Takaro TK, Brook JR, Chari R, Field CJ, et al. Exposure to household furry pets influences the gut microbiota of infant at 3-4 months following various birth scenarios. Microbiome. 2017;5(1):40.
47. Laursen MF, Zachariassen G, Bahl MI, Bergstrom A, Host A, Michaelsen KF, et al. Having older siblings is associated with gut microbiota development during early childhood. BMC Microbiol. 2015;15:154.
48. Martin R, Makino H, Cetinyurek Yavuz A, Ben-Amor K, Roelofs M, Ishikawa E, et al. Early-life events, including mode of delivery and type of feeding, siblings and gender, shape the developing gut microbiota. PLoS One. 2016;11(6):e0158498.
49. Mulligan CM, Friedman JE. Maternal modifiers of the infant gut microbiota: metabolic consequences. J Endocrinol. 2017;235(1):R1–r12.
50. Bengmark S. Nutrition of the critically ill—a 21st-century perspective. Nutrients. 2013;5(1):162–207.
51. Wu GD, Chen J, Hoffmann C, Bittinger K, Chen YY, Keilbaugh SA, et al. Linking long-term dietary patterns with gut microbial enterotypes. Science. 2011;334(6052):105–8.
52. Gibson GR, Hutkins R, Sanders ME, Prescott SL, Reimer RA, Salminen SJ, et al. Expert consensus document: the International Scientific Association for Probiotics and Prebiotics (ISAPP) consensus statement on the definition and scope of prebiotics. Nat Rev Gastroenterol Hepatol. 2017;14(8):491–502.
53. Belorkar SA, Gupta AK. Oligosaccharides: a boon from nature's desk. AMB Express. 2016;6(1):82.
54. Collins S, Reid G. Distant site effects of ingested prebiotics. Nutrients. 2016;8(9):E523.
55. de Oliveira Leite AM, Miguel MA, Peixoto RS, Rosado AS, Silva JT, Paschoalin VM. Microbiological, technological and therapeutic properties of kefir: a natural probiotic beverage. Braz J Microbiol. 2013;44(2):341–9.
56. Park KY, Jeong JK, Lee YE, Daily JW 3rd. Health benefits of kimchi (Korean fermented vegetables) as a probiotic food. J Med Food. 2014;17(1):6–20.
57. Giacosa A, Morazzoni P, Bombardelli E, Riva A, Bianchi Porro G, Rondanelli M. Can nausea and vomiting be treated with ginger extract? Eur Rev Med Pharmacol Sci. 2015;19(7):1291–6.
58. Haniadka R, Rajeev AG, Palatty PL, Arora R, Baliga MS. Zingiber officinale (ginger) as an anti-emetic in cancer chemotherapy: a review. J Altern Complement Med. 2012;18(5):440–4.
59. Lazzini S, Polinelli W, Riva A, Morazzoni P, Bombardelli E. The effect of ginger (Zingiber officinalis) and artichoke (Cynara cardunculus) extract supplementation on gastric motility: a pilot randomized study in healthy volunteers. Eur Rev Med Pharmacol Sci. 2016;20(1):146–9.
60. Miron A, Hancianu M, Aprotosoaie AC, Gacea O, Stanescu U. Contributions to chemical study of the raw polysaccharide isolated from the fresh pressed juice of white cabbage leaves. Rev Med Chir Soc Med Nat Iasi. 2006;110(4):1020–6.
61. Szaefer H, Krajka-Kuzniak V, Licznerska B, Bartoszek A, Baer-Dubowska W. Cabbage juices and indoles modulate the expression profile of AhR, ERalpha, and Nrf2 in human breast cell lines. Nutr Cancer. 2015;67(8):1342–54.
62. Zhgun AA, Aloiants GA. Effect of dehydrated cabbage juice on the secretory, acid- and pepsin-forming function of the stomach. Voenno-meditsinskii zhurnal. 1971;(4):36–8.
63. Wolffenbuttel E. Vitamin U (anti-ulcer factor), contained in cabbage juice, etc., as a possible means of correcting late dumping syndrome. Rev Bras Med. 1958;15(3):215–6.
64. Kh B, G'Labov T. Effect of juices of fresh and dried cabbage on secretory and motor functions of the stomach. Vopr Pitan. 1957;16(4):19–26.
65. Cheney G, Waxler SH, Miller IJ. Vitamin U therapy of peptic ulcer; experience at San Quentin Prison. Calif Med. 1956;84(1):39–42.
66. Cheney G. Vitamin U therapy of peptic ulcer. Calif Med. 1952;77(4):248–52.
67. Cheney G. Rapid healing of peptic ulcers in patients receiving fresh cabbage juice. Calif Med. 1949;70(1):10–5.
68. Hill C, Guarner F, Reid G, Gibson GR, Merenstein DJ, Pot B, et al. Expert consensus document. The International Scientific Association for Probiotics and Prebiotics consensus statement on the scope and appropriate use of the term probiotic. Nat Rev Gastroenterol Hepatol. 2014;11(8):506–14.
69. Bengum PS, Madhavi C, Rajagopal S, Viswanath B, Razak MA, Venkataratnamma V. Probiotics as functional foods: potential effects on human health and its impact on neurological diseases. Int J Nutr Pharmacol Neurol Dis. 2017;7(2):23–33.

Centrality of the GI Tract to Overall Health and Functional Medicine Strategies for GERD, IBS, and IBD

Ronald L. Hoffman

24.1 Impact of the GI Tract on Immunity – 380

24.2 GI Impact on the Brain – 381

24.3 GI Endocrine Effects – 381

24.4 GI Impact on Metabolism – 381

24.5 GERD – 382

24.6 Diet for GERD – 382

24.7 Supplements for GERD – 383

24.8 Irritable Bowel Syndrome – 383

24.9 GI-IBD – 385

References – 387

© Springer Nature Switzerland AG 2020
D. Noland et al. (eds.), *Integrative and Functional Medical Nutrition Therapy*,
https://doi.org/10.1007/978-3-030-30730-1_24

"All disease begins in the gut." This quote is frequently attributed to the ancient Greek physician Hippocrates nearly 2500 years ago. Traditional medical systems acknowledged the primacy of the gut as many remedies for systemic disorders involved purgatives to expel toxins and pernicious humors.

With the Scientific Revolution, the emergence of a rationalistic model ushered in insights about the process of digestion and nutrient assimilation. This resulted in a more reductionist model of the "alimentary canal" as a food tube with little function except as a mere conduit for nutrients and waste products (see ◘ Fig. 24.1).

Nevertheless, the early twentieth century saw a reinstatement of a more global appreciation of the role the GI tract plays in overall physical and mental well-being [1, 2].

The pioneering work of Eli Metchnikoff highlighted the involvement of intestinal microbes in promoting or attenuating "autointoxication." This was accompanied by popular reform movements, such as that of John Harvey Kellogg calling for enemas, fiber, and fasting to avert "toxemia." Radical surgeons even proposed colectomies to abrogate the "autointoxication" that was a presumed cause of "melancholia." [3]

The pendulum soon swung back with a firm repudiation of intestinal toxicity as an imaginary hypothesis. In 1912, famed pediatric surgeon Hastings Gilford stated that "autointoxication, the term as we commonly use it, is significant of nothing but a sort of mental flatulency on the part of the user as a substitute for thought." [4]

Consequently, investigation of the role of the GI tract as a determinant of overall health was kept in abeyance for much of the twentieth century, until a recent resurgence of interest in the microbiome led to a greater appreciation of its role in mediating metabolism, immunity, brain health, and endocrine function.

Microbiota lie within the anatomical gut. They consist of communities of bacteria, fungi, and viruses. Rather than being casual colonizers, the constituents of the microbiome elaborate essential nutrients, provide immunological signaling, perform detoxifying functions, influence metabolism, and influence neurotransmitters and hormones (see ◘ Fig. 24.2).

The essentiality of intestinal flora that makes up the microbiome has been underscored by experiments in which germ-free animals encounter numerous liabilities. The composition of the microbiome is, to some extent, genetically programmed; its proper acquisition is encouraged by inoculation via vaginal birth and supported by humoral factors secreted in breast milk. Western lifestyle is thought to pose serious challenges to the integrity of the microbiome (see ◘ Table 24.1, "Factors that undermine the microbiome").

To turn the conventional paradigm on its head, the GI tract is not merely a conduit for feces, but rather a carefully engineered housing for the microbiome, which itself may well deserve the status of an organ, with multiple functions.

24.1 Impact of the GI Tract on Immunity

The intestinal microbiota are important constituents of the human immune system and are essential for protection against pathogens. Evidence supporting this comes from studies showing that animals bred to be germ-free are highly susceptible to infection. The presence of commensal microbes provides necessary stimulation to the intestinal immune system which, via structures like Peyer's patches, regulates local and systemic immune responses via the elaboration of immunoglobulins [5].

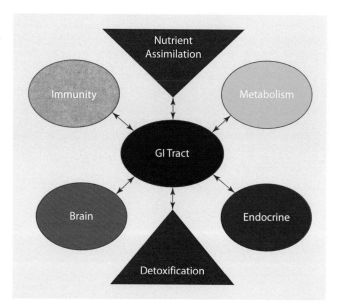

◘ Fig. 24.2 Broader understanding of the GI tract's bidirectional systemic relationships

◘ Fig. 24.1 Mechanistic model of GI tract

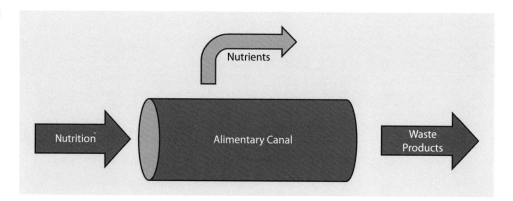

Clinical studies in humans suggest a robust effect of probiotic supplementation on infectious disease susceptibility. Frequency and duration of upper respiratory infections were reduced with oral probiotic administration [6].

It has also been demonstrated that human microbial dysbiosis is associated with production of pathogenic cytokines, which suggests a causative role in inflammatory and autoimmune disorders [7]. Researchers revealed that the presence of filamentous bacteria in young mice with an engineered predisposition to autoimmunity caused spontaneous generation of antibodies to cell nuclei – a feature of autoimmune disorders like lupus and systemic sclerosis [8]. Multiple sclerosis patients were found to have reduced levels of good bacteria ordinarily derived from diets high in healthy fiber [9]. Children with MS had different intestinal bacteria than those without MS [10].

24.2 GI Impact on the Brain

The gut-brain connection is now the subject of intense investigation. A well-acknowledged example of the impact of the GI tract on brain function is the phenomenon of hepatic encephalopathy in patients with advanced liver cirrhosis. Asterixis and delirium are relieved by administration of an osmotic laxative (lactulose) and luminally active antibiotics (neomycin and more recently rifaximin). Lately, probiotics have been shown to prevent hepatic encephalopathy in cirrhotics [11].

Speculation has arisen over the relationship between dysbiosis and neurodevelopmental disorders like autism. It has been demonstrated that children with autism have different bacteria and that their microbiota may modulate behavioral and physiological abnormalities [12]. While psychological stress has been found to alter the microbiome, the authors conclude that bacterial anomalies are the cause, not the consequence of, behavioral disorders [13].

Additional evidence has been adduced for the role of the microbiome in Parkinson's disease, Alzheimer's disease, depression, and anxiety. In fact, clinical trials of probiotic administration have been demonstrated to impact mood [14].

Table 24.1 Factors that undermine the microbiome

Antibiotics, antimicrobials (intentionally administered or from food or water)	Artificial sweeteners
Chlorinated, fluoridated water	Drugs (PPIs, H2 blockers)
C-section	Stress
Lack of breastfeeding	Sedentary lifestyle
Refined carbohydrates	Western hygiene
Inadequate fiber and polyphenols	GMOs?
Chemicalized food (e.g., emulsifiers)	

24.3 GI Endocrine Effects

The GI tract influences endocrine function in many ways. The systemic immune dysregulation that underlies celiac disease—and even non-celiac gluten sensitivity—has been shown to be associated with thyroid autoimmunity, with consequent hypothyroidism [15]. In a reciprocal fashion, the intestinal stasis that is sometimes a feature of hypothyroidism has been demonstrated to reinforce small intestinal bacterial overgrowth (SIBO), perpetuating a vicious cycle wherein dysbiosis generates pathological cytokines that promote thyroid autoimmunity.

Bacterial action is, in part, responsible for conjugation and enterohepatic processing of endogenous sex hormones and xenobiotics (sometimes referred to as "gender benders"). Hence, dysbiosis may be linked to such conditions as estrogen dominance, menstrual abnormalities, and reproductive system cancers [16]. The impact of the microbiome may account for the demonstrated ability of oral administration of fiber and especially lignans from flax and other plant sources to attenuate hormone imbalances.

Lipopolysaccharides (LPS) from bacteria have also been shown to inhibit growth hormone (GH), which may help account for the growth retardation seen in children with inflammatory bowel disease—which is sometimes treated with growth hormone injections.

It is even thought that the intestinal microbiome exerts an effect on skeletal fitness via its impact on sex hormones, glucocorticoids, serotonin, PTH, and cytokines [17].

24.4 GI Impact on Metabolism

In a series of remarkable experiments, it was demonstrated that propensity to obesity can be transferred from a fat rodent to a thin rodent via fecal inoculation; the converse also occurs when flora from lean animals are transferred to fat animals [18]. This underscores the notion that disruption of normal intestinal flora (dysbiosis) can predispose to metabolic derangement—even diabetes and associated comorbidities.

There is evidence that the relative balance of two major bacterial phyla within the digestive tract—*Firmicutes* and *Bacteroidetes*—is a determinant of susceptibility to weight gain [19]. This has prompted speculation that the global obesity epidemic is fueled not just by caloric excess and sedentary lifestyle but also by harmful environmental influences on the microbial diversity that characterizes a healthy microbiome.

While this phenomenon has not been definitively confirmed in humans, there are anecdotal accounts of individuals who have received fecal transplants for *Clostridium difficile* infections and became overweight after the donors were not properly screened for obesity [20].

Additionally, while it was once thought that the weight loss in gastric bypass surgery was mechanistically attributable to deliberately induced malabsorption and/or reduced gastric volume, new evidence has emerged that alterations in the GI microbiome may be, at least in part, responsible for altered metabolic disposition of consumed nutrients. Transfer of gut microbiota from Roux-en-Y-treated mice resulted in weight loss in surgically naive mouse recipients, suggesting a robust effect of microbial alteration on weight [21].

How does the microbiome exert its influence on metabolism? For one, nutrient extraction may be dependent on microbiota composition. For another, byproducts of microbial degradation of nutrients may play a role in signaling complex regulatory pathways, promoting or attenuating appetite and weight gain via insulin, adiponectin, leptin, or intestinal peptides [22]. Additionally, the presence of certain organisms may promote inflammation, an acknowledged contributor to metabolic syndrome [23].

The relationship of the microbiome to metabolism is now the subject of intense research. NIH funds were recently authorized for a study to see if the microbiome is linked to diabetes risk in Hispanic Americans, a population that has seen obesity rates skyrocket secondary to cultural changes [24].

24.5 GERD

For nearly 80 years, it has been unchallenged scientific dogma that heartburn (more recently dubbed gastroesophageal reflux disease or GERD) is a matter of unchecked acid secretion. Tens of millions of Americans have chewed calcium pills, quaffed first-generation acid blockers (Tagamet, Zantac, Pepcid), and now are reliant on even more powerful proton pump inhibitors (PPIs) like Prilosec, Prevacid, Nexium, Aciphex, and Protonix.

But according to a revolutionary new study, we may have been barking up the wrong tree: GERD is now thought to be caused by inflammation, not just hyperacidity. Researchers at the University of Texas Southwestern Medical Center and Dallas VA Medical Center have proven that GERD is actually an inflammatory response prompted by the secretion of pro-inflammatory cytokines [25].

Researchers "burned" the intestinal linings of mice with acid equivalent to the pH of refluxed gastric juice. They discovered that the resulting damage was not consistent with the typical pathology of GERD. Rather, the acid triggered a slow-developing inflammatory cascade that took weeks to resemble the usual findings of GERD.

This doesn't mean that acid suppression is without value in treating GERD. But it does suggest that targeting inflammation would be a better way of treating and preventing it, especially since long-term PPI use has so many side effects (see ▶ Box 24.1).

Additionally, many GERD patients obtain only incomplete or partial relief from their antacid drugs. Why should this matter to the estimated 20% of Americans suffering from upper GI pain? While acid blockers offer temporary relief, they don't get to the bottom of the problem. Additionally, chronic use may engender rebound hyperacidity. Therefore millions of Americans remain reliant on them, sometimes for their entire lifetimes.

Box 24.1 Conditions associated with chronic proton pump inhibitor use
- Nutrient deficiencies
- Osteoporosis
- Susceptibility to gastroenteritis and *Clostridium difficile*
- Coronary artery disease/MI
- Stroke
- Dementia
- Pneumonia
- Irritable bowel syndrome
- Nausea, rash, headache

24.6 Diet for GERD

Lifestyle interventions have long been fundamental to management of GERD symptoms. Patients are exhorted to lose weight, per likelihood that intra-abdominal pressure associated with excess girth may prompt reflux. They should eat smaller meals and refrain from eating before recumbency at bedtime. They should elevate the head of their beds, but not simply use extra pillows, which can increase pressure on the abdomen. Tight, constricting clothing should be avoided. Smoking is discouraged. Use of aspirin and NSAIDs, which have an erosive effect on the esophageal lining, is to be minimized.

Traditional GERD diets entail elimination of classic food and beverage precipitants:
1. Acidic foods and beverages like coffee, citrus, tomatoes, and vinegars
2. Carbonated beverages, which induce belching
3. Fried and fatty foods, which delay gastric emptying
4. Spicy food
5. Peppermint candies or gum, which relax the lower esophageal sphincter
6. Alcoholic beverages

But, in clinical practice, these measures often fail to curtail GERD symptoms. The reason may be that some patients have idiosyncratic reactions to specific "innocuous" foods, like dairy, wheat, or eggs. These intolerances are unlikely to be revealed by allergy skin testing or the IgE RAST. Rather, they must be unmasked via elimination and reintroduction.

Additionally, many patients with GERD react to the ubiquitous food additive MSG, often overlooked as a harmless "natural flavoring" in processed foods. MSG provokes gastric acid hypersecretion [26].

A promising avenue for patients unresponsive to the usual diet measures for GERD is the very low-carbohydrate (VLC) diet. In a small sample of patients, the VLC diet demonstrated symptom relief and reduced distal esophageal acid exposure [27]. While the authors claim the mechanism for this improvement is unknown, it is likely that carbohydrate restriction reduces microbial proliferation in the gut, as in IBS.

24.7 Supplements for GERD

DGL (Deglycyrrhizinated Licorice) While DGL has not been formally studied for GERD, studies in the 1980s demonstrated its curative effects in gastritis. While licorice has demonstrable anti-inflammatory effects, it contains a substance, glycyrrhetinic acid, which can cause pseudoaldosteronism, a condition where the body retains sodium, loses potassium, and accumulates fluid volume, sometimes resulting in edema and hypertension.

Melatonin Considerable research validates the benefits of melatonin in GERD. Melatonin [28] is best known as a sleep aid and circadian rhythm regulator. In the body, it is manufactured by the pineal gland, located deep within the brain. Often overlooked is that the gastrointestinal tract contains up to 400 times more melatonin than is secreted by the pineal gland. Research demonstrates that melatonin has the ability to suppress excess acid production. It also shields the GI lining from the destructive effects of free radicals caused by stress, toxic agents, or ulcer-causing drugs like NSAIDs.

A recent study showed that melatonin outperformed famotidine and boosted the effectiveness of omeprazole in alleviating symptoms and preventing tissue damage in patients with GERD [29]. The researchers summarized: "From the results of our study, it can be concluded that melatonin could be used in the treatment of GERD, either alone or in combination with omeprazole. The combination therapy of both melatonin and omeprazole is preferable, as melatonin accelerates the healing effect of omeprazole and therefore shortens the duration of treatment and minimizes its side effects." The suggested dose of melatonin for GERD is 3–6 mg at bedtime.

Mucilaginous Substances Marshmallow, plantain, and apple pectin are mucilaginous substances that soothe the esophageal mucosa; they belong to a category of herbs known for demulcent properties.

Limonene While its method of action is poorly understood and there is a dearth of human clinical studies, d-limonene, a terpene extract of citrus fruit, has been advanced as a potential ameliorant for GERD [30].

Alginate A "raft-forming agent" in the presence of gastric acid, alginate forms a gel. In the presence of bicarbonates added to the formula, alginate forms a pH buffer layer that offers protection to the esophageal mucosa.

STW 5 (Iberogast) A popular German formulation used extensively in Europe for treatment of dyspepsia, Iberogast is a liquid combination of the herbs *Iberis amara*, Angelica, chamomile, caraway fruit, St. Mary's thistle, balm leaves, peppermint leaves, celandine, and licorice root. Together, they exert anti-inflammatory, antibacterial, and pro-motility effects.

The latter makes Iberogast particularly helpful for GERD. By including certain herbal "bitters" that are traditional digestive aids, Iberogast promotes optimal gastric emptying and intestinal transit, alleviating the stuck feeling that many GERD sufferers experience. Due to its dual antibacterial and pro-motility effects, Iberogast is ideal for sufferers of SIBO (small intestinal bacterial overgrowth), a frequent contributor to the reflux and sour eructation of heartburn sufferers.

Aloe Another traditional remedy for gastrointestinal ailments, aloe is most familiar as a home first-aid treatment for burns—hence its application to esophageal inflammation.

For aloe to work best, it is preferable to use the gel form, in generous amounts, away from food. I generally suggest ¼ cup of a purified aloe gel product or fresh aloe juice 3 or 4 times daily away from meals.

In a recent study, aloe syrup was found to be comparable to Zantac and Prilosec in relieving symptoms of GERD and was well tolerated [31].

Probiotics No formal studies have been done on administering probiotics for GERD. But it stands to reason that easing dysbiosis might relieve out-of-control bacterial proliferation which can lead to upper intestinal gas and upward reflux of stomach contents.

Acidify Among the more controversial theories about GERD is that it may be the paradoxical result, not of too much stomach acid, but of too little. This condition, called hypochlorhydria, is known to be common in individuals over 50, especially women.

According to this theory, popularized by Jonathan Wright, MD, and others, lack of stomach acid results in poor motility, and incompletely digested food remains in the stomach, causing reflux [32]. This theory seems logical, as adequate acidification of the stomach can be a signal for gastric emptying, enabling food to transit to the small intestine.

On the other hand, the strategy can backfire: Some GERD sufferers report worsened pain when they take acid supplements like betaine hydrochloride as a digestive aid prior to meals. Therefore, a cautious empirical trial of a capsule or two of betaine HCl prior to meals (some patients prefer a tablespoon or two of apple cider vinegar) might be warranted in some cases of GERD, if it eases symptoms.

24.8 Irritable Bowel Syndrome

Irritable bowel syndrome (IBS) was once thought to be mostly psychogenic in origin, because there were no serological markers or definitive pathological changes to set it apart from distinct organic conditions like inflammatory bowel disease (IBD). We now know that there are a variety of measurable factors that characterize it.

While essentially a diagnosis of exclusion in the absence of other disease entities deemed more "serious," IBS can roughly be categorized into diarrhea predominant (IBS-D) or constipation predominant (IBS-C), with a considerable proportion of cases occupying the middle ground where diarrhea may alternate with constipation with associated bloating, urgency, tenesmus, mucus in the stool, and colicky pain (mixed type or IBS-M).

Recently, considerable interest has been generated about a subtype of IBS, dubbed postinfectious, which may be the long-term sequel to a bout of gastroenteritis. Symptoms persist, despite resolution of the initial infection, with no discernible traces of the implicated pathogen. The theory is that the infection may set the stage for prolonged dysbiosis; alternatively, there may be low-grade microscopic colitis, with lymphocyte aggregations seen on biopsy, but without the characteristic ulcerations or granulations pathognomonic of IBD. Indeed, a new blood test has been introduced to identify postinfectious IBS [33].

Much has been written about visceral hypersensitivity, genetically determined or acquired, in which feedback from the enteric plexus is amplified by aberrant brain circuits. From a conventional medicine standpoint, this argues for the use of bowel antispasmodics or antidepressant/anxiolytic medications. On the other hand, adopting a holistic perspective, natural modalities such as yoga, relaxation and meditation, biofeedback, hypnotherapy, and acupuncture might be invoked.

Fiber is often tapped by gastroenterologists as a panacea for IBS, but it is uncertain whether refined forms of soluble or insoluble fiber confer the benefits of plant polyphenols, present in fresh fruits and vegetables and even coffee, chocolate, spices, and tea, which may be vital substrates for cultivation of a healthy intestinal flora. For many Westerners on diets bereft of fiber, adding dietary or supplemental fiber may go a long way toward resolving digestive issues. On the other hand, overzealous fiber replacement may exacerbate IBS symptoms.

Of late, the concept of SIBO (small intestinal bacterial overgrowth) has been popularized as way of addressing IBS-D. One reason for the embrace and commercialization of this paradigm has been the introduction of the luminally active antibiotic rifaximin, which can curtail bacterial proliferation in the small intestine. But results can be temporary and equivocal, especially if patients do not address dietary precipitants of IBS.

Of course, celiac disease must be ruled out, but even in the absence of serological markers or biopsy-proven anatomical changes to the villi, non-celiac gluten sensitivity can be a major contributor to IBS [34]. While gluten elimination is mostly helpful for IBS-D, the presence of opiate-like gliadorphins in gluten may sometimes account for stubborn cases of constipation. Similarly, casomorphins from dairy products can trigger mu receptors in the colon, slowing intestinal transit.

Another way of defining optimal diets for IBS sufferers is by means of allergy testing. Conventional skin testing and IgE RAST testing are not useful for identifying food precipitants of IBS, but one study has demonstrated utility of IgG food allergy testing [35].

A useful accoutrement to antimicrobial treatment of SIBO, as well as a worthwhile starting point to test the efficacy of an elimination diet for IBS symptoms even prior to initiation of antimicrobial therapy, is the low-FODMAP (fermentable oligosaccharides, disaccharides, monosaccharides, and polyols) diet. Rather than a diet per se, it proposes a trial elimination of potentially problematic carbohydrates and sugar alcohols (see Table 24.2) [36, 37]. These are thought to provoke symptoms due to partial indigestibility (e.g., disaccharide intolerance), osmotic laxative action (e.g., polyols), or as a substrate for gas-generating fermentation. A recent study confirmed that IBS-D patients assigned to a low-FODMAP diet for 4 weeks obtained statistically significant symptom relief [38].

The low-FODMAP diet is not meant to be a lifelong prescription that requires rigid adherence; rather, it is a means for patients and nutritionists to identify potential food pre-

Table 24.2 Common foods high in FODMAPs

Fructose	Lactose	Fructans	Mannitol	Sorbitol	Galactans
Fruit: Apples, mango, pear, watermelon *Vegetables*: Asparagus, artichokes, sugar snap peas *Others*: Agave, high-fructose corn syrup, honey	Dairy (cow, goat, sheep)	*Fruit*: Custard apples, white peaches, nectarines, persimmon, watermelon *Vegetables*: Artichokes, garlic, leek, onion, spring onion (white part only), shallot *Grains/cereals*: Barley, rye, wheat-based food products *Nuts and legumes*: Cashews, pistachios, chickpeas, legumes, lentils *Others*: Fructo-oligosaccharides, inulin	*Fruit*: Stone fruits, peach, watermelon *Veggies*: Cauliflower, mushrooms, snow peas *Others*: Mannitol	*Fruit*: Apples and stone fruits; sugar-free candies and gum *Others*: Sorbitol	*Legumes*: Chickpeas, lentils, legumes (e.g., kidney beans, soy beans)

Based on data from Refs. [36, 37]

cipitants. Foods originally proscribed on the low-FODMAP diet that are well tolerated may be reintegrated.

Natural antimicrobials have applicability to SIBO because of their broad spectrum of action that discourages antibiotic and fungal resistance, their tolerability, relative affordability, and potential for long-term administration. These include berberine from Oregon grape root; aromatic oils from oregano, thyme, sage, and rosemary; olive leaf polyphenols; gentian; and allicin-rich garlic extracts.

Enteric-coated peppermint oil is well studied for relieving the symptoms of "spastic colon," a term formerly used for a common manifestation of IBS. In time-release form, the peppermint is released past the gastroesophageal junction, where it might exert a relaxing effect, prompting reflux. Instead, the peppermint acts on the intestinal smooth muscle, where it slows peristalsis by delaying the action potential of muscle fibers. In addition, peppermint has broad-spectrum antimicrobial effects, helping to alleviate SIBO.

Certain botanical agents possess pro-motility effects, which can overcome constipation, but also relieve the stasis which perpetuates colonization of the small intestine via translocation of inappropriate colonic bacteria. These include the carminative oils which are said to soothe the gastrointestinal tract, stimulate appetite and the release of digestive enzymes, and promote peristalsis and gastric emptying. They include fennel, cardamom, cumin, caraway, lemon balm, as well as many other herbs that are said to possess carminative properties.

Chamomile has a long tradition of use in treatment of gastroenteritis and especially IBS-D. It is thought to impart soothing and antispasmodic effects, with a mild CNS sedative component [39].

A formula consisting of nine herbs (including some of the aforementioned), STW 5 (Iberogast®), has been approved and used safely for five decades in European countries for the treatment of functional dyspepsia and IBS [40]. For constipation, a gradual ramp-up of magnesium citrate can prove helpful without the dependency of herbal laxatives that contain anthraquinones. Prebiotics from kiwi fruit, acacia, and fructo-oligosaccharides (FOS) may help to alleviate constipation, but care must be taken lest they produce uncomfortable gas and bloating (they are specifically contraindicated on the low-FODMAP diet).

Hydrochloric acid and digestive enzymes may play a role in relieving IBS symptoms by facilitating complete breakdown of food constituents.

Administration of certain strains of probiotics is documented to alleviate IBS-C [41]. Less consistent results have been seen in IBS-D. In the latter, it may be that certain species exert competitive inhibition or possess antimicrobial properties that discourage the proliferation of harmful flora. Alternatively, probiotics may exert a direct or indirect ameliorative effect on low-grade chronic intestinal immune activation that characterizes some forms of IBS. A recent pilot study showed alleviation of each of the cardinal symptoms of IBS (abdominal pain/discomfort, distention/bloating, and difficult defecation) with *Bifidobacterium* infantis; other studies have supported benefits of lactobacilli as well as the probiotic "cocktail" VSL#3® [42].

For acute diarrhea, the use of the soluble fiber pectin, along with adsorbent clays like kaolin, may rid the intestine of irritating bacterial endotoxins.

New discoveries have led us to appreciate the role that histamine may play in provoking intestinal symptoms. The intestinal lining is richly populated with mast cells that release histamine in combination with zonulin, a substance that disrupts gap junctions of intestinal epithelial cells and alters permeability. This may account for extraintestinal manifestations of not just celiac disease but also other less classic forms of food intolerance or dysbiosis.

Just as it does in the nasal passages, histamine produces hyperemia, swelling, mucous secretion, and pain in the GI tract. Histamine-mediated intestinal symptoms may be attenuated by a low-histamine diet and by administration of diamine oxidase (DAO), now available as an OTC supplement. DAO, normally produced by the intestinal epithelial cells, acts to metabolize excess histamine. The bioflavonoid quercetin also can attenuate the histamine response.

Since microscopic colitis is implicated in certain subtypes of IBS, it has been proposed that IBS sometimes represents a *forme fruste* of IBD; therefore some of the following strategies employed for IBD may be equally warranted for stubborn IBS symptoms.

24.9 GI-IBD

The term "colitis" comprises several types of inflammatory intestinal disorders, as distinguished from irritable bowel syndrome (IBS), a "functional" disorder of motility in which gross pathological changes are not apparent. The main types of colitis are ulcerative colitis (UC) (and its subtype ulcerative proctitis) and Crohn's disease (CD), which have different characteristic appearances and radiological and blood testing presentations. They are collectively referred to as inflammatory bowel disease (IBD).

There is now evidence that, on closer examination, a fairly high percentage of IBS patients actually have "microscopic colitis" in which intestinal surface cells show signs of damage, helping us to appreciate that the colitis spectrum may be wider than previously thought.

In addition, while CD and UC have not generally been thought of as infectious diseases, one form of colitis, pseudomembranous colitis, is known to be caused by a bacterium, *C. difficile*.

More researchers are coming to accept that imbalances in intestinal flora may be at the root of IBD [43]. Immune cells in the walls of the intestinal tract are engaged in continual "cross-talk" with the trillions of bacteria that inhabit our gut. When beneficial bacteria are suppressed and harmful bacteria proliferate, the intestinal defense system may go into "overdrive," resulting in an exaggerated immune response and triggering autoimmunity and consequent inflammation.

This disordered state of intestinal microbes is sometimes referred to as "dysbiosis." Evidence that dysbiosis might be a contributing factor in IBD comes from several sources. Frequent administration of antibiotics is known to be a risk factor for IBD, and UC and CD patients sometimes suffer exacerbations of their disease after antibiotics are prescribed for other conditions.

Also, use of powerful acid-blocking medications increases the risk for IBD, probably because it alters the intestinal flora. Breastfeeding, perhaps because of its "priming" effects on the intestinal immune system, appears to be protective against IBD [43].

In some studies, probiotics appear to ameliorate the symptoms of IBD [44], but no one is yet sure which probiotics work best, or in what form or dosage. Promising research [45] suggests that fecal flora harvested from healthy donors can help sufferers of IBD, but human trials are carefully regulated by the FDA and results have been mixed, with case reports of adverse consequences such as inadvertent transmission of cytomegalovirus resulting in acute exacerbation [46].

Commonly used medications sometimes play a role. Nonsteroidal anti-inflammatory drugs (NSAIDs) may set the stage for IBD by damaging the intestinal surface.

Additionally, stress provokes IBD. Anxiety and depression are associated with increased CD activity, and signals from the brain have been shown to influence inflammation by altering intestinal barrier function and even leading to changes in the composition of intestinal flora.

Whether because of overdependence on pharmacological drugs, refined foods, or other factors, IBD incidence is on the rise in developed countries, affecting as many as 1.4 million Americans, nearly 0.5% of the population. Some point to the "hygiene hypothesis," which proposes that our conquest over unsanitary conditions has increased the likelihood that our idle immune systems will inappropriately target our own tissues in a misguided attack. Supporting this notion is research suggesting that introduction of the nonpathogenic helminth *Plasmodium ovale* into the GI tracts of IBD sufferers can induce remission via immunological "decoying" [47].

The role of diet cannot be underestimated. Major changes in the Western diet have accompanied unprecedented rises in the incidence of IBD over the last 75 years. Introduction of refined and sugar- and chemical-laden foods has probably fueled the epidemic. Globalization of the food supply has rapidly introduced dairy products, cereal grains, potatoes, tomatoes, legumes, and other novel dietary items into the diets of populations not accustomed to these foods. Ubiquitous trans fats and refined omega-6 vegetable oils are thought to promote inflammation.

Certain food additives have been implicated in the causation and perpetuation of IBD. "Microparticles," such as titanium dioxide, are frequently added to food and have razor-like effects on the intestinal lining. Carrageenan, a "natural" seaweed derivative frequently added to foods to improve texture and "mouth feel," has been shown to inflame the intestinal lining in susceptible persons. Emulsifiers have recently been implicated.

This has led some to propose a return to a "paleo diet" to ameliorate the symptoms of IBD [48]. This ancestral diet highlights organic or grass-fed meat, chicken, and eggs; fish and shellfish; fresh, organic fruits and vegetables; and nuts and seeds. Conspicuously absent are grains, dairy products, legumes, nightshade family vegetables, processed and refined foods, and sugar.

Alternatively, some advocate the specific carbohydrate diet (SCD) for IBD. This grain-free, animal protein, and fruit/vegetable diet has many similarities to the paleo diet, except that it more stringently emphasizes starch avoidance while permitting fermented dairy products for those not specifically intolerant to cow's milk [49] (see ▶ Boxes 24.2 and 24.3). The goal of the SCD is to "break the vicious cycle" of microbial proliferation by denying pathogenic bacteria their preferred substrate of partially indigestible carbohydrate—much the same as the rationale for the low-FODMAP diet, with which the SCD shares similarities.

> **Box 24.2 Specific carbohydrate DIET (SCD)**
> - No disaccharides: lactose, sucrose, high-fructose corn syrup
> - No grains
> - No agar, carrageenan, or seaweed derivatives
> - No starchy legumes (some permitted)
> - No starchy vegetables (potatoes, sweet potatoes, parsnips, corn, etc.)
> - No processed meats
> - Based on data from Ref. [49]

> **Box 24.3 Foods allowed on SCD**
> - Unprocessed meats, eggs, poultry, fish
> - Most non-starchy vegetables
> - Most fruits and juices
> - Plain yogurt and lactose-free natural cheese
> - Nuts and nut flours for baked goods
> - Pulses, navy, string, lima beans
> - Oils, light tea and coffee, distilled alcohols, honey
> - Based on data from ▶ www.scdiet.org

A recent first-of-its-kind study validated the benefits of the SCD in children with ulcerative colitis and Crohn's disease. At the end of 12 weeks of strict adherence to the SCD, 8 of 10 patients who finished the study showed significant improvements and achieved remission from the dietary treatment alone [50]. On the other hand, stringent limitation of oligosaccharides may, at least in theory, have adverse consequences for IBD sufferers. The selfsame banned fermentable starches, upon bacterial degradation, yield beneficial short-chain fatty acids (SCFAs) like butyrate, which nourish colonic mucosa [51]. Indeed, some human trials have shown benefits of rectal instillation of sodium butyrate for proctitis and topical application for pouchitis.

Administration of oral butyrate supplements seems to be less efficacious than encouraging endogenous production via diet. Prebiotics that promote SCFAs include resistant starches

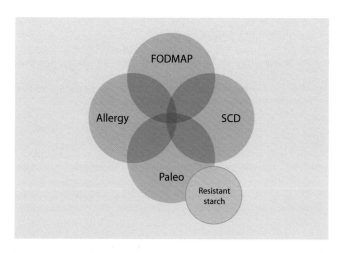

Fig. 24.3 GI diet similarities

Table 24.3	Nutraceuticals for IBD
Curcumin	EGCG (green tea polyphenol)
Boswellin	N-acetyl glucosamine (NAG)
Conjugated linolenic acid (CLA)	Melatonin
Phosphatidylcholine	Propionyl-L-carnitine (GPLC)
Fish peptides	Omega-3
Aloe	

like cooked, then re-cooled potatoes and unmodified potato starch powder (see Fig. 24.3).

There is no question that nutritional deficiencies play a prominent role in IBD. Malabsorption, diarrhea, and GI blood loss are common features of IBD, and therefore deficiencies of B vitamins, fat-soluble vitamins, and essential fatty acids—and key minerals such as magnesium, zinc, and selenium—are extremely common.

But supplementation is not just a matter of repletion of missing nutrients in IBD; certain vitamins and minerals have therapeutic effects beyond just staving off deficiency. One such nutrient is fish oil, which some studies indicate may suppress intestinal inflammation [52]. High doses are required, up to 9 g per day, delivered via enteric-coated capsules that open downstream to deliver omega-3 fatty acids directly to the intestinal epithelium.

Another is vitamin D, which studies suggest might be therapeutic at doses higher than required to simply ward off deficiency [53]. One clue comes from the finding that IBD is less commonly found at equatorial latitudes, where plentiful sun exposure seems to confer protection via production of vitamin D in the skin by UV rays. While a dose of 400 international units has traditionally been used to supplement vitamin D, doses in the thousands seem to be required to rein in inflammation and autoimmunity.

Iron is frequently low in patients with IBD because of malabsorption and gastrointestinal blood loss. Gastroenterologists, therefore, frequently prescribe potent iron supplements to their patients. But concern has arisen over the potentially pro-inflammatory and constipating effects of common iron supplements like ferrous sulfate [54]. Where possible, iron repletion via iron-rich foods like red meat and organ meats is preferable; if refractory to replacement with diet, parenteral iron may be indicated.

Natural products that have documented efficacy in IBD included curcumin [55], Boswellin [56], conjugated linolenic acid (CLA) [57], phosphatidylcholine [58], fish peptides derived from hydrolyzed white fish [59], aloe vera gel [60], EGCG from green tea [61], N-acetyl glucosamine [62], melatonin [63], and enteric-coated propionyl-L-carnitine [64] (see Table 24.3).

A traditional naturopathic remedy for bowel disorders is Bastyr (Robert's) Formula B, consisting of marshmallow, wild indigo, geranium, goldenseal, slippery elm, ginger, okra powder, niacinamide, and duodenum powder.

Probiotics may play a role in ameliorating IBD. Long-standing European experience with *Escherichia coli* Nissle 1917 (Mutaflor®) suggests benefits, but the formulation is not approved for importation to the United States. VSL#3®, a high-potency probiotic mixture, has been the object of several trials, including some that were double-blind placebo-controlled, with significant impact on improved stool frequency, self-reports of pain, physician assessment, and endoscopic scores [65].

Comfrey, a once-preferred gastrointestinal healing herb, has now fallen into disfavor because of its high content of hepatotoxic pyrrolidine alkaloids.

While there is evidence that administration of a wide variety of oral supplements and nutraceuticals may be justified in IBD, caution must be exercised. Many patients suffer from chronic diarrhea, cramping, and abdominal pain, or may have narrowing of segments of the intestine. A gradual introduction may be warranted, with supplements added only as tolerated.

As documented above, GI disorders are exceptionally amenable to natural interventions. These functional medicine strategies possess several advantages: (1) Preferential use of diet modification and nutraceuticals in lieu of pharmaceuticals generally minimizes serious side effects. (2) By comprehensively addressing the *causes* of GI complaints, natural approaches provide more than just transient, symptomatic relief. (3) Because of the centrality of the microbiome to overall health, and its role in the pathogenesis of what were previously thought to be "remote" diseases, functional medicine strategies contribute to long-term overall health optimization.

It is high time that such approaches take their rightful place in modern medicine's armamentarium. The pace of ongoing research guarantees that there will be an ever-expanding evidence base rationalizing their deployment.

References

1. Bested AC, et al. Intestinal microbiota, probiotics and mental health: from Metchnikoff to modern advances: part I – autointoxication revisited. Gut Pathogens. 2013;5:5.

2. Satterlee GR, et al. Symptomatology of the nervous system in chronic intestinal toxemia. JAMA. 1917;69(17):1414–8.
3. Gilford H. The relations of biology to pathology. Br Med J. 1912;1:279–83.
4. Tappenden K, et al. The physiological relevance of the intestinal microbiota—contributions to human health. JACN. 2007;26(6):679S–83S.
5. Popova M, et al. Beneficial effects of probiotics in upper respiratory tract infections and their mechanical actions to antagonized pathogens. J Appl Micro. 2012;113(6):1305–18.
6. Schirmer M, et al. Linking the human gut microbiome to inflammatory cytokine production capacity. Cell. 2016;167(4):1125–36.
7. Van Praet JT, et al. Commensal microbiota influence systemic autoimmune responses. EMBO J. 2015;34:466–74.
8. Chen J, et al. Multiple sclerosis patients have a distinct gut microbiota compared to healthy controls. Sci Rep. 2016;6:28484.
9. Tremlett H, et al. Gut microbiota in early pediatric multiple sclerosis: a case–control study. Eur J Neurol. 2016;23(8):1308–21.
10. Lunia MK. Probiotics prevent hepatic encephalopathy in patients with cirrhosis. Clin Gastroenterol Hepatol. 2014;12(6):1003–8.
11. Hsiao EY, et al. Microbiota modulate behavioral and physiological abnormalities associated with neurodevelopmental disorders. Cell. 2013;155:1451.
12. Steenbergen L, et al. A randomized controlled trial to test the effect of multispecies probiotics on cognitive reactivity to sad mood. Brain Behav Immun. 2015;48:258–64.
13. Foster JA, Rinaman L, Cryan JF. Stress & the gut-brain axis: regulation by the microbiome. Neurobiol Stress. 2017;7:124–36.
14. Chin LC, et al. Celiac disease and autoimmune thyroid disease. Clin Med Res. 2007;5(3):184–92.
15. Kunc M, et al. Microbiome impact on metabolism and function of sex, thyroid, growth and parathyroid hormones. Acta Biochim Pol. 2016;63(2):189–201.
16. Charles F, et al. The intestinal microbiome and skeletal fitness: connecting bugs and bones. Clin Immunol. 2015;159(2):163–9.
17. Bell DS, et al. Changes seen in gut bacteria content and distribution with obesity: causation or association? Postgrad Med. 2015;127(8):863–8.
18. Herbert T, et al. Gut microbiome, obesity, and metabolic dysfunction. J Clin Invest. 2011;121(6):2126–32.
19. Koliada A, Syzenko G, Moseiko V, et al. Association between body mass index and Firmicutes/Bacteroidetes ratio in an adult Ukrainian population. BMC Microbiol. 2017;17:120.
20. Alang N, et al. Weight gain after fecal microbiota transplantation. Open Forum Infect Dis. 2015;2(110):ofv004.
21. Liou AP, et al. Conserved shifts in the gut microbiota due to gastric bypass reduce host weight and adiposity. Sci Transl Med. 2013;5(178):178ra41.
22. Lean MEJ, et al. Altered gut and adipose tissue hormones in overweight and obese individuals: cause or consequence? Int J Obesity. 2016;40(4):622–32.
23. Tilg H, et al. Gut microbiome, obesity, and metabolic dysfunction. J Clin Invest. 2011;1121(6):2126–32.
24. Kaplan RC, Burk RD, Knight R. Epidemiology of the gut microbiome, prediabetes and diabetes in latinos. NIH Grant, 8 July 2016.
25. Dunbar K, et al. Association of acute gastroesophageal reflux disease with esophageal histologic changes. JAMA. 2016;315(19):2104–12.
26. Falalieiva TM, et al. Effect of long-term monosodium glutamate administration on structure and functional state of the stomach and body weight in rats. Fizio Zh. 2010;56(4):102–10.
27. Austin GL, et al. A very low-carbohydrate diet improves gastroesophageal reflux and its symptoms. Dig Dis Sci. 2006;51(8):1307–12.
28. Kandil TS, Mousa AA, El-Gendy AA, Abbas AM. The potential therapeutic effect of melatonin in gastro-esophageal reflux disease. BMC Gastroenterol. 2010;10:7.
29. TS K. The potential therapeutic effect of melatonin in gastro-esophageal reflux disease. BMC Gastroenterol. 2010;10:7.
30. Wilkins JS Jr. Method for treating gastrointestinal disorders. U.S. Patent (6,420,435). 16 July 2002.
31. Panahi Y, et al. Efficacy and safety of Aloe vera syrup for the treatment of gastroesophageal reflux disease: a pilot randomized positive-controlled trial. J Trad Chi Med. 1944;35(6):632–6.
32. Bredenoord AJ, et al. Symptomatic gastro-oesophageal reflux in a patient with achlorhydria. Gut. 2006;55(7):1054–5.
33. Barlow GM, et al. A definitive blood test for postinfectious irritable bowel syndrome? Expert Rev Gastroenterol Hepatol. 2016;10(11):1197–9.
34. Barmeyer C, Schumann M, Meyer T, et al. Long-term response to gluten-free diet as evidence for non-celiac wheat sensitivity in one third of patients with diarrhea-dominant and mixed-type irritable bowel syndrome. Int J Color Dis. 2016;32(1):29–39. doi: https://doi.org/10.1007/s00384-016-2663-x [published Online First: Epub Date].
35. Drisko J, et al. Treating irritable bowel syndrome with a food elimination diet followed by food challenge and probiotics. J Am Coll Nutr. 2006;25(6):514–22.
36. Mullin GE, Shepherd SJ, Chander Roland B, Ireton-Jones C, Matarese LE. Irritable bowel syndrome: contemporary nutrition management strategies. JPEN J Parenter Enteral Nutr. 2014;38(7):781–99. doi: https://doi.org/10.1177/0148607114545329 [published Online First: Epub Date].
37. Pasqui F, Poli C, Colecchia A, Marasco G, Festi D. Adverse food reaction and functional gastrointestinal disorders: role of the dietetic approach. J Gastrointestin Liver Dis. 2015;24(3):319–27. doi: https://doi.org/10.15403/jgld.2014.1121.243.paq [published Online First: Epub Date].
38. Eswaran SL, et al. A randomized controlled trial comparing the low FODMAP diet vs. modified NICE guidelines in US adults with IBS-D. Am J Gastroenterol. 2016;111:1824.
39. Srivastava JK, et al. Chamomile: a herbal medicine of the past with bright future. Mol Med Rep. 2010;3(6):895–901.
40. B O. STW 5 (Iberogast®)—a safe and effective standard in the treatment of functional gastrointestinal disorders. Wien Med Wochenschr. 2013;163(3–4):65–72.
41. Kim SE, Choi SC, Park KS, et al. Change of fecal flora and effectiveness of the short-term VSL#3 probiotic treatment in patients with functional constipation. J Neurogastroenterol Motil. 2015;21:111–20.
42. Quigley E. Probiotics in irritable bowel syndrome: an immunomodulatory strategy? J Am Coll Nutr. 2007;26(6):684S–90S.
43. Tamboli CP, Neut C, Colombel JF. Dysbiosis in inflammatory bowel disease. Gut. 2004;53(1):1–4.
44. Jones JL, Foxx-Orenstein AE. The role of probiotics in inflammatory bowel disease. Dig Dis Sci. 2007;52:607–11.
45. Lopez J, Grinspan A. Fecal microbiota transplantation for inflammatory bowel disease. Gastroenterol Hepatol (N Y). 2016;12(6):374–9.
46. Hohlmann EL. Case records of the Massachusetts General Hospital. Case 25-2014. A 37-year-old man with ulcerative colitis and bloody diarrhea. N Engl J Med. 2014;371(7):668–75.
47. Summers RW, et al. Trichuris suis therapy in Crohn's disease. Gut. 2005;54(1):87–90.
48. Cordain L. The pathologic role of dietary lectins in autoimmunity. British J Nutr. 2000;83:207–17.
49. Gottschall E. Breaking the vicious cycle. Ontario: Kirkton Press; 1994.
50. Suskind DL, et al. Clinical and fecal microbial changes with diet therapy in active inflammatory bowel disease. J Clin Gastr. 2016;52(2):155–63.
51. Bassaganya-Riera J, et al. Soluble fibers and resistant starch ameliorate disease activity in interleukin-10-deficient mice with inflammatory bowel disease. J Nutr. 2011;141(7):1318–25.
52. Turner D, et al. Maintenance of remission in inflammatory bowel disease using omega-3 fatty acids (fish oil): a systematic review and meta-analyses. Inflamm Bowel Dis. 2011;17(1):336–45.
53. Litwack G. editor. Vitamin D and inflammatory bowel disease. In: Vitamins and the immune system. 2011. p. 367–77.

54. Carrier J, et al. Effect of oral iron supplementation on oxidative stress and colonic inflammation in rats with induced colitis. Aliment Pharmacol Ther. 2001;15(12):1989–99.
55. Holt PR, et al. Curcumin therapy in inflammatory bowel disease: a pilot study. Dig Dis Sci. 2005;50(11):2191–3.
56. Gerhardt H. Therapy of active Crohn's disease with Boswellia serrata extract H 15. Z Gastroenterol. 2001;39(1):11–7.
57. Tech V. Novel therapy discovered for Crohn's disease. Science Daily, 19 March 2012.
58. Stremmel W, et al. Retard release phosphatidylcholine benefits patients with chronic active ulcerative colitis. Gut. 2005;54(7):966–71.
59. Nichols TW, et al. Improvements in mucosal integrity and function in IBD patients and reduction in GI symptoms in HIV patients with dietary peptides from hydrolyzed white fish. Regul Pept. 2002;108(1):31.
60. Langmead L, et al. Randomized double-blind placebo-controlled trial of oral aloe vera gel for active ulcerative colitis. Aliment Pharmacol Ther. 2004;19(7):739–47.
61. Abboud PA, et al. Therapeutic effect of epigallocatechin-3-gallate in a mouse model of colitis. Eur Jl Pharmacol. 2008;579(1–3):411–7.
62. Salvatore S, et al. A pilot study of N-acetyl glucosamine, a nutrient substrate for glycosaminoglycan synthesis in pediatric chronic inflammatory bowel disease. Aliment Pharmacol Ther. 2000;14(12):1567–79.
63. Chojnacki C, et al. Evaluation of melatonin effectiveness in the adjuvant treatment of ulcerative colitis. J Physiol Pharmacol. 2011;62(3):327–34.
64. Merra G, et al. Propionyl-l-carnitine hydrochloride for treatment of mild to moderate colonic inflammatory disease. World J Gastroenterol. 2012;18(36):5065–70.
65. Tursi A, et al. Treatment of relapsing mild-to-moderate ulcerative colitis with the probiotic VSl#3® as adjunctive to a standard pharmaceutical treatment. Am J Gastroenterol. 2010;105(10):2218–27.

The Microbiome and Brain Health

Sharon L. Norling

25.1 Introduction – 392

25.2 Dysbiosis – 392

25.3 Probiotics – 393

25.4 Second Brain – 394

25.5 Mood Disorders – 395

25.6 Schizophrenia – 396

25.7 Bipolar Disorder – 396

25.8 General Discussion of Mood Disorders – 396

25.9 Psychobiotics – 397

25.10 Brain-Derived Neurotrophic Factor (BDNF) – 398

25.11 Neurological Disorders – 399

25.12 Alzheimer's Disease – 399

25.13 Autism – 400

25.14 Parkinson's Disease – 400

25.15 Aging – 400

25.16 Fecal Transplants – 402

25.17 Food Sources – 402

25.18 Summary – 403

References – 403

© Springer Nature Switzerland AG 2020
D. Noland et al. (eds.), *Integrative and Functional Medical Nutrition Therapy*,
https://doi.org/10.1007/978-3-030-30730-1_25

25.1 Introduction

Strange but true, our bodies have more bacteria than we have human cells. The microbiome is inhabited by 100 trillion microorganisms, about 10 times the number of cells in the human body. Of note, it weighs 2–6 pounds. A dysfunction in this system significantly affects mental, emotional, and physical health [1]. It is estimated that the microbiome contributes 150–360 times more bacterial genes than human genes [2, 3].

The word microbiome is defined as the collection of microbes or microorganisms that inhabit an environment. It includes communities of symbiotic, commensal, and pathogenic bacteria, fungi, and viruses. The clusters of bacteria from different regions of the body are known as microbiota: skin microbiota, oral microbiota, vaginal microbiota, and GI microbiota. Scientists have recently come to understand that our GI microbiota, as a whole, determines whether pathogens in the gut coexist peacefully or cause disease [4].

Bacteria are distributed throughout the intestine, with the major concentration of microbes and metabolic activity found in the large intestine: It is commonly known as the intestinal microbiome. The intestinal microbiota are made up of 1000 species comprised of up to 130 trillion microbes, 70% of which cannot be cultivated in a microbiology laboratory [5, 6].

The gut microbiome is essential for the host for digestion, including the breakdown of dietary fibers and complex carbohydrates, the synthesis of vitamins, and the production of short-chain fatty acids. The balance and makeup of the microbiota play a critical role in metabolic functions [7]. The intestinal microbiota is essential to nutrient metabolism, opportunistic pathogen defense immune system development, and intestinal barrier function. About 70% of our immune cells reside in the GI tract. The development of the intestinal immune system is largely dependent upon exposure to microorganisms.

Scientists are conducting research which documents the significant impact of the gut on brain health. The National Institutes of Health devote millions of research dollars to understand these complex interactions. More than 90% of over 4000 articles on the microbiome were published in PubMed in the past 5 years [8]. While some physicians and scientists see probiotics as the "new magic medicine," it is far from new. Élie Metchnikoff received the Nobel Prize in Physiology or Medicine in 1908 for his work in discovering probiotics and phagocytes.

In the exciting area of research, one of the open questions is how chronic inflammation might be initiated and maintained in illnesses such as depression, and what the gut has to do with this. Emerging studies show that the normally very selective intestinal barrier may be compromised in depression (and in numerous conditions where depression is often a hallmark symptom) [9–11].

Psychological stress and exhaustive exercise have been shown to increase the permeability of the intestinal barrier [12–14]. However, a Westernized diet high in fat and sugar has also been shown to cause a more porous intestinal lining, the consequences of which include systemic access to food antigens, environmental toxins, and structural components of microbes, such as lipopolysaccharide endotoxin (LPS) [15]. The latter agent, LPS, is particularly important regarding depression; even relatively small elevations in systemic LPS levels have been shown to provoke depressive symptoms and disturb blood glucose control [16–22]. Endotoxins such as LPS can decrease the availability of tryptophan and zinc, thereby negatively influencing neurotransmission [23, 24]. Moreover, systemic LPS can elevate inflammation and oxidative stress. Traditional dietary practices have completely divergent effects of blood LPS levels; significant reductions (38%) have been noted after a one-month adherence to a prudent (traditional) diet, while the Western diet provokes LPS elevations [25]. These and other findings help establish mechanisms whereby the LPS-lowering, antioxidant, and anti-inflammatory properties of broad traditional dietary practices, as well as specific components within them, can help provide mood support. Indeed, when the limitation of intestinal absorption is overcome, individual phenolic structures have been shown, at least experimentally, to curb the breakdown of central neurotransmitters, mimicking the proposed mechanistic properties of some primary antidepressant medications [26, 27].

25.2 Dysbiosis

The term "dysbiosis" refers to situations where microbial composition and functions are shifted from their normal beneficial state to another that may be harmful to one's health. The microbiota dysbiosis may negatively impact CNS functioning through various pathways known as the brain-gut axis [28–30]. The brain-gut axis is impacted by intestinal permeability. Another name used for intestinal permeability is leaky gut (see ▶ Box 25.1).

Box 25.1 Intestinal Permeability (Leaky Gut)
– Modulation of local and systemic inflammation
– Decreased digestion and absorption of nutrients
– Decreased brain-derived neurotrophic factor (BDNF)
– Increased small intestinal bacterial overgrowth (SIBO)

Dysbiosis of the gut microbiota has profound effects on immune function and leads to inflammation. One common cause of dysbiosis is the Western diet as it affects both immune homeostasis and the gastrointestinal tract. For example, a high-sugar-content diet allows for the overgrowth of *Clostridium difficile* and *Clostridium perfringens* by increasing bile secretion. This dysbiosis can lead to metabolic syndrome, diabetes, obesity, celiac disease, inflammatory bowel disease, and irritable bowel syndrome.

One of the most important roles of the microbiota is the development and sustainment of local and systemic immunity through homeostasis of its environment. This has been demonstrated in studies showing the expansion of T and B cells in mesenteric lymph nodes and Peyer's patches, including CD4[+] cells and FOXP3 T_{reg} cells [31, 32]. An altered gut microflora has been reported in patients with rheumatoid arthritis and other autoimmune diseases as well as in allergic disease, implying that the normal gut microflora constitutes an ecosystem responding to inflammation in the gut and elsewhere in the human body.

25.3 Probiotics

Gut microbiota may be modulated with the use of probiotics, antibiotics, and fecal microbiota transplants as a prospect for therapy in microbiota-associated diseases. Probiotics can stabilize the gut microbiota and the intestinal permeability barrier, and they can improve the degradation of enteric antigens and alter their immunogenicity by countering the progression of inflammation.

In addition, probiotics can improve the immunological barrier, in particular the intestinal immunoglobulin A response. They are also able to mediate the balance between pro- and anti-inflammatory cytokines, thereby quelling the inflammation [33–36]. Probiotics can decrease inflammation.

Probiotics are on a dramatic growth curve. Probiotics are live microbial food supplements or components of bacteria, which have been shown to have beneficial effects on human health. It is estimated that the global probiotics market will exceed $46 billion by 2022 [37]. We are undoubtedly in an exciting period for probiotics. Probiotic products are entering the marketplace in all shapes and forms including yogurt, juices, and even chocolate. Probiotics are being adapted to almost everything that is ingestible. This raises the question as to whether or not these new food sources are just trendy or if they can effectively provide benefit. In terms of labeling probiotic products, the only standard required by the FDA for foods, supplements, and medical foods is that the labeling be truthful and not misleading. Sometimes, there is a disconnect between what's studied in published papers and what is actually available commercially on the market, but there are some well-validated, properly labeled products out there and those are the ones that should be utilized [38].

A probiotic should be able to survive the gastric barrier and bile acids to make it through the gastrointestinal tract (GIT) and reach the colon. It should have the ability to colonize the intestinal tract to perform its functions as a probiotic. A probiotic should be able to improve mucosal immunity and support barrier function. It should be able to competitively exclude pathogenic microbes. *Bacillus subtilis* are spore-forming bacteria and is one of the species of the *Bacillus* genus that fit these criteria; these spores survive the transit through the stomach and small intestines to reach the microbiome in the colon where they germinate, proliferate, and then resporulate [39–45].

Studies have shown that orally ingested *B. subtilis* spores are immunogenic and can disseminate to the Peyer's patches and mesenteric lymph nodes [46–52]. In this way, they offer some advantages over the more common *Lactobacillus*. The average over-the-counter probiotic typically has Lactobacilli and Bifidobacter organisms in various amounts, with some trying to outdo the other with larger and larger amounts of organisms. The gastric survival of these has been shown to be low as many are not acid-stable. The probiotic paradox is that dead cells can generate responses from the GIT, mainly immunomodulatory rather than antimicrobial [53].

The GI tract, with changes in pH, bile salts, and digestive enzymes, may require large numbers of probiotics to be consumed so that adequate numbers reach the lower GI tract. Most probiotics today emphasize the number of live bacteria. Lahtinen [54] suggests a product of fermentation may not require the probiotic to be viable when consumed and the probiotic will still be beneficial.

There is no consensus to the minimum number of bacteria needed to produce a beneficial effect on health. Studies such as McNulty and colleagues have shown that probiotics need to be consumed every day because they do not colonize the gut [44].

Probiotics reinforce the various lines of gut defense: immune exclusion, immune elimination, and immune regulation. Probiotics also stimulate nonspecific host resistance to microbial pathogens and thereby aid in their eradication. There is evidence for the use of probiotics for reducing anxiety in individuals with altered GI function, in addition to lowering systemic inflammatory cytokines and decreasing oxidative stress [55]. Messaoudi et al. found *Lactobacillus helveticus* combined with *Bifidobacterium longum* decreased anxiety in humans accompanied with a reduction in urinary free cortisol in treated humans [56]. *Lactobacillus rhamnosus* has been found to decrease anxiety and alters expression of GABA receptors in the brain, a receptor known to play a role in anxiety disorders [57]. Probiotics have also been promising as a potential preventative strategy for depression in human trials [58]. Moreover, there is evidence that the concentration of fatty acids, namely, arachidonic acid and docosahexaenoic acid, is increased in mice that received a strain of *Bifidobacterium breve* [59]. It is known that both fatty acids contribute to several neurodevelopmental processes, including neurogenesis and neurotransmission, and can impact such cognitive functions as learning and memory [60].

There is quite an amount of research evidence supporting the effectiveness of probiotics; however, the mechanisms of action are less clear. They may competitively exclude gut pathogens by repairing leaky gut or producing antimicrobial properties. Additionally, they may stimulate the immune system by increasing the production of anti-inflammatory cytokines [61]. They also communicate with the CNS via the vagus nerve and impact brain function.

Probiotics protect the inflamed intestinal epithelium. Competition for binding sites and inhibition of pathogen growth stimulate the immune system, including the stimulation of anti-inflammatory cytokines and enhancement of barrier function [62].

Regardless, multiple studies confirm the presence of a microbiota-gut-brain axis, and thus it is clear that probiotics can modulate aspects of brain function and brain health. Further research is needed to identify additional organisms and elucidate the underlying mechanism of actions.

While the focus of this chapter is intestinal microbiome, many different health targets have been studied with probiotics. One probiotic should not be expected to do all of these things, however. In general, the best evidence for probiotic health effects is in the area of decreasing duration or incidence of certain diarrheal diseases and in immune enhancement. Emerging evidence exists for probiotics in the areas of dental caries; prevention of allergy, intestinal infections, vaginal infections, and colds and respiratory infections; improved growth parameters in undernourished children; and improvements in quality-of-life indicators [63].

25.4 Second Brain

People may refer to the enteric nervous system (ENS) as an intuitive feeling, describing its actions as a "gut response" or "gut decision." So it is no surprise to find that the GI tract releases neurotransmitters that affect brain health and decision-making. The ENS is embedded in the wall of the digestive tract and is localized by the myenteric plexus and submucosal plexus. This system contributes to GI motility, nutrient handling, gastric acid secretion, and immune function.

As researchers gain a deeper understanding of ENS, this second brain located in the GI tract, it is clear that it is filled with neurotransmitters [64]. Together, the gut-brain connection significantly determines mental states and plays key roles in certain diseases throughout the body. While the second brain has an important role in brain health, it does not appear to participate in conscious thoughts or decision-making. "The second brain does not help with the great thought processes…religion, philosophy, and poetry is left to the brain in the head," says Michael Gershon, chairman of the Department of Anatomy and Cell Biology at New York–Presbyterian Hospital/Columbia University Medical Center, an expert in the field of neurogastroenterology, and the author of the Second Brain [64].

A dysfunction in the gut-brain axis is linked to neuropsychological, metabolic disorders such as obesity, immune, and endocrine disorders [65]. Dysbiosis has been linked to depression and autism spectrum disorder (ASD) and GI disorders including IBD and IBS [66] (see ▶ Box 25.2).

Box 25.2   Gut ➡ Brain
- Influences from the gut affecting the brain
- Dysbiosis
- Antibiotics and infections
- Environmental influences:
- Toxins
- Food and gluten sensitivities
- Genetic predisposition
- Nutrition and physical activity
- GI neurotransmitters
- Inflammatory cytokines
- Mode of delivery at birth

It is clear that the gut microbiota can be a key regulator of mood, cognition, pain, and obesity [67]. The microbiota-gut-brain (MGB) axis and the neuroimmune system provide understanding and management of anxiety, ADHD, autism, cytokines, depression, stress, and neuroimmune conditions [68]. Major depression and anxiety states are common in patients presenting with IBS.

Immune cells produce neurotransmitters. There are several mechanisms that facilitate neuroimmune interactions and the bidirectional communication. These include the anatomical proximity between immune cells and nerve cells, the receptors for neurotransmitters on immune cells, and for immune mediators on nerves and the intracellular signaling pathways that modulate nerve and immune phenotype expression and function [69]. Chronic active inflammation in the GI tract leads to neuroimmune plasticity, resulting in remodeling of both the neural and immune systems.

Gut microbiota and some probiotics can regulate immune functions. Benefits may be anti-inflammatory actions of certain bacteria and a capacity to affect HPA axis activity. Most physical and mental diseases have inflammation as their root cause. Much of our immune system, about 70%, is located in the gut microbiome [68].

When food enters the mouth and passes through the digestive system, it sends the brain a multitude of interacting signals that are filled with sensory, nutritive, and other information. According to Cenit [70], gluten sensitivity is a result of dysbiosis, not a genetic disorder. This may explain, in part, why patients who test positive for the celiac gene may not be gluten-sensitive while those patients who test negative for the celiac gene may have a gluten sensitivity. Dysbiosis is most prominent in the digestive tract or on the skin, but it can also occur on any exposed surface or mucous membrane. Dysbiosis creates inflammatory factors responsible for developing insulin resistance and body weight gain [70] (see ◘ Fig. 25.1). These conditions, in turn, create inflammation which can affect brain function and brain health.

Most information received by the brain relating to GI contents and activity is generally transmitted via vagal afferent feedback. The other enteric nervous system source is the

○ Fig. 25.1 Gut–brain interactions

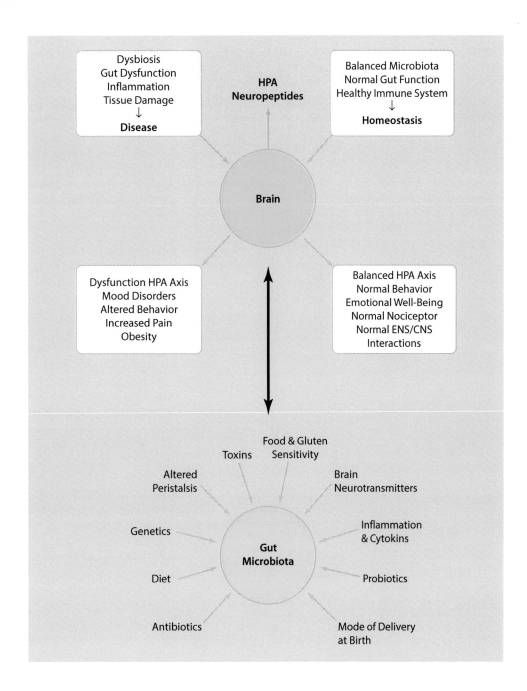

spinal cord afferents, which mediate GI pain rather than providing feedback to the brain relating to nutrient contents.

25.5 Mood Disorders

Conversely, visceral pain can affect central pain perception, mood, and behavior. Communication between the central nervous system (CNS) and ENS involves neural pathways as well as immune and endocrine mechanisms [71] (see ▶ Box 25.3).

Box 25.3 Brain ➡ Gut
- Influences from the brain affecting the gut microbiota [72]
- The hypothalamus-pituitary-adrenal (HPA) axis
- The CNS regulating areas of satiety
- Neuropeptides released from sensory nerve fibers

Chronic psychological stress is associated with a greater risk of depression, cardiovascular disease, diabetes, autoimmune diseases, upper respiratory infections, and poorer wound healing [73]. Stress induces leaky gut and increases mucosal immune response, which in turn alters the composition of the microbiome and leads to enhanced HPA drive. The hypothalamus is an integral part of the limbic system referred to as the "emotional" brain. The limbic system anatomically consists of the hypothalamus, amygdala, medial thalamus, and anterior cingulate cortex (ACC). Psychosocial factors can influence digestive function, symptom perception, illness behavior, and outcome [74].

Lackner and colleagues [75] reported a series of six individuals with irritable bowel syndrome (IBS) and six controls. Patients with IBS were treated with a 10-week course of cognitive therapy. Treatment was associated with a significant reduction in anxiety and digestive symptoms. PET scans showed reduced activity in the region of the left amygdala and right ACC following therapy [75].

25.6 Schizophrenia

The development of schizophrenia can be linked through the role of microbiota in regulating brain development, immune function, and metabolism. The evidence suggests possible microbiota alterations in schizophrenia resulting in structural damage to the GI tract, and a heightened immune response to infections, food antigens, and gluten sensitivity (see ▶ Box 25.4). Evidence that patients with schizophrenia may have altered microbiota [76]:

> **Box 25.4 Altered Microbiota in Psychiatric Disorders**
> – Structural damage to the GI tract in schizophrenia
> – Abnormal response to infections pathogens in schizophrenia
> – Abnormal response to food antigens in schizophrenia
> – Sensitivity to gluten and bovine casein

A recent study on the oral microbiome and schizophrenia has shown large differences between case and control individuals in terms of bacterial, viral, and fungal composition [77]. Schizophrenics had increased levels of lactic acid bacteria. Individuals with schizophrenia had decreased levels of many nonpathogenic bacteria and increased levels of intestinal immune activation as indicated by antibodies to food and intestinal antigens. The microbiome was significantly altered by probiotic therapy with a tendency toward normalization in the case individuals following treatment.

Major depressive disorders, bipolar disorder, and schizophrenia have been associated with immune responses, supporting the concept that mood disorders are related to inflammatory cytokines (see ▶ Box 25.5). The findings from clinical studies demonstrate an upregulated immune and inflammatory status in patients with schizophrenia [78] and a correlation between the level of inflammatory markers and severity of clinical symptoms [79]. It has been suggested that the uncontrolled neuroinflammation by pro-inflammatory cytokines is involved in the pathogenesis of schizophrenia [74]. Chronic macrophage activation and secretion of interleukin-2 and interleukin-2 receptors have been proposed as the basic biological mechanism of schizophrenia in earlier papers [74]. For example, the protozoa *Toxoplasma gondii* is known to cause major perturbation to the gut microbiota and is a recognized environmental risk factor for schizophrenia [74]. More recently, a chlorovirus (family *Phycodnaviridae*) has been identified in humans that affects cognitive function relevant to schizophrenia in animal models [74].

> **Box 25.5**
> Inflammation ➡ Cytokines ➡ HPA ➡ Mood disorders

N-methyl-D-aspartate (NMDA) receptor hypofunction is believed to be central to the pathophysiology of schizophrenia, as NMDA receptor antagonists produce schizophrenia-like symptoms while agents that enhance NMDA receptor (R) function reduce negative symptoms and improve cognition [74]. Variation in brain-derived neurotrophic factor (BDNF) expression is believed to play a role in the molecular mechanism underlying cognitive dysfunction in schizophrenia [74]. Given that normal development of the microbiota is necessary to stimulate brain plasticity through the appropriate expression of BDNF and NMDA receptors, it is possible that altered microbiota may contribute to the NMDA receptor dysfunction seen in schizophrenia [74].

25.7 Bipolar Disorder

Severance and colleagues [80] recently measured serological surrogate markers of bacterial translocation (soluble CD_{14} (sCD14) and lipopolysaccharide binding protein (LBP)) in bipolar subjects and schizophrenia subjects compared to controls. In bipolar disorder, sCD_{14} levels were significantly correlated with anti-tissue transglutaminase IgG ($r^2 = 0.037$, $P < 0.001$). The authors concluded that these bacterial translocation markers produced discordant patterns of activity that may reflect an imbalanced, activated innate immune state. Whereas both markers may upregulate following systemic exposure to gram-negative bacteria, autoimmunity, and non-lipopolysaccharide-based monocyte activation, and metabolic dysfunction may also contribute to the observed marker profiles [80].

25.8 General Discussion of Mood Disorders

Poor mental health has been associated with an increased likelihood of eating unhealthy foods [81]. Different diets create different gut flora. One study showed that rural African children eating a polysaccharide-rich diet had more Bacteroidetes and diversity than European Union children eating the Western diet [82]. In a 2014 study in *Nature*, it was indicated that these changes can happen in the human gut—within three or 4 days of a big shift in what one eats! [83] A variety of population studies have linked traditional dietary patterns with lowered risk of anxiety or depression [74]. More recent prospective investigations have shown that traditional healthy dietary patterns are associated with a 25–30% lower risk of depression [84, 85].

Microbes produce neurotransmitters and the gut microbiota produces three-fourths of the neurotransmitters in the body. More than 50% of the body's dopamine and 95% of your body's serotonin are produced in your gut, along with about 30 other neurotransmitters [86, 87]. These molecules are critical for signaling between cells of the nervous system. Dopamine and serotonin in the brain and the GI have both been shown to be involved in the regulation of eating behavior [88].

A decade ago, prior to the scientific hypothesis of Logan et al., the notion that the intestinal manipulation of the gut microbiota could provide therapeutic value to relieve depression and fatigue was, at the very least, outlandish [89, 90]. The mechanisms of probiotics and fermented foods are complex and may be related to their role in making neurotransmitters, decreasing oxidative stress, or improving cellular repair [91]. How can this happen? ▶ Box 25.6 shows mechanisms proposed by Logan and others that have been subject of study [89–91].

> **Box 25.6 Association of Gut Microbiota and Therapeutic Value in Mood Disorders**
> — Direct protection of the intestinal barrier
> — Influence on local and systemic antioxidant status, reduction in lipid peroxidation
> — Direct, microbial-produced neurotransmitter production, for example, gamma-aminobutyric acid (GABA)
> — Indirect influence on neurotransmitter or neuropeptide production
> — Prevention of stress-induced alterations to overall intestinal microbiota
> — Direct activation of neural pathways between gut and brain
> — Limitation of inflammatory cytokine production
> — Modulation of neurotrophic chemicals, including brain-derived neurotrophic factor (BDNF)
> — Limitation of carbohydrate malabsorption
> — Improvement of nutritional status, for example, omega-3 fatty acids, minerals, and dietary phytochemicals
> — Limitation of small intestinal bacterial overgrowth
> — Reduction of amine or uremic toxin burden
> — Limitation of gastric or intestinal pathogens (e.g., *Helicobacter pylori*)
> — Analgesic properties

Microbiota influence brain chemistry and consequently behavior. *Clostridium difficile* (CD), the hospital-based gut infection that kills 14,000 people each year in the USA, is associated with depression and dementia [92]. Two antidepressants, mirtazapine (Remeron) and fluoxetine (Prozac), are linked to a nearly 50% increased risk for CD infection [92].

Stress induces a dysbiosis, which in turn can trigger anxiety and depression [93–95]. Commensal bacteria modulate brain biochemistry and behavior through the vagus nerve, affecting neurotransmitters. Prolonged stress triggers unfavorable shifts in bacterial composition and diversity. Populations of beneficial microbes die off and dysbiosis flourishes [93–95]. More than one-third of people with depression have "leaky gut," the permeability of the gut lining that allows bacteria to enter the bloodstream (see ▶ Box 25.7).

> **Box 25.7 Serotonin's Role in the Gut [96]**
> — Motility patterns and gastric emptying
> — Secretion
> — Immune system
> — Pain and discomfort
> — Nausea and vomiting
> — Alters microbiome
> — Circulating 5-HT has the potential to impact many other tissues
> — Promotes homeostasis
> — Influences bone development
> — Receptor sites on immune cells B and T lymphocytes
> — Mast cells, macrophage, and T-cells synthesize 5-HT

25.9 Psychobiotics

The close relationships between gut microbiota, health, and disease have led to a great interest in using probiotics and/or prebiotics to prevent or treat disease [97]. The influence of probiotics on moods is significant. Psychobiotics is an emerging class of probiotics of relevance to psychiatry and mood disorders. A psychobiotic is a live organism that produces a health benefit in patients suffering from mood disorders. Such mind-altering probiotics act via their ability to produce various biologically active compounds [98].

Moreover, preliminary placebo-controlled human studies have shown that oral probiotic microbes can decrease anxiety, diminish perceptions of stress, and improve mental outlook [99]. Michael Messaoudi and colleagues from France evaluated *Lactobacillus helveticus* and *Bifidobacterium longum* combination probiotic, which was orally administered for 1 month in a one-month placebo-controlled study [100]. Among the otherwise healthy adults, significant improvements in depression, anger, anxiety, and lower levels of the stress hormone cortisol versus placebo were noted.

Bacteria are capable of producing and delivering GABA and serotonin, which are on the brain-gut axis [101]. Microbes that actively secrete GABA in the gut are strains of *Lactobacillus* and *Bifidobacterium*. Psychobiotics produce psychotropic effects on behavior, affecting the HPA axis and neurochemicals in the brain [98]. As cited by John F. Cryan, PhD, "two varieties of *Bifidobacteria* were more effective than escitalopram (Lexapro) at treating anxious and depressed behavior in a lab mouse strain known for pathological anxiety" [98].

Alterations in microbiota influence stress-related behaviors. GI tract bacteria, including commensal, probiotic, and pathogenic bacteria, can activate neural pathways and CNS signaling systems [102]. The MGB axis may provide novel approaches for prevention and treatment of mental illness, including anxiety and depression [102].

Tests revealed that chronic ingestion of probiotic *Lactobacillus plantarum* significantly reduced anxiety-like and depression-like behaviors [103]. It also reduced Lambert-Eaton Myasthenic Syndrome-induced elevation of serum corticosterone.

Lactobacillus plantarum also reduced inflammatory cytokine levels and increased anti-inflammatory cytokine levels in the serum. Furthermore, the dopamine levels in the brain were significantly increased. The psychotropic properties of certain bacteria have great potential for improving stress-related symptoms [103].

Lactobacillus rhamnosus is a bacterial strain that has been shown to reduce anxiety and depression in anxious mice [104]. *L. rhamnosus* markedly increased GABA levels. A number of microbes can produce other neurotransmitters, such as norepinephrine, serotonin, and dopamine. *Bifidobacterium infantis*, taken as a probiotic, alters serotonin levels just like Prozac but without the undesirable side effects [105].

The National Institute of Mental Health labels anxiety as a learning deficit because the brain is unable to learn to discriminate between dangerous and benign situations [106]. Recent research has led scientists to believe that the protein is an essential ingredient in combating anxiety. Scientists think this is due to the fact that it helps the brain learn to essentially work around the fear and create positive memories. In addition, higher levels of the protein ramp up levels of serotonin, which calms the brain down and increases the sense of safety [106].

Numerous studies have shown capability of bacteria in producing neurotransmitters and neuromodulators, including gamma-aminobutyric acid (GABA), norepinephrine, serotonin, dopamine, and acetylcholine [107]. Further, findings from germ-free animals indicated a decreased level of brain-derived neurotrophic factor (BDNF), important neurotrophic factor in neuronal growth and survival, and a reduced expression of some subunits of N-methyl-D-aspartate (NMDA) receptors involved in most abundant neurotransmission in the brain. The glutamatergic NMDA receptors, involved in excitatory neurotransmission of brain waves, engage the neural circuits involved in learning and memory [107]. GABA is the major inhibitory neurotransmitter in the CNS. Dysfunctions in GABA signaling are linked to anxiety and depression, defects in synaptogenesis, and cognitive impairments [107]. From these considerations, it can be concluded that, at least through contributing in neurotransmitter synthesis or receptor expression, probiotics might adjust the brain activity.

Accumulating evidence from experimental studies supports the hypothesis that—via affecting inflammation, endocrine system, and neurotransmission—the gut microbiome takes a crucial role in the CNS function [107]. Accordingly, it is suggested that dysfunction of the neuroendocrine system, behavior, and cognition is correlated with gut microbiota dysbiosis [74]. These considerations led to establishing the term "psychobiotics" to highlight the potential effects of probiotics in treatment of mental disorders [107].

Consistently, in a meta-analysis study, Kasińska and Drzewoski [108] reported a reduced homeostatic model of assessment for insulin resistance (HOMA-IR) and insignificant fasting plasma glucose (FGP) in probiotic-treated subjects [107]. Mazloom and colleagues [109] also reported that probiotic supplementation had no significant effect on fasting blood glucose, markers of insulin metabolism, and lipid profiles [107]. Consumption of symbiotic bread containing the heat-resistant probiotic *Lactobacillus sporogenes* (1×10^8 CFU/g) for 8 weeks also decreased the serum triglyceride and VLDL concentrations in patients with type 2 diabetes [107]. Probiotics have been shown to modulate many chronic illnesses. Decreasing inflammation and improving chronic disease have a significant impact on brain health.

25.10 Brain-Derived Neurotrophic Factor (BDNF)

In the past, it was believed that the brain was hardwired with a finite number of neurons. If a brain was damaged or diseased, there was little that could be done to reverse the injury. However, today we know that the brain is neuroplastic and significant regeneration can occur with a wide variety of therapies. Brain-derived neurotrophic factor (BDNF) is a protein produced inside nerve cells [106]. Although neurotransmitters like dopamine and serotonin are important in helping the brain function because they carry the signals of neurons, the protein BDNF builds and maintains the brain circuits which allow the signals to function [106]. Therapies that may be used include treating the root cause, cytokine signaling, increasing BDNF, lifestyle factors, and suppressing inflammation in the body. Increasing the expression of BDNF in the brain can be done by supplementing the excitatory neurotransmitter, glutamate, and increasing physical activity [106].

BDNF improves the function of neurons, encourages their growth, and strengthens/protects them against premature cell death [106]. It also binds to receptors at the synapses to improve signal strength between neurons. Essentially, the more BDNF in the brain, the better the brain works. Exposure to stress and the stress hormone cortisone and depression have been shown to decrease the expression of BDNF. Various studies have shown possible links between BDNF and conditions such as depression [110, 111], schizophrenia [63], obsessive-compulsive disorder [64], Alzheimer's disease [65], Huntington's disease [66], Rett syndrome [67], and dementia [68]. Ratey pointed out that a study of 30 depressed people found they all had low levels of BDNF [106].

25.11 Neurological Disorders

Encased in bone— our skull—the brain is our most protected organ. It weighs 3 pounds (1.36 kg), or approximately 2% of our body weight, yet it uses 20% of our body's oxygen and calories. Within the brain are approximately 100 billion neurons, with approximately 1000 synaptic connections per cell. In the visual cortex, there are approximately 12,000 synapses per cell, and in the prefrontal cortex, this increases to 80,000 synapses. The transmission speed along axons is measured at 200 miles per hour (322 km/h) [116].

Coursing throughout the brain to reach each neuron are capillaries, more than 400 miles (644 km) of them. These capillaries are not your ordinary ones; these are lined with specialized endothelial cells (ECs) and tight junctions, forming the blood-brain barrier (BBB). The BBB functions to help vital brain nutrients pass into the brain, protects the brain from plasma components, preserves the homeostasis of the brain, and directs inflammatory cells in response to an intracerebral event [74].

Where there is an intracerebral hemorrhage, ischemic injury, trauma, or neurodegenerative progression, plasma components, including red blood cells and leukocytes, cross the BBB and produce neurotoxins that may lead to abnormal synaptic and neuronal functions [74]. Tight junctions help the BBB's permeability to blood-borne solutes, which is dependent on their molecular weight and lipophilic and hydrophilic characteristics.

The consequences of central nervous system (CNS) inflammation are increased permeability of the BBB and leukocyte infiltration of the brain, both found in multiple sclerosis (MS). Major factors in the progression of MS are CD4+ interleukin-17 (IL-17)-producing T lymphocytes (Th_{17}) with IL-17 known to disrupt tight junctions in the ECs [66].

Further, T lymphocyte cerebral infiltration has been shown in patients with Alzheimer's disease (AD) and Parkinson's disease (PD). These again increase the permeability of the BBB [62, 99, 103–105]. Mounting evidence, both clinical and experimental, demonstrates the importance of BBB and any abnormalities of its functions, as well as their role in contributing to the progression of a number of inflammatory, infectious, and neurodegenerative diseases, especially in our aging population.

Acute stress increases GI and BBB permeability through activation of mast cells (MCs), which express high-affinity receptors for cortisol-releasing hormone (CRH). Chronic stress disrupted the intestinal barrier through MC activation and permitted penetration of luminal antigens, microflora metabolites, toxins, and lipopolysaccharide (LPS) in the systemic circulation and the CNS [39]. A molecule, called "microRNA-155," is responsible for cleaving epithelial cells to create microscopic gaps that let material through [40].

25.12 Alzheimer's Disease

The microbiota is a dynamic ecosystem which is influenced by several factors including genetics, diet, metabolism, age, geography, antibiotic treatment, and stress [46]. In addition, animal studies imply the necessity of an optimal function of what is known as the microbiome-gut-brain axis (MGB) in the behavioral as well as electrophysiological aspects of brain action [47].

There are correlations between microglial immunoreactivity and neuronal viability in Alzheimer's disease (AD) brain tissue. Immunohistochemical analysis demonstrated AD brain tissue expressed areas of diffuse fibrinogen indicative of a weakened BBB [41].

More than 40 million individuals throughout the world suffer from Alzheimer's, a disease for which there is no specific treatment. We know that Alzheimer's is an inflammatory disease. It is also known that gut bacteria can increase inflammation and probiotics can decrease inflammation.

Alzheimer's disease (AD) is recognized as one of the most common forms of senile dementia [43]. AD begins with memory loss of recent events (short-term memory impairment) and finally robs patients of their sense of self [44]. Increased biomarkers of oxidative stress [74], inflammation, and chronic neuroinflammation are reported to be associated with many neurodegenerative disorders of the central nervous system (CNS) including AD [45]. See ▶ Box 25.8 for emerging information on promotion of AD [43–45, 74].

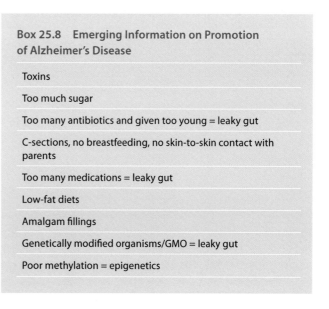

Box 25.8 Emerging Information on Promotion of Alzheimer's Disease

Toxins

Too much sugar

Too many antibiotics and given too young = leaky gut

C-sections, no breastfeeding, no skin-to-skin contact with parents

Too many medications = leaky gut

Low-fat diets

Amalgam fillings

Genetically modified organisms/GMO = leaky gut

Poor methylation = epigenetics

There is a preliminary research on the effect of probiotics on the prognosis of cognition [48]. However, data on the effects of probiotics on improving cognitive disorders are scarce

[49]. Gareau reported that intestinal dysbiosis in germ-free animals (containing no microbiota), bacterial infection with an enteric pathogen, and administration of probiotics can modulate cognitive behaviors, including learning and memory [50].

Postmortem analysis has shown lowered levels of BDNF in the brain tissues of people with Alzheimer's disease, although the nature of the connection remains unclear. Studies suggest that neurotrophic factors have a protective role against amyloid beta toxicity [112].

Researchers took a group of elderly Alzheimer's patients and studied them for 12 weeks [113]. Each participant underwent a test for mental function called the mini-mental status exam (MMSE), a standardized cognitive assessment used worldwide. They also underwent a blood test called highly sensitive C-reactive protein (hs-CRP), a powerful marker of inflammation. These tests were then repeated after 12 weeks.

The results were impressive [113]. The placebo group showed an increase in the inflammatory marker by 45%, while the probiotic group decreased the CRP by 18%. Over the 12 weeks, the placebo patients declined mentally. Their MMSE dropped from 8.47 to 8.00. The probiotic group improved their MMSE scores from 8.67 to 10.57. This study demonstrated that the probiotic administration for 12 weeks has favorable effects on mini-mental state examination (MMSE) score, malondialdehyde (MDA), serum high-sensitivity C-reactive protein (hs-CRP), markers of insulin metabolism, and triglyceride levels of the AD patients. These considerations led to establishing the term "psychobiotics" to highlight the potential effects of probiotics in the treatment of mental disorders [60]. Consistently, in a meta-analysis study, Kasińska and Drzewoski [60] reported a reduced homeostatic model of assessment for insulin resistance (HOMA-IR) and insignificant fasting plasma glucose (FGP) in probiotic-treated subjects. Mazloom et al. [109] also reported that probiotic supplementation had no significant effect on fasting blood glucose, markers of insulin metabolism, and lipid profiles.

The message here is that inflammation, which initiates or exacerbates all chronic illnesses including Alzheimer's disease, is dependent on the diversity and health of our gut bacteria. This, in turn, has major implications for brain health and function.

25.13 Autism

Autism is a condition in which intestinal microbiota is implicated. The onset of autism is often accompanied by intestinal dysfunction [79–81]. The first description of an association between autism and gastrointestinal syndrome began in 1971, with a report that 6 out of 15 autism patients had changed fecal character and defecation frequency [81].

Serotonin seeping from the second brain might even play some part in autism, the development disorder often first noticed in early childhood. Gershon has discovered that the same genes involved in synapse formation between neurons in the brain are involved in the alimentary synapse formation. "If these genes are affected in autism," he says, "it could explain why so many kids with autism have GI motor abnormalities" in addition to elevated levels of gut-produced serotonin in their blood [64]. About 95% of serotonin is produced by probiotics in the GI tract.

Food allergies, eczema, and asthma are associated with behavioral problems and neuropsychiatric disorders, including ADHD. Many children with autism spectrum disorders (ASD) present with GI systems and altered GI flora. ASD may involve brain inflammation. About 30% of children with ASDs have autoantibodies against brain proteins [42].

25.14 Parkinson's Disease

A high incidence rate of constipation is found in Parkinson's disease (PD) patients. Constipation can precede the onset of motor symptoms by more than 10 years [74].

Probiotics can efficiently reverse the impaired spatial learning and memory as well as synaptic transmission in diabetes mellitus [51]. It is demonstrated that other brain-related disorders, such as multiple sclerosis [52] and stress, are also influenced by probiotics [53]. In the levels of molecular mechanism, the microbiome is known to play a pronounced role in synaptic transmission. Numerous studies have shown capability of bacteria in producing neurotransmitters and neuromodulators including gamma-aminobutyric acid (GABA), norepinephrine, serotonin, dopamine, and acetylcholine [54]. Further, findings from germ-free animals indicated a decreased level of brain-derived neurotrophic factor (BDNF), important neurotrophic factor in neuronal growth and survival, and a reduced expression of some subunits of N-methyl-D-aspartate (NMDA) receptors [55]. GABA is the major inhibitory neurotransmitter in the CNS. Dysfunctions in GABA signaling are linked to anxiety, depression, and defects in synaptogenesis and cognitive impairments [57]. Also, the glutamatergic NMDA receptors are involved in the most important excitatory neurotransmission in the brain [58] and are engaged in the neural circuits involved in learning and memory. From these considerations, by contributing in neurotransmitter synthesis or receptor expression, probiotics might adjust the brain activity.

Accumulating evidence from experimental studies supports the hypothesis that via affecting inflammation, endocrine system, and neurotransmission, the gut microbiome takes a crucial role in the CNS function [59]. Accordingly, it is suggested that dysfunction of the neuroendocrine system, behavior, and cognition is correlated with gut microbiota dysbiosis [74].

25.15 Aging

We all know the brain changes as we age, as does the gastrointestinal tract. These changes, as well as a modification in lifestyle and decreased nutrition, increase the risk for infections, mental changes, physical impairments, and diminished cognitive abilities. The compromised microbiota in the elderly is associated with increased inflammation, also called inflammaging [11, 56, 74]. Inflammaging can also lead to decreased gut motility and constipation. Studies have illustrated an association between gut flora composition and cognitive processes, such as learning and memory. Additionally, intestinal microbiota contributes to the early development of normal social and cognitive behaviors [74].

Changes occurring in the microbiota during aging can have an impact on host health [74]. Notably, there is correlation between the relationship of falling microbial diversity and worsening frailty scores in elderly individuals. Additionally, drugs such as antibiotics, antacids, and H_2 receptor blockers can have profound negative effects on gut microbiota [74], all of which are more commonly used in the elderly as compared to a younger population. More than three-quarters of those aged 65 or older use at least one prescription medication, which can lead to unfavorable side effects on the GI tract [52]. For example, the use of opioids is linked to constipation [53], and the use of nonsteroidal anti-inflammatory drugs is linked with dysfunction in the GI defense system [54]. Furthermore, proton pump inhibitors, which are used to treat peptic ulcers, can trigger bacterial overgrowth of the upper GI tract through changes in the pH [17]. The use of broad-spectrum antibiotics can cause widespread disturbance in gut microbial composition and increase risk of *Clostridium difficile* infection, which in turn can reduce the biodiversity of gut microbiota [55]. As such, these medications can have implications on the microbiota-gut-brain-axis and influence CNS function.

Immunosenescence with chronic low-grade inflammation is characteristic in older age. This phenomenon, known as inflammaging, is distinguished by inflammation mediated by NK-kB and a loss of naïve CD4 T cells [56, 115]. A weakening of cell-mediated responses and reduction of the T-cell receptor repertoire have been shown in animal models upon aging [57]. In addition, clonal expansion of specific immune cells is impaired due to compromised cell division caused by a progressive shortening of telomeres through aging [58]. Notably, the GI microbiota of the elderly express a pro-inflammatory phenotype, as evidenced by a reduction in vitamin B12 synthesis and microbial reductase activity together with an increased incidence of DNA damage and immune compromise [50].

Indeed, inflammation is associated with a number of age-related diseases and neurodegenerative disorders, such as Parkinson's disease, Alzheimer's disease, multiple sclerosis, and motor neuron disease [60]. Claesson et al. showed that gut microbial composition and inflammation are directly linked to health outcomes [49]. They analyzed the microbiota composition and health outcomes in 178 patients living in different residences. Statistical analysis illustrated a clear difference in the microbial profiles depending on residence location, which significantly correlated with measures of nutrition, inflammation, frailty, and comorbidity. A decrease in community-associated microbiota and an increase in pro-inflammatory cytokines, as detected in long-stay hospital patients, were linked to increased frailty.

There is a decline in immune function in aging, and so microbiota-brain communication may be altered in the elderly, leading to changes in behavior. Supporting this is the fact that behavioral changes are observed in those with systemic infections as, similar to the gut, there is reciprocal communication between the CNS and the immune system [65]. Moreover, effects of gut microbiota and probiotics on inflammatory cytokines can have a direct effect on the brain.

Research groups have shown that gut microbes can also regulate the permeability of the blood–brain barrier (BBB), a highly selective barrier essential in protecting the brain from potential toxins [68]. Given that the function of both the BBB and microglia may deteriorate through age, these mechanisms are particularly relevant with regard to the role of microglia in elderly cognitive decline [69, 70]. Several epidemiological studies in elderly subjects have found links between diet and cognitive function [100]. In addition, chronic inflammation and a transformation in gut microbiota through age parallel a decline in cognitive function. The development of therapeutic strategies for diseases characterized by cognitive decline through alteration of the microbiota-gut-brain axis is an appealing possibility, particularly in the context of a global aging population [115]. However, much work still remains before this becomes a reality.

Decreased microbial diversity correlated with increased frailty, decreased diet diversity, health parameters, and increased levels of inflammatory markers. Individuals living in a community had the most diverse microbiota and were healthier as compared to those in short- or long-term residential care [74]. Firmicutes population drops with aging, while Bacteroidetes become the new dominant populations [74]. With a decline in Firmicutes, an increase in inflammatory diseases is found; this decline causes a decrease in immune response and reduced efficiency to digest. It was found that people at age 70 on average are still seen to have a similar biome of that of a 20- to 30-year-old; substantial changes are only seen within people who are older than 80 [74].

Deterioration in dentition, salivary function, digestion, and intestinal transit time may affect the intestinal microbiota with aging [74]. Aging is seen to negatively impact the microbiome, and the microbiome must be consistently maintained with a healthy diet to minimize the effects of aging [74].

In aging, BDNF is important as neuron morphology is critical in behavioral processes like learning and motor skills development [74]. BDNF levels have been shown to decrease in tissues with aging [74]. Studies using human subjects have found that hippocampal volume decreases with decreasing plasma levels of BDNF [74]. Like many other chemicals in the human body, aging decreases BDNF levels. This is why it takes the elderly longer to do complex tasks.

The relationship of BDNF to chronic illnesses and aging can be modified through a wide range of activities, nutrition, a healthy microbiome, and probiotics. Perlmutter rec-

Table 25.1 Strains of probiotics and which foods contain them

Lactobacillus plantarum	Sauerkraut, pickles, brined olives, kimchi, Nigerian ogi, sourdough, fermented sausage, stockfish, and some cheeses (such as cheddar)
Lactobacillus acidophilus	Yogurt kefir, miso, and tempeh
Lactobacillus brevis	Pickles, sauerkraut, and beer hop
Bifidobacterium lactis	Yogurt, miso, tempeh, pickled plum, pickles, kimchi, and many other forms of fermented and pickled fruits/vegetables that have not gone through the manufacturing process
Bifidobacterium longum	Yogurt, milk, fermented dairy foods, sauerkraut, and soy-based products

ommends five strains of probiotics to increase BDNF levels in the brain: *Lactobacillus plantarum, Lactobacillus acidophilus, Lactobacillus brevis, Bifidobacterium lactis,* and *Bifidobacterium longum* [114]. Table 25.1 lists the common foods to find these strains.

25.16 Fecal Transplants

While probiotics and antibiotics are generally known and available to treat gastrointestinal dysbiosis, fecal microbiota transplantation (FMT) has been rediscovered as a "new" way to restore gut microbiota. FMT is the administration of a solution of fecal matter from a donor into the intestinal tract of a recipient in order to directly change the recipient's gut microbial composition and confer a health benefit [74]. An early version of this practice was first documented in fourth-century China as "Yellow Soup." In some countries, maternal feces is inserted into the newborn's mouth to "jump-start" the colon. On June 17, 2013, the FDA approved FMT procedures for recurrent *Clostridium difficile* infection (CDI). There were zero documented serious side effects and there was a 92–95% success rate [74]. This strong interest is particular to patients with chronic gastrointestinal infections and inflammatory bowel diseases. Carefully screened donor stool is mixed with a saline solution. The solution is introduced into the GI tract via an NG tube, fecal enema, or oral capsules or during a colonoscopy. The "good" bacteria multiply and help flush out the *C. diff.* Bacteria. FMT reestablishes a balanced intestinal microbiota and results in impressive cure rates in patients with recurrent CDI. Standardization of FMT protocols and a randomized controlled trial are needed. FMT is likely to achieve widespread therapeutic benefit for a variety of diseases in the future [74].

FMT is currently approved by the FDA for the treatment of *Clostridium difficile* infection (CDI) with an IND (Investigational New Drug) application. FMT has also shown benefit in treatment of inflammatory bowel disease (IBD), irritable bowel syndrome (IBS), and ulcerative colitis, all of which affect brain health [117]. Additionally, FMT has been linked to normalization of multiple sclerosis symptoms and improvement of chronic fatigue syndrome [74]. Neurological improvement was reported in one patient with Parkinson's disease [74].

25.17 Food Sources

Mood disorders can make everything in life seem impossible. Mood-altering pharmaceutical drugs are frequently prescribed. One out of every five adult Americans now uses these drugs in spite of their side effects and addictive properties [74].

It is important to look at other natural options that are effective to balance neurotransmitters and improve brain health. The microbiome and probiotics have been discussed. Removing sugar, refined grains, and processed foods is a good beginning. Specific tests for food allergies and gluten sensitivities are important to identify which foods need to be eliminated from the diet to decrease inflammation and the risk of autoimmune disease.

Eating fresh, organic, non-GMO whole foods are essential to a healthy diversified microbiome and brain health. Adding fermented foods is beneficial. As Hippocrates famously said, "Let food be thy medicine and medicine be thy food." The diversity of whole foods containing phytochemical, antioxidants, and anti-inflammatory properties has a significant impact on the gut microbiome and brain health. The lipopolysaccharide endotoxin (LPS) is particularly important in provoking depressive symptoms and disturbing blood sugar control [118]. Traditional dietary practices have shown a 38% reduction of LPS after 1 month, while the Western diet increases LPS elevation [70]. The health benefits of eating live beneficial bacteria are becoming more evident. Foods are now being produced that contain probiotic bacteria.

Fermentation of foods and beverages is an ancient practice. Superficially, it would seem obvious, given the brain's dependence upon nutrients for its structure and function (including the micronutrients and non-nutrient dietary antioxidants), that nutrition should be a target of research in mental health [66].

Fermented foods and foods that contain live bacteria have been used throughout the world for centuries. Foods are fermented to extend shelf life, improve taste, and increase nutrient value. Fermented foods are known to increase specific nutrients and the phytochemical content of foods. Research consistently shows that a diet consisting of traditional whole foods improves mental health. Fermented foods also support the growth of microbes such as *Bifidobacteria* and *Lactobacillus sp.* which, by themselves, have been shown to produce neurotransmitters [66, 115].

Traditional fermented whole foods are instrumental in the complex systems of mental, emotional, and brain health [66]. Research has shown this influence may be by virtue of microbial actions, antioxidant and anti-inflammatory activity, reduction of intestinal permeability and decreasing LPS, improved glycemic control, nutritional support, lifestyle changes, and minimizing environmental toxins.

In addition to probiotics and fermented foods, other foods have also shown antidepressant properties, such as soy foods, cocoa, turmeric, green tea, coffee, blueberries, pomegranate, and honey [29–38]. Specific nutrients, such as magnesium, zinc, vitamin C, folic acid, and vitamin B12, have shown to decrease depressive symptoms [39–42]. Probiotics and the overall profile of the intestinal microbiota can influence tissue levels of mood-regulating minerals, such as magnesium and zinc [85, 86].

Since fatty acids seem to play an important role in shaping the gut microbiota metabolism, it is unsurprising that they have been considered as a dietary means to impede cognitive decline in aging. Human and animal studies on omega-3 polyunsaturated acids (n-3 PUFAs) have shed light on their neuroprotective roles through pathways of synaptic plasticity, neuroinflammation, and oxidative stress [106]. Recently, research has further suggested that through adult hippocampal neurogenesis, dietary choices, such as ingestion of n-3 PUFAs, can modulate brain structure and function throughout life [119].

25.18 Summary

The gut microbiota is a complex ecosystem which, in a healthy state, supports the major functions of the host by influencing metabolism, modulating the immune system, and protecting against pathogens.

The bidirectional communication and interaction between the ENS and CNS can support the body or cause disease depending on the diversity and health of the microbiome. The diseases not only affect the GI tract, through conditions like IBD, IBS, and colitis, or the CNS with mood disorders and neurological conditions, but they also have significant systemic impact and dysfunction.

Nagpal et al. clearly show that the gut microbiota is a promising target for preventative and therapeutic treatment of brain and mood disorders [115]. In the future, we may be able to use individual microbiome profiles in clinical practice to evaluate the current state and function of the gut. The use of prebiotics and probiotics, as well as fecal transplants, gives the clinicians a unique tool to treat chronic conditions.

UCLA's Mayer is doing work on how the trillions of bacteria in the gut "communicate" with enteric nervous system cells, which they greatly outnumber. His work with the gut's nervous system has led him to think that in the coming years, psychiatric treatment will expand to encompass the second brain in addition to the one atop the shoulders [74].

In terms of the potential treatment strategies, there is currently very limited research of the long-term effects of probiotics in human populations. Investigation of probiotics is further hindered by the varying quality control and efficacy of such a large family of compounds. Another area of research that is needed is the dose and the timing of probiotics and food. Carefully designed longitudinal clinical studies examining the microbial profiles of phenotype patients throughout life and the long-term effects of factors disturbing its compositions (e.g., via probiotics, antibiotics, dietary interventions) will be of great value to expand our knowledge of a promising area of research.

In 1907, Élie Metchnikoff probably never knew how exciting his discovery of the beneficial bacteria and phagocytes would be. Future research holds the potential of uncovering the intriguing connections between the gut microbiota and a multitude of chronic illnesses and neurological conditions.

» No disease that can be treated by diet should be treated with any other means. –Maimonides.

References

1. MacGill, Markus. Gut microbiota: definition, importance, and medical uses. Medical News Today, MediLexicon International, 26 June 2018. www.medicalnewstoday.com/articles/307998.php.
2. Backhed F, Ley RE, Sonnenburg JL, Peterson DA, Gordon JI. Host-bacterial mutualism in the human intestine. Science. 1915–1920;2005:307.
3. Backhed F, Roswall J, Peng Y, Feng Q, Jia H, Kovatcheva-Datchary P. Dynamics and stabilization of the human gut microbiome during the first year of life. Cell Host Microbe. 2015;17:690.
4. Bull MJ, Plummer NT. Part 1: the human gut microbiome in health and disease. Integr Med A Clin J. 2014;13(6):17–22.
5. Eckburg PB, Bik EM, Bernstein CN, et al. Diversity of the human intestinal microbial flora. Science. 2005;308(5728):1635–8.
6. Campbell AW. Your gut: no, not that one, this one. Alt Ther Health Med. 2014;20(2):9–10.
7. Tappenden KA, Deutsch AS. The physiological relevance of the intestinal microbiota—contributions to human health. J Am Coll Nutr. 2007;26(6):679S–83S.
8. Khanna S, Tosh PK. A clinician's primer on the role of the microbiome in human health and disease. Mayo Clin Proc. 2014;89:107–14.
9. Maes M, Kubera M, Leunis JC, Berk M, Geffard M, Bosmans E. In depression, bacterial translocation may drive inflammatory responses, oxidative and nitrosative stress (O&NS), and autoimmune responses directed against O&NS-damaged neoepitopes. Acta Psychiatr Scand. 2013;127:344–54.
10. Maes M, Kubera M, Leunis JC, Berk M. Increased IgA and IgM responses against gut commensals in chronic depression: further evidence for increased bacterial translocation or leaky gut. J Affect Disord. 2012;141:55–62.
11. Maes M, Kubera M, Leunis JC. The gut-brain barrier in major depression: intestinal mucosal dysfunction with an increased translocation of LPS from gram negative enterobacteria (leaky gut) plays a role in the inflammatory pathophysiology of depression. Neuro Endocrinol Lett. 2008;29:117–24.
12. Alonso C, Guilarte M, Vicario M, Ramos L, Rezzi S, Martínez C, Lobo B, Martin FP, Pigrau M, González-Castro AM, Gallart M, Malagelada JR, Azpiroz F, Kochhar S, Santos J. Acute experimental stress evokes a differential gender-determined increase in human intestinal macromolecular permeability. Neurogastroenterol Motil. 2012;24:740–6.
13. van Wijck K, Lenaerts K, Grootjans J, Wijnands KA, Poeze M, van Loon LJ, Dejong CH, Buurman WA. Physiology and pathophysiology of splanchnic hypoperfusion and intestinal injury during exercise: strategies for evaluation and prevention. Am J Physiol Gastrointest Liver Physiol. 2012;303:G155–68.

14. Li X, Kan EM, Lu J, Cao Y, Wong RK, Keshavarzian A, Wilder-Smith CH. Combat-training increases intestinal permeability, immune activation and gastrointestinal symptoms in soldiers. Aliment Pharmacol Ther. 2013;37:799–809.
15. Cani PD, Neyrinck AM, Fava F, Knauf C, Burcelin RG, Tuohy KM, Gibson GR, Delzenne NM. Selective increases of bifidobacteria in gut microflora improve high-fat-diet-induced diabetes in mice through a mechanism associated with endotoxaemia. Diabetologia. 2007;50:2374–83.
16. Cani PD, Amar J, Iglesias MA, Poggi M, Knauf C, Bastelica D, Neyrinck AM, Fava F, Tuohy KM, Chabo C, Waget A, Delmée E, Cousin B, Sulpice T, Chamontin B, Ferrières J, Tanti JF, Gibson GR, Casteilla L, Delzenne NM, Alessi MC, Burcelin R. Metabolic endotoxemia initiates obesity and insulin resistance. Diabetes. 2007;56:1761–72. https://doi.org/10.2337/db06-1491. [PubMed] [Cross Ref].
17. Reichenberg A, Yirmiya R, Schuld A, Kraus T, Haack M, Morag A, Pollmächer T. Cytokine-associated emotional and cognitive disturbances in humans. Arch Gen Psychiatry. 2001;58:445–52. https://doi.org/10.1001/archpsyc.58.5.445. [PubMed] [Cross Ref].
18. Prager G, Hadamitzky M, Engler A, Doenlen R, Wirth T, Pacheco-López G, Krügel U, Schedlowski M, Engler H. Amygdaloid signature of peripheral immune activation by bacterial lipopolysaccharide or staphylococcal enterotoxin B. J Neuroimmune Pharmacol. 2013;8:42–50.
19. Grigoleit JS, Kullmann JS, Wolf OT, Hammes F, Wegner A, Jablonowski S, Engler H, Gizewski E, Oberbeck R, Schedlowski M. Dose-dependent effects of endotoxin on neurobehavioral functions in humans. PLoS One. 2011;6:e28330.
20. Kullmann JS, Grigoleit JS, Lichte P, Kobbe P, Rosenberger C, Banner C, Wolf OT, Engler H, Oberbeck R, Elsenbruch S, Bingel U, Forsting M, Gizewski ER, Schedlowski M. Neural response to emotional stimuli during experimental human endotoxemia. Hum Brain Mapp. 2013;34:2217–27.
21. Dellagioia N, Devine L, Pittman B, Hannestad J. Bupropion pretreatment of endotoxin-induced depressive symptoms. Brain Behav Immun. 2013;31:197–204.
22. Benson S, Kattoor J, Wegner A, Hammes F, Reidick D, Grigoleit JS, Engler H, Oberbeck R, Schedlowski M, Elsenbruch S. Acute experimental endotoxemia induces visceral hypersensitivity and altered pain evaluation in healthy humans. Pain. 2012;153:794–9.
23. Dobos N, de Vries EF, Kema IP, Patas K, Prins M, Nijholt IM, Dierckx RA, Korf J, den Boer JA, Luiten PG, Eisel UL. The role of indoleamine 2,3-dioxygenase in a mouse model of neuroinflammation-induced depression. Alzheimers Dis. 2012;28:905–15.
24. Pekarek RS, Beisel WR. Effect of endotoxin on serum zinc concentrations in the rat. Appl Microbiol. 1969;18:482–4.
25. Pendyala S, Walker JM, Holt PR. A high-fat diet is associated with endotoxemia that originates from the gut. Gastroenterology. 2012;142:1100–1.
26. Yu Y, Wang R, Chen C, Du X, Ruan L, Sun J, Li J, Zhang L, O'Donnell JM, Pan J, Xu Y. Antidepressant-like effect of trans-resveratrol in chronic stress model: behavioral and neurochemical evidences. Psychiatry Res. 2013;47:315–22.
27. Pathak L, Agrawal Y, Dhir A. Natural polyphenols in the management of major depression. Expert Opin Investig Drugs. 2013;22:863–80.
28. Sekirov I, Russell SL, Antunes LC, Finlay BB. Gut microbiota in health and disease. Physiol Rev. 2010;90(3):859–904. Berg AM, Kelly CP, Farraye FA. Clostridium difficile infection in the inflammatory bowel disease patient. Inflamm Bowel Dis. 2013;19(1):194–204.
29. Pendyala S, Walker JM, Holt PR. A high-fat diet is associated with endotoxemia that originates from the gut. Gastroenterology. 2012;142(5):110–1101.
30. Ley RE. Obesity and the human microbiome. Curr Opin Gastroenterol. 2010;26(1):5–11.
31. Hrncir T, Stepankova R, Kozakova H, Hudcovic T, Tlaskalova-Hogenova H. Gut microbiota and lipopolysaccharide content of the diet influence development of regulatory T cells: studies in germ-free mice. BMC Immunol. November 2008;9:65.
32. Campbell A. Autoimmunity and the gut. Autoimmune Dis. 2014;2014:152428.
33. Isolauri E, Kirjavainen PV, Salminen S. Probiotics: a role in the treatment of intestinal infection and inflammation. Gut. 2002;50(Suppl 3):III54–9.
34. Malin M, Suomalainen H, Saxelin M, Isolauri E. Promotion of IgA immune response in patients with Crohn's disease by oral bacteriotherapy with Lactobacillus GG. Ann Nutr Metab. 1996;40(3):137–45.
35. Brint EK, MacSharry J, Fanning A, Shanahan F, Quigley EM. Differential expression of toll-like receptors in patients with irritable bowel syndrome. Am J Gastroenterol. 2011;106(2):329–36.
36. del MM G, Leonardi S, Maiello N, Brunese FP. Food allergy and probiotics in childhood. J Clin Gastroenterol. 2010;44(Suppl 1):S22–5.
37. Probiotics market to reach us$ 46 billion by 2022. Market watch press release online; 2018 Aug 22 [cited 2018 Sep 30]. Available from: https://www.marketwatch.com/press-release/probiotics-market-to-reach-us-46-billion-by-2022-2018-08-22/.
38. Sanders ME, Levy DD. The science and regulations of probiotic food and supplement product labeling. Ann N Y Acad Sci. 2011;1219:E1–E23.
39. Begley M, Hill C, Gahan CG. Bile salt hydrolase activity in probiotics. Appl Environ Microbiol. 2006;72(3):1729–38.
40. Isolauri E, Kirjavainen PV, Salminen S. Probiotics: a role in the treatment of intestinal infection and inflammation. Gut. 2002;50(Suppl 3):III54–9.
41. Hong HA, Duc le H, Cutting SM. The use of bacterial spore formers as probiotics. FEMS Microbiol Rev. 2005;29(4):813–35.
42. Hong HA, Khaneja R, Tam NM, et al. Bacillus subtilis isolated from the human gastrointestinal tract. Res Microbiol. 2009;160(2):134–43.
43. Mazza P. The use of Bacillus subtilis as an antidiarrhoeal microorganism. Boll Chim Farm. 1994;133(1):3–18.
44. Mcnulty NP, et al. The impact of a consortium of fermented milk strains on the gut microbiome of gnotobiotic mice and monozygotic twins. Sci Transl Med. 2011;3(106):106ra106. https://doi.org/10.1126/scitranslmed.3002701.
45. Kalliomäki M, Salminen S, Arvilommi H, Kero P, Koskinen P, Isolauri E. Probiotics in primary prevention of atopic disease: a randomised placebo controlled trial. Lancet. 2001;357(9262):1076–9.
46. Begley M, Hill C, Gahan CGM. Bile salt hydrolase activity in probiotics. Appl Environ Microbiol. 2006;72:1729–38.
47. Isolauri E, Kirjavainen PV, Salminen S. Probiotics: a role in the treatment of intestinal infection and inflammation? Gut. 2002;50:iii54–9.
48. Hong HA, Duc LH, Cutting SM. The use of bacterial spore formers as probiotics: table 1. FEMS Microbiol Rev. 2005;29:813–35.
49. Hong HA, Khaneja R, Tam NM, Cazzato A, Tan S, Urdaci M, Brisson A, Gasbarrini A, Barnes I, Cutting SM. Bacillus subtilis isolated from the human gastrointestinal tract. Res Microbiol. 2009;160:134–43.
50. Mazza P. The use of Bacillus subtilis as an antidiarrhoeal microorganism. Boll Chim Farm. 1994;133(1):3–18.
51. Gibson GR, Rouzaud G, Brostoff J, Rayment N. An evaluation of probiotic effects in the human gut: microbial aspects. Final technical report for Food Standards Agency (FSA) project ref 2005;G01022. BioBaia Website http://www.biogaia.com/study/evaluation-probiotic-effects-human-gut. Accessed 7 Oct 2018.
52. Kalliomäki M, Salminen S, Arvilommi H, Kero P, Koskinen P, Isolauri E. Probiotics in primary prevention of atopic disease: a randomised placebo-controlled trial. Lancet. 2001;357:1076–9.
53. Adams CA. The probiotic paradox: live and dead cells are biological response modifiers. Nutr Res Rev. 2010;23:37–46.
54. Lahtinen SJ. Probiotic viability – does it matter? Microb Ecol Health Dis. 2012;23 https://doi.org/10.3402/mehd.v23i0.18567.
55. Logan AC, Katzman M. Major depressive disorder: probiotics may be an adjuvant therapy. Med Hypotheses. 2005;64:533–8.

56. Messaoudi M, Lalonde R, Violle N, Javelot H, Desor D, Nejdi A, Bisson J-F, Rougeot C, Pichelin M, Cazaubiel M, et al. Assessment of psychotropic-like properties of a probiotic formulation (Lactobacillus helveticus R0052 and Bifidobacterium longum R0175) in rats and human subjects. Br J Nutr. 2011;105:755–64.
57. Bravo JA, Forsythe P, Chew MV, Escaravage E, Savignac HM, Dinan TG, Bienenstock J, Cryan JF. Ingestion of Lactobacillus strain regulates emotional behavior and central GABA receptor expression in a mouse via the vagus nerve. Proc Natl Acad Sci U S A. 2011;108:16050–5.
58. Steenbergen L, Sellaro R, van Hemert S, Bosch JA, Colzato LS. A randomized controlled trial to test the effect of multispecies probiotics on cognitive reactivity to sad mood. Brain Behav Immun. 2015;48:258–64.
59. Wall R, Marques TM, O'Sullivan O, Ross RP, Shanahan F, Quigley EM, Dinan TG, Kiely B, Fitzgerald GF, Cotter PD, et al. Contrasting effects of Bifidobacterium breve NCIMB 702258 and Bifidobacterium breve DPC 6330 on the composition of murine brain fatty acids and gut microbiota. Am J Clin Nut. 2012;95:1278–87.
60. Yurko-Mauro K, McCarthy D, Rom D, Nelson EB, Ryan AS, Blackwell A, Salem N, Stedman M. MIDAS investigators beneficial effects of docosahexaenoic acid on cognition in age-related cognitive decline. Alzheimers Dementia. 2010;6:456–64.
61. Gareau MG, Sherman PM, Walker WA. Probiotics and the gut microbiota in intestinal health and disease. Nat Rev Gastroenterol Hepatol. 2010;7:503–14.
62. Laukoetter MG, Nava P, Nusrat A. Role of the intestinal barrier in inflammatory bowel disease. World J Gastroenterol. 2008;14:401.
63. Rizos EN, Rontos I, Laskos E, Arsenis G, Michalopoulou PG, Vasilopoulos D, Gournellis R, Lykouras L. Investigation of serum BDNF levels in drug-naive patients with schizophrenia. Prog Neuro-Psychopharmacol Biol Psychiatry. 2008;32(5):1308–11. https://doi.org/10.1016/j.pnpbp.2008.04.007.
64. Hadhazy A (2010) Think twice: how the Gut's "second brain" influences mood and well-being. In: Scientific American. https://www.scientificamerican.com/article/gut-second-brain/. Accessed 7 Oct 2018.
65. Tougas G. The autonomic nervous system in functional bowel disorders. Gut. 2000;47:78iv–80.
66. Foster J, Zhou L. Psychobiotics and the gut–brain axis: in the pursuit of happiness. Neuropsychiatr Dis Treat. 2015;11:715.
67. Burokas A, Moloney RD, Dinan TG, Cryan JF. Microbiota regulation of the mammalian gut–brain axis. Adv Appl Microbiol. 2015;91:1–62.
68. Dinan TG, Cryan JF. Melancholic microbes: a link between gut microbiota and depression? Neurogastroenterol Motil. 2013;25:713–9.
69. Shea-Donohue T, Urban JF. Neuroimmune modulation of gut function. Gastrointest Pharmacol Handb Exp Pharmacol. 2016;239:247–67.
70. Cenit M, Olivares M, Codoñer-Franch P, Sanz Y. Intestinal microbiota and celiac disease: cause, consequence or co-evolution? Nutrients. 2015;7:6900–23.
71. Tougas G. The autonomic nervous system in functional bowel disorders. Gut. 2000;47:78iv–80.
72. Petra AI, Panagiotidou S, Hatziagelaki E, Stewart JM, Conti P, Theoharides TC. Gut-microbiota-brain axis and its effect on neuropsychiatric disorders with suspected immune dysregulation. Clin Ther. 2015;37:984–95.
73. Cohen S, Janicki-Deverts D, Doyle WJ, Miller GE, Frank E, Rabin BS, Turner RB. Chronic stress, glucocorticoid receptor resistance, inflammation, and disease risk. Proc Natl Acad Sci U S A. 2012;109:5995–9.
74. Drossman D, Creed F, Olden K, Svedlund J, Toner B, Whitehead W. Psychosocial aspects of the functional gastrointestinal disorders. In: Drossman D, Corazziari E, Talley N, Thompson W, Whitehead W, editors. Romm II. The functional gastrointestinal disorders: diagnosis, pathophysiology and treatment; a multinational consensus. 2nd ed. Degnon and Associates: McLean; 2000. p. 157–245.
75. Lackner JM, Lockwood A, Coad ML, et al. Alterations in GI symptoms, psychological status, and brain functioning following participation in cognitive therapy for IBS. Gastroenterology. 2004;126:A-477.
76. Cammarota G, Ianiro G, Bibbò S, Gasbarrini A. Gut microbiota modulation: probiotics, antibiotics or fecal microbiota transplantation? Intern Emerg Med. 2014;9(4):365–73.
77. Yolken R, Dickerson F. The microbiome-the missing link in the pathogenesis of schizophrenia. J Schizophr Res. 2014;153:S16.
78. Song X, Fan X, Song X, Zhang J, Zhang W, Li X, et al. Elevated levels of adiponectin and other cytokines in drug naive, first episode schizophrenia patients with normal weight. J Schizophr Res. 2013;150(1):269–73.
79. Hope S, Ueland T, Steen NE, Dieset I, Lorentzen S, Berg AO, et al. Interleukin 1 receptor antagonist and soluble tumor necrosis factor receptor 1 are associated with general severity and psychotic symptoms in schizophrenia and bipolar disorder. J Schizophr Res. 2013;145(1–3):36–42.
80. Severance EG, Gressitt KL, Yang S, Stallings CR, Origoni AE, Vaughan C, et al. Seroreactive marker for inflammatory bowel disease and associations with antibodies to dietary proteins in bipolar disorder. Bipolar Disord. 2013;16(3):230–40.
81. Oliver G, Wardle J, Gibson EL. Stress and food choice: a laboratory study. Psychosom Med. 2000;62(6):853–65.
82. Udem S. Faculty of 1000 evaluation for impact of diet in shaping gut microbiota revealed by a comparative study in children from Europe and rural Africa. F1000 – post-publication peer review of the biomedical literature. 2010.
83. David LA, Maurice CF, Carmody RN, Gootenberg DB, Button JE, Wolfe BE, et al. Diet rapidly and reproducibly alters the human gut microbiome. Nature. 2013;505(7484):559–63.
84. Drossman DA, Camilleri M, Mayer EA, Whitehead WE. AGA technical review on irritable bowel syndrome. Gastroenterology. 2002;123(6):2108–31.
85. Tougas G. The autonomic nervous system in functional bowel disorders. Gut. 2000;47(90004):78iv–80.
86. Eisenhofer G, Åneman A, Friberg P, Hooper D, Fåndriks L, Lonroth H, et al. Substantial production of dopamine in the human gastrointestinal tract. J Clin Endocrinol Metab. 1997;82(11):3864–71.
87. Kim D-Y, Camilleri M. Serotonin: a mediator of the brain-gut connection. Am J Gastroenterol. 2000;95(10):2698–709.
88. Koopman KE, Roefs A, Elbers DCE, Fliers E, Booij J, Serlie MJ, et al. Brain dopamine and serotonin transporter binding are associated with visual attention bias for food in lean men. Psychol Med. 2016;46(08):1707–17.
89. Logan AC, Rao AV, Irani D. Chronic fatigue syndrome: lactic acid bacteria may be of therapeutic value. Med Hypotheses. 2003;60(6):915–23.
90. Logan AC, Katzman M. Major depressive disorder: probiotics may be an adjuvant therapy. Med Hypotheses. 2005;64(3):533–8.
91. Selhub EM, Logan AC, Bested AC. Fermented foods, microbiota, and mental health: ancient practice meets nutritional psychiatry. J Physiol Anthropol. 2014;33(1):2.
92. Rogers MA, Greene MT, Young VB, Saint S, Langa KM, Kao JY, et al. Depression, antidepressant medications, and risk of Clostridium difficile infection. BMC Med. 2013;11(1):121.
93. Bravo JA, Forsythe P, Chew MV, Escaravage E, Savignac HM, Dinan TG, et al. Ingestion of Lactobacillus strain regulates emotional behavior and central GABA receptor expression in a mouse via the vagus nerve. Proc Natl Acad Sci U S A. 2011;108(38):16050–5.
94. Palma GD, Collins SM, Bercik P, Verdu EF. The microbiota-gut-brain axis in gastrointestinal disorders: stressed bugs, stressed brain or both? J Physiol. 2014;592(14):2989–97.
95. Barrett E, Ross R, O'Toole P, Fitzgerald G, Stanton C. γ-Aminobutyric acid production by culturable bacteria from the human intestine. J Appl Microbiol. 2012;113(2):411–7.

96. Mawe GM, Hoffman JM. Serotonin signalling in the gut—functions, dysfunctions and therapeutic targets. Nat Rev Gastroenterol Hepatol. 2013;10(8):473–86.
97. Prakash S, Rodes L, Coussa-Charley M, Tomaro-Duchesneau C, et al. Gut microbiota: next frontier in understanding human health and development of biotherapeutics. Biologics Targ Ther. 2011;5:71.
98. Dinan TG, Stanton C, Cryan JF. Psychobiotics: a novel class of psychotropic. Biol Psychiatry. 2013;74(10):720–6.
99. Bested AC, Logan AC, Selhub EM. Intestinal microbiota, probiotics and mental health: from Metchnikoff to modern advances: part III – convergence toward clinical trials. Gut Pathog. 2013;5(1):4.
100. Messaoudi M, Lalonde R, Violle N, Javelot H, Desor D, Nejdi A, et al. Assessment of psychotropic-like properties of a probiotic formulation (Lactobacillus helveticus R0052 and Bifidobacterium longum R0175) in rats and human subjects. Br J Nutr. 2010;105(05):755–64.
101. Evrensel A, Ceylan ME. The gut-brain axis: the missing link in depression. Clin Psychopharmacol Neurosci. 2015;13(3):239–44.
102. Foster JA, K-AM N. Gut–brain axis: how the microbiome influences anxiety and depression. Trends Neurosci. 2013;36(5):305–12.
103. Liu Y-W, Liu W-H, Wu C-C, Juan Y-C, Wu Y-C, Tsai H-P, et al. Psychotropic effects of Lactobacillus plantarum PS128 in early life-stressed and naïve adult mice. Brain Res. 2016;1631:1–12.
104. Collins SM, Bercik P. Intestinal bacteria influence brain activity in healthy humans. Nat Rev Gastroenterol Hepatol. 2013;10(6):326–7.
105. Nature's Bounty: The Psychobiotic Revolution [Internet]. Psychology Today. Sussex Publishers; [cited 2018Oct21]. Available from: https://www.psychologytoday.com/us/articles/201403/natures-bounty-the-psychobiotic-revolution.
106. What is BDNF and what does it do? [Internet]. Examined existence. [cited 2018Oct21]. Available from: https://examinedexistence.com/what-is-bdnf-and-what-does-it-do/.
107. Akbari E, Asemi Z, Kakhaki RD, Bahmani F, Kouchaki E, Tamtaji OR, et al. Effect of probiotic supplementation on cognitive function and metabolic status in Alzheimer's disease: a randomized, double-blind and controlled trial. Front Aging Neurosci. 2016;8:256.
108. Kasińska MA, Drzewoski J. Effectiveness of probiotics in type 2 diabetes: a meta-analysis. Pol Arch Int Med. 2015;125(11):803–13.
109. Mazloom Z, Yousefinejad A, Dabbaghmanesh MH. Effect of probiotics on lipid profile, glycemic control, insulin action, oxidative stress, and inflammatory markers in patients with type 2 diabetes: a clinical trial. Iran J Med Sci. 2013;38:38–43.
110. Dwivedi Y. Brain-derived neurotrophic factor: role in depression and suicide. Neuropsychiatr Dis Treat. 2009;5:433.
111. Brunoni AR, Lopes M, Fregni F. A systematic review and meta-analysis of clinical studies on major depression and BDNF levels: implications for the role of neuroplasticity in depression. Int J Neuropsychopharmacol. 2008;11(8):1169–80.
112. Mattson MP. Glutamate and neurotrophic factors in neuronal plasticity and disease. Ann N Y Acad Sci. 2008;1144:97–112. https://doi.org/10.1196/annals.1418.005.
113. Akbari E, Asemi Z, Daneshvar Kakhaki R, Bahmani F, Kouchaki E, et al. Effect of probiotic supplementation on cognitive function and metabolic status in Alzheimer's disease: a randomized, double-blind and controlled trial. Front Aging Neurosci. 2016;8:256. https://doi.org/10.3389/fnagi.2016.00256.
114. Perlmutter D. Brain maker: the power of gut microbes to heal and protect your brain – for life. Boston: Little, Brown and Company USA; 2015.
115. Nagpal R, Mainali R, Ahmadi S, Wang S, Singh R, Kavanagh K, Kitzman D, et al. Gut microbiome and aging: physiological and mechanistic insights. Nutr Healthy Aging. 2018;4(4):267–85.
116. Campbell AW. The blood-brain barrier. Alt Ther. 2016;22(2):6.
117. Tennant, McKenna. Fecal microbiota transplantation: the future of Feces. Yale Global Health Review, no 6, 2016.
118. Selhub EM, Logan AC, Bested AC. Fermented foods, microbiota, and mental health: ancient practice meets nutritional psychiatry. J Physiol Anthropol. 2014;33:2.
119. Leung K, Thuret S. Gut microbiota: a modulator of brain plasticity and cognitive function in ageing. Healthcare. 2015;3:898–916.

The Role of Nutrition in Integrative Oncology

Cynthia Henrich

26.1 Introduction – 409
26.1.1 What Is Cancer? – 409
26.1.2 Types of Cancer – 409
26.1.3 Statistics of Cancer – 409

26.2 Mechanisms of Oncogenesis and Metastases – 411

26.3 The Hallmarks of Cancer – 411

26.4 Current Conventional Medical Treatment – 411
26.4.1 Chemotherapy and Targeted Therapies – 411
26.4.2 Radiation – 413
26.4.3 Surgery – 414

26.5 Terrain Versus Cancer – 414

26.6 Optimization and Protection – 414

26.7 Synergistic Therapies – 414

26.8 Emotional Aspect of Cancer – 415

26.9 Lifestyle Detoxification – 416
26.9.1 Avoid or Remove? – 416
26.9.2 Sepsis – 417
26.9.3 Integrative and Functional Nutrition – 417
26.9.4 Liquid Oral or Enteral Nutrition – 418
26.9.5 Tea Therapy – 418
26.9.6 Turmeric – 418
26.9.7 Ginger – 419
26.9.8 Dandelion – 420
26.9.9 Methyl-Rich Foods – 420
26.9.10 Melatonin-Rich Foods – 421
26.9.11 Honokiol – 421
26.9.12 Limonene – 421
26.9.13 Cruciferous Vegetables – 421

© Springer Nature Switzerland AG 2020
D. Noland et al. (eds.), *Integrative and Functional Medical Nutrition Therapy*,
https://doi.org/10.1007/978-3-030-30730-1_26

26.9.14 Seeds – 422
26.9.15 Medicinal Mushrooms – 423
26.9.16 Bee Products – 423
26.9.17 Iodine Sufficiency – 423
26.9.18 Seaweed and Algae – 423
26.9.19 Fiber, Prebiotics, and Short-Chain Fatty Acids – 424
26.9.20 Modified Citrus Pectin and Pectin – 424
26.9.21 Fermented Foods and Probiotics – 425
26.9.22 Artemisia – 425
26.9.23 Carotenoid- and Polyphenol-Rich Foods – 425
26.9.24 Essential Oils – 425

26.10 Dietary Interventions – 426
26.10.1 Seasonal Eating and Living – 426
26.10.2 Autoimmune Paleo Diet – 426
26.10.3 Vegetarian Diet – 427
26.10.4 Ketogenic Diet – 427

26.11 Integrative Therapies Adjunctive to Nutrition Therapy – 429
26.11.1 Hyperthermia – 429
26.11.2 Mindfulness Practices or Meditation – 429
26.11.3 Movement and Exercise – 430
26.11.4 Massage or Touch Therapy – 430
26.11.5 Posttreatment: Creating an Empowerment Plan – 430

26.12 Conclusion – 431

References – 431

The Role of Nutrition in Integrative Oncology

Learning Objectives
- Understand conventional cancer modeling and treatments and their limitations and failure points.
- Create a repertoire of integrative therapies.
- Recognize the complexity of creating an integrative cancer protocol.
- Fit together the puzzle pieces of human metabolism, cancer cell metabolism, tumor microenvironment, and integrative therapies.
- Learn to use nutritional therapies to complement conventional cancer care.

26.1 Introduction

26.1.1 What Is Cancer?

Cancer is not a singular disease but a group of highly complex diseases characterized by unregulated cell proliferation that share pathways and molecular and metabolic mechanisms for oncogenesis, invasive growth, and metastases.

26.1.2 Types of Cancer

Cancer can be grouped into two main categories: solid tumors and blood cancers. Within those two groups, there is differentiation based on the type of cell in which the cancer originates.

Solid Tumors: Abnormal Mass of Tissue – Benign or Malignant
- *Carcinomas* originate in the epithelial tissues, including the skin and tissues that line and cover the internal organs and body cavities.
- *Sarcomas* develop in connective tissues, including the bones, cartilage, tendons, muscle, and other fibrous tissues.
- *Central nervous system cancers* develop in the brain and spinal cord.
- *Lymphomas* begin in the lymphatic system, including the lymphatic glands and vessels, the spleen, and the bone marrow.

Hematological Cancers: Blood-Forming Tissue Cancers
- *Leukemias* originate in the blood and bone marrow.
- *Multiple myeloma*
- *Lymphoma*

26.1.3 Statistics of Cancer

Despite medical science shining a tremendous amount of light on the growth mechanisms of these cancers, we have not yet made significant strides in conquering this disease. According to the American Cancer Society, in 2019, there will be an estimated 1,762,450 new cancer cases diagnosed, more than 4800 new cases daily, and approximately 606,880 cancer deaths [1] (▶ Boxes 26.1 and 26.2).

> **Box 26.1 Statistics List**
> - Three most common cancers diagnosed in men: prostate, lung, and colorectal
> - Three most common cancers diagnosed in women: breast, lung, and colorectal
> - Second most common cause of death among children aged 1–14 years in the US is cancer
> - Lifetime probability of being diagnosed with cancer: 39.3% for men and 37.7% for women
> - Based on data from Ref. [1]

> **Box 26.2 5-Year Relative Survival Rate**
> - Highest:
>   - Prostate 98%
>   - Melanoma of the skin 92%
>   - Female breast cancer 90%
> - Lowest:
>   - Pancreas 9%
>   - Liver 18%
>   - Esophagus 19%
>   - Lung 19%
> - Based on data from Ref. [1]

One out of every two males and one out of every three females can expect to be diagnosed with cancer within their lifetimes. Certain cancers have seen significant improvement in relative survival and cure rates, such as childhood leukemia, while the incidence of other cancers and their mortality rates are increasing. The American Association for Cancer Research released some unexpected projections for a shift in cancer statistics:

"Cancer incidence and deaths in the United States were projected for the most common cancer types for the years 2020 and 2030 based on changing demographics and the average annual percentage changes in incidence and death rates. Breast, prostate, and lung cancers will remain the top cancer diagnoses throughout this time, but thyroid cancer will replace colorectal cancer as the fourth-leading cancer diagnosis by 2030, and melanoma and uterine cancer will become the fifth and sixth most common cancers, respectively. Lung cancer is projected to remain the top cancer killer throughout this time period. However, pancreas and liver cancers are projected to surpass breast, prostate, and colorectal cancers to become the second and third leading causes of cancer-related deaths by 2030, respectively" [2].

These statistical analyses are done to determine the allocation of research funding as well as the relative effectiveness of the conventional routes of cancer treatment: chemotherapy, radiation, surgery, hormone therapy, immunotherapy, and targeted therapies.

26.1.3.1 The Functional, Integrative, Holistic, and Nutritional Perspective

A deeper holistic look at these statistics can illustrate shifts in causality, such as toxic exposures, infection [3], and dietary deficiencies [4]. This perspective is an opportunity to create change.

The statistics do not tell the story of the bio-individual aspects of the whole person. Lifestyle, exposure to toxins and pathogens, diet, emotional wellness, use of supplements, physical activity, exposure to EMF radiation, preconception history, all of these affect the molecular and epigenetic changes that can initiate or prevent the growth of cancer in the body.

On his deathbed, Louis Pasteur famously recanted his germ theory and said, "Bernard is correct. I was wrong. The germ is nothing. The terrain is everything." Claude Bernard's work aligned with that of Antoine Bechamp who originated the terrain theory [5]. Bechamp felt that disease resulted when microbes "changed form, function and toxicity" based on the terrain of the host.

> The constancy of the internal environment is the condition for a free and independent life. –Claude Bernard [5]

Cancer is the same. Conventional treatment focuses on the "germ or microbe" (tumor). Integrative care focuses on both the germ and the terrain (the health of the person).

Conventional therapy is effective at surgical debulking cytoreduction (surgically reducing tumor bulk), with or without chemotherapy and/or radiation. It does not address protecting healthy cells during treatment. Unfortunately, many therapies leave the patient with the risk of secondary cancers due to the toxicity of treatment, depression of the immune system causing a host of issues, the most dangerous of which is sepsis, as well as the risk of recurrence or metastasis of the original cancer in an enriched state. Enriched cancer cells arise when tumor bulk is reduced, but the stem cells are not eradicated. These remaining "enriched" stem cells then become resistant to future treatment, akin to MRSA in the bacterial world.

The ability of the body to prevent the growth of cancer and the potential of the immune system to conquer it are our most overlooked weapon in the war on cancer. Factors that can be optimized to create a body that can effectively engage in this battle include lifestyle, diet, honoring bio-individual needs, consuming anti-inflammatory foods that nourish the tissues and cells, reducing the toxic burden of the body, and attending to emotional and energetic imbalances [4, 6]. This improves clinical outcome and quality of life and aids in the management of conventional treatment. These are exactly the efficacy endpoints of any cancer treatment.

When considering how to utilize nutrition and supplementation in complementary and adjunctive cancer therapy, several things should be considered:

- Anti-inflammatory diet: Any biologically contraindicated foods should be avoided, such as gluten, dairy, eggs, lectins, nightshades, or other common allergens and inflammatory triggers, based on the needs of the individual.
- Targeted supplementation based on a nutritional and clinical assessment of any nutrient deficits (e.g., vitamin D, zinc, methyl nutrients, minerals) is important to optimize immune function until goal test markers are reached and maintained.
- Herbal therapies: Certain herbs and roots have gained a reputation as cancer fighters but should be examined in the setting of any underlying disease as well as drug regimens. For instance, turmeric/curcumin [6] being contraindicated when used in combination with certain chemotherapy drugs such as camptothecin, mechlorethamine, doxorubicin, or cyclophosphamide in the treatment of breast cancer [7–10].
- Detoxification support: Cytochrome pathways (discussed below) should be checked for inhibition or induction potential of food, herb, and drug combinations. Some of these situations are synergistic, while others are potentially harmful.

Moving forward, our goal should be to utilize the ability of certain foods and herbs to positively impact quality of life, sensitize cancer cells, protect healthy cells, and manipulate the tumor microenvironment to the benefit of the patient. If a substance can increase the amount of time that a chemotherapeutic agent is held in the body, while positively affecting disease markers, it may be possible that the drug dose can be reduced while simultaneously creating a positive patient outcome due to reduced side effects of the drug. This approach may make dose reductions of both chemo and radiation possible without a loss of efficacy, consequently allowing for potentially fewer side effects, better quality of life, and extension of treatment potential by resolving dose-limiting toxicities and time to resistance.

Phytonutrition is the therapeutic potential of the colorful pigments in plants to act as modifiers of physiological function with growing evidence related to modulation of carcinogenesis [10]. For the nutritionist, understanding the functional parameters that impact the absorption and utilization of these compounds will allow for their use to be optimized in specific interventions. Although most plants contain significant amounts of phytochemicals, certain groups are higher in lycopenes, such as red and pink fruits and vegetables, or the isothiocyanates found in cruciferous vegetables. An awareness of the evidence-based research specific to the impact of specific phytonutrients on certain cancers can create a significant impact on dietary recommendations for patients. This can enable the creation of integrative food- and botanical-based nutrition protocols for prevention and adjunctive therapies. The same theory can be applied to the metabolic function of fats and proteins and their physiological effects. Whole food and herbs can truly perform as complementary medicine [10].

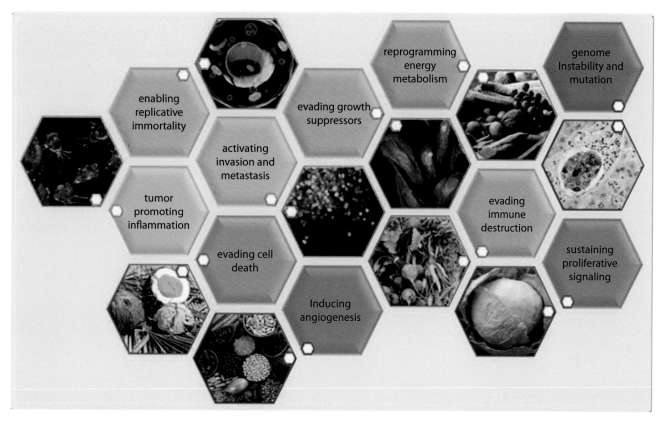

● Fig. 26.1 Graphic Hallmarks of Cancer (Courtesy of Cynthia Henrich)

26.2 Mechanisms of Oncogenesis and Metastases

The wealth of research on the mechanisms of cancer can be used to reverse engineer similar effects utilizing diet, lifestyle, and natural therapies either alone or integrated with conventional therapies. This integration can improve quality and length of life for patients while simultaneously contributing to their whole-body wellness. First, we must examine the underlying framework of mechanisms that enable the transformation of a healthy cell into a malignant cell capable of metastasizing to distant sites. According to the conventional model, the mechanisms are as follows and should each be viewed as an opportunity to affect the outcome with integrative strategies:

26.3 The Hallmarks of Cancer

Although cancer is a collection of highly complex diseases, cancer cells demonstrate a set of distinct traits across tumor types that enable them to reproduce and metastasize, according to conventional theories. Drs. Hanahan and Weinberg began to elucidate and expand on these traits [11]. Research has emerged of late that cancer cell metabolism and stem cells are not following the previous dictums and challenging these hallmark traits. Tumors exhibit another dimension of complexity in that they contain a repertoire of recruited, apparently normal cells that contribute to creating the tumor microenvironment (● Fig. 26.1) [11].

26.4 Current Conventional Medical Treatment

Chemotherapy, radiation, surgery, immunotherapy, hormone therapy, and targeted therapies have been designed to interrupt some point in the process of transformation of the cell into malignancy or to prevent that event from recurring. To offer individualized treatment options, histological compatibility and oncotyping tests have been developed to examine the genetics of a patient's tumor and their best response to treatment options. This can help to identify growth mechanisms and potentially chart a course for drug therapy.

26.4.1 Chemotherapy and Targeted Therapies

According the SEER*Rx Interactive Antineoplastic Drugs Database on the National Cancer Institute website, there are currently 1867 drugs that can be combined into 510 potential regimens for conventional cancer treatment [12].

Gaining an understanding of the classes of chemotherapeutic agents and their typical side effects, such as impact on platelets and cytochrome pathways, can make it easier to determine functional and genomic interactions with supplements and food. This will enable the practitioner to more carefully construct a dietary and botanical protocol that is most useful to the patient. There are several types of chemotherapy, each of which can be administered via oral, intravenous, intraperitoneal, transdermal, or implantable routes depending on the drug and cancer:

26.4.1.1 Alkylating Agents

Alkylating agents were one of the earliest classes of drugs developed to treat cancer. They work by chemically altering the DNA of a cancer cell by adding an alkyl group to directly modify DNA bases or form cross-links to prevent replication and/or induce apoptosis. These drugs are particularly devastating to the rapidly dividing cells that line the gut. Tremendous damage to the mucosal barrier impacts digestion and absorption, as well as frequently inducing side effects such as diarrhea and dehydration that can limit or end a course of chemotherapy [13].

Alkylating agents are most active in the resting phase of the cell. These types of drugs are cell-cycle nonspecific. There are several types of alkylating agents used in chemotherapy treatments:

- Mustard gas derivatives: Mechlorethamine, cyclophosphamide, chlorambucil, melphalan, and ifosfamide
- Ethylenimines: Thiotepa and hexamethylmelamine
- Alkyl sulfonates: Busulfan
- Hydrazines and triazines: Altretamine, procarbazine, dacarbazine, and temozolomide
- Nitrosoureas: Carmustine, lomustine, and streptozocin. Nitrosoureas are unique because, unlike most types of chemo treatments, they can cross the blood–brain barrier. They can be useful in treating brain tumors.
- Metal salts: Carboplatin, cisplatin, and oxaliplatin (► Box 26.3).

Box 26.3   Nutritional Therapies That Can Be Used to Preserve the Mucosal Barrier Function During Treatment
- Glutamine (cabbage is rich in the amino acid L-glutamine)
- Zinc carnosine
- Mastica
- Butyrate (ghee is rich in butyric acid)
- Immunoglobulin-rich colostrum
- Probiotic and prebiotic foods
- Mushrooms such as Cordyceps and Ganoderma
- Plant-based protein powders
- Elemental diet therapy
- Turmeric
- Ginger
- Herbs such as slippery elm, licorice root, aloe, marshmallow root and chamomile

---

26.4.1.2 Plant Alkaloids

Plant alkaloids are chemotherapy treatments derived from certain types of plants. The vinca alkaloids are made from the periwinkle plant (*Catharanthus roseus*). The taxanes are made from the bark of the Pacific Yew tree (taxus). The vinca alkaloids and taxanes are also known as antimicrotubule agents. The podophyllotoxins are derived from the mayapple plant. Camptothecin analogs are derived from the Asian "Happy Tree" (*Camptotheca acuminata*). Podophyllotoxins and camptothecin analogs are also known as topoisomerase inhibitors, which are used in certain types of chemotherapy. The plant alkaloids are cell-cycle specific. This means they attack the cells during various phases of division.
- Vinca alkaloids: Vincristine, vinblastine, and vinorelbine
- Taxanes: Paclitaxel and docetaxel
- Podophyllotoxins: Etoposide and teniposide
- Camptothecin analogs: Irinotecan and topotecan

26.4.1.3 Antitumor Antibiotics

Antitumor antibiotics are chemotherapeutic treatments made from natural products produced by species of the soil fungus, *Streptomyces*. These drugs act during multiple phases of the cell cycle and are considered cell-cycle specific. This class of drugs will have a significant impact on the diversity of the microbiome that must be addressed during and post-treatment. There are several types of antitumor antibiotics:
- Anthracyclines: Doxorubicin, daunorubicin, epirubicin, mitoxantrone, and idarubicin
- Chromomycins: Dactinomycin and plicamycin
- Miscellaneous: Mitomycin and bleomycin

26.4.1.4 Antimetabolites

Antimetabolites are types of chemotherapy treatments that are very similar to normal substances within the cell. When the cells incorporate these substances into the cellular metabolism, they are unable to divide. Antimetabolites are cell-cycle specific. They attack cells at very specific phases in the cycle. Antimetabolites are classified according to the substances with which they interfere.
- Folic acid antagonist: Methotrexate
- Pyrimidine antagonist: 5-fluorouracil, floxuridine, cytarabine, capecitabine, and gemcitabine
- Purine antagonist: 6-mercaptopurine and 6-thioguanine
- Adenosine deaminase inhibitor: Cladribine, fludarabine, nelarabine, and pentostatin

26.4.1.5 Topoisomerase Inhibitors

Topoisomerase inhibitors are types of chemotherapy drugs that interfere with the action of topoisomerase enzymes (topoisomerase I and II). During the process of chemotherapy treatments, topoisomerase enzymes control the manipulation of the structure of DNA necessary for replication.
- Topoisomerase I inhibitors: Irinotecan and topotecan
- Topoisomerase II inhibitors: Amsacrine, etoposide, etoposide phosphate, and teniposide

26.4.1.6 Miscellaneous Antineoplastics

Several useful types of chemotherapy drugs are unique:
- Ribonucleotide reductase inhibitor: Hydroxyurea
- Adrenocortical steroid inhibitor: Mitotane
- Enzymes: Asparaginase and pegaspargase
- Antimicrotubule agent: Estramustine
- Retinoids: Bexarotene, isotretinoin, tretinoin (ATRA)

Targeted therapies have been designed to target pathways outside of the common feature of cancer cells to be able to divide rapidly. Because many normal cells in the human body also divide rapidly, multiple side effects may unfortunately occur that can limit treatment and cause significant damage or morbidity to the person [14, 15].

The different types of targeted therapies are defined in three broad categories. Some targeted therapies focus on the internal components and function of the cancer cell. The targeted therapies use small molecules that can get into the cell and disrupt cellular function, resulting in apoptosis. There are several types of targeted therapy that focus on the inner parts of the cells, while other targeted therapies focus on receptors that are on the outside of the cell known as monoclonal antibodies.

Antiangiogenesis drugs target the blood vessels that supply oxygen to the cells, ultimately causing the cells to starve. Specifically, this class of drugs is designed to stop the formation of new blood vessels. There is some research indicating that this particular hallmark of cancer may need to be revisited because certain tumors seem to have the capability to co-opt existing blood vasculature. This can account for resistance to antiangiogenic drugs and may distinguish which tumors may be susceptible to therapy [16].

Monoclonal antibodies are a targeted cancer therapy that use an antibody to specifically target an antigen on the surface of a cancer cell. This lock-and-key function causes less toxicity to healthy cells but is limited to the cancers in which antigens and their respective antibodies have been identified. One way the immune system attacks foreign substances in the body is by making large numbers of antibodies. An antibody is a protein that sticks to a specific protein called an antigen. Antibodies circulate throughout the body until they find and attach to the antigen. Once attached, they can recruit other parts of the immune system to destroy the cells containing the antigen. The following are monoclonal antibodies: alemtuzumab, gemtuzumab ozogamicin, rituximab, trastuzumab, and ibritumomab tiuxetan.

Other targeted therapies include the following:
- Signal transduction inhibitors:
  - Imatinib mesylate (protein-tyrosine kinase inhibitor)
  - Gefitinib (epidermal growth factor receptor tyrosine kinase inhibitor, EGFR-TK)
    - Tyrosine kinases are a family of enzymes which catalyze phosphorylation and regulate biological processes like growth, differentiation, metabolism, and apoptosis of cells.
  - Cetuximab (epidermal growth factor receptor)
  - Lapatinib (epidermal growth factor receptor (EGFR))
  - Human epidermal receptor type 2 (HER2) tyrosine kinase inhibitor
    - Anti-EGFR therapies can inhibit the activation of the EGFR signaling pathway in cancer cells which has been linked to increased cell proliferation, angiogenesis, metastasis, and decreased apoptosis.
- Biologic response modifier agent: Denileukin diftitox
- Proteasome inhibitor: Bortezomib [17]

26.4.2 Radiation

Radiation therapy uses high-energy X-rays, gamma rays, electron beams, or protons to make small breaks in the DNA of cells or to create free radicals within the cells that will prevent replication and cause cancer cell death. The radiation may be delivered as external beam radiation therapy or as internal radiation therapy, also known as brachytherapy, via radioactive material placed in the body near the tumor. Systemic radiation therapy uses radioactive substances, such as radioactive iodine, to travel throughout the body to kill cancer cells. Approximately half of all cancer patients receive some type of radiation therapy sometime during the course of their treatment.

Radiation use is generally limited by late-onset adverse events that can occur during treatment or even months or years later. Secondary cancers, inflammatory conditions such as pneumonitis and proctitis, and radiation fibrosis, an irreversible condition characterized by increased connective tissue stiffness and loss of tissue function at the irradiated site, are a few of the typical outcomes of radiation therapy.

It may be possible to predict who may be at risk of radiation-induced fibrosis via genetic markers and methylation status. The application of this research may include dietary modification and supplement use to optimize methylation status, restore nutritional insufficiencies, monitor hypercoagulation (fibrinogen, platelets, and d-dimer markers), and mitochondrial protection [18].

Radiation has been found to create more cancer stem cells (CSC) that are resistant to treatment. Recent evidence indicates that radiation converts non-stem cancer cells into cancer stem cells, which exhibit similar radioresistance to intrinsic CSCs (cancer stem cells which are resistant to treatment and cause relapse and metastasis) [19]. A possible mechanism of cancer resistance to radiation may be through CSCs that should be killed by radiotherapy but are able to survive radiotherapy. After radiation-induced stresses disappear, these newly generated CSCs, together with the intrinsic CSCs, contribute to the relapse and metastasis of cancer [19].

An alternative to conventional radiation therapy is proton therapy. Proton therapy is an advanced form of radiation therapy that uses protons—positively charged atomic particles—instead of the photons used in standard X-ray radiation therapy. With proton therapy, doctors can precisely target the tumor while minimizing damage to the surrounding healthy tissue. Unlike X-ray radiation, such as intensity-modulated radiation therapy [IMRT], protons deposit much of their radiation directly in the tumor and then stop. For example, in prostate cancer, proton therapy eliminates about 60% of excess radiation delivered to healthy tissues surrounding the prostate compared to IMRT. Proton therapy can treat recurrent tumors in patients who have already received traditional forms of radiation, reduce the incidence of short- and long-term side effects, and incur a lower incidence of secondary tumors. This may potentially allow the use of higher, more effective doses of radiation.

26.4.3 Surgery

Surgical resection of a tumor with clear margins and negative lymph findings is the single best predictor of survival. Many conventional therapies are aimed at reducing tumor bulk in order to perform surgery. If tumor debulking surgery is not optimal, vascular endothelial growth factor (VEGF) may increase, which could increase angiogenesis.

26.5 Terrain Versus Cancer

Perhaps the most overlooked aspect in standard cancer care is the biological and emotional condition of the person. Science is doing an excellent job at uncovering pathways, genetics, and mechanisms of the genesis and growth of cancer, but because the focus is on drug development, the epigenetic impact of nutrition and lifestyle on these processes is largely ignored. The research, however, is extremely useful in terms of reverse-engineered nutrition and lifestyle solutions.

Consider the old argument of germ versus terrain referring to pathogenic infectious organisms and apply the concept to oncology. Enormous potential exists in the unexplored realm of cancer versus terrain. Use the information gleaned from the laser-focused study of cancer cells and tumors to reduce the manifestation of cancer by optimizing the cellular processes of the body through intensive bio-individually appropriate nourishment and detoxification. Imagine accounting for this within the framework of cancer statistics. We account for age, but none of the other driving factors that enable cancer to develop; or in the survivors, where radical lifestyle changes may dramatically shift where they fit in (or now defy) the survival statistics.

It is the epitome of the poem, "The Road Not Taken," by Robert Frost:

> *Two roads diverged in a wood, and I—*
> *I took the one less traveled by,*
> *And that has made all the difference.*

Dietary epigenetics, medical nutrition therapy, lifestyle medicine, circadian medicine, and detoxification protocols to optimize the human terrain will shift the entire way that we approach cancer prevention, treatment, and survival. These methods will enable practitioners to optimize the wellness of the patient, thereby creating a strong foundation for the body to withstand treatment, balance immune function, and tailor a dietary strategy that is ideal for the patient throughout different stages of treatment and recovery.

26.6 Optimization and Protection

The inclusion of the appropriate balance of macronutrients, saturation of micronutrients, and exclusion of harmful and toxic anti-nutrients is the core of cellular optimization. This allows the enzyme-driven processes of methylation, acetylation, phosphorylation, ubiquitination, and sumoylation to appropriately regulate cellular function and epigenetic expression, thereby enabling optimal mitochondrial metabolism.

Mitochondria are the primary source of energy for the body and dealing with mitochondrial metabolism can be thought of as two sides of a coin: optimize function and prevent dysfunction. The approach to this is multifaceted: provide the necessary nutrients, precursors, and cofactors; optimize liver function and detoxification; and determine bio-individual requirements for certain vitamins, conditionally essential nutrients, amino acids, and fatty acids to ensure optimal enzymatic function and genetic expression.

The importance of optimal mitochondrial function cannot be overstated. It is implicated in stem cell self-renewal, cardiomyocyte regeneration, neural stem cell survival [20], skeletal and heart muscle function and repair, neurological inflammation [21], cellular senescence [22], telomere shortening [23], oncogene initiation, and more.

26.7 Synergistic Therapies

Synergistic therapy in integrative oncology means that a particular substance or action will protect healthy cells while enabling another therapy to work more effectively. This can potentially enable the reduction of a chemotherapy drug dose through enhanced chemosensitivity, thereby reducing toxicity and side effects, or even dramatically shift an outcome by tapping into the power of diet on biology.

Synergistic therapies [24, 25] can be administered or performed concurrently with standard therapies, or they can be cycled throughout the regimen based on drug metabolism, pharmacokinetics, and pharmacogenetics. Each individual therapy, food, herb, vitamin or mineral, exercise, or mindfulness practice needs to be assessed on an individual basis, taking into consideration the regimen being administered as well as the bio-individual nature of the person. For instance, certain genetic markers or liver metabolism factors can create a need for optimization of methylation or adjustment for more rapid clearance of a substance through the cytochrome pathways. There is more and more information correlating phytonutrients and supplements with these metabolic pathways. It is helpful to check which pathway may be affected and cross-check that with the regimen and the patient.

Cytochrome P450 enzymes are essential for the metabolism of many medications [26]. Although this class has more than 50 enzymes, six of them metabolize 90% of drugs, with the two most significant enzymes being CYP3A4 and CYP2D6. Genetic variability (polymorphism) in these enzymes may influence a patient's response to commonly prescribed drug classes, including beta-blockers and antidepressants. Cytochrome P450 enzymes can be inhibited or induced by drugs, resulting in clinically significant drug-drug interactions that can cause unanticipated adverse reactions or therapeutic failures [26].

Drug metabolism is not limited to liver biotransformation but also occurs through the kidney, intestinal cells, lungs, central nervous system, gastric mucosa, prostate, and, in some cases, even the skin.

Because many regimens include premedication to alleviate side effects or to act as adjuvant therapy, the metabolism/drug interactions can become even more complicated. Referring to a clinical pharmacology P450 drug interaction table such as the one from the Indiana University School of Medicine can be helpful, but the complex nature of multidrug metabolism can make the use of therapeutic doses of food or supplements an incredibly complex situation [27].

26.8 Emotional Aspect of Cancer

Psycho-oncology explores how trauma and prolonged chronic stress can play a part in oncogenesis. It is extremely common when working with cancer patients for them to share that an emotional conflict or trauma occurred approximately 18–24 months prior to their diagnosis. The processing of a trauma of this nature can impact deep sleep patterns and the production of melatonin. Melatonin is primarily synthesized by the pineal gland and is a natural antioxidant with immune-enhancing properties. Melatonin production declines with age and may be one of the major contributors to immunosenescence and cancer genesis [28].

Among the various functions attributed to melatonin in the control of the immune system, antitumor defense assumes a primary role [28]. The nighttime physiological surge of melatonin in the blood or extracellular fluid has been suggested to serve as a natural restraint for tumor initiation, promotion, and/or progression. Melatonin was demonstrated to be oncostatic for a variety of tumor cells like breast carcinoma, ovarian carcinoma, endometrial carcinoma, human uveal melanoma cells, prostate tumor cells, and GI tumors [28].

Melatonin is also involved in genetic expression, primarily with the genes controlling the cell cycle, adhesion, and transport. It specifically impacts mitochondrial gene expression. Melatonin increases the production of cytokines IL-2, IL-6, IL-12, and IFN-gamma from lymphocytes and monocytes, which then increases the amount of natural killer cells.

The waterfall effect of stress reaches well past melatonin. As cortisol levels remain inappropriately elevated over time, stress is placed on the adrenal glands and adrenaline production. Adrenaline is required to stimulate the G-protein for the oxidative phosphorylation cycle of mitochondria, producing ATP by converting glucose in the Krebs citric acid cycle. As adrenaline production is depleted, the conversion of glucose to ATP is inhibited, and instead the glucose ferments and creates lactic acid as a by-product inducing cellular DNA and mitochondrial mutations.

Circadian rhythms and traditional Chinese medicine (TCM) both correlate time with physical balance and energy [29]. As described, in TCM, there are 14 major meridians that conduct the flow of Qi throughout the body. Twelve of these meridians make up the 24-hour clock, representing 2 hours each. The energy is constantly flowing through all of these meridians throughout the 24 hours, with each meridian having a 2-hour period of time as the primary meridian. The meridian reflects the energy of its associated organ, but also it reflects other processes in us: thoughts, emotions, colors, sounds, seasons, and even spiritual aspects [29]. The meridians are coupled in pairs of Yin (receiving energy) and Yang (expressing energy) of one of the five elements—earth, metal, water, wood, and fire [29].

So, for instance, recurring insomnia at the same time each night, or a physical complaint at the same time each day, may correlate to an emotion or issue that is causing a disruption or stagnation in the life force energy or chi. Examples include the following:
- Liver-gallbladder: anger, resentment, and frustration
- Heart-small intestine: emotional instability, discontent, lack of focus, and heartbreak
- Spleen-stomach: worry and lack of empathy
- Lung-large intestine: sadness and unresolved grief
- Kidney-urinary bladder: fear, abandonment, and anxiety

Often, the emotion of the initiating trauma will line up with an illness manifesting in the corresponding organ pair. This type of work is a beautiful opportunity to walk the border between physiology and emotion, navigating the Rivers of Qi.

If you can recall the moments of childhood where hours were spent so busy, happy, and content that one might forget to eat unless called, you can see that emotional balance and joy are sustenance in their own right. Addressing these emotional needs or imbalances by simply acknowledging their existence is often enough to create awareness and initiate the healing process. Nourishing the spirit can further that healing (▶ Box 26.4).

> **Box 26.4 Nutritional and Lifestyle Strategies for Emotional Support**
> - Optimize sleep
>   - Melatonin and foods rich in melatonin
>   - Theanine (found in green tea)
>   - Magnesium
>   - Apply a blue light filter on devices to mimic sundown.
>   - Sleep in a completely dark room.
>   - No caffeine
> - Mindfulness and meditation
> - Yoga
> - Deep breathing
> - Acupuncture
> - Exercise
> - Spiritual practices
> - Clean diet free of chemicals

26.9 Lifestyle Detoxification

From a holistic perspective, what you avoid and the relationships that influence you are as important as what you include in your life. Every action, product, food, movement, emotion, and breath is an opportunity to positively impact your wellness. Sharing this concept with patients will empower them to gain a sense of control and to participate in their healing process.

Detoxification is such an important aspect of wellness, both in the material and emotional world. Removing existing toxins and avoiding potential toxins reduce the environmental burden, making it easier for the body to focus on healing. Cleaning up the diet, personal and household products, and the emotional realm are the first steps toward wellness (see ▶ Chap. 13).

26.9.1 Avoid or Remove?

Antigenic triggers Based on the bio-individual needs of a patient, eliminate any potential allergens that might not have been previously considered: gluten, dairy, corn, eggs, soy, and nightshades.

Mold Any mold exposure should be remediated or avoided. This is extremely toxic and immunosuppressive.

Amalgam fillings Although controversial and especially when considered during cancer treatment, amalgam fillings may be removed by a biological dentist who uses the right precautions and detoxification procedures. The decision may depend on the degree of deterioration of the fillings.

Plastics as endocrine disruptor compounds (EDCs) Plastic wrap, plastic water bottles, and even the bisphenol-free plastics have been found to leach EDCs. Patients should be advised to never heat food in those containers but rather transfer it to a safe container for heating.

Chemicals Scented products, such as air fresheners and dryer sheets, cleaning products, beauty and personal products, detergents, and soaps that are commonly found in most households should be avoided. Styrofoam in packaging, drinking cups, and food containers should be avoided. Pesticides, fertilizers, and herbicides are commonly applied to crops, and patients undergoing treatment should be encouraged to purchase organic, if possible. Check labels of packaged food and personal products to eliminate chemical exposures, preservatives, and artificial ingredients and sweeteners. Perfluorinated alkylated substances (PFAS) such as in Teflon-coated clothing or cookware, fire retardant-coated clothing, furniture, and carpeting should be avoided.

EMF exposure Be conscious of the use of electronic devices in the evening as it will impact melatonin production. Leave your phone outside of your bedroom or set to airplane mode overnight. Do not carry the device close to your body and use air tube headsets to reduce EMF exposure to the head and neck area.

Microbiome disruptors Meat, fish, or animal products that are laden with hormones, antibiotics, and chemically altered feed have the potential to disrupt not only the animal microbiome but also the consumers. Fat stores from the animals contain even higher concentrations as these chemicals are lipophilic and of great concern.

High-glycemic foods Fast food, processed food, foods high in sugar, alcohol, and high-fructose corn syrup are not nutritious options for patients undergoing cancer treatment. Processed foods of low-nutrient density, such as high calorie, low nutrient density liquid enteral food and other chemically laden foods, sugar, undesirable fats, synthetic vitamins, and casein-filled anti-nutritive beverages, should be avoided and alternate recommendations made.

IV nutrition Intravenous nutrition is the fastest, most assured way of reaching clinically effective levels of the administered substances in the body. Deficiencies of stomach acid and enzymes or inflammatory gut conditions may inhibit the absorption of a variety of nutrients. IV nutrition bypasses the digestive tract and delivers the nutrients directly into the bloodstream for more direct access to the organs and tissues in need. It can also work to increase energy levels and reduce side effects and is excellent for recovery posttreatment. Careful follow-up of vitamin and mineral levels are needed to ensure safety.

Linus Pauling, one of the fathers of orthomolecular medicine and the Linus Pauling Institute of Oregon State University [30], have laid the foundation for evidence-based research on micronutrients. Several physicians and clinics have contributed vast stores of knowledge to this story, including Jeanne Drisko, MD [31], Dwight McKee, MD [32], Paul Anderson, ND and Mark Stengler, ND [33], and Keith Block, MD [34].

The NIH National Cancer Institute, Office of Cancer Complementary and Alternative Medicine (OCCAM) has references for clinical trials as well as specific drug integration [35]. They report that laboratory studies of high-dose vitamin C have shown decreased cell proliferation in prostate, pancreatic, hepatocellular, colon, mesothelioma, and neuroblastoma cell lines and that patients report improved quality of life and decreased cancer-related side effects. Refer to the Human Clinical Studies page for more examples and ongoing clinical trials [36].

Intravenous therapies may include:
- Ascorbic acid (a G6PD test should be run prior to administration as fatal hemolysis can occur if a patient has glucose-6-phosphate dehydrogenase deficiency; also contraindicated is renal insufficiency) [37]
- Glutathione
- B vitamins
- Magnesium
- Potassium
- Lipoic acid
- L-carnitine
- Minerals

Mixtures of various ingredients known as cocktails are also created to specifically target particular issues.

One combination of alpha lipoic acid (ALA), low-dose naltrexone (LDN), and vitamin D has been shown in case reports to have potential promise in the treatment of stage 4 adenocarcinoma of the pancreas [38, 39]. The authors described the long-term survival of a man with pancreatic cancer with metastases to the liver, treated with intravenous ALA and oral LDN without any adverse effects, and was alive and well 78 months after initial presentation. Three additional pancreatic cancer case reports were presented. At the time of the report, the first patient was alive and well 39 months after presenting with adenocarcinoma of the pancreas with metastases to the liver. The second patient who presented with the same diagnosis was treated with the ALA/LDN protocol, and after 5 months of therapy, PET scan demonstrated no evidence of disease. The third patient, in addition to his metastatic pancreatic cancer, had a past medical history of B-cell lymphoma and prostate adenocarcinoma. After 4 months of the ALA/LDN protocol, the CT/PET scan demonstrated no evidence of disease [38].

26.9.2 Sepsis

Sepsis is a common end-of-life occurrence in oncology [40]. Marik et al. will soon publish in *Chest* their own small before-and-after unblinded cohort study, born of an anecdote that should intrigue any intensivist: Three patients with "fulminant sepsis … almost certainly destined to die" from shock and organ failure, were infused with vitamin C and moderate-dose hydrocortisone out of desperation. All three patients recovered quickly and left the ICU in days "with no residual organ dysfunction."

Inspired by that experience, they went on to enroll and treat 47 septic patients with a cocktail of 1.5 g vitamin C IV every 6 hours, hydrocortisone 50 mg IV every 6 hours, and thiamine 200 mg IV every 12 hours (thiamine inhibits oxalate production and has potential benefits in septic shock). Controls were 47 patients matched in baseline characteristics. Hospital mortality was 4 of 47 (8.5%) in those treated with the cocktail, compared to 19 of 47 (40%) in those not. Vasopressors were weaned off in all cocktail-treated patients, usually in <24 hours (vs. 4 days for the controls). Renal function reportedly improved in all patients with acute kidney injury [41].

Intravenous nutrition therapy should be a first-line as well as an integrative therapy consideration for every patient. Riordan Clinic has developed a compendium of evidence-based research when deciding on treatment [42].

26.9.3 Integrative and Functional Nutrition

The impact of good nutrition in cancer prevention or as part of an integrative treatment program is well established. Food-based nutrients are capable of preventing cellular DNA damage from carcinogenic environmental factors. The correct dietary choices can ensure proper detoxification, immune system function, and metabolic function, reduce inflammation, increase longevity and quality of life, and minimize side effects of conventional treatment. The challenge lies more in breaking down the information so as not to fall into the trap of a one-target, one-compound solution mind-set.

The tapestry of synergistic nutrient and herbal benefits contained in food is woven from:
- Essential vitamins and minerals
- Fatty acids
- Fiber
- Thousands of phytochemicals including:
  - Terpenes
  - Polyphenols
  - Carotenoids
  - Isoflavones
  - Flavonoids
  - Chlorophylls
  - Alkaloids
  - Lignans
  - Phytosterols
  - Sulforaphanes
  - Stilbenes
  - Catechins
  - Tannins
- Folate
- Inositol
- Sulfides

Each food and nutrient, as well as bio-individual physical sufficiency of nutrients, either influences or is metabolized through detoxification pathways, including phase I CYP P450 isoenzymes, phase II conjugation enzymes, Nrf2 signaling, and metallothionein [43]. CYP450 enzymes are essential for the production of cholesterol, steroids, prostacyclins, and thromboxane A_2. Within the population, there is genetic variability or polymorphisms of CYP450 enzymes, with about 50% of the population having normal function and the rest having altered rapid to very low function. So ultimately, we need to consider the patient's particular metabolism, often referred to as normal, rapid, or slow metabolizers, layered with the effect of nutrients, and then the effect of drug regimens. Viewed this way, it is an extremely complex undertaking [43].

Super doses or consumption of certain nutrients, either via food or food-based supplements, will influence specific enzymes in a dose-dependent manner. They may increase or decrease the function of a pathway or act as adaptogens, balancing the function of an enzyme. Consumption of bioactive supplements in this manner may or may not be therapeutic, while food forms are often better tolerated and less potentially problematic. A whole foods-intensive diet is the safest and most important way to optimize biofunction [44, 45].

26.9.4 Liquid Oral or Enteral Nutrition

A simple glance at the label of any commercial oral or enteral feeding product will explain why they should be avoided: Sugar, corn syrup, sodium caseinate, corn maltodextrin, soy protein isolate, corn oil, and canola oil form a who's-who list of what not to consume. These ingredients are generally genetically modified and contribute to the toxic burden of the body. Sugar and corn syrup are two of the preferred energy sources for cancer and contribute to obesity, one of the leading causes of cancer, as well as nonalcoholic fatty liver disease. Although high in calories, the only vaguely nutritive value is added synthetic vitamins. The added inflammatory and endocrine-disrupting potential of filler ingredients is in direct opposition to the needs of the cancer patient. There are far healthier products on the market made of real organic high-quality ingredients. The Oley Foundation is a fantastic resource for real food options for blenderized or tube feeding needs [46].

Batch cooking and preparation are extremely helpful since patients are generally exhausted during times that require the greater food absorption and digestive rest that liquid nutrition provides. True nourishing support can be found with the following: freshly made vegetable and fruit juice blends (low sugar), smoothies (made with low-sugar ingredients and greens), blended soups, protein powders (pea, hemp, grass-fed beef protein, grass-fed whey, nutritional yeast, egg white powder as tolerated), organic whole fruit and vegetable and grass powders, and bone broths and collagen support sourced from grass-fed, free-range, organic animals.

26.9.5 Tea Therapy

Tea is a simple, yet powerful, way to provide daily support for detoxification, reduction of inflammation, and overall cancer care. Tea is a gentle way to enter the herbal realm. As always, choose high-quality teas that are not contaminated with heavy metals, fluoride, or pesticides. Tea contains polyphenols that include the catechins, alkaloids (caffeine, theophylline, and theobromine), amino acids, chlorophyll, and minerals.

From the NIH National Cancer Institute [47], among their many biological activities, the predominant polyphenols in green tea—EGCG, EGC, ECG, and EC—and the theaflavins and thearubigins in black teas have antioxidant activity. These chemicals, especially EGCG and ECG, have substantial free radical scavenging activity and may protect cells from DNA damage caused by reactive oxygen species.

Tea polyphenols have also been shown to inhibit tumor cell proliferation and induce apoptosis in laboratory and animal studies [47]. In other laboratory and animal studies, tea catechins have been shown to inhibit angiogenesis and tumor cell invasiveness. In addition, tea polyphenols may protect against damage caused by ultraviolet (UV) B radiation, and they may modulate immune system function. Furthermore, green teas have been shown to activate detoxification enzymes, such as glutathione S-transferase and quinone reductase, that may help protect against tumor development. Although many of the potential beneficial effects of tea have been attributed to the strong antioxidant activity of tea polyphenols, the precise mechanism by which tea might help prevent cancer has not been established.

Chamomile tea is a potent source of apigenin [48]. Apigenins appear to play a role in gene regulation, essentially reprogramming cancer cells so they act more like normal cells and die on schedule. One of the most striking discoveries in this body of research is that apigenins could potentially stop the spread of breast cancer. The latest research suggests apigenin binds to one of three types of proteins, each with a specific function. One of those proteins is called hnRNPA2. In a healthy hnRNPA2 protein, only one type of splicing takes place [48].

However, in cancer cells, two types of splicing take place. This abnormal splicing is a factor in about 80% of all cancers. Splicing is important because it prompts the production of mRNA, or messenger RNA, which then carries out instructions regarding gene activation. When apigenin connects with the hnRNPA2 protein in breast cancer cells, it changes the protein from two splices back to a single-splice setup. And with splicing back to normal, cells are able to die on schedule—or, at the very least, they become more vulnerable to the effects of chemotherapy drugs [48].

Herbal tea blends are available to gently cleanse the liver, support the mucosal health of the digestive system, support throat and esophageal health, relieve nausea, and more.

26.9.6 Turmeric

Turmeric is a member of the ginger family, and the rhizome has been used for thousands of years for its healing properties. Although the active constituent, curcumin, is the focus of much research, there are a multitude of compounds identified in the rhizome, and some research on curcumin-free turmeric indicates that these non-curcumin compounds, as well as the whole plant synergy, may hold equivalent efficacy [49].

Turmeric compounds include the yellow-colored polyphenols curcumin (diferuloylmethane), demethoxycurcumin (curcumin II), and bisdemethoxycurcumin (curcumin III), as well as turmerin, turmerone, elemene, furanodiene, curdione, bisacurone, cyclocurcumin, calebin A, and germacrone [49]. Studies have indicated that turmeric oil, present in turmeric, can enhance the bioavailability of curcumin. Dietary turmeric is especially important because consistent consumption in food overcomes some of the bioavailability issues of turmeric due to its hydrophobic nature, poor absorption, and metabolism. Different methods of delivery of curcumin have been studied, including using in combination with heat, fat, and the piperine compound from pepper. Much of these challenges are overcome naturally when turmeric is incorporated consistently in the diet [50]. Turmeric milk (turmeric mixed in warm milk) has been used for thousands of years as a tonic in India.

Evidence-based research indicates that turmeric possesses anti-inflammatory, hypoglycemic, antioxidant, wound-healing, antimicrobial, anticancer, and immune-supportive properties that support overall wellness [49, 50]. The polyphenol curcumin exhibits anticancer effects through a multitude of pathways, including the modulation of the cell cycle, by binding directly and indirectly to molecular targets including transcription factors (NF-kB, STAT3, β-catenin, and AP-1), growth factors (EGF, PDGF, and VEGF), enzymes (COX-2, iNOS, and MMPs), kinases (cyclin D1, CDKs, Akt, PKC, and AMPK), and inflammatory cytokines (TNF, MCP, IL-1, and IL-6).

There is upregulation of proapoptotic (Bax, Bad, and Bak) and downregulation of antiapoptotic proteins (Bcl(2) and Bcl-xL) [51]. Additionally, curcumin can suppress the activation of dendritic cells (DCs) by modulating the JAK/STAT/SOCS signaling pathway to restore immunologic balance in an experimental colitis model, colitis being a risk factor for colon cancer [52]. Because curcumin exhibits numerous mechanisms of cell death, it is possible that cells may not develop resistance [53].

Most critically of all, turmeric is proving to be an effective weapon against cancer stem cells in many types of cancer. Cancer stem cells (CSCs), as discussed above, are a subpopulation of tumor cells that possess the stem cell properties of self-renewal and differentiation. These cells have been identified in many solid tumors including breast, brain, lung, prostate, testis, ovary, colon, skin, and liver and also in acute myeloid leukemia. The CSC theory clarifies not only the issue of tumor initiation, development, metastasis, and relapse but also the ineffectiveness of conventional cancer therapies. Treatments directed against the bulk of the cancer cells may produce striking responses, but they are unlikely to result in long-term remissions if the rare CSCs are not targeted [54].

Curcumin was tested in a group of 25 advanced pancreatic cancer patients and found to induce tumor regression, induce tumor stability for 18 months, and reduce clinically relevant biomarkers, all without toxicity [55]. It was found to be safe for use in gemcitabine-resistant patients with pancreatic cancer [56]. Curcumin reduced the number and size of polyps found in the precancerous condition familial adenomatous polyposis (FAP). Curcumin reduces clinically relevant biomarkers in colorectal cancer that is refractory to chemotherapy. Pretreatment with curcumin considerably reduced the dose of cisplatin and radiation required to inhibit the growth of cisplatin-resistant ovarian cancer cells [57]. Curcumin combined with oxaliplatin may enhance efficacy of the latter in both p53wt and p53 mutant colorectal tumors [58].

Curcumin has been tested for safety and potential effect with the maximum tolerable dose range between 4 and 8 g per day. Bear in mind that combining curcumin with piperine will dramatically alter its bioavailability, thus changing that limit. All concurrent drug therapies should be checked for cytochrome isoenzyme activity to prevent toxicity and ensure efficacy [59].

26.9.7 Ginger

One of the ancient remedies for ailments ranging from stomach upset to aches and pains, the ginger rhizome now has a growing body of evidence of efficacy in cancer. There are many active compounds in this root, some of which are in the layer just under the peel, so when using the young root, it is best to scrub it, as opposed to peeling it, if making a tea, smoothie, or juice. If the root is harvested later, past the 9-month mark, the skin will be tougher and may need to be removed. The older roots yield higher amounts of ginger oil [60, 61].

The anticancer activity of ginger has many mechanisms including reducing inflammation, upregulating tumor-suppressor genes, cell-cycle arrest, inducing apoptosis via the mitochondrial pathway and production of ROS, inactivation of VEGF pathways, and modulation of signaling molecules such as NF-kB, STAT3, MAPK, PI3K, ERK1/2, Akt, TNF-α, COX-2, cyclin D1, cdk, MMP-9, survivin, cIAP-1, XIAP, Bcl-2, caspases, and other cell growth regulatory proteins [61] all essentially affecting a multitude of targets in the various stages of oncogenesis, angiogenesis, and metastasis. Additionally, ginger has been found to be neuroprotective, antiemetic, antimicrobial, antifungal, and protective of the liver, gastrointestinal system, and lungs [61, 62]. The main compounds of interest in ginger are terpenes, including zingiberene, β-bisabolene, α-farnesene, β-sesquiphellandrene, α-curcumene, zerumbone, and phenols including gingerol, paradol, and shogaol.

Ginger can be taken internally in the following forms: powdered spice, tea, pickled ginger, candied ginger, preserved ginger, ginger oil, or raw root, and it has been used in skin cancer research as a topical preparation. Ginger can stimulate appetite and enhance digestion. *Caution is advised* in patients on anticoagulants or with bleeding disorders [62].

Ginger has shown efficacy in the following cancers: skin, breast, lung, gastrointestinal, bile duct, pancreatic, ovarian, cervical, renal, prostate, liver, and brain. Ginger has been found helpful in ameliorating side effects of conventional treatment with radiation, doxorubicin, and cisplatin, as well as working synergistically with gemcitabine in pancreatic cancer by sensitizing tumor cells and suppressing tumor growth.

Interestingly, ginger combined with Gelam honey was found to produce more tumor cellular apoptosis than the use of 5-fluorouracil alone in the treatment of colorectal cancer cells [63]. The component zerumbone has been found to enhance the radiosensitivity of colon cancer cells. Beyond inducing apoptosis, autophagy, and modulating angiogenesis in ovarian cancer, ginger compounds may help to overcome chemotherapeutic drug resistance [64]. Perhaps the most encouraging research on ginger comes in the form of its ability to address stem and dendritic cells [65]. The ginger compound 6-shogaol was found to inhibit both breast cancer monolayer cells and stem cell spheroids at doses that are not toxic to normal cells [65, 66].

Also significant is the potential of 6-shogaol to decrease cancer development, progression, and metastasis by inhibiting the production of tumor-associated dendritic cells. The study reported tumor-associated dendritic cells (TADCs) facilitate lung and breast cancer metastasis in vitro and in vivo by secreting inflammatory mediator CC-chemokine ligand 2 (CCL2), but it is also the first to reveal that 6-shogaol can decrease cancer development and progression by inhibiting the production of TADC-derived CCL2. Human lung cancer A549 and breast cancer MDA-MB-231 cells increase TADCs to express high levels of CCL2, which increase cancer stem cell features, migration, and invasion, as well as immunosuppressive tumor-associated macrophage infiltration [66].

6-shogaol decreases cancer-induced upregulation of CCL2 in TADCs, preventing the enhancing effects of TADCs on tumorigenesis and metastatic properties in A549 and MDA-MB-231 cells. A549 and MDA-MB-231 cells enhance CCL2 expression by increasing the phosphorylation of signal transducer and activator of transcription 3 (STAT3), and the activation of STAT3 induced by A549 and MDA-MB-231 is completely inhibited by 6-shogaol. 6-shogaol also decreases the metastasis of lung and breast cancers in mice. 6-shogaol exerts significant anticancer effects on lung and breast cells in vitro and in vivo by targeting the CCL2 secreted by TADCs. Thus, 6-shogaol may have the potential of being an efficacious immunotherapeutic agent for cancers [66].

Ginger in all of its forms deserves a place of honor in our herbal healing arsenals for virtually all aspects of wellness, but specifically for cancer prevention and treatment.

26.9.8 Dandelion

Bitter greens, especially dandelion greens, are plentiful in the spring and have been used traditionally for centuries to stimulate bile and aid in phase 2 liver detoxification [67].

Dandelion flowers have antibacterial properties in their pollen; contain a terpene known as helenin, which is a powerful antifungal agent [68]; and are rich in vitamins A and B12. Dandelion leaves and flowers were found to be protective against UVB exposure and cellular senescence, suggesting potential use for UV sun protection and antiaging applications [69]. Perhaps the most powerful part of the dandelion is the root. The German Commission E has approved dandelion root for the treatment of disturbances in bile flow, stimulation of diuresis, loss of appetite, and dyspepsia.

The University of Windsor in Ontario is currently running the Dandelion Project, a clinical trial using dandelion root extract (DRE) for terminal cancer cases. Two refractory hematological malignancies experienced unusual responses to DRE [70]. DRE was shown to have the potential to induce apoptosis and autophagy in human pancreatic cancer cells with no significant effect on noncancerous cells [71].

Dandelion extracts have been used in traditional Native American medicine and traditional Chinese medicine to manage gastrointestinal and liver disorders. Upon study, DRE was found to be selectively cytotoxic to cultured leukemia cells [72].

Chronic myelomonocytic leukemia (CMML) is aggressive and highly resistant to treatment. DRE holds promise as a potential treatment [73]. DRE was shown to have selective efficacy in inducing two forms of programmed cell death in highly aggressive and resistant CMML cell lines. Rapid activation of caspase-8 not only activated the extrinsic pathway of apoptosis but also triggered pro-death autophagy selectively in these cells, suggesting that this extract has components that enhance its selective efficacy in targeting CMML cells. These results indicate that within the vast array of available natural products and compounds, there are nontoxic alternatives to conventional chemotherapy that are safe and effective [73].

And one of the most common cancers, colon, may also benefit from DRE treatment [74]. Phytochemical analyses of the extract showed complex multicomponent composition of the DRE, including some known bioactive phytochemicals such as α-amyrin, β-amyrin, lupeol, and taraxasterol. This suggested that this natural extract could engage and effectively target multiple vulnerabilities of cancer cells. Therefore, DRE could be a nontoxic and effective anticancer alternative, instrumental for reducing the occurrence of cancer cells drug resistance [74].

Dandelion root tea and extract are readily available and should be considered in all stages of treatment.

26.9.9 Methyl-Rich Foods

Methyl-rich foods are epigenetically powerful and include those that are high in folate, methionine, and choline, such as meat, fish, eggs, dairy, citrus, leafy greens, strawberries, and mushrooms [75].

Chemicals that enter our bodies can also affect the epigenome. One example, bisphenol A (BPA), is a compound used to make polycarbonate plastic. It is in many consumer products, including water bottles and linings of tin cans. Controversial reports questioning the safety of BPA have prompted some manufacturers to stop using the chemical. In the laboratory, BPA appears to reduce methylation of the agouti gene in mice. Agouti mice studied show yellow mothers give birth to pups with a range of coat colors from yellow to brown. When mothers were fed BPA, their babies were more likely to be yellow and obese. However, when mothers were fed BPA along with methyl-rich foods, the offspring were more likely to be brown and healthy. The maternal nutrient supplementation had counteracted the negative effects of exposure [76].

From there, it is easy to see the cascade effect of how attention to methylation and epigenetic programming can mitigate daily exposure to toxins, thereby reducing cancer risk.

26.9.10 Melatonin-Rich Foods

We have previously discussed the critical role of melatonin on the immune system. The decline in the production of melatonin with age has been suggested as one of the major contributors to immunosenescence and the possible development of neoplastic diseases [77]. To further complicate matters, melatonin metabolism from dietary sources can compete in the liver with other compounds, can be impacted by hormones, is affected by cytochrome enzymes, and is subject to other clinical deficiencies, such as pyroluria [78]. Pyrrole disorder, sometimes referred to as the mauve factor, is consistently associated with higher incidence of mental health issues, such as ADHD, bipolar disorder, and schizophrenia, all of which can be mediated with vitamin and mineral therapy, thus impacting endogenous melatonin production. There can also be a focus on foods that contain the necessary B6 and zinc and cofactors such as folate, B12, and manganese [79]. Also, as with any phytochemical, plant levels are dependent on growing conditions and harvesting [80].

Some excellent dietary choices for melatonin are tart cherry juice, tart cherries, walnuts, mustard seed, tomatoes, chilies, almonds, pineapple, bananas, oranges, fenugreek seeds, sunflower seeds, corn, rice, ginger root, peanuts, grains, corn, asparagus, mint, black and green tea, broccoli, pomegranate, olives, Brussel sprouts, and cucumbers [81–84].

26.9.11 Honokiol

Honokiol is a phytochemical isolated from the bark and leaves of the magnolia tree [85]. Honokiol has been demonstrated to possess anticarcinogenic, anti-inflammatory, antioxidative, and antiangiogenic properties, as well as an inhibitory effect on malignant transformation of papillomas to carcinomas in vitro and in vivo animal models without any appreciable toxicity. Honokiol affects multiple signaling pathways in molecular and cellular targets, including nuclear factor-κB (NF-κB), STAT3, epidermal growth factor receptor (EGFR), cell survival signaling, cell cycle, cyclooxygenase, and other inflammatory mediators [85].

Honokiol can be used in combination with oxaliplatin in the chemotherapy of colon cancer. This combination allows a reduction in oxaliplatin dose, thereby reducing its adverse effects. It may also enhance the chemotherapeutic effect of oxaliplatin for this disease [86]. Honokiol inhibits migration of non-small cell lung cancer cells [87].

Molecular docking analysis of honokiol in EGFR binding site indicated that the chemotherapeutic effect of honokiol against head and neck squamous cell carcinoma (HNSCC) is mediated through its firm binding with EGFR, which is better than that of gefitinib, a commonly used drug for HNSCC treatment [88]. Furthermore, honokiol shows effectiveness against several multidrug-resistant tumor cells [89].

26.9.12 Limonene

Limonene is a bioactive monoterpene found in the peel of citrus fruit. Even though it was proven to completely regress mammary tumors in rats over 25 years ago [90], human clinical trials are lacking. However, a recent study on the metabolomics of limonene is very promising and suggests that limonene's activity is likely through a general systemic effect rather than through a specific target [91].

Limonene is distributed extensively to human breast tissue when eaten and results in reduced breast tumor cyclin D1 expression that may lead to cell-cycle arrest and reduced cell proliferation. Overexpression of cyclin D1 promotes the transition of cells out of the G1 and into the cell cycle and is commonly overexpressed and deregulated early in breast tumorigenesis in humans [92]. This is notable because cyclin D1 is a key regulator of cell proliferation. It also controls other aspects of the cell fate, such as cellular senescence, apoptosis, and tumorigenesis [93].

D-limonene enhanced the antitumor effect of docetaxel against prostate cancer cells without being toxic to normal prostate epithelial cells. This combined beneficial effect could be through the modulation of proteins involved in the mitochondrial pathway of apoptosis. D-limonene could be used as a potent nontoxic agent to improve the treatment outcome of hormone-refractory prostate cancer with docetaxel [94].

26.9.13 Cruciferous Vegetables

Broccoli, broccoli sprouts, cabbage, watercress, kale, and all cruciferous vegetables are high in isothiocyanates, such as sulforaphane, which have been found to not only prevent cancer but to reduce and slow the growth of existing tumors in bladder, breast, colorectal, endometrial, gastric, lung, ovarian, pancreatic, prostate, and renal cancers [95].

Sulforaphane induces apoptosis in human prostate cancer by initiating ROS generation [95]. Although sulforaphane garners much of the anticancer attention, there are other compounds working synergistically to power this potential, such as benzyl isothiocyanate (BITC). The three processes that are impacted by BITC are ROS production, autophagy, and apoptosis. When a cruciferous vegetable is chewed or macerated, the enzyme myrosinase is released from a cellular compartment to hydrolyze the glucosinolates, producing ITCs and other products. There are nearly 120 identified ITCs, with sulforaphane and BITC simply being very well studied [96].

The referenced studies explore direct anticancer properties, but there are other factors to be considered. Cabbage, for instance, contains a significant amount of the amino acid L-glutamine, as well as vitamins C and K, folate, B6, calcium, magnesium, manganese, selenium, chlorophyll, flavonoids, fiber, carotenoids, indole-3-carbinol, lignans, phytosterols, and sulfur bioactives (other than glucosinolates).

Glutamine is a rich component of cruciferous vegetables and is the most abundant amino acid in the body and conditionally essential, meaning that normally the body produces enough but can utilize much more under conditions of stress. Adding glutamine to parenteral nutrition prevents deterioration of gut permeability and preserves mucosal structure [97]. Glutamine has been shown to prevent muscle wasting, one of the confounding factors of cancer morbidity.

Adding glutamine to the nutrition of clinical patients, enterally or parenterally, may reduce morbidity. Several excellent clinical trials have been performed to prove efficacy and feasibility of the use of glutamine supplementation in parenteral and enteral nutrition. The increased intake of glutamine has resulted in lower septic morbidity in certain critically ill patient populations [98]. It has been found that the use of glutamine supplementation is safe and can diminish risks of high-dose chemotherapy and radiation. There is some evidence that adequate glutamine availability can beneficially affect outcome, especially in patients undergoing bone marrow transplantation [99].

Glutamine supplementation was found to protect against radiation-induced mucositis, anthracycline-induced cardiotoxicity, and paclitaxel-related myalgias/arthralgias. Glutamine may prevent neurotoxicity of paclitaxel, cisplatin, oxaliplatin, bortezomib, and lenalidomide and is beneficial in the reduction of the dose-limiting gastrointestinal toxic effects of irinotecan and 5-FU-induced mucositis and stomatitis. Dietary glutamine reduces the severity of the immunosuppressive effect induced by methotrexate and improves the immune status of rats recovering from chemotherapy. In patients with acute myeloid leukemia requiring parenteral nutrition, glycyl-glutamine supplementation could hasten neutrophil recovery after intensive myelosuppressive chemotherapy [100].

There are individual genetic influences to be considered that may affect the metabolism of isothiocyanates. The Linus Pauling Institute discusses this in detail [30, 101]. Once absorbed, glucosinolate-derived isothiocyanates (such as sulforaphane) are promptly conjugated to glutathione by a class of phase II detoxification enzymes known as glutathione S-transferases (GSTs). This mechanism is meant to increase the solubility of isothiocyanates, thereby promoting a rapid excretion in the urine. Isothiocyanates are thought to play a prominent role in the potential anticancer and cardiovascular benefits associated with cruciferous vegetable consumption. Genetic variations in the sequence of genes coding for GSTs may affect the activity of these enzymes. Such variations have been identified in humans. It has been proposed that a reduced GST activity in these individuals would slow the rate of excretion of isothiocyanates, thereby increasing tissue exposure to isothiocyanates after cruciferous vegetable consumption. GSTs are involved in "detoxifying" potentially harmful substances like carcinogens, suggesting that individuals with reduced GST activity might also be more susceptible to cancer [101].

Finally, induction of the expression and activity of GSTs and other phase II detoxification/antioxidant enzymes by isothiocyanates is an important defense mechanism against oxidative stress and damage associated with the development of diseases like cancer and cardiovascular disease. The ability of sulforaphane to reduce oxidative stress in different settings is linked to activation of the nuclear factor E2-related factor 2 (Nrf2)-dependent pathway. Some, but not all, observational studies have found that GST genotypes could influence the associations between isothiocyanate intake from cruciferous vegetables and risk of disease (see ► Chap. 17).

26.9.14 Seeds

Seeds such as sunflower, pumpkin, chia, hemp, flax, sacha inchi, and sesame are very rich in essential fatty acids, vitamin E, B vitamins, and minerals like selenium, phosphorus, iron, copper, zinc, and potassium. Some seeds contain a significant amount of protein. Hemp seeds in particular are a complete source of all 20 known amino acids and are 25% protein, higher than many other seeds, including chia and flax.

The benefits of lignans have been called into question for hormone-linked cancers due to their phytoestrogenic action. However, the Oncology Nutrition Practice Group of the Academy of Nutrition and Dietetics shares that lignans are the type of phytoestrogens in flaxseeds, and they may change estrogen metabolism [102]. In postmenopausal women, lignans may cause the body to produce less active forms of estrogen and is believed to potentially reduce breast cancer risk. There is evidence that adding ground flaxseeds into the diet decreases cell growth in breast tissue as well. Some cell and animal studies have shown that two specific phytoestrogens found in lignans, enterolactone and enterodiol, may help suppress breast tumor growth. Researchers do not yet know if these results will apply to women with breast cancer, but this approach—adding flaxseed to the diet—looks promising. And several studies in women have shown that higher intake of lignans, the key phytoestrogen in flaxseeds, is associated with reduced risk of breast cancer [102].

Further, lignans in the diet are associated with less aggressive tumor characteristics in women who have been diagnosed with breast cancer. In other words, women who have already been eating lignans at the time of diagnosis seem to have tumors that are less aggressive [102].

26.9.15 Medicinal Mushrooms

Mushrooms have been used medicinally since ancient times. The chief medicinal uses of mushrooms to date are antioxidant, antidiabetic, hypocholesterolemic, antitumor, anticancer, immunomodulatory, anti-allergic, nephroprotective, and antimicrobial [103]. The mushrooms known to have anticancer effects belong to the genus *Phellinus, Pleurotus, Agaricus, Ganoderma, Clitocybe, Antrodia, Trametes, Cordyceps, Xerocomus, Calvatia, Schizophyllum, Flammulina, Suillus, Inonotus, Inocybe, Funlia, Lactarius, Albatrellus, Russula,* and *Fomes*. The anticancer compounds play crucial roles as reactive oxygen species promoters, mitotic kinase inhibitors, antimitotic, angiogenesis inhibitors, and topoisomerase inhibitors, leading to apoptosis and eventually halting cancer proliferation [103].

A particular mushroom compound, Active Hexose Correlated Compound (AHCC), acts as a biological response modifier [104]. In vivo and human clinical trials have shown that AHCC modifies both the innate and adaptive immune responses by increasing the production of cytokines, increasing the activity of NK cells by as much as 300–800%, increasing populations of macrophages and in some cases doubling it, increasing the number of dendritic cells, and increasing the number of T cells by as much as 200%.

Data from the treatment of more than 100,000 individuals with various types of cancer have shown AHCC treatment to be of benefit in 60% of cases. AHCC is particularly effective for liver, lung, stomach, colon, breast, thyroid, ovarian, testicular, tongue, kidney, and pancreatic cancers [105]. In addition to being able to fight cancer directly, AHCC also alleviates many of the side effects of chemotherapy [106]. Chemotherapy is replete with side effects, which range from psychologically distressing to life-threatening. Doctors noticed that chemotherapy patients taking AHCC did not lose hair. Subsequently, an in vivo study found that mice treated with AHCC were protected from chemically induced hair loss. Clinical studies in Korea and Japan have indicated that AHCC remarkably improves symptoms of nausea and vomiting in cancer patients.

Chemotherapy inhibits bone marrow function, which can be life-threatening. AHCC has been shown to raise the white blood cell count of cancer patients by about 30% [106]. One of the major drawbacks of chemotherapy is the demise of healthy cells in addition to cancer cells. An in vivo study found that rats given chemotherapy without AHCC experienced large increases in liver enzymes while those given chemotherapy plus AHCC had normal levels of liver enzymes.

Importantly, in vitro and animal studies show that AHCC has anticancer effects, not just effects on chemotherapy side effects [107]. In cisplatin-treated mice, AHCC increased its antitumor activity while reducing side effects and showed synergistic effects with gemcitabine in pancreatic cancer cells. A prospective study showed early evidence that AHCC improves prognosis after curative resection of hepatocellular carcinoma. AHCC also reduced chemotherapy-associated adverse effects in patients with advanced cancer, and in those with pancreatic ductal adenocarcinoma [107].

26.9.16 Bee Products

Both integrative and conventional medicine support the use of bee products as elixirs for virtually every health condition, including cancer [108–110]. Propolis, honey, royal jelly, pollen, and even venom have been found to be antimicrobial, anti-inflammatory, antioxidant, immune supportive, antiparasitic (parasites are an often-overlooked initiator of cancer), antidepressant, capable of inducing apoptosis, helpful in increasing energy levels, and wound healing.

26.9.17 Iodine Sufficiency

Although all parts of the body contain iodine, many tissues and organs in our body concentrate iodine [111]. Organs that have a concentration function include thyroid, stomach mucosa, salivary glands, breast tissue, thymus, skin, placenta, ovaries, uterus, prostate, pancreas, choroid plexus, and the ciliary body of the eye. The role of iodine in these tissues is believed to include antioxidant, anti-inflammatory, antiproliferative, antibacterial, proapoptotic, and pro-differentiating effects [112].

Research suggests that the therapeutic form of iodine in breast cancer is molecular iodine (I_2) and that I_2 is capable of inducing apoptosis in human breast cancer cells through mitochondrial-mediated pathways [112, 113]. Iodine may also affect the binding of estrogen receptors, thereby having an antiestrogenic effect on gene expression. Gastric cancers, such as stomach cancer and Barrett's esophagus, were found to have decreased or absent iodine levels in gastric tissues [113]. The subject of iodine and companion minerals, such as selenium, for endocrine health and breast cancer prevention and treatment has been a topic of popular publications [114, 115]. Iodine-rich foods include seaweeds, sea vegetables, spirulina, and chlorella, with lesser amounts in cranberries, baked potatoes, pastured eggs, organic dairy, shrimp, wild caught fish, and navy beans [116, 117].

26.9.18 Seaweed and Algae

Spirulina and chlorella are nutrient-dense chlorophyll-rich blue-green algae [118–121]. Part of their benefit is that there is a significant amount of nutrition available in powder or tablet forms, which is perfect for when appetite is waning but nutrient needs are high. Rich in phytonutrients, B-complex vitamins, carotenoids, iron, zinc, magnesium and other minerals, protein, amino acids, and fatty acids, they earn the rank of superfoods. Chlorella can increase excretion and decrease absorption of cadmium, lead, mercury, and other heavy metals, as well as radioactive particles [118–121].

Chlorella supports several aspects of immune function, including enhancing NK cell activity [122]. One preliminary study reported that patients taking chlorella suggested that the cellular components and functions of the immune system remained at near-normal levels and were less adversely affected when patients were undergoing chemotherapy and/or taking immunosuppressive medications such as steroids [122].

Fungal overgrowth is a common occurrence in oncology patients. The immunostimulatory effects of spirulina were reported to support clearance of systemic fungal conditions in a mouse model by increasing levels of tumor necrosis factor alpha and interferon gamma [123]. Phycocyanin, a component of spirulina, was found to induce apoptosis and autophagic cell death in pancreatic cancer cells [124]. One report indicates that the phycocyanin compound was able to affect multiple targets of oncogenesis in a variety of cancer types [125].

Seaweed is another nutritional powerhouse [126]. As with the algae, seaweed should be sourced carefully to obtain contamination-free products. Seaweed is a rich source of vitamins and minerals and is naturally high in iodine. Sea vegetables contain fucoidan, a polysaccharide studied for its ability to inhibit tumor development, induce cancer cell death, inhibit metastasis, protect against toxicity associated with chemotherapy and radiation, and act synergistically with certain chemotherapy regimens [126].

26.9.19 Fiber, Prebiotics, and Short-Chain Fatty Acids

A high-fiber diet decreases the risk of metabolic syndrome, thereby addressing one of the major risk factors for cancer [127]. The decreased transit time, dilution of fecal carcinogens, and increased microbial diversity of a high-fiber diet contribute to a decrease in cancer rates, including proximal and distal colon cancer [128], colorectal adenoma [129], breast cancer [130], and renal cell carcinoma [131].

The short-chain fatty acids (SCFAs), butyrate, propionate, and acetate, are generated in the large intestine through bacterial fermentation of dietary fiber and resistant starches [132–135]. Additionally, a rich source of butyrate can be found in bovine milk and milk-fat products such as butter and ghee. Insoluble fiber such as oat bran, wheat bran, and psyllium, as well as resistant starches found in cooked and cooled potatoes or rice, legumes, green bananas and plantains, or Jerusalem artichokes, supports the bacterial fermentation process. These foods act as prebiotics or nourishment for the healthy bacterial species that generate the SCFAs and serve to increase bacterial diversity, increase insulin sensitivity, reduce inflammation, and increase metabolism.

Butyrate has direct and specific anticancer potential [132–135], is known to induce autophagy and apoptosis of cancer cells, inhibits the growth of cancer cells in the bowel, supports the cancer-fighting effect of photodynamic therapy on astrocytoma cells, and increases the sensitivity of resistant cancer cells to irinotecan in the treatment of colon cancer. Propionate is mainly taken up by the liver and has been found to reduce systemic inflammation and counteract malignant cell proliferation in liver tissue [136].

26.9.20 Modified Citrus Pectin and Pectin

Modified citrus pectin (MCP) is a complex water-soluble indigestible polysaccharide obtained from the peel and pulp of citrus fruits and modified by means of high pH and temperature treatment. MCP has been found to sensitize prostate cancer cells to radiation therapy, overcome radioresistance, and reduce clinical radiation dose [137]. In addition, MCP has antimetastatic properties for multiple malignancies, including melanomas, and inhibits tumor angiogenesis and galectin-3, a regulator of cancer cell apoptosis which can increase the sensitivity of many cancer cell lines to various chemotherapies [138].

Pectin is a complex polysaccharide found in plant cell walls that acts as a natural gelling agent. When it interacts with acid, it produces compounds including arabinogalactans and galactans, which can bind to and inhibit galectin-3 as noted above. Pectin is also an excellent binding agent that may remove heavy metals such as mercury, lead, and aluminum, as well as radioactive particles from the body. It has been used in the Chernobyl region as recommended by the BELRAD Institute as part of food rations for children to effectively reduce the body burden of cesium [139].

MCP is protective against radiation-induced mucositis and works synergistically with doxorubicin to significantly increase the rate of apoptosis and limit the rate of cell-cycle division in prostate cancer cells [140]. MCP was also reported to be immune supportive and anti-inflammatory, protecting intestinal stem and epithelial cells against inflammation, especially following bacterial infection, by increasing tight-junction proteins and thus preserving the mucosal barrier [140].

Low-molecular-weight citrus pectin (LCP) effectively inhibits the growth and metastasis of gastrointestinal cancer cells, while demonstrating synergistic tumor-suppressor activity with 5-fluorouracil (5-FU) [141] against gastrointestinal cancer cells [142]. LCP can enhance the ability of oxaliplatin to inhibit cell proliferation and induce apoptosis, which may be associated with the activation of mitochondrial apoptosis pathways [143].

MCP affects numerous rate-limiting steps in cancer metastasis and works synergistically with chemotherapy and radiation treatment. It has been shown to be effective both in vitro and in vivo against prostate carcinoma, colon carcinoma, breast carcinoma, melanoma, multiple myeloma, and hemangiosarcoma. This is known as "one bullet with multiple targets" in oncology [144].

26.9.21 Fermented Foods and Probiotics

Fermented probiotic-rich foods and supplements support the health of the gastrointestinal system. The gut microbiome affects metabolism, nutrient absorption, and immune function. Certain strains can diminish the incidence of postoperative inflammation in cancer patients and relieve diarrhea resulting from radiation and chemotherapy. Some components of the immune response, including phagocytosis, natural killer cell activity, and mucosal immunoglobulin A production, can be improved by some probiotic bacteria [145]. *Saccharomyces boulardii* is an effective strain to prevent and resolve *Clostridium difficile* infection, antibiotic-associated diarrhea, and enteral nutrition-related diarrhea, all of which are concerns with immunocompromised oncology patients [146].

A review article on immunomodulation related to probiotics states the immune properties of the digestive mucosa are assisted by the gut-associated lymphoid tissue (GALT) [147]. The epithelial lining of the intestine comprises the primary physical barrier where immune responses are initiated. This intestinal barrier protects the host from the ingested external environment with a protective mucous layer, secretory IgA, and the tight junctions. Bacteria within the intestinal lumen interacts with intestinal epithelial and dendritic cells, macrophages, and lymphocytes. Probiotics modulate the immune system through the induction of cytokines from epithelial cells, and this is reported to be strain-specific [147].

There are high-quality nutraceutical probiotic blends tailored to specific needs that are strongly recommended for consideration for cancer prevention and during treatment. Good probiotic food choices include organic whole milk yogurt (if casein/animal milk is tolerated), organic kefir (if casein/animal milk is tolerated), fermented foods such as kimchi, sauerkraut, miso, and kombucha. If the patient is severely immunocompromised, each probiotic strain and food choice should be carefully researched to determine if it is appropriate for use.

26.9.22 Artemisia

Artemisia or wormwood plant contains powerful antiparasitic, antimicrobial, antifungal, and anti-inflammatory compounds, particularly artemisinin [148]. It is another multifaceted healer particularly in parasitic infections. Artemisinin has been found to reduce or prevent the formation of breast tumors [148], rapidly induce apoptosis in cancer cells [149], and recognized as an important worldwide medicinal plant [150, 151]. The entire plant, with all of its active compounds, is reported to be more effective than the artemisinin isolate [152]. Another study shows a synergistic cytotoxicity when artemisinin was used with butyric acid [153].

Lai and Singh have shown that with high concentration of iron in many types of cancer cells, it is possible to use this as a point of entry to induce selective apoptosis in leukemia cells by combining holotransferrin with Artemisia compounds [154]. This was also reported in radiation-resistant breast cancer cells [155], in drug-resistant small-cell-lung cancer cells [156], as well as a therapeutic effect against glioblastoma [157].

26.9.23 Carotenoid- and Polyphenol-Rich Foods

Carotenoids are red, orange, yellow, and green pigments found in plants and algae and include carotenes, xanthins, lutein, and lycopene. Partially functioning as antioxidants, certain carotenoids convert to vitamin A in the body, although that conversion is dependent on digestive function and absorption as well as thyroid function. Maceration or processing, cooking, and the presence of dietary fat increase the absorption of dietary carotenoids. Carotenoid-rich foods include carrots, sweet potatoes, peppers, dark leafy greens and lettuce, tomatoes, melons, cruciferous vegetables, squash, pumpkin, and citrus [158].

Polyphenols are responsible for the vibrant colors of fruit, berries, vegetables, and spices and add to their bitterness, astringency, flavor, and aroma. These phytochemicals are created by the plant in response to stimuli such as UV radiation, pathogens, bugs, and harsh climate [159]. In the human body, polyphenols help to reduce oxidative damage and inflammation, protect the skin from UV radiation, are neuroprotective, and help protect DNA from free radical damage and reverse epigenetic markers in the DNA, thereby preventing the development of cancer and reducing the growth of tumors.

The European Journal of Clinical Nutrition created a list of the 100 richest dietary sources of polyphenols. The richest sources were various spices and dried herbs, tea, some juices, whole grains, fruits, seeds and nuts, vegetables, cocoa products, and some darkly colored berries [160]. The bioavailability of polyphenols is increased when consumed with dietary fat.

Polyphenols found in pomegranate, green tea, broccoli, and turmeric have demonstrated significant anticancer effects involving angiogenesis, apoptosis, and proliferation, specifically in prostate cancer [161]. They can inhibit cancer cell growth by interacting with multiple signaling pathways in multiple cancers, including breast, gastrointestinal, prostate, pancreatic, lung, and ovarian, as well as refractory cancers and hormonally responsive cancers [162].

26.9.24 Essential Oils

Essential oils (EO) are aromatic volatile organic compounds that are the very olfactory essence of a plant [163]. They are typically extracted through steam distillation or cold-pressing. Their chemistry is extremely complex, consisting of hundreds of unique chemical compounds or constituents that depend on everything from climate to the soil the plant is grown in to the sounds that the plant "hears" while growing. Drought will yield a plant that is higher in certain constituents that the plant expressed in order to survive those conditions. Ancient writings and traditions tell us that essential oils have been used for thousands of years and, alongside herbs, were perhaps man's first medicine, both for physical and spiritual purposes. Modern science seems to be on the cusp of uncovering the biochemistry of what has been used for millennia.

The last decade has brought scientific discoveries indicating that odor receptors are not found solely in the nose but throughout the body—in the liver, heart, kidneys, muscle tissue, skin, even sperm—where they are responsible for initiating important physiological functions [163]. When we inhale a fragrance, those molecules are recognized first by the chemical receptors that line the olfactory membranes in our nose. Olfactory receptors behave like a lock-and-key system, with an odor molecule the key to the receptor's lock. When the right molecule lines up with the right receptor, it sets in motion an elaborate choreography of biochemical reactions. These signals are transmitted to the limbic system and the olfactory sensory nerves at the base of the brain. From there, physiological functions can be initiated.

The constituents of EOs have been found to be antibacterial, antiviral, antifungal, anti-inflammatory, analgesic, antiemetic, anxiolytic, and antidepressive [164]. EOs and their constituents act by multiple pathways and mechanisms involving apoptosis, cell-cycle arrest, antimetastatic and antiangiogenic, increased levels of reactive oxygen and nitrogen species (ROS/RNS), and DNA repair modulation to show their antiproliferative activity in the cancer cell. Reports also discuss tumor-suppressor proteins (p53 and Akt), transcription factors (NF-κB and AP-1), MAPK pathway, and detoxification enzymes like SOD, catalase, glutathione peroxidase, and glutathione reductase [164].

The use of essential oils is making great strides in oncology as aromatherapy, massage oils, and primary treatment. This is supported by in vitro and in vivo clinical research [165, 166]. Be cautious as cold-pressed citrus oils can cause a photosensitive reaction [166]. Research also points to benefits associated with components of the Boswellia species, or frankincense. The positive outcomes have been attributed to boswellic acids, but this most recent case study of bladder cancer suggests that the hydrodistillates may also have significant effect [167].

EOs are analyzed and categorized via gas chromatography. They consist of the hydrocarbons, including the terpenes, such as pinene, beta-caryophyllene, myrcene, limonene, chamazulene, and farnesene, and the oxygenated compounds, which are mainly esters, aldehydes, ketones, alcohols, phenols, and oxides. Each plant's essential oil contains a unique profile of a combination of these constituents, and thus its own particular biochemical potential application.

26.10 Dietary Interventions

Dietary interventions for cancer treatment are abundant, but common sense should be the ruling factor. Low-carbohydrate, vegetable-fruit phytonutrient-rich and nutrient-dense whole foods are fundamentals for all oncology dietary recommendations. Certain diets may be very beneficial for a period of time for detoxification or healing but may not be supportive for the long term. Think of it in the same terms that you would with conventional treatment: acute and maintenance interventions. Tailored targeted care should be prescribed depending on the patient's current treatment. Remain flexible and responsive to the whole person, and avoid being locked into a doctrine or belief that does not serve the patient's needs.

26.10.1 Seasonal Eating and Living

Humans are creatures of light and dark seasons and respond best when caring for themselves within the parameters of their location on the earth, with a nod to their heritage and genetics. For instance, in Ayurveda, seasonally balanced eating based on the seasons and a person's individual balance of dosha is essential: bitter greens in the spring to cleanse and detoxify the liver, gallbladder, and pancreas; cooling fruit and citrus in the summer to balance heat-induced lethargy; root vegetables, nuts, fats, meat, and hibernation foods in the autumn to conserve energy; and tonics in the winter to support the kidney, bladder, and heart in the darker months of contemplation and rest (see ► Chap. 45). Despite modern-day conveniences and artificial light and stimulation, at our core, we are disrupted when we do not remain in sync with nature. Simply contemplate the effect of blue light at night on our endocrine health or the stress of technology on our emotional health (see ► Chap. 34). Harmony with our natural surroundings and striving to balance what is lacking in our environment if we are city dwellers, should be our goal. This theory is applicable at all times for all people.

26.10.2 Autoimmune Paleo Diet

This approach is an effective strategy for reducing inflammation and promoting gut healing. The AIP diet removes most potentially inflammatory and typically allergic foods and focuses on nutrient-dense foods: vegetables, limited fruit, proteins including organ meats, fermented foods, and bone broths. However, in an oncology setting, it is important to not be overly focused on proteins as too much protein stimulates the mTOR pathway and contributes to the growth of cancer [168].

The mammalian target of rapamycin (mTOR) is an ancient molecular-signaling pathway that, when inhibited, will upregulate maintenance and repair, increase longevity, and impede the growth of cancer. When stimulated, it will allocate nutrients for growth, replication, and reproduction.

The Role of Nutrition in Integrative Oncology

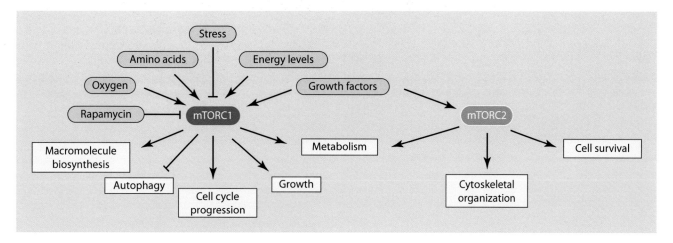

Fig. 26.2 mTOR: master regulator of the nutrient-sensing hormones insulin, leptin, and insulin-like growth factor (IGF). (Adapted from Laplante and Sabatini [168]. With permission from Elsevier)

The human biological imperative is geared toward reproduction to keep the species alive, as opposed to longevity. mTOR is a master regulator of the nutrient-sensing hormones insulin, leptin, and insulin-like growth factor (IGF), determining where and how nutrition will be used. Reducing protein intake reduces IGF, which in turn increases life span [169]. Glucose and amino acids are nutrients and fuel necessary for replication and reproduction. If glucose, amino acids, insulin, and growth factors like IGF are kept low, mTOR is suppressed, thereby allowing the upregulation of the genetic expression of maintenance and repair. mTOR also plays an important role in autophagy. A similar process is known as mitophagy, where damaged mitochondria are removed and replaced with new, healthy ones, and this process is also largely regulated by mTOR (◘ Fig. 26.2).

There is new evidence that protein restriction may be even more important than the restriction of net carbohydrates [170]. This theory has in fact been tested and found to hold true. One study reported that longevity and health were optimized when protein was replaced with carbohydrate. The authors reported that longevity may be extended in ad libitum-fed animals by manipulating the ratio of macronutrients to inhibit mTOR activation [170]. However, the conservation of protein may be even more important than the manipulation of carbohydrates. There is general agreement in oncology that the ideal amount of protein daily should not exceed 1 g per kilogram of lean body mass [171].

26.10.3 Vegetarian Diet

A vegetarian diet is a nourishing option as long as it involves consuming primarily vegetables and fruit, with a healthy dose of healthy fats. Unfortunately, many times, a vegetarian diet focuses on refined starches, like pasta, and other foods devoid of nutritive value that impart no healthful benefits, verging on a junk food diet at its most extreme. When a vegetarian diet consists of organic vegetables and fruit with enough healthy fats to ensure the absorption of the phytochemicals, it provides enough protein to satisfy biological requirements and, with attention to bio-individual needs, can be essentially autoimmune in nature [172]. Vegetarianism is really a spectrum of diets that is based on eating only plant-based foods but may include certain food groups such as dairy, eggs, and fish or may avoid cooked foods, as in the raw diet.

26.10.4 Ketogenic Diet

The research around the ketogenic diet and cellular metabolism is compelling (see ► Chap. 22). An ideal ketogenic diet is rich in healthy fats (70%), sufficient in clean proteins (25%), and low in carbohydrates (5%). This typically induces a state of nutritional ketosis where the body burns fat as fuel instead of sugar-glucose-carbohydrates, generating fewer reactive oxygen species (ROS) and secondary antioxidants, thereby lowering inflammation [173]. The effect of a ketogenic diet on the mitochondrial energy metabolism of cells may indicate that an additional hallmark of altered metabolism should be added as a hallmark of cancer as described by Ward and Thompson in 2012 as altering cell signaling and blocking cellular differentiation. Altered metabolism is a direct response to growth factor signaling [174]. Most cancer cells and proliferating normal cells still derive a significant fraction of their ATP through oxidative phosphorylation. Oxidative phosphorylation is the foundation of energy production within the ketogenic diet [174].

> Data now support the concepts that altered metabolism results from active reprogramming by altered oncogenes and tumor suppressors and that metabolic adaptations can be clonally selected during tumorigenesis. Altered metabolism should now be considered a core hallmark of cancer [174].

In normal cellular metabolism, cells convert glucose or glycogen stored in the liver into pyruvate through glycolysis. In an aerobic environment, this will result in the production of ATP via the Krebs citric acid cycle (CAC) or tricarboxylic acid cycle (TCA) through the electron transport chain, known as oxidative mitochondrial phosphorylation. Normal cells are able to adapt to utilize glutamine and ketones for energy production in a reduced glucose or fasting environment.

In the early twentieth century, German biochemist Otto Warburg noted that cancer cells produced lactate as a by-product, essentially a fermentation process in which the focus is on proliferation as opposed to energy production. Cancer cells are found to behave this way whether in an aerobic or anaerobic environment. Warburg believed this was due to damaged mitochondria [175]. The Warburg effect is the basis behind the PET scan, which delineates potentially cancerous areas based on uptake of a glucose-based tracer. Altered metabolism stems from damaged mitochondria, results from reprogramming of signaling pathways to sustain proliferation, and evades growth suppression and/or cell death [174].

Revisiting this consideration of mitochondrial cellular metabolism provides for a different potential application of diet in oncology. Restricting glucose to induce ketosis can theoretically optimize mitochondrial function, therefore directly impacting the malignancy and transformation of cells. In contrast to the nuclear somatic mutation, research across a broad range of tumor types, animal species, and experimental techniques has shown that normal mitochondria can suppress tumorigenicity. Nuclear transfer experiments suggest that tumorigenesis arises more from mitochondrial defects than from somatic mutations in the nuclear genome [176].

Ketogenic diets may act as an adjuvant cancer therapy by increasing the oxidative stress inside cancer cells and may also potentiate radiation and chemotherapies. A small trial of 10 patients with advanced cancer found that a ketogenic diet was able to halt or regress disease based on the extent of ketosis [177].

The combination of a ketogenic diet with hyperbaric oxygen therapy (HBOT) was found to have a synergistic effect and to inhibit tumor progression and prolong survival in animal studies [178]. A paleo-ketogenic diet alone has been shown to halt the progression of and partially regress a soft palate tumor in a patient over a 20-month period [179]. In many ways, a ketogenic diet mimics the effects of calorie restriction, such as reducing inflammation, decreasing side effects of treatment, potentiating standard therapies, and reducing cachexia [180].

26.10.4.1 Bio-individual Diets and Protocols

A bio-individual protocol entails assessing all aspects of the patient's condition, as discussed within this section, and then crafting a program of diet, supplements, detoxification, exercise, and emotional support that will address both their terrain and cancer. Some colon cancer patients have had colitis their whole lives, while others have been chronically constipated. Some lung cancer patients were heavy smokers, others never smoked. Some patients have deep emotional considerations to account for, others are well adjusted. Each of these nuances needs to be accounted for and addressed. It's really a marriage of ancient medicine with today's science.

Many clinics and physicians around the country are embracing these integrative strategies with successful outcomes, including case studies of work originally from Dr. William Kelley and subsequently from Drs. Nicholas Gonzalez and Linda Isaacs. Their approach is based on the pancreatic proteolytic enzyme research of the Scottish embryologist John Beard, who worked at the University of Edinburgh at the turn of the twentieth century [181].

This is an excellent summation of this concept, taken from an interview done by Dr. Conners with Dr. Gonzalez [182]:

> We divide patients into different metabolic categories, depending on each patient's particular genetic, biochemical and physiological make-up. In this model, patients with solid epithelial tumors, such as tumors of the lung, pancreas, colon, prostate, uterus, etc. do best on a largely plant-based diet. Such patients have a metabolism that functions most efficiently with a specific combination of nutrients that are found in fruits, vegetables, nuts, whole grains and seeds, and with minimal to no animal protein.
>
> On the other hand, patients with the blood or immune-based malignancies such as leukemia, myeloma and lymphoma do best on a high-animal protein, high-fat diet. Such patients do extremely well with a diet based on animal products with minimal to moderate amounts of plant-based foods, the particular design of the diet again depending on the individual patient's metabolic make-up. We find patients with pancreatic cancer always do best with a largely plant-based diet that emphasizes fruits, vegetables and vegetable juice, nuts, seeds and whole grains. Allowed protein includes fish one to two times a week, one to two eggs daily and yogurt daily, but no other animal protein. In our therapy, we use diets specifically because of the effect of food on the autonomic nervous system. This system consists of the sympathetic and parasympathetic branches and ultimately controls all aspects of our physiology, including immune function, cardiovascular activity, endocrine function and the entire action of our digestive system. The sympathetic and parasympathetic systems have opposing actions on the target organs and so can adjust our physiology depending on needs and demands, enabling our bodies to react to any situation, condition or stress. We believe disease, whatever the form, occurs because there is an imbalance in autonomic function. For example, we find solid tumors, such as tumors of the breast, lung, pancreas, colon, uterus, ovaries and liver, occur only in patients who have an overly strong sympathetic

nervous system and a correspondingly weak, ineffective parasympathetic nervous system. We believe that blood-based cancers, such as leukemia, lymphoma and multiple myeloma, only occur in patients that have an overly developed parasympathetic nervous system, and a correspondingly weak sympathetic nervous system. Previous research, such as Dr. Francis Pottenger's research during the 1920s and 1930s proposed that much if not all disease has autonomic imbalance as at least one of the major causes.

We have found that specific nutrients and foods have specific, precise and predictable effects on the autonomic nervous system. For example, a vegetarian diet emphasizes fresh fruits and vegetables, particularly leafy greens, and contains large doses of minerals such as magnesium and potassium. It has been shown in many studies that magnesium suppresses sympathetic function, while potassium stimulates parasympathetic activity. Furthermore, a largely vegetarian diet tends to be very alkalinizing, and the neurophysiologic research documents that in an alkalinizing environment, sympathetic activity is reduced, and parasympathetic activity increased. So, whatever other effect a vegetarian diet has, in terms of autonomic nervous system function, such a diet will reduce sympathetic activity and stimulate the parasympathetic system.

A meat diet is loaded with minerals such as phosphorus and zinc, which tend to have the opposite effect. A high-meat diet stimulates the sympathetic system and tones down parasympathetic activity. Furthermore, such a diet is loaded with sulfates and phosphates that in the body are quickly converted into free acid that in turn stimulates the sympathetic nervous system while suppressing parasympathetic activity.

So, by the careful use of diet, we are able to effect major changes in autonomic function and bring about balance in a dysfunctional nervous system. We find, further, as the autonomic system comes into greater harmony and balance, when the autonomic branches are equally strong, all systems – from the immune system to the cardiovascular system – work better regardless of the underlying problem. In essence, we are using diet to bring about greater physiological efficiency. For cancer patients, long experience has taught us that it is not enough to load patients with enzymes; the question of autonomic imbalance must also be addressed. In terms of pancreatic patients specifically, a plant-based diet provides all the nutrients to correct autonomic dysfunction. –Dr. Nicholas Gonzalez

Essentially, no one diet is perfect for everyone or all conditions; there are just far too many factors to take into consideration. But the prevailing concept of eating clean whole foods in moderation will always be sensible. Conquering the challenge of cancer and enabling healing are being accomplished on a field full of players who have opened their minds to the newer science of cancer as a metabolic disease rather than genetic in order to remain focused on the patient and every potential avenue to serve that patient on their cancer journey.

26.11 Integrative Therapies Adjunctive to Nutrition Therapy

26.11.1 Hyperthermia

Hyperthermia has been used with an intent to cure tumors for at least 4000 years and as a tool for the destruction of tumor masses well before that [183]. Hyperthermia effectively stimulates metabolic activity, triggering the release of stored toxins through detoxification pathways. Hyperthermia helps your body to mimic a fever response, which is what it uses normally to fight infection and cancer. Therapies include whole-body hyperthermia, hyperthermic intraperitoneal chemotherapy, radio-frequency ablation, regional hyperthermia, and the most recent advance, laser interstitial thermal therapy. Laser interstitial thermal therapy (LITT) is an emerging technique being used to treat primary and metastatic brain tumors by implanting a laser catheter into the tumor and heating it to temperatures high enough to kill the tumor [184]. Hyperthermia or infrared therapy can extend to patients' homes. Excellent options exist to provide support on a daily basis and for the long term, such as home infrared heating pad devices or an in-home infrared sauna.

26.11.2 Mindfulness Practices or Meditation

Mindfulness meditation is defined as the nonjudgmental awareness of experiences in the present moment [185]. The components through which mindfulness meditation exerts its effects are regulation of attention, body awareness, self-awareness, and regulation of emotion.

Emotions hold the power to manifest wellness or illness. Ancient practices of medicine tell us that improperly processed emotions create stagnation and block energy meridians in our body. Because humans are electrical beings, these blocked channels can result in physical illness. The hormonal stress response to prolonged stress, anger, hopelessness, fear, and chronic fatigue can contribute to immune suppression. Learning to live life as a meditation from moment to moment is one of the keys to a peaceful life.

Mindfulness meditation aids in lowering blood pressure, improves immune system and brain function, and minimizes pain sensitivity [186]. It can improve mood and reduce distress and distractive thoughts and behaviors [187].

26.11.3 Movement and Exercise

The human body was created to be in motion. In ancient times, movement was essential for survival, but the conveniences of modern life have drastically reduced that dependence. Our lymph system, which is essentially our immune system, is dependent on our physical movement to stimulate flow. The value of motion is important to our very well-being. The joy of moving our body, from a simple stretch or deep breath to a trail race, should be something that you enjoy as opposed to a dreaded task.

This becomes more complicated when experiencing treatment-induced fatigue or related to the cancer itself. At that point, the goal is to incorporate even the simplest movements. The benefits of physical activity have been connected to every aspect of our health, from our brain to our bone density. Exercise has been shown to reduce stress, improve sleep, prevent cognitive decline, alleviate anxiety, boost memory and ability to learn new things, enable neurogenesis (the creation of new brain cells), improve productivity and creativity, increase strength and flexibility, increase bone density, reduce the risk of cardiovascular disease, metabolic syndrome, and type 2 diabetes, reduce the risk of cancer, and increase longevity [188].

During treatment, proceed with care. The NCCN Clinical Practice Guidelines in Oncology (NCCN Guidelines®) for Cancer-Related Fatigue advise starting slowly and progressing incrementally [189]. Depending on fitness and comfort level, some people may want to start with a 10-minute walk around the block; others may find they can exercise for 20 minutes (or longer) initially, but the goal should be at least 30 minutes of aerobic exercise 5 days a week or more if tolerated. Caution is advised to find the level that fits with the circumstances.

The guidelines give practical suggestions for planning exercise such as if the patient doesn't have the energy to exercise a full half hour, break it up; try three 10-minute walks during the day; make exercise enjoyable; recruit a walking partner or listen to music with headphones while on a recumbent bike or treadmill; dress comfortably and drink plenty of water; consider yoga and tai chi which, while not aerobic, integrate movement and meditation and enhance wellness; and look for programs designed for cancer patients. Some health clubs and hospitals offer exercise classes that address the challenges and needs of people with cancer; avoid swimming pools as they can expose the patient to bacteria that may cause infections and the chlorine may irritate skin; and don't encourage exercise if the patient is not feeling well or running a fever [189].

26.11.4 Massage or Touch Therapy

Touch therapy is one of the oldest therapies known to man. From the lightest touch to deep tissue massage, the benefits range from relief of muscular pain to the opening of the energy meridians of the body to facilitate healing. Back massage given during chemotherapy significantly reduces anxiety and acute fatigue [190]. Massage therapy can be supportive in relieving lymphedema due to surgery or treatment. Gentle massage is used to help move lymph out of the swollen area and into areas with working lymph vessels [191].

Postsurgical scar formation can benefit from massage therapy in several ways. A valuable resource can be found to describe forms of massage therapy [192].

- Manual lymph drainage optimizes lymphatic circulation and drainage around the injured area.
- Myofascial release helps ease constriction of the affected tissue.
- Deep transverse friction can prevent adhesion formation and rupture unwanted adhesions.
- Lubrication of the scar helps soften and increase its pliability.
- Stretching aids in increasing range of motion.
- Heat application helps the pliability and flexibility of the scar.

26.11.5 Posttreatment: Creating an Empowerment Plan

Cancer treatment is a tug of war between fear and empowerment. Much of our power is handed over to physicians and the system during the duration of therapy, and when that time comes to a close, there is a big void in a patient's life. Oftentimes, fear and uncertainty fill this void, and several outcomes are a result of this phase. Some patients will return to their old ways with a misguided sense of the threat being over, some will remain stagnant in survivor groups reliving the trauma in a sense, some will throw caution to the wind and adopt habits that make them potentially more at risk for recurrence, and some will want to know how to lock the door in all ways possible to their cancer returning. This is a tremendous opportunity to positively impact and empower each and every one of these people and to arm them with holistic strategies to optimize their future. Giving patients strategies to keep themselves healthy, to rebuild their bodies, and to nourish and detoxify themselves can powerfully shift both their physical and mental well-being. Creating a plan for living should be the last step before releasing patients from treatment. Here are some ideas that the author incorporates as a foundation for an empowerment plan and can be modified depending on the particular patient and their mind-set and prognosis: Achieve and maintain optimal weight and body mass, detoxify, restore and optimize micronutrient levels in the body through food and supplements, and employ mental well-being strategies to manage fear, stress, worry, and body image and function.

Nutrition ultimately is about nourishing the body and the soul, and cancer treatment and posttreatment periods are incredibly powerful times to positively impact a patient's life. Take full advantage of that potential on every level to affect

> **Box 26.5 Sample IFMNT Nutrition Care Process**
> - *Stage 4 adenocarcinoma of the lung – non-smoker*
>   - *Assessment Data:* 63 yo female, BMI 23.5, osteoporosis, mild central adiposity, hx constipation, cough, mast cell allergy, with high-stress lifestyle for many years. Abnormal immunological findings in serum/+viral load/nutritional insufficiencies: vitamin A, zinc, vitamin D with genomic homozygous VDR, high Cu:Zn Ratio 2.1 (goal 0.9–1.0), anemia, macrocytic RBC, low IgA, low NK cell count and activity. Presented to ER with shortness of breath and X-ray revealed large masses in lung and liver, 4 liters fluid removed from pleural cavity. Diagnosis: Stage 4 adenocarcinoma of the lung cancer. Further testing ROS1+ (rare genetic mutation statistically does not respond to immunotherapy).
>   - *Nutrition Diagnosis:* Multiple nutritional insufficiencies/deficiencies with compromised immune system secondary to poor diet history and confirmed through laboratory testing.
>   - *Interventions:* Refer to integrative doctor for assessment. IV Rx: Antiviral plan with oral artemisinin, Lauricidin, Vit D3, zinc, copper−/iron-free multivitamin/mineral with bioactive forms of B vitamins, vitamin C, quercetin, and sulforaphane (broccoli sprout) with ketogenic diet. Educated in application of ketogenic diet with monitoring and logging to maintain nutritional ketosis with intermittent fasting and use of exogenous ketones. Follow up regarding nutrition plan and ketogenic diet weekly × 4 weeks and then monthly.
>   - *Monitoring and Evaluation:* Weekly monitoring of diet history, lab findings, and urine ketone log. Goal to maintain average of 1.5–3.0 ketones, healthy body mass, and no muscle cramps. IVs: Weekly high-dose vitamin C, artesunate, and immune support nutrients.
>   - Follow-Up 1 Year: "I feel amazing, better than I have in a long time." Cancer markers normal, CT/PET no evidence of disease.

the shift that is so desperately needed in this realm. Push the currently established boundaries until integrative care is the norm. The science is there; it needs only an army to carry out the mission. Be part of that army! (▶ Box 26.5)

26.12 Conclusion

This summary shares a case study that illustrates the potential of functional integrative care:

A patient called to ask how to submit imaging results of her cancer prior to an appointment. All of her focus was on her tumor, its size, and location. We explained to the patient that the primary driver of our course of action together would be the *whole* person and not specifically the imaging results. Integrated functional nutrition therapy will take into account the location of the cancer as it relates to the history of the patient and its possible emotional aspect but most importantly will create a set of tools that will support the metabolism, immune system, and mindfulness of the patient.

This illustration best summarizes the idea that we must focus on the terrain before the germ, the person before the cancer. It also brings to light the fear and disempowerment experienced by patients and the potential to radically improve the level of their care through education and the implementation of an integrative plan.

The majority of cancers arise from lifestyle issues and environment, which impact immune function, create inflammation, and/or initiate epigenetic triggers that allow cancer to develop. It is critical to explore every avenue available that can best optimize the response of the patient throughout treatment by bolstering their whole person and creating evidence-based synergies where available with conventional therapies.

Each person will present with a unique history, experience, mind-set, and disease progression that will help to define which options may support their best outcome. Each supportive strategy will provide results, but each patient may respond better to a stronger focus on different options.

Strategies discussed above should be employed in such a manner that they will optimize and complement any conventional treatment the patient is undergoing. Tremendous clinical data supporting the effectiveness of each of the abovementioned therapies has been presented, dictating their value in both stand-alone and integrative applications. The creation of a patient-centered team empowered with holistic, restorative, functional integrated modalities will define the future of cancer treatment.

References

1. Siegel RL, Miller KD, Jemal A. Cancer statistics, 2019. CA Cancer J Clin. 2019;69(1):7–34.
2. Rahib L, Smith BD, Aizenberg R, Rosenzweig AB, Fleshman JM, Matrisian LM. Projecting cancer incidence and deaths to 2030: the unexpected burden of thyroid, liver, and pancreas cancers in the United States. Cancer Res. 2014;74:2913–21.
3. Gannon OM, Antonsson A, Bennett IC, Saunders ND. Mini-review: viral infections and breast cancer – a current perspective. Cancer Lett. 2018;420:182–9.
4. Minihane AM, Vinoy S, Russel WR, Baka A, Roche HM, Tuohy KM, et al. Low-grade inflammation, diet composition and health: current research evidence and its translation. Br J Nutr. 2015;114: 999–1012.
5. Olmsted J, Harris E. Claude Bernard and the experimental method in medicine. New York: Henry Schuman; 1952.
6. Chang K, Pei-Shih P, Lin P, Hou C, Chen L, Tsai Y, et al. Curcumin upregulates insulin-like growth factor binding protein-5 (IGFBP-5) and C/EBPα during oral cancer suppression. Int J Cancer. 2010;127(1):9–20.
7. Ames BN. Micronutrients prevent cancer and delay aging. Toxico Let. 1998;102–103:5–18.
8. Ames BN. Low micronutrient intake may accelerate the degenerative diseases of aging through allocation of scarce micronutrients by triage. Proc Nat Acad Sci. 2006;103(47):17589–94.
9. Ames DN. Prolonging healthy aging: longevity vitamins and proteins. PNAS. 2018;115(43):10836–44.

10. Vidaček NS, Nanić L, Ravlić S, Sopta M, Gerić M, Goran M, Gajski G, et al. Telomeres, nutrition, and longevity: can we really navigate our aging? J Gerontol Series A. 2018;73(1):39–47.
11. Hanahan D, Weinberg R. Hallmarks of cancer: the next generation. Cell. 2011;144(5):646–74.
12. SEER*Rx Interactive Antineoplastic Drugs Database [Internet]. SEER. 2018 [cited 2019 Jun 16]. Available from: https://seer.cancer.gov/seertools/seerrx/?rx_type=drug.
13. Cleveland Clinic Cancer. Types of chemotherapy – What is chemotherapy? – Chemocare [Internet]. Chemocare.com. 2002 [cited 2019 Jun 15]. Available from: http://chemocare.com/chemotherapy/what-is-chemotherapy/types-of-chemotherapy.aspx.
14. CTCA A. Why does immunotherapy work for some but not others? [Internet]. Cancer Treatment Centers of America. Cancer Treatment Centers of America (CTCA) Comprehensive Cancer Care Network; 2017 [cited 2019 Jun 16]. Available from: http://www.cancercenter.com/discussions/blog/why-does-immunotherapy-work-for-some-but-not-others/.
15. Cleveland Clinic Cancer. About immunotherapy (Biologic response modifiers – colony-stimulating factors & tumor vaccines) – What is chemotherapy? – Chemocare [Internet]. Chemocare.com. 2019 [cited 2019 Jun 16]. Available from: http://chemocare.com/chemotherapy/what-is-chemotherapy/about-immunotherapy.aspx.
16. Pezzella F, Harris AL, Tavassoli M, Gatter KC. Blood vessels and cancer much more than just angiogenesis. Cell Death Disc. 2015;1:15064.
17. Cleveland Clinic Cancer. Targeted therapy: monoclonal antibodies, anti-angiogenesis, and other cancer therapies. Chemocare. www.chemocare.com/chemotherapy/what-is-chemotherapy/targeted-therapy.aspx.
18. Weigel C, Veldwijk MR, Cakes CC, Seibold P, Slynko A, Liesenfeld D, et al. Epigenetic regulation of diacylglycerol kinase alpha promotes radiation-induced fibrosis. Nat Commun. 2016;7:10893.
19. Li F, Zhou K, Gao L, Bin Zhang B, Li W, Yan W, et al. Radiation induces the generation of cancer stem cells: a novel mechanism for cancer radioresistance. 2016. Oncol Lett;12(5):3059–65.
20. Holmström K, Kostov RV, Dinkova-Kostova AT. The multifaceted role of Nrf2 in mitochondrial function. Cur Opin Tox. 2016;1:80–91.
21. Sadeghian M, Mastrolia V, Rezaei Haddad A, Mosley A, Mullali G, Schiza D, et al. Mitochondrial dysfunction is an important cause of neurological deficits in an inflammatory model of multiple sclerosis. Sci Rep. 2016;6:33249. https://doi.org/10.1038/srep33249.
22. Ziegler DV, Wiley CD, Velarde MC. Mitochondrial effectors of cellular senescence: beyond the free radical theory of aging. Aging Cell. 2015;14(1):1–7.
23. Passos JF, Saretzki G, Zglinicki T. DNA damage in telomeres and mitochondria during cellular senescence: is there a connection? Nucleic Acids Res. 2007;35(22):7505–13.
24. Carlson LE, Subnis UB, Piedalue KL, Vallerand J, Speca M, Lupichuk S, et al. The ONE-MIND Study: rationale and protocol for assessing the effects of ONlinE MINDfulness-based cancer recovery for the prevention of fatigue and other common side effects during chemotherapy. Eur J Cancer Care (Engl). 2019;28:e13074.
25. Tian J, Chen GL, Zhang HR. Sleep status of cervical cancer patients and predictors of poor sleep quality during adjuvant therapy. Sup Care in Cancer. 2015;23(5):1401–8.
26. Lynch T, Price A. The effect of cytochrome P450 metabolism on drug response, interactions, and adverse effects. Am Fam Physician. 2007;76(3):391–6.
27. Flockhart DA. Drug interactions: cytochrome P450 drug interaction table. Indiana University School of Medicine. 2007. https://drug-interactions.medicine.iu.edu. Accessed 1 June 2019.
28. Srinivasan V, Pandi-Perumal SR, Brzezinski A, Bhatnagar KP, Cardinali DP. Melatonin, immune function and cancer. Recent Pat Endocr Metab Immune Drug Discov. 2011;5(2):109–23.
29. Willard Terry. Insomnia: waking at the same time most nights. Terry Willard. 3 Jan 2013. www.drterrywillard.com/insomnia-waking-at-the-same-time-most-nights/.
30. Linus Pauling Institute [Internet]. 2019 [cited 2019 Jun 16]. Available from: http://lpi.oregonstate.edu/.
31. Drisko JA, Serrano OK, Chen Q, et al. Treatment of pancreatic cancer with intravenous vitamin C: a case report. Anti-Cancer Drugs. 2018;29(4):373–9.
32. Stargrove MB, Treasure J, McKee D. Herb, nutrient and drug interactions: clinical implications and therapeutic strategies. St. Louis: Mosby; 2007.
33. Paul Anderson. Outside the box cancer therapies alternative therapies that treat and prevent Cancer. Dr. Mark Stengler and Dr. Paul Anderson. https://markstengler.com/books/new-book-outside-box-cancer-therapies/
34. Block KI. The circadian system and cancer: it's about time! Integr Cancer Ther. 2018;17(1):3–4.
35. NIH Nat Cancer Inst. Human clinical trials. NIH Nat Cancer Inst. https://www.cancer.gov/about-cancer/treatment/cam/hp/vitamin-c-pdq/.
36. High-Dose Vitamin C [Internet]. National Cancer Institute. Cancer.gov; 2010 [cited 2019 Jun 16]. Available from: http://www.cancer.gov/about-cancer/treatment/cam/hp/vitamin-c-pdq#section/_16.
37. Padayatty SJ, Sun AY, Chen Q, Espey MG, Drisko J, Levine M. Vitamin C: intravenous use by complementary and alternative medicine practitioners and adverse effects. PLoS One. 2010;5(7):e11414.
38. Berkson BM, Rubin DM, Berkson AJ. Revisiting the ALA/N (alpha-lipoic acid/low-dose naltrexone) protocol for people with metastatic and nonmetastatic pancreatic cancer: a report of 3 new cases. Integr Cancer Ther. 2009;8(4):416–22.
39. Chiang KC, Yeh CN, Chen TC. Vitamin D and pancreatic cancer—an update. Cancers. 2011;3(1):213–26.
40. Centers for Disease Control and Prevention. National Center for Emerging and Zoonotic Infectious Diseases. Cancer, Infection and Sepsis Fact Sheet: A Potentially Deadly Combination Every Cancer Patient Should Know About. [Internet]. 2016 [cited 2019 Jun 16]. Available from: https://www.cdc.gov/sepsis/pdfs/cancer-infection-and-sepsis-fact-sheet.pdf.
41. Could vitamin C save lives in sepsis? These hospitals aren't waiting for proof. [Internet]. PulmCCM. 2017 [cited 2019 Jun 16]. Available from: http://pulmccm.org/main/2017/critical-care-review/vitamin-c-save-lives-sepsis/.
42. IVC Protocol Vitamin C Research | Riordan Clinic [Internet]. Riordan Clinic. 2014. Available from: https://riordanclinic.org/research-study/vitamin-c-research-ivc-protocol.
43. Lynch T, Price AL. The effect of cytochrome P450 metabolism on drug response, interactions, and adverse effects. Am Fam Physician. 2007;76(3):391–6.
44. Hodges RE, Minich D. Modulation of metabolic detoxification pathways using foods and food-derived components: a scientific review with clinical application. J Nutr Metab. 2015;2015:760689. Published online 2015 Jun 16.
45. Theoharides TC, Kavalioti M. Stress, inflammation and natural treatments. J Biol Regul Homeost Agents. 2018;32:1345–7.
46. Fessler T. Home Tube feeding with blenderized foods – Oley Foundation [Internet]. Oley.org. 2018 [cited 2019 Jun 16]. Available from: http://oley.org/default.asp?page=HomeTF_BlenderFoods.
47. Tea and Cancer Prevention [Internet]. National Cancer Institute. Cancer.gov; 2010 [cited 2019 Jun 16]. Available from: https://www.cancer.gov/about-cancer/causes-prevention/risk/diet/tea-fact-sheet.
48. Euler L. Clobber Cancer with a Daily Cup of This… [Internet]. Cancer Defeated. 2017 [cited 2019 Jun 16]. Available from: https://www.cancerdefeated.com/clobber-cancer-with-a-daily-cup-of-this-2/1220/.
49. Aggarwal BB, Yuan W, Li S, Gupta SC. Curcumin-free turmeric exhibits anti-inflammatory and anticancer activities: identification

of novel components of turmeric. Mol Nutr Food Res. 2013;57(9):1529–42.
50. Prasad S, Tyagi AK, Aggarwal BB. Recent Developments in Delivery, Bioavailability, Absorption and Metabolism of Curcumin: the Golden Pigment from Golden Spice. Cancer Res Treat. 2014;46(1):2–18. Published online 2014 Jan 15.
51. Shehzad A, Lee J, Lee YS. Curcumin in various cancers. Biofactors. 2013;39(1):56–68. https://doi.org/10.1002/biof.1068. Epub 2013 Jan 10
52. Zhao HM, Xu R, Huang XY, Cheng SM, Huang MF, Yue HY, et al. Curcumin suppressed activation of dendritic cells via JAK/STAT/SOCS signal in mice with experimental colitis. Front Pharmacol. 2016;7:455. eCollection 2016
53. Ravindran J, Prasad S, Aggarwal BB. Curcumin and cancer cells: how many ways can curry kill tumor cells selectively? AAPS J. 2009;11(3):495–510. Published online 2009 Jul 10
54. Soltanian S, Matin MM. Cancer stem cells and cancer therapy. Tumor Biol. 2011;32(3):425–40.
55. Gupta SC, Patchva S, Aggarwal BB. Therapeutic roles of curcumin: lessons learned from clinical trials. AAPS J. 2013;15(1):195–218.
56. Kanai M, Yoshimura K, Asada M, Imaizumi A, Suzuki C, Matsumoto S, Nishimura T, Mori Y, Masui T, Kawaguchi Y, Yanagihara K, Yazumi S, Chiba T, Guha S, Aggarwal BB. A phase I/II study of gemcitabine-based chemotherapy plus curcumin for patients with gemcitabine-resistant pancreatic cancer. Cancer Chemother Pharmacol. 2011;68(1):157–64.
57. Yallapu MM, Maher DM, Sundram V, Bell M, Jaggi M, Chauhan SC. Curcumin induces chemo/radio-sensitization in ovarian cancer cells and curcumin nanoparticles inhibit ovarian cancer cell growth. J Ovarian Res. 2010;3:11.
58. Howells LM, Mitra A, Manson MM. Comparison of oxaliplatin- and curcumin-mediated antiproliferative effects in colorectal cell lines. Int J Cancer. 2007;121(1):175–83.
59. Volak LP, Ghirmai S, Cashman JR, Court MH. Curcuminoids inhibit multiple human cytochromes P450 (CYP), UDP-glucuronosyltransferase (UGT), and sulfotransferase (SULT) enzymes, while piperine is a relatively selective CYP3A4 inhibitor. Drug Metab Dispos. 2008;36(8):1594–605 . Published online 2008 May 14. https://doi.org/10.1124/dmd.108.020552.
60. Prasad S, Tyagi AK. Ginger and its constituents: role in prevention and treatment of gastrointestinal cancer. Gastroenterol Res Pract. 2015;2015(142979):11pp.
61. Rahmani AH, Al Shabrmi FM, Aly SM. Active ingredients of ginger as potential candidates in the prevention and treatment of diseases via modulation of biological activities. Int J Physiol Pathophysiol Pharmacol. 2014;6(2):125–36. Published online 2014 Jul 12. PMID: 25057339.
62. Bode AM, Dong Z. Chapter 7: The amazing and mighty ginger. In: Benzie IFF, Wachtel-Galor S, editors. Herbal medicine: biomolecular and clinical aspects. 2nd ed. Boca Raton: CRC Press/Taylor & Francis; 2011.
63. Hakim L, Alias E, Makpol S, Ngah WZW, Morad NA, Yusof YAM. Gelam honey and ginger potentiate the anti-cancer effect of 5-FU against HCT 116 colorectal cancer cells. Asian Pac J Cancer Prev [Internet]. 2014;15(11):4651–7. https://doi.org/10.7314/APJCP.2014.15.11.4651.
64. Rhode J, et al. Ginger inhibits cell growth and modulates angiogenic factors in ovarian cancer cells. BMC Complement Altern Med BioMed Central. 2007;7:44.
65. Ray A, Vasudevan S, Sengupta S. 6-Shogaol inhibits breast cancer cells and stem cell-like spheroids by modulation of notch signaling pathway and induction of autophagic cell death. PLoS One. 2015;10(9):e0137614.
66. Hsu YL, Hung JY, Tsai YM, Tsai EM, Huang MS, Hou MF, Kuo PL. 6-shogaol, an active constituent of dietary ginger, impairs cancer development and lung metastasis by inhibiting the secretion of CC-chemokine ligand 2 (CCL2) in tumor-associated dendritic cells. J Agric Food Chem. 2015;63(6):1730–8. https://doi.org/10.1021/jf504934m. Epub 2015 Feb 9
67. *Taraxacum Officinale* DANDELION. http://www.anaturalhealingcenter.com/documents/Thorne/monos/Taraxicum%20mono.pdf. Accessed 6 June 2019.
68. Picman A. Antifungal activity of helenin and isohelenin. Biochem Syst Ecol. 1983;11(3):183–6.
69. Yang Y, Li S. Dandelion extracts protect human skin fibroblasts from UVB damage and cellular senescence. Oxid Med Cell Longev. 2015;2015:619560. Published online 2015 Oct 20.
70. Hamm C, Kanjeekal SM, Gupta R, Ng W. Dandelion root and chronic myelomonocytic leukemia. Blood. 2013;122(21):5216.
71. Ovadje P, Chochkeh M, Akbari-Asl P, Hamm C, Pandey S. Selective induction of apoptosis and autophagy through treatment with dandelion root extract in human pancreatic cancer cells. Pancreas. 2012;41(7):1039–47.
72. Ovadje P, Chatterjee S, Griffin C, Tran C, Hamm C, Pandey S. Selective induction of apoptosis through activation of caspase-8 in human leukemia cells (Jurkat) by dandelion root extract. J Ethnopharmacol. 2011;133(1):86–91.
73. Ovadje P, Hamm C, Pandey S. Efficient induction of extrinsic cell death by dandelion root extract in human chronic myelomonocytic leukemia (CMML) cells. PLoS One. 2012;7(2):e30604. Published online 2012 Feb 17.
74. Ovadje P, Ammar S, Guerrero JA, Arnason JT, Pandey S. Dandelion root extract affects colorectal cancer proliferation and survival through the activation of multiple death signalling pathways. Oncotarget. 2016;7:73080–100.
75. Niculescu MD, Zeisel SH. Diet, methyl donors and DNA methylation: interactions between dietary folate, methionine and choline. J Nutr. 2002;132(8 Suppl):2333S–5S.
76. Nutrition & the Epigenome [Internet]. Utah.edu. 2013 [cited 2019 Jun 16]. Available from: https://learn.genetics.utah.edu/content/epigenetics/nutrition/.
77. Srinivasan V, Pandi-Perumal SR, Brzezinski A, Bhatnagar KP, Cardinali DP. Melatonin, immune function and cancer. Recent Pat Endocr Metab Immune Drug Discov. 2011;5(2):109–23.
78. Peuhkuri K, Sihvola N, Korpela R. Dietary factors and fluctuating levels of melatonin. Food Nutr Res. 2012;56. https://doi.org/10.3402/fnr.v56i0.17252 . Published online 2012 Jul 20.
79. Mikirova N. Clinical test of pyrroles: usefulness and association with other biochemical markers. Clinical Medical Reviews and Case Reports [Internet]. 2015 Apr 30 [cited 2019 Jun 16];2(4). Available from: https://riordanclinic.org/wp-content/uploads/2015/04/CMRCR-2-027-Pyrroles.pdf.
80. Arnao MB, Hernández-Ruiz J. The physiological function of melatonin in plants. Plant Signal Behav. 2006;1(3):89–95.
81. Foods With Melatonin – Don't want to take synthetic melatonin? [Internet]. Immunehealthscience.com. 2010 [cited 2019 Jun 16]. Available from: http://www.immunehealthscience.com/foods-with-melatonin.html.
82. Getting melatonin from your diet [Internet]. Natural Health News. 2016 [cited 2019 Jun 16]. Available from: http://www.nyrnaturalnews.com/article/getting-melatonin-from-your-diet/.
83. Howatson G, Bell PG, Tallent J, Middleton B, McHugh MP, Ellis J. Effect of tart cherry juice (Prunus cerasus) on melatonin levels and enhanced sleep quality. Eur J Nutr. 2012;51(8):909–16.
84. Johns NP, Johns J, Porasuphatana S, Plaimee P, Sae-Teaw M. Dietary intake of melatonin from tropical fruit altered urinary excretion of 6-sulfatoxymelatonin in healthy volunteers. J Agric Food Chem. 2013;61(4):913–9.
85. Prasad R, Katiyar SK. Honokiol, an active compound of Magnolia Plant, inhibits growth, and progression of cancers of different organs. Adv Exp Med Biol. 2016;928:245–65.
86. Hua H, Chen W, Shen L, Sheng Q, Teng L. Honokiol augments the anti-cancer effects of oxaliplatin in colon cancer cells. Acta Biochim Biophys Sin Shanghai. 2013;45(9):773–9.

87. Singh T, Katiyar SK. Honokiol inhibits non-small cell lung cancer cell migration by targeting PGE2-mediated activation of β-catenin signaling. PLoS One. 2013;8(4):e60749.
88. Singh T, Gupta NA, Xu S, Prasad R, Velu SE, Katiyar SK. Honokiol inhibits the growth of head and neck squamous cell carcinoma by targeting epidermal growth factor receptor. Oncotarget. 2015;6(25):21268–82.
89. Saeed M, Kuete V, Kadioglu O, Börtzler J, Khalid H, Greten HJ, et al. Cytotoxicity of the bisphenolic honokiol from Magnolia officinalis against multiple drug-resistant tumor cells as determined by pharmacogenomics and molecular docking. Phytomedicine. 2014;21(12):1525–33. Epub 2014 Aug 28. https://www.ncbi.nlm.nih.gov/pubmed/25442261.
90. Haag JD, Lindstrom MJ, Gould MN. Limonene-induced regression of mammary carcinomas. Cancer Res. 1992;52(14):4021–6.
91. Miller JA, Pappan K, Thompson PA, Want EJ, Siskos A, Keun HC, et al. Plasma metabolomic profiles of breast cancer patients after short-term limonene intervention. Cancer Prev Res (Phila). 2015;8(1):86–93. Published online 2014 Nov 11. https://www.ncbi.nlm.nih.gov/pmc/articles/PMC4289656/.
92. Miller JA, Lang JE, Ley M, Nagle R, Hsu CH, Thompson PA, et al. Human breast tissue disposition and bioactivity of limonene in women with early-stage breast cancer. Cancer Prev Res (Phila). 2013;6(6):577–84.. Epub 2013 Apr 3. https://www.ncbi.nlm.nih.gov/pubmed/23554130.
93. Roué G, Pichereau V, Lincet H, Colomer D, Sola B. Cyclin D1 mediates resistance to apoptosis through upregulation of molecular chaperones and consequent redistribution of cell death regulators. Oncogene. 2008;27(36):4909–20. https://doi.org/10.1038/onc.2008.126. Epub 2008 Apr 28
94. Rabi T, Bishayee A. d-Limonene sensitizes docetaxel-induced cytotoxicity in human prostate cancer cells: Generation of reactive oxygen species and induction of apoptosis. J Carcinog. 2009;8:9.
95. Singh SV, Srivastava SK, Choi S, Lew KL, Antosiewicz J, Xiao D, et al. Sulforaphane-induced cell death in human prostate cancer cells is initiated by reactive oxygen species. J Biol Chem. 2005;280(20):19911–24. Epub 2005 Mar 11. https://www.ncbi.nlm.nih.gov/pubmed/15764812.
96. Lin JF, Tsai TF, Yang SC, Lin YC, Chen HE, Chou KY, et al. Benzyl isothiocyanate induces reactive oxygen species-initiated autophagy and apoptosis in human prostate cancer cells. Oncotarget. 2017;8(12):20220–34. https://doi.org/10.18632/oncotarget.15643. Published online 2017 Feb 23. PMCID: PMC5386757 PMID: 28423628 https://www.ncbi.nlm.nih.gov/pmc/articles/PMC5386757/.
97. van der Hulst RRWJ, von Meyenfeldt MF, Deutz NEP, Soeters PB, Brummer RJM, von Kreel BK, et al. Glutamine and the preservation of gut integrity. Lancet. 1993;341(8857):1363–5.
98. Boelens PG, Nijveldt RJ, Houdijk APJ, Meijer S, van Leeuwen PAM. Glutamine alimentation in catabolic state. J Nutr. 2001;131(9):2569S–77S.
99. Kuhn KS, et al. Glutamine as indispensable nutrient in oncology: experimental and clinical evidence. Eur J Nutr. 2010;49(4):197–210. U.S. National Library of Medicine.
100. Gaurav K, et al. Glutamine: a novel approach to chemotherapy-induced toxicity. Indian J Med Paediatr Oncol Off J Indian Soc Med Paediatr Oncol Medknow Publications & Media Pvt Ltd. 2012;33(1):13–20.
101. Cruciferous Vegetables [Internet]. Linus Pauling Institute. 2019 [cited 2019 Jun 16]. Available from: http://lpi.oregonstate.edu/mic/food-beverages/cruciferous-vegetables#disease-prevention-genetic-influences.
102. Dyer D, DiCioccio A. Flaxseeds and Breast Cancer [Internet]. Oncologynutrition.org. 2013 [cited 2019 Jun 16]. Available from: http://www.oncologynutrition.org/erfc/hot-topics/flaxseeds-and-breast-cancer/.
103. Patel S, Goyal A. Recent developments in mushrooms as anti-cancer therapeutics: a review. 3 Biotech, Springer Berlin Heidelberg. 2012;2(1):1–15.
104. Human Clinical Studies. AHCC Research Association. www.ahccresearch.org/pages/cs. Accessed 6 June 2019.
105. Smith JA, Mathew L, Gaikwad A, et al. From Bench to Bedside: evaluation of AHCC supplementation to modulate the host immunity to clear high-risk human papillomavirus infections. Front Oncol. 2019;9:173.
106. Ito T, Urushima H, Sakaue M, Yukawa S, Honda H, Hirai K, et al. Reduction of adverse effects by a mushroom product, active hexose correlated compound (AHCC) in patients with advanced cancer during chemotherapy—the significance of the levels of HHV-6 DNA in saliva as a surrogate biomarker during chemotherapy. Nutr Cancer. 2014;66(3):377–82.
107. AHCC. Memorial Sloan Kettering Cancer Center. www.mskcc.org/cancer-care/integrative-medicine/herbs/ahcc.
108. Bee Healthy Farms. Documents on SlideShare. www.slideshare.net/jflariviere/documents. Accessed 6 June 2019.
109. Premratanachai P, Chanchao C. Review of the anticancer activities of bee products. Asian Pac J Trop Biomed, Asian Pacific Tropical Medicine Press. 2014;4(5):337–44.
110. Fratellone Patrick M. Apitherapy products for medicinal use. OMICS International, OMICS International, 6 Nov. 2015. https://www.longdom.org/open-access/apitherapy-products-for-medicinal-use-2155-9600-1000423.pdf
111. Patrick L. Iodine: deficiency and therapeutic considerations. Altern Med Rev. 2008;13(2):116–27.
112. Brownstein D. Iodine: why you need it, why you can't live without it. 5th ed. West Bloomfield: Medical Alternative Press; 2014.
113. Kaczor T. Iodine and cancer: a summary of the evidence to date. Nat Med J. 2014;6(6)
114. Dach Jeffrey. Iodine treats breast cancer, overwhelming evidence. Jeffrey Dach MD, 17 Apr 2019. jeffreydachmd.com/iodine-treats-breast-cancer/.
115. Brownstein, David. The epidemic of breast cancer and thyroid disorders: the common link. Dr Brownstein, 7 June 2015. www.drbrownstein.com/the-epidemic-of-breast-cancer-and-thyroid-disorders-the-common-link/.
116. Hodges, Edmond. NaturoDoc Library. NaturoDoc, 7 Aug 2018. www.naturodoc.com/chlorella.htm.
117. Crawford, Krystal. The harmful deficiency that affects a staggering 72% of the population. Dr. Axe, 19 Mar 2018. draxe.com/iodine-deficiency/.
118. Queiroz MLS, et al. Protective effects of Chlorella Vulgaris in lead-exposed mice infected with Listeria Monocytogenes. Int Immunopharmacol, U.S. National Library of Medicine. 2003;3(6):889–900.
119. Ogawa K, et al. Evaluation of chlorella as a decorporation agent to enhance the elimination of radioactive strontium from body. PLoS One, Public Library of Science. 2016;11(2):e0148080.
120. Shim JA, et al. Effect of Chlorella intake on Cadmium metabolism in rats. Nutr Res Pract, The Korean Nutrition Society and The Korean Society of Community Nutrition. 2009;3(1):15–22.
121. Sears ME. Chelation: harnessing and enhancing heavy metal detoxification–a review. TheScientificWorldJournal, Hindawi Publishing Corporation. 2013;2013:219840.
122. Kwak JH, et al. Beneficial immunostimulatory effect of short-term Chlorella supplementation: enhancement of natural killer cell activity and early inflammatory response (randomized, double-blinded, placebo-controlled trial). Nutr J, BioMed Central. 2012;11:53.
123. Soltani M, et al. Evaluation of protective efficacy of Spirulina platensis in Balb/C mice with candidiasis. J De Mycologie Medicale, U.S. National Library of Medicine. 2012;22(4):329–34.
124. Liao G, Gao B, Gao Y, Yang X, Cheng X, Phycocyanin Inhibits OY. Tumorigenic potential of pancreatic cancer cells: role of apop-

125. Wang Zhujun, Xuewu Zhang. Inhibitory effects of small molecular peptides from Spirulina (Arthrospira) platensis on cancer cell growth. Food Funct, U.S. National Library of Medicine. 2016. Abstract
126. Atashrazm F, Lowenthal RM, Woods GM, Holloway AF, Dickinson JL. Fucoidan and cancer: a multifunctional molecule with anti-tumor potential. Mar Drugs. 2015;13(4):2327–46. https://doi.org/10.3390/md13042327.
127. Wei B, Liu Y, Lin X, Fang Y, Cui J, Wan J. Dietary fiber intake and risk of metabolic syndrome: a meta-analysis of observational studies. Clin Nutr. 2018;37(6 Pt A):1935–42.
128. Ma Yu, et al. Dietary Fiber intake and risks of proximal and distal colon cancers: a meta-analysis. Medicine, Wolters Kluwer Health. 2018.
129. Ben Qiwen, et al. Dietary fiber intake reduces risk for colorectal adenoma: a meta-analysis. Gastroenterology, U.S. National Library of Medicine. 2014.
130. Aune, D, et al. Dietary fiber and breast cancer risk: a systematic review and meta-analysis of prospective studies. Ann Oncol Off J Eur Soc Med Oncol, U.S. National Library of Medicine. 2012.
131. Huang, Tian-bao, et al. Dietary fiber intake and risk of renal cell carcinoma: evidence from a meta-analysis. Med Oncol (Northwood, London, England), U.S. National Library of Medicine. 2014.
132. Aoe, Seiichiro, et al. Effect of wheat bran on fecal butyrate-producing bacteria and wheat bran combined with barley on *Bacteroides* abundance in Japanese Healthy Adults. Nutrients, MDPI. 2018.
133. Zeng, Huawei, et al. Secondary bile acids and short chain fatty acids in the colon: a focus on colonic microbiome, cell proliferation, inflammation, and cancer. Int J Mol Sci, MDPI. 2019.
134. Mcnabney, Sean M, Tara M Henagan. Short chain fatty acids in the colon and peripheral tissues: a focus on butyrate, colon cancer, obesity and insulin resistance. Nutrients, MDPI. 2017.
135. Borycka-Kiciak, Katarzyna, et al. Butyric acid – a well-known molecule revisited. Przeglad Gastroenterologiczny, Termedia Publishing House. 2017.
136. Bindels, L B, et al. Gut microbiota-derived propionate reduces cancer cell proliferation in the liver. Br J Cancer, Nature Publishing Group. 2012.
137. Conti, Sefora, et al. Modified citrus pectin as a potential sensitizer for radiotherapy in prostate cancer. Integr Cancer Ther, SAGE Publications. 2018.
138. Glinsky, Vladislav V, Avraham Raz. Modified citrus pectin anti-metastatic properties: one bullet, multiple targets. Carbohydr Res, U.S. National Library of Medicine. 2009.
139. Sureban, Sripathi M, et al. Dietary pectin increases intestinal crypt stem cell survival following radiation injury. PLoS One, Public Library of Science. 2015.
140. Prostate Cancer, Nutrition, and Dietary Supplements (PDQ®): Integrative, Alternative, and Complementary Therapies – Health Professional Information [NCI]. *Prostate Cancer, Nutrition, and Dietary Supplements (PDQ®): Integrative, Alternative, and Complementary Therapies – Health Professional Information [NCI]*. https://www.uofmhealth.org/health-library/ncicdr0000719335. https://www.ncbi.nlm.nih.gov/books/NBK83984/#CDR0000719565__61pomi%20t
141. Fluorouracil [Internet]. Wikipedia. Wikimedia Foundation; 2019. Available from: https://en.wikipedia.org/wiki/Fluorouracil.
142. Wang S, et al. Chemoprevention of low-molecular-weight citrus pectin (LCP) in gastrointestinal cancer cells. Int J Biol Sci, Ivyspring International Publisher. 2016.
143. Lu W, et al. Influence of oxaliplatin combined with LCP on proliferation and apoptosis of colon cancer cell line HT29 [Abstract in English]. Zhonghua Wei Chang Wai Ke Za Zhi Chinese J Gastrointest Sur, U.S. National Library of Medicine. 2013.
144. Glinsky V, et al. Modified citrus pectin anti-metastatic properties: one bullet, multiple targets. Carbohydr Res, U.S. National Library of Medicine. 2009.
145. Lomax, AR, Calder, PC. Probiotics, immune function, infection and inflammation: a review of the evidence from studies conducted in humans. Curr Pharm Des, U.S. National Library of Medicine. 2009.
146. Kelesidis T, Pothoulakis C. Efficacy and safety of the probiotic Saccharomyces Boulardii for the prevention and therapy of gastrointestinal disorders. Ther Adv Gastroenterol, SAGE Publications. 2012.
147. McCulloch F. Anti-inflammatory probiotics: immunomodulation in the gut and beyond. Naturopathic Doctor News and Review, Naturopathic Doctor News and Review. 2018.
148. Lai H, Singh N. Oral artemisinin prevents and delays the development of 7,12-dimethylbenz[a]anthracene (DMBA)-induced breast cancer in the rat. Cancer Lett, Elsevier. 2005.
149. Singh N, Lai H. Artemisinin induces apoptosis in human cancer cells. Anticancer Res. 2004.
150. Efferth T, et al. Cytotoxic activity of secondary metabolites derived from Artemisia Annua L. towards cancer cells in comparison to its designated active constituent artemisinin. Phytomedicine, Urban & Fischer. 2011.
151. Efferth T, et al. Cytotoxic activity of secondary metabolites derived from Artemisia Annua L. towards cancer cells in comparison to its designated active constituent artemisinin. Phytomedicine Int J Phytother Phytopharmacol, U.S. National Library of Medicine. 2011.
152. Ferreira JFS, et al. Flavonoids from Artemisia Annua L. as antioxidants and their potential synergism with artemisinin against malaria and cancer. MDPI Mol Div Preserv Int. 2010.
153. Singh NP. Synergistic cytotoxicity of artemisinin and sodium butyrate on human cancer cells. Anticancer Res. 2005.
154. Lai H, Singh N. Selective cancer cell cytotoxicity from exposure to dihydroartemisinin and holotransferrin. Cancer Lett, U.S. National Library of Medicine. 1995.
155. Singh N, Lai H. Selective toxicity of dihydroartemisinin and holotransferrin toward human breast cancer cells. Life Sci, U.S. National Library of Medicine. 2001.
156. Sadava D, et al. Transferrin overcomes drug resistance to artemisinin in human small-cell lung carcinoma cells. Cancer Lett, U.S. National Library of Medicine. 2002.
157. Kim SH, et al. Therapeutic effects of Dihydroartemisinin and transferrin against glioblastoma. Nutr Res Pract, The Korean Nutrition Society and the Korean Society of Community Nutrition. 2016.
158. Higdon J, et al. Carotenoids: α-carotene, β-carotene, β-cryptoxanthin, lycopene, lutein, and zeaxanthin. Linus Pauling Institute. 2019.
159. Ramalingum N, Mahomoodally, MF. The therapeutic potential of medicinal foods. Adv Pharmacol Sci, Hindawi Publishing Corporation. 2014.
160. Pérez-Jiménez, J, et al. Identification of the 100 richest dietary sources of polyphenols: an application of the phenol-explorer database. Eur J Clin Nutr, U.S. National Library of Medicine. 2010.
161. Thomas, R, et al. A double-blind, placebo-controlled randomised trial evaluating the effect of a polyphenol-rich whole food supplement on PSA progression in men with prostate cancer–the U.K. NCRN Pomi-T Study. Prostate Cancer Prostatic Dis, Nature Publishing Group. 2014.
162. Fantini M, et al. In vitro and in vivo antitumoral effects of combinations of polyphenols, or polyphenols and anticancer drugs: perspectives on cancer treatment. Int J Mol Sci, MDPI. 2015.
163. Ferrer I, et al. Olfactory receptors in non-chemosensory organs: the nervous system in health and disease. Front Aging Neurosci, Frontiers Media S.A. 2016.
164. Gautam N, et al. Essential oils and their constituents as anticancer agents: a mechanistic view. BioMed Res Int, Hindawi Publishing Corporation. 2014.
165. Hunt KK, et al. Elafin, an inhibitor of elastase, is a prognostic indicator in breast cancer. Breast Cancer Res, BioMed Central, 15 Jan 2013. breast-cancer-research.com/content/15/1/R3.
166. Mori M, et al. Inhibition of elastase activity by essential oils in vitro. J Cosmet Dermatol, U.S. National Library of Medicine. 2002.

167. Xia D, et al. Cancer chemopreventive effects of Boswellia Sacra gum resin hydrodistillates on invasive urothelial cell carcinoma: report of a case. Integr Cancer Ther, SAGE Publications, 2017.
168. Laplante M, Sabatini DM. mTOR signaling in growth control and disease. Cell. 2012;149(2):274–93.
169. Ali S, Garcia J. M: sarcopenia, Cachexia and aging: diagnosis, mechanisms and therapeutic options - a mini-review. Gerontology. 2014;60:294–305.
170. Solon-Biet, SM, et al. The ratio of macronutrients, not caloric intake, dictates cardiometabolic health, aging, and longevity in ad libitum-fed mice. Cell Metabol, U.S. National Library of Medicine. 2014.
171. Escott-Stump S. Nutrition & diagnosis-related care. 8th ed. Philadelphia: Wolters Kluwer; 2015.
172. Campbell TC. The China Study – T. Colin Campbell Center for Nutrition Studies. Center for Nutrition Studies, 2005. nutritionstudies.org/the-china-study/.
173. Branco AF, et al. Ketogenic diets: from Cancer to mitochondrial diseases and beyond. Eur J Clin Invest, U.S. National Library of Medicine. 2016.
174. Ward PS, Thompson CB. Metabolic reprogramming: a cancer hallmark even Warburg did not anticipate. Cancer Cell. 2012;21(3):297–308.
175. Otto AM. Warburg effect(s)-a biographical sketch of Otto Warburg and his impacts on tumor metabolism. Cancer Metab. 2016;4:5. Published 2016 Mar 8.
176. Seyfried TN. Cancer as a mitochondrial metabolic disease. Front Cell Dev Biol, Frontiers Media S.A. 2015.
177. Fine EJ, et al. Targeting insulin inhibition as a metabolic therapy in advanced cancer: a Pilot Safety and Feasibility Dietary Trial in 10 patients. Nutrition (Burbank, Los Angeles County, Calif.), U.S. National Library of Medicine. 2012.
178. Poff AM, et al. The ketogenic diet and hyperbaric oxygen therapy prolong survival in mice with systemic metastatic cancer. PLoS One, Public Library of Science. 2013.
179. Tóth C, Clemens Z. Halted progression of soft palate cancer in a patient treated with the paleolithic ketogenic diet alone: a 20-months follow-up. Am J Med Case Rep, Science and Education Publishing. 2016.
180. O'Flanagan CH, et al. When less may be more: calorie restriction and response to cancer therapy. BMC Med, BioMed Central. 2017.
181. Gonzalez N. Case reports introduction. Dr. Nicholas Gonzalez. www.dr-gonzalez.com/cases_intro.htm.
182. Conners K, Hamilton K. Cancer and enzyme therapy. Cancer and Enzyme Therapy, 21 May 2019. www.connersclinic.com/cancer-and-enzyme-therapy/.
183. Glazer ES, Curley, SA. The ongoing history of thermal therapy for cancer. Surg Oncol Clin North Am, U.S. National Library of Medicine. 2011.
184. Laser Interstitial Thermal Therapy (LITT). MD Anderson Cancer Center. www.mdanderson.org/treatment-options/laser-interstitial-thermal-therapy.html.
185. Hölzel BK, et al. How does mindfulness meditation work? Proposing mechanisms of action from a conceptual and neural perspective. SAGE J. 2011.
186. Zen meditation: thicker brains fend off pain. ScienceDaily, University of Montreal. 2010.
187. Jain S, et al. A randomized controlled trial of mindfulness meditation versus relaxation training: effects on distress, positive states of mind, rumination, and distraction. Ann Behav Med A Publication of the Society of Behavioral Medicine, U.S. National Library of Medicine. 2007.
188. Physical Activity and Cancer. National Cancer Institute. www.cancer.gov/about-cancer/causes-prevention/risk/obesity/physical-activity-fact-sheet.
189. Exercising During Cancer Treatment. National Comprehensive Cancer Network. www.nccn.org/patients/resources/life_with_cancer/exercise.aspx.
190. Karagozoglu S, Kahve M. Effects of back massage on chemotherapy-related fatigue and anxiety: supportive care and therapeutic touch in cancer nursing. Appl Nurs Res ANR, U.S. National Library of Medicine. 2013.
191. Lymphedema (PDQ®)–Patient Version. *National Cancer Institute*, PDQ® Supportive and Palliative Care Editorial Board. PDQ Lymphedema. Bethesda: National Cancer Institute. 2019.
192. Cutler N. Six ways to remove scar tissue with massage. Massage Professionals Update, Institute for Integrative Healthcare Studies. 2019.

Further Reading and Research

Cancer as a Metabolic Disease: On the Origin, Management, and Prevention of Cancer by Thomas Seyfried (2012)

Fat as Fuel by Dr. Joseph Mercola (2017).

After Cancer Care by Drs. Gerald M. Lemole, Pallav K. Mehta and Dwight L. McKee.

The Metabolic Approach to Cancer by Dr. Nasha Winters, ND and Jess Higgins Kelley.

Beating Cancer with Nutrition by Patrick Quillin, PhD.

Herb, Nutrient, and Drug Interactions: Clinical Implications and Therapeutic Strategies by Drs. Mitchell Bevel Stargrove, Jonathan Treasure and Dwight McKee.

Conquering Cancer Volumes One and Two by Dr. Nicholas Gonzalez.

Healing with Whole Foods by Dr. Paul Pitchford.

Iodine: Why You Need It, Why You Can't Live Without It by Dr. David Brownstein.

What Your Doctor May Not Tell You About Breast Cancer by Dr. John Lee, David Zava, PhD, and Virginia Hopkins.

The Linus Pauling Institute at Oregon State University http://lpi.oregonstate.edu/.

The Weston A. Price Foundation https://www.westonaprice.org/.

Oncology Nutrition, a dietetic practice group of the Academy of Nutrition and Dietetics. http://www.oncologynutrition.org/.

Keto for Cancer: Ketogenic Metabolic Therapy as a Targeted Nutritional Strategy: A Comprehensive Guide for Patients and Practitioners by Miriam Kalamian.

Get Started with the Ketogenic Diet for Cancer. (free eBook) http://www.dietarytherapies.com/support.html.

The Cancer-Fighting Kitchen, Second Edition: Nourishing, Big-Flavor Recipes for Cancer Treatment and Recovery by Rebecca Katz and Mat Edelson.

Tripping Over the Truth: The Return of the Metabolic Theory of Cancer Illuminates a New and Hopeful Path to a Cure by Travis Christofferson.

Molecules of Emotion: The Science Behind Mind-Body Medicine by Candace B. Pert

Ketofast: Rejuvenate your health with a step-by-step guide to timing your ketogenic meals by Dr. Joseph Mercola.

The Microenvironment of Chronic Disease

Steven Gomberg

27.1 Introduction – 438

27.2 The Microenvironment in Malignancies – 438

27.3 Components of Tumor Microenvironments – 438

27.4 Cellular Components – 438

27.5 Molecular Components – 439

27.6 Nutritional Influences on the Tumor Microenvironment – 439
27.6.1 Influences on the Extracellular Matrix – 439
27.6.2 Angiogenesis – 440

27.7 Inflammatory Influences – 440

27.8 The Microenvironment in Chronic Autoimmune Disease and Similarities to Cancer – 440

27.9 Nutritional Influences on the Autoimmune Microenvironment – 441

27.10 The Microenvironment of Arthritic Conditions – 441

27.11 Nutritional Influences on the Microenvironment in Arthritic Conditions – 442

27.12 The Blood as a Potential Reflective Microenvironment – 442

27.13 Conclusion – 443

References – 444

© Springer Nature Switzerland AG 2020
D. Noland et al. (eds.), *Integrative and Functional Medical Nutrition Therapy*,
https://doi.org/10.1007/978-3-030-30730-1_27

27.1 Introduction

The term microenvironment related to disease states has many possible meanings. Typically, conventional designations about biological microenvironments distinguish a smaller part of the larger organism that is related to, yet distinct from, the larger macro environment. In this context, we will be referring to the milieu involving a neoplastic site, an inflammatory site such as a joint or organ system, or even the blood. In the diseased state, various pathologies can alter localized physiological processes, creating unique encapsulated microenvironments. These microenvironments nurture disease progression locally and have the potential to affect other organ systems in the larger context of the organism. This chapter will introduce several disease state microenvironments and the factors leading to the formation of these environments. Nutritional considerations will be addressed that may lead to the initiation or progression of these disease states, as well as the nutritional interventions that may be appropriate for their reversal. The microenvironments of tumors, autoimmune-affected organ systems, and arthritic conditions will be discussed. This chapter will also include a discussion of the blood as a physiological microenvironment that can be reflective of pathological changes elsewhere in the body through the examination of live red and white blood cell morphology and pathology.

27.2 The Microenvironment in Malignancies

Microenvironment alteration is an area of great interest in cancer disease progression. Microenvironmental changes occur in a variety of malignancies and in specific stages of disease progression [1]. Malignancies insidiously alter their own environments through a variety of processes to increase the potential for proliferation and metastasis [2, 3]. In doing so, malignancies contribute to relentless progression and cause secondary pathologies in local environments [4]. The alteration of malignant microenvironments relates not only to aberrantly mutated cells but also to other cellular tissues in the vicinity that have been negatively affected by the inflammatory changes. There is collateral damage to the normal cells in the local microenvironment. These include blood vessels and vascular cell precursors, stromal cells, and cellular components of the immune system, which form a heterogeneous matrix [5]. The tumor manipulates this matrix to avoid rejection by the immune system and increase its metastatic potential [6]. In addition, the largely anaerobic microenvironment of tumors increases the potential for localized infection [7]. Research has revealed a variety of pathways related to the interactions between tumor cells and their localized environment, creating these pathological and physiological changes, and has identified ways in which diet can affect this progression [8]. In addition, specific nutritional compounds, including phytochemicals and their derivatives, can alter the various aspects of the tumor microenvironment, including angiogenesis, metastatic potential, and the tumor load [9].

27.3 Components of Tumor Microenvironments

The tumor microenvironment (TME) in cancer is defined as the various places a tumor can exist, including blood vessels, lymphocytes, immune cells, fibroblasts and signaling molecules, and the extracellular matrix [10, 11]. Tumors develop a milieu in which they can modulate pathological processes contributing to their progression and metastasis, as well as the induction of peripheral immune intolerance [12]. This is done largely through the process of cell signaling and gives cancer cells the ability to express different morphology, metabolic features, motility, gene expression, proliferation, metastatic and angiogenic potential, and immune adaptivity. These processes stem from the phenomenon of tumor heterogeneity [13, 14].

Although cancer cell antigens initially trigger a CD8+ T-cell response, there is good evidence that possible immunoediting and TME immune suppression can lead to a failure of a systemic immune response aimed at controlling the prevalence and growth of a tumor [15].

27.4 Cellular Components

The tumor connective tissue or stroma in carcinomas typically sits beneath the epithelial cells that it is derived from and contains the extracellular matrix, including fibroblasts and various immune cells. The cancer stroma is prone to inflammatory response when challenged by intrusion of other normal immune components, such as tumor-infiltrating lymphocytes. This inflammation can encourage angiogenesis, increase cell cycling, and discourage apoptosis and taken together may enhance tumor growth [16].

Normal cellular fibroblasts typically produce collagen and other fibrotic components of connective tissue. However, in malignancies, cancer cells can pirate the function of normal fibroblasts, directing their function toward carcinogenesis [17]. These carcinogenesis-associated fibroblasts (CAFs) perform a variety of different cancer-inducing functions, including the secretion of such proangiogenic factors as platelet-derived growth factor (PDGF), basic fibroblast growth factors (bFGFs), and vascular endothelial growth factor (VEGF). CAFs can also shuttle lactate to cancer cells as they are capable of aerobic glycolysis [18]. CAFs are also associated with extracellular matrix (ECM) remodeling, a process by which collagen fibers in the matrix are malformed and release matrix metalloproteinases, enzymes involved in the degradation of the extracellular matrix [19]. This ECM remodeling allows for cancer cells to escape the local environment by destabilizing the proteins in the TME. CAFs inhibit T-cell penetration into the tumor microenvironment by increasing the density of the stromal matrix in the microenvironment.

Chemokine (C-X-C motif) ligand 2 (CXCL2) is a small cytokine belonging to the CXC chemokine family that is also called macrophage inflammatory protein 2-alpha (MIP2-alpha), growth-regulated protein beta (Gro-beta), and Gro oncogene-2 [12]. Biosynthesis of CXC motif cytokines (CXCL12) targets the most promising actionable treatments for anticancer therapy [20]. CAFs such as pericytes also increase the potential for metastasis through angiogenesis promotion and in shielding tumors from anti-VEGF therapies [21].

In the TME, other cellular components responsible for inflammatory immune response are upregulated. One group of cells that has a very strong correlation between the inflammatory response and cancer are tumor-associated macrophages (TAMs). TAMs are one of the myeloid-derived suppressor (MDSC) cells, a variety of tumor-infiltrating immune cells, all of which have well-established roles in the progression of cancer. These cells have various attributes. They can repress T-cell response in general, and they tend to gravitate toward the necrotic areas of a tumor where they mask it from the immune system through the secretion of interleukin 10 [3]. Activated TAMs can support tumor growth by secreting various factors such as epidural growth factor (EGF) [22]. TAMs also secrete a variety of angiogenesis-promoting factors such as VEGF [23], PDGF, and transforming growth factor beta (TGFβ) [24]. TAMs also slowly upregulate nuclear factor kappa beta (NFκβ) expression, which leads to enduring inflammation [25]. Overall, many studies show that infiltrating TAMs are indicative of a poor prognosis in malignancies, especially breast cancer and certain lymphomas [26, 27].

27.5 Molecular Components

Dysregulation of molecular pathways is characteristic of cancer in general, but it is also increased in the established cancer microenvironment due to chemical imbalances occurring secondary to overstimulation of cells by inflammatory chemokines, cytokines, and other substances [28]. Several pathways that respond to nutritional interventions are involved in this disrupted and divergent cell-signaling process. One such substance is the tumor protein p53, which is a crucial component of tumor suppression in multicellular organisms. p53 has various cellular functions, such as activation of DNA repair, restriction of cell cycling in cases where DNA is damaged, and initiation of apoptosis if DNA damage is irreparable [29]. In normal tissues, p53 levels are difficult to detect, being present in actively proliferating cells but not in resting cells [30]. In most human cancers, p53 is found in increased amounts and generally found to be mutated with deranged function [31]. Of all the genes associated with cancer suppression, p53 is the most commonly mutated. A 2002 study showed various lifestyle and dietary components associated with p53 mutations [32]. A high glycemic load, consumption of fast food, trans-fat consumption, and consumption of red meats were associated with increased p53 mutation as opposed to control groups [32]. Other studies have linked alcohol consumption with the increased likelihood of p53 mutations in breast cancer [33].

p53 can bind with T-cell antigens, forming complexes that inactivate immune molecules, and mutated p53 can activate tumor development through several different pathways [31]. Overall, the cellular role of p53 is quite broad and includes involvement in energy metabolism, cell senescence, immune response, cellular motility and migration potential, cell signaling and communication, cell cycling regulation, and apoptosis. In addition to its intracellular effects, there is growing evidence that in its normal state, p53 can suppress angiogenesis by increasing collagen-derived antiangiogenic molecules in the extracellular matrix by transcription suppression of VEGF and through the stimulation of hypoxia-inducible factor 1 alpha (Hif1α) degradation [34].

NFκβ is another intracellular transcription factor that binds specific DNA sequences involved in several intracellular processes, including regulating cellular growth and development, subsequent apoptosis, and immune and inflammatory responses. Dysregulation of NFκβ has been linked to a variety of diseases including cancer, viral infections, septic shock, and inflammatory and immune diseases [35]. NFκβ is an important regulator of these processes as it is rapidly acting – no additional proteins need to be synthesized for it to become active. NFκβ typically resides in the cellular cytoplasm, but when activated, it moves to the cell nucleus to initiate the expression of genes with specific binding sites. This can then lead to cellular proliferation, an inflammatory or immune response, or an antiapoptotic response [36].

Typical initiators of NFκβ activity include reactive oxygen species (ROS), tumor necrosis factor alpha (TNFα), interleukin 1 beta (IL-1β), bacterial by-products, certain drugs, and ionizing radiation [34]. As a potent upregulator of inflammation, NFκβ has a direct effect on the TME. NFκβ is a known inducer of such inflammatory tumor-promoting cytokines as interleukin 6 (IL-6), as well as progenitors for such inflammatory cytokines as cyclooxygenase 2 (COX-2). It also leads to low-level prolonged smoldering inflammation that is characteristic of the TME [37]. NFκβ is also associated with another DNA transcription factor known as signal transducer and activator of transcription 3 (STAT3). STAT3 and NFκβ engage in a variety of crosstalk, mediating the release of inflammatory cytokines in the TME. STAT3-mediated acetylation causes NFκβ to remain active in the cell nucleus longer and prolongs inflammatory response, which results in increased activity of TME inflammatory cytokines [38].

27.6 Nutritional Influences on the Tumor Microenvironment

27.6.1 Influences on the Extracellular Matrix

Many nutritional compounds can have an impact on the various aspects of the destabilized TME. One compound that helps to stabilize the extracellular matrix is epigallocatechin gallate (EGCG), derived from green tea. In vitro studies have shown that EGCG can inhibit both matrix metalloproteinases (MMP) 2 and 9 [39]. Vitamin D levels were inversely associated with circulating MMP-2 and MMP-9 in several studies and therefore may be of benefit as well [40]. Resveratrol may also be useful in downregulating both MMP-2 and MMP-9, as shown in studies on chondrosarcoma and fibrosarcoma cells [41]. Curcumin, one of the pigment compounds of turmeric, has been shown to inhibit and regulate MMPs in a variety of diseases [42], as well as acting with proanthocyanidins to downregulate collagenase, which also degrades the ECM in the TME [43]. The anti-inflammatory effects of omega-3 fatty acids also seem to decrease circulating MMPs in studies in MS patients [44]. The botanical yarrow achilleifolia has also been shown to downregulate MMP-2 and MMP-9 in the inflammatory response [45].

27.6.2 Angiogenesis

The inhibition of the angiogenic cascade or signaling switch is also a potential target of many nutritional compounds. As previously discussed, TAMS promote a variety of angiogenesis-promoting factors such as VEGF, PDGF, bFGF, TGFβ, and EGR. Nutritional inhibitors of VEGF include EGCG [46], resveratrol, curcumin, and proanthocyanidins. Other natural compounds isolated from botanicals having a direct effect on VEGF include artemisinin (extracted from Artemisia annua), baicalin (extracted from Scutellaria baicalensis), silymarin (extracted from milk thistle), and honokiol (extracted from magnolia bark) [47]. Dietary flavonoids and Vitamin E have also shown to be effective modulators of VEGF [48]. PDGF expression can be affected by ellagic acid, a polyphenol found in fruit and nuts. Ellagic acid also simultaneously suppresses VEGF expression [49]. EGCG influences PDGF receptors, modifying their propensity to be stimulated by PDGF and can consequently inhibit PDGF effects through cell signaling [50]. bFGF can be affected by stabilizing the ECM overall, which reduces its availability through motility inhibition. TNFα stimulates production of bFGF, so agents that inhibit TNFα can also decrease bFGF. Nettle leaf extract has been shown in studies to be a potent TNFα inhibitor [51]. TGFβ is typically a tumor inhibitor initially, but in the hypoxic TME, tumor cells become resistant to it and proliferate due to its presence. Inhibitors of TGFβ include curcumin, emodi (a phytochemical found in rhubarb and Japanese knotweed) [52], and resveratrol [53]. TAMs also increase EGF, which itself promotes tumor growth. Retinol has been shown in studies to decrease secretion of EGF in endothelial cells [54].

As previously discussed, CAFs increase lactic acid shuttling and contribute to the hypoxic environment in the TME, which can promote angiogenic signaling of VEGF via protein kinase C (PKC). This produces a high-insulin environment, stimulating hypoxic cells to produce more lactic acid, which contributes to cell signaling. Improving insulin metabolism and decreasing insulin resistance by decreasing dietary omega-6 linoleic and arachidonic fatty acids and increasing omega-3 fatty acids may decrease activation of PKC and improve this aspect of angiogenesis activation [55, 56].

Both p53 and Nfκβ have been extensively researched as players in the progression of carcinogenesis and angiogenesis. Consequently, many nutritional substances have been identified as modulators of these two upregulators of the TME. p53 functionality is closely related to redox status since the p53 protein contains numerous cysteine particles that are sensitive to antioxidants, so increased levels of antioxidants can be important in the prevention of p53 dysregulation. Zinc deficiency also seems to be associated with increased oxidative damage to DNA and p53 upregulation and mutations [57]. Other natural dietary compounds that influence p53 mutations include vitamin E, retinoic acid, quercetin, and folate, which all may inhibit the expression of mutated p53 [56].

27.7 Inflammatory Influences

As previously mentioned, the transcription factor NFκβ initiates a variety of cascades in the TME, notably those leading to inflammation. There are many nutritional supplements that can inhibit the NFκβ activation process. Dietary components with anti-inflammatory and antioxidant properties that inhibit NFκβ include curcumin, quercetin, EGCG, and 6-gingerol, extracted from ginger. Zinc, in addition to its effects on p53 mutations, influences NFκβ levels and associated VEGF expression relative to its depletion [58] and thus may play a role in regulating NFκβ and modulating copper levels resulting from an acute-phase reaction and low zinc status [59].

Normalization of the stromal density in a TME may be facilitated by the inhibition of CXCL12. Diindolylmethane, a brassican compound isolated from cruciferous vegetables, was shown to downregulate CXCL12 in breast cancer patients [60].

27.8 The Microenvironment in Chronic Autoimmune Disease and Similarities to Cancer

The environments of cancer and autoimmune disease share common characteristics. Specifically, both show hallmarks of an upregulated immune response, leading to inflammation and subsequent tissue damage (see ▶ Chap. 19). The divergent point of the two disorders relates to how the pathology in the microenvironment progresses. In malignancy, the tumor microenvironment induces immunosuppression where tumor cells are unrecognized by the immune system. There is mixed immune response in malignancy. The tumor cells have an increased anti-inflammatory response but are invisible to specific antibodies, while the surrounding microenvironment is pro-inflammatory. Tumors in a sense are wounds that do not heal, as the tumor reorients the immune system's healing response and uses it to drive its own progression and self-promotion.

In contrast, the microenvironment in autoimmune disease shows an activated inflammatory response that escalates in response to a lack of self-tolerance. The focus of the inflammatory response is the host tissue itself. There is expression of antibodies that are misrecognized, leading to further attack by the immune system. This self-perpetuating cycle results in permanent tissue damage over a long period. Typically in the microenvironment of autoimmune disease, the immune infiltrate is comprised primarily of either Th1 cells, and their inflammatory cytokines such as interleukin 2 and 17, or Th2 cells, and their anti-inflammatory cytokines such as interleukin 4 and 10. One interesting characteristic of autoimmune diseases is the diminished presence or impairment of CD4+ and CD25+ T regulatory cells, which have come to be thought of as an essential component of immune modulation and homeostasis [61].

Another commonality between the microenvironments of cancer and autoimmune disease is the propensity toward a hypoxic state. In autoimmune disease, the hypoxic environment is established largely due to immune cell infiltration, which increases the demand for oxygen over the available supply. In some autoimmune disorders, such as multiple sclerosis, the increased migration of lymphocytes and macrophages also increases the secretion of proangiogenic factors such as VEGF and MMPs. Although autoimmunity and cancer represent opposing ways in which the immune system can become dysfunctional, there is considerable overlap in terms of the potential targets for nutritional intervention.

Because of the propensity for inflammation, many of the cancer microenvironment anti-inflammatory strategies that downregulate proangiogenic factors, such as VEGF, could be considered appropriate. There are some different factors to consider, however. While there is good understanding of the role of macrophages and other immune cells in the cancer microenvironment, certain assumptions about the roles of specific immune cells in the microenvironment of autoimmune disease have led to research with conflicting conclusions. For example, there is an assumption that m1 macrophages would be present in Th1-type autoimmune diseases such as rheumatoid arthritis, Hashimoto's thyroiditis, and multiple sclerosis, but the research data is somewhat controversial [62, 63]. Conversely, the role of autoantibodies in autoimmune disease is well understood, but there is little understanding of their significance in cancer progression. Disease states such as hepatocellular carcinoma induced by hepatitis C or liver cirrhosis are proving to be helpful in delineating the role of the immune system in cancer progression [64].

27.9 Nutritional Influences on the Autoimmune Microenvironment

Hypoxia and proangiogenic factors are two components that cancer and autoimmune disease share. Consequently, suppressing factors that contribute to a hypoxic environment, such as VEGF and Hif1a, can be important in terms of destabilizing the autoimmune microenvironment (AIM). Dietary modulators of VEGF have been previously discussed in the section on cancer and include anti-inflammatory plant compounds that may also affect cytokine expression in autoimmune disease, such as curcumin, EGCG, proanthocyanidins, and resveratrol [46, 47]. Modulation of cytokine activity, both pro-inflammatory and anti-inflammatory, and transcription factors, such as NFκβ, are of interest in terms of the autoimmune microenvironment (AIM). In some autoimmune diseases, macrophages release TNFα, stimulating the migration of NFκβ into the cell nucleus, which in turn will upregulate inflammation and code for hyperproliferative cytokines such as interleukin 1, 6, and 8 [65]. Nutritional compounds that regulate TNFα and NFκβ include curcumin, quercetin, B6, and EGCG [66–69].

One well-known factor in autoimmune disease is the connection to pro-inflammatory "Western" diets high in fat, sugar, processed food, protein, and salt and the effect on CD4+ T cells in autoimmune disease [70]. Obesity tends to impact immunity through Treg cells and can promote a predominantly Th17 environment with upregulated IL6, which continues to signal immune response [70]. Consequently, the management of obesity and abnormal insulin response can be crucial in managing autoimmune disorders, especially of the inflammatory kind [71]. Vitamin D also seems to have a special role in the initiation and progression of autoimmune diseases. Vitamin D has been shown to increase the release of anti-inflammatory IL4 from immune cells and stimulate the production of Treg cells, which prevent the immune system from initiating autoimmune response by inducing self-tolerance [71].

27.10 The Microenvironment of Arthritic Conditions

Arthritic conditions, both inflammatory and noninflammatory, have an obvious inherent potential for microenvironmental changes and pathology. The microenvironment of the synovial capsule is an important target. It was previously assumed that the tendency for inflammation in arthritic conditions was isolated to those associated with autoimmune dysregulation, such as rheumatoid arthritis and psoriatic arthritis. However, research has shown that even in osteoarthritis, inflammation is a subclinical event that can lead to degradation of the joint [72]. One study examined the activation and infiltration of CD5+ cells in both rheumatoid and osteoarthritis and found them to be comparable [72]. Chondrocytes in an inflammatory synovial environment seem to be less susceptible to selective cell death and apoptosis and increase the inflammatory environment by secreting additional inflammatory cytokines such as MCP-1 and MIF [73]. Once tissue in the superficial zone of articular cartilage is damaged by inflammation, the mesenchymal stem cells present in it lose the ability to regenerate, and tissue damage becomes permanent. Inflammatory cytokines such as IL2, which upregulate Nfκβ, lead to low-level persistent inflammation [74]. Again, we see the prevalence of an inflammatory cascade in a disease microenvironment becomes self-sustaining and difficult to reverse. Even in supposed noninflammatory arthritic conditions, changes in the synovial microenvironment affect the regeneration of the joint capsule. One study showed that changes in the synovial ME in osteoarthritis changed the RNA sequencing in joint chondrocytes, resulting in downstream changes in the joint cartilage formation [75]. Another interesting aspect of arthritic conditions and microenvironmental concerns is the correlation between the gut microenvironment in inflammatory bowel disease and the development of enteropathic arthropathies. One etiological theory proposed is that lymphocytes or macrophages from the gut may migrate to the synovium and bind to the synovial tissue vessels, dependent on the development of adhesion molecules such as vascular adhesion protein 1 (VAP1). This in turn leads to the development of inflammatory arthritis in the affected joint [76].

Another proposed mechanism relates to mesenchymal cells present in both the gut and joint being predisposed to stimulation with simultaneous overexpression of TNF leading to TNF-mediated inflammation [77]. The development of loss of tolerance of normal bacterial flora in inflammatory bowel disease (IBD) seems to be the initiator of this cascade of events leading to IBD-centered arthropathies [78]. Also, activation of toll-like receptors on immune cells by certain bacteria in the intestinal environment, such as overabundant *Prevotella copri* in RA, can lead to the upregulation of inflammatory cytokines in the joint, such as Nfκβ and IL1r [79].

27.11 Nutritional Influences on the Microenvironment in Arthritic Conditions

Dietary factors that help mediate low-level inflammation are important in inflammatory response but also in mediating tissue damage in the joint capsule. Consequently, nutritional supplements that downregulate inflammatory cytokines, TNF, and Nfκβ – such as curcumin, quercetin, polyunsaturated fatty acids (PUFA) like fish oil, and EGCG – are important. PUFAs, such as those contained in krill oil, reduce cytokine infiltration into the joint capsule [80]. Collagen, in addition to being an important constituent of the joint capsule and being necessary for its regeneration, also seems to assist in regulating T-cell function so they do not become activated and attack the joint cartilage [81]. Vascular remodeling has been shown to be an important factor in the potential infiltration of macrophages and T cells into the joint capsule, so the downregulation of proangiogenic factors through the use of proanthocyanidins and curcumin could be useful [82].

Because of the role of the gut microbiome in inflammation in arthritic conditions, optimizing the gut and decreasing the load of pathogenic bacteria could possibly have a significant effect, especially in RA. Increased levels of *Prevotella copri* have been observed in the gut microbiota of patients with RA, and it has been postulated that lowering the levels of *Prevotella* and optimizing the diversity of the microbiota in RA patients could be of clinical significance [83].

27.12 The Blood as a Potential Reflective Microenvironment

One microsystem of the body that is relatively easy to explore is that of the blood. The blood microenvironment can be observed through a microscope, most effectively in darkfield and phase-contrast viewing. Due to the fluid nature of blood, one can experience the living dynamics of the blood environment by observing it soon after obtaining a sample, without having to stain the blood or through time elapse, which could cause the sample to degrade. Dark-field illumination, which condenses light by blocking light that passes through the specimen and allowing only for oblique rays from the circumference of a specially designed condenser, allows for the viewing of objects that are unstained and absorb little or no light, such as bacteria and cell wall structures. Dark-field microscopy is less effective in determining intracellular dynamics [84–86]. Phase-contrast illumination allows light to be passed through the sample at a variety of angles, offering an almost three-dimensional aspect to sample viewing. In this way, the cellular structure and viability of red blood cells and white blood cells can be evaluated, and extrapolations can be made as to the nature of the blood microenvironment, the plasma.

The test itself is controversial, and there is limited evidence for the validity of the interpretations of dark-field and phase-contrast live blood examinations. But it is important that practitioners are educated about the practice, in case patients ask about it or bring in results from a test.

Practitioners of live blood cell analysis assert that the evaluation can be seen as more qualitative rather than quantitative. Since the blood is observed immediately after the sample is obtained, the dynamics of both the cells themselves and the plasma are visible. For example, in conventional hematology, it is possible to quantify the numbers of various types of white blood cells such as neutrophils, macrophages, and basophils, but it is difficult to determine the viability of these cells. Observing them live in the blood can lead to postulations about the actual nature of potential immune response by assessing their viability. For example, although there may be many white cells present in a sample, they may exhibit poor phagocytic activity when observed in a live microscopic evaluation.

There are crucial considerations for sample preparation so that the possibility of artifacts remains minimal. Most often, comparing two samples from a patient as well as various views from a sample can reveal any inconsistencies arising from the sample preparation (e.g., abnormalities in slides or slide covers, air bubbles, or exogenous microorganisms). This section will briefly review some of the potential findings in live blood and the potential reflective clues they may have related to nutritional status. In terms of clinical relevance, findings in the live blood should generally not be considered diagnostic per se, but they may be useful in prioritizing other diagnostic testing or comparing to other diagnostic findings when considering a potential diagnosis.

Live blood cell analysis is most useful in enhancing existing findings arising from a conventional complete blood count (CBC) or, in some cases, findings in the metabolic panel (CMP). This relates to observing the morphology and activity of the blood cells within the microenvironment of the plasma. Cell wall structure can be observed, and phase-contrast viewing allows for some intracellular observations to be made. One example relates to relative cell size. Variabilities in cell size are easily visible in the live blood, and since anisocytosis is almost always associated with the various mechanisms leading to anemias, it can validate inferences about iron status or B vitamin levels, specifically B12 and folate [87]. In addition, a great variation in the sizes of RBCs can be interpreted as oxidative stress due to excessive homocysteine or, in cases of elevated oxidant catalysts such as mercury or excess iron, within the RBCs themselves [88]. In this case, the majority of cells tend toward macrocytosis. Another observation discernible in the live blood relates to cell membrane structure and viability. Many different abnormalities in the cell wall structure are visible under the microscope and include poikilocytosis, schistocytosis, and acanthocytosis. The blood cells appear with surface corrugations or spiculations. Two inferences can be made: potential oxidative stress causing the cell membrane to rupture and poor formation of the cellular membrane due to suboptimal essential fatty acid status subsequent to insufficient intake or metabolism.

White blood cell morphology is easily determined in either phase-contrast or dark-field viewing and can often enhance an understanding of the quantitative measurement available with a blood draw. For example, a CBC may show adequate neutrophils but may be hyper-segmented under microscopic viewing and therefore suggest B12 deficiency [89]. Reduced motility of various leukocytes may result from immunodeficiency disease processes [90]. Another example is related to the status of basophil activity. In conventional analysis, basophils may appear to quantitatively be within range, but microscopic evaluation may show excessive degranulation of the basophils, which would lead to an assumption of increased histamine activity [91].

Viewing the live blood under either phase-contrast or dark-field illumination can also show abnormalities in the dynamics of the red blood cells. One such phenomenon is the observation of rouleaux or stacking of the red blood cells. From a conventional perspective, rouleaux may be caused by acute-phase protein elevations in the blood, leading to a change in the overall pH of the blood microenvironment. Many times, it can be associated with increased fibrinogen or globulin levels, and often it is seen accompanied by spicules in the blood, also indicative of elevated fibrinogen [92]. In any case, the shift in valence of the serum pH can alter the potential of red blood cells to easily move through the peripheral vasculature, causing problems with circulation [93]. Often, improving protein digestion or the administration of proteolytic enzymes can reverse this condition in a short period of time.

Outside of conventional hematological findings that can be observed in the live blood, there are often observed findings that have a variety of potential interpretations. These observations are more often than not viewed as artifacts by conventional science. From the standpoint of certain schools of alternative medicine, these findings show the propensity of the blood microenvironment to be more conducive to various pathological processes [87, 94]. Several microbiologists in the past, including Gunther Enderlein and Gaston Naessens, have embraced this theory and extrapolated on the potential meaning of these findings in the live blood microenvironment [95, 96]. One example is the formation of pteroharps, or platelets spreading like wings of butterflies, in the blood. Platelets are observed to have spread out on the slide in a formation of a shadowy background. This does not occur in all samples and is variably interpreted to mean either a change in the blood valence due to an abnormal blood pH, similar to the rouleaux formation in the red blood cells, or it is interpreted to be a manifestation of an L-form bacteria without a cell wall. This finding is often associated with leaky gut syndrome and intestinal flora imbalances. Another typical finding in live blood is the presence of thecits, which are often thought to be corpuscles broken off other cells. An alternative interpretation is that these are a cell wall-deficient form of bacteria often correlated with intestinal dysbiosis.

27.13 Conclusion

The human body is complex and composed of a variety of structures and systems that are potential arenas for pathologies that can alter the local biological terrain and then have the potential to affect the organism. The understanding of disease microenvironments, therefore, is important not only in uncovering the process of systemic pathology but also in targeting nutritional interventions that can undermine the root of pathological processes before systemic disease occurs.

> **Case Study: Cervical Cancer**
>
> A 42-year-old female patient was diagnosed with stage IVb cervical cancer 1 month prior to initial integrative medicine consult. The patient was diagnosed after an abnormal PAP smear and biopsy with follow-up radiologic evaluation for staging. Multiple metastatic lesions were identified including the rectum and the supraclavicular lymph nodes. The patient otherwise was generally healthy, slightly obese, and only complained of chronic neck pain attributed to a previous motor vehicle accident. She elected to undergo conventional chemotherapy consisting of combined paclitaxel and platin-based therapy administered every 3 weeks for 6 cycles. The patient sought additional supportive integrative approaches throughout her therapy. Since antiangiogenic therapies such as bevacizumab were not used in the treatment of cervical cancer at this time, integrative treatment interventions focused on inhibition of angiogenesis in addition to standard anti-inflammatory and immune-modulating natural protocols. The patient was given an extract of curcumin, EGCG, and baicalin targeted to affect VEGF, PGGF, and p53 expression. In addition, she received high-dose resveratrol, which has been shown to inhibit STAT3 signaling. After the six cycles of treatment, the patient was sent for follow-up radiological exam and had negative CT/PET scan results. Conventional treatment was discontinued, but she remained on botanical and nutritional protocols for several years afterward. No adverse events were identified with this protocol. To date, 9 years after initial treatment, the patient remains disease-free. This case points to potential benefits possible when conventional cancer therapies are combined with dietary and medicinal plant interventions.

References

1. Pottier C, Wheatherspoon A, Roncarati P, Longuespée R, Herfs M, Duray A, Delvenne P, Quatresooz P. The importance of the tumor microenvironment in the therapeutic management of cancer. Expert Rev Anticancer Ther. 2015;15(8):943–54. https://doi.org/10.1586/14737140.2015.1059279.
2. Seyfried TN. Cancer as a metabolic disease: on the origin, management, and prevention of cancer. Hoboken: Wiley; 2012.
3. Hanahan D, Weinberg RA. Hallmarks of Cancer: the next generation. Cell. 2011;144(5):646–74. https://doi.org/10.1016/j.cell.2011.02.013.
4. Quail DF, Joyce JA. Microenvironmental regulation of tumor progression and metastasis. Nat Med. 2013;19(11):1423–37. https://doi.org/10.1038/nm.3394.
5. Pattabiraman DR, Weinberg RA. Tackling the cancer stem cells — what challenges do they pose? Nat Rev Drug Discov. 2014;13:497–512. https://doi.org/10.1038/nrd4253.
6. Coussens LM, Werb Z. Inflammation and cancer. Nature. 2002;420(6917):860–7. https://doi.org/10.1038/nature01322.
7. Cummins J, Tangney M. Bacteria and tumours: causative agents or opportunistic inhabitants? Infect Agent Cancer. 2013;8:11. https://doi.org/10.1186/1750-9378-8-11.
8. Sheeba CJ, Marslin G, Revina AM, Franklin G. Signaling pathways influencing tumor microenvironment and their exploitation for targeted drug delivery. Nanotechnol Rev. 2013;3(2):11–22. https://doi.org/10.1515/ntrev-2013-0032.
9. Béliveau R, Gingras D. Role of nutrition in preventing cancer. Can Fam Physician. 2007;53(11):1905–11.
10. NCI Dictionary of Cancer Terms, National Cancer Institute. https://www.cancer.gov/publications/. https://www.cancer.gov/publications/dictionaries/cancer-terms?cdrid=561725. Last accessed 8 May 2018.
11. Meehan K, Vella LJ. The contribution of tumor-derived exosomes to the hallmarks of cancer. Crit Rev Clin Lab Sci. 2016;53(2):121–31.
12. Whiteside TL. The tumor microenvironment and its role in promoting tumor growth. Oncogene. 2008;27(45):5904–12. https://doi.org/10.1038/onc.2008.271.
13. Hassan T, Afshinnekoo E, Wu S, Mason CE. Genetic and epigenetic heterogeneity and the impact on cancer relapse. Exp Hematol. 2017;54:26–30. https://doi.org/10.1016/j.exphem.2017.07.002.
14. Catenacci DV. Next-generation clinical trials: novel strategies to address the challenge of tumor molecular heterogeneity. Mol Oncol. 2015;9(5):967–96. https://doi.org/10.1016/j.molonc.2014.09.011.
15. Dunn GP, Old LJ, Schreiber RD. The development of immune resistant variants of cancer cells post anti-tumor response of the immune system in "The Three Es of Cancer Immunoediting". Annu Rev Immunol. 2004;22(1):329–60. https://doi.org/10.1146/annurev.immunol.22.012703.104803. PMID 15032581
16. Kumar KJ, Surh Y-J. Inflammation: gearing the journey to cancer. Mutat Res. 2008;659(1–2):15–30.
17. Räsänen K, Vaheri A. Activation of fibroblasts in cancer stroma. Exp Cell Res. 2010;316(17):2713–22. https://doi.org/10.1016/j.yexcr.2010.04.032.
18. Hanahan D, Coussens LM. Accessories to the crime: functions of cells recruited to the tumor microenvironment. Cancer Cell. 2012;21(3):309–22.
19. Weber CE, Kuo PC. The tumor microenvironment. Surg Oncol. 2012;21(3):172–7. https://doi.org/10.1016/j.suronc.2011.09.001.
20. Guo F, Wang Y, Liu J, Mok SC, Xue F, Zhang W. CXCL12/CXCR4: a symbiotic bridge linking cancer cells and their stromal neighbors in oncogenic communication networks. Oncogene. 2016;35(7):816–26. https://doi.org/10.1038/onc.2015.139.
21. Franco M, Roswall P, Cortez E, Hanahan D, Pietras K. Pericytes promote endothelial cell survival through induction of autocrine VEGF-A signaling and Bcl-w expression. Blood. 2011;118(10):2906–17. https://doi.org/10.1182/blood-2011-01-331694.
22. Biswas SK, Gangi L, Paul S, Schioppa T, Saccani A, Sironi M, Bottazzi B, Doni A, Vincenzo B, Pasqualini F, Vago L, Nebuloni M, Mantovani A, Sica AA. A distinct and unique transcriptional program expressed by tumor-associated macrophages (defective NF-κB and enhanced IRF-3/STAT1 activation). Blood. 2005;107(5):2112–22. https://doi.org/10.1182/blood-2005-01-0428.
23. Gonzalez FJ, Vicioso L, Alvarez M, Sevilla I, Marques E, Gallego E, Alonso L, Matilla A, Alba E. Association between VEGF expression in tumour-associated macrophages and elevated serum VEGF levels in primary colorectal cancer patients. Cancer Biomark. 2007;3(6):325–33.
24. Hong WK, Holland JF. Holland-Frei cancer medicine, vol 8. 8th ed. PMPH-USA; 2010:149–69.
25. Lawrence T. Macrophages and NF-κB in cancer. Curr Top Microbiol Immunol. 2011;349:171–84. https://doi.org/10.1007/82_2010_100.
26. Mathias RA, Gopal SK, Simpson RJ. Contribution of cells undergoing epithelial–mesenchymal transition to the tumour microenvironment. J Proteome. 2013;78:545–57. https://doi.org/10.1016/j.jprot.2012.10.016.
27. Yang M, Chen J, Su F, Yu B, Su F, Lin L, Liu Y, Huang JD, Song E. Microvesicles secreted by macrophages shuttle invasion-potentiating microRNAs into breast cancer cells. Mol Cancer. 2011;10:117. https://doi.org/10.1186/1476-4598-10-117.
28. Blaylock RL. Cancer microenvironment, inflammation and cancer stem cells: a hypothesis for a paradigm change and new targets in cancer control. Surg Neurol Int. 2015;6:92.

29. Guo G, Cui Y. New perspective on targeting the tumor suppressor p53 pathway in the tumor microenvironment to enhance the efficacy of immunotherapy. J Immunother Cancer. 2015;3:9. https://doi.org/10.1186/s40425-015-0053-5.
30. Rogel A, Popliker M, Webb C, Oren M. p53 cellular tumor antigen: analysis of mRNA levels in normal adult tissues, embryos, and tumors. Mol Cell Biol. 1985;5(10):2851–5.
31. Rivlin N, Brosh R, Oren M, Rotter V. Mutations in the p53 tumor suppressor gene important milestones at the various steps of tumorigenesis. Genes Cancer. 2011;2(4):466–74. https://doi.org/10.1177/1947601911408889.
32. Slattery ML, Curtin K, Ma K, Edwards S, Schaffer D, Anderson K, Samowitz W. Diet activity, and lifestyle associations with p53 mutations in colon tumors. Cancer Epidemiol Biomark Prev. 2002;11(6):541–8.
33. Freudenheim JL, Bonner M, Krishnan S, Ambrosone CB, Graham S, McCann SE, Moysich KB, Bowman E, Nemoto T, Shields PG. Diet and alcohol consumption in relation to p53 mutations in breast tumors. Carcinogenesis. 2004;25(6):931–9.
34. Hammond EM, Amato J. Giaccia hypoxia-inducible Factor-1 and p53: friends, acquaintances, or strangers? Clin Cancer Res. 2006;12(17):5007–9. https://doi.org/10.1158/1078-0432.CCR-06-0613.
35. Gilmore TD. Introduction to NF-kappaB: players, pathways, perspectives. Oncogene. 2006;25(51):6680–4. https://doi.org/10.1038/sj.onc.1209954. PMID 17072321
36. Nelson DE, Ihekwaba AE, Elliott M, Johnson JR, Gibney CA, Foreman BE, Nelson G, See V, Horton CA, Spiller DG, Edwards SW, McDowell HP, Unitt JF, Sullivan E, Grimley R, Benson N, Broomhead D, Kell DB, White MR. Oscillations in NF-kappaB signaling control the dynamics of gene expression. Science. 2004;306(5696):704–8. https://doi.org/10.1126/science.1099962.
37. Karin M. NF-kB as a critical link between inflammation and cancer. Cold Spring Harb Perspect Biol. 2009;1(5):a000141. https://doi.org/10.1101/cshperspect.a000141.
38. Hoesel B, Schmid JA. The complexity of NF-κB signaling in inflammation and cancer. Mol Cancer. 2013;12:86. https://doi.org/10.1186/1476-4598-12-86.
39. Roomi MW, Kalinovsky T, Monterrey J, Rath M, Niedzwiecki A. In vitro modulation of MMP-2 and MMP-9 in adult human sarcoma cell lines by cytokines, inducers and inhibitors. Int J Oncol. 2013;43(6):1787–98. https://doi.org/10.3892/ijo.2013.2113.
40. Timms PM, Mannan N, Hitman GA, Noonan K, Mills PG, Syndercombe-Court D, Aganna E, Price CP, Boucher BJ. Circulating MMP-9, vitamin D and variation in the TIMP-1 response with VDR genotype: mechanisms for inflammatory damage in chronic disorders? QJM. 2002;95(12):787–96.
41. Gweon EJ, Kim SJ. Resveratrol attenuates matrix metalloproteinase-9 and -2-regulated differentiation of HTB94 chondrosarcoma cells through the p38 kinase and JNK pathways. Oncol Rep. 2014;32(1):71–8. https://doi.org/10.3892/or.2014.3192.
42. Kumar D, Kumar M, Saravanan C, Singh SK. Curcumin: a potential candidate for matrix metalloproteinase inhibitors. Expert Opin Ther Targets. 2012;16(10):959–72. https://doi.org/10.1517/14728222.2012.710603.
43. Vayalil PK, Mittal A, Katiyar SK. Proanthocyanidins from grape seeds inhibit expression of matrix metalloproteinases in human prostate carcinoma cells, which is associated with the inhibition of activation of MAPK and NFkB. Carcinogenesis. 2004;25(6):987–95.
44. Shinto L, Marracci G, Baldauf-Wagner S, Strehlow A, Yadav V, Stuber L, Bourdette D. Omega-3 fatty acid supplementation decreases matrix metalloproteinase-9 production in relapsing-remitting multiple sclerosis. Prostaglandins Leukot Essent Fatty Acids. 2009;80(2–3):131–6. https://doi.org/10.1016/j.plefa.2008.12.001.
45. Benedek B, Kopp B, Melzig MF. Achillea millefolium L. s.l. – is the anti-inflammatory activity mediated by protease inhibition? J Ethnopharmacol. 2007;113(2):312–7.
46. Sartippour MR, Shao ZM, Heber D, Beatty P, Zhang L, Liu C, Ellis L, Liu W, Go VL, Brooks MN. Green tea inhibits vascular endothelial growth factor (VEGF) induction in human breast cancer cells. J Nutr. 2002;132(8):2307–11.
47. Sagar SM, Yance D, Wong RK. Natural health products that inhibit angiogenesis: a potential source for investigational new agents to treat cancer—Part 1. Curr Oncol. 2006;13(1):14–26. PMC1891166
48. Schindler R, Mentlein R. Flavonoids and vitamin E reduce the release of the angiogenic peptide vascular endothelial growth factor from human tumor cells. J Nutr. 2006;136(6):1477–82.
49. Labrecque L, Lamy S, Chapus A, Mihoubi S, Durocher Y, Cass B, Bojanowski MW, Gingras D, Béliveau R. Combined inhibition of PDGF and VEGF receptors by ellagic acid, a dietary-derived phenolic compound. Carcinogenesis. 2005;26(4):821–6.
50. Ahn HY, Hadizadeh KR, Seul C, Yun YP, Vetter H, Sachinidis A. Epigallocatechin-3 gallate selectively inhibits the PDGF-BB-induced intracellular signaling transduction pathway in vascular smooth muscle cells and inhibits transformation of sis-transfected NIH 3T3 fibroblasts and human glioblastoma cells (A172). Mol Biol Cell. 1999;10(4):1093–104.
51. Obertreis B, Ruttkowski T, Teucher T, Behnke B, Schmitz H. Ex-vivo in-vitro inhibition of lipopolysaccharide stimulated tumor necrosis factor-alpha and interleukin-1 beta secretion in human whole blood by extractum urticae dioicae foliorum. Arzneimittelforschung. 1996;46(4):389–94.
52. Thacker PC, Karunagaran D. Curcumin and emodin downregulate TGF-β signaling pathway in human cervical cancer cells. PLoS One. 2015;10(3):e0120045. https://doi.org/10.1371/journal.pone.0120045.
53. Kuroyanagi G, Otsuka T, Yamamoto N, Matsushima-Nishiwaki R, Kozawa O, Tokuda H. Resveratrol suppresses TGF-β-induced VEGF synthesis in osteoblasts: inhibition of the p44/p42 MAPKs and SAPK/JNK pathways. Exp Ther Med. 2015;9(6):2303–10. https://doi.org/10.3892/etm.2015.2389.
54. Miller LA, Cheng LZ, Wu R. Inhibition of epidermal growth factor-like growth factor secretion in tracheobronchial epithelial cells by vitamin A. Cancer Res. 1993;53(11):2527–33.
55. Thome JL, Campbell MJ. Nuclear receptors and the Warburg effect in cancer. Int J Cancer. 2015;137(7):1519–27. https://doi.org/10.1002/ijc.29012.
56. Boik J. Natural compounds in cancer therapy in xancer and natural medicine: a textbook of basic science and clinical research. Princeton: Oregon Medical Press; 1996.
57. Ho E, Courtemanche C, Ames BN. Zinc deficiency induces oxidative DNA damage and increases P53 expression in human lung fibroblasts. J Nutr. 2003;133:2543–8.
58. Golovine K, Uzzo RG, Makhov P, Crispen PL, Kunkle D, Kolenko VM. Depletion of intracellular zinc increases expression of tumorigenic cytokines VEGF, IL-6 and IL-8 in prostate cancer cells via NF-κB dependent pathway. Prostate. 2008;68(13):1443–9.
59. Antoniades V, Sioga A, Dietrich EM, Meditskou S, Ekonomou L, Antoniades K. Is copper chelation an effective anti-angiogenic strategy for cancer treatment? Med Hypotheses. 2013;81(6):1159–63. https://doi.org/10.1016/j.mehy.2013.09.035.
60. Hsu EL, Chen N, Westbrook A, Wang F, Zhang R, Taylor RT, Hankinson O. CXCR4 and CXCL12 downregulation: a novel mechanism for the chemoprotection of 3,3′-diindolylmethane for breast and ovarian cancers. Cancer Lett. 2008;265(1):113–23.
61. Rahat MA, Shakya J. Parallel aspects of the microenvironment in cancer and autoimmune disease. Mediators Inflamm. 2016;2016:4375120, 17 pages. https://doi.org/10.1155/2016/4375120. Last accessed May 8, 2018.
62. Palacios BS, Estrada-Capetillo L, Izquierdo E, Criado G, Nieto C, Municio C, González-Alvaro I, Sánchez-Mateos P, Pablos JL, Corbí AL, Puig-Kröger A. Macrophages from the synovium of active rheumatoid arthritis exhibit an activin a-dependent proinflammatory profile. J Pathol. 2015;235(3):515–26. https://doi.org/10.1002/path.4466.
63. Donlin LT, Jayatilleke A, Giannopoulou EG, Kalliolias GD, Ivashkiv LB. Modulation of TNF-induced macrophage polarization by synovial fibroblasts. J Immunol. 2014;193(5):2373–83.

64. Le Naour F, Brichory F, Misek DE, Bréchot C, Hanash SM, Beretta L. A distinct repertoire of autoantibodies in hepatocellular carcinoma identified by proteomic analysis. Mol Cell Proteomics. 2002;1(3):197–203.
65. Kinne R, et al. Macrophages in rheumatoid arthritis. Arthritis Res. 2000;2(3):189–202.
66. Bright JJ. Curcumin and autoimmune disease. Adv Exp Med Biol. 2007;595:425–51.
67. Li Y, Yao J, Han C, Yang J, Chaudhry MT, Wang S, Liu H, Yin Y. Quercetin, inflamm immunity. Nutrients. 2016;8(3):167.
68. Huang SC, Wei JC, Wu DJ, Huang YC. Vitamin B(6) supplementation improves pro-inflammatory responses in patients with rheumatoid arthritis. Eur J Clin Nutr. 2010;64(9):1007–13. https://doi.org/10.1038/ejcn.2010.107.
69. Yang F, Oz HS, Barve S, de Villiers WJ, McClain CJ, Varilek GW. The green tea polyphenol (−)-epigallocatechin-3-gallate blocks nuclear factor-κB activation by inhibiting IκB kinase activity in the intestinal epithelial cell line IEC-6. Mol Pharmacol. 2001;60(3):528–33.
70. Manzel A, Muller DN, Hafler DA, Erdman SE, Linker RA, Kleinewietfeld M. Role of "Western Diet" in inflammatory autoimmune diseases. Curr Allergy Asthma Rep. 2014;14(1):404.
71. Prietl B, Pilz S, Wolf M, Tomaschitz A, Obermayer-Pietsch B, Graninger W, Pieber TR. Vitamin D supplementation and regulatory T cells in apparently healthy subjects: vitamin D treatment for autoimmune diseases? Isr Med Assoc J. 2010;12(3):136–9.
72. Rollín R, Marco F, Jover JA, Fernandez-Gutierrez B. Early lymphocyte activation in the synovial microenvironment in patients with osteoarthritis: comparison with rheumatoid arthritis patients and healthy controls. Rheumatol Int. 2008;28(8):757–64. https://doi.org/10.1007/s00296-008-0518-7.
73. Röhner E, Matziolis G, Perka C, Füchtmeier B, Gaber T, Burmester GR, Buttgereit F, Hoff P. Inflammatory synovial fluid microenvironment drives primary human chondrocytes to actively take part in inflammatory joint diseases. Immunol Res. 2012;52(3):169–75. https://doi.org/10.1007/s12026-011-8247-5.
74. Buhrman C. Curcumin mediated suppression of Nfκβ promotes chondrogenic differentiation of mesenchymal stem cells in a high-density co-culture microenvironment. Arthritis Res Ther. 2010;12(4):R127.
75. Lewallen EA, Bonin CA, Li X, Smith J, Karperien M, Larson AN, Lewallen DG, Cool SM, Westendorf JJ, Krych AJ, Leontovich AA, Im HJ, van Wijnen AJ. The synovial microenvironment of osteoarthritic joints alters RNA-seq expression profiles of human primary articular chondrocytes. Gene. 2016;591(2):456–64. https://doi.org/10.1016/j.gene.2016.06.063.
76. Salmi M, Jalkanen S. Human leukocyte subpopulations from inflamed gut bind to joint vasculature using distinct sets of adhesion molecules. J Immunol. 2001;166:4650–7.
77. Armaka M, Apostolaki M, Jacques P, Kontoyiannis DL, Elewaut D, Kollias G. Mesenchymal cell targeting by TNF as a common pathogenic principle in chronic inflammatory joint and intestinal diseases. J Exp Med. 2008;205:331–7. https://doi.org/10.1084/jem.20070906.
78. Cox CJ, Kempsell KE, Gaston JS. Investigation of infectious agents associated with arthritis by reverse transcription PCR of bacterial rRNA. Arthritis Res Ther. 2002;5(1):R1–8.
79. Anderson KV. Toll signaling pathways in the innate immune response. Curr Opin Immunol. 2000;12(1):13–9.
80. Deutsch L. Evaluation of the effect of Neptune Krill Oil on chronic inflammation and arthritic symptoms. J Am Coll Nutr. 2007;26(1):39–48.
81. Bagchi D, Misner B, Bagchi M, Kothari SC, Downs BW, Fafard RD, Preuss HG. Effects of orally administered undenatured type II collagen against arthritic inflammatory diseases: a mechanistic exploration. Int J Clin Pharmacol Res. 2002;22(3–4):101–10.
82. García Pérez S, Krausz S, Ambarus CA, Baeten DL, Tak PP, Reedquist KA. The rheumatoid arthritis synovial microenvironment promotes differentiation of monocytes into pro-angiogenic macrophages responsive to angiopoietin signaling. Ann Rheum Dis. 2012;71:A80–1. https://doi.org/10.1136/annrheumdis-2011-201238.20.
83. Kim D, Kim WU. Editorial: can prevotella copri be a causative pathobiont in rheumatoid arthritis? Arthritis Rheumatol. 2016;68(11):2565–7.
84. Mehta V, Saurav K, Balachandran C. Dark ground microscopy. Indian J Sex Transm Dis. 2008;29:105–6.
85. Omoto CK. Using darkfield microscopy to enhance contrast: an easy and inexpensive method. Department of Genetics and Cell Biology Washington State University. https://public.wsu.edu/~omoto/papers/darkfield.html. Last accessed 8 May 2018.
86. Caprette, DR. Dark field viewing experimental biosciences: resources for introductory and intermediate level laboratory courses. Rice University Aug 10, 2012.
87. Denk S. Understanding live blood under the microscope. Biomedx. 1998.
88. McQueen, Sam. The basic 100- a health model interpretation of clinical chemistry parameters. Institute for Health Realities; 2006.
89. Keegan S, Jacinta A, Gruner T. Fresh capillary blood analysis using darkfield microscopy as a tool for screening nutritional deficiencies of iron and cobalamin (vitamin B12): a validity study. Adv Integr Med. 2016;3(1):15–21. https://doi.org/10.1016/j.aimed.2016.01.001.
90. Kummrow A, Frankowski M, Bock N, Werner C, Dziekan T, Neukammer J. Quantitative assessment of cell viability based on flow cytometry and microscopy. Cytometry A. 2013;83(2):197–204. https://doi.org/10.1002/cyto.a.22213.
91. Hastie R. The antigen-induced degranulation of basophil leucocytes from atopic subjects, studied by phase-contrast microscopy. Clin Exp Immunol. 1971;8(1):45–61.
92. Wagner C, Steffen P, Svetina S. Aggregation of red blood cells: from rouleaux to clot formation. arXiv:1310.1483v1 [physics.bio-ph]. 2013:1–13.
93. Laufman H. The significance of the blood sludge phenomenon. Quarterly Bulletin of the Northwest University Medical School. 1950.
94. Coyle M, editor. Advanced applied microscopy for nutritional evaluation and correction volume 1. Nu-life sciences 2004.
95. Grace S. An open letter on pleomorphic microbiology unbundling the enderlein legacy. Natural Philosophy Research Group. info@ecobiotics.com. 2001.
96. Lee BC, Yoon JW, Park SH, Yoon SZ. Toward a theory of the primo vascular system: a hypothetical circulatory system at the subcellular level. Evid Based Complement Alternat Med. 2013;2013:961957. https://doi.org/10.1155/2013/961957.

Chronic Pain

Jena Savadsky Griffith

28.1 Introduction – 448
28.1.1 Mechanisms of Pain (See Fig. 28.2) – 449
28.1.2 Plant Compounds – 460

References – 465

© Springer Nature Switzerland AG 2020
D. Noland et al. (eds.), *Integrative and Functional Medical Nutrition Therapy*,
https://doi.org/10.1007/978-3-030-30730-1_28

28.1 Introduction

Aristotle described pain as emotional, a passion of the soul, and the heart as the center of sensation. Later, Descartes asserted that pain was outside of the mind, a disturbance to the machine (body) that passed along nerve fibers through a pathway that led to the brain. The gate control theory introduced the brain as deeply involved in pain processing and perception [1]. In 1974, the International Association for the Study of Pain (IASP) essentially married the mind, body, and soul, weaving all theories into its current and finalized definition: "an unpleasant sensory and emotional experience associated with actual or potential tissue damage [2]." Although our understanding of pain is still in evolution, this integrative definition includes an emotional response, which, depending on the individual physiology, behavior and life experiences, will be perceived and then expressed differently for each person [3]. Even though pain and emotion travel similar pathways, this aspect of pain as personal experience is often ignored; it cannot convey what pain is or what it feels like to each individual [4]. Importantly, it does recognize that pain can be felt without tissue damage, as is the case with many chronic pain sufferers.

Thus, chronic pain is a complex multifactorial, biopsychosocial experience [5]. Chronic pain, as opposed to acute pain, exists for at least 3–6 months or beyond typical healing allowance [2]. It is a disease in and of itself, as in musculoskeletal pain or headaches, or it can be associated with other chronic conditions, such as diabetic or chemotherapy-induced neuropathy, obesity, rheumatoid arthritis (RA) or post-traumatic stress disorder (PTSD). As a central product of inflammation, pain intersects most chronic diseases today. According to the 2011 Institute of Medicine Report, there are more than 100 million people in the United States living with chronic pain, which is more than those affected by diabetes, heart disease, and cancer combined [6] (Fig. 28.1). It is the primary reason for visiting physicians and a leading cause of total disability, costing up to $635 million per year [7]. Affecting a total of 1.5 billion people worldwide, chronic pain is an increasing public health burden in most developed countries and afflicts people throughout the lifespan [8]. It is estimated that 25–46% of children and young adults experience chronic pain, most often in the form of headache or abdominal or musculoskeletal pain [9]. Of those surveyed, more than half of older adults, 53%, reported pain within the last month [10]. Pain in the elderly is often undertreated. It significantly affects quality of life and frequently crosses over into insomnia, depression and anxiety. Chronic pain not only becomes the central part of life for the sufferer, it impacts family, friends and the communities in which they live.

Most patients suffering with chronic pain are treated with pharmacology, local injection or surgery. Although these therapies play an important and necessary role, there is no epidemiological evidence that they have yet to alter the course or make a significant difference in patient outcomes, disability or cost [11, 12]. With its bio-individual manifestations, pain is ideally suited for an integrative and functional

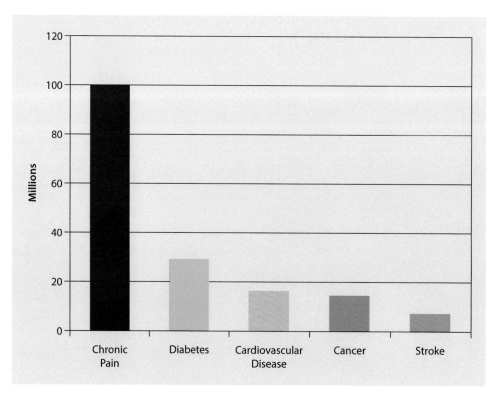

Fig. 28.1 Prevalence of Chronic Pain. (Courtesy of Jena Savadsky Griffith, RDN, IHC; Based on data from Pain [6], Diabetes [236], Heart Disease, Stroke [237], Cancer [238])

approach. The patient often needs multidisciplinary medical support and expertise in all aspects of life that are physical, emotional, and spiritual. A 2012 report on the practice of integrative medicine across America outlines significant strides being made in chronic pain, with nutrition at the therapeutic forefront [13]. In the first Multidisciplinary Pain Research workshop in Rome, 2016, researchers stated that patients with chronic pain should undergo nutritional assessment and counseling at the onset of treatment and that including nutrition in personalizing pain medicine is formidable and will improve any analgesic therapy, patient compliance and quality of life [14].

28.1.1 Mechanisms of Pain (See Fig. 28.2)

28.1.1.1 Nociceptive Pain

Nociceptive, or acute, pain results when noxious stimuli activate nociceptors in the peripheral nervous system (PNS) and serves an important protective function. This activation causes depolarization via first-, second- and third-order neurons in C fibers or A delta fibers that lead to the dorsal horn of the spinal cord and then up through the brainstem [5]. The thalamus then relays this nociceptive information to the higher cortical regions that include the amygdala, hypothalamus, periaqueductal grey, basal ganglia and regions of the cerebral cortex where the pain is processed, perceived and localized [15]. Pain is mediated via descending pathways by both excitatory (glutamate) and inhibitory (gamma-aminobutyric acid - GABA) neurotransmitters that either increase or decrease pain perception [5, 15]. Dysfunction can occur at all levels of ascending and descending pathways, resulting in chronic pain. The sequence of events in pain signaling involves four processes: transduction, transmission, modulation and perception [16].

28.1.1.2 Neuropathic Pain

Neuropathic pain is a complex multifactorial chronic pain state caused by a lesion, injury or dysfunction of either the peripheral or central nervous system (CNS). Accompanied by tissue damage or injured nerve fibers, these fibers misfire and send incorrect signals to various pain centers. These changes may not quickly resolve, as in diabetic neuropathy or postherpetic neuralgia. Thus, the impact of injury includes both damage to the fiber itself and the areas surrounding the injury. With neuropathic pain, neurons may continue to fire, leading to spontaneous pain and pain upon movement [5]. There may be inhibition impairment and alterations in central pain processing, which is key in chronic pain. Neuropathic pain can manifest in patients as shooting, burning, tingling or numbness.

28.1.1.3 Central Sensitization

It is now known that most patients with chronic pain have some degree of alteration in central nervous system processing called central sensitization (CS). CS is defined as a state in which the CNS is hyperactive or magnifies sensory input in multiple organ systems [5]. This hypersensitivity involves neuronal and microglial plasticity in response to activity, inflammation or neural injury [17]. The ensuing dysregulation produces allodynia (a greater-than-normal response to non-painful stimuli), hyperalgesia (heightened response to painful stimuli) and/or enables non-injured tissue to produce pain [18]. Localized pain can become generalized pain and manifest in joints or muscles. Pathophysiologically, signaling persists past the point of tissue healing and creates a sensory illusion where pain may be experienced with low or even absent sensory stimuli. These processes can either facilitate or inhibit pain transmission. This abnormal connectivity is driven by imbalances in concentrations of CNS neurotransmitters that control sensory processing, sleep, alertness, affect and memory [19]. Further, as a result of MR morphometry, it is now known that the cerebral cortex is reorganized and gray matter is decreased, creating a common "brain signature" in chronic pain sufferers [20].

Although an injury may be present with CS, it is not necessarily pathology of the tissue detected on a scan, but a functional pathology. Central sensitization is the underlying pathophysiology associated with fibromyalgia, irritable bowel syndrome (IBS), tension headaches, migraines, sleep disturbances, restless leg syndrome, chronic back and neck pain, osteoarthritis (OA), rheumatoid arthritis (RA), interstitial cystitis, and temporomandibular syndrome (TMJD). Importantly, CS is also implicated in anxiety, depression and PTSD [5, 21–23]. Whether or not chronic pain begins with a specific acute injury, as pain becomes chronic, multiple mechanisms overlap for many (Fig. 28.3). That is, the pain appears to be caused by a complex mixture of nociceptive, neuropathic and psychological factors with central sensitization playing a dominant role [24].

28.1.1.4 Psychological Pain

Chronic pain often exists in the absence of any physiological cause or injury. Although patients should never be told that the pain is "in their head," a new paradigm concludes that CNS processing may be altered by emotional states and traumatic events, in childhood or adulthood, much like a physical injury, into what is called psychological pain [25]. A history of trauma and adverse early adulthood experiences co-exists with many pain syndromes [26, 27]. Pain is experienced as more severe than expected, and emotional processing seems out of proportion to physical pathology, if any exists. The nervous system is sensitized over weeks, months or years. More subtly, chronic pain may develop because of unresolved emotional issues. Although those experiencing trauma have good psychosocial adjustment, experiencing pressure or an unrelated stressful incident at a later time may recall the initial trauma.

28.1.1.5 Psychosocial Mediators

Most of the study and treatment of pain has been based on the linear concepts of physical medicine. Pain was simply nociception. In the 1960s, pain was recognized as a multidimensional experience with sensory (location, quality and

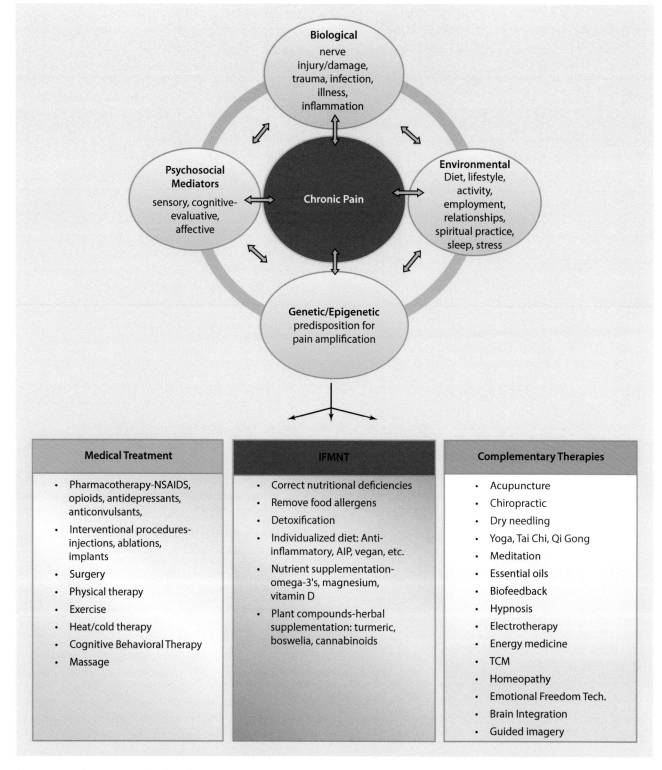

Fig. 28.2 Chronic Pain Pathophysiology and Management Algorithm. (Courtesy of Jena Savadsky Griffith, RDN, IHC)

intensity), affective (degree of unpleasantness) and cognitive-evaluative (attitude and beliefs about pain) components [28]. It integrated the common denominator of the individual and the embodiment of a person's history, energies and perceptions. Significantly, biopsychosocial factors provide a context for each person's pain experience. The degree of pain felt depends on disposition, mood, history of pain reactivity, attention, motivations, cognition, social influences, community, health and nutrition status and myriad other variables unique to that individual [12, 29].

Nociceptive (peripheral)	Neuropathic (peripheral)	Centralized Pain (central sensitization)
• Inflammation or mechanical tissue damage • NSAID, opioid responsive • Procedures	• Damage/dysfunction of peripheral nerves • Responds to NSAIDs, opioids, TCAs, neuroactive compounds	• Central disturbance in pain processing (hyperalgesia/allodynia) • Treated with neuroactive compounds to affect neurotransmitters
Examples Acute pain due to injury Osteoarthritis Rheumatoid Arthritis Cancer-related pain Post-operative pain	Examples Diabetic neuropathic pain Post herpetic neuralgia Trigeminal neuralgia Phantom limb pain Complex regional pain syndrome	Examples Fibromyalgia Irritable bowel syndrome TMJD Tension headache Low back pain

← MIXED PAIN STATES →
-Any combination may be present-

Fig 28.3 Mechanisms of pain. (Courtesy of Dr. Daniel Clauw, University of Michigan. NSAIDs nonsteroidal anti-inflammatory drugs, TMJD temporomandibular joint disorder, TCAs tricyclic antidepressants)

28.1.1.6 Sleep, Depression and Chronic Pain

Pain can trigger a cascade of events and often has a cyclic relationship with anxiety, depression, sleep disturbance, PTSD, obesity, and inactivity. Connections with pain, depression and anxiety are expected as they share similar biological pathways and neurotransmitters with overlapping pathophysiology in the central nervous system [19, 30]. Those in chronic pain are four times more likely to have comorbid anxiety or depression than pain-free primary care patients [31], with studies assessing comorbidity from 30% to 60% [30]. Similarly, up to 75% of those with depression in primary care settings complain of chronic pain, including headache, stomach pain, neck and back pain and idiopathic pain [31].

Sleep complaints are present in 67–88% of chronic pain conditions [32]. Most studies confirm its reciprocal relationship, with pain causing sleep disturbances and insomnia causing increased pain sensitivity. Loss of sleep increases inflammation, alters metabolism, depresses immunity and does not allow the body time to rest and replenish energy stores. Released along with glutamate during pain signaling is neurotransmitter substance P. During deeper stages of sleep, substance P is naturally decreased; therefore, inadequate sleep amplifies pain as more substance P remains [33]. Nutrition and lifestyle intervention should address inflammatory aspects of diet, coffee intake (see hydration), potential insulin resistance, supplementation, use of NSAIDs, stabilization of the gut microbiome and exercise.

28.1.1.7 Cancer-Related Pain

Due to the breadth and depth of cancer-related pain as its own entity, the chronic pain conditions discussed in this chapter are non-cancer-related. Cancer is a leading cause of chronic pain and presents challenges to both patient and practitioner. The American Cancer Society's most recent figures estimate 30% of newly diagnosed cancer patients have pain, 30–50% of patients undergo active treatment experience pain, and 70–90% of patients with advanced disease experience pain [34, 35]. Pain in cancer patients may be related to the cancer itself (metastatic bone disease or local tissue damage), to procedures (post-surgery, injections) or it can be related to treatment (radiation, chemotherapy) [12, 35, 36]. In addition, patients may experience nociceptive, neuropathic pain or both. Pain is one of the most feared aspects of cancer for both patients and their families; however, there are many diet and lifestyle interventions that can relieve patients of anxiety and stress, increase nutritional status, and empower patients to play an active role in their own care.

28.1.1.8 Nutrition and Chronic Pain

Eating is a daily practice that can contribute to or detract from health. According to the Centers for Disease Control and Prevention, 70% of all deaths in the United States result from chronic preventable diseases that are attributed to diet and lifestyle [37]. The current standard diet may be the most serious threat to public health [38]. Deficiencies, toxicities and sensitivities within any one or multiple biochemical and signaling processes can contribute to chronic pain conditions. While research is just beginning to address diet and pain, the connections are undeniable and individual dietary roadmaps can be created. There is evidence that specific foods, nutrients, diets and herbs can provide pain relief while other foods and dietary patterns contribute to increased pain. Regardless of the chronic pain label, from diabetic neuropathy to migraines to fibromyalgia, addressing food and its ingestion, absorption, digestion and elimination is an empowering and worthwhile first line of defense when a patient presents with chronic pain. Nutrition and diet may be the most appropriate initial recommendation for pain sufferers to activate a healing response [12].

28.1.1.9 Obesity

Chronic pain and nutrition status are linked in direct and indirect ways. Nutrition is clearly implicated in diabetes, obesity-related musculoskeletal pain, or where a food trigger may be at play, such as in migraines or IBS. Obesity and pain coexist and have negative reciprocal effects. As an epidemic with 150 million people, the treatment of obesity and chronic pain should be considered as synergistic [14]. As BMI increases, so does pain. That includes less obvious conditions as neuropathic pain and headaches [39]. In a survey of one million Americans, chronic pain rates were 68–254% higher in obese individuals compared to those who were non-obese [40]. Women who are overweight or obese have a 60–70% greater risk of developing fibromyalgia [41]. Extra weight can compress and cause body malalignment that leads to pain in the back, neck, knee, foot and hip [42]. Alternately, each pound of weight loss resulted in a four-pound reduction in stress on the knee per step, in a study of obese adults with OA [43]. Further, in the Framingham study, an 11-pound weight loss was associated with a 50% reduction in the risk of symptomatic knee arthritis [44]. Although most trials have been done with OA and knee pain, weight loss reduced pain and increased function in those with fibromyalgia and low back pain as well [45, 46]. Importantly, obesity and metabolic imbalance can be a consequence of chronic pain. Its continuance often leads to a decline in physical activity, trouble sleeping and use of food as coping mechanism [47, 48]. Chronic pain patients may have reduced ability to experience pleasure from food and thereby increase consumption, especially regarding "sugary and fatty foods [49]." Addressing and modifying the diet in the clinical setting is key in the management of pain chronification and its further progression to metabolic dysfunction.

28.1.1.10 Inflammation

Greater than a function of overload, however, obesity is more often related to pain as a product of systemic inflammation. Increased pain in non-weight bearing regions include the spine, neck and upper extremities, as well as conditions including fibromyalgia, migraine and headaches [42, 50–52]. Markers of inflammation, including C-reactive protein (CRP), tumor necrosis factor alpha (TNFα) and various cytokines, are elevated in progression of OA and low back pain, particularly in the obese population [39, 52–54]. For almost every type of chronic pain, whether it is biochemical, structural or stress-induced, systemic, local tissue or neural inflammation is an underlying or contributing factor [55–62]. By definition, pain is one of the four manifestations of inflammation [63]. All pain states—nociceptive, neuropathic, and mixed pain related—are associated with inflammation. Inflammatory mediators act to exacerbate or inhibit pain. An increase in proinflammatory cytokines is linked to the maintenance and induction of neuropathic pain, and these cytokines and free radicals produced at the site of injury may be involved with nociceptor sensitization and maintenance of nerve-injury-induced pain [64]. In the absence of injury, nerve inflammation causes neuropathic-like pain and induces spontaneous activity away from the generation site [65].

Designed to protect, inflammation is the body's immune response to pathogens, injury, or contaminants. As medicine looks less at symptoms of disease and more toward systems, inflammation gets more attention as a root cause and has become a common pathway for many of the chronic conditions that exist today, including persistent pain. A specific food can trigger an inflammatory response. Repeated consumption of inflammatory foods or dietary pattern can keep the body and immune system in a hyperactive inflammatory state. The current Western diet of highly processed, high-calorie foods, lower nutrient value, less fiber and suboptimal antioxidant power presents a heavy inflammatory load on all systems of the body and is central to the pathogenesis of chronic disease [38, 62]. The ideal balance of pro-inflammatory omega-6 essential fatty acids (EFAs) to anti-inflammatory omega-3 EFAs is 3:1; however, our standard American diet (SAD) supplies 20:1 and, for some, up to as much as 50:1 [66]. This often leaves a person profoundly burdened, malnourished, deficient and vulnerable to pain and pathogens. A ratio of 2–3:1 of omega-6 to omega-3 suppressed inflammation in patients with rheumatoid arthritis [67]. In response to chronic pain, an anti-inflammatory diet is a logical step. (see ▶ Chap. 23 and below ▶ Sect. 28.1.1.16)

28.1.1.11 Gut Microbiome

A normal inflammatory pain response requires a healthy gut microbiome. Although in its infancy, research continues to confirm Hippocrates' long-held edict that health begins and ends in the gut. Home to 100 million microbial cells, the gut microbiome not only facilitates nutrient absorption but has come to be known as the intrinsic regulator of the immune system, where 70–80% of it resides [68–70]. In a 2015 study, patients with RA had similarly altered gut and oral microbiomes that normalized after treatment [71]. A study of 42 fibromyalgia patients found that 100% of them had small intestinal bacterial overgrowth (SIBO—bacterial translocation from the large to the small intestine) [72]. Additionally, IBS was present in 30–70% of fibromyalgia patients [73]. This points to the disruption of a healthy microbiome and dysbiosis as an underlying feature for many with chronic pain and other inflammatory conditions.

With IBS, it is easier to make a direct food–gut connection as the pain resides in the abdomen. Affecting almost 15% of the population, it is the primary reason for visits to a gastroenterologist and second most common reason for work absence [74]. Etiology of IBS is, like chronic pain, as diverse as the people afflicted. While altered gut bacteria and infection are often causes, most attribute their condition to food sensitivities which are found in 50–84% of IBS patients [75]. The most common trigger foods are corn, wheat, coffee, eggs, tea, citrus and dairy foods. Followed by food sensitivity and lactose tolerance testing, and stool analysis, patients were put on an elimination diet for 1 year. In addition to many other symptomatic improvements, pain was reduced by 90%

[76]. It is common, however, to overlook the gut-brain connection that exists in IBS and chronic pain. As a regulator of the digestive process, the enteric nervous system (ENS) has its own pain receptors, nerves and neurotransmitters often identical to those in the CNS. Further, as the ENS has bidirectional communication with the brain, anxiety or stress can alter gut bacteria or altered bacteria can result in stress or anxiety. This confirms the central sensitization pathophysiological framework of life experience + psychological response + physiology that underlies the basis of chronic pain. Analysis of the gut microbiome has identified patterns of altered gut flora in patients with chronic pain conditions that also include complex regional pain syndrome, chronic fatigue syndrome, restless leg syndrome and comorbid conditions of anxiety, depression and obesity [77–79].

Diet can detract from or enrich the diversity of the microbiome. In general, a varied, fibrous whole-foods-rich diet that includes fruits, vegetables, nuts, quality protein, whole grains (if non-reactive), healthy fats that include omega-3 fatty acids, fermented foods and sufficient hydration, and is tailored to the individual goes far in restoring and maintaining health. Additional nutritional tools that can therapeutically increase gut health are prebiotics and probiotics [79, 80]. Detractors may be refined sugars, processed foods and hydrogenated vegetable oils/transgut microbiome fats, additives, preservatives, artificial sweeteners and high fructose corn syrup, in addition to multitude of other lifestyle factors that include alcohol, smoking, NSAIDs, antibiotics, acid blockers, other pharmaceuticals, chronic stress, cesarean sections and widespread use of antiseptics and sanitizers [80–85].

Thus, disruption of the gut microbiome does not only cause local distress but can affect other biological systems. Chronic pain patients may also present without traditional symptoms of gastrointestinal (GI) distress, leading clinicians to potentially miss an integral piece. For example, it is estimated that 57% of those with neurological dysfunction of unknown cause have gluten sensitivity [86]. In addition, 20 fibromyalgia patients, 17 of which had IBS and 8 with migraines, all testing negative for celiac disease, went on a gluten-free diet for at least 6 months. All 20 participants reported dramatic improvement in widespread chronic pain, with 15 indicating they were pain-free [87].

28.1.1.12 Food Sensitivities

Food sensitivities also play a direct role in migraine headaches, the second most common pain condition and one of the most prevalent pain disorders in the world. Overlapping many other pain conditions, migraines have genetic, hormonal, immune, environmental and potentially mitochondrial influences and are usually activated in response to a trigger [88, 89]. Overlooked causes include dehydration [90], altered oral and gut microbiome [91], celiac disease [92], gluten sensitivity, hypothyroid [93] or hyperinsulinemia [94]. As the most well-known trigger, food's association with migraines has been studied for over a century. Much research follows similar patterns to the 1930 study which resulted in 53% of migraine patients resolving headaches after avoidance of all allergenic foods with partial improvement shown in an additional 38%. Further, most research indicates a wide range of food as causation [95, 96]. Children with severe, frequent migraines were put on an elimination diet that consisted of one meat (lamb or chicken), one carbohydrate (rice or potato), one fruit (banana or apple), one vegetable, water, and vitamin supplements for 3–4 weeks. Children who did not improve were given a second elimination diet. On both diets, 88.6% recovered completely. Fifty-five foods provoked migraines. The most common were milk, egg, chocolate, orange and wheat. Other causes of migraines may be amines, nitrates, histamine, monosodium glutamate and artificial sweeteners [96, 97].

Trigger foods can only be identified if eliminated and then reintroduced. Although time consuming, individualized approaches where patients keep meticulous food journals (with symptoms) is an invaluable tool that helps patients understand the effect of food on their pain. Nutritional deficiencies linked with migraines and other forms of chronic pain include vitamin D, magnesium, vitamins B2, B6, B12, folic acid, COQ10 and alpha lipoic acid. (see Nutrients).

28.1.1.13 Oral Function

Affecting 7% of the population, oral facial pain (OFP) has a direct and reciprocal effect on nutritional status with risk for malnutrition. Due to decrease in function and chewing ability, common dietary advice suggests altered consistencies. However, simply recommending a soft diet may further contribute to deficiencies and malnutrition [98]. Offering customized dietary guidance based on an individual's challenges that may include protein (powders), antioxidant (fruit and vegetables), omega-3 and fiber rich (nuts, flax seeds) smoothies will potentially ensure maintenance of nutritional status, if not improvement in the pain condition.

28.1.1.14 Nutrition Assessment

As the first step, a complete nutritional assessment and interview of the chronic pain sufferer is crucial and improves outcomes and patient satisfaction [99]. All systems need to be reviewed and should include a comprehensive diet history, current intake, nutrition-focused physical findings and functional labs. An integrative assessment includes current and past medical and social health history, current work, stressors, role of exercise, sleep quality, relationships, current familial support and spiritual practice. It is a "how are you?" as opposed to "where is your pain?" Many chronic pain patients arrive angry, depressed, hopeless and have experienced losses in work and life that had previously given them meaning and purpose. Because of chronic pain's pervasive effect on the psychological and social aspects of life, listening, understanding and giving attention to the pain patient's story is the goal and builds trust [100]. Due to the multifactorial nature of pain, a thorough, thoughtful and compassionate patient-centered assessment cannot be understated. Patient participation, investment and co-creation of goals is an integral part of the relationship and is also necessary for a positive outcome. Asking open-ended questions and motivational interviewing is an optimal approach [101]. As caring for the chronic pain patient can be challenging, the health of the clinician must also be a priority. Introducing a more healthful diet is easier and more acceptable if the clinician can speak to its benefits from personal experience. There are myriad assessment tools that measure pain intensity, its effect on function, mood and specific types of sensation. As mentioned, there are many secondary factors and pain comorbidities including mental health, which encourages an interdisciplinary team approach. Although reduction of pain is clearly a goal, the overall objective is to increase wellbeing and function.

28.1.1.15 Functional Labs to Assess Nutritional Status

Deciding what testing is needed for a patient is an art, a skill and different for each individual. Tests cannot reveal everything, and they should not be solely relied upon, but together with interviews and questionnaires, they can help uncover integral pieces of the puzzle to create a treatment plan to optimize health. As many chronic illnesses begin in the gut, GI tests are often used. Below is a suggested list representative of baseline laboratory testing that is tailored to the individual patient:

- Organic acids test
- Comprehensive stool testing
- Immunologic: Food allergy/sensitivity/intolerance testing
- Glucose tolerance tests
- Candida antibody tests
- SIBO tests
- Methylation testing
- Hormone testing
- Vitamin D status
- Nutrient status testing
- Amino acid testing

28.1.1.16 Therapeutic Diets

As most pain patients view their situation as hopeless, diet can be a powerful and potentially transformative intervention [102]. Most research involving food and a certain disease state focuses on one isolated nutrient. The power of food, however, comes from its synergistic effect: a naturally coordinated and complex array of phytochemicals and nutrients combined, not only with other foods of high biological value, but how it interacts with the biology, genetics and biography of each individual. A shift in the patient's dietary pattern brings the most lasting change. Although much of the research linking dietary patterns and chronic pain has focused on different forms of arthritis [103], as outlined below, many diets have been successful in improving quality of life and ameliorating pain, at times, to resolution. Further, as the etiology of pain is steeped in bio-individuality, the dietary intervention should follow suit, and be tailored to the individual, including personal preferences, traditions and culture. The intervention may be as individual as the biology and biography of the patient in which it resides.

The *Elimination Diet* is universally accepted as the standard for identifying and eliminating potential food intolerances and is a frequent beginning diet. Although there are many versions, suspect foods are eliminated from the diet for a certain period, usually 3–12 weeks, followed by a reintroduction and food challenge phase [104, 105]. After the set period of elimination, foods are brought in one at a time, every 4 days, eating the food often. Accounting for 90% of all food reactions, the top eight allergens are milk, eggs, peanuts, nuts, wheat/gluten, soy, fish and shellfish [106]. Foods that are eaten most often that patients "cannot live without" are usually the most problematic. Many start out eliminating refined sugar, dairy, wheat and processed foods. Removing these foods potentially translates to less activation of innate immune pathways, less inflammation and less pain perception.

Celiac disease testing is recommended for pain patients due to the many non GI manifestations of the disease, but gluten-free diets have been used successfully without having a celiac diagnosis. Migraines, fibromyalgia, different forms of arthritis, IBS, chronic musculoskeletal pain and various idiopathic and identified neuropathies have been substantially improved on *gluten-free* diets [86, 87, 107]. There are ongoing trials to study gluten's effects on low back pain [108].

If the root cause of pain is chronic inflammation, *the anti-inflammatory diet* (AID) is a reasonable option. Considered to be the hallmark of integrative pain management, the AID is followed by many without chronic pain or illness, as it complements a healthy lifestyle. There are several versions, but the concepts are similar and strictness can be adjusted. The AID emphasizes high amounts of fruits and vegetables, whole grains, legumes, soy, fish, herbs and spices. Each dietary component addresses potential inflammatory markers that influence pain. High intake of antioxidants replete with vitamins and minerals modulate inflammatory cytokines, C-reactive protein production and temper damaging oxidative stress. In particular, evidence exists that removal of free radicals—nitric oxide, superoxide, and peroxynitrite—can prevent and reverse inflammatory pain, neuropathic pain, morphine-induced hyperalgesia, and tolerance [14]. In the British Cohort study, low intake of fruits and vegetables contributed to chronic widespread pain [109]. Soy foods may be beneficial in neuropathic pain and osteoarthritis [109, 110]. Broccoli, grapes, fish oil, ginger, green tea and tomatoes were found to reduce inflammation and pain [111]. Inclusion of fish and fish oil supplementation maximizes the omega-3 anti-inflammatory prostaglandins, while mitigating the competing omega-6 arachidonic cascade that is highly influential in modulating pain. (see ◘ Fig. 28.4). A personalized anti-inflammatory diet can reduce the prevalence of many of the chronic conditions associated with pain including diabetes, obesity and autoimmune disease.

The *Mediterranean diet*, a mainstay in the nutrition world, is also recommended to pain patients and has been successful in those with RA [110–112]. Similar to the anti-

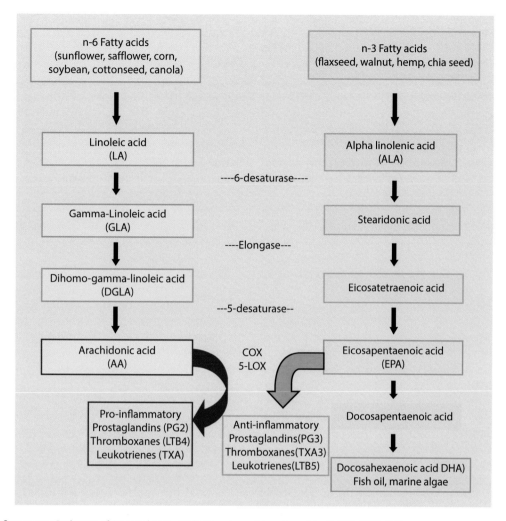

Fig. 28.4 Inflammatory Pathways of Essential Fatty Acids. (Courtesy of Jena Savadsky Griffith, RDN, IHC)

inflammatory diet, it is plant-heavy and features fruits, vegetables, whole grains, fish, less meat, legumes, nuts, olive oil, red wine and exercise. When more attention is given to a healthful eating plan, general health will improve, pain notwithstanding.

Based on the principle of the evolution discordance theory, genetic adaptation has not been able to adjust to our diet since the appearance of agriculture 10,000 years ago, and, more importantly, since the industrial revolution. Accordingly, food processing has altered these dietary indicators: (1) glycemic load, (2) fatty acid composition, (3) macronutrient composition, (4) micronutrient density, (5) acid-base balance, (6) sodium/potassium ratio, and (7) fiber content [38]. Considering we remain similar to Paleolithic ancestors, the principle of the *Paleo Autoimmune Protocol* (AIP) follows that their dietary pattern should still be relevant. Based on the work of Boyd Eaton, the paleo diet includes vegetables, fruits, nuts, roots, meat and organ meats and excludes dairy, grains, sugar, processed oils, alcohol and coffee [113]. Although open to interpretation, the AIP is generally an austere version of the diet, retaining only the most nutrient dense foods while excluding anything potentially inflammatory to even the smallest subset of people. In the age of information, research has yet to catch up with all who have successfully adopted AIP. While the scientific literature analyzing the efficacy of a paleo diet is small, it is unanimous [114]. In a 2017 study of 70 postmenopausal women who submitted to a paleo versus a prudent diet (low-fat, low-calorie, low-salt), both groups lost weight, but those on the paleo diet had greater reduction in inflammatory markers [115]. Type 2 diabetes patients following a paleo diet versus the diet suggested by the American Diabetes foundation (low fat dairy, whole grains, legumes) showed significantly greater glucose control and lipid profile [116]. Diagnosed with MS and confined to a wheelchair, Dr. Terry Wahls began researching brain and mitochondrial health, adopted a version of the AIP, added supplements, lifestyle changes and ended her chronic pain, fatigue and migraines. Her continued clinical trials repeating her process have been successful and have large implications for chronic pain conditions [117]. Critics of the diet suggest it has inadequate calcium supply and maintain that digestive abilities today are different than Paleolithic humans [118].

Vegan, *vegetarian* and *FODMAP* (fermentable oligo-, di-, monosaccharides and polyols) diets have also shown efficacy in chronic pain conditions [119, 120]. Fasting is practiced worldwide for cultural, religious and health reasons. There is ample empirical and observational evidence that medically supervised therapeutic fasting from 7 to 21 days confers benefits for rheumatic diseases and chronic pain syndrome [121]. After a 7-day juice fast in an integrative medicine clinic, 952 patients with various chronic pain syndromes significantly decreased their pain and improved quality of life [122]. Observational data and experimental research also support caloric restriction, intermittent fasting and a ketogenic diet [123]. Fasting mechanisms include increased production of neurotrophic factors, reduced oxidative mitochondrial stress, autophagy and neuroendocrine activation [124].

The nightshade vegetables and *oxalate*-containing foods may also be implicated in chronic pain. From the *Solanaceae* family, the nightshades include more than 2000 species, but most notably, tomatoes, potatoes, eggplant, peppers of all kinds and tobacco. Mostly linked with arthritis, the nightshades contain specific alkaloids—solanine in potato and eggplant, tomatine in tomato, nicotine in tobacco and capsaicin in garden peppers—that inhibits neurotransmitter enzyme, cholinesterase. When these alkaloids accumulate in the body, with help from other cholinesterase inhibitors such as caffeine, the result may be a paralytic-like muscle spasm, aches, pains, tenderness, inflammation and stiff body movements. Two surveys of arthritis patients adhering strictly to the "no nightshade" diet reported 72–94% had complete or substantial relief from the diet [125].

Foods high in *oxalate* may also cause inflammation and pain. Oxalate is found primarily in plant foods, with highest amounts in spinach, rhubarb, almonds and yams. Ordinarily, oxalate is routinely metabolized, minimally absorbed and excreted. However, compromised gut health causes oxalates to bind with calcium and form crystals that can lodge not only in the kidneys, but within body tissues causing pain. Excess oxalate also leads to oxidative damage and depletion of glutathione [126, 127]. The connection of pain and high oxalates is the basis for the *Pain Project* where participants have gotten relief from fibromyalgia, vulvodynia, interstitial cystitis and pelvic floor dysfunction after following a low oxalate diet [128]. While restoring gut health is of primary concern, experimenting with the elimination of these foods may be a next step if other therapeutic diets are not yielding expected results.

28.1.1.17 Nutrients and Supplementation in Pain Management

Correcting deficiencies and alleviating inflammation are at the core of most evidence-based supplementation for chronic pain. Supplementation of high doses is often used, but importantly, use of supplements has significant advantages and is universally preferred as the side effects are drastically reduced versus pain pharmacotherapy. Changing little over the last 40 years, micronutrient recommended dietary allowances (RDAs) are population statistics, represent rough estimates and are, to some extent, arbitrary figures. Therefore, the optimum level may be on a continuum well above or even below the RDA [129] (See Table 28.1).

For their role in scavenging reactive oxidant species and suppressing inflammation, antioxidant power is the backbone of a healthy diet, for those with pain and without. Further, concentrated nutrients that complement a healthy antioxidant, fiber-rich diet can increase efficacy of dietary nutrition intervention and further mitigate risks of medications. Due to the nature of scientific research and the study of nutrients in isolation, several supplements have been found to dramatically decrease the intensity of pain and increase function. With any neuropathy, it is recommended that clinicians check for "nutritional neuropathies" or those due to deficiency of specific nutrients, including thiamine, vitamin B12, vitamin E, vitamin B6, niacin, and copper and following bariatric surgery [130].

Bio-individuality is foundational for supplementation. Sufficiency for one patient may be deficiency or toxicity for another. Patient involvement, testing, tracking, individual preferences, current state of microbiota, absorption, excretion, skin color, age, other medical conditions, and medications all factor in to finding the optimal diet and nutrients in order to move toward reducing pain and restoring function.

28.1.1.18 Omega-3 Essential Fatty Acids (See Fig. 28.4)

Due to their established role in inflammatory pathways, and therefore pain control, long chain omega-3 (n-3) polyunsaturated fatty acids (PUFAs) are popular in pain management and one of the most effective natural anti-inflammatory agents available. Higher levels of n-3s that include DHA, EPA, and vegetable-sourced ALA are associated with lower levels of inflammatory mediators, anti-nociception, and greater emotional/cognitive functioning. Competing omega-6 (n-6) PUFAs are associated with inflammation, nociception and greater psychological distress [131]. Additionally, when EPA and DHA are metabolized by cyclooxygenase and lipoxygenase, they are converted into powerful resolvins and protectins that cause inflammatory recovery and suppression. Although both are essential, ratios of n-6:n-3 have increased substantially over and above the recommended 3:1 to about 20:1 or higher, causing greater inflammation within cell membranes and a biochemical susceptibility to excitotoxicity, cell damage, central sensitivity and chronic pain [132]. Omega-3s alleviated chronic nonsurgical back and neck pain after taking 1200–2400 mg for 8 weeks, with 59% discontinuing their use of prescription NSAIDs [133]. Similar results were found for RA with 2.7 g/day for 3 months [134], and after a dietary intervention of high n-3s and, importantly, a low intake of n-6s, chronic headache sufferers reduced headache pain, altered antinociceptive lipid mediators, and improved quality of life [135]. Although the mechanism is hypothesized, animal studies have found n-3s to be protective and alleviate neuropathic pain, post-traumatic brain injury, stroke, ALS, diabetes, lupus and other neuroinflammatory conditions [136]. Five

Table 28.1 Summary of dietary supplements used in the treatment of chronic pain

Nutrient/supplement/functional food	Pain states	Known action	Suggested dosage	Considerations
Omega-3	All	Anti-inflammatory	2.7 g/day EPA and DHA	Anticoagulant properties
Omega-6, borage/EPO		Anti-inflammatory, anti-thrombotic	500–1800 mg/day	
Cod liver oil (A,D,omega-3)	All	Immunity, inflammation, bone health	1 tsp/day	
Vitamin D	All	Immunity, inflammation, bone health	Level dependent 1000–2000 iu/day	Personalization, with A, K2, magnesium
Magnesium	All	Inflammatory mediator, muscle function, energy production, immunity	400 mg/day or more	Cofactors B6 and boron
Vitamin B1/thiamine	FM, alcoholic neuropathy	Mitochondrial function	600–1800 mg/day	
Vitamin B2/riboflavin	Headaches	Mitochondrial function	400 mg/day	
Vitamin B12	Back, neck, neuropathy, neuralgia	Neuroprotection, regeneration	1000 mcg/day	Cofactor folate
Bee pollen		Anti-inflammatory, antioxidant, rich nutrient profile including B complex	½–1 tsp.	Begin slowly to ascertain tolerance/allergy
Quercetin	RA, inflammatory pain, allergy headache, gout, spinal cord injury	Antioxidant, anti-inflammatory, analgesic, antihistamine	500–1000 mg/day	Often formulated with Bromelain, Contraindicated for kidney disease
Turmeric	Arthritis, JP, FM, IBS, DN, PMS, gout	Anti-oxidant, anti-inflammatory, analgesic, anti-arthritic	Average of 400–600 mg/3×/day	Anticoagulant properties
Ginger	RA, OA, migraine, JP, dysmenorrhea	Anti-inflammatory	200–500 mg/day	
Boswellia	RA, OA, dysmenorrhea	Anti-inflammatory, analgesic	300–500 mg/2–3×/day	
Willowbark	Headache, back, myalgia, RA, OA, dysmenorrhea, gout	Anti-inflammatory, analgesic	240 mg	Not for use with children, peptic ulcers or any condition where aspirin is contraindicated
Butterbur	Migraine/headache	Anti-inflammatory, antispasmodic	50–150 mg/day, age dependent	Extracted to remove hepatotoxic alkaloids
Feverfew	Migraines, RA, dysmenorrhea	Anti-inflammatory, inhibit platelet aggregation	6.25 mg 3×/day. (CO2 extracted)	Not with anticoagulants or during pregnancy

FM Fibromyalgia, *JP* Joint pain, *RA* Rheumatoid Arthritis, *OA* Osteoarthritis

patients with chronic pain resulting from burns, spinal stenosis, carpal tunnel syndrome, fibromyalgia, and cervical disc herniation were given 2400–7200 mg of EPA/DHA for at least 7 months, with all reporting significant pain reduction [137]. Fish oil also appears to attenuate the negative cellular and behavioral effects of opioids [138]. A lower dietary intake of omega-6 fatty acids, however, will necessitate a lower need for omega-3s. Quality sourcing of all essential fatty acids and understanding the potential for interaction with other anticoagulants (herbal and pharmaceutical) is important in clinical application.

Discovered in 2008, unsaturated fatty acid palmitoleic acid, or Omega-7, has also been found to have potential anti-inflammatory effects on metabolic diseases and muscle pain.

Potentially controlled by mTOR (mechanistic target of rapamycin) signaling, it is found in macadamia nuts, sea buckthorn oil, olive oil and in certain fish, such as salmon and anchovy [139].

28.1.1.19 Omega 6 Fatty Acids

While omega-3s understandably receive all of the anti-inflammatory notoriety, omega-6 fatty acids are in fact, essential and offer their own, often overshadowed, benefits for chronic pain and inflammation. Gamma linolenic acid (GLA) specifically, has proven anti-inflammatory benefits. Found in plentiful amounts in our food supply, linoleic acid (LA) from seed oils (sunflower, soy, corn) is the starting point for the synthesis of omega 6, mirroring the processes of omega 3, using the same or similar enzymes to form arachidonic acid:

LA → (GLA) → dihomo-gamma-linolenic acid (DGLA) → arachidonic acid (AA) (See Fig. 28.4).

DGLA is anti-inflammatory and anti-thrombotic and counteracts the inflammatory actions of AA. However, DGLA must be synthesized in the pathway from LA and GLA with all enzymes functioning, specifically, delta-6-desaturase, which naturally decreases with aging. If the conversion process is compromised, AA, in oversupply from the diet, flourishes, creating greater inflammation. Regarding chronic pain states, this reduced capacity to create DGLA has been associated with rheumatoid arthritis, diabetes and associated neuropathy, dermatitis/eczema, premenstrual syndrome, obesity, cancer and cardiovascular disease.

28.1.1.20 Vitamin D/Vitamin A

Extensive research concludes a worldwide deficiency of vitamin D, from 40% to 75% of the population, greater in the northern hemisphere and in those with pain. Known for its most direct role in osteomineral metabolism, vitamin D is more recently known for its contribution to immunity strength, cell growth, neuromuscular function and inhibition of inflammation [140, 141]. With vitamin D receptors in virtually every cell, it may be responsible for regulating more than 200 genes and has potential influence on multiple biologic systems of the body. It is also considered a neuroactive steroid that may modulate neuronal excitability. Some research shows indirect support for a role for vitamin D in nonspecific musculoskeletal pain, central sensitization (migraines and fibromyalgia), abdominal pain, knee pain, back pain and various pain comorbidities [142]. Although several studies have shown vitamin D supplementation to improve sleep, mood, pain levels and well-being, there is inconclusive evidence to implicate vitamin D deficiency in the etiology or maintenance of chronic pain or that supplementation alone will confer alleviation [142]. The most efficient source of vitamin D is the sun. Small amounts of vitamin D are found in several foods, including fatty fish, egg yolks, yogurt and mushrooms. The highest amount of food-sourced vitamin D is in cod liver oil at 450 iu/tsp and is often used, as it has a complementary amount of vitamin A. Pastured butter or ghee are additional sources of K [143]. If supplementation is used, vitamin D3 is the preferred form.

Testing should be done every 3 months. Vitamin D status is most often measured by storage 25(OH)D in the blood. The US lab reference is 30–74 ng/mL, while the Vitamin D Council recommends 40–80 ng/mL. However, based on a review of more than 1000 studies, the IOM recommends a more conservative range of 20–50 ng/mL, citing little to no benefit over 50 and increasing evidence that high levels cause harm. According to these definitions, the entire population may be deficient, and in fact, low OH (25) levels are found in both sick and healthy individuals. Therefore, a re-evaluation of this method of measurement is needed, as it does not consider the active form of vitamin D, the relationship between storage and active D, metabolic factors affecting its conversion (inflammation, infection, current mineral and nutritional status), genetics and other matters of bio-individuality.

In order for a nutrient, vitamin or mineral to become metabolically active, it often needs cofactors [143]. The metabolism of storage or pre-formed D, 25(OH)D (calcidiol) at every stage—in the skin, liver and kidney—into its active form 1,25(OH)2 (calcitriol) requires sufficient magnesium. In a state of hypomagnesemia, calcium is usually high. As Vitamin D absorbs intestinal calcium for use, with ample calcium, vitamin D may remain low. While magnesium has its own cofactors, attention should also be given to sodium and potassium to balance electrolytes. Further, vitamin A is D's biological antagonist. For vitamin D to be effective, sufficient, fat-soluble, vitamin A is needed; thus, vitamin A and vitamin D protect against the other's toxicity [144].

Recent studies also suggest that A and D also need vitamin K2, found in animal fats and fermented foods. Vitamin K2 ensures that calcium is appropriately transported to bones and teeth rather than to the soft tissues, contributing to disease. The many roles of vitamin D in the maintenance of health is a subject of continuing research. Whether low vitamin D is a physiological response or cause of disease has yet to be absolutely confirmed. In an integrative and functional setting, it is important to investigate the potential root cause before artificially raising vitamin D levels with supplementation.

28.1.1.21 Magnesium

Magnesium plays an essential role in hundreds of physiological processes and is critical to the electrical stability of cells. As it relates to pain, magnesium is instrumental in blocking the NMDA receptor, whose excessive stimulation is the primary mechanism of central sensitization. Low magnesium also enables excess substance P response, which triggers the entire inflammatory cascade [145]. Magnesium deficiency should be considered a public health crisis with some experts estimating current magnesium deficiency of the population as high as 80%. It quietly contributes to the process of inflammation and pain, therefore implicating all chronic disease and pain comorbidities including insulin resistance, obesity, depression, anxiety and insomnia. Magnesium is central in energy production, bone health, and muscle relaxation with deficiencies leading to cramping, spasms, myofascial tightness, and tightness in the smooth muscles of the GI tract.

Supplementation significantly alleviated pain in patients with migraines, fibromyalgia, neuropathic pain, chronic low back pain, IBS, diabetic neuropathy and postherpetic neuralgia [146–151]. Correcting magnesium deficiencies can potentially correct both the symptom and cause of chronic pain. With most magnesium stored in bone and soft tissue (99%) and the remaining under tight control, testing serum magnesium is inefficient and deficiency often goes unnoticed. A preferred test would be a Mag RBC test (magnesium red blood cell). Although not without its limitations, assessing RBC magnesium inside the cell as opposed to serum outside the cell will provide an earlier indicator of magnesium deficiency. An increase in refined foods, greater environmental and emotional stress, inflammation, infection, excess calcium, excess vitamin D, depleted soil mineralization, and drug interactions account for many of the contributing factors for deficiency. Cofactors include vitamin B6 and boron for absorption and proper utilization. Magnesium needs and levels fluctuate, depending on absorption, stress levels and myriad other factors with many deficiency symptoms included in a pain sufferer's profile: anxiety, insomnia, depression, fatigue, constipation, hypertension, muscle cramping, headaches, etc. As magnesium is a critical, water-soluble electrolyte needed daily, attention should also be given to the balance of sodium, chloride and potassium. Including mineral-rich sea salt and food sources of potassium (virtually all fruits and vegetables) is required to keep minerals regulated and/or from one mineral depleting the other. Dietary sources of magnesium include greens—one of the most missing foods in the modern diet—as well as nuts, seeds, and whole grains. Supplementation in the form of glycinate, malate, taurate, or threonate, 200–400 mg or higher, may be best for chronic pain sufferers; however, magnesium form may need to be personalized and started slowly. No less important is magnesium in oil form applied topically and Epsom salt baths.

28.1.1.22 B Vitamins

Essential for various metabolic processes, several B vitamin deficiencies are implicated in chronic pain. Vitamin B6 is a cofactor for the enzyme reaction that converts excitatory neurotransmitter glutamate into inhibitory neurotransmitter, GABA. Without sufficient B6, there is more glutamate and less GABA, which fuels excitotoxicity in the CNS [145]. Without sufficient B6, blood homocysteine levels increase, contributing to inflammation and potential toxic effects on CNS neurons. Elevated homocysteine is one potential mechanism for migraines [152]. Neuropathy due to B6 deficiency begins with numbness, paresthesias or burning pain in the feet which then ascends to affect the legs and eventually the hands. Although deficiencies are rare in the general population, chronic pain patients with inflammation and higher intake of NSAIDs will have disrupted B6 metabolism. Excess B6 can also cause neuropathy [130].

Fibromyalgia may have symptoms in common with thiamine deficiency; high doses of thiamine, 600–1800 mg/day, produced significant decreases in pain [153]. Alcohol intake should be assessed, as neuropathy can be the result of depleted thiamine. Riboflavin, vitamin B2, addresses the potential cause of migraines as aberrant mitochondrial metabolism. Daily use of 400 mg of B2 for 3 months resulted in a 50% reduction in headaches for the majority of patients [154].

Due to its integral role in neurological function and myelin integrity, vitamin B12 (methylcobalamin or hydroxycobalamin or adenosylcobalamin,) has been used for decades to treat pain. In other countries it has been labeled as an analgesic drug. Through supposed neuroprotective and regenerative actions, B12 administered orally and via injection has been successfully used to decrease pain in chronic low back pain, neck pain, diabetic or chemo-induced neuropathy and various forms of neuralgia [155]. Pain may be alleviated even with adequate levels of B12. Vitamin B12 supplementation of 1000 mcg, combined with a plant-based diet, improved pain in diabetic neuropathy patients compared to just supplementation alone [156].

A deficiency of B12 is linked to inadequate folate. Proper functioning of the folate cycle ensures formation of neurotransmitters, the pain signal messengers. As a critical cofactor for B12, inadequate folate can mask B12 deficiency; therefore, giving isolated folate without B12 is contraindicated. A sublingual dose of B12 at 1000 mcg is helpful. B12 may improve pain, insomnia and fatigue [157]. Assuming no genetic polymorphisms (MTHFR), taking a full range of B vitamins in combination may also be beneficial than one in isolation. Vitamins B1, B6 and B12 in combination with other medical therapy alleviated pain and increased mobility in those with back pain [158].

As the body does not store water-soluble B vitamins well, the need for them may be increased for the chronic pain population. Liver, poultry, shellfish and eggs have high amounts of B vitamins, especially B12. Plant sources include whole grains, potatoes, lentils and beans. Leafy greens are high in folate. An overlooked source of B-complex vitamins, including B12, is bee pollen, one of the richest sources of vitamins found in a single food in nature. If folate supplementation is used, L-methylfolate or folinic acid is preferred.

Other nutrients studied for their effect in the management of pain include coenzyme Q10, 5-HTP, glucosamine and chondroitin, MSM, alpha lipoic acid, melatonin and resveratrol.

28.1.1.23 Essential Nutrients

In a larger context, everything taken in, from food to oxygen to art, is essentially a nutrient, with its own specific energy. Experiences of life are absorbed in many ways that have physical, emotional and spiritual effects. Nutrients can be put on a continuum from inflammatory to nourishment, depending on individual, context and other known and unknown variables. Other overlooked and underestimated nutrients that increase pleasure, comfort, and joy while decreasing pain that are evidence-based and empirically substantiated include light, nature, music ("robustly relieves pain"), love, touch, community, dancing and laughter, hugging yourself and others, gratitude, and spiritual practice [159–167].

28.1.1.24 Hydration

Thirst is our own odometer, a signal from the hypothalamus that our blood volume is too high. However, for many, the sense of thirst becomes dulled. It is especially an issue within the vulnerable pain populations of the elderly and children. Among children and adolescents, water accounted for only 33% of fluid intake, with the remaining coming from high-calorie beverages [168]. As an underappreciated disease etiology, dehydration causes decreases in cognition and GI function, headaches, muscle and joint pain. Recently, even mild dehydration was found to increase pain sensitivity and pain perception [169]. Water provides lubrication for tissues, cartilage, discs and joints. Without water, there is slowed nutrient transportation, a buildup of metabolic waste and increased inflammation. Increased water intake reduced gout attacks by almost 50% [170]. There are several recommendations for water intake including the standard six to eight glasses or half the individual body weight in ounces, with needs fluctuating depending on activity, season, temperature, and energy intake.

In addition, many of the fluids commonly consumed today are diuretics—soft drinks, alcohol, tea and coffee. The most widely consumed legal drug in the world, coffee, is a CNS stimulant that is more of a cultural habit, than a hot beverage. Coffee as a recommendation for pain relief is controversial. Although the mechanism is unknown, caffeine is said to potentially reduce pain through its effect on adenosine receptors [171]. Conversely, caffeine triggers a stress response and causes various alterations in brain chemistry including neurotransmitter imbalances (serotonin, dopamine, GABA) that lead to irritability, mood issues, depression, headaches, insomnia and fatigue. It decreases melatonin [172] and can affect nutritional status by increasing insulin resistance, depleting vitamin B6 and inhibiting absorption of magnesium, B vitamins calcium, iron and vitamin D [173]. Although it is considered an antioxidant, there are individual sensitivities and varied genetic traits related to caffeine metabolism. Considering its range of parallel effects in the chronic pain population (sleep, anxiety, glucose control, dehydration), assessing function and pain without coffee/caffeine would be recommended.

28.1.2 Plant Compounds

28.1.2.1 Turmeric (*Curcuma longa*)

A long history of use in Ayurvedic medicine and extensive research over the past half century has confirmed that turmeric, with active ingredient *curcumin* (77%), has antioxidant, anti-plaque, anti-arthritic, anti-cancer and anti-inflammatory properties. In Sanskrit, turmeric is credited with 53 names, most likely due to its various mechanisms; however, research indicates that blocking the activation of NF-kB, the body's central inflammatory switch, along with other inflammatory mediators, may be its main pathway [63, 174]. Turmeric contains at least six different COX-2 inhibitors, the enzyme responsible for the inflammatory prostaglandin PG-E2, and may also deplete substance P. Studies indicate pain relief with turmeric in arthritis, diabetic neuropathy, gout, joint pain, IBS, PMS and fibromyalgia, in addition to positive influences on obesity, diabetes and depression [175, 176]. Curcumin was equal to and advantageous to ibuprofen in a 4 week study of OA of the knee [177]. The most comprehensive summary of turmeric benefits concluded that it outperforms many pharmaceuticals against chronic and debilitating diseases with virtually no adverse side effects [178]. Results have been found with as little as 200 mg/day or up to 8 g with an average dosage of 400–600 mg three times per day. In addition to capsule form, it can be added to food, taken as tea, and heated and made into a turmeric milk. Bioperine (black pepper) increases the bioavailability of curcumin by up to 2000% [179]. Turmeric's anticoagulant properties should be considered.

28.1.2.2 Ginger (*Zingiber officinale*)

Ginger is one of the most consumed dietary condiments and a member of the same plant family as turmeric (Zingiberaceae). It has been used alone and in combination with other herbs for thousands of years for its antioxidant anti-inflammatory properties. Ginger is effective in treating pain of RA, OA, migraine (in combination with feverfew), joint pain and dysmenorrhea. Historically used for OA and RA, ginger inhibits several inflammatory mediators, most notably enzymes 5-lipoxygenase and COX-2. Dosages range from 200 to 500 mg daily [180, 181].

28.1.2.3 Boswellia

Considered one of the most ancient and valued herbs in Ayurveda, gum resin is tapped from the boswellia tree then dried to extract the oil and solidify the resin. The resin has various terpenes and four major boswellic acids responsible for the inhibition of pro-inflammatory enzymes. The most potent is acetyl-11-keto-β-boswellic acid (AKBA) that specifically inhibits leukotrienes via 5-Lipoxygenase (5-LOX), a major inflammatory enzyme [182].

Boswellia has been used analgesically for RA, OA and dysmenorrhea. Pain from knee OA was reduced notably within 7 days, and there were significant improvements seen after 90 days of using boswellia standardized to contain 30% AKBA. Beyond its analgesic effects, patients had unexpected joint regeneration, assigning disease modification benefits [183]. Often taken with other herbs, a combination of boswellia and curcumin proved superior in effectiveness and tolerability compared to diclofenac for OA [63]. A later OA study proved similar results versus 200 mg celecoxib (celebrex) with a combination capsule of 700 mg curcumin and 300 mg boswellia extract. *Boswellia* products vary, but it is typically given as an extract standardized to contain 30–40% boswellic acids, 300–500 mg two or three times/day [63].

28.1.2.4 Willowbark

Dating back to the time of Hippocrates, patients were advised to chew on willow bark for pain and fever [184]. It is often called "nature's aspirin" because the active ingredient of willow bark is salicin, used to develop aspirin in the 1800s. Its analgesic and anti-inflammatory properties, however, come from salicin's combination with flavonoids and polyphenols within the bark. Although both are nonselective COX-1 and COX-2 inhibitors, its multicomponent synergy gives willow bark a broader mechanism of action than aspirin, without the GI damage. People have used the herb for headache, low back pain, myalgia, OA, RA, dysmenorrhea and gout. Patients with low back pain receiving 240 mg of willow bark had more significant relief than those receiving 120 mg or placebo [185]. Willow bark should not be used in children, in those with peptic ulcers or any other condition where aspirin is contraindicated [63].

28.1.2.5 Butterbur and Feverfew

Headaches are one of the most common types of chronic pain reported by Americans. After magnesium, both butterbur and feverfew, respectively, are commonly used in headache treatment and prevention, separately and in combination. Butterbur, *Petasites hybridus* root extract, is thought to inhibit inflammatory leukotriene production and have a potential antispasmodic effect on cerebral blood vessel walls [186]. Migraine frequency was reduced by 48% in those who took 75 mg of petasites BID [187]. Almost 80% of children and adolescents reported a reduction in migraine frequency of 50% by taking 50–150 mg/day, depending on age, for 4 months [188]. Butterbur has hepatotoxic alkaloids so the extract must be manufactured carefully to remove them.

Called medieval aspirin, the parthenolides within the leaves of feverfew are thought to be its active ingredient and mechanistic driver. These compounds may relieve migraines, RA and menstrual pain by inhibiting inflammation and platelet aggregation, which normalizes blood flow. Feverfew contains many medicinally active compounds including melatonin and an essential oil, chrysanthenyl acetate. Some evidence shows lack of efficacy which is thought to be due to varying plants and bioavailability, extracts used and dosage [186, 189]. Feverfew is not recommended during pregnancy or with anticoagulants. If taking for more than 1 week, stopping abruptly could produce rebound headaches and/or joint pain. Studies have used 100–325 mg up to four times daily for adults, standardized to contain 0.2–0.4% parthenolides. For feverfew supplements that are carbon dioxide extracted, 6.25 mg, 3 times daily, for up to 16 weeks was successfully used [189].

28.1.2.6 Cannabinoids

Despite political, legal and social issues surrounding its use and history, the addition of cannabinoids to the clinician's toolbox offers another approach and is evidential in chronic pain. Further, surveys report that due to lack of education and resources about cannabinoids, clinicians often do not bring up the topic [190]. Cannabis or medical marijuana is one of the most pharmacologically active plants known, with more than 400 chemical compounds, 85 of which are cannabinoids. The main cannabinoids by volume and notoriety are delta-9-tetrahydrocannabinol (THC), its primary psychoactive compound and cannabidiol (CBD), the non-psychoactive counterpoint [191]. Among its many properties, THC is a potent antioxidant, neuroprotective, anti-inflammatory and painkiller. Influencing many pathways, THC inhibits glutamate release, a central mechanism in most chronic pain conditions. CBD dampens THC's psychoactivity and elevates the plant's potency with its own complex anti-inflammatory and analgesic regulating effects. Possessing 20 times the antioxidant power of vitamin C and E, CBD notably inhibits TNFα. The study of cannabinoids led to the discovery of the endocannabinoid system, a group of endogenous cannabinoid receptors, CB1 and CB2, throughout the CNS and PNS that regulate homeostasis and have a significant role in the regulation of pain. Much like the bicarbonate buffer system regulates blood pH, the endocannabinoid system reacts to the environment and mediates response to mood, sleep, hormone regulation, pain and immunity [191, 192]. Notably, dietary omega-3 fatty acids reverse the dysregulation of the endocannabinoid system, improve insulin sensitivity and control body fat [66].

Current pain conditions may be a potential endocannabinoid deficiency. Although current research is hindered by its schedule one status by the US Drug Enforcement Administration, studies confirm that cannabis alleviates chronic pain conditions, neuropathic pain, headache, and untreatable pain, often with relief from stress/anxiety and insomnia, with minimal, if any side effects [193, 194]. Adverse effects are more often reported with synthetic versions [195]. Further, in 2014, the American Academy of Neurology provided a level A effectiveness to oral cannabis extract for spasticity related symptoms and pain [196]. Cannabinoids are delivered via smoking, vaporizers, sprays, tinctures and edibles. Medical marijuana is currently legal in 33 states and Washington, D.C., with only a few states approving treatment specifically for chronic, severe or intractable pain [197].

Interest in medicinal potential of CBD has catalyzed a rebirth of hemp in the USA and CBD hemp oil. Often confused, hemp and cannabis come from the same species, *Cannabis sativa*; however, hemp contains only trace amounts of THC. In order to be legal, it must have a THC content below 0.03%. There are varying hemp laws by state and much is imported [192, 198]. Securing its non-psychotropic status, it is considered a less controversial alternative while still maintaining the plant's benefits. Animal studies show CBD-induced suppression of chronic pain without tolerance. Researchers suggest these non-psychoactive components of cannabis may represent a novel class of therapeutic agents for the treatment of chronic pain [199]. Usual methods of delivery for CBD hemp oil include tincture and capsule form. Standards range; therefore attention must be given to a product's cultivation, harvest, extraction, exact constituents, testing, and manufacturing.

28.1.2.7 Medical Treatment and Nutritional Implications

A thorough physical, functional and psychological examination with detailed pain, medical and psychosocial history is vitally important to guide diagnosis and treatment. Additional testing is usually done to categorize the pain, determine its etiology and consider emotional and environmental factors. There is not one diagnostic test for chronic pain and, although helpful in certain cases, imaging does not always reveal a specific pathology or problem. Despite pain as one of the most universal symptoms in medicine, the assessment of a pain patient is challenging and takes time not often available. As pain is a subjective experience unable to be viewed by others, the evaluation of pain characteristics relies on self-reporting. Its complexity has further led to varying pain measurement scales and questionnaires that attempt to capture the patient, the pain and contributing and relieving factors. There are one-dimensional tools that that measure pain intensity, as in the Verbal Rating Scale (VRS), Numerical Rating Scale, Visual Analogue Scale (VAS) and Faces Pain Scale, which is helpful with children. More complex tools, such as the McGill Pain Questionnaire, distinguish nociceptive from neuropathic pain and attempt to decipher the pain dimension: sensory, affective and/or evaluative. Additional tools exist to assess function, substance abuse, opioid risk, emotional distress, potential for catastrophizing and pain-related fear and depression [200–203].

As complete analgesia is rarely achievable, the goal of medical treatment is to improve pain, and more importantly to optimize physical function and coping. One in three patients will get greater than 50% pain relief, which is regarded as an excellent result in chronic pain [200]. Pain medications fall into three general categories:

1. *Non-opioids*: aspirin, non-steroidal anti-inflammatories (NSAIDs), and acetaminophen (paracetamol).
2. *Opioids*: morphine, codeine, hydrocodone, oxycodone, methadone, fentanyl, etc. Although not true opioids biochemically, Tramadol and tapentadol are included as they work on the same receptors.
3. *Non-opioid or adjuvant analgesics*: Medications that may alleviate pain, although primary use is for another indication: antidepressants, anticonvulsants and antiarrhythmics, corticosteroids, muscle relaxants, local anesthetics, topical medications, adrenergic agonists, sympatholytics, and neuroleptics.

The use of NSAIDs remains a first line of defense for most classic physicians for mild to moderate joint, spine-related pain and headaches. Besides their well-known inhibition of prostaglandin synthesis via both COX-1 and COX-2 enzymes, their mechanism of action in producing analgesia is multifactorial, exhibiting both central and peripheral effects [63]. By blocking COX-1, which also normally acts to protect the gastrointestinal mucosa, nonselective NSAIDs and aspirin can cause significant GI and other tissue damage including gastritis, abdominal pain, ulceration, edema, hemorrhage, renal impairment, hypertension and death. This was the stimulus for the creation of such selective COX-2 NSAIDs as celecoxib (celebrex). However, more recently, all NSAIDs and in particular, COX-2, were found to increase risks for heart attack and stroke [204]. After 14 days, Ibuprofen was found to reduce testosterone production in males with implications for fertility, muscle mass and strength, fat distribution and mental health. Recommendations are to prescribe lowest possible doses for shortest duration. To mitigate GI damage from NSAIDs, many patients are given proton pump inhibitors (PPIs). Usually given for gastrointestinal reflux (GERD), PPIs are among the top ten most widely used drugs in the world with many patients remaining on them for long periods of time, without evidence-based indication [205]. However, these medications do not offer protection to the lower small intestine, but exacerbate NSAID-induced small intestinal lesions and increase the presence of pathogenic bacteria in the gut microbiome. The initial microbiotic insult from NSAID use decreases beneficial *lactobacilli* and impairs intestinal pH, permeability, mucosal production and immunity [206]. Following PPI treatment, there is an additional loss of beneficial and protective bacteria *Bifidobacterium* and *Lactobacillus*. Further, PPIs are associated with a deficiency of vitamin B12, iron, vitamin C, calcium, and magnesium [14, 204–207].

Due to the gastrotoxicity of NSAIDs, and PPIs when applicable, acetaminophen is recommended by the American College of Rheumatology for mild-to-moderate pain as first line therapy for OA of the knee and hip and lower back pain [208]. For its safety profile over NSAIDs, acetaminophen is also recommended for the elderly population [209]. Although too mild for most for most chronic pain patients and lacking in anti-inflammatory properties, it may have a sparing effect when taken with NSAIDs and opioids. Care should be taken regarding the potential hepatotoxicity of acetaminophen and its interaction with high doses of vitamin C [210].

Opiates are the mainstay of pain management for severe chronic pain. While opiates stem specifically from the alkaloids of the poppy plant (heroin, morphine, codeine), opioids are synthetic drugs that produce opiate-like effects and include oxycodone, Percocet, Demerol and methadone. Clinically, the term "opioid" now encompasses all natural and synthetic versions that act upon opioid receptors in the CNS, PNS, and immune system. The opioid system consists of three G-protein coupled receptors, Mu, Delta and Kappa, usually activated by endogenous opioid peptides like endorphins. Individual drugs interact with each receptor, accounting for the differences in analgesia and various side effects; however, the majority of activity occurs on the Mu receptor. Opioids are often called Mu agonists. The net effect of Mu receptor activation is inhibition of pain signal transmission to the brain and a decrease in the perception and experience of pain [11, 211].

Opioid receptors are also involved in other functions of the body, including modulation of reinforcement and reward mechanisms, which can sometimes drive the addictive pattern of use. Results of activation also include the most common side effects of constipation, respiratory depression, itching, sedation and nausea. With rates ranging from 15% to 90%, mitigating opioid-induced constipation is an important aspect of the management of a patient's quality of life. Increasing fiber, fluids, and activity is recommended, with fiber intake occurring several hours before or after taking the drug so as not to interfere with bioavailability [11].

Other side effects can be altered mood and cognitive function, dry mouth, urinary retention, sexual dysfunction, sleep disturbances, dizziness, hypotension and myoclonus (muscle twitching) [11]. Due to adverse side effects, approximately 20% discontinue therapy [212]. Often unrecognized, long-term opioid use also causes hormone suppression. Opioids action on receptor sites can lead to inhibitory or stimulatory effects on hormone release. Testosterone appears to be most often affected, leading to decreased libido and erectile dysfunction in men, oligomenorrhea or amenorrhea in women, and bone loss or infertility in both sexes [213].

Frequently misunderstood, tolerance and dependence are normal physiological responses to opioids. A more recently recognized opioid toxicity is induced hyperalgesia (OIH), where increasing the dose of the opioid exacerbates the pain [214].

While accepted for cancer-related pain, the use of opioids for chronic pain has been controversial due to concerns about side effects, long-term efficacy, functional outcomes, and the potential for drug abuse and addiction. These concerns have often led to the undertreatment of pain patients. Heeding criticism, physicians began prescribing more pain medications in the 1990s with the support from many professional pain societies. In response to the growing prevalence of chronic pain and its disabling effects, opioid prescriptions have increased to the point that they are the most commonly prescribed class of medications in the USA, with pharmacies dispensing 245 million prescriptions in 2014. The rate of death from opioid overdose has quadrupled in the past 15 years and nonfatal overdoses needing hospital care increased six times [215, 216]. While many opioids are diverted and improperly used, at the core of what is now called an opioid crisis is an increasing cohort of those in chronic pain, overreliance on a known analgesic, and limited alternatives for prescribing physicians [11, 217].

When pain cannot be controlled with medications and central sensitization cannot be prevented, there are various interventional pain procedures that can be used. These include nerve blocks, injections, ablative procedures, implants and neuromodulation.

28.1.2.8 Genetics

Genetics also has a role in the biopsychosocial matrix of an individual's perception and response to pain. Heritability can be explored across pain disease states or by the pain signaling process, adding another layer of complexity to its study. By examining pain behavior in female twins, genetic factors could account for 22–55% of the variance in chronic pain [218]. Although in early stages of research, various genes encoding for receptors are now known to play a role in pain sensitivity, reporting and susceptibility. Overall, the genetic contribution to chronic pain suggests small variations in a large number of single nucleotide polymorphisms (SNPs) that may represent different functional pathways [219, 220]. One of the most extensively studied "pain genes" is catechol-O-methyltransferase (COMT), an enzyme involved in the inactivation of dopamine, epinephrine, and norepinephrine (catecholamines) neurotransmission [221]. Sustained elevation of catecholamines, released from the adrenal glands in response to stress or severe pain, is associated with chronic pain conditions. Several single nucleotide polymorphisms (SNPs) of the gene that encodes COMT have been found to lower its activity, thereby increasing pain sensitivity. SNPs in the Mu opioid receptor gene, OPRM1, may contribute to individual differences in pain sensitivity. Further, mutations in OPRM1 and COMT combined may modulate opioid efficacy [222].

There are several genes that affect neurotransmitters. SNPs of the GCH1 gene (encoding GTP cyclohydrolase) has been shown to have a protective effect on lower back pain after surgery, less sensitivity to heat, ischemic and pressure pain and reduced incidence of low back pain. Polymorphisms of the serotonin transporter gene (*SLC6A4*), beta 2 adrenergic receptor ADRB2, and serotonin receptor 2A HTR2A are all associated with an increased risk for chronic widespread pain. Alterations in pain-related genes associated with ion channel function may also contribute to chronic pain susceptibility as found in sciatica, OA, phantom limb pain and discogenic disease. Called the neuropathic pain susceptibility gene, CACNG2, a protein involved in the transport of glutamatergic AMPA receptors, significantly affects susceptibility following nerve injury [221].

Although not related directly to pain signaling, genetic mutations in enzymes involved in the folate methylation pathway may be a factor in the pain population. Two common SNPs, C677T and A1298C have been identified in the MTHFR gene that codes for the methylenetetrahydrofolate reductase enzyme, also MTHFR. Mutations in this gene cause decreased activity of the enzyme, thereby disrupting methylation and blocking key detoxification pathways [220].

28.1.2.9 Epigenetics

In an experimental study of healthy adults, pain intensity ratings ranged from 0 to 100 (bottom and top of the scale) for an identical cold-water stimulus [222]. As experienced by each individual differently, the concepts of environment and bio-individuality are embedded within the definition of pain. Environmental toxins, medications, diet, and psychological stresses can alter epigenetic processes such as DNA methylation, histone acetylation, and RNA interference [223]. Epigenetic mechanisms modify DNA expression depending on information received from the environment. Early life experiences can positively or negatively leave their biological mark on the genome. Negative childhood experiences have been linked to mental health, drug addiction and obesity, all mediating factors in chronic pain and treatment [224]. There are also learned coping strategies and behaviors, learned reactions to pain and attention received as the result of pain at all ages. Familial patterns, lifestyle and diet are more of an epigenetic factor than genetic, and have a profound influence on susceptibility to disease. After nerve injury, more than 1000 genes may be activated or turned on with significant evidence pointing to epigenetic control from, at minimum, inflammation, immune response and opioid receptor function. What turns the acute pain into chronic is the biological terrain and environment it occurs within [225].

As an active mediator of inflammation, immunity and the microbiome, nutrition has significant epigenetic power. Nutrients can influence gene expression by directly inhibiting enzymes that catalyze DNA methylation or histone modifications or by altering the availability of substrates necessary for those enzyme reactions. Many of the components of an anti-inflammatory diet have positive epigenetic effects. Sulforaphanes like broccoli, Brussels sprouts and cabbage are known to increase histone acetylation. Butyrate, or butyric acid, is a short chain fatty acid found most prominently in butter, but also created by fermentation of dietary fiber in the intestines. It is a favored source of fuel for the colon, contributes to a healthy gut microbiome and increases histone acetylation [226]. Methyl donor nutrients include folate, vitamin B12, methionine, choline and betaine. Foods rich in folate include leafy green vegetables and citrus fruits. Sources of vitamin B12 are fish, meat and eggs. Choline is oxidized to form betaine, found in high concentration in beef liver, eggs and toasted wheat germ. The anti-inflammatory actions of vitamin D may be accredited to DNA methylation and histone modifications. Plant compounds curcumin and resveratrol, with known analgesic effects, have been found to positively alter epigenetic mechanisms [227].

28.1.2.10 Lifestyle

As discussed, psychological, social and environmental factors including activity, obesity, support, sleep and stress are foundational in the chronification of pain. Elevated BMI, past and current smoking and higher unemployment rate were associated with chronic pain in the 1958 British Cohort study. The study further concluded that chronic pain is a cause of unhealthy diet and lifestyles rather than a consequence [109]. Increased risk for chronic low back pain is also linked to weight gain and increased alcohol intake [228]. The musculoskeletal overload and systemic inflammation of obesity are significant contributors to the prevalence and severity of chronic pain. Exercise confers far-reaching, well-known beneficial effects including weight loss, lower inflammation and free radical reduction. Further, in the pain population, movement increases mental health, normalizes dysfunctional brain connectivity found in central sensitization, increases GABA production and reduces pain perception [229]. Fear of movement (FOM) has been increasingly recognized as a significant explanation for chronic pain development and is the basis for its recent characterization as a lifestyle disease. A cycle of fear, avoidance and catastrophizing leads to inactivity and physical deconditioning. In addition, study of the mesolimbic reward system reports that voluntary—as opposed to forced—exercise shows even greater analgesic effects [230].

Although the majority of low back pain (90%) resolves on its own in 3 months, it is often difficult to treat due to the psychosocial nature of its cause and is the most prevalent chronic pain condition [231]. With medical treatment often focusing on pain control instead of restoring function, modifiable lifestyle factors appear to make a difference and are studied often with this condition. Tai chi is an effective intervention regarding movement and posture in OA, low back pain, and fibromyalgia [232]. In a review of low back pain sufferers, yoga was found to have positive and significant effects on function and pain. Co-creating gradual exercise or movement intervention that is tailored to the interests, needs, abilities and fears of an individual will better prevent attrition and increase function, quality of life and recovery.

Contributing to its elusive treatment, low back pain is also linked to anxiety, depression, job dissatisfaction, poor body image and somatization. Lack of support, in relationships, family or work environments can lead to a perception of greater individual burden that can manifest physically. In addition to the physical demands of work, perceived lack of support from supervisors and lack of encouraging culture in the work unit are associated with an increased risk of intense low back pain and low back pain-related sick leaves [233].

Stress is linked to pain in multiple causative and consequential ways. In most chronic pain conditions, there is an ongoing reactivation of the stress response from pain or other stressors, that exhausts cortisol and the HPA axis, contributing to and/or causing pain and inflammation. Adding insult to the initial injury or event, the depletion of cortisol increases inflammation and creates vulnerability to pathogens, greater psychological stress and pain. Mind-body approaches that include progressive muscle relaxation, meditation, laughter, mindfulness-based approaches, hypnosis, guided imagery, yoga, biofeedback, and cognitive behavioral therapy have all shown efficacy in chronic pain [234].

The typical chronic pain patient has a heavy toxic burden that may include medications, decreased absorption, disturbed sleep and increased stress that compound and/or create conditions that take an acute injury and turn it into chronic pain. There are clear opportunities for clinicians to facilitate changes in lifestyle that have the potential to improve function, decrease inflammatory burden and create conditions for patients to flourish. As neuronal plasticity can cause central sensitization and chronic pain, diet and lifestyle changes can potentially decrease inflammation, desensitize the nervous system and reduce chronic pain.

28.1.2.11 Language of Pain

As mentioned, in the premiere definition of pain, it is described as a "negative experience." Further, all descriptive language for pain is in the context of weaponry and damage

[235]. Pain is viewed as something that must be killed, as it is a disability, a disorder and now a disease. If pain can only be negative, it can produce only negative effects and nothing positive may come out of it; the language, therefore, creates an inherent obstacle to recovery from pain conditions. As an integrative and functional medicine approach shifted its focus from treating disease to restoring health and wellness, for pain outcomes to change, the metaphors may need to change as well. If pain could be viewed as an invitation, a message to go deeper into its root cause, it may help patients look at pain not as something that must be destroyed, but as a potential doorway to physical, emotional and spiritual health.

References

1. Bonica JJ. History of pain concepts and pain therapy. Mt Sinai J Med. 1991;58(3):191–202. Available at: https://www.ncbi.nlm.nih.gov/pubmed/1875956. Accessed 12 May 2017.
2. Merskey H, Bogduk N. Classification of chronic pain: descriptions of chronic pain syndromes and definitions of pain terms. In: IASP task force on taxonomy. 2nd ed. Seattle: IASP Press; 1994. p. 209–14. Available at: https://www.iasp-pain.org/Taxonomy. Accessed 12 May 2017.
3. Lumley MA, Cohen JL, Borscz GS, et al. Pain and emotion: a biopsychosocial review of recent research. J Clin Psychol. 2011;67(9):942–68. https://doi.org/10.1002/jclp.20816.
4. Neilson S. Pain as metaphor: metaphor and medicine. Med Humanit. 2016;42(1):3–10. https://doi.org/10.1136/medhum-2015-010672. Accessed 13 May 2017.
5. Aronoff GM. What do we know about the pathophysiology of chronic pain? Implications for treatment considerations. Med Clin North Am. 2016;100:31–42. https://doi.org/10.1016/j.mcna.2015.08.004.
6. Institute of Medicine Report from the Committee on Advancing Pain Research, Care, and education: Relieving Pain in America, A Blueprint for Transforming Prevention, Care, Education and Research. The National Academies Press, 2011.
7. National Institute of Health. US Dept. of Health and Human Services. Pain Management Fact Sheet. Available at: https://report.nih.gov/nihfactsheets/ViewFactSheet.aspx?csid=57. Accessed 14 May 2017.
8. The American Academy of Pain Medicine. Facts and figures on pain. Available at: http://www.painmed.org/patientcenter/facts_on_pain.aspx#refer. Accessed 14 May 2017.
9. King S, Chambers CT, Huguet A. The epidemiology of chronic pain in children and adolescents revisited: a systematic review. Pain. 2011;152(12):2729–38. https://doi.org/10.1016/j.pain.2011.07.016.
10. Patel KV, Guralnik JM, Dansie EJ, Turk DC. Prevalence and impact of pain among older adults in the United States: findings from the 2011 National Health and Aging Trends Study. Pain. 2013;154(12):2649–57. Available at: https://www.ncbi.nlm.nih.gov/pubmed/24287107. Accessed 10 June 2017.
11. Rosenblum A, Marsch LA, Joseph H. Opioids and the treatment of chronic pain: controversies, current status and future directions. Exp Clin Psychopharmacol. 2008;16(5):405–16. https://doi.org/10.1037/a0013628. Available at: https://www.ncbi.nlm.nih.gov/pmc/articles/PMC2711509/. Accessed May 25, 2017.
12. Bonakdar RA, Sukkiennik AW. Integrative pain management. 1st ed. Oxford/New York: Oxford University Press; 2016.
13. Horrigan B, Lewis S, Abrams D, Pechura C. Integrative medicine in America: how integrative medicine is being practiced in clinical centers across the United States. Global Adv Health Med. 2012;1(3):18–94.
14. De Gregori M, Muscoli C, Schatman M, et al. Combining pain therapy with lifestyle: the role of personalized nutrition and nutritional supplements according to the SIMPAR feed your destiny approach. J Pain Res. 2016;9:1179–89. Available at: https://www.ncbi.nlm.nih.gov/pubmed/27994480. Accessed May 13, 2017.
15. Garland EL. Pain processing in the human nervous system: a selective review of nociceptive and biobehavioral pathways. Prim Care. 2012;39(3):561–71. https://doi.org/10.1016/j.pop.2012.06.013.
16. Benzon HT, Raja SN, Spencer LS, et al. Essentials of pain medicine. 3rd ed. Philadelphia, PA: Elsevier Saunders; 2011.
17. Latremoliere A, Woolf CJ. Central sensitization: a generator of pain hypersensitivity by central neural plasticity. J Pain. 2009;10(9):895–926. Available at: https://www.ncbi.nlm.nih.gov/pmc/articles/PMC2750819/. Accessed 25 May 2017.
18. Woolf CJ. Central sensitization: implications for the diagnosis and treatment of pain. Pain. 2011;152(3 Suppl):S2–15. https://doi.org/10.1016/j.pain.2010.09.030.
19. Clauw DJ. The development of treatments for pain. Advisory Committee Meeting Materials, Science Board of the Food and Drug Administration. Available at: https://www.fda.gov/downloads/AdvisoryCommittees/CommitteesMeetingMaterials/ScienceBoardtotheFoodandDrugAdministration/UCM489203.pdf. Accessed 26 May 2017.
20. May A. Chronic pain may change the structure of the brain. Pain. 2008;137(1):7–15.
21. Yunus MB. The prevalence of fibromyalgia in other chronic pain conditions. Pain Res Treat. 2012;2012(5):84573. Available at: https://www.ncbi.nlm.nih.gov/pmc/articles/PMC3236313/. Accessed 24 May 2017.
22. Bonavita V, De Simone R. The lesson of chronic migraine. Neurol Sci. 2015;36(Suppl 1):101–7. https://doi.org/10.1007/s10072-015-2175-4. Accessed May 24, 2017.
23. Hinwood M, Morandini J, et al. Evidence that microglia mediate the neurobiological effects of chronic psychological stress on the medial prefrontal cortex. Cereb Cortex. 2012;22(6):1442–54.
24. Gershwin ME, Hamilton ME. The pain management handbook: a concise guide to diagnosis and treatment. Totowa: Humana Press; 1998.
25. Davis B, Vanderah TW. A new paradigm for pain? J Fam Pract. 2016;65(598–600):602–5.
26. Schneiderhan J, Orizondo C. Chronic pain: how to approach these 3 common conditions. J Fam Pract. 2017;66(3):145-151–54-157.
27. Burger AJ, Lumley MA, Carty JN, et al. The effects of a novel psychological attribution and emotional awareness and expression therapy for chronic musculoskeletal pain: a preliminary, uncontrolled trial. J Psychosom Res. 2016;81:1–8.
28. Morone NE, Lynch CS, Greco CM, et al. The effects of mindfulness meditation on older adults with chronic pain: qualitative narrative analysis of diary entries. J Pain. 2008;9(9):841–8.
29. Simons L, Elman I, Borsook D. Psychological processing in chronic pain: a neural systems approach. Neurosci Biobehav Rev. 2014;39:61–78. https://doi.org/10.1016/j.neubiorev.2013.12.006.
30. Bair MJ, Wu J, Damush TM, et al. Association of depression and anxiety alone and in combination with chronic musculoskeletal pain in primary care patients. Psychosom Med. 2008;70(8):890–7.
31. Lépine JP, Briley MH. The epidemiology of pain in depression. Hum Psychopharmacol. 2004;19(Suppl 1):S3–7.
32. Finan PH, Goodin BR, Smith MT. The association of sleep and pain: an update and a path forward. J Pain. 2013;14(12):1539–52. https://doi.org/10.1016/j.pain.2013.08.007.
33. Field T, Diego M, Cullen C, et al. Fibromyalgia pain and substance P decrease and sleep improves after massage therapy. J Clin Rheumatol. 2002;8(2):72–6. https://www.ncbi.nlm.nih.gov/pubmed/8047572
34. American Cancer Society. Cancer Facts and Figures: Special Section: Cancer-related pain. Available at: https://www.cancer.org/content/dam/cancer-org/research/cancer-facts-and-statistics/annual-

cancer-facts-and-figures/2007/cancer-facts-and-figures-2007.pdf. Accessed 5 June 2017.
35. Lemay K, Wilson KG, Buenger U, et al. Fear of pain in patients with advanced cancer or in patients with chronic noncancer pain. Clin J Pain. 2011;27(2):116–24. https://doi.org/10.1097/AJP.0b013e3181f3f667.
36. International Association for the Study of Pain (IASP). Cancer Pain. Available at: https://www.iasp-pain.org/GlobalYear/CancerPain. Accessed 10 June 2017.
37. CDC, National Center for Chronic Disease Prevention and Health Promotion. The power of prevention: chronic disease. The public health challenge of the 21st century. 2009. Available at: https://www.cdc.gov/chronicdisease/pdf/2009-power-of-prevention.pdf. Accessed 26 May 2017.
38. Cordain L, Eaton SB, Sebastian A, et al. Origins and evolution of the Western diet: health implications for the 21st century. Am J Clin Nutr. 2005;81(2):341–54.
39. Okifuji A, Hare BD. The association between chronic pain and obesity. J Pain Res. 2015;8:399–408. Published online 2015 Jul 14. https://doi.org/10.2147/JPR.S55598.
40. Stone AA, Broderick JE. Obesity and pain are associated in the United States. Obesity (Silver Spring). 2012;20(7):1491–5.
41. Mork PJ. HUNT 1 and 2 study. Arthritis Care & Research. Eric Matteson, MD, chairman, department of rheumatology, Mayo Clinic, Rochester, Minn. Kyriakos A. Kirou, MD, rheumatologist, Hospital for Special Surgery, New York. National Fibromyalgia Association.
42. Anandacoomarasmy A, Fransen M, March L. Obesity and the musculoskeletal system. Curr Opin Rheumatol. 2009;9:88. Available at: https://www.ncbi.nlm.nih.gov/pubmed/19093327. Accessed 21 May 2017.
43. Messier SP, Gutekunst DJ, Davis C, et al. Weight loss reduces knee-joint loads in overweight and obese adults with knee osteoarthritis. Arthritis Rheum. 2005;52(7):2026–32. Available at: https://www.ncbi.nlm.nih.gov/pubmed/15986358. Accessed 26 May 2017.
44. Felson DT, Zhang Y, Anthony JM, et al. Weight loss reduces the risk for symptomatic knee osteoarthritis in women: the Framingham study. Ann Intern Med. 1992;116:535–9.
45. Senna MK, Sallam RA, Ashour HS, et al. Effect of weight reduction on the quality of life in obese patients with fibromyalgia syndrome: a randomized controlled trial. Clin Rheumatol. 2012;31(11):1591–7.
46. Roffey DM, Ashdown LC, Dornan HD, et al. Pilot evaluation of a multidisciplinary, medically supervised, nonsurgical weight loss program on the severity of low back pain in obese adults. Spine J. 2011;11(3):197–204.
47. Janke AE, Kozak AT. "The more pain I have, the more I want to eat": obesity in the context of chronic pain. Obesity (Silver Spring). 2012;20(10):2027–34.
48. Bousema EJ, Verbun JA, Seelen HA, et al. Disuse and physical deconditioning in the first year after the onset of back pain. Pain. 2007;130(3):279–86.
49. Geha P, Dearaujo I, Green B, et al. Decreased food pleasure and disrupted satiety signals in chronic low back pain. Pain. 2014;155(4):712–22. https://doi.org/10.1016/j.pain.2013.12.027.
50. Janke EA, Collins A, Kozak AT. Overview of the relationship between pain and obesity: what do we know? Where do we go next? J Rehabil Res Dev. 2007;44(2):245–62.
51. Hooper MM. Tending to the musculoskeletal problems of obesity. Cleve Clin J Med. 2006;73(9):839–45.
52. Yunus MB, Arlsan S, Aldag JC. Relationship between body mass index and fibromyalgia features. Scand J Rheumatol. 2002;31(1):27–31.
53. Sharif M, Shepstone L, Elson CJ, et al. Increased serum C reactive protein may reflect events that precede radiographic progression in osteoarthritis of the knee. Ann Rheum Dis. 2000;59:71–4.
54. Tam CS, Clement K, Baur LA, et al. Obesity and low grade inflammation: a pediatric perspective. Obes Rev. 2010;11:118–26.
55. Nicol A, Ferrante FM, Giron S, et al. The role of inflammation in the chronic pain paradigm. ASRA: Pain medicine meeting. 2015. Available at: https://www.asra.com/content/documents/736-the_role_of_central_inflammati.pdf. Accessed 28 May 2017.
56. Loggia ML, Chonde DB, Akeju O, et al. Evidence for brain glial activation in chronic pain patients. Brain. 2015;138(Pt 3):604–15. https://doi.org/10.1093/brain/awu377.
57. Tal M. A role for inflammation in chronic pain. Curr Rev Pain. 1999;3:440–6.
58. Kreutzberg GW. Microglia: a sensor for pathological events in the CNS. Trends Neurosci. 1996;19:312–8.
59. Watkins LR, Maier SF. Glia: a novel drug discovery target for clinical pain. Nat Rev Drug Discov. 2003;2:973–85.
60. Watkins LR, Milligan ED, Maier SF. Glial activation: a driving force for pathological pain. Trends Neurosci. 2001;24:450–5.
61. Grace PM, Hutchinson MR, Maier SF, Watkins LR. Pathologic pain and the neuroimmune interface. Nat Rev Immunol. 2014;14(4):217–31.
62. Minihane AM, Vinoy S, Russell WR. Low grade inflammation, diet composition and health: current research evidence and its translation. Br J Nutr. 2015;114(7):999–1012. https://doi.org/10.1017/S0007114515002093.
63. Maroon JC, Bost JW, Maroon A. Natural anti-inflammatory agents for pain relief. Surg Neurol Int. 2010;1:80. https://doi.org/10.4103/2152-7806.73804. Available at: https://www.ncbi.nlm.nih.gov/pmc/articles/PMC3011108/. Accessed 28 May 2017.
64. Tall J, Raja S. Dietary constituents as novel therapies of pain. Clin J Pain. 2004;20(1):19–26. https://www.researchgate.net/publication/8964906_Dietary_Constituents_as_Novel_Therapies_for_Pain.
65. Han C, Lee DH, Chung JM. Characteristics of ectopic discharges in a rat neuropathic pain model. Pain. 2000;84(2–3):253–61.2000.
66. Simopoulos AP. An increase in the omega-6/omega-3 fatty acid ratio increases risk for obesity. Nutrients. 2016;8(3):128.
67. Simopoulos AP. The importance of the omega-6/omega-3 fatty acid ratio in cardiovascular disease and other chronic diseases. Exp Biol Med (Maywood). 2008;233(6):674–88. https://doi.org/10.3181/0711-MR-311.
68. Wang Y, Kasper LH. The role of microbiome in central nervous system disorders. Brain Behav Immun. 2014;38:1–12. https://doi.org/10.1016/j.bbi.2013.12.015.
69. Guinane CM, Cotter PD. Role of the gut microbiota in health and chronic gastrointestinal disease: understanding a hidden metabolic organ. Ther Adv Gastroenterol. 2013;6(4):295–308. https://doi.org/10.1177/1756283X13482996.
70. Belkaid Y, Hand T. Role of the microbiota in immunity and inflammation. Cell. 2014;157(1):121–41. https://doi.org/10.1016/j.cell.2014.03.011.
71. Rogers G. Germs and joints: the contribution of the microbiome to rheumatoid arthritis. Nat Med. 2015;21(8):839–41. Available at: https://www.researchgate.net/publication/280869729_Germs_and_joints_The_contribution_of_the_human_microbiome_to_rheumatoid_arthritis. Accessed 28 May 2017.
72. Pimentel M, Wallace D, Hallegua DS, et al. A link between irritable bowel syndrome and fibromyalgia may be related to findings on lactulose breath testing. Ann Rheum Dis. 2004;63(4):450–2.
73. Wallace DJ, Hallegua DS. Fibromyalgia: the gastrointestinal link. Curr Pain Headache Rep. 2004;8(5):364–8.
74. Canavan C, West J, Card T. The epidemiology of irritable bowel syndrome. Clin Epidemiol. 2014;6:71–80. https://doi.org/10.2147/CLEP.S40245.
75. Mansueto P, D'Alcamo A, Seidita A, et al. Food allergy in irritable bowel syndrome: the case of non-celiac wheat sensitivity. World J Gastroenterol. 2015;21(23):7089–109. https://doi.org/10.3748/wjg.v21.i23.7089.
76. Drisko J, Bischoff B, Hall M, et al. Treating irritable bowel syndrome with a food elimination diet followed by food challenge and probiotics. J Am Coll Nutr. 2006;25(6):514–22.

77. Cashman MD, Martin DK, Dhillon S, et al. Irritable bowel syndrome: a clinical review. Curr Rheumatol Rev. 2016;12(1):13–26. Available at: https://www.ncbi.nlm.nih.gov/pubmed/26717952. Accessed 27 May 2017.
78. Fukudo S. Stress and visceral pain: focusing on irritable bowel syndrome. Pain. 2013;154(Suppl 1):S63–70. https://doi.org/10.1016/j.pain.2013.09.008.
79. Galland L. The gut microbiome and the brain. J Med Food. 2014;17(12):1261–72. https://doi.org/10.1089/jmf.2014.7000.
80. Lin CS, Chang CJ, Lu CC, et al. Impact of the gut microbiota, prebiotics, and probiotics on human health and disease. Biom J. 2014;37(5):259–68. https://doi.org/10.4103/2319-4170.138314.
81. Blaser M. Missing microbes: how killing bacteria creates modern plagues. London, England: Oneworld Publications; 2014.
82. Langdon A, Crook N, Dantas G. The effects of antibiotics on the microbiome throughout development and alternative approaches for therapeutic modulation. Genome Med. 2016;8:39. https://doi.org/10.1186/s13073-016-0294-z.
83. Noble EE, Hsu TM, Kanoski SE. Gut to brain dysbiosis: mechanisms linking western diet consumption, the microbiome, and cognitive impairment. Front Behav Neurosci. 2017;11:9. https://doi.org/10.3389/fnbeh.2017.00009.
84. Suez J, Korem T, Zeevi D, et al. Artificial sweeteners induce glucose intolerance by altering the gut microbiota. Nature. 2014:1–6. Available at: https://www.nature.com/nature/journal/vaop/ncurrent/pdf/nature13793.pdf. Accessed 28 May 2017.
85. Abou-Donia MB, El Masry EM, Abdel-Rahman AA. Splenda alters gut microflora and increases intestinal p-glycoprotein and cytochrome p-450 in male rats. J Toxicol Environ Health A. 2008;71(21):1415–29. https://doi.org/10.1080/15287390802328630.
86. Hadjivassiliou M, Grünewald RA, Davies-Jones GAB. Gluten sensitivity as a neurological illness. J Neurol Neurosurg Psychiatry. 2002;72:560–3. Available at: http://jnnp.bmj.com/content/72/5/560.citation-tools. Accessed 28 May 2017.
87. Isasi C, Colmenero I, Casco F, et al. Fibromyalgia and non-celiac gluten sensitivity: a description with remission of fibromyalgia. Rheumatol Int. 2014;34(11):1607–12. https://doi.org/10.1007/s00296-014-2990-6.
88. Alpay K, Ertaş M, Orhan EK, Üstay DK, Lieners C, Baykan B. Diet restriction in migraine, based on IgG against foods: a clinical double-blind, randomised, cross-over trial. Cephalalgia. 2010;30(7):829–37. https://doi.org/10.1177/0333102410361404.
89. Hall T, Briffa K, Hopper D. Clinical evaluation of cervicogenic headache: a clinical perspective. J Man Manip Ther. 2008;16(2):73–80.
90. Blau JN. Water deprivation: a new migraine precipitant. Headache. 2005;45(6):757–9. Available at: https://www.ncbi.nlm.nih.gov/pubmed/15953311. Accessed 27 May 2017.
91. Gonzalez A, Hyde E, Sangwan N, et al. Migraines are correlated with higher levels of nitrate-, nitrite-, and nitric oxide-reducing oral microbes in the American gut project cohort. mSystems. 2016;1(5):e00105–16. https://doi.org/10.1128/mSystems.00105-16.
92. Dimitrova AK, Ungaro RC, Lebwohl B, et al. Prevalence of migraine in patients with celiac disease and inflammatory bowel disease. Headache. 2013;53(2):344–55. https://doi.org/10.1111/j.1526-4610.2012.02260.x.
93. Lisotto C, Mainardi F, Maggioni F, et al. The comorbidity between migraine and hypothyroidism. J Headache Pain. 2013;14(Suppl 1):P138. https://doi.org/10.1186/1129-2377-14-S1-P138.
94. Gruber HJ, Bernecker C, Pailer S. Hyperinsulinemia in migraineurs is associated with nitric oxide stress. Cephalalgia. 2010;30(5):593–8. https://doi.org/10.1111/j.1468-2982.2009.02012.x.
95. Zaeem Z, Zhou L, Dilli E. Headaches: a review of the role of dietary factors. Curr Neurol Neurosci Rep. 2016;16(11):101.
96. Gaby A. The role of hidden food allergy/intolerance in chronic disease. Alt Med Rev. 1998;3(2):90–100. Available at: http://www.altmedrev.com/publications/3/2/90.pdf. Accessed 29 May 2017.
97. Patel RM, Sarma R, Grimsley E. Popular sweetener sucralose as a migraine trigger. Headache. 2006;46(8):1303–4.
98. Durham J, Touger-Decker R, Nixdorf DR, et al. Oro-facial pain and nutrition: a forgotten relationship? J Oral Rehabil. 2015;42:75–80. https://doi.org/10.1111/joor.12226.
99. Platt FW, Gaspar DL, Coulehan JL. "Tell me about yourself": the patient-centered interview. Ann Intern Med. 2001;134(11):1079–85. Available at: http://annals.org/aim/article/714557/tell-me-about-yourself-patient-centered-interview. Accessed 29 May 2017.
100. Rakel D, Weil A. Philosophy of integrative medicine. In: Rakel D, editor. Integrative medicine. Philadelphia: Elsevier; 2012. p. 2–11.
101. Habib S, Morrissey S, Helmes E. Preparing for pain management: a pilot study to enhance engagement. J Pain. 2005;6(1):48–54.
102. Cheatle MD. Depression, chronic pain, and suicide by overdose: on the edge. Pain Med. 2011;12(Suppl 2):S43–8. https://doi.org/10.1111/j.1526-4637.2011.01131.x.
103. Choi HK. Dietary risk factors for rheumatic diseases. Curr Opin Rheumatol. 2005;17(2):141–6.
104. Mahan LK, Escott-Stump S, Raymond JL. Krause's food and the nutrition care process. 13th ed. Philadelphia: Elsevier Saunders; 2012.
105. Denton C. The elimination diet. Minn Med. 2012;95(12):43–4. Available at: https://www.ncbi.nlm.nih.gov/pubmed/23346726. Accessed 28 May 2017.
106. U.S. Food and Drug Administration (FDA). Food allergies: what you need to know. Available at: https://www.fda.gov/food/resourcesforyou/consumers/ucm079311.htm. Accessed 28 May 2017.
107. Rodrigo L, Blanco I, Bobes J, et al. Clinical impact of a gluten-free diet on health-related quality of life in seven fibromyalgia syndrome patients with associated celiac disease. BMC Gastroenterol. 2013;13:157. https://doi.org/10.1186/1471-230X-13-157.
108. Mansueto P. Inflammatory diet and gluten-free diet. Available at: https://clinicaltrials.gov/ct2/show/NCT03017716. Accessed 28 May 2017.
109. Vandenkerkhof EG, Macdonald HM, Jones GT. Diet, lifestyle and chronic widespread pain: results from the 1958 British birth cohort study. Pain Res Manag. 2011;16(2):87–92. Available at: https://www.ncbi.nlm.nih.gov/pubmed/21499583. Accessed 29 May 2017.
110. Kjeldsen-Kragh J. Mediterranean diet intervention in rheumatoid arthritis. Ann Rheum Dis. 2003;62:193–5. https://doi.org/10.1136/ard.62.3.193.
111. Totsch SK, Waite ME, Sorge RE. Dietary influence on pain via the immune system. Prog Mol Biol Transl Sci. 2015;131:435–69. https://doi.org/10.1016/bs.pmbts.2014.11.013.
112. Gonzalez C, Rodriguez-Romero B, Carballo-Costa L. Importance of nutritional treatment in the inflammatory process of rheumatoid arthritis. Nutr Hosp. 2014;29(2):237–45. https://doi.org/10.3305/nh.2014.29.2.7067.
113. Eaton SB, Eaton SB III, Konner MJ. Paleolithic nutrition revisited: a twelve-year retrospective on its nature and implications. Euro J Clin Nut. 1997;51:207–16.
114. Spreadbury I. Comparison with ancestral diets suggests dense acellular carbohydrates promote an inflammatory microbiota, and may be the primary dietary cause of leptin resistance and obesity. Diabetes Metab Syndr Obes. 2012;5:175–89. https://doi.org/10.2147/DMSO.S33473.
115. Blomquist C, Alvehus M, Buren J. Attenuated low-grade inflammation following long-term dietary intervention in postmenopausal women with obesity. Obesity (Silver Spring). 2017;25(5):892–900. https://doi.org/10.1002/oby.21815.
116. Masharani U, Sherchan P, Schloetter M. Metabolic and physiologic effects from consuming a hunter-gatherer (Paleolithic)-type diet in type 2 diabetes. Eur J Clin Nutr. 2015;69(8):944–8. https://doi.org/10.1038/ejcn.2015.39.
117. Bisht B, Darling WG, Grossmann RE, et al. A multimodal intervention for patients with secondary progressive multiple sclerosis: feasibility and effect on fatigue. J Altern Complement Med. 2014;20(5):347–55. https://doi.org/10.1089/acm.2013.0188.

118. Tarantino G, Citro V, Finelli C. Hype or reality: should patients with metabolic syndrome-related NAFLD be on the hunter-gatherer (paleo) diet to decrease morbidity? J Gastrointestin Liver Dis. 2015;24(3):359–68. https://doi.org/10.15403/jgld.2014.1121.243.gta.

119. Donaldson MS, Speight N, Loomis S. Fibromyalgia syndrome improved using a mostly raw vegetarian diet: an observational study. BMC Complement Altern Med. 2001;1(7). Available at: https://www.ncbi.nlm.nih.gov/pubmed/11602026. Accessed 30 May 2017.

120. Nenonen MT, Helve TA, Rauma AL, et al. Uncooked, lactobacilli-rich, vegan food and rheumatoid arthritis. Br J Rheumatol. 1998;37(3):274–81. Available at: https://www.ncbi.nlm.nih.gov/pubmed/?term=vegan+diets+and+chronic+pain. Accessed 30 May 2017.

121. Michalsen A, Li C. Fasting therapy for treating and preventing disease-current state of evidence. Forsch Komplementmed. 2013;20(6):444–53. https://doi.org/10.1159/000357765.

122. Michalsen A, Hoffman B, Moebus S. Incorporation of fasting therapy in an integrative medicine ward: evaluation of outcome, safety, and effects on lifestyle adherence in a large prospective cohort study. J Altern Complement Med. 2005;11(4):601–7. https://www.ncbi.nlm.nih.gov/pubmed/16131283.

123. Masino SA, Ruskin DN. Ketogenic diets and pain. J Child Neurol. 2013;28(8):993–1001. https://doi.org/10.1177/0883073813487595.

124. Longo VD, Mattson MP. Fasting: molecular mechanisms and clinical applications. Cell Metab. 2014;19(2):181–92. https://doi.org/10.1016/j.cmet.2013.12.008.

125. Childers NF, Margoles MS. An apparent relation of nightshades (*Solanaceae*) to arthritis. Jour Neuro Ortho Med Surg. 1993;12:227–31. Available at: http://www.noarthritis.com/research.htm. Accessed 30 May 2017.

126. Lorenz EC, Michet CJ, Milliner DS, Lieske JC. Update on oxalate crystal disease. Curr Rheumatol Rep. 2013;15(7):340. https://doi.org/10.1007/s11926-013-0340-4.

127. Miller AW, Oakeson KF, Dale C. Effect of dietary oxalate on the gut microbiota of the mammalian herbivore *neotoma albigula*. Appl Environ Microbiol. 2016;82(9):2669–75. https://doi.org/10.1128/AEM.00216-16.

128. Solomons C. The pain project. Available at: http://www.thevpfoundation.org/effective_treatment.htm. Accessed 30 May 2017.

129. Kennedy DO. B vitamins and the brain: mechanisms, dose and efficacy—a review. Nutrients. 2016;8(2):68. https://doi.org/10.3390/nu8020068.

130. Hammond N, Wang Y, Dimachkie M, Barohn R. Nutritional neuropathies. Neurol Clin. 2013;31(2):477–89. https://doi.org/10.1016/j.ncl.2013.02.002.

131. Sibile KT, King C, Garrett TJ, et al. Omega-6: omega-3 PUFA ratio, pain, functioning and distress in adults with knee pain. Clin J Pain. 2017; https://doi.org/10.1097/AJP.0000000000000517.

132. Ramsden CE, Ringel A, Majchrzak-Hong SF. Dietary linoleic acid-induced alterations in pro- and anti-nociceptive lipid autacoids: implications for idiopathic pain syndromes? Mol Pain. 2016;12 https://doi.org/10.1177/1744806916636386.

133. Maroon JC, Bost JW. Omega-3 fatty acids (fish oil) as an anti-inflammatory: an alternative to nonsteroidal anti-inflammatory drugs for discogenic pain. Surg Neurol. 2006;65(4):326–31. Available at: https://www.ncbi.nlm.nih.gov/pubmed/16531187.

134. Goldberg RJ, Katz J. A meta-analysis of the analgesic effects of omega-3 polyunsaturated fatty acid supplementation for inflammatory joint pain. Pain. 2007;129(1–2):210–23. https://www.ncbi.nlm.nih.gov/pubmed/17335973.

135. Ramsden CE, Faurot KR, Zamora D. Targeted alteration of dietary n-3 and n-6 fatty acids for the treatment of chronic headaches: a randomized trial. Pain. 2013;154(11):2441–51. https://doi.org/10.1016/j.pain.2013.07.028.

136. Trepanier MO, Hopperton KE, Orr SK. N-3 polyunsaturated fatty acids in animal models with neuroinflammation. Eur J Pharmacol. 2016;785:187–206. https://doi.org/10.1016/j.ejphar.2015.05.045.

137. Ko GD, Nowacki NB, Arseneau L, et al. Omega-3 fatty acids for neuropathic pain. Clin J Pain. 2010;26:168–72.

138. Hakimian J, Minasyan A, Zhe-Ying L. Specific behavioral and cellular adaptations induced by chronic morphine are reduced by dietary omega-3 polyunsaturated fatty acids. PLoS One. 2017;12(4):e0175090. https://doi.org/10.1371/journal.pone.0175090.

139. Yin G, Wang Y, Cen XM. Identification of palmitoleic acid controlled by mTOR signaling as a biomarker of polymyositis. J Immunol Res. 2017;2017 https://doi.org/10.1155/2017/3262384.

140. Plotnikoff GA, Quigley JM. Prevalence of severe hypovitaminosis D in patients with persistent, nonspecific musculoskeletal pain. Mayo Clin Proc. 2003;78(12):1463–70. Available at: https://www.ncbi.nlm.nih.gov/pubmed/14661675. Accessed 30 May 2017.

141. Mangin M, Sinha R, Fincher K. Inflammation and vitamin D: the infection connection. Inflamm Res. 2014;63(10):803–19. https://doi.org/10.1007/s00011-014-0755-z.

142. Shipton EE, Shipton EA. Vitamin D deficiency and pain: clinical evidence of low levels of vitamin D and supplementation in chronic pain states. Pain Ther. 2015;4(1):67–87. https://doi.org/10.1007/s40122-015-0036-8.

143. National Institutes of Health, Office of Dietary Supplements. Vitamin D Fact Sheet. https://ods.od.nih.gov/factsheets/VitaminD-HealthProfessional/

144. Masterjohn C. Vitamin D toxicity redefined: vitamin K and the molecular mechanism. Med Hypotheses. 2007;68(5):1026–34. https://www.ncbi.nlm.nih.gov/pubmed/17145139.

145. Holton K. The role of diet in the treatment of fibromyalgia. Pain. 2016;6(4):317–20. https://doi.org/10.2217/pmt-2016-0019.

146. Gröber U, Schmidt J, Kisters K. Magnesium in prevention and therapy. Nutrients. 2015;7(9):8199–226. https://doi.org/10.3390/nu7095388.

147. Srebro D, Vuckovic S, Milovanovic A. Magnesium in pain research: state of the art. Curr Med Chem. 2016. https://www.ncbi.nlm.nih.gov/pubmed/27978803.

148. Yousef AA, Al-deeb AE. A double-blinded randomised controlled study of the value of sequential intravenous and oral magnesium therapy in patients with chronic low back pain with a neuropathic component. Anaesthesia. 2013;68(3):260–6. https://doi.org/10.1111/anae.12107.

149. Bagis S, Karabiber M, As I. Is magnesium citrate treatment effective on pain, clinical parameters and functional status in patients with fibromyalgia? Rheumatol Int. 2013;33(1):167–72. https://doi.org/10.1007/s00296-011-2334-8.

150. Kim YH, Lee PB, Oh TK. Is magnesium sulfate effective for pain in chronic post herpetic neuralgia patients comparing with ketamine infusion therapy? J Clin Anesth. 2015;27(4):296–300. https://doi.org/10.1016/j.jclinane.2015.02.006.

151. Rondon LJ, Privat AM, Daulhac L. Magnesium attenuates chronic hypersensitivity and spinal cord NMDA receptor phosphorylation in a rat model of diabetic neuropathic pain. J Physiol. 2010;588(Pt 21):4205–15. https://doi.org/10.1113/jphysiol.2010.197004.

152. Menon S, Nasir B, Avgan N. The effect of 1 mg of folic acid supplementation on clinical outcomes in female migraine with aura patients. J Headache Pain. 2016;17(1):60. https://doi.org/10.1186/s10194-016-0652-7.

153. Costantini A, Pala MI, Tundo S, et al. High-dose thiamine improves the symptoms of fibromyalgia. BMJ Case Reports. 2013;2013 https://doi.org/10.1136/bcr-2013-009019.

154. Schoenen J, Jacquy J, Lenaerts M. Effectiveness of high-dose riboflavin in migraine prophylaxis. A randomized controlled trial. Neurology. 1998;50(2):466–70. Available at: https://www.ncbi.nlm.nih.gov/pubmed/9484373. Accessed 31 May 2017.
155. Abyad A. Update on the use of vitamin B12 in management of pain. Middle East J Int Med. 2016;9(2). http://me-jim.com/August2016/VitaminB.pdf.
156. Bunner AE, Wells CL, Gonzales J. A dietary intervention for chronic diabetic neuropathy pain: a randomized controlled pilot study. Nutr Diabetes. 2015;5:e158. https://doi.org/10.1038/nutd.2015.8.
157. Tick H. Nutrition and pain. Phys Med Rehabil Clin N Am. 2014;26(2015):309–20. https://doi.org/10.1016/j.pmr.2014.12.006.
158. Geller M, Mibielli MA. Nunes CP. Comparison of the action of diclofenac alone versus diclofenac plus B vitamins on mobility in patients with low back pain. J Drug Assess. 2016;5(1):1–3.
159. Malenbaum S, Keefe FJ, Williams A. Pain in its environmental context: implications for designing environments to enhance pain control. Pain. 2008;134(3):241–4. https://doi.org/10.1016/j.pain.2007.12.002.
160. Hsieh C, Kong J, Kirsch I, et al. Well-loved music robustly relieves pain: a randomized, controlled trial. PLoS One. 2014;9(9):e107390. https://doi.org/10.1371/journal.pone.0107390.
161. Fishman E, Turkheimer E, DeGood DE. Touch relieves stress and pain. J Behav Med. 1995;18(1):69–79. Available at: https://www.ncbi.nlm.nih.gov/pubmed/7595953. Accessed 16 June 2017.
162. Strean WB. Laughter prescription. Can Fam Physician. 2009;55(10):965–7.
163. Mayor S. Community program reduces depression in patients with chronic pain, study shows. BMJ. 2016;353:i3352. Available at: http://www.bmj.com/content/353/bmj.i3352. Accessed 16 June 2017.
164. Tarr B, Launay J, Cohen E. Synchrony and exertion during dance independently raise pain threshold and encourage social bonding. Biol Lett. 2015; https://doi.org/10.1098/rsbl.2015.0767.
165. Gallace A, Torta DM, Moseley GL. The analgesic effect of crossing the arms. Pain. 2011;152(6):1418–23. https://doi.org/10.1016/j.pain.2011.02.029.
166. Hill PL, Allemand M, Roberts BW. Examining the pathways between gratitude and self-rated physical health across adulthood. Personal Individ Differ. 2013;54(1):92–6. https://doi.org/10.1016/j.paid.2012.08.011.
167. Hill P, Pargament KI. Advances in the conceptualization and measurement of religion and spirituality: implications for physical and mental health research. Am Psychol. 2003;58(1):64–74. https://doi.org/10.1037/0003-066X.58.1.64.
168. Patel AI, Hampton KE. Encouraging consumption of water in school and child care settings: access, challenges, and strategies for improvement. Am J Public Health. 2011;101(8):1370–9. https://doi.org/10.2105/AJPH.2011.300142.
169. Perry BG, Bear TLK, Lucas SJE. Mild dehydration modifies the cerebrovascular response to the cold pressor test. Exp Physiol. 2016;101(1):135–42. http://onlinelibrary.wiley.com/doi/10.1113/EP085449/pdf.
170. Neogi T, Chen C, Chaisson, et al. Drinking water can reduce the risk of recurrent gout attacks. Arthritis Rheum. 2009;60:S762–3. https://doi.org/10.1002/art.27110.
171. Baratloo A, Rouhipour A, Forouzanfar MM, Safari S, Amiri M, Negida A. The role of caffeine in pain management: a brief literature review. Anesthesiol Pain Med. 2016;6(3):e33193. https://doi.org/10.5812/aapm.33193.
172. Shilo L, Sabbah H, Hadari R. The effects of coffee consumption on sleep and melatonin secretion. Sleep Med. 2002;3(3):271–3. Available at: https://www.ncbi.nlm.nih.gov/pubmed/14592218. Accessed 18 June 2017.
173. Wolde T. Effects of caffeine on health and nutrition: a review. Food Sci Qual Manag. 2014;30:59–64. Available at: https://www.researchgate.net/publication/279923885_Effects_of_caffeine_on_health_and_nutrition_A_Review. Accessed 18 June 2017.
174. Prasad S, Aggarwal BB. Herbal medicine: biomolecular and clinical aspects, 2nd Ed. Chapter 13: turmeric, the Golden Spice: CRC Press/Taylor & Francis; 2011. Available at: https://www.ncbi.nlm.nih.gov/books/NBK92752/. Accessed 31 May 2017.
175. Gupta SC, Patchva S, Aggarwal BB. Therapeutic roles of curcumin: lessons learned from clinical trials. AAPS J. 2013;15(1):195–218. https://doi.org/10.1208/s12248-012-9432-8.
176. Aggarwal BB, Harikumar KB. Potential therapeutic effects of curcumin, the anti-inflammatory agent, against neurodegenerative, cardiovascular, pulmonary, metabolic, autoimmune and neoplastic diseases. Int J Biochem Cell Biol. 2009;41(1):40–59. https://doi.org/10.1016/j.biocel.2008.06.010.
177. Kuptniratsaikul V, Dajpratham P, Taechaarpornkul W, et al. Efficacy and safety of *Curcuma domestica* extracts compared with ibuprofen in patients with knee osteoarthritis: a multicenter study. Clin Interv Aging. 2014;9:451–8. https://doi.org/10.2147/CIA.S58535.
178. Duke JA. The garden pharmacy: turmeric, the queen of COX-2-inhibitors. Alternative Compl Ther. 2007;13(5):229–34.
179. Shoba G, Joy D, Joseph T. Influence of piperine on the pharmacokinetics of curcumin in animals and human volunteers. Planta Med. 1998;64(4):353–6.
180. Bode AM, Dong Z. Herbal medicine: biomolecular and clinical aspects, 2nd Ed. Chapter 7: the amazing and mighty ginger: CRC Press/Taylor & Francis; 2011. Available at: https://www.ncbi.nlm.nih.gov/books/NBK92775/. Accessed 31 May 2017.
181. Cady RK, Schreiber CP, Beach ME, et al. Gelstat migraine (sublingually administered feverfew and ginger compound) for acute treatment of migraine when administered during the mild pain phase. Med Sci Monit. 2005;11:165–9.
182. Siddiqui MZ. *Boswellia Serrata*, a potential antiinflammatory agent: an overview. Indian J Pharm Sci. 2011;73(3):255–61. https://doi.org/10.4103/0250-474X.93507.
183. Sengupta K, Krishnaraju AV, Vishal AA, et al. Comparative efficacy and tolerability of 5-Loxin and Aflapin against osteoarthritis of the knee: a double blind, randomized, placebo controlled clinical study. Int J Med Sci. 2010;7:366–77.
184. Levesque H, Lafont O. Aspirin throughout the ages: a historical review. Rev Med Interne. 2000;21(Suppl 1):8s–17s. https://www.ncbi.nlm.nih.gov/pubmed/10763200?dopt=Abstract.
185. Chrubasik S, Eisenberg E, Balan E. Treatment of low back pain exacerbations with willow bark extract: a randomized double-blind study. Am J Med. 2000;109(1):9–14.
186. Taylor FR. Lifestyle changes, dietary restrictions and nutraceuticals in migraine prevention. Tech Reg Anesth Pain Manag. 2009;13:28.
187. Lipton RB, Gobel H, Einhaupl KM. Petasites hybridus root (butterbur) is an effective preventive treatment for migraine. Neurology. 2004;63(12):2240–4.
188. Pothmann R, Danesch U. Migraine prevention in children and adolescents: results of an open study with a special butterbur root extract. Headache. 2005;45(3):196–203.
189. Pareek A, Suthar M, Rathore GS, et al. Feverfew (*Tanacetum parthenium* L.): a systematic review. Pharmacogn Rev. 2011;5(9):103–10. https://doi.org/10.4103/0973-7847.79105.
190. Fitzcharles MA, Ste-Marie PA, Clauw DJ. Rheumatologists lack confidence in their knowledge of cannabinoids pertaining to the management of rheumatic complaints. BMC Musculoskelet Disord. 2014;15:258. https://doi.org/10.1186/1471-2474-15-258.
191. Atakan Z. Cannabis, a complex plant: different compounds and different effects individuals. Ther Adv Psychopharmacol. 2012;2(6):241–54. https://doi.org/10.1177/2045125312457586.
192. Russo EB. Cannabinoids in the management of difficult to treat pain. Ther Clin Risk Manag. 2008;4(1):245–59.
193. Boychuk DG, Goddard G, Mauro G, et al. The effectiveness of cannabinoids in the management of chronic nonmalignant neu-

193. ropathic pain: a systematic review. J Oral Facial Pain Headache. 2015;29(1):7–14.
194. Baron EP. Comprehensive review of medicinal marijuana, cannabinoids, and therapeutic implications in medicine and headache: what a long strange trip it's been…. Headache. 2015;55(6):885–916.
195. van Amsterdam J, Brunt T, van den Brink W. The adverse health effects of synthetic cannabinoids with emphasis on psychosis-like effects. J Psychopharmacol. 2015;29(3):254–63. https://doi.org/10.1177/0269881114565142.
196. Yadav V, Bever C Jr, Bowen J, et al. Summary of evidence-based guideline: complementary and alternative medicine in multiple sclerosis: report of the guideline development subcommittee of the American Academy of Neurology. Neurology. 2014;82(12):1083–92.
197. Procon.org. 33 legal marijuana states and DC. Available at: http://medicalmarijuana.procon.org/view.resource.php?resourceID=000881. Accessed 25 Jan 2019.
198. National Conference of State Legislatures. State Industrial Hemp Statutes: May 16, 2017. Available at: http://www.ncsl.org/research/agriculture-and-rural-development/state-industrial-hemp-statutes.aspx. Accessed 3 June 2017.
199. Xiong W, Cui T, Chenk K. Cannabinoids suppress inflammatory and neuropathic pain by targeting α3 glycine receptors. J Exp Med. 2012;209(6):1121–34. https://doi.org/10.1084/jem.20120242.
200. Serpell M, editor. Handbook of pain management. London: Current Medicine Group; 2008.
201. Melzack R. The McGill pain questionnaire: major properties and scoring methods. Pain. 1975;1(3):277–99.
202. Center for Substance Abuse Treatment. Managing Chronic Pain in Adults With or in Recovery From Substance Use Disorders. Rockville (MD): Substance Abuse and Mental Health Services Administration (US); 2012. (Treatment Improvement Protocol (TIP) Series, No. 54.) Appendix B, Assessment Tools and Resources. Available at: https://www.ncbi.nlm.nih.gov/books/NBK92056/.
203. Dansie EJ, Turk DC. Assessment of patients with chronic pain. Colvin L, Rowbotham DJ, eds. Br J Anaesth 2013;111(1):19–25. https://doi.org/10.1093/bja/aet124.
204. Arfe A. Scotti L, Varas-Lorenzo C, et al. Non-steroidal anti-inflammatory drugs and risk of heart failure in four European countries: nested case-control study. BMJ. 2016;354:i4857. https://doi.org/10.1136/bmj.i4857.
205. Imhann F, Bonder MJ, Vich Vila A, et al. Proton pump inhibitors affect the gut microbiome. Gut. 2015;65:740. https://doi.org/10.1136/gutjnl-2015-310376.
206. Montenegro L, Losurdo G, Licinio R. Non steroidal anti-inflammatory drug induced damage on lower gastrointestinal tract: is there involvement of the microbiota? Curr Drug Saf. 2014;9(3):196–204. Available at: https://www.ncbi.nlm.nih.gov/pubmed/24809527/. Accessed 10 June 2017.
207. Leonard J, Marshall JK, Moayyedi P. Systematic review of the risk of enteric infection in patients taking acid suppression. Am J Gastroenterol. 2007;102(9):2047–56.
208. Abramson SB, Weaver AL. Current state of therapy for pain and inflammation. Arthritis Res Ther. 2005;7(Suppl 4):S1–6. https://doi.org/10.1186/ar1792.
209. Reid MC, Eccleston C, Pillemer K. Management of chronic pain in older adults. BMJ. 2015;350:h532. https://doi.org/10.1136/bmj.h532.
210. Pronsky ZM. Food medication interactions. 17th ed; 2012.
211. Pasternak GW, Pan Y-X. Mu opioids and their receptors: evolution of a concept. Sibley DR, editor. Pharmacol Rev 2013;65(4):1257–1317. https://doi.org/10.1124/pr.112.007138.
212. Moore RA, McQuay HJ. Prevalence of opioid adverse events in chronic non-malignant pain: systematic review of randomized trials of oral opioids. Arthritis Res Ther. 2005;7(5):R1046–51.
213. Vuong C, Van Uum SHM, O'Dell LE, Lutfy K, Friedman TC. The effects of opioids and opioid analogs on animal and human endocrine systems. Endocr Rev. 2010;31(1):98–132. https://doi.org/10.1210/er.2009-0009.
214. Lee M, Silverman SM, Hansen H. A comprehensive review of opioid-induced hyperalgesia. Pain Physician. 2011;14:145–61.
215. Levy B, Paulozzi L, Mack KA, Jones CM. Trends in opioid analgesic-prescribing rates by specialty, U.S., 2007-2012. Am J Prev Med. 2015;49:409–13.
216. CDC, National Center Health Statistics. Therapeutic drug use. Available at: https://www.cdc.gov/nchs/fastats/drug-use-therapeutic.htm. Accessed 25 Jan 2019.
217. Nora D, Mclellan AT. Opioid abuse in chronic pain—misconceptions and mitigation strategies. Longo DL, editor. N Engl J Med 2016; 374:1253–1263. https://doi.org/10.1056/NEJMra1507771.
218. James S. Human pain and genetics: some basics. Br J Pain. 2013;7(4):171–8. https://doi.org/10.1177/2049463713506408.
219. Zorina-Lichtenwalter K, Meloto CB, Khoury S. Genetic predictors of human chronic pain conditions. Neuroscience. 2016;338:36–62. https://doi.org/10.1016/j.neuroscience.2016.04.041.
220. Young EE, Lariviere WR, Belfer I. Genetic basis of pain variability: recent advances. J Med Genet. 2012;49(1) https://doi.org/10.1136/jmedgenet-2011-100386.
221. Smith SB, Reenilä I, Männistö PT, et al. Epistasis between polymorphisms in COMT, ESR1, and GCH1 influences COMT enzyme activity and pain. Pain. 2014;155(11):2390–9. https://doi.org/10.1016/j.pain.2014.09.009.
222. Fillingim R, Wallace M, Herbstman D, Ribeiro-Dasilva M, Staud R. Genetic contributions to pain: a review of findings in humans. Oral Dis. 2008;14(8):673–82. https://doi.org/10.1111/j.1601-0825.2008.01458.x.
223. Liang L, Lutz BM, Bekker A, Tao Y-X. Epigenetic regulation of chronic pain. Epigenomics. 2015;7(2):235–45. https://doi.org/10.2217/epi.14.75.
224. Notterman DA, Mitchell C. Epigenetics and understanding the impact of social determinants on health. Pediatr Clin N Am. 2015;62:1227–40.
225. Buchheit T, Van de Ven T, Shaw A. Epigenetics and the transition from acute to chronic pain. Pain Med. 2012;13(11):1474–90. https://doi.org/10.1111/j.1526-4637.2012.01488.x.
226. McGowan PO, Meaney MJ, Szyf M. Diet and the epigenetic (re)programming of phenotypic differences in behavior. Brain Res. 2008;1237:12–24. https://doi.org/10.1016/j.brainres.2008.07.074.
227. Reuter S, Gupta SC, Park B, Goel A, Aggarwal BB. Epigenetic changes induced by curcumin and other natural compounds. Genes Nutr. 2011;6(2):93–108. https://doi.org/10.1007/s12263-011-0222-1.
228. Parreira PCS, Maher CG, Ferreira ML. A longitudinal study of the influence of comorbidities and lifestyle factors on low back pain in older men. Pain. 2017;158:1571. https://doi.org/10.1097/j.pain.0000000000000952.
229. Paley CA, Johnson MI. Physical activity to reduce systemic inflammation associated with chronic pain and obesity: a narrative review. Clin J Pain. 2016;32(4):365–70. https://doi.org/10.1097/AJP.0000000000000258.
230. Senba E, Katsuya K. A new aspect of chronic pain as a lifestyle-related disease. Neurobiol Pain. 2017;1:6–15. https://doi.org/10.1016/j.ynpai.2017.04.003.
231. Holtzman S, Beggs RT. Yoga for chronic low back pain: a meta-analysis of randomized controlled trials. Pain Res Manag. 2013;18(5):267–72.

232. Peng PW. Tai chi and chronic pain. Reg Anesth Pain Med. 2012;37(4):372–82. https://doi.org/10.1097/AAP.0b013e31824f6629.
233. Eriksen W, Bruusgaard D, Knardahl S. Work factors as predictors of intense or disabling low back pain; a prospective study of nurses' aides. Occup Environ Med. 2004;61:398–404. https://doi.org/10.1136/oem.2003.008482.
234. Hassed C. Mind-body therapies—use in chronic pain management. Aust Fam Physician. 2013;42(3):112–7. Available at: https://www.ncbi.nlm.nih.gov/pubmed/23529519. Accessed 19 June 2017.
235. Neilson S. Pain as metaphor: metaphor and medicine. Med Humanit. 2015; https://doi.org/10.1136/medhum-2015-010672. http://mh.bmj.com/content/early/2015/08/07/medhum-2015-010672.
236. American Diabetes Association. Statistics about diabetes: prevalence, 2012. http://www.diabetes.org/diabetes-basics/diabetes-statistics/
237. Heart disease and stroke statistics—2011 update: a report from the American Heart Association. Circulation. 2011;123:e18–e209, page 20. http://circ.ahajournals.org/content/123/4/e18.full.pdf.
238. American Cancer Society, Prevalence of cancer: 2014. http://www.cancer.org/docroot/CRI/content/CRI_2_6x_Cancer_Prevalence_How_Many_People_Have_Cancer.asp.

Nutrition and Behavioral Health/Mental Health/Neurological Health

Ruth Leyse Wallace

29.1 Introduction – 474
29.1.1 Spectrums of Psychiatric Disorders – 474
29.1.2 Serious Mental Illness and Genetics – 474
29.1.3 Depression – 475
29.1.4 Bipolar Disorder – 475
29.1.5 Schizophrenia – 476

29.2 Other Conditions: Conditions that May Be Accompanied by Changes in Mental Status – 477
29.2.1 Brain-Derived Neurotropic Factor (BDNF) – 477
29.2.2 Celiac Disease – 477
29.2.3 Metabolic Syndrome – 477
29.2.4 Hyponatremia/Water Intoxication – 477
29.2.5 Caffeine Intoxication – 477
29.2.6 Food Insecurity – 477

29.3 The Effect of Micro- and Macronutrients on Mental/Behavioral/Neurological Health – 478
29.3.1 Lipids, Glucose, Amino Acids – 478
29.3.2 Vitamins and Minerals – 479
29.3.3 Minerals – 483
29.3.4 Potentially Toxic Minerals – 486

29.4 Drugs: Nutrition and Mental Status – 486

29.5 Summary Statement – 487

References – 488

© Springer Nature Switzerland AG 2020
D. Noland et al. (eds.), *Integrative and Functional Medical Nutrition Therapy*,
https://doi.org/10.1007/978-3-030-30730-1_29

29.1 Introduction

The brain, nerves, neurotransmitters, and genes are constantly impacted by numerous micro- and macronutrients. Gene activity may be turned off or on by a vitamin or mineral; nerves and cell membranes are built of essential fatty acids; amino acids are the building blocks of neurotransmitters. Brain structure is 80% lipid, and current evidence and theory indicate, in contrast to the traditional view, that ketones also play vital roles as metabolic and signaling mediators, even when glucose is abundant [1]. It is known that brain, blood, and cell content reflects the nutrient intake from the diet or supplements. There are communication pathways from brain to gut and gut to brain [2]. How could mental processes *not* be affected by nutritional status? [3] The question asked by nutritionists, molecular biologists, psychiatrists, geneticists, and other scientists is: "How?" Exactly how do nutrients affect these processes, individually or acting together?

The answers are not all in. Some studies focus on mood, which does not necessarily translate into clinical outcomes; epidemiological and tissue studies, meanwhile, struggle to disentangle the respective roles of nutrients in the body or brain [4]. Despite these challenges, we understand the connection between nutrition and mental health more now than in the past. The Nutrition Care Process, more than ever, needs to include assessment of nutritional factors that are influencing mental status.

In 2006 a poll of 1000 Americans indicated that more Americans feared loss of their mental capacity (62%) than feared loss of their physical capacity (29%) [5]. In 2014, there were an estimated 9.8 million adults aged 18 or older in the United States with serious mental illness (SMI)—4.2% of all US adults [6]. There were an estimated 43.6 million adults in the United States with any mental illness (AMI) in the past year—18.1% of all US adults [7]. The Surgeon General's Report of 1999 states that mental illness is the second leading cause of disability and premature mortality, comparable to heart disease and cancer [8–10].

29.1.1 Spectrums of Psychiatric Disorders

With respect to mental status and functioning, the human condition is wide-ranging and variable through time for each individual. If functioning varies in certain ways, or to such a degree that it disproportionately affects function and relationships, it may be diagnosed and treated to improve the quality of life of the individual. The ranges and degrees of differentiation of conditions may be expressed in a spectrum or continuum.

Mood disorders are a group of conditions that vary in severity and direction of the emotional states (◘ Fig. 29.1). Each disorder has certain criteria for diagnosis. This spectrum differs from the Spectrum for Psychotic Disorders and the Autism Spectrum.

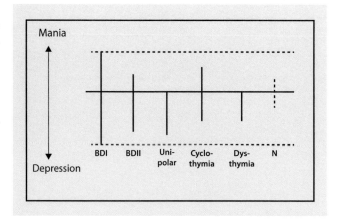

◘ **Fig. 29.1** Mood disorder spectrum. (Adapted with permission from Nemeroff [11]). *BDI* Bipolar Disorder I and II, *N* Normal

In the past, the Autism Spectrum encompassed the diagnoses of autism, Asperger Syndrome, Pervasive Developmental Disorder Not Otherwise Specified, and Childhood Disintegrative Disorder. The DSM-5 combined these into one category: "Autism Spectrum Disorder." Features of these disorders include social deficits and communication difficulties, stereotyped or repetitive behaviors and interests, sensory issues, and in some cases, cognitive delays [12]. (See also ► Chap. 30 for Intellectual and Developmental Disorders).

The spectrum of psychotic disorders includes schizophrenia, schizoaffective disorder, delusional disorder, schizotypal personality disorder, schizophreniform disorder, and brief psychotic disorder, as well as psychosis associated with substance use or medical conditions [13].

This chapter will provide in-depth discussion of the effect of nutrients on mental status and the major psychiatric diagnoses of depression, bipolar disorder, and schizophrenia.

29.1.2 Serious Mental Illness and Genetics

Schizophrenia, bipolar disorder, and severe depression have been shown to be genetically linked. A recent study determined that one of two genes code for the cellular mechanism for regulating the flow of calcium into neurons. Variation in one of these genes, CACNA1C, is known to impact brain activity involved in emotion, thinking, attention, and memory—functions disrupted in mental illnesses [14].

An earlier study found a gene common to all three; it appears to influence which symptoms occur [15]. This genetic abnormality results in removal of eicosapentaenoic acid (EPA), docosahexaenoic acid (DHA), arachadonic acid (AA), and gamma linolenic acid (GLA) from phospholipid structure. An intake ratio of 3 EPA:1 DHA is advantageous for treatment. This decreases the overactivity of one of the phospholipase enzymes that breaks down phospholipids' structures.

Table 29.1 Selected genes, interventions, and mental health

Gene	Effect	Potential nutrition intervention	Link to mental health
ACE	Produces an enzyme to help regulate the body's response to sodium intake. People with certain variants are at greater risk for high blood pressure when eating lots of salt. 7 in 10 people are at risk	Assess and advise re recommended sodium intake. Assess fluid intake and interaction between excessive fluid intake and Na intake	Individuals on lithium need to maintain consistent Na intake. Polydipsia, water intoxication, and/or hyponatremia has been noted in psychiatric populations
GLUT2	Helps regulate glucose, or sugar. People with certain variants prefer sweet foods and drinks. 1 in 5 people is at risk	Assess and advise re keeping % of kcal from low-nutrient sources. Assure adequate intake of Cr, Mg, and B-1, needed for CHO metabolism	Psychotropic drugs may increase craving for sweets
CYP1A2	Gene variants result in slow or fast metabolism of caffeine. Heavy coffee drinkers who are slow caffeine metabolizers retain more of the stimulant in the body	Assess and advise re caffeine intake. High residual caffeine increases risk of having a heart disease, developing diabetes, and hypertension, already increased in individuals with mental health disorders	Caffeine may increase stimulation of individuals in manic phase of bipolar disorder. Patients with eating disorder may use caffeine beverages to manage food intake
TPH2 Tryptophan hydroxylase (Patrick and Ames [16])	Regulates serotonin synthesis and release	Assess vit D level: Goal ≥30 ng/dl	Support serotonin synthesis and release/activation
		EPA, DHA intake Supplement: 2 g EPA and 1 g DHA from fish oil	"
		Exercise	Promote tryptophan entry into the brain

In July 2015, Jonathan Flint found two genetic markers reproducibly linked to depression. One was unknown; one was located next to a gene linked to energy production in mitochondria (Table 29.1) [17].

29.1.3 Depression

Depression may present with a variety of non-specific symptoms during a nutritional assessment—change in weight or appetite, fatigue, indecisiveness, or impaired concentration, as well as gastrointestinal symptoms and abdominal pain. Partial and acute vitamin deprivation studies suggest that the earliest impairments occur in measures of mood rather than mental performance [18]. Some researchers report the presence of inflammation underlying depression. Polyunsaturated fatty acids (PUFAs), folic acid, and probiotics are three supplements that target inflammation [19]. Evidence suggests a dysregulation in serotonin neurotransmission is central to the pathophysiology and treatment of Major Depressive Disorder (MDD).

Adolescents with Selective Serotonin Reuptake Inhibitor- (SSRI)-resistant depression exhibited deficits in red blood cell DHA compared with healthy adolescent controls, and it was shown that supplementation with fish oil was effective and well-tolerated. Two dose levels were compared: dose one was 2.4 g/day (EPA 1.6 g + DHA 0.8 g; 4 capsules/day) and dose two was 16.2 g/day (EPA, 10.8 g + DHA 5.4 g; 2 tbsp/day). Baseline-to-endpoint EPA and DHA levels increased, and depression symptom severity scores decreased, in both low-dose and high-dose groups [20].

A study of nutritional status in a group of depressed individuals showed intakes of vitamins A, B1, B2, B6, folate, C, sodium, potassium, Mg, Ca, P, Fe, and Zn and fiber ($p < 0.05$) were lower in the depression group. Fasting blood glucose levels, serum vitamins B12, and folic acid ($p < 0.05$) in the depression group were lower than controls. Meanwhile, median levels of body weight, waist circumference, hip circumference, and waist-to-hip ratio ($p < 0.05$) were significantly higher in the depression group. Serum insulin and Homeostatic Model Assessment (HOMA) levels of the two groups were similar [21].

Following a systematic search of PubMed, CINAHL, Cochrane Library, and Web of Science for clinical trials using adjunctive nutrients for depression, researchers concluded current evidence supports adjunctive use of S-adenosylmethionine (SAMe), methylfolate, omega-3, and vitamin D with antidepressants to reduce depressive symptoms [22].

29.1.4 Bipolar Disorder

Bipolar disorder (BD) is 80% heritable [23]. Folate-sensitive fragile genetic sites were found to be more frequent in patients with BD than in controls [24]. Patients and relatives had specific alterations in methylene tetrahydrofolate reductase (MTHFR) and were shown to have elevated homocysteine and be low in folate and vitamin B12 [25]. Normalizing these levels may decrease the symptoms of bipolar disorder.

Altered metabolic end-products suggest that individuals with BD have metabolic differences in the metabolism of PUFA when compared to control subjects. Patients with bipolar disorder have lower intake of Se, EPA, DHA, AA, and DPA as determined using fasted plasma samples & 7-day diet records [26]. A review of small trials reported that fish oil can enhance the effects of standard treatments for bipolar disorder and yield improvement in their bipolar depression symptoms [27].

Nierenberg et al. comment that data implicate mitochondrial dysfunction as an important component of the pathophysiology of bipolar disorder. Biological processes that may be dysregulated in BD are monoamine activity, immune and inflammatory processes, and oxidative stress [28]. Several tolerable and readily available mitochondrial modulators (MM) may be potential treatments, including N-acetyl-cysteine (NAC), acetyl-L-carnitine (ALCAR), (SAMe), coenzyme Q(10) (CoQ10), alpha-lipoic acid (ALA), creatine monohydrate (CM), and melatonin. Targets of treatment include mitochondrial dysfunction, oxidative stress, altered brain energy metabolism, and the dysregulation of multiple mitochondrial genes in patients with bipolar disorder. They conclude that clinical trials of individual MMs as well as combinations are warranted [29].

Lithium salts have been in use for the treatment of bipolar disorder for more than 50 years. There is a narrow therapeutic range between effectiveness and toxicity for lithium treatment. Target lithium blood levels are generally 0.8–1.1 mmol/L. Levels above 1.5 meq/L may result in marked tremor, nausea, diarrhea, or blurred vision. Levels above 2.0 meq/L have been associated with life-threatening side effects, such as neurotoxicity, delirium, and encephalopathy [30]. Reported cellular targets for Li (+) action involve Mg (2+)-activated enzymes, which are inhibited by Li(+) [31]. Sodium intake affects renal clearance by 30–50%. A consistent intake of Na supports a stable blood level of the drug [32].

Self-reported data from 348 patients in the United States diagnosed with bipolar disorder indicated the group had tried over 40 different supplements, and the long-term users took 19 different supplements. The most commonly taken supplements for both groups were fish oil, B vitamins, melatonin, and multivitamins [33].

29.1.5 Schizophrenia

In 2015 a proposal was made to drop the term "schizophrenia," with its connotation of hopeless chronic brain disease, and replace it with a term like "psychosis spectrum syndrome." Japan and South Korea have already abandoned this term [34]. Past classifications did not acknowledge the continuity between schizophrenia and other psychotic disorders, including schizoaffective disorder, delusional disorder, bipolar disorder with psychotic features, and depression with psychotic features. The DSM-5 terminology is "schizophrenia spectrum and other psychotic disorders" [35].

Folate concentrations have been inversely related to the Scale for Assessment of Negative Symptoms total score [36]. Negative symptoms associated with schizophrenia are social isolation, lack of initiative, socially awkward behavior, and discomfort when interacting with people. Positive symptoms are psychosis, hallucinations, delusions, secondary agitation, and thought disorder. Cognitive symptoms, which include memory, decision making, and problem solving, are a third type of impairment [37].

Studies show that phospholipids and essential fatty acids are depleted in red cell membranes of schizophrenic patients compared with control subjects, particularly arachidonic acid (AA) and DHA [38]. A healthy ratio decreases the overactivity of a phospholipase enzyme and normalizes the removal of EPA, DHA, AA, and GLA from phospholipid structure. Studies on the effect of diet and membrane changes on tardive dyskinesia have had mixed results.

Another 4-week, double-blind study of patients experiencing an acute episode of schizophrenia or schizoffective disorder found that baseline levels of PUFA affected the response to supplements of ethyl-EPA, vitamin E and/or vitamin C, given in divided doses. Symptoms became worse in individuals with a low baseline level of PUFA after taking a supplement of either ethyl-EPA (500 mg) OR vitamins (364 mg/day vitamin E or 1000 mg/day slow-release vitamin C). Use of the vitamin supplement increased the dose of medication needed for alleviating symptoms. Symptoms that worsened included positive symptoms of suspiciousness/persecution, delusions, hallucinations, paranoia, and hostility. Giving all three supplements neutralized this effect. These effects were not present in patients with a high baseline level of PUFA. Low serum α-tocopherol was the best predictor of low PUFA baseline levels. A possible explanation for this outcome is that, at high doses, vitamin E acts as a pro-oxidant in an environment of insufficient antioxidant capacity. At low baseline levels, and only when not combined with vitamins E & C, EPA increases RBC arachidonic acid, which increases positive symptoms (hallucinations, paranoia, delusions). *Knowledge of baseline PUFA needs to inform recommendations for supplements of vitamin C and E during acute episodes of schizophrenia* [39].

A comparison of diet histories of patients with schizophrenia and a matched group from the Health and Nutrition Examination Survey (NHANES) population led investigators to suggest that the higher proportion of obesity in the schizophrenic group was related to factors such as medication side effects and reduced physical activity rather than diet [40]. Another study of the intake of fatty acids and antioxidants suggested that membrane changes may be caused by abnormal membrane phospholipid metabolism rather than intake.

One explanation for the increased risk of obesity, metabolic syndrome, and diabetes is fat storage; computed tomography showed that with equal total body fat and equal subcutaneous fat, there was three times the amount of visceral fat stored by schizophrenic patients. The results of patients who were taking antipsychotics were no different than those who were not.

29.2 Other Conditions: Conditions that May Be Accompanied by Changes in Mental Status

29.2.1 Brain-Derived Neurotropic Factor (BDNF)

Peripheral BDNF has been positively correlated with lifetime depression episodes, psychiatric hospitalizations, and suicide attempts. Brain-derived neurotropic factor (BDNF) has been proposed as a biomarker for bipolar disorder, individuals with BD having lower levels of BDNF than controls. The BDNF gene provides instructions for making a protein found in the brain and spinal cord. This protein promotes the survival of neurons during growth, maturation, differentiation, and plasticity. The BDNF protein is found in regions of the brain that control eating, drinking, and body weight; it likely contributes to the management of these functions. The negative correlation between BDNF and number of mood episodes was moderated by impaired glucose metabolism [41].

29.2.2 Celiac Disease

Neurologic and psychiatric complications seen in gluten-sensitive patients may be the primary presentation in patients suffering from this disease. Data suggests that up to 22% of patients with celiac disease (CD) develop neurologic or psychiatric dysfunction and as many as 57% of people with neurological dysfunction of unknown origin test positive for anti-gliadin antibodies [42].

29.2.3 Metabolic Syndrome

The rate of metabolic syndrome was found to be higher in individuals with bipolar disorder (33%) and schizophrenia (47%) compared to matched NHANES controls (17% and 11%, respectively) [43]. Risk of metabolic syndrome was also associated with higher depression scores and abdominal obesity [44]. Nutrition care includes monitoring for indications of the development of metabolic syndrome, and intervening to prevent long-term consequences of the syndrome [45, 46]. (See also ▶ Chap. 46.)

The combined effect of depression and metabolic symptoms was greater than the sum of the individual effects: An individual with both depression and metabolic risk factors is more than six times more likely to develop diabetes [47].

Weight-loss programs designed for those with mental illness have been successful [48]. The behavior change process is facilitated through the use of self-monitoring, goal setting, and problem solving. Studies suggest that behavioral treatment produces weight loss of 8–10% during the first 6 months of treatment [49].

29.2.4 Hyponatremia/Water Intoxication

Psychotropic medications often increase dry mouth and thirst, which may increase fluid intake. Water intoxication may lead to hyponatremia. Symptoms of water intoxication in the early stages include nausea and vomiting, headache, or changes in mental state such as confusion or disorientation. Low serum sodium (hyponatremia) < 136 mEq/L is an osmotic shift of water into brain cells causing edema in the brain, caused by an excess of water relative to sodium. Signs and symptoms of acute hyponatremia include headache, confusion, and stupor. More severe symptoms can also occur: muscle weakness, spasms, or cramps; seizures; unconsciousness; coma. To prevent hyponatremia, avoid drinking more than 1 L of fluid per hour [50].

29.2.5 Caffeine Intoxication

Assessment for caffeine intoxication needs to include a variety of potential sources: prescription and over-the-counter medications, coffee, tea, soft drinks, caffeine shots, or chocolate. Signs and symptoms may involve (1) the central nervous system (CNS): headache, lightheadedness, anxiety, agitation, tremulousness, perioral and extremity tingling, confusion, psychosis, or seizures; (2) the gastrointestinal (GI) system: nausea and vomiting, abdominal pain, diarrhea, bowel incontinence, or anorexia; or (3) the cardiovascular system: palpitations or racing heart rate, or chest pain [51].

Wang, Woo and Bahk [52] and others [53] have discussed the possibility that caffeine could induce de novo psychotic and manic symptoms as well as aggravate previous disorders. Single doses of caffeine up to 200 mg (about 3 mg/kg for a 70-kg adult) are not a safety concern [54, 55].

29.2.6 Food Insecurity

Food-insufficient adolescents were significantly more likely to have had dysthymia, thoughts of death, and a desire to die and to have attempted suicide. Data from the NHANES III survey found that 5% of 15- and 16-year-old adolescents reported they had attempted suicide, and 38.8% reported at least one suicidal symptom [56, 57]. "Of all the top ten causes of death in the United States, only suicide is on the rise," said Charles Nemeroff, MD, at the Congressional Neuroscience Caucus briefing on June 29, 2016 [10].

29.3 The Effect of Micro- and Macronutrients on Mental/Behavioral/Neurological Health

Assessment vitamin, mineral, essential fatty acid, or amino acid status is not common, and methods are not universally accepted. These assessments are also an expense insurance companies are not willing to support. However, it is clear that a deficiency of many nutrients will effect mental status and that the majority of Americans do not meet the 2015–2020 guidelines for a healthy diet [58]. Less than 25% of Americans eat the recommended amounts of fruits, vegetables, or dairy foods. Who knows how many Americans experience fatigue, irritability, difficulty concentrating, altered cognition, anxiety, low-grade inflammation, or other effects of poor nutritional status?

29.3.1 Lipids, Glucose, Amino Acids

29.3.1.1 Essential Fatty Acids

Substantial evidence implicates a dietary deficiency in long-chain omega-3 (LCn-3) fatty acids, EPA and DHA, in the pathophysiology of psychiatric disorders, including MDD, bipolar disorder, schizophrenia, and attention deficit hyperactivity disorder (ADHD). A deficiency may coincide with, and may precede, the initial onset. Case–control studies have consistently observed low erythrocyte eicosapentaenoic acid EPA and/or DHA levels in patients with these disorders [59]. Deficiency of essential fatty acids (EFA) has been linked to violence, hostility and aggression, and suicide [60].

It is wise to remember that research in these areas is complex and fairly recent. A 2015 Cochran Review of research on the use of fatty acids for depression reports the quality of the evidence for all outcomes as low to very low. The number of studies and number of participants were low, and the majority of studies were judged to be at high risk of bias and also suggested a likely bias toward a positive finding for n-3 PUFAs [61].

If n-3 fatty acids are replaced in the diet by n-6 fatty acids the ratios in cell and brain tissues also change. Changes in fatty acid content of cell membranes may increase vulnerability to depression.

Messamore and McNamara report results that indicate the majority (75%) of psychiatric patients exhibit whole-blood EPA + DHA levels at ≤4% of total fatty acid composition. Twenty-five percent of the general population has this level of total fatty acids. Corrective treatment with fish oil-based products has resulted in improvements in psychiatric symptoms without notable side effects. This provides a rationale for screening for and treating omega-3 fatty acid deficiency in patients with psychiatric illness [59, 62].

Even a short course (2 weeks) of a nutritional supplement containing EPA has reduced the rates of new-onset depression to 10% [63]. In addition, both EPA and DHA delayed the onset of depression, and both treatments were well tolerated, with no serious side effects.

Use of EFA as an adjunct has allowed reduced doses of other medications and has reduced the relapse rate in bipolar disorder [59].

Total normal brain and hippocampus volumes have been directly associated with levels of omega-3 fatty acids in a study of more than 1000 postmenopausal women [64].

After a review of evidence, essential fatty acids were accepted in 2006 by the American Psychiatric Association Committee on Research on Psychiatric Treatments, the Council on Research, and the Joint Reference Committee of the American Psychiatric Association (APA) as a treatment option for depression. They concluded that evidence of epidemiologic and tissue compositional studies support a protective effect of n-3 EFA intake for unipolar and bipolar depression. Bipolar disorder and major depression were more influenced by supplementation than mild-to-moderate depression. They also felt use of EPA and DHA appear to have negligible risk. They concluded that the evidence for n-3 fatty acid's protective effect for schizophrenia was less conclusive [65, 66]. The APA adopted the consensus recommendations of the American Heart Association for an EPA + DHA dose of total 1 g/day in patients with MDD.

Controlled intervention studies have in general found that daily doses in the range of 0.2–2 g/day of EPA + DHA may be effective for the treatment of mood symptoms. Emerging evidence also suggests that a larger ratio of EPA to DHA may be more effective for treating symptoms of depression. Edward Kane recommends a ratio of 4:1 LA:ALA (4 parts n-6 [linoleic acid, or LA] to 1 part n-3 [a-linolenic acid, or ALA] [67]. The US Food and Drug Administration (FDA) consider LCn-3 fatty acids doses up to 3 g/day to be generally regarded as safe.

Use of fish oil and krill oil products, when matched for dose and EPA + DHA content, yielded similar plasma and RBC levels of EPA + DHA, indicating comparable bioavailability irrespective of formulation [68].

Reference values for serum essential fatty acids and others may be found at Mayo Clinic Medical Laboratories in Rochester at ► http://www.mayomedicallaboratories.com/test-catalog/Clinical+and+ Interpretive/82426. The American Medical Association current procedural terminology (CPT) codes for blood fatty acid analysis by Gas Liquid Chromatography are (GLC) (82725, 82544, 82492), and those for GLC testing in general are (82541, 82542).

29.3.1.2 Cholesterol

Cholesterol-reducing diets often reduce the n-3 fatty acids in the diet and lower DHA, which is an essential part of the neuronal cell membranes. The brain synthesizes its own cholesterol, and there is little correlation between blood levels of cholesterol and neurotransmitter function [69, 70]. Exploratory correlations (but not planned statistical comparisons) of clinical variables revealed that mania severity and suicidality were positively correlated with UE:E EPA ratio (unesterified: esterified EPA), and that several plasma levels and ratios correlated with panic disorder and psychosis [71].

Although there is controversy about cholesterol becoming too low, total blood cholesterol levels below normal/desired levels have been linked to depression and suicide [72], especially in the young and the elderly. Multivariate Cox regressions predicting suicide, homicides, and accidents suggested that three predictor variables were significant: low total cholesterol, morbid depression, and anti-social personality disorder. The interaction between low cholesterol and morbid depression ($p < 0.005$) indicated individuals with both at baseline were ~7 times more likely to die from external mortality. Patients presenting with low cholesterol and morbid depression in clinical practice may warrant clinical attention and surveillance [73]. In older (80-year-old) males, total cholesterol levels of <160 mg/dL have been associated with depression [74]. Women with non-medicated LDL levels <100 were at increased risk of depression [75].

29.3.1.3 Amino Acids

Amino acids are precursors of neurotransmitters (◘ Table 29.2). The nine essential amino acids are histidine, isoleucine, leucine, lysine, methionine, phenylalanine, threonine, tryptophan, and valine. A non-protein amino acid, homocysteine, can be produced from methionine or converted into cysteine. Hyperhomocysteinemia is a risk factor for brain atrophy, cognitive impairment, and dementia. Plasma concentrations of homocysteine can be lowered by dietary intake of B vitamins.

29.3.1.4 Carbohydrates: Glucose

Possible mental symptoms of hypoglycemia (<70 mg/dL) include: changes in behavior or personality, irritability or nervousness, argumentativeness, and trouble concentrating. Signs and symptoms of hyperglycemia include fatigue, headaches, constipation, diarrhea, and nerve damage. Fasting hyperglycemia is defined as blood glucose higher than 130 mg/dL after not eating or drinking for at least 8 hours. Postprandial hyperglycemia is defined as blood glucose over 180 mg/dL 2 hours after you eat [82].

29.3.2 Vitamins and Minerals

29.3.2.1 Vitamins

Vitamins affect energy metabolism in all cells, the synthesis of neurotransmitters, the activation or inhibition of genes, the repair of DNA, and the function and atrophy of the brain [83]. Below is information regarding the effects on mental status of selected essential vitamins.

Vitamin B1 (Thiamin)

Thiamin deficiency often presents as changes in mental status (alertness, attention, memory) and gait/mobility. Failure to suspect a vitamin deficiency can lead to permanent cognitive and physical disabilities that may necessitate lifelong care [84]. Wernicke's encephalopathy is an acute neuropsychiatric condition caused by an insufficient supply of thiamin to the brain. If undiagnosed or inadequately treated, it is likely to proceed to Korsakoff's Syndrome. Wernicke-Korsakoff encephalopathy (WKE) has historically been associated with alcoholism [85]; but it has more recently been recognized in patients who have anorexia nervosa, a history of bariatric surgery, cancer [86], or who consume a diet of "empty calories" which result in a high calorie/thiamin ratio [87].

Thiamin diphosphate acts as a rate-limiting co-factor for a number of thiamin-dependent enzymes involved in the degradation of branched chain amino acids. Deficiency can lead to exaggeration of emotional and stress reflexes, altered mitochondrial activity, impaired oxidative metabolism, and eventually selective neuronal death [88, 89]. Thiamin and its phosphate esters may regulate gene expression by influencing mRNA structure and in DNA repair [90]. Chronic alcoholism can impair the functions of gene mutations. Excessive alcohol intake keeps SCL19A2 and SCL19A3 turned off [91].

Vitamin B2 (Riboflavin)

An autosomal recessive disorder, Brown–Vialetto–Van Laere Syndrome 2 (BVVLS2) is a progressive neurodegenerative disease that is a consequence of severe riboflavin deficiency, related to disruption of riboflavin transporters. Treatment of a 20-month-old female with BVVLS2 was initiated with 10 mg/kg/day dose of riboflavin and titrated up to 70 mg/kg/day over 2 months. The child tolerated the riboflavin therapy well. Symptoms began improving in a week with widespread improvement in 2–3 months. No new symptoms were observed [92].

Vitamin B3 (Niacin)

A niacin deficiency, or decreased tryptophan (Trp) availability to the brain, has been linked to depression and anxiety through a decrease in serotonin. Pellagra, the niacin-deficiency disease, is known by the "four Ds": dermatitis, diarrhea, dementia, and death. Possible symptoms include depression, anxiety, vertigo, memory loss, paranoia, psychotic symptoms, and aggression [93].

Pellagra was present in 27% of patients who died during hospitalization for alcohol dependence. It has been suggested that niacin deficiency be included in the differential diagnosis of alcoholic withdrawal syndrome [94].

Vitamin B6 (Pyridoxine)

High total intakes of vitamins B6 and B12 are reported as protective against depressive symptoms over time. For every 10 mg increase in the intake of vitamin B6 and for every 10 μg increase in vitamin B12 the risk of developing symptoms of depression were decreased by 2% per year [95]. Low concentrations of vitamin B6, but not of folate or homocysteine, have been associated with lower mini-mental state examination (MMSE) scores and lower attention and executive function scores in a multivariate analysis [96].

Beyond its role as a necessary cofactor in the folate cycle, the role of vitamin B6 in amino acid metabolism makes it a rate-limiting cofactor in the synthesis of neurotransmitters such as dopamine, serotonin, γ-aminobutyric acid (GABA),

Table 29.2 Amino acids and neurotransmitters

Neurotransmitter	Related nutrients	Effect on the brain	Effect on mood	Tissue secreting	Other	Food sources of neurotransmitter precursors
Serotonin—A monoamine	Tryptophan, with B6, B12, and folic acid	Serotonin levels: 1. increase with dietary carbohydrate; 2. increase with omega-3 fatty acids; 3. decrease with high-protein diet	Increased serotonin levels: 1. improve mood; 2. increase pain tolerance and sleep; 3. decrease aggression and cravings	80–90% is secreted by the gastrointestinal tract; serotonin is synthesized by the brain and neurons and secreted into synapses between neurons	Other amino acids, especially branched-chain amino acids (leucine, isoleucine, and valine) compete with tryptophan to cross the blood/brain barrier; serotonin synthesis is sensitive to the availability of tryptophan from meals [76]	1. Foods high in tryptophan: meats, milk, yogurt, and eggs 2. Foods high in carbohydrates: grains, potatoes, fruit, and milk
Dopamine—A monoamine	Tyrosine and phenylalanine; along with B-12, folic acid, magnesium	Tyrosine blood levels rise with a high-protein diet; dopamine synthesis is insensitive to the availability of tyrosine from meals due to activity of tyrosine hydroxylase	1. Sufficient dopamine increases tolerance, mood, alertness, cognition, and problem solving 2. Dopamine depletion results in blunted pleasure response, clouded thinking, lowered motivation, and lowered vigilance in patients with schizophrenia [77]	Produced in the brain and released by the hypothalamus	Stress depletes tyrosine from the blood	1. Sources of L-phenylalanine: beef, poultry, pork, fish, dairy products, eggs, and soy products such as soybean flour and tofu and almond 2. The artificial sweetener aspartame is also high in phenylalanine 3. Sources of tyrosine include the following: oatmeal, soybeans, skim milk, beef, chicken, brown rice, and almonds

Norepi-nephrine—A mono-amine	Tyrosine with B12, folic acid, magnesium	Norepinephrine increases with a high-protein diet	Low nor-epinephrine levels associated with irritability, depression, moodiness	Norepinephrine is synthesized and released by neurons in the brain and central nervous system and also by the sympathetic ganglia located near the spinal cord or in the abdomen and is released directly into the bloodstream by the adrenal glands	1. Too much results in fear, depression, compulsivity, mood swings, mania, and addiction
2. Too little results in depression and paranoia	Sources of tyrosine include the following: oatmeal, soybeans, skim milk, beef, chicken, brown rice, and almonds
Histamine	Histidine, one of the essential biogenic amines; responses, such as inflammation, gastric acid secretion, neurotransmission, and immune modulation	Histamine modulates neurotransmitters in the brain, stress-related release of hormones from the pituitary and central aminergic neurotransmitters [78] Diamine oxidase breaks down histamine to prevent buildup in plasma; deficiency may be genetic or acquired Alcohol attacks diamine oxidase and histamine builds up	Promotes/controls feeding behavior, wakefulness [79]	H_3 subtype, focused on as the therapeutic target for cognitive dysfunctions; expressed preferentially in neurons and modulates neurotransmitter release [80]. Produced by the granules in mast cells and basophils as part of a local immune response. Found in connective tissue; also synthesized in the brain; acts as a neurotransmitter. Stored and released in the entero-chromaffin-like (ECL) cells of the stomach	Serum: N = diamine oxidase activity ≥80 HDU/ml (HDU = high-dependency unit) Symptoms of excess histamine include migraine headaches, gastrointestinal disorders, chronic fatigue, and ADHD	Sources of histidine include the following: beef, pork, chicken, turkey, peanut butter, cheese, eggs, sesame seeds, and sunflower seeds Diamine oxidase is available on non-prescription basis Avoidance of alcohol advised

Adapted with permission from Leyse-Wallace [81]

noradrenaline, and the hormone melatonin. Even mild deficiency results in down-regulation of GABA and serotonin synthesis [93].

A 2012 report from the CDC demonstrated that 10.5% of the entire US population were biochemically deficient in vitamin B6 [97]. Groups most likely to be deficient in B6 are women of reproductive age, especially current and former users of oral contraceptives; male smokers; non-Hispanic African-American men; and men and women over age 65 [98].

Sensory neuropathy typically develops at doses of pyridoxine in excess of 1000 mg/day with pain and numbness of the extremities and in severe cases, difficulty walking. Since placebo-controlled studies have generally failed to show therapeutic benefits of high doses of pyridoxine, there is little reason to exceed the UL of 100 mg/day [99].

Vitamin B12

Mental signs of a B12 deficiency include agitation, apathy, emotional instability, psychosis, irritability, memory problems, mood swings, sleep disturbance, personality changes, suspiciousness/paranoia, and trouble concentrating. Psychiatric symptoms of B12 deficiency may present before hematologic findings [100]. Seventy to ninety percent of persons with a clinically relevant B12 deficiency have neurological disorders such as numbness and tingling in the hands and feet; in about 25% of cases these are the only clinical manifestations of the B12 deficiency [101]. Damage to the CNS may become irreversible. A review by the *Cochrane Group* confirmed orally administered cyanocobalamin with initial doses of 1000–2000 μg daily, then weekly, is just as effective as parenteral administration [101]. Oral supplements have been confirmed to be as effective as injections.

Vitamin B12 functions as a cofactor for methionine synthase and L-methylmalonyl-CoA mutase. Methionine synthase catalyzes the conversion of homocysteine to methionine. Methionine is required for the formation of SAMe, a universal methyl donor for almost 100 different substrates, including DNA, RNA, hormones, proteins, and lipids [102]. A deficiency in vitamin B12 causes an accumulation of homocysteine in the blood and may decrease the ability to metabolize neurotransmitters.

Elderly patients judged to have normal nutritional status using the Mini Nutritional Assessment (MNA) were found to be low in vitamin B12 [103]. B12 and folate have independently been shown to predict positive affect [104].

Vitamin C

Prior to the development of scurvy, vitamin C deficiency is accompanied by symptoms of mood disturbance, personality changes, and changes in scores on the Minnesota Multiphasic Personality Inventory (MMPI). Psychological symptoms occur earlier and with less severe depletion than physical signs and symptoms. Kinsman reports personality changes occurred at levels of whole blood of 1.21–1.17 mg/100 mL. Physical signs/symptoms occurred at 0.67–0.14 mg/100 mL ascorbic acid in whole blood [105]. In 2010, based on data from 2003 to 2004, 7% of the US population was determined to be biochemically deficient in ascorbic acid [106]. Tobacco smoking is frequent in mental health populations, which increases the need for vitamin C and also folic acid.

Short-term therapy for hospitalized patients with vitamin C raised plasma vitamin C levels, improved mood disturbance by 71%, and reduced psychological distress by 51% [107]. A number of reports describe observing signs of vitamin C deficiency in patients being treated for anorexia nervosa [108–111]. Low intakes of vitamin C, as well as other nutrients, are reported by people experiencing food insecurity [112].

Polymorphisms in the vitamin C transporter genes have been shown to compromise genes encoding sodium-dependent ascorbate transport proteins [113].

Folate

Low folate has been associated with changes in cognition affecting memory, attention, and language. Three weeks of a folate-deficient diet (5 μg/day) was reported to produce a decrease of serum levels from 7 ngs/mL to below 3 ngs/mL. In the 18th week, mental changes including insomnia, irritability, fatigue, and forgetfulness occurred, as well as anemia and red cell macrocytosis. Oral folate replacement of 250 μg caused remission of symptoms [114, 115].

There are contradictory opinions in the literature regarding interpretation of studies involving folic acid and depression [116, 117]. A group that received selective serotonin reuptake inhibitors and selective norepinephrine reuptake inhibitors (SSRI/SNRI), combined with 7.5–15 mg/day of L-methylfolate, was more effective than monotherapy with only an SSRI/SNRI [118].

Taking folic acid does not appear to improve the antidepressant effects of lithium in people with bipolar disorder. However, taking folate with the medication valproate improves the effects of valproate. There is conflicting evidence about the role of folic acid in age-related decline in memory and thinking skills. However, other research suggests no benefit [119].

A 16-week study showed that 2 mg of folic acid plus 400 μg of vitamin B12 improved negative symptoms of schizophrenia significantly. However, improvements were noted only when genotype was taken into account. Only patients homozygous for the 484T allele of FOLH1(rs202676) demonstrated significantly greater benefit with active treatment [120].

The tolerable upper intake of folic acid for healthy adults is 1000 mcg/day. Folate interacts with a large number of common and specialty drugs. Doses over 1000 mcg/day should be assessed for potential interactions. Excess folate may cause low blood pressure and may lower blood glucose. The Mayo Clinic website lists many herbs, supplements, and prescription and over-the-counter drugs that may interact with folic acid. For the entire list, see: ▶ http://www.mayoclinic.org/drugs-supplements/folate/safety/hrb-20059475.

Vitamin A

The retinoids are a family of compounds derived from vitamin A or pro-vitamin A carotenoids. They are necessary for the function of the brain and central nervous system, and new discoveries point to a central role ranging from neuroplasticity to neurogenesis. Retinoic acid is important for a regular pattern of sleep [121].

Isotretinoin is a vitamin A-derived medication that is associated with significant adverse effects including arthralgia myalgia, nosebleeds, headache, dyslipidemia, liver dysfunction, and depression. Case reports describe depression, suicidality, psychosis, violence, and aggression developing in conjunction with isotretinoin treatment (13-cis retinoic acid/Acutane) used in treating severe acne [122]. Evidence suggests probable clinical exacerbation of bipolar mood disorder and possible links to psychosis. Genetic polymorphisms in the retinoic acid receptor alpha (RARA), one of the main targets of isotretinoin, showed an association with the adverse effects of oral isotretinoin therapy [123].

Vitamin D

Low levels of vitamin D are found frequently in those with mental illness, especially schizophrenia [124–127]. One study showed that vitamin D status was associated with more negative symptoms in first-episode schizophrenia. Mixed results are reported in the relationship between vitamin D levels and depression. One study observed a trend toward a greater decrease in the Beck Depression Inventory in the vitamin D treatment group than in the placebo group [128]. One study showed that 58% of individuals who attempted suicide were deficient in vitamin D [129]. Another found initially low vitamin D levels were associated with clinically significant depressive symptoms across follow-up [130].

Vitamin D is now considered a potent neuroactive/neurosteroid hormone, critical to brain development and normal brain function, and valued for its anti-inflammatory property. A vitamin D ligand-receptor has been found throughout the body including the central nervous system [131].

A proposal by Patrick and Ames [16] in 2015 describes evidence that insufficient levels of vitamin D and essential fatty acids may explain an underlying mechanism contributing to psychiatric disorders and depression through a vitamin D receptor and the enzyme tryptophan hydroxylase (TPH2). Optimal intake of these nutrients may prevent and modulate the severity of brain dysfunction and alter behavior. Tryptophan depletion in the brain reduces serotonin levels, which results in favoring short-term over long term rewards, impulsiveness, antisocial aggressive behavior, feelings of anger, and incidence of self-injury. They comment that exercise is a helpful intervention because muscles preferentially absorb branch-chain amino acids, which compete with tryptophan for transporters across the blood-brain barrier. Exercise facilitates more tryptophan crossing the brain-blood barrier for serotonin synthesis.

Vitamin K

The vitamin K-dependent protein GAS6 is involved in chemotaxis, mitogenesis, cell growth, and re-myelination of nerves. Low levels of vitamin K have been associated with dementia and alterations in cognition, although much of the research on vitamin K and the nervous system has been on animals. One area of future research is the relative deficiency state of vitamin K induced by use of the drug Warfarin and its potential effect on cognition.

29.3.3 Minerals

Some minerals are essential for health and functioning. Some are generally toxic to humans. (See also ◘ Table 29.3) Some are essential, but may be toxic at extreme levels of intake. Either inadequate or toxic levels of minerals may affect mental functioning.

29.3.3.1 Nutrient Minerals
Copper

High concentrations of copper are found in the brain. Recent studies have found that copper helps brain cells communicate with each other by acting as a brake when it is time for neural signals to stop [133]. Some patients with Alzheimer's Disease had normal serum copper, but higher levels were found in cerebrospinal fluid [134]. In cases of copper deficiency, dopamine-hydroxylase (DBH) activity may decrease and result in elevated ratios of dopamine to norepinephrine. The balance between copper and zinc is important in maintaining neurotransmitter levels and function. Walsh has associated overload of copper and copper-zinc imbalance with schizophrenia, bipolar disorder, ADHD, and postpartum depression [135].

Wilson's disease is an autosomal recessive disorder resulting in copper accumulation in basal ganglia and neuronal loss. One third of patients present with psychiatric symptoms such as depression, labile mood, impulsiveness, disinhibition, self-injurious behavior, or psychosis.

Free plasma copper (in contrast to bound plasma copper) shows a significant inverse relationship with mini-mental state examination (MMSE) and attention-related neuropsychological test scores. About 95% of patients with the neurological presentation manifest the Kayser-Fleischer ring—a primary site of copper deposition in Wilson's disease.

A case of a male diagnosed with celiac disease and a hemoglobin of 4.9 g/dL, which was non-responsive to a gluten-free diet or supplements of iron, folic acid, and B12, was found to be low in copper. Two months of copper supplementation normalized his neutropenia [136].

Lithium

Lithium inhibits the release of thyroid hormone (TH), and lithium treatment has been associated with the development of goiter within weeks to months of initiating lithium ther-

Table 29.3 Potentially toxic doses of drug of minerals [132]

Mineral	Blood reference level	DRI: adults, M/F	Potentially toxic dose
Calcium	Serum Ca Hypocalcemia ≤8.0–8.5 mg/dL, Hypercalcemia: S/S > 11.5 mg/dL, Critical value: >12.0 mg/dL, A Medical emergency >15.0 mg/dL	1000–1300 mg/day	12,000 mg
Phosphorous	Serum phosphorous, hypophosphatemia 1.5–2.4 mg/dL moderately low; not usually associated with clinical signs and symptoms <1.5 mg/dL may result in muscle weakness, hemolysis of red cells, coma, bone deformity, and impaired bone growth <1.0 mg/dL are potentially life-threatening Values are lowest in the morning, peak first in the late afternoon, and peak again in the late evening	700 mg/day	12,000 mg
Magnesium	S/S Mg deficiency ≤1.0 mg/dL. Levels ≥9.0 mg/dL may be life-threatening	320–420 mg/day	6000 mg
Iron	Serum ferritin Males, 24–336 mcg/L Females, 11–307 mcg/L May be elevated inflammation, liver disease, chronic infection, autoimmune disorders, and malignancy May be elevated in iron storage disorders such as hemochromatosis, hemolytic anemia, and sideroblastic anemia and in those who have had multiple blood transfusions Native Africans, African Americans, and Asians may have higher mean concentrations of serum ferritin	8–18 mg/day	100 mg
Zinc	Normal urinary excretion of zinc—300 to 600 mcg/g creatinine/day Serum zinc—0.66 to 1.10 mcg/mL High urine/low serum zinc may be caused by hepatic cirrhosis, neoplastic disease, or increased catabolism High urine/normal or elevated serum zinc indicates a large dietary source, usually in the form of high-dose vitamins Low urine/low serum zinc may be caused by dietary deficiency or loss through exudation or gastrointestinal losses	8–11 mg/day	500 mg
Copper	16 years ± Urine 15–60 mcg/L Low urine copper levels: malnutrition, hypoproteinemias, malabsorption, and nephrotic syndrome Low serum copper associated with Wilson disease Excess use of denture cream containing zinc can cause hypocupremia Increased zinc consumption interferes with normal copper absorption	900 µg/day	100 mg
Fluoride	Plasma N = 0.0–4.0 mcmol/L Plasma fluoride >4 mcmol/L = excessive exposure Plasma fluoride >4 mcmol/L = adverse effects W/ fluoride-treated water-plasma fluoride 1–4 mmol/L W/O fluoride-treated water-plasma fluoride <1 mcmol/L	3–4 mg/day	4–20 mg
Iodine	Serum N = 40–92 ng/mL 80–250 ng/mL may indicate hyperthyroidism I ≥ 250 ng/mL may indicate iodine overload	120–150 µg/day	2 mg
Selenium	Serum N adult = 70–150 ng/mL (0.15 parts per million) Children require less circulating selenium than adults Se deficiency serum <40 ng/mL; associated with loss of glutathione peroxidase activity Usual treatment—Selenium supplementation to raise serum concentration > 70 ng/mL	55 µg/day	1 mg

apy. Goiter in patients receiving lithium therapy ranges from 20% of patients in iodine-replete areas to 87% in patients residing in iodine-deficient areas. Up to a third of patients on lithium therapy who develop goiter also may develop hypothyroidism, which usually remains subclinical, although it could conceivably affect weight status [137].

Magnesium

Possible causes of hypermagnesemia include laxative abuse, diuretics, lithium intoxication, and hypothyroidism; causes of hypomagnesemia, meanwhile, can include gastrointestinal diseases (Crohn's, ulcerative colitis, etc.), renal disease, vomiting, laxative abuse, and alcohol abuse [138].

Chronic magnesium deficiency may be related to numerous physical and mental disorders, including chronic fatigue, depression, irritability, disorientation, depression, and psychosis. Magnesium ions may block synaptic transmission by interfering with the release of acetylcholine. Consumption of alcohol, even in modest amounts, can double or even quadruple the excretion of magnesium. Some over-the-counter and prescription drugs, such as proton pump inhibitors, can lower body magnesium levels.

Mg is essential for the proper activity of DNA polymerase I, RNA polymerase, and DNA helicase. Its deficiency results in errors during DNA replication, transcription, and translation.

Supplements in the form of magnesium oxide, chloride, or lactate provide 60–84 mg magnesium per tablet. Sustained-release forms may reduce the potential diarrhea from magnesium supplementation.

Manganese

A U-shaped curve describes the need and effect of manganese in humans. Manganese (Mn) toxicity is a greater concern than deficiency. Manganese toxicity can result in a permanent neurological disorder such as tremors, difficulty walking, and facial muscle spasms. The toxicity syndrome, manganism, may be preceded by psychiatric symptoms, such as irritability, aggressiveness, and even hallucinations [139]. Iron deficiency has been known to increase manganese accumulation in the brain.

In the brain, the manganese-activated enzyme glutamine synthetase converts the amino acid glutamate to glutamine. Glutamate, an excitotoxic neurotransmitter, is a precursor to an inhibitory neurotransmitter, γ-aminobutyric acid (GABA). Expression of genes implicated in manganese, iron, and zinc homeostasis has been shown to increase with vitamin D3 treatment. It is possible that the control of systemic levels of zinc and manganese is regulated by vitamin D3 in the intestine [140]. With vitamin D treatment, the gene encoding the zinc and manganese transporter, ZnT10, had about a 15-fold increase in expression. Vitamin D3 also increases expression of the SLC30A10 gene. Mutations of SLC30A10 can cause manganese intoxication.

Selenium

Small amounts of selenium (Se) are essential; excess amounts of selenium are toxic. There is greater uptake of selenium in the brain than in other tissues. At times of deficiency, the brain retains selenium to a greater extent than other areas of the body.

The function of selenium in the brain is related to functions of selenium-dependent enzymes in the oxidative damage-protective system: polyunsaturated fatty acids (PUFA) are protected from peroxidation by selenium-dependent glutathione peroxidase (Se-GPX). Se-GPX activity occurs in myelin tissue.

Energy-adjusted intakes of selenium have been found to be a predictor of high scores for depression on the Beck Depression Inventory in Iran [141]. An analysis for 35 serum trace elements showed a different profile in schizophrenic patients and controls in China. Concentrations of selenium, zinc, and cesium were significantly lower in schizophrenia [142].

Serum selenium concentrations decline with age. Marginal or deficient selenium concentrations might be associated with age-related declines in brain function, possibly due to decreases in selenium's antioxidant activity [143].

Supplementation with a selenium-rich Brazil nut has shown potential in reducing cognitive decline in patients with mild cognitive impairment (MCI) patients, and could prove to be an effective nutritional approach early in the disease process to slow decline. Response to intake of Brazil nut intake by mRNA appears to be influenced by a number of SNPs and genotypes [144, 145]. Brazil nuts contain very high amounts of selenium (68–91 mcg per nut) and could possibly cause selenium toxicity if consumed regularly.

A strong garlic-like odor is usually present in acute, subtoxic, and chronic selenium poisoning, attributed to dimethylselenide. Neurologic symptoms of acute toxicity may include tremor, peripheral neuropathy, muscle spasms, restlessness, confusion, delirium, and coma. The most common sign of chronic selenium poisoning are nail changes: Nails become brittle, with white spots and longitudinal streaks. Other signs include nausea, vomiting, diarrhea, fatigue, and skin lesions. As selenosis progresses, decreased cognitive function, weakness, paralysis, and death occur [146]. The upper limit for a maximum safe daily dietary intake, based on the amounts of selenium that are associated with hair loss and nail brittleness, has been estimated at 400 μg for selenium from food and supplements combined [143].

Zinc

Zinc deficiency has been shown to induce depression-like and anxiety-like behaviors, and zinc supplementation has improved the efficacy of antidepressant drugs in depressed patients. Zinc deficiency may cause decreased appetite, change in the sense of taste, and weight loss. Data may indicate a role for zinc deficiency in the development of mood disorders, but also show that zinc may also be important in their treatment [147].

Zinc is found in in many body systems, participating in the activity of over 200 enzymes. Zinc functions in the incorporation of thymine into DNA, one of the four bases forming the structure of DNA. Zinc deficiency leads to both primary and secondary alterations in brain development and growth and is one of the causes of primary neural tube defect. Indirect effects of zinc deficiency occur in the lipid content of the developing brain.

Changes in brain zinc concentration are observed in Alzheimer's disease, Down syndrome, epilepsy, multiple sclerosis, retinal dystrophy, and schizophrenia. Recent investigations into molecular mechanisms suggest a role for zinc in the regulation of neurotransmitter systems, antioxidant mechanisms, neurotrophic factors, and neuronal precursor cells.

29.3.4 Potentially Toxic Minerals

29.3.4.1 Cadmium

In a review of heavy metals, Wu et al. explained Cadmium (Cd) toxicities are expressed through depletion of glutathione and bonding to sulfhydryl groups of proteins. Cadmium indirectly generates reactive oxygen species (ROS) by its ability to replace iron and copper. Lead becomes toxic to organisms through the depletion of antioxidants [148].

The most significant exposure to cadmium comes from cigarette smoke. Occupational exposure to cadmium has led to (a) slowing of visuomotor functioning on neurobehavioural testing; (b) increase in complaints of peripheral neuropathy; (c) altered equilibrium; and (d) lowered ability to concentrate, which were dose-dependent with cadmium exposure. Environmental exposure to cadmium has been shown marginally significant in relation to the development of Alzheimer's [149].

Sources of cadmium for non-smokers are terrestrial foods (98%), aquatic foods (1%), and water (1%). The content of terrestrial foods varies significantly related to the type of food crop grown, agricultural practices, and the atmospheric deposition of cadmium onto exposed plant parts.

With an absorption rate of 5% from ingestion, the average person is believed to retain about 0.5–1.0 μg of cadmium per day from food. The World Health Organization (WHO) established a provisional tolerable *weekly* intake (PTWI) for cadmium at 7 μg/kg of body weight. For a 150-pound (68 kg) person, this amounts to 477 μg/week, or 68 μg/day [150].

29.3.4.2 Mercury

Methyl mercury is more than 95% absorbed and can accumulate if rates of consumption exceed rates of excretion. It has a strong affinity for sulfhydryl groups in tissues and accumulates to a greater concentration in brain, muscle, and kidney. Methyl mercury easily crosses the blood-brain barrier, where it is biotransformed into inorganic mercury. The brain: blood concentration ratio after the initial distribution is between 10:1 and 5:1. Once in the central nervous system, methyl mercury can be demethylated to inorganic mercury. Inorganic mercury has a half-life of years in brain tissue [151, 152].

Since fish consumption is promoted for a healthy diet and excess fish intake may carry contaminants such as mercury, fish consumption should be part of diet histories and comprehensive health screenings to identify those at risk for mercury accumulation. The symptoms of excess methyl mercury include fatigue, headache, decreased memory, decreased concentration, muscle or joint pain, prickly sensations, and problems with fine motor coordination, speech, sleep, and walking. Although fish is the main source of mercury, additional routes of exposure to mercury are mercury vapor in certain occupations and some medications, herbs, remedies, and vaccines [153].

A general medical practice in Seattle reported out of 700 individuals who were screened, 123 needed laboratory evaluation. One hundred three (89%) of these had had mercury levels ≥5.0 μg/L; 63 (54%) had levels ≥10 μg/L; 19 (16%) had levels ≥20 μg/L, and four had levels >50 μg/L. After patients were advised to omit mercury sources or fish, follow-up found mercury levels declined rapidly in the first 3 weeks, followed by a slower reduction over time. All but two patients reduced their level to <5.0 μg/L by 41 weeks [154].

> Mercury Reference Dose (RfD): no more than 0.1 μg Hg/kg body weight/day.
> Acceptable levels:
> Whole blood mercury level < 5.0 μg/L.
> Hair level < 1.0 μg/g.

Consumer Reports lists these types of seafood as containing the lowest mercury content: shrimp, scallops, sardines, oysters, squid, tilapia, and wild and Alaskan salmon. The publication estimates that an adult could safely eat 36 ounces per week and children could eat 18 ounces per week of these fish [155].

29.4 Drugs: Nutrition and Mental Status

Weight gain is one effect of a number of psychotropic drugs. (see ◘ Table 29.4) This, in turn, is a frequent cause of patients discontinuing medication. Nutrition counseling incorporating diet and exercise principles has been successful with psychiatric patients, who have shown interest and motivation with multidisciplinary approaches [158].

- Drug: Nutrient Interactions include:
  - Protein pump inhibitors (Prilosec, Prevacid), H2 receptor antagonists (Tagamet, Pepcid, Zantac), and metformin reduce absorption of vitamin B12 and may induce a deficiency.
  - Antidepressant medications:
    - Taking vitamin B6 supplements may improve the effectiveness of some tricyclic antidepressants, such as nortriptyline (Pamelor), amitriptyline (Elavil), desipramine (Norpramin), and imipramine (Tofranil).
    - Monoamine oxidase inhibitors (MAOIs) may reduce blood levels of vitamin B6. Examples of

Table 29.4 Nutrition and psychotropic medications [156, 157]

Medication type	Generic Name (trade/brand name)	Side effects
Antidepressants: Selective serotonin reuptake inhibitors (SSRIs) are often prescribed for depression, obsessive compulsive disorder, social anxiety disorder, and occasionally eating disorders	Fluoxetine (Prozac) Sertraline (Zoloft) Paroxetine (Paxil) Fluvoxamine (Luvox) Citalopram (Celexa)	Weight gain Gastrointestinal symptoms Avoid tryptophan supplements St. John's wort
Antidepressants: Tricyclic	Amitriptyline (Elavil) Amoxapine (Asendin) Nortriptyline (Aventyl, Pamelor) Doxepin (Adapin, Sinequan) Imipramine (Janimine, Tofranil) Desipramine (Norpramin) Clomipramine (Anafranil) Protriptyline (Vivactil)	Carbohydrate cravings, slowing of metabolism, and appetite stimulation Weight gain may be dose-dependent and related to duration of therapy The greatest weight gain is from imipramine and amitriptyline. Minimal weight gain occurs with protriptyline and desipramine
Antipsychotic	Thioridazine (Mellaril) Novo-Ridazine Haloperidol (Haldol, Peridol) Molindone (Moban) Chlorpromazine (Thorazine)	Appetite changes, elevated cholesterol, glucose up or down, weight increase Take Mg, Ca, and iron supplement separately by 4 hours May increase riboflavin need
Atypical antipsychotics	Clozapine (Clozaril) Olanzapine (Zyprexa) Sertindole (Serlect) Risperidone (Risperdal) Quetiapine (Seroquel) Ziprasidone (Zeldox)	Greatest association with weight gain with clozapine and olanzapine
Antiseizure/anticonvulsants Phenytoin, Depakote anticonvulsants/putative mood stabilizers May be prescribed for seizure disorders, bipolar disorder, and selected forms of depression	Divalproex (Depakote) Carbamazepine (Tegretol) Lamotrigine (Lamictal) Gabapentin (Neurontin) Topiramate (Topamax)	Initial weight gain, esp if patient at or below normal BMI May cause hyperinsulinemia; increased appetite May need Ca, Vit D, B-1, carnitine supplement; folate suppl frequently Rx; folic acid is an antagonist of phenytoin (dilantin), phenobarbitol, methotrexate, and other medications, appetite change
Lithium		Requires consistent fluid and sodium intake; moderate caffeine intake

Monoamine oxidase inhibitors
  Nonselective irreversible type
  Reversible (RIMA) types
	Isocarboxazid (Marplan), Phenelzine (Nardil), Tranylcypromine (Parnate)
	Moclobemide (Manerix) and Toloxatone (Humoryl)
		Limit foods w/ tyramine with nonselective MAOI: chocolate, aged and mature cheeses, smoked and aged/fermented meats, hot dogs, some processed lunch meats, fermented soy products, and draft beers

MAOIs include phenelzine (Nardil) and tranylcypromine (Parnate).
- Phenytoin (Dilantin used to treat seizures): Vitamin B6 makes phenytoin less effective [159].
- Riboflavin intake/status: Some drugs may lower riboflavin levels, but riboflavin may improve the effects of antidepressants: imipramine (Tofranil), desipramine (Norpramin), amitriptyline (Elavil); nortriptyline (Pamelor), phenothiazines (Thorazine), phenytoin (Dilantin).

29.5 Summary Statement

Nutritional status and deficient or excess nutrients have meaningful effects on mental status, through enzymes, neurotransmitters, and metabolism; they interact with many drugs; and they influence genetic transcription. Thorough nutritional assessment and targeted interventions have the potential to improve the course of treatment and quality of life for individuals suffering from altered mental status.

References

1. Puchalska P, Crawford PA. Multi-dimensional roles of ketone bodies in fuel metabolism, signaling, and therapeutics. Cell Metab. 2017;25(2):262–84. https://doi.org/10.1016/j.cmet.2016.12.022.
2. Koloski NA, Jones M, Talley NJ. Evidence that independent gut-to-brain and brain-to-gut pathways operate in the irritable bowel syndrome and functional dyspepsia: a 1-year population-based prospective study. Aliment Pharmacol Ther. 2016;44(6):592–600. https://doi.org/10.1111/apt.13738.
3. Lim SY, Kim EJ, Kim A, et al. Nutritional factors affecting mental health. Clin Nutr Res. 2016;5(3):143–52. https://doi.org/10.7762/cnr.2016.5.3.143.
4. Hallahan B, Ryan T, Hibbeln JR, et al. Efficacy of omega-3 highly unsaturated fatty acids in the treatment of depression. Br J Psychiatry. 2016;209(3):192–201. https://doi.org/10.1192/bjp.bp.114.160242.
5. Parade Publications. What Americans think about aging and health. *Parade Magazine*. New York, 5 Feb 2006. p. 11.
6. National Institute of Mental Health (NIH). Serious mental illness (SMI) among U.S. adults. Retrieved from: https://www.nimh.nih.gov/health/statistics/prevalence/serious-mental-illness-smi-among-us-adults.shtml.
7. National Institute of Mental Health (NIH). Any mental illness (AMI) among U.S. adults. Retrieved from: https://www.nimh.nih.gov/health/statistics/prevalence/any-mental-illness-ami-among-us-adults.shtml.
8. National Association of Social Workers (NASW). BULLETIN: groundbreaking surgeon general's report on mental health. Dec 1999. Retrieved from: http://www.naswdc.org/practice/behavioral_health/surgeon_gen.asp.
9. U.S. Department of Health and Human Services. Mental health: a report of the surgeon general. Rockville, 1999. Retrieved from: https://profiles.nlm.nih.gov/ps/access/NNBBHS.pdf.
10. Charles B. Nemeroff headlines capitol hill hearings on depression, suicide and opioid addiction. The American College of Neuropsychopharmacology (ACNP) Liaison Committee, co-chaired by Charles B. Nemeroff, M.D., Ph.D., Leonard M. Miller Professor and Chair of the Department of Psychiatry and Behavioral Sciences, presented a briefing on depression and suicide titled "A Precision Medicine Approach to Mental Illness," at the Congressional Neuroscience Caucus on June 29. Retrieved from: http://med.miami.edu/news/dr.-charles-b.-nemeroff-headlines-capitol-hill-hearings-on-depression-suici.
11. Nemeroff CB. Psychiatric disorders are brain diseases: the pathway to advancing treatment and eliminating stigma. 2015. http://nami-ofmiami.org/wp-content/uploads/sites/50/2015/05/Nemeroff-Psychiatric-Disorders-are-Brain-Diseases-NAMI-May-2015-Copy.pdf.
12. WebMD. What are the types of autism spectrum disorders? Retrieved from: http://www.webmd.com/brain/autism/autism-spectrum-disorders#1.
13. Nemade R, Dombeck M. Formal DSM schizophrenia spectrum and other psychotic disorders diagnoses. AMHC. http://www.amhc.org/poc/view_doc.php?type=doc&id=8815&cn=7.
14. Smoller J, Kendler K, Craddock N, Cross-Disorder Group of the Psychiatric Genomics Consortium, et al. Identification of risk loci with shared effects on five major psychiatric disorders: a genome-wide analysis. Lancet. 2013;381(9875):1371–9. Published online 2013 Feb 28. https://doi.org/10.1016/S0140-6736(12)62129-1.
15. Jones, JW and Sidwell M. Essential fatty acids and treatment of psychiatric diseases. Orig Internist, Inc. 2001 March, 8:5.
16. Patrick RP, Ames B. Vitamin D and the omega-3 fatty acids control serotonin synthesis and action, part 2: relevance for ADHD, bipolar disorder, schizophrenia, and impulsive behavior. FASEB J. 2015;29(6):2207–22. https://doi.org/10.1096/fj.14-268342.
17. Flint J. First robust genetic links to depression emerge. Nature. 2015;523(7560):268–9. Reported by Heidi Ledford.
18. Haller J. Biokinetic parameters of vitamins a, B-1, B-2, B-6, E, K and carotene in humans. In: Lieberman HR, Kanarek RB, Prasad C, editors. Nutritional Neuroscience. New York: CRC Taylor and Francis; 2005. p. 229.
19. Hastings CN, Sheridan H, Pariante CM, Mondelli V. Does diet matter? The use of polyunsaturated fatty acids (PUFAs) and other dietary supplements in inflammation-associated depression. Curr Top Behav Neurosci. 2016;31:321–38. http://www.ncbi.nlm.nih.gov/pubmed/27431396.
20. McNamara RK. Detection and treatment of long-chain omega-3 fatty acid deficiency in adolescents with SSRI-resistant major depressive disorder. PharmaNutrition. 2014;2(2):38–46. 49885 (8/2014).
21. Kaner G, Soylu M, Yuksel N, et al. Evaluation of nutritional status of patients with depression. Biomed Res Int. 2015;2015:521481. https://doi.org/10.1155/2015/521481.
22. Sarris J, Murphy J, Mischoulon D, et al. Adjunctive nutraceuticals for depression: a systematic review and meta-analyses. Am J Psychiatry. 2016;173(6):575–87. https://doi.org/10.1176/appi.ajp.2016.15091228.
23. McGuffin P, Rijsdijk F, Andrew M, Sham P, Katz R, Cardno A. The heritability of bipolar affective disorder and the genetic relationship to unipolar depression. Arch Gen Psychiatry. 2003;60:497–502.
24. Demirhan O, Tastemir D, Sertdemir Y. The expression of folate sensitive fragile sites in patients with bipolar disorder. Yonsei Med J. 2009;50(1):137–41.
25. Ozbeck Z, Kuckkali CI, Ozkok E, Orhan N, Aydin M, Kilic G, Sazci A, Kara I. Effect of the methylenetetrahydrofolate reductase gene polymorphisms on homocysteine, folate and vitamin B12 in patients with bipolar disorder and relatives. Prog Neuro-Psychopharmacol Biol Psychiatry. 2008;32(5):1331–7.
26. Evans SJ, Ringrose RN, Harrington GJ, et al. Dietary intake and plasma metabolomic analysis of polyunsaturated fatty acids in bipolar subjects reveal dysregulation of linoleic acid metabolism. J Psychiatr Res. 2014;57:58–64. https://doi.org/10.1016/j.jpsychires.2014.06.001.
27. NYU Langone University. Mental & behavioral health. Retrieved from: http://nyulangone.org/conditions/areas-of-expertise/mental-behavioral-health.
28. Lopresti AL, Jacka FN. Diet and bipolar disorder: a review of its relationship and potential therapeutic mechanism of action. J Altern Complement Med. 2015;21(12):733–9. https://doi.org/10.1089/acm.2015.0125.
29. Nierenberg AA, Kansky C, Brennan BP, et al. Mitochondrial modulators for bipolar disorder: a pathophysiologically informed paradigm for new drug development. Aust N Z J Psychiatry. 2013;47(1):26–42. https://doi.org/10.1177/0004867412449303. Epub June 18, 2012. http://www.ncbi.nlm.nih.gov/pubmed/22711881.
30. Center for Quality Assessment and Improvement in Mental Health (CQAIMH). Bipolar disorder: monitoring lithium serum levels. Retrieved from: http://www.cqaimh.org/measure_ls.html.
31. Mota de Freitas D, Castro MM, Geraldes CF. Is competition between Li+ and Mg2+ the underlying theme in the proposed mechanisms for the pharmacological action of lithium salts in bipolar disorder? Acc Chem Res. 2006;39(4):283–91.
32. Pronsky ZM. Food medication interactions. 15th ed. Birchrunville; 2008. p. 182.
33. Bauer M, Glenn T, Conell J, et al. Common use of dietary supplements for bipolar disorder: a naturalistic, self-reported study. Int J Bipolar Disord. 2015;3(1):29. https://doi.org/10.1186/s40345-015-0029-x.
34. van Os J. 'Schizophrenia' does not exist. BMJ. 2016;352:i375. https://doi.org/10.1136/bmj.i375.
35. Zupanick CE. The new DSM-5: schizophrenia spectrum and other psychotic disorders. Aroostook Mental Health Services. http://www.amhc.org/1418-dsm-5/article/51960-the-new-dsm-5-schizophrenia-spectrum-and-other-psychotic-disorders.

36. Goff DC, et al. Folate, homocysteine, and negative symptoms in schizophrenia. Am J Psychiatry. 2004;161(9):1,705–8.
37. Coyle J. Treating the negative symptoms of schizophrenia. Medscape Psychiatry Mental Health. 2006;11(2):2.
38. Peet M, Laugharne JD, Mellor J, Ramchand CN. Essential fatty acid deficiency in erythrocyte membranes from chronic schizophrenic patients and the clinical effects of dietary supplementation. Prostaglandins Leukot Essent Fat Acids. 1996;55(1–2):71–5.
39. Bentsen H, Osnes K, Refsum H, Solberg DK, Bøhmer T. A randomized placebo-controlled trial of an omega-3 fatty acid and vitamins E+C in schizophrenia. Transl Psychiatry. 2013;3:e335. https://doi.org/10.1038/tp.2013.110.
40. Henderson D, Borba C, Daley T, Boxill R, Nguyen D, Culhane M, Louie P, Cather C, Evins AE, Freudenreich O, Taber S, Goff D. Dietary intake profile of patients with schizophrenia. Ann Clin Psychiatry. 2006;18(2):99–105.
41. Mansur RD, Santos CM, Rizzo LB, et al. Brain-derived neurotrophic factor, impaired glucose metabolism and bipolar disorder course. Bipolar Disorder. 2016;18(4):373–8. https://doi.org/10.1111/bdi.12399.
42. Jackson JR, Eaton WW, Cascella NG, Fasano A, Kelly DL. Neurologic and psychiatric manifestations of celiac disease and gluten sensitivity. Psychiatry Q. 2012;83(1):91–102. https://doi.org/10.1007/s11126-011-9186-y.
43. Bly MJ, Taylor SF, Dalack G, et al. Metabolic syndrome in bipolar disorder and schizophrenia: dietary and lifestyle factors compared to the general population. Bipolar Disord. 2014;16(3):277–88. https://doi.org/10.1111/bdi.12160. Epub Dec 13, 2013. http://www.ncbi.nlm.nih.gov/pubmed/24330321.
44. Silarova B, Giltay EJ, Dortland AVR, et al. Metabolic syndrome in patients with disorder: comparison with major depressive disorder and non-psychiatric control. J Psychosom Res. 2015;78(4):391–8. https://doi.org/10.1016/j.jpsychores.2015.02.010.
45. Lisi Donna M. Diabetes and the psychiatric patient. U.S. Pharmacist. Oct 2016. Medscape. http://www.medscape.com/viewarticle/733705_6.
46. Nasrallah HA. Why are metabolic monitoring guidelines being ignored? Curr Psychiatr Ther. 2012;11(12):4–5. http://www.currentpsychiatry.com/home/article/why-are-metabolic-monitoring-guidelines-being-ignored/6ed39aa482a23d88a558bc0350ff7095.html.
47. Schmitz N, Deschênes SS, Burns RJ, et al. Depression and risk of type 2 diabetes: the potential role of metabolic factors. Mol Psychiatry. 2016;21(12):1726–173. http://www.nature.com/mp/journal/vaop/ncurrent/full/mp20167a.html.
48. Daumit GL, Dickerson FB, Wang N-Y, et al. A behavioral weight-loss intervention in persons with serious mental illness. New EngJ Med. 2013;368:1594–602.
49. Foster GD, Makris AP, Bailer BA. Behavioral treatment of obesity. Am J Clin Nutr. 2005;82(Suppl):230S–5S. http://ajcn.nutrition.org/content/82/1/230S.full.pdf+html.
50. Merck Manual (Professional Version). Hyponatremia. Retrieved from: http://www.merckmanuals.com/professional/endocrine-and-metabolic-disorders/electrolyte-disorders/hyponatremia.
51. Yew D. Caffeine toxicity. June 6, 2017. Medscape. Retrieved from: http://emedicine.medscape.com/article/821863-overview.
52. Wang HR, Woo Y, Bahk WM. Caffeine-induced psychiatric manifestations: a review. Int Clin Psychopharmacol. 2015;30(4):179–82.
53. Hedges DW, Woon FL, Hoopes SP. Caffeine-induced psychosis. CNS Spectr. 2009;24(3):127–9.
54. European Food Safety Authority report on caffeine. Published: May 27, 2015. http://www.efsa.europa.eu/en/efsajournal/pub/4102.htm.
55. Harrison-Dunn A-R. EFSA: 400mg of caffeine a day is safe. NUTRAingredients.com, 16 Jan 2015. Retrieved from: http://www.nutraingredients.com/Regulation-Policy/EFSA-400mg-of-caffeine-a-day-is-safe?utm_source=newsletter_special_edition&utm_medium=email&utm_campaign=16-Jan-2015.
56. Cook JT, Frank DA. Food security, poverty, and human development in the United States. Ann N Y Acad Sci. 2008;1136:193–209. https://doi.org/10.1196/annals.1425.001.
57. Alaimo K, Olson CM, Frongillo EA. Family food insufficiency, but not low family income, is positively associated with dysthymia and suicide symptoms in adolescents. J Nutr. 2002;132(4):719–25.
58. U.S. Department of Health and Human Services, U.S. Department of Agriculture. 2015–2020 dietary guidelines for Americans. 8th Edition. 2015. Dietary intakes compared to recommendations. Percent of the U.S. population ages 1 year & older who are below, at, or above each dietary goal or limit. Chapter 2, Figure 2-1: page 39. https://health.gov/dietaryguidelines/2015/resources/2015-2020_Dietary_Guidelines.pdf.
59. Messamore E, McNamara RK. Detection and treatment of omega-3 fatty acid deficiency in psychiatric practice: rationale and implementation. Lipids Health Dis. 2016;15:25. https://doi.org/10.1186/s12944-016-0196-5.
60. Hibbeln J. Brain futures 2015. Modern fats and the modern mind. Mental Health Assoc of Maryland. https://www.brainfutures2015.org/wp-content/uploads/2015/12/Hibbeln-Joseph-4.pdf.
61. Appleton KM, Sallis HM, Perry R, et al. Omega-3 fatty acids for depression in adults. Cochrane Database Syst Rev. 2015;11:CD004692. https://doi.org/10.1002/14651858.CD004692.pub4.
62. McNamara RK, Strawn JR. Role of long-chain omega-3 fatty acids in psychiatric practice. PharmaNutrition. 2013;1(2):41–9. https://doi.org/10.1016/j.phanu.2012.10.004.
63. Su K-P, Lai H-C, Yang H-T, et al. Omega-3 fatty acids in the prevention of interferon-alpha-induced depression: results from a randomized, controlled trial. Biol Psychiatry. 2014;76(7):559–66. https://doi.org/10.1016/j.biopsych.2014.01.008.
64. Pottala J, Yaffe K, Robinson JG, et al. Omega-3 fatty acids linked to brain volume. Neurology. 22 Jan 2014. http://www.medscape.com/viewarticle/819632, https://doi.org/10.1212/WNL.0000000000000080.
65. Freeman MP, et al. Omega-3 fatty acids: evidence basis for treatment and future of research in psychiatry. J Clin Psychiatry. 2006;67(12):1954–67.
66. Martins JG. EPA but not DHA appears to be responsible for the efficacy of omega-3 long chain polyunsaturated fatty acid supplementation in depression: evidence from a meta-analysis of randomized controlled trials. Am Coll Nutr. 2009;28(5):525–42.
67. Kane E. The remarkable 4:1 fatty acid ratio and the brain. BodyBio Bulletin. Apr 2015. http://www.bodybio.com/BodyBio/docs/Remarkable-FattyAcid.pdf.
68. Yurko-Mauro K, Kralovec J, Hall EB, et al. Similar eicosapentaenoic acid and docosahexaenoic acid plasma levels achieved with fish oil or krill oil in a randomized double-blind four-week bioavailability study. Lipids Health Dis. 2015;14:99. http://www.lipidworld.com/content/14/1/99. https://doi.org/10.1186/s12944-015-0109-z.
69. Zhang J, Liu Q. Cholesterol metabolism and homeostasis in the brain. Protein Cell. 2015;6(4):254–64. https://doi.org/10.1007/s13238-014-0131-3.
70. Vance JE. Dysregulation of cholesterol balance in the brain: contribution to neurodegenerative diseases. Dis Model Mech. 2012;5:746–55. https://doi.org/10.1242/dmm.010124.
71. Saunders EF, Reider A, Singh G, et al. Low unesterified:esterified eicosapentaenoic acid (EPA) plasma concentration ratio is associated with bipolar disorder episodes, and omega-3 plasma concentrations are altered by treatment. Bipolar Disord. 2015;17(7):729–42. https://doi.org/10.1111/bdi.12337. Epub Oct 1, 2015. http://www.ncbi.nlm.nih.gov/pubmed/26424416.
72. Troisi A. Low cholesterol is a risk factor for attentional impulsivity in patients with mood symptoms. Psychiatry Res. 2011;188(1):83–7.
73. Boscarino J, Erlich PM, Hoffman SN. Low serum cholesterol and external-cause mortality: potential implications for research and surveillance. J Psychiatr Res. 2009;43(9):848–54.
74. Morgan RE, Palinkas LA, Barrett-Connor EL, Wingard DL. Plasma cholesterol and depressive symptoms in older men. Lancet. 1993;341:75–9.

75. Persons JE, Robinson JG, Coryell WH, et al. Longitudinal study of low serum LDL cholesterol and depressive symptom onset in postmenopause. J Clin Psychiatry. 2016;77(2):212–20. https://doi.org/10.4088/JCP.14m09505.
76. Best J, Nijhout HF, Reed M. Serotonin synthesis, release and reuptake in terminals: a mathematical model. Theor Biol Med Model. 2010;7:34. https://doi.org/10.1186/1742-4682-7-34.
77. Voruganti LN, Awad AG. Subjective and behavioural consequences of striatal dopamine depletion in schizophrenia – findings from an in vivo SPECT study. Schizophr Res. 2006;88(1–3):179–86.
78. Brown RD, Stevens DR, Haas HL. The physiology of brain histamine. Prog Neurobiol. 2001;63(6):637–72.
79. Passani M, Pertti Panula B, Lin J-S. Histamine in the brain. Front Syst Neurosci. 2014;8:64. https://doi.org/10.3389/fnsys.2014.00064.
80. Ichikawa A, Tanaka S. Histamine biosynthesis and function. 2012. Wiley Online Library doi: https://doi.org/10.1002/9780470015902.a0001404.pub2
81. Leyse-Wallace R. "Nutrition and the brain/CNS". Brain data and dogma: expanding MNT to increase fiscal reimbursement. Handout for the Pre-FNCE Workshop of the Behavioral Health Nutrition Dietetics Practice Group of the Academy of Nutrition and Dietetics; Oct 2015. Nashville.
82. WebMD. High blood sugar and diabetes. Retrieved from: http://www.webmd.com/diabetes/guide/diabetes-hyperglycemia.
83. Smith AD, Smith SM, de Jager CA, Whitbread P, Johnston C, et al. Homocysteine-lowering by B vitamins slows the rate of accelerated brain atrophy in mild cognitive impairment: a randomized controlled trial. PLoS One. 2010;5(9):e12244. https://doi.org/10.1371/journal.pone.0012244.
84. McCormick LM, Buchanan JR, Onwuameze OE, et al. Beyond alcoholism: Wernicke-Korsakoff syndrome in patients with psychiatric disorders. Cogn Behav Neurol. 2011;24(4):209–16. https://doi.org/10.1097/WNN.0b013e31823f90c4.
85. Scalzo SJ, Bowden SC, Ambrose ML, et al. Wernicke-Korsakoff syndrome not related to alcohol use: a systematic review. J Neurol Neurosurg Psychiatry. 2015;86(12):1362–8. https://doi.org/10.1136/jnnp-2014-309598.
86. Isenberg-Grzeda E, Alici Y, Hatzoglou V, et al. Nonalcoholic thiamine-related encephalopathy (Wernicke-Korsakoff syndrome) among inpatients with cancer: a series of 18 cases. Psychosomatics. 2016;57(1):71–81. https://doi.org/10.1016/j.psym.2015.10.001.
87. Wijnia JW, Oudman E, Bresser EL, et al. Need for early diagnosis of mental and mobility changes in Wernicke encephalopathy. Cogn Behav Neurol. 2014;27(4):215–21. https://doi.org/10.1097/WNN.0000000000000041.
88. Thomson AD, Guerrini I, Marshall EJ. The evolution and treatment of Korsakoff's syndrome: out of sight, out of mind? Neuropsychol Rev. 2012;22(2):81–92. https://doi.org/10.1007/s11065-012-9196-z.
89. Lonsdale D. Thiamine and magnesium: keys to disease. Med Hypotheses. 2015;84:129–34.
90. Kowalska E, Kozik A. The genes and enzymes involved in the biosynthesis of thiamin and thiamin diphosphate in yeasts. Cell Mol Biol Lett. 2008;13(2):271–82.
91. GB HealthWatch. Thiamin. Retrieved from: https://www.gbhealthwatch.com/Nutrient-Thiamin-Genes.php.
92. Petrovski S, Shashi V, Petrou S, et al. Exome sequencing results in successful riboflavin treatment of a rapidly progressive neurological condition. Cold Spring Harb Mol Case Stud. 2015;1:a000257. https://doi.org/10.1101/mcs.a000257.
93. Kennedy DO. B vitamins and the brain: mechanisms, dose and efficacy—a review. Nutrients. 2016;8:68. https://doi.org/10.3390/nu8020068.
94. Oldham MA, Ivkovic A. Pellagrous encephalopathy presenting as alcohol withdrawal delirium: a case series and literature review. Addict Sci Clin Pract. 2012;7:12. http://www.ascpjournal.org/content/7/1/12.
95. Skarupski KA, Tangney C, Li H, et al. Longitudinal association of vitamin B-6, folate, and vitamin B-12 with depressive symptoms among older adults over time. Amer J Clin Nutr. 2010;92(2):269–70.
96. Moorthy D, Peter I, Scott TM, et al. Status of vitamins B-12 and B-6 but not of folate, homocysteine, and the methylenetetrahydrofolate reductase C677T polymorphism are associated with impaired cognition and depression in adults. J Nutr. 2012;142(8):1554–60. https://doi.org/10.3945/jn.112.161828.
97. CDC. Second national report on biochemical indicators of diet and nutrition in the US population. Hyattsville: U S Department of Health and Human Services, Centers for Disease Control and Prevention; 2012.
98. Morris MS, Picciano MF, Jacques PF, Selhub J. Plasma pyridoxal 5-phosphate in the US population: the National Health and Nutrition Examination Survey, 2003–2004. Am J Clin Nutr. 2008;87(5):1446–54.
99. Delage B (updating author), Gregory JF (reviewer). Vitamin B-6. *Linus Pauling Institute Micronutrient Information Center*. Oregon State University. 2014. http://lpi.oregonstate.edu/mic/vitamins/vitamin-B6.
100. Ramsey D. Vitamin deficiencies and mental health: how are they linked? Curr Psychiatr Ther. 2013;12(1):37–44.
101. Grober U, Kisters K, Schmidt J. Neuroenhancement with vitamin B12—Underestimated neurological significance. Nutrients. 2013;5(12):5031–45. https://doi.org/10.3390/nu5125031.
102. National Institutes of Health (NIH). Vitamin B12. https://ods.od.nih.gov/factsheets/VitaminB12-HealthProfessional/ updated 2/11/16.
103. Araujp DA, Noronha MB, Cunha NA, et al. Low serum levels of vitamin B12 in older adults with normal nutritional status by mini nutritional assessment. Eur J Clin Nutr. 2016;70(7):859–62. https://doi.org/10.1038/ejcn.2016.33.
104. Edney LC, Burns NR, Danthiir V. Subjective well-being in older adults: folate and vitamin B12 independently predict positive affect. Br J Nutr. 2015;114(8):1321–8. https://doi.org/10.1017/S0007114515002949.
105. Kinsman RA, Hood J. Some behavioral effects of ascorbic acid deficiency. Am J Clin Nutr. 1971;24:455–64.
106. Report of the dietary guidelines advisory committee on the dietary guidelines for Americans, 2010 section D2–23. Retrieved from: http://www.cnpp.usda.gov/DGAs2010-DGACReport.htm.
107. Wang Y, Liu XJ, Robotaille L, et al. Effects of vitamin C and vitamin D administration on mood and distress in acutely hospitalized patients. Am J Clin Nutr. 2013;98(3):705–11. https://doi.org/10.3945/ajcn.112.056366.
108. Andre R, Gabielli A, Laffitte E, Kherad O. Atypical scurvy associated with anorexia nervosa. Ann Dermatol Venereol. 2016;144(2):125–9. pii: S0151-9638(16)30324-6. https://doi.org/10.1016/j.annder.2016.06.005.
109. Christopher K, Tammaro D, Wing EJ. Early scurvy complicating anorexia nervosa. South Med J. 2002;95(9):1065–6.
110. Strumia R. Dermatologic signs in patients with eating disorders. Am J Clin Dermatol. 2005;6(3):165–73.
111. Phillipp E, Pirke K-M, Seidl M, Tuschl RJ, Fichter MM, Eckert M, Wolfram G. Vitamin status in patients with anorexia nervosa and bulimia nervosa. Int J Eat Disord. 1988;8(2):209–18.
112. Davison KM, Kaplan BJ. Food insecurity in adults with mood disorders: prevalence estimates and associations with nutritional and psychological health. Ann General Psychiatry. 2015;14:21. https://doi.org/10.1186/s12991-015-0059-x.
113. Shaghaghi MA, Kloss O, Eck P. Genetic variation in human vitamin C transporter genes in common complex diseases. Adv Nutr. 2016;7:287–98.
114. Herbert V. Experimental nutritional folate deficiency in man. Trans Assoc Am Physicians. 1962;75:307–20.

115. Thornton WE, Thornton BP. Folic acid, mental function and dietary habits. J Clin Psychiatry. 1978;39(4):315–22.
116. Melong J, Gardner D. Women with depression should be offered folic acid. Can Fam Physician. 2011;57(9):993–6. http://www.ncbi.nlm.nih.gov/pmc/articles/PMC3173416/.
117. Coppen A, Bailey J. Enhancement of the antidepressant action of fluoxetine by folic acid: a randomised, placebo controlled trial. J Affect Disord. 2000;60(2):121–30.
118. Ginsberg LD. L-methylfolate effective in the treatment of a MDD. APA 2010, Poster NR3–46. Psychiatric Dispatches: News from the 163rd Annual Meeting of the American Psychiatric Association. Primary Psychiatry. July 1, 2010.
119. RxList. Folic acid. Retrieved from: http://www.rxlist.com/folic_acid/supplements.htm.
120. Roffman JL, Lamberti JS, Achtyes E, et al. Randomized multicenter investigation of folate plus vitamin B12 supplementation in schizophrenia. JAMA Psychiat. 2013;70(5):481–9. https://doi.org/10.1001/jamapsychiatry.2013.900.
121. Ransom J, Morgan PJ, McCaffery PJ, Stoney PN. The rhythm of retinoids in the brain. J Neurochem. 2014;129(3):366–76. https://doi.org/10.1111/jnc.12620.
122. Bremner J. Douglas and Peter McCaffery. The neurobiology of retinoic acid in affective disorders. Prog Neuro-Psychopharmacol Biol Psychiatry. 2008;32(2):315–31. https://doi.org/10.1016/j.pnpbp.2007.07.001.
123. Alzoubi KH, OF Khabour, Hassan RE, et al. The effect of genetic polymorphisms of RARA gene on the adverse effects profile of isotretinoin-treated acne patients. Int J Clin Pharmacol Ther. 2013;51(8):631–40. https://doi.org/10.5414/CP201874.
124. Nerhus M, Berg AO, Dahl SR, et al. Vitamin D status in psychotic disorder patients and healthy controlsDOUBLEHYPHENThe influence of ethnic background. Psychiatry Res. 2015;230(2):616–21. https://doi.org/10.1016/j.psychres.2015.10.015.
125. Graham KA, Keefe RS, Lieberman JA, Calikoglu AS, Lansing KM, Perkins DO. Relationship of low vitamin D status with positive, negative and cognitive symptom domains in people with first-episode schizophrenia. Early Interv Psychiatry. 2015;9:397–405.
126. Crews M, Lally J, Gardner-Sood P, et al. Vitamin D deficiency in first episode psychosis: a case-control study. Schizophr Res. 2013;150(2–3):533–7. https://doi.org/10.1016/j.schres.2013.08.036.
127. Belvederi Murri M, Respino M, Masotti M, et al. Vitamin D and psychosis: mini meta-analysis. Schizophr Res. 2013;150(1):235–9. https://doi.org/10.1016/j.schres.2013.07.017.
128. Sepehrmanesh Z, Kolahdooz F, Abedi F, et al. Vitamin D supplementation affects the beck depression inventory, insulin resistance, and biomarkers of oxidative stress in patients with major depressive disorder: a randomized, controlled clinical trial. J Nutr. 2015;146(2):243–8. https://doi.org/10.3945/jn.115.218883.
129. Grudet C, Malm J, Westrin A, Brundin L. Suicidal patients are deficient in vitamin D, associated with a pro-inflammatory status in the blood. Psychoneuroendocrinology. 2014;50:210–9. https://doi.org/10.1016/j.psyneuen.2014.08.016.
130. Kerr DCR, Zava DT, Piper WT, et al. Associations between vitamin D levels and depressive symptoms in healthy young adult women. Psychiatric Res. 2015;227(1):46–51.
131. Chiang M, Natarajan R, Fan X. Vitamin D in schizophrenia: a clinical review. Evid Based Ment Health. 2016;19(1):6–9.
132. Christopher BJ. Nutrification of Foods. In: Shils ME, Olson JA, Shike M, editors. Modern nutrition in health and disease. Philadelphia: Lea and Febiger; 1994. 1,582.
133. Krishnamoorthy L, Cotruvo JA, Chan J, et al. Copper regulates cyclic-AMP-dependent lipolysis. Nat Chem Biol. 2016;12(8):586–92. https://doi.org/10.1038/nchembio.2098.
134. Desai V, Kaler SG. An increased concentration of copper in cerebrospinal fluid with normal plasma copper concentrations has been noted in some patients with Alzheimer disease: role of copper in human neurological disorders. J Clin Nutr. 2008;88(suppl):855S–8S.
135. Walsh WJ. Nutrient power: heal your biochemistry and heal your brain. New York: Skyhorse Publishing; 2012. p. 18–20; 141.
136. Khera D, Sharma B, Singh K. Copper deficiency as a cause of neutropenia in a case of coeliac disease. BMJ Case Rep. 2016;15:2016. https://doi.org/10.1136/bcr-2016-214874.
137. Kraszewska A, Chlopocka-Wozniak M, Abramowicz M, et al. A cross-sectional study of thyroid function in 66 patients with bipolar disorder receiving lithium for 10-44 years. Bipolar Disord. 2014;17(4):375–80. https://doi.org/10.1111/bdi.12275.
138. Devkota BP. Magnesium. 2014. Medscape. Retrieved from: http://emedicine.medscape.com/article/2088140-overview#a2.
139. Manganese. Micronutrient information Center. Linus Pauling Institute. Oregon State University. Reviewed in 2010 by Michael Aschner and Gray EB Stahlman. http://lpi.oregonstate.edu/mic/minerals/manganese
140. Claro da Silva T, Hiller C, Gai Z, Kullak-Ublick GA. Vitamin D3 transactivates the zinc and manganese transporter SLC30A10 via the vitamin D receptor. J Steroid Biochem Mol Biol. 2016;163:77–87. https://doi.org/10.1016/j.jsbmb.2016.04.006.
141. Banikazemi Z, Mirzaei H, Mokhber N, Mobarhan MG. Selenium intake is related to beck's depression score. Iran Red Crescent Med J. 2016;18(3):e21993. https://doi.org/10.5812/ircmj.2199.
142. Cai L, Chen T, Yang J, et al. Serum trace element differences between schizophrenia patients and controls in the Han Chinese population. Sci Rep. 2015;5:15013. https://doi.org/10.1038/srep15013. www.nature.com/scientificreports/.
143. National Institutes of Health (NIH). Selenium. Retrieved from: https://ods.od.nih.gov/factsheets/Selenium-HealthProfessional/.
144. Cardoso BR, Busse AL, Hare DJ, et al. Pro198Leu polymorphism affects the selenium status and GPx activity in response to Brazil nut intake. Food Funct. 2016;7(2):825–33. https://doi.org/10.1039/c5fo01270h.
145. Karunasinghe N, Han DY, Zhu S. Serum selenium and single-nucleotide polymorphisms in genes for selenoproteins: relationship to markers of oxidative stress in men from Auckland, New Zealand. Genes Nutr. 2012;7(2):179–90. https://doi.org/10.1007/s12263-011-0259-1.
146. Nuttall KL. Evaluating selenium poisoning. Ann Clin Lab Sci. 2006;36(4):409–20.
147. Cope EC, Levenson CW. Role of zinc in the development and treatment of mood disorders. Curr Opin Clin Nutr Metab Care. 2010;13(6):685–9. https://doi.org/10.1097/MCO.0b013e32833df61a.
148. Wu X, Cobbina SJ, Mao G, et al. A review of toxicity and mechanisms of individual and mixtures of heavy metals in the environment. Environ Sci Pollut Res Int. 2016;23(9):8244–59. https://doi.org/10.1007/s11356-016-6333-x.
149. Min J-Y, Min K-B. Blood cadmium levels and Alzheimer's disease mortality risk in older US adults. Environ Health. 2016;15(1):69. https://doi.org/10.1186/s12940-016-0155-7.
150. International Cadmium Association (ICdA). Cadmium exposure and human health. Retrieved from: http://www.cadmium.org/environment/cadmium-exposure-and-human-health.
151. Consortium for Interdisciplinary Environmental Research. Information sources for health professionals. Stony Brook University, StonyBrook, New York. http://www.stonybrook.edu/commcms/gelfond/physicians/info.html.
152. Rooney JP. The retention time of inorganic mercury in the brain – a systematic review of the evidence. Toxicol Appl Pharmacol. 2014;274(3):425–35. https://doi.org/10.1016/j.taap.2013.12.011.
153. Saper RB, Kales SN, Paquin J, et al. Heavy metal content of ayurvedic herbal medicine products. JAMA. 2004;292(23):2868–73.
154. Hightower JM, Moore D. Mercury levels in high-end consumers of fish. Environ Health Perspect. 2003;111(4):604–8.

155. Consumer Reports. Choose the right fish to lower mercury risk exposure. Aug 2014. Retrieved from: http://www.consumerreports.org/cro/magazine/2014/10/can-eating-the-wrong-fish-put-you-at-higher-risk-for-mercury-exposure/index.htm.
156. Pronsky ZM. Food medication interactions. 15th ed. Birchrunville: Food Medication Interactions; 2008.
157. Lasslo-Meeks M. Weight gain liabilities of psychotropic and seizure disorder medications. SCAN's Pulse. 2003;22(2):7–12. SCAN is a Practice Group of The Academy of Nutrition and Dietetics.
158. Schwartz TL, Meszaros ZA, Khan R, Nihalani N. How to control weight gain when prescribing antidepressants. Curr Psychiatr Ther. 2007;6(5):43–53. http://www.currentpsychiatry.com/home/article/how-to-control-weight-gain-when-prescribing-antidepressants/409757c0a7b3e1e2a0adc3d31cd13e52.html.
159. University of Maryland Medical Center. Vitamin B6 (Pyridoxine). Retrieved from: http://umm.edu/health/medical/altmed/supplement/vitamin-b6-pyridoxine.

Neurodevelopmental Disorders in Children

Mary Anne Morelli Haskell

30.1 Introduction – 495

30.2 Children Are Not Little Adults – 495

30.3 Toxins and Toxicants – 495

30.4 Family History and Genetics – 497

30.5 Lifestyle Factors – 497

30.6 Autism Spectrum Disorders (ASD) – 498

30.7 History – 499

30.8 Gastrointestinal Symptoms – 499

30.9 Mitochondrial Issues in Autism – 500

30.10 Biomedical Assessment for Autism – 500

30.11 Functional Lab Assessment for Autism – 500

30.12 Family History and Autism – 501

30.13 Polymorphisms in Autism – 501

30.14 Genes Affecting GI Health – 502

30.15 Folate and Methylation Genes – 502

30.16 Transsulfuration Pathway Genes – 502

30.17 Neurotransmitter Genes – 502

30.18 Genes Affecting Detoxification – 502

30.19 Nutritional Support – 502

© Springer Nature Switzerland AG 2020
D. Noland et al. (eds.), *Integrative and Functional Medical Nutrition Therapy*,
https://doi.org/10.1007/978-3-030-30730-1_30

30.20 ADHD – 505
30.20.1 Lifestyle and Environmental Factors – 505

30.21 Dr. Amen's Seven Types of ADHD [87] – 505
30.21.1 General Dietary and Lifestyle Strategies for ADHD – 506

30.22 Seizure Disorders – 508
30.22.1 Contraindications to the Ketogenic Diet – 509
30.22.2 Ketogenic Diet Labs – 509
30.22.3 Supplements for the Ketogenic Diet – 509

30.23 PANDAS and PANS – 510
30.23.1 NIMH Guidelines for Diagnosing PANDAS – 510
30.23.2 NIMH Guideline for Diagnosing PANS [108] – 511

30.24 Conclusion – 513

References – 513

Learning Objectives
- To support children with neurodevelopmental and neuro-immune disorders in reaching their optimal potential through lifestyle teaching, individualized diet, targeted supplements, and other modalities based on symptoms and biomedical testing.
- To understand the use of biomarkers and genetic individuality to support vulnerable areas of their biochemistry as evidenced by related dysfunctions in multiple systems to help these children thrive.

30.1 Introduction

The number of children with neurodevelopmental and neuro-immune issues is increasing at an alarming rate. Changes in obstetrics, farming, food production, health care, and the sheer number of chemicals, pesticides, and toxins in our daily life contribute to the rapid growth in neurologic, developmental, and autoimmune issues our children face. The importance of understanding these changes helps us support the healing of the gut, the immune system, and the neurologic system, including the brain, with therapeutic nutrition. Many factors need to be assessed to help these children to reach their optimal potential. The aim of this chapter is to highlight some of the current research relevant to helping children with neurodevelopmental disorders using autism spectrum disorder, ADHD, PANDAS, and seizure disorders as prototypes.

30.2 Children Are Not Little Adults

Understanding nutrition for growth and development, in addition to harnessing nutrition to mitigate toxic exposures prenatally and postnatally, is essential in caring for our children. Children are not "little adults." They differ from adults in their anatomy, physiology, musculoskeletal system, and organ systems as they are in the process of maturing. Children have a higher metabolic rate with an increased oxygen and caloric need. Newborns have a higher fluid-to-solid ratio: 75–80% water versus 60% in adults. Control of electrolyte status is less effective until 12–18 months of age, and children are more susceptible to malnutrition and electrolyte imbalance beyond 18 months. The daily water exchange rate is higher in children. Small children are more susceptible to head injury [1] due to thinner cranial vault bones and a large head-to-torso ratio and the length of time air-filled sinuses take to develop. At birth much of the skeleton is cartilage, while the flat bones of the skull, face, and jaw are composed of membranous bone for safe passage of the baby. Until puberty, bones are softer and more easily bent and fractured. The brain is immature at birth and completes about 60% of its development in the first 2 years of life. Because the brain is growing so rapidly during this time, adequate healthy fats, protein, energy, iron, zinc, copper, iodine, selenium, choline, folate, and vitamin A are important for development of the brain and the nervous system [2].

The developing immune system is influenced before birth by the fetal-maternal interface via the placenta and by the type of delivery. It turns out that there are more immune communications between mom and fetus than previously thought [3]. With a vaginal delivery, the infant is colonized by mom's vaginal and intestinal flora. However, during a C-section, this contact with vaginal microflora is absent and environmental flora plays a more important role in colonizing the infants' microbiome. The circumstances of the C-section also have an influence; a planned C-section, with no labor or rupture of membranes, has less microbial exposure, while an emergent C-section after a long labor with rupture of the membranes will introduce more maternal flora [4].

Atopic disease tends to be more common in children with C-section births than vaginal births, especially if there is a family history of allergy. The lack of exposure to maternal vaginal and intestinal flora is associated with alterations in the neonatal gut flora and cytokine responses which can lead to changes in the Th1/Th2 helper cells that favor atopy [5]. You may notice immune differences in siblings related to their type of delivery and exposures prenatally and postnatally. The composition of the early microbiota is an important factor in the health of the immune system [6]. The initial colonization of the infant gut is affected by genetics, prenatal exposures, antibiotics, type of delivery, and diet composition of breast milk or formula (◘ Fig. 30.1).

Malamitsi-Puchner et al. found vaginal delivery promoted cytokines important to neonatal immune health. Exclusive breastfeeding, when possible, supports a healthy microbiota in infants by seeding and selecting for specific populations of bacteria. C-sections, antibiotics, and formula feeding may alter the development of the microbiome and have been linked to increased metabolic and immune issues [7].

30.3 Toxins and Toxicants

Children and fetuses are more vulnerable to toxins and toxicants. There is no known safe level of neurotoxins for an embryo and developing fetus. Due to trans-placental exposures and communication, the fetus is exposed to everything the mother takes in [8]. Of the 287 chemicals the Environmental Working Group detected in umbilical cord blood in 2004 [9]:
- 180 cause cancer in humans or animals.
- 217 are toxic to the brain and nervous system.
- 208 cause birth defects or abnormal development in animal tests.

After birth, infants and children experience greater exposure to chemicals pound-for-pound than do adults. They also have an immature and porous blood-brain barrier, which allows greater chemical/toxic exposures to reach their developing brain. In addition, their neurological, immunological, and detoxification organs are not yet mature. The liver, the main detoxifying organ, is more efficient by age 7, but it continues to mature until late adolescence and beyond. Children are

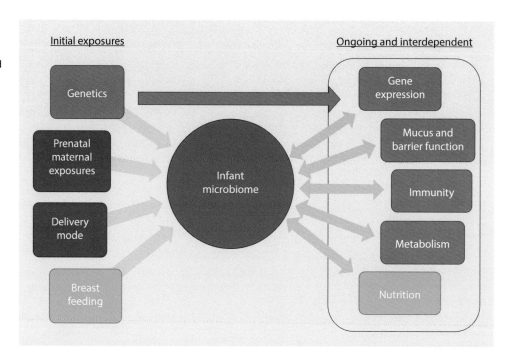

Fig. 30.1 Colonization of infant microbiome effected by initial and ongoing issues. (Reprinted from Houghteling and Walker [6]. With permission from Wolters Kluwer Health, Inc.)

exposed to more pesticides and toxins because they spend more time at ground level and through their exploratory and hand-to-mouth behaviors. Pregnant and breastfeeding mothers would be wise to avoid toxic exposures in food and the environment. The father's exposure history is also important to know. Has dad been exposed to work-related toxins, medications, alcohol, recreational drugs, or infectious agents?

There is an increasing epidemic of learning difficulties, as well as ASD and ADHD, because of the ever-increasing impact of environmental toxins on the development of children. Heavy metals can have a direct toxic effect reducing the ability to absorb essential nutrients. Mercury is considered the most toxic heavy metal. It affects the nervous system and alters brain function, leading to tremors, memory issues, irritability, and a host of other issues [10]. Mercury (Hg) blocks the absorption of iron, zinc, and selenium; however rat studies suggest adequate zinc can help decrease the toxicity of Hg exposure [11]. Zinc supports homeostasis, immune function, and apoptosis and reduces oxidative stress [12]. Aluminum displaces magnesium, calcium, and iron which can affect intercellular communication, cell growth, and secretory functions. Aluminum overload leads to inflammation and neurotoxicity [13]. Heavy metals and pesticides have a synergistic effect in humans and animals [14]. Likewise, an abundance of nutrient minerals can help to protect the body from absorbing as many toxic elements [15].

The enormous increase in electromagnetic frequency (EMF) waves, such as radio, television, cell phones, and Wi-Fi, is a vital concern for health. Radiofrequency radiation covers a wide range of frequencies from 900 MHz (megahertz) to greater than 15 GHz (gigahertz) for current 5G (5th-generation mobile networks). Children are disproportionately affected by cell phone radiation and EMF, and it is of greater concern for the developing fetus, infants, and children whose brains are thought to be more vulnerable because of smaller head size. The distance to the center of the brain is shorter because the skull and ears are thinner, allowing more radiation to reach critical structures. Their brains contain more fluid and therefore absorb more microwave radiation [16]. As body organs and systems are developing, there is increasing risk augmenting damage from previous EMF exposure in utero.

There is evidence that these frequencies have many effects down to the cellular level, adversely affecting mitochondria, causing oxidative damage from reactive oxygen species, and affecting cell membrane permeability [17]. Oxidative stress has been demonstrated in the brains and livers of developing rats exposed to EMF related to decreased glutathione peroxidase and antioxidant concentrations. Negative effects have been noted in rat testes and human reproduction. EMF exposure affects a wide range of human body systems with symptoms including headache, sleep issues, fatigue, skin problems, attention, and memory issues, as well as neuropsychiatric effects including depression and presumably autism [18]. The neuropsychiatric effects appear to be from activation of voltage-gated Ca^{2+} channels (VGCCs) in the outer membrane of cells in the brain, allowing excessive neurotransmitter and neuroendocrine release [19]. Excess calcium in the cells increases oxidative stress and ultimately affects mitochondrial function and DNA.

> Because the World Health Organization considers wireless radiation a possible human carcinogen, wireless radiation does not belong in schools with young children.
> – Anthony B. Miller, MD, PhD
> expert advisor to WHO, Professor Emeritus, University of Toronto
> The C4ST Women's College Hospital Symposium 9/12/14
> ► http://c4st.org/dr-anthony-miller/

30.4 Family History and Genetics

The field of genetics is changing daily. Going beyond genetic determinism, epigenetics has shown us that factors such as diet, nutritional status, lifestyle, environmental toxicants and toxins, including air and water pollution, food processing, agricultural practices, and extremes of stress can affect an individual as well as future generations [20, 21]. Epigenetic marks can change for the better in response to a positive attitude, healthy diet and lifestyle, and limiting exposure to the increasing number of environmental toxins and toxicants.

"Epi" means "above." Epigenetics is "control above the genes." It denotes a change in gene expression (phenotype) caused by an external modification to DNA by the environment that turns genes on and off without a change to the DNA sequence (genotype). These changes are potentially reversible and preventable. The cell membrane picks up environmental signs that control the reading and expression of genes. The embryo stage of development lays down the genetic code, and from the fetal stage on, there is epigenetic modification.

Our genes work in concert with biochemical chain reactions. Nutrition and tailored supplementation are profound shapers of epigenetic changes and the subsequent effect on metabolic processes. A gene with a single nucleotide polymorphism (SNP) variation may mean the gene efficiency is altered; it could be working at a decreased or increased efficiency or may have lost regulatory controls. Related to how an enzyme functions, upregulation or downregulation can have far-reaching effects. Metabolic pathways require nutrient cofactors, which are vitamins and minerals required for enzyme function. If there is an adverse effect related to a SNP, supplementation may help mitigate the effect by stimulating the metabolic pathway affected. A deficiency in cofactors may limit the efficiency of a metabolic pathway. Supplying cofactors (vitamins and minerals) through supplementation may support gene function through epigenetic changes and/or metabolic pathways; SNPs may be indirectly supported by nutrients that modulate pathways through other mechanisms.

The tricky part is that the genes work in conjunction with one another, so the knowledge of a single SNP does not give a complete roadmap as effects vary for each person depending on their individual genetic makeup. For instance, if a methylation pathway is affected by a methylenetetrahydrofolate reductase (MTHFR) SNP, it may be under- or overmethylated and interact with other genes in this important pathway. (See ▶ Chap. 18 on Methylation.) Biomarkers, such as homocysteine levels or actual genetic testing if available, are useful to determine exposure, effect of exposure, disease progression, and susceptibility factors, as well as which SNPs may need support. Genetic SNPs need only be addressed if they need support, based on symptoms and biomarkers. Always treat the patient, not a lab test or SNPs. To understand the workings of the genes and their integration with biomarkers, it will take continued research and the help of artificial intelligence tools, since there are more than 20,000 genes, of which only a fraction are well-understood.

30.5 Lifestyle Factors

The basics of health are the same for all children:
- Remove foods containing additives and food triggers from the diet: processed foods, artificial colors, flavors, preservatives, GMO foods, pesticides, nitrites, artificial sweeteners, corn syrup, high fructose corn syrup, trans fats, MSG (monosodium glutamate), etc.
- Remove sources of environmental toxicants from the home: pesticides, chemical cleaners, air fresheners, fabric softeners, plastics, outgassing furniture and carpets, as well as any sources of mold. Don't forget EMF pollution and the effects of "smart meters." [22] Interestingly, many parents find their children sleep better when Wi-Fi is turned off at night. Avoid electronics in the bedroom.
- Pure water is essential. Additives to municipal systems can also be absorbed through bathing. Use multi-stage or whole-house filters to remove species such as asbestos, fluoride, arsenic, and other well-known biotoxins including drug residues. Because water is a source of minerals in its natural state, insure adequate minerals in pure spring water or the diet.
- Basic healthy, whole-foods diet: fats, protein, carbohydrates, fruits, and vegetables.
- Basic dietary supplements as needed: fatty acids, enzymes, probiotics, and appropriate multivitamin/mineral.
- Determine any special diet or dietary needs, including any food sensitivities or allergies.
- Increase nutrient-dense foods: eggs; meat from well-raised, pastured chickens and animals; organic liver [23]; sweet potatoes; leafy greens; beans; legumes; nuts; seeds; some grains; seaweed; blackstrap molasses; organic bone or vegetable broth; balanced smoothies or fresh juices; and fermented foods as tolerated [24].
- Repair the gut: leaky gut, yeast overgrowth, parasites, dysbiosis, and balance intestinal flora (see ▶ Chaps. 23 and 24).
- Targeted supplements: as indicated for immune support, reducing inflammation, healing the gut, and genetic support.
- Support detox pathways: optimize elimination through healthy bowel function and liver function and consider mild chelators and binders.

Sleep is crucial for children [25]. Babies sleep 20 hours a day to support the metabolic demands of their rapidly growing brain. Most toddlers and preschoolers need 11–14 hours of sleep. Insufficient sleep can add stress to an already challenged nervous system and can affect the neuroendocrine system by elevating stress hormones. In fact, a study of adolescents has found that sleep deprivation can mimic ADHD, including hyperactivity, impulsivity, poor attention, and other behavior problems [26]. Parents can visibly recognize the behavioral effects of sleep deprivation in young children, which have been evidenced in studies as affecting cognitive function as well as behavior [27].

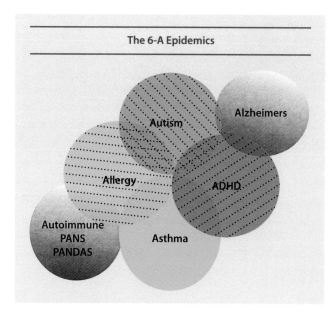

● Fig. 30.2 Epidemics of developmental disabilities related to oxidative stress and inflammation

The 6-A Epidemics are rising at an alarming rate and have oxidative stress mechanisms and inflammation in common. They are influenced by environmental toxins, heavy metals, chemicals, infection, and damaged food. According to the Centers for Disease Control and Prevention report in 2008, one in six children in the United States has a developmental disability. At the same time, Alzheimer's disease is also increasing at an alarming rate. It is the sixth leading cause of death in the United States, having increased 55% between 1999 and 2014. What in our environment, lifestyle, and diet are causing these dramatic increases in neurologic diseases in children and adults? (● Fig. 30.2).

30.6 Autism Spectrum Disorders (ASD)

Autism spectrum disorder (ASD) is one of the most confounding neurodevelopmental disorders in children. The CDC data from November 2015 shows that 1 in 45 children is now diagnosed with autism. 2.2% of children between ages 3 and 17 are currently affected by ASD. In the 1980s, the rate was 1 in 10,000. Prior to 1960, most cases of autism showed signs at birth. However regressive autism now represents 80% of the cases where children appear to be developing normally and begin losing developmental skills, most often between 15 and 24 months [28]. The incidence is four to five times higher in males, and families with one child with ASD have a 2–18% chance of having a second affected child. Although diagnosis and awareness have improved, teachers, grandparents, and pediatricians will tell you these children are truly different and each child is unique in their individual presentation. ASD is a heterogeneous condition with many variations in presentation. Genetic factors are involved, as well as environmental and other epigenetics factors, but genes do not create epidemics, though they may increase sensitivity to developing ASD. It has been aptly said that genetic vulnerability loads the gun and environmental toxins pull the trigger. This is an epidemic with high levels of special needs, devastating families and taxing school systems.

Early signs that parents notice are the absence of big smiles or warm joyful expressions by 6 months. No sharing of sounds, smiles, and facial expressions by 9 months. No babbling, pointing, or waving by 12 months. No words by 16 months. No two-word sentences by 24 months. Frequently early development appears on track and parents later notice a regression such as blank stares or loss of eye contact or words which can occur over weeks or months. Persistent regression is never normal in childhood development.

DSM-V criteria for autistic disorder describes a behavioral disorder with:
- Impaired social interaction
- Impaired social communication
- Restricted repertoire of activities and interests

Assessment of severity is based on social communication impairments and restricted repetitive patterns of behavior. This disorder affects the essence of a child's ability to communicate, rephrase, and use language, as well as their sense of self. Recovery rates are estimated to be 3–25% with traditional treatment. The earlier interventions are begun, the better the outcome, preferably before the age of 3. Changes are slower after the age of 8. However, with proper biomedical care, continued improvements are possible into adulthood, though they are more modest. Some children will be able to lose the diagnosis of autism and be mainstreamed in school.

Autism, once thought to be a behavioral disorder, is a **biological/neurological** disorder involving systemic inflammation and changes in the brain [29]. Oxidative stress, mitochondrial dysfunction, and immune dysregulation/inflammation have been recognized using peripheral biomarkers from blood and urine tests. Recent studies also report these abnormalities in brain tissue in individuals with ASD.

Associated conditions include:
- Seizures (30% +)
- Cognitive deficits or intellectual disabilities
- Savant skills
- Sensory dysfunction and/or impaired sensory-motor integration
- Motor dysfunction, including motor planning, delayed motor milestones, toe walking, etc.

Autism is not one disease but is different in each child. Many different systems can be affected. Kenneth A. Bock, MD [30], suggests there are subgroups of autism. One or more areas may be affected. Looking at symptoms and biomarkers can lead to effective interventions:
- Gastrointestinal: 50–70% have indigestion, constipation, loose stools, or abdominal pain
- Allergies/sensitivities (50% +)
- Immune deficiency/autoimmune/dysregulation
- Infections: viral, bacterial, fungal, and parasitic
- Metabolic: enzyme, mitochondrial dysfunction

- Nutritional imbalances
- Hormonal imbalances
- Toxic injury

Our food supply is compromised by the overuse of herbicides and pesticides [31]. The toxicity of commonly used glyphosate, originally patented as an antibiotic is compounded by the adjuvants in the herbicide that enhance bioaccumulation. This affects plants grown for human consumption as well as meat, eggs, dairy and farmed fish and our entire food chain, as animals and farmed fish are fed GMO feed. Many environmental factors come into play and affect homeostasis in multiple body systems. This creates physiological stresses that can include increased oxidative stress, heightened inflammatory responses, gut dysfunction, neurologic damage, and immune dysregulation.

The interaction of genes and environmental toxins and toxicants in the prenatal, perinatal, and postnatal period are contributing to the rise in ASD. Maternal immune activation can alter brain development and has also been linked to ASD [32, 33]. Neuro-inflammation is involved in autistic behavior. Findings suggest microglia activation or dysfunction can affect neurodevelopment [34]. Increased levels of inflammatory cytokines in patients with ASD have been reported, as well as the presence of serum antibodies against the central nervous system and maternal IgG reactive to fetal brain tissue [35].

30.7 History

A thorough history and timeline are essential. Was pregnancy easily achieved or was intervention needed? Were the parents excited about the pregnancy? Was the mother in a supportive relationship? What was her stress level? Were there drug or alcohol issues with either parent? What was the geographic location? Were there environmental exposures? Were mom or dad exposed to toxins at home or on the job? Were transgenerational epigenetic factors present, such as parental exposure to diethylstilbestrol, war related issues, etc?

Were both parents in good health? Any illnesses or autoimmune conditions? Was mom adrenally depleted or diagnosed with low thyroid? Thyroid hormones are critical to fetal development. Fetal thyroid function does not begin until 14 weeks of gestation. Iodine deficiency can lead to mental retardation. Inadequate thyroid hormone disrupts neuronal migration into the cortical layers of the brain in the fetus. Children with ASD can have defects of neurogenesis and neuronal migration. Maternal hypothyroidism can increase the risk of placental abruption, preterm labor, and postpartum hemorrhage. Evidence suggests that low maternal T4 levels can affect cortical development of the fetus [36]. Screening for congenital hypothyroidism between days 2 and 4 is critical as specimens prior to 48 hours may be falsely elevated.

What medications, vaccines, Rhogam (a necessary medication but potential risk factor as it can contain a mercury preservative), epidurals, Pitocin, etc., did mom receive? Was birth by vaginal birth or C-section? How long was labor? A C-section after 36 hours of labor is a different stress than a planned C-section. How was the birth? Was there birth trauma? Were antibiotics given to mom during labor or baby? How did baby look at birth? Potential risk factors for birth trauma include an uncommon fetal presentation, umbilical cord issues, fetal distress, birth injury, maternal hemorrhage, low birth weight, multiples and small size for gestational age [37]. Was medical intervention needed? Was the child taken to the neonatal intensive care unit (NICU)? Did the infant need more than 3 days in a NICU? Were there issues with colic, sleep, feeding, or gastroesophageal reflux? Was the sequential progression of development timely? Did any significant events happen around the time regression or developmental issues were first noticed? Finding the multiple risk factors and events that led to this child to develop autism can help us unravel the puzzle that leads to improvement in their lives and the lives of their families.

These are all very important questions to help in charting a course of treatment and unraveling contributing factors. Parents can do a timeline listing developmental events, vaccines (note that reactions may not be severe or immediate), illnesses, medications, stressful events, and toxic exposures such as black mold or pesticides, to aid in pinpointing where insults may have occurred.

Prenatal factors associated with ASD are under study. Several factors that are suspected to be contributors are gestational or pre-gestational diabetes mellitus, some maternal infections and other causes of maternal immune activation, valproic acid, SSRIs, organophosphates, pesticides, air pollutants, alcohol, and heavy metals. Diabetes mellitus affects fetal growth, which can increase complications during pregnancy and delivery. Diabetes can also affect fine and gross motor development and increase the rate of ASD, ADHD, and learning difficulties. It is thought maternal diabetes may cause increased fetal oxidative stress or epigenetic changes. Tight control lessens the effects of diabetes on the fetus, but does not eliminate risk [38].

During pregnancy, the demands for folate increase to support the growth and development of the fetus. Mothers with MTHFR and dihydrofolate reductase (DHFR) polymorphisms do not process folic acid well and are better supported with foods high in folate and vitamins that contain 5-MTHF (methylated folate) [39]. Both of these polymorphisms are common in parents and patients with ASD.

30.8 Gastrointestinal Symptoms

Children with ASD have a greater risk of gastrointestinal issues, especially constipation, chronically loose stools, diarrhea, gas, bloating, and abdominal pain [40]. Children with intestinal inflammation may also have sleep disturbances, behavioral issues, or unusual posturing [41]. Many adopt a posture that puts pressure on the lower abdomen to lessen pain. Those who are unable to effectively defecate on the toilet may use a squatting posture to produce a Valsalva maneuver. GI dysfunction can worsen symptoms through pain, stress, and discomfort, resulting in such issues as agitation, anxiety, aggression, self-injury, brain fog and sleep issues.

GI symptoms can occur via different mechanisms. Pathways affecting the gut-brain axis can affect behavior and cognition. Studies have shown in children and mice that having a gluten-free diet lessens GI symptoms and improves behavior [42–44]. Gluten, a protein in some grains, can cause inflammation by multiple means including allergy, intolerance, gluteomorphin, celiac, non-celiac gluten intolerance, or contaminants in the grain. Multiple mechanisms by which gluten may cause issues can make definitive testing challenging. Modern wheat has been hybridized to produce a higher gluten content and is often desiccated with glyphosate. Some children with GI symptoms have a lymphocytic enterocolitis with dysregulated intestinal mucosal immunity and increased cytokines. Therapeutic diets can decrease inflammation [45].

Pathophysiological mechanisms that may link ASD and GI symptoms include [30]:
- Intestinal inflammation (with or without autoimmunity)
- Gluten-related issues, such as celiac, wheat allergy, non-celiac gluten sensitivity, and gluteomorphin metabolites
- Functional abdominal pain
- GI dysmotility with dysautonomia
- Gastroesophageal reflux
- Gut microbiota dysregulation affecting gastrointestinal permeability, mucosal immune function, intestinal motility, and sensitivity [46]

30.9 Mitochondrial Issues in Autism

Studies have shown children with autism to have mitochondrial issues including mitochondrial dysfunction, mtDNA (mitochondrial DNA) over-replication, and deletions more often than neurotypical children [47]. Children with mitochondrial dysfunction can show nonspecific symptoms such as anxiety and GI abnormalities as abdominal pain, constipation, reflux, nausea, and vomiting; dizziness, headache and fatigue. They may have a family history of mitochondrial disease; neurodevelopmental regression (which should be a very rare event in children); seizures; fatigue/lethargy; ataxia; motor delay; or cardiomyopathy. Oxidative stress and inflammation adversely affect mitochondria [48–50]. Robert Naviaux, MD, a mitochondrial researcher at UCSD, has proposed the cell danger response (CDR) hypothesis that could prove to be the root cause of autism, ADHD, asthma, atopy, allergies, epilepsy, Alzheimer's disease, bipolar disorder, and other chronic issues [51]. He describes it as universal metabolic response to stress or injury and suggests that these issues trace back to an abnormal persistence of a normal, alternative functional state of the extracellular ATP signaling, causing a loss of ATP. The CDR is protective until these changes become sustained. When they occur during pregnancy or the first 3 years of life, they can alter the trajectory of normal development [52].

Until we find more specific ways to support mitochondria, lifestyle, nutrition, and supplementation are helpful. Nutrients supportive of mitochondrial function are L-carnitine, coenzyme Q10, and vitamins C, D, E, and B5 [53]. CoQ10 is needed to produce ATP, B1, B2, B3, B5, and antioxidants [54].

30.10 Biomedical Assessment for Autism

There is currently a search underway for nutritional and biomarkers in ASD [55]. Many tests are useful in teasing out metabolic and nutritional issues in autism. Alterations in metabolic pathways vary according to exposures and genetic individuality.

However, collecting blood and other bodily fluids can be quite traumatic for these children. An organic acid and urine peptide test is a good place to start as urine is usually the least traumatic body fluid to collect.

William Walsh, PhD, of the Walsh Research Institute, found 85% of children on the autistic spectrum were zinc-deficient and many had an increased copper-to-zinc ratio (should be 1:1). Another study indicated children with ASD and GI disease showed less stimming and hyperactivity with zinc supplementation [56]. Optimizing zinc can improve appetite, as taste buds are more efficient with adequate zinc. Zinc can also lessen anxiety. Improving the zinc/copper balance can take 2 months to show improvement. There may be a brief period of worsening symptoms as the copper is detoxed.

30.11 Functional Lab Assessment for Autism

1. Urine for organic acid testing: to evaluate for mitochondrial function via Krebs cycle and amino acid metabolites, neurotransmitter metabolites, and microbial overgrowth – yeast, bacteria, clostridia markers, nutritional and antioxidant issues, fatty acid metabolism and oxalates as well as detox indicators related to glutathione, ammonia, and GI bacteria.
2. Urine peptide testing: gluteomorphin and casomorphin – neuropeptides can affect cognitive function, speech, and auditory integration.
3. Blood chemistry: metabolic panel, CBC, magnesium and selenium, vitamin D, vitamin C, fat-soluble vitamins, ferritin, total iron-binding capacity (TIBC), serum lead, cholesterol, RBC folate, RBC zinc, copper, and ceruloplasmin.
4. Markers of immune function and inflammation: Activated T- and B-cell subsets, immunoglobulins, cytokines, ANA, ESR, C-reactive protein, anti-casein, anti-gluten, and anti-soy IgG. Viral/bacterial markers if indicated.
5. Genetic screening: Genetic sequencing can be done by saliva or blood, which is still very expensive. It is difficult to test the entire genome through saliva unless it is possible to collect several milliliters; however several

companies can test a limited number of genes from a paper dot saturated with saliva. Some polymorphisms can be supported with supplementation. CGH (comparative genomic hybridization) array identifies small deletions and duplications to check for genetic anomalies. Note that FMR1 gene, which codes for fragile X mental retardation protein, has symptoms that can overlap with autism.

6. Hormones: thyroid function – FT3, FT4, and TSH. Low thyroid can cause fatigue and brain fog. AM cortisol or salivary cortisol. Low cortisol can be related to transition difficulties, tantrums, and anxiety.
7. Allergy: IgE, IgG – food sensitivity and delayed reactions can contribute to behavior issues. A large number of IgG food allergies indicate a leaky gut. How a child acts after eating a food should be noted. This information and appropriate laboratory findings will help to decide which foods are most important to eliminate. Diet should be as varied as possible. In some children, removing allergenic foods may cause withdrawal and worsening behavior for up to 2 or 3 weeks.
8. Oxidative stress: can be evaluated by urine for lipid peroxides indicating oxidative damage to cell membranes and 8-OHdG-oxidative damage to DNA. Blood markers: lipid peroxides, nitrotyrosine, transferrin, 8-OHdG, glutathione, cysteine/cystine ratio, and the enzymes glutathione peroxidase and superoxide dismutase. Oxidative stress is common in neuropsychiatric disorders [57].
9. Mitochondrial dysfunction (MD): fasting morning labs – lactate, pyruvate, free and total carnitine, acylcarnitine panel, ubiquinone, ammonia, CK, AST/ALT, CO2, glucose, and mitochondrial DNA sequencing. ASD has been associated with elevations in lactate-to-pyruvate ratio, CK, ammonia, AST, and decreased carnitine. Some of the newer biomarkers include buccal cell enzymology and mitochondrial antibodies [58].
10. Autoimmune testing: myelin basic protein autoantibodies indicate inflammatory demyelination, cerebellar antibodies. Anti-neuronal antibodies have implicated in severe autism [59].
11. Toxins: urine porphyrins – an oxidized metabolite of heme biosynthesis associated with genetic disorders, metabolic issues, oxidative stress, and toxic exposures to metals like mercury, lead, and arsenic or chemicals. Because porphyrins are assembled in the mitochondria, this gives information on the health of the mitochondria. Patterns of porphyrin elevations can indicate clues to specific toxicities. For example, coproporphyrin or precoproporhyrins can be associated with mercury exposure. Several labs have toxic panels that test for alkylphenols, organochlorines, organophosphates, plasticizers, PCBs, and volatile organic compounds [60].
12. Kryptopyrrole quantitative urine: tests for pyroluria, an inherited disorder that robs the body of zinc and B6. This condition may be found in ASD, ADHD, and behavior disorders, including obsessive-compulsive disorder, anxiety and depression [61]. Symptoms include poor anger control, mood swings, social anxiety, infections, sleep issues, absence of dreams, and light, sound, and tactile sensitivities. Greater than 10 mcg/dl may indicate borderline pyroluria, and greater than 20 mcg/dl is positive for pyroluria. In pyrrole disorder, B6 and zinc are rendered inactive by the build-up of HPL (hydroxyhemopyrrolin-2-one) involved in the synthesis of heme. The result is a major deficiency of B6 and zinc. Individualized doses of B6, P5P, zinc, and possibly manganese and evening Gamma linolenic acid may be helpful. This condition should be treated prior to detox. If pyrroles are an issue, results with supplementation can usually be seen in 4 weeks.

30.12 Family History and Autism

Neurologic and autoimmune issues commonly run in families [62]. Neurologic disorders found in families with autism include Asperger's, Tourette's, ADHD, depression, schizophrenia, and bipolar disorder. Autoimmune conditions in families with autism include lupus, Hashimoto's thyroiditis, type 1 diabetes, chronic fatigue, rheumatoid arthritis, fibromyalgia, and Crohn's disease [63]. A common factor linking these issues may include impaired transsulfuration and methylation pathways.

30.13 Polymorphisms in Autism

Autism has a strong genetic component. ASD is a heterogeneous disorder with marked genotypic and phenotypic variability and complexity [64]. However, genes do not cause disease or epidemics. A gene codes for how to make a protein. The protein and RNA molecules interact with one another in dance of dizzying complexity. The genome can provide many answers, as we learn its language and complex interactions.

Many genes have been associated with autism, but it is genetically heterogeneous and has interactions with environmental factors [64]. Monozygotic twin discordance studies where only one twin is affected suggest a role for non-genetic factors [65].

The CHD8 gene is one of the few genes consistently related to a small subtype of autism that occurs early in development consisting of ASD, macrocephaly with a prominent forehead, wide-set eyes and a pointed chin, GI complaints, and marked sleep dysfunction [66]. This disruption is interesting because of the comorbidities between brain development and enteric innervation.

It is thought that common gene variants contribute to the risk of developing autism. Many of these polymorphisms can be positively influenced by nutritional interventions [67] and supporting biochemical pathways, such as methylation, sulfation, and neurotransmitter synthesis metabolism. Genes considered clinically useful by integrative practitioners include those related to gut health, cellular energy, folate metabolism, methylation detox, detoxification and the methionine cycle.

30.14 Genes Affecting GI Health

HLA (human leukocyte antigen) genes may increase the risk of gluten intolerance, peanut allergy, and celiac disease.

MCM6 (minichromosome maintenance complex 6) controls the activity of LCT gene, which codes for lactase. Variants in the MCM6 gene improve lactose tolerance and lack of variants increase the risk of lactose intolerance.

FUT2 (fucosyltransferase 2), involved in methylation pathway, can lead to lower levels of bifidobacteria and prebiotics [68]. Disruptions in these genes can affect intestinal health and contribute to GI comorbidity, as well as affect the strength of the immune system.

30.15 Folate and Methylation Genes

MTHFR (methylenetetrahydrofolate reductase) is the enzyme that activates folates to 5-MTHF needed for the re-methylation of homocysteine to methionine. In studies of Caucasian and Hispanic populations, it is estimated that 40–50% of the population has at least MTHFR SNP [69], but it is estimated 90% of children with autism have a SNP in one of the MTHFR genes and tend to undermethylate. The incidence of MTHFR and MTRR SNPs vary regionally and among different ethnic groups [70]. These SNPs affect detoxification, immune function, arterial health, vascular function, and neurocognitive and mental health.

Physical signs associated with folate SNPs include midline defects such as tongue, lip ties, spina bifida [71], hypospadias, cleft palate, Mongolian spots, sacral dimples, umbilical hernia, and impaired detoxification. This author has found many babies with tongue or lip ties have at least one MTHFR gene SNP and some babies show multiple midline issues such as tongue tie, hypospadias, sacral dimple and midline birth marks.

MTHFR, FOLR (folate receptor), and DHFR (dihydrofolate reductase) gene variants may require more folate-rich food or folate supplements. However, in people who overmethylate, there can be difficulty with methyl donors which can manifest as irritability, behavioral issues, and/or hyperactivity. Methylation is affected by numerous genes, nutrition, and environment. With these SNPs, avoid folic acid in enriched processed foods and supplements as unmetabolized folic acid can lead to immune dysfunction. The MTHFR polymorphism calls for methylfolate instead of folic acid and methylcobalamin rather than cyanocobalamin. Adeno sylcobalamin or hydroxycobalamin may work better depending on MTR (methionine synthase) and MTRR (methionine synthase reductase) methionine cycle genes.

RFC: reduced folate carrier polymorphisms in the mother can affect embryogenesis and organogenesis by compromising intrauterine availability of folate products and can disrupt normal neurodevelopment.

TCN2: transcobalamin II gene encodes a protein that transports vitamin B12 from blood into cells. One variant is associated with low serum B12 and high homocysteine. Decreased brain levels of B12 can be found in autism.

30.16 Transsulfuration Pathway Genes

CBS: cystathionine beta synthase regulates homocysteine to cystathionine to the transsulfuration pathway. CBS mutations that upregulate this enzyme can cause excess ammonia, taurine, and sulfur. A diet high in sulfur and related supplements may have a negative effect with these SNPs. The active form of B6, pyridoxal-5-phosphate can work better for those with genetic variations that slow CBS activity.

30.17 Neurotransmitter Genes

COMT: catechol-O-methyltransferase helps break down catecholamines, and the neurotransmitters dopamine, norepinephrine, and epinephrine, as well as some chemicals, toxins, and estrogen. COMT affects parts of the prefrontal cortex which is involved with personality, controlling behaviors, short-term memory, planning, abstract thinking, and emotion. COMT++ individuals may have difficulty breaking down the above chemicals and may have difficulty with methyl donors, leading to hyperactivity, behavior issues, anxiety, and irritability.

GABRB3: affects the GABA receptor and has been noted in some cases of autism [72].

30.18 Genes Affecting Detoxification

GST: glutathione S-transferase detoxifies some products of oxidative stress and toxins [73].

GSTM1: (liver) and GSTP1 (lungs and brain) – responsible for making the master antioxidant glutathione [73].

PON1: Paraoxonase aids in clearing pesticides and other toxins [73].

SOD: Superoxide dismutase is an antioxidant enzyme involved in Phase 2 liver detox that converts reactive oxygen species to hydrogen peroxide to quell free radical damage [74].

CYP2C9: There are 60 CYP (cytochrome P450) genes that aid the liver in clearing toxic substances in Phase 1 liver detox [74].

Variants in CYP, Pon1, SOD, GSTM1, and glutathione can decrease the effectiveness of detoxification and lead to environmental and chemical sensitivities. SNPs are common in these genes with autism [74].

30.19 Nutritional Support

In *Changing the Course of Autism,* author Bryan Jepson states, "Diet changes often result in rapid improvements in both neurological and GI function, leading to better absorption of nutrients, decreased GI inflammation and decreased immune system activation: subsequent improvement in sleep, bowel function, mood and immunity follow" [75].

A gluten-free/casein-free diet is a common starting point. If a GF/CF diet seems like it would be difficult for the family, a urine peptide test for casomorphin and gluteomorphin is helpful to convince the family if a GF/CF diet is worth the effort. If one or both tests are positive and the family is willing to implement the diet, improved behavior is often seen [76]. Studies on GF/CF diets are inconclusive but parents often report improvements in learning and language. A gluten-free diet could foster inadequate B vitamins, iron, and zinc as well as lower fiber intake; however, these deficiencies are easily overcome with a whole-foods diet and supplementation when needed.

A recent study showed improvement in 67 children and adults on a healthy gluten-free, casein-free, soy-free (HGSCF) diet with sequentially added nutritional supplements. The results suggest that dietary and nutritional intervention can improve nutritional status, non-verbal IQ, and other symptoms of autism [77].

In susceptible children, dietary intervention is to reduce gut inflammation so the gut can heal. For instance, a gluten-free/casein-free diet is worth trying for at least 3 months [78]. Though studies are not clear that this is a helpful intervention, parents often report improvement. Signs that a GF/CF diet could be helpful include a distant, "spaced out" effect, picky eater, diarrhea/constipation, abdominal distention, and speech and auditory issues. Gluten and casein contain high levels of glutamate and aspartate, both excitatory neurotransmitters. Glutamate is involved in memory and learning. Glutamate converts to GABA, which is calming and helps with speech. However too much glutamate inhibits GABA conversion and creates neurological inflammation [79]. A healthy gut with a flourishing microbiome and good nutrition are key for a healthy brain.

Addressing biochemical individuality in diet is key. Clues can be assessed using food cravings, diet diary, food frequency, reactions to food, symptoms, biomarkers and lab results, and genetics. After several months on a GF/CF diet, the diet can be fine-tuned, changed, or layered with other diets. The Specific Carbohydrate Diet (SCD) is helpful for bowel inflammation, continued diarrhea, and gut dysbiosis. The Gut and Psychology Syndrome (GAPS) diet was derived from the SCD. It eliminates grains, dairy, sugars, and starchy carbs. It uses fresh meat and fish and bone broths to help heal the gut. The Body Ecology Diet is helpful for yeast overgrowth, digestive discomfort, excess gas, and yeast symptoms. A Low Oxalate Diet can be helpful for discomfort in the urinary or GI tract, body pain, continued stimming,

Box 30.1 Signs of Phenol Sensitivity
- Sulfate in blood and urine
- Reactions to phenol/salicylates/Tylenol
- Cravings for phenolic foods—dark fruits, apples, onions, potatoes, curly kale, cabbage
- Family history of autoimmune/neurologic issues
- Phase 2 liver detox test for sulfation

excess masturbation, and bowel issues. Oxalates cause gut inflammation and often show up high on urine organic acid tests. Calcium citrate can be used to lower oxalates with meals. Many healthy foods contain oxalates such as spinach, beets, berries, beans, nuts, and chocolate so just high oxalate foods are limited. Oxalates can be removed by soaking grains, nuts, and beans [79].

Many whole foods have a high phenolic content with many protective properties and antioxidants. However, children with autism tend to have low sulfate levels and phenol sulfotransferase (PST), leading to a decreased capacity to detoxify phenols [80]. Dysbiosis and heavy metals can also adversely affect sulfation leading to phenol sensitivity. Common symptoms of phenol intolerance include dark circles under the eyes, red face/ears, head banging, self-injurious behavior, behavior issues, hives, inappropriate laughter, diarrhea, and stomach aches [79] (▶ Box 30.1).

Finding the best foods and diet takes patience. Whenever possible, expand healthy food choices and rotate foods to lessen food reactions. Textures are often an issue. How food is prepared can alter textures to be more acceptable.

Nutrient support: Children with autism have brain dysfunction. These children operate in narrow range. Add supplements very slowly and monitor the results. A vitamin/mineral supplement designed for sensitive children can be added. Balanced essential fatty acid supplementation and healthy fats in the diet support brain health. Digestive enzymes can be helpful. For children with phenol sensitivity, No-Fenol and molybdenum can help process phenols [79].

Carnitine helps transport fatty acids in cells, needed for fatty acid metabolism and energy production in the heart and muscle. Probiotics support gut health and *Saccharomyces* helps reduce clostridia overgrowth [81]. Vitamin D can have an immunomodulatory on T helper cells and T regulatory cells [82].

Epsom salt baths or foot soaks work for some children. Epsom salt is magnesium sulfate, which can be absorbed through the skin. Start with 1 teaspoon for sensitive children and work up to ½ to 2 cups per bath for 20 minutes depending on the size of the child. Some parents report their children sleep better and are more relaxed with Epsom salt baths.

Case Study: ASD

Jack is a 2-year-old boy with developmental and speech delay and recently diagnosed with autism. He has decreased eye contact, aggressive behavior, and difficulty with transitions and engages in solitary play.

Prenatal History: Second pregnancy, which was a pleasant surprise. Mom was in good health but suffered severe nausea for the first month. At 26 weeks of gestation, Mom had a TDaP vaccine and the baby stopped moving for several weeks.

Mom had severe redness and swelling at the injection site and had difficulty lifting her arm for months. In the third trimester, she developed severe depression, which persisted postpartum.

Birth History: Labor began spontaneously at 41 weeks and lasted 21 hours. No medications were used, but mom pushed for 2 hours. Birth weight was 7.5 pounds and Apgar scores were 8 and 9. Baby's head shape was crooked. Baby was alert and nursed immediately but ineffectively.

Milk supply was low, possibly due to a history of breast reduction surgery. He initially lost 20 ounces in the first week. He breastfed for 2 months. He preferred to bottle feed and was intolerant to cow dairy and soy formula, so he was given a hypoallergenic formula and then a non-GMO sensitive formula. He spit up after every feeding. Spitting up is most often irritation of the vagus nerve (CNX) from birth trauma but can also be in response to a food that mom is eating or formula, most often cow dairy. Rarely it can be due to a congenitally lax sphincter. Difficulty latching is also a structural issue as birth trauma can cause compression of the hypoglossal nerve (CN XII), which does not have a canal yet, but traverses through a slit in the occipital condyles. He did not have a tongue tie, which is sometimes related to a MTHFR SNP.

Past Medical History: At 2 months, he had DTaP vaccine. At 3 months, he had a fever of 103 and was hospitalized for late-onset Group B strep. At 8 months, he had a second DTaP and became lethargic and miserable and lost his appetite. At 9 months, he developed repetitive play, spinning wheels of trucks, and stopped interacting in play.

Diet History: Jack has a poor appetite and is a picky eater. His diet consists of oatmeal bars, crackers, yogurt, meat, chicken, nuts, a few fruits and veggies, as well as two bottles daily of a pediatric meal replacement beverage. The meal replacement contains water, sugar, corn maltodextrin, milk protein concentrate, high oleic, safflower oil, canola oil, and soy protein isolate. Less than 0.5% of the following: short-chain fructooligosaccharides, natural and artificial flavor, cellulose gel, potassium chloride, magnesium, phosphate, potassium citrate, calcium phosphate, tuna oil, calcium carbonate, potassium, salt, cellulose gum, choline chloride, ascorbic acid, soy lecithin, monoglycerides, potassium hydroxide, m-inositol, carrageenan, taurine, ferrous sulfate, dl-alpha-tocopheryl, acetate, L-carnitine, zinc sulfate, calcium pantothenate, niacinamide, manganese sulfate, thiamine chloride hydrochloride, pyridoxine hydrochloride, riboflavin, lutein, cupric sulfate, vitamin A palmitate, folic acid, chromium chloride, biotin, potassium iodide, sodium selenate, sodium molybdate, phylloquinone, vitamin D3, and cyanocobalamin. Corn maltodextrin, soy, safflower, and canola oil are commonly from GMO crops. He prefers sweet and salty foods.

Family History: postpartum depression, depression, and anxiety.

Physical Exam Positives: well-developed child with pale skin, dark circles under his eyes, aphthous ulcers, distended abdomen, + gas, bloating, mild perioral rash, poor balance, lumbosacral area is compressed, walks with his left leg externally rotated, occipital and frontal bones show molding from birth, and head feels compressed, particularly in the occipital area. Restrictions in the occipital area can reflect on the health of the cerebellum, which is involved with motor skills, balance, and muscle memory. Poor motor skills affect social skills and communication skills.

Labs: Gluten/Casein Peptides Test – positive for both indicating a GF/CF diet should be very helpful.

In this case, both gluten and casein are being metabolized in gluteomorphin and casomorphin peptides, having opiate-like effects causing behavioral issues, brain fog, and addictive symptoms to these foods. Organic acid test showed high furancarbonylglycine, the only yeast and fungal marker that was high. Several mitochondrial markers were high, including succinic acid which may relate to a relative deficiency of riboflavin and/or CoQ10 which are cofactors for succinic dehydrogenase in the Krebs cycle.

High malic acid supports the use of niacin and CoQ10 and high aconitic acid requires glutathione to support the enzyme aconitase. High quinolinic acid/5-HIAA ratio suggests neural excitotoxicity, common in autism. Inflammation from infection or immune overstimulation, increased cortisol, or exposure to phthalates all can increase this ratio. Carnitine, which mediates the transfer of fatty acids across the membrane of the mitochondria, melatonin, turmeric, and garlic, may reduce injury to the brain by quinolinic acid.

MAO B + (an X-linked gene): associated with aggression, poor breakdown of serotonin, COMT ++ affecting methylation, neurotransmitter degradation, MTHFR C677T +-, VDR + -.

MAO-B interacts with COMT and VDR. MAO-B may present clinically with anxiety, allergy, OCD (obsessive-compulsive disorder), and tics. MAO-B++ tends to have high amines, dopamine, and PEA (phenylethylamine), whereas MAO-B - tends to have low dopamine and MAO-B + - can have up-and-down dopamine.

Treatment: Parents began a GF/CF diet and he continued his basic supplements. He began a program of osteopathic manipulative treatment aimed at releasing restricted areas of the body and improving blood flow and glymphatic drainage in the brain. (Glymphatics are essentially the lymphatics of the brain.) Within 3 weeks he began saying more words, developed better eye contact, and began interacting with his siblings more (although not always peacefully). GSH cream and ascorbyl palmitate vitamin C were added to support detoxification. Glutathione is a major antioxidant and reduces oxidative stress from free radicals, lipid peroxides, and heavy metals. Ascorbyl palmitate vitamin C is lipid-soluble and thought to be most helpful for the brain [83].

Parents were instructed to do a 20-minute Epsom salt baths twice a week and "clay play" with bentonite clay. Clay play is best done outside because plants love it but plumbing pipes do not. He continued progressive therapeutic measures with nutrient support, speech therapy, and occupational therapy (OT) and applied behavioral analysis (ABA) and continues to make progress.

30.20 ADHD

ADHD is defined as inattention, hyperactivity, forgetfulness, distractibility, and impulsive behavior, although symptoms and causes vary for each individual child. ADHD can trigger intense emotions that overwhelm the brain. These children tend to have low frustration tolerance, hot tempers, excitability, social anxiety, and difficulty with relationships. According to CDC reports in 2016, one in nine children age 2–17 in the United States has been diagnosed with ADHD. It is exploding in incidence and there is no good answer for how to treat it or prevent it. A thorough assessment by a team is needed.

Treatment should include diet modification, biomedical treatment, and nutritional supplementation as indicated, along with behavior modification, lifestyle change, and counseling. Medication can be helpful for some children, but should never be the only treatment as it has not been shown to necessarily have long-term benefits.

Shaw found that children with ADHD are behind their classmates by about 3 years in brain development, although they had precocious motor cortex development [84]. On the positive side, these children can be fun, smart, creative, and driven and, as adults, often run large companies.

One study shows the brain maturation delay theory to include subcortical structures such as the nucleus accumbens (motivation, reward circuit, involves dopamine and serotonin), amygdala (processing of memory, emotional reactions, decision-making), caudate (procedural and associative learning and inhibitory control of action), hippocampus (short-term to long-term memory), and putamen (influences learning, uses GABA and acetylcholine) [85].

ADD/ADHD is not a single issue and it is difficult to define pathology. It tends to run in families and several genes are often involved. Environmental issues are an increased issue. Similar symptoms can also be seen with other issues such as thyroid issues, lead toxicity, head injury, divorce, environmental toxins, food additives, fetal alcohol syndrome, fragile X, mitochondrial issues, developmental delays, anxiety, and food allergies.

30.20.1 Lifestyle and Environmental Factors

ADHD may be caused by or made worse by toxic exposure, trauma, food allergies/sensitivities, yeast overgrowth, and nutritional deficiencies.
  Biomedical testing:
- Organic acid testing:
  1. Evaluates biomarkers for *Candida* overgrowth, which can affect brain function and sensory issues and contribute to self-stimulatory behavior, bacterial overgrowth, and neurotransmitter metabolites
  2. Clostridia toxic metabolite markers
- Lab tests: cholesterol panel, copper/zinc ratio, ferritin, vitamin D, thyroid testing, and streptococcus antibodies. Copper/zinc ratio: Elevated copper with depressed zinc can contribute to ADHD, hyperactivity, and rage behavior [86].
- Comprehensive Fatty Acids Test
- Comprehensive stool analysis
- Gluten/Casein Peptides Test – urine
- IgE/IgG food allergy/sensitivity testing including candida
- Metals hair test or toxin testing histamine levels normal range: 40–70 ng/ml. Histamine is broken down by methyl groups, so high histamine can relate to low neurotransmitters and under-methylation, depression, allergies, OCD, and hyperactivity. Low histamine can relate to high neurotransmitters and over-methylation, frustration, cries easily.
- Gene testing: MTHFR SNPs are common. Dopamine D4 receptor gene (DRD4), dopamine D5 receptor gene, COMT, dopamine transporter gene (SLC6A3 or DAT1), and MAO genes are some genes thought to be involved. Dopamine is often low.

30.21 Dr. Amen's Seven Types of ADHD [87]

Dr. Amen's clinical experience and SPECT brain scan imaging led him to delineate seven types of ADHD.
1. Classic ADHD: Inattentive, distractible, hyperactive, disorganized, and impulsive. Energetic, fun, loves to play, often walked early. Difficulty with time, getting homework done efficiently.

Cause: Brain activity is normal at rest but decreases during concentration. There is decreased blood flow to the prefrontal cortex, cerebellum, and basal ganglia. Dopamine supports attention span, focus, and motivation and is likely deficient.

Treatment: Stimulating supplements such as rhodiola, green tea, L-tyrosine, and high EPA fish oil. Physical activity especially before school and homework. Stimulant medications are sometimes used.
2. Inattentive ADD: Short attention span, distractible, disorganized, procrastinates, and daydreams.

Cause: Dopamine is deficient.

Treatment: High-protein, lower-carb diets are helpful as is regular exercise. For types 1 and 2, L-tyrosine can help build dopamine. It is best taken on an empty stomach. It can increase the brain level of phenylethylamine (PEA), a mild stimulant that can increase motivation. PEA is byproduct of the amino acid phenylalanine. (Chocolate contains PEA, explaining its longtime popularity.)
3. Over-focused ADD: Classic ADD with signs of having difficulty shifting attention and negative thought patterns or behaviors.

Cause: Dopamine and serotonin deficiencies, over-activity in the anterior cingulate gyrus, reducing flexibility.

Treatment: L-tryptophan, 5-HTP (can reduce depression), saffron, and inositol (helps boost alertness, focus, mood, and mental clarity). Avoid a high-protein diet if it triggers mean behavior. Neurofeedback can be helpful to train the brain to recruit focus brain waves.

4. Temporal lobe ADD: Classic ADD as well as learning, memory, and behavioral problems such as quick anger, aggression, anxiety, and mild paranoia.

Cause: Changes in the temporal lobe, decreased activity in the prefrontal cortex.
    Treatment: Amino acid GABA is used to calm neural activity and inhibit nerve cells from over-firing or erratic firing, magnesium for anxiety and irritability, and gingko or vinpocetine for learning and memory problems. Sometimes anticonvulsant medications are used to help with mood stability.
5. Limbic ADD: This type of ADD can include depression, moodiness, negativity, irritability, social isolation, and low self-esteem. Depression tends to be cyclical, whereas ADD symptoms tend to be more constant.

Cause: Decreased prefrontal cortex activity, hyperactivity in the deep limbic (emotional center) center of the brain. Painful emotions can occur when the deep limbic system is inflamed. The deep limbic system also processes the sense of smell and affects sleep, appetite, social awareness, motivation, and drive.
    Treatment: Diet and aerobic exercise. Amino acid supplements, DL-phenylalanine and L-tyrosine. DL-phenylalanine may cause anxiety, jitteriness, and hyperactivity in children.
6. Ring of fire ADD: Includes ADD symptoms with sensitivity to noise, light, clothes, touch, rigid thinking, grandiose thoughts, and unpredictable behavior. Symptoms tend to be consistent, as opposed to bipolar disorder where symptoms cycle.

Cause: Hyperactive brain especially the cingulate gyrus, parietal lobes, temporal lobes, and prefrontal cortex.
    Treatment: Calming supplements GABA, 5HTP, and L-tyrosine can be helpful. Stimulant medication may worsen symptoms, if used alone. Diet to reduce allergies and inflammation and exercise are helpful.
7. Anxious ADD: Includes ADD symptoms with signs of anxiety, headaches from stress, difficulty with social interactions, insecurity, and self-consciousness.

Cause: Over-activity in the basal ganglia affecting dopamine.
    Treatment: Inositol, L-theanine, Holy Basil, and magnesium promote relaxation and focus. A combination of *Magnolia officinalis* and *Phellodendron amurense* has anti-stress benefits and reduces cortisol [88]. Physical exercise and reducing electronic overstimulation have calming effects. Stimulant medications can increase anxiety and sleep issues as well as cause these kids to become more mechanical. If stimulant medication is needed, try the lowest dose possible.

30.21.1 General Dietary and Lifestyle Strategies for ADHD

Many children with ADHD have nutritional or metabolic imbalances, dietary issues, food allergies or sensitivities, toxic exposures, yeast overgrowth, gastrointestinal imbalances, or chronic infections that can be improved with nutrition. Diet alone will not treat attention deficit issues.

- A clean whole-foods diet is essential and can reduce pesticide accumulation [89]. This is important for all children. Encourage parents to purchase the best food they can afford.
- Avoid processed foods as much as possible. Help children recognize the effects of what they eat and the possible consequences.
- Reduce high-sugar foods and starchy carbohydrates. These foods increase insulin levels triggering hypoglycemia. This can lead to the brain secreting increased glutamate, a chemical messenger which becomes an excitotoxin in excess. Too much glutamate affects mood by causing agitation, anger, anxiety, depression, and panic attacks.
- Free glutamate is a component of many food additives such as MSG, yeast extract, calcium caseinate, etc. High-sugar diets also increase chronic inflammation. MSG and hydrolyzed vegetable protein can decrease dopamine levels.
- High glycemic diets impair delta 5 and delta 6 dehydrogenase, enzymes involved in EPA and DHA production needed for brain health. Hydrogenated oils impair cell membrane and brain function.
- Avoid allergens. The top seven are soy, wheat, cow dairy (goat milk does not contain A1 casein, which can cross-react with gluten and may be better tolerated), peanuts, tree nuts, eggs, and shellfish. Watch for food family allergies. For instance, those with latex allergies might need to avoid avocados, bananas, kiwis, and papayas.
- Gluten sensitivity can affect the nervous system and the brain. For children with a gluten intolerance, a GF diet can improve behavior and attention. Many patients with celiac disease have ADHD symptoms that improve with a GF diet.
- Avoid food additives and genetically modified ingredients. Blue #1 and #2, Green #3, Orange B, Red #3 and #40, Yellow #5 and #6, and sodium benzoate are among the many food additives that can affect focus and behavior [90]. Artificial coloring can be found in sports drinks, candy, cake, chewable vitamins, and toothpaste. Glyphosate, the active ingredient in Monsanto's Roundup herbicide, was originally licensed as an antibiotic. It destroys the microbiome of the soil and the human gut. In addition, it limits the body's ability to detoxify and blocks vitamin D. GMO foods have been modified to survive direct application of herbicides; hence eating GMO foods means ingesting glyphosate and other toxic components of the herbicide.
- Nitrites used in some cured meats such as lunchmeat, bacon, ham, sausages, and hotdogs can cause restlessness and rapid heart rate as well as increased risk of type 1 diabetes, IBS, and cancer. Artificial sweeteners can affect cognitive function and emotional balance.

- Omega-3 fatty acids can be helpful in reducing hyperactivity, behavioral issues, and learning problems. They support the executive function centers of the brain. DHA and EPA from clean fish or fish oil are most helpful as not all people can convert ALA from vegetarian sources. Wild-caught salmon also contains B6. Avoid damaged fats in refined processed foods.
- Support healthy bowel flora and healthy, daily bowel movements. A toxic gut can increase behavior symptoms, ADHD, depression, and autism.
- B-6 at 15–30 mg/kg may be helpful for focus with a B complex, EPA/DHA, multi-mineral, and probiotics.
- Vitamin D supplementation may improve cognitive function and attention and lessen hyperactivity and impulsivity in children with ADHD who are vitamin D deficient [91].
- Inositol, Bacopa, and theanine can be relaxing.
- Rhodiola rosea can improve focus. There are several good supplement formulas for improving focus; however not all are user-friendly for the pediatric population as they require cooperation and being able to swallow capsules.
- Sensory integration issues can compound ADHD and interfere with learning. Sensory integration refers to how our senses work together to process and organize incoming information.
- Exercise can enhance dopamine and norepinephrine production, which can help improve attention and focus. Regular exercise can help relieve stress, regulate hyperactivity, and improve concentration. Outdoor activity is important, including exposure to natural light. Avoid "nature-deficit disorder," a term coined by author Richard Louv [92]. Exercise before school or on the way to school as well as before and during homework is helpful. Incorporate cross-body exercises such as running, skipping, marching, or Simon Says using cross-lateral motions.
- Downtime is important too: allow time for free play, building with Legos or blocks, drawing, playing in the dirt, petting the cat, etc.
- Neurofeedback can help the frontal lobes learn to stay in a focused beta state. Studies show that the results can have lasting effects. Cognitive behavioral therapy can improve skills and habits for executive function and may improve emotional and social self-regulation.
- Essential oils: vetiver and cedarwood can calm and improve focus. Rosemary and peppermint can improve alertness and enhance memory. Frankincense brings clarity and higher cognitive function. Ylang ylang and lavender are calming.
- Sleep issues are quite common in children with ADHD. They may have difficulty falling asleep, restless sleep, or difficulty waking. Areas of the brain that regulate attention also regulate sleep. Dysregulation of the serotonin system can affect both these areas. Their internal circadian clock may be off, not releasing melatonin at the best time in the evening. This faulty signaling can contribute to difficulty with schedules and keeping track of time. The COMT gene is responsible for breaking down dopamine which helps regulate sleep. Children with COMT gene snips may have more difficulty with sleep. Medication and caffeine can also affect sleep. A regular bedtime every day, a comfortable bed, and a good sleep environment facilitate restful sleep.
- Melatonin can help with a faulty circadian clock when taken an hour before bedtime. Children's dosages are based on body weight and metabolism. GABA works for some, and sleep tea or essential oils like lavender and rosemary can be helpful. Benadryl (diphenhydramine) causes drowsiness but has been recently implicated in dementia when used regularly so it's not a good idea in children with brain challenges.
- No computers, phones, or TV in the bedroom. Turn off Wi-Fi, which is low-level microwave radiation, at night (or better yet, use wired computers). Radiation from wireless devices can delay deep non-REM sleep, which can impact learning and memory, as well as increase oxidative stress. Electromagnetic frequencies (EMF)/radiofrequency radiation (RFR) can also adversely affect immune and metabolic function. Avoid placing beds near walls that have smart meters outside.
- Avoid blue light from computers, TV, iPads, and iPhones for 1–2 hours before bed. Blue light signals the pineal gland to shut off melatonin production.
- Digesting snacks too close to bedtime can keep children awake. Warm milk, turkey, or an apple contains tryptophan, which can facilitate sleep. Avoid GMO Arctic apples.
- Parenting is a tough job. This is particularly true for parents with ADHD parenting children with ADHD. They can thrive on turmoil in relationships unconsciously because it stimulates their brain and makes them more alert. Retraining patterns of behavior or behavior modification can be a big help for families.
- Some teens with ADHD self-medicate with marijuana, often to reduce anxiety. Marijuana has negative effects on attention, memory, and learning that can last for days to weeks. The hippocampus region of the brain is important for memory, learning, and the integration of sensory experience with emotions and motivations as well as converting information into short-term memory. THC (delta-9-tetrahydrocannabinol) suppresses the way the information processing system of the hippocampus works [93].
- Executive function support is helpful. Watches and timers help with time periods needed for tasks. Break up lengthy tasks into small steps. Support emotional regulation and provide positive motivation.
- Finding the best teacher/school is essential. These children need strategies for emotional regulation, support planning for the future, externalized motivation, and encouragement.

Biomedical treatments, nutritional supplementation, and diet are an important adjunct to behavior modification, lifestyle changes, executive function tutoring, occupational therapy,

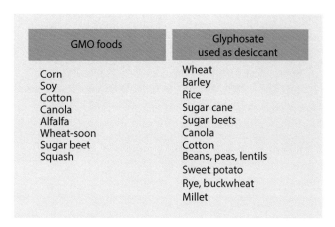

□ Fig. 30.3 Recommended foods to avoid to lessen oxidative stress and inflammation

neurofeedback, and, if needed, medication. Underlying issues need to be addressed and medications should never be the only intervention because it does not provide long-term improvements in attention and learning (□ Fig. 30.3).

30.22 Seizure Disorders

Seizures have many different causes and can occur at any age. The term epilepsy encompasses various types of seizure disorders thought to be caused by abnormal electrical signals in the brain. Seizures in newborns can be triggered by anoxia, infection, bleeding, metabolic imbalances and drug withdrawal. Structural damage to the brain can occur in the perinatal period and during birth from vascular or cerebral trauma; later from head injury, sports injuries, bike or motor vehicle accidents, cerebrovascular accidents, as well as from intracranial tumors, metabolic diseases, toxins, vaccines, and drugs. This damage may be visible or more often invisible and most often is not seen on CT or MRI studies. Electroencephalogram (EEG) is used to look for seizure activity in the brain, but in patients with observed seizures, it is sometimes difficult to capture the activity on EEG. For seizures resulting from a traumatic cause, cranial osteopathy or cranial sacral therapy may be helpful. Seizures impacting the life of a child are most commonly treated with anticonvulsant medications. Surgery is used only if there is a specific identifiable focus. Seizures are driven by inflammation and excitotoxicity. Supportive care for children with seizure disorders includes diet, vitamins, supplements, herbs, homeopathy, mind-body techniques, and manual medicine.

A vitamin B6 (pyridoxine) deficiency can cause or worsen seizures, especially in infants and newborns [94]. B6 can sometimes help older children also. Sodium, calcium, and magnesium are necessary for stable electrical activity of the brain. Low magnesium levels can trigger seizures and may lead to low calcium levels. Low blood sugar can trigger seizures and is most common in children with type 1 diabetes taking insulin.

Low sodium can be caused by diuretics, anti-seizure medications such as carbamazepine and oxcarbazepine, or excessive water intake.

Rare inherited metabolic disorders such as pyridoxine-dependent seizures, biotinidase deficiency, and folinic acid-responsive seizures respond to supplying the missing nutrient. Pyridoxine-dependent epilepsy is not responsive to antiepileptic medication, but responds clinically and with improvements on EEG to large daily supplementation of B6. The folate receptor 1 (FOLR1) gene may also be involved.

A SNP in this gene can lead to cerebral folate deficiency that may affect social interactions, developmental delay, and seizure control.

Biotinidase deficiency can be profound or partial, depending on how severe the enzymatic defect is [95]. Symptoms most often occur between 3 and 6 months with seizures, hypotonia, developmental delay, skin issues, and respiratory problems. There are four biotin-dependent carboxylases that can be treated with biotin. This issue may be picked up on the newborn screening exam.

The ketogenic diet, developed by Russell Wilder at the Mayo Clinic in 1921, is now considered an important therapeutic option for epilepsy as well as infantile spasms [96] and is being investigated for children with autism and Rett syndrome. Despite the success in resolving or lessening seizures it provided for many children, it remained little used until it became popular in the 1990s, with the Charlie Foundation for Ketogenic Therapies. The ketogenic diet involves changing the body's primary energy source from glucose to ketones. The breakdown products of ketones, beta-hydroxybutyric acid, cross the blood-brain barrier and are used by the brain for energy. The ketogenic diet helps reduce microglial activation, brain inflammation, and excitotoxicity. Studies have shown that half of children will have a 50% improvement in seizures and about one third show a 90% improvement. A small percentile of children become seizure-free [97]. This diet is sometimes initiated under hospital conditions. "The diet is difficult," according to one mother, "but not nearly as difficult as watching your child have a seizure." It requires a lot of prep work, counting carbs, weighing food, and making substitutions that suit a child's taste, but the result can be eventual resolution of seizures in some cases or a reduction in seizures.

If the diet is successful, there are fewer emergency room visits and doctor visits and less medication which often translates to improved development. Some children with seizures prefer high-fat foods, and one study from 2015 suggests this may be helpful in predicting which children will have success with the ketogenic diet. The ketogenic diet should be managed by a neurologist and dietitian. This is a medical diet and needs supervision for side effects, efficacy, medication adjustment, and weight gain. It is initiated by fasting during a brief hospital stay.

Several other versions of the ketogenic diet that are used for seizures include the modified Atkins diet (MAD), the medium-chain triglyceride (MCT) diet, and the low glycemic index treatment (LGIT) [98]. The efficacy of the MAD diet is good and it is easier to manage, although stricter ketosis works better for some children. The modified Atkins diet

> **Box 30.2 Conditions Likely to Respond to Ketogenic Diet**
> - GLUT-1 deficiency syndrome (SLC2A1 gene)
> - Mitochondrial disease
> - Children receiving tube feeding or mostly formula
> - Pyruvate dehydrogenase deficiency
> - Myoclonic-astatic epilepsy (Doose syndrome) often presents ages 3–5 years with head drop seizures
> - Dravet syndrome (severe myoclonic epilepsy of infancy (SCN1A gene)
> - Infantile spasms (West syndrome)-1 study showed 2/3 have greater than 90% reduction in spasms
> - Tuberous sclerosis complex—may be more long-term treatment
> - Rett syndrome

(MAD) can be started outside the hospital and does not require fasting to initiate. It should be under the supervision of the neurologist.

The advantages of the modified Atkins diet (MAD) are:
- No fluid or calorie restrictions or limitations.
- There are no restrictions on proteins as fats do not have to be weighed or measured. Typically 35% of calories for a patient on the MAD come from protein.
- Foods do not have to be weighed and measured, but carbohydrate counts are limited to 15–20 g a day.
- The MAD is easier to manage outside the home and in restaurants and is easier for the family.
- Family members can eat the same way.

MAD has a similar efficacy as the ketogenic diet (40–50% with greater than 50% seizure reduction, including approximately 15% seizure-free). It works for men and women equally and is being used in children, adolescents, and adults. Like the ketogenic diet, it is most commonly used for patients with daily seizures who have not fully responded to medications. It is under study for regions of the world with limited resources for which the classic ketogenic diet would be too difficult and/or time-consuming.

The ketogenic diet is worth trying for all seizure types and severities, in all ages and sizes (▶ Box 30.2).

30.22.1 Contraindications to the Ketogenic Diet

- Carnitine deficiency (primary)
- Carnitine palmitoyltransferase (CPT) I or II deficiency
- Carnitine translocase deficiency
- B-oxidative defects
- Pyruvate carboxylase deficiency
- Porphyria

Ketogenic diets do not work as well if there are issues maintaining adequate nutrition, a surgical focus for the seizures, or parental or caregiver noncompliance.

One challenge of the ketogenic diet has been that the recommended fat sources were not always the healthiest and contained damaged fats. By using healthy fat sources such as grass-fed butter, organic ghee, nut butters, coconut and MCT oil, olive oil, flaxseed, omega-3 fatty acids, and homemade mayonnaise or mayonnaise made from healthier fat sources, the chances for brain healing on a cellular level are improved.

Food additives such as colorings, preservatives, monosodium glutamate (MSG), and artificial sweeteners including aspartame, saccharin, and phenylalanine can trigger seizures in some people. Some ketogenic formulas contain aspartame or other artificial sweeteners. Substitute healthier sweeteners such as stevia for the artificial sweeteners often recommended.

MCT is found in coconuts, in coconut and palm kernel oil, and in smaller amounts in butter and high-fat dairy products from grass-fed cows and goats. There are four types of MCTs. Coconut oil contains all types.
- 6 carbons (C6), caproic acid
- 8 carbons (C8), caprylic acid
- 10 carbons (C10), capric acid
- 12 carbons (C12), lauric acid – known for antibacterial, antimicrobial, and antiviral properties

Fractionated coconut oil contains primarily C8 and C10, which support brain health and curb appetite.

MCTs have antioxidant and anti-inflammatory properties as they produce fewer oxygen species when metabolized to ATP. Most MCT oil products contain C8 and C10 fats. C8 (caprylic acid) converts to ketones more efficiently than C10. Ketones make great fuel for mitochondria and support their functioning. Some companies make MCT oil with a higher C8 content, which may be better tolerated. High C8 MCT oil is best given in the morning as it wakes the brain up. Dose is important as too much MCT oil can cause diarrhea in some children.

30.22.2 Ketogenic Diet Labs

Check labs prior to starting the diet and at 3 and 6 months and then every 6 months thereafter.

Urinalysis
After an 8-hour fast:
- CBC with differential
- Complete metabolic profile (SMA-20)
- Complete lipid profile (fasting)
- Selenium level
- Carnitine profile (total and free)
- 1,25-OH-vitamin D level
- A red cell fatty acid profile can help guide choices of fats

30.22.3 Supplements for the Ketogenic Diet

- Multivitamins: Because the ketogenic diet eliminates grains and limits fruits and vegetables, it is essential that children take a complete, low-carb multivitamin.

- Calcium and vitamin D: Many epilepsy medications affect nutritional status including calcium metabolism and subsequent bone loss. In addition, the diet is low in calcium sources. Heavy cream, often used in the ketogenic diet, is low in calcium. Consider the need for calcium supplements and vitamin D.
- Oral citrates can help prevent kidney stones by alkalinizing the urine and solubilizing urine calcium.
- Constipation is a common side effect of the ketogenic diet. Food choices such as MCT oil, avocados, and high-fiber vegetables like cucumbers, asparagus, and zucchini are helpful, as well as sufficient fluids, and exercise. Magnesium citrate is used to relieve constipation but check for drug interactions.
- Check nutrient depletions related to anticonvulsants and other medications for all patients. For instance, phenytoin (Dilantin), carbamazepine (Tegretol), valproic acid (Depakote), topiramate (Topamax), and gabapentin (Neurontin) can deplete B6, folate, and B12.
- Carnitine is needed for fatty acid to be transported into the mitochondria during the breakdown of fats. Fatigue and lethargy can be symptoms of carnitine deficiency. Supplement if deficiency or symptoms occur.
- Selenium: an antioxidant nutrient involved in reducing oxidative stress and preventing cardiomyopathy. Food sources include Brazil nuts, tuna, beef, chicken, turkey, and grains. Check the multivitamin for adequate amounts.

Supplements to improve seizure threshold:
- Curcumin, quercetin, carnosine, acetyl L-carnitine, resveratrol, and bioflavonoids are anti-inflammatory, potentially reducing microglial activation and excitotoxicity.
- DHA, omega-3 oils play a role in brain function, growth, and development and are anti-inflammatory and reduce excitotoxicity. Studies show a decrease in seizures in children taking omega-3s (EPA-DHA) [99]. Some children are low in arachidonic, an omega-6. Best to test with a red blood cell fatty acid test.
- Magnesium citrate is anti-inflammatory and blocks one of the main glutamate receptors in seizures. Magnesium has been shown to raise seizure threshold levels. Some forms of magnesium may be more beneficial [100]. Magnesium L-threonate crosses the blood-brain barrier and magnesium glycinate is well absorbed and less likely to cause diarrhea. However, magnesium levels are difficult to assess. An RBC magnesium test is more accurate than serum.
- Melatonin supports healthy sleep cycles and has anti-inflammatory and antioxidant properties.
- One pilot study showed that correcting vitamin D deficiency improved seizure control in drug-resistant epilepsy [101, 102].
- Grape seed extract contains oligomeric proanthocyanidins, which help protect the hippocampus. It reduces stress and supports mitochondrial function [103] (▶ Box 30.3).

Box 30.3 Medications That May Increase Seizure Frequency
- Antihistamines
- Insulin and diabetes medications
- Oxytocin
- Antidepressants
- Clomipramine
- Clozapine
- Lithium
- Fluoroquinolone antibiotics
- Methylphenidate
- Metronidazole or Tinidazole

There are herbal combinations and supplements in various cultures that have anticonvulsant properties. Also, there are herbal remedies and supplements that can have proconvulsant effects or interact with antiepileptic medications [104].

General diet guidelines include avoiding artificial sweeteners and food additives, excitotoxic food additives such as MSG, aspartame, l-cysteine, carrageenan, hydrolyzed proteins, soy proteins and isolates, soy sauce, autolyzed yeast, and caseinates. These foods can contribute to already overstimulated glutamate receptors, which can trigger seizures. Cheese, milk concentrates, and mushrooms are high in glutamate.

30.23 PANDAS and PANS

PANDAS (Pediatric Autoimmune Neuropsychiatric Disorders Associated with Streptococcal Infections) and PANS (Pediatric Autoimmune Acute-Onset Neuropsychiatric Syndrome) strike fear in the heart of parents who have seen their children change overnight, despite the hot debates about the existence of PANDAS for years.

PANDAS was described in the 1990s at the National Institute of Mental Health (NIMH) by Drs. Susan Swedo, Henrietta Leonard, et al., who observed cases of OCD that began abruptly, with dramatic symptoms rather than coming on gradually over weeks or months which is typical of OCD [105].

PANDAS began as a clinical diagnosis made when children and adolescents develop an *abrupt* obsessive-compulsive disorder and/or tic disorders following a strep infections. More comprehensive tests to aid diagnosis are now available.

30.23.1 NIMH Guidelines for Diagnosing PANDAS

- Presence of clinically significant obsessions, compulsions, and/or tics.
- Unusually abrupt onset of symptoms or a relapsing-remitting course of symptom severity.
- Pre-pubertal onset. Symptoms of the disorder first become evident between 3 years of age and puberty.

- Association with Group A streptococcal (GAS) infection, repeat strep, rheumatic, or scarlet fever.
- Note: In PANDAS, GAS infections may be present without apparent pharyngitis (i.e., no complaints of a sore throat).
- Association with other neuropsychiatric symptoms.

Other common symptoms listed by NIMH include:
- Parents can usually remember the day their child's behavior changed. PANS and PANDAS is characterized by an abrupt onset of obsessive-compulsive disorders and/or tics.
- Severe separation anxiety.
- Generalized anxiety.
- Motoric hyperactivity, abnormal movements, and a sense of restlessness.
- Sensory abnormalities (sensitivity to light or sounds), distortions of visual perceptions, and, occasionally, visual or auditory hallucinations.
- Difficulties concentrating and loss of academic abilities, particularly in math and handwriting.
- Increased urinary frequency or sense of urgency and/or a new onset of bedwetting.
- Irritability (sometimes with aggression) and emotional lability. Abrupt onset of depression can also occur, with thoughts about suicide.
- Developmental regression, including temper tantrums and "baby talk."

It is hypothesized that PANDAS is related to immune dysfunction at one or more levels [106]:
- Cross-reactive "anti-brain" antibodies trigger OCD, tics, and neuropsychiatric symptoms [106].
- Inflammation of the basal ganglia which are involved with movement and behavior.
- Disruption of chemokines or cytokines altering central nervous system function.

This emphasizes the link between autoimmunity and inflammation with neuropsychiatric disease.

30.23.1.1 Laboratory Testing for PANDAS
- Throat culture GABHS positive.
- Elevated antistreptolysin titers (ASO): two–four-fold rise in titer from initial infection in 1–4 weeks.
- Elevated anti-DNase B: increases from initial infection in 6–8 weeks 2- to four-fold.
- Cunningham panel: Elevated anti-neuronal antibodies such as dopamine D1 receptor, dopamine D2L receptor, lysoganglioside, tubulin, and elevated CaM kinase II – results in neuronal excitation and increased dopamine transmission [107].
- Increased basal ganglia volume on volumetric MRI.
- Elevated B-cell antigen D8/17, a possible marker of rheumatic fever, OCD, or Sydenham's chorea.
- CMP, ANA, CRP, thyroid panel and antibodies, sed rate, ferritin, immunoglobulins, vitamin D, B12, folate, gluten sensitivity and cross-reactivity, organic acid testing, genetic testing, and stool testing.
- When Group A streptococcal infection is not found, test for other infections.

Note: Titers from prior strep infection may remain elevated for months. Exacerbations of PANDAS after an initial strep infection may be caused by non-strep triggers.

Other support for PANDAS includes a positive family history of OCD, tic disorders, and autoimmunity as well as a response to antibiotics or immunomodulatory therapy such as IVIG and plasmapheresis to remove autoantibodies (Table 30.1).

Table 30.1 Factors contributing to exacerbations of PANDAS/PANS

Infectious	Other
Group A beta hemolytic strep	Psychosocial stress
Colds, Coxsackie A and B	Nutrition
Sinusitis, EBV	Medication
Mycoplasma pneumonia	Genetics
Influenza, CMV	Allergies
Varicella zoster	Chlorine
Herpes simplex	
Lyme, Babesia, *Bartonella*	
HPA axis dysfunction	

PANS (Pediatric Acute-Onset Neuropsychiatric Syndrome) may begin with an infection other than strep, environmental toxins, or immune dysfunction

30.23.2 NIMH Guideline for Diagnosing PANS [108]

1. Abrupt, dramatic onset of obsessive-compulsive disorder
2. At least two additional neuropsychiatric symptoms:
   - Anxiety, separation anxiety
   - Emotional liability: extreme mood swings, depression
   - Irritability, aggression, or severe oppositional behaviors
   - Behavioral (developmental) regression
   - Deterioration in school performance
   - Sensory or motor abnormalities
   - Sleep disturbances, bedwetting, or urinary frequency

Treatment for PANDAS and PANS: Depends on the severity of the illness and may include cognitive behavioral therapy (CBT) and antibiotics. In severe cases, IV immunoglobulin (IV IG) and plasmapheresis are used for immunomodulation to reduce offending antibodies. Address all basic nutrition, immune, and genetic issues. Remove excitotoxins, sugar, processed foods, junk food, hydrogenated, and trans fats and allergic foods.

Support fatty acids and consider a red cell fatty acid test. Steroids can reduce symptoms, but they tend to return after stopping the drug. *Streptococcus salivarius* probiotics are helpful for some children. Selective serotonin reuptake inhibitor medications are sometimes used for OCD symptoms, but are often not be well tolerated. Occasionally a tonsillectomy is helpful if the tonsils are large, which can cause sleep apnea, and become a reservoir for strep. In susceptible children, aggressively treating infection may be helpful. Antibiotic prophylaxis has pros and cons as gut immunity is severely affected. Maintain serum 25-hydroxyvitamin D above 30 ng/mL. VDR polymorphism may need higher levels.

Case Study: PANDAS

Joe is an 11-year-old male who developed a severe sore throat and flu-like symptoms. Several days into the illness, he became severely agitated with aggressive and obsessive-compulsive behaviors. His mother and grandmother have a history of anxiety and OCD behavior.

Birth and previous medical history: Term birth via C-section for breech position and early preeclampsia. Apgar score was 9 and 10; he was nursed immediately and received hepatitis B vaccine on the first day of life. At 4 months of age, he received vaccines for seven diseases. He cried inconsolably with a high-pitched cry for 6 hours, was limp with no eye contact, and was taken to the emergency room for evaluation.

At age 3, he began stuttering and received speech therapy. At age 5, he had severe flu-like illness with sore throat, fever, and extreme fatigue with a negative throat culture.

He developed fear of germs, repeatedly washed his hands, and refused to come out of his room. Gradually he improved, but never to his previous baseline. He would intermittently have vocal tics.

Labs: Cunningham panel showed elevated dopamine D1 titers which can be associated with aggression. Tubulin and lysoganglioside titers were mildly elevated CaM kinase II. Vitamin D level was 29. Gene polymorphisms related to methylation which affect mood and can cause anxiety include CBS++, COMT++, MAO-A+, and MTHFD ++.

CBS genes convert homocysteine to cystathionine and are involved in transsulfuration and glutathione synthesis, ridding the body of excess sulfur amino acids as well as part of the methylation cycle. Ammonia can be a byproduct of sulfur metabolism, which can adversely affect brain function.

His SNP is related to high homocysteine and reduced CBS activity. Helpful nutrients include N-acetyl L-carnitine, B6, phosphatidylcholine, and phosphatidylserine.

MAO-A (monoamine oxidase A) is an enzyme that breaks down neurotransmitters such as noradrenaline, adrenaline, serotonin, and dopamine. It is located on the X chromosome, so men only carry one copy. It is important in regulating mood and can be linked to anxiety, OCD, and an increased risk of aggression and violence; hence it is referred to as the "warrior gene." This enzyme needs B2 (riboflavin). It may cause an intolerance of methylfolate.

In addition, adequate B12 is needed before supporting folate. With issues in CBS, homocysteine may create excess glutamate increasing anxiety.

COMT genes break down neurotransmitters and catecholamines including dopamine and norepinephrine.

COMT is involved in the prefrontal cortex, which is involved with personality, planning, emotion, behavior control, and abstract thinking. With a double COMT SNP, enzyme activity to break down dopamine can be reduced three–four-fold. Methylation SNPs can further alter the function of COMT genes. Supportive nutrients include low-dose lithium, magnesium, and B12. SAMe is sometimes helpful for some children but may increase dopamine for others.

MTHFD1 is an enzyme involved in the conversion of dietary folate.

Treatment in brief: He was treated with azithromycin and his behavior symptoms dramatically improved for a time. Then he refused to go out of the house again for fear of germs. A course of augmentin was given. His appetite was poor. He was started on a small dose of a multivitamin with hydroxocobalamin as well as lithium orotate 5 mg to support B12 and folate uptake into cells. Anti-inflammatory supplements, colostrum, healthy fats, and antiviral herbs were added and nutrients increased. Intestinal dysbiosis was treated. He received cognitive behavioral therapy. As he began to improve slowly, his appetite improved and he returned to school on a modified schedule. The Cunningham panel had returned to normal when re-checked a year later.

As we learn to deal with the increase in neuropsychiatric autoimmune disorders, all the basics of a functional medicine approach need to be addressed. For autoimmune disease to be present, there need to be a genetic predisposition, an environmental or infectious trigger, and intestinal permeability. We must have an appreciation of these interactions between the behavioral, neural, endocrine, and immune processes. The brain and the immune system are linked by the autonomic nervous system and the neuroendocrine system. Neurotransmitters and cytokines "talk" [109].

Factors include:
— Infections: bacterial, viral, and fungal
— Toxins, heavy metals, and environmental allergies and sensitivities
— GI issues: dysbiosis, intestinal hyperpermeability, food allergies, and sensitivities
— Nutritional issues
— Hormonal and immunological imbalances
— Stress, emotions, sleep, and lifestyle issues
— Neurotransmitter imbalances

The number of children suffering from developmental and mental illness, depression, anxiety, OCD, oppositional defiant disorder (ODD), ASD, ADHD, and PANS and PANDAS is increasing. It is imperative that we look beyond diagnosis and medication to address the underlying causes of these issues.

30.24 Conclusion

Neurodevelopmental and neuro-immune-psychiatric issues are increasing dramatically in children. Understanding the factors that have led to this increase allows us to work upstream to create preventative solutions and find interventions to help each child reach their individual potential. Children are not little adults. Understanding the differences in their metabolism, physiology, and body composition as well as the impact of toxic exposures prenatally and postnatally is crucial. Health care has become infinitely more complicated by chemicals, a damaged food supply, EMFs, and the condition of our air, soil, and water. For all of these diverse issues, healthy lifestyle including an individualized whole-foods diet, appropriate supplementation as indicated, exercise, sleep, stress management, and avoiding neurotoxins is important. Autism has increased from 1 in 10,000 to 1 in 45, with the majority now presenting as regressive autism. This suggests that environmental insults in utero and/or after birth involve genetic and epigenetic factors leading to oxidative stress, brain inflammation, and out-of-balance biochemistry. Autism is a treatable disorder requiring a thorough history and timeline of symptoms, diet history, functional lab assessment, and supporting biochemical individuality to improve the function of the body and the brain. This is true for ASD, ADHD, sensory deficits, learning disabilities, depression, developmental delay, behavior disorders, and other neurodevelopmental disorders. Seizures, even those resistant to medications, often respond to a ketogenic or modified ketogenic diet that can help heal the brain. Certain nutrients and a healthy lifestyle can improve seizure threshold as well as neurodevelopmental outcomes. Avoiding excitotoxic food chemicals can also help lower seizure threshold. Neuropsychiatric autoimmune disorders present the challenge of learning how multiple systems interact, as cross-reactive anti-brain antibodies can trigger OCD, tics, and neuropsychiatric symptoms. Controlling risk factors before, during, and after pregnancy as well as cleaning up our environment and food supply is so important for future generations.

In summary, this chapter reviews an important group of neurodevelopmental problems that are increasing in children in epidemic proportions. Better answers are needed through understanding the interplay of various systems in the body including but not limited to neuropsychiatric, immunologic, gastrointestinal, genetic, and/or acquired inborn errors of metabolism with our current environment via lab evaluation and clinical assessment with appropriate treatments. Parents and caregivers are essential members of the treatment team alongside integrative and functional medicine practitioners.

References

1. Alexiou GA, Sfakianos G, Prodromou N. Pediatric head trauma. J Emerg Trauma Shock. 2011;4(3):403.
2. Georgieff M. Nutrition and the developing brain: nutrient priorities and measurements 1'2'3'. Am J Clin Nutr. 2007;85(5):614s–20s.
3. Ivarsson MA. Differentiation and functional regulation of human fetal NK cells. J Clin Invest. 2013;123(9):3889–901.
4. Neu J. The microbiome and its impact on disease in the preterm patient. Curr Pediatr Rep. 2013;1(4):215–21.
5. Kolokotroni O, et al. Asthma and atopy in children born by caesarean section: effect modification by family history of allergies-a population based cross-sectional study. BMC Pediatr. 2012;12:179.
6. Houghteling P, Walker W. Why is initial bacterial colonization of the intestine important to the infant's and child's health? J Pediatr Gastroenterol Nutr. 2015;60(3):294–307.
7. Mueller NT. The infant microbiome development: mom. Trends Mol Med. 2015;21(2):109–17.
8. Grandjean P, Landrigan PJ. Neurobehavioral effects of developmental toxicity. Lancet Neurol. 2014;13(3):330–8.
9. Environmental Working Group. Body burden: the pollution in newborns. 14 July 2005. Accessed: https://www.ewg.org/research/body-burden-pollution-newborns.
10. Jaishankar M, et al. Toxicity, mechanism and health effects of some heavy metals. Interdiscip Toxicol. 2014;7(2):60–72.
11. Mesquita M, et al. Effects of zinc against mercury toxicity in female rats 12 and 48 hours after HgCl2 exposure. EXCLI J. 2016;15:256–67.
12. Chasapis CT, et al. Zinc and human health: an update. Arch Toxicol. 2012;86(4):521–34.
13. Fulgenzi A, et al. Aluminium involvement in neurotoxicity. Biomed Res Int. 2014;2014:758323.
14. Singh N, Gupta VK, Kumar A, Sharma B. Synergistic effects of heavy metals and pesticides in living systems. Front Chem. 2017;5:70.
15. Chen C, et al. The roles of serum selenium and selenoproteins on mercury toxicity in environmental and occupational exposure. Environ Health Perspect. 2006;114(2):297–301.
16. Gandhi OP, Morgan LL, de Salles AA, Han Y-Y, Herberman RB, Davis DL. Exposure limits: the underestimation of absorbed cell phone radiation, especially in children. Electromagn Biol Med. 2012;31:34–51. https://doi.org/10.3109/15368378.2011.622827.
17. Hardell L. World Health Organization, radiofrequency radiation and health-a hard nut to crack (review). Int J Oncol. 2017;51(2):405–13.
18. Carlo GL, Mariea TJ. Wireless radiation in the aetiology and treatment of autism: clinical observations and mechanisms. J Austr Coll Nutr Environ Med. 2007;26(2):3–7.
19. Kaplan S, Davis D. Controversies on EMF in neurobiology of organisms. J Chem Neuroanat. 2016;75(PtB):41–2.
20. Yehuda R, et al. Holocaust exposure induced intergenerational effects on FKBP5 methylation. Biopsych. 2016;80(5):372–80.
21. University of Southampton. Risk of obesity influenced by changes in our genes. ScienceDaily. 26 Apr 2017. www.sciencedaily.com/releases/2017/04/170426093316.htm.
22. Sage C, Burgio E. Electromagnetic fields, pulsed radiofrequency radiation, and epigenetics: how wireless technologies may affect childhood development. Child Dev. 2018;89:129–36.
23. Lane M, Robker RL, Robertson SA. Parenting from before conception. Science. 2014;345(6198):756–60.
24. Matthews J. Nourishing hope for autism: nutrition and diet guide for healing our children. San Francisco: Healthful Living Media; 2008. p. 143–4.
25. Owens LJ, France KG, Wiggs L. Behavioural and cognitive-behavioural interventions for sleep disorders in infants and children: a review. Sleep Med Rev. 1999;3(4):281–302.
26. Shannon Scott M. Kids and sleep loss. In: Mental health for the whole child. New York, London: W. W. Norton & Company; 2013. p. 33.
27. Astill RG, et al. Sleep, cognition, and behavioral problems in school-age children: a century of research meta-analyzed. Psychol Bull. 2012;138(6):1109–38.
28. Stefanatos G. Regression in autistic spectrum disorders. Neuropsychol Rev. 2008;18(4):305–19.
29. Rossignol DA, Frye RE. Evidence linking oxidative stress, mitochondrial dysfunction, and inflammation in the brain of individual with

autism. Front Physiol. 2014;5:150. Review Article. Published: 22 Apr 2014. https://doi.org/10.3389/fphys.2014.00150.
30. Bock K. Healing the new childhood epidemics: autism, ADHD, asthma, and allergies: the groundbreaking program for the 4-A disorders. New York: Ballantine Publishing; 2008.
31. Curl C, Fenske R, Elgethun K. Organophosphorus pesticide exposure of urban and suburban preschool children with organic and conventional diets. Env Health Perspect. 2003;111:377–82.
32. Bauman MD, et al. Maternal antibodies from mothers of children with autism alter brain growth and social behavior development in the rhesus monkey. Transl Psychiatry. 2013;3:278.
33. Coren LA, et al. Maternal mid-pregnancy autoantibodies to fetal brain protein: the early markers for autism study. Biol Psychiatry. 2008;64(7):583–8.
34. Takano T. Role of microglia in autism: recent advances. Dev Neurosci. 2015;37(3):195–202.
35. Estes ML, McAllister AK. Maternal Th17 cells take a toll on baby's brain. Science. 2016;351(6276):919–20.
36. Gustavo C, et al. Association of gestational maternal hypothyroxinemia and increased autism risk. Ann Neurol. 2013;74(5):733–42.
37. Gardener H, Spiegelman D, Buka SL. Perinatal and neonatal risk factors for autism: a comprehensive meta-analysis. Pediatrics. 2011;128(2):344–55.
38. Ornoy AL, Weinstein-Fudim L, Ergaz Z. Prenatal factors associated with autism spectrum disorder (ASD). Reprod Toxicol. 2015;56:155–69.
39. Greenberg JA. Folic acid supplementation and pregnancy: more than just neural tube defect prevention. Rev Obstet Gynecol. 2011;4(2):52–9.
40. Wasilewska J, Klukowski M. Gastrointestinal symptoms and autism spectrum disorder: links and risks-a possible new overlap syndrome. Pediatric Health Med Ther. 2015;6:153–66.
41. Mumper EA. Call to action: recognizing and treating medical problems of children with autism. N Am J Med Sci. 2012;5(3):180–4.
42. Ghalichi F, et al. Effect of gluten free diet on gastrointestinal and behavioral indices for children with autism spectrum disorders: a randomized clinical trial. World J Pediatr. 2016;12(4):436–42.
43. Hsiao EY, et al. Microbiota modulate behavioral and physiological abnormalities associated with neurodevelopmental disorders. Cell. 2013;155(7):1451–63.
44. Li Q, Zhou JM. The microbiota-gut-brain axis and its potential therapeutic role in autism spectrum disorder. Neuroscience. 2016;324:131–9.
45. Ashwood P, Anthony A, Torrente F, Wakefield AJ. Spontaneous mucosal lymphocyte cytokine profiles in children with autism and gastrointestinal symptoms: mucosal immune activation and reduced counter regulatory interleukin-10. J Clin Immunol. 2004;24(6):664–73.
46. Bock K, Sptauth C. Healing the new childhood epidemics. In: Ballantine Books. New York; 2008. p. 249.
47. Giulivi C, et al. Mitochondrial dysfunction in autism. JAMA. 2010;304(21):2389–96.
48. Rose S, et al. Evidence of oxidative damage and inflammation associated with low glutathione redox status in the autism brain. Transl Psychiatry. 2012;2:e134.
49. Rossignol DA, Frye RE. A review of research trends in physiological abnormalities in autism disorders; immune dysregulation, inflammation, oxidative stress, mitochondrial dysfunction and environmental toxicant exposures. Mol Psychiatry. 2012;17(4):389–401.
50. Rossignol DA, Frye RE. Mitochondrial dysfunction in autism spectrum disorders: a systematic review and meta-analysis. Mol Psychiatry. 2012;17(3):290–314.
51. Naviaux RK. Metabolic features of the cell danger response. Mitochondrion. 2014;16:7–17. 6 Aug 2013.
52. Naviaux RK. Antipurinergic therapy for autism-an in-depth review. Mitochondrion. 2018;43:1–15. 7 Dec 2017.
53. Kelley R. Evaluation and treatment of patients with autism and mitochondrial disease. Kennedy Krieger Institute; June 13, 2009, Baltimore, MD.
54. Haas R, et al. Supporting children with mitochondrial disorders. ICAN. 2014;6(3):160–3.
55. Esparham AE, et al. Nutritional and metabolic biomarkers in autism spectrum disorders: an exploratory study. Integr Med (Encinitas). 2015;14(2):40–53.
56. Russo AJ. Increased copper in individuals with autism normalizes post zinc therapy more efficiently in individuals with concurrent GI disease. Nutr Metab Insights. 2011;4:49–54.
57. Tsaluchidu S, et al. Fatty acids and oxidative stress in psychiatric disorders. BMC Psychiatry. 2008;8(Suppl 1):S5.
58. Rose S, et al. Clinical and molecular characteristics of mitochondrial dysfunction in autism spectrum disorder. Mol Diagn Ther. 2018;22:571–93.
59. Piras IS, et al. Anti-brain antibodies are associated with more severe cognitive and behavioral profiles in Italian children with autism spectrum disorder. Brain Behav Immun. 2014;38:91–9.
60. Shaw W. Elevated urinary glyphosate and clostridia metabolites with altered dopamine metabolism in triplets with autistic spectrum disorder or suspected seizure disorder: a case study. Integr Med. 2017;16(1):50–7.
61. Jackson J, et al. Urine pyrroles and other orthomolecular tests in patients with ADD/ADHD. J Orthomol Med. 2010;25(1):39–42.
62. Atladottir HO, et al. Association of family history of autoimmune diseases and autism spectrum disorders. Pediatrics. 2009;124(2):687–94.
63. Wu S, et al. Family history of autoimmune diseases is associated with an increased risk of autism in children: a systematic review and meta-analysis. Neurosci Biobehav Rev. 2015;55:322–32.
64. Geschwind D. Autism: many genes, common pathways? Cell. 2008;135(3):391–5.
65. Wong CC. Methylomic analysis of monozygotic twins discordant for autism spectrum disorder and related behavioral traits. Mol Psychiatry. 2014;19(4):495–503.
66. Bernier R, et al. Disruptive CHD8 mutations define a subtype of autism early in development. Cell. 2014;158(2):263–76.
67. Elsamanoudy A, et al. The role of nutrition related genes and nutrigenetics in understanding the pathogenesis of cancer. J Microsc Ultrastruct. 2016;4:115–22.
68. Wacklin P, et al. Secretor genotype (FUT2 gene) is strongly associated with the composition of Bifidobacteria in the human intestine. PLoS One. 2011;6(5):e20113.
69. Peng F, Labelle LA, Rainey B, Tsongalis GJ. Single nucleotide polymorphisms in the methylenetetrahydrofolate reductase gene are common in US Caucasian and Hispanic American populations. Int J Mol Med. 2001;8(5):509–11.
70. Wang X, et al. Geographical and ethnic distributions of the MTHFR C677, A1298C and MTRR A66G gene polymorphisms in Chinese populations; a meta-analysis. PLoS One. 2016;11(4):e0152414.
71. Zhang T, et al. Genetic variants in the folate pathway and the risk of neural tube defects: a meta-analysis of the published literature. PLoS One. 2013;8(4):e59570.
72. Chen CH, et al. Genetic analysis of GABRB3 as a candidate gene of autism spectrum disorders. Mol Autism. 2014;5:36.
73. NIH genetics home reference for gene descriptions. https://ghr.nlm.nih.gov/.
74. Kovac J, et al. Rare single nucleotide polymorphisms in the regulatory regions of the superoxide dismutase genes in autism spectrum disorder. Autism Res. 2014;7:138–44.
75. Jepson B, Johnson J. Changing the course of autism: a scientific approach for parents and physicians. Boulder: Sentient Publications; 2007.
76. Cruchet S. Truths, myths, and needs of special diets: attention-deficit/hyperactivity disorder, autism, non-celiac gluten sensitiv-

ity, and vegetarianism. Ann Nutr Metab. 2016;68(Suppl 1):43–50. Published online: 30 June 2016.
77. Adams J, et al. Comprehensive nutritional and dietary intervention for autism spectrum disorder-a randomized, controlled 12-month trial. Nutrients. 2018;10(3):369.
78. Whiteley P, et al. Gluten- and casein-free dietary intervention for autism spectrum conditions. Front Hum Neurosci. 2012;6:344.
79. Matthews J. Nourishing hope for autism. 1 Nov 2008.
80. Waring RH, et al. Biochemical parameters in autistic children. Dev Brain Dysfunc. 1997;10:40–3.
81. Buts JP, Cortjier G, Delmee M. Saccharomyces boulardii for Clostridium difficile-associated enteropathies in infants. J Pediatr Gastroenterol Nutr. 1993;16(4):419–25.
82. Mostafa GA, Al-Ayadhi LY. Reduced serum concentrations of 25-hydroxyvitamin D in children with autism: relation to autoimmunity. Licensee BioMed Central Ltd 2012. J Neuroinflammation. 2012;9:201.
83. Pokorski M. Stability of ascorbyl palmitate molecule in the rat brain. J Physiol Pharmacol. 2005;56(Suppl 4):197–201.
84. Shaw P, et al. Attention-deficit/hyperactivity disorder is characterized by a delay in cortical maturation. PNAS. 2007;104(49):19649–54.
85. Hoogman M, et al. Subcortical brain volume differences in participants with attention deficit hyperactivity disorder in children and adults: a cross-sectional mega-analysis. Lancet Psychiatry. 2017;4(4):310–9. Published: 15 Feb 2017.
86. Walsh WJ. Nutrient power heal your biochemistry and heal your brain. Ontario: Skyhorse Publishing; 2012, 2014. p. 128.
87. Amen DG. All-new revised edition healing ADD, the breakthrough program that allows you to see and heal the 7 types of ADD. 2013.
88. Talbott SM, et al. Effect of magnolia officinalis and phellodendron amurense (Relora) on cortisol and psychological mood state in moderately stressed subjects. J Int Soc Sport Nutr. 2013;10:37.
89. Bouchard M, et al. Attention-deficit/hyperactivity disorder and urinary metabolites of organophosphate pesticides. Pediatrics. 2010;125(6):e2170–7.
90. Arnold LE, et al. Artificial food colors and attention-deficit/hyperactivity symptoms: conclusions to dye for. Neurotherapeutics. 2012;9(3):599–609.
91. Elshorbagy H, et al. The impact of vitamin D supplementation on attention-deficit hyperactivity disorder in children. Ann Pharmacother. 2018;52(7):623–31. online ISSN: 1542-6270.
92. Louv R. Last child in the words: saving our children from nature-deficit disorder. In: Algonquin Books. Chapel Hill; 2005.
93. Medina KL, Schweinsburg AD, Cohen-Zion M, Nagel BJ, Tapert SF. Effects of alcohol and combined marijuana and alcohol use during adolescence on hippocampal volume and asymmetry. Neurotoxicol Teratol. 2007;29(1):141–52.
94. Wang HS. B6 related epilepsy during childhood. Chang Gung Med J. 2007;30(5):369–401.
95. Bhardwaj P, Kaushal RK, Chandel A. Biotinidase deficiency: a treatable cause of infantile seizures. J Pediatr Neurosci. 2010;5(1):82–3.
96. Kossoff E, et al. Ketogenic diets: an update for child neurologists. J Child Neuro. 2009;24(8):979–88.
97. Kossoff E, et al. The ketogenic and modified Atkins diets: treatments for epilepsy and other disorders. 6th ed. New York: Demos Health; 2016.
98. Martin K, et al. Ketogenic diet and other dietary treatments for epilepsy. Cochrane Database Syst Rev. 2016;(2):CD001903.
99. Diala A. Fish oil intake and seizure control in children with medically resistant epilepsy. N Am J Med Sci. 2015;7(7):317–21.
100. Osborn KE, et al. Addressing potential role of magnesium dyshomeostasis to improve treatment efficacy for epilepsy: a reexamination of the literature. J Clin Pharmacol. 2016;56(3):260–5.
101. Pendo K. Vitamin D3 for the treatment of epilepsy: basic mechanisms, animal models, and clinical trials. Front Neurol. 2016;7:218.
102. Hollo A. Correction of vitamin D deficiency improves seizure control epilepsy: a pilot study. Epilepsy Behav. 2012;24(1):131–3. 11 Apr 2012.
103. Zhen J. Effects of grape seed proanthocyanidin extract on pentylenetetrazole-induced kindling and associated cognitive impairment in rats. Int J Mol Med. 2014;34(2):391–8.
104. Tyagi A, Delanty N. Herbal remedies, dietary supplements, and seizures. Epilepsia. 2003;44(2):228–35. Blackwell Publishing, Inc. 2003 International League against Epilepsy.
105. Swedo S, et al. Pediatric autoimmune neuropsychiatric disorders associated with streptococcal infections: clinical description of the first 50 cases. Am J Psychiatry. 1998;155:2.
106. Kirvan CA, et al. Mimicry and autoantibody-mediated neuronal cell signaling in Sydenham Chorea. Nat Med. 2003;9(7):914–20.
107. Cox CJ, Zuccolo AJ, Edwards EV, Mascaro-Blanco A, Alvarez K, Stoner J, Kiki C, Cunningham MW. Antineuronal antibodies in a heterogeneous group of youth and young adults with tics and obsessive-compulsive disorder. Child Adolesc Psycopharm. 2015;25:76–85.
108. Chang K, et al. Clinical evaluation of youth with pediatrics acute-onset neuropsychiatric syndrome (PANS): recommendations from the 2013 PANS Consensus Conference. J Child Adolesc Psychopharmacol. 2015;25(1):3–13.
109. Alder R, Cohen N, Felten D. Psychoneuroimmunology: interactions between the nervous system and the immune system. Lancet. 1995;345:99–103.

Nutritional Influences on Hormonal Health

Filomena Trindade

31.1 Introduction to Hormonal Health – 518

31.2 Insulin – 518
31.2.1 Insulin Resistance – 518
31.2.2 Impaired Glucose Tolerance – 518
31.2.3 Prediabetes – 519
31.2.4 Diabetes Mellitus Type 2 – 519
31.2.5 Metabolic Syndrome Vs. Insulin Resistance – 519
31.2.6 Identifying Patients at Risk for Insulin Resistance – 519
31.2.7 Protocol – 519
31.2.8 Staging the Patient – 520
31.2.9 Laboratory Evaluation – 520
31.2.10 Treatment – 521
31.2.11 A Final Word on Insulin Resistance – 523

31.3 Adrenal Dysfunction – 524
31.3.1 What Is Adrenal Dysfunction? – 524
31.3.2 What Causes Adrenal Dysfunction? – 525
31.3.3 Assessment and Treatment of Adrenal Dysfunction – 525
31.3.4 A Final Word on Adrenal Dysfunction – 526

31.4 Thyroid – 526
31.4.1 Assessment – 527
31.4.2 Treatment – 527

31.5 Sex Steroid Hormones – 528

31.6 Conclusion – 529

References – 529

© Springer Nature Switzerland AG 2020
D. Noland et al. (eds.), *Integrative and Functional Medical Nutrition Therapy*,
https://doi.org/10.1007/978-3-030-30730-1_31

31.1 Introduction to Hormonal Health

An integrative or functional medicine approach to illness involves identifying the root cause(s) and personalizing treatment plans to address those underlying causes. In this way, the approach helps patients avoid and reverse illness as they move toward optimal health. This is a giant leap forward from conventional drug-based symptom suppression and disease management approaches. Fortunately, combining integrative/functional and conventional approaches to medicine is not entirely incompatible. It is wise to integrate the best of conventional laboratory analysis, diagnostic imaging methods, and appropriate treatments, along with cutting-edge nutritional, lifestyle, biochemical, and genetically driven interventions to create the best outcomes possible for patients. This approach of integrating the best of both paradigms has been useful in developing and personalizing treatment of hormonal imbalance.

A recent report highlighted the fact that endocrine active chemicals (EACs) or endocrine disruptors may cause a variation in normal hormone function [1]. It has been found that even small alterations in hormone concentrations, especially during embryonic development, can have significant and lasting effects. Further, some EACs have the potential to impact human health at lower doses than those proposed in traditional toxicity testing by regulatory agencies, which means that some effects may be overlooked [1].

A protocol has been developed to address hormonal imbalances in a systematic way and includes nutritional influences in hormone metabolism. To be effective, an ordered evaluation followed by systematic balancing is key, as will be detailed below. Achieving optimal hormonal health happens as hormones are balanced individually and then as they begin to work in concert with each other. Hormones are like a symphony in that each hormone must not only be balanced or fine-tuned but also working optimally with other hormones to make beautiful music. In the author's experience, the orderly approach in hormone balancing begins with insulin, followed by the hypothalamic pituitary adrenal axis (hereinafter referred to as Adrenals), thyroid, and sex steroid hormones (estrogens, progesterone, and testosterone) with a final step optimizing *estrogen metabolism* in both men and women.

31.2 Insulin

Insulin is a major hormone and one of critical importance. It not only affects all other hormones, but if there is decreased insulin sensitivity or insulin resistance, it is postulated to contribute to cardiovascular disease, premature death, cancer, and dementia, to name a few [2]. Moreover, insulin resistance is a major cause of premature aging and death in both the developed and developing worlds. It is arguably our biggest global health epidemic. Human history has never before seen as many new cases of type 2 diabetes diagnosed, particularly in children [3]. The diagnosis of adult-onset or type 2 diabetes in children has climbed astronomically, increasing more than 1000 percent over the past two decades. A mere 15 years ago, only 3 percent of all diabetes cases in children were of the type 2 variety; now that number has climbed to 50 percent. Furthermore, what is shocking is that 40 percent of children are now overweight and two million are morbidly obese (exceeding the 99th percentile for weight) [4].

It is known that people with diabetes suffer vascular disease at much higher rates [5]. It is estimated that diabetics are four times more likely to die of heart disease than their nondiabetic counterparts. Furthermore, the rate of cerebrovascular accidents (CVAs) is three to four times higher in diabetics. Unfortunately, these risks do not only apply to the fully diabetic patient. Of concern, women are at an ever-increasing risk for cardiovascular disease, especially when insulin resistance is present, and women in perimenopause or menopause who suffer more hot flashes and night sweats are more likely to have undiagnosed insulin resistance [6].

The risk for cardiovascular disease also encompasses those who are considered prediabetic. Patients with prediabetes are four times more likely to die of heart disease and have a 21 percent higher risk of stroke [7]. Therefore, the prefix "pre" in prediabetes is arguably not really "pre"-disease. People with prediabetes are diseased secondary to the elevated degree of insulin resistance. Therefore, it is not the diagnosis of diabetes alone that renders the patient at high risk for major life-threatening disease but, rather, the state of insulin resistance. There is a critical need for clinicians to recognize and address insulin resistance and understand the continuum of insulin resistance. The continuum refers to the progression from insulin resistance to impaired glucose tolerance to prediabetes to diabetes. Treatment of this continuum begins with detection of early insulin resistance and of learning the tools to apply to reverse this state in order to avoid the consequences and progression to diabetes.

31.2.1 Insulin Resistance

Simply put, insulin resistance is a condition where insulin becomes less effective at lowering blood sugar [5–8]. It is the inability of insulin to facilitate glucose uptake from the blood into the cell because of poor insulin binding at the cells' receptors. The body responds to this by stimulating the pancreas to produce more insulin to ensure that glucose can leave the bloodstream and enter the cells. Initially, the body's response is effective in that the excess insulin is able to keep blood sugar in the normal range. The first sign of imbalance lies in the increased levels of insulin production.

31.2.2 Impaired Glucose Tolerance

Impaired glucose tolerance occurs when glucose rises above the normal range, despite the pancreas continuing to overproduce insulin [5–9]. The rise can be seen after meals (postprandial) or at in the fasting state. In many patients at this stage along the continuum, fasting glucose may be normal, but you will see an elevated glucose level after a meal or after a high glycemic load. Levels are elevated but not enough to qualify the patient as prediabetic. These are the patients in which there is an elevated hemoglobin A1c (HgbA1c) despite a normal fasting glucose. Simply put, impaired glucose tolerance is when the pancreas and its insulin production are no longer able to keep blood sugar normal after a glucose challenge.

31.2.3 Prediabetes

As the pancreas continues to overproduce insulin, its ability to keep fasting blood sugar in the normal range is impaired as evidenced by a fasting blood sugar over 100 mg/dL to 125 mg/dL [5, 8, 9]. This should be considered an advanced state of disease because close behind prediabetes is the end of the continuum with type 2 DM. Therefore, prediabetes is clearly a state of major dysfunction and it is not "pre"-disease. It is a disease in and of itself.

31.2.4 Diabetes Mellitus Type 2

Diabetes mellitus type 2 is defined as fasting blood sugar of 126 mg/dL or higher or a random blood sugar of over 200 mg/dL on more than one occasion [5, 8, 9].

31.2.5 Metabolic Syndrome Vs. Insulin Resistance

The synergistic and aggregate effects of insulin resistance, obesity, hypertension (HTN), and dyslipidemia all form a construct called the metabolic syndrome [7]. Metabolic syndrome was initially coined "Syndrome X" by Gerald M. Reaven as "a constellation of lipid and non-lipid risk factors of metabolic origin." Metabolic syndrome was formally recognized by the Adult Treatment Panel III (ATP III) of the National Cholesterol Education Program (NCEP) 2004 update and identifies the significance of cardiovascular disease (CVD) risk factors [8]. Risk factors for metabolic syndrome include visceral adiposity, high body mass index (BMI), insulin resistance, and high insulin. In order to have metabolic syndrome, a patient must have three out of the following five characteristics present: increased waist circumference (greater than 40 inches for men and 35 inches for women), hypertension (blood pressure greater than 130/85 mmHg), elevated triglycerides (greater than 150 mg/dL), elevated fasting blood sugar (greater than 100 mg/dL), and decreased HDL (40 mg/dL in men and 50 mg/dL in women).

Focus has recently shifted to insulin resistance as the dominant and independent predictor of age-related diseases [10, 11]. A normal blood sugar level is less than 100 mg/dL; however, we know that risks extend below that threshold. Studies have shown that a blood sugar greater than 87 mg/dL equals a progressive increase of type 2 diabetes. Further, the lowest risk of progression from insulin resistance to type 2 diabetes is a fasting blood sugar less than 75 mg/dL.

In the United States, we have seen a rise of obesity in epidemic proportions over the past 20 years [2, 3, 5, 8, 10]. Among the American adults who are simply overweight, more than 70 percent already have prediabetes because of undiagnosed and untreated insulin resistance, with significant increased risk of cardiovascular disease and death. The unfortunate fact is that most are unaware of their condition and its risks. The awareness problem is further aggravated by the fact that there are no current national screening guidelines or reimbursement for healthcare providers to diagnose and treat insulin resistance. Guidelines and reimbursement are only available once diabetes develops. In addition, healthcare practitioners are not properly trained to diagnose or treat the largest epidemic in the United States. In sum, it is well-known that obesity increases insulin resistance and leads to diabetes. Conversely, the most important *cause* of obesity is insulin resistance that often goes undiagnosed.

31.2.6 Identifying Patients at Risk for Insulin Resistance

To determine if patients have insulin resistance, it is not as simple as looking at body habitus. According to the National Health and Nutrition Evaluation Survey (NHANES) correlation study from 1999 to 2004, up to 25 percent of normal weight patients (BMI less than 25) may be insulin-resistant [11]. Further, some patients may be insulin-resistant and still have normal fasting insulin levels, normal HgbA1c, and normal fasting glucose but are on the continuum to develop type 2 diabetes and metabolic syndrome. Important for clinicians, the underlying causes of type 2 diabetes, metabolic syndrome, and insulin resistance are reversible and preventable [5, 7] by using nutritional and lifestyle interventions. The question becomes how do we identify the patients at risk? This is best addressed by looking at the confluence of genes and metabolism, both of which underlie the abnormal patterns of function seen in hyperinsulinemia. Clinicians need to determine why and how a patient became insulin-resistant. How and where do we start? What is the root cause? How do we identify and treat insulin resistance?

31.2.7 Protocol

The author has developed a treatment protocol for insulin resistance that incorporates integrative and functional medicine principles. This starts with collecting a thorough and detailed history and conducting a complete physical examination. In the process of data gathering, it is imperative to look for basic imbalances in digestion/absorption/barrier integrity, detoxification pathways, hormone/neurotransmitter health, immune/inflammatory process, and mind/spirit equilibrium imbalance. The goal is to get to the underlying cause of illness. This allows the practitioner to connect the dots between the history and the physical exam to find clues leading to the root cause or foundation of the dysfunction. Once the investigation into development of insulin resistance has proceeded and potential underlying cause(s) are identified, laboratory testing can be ordered and evaluated. Patients are staged along the continuum of insulin resistance at this juncture. Individualized treatment protocols are prescribed that incorporate certain basic standards (see Table 31.1).

Table 31.1	Insulin resistance assessment protocol
Poor-quality diet	Processed foods High in sugar and high fructose corn syrup Trans, hydrogenated, and saturated fats Polyunsaturated omega-6 oils (except gamma linoleic acid, GLA) Low in vegetables, fruits, and antioxidants Low in fiber
Hydration	Assess hydration status and source/content of water and liquids Can be cause of insulin resistance, i.e., look for sugar and high fructose corn syrup drinks
Food allergies/sensitivities	Modify and personalize elimination diet/ketogenic diet and/or functional testing for allergies and sensitivities
Oxidative stress and/or mitochondrial dysfunction	Assess clinically and with laboratory evaluation
Stress and adrenal dysfunction	Lack of exercise Poor sleep
Hormone imbalance	Use history and physical exam, as well as laboratory evaluation
Toxins	Heavy metals, persistent organic pollutants, endocrine disruptors volatile solvents, dirty electricity, electromagnetic radiation, suboptimal estrogen metabolism, etc.
Dysbiosis	Assess with comprehensive stool analysis
Infections	Occult esp. dental Gastrointestinal Viral, bacterial, parasitic, and fungal
Nutrient deficiencies	Including vitamins, minerals, antioxidants, amino acids, methylating factors, and essential fatty acids Assess clinically with history and physical and with functional and/or conventional testing
Prescription drugs	Prior or current use of statins or other cholesterol-lowering drug use, proton pump inhibitors, or beta blockers
Hyperinsulinemia causing insulin resistance; beta cell dysfunction	Review all above to look for root cause

© Copyright 2013 Filomena Trindade MD, MPH. All rights reserved. Reprinted with permission of the author

31.2.8 Staging the Patient

Staging facilitates understanding of where the patient falls along the continuum of insulin resistance. It is the stage that determines the treatment and predicts how likely they are to progress to type 2 diabetes. In the continuum of insulin resistance, with wellness on one end and illness on the other, the patient progresses from insulin-sensitive, to insulin-resistant, to impaired glucose tolerance, followed by prediabetes, and then to type 2 diabetes.

The three stages and their components are described in detail below in the "Laboratory Evaluation" section. Stage 1 is when adiponectin (an adipose-derived protein) starts to decrease [11]. In stage 2, insulin is already starting to increase, and in stage 3, proinsulin (the prohormone precursor to insulin) increases [11]. Not everyone moves through these stages smoothly, as there may be variations in timing and presentation. We also know that there may be some patients who fall between type 1 and type 2 diabetes, related to an underlying autoimmune disease [12, 13]. Overall, these stages are valuable as each stage will direct proper treatment. For instance, if a patient is found to have insulin resistance and hyperlipidemia, it is the insulin resistance that is driving the hyperlipidemia and the progression to glucose intolerance and prediabetes. Consequently, we treat insulin resistance first.

31.2.9 Laboratory Evaluation

Laboratory evaluation is an important tool to help guide assessment of the patient's stage along the continuum of insulin resistance. Although not all insulin-resistant patients develop diabetes, it is insulin resistance that is the underlying problem driving metabolic syndrome and chronic disease, such as cardiovascular risk. Consequently, it is very important to identify those with insulin resistance early and promptly treat them to halt and/or reverse disease progression. Laboratory assessment is critical.

Most test result ranges in conventional laboratory analysis are very broad. This occurs because the lab is aggregating multiple patients' ranges who are considered normal although disease may be present at an early stage; these ranges are not functional ranges. A general guideline to determine abnormality is to look at the upper third and fourth quartile when assessing conventional lab values for glucose and insulin while looking at the first and second quartiles when assessing micronutrient levels. This gives a more functional/integrative medicine reference range that may guide intervention. It is important to understand that although the value may be in a normal range according to the conventional lab, early disease and dysfunction may be present. This helps when analyzing results for patients who can only afford to undergo conventional insurance-covered laboratory evaluations and do not have access to more extensive laboratory testing.

31.2.9.1 Hemoglobin A1c

Hemoglobin A1c (HbA1c) was originally developed to determine if diabetic patients were being compliant with their glucose-lowering medication [8, 14]. Put simply, it measures the amount of glucose bound to the hemoglobin molecule and reflects levels of glucose for the previous 90 to 120 days. It was not developed as a screening tool, although it is currently being used as one.

With this in mind, what does it mean if a HbA1c is elevated? Where does the elevated HbA1c place the patient on the continuum? HgbA1c is an independent risk factor for cardiovascular mortality: "The predictive value of HgbA1c for total mortality was stronger than that documented for cholesterol concentration, body mass index and blood pressure." [14]. If HbA1c is 5.4 percent or higher, the patient has impaired glucose tolerance and increased risk. If the HbA1c is increased, the patient is having episodes of hyperglycemia. However, it might only be elevated postprandially and not in the fasting state. Consequently, patients must check the postprandial blood sugars both in the laboratory and at home. With insulin resistance it is imperative to preserve pancreatic beta cell function and aggressively address the underlying cause of the dysfunction.

31.2.9.2 HOMA-IR

The homeostatic model assessment for insulin resistance (HOMA-IR) is a calculation based on plasma levels of fasting glucose and insulin [15, 16]. It is used to assess insulin sensitivity. There may be better ways of assessing insulin resistance.

31.2.9.3 Adiponectin, Insulin, and Proinsulin: Staging the Progression of Insulin Resistance to Prediabetes and Type 2 Diabetes

Adiponectin is a protein hormone that is produced in fat cells and helps regulate glucose levels and fatty acid breakdown [17]. Adiponectin is protective against atherosclerosis, modulates fat tissue production and breakdown, promotes insulin sensitivity, and decreases hepatic glucose and lipid production. Biologic adiponectin levels decrease prior to increases in insulin levels; therefore it can be considered a marker of insulin sensitivity. Other markers assessed by laboratory evaluation for insulin sensitivity include insulin, proinsulin, HbA1c, fasting, 30 minute, 1 hour and 2 hour glucose levels, a 30 minute, 1 hour and a two-hour insulin level after a 75 gram glucose load. Both conventional and specialty laboratories can provide these tests. While a specialty lab plots the results in graphical form with functional ranges, conventional labs give numerical values that can be plotted in graphical form by the practitioner. With optimal control and regulation of glucose, adiponectin will be normal, fasting insulin will be 5–7 mIU/L, HbA1c will be 5.3 percent or less, and fasting glucose will be less than 75 mg/dL. The risk of diabetes significantly increases when the fasting blood glucose is greater than 87 mg/dL [6].

In *stage 1* insulin resistance, adiponectin is declining while fasting insulin remains normal [11, 17]. Postprandial insulin might be elevated, but the primary determining factor of stage 1 is a normal fasting insulin. Proinsulin, glucose readings, HOMA-IR, and HbA1c are all usually normal, although HOMA-IR could be slightly elevated. Adiponectin is the key factor that needs to be restored. Low adiponectin is a marker for insulin resistance, making the patient more at risk for dyslipidemia. Decreasing adiponectin also tends to be associated with increasing inflammatory markers. With that, there is increased risk for vascular injury and increased risk of progression along the continuum toward type 2 diabetes [11, 17].

Treatment includes diet and lifestyle changes. Focus is on body composition and making sure patients are eating a healthful diet. Experts recommend a Mediterranean-type, low glycemic load diet, but the diet needs to be personalized to the individual [18].

In *stage 2* insulin resistance, adiponectin levels are decreasing, insulin is starting to increase, while proinsulin remains normal [11, 17]. There may be early beta cell impairment. In stage 2, HOMA-IR is higher than normal and may be associated with a high postprandial glucose. At this stage, fasting glucose may be borderline elevated while random and postprandial sampling of glucose may be mildly elevated but at times be normal. Stage 2 treatment usually consists of diet and lifestyle changes and supplementation. Pharmacotherapy should not be started at this stage, but rather in early stage 3 and even then, there is the potential for reversal with diet, nutrition, lifestyle modification and nutraceutical support. Again, this is case dependent.

In *stage 3* insulin resistance, proinsulin is increasing, indicating the pancreas is struggling to keep glucose normal. The pancreas is attempting to excrete insulin as fast as it is manufactured and before the final cleavage has happened. In stage 3 there is low adiponectin, high HOMA-IR, high insulin, elevated proinsulin, possibly high glucose, and high HbA1c [11, 17]. There are some patients who can maintain normal or close to normal fasting glucose even in the face of these abnormalities. However, with glucose challenge testing, the post-challenge glucose level will be high. In most cases, the postprandial glucose will elevate with a high glycemic load meal. Despite these findings, the patient may not be classified diabetic. This group might not even fulfill the established criteria for prediabetes, as the American Diabetes Association definition for prediabetes is a fasting glucose level of 100–125 mg/dl [8]. The overarching treatment aim in stage 3 is to preserve beta cell function. Having patients check postprandial glucose levels also helps the patient learn what foods increase their glucose level.

31.2.10 Treatment

▶ Box 31.1 presents the approach for the treatment of insulin resistance. It will be necessary to address and re-address these factors both at baseline and throughout the course of treatment.

> **Box 31.1 Insulin Resistance Treatment Protocol**
> - Foundation is nutritional support: Wholesome food that is fresh, whole, unprocessed, organic, colorful, fermented, high in omega-3 fatty acids
> - Identify the underlying functional imbalances, prioritize, and address
> - Personalize an elimination diet or modified ketogenic diet according to individual patient needs
> - Lifestyle modification: Individualized stress reduction, alcohol consumption, smoking cessation, sleep
> - Exercise: Tailor to individual patient needs
> - Nutritional supplementation: Personalized based on clinical exam and laboratory testing
> - Mind-body-spirit connection: Find and foster purpose and meaning in life as well as gratitude, and explore history of trauma that may be preventing compliance and development of treatment protocol
> - Assess possible need for pharmacological treatment based on functional testing, patient needs, and response to above approach
>
> © Copyright 2012 Filomena Trindade MD, MPH. All rights reserved. Reprinted with permission of the author

31.2.10.1 Dietary Management

Dietary management with healthful whole foods for patients with insulin resistance results in decreasing insulin stimulation and increasing insulin sensitivity [19–26]. Dietary modifications to decrease insulin release always begin with an initial assessment to explore the patient's intake of fiber, good and bad fats, types and quantity of protein, and simple and complex carbohydrates, as noted above. Education of patients about the quality of dietary components and helping them understand glycemic load is imperative at this early stage. Crucial is the fact that food components are important signaling molecules that are not often found in processed foods [27]. Many phytochemicals work as tissue-specific kinase response modulators (SKRMs). Kinases are a group of enzymes with multiple functions that include facilitating insulin function. SKRMs are going to affect cellular communication and cellular signaling. Some dietary phytochemicals that modulate these pathways include berberine, cinnamon, ginseng, quercetin, resveratrol, green tea extract, and hops extract.

Components of this type of diet include organic, fresh, whole, unprocessed, colorful food. It is important to make sure patients have protein at every meal and snack. Emphasize elimination of sugar, trans fats, some saturated fats, and food allergens. Patients need education regarding inflammatory foods, including gluten, dairy, soy, corn, and nightshades. An elimination diet may be helpful to identify and remove any dietary source of inflammation. Also address the areas of dysfunction or imbalance in laboratory assessment and prioritize. Depending where the patient is on the continuum, it may be necessary to start nutraceuticals that increase cellular responsiveness to insulin, including chromium, alpha-lipoic acid, magnesium, CoQ10, and protein kinase modulators, to name a few [27–35]. See ◘ Table 31.2.

31.2.10.2 Exploring Detoxification

Body burden of environmental toxins is known to be another underlying problem triggering insulin resistance [36–39]. Toxic endocrine disruptors affect estrogen and estrogen metabolism and are known to result in the etiology of type 2 diabetes mellitus [36–38]. Low-dose organochlorine pesticides and polychlorinated biphenyls predict obesity, dyslipidemia, and insulin resistance among people free of diabetes [36–38]. Heavy metals may also be a problem and can contribute to insulin resistance [39]. The liver enzyme, GGTP, in the high normal range is associated with insulin resistance. It may be helpful to monitor GGTP and consider it a predictor if it is 30 IU/L or higher. If it is over 40 IU/, then glutathione production needs to be improved.

31.2.10.3 Exercise

As exercise is an important tool for intervention in insulin resistance, it is critical to assess how much a patient is moving. It is imperative to design an appropriate exercise regimen for each patient that incorporates aerobic exercise and weight training and re-evaluate adherence to the program at regular intervals with change as needed. It is wise to review the published studies with patients and give them the choice between diet, lifestyle, and physical activity versus medications with their potential side effects. Typically, in the author's practice, patients choose diet and lifestyle changes for management. Patients need to be informed that if they do not change their diet and lifestyle, the outcome will likely be increased cardiovascular disease and progression to type 2 diabetes. Are they willing to take those risks? When explaining how diet, nutrition, exercise, and lifestyle modifications have been proven superior to prescription medications [40], patients express amazement and choose to succeed with this approach.

According to the World Health Organization, adults 18–64 should do at least 150 minutes of moderate-intensity aerobic physical activity throughout the week or at least 75 minutes of vigorous-intensity aerobic physical activity throughout the week or an equivalent combination of moderate- and vigorous-intensity activity [41]. Aerobic activity should be performed in episodes of at least 10 minutes' duration. For additional health benefits, adults should increase their moderate-intensity aerobic physical activity to 300 minutes per week or engage in 150 minutes of vigorous-intensity aerobic physical activity per week or an equivalent combination. Muscle-strengthening activities should be done involving major muscle groups on 2 or more days a week. Exercise alters skeletal muscle metabolism and improves glucose uptake, reduces low-density lipoprotein, raises HDL, lowers blood pressure, and reduces inflammation and oxidative stress [42]. In a Diabetes Prevention Program Research Group 2002 study, 3234 people with prediabetes were randomized to placebo, metformin, or lifestyle modification (≥7% weight loss and ≥ 150 min/week of physical activity) for 2.8 years [40]. Compared to the placebo group, lifestyle intervention decreased incidence of type 2 DM by 58 percent. Metformin (850 mg BID) decreased type 2 diabetes mellitus by only 31 percent.

Table 31.2 Micronutrient recommendations for insulin resistance

Chromium [35]	Generally, give 400 mcg/daily if insulin-resistant Likely most effective if deficient but difficult to test
Vitamin D [32]	Test 25(OH)D and supplement as appropriate (or supplement 2000–5000) IU/daily Vitamin D levels should be 50–80 ng/ml, depending on whether other medical conditions are present A study showed a positive correlation of 25(OH)D concentration with insulin sensitivity and a negative effect of hypovitaminosis D on beta cell function Subjects with hypovitaminosis D are at higher risk of insulin resistance and the metabolic syndrome Increasing 25(OH)D from 10–30 ng/mL can improve insulin sensitivity by 60%
CoQ10 [28, 29]	Supplement in patients with metabolic syndrome, insulin resistance, hypertension, or mitochondrial dysfunction Dose depends on functional levels: Optimize to >2 mcg/mL (plasma) In a patient with high oxidative stress or mitochondrial dysfunction, may need to supplement at a much higher dose 120 mg/day of coenzyme Q10 improves glycemic control and blood pressure in NIDDM 200 mg of CoQ10 daily improved HgA1C and blood pressure in NIDDM patients
Alpha-lipoic acid [30, 31]	Doses of 600 to 1800 mg/day may improve insulin sensitivity 600–1200 mg/day of ALA may improve microcirculation and diabetic polyneuropathy
Magnesium [33, 34]	Epidemiological studies show that high daily mg intake is predictive of a lower incidence of NIDDM Poor intracellular mg concentration is found in NIDDM and in hypertensive patients Daily mg administration in NIDDM patients and in insulin-resistant patients restores intracellular mg concentration and contributes to improve insulin sensitivity and glucose uptake Magnesium supplementation has been shown to improve insulin sensitivity Patients have a hard time absorbing magnesium intracellularly Buccal swab is the best way to assess intracellular magnesium Patients with low magnesium are at a slightly higher risk for atrial fibrillation and arrhythmias Many of these patients may also need potassium. Magnesium glycinate may have less effect on the gut causing loose bowels, and it is generally well absorbed Starting dose 200–400 mg of a chelated magnesium and increase to bowel tolerance if normal kidney function is established

31.2.10.4 Stress and Autonomic Dysfunction

Stress and autonomic dysfunction have a powerful impact on progression of insulin resistance and contribute to cardiovascular disease. To assess autonomic dysfunction, an important tool to incorporate is heart rate variability [43, 44]. As heart rate variability lowers, there is progression from insulin resistance to impaired glucose tolerance to diabetes. Of interest in patients with coronary artery disease, lower heart rate variability has also been correlated with impaired pancreatic function. Visceral fat (not subcutaneous fat) has also been associated with lower heart rate variability as well. In patients with either diabetes or a family history of diabetes, those genetically predisposed to NIDDM have more visceral fat and lower insulin sensitivity compared with those unaffected individuals [45]. Knowing these connections, the dysregulation of the autonomic nervous system can now be linked to visceral adiposity and insulin resistance.

31.2.10.5 Sleep

Sleep is another important factor in the management of insulin resistance. Lack of sleep can affect insulin resistance in several different ways as it can cause inflammatory-related insulin resistance, changes in glucose metabolism, altered appetite, and hypothalamic-pituitary-adrenal axis dysfunction [46]. Sleep deprivation may lead to insulin resistance and subsequently to diabetes mellitus. For those with metabolic syndrome, recommended average sleep times range from 7 to 9 hours per night and ideally should be 8 hours [46]. It is also critical to ask patients about the quality of their sleep. Are they waking up multiple times? Are they feeling rested when they wake up? Is there snoring? Obstructive sleep apnea can affect insulin resistance and metabolic syndrome through poor oxygenation. In addition, many patients with obstructive sleep apnea have undiagnosed insulin resistance.

31.2.11 A Final Word on Insulin Resistance

We have established that insulin resistance is a disease contributing to numerous chronic and life-threatening conditions and affects countless patients, many of whom go undiagnosed. The good news is that insulin resistance, as well as the progression to diabetes, can not only be prevented but can be treated and reversed. Therefore, it is our responsibility as clinicians to track down early cases of insulin resistance before they advance into later stages along the continuum and to reverse whatever damage may have occurred. Likewise, it is our duty to offer an effective protocol to those who already suffer from a clinical diagnosis of prediabetes or diabetes.

Integrative and functional medicine tools help identify the underlying dysfunctions and the root causes in a personalized manner. Using clinical evaluation, as well as laboratory measurements including adiponectin, fasting insulin, proinsulin, HgbA1c, postprandial insulin, and glucose, the clinician can assess where the patient lies along the continuum of insulin resistance. Once the stage of insulin resistance has been identified, a personally tailored treatment plan is prescribed to reverse beta cell dysfunction and prevent further progression to diabetes. As needed, treatment options may address any of the following areas of a patient's life: diet, sleep, stress, exposure to toxins and ability to detoxify, exercise, and/or micronutrient consumption. In conclusion, by way of acknowledging the role of insulin resistance in the bigger picture of obesity, diabetes, and cardiovascular disease, we are able to recognize how vitally important it is to incorporate the effective diagnosis and treatment of insulin resistance into clinical practice.

31.3 Adrenal Dysfunction

We live in a world that is increasingly stressful and often coupled with sedentary lifestyle practices. Many people eat processed foods lacking in nutritive quality at most meals. Each day we are exposed to a skyrocketing number of synthetic chemicals in our personal care products, the food we eat, the water we drink, and the air we breathe. Our minds are constantly distracted by relentless electronic stimulation. Societies have collectively induced self-psychological and economic slavery. This is our world. The shocking part is this way of life is accepted and considered normal and the dysfunction may be difficult to perceive, much less find successful life balance. In this environment, feeling anxious and/or depressed is an understandable reaction to the overwhelming stimuli of a typical modern lifestyle.

31.3.1 What Is Adrenal Dysfunction?

The adrenal glands are bean-shaped organs situated above the kidneys with secretions that are required for maintenance of life [47]. They are made up of an outer region or cortex and an inner region or medulla. The hormones of the adrenal cortex are steroids and act at the level of the genome to regulate expression of genes for operation of important processes. There are three major categories of adrenal steroid hormones, mineralocorticoids, glucocorticoids, and androgens, and all are essential for maintaining internal equilibrium [47]. The adrenal medulla produces two important hormones, epinephrine and norepinephrine, that play a critical role in the response to stress. The catecholamines affect the cardiovascular, respiratory, excretory, and skeletal systems and equip the body for fright, fight, or flight. An important function of the adrenal glands is to provide dynamic balance or homeostasis of the hormonal systems whether in rest or stress.

The job of the adrenals when it comes to homeostasis during stressful events is to handle the disruption in hormones. Their handling of stress is akin to managing a bank account – it's your *adrenal* bank account. If you have a lot of adrenal reserve, it means you can spend on the account. Thus, many people choose to spend frivolously early in life when reserves are naturally higher. As we grow older, however, it behooves us to pay more attention to the adrenal savings account and work to replenish any debt or accumulated depletion. No matter what age, we must learn to manage our adrenal bank account and live within our means. That means choosing diet, lifestyle, hobbies, and sleep patterns that work to regenerate what has been lost. If we do not do this, there are consequences.

There is a threshold of stress that our adrenals can handle within a healthy range. Beyond that threshold, however, damage is done. This threshold setpoint differs for each of us and is very much like the stress on a violin string. Insufficient tension produces a dull, raspy sound. Too much tension makes a shrill, annoying noise or snaps the string. However, just the right degree of pressure can create a magnificent tone. Similarly, we all need to find the proper level of stress that allows us to perform optimally and make melodious music as we go through life.

A problem arises for many when they overspend their adrenal reserves and find themselves broke. In other words, we reach a level at which we can no longer adapt to further stress and have lost adrenal resilience. When this occurs, the body can no longer cope with stress as it once had and the effects are deleterious. Under normal circumstances, once an acute threat has passed, the stress response becomes inactivated and levels of stress hormones return to baseline levels [47]. This translates into resilience (see ◘ Table 31.3). In this case, we are using the normal relaxation response to recover from stress. However, if there is unremitting stress, we can then reach a point where adaptation can no longer occur and adrenal dysfunction ensues. The hypothalamic-pituitary-adrenal axis responds to the unremitting stress by downregulation at the level of the adrenal glands, and that is what we are calling Adrenal dysfunction. The actual mechanisms involved are not yet fully understood. Adrenal dysfunction is composed of three stages which were described by Hans Selye in the 1930s, although some controversy surrounds this classification [48]. Selye called the series of stages "the general adaptation syndrome," described in ◘ Table 31.3.

◘ Table 31.3 Stages of adrenal response

Stages of adrenal response	
Stage 1: Arousal	Both cortisol and DHEA increase with episodic stress, but recovery returns to baseline This may or may not be asymptomatic
Stage 2: Adaptation	Cortisol is chronically elevated and DHEA declines Symptoms of stress arise such as anxiety attacks, mood swings, depression
Stage 3: Exhaustion	Adrenal insufficiency yields low cortisol and DHEA levels Patients present with depression and severe fatigue

Based on data from Ref. [48]

Modern life poses ongoing stressful events that are not short-lived. When we lose our resilience, we progress to stage 2 adrenal dysfunction or "the adaptation phase." If we continue to borrow from our account as our savings dwindle, then we progress further to stage 3 (◘ Table 31.3). At this point, stress becomes chronic. This continual exposure to stress puts us at risk for many other health problems. Patients at this stage often struggle with severe fatigue or chronic fatigue syndrome, fibromyalgia, hair loss, premature hormone loss or hormone imbalance, weight gain, autoimmune disorders, suppression of the immune system, arthritis, anxiety, and depression [49–51].

Adrenal dysfunction, when at the severe maladaptation stage, is generally referred to as adrenal fatigue. The term adrenal fatigue has been inaccurately used to refer to any of the stages of adrenal dysfunction. Correctly, it should strictly refer only to the third stage of adrenal dysfunction. Patients may present with a variety of symptoms of adrenal gland dysfunction that include fatigue, low energy, weakness, moodiness, anxiety or depression, muscle and bone loss, hormonal imbalance, insomnia, and skin problems, among others. Also common is an overreaction to stress in which a minor event causes a disproportionate reaction of anger, irritation, or sadness.

31.3.2 What Causes Adrenal Dysfunction?

Chronic unremitting stress is the leading cause of adrenal dysfunction and the term stress can refer to physical *or* psychological stress [49–51]. For example, physical stress could include environmental toxin exposure, infections, nutrient deficiencies, hormone imbalances, allergies, and certain prescription medications. A nutrient-poor diet high in sugar, simple carbohydrates, food preservatives, alcohol, or stimulants such as caffeine leads to adrenal dysfunction. These foods can cause the blood sugar to spike, and in response, the adrenal cortex may produce excess cortisol, which over time can lead to hypothalamic pituitary adrenal dysfunction. Psychological stress could be related to events such as divorce, an unhappy marriage, a stressful job, difficult family life, lack of social or familial support or history of past or current abuse (physical, sexual or psychological). It could also come in the form of perceived stress. For instance, reliving past trauma or worrying about what people may be thinking takes the same physiologic toll as an immediate stressful event. The body doesn't know the difference between psychological stress – real or perceived – and a physical emergency. To the body, stress is stress, whether physical or psychological or both, and both are equally detrimental.

31.3.3 Assessment and Treatment of Adrenal Dysfunction

One useful laboratory evaluation for adrenal dysfunction is a four-point or 6-point salivary cortisol test. The latter is the 4 point salivary cortisol throughout the day with the Cortisol Awakening Response (CAR). Salivary cortisol concentrations reflect the free, biologically active fraction of cortisol in the plasma, and tests of it provide results that are of greater diagnostic significance than plasma total concentrations alone [52, 53].

The author has developed a three-legged stool approach with an optional fourth leg, if needed, to treat adrenal dysfunction. At the most basic level, the first leg of the approach covers modifications in nutrition and lifestyle. Eating organically grown, phytochemically dense foods that are devoid of fillers and preservatives is the foundation for recovery. Meals need to have a low glycemic load and good sources of clean protein. The majority of the plate, about two-thirds, must be composed of vegetables and a few servings of low-glycemic-load fruits. Chewing food appropriately must be emphasized and hydrating well with clean water is critical. The first leg also includes the lifestyle component that incorporates exercise and dealing with trauma. Exercise recommendations should be appropriate for the stage of adrenal dysfunction. A person in stage 1 adrenal dysfunction with high cortisol levels would need more of an aerobic program, whereas someone in stage 3 needs more gentle movement that will not cause further stress to the system. It is important to address the level of stress and discuss how to best modify it. It is crucial to emphasize hobbies that are enjoyable and a relaxation protocol that works for the individual patient. Further, trauma or history of abuse must be explored. According to Bethel Van der Kolk in his book: The Body Keeps the Score (brain, mind, and body in the healing of trauma) we must address and heal trauma in order to restore health. This book reviews the scientifically proven ways to heal trauma by rewiring the brain including eye movement desensitization and reprogramming (EMDR), neurofeedback, emotional freedom technique (EFT) among others. It is an essential read inorder to help patients fully recover from adrenal or HPA axis dysfunction.

The second leg involves tailored nutraceuticals that help heal and reverse the adrenal dysfunction. This includes nutraceuticals such as B-complex with B6 in the activated or P-5-P form, 5-MTHF, biotin, B5 or pantothenic acid, omega-3 fatty acids, vitamin C, magnesium, and zinc [54–68]. Pantothenic acid has the added benefit that it can enter the steroidogenic pathway early at the level of pregnenolone. This is particularly important for those patients who don't want to use hormonal replacement as higher doses of pantothenic acid, up to 1500 mg per day, may be used. Zinc is an important mineral that helps with healing. See ◘ Table 31.4.

The third leg of the three-legged stool consists of adaptogens. Adaptogens are herbs and botanicals that help the body adapt to the stressors and may facilitate communication between the adrenals and the higher centers of the brain [69, 70]. Examples include *Rhodiola rosea* (tyrosol, salidroside, rosiridin), *Eleutherococcus senticosus* (eyringin, eleutheroside E), *Schisandra chinensis* (schizandrin), *Panax ginseng* (ginsenosides), and *Withania somnifera* (sitoindosides). For the reader interested in greater detail, please explore the references provided.

Table 31.4 Nutraceuticals for adrenal dysfunction

Pantothenic acid (B5)	Stimulates ability of adrenal cells to secrete progesterone and corticosterone 500–1500 mg/day Enters early in steroidogenic pathway Can lead to increase in downstream adrenal hormones. Induces adrenal hyperresponsiveness to ACTH stimulation
B6 (ideally P5P)	50–100 mg/day Higher need due to stress. A co-factor in energy production
Folate (5MTHF)	400–800 mcg/day Co-factor in energy production
Magnesium	400–600 mg/day of chelated form Co-factor in many reactions of the Krebs cycle. Low magnesium increases the release of stress hormones. Magnesium decreases nocturnal ACTH secretion
Zinc	The glucocorticoid receptor, like all steroid receptors, is a zinc-finger transcription factor >300 human enzymes require zinc 30–50 mg/day balance with copper (10–20:1 ratio)
Vitamin C	Preferentially concentrated in adrenal glands and lost in times of stress from adrenal cortex
Omega 3	Blunts the stress response. Decreases cortisol and epinephrine 1000–3000 mg depending on co-existing pathologies
Vitamin D	2000 IU/day Reduces latent virus reactivation with increased stress

Based on data from Ref. [54–68]

An optional fourth leg may be necessary at times and includes hormonal supplementation. Hormonal supplementation may help the adrenals recover more quickly and may assist both the brain and adrenals to re-establish their balance and connection with one another [71–73]. The most common hormones that may be considered are DHEA and pregnenolone, but these are recommended in very low doses. It is strongly recommended that hormones only be used when the other three legs are in place and more support is needed. Fortunately, in many cases, the first three legs are satisfactory interventions to bring the hypothalamic-pituitary-adrenal axis back to balance.

31.3.4 A Final Word on Adrenal Dysfunction

The two most important aspects of addressing adrenal dysfunction are to stage the patient and understand the root cause of the dysfunction. Then and only then can a personalized approach utilizing functional and integrative medicine tools be tailored to the individual needs of each patient. With focus and attention to the nutritional content of the diet, lifestyle modification, and supplementation with vitamins, minerals, adaptogens, and possibly hormones, the good news is that it is possible to achieve a complete recovery from adrenal dysfunction.

31.4 Thyroid

A brief review of the physiology of thyroid function is essential here. The hypothalamus produces thyroid-releasing hormone (TRH), which signals the pituitary to release thyroid-stimulating hormone (TSH), which then acts on the thyroid gland to make 95 percent thyroxine (T4) and 5 percent triiodothyronine (T3) [74]. The liver and kidneys are responsible for converting T4 to T3. Free T3 is the active form of the hormone. When properly functioning, thyroid hormone keeps metabolism in control.

The thyroid is an endocrine organ that is susceptible to nutritional and environmental challenges [36, 37]. There are many disruptors of thyroid hormone function, including food sensitivities and/or food allergies, toxins, infections, nutrient deficiencies, stress, sleep deprivation, pharmaceutical drugs, adrenal dysfunction, oxidative stress, stress in general, and changes in the gut microbiota. These disruptors are extremely important because suboptimal thyroid function has far-reaching consequences.

Some consequences of low thyroid hormone include fatigue, weight gain, depression cardiovascular dysfunction (including dyslipidemias and atherogenesis), glucose intolerance or insulin resistance, and poor pregnancy outcomes [75, 76]. Inhibitors or promoters of proper thyroid hormone function will be divided according to the location of operation. First, at the level of the thyroid gland or at the conversion of T4 to reverse T3 (RT3) or to T3, there are factors that contribute to or inhibit proper thyroid function. Factors contributing to proper production of thyroid hormones include the nutrients iron, iodine, tyrosine, zinc, selenium, vitamin A, vitamin E, vitamin B2, vitamin B3, vitamin B6, vitamin C, and vitamin D [77–84]. Factors that inhibit proper production of thyroid hormone include stress, infection, trauma, radiation, medications, fluoride and bromine (antagonist to iodine), toxins such as pesticides, Hg, Cd, Pb, and autoimmune diseases such as celiac disease [36, 37]. Minerals important for conversion of T4 to T3 by deiodinase enzyme include selenium and zinc. Factors that increase conversion of T4 to RT3 include stress, trauma, low-calorie diets, inflammation and cytokines, toxins, infections, liver and/or kidney dysfunction, and certain medications. Factors that improve cellular sensitivity to thyroid hormones include vitamin A, exercise, and zinc. Factors that promote or inhibit conversion of T4 to T3 by a deiodinase enzyme are provided in Table 31.5.

Table 31.5 Factors that facilitate or inhibit deiodinase conversion of T4 to T3

Facilitate	Micronutrients
	  Vitamins: A, E, riboflavin, etc.
	  Minerals: selenium, potassium, iodine, iron, zinc
	Hormones
	  Cortisol
	  Growth hormone
	  Testosterone
	  Insulin
	  Glucagon
	  Melatonin

Inhibit	Micronutrient excesses or deficiencies
	  Selenium deficiency
	  Iodine
	  Other micronutrient deficiencies
	Hormones
	  Excess estrogen
	  Oral contraceptives
	  Prescription medications
	  Excess cortisol, catecholamines
	Chronic illness (cytokines, free radicals)
	Long-term hospitalization
	Compromised liver or kidney function
	Poor-quality diets
	  Deficient protein
	  Excess sugar
	Environmental toxicity
	  Heavy metals
	  Herbicides, pesticides
	  Polycyclic aromatic hydrocarbons

Table 31.6 Assessment of abnormal thyroid function

Symptoms	Memory and concentration problems Diffuse headache, migraines Depression; melancholia Constipation: hard or pellet-like bowel movements and decreased frequency Low libido Reactive hypoglycemia Hypotension Hypoglycemia Poor tolerance to stress and exercise Fatigue Hair loss Poor concentration Cold extremities Headaches		
Physical exam	Dry skin, elbow keratosis, brittle nails Diffuse hair loss Puffy face, swollen eyelids Edema in legs, feet, hands Elevated cholesterol, generally LDL Easy bruising Prolonged Achilles tendon reflex Keratoderma Enlarged thyroid gland		
Laboratory testing and ranges	TSH Free T4 Free T3 Total T3 Anti-peroxidase antibodies Anti-thyroglobulin antibodies Anti TSH receptor antibodies rT3 Total T3 (TT3) Free T3/free T4 TT3/rT3 CBC and ferritin RBC selenium RBC zinc 25-OH vitamin D Serum vitamin A Urinary fasting morning spot iodine Celiac panel Toxic metals as indicated by history and physical exam Food sensitivity testing	*Functional ranges*	*Conventional ranges*
		TSH: 0.4–2/2.5 mIU/L	(0.4–5.5)
		Free T4: 15–23 pmol/L	(9–23)
		Free T3: 5–7 pmol/L	(3–7)
		Total T3: 120–181 ng/dl	(76–181)
		Thyroid antibodies: WNL	
		rT3: 11–18 ng/dl	(11–31)
		FT3/FT4: >0.33 TT3/RT3: >6	

31.4.1 Assessment

For overall assessment see Table 31.6.

When evaluating laboratory results for thyroid testing, it is important to realize that most conventional laboratories have not adopted the optimal reference ranges that have been promoted by current literature and used in integrative/functional practices. Instead they use standard reference ranges that have been developed by tabulating results from multiple patient samples who were considered normal without regard to level of dysfunction. In appropriate circumstances, red blood cell heavy metal testing may be done for acute ongoing exposure, and urinary heavy metals may be used for assessment of total body load. Urinary testing can be done with or without provocation with a chelator. However, if ordering a provoked 24-hour or 6-hour urinary test, make sure patient will tolerate this challenge. A chronically ill patient may not be able to tolerate provocation.

31.4.2 Treatment

Treatment should begin by first focusing on lifestyle modifications and supplementation. After this has been optimized, the final intervention may include hormonal therapy, if necessary. Special attention is directed to removing the offending cause or causes and supplementing the deficiencies as appropriate. Recommendations include a dietary plan that is low in reactive foods. Choices for the plan include foods that are high in pre- and probiotic content and high in phytonutrients and balanced in the omega-6/omega-3 ratio. The plan

Table 31.7 Thyroid function and the relationship to other hormones

Hormone	
Adrenal	High adrenal activity impairs 5' deiodinase leading to higher T4, lower T3, normal or elevated TSH Low adrenal activity may result in lower T4, higher T3, normal or elevated TSH Excess adrenaline can desensitize T3 receptors leading to T3 resistance, higher T3, despite symptoms of hypothyroid With excess adrenaline, the body compensates by lowering T4 leading to symptoms of hypothyroid but often the patient intolerant of thyroid supplementation (always balance adrenals first)
Sex steroid hormones	Hypothyroidism stimulates CYP3A4 leading to increased production of 16αOHE1 Hypothyroid decreases concentration of SHBG which causes more bioavailable E2 and testosterone Hyperthyroid increases SHBG leading to less bioavailable E2 and testosterone Hypothyroidism associated with less deactivation of cortisol to cortisone (hyperthyroidism leads to the opposite)

should strive to reduce inflammation by reducing saturated fats, damaged fats, and trans-fatty acids. Critical for the patient is ensuring sufficient sleep, exercise, hydration, and stress reduction techniques. Evaluating for adequate social and community support is key.

Key nutrients to consider supplementing include selenium, zinc, iron or ferritin, iodine, vitamin D, and vitamin A [77–84]. Zinc and selenium are especially important minerals to consider. For example, a low T3/T4 ratio may be related to impaired zinc and/or selenium status. Remember to begin with food first; Brazil nuts are particularly high in selenium and zinc and may be recommended before supplements. When needed, supplementation has been associated with modest changes in thyroid hormones and with a normalization of T4 and RT3 plasma levels. Iron deficiency impairs thyroid hormone synthesis by reducing the activity of heme-dependent thyroid peroxidase. Evaluate ferritin levels by laboratory testing instead of iron alone and supplement with ferritin when levels are low. Supplementation is recommended at levels of 80–100 mg/ml in women and slightly higher in men. Further, iron-deficiency anemia blunts and iron supplementation improves the efficacy of iodine supplementation [77–84]. Typical amounts to consider are selenium, 200–400 mcg; zinc, 15–30 mg; vitamin D, 2000 IU; vitamin A, 2000 IU; iodine, 150 mcg; and iron, 15–20 mg (in a menstruating woman). Important points to consider for thyroid function and the relationship to other hormones are listed in Table 31.7.

31.5 Sex Steroid Hormones

Sex steroid hormones for this discussion will refer broadly to estrogen, progesterone, and testosterone. In the author's experience, if insulin sensitivity, adrenal function, and thyroid status have been addressed, the sex hormones may no longer need special attention or additional hormone replacement. In other words, there may be imbalances in sex hormones due to problems with insulin sensitivity, adrenal function, or thyroid status. It is important to assess sex steroid hormones by laboratory testing since abnormalities may be present in men and women of all ages and not just in the perimenopause, menopause, or andropause stages. ► Box 31.2 details the approach to the sex steroid hormone assessment.

Box 31.2 Assessment of Sex Steroid Hormone Balance
- Thorough history including past medical history, childhood and family history, hobbies
- Evaluation of insulin, adrenals, thyroid, sex hormones (including ovarian function), steroidogenic cascade, and estrogen metabolism
- Balance all hormones in specific order as outlined
- Consider estrogen metabolism/genomics
- Consider the effect of hormone balancing before drugs
  - Food and lifestyle management first
- Assessment of impact of hormones in the individual
- Decision to give hormones and informed consent
- Consider best route of delivery
- Regular reassessment of hormone levels and symptoms

It is strongly cautioned not to begin bioidentical hormone replacement without adequate evaluation first. Since women with increased or persistant vasomotor symptoms (VMS) are at higher risk for cardiovascular disease due to insulin resistance [6, 85], it is imperative to screen for insulin resistance prior to balancing the sex hormones. Much has been written about hormone replacement therapy with respect to potential increased risk of heart disease and breast cancer [6, 85–88]. It has been stated that hormone replacement therapy should be considered in these women to avoid these increased risks [85]. If VMS are significantly associated with insulin resistance, it may also be postulated that women with insulin resistance would have a greater need for menopausal hormone treatment than those without insulin resistance. In this regard, there is the possibility of a beneficial effect with replacement of menopausal hormones in patients with insulin resistance.

In women beginning in pre- and perimenopause, progesterone and testosterone levels decline before estrogen levels decline. For menstruating women, it is helpful to measure progesterone in the luteal phase on days 19–21 of the cycle and the estrogens during the follicular phase on days 10–12 to evaluate the respective peaks. Further, if we focus on optimizing estrogen metabolism for both men and women, theoretical risks from hormone replacement may be mitigated [89]. Daily unremitting stress is associated with early decline in progesterone and estrogen in younger women. Because of its position at the beginning of the steroidogenic pathway, progesterone may be diverted down the pathway to the stress hormone, cortisol. This is related to the physiologic need to respond to stress for survival, which is immediate and pressing. In addition, environmental toxins compete with our innate hormones because they are endocrine disruptors and/or xenoestrogens that disrupt hormone signaling. In men, testosterone levels are declining at much younger ages related to exposure to environmental toxins [90, 91].

As emphasized above, if there is an imbalance in baseline insulin sensitivity, adrenal/HPA axis function, or with thyroid gland, begin by repairing these baseline abnormalities. However, if after fixing the root causes sex hormone replacement is needed, studies show that transdermal bioidentical estrogen and testosterone, as well as micronized oral or transdermal progesterone, are generally safe and may not increase a woman's risk of heart disease or breast cancer if prescribed appropriately [92–98]. It is necessary to measure hormone levels and evaluate estrogen metabolism and then follow these levels on a regular basis if hormone replacement is begun to ensure safety. Use non-synthetic hormones. Progestins are an example of synthetic hormones which are associated with increased cardiovascular risks, as demonstrated in the Women's Health Initiative study [86, 96].

Hormone level baseline assessment should include progesterone, testosterone (both total and free), sex hormone-binding globulin (SHBG), albumin, estradiol, estrone, and estriol. With values for SHBG, albumin, and total testosterone, the bioavailable testosterone can be calculated in men and women. The testosterone is tightly bound to SHBG, while testosterone is not tightly bound to albumin; thus the bioavailable amount is important to consider. Follow-up laboratory evaluation is critical any time hormone replacement is prescribed. Testing should also include downstream metabolites.

When identifying declining levels of progesterone early in the peri-menopause, supporting adrenal function and adding chasteberry or vitex may be sufficient to abate symptoms and restore function [99–101]. Low testosterone levels can be enhanced in women with low doses of sublingual DHEA such as 5–10 mg twice daily along with detoxification support, particularly in the conjugation pathways of methylation and glutathione. DHEA, when supplemented in men, tends to go downstream to estrogens, while in women it is converted to testosterone due to the activity of the different enzymes in men and women. In men, addition of zinc and general detoxification are recommended. In addition, flaxseed meal in both men and women at two tablespoons per day can enhance detoxification and help with hormonal imbalance symptoms. The extract of Rheum rhaponticum (ERr 731), also known as Estrovera in the United States, Canada, and South Africa, has been used for menopausal symptoms with success [89, 102–104]. In addition to helping with hot flashes, it also helps with the anxiety and mood changes that often accompany hormonal imbalances [105]. Estrovera, or Err 731, binds to estrogen receptor beta, which may be protective for hormone-related cancers. The central functions relevant to climacteric complaints are proposed to be mediated via estrogen receptor beta (ERbeta) activation [106].

To allow for smooth estrogen metabolism, supporting phase 2 conjugation detoxification enzymes, especially methylation with activated B vitamins and supplements such as alpha-lipoic acid, N-acetyl cysteine, selenium and liposomal glutathione, can be helpful. Studies have shown that reducing the levels of estrogen-DNA adducts could prevent the initiation of human cancer [107–109]. The dietary supplements N-acetylcysteine and resveratrol also inhibit formation of estrogen-DNA adducts in cultured human breast cells and in women [110]. These results suggest that the two supplements offer an approach to reducing the risk of developing various prevalent types of human cancer.

31.6 Conclusion

Addressing hormonal issues in a systematic way is critical to discovering the root cause of hormonal imbalance. This is most effective when addressing insulin dysfunction first, followed by adrenal support, followed by addressing thyroid issues, and lastly balancing the sex steroid hormones along with their metabolism. This approach allows the practitioner to fix underlying imbalances without resorting to hormone replacement in many, if not most, cases.

References

1. Application of Systematic Review Methods in an Overall Strategy for Evaluating Low-Dose Toxicity from Endocrine Active Chemicals. Committee on Endocrine-Related Low-Dose Toxicity; Board on Environmental Studies and Toxicology; Division on Earth and Life Studies; National Academies of Sciences, Engineering, and Medicine, 2017. http://nap.edu/24758.
2. Saltiel AR, Olefsky JM. Inflammatory mechanisms linking obesity and metabolic disease. J Clin Invest. 2017;127(1):1–4.
3. Ludwig DS, Ebbeling CB. Type 2 diabetes mellitus in children: primary care and public health considerations. JAMA. 2001;286(12):1427–30.
4. Pinhas-Hamiel O, Zeitler P. The global spread of type 2 diabetes mellitus in children and adolescents. J Pediatr. 2005;146:693–700.
5. Hyman M. The blood sugar solution: the Ultrahealthy program for losing weight, preventing disease, and feeling great now! 1st ed. New York: Little, Brown and Company; 2012. p. 7–10.
6. Tuomikoski P, Savolainen-Peltonen H. Vasomotor symptoms and metabolic syndrome. Maturitas. 2017;97:61–5.
7. Reaven GM. Insulin resistance, the insulin resistance syndrome, and cardiovascular disease. Panminerva Med. 2005;47(4):201–10.
8. Huang PL. A comprehensive definition for metabolic syndrome. Dis Model Mech. 2009;2(5–6):231–7. https://doi.org/10.1242/dmm.001180.
9. Tirosh A, Shai I, Tekes-Manova D, Israeli E, Pereg D, Shochat T et al, Israeli Diabetes Research Group. Normal fasting plasma glucose levels and type 2 diabetes in young men. N Engl J Med. 2005;353(14):1454–62.

10. Wildman RP, Muntner P, Reynolds K, McGinn AP, Rajpathak S, Wylie-Rosett J, et al. Obese without cardiometabolic risk factor clustering and the normal weight with cardiometabolic risk factor clustering: prevalence and correlates of 2 phenotypes among the US population (NHANES 1999-2004). Arch Intern Med. 2008;168(15):1617–24.
11. Pfutzner A, Weber MM, Forst T. A biomarker concept for assessment of insulin resistance, beta-cell function and chronic systemic inflammation in type 2 diabetes mellitus. Clin Lab. 2008;54(11–12): 485–90.
12. Gabriel CL, Smith PB, Mendez-Fernandez YV, Wilhelm AJ, Ye AM, Major AS. Autoimmune-mediated glucose intolerance in a mouse model of systemic lupus erythematosus. Am J Physiol Endocrinol Metab. 2012;303(11):E1313–24. https://doi.org/10.1152/ajpendo.00665.2011.
13. Fasano A. Surprises from celiac disease. Sci Am. 2009;301(2): 54–61.
14. Khaw KT, Wareham N, Luben R, Bingham S, Oakes S, Welch A, et al. Glycated hemoglobin, diabetes, and mortality in men in Norfolk cohort of European prospective investigation of cancer and nutrition (EPIC-Norfolk). BMJ. 2001;322(7277):15–8.
15. Ahuja V, Kadowaki T, Evans RW, Kadota A, Okamura T, El Khoudary SR, Fujiyoshi A, Barinas-Mitchell EJ, Hisamatsu T, Vishnu A, Miura K, Maegawa H, El-Saed A, Kashiwagi A, Kuller LH, Ueshima H, Sekikawa A, ERA JUMP Study Group. Comparison of HOMA-IR, HOMA-β% and disposition index between US white men and Japanese men in Japan: the ERA JUMP study. Diabetologia. 2015;58(2):265–71. https://doi.org/10.1007/s00125-014-3414-6.
16. Wallace TM, Levy JC, Matthews DR. Use and abuse of HOMA modeling. Diabetes Care. 2004;27(6):1487–95. https://doi.org/10.2337/diacare.27.6.1487.
17. Stern JH, Rutkowski JM, Scherer PE. Adiponectin, Leptin, and fatty acids in the maintenance of metabolic homeostasis through adipose tissue crosstalk. Cell Metab. 2016 May 10;23(5):770–84. https://doi.org/10.1016/j.cmet.2016.04.011.
18. Fragopoulou E, Panagiotakos DB, Pitsavos C, Tampourlou M, Chrysohoou C, Nomikos T, Antonopoulou S, Stefanadis C. The association between adherence to the Mediterranean diet and adiponectin levels among healthy adults: the ATTICA study. J Nutr Biochem. 2010 Apr;21(4):285–9. https://doi.org/10.1016/j.jnutbio.2008.12.013.
19. Chaput JP, Tremblay A, Rimm EB, Bouchard C, Ludwig DS. A novel interaction between dietary composition and insulin secretion: effects on weight gain in the Quebec Family Study. Am J Clin Nutr. 2008;87(2):303–9.
20. Davis DR, Epp MD, Riordan HD. Changes in USDA food composition data for 43 garden crops, 1950 to 1999. J Am Coll Nutr. 2004;23(6):669–82.
21. Luo Z, Zhang Y, Li F, He J, Ding H, Yan L, et al. Resistin induces insulin resistance by both AMPK-dependent and AMPK-independent mechanisms in HepG2 cells. Endocrine. 2009;36(1):60–9.
22. McGarry JD. Glucose-fatty acid interactions in health and disease. Am J Clin Nutr. 1998;67(3 Suppl):500S–4S.
23. Meydani M. A Mediterranean-style diet and metabolic syndrome. Nutr Rev. 2005;63(9):312–4.
24. Esposito K, Marfella R, Ciotola M, Di Palo C, Giugliano F, Giugliano GD, et al. Effect of a Mediterranean-style diet on endothelial dysfunction and markers of vascular inflammation in the metabolic syndrome: a randomized trial. JAMA. 2004;292(12):1440–6.
25. USDA National Agriculture Library: nutrient changes over time: frequently asked questions. [Internet 2017]. Available from: http://www.nal.usda.gov/fnic/faq.
26. Fung TT, Malik V, Rexrode KM, Manson JE, Willett WC, Hu FB. Sweetened beverage consumption and risk of coronary heart disease in women. Am J Clin Nutr. 2009;89(4):1037–42.
27. Upadhyay S, Dixit M. Role of polyphenols and other phytochemicals on molecular signaling. Oxid Med Cell Longev. 2015, Article ID 504253, 15 pages. https://doi.org/10.1155/2015/504253.
28. Singh RB, Niaz MA, Rastogi SS, Shukla PK, Thakur AS. Effect of hydrosoluble coenzyme Q10 on blood pressures and insulin resistance in hypertensive patients with coronary artery disease. J Hum Hypertens. 1999;13(3):203–8.
29. Hodgson JM, Watts GF, Playford DA, Burke V, Croft KD. Coenzyme Q10 improves blood pressure and glycaemic control: a controlled trial in subjects with type 2 diabetes. Eur J Clin Nutr. 2002;56(11): 1137–42.
30. Jacob S, Ruus P, Hermann R, Tritschler HJ, Maerker E, Renn W, et al. Oral administration of RAC-alpha-lipoic acid modulates insulin sensitivity in patients with type-2 diabetes mellitus: a placebo-controlled pilot trial. Free Radic Biol Med. 1999;27(3–4):309–14.
31. Haak E, Usadel KH, Kusterer K, Amini P, Frommeyer R, et al. Effects of alpha-lipoic acid on microcirculation in patients with peripheral diabetic neuropathy. Exp Clin Endocrinol Diabetes. 2000;108(3): 168–74.
32. Chiu KC, Chu A, Go VL. Saad MF Hypovitaminosis D is associated with insulin resistance and beta cell dysfunction. Am J Clin Nutr. 2004;79(5):820–5.
33. Barbagallo M, Dominguez LJ, Galioto A, Ferlisi A, Cani C, Malfa L, et al. Role of magnesium in insulin action, diabetes and cardio-metabolic syndrome X. Mol Asp Med. 2003;24(1–3):39–52.
34. Guerrero-Romero F, Tamez-Perez HE, González-González G, Salinas-Martínez AM, Montes-Villarreal J, Treviño-Ortiz JH, et al. Oral magnesium supplementation improves insulin sensitivity in non-diabetic subjects with insulin resistance. A double-blind placebo-controlled randomized trial. Diabetes Metab. 2004;30:253–8.
35. Kim CW, Kim BT, Park KH, Kim KM, Lee DJ, Yang SW, et al. Effects of short-term chromium supplementation on insulin sensitivity and body composition in overweight children: randomized, double-blind, placebo-controlled study. J Nutr Biochem. 2011;22(11): 1030–4.
36. Diamanti-Kandarakis E, Bourguignon JP, Giudice LC, Hauser R, Prins GS, Soto AM, Zoeller RT, Gore AC. Endocrine-disrupting chemicals: an Endocrine Society Scientific Statement. Endocr Rev. 2009;30940:293–342. https://doi.org/10.1210/er.2009-0002.
37. Zoeller RT, Brown TR, Doan LL, Gore AC, Skakkebaek NE, Soto AM, Woodruff TJ, Vom Saal FS. Endocrine-disrupting chemicals and public health protection: a statement of principles from the Endocrine Society. Endocrinology. 2012;153(9):4097–110. https://doi.org/10.1210/en.2012-1422.
38. Alonso-Magdalena P, Quesada I, Nadal A. Endocrine disruptors in the etiology of type 2 diabetes mellitus. Nat Rev Endocrinol. 2011;7(6):346–53.
39. Kim KN, Park SJ, Choi B, Joo NS. Blood mercury and insulin resistance in nondiabetic Koreans (KNHANES 2008-2010). Yonsei Med J. 2015;56(4):944–50.
40. Knowler WC, Barrett-Connor E, Fowler SE, Hamman RF, Lachin JM, Walker EA. Nathan DM; diabetes prevention program research group. Reduction in the incidence of type 2 diabetes with lifestyle intervention or metformin. N Engl J Med. 2002;346(6):393–403.
41. WHO: Global Strategy on Diet, Physical Activity and Health. Physical activity and adults. [Internet 2018]. Accessed 12/22/2018: https://www.who.int/dietphysicalactivity/factsheet_adults/en/.
42. Klein S, Sheard NF, Pi-Sunyer X, et al. Weight management through lifestyle modification for the prevention and management of type 2 diabetes: rationale and strategies. A statement of the American Diabetes Association, the north American Association for the Study of Obesity, and the American Society for Clinical Nutrition. Am J Clin Nutr. 2004;80(2):257–63.. Review
43. Lampert R, Bremner JD, Su S, Miller A, Lee F, Cheema F, Goldberg J, Vaccarino V. Decreased heart rate variability is associated with higher levels of inflammation in middle-aged men. Am Heart J. 2008;156(4):759.el–7.
44. Kataoka M, Ito C, Sasaki H, Yamane K, Kohno N. Low heart rate variability is a risk factor for sudden cardiac death in type 2 diabetes. Diabetes Res Clin Pract. 2004;64(1):51–8.
45. American Diabetes Association. Diagnosis and classification of diabetes mellitus. Diabetes Care. 2009;32(Suppl 1):S62–7. https://doi.org/10.2337/dc09-S062.

46. Aldabal L, Bahammam AS. Metabolic, endocrine, and immune consequences of sleep deprivation. Open Respir Med J. 2011;5:31–43.
47. Goodman HM. Chapter 4 – adrenal glands. In: Basic medical endocrinology. 4th ed. Elsevier; 2009. p. 61–90. https://doi.org/10.1016/B978-0-12-373975-9.X0001-8
48. Jackson M. Chapter 1 Evaluating the role of Hans Selye in the modern history of stress. In: Cantor D, Ramsden E, editors. Stress, shock, and adaptation in the twentieth century: Rochester Studies in Medical History. Rochester: University of Rochester Press; 2014.
49. Brady KT, Sinha R. Co-occurring mental and substance use disorders: the neurobiological effects of chronic stress. A J Psych. Published Online: 1 Aug 2005. https://doi.org/10.1176/appi.ajp.162.8.1483.
50. McEwen BS. Protection and damage from acute and chronic stress: allostasis and allostatic overload and relevance to the pathophysiology of psychiatric disorders. Ann N Y Acad Sci. 2009;1032(1):1–7. https://doi.org/10.1196/annals.1314.001.
51. A to Z Guide: WebMD. Adrenal fatigue: is it real? [Internet WebMD Medical Reference Reviewed by William Blahd, MD on January 23, 2017]. Accessed 12/28/2018. https://www.webmd.com/a-to-z-guides/adrenal-fatigue-is-it-real#1.
52. Evans PJ, Peters JR, Dyas J, Walker RF, Riad-Fahmy D, Hall R. Salivary cortisol levels in true and apparent hypercortisolism. Clin Endocrinol. 1984;20(6):709–15.
53. Peters JR, Walker RF, Riad-Fahmy D, Hall R. Salivary cortisol assays for assessing pituitary-adrenal reserve. Clin Endocrinol. 1982;17(6):583–92.
54. Mccabe D, Colbeck M. The effectiveness of essential fatty acid, B vitamin, vitamin C, magnesium and zinc supplementation for managing stress in women: a systematic review protocol. JBI Database System Rev Implement Rep. 2015;13(7):104–18.
55. Vinogradov V, Shneider A, Senkevich S. Thiamine cardiotropism. Cor Vasa. 1991;33(3):254–62.
56. Nakagawasai O, Yamadera F, Iwasaki K, Arai H, Taniguchi R, Tan-No K, et al. Effect of kami-untan-to on the impairment of learning and memory induced by thiamine-deficient feeding in mice. Neuroscience. 2004;125(1):233–41.
57. Peters EM, Anderson R, Nieman DC, Fickl H, Jogessar V. Vitamin C supplementation attenuates the increases in circulating cortisol, adrenaline and anti-inflammatory polypeptides following Ultramarathon running. Int J Sports Med. 2001;22(7):537–43.
58. Seelig MS. Consequences of magnesium deficiency on the enhancement of stress reactions; preventive and therapeutic implications (a review). J Am Coll Nutr. 1994;13(5):429–46.
59. Delarue J, Matzinger O, Binnert C, Schneiter P, Chioléro R, Tappy L. Fish oil prevents the adrenal activation elicited by mental stress in healthy men. Diabetes Metab. 2003;29(3):289–95.
60. Moriguchi T, Greiner RS, Salem N. Behavioral deficits associated with dietary induction of decreased brain Docosahexaenoic acid concentration. J Neurochem. 2008;75(6):2563–73.
61. Hutchins H, Vega C. Omega-3 fatty acids: recommendations for therapeutics and prevention. MedGenMed. 2005;7(4):18.
62. Zwart SR, Mehta SK, Ploutz-Snyder R, et al. Response to Vitamin D supplementation during Antarctic winter is related to BMI, and supplementation can mitigate Epstein-Barr virus reactivation. J Nutr. 2011;141(4):692–7.
63. Human Vitamin and Mineral Requirements. Ch. 16 Zinc: zinc metabolism and homeostasis. FAO/WHO-FAO Corporate Document Repository. Retrieved 6-9-15 from http://www.fao.org/docrep/004/y2809e/y2809e0m.htm
64. Deans E. Zinc: an antidepressant, the essential mineral for resiliency. Psychology Today: Evolutionary Psychiatry. 2013. Accessed: https://www.psychologytoday.com/us/blog/evolutionary-psychiatry/201309/zinc-antidepressant
65. Prasad A. Zinc: role in immunity, oxidative stress and chronic inflammation. Curr Opin Clin Nutr Metab Care. 2009;12(6):646–52.
66. Frederickson CJ, et al. Importance of zinc in the central nervous system: the zinc-containing neuron. J Nutr. 2000;130(5):1471S–83S.
67. Yoshida E, Fukuwatari T, Ohtsubo M, Shibata K. High-fat diet lowers the nutritional status indicators of pantothenic acid in weaning rats. Biosci Biotechnol Biochem. 2010;74(8):1691–3.
68. Jaroenporn S, Yamamoto T, Itabashi A, et al. Effects of pantothenic acid supplementation on adrenal steroid secretion from male rats. Biol Pharm Bull. 2008;31(6):1205–8.
69. Panossian A, Wikman G, Kaur P, Asea A. Adaptogens exert a stress-protective effect by modulation of expression of molecular chaperones. Phytomedicine. 2009;16(6–7):617–22.
70. Panossian A, Wikman G, Kaur P, Asea A. Adaptogens stimulate neuropeptide Y and Hsp72 expression and release in neuroglia cells. Front Neurosci. 2012;6:6.
71. Morales J, et al. Effects of replacement dose of DHEA in men and women of advancing age. J Clin Endo Metab. 1994;78(6):1360–7.
72. Vallée M, et al. Role of pregnenolone, dehydroepiandrosterone and their sulfate esters on learning and memory in cognitive aging. Brain Res Brain Res Rev. 2001;37(1–3):301–12.
73. Zhai G, et al. Eight common genetic variants associated with serum DHEAS levels suggest a key role in ageing mechanisms. PLoS Genet. 2011;7(4):e1002025. https://doi.org/10.1371/journal.pgen.1002025.
74. Ross DS. Thyroid hormone synthesis and physiology. Up to date [Internet February 23, 2018]. Accessed 12/20/2018: https://www.uptodate.com/contents/thyroid-hormone-synthesis-and-physiology.
75. Surks MI. Clinical manifestations of hypothyroidism. Up to Date. [Internet October 31, 2018]. Accessed 12/28/2018: https://www.uptodate.com/contents/clinical-manifestations-of-hypothyroidism.
76. Lauritano EC, Bilotta AL, Gabrielli M, et al. Association between hypothyroidism and small intestinal bacterial overgrowth. J Clin Endocrinol Metab. 2007;92:4180.
77. Berger MM, et al. Influence of selenium supplements on the post-traumatic alterations of the thyroid axis: a placebo-controlled trial. Intensive Care Med. 2001;27(1):91–100.
78. Olivieri O, et al. Selenium, zinc, and thyroid hormones in healthy subjects: low T3/T4 ratio in the elderly is related to impaired selenium status. Biol Trace Elem Res. 1996;51(1):31–41.
79. Bremner AP, et al. Significant association between thyroid hormones and erythrocyte indices in euthyroid subjects. Clin Endocrinol. 2012;76(2):304–11.
80. Zimmermann MB, Köhrle J. The impact of iron and selenium deficiencies on iodine and thyroid metabolism: biochemistry and relevance to public health. Thyroid. 2002;12(10):867–78.
81. Eftekhari MH, et al. Effect of iron repletion and correction of iron deficiency on thyroid function in iron-deficient Iranian adolescent girls. Pak J Biol Sci. 2007;10(2):255–60.
82. Howdeshell KL. A model of the development of the brain as a construct of the thyroid system. Environ Health Perspect. 2002;110(Suppl 3):337–48.
83. Kivity S, et al. Vitamin D and autoimmune thyroid diseases. Cell Mol Immunol. 2011;8(3):243–7.
84. Feart C, et al. Aging affects the retinoic acid and the triiodothyronine nuclear receptor mRNA expression in human peripheral blood mononuclear cells. Eur J Endocrinol. 2005;152(3):449–58.
85. Kwon DH, Lee JH, Ryu KJ, Park HT, Kim T. Vasomotor symptoms and the homeostatic model assessment of insulin-resistance in Korean postmenopausal women. Obstet Gynecol Sci. 2016;59(1):45–9.
86. Writing Group for the Women's Health Initiative Investigators. Risks and benefits of Estrogen plus progestin in healthy postmenopausal women principal results from the Women's Health Initiative randomized controlled trial. JAMA. 2002;288(3):321–33. https://doi.org/10.1001/jama.288.3.321.
87. Lobo RA. Where are we 10 years after the Women's Health Initiative? J Clin Endocrinol Metabol. 2013;98(5):1771–80. https://doi.org/10.1210/jc.2012-4070.
88. Gurney EP, Nachtigall MJ, Nachtigall LE, Naftolin F. The Women's Health Initiative trial and related studies: 10 years later: a clinician's view. J Steroid Biochem Mol Biol. 2014;142:4–11.

89. Chang JL, Montalto MB, Heger PW, Thiemann E, et al. Rettenberger R, Rheum rhaponticum extract (ERr 731): Postmarketing data on safety surveillance and consumer complaints. Integr Med (Encinitas). 2016;15(3):34–9.
90. Castro B, Sánchez P, Torres JM, Preda O, del Moral RG, Ortega E. Bisphenol A exposure during adulthood alters expression of aromatase and 5α-reductase isozymes in rat prostate. PLoS One. 2013;8(2).
91. Joensen UN, Veyrand B, Antignac JP, Blomberg Jensen M, Petersen JH, Marchand P, Skakkebæk NE, Andersson AM, Le Bizec B, Jørgensen N. PFOS (perfluorooctanesulfonate) in serum is negatively associated with testosterone levels, but not with semen quality, in healthy men. Hum Reprod. 2013;28(3):599–608. https://doi.org/10.1093/humrep/des425. Epub 2012 Dec 18. Erratum in: Hum Reprod. 2014 Jul;29(7):1600.
92. Stephensen K, et al. The effects of compounded bioidentical transdermal hormone therapy on hemostatic, inflammatory, immune factors, cardiovascular biomarkers, quality – of-life measures; and health outcomes in postmenopausal women. Int J Pharm Compd. 2013;17(1):74–85.
93. Santen R, et al. Postmenopausal hormone therapy: an Endocrine Society Scientific Statement. JCEM. 2010;95(1):7.
94. Simon JA. What's new in hormone replacement therapy: focus on transdermal estradiol and micronized progesterone. Climacteric. 2012;15(Suppl 1):3–10.
95. Fournier A, Berrino F, Riboli E, Avenel V, Clavel-Chapelon F. Breast cancer risk in relation to different types of hormone replacement therapy in the E3N-EPIC cohort. Int J Cancer. 2005;114(3):448–54.
96. Casanova G, Spritzer PM. Effects of micronized progesterone added to non-oral estradiol on lipids and cardiovascular risk factors in early postmenopause: a clinical trial. Lipids Health Dis. 2012;11:133.
97. Morgantaler A, et al. Testosterone therapy and cardiovascular risk: advances and controversies. Mayo Clin Proc. 2015;90(2):224–51.
98. L'hermite M, Simoncini T, Fuller S, Genazzani AR. Could transdermal estradiol + progesterone be a safer postmenopausal HRT? A review Maturitas. 2008;60(3–4):185–201.
99. Roemheld-Hamm B. Chasteberry. American Family Physician. [Internet September 1, 2005]. Accessed 1/12/2019: https://www.aafp.org/afp/2005/0901/p821.html.
100. Romm A. Balance your hormones: 5 ways vitex can help your hormones. [Internet October 27, 2018]. Accessed 1/12/2019: https://avivaromm.com/vitex-hormones/.
101. Warren MP, Ramos RH. Chapter 32 – Alternative therapies to hormone replacement therapy. In: Lobo RA, Kelsey J, Marcus R. Menopause: biology and pathobiology. Academic Press; 2000. p. 459–80. https://doi.org/10.1016/B978-012453790-3/50033-0.
102. Kaszkin-Bettag M, Richardson A, Rettenberger R, Heger PW. Long-term toxicity studies in dogs support the safety of the special extract ERr 731 from the roots of Rheum rhaponticum. Food Chem Toxicol. 2008;46(5):1608–18.
103. Kaszkin-Bettag M, Beck S, Richardson A, Heger PW, Beer AM. Efficacy of the special extract ERr 731 from rhapontic rhubarb for menopausal complaints: a 6-month open observational study. Altern Ther Health Med. 2008;14(6):32–8.
104. Hasper I, Ventskovskiy BM, Rettenberger R, Heger PW, Riley DS, Kaszkin-Bettag M. Long-term efficacy and safety of the special extract ERr 731 of Rheum rhaponticum in perimenopausal women with menopausal symptoms. Menopause. 2009;16(1):117–31.
105. Kaszkin-Bettag M, Ventskovskiy BM, Kravchenko A, Rettenberger R, Richardson A, et al. The special extract ERr 731 of the roots of Rheum rhaponticum decreases anxiety and improves health state and general well-being in perimenopausal women. Menopause. 2007;14(2):270–83.
106. Wober J, Möller F, Richter T, Unger C, Weigt C, et al. Activation of estrogen receptor-beta by a special extract of Rheum rhaponticum (ERr 731), its aglycones and structurally related compounds. J Steroid Biochem Mol Biol. 2007;107(3–5):191–201.
107. MARIE-GENICA Consortium on Genetic Susceptibility for Menopausal Hormone Therapy Related Breast Cancer Risk. Genetic polymorphisms in phase I and phase II enzymes and breast cancer risk associated with menopausal hormone therapy in postmenopausal women. Breast Cancer Res Treat 2010;119(2):463–474.
108. Cavalieri E, Chakravarti D, Guttenplan J, Hart E, Ingle J, et al. Catechol estrogen quinones as initiators of breast and other human cancers: implications for biomarkers of susceptibility and cancer prevention. Biochim Biophys Acta. 2006;1766(1):63–78.
109. Cavalieri E, Rogan E. Catechol quinones of estrogens in the initiation of breast, prostate, and other human cancers: keynote lecture. Ann N Y Acad Sci. 2006;1089:286–301.
110. Cavalieri EL, Rogan EG. Depurinating estrogen-DNA adducts, generators of cancer initiation: their minimization leads to cancer prevention. Clin Transl Med. 2016;5(1):12.

Nutritional Influences on Reproduction: A Functional Approach

Brandon Horn and Wendy Yu

32.1 Introduction – 534

32.2 Microenvironments and Complexity – 534

32.3 Examining the Structural Layer – 536

32.4 Examining the Functional Layer – 536
32.4.1 Energy Production – 536
32.4.2 Energy Utilization – 537
32.4.3 Hormonal Factors – 537
32.4.4 Nutrient Factors – 539
32.4.5 Reproductive Toxicology – 540
32.4.6 Reproductive Immunology – 544
32.4.7 Exercise – 546

32.5 Examining the Informational Layer – 547
32.5.1 Genetics – 547
32.5.2 Innate Genetic Issues – 547
32.5.3 Neural Tube Defects and Folate Supplementation – 548
32.5.4 Risk Factors of Folic Acid – 548

32.6 Acquired Genetic Issues – 549

32.7 Summary – 549

References – 550

© Springer Nature Switzerland AG 2020
D. Noland et al. (eds.), *Integrative and Functional Medical Nutrition Therapy*,
https://doi.org/10.1007/978-3-030-30730-1_32

32.1 Introduction

Reproduction is one of the most basic biological processes, yet it is also among the most mysterious. On one level, human reproduction merely requires the fusion of two haploid gametes, forming a diploid zygote. This zygote must then divide 40 times, amassing 1.4 trillion cells in the process (assuming the average weight of a newborn is 3250 g) [1]. The resulting conglomeration of cells is then ready to enter the world as a human infant. This newborn, however, is not merely a mass of molecules. It is a highly organized and immensely complicated network of other living organisms. Some of these organisms are endogenously derived (the 1.4 trillion human cells) [2], while others are acquired exogenously. Of those acquired, some are inherited (mitochondria), some are acquired within the womb, and still others are acquired as the fetus is exposed to the environment during and following childbirth [3].

Within the microenvironment, each of these cells has its own nutritional requirements, need for waste management, and defense against invaders. They all serve some unique purpose within the larger community that is a human organism. For endogenously derived cells, each cell's purpose is expressed through its phenotype. A cell that migrates to the heart will develop phenotypical characteristics of a cardiac cell, whereas a cell that migrates to the liver will develop characteristics of a hepatic cell. Despite clear phenotypical differences, both cells are genetically identical. The environment determines a cell's ultimate phenotypical expression, whether in the promotion of physiology or pathology.

The environment, therefore, is one of the key determinative factors in the health of an organism. Nutrients, as largely exogenous molecular influences, make up an important part of this environment, and variations in the type and quantity of such influences can significantly alter physiology. It is within this context that we will explore the dynamics of reproduction. The goal of this chapter is to help the reader understand the complexities inherent in reproduction and to create effective strategies for determining an appropriate nutritional intervention. This chapter is not, nor is it intended to be, a comprehensive catalogue of nutritional influences on reproduction.

32.2 Microenvironments and Complexity

Reproduction is a biological imperative and our bodies have evolved to survive and reproduce. As a result, the successful treatment of reproductive issues often has more to do with removing impediments to reproduction than directly facilitating conception. While both strategies merit consideration within any given case, it should be noted that the direct facilitation of conception through assisted reproductive technologies (ART), such as in vitro fertilization (IVF), even in the youngest and most fertile population, has a success rate of less than 50%. By the time a woman is 43 years old, she has a less than 4% chance of achieving a successful pregnancy through ART [4].

Interestingly, when ART techniques are combined with other modalities designed to address factors influencing the uterine environment, significant increases in pregnancies and live birth rates have been reported [5–7]. As we will explore, the uterine and ovarian microenvironments are essential factors in reproduction. The status of these microenvironments, however, is intimately entangled with the status of the relative macroenvironment of the organism. Indeed, many reproductive pathologies are merely one aspect of a larger clinical picture. For example, endometriosis is a condition that affects 10–15% of reproductive-age women and is associated with reduced fertility [8]. Researchers noted a significant association between endometriosis and pigmentary traits, including an increased number of cutaneous nevi, freckles, and those with sun-sensitive skin [9]. While the precise relationship between pigment and endometriosis has yet to be elucidated, it has been shown that women with endometriosis have a higher risk of melanoma [10]. Furthermore, both melanoma and endometriosis have been associated with environmental toxins, including PCB and organochlorine exposure [11]. One's ultimate toxic burden, while strongly correlated with the dose and type of toxic exposure, is also largely influenced by one's detoxification capacity for such toxins. In the case of PCBs, biotransformation occurs via the cytochrome P450 1A1 enzyme (CYP1A1) [12]. It was found that individuals with genetic polymorphisms in the gene that codes for CYP1A1 (A4889G) have significantly higher risks of cancer [13] and endometrial hyperplasia [14]. Similarly, genetic polymorphisms in other genes, such as the GSTM1 gene involved in the regulation of glutathione S-transferase (and thus protection from PCBs), were associated with an increased risk of endometriosis [15]. Because GSTM1 is also involved in detoxification of electrophilic compounds including carcinogens, therapeutic drugs, and other environmental toxins and products of oxidative damage, any findings related to issues in glutathione-mediated detoxification pathways warrant an inquiry into GSTM1 and other polymorphisms. In the case of GSTM1, the frequency of the particular polymorphism rs366631 in the general population is 50–78% (depending on race) [16]. Therefore, it is more likely than not that a patient presenting with endometriosis will develop a number of seemingly unrelated pathologies.

Endometriosis has also been shown to have a strong inflammatory component in both its pathogenesis and the resulting pathophysiology. Consequently, endometriosis has been associated with an increased risk of cardiovascular disease [17]. Vascular insult and atheromatous plaque progression are commonly found with endometriosis-associated coronary heart disease and can also be found in the general population of women with unexplained infertility [17–18]. While there are many systems affected by inflammation, an attempt to uncover the source of such inflammation should be made. As inflammation is an immune-mediated response, it comes as no surprise that endometriosis has been associated with infectious agents. Less intuitive is that infectious influence is not restricted to pathogens local to the reproductive tract. Take *Helicobacter pylori* (*H. pylori*), for example. *H. pylori* is not known to infect the genital tract or associated nervous or glandular structures, yet it is significantly more prevalent among infertile men and women than among healthy controls [20]. Extra-reproductive pathogens influencing reproduction are not limited to *H. pylori*. Periodontal disease, for example, was also significantly associated with endometriosis [20]. It has been subsequently hypothesized that systemic autoimmune disorders and possibly endometriosis may be an inflammatory response to seeded periodontal bacterial pathogens [20]. A variety of local pathogens have also been associated with an increased risk of developing endometriosis. Some of these include the human papillomavirus, *Escherichia coli*, *Chlamydia trachomatis*, and *Mycoplasma genitalium* [21–22]. In addition to pathogens, inflammation can be induced by autoimmunity. In women with endometriosis, the presence of autoantibodies is a common feature [23].

Aside from the cardiovascular, toxicological, reproductive, and immune factors, endometriosis has perhaps most strongly been associated with endocrine pathology. Indeed, estrogen is often considered to be one of the most important factors in the pathogenesis of endometriosis [24]. What is less well known is that progesterone resistance may play an equally important role [25]. Progesterone resistance can result from various etiologies, one of the more important of which is via modulation of the endocannabinoid system (ECS). The cannabinoid receptor type 1 (CB1-R), for example, has been found to have essential reproductive functions [26] and appears to be responsible for progesterone's anti-inflammatory actions. Indeed, within the reproductive tract, the uterus contains the highest concentration of CB1-R ligands, and the activation of CB1-R seems to be directly involved in implantation. Altering the endocannabinoid system by inhibiting progesterone (a CB1-R ligand), or through direct exogenous ligand exposure (via cannabis use), resulted in reproductive malfunctions [25].

Other hormones have been associated with endometriosis as well. For example, low levels of vitamin D were found in women with endometriosis and associated pelvic pain [27]. Excessive exposure to thyroid hormone, for example, in Graves' disease, was also associated with endometriosis [28], as were genetic polymorphisms in the renin-angiotensin system [29].

Emotional factors were also found associated with endometriosis, with stress contributing to the development and severity [30]. Interestingly, endometriosis patients with high stress had lower salivary cortisol than controls, leading researchers to postulate that the hypocortisolism may be an adaptive response [31].

While there are many other factors that can influence endometriosis and reproduction in general, this example illustrates the importance of approaching infertility as a "nonrandom" occurrence. Different infertility etiologies are genetically and clinically linked with other diseases, and researchers suggest that we should be treating infertility problems using phenomics, rather than viewing infertility as an isolated and exclusive disease of the reproductive system or HPO/HPA axis [32].

As the previous example illustrates, there are many etiological and pathophysiological factors involved in reproductive pathology. Many of these factors overlap and form an interconnected web. Organizing the masses of information can be challenging. One useful model has its roots in Chinese medicine, where they describe all phenomena as some combination of three levels: material/structural, energy/function, and informational (which includes psychospiritual factors). While all phenomena will have some interaction with each of these factors, there tends to be a dominant factor. To illustrate this concept, let us examine a common reproductive pathology: uterine leiomyomas (fibroids).

A fibroid is a structural anomaly that can inhibit reproduction through a number of potential mechanisms, the primary of which is usually some form of physical obstruction. While hormone modulation and potentially certain phytochemicals can inhibit the growth and even regress the tumor [33], many of the therapeutic targets used to accomplish this are antagonistic to reproduction. Estrogen and progesterone, for example, are known growth factors for fibroids. Downregulation of ovarian steroid hormones using a GnRH agonist often results in tumor regression [34]; however, reproduction is not possible under such conditions and the therapy is only approved in the short term. Other hormonal and potentially nutraceutical therapies, such as epigallocatechin gallate (EGCG), work by modulating aromatase or directly inhibiting cellular reproduction [35–36]. In cases of infertility, inhibiting cellular reproduction would be counterproductive. As a result of a lack of options, fibroids suspected of causing infertility are often surgically excised, yet the 8-year recurrence rate for laparoscopic myomectomies is 76% [37]. Such a high recurrence rate may be the result of the failure to address antecedents or triggering events. Indeed, physical activity [38], lower BMI [39], intake of fruit, intake of preformed vitamin A [40], and sufficient sunlight exposure [41] have each been independently associated with lower incidence of fibroids. In animal models, activated vitamin D3 (1,25(OH)2D3) shrank uterine fibroids [42], lending further weight to the epidemiological data [41]. Genetic factors have also been found to play a role [429].

While a number of nutrients have been explored in the treatment of fibroids, human studies are lacking. As a general proposition, correlations between animal studies and human studies are largely disappointing. A review of the highest impact animal studies found only 10% resulted in a treatment approved for use in patients [43]. Thus, animal studies are not reliable predictors of human response. In addition, and as we'll see later in the chapter, many nutritional supplements seem to have dose-dependent biphasic effects: promoting or inhibiting the same function. Genistein, for example, was found either to proliferate or inhibit human uterine leiomyoma cells, depending on the dose [44].

In summary, uterine fibroids have a structural component, being the mass itself and the potential to physically obstruct important aspects of reproduction. They also have a functional component, which covers the etiology and pathophysiology of the fibroid and includes its antecedents, triggering events, and mediators. There are also psychological and informational aspects, including possible genetic and stress components.

As the aforementioned example illustrates, structural abnormalities are often the result of functional or informational problems. Therefore, when dealing with structural issues, an attempt should always be made to identify and address any confounding informational and functional issues.

32.3 Examining the Structural Layer

Structure dictates function. While this is a mantra of biology, as we will discuss, the converse is also true. Structural issues can present on a macroscale, such as a uterine fibroid, or on a microscale, such as improper protein structures or folding. Within reproductive pathophysiology we can roughly divide structural issues into direct and indirect.

Direct structural issues are gross anatomical abnormalities that can change how the reproductive process functions. Hydrosalpinx, ovarian cysts, endometriomas, fibroids, and polyps are all examples of masses that have potential for causing direct structural issues. Within male fertility, these may relate to varicoceles and other tumors, as well as morphological defects in sperm.

Indirect structural issues occur where structural damage or morphological anomalies interfere with cellular communication, often leading to over- or underproduction of hormones or other ligands. This form of structural issue has significant overlap with the energy/function level; however, if the structural level is not addressed, it is very difficult to correct the functional aspects. Therefore, the ideal therapeutic target is usually the structural abnormality and its antecedents. An example of this is a Leydig cell tumor (LT), which can secrete sex hormones resulting in reproductive problems. In men, an LT can lead to feminization, loss of libido, erectile dysfunction, and gynecomastia. In women, they can lead to masculinization, anovulation, amenorrhea, clitoromegaly, and increase in musculature. Treatments aimed at inhibiting the effects of estrogen or testosterone are of limited benefit and destruction of the tumors is necessary to stop the process [45].

Not all tumors, however, should be treated surgically. While a common procedure, aspiration of ovarian cysts has not been found to be effective at increasing fertility outcomes [46] and the "effects of surgical treatment are often more harmful than the cyst itself to the ovarian reserve" [47].

While tumors are one of the more common indirect structural issues, tissue damage is another. Head injuries, for example, can cause damage to the hypothalamus, resulting in hypopituitarism with elevated prolactin levels [48]. Scar tissue from abdominal surgeries or endometriosis can lead to pelvic adhesions and even infertility [49]. The formation of masses in the blood (i.e., clots) are also a common cause of reproductive problems [50]. These can inhibit blood flow or lead to ischemia and are a well-known risk factor in recurrent miscarriage and suspected factors in implantation failure [50]. Clots are an interesting subset of structural issues because nutritional intervention can be very effective. Nutraceuticals such as nattokinase, lumbrokinase, *Desmodium styracifolium*, and *Sargassum fulvellum* have strong fibrinolytic activity [51]. Many Chinese herbs have also been shown to inhibit a number of prothrombotic pathways [51]. Dietary factors can also influence clotting risk, with red and processed meats tending to increase thrombotic risk and fruit, vegetables, and fish decreasing risk [52]. Unfortunately, most studies incorporating meat are poorly designed. Red meat from grass-fed cows, for example, has a completely different nutritional profile than the same meat from grain-fed cows. An analysis done by the Weston A. Price Foundation with the University of Illinois found that "[t]he ratio of omega-6 to omega-3 fatty acids was over 16:1 for the grain-fed tallow but only 1.4:1 for the grass-fed tallow." It should be noted that a higher n6:n3 ratio has been found to be associated with higher thrombotic risk [53]. Therefore, studies discussing "meat" (or any other class of foods without specifying type or source) should be approached with significant skepticism.

Structural impingement of nerves or blood vessels may also result in altered function of a variety of systems. While high-quality data are lacking, biological mechanisms exist whereby spinal or other nerve impingement may negatively impact sexual and reproductive function [54]. Where such patho-mechanisms are suspected, manual manipulation such as massage, chiropractic, or osteopathic manipulation should be explored.

Finally, we have structural damage to DNA. The cause of such damage is often reactive oxygen species (ROS) or reactive nitrogen species (RNS). As will be discussed later in the chapter, this can cause damage leading to poor oocyte quality or oocyte destruction.

32.4 Examining the Functional Layer

Function is life. It is what separates an organism from an inanimate object. The term "function" is a concept. The means of executing that concept is energy. Therefore, energy is one of the most important and fundamental requirements for life and, as a consequence, for reproduction. Indeed, the body seems to have evolved to suppress reproductive function in response to energy deficits [55]. The energy cycle has four basic aspects: production, storage, delivery, and utilization. While energy storage and delivery can be important factors in reproduction, we will focus this discussion on production and utilization.

32.4.1 Energy Production

Efficient energy production occurs primarily through oxidative phosphorylation within the mitochondria. Indeed, the estimated number of mitochondria per cell can give us some idea of the energy requirements of the function of a given tissue. For example, a skeletal myocyte has been estimated to contain over 3500 mitochondria [56], whereas a cardiac myocyte, which is active all the time, contains almost 7000. In contrast, the number of mitochondria in an oocyte (mature egg cell) is close to 800,000 [57]. The mitochondrial requirements of an oocyte are therefore orders of magnitude higher than other cells. It comes as little surprise therefore, that one of the most fundamental sources of embryo failure is the inability to produce sufficient energy [58].

While the quantity of mitochondria is certainly important for energy production, the quality of energy production is equally important. For proper function, mitochondria need both substrate and cofactors. With aerobic respiration, the substrates are pyruvate and oxygen with pyruvate being primarily a glucose derivative resulting from glycolysis. Glycolysis, however, occurs within the cells and therefore glucose must efficiently enter such cells. Suboptimal glucose transport is thus correlated with poor fertility outcomes. Polycystic ovary syndrome (PCOS), for example, is a condition that affects up to 20% of women [59] and is highly associated with insulin resistance [60]. While women with PCOS have an ordinary number of primordial follicles, the number of primary and secondary follicles is significantly increased. However, despite the large quantity of follicles, the growth of such follicles often arrests prematurely. Without a dominant follicle, anovulation results. For those women with PCOS that do ovulate and achieve pregnancies, the incidence of spontaneous abortion is as high as 73% [59]. Improved control of glucose, however, may result in significantly improved pregnancy outcomes [61].

Oxygen is essential for human life and reproduction. Oxygen utilization level within the oocyte was found to be a predictor of reproductive success [62]. Consequently, the availability, perfusion, transport, and utilization of oxygen should be examined in the context of reproduction. For example, pulmonary issues can predispose women to fertility difficulties [63], as can issues with oxygen transport [64]. Once oxygen is in the cell, the mitochondria require a number of nutrient cofactors for proper function. Electron transport chain (ETC) cofactors include nicotinamide, thiamine, riboflavin, succinate, and coenzyme Q10 (CoQ10) [65].

The aerobic metabolism of energy is a delicate balance. The reduction of oxygen produces ROS, including superoxide and peroxide anions, which can damage DNA and cellular structures and is associated with fertility decline [66]. At the same time, ROS and RNS are important facilitators of ovulation, folliculogenesis, implantation, and even ovarian steroidogenesis [67]. A study measuring follicular fluid ROS levels found increased ROS was a marker predictive of in vitro fertilization (IVF) success [68]. Optimal ROS levels change depending on the location and timing of the reproductive cycle and, as such, the indiscriminate use of antioxidants could be detrimental to fertility [69].

32.4.2 Energy Utilization

While energy production is essential, of equal importance is its utilization. We can broadly classify the use of energy through the related concepts of communication and interaction. Communication allows for interaction and, while a very broad topic, we will narrow our focus to how the way we interact and process our environment influences reproduction.

The process of communication deals with signal distribution, be it electrical, chemical, or otherwise. The nervous system, as a mediator of brain signaling, is intimately involved in almost every aspect of reproduction [70]. Insults to the nervous system can inhibit proper reproductive function, whether from disease process, emotions, or exposure to environmental or immunological challenges [71–75]. Chemical signaling, whether endocrine, paracrine, or autocrine, is indispensable for human reproduction [76–78]. Disruption in chemical signaling can also come from disease process, emotions, and exposure to environmental and immunological factors [79–81]. In addition, there seem to be other communication and delivery mechanisms that are more amorphous, such as quantum entanglement within neural networks [82] and the meridian theory of Chinese medicine which posits a network of channels covering the entire body. While research is ongoing, there have been a number of recent studies affirming that biophysical characteristics at acupuncture points are in fact different from non-acupuncture points [83]. The modulation of these communication routes via acupuncture has been found to improve a variety of fertility issues, including improvement in markers of poor ovarian reserve [84]; premature ovarian failure [85]; ovulation frequency in women with PCOS [86]; sperm count, motility, and morphology [87–89]; and so forth.

There are many physiological and pathological intricacies to the facilitation and inhibition of the process of communication and transformation. Some of the more important facilitating molecules include hormones and various nutrient cofactors.

32.4.3 Hormonal Factors

The primary role of hormones is to carry particular messages from one site to another. Sex hormones, including progesterone, androgens, and estrogens, are essential to reproduction, but there are many other hormones involved as well. Some of these include gonadotropin-releasing hormone (GnRH), luteinizing hormone (LH), follicle-stimulating hormone (FSH), prolactin (PRL), oxytocin, inhibin, activin, prostaglandin F2a, PGE2, human chorionic gonadotropin (hCG), placental lactogen, relaxin, thyroid hormones, DHEA, insulin, anti-Mullerian hormone, albumin, SHBG, cortisone, melatonin, and angiotensin. While a thorough investigation of the interactions of these hormones is beyond the scope of this survey, we will review selective interactions to further demonstrate the role of systems biology within reproduction.

During the follicular phase, the hypothalamus sends low-frequency, high-amplitude pulses of GnRH to the anterior pituitary, which responds by secreting FSH. FSH then stimulates the aromatization of androgens within the granulosa cells to create more estrogen. However, LH is just as important as it stimulates the theca cells to produce the androgens that are aromatized [90]. FSH also stimulates follicular growth. As follicles grow, they produce more estrogen. Estrogen, in turn, helps develop the uterine lining, stimulates vaginal lubrication, upregulates expression of progesterone receptors, and increases vasopressin and oxytocin production in preparation for ovulation. With respect to FSH, estrogen has a bimodal effect, first inhibiting FSH through a negative feedback loop until E2 levels reach 2–4 times the early follicular level. At that point, the response of the hypothalamus switches to a positive feedback that results in a surge of gonadotropin production, though a number of factors, including insufficient oxytocin levels and elevated melatonin levels, may inhibit the LH surge [91]. The LH surge prevents further growth of nondominant follicles and causes final oocyte maturation to begin, a process involving the renin-angiotensin system. As a result, the use of ACE inhibitors may interfere with this process [92].

Following the LH peak, the outer granulosa layer stops aromatizing androgens and, with the help of other hormones like melatonin [93] and angiotensin and renin [94], switches to progesterone synthesis. During ovulation, the follicle is ruptured via the degradation of the follicular wall by enzymes such as collagenase, plasminogen activator, and gelatinase. After release of the oocyte, the follicle (corpus luteum) primarily secretes oxytocin, progesterone, estrogen, and inhibin.

Oxytocin (OT) is believed to facilitate sperm transport to the proper fallopian tube [95], is anxiolytic, and seems to be a cofactor in angiogenesis [96]. Progesterone converts the endometrium to the secretory stage, thickens cervical mucus to block sperm transit, and decreases the maternal immune response. Progesterone also has anxiolytic effects, perhaps to help minimize the negative impact of stress on implantation [97], and helps to control oxytocin levels. While OT seems useful to facilitate sperm transport, continued myometrial contractions may interfere with implantation. Indeed, the use of OT antagonists following IVF was associated with increased implantation rates [98].

Like OT, PRL has important bimodal functions in reproduction. While PRL is necessary for embryo implantation and can rescue stress-threatened pregnancies by preventing progesterone degradation [99], its presence in sufficient quantities at the wrong time can inhibit ovulation through suppression of GnRH. Perhaps this is how vitex agnus-castus (VAC) improves fertility in some women, as it was found to lower PRL levels equivalent to the dopamine agonist bromocriptine [100]. However, caution should be exercised in administering VAC to eu-PRL individuals, as low prolactin may have unwanted effects on fertility.

Melatonin is present in significant amounts in follicles [93] and peaks during the luteal phase. The cyclical nature of melatonin throughout the reproductive cycle is essential as prematurely elevated levels may suppress ovulation [101]. In the luteal phase, melatonin plays a number of important roles, particularly in the maintenance of progesterone levels and in the reduction of oxidative stress. Indeed, melatonin's antioxidant effects are strong enough to help prevent premature ovarian failure in chemotherapy-treated females [102] and to significantly extend the reproductive life of hens [103]. As melatonin levels decline with age, the likelihood of poor oocyte quality due to oxidative damage increases significantly [104]. In pregnancy, melatonin readily crosses the placental barrier and has an essential role in protecting the fetus from oxidative damage [104]. It also appears to be a crucial component in proper fetal development, particularly in the prevention of neurodevelopmental disorders. Low levels of melatonin are associated with autism spectrum disorders (ASD) [104]. As melatonin secretion occurs mainly at night in the absence of light, it is essential to reinforce the importance of proper sleep hygiene. Minimization of nocturnal light is important to maintain proper melatonin levels. Exposures to as little as 285 lux for 2 hours were sufficient to negatively impact melatonin levels [105], though filtering out blue light via orange-tinted safety glasses seems protective [106].

Following ovulation, the oocyte is released into the peritoneal cavity where the fimbriae at the end of the nearest fallopian tube facilitate its entrance. Propulsion of sperm and oocytes within the fallopian tubes is accomplished through a complex interaction between peristaltic waves, ciliary activity, and the flow of tubal secretions [107]. Fertilization typically occurs in the ampulla of the fallopian tube, with sperm able to live for up to 7 days waiting for the arrival of the oocyte [108]. While it has often been assumed that ovulation only occurs once per cycle, multiple batches of maturing eggs have been observed in nearly all human subjects and multiple ovulations per cycle in 10% [109]. This perhaps explains why the rhythm method of contraception tends to be unreliable [110], though there are other factors, such as pheromones, that can trigger ovulation [111] and could influence rhythm method failures.

Upon fertilization, it takes the embryo ~4–6 days to travel from the fallopian tube into the uterus where it spends ~3 days in the uterine cavity prior to implantation. During this time, the embryo is nourished solely through fluids in the uterine cavity. Therefore, fat-soluble vitamins and iron are particularly important [112]. The endometrium then absorbs the fluids to allow contact between the blastocyst and the uterus. Failure of fluid absorption (endocytosis) significantly increases implantation failure [113]. The invasion of the embryo into the uterus then begins. This is an inflammatory process and the use of anti-inflammatory medication may be counterproductive. For example, dexamethasone administration for 4 days inhibited implantation in 80% of rats given E2. However, injecting two doses of histamine reversed these effects. [114]. Therefore, eating fermented or other foods high in histamine may help facilitate implantation. Similarly, PGE2 reversed dexamethasone-induced implantation inhibition [114].

There are many hormones involved in implantation, including thyroid hormone, E2, P4, PRL, CRH, melatonin, and PGE2. [115–118]. If hCG or other signals are not received indicating implantation, hormone levels drop, triggering menstruation. If hCG is detected, the body begins preparation for a pregnancy.

From a treatment standpoint, there are many dietary strategies to modulate hormones. One common strategy is the use of natural aromatase inhibitors, such as flavonoids. In particular, chrysin, naringin, naringenin, hesperetin, eriodictyol, and apigenin [119–120] are among the compounds shown to have significant inhibitory effects on aromatase. Tea polyphenols can also inhibit aromatase, with black tea having significantly more potent effects than green tea [119]. A variety of other phytocompounds, particularly from a variety of herbal medicines and foods, are also effective [119]. Genistein and quercetin, on the other hand, induced aromatase with a 2.5- and 4-fold induction, respectively [121]. Aromatase modulation should be attempted only relative to the patient. In diseases often characterized by estrogen-dominant states such as endometriosis, aromatase inhibition may be therapeutically indicated. However, in cases of androgen dominance, such as PCOS, this would be counterproductive. Strongly inhibiting aromatase may result in poor reproductive outcomes, and there is some evidence that it could induce autoimmune disease [122]. Indeed, estrogen seems to be an important regulatory hormone for development of peripheral T regulators and therefore can help maintain immune tolerance. Estrogen was also found to be an important factor in embryo implantation [123].

Cortisol is released by the adrenal glands in response to stress and low blood sugar. Increased cortisol levels have been associated with untimely increases in gonadotropin and progestin levels during the follicular phase, impending ovulation and implantation [97]. It has also been associated with early pregnancy loss [124]. During pregnancy, elevated cortisol is associated with intrauterine growth restriction, preterm labor, preeclampsia, and chorioamnionitis [125]. In contrast, cortisol was also found to have beneficial effects on stimulating hCG secretion and promoting trophoblast growth [125]. The adverse effects of cortisol are unlikely to stem from normal physiological secretion, but rather from stress-induced production. Cortisol levels are upregulated by caffeine, stress, and sleep deprivation. They are downregulated by relaxation, massage, intercourse, magnesium, vitamin C, n3 fatty acids, and having fun or laughing. Estrogen also increases cortisol levels [97].

32.4.4 Nutrient Factors

Nutrition is one of the most important factors in the survival and reproduction of any organism. Indeed, nutrition provides substrates for both structure and function. In reproduction, proper nutrition has been found to be essential for fertility and proper fetal development [126]. To begin with, nutrition provides the substrates for energy production. Reproduction requires a tremendous amount of energy expenditure. It is estimated that over a normal gestational period with normal fetal weight gain, the caloric requirement is 78,000 kcal [55]. Calories, however, are not a sufficient metric; the type of caloric intake is also important. Numerous food substances have been found to impair fertility. For example, trans-fatty acid intake was inversely associated with sperm count [127]. Replacement of 1% energy of saturated fatty acids with 1% energy of long-chain n3-PUFAs more than doubled live birth rates [128]. Caffeine consumption (coffee, tea, sodas, etc.) in both men and women was associated with a greater than 70% increase in the risk of pregnancy loss [129] and alcohol more than doubled the risk of miscarriage [130]. Study results on alcohol and caffeine, however, have not been consistent [131]. Sugar consumption reduced live birth rates [132] and adding just one serving of poultry per day resulted in a 32% higher chance of developing ovulatory infertility [133–134]. Interestingly, no other meats had this effect, though both meat and fish have toxicological concerns that are known to negatively impact fertility [131].

In contrast, proper nutrition has consistently been found to increase fertility [131]. High-quality food should be free of pesticides and as natural as possible. Organic foods avoid endocrine damage from pesticide residues. Indeed, such residues were associated with fertility problems including reduced semen quality in men and a lower probability of pregnancy and live birth in women [7, 135]. The Mediterranean diet was examined in relationship to fertility and found to significantly improve live birth rates [136] and reduced the risk of preeclampsia [137]. On the other hand, caution is warranted for patients on restrictive diets. Carbohydrate restriction in the form of avoiding whole grains may be counterproductive, as whole grain intake was associated with a 51% higher probability of live birth [138]. Consumption of fruits and vegetables was positively correlated with healthy birth weight [139], and tomato juice, in particular, improved sperm motility [140]. Consistent use of a prenatal multivitamin resulted in a 55% reduction in the risk of miscarriage [129], stillbirths [141], preeclampsia, preterm labor, and small gestational-age babies [142]. Intake of iron, zinc, folic acid, calories, and protein was also found to be higher in fertile women, specifically in the form of meat, nuts, dried fruit, and green leafy vegetables [143].

Specific nutrients have also been found to exert important influences on fertility; however, whether a nutrient is beneficial or detrimental is often dose-dependent. For example, levels of iron (Fe), copper (Cu), and magnesium (Mg) were found to be lower in males with sperm motility issues, yet excess Fe and Cu (especially if a person is positive for hemochromatosis or Wilson's disease) both impaired sperm parameters and could lead to impotence [144, 145]. Zinc (Zn) levels were found to be significantly higher in fertile men than in non-fertile men, and seminal Zn levels were positively correlated with improved sperm count and morphology [146]. High levels of Zn, however, produced the opposite effect [147]. Selenium was found to reduce the risk of miscarriage, preeclampsia, preterm labor, and gestational diabetes [142]. Deficiencies in selenium have also been associated with reduced sperm quality. However, too much selenium reduced sperm quality [148–149].

Vitamin D, technically a hormone, regulates calcium, magnesium, and phosphate homeostasis and plays an important role in fertility. Low levels of vitamin D are associated with a decrease in sperm count and motility, along with an increase in abnormal sperm morphology. High vitamin D levels were found associated with poor sperm count, motility, and morphology [150]. Furthermore, elevated vitamin D levels in follicular fluid reduced fertilization rate and embryo quality [150], and higher values of vitamin D were associated with a lower possibility of achieving pregnancy [150]. Indeed, the activated form of vitamin D (1,25(OH)2D3) is transferred through the placental barrier, where it can leach out embryonic skeletal calcium stores, potentially compromising postnatal survival [150]. A significant portion of vitamin D's effects on fertility is thought to be its influence on calcium homeostasis. One study examining vitamin D-deficient subfertile mice found that calcium supplementation restored fertility [151]. Indeed, calcium plays a major role in fertility as calcium ion oscillations initiate embryogenesis [152].

Magnesium (Mg) also has important influences on fertility. Mg ions were found to preserve fertility on exposure to ionizing radiation [153]. Women supplementing with 300 mg of Mg were found to have fewer instances of intrauterine growth retardation, preterm labor, preeclampsia, gestational diabetes mellitus, leg cramping, pregnancy-induced hypertension, and abnormal weight gain. The Cesarean delivery rate was 67% lower, and children born to these women had higher Apgar scores when compared to women taking only 100 mg of Mg daily [154].

Maternal intake of folic acid during pregnancy was found to lower the risk of autism spectrum disorder (ASD) [155], neural tube defects, cardiac malformations, and cleft palate [156]. However, one report found that excessive levels of folate and vitamin B12 in the blood increased autism risk by 17,600%. Excess levels were defined as more than 59 nanomoles per liter of folate and more than 600 picomoles per liter of B12. While this has not yet been confirmed, a bimodal effect of folic acid would be commensurate with other nutrients [157].

Vitamin C (AA) can also help protect the reproductive system from oxidative stress. It is also important during follicular growth [158] and to help prevent reproductive damage from lead and other toxins [159]. AA was found to protect the fetus from epigenetic damage due to maternal smoking [160] and to help reverse testicular damage in diabetic rats [161]. A dose of 500 mg/day of time-release vitamin C was found to almost double pregnancy rates [162].

While many other nutrients may influence reproduction, the above examples make it clear that many nutrients can either improve or impair fertility. Clinicians should therefore consider careful evaluation of nutrient status prior to commencement of a supplement regimen in support of reproductive health. In addition to supporting physiology through proper nutrition, avoidance or neutralization of inhibitory factors is of equal importance. These factors often arise as a result of environmental exposures. Indeed, in modern times where nutrients are relatively easily accessed, toxicological concerns have become increasingly prominent.

32.4.5 Reproductive Toxicology

There are over 80,000 chemicals registered for use in the United States, and 2,000 new ones introduced each year [163]. While many chemicals have been found to cause reproductive harm, most of the chemicals in use have never been tested. According to the *New York Times*,

» MANY Americans assume that the chemicals in their shampoos, detergents and other consumer products have been thoroughly tested and proved to be safe. This assumption is wrong. Unlike pharmaceuticals or pesticides, industrial chemicals do not have to be tested before they are put on the market… [Furthermore,] companies are not required to provide any safety data when they notify the [EPA] about a new chemical, and they rarely do it voluntarily … If the EPA does not take steps to block the new chemical within 90 days or suspend review until a company provides any requested data, the chemical is, by default, given a green light… [164].

In such an environment, a toxicological evaluation becomes essential. Official determinations of safe levels of potentially toxic substances may err by orders of magnitude. This is the result of most toxicity studies examining isolated elements. Real-world exposures are rarely isolated and substances considered harmless at low levels can become highly toxic when combined. For example, experiments where mercury was administered at a level that kills 1% of a test population (LD1) and then combined with an amount of lead that kills 0.05% of a test population (1/20 of LD1) killed 100% of the animals [165]. This is one of the reasons for adverse event reporting at dosages deemed safe in controlled trials.

A second important toxicological principle is that route of administration can exponentially increase the toxicity of various substances. For example, safe levels of GI exposure can be highly toxic if given parenterally. A study looking at the LD50 for cholinesterase inhibitors found the LD50 for oral Ciodrin was more than 40 times higher than the LD50 for intravenous and more than 15 times higher than subcutaneous administration [166]. Finally, toxicity dosage is not linear. Researchers examining aluminum oxyhydroxide (AlO(OH)), the primary adjuvant for vaccines, found that "[n]eurobehavioural changes, including decreased activity levels and altered anxiety-like behaviour, were observed compared to controls in animals exposed to 200 μg Al/kg but **not** [emphasis added] at 400 and 800 μg Al/kg." They concluded that "[Al] injected at low dose in mouse muscle may selectively induce long-term Al cerebral accumulation and neurotoxic effects; and that 'the dose makes the poison' rule of classical toxicology appears to be overly simplistic" [167].

It should also be mentioned that, for obvious reasons, most controlled toxicological studies (including the ones we will be citing below) are done in nonhuman, animal models and extrapolated to humans despite the imperfect correlation.

Toxins are ubiquitous in the environment with common sources including food, air, water, electronics, cleaning supplies, pesticides, insecticides, pharmaceuticals, nutraceuticals, and medical procedures. Reproductive toxins in food can arise "naturally" or through contamination with food additives. For example, grains naturally attract mold growth. As such, cereal grains designated for consumption are frequently contaminated with mycotoxins such as zearalenone that are highly toxic to the reproductive system, lead to the degeneration of oocytes [168], and inhibit sperm viability [169]. Rice, particularly the bran, has been shown to contain high levels of arsenic (discussed further below) [170], a compound that is highly toxic to the reproductive system. Arsenic can cross the placental barrier, inducing developmental toxicity and even fetal death [171]. It also damages male fertility, reducing testis size and interfering with spermatogenesis and steroidogenesis [172], though quercetin may ameliorate some of the negative impact on the male reproductive system [173]. The bran of rice contains 10× the arsenic levels of polished rice; thus brown rice has much higher arsenic levels than white rice. In addition, proper preparation of rice can significantly reduce arsenic levels. Cooking with a water-to-rice ratio of 12:1 was found to reduce arsenic levels by 57% [174].

Other common metals that have significant reproductive toxicity include cadmium, mercury, lead, and aluminum. Cadmium (Cd) is found in various foods such as shellfish, mushrooms, cereal grains, chocolate, potatoes, and other vegetables (usually acquired from the soil), as well as cigarettes, gasoline combustion, fertilizers, PVC, ceramic glazes, textiles, and paints [175]. A combination of zinc and selenium was found to ameliorate Cd damage to ovarian tissue. Vitamin C, retinol and blueberries also significantly reduced the negative impact of Cd exposure. Low-protein diets on the other hand increase cadmium toxicity, presumably due to lower access to cysteine – necessary for metallothionein synthesis [176–178].

Mercury is commonly found in seafood, fluorescent light bulbs, vaccines, contact lens solutions, makeup, and skin creams. Dental amalgams contain 50% elemental mercury and may produce a significant amount of vapor, as it is consistently released from amalgams and absorbed by the human body [179]. Modern amalgams are higher in copper, causing still higher mercury emissions [180]. While it was previously believed that mercury released from dental amalgams was not harmful, new findings dispute this [181]. Heat can increase the release of mercury vapor, as can electromagnetic fields [181]. Exposure of patients, for example, to radiofrequency emissions from conventional Wi-Fi devices increased mercury release from dental amalgams [182]. Consequently, researchers have urged pregnant women with dental amalgams to minimize their exposure to electromagnetic fields [181].

Mercury is highly toxic to the reproductive system and all mercury compounds can cause fetal harm [183]. Methylated mercury easily crosses the placental barrier and denatures DNA. It can also induce chromosomal aberrations, increasing aneuploidy incidence [183]. Garlic is a relatively safe compound that effectively inhibits the embryotoxicity of methylmercury [184] and, given its safety and efficacy profile, should be considered prophylactically for persons of reproductive age with potential mercury exposure. A variety of other compounds have also been found to attenuate the toxic effects of mercury compounds and should be considered both in treatment and prophylaxis. Examples include selenium [185], cysteine and methionine [178], alpha-lipoic acid [186], niacin [187], chrysin [188], and vitamin C [189]. Interestingly, morphine [190] and prolactin [191] were also found to have protective effects.

Arsenic (As) is commonly found in water, seafood, chicken, pork, wine, tobacco, rice, nutritional supplements, and wood. It is highly embryotoxic and tends to cross the placental barrier, inducing irreversible defects in the developing fetus and embryo. It also damages testicular function, resulting in reduced count and motility [192]. Arsenic also inhibits steroidogenic enzymes 3β- and 17β-hydroxysteroid dehydrogenase, whose activity is essential in the production of progesterone, testosterone, and estrogen [193]. A number of compounds have been found to inhibit arsenic-induced toxicity. These compounds include vitamin C [194]; vitamin E [195]; alpha-lipoic acid [196]; vitamin B12; carotenoids; flavonoids such as quercetin, naringenin, and silymarin; and organic forms of calcium, iron(II), and zinc. Of note, the inorganic forms of these metals can be problematic in combination with arsenic [197]. Amino acids, such as cysteine and methionine, were all found to mitigate arsenic's toxic effects. Selenium, specifically, decreases teratogenicity of As [178].

Lead (Pb) is commonly found in soil, dust, drinking water, paints in older homes, crystal, some ceramic glazes, imported foods, batteries, chocolate, and nutritional supplements [198]. Pb has been associated with sterility, spontaneous abortions, stillbirths, and neonatal deaths [197]. It can also reduce semen morphology, motility, count, and volume [199]. Fetal exposure to Pb can result in impaired cognitive development and was even associated with an increase in criminal arrests in early adulthood. For every 5 $\mu g\ dL^{-1}$ increase in blood Pb levels by 6 years of age, the risk of being arrested for a violent crime as a young adult increased by almost 50% [200]. Omega-3 fatty acids were found to significantly decrease brain damage due to lead exposure [201]. As Pb mimics calcium, vitamin D increases Pb absorption and its deposition into the bone and the kidneys [178]. Phosphorus, on the other hand, decreases bone retention of Pb [178]. Zinc supplementation decreases Pb GI absorption and tissue accumulation [178].

Aluminum (Al) is the third most abundant element in the earth's crust, yet it is highly neurotoxic [202]. It is found in a wide variety of commonly used products such as baking powder, sugar, and salt (as an anticaking agent), a bleaching agent in white flour, antiperspirants, lotions, infant formulas, parenteral nutrition (TPN) [203], antacids, and cheeses [204]. It is also contained in vaccines in biologically significant amounts [167] routinely recommended for pregnant women [205–206]. Following prenatal exposure, Al transplacentally traverses and accumulates in the fetal tissues in amounts that can adversely influence fetal development [192]. Pregnancy enhances susceptibility to Al toxicity. In developmental studies, Al has been shown to cause bone abnormalities, fetal internal hemorrhaging, neurodegeneration, and immune system activation [192, 202, 207]. Silica has been shown to chelate Al [208]. In one study, the consumption of 1 L of silicon-rich mineral water daily for 12 weeks significantly reduced the body burden of Al and improved cognitive function in Alzheimer's disease patients [209]. In addition, vitamin E and selenium [210], vitamin C [211], green tea leaf extract [212], and curcumin have all been shown to reduce the toxic effects of Al.

Organophosphates (OPs) and carbamates (CMs) are the most commonly used pesticides in agriculture and have been shown to induce reproductive toxicity, developmental toxicity, and endocrine disruption [192]. OPs are irreversible acetylcholinesterase (AChE) inhibitors and operate similarly to nerve gases, such as sarin. AChE inhibitors have also been shown to disrupt cell replication and differentiation. OPs and CMs are also potent endocrine disruptors and were found to disrupt normal sexual differentiation, ovarian function, sperm production, and pregnancy [192]. OPs and CMs are common residues on a variety of foods, including peaches, apples, nectarines, popcorn, pears, corn, and grapes [213]. Not surprisingly, an organic diet was found to significantly lower dietary exposure to OPs [214]. Quercetin has been shown to protect against multi-OP pesticides [215]. Vitamins C and E [216] and lycopene [217] also have protective effects.

Polychlorinated aromatic hydrocarbons (PAHs) are a large group of chemicals that are byproducts of combustion. Typical sources include combustion of coal, wood, and biofuels for cooking, cigarette smoke, and vehicle emissions. Food preparation is a common source as well, with PAHs resulting from a variety of food processing techniques, including curing, drying, smoking, roasting, grilling, barbecuing, and refining [218]. PAHs have a broad range of anti-fertility actions. For example, PAH exposure was found to destroy oocytes [219] and damage sperm DNA [220]. It also has toxic effects on embryonic development [192]. Low-dose repeated exposures were found to be substantially more toxic than a single high-dose exposure [219], thus emphasizing the importance of limiting dietary and environmental exposures. PAHs are also xenosteroids and can significantly influence a wide variety of serum hormone levels, including estrogen, progesterone, androgens, PRL, and LH [192]. Ellagic acid, epigallocathechingallate, chlorophyllin, and benzyl isothiocyanate were found to ameliorate toxicity from PAHs, though the protective effect depended on the type of PAH [221]. Vitamin C had the paradoxical effect of increasing PAH toxicity by interfering with superoxide dismutase in one study [222] and, along with vitamin E, demonstrated antigenotoxic effects in another [223].

There are many other common chemicals that negatively impact fertility. Polychlorinated biphenyls (PCBs) and dioxins are highly toxic persistent organic pollutants (POPs). Dioxins are the primary toxic constituent of Agent Orange, for example. They are lipophilic byproducts of combustion (wood, coal, petroleum, etc.) and the manufacture of chlorine products, such as bleach and paper manufacturing [224]. Over 90% of human exposure is through food [225], with approximately 41% of exposure from dairy products, 38% from beef ingestion, and an overall load of 98% from animal products [226]. Dioxins can be found in much smaller amounts in hygiene products such as toilet paper and diapers [227], though the contribution to disease at such low concentrations has not been confirmed. Dioxins can lead to ovarian dysfunction and have been implicated in the pathophysiology of endometriosis [228]. Transformation of the aryl hydrocarbon receptor (AhR) is the initial step of dioxin toxicity [229]. Catechins, flavonols, and flavones are natural antagonists of AhR. Chlorophylls, lutein, quercetin, carotenoids, and green tea catechins also inhibit AhR transformation. Among foods, AhR transformation inhibition was found with spinach, citrus fruits (particularly lime, grapefruit, and lemon), sage, rosemary, oregano, peppermint, clove, olive oil [230], coffee [231], and cacao [232]. However, a number of these components had biphasic activity: acting as an AhR antagonist at low doses, while high doses had agonist (and therefore potentially toxic) effects [231–232]. The biphasic activity of many substances serves as a reminder to utilize caution with aggressive dosing.

PCBs are similarly lipophilic and environmentally persistent and travel up the food chain. They are reproductive toxins that can induce a reduction in the number of ovarian follicles and may induce premature ovarian failure [233]. Semen abnormalities have also been induced by PCBs [233]. A study on mink found PCBs from fish ingestion caused complete, though reversible, reproductive failure [234]. Placental transfer also occurs with PCBs but not readily. Lactation, however, can be a significant source of PCB exposure for the neonate [235]. PCBs were widely used in electronic equipment manufacturing and, prior to their use being banned in 1979, were ubiquitous. Fortunately, environmental levels have begun to decline in recent years; however, seafood, meat, and dairy products are still significant sources of exposure. Vitamins E and C [236] and quercetin [237] were found to be protective. EGCG was found to inhibit PCBs xenoestrogenic effects as well [238]. However, as with dioxins, many of the substances that ameliorated negative effects from PCBs had biphasic effects and could inhibit or promote toxicity depending on circumstances [239].

Phthalates (PHTs) are another group of lipophilic and ubiquitous compounds used to make plastics flexible and as lubricants. Unlike PCBs, PHT is currently used everywhere. Some examples include PVC pipes, paints, children's toys, sex toys, food containers, sandwich bags, infant care products, electronics, and cosmetics. While the primary route of exposure is through food, considerably higher levels of exposure come from hospital environments [240]. For example, large amounts of PHT are leached from intravenous (IV) tubing [241] and PVC bags housing IV solution [242]. It is also added to the enteric coating in pharmaceutical pills [243]. From a reproductive standpoint, PHTs easily cross the placenta and act as anti-androgens, reducing testosterone and negatively impacting male reproductive development. In females, they also act as endocrine disruptors, inducing PCOS [244], endometriosis, adenomyosis, and fibroids [245], interfering with steroidogenesis, and inhibiting ovulation by suppressing the LH surge. Interestingly, they were found to act as antagonists to the G protein-coupled cannabinoid 1 (CB1) receptor [246]. Endocannabinoids are important in the regulation of successful reproduction. Plasma levels of anandamide (AEA, an endocannabinoid ligand) fluctuate throughout the menstrual cycle and pregnancy. Successful implantation and pregnancy seem to require lower levels of AEA, whereas higher levels are found during the follicular phase, and a surge of AEA occurs during active labor. Not surprisingly, the use of exocannabinoids (such as marijuana) has been found to increase the risk of adverse pregnancy outcomes [247]. CB1 receptors have been found in the brain tissue of fetuses and are important contributors to neuronal diversity during early CNS development [248]. Oxidative stress was found to play a critical role in how PHT induces neurotoxic effects [249]. Selenium was found to protect the male reproductive system from PHT damage [250], as was vitamin E [251]. Paradoxically, administration of vitamin C or resveratrol increased PHT-induced oxidative damage [252]. However, if vitamin C was combined with vitamin E, it significantly inhibited testicular injury [253]. Finally, PHTs (and bisphenol A) were found to disrupt circulating vitamin D levels in pregnancy [254]. As adequate vitamin D is essential for proper reproduction, pregnancy, and fetal development, levels should be monitored closely. Significant protective effects of vitamin D during pregnancy were found where serum 25(OH)D exceeded 30 ng/ml [255].

In addition to material-based functional interference, we are also exposed to large amounts of electromagnetic radiation (EMR) with potentially significant implications for reproduction. The electromagnetic spectrum can roughly be divided into ionizing and nonionizing radiation at approximately 2417 THz/124 nm. This falls in the upper ultraviolet spectrum (UVC). UVB, the wavelengths responsible for tanning and necessary for vitamin D production (280–320 nm) [256], are too long for ionization, yet UVB radiation can lead to DNA damage, leading to apoptotic keratinocytes and increasing cancer risk [257].

While the reproductive toxicity of ionizing radiation is well described [258–259], the growing body of literature demonstrating toxic effects of nonionizing radiation has not received widespread acknowledgment. Nonionizing radiation (NIER) is a ubiquitous, potentially deleterious modern environmental influence that may greatly impact both female and male reproduction and overall health. A review of the scientific literature regarding immunological effects of NIER stated: "it must be concluded that the existing public safety limits are inadequate to protect public health, and that new public safety limits, as well as limits on further deployment of untested technologies, are warranted" [260]. NIER is absorbed by living tissue [261], leading to a variety of adverse effects including reproductive, immune, neurological, psychiatric, and oncologic effects [260, 262–265].

From a reproductive standpoint, NIER has been found to cause a variety of disturbances including increased miscarriage rates [262], preterm births [266], fetal growth abnormalities [267], neurodevelopmental issues [268], disturbances in semen parameters [269], and hormonal disruption [269]. The most common sources of exposure are from cell towers, power lines, transformers, power meters, mobile phones and watches, wireless routers, cordless telephones, and microwave ovens [270]. Common mobile phone usage involves leaving phones in pockets, on belt clips, in bras, and resting them on one's abdomen while pregnant. The unique location of the testes and temperature sensitivity makes them more susceptible to NIER penetration and damage. In males, NIER has been shown to reduce seminiferous tubule diameters [271–272] and tunica albuginea thickness [273]. It has also been found to induce changes consistent with oxidative damage, including reductions in CAT and SOD activity [272], lipid peroxidation, and DNA damage [273]. NIER can also decrease spermatozoa formation [271], impair motility [274], create sperm aggregation [275], and decrease fertilization potential [276].

In females, NIER induces oxidative stress and increases apoptosis in endometrial tissue [277–278] with significant reductions in ovarian primordial follicles [279–280] and increases in follicular atresia, apoptosis, and stromal fibrosis [280].

In pregnancy, exposure to 900 MHz frequencies (common in cordless and mobile phones) during pregnancy in rats resulted in male offspring with a higher apoptotic index, greater DNA oxidation levels, lower sperm motility and vitality, and altered seminiferous tubules [281]. Female offspring with in utero exposure to NIER had a decreased number of ovarian follicles [282], broken oocyte nests, binucleate primordial follicles, loose stromal cells, cytoplasmic vacuolization, greater follicular atresia, and chromatin condensation [283].

Some natural substances have been examined for their ability to counteract NIER-related damage. *Cordyceps*, a type of fungus rich in selenium, was shown to reduce testicular damage by increasing sperm formation and decreasing apoptotic spermatogenic cells [271]. Melatonin was found to prevent DNA damage in male germ cells [284], and vitamins C and E protected the testes from oxidative damage with increased total glutathione and glutathione peroxidase levels and reduced lipid peroxidation values [285]. It should also be noted that, like many nutrients, NIER may have a biphasic activity that is dose-dependent. Some studies have found NIER improved reproductive outcomes. For instance, rats exposed to NIER at 900 MHz had increased sperm motility and testosterone levels, improved sperm morphology, and lowered tail abnormalities, though it also had the result of increasing precocious puberty [286]. Lastly, of particular importance is a review of the mobile phone safety literature. Panagopoulos et al. [287] observed that 50% of the studies using simulated mobile signals found no effects, in sharp contrast to studies using real exposures to mobile signals which resulted in an almost 100% effect rate. The reader should note it is inappropriate to use simulated exposures for safety data when sources of real-world exposures are available.

Glyphosate (GLY) is a now ubiquitous, broad-spectrum herbicide brought to the consumer market in Roundup. Exposure has been linked to a wide range of health problems including reproductive problems, autism, Parkinson's disease, Alzheimer's disease, and cancer [288]. In males, it has been found to impair sperm motility [289–290] and concentration [290–291], cause testicular lesions [290], and impair mitochondrial function [289]. When combined with soy, GLY was associated with a decrease in testosterone levels, degeneration of Sertoli and Leydig cells, and a decrease in seminiferous tubule diameter [292]. In females, GLY resulted in changes to granulosa cells [293], abnormal uterine development, and increased risk for neoplasias [294]. GLY reduced fecundity by inducing endometrial morphological abnormalities and decreasing the expression of estrogen and progesterone receptors [295]. It also disrupted steroidogenic acute regulatory (StAR) protein expression [296]. Offspring had higher levels of oxidative damage, lipid peroxidation, lower levels of glutathione [297], and skeletal alterations affecting the cephalic and neural crests [298–299].

Perinatal exposure to GLY negatively impacted the set point of the hypothalamus-pituitary-thyroid axis [300]. Neuropsychiatric effects were commonly found with predispositions to attention-deficit hyperactivity disorder [301], anxiety [302], depression, reduced locomotor activity [302], and glutamate excitotoxicity in the hippocampus persisting throughout adulthood [303–304].

Finally, a curious study found GLY-induced dysbiosis, characterized by reduced *Firmicutes* and increased *Bacteroides* ratio in female rat gut microbiota, but not in males [305]. *Bacteroides* can be commensal, mutualist, or pathobiont [306]. Increased levels can lead to reproductive pathogenic states, such as antibiotic-resistant bacterial vaginosis [307]. Changes in vaginal microflora by pathogenic organisms can lead to infertility [308]. Women with asymptomatic vaginosis and women with bacterial vaginosis both had higher incidences of infertility [308]. Low levels of microflora, *Firmicutes*, may contribute to infertility. In a study comparing normal vaginal flora to women with infertility, it was found that healthy women's flora contained on average 27.8% *Firmicutes*. Infertile women averaged only 3.5%. Therefore, GLY may increase risk of infertility through modulation of the female microbiome. In addition, Group B Strep, Gram variable coccobacilli, and *Trichomonas vaginalis* were found in infertile women but not in healthy women. Furthermore, 25% of a healthy woman's flora consisted of diphtheroids and *Micrococcus*, whereas infertile women had none of these organisms. While *E. coli* and *Enterococcus* spp. were found in both groups, there were two to four times as many in infertile women.

GLY has multiple targets of interference, including oxidative damage [297]; excitotoxicity [304]; microbiome disruption [305]; inhibition of aromatase [309]; depletion of iron, cobalt, molybdenum, copper, tryptophan, tyrosine, methionine, and selenomethionine [310]; disruption of the intestinal barrier [311]; and breakdown of tight junctions [312]. There is a paucity of studies on mitigation of GLY toxicity. Lignite extract was recently shown to inhibit GLY-induced intestinal permeability [312]; however, the study was conducted by a nutraceutical company on its own product and should be independently verified. Despite the lack of data, evaluating patients on the above targets and addressing individual shortcomings may prove clinically useful in GLY mitigation.

While presented here, it should be noted that the evidence for reproductive toxicity from GLY, and the toxicity of GLY and other chemicals in general, is the subject of intense debate and controversy [313]. The topic has also become a wider political debate involving the safety of genetically modified organisms (GMOs) as a whole. There are many studies establishing the safety of GLY and GMOs [314]. However, the weight and reliability of evidence should be approached with caution [315]. With GMOs and GLY having very significant public health implications and concerns, the precautionary principle should be evoked [316].

32.4.6 Reproductive Immunology

"One of the most remarkable aspects of reproductive biology is the simple fact that a healthy woman with a fully functional immune system can successfully carry a semiallogeneic pregnancy to full term without immune rejection" [317]. Indeed, tolerance of the rapid growth of an invasive mass, from implantation through fetal development, has been described as "pseudomalignant" or even a "physiological metastasis" [318]. A number of immune cells, such as natural killer cells (NKs), regulatory T cells (Tregs), effector T cells (Teffs), and dendritic cells (DCs) populate the maternal-fetal interface [318]. For many years, it was believed that recurrent pregnancy loss often resulted from an overactive maternal immune system. However, results obtained by direct suppression of the maternal immune system were disappointing [317]. In contrast, studies examining the role of Tregs, and particularly peripheral Tregs (pTregs), have suggested that pTregs play a major role in governing fetal immune tolerance [317] and are believed to be the primary suppressors of the immune response [319]. The ability of Tregs to effect immune tolerance is dependent on the maintenance of low levels of L-tryptophan (TRP) locally [320]. This is accomplished through the metabolism of TRP by the iron-dependent enzyme indoleamine 2,-3-dioxygenase (IDO) expressed by DCs. Perhaps paradoxically, free iron prevents TRP degradation [321], potentially resulting in reduced fetal tolerance.

The expression of IDO is enhanced by PGE2 [322], a downstream metabolite of arachidonic acid (AA) requiring cyclooxygenase 2 (COX-2) for conversion. Inhibition of COX-2 has been found to constrain IDO activity [323]. Therefore COX-2 inhibitors, including NSAIDs such as aspirin (which irreversibly binds to COX-2 [324]), should be used with caution in pregnant women or those trying to conceive. Aspirin is prescribed to many women undergoing in vitro fertilization (IVF) under the hypothesis that it may improve blood flow to the implantation site; however, studies have not found a clear benefit [325] and paradoxically it may prevent successful pregnancy. COX-2 induction was found to be crucial in implantation [326]. In mice, for example, deletion of the gene encoding for COX-2 (Ptgs2) resulted in complete female infertility [327], as did the removal of the PGE2 receptor (EP2) [328].

The facilitation of Treg activity has multiple targets. Some include increasing dietary AA (e.g., chicken, eggs, beef, turkey, and pork [329]), facilitating AA's conversion to PGE2 (via COX-2), inducing COX-2 directly, or minimizing the use of COX-2 inhibitors. One can also directly facilitate Treg activity, for example, through the use of short-chain fatty acids (such as butyrate), although medium-chain fatty acids (such as coconut oil) and long-chain fatty acids (such as DHA) inhibit Treg activity [330–331]. It is noteworthy that plants and animals contain multiple chemical compounds often having ligands for multiple targets that can produce seemingly contradictory results. For example, butyrate can inhibit COX-2 [332], suggesting that it may downregulate Treg production, yet it was shown to increase pTregs specifically [333]. Seafood can be a significant source of AA, yet it shuttles conversion of AA to PGE3 instead of PGE2 [334], making EPA and DHA supplementation potentially counterproductive in cases of immunological infertility.

A number of other substances can modulate Treg activity. For example, vitamins A (as retinoic acid) and D were found to increase Tregs, as was gluten [335]. In general, where increased Tregs are found, it is often unclear whether the substance is increasing Tregs directly or whether the Tregs are increasing as a response to control an inflammatory immune response induced by the substance. In the case of gluten, this may be a worthwhile evaluation. Those on a gluten-free diet were found to have significantly decreased percentages of circulating Tregs [335], suggesting that arbitrarily restricting gluten may turn out to be counterproductive to fertility.

Another important substance in optimizing Treg function is folate. Researchers found that Tregs express high levels of folate receptor 4 and folate depletion via methotrexate (MTX) resulted in significantly reduced numbers of Tregs [432]. As MTX inhibits dihydrofolate reductase (DHFR), those with genetic polymorphisms in the DHFR and/or MTHFR genes may also experience poor Treg survival. Therefore, genetic and functional evaluations of folate status can be important factors in treating infertility.

Commensal microbes have also been found to facilitate Treg activity. Probiotic mixtures such as *Lactobacillus acidophilus* and *Bifidobacterium longum* were found to increase immune-modulating activities of Tregs by increasing expression of FoxP3, an important component of immune tolerance. Other organisms were also found to increase FoxP3 expression, including *Bifidobacterium infantis 35624*, *Lactobacillus rhamnosus* GG (ATCC 53103) and *Bifidobacterium lactis* (Bb-12), *Lactobacillus salivarius* and *Lactobacillus reuteri* (ATCC 23272), a probiotic mixture called VSL#3, and *Helicobacter pylori* [335].

Not all organisms that facilitate Treg activity are beneficial for fertility. The immune interaction with microbes is complex and the same microbe can be pathogenic in some instances and beneficial in others. *H. pylori*, for example, may upregulate Treg efficacy to prevent its own destruction. As a consequence, it may have a protective role in some autoimmune diseases [336] and in reducing stroke risk [337], while increasing susceptibility to gastric cancer [338] and infertility [19]. In the case of infertility, anti-*H. pylori* antibodies have been found in the cervical mucus and there is evidence that cross-reaction with sperm may occur inhibiting sperm progression and thus fertility [339].

The reproductive microbiota play an exceedingly important role in reproduction. Endometrial microbiota consisting of less than 90% *Lactobacillus* species was found to significantly decrease implantation [60.7% vs 23.1%], pregnancy [70.6% vs 33.3%], ongoing pregnancy [58.8% vs 13.3%], and live birth [58.8% vs 6.7%] rates [340]. A vaginal microbiome composed solely of *Lactobacillus* seems to produce the highest chances of a successful IVF outcome [341]. In contrast, the isolation of *Staphylococcus* species or *Enterobacteriaceae* from the cervix or vagina resulted in substantially lower pregnancy rates [342]. Likewise, a preponderance of *Clostridium perfringens, Bacteroides* sp., *Staphylococcus aureus*, and *S. epidermis* were associated with fewer pregnancies [341].

Chlamydia trachomatis (CT) infection is the most prevalent sexually transmitted disease in the world [343], affecting 1.6 million people in the United States annually [430]. Infertility is estimated to result in 3% of infected women and 2% will have adverse pregnancy outcomes [345]. It can also cause male sterility [343], though the percentage of the population affected is unclear due to poor testing methodology. Non-pharmaceutical interventions show significant potential in treating CT infections and can be used alone or to potentiate standard antibiotic therapies [344]. Polyphenols, such as flavones (apigenin, luteolin, acacetin), flavonols (quercetin, myricetin, morin, rhamnetin), natural coumarins (methoxypsoralen), and gallates, are all highly active chlamydia inhibitors [344]. Lipid compounds such as capric acid, monocaprin, and lauric acid caused a greater than 10,000-fold reduction in the infectivity titer with monocaprin having a greater than 100,000-fold inactivation rate [344]. Peptides including melittin, isolated from bee venom, and cecropin, isolated from cecropia moths, also exhibit anti-CT properties. Vaginal lactobacilli such as *L. vaginalis, L. brevis, L. gasseri*, and *L. crispatus* had strong inhibitory effects on CT as well, with *L. crispatus* having the highest efficacy [344].

Many other infections can lead to infertility. Indeed, 15% of male factor infertility is estimated to be due to infectious agents [19]. Some of these agents include protozoa such as *Toxoplasma gondii*, which can cause endometritis, impaired folliculogenesis, ovarian and uterine atrophy, amenorrhea, and reduced semen quality [431]. Mycoplasma species, particularly *M. hominis* and *M. genitalium*, *Ureaplasma urealyticum* [346], *Neisseria gonorrhoeae*, *Gardnerella vaginalis*, *Trichomonas vaginalis* [347], *Candida albicans*, *Listeria monocytogenes* [348], and others, have also been implicated in impairing fertility and have been associated with increased miscarriage rates [349]. Other common organisms that were associated with increased miscarriage rates include human herpesvirus, *Cytomegalovirus* [350], *Chlamydia trachomatis* [351], *Mycoplasma hominis* [346], and *Peptostreptococcus* spp. [352].

While microbial diversity in the gut is believed to confer health benefits [353], bacterial diversity in the vagina was found to be associated with vaginal inflammation and reduced implantation [354]. Normal vaginal microbiota is dominated by *Lactobacillus* species, especially *L. crispatus*, *L. iners*, *L. gasseri*, and *L. jensenii* [355]. The proportions of various species differ among various ethnic groups and also change during pregnancy, with an increase in *Lactobacillus* species and a decrease in anaerobic species as pregnancy progresses [356].

Abnormal vaginal microbiota (AVM) composition, such as in bacterial vaginosis (BV), aerobic vaginitis (AV), or abnormal vaginal flora (AVF), has been implicated in a variety of adverse fertility and pregnancy outcomes, including increased risk of preterm birth. While antibiotics can eradicate BV in pregnancy, the increased risk of preterm birth is not altered [357]. The destruction of beneficial bacteria may also have adverse consequences on the health of the baby, and even a short course of antibiotics can have lasting adverse consequences on the gut microbiome [358].

A review found that prophylactic antibiotics during the second and third trimesters, though reducing infection, do not reduce adverse pregnancy outcomes and morbidity. Instead, they result in short- and longer-term harm for children of mothers exposed to antibiotics, including neonatal death [359–360]. In contrast, prophylaxis and treatment of AVM with *Lactobacillus* species (*Lactobacillus rhamnosus* HN001 and *Lactobacillus acidophilus* GLA-14) were found to effectively inhibit both AV and BV with a 100% inhibition of all studied pathogens (*S. aureus*, *E. coli*, *G. vaginalis*, and *A. vaginae*) after 48 hours [361]. Two studies suggested that the use of oral fermented milk or vaginally administered yogurt reduced the risk of preterm birth [362]. In contrast to antibiotics, administration of probiotics during pregnancy has been associated with increased resistance to respiratory infections, reduced IgE-associated disease burden, and reduced incidence of atopic eczema in children. In mothers, a reduction in blood glucose and increase in glucose tolerance have been observed [362].

Other strategies to reduce microbial and immunological burden include aversions during pregnancy. Indeed, up to 80% of women experience some form of food aversion, nausea, and/or vomiting in the first trimester [363–364], with nausea and vomiting being associated with reduced risk of pregnancy loss [365]. Meat (which includes poultry [366] and fish), for example, is a common aversion, as are coffee, sour foods, beans, and rice [364]. As discussed previously, meats are high sources of TRP, which needs to be kept at lower levels, at least locally to maintain immune tolerance of the fetus. TRP is also necessary for the replication of many intracellular microorganisms such as *Chlamydia trachomatis* [321], and an aversion to sources of high TRP may be essential for suppressing the replication of such organisms. Beans are high in iron, which, as discussed previously, may impair TRP breakdown. Sour fruits, such as citrus, often contain COX-2 inhibitors [367–369] and rice may contain significant levels of arsenic, which has been associated with spontaneous pregnancy loss [370]. Thus, food aversions, nausea, and vomiting in early pregnancy may play an important physiological defense mechanism for maintaining a healthy pregnancy. Therapies to suppress these reactions, whether pharmaceutical or natural, should be approached with caution.

While the interface between immunology and infertility is vast, when dealing with fertility challenges, immunology must be carefully considered. Testing for the presence of pathogenic microorganisms and addressing them with antimicrobials may be insufficient or even counterproductive. While antimicrobial therapy may be necessary in some instances, consideration for the totality of the microbial terrain and the body as a whole is essential for optimal reproductive outcomes.

32.4.7 Exercise

Immune modulation can also come from exercise. Low-intensity exercise reduces cortisol levels, whereas moderate- to high-intensity exercise increases them [371]. Indeed, athletic women are much more likely to experience reproductive cycle dysfunction [372]. A dose-response relationship in nonobese women between vigorous physical activity (of 5 hours or more per week) and delayed time to conceive was found [373]. On the other hand, for overweight or obese women with infertility, a combination of reduced calorie diet and aerobic exercise significantly improved pregnancy rates [374]. Interestingly, the improved chances of pregnancy in these women held true for natural pregnancies only. For IVF, diet and exercise did not seem to change pregnancy chances significantly [357]. During pregnancy, the amount of exercise was proportional to the risk of miscarriage, with hazard ratios (HR) increasing, beginning with 75 minutes of exercise per week (HR 1.8) and increasing to an HR above 3 for 420+ minutes per week through week 18 of pregnancy [375]. After week 18, those exercising had reduced risk of miscarriage. It is worth noting that even those engaging in low-impact exercises (dancing, walking, hiking, aerobic) had twice the risk of miscarriage. Those engaged in high-impact exercise (jogging, ball games, and racket sports) had 3.6x the risk of miscarriage. Swimming was the one sport that did not seem to increase miscarriages, though swimming is often a source of chlorine exposure which was found to increase risk of birth defects [376]. For men, vigorous exercise was associated with reduced fertility rates, particularly in endurance athletes [372]; however, moderate exercise does not seem to have any negative impact.

32.5 Examining the Informational Layer

Information is what gives meaning. Information is not only messages themselves, but also how messages are interpreted and processed. There is significant overlap between the informational, functional, and physical layers because information is what drives everything. Any physical structure requires information to be built. Any function requires information to instruct it. Therefore, information is central to all physiological processes. It has been assumed that the brain is the central repository of information and thinking; however, recent advances in quantum effects are questioning this assumption [377–378]. Whether the mind is local or nonlocal to the brain, thoughts and attitude can significantly influence reproduction. For example, a woman's level of positive expectation related to motherhood and a man's quality of integration between the wish for a child and sexual relationships were found to be significant predictors of a couple's fertility status [379]. The effects of psychological stress on fertility outcomes have been studied during IVF cycles, with lower perceived stress at the time of embryo transfer being associated with improved pregnancy rates [380]. Women with higher anxiety and depression scores on the day of pregnancy detection were less likely to be pregnant [6], whereas lower levels of norepinephrine and cortisol at the time of oocyte retrieval as well as lower levels of cortisol at the time of pregnancy test were associated with favorable pregnancy outcomes [6]. These findings suggest that therapy and stress reduction techniques may provide important benefits to infertile couples. Indeed, a meta-analysis found psychological interventions improved pregnancy rates, though they ironically did not improve mental health measures [381].

32.5.1 Genetics

Genes are a structural source of information that can be either a direct or indirect source of infertility. Direct genetic issues generally fall into two major categories: inborn and acquired. Inborn faults span a wide variety of genetic syndromes, including fragile X syndrome, Kartagener's syndrome, myotonic dystrophy, Noonan syndrome, Klinefelter syndrome, 47,XYY syndrome, Y chromosome microdeletions, and polymorphisms in many genes, including DAX1, CBX2, NR5A1, WNT4, GNRH1, DAX1, KISS1, MTHFR [382–383], and so on. Some of these are directly responsible for infertility and some indirect. An example of a direct genetic cause of infertility is Klinefelter syndrome (KS). KS is a relatively common (1 in 500 men) condition characterized by additional X chromosomes. This leads to azoospermia (absence of sperm in semen). As a consequence, female fertilization is not possible without assisted reproductive technologies, specifically IVF-ICSI.

32.5.2 Innate Genetic Issues

An example of how a common genetic issue can have a wide range of influences indirectly leading to infertility can be found with a gene involved in folate metabolism. Folate is an essential nutrient for both fertility and pregnancy. Women with the highest intake of folic acid reduced the chances of anovulatory infertility by 59% [384]. While fortification of foods with folic acid is controversial due to potential associations with cancer and the masking of vitamin B12 deficiency [385], in terms of pregnancy, folic acid fortification has been called one of the "most significant public health measures for the prevention of pregnancy-related disorders" [386]. Significant sources of folates include edamame, arrowroot, chickpeas, peanuts, lentils, wheat germ, sunflower seeds, spinach, avocado, broccoli, artichokes, and asparagus.

One of folate's primary functions is to help run the methionine cycle that generates S-adenosyl-L-methionine (SAM): the primary methyl donor for most biological methylation reactions [387]. The rate-limiting step in the methyl cycle is controlled by the enzyme 5-methylenetetrahydrofolate reductase (MTHFR), which converts 5, 10-methylenetetrahydrofolate into 5-methyltetrahydrofolate (5MTHF). A gene of the same name codes for the MTHFR enzyme. There are a number of polymorphisms that occur in this genetic pathway, the most researched of which is the C665T transition (previously known as C677T). Women with a C665T mutation had reduced numbers of oocytes, lower serum estradiol concentrations at ovulation [384], higher risk of recurrent implantation failure [388], higher risk of neural tube defects [389], and a 6.3-fold increase in early miscarriage risk [390]. Paternal MTHFR mutation is also a risk factor for recurrent pregnancy loss [391]. Homozygotes have a nine-fold increase in the odds of idiopathic male factor infertility [392].

MTHFR polymorphisms are extremely common with around 50% of whites and Hispanics and 12% of African Americans having one or more mutations. Consequently, MTHFR-related methylation issues may be encountered frequently in a clinical setting. Methylation is essential to life as it is one of the primary tools of epigenetic control of our genome. It enables the body to silence a gene without modifying the structural composition. Methylation is also necessary for immune modulation, histamine metabolism, energy production, cellular repair, detoxification, and a number of other processes. Polymorphisms in the MTHFR gene can reduce the efficiency of methylation, causing a number of both direct and indirect consequences. As methylfolate plays an important part in the metabolism of homocysteine to methionine, an MTHFR polymorphism, particularly in the absence of adequate levels of vitamins B2 [393], B6, B12 and folic acid [394], frequently leads to elevated homocysteine (Hcy) levels. From a reproductive standpoint, elevated Hcy has been associated with poor oocyte quality [395], unexplained female sterility [396], fertilization failure [397], recurrent pregnancy loss [398], NMDA excitotoxicity (which can impair early brain development) [399–400], oxidative damage [401], premature ovarian failure [402], and paternal-cause recurrent pregnancy loss [391]. Therefore, it is prudent to test Hcy levels in patients with known or suspected methylation defects.

Correcting an MTHFR deficiency can have far-reaching effects. For example, vitamin B2 was shown to nullify the negative influence of MTHFR polymorphisms on Hcy levels [393], though its impact on other MTHFR related issues remains to be investigated. Currently, the most logical and popular strategy of bypassing an MTHFR deficiency is supplementation with 5MTHF, the metabolic end product of the MTHFR enzyme. While studies have indicated 5MTHF supplementation has theoretical benefits over folic acid supplementation [403], clinicians frequently prescribing 5MTHF have described severe negative reactions that have yet to appear in the medical literature [404]. While the precise mechanism of these reactions is unclear, there are reasons to believe it may involve methyl trapping [403].

Conversion of homocysteine to methionine is facilitated by methionine synthase (MS). MS requires methylated vitamin B12 in order to convert Hcy to methionine. Under normal circumstances, vitamin B12 is methylated by 5MTHF. 5MTHF then loses its methyl group, becomes THF, and is further metabolized into other non-methylated folates required for other physiological processes including proper DNA and RNA synthesis. If all the cobalamin molecules are pre-methylated (e.g., when supplementing with methylcobalamin, a popular form of vitamin B12), 5MTHF will not be able to give its methyl group to cobalamin and consequently will be unable to form the other folates necessary for proper physiology. One strategy to circumvent this problem is to avoid combining 5MTHF and methylcobalamin. Instead, supplementation with the most common natural forms of cobalamin, adenosylcobalamin or hydroxocobalamin [405], may be considered, as these are unmethylated and will still require methyl groups from 5MTHF to run the methionine cycle. If methylcobalamin is required, then supplementation with the so-called "synthetic" folate (folic acid) will provide the other folates to varying degrees [403] and potentially avoid some of the reactions to 5MTHF.

32.5.3 Neural Tube Defects and Folate Supplementation

Appropriate access to folate is critical for proper fetal development. One of the great successes of targeted nutritional intervention has been in the reduction of neural tube defects (NTDs) via folic acid supplementation. NTDs are a group of birth defects related to the brain, spine, or spinal cord, with spina bifida and anencephaly being the most common presentations. Suspicions of the NTD-folate relationship began in 1964 when Hibbard demonstrated an association between folate deficiency and NTDs [406]. Red cell folate (RCF) was subsequently found to be an independent risk factor for NTDs [407]. While folic acid has extensive data conclusively demonstrating its efficacy in significantly reducing the incidence of NTDs [406], other forms of folate may also be effective.

The three most widely available forms of folate are folic acid (FA), folinic acid (FLA), and 5MTHF. FA is an oxidized form of folate that is often mischaracterized as an "unnatural" or "synthetic" folate. FA, in fact, has been isolated from spinach, chickpeas, tomatoes, green beans, and cabbages [408]. From the standpoint of reproduction, the main advantage of FA is the volume of research establishing safety and efficacy, particularly in the prevention of neural tube defects (NTDs). Studies have found FA alone reduces the risk of NTDs by up to 70% [409], though this was at a dosage of 4 mg/day. More moderate dosages (800 µg) combined with other vitamins and minerals, including vitamins B12, B6, and B2, reduced the risk by up to 90% [410]. The general consensus of recommended periconceptional daily folate intake ranges from 400 to 800 µg [409–410], with an assumption that an additional 200 µg of folates will come from the diet. For women with a previous history of NTDs, the daily recommendation is 4 mg [409].

People with MTHFR C665T polymorphisms are at an increased risk for NTDs [403], presumably due to a 30–70% decrease in the activity of MTHFR and therefore in the production of 5MTHF. In Europe, depending on the area, this polymorphism affects one third to half of the population [411]. It is therefore a significant public health concern. Consequently, supplemental 5MTHF has been hypothesized to be superior to folic acid and is being recommended as a substitute due to its perceived ability to bypass MTHFR [412]. Unfortunately, the mechanism by which folic acid protects against NTDs is unknown and only 13% of NTDs could be attributed to the MTHFR C665T mutation [413]. Recommendations for substituting FA with 5MTHF are based largely on serum and/or RBC folate levels achieved by 5MTHF administration, rather than from studies on actual NTD incidence. Such assumptions, while theoretically sound, do not account for potentially unique properties of the different folates. For example, FA was found to have free radical scavenging properties [414]. As oxidative damage is a well-established risk factor influencing the incidence of NTDs [415], the unique antioxidant properties of FA may be a significant contributing factor in its efficacy. In such a case, prescribing other folates could paradoxically result in a significant increase in NTD incidence. While future studies may establish 5MTHF to be equivalent or even superior to FA in reducing NTD rates, it is not supported by the current state of evidence.

32.5.4 Risk Factors of Folic Acid

While FA is effective in reducing NTD incidence, its use is not without risk. Unlike 5MTHF, FA can mask symptoms of a vitamin B12 deficiency, especially in dosages above

1 mg [409]. Vitamin B12 deficiencies can result in permanent neurological damage. Consequently, testing serum B12 and methylmalonic acid may be prudent when supplementing with FA, particularly in higher doses. Co-administration of vitamin B12 should also be considered, taking into consideration the potential risk of methyl trapping if methylated B12 is used.

A more recent concern with supplemental FA has surfaced with the discovery of unmetabolized folic acid (UMFA) in serum. This occurs when supplemental FA is used in concentrations above 200 μg [403]. The significance of UMFA is currently unclear. One group, for example, found UMFA levels associated with lower natural killer cell (NK) cytotoxicity [416]. However, this association was only found in a subset of older women (60–75 years old). Interestingly, the researchers also found a decrease in NK cytotoxicity for a subset of subjects with elevated serum levels of 5MTHF. It is unclear at this time what the significance of these findings is. While UMFA may ultimately prove to confer certain risks, the greater issue may be with too much folate. What constitutes "too much" is an individual analysis. Where metabolism of folates is genetically impaired (e.g., in cases of DHFR or MTHFR deficiencies), the use of FLA or 5MTHF has theoretical benefits, but lacks clinical data.

As clinical decisions at times precede evidence, in cases where a clinician has concerns regarding DHFR and/or MTHFR deficiencies and folic acid is not producing desired biochemical changes, one may consider the use of other folates. Dosages of up to 200 μg of 5MTHF are theoretically equivalent to the first 200 μg of FA (which is completely converted to 5MTHF prior to absorption). Where DHFR polymorphisms exist, FLA may be a second option to bypass the need for DHFR (see ◘ Fig. 32.1). However, it should be emphasized that, at the time of this writing, neither 5MTHF nor FLA have been sufficiently established in studies where NTDs are the outcome measure [409]. In mouse models, FA and FLA achieved different effects in preventing NTDs and dosage data on FLA in humans for such purposes is lacking [409]. Regardless, monitoring homocysteine levels, SAM/SAH ratios, and whole blood histamine may be prudent when supplementing MTHFR-deficient patients.

32.6 Acquired Genetic Issues

Acquired genetic issues are typically in the form of DNA damage, which results from primarily endogenous insults [417]. Reactive oxygen species (ROS), aldehydes from LPO, some estrogen metabolites, methylation agents, reactive nitrogen species, and reactive carbonyl species are some of the common pathways to DNA damage, with the most studied being ROS and the resulting oxidative damage [417]. One of the most important endogenous sources of ROS is mitochondrial, via the electron transport chain and nitric oxide synthase reaction. ROS are implicated in a variety of reproductive issues. Oxidative damage to sperm DNA was found to be an important contributory factor in the etiology of male infertility [418] and a likely cause of genetic diseases in offspring [419]. ROS, while important facilitators of oocyte maturation, ovulation, and implantation, correspond to poor embryo quality where elevated levels were detected in follicular fluid [420]. ROS further leads to spontaneous aneuploidy [421] and contributes to cellular aging in general [422]. Where exposure to elevated levels of ROS is likely, such as in chronic inflammatory conditions, the use of antioxidants may be of benefit [423]. Antioxidants, such as N-acetylcysteine; melatonin; vitamins A, C, and E; folic acid; myoinositol; zinc; and selenium, had only weak evidence of improving fertility outcomes [423]. In such studies, it is rare to find an assessment of the patients' antioxidant status prior to treatment. There is often an assumption that antioxidants are "good" and oxidative stress is "bad." As a result, antioxidants may be overemphasized, leading to what has been termed in the scientific literature "antioxidative stress" (AOS) [422]. AOS may interfere with the immune system's ability to fight bacteria and essential defensive mechanisms for removal of damaged cells, including those that are precancerous and cancerous [422]. Therefore, individualized evaluations of the ratio of ROS to antioxidants should be encouraged prior to supplementation.

Another common etiology of acquired genetic issues that can have profound effects on reproduction is epigenetics [424]. In essence, epigenetics is a modification in the function of specific genes that does not modify the DNA sequence [425]. Epigenetic changes can be stimulated by a variety of factors including social interactions, psychological state, diet, seasons, diurnal fluctuations, disease exposure, toxic chemicals, drugs, financial status, exercise, microbiome, therapeutic drugs, and alternative medicine [426].

Epigenetic changes involve histone modification through a variety of mechanisms, the most studied of which is methylation [427]. Such epigenetic modifications are heritable [424]. For example, paternal diet was found to influence insulin secretion, glucose tolerance, and lipid profiles in offspring [425]. There is now increasing evidence suggesting that environmental factors can have negative effects on epigenetic processes controlling implantation, placentation, and fetal growth [428]. Epigenetic factors that negatively impact fertility can arise directly, such as by not having enough access to methyl groups to methylate DNA, or indirectly, such as through modification of environmental factors, both of which have been discussed previously. Addressing these causes requires a careful examination of the totality of a person's life. However, one of the primary controllable influences on our genes and reproductive health is nutrition.

32.7 Summary

In summary, reproduction is perhaps the most important biological mandate presenting unique challenges in modern times. Recognizing and analyzing the interplay of biological

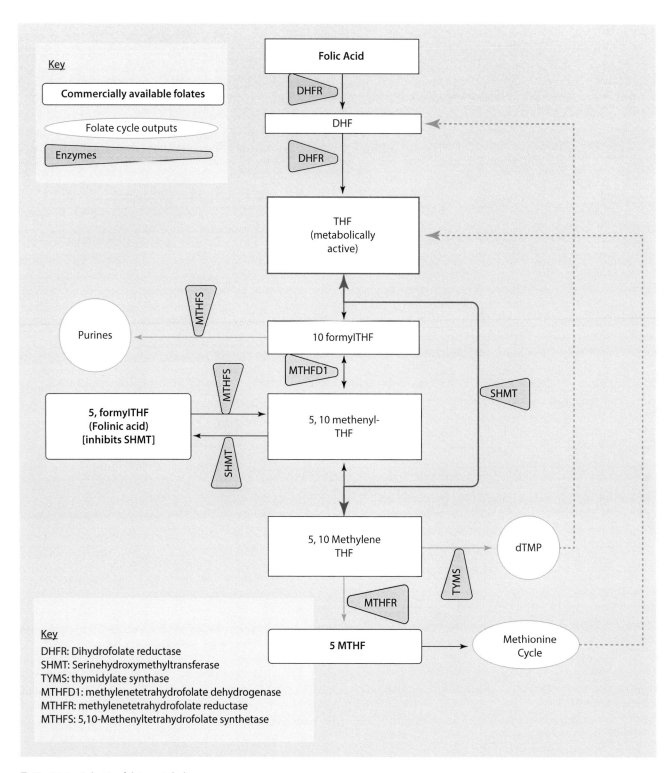

● Fig. 32.1 Selective folate metabolism

systems is essential to the optimization of clinical results. Furthermore, generalized statistical modeling that fails to account for a patient's unique combination of genetic, environmental, social, and psychological factors can contribute to the persistence or iatrogenic aggravation of reproductive conditions. Considering these factors will both improve a patient's overall health and optimize their reproductive efficacy.

References

1. Sender R, Fuchs S, Milo R. Revised estimates for the number of human and bacteria cells in the body. PLoS Biol. 2016;14(8):e1002533.
2. Trifonov EN. Vocabulary of definitions of life suggests a definition. J Biomol Struct Dyn. 2011;29(2):259–66.
3. Walker RW, Clemente JC, Peter I, Loos RJF. The prenatal gut microbiome: are we colonized with bacteria in utero? Pediatr Obes. 2017;12:3–17.

4. Society for Assisted Reproductive Technology. Final CST data for 2105. [Internet]; 2015 [cited 2018 May 8]. Available from: https://www.sartcorsonline.com/rptCSR_PublicMultYear.aspx?reportingYear=2015.
5. Manheimer E, Zhang G, Udoff L, Haramati A, Langenberg P, Berman BM, Bouter LM. Effects of acupuncture on rates of pregnancy and live birth among women undergoing in vitro fertilisation: systematic review and meta-analysis. BMJ. 2008;336(7643):545–9.
6. An Y, Sun Z, Li L, Zhang Y, Ji H. Relationship between psychological stress and reproductive outcome in women undergoing in vitro fertilization treatment: psychological and neurohormonal assessment. J Assist Reprod Genet. 2012;30(1):35–41.
7. Chiu Y-H, Williams PL, Gillman MW, Gaskins AJ, Mínguez-Alarcón L, Souter I, et al. Association between pesticide residue intake from consumption of fruits and vegetables and pregnancy outcomes among women undergoing infertility treatment with assisted reproductive technology. JAMA Intern Med. 2018;178(1):17–26.
8. Santulli P, Marcellin L, Tosti C, Chouzenoux S, Cerles O, Borghese B, et al. MAP kinases and the inflammatory signaling cascade as targets for the treatment of endometriosis? Expert Opin Ther Targets. 2015;19(11):1465–83.
9. Kvaskoff M, Mesrine S, Clavel-Chapelon F, Boutron-Ruault M-C. Endometriosis risk in relation to naevi, freckles and skin sensitivity to sun exposure: the French E3N cohort. Int J Epidemiol. 2009;38(4):1143–53.
10. Viganò P, Somigliana E, Panina P, Rabellotti E, Vercellini P, Candiani M. Principles of phenomics in endometriosis. Hum Reprod Update. 2012;18(3):248–59.
11. Farland LV, Lorrain S, Missmer SA, Dartois L, Cervenka I, Savoye I, Mesrine S, Boutron-Ruault MC, Kvaskoff M. Endometriosis and the risk of skin cancer: a prospective cohort study. Cancer Causes Control. 2017;28(10):1011–9.
12. Laden F, Ishibe N, Hankinson SE, Wolff MS, Gertig DM, Hunter DJ, Kelsey KT. Polychlorinated biphenyls, cytochrome P450 1A1, and breast cancer risk in the Nurses' Health Study. Cancer Epidemiol Biomark Prev. 2002;11(12):1560–5.
13. Singh N, Mitra AK, Garg VK, Agarwal A, Sharma M, Chaturvedi R, et al. Association of CYP1A1 polymorphisms with breast cancer in North Indian women. Oncol Res. 2007;16(12):587–97.
14. Esinler I, Aktas D, Alikasifoglu M, Tuncbilek E, Ayhan A. CYP1A1 gene polymorphism and risk of endometrial hyperplasia and endometrial carcinoma. Int J Gynecol Cancer. 2006;16(3):1407–11.
15. Zhu H, Bao J, Liu S, Chen Q, Shen H. Null genotypes of GSTM1 and GSTT1 and endometriosis risk: a meta-analysis of 25 case-control studies. PLoS One. 2014;9(9):e106761.
16. Huang RS, Duan S, Kistner EO, Zhang W, Bleibel WK, Cox NJ, et al. Identification of genetic variants and gene expression relationships associated with pharmacogenes in humans. Pharmacogenet Genomics. 2008;18(6):545–9.
17. Mu F, Edwards JR, Rimm EB, Spiegelman D, Missmer SA. Endometriosis and risk of coronary heart disease. Circ Cardiovasc Qual Outcomes. 2016;9(3):257–64.
18. Verit FF, Zeyrek FY, Zebitay AG, Akyol H. Cardiovascular risk may be increased in women with unexplained infertility. Clin Exp Reprod Med. 2017;44(1):28.
19. Pellati D, Mylonakis I, Bertoloni G, Fiore C, Andrisani A, Ambrosini G, Armanini D. Genital tract infections and infertility. Eur J Obstet Gynecol Reprod Biol. 2008;140(1):3–11.
20. Kavoussi SK, West BT, Taylor GW, Lebovic DI. Periodontal disease and endometriosis: analysis of the national health and nutrition examination survey. Fertil Steril. 2009;91(2):335–42.
21. Heidarpour M, Derakhshan M, Derakhshan-Horeh M, Kheirollahi M, Dashti S. Prevalence of high-risk human papillomavirus infection in women with ovarian endometriosis. J Obstet Gynaecol Res. 2016;43(1):135–9.
22. Lin W-C, Chang CY-Y, Hsu Y-A, Chiang J-H, Wan L. Increased Risk of endometriosis in patients with lower genital tract infection: a nationwide cohort study. Medicine. 2016;95(10):e2773.
23. Lang GA, Yeaman GR. Autoantibodies in endometriosis sera recognize a Thomsen–Friedenreich-like carbohydrate antigen. J Autoimmun. 2001;16(2):151–61.
24. Kitawaki J, Kado N, Ishihara H, Koshiba H, Kitaoka Y, Honjo H. Endometriosis: the pathophysiology as an estrogen-dependent disease. J Steroid Biochem Mol Biol. 2002;83(1–5):149–55.
25. Resuehr D, Glore DR, Taylor HS, Bruner-Tran KL, Osteen KG. Progesterone-dependent regulation of endometrial cannabinoid receptor type 1 (CB1-R) expression is disrupted in women with endometriosis and in isolated stromal cells exposed to 2,3,7,8-tetrachlorodibenzo-p-dioxin (TCDD). Fertil Steril. 2012;98(4):948.
26. Komorowski J, Stepień H. The role of the endocannabinoid system in the regulation of endocrine function and in the control of energy balance in humans. Postepy Hig Med Dosw (Online). 2007;61:99–105.
27. Anastasi E, Fuggetta E, Vito CD, Migliara G, Viggiani V, Manganaro L, et al. Low levels of 25-OH vitamin D in women with endometriosis and associated pelvic pain. Clin Chem Lab Med (CCLM). 2017;55(12):e282.
28. Yuk J-S, Ji H-Y, Lee J, Park W. Graves' disease is associated with endometriosis: 3-year population-based cross-sectional study. J Med (Baltimore). 2016;95(10):e2975.
29. Hsieh Y-Y, Lee C-C, Chang C-C, Wang Y-K, Yeh L-S, Lin C-S. Angiotensin I-converting enzyme insertion-related genotypes and allele are associated with higher susceptibility of endometriosis and leiomyoma. Mol Reprod Dev. 2007;74(7):808–14.
30. Cuevas M, Flores I, Thompson KJ, Ramos-Ortolaza DL, Torres-Reveron A, Appleyard CB. Stress exacerbates endometriosis manifestations and inflammatory parameters in an animal model. Reprod Sci. 2012;19(8):851–62.
31. Petrelluzzi KFS, Garcia MC, Petta CA, Grassi-Kassisse DM, Spadari-Bratfisch RC. Salivary cortisol concentrations, stress and quality of life in women with endometriosis and chronic pelvic pain. Stress. 2008;11(5):390–7.
32. Tarin JJ, García-Pérez MA, Hamatani T, Cano A. Infertility etiologies are genetically and clinically linked with other diseases in single meta-diseases. Reprod Biol Endocrinol. 2015;13:31.
33. Islam MS, Akhtar MM, Ciavattini A, Giannubilo SR, Protic O, Janjusevic M, et al. Use of dietary phytochemicals to target inflammation, fibrosis, proliferation, and angiogenesis in uterine tissues: promising options for prevention and treatment of uterine fibroids? Mol Nutr Food Res. 2014;58(8):1667–84.
34. Ali M, Chaudhry ZT, Al-Hendy A. Successes and failures of uterine leiomyoma drug discovery. Expert Opin Drug Discovery. 2017;13(2):169–77.
35. Parsanezhad ME, Azmoon M, Alborzi S, Rajaeefard A, Zarei A, Kazerooni T, et al. A randomized, controlled clinical trial comparing the effects of aromatase inhibitor (letrozole) and gonadotropin-releasing hormone agonist (triptorelin) on uterine leiomyoma volume and hormonal status. Fertil Steril. 2010;93(1):192–8.
36. Roshdy E, Rajaratnam V, Maitra S, Sabry M, Allah A, Al-Hendy A. Treatment of symptomatic uterine fibroids with green tea extract: a pilot randomized controlled clinical study. Int J Women's Health. 2013;5:477–86.
37. Kotani Y, Tobiume T, Fujishima R, Shigeta M, Takaya H, Nakai H, et al. Recurrence of uterine myoma after myomectomy: open myomectomy versus laparoscopic myomectomy. J Obstet Gynaecol Res. 2018;44(2):298–302.
38. Baird DD, Dunson DB, Hill MC, Cousins D, Schectman JM. Association of physical activity with development of uterine leiomyoma. Am J Epidemiol. 2007;165(2):157–63.
39. Mcwilliams M, Chennathukuzhi V. Recent advances in uterine fibroid etiology. Semin Reprod Med. 2017;35(02):181–9.
40. Wise LA, Radin RG, Palmer JR, Kumanyika SK, Boggs DA, Rosenberg L. Intake of fruit, vegetables, and carotenoids in relation to risk of uterine leiomyomata. Am J Clin Nutr. 2011;94(6):1620–31.
41. Baird DD, Hill MC, Schectman JM, Hollis BW. Vitamin D and risk of uterine fibroids. Epidemiology. 2013;24(3):447–53.

42. Halder SK, Sharan C, Al-Hendy A. Faculty of 1000 evaluation for 1,25-dihydroxyvitamin D3 treatment shrinks uterine leiomyoma tumors in the Eker rat model. Biol Reprod. 2012;86(4):116.
43. Worp HBVD, Howells DW, Sena ES, Porritt MJ, Rewell S, Ocollins V, et al. Can animal models of disease reliably inform human studies? PLoS Med. 2010;7(3):e1000245.
44. Moore A, Castro L, Yu L, Zheng X, Di X, Sifre M, et al. Stimulatory and inhibitory effects of genistein on human uterine leiomyoma cell proliferation are influenced by the concentration. Hum Reprod. 2007;22(10):2623–31.
45. Zhang HY, Zhu JE, Huang W, Zhu J. Clinicopathologic features of ovarian Sertoli-Leydig cell tumors. Int J Clin Exp Pathol. 2014;7(10):6956–64.
46. McDonnell R, Marjoribanks J, Hart RJ. Ovarian cyst aspiration prior to in vitro fertilization treatment for subfertility. Cochrane Database Syst Rev. 2014;(12):CD005999.
47. Legendre G, Catala L, Morinière C, Lacoeuille C, Boussion F, Sentilhes L, et al. Relationship between ovarian cysts and infertility: what surgery and when? Fertil Steril. 2014;101(3):608–14.
48. Strauss JF, Barbieri RL. Yen and Jaffe's reproductive endocrinology: physiology, pathophysiology, and clinical management. 5th ed. Philadelphia: Saunders/Elsevier; 2004. p. 545.
49. Buțureanu SA, Buțureanu TA. Pathophysiology of adhesions. Chirurgia (Bucur). 2014;109(3):293–8.
50. Martinez-Zamora MA, Creus M, Tassies D, Reverter JC, Civico S, Carmona F, et al. Reduced plasma fibrinolytic potential in patients with recurrent implantation failure after IVF and embryo transfer. Hum Reprod. 2011;26(3):510–6.
51. Chen C, Yang F-Q, Zhang Q, Wang F-Q, Hu Y-J, Xia Z-N. Natural products for antithrombosis. Evid Based Complement Alternat Med. 2015;2015:1–17.
52. Cushman M. Epidemiology and risk factors for venous thrombosis. Semin Hematol. 2007;44(2):62–9.
53. Kalogeropoulos N, Panagiotakos DB, Pitsavos C, Chrysohoou C, Rousinou G, Toutouza M, et al. Unsaturated fatty acids are inversely associated and n-6/n-3 ratios are positively related to inflammation and coagulation markers in plasma of apparently healthy adults. Clin Chim Acta. 2010;411(7–8):584–91.
54. Azadzoi KM, Siroky MB. Neurologic factors in female sexual function and dysfunction. Korean J Urol. 2010;51(7):443–9.
55. Skrzypek M, Wdowiak A, Marzec A. Application of dietetics in reproductive medicine. Ann Agric Environ Med. 2017;24(4):559–65.
56. Miller FJ, Rosenfeldt FL, Zhang C, Linnane AW, Nagley P. Precise determination of mitochondrial DNA copy number in human skeletal and cardiac muscle by a PCR-based assay: lack of change of copy number with age. Nucleic Acids Res. 2003;31(11):e61.
57. Barritt J, Kokot M, Cohen J, Steuerwald N, Brenner C. Quantification of human ooplasmic mitochondria. Reprod Biomed Online. 2002;4(3):243–7.
58. Kloc M, editor. Oocytes – maternal information and functions. Springer International Publishing; 2017. https://www.springer.com/gp/book/9783319608549.
59. Sirmans S, Pate K. Epidemiology, diagnosis, and management of polycystic ovary syndrome. Clin Epidemiol. 2014;6:1–13.
60. Marshall JC, Dunaif A. Should all women with PCOS be treated for insulin resistance? Fertil Steril. 2012;97(1):18–22.
61. Almalki M, Buhary B, Almohareb O, Aljohani N, Alzahrani S, Elkaissi S, et al. Glycemic control and pregnancy outcomes in patients with diabetes in pregnancy: a retrospective study. Indian J Endocrinol Metabol. 2016;20(4):481–90.
62. Tejera A, Herrero J, Santos MDL, Garrido N, Ramsing N, Meseguer M. Oxygen consumption is a quality marker for human oocyte competence conditioned by ovarian stimulation regimens. Fertil Steril. 2011;96(3):618–23.
63. Gade EJ, Thomsen SF, Lindenberg S, Backer V. Fertility outcomes in asthma: a clinical study of 245 women with unexplained infertility. Eur Respir J. 2016;47(4):1144–51.
64. Darwish OA, El-nagar M, Eid EE, El-sherbini AF. Interrelationships between anemia and fertility patterns in rural Egyptian women. Egypt Popul Fam Plann Rev. 1979;13(1–2):1–28.
65. Orsucci D, Filosto M, Siciliano G, Mancuso M. Electron transfer mediators and other metabolites and cofactors in the treatment of mitochondrial dysfunction. Nutr Rev. 2009;67(8):427–38.
66. Bruin JD, Dorland M, Spek E, Posthuma G, Haaften MV, Looman C, et al. Ultrastructure of the resting ovarian follicle pool in healthy young women. Biol Reprod. 2002;66(4):1151–60.
67. Agarwal A, Gupta S, Sharma RK. Role of oxidative stress in female reproduction. Reprod Biol Endocrinol. 2005;3:28.
68. Attaran M, Pasqualotto E, Falcone T, Goldberg JM, Miller KF, Agarwal A, Sharma RK. The effect of follicular fluid reactive oxygen species on the outcome of in vitro fertilization. Int J Fertil Womens Med. 2000;45(5):314–20.
69. Ruder EH, Hartman TJ, Blumberg J, Goldman MB. Oxidative stress and antioxidants: exposure and impact on female fertility. Hum Reprod Update. 2008;14(4):345–57.
70. National Research Council (US) Committee on Neurotoxicology and Models for Assessing Risk. Environmental neurotoxicology. Washington, D.C.: National Academies Press (US); 1992.
71. Cavalla P, Rovei V, Masera S, Vercellino M, Massobrio M, Mutani R, et al. Fertility in patients with multiple sclerosis: current knowledge and future perspectives. Neurol Sci. 2006;27(4):231–9.
72. Kalantaridou SN, Makrigiannakis A, Zoumakis E, Chrousos GP. Stress and the female reproductive system. J Reprod Immunol. 2004;62(1–2):61–8.
73. Sanders KA, Bruce NW. A prospective study of psychosocial stress and fertility in women. Hum Reprod. 1997;12(10):2324–9.
74. Smeenk J, Verhaak C, Vingerhoets A, Sweep C, Merkus J, Willemsen S, et al. Stress and outcome success in IVF: the role of self-reports and endocrine variables. Hum Reprod. 2005;20(4):991–6.
75. Lei H-L, Wei H-J, Ho H-Y, Liao K-W, Chien L-C. Relationship between risk factors for infertility in women and lead, cadmium, and arsenic blood levels: a cross-sectional study from Taiwan. BMC Public Health. 2015;15:1220.
76. Schug TT, Johnson AF, Birnbaum LS, Colborn T, Guillette LJ, Crews DP, et al. Minireview: endocrine disruptors: past lessons and future directions. Mol Endocrinol. 2016;30(8):833–47.
77. Evers AS. Paracrine interactions of thyroid hormones and thyroid stimulation hormone in the female reproductive tract have an impact on female fertility. Front Endocrinol. 2012;3:50.
78. Subirán N, Candenas L, Pinto FM, Cejudo-Roman A, Agirregoitia E, Irazusta J. Autocrine regulation of human sperm motility by the met-enkephalin opioid peptide. Fertil Steril. 2012;98(3):617–25.
79. Barrett ES, Sobolewski M. Polycystic ovary syndrome: do endocrine-disrupting chemicals play a role? Semin Reprod Med. 2014;32(3):166–76.
80. Roney JR, Simmons ZL. Elevated psychological stress predicts reduced estradiol concentrations in young women. Adapt Hum Behav Physiol. 2014;1(1):30–40.
81. Purohit A, Newman SP, Reed MJ. The role of cytokines in regulating estrogen synthesis: implications for the etiology of breast cancer. Breast Cancer Res. 2002;4(2):65–9.
82. Cui Y, Shi J, Wang Z. Complex rotation quantum dynamic neural networks (CRQDNN) using complex quantum neuron (CQN): applications to time series prediction. Neural Netw. 2015;71:11–26.
83. Li J, Wang Q, Liang H-L, Dong H-X, Li Y, et al. Biophysical characteristics of meridians and acupoints: a systematic review. Evid Based Compl Alt Med. 2012;2012:793841.
84. Wang Y, Li Y, Chen R, Cui X, Yu J, Liu Z. Electroacupuncture for reproductive hormone levels in patients with diminished ovarian reserve: a prospective observational study. Acupunct Med. 2016;34(5):386–91.
85. Chen Y, Fang Y, Yang J, Wang F, Wang Y, Yang L. Effect of acupuncture on premature ovarian failure: a pilot study evidence based complementary and alternative medicine. Evid Based Compl Alt Med. 2014;2014:718675.
86. Johansson J, Redman L, Veldhuis PP, Sazonova A, Labrie F, Holm G, et al. Acupuncture for ovulation induction in polycystic ovary syndrome: a randomized controlled trial. Am J Physiol Endocrinol Metabol. 2013;304(9):E934–43.

87. Siterman S, Eltes F, Schechter L, Maimon Y, Lederman H, Bartoov B. Success of acupuncture treatment in patients with initially low sperm output is associated with a decrease in scrotal skin temperature. Asian J Androl. 2009;11(2):200–8.
88. Gurfinkel E, Cedenho AP, Yamamura Y, Srougi M. Effects of acupuncture and moxa treatment in patients with semen abnormalities. Asian J Androl. 2003;5(4):345–8.
89. Sherman S, Eltes F, Wolfson V, Zabludovsky N, Bartoov B. Effect of acupuncture on sperm parameters of males suffering from subfertility related to low sperm quality. Arch Androl. 1997;39(2):155–61.
90. Kumar P, Sait S. Luteinizing hormone and its dilemma in ovulation induction. J Hum Reprod Sci. 2011;4(1):2–7.
91. Evans J. GnRH and oxytocin have nonidentical effects on the cellular LH response by gonadotrophs at pro-oestrus. J Endocrinol. 1999;163(2):345–51.
92. Yoshimura Y, Koyama N, Karube M, Oda T, Akiba M, Yoshinaga A, et al. Gonadotropin stimulates ovarian renin-angiotensin system in the rabbit. J Clin Investig. 1994;93(1):180–7.
93. Nakamura Y, Tamura H, Takayama H, Kato H. Increased endogenous level of melatonin in preovulatory human follicles does not directly influence progesterone production. Fertil Steril. 2003;80(4):1012–6.
94. Carrera MP, Ramírez-Expósito MJ, Valenzuela MT, Dueñas B, García MJ, Mayas MD, et al. Renin-angiotensin system-regulating aminopeptidase activities are modified in the pineal gland of rats with breast cancer induced by N-methyl-nitrosourea. Cancer Investig. 2006;24(2):149–53.
95. Kunz G, Beil D, Deininger H, Wildt L, Leyendecker G. The dynamics of rapid sperm transport through the female genital tract: evidence from vaginal sonography of uterine peristalsis and hysterosalpingoscintigraphy. Hum Reprod. 1996;11(3):627–32.
96. Cattaneo MG, Chini B, Vicentini LM. Oxytocin stimulates migration and invasion in human endothelial cells. Br J Pharmacol. 2008;153(4):728–36.
97. Nepomnaschy PA, Welch K, Mcconnell D, Strassmann BI, England BG. Stress and female reproductive function: a study of daily variations in cortisol, gonadotrophins, and gonadal steroids in a rural Mayan population. Am J Hum Biol. 2004;16(5):523–32.
98. Decleer W, Osmanagaoglu K, Devroey P. The role of oxytocin antagonists in repeated implantation failure. Facts Views Vis Obgyn. 2012;4(4):227–9.
99. Douglas AJ. Baby on board: do responses to stress in the maternal brain mediate adverse pregnancy outcome? Front Neuroendocrinol. 2010;31(3):359–76.
100. Die MV, Burger H, Teede H, Bone K. Vitex agnus-castus extracts for female reproductive disorders: a systematic review of clinical trials. Planta Med. 2012;79(7):562–75.
101. Fernando S, Rombauts L. Melatonin: shedding light on infertility? – A review of the recent literature. J Ovarian Res. 2014;7(1):98.
102. Jang H, Hong K, Choi Y. Melatonin and fertoprotective adjuvants: prevention against premature ovarian failure during chemotherapy. Int J Mol Sci. 2017;18(6):1221.
103. Jia Y, Yang M, Zhu K, Wang L, Song Y, Wang J, et al. Melatonin implantation improved the egg-laying rate and quality in hens past their peak egg-laying age. Sci Rep. 2016;6(1):39799.
104. Reiter RJ, Tan D-X, Manchester LC, Paredes SD, Mayo JC, Sainz RM. Melatonin and reproduction revisited. Biol Reprod. 2009;81(3):445–56.
105. Aoki H, Yamada N, Ozeki Y, Yamane H, Kato N. Minimum light intensity required to suppress nocturnal melatonin concentration in human saliva. Neurosci Lett. 1998;252(2):91–4.
106. Figueiro MG, Wood B, Plitnick B, Rea MS. The impact of light from computer monitors on melatonin levels in college students. Neuro Endocrinol Lett. 2011;32(2):158–63.
107. Lyons RA, Saridogan E, Djahanbakhch O. The reproductive significance of human fallopian tube cilia. Hum Reprod Update. 2006;12(4):363–72.
108. Poblete FA. The probability of conception on different days of the cycle with respect to ovulation: an overview. Adv Contracept. 1997;13(2–3):83–95.
109. Baerwald AR, Adams GP, Pierson RA. A new model for ovarian follicular development during the human menstrual cycle. Fertil Steril. 2003;80(1):116–22.
110. Kumar RS, Narayanan SN, Paval J. High contraceptive failure rate of rhythm method: possible involvement of pheromones. Med Hypotheses. 2009;73(6):1079.
111. Stern K, Mcclintock MK. Regulation of ovulation by human pheromones. Nature. 1998;392(6672):177–9.
112. Boron W, Boulpaep E. Medical physiology: a cellular and molecular approach. Oxford: Elsevier; 2004. ISBN: 1-4160-2328-3
113. He R-H, Gao H-J, Li Y-Q, Zhu X-M. The associated factors to endometrial cavity fluid and the relevant impact on the IVF-ET outcome. Reprod Biol Endocrinol. 2010;8(1):46.
114. Johnson DC, Dey SK. Role of histamine in implantation: dexamethasone inhibits estradiol-induced implantation in the rat. Biol Reprod. 1980;22(5):1136–41.
115. Colicchia M, Campagnolo L, Baldini E, Ulisse S, Valensise H, Moretti C. Molecular basis of thyrotropin and thyroid hormone action during implantation and early development. Hum Reprod Update. 2014;20(6):884–904.
116. Bagchi IC, Li Q, Cheon YP. Role of steroid hormone-regulated genes in implantation. Ann N Y Acad Sci. 2001;943(1):68–76.
117. Stefanoska I, Krivokuća MJ, Vasilijić S, Ćujić D, Vićovac L. Prolactin stimulates cell migration and invasion by human trophoblast in vitro. Placenta. 2013;34(9):775–83.
118. Makrigiannakis A, Zoumakis E, Kalantaridou S, Coutifaris C, Margioris AN, Coukos G, et al. Corticotropin-releasing hormone promotes blastocyst implantation and early maternal tolerance. Nat Immunol. 2001;2(11):1018–24.
119. Balunas M, Su B, Brueggemeier R, Kinghorn A. Natural products as aromatase inhibitors. Anti Cancer Agents Med Chem. 2008;8(6):646–82.
120. Lephart ED. Modulation of aromatase by phytoestrogens. Enzyme Res. 2015;2015:1–11.
121. Sanderson JT, Hordijk J, Denison MS, Springsteel MF, Nantz MH, van den Berg M. Induction and inhibition of aromatase (CYP19) activity by natural and synthetic flavonoid compounds in H295R human adrenocortical carcinoma cells. Toxicol Sci. 2004;82(1):70–9.
122. Zarkavelis G, Kollas A, Kampletsas E, Vasiliou V, Evripides Kaltsonoudis E, Drosos A, et al. Aromatase inhibitors induced autoimmune disorders in patients with breast cancer: a review. J Adv Res. 2016;7(5):719–26.
123. Tai P, Wang J, Jin H, Song X, Yan J, Kang Y, et al. Induction of regulatory T cells by physiological level estrogen. J Cell Physiol. 2007;214(2):456–64.
124. Nepomnaschy PA, Welch KB, McConnell DS, Low BS, Strassmann BI, England BG. Cortisol levels and very early pregnancy loss in humans. Proc Natl Acad Sci U S A. 2006;103(10):3938–42.
125. Michael AE, Papageorghiou AT. Potential significance of physiological and pharmacological glucocorticoids in early pregnancy. Hum Reprod Update. 2008;14(5):497–517.
126. Shaum KM, Polotsky AJ. Nutrition and reproduction: is there evidence to support a "fertility diet" to improve mitochondrial function? Maturitas. 2013;74(4):309–12.
127. Chavarro JE, Minguez-Alarcon L, Mendiola J, Cutillas-Tolin A, Lopez-Espin JJ, Torres-Cantero AM. Trans fatty acid intake is inversely related to total sperm count in young healthy men. Hum Reprod. 2014;29(6):1346–7.
128. Chiu Y-H, Karmon AE, Gaskins AJ, Arvizu M, Williams PL, Souter I, et al. Serum omega-3 fatty acids and treatment outcomes among women undergoing assisted reproduction. Hum Reprod. 2017;33(1):156–65.
129. Louis GMB, Sapra KJ, Schisterman EF, Lynch CD, Maisog JM, Grantz KL, et al. Lifestyle and pregnancy loss in a contemporary cohort of

129. women recruited before conception: the LIFE Study. Fertil Steril. 2016;106(1):180–8.
130. Avalos LA, Roberts S, Kaskutas LA, Block G, Li D-K. Volume and type of alcohol during early pregnancy and the risk of miscarriage. Subst Use Misuse. 2014;49(11):1437–45.
131. Gaskins AJ, Chavarro JE. Diet and fertility: a review. Am J Obstet Gynecol. 2018;218(4):379–89.
132. Machtinger R, Gaskins AJ, Mansur A, Adir M, Racowsky C, Baccarelli AA, et al. Association between preconception maternal beverage intake and in vitro fertilization outcomes. Fertil Steril. 2017;108(6):1026–33.
133. Sharma R, Biedenharn KR, Fedor JM, Agarwal A. Lifestyle factors and reproductive health: taking control of your fertility. Reprod Biol Endocrinol. 2013;11(1):66.
134. Chavarro JE, Rich-Edwards JW, Rosner BA, Willett WC. Protein intake and ovulatory infertility. Am J Obstet Gynecol. 2008;198(2):210.e1–7.
135. Chiu YH, Afeiche MC, Gaskins AJ, Williams PL, Petrozza JC, Tanrikut C, Hauser R, Chavarro JE. Fruit and vegetable intake and their pesticide residues in relation to semen quality among men from a fertility clinic. Hum Reprod. 2015;30(6):1342–51.
136. Karayiannis D, Kontogianni MD, Mendorou C, Mastrominas M, Yiannakouris N. Adherence to the Mediterranean diet and IVF success rate among non-obese women attempting fertility. Hum Reprod. 2018;33(3):494–502.
137. Torjusen H, Brantsaeter AL, Haugen M, Alexander J, Bakketeig LS, Lieblein G, et al. Reduced risk of pre-eclampsia with organic vegetable consumption: results from the prospective Norwegian Mother and Child Cohort Study. BMJ Open. 2014;4(9):e006143.
138. Gaskins AJ, Chiu Y-H, Williams PL, Keller MG, Toth TL, Hauser R, et al. Maternal whole grain intake and outcomes of in vitro fertilization. Fertil Steril. 2016;105(6):1503.
139. Mikkelsen TB, Osler M, Orozova-Bekkevold I, Knudsen VK, Olsen SF. Association between fruit and vegetable consumption and birth weight: a prospective study among 43,585 Danish women. Scand J Public Health. 2006;34(6):616–22.
140. Yamamoto Y, Aizawa K, Mieno M, Karamatsu M, Hirano Y, Furui K, Miyashita T, Yamazaki K, Inakuma T, Sato I, Suganuma H, Iwamoto T. The effects of tomato juice on male infertility. Asia Pac J Clin Nutr. 2017;26(1):65–71.
141. Balogun OO, da SK L, Ota E, Takemoto Y, Rumbold A, Takegata M, Mori R. Vitamin supplementation for preventing miscarriage. Cochrane Database Syst Rev. 2016;(5):CD004073.
142. Perkins AV, Vanderlelie JJ. Multiple micronutrient supplementation and birth outcomes: the potential importance of selenium. Placenta 2016;48 Suppl 1:S61–5.
143. Dhanashree N, Anuradha S, Ketan S. Effect of diet and nutrient intake on women who have problems of fertility. Int J Pure App Biosci. 2016;4(4):198–204.
144. Hashemi MM, Behnampour N, Nejabat M, Tabandeh A, Ghazi-Moghaddam B, Joshaghani HR. Impact of seminal plasma trace elements on human sperm motility parameters. Rom J Intern Med. 2018;56(1):15.
145. Tvrda E, Peer R, Sikka SC, Agarwal A. Iron and copper in male reproduction: a double-edged sword. J Assist Reprod Genet. 2014;32(1):3–16.
146. Colagar AH, Marzony ET, Chaichi MJ. Zinc levels in seminal plasma are associated with sperm quality in fertile and infertile men. Nutr Res. 2009;29(2):82–8.
147. Riffo M, Leiva S, Astudillo J. Effect of zinc on human sperm motility and the acrosome reaction. Int J Androl. 1992;15(3):229–37.
148. Zhou J-C, Zheng S, Mo J, Liang X, Xu Y, Zhang H, et al. Dietary selenium deficiency or excess reduces sperm quality and testicular mRNA abundance of nuclear glutathione peroxidase 4 in rats. J Nutr. 2017;147(10):1947–53.
149. Hawkes WC, Turek PJ. Effects of dietary selenium on sperm motility in healthy men. J Androl. 2001;22(5):764–72.
150. Laganà AS, Vitale SG, Frangež BH, Vrtačnik-Bokal E, D'Anna R. Vitamin D in human reproduction: the more, the better? An evidence-based critical appraisal. Eur Rev Med Pharmacol Sci. 2017;21(18):4243–51.
151. Uhland AM, Kwiecinski GG, Deluca HF. Normalization of serum calcium restores fertility in vitamin d-deficient male rats. J Nutr. 1992;122(6):1338–44.
152. Hachem A, Godwin J, Ruas M, Lee HC, Buitrago MF, Ardestani G, et al. PLCζ is the physiological trigger of the Ca 2 oscillations that induce embryogenesis in mammals but conception can occur in its absence. Development. 2017;144(16):2914–24.
153. Kiss SA, Kiss I. Effect of magnesium ions on fertility, sex ratio and mutagenesis in Drosophila melanogaster males. Magnes Res. 1995;8(3):243–7.
154. Tarjan A, Zarean E. Effect of magnesium supplement on pregnancy outcomes: a randomized control trial. Adv Biomed Res. 2017;6(1):109.
155. Wang M, Li K, Zhao D, Li L. The association between maternal use of folic acid supplements during pregnancy and risk of autism spectrum disorders in children: a meta-analysis. Mol Autism. 2017;8:51.
156. Grajecki D, Zyriax B-C, Buhling KJ. The effect of micronutrient supplements on female fertility: a systematic review. Arch Gynecol Obstet. 2012;285(5):1463–71.
157. Bloomberg School of Public Health. Too much folate in pregnant women increases risk for autism, study suggests [Internet]. John Hopkins University; 2016 May 11. Available from: https://www.jhsph.edu/news/news-releases/2016/too-much-folate-in-pregnant-women-increases-risk-for-autism-study-suggests.html.
158. Murray AA, Molinek MD, Baker SJ, Kojima FN, Smith MF, Hillier SG, Spears N. Role of ascorbic acid in promoting follicle integrity and survival in intact mouse ovarian follicles in vitro. Reproduction. 2001;121(1):89–96.
159. Nam SM, Chang B-J, Kim J-H, Nahm S-S, Lee J-H. Ascorbic acid ameliorates lead-induced apoptosis in the cerebellar cortex of developing rats. Brain Res. 2018;1686:10–8.
160. Shorey-Kendrick LE, Mcevoy CT, Ferguson B, Burchard J, Park BS, Gao L, et al. Vitamin C prevents offspring dna methylation changes associated with maternal smoking in pregnancy. Am J Respir Crit Care Med. 2017;196(6):745–55.
161. Aguirre-Arias MV, Velarde V, Moreno RD. Effects of ascorbic acid on spermatogenesis and sperm parameters in diabetic rats. Cell Tissue Res. 2017;370(2):305–17.
162. Crha I, Hrubá D, Ventruba P, Fiala J, Totusek J, Visnová H. Ascorbic acid and infertility treatment. Cent Eur J Public Health. 2003;11(2):63–7.
163. National Toxicology Program. About NTP [Internet]. U.S. Department of Health and Human Services. Available from: https://ntp.niehs.nih.gov/about/.
164. The New York Times. Think those chemicals have been tested? [Internet]. Ian Urbina: The New York Times; 2013 Apr 13. Available from: https://www.nytimes.com/2013/04/14/sunday-review/think-those-chemicals-have-been-tested.html.
165. Schubert J, Riley EJ, Tyler SA. Combined effects in toxicology-a rapid systematic testing procedure: cadmium, mercury, and lead. J Toxicol Environ Health. 1978;4(5–6):763–76.
166. Natoff IL. Influence of the route of administration on the toxicity of some cholinesterase inhibitors. J Pharm Pharmacol. 1967;19(9):612–6.
167. Crépeaux G, Eidi H, David M-O, Baba-Amer Y, Tzavara E, Giros B, et al. Non-linear dose-response of aluminium hydroxide adjuvant particles: selective low dose neurotoxicity. Toxicology. 2017;375:48–57.
168. Minervini F, Giannoccaro A, Fornelli F, Dell'Aquila ME, Minoia P, Visconti A. Influence of mycotoxin zearalenone and its derivatives (alpha and beta zearalenol) on apoptosis and proliferation of cultured granulosa cells from equine ovaries. Reprod Biol Endocrinol. 2006;4:62.

169. Benzoni E, Minervini F, Giannoccaro A, Fornelli F, Vigo D, Visconti A. Influence of in vitro exposure to mycotoxin zearalenone and its derivatives on swine sperm quality. Reprod Toxicol. 2008;25(4):461–7.
170. Meharg AA, Lombi E, Williams PN, Scheckel KG, Feldmann J, Raab A, et al. Speciation and localization of arsenic in white and brown rice grains. Environ Sci Technol. 2008;42(4):1051–7.
171. Golub MS, Macintosh MS, Baumrind N. Developmental and reproductive toxicity of inorganic arsenic: animal studies and human concerns. J Toxicol Environ Health B Crit Rev. 1998;1(3):199–237.
172. Kim Y-J, Kim J-M. Arsenic toxicity in male reproduction and development. Dev Reprod. 2015;19(4):167–80.
173. Jahan S, Rehman S, Ullah H, Munawar A, Ain QU, Iqbal T. Ameliorative effect of quercetin against arsenic-induced sperm DNA damage and daily sperm production in adult male rats. Drug Chem Toxicol. 2015;39(3):290–6.
174. Carey M, Jiujin X, Farias JG, Meharg AA. Rethinking rice preparation for highly efficient removal of inorganic arsenic using percolating cooking water. PLoS One. 2015;10(7):e0131608.
175. Agency for Toxic Substances & Disease Registry. Cadmium toxicity: where is cadmium found? [Internet]. Case Studies in Environmental Medicine (CSEM). Available from: https://www.atsdr.cdc.gov/csem/csem.asp?csem=6&po=5.
176. Izaguirry AP, Soares MB, Vargas LM, Spiazzi CC, Brum DDS, Noremberg S, et al. Blueberry (Vaccinium ashei Reade) extract ameliorates ovarian damage induced by subchronic cadmium exposure in mice: potential δ-ALA-D involvement. Environ Toxicol. 2015;32(1):188–96.
177. Nad P, Massanyi P, Skalicka M, Korenekova B, Cigankova V, Almasiova V. The effect of cadmium in combination with zinc and selenium on ovarian structure in Japanese quails. J Environ Sci Health A. 2007;42(13):2017–22.
178. Peraza MA, Ayala-Fierro F, Barber DS, Casarez E, Rael LT. Effects of micronutrients on metal toxicity. Environ Health Perspect. 1998;106:203–16.
179. Brownawell AM, Berent S, Brent RL, Bruckner JV, Doull J, Gershwin EM, et al. The potential adverse health effects of dental amalgam. Toxicol Rev. 2005;24(1):1–10.
180. Bengtsson UG, Hylander LD. Increased mercury emissions from modern dental amalgams. Biometals. 2017;30(2):277–83.
181. Mortazavi G, Mortazavi S. Increased mercury release from dental amalgam restorations after exposure to electromagnetic fields as a potential hazard for hypersensitive people and pregnant women. Rev Environ Health. 2015;30(4):287.
182. Paknahad M, Mortazavi SMJ, Shahidi S, Mortazavi G, Haghani M. Effect of radiofrequency radiation from Wi-Fi devices on mercury release from amalgam restorations. J Environ Health Sci Eng. 2016;14(1):12.
183. Mailhes JB. Methylmercury effects on Syrian hamster metaphase II oocyte chromosomes. Environ Mutagen. 1983;5(5):679–86.
184. Lee JH, Kang HS, Kang J. Protective effects of garlic juice against embryotoxicity of methylmercuric chloride administered to pregnant Fischer 344 rats. Yonsei Med J. 1999;40(5):483–9.
185. Spiller HA. Rethinking mercury: the role of selenium in the pathophysiology of mercury toxicity. Clin Toxicol. 2017;56(5):313–26.
186. Yang T, Xu Z, Liu W, Xu B, Deng Y. Protective effects of Alpha-lipoic acid on MeHg-induced oxidative damage and intracellular Ca2 dyshomeostasis in primary cultured neurons. Free Radic Res. 2016;50(5):542–56.
187. Paula ESD, Carneiro MFH, Grotto D, Hernandes LC, Antunes LMG, Barbosa F. Protective effects of niacin against methylmercury-induced genotoxicity and alterations in antioxidant status in rats. J Toxic Environ Health A. 2016;79(4):174–83.
188. Manzolli ES, Serpeloni JM, Grotto D, Bastos JK, Antunes LMG, Barbosa F, et al. Protective effects of the flavonoid chrysin against methylmercury-induced genotoxicity and alterations of antioxidant status, in vivo. Oxidative Med Cell Longev. 2015;2015:1–7.
189. Mozhdeganloo Z, Jafari AM, Koohi MK, Heidarpour M. Methylmercury-induced oxidative stress in rainbow trout (Oncorhynchus mykiss) liver: ameliorating effect of vitamin C. Biol Trace Elem Res. 2015;165(1):103–9.
190. Costa-Malaquias A, Almeida MB, Monteiro JRS, Macchi BDM, Nascimento JLMD, Crespo-Lopez ME. Morphine protects against methylmercury intoxication: a role for opioid receptors in oxidative stress? PLoS One. 2014;9(10):e110815.
191. Silva-Pereira L, Rocha CD, Cunha L, Costa ED, Guimarães A, Pontes T, et al. Protective effect of prolactin against methylmercury-induced mutagenicity and cytotoxicity on human lymphocytes. Int J Environ Res Public Health. 2014;11(9):9822–34.
192. Ramesh CG, editor. Reproductive and developmental toxicology. London: Academic Press, Elsevier Science; 2011. p. 1310.
193. Jana K, Jana S, Samanta PK. Effects of chronic exposure to sodium arsenite on hypothalamo-pituitary-testicular activities in adult rats: possible an estrogenic mode of action. Reprod Biol Endocrinol. 2006;4:9.
194. Karasavvas N. Vitamin C protects HL60 and U266 cells from arsenic toxicity. Blood. 2005;105(10):4004–12.
195. Mittal M, Flora SJS. Vitamin E supplementation protects oxidative stress during arsenic and fluoride antagonism in male mice. Drug Chem Toxicol. 2007;30(3):263–81.
196. Ghosh S, Mishra R, Biswas S, Bhadra RK, Mukhopadhyay PK. α-lipoic acid mitigates arsenic-induced hematological abnormalities in adult male rats. Iran J Med Sci. 2017;42(3):242–50.
197. Chakrabarty N, editor. Arsenic toxicity: prevention and treatment. Boca Raton: CRC Press; 2016.
198. New York State. Sources of lead [Internet]. Department of Health; 2010. Available from: https://www.health.ny.gov/environmental/lead/sources.htm.
199. Apostoli P, Kiss P, Porru S, Bonde JP, Vanhoorne M. Male reproductive toxicity of lead in animals and humans. ASCLEPIOS Study Group. Occup Environ Med. 1998;55(6):364–74.
200. Sanders T, Liu Y, Buchner V, Tchounwou P. Neurotoxic effects and biomarkers of lead exposure: a review. Rev Environ Health. 2009;24(1):15–45.
201. Singh PK, Nath R, Ahmad MK, Rawat A, Babu S, Dixit RK. Attenuation of lead neurotoxicity by supplementation of polyunsaturated fatty acid in Wistar rats. Nutr Neurosci. 2015;19(9):396–405.
202. Couette M, Boisse M-F, Maison P, Brugieres P, Cesaro P, Chevalier X, et al. Long-term persistence of vaccine-derived aluminum hydroxide is associated with chronic cognitive dysfunction. J Inorg Biochem. 2009;103(11):1571–8.
203. Sánchez HA, Tejada-González P, Arteta-Jiménez M. Aluminium in parenteral nutrition: a systematic review. Eur J Clin Nutr. 2013;67(3):230–8.
204. Yokel RA, Hicks CL, Florence RL. Aluminum bioavailability from basic sodium aluminum phosphate, an approved food additive emulsifying agent, incorporated in cheese. Food Chem Toxicol. 2008;46(6):2261–6.
205. United States Food and Drug Administration. Vaccines. Adacel prescribing information. FDA. [Internet]; 2018 [cited 2018 May 8]. Available from: https://www.fda.gov/downloads/BiologicsBloodVaccines/Vaccines/ApprovedProducts/UCM142764.pdf.
206. United States Food and Drug Administration. Vaccines. Boostrix prescribing information. FDA. [Internet]; 2018 [cited 2018 May 8]. Available from: https://www.fda.gov/downloads/BiologicsBloodVaccines/UCM152842.pdf.
207. Shaw CA, Li D, Tomljenovic L. Are there negative CNS impacts of aluminum adjuvants used in vaccines and immunotherapy? Immunotherapy. 2014;6(10):1055–71.
208. Rondeau V, Jacqmin-Gadda H, Commenges D, Helmer C, Dartigues J-F. Aluminum and silica in drinking water and the risk of Alzheimer's disease or cognitive decline: findings from 15-year follow-up of the PAQUID cohort. Am J Epidemiol. 2008;169(4):489–96.

209. Davenward S, Bentham P, Wright J, Crome P, Job D, Polwart A, Exley C. Silicon-rich mineral water as a non-invasive test of the 'aluminum hypothesis' in Alzheimer's disease. J Alzheimers Dis. 2013;33(2):423–30.
210. El-Demerdash FM. Antioxidant effect of vitamin E and selenium on lipid peroxidation, enzyme activities and biochemical parameters in rats exposed to aluminium. J Trace Elem Med Biol. 2004;18(1):113–21.
211. Olajide OJ, Yawson EO, Gbadamosi IT, Arogundade TT, Lambe E, Obasi K, et al. Ascorbic acid ameliorates behavioural deficits and neuropathological alterations in rat model of Alzheimer's disease. Environ Toxicol Pharmacol. 2017;50:200–11.
212. Jelenković A, Jovanović MD, Stevanović I, Petronijević N, Bokonjić D, Živković J, et al. Influence of the green tea leaf extract on neurotoxicity of aluminium chloride in rats. Phytother Res. 2013;28(1):82–7.
213. The Environmental Working Group. Overexposed organophosphate insecticides in children's food [Internet]. Wiles R, Davies K, Campbell C: Washington D.C.; 1998 Jan 29. Available from: https://www.ewg.org/research/overexposed-organophosphate-insecticides-childrens-food/download#.WlwtX1Q-dTY.
214. Lu C, Toepel K, Irish R, Fenske RA, Barr DB, Bravo R. Organic diets significantly lower children's dietary exposure to organophosphorus pesticides. Environ Health Perspect. 2006;114(2):260–3.
215. Li S, Cao C, Shi H, Yang S, Qi L, Zhao X, et al. Effect of quercetin against mixture of four organophosphate pesticides induced nephrotoxicity in rats. Xenobiotica. 2015;46(3):225–33.
216. Eroğlu S, Pandir D, Uzun FG, Bas H. Protective role of vitamins C and E in dichlorvos-induced oxidative stress in human erythrocytes in vitro. Biol Res. 2013;46(1):33–8.
217. El-Saad AA, Ibrahim M, Hazani A, El-Gaaly G. Lycopene attenuates dichlorvos-induced oxidative damage and hepatotoxicity in rats. Hum Exp Toxicol. 2015;35(6):654–65.
218. Tongo I, Ogbeide O, Ezemonye L. Human health risk assessment of polycyclic aromatic hydrocarbons (PAHs) in smoked fish species from markets in Southern Nigeria. Toxicol Rep. 2017;4:55–61.
219. Borman S, Christian P, Sipes I, Hoyer P. Ovotoxicity in female fischer rats and b6 mice induced by low-dose exposure to three polycyclic aromatic hydrocarbons: comparison through calculation of an ovotoxic index. Toxicol Appl Pharmacol. 2000;167(3):191–8.
220. Han X, Zhou N, Cui Z, Ma M, Li L, Cai M, et al. Association between urinary polycyclic aromatic hydrocarbon metabolites and sperm DNA damage: a population study in Chongqing, China. Environ Health Perspect. 2010;119(5):652–7.
221. Lagerqvist A, Håkansson D, Frank H, Seidel A, Jenssen D. Structural requirements for mutation formation from polycyclic aromatic hydrocarbon dihydrodiol epoxides in their interaction with food chemopreventive compounds. Food Chem Toxicol. 2011;49(4):879–86.
222. Jarabak R, Harvey RG, Jarabak J. Redox cycling of polycyclic aromatic hydrocarbon o-quinones: reversal of superoxide dismutase inhibition by ascorbate. Arch Biochem Biophys. 1997;339(1):92–8.
223. Gajecka M, Kujawski LM, Gawecki J, Szyfter K. The protective effect of vitamins C and E against B(a)P-induced genotoxicity in human lymphocytes. J Environ Pathol Toxicol Oncol. 1999;18(3):159–67.
224. Institute of Medicine (US) Committee on the Implications of Dioxin in the Food Supply. Dioxins and dioxin-like compounds in the food supply: strategies to decrease exposure. Washington, D.C.: National Academies Press (US); 2003.
225. United States Environmental Protection Agency. Learn about dioxin [Internet]. Available from: https://www.epa.gov/dioxin/learn-about-dioxin.
226. U.S. EPA. EPA's reanalysis of key issues related to dioxin toxicity and response to NAS comments (Volume 1) (Interagency Science Discussion Draft). U.S. Environmental Protection Agency, Washington, D.C., EPA/600/R-10/038D, 2012. Figures are from 4/94 data, newer data was not conveniently presented.
227. Devito MJ, Schecter A. Exposure assessment to dioxins from the use of tampons and diapers. Environ Health Perspect. 2001;110(1):23–8.
228. Anger DL, Foster WG. The link between environmental toxicant exposure and endometriosis. Front Biosci. 2008;13:1578–93.
229. Fukuda I, Sakane I, Yabushita Y, Kodoi R, Nishiumi S, Kakuda T, et al. Pigments in green tea leaves (camellia sinensis) suppress transformation of the aryl hydrocarbon receptor induced by dioxin. J Agric Food Chem. 2004;52(9):2499–506.
230. Kalaiselvan I, Samuthirapandi M, Govindaraju A, Malar DS, Kasi PD. Olive oil and its phenolic compounds (hydroxytyrosol and tyrosol) ameliorated TCDD-induced heptotoxicity in rats via inhibition of oxidative stress and apoptosis. Pharm Biol. 2015;54(2):338–46.
231. Amakura Y, Tsutsumi T, Sasaki K, Nakamura M, Yoshida T, Maitani T. Influence of food polyphenols on aryl hydrocarbon receptor-signaling pathway estimated by in vitro bioassay. Phytochemistry. 2008;69(18):3117–30.
232. Mukai R, Fukuda I, Nishiumi S, Natsume M, Osakabe N, Yoshida K-I, et al. Cacao polyphenol extract suppresses transformation of an aryl hydrocarbon receptor in C57BL/6 mice. J Agric Food Chem. 2008;56(21):10399–405.
233. Foster WG. The reproductive toxicology of great lakes contaminants. Environ Health Perspect. 1995;103:63.
234. Aulerich RJ, Ringer RK. Current status of PCB toxicity to mink, and effect on their reproduction. Arch Environ Contam Toxicol. 1977;6(1):279–92.
235. Lorber M, Toms L-ML. Use of a simple pharmacokinetic model to study the impact of breast-feeding on infant and toddler body burdens of PCB 153, BDE 47, and DDE. Chemosphere. 2017;185:1081–9.
236. Takhshid MA, Tavasuli AR, Heidary Y, Keshavarz M, Kargar H. Protective effect of vitamins E and C on endosulfan-induced reproductive toxicity in male rats. Iran J Med Sci. 2012;37(3):173–80.
237. Selvakumar K, Bavithra S, Suganthi M, Benson CS, Elumalai P, Arunkumar R, et al. Protective role of quercetin on pcbs-induced oxidative stress and apoptosis in hippocampus of adult rats. Neurochem Res. 2011;37(4):708–21.
238. Baker KM, Bauer AC. Green tea catechin, EGCG, suppresses PCB 102-induced proliferation in estrogen-sensitive breast cancer cells. Int J Breast Cancer. 2015;2015:1–7.
239. James MO, Sacco JC, Faux LR. Effects of food natural products on the biotransformation of PCBs. Environ Toxicol Pharmacol. 2008;25(2):211–7.
240. Erythropel HC, Maric M, Nicell JA, Leask RL, Yargeau V. Leaching of the plasticizer di(2-ethylhexyl)phthalate (DEHP) from plastic containers and the question of human exposure. Appl Microbiol Biotechnol. 2014;98(24):9967–81.
241. Bagel-Boithias S, Sautou-Miranda V, Bourdeaux D, Tramier V, Boyer A, Chopineau J. Leaching of diethylhexyl phthalate from multilayer tubing into etoposide infusion solutions. Am J Health Syst Pharm. 2005;62(2):182–8.
242. Demoré B, Vigneron J, Perrin A, Hoffman MA, Hoffman M. Leaching of diethylhexyl phthalate from polyvinyl chloride bags into intravenous etoposide solution. J Clin Pharm Ther. 2002;27(2):139–42.
243. Mcpartland JM, Guy GW, Marzo VD. Care and feeding of the endocannabinoid system: a systematic review of potential clinical interventions that upregulate the endocannabinoid system. PLoS One. 2014;9(3):e89566.
244. Davis B, Maronpot R, Heindel J. Di-(2-ethylhexyl) phthalate suppresses estradiol and ovulation in cycling rats. Toxicol Appl Pharmacol. 1994;128(2):216–23.
245. Huang P-C, Tsai E-M, Li W-F, Liao P-C, Chung M-C, Wang Y-H, et al. Association between phthalate exposure and glutathione S-transferase M1 polymorphism in adenomyosis, leiomyoma and endometriosis. Hum Reprod. 2010;25(4):986–94.

246. Bisset KM, Dhopeshwarkar AS, Liao C, Nicholson RA. The G protein-coupled cannabinoid-1 (CB1) receptor of mammalian brain: inhibition by phthalate esters in vitro. Neurochem Int. 2011;59(5):706–13.
247. Paria B, Wang H, Dey S. Endocannabinoid signaling in synchronizing embryo development and uterine receptivity for implantation. Chem Phys Lipids. 2002;121(1–2):201–10.
248. Basavarajappa B, Nixon R, Arancio O. Endocannabinoid system: emerging role from neurodevelopment to neurodegeneration. Mini-Rev Med Chem. 2009;9(4):448–62.
249. Tseng I-L, Yang Y-F, Yu C-W, Li W-H, Liao VH-C. Correction: phthalates induce neurotoxicity affecting locomotor and thermotactic behaviors and AFD neurons through oxidative stress in caenorhabditis elegans. PLoS One. 2013;8(12):e82657.
250. Erkekoglu P, Giray BK, Kızılgün M, Rachidi W, Hininger-Favier I, Roussel A-M, et al. Di(2-ethylhexyl)phthalate-induced renal oxidative stress in rats and protective effect of selenium. Toxicol Mech Methods. 2012;22(6):415–23.
251. Wang Y, Chen B, Lin T, Wu S, Wei G. Protective effects of vitamin E against reproductive toxicity induced by di(2-ethylhexyl) phthalate via PPAR-dependent mechanisms. Toxicol Mech Methods. 2017;27(7):551–9.
252. Botelho GGK, Bufalo AC, Boareto AC, Muller JC, Morais RN, Martino-Andrade AJ, et al. Vitamin C and resveratrol supplementation to rat dams treated with di(2-ethylhexyl)phthalate: impact on reproductive and oxidative stress end points in male offspring. Arch Environ Contam Toxicol. 2009;57(4):785–93.
253. Ishihara M, Itoh M, Miyamoto K, Suna S, Takeuchi Y, Takenaka I, et al. Spermatogenic disturbance induced by di-(2-ethylhexyl) phthalate is significantly prevented by treatment with antioxidant vitamins in the rat. Int J Androl. 2000;23(2):85–94.
254. Johns LE, Ferguson KK, Cantonwine DE, Mcelrath TF, Mukherjee B, Meeker JD. Urinary BPA and phthalate metabolite concentrations and plasma vitamin D levels in pregnant women: a repeated measures analysis. Environ Health Perspect. 2017;125(8):087026.
255. Heyden E, Wimalawansa S. Vitamin D: effects on human reproduction, pregnancy, and fetal well-being. J Steroid Biochem Mol Biol. 2018;180:41–50.
256. Turnbull D, Parisi A, Kimlin M. Vitamin D effective ultraviolet wavelengths due to scattering in shade. J Steroid Biochem Mol Biol. 2005;96(5):431–6.
257. D'Orazio J, Jarrett S, Amaro-Ortiz A, Scott T. UV radiation and the skin. Int J Mol Sci. 2013;14(6):12222–48.
258. Ogilvy-Stuart AL, Shalet SM. Effect of radiation on the human reproductive system. Environ Health Perspect. 1993;101:109–16.
259. Biedka M, Kuźba-Kryszak T, Nowikiewicz T, Żyromska A. Fertility impairment in radiotherapy. Współczesna Onkologia. 2016;3:199–204.
260. Johansson O. Disturbance of the immune system by electromagnetic fields – a potentially underlying cause for cellular damage and tissue repair reduction which could lead to disease and impairment. Pathophysiology. 2009;16(2–3):157–77.
261. Wdowiak A, Mazurek PA, Wdowiak A, Bojar I. Effect of electromagnetic waves on human reproduction. Ann Agric Environ Med. 2017;24(1):13–8.
262. Li D-K, Chen H, Ferber JR, Odouli R, Quesenberry C. Exposure to magnetic field non-ionizing radiation and the risk of miscarriage: a prospective cohort study. Sci Rep. 2017;7(1):17541.
263. Kim JH, Yu D-H, Huh YH, Lee EH, Kim H-G, Kim HR. Long-term exposure to 835 MHz RF-EMF induces hyperactivity, autophagy and demyelination in the cortical neurons of mice. Sci Rep. 2017;7:41129.
264. Abdel-Rassoul G, El-Fateh OA, Salem MA, Michael A, Farahat F, El-Batanouny M, et al. Neurobehavioral effects among inhabitants around mobile phone base stations. Neurotoxicology. 2007;28(2):434–40.
265. Morgan LL, Miller AB, Sasco A, Davis DL. Mobile phone radiation causes brain tumors and should be classified as a probable human carcinogen (2A) (review). Int J Oncol. 2015;46(5):1865–71.
266. Sadeghi T, Ahmadi A, Javadian M, Gholamian SA, Delavar MA, Esmailzadeh S, et al. Preterm birth among women living within 600 meters of high voltage overhead Power Lines: a case-control study. Rom J Intern Med. 2017;55(3):145–50.
267. Vocht FD, Hannam K, Baker P, Agius R. Maternal residential proximity to sources of extremely low frequency electromagnetic fields and adverse birth outcomes in a UK cohort. Bioelectromagnetics. 2014;35(3):201–9.
268. Birks L, Guxens M, Papadopoulou E, Alexander J, Ballester F, Estarlich M, et al. Maternal cell phone use during pregnancy and child behavioral problems in five birth cohorts. Environ Int. 2017;104:122–31.
269. Schauer I, Al-Ali BM. Combined effects of varicocele and cell phones on semen and hormonal parameters. Wien Klin Wochenschr. 2018;130(9–10):335.
270. Macrì MA, Luzio DS. Biological effects of electromagnetic fields. Int J Immunopathol Pharmacol. 2002;15(2):95–105.
271. Miao X, Wang Y, Lang H, Lin Y, Guo Q, Yang M, et al. Preventing electromagnetic pulse irradiation damage on testis using selenium-rich cordyceps fungi. A Preclinical Study in Young Male Mice. OMICS. 2017;21(2):81–9.
272. Bin-Meferij MM, El-kott AF. The radioprotective effects of *Moringa oleifera* against mobile phone electromagnetic radiation-induced infertility in rats. Int J Clin Exp Med. 2015;8(8):12487–97.
273. Houston BJ, Nixon B, King BV, Iuliis GND, Aitken RJ. The effects of radiofrequency electromagnetic radiation on sperm function. Reproduction. 2016;152(6):R263.
274. Kamali K, Atarod M, Sarhadi S, Nikbakht J, Emami M, Maghsoudi R, et al. Effects of electromagnetic waves emitted from 3G wi-fi modems on human semen analysis. Urologia. 2017;84(4):209–14.
275. Yan J-G, Agresti M, Bruce T, Yan YH, Granlund A, Matloub HS. Effects of cellular phone emissions on sperm motility in rats. Fertil Steril. 2007;88(4):957–64.
276. Falzone N, Huyser C, Becker P, Leszczynski D, Franken DR. The effect of pulsed 900-MHz GSM mobile phone radiation on the acrosome reaction, head morphometry and zona binding of human spermatozoa. Int J Androl. 2011;34(1):20–6.
277. Oral B, Guney M, Ozguner F, Karahan N, Mungan T, Comlekci S, et al. Endometrial apoptosis induced by a 900-MHz mobile phone: preventive effects of vitamins E and C. Adv Ther. 2006;23(6):957–73.
278. Guney M, Ozguner F, Oral B, Karahan N, Mungan T. 900 MHz radio-frequency-induced histopathologic changes and oxidative stress in rat endometrium: protection by vitamins E and C. Toxicol Ind Health. 2007;23(7):411–20.
279. Bakacak M, Bostancı MS, Attar R, Yıldırım ÖK, Yıldırım G, Bakacak Z, et al. The effects of electromagnetic fields on the number of ovarian primordial follicles: an experimental study. Kaohsiung J Med Sci. 2015;31(6):287–92.
280. Türedi S, Hancı H, Çolakoğlu S, Kaya H, Odacı E. Disruption of the ovarian follicle reservoir of prepubertal rats following prenatal exposure to a continuous 900-MHz electromagnetic field. Int J Radiat Biol. 2016;92(6):329–37.
281. Odacı E, Hancı H, Yuluğ E, Türedi S, Aliyazıcıoğlu Y, Kaya H, et al. Effects of prenatal exposure to a 900 MHz electromagnetic field on 60-day-old rat testis and epididymal sperm quality. Biotech Histochem. 2015;91(1):9–19.
282. Gul A, Çelebi H, Uğraş S. The effects of microwave emitted by cellular phones on ovarian follicles in rats. Arch Gynecol Obstet. 2009;280(5):729–33.
283. Soleimani-Rad S, Roshangar L, Hamdi B, Khaki A. Effect of low-frequency electromagnetic field exposure on oocyte differentiation and follicular development. Adv Biomed Res. 2014;3(1):76.

284. Liu C, Gao P, Xu S-C, Wang Y, Chen C-H, He M-D, et al. Mobile phone radiation induces mode-dependent DNA damage in a mouse spermatocyte-derived cell line: a protective role of melatonin. Int J Radiat Biol. 2013;89(11):993–1001.
285. Al-Damegh M. Rat testicular impairment induced by electromagnetic radiation from a conventional cellular telephone and the protective effects of the antioxidants vitamins C and E. Clinics. 2012;67(7):785–92.
286. Nisbet HÖ, Nisbet C, Akaras A, Çevik M, Karayiğit MÖ. Effects of exposure to electromagnetic field (1.8/0.9 GHz) on testicular function and structure in growing rats. J Exp Clin Med. 2013;30(3):275.
287. Panagopoulos DJ, Johansson O, Carlo GL. Real versus simulated mobile phone exposures in experimental studies. Biomed Res Int. 2015;2015:1–8.
288. Malagoli C, Costanzini S, Heck JE, Malavolti M, Girolamo GD, Oleari P, et al. Passive exposure to agricultural pesticides and risk of childhood leukemia in an Italian community. Int J Hyg Environ Health. 2016;219(8):742–8.
289. Anifandis G, Amiridis G, Dafopoulos K, Daponte A, Dovolou E, Gavriil E, et al. The in vitro impact of the herbicide roundup on human sperm motility and sperm mitochondria. Toxics. 2017;6(1):2.
290. Owagboriaye FO, Dedeke GA, Ademolu KO, Olujimi OO, Ashidi JS, Adeyinka AA. Reproductive toxicity of Roundup herbicide exposure in male albino rat. Exp Toxicol Pathol. 2017;69(7):461–8.
291. Cai W, Ji Y, Song X, Guo H, Han L, Zhang F, et al. Effects of glyphosate exposure on sperm concentration in rodents: a systematic review and meta-analysis. Environ Toxicol Pharmacol. 2017;55:148–55.
292. Nardi J, Moras PB, Koeppe C, Dallegrave E, Leal MB, Rossato-Grando LG. Prepubertal subchronic exposure to soy milk and glyphosate leads to endocrine disruption. Food Chem Toxicol. 2017;100:247–52.
293. Perego MC, Schutz LF, Caloni F, Cortinovis C, Albonico M, Spicer LJ. Evidence for direct effects of glyphosate on ovarian function: glyphosate influences steroidogenesis and proliferation of bovine granulosa but not theca cells in vitro. J Appl Toxicol. 2016;37(6):692–8.
294. Schimpf MG, Milesi MM, Ingaramo PI, Luque EH, Varayoud J. Neonatal exposure to a glyphosate based herbicide alters the development of the rat uterus. Toxicology. 2017;376:2–14.
295. Ingaramo PI, Varayoud J, Milesi MM, Schimpf MG, Muñoz-De-Toro M, Luque EH. Effects of neonatal exposure to a glyphosate-based herbicide on female rat reproduction. Reproduction. 2016;152(5):403–15.
296. Walsh LP, Mccormick C, Martin C, Stocco DM. Roundup inhibits steroidogenesis by disrupting Steroidogenic Acute Regulatory (StAR) protein expression. Environ Health Perspect. 2000;108(8):769.
297. Almeida LLD, Teixeira ÁAC, Soares AF, Cunha FMD, Silva VAD, Filho LDV, et al. Effects of melatonin in rats in the initial third stage of pregnancy exposed to sub-lethal doses of herbicides. Acta Histochem. 2017;119(3):220–7.
298. Dallegrave E, Mantese FD, Coelho RS, Pereira JD, Dalsenter PR, Langeloh A. The teratogenic potential of the herbicide glyphosate-roundup in Wistar rats. Toxicol Lett. 2003;142(1–2):45–52.
299. Paganelli A, Gnazzo V, Acosta H, López Silvia L, Carrasco Andrés E. Glyphosate-based herbicides produce teratogenic effects on vertebrates by impairing retinoic acid signaling. Chem Res Toxicol. 2010;23(10):1586–95.
300. Souza JSD, Kizys MML, Conceição RRD, Glebocki G, Romano RM, Ortiga-Carvalho TM, et al. Perinatal exposure to glyphosate-based herbicide alters the thyrotrophic axis and causes thyroid hormone homeostasis imbalance in male rats. Toxicology. 2017;377:25–37.
301. de Araujo JS, Delgado IF, Paumgartten FJ. Glyphosate and adverse pregnancy outcomes, a systematic review of observational studies. BMC Public Health. 2016;16:472.
302. Gallegos CE, Bartos M, Bras C, Gumilar F, Antonelli MC, Minetti A. Exposure to a glyphosate-based herbicide during pregnancy and lactation induces neurobehavioral alterations in rat offspring. Neurotoxicology. 2016;53:20–8.
303. Cattani D, Cesconetto PA, Tavares MK, Parisotto EB, Oliveira PAD, Rieg CEH, et al. Developmental exposure to glyphosate-based herbicide and depressive-like behavior in adult offspring: implication of glutamate excitotoxicity and oxidative stress. Toxicology. 2017;387:67–80.
304. Cattani D, De Liz Oliveira Cavalli VL, Rieg CEH, Domingues JT, Dal-Cim T, Tasca CI, et al. Mechanisms underlying the neurotoxicity induced by glyphosate-based herbicide in immature rat hippocampus: involvement of glutamate excitotoxicity. Toxicology. 2014;320:34–45.
305. Lozano VL, Defarge N, Rocque L-M, Mesnage R, Hennequin D, Cassier R, et al. Sex-dependent impact of roundup on the rat gut microbiome. Toxicol Rep. 2018;5:96–107.
306. Wexler AG, Goodman AL. An insiders perspective: *Bacteroides* as a window into the microbiome. Nat Microbiol. 2017;2(5):17026.
307. Turovskiy Y, Noll KS, Chikindas ML. The etiology of bacterial vaginosis. J Appl Microbiol. 2011;110(5):1105–28.
308. Babu G. Comparative study on the vaginal flora and incidence of asymptomatic vaginosis among healthy women and in women with infertility problems of reproductive age. J Clin Diagn Res. 2017;11(8):DC18.
309. Richard S, Moslemi S, Sipahutar H, Benachour N, Seralini G-E. Differential effects of glyphosate and roundup on human placental cells and aromatase. Environ Health Perspect. 2005;113(6):716–20.
310. Samsel A, Seneff S. Glyphosate, pathways to modern diseases II: celiac sprue and gluten intolerance. Interdiscip Toxicol. 2013;6(4):159–84.
311. Vasiluk L, Pinto LJ, Moore MM. Oral bioavailability of glyphosate: studies using two intestinal cell lines. Environ Toxicol Chem. 2005;24(1):153–60.
312. Gildea JJ, Roberts DA, Bush Z. Protective effects of lignite extract supplement on intestinal barrier function in glyphosate-mediated tight junction injury. J Clin Nutr Diet. 2017;3:1.
313. Nature. International Weekly Journal of Science. Paper claiming GM link with tumours republished [Internet]. Barbara Casassus: Nature; 2014 Jun 24. Available from: https://www.nature.com/news/paper-claiming-gm-link-with-tumours-republished-1.15463.
314. Williams GM, Kroes R, Munro IC. Safety evaluation and risk assessment of the herbicide roundup and its active ingredient, glyphosate, for humans. Regul Toxicol Pharmacol. 2000;31(2):117–65.
315. Ioannidis JPA. Why most published research findings are false: authors reply to Goodman and Greenland. PLoS Med. 2007;4(6):e215.
316. Goldstein BD. The precautionary principle also applies to public health actions. Am J Public Health. 2001;91(9):1358–61.
317. Williams Z. Inducing tolerance to pregnancy. N Engl J Med. 2012;367(12):1159–61.
318. Nehar-Belaid D, Courau T, Dérian N, Florez L, Ruocco MG, Klatzmann D. Regulatory T cells orchestrate similar immune evasion of fetuses and tumors in mice. J Immunol. 2015;196(2):678–90.
319. Bonney EA. Immune regulation in pregnancy: a matter of perspective? Obstet Gynecol Clin N Am. 2016;43(4):679–98.
320. Kanellopoulos-Langevin C, Caucheteux SM, Verbeke P, Ojcius DM. Tolerance of the fetus by the maternal immune system: role of inflammatory mediators at the feto-maternal interface. Reprod Biol Endocrinol. 2003;1:121.
321. Krausse-Opatz B, Wittkop U, Gutzki FM, Schmidt C, Jürgens-Saathoff B, Meier S, et al. Free iron ions decrease indoleamine 2,3-dioxygenase expression and reduce IFNγ-induced inhibition of Chlamydia trachomatis infection. Microb Pathog. 2009;46(6):289–97.

322. Trabanelli S, Lecciso M, Salvestrini V, Cavo M, Očadlíková D, Lemoli RM, et al. PGE2-induced IDO1 inhibits the capacity of fully mature DCs to elicit an in vitro antileukemic immune response. J Immunol Res. 2015;2015:1–10.
323. Lee SY, Choi HK, Lee KJ, Jung JY, Hur GY, Jung KH, et al. The immune tolerance of cancer is mediated by IDO that is inhibited by COX-2 inhibitors through regulatory T cells. J Immunother. 2009;32(1):22–8.
324. Flower R. What are all the things that aspirin does? This fascinating but simple and cheap drug has an assured future. BMJ. 2003;327(7415):572–3.
325. Schisterman EF, Gaskins AJ, Whitcomb BW. Effects of low-dose aspirin in in-vitro fertilization. Curr Opin Obstet Gynecol. 2009;21(3):275–8.
326. Shah BH, Catt KJ. Roles of LPA3 and COX-2 in implantation. Trends Endocrinol Metab. 2005;16(9):397–9.
327. Wang H, Ma W-G, Tejada L, Zhang H, Morrow JD, Das SK, et al. Rescue of female infertility from the loss of cyclooxygenase-2 by compensatory up-regulation of cyclooxygenase-1 is a function of genetic makeup. J Biol Chem. 2003;279(11):10649–58.
328. Norman RJ, Wu R. The potential danger of COX-2 inhibitors. Fertil Steril. 2004;81(3):493–4.
329. Taber L, Chiu C-H, Whelan J. Assessment of the arachidonic acid content in foods commonly consumed in the American diet. Lipids. 1998;33(12):1151–7.
330. Bhutia YD, Ganapathy V. Short, but smart: SCFAs train T cells in the gut to fight autoimmunity in the brain. Immunity. 2015;43(4):629–31.
331. Yessoufou A, Plé A, Moutairou K, Hichami A, Khan NA. Docosahexaenoic acid reduces suppressive and migratory functions of CD4CD25 regulatory T-cells. J Lipid Res. 2009;50(12):2377–88.
332. Tong X, Yin L, Giardina C. Butyrate suppresses Cox-2 activation in colon cancer cells through HDAC inhibition. Biochem Biophys Res Commun. 2004;317(2):463–71.
333. Zhang M, Zhou Q, Dorfman RG, Huang X, Fan T, Zhang H, et al. Butyrate inhibits interleukin-17 and generates Tregs to ameliorate colorectal colitis in rats. BMC Gastroenterol. 2016;16:84.
334. Bost J, Maroon A, Maroon J. Natural anti-inflammatory agents for pain relief. Surg Neurol Int. 2010;1(1):80.
335. Navikas SI, Teimer R, Bockermann R. Influence of dietary components on regulatory T cells. Mol Med. 2012;18(1):95–110.
336. Hasni SA. Role of helicobacter pylori infection in autoimmune diseases. Curr Opin Rheumatol. 2012;24(4):429–34.
337. Chen Y, Segers S, Blaser MJ. Association between Helicobacter pylori and mortality in the NHANES III study. Gut. 2013;62(9):10.
338. Nagase H, Takeoka T, Urakawa S, Morimoto-Okazawa A, Kawashima A, Iwahori K, et al. ICOS+ Foxp3+ TILs in gastric cancer are prognostic markers and effector regulatory T cells associated with Helicobacter pylori. Int J Cancer. 2016;140(3):686–95.
339. Moretti E, Figura N, Collodel G, Ponzetto A. Can Helicobacter pylori infection influence human reproduction? World J Gastroenterol. 2014;20(19):5567–74.
340. Moreno I, Codoñer FM, Vilella F, Valbuena D, Martinez-Blanch JF, Jimenez-Almazán J, et al. Evidence that the endometrial microbiota has an effect on implantation success or failure. Am J Obstet Gynecol. 2016;215(6):684–703.
341. Hyman RW, Herndon CN, Jiang H, Palm C, Fukushima M, Bernstein D, et al. The dynamics of the vaginal microbiome during infertility therapy with in vitro fertilization-embryo transfer. J Assist Reprod Genet. 2012;29(2):105–15.
342. Selman H, Mariani M, Barnocchi N, Mencacci A, Bistoni F, Arena S, et al. Examination of bacterial contamination at the time of embryo transfer, and its impact on the IVF/pregnancy outcome. J Assist Reprod Genet. 2007;24(9):395–9.
343. Samplaski MK, Domes T, Jarvi KA. Chlamydial infection and its role in male infertility. Adv Androl. 2014;2014:1–11.
344. Brown M, Potroz M, Teh S-W, Cho N-J. Natural products for the treatment of chlamydiaceae infections. Microorganisms. 2016;4(4):39.
345. Paavonen J. Chlamydia trachomatis: impact on human reproduction. Hum Reprod Update. 1999;5(5):433–47.
346. Donders GG, Bulck BV, Caudron J, Londers L, Vereecken A, Spitz B. Relationship of bacterial vaginosis and mycoplasmas to the risk of spontaneous abortion. Am J Obstet Gynecol. 2000;183(2):431–7.
347. Kranjcic-Zec I, Dzamic A, Mitrovic S, Arsenijevic-Arsic V, Radonjic I. The role of parasites and fungi in secondary infertility. Med Pregl. 2004;57(1–2):30–2.
348. Li W, Chang Y, Liang S, Zhong Z, Li X, Wen J, et al. NLRP3 inflammasome activation contributes to Listeria monocytogenes-induced animal pregnancy failure. BMC Vet Res. 2016;12(1):36.
349. Cicinelli E, Matteo M, Tinelli R, Pinto V, Marinaccio M, Indraccolo U, et al. Chronic endometritis due to common bacteria is prevalent in women with recurrent miscarriage as confirmed by improved pregnancy outcome after antibiotic treatment. Reprod Sci. 2013;21(5):640–7.
350. Giakoumelou S, Wheelhouse N, Cuschieri K, Entrican G, Howie SEM, Horne AW. The role of infection in miscarriage. Hum Reprod Update. 2016;22(1):116–33.
351. Donders GG, Calsteren VK, Bellen G, Reybrouck R, Bosch VT, Riphagen I, Lierde VS. Predictive value for preterm birth of abnormal vaginal flora, bacterial vaginosis and aerobic vaginitis during the first trimester of pregnancy. BJOG. 2009;116(10):1315–24.
352. Moberg PJ, Gottlieb C, Nord CE. Anaerobic bacteria in uterine infection following first trimester abortion. Eur J Clin Microbiol. 1982;1(2):82–6.
353. Mosca A, Leclerc M, Hugot JP. Gut microbiota diversity and human diseases: should we reintroduce key predators in our ecosystem? Front Microbiol. 2016;7:455.
354. Jespers V, Kyongo J, Joseph S, Hardy L, Cools P, Crucitti T, et al. A longitudinal analysis of the vaginal microbiota and vaginal immune mediators in women from sub-Saharan Africa. Sci Rep. 2017;7:11974.
355. Ravel J, Gajer P, Abdo Z, Schneider MG, Koenig SSK, McCulle SL, et al. Vaginal microbiome of reproductive-age women. Proc Natl Acad Sci U S A. 2011;108(Suppl 1):4680–7.
356. Romero R, Hassan SS, Gajer P, Tarca AL, Fadrosh DW, Bieda J, et al. The vaginal microbiota of pregnant women who subsequently have spontaneous preterm labor and delivery and those with a normal delivery at term. Microbiome. 2014;2(1):18.
357. Brocklehurst P, Gordon A, Heatley E, Milan SJ. Antibiotics for treating bacterial vaginosis in pregnancy. Cochrane Database Syst Rev. 2013;(1):CD000262.
358. The Guardian. Giving antibiotics to babies may lead to obesity, researchers claim [Internet]. Ian Sample: The Guardian; 2014 Aug 14. Available from: https://www.theguardian.com/science/2014/aug/14/antibiotics-babies-toddlers-obesity-fat.
359. Flenady V, Hawley G, Stock OM, Kenyon S, Badawi N. Prophylactic antibiotics for inhibiting preterm labour with intact membranes. Cochrane Database Syst Rev. 2013;(12):CD000246.
360. Kuperman AA, Koren O. Antibiotic use during pregnancy: how bad is it? BMC Med. 2016;14:91.
361. Bertuccini L, Russo R, Iosi F, Superti F. Effects of Lactobacillus rhamnosus and Lactobacillus acidophilus on bacterial vaginal pathogens. Int J Immunopathol Pharmacol. 2017;30(2):163–7.
362. Sanz Y. Gut microbiota and probiotics in maternal and infant health. Am J Clin Nutr. 2011;94(Suppl-6):2000S–5S.
363. Bayley TM, Dye L, Jones S, Debono M, Hill AJ. Food cravings and aversions during pregnancy: relationships with nausea and vomiting. Appetite. 2002;38(1):45–51.
364. Steinmetz AR, Abrams ET, Young SL. Patterns of nausea, vomiting, aversions, and cravings during pregnancy on Pemba Island, Zanzibar, Tanzania. Ecol Food Nutr. 2012;51(5):418–30.

365. Hinkle SN, Mumford SL, Grantz KL, Silver RM, Mitchell EM, Sjaarda LA, et al. Association of nausea and vomiting during pregnancy with pregnancy loss. JAMA Intern Med. 2016;176(11):1621–7.
366. Hook EB. Dietary cravings and aversions during pregnancy. Am J Clin Nutr. 1978;31(8):1355–62.
367. Shirasaka Y, Shichiri M, Mori T, Nakanishi T, Tamai I. Major active components in grapefruit, orange, and apple juices responsible for OATP2B1-mediated drug interactions. J Pharm Sci. 2013;102(9):3418–26.
368. Chao C-L, Weng C-S, Chang N-C, Lin J-S, Kao S-T, Ho F-M. Naringenin more effectively inhibits inducible nitric oxide synthase and cyclo-oxygenase-2 expression in macrophages than in microglia. Nutr Res. 2010;30(12):858–64.
369. Yang H-L, Chen S-C, Kumar KJS, Yu K-N, Chao P-DL, Tsai S-Y, et al. Antioxidant and anti-inflammatory potential of hesperetin metabolites obtained from hesperetin-administered rat serum: an ex vivo approach. J Agric Food Chem. 2011;60(1):522–32.
370. Bloom MS, Fitzgerald EF, Kim K, Neamtiu I, Gurzau ES. Spontaneous pregnancy loss in humans and exposure to arsenic in drinking water. Int J Hyg Environ Health. 2010;213(6):401–13.
371. Hill EE, Zack E, Battaglini C, Viru M, Viru A, Hackney AC. Exercise and circulating cortisol levels: the intensity threshold effect. J Endocrinol Investig. 2008;31(7):587–91.
372. Olive DL. Exercise and fertility: an update. Curr Opin Obstet Gynecol. 2010;22(4):259–63.
373. Wise LA, Rothman KJ, Mikkelsen EM, Sørensen HT, Riis AH, Hatch EE. A prospective cohort study of physical activity and time-to-pregnancy. Fertil Steril. 2012;97(5):1136–42.
374. Best D, Avenell A, Bhattacharya S. How effective are weight-loss interventions for improving fertility in women and men who are overweight or obese? A systematic review and meta-analysis of the evidence. Hum Reprod Update. 2017;23(6):681–705.
375. Madsen M, Jørgensen T, Jensen M, Juhl M, Olsen J, Andersen P, et al. Leisure time physical exercise during pregnancy and the risk of miscarriage: a study within the Danish National Birth Cohort. J Reprod Immunol. 2011;90(2):156–7.
376. Hwang B-F, Jaakkola JJ, Guo H-R. Water disinfection by-products and the risk of specific birth defects: a population-based cross-sectional study in Taiwan. Environ Health. 2008;7(1):23.
377. Jedlicka P. Revisiting the quantum brain hypothesis: toward quantum (neuro)biology? Front Mol Neurosci. 2017;10:366.
378. Hameroff SR, Craddock TJ, Tuszynski JA. Quantum effects in the understanding of consciousness. J Integr Neurosci. 2014;13(2):229–52.
379. Stoleru S, Teglas J, Fermanian J, Spira A. Psychological factors in the aetiology of infertility: a prospective cohort study. Hum Reprod. 1993;8(7):1039–46.
380. Balk J, Catov J, Horn B, Gecsi K, Wakim A. The relationship between perceived stress, acupuncture, and pregnancy rates among IVF patients: a pilot study. Complement Ther Clin Pract. 2010;16(3):154–7.
381. Database of Abstracts of Reviews of Effects (DARE): quality-assessed reviews. York (UK): Centre for Reviews and Dissemination (UK); 1995. https://www.ncbi.nlm.nih.gov/books/NBK285222/.
382. Zorrilla M, Yatsenko AN. The genetics of infertility: current status of the field. Curr Genet Med Rep. 2013;1(4):247–60.
383. Park JH, Lee HC, Jeong Y-M, Chung T-G, Kim H-J, Kim NK, et al. MTHFR C677T polymorphism associates with unexplained infertile male factors. J Assist Reprod Genet. 2005;22(9–10):361–8.
384. Thaler C. Folate metabolism and human reproduction. Geburtshilfe Frauenheilkd. 2014;74(09):845–51.
385. Crider KS, Bailey LB, Berry RJ. Folic acid food fortification – its history, effect, concerns, and future directions. Nutrients. 2011;3(3):370–84.
386. Tamura T, Picciano MF. Folate and human reproduction. Am J Clin Nutr. 2006;83(5):993–1016.
387. Arruda ITSD, Persuhn DC, Oliveira NFPD. The MTHFR C677T polymorphism and global DNA methylation in oral epithelial cells. Genet Mol Biol. 2013;36(4):490–3.
388. Choi Y, Kim JO, Shim SH, Lee Y, Kim JH, Jeon YJ, et al. Genetic variation of Methylenetetrahydrofolate Reductase (MTHFR) and Thymidylate Synthase (TS) genes is associated with idiopathic recurrent implantation failure. PLoS One. 2016;11(8):e0160884.
389. Methylenetetrahydrofolate Reductase (MTHFR, 1p36.3). Encyclopedic dictionary of genetics, genomics and proteomics. Updated on 2018 May 6.
390. Mukhopadhyay R, Saraswathy KN, Ghosh PK. MTHFR C677T and factor V Leiden in recurrent pregnancy loss: a study among an endogamous group in North India. Genet Test Mol Biomarkers. 2009;13(6):861–5.
391. Govindaiah V, Naushad SM, Prabhakara K, Krishna PC, Devi ARR. Association of parental hyperhomocysteinemia and C677T methylenetetrahydrofolate reductase (MTHFR) polymorphism with recurrent pregnancy loss. Clin Biochem. 2009;42(4–5):380–6.
392. Irfan M, Ismail M, Beg MA, Shabbir A, Kayani AR, Raja GK. Association of the MTHFR C677T (rs1801133) polymorphism with idiopathic male infertility in a local Pakistani population. Balkan J Med Genet. 2016;19(1):51–62.
393. Ocal P, Ersoylu B, Cepni I, Guralp O, Atakul N, Irez T, et al. The association between homocysteine in the follicular fluid with embryo quality and pregnancy rate in assisted reproductive techniques. J Assist Reprod Genet. 2012;29(4):299–304.
394. García-Minguillán CJ, Fernandez-Ballart JD, Ceruelo S, Ríos L, Bueno O, Berrocal-Zaragoza MI, Molloy AM, Ueland PM, Meyer K, Murphy MM. Riboflavin status modifies the effects of methylenetetrahydrofolate reductase (MTHFR) and methionine synthase reductase (MTRR) polymorphisms on homocysteine. Genes Nutr. 2014;9(6):435.
395. Lonn E, Yusuf S, Arnold MJ, Sheridan P, Pogue J, Micks M, McQueen MJ, Probstfield J, Fodor G, Held C, Genest J Jr. Heart outcomes prevention evaluation (HOPE) 2 investigators. Homocysteine lowering with folic acid and B vitamins in vascular disease. N Engl J Med. 2006;354:1567–77.
396. Duva M, Micco PD, Strina I, Alviggi C, Iannuzzo M, Ranieri A, et al. Hyperhomocysteinemia in women with unexplained sterility or recurrent early pregnancy loss from Southern Italy: a preliminary report. Thromb J. 2007;5(1):10.
397. Jerzak M, Putowski L, Baranowski W. Homocysteine level in ovarian follicular fluid or serum as a predictor of successful fertilization. Article in Polish. Ginekol Pol. 2003;74(9):949–52.
398. Nelen W. Homocysteine and folate levels as risk factors for recurrent early pregnancy loss. Obstet Gynecol. 2000;95(4):519–24.
399. Choudhury S, Borah A. Activation of NMDA receptor by elevated homocysteine in chronic liver disease contributes to encephalopathy. Med Hypotheses. 2015;85(1):64–7.
400. Aida T, Ito Y, Takahashi YK, Tanaka K. Overstimulation of NMDA receptors impairs early brain development in vivo. PLoS One. 2012;7(5):e36853.
401. Liu C-C, Ho W-Y, Leu K-L, Tsai H-M, Yang T-H. Effects of S-adenosylhomocysteine and homocysteine on DNA damage and cell cytotoxicity in murine hepatic and microglia cell lines. J Biochem Mol Toxicol. 2009;23(5):349–56.
402. Hou N, Chen S, Chen F, Jiang M, Zhang J, Yang Y, et al. Association between premature ovarian failure, polymorphisms in MTHFR and MTRR genes and serum homocysteine concentration. Reprod Biomed Online. 2016;32(4):407–13.
403. Pietrzik K, Bailey L, Shane B. Folic acid and L-5-Methyltetrahydrofolate: comparison of clinical pharmacokinetics and pharmacodynamics. Clin Pharmacokinet. 2010;49(8):535–48.
404. MTHFR.Net. Methylfolate side effects [Internet]. Dr Lynch: MTHFR Mutations; 2012 Mar 1. Available from: http://mthfr.net/methylfolate-side-effects/2012/03/01/.

405. Farquharson J, Adams JF. The forms of vitamin B12 in foods. Br J Nutr. 1976;36(01):127–36.
406. Hoffbrand AV, Weir DG. The history of folic acid. Br J Haematol. 2001;113(3):579–89.
407. Kirke PN, Molloy AM, Daly LE, Burke H, Weir DG, Scott JM. Maternal plasma folate and vitamin B_{12} are independent risk factors for neural tube defects. Q J Med. 1993;86(11):703–8.
408. Delchier N, Herbig A-L, Rychlik M, Renard CM. Folates in fruits and vegetables: contents, processing, and stability. Compr Rev Food Sci Food Saf. 2016;15(3):506–28.
409. Cavalli P. Prevention of neural tube defects and proper folate periconceptional supplementation. J Prenat Med. 2008;2(4):40–1.
410. Czeizel AE, Bártfai Z, Bánhidy F. Primary prevention of neural-tube defects and some other congenital abnormalities by folic acid and multivitamins: history, missed opportunity and tasks. Ther Adv Drug Saf. 2011;2(4):173–88.
411. Botto LD, Yang Q. 5, 10-methylenetetrahydrofolate reductase gene variants and congenital anomalies: a huge review. Am J Epidemiol. 2000;151(9):862–77.
412. Obeid R, Holzgreve W, Pietrzik K. Is 5-methyltetrahydrofolate an alternative to folic acid for the prevention of neural tube defects? J Perinat Med. 2013;41(5):469.
413. Crino P. Faculty of 1000 evaluation for association of the maternal MTHFR C677T polymorphism with susceptibility to neural tube defects in offsprings: evidence from 25 case-control studies. F1000 – post-publication peer review of the biomedical literature. 2012.
414. Joshi R, Adhikari S, Patro B, Chattopadhyay S, Mukherjee T. Free radical scavenging behavior of folic acid: evidence for possible antioxidant activity. Free Radic Biol Med. 2001;30(12):1390–9.
415. Yuan Y, Zhang L, Jin L, Liu J, Li Z, Wang L, et al. Markers of macromolecular oxidative damage in maternal serum and risk of neural tube defects in offspring. Free Radic Biol Med. 2015;80:27–32.
416. Troen AM, Mitchell B, Sorensen B, Wener MH, Johnston A, Wood B, et al. Unmetabolized folic acid in plasma is associated with reduced natural killer cell cytotoxicity among postmenopausal women. J Nutr. 2006;136(1):189–94.
417. Bont RD. Endogenous DNA damage in humans: a review of quantitative data. Mutagenesis. 2004;19(3):169–85.
418. Shen HM, Chia SE, Ong CN. Evaluation of oxidative DNA damage in human sperm and its association with male infertility. J Androl. 1999;20(6):718–23.
419. Cocuzza M, Sikka SC, Athayde KS, Agarwal A. Clinical relevance of oxidative stress and sperm chromatin damage in male infertility: an evidence based analysis. Int Braz J Urol. 2007;33(5):603–21.
420. Opuwari CS, Henkel RR. An update on oxidative damage to spermatozoa and oocytes. Biomed Res Int. 2016;2016:1–11.
421. Wang C-Y, Liu L-N, Zhao Z-B. The role of ROS toxicity in spontaneous aneuploidy in cultured cells. Tissue Cell. 2013;45(1):47–53.
422. Poljsak B, Šuput D, Milisav I. Achieving the balance between ROS and antioxidants: when to use the synthetic antioxidants. Oxidative Med Cell Longev. 2013;2013:1–11.
423. Vitamins and minerals for subfertility in women. Cochrane Database of Systematic Reviews: Plain Language Summaries [Internet]. 2017 July 28.
424. Das L, Parbin S, Pradhan N, Kausar C, Patra SK. Epigenetics of reproductive infertility. Front Biosci (Schol Ed). 2017;9:509–35.
425. Stuppia L, Franzago M, Ballerini P, Gatta V, Antonucci I. Epigenetics and male reproduction: the consequences of paternal lifestyle on fertility, embryo development, and children lifetime health. Clin Epigenet. 2015;7(1):120.
426. Kanherkar RR, Bhatia-Dey N, Csoka AB. Epigenetics across the human lifespan. Front Cell Dev Biol. 2014;2:49.
427. Handy DE, Castro R, Loscalzo J. Epigenetic modifications: basic mechanisms and role in cardiovascular disease. Circulation. 2011;123(19):2145–56.
428. Dada R, Kumar M, Jesudasan R, Fernández JL, Gosálvez J, Agarwal A. Epigenetics and its role in male infertility. J Assist Reprod Genet. 2012;29(3):213–23.
429. Eggert SL, Huyck KL, Somasundaram P, Kavalla R, Stewart EA, Lu AT, et al. Genome-wide linkage and association analyses implicate FASN in predisposition to uterine leiomyomata. Am J Hum Genet. 2012;91(4):621–8.
430. United States Centers for Disease Control and Prevention. Reported STDs in the United States. CDC. [Internet]; 2016 [cited 2018 May 8]. Available from: https://www.cdc.gov/nchhstp/newsroom/docs/factsheets/std-trends-508.pdf.
431. Shiadeh MN, Niyyati M, Fallahi S, Rostami A. Human parasitic protozoan infection to infertility: a systematic review. Parasitol Res. 2015;115(2):469–77.
432. Kunisawa J, Hashimoto E, Ishikawa I, Kiyono H. A pivotal role of vitamin B9 in the maintenance of regulatory T cells in vitro and in vivo. PLoS One. 2012;7(2):e32094.

Lifestyle Patterns of Chronic Disease

Sarah Harding Laidlaw

33.1 History and Definition of Chronic Disease – 564

33.2 Underlying Origins of Chronic Disease – 564

33.3 Primary Origins – 564
33.3.1 Genetic – 564
33.3.2 Epigenetic – 564
33.3.3 Preconception and Intrauterine – 565
33.3.4 Early Life Experiences – 565

33.4 Secondary Origins – 565
33.4.1 Stress – 565
33.4.2 Smoking – 566
33.4.3 Alcohol – 566
33.4.4 Sleep – 566
33.4.5 Social Determinants – 567
33.4.6 Physical Activity – 568
33.4.7 Dietary Patterns – 568

33.5 Patterns Difficult to Control on an Individual Basis – 569
33.5.1 Food Insecurity – 569
33.5.2 Environmental Toxins – 570
33.5.3 Legislative Policy and Chronic Disease – 571

33.6 Tertiary Origins – 571

References – 574

© Springer Nature Switzerland AG 2020
D. Noland et al. (eds.), *Integrative and Functional Medical Nutrition Therapy*,
https://doi.org/10.1007/978-3-030-30730-1_33

33.1 History and Definition of Chronic Disease

Chronic disease is a relatively new phenomenon that began around the time of the industrial revolution when machines took over the work of men, with recognition beginning after World War II. Unhealthy lifestyle behaviors are at the root of the global burden of noncommunicable diseases and account for about 63% of all deaths [1]. Common and costly, most chronic diseases are preventable, noting that 60–70% of primary care visits in developed countries are for preventable conditions [2].

According to the World Health Organization (WHO), "Noncommunicable diseases (NCDs), also known as chronic diseases, are not passed from person to person. They are of long duration and generally slow progression. The four main types of noncommunicable diseases are cardiovascular diseases (like heart attacks and stroke), cancers, chronic respiratory diseases (such as chronic obstructive pulmonary disease and asthma) and diabetes." [3]

There has been a significant shift to noncommunicable diseases, as noted in the 2010 Global Burden of Disease (GBD) estimates. Years lived with disability (YLD) as a result of communicable, maternal, neonatal, and nutrition-related diseases decreased by 19.5% between 1990 and 2010, while YLD due to chronic disease increased: cardiovascular diseases by 17.7%; chronic respiratory disease by 8.5%; neurological conditions by 12.2%; diabetes by 30.0%; and mental and behavioral illnesses by 5.0% [4].

33.2 Underlying Origins of Chronic Disease

Causes of chronic disease can be loosely categorized into primary, secondary, and tertiary management groupings characterized by the level of control one has over them. With primary causes, there is little control as they have occurred either transgenerationally or within the womb. Secondary causes are those that there is some control over during one's lifetime such as diet, smoking, or exercise. The tertiary category focuses on preventing or managing the chronic disease based on what is known about the individual by identifying such risk markers as low serum vitamin D levels, chronic inflammatory markers, a low nutrient-dense diet, high intake of trans fats, and genomic risk (e.g., celiac disease) and addressing post-diagnosis complications of these conditions.

33.3 Primary Origins

33.3.1 Genetic

A person's genes are one factor in determining whether chronic illness develops early in life, in later years, or if at all. Genetic predisposition to a chronic disease involves the presence of one or more gene mutations and or combination of alleles. The presence of gene mutations or alleles may not be abnormal, but environmental influences and multiple interactions may determine disease occurrence. The risk of developing chronic disease is influenced by genetics and exposures to gene-altering conditions and interactions with these exposures. Whether our inherited genomes are the primary causes of chronic diseases or just part of the equation is still being investigated. A study of monozygotic twins from Western European countries was used to estimate population attributable factors of 28 chronic diseases. Of 1.53 million deaths in 2000, 0.25% was attributable to genetics plus shared exposures. Genetic population attributable factors ranged from 3.4% for leukemia to 48.6% for asthma. Cancers had the lowest population attributable factors (median, 8.26%), while neurological (median, 26.1%) and lung (median, 33.6%) diseases had the highest population attributable factors [5].

The increasing incidence of obesity is a public health concern that has far-reaching consequences. It is associated with the development of several of the chronic diseases that are impacting health and healthcare globally. Genetic links to obesity have been identified and, combined with sensitivity to environmental cues, may provide at least one explanation for how genetic risk for obesity may promote unhealthy eating behaviors. Using well-known obesity gene FTO rs9939609 SNP in a sample of 78 children, ages 9–12 years, Rapuano et al. observed that children at risk for obesity exhibited a stronger response to food commercials than children not at risk. These results suggest that children who are genetically at risk for obesity respond to reward signals stronger than those not at risk, which in turn may lead to lifelong unhealthy eating behaviors [6]. In addition, this polymorphism has been related to increased BMI and obesity throughout life, as well as type 2 diabetes, and influences food intake and food choice but not energy expenditure [7].

33.3.2 Epigenetic

Recognition of the origins of chronic disease during early periods of development may provide the key to understanding, preventing, and treating chronic diseases in the future. It is now well understood that early life events play a major role in the development of chronic disease later on. The four critical periods for the development of obesity may very well be the same for other chronic, non-obesity-related disease. These periods of development are prenatal, early childhood, adolescence, and pregnancy [8].

The social determinant of health, defined as any nonmedical factor directly influencing health, including values, attitudes, knowledge and behaviors, family, neighborhood, and social network, has recently attracted interest among researchers. Social disadvantage is compellingly related to health throughout life [9]. Environmental factors, including diet, nurturing, and socioeconomic status, are closely linked to one's genetic code and have the propensity to leave biochemical marks on the genome. Factors such as physical and geographic characteristics of neighborhoods including food deserts, air quality, lead paint exposure, racial disparity, or inequality influence health, including neuroendocrine, developmental, immunologic, and vascular pathways. Information can provide an adaptive advantage or a disadvantage to offspring depending on timing and duration of the interaction between the individual and his or her environment. Early childhood social disadvantage has been shown to have an impact on later child and adult health, socioeconomic well-being, and cognitive ability. Under the right circumstances, as listed above, as well as maternal behavior, environmental signals may result in patterns of epigenetic marks in offspring that reflect those of the parent. Evidence supports the direct impact of early life experiences on epigenetic patterns in specific loci on the genome. Negative early life experiences have been linked to differences in epigenetic patterns for genes related to drug addiction, mental health, and obesity [10].

Telomeres, the crucial part of the human cell that influences aging, have also been shown to be associated with environmental and social stress. Although telomere shortening may be related to normal aging, accelerated telomere shortening has been linked to a number of acquired disease processes, including inflammation-connected cancers and coronary artery disease. Some research has proposed that stress increases the oxidative burden on the cell, damaging and reducing telomere length [11].

33.3.3 Preconception and Intrauterine

The intrauterine period appears to be one of the most sensitive times for formation of epigenetic variability influencing development and risk for a range of diseases that can develop later in life [12]. Obesity, cardiovascular disease, type 2 diabetes, stroke, metabolic syndrome, and osteoporosis later in life all appear to be related to in utero influences that result from small size and thinness at birth. Studies reviewed by Barker demonstrated that retarded growth in utero and infancy is strongly related to cardiovascular disease mortality and some of its known risk factors, proposing three explanations for the findings:
1. Birth weight is merely a marker for adverse environmental influences that appear later in life.
2. Genetic influences early in life shown as growth failure later appear as degenerative diseases, implying that genes related to low birth weight are related to adult-onset degenerative diseases.
3. Most likely, the relationship between early retarded growth and adult disease is due to long-term effects on physiology and metabolism caused by an unfavorable environment during critical periods of development [13].

The well-known Dutch Winter Famine study further revealed the long-term health outcomes of approximately 40,000 children born to mothers with gestational famine during pregnancy as a result of Nazi blockage of food supply lines during World War II. Early gestation appeared to be the most vulnerable time for developing fetuses. Those conceived during this period were found to be at higher risk for impaired glucose tolerance, obesity, schizophrenia, and depression and had an increased atherogenic lipid profile that was more than twice normal. They were more responsive to stress and were at double the rate for coronary heart disease and performed worse on cognitive tests. In addition, changes in reproductive function including earlier menopause, breast cancer, and changes in insulin-like growth factor were observed [14–16].

33.3.4 Early Life Experiences

Childhood abuse and neglect as well as maternal nurturing have been shown to have an adverse effect on adult cognition and development of psychiatric disorders. Through rodent studies, researchers have been able to demonstrate that rats who were neglected in early postnatal periods exhibited methylation of the brain-derived neurotrophic factor gene (BDNF) throughout the lifespan. As they grow up, these rodents mistreat their own offspring, and these offspring also have significant DNA methylation [17].

As far as other chronic illnesses, most data come from studies of prenatal famine during the Dutch Winter Famine. Although not as widely studied, famine postnatally is also associated with a higher incidence of coronary heart disease in women and their offspring. Children small at birth exhibited higher levels of adipose tissue as compared to lean tissue, and they exhibited more insulin resistance. Small size at birth, low weight gain during infancy, and a rapid increase in body weight/BMI during childhood are associated with metabolic syndrome, hypertension, and coronary heart disease in adulthood [18].

A recent study that looked at the children of 5034 mothers in New York State who delivered between 2008 and 2010 assessed five developmental domains (fine motor, gross motor, communication, personal-social functioning, and problem-solving ability). The study of maternal and paternal obesity, believed to be the first of its kind, established the relationship between maternal obesity and delays in fine motor development, while paternal obesity associated with delays in personal-social development. The results regarding child development were not the only facts of interest in this study, as maternal obesity was also related to paternal obesity and lower socioeconomic status. Additionally, obesity was associated with an increased likelihood of smoking, being diagnosed with gestational diabetes or hypertension, and lower likelihood of alcohol intake, multivitamin use, and fish oil supplementation during pregnancy [19].

33.4 Secondary Origins

33.4.1 Stress

Stress affects people of all ages. Life is full of stress, and it is known to have both positive and negative effects on health. When stress helps the body overcome lethargy or enhance performance, it can be positive. Short-term stress can boost the immune system, but long-term, chronic stress resulting in fatigue or inability to cope is a promoter of chronic disease and early aging. Environmental (experiential), psychological (emotional), and biological (physiological) stress can have different results on and between individuals. Chronic stress is inflammatory due to injury or infection, poor diet, quality of life including socioeconomic factors, smoking, and emotions. Increased stress can result in oxidative stress that depletes nutrient stores or increases nutrient requirements as a result of damage to the body from disease or injury. Emotional stress is a major contributing factor to six of the leading causes of death in the United States: cancer, coronary heart disease, accidental death, cirrhosis of the liver, respiratory diseases, and suicide. Chronic stress has an immunosuppressive effect that interferes with the body's ability to elicit a prompt and effective immune response. Longer exposure to stress also makes it less likely that the body can shift from adaptive response (as in fight-or-flight response) and shifts to detrimental changes at the cellular level and then broader immune system function [20].

33.4.2 Smoking

It is widely accepted that smoking is a leading cause of morbidity and mortality, not only in the United States but worldwide. Tobacco use is a modifiable risk factor for numerous chronic diseases including cancer, cardiovascular disease, diabetes, inflammatory disease, and pulmonary-related illnesses. A study conducted by Campos et al., in Brazil, found that one in ten persons with chronic disease smoked, lightly but daily, and also lived with someone who smoked. Most patients had either hypertension, diabetes, or chronic kidney disease, all comorbid conditions that are negatively impacted by cigarette smoke [21].

Smoking is considered a risk factor for chronic obstructive pulmonary disease (COPD), although not all smokers develop the disease. Use of tobacco products in persons with reflux disease is contraindicated due to its lower esophageal sphincter pressure lowering effect and decreased salivation. Additionally, smoking increases the risk of esophageal and lung cancer. Cigarette smoking also exerts a toxic effect on osteoblasts, increasing the risk for and incidence of osteoporosis [22].

There is no risk-free level of exposure to smoke. Secondhand smoke, as well as smoking itself, no matter how much, leads to rapid and sharp increase in endothelial dysfunction by injuring epithelium, impairing epithelial vasodilation, and triggering inflammation. Powerfully addictive, cigarette smoking or low-level exposure via secondhand smoke leads to inhalation of toxic chemicals burned during combustion, resulting in DNA damage, inflammation, and oxidative stress. Smoking or exposure to smoke during reproductive years can affect fertility and conception and impact fetal and child development. Tobacco smoke exposure is associated with chromosomal damage or DNA damage resulting in reduced sperm quality, possibly leading to birth defects or early fetal demise. Smoking may also be related to inflammation in the mother as well as decreased nutrient absorption, decreased nutrient uptake, and reduced levels of necessary embryonic tissue nutrients, such as multivitamins and folate. In addition, evidence is consistent that carbon monoxide from smoking leads to lower birth weights and may be linked to cognitive and neurobehavioral deficits in the infant/child.

Residual nicotine and other chemicals left from smoking, known as thirdhand smoke, can react with indoor pollutants to create a toxic mix of cancer-causing compounds. These compounds pose risk to non-smokers and young children. A relatively new concept, the residuals from thirdhand smoke, cannot be removed easily from fabrics and surfaces or from airing out a room or restricting smoking to one location. Inhaling, touching, or swallowing items exposed to thirdhand smoke may place non-smokers and children at risk of smoking-related diseases. This can be especially problematic for infants and young children who have a tendency to put things in their mouths. [22, 23]

33.4.3 Alcohol

According to the Centers for Disease Control and Prevention, alcohol is responsible for 88,000 deaths in the United States each year and is identified as single condition for 25 diseases and chronic conditions in the *International Classification of Disease* (ICD)-10 manual. Alcohol has been a part of human culture since history was first recorded. It has been noted to have both beneficial and detrimental effects on health, depending on the amount consumed. However, long-term excessive intake has been linked to tumors, neuropsychiatric conditions, and many cardiovascular and digestive diseases, while short-term health risks include violence, injury, risky sexual behavior, and risk of miscarriage or preterm birth. Excessive alcohol consumption has been identified as responsible for 1 in 10 deaths among working age adults ages 20–64 in 2010 and involves binge drinking (for women, 4 or more drinks during a single occasion and 5 for men), heavy drinking (8 or more drinks per week for women and 15 or more for men), and any drinking by pregnant women or people younger than age 21 [24–27].

Some people use alcohol to reduce stress or believe it is a mood elevator. Social influences and norms can also prompt a person to drink, and, in some instances, too much. Low levels of parental involvement and support have been shown to increase the likelihood of adolescent alcohol use and other problem behaviors, including smoking. A family history of alcoholism has been noted in the development of alcoholism, and studies have clearly demonstrated that alcoholism is a genetically complex disease, influenced by multiple genes interacting with each other and with environmental factors to cause disease [24–27].

33.4.4 Sleep

Insufficient sleep alters the endocrine regulation of hunger, appetite, and energy expenditure. In the short term, lack of sleep inhibits leptin production, suggesting a potential mechanism for the early development of obesity related to increased fat mass [27, 28]. Sleep deprivation, shift work, and exposure to bright light at night can have an impact on the body's circadian rhythm and increase the probability of developing obesity. A prospective cohort study of 8234 children aged 7 years and a subsample of 909 children (children in focus) with data on additional early growth-related risk factors for obesity in Great Britain revealed a relationship between lack of sleep and obesity. Amount of sleep in children at 30 months of age was independently correlated with obesity at 7 years of age. Those children with the lowest amount of sleep (<10.5 hours and 10.5–10.9 hours) before age 7 were more likely to be obese than those recording 12 hours of sleep or more [29].

Obesity, in and of itself, is a risk factor for developing sleep apnea, which can further exacerbate chronic illness. The presence of sleep apnea in obese individuals is estimated to be up to 45% of subjects observed. A review of six large studies conducted worldwide suggests that 25% of individuals with a BMI between 25 kg/m^2 and 28 kg/m^2 have at least some mild sleep apnea. Independent of obesity, several cardiometabolic abnormalities have been identified in persons with sleep apnea, including glucose intolerance and insulin resistance, both of which are precursors to diabetes and cardiovascular disease. In addition, persons with obstructive sleep apnea, have exhibited a heightend systemic inflammatory state as indicated by increased cytokines, serum amyloid- A production and in some, elevations in C-reactive protein [28].

33.4.5 Social Determinants

The Commission on Social Determinants of Health defines social determinants of health as economic and social conditions that influence the health of people and communities. Money, power, and resources that people have, which are often influenced by policy choices, form these environments [30]. The factors determining the social determinants of health are related to health outcomes and include:
- Early childhood development
- Education level
- Employment and retaining a job
- Type of work a person does
- Food security
- Access to healthcare and quality of service
- Housing, including neighborhood quality and safety
- How much money a person earns
- Discrimination and social support (CDC, Social Determinants of Health)

They are the conditions in which individuals are born, grow, live, work, and age. Notterman (2015) refers to the social determinants of health as any nonmedical factor directly influencing health, such as values, attitudes, knowledge, and behavior [10]. Family, neighborhood, and social network context are external sources of influence. Social disadvantage, risk exposure, and social inequalities are recognized as having causal roles impacting health across the lifespan [9].

Bharmal et al. have proposed main categories of upstream factors that provide a better understanding of how these macro factors lead to poor health and inequities. Greater social disadvantage appears to be dose-responsive and is associated with poorer health. This social disadvantage approach is similar to, but elaborates on, those defined by the Commission on Social Determinants of Health:
1. Neighborhood conditions that can influence health through physical conditions such as air, playgrounds, and safety.
2. Working conditions with the ability to influence physical, social, and psychosocial aspects of health from injury, sedentariness, obesity and obesity-related chronic diseases, ventilation and air quality, imbalance of work efforts and rewards, and support among and of coworkers.
3. Education, which has been linked to increased knowledge, resulting in better health and healthy behaviors. It shapes employment opportunities that determine economic resources influencing health, and it influences health through psychological and social factors with a perception of greater control, higher social status, and increased social support.
4. Attainment of income and wealth while income inequality is linked with poor health due to social erosion and lack of social cohesion.
5. Ethnicity and racism affects individual opportunities based on race or ethnicity and perpetuates social disadvantage in segregated neighborhoods, unsafe housing, and low quality/poor educational opportunities.
6. Stress, as described above, may also be a causal factor of chronic disease through accumulated injury from stressful socioeconomic and environmental situations that trigger the release of cortisol, cytokines, and other substances that can damage the immune system, vital organs, and physiological systems that lead to more rapid onset – or progression to – chronic diseases, such as cardiovascular disease [9].

In addition to the social disadvantage theory, Bharmal also describes a life course approach to health that explores three life periods that are vulnerable to the effects of the risk and duration of exposure to social determinants of health.
1. Adverse health effects of early childhood experiences that are related to social disadvantage that affect children's cognitive, behavioral, and physical development, in turn affecting lifelong health.
2. Intergenerational transfer of advantage has been studied over at least two decades. Children of socially disadvantaged parents have limited educational choices and are less healthy, which translates into decreased advantage for good health and social advantage as adults. Research on gene and environment interactions proposes that intergenerational transmission of social advantage may be partly explained by changes in gene expression that are passed on to future generations.

3. In animal studies, a potential causal link of epigenetics suggests that social status can regulate the expression of genes that control the immune system. Perceived stress, occupational status, educational achievement, work schedules, and intimate partner violence have been linked with telomere shortening, resulting in cellular aging [9].

33.4.6 Physical Activity

Lifestyle shifts, including changes in access to healthcare, migration from more rural to urban areas, transportation accessibility, frequent job changes, access to more electronic devices including television, concern regarding safety in communities, and chronic disease are becoming more prominent in younger individuals, leaving them encumbered with poor health for the rest of their lives. These shifts have resulted in the prevalence of physical inactivity, a change that has been identified as being partly responsible for the earlier onset of chronic disease. People are less physically active today than they have been in past generations, with the number of daily steps taken approximately 50–70% less since industrialization and the introduction of powered machinery [31].

A sedentary lifestyle over years can lead to primary decrease in functionality, obesity, cardiovascular disease, type 2 diabetes, and premature death. Furthermore, individuals with a chronic disease are more likely to become less physically active, if they were not already, which leads to deconditioning and further decreases in the ability to perform physical activity. If this cycle is not halted, the likelihood for suboptimal quality of life and long-term poor health increases significantly.

In order for people to enjoy health benefits, WHO recommends they get at least 600 MET minutes (the physiological measure expressing the energy cost of physical activities) of total activity a week. Moderate-intensity activity such as brisk walking 150 minutes/week or 75 minutes of vigorous physical activity such as running meets these guidelines. A review of literature citing the association between physical activity and the risk of breast cancer, colon cancer, diabetes, ischemic heart disease, and ischemic stroke was completed by Kyu et al., in order to quantify the dose-response association between disease risk and physical activity. They concluded that people who achieved total physical activity levels several times the WHO recommendations experienced a significant reduction in the risk of the five diseases. Most health gains were observed at levels of 3000–4000 MET minutes/week [32].

With the exception of genetic causes, cardiovascular disease is highly correlated to physical inactivity. Daily physical activity reduces the risk of cardiovascular disease, and it improves functional capacity. Diabetes affects more than 24 million Americans, with another 60% having prediabetes and at greater risk for developing type 2 diabetes. An increase in physical inactivity has been linked to the rising rates of type 2 diabetes with increased blood lipids, blood pressure, and glucose intolerance. Obesity, linked to many chronic diseases including heart disease, diabetes, osteoarthritis, and cancer, is directly linked to physical inactivity. When excess energy (calories) is consumed and not expended, the surplus is stored as adipose tissue. Daily physical activity aids in weight loss, balancing the intake output equation, and in improvements of functional capacity. Physical inactivity has also been associated with certain types of cancer, while epidemiological studies show that increased physical activity is associated with a decreased risk of breast, colon, and prostate cancers [33].

33.4.7 Dietary Patterns

Changes in the way people eat and live are increasingly impacting the health and development of chronic disease. Diet-related chronic diseases have increased as the American diet and lifestyle has changed. A nutritional transition is occurring, whereby traditional diets are being replaced by diets higher in processed foods, animal fats, and sugar. Rising incomes have increased the demand for, and prestige of, large amounts of animal proteins, as well as generous sized meals. This transition, combined with physical inactivity, is increasing the burden of chronic disease by promoting overconsumption and underexercising (see ▶ Chap. 2).

The standard American diet (SAD) is one high in processed foods, fat, sugar, and animal products and is consumed by a majority of the population. It contains a low amount of fruit and vegetables, which is linked to a lower intake of nutrient-dense foods that contain phytonutrients, antioxidants, fiber, and other health-conferring properties. This shift in diet began after World War II and rapidly escalated with convenience foods, fast foods, and sugar-containing beverages, being the norm for families who once relied on a stay-at-home mom to prepare meals. Now working mothers with limited time and a hungry family find they must rely on options that are quick and filling.

A traditional diet, once the "norm," is one centered on plant-based foods with small amounts of animal proteins, with sweets and desserts reserved for special occasions. These plant-based foods maximize nutrients, while the SAD diet minimizes them. Traditional diets are also inspired by rich cultural and historical cuisines around the globe. Unfortunately, many of the healthier traditional diets are now threatened by exposure to Western-style foods with fast food and convenience foods permeating all corners of the globe. Eating a traditional diet can bestow benefits that are health-promoting. These diets include the diet popular in healthcare – the Mediterranean diet with its emphasis on plant-based foods, legumes, healthy fats, low fat dairy, eggs and poultry in small amounts, fish as the predominant source of protein, and red meat and sweets once a week or less. Other less well-known or practiced traditional diets include the African Heritage diet, Latin American diet, Asian diet, and more widely seen and accepted vegetarian and vegan diets. Of course, the foods and the preparation techniques of these diets can and have been changed to meet the American taste, but if eaten as intended, they can and will confer health benefits on the consumer.

33.5 Patterns Difficult to Control on an Individual Basis

33.5.1 Food Insecurity

According to the United States Department of Agriculture (USDA), "Food security means access by all people at all times to enough food for and active, healthy lifestyle." Food insecurity, on the other hand, is defined "as a lack of consistent access to enough food for an active and healthy life and or a household-level economic and social condition of limited or uncertain access to adequate food." It is further defined as low or very low food security (◘ Fig. 33.1). Both low and very low food security households with children reported similar patterns of lack of food or uncertain access to it. According to the USDA, "The defining characteristic of

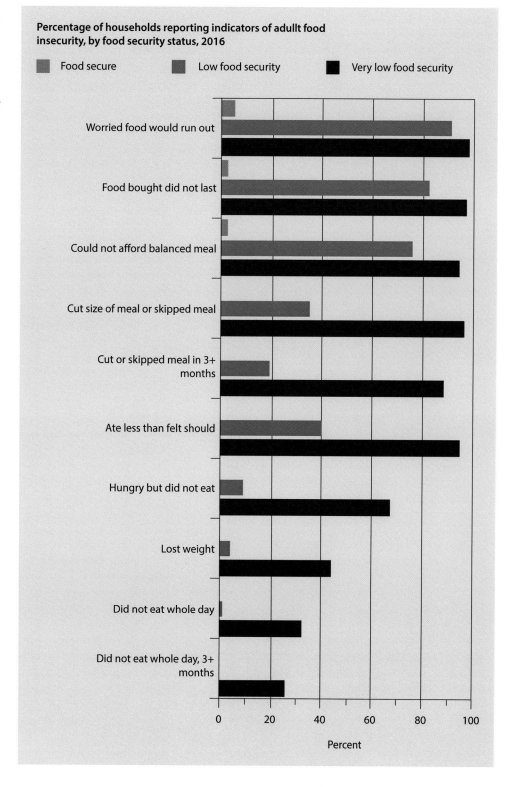

◘ Fig. 33.1 Percentage of households reporting indicators of adult food insecurity, by food security status, 2016. (Source: USDA, Economic Research Service, using data from the December 2016 Current Population Survey Food Security Supplement)

very low food security is that, at times during the year, the food intake of household members is reduced and their normal eating patterns are disrupted because the household lacks money and other resources for food [34]."

In 2015, some 42.2 million Americans lived in 15.8 million food-insecure households, with about 6.3 million of these people experiencing very low food security. Households with children reported food insecurity at a significantly higher rate than those without children and especially so in households headed by single mothers [35]. Data from several studies have demonstrated that food insufficiency is associated with poor dietary intake, inadequate micronutrient intake, and poor diet quality. Over time, these factors can contribute to the development of chronic disease including osteoporosis, sarcopenia, and anemia. Analysis of data from the National Health and Nutrition Examination Survey (NHANES) showed that of the 15,199 persons, approximately 1256 reported some level of food insecurity and experienced 1 of 3 chronic diseases – diabetes, hypertension, or chronic kidney disease. In the review of this data, food-insecure individuals were more likely to have diabetes [36]. Food-insecure adults with diabetes were observed to have poorer glycemic control than food-secure individuals. They also have a higher prevalence of metabolic syndrome and have diets of lower quality that contribute to risk of cardiometabolic disease. In addition, food insecurity was associated with several individual chronic disease risk factors including smoking, higher concentrations of HgBA1c, higher levels of C-reactive protein, and greater incidence of obesity among women. It has been suggested that financially strapped families may purchase cheap energy-dense foods that are high in calories and highly processed [37].

33.5.2 Environmental Toxins

The environmentally based contributors to chronic disease risk are, although poorly understood, known to have a cumulative and dose-responsive effect. Most Americans have detectable levels of an extensive assortment of environmental chemicals in their bodies, including many with known endocrine-disrupting, neurotoxic, and carcinogenic activities. In animal models, these are biologically and toxicologically relevant to the development of chronic diseases such as cardiovascular disease, diabetes, hypertension, obesity, and cancer [38].

Among environmental hazards are air pollution, including smoking and occupational exposure, polycyclic aromatic hydrocarbons, heavy metals, and bioaerosols linked to chronic disease through epigenetic processes, including gene methylation and histone modification. Although epigenetic modification and transgenerational inheritance of disease risk have been demonstrated in animals, definitive evidence for heritable environmentally induced epigenetic changes in humans is limited. However, there are epidemiologic studies that establish a relationship between maternal and grandmaternal smoking and the risk for asthma in children [4].

Endocrine-disrupting chemicals (EDC), widely present in the environment, are natural or synthetic compounds which, through environmental or inappropriate developmental exposures, alter the hormonal and homeostatic systems that enable the body to communicate with and respond to the environment. The most commonly studied EDCs are bisphenol A (BPA), phthalates, atrazine (ATR), polychlorinated biphenyls (PCB) and polybrominated diphenyl ethers, and DDT and DDE. BPA has long been used in a wide variety of manufacturing, food packaging, toys, and other uses and is produced and used in larger amounts than any other chemical. Resins of BPA are found in foods and beverages regularly consumed. Phthalates and their esters are used as liquid plasticizers and found in everyday products such as plastics, coating, cosmetics, and medical tubing. Atrazine is an herbicide used for control of broadleaf grasses and has been the major herbicide used worldwide since 1959; it and its metabolites are common groundwater contaminants. p,p'-Dichlorodiphenyltrichloroethane (DDT) is an industrial and household synthetic insecticide. It was used extensively until 1972, when it was banned due to its effects on human health and the environment.

Dichlorodiphenyldichloroethylene (DDE) and dichlorodiphenyldichloroethane (DDD) are metabolites of DDT and have been linked to endocrine-related chronic diseases including endometrial, pancreatic and breast cancer, testicular tumors, and type 2 diabetes [38].

Toxins during gestation have been associated with adverse outcomes at birth for decades. Children exposed in utero to thalidomide were born with limb deformities, while those born to alcoholic mothers experience fetal alcohol syndrome. Insufficient folate in the maternal diet may result in spina bifida. Methyl mercury exposure has resulted in Minamata disease, and diethylstilbestrol (DES)-exposed children of both sexes may have multiple reproductive disorders, certain cancers, cryptorchidism, and other diseases. Increased disease risk is now being observed in the grandchildren of DES-treated women, as well as examples of disease in those children with transgenerational exposure to BPA, PCBs, pesticides, and other environmental toxins [38].

Air pollution, either outdoor or indoor, is known to contribute to respiratory diseases, cancers, cardiovascular disease, and adverse pregnancy outcomes. Air pollutants have been categorized into four different groups: (1) gaseous pollutants such as sulfur dioxide, carbon dioxide, ozone, or volatile organic compounds; (2) persistent organic pollutants including dioxin; (3) heavy metals such as lead and mercury; and (4) particulate matter.

All kinds of air pollution can affect the respiratory system, from itchy throat and bronchoconstriction to asthma, emphysema, and lung cancer. Carbon monoxide affects the organs that consume a lot of oxygen, such as the heart and brain, resulting in impaired concentration, slow reflexes, and confusion. Particulate matter has the ability to interfere with blood coagulation and lung inflammation and can lead to myocardial infarction or angina. Dioxin has been shown to result in ischemic heart disease and adverse effects on the nervous system by decreasing nerve conduction velocity and impaired mental development in children. Heavy metals such as lead, mercury, and arsenic have been linked to neuropathies, due to neurotoxicity, resulting in memory disturbances, sleep disorders, anger, fatigue, and other problems. Heavy metals have been associated with kidney damage and disease and dioxin with liver cell damage. During pregnancy, pollutants can lead to spontaneous abortion, reduced fetal growth, and prematurity, while lead exposure is known to lead to congenital malformations and lesions on the developing brain, interfering with infant motor and cognitive functions. Dioxins are endocrine disruptors and affect the growth and development of the fetal nervous system [39].

33.5.3 Legislative Policy and Chronic Disease

33.5.3.1 Asbestos

It is well-known that exposure to asbestos results in chronic lung disease, including lung cancer, mesothelioma, and asbestosis or fibrosis of the lung. Although its use in products is limited, its effects will be felt well into the twenty-first century. Workers who are exposed to asbestos are five times more likely to develop lung-related diseases than nonsmokers not exposed, but the disease risk rises to 50 times more likely for those who smoke. Asbestos has been seen as the principal cause of occupational cancer in the United States and a significant cause of disease and disability from noncancer-related disease. Estimates indicate that the cumulative total number of asbestos-associated deaths in the United States may exceed 200,000 by the year 2030 [40–42].

In the 1970s, some uses of asbestos were banned or otherwise regulated, and its importation from other countries was dramatically reduced in light of growing evidence concerning its connections to serious health conditions. Continued research has demonstrated the adverse effects of asbestos, and, in 2002, the United States discontinued mining it. Still, asbestos remains in use in some products today and still exists in many old buildings. As of this writing, although lawmakers have attempted to introduce legislation banning asbestos use, measures to do so have failed. Thus, there is no comprehensive law governing its use in the United States [41].

33.5.3.2 Lead

Lead has been used since the beginning of time and continues to be found in man-made and natural products, so its complete elimination is virtually impossible. The use of lead in paint, gasoline, pipes, some construction materials, and cosmetics has been banned for a number of years. However, lead from gasoline emissions and other deposits from years ago remain in the soil and will not break down or go away. Lead is still in use in lead-acid batteries, ammunition, cable coverings, piping, in the manufacturing of brass and bronze, as bearing metal in machinery, and in sheet lead. Lead compounds are used in matches, some paints and ceramics, soldering and building materials, and photography and as catalysts to help chemical reactions [40, 43].

Lead poisoning in children has brought significant attention to the topic since the 1970s. Children exposed to lead, primarily from paint chips and dust in the environment, are at greater risk due to their developing brain and rapid uptake of nutrients. Children exposed to lead often exhibit cognitive issues, including behavioral and learning problems, stunted growth, and mental retardation. Studies looking at the relationship of lead to cancer are mixed; however lead levels in adults can lead to headaches, mood swings, and problems with memory and damage peripheral nerves and kidney and reproductive function [43].

33.5.3.3 Arsenic

Arsenic exposure can occur through such occupations as mining and smelting, the burning of arsenic-rich coal, and from drinking contaminated water. Arsenic can enter the water through soil or as a result of industrial waste contamination and burning of fossil fuels, through the use of agricultural pesticides and food additives, or through natural decay of minerals in weathered rocks and soils. Arsenic in drinking water is a concern throughout the world as it can cause morbidity and mortality from a number of diseases, including skin lesions, diabetes, peripheral neuropathy, gastrointestinal symptoms, high blood pressure, and cardiovascular disease [40]. The Environmental Protection Agency (EPA) updated arsenic standards in the early 2000s after considerable delay, and while levels have declined, it remains present. The most common source of arsenic in drinking water appears to be from private water wells, although contaminants may be present in larger municipalities where the main source of drinking water is from wells [44].

33.6 Tertiary Origins

Although not an *origin* of chronic disease, tertiary is more aptly defined as the management of the chronic disease and prevention of further complications, based on what is known about the individual. This approach is central to the expanding field of lifestyle medicine, whereby the clinician uses risk factor markers to assess for and identify chronic disease. Lifestyle medicine focuses on the prevention and treatment of chronic disease caused by lifestyle factors including physical inactivity, poor diet, tobacco use, and stress and psychosocial factors. Below is a table of chronic diseases with their risk factor markers, potential origins that may help in identifying the approach appropriate for the identified chronic disease (◘ Table 33.1).

Noncommunicable, chronic diseases have surpassed communicable and infectious diseases as the cause for morbidity and mortality worldwide. More than one third of Americans have been identified as obese, and diseases related to obesity – cancer, heart disease, type 2 diabetes, and stroke – are among the leading preventable causes of morbidity and death. New approaches to identify and treat these conditions are needed to reduce the incidence of these diseases and to prevent their continued explosion. Lifestyle medicine is

Table 33.1 Chronic diseases, risk factor markers, and their potential origins

Disease	Risk factor markers	Primary origin [45]	Secondary origin
Cardiometabolic			
Obesity	Weight		Diet
Diabetes Heart disease	Glucose HgB A1c Lipids Homocysteine Blood pressure	Type 1 diabetes – HLA-DQA1, HLA-DQB1, and HLA-DRB1 genes	Smoking Stress Psychosocial Inactivity
Cancer	Obesity	BRCA1	Diet
	Diabetes	BRACA2	Chemical exposure
	Inflammation	Multiple tumor markers	Smoking
Hormonal and endocrine			
Thyroid	Hyperthyroid Hypothyroid Low vitamin D		Iodine intake Drugs Infections Chemicals
Hashimoto's thyroiditis	TSH Antithyroid antibodies Free T-4 test	HLA genes CTLA-4 gene PTPN22 gene VDR gene Cytokine genes Female Postpartum Fetal microchimerism Human leukocyte antigen (HLA) complex	Change in sex hormones Excess iodine intake Viral infections Ionizing radiation
PCOS	Thyroid function tests Prolactin Free androgen index Fasting glucose Fasting insulin Elevated blood pressure Weight Miscarriage Dysmenorrhea Hirsutism Acne	Possible multiple genes with relationship to PCOS	Stress/anxiety Low socioeconomic status
Adrenal (Cushing disease)	Weight Blood pressure Moon face Cortisol levels	Does not have a clear pattern of inheritance Somatic genetic mutations	Prolonged stress and/or depression
Autoimmune			
Allergies and intolerances	Physical measures – NFPA IgE IgG	Ethnicity Genetic factors are yet unknown	Environmental smoke Diet/nutrition Sleep Air pollution
Celiac disease	Type 1 diabetes Weight loss Skin rash	Variants of the HLA-DQA1 and HLA-DQB1 genes Possibly changes in other genes	Diet
Rheumatoid arthritis	Red, swollen, warm joints Anemia	HLA complex Various genes Sex Age Family history	Smoking Obesity Environmental pollutants Viral/bacterial infections

Table 33.1 (continued)

Disease	Risk factor markers	Primary origin [45]	Secondary origin
Systemic lupus erythematosus	Joint pain Fatigue Weight loss	Immune function-related genes	Drugs Viral infections Diet Stress Chemical exposures Sunlight
Gastrointestinal			
Inflammatory bowel disease/ulcerative colitis	Weight (loss) GI function	Variations in various genes related to protective function of intestine	Abnormal immune response to bacteria and toxins
Crohn's disease	Weight (loss) Anemia	Variations in the ATG16L1, IRGM, and NOD2 genes IL23R gene Variations in certain regions of chromosome 5 and chromosome 10	Cigarette smoking
Gastroesophageal reflux (GERD) Barrett's esophagus	Weight (obesity) Non-cardiac chest pain Dental erosion Hiatal hernia	Polygenic basis overlapping with Barrett's esophagus, but incompletely understood [46]	Alcohol Large meals Viral infections COPD
Ulcers	Weight (loss) Appetite loss	Many individuals with peptic ulcers have close relatives with ulcers, suggesting that genetic factors may be involved but has yet to be fully elucidated	*H. pylori* infection Gastritis Aspirin use Stress/severe illness Tobacco use Alcohol Stress
Developmental			
Attention deficit hyperactivity disorder (ADHD)/attention deficit disorder (ADD)	Behavioral problems	Research suggests there may be a genetic component as it appears to run in families	Exposure to lead Smoking and alcohol use during pregnancy
Autism spectrum disorder	Impaired social interaction Poor communication skills Repetitive behavior	It is believed that 50% of the risk of ASD is due to genetic variation, but exact genes are yet positively identified [47]	Drugs taken during pregnancy Environmental toxins Alcohol
Neurological			
Parkinson's disease	Tremor Weight (loss)	Mutations in LRRK2, PARK2, PARK7, PINK1, or SNCA genes. Effect of changes not fully understood	Free radical production
Alzheimer's disease	Memory loss Weight (loss) Malnutrition	Gene mutation passed from parent to child – APP, PSEN1, or PSEN2 genes	Poor diet (lack of folic acid, phytonutrients, naturally occurring nitrates, b vitamins) in middle age
Amyotrophic lateral sclerosis (ALS)	Dysphagia Muscle weakness, twitching, and cramping	Mutations in the C9orf72, SOD1, TARDBP, and FUS genes	Secondary origins do not appear to affect the development of ALS
Multiple sclerosis (MS)	Muscle weakness Tremors Difficulty walking Vision difficulties	Variations in numerous genes are thought to cause MS – HLA-DRB1, IL7R	Vitamin D intake Smoking Exposure to Epstein-Barr virus
Muscular dystrophies (Duchenne and Becker)	Progressive muscle weakness Muscle wasting	Mutations in the DMD X-linked recessive pattern carried by the mother	Secondary origins do not appear to affect the development of muscular dystrophies

becoming the recommended approach to preventing and treating chronic disease. This approach considers lifestyle intervention including physical activity, nutrition, stress reduction, and avoidance of excess alcohol and smoking as the preferred interventions. Although a challenge, given the current state of health worldwide, this approach, rather than a curative approach, is the one with the most ability to impact health over the long term.

References

1. Kushner RF, Sorensen KW. Lifestyle medicine: the future of chronic disease management. Curr Opin Endocrinol Diabetes Obes. 2013;20:389–95.
2. Egger GJ, Binns AF, Rossner SR. The emergence of "lifestyle medicine" as a structured approach for management of chronic disease. Med J Aust. 2009;190(3):143–5.
3. Noncommunicable Diseases. World Health Organization website. http://www.who.int/topics/noncommunicable_diseases/en/. Accessed 29 Nov 2016.
4. Sly P, Carpenter DO, van den Berg M, et al. Health consequences of environmental exposures: causal thinking in global environmental epidemiology. Ann Global Health. 2016;82(1):c3–9. https://doi.org/10.1016/j.aogh.2016.01.004.
5. Rappaport SM. Genetic factors are not the major cause of chronic diseases. PLoS One. 2016;11(4):e0154387. https://doi.org/10.1371/journal.pone.0154387.
6. Rapuano, KM, Zieselman AL, Kelley WM, et al. Genetic risk for obesity predicts nucleus accumbens size and responsivity to real-world food choices. PNS Early Edition. 2016. Available at www.pnas.org/cgi/doi/10.1073/pnas.1605548113. Accessed 28 Dec 2016.
7. Ahmad T, Lee IM, Pare G, et al. Lifestyle interaction with fat mass and obesity-associated (FTO) genotype and risk of obesity in apparently healthy U.S. women. Diabetes Care. 2011;34(3):675–80.
8. Dietz WH. Overweight in childhood and adolescence. NEJM. 2004;350(9):855–7.
9. Bharmal N, DeRose KP, Felician M, Weden MM. Understanding the upstream social determinants of health. RAND health. RAND social determinants of health interest Group 2015. Available at https://www.rand.org/content/dam/rand/pubs/working_papers/WR1000/WR1096/RAND_WR1096.pdf. Accessed 26 Dec 2016.
10. Notterman DA, Mitchell C. Epigenetics and understanding the impact of social determinants of health. Pediatr Clin N Am. 2015;62:1227–40.
11. Shammas MA. Telomeres, lifestyle, cancer and aging. Curr Opin Clin Nutr Metab Care. 2011;14(1):28–34. https://doi.org/10.1097/MCO.0b013e32834121b1.
12. Gluckman PD, Hanson MA, Cooper C, Thornburg KL. Effect of in utero and early-life conditions on adult health and disease. NEJM. 2008;359:61–73.
13. Barker DJP, Marytn CN. The maternal and fetal origins of cardiovascular disease. J Epidemiol Community Health. 1992;46(1):8–11.
14. Rosebloom TJ, Painter RC, an Abeelen AFM, Veenendaal MVE, de Rooij SR. Hungry in the womb: what are the consequences? Lessons from the Dutch famine. Maturitas. 2011;70(2):141–5.
15. Lumey LH. Reproductive outcomes in women prenatally exposed to undernutrition: a review of findings from the Dutch famine birth cohort. Proc Nutr Soc. 1998;57:129–35.
16. Kyle UG, Pichard C. The Dutch famine of 1944-1945: a pathophysiological model of long-term consequences of wasting disease. Curr Opin Clin Nutr Metab Care. 2006;9(4):388–94.
17. Roth TL, Lubin FL, Funk AJ, Sweatt JD. Lasting epigenetic influence of early-life adversity on the BDNF gene. Biol Psychiatry. 2009;65(9):760–9.
18. Van Ableelen AFM, Elias SG, Bossyut PMM, et al. Cardiovascular consequences of famine in the young. Eur Heart J. 2012;33(4):538–45.
19. Yeung EH, Sundaram R, Ghassabian A, Xie Y, Louis GB. Parental obesity and early childhood development. Pediatrics. 2017;139:e20161459. https://doi.org/10.1542/peds.2016-1459.
20. Salleh RM. Life event, stress and illness. Malaysian J Med Soc. 2008;15(4):9–18.
21. Campos TS, Richter KP, Cupertino AP, et al. Cigarette smoking among patients with chronic disease. Int J Cardiol. 2014;174(3):808–10.
22. Centers for Disease Control and Prevention (US); National Center for Chronic Disease Prevention and Health Promotion (US); Office on Smoking and Health (US). How Tobacco Smoke Causes Disease: The Biology and Behavioral Basis for Smoking-Attributable Disease: A Report of the Surgeon General. Atlanta (GA): Centers for Disease Control and Prevention (US); 2010. Available from: https://www.ncbi.nlm.nih.gov/books/NBK53017/. Accessed 18 Dec 2016.
23. Hays JT. What is thirdhand smoke, and why is it a concern? Mayo Foundation for Medical Education and Research (MFMER). July 13th, 2017. Retrieved from: https://www.mayoclinic.org/healthy-lifestyle/adult-health/expert-answers/third-hand-smoke/faq-20057791.
24. At a Glance. 2016 Excessive Alcohol Use. CDC Fact Sheet. Available at https://www.cdc.gov/chronicdisease/resources/publications/aag/alcohol.htm. Accessed 5 Jan 2017.
25. Fact Sheets – Alcohol use and your health. CDC Fact Sheet. Available at https://www.cdc.gov/alcohol/fact-sheets/alcohol-use.htm. Accessed 5 Jan 2017.
26. Shield KD, Parry CP, Rehm J. Focus on: chronic diseases related to alcohol use. National Institute on Alcohol Abuse and Alcoholism Available at https://pubs.niaaa.nih.gov/publications/arcr352/155-173.htm. Accessed 5 Jan 2017.
27. Alcohol Research and Health. Why do some people drink too much? Available at https://pubs.niaaa.nih.gov/publications/arh24-1/17-26.pdf. Accessed 5 Jan 2017.
28. Romero-Corral A, Caples SM, Lopez-Jimenez F, Sommers V. Interactions between obesity and sleep apnea. Chest. 2010;137(3):711–9. https://doi.org/10.1378/chest.09-0360. Accessed 19 Dec 2016.
29. Reilly JJ, Armstrong J, Dorosty AR, et al. Early life risk factors for obesity in childhood: a cohort study. BMJ. https://doi.org/10.1136/bmj.38470.670903.E0. (published 20 May 2005). Accessed 19 Dec 2016.
30. Centers for Disease Control and Prevention. NCHHSTP social determinants of health. Available at https://www.cdc.gov/nchhstp/socialdeterminants/faq.html. Accessed 19 Dec 2016.
31. Booth FW, Roberts CK, Laye MJ. Lack of exercise is a major cause of chronic disease. HIH Public Access. Available at https://www.ncbi.nlm.nih.gov/pmc/articles/PMC4241367/. Accessed 28 Dec 2016.
32. Kyu HH, Bachman VF, Alexander LT, et al. Physical activity and risk of breast cancer, colon cancer, diabetes, ischemic heart disease, and ischemic stroke events: systematic review and dose-response meta-analysis for the Global Burden of Disease Study 2013. BMJ. 2016;354:i3857. Available at https://doi.org/10.1136/bmj.i3857. Accessed 28 Dec 2016.
33. Durstine JL, Gordon B, Wang Z, Luo X. Chronic disease and the link to physical activity. J Sport Health Sci. 2013;2:3–11.
34. USDA. Definitions of food security. Available at https://www.ers.usda.gov/topics/food-nutrition-assistance/food-security-in-the-us/definitions-of-food-security.aspx. Accessed 31 Dec 2016.
35. Hunger and poverty statistics. http://www.feedingamerica.org/hunger-in-america/impact-of-hunger/hunger-and-poverty/hunger-and-poverty-fact-sheet.html. Accessed 31 Dec 2016.
36. Terrill A, Vargas R. Is food insecurity associated with chronic disease and chronic disease control? Ethnicity Dis. 2009;19:S3–3–6.
37. Ford ES. Food security and cardiovascular disease risk among adults in the United States: findings from the National Health and Examination Survey, 2003-2008. Prev Chronic Dis. 2013;10:130244. https://doi.org/10.5888/pcd10.130244. Accessed January 4, 2017.

38. Gore AC, Chapell VA, Fenton SE, et al. EDC-2: the Endocrine Society's second scientific statement on endocrine-disrupting chemicals. Endocrine Rev. 2015;36:E1–E150. https://doi.org/10.1210/er.2015-1010.
39. Kampa M, Castanas E. Human health effects of air pollution. Environ Pollution. 2008;151:362–7.
40. Prüss-Ustün A, Vickers C, Haefliger P, Bertollini R. Knowns and unknowns on burden of disease due to chemicals: a systematic review. Environ Health. 2011;10(9). Available at http://ehjournal.biomedcentral.com/articles/10.1186/1476-069X-10-9. Accessed 5 Jan 2017.
41. Asbestos.com. National asbestos regulations. Available at https://www.asbestos.com/legislation/. Accessed 5 Jan 2017.
42. ATSDR Case Studies in Environmental Toxicity: Asbestos toxicity. Available at https://www.atsdr.cdc.gov/csem/asbestos_2014/docs/asbestos.pdf. Accessed 5 Jan 2017.
43. American Cancer Society. Lead: what is lead? Available at http://www.cancer.org/cancer/cancercauses/othercarcinogens/athome/lead. Accessed 5 Jan 2017.
44. Environmental Protection Agency (EPA) Drinking water regulations and contaminants. Available at https://www.epa.gov/dwregdev/drinking-water-regulations-and-contaminants. Accessed 9 Jan 2017.
45. Your guide to understanding genetic conditions. Genetics home reference. National Institutes of Health, National Health Library. Available at https://ghr.nlm.nih.gov/. Accessed 8 Jan 2016.
46. Garahkhani P, Tung J, Hinds D, et al. Chronic gastroesophageal reflux disease shares genetic background with esophageal adenocarcinoma and Barrett's esophagus. Hum Mol Genet. 2016;25(4):828–35. https://doi.org/10.1093/hmg/ddv512.
47. Yoo H. Genetics of autism spectrum disorder: current status and possible clinical applications. Exper Neurobiology. 2015;24(4):257–72. https://doi.org/10.5607/en.2015.24.4.257.

Additional Resources

Food Security and Health. Policy brief by the scientific advisory board of the UN Secretary General. December 28, 2016.

Lung Cancer Fact Sheet. Available at http://www.lung.org/lung-health-and-diseases/lung-disease-lookup/lung-cancer/learn-about-lung-cancer/lung-cancer-fact-sheet.html. Accessed 7 Jan 2017.

National Institute of Environmental Health Sciences. Air pollution. Available at https://www.niehs.nih.gov/health/topics/agents/air-pollution/. Accessed 4 Jan 2017.

WHO/FAO Expert Consultation. Diet, nutrition and the prevention of chronic disease. 2003. Available at http://www.who.int/dietphysicalactivity/publications/trs916/en/. Accessed 31 Dec 2016.

Circadian Rhythm: Light-Dark Cycles

Corey B. Schuler and Kate M. Hope

34.1 Introduction – 579

34.2 Background of Scientific Relevance to Metabolism and Nutrition Status – 579

34.3 Central and Peripheral Circadian Clocks – 580

34.4 Circadian Rhythm and Disruption of the Sleep-Wake Cycle – 581

34.5 Types of Circadian Rhythm Sleep Disorders (CRSDs) – 581

34.6 Treatment for Circadian Rhythm Sleep Disorders – 581
34.6.1 Light, Melatonin, and Vitamin B12 – 581

34.7 Nutrition and the Circadian System – 582
34.7.1 The Circadian Clock System Prepares the Body for Feeding – 583
34.7.2 Coupling Between Metabolism and Clocks – 583
34.7.3 Eating Patterns: The Feeding/Fasting Cycle – 583

34.8 Time-of-Day Restricted Feeding – 584

34.9 Circadian Clock Genes – 585

34.10 Pathophysiology Mechanisms and Relevance to Chronic Disease Risk – 586
34.10.1 Metabolism and Obesity – 586
34.10.2 Cancer – 586
34.10.3 Diabetes – 587
34.10.4 Cardiovascular Issues – 587
34.10.5 Renal Function – 588
34.10.6 Endocrine System – 588
34.10.7 Immune System – 588
34.10.8 Reproductive System – 588
34.10.9 Sleep Hygiene – 589

© Springer Nature Switzerland AG 2020
D. Noland et al. (eds.), *Integrative and Functional Medical Nutrition Therapy*,
https://doi.org/10.1007/978-3-030-30730-1_34

34.11 Toxin-Related Influences – 589

34.12 Key Lifestyle Influences – 589

34.13 Chronotypes and Personalized Nutrition – 590

34.14 Conclusion – 590

References – 591

34.1 Introduction

Chronobiology is the study of how biological clocks control and regulate virtually every function of life. Most living organisms, including humans, have an endogenous biological clock that operates on a near 24-hour cycle, in concert with Earth's daily rotation and the lunar cycles. This is called the circadian rhythm (from the Latin "circa diem," which means "approximately a day"). This intrinsic clock regulates a wide variety of functions, including sleep, arousal, feeding, and a myriad of metabolic processes. Indeed, it appears that nearly all biological mechanisms are regulated by the circadian clock system, including blood pressure, body temperature, cell division, immune function, digestion, and hormone secretion [1].

In humans, the master biological clock is located in the suprachiasmatic nucleus (SCN) within the anterior hypothalamus of the brain. The SCN receives photic input from the retina each day, which serves to reset the clock. The master clock then coordinates the peripheral clocks which are present in nearly all cells of the body, by way of a common set of clock genes that work in various negative feedback loops to create the circadian oscillations that regulate metabolism [1–3]. Functional genomics have established that these circadian clocks govern a large portion of the human genome.

Though circadian rhythms are endogenous in nature and regulated by the clock genes, they can be entrained or adjusted by zeitgebers (from the German, "time giver"), which are external cues from the environment. Zeitgebers include light, temperature, redox cycles, diet, exercise, and social contacts [4, 5]. Disruptions in the circadian coordination, whether environmental or genetic, can cause metabolic imbalances, leading to sleep issues, which may eventually affect the nervous, endocrine, cardiovascular, and immune systems. This can result in dyslipidemia, obesity, hypertension, inflammation, and mental health disorders [6]. Growing evidence also shows support for the idea that circadian disruption from shift work, chronic jet lag, or artificial light exposure at night may be risk factors for the development of cancer [7–9].

Circadian rhythm sleep disorders (CRSDs) or circadian sleep disorders are clinically diagnosed disorders that arise from abnormalities in the length, timing, or alignment to the day-night cycle. These include delayed sleep phase syndrome (DSPS), advanced sleep phase syndrome (ASPS), non-24-hour sleep-wake disorder (Non-24), irregular sleep-wake disorder (ISWD), shift work disorder, and jet lag disorder [10].

The gold-standard measures of circadian rhythm in humans have been core body temperature and salivary or plasma melatonin levels [4]. The SCN regulates the secretion of the hormone melatonin (N-acetyl-5-methoxytryptamine) by the pineal gland in response to the environmental light-dark cycle, inducing sleep. Abnormal oscillations in melatonin secretion can affect sleep patterns [11]. Bright light exposure and melatonin therapy have been shown to be effective treatments for circadian rhythm sleep disorders [12, 13].

The circadian rhythm is tightly connected to energy homeostasis and is affected by the timing of meals, as well as food composition and components. The clocks control energy metabolism, and the metabolic states, in return, can influence the circadian clocks. In many countries where meal timing and sleep habits are not in tune with the natural circadian rhythm, metabolic disease can arise [14]. The principles of chrononutrition, which studies the relationship between the circadian system and metabolism, have been shown to improve human weight loss and will likely benefit people who suffer from metabolic disease, as well as the general public [15].

34.2 Background of Scientific Relevance to Metabolism and Nutrition Status

As the saying goes, "You are what you eat," however, research indicates that we may need to add a companion axiom that states, "You are *when* you eat," as well. Food has been shown to be a strong driver of circadian rhythms, and a disruption in metabolism can occur if food consumption is not in phase with the light-dark cycle [5]. The term "chrononutrition" refers to the timing of food ingestion in coordination with the daily circadian rhythm. Time-specific food intake has been shown to have distinct consequences on physiology, and the circadian clock system provides a critical connection between nutrition and an organism's homeostasis [16]. Meal timing can affect the sleep-wake cycle, core body temperature, performance, alertness, and hormone levels, and it appears to be connected to obesity and metabolic pathologies [16].

In addition to the timing, the type and quality of macronutrients and micronutrients have the capability of functioning as zeitgebers by modulating clock proteins or nuclear receptors. The use of targeted nutrients based on SNPs (single nucleotide polymorphisms) in human clock genes, as well as a person's specific chronotype, has the potential for clinical utilization [5]. Nutrients can either disrupt or restore the synchronization between the peripheral clocks and the suprachiasmatic nucleus (SCN), also known as the central pacemaker in the brain.

The SCN is a network of neurons. It is a wing-shaped structure located within the hypothalamus gland that sits directly above the optic chiasm. This neural network consists of thousands of neurons, and it sends out signals that keep the rest of the body on a near 24-hour schedule.

The role of the microbiome is being explored for its role in circadian rhythm. A large portion of the body's microbiota is located in the gastrointestinal tract, a site that has been shown to have a powerful circadian clock. The microbiota participates in the daily cycles of digestion, and changes in its levels and composition can have an impact on the system-wide cyclic homeostasis of the body [16]. An appreciation of the circadian system has many implications for nutritional science and may ultimately help reduce the burden of chronic disease [17].

34.3 Central and Peripheral Circadian Clocks

Even without external cues from the environment, humans maintain a sleep-wake rhythm that is very close to 24 hours. In mammals, the central circadian clock that regulates this rhythm is the SCN. This master pacemaker regulates rhythms such as the sleep-wake cycle, the autonomic nervous system, body temperature, gene expression, and hormone secretion, including melatonin [17]. There is also evidence that the cell cycle is regulated via the day-night cycle, and numerous studies have shown that cell division is governed by a circadian variation in mammalian tissues, including the epithelia of the tongue, skin, oral mucosa, intestine, esophagus, stomach, duodenum, jejunum, rectum, as well as the bone marrow, and pancreas [3].

However, without time cues, the internal clock's period is actually about 24.2 hours, so it must be reset or entrained every day to the 24-hour day, to keep the organism in sync with the external world [17]. The SCN is entrained by light cues that enter the retina. The retina transfers information about light and darkness to the SCN, which then synchronizes a phase of rhythms with the external environment. This is orchestrated through a monosynaptic pathway from photosensitive cells in the retina that take photic input to the SCN. Then, by way of a multisynaptic pathway, information is transferred from the SCN to the pineal gland. The pineal gland synthesizes melatonin, a neurotransmitter-like substance that regulates circadian rhythms and sleeping patterns.

The central circadian clock also sends signals to peripheral clocks throughout the body. The peripheral clocks, or local clocks, are found in every cell, tissue, and organ [2, 3]. While the central clock is entrained by light, the peripheral clocks are entrained by feeding cues and dominate local metabolic rhythms including glucose, lipid homeostasis, and hormonal secretion. Nutrients and meal timing can affect the clock system. Thus, a current view for the adaptive significance of circadian clocks is their basic role in coordinating metabolism [18] (◘ Fig. 34.1).

The master circadian clock is located in the suprachiasmatic nucleus (SCN) in the hypothalamus. This master clock is entrained to the 24-hour day cycle by light which enters the retina, and it dominates activity rhythms. The central circadian clock sends signals to peripheral clocks located in cells throughout the body. The peripheral clocks are entrained by feeding cues that dominate local metabolic rhythms. Both nutrients and meal timing can affect the clock system. Chrononutrition has two facets: (1) nutrients and substances, for example, caffeine, can alter the periodicity of the clock, and high-fat diets can affect the rhythm of lipogenesis, circulating lipids, and feeding behavior and (2) meal-timing can affect the clock system, for example, skipping breakfast or nocturnal eating can increase the risk of obesity, whereas regular or time-restricted eating can synchronize the clock system and offset metabolic disorders.

Synchronicity between the central and peripheral clocks appears to be important for metabolic health. It is shown that workers on the night shift or a rotating schedule have a greater risk for type 2 diabetes, obesity, and cardiovascular disease. If the light-entrainable pacemaker of the SCN is out of sync with the food-entrainable oscillators of the peripheral tissues, then there may be a disruption in the energy balance [5].

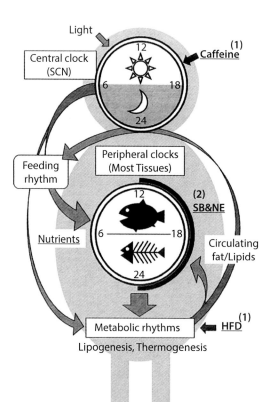
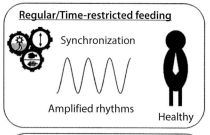
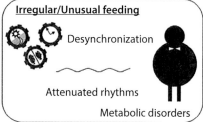

◘ Fig. 34.1 The circadian clock system and chrononutrition. (Reprinted from Oike et al. [14]. With permission from Springer Nature)

34.4 Circadian Rhythm and Disruption of the Sleep-Wake Cycle

Identifying circadian rhythm sleep disorders or functional versions of these disorders can be an important part of nutritional management. Medical care often overlooks functional sleep disorders and is not well-equipped to support sleep disorders non-pharmacologically.

Initial assessment can identify clues to the functional or pathologic state. Either on intake documents or initial interview, questions regarding diurnal preference (is the client a "night owl" or "morning lark"?) and whether or not they nap may be important regardless of the presenting complaint or health goal. Excessive daytime sleepiness and sleep deprivation are two common complaints in nutrition practice. Not only do these result in metabolic consequences but also poor decision-making as it pertains to healthy choices.

34.5 Types of Circadian Rhythm Sleep Disorders (CRSDs)

As a practical note when speaking about circadian rhythm, the term *delayed* is used to mean a later-than-usual onset of sleep. If normal onset of sleep is 10 p.m., then a delayed onset may mean an 11 p.m. or later sleep onset. The term *advanced* suggests the opposite and is used to mean an earlier-than-usual onset of sleep. This is important when evaluating therapeutics and methods that shift the circadian pattern and is relevant in CRSD nomenclature.

Delayed Sleep Phase Syndrome (DSPS) is thought to be the most common and similarly most understood CRSD. Typically, the delay is 3–6 hours. These clients are often awake until 2–6 a.m. and wake up between 10 a.m. and 1 p.m. or at least would if allowed. When a conventional wake time in those with DSPS is enforced, they often suffer from sleep deprivation [19]. Prevalence of DSPS in those with chronic insomnia is 5–10% but only 0.13–0.17% of the general population [20–22]. Polymorphisms in circadian rhythm genes, such as Human Per3 (hPer3), arylalkylamine N-acetyltransferase, HLA, and Clock, may play a role is DSPS [23–27].

Advanced Sleep Phase Syndrome (ASPS) can be clinically overlooked as a CRSD. Individuals with ASPS wake up early without much effort. Following their circadian pattern, they may choose sleep onset at 6–9 p.m. However, social pressures may delay their preferred bedtime and cause chronic sleep deprivation. Advancement in the phase of sleep-wake cycle is commonly seen with aging [28]. Prevalence is thought to be exceedingly rare [22]. Polymorphisms in circadian clock gene hPer2 have been identified [29].

Non-24-Hour Sleep Pattern is a CRSD that is primarily reported in blind individuals with no light perception. Delays or advances may result since the SCN does not respond to dark-light cycles. Up to 70% of blind individuals complain of chronic sleep disruption [30].

Irregular Sleep-Wake Pattern as a CRSD includes those with poor sleep hygiene or those who voluntarily have irregular cycles. Shift work and jet lag are two examples of this. Not all shift workers have sleep difficulty as the ability to cope with shifting circadian patterns varies between individuals [31]. Similarly, not all travelers suffer from jet lag. Eastward travel is often more problematic than westward travel. Irregular sleep-wake pattern also can occur in those with developmental conditions, brain injury, and dementia and those in institutional care [32–35].

Sleep Apnea is not recognized as a circadian rhythm sleep disorder per se. However, it can dramatically affect circadian patterns, may be affected by diurnal patterns, is a common sleep disorder, and can be modified, in part, by nutritional interventions. Poor sleep quality causes daytime sleepiness and affects both wake timing and normal somnolence. Sleep apnea severity is described by the average number of respiratory events per hour of sleep (frequency), but also the duration of each event can be monitored. At a circadian phase corresponding to early morning, apnea durations are longest and frequency was low. In the late afternoon and early evening, event durations are shortest and frequency is high [36]. This may be due to the concept that respiratory plasticity or neuronal control of respiration is modulated by a circadian rhythm [37]. Also, upper airway muscle activity, which helps maintain upper airway patency, is also modulated by a circadian rhythm [38, 39]. Prescribing a sleep schedule and possibly promoting daytime napping may be involved in managing sleep apnea as it pertains to breathing stability [37]. Addressing excess body weight and inflammation are key nutritional interventions to support individuals with sleep apnea [40, 41].

34.6 Treatment for Circadian Rhythm Sleep Disorders

34.6.1 Light, Melatonin, and Vitamin B12

Treatments are limited for circadian rhythm sleep disorders because of a lack of ability to assess circadian function in a practical and meaningful way, and there are limited randomized controlled clinical trials, especially in regard to functional alterations. In various combinations of dosing and timing, light, melatonin, and vitamin B12 routinely appear to support circadian rhythm.

Melatonin is not corrective and cannot be relied upon as a sole tool in the management of DSPS as treatment shows a wide variability among subjects [42–45]. Dosage trial and error is often required over the course of several months. The lowest effective dose is recommended.

Vitamin B12 likely acts to enhance the entraining light signal to the circadian pacemaker (SCN). Whether methylcobalamin is the right form is debatable. Cyanocobalamin, hydroxocobalamin, and adenosylcobalamin may be converted to methylcobalamin. If a clinician is tracking homocysteine as a functional marker, methylcobalamin would have to be used in order to note treatment effectiveness ◻ Table 34.1.

Table 34.1 Dosing light, melatonin, and vitamin B12 for disordered sleeping

Condition	Light	Melatonin	Vitamin B12
Delayed sleep phase syndrome	Early morning bright light (2500 lux) for 2 hours	Moderate dose (3–6 mg) between 5 p.m. and 7 p.m. for 1 month followed by moderate dose at 7:30 p.m. for two additional months	Dosed according to clinical need (1.5–4.5 mg methylcobalamin)
Advanced sleep phase syndrome	Early evening bright light (2500 lux) for 2 hours	Low dose (0.3–0.9) early morning	
Non-24-hour sleep pattern	If sighted or blind with light perception, bright light (2500 lux) for 2 hours	High dose (10 mg) before bed for entrainment Low dose (0.5 mg) at night	Dosed according to clinical need (1.5–4.5 mg methylcobalamin)
Irregular sleep-wake pattern shift work disorder	Bright light (1200–10,000 lux) for 3–6 hours during the night shift or 20 minutes per hour	Moderate–high dose (3–10 mg) at bedtime after night shift for 2–6 days	
Irregular sleep-wake pattern jet lag (eastbound)	Avoid bright light in the morning; get as much as possible in the afternoon	Moderate dose (2–5 mg) at bedtime	
Irregular sleep-wake pattern jet lag (westbound)	Stay awake while it is daylight at the destination; sleep when it gets dark		
Irregular sleep-wake pattern (other)	Early morning bright light (2500 lux) for 2 hours	Low–high dose (0.3–10 mg) before bed	Dosed according to clinical need (1.5–4.5 mg methylcobalamin)
Sleep apnea	Need determined by chronotype	Moderate dose (2–5 mg) at bedtime	

Some nutritional interventions may be applicable to virtually all aberrant circadian patterns. Assessment for the applicability in individual patients of these interventions may provide a useful clinical roadmap. Inflammatory signals have profound effects on circadian clocks. Essential fatty acid balance and natural substances that target NF-KappaB and AMPK (5′ adenosine monophosphate-activated protein kinase) can be considered [46, 47]. As examples, omega-3 fatty acids such as eicosapentaenoic and docosahexaenoic acids (EPA and DHA), turmeric, and berberine, respectively, can be considered. Red blood cell fatty acid analysis, clinically actionable markers of inflammation including C-reactive protein and erythrocyte sedimentation rate (ESR), and blood sugar dysregulation markers such as hemoglobin A1c, fasting glucose, and fasting insulin can be used to identify these opportunities for intervention. Supplements of and foods rich in magnesium and the other B vitamins beyond B12 can be included in the management of individuals with circadian rhythm sleep disorders [48, 49]. The type and dose of each require individualized clinical evaluation and a trial of therapy. Dietary management of circadian rhythm disorders may include the inclusion of phytonutrients such as polyphenol- and flavonoid-rich vegetables and spices [50, 51]. Additionally, nutrition professionals must consider macronutrient choices that support glycemic balance [52, 53].

34.7 Nutrition and the Circadian System

The circadian clock directs when we are active, when we sleep, and, to a great extent, when we eat. This system maintains the rhythm in the endocrine and metabolic pathways required for our body's homeostasis. A wide range of metabolic pathways are controlled by the clock system, whose daily fluctuations are affected by the food we ingest [54]. Today's modern lifestyle, which includes artificial lighting and ready-made, easy-to-access snacks, may place an evolutionary stress on our bodies that can impact our diet and health. The complex circadian system optimizes behavior and physiology according to the time of day and is organized with a central clock in the SCN primarily entrained by light. Patients are commonly exposed to less light during the day and more light at night because of artificial lighting. This further impairs the circadian system and disrupts sleep. Disturbed sleep thus results in widespread deleterious effects on metabolic health. As an example, sleep disruption promotes increased energy intake, reduced energy expenditure, insulin resistance in some individuals, and an increased propensity to make less healthy food choices.

The human biological clock can be disrupted or adjusted, depending on environmental factors, including not only the pattern of day and night, but the rhythmic intake of food [16].

34.7.1 The Circadian Clock System Prepares the Body for Feeding

The circadian clock system readies the body for daytime feeding. Gastrointestinal motility and gastric emptying peak in the morning, and animal studies have shown that the clock regulates the action of bile acids and nutrient transporters which optimize digestion [17]. In addition, the gut microbiota exhibit daily circadian oscillations, which are time-of-day specific in order to perform certain functions. The intestinal cells have a circadian clock that scientists are now investigating in order to discover its physiological pathway. The gut clock reacts to the rhythmic intake of food, and it has been shown that the microbiota participate in food processing and react with the intestinal clock [16]. Energy metabolism is enhanced by the gut microbes during the active phase, while detoxification occurs more often in the resting phase [17, 55]. There is a complex connection between the circadian rhythm and the microbiome, and a disruption in the clock system can have a disorganizing effect on the gut microbes [56]. This information lends itself to recommendations of avoiding high-fat meals in the evening and also ensuring micronutrient-rich evening meals to support detoxification during rest (◘ Fig. 34.2).

The intestinal cells have a circadian clock that research is showing is connected to the central clock in the SCN. The gut clock responds to the rhythmic intake of food, and it has been demonstrated that the microbiota participate in food processing and react to the intestinal clock. The action is bidirectional, in that the gut clock needs the microbiota to function correctly, and the gut bacterial levels fluctuate according to the gut cycles. Thus, the circadian clock is an integral player between food and the intestinal system.

The circadian clock also controls rhythms in the blood concentrations of many nutrients, for example, glucose and lipids [57]. Due to the clock-controlled regulation of the gastrointestinal system, practitioners may want to consider the timing of nutritional testing. There has been some interesting research with rodent models and the timing of food intake and allergies. It was shown that exposure to food antigens during the resting phase produced more severe food allergy symptoms [58].

The circadian system regulates energy storage in the body's tissues. To a large extent, appetite is clock-controlled. There is a bimodal daily peak of insulin within the active phase, and diet-induced thermogenesis is rhythmically regulated and peaks in the morning. Hunger is typically lower in the morning and higher at night. This may be of importance when discussing obesity, since delayed bedtimes allow more time for food consumption when the appetite is higher. Some suggest that consuming a greater percentage of food in the morning may encourage a negative energy balance and a decrease in body mass [17, 59]. This works to the extent that a patient does, in fact, reduce portions in the evening rather than simply consuming more in the morning and allowing natural appetite and satiety to control excess eating at night. Encouraging self-awareness of a satiety index may be a central principle in medical nutrition therapy when these tools are used.

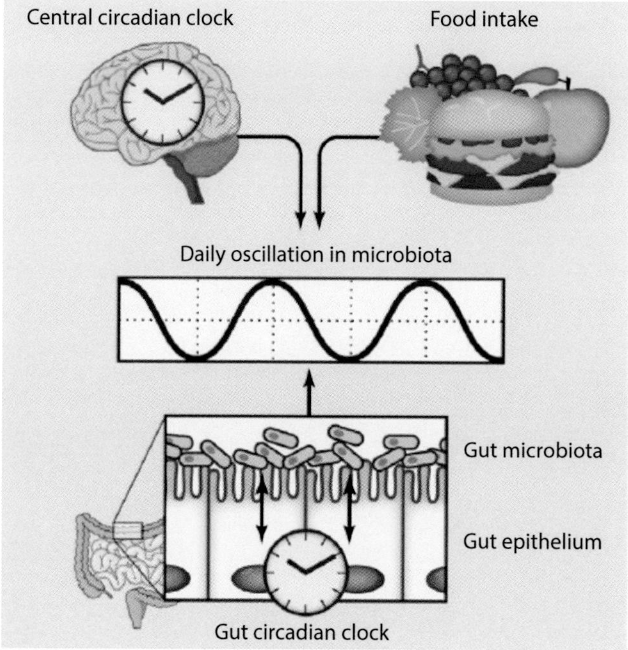

34.7.2 Coupling Between Metabolism and Clocks

The light-dark cycle entrains the central clock in the SCN and governs activity-related rhythms such as the sleep-wake cycle, core body temperature, the autonomic nervous system, and melatonin production, whereas the peripheral clocks, present in most tissues including the brain, are entrained by the feeding and fasting cycles. The peripheral clocks oversee local physiological processes such as glucose and lipid levels, hormone secretion, digestive functions, and the immune response [14, 60].

Recent studies have shown that the peripheral circadian oscillations respond differently to dietary cues than those of the SCN and that an irregular feeding schedule can cause a decoupling between the SCN and the peripheral system. A desynchronization or decoupling among the clocks is thought to result in the development of cancer, metabolic, and psychiatric disorders [14].

34.7.3 Eating Patterns: The Feeding/Fasting Cycle

The circadian system primes organisms to feed at specific times, and if feeding occurs outside the normal boundaries, negative health consequences may result [17, 61]. Studies in

◘ Fig. 34.2 The connection between the microbiota and the gut circadian clock. (Reprinted from Asher and Sassone-Corsi [16]. With Elsevier)

both humans and animals suggest that the timing of meals may be related to diabetes, obesity, and impaired cognitive function via the peripheral clock network and metabolism [17, 61]. In humans, skipping meals, especially breakfast, has been shown to be associated with obesity and other metabolic issues. On the flip side, eating later at night can disrupt the rhythm of meal timing and sleep onset and has also been associated with weight gain. Though conclusive data is somewhat limited, there have been many studies on the effects of the fasting/feeding cycle on the circadian rhythm. Rodent studies show that fasting during the early "activity" phase (which for them is nighttime, since rodents are nocturnal) results in an increase in both body weight and hepatic lipid deposition. This early nocturnal fasting disrupts the peripheral clock system and results in de novo lipid synthesis and a tendency toward obesity. This is akin to skipping breakfast in humans [62].

There have been a number of studies that show the importance of eating a meal upon the beginning of the active phase to prevent obesity. Thus, one could make an argument for eating on a regular schedule, during the normal "active phase" (or light phase for humans) for proper peripheral clock management and optimal health [62]. Human studies have compared isocaloric weight-loss groups and have found there was a greater improvement in metabolic markers in the group given the larger breakfast and a smaller dinner, rather than vice versa. Other studies have found a correlation between earlier meal times and a decrease in serum lipid levels. Evidence also shows that the consumption of breakfast among adolescents is inversely associated with weight gain [63]. In addition, time-delayed eating, as in nighttime eating, is positively associated with an elevated body mass index (BMI) [14, 16]. In studies of both rats and humans, it was found that there was an affinity toward higher fat foods in the evening compared to morning. Though studies are not conclusive, this preference may indicate an association between late-night feeding and obesity [16].

34.8 Time-of-Day Restricted Feeding

The eating pattern of humans in modern societies typically consists of three meals and a few snacks every day. This is abnormal from an evolutionary perspective, where research shows hunter-gatherers eating intermittently. The body evolved to take up and store glucose (liver glycogen stores) and longer-lasting energy substrates, such as fatty acids in adipose tissue, and was designed for a diurnal rhythm of activity/rest and feeding/fasting. Modern lifestyle, with artificial light and erratic eating patterns, can upset our biological circadian system [64].

The timing of meals seems to be a significant zeitgeber for the oscillations in the peripheral clocks and could be seen as an important factor in regulating metabolism. Though the mechanisms by which an irregular feeding time affects the metabolism are not completely known, the evidence suggests that differential meal timing may alter the circadian cycle independent of lighting conditions. For the most part, the SCN clock is controlled by the light-dark cycle, whereas the peripheral clock rhythms are affected by feeding cycles. Time-of-day restricted feeding (TRF), where food availability is restricted to a smaller window during the day, has been shown to entrain the circadian rhythm in peripheral tissues without disrupting the clock rhythm of the SCN [62, 65]. The effect of restricted feeding is not the same as caloric restriction, which can affect the phase of circadian rhythms in the central pacemaker [5].

A number of studies show the beneficial effects of restricting food intake to a discrete window of time within the daily active cycle (daylight for humans). There have been several human epidemiological studies that exhibited a correlation between eating pattern and obesity. When food is restricted to a smaller time frame within the active period, increase in glucose tolerance, reduction in insulin resistance, and decrease in serum lipid levels have been shown. This has implications for reducing the risk of developing obesity [16, 66] (◘ Fig. 34.3).

There is a wide body of evidence that shows that when food is restricted to a discrete window of time within the daily cycle, various metabolic processes are improved. When food is restricted to a smaller time frame within the active period, an increase in glucose tolerance, a reduction in insulin resistance, and a decrease in serum lipid levels have been shown. This has implications for reducing the risk of developing obesity.

In a study of overweight and obese women, those who consumed a larger percentage of daily calories earlier in the day lost more weight than those eating later in the day [67]. These findings are consistent with others that show that earlier lunch consumption is associated with greater weight loss after 20 weeks [17].

Restricting the delivery of nutrients in a rhythmic fashion may help to reset the peripheral clocks, which could potentially have a positive impact on the immune system. Some

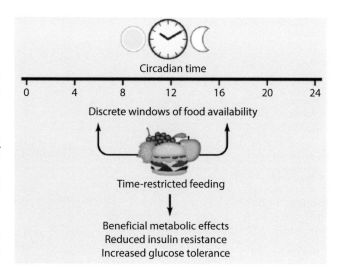

◘ Fig. 34.3 The benefits of time-of-day restricted feeding (TRF). (Reprinted from Asher and Sassone-Corsi [16]. With Elsevier)

emerging research with both human and animal models shows improved health with intermittent energy restriction of as little as 16 hours (an 8-hour eating window). Here there is a shift to fat metabolism and ketone production, which can stimulate adaptive cellular stress response and repair molecular damage [64]. Circadian dysregulation can induce adverse health effects as well as hamper recovery. When dealing with ill patients in the ICU or those who are on enteral feeding, restoring their day-night cycle might help realign their biological rhythms during the recovery phase to improve outcomes [2].

Practitioners who develop nutritional plans for clients should not ignore the impact of the circadian rhythm on nutrition and health. The timing of meals has shown to affect body weight and the risk of obesity in humans [14, 68, 69]. Though not conclusive, it appears late meals and skipping breakfast encourage a gain in body weight and propensity toward obesity. Eating a larger breakfast and a smaller dinner may lead to better markers for fasting glucose, insulin, ghrelin, as well as improved mean hunger and satiety scores [14, 67]. Early mealtimes are also shown to reduce triglycerides and LDL levels in serum [65]. In addition, restricting the consumption of food to a smaller window during the active period may be beneficial to health [64].

Intermittent fasting (IF) is popular in the health and fitness field today. Practitioners often recommend a 12- to 16-hour fast, and many studies show a positive effect on weight loss [70–72]. However, breakfast is the meal most often skipped. Though positive weight loss may occur when breakfast is removed, evidence indicates that for overall health and weight management, it is better to eat breakfast [63, 73]. This is keeping with the idea that it is more advantageous to consume one's daily calories within the active (daylight) cycle [16]. Recent evidence demonstrates that when the first meal of the day is missed, neuroendocrine parameters may be negatively affected. In one study, levels of testosterone and IGF-1 were shown to be reduced, as was leptin [74]. Another study in men found that omitting breakfast led to an increased risk of type 2 diabetes [73]. It is worth speculating if perhaps the benefits of intermittent fasting could be achieved without the deleterious effects if the fasting window began in the early evening and breakfast was consumed on a regular basis.

34.9 Circadian Clock Genes

Within the neurons of the SCN, there are a number of genes that are activated or inhibited in a regular pattern over the course of a day. The fluctuations act as molecular gears of the biological clock which regulates the 24-hour cycle. These molecular oscillations work in a negative feedback fashion, whereby the protein product of a gene turns off production of more protein.

At the core of the molecular network are transcription factors that drive expression and regulate the biological function of the master circadian genes, CLOCK and BMAL1. The transcription of a large number of clock-controlled genes also directs the transcription of their own repressors, period (PER) and cryptochrome (CRY), which create a self-regulated feedback loop. During daytime, the *per* and *cry* genes create the PER and CRY repressors, which accumulate over the day. This eventually inhibits the CLOCK-BMAL1-driven transcription of *per* and *cry*. Then the degradation of PER and CRY allows the CLOCK-BMAL1 transcription to proceed once again. This creates the cycling in the circadian gene expression [16, 75]. Additional clock genes or clock-controlled genes, such as Rev-erbα/ß, Rorα/ß, Dpb, Dec1/2, CK1ε/δ, and NPAS2, also work to sustain mammalian circadian clocks [14] (◘ Fig. 34.4).

CLOCK and BMAL1 are transcriptional activators that dimerize to stimulate the expression of many clock-controlled genes (CCGs) with E-box promoter elements in their promoters. CLOCK:BMAL1 also stimulate the expression of the Period (Per) and Cryptochromes (Cry) gene families. PER and CRY protein levels become elevated during the night. They dimerize and translocate to the nucleus to repress CLOCK:BMAL1-mediated transcription. PERs and CRYs undergo a number of posttranslational modifications that cause their degradation, which then allows them to start a

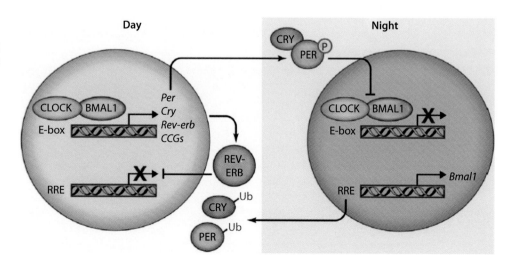

◘ Fig. 34.4 The molecular organization of the circadian clock. (Reprinted from Asher and Sassone-Corsi [16]. With Elsevier)

new circadian cycle. Another loop involves the proteins REV-ERBa/b, whose levels increase during the day and bind to specific responsive promoter elements (RRE), inhibiting Bmal1 transcription. At night, REV-ERBa protein levels decrease, allowing Bmal1 transcription to take place. These regulatory loops operate in most cells and control a large percentage of the human genome.

Disruption in the feeding/fasting cycle can impact the circadian clock genes. In mice, early nocturnal fasting (which correlates with skipping breakfast in humans) alters lipid metabolism and the peripheral clocks. It increases caloric intake and alters lipogenesis at the levels of transcriptional and metabolic cycling, changing the expression of clock genes such as *CLOCK*, *BMAL1*, *Cry1*, and *Per2* [62]. Single nucleotide polymorphisms (SNPs) in CLOCK have been associated with non-alcoholic steatohepatitis, metabolic syndrome, small dense low-density lipoprotein levels, obesity, and diabetes. One of the most studied associations is that of obesity. So far, eight common CLOCK SNPs have been associated with obesity, and three of them are linked with energy intakes [17, 76].

Disruption of the molecular clock can produce metabolic disturbances, a fact which is supported by some observational studies [7, 9, 57]. Nutritional interventions, such as caloric restriction, may hold promise in mediating some beneficial effects [17]. Rodent models suggest that a calorie-restricted diet reprograms the circadian clocks both transcriptionally and posttranscriptionally. It appears that the transcription factor BMAL1 is important for the reprogramming of the clock [77].

Further study of clock genes and their alterations will expand our knowledge of the molecular workings of the circadian clock that could lead to potential therapies for treating metabolic, sleep, and neuropsychiatric disorders. Of particular interest is identifying the at-risk CLOCK genotypes, who may be more susceptible to overeating and gaining weight when they experience shorter sleep time [76].

34.10 Pathophysiology Mechanisms and Relevance to Chronic Disease Risk

34.10.1 Metabolism and Obesity

A large body of evidence shows that cellular metabolism is intimately interconnected with the circadian clock [16, 64, 78]. Some say it may be due to an evolutionary advantage that separates the anabolic and catabolic reactions of the body during different times of the day, such as gluconeogenesis and glycolysis. This way, the metabolic cycles are aligned to the sleep-wake cycle. The circadian clock mechanism influences homeostasis over a wide range of processes including glucose and lipid metabolism, body temperature, endocrine hormone secretion, and cardiovascular health [64].

Numerous studies have focused on the connection between the scheduling of meals and such pathologies as obesity and diabetes [60, 62, 63]. It can be conjectured that "optimal" feeding schedules may also provide health benefits beyond these conditions, such as preventing morbidity and reducing mortality in relation to other pathologies such as aging [16].

There is a daily cyclic control over glucose metabolism [64, 79]. Glucose tolerance and the action of insulin vary throughout the day. Glucose tolerance is reduced in the evening, as opposed to the morning hours, because insulin secretion is lower, and insulin sensitivity is diminished during the nighttime. Glucose levels peak right before the "active period" or daytime hours [64]. Insulin sensitivity has a bimodal peak during the daytime phase, and appetite is actually lowest in the morning [17, 59, 79]. Thermogenesis also has a circadian rhythm that is induced by the diet and peaks in the morning. Nutrition professionals may want to recommend a larger percentage of dietary energy during the early hours of the active phase. This will encourage a negative energy balance, which is the principle determining factor of reducing body mass [17, 80].

Poor-quality sleep and reduced sleep duration are linked with impairment of glucose tolerance, along with reduced insulin response following a glucose challenge. Studies with shift workers have revealed that, with a disruption in the light-dark cycle, the risks of metabolic dysfunction, cardiovascular disease, stroke, obesity, and cancer all increase. Evidence also shows a rise in BMI when there is a discrepancy between the natural circadian clock and one's social clock. In a compelling clinical study, volunteers were put on a cycle that was shifted 12 hours from their normal sleep-wake cycle. The subjects exhibited increased glucose, decreased leptin, and elevated blood pressure. Their postprandial glucose response was similar to a prediabetic patient [64, 79].

In mouse studies, feeding at the incorrect time in the light-dark cycle can elevate the obesogenic effects of a high-calorie diet, due to the desynchronization of the hormonal and molecular rhythms in maintaining energy balance [64, 81]. While feeding a high-fat diet during the rest phase leads to weight gain, restricting feeding to the active phase leads to an improvement in obesity and metabolic disorders.

Directing therapy toward weight loss in obese and overweight patients may improve sleep, such as is the case with sleep apnea [40]. The potential positive effects of normalizing body composition include improved glucose tolerance, normalized lipid markers and reduced blood pressure, reduced inflammation, and reduced stress to the circadian system. Obesity and related metabolic sequelae, including cardiovascular disease, accelerated aging, diabetes, cancer, and neurodegenerative diseases, are conditions of increased oxidative stress. It has been proposed that oxidative stress and circadian rhythm be supported simultaneously when one or both clinical challenges are identified [82].

34.10.2 Cancer

In human and rodent studies, abnormal circadian rhythms have been associated with a greater risk of both the development and progression of cancer. A disruption in the light-dark cycle can result in poorer outcomes for cancer patients. Faster tumor growth and shorter survival times are seen when the rhythm is unbalanced. Shift work, with the attendant disruption of the natural circadian rhythm, is a significant and independent risk factor for the development of a variety of cancers. Studies in women with breast cancer show that those whose cortisol rhythms were off due to circadian disruption had a poorer survival rate [7]. It is believed that the prolonged disturbance of the circadian clock system may lead not only to cancers of the breast and prostate but to other types of cancers, such as ovarian, kidney, brain, colorectal, lung, head and neck, pancreatic, and hematological cancers [9].

Interventions that target the circadian timing system hold promise for potential therapies that may help improve the quality of life and survival in cancer patients [9]. In addition, the altered expression of various clock genes may be an important mechanism contributing to the progression of cancer and is an area of further research interest. Practitioners can work with their cancer patients to regulate the timing of meals in concert with the natural circadian rhythm. An organized feeding routine within the normal rhythm of the light-dark cycle will help maintain homeostasis, which may lead to more favorable outcomes.

34.10.3 Diabetes

Evidence is accumulating that inadequate sleep may be a risk factor for type 2 diabetes. Short duration of stage 3 and 4 NREM (non-rapid eye movement) sleep has been shown to increase insulin resistance. In clinical studies, decreased glucose tolerance was observed in adults after several nights of sleeping only 4–5 hours per night. These preliminary findings indicate that there may be a relationship between the time spent in slow-wave, deep sleep, and the circadian rhythm of glucose metabolism [83].

Studies involving circadian clock genes suggest a role for the clock proteins in diabetes. In animal models, circadian desynchronization results in a faster onset of type 1 diabetes. In this case, the overexpression of human islet amyloid polypeptide leads to pancreatic beta-cell failure. Other rodent studies show that when the SCN is ablated, the circadian rhythm of glucose uptake is lost, leading to hepatic insulin resistance [84–86].

Other investigations show that constant light exposure disrupts the circadian rhythmicity, resulting in insulin resistance [84, 87]. Melatonin is regulated by the light-dark cycle, and evidence for its role in diabetes comes from research that identified a SNP near the melatonin receptor 1B gene. This gene encodes the melatonin 2 receptor, which is linked with increased fasting glucose and a greater incidence of type 2 diabetes in white Europeans and Asians [84, 88, 89]. Melatonin is secreted in a diurnal fashion and influences insulin secretion. The circadian secretion of insulin from the pancreas is also clock-driven. Evidence suggests a connection between melatonin and the control of glucose-stimulated insulin secretion. The dysregulation of the circadian secretion of melatonin may be a key to the development of type 2 diabetes [84, 90].

In humans, a study demonstrated that two SNPs in the clock gene, BMAL1, led to a greater risk of gestational diabetes mellitus in women. Other recent work has illustrated a role for Cry1 and Cry2 in diabetes. In addition, there is ongoing research exploring the use of Cry1/2 agonists to treat the disease. Continued exploration into the physiological significance of the gene interactions may lead to therapies that successfully target the circadian clock for the treatment of diabetes and other metabolic disorders [84].

34.10.4 Cardiovascular Issues

Sleeping for less than 7 hours per night is associated with a higher risk of cardiovascular disease and poor cardiovascular health outcomes, such as hypertension, hypercholesterolemia, myocardial infarction, and cerebrovascular events [83]. Several studies in both rodents and humans have shown the importance of the circadian clock in cardiovascular disease, due to different cardiovascular phenotypes associated with circadian disruption or desynchronization [84, 91–93]. The desynchronizing of the clock system is often caused by rotating shift work or chronic jet lag [84]. There is mounting evidence that shows the importance of maintaining healthy circadian function for cardiovascular health. An area of active investigation is the regulation of blood pressure. Circadian clock proteins regulate the "dipping" of blood pressure at night (by about 10%) and its rise in the morning. The mechanism is unknown, but many people with hypertension, type 2 diabetes, and chronic kidney disease do not experience this effect. It is likely that a combination of sympathetic nervous system activity, hormone signaling, nitric oxide, and sodium reabsorption together contributes to the circadian control of blood pressure [84].

As with blood pressure, heart rate decreases at night and rises during the day. The exact mechanism for the circadian control of the heart rate is not yet fully understood, but peroxisome proliferator-activated receptor-gamma has been suggested, by way of an unknown pathway involving BMAL1 in the endothelium. It has been demonstrated that the circadian clock controls numerous intracellular calcium channels that are important for heart rate regulation. Myocardial repolarization also appears to be controlled by circadian rhythm [84].

The incidence of heart attack and stroke peaks in the morning, attributing to the rise in blood pressure and heart rate occurring at that time. Human studies show that infarct size and myocardial injury both display circadian oscillations, with the highest damage early in the day [84]. Cardiac remodeling occurs during sleep, and a desynchronization leads to an increased risk of cardiovascular events. Studies with shift workers who have rotating schedules show an increased risk for cardiac issues [94–98]. Those who suffer from chronic jet lag may have similar risks due to desynchronization. Therefore, it is important to maintain a healthy circadian rhythm for cardiovascular health [84].

34.10.5 Renal Function

Many processes in the kidney follow a circadian rhythm including renal blood flow, glomerular filtration rate, as well as potassium and sodium excretion. The clock gene, Per1, is involved with the epithelial sodium channel (ENaC) in the renal collecting duct, which is a key mediator of sodium reabsorption and blood volume control by the kidney. Per1 also regulates multiple genes involved in sodium reabsorption [99]. In one human study, it was found that Per1 was upregulated 1.7-fold in the renal medulla of kidneys in patients with hypertension, compared with normotensive controls [100]. Thus, it is suggested that Per1 may play a role in blood pressure control in humans.

34.10.6 Endocrine System

The circadian clock is a key regulator of the endocrine system, and conversely, the endocrine system plays a role in the synchronization and regulation of the peripheral circadian clocks. A mutual physiologic interaction between the circadian clock system and the hypothalamic-pituitary-adrenal axis (HPA axis) is orchestrated by the Clock/Bmal1 transcription factors, along with the glucocorticoids and glucocorticoid receptors. This serves to manage the adaptations not only to the day-night cycles but to the variety of stress stimuli the body is exposed to on a daily basis. The circadian clock system and the body's stress response system are both crucial for survival, and they communicate with each other on many levels to constantly adjust the various physiological activities. Dysregulation in either system may lead to similar pathologies [101].

The glucocorticoids, steroid hormones produced in the adrenal glands, are involved in reducing inflammation, and they play a part in stress response, metabolism, and the function of the neurological and cardiovascular systems. Glucocorticoids are thought to synchronize the circadian clock, as studies demonstrate that they can entrain the peripheral circadian oscillators [102]. Glucocorticoid levels are tightly regulated, and they express a daily rhythm, with their highest levels occurring in the early morning and their lowest in the late evening (in diurnal animals). The SCN, under the control of daily light input, regulates the rhythmic secretion of glucocorticoids from the adrenal glands by influencing the activity of the HPA axis. The glucocorticoids, in turn, phase-delay the peripheral clocks via the clock genes. This process is important for the body's adaptation and response to stress. A regulatory feedback loop is enacted by the peripheral clocks through the interaction of Clock/Bmal1 and the glucocorticoid receptors of the central clock on the HPA axis [101].

Numerous in vitro studies show that synthetic glucocorticoids can synchronize cell cultures via the circadian gene expression. In animal models, the administration of hydrocortisone synchronized circadian gene expression in the lung, pineal gland, salivary gland, cornea, and liver [102–104].

34.10.7 Immune System

The circadian system is an important component of the hypothalamic-immune communication. Disruption of this rhythmic process has been linked to immune dysregulation, and studies show that the circadian clock genes are involved in the immune response. Interactions between the HPA axis and the circadian clock system are seen in the regulation of immune function. The circadian oscillation of numerous cytokines has been observed in humans, including interleukin-1 beta (IL-1beta), IL-6, interferon (IFN)-gamma, and TNF-alpha in T- and B-lymphocytes and natural killer cells. In animal models, those with clock gene mutations display dysfunctional immune systems, including disrupted circadian rhythm in leukocytes and cytokine production, altered response to lipopolysaccharide (LPS)-induced endotoxic shock, and the defective development of B-lymphocytes [101]. Activation of the HPA axis and the resulting release of glucocorticoids strongly influence immune activity and the inflammatory response [101].

Sleep deprivation can evoke an acute stress response within an organism, and sleep loss can be a risk factor for impaired immune function. A number of studies have indicated that under normal sleep-wake and lighting conditions, circulating blood cell populations follow a diurnal rhythm. If sleep-deprived, however, granulocyte levels and their rhythmicity are disrupted, and these changes reflect the body's immediate immune response when exposed to a stressor [105]. In a rodent model of chronic jet lag, research demonstrated that natural killer cell activity is compromised [84, 106].

34.10.8 Reproductive System

Recent studies have added to the knowledge of molecular circadian rhythm activity in the reproductive system. Most of the work has explored the role of the central clock system in the female reproductive cycle; however, newer work has also revealed the existence of a testicular clock [84]. Daily and seasonal environmental signals impact fertility through the clock system, as it coordinates the various processes involved in sex hormone production, ovulation, spermatogenesis, mating, embryo development, and childbirth [107, 108].

In females, the reproductive clock is involved in the neurons that express gonadotropin-releasing peptide hormone (GnRH). GnRH is responsible for releasing follicle-stimulating hormone and luteinizing hormone from the anterior pituitary gland. GnRH displays a circadian pattern. The central SCN clock appears to have control over the peripheral ovarian clock. Animal models show the oscillation of clock genes in the ovaries, and those with mutations in BMAL1 and CLOCK show decreased fertility. It is interesting to note that the existence of the circadian clock in the placenta has been found; however, the significance remains to be seen [84].

Circadian clock gene expression has been discovered in the male reproductive system, and it is demonstrated that the proteins CLOCK and BMAL1 are involved in the development and physiology of the chromatoid body in the male germ cells, which are crucial to spermatogenesis [107]. In rodent studies, BMAL1 was shown to be necessary for fertility and proper testosterone production. Other animal models show that clock gene mutations influence reproductive function [107, 109, 110].

34.10.9 Sleep Hygiene

The concept of proper sleep hygiene is currently attracting a good deal of attention in the health field. Many professionals recommend consistent daily rituals, including an ideal bedtime environment that pays attention to all the senses.

Human circadian rhythms were entrained by the light-dark pattern of the solar day. Since the development of electric lights, exposure to nighttime lighting makes it more difficult for the body to synchronize biological processes [111]. The natural effect of darkness on the initiation and maintenance of sleep may be disrupted by the bright lights from modern media displays, such as phones, tablets, televisions, and computers. Both bright lights and stimulating content curb the natural decrease in body temperature and heart rate, thereby affecting the inclination to sleep. Research shows exposure to bright light at night can decrease melatonin production and circumvent the physiologic sleep processes [83]. Lowering lights at night, turning off displays, and sleeping in a darkened room can encourage the natural sleep cycle [83].

The brain processes sounds even while the body sleeps. Some sounds can be disruptive, but some may be soothing and may help induce sleep [83]. Earplugs can help protect from disruptive noise, while on the other hand, a sound machine producing soothing nature sounds like a waterfall may help some people fall asleep more quickly [112].

Aromatherapy offers a safe, cost-effective intervention that can aid in decreasing some of the negative impacts of lack of sleep. Natural essential oils that contain sedative or hypnotic properties are encouraging as a sleep therapy. Numerous studies, including some randomized controlled trials, have shown that essential oils, especially lavender, can improve sleep issues [113–116]. Exercise is a healthy activity that can reduce the risk of cardiovascular disease and other chronic issues. The National Sleep Foundation states that exercise is a non-pharmacological intervention that can improve the quality of sleep. However, the timing of exercise may affect sleep quality. Studies have shown that moderate aerobic exercise in the morning, as opposed to later in the evening, results in longer periods of deep sleep and shorter sleep-onset latency. In addition, morning exercise is demonstrated to optimize blood pressure changes that lead to an improvement in sleep architecture [117]. Information is lacking regarding the timing of meals and sleep quality, except that, as discussed in an earlier section, later bedtimes result in an increased food intake during the late-night hours. Current studies are finding that sleep restriction may lead to greater desire and consumption of palatable food through the endocannabinoid receptor system [118]. Though the study of incorporating sleep hygiene into weight loss programs is in its infancy, research has demonstrated better weight management when a sleep component is added [119–121]. This may provide some insight to nutritionists and health professionals who are helping patients with weight issues [83].

There is growing consideration that keeping in tune with the body's natural circadian rhythm is advantageous for health. Developing a regular daily routine that includes optimal meal timing, exercise, and sleep habits may contribute to proper weight management and ward off chronic disease.

34.11 Toxin-Related Influences

Caffeine, while the most common drug used to combat daytime sleepiness, disrupts circadian rhythm. In fact, caffeine can shift sleep onset forward by as much as 45–60 minutes even when consumed 6 hours prior to a scheduled sleep time. Later consumption can have even more profound effects on sleep time [122]. Those with circadian pattern concerns should be counseled to reduce or eliminate caffeine or, at minimum, limit use to only early in the day. While medical and nutrition professionals may take this information for granted, consumption statistics suggest that patients are using caffeine at all times of the day. A population-based study of adults estimated that 90% of individuals consume caffeine in the afternoon between noon and 6 p.m. and 68.5% of people consume caffeine in the evening between 6 p.m. and midnight [123]. Forward shifts of sleep time and artificial waking times can further lead to reliance upon caffeine. The effects of caffeine are quickly adapted to thus further increasing the dose required for a cognitive benefit [124, 125].

Of particular interest in circadian patterns is the influence of heavy metals, such as cadmium. Cadmium changes the expression of various circadian rhythm-related genes, including Bmal1, Per1, Per2, Cry1, and Cry2, and disrupts virtually every hormone secreted by the pituitary gland in animal models, even at low doses via drinking water. Melatonin has been shown to reduce these toxic effects including disruption of prolactin, luteinizing hormone, thyroid stimulating hormone, and corticosterone [126]. Melatonin's protective effect may be derived from supplemental use or from endogenous production. However, if endogenous melatonin is shunted toward cadmium amelioration, less is available for other functions, in particular its key role in circadian control. Melatonin serves as a suicidal antioxidant and is not recycled by additional antioxidants as is the case with vitamins C and E, coenzyme Q10, and others.

34.12 Key Lifestyle Influences

Alcohol affects circadian patterns with both acute and chronic use [127]. Ethanol impairs sleep, and sleep problems often contribute to relapse drinking [128–132]. In one study, healthy subjects who consumed a significant amount of alcohol over the course of a day did not see changes in cortisol levels but did see increases in testosterone [133]. In a similar binge replication study, alcohol dramatically decreased nocturnal thyroid stimulating hormone (TSH) secretion [134].

If cessation is suggested, a significant time period of 1 week or more should be used to determine the effectiveness of cessation on circadian patterns since alcohol withdrawal can have its own set of repercussions to diurnal chemical secretion and its effects. Of particular interest is that when alcoholics abstain from alcohol, a delay in the nocturnal rise of melatonin occurs [135]. Thus, during cessation, supplemental melatonin may support sleep and reduce the propensity for relapse that often occurs as subjects rely on alcohol as a sleep aid. In non-addicted individuals, alcohol also shows a suppression of melatonin secretion in a dose-dependent manner [136].

This is not the only place in the scientific literature where melatonin and alcohol intersect. Of specific interests to nutrition professionals, melatonin may counteract some of the effects that alcohol has on duodenal paracellular permeability [137]. Intestinal hyperpermeability, while not exclusively related to circadian patterns, is implicated in a variety of chronic health conditions. Alcohol consumption appears to decrease endogenous daytime glutamate signaling to the SCN, which may or may not be a clinically actionable mechanism [127]. Brain-derived neurotrophic factor (BDNF) can be increased via a Mediterranean-style diet [138], physical activity [139], intermittent fasting [140], and caloric restriction [141, 142]. Per2 abnormalities may predispose individuals to alcohol dependence [143]. Of interest for preventative models of care, those with abnormal circadian patterns may be good candidates for screening and preventative counseling for alcohol or other drug use and abuse.

Nutrition professionals should fully investigate alcohol consumption patterns among those with known circadian and/or sleep challenges and thyroid disorders. Those with gastrointestinal disorders may additionally benefit from such investigation. Some patients will benefit from abstinence even in the absence of addiction.

34.13 Chronotypes and Personalized Nutrition

With today's interest in personalized nutrition, understanding an individual's chronotype may be helpful in developing a personal diet and lifestyle plan. A person is classified by chronotype according to preference for when to sleep and when to be active. If laboratory investigations of an individual's internal time clock are not practical, a questionnaire may be used to estimate chronotype. These include the Morningness-Eveningness Questionnaire and the Munich Chronotype Questionnaire Test [144, 145]. If a person's chronotype influences physiological processes, then it may be important to consider chronotype when developing a nutritional plan. Recent studies are beginning to explore the role of clock gene SNPs in response to diet. Knowledge of gene variants in the circadian clock system may help practitioners design successful personalized nutrition [17].

34.14 Conclusion

Circadian rhythm disorders and disruptions in sleep habits may be overlooked or dismissed in clinical practice. The evidence of detrimental effects on metabolic function and dietary choices are accumulating. Emphasis on supporting circadian system function and addressing sleep disruption may significantly enhance health and productivity for many individuals.

Biochemical assessment of individuals who may have circadian rhythm dysfunction includes evaluation of the HPA axis via saliva or urinary parameters; inflammation markers such as C-reactive protein (CRP) or erythrocyte sedimentation rate (ESR); DHEA-S; glycemic handling markers, including hemoglobin A1c, fasting glucose, and insulin; and B-vitamin status as estimated by homocysteine, methylmalonic acid, and a complete blood count. Advanced biochemical assessment may include heavy metal testing, such as a provoked urinary sample or urinary porphyrins, genetic analysis, and red blood cell fatty acids. Markers for gastrointestinal microbiotic dysbiosis and antioxidant status may further inform treatment strategies and promote treatment adherence.

Included in an interview in such a case should be the exploration of the regularity or irregularity of sleep and wake times, as well as sleep hygiene, including bedtime routine, light and sound exposures, electromagnetic frequency (EMF) exposure [146, 147], coping mechanisms, and stress influences, such as the Holmes and Rahe Personal Stress Inventory. Dietary intake assessment should include timing of meals to determine the normal eating window. Clinical assessment may include stress, depression and anxiety screening questionnaires, and sleep questionnaires such as the Pittsburgh Sleep Quality Index (PSQI).

Integrative physicians and nutrition professionals wishing to optimize the circadian system may consider emphasizing eating upon arising, minimizing the eating window on a daily basis to 8–12 hours, cautioning against high-fat evening meals, prudent use of melatonin and vitamin B12 as needed, and moderation or abstinence of alcohol and caffeine, especially in the last 6–8 hours before bed. Antioxidant- and mineral-rich foods should be promoted, and supplemental use of bioavailable turmeric, berberine, and other botanicals may offer additional support given the outcomes of inflammatory and glycemic markers.

Increasing genetic and genomic screening may further inform clinicians how aggressive these and other interventions should be. Chrononutrition is still a promising yet underdeveloped science and clinical practice. Best practices are still evolving, and further intervention strategies await.

Satisfactory chronic disease care on the whole is an unmet medical need. While there are numerous contributing factors to chronic disease, the elements of sleep, circadian rhythm, and the stress response (HPA axis dysregulation) form a triad of independent, yet intimately related, components [148]. Each of these aspects can disrupt the other, and they may be impossible to treat independently. HPA axis dysregulation can cause circadian rhythm imbalances and/or sleep disruption. Circadian rhythm imbalances can cause sleep disruption and/or HPA axis dysregulation. Sleep disruption can cause HPA axis dysregulation and/or circadian rhythm imbalances. If a patient is being treated for a chronic disease, sleep quality and quantity should be evaluated from both a current and a historical perspective. Supporting a balanced and natural circadian rhythm is a favorable preventive medical approach to cancer, diabetes, cardiovascular disease, neurodegenerative conditions, and other chronic disease manifestations [149–154]. Considering that the natural circadian cycle with nutritional and lifestyle approaches does not necessarily fall under the role of any specific medical specialty, nutritional professionals, integrative medicine practitioners, and those in primary care can screen for and identify interruptions in circadian patterns and poor sleep as key health determinants as well as a component of chronic disease management.

The circadian clock shows promise for the development of new therapies, both pharmacologic and nutritional, that address a variety of disorders. The day-night rhythm plays a key role in multiple physiological systems and may prove to be a target for diseases that involve multiple organ systems. Understanding the molecular interactions and exploring the adverse metabolic consequences of circadian misalignment will help formulate therapies for metabolic, behavioral, and immune diseases.

References

1. Feillet C, van der Horst GT, Levi F, Rand DA, Delaunay F. Coupling between the circadian clock and cell cycle oscillators: implication for healthy cells and malignant growth. Front Neurol. 2015;6:96.
2. Sunderram J, Sofou S, Kamisoglu K, Karantza V, Androulakis IP. Time-restricted feeding and the realignment of biological rhythms: translational opportunities and challenges. J Transl Med. 2014;12:79.
3. Grechez-Cassiau A, Rayet B, Guillaumond F, Teboul M, Delaunay F. The circadian clock component BMAL1 is a critical regulator of p21WAF1/CIP1 expression and hepatocyte proliferation. J Biol Chem. 2008;283(8):4535–42.
4. Monk TH. Enhancing circadian zeitgebers. Sleep. 2010;33(4):421–2.
5. Ribas-Latre A, Eckel-Mahan K. Interdependence of nutrient metabolism and the circadian clock system: importance for metabolic health. Mol Metab. 2016;5(3):133–52.
6. Grechez-Cassiau A, Fillet C, Guerin S, Delaunay F. The hepatic circadian clock regulates the choline kinase α gene through the BMAL1-REV-ERBα axis. Chronobiol Int. 2015;32:6.
7. Innominato PF, Focan C, Gorlia T, et al. Circadian rhythm in rest and activity: a biological correlate of quality of life and a predictor of survival in patients with metastatic colorectal cancer. Cancer Res. 2009;69(11):4700–7.
8. Siffroi-Fernandez S, Dulong S, Li X-M, et al. Functional genomics identify Birc5/Survivin as a candidate gene involved in the chronotoxicity of cyclin-dependent kinase inhibitors. Cell Cycle. 2014;13(6):984–91.
9. Litlekalsoy J, Rostad K, Kalland KH, Hostmark JG, Laerum OD. Expression of circadian clock genes and proteins in urothelial cancer is related to cancer-associated genes. BMC Cancer. 2016;16:549.
10. Merck Manual Professional Version. Circadian rhythm sleep disorders. Accessed at https://www.merckmanuals.com/professional/neurologic-disorders/sleep-and-wakefulness-disorders/circadian-rhythm-sleep-disorders.
11. Etain B, Dumaine A, Bellivier F, et al. Genetic and functional abnormalities of the melatonin biosynthesis pathway in patients with bipolar disorder. Hum Mol Genet. 2012;21(18):4030–7.
12. Zisapel N. Circadian rhythm sleep disorders: pathophysiology and potential approaches to management. CNS Drugs. 2001;15(4):311–28.
13. Okawa M, Mishima K, Nanami T, et al. Vitamin B12 treatment for sleep-wake rhythm disorders. Sleep. 1990;13(1):15–23.
14. Oike H, Oishi K, Kobori M. Nutrients, clock genes and chrononutrition. Curr Nutr Rep. 2014;3:204–12.
15. Johnston JD, Ordovas JM, Scheer FA, Turek FW. Circadian rhythms, metabolism, and chrononutrition in rodents and humans. Adv Nutr. 2016;7:399–406.
16. Asher G, Sassone-Corsi P. Time for food: the intimate interplay between nutrition, metabolism, and the circadian clock. Cell. 2015;161:84–92.
17. Potter GD, Cade JE, Grant PJ, Hardie LJ. Nutrition and the circadian system. Br J Nutr. 2016;116(3):434–42.
18. Mohawk JA, Green CB, Takahashi JS. Central and peripheral circadian clocks in mammals. Annu Rev Neurosci. 2012;35:445–62.
19. Regestein QR, Monk TH. Delayed sleep phase syndrome: a review of its clinical aspects. Am J Psychiatry. 1995;152(4):602–8.
20. Weitzman ED, Czeisler CA, Coleman RM, et al. Delayed sleep phase syndrome. A chronobiological disorder with sleep-onset insomnia. Arch Gen Psychiatry. 1981;38(7):737–46.
21. Ando K, Kripke DF, Ancoli-Israel S. Estimated prevalence of delayed and advanced sleep phase syndromes. Sleep Res. 1995;24:509.
22. Schrader H, Bovim G, Sand T. The prevalence of delayed and advanced sleep phase syndromes. J Sleep Res. 1993;2(1):51–5.
23. Archer SN, Robilliard DL, Skene DJ, et al. A length polymorphism in the circadian clock gene Per3 is linked to delayed sleep phase syndrome and extreme diurnal preference. Sleep. 2003;26(4):413–5.
24. Hohjoh H, Takasu M, Shishikura K, et al. Significant association of the arylalkylamine N-acetyltransferase (AA-NAT) gene with delayed sleep phase syndrome. Neurogenetics. 2003;4(3):151–3.
25. Iwase T, Kajimura N, Uchiyama M, et al. Mutation screening of the human clock gene in circadian rhythm sleep disorders. Psychiatry Res. 2002;109(2):121–8.
26. Ebisawa T, Uchiyama M, Kajimura N, et al. Association of structural polymorphisms in the human period3 gene with delayed sleep phase syndrome. EMBO Rep. 2001;2(4):342–6.
27. Takahashi Y, Hohjoh H, Matsuura K. Predisposing factors in delayed sleep phase syndrome. Psychiatry Clin Neurosci. 2000;54(3):356–8.
28. Bliwise DL, King AC, Harris RB, Haskell WL. Prevalence of self-reported poor sleep in a healthy population aged 50-65. Soc Sci Med. 1992;34(1):49–55.
29. Toh KL, Jones CR, He Y, et al. An hPer2 phosphorylation site mutation in familial advanced sleep phase syndrome. Science. 2001;291(5506):1040–3.
30. Martens H, Endlich H, Hildebrandt G. Sleep-wake distribution in blind subjects with and without sleep complaints. Sleep Res. 1990;9:398.
31. Akerstedt T, Torsvall L. Shiftwork. Shift dependent Well-being and individual differences. Ergonomics. 1981;24:265–73.
32. Witting W, Kwa IH, Eikelenboom P, Mirmiran M, Swaab DF. Alterations in the circadian rest-activity rhythm in aging and Alzheimer's disease. Biol Psychiatry. 1990;27(6):563–72.
33. Hoogendijk WJ, van Someren EJ, Mirmiran M, et al. Circadian rhythm-related behavioral disturbances and structural hypothalamic changes in Alzheimer's disease. Int Psychogeriatr. 1996;8(Suppl 3):245–52; discussion 269–72.

34. van Someren EJ, Hagebeuk EE, Lijzenga C, et al. Circadian rest-activity rhythm disturbances in Alzheimer's disease. Biol Psychiatry. 1996;40(4):259–70.
35. Pollak CP, Stokes PE. Circadian rest-activity rhythms in demented and nondemented older community residents and their caregivers. J Am Geriatr Soc. 1997;45(4):446–52.
36. Butler MP, Smales C, Wu H, et al. The circadian system contributes to apnea lengthening across the night in obstructive sleep apnea. Sleep. 2015;38(11):1793–801.
37. Mateika JH, Syed Z. Intermittent hypoxia, respiratory plasticity and sleep apnea in humans: present knowledge and future investigations. Respir Physiol Neurobiol. 2013;188(3):289–300.
38. Martin A, Carpentier A, Guissard N, van HJ, Duchateau J. Effect of time of day on force variation in a human muscle. Muscle Nerve. 1999;22:1380–7.
39. Zhang X, Dube TJ, Esser KA. Working around the clock: circadian rhythms and skeletal muscle. J Appl Physiol. 2009;107:1647–54.
40. Mitchell LJ, Davidson ZE, Bonham M, et al. Weight loss from lifestyle interventions and severity of sleep apnoea: a systematic review and meta-analysis. Sleep Med. 2014;15(10):1173–83.
41. Entzian P, Linnemann K, Schlaak M, Zabel P. Obstructive sleep apnea syndrome and circadian rhythms of hormones and cytokines. Am J Respir Crit Care Med. 1996;153(3):1080–6.
42. Armstrong SM, McNulty OM, Guardiola-Lemaitre B, Redman JR. Successful use of S20098 and melatonin in an animal model of delayed sleep-phase syndrome (DSPS). Pharmacol Biochem Behav. 1993;46(1):45–9.
43. Oldani A, Ferini-Strambi L, Zucconi M, et al. Melatonin and delayed sleep phase syndrome: ambulatory polygraphic evaluation. Neuro Rep. 1994;6(1):132–4.
44. Kamei Y, Hayakawa T, Urata J, et al. Melatonin treatment for circadian rhythm sleep disorders. Psychiatry Clin Neurosci. 2000;54(3):381–2.
45. Nagtegaal JE, Laurant MW, Kerkhof GA, et al. Effects of melatonin on the quality of life in patients with delayed sleep phase syndrome. J Psychosom Res. 2000;48(1):45–50.
46. Kim SM, Neuendorff N, Chapkin RS, Earnest DJ. Role of inflammatory signaling in the differential effects of saturated and polyunsaturated fatty acids on peripheral circadian clocks. EBioMedicine. 2016;7:100–11.
47. Jordan SD, Lamia KA. AMPK at the crossroads of circadian clocks and metabolism. Mol Cell Endocrinol. 2013;366(2):163–9.
48. Held K, Antonijevic IA, Künzel H, et al. Oral Mg(2+) supplementation reverses age-related neuroendocrine and sleep EEG changes in humans. Pharmacopsychiatry. 2002;35(4):135–43.
49. Beydoun MA, Gamaldo AA, Canas JA, et al. Serum nutritional biomarkers and their associations with sleep among US adults in recent national surveys. PLoS One. 2014;9(8):e103490.
50. Bladé C, Aragonès G, Arola-Arnal A, et al. Proanthocyanidins in health and disease. Biofactors. 2016;42(1):5–12.
51. Chung S, Yao H, Caito S, et al. Regulation of SIRT1 in cellular functions: role of polyphenols. Arch Biochem Biophys. 2010;501(1):79–90.
52. Ford ES, Wheaton AG, Chapman DP, et al. Associations between self-reported sleep duration and sleeping disorder with concentrations of fasting and 2-h glucose, insulin, and glycosylated hemoglobin among adults without diagnosed diabetes. J Diabetes. 2014;6(4):338–50.
53. Herrmann TS, Bean ML, Black TM, Wang P, Coleman RA. High glycemic index carbohydrate diet alters the diurnal rhythm of leptin but not insulin concentrations. Exp Biol Med (Maywood). 2001;226(11):1037–44.
54. Abbondante S, Eckel-Mahan KL, Ceglia NJ, Baldi P, Sassone-Corsi P. Comparative circadian metabolomics reveal differential effects of nutritional challenge in the serum and liver. J Biol Chem. 2016;291(6):2812–28.
55. Thaiss CA, Zeevi D, Levy M, et al. Transkingdom control of microbiota diurnal oscillations promotes metabolic homeostasis. Cell. 2014;159:514–29.
56. Leone V, Gibbons SM, Martinez K, et al. Effects of diurnal variation of gut microbes and high fat feeding on host circadian clock function and metabolism. Cell Host Microbe. 2015;17(5):681–9.
57. Morgan L, Arendt J, Owens D, et al. Effects of the endogenous clock and sleep time on melatonin, insulin, glucose and lipid metabolism. J Endocrinol. 1998;157:443–51.
58. Tanabe K, Kitagawa E, Wada M, et al. Antigen exposure in the late light period induces severe symptoms of food allergy in an ova-allergic mouse model. Sci Rep. 2015;5:14424.
59. Scheer FA, Morris CJ, Shea SA. The internal circadian clock increases hunger and appetite in the evening independent of food intake and other behaviors. Obesity (Silver Spring). 2013;21:421–3.
60. Hatori M, Vollmers C, Zarrinpar A, et al. Time-restricted feeding without reducing caloric intake prevents metabolic diseases in mice fed a high-fat diet. Cell Metab. 2012;5:848–60.
61. Xu K, DiAngelo JR, Hughes ME, et al. The circadian clock interacts with metabolic physiology to influence reproductive fitness. Cell Metab. 2011;13:639–54.
62. Yoshida C, Shikata N, Seki S, Koyama N, Noguchi Y. Early nocturnal meal skipping alters the peripheral clock and increases lipogenesis in mice. Nutr Metab (Lond). 2012;9(1):78.
63. Timlin MT, Pereira MA, Story M, Neumark-Sztainer D. Breakfast eating and weight change in a 5-year prospective analysis of adolescents: project EAT (Eating Among Teens). Pediatrics. 2008;121(3):e638–45.
64. Mattson MP, Allison DB, Fontana L, et al. Meal frequency and timing in health and disease. Proc Natl Acad Sci U S A. 2014;111(47):16647–53.
65. Marcheva B, Ramsey KM, Peek CB, et al. Circadian clocks and metabolism. Handb Exp Pharmacol. 2013;217:127–55.
66. Yoshizaki T, Tada Y, Hida A, et al. Effects of feeding schedule changes on the circadian phase of the cardiac autonomic nervous system and serum lipid levels. Eur J Appl Physiol. 2013;113(10):2603–11.
67. Jakubowicz D, Barnea M, Wainstein J, Froy O. High caloric intake at breakfast vs. dinner differentially influences weight loss of overweight and obese women. Obesity (Silver Spring). 2013;21(12):2504–12.
68. Garaulet M, Gomez-Abellan P, Alburquerque-Bejar JJ, et al. Timing of food intake predicts weight loss effectiveness. Int J Obes. 2013;37(4):604–11.
69. Wang JB, Patterson RE, Ang A, et al. Timing of energy intake during the day is associated with the risk of obesity in adults. J Hum Nutr Diet. 2014;27(Suppl 2):255–62.
70. Varady KA, Bhutani S, Church EC, et al. Short-term modified alternate-day fasting: a novel dietary strategy for weight loss and cardioprotection in obese adults. Am J Clin Nutr. 2009;90:1138–43.
71. Bhutani S, Klempel MC, Kroeger CM, et al. Alternate day fasting and endurance exercise combine to reduce body weight and favorably alter plasma lipids in obese humans. Obesity. 2013;21:1370–9.
72. Anson RM, Guo Z, de Cabo R, et al. Intermittent fasting dissociates beneficial effects of dietary restriction on glucose metabolism and neuronal resistance to injury from calorie intake. Proc Natl Acad Sci U S A. 2003;100:6216–20.
73. Mekary RA, Giovannucci E, Willet WC, et al. Eating patterns and type 2 diabetes risk in men: breakfast omission, eating frequency, and snacking. Am Soc Nutr. 2012;95(5):1182–9.
74. Moro T, Tinsley G, Bianco A, et al. Effects of eight weeks of time-restricted feeding (16/8) on basal metabolism, maximal strength, body composition, inflammation, and cardiovascular risk factors in resistance-trained males. J Transl Med. 2016;14:290.
75. Takahashi JS, Hong H-K, Ko CH, McDearmon EL. The genetics of mammalian circadian order and disorder: implications for physiology and disease. Nat Rev Genet. 2008;9(10):764–75.
76. Valladares M, Obregon AM, Chaput JP. Association between genetic variants of the clock gene and obesity and sleep duration. J Physiol Biochem. 2015;71(4):855–60.
77. Patel SA, Velingkaar N, Makwana K, Chaudhari A, Kondratov R. Calorie restriction regulates circadian clock gene expression through BMAL1 dependent and independent mechanisms. Sci Rep. 2016;6:25970.

78. Eckel-Mahan KL, Patel VR, de Mateo S, et al. Reprogramming of the circadian clock by nutritional challenge. Cell. 2013;155(7):1464–78.
79. Scheer FA, Hilton MF, Mantzoros CS, Shea SA. Adverse metabolic and cardiovascular consequences of circadian misalignment. PNAS. 2009;106(11):4453–8.
80. Gibbs M, Harrington D, Starkey S, Williams P, Hampton S. Diurnal postprandial responses to low and high glycaemic index mixed meals. Clin Nutr. 2014;33(5):889–94.
81. Kohsaka A, Laposky AD, Ramsey KM, et al. High-fat diet disrupts behavioral and molecular circadian rhythms in mice. Cell Metab. 2007;6(5):414–21.
82. Wilking M, Ndiaye M, Mukhtar H, Ahmad N. Circadian rhythm connections to oxidative stress: implications for human health. Antioxid Redox Signal. 2013;19(2):192–208.
83. Golem DL, Martin-Biggers JT, Koenings MM, Davis KF, Byrd-Bredbenner C. An integrative review of sleep for nutrition professionals. Adv Nutr. 2014;5(6):742–59.
84. Richards J, Gumz ML. Mechanism of the circadian clock in physiology. Am J Physiol Regul Integr Comp Physiol. 2013;304(12):R1053–64.
85. la Fleur SE, Kalsbeek A, Wortel J, et al. A daily rhythm in glucose tolerance: a role for the suprachiasmatic nucleus. Diabetes. 2001;50:1237–43.
86. Coomans CP, van den Berg SA, Lucassen EA, et al. The suprachiasmatic nucleus controls circadian energy metabolism and hepatic insulin sensitivity. Diabetes. 2012;62:1102–8.
87. Coomans CP, van den Berg SA, Houben T, et al. Detrimental effects of constant light exposure and high-fat diet on circadian energy metabolism and insulin sensitivity. FASEB J. 2013;27(4):1721–32.
88. Bouatia-Naj N, Bonnefond A, Cavalcanti-Proenca C, et al. A variant near MTNR1B is associated with increased fasting plasma glucose levels and type 2 diabetes risk. Nat Genet. 2008;41:89–94.
89. Liu C, Wu Y, Li H, et al. MTNR1B rs10830963 is associated with fasting plasma glucose, HbA1C and impaired beta-cell function in Chinese Hans from Shanghai. BMC Med Genet. 2010;11:59.
90. Peschke E, Bahr I, Muhlbauer E. Experimental and clinical aspects of melatonin and clock genes in diabetes. J Pineal Res. 2015;59(1):1–23.
91. Martino TA, Tata N, Belsham DD, et al. Disturbed diurnal rhythm alters gene expression and exacerbates cardiovascular disease with rescue by resynchronization. Hypertension. 2007;49:1104–13.
92. Martino TA, Tata N, Simpson JA, et al. The primary benefits of angiotensin-converting enzyme inhibition on cardiac remodeling occur during sleep time in murine pressure overload hypertrophy. J Am Coll Cardiol. 2011;57(20):2020–8.
93. Durgan DJ, Young ME. The cardiomyocyte circadian clock: emerging roles in health and disease. Circ Res. 2010;106(4):647–58.
94. Wang A, Arah OA, Kauhanen J, Krause N. Shift work and 20-year incidence of acute myocardial infarction: results from the Kuopio Ischemic Heart Disease Risk Factor Study. Occup Environ Med. 2016;73(9):588–94.
95. Wang A, Arah OA, Kauhanen J, Krause N. Work schedules and 11-year progression of carotid atherosclerosis in middle-aged Finnish men. Am J Ind Med. 2015;58(1):1–13.
96. Kantermann T, Duboutay F, Haubruge D, et al. Atherosclerotic risk and social jetlag in rotating shift-workers: first evidence from a pilot study. Work. 2013;46(3):273–82.
97. Fujino Y, Iso H, Tamakoshi A, et al. A prospective cohort study of shift work and risk of ischemic heart disease in Japanese male workers. Am J Epidemiol. 2006;164(2):128–35.
98. Lieu SJ, Curhan GC, Schernhammer ES, Forman JP. Rotating night shift work and disparate hypertension risk in African-Americans. J Hypertens. 2012;30(1):61–6.
99. Richards J, Greenlee MM, Jeffers LA, et al. Inhibition of αENaC expression and ENaC activity following blockade of the circadian clock-regulatory kinases CK1δ/ε. Am J Physiol Renal Physiol. 2012;303(7):F918–27.
100. Marques FZ, Campain AE, Tomaszewski M, et al. Gene expression profiling reveals renin mRNA overexpression in human hypertensive kidneys and a role for microRNAs. Hypertension. 2011;58(6):1093–8.
101. Nader N, Chrousos GP, Kino T. Interactions of the circadian CLOCK system and the HPA axis. Trends Endocrinol Metab. 2010;21(5):277–86.
102. Pezuk P, Mohawk JA, Wang LA, et al. Glucocorticoids as entraining signals for peripheral circadian oscillators. Endocrinology. 2012;153(10):4775–83.
103. Balsalobre A, Brown SA, Marcacci L, et al. Resetting of circadian time in peripheral tissues by glucocorticoid signaling. Science. 2000;289:2344–7.
104. Son GH, Chung S, Kim K. The adrenal peripheral clock: glucocorticoid and the circadian timing system. Front Neuroendocrinol. 2011;32:451–65.
105. Ackermann K, Revell VL, Lao O, et al. Diurnal rhythms in blood cell populations and the effect of acute sleep deprivation in healthy young men. Sleep. 2012;35(7):933–40.
106. Logan RW, Zhang C, Murugan S, et al. Chronic shift-lag alters the circadian clock of NK cells and promotes lung cancer growth in rats. J Immunol. 2012;188(6):2583–91.
107. Peruquetti RL, de Mateo S, Sassone-Corsi P. Circadian proteins CLOCK and BMAL1 in the chromatoid body, a RNA processing granule of male germ cells. PLoS One. 2012;7(8):e42695.
108. Kennaway DJ, Boden MJ, Varcoe TJ. Circadian rhythms and fertility. Mol Cell Endocrinol. 2012;349:56–61.
109. Alvarez JD, Hansen A, Ord T, et al. The circadian clock protein BMAL1 is necessary for fertility and proper testosterone production in mice. J Biol Rhythm. 2008;23(1):26–36.
110. Boden MJ, Kennaway DJ. Circadian rhythms and reproduction. Reproduction. 2006;132:379–92.
111. Bedrosian TA, Nelson RJ. Timing of light exposure affects mood and brain circuits. Transl Psychiatry. 2017;7(1):e1017.
112. Farokhnezhad Afshar P, Bahramnezhad F, Asgari P, et al. Effect of white noise on sleep in patients admitted to a coronary care. J Caring Sci. 2016;5(2):103–9.
113. Lillehei AS, Halcón LL, Savik K, et al. Effect of inhaled lavender and sleep hygiene on self-reported sleep issues: a randomized controlled trial. J Altern Complement Med. 2015;21(7):430–8.
114. Perl O, Arzi A, Sela L, et al. Odors enhance slow-wave activity in non-rapid eye movement sleep. J Neurophysiol. 2016;115(5):2294–302.
115. Dyer J, Cleary L, McNeill S, Ragsdale-Lowe M, Osland C. The use of aromasticks to help with sleep problems: a patient experience survey. Complement Ther Clin Pract. 2016;22:51–8.
116. Kasper S, Anghelescu I, Dienel A. Efficacy of orally administered Silexan in patients with anxiety-related restlessness and disturbed sleep--a randomized, placebo-controlled trial. Eur Neuropsychopharmacol. 2015;25(11):1960–7.
117. Fairbrother K, Cartner B, Alley JR, et al. Effects of exercise timing on sleep architecture and nocturnal blood pressure in prehypertensives. Vasc Health Risk Manag. 2014;10:691–8.
118. Hanlon EC, Tasali E, Leproult R, et al. Sleep restriction enhances the daily rhythm of circulating levels of Endocannabinoid 2-Arachidonoylglycerol. Sleep. 2016;39(3):653–64.
119. Clifford LM, Beebe DW, Simon SL, et al. The association between sleep duration and weight in treatment-seeking preschoolers with obesity. Sleep Med. 2012;13(8):1102–5.
120. Taheri S. The link between short sleep duration and obesity: we should recommend more sleep to prevent obesity. Arch Dis Child. 2006;91:881–4.
121. Chaput JP, Despres JP, Bouchard C, et al. Longer sleep duration associates with lower adiposity gain in adult short sleepers. Int J Obes. 2012;36(5):752–6.
122. Drake C, Roehrs T, Shambroom J, Roth T. Caffeine effects on sleep taken 0, 3, or 6 hours before going to bed. J Clin Sleep Med. 2013;9(11):1195–200.

123. Penolazzi B, Natale V, Leone L, Russo PM. Individual differences affecting caffeine intake. Analysis of consumption behaviours for different times of day and caffeine sources. Appetite. 2012;58:971–7.
124. Zwyghuizen-Doorenbos A, Roehrs TA, Lipschutz L, Timms V, Roth T. Effects of caffeine on alertness. Psychopharmacology. 1990;100:36–9.
125. Snel J, Lorist MM. Effects of caffeine on sleep and cognition. Prog Brain Res. 2011;190:105–17.
126. Jiménez-Ortega V, Cano Barquilla P, Fernández-Mateos P, Cardinali DP, Esquifino AI. Cadmium as an endocrine disruptor: correlation with anterior pituitary redox and circadian clock mechanisms and prevention by melatonin. Free Radic Biol Med. 2012;53(12):2287–97.
127. Prosser RA, Glass JD. Assessing ethanol's actions in the suprachiasmatic circadian clock using in vivo and in vitro approaches. Alcohol. 2015;49(4):321–39.
128. Landolt HP, Gillin JC. Sleep abnormalities during abstinence in alcohol-dependent patients: aetiology and management. CNS Drugs. 2001;15:413–25.
129. Roehrs T, Petrucelli N, Roth T. Sleep restriction, ethanol effects and time of day. Human Psychopharm. 1996;11:199–204.
130. Roehrs T, Roth T. Sleep, sleepiness, sleep disorders and alcohol use and abuse. Sleep Med Rev. 2001;5:287–97.
131. Clark CP, Gillin JC, Golshan S, et al. Increased REM sleep density at admission predicts relapse by three months in primary alcoholics with a lifetime diagnosis of secondary depression. Biol Psych. 1998;43:601–7.
132. Brower KJ, Aldrich MS, Robinson EAR, Zucker RA, Greden JF. Insomnia, self-medication, and relapse to alcoholism. Am J Psychiat. 2001;158:399–404.
133. Danel T, Vantyghem M-C, Touitou Y. Responses of the steroid circadian system to alcohol in humans: importance of the time and duration of intake. Chronobiol Int. 2006;23:1025–34.
134. Danel T, Touitou Y. Alcohol decreases the nocturnal peak of TSH in healthy volunteers. Psychopharmacology. 2003;170:213–4.
135. Kuhlwein E, Hauger RL, Irwin MR. Abnormal nocturnal melatonin secretion and disordered sleep in abstinent alcoholics. Biol Psych. 2003;54:1437–43.
136. Rojdmark S, Wikner J, Adner N, Andersson DEH, Wetterberg L. Inhibition of melatonin secretion by ethanol in man. Metab Clin Exp. 1993;42:1047–51.
137. Sommansson A, Saudi WS, Nylander O, Sjöblom M. Melatonin inhibits alcohol-induced increases in duodenal mucosal permeability in rats in vivo. Am J Physiol Gastrointest Liver Physiol. 2013;305(1):G95–105.
138. Sánchez-Villegas A, Galbete C, Martinez-González MA, et al. The effect of the Mediterranean diet on plasma brain-derived neurotrophic factor (BDNF) levels: the PREDIMED-NAVARRA randomized trial. Nutr Neurosci. 2011;14(5):195–201.
139. Saucedo Marquez CM, Vanaudenaerde B, Troosters T, Wenderoth N. High-intensity interval training evokes larger serum BDNF levels compared with intense continuous exercise. J Appl Physiol (1985). 2015;119(12):1363–73.
140. Cherif A, Roelands B, Meeusen R, Chamari K. Effects of intermittent fasting, caloric restriction, and Ramadan intermittent fasting on cognitive performance at rest and during exercise in adults. Sports Med. 2016;46(1):35–47.
141. Araya AV, Orellana X, Espinoza J. Evaluation of the effect of caloric restriction on serum BDNF in overweight and obese subjects: preliminary evidences. Endocrine. 2008;33(3):300–4.
142. Contestabile A. Benefits of caloric restriction on brain aging and related pathological states: understanding mechanisms to devise novel therapies. Curr Med Chem. 2009;16(3):350–61.
143. Spanagel R, Pendyala G, Abarca C, et al. The clock gene Per2 influences the glutamatergic system and modulates alcohol consumption. Nature Med. 2005;11:35–42.
144. Horne JA, Ostberg O. A self-assessment questionnaire to determine morningness-eveningness in human circadian rhythms. Int J Chronobiol. 1976;4(2):97–110.
145. Roenneberg T, Wirz-Justice A, Merrow M. Life between clocks: daily temporal patterns of human chronotypes. J Biol Rhythm. 2003;18(1):80–90.
146. Lewczuk B, Redlarski G, Żak A, Ziółkowska N, Przybylska-Gornowicz B, Krawczuk M. Influence of electric, magnetic, and electromagnetic fields on the circadian system: current stage of knowledge. Biomed Res Int. 2014;2014:169459. https://doi.org/10.1155/2014/169459.
147. Redlarski G, et al. The influence of electromagnetic pollution on living organisms: historical trends and forecasting changes. Biomed Res Int. 2015;2015:234098. *PMC*. Web. 3 Oct. 2018.
148. Garbarino S, Lanteri P, Durando P, Magnavita N, Sannita WG. Co-morbidity, mortality, quality of life and the healthcare/welfare/social costs of disordered sleep: a rapid review. Int J Environ Res Public Health. 2016;13(8). pii: E831.
149. Barone DA, Chokroverty S. Neurologic diseases and sleep. Sleep Med Clin. 2017;12(1):73–85.
150. Armstrong TS, Shade MY, Breton G. Sleep-wake disturbance in patients with brain tumors. Neuro Oncol. 2016. pii: now119.
151. Whitehead LC, Unahi K, Burrell B, Crowe MT. The experience of fatigue across long-term conditions: a qualitative meta-synthesis. J Pain Symptom Manage. 2016;52(1):131–43.e1.
152. McHill AW, Wright KP. Role of sleep and circadian disruption on energy expenditure and in metabolic predisposition to human obesity and metabolic disease. Obes Rev. 2017;18(Suppl 1):15–24.
153. Forrestel AC, Miedlich SU, Yurcheshen M, Wittlin SD, Sellix MT. Chronomedicine and type 2 diabetes: shining some light on melatonin. Diabetologia. 2017;60(5):808–22.
154. Touitou Y, Reinberg A, Touitou D. Association between light at night, melatonin secretion, sleep deprivation, and the internal clock: health impacts and mechanisms of circadian disruption. Life Sci. 2017;173:94–106.

Nutrition with Movement for Better Energy and Health

Peter Wilhelmsson

35.1 Background – 596
35.1.1 Integrating Nutrition and Movement – 596

35.2 Physiological Mechanisms Associated with Exercise and Nutrition – 597
35.2.1 Oxygen Plays a Key Role – 597
35.2.2 Nutrients Play a Key Role – 600
35.2.3 MET: Powerful Prevention and Treatment of Chronic Disease – 602

35.3 MET and Nutrients for People with Various Conditions – 606
35.3.1 Commonalities Across Conditions – 606
35.3.2 Sarcopenia – 608

References – 610

> Exercise is king. Nutrition is queen. Put them together and you've got a kingdom. –Jack LaLanne

35.1 Background

Inactivity and poor nutrition are major drivers of chronic disease risk, suffering, and early mortality. Physical inactivity is the fourth-leading risk factor for global mortality (6% of deaths), trailing only hypertension (13%), tobacco use (9%), and hyperglycemia (6%). Approximately 3.2 million annual deaths are linked to insufficient physical activity. In fact, inactivity is a greater mortality risk than being overweight or obese (5% of global mortality).

In 2008, 31 percent of adults worldwide – slightly more women than men – were considered inactive. In the American and European-Mediterranean populations, about half of the women and 40 percent of men were insufficiently active. Southeast Asia had the lowest percentages of inactive populations, with men at 15 percent and women at 19 percent [1].

As will be highlighted in this chapter, most health conditions can be improved with movement and exercise training (MET) or a combination of MET and other lifestyle medicine therapies. One challenge in the public health arena, preventive medicine, and disease-oriented medicine is how to inspire, educate, organize, and follow up on lifestyle medicine therapies. People will always look for quick fixes, but how do we educate, inspire, and motivate them to establish and maintain lifestyle habits that promote health instead of disease?

Inactivity, poor food choices, alcohol abuse, smoking, and poor stress management are commonly listed as drivers of chronic disease. The functional medicine model adds environmental toxins as another risk factor. The more we can identify and rectify these causes of chronic disease, the more momentum is created and the more success we will have with individual patients. As a result, we will see lower healthcare costs.

35.1.1 Integrating Nutrition and Movement

In the same way, one should not separate the soul and body and one should not separate movement and nutrition. They influence each other. You cannot treat nutrient deficiencies with exercise, and you cannot treat inactivity with nutrition. Both are needed all the time.

Administering the correct dosages of movement and exercise training (MET) will make everything else work better. Even dietary supplements like fish oil work better when combined with movement and exercise [2]. This 2014 study in *Neuropsychologia* showed a difference in cognitive function between patients who took omega-3 supplements and exercised. They also saw that the omega-6 to omega-3 ratio improved by adding exercise to omega-3 intake. This shows the positive potential of combining the intake of nutrients with movement and exercise. The transport, uptake, and function of nutrients with or without different levels of exercise and oxygenation of the cells may differ, according to the previous study.

The addition of consistent MET will complement and make other lifestyle habits or therapies more effective, whether they are stress management, dietary habits, supplement intake, or even use of pharmaceutical drugs. Our physiology depends on movement for basic and optimal function. Daily movement and exercise provide good circulation to facilitate normal physiological and biochemical functions.

People often underestimate the power of changing small habits, such as drinking a couple of glasses of water every morning, using public transportation, using a standing workplace, eating a salad every day, taking a daily walk, or just doing a little more daily stretching and activity. In the public health arena, larger populations making a few small, positive changes produce tremendous health effects, healthcare savings, and increased productivity. According to a study in Australia, walking and moving for 35 minutes per day has a positive effect on health and survival. The study showed people in the city of Melbourne alone had 272 fewer deaths per year, 903 fewer cases of chronic disease and a savings of $12.2 million when they used public transport instead of taking a car [3]. This and other studies show how everyday exercise and movement, even walking to the subway or to work every day, has a positive effect on health.

35.1.1.1 Movement and Detoxification

We are surrounded and affected by daily toxins. In the past 80 years, we have been subjected to more than 100,000 new chemicals [4]. At least 62,000 of these chemicals have shown to be transported by our cell membranes and enter our cells. Most of these are toxic to our cells and mitochondria, and they can affect and disturb our hormones. These chemicals are everywhere, in air, water, food, cosmetics, pharmaceuticals, personal care products, cleaning products, fuel, receipts, plastic bottles, and more.

Why is MET so important for detox? Exercise increases thyroid activity, whole-body metabolism, perspiration, lymphatic circulation, and bowel movement so that all these functions become more efficient in cleaning out waste.

- MET activates the lymphatic system and helps with filtration of toxins through the lymph nodes and contributes to a better cleanse of interstitial tissue fluid.
- MET increases sweating, which carries toxins from the inside of the body to the outside of the body, where it evaporated or wiped off by clothes, towels, etc. The more you move and sweat, the more you get rid of environmental toxins which relieve the kidneys and liver for the detoxification process.
- MET increases the activities of autophagosomes and the "cleaning process" that helps to break down and remove or recycle the old worn-out cell components. The same applies to "urban workers" in the brain which are primarily active at night thanks to the movements through the lymphatic system of the brain, the so-called glymphatic system (lymphatic system in glial cells). Activity during the daytime combined with a good night of sleep has shown to optimize this glymphatic cleaning process, that is so valuable for removing debris from the brain.

- Increased oxygenation of cells occurs by exercise and makes the combination of circulation and nourishment to the citric acid cycle help bind and displace toxins.
- The balance of redox status. Consistent moderate MET helps maintain a healthy redox balance. Both under-exercise and overtraining create problems with redox balance and an abundance of free radicals, accelerating the aging process and increasing the risk of developing chronic disease. Exercise induces increases in superoxide dismutase and other markers. MET and aerobic capacity are important factors in maintaining balance in redox status [5].

To further optimize the expelling of toxins through sweating, make sure you shower directly after you work out, and even better, shower and take a 15–30-minute post-workout sauna or hot bath.

35.1.1.2 Sitting Is the New Smoking

Over the last 10 years, many studies have concluded that excessive sitting is a major risk for mortality or chronic disease, independent from MET. Many people who have transitioned from a sitting work desk to a standing work desk can attest to that it is one of the easiest changes one can make to decrease the risk of mortality and chronic disease. A meta-analysis in 2015 concluded that sedentary time, primarily hours of sitting time, was independently associated with a greater risk for all-cause mortality [6]. Even if someone gets the widely recommended 30 minutes of daily exercise, being active in the remaining approximate 6500 waking minutes of the week makes a greater difference. An international mobile health (mHealth) program for studying obesity, inactivity, and sitting showed improvement when these people could stand and move more [7]. In fact, reduced sitting time may have a positive impact on psychological and physical health [8].

Sitting and watching TV is more dangerous than just sitting and working because watching TV is often associated with snacking on processed foods. Every hour of TV viewing after the age of 25 takes 22 minutes off a person's life [9, 10]. Sitting with one's upper back hunched forward also restricts the flow of oxygen and induces shallow breathing. Shallow breathing and stress breathing are common in modern societies. Deep breathing exercises contribute to increased oxygenation of the tissues, improvement in blood and lymphatic flow, and better digestive functions. Aerobic exercise improves the heart's blood-pumping capacity, as well as the capacity of the lungs. Both aerobic and anaerobic exercises have been shown to increase our capacity to better quench oxidative stress and improve redox balance [11]. Standing, walking, and MET with good posture increase the vital flow of oxygen to tissues. Transitioning to a standing desk is a simple change to decrease sedentary time. At a standing work desk, it is easier to move, fidget, stretch, and maintain good posture, all of which improve oxygenation to the tissues. Transition to better sitting positions at work is a good first step. Standing and working or walking and working are ideal (◘ Fig. 35.1).

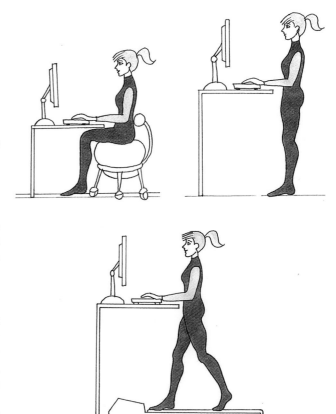

◘ Fig. 35.1 Sitting upright, standing, and walking at workplace

35.2 Physiological Mechanisms Associated with Exercise and Nutrition

35.2.1 Oxygen Plays a Key Role

35.2.1.1 Oxygen, Nutrition, and Movement

We eat 4 pounds of food and drink 4 quarts of fluid a day, but in that same time, we take in 7.5 pounds of oxygen and produce waste from the burning of oxygen in biochemical reactions. About 3% of this oxygen intake ends up as free radicals due to by-products from energy production. These free radicals are complexed by our own production and even by the external reserve of antioxidants we ingest from foods, superfoods, and supplements.

There is a tendency for oxygen deprivation in the modern world, primarily because of our sedentary lifestyle. Boosting the production and transport of oxygen stimulates the synthesis and activity of white blood cells, resulting in a stronger immune system. Increasing oxygen levels may be the principal method to combat damage caused by aging and to discourage the growth of abundant number of anaerobic bacteria, yeast, and cancer cells. Production of energy and the use of oxygen are controlled tightly and influenced a great deal by MET. Oxygenation to muscle cells can increase nearly 15 times through intensive training as compared to a resting state, such as sitting. In a resting state, the body uses about 0.21 L of oxygen per minute, compared to the approximate 2.5 L/minute rate from brisk walking and the 3.0 L/minute rate from running. The total cardiac output of blood in a resting state is approximately 5 L/minute, but that rapidly increases to about 25 L/minute when engaging in vigorous MET.

The heart must work against gravity to pump blood up to the brain, the body's largest consumer of oxygen. One of the main causes of mental deterioration with age is hardening arteries and high blood pressure, both of which result from decreased nitric oxide production/supply, oxidized lipids, chronic inflammation, autoimmunity, or other sources of endothelial dysfunction. Years of inactivity, especially coupled with poor nutrition habits, lead to endothelial dysfunction and a decreased oxygen supply to the brain. Thus, a major step in reducing mental deterioration over time may be consistently increasing the oxygen to the brain. In addition, exercise has been shown to increase brain plasticity through the increased production and expression of brain-derived neurotrophic factor (BDNF) [12].

Aerobic exercise increases the efficiency of the lungs in transporting oxygen into the blood and trains the heart to increase oxygen-rich blood to the body. Combining aerobic and resistance training lowers blood sugar levels and leads to a rise in the number of mitochondria in muscle tissue. Improved insulin sensitivity and glucose transport mean less glycation. Oxygen increases insulin sensitivity, particularly in muscle cells, efficiently transporting glucose as an easy fuel source for production of ATP energy for the muscles. A combination of aerobic and resistance exercise also increases the oxygen-carrying efficiency of hemoglobin and speeds up the general metabolism. It keeps the glucose, nutrients, and oxygen in the blood moving quickly through the system and saturating the body's tissues and cells.

Exercise training increases the transport and use of glucose in the cells and mitochondria by as much as 20 times. Inactivity contributes to insulin resistance by continually having too much glucose in the blood that must be transported by insulin into the cells or stored as glycogen or fat. Too much glucose creates an oxygen deprivation in which disease and pathogens can take root. Red blood cells contain hemoglobin molecules with an iron atom. A lack of red blood cell function means the body is deprived of oxygen. Hence, MET together with a low-glycemic, high-fiber, and mineral-rich diet build strong hemoglobin, increase blood flow, and ensure a highly oxygenated body. Proper oxygenation and adequate nutrient status ensure that all tissues are healthy. A copper and/or zinc deficiency increases glycation of hemoglobin and lowers the thyroid function, which is copper- and zinc-dependent [13].

35.2.1.2 Oxygen, ATP Production, and Nitric Oxide

A good example of the synergy of movement and nutrition is how all our cells produce adenosine triphosphate (ATP), our vital energy currency. To create ATP, the mitochondria must have access to oxygen. Hence, a major task for our respiratory and circulatory system is to deliver oxygen to the tissues so that our mitochondria can access it for energy. A basic prerequisite for health is a proper supply of nutrients and oxygen to our mitochondria for energy production and the disposal or recycling of waste products through aerophagia or biotransformation through, for example, the cytochrome p450 system. But this flow of life can easily be sabotaged, and one may become limited or incapacitated.

However, we can produce ATP energy units through the anaerobic system, which does not need oxygen. This system can produce energy for a short or limited time, but is not as effective as the aerobic system and creates excess lactic acid in the process. It may be viewed as a Plan B system for long-term energy production, but it is used as a Plan A system for sprint athletes and short interval exercise training.

Both physically fit and unfit people feel a burning pain when they do intense exercise that their body is unaccustomed to. A well-trained athlete, by contrast, is more efficient at converting excess lactic acid to glucose so the excess lactate content and the workout "burn" is reduced.

Good production of nitric oxide (NO) allows the endothelial tissue to be softer and more flexible. This enables more oxygen to reach the muscle cells for growth, function, and good ATP production. The enzyme nitric oxide synthase (NOS) produces nitric oxide by regulation of the protein caveolin. Nitric oxide (NO) is formed primarily in endothelial tissue.

Nitrate-rich vegetables and the body's production of nitric oxide and arginine are important precursors to form NO. Endothelial nitric oxide synthase (eNOS) keeps the cardiovascular system healthy. Arginine from food or supplements helps to synthesize eNOS, and citrulline keeps it circulating in the blood. Optimal NO production supports relaxed and healthier blood vessels, enabling better transport of oxygen to the cells. NO also has an anti-inflammatory effect and helps prevent blood clots [14]. Many endurance athletes drink beetroot juice or take nitrate-rich supplements to improve their performance through increased blood flow and oxygenation of muscle tissues.

Production of NO decreases with age, and most people eat few nitrate-rich plants, such as arugula, spinach, kale, pomegranate seeds, and beetroot. Furthermore, when we are out of balance, we can produce too much of a toxic form of NO called peroxynitrite. Peroxynitrite increases the risk of cardiovascular disease and diabetes, additionally poisons our mitochondria, and reduces the production of ATP [15]. Poor NO production, along with other factors like a slow metabolism, creates a poor blood circulation.

> **Box 35.1 Common Symptoms Associated with Poor Circulation**
> - Coldness in extremities, regardless of environment but worse when it is cold outside
> - Headaches, including migraines
> - Brain fog and poor concentration
> - Varicose veins
> - Swelling and water retention, especially in the ankles, neck region, and upper arms
> - Fatigue
> - Numbness in extremities that is not related to a pinched nerve

35.2.1.3 The Glymphatic System

A study published in a 2015 issue of *Nature* changed our understanding of anatomy and physiology by introducing a system of lymphatic vessels that circulate to the brain that had not been widely known or proven [16]. This study suggests we have underestimated the effects of the lymphatic system in health and disease, particularly brain health. The lymphatic system has been shown to play a major role in the immune system that influences neurological diseases. It has also been shown to be vital to brain health by increasing transport of oxygen and nutrients to the brain cells and by helping to carry away waste products that are a natural consequence of normal brain functioning. "In Alzheimer's, there are accumulations of big protein chunks in the brain," lead author Jonathan Kipnis said in a press release [17]. "We think they may be accumulating in the brain because they're not being efficiently removed by these vessels." In other words, increased lymphatic circulation and drainage by the glymphatic channels of the brain are important to brain health and prevention of cognitive impairment diseases. Old and new illustrations of the lymphatic system show a stark contrast of how limited our knowledge has been of the vast functions of the lymphatic-glymphatic system. The following pictures tell a thousand words (Fig. 35.2):

The new discovery emphasizes the importance of exercise and breathing in stimulating the flow of lymphatic and glymphatic fluids to and through the brain, increasing oxygenation of tissues and drainage of metabolic by-products.

There is no central pump for the lymphatic system. Lymphatic fluid is pumped through both movement and breathing. Diaphragm movement during deep breathing and physical activity are important stimuli for the lymphatic system. Adequate MET support a well-functioning lymph system. Massage, skin brushing, saunas, and other therapies also contribute to lymphatic flow. The lymphatic system's movement is so important to do abdominal breathing versus shallow chest breathing. Another reason is that abdominal breathing keeps you calmer and makes sure you spend more time during the day in the parasympathetic dominant state. This ensures that you are not in a sympathetic-dominant-stimulated state of stress for a long time.

Both abdominal breathing and increased circulation through exercise oxygenate muscle cells via the citric acid cycle. MET stimulates the movement of fluids in the lymphatic vessels and helps to transport waste from the cells. Evidence from the 2015 study in *Nature* suggests breathing and movement have a greater impact on brain health than previously thought. The reason this was overlooked is that the tiny lymphatic vessels are so intimately wrapped in the blood vessels, which move from the throat and sinuses through the brain. Drainage of waste by the glymphatic system works together with movement of the cranial spinal fluid. This extra movement of waste by the glymphatic system may partly explain why regular exercise has a protective effect against dementia and neurological diseases.

Poor lymphatic flow and blood perfusion due to inactivity contribute to crippled ATP energy production. The poor

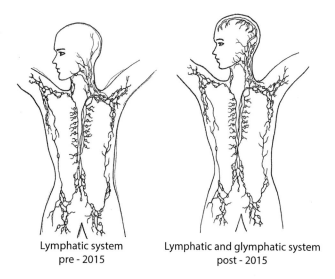

 Fig. 35.2 Recent discovery in glymphatic anatomy and drainage

lymphatic flow hampers the filtering of waste products and extracellular debris. Restoring and ensuring healthy blood and lymphatic circulation through MET is key to all physiological, psychological, and biochemical functions.

35.2.1.4 Pathophysiology Resulting from Lack of Oxygen and Energy

Nothing is more fundamental to life than respiration – the intake of oxygen and release of carbon dioxide. Healthy red blood cells (erythrocytes) absorb oxygen from the lungs for delivery to cellular tissues by a vast capillary network throughout the body. To support this process, the bone marrow produces a continuous supply of new blood cells, giving birth to two million new red blood cells every second. Each day, the body produces more than 200 billion new red blood cells, and at any given moment, there are more than 25 trillion blood cells in circulation.

Good blood circulation enriches the blood and cells with oxygen, transports nutrients throughout the body, helps manage waste production, improves muscle recovery, and accelerates healing. When circulation is compromised by aging, sitting too much, inactivity, disease, or other factors, the muscles may atrophy, and numbness and tingling can occur. Poor circulation increases the risk for serious health threats, including stroke, hypertension, and kidney failure.

Over time, consistent MET strengthens the vessels and the heart. One reward of consistent MET is a stronger heart and a lower pulse. A pulse below 62 beats per minute and a blood pressure under 120/80 are generally considered protective signs that decrease your risk for chronic disease, especially cardiovascular disease. Even your risk for psychological disturbances, substance abuse and violent behavior are reduced by MET, according to a 2016 Karolinska Institute study conducted on one million Swedish men [18].

A strong heart and healthy endothelial tissue contribute to excellent blood flow, requiring less work to transport nutrients, carry away waste products and oxygenate the cells. Hypertension is still considered the largest risk factor for cardiovascular diseases and the combination of a healthy diet, an appropriate MET program and improved stress reduction may improve health and decrease mortality risk.

35.2.2 Nutrients Play a Key Role

35.2.2.1 Key Nutrient Patterns for Better Energy and Health

Macronutrients, carbohydrates, amino acids, or fats are needed for conversion into fuel for the citric acid cycle. Nutrients such as citric acid, malic acid, and fumaric acid are needed for the cycle to create energy, which will enter the final stage of energy production, oxidative phosphorylation. B vitamins and metabolites NAD and NADH are needed as cofactors in the citric acid cycle. Magnesium and coenzyme Q10 are important for oxidative phosphorylation to work. Antioxidants are needed in the body to take care of the free radicals produced as a waste product from the production of energy. Nitrates are needed for optimal production of nitric oxide, which provides optimum oxygenation to the cells. B vitamins and other nutrients may be needed to ensure that the methylation and other cellular functions work and keep our mitochondria, DNA, and RNA in balance. Oxygen is needed for the whole process of aerobic energy production to work at all.

For blood to be able to carry oxygen to the cells, it needs hemoglobin. In adult men, hemoglobin concentrations can range from 13.5 to 18 g/100 ml of blood. In adult women, the concentrations vary from 11.5 to 16.4 g. The following factors are among the most important reasons that affect the amount of hemoglobin in erythrocytes: nutrients such as iron, vitamin B12, folic acid, and the availability of nitrates, proteins, fatty acids, elevation, and oxygen supply. The most important factor, however, is movement and exercise. Well-trained elite athletes usually have very high hemoglobin levels and thus have an excellent transport of oxygen to the muscle cells.

Through the combination of MET and nutrition, as well as certain genetic and epigenetic factors, you reach optimal transport of oxygen to the muscle cells. With moderate to intense MET, you can produce at least 30 times as much ATP compared to a resting state. During intense exercise, such as sprint intensity training (SIT), you can use more than 1000 times more ATP than in the resting state. After 10–15 seconds, the normal ATP production takes place through anaerobic and aerobic combination, instead of only via anaerobic metabolism.

35.2.2.2 Nutrients for Oxygen Transport

Here are some of the ways that nutrients and oxygen work hand in hand:
- The B vitamins B12 and folate are necessary to produce red blood cells.
- Iron is important for the transport of nutrients and oxygen through healthy hemoglobin transport in the blood.
- Vitamin C is essential to the uptake and utilization of iron.
- Vitamin E helps transport oxygen to the blood cells.
- Vitamin K2 increases the oxygen-carrying capacity of the hemoglobin in red blood cells. Vitamin K2 and nitrates in foods, amino acids arginine, and citrulline and nitrates eaten in a plant-based diet through the intake of spinach, red beets, pomegranates, arugula, and other nitrate- and chlorophyll-rich greens all aid in the status of healthy endothelial cells. This increases blood flow and oxygen transport. Inactivity, lack of chlorophyll and minerals, lower oxygen levels in the air, and air pollution all contribute to low oxygen.
- Magnesium enhances the binding of oxygen to heme proteins and gives red blood cells the flexibility to enter tiny capillaries. Magnesium also stimulates the movement of oxygen atoms from the bloodstream to the cells and prevents blood vessels from constricting and platelets from sticking together. Calcium and magnesium aid in the transport of oxygen into the cell. Magnesium is also needed to shift calcium in and out of cells, thereby providing the correct pH for high-cellular oxygen. Magnesium-deficient individuals use more oxygen during physical activity, which in turn produces more free radicals.
- Antioxidants found in foods, phytochemicals, amino acids, vitamins, and minerals help the body use oxygen better and protect against damage by free radicals formed when excess oxygen is produced in intense exercise or overexertion. Copper, manganese, and zinc are used by the body to synthesize superoxide dismutase, a potent antioxidant enzyme. This is extremely important in the body's defense against oxidation and the accumulation of excess glycation, so-called aggregated glycation end products which contribute to systemic oxidative stress and disturb the proper function of proteins.

35.2.2.3 Nutrients for ATP Production

What can affect or limit ATP production? Energy production will gradually become less efficient with aging, especially after one is in their 40s. Infections, such as mycoplasma infections and Lyme disease, meningitis, heavy metal poisoning, and poisoning from chemicals or plastics, can all limit ATP energy production. Lack of oxygen and ATP-rich nutrients, such as, CoQ10, B2, NADH (niacin metabolite), B5, B6, and magnesium, may also limit energy production. The bases for protecting and optimizing the mitochondria are lifestyle factors, access to optimal amounts of oxygen, macro- and micro-nutrients and minimal exposure, and storage of environmental toxins.

It has been found that the coenzyme Q10, the active form of folate (5-methyltetrahydrofolate, 5-MTHF), grape seed extract, lipoic acid, melatonin, artichoke, the herb selfheal (*Prunella vulgaris*), and naturally occurring polyphenols and other phytochemicals present in beetroot, olives, olive oil, tomatoes, cocoa, ginkgo biloba, pomegranate, and hawthorn berries all contribute to normal NO production. These promote good circulation and oxygenation and also provide protection from the overproduction of peroxynitrite. Nutrition for better absorption of oxygen can be found in nitrate-rich vegetables (e.g., arugula, kale, beets, etc.), arginine, citrulline, CoQ10, magnesium bisglycinate, B-complex vitamins, and vitamin E complex.

> **Box 35.2 Twelve Tips for Better Mitochondrial Health**
> 1. Regular physical activity equivalent to at least 12,000 steps per day.
> 2. A plant-based diet with a low-glycemic index.
> 3. Avoid gluten and possibly other substances/foods that you may be allergic or oversensitive to.
> 4. Frequent mild detox and detox cures for a few weeks, several times a year.
> 5. Abdominal breathing and, if possible, treatment for the lymphatic system through massage, sacro-cranial treatment, dry brushing, saunas, and relaxing Epsom salt or herbal baths.
> 6. Extra intake of nitrates from arugula, spinach, beets, pomegranate, and supplements of nitrates from concentrated red beet juice or juice-concentrated powders or capsules.
> 7. Good sleep, staying calm, and using a stress management program as an everyday part of life.
> 8. Use clean-toxic-free products in terms of hygiene, cosmetics, food, etc. and gradually chelate accumulated heavy metals and chemicals in tissues by eating organic veggies, sprouts and regularly using spices such as tumeric, cayenne, cilantro, rosemary, oregano and basil. Regular use of green superfood powders such as chlorella, spirulina and wheat grass is a great way to gentle move out toxins from the body.
> 9. Mitochondrial-supportive supplements, such as CoQ10, ribose, L-carnitine, acetyl L-carnitine, resveratrol, vitamin B complex, magnesium, and lipoic acid.
> 10. Extra methylation supplements if necessary and if there are gene deviation (DNA testing) and elevated homocysteine levels above 8 from blood testing: Vitamin B6, B12, folic acid, choline, betaine, trimethylglycine, or SAMe supplementation will most likely improve homocystein levels and methylation functions.
> 11. Acupuncture for better energy.
> 12. Adaptogenic herbs: *Astragalus*, ashwaganda, eleuthrococcus root, arctic root, ginseng.
> The addition of nutrients, herbs, or low-dose aspirin that counteract blood viscosity. The most common blood thinning supplements are omega-3 fatty acids, garlic, gingko, and meadowsweet. These are especially valuable for elderly people, or those at cardiovascular risk.

35.2.2.4 Unique Considerations for Cardiometabolic Syndrome

Cardiometabolic syndrome is a term which recognizes the common development of cardiovascular diseases due to an etiology of the combination of insulin resistance, obesity, and inflammation. From a functional medicine perspective, we would add other contributing factors such as inactivity, nutritional imbalances, and toxicity.

Insulin resistance and glycemic control are influenced by many mechanisms. MET seems to be one of the most effective strategies in helping the body maintain healthy blood sugar. MET stimulates 5'-AMP-activated kinase that influences GLUT4 glucose transport of glucose inside the cell.

Another way it helps is by influencing insulin pathways and modulating inflammation-signaling pathways. Modulation of these pathways is dose-dependent [19]. Aside from inactivity, other lifestyle factors reported to impact insulin resistance are xenoestrogens, obesity, stress, and dietary factors.

Even though one can approach the challenges of insulin resistance with weight reduction, biotransformation/detoxification of chemicals, caloric or dietary restrictions, or use of supplements such as chromium and cinnamon extract, MET or pharmaceutical therapies such as metformin, a more ideal approach is to use a combination of lifestyle therapies to affect improvement either before or in conjunction with food supplement or pharmaceutical support. A simple but consistent lifestyle support program has been shown to be very effective in affecting positive change. One 2002 study showed better improvement with lower costs compared to taking the pharmaceutical metformin [20].

A healthy and fit person can really enjoy a great level of aerobic fitness by doing 45–120 minutes of MET, but this dose is too much, of course, for one who is not well-trained, or for a person with pathology or strong risks for problems, such as those who are obese, take lots of medications, or have past or current cardiovascular diseases. A study conducted at Mayo Clinic in 2012 found that, for cardiovascular patients, the optimal dose is more likely to be 30 minutes a day of moderate training instead of the 60–90 minutes for the very fit. For these cardiovascular patients, too much oxidative stress is initiated with 60-minute moderate or vigorous MET. The strongest risks are at a sustained effort of vigorous exercise such as long interval training [21].

> **Box 35.3 Improvement of Cardiometabolic Markers**
> The following cardiometabolic markers have been shown to be improved by regular exercise training:
> - Lipid, inflammation, and glucose profile:
>   - Trigylceride (Tg)
>   - HDL
>   - LDL
>   - Fasting glucose, glucose stress test
>   - Fasting insulin, insulin stress test
>   - HbA1C (long-term sugar management)
>   - Hs CRP
>   - Uric acid
>   - Homocysteine
>   - Fibrinogen
>   - Lp_PLA2
> - VAP advanced lipid profile:
>   - Apo B, Apo A1 and or Apo B/A1 quota
>   - VLDL 3
>   - HDL 2
>   - Lp (a)
>   - GGT (oxidative stress/toxicity marker)

The level of intensity of aerobic training estimated to improve cardiorespiratory fitness has been 69% exertion of maximum heart rate (max HR). This is considered to be moderate training pace, where you can carry on a strained conversation, or a 6–7 out of 10 on the scale of exertion, a 13–15 on the Borg scale of 1–20 of exertion. This level of exertion is also considered the minimal effective dose for improving cardiorespiratory fitness. It is also a rate where you are burning about 45% fat and 55% carbohydrates as fuel.

> **Box 35.4 Functional Medicine Causative or Contributing Factors**
> - Insulin resistance
> - Toxicity
> - Oxidative stress
> - Inflammation
> - Endothelial dysfunction, dyslipidemia
> - Hormonal
> - Genetic, epigenetic factors
> - Dysimmunity, autoimmunity
> - Energy, mitochondrial dysfunction
> - Microbiome
> - Structural
> - Tissue Perfusion—blood and lymph and maintenance of muscle mass
> - Emotional, mental, spiritual

In addition to the challenges of the pathophysiology of the combination of insulin resistance and cardiovascular irregularities or dysfunction, a functional medicine-trained nutritionist will also investigate other underlying processes, which may be important parts of the etiology or contribute to the problems. Most of these processes will be influenced by increases in various types of physical activity. I have even added tissue perfusion, increased cardio conditioning, and increased muscle mass as a solution to one of these functional medicine links in the causal chain of chronic disease. Regardless of the mechanisms and processes, people are most concerned about the length and quality of their lives. The bottom line is that MET gives you the most "bang for your buck." Better aerobic conditioning gives you not only a longer life but a better quality of life.

- Among other things, VO_2 max is correlated with both lowered morbidity and mortality [22].
- Chromosome telomere length, which is a predictor of health and longevity, is spared for those who have a better VO_2 max [23].
- Telomere length (mitochondrial mitogenesis) is increased by exercise [24].
- Epigenetic studies show that activity helps modify DNA signaling [25].

35.2.3 MET: Powerful Prevention and Treatment of Chronic Disease

Not only is MET vital for prevention of disease and rehabilitation of injuries; it is often very powerful for alleviating suffering, increasing quality of life, restoring function, and assisting in improving states of pathophysiology. The following are only a few of the many chronic disease states where MET has shown to be a valuable in conjunction with other treatments.

35.2.3.1 Types of MET

Integrated standing, movement, and exercise training (ISMET) is a vital part of a holistic and integrative lifestyle strategy to promote healthy aging and decrease the ravages of chronic disease. Standing, movement, and exercise training maintain or upgrade normal physiological and biochemical functions, thereby preventing breakdown of function and development of progressive pathology.

ISMET should be part of a holistic lifestyle prescription and support system like that which has been repeatedly implemented by Dr. Dean Ornish. In several of his studies, where movement and exercise training (MET), meditation, nutrition, supplementation, and psychosocial support are implemented as an integrative system, intensive lifestyle changes may affect the progression of prostate cancer [26]. Depending on the patient-doctor/lifestyle coach interaction, a comprehensive lifestyle program can be suggested and negotiated, with the proper support system, inspiration, and knowledge to implement the program. In integrative and functional medicine, the increasing model has shown that it takes a team or network of professionals or lay coaches to help the patient turn their life around and integrate new habits of health into their lives.

ISMET is as an important part of these regenerative changes and strategies as dietary changes or psychosocial changes. These types of integrated comprehensive programs have been shown to often be effective in various types of chronic disease [27].

Although setting a goal of walking the equivalent of more than 10,000 steps daily is admirable, and will improve your health or disease status, setting more specific goals for whole body training is even better. The integrative model I am recommending with the acronym, ISMET, involves improvements in the following areas:

1. More standing, less sitting, the 12–12 rule (12 hours sleeping and sitting, 12 hours standing, moving or exercising)
2. More habits of incidental movement
3. Walking and play, hobbies, and extra activities
4. Exercise training, both planned and spontaneous

The planned or habitual MET can be a combination of aerobic and anaerobic activities and can include a combination of dancing, playing, endurance training, interval training, strength/resistance training, and flexibility and balance training (i.e., yoga, stretching). If possible, spend some of this time in a natural environment, such as beaches or paths through wooded areas or near lakes. Nature deficit disorder, as identified in 2005 by the author Richard Louv in the book, *Last Child in the Woods*, emphasized the added bonus of health benefits and joy in playing, moving, and exercising in a natural environment.

> **Box 35.5 Good Tips for Starters, from Inactivity to Active Movers**
> Some good tips for those who are starting to exercise:
> - Start slow and start small.
> - Work out at a time of the day when you have more energy.

- Find activities that bring you joy or connection with people, your tribe.
- Be comfortable and use the right clothing, shoes, and gear.
- Reward yourself for successes and consistency.
- Make exercise a social activity with friends or community groups. Community exercise groups are effective. This has been demonstrated repeatedly in lifestyle support groups, such as those formed at Saddleback Church, called The Daniel Plan (▶ www.danielplan.com).
- Use apps, technology, and tracking devices to see improvements and for coaching.
- Keep gradually increasing the MET dose as your motivation and goals grow, but be careful, gain knowledge and experience before doing extreme sports. Extreme exercise, such as triathlon training and competition, can easily cause too much oxidative stress, gastrointestinal disruption, inflammation, injury, hormonal disturbances, anemia, eating disorders, osteopenia, etc.

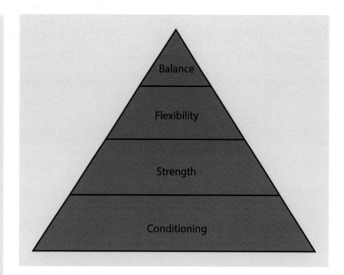

Fig. 35.3 MET pyramid

35.2.3.2 Key Exercise Patterns for Better Energy and Health

Traditional exercise trainings, such as yoga, tai chi, and qigong, have proven through centuries of use to be important for both the health and well-being of individuals and cultures. One of several thousand studies on these ancient practices shows increased quality of life for the elderly [28]. These practices not only involve movement but also concentration, balance, stretching, meditation, breathing, and stress management. These are an excellent complement to aerobic conditioning training, which is the base of the MET pyramid.

> **Box 35.6 ET Training Pyramid**
> Cardio or aerobic training is the base of the pyramid of training. It is important that everyone has a basic level or energy and endurance. The next three levels are maintaining the goals of having strength, flexibility, and balance. These levels provide great fitness focuses for goals to help prevent sarcopenia, accidents, falls, and development of chronic disease (◘ Fig. 35.3).

Many people struggle with scheduling movement and MET into their daily lives; but by simply increasing your amount of incidental activities and planning regular MET sessions, you can enjoy more movement and health. All it takes are a few small changes of habit, like parking farther from work, taking public transportation, bicycling more, walking to nearby neighbors' homes or stores, walks and talks with colleagues or friends, taking the stairs regularly, more play with the kids or grandkids, walking somewhere in the building every hour at your workplace, and more house cleaning, yard work, or gardening. Such changes can add the equivalent of 2000–6000 steps per day. After a few weeks, it has become a lifestyle habit and you don't notice it. It is now on autopilot. This is especially important if you have an occupation, workplace, or home environment that is not conducive to movement.

Walking is a great place to start for an inactive person who wants to become active. One can always start low and go slow to avoid pain and injury. If one has a difficult time taking an extra 30–60-minute walk every morning or evening, taking 3–6 shorter walks will accomplish a similar result. Using a pedometer or step monitor is one of the easiest tracking tools for your health. Devices worn on the belt, wrist, or as a wristwatch can measure the amounts of steps the wearer takes per day. Minimal recommendations usually suggest 4000 steps of incidental activity and 3000 steps of walking, for a daily total of 7000. A 30-minute power walk will give you about 4000 steps.

According to many public health recommendations, you should create a habit where you spend at least 30 minutes a day walking 5 days per week to accumulate at least 15,000 weekly steps [29]. This recommendation, while modest, is deemed sufficient for creating better health and decreasing risks for chronic disease. A breakthrough book in this field, *Biomarkers*, was published in 1992 by Evans and Rosenberg at Tufts University. This book shows that even walking 30–60 minutes five times per week, when started by obese/overweight individuals, improves key biomarkers of health and disease. These improvements are easy to track and measure with blood tests, a tape measure and bio impedance calculations.

Exercise goals and types vary quite a bit depending on one's situation and personal wishes. To make progress with obesity, one most often needs to exercise more than those who are just looking for a minimal amount of exercise or steps for general health. Eating the right amounts and kinds of nutritious foods is also important.

As seen below, certain levels of movement and exercise training have positive effects on insulin resistance and metabolic resistance. However, consistency is the most important key. Even modest progress can be noticed when you are walking every day. The best form of exercise is that which you do consistently.

Fig. 35.4 The role of nutrition and exercise in the mechanisms of pathology or in pathophysiology. (Reprinted from Handschin and Spiegelman [39]. With permission from Springer Nature)

Box 35.7 Step Goals
The following is a guideline for amounts of steps to aim for each day, on average. This includes incidental activities, walking, play, and planned movement and exercise training (MET):
- Inactivity contributing to chronic disease (<5500 steps daily)
- Some improvements in immunity functions and health markers (>7000 steps daily)
- Metabolic syndrome/insulin resistance improvement (>8000 steps day)
- BMI-defined weight status improvement (>10,000 steps daily)
- Optimal daily steps matching our physiological needs and our historical past (>14,000 steps daily)

Taking 7000–10,000 steps per day from incidental activities, movement, and walks should be seen as the minimal amount of effort for combined daily movement and planned exercise time. If you do a moderate-to-intense gym workout, dancing, aerobics, or strength training, you can count that as 2000–4000 equivalent steps per half hour, depending on the level of intensity/challenge.

A more optimal number of steps per day would be 12,000–20,000 steps with combined, incidental movement, play, walks, and MET (planned movement and exercise training). The more natural and healthy cultures on the planet, the "Blue Zone" communities, have a daily average of 18,000–24,000 steps movement. This also reflects the amount of movement many of our ancestors had many centuries ago. These goals should also be an excellent complement to the habit of sitting less.

My recommendation is the 12–12 rule. Out of your daily 24 hours, spend 12 of them sleeping and sitting, and spend the other 12 moving and standing. Moving more at work and home, exercising regularly and using a standing work desk are some measures that can help you reach the 12–12 balance for greater health. MET should be put in a context of good technique, body posture, and lots of standing during the day. I refer to this healthy habit as integrated standing, movement, and exercise training (ISMET).

> **Box 35.8 Get the Blood Flowing and Keep It Flowing**
> - More standing and less sitting. No more than 4 hours per day sitting, 7–8 hours of sleeping, and the rest of the time should be standing or moving. The 12–12 rule of ISMET.
> - Movement improvement, incidental activity producing >6000 steps daily.
> - Planned exercise at least 4 hours a week, whether at the gym, dance studio, in the outdoors, or at home looking and following a DVD/online workout program. For two to four sessions per week, transition from walking to power walking, Nordic walking (walking with poles), cardio workouts, running, biking, swimming, spin cycling, or cross training for a more strenuous workout.
> - Fidget and move at a standing desk. Taking a break every hour for some walking, doing squats, stretching, yoga poses, or having a walk-and-talk meeting with a colleague.
> - Lots of little movements and stretching throughout the day. Keeping all body parts moving every day. A good stretching program like yoga helps stimulate blood flow and balance to all parts of the body.
> - Consuming more spices like cayenne, cardamom, ginger, garlic, rosemary, sage, and pepper to stimulate and assist in increased blood flow and perfusion of the tissues.

The best types of ISMET exercise programs are the ones you do consistently. It is ideal if your planned exercise training contains some of the following ingredients:
- Personal health and fitness goals.
- Social support, exercising with friends.
- Competition, if it challenges you and motivates you. The FITTPRO model, found at ▸ www.exerciseismedicine.org, can give basic guidelines about frequency (5–7× weekly), intensity (50–70% mHR), type of exercise (mixing aerobic, resistance, and strength), duration (30–60 minutes), and suggestions for increasing these factors by 10 percent per week.
- Overcoming or improving states of illness, personalized goals, and programs.

> **Box 35.9 Types of Exercise Training**
> The following flow chart can depict basic types of training that can be combined and coordinated for whole-body MET. These play different roles in building physical and psychological health and resilience. Together with my 12–12 recipe for more standing and movement, these areas illustrate the ISMET model:
> - Aerobic training builds endurance and cardiovascular and psychological health.
> - Anaerobic training such as sprint intensity training (SIT) or high-intensity interval training (HIIT) builds and maintains muscle mass, speed, stimulating hormone production, and a healthy metabolism.
> - Flexibility and balance training, such as tai chi and yoga, balances muscles and enhances breathing, flexibility, and balance.
> - Strength and resistance training builds strength and power, stimulating hormone production and protects from sarcopenia.
> - Play, gardening, dance, and sports activities can balance other times of training and bring extra fun and variation to the mix of MET.

Physical therapy and targeted training are specialized trainings to achieve specific physical goals or target specific muscles, muscle groups, or function.

Moderate aerobic exercise training, high-intensity interval training (HIIT), and sprint intensity training (SIT) have been shown to be effective in improving the symptoms and progression of chronic diseases. Different forms, doses, and combinations of these are appropriate for different individuals and needs. HIIT can be twice as effective as moderate-intensity aerobic training for people with lifestyle-induced diseases [30].

Dr. Phil Maffatone, who has trained over 10,000 amateur and elite athletes, recommends that at least 80% of your training time should be in the low-moderate fat-burning endurance zone, which he basically calculates with 180 minus your age. When you do your MET, you can mix up the above types of training, so that each of these types is done each week. When time is an issue, you can get a great workout in 15–20 minutes with HIIT or SIT workouts.

> **Box 35.10 Training Zones**
> The number 220, minus your age, is the most common way to, on a gross level, estimate the maximum pulse/heart rate (max HR) or the liverpool model for calculating your max HR which is often a little more accurate. Take 217 minus your age times 0.85. The following are the amounts of effort and heart rate that give different kinds of advantage:
> - 50–75% of your max HR for aerobic conditioning and endurance training (55–65% of this is the max fat burning zone). Long MET sessions of 1–3 hours
> - 75–95% max HR for medium to hard training, where sprinting and intervals are done. Short-duration MET training. 15–60 minutes including warm-ups and cool down

> **Box 35.11 Training Intensity**
> Training intensity guidelines vary but are also, for the most part, similar. ACSMs (American College of Sports Medicine) guidelines and other organizations (▶ www.acefitness.org) can be summarized as the following:
> - 220 minus age and then the following level of effort (Borg scale of 1–20 and percentage of max heart rate):
>   - Low effort, Borg scale: 9–12: 50–63% of max HR (you can sing)
>   - Moderate: Borg scale: 13–14, 64–76% of max HR (you can talk)
>   - High/intense effort: Borg scale: 16–18, 77–93% max HR (you cannot talk)

35.2.3.3 Fitness and Lifestyle Assessments

Fitness assessments are offered in many training centers and gyms. There are also plenty of these available online or in books or clinic brochures. One of the simplest is doing the following four easy tests that can assess your fitness, but these fitness tests and others should be combined with anthropometric testing, a physical exam and lab analysis for a more comprehensive personal analysis.
1. The push-up test (measures muscular strength and endurance)
2. The crunch test (measures abdominal strength and endurance)
3. The 3-minute step test (measures aerobic fitness)
4. The 1-mile walk test (measures aerobic fitness)

Nutritionists, dieticians, and lifestyle coaches can use clinical questionnaires to evaluate the ISMET status of the client/patient and then follow up with coaching, either online or via office visits. There are many online questionnaires, including ones accessible in the functional medicine tool kit for those who become members of the institute for functional medicine (▶ www.functionalmedicine.org).

35.2.3.4 Integrative Lifestyle Therapies

Integrated standing, movement, and exercise training (ISMET) is a vital part of a holistic and integrative lifestyle strategy to promote healthy aging and decrease the ravages of chronic disease. Standing, movement, and exercise training maintain or upgrade normal physiological and biochemical functions, thereby preventing breakdown of function and development of progressive pathology. ISMET can be part of an integrative system of overall lifestyle changes that include meditation, nutrition, supplementation, and psychosocial support.

It would be great if we can expand our recommendations from nutrition programs to comprehensive lifestyle programs, such as the ones that Dr. Dean Ornish has repeatedly shown are very powerful [31].

Incorporating an integrated lifestyle program that includes a personalized ISMET strategy will be more common in the days and years to come as the science and experiences emerge that confirm the efficacy and power of patient/consumer driven self-care and healthcare. The convenience of the information and technology revolution will both bring increased temptations that distract from ISMET, while making available technology – such as wearable-monitoring devices – that will drive and increase the efficacy of personalized lifestyle medicine programs. As the science and monitoring technology evolve, we will be able to prove to ourselves and others the value of ISMET and lifestyle medicine.

35.2.3.5 Both Undertraining and Overtraining Can Lead to Higher Disease Risk

The effects of physical activity on markers on specific IGF-1 markers for disease prevention can be seen in a study from Poland conducted in 2015. The study demonstrated that MET has a positive effect on the balance of IGF-1 and IGFBP in the blood, therefore contributing to prevention of disease through the proper regulation of these important signaling molecules. The majority of these studies indicate that mechanical loading is a key mechanism linking IGF-1/IGFBPs concentration and selected chronic diseases development. The duration and intensity of physical activity have a significant impact on IGF-1 and IGFBP serum. The highest concentration of IGF-1 in serum was after eccentric training. "Overtraining" increases unfavorable and unbound IGF-1 levels and contributes to the increased incidence of hormone-cancer and osteoarthritis. Irregularity of the GH/IGF-1 axis may affect the development of rheumatic diseases, metabolic syndrome, and cardiovascular diseases [32].

35.2.3.6 Single Exercise Training, Combined Exercise Training, and Lifestyle Programs

Many studies show examples of improved insulin resistance, decreased metabolic syndrome and decreased diabetes markers, and pathology with exercise training. However, these studies can also be looked at comparing the effects of endurance training, strength training, and total lifestyle programs. One study with 250 adults showed benefits with both endurance training and strength training, but the improvements were even greater when these two were combined. Even more improvement came after combining endurance and strength training and using a more comprehensive exercise, diet, and lifestyle program [33].

Several lifestyle intervention programs have demonstrated that a comprehensive program including diet and exercise is much more cost-effective and successful than pharmaceutical intervention with the drug metformin for the treatment of diabetes [34, 35].

35.3 MET and Nutrients for People with Various Conditions

35.3.1 Commonalities Across Conditions

35.3.1.1 Therapeutic MET for Rehabilitation (TMETR)

Most hospitals and many clinics and private practices have programs for rehabilitation of sports injuries; auto-, home-, or work-related accidents; war casualties; and various forms of physical limitations or handicaps. This can involve the use of many physical modalities, machines, body work, exercise prescriptions, and surgical procedures that are used by many different types of trained professionals. That which is in its infancy, however, is the aggressive use of TMETR for chronic disease unrelated to acute accidents or injuries.

Although the American College of Sports Medicine (► www.acsm.org) and institutions in many countries, including my own country of Sweden, have some guidelines (► www.fyss.se) for TMETR for specific chronic conditions, these are seldom used in the clinical practice. Primarily, as with personalized nutrition programs, there is no or very little training of doctors, nurses, dietitians, health coaches, chiropractors, physical therapists, or acupuncturists, concerning the use of exercise prescriptions. The field of TMETR for chronic disease states is still very young, especially compared to the use of TMETR in physical rehabilitation from injuries. Until the value of TMETR for chronic disease application is further studied, used, and incorporated into health professional training, the use of TMETR will be an underutilized powerful asset in the similar way personalized nutrition has been.

I will present information that introduces and validates TMETR not only for injuries but primarily for chronic disease states. In the functional medicine model, as eloquently stated by one of the pioneers in medicine, William Osler, "The good physician treats the disease, the great physician treats the patient who has the disease." We are more interested in understanding and caring for the patient than for the disease. "The patient is more important than the disease the patient has." Following some guidelines of aerobic and strength training for a disease state is helpful, but ultimately that information and recommendation must be adjusted to each patient. So TMETR must ultimately be personalized to the living situation, family and support group, limitations, possibilities, socioeconomic status, motivations, goals, and aspirations of the unique person with the disease.

> Box 35.12 Seven Great Goals for TMETR Is to Restore Health
> 1. Personalize, communicate, and coach the TMETR.
> 2. Alleviate suffering.
> 3. Halt and reverse pathology.
> 4. Restore and improve function.
> 5. Improve quality of life.
> 6. Build resilience/organ reserve/organ capacity.
> 7. Reduce risk for chronic disease reoccurrence.

35.3.1.2 MET and Physical Therapies for Injuries

MET and correct physical therapy is a very important adjunct to knee surgeries and other medical procedures. Many times, the correct use of MET and physical therapy can substitute for surgery and bring as good as or better results than surgery and at a much lower cost. According to medical investigator Norman Swan, about 20% percent of knee replacement costs (over $1 billion) incurred by the Australian healthcare system were unnecessary [36].

35.3.1.3 Unique Considerations for Inflamm-Aging, Cardiometabolic Disease, Dementia, Sarcopenia and NAFLD

In the following sections, we will investigate four chronic complex conditions that cause a great deal of suffering and a huge strain on our healthcare system. These are also models of how an integrative lifestyle program, including diet, MET, and food supplementation, can be combined to achieve very good results. These four conditions are cardiometabolic syndrome, dementia/Alzheimer's, sarcopenia, and nonalcoholic fatty liver disease (NAFLD). Their common etiology is a combination of underlying causes such as toxicity, chronic insulin resistance, toxicity, and inflammation that are causative or a result of these problems. If we scratch below the surface of these conditions, we will find the same culprits that are causative: insulin resistance, inflammation, toxicity and obesity, inactivity, poor stress management, poor food choices, cigarette smoking or toxins in the environment, and other lifestyle factors. These factors potentiated by genetic and epigenetic factors often lead to years of suffering from chronic diseases that transgress across body systems and disease. First, however, let us examine the process of chronic inflammation on disease etiology and progression.

Inflammation control: moderate MET decreases the risk of chronic inflammation. Moderate endurance and strength training have been shown to decrease chronic inflammation, and flexibility and balance exercise practices, such as yoga, have been shown to lower inflammation markers. Ambarish Vijayaraghava and his research colleagues revealed in a fascinating study that, "Regular practice of yoga lowers basal TNF-α and IL-6 levels. It also reduces the extent of increase of TNF-α and IL-6 to a physical challenge of moderate exercise and strenuous exercise" [37].

The nutrient and oxygen needs of the mitochondria are very important also to prevent mitochondrial inflamm-aging, a chronic low-grade inflammation leading to dysfunctional mitochondria and oxidative stress. The microRNAs are involved in this process, which is hypothesized in an article in *Experimental Gerontology* from 2014 [38].

This chronic inflamm-aging in the mitochondria, which occurs in some hard-training athletes, can perhaps account for damage to heart and muscle cells from overtraining. Intense exercise for long periods of time increases the risk of chronic inflammation [39].

Elite training runs the risk of both injuries and chronic inflammation, if the training, sleep, and nutrition protocols are not carefully planned and monitored. Antioxidant and anti-inflammatory foods and supplements can balance this and protect from meta-inflammation or injuries.

The big culprit in creating chronic inflammation in our society is not overtraining but inactivity, in combination with inflammation-stimulating processed foods. The following illustration shows how inactivity and obesity contribute to inflammation (◘ Fig. 35.4).

> **Box 35.13 Impact of Sarcopenia**
> - On an individual level:
>   - Sarcopenia greatly diminishes independence and quality of life.
>   - Sarcopenia reduces individual health and happiness.
>   - Sarcopenia increases the processes of inflammation and insulin resistance that underlie the progression of many types of diseases.
> - On a societal or public health level:
>   - Sarcopenia increases the risk of all-cause mortality and disability.
>   - Sarcopenia puts every patient who has surgery at greater risk of complications.

35.3.2 Sarcopenia

Sarcopenia, the loss of muscle tissue that is primarily a by-product of the aging process and inactivity, contributes to obesity, diabetes, dementia, and the cardiometabolic syndrome. The main cause of sarcopenia is physical inactivity, especially lack of resistance training. Muscle loss and fat gain is reversed by fitness and resistance training.

Sarcopenia and the aging process are associated with less efficient function of the mitochondria, the effects of which were well summarized in an article by Marita Rippo et al. in Experimental Gerontology in 2014. "Mitochondria are intimately involved in the aging process. The decline of autophagic clearance during aging affects the equilibrium between mitochondrial fusion and fission, leading to a build-up of dysfunctional mitochondria, oxidative stress, chronic low-grade inflammation and increased apoptosis rates, the main hallmarks of aging" [40].

> **Box 35.14 Multifactorial Causes of Sarcopenia**
> Like most chronic conditions associated with aging, sarcopenia has some multifactorial causes associated with the following imbalances:
> - Hormonal changes (decreased levels of testosterone, human growth hormone, DHEA, etc.)
> - Chronic inflammation
> - Ectopic fat deposition
> - Decreased satellite cell health
> - Blunted responses to anabolic stimuli
> - Decreased metabolism
> - Inactivity

Without MET, especially strength training, both muscle mass and strength decline every year after the age of 50, and it accelerates after age 70. Muscle mass declines at the average rate of 0.9% in men and 0.65% per year in women after age 75, doubling the rate of loss per year from 10 to 15 years earlier. Loss of muscle mass is usually more pronounced in the lower extremities, an influential factor in age-related functional impairment there. This often results in movement restriction, poor quality of life, disability, and mortality risk. At age 75, the loss of strength per year, without extra strength training, is two to five times faster than loss of muscle mass [41].

Strength training or resistance exercise, the act of loading muscle against an external force, remains the most effective form of exercise in counteracting decreased function and sarcopenia. Because aging is associated with declines in hormones, muscle mass, muscle function, and aerobic capacity, more resistance training is needed [42].

Per figures from the Journal of the American Geriatric Society, "The estimated direct healthcare cost attributable to sarcopenia in the United States in 2000 was $18.5 billion ($10.8 billion in men, $7.7 billion in women), which represented about 1.5% of total healthcare expenditures for that year. The excess healthcare expenditures were $860 for every man with sarcopenia and $933 for every sarcopenic woman. A 10% reduction in sarcopenia prevalence would result in savings of $1.1 billion (dollars adjusted to 2000 rate) per year in U.S. healthcare costs" [43].

35.3.2.1 Unique Considerations for Dementia

Alzheimer's has been referred to as type 3 diabetes by some researchers and doctors because of the similar underlying pathophysiology and processes that lead to both diabetes type 2 and dementia/Alzheimer's disease. Insulin resistance, toxicity, inactivity, obesity, oxidative stress, and obesity have been closely correlated with both diabetes and dementia.

MET has been shown to be one of the best strategies to decrease cognitive decline [44]. Reductions of mean blood flow values are significantly correlated with the severity of dementia [45]. Increased oxygenation through MET has many advantages for keeping brain tissue healthier. Recent results show that MET also helps move glymphatics, which carry away waste products and cellular debris that accumulate from an aging brain. MET has also specifically been shown to affect hippocampal-dependent cognition and to increase blood perfusion to these areas [46]. High-flavanol cocoa has also been shown to improve memory by stimulating the input region of the hippocampus, the dentate gyrus [47]. In addition to a bigger and healthier hippocampus, MET stimulates the production of brain-derived neurotrophic factors (BDNFs) that appear to influence energy metabolism, appetite, and important aspects of neurocognitive function [48]. Not only that, but a study from Karolinska Institute in Stockholm has shown that there is indeed an antidepressant effect from running that is associated with an increase of hippocampal cells [49]. In addition to the benefits to the hippocampus, different types of MET affect different parts of the brain. According to a Canadian study of elderly women, resistance training had a more profound effect on staving off cognitive decline compared to exercise programs designed to tone and balance the body [50].

According to an article by Teal Burrell, different types of exercises affect different parts of the brain to a lesser and stronger degree [51]. Neuroimaging techniques such as

MRI and SPEC scans confirm the positive influences of increased blood flow to the brain. Some of these effects include improved performance across cognitive domains and core functions that improve executive decisions [52]. Dr. Daniel Amen, a neurologist who has done more than 120,000 SPEC scans, affirms the strong positive effects of MET and other lifestyle improvements as highly restorative for damaged or low-functioning brains. MET gives us better cognition and protection against dementia, and there is a dose-response relationship [53, 54]. Not only does MET sharpen the aging mind, it keeps people happier in their golden years. The elderly who do MET have less depression, better moods, higher self-esteem, and self-confidence. They also sleep better, enjoy more energy, and have better resilience [55].

35.3.2.2 Unique Considerations for Sarcopenia

Consistent exercise training is the primary treatment to prevent and affect sarcopenia. A balance of endurance and primarily strength exercises at least 3 days per week is usually effective. It is never too late to start. Older people get just as much out of MET as younger people [56]. Casual walks and light physical activity will not do much for reversing sarcopenia. A study on the elderly in Iceland illustrated that moderate to vigorous physical activity is needed to be done consistently at least three times per week for effective results on the elderly [57].

The older one is the more important it is to engage in muscle sparing and muscle building. In a 1994 study presented in *The New England Journal of Medicine*, high-intensity resistance exercise training is shown to be a feasible and effective means of countering muscle weakness and physical frailty in very elderly people. In contrast, multi-nutrient supplementation without concomitant exercise does not reduce muscle weakness or physical frailty [58].

The best approach is to engage in endurance training, strength training, and yoga or stretching for flexibility, balance, and core strength. When possible, Nordic walking is preferable to normal walking. Walking at a more strained rate with poles helps activate more muscles, your metabolism, and more hormones. Taking a 30–60-minute walk daily is great, but make it a power walk or Nordic-power walk every other day. If you swim, bike, row, or run, make every other session a moderate to vigorous one and make sure to warm up before and after. Three sessions of 45–60 minute moderate to vigorous workouts per week that include strength training and HIIT or SIT seem to be a great goal to shoot for, since it both deals with the ravages of sarcopenia and puts the brakes on brain aging. Remember though the 12–12 goal: 12 hours a day sleeping and sitting and 12 hours standing and moving.

The combination of exercise training and supplementation reinforces the effects each can have on improving strength, balance, and speed. Protein, creatine, vitamin D, and calcium together showed a positive effect [59]. Sufficient protein drinks during or after the workout will help build muscle. Use branched-chain amino acids or whey protein during and after your 2–3 toughest weight training, SIT, or HIIT sessions. A plant-based healthy diet and the daily intake of omega-3 supplements and good multivitamin and mineral supplements are a great complement to your exercise plan for decreasing the risk of inflammation and injury.

35.3.2.3 Unique Considerations for Nonalcoholic Fatty Liver Disease NAFLD

Like many other recognized lifestyle diseases, such as diabetes, cardiovascular disease, and dementia, nonalcoholic fatty liver disease is increasing in epidemic proportions where cultures have adopted the modern lifestyle of inactivity and indulgence in junk foods. These problems have seen a dramatic increase in the last 20 years all over the world in places that have adopted the Western lifestyle. In the United States alone, the increase has been dramatic, increasing by 170% in the brief period of 2004–2013 [60].

The 2004 documentary film, *Super Size Me*, by Morgan Spurlock demonstrated the power of junk foods and healthy foods on our health outcomes and specifically on the increase and decrease of liver enzymes. This movie brought the message loud and clear that the intake of substances other than alcohol can elevate your liver enzymes and threaten your health. In the movie, the host was shocked that eating three fast food meals daily resulted in elevated liver enzymes within a few weeks, starting on the path of liver pathology. A healthy plant-based diet reversed this progression within several months. Not only has a plant-based diet been demonstrated to reverse fatty liver damage, so have many herbs, phytonutrients, and exercise training. Nonalcoholic fatty liver disease (NAFLD) and nonalcoholic steatohepatitis (NASH) have been shown to be very responsive to diet, exercise training, and nutritional supplements. Lifestyle therapies have been shown to have a positive effect on decreasing the progression of this common and rising chronic disease process.

There are studies showing the ability of herbs and nutrients in protecting liver cells. Isoflavone-rich supplements alone have been shown to protect against fatty liver disease through various pathways which modulate fructose production, fatty acid β-oxidation, lipid synthesis, and oxidative stress [61]. In addition, studies show there can be beneficial effects of probiotics and prebiotics in protecting liver cells from development of NAFLD by regulating the intestinal barrier function [62].

Different kinds of MET, whether endurance, strength, HIIT, or flexibility and balance, have been shown to have positive effects on degeneration due to the improved milieu surrounding and within the tissue. The increase in oxygen and nutrients are used for restoration and regeneration of liver cells and tissue. One study by Takahashi shows that resistance exercise comprising squats and push-ups helps to improve the characteristics of metabolic syndrome in patients with nonalcoholic fatty liver disease [63].

Endurance training, such as aerobic swimming training, can prevent NAFLD via the regulation of fatty acid transport-, lipogenesis-, and β-oxidation-associated genes. In addition, the benefits from aerobic swimming training were achieved partly through the PANDER-AKT-FOXO1 pathway [64].

35.3.2.4 Unique Considerations for Spinal Cord Injuries

The practice and science of the rehabilitation of spinal cord injuries, accidents, and traumas have been investigated thoroughly in the last century. The advances during the last decade of advanced stem cell and other modern techniques are a big boost to the tried-and-tested combination of physical therapy, MET, therapeutic modalities, and surgery to bring relief for spinal cord injuries, to help accident-injured patients, or to work athletes back into shape. Many studies have been done over the years to show the advantage of the value of combining physical therapy, movement, and pulse current for injury and trauma rehab [65].

Another study shows the value of MET for not only helping with a spinal cord injury but the other pathologies or risks that can accompany the injury. The increased risk of cardiovascular - and other chronic disease in persons inflicted with spinal cord injuries can be improved by consistent MET [66].

The goals of physical therapy and MET should be not only to improve function after accidents, handicaps, or injuries but also to decrease the risk of chronic diseases that can increase in the future.

35.3.2.5 Unique Considerations for Rheumatoid Arthritis RA

MET especially resistance and strength training, has been shown to be a safe and effective means of restoring muscle mass and functional capacity in patients with established and chronic rheumatoid arthritis (RA). The authors of a study of the effects of resistance training for appropriate RA patients recommend programs similar to theirs be included in disease management [67].

35.3.2.6 Unique Considerations for Cancer

Undertaking regular exercise in carbohydrate-restricted states may be a practical approach to achieve the physiological benefits of consistent p53 signaling [68].

HIIT can be performed by female cancer survivors without adverse health effects. Here, HIIT and LMIE (low-to-moderate-intensity exercise) both improved work economy, quality of life and cancer-related fatigue, body composition, or energy expenditure [69].

Studies have shown that the single or combined use of MET and the nutrient supplement resveratrol may improve the effects of chemotherapy and protection from some of the cardiotoxicity of the agent dexrazoxane, (DOX) [70].

MET helps cancer survivors control body weight, improve aerobic capacity, reduce fatigue, and improve their gait, balance, and mood [71].

35.3.2.7 Unique Considerations for Parkinson's

More and more Parkinson's (PD) patients are turning to the gym and their bicycles to improve their disease symptoms and the quality of their lives. Results can be astounding and powerful sometimes, as reported by many doctors and patients and in a study reported in the Clinical Journal of Sports Medicine. "The results of the present research synthesis support the hypothesis that patients with PD improve their physical performance and activities of daily living through exercise. Future studies should include the development of standardized exercise programs specific for problems associated with PD as well as standardized testing methods for measuring improvements in PD patients" [72].

35.3.2.8 Unique Considerations for MS

It is not only walking that is important for multiple sclerosis (MS) patients but walking downhill may be even more helpful. The stimulation and firing of certain nerves and muscle groups during downhill walking seem to have a powerful effect. According to a study reported in International Journal of MD Care 2016 "After the intervention, significant improvement was found in the downhill group versus the uphill group in terms of fatigue, mobility, and disability indices; functional activities; balance indices; and quadriceps isometric torque ($P < 0.05$). The results were stable at 4-week follow-up" [73].

35.3.2.9 Unique Considerations for Schizophrenia

MET seems to be very beneficial for those afflicted by schizophrenia, especially resistance training, according to a study from Brazil. "In this sample of patients with schizophrenia, 20 weeks of resistance or concurrent exercise program improved disease symptoms, strength, and quality of life [74].

References

1. World Health Organization. Global health risks: mortality and burden of disease attributable to selected major risks. Geneva, 2009 [updated 2016; cited 2016 Dec 5]. Available from: http://www.who.int/dietphysicalactivity/factsheet_inactivity/en/.
2. Leckie RL, Manuck SB, Bhattacharjee N, Muldoon MF, Flory JM, Erickson KI. Omega-3 fatty acids moderate effects of physical activity on cognitive function. Neuropsychologia. 2014;59:103–11.
3. Beavis MJ, Moodie M. Incidental physical activity in Melbourne, Australia: health and economic impacts of mode of transport and suburban location. Health Promot J Austr. 2014;25(3):174–81.
4. Lowry J. We are walking, talking toxic wastes sites. Scripps Howard News Service. Seattle Post Intelligencer. 22 March 2001.
5. Djordjevic D, Cubrilo D, Macura M, Barudzic N, Djuric D, Jakovljevic V. The influence of training status on oxidative stress in young male handball players. Mol Cell Biochem. 2011;351(1–2):251–9.
6. Aviroop B, et al. Sedentary time and its association with risk for disease incidence, mortality and hospitalization in adults. Ann Intern Med. 2015;162(2):123–32.

7. Ganesan AN, Louise J, et al. International mobile-health intervention on physical activity, sitting, and weight: The Stepathlon Cardiovascular Health Study. J Am Coll Cardiol. 2016;67(21):2453–63.
8. Rebar AL, Duncan MJ, Short C, Vandelanotte C. Differences in health-related quality of life between three clusters of physical activity, sitting time, depression, anxiety, and stress. BMC Public Health. 2014;14:1088.
9. Sun JW, Zhao LG, Yang Y, Ma X, Wang YY, Xiang YB. Association between television viewing time and all-cause mortality: a meta-analysis of cohort studies. Am J Epidemiol. 2015;182(11):908–16.
10. Basterra-Gortari FJ, Bes-Rastrollo M, Gea A, et al. Television viewing, computer use, time driving and all-cause mortality: the SUN cohort. J Am Heart Assoc. 2014;3(3):e000864.
11. Park S-Y, Kwak Y-S. Impact of aerobic and anaerobic exercise training on oxidative stress and antioxidant defense in athletes. J Exerc Rehabil. 2016;12(2):113–8.
12. Gomez-Pinilla F, et al. Exercise impacts brain-derived neurotrophic factor plasticity by engaging mechanisms of epigenetic regulation. Eur J Neurosci. 2011;33(3):383–90.
13. Aihara K, et al. Zinc, copper, manganese, and selenium metabolism in thyroid disease. Am J Clin Nutr. 1984;40(1):26–35.
14. Mikolai J. Vascular biology, endothelial function, and natural rehabilitation, part 1: the nitric oxide pathway. Townsend Letter. May 2014. Issue # 370: 52–7.
15. Bertoluci MC, et al. Endothelial dysfunction as a predictor of cardiovascular disease in type 1 diabetes. World J Diabetes. 2015;6(5):679–92.
16. Louveau A, et al. Structural and functional features of central nervous system lymphatics. Nature. 2015;523(7560):337–41.
17. Barney J. They'll have to rewrite the textbooks. [Internet] Press release 2016, UVA Today. [cited 2016 Dec 10]. Available from: https://news.virginia.edu/illimitable/discovery/theyll-have-rewrite-textbooks.
18. Latvala A, Kuja-Halkola R, et al. Association of Resting Heart Rate and Blood Pressure in late adolescence with subsequent mental disorders: a longitudinal population study of more than 1 million men in Sweden. JAMA Psychiat. 2016;73(12):1268–75.
19. Röhling M, Herder C, Stemper T, Müssig K. Influence of acute and chronic exercise on glucose uptake. J Diabetes Res. 2016;2016:2868652.
20. Knowler WC, Barrett-Connor E, Fowler SE, Hamman RF, Lachin JM, Walker EA, et al. Reduction in the incidence of type 2 diabetes with lifestyle intervention or metformin. N Engl Med. 2002;346(6):393–403.
21. O'Keefe James H, Harshal R, et al. Potential adverse cardiovascular effects from excessive endurance exercise. Mayo Clin Proc. 2012;87(6):587–95.
22. Buscemi S, et al. Relationships between maximal oxygen uptake and endothelial function in healthy male adults: a preliminary study. Acta Diabetol. 2013;50(2):135–41.
23. Cherkas LF, Hunkin JL, et al. The association between physical activity in leisure time and leukocyte telomere length. Arch Intern Med. 2008;168(2):154–8.
24. Ljubicic V, Joseph AM, Saleem A, et al. Transcriptional and post-transcriptional regulation of mitochondrial biogenesis in skeletal muscle: effects of exercise and aging. Biochem Biophys Acta. 2010;1800(3):223–34.
25. Lindholm Marlene E, et al. An integrative analysis reveals coordinated reprogramming of the epigenome and the transcriptome in human skeletal muscle after training. Epigenetics. 2014;9(12):1557–69.
26. Ornish D, Weidner G, Fair WR, Marlin R, Pettengill EB, Raisin CJ, et al. Intensive lifestyle changes may affect the progression of prostate cancer. J Urol. 2005;174(3):1065–9; discussion 1069–70
27. Khaw KT, Wareham N, Bingham S, Welch A, Luben R, Day N. Combined impact of health behaviours and mortality in men and women: the EPIC-Norfolk prospective population study. PLoS Med. 2008;5(1):e12.
28. Suzuki S, Uematsu A, Shimazaki T, Kobayashi H, Nakamura M, Hortobagyi T. Qigong exercise improves Japanese old Adults' quality of life: 174 board #11 June 1, 9: 30 AM – 11: 00 AM. Med Sci Sports Exerc. 2016;48(5 Suppl 1):32.
29. Marshall SJ, Levy SS, Tudor-Locke CE, et al. Translating physical activity recommendations into a pedometer-based step goal: 3000 steps in 30 minutes. Am J Prev Med. 2009;36(5):410–5.
30. Weston Kassia S, et al. High-intensity interval training in patients with lifestyle-induced cardiometabolic disease: a systematic review and meta-analysis. Br J Sports Med. 2014;48:1227–34.
31. Khaw KT, Wareham N, Bingham S, et al. Combined impact of health behaviors and mortality in men and women: the EPIC-Norfolk prospective population study. PLoS Med. 2008;5(1):e12.
32. Marta M, Danuta S. Effect of physical activity on IGF-1 and IGFBP levels in the context of civilization diseases prevention. Rocz Panstw Zahl Hig. 2016;67(2):105–11.
33. Cheng YJ, Gregg EW, et al. Muscle – strengthening activity and its association with insulin sensitivity. Diabetes Care. 2007;30(9):2264–70.
34. Herman WH, Hoerger TJ, et al. The cost-effectiveness of lifestyle modification or metformin in preventing type 2 diabetes in adults with impaired glucose tolerance. Ann Intern Med. 2005;142(5):323–32.
35. Li G, Zhang P, et al. The long–term effect of lifestyle interventions to prevent diabetes in the China Da qing diabetes prevention study: a 20-year follow-up study. Lancet. 2008;371(9626):1783–9.
36. Swan Norman. Patients at risk and billions of dollars being wasted because of tests, scans and procedures that don't work. 10 Oct 2016. www.abc.net.au/news.
37. Vijayaraghava A, et al. Effect of yoga practice on levels of inflammatory markers after moderate and strenuous exercise. J Clin Diagn Res. 2015;9(6):CC08–12.
38. Rippo MR, et al. MitomiRs in human inflamm-aging: a hypothesis involving miR.181a and miR-146a. Exp Gerontol. 2014;56:154–63.
39. Handschin C, Spiegelman BM. The role of exercise and PGC1 alpha in inflammation and chronic disease. Nature. 2008;454(7203):463–9.
40. Rippo MR, Olivieri F, Monsurrò V, Prattichizzo F, Albertini MC, Procopio AD. MitomiRs in human inflamm-aging: a hypothesis involving miR-181a, miR-34a and miR-146a. Exp Gerontol. 2014;56:154–63.
41. Mitchell WK, et al. Sarcopenia, dynapenia, and the impact of advancing age on human skeletal muscle size and strength; a quantitative review. Front Physiol. 2012;3:260.
42. Brook MS, Wilkinson DJ, et al. Skeletal muscle homeostasis and plasticity in youth and ageing: impact of nutrition and exercise. Acta Physiol (Oxf). 2016;216(1):15–41.
43. Janssen I, Shepard DS, Katzmarzyk PT, Roubenoff R. The healthcare costs of sarcopenia in the United States. J Am Geriatr Soc. 2004;52(1):80–5.
44. Muscari A, Giannoni C, Pierpaoli L, Berzigotti A, Maietta P, Foschi E, et al. Chronic endurance exercise training prevents aging-related cognitive decline in healthy older adults; a randomized controlled trial. Int J Geratr Psychiatry. 2010;25(10):1055–64.
45. Tachibana H, Meyer JS, Kitagawa Y, Rogers RL, Okayasu H, Mortel KF. Effects of aginine on cerebral blood flow in dementia. J Am Geriatr Soc. 1984;32(2):114–20.
46. Alfini AJ, Weiss LR, Leitner BP, Smith TJ, Hagberg JM, Smith JC, et al. Hippocampal and cerebral blood flow after exercise cessation in master athletes. Front Aging Neurosci. 2016;8:184.
47. Small Scott A. Enhancing dentate gyrus function with dietary flavanols improves cognition in older adults. Nature Neurosci. 2014;17:1798–803.
48. Toshio M, Yasunori A. Effect of exercise and nutrition upon lifestyle-related disease and cognitive Function. J Nutr Sci Vitaminol. 2015;61:S122–4.

49. Bjornebekk A, Mathe AA, Brene S. The antidepressant effect of running is associated with increased hippocampal cell proliferation. Int J Neuropsychopharmacol. 2005;8(3):357–68.
50. Liu-Ambrose T, et al. Resistance training and executive functions: a 12-month randomized controlled trial. Arch Intern Med. 2010;170(2):170–8.
51. Burrell Teal. Five different physical exercises that affect the brain in very different ways. www.preventdisease.com. 14 June 2016.
52. Chelsea WN. Brain activation during dual-task processing is associated with cardiorespiratory fitness and performance in older adults. Front Aging Neurosci. 2015;7:154.
53. Geda YE, Roberts RO, Knopman DS, Christianson TJ, Pankratz VS, Ivnik RJ, et al. Physical exercise, aging, and mild cognitive impairment. A population-based study. Arch Neurol. 2010;67(1):80–6.
54. Vidoni ED, Johnson DK, Morris JK, Van Sciver A, Greer CS, Billinger SA, et al. Dose-response of aerobic exercise on cognition: a community-based, pilot randomized controlled trial. PLoS One. 2015;10(7):e0131647.
55. Sjösten N, Kivelä SL. The effects of physical exercise on depressive symptoms among the aged: a systematic review. Int J Geriatr Psychiatry. 2006;21(5):410–8.
56. Landi F, Marzetti E, Martone AM, Bernabei R, Onder G. Exercise as a remedy for sarcopenia. Curr Opin Clin Nutr Metab Care. 2014;17(1):25–31.
57. Mendes R, Sousa N, Themudo-Barata J, Reis V. Impact of a community-based exercise programme on physical fitness in middle-aged and older patients with type 2 diabetes. Gac Sanit. 2016;30(3):215–20.
58. Fiatarone MA, O'Neill EF, Ryan ND, Clements KM, Solares GR, Nelson ME, et al. Exercise training and nutritional supplementation for physical frailty in very elderly people. N Engl J Med. 1994;330:1769–75.
59. Vásquez-Morales A, Wanden-Berghe C, Sanz-Valero J. Exercise and nutritional supplements: effects of combined use in people over 65 years; a systematic review. Nutri Hosp. 2013;28(4):1077–84.
60. Wong RJ, Aguilar M, Cheung R, et al. Nonalcoholic steatohepatitis is the second leading etiology of liver disease among adults awaiting liver transplantation in the United States. Gastroeneterology. 2015;148(3):547–55.
61. Qiu LX, Chen T. Novel insights into the mechanisms whereby isoflavones protect against fatty liver disease. World J Gastroenterol. 2015;21(4):1099–107.
62. Miura K, Ohnishi H. Role of gut microbiota and toll-like receptors in nonalcoholic fatty liver disease. World J Gastroenterol. 2014;20(23):7381–91.
63. Takahashi A, Abe K, Usami K, et al. Simple resistance exercise helps patients with non-alcoholic fatty liver disease. Int J Sports Med. 2015;36(10):848–52.
64. Wu H, Jin M, Han D, Zhou M, Mei X, Guan Y, Liu C. Protective effects of aerobic swimming training on high-fat diet induced nonalcoholic fatty liver disease: regulation of lipid metabolism via PANDER-AKT pathway. Biochem Biophys Res Commun. 2015;458(4):862–8.
65. Johnston TE, Modlesky CM, Betz RR, Lauer RT. Muscle changes following cycling and/or electrical stimulation in pediatric spinal cord injury. Arch Phys Med Rehabil. 2011;92(12):1937–43.
66. Warburton Darren ER, et al. Cardiovascular health and exercise rehabilitation in spinal cord injury. Top Spinal Cord Inj Rehabil. 2007;12(1):98–122.
67. Lemmey Andrew B, Casanova F, et al. Effects of high-intensity resistance training in patients with rheumatoid arthritis: a randomized controlled trial, exercise and RA. Arthritis Rheum. 2009;61(12):1726–34.
68. Bartlett JD, Close GL, et al. The emerging role of p53 in exercise metabolism. Sports Med. 2014;44(3):303–9.
69. Schmitt J, et al. A 3 week multimodal intervention involving high intensity interval training in female cancer survivors: a randomized controlled trial. Physiol Rep. 2016;4(3):e12693.
70. Pedersen L, et al. Voluntary running suppresses tumor growth through epinephrine- and IL-6-dependent NK cell mobilization and redistribution. Cell Metab. 2011;23(3):554–62.
71. ACSM. Exercise management for persons with chronic disease and disabilities. 2009:214–6. ACSM Roundtable of Exercise Guidelines for Cancer Survivors. 2012.
72. Crizzle AM, Newhouse IJ. Is physical exercise beneficial for persons with Parkinson's disease? Clin J Sport Med. September 2006;16(5):422–5.
73. Samaei A, et al. Uphill and downhill walking in multiple sclerosis: a randomized controlled trial. Int J MS Care. 2016;18(1):34–41.
74. Silva BA e, et al. A 20-week program of resistance or concurrent exercise improves symptoms of schizophrenia: results of a blind, randomized controlled trial. Rev Bras Psiquiatr. 2015;37:271–9.

Resources

Dean W, English J. Krebs Cycles intermediates: maximizing your body's performance. Nutrition Review. 22 April, 2013.

Egger G, Binns A, Rossner S. Lifestyle medicine textbook. Australia: McGraw Hill; 2008.

Haines A. IFM conference lectures, San Diego, May 13–14, 2016.

Higgins J. Exercise is the best medication we have! Go get it! ACSM Fit Society. 2014;16(2)

Pizzorno J, Katzinger J. Clinical pathology, a functional perspective. Coquitlam: Mind Publishing; 2012.

Rosenberg I, Evans W. Biomarkers. New York: Simon & Schuster; 1991.

Wilhelmsson P. Snabbare, starkare och friskare med Integrativ Idrottsnutrition. Sweden: Örtagården; 2016.

Wisloff U, et al. Cardiovascular risk factors emerge after artificial selection for low aerobic capacity. Science. 2005;307:418–20.

www.ecotherapyheals.org

www.ecsm.org. ACSMs Exercise Management for Persons with Chronic Diseases and Disabilities 3rd edition.

www.ideafit.com/fitness-library/hiit-vs-continuous-endurance-training-battle-of-the-aerobic-titans

www.richardlouv.org

www.shinrin-yoku.org

Mental, Emotional, and Spiritual Imbalances

Muffit L. Jensen

36.1 Mind-Body Techniques – 614

36.2 Six Common Mind-Body Techniques – 614
36.2.1 Neuro Emotional Technique (NET) – 614
36.2.2 Biofeedback – 615
36.2.3 Neurofeedback – 615
36.2.4 Cognitive Behavioral Therapy – 615
36.2.5 Emotional Freedom Technique (EFT) – 616
36.2.6 Hypnotherapy – 616

36.3 Summary – 616

References – 617

After careful review of the Functional Medicine Matrix (IFM Matrix) and the Medical Symptoms Questionnaire (MSQ), you have determined that part of the patient's clinical picture has an emotional or mental component. Along with the nutritional treatment plan, the patient may require another form of care or a multidisciplinary approach. If this is the case, it is time to consider some form of psychotherapy or a type of complementary or alternative mind-body medicine. In order to understand how the brain affects the body, creating or potentiating illness, we must examine some of the research.

Research of modern mind-body medicine dates back to the 1960s, although the concept has been recognized as far back as Hippocrates.

» Disease is not an entity, but a fluctuating condition of the patient's body, a battle between the substance of the disease and the natural self-healing tendency of the body. –Hippocrates

A psychiatrist, George Solomon, began to investigate the impact that emotions had on inflammation and the immune system in general. This new field was called psychoneuroimmunology [1].

Dr. Candace Pert (1946–2013) was a significant contributor to the emergence of mind-body medicine as an area of legitimate scientific research in the 1980s. She has been called "The Mother of Psychoneuroimmunology" and "The Goddess of Neuroscience" by her many fans. To her colleagues, she was an internationally recognized neuroscientist and pharmacologist who published more than 250 research articles and many books on how the brain, endocrine, immune system, and many other organs and systems interact to create disease processes or with the correct interventions create wellness through peptide/receptor cellular communication.

"A feeling sparked in our mind or body will translate as a peptide being released somewhere. Brain, organs, tissues, skin, muscle and endocrine glands all have peptide receptors on them and can retrieve, repress or store emotional information (memories) and or behaviors," she wrote. "You can access emotional memory anywhere in the peptide/receptor network in any number of ways. I think unexpressed emotions are literally lodged in the body [2]."

Emotions, largely ignored within the traditional confines of science and medicine, are actually the key to understanding psychoneuroimmunology and its emerging picture of how the body and mind affect each other. By examining and treating the triad of mind, body, and spirit, we can close the gap and acquire a positive result that of whole wellness (◘ Fig. 36.1).

36.1 Mind-Body Techniques

Mind-body techniques aid in the management of pain and many other diseases because they reduce anxiety, encourage relaxation, improve coping skills, and ease tension. Managing stress will decrease stress hormones and increase good emotion-modulating neuropeptides. The goals of mind-body

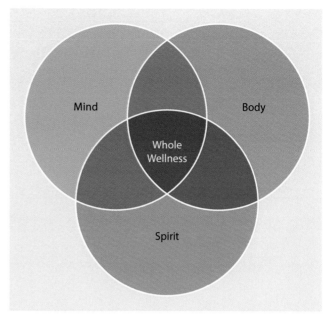

◘ Fig. 36.1 The overlap of integrating mind, body, and spirit to create whole wellness

techniques are to reduce stress hormone levels and to reset emotional cellular memory, so that the body is better able to fight off illness.

Some conditions that mind-body techniques may be helpful for include:
- Anxiety
- Asthma
- Autoimmune disorders
- Cancer
- Coronary heart disease
- Depression
- Fibromyalgia
- General eating disorders
- Hypertension
- Insomnia
- Menopause
- Mental health issues
- Obesity
- Pain (acute and chronic)
- Poststroke
- Stomach and intestinal problems
- Weight loss

36.2 Six Common Mind-Body Techniques

36.2.1 Neuro Emotional Technique (NET)

36.2.1.1 What Is It?

NET uses a methodology of finding and removing neurological imbalances related to the physiology of unresolved stress. NET is a tool that can help improve many behavioral and physical conditions.

36.2.1.2 How Does It Work?

NET is based on the physiological foundations of stress-related responses. As discovered in the late 1970s, emotional responses are composed of neuropeptides (amino acid chains) and their receptors, which lie on neurons and other cells of remote tissues in the body. The neuropeptides are ejected from the neuron and carry the encoded information to other sites within the body. These neuropeptides are in a category of neurochemicals known as Information Substances (IS), which are released at times of stress-related arousal and become attached to remotely positioned neuroreceptors. Significantly, this process also happens when a person recalls to memory an event in which a stress originally occurred. This is a key factor in the NET treatment. Thus, the physiological status of the body is emotionally replicating a similar physiological state that was found in the original conditioning event by the process of remembering. It is at this point that the practitioner applies the principles of the technique to then "reset/erase" the original event through the nervous system. This technique is highly effective, although currently taught to chiropractors, acupuncturists, and psychotherapists only. This technique is worth a co-referral to a trained practitioner within the interdisciplinary team.

The ONE Research Foundation was established in 1993 by Neuro Emotional Technique (NET) founder Scott Walker, D.C., to raise funds for research to demonstrate empirically the validity of the BioPsychoSocial (BPS) model as an effective methodology for intervening in a broad spectrum of physical, emotional, and psychological conditions. The ONE Research Foundation (ONE) is comprised of individuals who are committed to the natural noninvasive healing of the mind and body. ONE is a nonprofit 501(c)(3) dedicated to making these natural methods available to all as the standard of care [3].

36.2.2 Biofeedback

36.2.2.1 What Is It?

Biofeedback is a noninvasive drug-free form of treatment. It is a process that enables an individual to learn how to change physiological activity for the purposes of improving health and performance. Precise instruments measure physiological activity such as heart function, breathing, muscle activity, and skin temperature. These instruments rapidly and accurately "feedback" information to the user. The presentation of this information, often in conjunction with changes in thinking, emotions, and behavior, supports desired physiological changes. Over time, these changes can endure without continued use of an instrument.

36.2.2.2 How Does It Work?

The therapist attaches sensors or electrodes to the body. These sensors provide a variety of readings – feedback – which is displayed on equipment, usually a television monitor, for the patient to see. With this information, patients can learn to make changes so subtle that at first, they cannot be consciously perceived. With practice, the new responses and behaviors can help to bring relief and improvement to a variety of disorders.

36.2.3 Neurofeedback

36.2.3.1 What Is It?

Neurofeedback is a natural, self-regulating approach that helps restore the brain's ability to function in the way it was designed to function. Also known as EEG biofeedback, brain training and brain computer interface (BCI), neurofeedback utilizes the operant conditioning learning model to achieve self-regulation. Additionally, monitoring the brain's bioelectrical system also advises us of chemical activity within the brain. It is a safe, noninvasive, painless learning procedure during which sensors are placed on the surface of the patient's head.

36.2.3.2 How Does It Work?

A neurofeedback session involves the client sitting in a comfortable chair in a quiet room. He or she will have electrodes placed on the area of the scalp where the training will occur, which is determined by the QEEG interpretation. The sensors record brain electrical activation levels and enable participants to learn to improve mental performance, normalize behavior, and/or stabilize mood. The information is displayed on a computer screen, together with sounds, which change according to the brain's activity levels. Therefore, the patient can read, understand, and influence his or her brainwave activity. Once the patient learns to access and regulate the brain more effectively, symptoms begin to improve or performance optimizes. The visual feedback on the computer screen and auditory feedback through speakers or headphones are positive reinforcement. Creating these reward parameters involves skill and training. It is perhaps the biggest breakthrough in noninvasive medicine in the last 50 years.

Medical conditions that have been found to benefit most from neurofeedback assessment and treatments include autism, ADHD/ADD, mood disorders, postanesthesia, head injury, PTSD, sleep interruption, and stroke rehabilitation [4].

36.2.4 Cognitive Behavioral Therapy

36.2.4.1 What Is It?

Cognitive behavioral therapy is a psychosocial intervention that is among the most widely used evidence-based practices for treating mental disorders. Guided by empirical research, CBT focuses on the development of personal coping strategies that target solving current problems and changing unhelpful patterns in cognitions, behaviors, and emotional regulation. It was originally designed to treat depression and is now used for a number of mental health conditions.

36.2.4.2 How Does It Work?

Cognitive behavioral therapy requires the patient and therapist to work as a team, collaborating to solve problems. Rather than waiting for problems to get better after talking about them repeatedly from week to week, patients can take an active role in their own treatment, using self-help assignments and CBT tools between sessions to speed up the process of change. Each session is focused on identifying ways of thinking differently and unlearning unwanted reactions.

In adults, CBT has been shown to have effectiveness and a role in the treatment plans for anxiety disorders, depression, eating disorders, chronic low back pain, personality disorders, psychosis, schizophrenia, and substance use disorders, in the adjustment, depression, and anxiety associated with fibromyalgia and with postspinal cord injuries [5].

36.2.5 Emotional Freedom Technique (EFT)

36.2.5.1 What Is It?

Emotional freedom technique, also known as tapping. EFT combines the physical benefits of acupuncture with the cognitive benefits of the traditional talk therapy for a much faster result. Most aspects of our lives are not working the way we would like due to a disruption in the body's energy system. This may be from a past trauma or memories which cause negativity, anxiety, stress, physical pain, and unresolved emotional issues. EFT is an amazing healing tool and can help reset the subconscious mind and eliminate old beliefs that patients may not realize they are holding onto.

36.2.5.2 How Does It Work?

According to the EFT manual, the procedure consists of the participant rating the emotional intensity of their reaction on a Subjective Units of Distress Scale (SUDS) (a Likert scale for subjective measures of distress, calibrated 0–10) and then repeating an orienting affirmation while rubbing or tapping specific points on the body. It's a similar approach to acupuncture, but instead of needles, we tap on the main energy spots on our body, called meridians, to release negative emotions. Some practitioners incorporate eye movements or other tasks. The emotional intensity is then rescored and repeated until no changes are noted in the emotional intensity. Proponents of EFT and other similar treatments believe that tapping/stimulating acupuncture points provides the basis for significant improvement in psychological problems [6].

36.2.6 Hypnotherapy

36.2.6.1 What Is It?

Hypnotherapy is a form of psychotherapy that is used to create subconscious change in a patient in the form of new responses, thoughts, attitudes, behaviors, or feelings. It is undertaken with a subject in hypnosis.

36.2.6.2 How Does It Work?

The therapist consults with clients to determine the nature of their problems and prepares them to enter a hypnotic state by explaining how hypnosis works and what they will experience. He or she will then induce the client into a hypnotic state using individualized methods and techniques of hypnosis based on interpretation of test results and analysis of the client's problem. A person who is hypnotized displays certain unusual behavior characteristics and propensities, compared with a non-hypnotized subject, most notably heightened suggestibility and responsiveness. The goal of hypnotherapy is to increase motivation or alter behavior patterns. The therapist may train the client in self-hypnosis conditioning.

36.3 Summary

The above techniques are only a few of the many wonderful complementary techniques available. See ◘ Fig. 36.2 for other suggestions. Although some modalities do not have a strong evidence base, they have historical recognition of value, some as far back as thousands of years in traditional medical practice. Just because they are not researched in a RCT format doesn't mean that they are not highly effective. The way we live, how we think, what we eat, and how we feel affect our health. The value of these therapies as part of the patient's treatment plan is truly worthwhile. You may be asking at this point, "what technique should I choose for the patient?" The simple answer is: the best technique is the one that the patient likes and will do. Always remember, "there is a cork for every bottle."

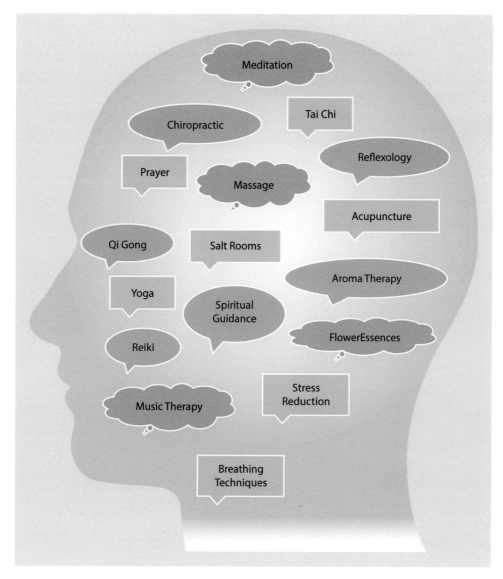

Fig. 36.2 Further suggestions for mind-body techniques not discussed that are commonly used in integrative medicine

References

1. Solomon GF, Moos RH. Emotions, immunity, and disease: a speculative theoretical integration. Arch Gen Psychiatry. 1964;11:657–74.
2. Pert C. Molecules of emotion: the science behind mind-body medicine (paperback). New York: Simon & Schuster; 1999.
3. NetMindBody. Home page. https://www.netmindbody.com/. Accessed March 2019.
4. Psychology Today. Neurofeedback. 2019. https://www.psychologytoday.com/us/therapy-types/neurofeedback. Accessed March 2019.
5. *Cognitive behavioral therapy*. (n.d.). Retrieved 19 July 2017, from Wikipedia: The Free Encyclopedia: http://en.wikipedia.org/wiki/Cognitive_behavioral_therapy.
6. *Hypnotherapy*. (n.d.). Retrieved 19 July 2017, from Wikipedia: The Free Encyclopedia: http://en.wikipedia.org/wiki/Hypnotherapy.

IFMNT Nutrition Care Process

Contents

Chapter 37 The IFMNT Practitioner – 621
Robin L. Foroutan

Chapter 38 The Patient Story and Relationship-Centered Care – 633
Leigh Wagner

Chapter 39 The Nutrition-Focused Physical Exam – 637
Mary R. Fry

Chapter 40 Modifiable Lifestyle Factors: Exercise, Sleep, Stress, and Relationships – 695
Margaret Christensen

Chapter 41 Developing Interventions to Address Priorities: Food, Dietary Supplements, Lifestyle, and Referrals – 715
Aarti Batavia

Chapter 42 Therapeutic Diets – 743
Tracey Long and Leigh Wagner

Chapter 43 Dietary Supplements: Understanding the Complexity of Use and Applications to Health – 755
Eric R. Secor

Chapter 44 Clinical Approaches to Monitoring and Evaluation of the Chronic Disease Client – 769
Cynthia Bartok and Kelly Morrow

Chapter 45 Ayurvedic Approach in Chronic Disease Management – 783
Sangeeta Shrivastava, Pushpa Soundararajan, and Anjula Agrawal

The IFMNT Practitioner

Robin L. Foroutan

37.1 Who Is the IFMNT Practitioner? – 622

37.2 What About Their Approach Sets Them Apart? – 622

37.3 The Functional Medicine Matrix – 624

37.4 Opportunities for Education (Certification, Credentialing, and Other Accredited Programs) – 624

37.5 Major Advocate for Development of IFMNT: Dietitians in Integrative and Functional Medicine (DIFM) – 625

37.6 Standards of Practice and Professional Performance – 627

37.7 The IFMNT Radial – 628
37.7.1 Key Area: Food, Lifestyle, Environment – 628
37.7.2 Key Area: Nutrition Physical Signs and Symptoms – 628
37.7.3 Key Area: Biomarkers – 629
37.7.4 Key Area: Metabolic Pathways & Networks – 630
37.7.5 Key Area: Systems – 630
37.7.6 Precipitating Factors – 630
37.7.7 IFMNT Radial Summary – 630

37.8 Summary – 630

References – 631

37.1 Who Is the IFMNT Practitioner?

Integrative and Functional Medical Nutrition Therapy (IFMNT) is a term championed by the Executive Committee of Dietitians in Integrative and Functional Medicine (DIFM), a dietetic practice group of the Academy of Nutrition and Dietetics. DIFM and others in the Integrative and Functional Medicine arena have used IFMNT as a specific approach to medical nutrition therapy (MNT) that incorporates both Integrative and Functional Medicine methodology to deliver a "food as medicine" approach to nutrition care. This style of care requires education and training beyond the core nutrition curriculum that exists for traditionally trained medical and allied healthcare practitioners, such as the required educational competencies for registered dietitians-nutritionists as determined by the Accreditation Council for Education in Nutrition and Dietetics (ACEND), the accrediting body of the Academy of Nutrition and Dietetics.

The basic tenet of this emerging area of healthcare centers on optimizing nutritional status and correcting underlying system imbalances by establishing nutritional therapies that enable the body to function optimally. Because IFMNT interventions are nutrition-based therapies rooted in dietary interventions and dietary and herbal supplements, the IFMNT-trained nutrition professional is poised to play a critical role in this medical paradigm. This version of diet therapy is relatively new, and much of the growth and momentum of Integrative and Functional Medicine, including IFMNT, grew as a result of consumer demand. In response to this call for a more thoughtful and natural approach to healing, paired with a growing body of evidence lending credence to IFMNT practices, an increasing number of medical professionals are seeking out opportunities to become trained and proficient in IFMNT.

Among dietitians and nutritionists, interest in this specialty area of nutrition and health is high, as demonstrated by the exponential growth of DIFM. Yet the information and specific training necessary to deliver IFMNT have not yet widely trickled down to the student level in most accredited dietetics programs [1, 2]. In recent years, many more opportunities for learning have become available to nutrition professionals, a trend that is sure to continue across disciplines.

Dietitians and certified nutritionists are not the only IFMNT practitioners. IFMNT practices have been employed and embraced by many other healthcare providers, including medical doctors, naturopathic doctors, osteopaths, physician assistants, nurses and nurse practitioners, chiropractors, and licensed acupuncturists. Organizing bodies, such as the Holistic Nurses Association, exist across disciplines and serve to bring together those practicing under similar philosophies. Because many different types of medical professionals are practicing Integrative or Functional Medicine or are interested in offering these services to their patients, the IFMNT provider can function independently in a private practice setting or as part of an Integrative/Functional Medicine team. In fact, these interdisciplinary teams have started to revolutionize medical institutions in ways that were difficult to imagine only a decade ago.

Integrative and Functional Medicine has become better accepted by both conventional institutions and health professionals in recent years, further encouraging the growth and interest of the Integrative-Functional Medical paradigm, of which IFMNT is an important part. One example is the recent opening of the Cleveland Clinic's Center for Integrative & Lifestyle Medicine, with its expressed dedication to address "the increasing demand for integrative healthcare by researching and providing access to practices that address the physical as well as lifestyle, emotional, and spiritual needs of patients." Similarly, New York Presbyterian created the Integrative Health and Wellbeing program in conjunction with Weill Cornell Medical Center, which employs an interdisciplinary team including medical doctors, registered dietitian nutritionists, licensed acupuncturists, and nurses. These types of institutions have the potential to measurably change the kind of health and medical interventions available to the public on a large scale.

37.2 What About Their Approach Sets Them Apart?

In contrast to a conventional, symptom- or disease-centered approach to medicine, Integrative and Functional Medicine practitioners offer a different approach; the practitioner works in partnership with the patient to identify the underlying causes and metabolic imbalances driving the symptoms and disease state. The *integrative and functional medical model* uses evidence-based research to provide patient-centered care in order to create a unique personalized care plan based on each person's specific environment, lifestyle, and genes [3]. A major component of environment is a person's diet, and nutrient status is a recognized determining factor for health. Thus, the foundation of any integrative and functional intervention rests firmly on dietary habits and optimizing nutritional status. This is where IFMNT comes in. IFMNT represents the same kind of patient-centered approach for the nutrition professional. This approach moves away from a one-size-fits-all methodology to nutrition and wellness, acknowledging and harnessing the body's innate ability to heal itself, given the right conditions. An IFMNT practitioner recognizes that optimal health is more than simply the absence of disease.

To provide patient-centered care, an IFMNT practitioner gathers information by listening to the patient's story, looking for hints and clues about the etiology of the underlying imbalances that trigger symptoms. The focus is on the whole person, who is more than simply a patient with a disease. Establishing a *therapeutic relationship* between patient and practitioner is important and promotes healing (see ▶ Chap. 6). An individual's beliefs, attitudes, and motivations are all considered relevant to physical and emotional well-being and are typically explored as part of the "story." The mind-body-spirit connection is not only acknowledged but honored by IFMNT, (see ▶ Chap. 36) as well as physical, social, and lifestyle factors unique to each person. In fact, the concept of individuality is at the forefront of IFMNT. Voluntary actions such as decision-making, emotional responses, stress responses, and lifestyle habits are taken into account, yet so are involuntary actions within the microenvironment of the body, such as nutrient metabolism and absorption, cellular functions, and interactions between organ systems, largely influenced by genetic variability and epigenetic influences.

Aside from an individual's dietary and lifestyle choices, which influence overall health, people have unique metabolic patterns based on genetics and epigenetic influences that affect nutrient requirements. The interactions between diet, genes, and environment are considered and evaluated as part of an IFMNT practitioner's approach. Even a minor imbalance in the body can trigger a cascade of biological effects that may eventually lead to presentation of symptoms, compromised health, or chronic disease. For example, chronically low-normal vitamin D may contribute to osteoporosis, poor immune function, and other consequences. Chronically low-normal vitamins B12, B6, and folate can result in hyperhomocysteinemia, which increases the risk for cardiovascular disease and blood clots. Inadequate vitamin B12 status is also linked with fatigue, depression, and anxiety disorders. This is why IFMNT practitioners utilize a wide range of assessment tools and measures, including a nutrition-focused physical examination and laboratory data, in an effort to determine nutrient status. These same tools also help to identify key system imbalances that may contribute to symptoms and influence metabolism and health. Rather than suppress a metabolic function to mask a symptom, the goal of IFMNT is to restore optimal metabolic function to promote health and healing. Therefore, the focus is on true healing: the reversal of the metabolic disruptions that influence a disease process and create symptoms, not just managing or masking symptoms. This focus is at the heart of IFMNT.

In order to address underlying metabolic disturbances at play, an IFMNT practitioner is well-versed in biochemical processes and nutritional biochemistry. These metabolic pathways are complex and reliant on nutrient cofactors to operate optimally. Even slight nutrient deficiencies or "chronic insufficiencies" left uncorrected over a long period of time can disrupt metabolic function. According to the triage theory developed by Bruce N. Ames, when faced with chronic nutrient deficiency or insufficiency, the body must prioritize available micronutrients to support the functions most critical to survival and reproduction. This prioritization occurs at the expense of the functions that protect against future cellular damage and aging. [4, 15]. This is how chronic insufficiencies can contribute to accelerated aging, chronic disease, physiological stress, and chronic inflammation over years. The effects of nutrients, and thus nutrient deficiencies and chronic insufficiencies, in metabolic pathways are important for an IFMNT practitioner to understand (see ▶ Chap. 46).

Chronic Insufficiency versus Nutrient Deficiency

Overt nutrient deficiencies resulting in long latency deficiency diseases are a public health concern, yet these deficiency diseases (such as rickets, beriberi, and scurvy, the deficiency diseases of vitamin D, thiamin, and vitamin C, respectively) are rare in the United States and Canada [5]. However, established lower limits for what is considered "within normal limits" for specific nutrients may fall well outside the *optimal range* for ideal metabolic functioning. In fact, suboptimally low levels of these same vitamins (vitamin D, thiamin, and vitamin C) are often seen in clinical practice. Lab values that fall on the high or low edges of normal limits are typically overlooked in conventional medicine. Practitioners may be inadvertently missing opportunities to optimize nutrition status, allowing nutrient chronic low-level insufficiencies to contribute to long-term chronic disease risk [6].

Because nutrients serve as cofactors for all biochemical processes in the body, IFMNT acknowledges that chronically low or suboptimal levels of certain nutrients may indeed take a toll on human health by inhibiting or slowing down metabolic processes. While chronic nutrient insufficiencies may trigger minor signs and symptoms of metabolic imbalances – even in the absence of overt disease states – diminished nutrient status over the long term can trigger progression into chronic diseases. The integrative and functional medical model maintains that there are consequences to insufficient nutrient status over time, such as mitochondrial decay and increased DNA damage [4]. Therefore, IFMNT practitioners may favor optimal ranges when evaluating nutrient laboratory data, rather than relying on conventional laboratory reference ranges, which are based on plus or minus two standard deviations of levels observed in the general population. In line with recognizing genetic individuality and person-centered care, functional lab testing may be used in addition to conventional blood chemistries to provide more specific information as to whether nutrient cofactors are present in levels that are optimal for a specific individual.

Laboratory Data for the IFMNT Practitioner
- Conventional biomarkers may include:
  - Complete blood count (CBC)
  - Complete metabolic panel (CMP)
  - Serum nutrient status (e.g., vitamins, minerals)
  - Red blood cell nutrient status (e.g., magnesium, zinc)
  - Kidney and liver function tests
  - Cholesterol panels
  - Homocysteine levels
  - Ammonia levels
  - Uric acid levels
  - Heavy metals

- Fatty acids panel
- Amino acids panel
- IgE food allergy panels
- Genetic information may include:
  - Single nucleotide polymorphisms (e.g., MTHFR, VDR, COMT)
  - Haplotypes (e.g., celiac disease haplotpes HLA-DQ2 and HLA-DQ8)
- Functional testing may include:
  - Digestive stool analyses
  - Microbial stool analyses
  - Organic acids panel
  - White blood cell micronutrient status
  - Methylmalonic acid (MMA)
  - Formiminoglutamic acid (FIGLU)
  - Non-IgE food sensitivity panels

Once the subjective and laboratory information is gathered, an IFMNT practitioner may use both conventional and integrative therapies, as appropriate. Integrative and functional approaches are generally employed to focus on identifying the root causes of symptoms and diseases to focus on preventive or restorative care. Traditional, non-Western, approaches may also be used, such as principles of Traditional Chinese Medicine or Ayurveda. The IFMNT practitioner will make use of any evidence-based approach, whether from conventional or alternative medicine practices. Modalities such as food elimination diets and dietary supplements (e.g., vitamins, minerals, herbs, botanicals, probiotics) are commonly used, as are metabolic detoxification support and digestive support programs. Because IFMNT practitioners recognize the value of mind-body interventions, they may also recommend other modalities such as yoga, Qigong, meditation, guided imagery, sound healing, and healing touch. Whatever the treatment plan, IFMNT practitioners use their vast knowledge of the dynamic interaction between biological systems and biochemical pathways, nutrient status, biochemical individuality, genetics, and epigenetics to create an effective, holistic nutrition care plan [7].

37.3 The Functional Medicine Matrix

"Functional Medicine," as a concept, was proposed by Jeffrey Bland, PhD, in the early 1980s. Through his vision and leadership, the Institute of Functional Medicine (IFM) emerged. As with Integrative Medicine, Functional Medicine seeks to address the underlying causes of disease states and uses a systems biology-oriented approach that recognizes the interconnectedness and interplay between the various organ systems, while aiming to foster a therapeutic relationship between patient and practitioner [3]. Particular attention is paid to acknowledging each individual's biochemical uniqueness in order to personalize medical care. Functional Medicine practitioners consider the complex interactions between genetic predispositions, environmental factors, diet, and lifestyle to assess, prevent, and treat chronic disease.

IFM represents one of the major leaders educating healthcare providers in the theories and practices of Integrative and Functional Medicine. There are several qualified organizations that offer educational programs, certification courses, and publications in Functional Medicine (see Resources). As such, IFM developed the first assessment tool, the *Functional Medicine Matrix*. The Matrix serves as architecture for organizing the complexity of each case to synthesize information and create target interventions for patient care. It offers a unique system for homing in on clinically relevant information that is part of the patient's "story" and helps the practitioner organize the information to help clarify the complex interactions between symptoms that are characteristic of chronic disorders. It provides a framework to aid the practitioner in identifying and interpreting the various factors involved in promoting health and preventing disease in each individual (Fig. 37.1).

The Functional Medicine Matrix is divided into three main parts: (1) the patient's story, (2) an assessment of lifestyle factors, and (3) physiological assessments, observations, and other signs of clinical imbalances. The patient's story is further delineated into three subsections: antecedents and predispositions, the "triggering event(s)," and mediators that appear to contribute to the patient's medical condition, health crisis, or current set of symptoms. Lifestyle assessments, referred to as fundamental lifestyle factors, include dietary habits, exercise, sleep, relationships, and life meaning and purpose, with the latter including attitudes and core beliefs. Biological and physiological assessments are represented in the largest section, "Clinical Imbalances Found in the Functional Organizing Systems." This section is where systems data are aggregated and analyzed, including information such as inflammatory biomarkers, body composition, lab test results, toxicity/detoxification, and physiological imbalances. Integrated throughout is the understanding that the mind-body-spirit connection influences each individual patient and must be honored. The Functional Medicine Matrix represents one of the first assessment tools for Functional Medicine practitioners.

37.4 Opportunities for Education (Certification, Credentialing, and Other Accredited Programs)

For many years, RDNs and other nutrition professionals were primarily self-taught in the concepts and practices of Integrative and Functional Medicine, as there had been a lack of accredited programs that offered comprehensive education around IFMNT. With the exception of Bastyr University in Seattle, WA, the University of Kansas Medical Center in

Fig. 37.1 Functional Medicine Matrix. (Used with permission from The Institute for Functional Medicine ©2015)

Kansas City, KS, and the Certified Nutrition Specialist (CNS) credential, recognized and accredited programs were in short supply. In the 1990's, other trailblazers developed programs for healthcare professionals with a focus on nutrition, such as The University of Arizona Andrew Weil Center for Integrative Medicine and IFM. Fortunately, the tides have shifted, and several accredited programs now offer advanced degrees in IFMNT, as well as a multitude of certifications and other training courses (Table 37.1).

37.5 Major Advocate for Development of IFMNT: Dietitians in Integrative and Functional Medicine (DIFM)

A subgroup of the Academy of Nutrition and Dietetics (formerly the American Dietetic Association), Dietitians in Integrative and Functional Medicine (DIFM) had been previously known as Nutrition in Complementary Care. It was originally established in 1998 by a group of like-minded dietitians interested in expanding their skills in topics such as functional foods, dietary supplements, and alternative medical systems, such as Traditional Chinese Medicine and Ayurvedic Medicine. In 2009, the group changed its name to DIFM to better reflect its role in the emerging Integrative and Functional Medicine community. As of 2019, DIFM has more than 5,500 members consisting of RDNs, dietetic technicians (DTRs), dietetic interns, and students interested in practicing an integrated and personalized approach to health and healing.

> **Dietitians in Integrative and Functional Medicine (DIFM) Mission and Vision [8]**
> - *Vision:* Optimize health and healing with integrative and functional nutrition
> - *Mission:* Empower members to be leaders in integrative and functional nutrition
> - *Values:* Innovation, integrity, and compassion

DIFM's contribution to the field of IFMNT has been vast. It offers members opportunities for continuing education and rolled out a certification program in 2017.

Table 37.1 List of IFMNT Training Programs

Organization	Educational Opportunities
Academic Programs and Degrees	

Bastyr University	*Bachelors of Science degrees in:*
	  Nutrition
	  Nutrition & Culinary Arts
	  Nutrition & Exercise Science
	  Combined program in Nutrition and Didactic Program in Dietetics
	  Herbal Sciences
	Masters of Science degrees in:
	  Nutrition Research
	  Combined program in Nutrition and Didactic Program in Dietetics
	  Nutrition and Clinical Health Psychology
	  Non-degree Dietetic Internship

Maryland University of Integrative Health	Doctoral Degree in Integrative Nutrition (online) Masters Degree in Nutrition and Integrative Health
Rutgers University	Masters of Science (MS) Degree in Health Science Integrative Health and Wellness
University of Bridgeport	Masters of Science in Nutrition (online) Dual degree, Masters of Science in Nutrition and Doctor of Naturopathic Medicine Bachelors of Science in Nutrition
University of Kansas School of Health Professions, Departments of Integrative Medicine and Dietetics and Nutrition	*Non-degree*: graduate certificate: Dietetics and Integrative Medicine (DIM) 12 hour certificate (online) *Degree*: Dietetic Internship Fellowship and MS in Dietetics & Nutrition with Integrative Nutrition emphasis
Certificate, Accreditation and Training Programs	
Academy of Nutrition and Dietetics	Integrative and Functional Nutrition Certificate of Training Program (online)
Arizona Center for Integrative Medicine	iHELP Program
Bauman College	Nutrition Consultant Training Program (online option)
Certified Nutrition Specialist (CNS)	Board for Certification of Nutrition Specialists (BCNS) American College of Nutrition (CAN) (MS or PhD eligible)
Certified Clinical Nutritionist (CCN)	Clinical Nutrition Certification Board (CNCB) International and American Association of Clinical Nutritionists (IAACN)
Duke University	Integrative Health Coaching Certification
Institute for Functional Medicine	Institute for Functional Medicine Certification Program (IFMCP) Applying Functional Medicine in Clinical Practice (AFMCP) Introduction to Functional Nutrition Course (online, free)
Integrative and Functional Nutrition Academy (IFNA)	Integrative and Functional Nutrition Certified Practitioner (IFNCP) Advanced Practice Credential (online) IFNCP Certificate of Training (online)
Next Level Functional Nutrition IFMNT	Integrative and Functional Nutrition Foundations Course (online) IFMNT Certificate of Training Program (online) Advanced Studies (online)
Rutgers University	Graduate certificates available in Complementary and Alternative Medicine Graduate certificate in Health Coaching Graduate Certificate in Integrative Medicine Research
University of Kansas School of Health Professions, Departments of Integrative Medicine and Dietetics and Nutrition	Graduate Certificate: Dietetics and Integrative Medicine (DIM) 12 hour certificate (online)
University of Miami	Integrative and Complementary Academic Medicine: Clinical Nutrition Conference Series

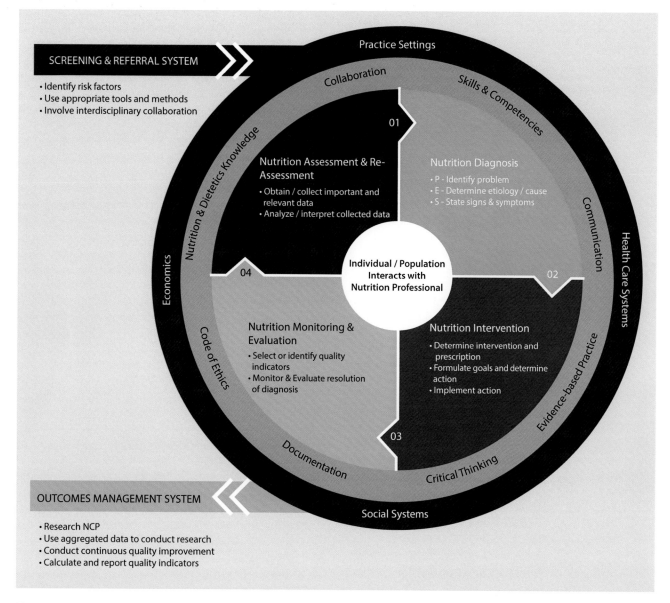

Fig. 37.2 Nutrition Care Process and model. (Reprinted from Swan et al. [13]. With permission from Elsevier)

37.6 Standards of Practice and Professional Performance

In 2011, DIFM developed and published core competencies as part of their Standards of Process (SOP) and Standards of Professional Performance (SOPP), establishing an official framework to guide the practice and performance of registered dietitian nutritionists (RDNs) practicing IFMNT. These were updated and revised in 2019. The SOP for IFMNT builds upon the previously established four steps of the *Nutrition Care Process*, which provides a general framework for the nutrition professional to personalize care and tailor interventions to each individual. First described by the Academy of Nutrition and Dietetics and published as part of the Nutrition Care Model, the four steps of the Nutrition Care Process (NCP) are nutrition assessment, diagnosis, intervention, and, finally, monitoring and evaluation. These steps are designed to facilitate safe and effective nutrition services, promote evidence-based practice, and serve as a framework for critical thinking and decision-making for all RDNs in all settings [7, 9, 14] (◘ Fig. 37.2).

Also previously described by the Academy of Nutrition and Dietetics are the Standards of Professional Performance (SOPP), which detail competency levels for credentialed RDNs and standards of professional behavior. The SOPP are designed to serve as guides for self-evaluation and to assess one's own individual performance. They also serve as tools to determine additional education and skills necessary to advance one's individual level of practice [10]. These tools allow the RDN to continually evaluate his and her own overall competency by the self-evaluation of skills, learning, training, education, and existing knowledge, as well as responsibility and accountability in the practice of nutrition and dietetics [11].

> **Steps in the Nutrition Care Process**
> The Nutrition Care Process includes the following steps [9]:
> - *Nutrition assessment*: Collection and documentation of information, including health history, biochemical data, labs and medical test results, previous medical evaluations, information from a nutrition-focused physical, anthropometric data, and diet record
> - *Diagnosis*: Information collected from the nutrition assessment is synthesized in the selection of appropriate nutrition diagnoses
> - *Intervention*: The RDN will determine the appropriate nutrition intervention to address the root cause of the problem with the goal of alleviating signs and symptoms of the previously determined diagnoses
> - *Monitoring/evaluation*: Monitoring and evaluation are used to determine if the patient is making progress toward or has achieved established goals

The SOPP are divided into several specific standards, with each standard describing a focus area or domain meant to serve as a guide for self-evaluation and professional development. The six domains of professional performance published in the 2019 SOPP are as follows: quality in practice, competence and accountability, provision of services, application of research, communication and application of knowledge, and utilization and management of resources. These standards are to be used by RDNs as a guide for self-evaluation and demonstrating competence, and as a roadmap to expand their practice and identify areas for professional development. Each focus area standard has several indicators or quantifiable actions that explain how each standard might be used. The DIFM-developed SOP/SOPP can help determine whether an RDN is qualified to provide IFMNT and clearly describes the skills, knowledge, education, and competencies required to safely and effectively provide quality IFMNT. It further allows RDNs to categorize themselves as one of three competency levels of practice: competent practitioner, proficient practitioner, and expert practitioner [14].

In order to further support IFMNT practitioners, expert RDNs Kathie Swift, Diana Noland and Elizabeth Redmond developed the IFMNT Radial, which was published as part of the 2011 SOP/SOPP and revised in 2018. The Radial customized the NCP for the IFMNT practitioner and incorporates the key factors at the core of IFMNT, facilitating the evaluation of complex information using the NCP [14]. It represents the first of such tools developed specifically for the IFMNT practitioner.

37.7 The IFMNT Radial

Published within the SOP/SOPP in 2011 and updated in 2018, three expert IFMNT practitioners established a visual conceptual framework to aid RDNs in implementing IFMNT in their practice. This patient-/client-centered model for critical thinking facilitates the shared decision-making process between patient and practitioner and offers a specialized version of the NCP for the IFMNT practitioner. Its circular architecture signifies the multidirectional motion and interrelationships between all aspects of health, imbalances, and disease states.

At the center of the Radial is the NCP, as the focal point of any IFMNT practice is to provide personalized nutrition care. The steps in providing this care are assessment, diagnosis, intervention, monitoring, and evaluation. The imagery in the center circle depicts food as a determining factor in health and disease. Surrounding the center circle are five circles representing each of the five key factors to evaluate a person's health: Diet, lifestyle and environment; signs and symptoms; biomarkers; metabolic pathways; and biological systems to consider. Since IFMNT recognizes that each of these areas is influenced by a person's biochemical and genetic uniqueness, the imagery of DNA and microbiota strands link the five key areas together signifying that each of these areas is interconnected, with any imbalance of one potentially affecting the others. Additional influences to be considered include precipitating factors, including stress and belief systems, toxic exposure, pathogens and infections, and allergens and intolerances. These potentially antagonistic factors all affect metabolism and overall health [7, 14] (◘ Fig. 37.3).

37.7.1 Key Area: Food, Lifestyle, Environment

The food, lifestyle and environment area represents the patient's "story" and a subjective description of overall habits, which an IFMNT practitioner must consider as part of the NCP. Since food choices are an important determinant of health and disease, food preferences, access to foods, culture, and traditions are major factors to be discussed and understood by the practitioner. It also includes the health of those around the patient, how often he or she exercises and moves, sources and levels of stress, and the coping methods each person uses. It may include his or her access to nature, fresh air, and the amount of sunlight they are exposed to on a regular basis, as well as the kinds of relationships they have in their lives. Tools utilized might include a questionnaire or nutrition intake survey, food frequency questionnaire, or a food record. This is where the practitioner might begin to determine various triggers, antecedents, or precipitating factors related to the present health concern [7, 12].

37.7.2 Key Area: Nutrition Physical Signs and Symptoms

A major feature of IFMNT is the *nutrition-focused physical exam*, which enables the practitioner to assess and evaluate signs and symptoms. Visual symptoms can be clinically relevant, such as skin color, texture and moisture, and the appearance of the tongue and nails. Other tools employed here

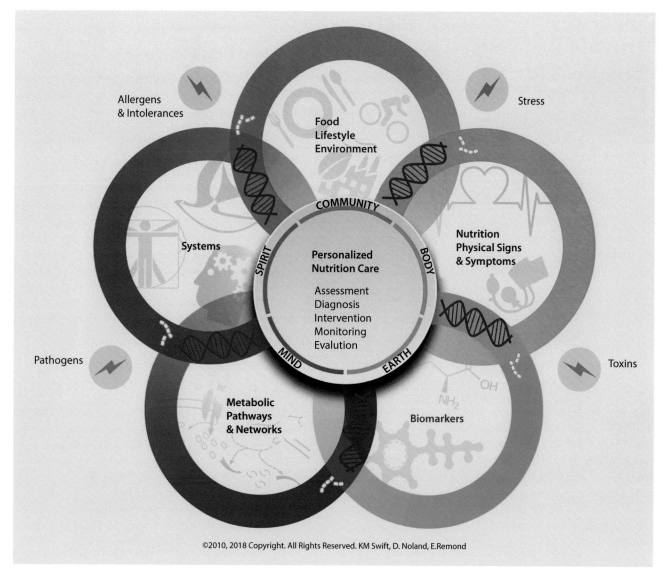

Fig. 37.3 Integrative and Functional Medical Nutrition Therapy (IFMNT) Radial. (Courtesy of Kathie Madonna Swift, Diana Noland, Elizabeth Redmond)

might include a multi-system questionnaire or other clinical signs and symptoms as described by the patient. From this information, the IFMNT practitioner can glean a systems analysis. In other words, this key area helps to shed light on body systems that may require support, such as digestive, endocrine, immune, nervous, reproductive, or detoxification/biotransformation systems [7, 12].

37.7.3 Key Area: Biomarkers

Biomarkers are of considerable importance to any practitioner, including those practicing IFMNT. Biological measurements, such as anthropometrics, vital signs, and laboratory data, are used to help shed light on abnormalities reported by the patient or identified as part of the nutrition-focused physical exam. Conventional biomarkers, such as blood and urine tests, can be used to evaluate nutrient status and look for nutrient insufficiencies and deficiencies that may be impacting health, as does more advanced functional testing for macro- and micronutrient status. These and other laboratory measurements are critical to IFMNT practice because they seek information on how well metabolic pathways are functioning. Each of these pathways is driven by complex molecular cascades that are largely reliant on nutrient cofactors that can be measured either with conventional or functional testing. Other influencers of these metabolic pathways include things like accumulated toxins, heavy metal interference, digestion and absorption issues, dysbiosis, and single nucleotide polymorphisms that can be measured using specialty labs and functional diagnostic testing [7, 12].

37.7.3.1 Examples of Conventional Tests
- Waist-hip ratio
- BMI
- Blood tests (e.g. complete blood count (CBC), comprehensive metabolic panel (CMP), vitamin and mineral

levels, immunoglobulin markers, cholesterol panel, homocysteine, thyroid panel, iron panel, food allergy panels, MTHFR, high-sensitivity C-Reactive Protein (CRP), hemoglobin A1C, sedimentation rate)
- Urinalysis (e.g. iodine levels, selenium levels, bacteria count, white blood cells, red blood cells, glucose, ketones)

37.7.3.2 Examples of Functional Tests
- Organic acids testing
- Genomics (e.g., MTHFR, VDR, GST, COMP, 23andMe, lipoprotein(a), detoxification panel, apoE)
- Digestive and microbial stool analyses
- Provoked urinary heavy metal testing
- Food sensitivity testing
- Salivary cortisol testing
- Salivary hormone testing
- Dried urine hormone testing

37.7.4 Key Area: Metabolic Pathways & Networks

Metabolic pathways are the biological machinery of the body, and overall health depends solely on how well these pathways run. The metabolic pathways of the body are driven by enzymes, the biological catalysts that trigger the conversion of one molecule to another. These enzymes are dependent on micronutrient cofactors, which can be measured and assessed by examining biomarkers. Even a subclinical micronutrient insufficiency can have a disruptive effect on metabolic pathways because the body prioritizes available micronutrients for the most important biological functions. Micronutrient insufficiencies may be insignificant in the short term but over time can contribute to the development of chronic conditions, poor antioxidant status, increased oxidative stress, premature aging, and a compromised ability to heal from injury or illness. Understanding nutritional biochemistry allows the IFMNT practitioner to better understand and recognize micronutrient imbalances, suboptimal metabolic processes, and the potential effects on the present state of health [7, 12].

This area focuses attention on assessing the integrity of the metabolic pathways to identify any core imbalances that may be present. By reviewing biochemical markers and pathway intermediaries, the IFMNT practitioner gathers information on metabolic pathway functionalities, which can then be interpreted within the context of the patient's current health status [12].

37.7.5 Key Area: Systems

The systems area serves to summarize and synthesize the findings from the other key areas and clarifies their overall effect on health status. By using the Radial, the IFMNT practitioner can identify the various causes that contribute to a specific malady, as well as identifying core imbalances that may be the root cause of several different symptoms. In this way, the "systems" area clarifies the opportunities for intervention and allows the practitioner to prioritize each inbalance and intervention.

Major system imbalances may include disruption in cellular integrity, digestion, detoxification, energy metabolism, oxidative stress and inflammation, nutritional status, immune system, endocrine systems, cardiovascular, and neurological systems [7].

37.7.6 Precipitating Factors

Linking each circle of the Radial are microbial and DNA strands with examples of precipitating factors that influence overall health, resilience, and predisposition to illness. In this way, all areas of the Radial are interconnected and influenced by each person's unique biochemistry and genetics. Each person's individual genome, epigenome and microbiome influences their health and biology in unique and complicated ways. Some of these factors include stress, toxic exposure, infection and pathogens, and allergies and intolerances [14].

Once the IFMNT practitioner has made a comprehensive assessment, including all relevant aspects of the Radial, he or she can develop a treatment plan that prioritizes the areas where the most important system imbalances lie. In terms of the Radial, this will be the circles with the greatest number of dysfunctions, specifically those that are triggers of the patient's primary symptoms and health concerns [12].

37.7.7 IFMNT Radial Summary

The IFMNT Radial provides a framework for the IFMNT practitioner to be thorough and organized in prioritizing information. It also helps the practitioner identify where further investigation is warranted. Organizing findings in this way can help the practitioner examine clinical findings into key imbalances, further clarifying underlying causes of disease and symptoms from a systems biology approach. In this way, the Radial is "a model for critical thinking that embraces both the science and the art of personalized nutrition care with considerations of multiple conventional or complementary care disciplines" [7].

37.8 Summary

An increase in demand for a more natural, holistic approach to healing, as well as the growing body of evidence supporting the safety and efficacy of Integrative and Functional Medicine practices, has driven the expansion of Integrative and Functional Medicine practitioners. Because nutrition and diet modification are the foundation of Integrative and Functional Medicine interventions, the growth in IFMNT-trained practitioners and those seeking training has also grown exponentially. Many healthcare providers have embraced IFMNT practices across disciplines, bringing nutrition and MNT as a modality for healing to the forefront of Integrative and Functional Medicine.

References

1. Swift KM. The changing landscape of nutrition and dietetics: a specialty group for integrative and functional medicine. Integr Med. 2012;11(2):19–20.
2. Wagner LE, Evans RG, Noland D, Barkley R, Sullivan DK, Drisko J. The next generation of dietitians: implementing dietetics education and practice in integrative medicine. J Am Coll Nutr. 2015;34(5):430–5.
3. The Institute for Functional Medicine. Functional Medicine determines how and why illness occurs and restores health by addressing the root causes of disease for each individual. https://www.functionalmedicine.org/Patients/WhatisFM/. Accessed 28 March 2017.
4. Ames BN. Prevention of mutation, cancer, and other age-associated diseases by optimizing micronutrient intake. J Nucleic Acids. 2010;2010:725071. https://doi.org/10.4061/2010/725071.
5. Centers for Disease Control. CDC's Second Nutrition Report: a comprehensive biochemical assessment of the nutrition status of the U.S. population. https://www.cdc.gov/nutritionreport/pdf/4page_%202nd%20nutrition%20report_508_032912.pdf. Accessed 28 March 2017.
6. Heaney RP. Long-latency deficiency disease: insights from calcium and vitamin D. Am J Clin Nutr. 2003;78(5):912–9.
7. Ford D, Raj S, Batheja RK, et al. American Dietetic Association: standards of practice and standards of professional performance for registered dietitians (competent, proficient, and expert) in integrative and functional medicine. J Am Diet Assoc. 2011;111(6):902–913.e1–23. https://doi.org/10.1016/j.jada.2011.04.017.
8. Dietitians in integrative and functional medicine: annual report 2015–2016. http://integrativerd.org/wp-content/uploads/2012/04/DIFM_DPG_Annual_Report_2015-2016_FINALIZED.pdf. Accessed 26 March 2017.
9. eatrightPRO. Nutrition care process. 2017. http://www.eatrightpro.org/resources/practice/nutrition-care-process. Accessed 20 March 2017.
10. eatrightPRO. Standards of practice. 2017. http://www.eatrightpro.org/resources/practice/quality-management/standards-of-practice. Accessed 20 March 2017.
11. Focus area standards for CDR specialist credentials. J Acad Nutr Diet. 2017. http://www.andjrnl.org/content/credentialed. Accessed 20 March 2017.
12. Dean S. Introduction to Integrative and Functional Nutrition, presented at CDA Annual Conference, 2015. http://www.dietitian.org/d_cda/docs/annual_mtg_2015/conference_handouts/thu4-9-15/Dean_IntroductionIFN.pdf. Assessed 5 Feb 17.
13. Swan WI, Vivanti A, Hakel-Smith NA. Nutrition care process and model update: toward realizing people-centered care and outcomes management. J Acad Nut Diet. 2017;117(12):2003–14.
14. Diana Noland, Sudha Raj, (2019) Academy of Nutrition and Dietetics: Revised 2019 Standards of Practice and Standards of Professional Performance for Registered Dietitian Nutritionists (Competent, Proficient, and Expert) in Nutrition in Integrative and Functional Medicine. Journal of the Academy of Nutrition and Dietetics 119 (6):1019-1036.e47
15. Bruce N. Ames, (2018) Prolonging healthy aging: Longevity vitamins and proteins. Proceedings of the National Academy of Sciences 115 (43):10836-10844

The Patient Story and Relationship-Centered Care

Leigh Wagner

38.1 Patients, Providers, the Therapeutic Approach, and Relationship-Centered Care [1] – 635
38.1.1 "PEECE" in Practice – 635
38.1.2 The Patient's Story and the Institute for Functional Medicine Matrix – 635
38.1.3 Patient Example – 636

References – 636

© Springer Nature Switzerland AG 2020
D. Noland et al. (eds.), *Integrative and Functional Medical Nutrition Therapy*,
https://doi.org/10.1007/978-3-030-30730-1_38

A defining feature of integrative and functional medicine is the careful collection of the patient's story. The patient's story builds on the conventional (acute) medical model of gathering a patient's past medical history, history of present illness, review of systems, and accounts for the complexity of chronic disease. This information can illustrate to the clinician and the patient how the patient's life events and health story have contributed to their health status at the time of the consultation.

The importance of the patient story is emphasized by the idea that integrative and functional medicine practitioners treat each patient from the perspective of the "N of 1," meaning that the practitioner views the patient as an individual and takes into account the nuances of the patient's metabolism, lifestyle, genetics, stressors, and environment. Thus, each patient is considered his or her own control when evaluating the effectiveness of a therapy or treatment.

Gathering a useful patient story requires a clinician to ask careful questions that the patient might find culturally and personally sensitive [1]. The way a practitioner performs the patient interview can determine whether patients feel comfortable enough to share their stories and follow the recommended plan of care. The practitioner's interviewing skill can be honed over time. It is also important to keep in mind that the story will continue to develop as the patient works with the practitioner.

There are several goals of gathering a comprehensive patient story: (1) identifying antecedents, triggers, and mediators that impact the patient's health; (2) establishing a therapeutic relationship between practitioner and patient by building trust and rapport with the patient; (3) having a starting point and reference for the provider and the patient to build on during future appointments; (4) identifying any barriers or challenges that the patient might have (e.g., poor community or family support, physical, psychological, or emotional challenges); (5) creating a sense of hopefulness in the patient and the provider that there are therapeutic options for a patient who may have lost hope in any other medical options; and (6) informing the patient of the intervention plan.

The Institute for Functional Medicine (IFM) Matrix illustrates the complex and comprehensive information gathered from the patient (Fig. 38.1). The IFM Matrix is a guide for learning about the patient's health story from the standpoint of mechanism history. This means that the practitioner asks about the patient's health from the perspective of chronic

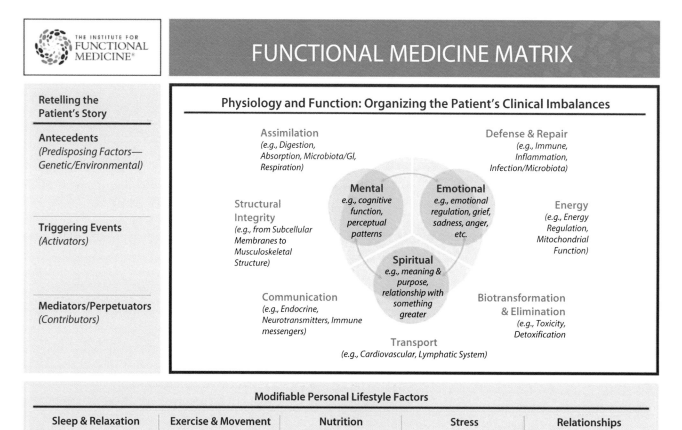

Fig. 38.1 Functional Medicine Matrix. (Used with permission from The Institute for Functional Medicine ©2015)

disease mechanisms and core imbalances. In other words, the practitioner can fill in the components of the IFM matrix as the patient tells his or her story.

The patient story is both a narrative and a means to note patterns that can provide clues to metabolic imbalances. It also informs the patient's intervention plan.

38.1 Patients, Providers, the Therapeutic Approach, and Relationship-Centered Care [1]

There are distinct roles that providers and patients take with respect to the therapeutic approach to healing. Patients are expected to take an active role in their own healing. One of the most important aspects of the provider and patient is the therapeutic relationship between the two parties. Relationship-centered care lowers overall healthcare costs [2, 3].

Providers see themselves as a navigator for the patient, who is the pilot of their own healthcare. Providers also take into account the beliefs and cultural nuances of each patient and understand the healing journey is unique to each individual [1]. Providers help patients overcome barriers to wellness and healing. In addition to educating, mentoring, and coaching patients, providers typically live their own therapeutic, healthy lifestyle; they do this for various reasons: [1] to empathize with the changes the patient is being asked to make, [2] to live a healthy lifestyle to be able to best care for themselves and their patients, and [3] to best guide patients on their healing journey.

The therapeutic approach to healing involves a holistic mind-body-spirit view. The mind accounts for the patient's mental health and knowledge base, especially regarding the understanding of his or her own lifestyle choices and their impact on their health outcomes. The body is probably where conventional healthcare spends most of its focus, checking anthropometrics (height, weight, body mass index), blood pressure, cholesterol levels, and other measurable changes. Integrative and functional practitioners also incorporate mind-body and spiritual aspects into the care plan. Spirituality can encompass religion, spiritual practices, mindfulness and/or meditation, and even the patient's community support. Each one of these aspects of the mind, body, and spirit is acknowledged as contributing to the patient's well-being, which the provider often conveys to the patient.

The therapeutic approach to healing also includes evidence-based practice that encompasses three facets [4]:
- The patient's goals and vision for health.
- Evidence in the scientific literature.
- The practitioner's clinical experience.

Combined, these three aspects of evidence-based healthcare can be weighed through gathering the patient's story and through the therapeutic relationship. Evidence is also collected from a variety of sources.

Ultimately, compassion and support are critical to the healing encounter between provider and patient. Whether or not a cure is possible for a patient's condition, the provider can always give encouragement and help guide the patient toward healing and wellness, if the patient sees that as an important goal.

38.1.1 "PEECE" in Practice

Rakel and Fortney describe the concept of "PEECE" in clinical practice in their chapter on "The Healing Encounter" in the textbook "Integrative Medicine" [1]. The acronym stands for Positive prognosis, Empathy, Empowerment, Connection, and Education. Positive prognosis refers to the concept of hope and enhancing positive patient expectations. Empathy and connection enhance the feeling of being cared for by the practitioner. Empowerment and education give the patient motivation to seek health-promoting behaviors.

38.1.2 The Patient's Story and the Institute for Functional Medicine Matrix

One method of organizing and seeing the influence of the patient's story on his or her own health is by using the Institute for Functional Medicine's (IFM) Matrix [5] (◘ Fig. 38.1). The IFM Matrix is a tool that can help the clinician record and visualize the impact of the patient's story on his or her health. On the left perimeter of the matrix, the clinician is prompted to ask the patient to retell his or her life and health story. The clinician can record this in a timeline format or can make notes in each of these categories:
- Antecedents (predisposing factors – genetic/environmental).
- Triggering events (activators).
- Mediators/perpetuators (contributors).

In an IFM presentation, Dr. Dan Lukaczer, ND, [6] describes antecedents as "factors, genetic or acquired, that predispose [an] individual to an illness." He defines triggers as "factors that provoke the symptoms and signs of illness" and mediators/perpetuators as "factors, biochemical or psychosocial, that contribute to pathological changes and dysfunctional responses." Examples of antecedents, triggers, and mediators include stress, infections (chronic or acute), toxic exposures, foods, and allergens, sensitivities, and/or intolerances.

The clinician can use the central part of the matrix to record, organize, and visualize the patient's clinical imbalances: assimilation, structural integrity, communication, transport, biotransformation and elimination, energy, defense, and repair. In the middle of the matrix, the practitioner is reminded of the importance of the patient's mental, emotional, and spiritual health. Along the bottom border of the matrix, the clinician is reminded of the modifiable lifestyle factors that should be reviewed with the patient to find out where excesses or deficiencies might appear. These lifestyle factors include sleep, relaxation, exercise, movement, nutrition, stress, and relationships. While listening to the patient's story, the clinician can fill in the information on the matrix to create a visual to work from and create a personalized plan of care for the patient.

38.1.3 Patient Example

A 30-year-old woman visits with persistent hives. She has seen her primary care physician, a dermatologist, and an allergist and has tried several creams and medications. She has some relief of her symptoms, but she still has hives. She sought out a consultation from an integrative and functional dietitian, who asked her about the onset of her symptoms (prior to and during her two pregnancies) and her triggers (stress and pressure on her skin). She said the hives would persist for weeks, and several would develop while others were healing. After getting the patient's medical background, she shared that she hates and avoids all fish and seafood (antecedent/mediator) and that her son has skin rashes when he drinks cow's milk. She also says that the symptoms persisted after the birth of her second child. A mechanistic understanding of the essential fatty acid/eicosanoid pathway helps illustrate how the patient's story may give clues for laboratory testing for the patient and a means to intervene. The dietitian recommended nutritional lab testing: high-sensitivity C-reactive protein (hsCRP), vitamins D, B12, folate, and an omega-3 index. The labs showed that she was, in fact, deficient in omega-3 fats and vitamin D was deficient at 17 ng/dl. The patient's story helped inform what was driving the hives to persist; it led the practitioner to consider the mechanisms of inflammation and immune dysregulation, which can be affected by the eicosanoid cascade (omega fatty acids) and vitamin D's effect on the immune system. The family history of hives in her son following milk ingestion advises the practitioner to also check food sensitivities in the patient. After successful reintroduction of deficient micronutrients and food elimination, the hives resolved.

References

1. Rakel D, editor. Integrative medicine. 3rd ed. Philadelphia: Elsevier; 2012.
2. De Maeseneer JM, De Prins L, Gosset C, Heyerick J. Provider continuity in family medicine: does it make a difference for total health care costs? Ann Fam Med. 2003;1(3):144–8.
3. Safran DG, Miller W, Beckman H. Organizational dimensions of relationship-centered care. J Gen Intern Med. 2006;21 Suppl 1:S9–15.
4. Spring B. Evidence-based practice in clinical psychology: what it is, why it matters; what you need to know. J Clin Psychol. 2007;63(7): 611–31.
5. Jones DS, editor. Textbook of functional medicine. Gig Harbor: The Institute for Functional Medicine; 2010.
6. Lukaczer D, editor. Clinical integration and the Functional Medicine Matrix part 1. Federal Way, WA: Institute for Functional Medicine; 2015.

The Nutrition-Focused Physical Exam

Mary R. Fry

39.1 Background/Introduction – 638

39.2 Conclusion – 691

References – 692

Learning Objectives

- Recognize the importance of the nutrition physical exam for both nutrition professionals (dietitians and nutritionists) and medical providers.
- Identify signs of nutritional status and hydration from physical examination.
- Relate pertinent physical examination findings to their nutritional (and medical) causes.
- Compare and contrast anthropometric measures and their clinical utility.

39.1 Background/Introduction

The nutrition-focused physical assessment (NFPA), also referred to as the nutrition-focused physical exam (NFPE), is a vital part of a nutritional assessment and forms an integral part of the nutrition care process (NCP) model, which is comprised of the following separate and consecutive steps: assessment, diagnosis, intervention, monitoring, and evaluation [1]. The nutrition progress note for the NCP model is correspondingly referred to as an "ADIME" note. This model was devised, in part, to ensure a comprehensive nutrition assessment. A nutrition-focused physical exam forms an integral part of the NCP model assessment step, which itself is comprised of five domains:
- Food/nutrition-related history
- Medical/psychological or behavioral history
- Anthropometric measurements
- Biochemical data/tests/procedures
- Nutrition-related physical findings

Findings from each of these five domains are required to systemically analyze and integrate the findings from each portion of the NCP model in order to accurately determine nutritional status [1–3].

Furthermore, each of the steps listed above (and their domains) informs the subsequent step and actions to be taken by the nutrition care professional. For example, dietary and medical history should focus on the physical exam and subsequent laboratory assessments, which can then result in a clear and accurate working nutritional diagnosis (or diagnoses). This diagnosis then informs and guides clinical intervention and the monitoring/evaluation thereof [2–5].

Despite the critical role that the nutrition-focused physical exam plays in a comprehensive nutritional assessment, it did not become a part of a dietitian's (or nutritionist's) practice or education until the 1990s, a full decade after the concept for integrating it into the nutritional assessment was proposed [6–9]. Despite nearly three decades passing since its integration into practice, a vast majority of dietitians still do not readily incorporate it into a routine patient assessment [5].

A prospective Internet-based survey of registered dietitians in the USA reported the following as factors limiting the incorporation of the NFPE into practice: workload (70.3%), lack of prior education (42.4%), and training (43.5%) [9]. Regarding workload or time, 52.3% of US dietitians surveyed by Mackle et al. [10] cited time as a limiting factor to using the NFPE, and 25.3% cited confidence in their ability to competently perform the NFPE.

The majority of those surveyed reported that they did not receive sufficient training to competently perform the NFPE on their patients, with the exception of the following assessments: height (96.4% of those surveyed reported feeling competent to measure), weight (97.4% of those surveyed reported feeling competent to measure), waist circumference (78.0% of those surveyed reported feeling competent to measure), and skinfold measurements with a caliper (65.2% of those surveyed reported feeling competent to measure) [9].

To date, similar analyses have not been conducted for nutritionists in practice, but similar practice patterns and competencies in the NFPE are likely, based on a comparison of the average curriculum across these professions. As it happens, the practice of integrative nutrition may be even more challenged in the realm of the NFPE, with a wider variety of functional tests to learn about and gain competency in, and with fewer opportunities for hospital-based internships (where a variety of NFPE skills are generally more likely to be practiced).

On the medical front, physicians' use of the physical exam is similarly threatened or arguably underutilized. In the "Physician's Perspective," published in *The New England Journal of Medicine* [11], Dr. Jauhar attributes the demise of the physical exam in medicine to a discomfort with uncertainty and a fear of subjective observation. Technological methods of diagnosis are increasingly seen as more certain and legally defensible. Rathe's commentary on the "Complete Physical" [12] concurs, stating that failure to diagnose (accurately) commonly results in medical malpractice suits.

However, proponents of the physical exam emphasize its importance in supporting the development of trust, empathy, and relationship-building. Verghese and Horwitz [13] contend that physicians skilled in physical exams order fewer unnecessary tests and make better use of the information gained from these tests than counterparts who eschew or limited their use of the physical exam.

As to a physician's knowledge of nutrition (since the emphasis of this chapter is on the nutrition-focused physical exam), a recent review of the status of nutrition education in US medical schools reports that "the amount of nutrition education medical students receive continues to be inadequate" [14]. The survey conducted in this study questioned nutrition educators at medical schools directly (109 surveyed, with 105 completing the survey) and found that only 27% of the schools surveyed provided the minimum 25 hours of nutrition instruction recommended by the National Academy of Sciences. Moreover, medical residents, recent graduates, and practicing physicians surveyed in other studies reported feeling ill-prepared to provide nutritional counsel and care [14].

The NFPE is a full head-to-toe exam of a patient in which their appearance and function are assessed in order to detect any signs of malnutrition, nutrient deficiencies, or nutrient toxicities [5]. In the realm of integrative and functional nutrition, this can be expanded to include signs of suboptimal organ system function and imbalances (which will be covered extensively in the tables associated with this chapter).

The goal of a NFPE is to identify factors that may impede normal dietary intake and habits that may impact nutritional status and/or that may reflect conditions that are nutritional in nature [8] or to examine the fat, muscle, fluid, and micronutrient status of a patient and whether or not it is outside of normal levels as a result of inflammation, illness, and/or excess or deficient nutrient intake [15]. Dietitians and nutritionists are not qualified/permitted to diagnose medical conditions; however, they are qualified to make a nutritional diagnosis, which is defined as a condition that can be resolved or improved by nutrition intervention [16]. The NFPE is designed to assess and diagnose nutritional conditions and is to be used to inform other domains of the nutritional assessment as well as to guide and inform that monitoring and evaluation of a patient's case, enabling the nutritionist to gauge the patient's response to the intervention [1, 4]. The NFPE can bring medicine, nutritional sciences, food science, and technology together to address suspected nutrition-related health conditions and imbalances [4].

The components of the physical exam were loosely defined originally and ranged from body composition measurements (height, weight, percentage body fat, assessment of muscle tissue) to vital signs, visual assessment for clinical signs of nutrient deficiencies, and oral, cardiac, pulmonary, abdominal, and dermatological examination [8]. A more current working definition of the NFPE for dietitians derived from Touger-Decker [8] is the measurement of vital signs and body composition (height, weight, and bioelectrical impedance analysis, among others); inspection for clinical signs of nutrient deficiencies; auscultation of heart, lung, and abdominal signs; and examination of the head, neck, oral cavity, skin, and cranial nerves. Touger-Decker goes on to differentiate NFPE skills into those expected of an entry-level practitioner: vital signs (pulse, blood pressure); head, neck, and oral examination, including extraoral (lips, eyes, nose, ears) and intraoral (hard and soft tissue, tongue, teeth, and saliva); recognition of the components of lymph node and cranial nerve examination; dermatologic examination for both nutrient deficiencies and edema; body composition analysis (bioelectrical impedance analysis and calipers); cardiac and pulmonary auscultation; and an abdominal exam. For a more advanced practitioner, competence in performing all of the aforementioned techniques is expected, along with the ability to *conduct* lymph and cranial nerve exams versus just recognizing their components. Esper [1] describes the NFPE as a system-based examination, which is similar to the physical exam performed by a licensed healthcare provider. It is outlined: general inspection, vitals, skin, nails, head/hair, eyes/nose, mouth, neck/chest, abdomen, and musculoskeletal. Examination techniques employed are similar to those utilized in a physical exam performed by a licensed healthcare provider and include inspection, palpitation, and auscultation (note that percussion is seldom performed by dietitians/nutritionists) [17].

Medical nutrition therapy (MNT) involves the assessment of a patient with one or more medical conditions which can place an individual at nutritional risk. MNT has been shown to decrease the duration of hospital stays and decrease the need for/use of more costly medical interventions and can, in some cases, obviate the need for hospitalization [2]. Integrative and functional medical nutrition therapy (IFMNT) takes dietetics and medicine to another level with the focus on optimizing nutritional status to optimize health and, in turn, reduce the risk of nutrition-related disease [18].

The IFMNT model is an extension of the NCP model. The IFMNT approach is encapsulated in the following radial, with the NCP's ADIME at its core and an array of factors to consider radiating out from this. In this chapter, a conventional dietetics approach to the NFPE will be covered as part of a more extensive and integrated IFMNT approach. The NFPE of an integrative nutritionist is more comprehensive and sensitive (to detect early and subtle changes that reflect suboptimal nutrient status and health) than that classically taught to dietitians. Note that in the IFMNT radial below, at least three of the five factors (biomarkers, systems, signs and symptoms) influencing nutritional status rely on NFPE findings [19] (◘ Fig. 39.1).

To thoroughly perform this exam, each body region is covered below, along with possible pertinent abnormal findings, for reference. Note that the scope and techniques described below extend beyond a typical NFPE performed by a dietitian as functional signs and techniques are included (as are, in some cases, more medically oriented techniques and findings). These additional examination tools and techniques can aid in detecting early signs of nutrient deficiency or toxicity and functional imbalances that could be caused by or contribute to changes in nutritional status.

It is my hope that this chapter revivifies the physical exam for dietitians, nutritionists, and other healthcare providers and, in so doing, fosters more comprehensive and qualitative nutritional and nutrition-oriented medical care and enhances the patient-provider relationship (◘ Table 39.1).

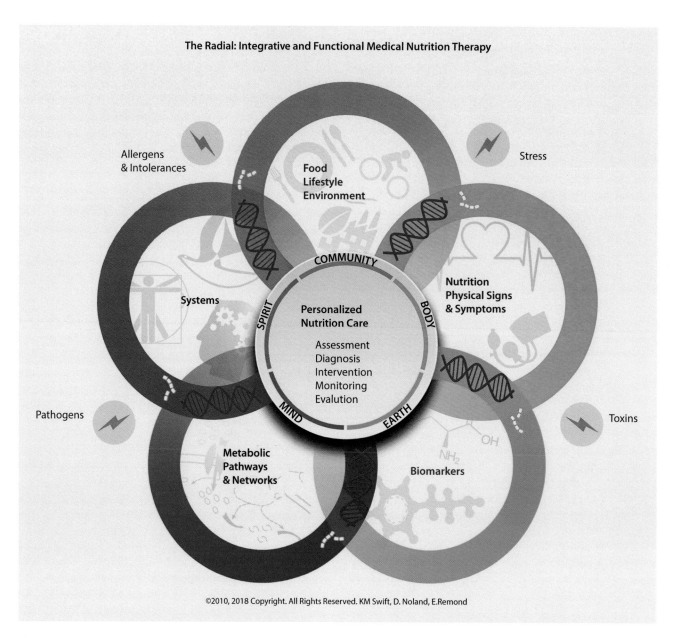

■ Fig. 39.1 Integrative and Functional Medical Nutrition Therapy (IFMNT) Radial. These factors influence one's function (physiologic) as well as one's nutritional status. Further, they provide a common language and bridge between nutritionists/dietitians and medical providers (particularly those practicing functional medicine), thus facilitating collaboration and coordination of care. (Courtesy of Kathie Madonna Swift, Diana Noland, Elizabeth Redmond)

Table 39.1 Nutrition-focused physical exam reference table

Region examined	Nutrition-associated examination findings (and select examination/test techniques)		Nutritionally related cause/association
General (also referred to as "Global Examination") [2]	*This section of the physical exam is devoted to a general survey of the patient for alertness, affect, speech, coordination body size (overweight/obese or wasting), and habitus (i.e., build – endo-, meso-, or ectomorphic) and body movements. This is where your first impressions of the patient/client are formed. You are looking to assess their overall state of health, are they currently ill or frail, or fit or robust? Are they showing signs of mental or physical distress? What is their level of consciousness and alertness? How are they groomed? Do you detect any unusual odor? [16] This global exam will drive what you focus on for the remainder of your exam, so pay careful attention to these clinical findings.*		
	Grooming [20]	Poorly groomed (unkempt, unclean clothing, and personal hygiene [grooming/care of the hair, skin, teeth, mouth, overall body])	Dementia Psychiatric disorders (psychosis, schizophrenia in particular) Homelessness or poor living conditions Essential fatty acid (EFA) imbalance Vitamin B_{12} status insufficiency
	Clothed inappropriately for the season [20]	Cold intolerance	Anemia (iron-deficient) Low thyroid function
	Shoes untied or slippers [20]	Pedal edema	See "Edema" section below
	Toes of shoes cut out [20]		Gout or foot ulcers Neuropathy with burning feet "Burning feet syndrome" – thiamine deficiency (accompanied by peripheral neuropathy)
	Body piercings or tattoos [21]		May be at increased risk of hepatitis C
	Odor	Body	Poor hygiene (see "Grooming" above) Disturbances in skin microflora balance [22] Diet (spices (cumin), garlic, and onions) Stress [23] Steroid hormone imbalances SNPA ABC11 538G/A or G/G polymorphism associated with strong axillary odor with perspiration and staining of clothes (axillary osmidrosis) [24] Trimethylaminuria (fish malodor syndrome) is a genetic metabolic disorder that results in an odor of decaying fish [25] Liver disease (ammonia) Kidney disease (uremia) Chronic genital infections
		Breath [21, 26]	Alcohol (evaluate for macronutrient status – energy, protein, and micronutrient status: iron, thiamine, riboflavin, pyridoxine, niacin, folate) Fruity/acetone odor – diabetes – due to ketosis [27] Garlic odor with selenium toxicity Putrid – may have bacterial overgrowth in GI Uremia with renal disease
	Pain	Wincing or painful grimace, increased perspiration, protecting the affected area [21]	Examine/evaluate the painful region or symptoms further for the etiology or origins thereof
	Alertness	Evaluate by asking/observing the patient's awareness of their environment and their attention span	Allergies or suboptimal liver function/toxicity Fatigue [28] Medication [28] Narcolepsy [28] Attention deficit disorder/attention deficit hyperactivity disorder [28] Dementia [28] Schizophrenia [28]

(continued)

◻ Table 39.1 (continued)

Region examined	Nutrition-associated examination findings (and select examination/test techniques)		Nutritionally related cause/association
	Orientation	Oriented to person, time, and place?	Essential fatty acid (EFA) imbalance Vitamin B_{12} insufficiency Phosphorus and niacin status for insufficiency
	Affect	Flat	Iron and vitamin B_{12} insufficiency Dysthymia/depression Hypothyroidism/suboptimal thyroid function Schizophrenia
	Speech	Dentures click when talking	May be related to weight loss
		Pressured and/or rapid speech	Hypomanic/manic (bipolar disorder) Rule out hyperthyroidism or thyroid storm as cause
	Skin color	*To evaluate pigmentation, check the conjunctiva, nail beds, palms (loss of crease in severe anemia), lips, and tongue*	
		Hyperpigmentation	Niacin deficiency (with peeling in sun-exposed areas), folate, and/or B_{12} deficiency [29], protein energy malnutrition (when on extremities) [30] Addison's disease
		Hypopigmentation In dark-skinned individuals, examination of the palms and soles may be helpful	Vitiligo, pallor, and anemia (iron, folate, or B12 deficiency [29], arterial insufficiency) B_{12} deficiency (associated with vitiligo) Copper deficiency (hypopigmentation on the neck, axilla, and/or trunk in Menkes disease) [30]
		Pallor	Anemia
		Erythema (redness) [16, 31]	Facial redness, suspect: fever, menopause, rosacea, hyperthyroidism, niacin supplementation, caffeine, spicy food intake, or possibly inflammation Redness in other locations of the body may result from inflammation, injury, rash or infection, or allergy (red ears) Riboflavin, zinc, and EFA deficiency [30]
		Yellow (jaundice) To see jaundice in the lips, blanch with a glass slide or small flat glass implement It is typically first noted in the eyes and then seen in the face, mucous membranes, and eventually all of the skin	Results from excess amounts of bilirubin in circulation as a result of increased hemolysis, liver disease, or obstruction of the flow of bile to the intestine [16] If it spares the sclera (white part of the eyes), it is carotenemia and not jaundice It can occur in pernicious anemia (vitamin B_{12} deficiency) [16, 29, 31]
		Orange	Carotenemia – look at the palms, soles, and face. Results from excess consumption of carotenoids
		Brown [21] Melasma (also known as chloasma or the "mask of pregnancy") is characterized by dark, discolored patches on the skin (mainly facially) and is generally seen in women of reproductive age) Acanthosis nigricans – velvety brown-to-black markings in areas of body creases (skin folds, armpits, over joints in the fingers and toes) [16]	Sun exposure Melasma can occur in pregnancy (and with sun exposure, use of oral contraceptives) Acanthosis nigricans can occur with obesity and metabolic syndrome, dyslipidemia, hypertension, Addison's disease, pituitary disorders, hypothyroidism, polycystic ovarian syndrome (PCOS), and administration of growth hormones, oral contraceptives [32], with excess niacin supplementation [30]. Its presence in children heralds the development of metabolic syndrome. It can also be seen in those with lymphoma and cancers of the genitourinary or gastrointestinal tract or from medications and nutrient supplementation (nicotinic acid) [16]. Excess niacin intake (may be accompanied by blistering) Hemochromatosis (iron storage disorder)

The Nutrition-Focused Physical Exam

Table 39.1 (continued)

Region examined	Nutrition-associated examination findings (and select examination/test techniques)		Nutritionally related cause/association
		Gray	Iron toxicity [30]
		Cyanosis [16, 31]	Decreased oxygenation of the blood (note that testing for oxygen saturation is generally considered a more accurate of blood oxygenation) Central cyanosis: cyanotic lips, buccal mucosa, tongue – see in COPD, heart disease, abnormal hemoglobins Peripheral cyanosis: cyanotic nail beds can occur with exposure to cold, anxiety, CHF, Raynaud's disease, and venous obstruction Hemoglobinopathy
		Purpura (bruising)	Vitamin C/K insufficiency If bruising readily with trauma, may be due to essential fatty acid (EFA) insufficiency [30] Easy bruising – can be due to platelet disorders (aplastic anemia), liver failure, hypersplenism (splenomegaly), autoimmune disorders, alcoholism, and hematologic malignancies
		Use of makeup or cosmetics	May affect accuracy of skin assessment (hide skin tone, lesions, acne) Tip: inquire about brand of cosmetics (ensure lowest toxin cosmetics are used) [33]
	Edema (peripheral) [30, 34]	Positive pitting test: Depressing skin over suspected edematous area and then removing pressure yields a pit that remains for over 5 seconds	Protein, energy, zinc, and thiamine status for insufficiency Low serum albumin [3]
	Weight loss, wasting, or cachexia [21, 26] (Sarcopenia is referenced later in this table)	Check temporal lobes and cheeks – well-demarcated bony prominences and veins [3, 29]	Protein, fat, and energy insufficiency Malnutrition (a risk factor for infection and poor wound healing) Cancer Dysphagia Vomiting/diarrhea Malabsorption HIV/AIDS (lipodystrophy)
	Overweight/ obesity [20, 21]	BMI Waist-hip ratio (WHR) Waist-to-height ratio (WHtR) Skin caliper testing Bioelectric impedance analysis (See "Aanthropometrics" table following this table)	Excess caloric intake/insufficient physical activity/endocrinological imbalance (e.g., low thyroid function)/genetics Assess for toxicity (particularly dioxins and heavy metals) Tip: educate the patient about risks of excess weight, diabetes, heart disease, stroke, hypertension, osteoarthritis, sleep apnea, some types of cancer, inflammation, and dysbiosis
		Central obesity [34] "Apple shape" Adiposity is concentrated in the upper half of the body (neck, cheeks, shoulder, chest, upper abdomen) More common in men	Increased risk of metabolic syndrome (hypertension, insulin resistance, inflammation, cardiovascular disease, dyslipidemia, sleep apnea) with increased visceral adiposity

(continued)

Table 39.1 (continued)

Region examined	Nutrition-associated examination findings (and select examination/test techniques)		Nutritionally related cause/association
		Peripheral obesity [34] "Pear" shape Adiposity is concentrated in the lower half of the body (lower abdomen, pelvic girdle, buttocks, and thighs) More common in women	Carries health risks of obesity, but lower health risks than central obesity
	Uncoordinated gait (ataxia)		Thiamine or copper deficiency [1] Mitochondrial dysfunction [20] Essential fatty acid imbalance [20] Toxin burden [20]
	For additional gait disturbances, see the "Neurological" section later in this table		
	Amputation [26]		Possible diabetes complications Possible history of severe infection Possible history of traumatic injury/accident Amputations will affect calculations of weight and require calculation of an adjusted weight
	Body movements [4]	Can the individual feed independently (i.e., do they have limited use of hands or inability to sit upright)?	Lack of ability to independently feed must be addressed to ensure sufficient nutrient intake
	Posture [20]	Dowager's hump/stooped posture	Bone loss (osteomalacia or osteoporosis) Vitamin A and D, magnesium, or calcium insufficiency Tip: consider further testing (DEXA and N-telopeptides cross-links (NTx) to monitor response to interventions) [35] Depression Parkinson's disease [36]
		Preference to sit erect, neck veins distended	Left-sided heart failure [21]
		Preference to lean forward, arms crossed	Chronic obstructive pulmonary disease (COPD) [21]
		Fetal position	May be due to pancreatitis [34]
		Bend toward the side of the lesion	May be due to renal or perirenal abscesses [34]
		Still: avoid movement as it provokes pain	May be due to peritonitis [34]
		Restless	May be due to intestinal obstruction [34]
	Tanner stage	Evaluate if developmental age is consistent with chronological age	Assess for signs of macronutrient and micronutrient deficiencies that may have delayed growth and development Assess for developmental delays that may have affected food and nutrient intake [16]
Bates – mini	Breathing	Labored Rapid, shallow Wheezing	(See "Respiratory" section of table below)

Table 39.1 (continued)

Region examined	Nutrition-associated examination findings (and select examination/test techniques)		Nutritionally related cause/association
	Subjective global assessment (SGA) of nutritional status [16, 26, 34]: Tool to assess malnutrition based on patient's history and physical exam. It is also a powerful predictor of postoperative complications. Assesses recent loss of body weight, changes in the usual diet, the presence of significant GI symptoms and the patient's functional capacity, and three elements of the physical exam Assess the following using a scale of normal (0), mild (1+), moderate (2+), and severe (3+) **History assesses:** Recent loss of body weight Changes in the usual diet Presence of significant GI symptoms Functional capacity **Physical exam assesses:** **Subcutaneous fat loss** – assess the following sites for a lack of fullness (i.e., the skin fits loosely over the tissues): triceps, midaxillary line (costal margin), palmar areas of the hand, interosseous muscles (between the thumb and forefinger as an example), and deltoid region of the shoulder –Assess using a scale of normal (0), mild (1+), moderate (2+), and severe (3+) **Muscle wasting** – assess by palpation (and inspection at a distance, as helpful). Examine the quadriceps and deltoids. "Squared off" shoulders are visible with muscle wasting and subcutaneous fat loss **Loss of fluid from the intravascular space**- ankle or sacral edema, ascites. Assess for edema via palpation (pitting test) **Score/classes:** Class A (well-nourished) Class B (moderately malnourished) Class C (severely malnourished)		
Vital signs	*The utility of a vital sign is to give the clinician an early indication of ill health or imbalances. They are the signs of life* [16] *or "vitality." They serve to establish a baseline and are a helpful tool for monitoring and evaluation thereafter*		
	Temperature *Average normal temperature is 97.7 degrees Fahrenheit (or 36.5 degrees Celsius). (McGee)*	Elevated	Confers increased food and fluid requirements [3]. Can indicate infectious disease, trauma (surgery/injury), malignancies, immune disorders, blood disorders (hemolytic anemia), infarctions, and medications. Fever with chills (shaking) – accurate sign of bacteremia [31]
		Depressed	Hypothyroidism or Wilson's temperature syndrome, hypoglycemia, insomnia, severe stress/shock, exposure to cold, anemia, diabetes, Addison's disease, infection, liver/kidney failure, paralysis, excess alcohol intake, starvation, medications [21] Note that a depressed body temperature can lead to deleterious consequences metabolically (enzymes do not function optimally at lower body temperature) Wide variability of temperature can be seen in adrenal insufficiency [16, 31]
	Pulse	*Measure radial pulse for at least 15 seconds to obtain (30 seconds is preferred). A number of different qualities of the pulse are measured/observed in Ayurvedic and Chinese medicine diagnostics.* **Normal sinus rate = 50–95 beats per minute (bpm)**	
		Elevated (tachycardia) >100 bpm	Anemia, hyperthyroidism, with elevated body temperature, anxiety, pain, exercise, medications (antidepressants and others) [3, 31] Prognostic: can portend increased complications and poorer survival rate across many clinical disorders Magnesium, calcium, and/or niacin insufficiency may be a cause of tachycardia
		Depressed (bradycardia) <50 bpm	Exercise/training (i.e., improved conditioning), heart disease, hypothyroidism, medications [3]
	Irregular heart rate/rhythm may be due to potassium deficiency (can be seen with diuretics), magnesium deficiency, dehydration, or allergies		

(continued)

Table 39.1 (continued)

Region examined	Nutrition-associated examination findings (and select examination/test techniques)		Nutritionally related cause/association
	Respiratory rate	Measure of the number of breaths per minute Respiratory rate is the only vital sign under voluntary control (why measuring it without telling the patient is advised (so take before/after take patient/client's pulse) Take for 30–60 seconds to ensure accuracy Normal respiratory rate = 20 breaths/min (ranges from 16 to 25 breaths/min) [31]	
		Increased (tachypnea) >25 breaths/min	Dyspnea (difficult or labored breathing) with patient lying down that is relieved while sitting up – can be due to ascites, pleural effusion, morbid obesity, congestive heart failure (CHF), or severe pneumonia Rapid deep breathing (hyperventilation) – can occur with exercise, anxiety, metabolic acidosis, and brain stem injury [21] Can affect calorie and protein requirements as well as type of food eaten and quantity [3] – as eating may increase energy expenditure, fatigue, or dyspnea [16]
		Decreased (bradypnea) <8 breaths/min	Diabetic coma, medication-induced or due to increased intracranial pressure
		Kussmaul respirations – rapid and deep –seen in patients with metabolic acidosis (most other abnormal respiratory rates see shallow respirations). The rate can be fast, slow, or normal [21] Cheyne-Stokes breathing/periodic breathing – alternating periods of hyperpnea and apnea. Occurs when there is enhanced sensitivity to carbon dioxide. Seen in stable congestive heart failure, neurological disorders, hemorrhage, infarction, tumors, meningitis, and some head trauma (brain damage), from medications, and at high altitudes or during sleep (considered normal) With fluctuations of respiration come fluctuations in mental status Prognostic: in a patient with heart disease, this respiratory rate is associated with a poor prognosis	
	Blood pressure When evaluating hypertension in children and adolescents – use blood pressure for male/female by age and height percentiles table(s) [26]	Systolic blood pressure – maximal arterial pressure during systole (the phase of cardiac contraction) Diastolic blood pressure – minimal arterial pressure during diastole (the phase of cardiac relaxation) Normal blood pressure: 120–129/80–84 mm Hg <120/80 mm Hg Recommendation to take the mean of at least 2 seated blood pressure readings on 2 or more days before deciding on a diagnosis of hypertension [16] Common errors in measuring: wrong cuff size (kids), auscultatory gap, stethoscope pressure too firm, inappropriate level of the arm, feet not firmly on the floor, terminal digit preference (clinician's like to round to the nearest 0. 5 or another preferred number) [26, 31, 37] "White coat hypertension" – temporary increase in blood pressure when in a medical environment. Consider ABPM (ambulatory blood pressure monitoring) if persists. ABPM periodically measures and records blood pressure over 24-hour period (or longer)	
		Elevated (hypertension) > 120/80 mm Hg	"Masked hypertension" – blood pressure readings are normal in a medical visit but elevated in everyday life [16] Can be seen with blood loss, myocardial infarction and cardiac failure, irregular heart rate, medications [3] May be due to excess sodium intake (in those with a single nucleotide polymorphism (SNP) that codes for salt sensitivity) [38] Potassium, magnesium and calcium [37, 39], and vitamin D insufficiency [29, 30]
		Depressed (hypotension)	Medications Anorexia, electrolyte loss, or dehydration [16]

Table 39.1 (continued)

Region examined	Nutrition-associated examination findings (and select examination/test techniques)		Nutritionally related cause/association
		Postural or orthostatic hypotension: As a patient stands after lying supine, blood volume shifts to lower body. Normally blood pressure can stay stable during this shift (via compensatory mechanisms to increase cardiac output, heart rate, vascular resistance, and circulation of the blood to the body (from the heart)), which involves increased sympathetic outflow. If compensatory mechanisms fail, (decrease in systolic blood pressure of 20 mm Hg or a decrease in diastolic blood pressure of 10 mm Hg within 3 minutes of rising from sitting or supine position), a diagnosis of postural hypotension is made [40] Note that this may be accompanied by a feeling of dizziness or lightheadedness on standing, weakness or fatigue, and syncope Occasionally orthostatic dyspnea/chest pain can be seen [16, 40] Postprandial hypotension (decrease in systolic blood pressure of at least 20 mm Hg within 75 minutes of a meal) is related to orthostatic hypotension [40]	Associated with adrenal insufficiency, dehydration, blood loss, shock, anemia, pregnancy/postpartum, hypothyroidism, eating disorders, hypokalemia, acute hyperglycemia, AIDS, anxiety or panic disorder, eating disorders, prolonged bed rest, cardiovascular conditions (congestive heart failure, MI, arrhythmia, myocarditis, pericarditis, valvular disease, venous insufficiency), alcohol, medications (antiadrenergics, antiarrhythmics, antianginals, antidepressants, diuretics, sedatives, neuroleptics, antiparkinsonian agents, narcotics) [40] Can see more commonly in those >65 y.o. and in those who are frail [40], less physically active at work or who have multiple comorbidities (such as CHF, stroke, chronic kidney disease [41] Prognostic: associated with an increased risk of dementia [42] and may predict a poor prognosis in elderly patients with dementia [43] May be predictive of increased risk of mortality [44]
	Pulse pressure [16, 31]	Pulse pressure = systolic blood pressure (mm Hg) − diastolic blood pressure (mm Hg), e.g., blood pressure is 120/80 mm Hg; pulse pressure is 40	
		Elevated >40 The pulse pressure becomes elevated as the aorta becomes more stiff (may be due to high blood pressure or atherosclerosis)	Magnesium deficiency Severe iron-deficient anemia and in hyperthyroidism Prognostic: a high pulse pressure may be a strong predictor of heart problems and may indicate heart valve incompetence In older adults, a pulse pressure > 60 is a risk factor for cardiovascular disease, and elevated pulse pressures in older adults in general are considered a sensitive marker for carotid artery stenosis (can increase the risk of stroke, coronary heart disease, and sudden death) [16]
		Depressed <40	May indicate poor heart function
	Tip: interventions to address hypertension usually also address pulse pressure		
	Oxygen saturation [16, 31] Also known as "O_2 Sat" Factors affecting measurement: cold digit, hypotension, excessive ambient light, motion, discoloration of nail bed (if performing on hand) – can be due to bruising under the nail or darker nail polish Indicates how well the blood is becoming oxygenated by the lungs – more specifically, the degree of oxygenation of hemoglobin Can measure on finger/pinna (earlobe) or toe [measure in patient while supine and then raise the leg (keeping the O_2 Sat monitor clipped on the toe) 12–16 inches]. Hold there for 15 seconds. The O_2 Sat should not decline more than 2% Normal: 90–99% O_2 Sat (optimal 95–99%)		

(continued)

Table 39.1 (continued)

Region examined	Nutrition-associated examination findings (and select examination/test techniques)		Nutritionally related cause/association
		Increased	Not clinically significant
		Decreased <90%	Anemia, smoking (cigarettes), shock, pneumonia and other pulmonary diseases, congenital heart disease, chronic obstructive pulmonary disease (COPD), emphysema, alpha antitrypsin deficiency [45], obesity, hyperventilation syndrome, sleep apnea, high altitude, medications (narcotics, anesthetics) Clinical significance: negatively impacts wound healing (95% O_2 Sat needed for optimal wound healing)
		Decreased toe O_2 Sat (i.e., >2% change with raised leg test)	Indicative of compromise in health of blood vessels in legs (peripheral vascular disease, heart failure, stroke, diabetes, hypertension, hyperlipidemia, atherosclerosis, cardiac surgery) [37]
Skin	The skin is considered an organ of vicarious elimination – skin lesions and pathologies thus often lead us to examine impaired elimination in the body – due to digestive imbalances (constipation and other dysfunctions), toxicity, and other causes. Skin changes can also point to impaired absorption of nutrients or alterations in the utilization of nutrients. The nutrient requirements of the skin are dependent on those nutrients needed for the replacement of its cells and layers thereof [30]. With the skin turning over fairly rapidly, it is often a marker of fairly recent changes in nutritional status		
	Color (see above "General" section of this table)		
	Temperature (see above "General" section of this table)		
	Moisture	Decreased perspiration (hypohidrosis)	Thiamine deficiency [30] Medications, skin trauma or disorders, diabetes, alcoholism, neuropathies, Sjogren's syndrome [46]
		Increased perspiration (hyperhidrosis)	Thiamine deficiency [30] Medications, tumor, injury, diabetes, gout, menopause [46]
	Lesions	Examine for the following: Location and distribution (generalized or localized) – i.e., extensor or flexor surfaces, all over or only in specific regions (acne – typically face, upper chest, and back), psoriasis (knees, elbows (and maybe other areas such as scalp), *Candida* in skin folds Arrangement (linear/clustered/annular (in a ring)). A ring pattern can be ringworm. A unilateral, dermatomal pattern is suggestive of herpes zoster or shingles Type (macule, papule) – a macule is a flat, nonpalpable, circumscribed area of discoloration less than 1 cm in diameter (a freckle), a papule is a raised, palpable skin lesion <1 cm in diameter – may be pigmented or nonpigmented (e.g., nevus) Color Examine ABCDEs: (asymmetry, border (ragged/irregular vs. smooth), color (more than one shade), diameter (increased (> 6 mm)), elevation (not as specific a finding so often excluded))	
		Acne	Zinc, vitamin A excess [29, 30], omega-3 fatty acids, iodine/vitamin B_{12} insufficiency [30] May be related to higher glycemic load in diet/hyperinsulinemia [47] Correlated with consumption of dairy products [40] Pubertal hormonal changes, sex hormone imbalances Food allergy, gastric hydrochloric acid deficiency, blood sugar dysregulation
		Rosacea	Vitamin B_{12} intake [30] Food allergies and sensitivity to carbohydrates May be due to decreased gastric acid and pepsin production [29]

The Nutrition-Focused Physical Exam

Table 39.1 (continued)

Region examined	Nutrition-associated examination findings (and select examination/test techniques)		Nutritionally related cause/association
		Skin tags Pedunculated soft lesion (appear like grain of puffed rice cereal attached at one end) May be pigmented Typically found on the eyelid, neck, or axillae	Can occur due to friction of clothes/jewelry in some cases Diabetes, insulin resistance, elevated triglycerides, low LDL (low-density lipoprotein) cholesterol [48]
		Actinic keratosis Actinic keratoses (AK) are premalignant lesions that result from long-term exposure to sun (damage therefrom)	Diet may play a role in the development of AK, with moderate intake of oily fish (tuna, salmon, and sardines) and wine being protective [49] Prognostic: AK are strongly predictive of all forms of skin cancer
		Urticaria	May be due to food allergies Chronic form associated with autoimmune thyroid conditions [50] Approximately 30–50% of those with chronic urticaria are autoimmune-mediated [51]
		Petechiae	Vitamin C, K insufficiency
	Texture	Rough	Hypothyroidism Keratosis pilaris or follicular hyperkeratosis – thickening of the outermost layer of the epidermis producing small red bumps at the base of the hair follicles on the back of the arms and the anterior thighs Vitamin A, copper, zinc, and/or EFA insufficiency Excess vitamin A [29, 30] Pancreatic enzyme insufficiency (EFA, vitamin A malabsorption) [29]
		Scaling	Vitamin A, zinc, EFA insufficiency, or vitamin A excess
		Dry (xerosis)	Essential fatty acid, vitamin A, biotin, zinc insufficiency [29]
		Wrinkles	Protein and/or energy insufficiency [30] Increased wrinkle score associated with decreased bone density or fracture risk (suggests that collagen contributes to both) [32, 52] (Around the eyes) sleep deprivation
	Wounds and ulcers	Poorly healing	Decreased oxygenation of the blood [16] Diabetes, steroid use, malignancy, or AIDS [3] Zinc, copper, vitamin C, protein, and/or EFA, deficiency [3, 30] vitamin A [29]
		Scars	(Cue to inquire about surgery/injury)
		Keloids Hypertrophic growth of scar (grows larger than the original scar)	Etiology is unknown More common in individuals with darker skin Preliminary research suggests that keloids may be associated with insufficiency of EFAs, soluble fiber, and phytochemicals [53] Tip: Weigh risks/benefits of surgical procedures as there could be significant scarring/scar tissue with procedures
		Fissuring (dermatomalacia)	Vitamin A deficiency [30], niacin, zinc insufficiency [29]

(continued)

Table 39.1 (continued)

Region examined	Nutrition-associated examination findings (and select examination/test techniques)		Nutritionally related cause/association
	Rashes	Dermatitis, eczema	Macronutrient (protein and energy) insufficiency Micronutrient (riboflavin, niacin, zinc) insufficiency [3, 29, 30] EFA insufficiency [3, 29, 30] Vitamin B_{12} toxicity [30] Seborrheic form related to deficiencies of biotin, riboflavin, vitamin B_6, zinc, and EFAs [29, 30] Tip: Consider possible imbalances in GI function and/or the microbiome and evaluate for food allergies [29]
		Dermatitis herpetiformis	Gluten intolerance/gluten-sensitive enteropathy, celiac disease
		Psoriasis	Vitamin A or zinc insufficiency [29, 30] Clinical significance: considered sign of bowel toxemia and suboptimal liver health
	Hydration/turgor	To assess, lightly pinch the back of the patient/client's hand, over the sternum or on the inner thigh (using your thumb and forefinger). If the skin does not return to the normal tension/form within 3 seconds, there is decreased turgor [16] Note that this is less reliable in patients of increased age (the skin loses elasticity) Other signs of decreased hydration include dry mucous membranes, dry axillae, sunken eyes, and longitudinal tongue furrows [34]	Decreased turgor with insufficient riboflavin [29] Insufficient water/electrolytes Decreased turgor can result from the use of diuretic medication
	Edema	Due to excess fluid accumulation in the tissues due to ↑venous pressure (CHF), ↑vascular permeability (inflammation), ↓oncotic pressure (hypoalbuminemia), lymphatic obstruction (lymphedema), and deposition of additional tissue (myxedema (hypothyroidism), lymphedema (obesity) Pitting or non-pitting related to protein content of edema fluid. Low protein leads to more rapid pitting and recovery Pitting test (depressing over suspected edematous area and then removing pressure yields a pit that remains for over 5 seconds) Non-pitting edema – depression over 5 seconds leads to only slightly pitting and resumes to normal skin tone rapidly (rated as 1 + trace) to pitting edema (1.3–2.5 mm), 2–5 minutes to recover from (rated at 4+ severe)	Nephrotic syndrome, CHF Protein, energy, zinc, niacin, and thiamine insufficiency [30] Hypervitaminosis A Low serum albumin Evaluate for allergies

Table 39.1 (continued)

Region examined	Nutrition-associated examination findings (and select examination/test techniques)		Nutritionally related cause/association
	Vascularity	Capillary fragility Evaluate with blood pressure cuff: increase the venous pressure in the forearm with a blood pressure cuff and inspect the skin for petechial eruptions (draw a 1 inch circle 10 cm below the elbow crease before inflating the cuff to a pressure midway between the patient's systolic and diastolic pressure). Wait up to 5 minutes for any petechiae to form	If >10 broken capillaries (petechiae) in the circle, consider vitamin C deficiency [30]
	Pruritus	Itchiness	Parasites, allergies, chlorine exposure (topical), cancer, iron deficiency or niacin or vitamin A toxicity [29, 30] Hypothyroidism, liver disease, kidney failure
	Dermatographism	Writing on the skin (with blunt object), the "draw test" – produces localized hives that last 15–30 minutes	Associated with hypersensitivities/allergies, heat, stress, cold exposure (exaggerated histamine response)
Nails [16, 29, 30, 54, 55, 56]	Fingernails protect the distal ends of our fingers and toes. They are comprised of keratin (which is derived from the sulfur-rich amino acid cystine). Nails (both fingernails and toenails) can show signs of nutrient insufficiency/toxicity and exposure to toxins and systemic or localized disease or other conditions. A healthy nail structure requires sufficient protein and fat in the diet and sufficient hydration (to remain flexible). The nails are a sensitive, early marker of nutritional imbalances (with changes often preceding other clinical signs or laboratory findings). Clinical significance: Nail nutrient deficiencies often link to gastrointestinal health **Nail anatomy** – the nail plate is what one considers the "nail." The pinkish color of it is due to a vascular nail bed that is attached to the lunula (white moon of the nail). The cuticle extends from the nail fold and seals the space between the fold and plate from excess moisture. The "end" of the nail is referred to as the "free edge" Fingernails grow ~ 0.1 mm every 6–10 days and are completely renewed in 6 months (toenails grow more slowly) (at about 1/3 of the rate of fingernails). This timing can help to estimate the onset/exacerbation of medical conditions or diseases (measure the distance from the nail bed to the lesion/nail pathology and multiply out by 0.1 mm to approximate the days since the pathology started). With age, nails can lose luster and become more yellow and thickened (especially the toenails) **Examination** – be sure to ask if the patient bites or chews their nails and had an injury to the nails in the last 6 months or up to a year and if they recently/currently had (or have) polished nails		
	Shape or growth change		
Figure 39.2	Clubbing	Nail plates have exaggerated upward curve and curl around the fingertips Look for Schamroth's sign: Disappearance of diamond-shaped window between digits that is normally present when paired digits are juxtaposed (anteriorly) (Fig. 39.3)	Iodine deficiency Vitamin A excess Inflammatory GI disorders (IBD, celiac) Dysentery Fistulas Idiopathic pulmonary fibrosis Pulmonary malignancy Hodgkin's lymphoma Tuberculosis Asbestosis Chronic obstructive pulmonary disorder (COPD) Cirrhosis Congenital heart disease Endocarditis

(continued)

☐ Table 39.1 (continued)

Region examined	Nutrition-associated examination findings (and select examination/test techniques)		Nutritionally related cause/association
☐ Figure 39.4	Koilonychia (spoon nails)	Nail plates are depressed and appear flat or scooplike. This can be normal in children Water drop test [16] to evaluate for this nail change: Place a drop of water on the nail – if it does not drip off, the nail is flattened (early koilonychias)	Riboflavin, niacin, vitamin C, zinc, copper, chromium, selenium, protein (especially methionine and cysteine), and iron insufficiency Iron deficiency anemia Plummer-Vinson syndrome Malabsorption Hypochlorhydria Hemochromatosis Raynaud's disease Lupus (SLE) Nail-patella syndrome Thyroid disorders Rheumatic fever Syphilis Persistent occupational exposure of hands to petroleum-based solvents Trauma to the nail
☐ Figure 39.5	Onycholysis	The nail bed becomes separated from the nail plate	Iron deficiency anemia, pellagra (niacin deficiency) Cronkhite-Canada syndrome (low protein in the blood) – rare Psoriasis Eczema Infection (onychomycosis) Hyperthyroidism/thyrotoxicosis (referred to as "Plummer's nails") Sarcoidosis, amyloidosis, and/or connective tissue disorders Trauma to the nail
☐ Figure 39.6	Pitting	Few patterns: 1. Ridges across the nail plate 2. Red-brown spots ("oil drop sign") 3. Rows	Patterns correspond to: 1. Eczema 2. Psoriasis 3. Alopecia areata (autoimmune)
☐ Figure 39.7	Beau's lines	Depressed horizontal furrows across (transverse) the nail plate	Malnutrition Protein, calcium, niacin, zinc, vitamin A, and vitamin C insufficiency (May be seen in those with recent history of anorexia) Carpal tunnel syndrome Severe illness that disrupts nail growth (mumps, measles, myocardial infarction) Hypotension Hypocalcemia Uncontrolled diabetes Surgery Chemotherapy or immunosuppressive therapy Exposure to cold temperatures (Raynaud's disease) Trauma to the nail
	Unusually wide, square nail plates		Hormonal (endocrine) disorders
	Unusually long, narrow nail plates		Pituitary hormone deficiency Marfan syndrome
	Chronically chipped, sawtooth nails	Nail edge has sawtooth pattern from chipping	Malnutrition, vitamin deficiencies Radiation exposure Chemical exposure/damage

Table 39.1 (continued)

Region examined	Nutrition-associated examination findings (and select examination/test techniques)		Nutritionally related cause/association
Figure 39.8	Ridging of the nails	Often accompanies brittle nails	Mineral malabsorption Protein insufficiency Insufficient fat intake Iron deficiency (with central ridge(s)) B vitamin insufficiency (thiamine, riboflavin, niacin, pyridoxine, folate) Zinc, vitamin A insufficiency
	Onychomadesis	Spontaneous separation of the nail plate from the nail bed originating from the proximal end. The nail is then shed as it grows	Hypocalcemia Idiopathic After severe systemic illness Drug reactions Infections Cronkhite-Canada syndrome Trauma to the nail
	Acrodynia	Red tips with nail growth abnormalities	Mercury toxicity
	Onychorrhexis (Brittle nails)	Irregular, frayed and torn nail borders Called "onychoschizia" if the nail splits	Protein, EFA, iron, biotin, boron, folate, zinc, calcium, magnesium, silica, sulfur, zinc, vitamin A, carotenoid, vitamin C, vitamin D, and vitamin B_{12} insufficiency Selenium or vitamin A toxicity Anorexia Hyperthyroidism Hypochlorhydria Metabolic bone disease Cronkhite-Canada syndrome Tip: in postmenopausal women and older men, check bone density
	Absent lunula	No visible half-moon on nail	Zinc, protein insufficiency
	Color change		
Figure 39.9	Terry's nails	Most of the nail bed is white except for a pink band near the nail tip. Due to decreased vascularity and increased connective tissue in the nail bed Generally, the lunula is obliterated	Niacin, protein insufficiency Copper toxicity Cirrhosis Hepatic failure Diabetes mellitus Congestive heart failure Hyperthyroidism
	Azure lunula	Blue half-moons in the nail bed	Copper toxicity Wilson's disease (hepatolenticular degeneration) Silver poisoning (blue-gray) Quinacrine therapy Impaired circulation
	Yellow lunula	Yellow half-moons in the nail bed	Tetracycline therapy
	Brown/black lunula	Brown/black half-moons in the nail bed	Excess fluoride Hyperthyroidism (nail plate brown)
	Red lunula	Red half-moons in the nail bed	Cardiac failure
Figure 39.10	Half-and-half nails (Lindsay's nails)	The proximal half of the nail is white, and the distal half (horizontally) is pink or brown, and the lunula is obliterated	Niacin zinc insufficiency Renal failure (pathognomic)

(continued)

Table 39.1 (continued)

Region examined	Nutrition-associated examination findings (and select examination/test techniques)		Nutritionally related cause/association
Figure 39.11	Muehrcke's lines	Two white lines (arcuate) parallel to the lunula and separated from the nail bed with normal nail. Due to abnormality of vascular nail bed – so does not move with nail growth. Lines disappear transiently with blanch test	Malnutrition (protein deficiency) Acrodermatitis enteropathica (rare, inherited form of zinc deficiency) Hypoalbuminemia (pathognomic) Nephrotic syndrome Liver disease
Figure 39.12	Mees' lines (Reynolds/Aldrich lines)	Longitudinal white streaks from the cuticle to the free edge	Niacin, calcium, zinc insufficiency (hypocalcemia) Acrodermatitis enteropathica (zinc deficiency)) Renal failure Hodgkin's lymphoma Myocardial infarction, congestive heart failure Leprosy Malaria Sickle cell anemia Carbon monoxide poisoning Arsenic poisoning Thallium poisoning Chemotherapy (cancer)
	Hapalonychia	Nails thin causing the bending and breaking of the free edge and longitudinal fissures	Kwashiorkor Insufficient intake of protein and/or fat Vitamin A, pyridoxine, and vitamin C and D insufficiency Calcium, silica, iron, sulfur, and zinc insufficiency
	Pallor		Iron, vitamin B_{12} insufficiencies/anemia, selenium and riboflavin insufficiency
	Yellow nail	Nails grow more slowly and develop "heaped up"/thickened appearance and lunula disappear. Thought to be due to abnormal lymph drainage or leakage of protein as a result of increased microvascular permeability [16]	Rule out smoking and recent nail polish use Beta carotene excess Vitamin D insufficiency Long-term use of tetracycline Type 2 diabetes [48] Lymphedema Pleural effusion Bronchiectasis Sinusitis Rheumatoid arthritis Nephrotic syndrome Thyroiditis Tuberculosis Raynaud's disease Immunodeficiency Cancer (possibly) Tip: vitamin E may help as an intervention
	White nails		Selenium, zinc insufficiency
	Red nails		Malnutrition, lupus, carbon monoxide angioma, or polycythemia
	Brown/gray nails		Vitamin B_{12} insufficiency, diabetes mellitus, hyperthyroidism [54], or cardiovascular disease
	Blue nails		Peripheral cyanosis Copper toxicity
	Longitudinal striations (Trachyonychia)		Protein, calcium, pyridoxine, biotin, and vitamin C insufficiency Vitamin A excess Alopecia areata Vitiligo Atopic dermatitis Psoriasis

Table 39.1 (continued)

Region examined	Nutrition-associated examination findings (and select examination/test techniques)		Nutritionally related cause/association
Figure 39.13	Splinter hemorrhage	Longitudinal red streaks from nail bed toward proximal margin. Signify bleeding of the capillaries	Scurvy (vitamin C deficiency) Oral contraceptive use Pregnancy Chronic hypertension SLE Rheumatoid arthritis Psoriasis Peptic ulcer Subacute bacterial endocarditis Trichinosis Trauma to the nail
	Telangiectasia	Dilated blood vessels in the nail at the margin of the finger nails	Rheumatoid arthritis SLE Dermatomyositis Scleroderma
Figure 39.14	Punctate leukonychia	White spots or patches between the nail and the nail bed. Due to subungual air bubbles	If isolated dots, can be due to trauma to the nail (including repeated manicures) or zinc, selenium, and niacin insufficiency If on the entire nail, likely due to a congenital disorder
	Longitudinal melanonychia	Brown/black discoloration – streaks spread from the nail bed to the surrounding finger	Vitamin D, folate, and vitamin B_{12} insufficiency Malnutrition (protein deficiency) GI polyps (discoloration on the nail is in spots) Benign nevus Normal variant in darkly pigmented people Melanoma (large patch or collection of small freckles most commonly on the thumb and big toes)
	Other changes/evaluation		
Figure 39.15	Paronychia	Infection around the fingernail Can have redness, pus, swelling, and tenderness (Acute – *Staphylococcus* infection Chronic – fungal)	EFA, zinc, and vitamin C insufficiency [29]
	Hangnails		Zinc, vitamin C, folate, and protein insufficiency
	Ragged cuticles		Boron and iron insufficiency
Figure 39.16	Onychophagia	Nail biting	Stress Anxiety
	Onychotillomania	Nail picking	Stress Anxiety
	Capillary refill test	Technique: Nails must be unpolished. Apply pressure to the nail until it blanches/turns white. Remove pressure. Patient is to hold the hand above the heart, while the time it takes for blood to return to the tissues (nail turns pink) is measured Indicator of tissue perfusion and hydration	Blanch time > 2 seconds (i.e., decreased capillary refill): Dehydration Vitamin A and C insufficiency Peripheral vascular disease (arteriosclerosis, hypertension, diabetes mellitus, smoking, heart disease, abnormal cholesterol) Shock, hypothermia
Head	Examine the head, scalp, and hair		

(continued)

Table 39.1 (continued)

Region examined	Nutrition-associated examination findings (and select examination/test techniques)		Nutritionally related cause/association
Hair	Examine for quality, quantity, distribution, color, texture, and baldness. Look for hair loss diffusely or in patchy areas on the scalp. Be sure to inquire about any chemical processing, dying, or bleaching of hair which could affect the color and/or texture of the hair. Normal hair loss is considered to be 50–100 hairs per day		
	Hair loss [34, 57, 58]		Stress of serious illness or injury, parasites, heavy metal toxicity, oral contraceptives Protein, iron, zinc, or biotin insufficiency, anemia In women, hair loss can occur following (may be up to 2 months after) extreme stress from a major illness, from following a very low-calorie diet, from rapid weight loss (post-bariatric surgery), polycystic ovarian syndrome (PCOS), thyroid disease, or as a result of medication [16] Alopecia areata – autoimmune disorder (immune system attacks hair follicles) – patchy hair loss Tip: may respond to zinc and biotin supplementation [16]
	Baldness		Vertex baldness with hypertension or high cholesterol – possible marker for increased risk of coronary heart disease [59] In women, frontal baldness is associated with PCOS (usually seen with hirsutism)
	Sparse or thin hair		Protein, zinc, iron, biotin, linoleic acid, and/or manganese insufficiency [4, 16, 32] Vitamin A excess Hyperthyroidism
	Hair that is easy to pluck		Protein, zinc, and iron insufficiency
	Premature graying		PABA (para-aminobenzoic acid) deficiency Vitiligo Stress
	Dry/brittle		EFA and/or manganese insufficiency [29] Hypothyroidism
	Dandruff		EFA, fat-soluble vitamin insufficiency (due to biliary dysfunction) Calcium, pyridoxine, and other B insufficiencies Antioxidant insufficiency (selenium especially) Excess refined carbohydrate intake Hypochlorhydria
	Flag sign	Bands of depigmented hair can appear horizontally (transverse to the length of hair) during periods of inadequate protein intake When protein intake returns to normal, pigmented hair will grow again – creating the alternating band-like appearance	Kwashiorkor, protein deficiencies in ulcerative colitis, and other conditions
	Corkscrew hair	At the base of the hair follicle	Vitamin C insufficiency
	Menkes steely hair		Copper insufficiency (due to disordered copper metabolism) [29]
Scalp	Itchiness		Calcium insufficiency Skin rash (psoriasis or other)
	Psoriasis		Sulfur insufficiency Imbalanced omega-3 or omega-6 fatty acid ratio

Table 39.1 (continued)

Region examined	Nutrition-associated examination findings (and select examination/test techniques)		Nutritionally related cause/association
Face	Affect	See above under "General" section	
	Asymmetry		Stroke/neurological issue Bell's palsy – paralysis of muscles on one side of the face (with inability to control facial muscles) generally idiopathic, though may be due to diabetes, Guillain-Barre syndrome, Lyme disease, viral infection (recent URI), or other causes (tumor) Facial nerve (CN7) defect
	Facial nerve (CN7) examination: Ask patient to make a variety of facial expressions – raise eyebrows bilaterally, close the eyes, frown, smile, show their teeth, and puff out their cheeks {Bates}		
	Vertical creases near midline of the forehead		Consider duodenal ulcers
	Seborrheic dermatitis		Biotin, riboflavin, vitamin B_6, zinc, and EFA insufficiency Can be seen with diets high in refined sugars Can be due to food allergies (research supporting this in children) [60]
	Myxedema	Characterized by a relatively hard edema of subcutaneous tissue, with increased content of proteoglycans in the fluid. May also see swelling in the neck if goiter	Hypothyroidism
	Loss of lateral 1/3 of eyebrow		Hypothyroidism
	Dilated capillaries on cheeks and nose		Excess alcohol, gastric hydrochloric acid deficiency
	Flushing	May also be seen with red ears	Allergy
	Yellow skin		Pernicious anemia (vitamin B_{12} deficiency) [16, 29, 31]
	Brown, patchy pigmentation of the skin	Especially on cheeks with parotid enlargement and a "moon" facies	Protein-calorie deficiency
	Acne	See "Skin" section of table above	
	Acne rosacea	See "Skin" section of table above	
	Tics/abnormal movements	Evaluate for with smile test	Facial nerve (CN7) defect Tourette's syndrome
	Chvostek's sign	To test for this sign, percuss at the top of the cheek (just below cheek bone) with tip of the index or middle finger. Will see repeated contractions of the facial muscle if this sign is present	Calcium and magnesium deficiency Tetanus, tetany Hyperventilation (induces hypocalcemia)
	Trigeminal nerve examination (CN5) examination: Pain and light touch are examined across three zones of this nerve on the face (ophthalmic, maxillary, and mandibular). Motor function of CN5 is evaluated by feeling for contraction of temporal and masseter muscles as well as evaluation of the corneal reflexes		
	Hair growth	Excess in women (hirsutism) Frontal baldness and acne may be present with hirsutism	Hormonal imbalance (excess androgens). Cushing's, PCOS, elevated cholesterol, sleep apnea, insulin resistance, diabetes, tumors, and medications

(continued)

Table 39.1 (continued)

Region examined	Nutrition-associated examination findings (and select examination/test techniques)		Nutritionally related cause/association
Eyes	Eye physiology: The eye is a sphere enclosed by three layers – outermost is the sclera/cornea; then there are the choroid, ciliary body, and iris; and the innermost layer is the retina The eyeball is largely covered by the sclera (the "white" of the eye). Over the iris is the cornea – this is transparent so light can pass through the eye The eye is composed of two chambers of fluid separated by the lens. These carry nutrients and are transparent to allow light to pass through the eyes. The shape of these chambers changes to allow vision at different distances The iris is a radial muscle that controls the size of the pupil (allows it to dilate or constrict) In a nutrition physical exam, the eyes are examined largely with a penlight. Physicians and ophthalmologists typically use ophthalmoscopes to visualize the retina and can detect retinal hemorrhages and changes in the retinal arteries that occur with hypertension and diabetes		
	Eyelids	Ptosis	Neurological/muscular disease [21]
		Retracted lid	Neurological/muscular disease, mechanistic (includes atopic dermatitis and essential hypertension [61]
		Exophthalmos	Hyperthyroidism (Graves' disease)
		Angular palpebritis or blepharitis	Riboflavin, niacin, or pyridoxine insufficiency
		Periorbital edema	Allergies, local inflammation, myxedema, nephrotic syndrome/fluid-retaining state
		Dark circles under the eyes	Allergies (allergic shiners), dysbiosis, liver congestion with detoxification problems (phase I and II detoxification) [20], hypochlorhydria, pancreatic insufficiency, adrenal hypofunction, and mineral deficiency [57, 58] Fatigue, sleep deprivation [62]
		Dennie's lines (horizontal creases across the lower eyelids) [60]	Food allergies
		Meibomian gland dysfunction	May be due to oxidative stress Tip: may benefit from omega-3 fatty acid supplementation
	Lesions (lumps, swellings around the eyes)	Xanthelasma (Slightly raised, yellowish, and well-circumscribed plaques that appear along the nasal portion of one or both of the eyelids)	May accompany lipid disorders such as hypercholesterolemia [57]
	Conjunctiva	Conjunctivitis	Allergies, irritation, infection
		Inflammation	Vitamin A or riboflavin insufficiency
		Conjunctival pallor	Iron, folate, and B_{12} insufficiency
		Bitot's spots (foamy, superficial patches on the conjunctiva) [1]	Vitamin A deficiency
		Conjunctival xerosis (dryness)	Vitamin A or riboflavin deficiency (or environmental/chemical irritation) Diabetes [63]
	Pupils	Dilation	Allergy (especially dairy) [29]
		Swinging flashlight test Test of pupillary response to light Shine penlight first in one eye and then the other Normal response: as light swings from one eye to the other, each pupil should constrict briskly Optic nerve or retinal lesion: dilates upon exposure to light	Optic nerve (CN2) or retinal lesion

Table 39.1 (continued)

Region examined	Nutrition-associated examination findings (and select examination/test techniques)		Nutritionally related cause/association
		Nystagmus	Thiamine insufficiency [29] Medication, CNS diseases Inflammation of inner ear Albinism Astigmatism, myopia Defects in CN3, CN4, and CN6
	Lens and cornea	Corneal arcus	Vitamin A deficiency Dyslipidemia (middle-aged men) (DN)
		Corneal scar, ulcerations	Vitamin A deficiency
		Corneal xerosis (dull appearance)	Vitamin A deficiency
		Keratomalacia (cornea is soft)	Vitamin A deficiency
		Xerophthalmia	Vitamin A deficiency
		Pterygium	Epidemiologic evidence of an association between obesity and pterygium in women [64]
		Cataracts	Dysglycemia, vitamin C insufficiency May be marker of oxidative damage and ultimately coronary heart disease (CHD)
		Glaucoma	Hereditary Chronic subclinical inflammation May be associated with imbalances in oral microbiome [65]
		Kayser-Fleischer rings (Greenish-brown deposition of copper in annular ring around periphery of the cornea (where it meets the sclera/white of the eye))	Results from inherited accumulation of copper in the liver due to inherited ceruloplasmin defect Wilson's disease
	Iris	Iritis	Trauma, can also be seen with certain diseases, such as ankylosing spondylitis, Reiter syndrome, sarcoidosis, inflammatory bowel disease, and psoriasis. Infectious causes may include Lyme disease, tuberculosis, toxoplasmosis, syphilis, and herpes simplex and herpes zoster viruses
		Iris contraction test (A variation on the swing flashlight test: upon exposure to light in a darkened room, the pupil should contract immediately and remain contracted)	With adrenal insufficiency, the pupil will not remain contracted with exposure to light, but will dilate. This dilation will occur within 2 minutes and can last for 30–45 seconds before another contraction. It is best to time when the dilation occurs and how long it lasts. Subsequent retesting can then serve as a monitor for adrenal status as interventions are employed
	Optic (CN2) and oculomotor (CN3) examination: assess extraocular movements ("H" and "X" in space) and evaluation of pupillary response to light and accommodation		
		Ophthalmoplegia Weakness or paralysis of one or more of the extraocular muscles	Thiamine deficiency [29]
		Nystagmus	Thiamine insufficiency [29] Medication, CNS diseases Inflammation of inner ear Albinism Astigmatism, myopia Defects in CN3, CN4, and CN6

(continued)

Table 39.1 (continued)

Region examined	Nutrition-associated examination findings (and select examination/test techniques)		Nutritionally related cause/association
	Vision changes	Reduced visual acuity	Essential fatty acid deficiency [29] Optic nerve defect
		Reduced night vision	Vitamin A [29], zinc insufficiency
		Color blindness Inherited poor color vision usually affects both eyes, and the severity doesn't change over one's lifetime Examine with Ishihara or Hardy Rand and Rittler testing	Diabetes, glaucoma, macular degeneration, Alzheimer's disease, Parkinson's disease, chronic alcoholism, leukemia, and sickle cell anemia Medications can alter color vision, such as some drugs that treat heart problems, high blood pressure, erectile dysfunction, infections, nervous disorders, and psychological problems Ability to see colors deteriorates slowly as you age Exposure to some chemicals (carbon disulfide and fertilizers) may cause loss of color vision [66]
		Retinal field defect	Vitamin A, vitamin E deficiency [29]
		Macular degeneration Use Amsler grid to examine	Tip: lutein, zeaxanthin, and vitamin C may help prevent macular degeneration
		Visual contrast	Can use to assess loss of visual contrast sensitivity in macular degeneration, cataracts, and glaucoma and can help prevent falls in elderly due to loss of contrast sensitivity A number of factors can affect the ability to perceive visual contrast: nutritional deficiencies, alcohol consumption, medication or drugs, exposure to endogenous or exogenous toxins, including but not limited to neurotoxins, biotoxins, volatile organic compounds (VOCs), mycotoxins, mold, microbial, VOCs, parasites, cyanobacteria, dinoflagellates, heavy metals, Lyme disease Visual contrast testing is used to track the progress of patients undergoing treatment for Lyme disease Can measure neurotoxicity in those with M.S. Note: Loss of visual contrast is not diagnostic, but may warrant further evaluation for these conditions [67, 68]
		Photophobia	Zinc deficiency [29]
		Floaters	Those with nearsightedness are more prone to floaters (also those with diabetes and a history of cataract surgery) Vitamin C and K and bioflavonoid insufficiency
Ears	Diagonal earlobe crease		Highly associated with cardiovascular disease (in men)
	Tophi on external earlobe		Gout
	Hair in the ear canal		Associated with coronary artery disease
	Excess cerumen (wax) in the canal		EFA insufficiency or allergies
	Tinnitus		Cardiovascular disease, allergies, aspirin toxicity, B12 deficiency [69]
	Acoustic nerve (CN8) examination: assess for hearing by rubbing fingers together on either side of the ear at varying distances and asking the patient to localize as well as evaluating for conductive hearing loss (Weber and Rinne tests with tuning fork) [21]		
Nose	External nose	Examine for redness, dryness, rashes/skin abnormalities	

Table 39.1 (continued)

Region examined	Nutrition-associated examination findings (and select examination/test techniques)		Nutritionally related cause/association
		Rhinophyma A thickening and reddening of the skin on the nose. Can see broken blood vessels	Related to excess alcohol ingestion Hypochlorhydria May be related to acne rosacea
		Malar rash (also referred to as "butterfly rash" for the butterfly shape with the wings of the butterfly on the cheeks and the body over the bridge of the nose) It is a purplish, scaly rash	Systemic lupus erythematosus (SLE) Niacin deficiency
		Salute sign/allergic salute [60] Occurs from rubbing the nose (due to nasal drip/rhinitis) with an upward movement of the hand Allergic crease horizontal lines across lower portion of the nose from rubbing the nose with the hand (saluting)	Allergies (especially dairy)
		Nasolabial seborrhea (Seborrheic rash around the base of the nose and perioral region)	Pyridoxine, EFA deficiency [29]
		Nasolabial dyssebacia (a disorder of the sebaceous glands characterized by excess oil production, inflammation, exfoliation, and fissuring of the sebaceous glands (which appear moist and reddened) around the base of the nose and perioral region)	Riboflavin deficiency [29]
	Nasal canal/intranasal [34, 57]	Intranasal polyps	Allergies (chronic rhinitis) Salicylate sensitivity (often seen with asthma) [29]
		Epistaxis (nosebleeds)	Vitamin C, vitamin K, and flavonoid insufficiency [60] Allergies Local trauma Infection Dry air Hypertension or coagulation disorders Medications

(continued)

Table 39.1 (continued)

Region examined	Nutrition-associated examination findings (and select examination/test techniques)		Nutritionally related cause/association
	Smell Evaluate sense of smell using smell identification test [UPSIT]: Self-administered 40-item test with smell card booklet Smell discrimination test: Sniffin' Sticks [70]	Hyposmia – decreased ability to smell	Zinc deficiency, pernicious anemia May be due to nasal polyps, cadmium poisoning, smoking, asthma, allergy, diabetes, hypothyroidism, fibromyalgia, MS, schizophrenia, sarcoidosis, SLE, hepatic or renal failure, tumor Medications: zinc-based intranasal sprays; intranasal medications; antibiotics; antidepressants; antilipidemic; medications for hypertension, rheumatism, and hyperthyroidism; chemotherapeutic medications; and opioids [71] Smell disorders are present in a number of neurodegenerative illnesses and part of their early diagnosis (changes in the sense of smell are the first sign of idiopathic Parkinson's disease). Can also be a sign of early Alzheimer's disease [72] Olfactory (CN1) defect Prognostic: Decreased olfactory discrimination is a significant predictor of future cognitive decline [71]. Olfactory dysfunction in old age predicts mortality [73]. Lack of ability to identify smells related to overall health and one's life expectancy may be negatively associated with lowered olfaction [73]
		Anosmia – lack of ability to smell	Unilateral anosmia may be due to trauma Chronic infection or inflammation of the nasal passages
		Hyperosmia – increased ability to smell	Hyperosmia can occur with Addison's disease, head injury, multiple chemical sensitivity, before or during migraine attacks, and following rapid chronic withdrawal of drugs Depression often accompanies cases of hyperosmia [74]
		Peanut butter test Test patient with closed eyes, for their ability to detect the odor of peanut butter, one nostril at a time. Hold a container of peanut butter medially below the nostril (ideally it is not a very wide mouth (1 oz. is ideal) container) and move up 1 cm at a time during the patient's exhalation. If an odor is detected, note the distance between the patient's nostril and the top of the container with a 30-cm ruler	Left nostril impairment of odor detection in patients with probable Alzheimer's disease and with mild cognitive impairment This test may also be useful for monitoring disease progression (the asymmetry was greatest in early stages of the disease. With disease progression, the right nostril became more impaired – resulting in an overall decrease of asymmetry [75]
Sinuses	Examine externally – looking for pain, tenderness, redness, and swelling over the affected sinus(es)		
	Congestion or (post)nasal drip		Chronic allergies Yellow-green sputum suspicious of infection
	Dysbiosis		Can be due to hypochlorhydria and pancreatic enzyme deficiency
Mouth	Due to the rapid turnover of cells in the oral mucosa, a number of diseases, disorders, and imbalances have clues, or definitive diagnoses, which can be made from examining the mouth early in the course of disease (lead poisoning, eating disorders, nutritional deficiencies, etc.). Prior to starting the exam, it is advised to ask the patient about any recent changes in taste, burning sensations, pain, or bleeding of the gums or mucosa (Radler and Lister)		

Table 39.1 (continued)

Region examined	Nutrition-associated examination findings (and select examination/test techniques)		Nutritionally related cause/association
	Lips	Cheilosis, angular stomatitis Redness, cracking of one or both angles of the mouth	Riboflavin, niacin, pantothenic acid, pyridoxine, folate, B_{12}, iron, and or zinc deficiency [29, 57, 76] Infection (*Candida* in those with dentures, *Staph.* for those without) Poorly fitting dentures (saliva leaks to the mouth corners – Causing skin irritation)
		Cheilitis	Early sign of Crohn's disease Nutritional deficiencies of cheilosis and/or excess vitamin A
		Actinic cheilitis Usually on the lower lip	Excess exposure to sun (loss of pigment in the lips and may be scaly and thickened)
		Lip ulcers or sores	Cold sores: selenium deficiency [29] From herpes labialis (herpes simplex virus I (HSV-1) – cold sores) If you see lesions on or around the upper border of the lip – usually it means it is herpes. Herpes lesions can be precipitated by stress and adrenal overload and calcium and EFA deficiency: deficiency of lysine, exposure, vitamin C, and/or flavonoids [29, 58] Cancer (squamous cell carcinoma) – usually affects the lower lip (may be a scaly plaque, ulcer, or nodular). Seen in those with fair skin and history of prolonged sun exposure
		Enlarged lip	Trauma Angioedema (which can be secondary to food allergies)
		Pigmented lip	In Peutz-Jeghers syndrome, you will see multiple pigmented spots on the lips and oral mucosa and intestinal polyps. There is a slightly higher risk of gastrointestinal malignancies in this condition
	Palate	Visualize the roof of the mouth Assess the pharynx (pillars, tonsils, uvula) for movement, enlargement, or lesions Use the tongue blade in the middle 1/3 of the tongue. Traction forward and down and have the patient say "Aah" (elevates the soft palate and allows one to assess cranial nerve function… more on this later). Don't touch the back of the tongue or you can elicit the gag reflex Note that those in a state of parasympathetic stress will have an easy gag reflex (due to increased alkalinity with too much potassium relative to calcium)	
		Candida (thrush) White discharge on palate/ pharynx or mucosa	May be due to iron and/or vitamin C deficiency Medications (antibiotics, corticosteroids) Occurs with reduced immunity and malabsorption Can result from dry mouth Diabetes, HIV, cancer, renal failure [76]
		Blue-gray	Hemochromatosis
		Psoriasis May cause pain and discomfort and thus affect the types of foods consumed and quantities	May be an indication of the liver and GI imbalances (both associated with psoriasis)

(continued)

◨ Table 39.1 (continued)

Region examined		Nutrition-associated examination findings (and select examination/test techniques)	Nutritionally related cause/association
	Pharynx	Absence of the uvula	May be surgically removed (as may be done for obstructive sleep apnea)
		Swollen uvula	Pharyngitis
		Tonsillitis	May be due to allergies
		Enlargement of tonsils	Vitamin A and/or C and zinc deficiency [29] Allergies/sensitivities Chronic infection Sleep apnea
		"Crimson crescents" – bilateral reddening (a purplish red hue) of the pillars without any pain or sore throat	Chronic fatigue syndrome
		Warts on tonsils/pillars	HPV (human papillomavirus)

Glossopharyngeal nerve (CN9) and vagus (CN10) examination: Assess movement of the pharynx while the patient says "ah" and evaluate for gag reflex (bilaterally). Listen to the voice for hoarseness or a nasal tone and observe for any difficulty swallowing (CN10 lesion)

	Buccal mucosa	White spots	Oral thrush (*Candida*) – these can be scraped off – revealing red, raw (inflamed, bleeding) oral mucosa underneath. May also see squamous cell carcinoma, Koplik's spots (rubeola), or HIV (hairy leukoplakia) Worry more about hairy leukoplakia (precancerous) in those who smoke, chew tobacco, abuse alcohol, or have a history thereof
		Pigmented spots	Peutz-Jeghers syndrome, smokers' melanosis, malignant melanoma, Addison's disease, hemochromatosis
		Red spots: petechiae	Vitamin C, K, protein, and/or energy insufficiencies Thrombocytopenia Infectious mononucleosis, pyogenic granuloma, erythema migrans, Kaposi's sarcoma
		Xerostomia (dry mouth)	Zinc insufficiency
		Stomatopyrosis (painful inflamed mouth) and dysesthesia (burning mouth syndrome)	Iron, B12, folate, and/or magnesium insufficiency
		Pallor	Iron deficiency
		Ulcers/sores (aphthous ulcers [canker sores])	Vitamin B_{12}, folate insufficiency Allergies (including celiac disease) (More rarely seen with infection and cancer) Can recur with use of toothpaste containing sodium lauryl sulfate
	Gums	Gum hypertrophy, bleeding	Bioflavonoid and vitamin B_{12} insufficiency Poor brushing technique/use of a brush with hard bristles Vitamin C and D deficiency (with scurvy see petechiae (small red or purple spots caused by bleeding into the skin) on gums) Puberty, pregnancy Inflammatory bowel disease (IBD), uveitis, ankylosing spondylitis, peripheral arthritis, erythema nodosum [77] Leukemia Medications (phenytoin, cyclosporin A)

Table 39.1 (continued)

Region examined		Nutrition-associated examination findings (and select examination/test techniques)	Nutritionally related cause/association
		Gum recession Gingivitis Inflammation of the gum tissue Gingivitis with periodontitis (bacterial infection) is known as periodontal disease. In more advanced stages, larger gum pockets and teeth loosen and can fall out	Silica, calcium, CoQ_{10}, quercetin, antioxidant insufficiency Vitamin A and D, riboflavin, and niacin insufficiency (also see gingival tenderness with niacin deficiency) [78] Gingival inflammation significant predictor of cognitive decline [79] Oral inflammation and poor oral hygiene associated with hypertension in individuals <65 years of age [80] Strong links between chronic oral inflammation and a number of other health conditions (cardiovascular disease, stroke, diabetes, Alzheimer's and other dementias, and pancreatic cancer) [32, 78]
		Lead line A bluish-black line on the gums (~1 mm from the gum margin) that may indicate chronic lead poisoning. The line follows the contours of the gums and is absent where there are no teeth (it is produced by tartar-forming bacteria)	Lead or bismuth exposure
	Teeth	Long teeth	Gum recession
		Bruxism	Excess stress, allergies, parasites
		Tooth decay	Poor mineral absorption Vitamin C and B_{12}, fluoride, and phosphorus deficiency [29]
		Tooth discoloration with malposition and hypoplastic line across upper primary incisors (become yellow-brown in color)	Protein calorie malnutrition
		Mottled enamel, fluorosis	Excess fluoride, calcium insufficiency
		Tooth erosion	Excess consumption of fresh citrus or sugary sodas Bulimia (along posterior aspect of the teeth – particularly the incisors) GERD Swimming in chlorinated water (chronically) Frequent wine consumption (acids can erode the teeth)
	Dental materials Nickel and/or other metals used in restoration may lead to allergic reactions in some patients Mercury amalgams, root canals, implants, mixed noble- and base-metal crowns, partial dentures made of chrome cobalt, BPA-based resin fillings, night guards, and dentures may create oxidative stress in some patients Mercury amalgams in young adulthood may lead to increased risk of diabetes mellitus in later life [81]		
	Taste	**Nutrients involved in taste [34, 58]** *Vitamins:* A, E, riboflavin, niacin, pantothenic acid, pyridoxine, folate *Minerals:* Copper, iodine, iron, zinc Taste tests available for individual minerals (potassium, zinc, magnesium, copper, chromium, manganese, molybdenum, and selenium) may provide some initial information to guide physical exam and laboratory evaluation (note that these tests are generally considered less definitive than laboratory assays for these minerals, but are an affordable and rapid in-office assessment)	

(continued)

Table 39.1 (continued)

Region examined	Nutrition-associated examination findings (and select examination/test techniques)		Nutritionally related cause/association
		Dysgeusia Altered taste perception Differential diagnosis (Ddx): loss of smell, which can contribute to/cause loss of taste	Zinc and vitamin B$_{12}$ insufficiency Diabetes, hypothyroidism, or other metabolic conditions Cancer treatment (chemotherapy and/or radiation), postoperatively and/or from certain medications such as azithyromycin [17], amlodipine, metronidazole, tetracycline groups, statins, and a number of thyroid medications [82]
		Loss of taste (general) Be sure to evaluate for loss of smell (which can contribute to/cause loss of taste)	Zinc insufficiency
		Loss of taste for meat	Incomplete protein digestion (most likely cause thereof is hypochlorhydria) Zinc deficiency Pepsin and protease insufficiency
	Intraoral inflammation	Redness of mucosa and/or gums	Vitamin C insufficiency [1]
	Bitter strips to evaluate supertaster status Place the control strip on the tongue first and then the bitter strip on the mouth and evaluate for taste Supertasters are homozygous for the allele that detects this taste. They have more fungiform papillae Women, Asians, and African Americans are more likely to be supertasters, and about 25% of all Americans are classified as supertasters. (Being a supertaster may have served an evolutionary advantage in avoiding potentially toxic plant alkaloids) Clinical significance: Supertasters tend to have a higher risk of colon cancer, consume more salt, and be leaner than the average population. They may be more resistant to bacterial sinus infections, and they tend to be pickier eaters Finally, supertasters may pass the zinc tally taste test, but still be deficient in zinc on later blood analyses Tasters are more likely to be non-smokers and not in the habit of drinking coffee or tea and more likely to find green vegetables bitter. Taster status is found more commonly in Native Americans, Inuits, and Australian or New Guinea aboriginals		
	Breath odor (halitosis)	Disorders of oral cavity	Niacin insufficiency [29] Retained food, stomatitis, glossitis, periodontal disease, poorly cleaned dentures, xerostomia (dry mouth – decreased saliva)
		Disorders of the nose and sinuses	Atrophic rhinitis, chronic sinusitis, nasal septal perforation, ozena (atrophic disease of nose and turbinates), nasal septal perforation (can be due to cocaine use), and retained foreign bodies
		Disorders of the tonsils and pharynx	Recurrent tonsillar and adenoid infections, pharyngitis
		Disorders of the digestive organs	Achalasia, GERD, empyema Dysbiosis, pancreatic enzymes, and/or hydrochloric acid insufficiency [58]
		Disorders of the lungs	Abscesses, bronchiectasis, pneumonia, empyema (is the collection of pus in a body cavity – in this case the lungs ("pleural empyema"))
	Fruity/ammoniacal odor	Systemic causes	Diabetic ketoacidosis (fruity smell), fetor hepaticus, and uremia (ammoniacal odor)
		Psychiatric conditions	Odor of bad breath perceived by the patient but not the physician/health provider. May be a hallucination. (Some patients will extend this to overall smells emanating from their body and will isolate themselves to avoid others smelling them)

The Nutrition-Focused Physical Exam

Table 39.1 (continued)

Region examined	Nutrition-associated examination findings (and select examination/test techniques)		Nutritionally related cause/association
	Lymph glands (neck)	Lymphadenopathy (swollen glands)	Food allergies Infection Cancer
	Parotid gland	Enlargement	Thiamine deficiency [29] Protein insufficiency Bulimia

Tongue [34, 57]	The tongue is composed of skeletal muscle covered with mucosa. On the top (dorsum) of the tongue are three different types of taste buds/papillae:
	  Filiform
	  Fungiform (which cover the entire surface of the tongue)
	  Circumvallate (posterior dorsum of the tongue in semicircular arrangement). In addition to containing the taste buds, these papillae also increase the surface area of the tongue
	Visualization: Have the patient say "Ah" while protruding (and then curling up) the tongue so as to visualize both the dorsum of the tongue and the sublingual surface. Gauze can be used to examine the far interior and lateral aspects (place on tip and grasp and pull with the aid of the gauze)		

	Color	Pale	Anemia
		Red *Candida* (beefy red with white coat); glossitis – various shades of coloration	Glossitis: B vitamin insufficiency (thiamine, riboflavin, niacin, pyridoxine, folate, B_{12}) Magenta (riboflavin insufficiency) Fiery red (niacin insufficiency) With hyperkeratotic appearance – vitamin A insufficiency [83] Note: in folate deficiency, glossitis is accompanied by normal proprioception, but in glossitis due to vitamin B_{12} deficiency/pernicious anemia, abnormal proprioception is seen Iron deficiency anemia Severe protein-calorie nutrition and malabsorption Alcoholism Oral estrogen – likely due to hormone-induced depletion of B vitamins [83] If *Candida* is the cause, may co-occur with iron and/or vitamin C deficiency
	Coat	Thin If occurs with atrophied taste buds "slick or smooth tongue" or atrophic glossitis	The tongue can often be sore (indicates atrophy). Vitamin B deficiencies, iron deficiency, alcoholism, severe protein-calorie malnutrition, malabsorption May occur with hypochlorhydria and/or small intestinal bacterial overgrowth (SIBO)
		White, cheesy discharge (atop beefy red tongue)	Candidiasis, GI flora imbalance, AIDS
		Thin coat with parched appearance	Dehydration
	Ridges/furrow	Fissured tongue (also known as "scrotal tongue")	Niacin deficiency and gut-triggered autoimmune issues thought to be possible causes as well May occur with psoriasis (along with geographic tongue) and Sjogren's syndrome [83]
	Swelling	Can often be seen with teeth marks on the lateral border of the tongue	Poor digestion, allergies, hypothyroidism, dysbiosis, amyloidosis, Down syndrome [29] Can also be seen with thickened speech, snoring, and sleep apnea

(continued)

Table 39.1 (continued)

Region examined	Nutrition-associated examination findings (and select examination/test techniques)		Nutritionally related cause/association
	Lesions	Geographic tongue A form of atrophic glossitis that has scattered smooth red areas of denuded papillae alternating with normal patches of the tongue – lending the map-like appearance that changes over time (migratory). Considered benign and self-limited in conventional medical circles	B Vitamin and zinc insufficiency Allergies May occur with psoriasis along with fissuring [83] and with chemical sensitivity [32] Celiac disorder
		Hairy black tongue – the "hair" is elongated papillae on the tongue dorsum	May follow antibiotic therapy, smoking, and exposure to bismuth exposure Use of charcoal at high doses in cases of poisoning/overdose can also lead to this condition
		Hairy leukoplakia – multiple white, warty, and painless plaques. These are usually located on the lateral aspect of the tongue (and can also be seen on the inner mucosa of the cheeks). The plaques each have hair-like projections. Ddx – *Candida*: In *Candida*, the white coat can be scraped off, which is not the case with hairy leukoplakia	Insufficiencies of Vitamin A, riboflavin, niacin, pyridoxine, folate, B_{12} precancerous, HIV/AIDS
		Tongue biting	If with history of syncope (fainting), suspect seizures (generalized tonic-clonic), especially when biting is on the sides (lateral portions) of the tongue
		Varicose veins on sublingual (underside of the tongue) veins. Referred to as "caviar tongue"	May be normal variant of age, but in some patients, may indicate a chronic increase in right-sided pressures (e.g., CHF)
	Movement	Tremors or involuntary movement	Hypoglossal (CN12) nerve defect
	Hypoglossal (CN12) examination: assess the patient's ability to articulate; inspect the tongue on resting and with protrusion		
Jaw	Jaw function – assess for temporomandibular joint (TMJ) dysfunction (it can limit/change food intake) Palpate the patient's jaw (also opening of ear canal) as they open and close their mouth. Feel and listen for audible clicks over the jaw. Assess for chewing and swallowing		
	Movement	Clicking and misalignment (and often reports of pain) due to TMD/TMJ or TMJD (temporomandibular disorder, temporomandibular joint disorder)	TMD/TMJD correlated with increased psychological stress and may be associated with abdominal obesity and a lower BMI in women [84] Headaches, allergies, depression, fatigue, degenerative arthritis, fibromyalgia, autoimmune disorders, sleep apnea, and gastrointestinal complaints were shown to be more prevalent in those suffering from TMJD [85] and in those with rheumatoid arthritis, ankylosing spondylitis, and primary Sjogren's syndrome [86]

The Nutrition-Focused Physical Exam

Table 39.1 (continued)

Region examined	Nutrition-associated examination findings (and select examination/test techniques)		Nutritionally related cause/association
	Chewing and swallowing	Important to assess for both chewing and swallowing in elderly 3 oz. water swallow test If this test is passed, thin liquids and other food consistencies can be recommended without further dysphagia assessment [88] Dysphagia risk assessment for the community-dwelling elderly (DRACE): A valid and reliable tool for detecting latent risk of chewing and swallowing disorders in the elderly community-dwelling population [89]	Those with chewing and swallowing difficulties had significantly lower vitamin A, E, and manganese levels, and those with chewing difficulties had significantly lower magnesium and vitamin E [87] Vagus nerve defect (CN10) could interfere with swallowing
Neck[31,34]	Lymphatic vessels are located throughout the body in all tissues and organs. Lymphatic vessels accompany blood vessels – which helps to locate the nodes in some regions of the body. Lymphatic vessels function to collect lymph (ECF) and carry it to venous circulation. In transit to the venous system, lymph passes through lymph nodes – which serve to filter lymph fluid. Microbes, malignant cells, and other debris in the lymph nodes can lead to enlargement and/or hardening of the lymph nodes. If this enlargement or hardening is significant enough, the nodes can be palpable, and the diagnosis of "peripheral lymphadenopathy" is made. Note that a normal adult has around 400–450 lymph nodes in the body, though only ~100 of these are palpable (in the arm and underarm (axilla), in the leg and in the head and neck)		

		Lymphadenopathy
		When palpating a lymph node, do so lightly and assess for:
		  Size – >1 cm diameter (usually measured with calipers); considered significant and may be pathological (may see large nodes in intravenous (IV) drug users – often benign)
		  Consistency – Rock hard suggestive of malignancy (though Hodgkin's lymphoma nodes are usually rubbery). Note that cysts, being fluid filled, are softer, and swollen nodes without malignancy (infection or allergies) will generally be softer
		  Matting – individual nodes swell together to create larger nodes that may or may not be stuck to overlying skin or underlying tissue
		  Tenderness – usually tender in inflammation, less often with cancer
		In general:
		  Benign nodes are usually small, soft, non-tender, and discrete (not matted)
		Inflammatory nodes, tender, firm, and often matted.
		  Cancerous nodes usually large, non-tender, matted, and rock hard. "Sentinel" nodes are nodes that, when enlarged or abnormal to palpation, signal a specific condition
			Allergies, infection (systemic, may have generalized and chronic LA (lymphadenopathy); acute, may be isolated node(s) and resolve as the infection resolves), toxin exposure, Hodgkin's, lymphoma, leukemia, cancer, HIV, CT (connective tissue) disorders. Note that cysts and lipomas must be considered in the differential
			Significance by location:
			  Occipital (kids, childhood infections; adults, rare (maybe with scalp infection))
			  Posterior cervical – dandruff
			  Preauricular – conjunctivitis and lymphoma
			  Submandibular, submental – dental or cancer of oral or nasal region
			  Supraclavicular node enlargement on the left known as "Troisier's node" signifies spread of intra-abdominal or intrapelvic malignancies
			  Supraclavicular node enlargement on the right – lung and breast cancers

(continued)

Table 39.1 (continued)

Region examined	Nutrition-associated examination findings (and select examination/test techniques)		Nutritionally related cause/association
	Venous distension along the carotid artery		Fluid status overload [1, 3] – may be due to heart failure, hypoalbuminemia, retention of salt and water, venous, or lymphatic stasis
	Parotid gland enlargement		Protein deficiency/bulimia (bilateral) [3] Thiamine deficiency [29] Cyst, tumor, or hyperparathyroidism [3]
	Range of motion	Decreased range of motion can affect ability to properly masticate food [3]	
Thyroid [31, 34]	With palpation, note size, consistency (soft, firm, rubbery, or hard), texture (diffuse or nodular), tenderness, tracheal deviation (seen with asymmetrical goiter), and enlarged lymph nodes [over the thyroid] Rubbery consistency is palpable with Hashimoto's thyroiditis, and hardness is noted with cancer The size of the thyroid varies with the supply of iodine in a diet (less iodine leading to larger glands). Note that women's thyroid glands are usually larger and easier to palpate than men's. And the right lobe of the thyroid is often a little larger than the left		
		Goiter	Iodine deficiency/excess Can occur with hypothyroid, euthyroid, or hyperthyroid status
		Nodules	Iodine deficiency Hashimoto's thyroiditis Thyroid adenoma, cyst, cancer
Chest	**Cardiac exam:** *Normal:* "Lub dub" (S1, S2, or first and second heart sounds produced by the closure of the AV valves (tricuspid and mitral) and then the pulmonic and aortic valves (S2)). This explains why S1 is heard more in the lower chest and S2 in the upper Listen for abnormalities of the sound of S1 and S2 over each of the valves and any additional sounds *Clinical significance:* Auscultation can be used to detect a wide range of cardiac conditions using careful techniques of listening for S1 and S2 over each of the valves, using specific positions or techniques to accentuate suspected abnormalities and listening for adventitious sounds **Pulmonary exam:** Perform both anteriorly (patient lying on back) and posteriorly (patient seated with arms crossed (spreads scapula to better examine the lungs)). Work from head down – comparing sides as you go (asking them to take a deep breath in each time you place your stethoscope on a new site) *Palpation* Tenderness (can occur with costochondritis – inflammation of the costal cartilage) *Percussion* Determines if the lungs are full of air/fluid/solid masses *Auscultation* Compare sides with patient breathing through the mouth		
	Cardiac [31, 34, 57]	Mitral valve prolapse (MVP) [31, 34, 57]	Magnesium, carnitine, potassium, calcium, and niacin insufficiency Dehydration, allergies, medications
		Palpitations, arrhythmia	Thiamine, magnesium, coenzyme Q_{10} (CoQ_{10}), vitamin K, and calcium deficiency [29] Potassium or magnesium deficiency/excess, calcium, or phosphorus deficiency [3]
		Cardiomegaly	Selenium, thiamine [29]
		Tachycardia	Thiamine [3], CoQ_{10} deficiency [29] Dehydration [3] Organic heart disease, severe pulmonary disease, respiratory insufficiency, excessive alcohol consumption, or drug toxicity

Table 39.1 (continued)

Region examined	Nutrition-associated examination findings (and select examination/test techniques)		Nutritionally related cause/association
	Pulmonary inspection	Barrel chest	Thin elderly individuals Chronic obstructive pulmonary disease (COPD) such as emphysema
		Pigeon chest (bowed chest)	Rickets, Marfan syndrome Familial (normal)
		Funnel chest (hollow chest)	Can lead to arrhythmias or MVP
		Retraction of interspaces	Severe asthma, COPD
		Use of accessory muscles to breathe (sternocleidomastoid (SCM) and scalene) during inspiration and abdominal muscles in expiration. Normally the diaphragm is the only muscle used in breathing	Signifies COPD or respiratory muscle fatigue
		Reactive airways	Magnesium deficiency
	Respiratory rate and rhythm	Increases in respiratory rate lead to increases in caloric, and often fluid, requirements, and changes in respiratory rate can affect acid-base balance [3]	
		Pursed lip breathing	Emphysema (often the result of chronic cigarette smoking)
		Tachypnea (rapid breathing)	Physical exertion Heat stroke, shock, anxiety/panic attack Metabolic acidosis, diabetic ketoacidosis, pneumonia, cystic fibrosis, pulmonary embolism, heart failure, sepsis
		Bradypnea	Hypothyroidism Medications: sedatives/narcotics
		Apnea (absence of breathing 20 seconds while awake, 30 seconds while asleep)	Sleep apnea
	Peak expiratory flow	Measurement of air flowing out of the lungs See Fig. 39.17. Peak Expiratory Flow Rate Green, Yellow Zone and Red Zone – relative to peak flow	Asthma associated with a narrowed airway – using a peak flow meter gives information about how open the airways in the lung are (during an attack, the airways in the lungs slowly begin to narrow) Daily PFM readings for those with asthma are encouraged. This provides valuable information about the status of the airways and can help to determine triggers of asthma, how well an asthma management/treatment program is and when emergency medical care is needed. With an impending attack, the peak flow rates start to drop. This allows early changes in one's medication or routines to prevent worsening of asthma symptoms Decreased peak flow rates may be seen with EFA, magnesium insufficiency, with food/environmental allergy and if fewer methyl donors (folate, vitamin B_{12} insufficiency) available (as these are required for histamine metabolism) Prognostic: decreased peak expiratory flow rate predictive of increased mortality in those with COPD [90]
	Auscultation	Rales (crackles)	Associated with fluid in the alveoli (fluid overload), as seen in CHF

(continued)

Table 39.1 (continued)

Region examined	Nutrition-associated examination findings (and select examination/test techniques)		Nutritionally related cause/association
		Rhonchi (high pitched, continuous, clear with cough)	Obstruction
		Wheezes (high pitched, musical)	Due to narrowing of airways with chronic obstructive lung disease (including asthma). Hear on expiration

		Kussmaul (Deeper and more rapid breathing)	Diabetic ketoacidosis anion-gap acidosis "MAKE UP a List" [34]:
			    Methanol poisoning
			    Aspirin intoxication
			    Ketoacidosis
			    Ethylene glycol ingestion
			    Uremia
			    Paraldehyde administration
			    Lactic acidosis

Abdomen [21, 31, 34]	Generally, the 4-quadrant method is preferred as the 9-quadrant system often has organs occupying more than one quadrant Have the patient lie supine (on back) – pillow under the head and with knees bent (or pillow under) Ask if they need to void their bladder before having them lay down on the exam table Patient's arms should be at sides or folded over the chest. Ask the patient to point to any areas of pain and examine these last. While performing the examination, watch the patient's face for signs of discomfort. Make sure your hands and stethoscope are warm and your nails trimmed. If the patient is tense or ticklish, begin the palpation step with their hand under yours – you can then slip your hand under soon thereafter as they will usually calm quickly. As you carry out the four steps (inspection, auscultation, percussion, palpation), try to visualize the organs underneath		
	Inspection	Scars, bruising/ecchymosis	Sign of abdominal hemorrhage

		Distension	6 Fs:
			    Fluid (ascites)
			    Fat (obesity)
			    Flatulence (gas)
			    Fetus (pregnancy)
			    Feces (constipation)
			    Full-sized tumor (abnormal lesion) [16]

		Striae (stretch marks)	Considerable weight changes Pregnancy Cushing's syndrome can see pink-purple striae
		Scaphoid (sunken) – occurs with loss of subcutaneous fat	Insufficient caloric intake [3]
		Dilated veins	Hepatic cirrhosis obstruction inferior vena cava
		Rashes, lesions, nodules	Keratin and sebum can build up in the umbilicus – leading to a nodule which can be readily extracted Prognostic: Sister Mary Joseph Nodule – metastatic carcinoma of the umbilicus, usually from the stomach, colon ovary, or pancreas – often means only 10–11 months left of life
		Contour (rounded, protuberant, bulging flanks)	Protein deficiency [29] Gas Obesity Ascites (often accompanied by edema and hypoalbuminemia) [31]
		Symmetry	Asymmetry with enlarged organ or mass
		Caput medusae (abnormal venous networks on the abdominal wall)	Portal hypertension (usually seen in those with cirrhosis)
		Ascites (accumulation of fluid in the peritoneal cavity) – bulging flanks, flank dullness	Liver disease (severe), cancer, CHF, pancreatitis

Table 39.1 (continued)

Region examined	Nutrition-associated examination findings (and select examination/test techniques)		Nutritionally related cause/association
		Scratch test (Sergent's white line) Take the cap end of a pen and stroke it lightly across the abdomen (make about a 6-inch line) A normal response is to see a whitish line, which reddens within a few seconds. In adrenal insufficiency, the line remains white and may grow wider	If redness is delayed/absent when scratch near the umbilicus, evaluate for decreased adrenal function
	Auscultation	Follows inspection (prior to percussion and palpation) as these can alter the frequency of bowel sounds. This step of the examination is helpful in determining bowel motility (via the bowel sounds) and exploring vascular obstructions or abnormalities (such as bruits) Listen for bowel sounds using your stethoscope over all four quadrants. Bowel sounds usually occur on the order of 5–34 per minute. You may hear the stomach growling (prolonged gurgle of hyperperistalsis known as borborygmi). Listen for 2 minutes before deciding if absent If patient has high blood pressure, listen in epigastric region for bruits∗. (If arterial insufficiency is suspected in the legs, listen over the aorta, iliac arteries, and femoral arteries. And if the liver pathology is suspected, listen over the liver for hepatic bruits) ∗A bruit is caused by turbulent blood flow in an artery	
		Increased borborygmi	Diarrhea and in early small bowel obstruction
		Decreased borborygmi	Post-surgery, later stage bowel obstruction, paralytic ileus, peritonitis [3]
	Percussion	Helps to assess the amount and distribution of gas in the abdomen and to identify possible masses	
		Hepatomegaly	Protein deficiency or vitamin A excess Hepatitis Chronic alcohol abuse/cirrhosis, lymphadenopathy [31]
		Splenomegaly	Lymphadenopathy, infection (tropical diseases (malaria, typhoid)), HIV, jaundice, hepatocellular disease, cirrhosis, leukemias, lymphomas [31], hemolytic anemia
	Palpation	Tenderness to the left of the xiphoid at the fourth intercostal space	May be due to hiatal hernia syndrome
		Tender areas in the colon	Investigate for yeast overgrowth, dysbiosis, diverticulitis, colitis, cancer (particularly if accompanied by occult blood), and a variety of functional GI disorders Testing HCL levels may be indicated [57]
		Abdominal wall tenderness test (Carnett's test) [34, 91] Identify the area of maximal tenderness from palpation and apply enough pressure to the point to elicit a moderate amount of tenderness. Continue to apply pressure as the patient folds the arms on the chest and lifts the head and shoulders (partial sit-up). If this elicits increased tenderness at the site being palpated, it is a positive test (and thus peritonitis less likely)	Diabetic neuropathy elicits a positive test and will often occur with cutaneous hypersensitivity and weakness of the abdominal organs (producing bulging of the abdominal wall) In patients with chronic abdominal pain, a positive abdominal wall tenderness test decreases the probability that the pain has a visceral (i.e., organ-related) cause
		Hepatojugular reflex With patient in supine position, exert firm and sustained pressure on the lower liver edge (in and up toward the head at the lower right costal margin)	In CHF, will see filling of the neck veins – which characterizes a positive reflex test

(continued)

◼ Table 39.1 (continued)

Region examined	Nutrition-associated examination findings (and select examination/test techniques)		Nutritionally related cause/association
Musculoskeletal [21, 26, 31]	Inspect – for joint swelling, tenderness, and deformity Palpate – for joint warmth, tenderness, and crepitus Investigate – range of motion (ROM) of joint Note that joint pain within the joint (articular disease) often manifests in swelling and tenderness surrounding the joint and often limits ROM with both passive and active movements, whereas pain outside the joint (extra-articular disease) causes swelling and tenderness which is localized to specific regions within the joint and does not affect all aspects of a joint's ROM Musculature and skeletal system are related to nutritional status. In developing countries, it can be seen with marasmus (energy deficiency) and kwashiorkor (protein and energy deficiency) with frank malnutrition, whereas in developed countries, the presentation is generally subtler, with lower body weight, muscle atrophy and loss of subcutaneous fat, weakness, and laboratory abnormalities (low albumin, increased creatinine excretion)		
	Skull	Delayed closure of fontanelles in infants or abnormal softening of bones in the skull (craniotabes), frontal, or parietal bossing (swelling or thickening of the front and sides of the head)	Vitamin D and calcium deficiency [3, 29]
	Head/face	Muscle wasting (temporal)	Protein and calorie insufficiency [1, 3]
	Torso	Beading of ribs (prominent knobs at the base of the costochondral joints (also known as Rachitic Rosary). These knobs create appearance of beads under the skin of the rib cage)	Rickets, with bowed legs, pain in spine, pelvis and legs, thickened wrists and ankles, projection of the breastbone, delayed growth, and muscle weakness, due to vitamin D deficiency
		Subperiosteal hemorrhages (of the femur)	Children with vitamin C deficiency
		Scoliosis A lateral curvature of the spine	Structural (with vertebral rotation or thoracic deformity) or functional (unequal leg length without vertebral rotation or thoracic deformity) Some evidence to suggest that scoliosis is associated with methylation defects (methylenetetrahydrofolate reductase (MTHFR)) [92]
		Kyphosis A rounded thoracic convexity Dowager's hump	Osteoporosis (adults) If in adolescents, consider Scheuermann's disease (seen in tall and underweight adolescents). There is a hereditary component to this disease
		Lordosis Accentuation of the normal lumbar curve	Can develop in compensation to protuberant abdomen (pregnancy or marked obesity) or in compensation for flexion deformities of the hip
	Legs	Bowing of legs and/or knock knees	Rickets (vitamin D deficiency)
		Muscle wasting gastrocnemius, buttocks	Insufficient protein and calorie intake Thiamine, vitamin C, phosphorus, and calcium deficiency [3]
		Low skeletal muscle mass lower leg	Independently associated with osteoarthritis of the knee in obese individuals [93]
		Weakness, pain in calf muscles	Thiamine deficiency
		Epiphyseal enlargement	Vitamin D deficiency (if painless) or vitamin C deficiency (if painful)
	Arm(s)	Arm muscle area – assesses total amount of muscle or protein in the body (uses triceps skinfold measure)	(See ◼ Fig. 39.18. Anthropometrics Flow Chart)
		Deltoid muscle wasting	Protein and calorie insufficiency [1]

Table 39.1 (continued)

Region examined	Nutrition-associated examination findings (and select examination/test techniques)		Nutritionally related cause/association
	Spinal accessory (CN11) examination: Assess patient's ability to shrug shoulders and strength of sternocleidomastoid muscles with flexion as head turns against your hand [21]		
	Wrists and carpal joints	Ganglion (cystic, round, usually non-tender swelling along the tendon sheath on the dorsum of the wrist commonly). They can also develop elsewhere on hands, wrists, ankles, and feet	Not a known nutritional association. (Though homeopathic silica is often used to treat ganglion cysts.)
		Dupuytren contracture A fibrous thickening on the palm along the tendon of the fourth finger which keeps the finger partially flexed	Hereditary (in part), males over 40, dysglycemia, seizures (for reasons not yet understood) History of hand trauma, alcohol, and tobacco use may also predispose one to the condition
		Subluxation of the ulna	Chronic rheumatoid arthritis
		Epiphyseal enlargement	Deficiency of vitamin D (if painless) and of vitamin C (if painful)

		Grip strength
		Measure using a dynamometer and compare readings obtained to instrument's normative tables
			Hand grip strength reflects nutrient status and accurately predicts postoperative complications (Lower hand grip strength indicates poorer nutrient status [particularly protein and vitamin D [29]] and increased risk of post-op complications) and may be indicative of presence of sarcopenia
			Recommended cut-off values for sarcopenia [94]:
			  Caucasian:
			    <20 kg in women
			    <30 kg in men
			  Asian:
			    <18 kg in women
			    <26 kg in men
			Prognostic: There is also evidence to suggest that in postmenopausal women, grip strength is positively related to normal bone mineral density and may help to identify women who could benefit from additional bone density evaluation [95]
			Predictive of risk of cardiovascular disease, cardiovascular death, and all-cause death (across both genders) [96] and was a stronger predictor of cardiovascular and all-cause mortality than systolic blood pressure [97]
			Grip strength (post-stroke) of the unaffected side considered an independent predictor for short-term functional improvements [98]

	Fingers	Swan neck deformity, Boutonniere deformity, ulnar deviation	Rheumatoid arthritis
		Heberden's nodes Nodules on the distal interphalangeal joints (due to bony overgrowth of the joint). They are usually hard and are painless Bouchard's nodes are located on the proximal interphalangeal joints and are less common	Osteoarthritis (consider allergy to nightshade vegetables) [29]
		Callus on the back of the hand	Bulimia (from using fingers to stimulate the gag reflex in order to vomit)

(continued)

Table 39.1 (continued)

Region examined	Nutrition-associated examination findings (and select examination/test techniques)		Nutritionally related cause/association
		Gout	Associated with increased protein (purine-rich foods such as red meat and sardines), alcohol, medications (diuretics, low-dose aspirin, cyclosporine, end-stage renal disease), and metabolic syndrome
		Skin cracks/splits	Zinc and EFA deficiency [29]
	Global symptoms	Bone/joint pain	Vitamin A, C, and D deficiency [3, 29] Infection Degenerative joint changes Tumor(s) Multiple myeloma
		Muscle spasms	Magnesium insufficiency (Carpopedal) calcium, magnesium insufficiency [29]
		Muscle tenderness bilaterally	Thiamine insufficiency [3]
		Decreased muscle tone (especially of hips and pelvis)	Vitamin D insufficiency
	Sarcopenia	Defined as the loss of muscle mass and strength that occur with advanced age [99]	Vitamin D insufficiency (and thus likely those with VDR polymorphism) and protein (and hence those with hypochlorhydria, pancreatic insufficiency, or leaky gut are at increased risk [32]) Antioxidant insufficiency (vitamins C and E, carotenoids, and selenium) is linked to increased risk of sarcopenia in older adults [100] Considered to be due to chronic low-grade inflammation driven by decreased physical activity, anabolic hormone, cytokines, oxidative stress, and adipokines (with obesity) [101]
Neurological [34]	Mental status	Confabulation (making up stories) and disorientation	May be due to Korsakoff's psychosis (deficiency of thiamine resulting from alcoholism) or Wernicke's encephalopathy (precipitates the brain damage that leads to Korsakoff's psychosis)
		Acute disorientation	Phosphorus and niacin insufficiency [29]

	Reflexes	Reflexes are involuntary contractions of the muscles which are induced by specific stimuli
		We will focus on muscle stretch reflexes, in which a brisk stretch of the muscle induces the reflex via muscle spindles to the spinal cord (which then sends the signal back to the muscle and causes the reflexive contraction)
		Examining reflexes:
		  Have the patient seated on an examination table with their leg dangling and limbs hanging symmetrically
		  Sharply tap (with a reflex hammer) on the point where the muscle inserts distally on the bone. On tapping the pointed end or blunt end can be used. The pointed end is better in smaller locations. Hold the hammer between the thumb and index finger so it swings freely within the hand
		  Check reflexes bilaterally and compare. Use only enough force needed to generate a response. Can reinforce the knee reflex (i.e., accentuate if the reflex response is difficult to elicit) by asking the patient to clench their jaw and interlace their fingers
		  Reflexes typically tested include the biceps, triceps, brachioradialis, abdominal, patellar, ankle, and plantar [3]
		Rate the reflex response according to the following scale:
		  0 = absent
		  1 = hypoactive (diminished)
		  2 = normal
		  3 = brisk (increased)
		  4 = hyperactive
		For the Babinski reflex, stroke the lateral aspect of the foot with the metal end of the reflex hammer. A normal response is the absence of a reflex response. If dorsiflexion (the toe goes up), it is a positive sign
		Asymmetry of reflexes more suggestive of pathology	

Table 39.1 (continued)

Region examined	Nutrition-associated examination findings (and select examination/test techniques)		Nutritionally related cause/association

		Hypoactive (Achilles, patellar reflexes)	Hypothyroidism (Achilles)
			Nutrient deficiencies:
			  Potassium [3]
			  Thiamine, pyridoxine, vitamin B_{12}, or possibly with excess B_{12} (Achilles, patellar reflexes)
			  Respiratory alkalosis [3]

		Hyperactive	CNS disease Calcium deficiency/hypocalcemia
		Positive Babinski sign	Parasites, drug/alcohol intoxication, post-seizure (or upper motor neuron disease – such as amyotrophic lateral sclerosis (ALS)), vitamin B_{12} deficiency [102] (Positive sign is normal to see in infants)
	Touch	Examine for ability to perceive touch by touching the skin lightly over different dermatomes with cotton wisp covering as many dermatomes as possible A dermatome is the area of the skin innervated by the sensory fibers of a single nerve root If a nerve root is damaged – there will be loss of sensation across the whole dermatome [31]	
		Monofilament testing Apply pressure on the plantar side of the foot until the monofilament buckles Hold for 1 second and release	If a patient cannot sense this pressure, they have loss of protective association (seen in diabetes) Prognostic: increased risk of foot ulceration and amputation
		Paresthesias Abnormal sensation with "pins and needles" (numbness, tingling, pricking sensation) due to damage to (or pressure on) the peripheral nerves	Thiamine, iodine, vitamin B_6, vitamin B_{12}, vitamin E, omega-3 fatty acid, phosphorus insufficiency [29] Carpal tunnel syndrome, ulnar neuropathy, multiple sclerosis (MS), diabetes, hypothyroidism, alcoholic neuropathy, drug toxicity
	Temperature	Examine for ability to perceive temperature by touching the skin lightly over different dermatomes with tubes of warm/cold water or cold tuning fork and warm index finger of clinician (to discern difference) covering as many dermatomes as possible [31]	
	Pain	Examine for ability to perceive pain by touching the skin lightly over different dermatomes with sharp and dull object/stimuli (to discern difference) covering as many dermatomes as possible [31]	
	Sensation	Dermatomal testing discussed previously	
	Vibration	Test with a tuning fork (128 Hz) Strike the tuning fork forcefully against the heel of your palm and then apply the stem of the fork to the lateral malleolus (ankle bone) should perceive vibrations for ≥11 seconds (decreases by 2 seconds for every decade starting at 40 years) And ≥15 seconds (decreases by 2 seconds for every decade starting at 40 years) on the ulnar styloid	Decreased vibratory sense seen with exposure to heavy metals, neurotoxins [103] decreased antioxidant intake, methylation defects (riboflavin, B_6, folate, B_{12}) Thiamine, riboflavin, niacin, pantothenic acid, and B_{12} insufficiency Polyneuropathy of diabetes, Lyme disease, collagen vascular disease, allergy, and autoimmune conditions [3]
	Proprioception/position sense [31]	Grasp the lateral aspect of the toe or index finger between the thumb and forefinger. Have the patient close his/her eyes and move the digit up and down – pausing to ask what orientation they perceive the digit to be in	Loss of proprioception with peripheral neuropathy Thiamine [29], vitamin B_{12} deficiency Diabetes, MS, history of stroke, severe brain disease

(continued)

Table 39.1 (continued)

Region examined	Nutrition-associated examination findings (and select examination/test techniques)		Nutritionally related cause/association
	Neuropathy	Peripheral neuropathy Characterized by weakness, paresthesias, ataxia, decreased tendon reflexes, fine tactile sense, decreased vibratory sense, and (position sense)	EFA, vitamin E, thiamine, pyridoxine, vitamin B_{12} insufficiency, or excess pyridoxine [3, 29] Alcoholism (polyneuropathy) [36]
	Tremors [21, 31]	Resting (static) Tremor is most prominent at rest and may decrease or disappear with voluntary movement	Parkinson's disease Magnesium and pyridoxine insufficiency [29]
		Active tremors Postural: Active tremors appear when actively holding the affected part. It may worsen with intention. Essential tremor is the term used when it is a benign tremor. It is the most common movement disorder, and the etiology is unknown. Intention: The tremor is absent at rest, appears with activity, and worsens as one nears the target (e.g., picking up a mug)	Postural: hyperthyroidism, fatigue, anxiety, and also benign (familial) tremor Intention tremor: multiple sclerosis
	Balance	Romberg test – patient should stand with feet together and eyes open and then close both eyes for 60 seconds. Note that you will want to have an arm in front of and behind them to protect them from falling should they lose balance. Only very minimal swaying should occur unless ataxia (problems with coordination and nervous system function due to neurological lesion), cerebellar disorders, or an acoustic nerve lesion (CN8) is present	
		Positive Romberg	Diabetic neuropathy, Guillain-Barre, MS Ataxia Vascular injuries to the thalamus [36]
	Gait [36]	Walking is a very complex action that requires integration of many systems (motor, sensory, vestibular, and visual) so most abnormalities in the nervous system are visible in the gait. Look for symmetry versus asymmetry (muscle, joint, and pain disorders cause asymmetry, whereas abnormal limb control tends to be more symmetrical and due to central lesions) Disorders of the gait can be due to pain, joint immobility, muscle weakness, or abnormal control of the limbs Note that alcoholism/alcohol intoxication affects gait enormously (can lead to wide gait, poor tandem gait, and possibly leg ataxia)	
		Gait speed: Associated with increased survival in older adults, said to reflect health and functional status and to be a potentially useful clinical indicator of well-being [104] and survival [105] in older adults Measure via measured walk from standing start (over 8 feet [104] or 4 m [105] Patient to walk from start to finish at their usual speed "just as if… walking down the street to go to the store" [104] Time two walks and average May be less helpful for older adults who already report dependence in basic activities of daily living [105]	≤0.8 m/s associated with sarcopenia (other measures (of muscle mass and strength) need to be assessed to confirm/strengthen diagnosis [100] <0.6 increased likelihood of poor health/function >1.0 m/s associated with healthy aging [105]

Table 39.1 (continued)

Region examined	Nutrition-associated examination findings (and select examination/test techniques)		Nutritionally related cause/association
		Slow gait	Sarcopenia Depression Schizophrenia Prognostic: predictive of cognitive decline
		Rapid gait	Mania Hyperthyroidism
		Tandem gait	Schizophrenia, alcoholism [36]
		Wide gait	Thiamine and vitamin B_{12} insufficiency [29] Alcoholism
		Shorter stride length	Schizophrenia Cortical/basal ganglia disorders
		"Parkinsonian gait" Short steps (petit pas), reduced arm swing, stooped posture, anteropulsion/retropulsion, festination, postural instability	Parkinson's disease Depression (can resolve as depression resolves) Schizophrenia (mild Parkinsonian gait) Antipsychotic medications
		Ataxia Increased width of gait, arrhythmic steps, unsteadiness and impaired tandem gait, and loss of position sense	Cerebellar, pons or thalamus injury, possible cortical damage Alcohol or medications (benzodiazepines, anticonvulsants) Thiamine, vitamin E, vitamin B_{12}, and copper insufficiency [1, 29] Mitochondrial dysfunction, EFA imbalance, toxin burden [20]
		Antalgic gait	Degenerative joint disease (DJD), orthopedic injury
		Spastic gait with scissoring	Osteoarthritis, B_{12} deficiency, trauma, inflammation, tumor, demyelination disorder
		Abrupt onset of gait disturbance, selective disability, gait disturbance related to minor trauma	Psychogenic balance disorders
	Risk of falling	Stops walking when talking (difficulty performing both simultaneously confers increased risk of falling) Timed chair stands: times how long it takes for a patient to get up from a chair and sit again three times in a row (positive is >10 seconds) Get-up-and-go test: assesses mobility and predicts falls – positive is ≥12 seconds [106]	Polypharmacy (medications risk factor for falls in at least 30% of elderly exhibiting gait or balance disorders – especially psychiatric medications) Hypotension Positive get-up-and-go test with sarcopenia
	Coordination	Assessed by: Finger-to-nose test Rapid alternating movements Heel-to-shin tests	Cerebellar disease – clumsy/unsteady movement [21]

Anthropometrics				
Measure	Examination/measurement method	Optimal	Excess/elevated	Deficient/depressed
Stature/height	The simplest method to accurately measure height involves positioning the individual to be measured with their back against a flat vertical surface upon which a measuring stick or measuring tape has been affixed. Then a headboard is positioned at a right angle to the wall or vertical surface upon which their back rests and their stature can be obtained (ideally to the closest 1/8 of an inch) A stadiometer can also be used to measure height Some notes on ensuring accuracy with this measurement: If a wall is used, the baseboard should be minimal, and the individual being measured should be standing on a hard, uniform surface without carpet For those with difficulty walking or who have severe spinal curvature, measuring height by this method would be inaccurate, so their stature is best estimated from knee height (normative reference tables are published to interpret this measure) -Knee height correlates closely with stature An alternative method is to estimate stature from upper or lower arm length (though this is used less often) The knee and ankle of the left leg are typically measured (as the left leg was used by researchers in developing the equations used to determine normative values) The knee and ankle should be at a 90-degree angle to each other (verify this with a right triangle, square, or other device) The fixed blade of the sliding caliper is positioned under the heel of the left foot, and the moveable blade is placed on the anterior surface of the left thigh [26]	Used relatively (in other measures (such as body mass index and waist-to-height ratio)) and for growth chart assessment		
Length (aka recumbent length) Pediatrics	Obtained with subject lying down in face-up (supine) position. Usually limited in use for children ≤24 months of age or children who are unable to stand without assistance. Length forms the basis of growth charts up to 24 months of age, with height used in growth charts for children at 2 years and beyond To measure recumbent length accurately, a special measuring device is required, and two people are needed to accurately obtain the measurement. A stationary headboard and moveable footboard must be perpendicular to the backboard The length is measured from the foot board ("0" starts at the headboard), with the crown of the head held securely against the headboard, and a right angle made between the Frankfort horizontal plane and the backboard The soles of the feet are to be held firmly against the footboard at this time (toes pointing upward), and the child's shoulders and buttocks should be firmly touching the backboard. The length of the child can then be recorded to the nearest 1/8 of an inch. Make notes of any challenges or increased estimation based on the child moving about and not laying still during this procedure The rate of gain in length is indicative of nutritional adequacy [26]	Used relatively (in other measures (such as body mass index and waist-to-height ratio) and for growth chart assessment		

The Nutrition-Focused Physical Exam

Measure	Examination/measurement method	Optimal	Excess/elevated	Deficient/depressed
Head circumference {Pediatrics} [26]	To perform this measurement, with the child seated in their caregiver's lap and the lower edge of a non-stretchable, but flexible, measuring tape positioned just above the eyebrows and ears around the occipital prominence of the head. The tape should be tight enough to compress the hair. The measurement should be read to the nearest 1/8 of an inch. A duplicate reading is best to ensure an accurate reading	The measurement of head circumference is very important for screening in the first year of life, in particular. In the first 12 months of life, head circumference increases rapidly and slows by 36 months of age. Pediatric growth charts are used to evaluate head circumference for age as a means to monitor growth and development		
Weight [26]	Weight is one of the most important measurements in nutritional assessment. It is important in predicting caloric expenditure and in determining body composition. Electronic scales are preferred over balance beam scales (for accuracy and they also are generally lighter weight, more portable, and easier to use and have a higher weight capacity (best when weighing obese clients)). Many models of electronic scales can connect into computer networks, and information on weight, body mass, and stature can be directly recorded into a client's electronic chart/record. To weigh a child or adult correctly/accurately, ensure that both feet are firmly planted on the base of the scale (in the middle of the platform) without touching anything and with weight evenly distributed over both feet. Read the weight to within 0.2 lb. Any subsequent reading, if performed correctly, should agree within 0.2 lb. It is best to weigh both children and adults after voiding (after they have emptied the bladder) and dressed in an item of clothing of known weight, such as a gown. There are diurnal variations in weight of about 2 lbs. in children and 5 lbs. in adults, so it is best to record the time that an individual was weighed and to weigh them subsequently at approximately the same time of the day. Non-ambulatory persons require a bed scale or wheelchair scale. The individual is positioned into a weighing sling, which is then raised until the person is suspended over the bed. A chair scale can also be used to obtain weight (which has the person sitting upright in their chair while leaning against a backrest). And finally, there are wheelchair scales (upon which the wheelchair is rolled) for obtaining a person's weight. Note: Weight in adolescents more accurately reflects nutritional adequacy than height [17]	Optimal weight is 22.5–25 kg/m^2. The risk of death is lowest when the body mass index (BMI) is approximately 22.5–25.0 kg/m^2. For every 5 kg/m^2 increase in BMI, risk of death from all causes increases by 30%, from cardiovascular disease by 40%, from diabetic renal and hepatic diseases by 60–120%, and from neoplastic diseases by 10%. Usual body weight (UBW). Variations from UBW strongly linked to nutritional risk and health complications. Nutritional risk: >5% unexplained weight change in <1 month or >10% in 6 months. Percent usual body weight [107] =current body weight ×100 usual body weight. Percent weight (wt) change = *Present body weight – usual body weight* ×100 usual body weight or: % usual body weight – 100 = % change	Overweight. A body weight in excess of some reference point. The reference point is usually defined in terms of weight for height. Possible for very muscular person to be overweight (though not commonly the case). BMI ≥25 kg/m^2. Obesity. An excess amount of body fat relative to lean body mass usually expressed as a percentage of body weight that is adipose tissue. BMI ≥30 kg/m^2. Those with a BMI in the 30–35 kg/m^2 range died 2–4 years earlier. Those with a BMI in the 40–45 kg/m^2 range died 8–10 years earlier (comparable to the effect of cigarette smoking on lifespan)	Underweight. Defined as a BMI <18.5 kg/m^2. With a BMI <22.5 kg/m^2 (not even underweight), the risk of death increases, especially for smoking-related diseases such as respiratory diseases and lung cancer

Measure	Examination/measurement method	Optimal	Excess/elevated	Deficient/depressed
Body mass index (BMI) (Quetelet index)	Because of the difficulty of determining body composition and the ease of measuring height and weight, BMI is used to screen for overweight and obesity in children, adolescents, and adults Correlates with underwater weighing and dual-energy X-ray absorptiometry (DEXA) BMI = (W/H [2]), which is weight in kilograms divided by height in meters (squared) Calculations or nomograms available for interpretation or classification and online calculators are available for metric and nonmetric calculation BMI differs with race/ethnicity, sex, age, and musculature (athletes with increased muscle mass may have elevated BMI measure in the absence of elevated total body fat). These differences must be considered in interpreting the BMI	Child and adolescent: 5th–85th percentile on growth chart (BMI for age percentile charts) [17] Adult [17]: 22.5–25 kg/m²	Child and adolescent: Overweight 85th–95th percentile Obese: ≥ 95th percentile Adult [17]: Overweight: 25–29.9 kg/m² Obese: ≥30 kg/m²	Child and adolescent: <5th percentile Adult: Underweight<18.5 kg/m² BMI <23 in older adults (>65 y.o.) confers an increased risk of mortality [17]

Growth charts	Growth charts are used to assess the growth and development of infants, children, and adolescents (up to 20 years of age)
	They are based on growth data obtained from large numbers of healthy infants, children, and adolescents and are an important tool in assessing nutritional status, general well-being and growth, and development of individuals up to 20 years of age (For infants/young children less than 2 years of age, they are known as the growth standard (as opposed to a *growth reference* – which determines what is/exists and is comprised of data from all kids, not just healthy individuals)
	Note that growth reference data is all that we have available for the 2–20 years of age growth charts
	Growth charts are a helpful tool to monitor nutritional status in those undergoing medical treatment – allowing adjustments in nutritional intake (enteral or parenteral) to be made in a timely and effective manner
	These charts have historically been used to screen for malnutrition but are increasingly being used to screen for overweight/obesity now
	There are two sets of growth charts for both male and female:
	   Birth – 24 months of age and 2–20 years of age
	There are growth charts that give percentile curves for:
	   Weight-for-age
	   Length-for-age
	   Weight-for-length
	   Head circumference-for-age			

Waist circumference (WC) [107]	Where fat is distributed in the body may have a greater impact on health than total amount of fat in the body
	To measure:
	   Have the patient remove any outer clothing covering the abdomen and waist and place the measuring tape on bare skin
	   Obtain the measurement by measuring the distance around the narrowest area of the waist (between the lowest rib and the iliac crest)
	   The person being measured should stand erectly with the abdomen relaxed, arms at side, and feet together
	   The tape should be snug against the skin but not to the point where the skin is compressed
	   Take the measurement after the end of a normal expiration
	   Repeat 1–2 times to ensure accuracy and record to the nearest 0.1 inch
	   If the BMI >30, assuming central adiposity, WC is not necessary to perform [26]
	Waist circumference may be used to evaluate success of weight loss treatment
	Little predictive value in people with BMI ≥35 kg/m²
	May not apply with height <60 inches (5 ft.)
	Ethnic variation (Asian descent – WC > BMI for predicting disease risk)
	Ethnic variation – generally predictive for racial and ethnic groups in North America – those of Asian descent – WC > BMI for predicting disease risk
	Fat accumulation in the abdomen is linked to increased risk of type II diabetes and obesity-related diseases in general
	Recommended to be used as part of routine physical exam
			Increased waist circumference is an independent risk factor for disease [17]
			Male:
			   High risk >40 inches
			Female:
			   High risk >35 inches
			May not be a useful measure with height in excess of 60 inches or with BMI ≥35	

Measure	Examination/measurement method	Optimal	Excess/elevated	Deficient/depressed

Waist-to-hip ratio[17]	=Waist/hip circumference
	Abdominal fat is comprised of:
	  Subcutaneous fat
	  Visceral fat – Research that visceral fat (fat around a number of abdominal organs (such as the stomach, liver, and spleen)) is most strongly correlated with morbidity and mortality
	    Retroperitoneal fat (retroperitoneal fat = outside of and posterior to peritoneal cavity (which lines the surface of the abdominopelvic wall and contains most of the organs))
	  Most accurate measure of abdominal fat is via MRI or CT – but costly and less practical, so WHR is used to estimate abdominal adiposity
	(Hip measurement is taken at the widest part of the buttocks (i.e., hip circumference at largest point. The anatomic landmark for this is the greater trochanter.))
		Preferred ratio is to have the waist circumference less than the hip circumference (so the ratio <1 (due to the increased risk for hypertension with higher waist circumferences))
		Optimal/ideal:
		  <0.8 female
		  <0.9 male
			Increased abdominal fat is an independent risk factor for morbidity (sickness) and mortality
			World Health Organization (WHO):
			  >9 in men
			  >0.85 in women
			Benchmark for metabolic syndrome (predicts all cause and CVD mortality) [17]	

Waist-to-height ratio (WHtR)	WHtR = Waist circumference divided by height
	A measure of the distribution of adipose tissue [17]
	More sensitive than BMI as early warning of health risks and also considered to be more sensitive than the waist circumference in several different populations (due in part to the fact that it includes a measure of stature) [108]
	It is easier to measure/calculate than BMI and more affordable (no scale is required, just measuring tape for waist circumference and height [108]
	More accurate in adolescents and across different ethnicities
	WHtR > WC > BMI for detecting cardiometabolic risk across both genders [109, 110, 111, 112]
	The WHtR counteracts differences across ethnicities that the waist circumference and WHR cannot (such as the fact that health risks for Asians are increased with smaller levels of visceral fat than for Caucasians). With the average height of Asians being less than Caucasians, the WHtR can account for this
	WC and WHtR are superior to BMI in predicting diabetes and CVD risk (BMI does not assess the accumulation of visceral/abdominal fat) [109]
		WHtR = 0.5
		"Keep your waist circumference to less than half of your height" [109]
		General age-related ranges:
		  0.43–0.5 adults less than 40 y.o.
		  0.5–0.6 adults 40–50 y.o.
		  ≤0.6 adults over 50 y.o [17].
			(Apple and pear shapes)
			Overweight:
			  F: 0.49–0.54
			  M: 0.53–0.58
			Obese:
			  F: 0.54–0.58
			  M: 0.58–0.63
			Very obese:
			  F: >0.58
			  M: >0.63
			Higher values of WHtR relate to increased risk of metabolic syndrome and to obesity-related cardiovascular conditions [17, 108]
				(Chile shape)
				Slim:
				  F: 0.35–0.42
				  M: 0.35–0.43
				Underweight:
				  F: <0.35
				  M: <0.35

Body composition analysis – there are several methods for determining body composition [17, 26]
  Densitometry
    Underwater weighing
    Air displacement plethysmography
  Total body water
  Total body potassium
  Neutron activation analysis
  Creatinine excretion
  3-Methylhistidine
  Electrical conductance
    Bioelectrical impedance analysis (BIA)
  Infrared interactance
  Ultrasound
  Computed tomography (CT)
  Magnetic resonance imaging (MRI)
  Dual-energy X-ray absorptiometry
The most accurate and widely used will be discussed in more depth in this table hereafter

Measure	Examination/measurement method	Optimal	Excess/elevated	Deficient/depressed

Skinfold thickness [26, 107]	Most widely used method of estimating percent body fat
	As these measures assume subcutaneous fat comprises 50% of the total body fat, and the measures require considerable skill to perform correctly, they are not considered the most accurate method of determining percentage body fat. However, they do have some advantages:
	  Estimates energy reserves (fat and protein) in subcutaneous tissue
	  Measurements can be easily and quickly obtained
	  Done accurately, measures can correlate well with hydrostatic weighing
	  Inexpensive and portable equipment (vs. hydrostatic weighing)
	Typical sites measured include:
	  Chest
	  Triceps
	  Subscapular
	  Midaxillary
	  Suprailiac
	  Abdomen
	  Thigh
	  Medial Calf
	Calculations and/or nomograms are used to determine body fat percentage from one or more specific skinfold sites (sites chosen for measurement/calculation vary by gender and age)
	In hospitalized patients, the sum of:
	  Triceps and subscapular skinfold thicknesses (measured recumbently) are used as indicator of body's energy reserves
		Normative tables, equations, and nomograms available for interpretation [26]		

Arm muscle area (AMA) [26]	Arm muscle area is used to assess total amount of muscle in the body (uses triceps skinfold (TSF) measure and mid-arm circumference (MAC) for calculation/determination) Limitations with accurately measuring MAC and TSF limit use of this measure AMA is correlated with creatinine excretion (children) and total body muscle mass (adults)	Normative tables, equation, and nomograms available for interpretation		

Measure	Examination/measurement method	Optimal	Excess/elevated	Deficient/depressed

Bioelectrical impedance analysis (BIA) [17, 26]	Electrical current passed through the body to determine fat vs. fat-free mass. This current is harmless and cannot be felt by the subject
	Principle: The current is opposed by nonconducting tissues (fat and cell membranes) and transmitted by electrolytes in water (in fat-free tissues mainly)
	Opposition to alternating current (AC) = "Impedance"
	Accuracy – comparable to skinfold measurements (may be better)
	Particularly helpful in assessing body composition in those with edema or in patients who have had an amputation
	Weaknesses: assumes normal hydration
	With the weakness of dehydration skewing results, it is advised to have subjects drink plenty of water and refrain from exercising heavily for 4–6 hours and from consuming alcohol and/or taking diuretics for 24 hours prior to the test
	Depending on the device used, a BIA machine may measure any/all of the following:
	  Weight
	  Percent total fat
	  Fat mass and fat-free mass
	  Muscle mass
	  Bone mass
	  BMR
	  Metabolic age
	  Visceral fat rating
	  BMI
	  Physique rating
	For more specifics on these measures and their interpretations, consult the guide(s) of the specific machine(s) that you will use for BIA testing
		For more specifics on these measures and their interpretations, consult the guide(s) of the specific machine(s) that you will use for BIA testing
		BIA skeletal muscle mass (whole body) measure can be used to screen for sarcopenia [94] (in conjunction with handgrip strength and gait speed)
		$<8.87–10.76$ kg/m^2 men
		$<6.42–6.76$ kg/m^2 women		

Air displacement plethysmography (BOD-POD) [17, 26]	Body density is measured as a means to estimate body fat and fat-free mass. The BOD-POD device is considered to be an accurate method for determining body composition Patient enters a capsule/pod wearing tight-fitting swimsuit and cap to minimize measurement error It is relatively quick, easy, and affordable to test/obtain results and is preferred for children, the elderly, and those with disabilities over underwater weighing As it does not rely on body water content for determining body density and composition, it has potential utility in those with hydration imbalances and in adults with end-stage renal disease			
Dual-energy X-ray absorptiometry (DEXA) [17]	Measures fat, bone mineral, and fat-free soft tissue Used in hospital setting Considered easy to use. Patient needs to remain still for accurate measure so can be a problem for those with chronic pain and the elderly Affected by hydration status and thickness of tissues Emits low-level radiation			

Measure	Examination/measurement method	Optimal	Excess/elevated	Deficient/depressed
Measuring energy expenditure [26]	Measuring energy expenditure Direct calorimetry Indirect calorimetry Doubly labeled water Bicarbonate urea Estimating energy expenditure (REE) Harris-Benedict equation World Health Organization (WHO) NIH University of Vermont Mifflin St. Jeor Estimating energy expenditure – major determinant is fat-free mass (determine 70–80% of variance in REE, with the remaining 20–30% due to genetics). Note that the WHO equations do not include stature (it was not found to significantly improve the predictive ability of the equations) The resting energy expenditure equations must be increased to account for total energy expenditure, and this is achieved by multiplying the REE by an activity factor. In theory, REE includes the thermic effect of food (TEF) and total energy expenditure (TEE); however, in most clinical settings, allowance for additional energy needs from the TEF is not given In cases of disease, injury, and surgery, an additional factor can be included to estimate the 24-hour energy expenditure of these patients. The Estimated Energy Requirement (EER) developed by the National Academy of Sciences is defined as "average dietary intake that is predicted to maintain energy balance in a healthy person of defined age, gender, weight, height and level of physical activity consistent with good health." These equations apply only to people with healthy weight. For overweight, use the TEE equations			
Nutrition screening questionnaires				
Mini Nutritional Assessment [26] (MNA and MNA-SF)	Designed to efficiently identify elderly patients at nutritional risk who may then need a more extensive nutritional assessment The full MNA can be completed in 10–15 minutes The short form (MNA-SF) can be completed in less than 5 minutes			
Malnutrition Universal Screening Tool (MUST) [26]	Designed to identify adults who are malnourished, at risk of malnutrition, underweight, or obese It evaluates BMI, history of unintentional weight loss, and the presence of decreased food intake due to illness Can be completed in 3–5 minutes			

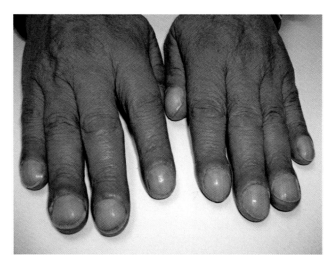

■ Fig. 39.2 Clubbing. (Reprinted from ► https://commons.wikimedia.org/wiki/File:Acopaquia.jpg. With permission from Creative Commons License 4.0: ► https://creativecommons.org/licenses/by-sa/4.0/)

■ Fig. 39.3 Schamroth's sign: disappearance of diamond-shaped window between digits that is normally present when paired digits are juxtaposed (anteriorly) seen in clubbing. (Reprinted with permission from ► https://www.flickr.com/photos/104346167@N06/17001286589)

The Nutrition-Focused Physical Exam

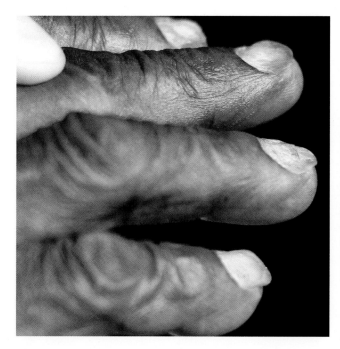

■ **Fig. 39.4** Koilonychia. (Reprinted from ► https://www.flickr.com/photos/coreyheitzmd/15023020192. With permission from Creative Commons License 2.0: ► https://creativecommons.org/licenses/by/2.0/)

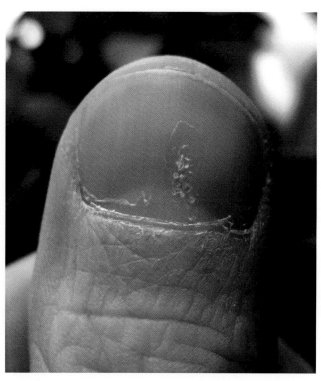

■ **Fig. 39.6** Nail pitting. (Reprinted from ► https://commons.wikimedia.org/wiki/File:Luszczyca_paznokcia.jpg. With permission from Creative Commons License 3.0: ► https://creativecommons.org/licenses/by-sa/3.0/deed.en)

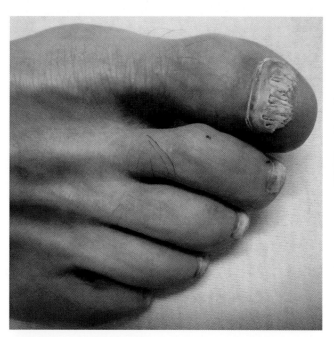

■ **Fig. 39.5** Onycholysis nail bed becomes separated from the nail plate. (Reprinted from ► https://commons.wikimedia.org/wiki/File:Oncymycosis.JPG. With permission from Creative Commons License 3.0: ► https://creativecommons.org/licenses/by-sa/3.0/deed.en)

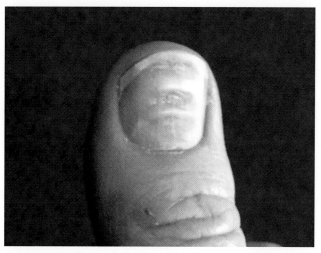

■ **Fig. 39.7** Beau's lines. (Reprinted from ► https://commons.wikimedia.org/wiki/File:Beau%27s_lines.JPG. With permission from Creative Commons License 4.0: ► https://creativecommons.org/licenses/by-sa/4.0/)

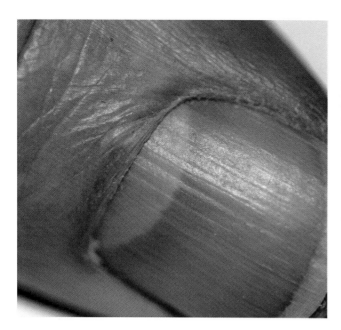

■ Fig. 39.8 Ridging of the nails. (Reprinted from ► https://commons.wikimedia.org/wiki/File:Lunula_07.jpg. With permission from Creative Commons License 3.0: ► https://creativecommons.org/licenses/by-sa/3.0/deed.en)

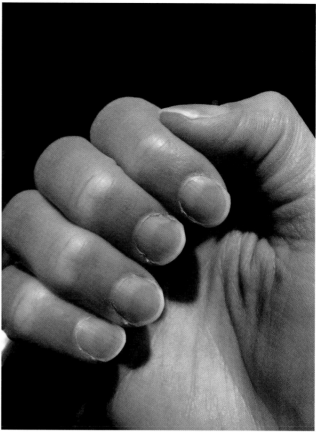

■ Fig. 39.10 Half-and-half nails (Lindsay's nails). (Reprinted from ► https://commons.wikimedia.org/wiki/File:Lindsays_Nails_2.jpg. With permission from Creative Commons License 4.0: ► https://creativecommons.org/licenses/by-sa/4.0/)

■ Fig. 39.9 Terry's nails. (Reprinted from ► https://commons.wikimedia.org/wiki/File:Terry%27s_nails.jpg. With permission from Creative Commons License 3.0: ► https://creativecommons.org/licenses/by-sa/3.0/deed.en)

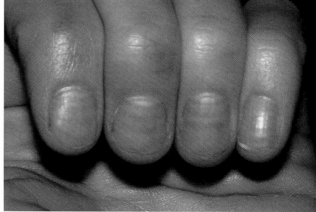

■ Fig. 39.11 Muehrcke's lines. (Reprinted with permission from ► https://commons.wikimedia.org/wiki/File:Muehrcke%27s_lines.JPG)

The Nutrition-Focused Physical Exam

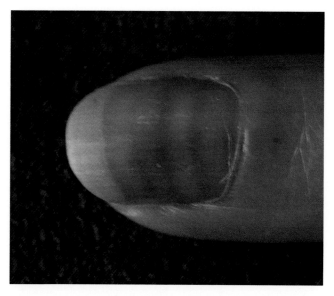

■ **Fig. 39.12** Mee's lines (Reynolds/Aldrich lines). (Reprinted from ► https://commons.wikimedia.org/wiki/File:Mee%27s_lines.JPG. With permission from Creative Commons License 3.0: ► https://creativecommons.org/licenses/by-sa/3.0/deed.en)

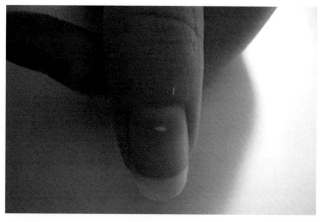

■ **Fig. 39.14** Punctate leukonychia. (Reprinted from ► https://en.wikipedia.org/wiki/Leukonychia. With permission from Creative Commons License 3.0: ► https://creativecommons.org/licenses/by-sa/3.0/deed.en)

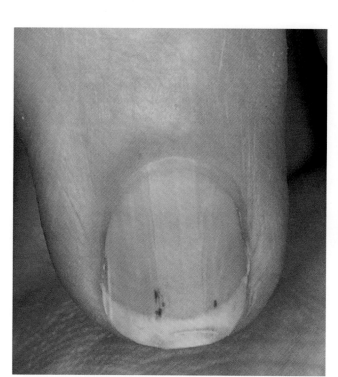

■ **Fig. 39.13** Splinter hemorrhage. (Reprinted with permission from ► https://commons.wikimedia.org/wiki/File:Splinter_hemorrhage.jpg)

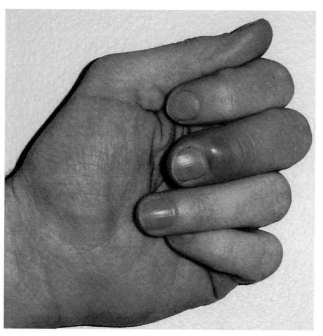

■ **Fig. 39.15** Paronychia. (Reprinted with permission from ► https://commons.wikimedia.org/wiki/File:Paronychia.jpg)

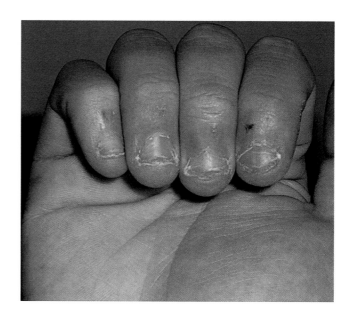

Fig. 39.16 Onychotillomania. (Reprinted with permission from https://commons.wikimedia.org/wiki/File:Nailbitebad.jpg)

Using a Peak Flow Meter (PFM),

- Set the sliding pointer on the meter to zero
- Hold the PFM by the handle
- Stand erect
- Remove chewing gum or food from mouth
- Take a full deep breath and seal both lips and tongue around the mouthpiece
- Exhale as quickly and as forcefully as you can
- Note the number where the sliding pointer stopped
- Reset the sliding pointer to zero
- Repeat 3 times
- If any coughing occurs in this process, repeat
- Record the highest reading of the 3 in a graph or log (This is the *peak flow*.)
- Use daily – ideally around the same time each day
- If a new meter is needed, be sure to determine peak flow for that meter

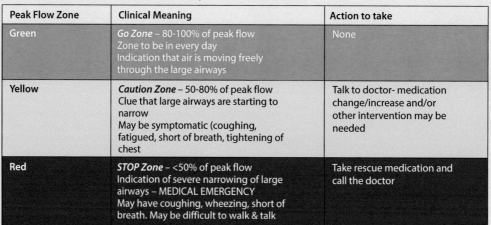

Peak Flow Zone	Clinical Meaning	Action to take
Green	*Go Zone* – 80-100% of peak flow Zone to be in every day Indication that air is moving freely through the large airways	None
Yellow	*Caution Zone* – 50-80% of peak flow Clue that large airways are starting to narrow May be symptomatic (coughing, fatigued, short of breath, tightening of chest	Talk to doctor- medication change/increase and/or other intervention may be needed
Red	*STOP Zone* – <50% of peak flow Indication of severe narrowing of large airways – MEDICAL EMERGENCY May have coughing, wheezing, short of breath. May be difficult to walk & talk	Take rescue medication and call the doctor

Predicted average peak expiratory flow (L/mins):

Age (yrs)	Height					
	65"		70"		75"	
30	489	617	502	637	513	655
40	483	620	496	641	507	659

Fig. 39.17 Peak expiratory flow rate (PEFR) technique

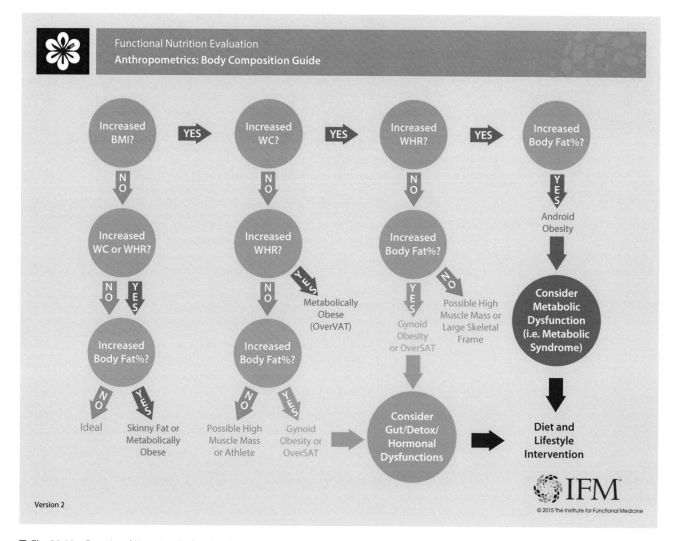

Fig. 39.18 Functional Nutrition Evaluation. Anthropometrics: Body Composition Guide. (Used with permission from The Institute for Functional Medicine ©2015)

39.2 Conclusion

In this chapter, we have covered an array of physical signs and cues that signal nutritional and medical imbalances, disorders, and diseases. By integrating this knowledge into practice, the dietitian/nutritionist or medical provider may benefit from the words of William Osler, a Canadian physician who emphasized the importance of the history and physical exam in making an accurate diagnosis:

> Learn to see, learn to hear, learn to feel, learn to smell and know that by practice alone can you become experts. [113]

Indeed, learning the language of the body takes time and presence. The more examinations one performs, the more the examiner builds a repertoire of knowledge and also an embodied sense of what is normal. This sense then allows one's antennae to more accurately pick up something abnormal [114]. It is no wonder that Sir Arthur Conan Doyle, the creator of iconic fictional detective and astute observer Sherlock Holmes, was a physician [115]!

The physical exam offers a unique opportunity to truly relate to a patient in a more intuitive and sensory manner, to connect with them through a kind of ritual that can initiate a patient-provider bond. Verghese entreats the examiner to perform the exam with skill and consideration so as to not violate the sacredness thereof and the transformation that this ritual can incite [13]. Furthermore, he and his colleagues introduce a conceptual framework for the utility of the physical exam organized into seven themes [116].

1. Diagnostic accuracy: Well-performed technique, practiced over time increases diagnostic accuracy.
2. Ongoing care and prognosis: Physical exam findings can guide therapy and prognostic assessments.
3. Patient contact: The physical exam can improve the patient-provider relationship through the ritual and touch that it involves.
4. Accessibility: Rapid assessment (often with minimal equipment) is possible with the physical exam
5. Pedagogic value: Teaches clinical reasoning and bedside manner and can also be used in patient education.

6. Cost: The physical exam may be more cost-effective than comparable technological tests.
7. Patient safety: Many physical examination procedures offer lower risk assessment than corresponding technological evaluations.

They contend that the physical exam may offer healing not only to the patient but to many pressing issues in healthcare today, such as misdiagnoses, escalating costs, decline in the provider-patient relationship, and provider burnout.

This chapter provides the knowledge needed to recognize and interpret a wide variety of abnormal findings. From what has been covered here, you may notice that a patient who presents with keratosis pilaris, fissuring eczema, ridging and missing lunulae of the nails, a tendency toward paronychia, thinning of the hair, photophobia, and a reduced sense of smell and taste may be suffering from deficiencies of zinc and possibly protein. With your counsel, the patient does a trial switch from a vegan diet to one with high-quality animal protein daily, and within a few months, partial lunulae are present on all of their nails, and, within 6 months, the lunulae are fully intact and accompanied by changes to the health of their skin and hair. These changes reflect to the examiner that their zinc and protein status are improving and with it, a return of greater health overall.

Lest we not touch on one important point before closing: in an era where dietitians, nutritionists, and healthcare providers are besieged with pressures and growing amounts of data to create and sift through, a little fun may be welcome, and physical diagnosis, and the satisfaction of arriving at a diagnosis solely (or largely) with one's wits and senses, while connecting more with one's patient [34], may be just what the doctor ordered.

References

1. Esper DH. Utilization of nutrition-focused physical assessment in identifying micronutrient deficiencies. Nutr Clin Pract. 2015;30(2):194–202.
2. Hammond K. Nutrition-focused physical assessment. Home Healthc Nurse. 1999;17(6):354–5.
3. Hammond KA. The nutritional dimension of physical assessment. Nutrition. 1999;15(5):411–9.
4. Litchford M. Putting the nutrition-focused physical assessment into practice in long-term care settings. Ann Longterm Care. 2013;21(11):38–41.
5. Dennett C. Nutrition-focused physical exams. Today's Dietitian. 2016;18(2):36.
6. Baker JP, Detsky AS, Wesson DE, Wolman SE, Stewart S, Whitewell J, et al. Nutritional assessment: a comparison of clinical judgement and objective measures. N Engl J Med. 1982;306(16):969–72.
7. Detsky AS, Smalley PS, Chang J. The rational clinical examination. Is this patient malnourished? JAMA. 1994;271(1):54–8.
8. Touger-Decker R. Physical assessment skills for dietetics practice. The past, the present, and recommendations for the future. Top Clin Nutr. 2006;21(3):190–8.
9. Stankorb SM, Rigassio-Radler D, Khan H, Touger-Decker R. Nutrition focused physical examination practices of registered dietitians. Top Clin Nutr. 2012;25(4):335–44.
10. Mackle TJ, Touger-Decker R, Maillet JO, Holland BK. Registered dietitians' use of physical assessment parameters in professional practice. J Am Diet Assoc. 2003;103:1632–8.
11. Jauhar S. The demise of the physical exam. N Engl J Med. 2006;354:548–51.
12. Rathe R. The complete physical. Am Fam Physician. 2003;68(7):1439–44.
13. Verghese A, Horwitz RI. In praise of the physical examination. It provides reason and ritual. BMJ. 2009;339:b5448.
14. Adams KM, Kohlmeier M, Zeisel SH. Nutrition Education in U.S. medical schools: latest update of a National Survey. Acad Med. 2010;85(9):1537–42.
15. Mordarski B. Nutrition-focused physical exam hands-on training workshop. J Acad Nutr Diet. 2016;116(5):868–9.
16. Litchford MD. Nutrition focused physical assessment: making clinical connections. Greensboro: CASE Software and Books; 2013.
17. Mahan LK, Raymond JL. Krause's food & the nutrition care process. 14th ed. St. Louis: Elsevier; 2017.
18. Ford D, Raj S, Batheja RK, DeBusk R, Grotto D, Noland D, et al. American Dietetic Association: standards of practice and standards of professional performance for registered dietitians (competent, proficient, and expert) in integrative and functional medicine. J Am Diet Assoc. 2011;111(6):902–11.
19. Dean S. Introduction to integrative and functional nutrition. Conference handout. California Academy of Nutrition & Dietetics (CAND) annual meeting and exhibition. 2015 Apr 9–11; Riverside. Available from: http://www.dietitian.org/d_cda/docs/annual_mtg_2015/conference_handouts/thu4-9-15/Dean_IntroductionIFN.pdf.
20. Noland D. The University of Kansas Medical Center Clinical Nutrition Seminar. Kansas City: Department of Dietetics & Nutrition and Department of Integrative Medicine; 2011. Courtesy of Diana Noland
21. Bickley LS, Szilagyi PG. Bates' pocket guide to physical examination and history taking. 7th ed. Philadelphia: Wolters Kluwer Health|Lippincott Williams & Wilkins; 2007.
22. James AG, Austin CJ, Cox DS, Taylor D, Calvert R. Microbiological and biochemical origins of human axillary odour. FEMS Microbiol Ecol. 2013;83:527–40.
23. Callewaert C, Lambert J, Van de Wiele T. Towards a bacterial treatment for armpit malodour. Exp Dermatol. 2017;26:388–91.
24. Ishikawa T, Toyoda Y, Yoshiura K, Nikawa N. Pharmacogenetics of human ABC transporter AVBCC11: new insights into apocrine gland growth and metabolite secretion. Front Genet. 2013;3(306):1–13.
25. Hur E, Gungor O, Bozkurt D, Bozgul SMK, Dusunur F, Caliskan H, et al. Trimethylaminuria (fish malodour syndrome) in chronic renal failure. Hippokratia. 2012;16(1):83–5.
26. Lee RD, Nieman DC. Nutritional assessment. 6th ed. New York: McGraw-Hill; 2013.
27. Musa-Veloso K, Likhoidi SS, Cunnane SC. Breath acetone is a reliable indicator of ketosis in adults consuming ketogenic meals. Am J Clin Nutr. 2002;76:65–70.
28. Sadock BJ, Sadock VA. Kaplan & Sadock's pocket handbook of clinical psychiatry. 4th ed. Philadelphia: Lippincott Williams & Wilkins; 2005.
29. Noland D, Huls A. Nutritional assessment by location table. University of Kansas Med Center Seminar; 2011.
30. nsight.org [Internet]. Federal Way: Institute of Functional Medicine. Functional Nutrition Evaluation. Skin Exam Quick Reference Guide;c2016 [cited 2016 Feb 7]. Available from: https://www.ifm.org/learning-center/skin-exam-n-sight-companion-guide/.
31. McGee S. Evidence-based physical diagnosis. 3rd ed. Philadelphia: Elsevier Saunders; 2012.
32. Newton C. Let me hear your body talk: the functional nutrition-focused physical exam: Integrative and Functional Nutrition Academy [Webinar]; 2016. http://integrativerd.org/wp-content/uploads/2014/06/DIFM-webinar-Body-Talk-2014-corrected.pdf

33. Suzuki D. Vancouver: David Suzuki Foundation;c2014 [cited Feb 26]. 'Dirty Dozen' cosmetic chemicals to avoid; [about 2 screens]. Available from: http://www.davidsuzuki.org/issues/health/science/toxics/dirty-dozen-cosmetic-chemicals/.
34. Mangione S, editor. Physical diagnosis secrets. Philadelphia: Hanley & Belfus Inc; 2000.
35. Brown SE. Better bones [Internet]. East Syracuse: Brown SE. 2015 Sep 14 – [cited 2017 Feb 23]. Available from: http://www.betterbones.com/bone-health-basics/whats-the-cause-of-your-osteoporosis/.
36. Sanders RD, Gilling PM. Gait and its assessment in psychiatry. Psychiatry (Edgmont). 2010;7(7):38–43.
37. nsight.org [Internet]. Federal Way: Institute of Functional Medicine. Functional Nutrition Evaluation. Blood Pressure Quick Reference Guide; c2016 [cited 2016 Feb 7]. Available from: https://www.ifm.org/learning-center/blood-pressure-n-sight-companion-guide/.
38. DeBusk R, Jaffe Y. It's not just your genes. Your diet. Your lifestyle. Your genes. San Diego: BKDR Publishing; 2006. p. 36–8.
39. National Institute of Health [Internet]. Bethesda: National Heart, Lung, and Blood Institute; [cited e2017 Feb 23]. Description of the DASH eating plan. Available from: https://www.nhlbi.nih.gov/health/health-topics/topics/dash.
40. Lanier JB, Mote MB, Clay EC. Evaluation and management of orthostatic hypotension. Am Fam Physician. 2011;84(5):527–36.
41. Zhu QO, Tan CSG, Tan HL, Wong RG, Joshi CS, Cuttilan RA, et al. Orthostatic hypotension: prevalence and associated risk factors among the ambulatory elderly in an Asian population. Singap Med J. 2016;57(8):444–51.
42. Wolters FJ, Mattace-Raso FUS, Koudstaal PJ, Hofman A, Ikram MA. Orthostatic hypotension and the long-term risk of dementia: a population-based study. PLoS Med. 2016;13(10):e1002143.
43. Oishi E, Sakata S, Tsuchihashi T, Tominaga M, Fujii K. Orthostatic hypotension predicts a poor prognosis in elderly people with dementia. Intern Med. 2016;55:1947–52.
44. Veronese N, De Rui M, Bolzetta F, Zambon S, Corti MC, Baggio G, et al. Orthostatic changes in blood pressure and mortality in the elderly: the Pro V.A. Study. Am J Hypertens. 2015;28(10):1248–56.
45. Ye Q, D'Urzo AD. Challenge of alpha 1-antitrypsin deficiency diagnosis in primary care. Can Fam Physician. 2016;62:899–901.
46. Mayo Clinic. Scottsdale: Mayo Foundation for Medical Education and Research; c1998–2017 [cited 2017 Feb 23]. Anhidrosis; [about 1 screen]. Available from: http://www.mayoclinic.org/diseases-conditions/anhidrosis/basics/risk-factors/con-20033498.
47. Bronsnick R, Murzaka EC, Rao BK. Diet in dermatology. Part I. Atopic dermatitis, acne, and nonmelanoma skin cancer. J Am Acad Dermatol. 2014;71:1039.e1–12.
48. Van Hattem S, Bootsma AH, Thio HB. Skin manifestations of diabetes. Cleve Clin J Med. 2008;75(11):772–87.
49. Hughes MC, Williams GM, Fourtanier A, Green AC. Food intake, dietary patterns, and actinic keratosis of the skin: a longitudinal study. Am J Clin Nutr. 2009;89:1246–55.
50. Nuzzo V, Tauchmanova L, Colasanti P, Zuccoli A, Colao A. Idiopathic chronic urticaria and thyroid autoimmunity. Experience of a single center. Dermatolendocrinol. 2011;3(4):225–8.
51. Godse KV. Urticaria and treatment options. Indian J Dermatol. 2009;54(4):310–2.
52. Pal L, Kidwai N, Glockenburg K, Schneller N, Altun T, Figueroa A. Skin wrinkling and rigidity are predictive of bone mineral density in early postmenopausal women. Proceedings of the Endocrine Society's 93rd annual meeting & expo; 2011 Jun 4–7; Boston. https://doi.org/10.1210/endo-meetings.2011.PART3.P23.P3-126.
53. Ferreira AC, Hochman B, Furtado F, Bonatti S, Ferreira LM. Keloids: a new challenge for nutrition. Nutr Rev. 2010;68(7):409–17.
54. Fawcett RS, Linford S, Stulberg DL. Nail abnormalities: clues to systemic disease. Am Fam Physician. 2004;69(6):1417–24.
55. Fitzgerald K. [Internet] Sandy Hook: Dr. Kara Fitzgerald Functional Medicine;c2017 [cited 2017 Feb 13]. Fingernails: a window into your metabolic soul. [about 1 screen]. Available from: http://www.drkarafitzgerald.com/2014/08/30/fingernails-a-window-into-your-metabolic-soul/.
56. Fitzgerald K. [Internet] Sandy Hook: Dr. Kara Fitzgerald Functional Medicine;c2017 [cited 2017 Feb 13]. Nutrition physical exam: tongue & nails suggest gut health. Available from: https://www.drkarafitzgerald.com/2015/07/09/nutrition-physical-exam-tongue-nails-suggest-gut-health/.
57. Marz RA. Medical nutrition from Marz. 2nd ed. Portland: OmniPress; 1997.
58. Weatherby D. Signs and symptoms analysis from a functional perspective. 2nd ed. Ashland: Bear Mountain Publishing; 2004.
59. Lotufo PA, Chae CU, Ajani UA, Hennekens CH, Manson JE. Male pattern baldness and coronary heart disease. The Physicians Health Study Arch Intern Med. 2000;160:165–71.
60. Gaby A. Nutritional medicine. 1st ed. Fritz-Perlberg: Concord; 2011.
61. Bartley GB. The differential diagnosis and classification of eyelid retraction. Trans Am Opthalmol Soc. 1995;93:371–89.
62. Sundelin T, Lekander M, Kecklund G, Van Someren EJW, Olsson A, Axelsson J. Cues of fatigue: effects of sleep deprivation on facial appearance. Sleep. 2013;36(9):1355–60.
63. Fuerst N, Langelier N, Massaro-Giordano M, Pistilli M, Stasi K, Burns C, et al. Tear osmolarity and dry eye symptoms in diabetics. Clin Opthalmol. 2014;8:507–15.
64. Nam GE, Kim S, Paik JS, Kim HS, Na KS. Association between pterygium and obesity status in a South Korean population. Medicine. 2016;95(50)
65. Astafurov K, Elhawy E, Ren L, Dong CQ, Igboin C, Hyman C, et al. Oral microbiome link to neurodegeneration in glaucoma. PLoS One. 2014;9(9):e104416.
66. Mayo Clinic. [Internet]. Scottsdale: Mayo Foundation for Medical Education and Research; c1998–2017 [cited 2017 Feb 23]. [about 1 screen]. Available from: http://www.mayoclinic.org/diseases-conditions/poor-color-vision/basics/causes/con-20022091.
67. Mercola.com. [Internet]. Dr. Joseph Mercola; c1997-2017 [cited 2017 Feb 21]. Effective strategies to identify and correct the inflammation caused by mold exposure. [About 1 screen]. Available from: http://articles.mercola.com/sites/articles/archive/2012/07/22/mold-and-other-chronic-diseases.aspx.
68. Biotoxinjourney.com c2016 [cited 2017 Feb 20] Are You Moldy? [About 1 screen]. Available from: http://biotoxinjourney.com/areyoumoldy/-VCSTest.
69. Singh C, Kawatra R, Gupta J, Awasthi V, Dungara H. Therapeutic role of vitamin B12 in patients of chronic tinnitus: a pilot study. Noise Health. 2016;18(81):93–7.
70. Freiherr J, Gordon AR, Alden EC, Ponting AL, Hernandez MF, Boesveldt S, et al. The 40-item Monell extended Sniffin' sticks identification test (MONEX-40). J Neurosci Methods. 2012;205(1):10–6.
71. Mackay-Sim A, Johnston AN, Owen C, Burne TH. Olfactory ability in the healthy population: reassessing presbyosmia. Chem Senses. 2006;31(8):763–71.
72. Sohrabi HR, Bates KA, Weinborn MG, Johnston ANB, Bahramian A, Taddei K, et al. Olfactory discrimination predicts cognitive decline among community-dwelling older adults. Transl Psychiatry. 2012;2:e118.
73. Wilson RS, Yu L, Bennett DA. Odor identification and mortality in old age. Chem Senses. 2010;36:63–7.
74. Hawkes C. Disorders of smell and taste. In: Office practice of neurology. 2nd ed. Philadelphia: Churchill Livingstone; 2003. p. 102–12.
75. Stamps JJ, Bartoshuk LM, Heilman KM. A brief olfactory test for Alzheimer's disease. J Neuro Sci. 2013;15:333.
76. Radler DR, Lister T. Nutrient deficiencies associated with nutrition-focused physical findings of the oral cavity. Nutr Clin Pract. 2013;28(6):710–2.
77. Lankarani KB, Sivandzadeh GR, Hassanpour S. Oral manifestation in inflammatory bowel disease: a review. World J Gastroenterol. 2013;19(46):8571–9.

78. nsight.org [Internet]. Federal Way: Institute of Functional Medicine. Functional Nutrition Evaluation. Dental Exam Quick Reference Guide; c2016 [cited 2016 Feb 7]. Available from: https://www.ifm.org/learning-center/dental-exam-n-sight-companion-guide/.
79. Stewart R, Weyant RJ, Garcia ME, Harris T, Launer LJ, Satterfield S, et al. Adverse oral health and cognitive decline: the health, aging and body composition study. J Am Geriatr Soc. 2013;61(2):177–84.
80. Darnaud C, Thomas F, Pannier B, Danckin N, Bouchard P. Oral health and blood pressure: the IPC cohort. Am J Hypertens. 2015;28(10):1257–61.
81. He K, Zun P, Liu K, Morris S, Reis J, Guallar E. Mercury exposure in young adulthood and incidence of diabetes later in life. The CARDIA Trace Element Study. Diabetes Care. 2013;36(6):1584–9.
82. Pugazhenthan T, Singh H, Kumar P, Hariharan BI. Dysgeusia going to be a rare or a common side-effect of amlodipine? Ann Med Health Sci Res. 2014;4(Suppl 1):S43–4.
83. Fitzgerald K. [Internet]. Sandy Hook: Dr. Kara Fitzgerald Functional Medicine. c2017; [cited 2017 Feb 25]. Introduction to Tongue Nutrient Diagnosis in Functional Medicine; [about 1 screen]. Available from: http://www.drkarafitzgerald.com/2015/09/11/introduction-to-tongue-nutrient-diagnosis-in-functional-medicine-2/.
84. Rhim E, Han K, Yun KI. Association between temporomandibular disorders and obesity. J Craniomaxillofac Surg. 2016;44:1003–7.
85. Hoffman RG, Kotchen JM, Kotchen TA, Cowley T, Dasgupta M, Cowley AW. Temporomandibular disorders and associated clinical comorbidities. Clin J Pain. 2011;27:268–74.
86. Keris EY, Yaman SD, Demirag MD. Temporomandibular joint findings in patients with rheumatoid arthritis, ankylosing spondylitis, and primary Sjögren's syndrome. J Investig Clin Dent. 2017;8(24)1–7.
87. Mann T, Heuberger R, Wong H. The association between chewing and swallowing difficulties and nutritional status in older adults. Aust Dent J. 2013;58:200–6.
88. Suiter DM, Leder SB. Clinical utility of the 3-ounce water swallow test. Dysphagia. 2008;23:244–50.
89. Miura H, Kariyasu M, Yamasaki K, Arai Y. Evaluation of chewing and swallowing disorders among frail community-dwelling elderly individuals. J Oral Rehabil. 2007;34:422–7.
90. Hansen EF, Vestbo J, Phanareth K, Kok-Jensen A, Dirksen A. Peak flow as a predictor of overall mortality in asthma and chronic obstructive pulmonary disease. Am J Resp Crit Care Med. 2001;163:690–3.
91. Hall MW, Sowden DS, Gravestock N, Greenbaum DS, Joseph JG, Freeman E. Abdominal wall tenderness tests. Lancet. 1991;337(8757):1606–7.
92. Muno T, Patel J, Badilla-Porras R, Kronick J, Mercimek-Mahmutoglu S. Severe scoliosis in a patient with severe methylenetetrahydrofolate reductase deficiency. Brain Dev. 2015;37:168–70.
93. Lee SY, Ro HJ, Chung SG, Kang SH, Seo KM, Kim DK. Low skeletal muscle mass in the lower limbs is independently associated with to knee osteoarthritis. PLoS ONE. 2016;11(11):e0166385.
94. Lee DC, Shook RP, Drenonutz C, Blair SN. Physical activity and sarcopenic obesity: definition, assessment, prevalence and mechanism. Future Sci OA. 2016;2(3):FS0127.
95. Kärkkäinen M, Rikkonen T, Kröger H, Sirola J, Tuppurainen M, Salovaara K. Physical tests for patient selection for bone mineral density measurements in postmenopausal women. Bone. 2009;44:660–5.
96. Gale CR, Martyn CN, Cooper C, Sayer AA. Grip strength, body composition and mortality. Intl J Epidemiol. 2007;36:228–35.
97. Leong DP, Teo KK, Rangarajan S, Lopez-Jaramillo P, Avezum A, Orlandi A, et al. Prognostic value of grip strength: findings from the prospective urban rural epidemiology (PURE) study. Lancet. 2015;386:266–73.
98. Youbin Y, Shim JS, Oh BM, Seo HG. Grip strength on the unaffected side as an independent predictor of functional improvement after stroke. Am J Phys Med Rehabil. 2017;96:616–20.
99. Robinson S, Cooper C, Sayer AA. Nutrition and sarcopenia: a review of the evidence and implications for preventive strategies. J Aging Res. 2012;2012:510801.
100. Cruz-Jentoft AJ, Baeyens JP, Bauer JM, Boirie Y, Cedarholm T, Landi F. Sarcopenia: European consensus on definition and diagnosis. Report of the European Working Group on Sarcopenia in Older People. Age Ageing. 2010;39:412–23.
101. Jensen GL. Inflammation: roles in aging and sarcopenia. JPEN J Parenter Enteral Nutr. 2008;32(6):656–9.
102. Greene L, Hartz C, Ma A. Vitamin B12 deficiency. Act early to prevent neurological damage. Adv NPs PAs. 2006;14(7):63.
103. Gilman S. Joint position sense and vibration sense: anatomical organisation and assessment. J Neurol Neurosurg Psychiatry. 2002;73:473–7.
104. Guralnik JM, Simonsick EM, Ferrucci L, Glynn RJ, Berkman LF, Blazer DO. A short physical performance battery assessing lower extremity function: association with self-reported disability and prediction of mortality and nursing home admission. J Gerontol. 1994;49(2):M85–94.
105. Studenski S, Perera S, Patel K, Rosano C, Faulkner K, Inzitari M. Gait speed and survival in older adults. JAMA. 2011;305(1):50–8.
106. CDC.gov National Center for Injury Prevention and Control. [Internet]. Atlanta: STEADI. 2016 Nov 23; [cited 2017 Feb 25]; The Timed Up and Go (TUG) Test [Download]. Available from: https://www.cdc.gov/steadi/pdf/tug_test-a.pdf.
107. Nelms M, Sucher KP, Lacey K. Nutrition therapy and pathophysiology. 3rd ed. Boston: Cengage; 2016.
108. Ashwell M, Hsieh SD. Six reasons why the waist-to-height ratio is a rapid and effective global indicator for health risks of obesity and how its use could simplify the international public health message on obesity. Int J Food Sci Nutr. 2005;56(5):303–7.
109. Ashwell M, Gunn P, Gibson S. Waist-to-Height Ratio is a better screening tool than waist circumference and BMI for adult cardiometabolic risk factors: systematic review and meta-analysis. Obes Rev. 2012;13:275–86.
110. Zeng Q, He Y, Dong S, Zhao X, Chen Z, Song Z, et al. Optimal cut-off values of BMI, waist circumference and waist:height ratio for defining obesity in Chinese adults. Br J Nutr. 2014;112:1735–44.
111. Browning LM, Hsieh DS, Ashwell M. A systematic review of waist-to-height ratio as a screening tool for the prediction of cardiovascular disease and diabetes: 0.5 could be a suitable global boundary value. Nutr Res Rev. 2010;23:247–69.
112. Yan W, Bingxian H, Yao H, Jianghong D, Jun C, Dongliang G, et al. Waist-to-height ratio is an accurate and easier index for evaluating obesity in children and adolescents. Obesity. 2007;15(3):748–52.
113. Osler W. Sir William Osler: aphorisms from his bedside teachings and writings. Springfield: Charles C. Thomas Publisher; 1961:129. Quoted from Thayer W. Osler, the teacher. Bull Johns Hopkins Hosp 1919;30:198
114. Kelly M, Tink W, Nixon K, Dornan T. Losing touch? Refining the role of physical examination in family medicine. Can Fam Physician. 2015;61:1041–3.
115. Arthur Conan Doyle. London: The Sir Arthur Conan Doyle Literary Estate; c2000-2017 [cited 2017 Feb 22] Biography; [about 1 screen]. Available from: http://www.arthurconandoyle.com/about-us.html.
116. Zaman J, Verghese A, Elder A. The value of the physical examination: a new conceptual framework. South Med J. 2016;109(12):754–7.

Modifiable Lifestyle Factors: Exercise, Sleep, Stress, and Relationships

Margaret Christensen

40.1 Introduction – 697

40.2 Gathering Lifestyle Data: The Big Picture – 697
40.2.1 Establishing the Therapeutic Relationship – 697

40.3 Gathering Oneself – 698

40.4 Utilizing Assessment Questionnaire Tools – 698
40.4.1 Benefits for Practitioners – 698
40.4.2 Benefit for Clients – 699
40.4.3 Insights and Change – 699
40.4.4 Storytelling and Healing – 699

40.5 How to Ask Questions: Establishing the Therapeutic Relationship – 699

40.6 Assessing Exercise/Movement – 700
40.6.1 Exercise: Assessing the Big Picture – 700
40.6.2 Questions to Ask: Exercise and Movement – 700
40.6.3 Assessing Movement at Home – 701
40.6.4 Assessing Movement at Work – 702
40.6.5 Resources to Motivate and Track Exercise – 703

40.7 Assessing Sleep and Restorative Activities – 703
40.7.1 Restorative Activities: The Big Picture – 703
40.7.2 Screening Questions for Sleep Apnea – 703
40.7.3 Questions to Assess Sleep Habits – 703
40.7.4 Other Restorative Activities – 704

40.8 Stress and Resilience – 704
40.8.1 Stress and Resilience: Assessing the Big Picture – 704
40.8.2 Tools for Assessment of Stress and Self-Care – 705
40.8.3 Stress: Important Questions to Ask – 705
40.8.4 Abuse and Neglect – 705

© Springer Nature Switzerland AG 2020
D. Noland et al. (eds.), *Integrative and Functional Medical Nutrition Therapy*,
https://doi.org/10.1007/978-3-030-30730-1_40

40.8.5	Assessing Anxiety and Depression – 706	
40.8.6	Basic Questions to Assess Stress Levels – 706	
40.8.7	Assessing Resilience – 707	
40.8.8	Resources for Assessing and Addressing Stress and Resilience – 707	

40.9 Assessing Relationships and Beliefs – 707
40.9.1 Relationships and Beliefs: The Big Picture – 707
40.9.2 Relationships, Social Networks, and Health Outcomes – 708
40.9.3 Beliefs – 709
40.9.4 Spirituality and Religion – 709
40.9.5 Questions to Ask About Relationships and Beliefs – 710

40.10 Assessing Readiness to Change – 710
40.10.1 The Nature of Change – 710
40.10.2 Smart Goal-Setting – 711

40.11 Conclusion – 712

References – 712

40.1 Introduction

The skilled IFMNT practitioner understands that developing a clear picture of a client's current life circumstance is imperative to helping the client implement desired functional nutritional and lifestyle prescriptions for the long term. Optimizing lifestyle change is the fundamental intervention of the integrative functional medicine approach. Initiating and sustaining that transformative change is dependent on many factors, including how the IFMNT practitioner establishes the therapeutic relationship and maintains knowledge of critical factors in the social milieu. Knowing how to obtain a comprehensive assessment of modifiable lifestyle factors in an insightful, educational, and compassionate manner can have lasting impact on the achievement and maintenance of health goals.

The detailed metabolic and physiologic pathways for nutritional influences on health and disease outcomes have been previously discussed in this textbook. The study of the science behind physiologic mechanisms and biochemical pathways is the science of medicine. Knowing how to engage the patient and how to ask questions is the art of medicine. So, how does the practitioner translate knowledge into action for the patient? What tools are available for assessment? For education and interventions? For ongoing support? What personal interaction skills are necessary for the practitioner? How do we assess the client's readiness to change? How do we support lifelong changes? Each area is discussed in (▶ Chaps. 6 and 9).

We begin by establishing a mutually beneficial dialogue with the client to understand where they are in their life at this moment. The bottom of the IFM Matrix tool emphasizes the role of current lifestyle areas as root factors influencing the complex interconnecting physiologic and biochemical factors of the Matrix above it. These areas represent the ongoing modifiable triggers and mediators of biologic function. As a teaching tool, the practitioner can explain how these five primary "roots" of the Functional Matrix "tree of symptoms" act as major influences on current health. The Matrix provides an excellent visual to explain complex phenomena in a simple manner (◘ Fig. 40.1).

At the center of these modifiable lifestyle factors is nutrition assessment, the purview of ▶ Chap. 9 on IFMNT NIBLETS. The four other important areas to assess, in a nonjudgmental and compassionate fashion, are past and current exercise and movement, sleep and restoration habits, stress levels and their perception (including resilience levels), as well as relationships and beliefs. Detailed physiological impacts of these areas on biochemical pathways have been previously discussed (see Part I, Section 2: ▶ Chaps. 7 and 8). In addition, as part of implementing any kind of program, it is valuable for the practitioner to know how to assess the client's readiness to change.

Most practitioners are very comfortable asking about exercise and sleep issues and have had training in asking basic stress and relationship questions, but many are not yet comfortable, nor have the tools, to help a client explore their deeper belief systems and underlying spiritual values. It is often that these very important underlying and unspoken codes are the "back story" that unconsciously governs sabotaging behavior. Positive beliefs and spiritual connections can be harnessed to engage not just immediate change but for transformative, inspirational, lifelong benefit.

40.2 Gathering Lifestyle Data: The Big Picture

40.2.1 Establishing the Therapeutic Relationship

For the practitioner, understanding in detail where the patient stands regarding current levels of stress and overwhelm can help direct priorities and frame realistic expectations for the client's current capabilities. In addition, interest from the clinician can serve to markedly assist in developing trust from the patient who feels heard. When the client has been validated for their obstacles and challenges and sees that the practitioner is interested in finding workable solutions to those challenges, they are then more willing to trust the clinician's guidance, want to actively cooperate, and are more likely to persevere in the face of stumbling blocks. In order to get optimal outcomes, the client must be mentally, emotionally, and physically capable of implementing the requested changes to their diet and lifestyle.

By evaluating each of the subject areas, the practitioner can prioritize and provide support tools to help the client implement change and move beyond limiting beliefs. If the practitioner doesn't understand the current circumstances and capabilities of the client, there is likely to be frustration and disillusionment on both sides, as well as wasted time and resources, no matter how brilliant a diagnostician is. There is no use telling a client to go on an elimination diet when they are in the midst of remodeling their kitchen or are drowning in stress from spending their hours in traffic driving kids to sports practice with no time for self-care. Meeting the client where they are is imperative.

> "Meeting the client where they are is imperative."

Identifying obstacles and opportunities in each area allows for individualized support tools to be recommended [see below] and can have profound impact on outcomes. For example, one patient may be a former athlete whose belief system is easily engaged in focusing on exercise as a leverage point, whereas another may be very committed to their faith practice, whose tenets may be used to support changes in eating habits. It's critical for the practitioner to help clients see new avenues for change, often from years of entrenched, unhealthy patterns. The lifestyle factors section on the Adult Intake Questionnaire [Male or Female] (▶ Chap. 46) is an

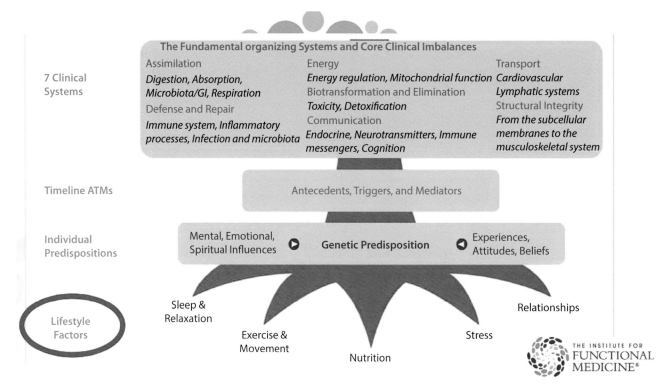

Fig. 40.1 Modified IFM tree. (Used with permission from The Institute for Functional Medicine ©2015)

excellent starting tool for basic assessments, pointing the practitioner to areas that may require more in-depth assessment with other questionnaires.

40.3 Gathering Oneself

The IFM acronym "GOTOIT" [1] is an organizational framework pneumonic that refers not only to gathering information and data on the client but to gathering one's self as a practitioner.

"GOTOIT"
- Gather
- Organize
- Tell
- Order
- Initiate
- Track

We must ask ourselves, "How well do I do as a practitioner regarding daily meditation or mindfulness-prayer practices, movement, rest, and restoration?" Does the clinician have a support system? What kind of personal growth and ongoing psychotherapeutic work has been done? Burnout is extremely common among today's physicians and nurses [2], affecting the clinician's ability to be present. The practitioner's energetic state can markedly influence the patient's receptivity. Calm, present, mindful clinicians, following their own advice, are much more likely to inspire confidence and participation by the client [3]. ▶ Chapter 46 has an assessment questionnaire, "Walk the Talk," specifically designed for practitioners to look at themselves. In this arena, the ancient wisdom teaching of "know thyself" connected to the biblical command of "physician heal thyself." (▶ Chap. 6) Both speak volumes!

Since studies have shown that 60–80% of office visits for physical complaints are, in reality, related to personal psychosocial stressors [4], the clinician must cultivate both internal personal resources and external network of resources for referral to the appropriate support mechanisms, such as psychotherapists, social workers, support groups, domestic abuse shelters, trainers, nutritionists, bodyworkers, lifestyle coaches, acupuncturists, chiropractors, biologic dentists, crisis centers, as well as clinical subspecialists. It takes a village, and creating health is a team sport!

40.4 Utilizing Assessment Questionnaire Tools

40.4.1 Benefits for Practitioners

Collecting lifestyle data can be done both with direct questioning and indirectly by using any of several different questionnaires and assessment tools. As direct questioning in all these different areas at one sitting is a daunting task, the use of questionnaire materials ahead of time can allow for the gathering of much data with minimal time. Many useful questionnaires are available through organizations (such as the IFM Toolkit, available on the IFM website, and referred to henceforth as "Toolkit") [5] as well as online resources.

Practitioners can initially utilize the answers on a modifiable intake form, such as the comprehensive Adult Medical Intake Questionnaire [male or female] found in the IFM Toolkit. Each lifestyle area is addressed with basic questions to help begin a dialogue. While reviewing the questionnaires with the client on intake, the opportunity often arises to segue directly into brainstorming of ideas for improvement or changes that can be made. Multiple examples are listed throughout this chapter and provide an opportunity to have a back-and-forth dialogue about what is possible.

For more in-depth and illuminative information, areas of concern can be investigated with additional self-administered questionnaires such as the sleep questionnaire, the DASS [Depression Anxiety Stress Scale] questionnaire or an exercise assessment, or self-care questionnaire (Appendix A2). The practitioner can easily uncover areas of concern while noticing opportunities for leveraging behaviors and beliefs to expand on what is already working. The questionnaires can additionally help triage the client's ongoing needs to the appropriate practitioner. For example, a busy physician may not have the time or expertise to counsel a client on an exercise prescription, but they recognize the need to refer to the lifestyle coach or to refer another client for counseling because of abuse issues.

40.4.2 Benefit for Clients

The benefit for the client from filling out these detailed questionnaires is that the process itself can have educational and therapeutic value. They may wonder why they're being asked about how they were born or whether they had ear infections while growing up. They may ask how answering that question will help them now. A question regarding whether a client was breastfed helps them to begin to understand the connection between early life nutrition and later outcomes such as autoimmune and gut function. It is often an eye-opening experience for the client to begin to see just how many layers of stress they have been dealing with, or how little time they have for self-care, or how having lots of antibiotics as a child may be impacting their health now. We may often be the first to ask about histories of childhood abuse, and they will be relieved and grateful. Many have new insights into what is really underlying their health issues and begin connecting dots from the past to be able to make the necessary changes moving forward. The questions themselves prepare the client to be more receptive to solutions the practitioner can provide.

40.4.3 Insights and Change

Questionnaire-induced insights may bring about shifts even before the practitioner meets with them. For example, after filling out a DASS, the client may purposefully seek to cut down on watching stressful television or the amount of alcohol they consume, or they may realize just how much their job is impacting their health and begin contemplating a change. Filling out the exercise questionnaire often gets them moving a little more before they come in for their visit. Answering questions about stress factors from the past, such as the Adverse Childhood Events (ACE) score can be illuminating. These questions can help a client to see and realize just how important current strategies and tools to help modulate the lifelong impact of those events become. An opportunity arises for discussing resiliency tools and strategies. The ACE questionnaire tool and resources can be found at the Centers for Disease Control and Prevention website (▶ cdc.gov) or for the layperson at ▶ ACEsTooHigh.com.

It is particularly valuable to focus on *whatever* a patient is doing right and build on the positives rather than focusing on the negatives. The use of the word "why" as in "Why aren't you exercising?" should be avoided as it creates defensiveness. Instead, when exploring obstacles and challenges, use a phrase such as, "Tell me about….". For example, "Tell me about what gets in the way of exercising or movement for you."

40.4.4 Storytelling and Healing

Just the act of telling their story to an attentive listener and being witnessed has therapeutic value for the patient and has been documented as a healing tool in and of itself [6]. Unfortunately, because of the time limitations within the conventional medical model, few clients have had an opportunity to tell their histories in a comprehensive way and to discuss their current lifestyle challenges in-depth. As the client is watching, the practitioner can draw arrows from the lower half of the Matrix, illustrating connections from lifestyle factors to the functional metabolic disturbances above – an excellent, concrete teaching strategy.

Once the client's information is gathered, the attentive and compassionate practitioner tells the patient's story back to them, organized in such a way that a path to healing can be illuminated through the modifiable factors, inspiring hope and confidence. Reiterating the client's history in this way lets them know they have been heard and often elicits astonished appreciation and a willingness to eagerly move forward. Not uncommonly, a sense of great relief and well-being happens for the client without the practitioner actually "doing" anything. Thus, with a thorough and caring intake assessment, done well, a therapeutic relationship begins.

40.5 How to Ask Questions: Establishing the Therapeutic Relationship

> The consciousness in which you gather your information is just as significant as the information you gather.
> – Monique Class, RN, MSN, IFMCP

When gathering lifestyle information, it is important to ask questions with empathy in a nonjudgmental and neutral way, as many of these areas are sensitive or may elicit guilt or shame response from the client's perceived lack of control or failure to have previously implemented changes. Focusing on what is already working and expanding those areas can often be more useful than just focusing on what a client is not doing "right." Again, one of the most important things a clinician can do to improve their credibility with clients is to assess and address each of these areas in themselves [7–9].

The following are eight practices that are found to promote healing relationships, outlined in an excellent article by psychologists Churchill and Schenck on "Healing Skills in Medical Practice" [10]. They are well worth printing out for clinicians to remind themselves how to create the most positive interactions.

1. *Do the little things*: Introduce yourself and your team, greet everyone, shake hands, smile, sit down, make eye contact, give your undivided attention, be human, and be personable.
2. *Take time and listen*: Be still, be quiet, be interested, and be present.
3. *Be open*: Be vulnerable, be brave, face the pain, and look for the unspoken.
4. *Find something to like, to love*: Take the risk, stretch yourself and your world, and think of your family.
5. *Remove barriers*: Practice humility, pay attention to power and its differentials, create bridges, be safe, and make welcoming spaces.
6. *Let the patient explain*: Listen for what and how they understand, listen for fear and anger, and listen for expectations and hopes.
7. *Share authority*: Offer guidance, get permission to take the lead, support patients' efforts to heal themselves, and be confident.
8. *Be committed and trustworthy*: Do not abandon, invest in trust, be faithful, and be thankful.

40.6 Assessing Exercise/Movement

40.6.1 Exercise: Assessing the Big Picture

It is well established that adequate and regular exercise has a great number of health benefits impacting every area of metabolism and physiology, as elucidated in earlier chapters. A person need not be familiar with intricate beneficial biochemical cellular impacts to intuitively understand that they feel better when moving. There are those clients who are already engaged in an excellent exercise regime and may not require much motivation, but unfortunately, the majority of clients present struggling to incorporate "exercise" into their lives. Currently 70% of the average American adult's time is sedentary, while the remaining 30% is only light activity [11]. The data is clear that chronic sitting is as deadly as smoking. Encouraging clients to replace any sedentary time with sleeping, standing, walking, or moderate physical activity has been shown to reduce morbidity and mortality [12].

So, what are the challenges faced by so many of us in implementing a regular exercise program? This is where the compassionate inquiry skills of the IFMNT practitioner are beneficial. For example, for many women in particular, "exercise" is a dirty word, something else added to the "should" list of too many things to do already, when they are exhausted, not well or don't enjoy it. Having tools to help assess and reframe their beliefs from the idea of more "work" or a "have-to" into one of compassionate self-care can be valuable. Identifying and engaging in opportunities to increase movement throughout the day, no matter if formal exercise or not, can be useful.

40.6.2 Questions to Ask: Exercise and Movement

Beyond the questions on the Adult Intake Questionnaire, the IFM Exercise History and Activities of Daily Living questionnaires are excellent tools to help develop a comprehensive assessment of overall daily movement, including formal exercise and general activity levels, enjoyability, and perceived roadblocks. Yet for some, filling out the extended Exercise History Questionnaire can sometimes feel overwhelming and shaming when inadequacies in this area are revealed.

It is always best to initially frame these questions in terms of "what are the motivations for and perceived benefits of the client's long-term health goals?" For example, start with: "Would you like better sleep, mood, weight loss, brain function, cardiovascular health, cancer prevention?" Beginning with the end in mind helps create a bridge for incremental change (see ◘ Fig. 40.2).

A comprehensive exercise intake assessment includes "FITT" questions about frequency, intensity level, type of exercise, and time done and covers cardiovascular activities, strengthening, stretching, and balance. Other pertinent areas to assess are:

- What types of activities do they prefer? Formal or informal?
- What do they like most about exercising?
- What do they like least?
- Do they prefer solo, partner, or group activities?
- Do they have a membership at any facility that includes movement?
- Do they have any of their own equipment?
- What obstacles do they have? Time constraints? Financial? Transport?
- What are some possible solutions?
- Best time of day?
- When do they have the most energy and time?
- What are their goals regarding balance, flexibility, fitness, and body composition?
- What kind of injuries or physical limitations with joint or musculoskeletal pain do they have?

> **What are your motivators for exercise? (Check all that apply)**
> ☐ Prevent cardiac disease and stroke ☐ Decrease stress
> ☐ Reduce blood pressure   ☐ Improve sleep
> ☐ Control blood glucose   ☐ Weight reduction
> ☐ Prevent bone loss   ☐ Increase mental alertness
> ☐ Increase energy   ☐ Better endurance
> ☐ Increase self esteem   ☐ Increase interest in sex
> ☐ Improve mood   ☐ Other _____
>
> THE INSTITUTE FOR FUNCTIONAL MEDICINE

Fig. 40.2 Motivators for exercise. (Used with permission from The Institute for Functional Medicine ©2015)

- What kind of medical conditions such as neurologic issues, breathing issues, balance, congestive heart failure, or angina limit their activities?

It is often helpful to walk a client through their day from morning to bedtime to help reveal where the obstacles and opportunities lie. Encouraging the patient to substitute any movement for screen time can be beneficial for every aspect of health [13, 14].

Explain at a metaphorical level that movement is about "taking the next right step" and "moving forward" in life. Literal movement is a concrete external manifestation of the internal changes the client wants to make. Movement can be reframed as a form of prayer or meditation, e.g. taking time to be mindful of each step, of the neighbor's gardens, and of the changing seasons or while mentally rehearsing a favorite spiritual verse or poem, dedicating themselves to a higher good. "My body is my temple and the home to my soul" is a wonderful mantra to repeat while walking.

If a client is feeling "stuck" in old patterns and hates the concept of "exercise," this attitude can be reframed into an opportunity to literally "create more movement" in their life. A simple strategy to encourage moving beyond inertia is to suggest putting tennis shoes and sweats right by the bed, getting up first thing without thinking, putting them on, and walking 10 minutes in one direction out the door and 10 minutes back. No gym required.

Additional strategies include asking, "Do you feel safe walking in your neighborhood? Do you have a friend you can go with?" For some, the opportunity to have "walk and talk" therapy time with a good friend or buddies cycling together can provide the motivation to get out and going. Having a client setting a goal of only watching their favorite rerun while on the treadmill, balance ball, or mini trampoline can provide incentive. Make it fun by encouraging the use of favorite dance music from the time a person was in his/her teens and 20s as a way to engage the joyful, high energy spirit that naturally makes the body want to move.

For the more sedentary or ill client who is having difficulty moving, utilizing the Activities of Daily Living Questionnaire is a gentler introduction. It provides not only a realistic assessment of physical capability but also an opportunity to suggest simple measures that can be incorporated to enhance what movement is already naturally occurring day to day. For example, chores can be arranged so a person can alternate downstairs and upstairs tasks, purposefully making as many round trips as able. Using breath and mindfulness techniques at the same time can help keep the person present to the miracle of movement.

For those with severe fibromyalgia, chronic fatigue syndrome, or other debilitating illness, an excellent functional impact questionnaire that includes activity can be found online at ▶ Myalgia.com or ▶ FIQR.info. These questionnaires can be downloaded, or better yet, online electronic calculators used, to make it very fast and easy to assess and follow a client's progress over time. The Toolkit handout of Lifestyle Modification for Chronic Pain and Fibromyalgia Syndromes can be supportive.

Inquire about the patient's experience with tai chi, Pilates, and yoga (which can be done at all intensity levels from restorative to intense) as all are good choices for core strengthening, flexibility, and balance. Many people can't stand the thought of a gym, whereas others find it energizing and improving of their self-esteem, but transportation and membership cost can be an obstacle [15]. Dancing, hiking, gardening, and yard work are all ways to move aerobically without a gym. Of course, running, cycling, weights, swimming, martial arts, and aerobics classes are excellent modalities. A balance ball is a very inexpensive tool to utilize for sitting both at home and work. Help clients look for lower cost resources such as a YMCA or recreation center and classes such as senior "SilverSneakers" or chair yoga classes for the over-65 crowd. "Pickleball" is the new rage for all ages, especially found in senior communities, combining tennis, badminton, and table tennis, all elements for multigenerational fun. Dance is an excellent form of exercise, from Zumba to ballroom dancing, country and western to Argentine tango, ballet to barre, involving balance and flexibility as well as aerobic activity. Dance classes can provide a social milieu as well. Encourage at least one activity a week in community for the importance of social connection.

40.6.3 Assessing Movement at Home

Unfortunately, given all the "convenience" appliances and foods of this day and age, people are moving far less and sitting more both at home and at work. They work in order to

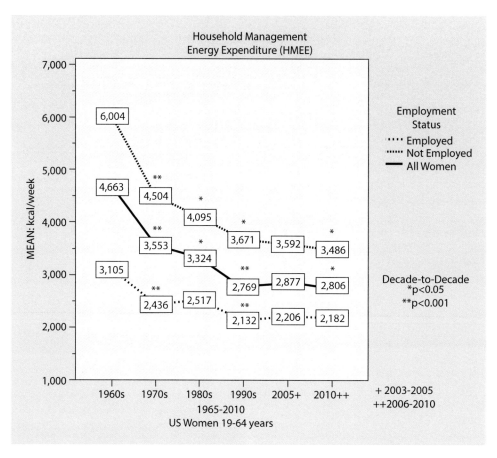

Fig. 40.3 Household energy expenditure. Depicts the decade to decade change in household management energy expenditure per week (HMEE/week) for all women and by employment status. (Reprinted from Archer et al. [16]. With permission from Creative Commons License)

pay for those conveniences and then pay more to go to the gym to workout. Yet, too often the thought of going to the gym or standing and preparing a real meal becomes exhausting for a burned-out working individual, especially with children. Commonly, many are recuperating from stressful days through sedentary time involved with watching television, sitting at a computer, or using handheld devices while eating a prepackaged microwaved meal. Unfortunately, these activities have become the sole source of "restorative" activities for a vast majority of the American population. Yet, for each additional hour of television watched per day, there is an 11% increase in all-cause mortality [14].

For those who are already quite sedentary, examine their written exercise intake or use the verbal interview to discover how much movement they are already engaged in with housework or yard work. Expanding on those areas can be useful ways to begin to increase caloric expenditure. Everyday chores can be a source of burning extra calories. The average 1950s housewife was burning 1000–3000 more calories a week with sweeping, doing laundry and walking to the local market, chopping and preparing daily meals, etc. [16]. Suggest listening to good music or podcasts while engaging the whole family in chopping vegetables, making meals and hand-washing dishes or while mindfully standing and ironing, turning those humdrum chores into activities of pleasure rather than boring repetitive tasks.

Encouraging singing along loudly and dancing while sweeping, vacuuming, going up and down the stairs multiple times to carry laundry, making the bed, and bending and moving while cleaning can turn mundane "have to's" into fun and playful opportunities to burn calories. Putting the kids in a stroller and going out for a walk after dinner is a time-honored way to substitute TV time for moving family time (Fig. 40.3).

Additionally, we are burning less calories from maintaining temperature regulation compared to even 40 years ago because of HVAC systems keeping the temperature at steady states so our bodies don't have to. Keeping indoor temperatures cooler has been shown to expend more calories from thermogenesis as well as exercising in the cold [17–20]. Shivering and sweating are simple ways to increase caloric expenditure whether moving bodies indoors or out. Have everyone turn thermostats down in winter and up in summer.

40.6.4 Assessing Movement at Work

One can assess if the client's job is stationary/sitting or if it naturally involves movement, such as being a nurse or retail worker required to spend most of their shifts on their feet. A number of wearable devices are now available, such as the

Fitbit, and many smartphones can track movements and provide positive feedback to the client to encourage incrementally increasing their goals [see Toolkit for a list of personal device options]. Engaging their competitive spirit with co-workers can also be motivating.

Portable standing desktop stations and chairs that raise from sitting to standing are available that can encourage the client to frequently change positions. Balance-ball chairs actively engage core muscles while sitting. Devices, such as a stationary pedal for underneath the desk, are engaging and easy. Utilizing simple at-work stretching exercises for desk sitting and setting a timer to remind to stand every 15–20 minutes can be useful. Getting up frequently to go to the water cooler or taking stairs instead of an elevator can easily add to a daily step count. Helping the patient to think creatively can assist them in their goals.

40.6.5 Resources to Motivate and Track Exercise

Once an assessment is made of the current activity level and the challenges and opportunities outlined, it is much easier to utilize a tool such as the IFM Exercise Prescription, which reiterates the FITT model of desired frequency, intensity, type, and time. High-intensity interval training [HIIT] such as the "7-minute workout" (many versions available online) has become a popular, efficient, and effective use of time for cardiovascular fitness. Again, the client is much more likely to have interest and be compliant if the practitioner is also practicing what they are preaching [12]. Tracking over time can be done with a form such as the IFM Exercise Goals and Tracking Journal, both found in the Toolkit.

Two excellent handouts for motivation which also can be found in the Toolkit include "The Power of Movement: Living an Active Lifestyle" and "Tips to Incorporate Mindful Daily Movement." Suggest a TV moratorium for 4 weeks and see how much time is freed up for all forms of self-care!

40.7 Assessing Sleep and Restorative Activities

40.7.1 Restorative Activities: The Big Picture

More than a third of US adults don't get enough quality sleep. According to recent data from the CDC, 50–70 million adults in the United States have chronic sleep and wakefulness disorders [21]. Sleep difficulties are associated with all chronic diseases, mental disorders, health-risk behaviors, and the underlying limitations of daily functioning, injury, and mortality. Insomnia is a known independent risk factor for depression and other psychiatric disorders. Most patients with mood disorders experience insomnia. Conversely, one-third to one-half of patients with chronic sleep problems have mood disorders leading to a vicious cycle [22].

Recent data on all causes of mortality in the United States show sleep apnea to be a major cofactor for many chronic diseases. Although sleep apnea affects more than 22 million Americans, 80% of cases are undiagnosed. Sleep deprivation contributes to many health challenges including metabolic syndrome, cardiovascular disease, inflammation, neuropsychiatric disorders, and type 2 diabetes, along with impaired memory [23–25]. Sleep disorders may also trigger immune system abnormalities, inducing autoantibody production and autoimmune disease: SLE, RA, etc. [26] (▶ Chap. 35).

Thus, screening for insomnia, narcolepsy symptoms, snoring/sleep apnea, and restless leg syndrome is paramount. Appropriate strategies to improve sleep quality can be implemented, and the need for further diagnostic testing such as a sleep study at an overnight sleep center can be recommended. Fortunately, home sleep study equipment is now available through a number of different vendors which may be less expensive, less disruptive, and more amenable to encouraging a client to be proactive in addressing their issues than going to a sleep center. For many with mild-to-moderate obstructive sleep apnea, a special dental appliance, created by a dentist trained in this area, can be an economical option in lieu of a CPAP machine, improving compliance.

40.7.2 Screening Questions for Sleep Apnea

Important screening questions to identify sleep disturbance pathology include:
- On average, how many hours of sleep do you get in a 24-hour period?
- Do you snore?
- During the past 30 days, for about how many days did you find yourself unintentionally falling asleep during the day?
- During the past 30 days, have you ever nodded off or fallen asleep, for even just a brief moment, while driving?

40.7.3 Questions to Assess Sleep Habits

A more extensive sleep questionnaire can include questions about sleep environment, including light, sound, temperature, pets etc., including the use of electronic devices and sources of EMFs that may interfere with sleep. Physiologic changes in the brain have been documented from electromagnetic fields generated by cell phones, electronic appliances, and household electrical meters positioned near bedrooms. These have been shown to either produce or exacerbate a wide spectrum of neuropsychiatric effects including depression [27].
Questions include:
- What time do you go to bed? What time to you wake up?
- Are you watching the news or stimulating TV before bed?
- Are you on your laptop or electronic device before bed?

- On average how long does it take you to fall asleep?
- How many times do you wake up? Why? Going to bathroom? Hot flashes? Mind racing?
- Are you able to fall back asleep?
- Are you refreshed when you wake up?
- Are there pets in the bedroom?
- Does your spouse snore?
- Is there a TV in the bedroom?
- Do sounds and/or lights wake you easily?
- How much alcohol and caffeine do you typically drink per day?

It can be useful to help the client to take time for a bedtime routine, beginning an hour prior to bed, as a form of enjoyable deep self-care, rather than another "have to." Consider putting on calming music and taking time for a hot Epsom salt bath with essential oils such as lavender and frankincense (known to have calming effects on the nervous system). Everyone needs time for good oral hygiene, and women in particular often need extra time to mindfully take care of their skin, face, and hair with nontoxic cleansers. Reading a good book, stretching, journaling, and utilizing meditation, prayer, and gratitude can be very simple mindful ways to peacefully wind down, bringing a sense of contentment and closure to an otherwise very hectic day. Counsel patients that bedtime begins an hour before sleep with routine preparations that may include bathing and story time for themselves, no different than the parent has a routine with their children!

40.7.4 Other Restorative Activities

Leading-edge science on neuroplasticity has shown many benefits of tools like mindfulness meditation, gentle yoga, tai chi, gratitude, and journaling practices in rewiring the circuitry leading to chronic activation of the hypothalamus-pituitary-adrenal (HPA) axis. It is important to ask, "Besides sleep, what other type of restorative activity do you engage in? How do you express yourself creatively? What brings you joy? Makes you laugh?" Many clients will get blank looks or even well up with tears when asked these questions, as they realize they have no answers. Particularly women, who are wrapped up in meeting their families and children's needs, find themselves last on their list.

As discussed in previous chapters, meditation and mindfulness-based stress reduction techniques are ancient practices revealed by modern-day technologies to significantly impact learning and memory, emotional regulation, perspective, cognitive restructuring and learning, immune function, and HPA axis regulation. It is known to increase brain-derived neurotrophic factor (BDNF), stimulate the prefrontal cortex, and facilitate neurogenesis. It has been shown to be one of the most cost-effective intervention techniques for improving quality of life, particularly in chronic health conditions [28–31]. Other meditative and restorative practices can be knitting or other needlework, fishing, hiking, gardening, tinkering with tools, and cooking. When mindful concentration is brought to the moment, no matter the activity, it has the capacity to rest the mind, body, and spirit, shifting physiology. Bringing mindfulness and breath focus, to whatever pursuit engaged in, can significantly reduce levels of distress for clinician and client alike.

Laughter is truly the best medicine and yet is something that is frequently lacking. We know it raises endorphins, lowers stress hormones, and modulates immune function, helping to mitigate the impact of stress [32, 33]. Ask, "what makes you laugh, and how often?" One suggestion could be that if they are going to watch TV, choose only something benign, like a cooking or auto show, or something funny that will make them laugh. Pets can also be a great resource in this area. Seeking out humor on purpose, especially during times of stress, can help alleviate its intensity.

Questions for Restorative Activities:
- What is your favorite way to relax?
- What do you do that is fun? Joyful? Playful?
- What makes you laugh? And how often?
- How do you express yourself creatively?
- What did you do for fun as a kid?

40.8 Stress and Resilience

40.8.1 Stress and Resilience: Assessing the Big Picture

We live in a highly stressed culture, rapidly moving from one activity to the next, focused on maximum productivity, minimal downtime with much hidden as well as overt violence. Our society doesn't value space, emptiness, silence, and nature, unfortunately to the detriment of our long-term health. Stress is a word that can have many different connotations for the client along with many different sources. Patients are usually focused on the external sources of stress such as work, finances, family life, and poor health, which may be both acute and chronic. Many people utilize sugar, alcohol, or TV and excessive electronics to manage their stress level, which inputs often only further exacerbate the total stressful load. Adequate sleep and restorative activities as discussed above can contribute significantly to building resilience from stress, as can other activities, which are outlined below.

For the clinician, it is imperative to understand the sources of the patient's stress, the current perceptions of their impact, and their level of resilience and coping strategies in order to target appropriate interventions. Does the patient have a strong internal locus of control with established resiliency factors such as optimism, perseverance, and a social support system, or are they focused on an external locus, blaming others, having chronic negative self-thoughts and feelings of hopelessness? How an individual responds to stress is extremely variable and has genetic, learned and environmental components whose internal physiologic and metabolic consequences are extensively discussed in Parts II and IV of this textbook. From a functional medicine standpoint, the practitioner aims to support the client by alleviating the total body burden of external stressors impacting the internal sources, offering them tools to help ameliorate both sides of the equation.

Unfortunately for many seeking functional medicine evaluations, the largest sources of stress are their chronic health issues, unresolved or worsened, often after years of seeking help from many different specialists with little improvement. They may be disillusioned, angry, and skeptical. Failure to resolve root causes of health issues has impacted their finances, relationships, self-esteem, level of anxiety, and depression and kept them in a downward spiral of hopelessness. Channeling anger and hopelessness into community and action can be important ways to transform those stressors.

Assessing how much stress a person feels on a daily/weekly/yearly basis can also be useful in understanding what testing to order and interventions to recommend. Every single physiologic node of the Matrix is impacted: hormones and neurotransmitters, inflammation, microbiome, etc. For example, depending on the presentation, the clinician may want to run a diurnal salivary cortisol test to see the impact on the HPA axis, or in another with severe IBS, a stool test to assess effect on the microbiome and intestinal permeability.

Focusing on implementing simple external coping strategies can be one of the most effective interventions. Often the least expensive and time-honored prescriptions are ones of community, laughter, downtime, adequate sleep, or mindful breathwork and meditation – all shown to benefit the internal physiology. Physically writing a prescription for self-care on a prescription pad and telling the client to put it on their bathroom mirror or refrigerator where they can see it every day can be an effective measure to engage their self-compassion and compliance.

40.8.2 Tools for Assessment of Stress and Self-Care

Excellent preliminary intake questions encompassing the common stress sources are found in the IFM Adult Intake Questionnaire and the Self-Care Assessment Questionnaire, both found in the Toolkit. These questions gently raise awareness in a nonjudgmental way and give the client insight into possibilities for change. Emphasize that there are no wrong or right answers. For more in-depth and insightful assessments, the DASS and ACE questionnaires (discussed below) can be essential tools for evaluating dimensions of emotional distress and root causes of chronic illness. All practitioners should be familiar with them.

40.8.3 Stress: Important Questions to Ask

A simple way to begin a conversation is: "On a scale of 1–10, with 10 being the highest amount of stress, where do you feel you are running on average today? For the last month? For the last year? What are your major sources of stress and can you tell me more?" Asking about work, family, social relationships, intimate relationships, children, financial stressors, and health are all important to get an overall picture of challenges and strengths.

What about their work life? What kind of work do they do? Are they satisfied with it? How are their relationships with co-workers or bosses? How much pressure do they feel and is there adequate time and support to get the work done they need? What kind of commute do they have? Driving in traffic can often be one of the biggest sources of chronic low-grade stress. Using tools such as listening to interesting podcasts, fun or inspirational music and books on tape while commuting, rather than cacophonous talk radio or stressful news, can be helpful.

Ask about their social life. Friends. Their significant other. Who takes care of the children if both parents work? What other health issues are other family members dealing with? Teenagers? Elderly parents? Are there major financial or legal stressors? For many working women, childcare issues and lack of time for self-care contribute to feelings of being majorly overwhelmed. For stay-at-home moms, acknowledge that raising children and caring for a home is a full-time job with its own set of stressors, not to be devalued. Men often shoulder their burdens internally and in isolation. In a culture that frowns upon asking for help, they tend to be minimizers on questionnaires and may require subtler questioning.

It may be helpful to ask whether there are ongoing issues such as alcoholism, addiction, or abuse. Does the person feel safe in their home? With their partner? In their neighborhood? One in every three women and one in every four men have been victim of violence in their lifetime, most often from an intimate partner [34]. These questions cannot be ignored.

40.8.4 Abuse and Neglect

Unfortunately, far too many of us, practitioners and patients alike, did not grow up in optimum circumstances, and some traumas can have lifelong health impacts if not acknowledged and addressed. Ask what things were like growing up. How stressed were their parents? What behaviors and coping skills were modeled? Were they allowed to speak their voice and offered comfort and appropriate support, or did dysfunction, verbal and physical abuse, neglect, and other forms of violence predominate? Did they grow up in an alcoholic family system? All practitioners should be familiar with administering the CDC's Adverse Childhood Events [ACEs] Study and scale, which looks at ten different types of childhood trauma, to understand the sobering long-term implications for personal and public health.

The ACE is an extremely useful tool for helping patients to understand how early childhood trauma can have long-term health impacts and can help to dissipate some of the shame and silence often associated with these events. Unfortunately, shame and guilt are extremely toxic emotions to the nervous and immune systems. Silence often leads to the perpetuation of violence and abuse from one generation to the next, no matter the socioeconomic status. Findings of the ACE study show:

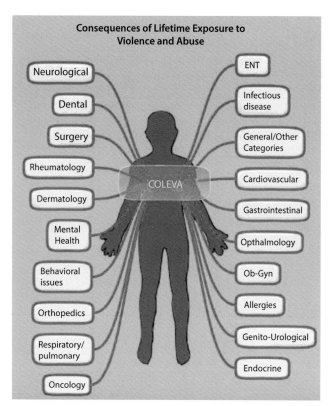

• Fig. 40.4 Consequences of lifetime exposure to violence and abuse. (Based on data from Academy on Violence & Abuse (▶ AVAHealth.org). Exposure to Violence and Abuse (COLEVA). (Retrieved from: ▶ https://www.avahealth.org/resources/ava_publications/ava_publications_new.html)

Childhood trauma is very common, even in employed white middle-class, college-educated people. There is a direct link between childhood trauma and adult onset of chronic disease, including heart disease, lung cancer, diabetes, and many autoimmune diseases, as well as depression, violence, being a victim of violence, suicide, and a host of other mental health issues. People usually experience more than one type of violence, rarely is it only sexual or verbal abuse. More types of trauma increased the risk of health, social, and emotional problems [34].

Another excellent interactive resource for understanding the documented impacts of violence and chronic childhood stress on all organ systems is from the COLEVA project: Consequences of Lifetime Exposure to Violence and Abuse, a project of the Academy on Violence and Abuse (▶ avahealth.org) shown in ◻ Fig. 40.4.

Whichever stressors are uncovered, the clinician can impart the positive, hopeful findings of neuroplasticity, including the possibility of rewiring the brain, to change PTSD and stress responses by cultivating resiliency tools such as mindfulness-based stress reduction (MBSR), mind-body breathwork, EMDR, HeartMath, and other therapeutic modalities. All clinicians should be familiar with the attributes of resiliency detailed below for themselves as well as patients and must have a network of resources to which they can refer, such as crisis hotlines, psychologists, 12-step recovery groups, and licensed social workers familiar with how to support the client, who may be revealing this information for the first time to a healthcare practitioner.

40.8.5 Assessing Anxiety and Depression

Anxiety and depression are some of the most common symptoms presented by clients as ongoing sources of chronic distress. While there are many different depression and anxiety assessment tools, including the Beck Depression Inventory, Hamilton Anxiety Depression Scale, and GAD-7, the use of the Depression Anxiety Stress Scale [DASS] developed by Lovington and Lovington in Australia adds sensing of the third axis of physiologic stress separated from anxiety and depression symptoms (see ▶ Chap. 29).

> The DASS is a 42-item questionnaire which includes three self-report scales designed to measure the negative emotional states of depression, anxiety and stress. Each of the three scales contains 14 items, divided into subscales of two to five items with similar content. The Depression scale assesses dysphoria, hopelessness, devaluation of life, self-deprecation, lack of interest/involvement, anhedonia, and inertia. The Anxiety scale assesses autonomic arousal, skeletal muscle effects, situational anxiety, and subjective experience of anxious affect. The Stress scale (items) is sensitive to levels of chronic non-specific arousal. It assesses difficulty relaxing, nervous arousal, and being easily upset/agitated, irritable/over-reactive and impatient. Respondents are asked to use four-point severity/frequency scales to rate the extent to which they have experienced each state over the past week [35]. The questionnaire, along with its interpretive guide, can be accessed through the Psychology Foundation of Australia website or can be found in the IFM Toolkit.

40.8.6 Basic Questions to Assess Stress Levels

- How stressed do you feel on a scale of 1–10?
- What are the major sources of your stress?
- How are things going for you at home? With work? With relationships? With finances?
- How much is your health contributing to your stress levels?
- Did you feel safe growing up?
- Do you feel safe at home now?
- Physical, emotional, and sexual abuse are unfortunately very common – is this an area for us to discuss?
- What "courageous conversation" are you not having about an issue that's bothering you?
- Is the level of anxiety or depression, grief, or sadness you feel impacting your quality of life?

40.8.7 Assessing Resilience

Studies show that recovery is possible from even extreme trauma and stressors, both past and future [36, 37]. The following list, adapted from the work of psychotherapist Elizabeth Scott, MS, explains how qualities found in resilient individuals can be purposefully cultivated and include [38]:

1. Developing a positive attitude – learning to view life's difficulties as challenges to overcome, rather than fears to be avoided, moving from self-pity and blame to positive self-talk and taking responsibility.
2. Developing emotional awareness – asking "Why are you upset?" "What needs to change?" Journal to express feelings.
3. Developing an internal locus of control – even if we can't control an external situation, we can control how we respond.
4. Cultivate optimism by maximizing strengths and minimizing weaknesses.
5. Rally social support and tap into community resources.
6. Maintain a sense of humor.
7. Exercise and get adequate rest.
8. Get in touch with spiritual side, in whatever form is meaningful.
9. Perseverance and determination not to give up.

Many resources from books, DVDs, personal devices, and apps are available to help clients cultivate resiliency and learn techniques to retrain emotional and physiologic reactivity. Perhaps some of the best-studied tools include those from the HeartMath Institute (▶ HeartMath.org), which have created many biofeedback programs focusing on heart rate variability and coherence for both personal and professional use. The Monroe Institute (▶ MonroeInstitute.org) offers many modalities focusing on using Hemi-Sync binaural sound technology to entrain brain waves into relaxed states. QEEG neurofeedback harnesses the power of the brain's electromagnetic fields to restore calm; both personal devices such as Muse and professional devices are available. Many apps can be downloaded to help learn mindfulness and mediation including Headspace, Calm, Insight Timer, and others. Well-known investigators and research pioneers in the fields of meditation, mindfulness, and trauma recovery have audio. A number of spiritual leaders, from all traditions, have online broadcasts that are easily downloaded to a portable device.

Ask your clients "How do you manage your stressors?" and explore what are their resources for support: Family? Friends? Community? Church? Pets? Nature? Exercise? Meditation? Prayer? Silence? Have they sought counseling? Support groups? When is the last time they had a vacation or long weekend? How are they sleeping? Do they get some exercise? How often do they laugh? What tools do they use to relax and have fun or express themselves? As discussed above, some of the most powerful and insightful question you can ask is: "What do you do that is fun, joyful, creative, or playful?" It is amazing how many clients will get tearful when you ask these questions when they recognize they don't have an answer.

Twelve-step programs, such as AA and Al-Anon, Overeaters Anonymous, etc., offer inexpensive and incredibly valuable support tools that provide a day-to-day plan for getting through the most difficult of circumstances. All practitioners should be familiar with them and are encouraged to attend open meetings to learn more. Remembering that joy, fun, humor, and creativity even in the midst of severe external circumstances can be what carry people through. They are potent healing modalities. Build on whatever they are doing well, look for positives, and expand them.

40.8.8 Resources for Assessing and Addressing Stress and Resilience

The Toolkit offers resources to share with clients on an array of topics, including:
- Relaxation response
- Strategies for transforming stress
- Mindfulness for sleep disorders and anxiety
- Integrated guide to visualization and imagery
- Meditation-getting started
- IFM gratitude journal
- Breathing techniques to soothe the soul, restoration prescription
- Cultivating self-awareness and mindfulness
- Restoration activities
- Health benefits of napping
- Tips to incorporate mindful movement daily

Other modalities to raise sympathetic tone may include bodywork such as massage, lymphatic therapy, craniosacral work, acupuncture, and laughter.

40.9 Assessing Relationships and Beliefs

40.9.1 Relationships and Beliefs: The Big Picture

Of critical importance for the practitioner to assess is the client's current state of relationships, whether at home, with extended family and friends, pets, work, and community, and including their support systems and beliefs. These factors are the "back story" which underlie and can either completely undermine or be the lynchpin for positive change. Those wishing to make significant lifestyle changes that deviate from the norm of their friends and family may encounter major obstacles in obtaining support or be directly undermined. At the same time, many feel isolated from a community or are far from family or perhaps they are purposely disengaged from a dysfunctional family system and need a source of community to support their goals. Unlike small communities and extended generations, where skills were passed from one generation to another, today's highly mobile and distant lifestyle can make many feel isolated and alone. Pets and animals can be an important source of comfort, relaxation, affection, and attachment for many. They may often be an incredibly important primary relationship in a person's life, whose well-being is tied to their own (▶ Chap. 6).

Beliefs encompass not only belief in themselves and their ability to make the required lifestyle changes but also the person's belief in whether they feel connected to something larger than themselves – a religious or spiritual source of sustenance, no matter the particulars. Developing knowledge of the client's stage of readiness to change, from pre-contemplative to ready to take action, can focus the efforts of clinician and client alike. By having an understanding of these forms of connectivity, along with the obstacles and opportunities presented, the clinician can help inform and guide the utilization of positive forces and known benefits of social networks in supporting the health goals of their patients.

40.9.2 Relationships, Social Networks, and Health Outcomes

Many studies have shown that a person's habits and state of health are closely tied to their relationships with family and community of friends. Family ties and social networks can markedly impact longevity and quality of life [39]. The famous Roseto study of an Italian-American community, published in the early 1960s, showed rates of death from heart disease were unexpectedly lower than surrounding neighborhoods, even though the theoretical cardiovascular risks from smoking and poor diets were the same. The factors that seemed to play the important role in longevity were close intergenerational family ties with a strong social network of like-minded values [40]. A recent study concluded that family aids most in longevity, as marriage bonds were found to be more important than friends in terms of survival [41]. However, what about overall health, weight, obesity, and nutritional and lifestyle choices? How do family and friends impact these factors?

A pivotal longitudinal study by Christakis, published in 2007, looked at the spread of obesity between friends and family and revealed that it is the network of friends, particularly the ones with the closest and strongest ties, which had the most powerful influence on gaining weight or not. Christakis concludes, "The spread of obesity in social networks appears to be a factor in the obesity epidemic. Yet the relevance of social influence also suggests that it may be possible to harness this same force to slow the spread of obesity. Network phenomena might be exploited to spread positive health behaviors, in part because people's perceptions of their own risk of illness may depend on the people around them... People are connected, and so their health is connected" [42].

It is the social support, advocacy, accountability, and supportive influences found in *groups*, not necessarily the directives of what a person is actually being told to do, that create the anticipated behavioral change. These factors have been utilized for a number of years in the layperson setting of peer support groups, ranging from divorce and grief support groups to Weight Watchers, smoking cessation, Alcoholics Anonymous, and other similar 12-step programs, and have shown the power of a collective consciousness in creating positive outcomes.

The IFMNT practitioner can harness this phenomenon to great advantage with the opportunity to do learning and group visits, also known as shared medical appointments (SMAs), which have been shown to enhance individual patient experience, improve population health, and reduce the cost of healthcare, three cofactors upon which some models of reimbursement are now based [43–46]. Coined the "Triple Aim" of improved "care, health, and cost," stakeholders have recently acknowledged the necessity of upgrading to the "Quadruple Aim," adding the fourth component of "provider satisfaction," as rates of physician, nursing, and clinical staff burnout are significantly high. Healthcare clinicians and staff, appropriately trained in the necessary group facilitation skills based on a team model of shared responsibilities, can significantly benefit from this group model, with overall improved personal and professional satisfaction [47, 48].

The SMA/group care model has been successfully implemented in many primary care settings and includes improved outcomes in diabetic and cardiovascular care, prenatal care, cancer, autoimmune, asthma, diabetes, ulcerative colitis, multiple sclerosis, HIV, pain management, menopause, insomnia, and stress [49–51]. An excellent review, published by Massachusetts General Hospital in 2012 for the hospital and outpatient clinic setting, outlines a practical framework for implementing the three basic goals of SMAs:
1. Access to medical care visits
2. Education for patients on their medical condition or disease
3. Enhancement of self-management skills for lifestyle and behavioral change, to promote self-management at home and consistency in follow-through on medical recommendations [51] (see ◘ Fig. 40.5)

An additional excellent resource for implementing group visits, focused specifically on the practitioner of functional medicine, has been developed by family practice physician Shilpa Saxena, MD, IFMCP and can be found through her website.

Ideally, impacting population health for the better begins with impacting prenatal and postnatal health. An excellent example of how relationships developed in early prenatal care, utilizing a group visit model, can be used to great advantage to achieve the Quadruple Aim, is reviewed in Strickland's 2016 article, "Centering Pregnancy: Meeting the Quadruple Aim in Prenatal Care" [52]. Originally developed in the early 1990s by certified nurse-midwife Sharon Rising and championed by the Centering Pregnancy Health Institute [CHI] to great advantage, "Centering Pregnancy is a group prenatal care model that engages pregnant women in their care, [resulting] in promising health and system outcomes...this innovative care model deliver[s] the Quadruple Aim of improved patient experience, quality of care, cost containment, and provider satisfaction" [53]. Strickland's article outlines how women involved in supporting one another through pregnancy and postnatal care had significantly

Types of Groups			
Main Focus of the Group	**ACCESS** To improve access to medical care and address direct medical needs	**EDUCATION** To improve health education and teaching skills for self management	**BEHAVIORAL CHANGE** To promote and enhance strategies for lifestyle and behavioral change
Examples of Groups by Focus	• Shared medical appointments • Group medical clinics, veterans administration hospital	• Diabetes self-management education groups by CDE diabetes nurse educators • Health coaching	• Medical group visits • Group psychotherapy • Patient peer-to-peer support groups

Fig. 40.5 Types of group visits or shared medical appointments. (Reprinted from with permission from Putting Group Visits into Practice- A Practical Guide to Implementation and Maintenance of Group Visits at Massachusetts General Hospital, Jan 2012. Retrieved from: ► http://www.massgeneral.org/stoecklecenter/assets/pdf/group_visit_guide.pdf)

reduced episodes of preterm birth, increased healthy birth weights, and markedly improved rates of breastfeeding, all known factors in long-term cost and health outcomes for later life. In addition, markedly improved clinician personal and professional satisfaction was a measured outcome in several studies referenced, bringing back a sense of hope, meaning, and rededication to their vocations [54].

40.9.3 Beliefs

Beliefs encompass everything from a person's thoughts about themselves and their capabilities, to their place in the world, to their connection to a spiritual source greater than themselves. Belief systems can be powerfully engaged to help the client make needed changes toward their health goals. Determining if a client has a sense of an internal locus of control in which they believe in themselves, versus an external locus, feeling chronically buffeted by outside forces over which they feel helpless, can direct the clinician in the language and resources necessary to engage and inspire that client. Ask what are they passionate about? What moves them? Is there an area they feel strongly about? How do they feel about their ability to implement and sustain necessary changes? What can help or hinder them?

40.9.4 Spirituality and Religion

Spirituality is the point of connection for the entire Patient's Story. Asking in a nonjudgmental, non-directed way whether or not a client has a regular spiritual or meditative practice can open a dialogue to engage God, truth, or a higher power as a personal resource for positive behavioral change. As discussed by Saguil et al., "The medical literature has become replete with articles purporting positive links between spirituality and such diverse conditions as cardiovascular disease, COPD, cancer, cirrhosis, rheumatoid arthritis, leukemia, AIDS, depression, adolescent risk behaviors, anxiety, and pain. In addition, multiple studies have reported that increasing levels of spirituality correlate with decreasing levels of medical utilization, healthcare costs, and death" [54].

In 2012, Dr. Harold Koenig of Duke University's Department of Spirituality and Medicine published an exhaustive review of more than 1200 papers on prayer, health outcomes, and longevity [54]. Other studies have shown even that most physicians believe in medical miracles [55]. The data is clear that a regular spiritual or meditation practice is associated with improved well-being and outcomes and that all practitioners should ask their patients about spiritual needs. Indeed, the Joint Commission on Accreditation and the Association of American Medical Colleges have directed all health institutions and medical schools to implement a short spiritual assessment as part of the medical history, to help caregivers deliver compassionate and appropriate care [56].

Although religious beliefs and practices play a large role in many people's lives, those who find themselves a-religious may feel a strong sense of spiritual connection through animals or nature or feel drawn to the embodied spirituality of yoga or tai chi practices or martial arts.

Engaging larger values in serving a greater cause can help motivate new behaviors. For various reasons, a client may have a hard time believing they can make changes for themselves, yet they would be willing to take steps to be of benefit others such as animals or planet or family or their children or grandchildren. For example, many people are aware of large issues such as terrorism or wars but feel powerless or hopeless to do anything about it. Helping them see how day-to-day choices, e.g., by becoming mindful of their spending habits and which foods they choose, are personal ways of impacting forces much greater than themselves. Some can be motivated to engage in frequenting a local community garden or farmer's market as a way to get exercise and healthy food, to spend quality time with family and community, and to know they are contributing to supporting local and organic economies.

Engaging their "inner child" can be a tool to use to help motivate. "If your adult-self were watching your child-self stay up late, not eat, punish yourself with negative words and thoughts, how would you re-parent yourself?" Supporting a client in awakening compassion, not blame toward their previous choices, helps to overcome shame and guilt and creates a hopeful path forward.

The Rev. Rick Warren, along with functional medicine leader Mark Hyman, MD, showed exactly the benefit of engaging small groups and beliefs utilizing "Food + Fitness + Faith + Focus + Friends" in their combined effort called "The Daniel Plan." Small prayer groups already found at the church were engaged in lifestyle changes grounded in biblical principles to help the entire congregation get on board. Regarding positive outcomes, as Dr. Hyman likes to joke, "A New England Jew meets a born-again Christian and an entire church loses 10 tons along the way!" Jewish Community Centers have utilized this concept to help create centers of service and community that include everything from childcare resources to mental health and counseling to exercise and religious study, as has the YMCA. Whether in Hare Krishna, Buddhist temples, or Muslim mosques, faith, fitness, and fellowship can go together. Some churches are educating their communities about the detrimental environmental and health consequences of genetically modified foods by using the adage that GMO stands for "God Move Over"!

As outlined in detail above, no matter what faith, or none, mindfulness meditation is a powerful tool that has shown extraordinary ability to create new neural pathways and can be combined with any daily activity, faith or healing practice. For example, the Christian concept of centering prayer uses mindfulness techniques and a repeated sacred word as points of focus. Fishing or knitting both use active forms of mindfulness. Respectfully exploring the importance of a patient's spiritual beliefs gives permission to actively engage this most powerful resource of healing.

40.9.5 Questions to Ask About Relationships and Beliefs

The practitioner should approach relationship and beliefs assessments in a nonjudgmental, non-threatening manner (without injecting their own beliefs unless asked) and using open-ended questions such as:
- How are things going at home with your spouse/partner?
- With your co-workers and boss?
- With your children/parents?
- With extended family?
- Do you have pets to whom you are connected?
- Do you feel safe and supported at home?
- Do you have close friends who know what is going on with you?
- Do you feel you have someone to talk to in times of need and to share triumphs?
- How supportive are your close friends and family of your health goals?
- Do you feel like your life has purpose and meaning?
- Do you have a regular spiritual or faith practice that is meaningful to you?
- What kind of extended support system do you have?
- Do you see a counselor or therapist?
- Any support groups to which you belong?
- Do you have a sense of belonging?
- Is there any activity you are engaged in that makes you feel part of a community?
- Do you have silence in your life?
- What wishes do you have for yourself, what do you value most about yourself and your work?
- What makes you feel most alive?
- Do you believe there is a higher purpose in the health challenges you have suffered?

The IFM Adult Intake Questionnaire [male/female] includes a section in which similar questions are easily covered, including stress assessments and relationships together and can be an initial resource to investigate areas of concern.

Finding a higher purpose and meaning in their health journey may be one of the key factors that can help inspire hope and enable clients to become willing to move through the necessary steps of readiness to change. Particularly for those who have been very sick and isolated for a long time, finding meaning in their suffering can be a source of hope for healing. Viktor Frankl, the famed psychotherapist, Nazi concentration camp survivor, and author of the classic, "*Man's Search for Meaning*," began a powerful psychotherapeutic approach, logotherapy, which arose out of his horrific experience [57]. He realized that those prisoners who were able to find a higher purpose or meaning in their suffering were much more likely to survive the bleak and dangerous hopelessness of the concentration camp than those who felt there was nothing to live for. For himself, it was imperative to do everything he could to survive, so he could be a witness to the horrors he experienced.

Can you help your client to be able to share his or her own process of healing? To take the tools and information they will be learning and turn around their experience to reach out and to help others out of their suffering? For all people, young or old, with devastating anxiety and/or depression, cancer or CFS, gastrointestinal and autoimmune issues, or neurologic disorders, all conditions being suffered by hundreds of thousands of their peers – what if they can become a voice for greater change as their challenges are turned around to be a beacon of light for so many others on similar paths? No matter who, telling their story through the lens of the hero's/heroine's journey can be a powerful tool for inspiration, enabling long-lasting health and well-being.

40.10 Assessing Readiness to Change

40.10.1 The Nature of Change

> Change is not just about integrating new ideas.... It's about transforming old ones! –Monique Class, MSN, IFMCP [58]

Change and transformation are the underlying mechanisms for improving lifestyle. So, what is change and how is it different from transformation? What can we do to foster change in patients (and ourselves)? Why is change difficult? What are stages that we should recognize? Where do patients get stuck? Where do practitioners get stuck? Why is it important to talk about spirituality with our patients to foster change, transformation, and healing? These are all important questions for practitioners to grasp when engaging their clients to make changes for optimal health outcomes.

We want to be able to encourage "change," which is a goal-oriented, past or future action and is measurable. At the same time, we would like to inspire "transformation," which can be defined as a significant alteration in behaviors due to insight from the subconscious mind and which is more likely to help sustain change for the long run.

Understanding the stages of change and where the patient is on their journey can be helpful to direct the appropriate education, inspiration, and interventions, to help a person meet their health goals. Prochaska, in his seminal work "Changing for Good," outlines the stages that occur not necessarily in a linear fashion, but more akin to moving up and down a spiral: the five stages include [59]:

1. Pre-contemplation: I don't even know or don't want to know that I need to do something.
2. Contemplation: I need to do something in the next 6 months.
3. Preparation: I am going to do something in the next month.
4. Action: I am actively pursuing changes.
5. Maintenance: I am continuing to have new behaviors ongoing.

The IFM Adult Questionnaire has an abbreviated version addressing these stages as part of the intake, assessing how willing and ready a person is to make changes in their diet, nutrients, fitness, tracking, sleep, etc. utilizing a scale of 1–5. If a patient is still in the pre-contemplative stage, they may require further education or questioning about their beliefs, values, social support, and perceived benefits to help move them forward, versus someone who is seeking out a practitioner purposefully because they are ready to take action. Ongoing maintenance is often a more difficult area and one that is influenced by all of the factors discussed above, including relationships, beliefs, social support, and barriers such as transport and cost.

40.10.2 Smart Goal-Setting

Once opportunities for change have been identified in each area, encouraging the client to set attainable goals in each area of lifestyle factors can be very helpful. Start with one simple goal in each area and add incrementally. Having the client write down their goals, using the SMART technique of naming a specific goal, that is measurable, action-oriented, realistic and timely can markedly improve engagement. A specific goal answers the questions who when, what, where and why. For example, "I will ride my bike three times a week for 15 minutes in the neighborhood, to help with stress relief and improve my metabolism." A measurable goal answers the question of "how?" As in, "I can keep track of my mediation time with my Headspace or Calm app, and work on adding an additional minute each week, until I've reached 30 minutes per day." Or, "I can write three gratitudes in my journal for what I have done well that day" (Fig. 40.6)

SMART Goal Component	Example
Specific State the desired outcome as explicitly as possible, and target a specific area for improvement. This is the "who, what, where, when, which, and why" of your goal.	I will walk at least five days per week in the evenings to help me reduce my waist size (in inches).
Measurable Identify the ways in which you will track your progress, and be as specific as possible. This is the "how" of your goal.	I will meditate for 30 minutes a day five times a week in order to lower my stress levels and blood pressure.
Action-oriented Start with small, achievable goals that are easily outlined into specific steps that will enable you to complete the goal. Then, as you meet those smaller goals, work up to intermediate goals and goals that are more difficult to achieve.	I will make an effort to move my body for at least 15 minutes three days a week, increasing my time each week by five minutes until I reach 30 minutes per day. I will add an extra day every two to four weeks until I reach 30-60 minutes for five days a week.
Realistic Create a goal that you are both willing and able to accomplish.	I will begin my bedtime ritual one hour before bedtime, which will help me fall asleep faster each night.
Timely Set a deadline or time for achieving your goal to help keep you motivated.	Over the next month, I will start eating breakfast every day. For the first week, I will make breakfast (or prepare it ahead the night before) twice per week. In the second week, I will make breakfast three times per week. In the third week, I will make breakfast five times per week. In the fourth week, I will make breakfast every day.

Fig. 40.6 Smart goal-setting. (Used with permission from The Institute for Functional Medicine ©2015)

40.11 Conclusion

As practitioners, by doing an in-depth, compassionate, and insightful assessment of all lifestyle factors, we can help inspire transformative change in the five fundamental modifiable lifestyle areas: nutrition, exercise and movement, stress and resilience, sleep and restoration, and relationships and beliefs. Gathering oneself as a practitioner, being present and mindful, and knowing how and what questions to ask, including the use of targeted questionnaires, can efficiently and compassionately engage the client in a therapeutic relationship.

Supporting transformative change by utilizing nonjudgmental questioning techniques and encouraging statements, such as "Your body has a remarkable ability to heal itself from the cellular to the spiritual levels. Everything we consume at all levels of body, mind and spirit become us, so changing and transforming what you are consuming, at all levels, helps to change and transform the body and your health," can be very empowering. And, we can ask, how can they take their suffering, their challenges, their learning, and their opportunities for transformation and find meaning?

Acknowledging the client by retelling their story with empathetic statements like, "Wow Mr. Jones, I can see you have really suffered and are feeling very frustrated and angered by how your complaints have been previously marginalized or trivialized, you have amazing fortitude to get you here! Let us see how we can help you find and build increasing sources of resilience to help you turn around," gives hope and engenders partnership and cooperation to encourage lifelong, positive lifestyle change.

References

1. The Institute for Functional Medicine. The functional medicine approach. Retrieved from: https://www.ifm.org/functional-medicine/what-is-functional-medicine/. Accessed 4 May 2019.
2. Peckham C. Medscape lifestyle report 2016: bias and burnout. Medscape Med News Rep. 2016 Jan 13. Slide 17. Accessed 29 Jan 2017.
3. Frank E, Dresner Y, Shani M, Vinker S. The association between physicians' and patients' preventive health practices. Can Med Assoc J. 2013;185(8):649–53.
4. Nerurkar A, Bitton A, Davis RB, Phillips RS, Yeh G. When physicians counsel about stress: results of a National Study. JAMA Intern Med. 2013;173(1):76–7. https://doi.org/10.1001/2013.jamainternmed.480.
5. The Institute for Functional Medicine. What is the IFM toolkit? Retrieved from: https://www.ifm.org/news-insights/toolkit-what-is-the-ifm-toolkit/. Accessed 4 May 2019.
6. http://www.naturalawakeningsmag.com/Natural-Awakenings/June-2014/The-Healing-Power-of-Story. Accessed 20 Jan 2017.
7. Rollin MC. The energetic heart: bioelectromagnetic interactions within and between people. Institute of HeartMath: Boulder Creek; 2003.
8. Rakel D, Barrett B, Zhang Z, et al. Perception of empathy in the therapeutic encounter: effects on the common cold. Patient Educ Couns. 2011;85(3):390–7. https://doi.org/10.1016/j.pec.2011.01.009.46.
9. Abramson S, Stein J, Schaufele M, Frates E, Rogan S. Personal exercise habits and counseling practices of primary care physicians: a national survey. Clin J Sport Med. 2000;10(1):40–8.
10. Churchill LR, Schenck D. Healing skills for medical practice. Ann Intern Med. 2008;149:720–4.
11. Owen N, Sparling PB, Healy GN, Dunstan DW, Matthews CE. Sedentary behavior: emerging evidence for a new health risk. Mayo Clin Proc. 2010;85(12):1138–41. https://doi.org/10.4065/mcp.2010.0444.
12. Berra K, Rippe J, Manson JE. Making physical activity counseling a priority in clinical practice, the time for action is now. JAMA. 2015;314(24):2617–8.
13. Stamatakis E, Rogers K, Ding D, et al. All-cause mortality effects of replacing sedentary time with physical activity and sleeping using an isotemporal substitution model: a prospective study of 201,129 mid-aged and older adults. Int J Behav Nutr Phys Act. 2015;12:121.
14. Dunstan DW, Barr EL, Healy GN, et al. Television viewing time and mortality: the Australian Diabetes, Obesity and Lifestyle Study (AusDiab). Circulation. 2010;121(3):384–91.
15. Murray J, Fenton G, Honey S, Bara AC, Hill KM, House A. A qualitative synthesis of factors influencing maintenance of lifestyle behaviour change in individuals with high cardiovascular risk. BMC Cardiovasc Disord. 2013;13:48.
16. Archer E, Shook RP, Thomas DM, Church TS, Katzmarzyk PT, Hébert JR, et al. 45-year trends in women's use of time and household management energy expenditure. PLoS One. 2013;8(2):e56620.
17. Yoneshiro T, Aita S, Matsushita M, et al. Recruited brown adipose tissue as an antiobesity agent in humans. J Clin Invest. 2013;123(8):3404–8.
18. Johnson F, Mavrogianni A, Ucci M, Marmot A, Batterham R, Vidal-Puig A. Could increased time spent in a thermal comfort zone contribute to population increases in obesity? Obes Rev. 2011;12(7):543–51.
19. Hansen JC, Gilman AP, Odland JA. Is thermogenesis a significant causal factor in preventing the "globesity" epidemic? Med Hypotheses. 2010;75:250–6.
20. Mavrogianni A, Johnson F, Ucci M, Marmot A, Wardle J, Oreszczyn T, Summerfield A. Historic variations in winter indoor domestic temperatures and potential implications for body weight gain. Indoor Built Environ. 2011;22(2):360–75.
21. CDC. Morbidity and mortality weekly report March 4, 2011;60(8). https://stacks.cdc.gov/view/cdc/29020. Accessed 9 Dec 2019.
22. Benca RM, Okawa M, Uchiyama M, Ozaki S, Nakajima T, Shibui K, Obermeyer WH. Sleep and mood disorders. Sleep Med Rev. 1997;1(1):45–56.
23. Briançon-Marjollet A, Weiszenstein M, Henri M, Thomas A, Godin-Ribuot D, Polak J. The impact of sleep disorders on glucose metabolism: endocrine and molecular mechanisms. Diabetol Metab Syndr. 2015;7:25.
24. Raison CL, Rye DB, Woolwine BJ, Vogt GJ, Bautista BM, Spivey JR, Miller AH. Chronic interferon-alpha administration disrupts sleep continuity and depth in patients with hepatitis C: association with fatigue, motor slowing, and increased evening cortisol. Biol Psychiatry. 2010;68(10):942.
25. Meier-Ewert HK, Ridker PM, Rifai N, Regan MM, Price NJ, Dinges DF, Mullington JM. Effect of sleep loss on C-reactive protein, an inflammatory marker of cardiovascular risk. J Am Coll Cardiol. 2004;43(4):678–83.
26. Sangle SR, Tench CM, D'Cruz DP. Autoimmune rheumatic disease and sleep: a review. Curr Opin Pulm Med. 2015;21(6):553–6.
27. Pall ML. Microwave frequency electromagnetic fields (EMFs) produce widespread neuropsychiatric effects including depression. J Chem Neuroanat. 2015;75(Pt B):43–51.
28. Manna A, et al. Neural correlates of focused attention and cognitive monitoring in meditation. Brain Res Bull. 2010;82(1–2):46–56.
29. Fell J, Axmacher N, Haupt S. From alpha to gamma: electrophysiological correlates of meditation related states of consciousness. Me Hypotheses. 2010;75(2):218–24.
30. Jung YH, et al. The effects of mind-body training on stress reduction, positive affect, and plasma catecholamines. Neurosci Lett. 2010;479(2):138–42.

31. Newberg AB, et al. Cerebral blood flow difference between long-term mediators' and non-mediators'. Consciou Cogn. 2010;19(4):899–905.
32. Bennett MP, Zeller JM, Rosenberg L, McCann J. The effect of mirthful laughter on stress and natural killer cell activity. Altern Ther Health Med. 2003;9(2):38–45.
33. Berk LS, Felten DL, Tan SA, Bittman BB, Westengard J. Modulation of neuroimmune parameters during the eustress of humor-associated mirthful laughter. Altern Ther Health Med. 2001;7(2):62–72, 74–6
34. Centers for Disease Control and Prevention (CDC). National intimate partner and sexual violence survey: 2010 summary report. Retrieved from: https://www.cdc.gov/violenceprevention/pdf/nisvs_report2010-a.pdf. Accessed Jan 2017.
35. Aces Too High News. Got Your ACE Score? Retrieved from: https://acestoohigh.com/got-your-ace-score/. Accessed Jan 2017.
36. Bonanno GA, Galea S, Bucciarelli A, Vlahov D. What predicts psychological resilience after disaster? The role of demographics, resources and life stress. J Consult Clin Psychol. 2007;75(5):671–82.
37. Southwick SM, Vythilingam M, Charney DS. The psychology of depression and resilience to stress. Annu Rev Clin Psychol. 2005;1:255–91.
38. verywell mind. Cope with stress and become more resilient. Retrieved from: https://www.verywell.com/cope-with-stress-and-become-more-resilient-3144889. Accessed 17 Jan 2017.
39. Compare A, Zarbo C, Manzoni GM, et al. Social support, depression, and heart disease: a ten year literature review. Front Psychol. 2013;4:384.
40. Stout C, Morrow J, Brandt EN Jr, Wolf S. Unusually low incidence of death from myocardial infarction: Study of an Italian American Community in Pennsylvania. JAMA. 1964;188(10):845–9.
41. American Sociological Association. Relationships with family members, but not friends, decrease likelihood of death. Retrieved from: http://www.asanet.org/press-center/press-releases/relationships-family-members-not-friends-decrease-likelihood-death. Accessed Jan 2017.
42. Christakis N, et al. The spread of obesity in a large social network over 32 year. N Engl J Med. 2007;357:370–9.
43. Wynn JD. The value of exceptional patient experience. N C Med J. 2016;77(4):290–2.
44. Berwick DM. A user's manual for the IOM's 'Quality Chasm' report. Health Aff (Millwood). 2002;21(3):80–90.
45. Berwick DM, Nolan TW, Whittington J. The triple aim: care, health, and cost. Health Aff (Millwood). 2008;27(3):759.
46. Medical Economics. Chronic disease: primary care group visits may be key bossting patient. Retrieved from: http://medicaleconomics.modernmedicine.com/medical-economics/content/tags/chronic-disease/primary-care-group-visits-may-be-key-boosting-patient. E-published July 2014. Accessed Jan 2017.
47. Bronson DL, Maxwell RA. Shared medical appointments: increasing patient access without increasing physician hours. Cleve Clin J Med. 2004;71(5):369, 370, 372, 374 passim.
48. Bodenheimer T, Sinsky C. From triple to quadruple aim: care of the patient requires care of the provider. Ann Fam Med. 2014;12(6):573–6.
49. Housden L, Wong ST, Dawes M. Effectiveness of group medical visits for improving diabetes care: a systematic review and meta-analysis. CMAJ. 2013;185(13):E635–44.
50. Bartley KB, Haney R. Shared medical appointments: improving access, outcomes, and satisfaction for patients with chronic cardiac diseases. J Cardiovasc Nurs. 2010;25(1):13–9.
51. Putting group visits into practice: a practical overview to preparation, implementation, and maintenance of group visits at Massachusetts General Hospital January 2012. http://www.massgeneral.org/stoecklecenter/assets/pdf/group_visit_guide.pdf. Accessed Jan 2016.
52. Strickland C. Centering pregnancy: meeting the quadruple aim in prenatal care. N C Med J. 2016;77(6):394–7.
53. McNeil DA, Vekved M, Dolan SM, Siever J, Horn S, Tough SC. A qualitative study of the experience of CenteringPregnancy group prenatal care for physicians. BMC Pregnancy Childbirth. 2013;13(Suppl 1):S6.
54. Koenig H. Religion, spirituality, and health: the research and clinical implications. ISRN Psychiatry. 2012;2012:278730, 33 pages.
55. Dossey L. Survey of physicians' views on miracles. The Louis Finkelstein Institute for Religious and Social Studies of the Jewish Theological Seminary, New York, December 2004.
56. Hodge D. A template for spiritual assessment: a review of the JCAHO requirements and guidelines for implementation. Soc Work. 2006;51(4):317–26.
57. Maria Marshall, Edward Marshall. Logotherapy revisited: review of the Tenets of Viktor E. Frankl's logotherapy. Ottawa Institute of Logotherapy. 2012.
58. Class M. Shifting the therapeutic encounter into an environment of insight. Lecture presented at Institute for Functional Medicine: Washington, D.C.; 2018.
59. Prochaska JO, Norcross JC, DiClemente CC. Changing for good. New York: Avon Books; 1995.

Developing Interventions to Address Priorities: Food, Dietary Supplements, Lifestyle, and Referrals

Aarti Batavia

41.1 Introduction – 716

41.2 Initial Interventions – 717
41.2.1 Food – 717
41.2.2 Whole Food Versus Specific Dietary Nutrient – 719
41.2.3 Meal Timings and Intervals – 720
41.2.4 Increased Intake of Vegetables with Moderate Fruit Intake – 721
41.2.5 Consumption of Organically Raised and Locally Grown Foods whenever Possible, with Low-Fat Animal Products (Low-Fat Dairy and Poultry) in Moderation – 722
41.2.6 Increased Intake of Fiber – 723
41.2.7 Removal of Potential or Documented Allergens and Excessive Alcohol – 724
41.2.8 Adequate Intake of Fatty Acids – 727
41.2.9 Adequate Water Intake – 730
41.2.10 Inclusion of Prebiotic and Probiotic Foods – 731
41.2.11 Lifestyle – 736
41.2.12 Referrals – 739

41.3 Monitoring and Evaluating Interventions – 739

References – 739

© Springer Nature Switzerland AG 2020
D. Noland et al. (eds.), *Integrative and Functional Medical Nutrition Therapy*,
https://doi.org/10.1007/978-3-030-30730-1_41

41.1 Introduction

Integrative and functional medical nutrition therapy is done within the context of therapeutic and healing partnerships. The purpose of such a relationship is to help the patient heal by identifying the influences and underlying mechanisms that initiated and continue to mediate the patient's illness. It is an equal investment of focus by both the dietitian/nutritionist and the patient to work together to identify where to apply the leverage of change – may it be nutritional or related to sleep, relationships, or stress management. Connection, rapport building, deep listening, reflection, presence, humility, vulnerability, compassion trust, and gratitude are all essential for healing to occur [1].

There are four essential components in integrative and functional medicine nutrition therapy (IFMNT), and they can be used by all healthcare professionals applying the functional medicine approach:
1. Listening to the patient story on the initial intake
2. Evaluating, prioritizing, and focusing on the patient's modifiable lifestyle factors
3. Organizing the patient's clinical imbalances by underlying causes into a systems biology matrix framework
4. Creating a therapeutic and healing partnership between the dietitian/nutritionist and patient to provide practical evidence-based interventions with timely monitoring and evaluation [2]

After gathering and working through the patient's history and laboratory findings; organizing the subjective and objective data on the functional medicine timeline and matrix; retelling the story back to the patient and acknowledging their goals, antecedents, triggers, and mediators; and identifying the clinical imbalances in the organization of physiological systems of the matrix, the dietitian/nutritionist comes to the most critical aspect of the encounter: intervention.

A completed matrix facilitates the review of common pathways, mechanisms, and mediators of disease and helps the nutrition expert select points of leverage for nutritional and lifestyle interventions. The nutrition practitioner may also find an opportunity to refer the patient to another healthcare practitioner for their medical expertise. For example, one might send a patient for vision evaluation if one complains of strained eye/poor vision or send a male patient to a physician specializing in bioidentical hormones if he has very low testosterone and cardiovascular or cognitive decline. However, even with the matrix as an aid to synthesizing and prioritizing information, it can be very useful to consider the impact of each variable at five different levels:
1. Whole body (the "macro" level)
2. Organ system
3. Metabolic or cellular
4. Subcellular/mitochondrial
5. Subcellular/gene expression [3]

It is important for the nutrition practitioner to identify therapies based on their potential impact on the most central imbalances of the particular patient. Evaluating interventions that are available at each of these five levels can help to identify a reasonably thorough set of strategies from which to choose. The following list incorporates only a few examples of various types of interventions within these five different levels.

1. *Whole-body interventions:* The human organism is a complex adaptive system with countless points of access. Therefore, interventions at one level will affect points of activity in other areas as well. For example, improving the patient's sleep will beneficially influence the immune response and melatonin levels, along with decreasing oxidative stress. Exercise reduces stress, improves insulin sensitivity, and improves detoxification. Reducing stress can reduce cortisol levels, improve sleep, improve emotional well-being, and reduce the risk of heart disease. Changing an individual's nutrition plan can have myriad effects on health, from reducing inflammation to reversing coronary artery disease or insulin resistance to even impacting cognitive decline.

2. *Organ system interventions:* These interventions are used more frequently in the acute presentation of illness and, in most cases, would require a referral to a non-nutrition professional. Examples include draining lesions; repairing lacerations; reducing fractures, pneumothoraces, hernias, or obstructions; or removing a stone to reestablish whole-organ function. There are many interventions that improve organ function. For example, continuous positive airway pressure (CPAP) machines improve inspiratory flow rate sleep quality and decrease daytime sleepiness and oxidative stress in individuals with obstructive sleep apnea syndrome (OSAS), thereby improving metabolic functions and cognition [4]. There is also evidence that effective management of OSAS not only has positive impacts on the individual but also infers positive benefits to their partners' sleep and daytime functioning [5].

3. *Metabolic or cellular interventions:* Cellular health can be addressed by ensuring the adequacy of MAPDOM status – minerals and fiber, antioxidants/phytonutrients, proteins, vitamins, fatty acids, and methylation factors in the diet or, if necessary, from supplementation. An individual's metabolic enzyme polymorphisms can profoundly affect his or her nutrient requirements. For example, adding conjugated linoleic acid (CLA) to the diet can affect body weight, alter the peroxisome proliferator-activated receptor (PPAR) system, and modulate the inflammatory response. However, in a person who is insulin resistant and/or has diabetes, adding CLA may induce hyperinsulinemia, which is detrimental. Altering the types and proportions of carbohydrates in the diet may reduce insulin secretion, increase insulin sensitivity, and essentially alter metabolism in the insulin-resistant patient.

4. *Subcellular/mitochondrial interventions:* There are several examples of mitochondrial nutrient-support interventions. Inadequate iron intake causes oxidants to leak

from the mitochondria, damaging mitochondrial function and mitochondrial DNA. Making sure there is sufficient iron helps to mitigate this concern. Ensuring sufficient antioxidants and cofactors for the at-risk individual must be considered in each part of the matrix. Carnitine, for example, is required as a carrier for the transport of fatty acids from the cytosol into the mitochondria and in improving the efficiency of beta oxidation of fatty acids and resultant ATP production. In patients who have lost significant weight, carnitine undernutrition can result in fatty acids undergoing omega oxidation, a far less efficient form of metabolism. Patients with low carnitine may also respond to riboflavin supplementation.

5. *Subcellular/gene expression interventions:* Many compounds interact at the gene level to alter cellular response, thereby affecting health and healing. Any intervention that alters NF-κB entering the nucleus, binding to DNA, and activating genes that encode inflammatory modulators such as IL-6 (and thus CRP), cyclooxygenase 2, IL-1, lipoxygenase, inducible nitric oxide synthase, TNF-alpha, or a number of adhesion molecules will impact many disease conditions. There are several ways to alter the environmental triggers or NF-κB, including altering emotional stress, exercising, lowering oxidative stress, and consuming adequate phytonutrients, antioxidants, alpha-lipoic acid, EPA, DHA, and GLA. Vitamin D has key roles in controlling gene expression, inflammation, bone metabolism, calcium homeostasis, endocrine and cardiovascular physiology, and healing [3].

Once the nutritional professional has ample experience using this model along with improved pattern-recognition skills, it will often lessen the need for extensive laboratory assessments. However, there will always be certain clinically perplexing cases that simply cannot be assessed without objective data, and, for most patients, there may be an irreducible minimum of laboratory assessments required to accumulate information. For example, in the clinical workup of patients with Alzheimer's disease, heavy metal toxicity, infection, or mold exposure may play an important role [6, 7]. Heavy metal body burden or mold toxicity cannot be sensibly assessed without laboratory studies. In such cases, it becomes critical for the nutrition practitioner to develop a network of capable, collaborative healthcare practitioners with whom to co-manage challenging patients and to whom referrals can be made for therapies outside the nutritional professional's own expertise. This will enrich patient care and strengthen the nutritionist/dietitian-patient relationship.

No two patients are treated identically in functional medicine, but some advice is applicable to nearly every patient. The following lifestyle recommendations also meet the criterion of helping to normalize and have a large impact on multiple systems, working as whole-body interventions and thus providing a foundation of health.

41.2 Initial Interventions

41.2.1 Food

Food is not only nutrition; it also plays a significant role in cultural and personal identity, along with shaping and expressing social relations. The nutrition professional should consider these factors while personalizing a nutrition plan for the patient. It is also crucial to consider family history, history of diseases and chronic conditions, physical activity, allergies and food intolerances, access to food, epigenetic markers, and food preferences.

This is an era when diets can be a lot like religion, and there are proponents of each food culture – from paleo to vegan or vegetarian. It is crucial that the nutrition practitioner has insights into the latest "diet trends," along with critically differentiating between an effective therapeutic food plan tailored for the patient utilizing evidence-based research and whatever diets are popular at the moment. In current times, the dietitian/nutritionist has a plethora of therapeutic diets to work with based on the patient's needs. Examples of therapeutic diets are Specific Carbohydrate Diet (SCD); fasting; the ketogenic plan; intermittent fasting; time-restricted feeding, low-glycemic plan; elimination food plans; low-histamine diet; the Dietary Approaches to Stop Hypertension (DASH) diet; gluten-free or casein-free diets; the Gut and Psychology Syndrome (GAPS) diet; fermentable oligosaccharide, disaccharide, monosaccharide, and polyol (FODMAPs) diet; a Mediterranean food plan; and the Zone diet. The nutritionist/dietitian also needs to be aware of the different effects of nutrients according to our genetic constitution (nutrigenetics) and how nutrients may affect gene expression (nutrigenomics). Dietary advice that is specific to individuals with a particular genotype is more effective at preventing chronic diseases than general recommendations about diet. The next chapter will go into the details about each of these therapeutic diets.

Personalized and precision medicine and nutrition can be considered as occurring at three levels: (1) conventional nutrition based on general guidelines for population groups by age, gender, and social determinants; (2) individualized nutrition that adds phenotypic information about the person's current nutritional status (e.g., anthropometry, biochemical and metabolic analysis, physical activity); and (3) genotype-directed nutrition based on rare or common gene variation. Research and appropriate translation into medical practice and dietary recommendations must be based on a solid foundation of scientific knowledge and practical solutions that the patient is able to follow and comply [8].

1. *Removal of processed foods: high-glycemic-index foods, high-fructose corn syrup, and trans or partially hydrogenated fats*

    The Western-style diet, also known as the meat-sweet diet or standard American diet, is characterized by an over-availability and overconsumption of food, along with high intakes of high-sugar desserts, sweetened drinks, high-fat foods, refined grains, red meat, and high-fat dairy products. The heat processing of these foods involves high levels of dietary advanced glycation end products (dAGEs). The National Health and Nutrition Examination Survey (NHANES) showed that 68.3% of those studied were considered overweight (BMI ≥25) and 33.9% were obese (BMI ≥30) [9]. Chronically increased intake of these foods correlates with higher concentration of fasting insulin, C-peptide, leptin, tissue-type plasminogen activator antigen (tPA), C-reactive protein (CRP), E-selectin (sSELE), intercellular adhesion molecule 1 (sICAM-1), vascular cell adhesion molecule 1 (sVCAM-1), and interleukin 6 (IL-6). This affects multiple metabolic functions, causing oxidative stress, a higher osteoporosis risk, and chronic kidney disease [10–12]. In general, processed foods impact each and every clinical imbalance in the matrix, causing changes in digestion and absorption, impairing insulin regulation, leading to immune system activation and vasculature to causing hormone biotransformation.

2. *Consumption of whole-food meals at regular intervals*
Epidemiological studies have shown that regular consumption of whole foods is associated with health promotion and reduced risks of developing several chronic diseases, as well as all-cause mortality. The underlying physiological mechanisms behind the protective effects of whole foods can most likely be ascribed to a concerted action of a wide variety of phytochemicals, phenolic compounds, carotenoids, vitamin E, sterols, and dietary fiber [13]. Plants are the only source of these nutrients, whereas animals are essentially devoid of them [14].

> - Emphasize whole grains such as oatmeal, brown rice, beans, quinoa, and whole grain or bean/lentil pasta over meat when making meal choices.
> - Limit your intake of processed foods, which are often pasteurized at high temperatures.

What Can You Do to Reduce High-AGE Foods in Your Diet?
- Eat more raw or steamed fruits and vegetables and eat meat less often for damage control. Fruits and vegetables are high in antioxidants and have anti-aging properties.
- Limit the amount of grilled, fried, or broiled foods in your diet.
- Lower the flame to low on a gas grill.
- Use "wet" cooking methods, such as stewing, boiling, braising, slow-cooking, or steaming, often because dry-heat cooking methods create more AGEs than moist ones. For example, a chicken breast broiled for 15 minutes contains more than five times as many AGEs as the same food boiled for 1 hour.
- Trim visible fat from meat before cooking at high temperatures. Cook with less fat.
- When you do grill, use acidic marinades including lemon juice and vinegar, which are thought to help to fight AGEs.
- Don't eat the browned or charred portions of cooked meats and cheeses.
- After cooking, cut off any portions of food that were charred.

Phytochemicals are a large group of chemical compounds naturally occurring in plants, fruits, vegetables, legumes, whole grains, nuts, seeds, fungi, herbs, and spices, conferring aroma, flavor, texture, and color. These compounds have been developed over thousands of years of evolution to defend organisms from the effects of free radicals, fungi, bacteria, and viruses. One would wonder as to why these phytochemicals have beneficial effects on humans if they are produced to protect plants. Humans have been ingesting plants as part of their diet throughout their evolutionary history, and our neurons have conserved many of the same signaling pathways that first evolved in insects and other herbivores that preceded humans in evolution. Examples include pathways that signal via SIRT1, AMPK, and Nrf2. Activation of one or more of these signaling pathways that evolved to defend cells against potentially toxic phytochemicals appears to be a major reason why ingestion of phytochemicals can protect us against injury and disease [15].

Phytochemicals can be classified into several major groups based on their chemical structure: alkaloids, sulfur-containing phytochemicals, terpenoids, and polyphenols [16]. Apart from imparting flavor, color, and texture, these phytochemicals perform a slew of functions – they decrease oxidative stress and neutralize free radicals, act as anti-inflammatory and antibacterial agents, reduce carcinogenic activity via inhibiting tumor growth, detoxify carcinogens, and slow down the progression of such degenerative diseases as osteoporosis, macular degeneration, and cataracts [14].

Phytochemicals have beneficial or stimulatory effects on animal cells, though they can be toxic when consumed in high amounts. The beneficial effects are an example of "hormesis." When cells and organisms are challenged with mild stress by some of the noxious phytochemicals present in the plants, they respond adaptively in ways that help them withstand more severe stress. Hormetic phytochemicals such as resveratrol, sulforaphane, curcumin, catechins, allicin, and hypericin are reported to activate adaptive stress response signaling pathways that increase cellular resistance to injury and disease [15].

Alkaloids from *Piper nigrum* (black pepper) exhibit anti-inflammatory activity via activating the Nrf pathway. The activation of Nrf2 has been shown to be a key step in the cellular response against several common diseases, including inflammation, aging, diabetes, cardiovascular disease, acute pulmonary injury, neurodegenerative diseases, and cancer [17]. Glycoalkaloids produced by eggplants (α-solamargine and α-solasonine), potatoes (α-chaconine and α-solanine), and tomatoes (α-tomatine) or their hydrolysis products (mono-, di-, and trisaccharide derivatives and the aglycones solasodine, solanidine, and tomatidine) have been shown to inhibit the growth of cancer cells in culture (in vitro), as well as tumor growth in vivo. In experimental studies, α-tomatine also contributed to protective immunity elicited by a malaria vaccine candidate [18].

Organosulfur compounds such as allicin, ajoene, and isothiocyanates have shown antibacterial activity against both gram-negative and gram-positive bacteria. Allicin is an organosulfur compound obtained from garlic when it is crushed, which causes the enzyme alliinase to convert alliin into allicin. This transformation occurs only in crushed plants; in intact plants, alliin and alliinase are kept in different cell compartments, coming into contact only when the cloves are broken. Allicin has also shown antibacterial activity against *P. aeruginosa*, *S. epidermidis*, *Burkholderia cepacia*, *S. agalactiae*, MRSA, and oral pathogens causing periodontitis and caries [16].

Terpenoids are a group of substances that occur in nearly every natural food. Their main subclasses discussed as beneficial to maintain and improve health are monoterpenes (like limonene, carvone, or carveol), diterpenes (including the retinoids), tetraterpenes (which include all different carotenoids like α- and β-carotene, lutein, lycopene, zeaxanthin, and cryptoxanthin), and terpenoid chromanols [19].

Tocotrienols and tocopherols (vitamin E) are terpenoid chromanols and tocotrienols that possess the ability to stimulate the killing of cancer cells selectively through apoptosis in order to reduce cancer cell proliferation, while leaving normal cells unaffected. One of the mechanisms by which tocotrienols are thought to suppress cancer is related to the isoprenoid side chain. Isoprenoids, along with selenium, have been shown to suppress the initiation, growth, and progression of prostate and other cancers. They are commonly found in almonds, peanuts, and walnuts, which may explain why diets rich in these foods such as the Mediterranean diet have consistently been shown to reduce the incidence of cancer. Much of the broad involvement of vitamin E in human metabolism is due to its role as the body's primary lipid-soluble antioxidant. Tocopherols and tocotrienols are part of the body's highly effective defense system, without which life could not exist. This defense system consists of a network of antioxidants, interacting with and supporting each other. Antioxidants such as vitamin C, coenzyme Q10, and glutathione are needed for effective recycling of tocopherols and tocotrienols. The unique power of both tocopherols and tocotrienols is their ability to break the chain reaction of lipid peroxidation by neutralizing peroxyl radicals to prevent the spread of free radical damage in cell membranes. Vitamin E is a generic term for at least eight structurally related molecules: α-tocopherol (αT), β-tocopherol, γ-tocotrienol (γTE), and δ-tocotrienol. Among them, αT is the predominant form of vitamin E in plasma and tissues and is the form that has drawn most attention in the past. Combinations of different forms of vitamin E may be superior to each alone [20].

Polyphenols are one of the largest secondary plant metabolites ubiquitously present in fruits and vegetables considered an integral part of the human diet. Polyphenols are classified as flavonoids and non-flavonoids, based on their chemical structure. Flavonoid-type polyphenols are divided into different subclasses: anthocyanins, anthocyanidins, flavones, flavanols, catechins, isoflavones, flavanones, and flavonols. Non-flavonoids are comprised of stilbenes, phenolic acids, lignans, and hydroxycinnamic acids [21].

Polyphenols have been studied for their potential involvement in many areas, including cardiovascular, diabetes, cancer, neurodegenerative conditions, inflammation, and microbial diseases. Their protective activity was firstly attributed to their free radical scavenger, antioxidant, and metal chelator properties and later on to their ability to inhibit or reduce different enzymes. The best studied stilbene is resveratrol, found in grapes and grape products, for its role in activation of sirtuin enzymes that increase longevity, reduction of TNF-α and TGF-β, and induction of neurotrophic factors such as brain-derived neurotrophic factor (BDNF) [22, 23]. The richest sources of lignans are flaxseed and sesame seeds, while minor sources are cereals, lentils, fruits (pears, prunes), and vegetables (garlic, asparagus, carrots) [22]. Sesamin, a lipid-soluble lignin, has shown to potentiate TNF-α-induced apoptosis, inhibit proliferation of a wide variety of tumor cells, and downregulate NF-κB activation induced by various inflammatory stimuli and carcinogens [24].

Fibers found in whole plant foods powerfully support the gastrointestinal, cardiovascular, and immune systems through multiple mechanisms and will be discussed further in this chapter. Yet more than 90% of adults and children in the United States do not get the minimum recommended amount of dietary fiber [14].

41.2.2 Whole Food Versus Specific Dietary Nutrient

Consumption of whole foods is strongly associated with reduced risk of chronic diseases. It is therefore reasonable for scientists to identify the bioactive compounds responsible and hope to find the "magic pill" to prevent chronic diseases. The critical question then is whether a purified phytochemical has the same health benefit as the phytochemical present in whole food or a combination of foods [25]. Summaries, which mostly represent meta-analyses, show no long-term benefit for vitamin supplements (Collin C) if no dietary and interventional changes are made. Dietary supplements do not have the same health benefits as a diet rich in whole grains, legumes, nuts, seeds, fruits and vegetables, because, taken alone, the individual antioxidants studied in clinical trials do not appear to have consistent preventive effects. Constituents delivered by foods taken directly from their biological environment have different effects from those formulated through technologic processing. The isolated pure compound either loses its bioactivity or may not behave the same way as the compound in whole foods [25, 26].

For example, numerous studies have shown that the risk of cancer is inversely related to the consumption of green and yellow vegetables and fruits. As beta-carotene is present in abundance in these vegetables and fruits, it has been extensively investigated as a possible cancer-preventive agent. However, the role of carotenoids as anticancer supplements has recently been questioned as a result of several clinical studies. In one study, the incidence of non-melanoma skin cancer was unchanged in patients receiving a beta-carotene supplement. In other studies, smokers gained no benefit from supplemental beta-carotene with respect to lung cancer [25].

Vitamin C supplementation also has been shown to lower the incidence of cancer and heart disease. A Cornell University study reported that phytochemical extracts from fruit have strong antioxidant and antiproliferative effects and proposed that the combination of phytochemicals in fruits and vegetables is critical to powerful antioxidant and anticancer activity. For example, the total antioxidant activity of phytochemicals in 1 g of apples with skin is equivalent to 83.3 mol vitamin C equivalents – that is, the antioxidant value of 100 g of apples is equivalent to 1500 mg of vitamin C. This is much higher than the total antioxidant activity of 0.057 mg of vitamin C (the amount of vitamin C in 1 g of apples with skin). In other words, vitamin C in apples contributed only <0.4% of total antioxidant activity. Thus, most of the antioxidant activity came from phytochemicals, not from the vitamin C. The natural combination of phytochemicals in fruit and vegetables is responsible for their potent antioxidant activity [27].

In another study, researchers compared the efficacy of lycopene supplementation with consumption of tomato products on intermediate cardiovascular risk factors, including oxidative stress, inflammation, endothelial function, blood pressure, and lipid metabolism. Tomato intake provided more favorable results on cardiovascular risk endpoints than did lycopene supplementation [28].

In short, the effect of the whole food is more powerful than the effect of a specific dietary nutrient. However, public confusion proliferates on this issue, incurring huge monetary and social costs [29]. It is important that the nutrition practitioner emphatically relays this message to their patients. Practitioners have often seen new patients coming to their first visit with a bagful of supplements. With all the information available on the Internet and individuals and companies claiming benefits of these supplements, a desperate patient might buy these supplements without correct guidance. Helping patients to understand the importance of whole foods then becomes one of the important objectives of nutrition education during that visit or subsequent visits, whichever seems apt for the practitioner.

41.2.3 Meal Timings and Intervals

We are increasingly moving away from regular meals as more meals are being eaten outside the home and family meals have been eroded by hectic schedules. The prevalence of irregular meal patterns is greater among adolescents than it was during previous decades. Irregular snacking has become more common in children and may have contributed to both the increasing prevalence of obesity in children as well as dyslipidemia in adolescents during the past few decades.

In a study conducted at Queen's Medical Centre in the United Kingdom, irregular meal frequency disturbed energy metabolism in healthy lean women, leading to a lower postprandial energy expenditure, higher degree of insulin resistance, or lower postprandial insulin sensitivity and higher fasting lipid profiles, thereby indicating a deleterious effect on these cardiovascular disease risk factors [30]. A more evenly spaced pattern of three meals per day with no snacks was more strongly associated with a lower prevalence of overweight and obesity and a nutritionally higher-quality diet [31]. Meanwhile, prolonged intervals between meals and skipping meals may trigger migraines and hypoglycemia in some individuals.

Physiological responses to what and how much we eat represent the foundation for translational as well as basic science aimed at preventing and treating metabolic diseases, including obesity and diabetes. However, the timing of food consumption independent of total caloric intake and macronutrient quality has emerged as a critical factor in maintaining metabolic health. Panda states that when healthy adults eat identical and isocaloric meals at breakfast, lunch, or dinner, the postprandial glucose rise is lowest after breakfast and highest after dinner, as if the dinner were twice the size of the breakfast. In addition, when healthy adults are given a constant glucose infusion over 24 hours, blood glucose rises at night and falls around dawn. This is likely because melatonin inhibits insulin release from pancreatic islets through the melatonin receptor and the evening rise in melatonin likely causes hyperglycemia. This indicates that in addition to what and how much we eat, *when we eat* helps determine the physiological response to nutrient availability [32, 33].

Almost every cell in the body has a circadian clock machinery, each with an approximately 24-hour period. Circadian rhythms are an integral part of physiology that seems to be essential for health. The circadian system is a master integrator of both the internal state of the organism and the organism's interaction with nutrition and ambient light. The suprachiasmatic nucleus (SCN) in the hypothalamus acts as a master clock to coordinate independent oscillators throughout the body and determine the period of the organism. Unlike peripheral oscillators, the SCN is composed of a network of neurons with intricate intercellular communication to produce robust outputs through both neural and humoral cues. In addition to this internal regulation, the SCN also receives external input, such as light, to help an organism coordinate with their environment. Nutrient consumption also has a large influence on biological rhythms, but it has a more direct effect on peripheral oscillators than the SCN. Together, light and nutrients coordinate internal biological rhythms with the environment.

The erratic lifestyle associated with modern society – aberrant eating and sleep patterns, inappropriate light exposure, jet lag, and shift work – contributes to circadian rhythm disruption. This disruption compromises multiple levels of physiology, metabolism, and inflammation and increases the risk for noninfectious chronic diseases such as metabolic disorder, diabetes, cardiovascular disease, dental caries, and cancer [34]. Light at night suppresses sleep and promotes extended wakefulness, thereby allowing ingestive behavior to continue late into the night. Therefore, room lighting (or darkness) is critical during sleep. It is important to keep bedrooms as dark as possible and avoid blue light before sleep. Conversely, maintaining a consistent daily eating and fasting rhythm and sleeping in a dark room maintains normal circadian physiology and can prevent or mitigate several of these chronic diseases [35, 36].

41.2.4 Increased Intake of Vegetables with Moderate Fruit Intake

Fruits and vegetables are rich sources of a variety of nutrients, including trace minerals, vitamins, and dietary fibers, along with other classes of biologically active compounds. These phytochemicals can have complementary and overlapping mechanisms of action, including modulation of cholesterol synthesis, reduction of platelet aggregation, detoxification enzymes, stimulation of the immune system and hormone metabolism, reduction of blood pressure, promote brain health and antibacterial, antioxidant, and antiviral effects [37].

41.2.4.1 Antioxidant Activity

The antioxidant defense system has both enzymatic and nonenzymatic components that prevent radical formation, eliminate damaged molecules, repair oxidative damage, remove radicals before damage can occur, and prevent mutations. Phytochemicals from fruit and vegetables can induce the expression of endogenous antioxidant enzymes along with metalloenzymes. For example, mitochondrial superoxide dismutase is a manganese-containing enzyme, and adequate consumption of vegetables and fruit can provide the necessary manganese [15]. Nonenzymatic components of the antioxidant defense system interrupt the free radical-initiated chain reaction of oxidation or scavenge and disable free radicals before they react with cellular components. Typically, the antioxidants vitamins C and E and beta-carotene have received the most attention. Two-week administration of carrot and tomato juices (330 mL/d) and spinach powder (10 g/d), added separately to a low-carotenoid diet, decreased endogenous lymphocyte DNA strand breaks and reduced DNA-based oxidation in 23 healthy men [38].

41.2.4.2 Modulation of Detoxification Enzymes

Detoxification, or drug-metabolizing, enzymes are essential for the biotransformation of many important endogenous compounds and xenobiotics. Phase I enzymes, such as cytochrome P450 (CYP)-dependent monooxygenase, catalyze oxidation, reduction, and hydroxylation reactions and convert hydrophobic compounds to reactive intermediate metabolites in preparation for their reaction with water-soluble moieties to improve excretion. Phase II enzymes – such as glucuronosyltransferases, sulfotransferases, and glutathione transferases – catalyze these conjugation reactions.

Numerous constituents of plant foods, including isothiocyanates, flavonoids, and allyl sulfides, are potent modulators of the CYP monooxygenases. However, the effects of some of these phytochemicals on CYPs are complex. They have the capacity to inhibit certain enzymes at high concentrations of the compound or to activate moderately the same enzyme at lower concentrations, while others may be competitive inhibitors of CYPs. Even when present at low concentrations and in combination with other compounds, their actions can be significant.

The capacity to conjugate metabolically activated intermediates and excrete them from the body is critical in protecting against many potential mutagens, and efforts have focused on determining how vegetable and fruit constituents can influence the phase II conjugating enzymes. In a series of small studies where individuals were supplemented with 300 g of brussels sprouts, the results showed the capacity of this cruciferous vegetable to affect specific glutathione transferase isoenzymes and increase plasma concentrations of glutathione transferase. Induction of these enzyme systems is rapid, and enzymatic activity rises and plateaus within 5 days of continued daily ingestion of a food with inducing capacity. Similarly, the enzyme activities drop rapidly when the food is removed from the diet.

Xenobiotics share many of the same metabolic pathways with drugs, relying on the same enzyme systems for their biotransformation. For example, compounds in cruciferous vegetables can alter drug metabolism by both inhibiting and inducing certain CYPs and possibly by inducing select conjugating enzymes. Lampe mentions a study in which feeding individuals a diet of cruciferous vegetables enhanced their ability to metabolize phenacetin and antipyrine (a pain-relieving drug) oxidatively. Consuming a diet containing brussels sprouts and cabbage for 1 week could also influence select conjugating enzyme systems, stimulating the conjugation of acetaminophen with glucuronide but not with sulfate.

41.2.4.3 Stimulation of the Immune System

Nutrients and other constituents of fruit and vegetables have the potential to affect almost all aspects of the immune system. Several of the vitamins associated with diets high in fruit and vegetables have been shown to improve immune status. Del Corno and his colleagues demonstrated the immune-protective role of polyphenol metabolites to human dendritic cells in markedly impairing the production of pro-inflammatory cytokines and chemokines in response to bacterial application and its role in the prevention of acute and chronic inflammatory diseases [39, 40].

41.2.4.4 Decreased Platelet Aggregation

Blood platelet aggregation is fundamental to a wide range of physiologic processes – e.g., normal blood coagulation, thrombosis, atherosclerosis, and tumor formation and metastasis. Activated platelets adhere to altered blood vessel walls, where they aggregate and promote the release of mitogenic factors that stimulate endothelial cell proliferation. A subsequent cascade of events leads ultimately to occluded vessels.

Lampe's article mentions two studies showing reductions in platelet aggregation of 16.4% and 58% with 9 g and 10 g of fresh garlic cloves, respectively. The antiplatelet effects are attributed to the allyl propyl disulfide, diallyl disulfide, and other sulfur compounds contained in the essential oil.

Mitogenic factors released by platelets, such as platelet-derived growth factor (PDGF), are thought to be important in the initiation and progression of atherosclerotic lesions and to respond to alterations in diet. High consumption of carotenoid-rich vegetables (carrots and spinach) and soy foods (tofu and textured soy protein) increased concentrations of serum PDGF in study participants, thereby demonstrating the anti-atherogenic role of carotenoids [38, 41].

41.2.4.5 Alterations of Cholesterol Metabolism

Isolated dietary fibers from vegetable and fruit sources, particularly pectins, have hypocholesterolemic action. The addition of pectin and fiber-containing foods in experimental diets lowered plasma cholesterol to varying degrees – a variety of vegetables (570 g/d) and fresh apples (600 g/d) by 4%, fresh carrots (200 g/d) by 11%, apples (350–400 g/d) by 8–11%, and prunes (100 g/d) by 5%. The reductions in cholesterol are probably due to different mechanisms specific to the type of fiber source and to different dietary fiber intake amounts. On the basis of the physiologic effects observed in humans, possible mechanisms include (1) increased excretion of fecal bile acids and neutral steroids; (2) altered ratios of primary to secondary bile acids; (3) increased fecal cholesterol and fatty acid excretion; (4) and indirect effects, such as high-fiber foods replacing fat- and cholesterol-containing foods in the diet [38].

41.2.4.6 Modulation of Steroid Hormone Concentrations and Hormone Metabolism

A vegetarian diet is associated with lower circulating concentrations of sex steroid hormones, increased fecal excretion of estrogens, and different hormonal profiles than those observed with consumption of an omnivorous diet. In the past, much of the response to such a diet has been attributed to the physiologic effects of dietary fiber. However, other constituents of vegetables and fruit also may influence metabolism of endogenous steroid hormones.

Effects of indole-3-carbinol (I3C), a constituent of cruciferous vegetables, on estrogen metabolism have been observed in men and women. In men, 500 mg I3C/d (equivalent to 300–500 g of cabbage) for 7 days increased estrone 2-hydroxylation from 29% to 46%. A randomized clinical trial compared the effects of 400 mg I3C, 20 g of cellulose, and placebo daily on 2-hydroxylation of estrogen in women. The mean ratio of 2-hydroxyestrone to estriol, a measure of 2-hydroxylation, increased 1.6-fold in the first month and remained at that level for the 3 months of the study, while no change from baseline was observed in the cellulose or placebo group [38].

It has been hypothesized that shifting the metabolism of 17β-estradiol toward 2OHE1, and away from 16OHE1, could decrease the risk of estrogen-sensitive cancers, such as breast cancer. Reding and her colleagues demonstrated that fruit consumption, specifically banana and citrus fruits, was positively associated with 2-OHE_1 concentrations and not associated with 16α-OHE_1, thereby suggesting that breast tissue exposure to estrogen metabolites may be influenced by diet [42].

41.2.4.7 Antibacterial and Antiviral Activity

Plants have developed sophisticated active defense systems against pathogens, one being the production of antibiotic compounds. Cranberry juice has long been advocated as a treatment for urinary tract infections in women. In a double-blind, randomized, placebo-controlled trial, Avorn et al. showed that 300 mL of cranberry juice per day for 6 months can influence bacterial flora in the urinary tract.

The protective effects of vegetables and fruit observed in epidemiologic studies are not observed with pharmacologic doses of plant foods or their constituents, but with intakes constituting part of a usual diet. Research on the effects of pharmacologic doses of individual plant food constituents may identify particular uses for selected compounds. However, it is unlikely that any one compound will be a magic bullet preventing a whole range of chronic diseases, and humans will still continue to eat food to nourish body and soul, so we must remain committed to consuming a variety of vegetables and fruit as part of a whole-food diet [38].

41.2.5 Consumption of Organically Raised and Locally Grown Foods whenever Possible, with Low-Fat Animal Products (Low-Fat Dairy and Poultry) in Moderation

Numerous studies have compared organic and conventional food production, in relation to both their nutritional value and their content of chemical residues. Reganold looked at 15 reviews or meta-analyses of the scientific literature comparing the nutrition of organic and conventional foods that were published in the preceding 15 years. Twelve of these studies found some evidence of organic food being more nutritious (for instance, having higher concentrations of vitamin C, total antioxidants and total omega-3 fatty acids, and omega-3 to omega-6 ratios). Whether or not these are nutritionally meaningful differences continues to be debated. The rest of the three studies concluded that there were no consistent nutritional differences between organic and conventional foods.

Organic farming systems produce lower yields compared with conventional agriculture. However, they are more profitable and environmentally friendly, and they deliver equally or more nutritious foods that contain less (or no) pesticide residues, compared with conventional farming.

Studies have found that children who eat conventionally produced foods have significantly higher levels of organophosphate pesticide metabolites in their urine than children who eat organically produced foods [43]. In 2012, the American Academy of Pediatrics reported that an organic diet reduces children's exposure to pesticides and provided resources for parents seeking guidance on which foods tend to have the highest pesticide residues [44].

Organic livestock are fed with organically produced feed that is free of pesticides and animal byproducts, and therefore it is supposed that there should be lower accumulation of chemical residues. An intriguing study by Hernandez and his colleagues found pollutants in both conventional and organic meat. This group examined 76 samples of organic and conventional beef, chicken, and lamb for 33 carcinogenic pollutants that are commonly found in nonorganic meat. They found pollutants not only in the conventional samples but also in the organic samples. In fact, the difference in the level of pollutants between both was minimal. Of the samples, they found that lamb – both nonorganic and organic – had the highest level of pesticides, with the organic samples actually containing more pollutants than the nonorganic. Both kinds of chicken had the lowest levels [45]. Another study found that conventional chicken and pork had a 33% higher risk for contamination with antibiotic-resistant bacteria compared to organic alternatives [43]. In short, the organic meat was far from being devoid of persistent organic pollutants.

A large meta-analysis by Dominika and colleagues iterated that there was a difference in the composition and quality of fat between conventional and organic meat, with organic meat having more unsaturated fat, including anti-inflammatory omega-3 s [46].

These studies demonstrate that environmental contamination by POPs is ubiquitous and human exposure is difficult to avoid. Even consumers who choose to consume organic food, which is theoretically healthier with better composition of fatty acids, have exposure rates to fat-soluble pollutants that can become even higher than those of consumers who eat conventional food. There needs to be a consistent effort to minimize environmental toxic pollutants and consume animal products in moderation [45, 47].

41.2.6 Increased Intake of Fiber

The World Health Organization (WHO) and Food and Agriculture Organization (FAO) state that dietary fiber is a polysaccharide with 10 or more monomeric units which is not hydrolyzed by endogenous hormones in the small intestine. Dietary fiber and whole grains contain a unique blend of bioactive components including resistant starches, vitamins, minerals, phytochemicals, and antioxidants. Dietary fiber is indigestible in the human small intestine but digested completely or partially fermented in the large intestine. It is examined in two groups: water-soluble and water-insoluble organic compounds.

Based upon their digestibility in the gastrointestinal (GI) tract, they are divided into two basic groups: nonstructural, nonfibrous polysaccharides or simple carbohydrates (starch, simple sugars, and fructans) and structural, non-starch polysaccharides (NSP) or complex carbohydrates that are resistant to digestion in the small intestine and require bacterial fermentation located in the large intestine (cellulose, hemicellulose, lignin, pectin, and beta-glucans). NSP can be further subdivided into the two general groups of soluble and insoluble.

Soluble fiber (pectins, gums, inulin-type fructans, and some hemicelluloses) dissolves in water, forming viscous gels, bypasses the digestion of the small intestine, and is easily fermented by the microflora of the large intestine. Insoluble fibers (lignin, cellulose, and some hemicelluloses) are not water-soluble and do not form gels, so their fermentation is severely limited [48, 49].

41.2.6.1 Health Benefits of Fiber
Cardiovascular

High levels of dietary fiber intake are associated with lower rates of coronary heart disease, stroke, and peripheral vascular disease, as well as lowering such major risk factors as hypertension, diabetes, obesity, and dyslipidemia.

The probable mechanisms behind dietary fiber and cardiovascular disease prevention may be as follows: First, soluble fibers have been shown to increase the rate of bile excretion, therefore reducing serum total and LDL cholesterol. Second, dietary fiber affects the ability to regulate energy intake, thus enhancing weight loss or maintenance of a healthier body weight. Third, short-chain fatty acid production, especially propionate, has been shown to inhibit cholesterol synthesis. Fourth, dietary fiber has been shown to lower risk of type 2 diabetes, either through glycemic control or reduced energy intake. Fifth, fiber has been shown to decrease proinflammatory cytokines such as interleukin-18, which may have an effect on plaque stability. Sixth, increased consumption of dietary fiber has been shown to decrease circulating levels of C-reactive protein (CRP), a marker of inflammation and a predictor for coronary heart disease.

Obesity

Dietary fiber's ability to decrease body weight or mitigate weight gain could be attributed due to several factors. First, fiber may significantly decrease energy intake. Second, soluble fiber, when fermented in the large intestine, produces glucagon-like peptide (GLP-1) and peptide YY (PYY). Both these gut hormones play a role in inducing satiety. Third, dietary fiber may decrease diet metabolizable energy, which is gross energy minus the energy lost in the feces, urine, and combustible gases. Also, as dietary fiber intake increases, the intake of simple carbohydrates tends to decrease. Both soluble and insoluble fiber may lead to weight loss.

Diabetes

Although other risk factors such as obesity, lack of physical activity, and smoking are precursors for type two diabetes, dietary factors also seem to play a significant role. Type 2 diabetes results from decreased insulin sensitivity and hyperglycemia. Several explanations have been proposed to understand the physiology behind the relationship between glycemic index and diabetes. First, simple carbohydrates are high in glycemic index and produce higher blood glucose levels. This chronic hyperglycemia may lead to the dysfunction of beta cells in the pancreas, thus decreasing insulin release. Second, because of the high glycemic load, tissues such as skeletal muscle, liver, and adipose become resistant to insulin.

There is an inverse relationship between dietary fiber and diabetes, independent of age and body weight. The mechanisms behind insoluble fiber are more peripheral and not limited to nutrient absorption. First, an accelerated secretion of glucose-dependent insulinotropic polypeptide (GIP) is observed directly after the ingestion of an insoluble fiber. GIP is an incretin hormone that stimulates postprandial insulin release. Second, insoluble fiber can result in a reduced appetite and food intake, and this may lead to a decreased caloric intake and BMI. Third, short-chain fatty acids, via fermentation, have been shown to reduce postprandial glucose response in insulin-sensitive tissues.

Gastrointestinal Conditions

Dietary fibers affect the entire gastrointestinal tract from the mouth to the anus. Soluble fibers usually delay gastric emptying, while insoluble fibers tend to decrease gastric transit time. In the small intestine, fibers can reveal responses of a wide variety of gastrointestinal hormones that serve as incretins to stimulate insulin release and affect appetite. They can bind bile acids, impede micelle formation, and thereby increase fecal excretion of cholesterol and bile acids. They are also effective in increasing total fecal volume and promoting regular bowel movements. In the colon, soluble fibers increase microbiota mass, with some acting as prebiotics to promote health-promoting bacteria bifidobacterium and lactobacilli.

Guar gum and other soluble fibers are associated with low levels of gastric acid production, which may protect from GERD. Evidence strongly indicates that high fiber intake is associated with lower prevalence of duodenal ulcer disease than lower fiber intake. Judicious use of soluble fibers may offer benefits to individuals who are in remission from ulcerative colitis. Inulin as a prebiotic has the ability to stimulate the growth of bifidobacterium while restricting the growth of potential pathogenic bacterial species such as *E. coli*, *Salmonella*, and *Listeria*. This could be beneficial in such disorders as ulcerative colitis and *C. difficile* infections. In one study, inulin decreased the risk of colon cancer by reducing colorectal cell proliferation, decreasing interleukin-2 release, and decreasing exposure to genotoxins.

Cancer

Cancer continues to be one of the top health concerns of populations worldwide. Insoluble fiber found in whole grains has direct effects in the colon by promoting laxation, decreasing transit time, and binding substances such as bile acids and carcinogens. Soluble fiber is acted upon by the gut microbiota, and the utilization results in changes to the numbers and types of bacteria and, more importantly, changes to their metabolic activities in terms of the formation of tumor promoters, genotoxins, and carcinogens.

Selective prebiotic fiber sources – such as resistant starches, inulin, and some oligosaccharides – act as a selective substrate for bacteria that produce specific short-chain fatty acids and can lower the intestinal pH and prevent the growth of pathogenic bacteria. They increase the number of bifidobacteria in the colon, and this and the reduced intestinal pH have a direct impact on carcinogenesis in the large intestine. The SCFA butyrate has shown to increase apoptosis of colonic tumor cell lines.

Immune System

Soluble fibers such as inulin and other oligofructoses stimulate growth of bifidobacteria and lactobacilli in the colon. These bacteria generate short-chain fatty acids and stimulate the immune system. Some of the other health benefits of these bacteria include protection from intestinal infection; activation of intestinal function; assistance in digestion and absorption, especially of calcium; lowering of intestinal pH for formation of acids after assimilation of carbohydrates; reduction of the number of potentially harmful bacteria; production of vitamins and antioxidants; bulking activity and production of fecal matter; and potential reduction in the risk for colorectal cancer [48].

41.2.7 Removal of Potential or Documented Allergens and Excessive Alcohol

One of the most powerful tools for a whole-body intervention is identifying food triggers, food allergens, and intolerances. This becomes an integral part of a nutritionist/dietitian's 5-R approach (see ◘ Table 41.1). Identification of food allergies is done through IgE testing for foods, whereas to identify non-IgE-related food sensitivities, there are a plethora of tests in the market. If the patient has monetary restrictions, a well-structured elimination plan focusing on MAPDOMs offers the patient the opportunity to experience a much healthier approach to eating. Many common ailments tend to resolve once the individual focuses on more plant-based and organic foods.

Developing Interventions to Address Priorities: Food, Dietary Supplements, Lifestyle, and Referrals

Table 41.1 Integrative 5-R approach

Interventions	Definition	Food	Dietary supplements	Lifestyle	Referrals
Remove	Remove stressors Remove what harms This step focuses on eliminating pathogenic bacteria, viruses, fungi, parasites, and other environmentally derived toxic substances from the gastrointestinal tract Eliminate or reduce ongoing toxic exposures in the home and/or workplace	Remove foods and beverages that are likely to contain toxins, food allergens, or antigenic challenges Elimination diets Top 10 antigens FODMAP Remove processed foods containing dyes or chemicals Avoid highly processed foods Avoid foods cooked at high temperatures – these are high in advanced glycated end products (AGEs) Eliminate refined flours Eliminate soda and sugary beverages Limit/ eliminate alcohol	Antimicrobial Anti-parasites Oil of oregano, olive leaf extract, garlic extract, berberine, undecylenic acid, grapefruit seed extract, monolaurin, mastic gum, caprylic acid, wormwood extract, clove extract, black walnut hull, artemisinin, cinnamon bark, pau d'arco inner bark extract, rosemary leaf extract, thyme oil, apple cider vinegar Chelating agents: modified citrus pectin, chlorella, activated charcoal Energetics: bioenergetic therapies (cancer – Rad-Tox from Apex)	Eliminate toxic elements and chemicals from food and environment Mold remediation HEPA filters Toxic relationships EMF Fix irregular sleep patterns: regularize sleep as per circadian rhythms Eliminate Teflon pans, plastic containers Smoking cessation	DC/DO/NUCA-C1/C2/ Correct structural stress Medical specialties – gastroenterology Reduction of hiatal hernia – osteopathic, chiropractic, or visceral manipulation Allergy and immunology, infectious disease, ILADS, Lyme-literate practitioner Functional medicine psychiatrist Referral for IV chelation
Replace	The second clinical site: Replace factors that may be lacking or limited	Optimize nutrition through whole foods Correct nutritional deficiencies Apple cider vinegar Papaya, pineapple	Digestive enzymes, betaine HCl, zinc for HCl production, lipotropic factors, pancreatic enzymes, diamine oxidase, DPP IV proteases Swedish bitters Gentian Prokinetics such as D-limonene ginger iberogast, Chinese herbs (TJ 43)	Mindful eating Chewing food Eating meals at regular timings Clean filtered water Sleep hygiene helps with melatonin production – important for digestive support HEPA/ULPA filters to reduce dust, molds, volatile organic compounds Indoor plants	Meyer's cocktail, Vitamin C Acupuncture

(continued)

Table 41.1 (continued)

Interventions	Definition	Food	Dietary supplements	Lifestyle	Referrals
Reinoculate	Reintroduction of desirable bacteria, or "probiotics," into the intestine to reestablish microfloral balance	Butyrate in food: pasture butter/buttermilk/SCFA Prebiotic foods: Asparagus Bananas Burdock root Chicory Chia seeds Fruit Garlic Leeks Onions Peas Sunchokes Probiotic foods: Buttermilk Jeruk Khimchi Kombucha Miso Nato Olives Raw pickles Sauerkraut Antibiotic-free animal produce	Butyrate inulin, larch-arabinogalactans, psyllium modified citrus pectin Probiotics *Saccharomyces boulardii* *Lactobacillus* species – *reuteri*, *casei*, *rhamnosus*, *acidophilus* *Streptococcus* species *Bifidobacterium* species: *infantis*, *lactis*, *longum*, *bifidum*	Sleep Emotional resilience Anger management Avoid negative self-talk Meditation, practicing loving kindness	Medical specialist for prescribing butyrate enemas, fecal transplantation
Repair	Provide nutritional support for regeneration and healing of the gastrointestinal mucosa	Whole-food nutrition Raw spinach, cabbage, okra, aloe vera Avoid highly processed foods Avoid high-temperature cooking processes – high advanced glycated end products	Glutamine N-acetyl glucosamine Fiber for SCFA (both soluble and insoluble) butyrate colostrum aloe, artemisia, boswellia, curcumin, geranium, licorice/DG, marshmallow, quercetin, rutin, slippery elm Essential fatty acids Antioxidants: Vitamins C, A, and E, CoQ10 Liposomal glutathione Pro-resolvins Alpha lipoic acid Zinc picolinate/carnosine Gamma oryzanol N-acetyl cysteine	Prayer/grace before meals Quiet environment Avoid microwave Forgiveness in relationships	Medical specialist for IV glutathione Psychotherapist Chiropractor Physical therapy

Table 41.1 (continued)

Interventions	Definition	Food	Dietary supplements	Lifestyle	Referrals
Rebalance	Modify attitude, diet, and lifestyle to promote healthier living	Whole clean foods appropriate for the individual Eat no later than 2–3 hours before bedtime Meal spacing Eating at regular intervals	As needed for maintenance Melatonin 5HTP GABA Adrenal support Magnesium Fiber Pro-resolvins Fatty acids Phosphatidylserine Ashwagandha	Healthy mind/healthy body/mindful eating Practice gratitude Movement Purpose Dinacharya: practice of daily routine as per Ayurveda HeartMath Acupuncture Massage Prayer Meditation Regular exercise Laughter: What makes you laugh? Steam/sauna Hydrotherapy Dry skin brushing	Yoga therapy Chiropractor Craniosacral practitioner Music therapy Water therapy Art Chakra balancing Guided imagery Hypnosis

The elimination diet excludes saturated and trans fats, empty calories, sugars, additives, hormones, and antibiotics in food along with toxic chemicals and xenobiotics. Alcohol needs to be eliminated, as ethanol depletes intestinal integrity and causes gut dysbiosis. Butyrate, a fermentation byproduct of gut microbiota, is also negatively altered following chronic alcohol consumption [50].

With this one therapeutic modality, the dietitian/nutritionist can reduce food antigenicity and inflammatory reactions, improve gut permeability, rebalance Th1 and Th2 cytokines, rebuild the brush border, and improve digestion and absorption. Implementation is simple but may not be easy for patients and their families. Some improvements stem from the reduction of the antigenic load at the level of the gut and the resultant T-helper cell Th1 vs. Th2 immune shifting. While the simple sensitivities may take hours to days to improve, immune system rebalancing may take weeks to decrease and eliminate antigens from the gut and show improved symptoms on a clinical scale.

No one likes to change dietary habits, but it is an essential part of the management of chronically ill patients. It is important to emphasize the importance of complete adherence to the plan. When patients are prescribed an elimination diet, they will typically start "negotiating" the terms, including alcohol. It is here where motivational interviewing skills will serve both the practitioner and the patient.

Elimination of dietary triggers helps with most chronic conditions, including mental health issues. A paper published by Karakula et al. reviews the relationship between gut microbiota, intestinal permeability, and dietary antigens in the etiopathology of schizophrenia [51]. Studies on the role of ketogenic diet in mental health have been on a rise [52, 53].

41.2.8 Adequate Intake of Fatty Acids

Fatty acids are constituents of cell membranes, act as an energy source, and have biological activities that act to influence tissue, metabolism, and function, along with responsiveness to hormonal and other signals. The biological activities may be grouped as regulation of membrane structure and function; transcription factor activity; regulation of intracellular signaling pathways and gene expression; and regulation of the production of bioactive lipid mediators. Fatty acids influence health and well-being, and aberrations in their profiles influence metabolic conditions such as cardiovascular diseases, neurovascular diseases, type 2 diabetes, inflammatory diseases, and cancer.

Different foods contain different amounts of fat and different types of fatty acids, and these may be affected by processing, storage, and cooking methods. Humans eating a varied diet will consume many different types of fatty acid each day. The pattern of fatty acid consumption varies from meal to meal, from day to day, and from season to season. Aging, geographical location, and cultural and religious practices also impact fatty acid intake. Fatty acids can be synthesized in the human body, either from non-lipid precursors such as glucose or from other fatty acids; exceptions to this are the so-called essential fatty acids, which must come from the diet.

41.2.8.1 Effects of Saturated Fatty Acids

Palm oil, coconut oil, cocoa butter, and animal-derived fats such as lard, tallow, and butter are rich sources of saturated fatty acids. The amounts of individual saturated fatty acids present vary among these sources. Many plant oils contain a significant amount of saturated fatty acids, particularly palmitic acid, as do fish and fish oils. Dairy fats contain some odd-chain saturated fatty acids. Saturated fatty acids are also synthesized de novo in humans, the precursor being acetyl-CoA, produced from carbohydrate or amino acid metabolism. Processed and refined carbohydrates, when consumed in excess, are also converted to fatty acids.

Saturated fatty acids influence cell signaling in many cell types through their roles in phospholipid, sphingolipid, ganglioside, and lipid raft structure and in covalent modification of proteins. Through effects on regulation of transcription factors involved in lipid metabolism and in inflammation (NF-κB), saturated fatty acids influence cholesterol, fatty acid, and triacylglycerol biosynthesis; lipoprotein assembly, secretion and clearance, and metabolism; and inflammatory processes. Hence, they play important roles in normal cellular and tissue metabolism and function but also influence factors involved in determining risk of cardiometabolic and other inflammatory diseases [54].

Based on the recent controversy with saturated fats, one could easily assume that the "butter/bacon is back" movement has won and that reducing saturated fat is no longer important for heart health. A few recently published studies conclude that there is no association between intake of saturated fats and cardiovascular risk [55, 56]. One of these studies was a meta-analysis of observational studies reporting associations of saturated fat with all-cause mortality, cardiovascular mortality, and stroke (de Souza). The researchers concluded that saturated fats are not associated with those outcomes. The other popular study was the Minnesota Coronary Experiment (MCE), a randomized control trial conducted with 9423 men and women [56]. The researchers concluded that replacing saturated fat in the diet with polyunsaturated fats had no impact on mortality, even though it reduced cholesterol. While these studies and arguments seem compelling, the vast majority of the research still links saturated fat intake with increased risk for cardiovascular risk. Critiques of the above meta-analysis show that the researchers based their conclusions on selective data and ignored the data that showed replacing saturated fat with high-quality carbohydrates – fruits, vegetables, and whole grains – reduced the risk of cardiovascular disease [57, 58].

41.2.8.2 Effects of MUFAs

Oleic acid is the most prevalent monounsaturated fatty acid (MUFA) in the human diet. Plant oils and animal-derived fats such as lard, tallow, and butter are good sources of oleic acid. Olive oil is an especially rich source, with oleic acid typically contributing about 70% of fatty acids present. In low-erucic acid rapeseed oil (aka canola oil), which is genetically engineered, oleic acid typically makes up about 60% of fatty acids. Palmitoleic acid is found in low amounts in many plants oils and animal fats, and it is quite abundant in macadamia oil and in fatty fish and fish oils. Both palmitoleic and oleic acids can be synthesized de novo in humans.

Cell membrane phospholipids contain some palmitoleic acid and significant proportions of oleic acid. Replacing saturated fatty acids with oleic acid has a small cholesterol- and LDL cholesterol-lowering effect with an inconsistent effect on HDL cholesterol. Oleic acid also renders LDL fairly resistant to oxidation and limits the formation of pro-atherogenic oxidized LDL. Although oleic acid is often called "anti-inflammatory," its effects on inflammatory processes are modest, and it seems likely that any anti-inflammatory effects of olive oil are due to polyphenolic substances rather than oleic acid. Oleic acid seems to improve glucose control and insulin sensitivity along with a small lowering effect on blood pressure.

Effects of oleic acid are observed when it is used to replace saturated fatty acids in the diet, and it is likely that these effects are due to partial removal of the deleterious biological effect of saturated fatty acids rather than to any specific molecular or cellular action of oleic acid. However, some studies show effects of oleic acid on transcription factors involved in lipid homeostasis. The effects seen when saturated fatty acids are replaced with oleic acid would be expected to lower risk of cardiovascular disease [54].

In terms of preventing cardiovascular events, the Prevención con Dieta Mediterránea (PREDIMED) study reported that a 4.8-year intervention with a Mediterranean diet that included extra-virgin olive oil – a source of oleic acid – reduced myocardial infarction, stroke, and cardiovascular death by 30% compared with the control diet [59].

Over the past decade, palmitoleic acid – found in macadamia nuts, avocado, and sea buckthorn – has received significant interest due to its description as a "lipokine." A lipokine is a lipid messenger released from adipose tissue having effects on other tissues such as the skeletal muscle, by increasing insulin sensitivity, and in the liver, where it reduces fat accumulation.

In conclusion, the two major dietary MUFAs appear to have a cardioprotective and anti-inflammatory role, especially when substituted for saturated fatty acids [54].

41.2.8.3 Effects of Omega-6 PUFAs

Linoleic acid is the most prevalent omega-6 (n-6) polyunsaturated fatty acids (PUFAs) in the human diet. Many nuts, seeds, and oils contain high proportions of linoleic acid. For example, linoleic acid comprises 75% of fatty acids in safflower oil and around 50–55% in corn oil, soybean oil, sunflower oil, and cottonseed oil. It also makes a significant contribution to vegetable oils that are rich in oleic acid (rapeseed, peanut, and olive oils) or palmitic acid (palm oil). Some seeds and seed oils contain metabolic products of linoleic acid such as γ-linolenic acid and are found in evening primrose oil, but these are not important dietary constituents.

The second most common dietary n-6 PUFA is arachidonic acid, which is found in foods of animal origin, such as meat and eggs. Linoleic acid is the metabolic precursor of γ-linolenic acid, dihomo-γ-linolenic acid, and arachidonic acid. Membrane phospholipids of human cells contain significant proportions of both linoleic and arachidonic acids and often contain some dihomo-γ-linolenic acid. Linoleic acid is an important constituent of ceramides, especially those found in the skin, and essential fatty acid deficiency results in breakdown of skin integrity and inability to prevent transdermal water loss. However, only a fairly low intake of linoleic acid is required to prevent essential fatty acid deficiency (1% of energy), and most diets provide intakes greatly in excess of this.

Linoleic acid lowers blood cholesterol and LDL cholesterol concentrations, particularly when it replaces the common saturated fatty acids. It has an effect in hepatic cholesterol metabolism by inducing the gene encoding cholesterol 7α-hydroxylase, which is the rate-limiting enzyme in the synthesis of bile acids from cholesterol. This upregulation increases the production of bile acids that are exported from the liver to the gallbladder, thereby lowering hepatic cholesterol and LDL. Linoleic acid was found to increase insulin sensitivity and activate NF-κB, but is not associated with increasing CRP, IL-6, or soluble TNF receptor and may thus have only a limited effect on inflammation in humans.

Arachidonic acid has a structural role in the brain. Studies of infants receiving formulas that combine arachidonic acid with the long-chain n-3 PUFA docosahexaenoic acid have found improvements in cognitive development. Free arachidonic acid has roles in cell signaling and can directly promote inflammation by acting via the NF-κB pathway. A major role of cell membrane arachidonic acid is as a substrate for the synthesis of eicosanoids, which include series 2 prostaglandins, thromboxanes, and leukotrienes. These are formed by metabolism of arachidonic acid by cyclooxygenase or lipoxygenase pathways. The resulting metabolites have many roles in platelet aggregation and blood clotting, inflammation and pain, regulation of the immune response and bone turnover, smooth muscle contraction, tumor cell proliferation, renal function, and cancer progression. Eicosanoid mediators are formed in a cell-specific manner and often have opposing effects to one another, thereby acting to ensure a controlled biological response [54]. It is well-known that n-6 and n-3 fatty acids share metabolic pathways and can potentially compete with each other, causing n-6 to interfere with potential cardiovascular benefits of the n-3. Studies have found that the combination of both types of fatty acids is associated with the lowest levels of inflammation and lowest risk for metabolic diseases [60].

41.2.8.4 Effects of n-3 PUFAs

The essential n-3 PUFA α-linolenic acid is synthesized in plants. Flaxseeds and flaxseed oil, walnuts, and chia seeds are rich sources of α-linolenic acid. Soybean oil and rapeseed oil contain about 7–10% α-linolenic acid. There is a metabolic pathway by which α-linolenic acid can be converted to eicosapentaenoic acid (EPA) and then on to docosapentaenoic acid (DPA) and docosahexaenoic acid (DHA). However, the conversion of α-linolenic acid to DHA is poor. There is direct competition for metabolism of n-6 and n-3 PUFAs as this pathway shares the same enzymes as the metabolism of linoleic acid. A typical dietary ratio of linoleic to α-linolenic acid in Western diets is 7–20. This may be one reason why metabolism of α-linoleic acid to EPA is poor in humans and that to DHA is extremely limited.

EPA, DPA, and DHA are found in seafood, especially in fatty fish, and in fish oil supplements. Unless a person is regularly consuming fatty fish or taking a fish oil supplement daily, α-linolenic acid will be the most common n-3 PUFA in the diet. Cell membranes contain very little α-linolenic acid, but most contain modest amounts of EPA and greater amounts of DPA and DHA. Membranes of the brain (gray matter) and eye (rod outer segments) contain high amounts of DHA. The structure of n-3 PUFAs, especially DHA, has a strong influence on the physical properties of membranes into which they are incorporated and on membrane protein function. As the result of effects on membrane-generated intracellular signals, EPA and DHA can modulate transcription factor activation and thereby expression of genes. The transcription factors affected by EPA and DHA include NF-κB, Ki-67, SREBPs, and PPAR-α and PPAR-γ. These effects are central to their role in controlling inflammation, fatty acid and triacylglycerol metabolism, and adipocyte differentiation.

The replacement of cell membrane arachidonic acid by EPA and DHA influences the pattern of lipid mediators produced. There is decreased production of eicosanoids from arachidonic acid and increased production of eicosanoids from EPA, especially of the selective proresolving mediators (SPM), which play an anti-inflammatory role and also influence the immune system. Therefore, EPA and DHA can influence inflammation, immune function, blood clotting, vasoconstriction, and bone turnover, along with several other processes. An adequate supply of DHA is essential for optimal neural, behavioral, and visual development of the infant. EPA and DHA have important roles in brain function throughout the life course.

These n-3 PUFAs are effective at lowering blood triglyceride concentrations. DHA can influence the concentrations of cholesterol-carrying lipoproteins and can cause a small increase in both LDL and HDL cholesterol. EPA and DHA also increase LDL particle size, rendering LDL less atherogenic. They also exert an effect on blood pressure, aldosterone secretion, generation of nitric oxide by the endothelium, vascular reactivity, and cardiac hemodynamics. EPA and DHA also reduce inflammation.

Evidence from prospective and case-control studies indicate that consumption of EPA and DHA reduces the risk of CVD outcomes. Three key mechanisms have been suggested to contribute to the therapeutic effect of EPA + DHA: first, altered cardiac electrophysiology, seen as lower heart rate, fewer arrhythmia, and increased heart rate; second, an antithrombotic action resulting from the altered pattern of production of eicosanoid mediators that control platelet aggregation; and lastly, due to the anti-inflammatory effect of EPA and DHA, which serve to stabilize atherosclerotic plaques, preventing their rupture. EPA and DHA may directly influence cancer cells and the tumor environment by inhibiting or slowing tumor initiation, invasion, and cell proliferation and by promoting tumor cell apoptosis [54].

Most human studies of the functional effects of EPA and DHA have used supplemental forms with differing ratios and dosages, where the fatty acids have been provided without any dietary change and without removal of other fatty acids from the diet [54]. While this has undoubtedly contributed to confusion among medical professionals, the current scientific literature provides a strong evidence that n-3 fatty acids reduce the risk of metabolic diseases, including cardiovascular risks and brain health [61]. This has encouraged national and international guidelines to collectively recommend that healthy adults consume at least 250 mg per day of long-chain omega-3 fatty acids to maintain cardiovascular health, with many organizations recommending higher amounts for those at greater risk of CVD and other inflammatory conditions [62, 63].

41.2.8.5 Role of Trans Fats

Industrially produced trans fats resulting from the hydrogenation of vegetable oils are generally found in snack foods and baked goods, such as cakes, muffins, and pies. These are identified as partially or fully hydrogenated vegetable oils on the label. A 2% increase in trans fat consumption is associated with a 23% increase in the incidence of cardiovascular disease. The presence of a trans double bond causes the physical properties of the fatty acid to be more like a saturated fatty acid than an unsaturated one. Therefore, incorporation of trans fatty acids into cell membranes causes the membranes to become less fluid, which then influences membrane protein function and interactions, in turn affecting cell-signaling processes. The trans fats adversely affect LDL particle number and inflammation markers such as CRP and IL-6. Because of these biological effects, trans fatty acids as a class appear likely to confer greater health risk than other fatty acid classes. The current dietary guidelines recommend that trans fats should be limited to less than 1% of energy or as low as possible [54, 64].

The functional medicine use of fatty acids, especially essential fatty acids, in nutrition intervention is an important therapeutic tool. Understanding the pathophysiology involved in inflammation and metabolic dysfunction allows the nutrition professional to manage patients in a vigilant and systematic manner.

41.2.9 Adequate Water Intake

Hydration is integral to health, and water trails only air among the most important substances in maintaining hemodynamics and functional reserve. The human body survives for only a few days without meeting essential water requirements but could sustain itself for several weeks without ingesting macronutrients or micronutrients. Water is involved in critical anatomical and physiological functions, such as in providing mass and structure to the cells, functioning as the medium and reagent for metabolic reactions. Intercellular water binds to proteins, carbohydrates, and nucleic acids to maintain their proper function, lubricates adjoining tissues, transports substrate and metabolic waste,

and is the primary means for dissipating excess body heat. When a person is dehydrated, the structure of critical cellular biomolecules is adversely affected, reducing their function. All systems are influenced by dehydration, resulting in reduced energy production, neurotoxicity, alteration in the function of concentration of substances within cells, pH changes, and altered enzyme function. Muscle cramping, muscle soreness, constipation, fatigue, and sensitivity to toxic substances all increase during states of dehydration, while mental clarity is reduced. In these cases, water becomes the limiting nutrient for support of functional physiology [65].

In a functional medicine assessment, the adequacy of water to maintain functional organ reserve should be a first-stage priority. Body composition measurements using bio-impedance analysis can provide information about both intra- and intercellular water status. Three simple indicators of hydration status include thirst, body weight, and urine osmolality. Thirst typically lags behind an acute change in hydration, developing after humans incur a 1% to 2% acute reduction in body mass. It is observed that endurance and cognitive performance decline at slightly less than 2% acute loss of body mass. The signal to initiate drinking behavior does not appear to be synchronized to prevent deterioration of function, and therefore thirst is commonly viewed as a less-than-ideal method of tracking hydration. Therefore, thirst should be used in combination with acute change in body weight and urine color to develop fluid intake strategies. Body weight and acute change in body weight following exercise and/or heat exposure are a relatively accurate and reliable measures of hydration status. Specific gravity or urine osmolality is also a predictor of hydration status, and elevations in osmolality or specific gravity of a urine sample suggest the onset of dehydration. For specific gravity, a value of greater than 1.020 (equivalent to an osmolality of approximately 900 mOsm/kg) categorizes an individual as dehydrated, and a value of 1.035 is considered frank dehydration.

Hydration has shown to impact various aspects of human health. Poor hydration status may be associated with compromised oral health, such as dental caries and erosion of dental enamel. Adequate fluid consumption has a positive effect on intestinal and hormonal health. Consumption of water has shown to increase stool weight and volume along with decreasing bowel transit time in some individuals. It has also been associated inversely with certain types of cancers. This is because water consumption results in accelerated transit of potential carcinogens through the intestinal tract and decreased time for exposure and dilution of carcinogens in the water phase (phase 2 detoxification – sulfation and glucuronidation) of the stool. It may also be because of greater intestinal excretion of hormones such as estrogen that are associated with cancer, because weak evidence of a relationship was seen between bowel motility and breast cancer. Enhanced hydration also leads to a reduced sympathovagal ratio, decreases the sympathetic nervous system drive to retain fluids, and increases the parasympathetic activation to excrete fluids, all of which could, in a state of dehydration, facilitate responses leading to hypertension. In individuals who are susceptible to nephrolithiasis, increased water ingestion reduces the risk of subsequent kidney stones, specifically calcium nephrolithiasis with stones composed primarily of calcium oxalate. The mechanism is thought to be a result of the dilution in concentration of materials that would precipitate stones on saturation. Continuous ingestion of water stimulates urine production and contributes to the dilution and excretion [66].

The WHO has declared that "all people, whatever their stage of development and their social and economic conditions, have the right to have access to an adequate supply of safe drinking water." Although presence of a public water distribution network is often an indicator of a suitable water supply, it should not be expected that the piped water quality is always adequate for human consumption [67]. Like the fish we eat – which are now significant sources of toxins such as mercury, cocaine, and antidepressants – our water can contain everything from pesticides to substances we add to water as a function of general public use (chlorine, fluoride, lead, and iron leaking from lead pipes, softening agents), to residues from farming practices, to bacterial or parasitic agents resulting from contamination or overuse. People are also exposed to disinfected drinking, shower, or bathing water containing at least 600 identified disinfectant by-products [68].

Citizens of a small town in Colombia had a significant prevalence of autoimmune thyroiditis, which was traced back to the contamination of drinking water with phenolic chemicals. Those chemicals adversely influenced the immune system, resulting in production of anti-thyroid antibodies. When the water supply was purified, the prevalence of autoimmune thyroiditis was significantly reduced in the community [69, 70].

In April 2014, the city of Flint, Michigan, switched to the Flint River as a temporary drinking water source – without implementing corrosion control – discontinuing the purchase of treated water from the Detroit Water and Sewer Department (DWSD). Ten months later, water samples collected from a Flint residence revealed progressively rising water lead levels – 104, 397, and 707 µg/L – coinciding with increasing water discoloration. An intensive follow-up monitoring event at this home investigated patterns of lead release by flow rate. All water samples contained lead above 15 µg/L, and several exceeded hazardous waste levels (>5000 µg/L). After analysis of blood lead data revealed spiking lead in the blood of Flint children in September 2015, a state of emergency was declared, and public health interventions (distribution of filters and bottled water) likely averted an even worse exposure event due to rising water lead levels. This incidence will have long-term health consequences in infants as well as in adults, including developmental delays, learning difficulties, mood disorders, miscarriages, joint pains, hypertension, and cognitive decline in adults [71].

The Safe Drinking Water Act (SDWA) is the federal law that protects public drinking water supplies throughout the nation. Under the SDWA, the Environmental Protection Agency (EPA) sets standards for drinking water quality and, with its partners, implements various technical and financial programs to ensure drinking water safety. Even with the law in place, more action needs to be taken with regard to the water quality. The EPA in January of 2017 initiated a peer review of draft scientific modeling approaches to inform the agency's evaluation of potential health-based benchmarks for lead in drinking water [72].

Pure water on this planet has become a diminished resource. All of these problems have significant health implications, ranging from neurological damage to respiratory problems to gastrointestinal ailments. Access to safe water for drinking, bathing, and normal household use is a very important part of a healthy life – one of which we are most aware when it no longer exists.

One of the smarter options is to recommend patients filter their own water and carry it with them in a stainless steel bottle. There are many filtering options. A few options would be to use simple carbon filters or a reverse-osmosis filter system, which puts the water through a multistep process to remove the toxins. There is an initial installation cost, but it is cheaper over the long run. Filtering bathing and washing water is also recommended. A practitioner may recommend a whole-house water filter with an additional drinking water filter as the best option.

> **Hydration Tips**
> - On a sedentary day, try to drink around 2 liters of water.
> - Start by drinking a glass of fresh water when you get up in the morning.
> - If you are not used to drinking water regularly, try initially replacing just one of your other drinks a day with fresh water, increasing your consumption as the weeks go by.
> - Drink a glass of water before and during each meal.
> - Hot water with fresh mint, lemon balm, or a piece of fruit in it – like lime, lemon, and orange – often helps those who want a hot drink.
> - Carry a bottle filled with water with you whenever you leave the house.
> - During exercise, drink at 10- to 15-minute intervals, or think of it as a full glass every 30 minutes. Drink slowly and drink early, as it's physically easier to do this when you are still feeling fresh.
> - Keep a check on your urine. As a general guide to hydration, it should be plentiful, pale in color, and odorless.

41.2.10 Inclusion of Prebiotic and Probiotic Foods

Nutrition plays a key role in the modulation of gut microbiota composition, and, in turn, the gut microbiota plays a critical role within the body. The gut microbiota is mainly involved in the development and growth of immunity; the promotion of barrier integrity; the prevention of antigens and pathogens from entering the mucosal tissues; the regulation of several fundamental metabolic pathways such as synthesizing vitamin K; the promotion of angiogenesis and enteric nerve function; and reductive reactions such as methanogenesis, acetogenesis, nitrate reduction, and sulfate reduction, to name a few.

Dysbiosis, on the other hand (quantitative and/or qualitative alterations of gut microbiota), impairs homeostasis in the immune system as well as metabolic pathways, leading to the development of several gut microbiota-related diseases. These include functional gastrointestinal diseases, inflammatory bowel disease, gastrointestinal malignancies, allergic diseases, intestinal infectious diseases, liver diseases, obesity and metabolic syndrome, diabetes mellitus, autism, and others.

The four major microbial phyla that represent over 90% of the bacterial component of the gut microbiota are *Firmicutes*, *Bacteroides*, *Proteobacteria*, and *Actinobacteria*. The majority of "good" bacteria harboring the human gut microbiota are represented by *Firmicutes* and *Bacteroides*. *Firmicutes* are subgrouped in *Clostridium*, whereas *Bacteroides* phyla are subgrouped with a great number of *Prevotella* and *Porphyromonas*. The human gut microbiota also includes viruses, especially phages, Eukarya as fungi, *Blastocystis*, Amoebozoa, and Archaea. Lactic acid bacteria (LAB) and bifidobacteria (*Actinobacteria*) are two important types of gut bacteria, which are present from birth or acquired from digested food. *Lactobacillus* and *Leuconostoc* spp. are the main lactic acid bacteria found in the human intestine. *Bifidobacterium* spp. are the predominant bacteria found among the first colonizers of newborns and persist at a low level in adults. Infants fed with breast milk had higher levels of *Bifidobacteria* spp., while infants fed with formula had higher levels of *Bacteroides* spp., *Clostridium coccoides*, and *Lactobacillus* spp. [73].

Diet has been known to have a strong influence on the composition of intestinal microbiota, and to confirm this hypothesis, researchers sequenced oral microbiota from skeleton teeth of people who lived over the various eras. The most significant changes in human gut microbiota occurred during two socio-dietary breakthroughs over the human history: the first one around 10,000 years ago, during the passage from the hunter-gatherer Paleolithic era to the farming Neolithic era, with a diet rich in carbohydrates, and the second one at the beginning of the industrialized period approximately two centuries ago, characterized by a processed flour and sugar diet. Studies show that there are differences between the intestinal microbiota of subjects fed with a standard American diet and that of subjects with a diet rich in fibers. Literature shows an increased prevalence of *Bacteroides* and *Actinobacteria* is positively associated with animal protein and a high-fat diet (more prevalent in Western countries), but is negatively associated with fiber intake, whereas *Firmicutes* and *Proteobacteria* show the opposite association. In contrast, the *Prevotella*-prevailing enterotype is associated with high consumption of carbohydrates and simple sugars, indicating a correlation with a carbohydrate-based diet more typical of agrarian societies and vegetarians [74].

41.2.10.1 Probiotics

Probiotics are live microorganisms that, when administered in adequate amounts, confer a health benefit on the host [75]. There exists a long history of human consumption of probiotics (particularly lactic acid bacteria and bifidobacteria) and prebiotics, either as natural components of food or as fermented foods.

The colonic microbiota is able to metabolize complex carbohydrates and oligosaccharides into short-chain fatty acids (SCFAs) such as butyrate, acetate, and propionate. These metabolites play a role in regulating intestinal pH. A small variation of acid concentrations may exert important consequences. A one-unit decrease in pH from 6.5 to 5.5 has been shown to have a profound selective effect on the colonic microbial population, with a prevalent decrease of *Bacteroides* spp. and a growth of butyrate-producing gram-positive bacteria. The decrease of pH caused by high concentration of SCFAs prevents the growth of potentially pathogenic bacteria, such as *E. coli* and other members of the *Enterobacteriaceae* family.

The most common sources of probiotics are yogurt, buttermilk, cheese, and kefir. Traditionally, yogurt has only one or two bacteria, whereas kefir tends to have several probiotic bacteria. Other foods produced by bacterial fermentation are Japanese miso, natto, tempeh, sauerkraut, sourdough, kimchi, kombucha, olives, and pickled vegetables.

41.2.10.2 Prebiotics

Different types of dietary fibers, complex carbohydrates (digestible and nondigestible) and oligosaccharides, act as prebiotics and exert a different influence on the composition of the gut microbiota and, consequently, on its fermentative metabolism. Prebiotics are nondigestible food ingredients that beneficially affect the host by selectively stimulating the growth and/or activity of one or a limited number of bacteria in the colon that can improve the host health. Prebiotics cannot be digested by pathogenic bacteria. They stimulate the growth of probiotic bacteria and allow them to grow predominantly. Prebiotics are usually polysaccharides or oligosaccharides. They are naturally found in such fruits and vegetables as asparagus, bananas, dandelion greens, eggplant, endives, garlic, honey, Jerusalem artichokes, jicama, leeks, legumes, onions, radishes, tomatoes, and whole grains. The commonly known prebiotics are inulin, fructo-oligosaccharides (FOS), galacto-oligosaccharides (GOS), soya-oligosaccharides, xylooligosaccharides, pyrodextrins, isomalto-oligosaccharides, and lactulose [76].

The Western diet is considered to cause intestinal dysbiosis and trigger local inflammation, leading to an increase of intestinal permeability. Studies show that a high-fat diet induces the proliferation of certain gram-negative bacteria which can ultimately result in increased intestinal lipopolysaccharide (LPS) and increased gut permeability. Increased gut permeability is associated with a decrease in *Bifidobacteria* spp., bacteria that are known to reduce LPS levels and also improve gut barrier function. Interestingly, prebiotic treatment with nondigestible carbohydrates increases bifidobacteria and reduces intestinal permeability, LPS concentrations, and metabolic endotoxemia [74].

The healthful effects of probiotics and prebiotics factor in their potential impact on the balance of the body's microflora and directly or indirectly in their enhancement of the function of the gut and systemic immune system. Tailoring the diet to include probiotic foods that the patient enjoys can have a significant positive impact in improving the patient's nutritional intake [77].

41.2.10.3 Translating Expertise into Intervention

Diet has an enormous impact on many aspects of our health, even beyond providing energy and essential nutrients. The circulating substrates derived from food have both direct and indirect actions to activate receptors and signaling pathways, in addition to providing fuel and essential micronutrients. Ultimately one can consider food as a cocktail of "hormones." A hormone is a regulatory compound produced in one organ that is transported in blood to either inhibit or stimulate specific cells in another part of the body. They exert their effects on target tissues by acting on cellular receptors to alter activity through intracellular signaling cascades or via nuclear receptors to regulate gene transcription. Although food is not produced in the body, its components travel through the blood, and nutrient substrates can act as signaling molecules by activating cell-surface or nuclear receptors [78]. If the body does not receive appropriate materials or if there is inappropriate signaling due to nutritional deficiencies – for example, B12 and homocysteine or excessive amounts of omega-6 foods – metabolic processes suffer and health decline.

As nutritional practitioners, we hold a valuable position in helping patients through one of the most critical modifying lifestyle factors, and that is food and nutrition. Planning and implementation are two components of nutrition intervention. The practitioner has to prioritize nutritional diagnosis, based on the severity of the problem, safety, patient/client need, the likelihood that the nutrition intervention will impact the problem, and the patient's perception of importance. The nutrition intervention is directed whenever possible at the etiology or the cause of the problem identified. If it is not possible to direct the nutrition intervention, the nutrition intervention is aimed at reducing the impact of the signs and symptoms of the problem. For example, if the etiology is related to food allergies or sensitivities, nutrition intervention is based on the elimination of trigger or offending foods. If excessive energy intake is caused by depression and initiation of a psychiatric prescription (changing or withholding prescription drugs is out of RDNs' scope of practice), the nutrition intervention is aimed at reducing the impact of the signs and symptoms of the problem – for example, weight gain or high blood glucose due to increased energy intake.

During the planning phase of intervention, the nutrition practitioner uses evidence-based guidelines along with institutional policies and procedures to reach medical nutrition therapy goals, desired behavior changes, and/or expected outcomes. The nutrition prescription identifies the specific nutritional interventional strategies and establishes the patient-focused goals to be accomplished. The goals should be measurable, achievable, and time-defined. The functional medicine nutrition professional must make sure that they are jointly set with the patient, keeping in mind not just the clinical indicators and biomarkers but also the socioeconomic, cultural, spiritual, and emotional aspects of healing.

Implementation of nutrition intervention should also be accompanied by data collection initiated during nutrition assessment, and practitioners should revise the intervention based on the patient's response. In most cases, the initial nutrition intervention consists of nutrition education, where the nutritional practitioner imparts knowledge and shares skills to help the patient manage and modify food choices and eating behaviors. Nutritional counseling, a supportive process, organically follows, either during the first visit or during follow-ups. This is a collaborative relationship to set priorities, evaluate goals, and fine-tune a nutrition plan to foster self-care and treat or manage an existing condition and promote health.

It is important to note that initiating multiple changes during one single visit may overwhelm indiviudals seeking help. Some patients are able to make many changes at a time, but most patients with long-term chronic issues need time. They might need to make just a few changes at a time. The key to successful nutrition intervention in the nutrition care process is helping patients understand the rationale behind prescribing an individualized food plan; keeping health goals and objectives at the forefront; and fostering self-efficacy by sharing recipes, menu-planning, and providing resources that will enable them to translate the nutrition consultation into practical application and behavioral change. This helps with compliance, builds trust, and fosters a healthy practitioner-patient relationship.

> **Recipe: Carrot Beet Probiotic Beverage (Carrot-Beet Kanji)**
> - *Ingredients*:
>   - Water – 8 cups
>   - Carrots (orange or purple) – 2 medium, peeled and julienned
>   - Beet – 1, peeled and julienned
>   - Green chilies – to taste, slit on one side (1 small, optional)
>   - Powdered mustard seeds – 1 1/2 tbsp
>   - Salt – 1 1/2 tsp (or to taste)
>   - Red chili powder (optional) – 1 tsp

- *Method*
  1. In a clean pitcher or bottle with a lid, preferably glass or ceramic, add all of the ingredients and mix well. Do not use plastic bottles or pitchers.
  2. Cover and keep the pitcher in the sun for 3–4 days, stirring at least once daily with a clean spoon.
  3. Once fermented, taste the kanji. When it's ready, it has a tangy and fermented taste. Taste it daily just to understand the change in flavor, its food chemistry in action! Store the kanji in the refrigerator.
  4. Serve chilled. Mix before serving. Carrots, beets, and green chilies can be eaten.

In North India, deep purple-colored carrot is fermented along with crushed mustard seed, hot chili powder, and salt for a few days to get a popular drink called kanji, which is considered to have high nutritional value and cooling and soothing properties. According to the *Indian Journal of Microbiology*, lactic acid bacteria (LAB) play an important role in the fermentation of vegetables, improving nutritive value, palatability, acceptability, microbial quality, and shelf life.

41.2.10.4 Dietary Supplements

The majority of adults in the United States take one or more dietary supplements either every day or occasionally (*ODS. NIH*). According to the Academy of Nutrition and Dietetics position paper on nutrition supplementation, "dietetics practitioners should position themselves as the first source of information on nutrient supplementation." In order to accomplish this, the registered dietitian/clinical nutritionist must keep up to date on the issues associated with regulation, safety, and efficacy, along with identifying supplements available in the market.

According to the *Food and Drug Administration* (FDA), a dietary supplement is a product intended for ingestion that contains a "dietary ingredient" intended to add further nutritional value to (supplement) the diet. A "dietary ingredient" may be one, or any combination, of the following substances:
- A vitamin
- A mineral
- An herb or other botanical
- An amino acid
- A dietary substance for use by people to supplement the diet by increasing the total dietary intake
- A concentrate, metabolite, constituent, or extract

Dietary supplements may be found in such forms as tablets, capsules, softgels, gelcaps, liquids, or powders. Some dietary supplements can help ensure that you get an adequate dietary intake of essential nutrients; others may help you reduce your risk of disease.

Dietary supplements are complex products. The FDA has established good manufacturing practices (GMPs) for dietary supplements to help ensure their identity, purity, strength, and composition. These GMPs are designed to prevent the inclusion of the wrong ingredient, the addition of too much or too little of an ingredient, the possibility of contamination, and the improper packaging and labeling of a product. The FDA periodically inspects facilities that manufacture dietary supplements.

In addition, several independent organizations such as US Pharmacopeia and NSF International offer quality testing for dietary supplement manufacturers and allow products that pass these tests to display their seals of approval. These seals of approval provide assurance that the product was properly manufactured, contains the ingredients listed on the label, and does not contain harmful levels of contaminants. These seals of approval do not guarantee that a product is safe or effective. NSF International also offers an NSF Certified for Sport Program so that athletes can be sure that their dietary supplements do not contain banned substances.

Several companies provide unsolicited post-market surveillance of dietary supplements to ensure quality and integrity in the supplement industry. Consumer Labs and the nonprofit Center for Science in the Public Interest are two such prominent entities.

Under the Dietary Supplement Health and Education Act of 1994 (DSHEA), supplement manufacturers are solely responsible for ensuring that their products are safe. There is no approval of claims, no product registration, and no formula standards. The FDA does pre-market review of new dietary ingredients (NDI). The term "new dietary ingredient" means a dietary ingredient that was not marketed in the United States in a dietary supplement before October 15, 1994. There is no authoritative list of dietary ingredients that were marketed in dietary supplements before October 15, 1994. Even if the NDI is reviewed by the FDA, there is no guarantee of safety. If a dietary supplement causes harm, it is up to the FDA to prove the supplement is unsafe and remove it from the market. Under the current legislation, it is a daunting task for the government to regulate the dietary supplement industry.

Very few supplements have been banned by the FDA in recent times. *Ephedra sinica*, aristolochic acid, and dimethylamylamine (DMAA) are on the list of banned supplements. Herbal products may pose a significant risk in some cases as they have multiple chemical constituents that have not been adequately characterized. *Hypericum perforatum*, also called St. John's wort, is one of the most problematic herbs as far as pharmacokinetics and safety go, along with *Schisandra chinensis* and black pepper. It is important that a nutritional professional recommending herbal products therapeutically seek additional herbal training.

41.2.10.5 Rationale for Recommending Supplements

Before recommending supplements, the functional medicine/nutrition practitioner must determine the clinical indicators for the use. Some of the factors that can help in the decision making are:

- *Filling nutritional gaps*: Does the patient have nutritional deficiencies evident from dietary analysis, signs and symptoms, or laboratory evaluations?
- *Increased nutritional needs*: Does the patient have an increased need for nutrients as a result of inborn errors of metabolism, absorption or transport defects, abnormal enzyme production, or excessive renal losses? The patient may have elevated needs of additional nutrients or enzymes secondary to acute or chronic illness, such as burns, traumatic brain injury, hypochlorhydria, small intestinal bacterial overgrowth, IBD, and pancreatic enzyme deficiency.
- *Maintenance*: Does the patient need supplements to maintain balance of nutrients? For instance, patients who are on calcium supplements for bone health often require magnesium to maintain an appropriate calcium-to-magnesium ratio.
- *Drug-nutrient interactions*: Is the patient on prescription drugs such as statins, antacids, or corticosteroids? If so, they might benefit from supplementation of CoQ10, B12, and calcium, respectively.
- *Genetic factors*: Does the patient have single-nucleotide polymorphisms and therefore increased needs of certain nutrients? For example, MTHFR hyperhomocysteinemia might require adequate folate and B12 intervention.
- The recommended daily allowances (RDAs) and adequate intakes establish nutrient intake levels for healthy individuals and may not address the specific needs of every individual. Individuals who are very active or participate in sports, or who have chronic inflammatory conditions, may benefit from dietary supplementation. There might be a need for supplementation even in women of childbearing age, pregnant women, and older adults to meet optimal nutrient needs.

41.2.10.6 Roles and Responsibilities

It is important to use an evidence-based approach when recommending supplements and screen for drug-nutrient interactions and over-supplementation to ensure safety.

It is the responsibility of the nutrition practitioner to make sure any products dispensed from the office are high-quality products and that the manufacturers comply with Good Manufacturing Practices to ensure safety and efficacy. Look for quality ingredients such as chelated minerals that can be more easily absorbed than mineral salts; active B vitamins, especially for patients with single nucleotide polymorphisms, for example, pyridoxal 5-phosphate instead of pyridoxine hydrochloride; and D-alpha tocopherol with mixed tocopherols versus the synthetic counterpart DL-alpha tocopherol. Just as the functional medicine nutrition practitioner encourages patients to choose clean eating and minimally processed foods, the practitioner should also recommend dietary supplements with minimal processing and excipients: binders, flavorings, colorings, high-fructose corn syrup, and so on. For patients with food allergies or sensitivities, ensuring allergenic-free supplements is critical.

Some of the responsibilities of the practitioner initiating the use of dietary supplements are as follows: to assess nutritional status of the patient to determine the likelihood of inadequate or excessive intake of minerals and vitamins; to evaluate for the potential benefit or harm of nutrient supplementation; to evaluate safety with regard to dosage and potential drug-nutrient interaction; to educate patients about the potential benefits and rationale behind using dietary supplements; to be informed of the research on the supplements suggested; and to be aware of the regulatory, legal, and ethical issues of recommending and selling supplements.

Whenever possible, it is advisable to validate the need for supplementation with laboratory testing and therefore collaboration of care. Referrals, and building a community of healthcare providers, become important in a functional medicine practice. Specific nutrient levels can be obtained through standard labs and specialty functional medicine testing such as organic acid tests, stool analysis, fatty acid profiles, hormone metabolites, and nutrigenomic testing.

The functional nutrition practitioner should be aware of potential adverse effects of dietary supplements and should monitor patients appropriately. While the use of dietary supplements usually has an excellent safety profile, side effects can occur, and high doses of certain nutrients such as B3, B6, B12, zinc, and vitamin D can have serious toxic effects. Sensitive individuals may have allergic reactions not only from the excipients, additives, colorings, and flavorings added but also from food-based ingredients and products containing herbs. Patients with advanced renal or hepatic insufficiency may develop toxicity from some nutrients. In such patients, supplements may be contraindicated, and they should be used only under the direction of practitioners who are proficient in management of serious cases.

The functional nutrition practitioner must update his/her knowledge on dietary supplements on a regular basis.

Supplement Education Resources
- Websites:
  - Office of Dietary Supplements: ▶ https://ods.od.nih.gov/
  - National Center for Complementary and Integrative Health: ▶ https://nccih.nih.gov/
  - Linus Pauling Institute Micronutrient Information Center: ▶ http://lpi.oregonstate.edu/mic
  - ▶ Drugs.com Drug Interactions Checker: ▶ https://www.drugs.com/drug_interactions.html
  - Medscape Drug Interaction Checker: ▶ http://reference.medscape.com/drug-interaction-checker
  - Natural Medicines Comprehensive Database: ▶ https://naturalmedicines.therapeuticresearch.com/
  - Consumer Lab: ▶ https://www.consumerlab.com/

- Books:
  - Mosby's Handbook of Herbs and Natural Supplements, 4th edition by Linda Skidmore-Roth, RN, MSN, NP
  - The Health Professional's Guide to Dietary Supplements by Shawn M. Talbot and Kerry Hughes
  - The Health Professional's Guide to Popular Dietary Supplements, 3rd edition, by Allison Sarubin Fragakis, MS, RD with Cynthia A. Thomson, PhD, RD
  - Nutritional Medicine by Alan Gaby

41.2.11 Lifestyle

Lifestyle is way of living. It is a set of habits and attitudes – physical, mental, social, and spiritual – that together constitute a mode of living for an individual or group.

At the base of the functional medicine matrix are modifiable personalized lifestyle factors such as sleep and relaxation, exercise and movement, nutrition and hydration, stress and resilience, and, last but not least, relationships and networks. Earlier in this chapter, we reviewed the nutrition and hydration aspects. This section will be an overview of the other lifestyle factors.

41.2.11.1 Sleep

Circadian rhythms are endogenous rhythms with a periodicity of approximately 24 hours, plus or minus 4 hours. These rhythms are dependent on an internal clock located in the suprachiasmatic nucleus (SCN) of the anterior hypothalamus. Each of the paired suprachiasmatic nuclei is composed of a group of about 10,000 interconnected neurons that give rise to circadian rhythms through specific neuronal gene expression patterns and by the rate at which they fire action potentials. In addition to the SCN, peripheral clocks have been identified in numerous tissues such as the liver, kidney, heart, skin, and retina, and these are capable of acting in an autonomous manner. The SCN subsequently synchronizes peripheral clocks with each other and thus aligns the entire circadian system to the external light-dark cycle.

The circadian system in human beings is a complex entity that starts in the eye and terminates in the pineal gland, which produces melatonin and is essential for functioning of the clock. In humans, melatonin is secreted during the dark phase of the light-dark cycle. Daytime melatonin levels are very low. Light is considered to be the most potent circadian synchronizer for humans, although non-photic time cues – such as meal times, physical activity, and social interaction – also play a part in synchronization of the circadian system.

Even low-intensity light, as emitted by recent technologies such as LEDs, computer screens or televisions, mobile phones, and tablets, is capable of acting on the clock, thus leading to a phase delay and a slowing of melatonin secretion. In addition to being exposed to light from a range of LED consoles throughout the day, many individuals use such devices at night. This results in circadian phase delay and is associated with a lack of sleep and is thought to underlie clock desynchronization disorders. This amounts to a type of chronic jet lag, also referred to as social jet lag – that is, a misalignment between the clock and social time – and can jeopardize health. Desynchronization manifests through atypical clinical symptoms, such as persistent fatigue, sleep disorders leading to chronic insomnia, poor appetite, and mood disorders that can cause depression, though some desynchronized people do not experience any of these clinical signs.

Chronic sleep deprivation results in drowsiness, decreased levels of attention and alertness that underlie a doubling in the risk of traffic accidents, cardiovascular risk (elevated blood pressure and lipids), GERD, dysregulation in glucose metabolism, insulin resistance (diabetes and obesity), macular degeneration, and cognitive and mental health disorders.

Educating and creating awareness in patients about the importance of sleep and encouraging them to have good sleep hygiene may help bring metabolic and physical changes and improve health and well-being along wih preventing societal harm [79, 80].

> **Strategies for Restful Sleep**
> - Avoid exposure to light up to 30 min prior to going to sleep.
> - Opt for a morning shift that starts before 7 a.m.
> - Keep the same sleep timing most days.
> - Avoid stimulants such as alcoholic, caffeinated, or sugary beverages.
> - Complete aerobic exercise 3 hours before bedtime.
> - Avoid reading stimulating materials at night.
> - Avoid arguments and difficult conversations before bedtime.
> - Avoid large meals before bedtime.
> - Take hot Epsom salt aromatherapy baths.
> - Sleep in a dark room.
> - Use HEPA or air purifiers/filters to clean air in your bedroom.
> - Use humidifiers if needed.
> - Avoid shift work or night work when pregnant.

41.2.11.2 Exercise and Movement

Exercise and movement may alter and therapeutically impact mental (improve memory, increase hippocampal size, and increase BDNF), emotional (increase mood, decrease depression, mood, and anxiety), and spiritual aspects (experience sense of purpose; enhance social network, peer support, and sense of community; nature-based exercises aka "dose of nature") in an individual – the very core of the functional medicine matrix [81–83].

Exercise and movement also have a therapeutic impact on each of the clinical imbalances in the function medicine matrix.

- **Assimilation:**
  - Exercise initiated early in life increases gut bacteria species which promotes psychological and metabolic health.
  - Exercise improves the *Bacteroidetes-Firmicutes* ratio and stimulates proliferation of bacteria which can modulate mucosal immunity and improve barrier function.
  - Exercise may also stimulate bacteria capable of producing substances such as SCFAs that protect against gastrointestinal disorders and colon cancer along with providing psychological health benefits [84].

- **Structural integrity:**
  - Increases muscle mass
  - Helps in the reduction of body fat
  - Increases peak oxygen intake
  - Decreases blood viscosity and fibrinogen

- **Communication:**
  - Increases insulin sensitivity, AMPK, adiponectin, NO, semen quality, and vagal tone [85, 86]
  - Reduces HbA1c, adrenergic activity, and estradiol [85, 87]

- **Transport:**
  - Increases maximal oxygen consumption, heart rate variability, flow-mediated dilatation, angiogenesis, endothelial function, HDL, and aerobic threshold
  - Decreases resting heart rate, blood pressure, LDL, and triglycerides [88]

- **Biotransformation and elimination:**
  - Improves bowel function
  - Increases skeletal muscle glucose uptake

- **Energy:**
  - Increases mitochondrial biogenesis and ATP production
  - Decreases fatigue

- **Defense and repair:**
  - Improved immune system
  - Decreased inflammation
  - Increase natural killer cell activity

The intensity and duration of exercise also determine energy consumption. Exercise performed for the same duration at different intensities results in more energy consumption during the higher-intensity segments. Exercise intensity also affects the secretion of hormones associated with energy substrate oxidation. Studies report that carbohydrates are the main energy substrate in high-intensity exercise, compared with lipids being utilized more in low-to-moderate intensity exercise. In recent times, high-intensity interval training (HIIT) has gained attention [89].

Many individuals may feel daunted by the thought of changing their lives and starting new, more active routine. A functional medicine practitioner/nutritionist can help provide motivation and support during this transition by helping patients to remember a few key tips:

1. Emphasize fun: Focus on what the patient enjoys and loves to do rather than exercise being a punishment.
2. Attach activity to habits: Taking a walk after dinner is a time-honored way to get moving. Any routine behavior can have a small activity bonus built in.
3. Involve others: Chances are that the patient's friends, family, and co-workers want to be more active too. It then becomes a community and peer-shared experience. Swap sitting at a coffee shop for walks; swap dinners for bowling or after-dinner walks.
4. Be inventive: Challenge your patient with more active living. Suggest ideas such as an extra lap around the grocery store perimeter, standing while watching television, getting off a bus earlier, using a standing desk at work, and stretching one part of the body when your patient texts someone.
5. Be forgiving: If the patient had one sedentary day, suggest them to not overwork the next day or punish themselves.
6. Track progress: Consider suggesting a pedometer app/device and having fun with it.

It is very important for functional medicine practitioners to walk the talk. Modeling healthy habits not only provides the functional medicine practitioner the deep knowledge of what is required of the patient but also increases the likelihood that you will counsel patients successfully – may it be providing recipes, healthy eating, exercising, or creating a work-life balance. Improving one's own health practices increases efficacy at lifestyle counseling. Practicing one's own preventative and healthy behaviors increases patient adherence to recommendations.

41.2.11.3 Stress and Resilience

Walter Cannon, professor and chairman of the Department of Physiology at Harvard Medical School, wrote extensively about stress, and his work during World War I helped to establish the groundwork for understanding the physiology of mind-body interactions [90]. Later, Hans Seyle reintroduced the term and he is now considered the "father of stress." Per Seyle, "stress is the nonspecific response of the body to any demand, and a stressor is any agent that produces stress at any time." Stressors can be loosely classified as direct physical systemic threats, e.g., starvation, cold, pain, hemorrhage, or "progressive threats," e.g., psychosocial threats such as defeat, separation, helplessness, and social isolation.

Allostasis is the ability to achieve homeostasis when a stress is applied to the system. Homeostasis is achieved via acute and chronic changes through the neuroendocrine axis (primarily the hypothalamic-pituitary-adrenal, or HPA, axis), the immune system, the autonomic nervous system, and the cardiovascular system. The HPA axis responds to input by involving the immune, cardiovascular, and metabolic systems. Acute stress induces changes that are generally thought to be adaptive and essential, whereas chronic stress is thought to induce a higher physiological price by creating an allostatic load, defined as a long-term effect of physiologic response to stress.

While acute stress heightens cognition, these same responses when activated chronically are highly deleterious. It may injure hippocampal cells via cortisol release in response to CRF and ACTH. Such injury could lead to dysfunction and atrophy of critically important memory and emotional brain structure [91].

Chronic or cumulative stress, representative of lifetime exposure to adversity, is also thought to exert its influence through a "chain of risk," wherein early adverse life events increase the risk for later exposures. The cumulative stress, in turn, impacts physical, behavioral, and mental outcomes such as hypertension, pain, physical disability, psychiatric disorders, drug and alcohol abuse, and other chronic illness. Such chronic stress may impact the autonomic nervous system (ANS) and can be noninvasively measured by methods such as heart rate variability (HRV). Lower indices of HRV and altered ANS function may contribute to cardiac mortality, decrease vagal tone, and increase cardiac workload and endothelial dysfunction [92].

Simply asking a patient "What are the main stressors in your life?" can be very useful. Together the functional medicine practitioner and patient can generate a list of all the stressors, so that they can be tracked and addressed.

Like nutrition and exercise, there is a role for personalization in one's approach to stress, which can be encompassed in the diversity of modalities from which an individual can choose to modify their behavior to cope with it. This may include such mind-body practices as relaxation response developed by Herbert Benson, mindfulness-based stress reduction (MBSR) from Jon Kabat-Zinn, vipassana meditation [93], yoga, Kirtan Kriya, squared breathing, and diaphragmatic breathing.

Studies in the area of mind-body medicine are now becoming more molecular-focused. For instance, Balasubramanian et al. compared salivary expression of 22 differentially expressed proteins in subjects after 20 minutes of yogic breathing and those who chose to read a text of their choice. The subjects who practiced the 20-minute yogic breathing technique had increased salivary secretion which regulates digestive, nervous, immune, and respiratory systems. The researchers also elucidated that yogic breathing could potentially stimulate salivary expression of the nerve growth factor, a trophic factor involved in the development, maintenance, and survival of the peripheral nervous system and cholinergic neurons of the CNS that are significantly reduced in Alzheimer's disease patients [94].

Certain individuals may be epigenetically and genetically more inclined to respond to stress. Children subjected to abuse have decreased hippocampal expression of glucocorticoid receptor and heightened stress responses. Early stressed offspring must learn to adapt to environmental challenges and maintain stability through change, a characteristic referred to as "resilience." In addition to epigenetic effects, a number of genetic variants have been identified to cause altered response to stress and mood disorders. For example, polymorphisms in catechol-O-methyltransferase enzyme have been implicated in mental illness with the presence of certain SNPs resulting in susceptibility to posttraumatic stress disorder.

In summary, a functional medicine practitioner/nutritionist needs to not only understand the nutrition intervention aspect but also have insight into psychosocial and genetic aspect of an individual in order to better help the patient. Functional medicine can be integral throughout a patient's life, from prevention to preclinical symptoms to disease manifestation and progression [95].

Relaxation Response

Herbert Benson, MD, elucidated the physiological underpinnings of the trophotropic center – a combination of the parasympathetic nervous system, somatic muscle relaxation, and cortical beta rhythm synchronization. First, he described what he called "the four states": awake, sleep, dream, and the fourth state. The difference between the first three states and the fourth state is that the first three happen spontaneously, while specific actions are required to enter into the fourth state or what Benson initially called "eliciting the relaxation response (RR)."

— Comfort: Sit easily in a chair or on the floor.
— Quiet: Be alone in a spot where you will not be disturbed, i.e., no texts, emails, cell phones, etc., while eliciting the relaxation response.
— Tool: Focus on a word, thought, breathing, sound, or short prayer.
— Attitude: When other thoughts enter your mind, refocus on your tool to the exclusion of everything else for 10–20 minutes twice a day.

What Is Squared Breathing?

Squared breathing is just four simple breath segments done to a count of four.

— Inhale 2 3 4
— Hold 2 3 4
— Exhale 2 3 4
— Hold 2 3 4

Focus on the breath and the count of four; repeat the same process until you reach a relaxed state.

41.2.12 Referrals

There might be cases in practice when nutritional therapies, use of supplements, and lifestyle modifications show no improvement or benefit or only slight improvements. For a functional medicine practitioner, it is imperative to get to the root cause of the problem. In such cases, look for toxicity and underlying infections such as molds and their toxins, silicone breast implants and other implants, parasites, viruses, or Lyme disease and its co-infections. In such cases, refer the patient to a practitioner proficient in those areas.

No single practitioner – and no single discipline – can cover all the viable therapeutic options. Interventions differ by training, specialty focus, licensure, and even by beliefs and ethnic heritage. Practitioners in any specialty can, to the degree allowed by their training and licensure, utilize a functional medicine approach, including integrating the matrix as a basic template for organizing and coupling knowledge and data. So it is important to seek out other providers who have also acquired some functional medicine training, building a team of practitioners in the community you work with, whether in-office or virtual. Treatment success can be affected by the cultural bias of the patient and the provider, so get to know your colleagues in other fields.

Educate yourself about what other disciplines have to offer. As a nutritional practitioner, you may want to build up a team that includes physicians, a chiropractor, a naturopathic or osteopathic physician, an acupuncture specialist, practitioners trained in traditional Chinese medicine, a yoga instructor, practitioners trained in Ayurveda, a personal trainer, a psychologist, a functional medicine psychiatrist, and so on. It is beyond the mission of this chapter to discuss the training, scope of practice, or therapeutics of the many different healthcare approaches available today, but many publications and websites do provide such information. Developing a network of capable, collaborative healthcare practitioners with whom you can co-manage challenging patients and to whom you can refer for therapies outside your own expertise will enrich the care your patients receive and will strengthen the clinician-patient relationship. Many patients do not tell one practitioner about care they are receiving from other practitioners. Keep those lines of communication open and alive. All referral relationships should be handled with the same professionalism and courtesy.

41.3 Monitoring and Evaluating Interventions

At every follow-up visit, the nutrition practitioner monitors the amount of progress made toward the patient's goals and expected outcomes and compared to the patient's previous status. The four categories of outcomes that the nutrition practitioner monitors and evaluates are food- and nutrition-related outcomes (intake of food and dietary supplements, knowledge and beliefs, food availability, physical activity, nutrition, and quality of life); biochemical data, medical tests, and procedure outcomes; anthropometric measurement; and nutrition-focused physical findings.

Follow-up visits give the patient and practitioner the time and space to monitor and evaluate for compliance of nutrition intervention and laboratory biomarkers; clarify patients' questions or concerns and provide support; identify positive and negative outcomes; change nutrition behaviors with foods and supplements; implement other new behaviors, including sleep, exercise, relaxation, relationships, and hydration; and reinforce motivation and a sense of well-being for the patient until the patient reaches his or her goal. The follow-ups can be spaced out for longer duration at regular intervals during the maintenance phase until the patient is discharged.

References

1. Churchill LR, Schenck D. Healing skills for medical practice. Ann Intern Med. 2008;149(10):720–4.
2. Hanaway P. Form follows function: a functional medicine overview. Perm J. 2016;20(4):16–109.
3. Jones D Hoffman L Quinn S 21st century medicine: a new model for medical education and practice [Internet]. Institute of Functional Medicine; 2010 [Cited 2017 Feb 2]. Available from: https://www.functionalmedicine.org/files/library/21st-century-medicine_001.pdf.
4. Maspero C, Ginannini L, Galbiati G, Rosso G, Farronato G. Obstructive sleep apnea syndrome: a literature review. Minerva Stomatol. 2015;64(2):97–109.
5. Luyster FS. Impact of obstructive sleep apnea and its treatments on partners: a literature review. J Clin Sleep Med. 2017;13(3):467–77.
6. Killian L, Starr J, Shiue I, Russ T. Environmental risk factors for dementia: a systematic review. BMC Geriatr. 2016;16:175. https://doi.org/10.1186/s12877-016-0342-y.
7. Harris S, Harris E. Herpes simplex virus type 1 and other pathogens are key causative factors in sporadic Alzheimer's disease. J Alzheimers Dis. 2015;48(2):319–53. https://doi.org/10.3233/JAD-142853.
8. Ferguson L, De Caterina R, Görman U, et al. Guide and position of the International Society of Nutrigenetics/Nutrigenomics on personalised nutrition: Part 1 – fields of precision nutrition. J Nutrigenetics Nutrigenomics. 2016;9:12–27. https://doi.org/10.1159/00445350.
9. Sampey B, Vanhoose A, Winfield H, et al. Cafeteria diet is a Robust Model of human metabolic syndrome with liver and adipose inflammation: comparison to high-fat diet. Obesity (Silver Spring, Md). 2011;19(6):1109–17. https://doi.org/10.1038/oby.2011.18.
10. Odermatt A. The Western-style diet: a major risk factor for impaired kidney function and chronic kidney disease. Am J Physiol Renal Physiol. 2011;301(5):F919–31. https://doi.org/10.1152/ajprenal.00068.2011.
11. Uribarri J, Woodruff S, Goodman S, et al. Advanced Glycation End Products in foods and a practical guide to their reduction in the diet. J Am Diet Assoc. 2010;110(6):911–16.e12. https://doi.org/10.1016/j.jada.2010.03.018.
12. Defagó Elorriaga N, Irazola V, Rubinstein A. Influence of food patterns on endothelial biomarkers: a systematic review. J Clin Hypertens (Greenwich). 2014;16(12):907–13. https://doi.org/10.1111/jch.12431.

13. Knudsen B, Norskov N, Bolvig A, Hedemann M, Laerke H. Dietary fibers and associated phytochemicals in cereals. Mol Nutr Food Res. 2017;61:1600518. https://doi.org/10.1002/mnfr.201600518.
14. Heaver J. Plant-based diets: a physician's guide. Perm J. 2016;20(3):93–101. https://doi.org/10.7812/TPP/15-082.
15. Murugaiyah V, Mattson M. Neuro-hormetic phytochemicals: an evolutionary- bioenergetics perspective. Neurochem Int. 2015;89:271–80. https://doi.org/10.1016/j.neuint.2015.03.009.
16. Barbieri R, Coppo E, Marchese A, Daglia M, et al. Phytochemicals for human disease: an update on plant-derived compounds antibacterial activity. Microbiol Res. 2017;196:44–68. https://doi.org/10.1016/j.micres.2016.12.003.
17. Ngo Q, Tran P, Tran M, et al. Alkaloids from Piper nigrum exhibit anti inflammatory activity via activating the Nrf2/HO1 pathway. Phytother Res. 2017;31:663. https://doi.org/10.1002/ptr.5780.
18. Friedman M. Chemistry and anti carcinogenic mechanisms of glycoalkaloids produced by eggplants, potatoes, and tomatoes. J Agric Food Chem. 2015;63(13):3323–37. https://doi.org/10.1021/acs.jafc.5b00818.
19. Wagner K, Elmada I. Biological relevance of terpenoids. Overview focusing on mono, di-and tetrapenes. Ann Nutr Metab. 2003;47(3–4):95–106.
20. Rabi T, Gupta S. Dietary terpenoids and prostate cancer. Front Biosci. 2008;13:3457–69.
21. Farzaei M, Abdollahi M, Rahimi R. Role of dietary polyphenols in the management of peptic ulcer. World J Gastroenterol. 2015;21(21):6499–517. https://doi.org/10.3748/wjg.v21.i21.6499.
22. Kim Y, Keogh JB, Clifton PM. Polyphenols and glycemic control. Nutrients. 2016;8(1):17. https://doi.org/10.3390/nu8010017.
23. Malhotra A, Bath S, Elbarbry F. An organ system approach to explore the antioxidative, anti-inflammatory, and cytoprotective actions of resveratrol. Oxid Med Cell Longev. 2015;2015:803971. https://doi.org/10.1155/2015/803971.
24. Harikumar K, Sung B, Tharakan S, et al. Sesamin manifests chemopreventive effects through suppression of nf-κb-regulated cell survival, proliferation, invasion and angiogenic gene products. Mol Cancer Res MCR. 2010;8(5):751–61. https://doi.org/10.1158/1541-7786.MCR-09-0565.
25. Liu R. Health benefits of fruit and vegetables are from additive and synergistic combinations of phytochemicals. Am J Clin Nutr. 2003;78(3 Suppl):517S–20S.
26. Jacobs D, Gross M, Tapsell L. Food synergy: an operational concept for understanding nutrition. Am J Clin Nutr. 2009;89(5):1543S–8S. https://doi.org/10.3945/ajcn.2009.26736B.
27. Liu R. Antioxidant activity of fresh apples. Nature. 2000;405(6789):903–4. https://doi.org/10.1038/35016151.
28. Burton B. Whole food versus supplement: comparing the clinical evidence of tomato intake and lycopene supplementation on cardiovascular risk factors. Adv Nutr. 2014;5(5):457–85. https://doi.org/10.3945/an.114.005231.
29. Campbell T. Untold nutrition. Nutr Cancer. 2014;66(6):1077–82. https://doi.org/10.1080/01635581.2014.927687.
30. Farsshchi H, Taylor M, Macdonald I. Beneficial metabolic effects of regular meal frequency on dietary thermogenesis, insulin sensitivity, and fasting lipid profiles in healthy obese women. Am J Clin Nutr. 2005;81(1):16–24.
31. McCrory M, Shaw A, Lee J. Energy and nutrient timing for weight control: does timing of ingestion matter? Endocrinol Metab Clin N Am. 2016;45(3):689–718. https://doi.org/10.1016/j.ecl.2016.04.017.
32. Panda S. Circadian physiology of metabolism. Science. 2016;354(6315):1008–15. https://doi.org/10.1126/science.aah4967.
33. Leung G, Huggins C, Ware R, Bonham M. Time of day difference in postprandial glucose and insulin responses: Systematic review and meta-analysis of acute postprandial studies. Chronobiol Int. 2019:1–16. https://doi.org/10.1080/07420528.2019.1683856.
34. Roestamadji R, Nastiti N, Surboyo M, Irmawati A. The risk of night shift workers to the glucose blood levels, saliva, and dental caries. Eur J Dent. 2019;13(3):323–9. https://doi.org/10.1055/s-0039-1697211.
35. Manoogian E, Panda S. Circadian rhythms, time restricted feeding, and healthy aging. Ageing Res Rev. 2017;39:59–67. https://doi.org/10.1016/j.arr.2016.12.006.
36. Jian P, Turek F. Timing of meals: When is as critical as what and how much. Am J Physiol Endocrinol Metab. 2017;312(5):E369–80. https://doi.org/10.1152/ajpendo.00295.
37. Rajaram S, Jones J, Lee G. Plant-based dietary patterns, plant foods, and age-related cognitive decline. Adv Nutr. 2019;10(4 Suppl):S422–36. https://doi.org/10.1093/advances/nmz081.
38. Lampe J. Health effects of vegetables and fruit: assessing mechanisms of action in human experimental studies. Am J Clin Nutr. 1999;70(3 Suppl):475S–90S.
39. Del Corno M, Scazzocchio B, Masella R, Gessani S. Regulation of dendritic cell function by dietary polyphenols. Crit Rev Food Sci Nutr. 2016;56(5):737–47. https://doi.org/10.1080/10408398.2012.713046.
40. Khan H, Sureda A, Belwal T, Çetinkaya S, Süntar İ, Tejada S, Devkota HP, Ullah H, Aschner M. Polyphenols in the treatment of autoimmune diseases. Autoimmun Rev. 2019;18(7):647–57. https://doi.org/10.1016/j.autrev.2019.05.001.
41. Lutz M, Fuentes E, Ávila F, Alarcón M, Palomo I. Roles of phenolic compounds in the reduction of risk factors of cardiovascular diseases. Molecules. 2019;24(2):E366. https://doi.org/10.3390/molecules24020366.
42. Reding K, Atkinson C, Westerlind K, et al. Fruit intake associated with urinary estrogen metabolites in healthy premenopausal women. Open J Prev Med. 2012;2(1):1. https://doi.org/10.4236/ojpm.2012.21001.
43. Reganold J, Wachter J. Organic agriculture in the twenty-first century. Nature Plants. 2016;2:15221. https://doi.org/10.1038/nplants.2015.221.
44. Forman J, Silverstein J. Organic foods: health and environmental advantages and disadvantages. Pediatrics. 2012;130(5):e1406–15. https://doi.org/10.1542/peds.2012-2579.
45. Hernandez A, Boada L, Mendoza Z, et al. Consumption of organic meat does not diminish the carcinogenic potential associated with the intake of persistent organic pollutants (POPs). Environ Sci Pollut Res Int. 2015;24(5):4261–73. https://doi.org/10.1007/s11356-015-4477-8.
46. Dominika S, Barabski M, Seal C, et al. Composition differences between organic and conventional meat: a systemic literature review and meta-analysis. Br J Nutr. 2016;115(6):994–1011. https://doi.org/10.1017/S0007114515005073.
47. Hurtado-Barroso S, Tresserra-Rimbau A, Vallverdú-Queralt A, Lamuela-Raventós RM. Organice food and the impact on human health. Crit Rev Food Sci Nutr. 2019;59(4):704–14. https://doi.org/10.1080/10408398.2017.1394815.
48. Otles S, Ozgoz S. Health effects of dietary fiber. Acta Sci Pol Technol Aliment. 2014;13(2):191–202.
49. Willis HJ, Slavin JL. The influence of diet interventions using whole, plan food on the gut microbiome: a narrative review. J Acad Nutr Diet. 2019; pii: S2212-2672(19)31372-3. https://doi.org/10.1016/j.jand.2019.09.017.
50. Cresci G, Glueck B, McMullen M, et al. Prophylactic tributyrin treatment mitigates chronic-binge alcohol induced intestinal barrier and liver injury. J Gastroenterol Hepatol. 2017;32:1587. https://doi.org/10.1111/jgh.13731.
51. Karakula H, Dzikowski M, Pelczarska A, Dzikowska I, Juchnowicz D. The brain-gut dysfunctions and hypersensitivity to food antigens in the etiopathogenesis of schizophrenia. Psychiatr Pol. 2016;50(4):747–60.
52. Sarnyai Z, Kraeuter AK, Palmer CM. Ketogenic diet for schizophrenia: clinical implication. Curr Opin Psychiatry. 2019;32(5):394–401. https://doi.org/10.1097/YCO.0000000000000535.

53. Brietzke E, Mansur RB, Subramaniapillai M, Balanzá-Martínez V, Vinberg M, González-Pinto A, Rosenblat JD, Ho R, McIntyre RS. Ketogenic diet as a metabolic therapy for mood disorders: Evidence and developments. Neurosci Biobehav Rev. 2018;94:11–6. https://doi.org/10.1016/j.neubiorev.2018.07.020.
54. Calder P. Functional roles of fatty acids and their effects on human health. JPEN J Parenter Enteral Nutr. 2015;39(1 Suppl):18S–32S. https://doi.org/10.1177/0148607115595980.
55. de Souza R, Mente A, Maroleanu A, et al. Intake of saturated and trans unsaturated fatty acids and risk of all cause mortality, cardiovascular disease, and type 2 diabetes: systematic review and meta-analysis of observational studies. Br Med J. 2015;351:h3978. https://doi.org/10.1136/bmj.h3978.
56. Ramsden C, Zamora D, Majchrzak-Hong S, et al. Reevaluation of the traditional diet-heart hypothesis: analysis of recovered data from Minnesota Coronary Experiment (1968–73). Br Med J. 2016;353:i1246. https://doi.org/10.1136/bmj.i1246.
57. Hooper L, Mann J. Observational studies are compatible with an association between saturated and trans fats and cardiovascular disease. Evid Based Med. 2016;21:37. https://doi.org/10.1136/ebmed-2015-110298.
58. Willett WC. Research review: old data on dietary fats in context with current recommendations. Harvard School of Public Health. https://www.hsph.harvard.edu/nutritionsource/2016/04/13/diet-heart-ramsden-mce-bmj-comments/.
59. Martínez-González M, Zazpe I, Razquin C, et al. Empirically derived food patterns and the risk of total mortality and cardiovascular events in the PREDIMED study. Clin Nutr. 2015;34:859–67. https://doi.org/10.1016/j.clnu.2014.09.006.
60. Pischon T, Hankinson SE, Hotamisligil GS, Rifai N, Willett WC, Rimm EB. Habitual dietary intake of n-3 and n-6 fatty acids in relation to inflammatory markers among US men and women. Circulation. 2003;108:155–60.
61. Dyall SC. Long–chain omega-3 fatty acids and the brain: a review of the independent and shared effects of EPA, DPA and DHA. Front Aging Neurosci. 2015;7:52. https://doi.org/10.3389/fnagi.2015.00052.
62. Lloyd-Jones D, Hong Y, Labarthe D, et al. Defining and setting national goals for cardiovascular health promotion and disease reduction: the American Heart Association's strategic impact goal through 2020 and beyond. Circulation. 2010;121:586–613. https://doi.org/10.1161/CIRCULATIONAHA.109.192703.
63. Agostoni C, Bresson J, Fairweather-Tait S, et al. Scientific opinion on dietary reference values for fats, including saturated fatty acids, polyunsaturated fatty acids, monounsaturated fatty acids, trans fatty acids, and cholesterol. EFSA Journal. 2010;8:1461–566.
64. Wang D, Li Y, Chiuve S, et al. Association of specific dietary fats with total and cause-specific mortality. J Am Med Assoc Int Med. 2016;176(8):1134–45. https://doi.org/10.1001/jamainternmed.2016.2417.
65. Hedaya R. Stress, spirituality, poverty, and community- effects on health. Textbook of functional medicine. Gig Harbor: Institute of Functional Medicine; 2010. p. 670.
66. Craig A, Lynn J. Hydration and health. Am J Lifestyle Med. 2011;5(4):304–15. https://doi.org/10.1177/1559827610392707.
67. Seyed A, Vali A, Mohammad M, Hamed B. Consumer perception and preference of drinking water sources. Electron Physician. 2016;8(11):3228–33. https://doi.org/10.19082/3228.
68. Ceretti E, Moretti M, Zerbini I, et al. Occurrence and control of genotoxins in drinking water: a monitoring proposal. J Public Health Res. 2016;5(3):769. https://doi.org/10.4081/jphr.2016.769.
69. Gaitan E, Wahner HW, Gorman CA, et al. Measurement of triiodothyronine in unextracted urine. J Lab Clin Med. 1975;86:538–46.. https://www.ncbi.nlm.nih.gov/pubmed/1151171
70. Gaitan E. Goitrogens. Bailliveres Clin Endocrinol Metab. 1988;2:683–702.
71. Pieper K, Tang M, Edwards M. Flint water crisis by interrupted corrosion control: investigating "ground zero" home. Environ Sci Technol. 2017;51:2007. https://doi.org/10.1021/acs.est.6b04034.
72. Use of lead free pipes, fittings, fixtures, solder and flux for drinking water. 2017. https://www.epa.gov/dwstandardsregulations/use-lead-free-pipes-fittings-fixtures-solder-and-flux-drinking-water. Accessed 2 Sept 2017.
73. Yu-Jie Z, Sha L, Ren-You G, Tong Z, Dong-Ping X, Hua-Bin L. Impacts of gut bacteria on human health and disease. Int J Mol Sci. 2015;16(4):7493–519. https://doi.org/10.3390/ijms16047493.
74. Bibbo S, Laniro G, Giorgio V, et al. The role of diet on gut microbiota composition. Eur Rev Med Pharmacol Sci. 2016;20(22):4742–9.
75. Linda V, Kaori S, Jia Z. Probiotics: a proactive approach to health. A symposium report. Br J Nutr. 2015;114(S1):S1–S15. https://doi.org/10.1017/S0007114515004043.
76. Bemmo L, Sahoo M, Jayabalan R, Zambou N. Honey, probiotics and prebiotics: review. Res J Pharm Biol Chem Sci. 2016;7(5):2428–38.
77. Kingston J, Radhika M, Roshini P, Raksha M, Murali H, Batra H. Molecular characterization of lactic acid bacteria recovered from natural fermentation of beetroot and carrot Kanji. Indian J Microbioc. 2010;50(3):292–8. https://doi.org/10.1007/s12088-010-0022-0.
78. Ryan K, Randy S. Food as a hormone. Science. 2013;339(6122):918–9. https://doi.org/10.1126/science.1234062.
79. Touitou Y, Reinberg A, Touitou D. Association between light at night, melatonin secretion, sleep deprivation, and the internal clock: health impacts and mechanism of circadian disruption. Life Sci. 2017;173:94–106. https://doi.org/10.1016/j.lfs.2017.02.008.
80. Walker MP. A societal sleep prescription. Neuron. 2019;103(4):559–62. https://doi.org/10.1016/j.neuron.2019.06.015.
81. Dethlefsen C, Pedersen K, Hojman P. Every exercise bout matters: linking systemic exercise responses to breast cancer control. Breast Cancer Res Treat. 2017;162(3):399–408. https://doi.org/10.1007/s10549-017-4129-4.
82. Sims-Gould J, Vazirian S, Li N, Remick R, Khan K. Jump step- a community based participatory approach to physical activity & mental wellness. BMC Psychiatry. 2017;17(1):319. https://doi.org/10.1186/s12888-017-1476-y.
83. Pretty J, Rogerson M, Barton J. Green Mind Theory: how brain-body-behaviour links to natural and social environments for healthy habits. Int J Environ Res Public Health. 2017;14(7):E706. https://doi.org/10.3390/ijerph14070706.
84. Monda V, Villano I, Messina A, et al. Exercise modifies the gut microbiota with positive health effects. Oxidative Med Cell Longev. 2017;2017:3831972. https://doi.org/10.1155/2017/3831972.
85. Kiobsted R, Woitaszweski J, Treebak J. Role of AMP- activated protein kinase for regulating post-exercise insulin sensitivity. EXS. 2016;107:81–126. https://doi.org/10.1007/978-3-319-43589-3_5.
86. Rosety M, Diaz A, Rosety J, et al. Exercise improved semen quality and reproductive hormone levels in sedentary obese adults. Nutr Hosp. 2017;34(3):603–7. https://doi.org/10.20960/nh.549.
87. Tharmaratnam T, Tabobondung T, Tabobondung T, Sivagurunathan S, Iskandar M. Exercise and oestrogens: aerobic high-intensity exercise promotes leg vascular and skeletal muscle mitochondrial adaptations in early postmenopause. J Physiol. 2017;595:6379–80. https://doi.org/10.1113/JP275063.
88. Simmonds MJ, Sabapathy S, Serre KR, et al. Regular walking improves plasma protein concentrations that promote blood hyperviscosity in women 65–74 yr with type 2 diabetes. Clin Hemorheol Microcirc. 2016;64(2):189–98.
89. Hyeon-Ki K, Karina A, Hiroko T, et al. Effects of different intensities of endurance exercise in morning and evening on lipid metabolism response. J Sports Sci Med. 2016;15(3):467–76.
90. Brown T, Fee E. Walter Bradford Cannon. Pioneer physiologist of human emotions. Am J Public Health. 2002;92(10):1594–5.

91. Khalsa D. Stress, meditation and Alzheimer's disease prevention: where the evidence stands. J Alzheimers Dis. 2015;48(1):1–12. https://doi.org/10.3233/JAD-142766.
92. Lampert R, Keri T, Hong K, Donovan T, Lee F, Sinha R. Cumulative stress and autonomic dysregulation in a community sample. Stress. 2016;19(3):269–79. https://doi.org/10.1080/10253890.2016.1174847.
93. Kakumanu R, Nair A, Venugopal R, Sasidharan A, Ghosh P, John J, Mehrotra S, Panth R, Kutty B. Dissociating meditation proficiency and experience dependent EEG changes during traditional Vipassana meditation practice. Biol Psychol. 2018;135:65–75. https://doi.org/10.1016/j.biopsycho.2018.03.004.
94. Balasubramanian S, Mintzer J, Wahlquist A. Induction of salivary nerve growth factor by yogic breathing: a randomized controlled trial. Int Psychogeriatr. 2015;27(1):168–70. https://doi.org/10.1017/S1041610214001616.
95. Minich D, Bland J. Personalized lifestyle medicine: relevance for nutrition and lifestyle recommendations. Sci World J. 2013;2013:129841. https://doi.org/10.1155/2013/129841.

Therapeutic Diets

Tracey Long and Leigh Wagner

42.1 Introduction – 744

42.2 Defining "Therapeutic Diets" – 744

42.3 Traditional/Historical Perspective on Therapeutic Diets – 744

42.4 Conventional Approaches to Therapeutic Diets (and the Nutrition Care Process) – 744

42.5 Integrative and Functional Approaches to Therapeutic Diets (and the Nutrition Care Process) – 744

42.6 IFMNT Assessment Tools – 744

42.7 Commonly Used IFMNT Therapeutic Diets – 745
42.7.1 Elimination Diets – 745
42.7.2 Low-Histamine Diet – 746

42.8 Where to Find Histamine – 747

42.9 How Can We Support Histamine Overload? – 748
42.9.1 Low-Carbohydrate High-Fat Diet – 748

42.10 How Do Diabetic Ketoacidosis and Nutritional Ketosis Differ? – 748

42.11 What Is LCHF? – 749

42.12 What Are the Benefits of LCHF? – 749
42.12.1 Carbohydrate Restriction in Cancer Therapy – 749

42.13 Who Should Avoid LCHF? – 749

42.14 Intermittent Fasting – 750

42.15 Benefits of Intermittent Fasting – 750

42.16 Concerns and Special Considerations for Intermittent Fasting – 751

42.17 Different Methods of Intermittent Fasting – 751

42.18 Summary and Conclusion – 752

References – 752

© Springer Nature Switzerland AG 2020
D. Noland et al. (eds.), *Integrative and Functional Medical Nutrition Therapy*,
https://doi.org/10.1007/978-3-030-30730-1_42

42.1 Introduction

We so often hear the saying by Hippocrates, "Let food be thy medicine and medicine be thy food," but do we really pause to consider the meaning and relevance of his words? "Food as medicine" is using the functional characteristics of foods that allow the body to heal and function optimally. Medicine as our food is the idea that how we eat can be used on a daily basis, as a lifestyle or preventive medicine-type approach that will keep us well or prevent disease. This chapter will summarize and examine therapeutic diets from an integrative and functional perspective. We will describe dietary approaches used therapeutically by practitioners, the evidence for their use, the application, and any caveats or cautions for their use.

42.2 Defining "Therapeutic Diets"

The term "therapeutic diets" is defined by the Academy of Nutrition and Dietetics as "a diet intervention ordered by a healthcare practitioner as part of the treatment for a disease or clinical condition manifesting an altered nutritional status, to eliminate, decrease, or increase certain substances in the diet [1]." We will use this working definition of therapeutic diets within the context of nutrition approaches for integrative and functional medical nutrition therapy (IFMNT).

42.3 Traditional/Historical Perspective on Therapeutic Diets

In early civilizations, particularly of India and China, ancient traditions included dietary approaches to healing. These were some of the earliest documented accounts of using food as medicine. India's ancient healing paradigm, Ayurveda, is one of the oldest medical systems in the world. Its name is derived from Sanskrit words "life science" or "life knowledge." Ayurveda takes the approach to health with the belief that all aspects of an individual's lives are interconnected (humans, health, and the universe) [2]. Ayurveda also uses individual constitutions or "doshas" that characterize an individual's tendencies to disease and specific remedies, especially dietary recommendations.

Similarly, and slightly more recently, traditional Chinese medicine (TCM) is an ancient medical system that also uses therapeutic diets as part of its foundation to health. TCM is similar to Ayurveda in viewing human health as connected to the larger environment and universe as a whole [3].

42.4 Conventional Approaches to Therapeutic Diets (and the Nutrition Care Process)

Conventional nutritional approaches to therapeutic diets have included several categories of diet modifications, including nutrient modifications (i.e., macronutrients, micronutrients, fiber), texture or consistency modifications (mechanical soft, puree, etc.), food allergies and intolerance modifications, and additional feedings (snacks and oral supplements). Each of these is recommended based on assessment of the patient and recommendation by the doctor, another healthcare provider, and/or a registered dietitian.

42.5 Integrative and Functional Approaches to Therapeutic Diets (and the Nutrition Care Process)

IFMNT is a specific approach to medical nutrition therapy (MNT) that expands on the assessment portion of the nutrition care process (NCP). The NCP involves assessment, diagnosis (nutritional diagnosis), intervention, monitoring, and evaluation. An expanded assessment requires specific tools (including time) as a means to investigate core nutritional imbalances in patients. The IFMNT practitioner has the benefit of additional time with the patient, often spending 60–90 minutes for an initial consultation and close (often frequent) monitoring of interventions.

42.6 IFMNT Assessment Tools

Prior to an initial appointment, a patient completes paperwork that includes prompts about their goals, lifestyle barriers and challenges, health history, family history, social history, nutrition and diet histories, environmental exposures history, physical activity, dental history, symptoms, dietary supplementation and medication histories, and other backgrounds. During the initial consultation (often 60–90 minutes), the IFMNT practitioner will ask about any questions identified from the patient's initial intake forms. Through the interview, the IFMNT practitioner collects the patient's story (see ▶ Chap. 38 for more on the patient's story) and fills in any gaps left in the paperwork.

During the initial appointment, the IFMNT practitioner assesses more than just the patient's nutritional status and metabolism; she also assesses their readiness to change, self-efficacy, and level of feasibility to make necessary changes. The IFMNT practitioner will be able to help troubleshoot any of the patient's challenges to ensure that if the patient would like to make changes, they will be able to access the support they need to make them. Sometimes this involves family, friends, and their community to serve as caregivers or support systems (▶ Box. 42.1).

> **Box 42.1 IFMNT Tools**
> – Time
> – The client's story and health timeline
> – IFM Matrix
> – IDU assessment tool
> – Intake assessment form
> – Health symptoms questionnaire
> – Fats and oils questionnaire
> – Nutrition physical exam
> – Body composition measurement
> – Laboratory testing (conventional and functional)
> – Motivational interviewing and stages of change model

42.7 Commonly Used IFMNT Therapeutic Diets

There are countless "diets" widely promoted on the Internet and social media that may or may not be based on evidence and many that are conventionally accepted and used: carbohydrate counting for diabetic patients (often matching carbohydrate intake to insulin dosage) [4]; a "cardiac diet," which includes lowering total and saturated fat intake [5], sodium levels, and egg yolks; calorie counting for weight loss; and a general Mediterranean diet for heart health. These are generally well accepted in the conventional medical world but in a broad sense lack the nuance of each individual patient. The IFMNT approach to therapeutic diets is based on the foundational concept of evaluating the patient's nutrition status and implementing an intervention that is personalized to that individual.

42.7.1 Elimination Diets

Elimination diets (also called "exclusion diets") [6] are both a diagnostic tool to identify food adverse food reactions and a broad term for a systematic dietary approach to mitigate symptoms of adverse food reactions or to alter disease progression. At the extreme end, celiac disease and food allergies are conditions that necessitate the avoidance of certain foods to protect the patient from life-threatening illness. Less severe, but also on the spectrum of adverse food reactions, are food sensitivities and intolerances, which also warrant the elimination of certain foods (e.g., lactose or fructose intolerance or non-celiac gluten sensitivity); these have a less severe but still detrimental impact on one's health and often quality of life. Another category of conditions that may warrant an elimination diet are autoimmune diseases like rheumatoid arthritis [7] or Crohn's disease [8].

Although there has been skepticism of sensitivities to foods like wheat and gluten outside of overt celiac disease, evidence is emerging that there can be measurable physical changes in immune activation and cellular damage in people with wheat sensitivity without celiac disease [9]. Alessio Fasano [10] has contributed significantly to the evidence around characterization of non-celiac gluten sensitivity.

A PubMed search for "elimination diet" shows that most recent publications were specifically related to therapeutic use for eosinophilic esophagitis (EoE) [11–31]. When searching for "exclusion diet," one may find that most publications refer to inflammatory bowel diseases [6, 32–38]. Terminology of elimination diets seems to vary by diagnosis or area of research. Other conditions studied for the therapeutic use of elimination diets include irritable bowel syndrome (IBS) [39, 40], autism spectrum disorder [41–43], migraine headache [44], and others.

The elimination diet is often used by individuals who suspect they have food intolerances. Food intolerances are adverse food reactions that are not overt food allergies (mediated by immunoglobulin E (IgE)). Food intolerances involve an adverse reaction to food due to a lack of adequate amounts of an enzyme to digest the food (e.g., lactose intolerance is a deficiency of the enzyme lactase) [45]. Food sensitivities are a non-IgE, immune-mediated reaction to a food when ingested. These are often less severe reactions like adverse digestive symptoms (stomach pain, bloating, nausea, breathing problems, eczema, brain fog, etc.). Non-celiac gluten sensitivity is a condition characterized by adverse symptoms (fatigue, digestive problems, brain fog, etc.) caused by ingestion of gluten without having classic characteristics of celiac disease [10]. Although the mechanism is not fully understood, some hypothesize that the adverse reaction may be due to histamine intolerance [46]. Further, evidence suggests that the pathophysiology of celiac disease is related to a response by the adaptive immune system, while non-celiac gluten sensitivity is a reaction by the innate immune system [47]. This phenomenon has been measured via fecal assays suggesting an immune reaction to gluten/dairy [48].

People with IBS often try to eliminate certain foods (e.g., gluten and dairy) in an attempt to alleviate symptoms [40]. A low FODMAP diet is another version of an elimination diet with compelling evidence that it contributes to amelioration of IBS symptoms [49].

42.7.1.1 Implementation of an Elimination Diet

Elimination diets are often followed for 4–12 weeks (the time period foods are withdrawn) with additional time to systematically reintroduce foods back into the diet. To begin, a person avoids one or several foods for the specified time period. During that time, the individual keeps a diary of their dietary intake and symptoms (see ◘ Fig 42.1) to ensure that suspected foods are completely removed and there aren't any hidden sources of the food(s) remaining in the diet. Symptoms are recorded, along with the time and food eaten, to note any patterns in food ingestion with elicited symptoms. At the end of the period of elimination, each of the eliminated foods is reintroduced ("challenged") back into the diet, one-by-one, to determine which (if any) foods elicit symptoms. If the person has an adverse response to the food, they are to eliminate it again, wait for the symptom(s) to subside, and try another new food to assess tolerance. After all foods are reintroduced or "challenged" back into the diet, then the person can wait another period of time before trying to reintroduce any remaining foods again.

42.7.1.2 Types of Elimination Diets

There are many dietary approaches that aren't necessarily referred to as "elimination diets" but may fall within the general umbrella of the term: foods are eliminated with the expectation that symptoms or signs of disease or illness may improve. Examples of elimination diets include (1) the Paleolithic or ancestral diet (Paleo diet) [50], (2) a low FODMAP diet (fermentable oligo-di-monosaccharides

Fig 42.1 Diet and symptom diary

Date/time	Location/activity	Food or beverage consumed	Amount (cup, etc)	Mood and symptoms

Reviewed by _____ Date/Time _____

Table 42.1 Potential vulnerabilities

Eliminated food	Macronutrient vulnerabilities	Micronutrient and phytonutrient vulnerabilities
Gluten [54]	Fiber, carbohydrate	Iron, folate, calcium, selenium, magnesium, zinc, niacin, thiamin, riboflavin, vitamins A and D
Dairy	Protein, fat	Calcium, potassium, phosphorus, vitamin A, vitamin D, vitamin B12, riboflavin, niacin
Soy [55]	Fiber, protein	Calcium, B-vitamins, iron, zinc
Egg [56]	Protein, fat/cholesterol	Choline, retinol (vitamin A), vitamin E, thiamin (B1), riboflavin (B2), niacin (B3), pantothenic acid (B5), pyridoxine (B6), folate (B9)
Nuts [57]	Mono- and polyunsaturated fatty acids, protein, fiber	Vitamins E and K, folate, magnesium, copper, potassium, selenium, carotenoids, antioxidants, phytosterols
Grains [58]	Fiber, carbohydrate	Folate, thiamin, iron, niacin, riboflavin, vitamin B6, zinc, sodium
Paleo (Free of grains, dairy, legumes) [50]	Fiber, carbohydrate	Folate, thiamin, iron, niacin, riboflavin, vitamin B6, vitamin B12, zinc, sodium, calcium, potassium, phosphorus, vitamins A and D
Low FODMAP [51]	Fiber, fat	Iodine, selenium
Specific carbohydrate diet [32] (study in pediatric population)	Fiber (due to rationale for the diet)	Vitamin D, calcium [32] (limited data and study in pediatric population)

and polyols): eliminating all categories of highly fermentable carbohydrates and reintroducing them back into the diet one FODMAP category at a time) [51], and (3) the specific carbohydrate diet [32]. Other approaches involve individualized elimination diets based on food sensitivity testing (IgG blood testing) to determine which foods to eliminate [52]. Even a low-histamine diet and a ketogenic diet (high fat, low carbohydrate) could be considered elimination diets, and they will be covered later in this chapter. Although there have been clinical trials on the implementation of elimination diets [53], there is still a lot to be learned about elimination diets and their indication and safe implementation.

42.7.1.3 Cautions for Elimination Diets

A major concern for the implementation of an elimination diet is the risk for nutritional deficiencies or insufficiencies, especially if followed for a long period of time. Table 42.1 summarizes the potential macro- and micronutrients that may be insufficient in someone following an elimination diet.

42.7.2 Low-Histamine Diet

A low-histamine diet is one of several specialized food patterns to assist patients with food sensitivities, food intolerances, and/or food allergies. Focusing on this specific diet is warranted, as this is a therapeutic diet that is seemingly more and more common. Many integrative and functional nutrition practitioners are noticing an increase in patients needing support for histamine intolerance.

Histamine belongs to a family of biogenic amines that are classified into monoamines and polyamines. Monoamines are derived from one ammonia molecule, NH3, where one of the hydrogen molecules is dropped and replaced with another chemical structure. Polyamines have more than one NH3 starter group. Five commonly known monoamines are histamine, serotonin, dopamine, norepinephrine, and epinephrine [59]. Another trace monoamine is tyramine, which can be yet another cause of biogenic amine-related food intolerance. To understand histamine, it is helpful to understand genetic and biochemical possibilities. Several boxed genetic notes are provided for more in-depth understanding.

> Genetics Note 1: the enzyme histidine decarboxylase (HDC) is made from the HDC gene. Its purpose is to convert the amino acid histidine into histamine. In some individuals, genetic variants may cause an increase or decrease in the enzyme function possibly resulting in too much or too little histamine. Active Vit B6 or P-5-P is a required cofactor of HDC [60]. Some fish are high in microbes that are capable of producing HDC that, when consumed by some individuals who do not clear histamine well, can contribute to histamine overload [61].

Histamine is most recognized for its role in symptoms associated with classic allergy in IgE-mediated immune system activation: hives, tissue swelling, nasal congestion, asthma, headaches, oral allergy symptoms, and gastrointestinal complaints. However, histamine has many beneficial roles in human function including neurotransmission, gastric acid secretion, inflammation and immune system support, and smooth muscle tone [60].

Those with histamine intolerance may be affected by a variety of sudden onset and seemingly unexplainable symptoms such as flushing, hives, rapid heartbeat, profuse sweating, nosebleeds, car sickness, migraines, itchiness, and more. For a comprehensive lists of symptoms of histamine overload, see references listed below, especially the books by Jarisch and Lynch.

42.8 Where to Find Histamine

Histamine intolerance may actually be an imbalance between histamine production and accumulation and the ability to degrade histamine. The body is capable of making histamine and storing it in specialized immune cells called mast cells. Mast cells are sentinel cells found primarily in mucosal tissue that help initiate an inflammatory response when a threat is detected. Releasing histamine is important for immune function, unless the mast cells are overactive. This is a phenomenon known as mast cell activation disorder (MCAD) [62]. Because mast cells live in the mucosal lining of the gut, if the gut lining is inflamed for any reason, mast cells may become activated and/or have the opportunity to migrate from the mucosa system-wide due to leaky mucosal tight junctions [63].

Histamine load may also come from food, especially food subject to spoiling or degradation as microbes degrade the amino acid histidine in food with the HDC enzyme and make histamine. Foods rich in protein are culprits, but so are foods and beverages allowed to ferment, such as wine, sauerkraut, and kombucha. Foods high in protein will be degraded by microbes after cooking, raising histamine content. While refrigeration will slow histamine production, only freezing will stop histamine content from increasing [64].

Allergies to food and the environment are the more obvious sources of histamine production. IgE-type allergy testing may be warranted to reduce load [64]. Elimination of antigens and/or treatment for allergic-type hyper-response may be needed.

The stress response may increase histamine production [65]. Assisting patients with stress management is key in regulating histamine load.

Estrogens (both naturally occurring and estrogen-mimicking "xenoestrogens") also contribute to the histamine load [66]. Assessing patients for estrogen dominance and genetic risk for estrogen dominance may be helpful. Lifestyle and dietary interventions to reduce estrogen involve avoiding exposure to plastics or heating food in plastics. Increasing cruciferous food intake is also supportive to decrease estrogen.

Emerging evidence suggests that specific species of gastrointestinal microbes produce histamine. Examples are *Lactobacillus casei*, *Lactobacillus delbrueckii*, and *Lactobacillus bulgaricus*. Testing for small intestine bacterial overgrowth may be a diagnostic tool in determining if dysbiosis may be an underlying contributor. Certain microbes are known to assist with histamine degradation such as spore-based probiotics, *Lactobacillus plantarum*, *Lactobacillus rhamnosus*, and *Bifidobacterium longum* [63].

> Genetics Note 2: the enzyme catechol-O-methyltransferase (COMT) made by the gene of same name is important for estrogen degradation. Individuals who are homozygous positive for variants may have estrogen dominance and benefit from estrogen clearing support. SAMe (methylation ability), magnesium, and vitamin C are cofactors for COMT [67].

Based on the above-described reasons that histamine may accumulate and individual sensitivity to histamine, it is best to describe histamine intolerance as "histamine overload." When the body is unable to handle the load, histamine can build up and eventually "spill over" the tolerable limit for the individual and create a histamine response.

42.9 How Can We Support Histamine Overload?

1. Decrease the dietary load. Limit very high histamine foods: red wine, champagne, aged cheeses, cured meat and fish, bone broth and fish stock, vinegar and fermented foods such as sauerkraut, and chocolate. A more comprehensive list of high and moderately high histamine foods is available in the references. Patients may benefit from suggested meals and meal plans based on lower histamine foods.
2. Avoid fish high in histamine unless very fresh or fresh and flash frozen.
3. Avoid or limit eating leftovers. Freeze leftovers for future reheating and consumption.
4. Assess genes that may influence the ability to degrade histamine. See boxed notes for DAO and HNMT genetics.
5. Support with probiotics (mentioned above) that are known to degrade histamine.
6. Identify possible allergens and eliminate and/or refer for treatment.
7. Assess for gut inflammation and/or SIBO and support as needed
8. Support with nutrients and/or supplements known to support degradation of histamine such as selenium, quercetin, vitamin C, stinging nettle, EGCG (primary catechin in green tea), and the DAO enzyme [68, 69].
9. Collaborate with other healthcare providers who may be able to offer support with prescription antihistamines. Be aware that many antihistamines decrease stomach acid secretion that may result in low vitamin B12. Consider assessing for B12 deficiency.
10. Assist with stress management techniques.

> Genetics Note 3: the enzyme diamine oxidase (DAO) is produced by the cells that line the gastrointestinal tract. DAO is a key enzyme that degrades dietary histamine in the extracellular space (especially the gastrointestinal tract). Individuals with significant genetic variants might be impaired at making adequate DAO. Even without genetic variants, a damaged gut lining may impair DAO production. DAO is available as a supplement that is recommended for use as 15 minutes prior to consuming high histamine foods. Vitamin B6 is a cofactor for DAO [68, 69].
>
> Genetics Note 4: the enzyme histamine N-methyltransferase (HNMT) degrades histamine inside the cell. It is primarily concentrated in the liver. Individuals with significant genetic variations may benefit from additional cofactor support to help speed up HNMT which is SAMe [68, 69].
>
> Genetics Note 5: the enzyme monoamine oxidase B (MAOB) provides an important step in the histamine degradation pathway. HNMT degrades histamine to N-methylhistamine, which then is further degraded by MAOB so the body can finally get rid of the histamine. If MAOB is not working well due to lack of cofactor and/or presence of genetic variants through feedback inhibition, HNMT slows down making histamine buildup. The easy fix is a trial of the cofactor vitamin B2 (riboflavin) as about 400 mg taken 2–5 times a week [68, 69].

42.9.1 Low-Carbohydrate High-Fat Diet

"There has been a dramatic resurgence of interest in the ketogenic diet during the past several years," as stated in a 1997 review article by Swink et al. [70]. This original study emphasized the utility of the KD in treating children with epilepsy starting in the 1920s, which then fell out of favor in the 1970s due to the invention of antiepileptic medications. The 1997 paper mentioned the return of interest in the KD in severe cases of pediatric epilepsy when medications did not work. The same research team published a follow-up review in 2007, *The ketogenic diet: one decade later*, where they acknowledged that this diet, with evidence of use as far back as 500 B.C., was maintaining its momentum after 10 years [71]. This food trend continues to thrive as numerous research teams are diving into its many health applications [70–78].

This summary of the KD will be referred to from this point on as the low-carb high-fat or low-carb healthy fat (LCHF) food pattern. Many health professionals are making the transition to using the term LCHF due to the negative and often misunderstood confusion between nutritional ketosis and diabetic ketoacidosis.

42.10 How Do Diabetic Ketoacidosis and Nutritional Ketosis Differ?

Diabetic ketoacidosis is an acute life-threatening condition usually only seen in type 1 diabetes and rarely in type 2 diabetes [79]. In diabetic ketoacidosis, there is too little insulin to control a sharp rise in glucose and ketones, a result of an acute (usually within 2–4 hours) onset of a catabolic state [80]. Other biomarkers include a pH less than 7.0, a bicarbonate less than 10 mEq/L, and an anion gap >15 mEq/L. Blood ketones are measured as beta-hydroxybutyrate with serum levels >8 mmol/L [80] or 15–25 mmol/L [72]. Notice that researchers do not agree on a standard level for beta-hydroxybutyrate, but the combination of the above markers is what is significant. This combined "perfect storm" can result in a critical threat warranting immediate medical treatment.

Nutritional ketosis is a mild form of ketosis where blood glucose is relatively low, blood pH remains within normal reference ranges, and blood ketone values usually are between 0.5 and 5 mmol/L [74]. Medically therapeutic ketosis for treating epilepsy ranges from 2.0 to 7.0 [81].

■ Table 42.2 Macronutrient percentages

Macronutrients as percent of total kcals	Standard American Diet [82]	Paleo diet [83]	Mild LCHF diet [72, 73]	Moderate LCHF diet [72, 73]	Medically therapeutic LCHF (epilepsy and Alzheimer's disease [70, 73]
Carbohydrate	50	22–40	20	10–15	7–8
Protein	15	19–35	20	15–20	12–13
Fat	35	28–47	65	70	79–80

42.11 What Is LCHF?

A LCHF food pattern is high in (healthy) fat, adequate or moderate in protein, and low in carbohydrates. Healthy fats are considered, in the context, to be fats derived from whole food sources (grass-fed meat and poultry, avocados, olives, cold-pressed oils, butter, etc.) versus fats derived from chemical and heat processing (seed and soy-based oils). This pattern is able to assist the human body in making ketone bodies for energy. The formation of ketones for energy occurs with fasting, prolonged exercise, and very low carb intake. Comparing the macronutrient percentages of LCHF variations with the standard American intake may be helpful as listed in ■ Table 42.2. The Paleo diet was included for additional reference as this food pattern, which emphasizes a whole food approach to food intake, is often used as a reference template for LCHF.

42.12 What Are the Benefits of LCHF?

Scientists are rapidly studying many possible benefits of a LCHF food pattern. Research covers a wide range of hypotheses from effects on exercise performance to metabolic derangement to chronic disease. Here is a summary of some of the current findings.

42.12.1 Carbohydrate Restriction in Cancer Therapy

First, note that not all cancers are related to diet, such as those associated with virus, environmental exposures, age, and genetic mutations [74]. However, when these types of cancer are diagnosed and treated, diet can be an important factor in outcomes.

Some common cancers have been linked to dietary influences. These cancers include breast, colon, some lung, prostate, and gallbladder/biliary cancers and endometrial adenocarcinoma. These cancers are linked to hyperinsulinemia based on data from people with diabetes mellitus (DM) [84–86]. A therapeutic diet could be one that promotes lower insulin and may be helpful in cancer prevention. Dr. Dawn Lemanne is recommending a moderate carbohydrate restriction of about 100 grams of net carbohydrates per day in breast cancer and in stage III colon cancer in those with BMI >25 [74, 75]. The breast cancer recommendations come from the large WHEL (Women's Healthy Eating and Lifestyle) and WINS (Women's Interventional Nutrition Study) trials [87, 88]. A LCHF food pattern may also increase lifespan of those with glioblastoma and metastatic glioblastoma [77, 78].

In addition to insulin, other biomarkers may be considered. An observational study of 269,391 participants in Korea over 2 years between 2002 and 2005 found that all-cause mortality, including cancer, was lower in those with higher blood lipids [89]. Certainly clinical trials are needed to verify this, but studies such as this support possible benefits to higher lipids which can occur in some people following LCHF.

The LCHF food pattern may support starving cancer cells of glucose. The energy demand of cancer cells dependent on glucose for energy is about twenty times greater than normal cells. This altered energy utilization and increased demand is known as the Warburg effect [76].

When should LCHF be avoided in cancer? Genetic mutations such as the BRAF V600E mutation result in dietary fat-fueled tumor growth [90]. Note that some cancers are learning to use other sources of fuel such as protein, amino acids, and fats instead of glucose [91–93].

- **Rationale for benefits of a LCHF dietary pattern**
1. Alzheimer's disease: LCHF resulting in mild to metabolically therapeutic level of ketone production is recommended by The Bredesen Protocol to End Alzheimer's for three reasons: (1) to provide an alternative energy source to brain cells, (2) to decrease neural inflammation, and (3) to increase brain-derived neurotrophic factor [73].
2. Cardiometabolic disease: LCHF may support reversing chronic diseases such as type 2 DM, blood lipid dysregulation, hypertension, obesity and overweight, and chronic inflammation and decrease the need for insulin in type 1 DM [72, 94].
3. LCHF has been shown to enhance athletic performance, especially in endurance athletes [95].
4. LCHF may simplify food intake as many find this food pattern more satiating with a reduction in cravings for carbohydrates and the need to eat frequently [72, 94].

42.13 Who Should Avoid LCHF?

Many people are finding benefit from increasing healthy fats while lowering carbohydrates. As more adopt this food pattern and research data accumulates, we may find additional reasons to avoid LCHF, but at this time the only recommended avoidance is for starting this food pattern during pregnancy [72, 94]. The most important consideration when adopting LCHF is a slow acclimation and progression to LCHF to assist the body to adapt to using fat for fuel over carbohydrates to avoid possible flu-like symptoms and muscle cramps [96].

42.14 Intermittent Fasting

Intermittent fasting (IF) is the conscious choice to abstain from food for health-promoting reasons such as spiritual, cleansing, or detoxing or as a method to ameliorate a disease. IF is also referred to as time-restricted feeding or periodic fasting. This section will refer to IF but could mean any of the three above titles. Intermittent fasting is not starvation. Starvation is a state of forcefully being deprived of food such as lack of availability (war or famine) or withholding food. Note: anorexia nervosa is considered a form of purposeful food refusal involving complex emotional, social, mental, and nutrition-related factors. Appropriately, this discussion does not include anorexia nervosa.

Fasting has no set duration. It is a cycle of consuming food followed by an abstinence period from food. The concept of IF represents cyclical and pre-planned periods of fasting followed by appropriate food intake. This could be simply avoiding food intake between meals, i.e., avoiding snacking between a standard breakfast and lunch. Other examples of fasting cycles are overnight, alternate-day, or extended-day fasts. The modified, shorter versions of fasting are showing promise with many health benefits and are more appealing to the general population over extended fasts. IF can be eucaloric or hypocaloric and still result in positive outcomes. The variability in fasting regimens allows the healthcare professional to individualize the needs of patients. Many patients benefit from individualized guidance on how to implement a fasting protocol based on health goals and reason(s) for fasting. Understanding the basics of fasting will provide guidance for patient support.

42.15 Benefits of Intermittent Fasting

A 2017 review by Patterson et al. [97] found that IF supports weight loss, improves metabolic health markers, and may influence other aspects of health, such as improved sleep circadian rhythm and microbiome biology. Some of the metabolic health markers include lipids, such as lower total cholesterol, lower LDL cholesterol, improved HDL cholesterol, lower triglycerides, and improved blood glucose and insulin. A more recent review found that most evidence related to IF and weight loss and improved lipid profiles are observational and suggested the need for more clinical studies to confirm these initial findings [98].

Regarding weight loss, IF is supportive as a manageable method for caloric restriction, promoting adipose thermogenesis, and altering the gut microbiome to increase metabolism via influences on adipose tissue [99–101]. Stockman et al., in a 2018 review, summarized findings in animal models showing that IF may reduce oxidative stress, improve cognition, and slow down the aging process [102]. This same review found that while the human clinical trials are small (many with fewer than 300 participants), they are showing evidence of sustainable weight loss and improved insulin sensitivity.

IF between meals and overnight is a newer dietary therapy for digestive problems such as inflammatory bowel disease (IBD), irritable bowel syndrome (IBS), and small intestine bacterial overgrowth (SIBO). The author (TL) uses IF for 12 hours overnight and 4–5 hours between meals for bowel rest with good clinical and anecdotal success for these gastrointestinal (GI) conditions. This concept is based on the work of Mark Pimentel, MD, the director of the GI Motility Program and Laboratory at Cedars-Sinai. He has authored many papers, and his book, *A New IBS Solution*, describes the migrating motor complex (MMC) function in the gastrointestinal tract [103]. The MMC is a wavelike pulse that has been shown to sweep microbes and food debris out of the small intestine helping to prevent SIBO. He describes the MMC as pulsing naturally about every 90 minutes when food is not present. He has found that when the MMC runs 10–12 times every 24 hours, IBS/SIBO treatment is optimized. This can only occur during intermittent fasting states instead of the conventional ideal of eating three meals and two to three snacks each day. Additionally, bowel rest has traditionally been used for IBD conditions for the past 40 years as a method to decrease inflammation. IF is a method to provide rest intermittently with nutritious food intake [104, 105].

Prominent clinicians are using varying methods of IF to support treatment of some chronic diseases. One example is Dr. Jason Fung, MD, Canadian nephrologist and head of the Intensive Dietary Management program to support weight loss and type 2 diabetes reversal. He teaches both low-carbohydrate diets and IF to patients. Dr. Fung has written two books on his methods noted in the references. He is collecting data in his clinic and shares many clinical and anecdotal stories about his success using his protocols to reverse diabetes, a disease that conventional wisdom has touted as being progressive, insidious, and irreversible [94, 106]. Because Dr. Fung's IF methods are individualized to each patient, such as 12-hour overnight fasting, between-meal fasting, and/or alternate-day or extended-day fasting, refer to his books for further examples and application. He shares clinical assessments to ensure safe implementation of IF.

Dale Bredesen, MD, neurologist and founder of MPI Cognition, studies Alzheimer's and other neurodegenerative diseases. A key dietary aspect of his protocol is IF. Dr. Bredesen recommends a 12-hour overnight fast on a daily basis, but those with increased genetic risk are instructed to extend their fast more. Those with the ApoE 3/4 genetics for Alzheimer's risk use a minimum 14-hour overnight fast, and those with the ApoE 4/4 genetic variants for Alzheimer's risk are instructed to use a minimum overnight fast of 16 hours. The goal is to produce a state of mild nutritional ketosis to help fuel the brain, optimize brain mitochondrial function, and reduce neural inflammation [73].

Also noteworthy is the research by Valter Longo, PhD, Professor in Gerontology and Professor in Biological Science, University of Southern California. His clinical animal trials have yielded the following findings [107, 108]:

- IF is chemo protective because it protects normal cells from treatment side effects. When starved, normal cells slow division, which is protective during cancer treatment. However, when cancer cells are starved, they continue dividing because their growth switch is broken in the "on" position making them more vulnerable to treatment.
- IF sensitizes tumor cells to chemotherapy treatment.
- IF slows tumor growth even without chemotherapy.

In 2010, a group of ten volunteer patients requested that Dr. Longo use them in a human trial with a model used in his rodent trials. At the time, he refused, citing many concerns that included unlikely IRB approval. The group of patients said they would attempt his animal model fasting techniques on their own, and he was invited to take notes. The results of this volunteer study were published in 2012 with the following findings [109, 110]:

- Fasting was well tolerated with some mild lightheadedness and weakness (temporary).
- Fasting reduced fatigue.
- Fasting reduced overall weakness.
- Fasting resulted in fewer gastrointestinal side effects.
- No adverse effects on tumor volume or serum tumor markers were identified.
- The proposed mechanism is that fasting enhances leptin sensitivity and lowers insulin (a growth hormone).

Note: this study included ten volunteers with various malignancies. They fasted 48–140 hours prior to chemotherapy and for 5–56 hours following.

42.16 Concerns and Special Considerations for Intermittent Fasting

Concerns and considerations for IF vary depending on the model. These are summarized below based on Dr. Jason Fung's book and on the clinical experience and application of IF by the author (TL) [106]. A literature review did not reveal any specific scientific studies on possible harm with IF, although basic understanding of physiology, psychology, and biochemistry provides background to address concerns and considerations.

- Making drastic, significant changes: Many patients get enthusiastic about fasting, which can lead to discouragement and failure. Proper assessment for readiness and appropriate stage is important. Starting small, like fasting overnight for 12 hours and/or helping patients restructure meals to feel satiated (often with higher healthy fat and higher fiber options), will help with a between-meal fast.
- Electrolyte imbalances: Some patients may experience symptoms such as dizziness, headaches, and muscle cramps. Patients should be carefully monitored and provided with electrolyte replacement if warranted, along with plenty of water during any fast [106]. Be aware for extended fast, usually beyond 3 days, risk of refeeding syndrome is possible [111].
- Coach patients about managing feelings of hunger. Hunger tends to come in waves so strategies to assist through these sporadic periods are important. Some strategies include drinking a glass of water or a cup of green tea, taking a walk, or engaging in an activity that requires concentration; staying busy during the fasting period can be helpful.
- Many overeat when breaking a fast, especially longer fasts. Patients need to be guided in breaking a fast, such as planning a small snack to start with or pretend he/she never fasted and eat a normal meal.
- Advise patients that they should not make an announcement about their fast. Many people will think fasting is extreme and may not be supportive. Suggest that patients only share this with a few supportive people.
- Heartburn can be a problem when breaking a fast. This can be avoided by consuming smaller meals after fasting.
- Some patients may need to take medication during a fast. Dr. Fung suggests that while fasting patients can eat a few pieces of lettuce before swallowing medications [106].
- Patients on blood sugar-lowering medication need to carefully monitor blood glucose levels and seek assistance on altering medication doses during fasting, if needed.
- Patients with diagnosed eating disorders should not fast. Each situation needs to be assessed. For example, a mother of teenage girls may want to avoid fasting due to risk of influencing possible disordered eating in her children.

42.17 Different Methods of Intermittent Fasting

The studies of IF have been difficult to review due to varying models. ◘ Table 42.3 below summarizes the most commonly used fasting models. Some fasts encourage water-only fasting with electrolytes, as needed. Other models, especially when combined with a low-carb high-fat protocol, allow for consumption of fat such as MCT oil or cream and/or bone broth during the fast. Beverages such as coffee or tea with added cinnamon and/or added fat are acceptable according to some providers [106]. Another type of fast called the 5:2 fast suggests continuing normal dietary intake for 5 days out of the week; the other 2 days are fasting days that allow for intake of 0–600 kcals from protein, fat, and non-starchy vegetables. A primary goal of many fasting models is a compressed eating window (CEW). Because eating food (some foods more than others) creates an inflammatory response, a CEW is often recommended as a method to decrease inflammation [104, 105]. A CEW may also support gastrointestinal conditions mentioned above like IBD, IBS, and SIBO.

Table 42.3 Types of intermittent fasting

Type of intermittent fasting	Description
Time-restricted fasting	Compressed eating window such as 12:12 (fasting for 12 hours followed by food intake occurring during the next 12 hours). Other examples are 14:10, 16:8, 18:6, and 20:4
Time-restricted fasting for cancer prevention and treatment (breast, colon, and glioblastoma) [112]	13-hour overnight fast: not eating from 6 p.m. until 7 p.m. the following day
Time-restricted fasting for Alzheimer's prevention and/or reversal [73]	3/12–16: (3-hour fast between dinner and bedtime with an overnight fast of 12–16 hours depending on ApoE 4 genetics)
5:2 fast	Eat normal intake for 5 days during the week. For two consecutive or nonconsecutive days, eat 0–600 kcals of protein, fat, and non-starchy vegetables
Alternate-day fasting [106]	For alternating days during the week, fast for 20–24 hours such as Monday, Wednesday, and Friday. For example, eat dinner on Sunday night at 7 p.m. and avoid eating again until 5 p.m. on Monday[a]
Extended fasting [106]	Fasting for 2–14 days. Note: fasting beyond 3 days may increase risk of refeeding syndrome
Extended fasting as an adjunct to cancer treatment [112]	Fasting 24–48 hours prior to chemotherapy and for as long as possible after as applied by Dr. Dawn Lemmane based on research by Dr. Valter Longo

[a]Coffee or tea with added fat such as MCT oil or cream and/or bone broth is considered acceptable or as a strict water-only fast. Electrolytes are recommended during alternate-day and extended-day fasts [106]

42.18 Summary and Conclusion

Therapeutic diets have been used for thousands of years for the purpose of promoting health and healing. The modern IFMNT approach involves applying the nutrition care process in order to personalize the therapeutic use of nutrition, food, and targeted supplements. The IFMNT practitioner should be aware of the nutritional vulnerabilities (potential nutrient deficiencies) of each individual diet and ensure that any dietary intervention provides adequate amounts of macronutrients, micronutrients, and phytonutrients, which can be achieved by foods and/or supplements.

References

1. Boyce B. CMS final rule on therapeutic diet orders means new opportunities for RDNs. J Acad Nutr Diet. 2014;114(9):1326–8.
2. Weber Wea. Ayurveda: In Depth: National Center for Complementary and Integrative Health; 2015 [Available from: https://nccih.nih.gov/health/ayurveda/introduction.htm.
3. Burke Aea. Traditional Chinese Medicine: In Depth: National Center for Complementary and Integrative Health; 2013 [updated October 2013. Available from: https://nccih.nih.gov/health/whatiscam/chinesemed.htm.
4. Gray A. Nutritional recommendations for individuals with diabetes. Endotext [Internet]: MDText. com, Inc.; 2015.
5. Satija A, Hu FB. Plant-based diets and cardiovascular health. Trends Cardiovasc Med. 2018;28(7):437–41.
6. Kakodkar S, Mutlu EA. Diet as a therapeutic option for adult inflammatory bowel disease. Gastroenterol Clin N Am. 2017;46(4):745–67.
7. Gianfranceschi P, Fasani G, Speciani AF. Rheumatoid arthritis and the drop in tolerance to foods: elimination diets and the reestablishment of tolerance by low-dose diluted food. Ann N Y Acad Sci. 1996;778:379–81.
8. Cohen SA, Gold BD, Oliva S, Lewis J, Stallworth A, Koch B, et al. Clinical and mucosal improvement with specific carbohydrate diet in pediatric Crohn disease. J Pediatr Gastroenterol Nutr. 2014;59(4):516–21.
9. Uhde M, Ajamian M, Caio G, De Giorgio R, Indart A, Green PH, et al. Intestinal cell damage and systemic immune activation in individuals reporting sensitivity to wheat in the absence of coeliac disease. Gut. 2016;65(12):1930–7.
10. Fasano A, Sapone A, Zevallos V, Schuppan D. Nonceliac gluten sensitivity. Gastroenterology. 2015;148(6):1195–204.
11. Chen JW, Kao JY. Eosinophilic esophagitis: update on management and controversies. BMJ (Clinical research ed). 2017;359:j4482.
12. Cianferoni A, Shuker M, Brown-Whitehorn T, Hunter H, Venter C, Spergel JM. Food avoidance strategies in eosinophilic oesophagitis. Clin Exp Allergy. 2019;49(3):269–84.
13. Gomez Torrijos E, Gonzalez-Mendiola R, Alvarado M, Avila R, Prieto-Garcia A, Valbuena T, et al. Eosinophilic esophagitis: review and update. Front Med. 2018;5:247.
14. Kinoshita Y, Ooouchi S, Fujisawa T. Eosinophilic gastrointestinal diseases - pathogenesis, diagnosis, and treatment. Allergol Int. 2019;68(4):420–9.
15. Kliewer KL, Cassin AM, Venter C. Dietary therapy for eosinophilic esophagitis: elimination and reintroduction. Clin Rev Allergy Immunol. 2018;55(1):70–87.
16. Munoz-Persy M, Lucendo AJ. Treatment of eosinophilic esophagitis in the pediatric patient: an evidence-based approach. Eur J Pediatr. 2018;177(5):649–63.
17. Nhu QM, Aceves SS. Medical and dietary management of eosinophilic esophagitis. Ann Allergy Asthma Immunol. 2018;121(2):156–61.
18. Nhu QM, Moawad FJ. New developments in the diagnosis and treatment of eosinophilic esophagitis. Curr Treat Options Gastroenterol. 2019;17(1):48–62.
19. Patel RV, Hirano I. New developments in the diagnosis, therapy, and monitoring of eosinophilic esophagitis. Curr Treat Options Gastroenterol. 2018;16(1):15–26.
20. Pesek RD, Gupta SK. Emerging drugs for eosinophilic esophagitis. Expert Opin Emerg Drugs. 2018;23(2):173–83.
21. Wilson JM, McGowan EC. Diagnosis and management of eosinophilic esophagitis. Immunol Allergy Clin N Am. 2018;38(1):125–39.
22. Akhondi H. Diagnostic approaches and treatment of eosinophilic esophagitis. A review article. Ann Med Surg. 2017;20:69–73.
23. Cotton CC, Eluri S, Wolf WA, Dellon ES. Six-food elimination diet and topical steroids are effective for eosinophilic esophagitis: a meta-regression. Dig Dis Sci. 2017;62(9):2408–20.

24. de Bortoli N, Penagini R, Savarino E, Marchi S. Eosinophilic esophagitis: update in diagnosis and management. Position paper by the Italian Society of Gastroenterology and Gastrointestinal Endoscopy (SIGE). Dig Liver Dis. 2017;49(3):254–60.
25. Hommeida S, Alsawas M, Murad MH, Katzka DA, Grothe RM, Absah I. The association between celiac disease and eosinophilic esophagitis: Mayo experience and meta-analysis of the literature. J Pediatr Gastroenterol Nutr. 2017;65(1):58–63.
26. McGowan EC, Platts-Mills TA. Eosinophilic esophagitis from an allergy perspective: how to optimally pursue allergy testing & dietary modification in the adult population. Curr Gastroenterol Rep. 2016;18(11):58.
27. Molina-Infante J, Gonzalez-Cordero PL, Arias A, Lucendo AJ. Update on dietary therapy for eosinophilic esophagitis in children and adults. Expert Rev Gastroenterol Hepatol. 2017;11(2):115–23.
28. Newberry C, Lynch K. Can we use diet to effectively treat esophageal disease? A review of the current literature. Curr Gastroenterol Rep. 2017;19(8):38.
29. Philpott H, Dellon ES. The role of maintenance therapy in eosinophilic esophagitis: who, why, and how? J Gastroenterol. 2018;53(2):165–71.
30. Philpott H, Kweh B, Thien F. Eosinophilic esophagitis: current understanding and evolving concepts. Asia Pac Allergy. 2017;7(1):3–9.
31. Sun MF, Gu WZ, Peng KR, Liu MN, Shu XL, Jiang LQ, et al. Eosinophilic esophagitis in children: analysis of 22 cases. Zhonghua Er Ke Za Zhi. 2017;55(7):499–503.
32. Braly K, Williamson N, Shaffer ML, Lee D, Wahbeh G, Klein J, et al. Nutritional adequacy of the specific carbohydrate diet in pediatric inflammatory bowel disease. J Pediatr Gastroenterol Nutr. 2017;65(5):533–8.
33. Gunasekeera V, Mendall MA, Chan D, Kumar D. Treatment of Crohn's disease with an IgG4-guided exclusion diet: a randomized controlled trial. Dig Dis Sci. 2016;61(4):1148–57.
34. Jian L, Anqi H, Gang L, Litian W, Yanyan X, Mengdi W, et al. Food exclusion based on IgG antibodies alleviates symptoms in ulcerative colitis: a prospective study. Inflamm Bowel Dis. 2018;24:1918.
35. Limketkai BN, Iheozor-Ejiofor Z, Gjuladin-Hellon T, Parian A, Matarese LE, Bracewell K, et al. Dietary interventions for induction and maintenance of remission in inflammatory bowel disease. Cochrane Database Syst Rev. 2019;2:Cd012839.
36. Penagini F, Dilillo D, Borsani B, Cococcioni L, Galli E, Bedogni G, et al. Nutrition in pediatric inflammatory bowel disease: from etiology to treatment. A systematic review. Nutrients. 2016;8(6):1–27.
37. Ruemmele FM. Role of diet in inflammatory bowel disease. Ann Nutr Metab. 2016;68(Suppl 1):33–41.
38. Wang G, Ren J, Li G, Hu Q, Gu G, Ren H, et al. The utility of food antigen test in the diagnosis of Crohn's disease and remission maintenance after exclusive enteral nutrition. Clin Res Hepatol Gastroenterol. 2018;42(2):145–52.
39. Casellas F, Burgos R, Marcos A, Santos J, Ciriza-de-Los-Ríos C, Garcia-Manzanares A, et al. Consensus document on exclusion diets in irritable bowel syndrome (IBS). Nutr Hosp. 2018;35(6):1450–66.
40. Werlang ME, Palmer WC, Lacy BE. Irritable bowel syndrome and dietary interventions. Gastroenterol Hepatol. 2019;15(1):16–26.
41. Endreffy I, Bjorklund G, Dicso F, Urbina MA, Endreffy E. Acid glycosaminoglycan (aGAG) excretion is increased in children with autism spectrum disorder, and it can be controlled by diet. Metab Brain Dis. 2016;31(2):273–8.
42. Kawicka A, Regulska-Ilow B. How nutritional status, diet and dietary supplements can affect autism. A review. Roczniki Panstwowego Zakladu Higieny. 2013;64(1):1–12.
43. Ly V, Bottelier M, Hoekstra PJ, Arias Vasquez A, Buitelaar JK, Rommelse NN. Elimination diets' efficacy and mechanisms in attention deficit hyperactivity disorder and autism spectrum disorder. Eur Child Adolesc Psychiatry. 2017;26(9):1067–79.
44. Mitchell N, Hewitt CE, Jayakody S, Islam M, Adamson J, Watt I, et al. Randomised controlled trial of food elimination diet based on IgG antibodies for the prevention of migraine like headaches. Nutr J. 2011;10:85.
45. Caminero A, Meisel M, Jabri B, Verdu EF. Mechanisms by which gut microorganisms influence food sensitivities. Nat Rev Gastroenterol Hepatol. 2019;16(1):7–18.
46. Schnedl WJ, Lackner S, Enko D, Schenk M, Mangge H, Holasek SJ. Non-celiac gluten sensitivity: people without celiac disease avoiding gluten-is it due to histamine intolerance? Inflamm Res. 2018;67(4):279–84.
47. Sapone A, Lammers KM, Casolaro V, Cammarota M, Giuliano M, De Rosa M, et al. Divergence of gut permeability and mucosal immune gene expression in two gluten-associated conditions: celiac disease and gluten sensitivity. BMC Med. 2011;9(1):23.
48. Carroccio A, Brusca I, Mansueto P, Soresi M, D'Alcamo A, Ambrosiano G, et al. Fecal assays detect hypersensitivity to cow's milk protein and gluten in adults with irritable bowel syndrome. Clin Gastroenterol Hepatol. 2011;9(11):965–71.e3.
49. Manning LP, Biesiekierski JR. Use of dietary interventions for functional gastrointestinal disorders. Curr Opin Pharmacol. 2018;43:132–8.
50. Masharani U, Sherchan P, Schloetter M, Stratford S, Xiao A, Sebastian A, et al. Metabolic and physiologic effects from consuming a hunter-gatherer (Paleolithic)-type diet in type 2 diabetes. Eur J Clin Nutr. 2015;69(8):944–8.
51. Staudacher HM, Ralph FSE, Irving PM, Whelan K, Lomer MCE. Nutrient intake, diet quality, and diet diversity in irritable bowel syndrome and the impact of the low FODMAP Diet. J Acad Nutr Diet. 2019; https://doi.org/10.1016/j.jand.2019.01.017.
52. Drisko J, Bischoff B, Hall M, McCallum R. Treating irritable bowel syndrome with a food elimination diet followed by food challenge and probiotics. J Am Coll Nutr. 2006;25(6):514–22.
53. Neuendorf R, Corn J, Hanes D, Bradley R. Impact of food immunoglobulin G-based elimination diet on subsequent food immunoglobulin G and quality of life in overweight/obese adults. J Altern Complement Med. 2019;25(2):241–8.
54. El Khoury D, Balfour-Ducharme S, Joye IJ. A review on the gluten-free diet: technological and nutritional challenges. Nutrients. 2018;10(10):1–25.
55. Rizzo G, Baroni L. Soy, soy foods and their role in vegetarian diets. Nutrients. 2018;10(1):1–51.
56. Rehault-Godbert S, Guyot N, Nys Y. The Golden egg: nutritional value, bioactivities, and emerging benefits for human health. Nutrients. 2019;11(3):1–26.
57. de Souza RGM, Schincaglia RM, Pimentel GD, Mota JF. Nuts and human health outcomes: a systematic review. Nutrients. 2017;9(12):1–23.
58. Papanikolaou Y, Fulgoni VL. Grain foods are contributors of nutrient density for American adults and help close nutrient recommendation gaps: data from the National Health and Nutrition Examination Survey, 2009-2012. Nutrients. 2017;9(8):1–14.
59. Purves D, Williams SM. Neuroscience. 2nd ed. Biogenic Amines Chapter: Sinauer Associates; 2001.
60. HDC gene - Genetics Home Reference [Internet]. U.S. National Library of Medicine. National Institutes of Health; [cited 2018May11]. Available from: https://ghr.nlm.nih.gov/gene/HDC.
61. Kanki M, Yoda T, Tsukamoto T, Baba E. Histidine decarboxylases and their role in accumulation of histamine in tuna and dried saury. Appl Environ Microbiol. 2007;73(5):1467–73.
62. Urb M, Sheppard DC. The role of mast cells in the defense against pathogens. PLoS Pathog. 2012;8(4):e1002619.
63. Krishnan K. How the microbiome shapes the systemic immune system in health and disease. Lecture presented at: Premier On-line Training with Susan Allen-Evenson and Kiran Krishnan; 2017.
64. Joneja JMV. Dealing with food allergies in babies and children. Boulder: Bull Pub. Co.; 2007.
65. Smolinska S, Jutel M, Crameri R, O'Mahony L. Histamine and gut mucosal immune regulation. Allergy. 2013;69(3):273–81.
66. Bonds RS, Midoro-Horiuti T. Estrogen effects in allergy and asthma. Curr Opin Allergy Clin Immunol. 2013;13(1):92–9.

67. Gogos J. COMT (catechol-O-methyltransferase). Wiley Encyclopedia of Molecular Medicine; 2002.
68. Jarisch R. Histamine intolerance histamine and seasickness. Berlin: Springer Berlin; 2014.
69. Lynch B. Dirty genes: a breakthrough program to treat the root cause of illness and optimize your health: HarperCollins Publishers; 2018.
70. Swink TD, Vining EP, Freeman JM. The ketogenic diet: 1997. Advances in Pediatrics. 1997;40:297–329.
71. Freeman JM, Kossoff EH, Hartman AL. The ketogenic diet: one decade later. Pediatrics. 2007;119(3):535–43.
72. Volek JS, Phinney SD, Kossoff E, Eberstein J, Moore J. The art and science of low carbohydrate living an expert guide to making the life-saving benefits of carbohydrate restriction sustainable and enjoyable. Lexington: Beyond obesity; 2011.
73. Bredesen DE. The end of Alzheimer's: the first program to prevent and reverse the cognitive decline of dementia. London: Vermilion; 2017.
74. Lemanne D. Carbohydrate restriction in cancer therapy. Lecture presented at: Low Carb Breck 2017; Colorado; 2017.
75. Meyerhardt JA, Sato K, Niedzwiecki D, Ye C, Saltz LB, Mayer RJ, et al. Dietary glycemic load and cancer recurrence and survival in patients with stage III colon cancer: findings from CALGB 89803. JNCI. 2012;104(22):1702–11.
76. Epstein T, Gatenby RA, Brown JS. The Warburg effect as an adaptation of cancer cells to rapid fluctuations in energy demand. PLoS One. 2017;12(9):e0185085.
77. Seyfried TN, Sanderson TM, El-Abbadi MM, Mcgowan R, Mukherjee P. Role of glucose and ketone bodies in the metabolic control of experimental brain cancer. Br J Cancer. 2003;89:1375–82.
78. Abdelwahab MG, Fenton KE, Preul MC, Rho JM, Lynch A, Stafford P, Scheck AC. The ketogenic diet is an effective adjuvant to radiation therapy for the treatment of malignant glioma. PLoS One. 2012;7:e36197.
79. What You Should Know About Diabetic Ketoacidosis [Internet]. WebMD. WebMD; [cited 2018May8]. Available from: https://www.webmd.com/diabetes/ketoacidosis.
80. Kitabchi AE, Fisher JN. Hyperglycemic crises: diabetic ketoacidosis (DKA) and hyperglycemic hyperosmolar state (HHS). Acute Endocrinol. 2008:119–47.
81. Vanitallie TB, Nufert TH. Ketones: metabolisms ugly duckling. Nutr Rev. 2003;61(10):327–41.
82. Appendix 7. Nutritional Goals for Age-Sex Groups Based on Dietary Reference Intakes and Dietary Guidelines Recommendations [Internet]. Chapter 4 - 2008 Physical activity guidelines. [cited 2018May9]. Available from: https://health.gov/dietaryguidelines/2015/guidelines/appendix-7/.
83. Cordain L. The Paleo Diet. Place of publication not identified: John Wiley & Sons Ltd; 2010.
84. Farooki A, Schneider SH. Increased cancer-related mortality for patients with type 2 diabetes who use sulfonylureas or insulin: response to Bowker et al. Diabetes Care. 2006;29(8):1989–90.
85. Zendehdel K. Cancer incidence in patients with type 1 diabetes mellitus: a population-based cohort study in Sweden. Cancer Spectrum Knowledge Environment. 2003;95(23):1797–800.
86. Renehan AG, Zwahlen M, Minder C, O'Dwyer ST, Shalet SM, Egger M. Insulin-like growth factor (IGF)-I, IGF binding protein-3, and cancer risk: systematic review and meta-regression analysis. Lancet. 2004;363(9418):1346–53.
87. Influence of a diet very high in vegetables, fruit, and fiber and low in fat on prognosis following treatment for breast cancer: the Women's healthy eating and living (WHEL) randomized trial. Breast diseases: a year book quarterly 2008;19(1):35–36.
88. Blackburn GL, Wang KA. Dietary fat reduction and breast cancer outcome: results from the Women's intervention nutrition study (WINS). Am J Clin Nutr. 2007;86(3):878S.
89. Jeong S-M, Choi S, Kim K, Kim S-M, Lee G, Son JS, et al. Association of change in total cholesterol level with mortality: a population-based study. PLoS One. 2018;13(4):e0196030.
90. Ketogenesis Drives BRAF-MEK1 Signaling in BRAFV600E-Positive Cancers. Cancer Discovery. 2015;5(9).
91. Lyssiotis C, Cantley L. Acetate fuels the cancer engine. Cell. 2015;160(3):567. https://doi.org/10.1016/j.cell.2015.01.021.
92. Cao MD, Lamichhane S, Lundgren S, Bofin A, Fjøsne H, Giskeødegård GF, Bathen TF. Metabolic characterization of triple negative breast cancer. BMC Cancer. 2014;14(1) https://doi.org/10.1186/1471-2407-14-941.
93. Wise DR, Thompson CB. Glutamine addiction: a new therapeutic target in cancer. Trends Biochem Sci. 2010;35(8):427–33. https://doi.org/10.1016/j.tibs.2010.05.003.
94. Fung J. Diabetes code prevent and reverse type 2 diabetes naturally. Carlton: Scribe Publications; 2018.
95. Volek JS, Phinney SD. The art and science of low carbohydrate performance. Beyond Obesity LLC: Berlín; 2012.
96. Harvey CJ, Schofield GM, Williden M, Mcquillan JA. The effect of medium chain triglycerides on time to nutritional ketosis and symptoms of keto-induction in healthy adults: a randomised controlled clinical trial. J Nutr Metab. 2018;2018:1–9. https://doi.org/10.1155/2018/2630565.
97. Patterson RE, Sears DD. Metabolic effects of intermittent fasting. Annu Rev Nutr. 2017;37(1):371–93.
98. Santos HO, Macedo RC. Impact of intermittent fasting on the lipid profile: assessment associated with diet and weight loss. Clin Nutr ESPEN. 2018;24:14.
99. Golbidi S, Daiber A, Korac B, Li H, Essop MF, Laher I. Health benefits of fasting and caloric restriction. Curr Diab Rep. 2017;17(12):123.
100. Kim K-H, Kim YH, Son JE, Lee JH, Kim S, Choe MS, et al. Intermittent fasting promotes adipose thermogenesis and metabolic homeostasis via VEGF-mediated alternative activation of macrophage. Cell Res. 2017;27(11):1309–26.
101. Haas JT, Staels B. Fasting the microbiota to improve metabolism? Cell Metab. 2017;26(4):584–5.
102. Stockman M, Thomas D, Burke J, Apovian CM. (2018). Intermittent fasting: is the wait worth the weight? Retrieved from https://rd.springer.com/article/10.1007/s13679-018-0308-9.
103. Pimentel M. A new IBS solution: bacteria, the missing link in treating irritable bowel syndrome. Sherman Oaks: Health Point Press; 2011.
104. Doig CM. Controlled trial of bowel rest and nutritional support in the management of Crohn's disease. J Pediatr Surg. 1989;24(9):945.
105. Mcintyre PB, Powell-Tuck J, Wood SR, Lennard-Jones JE, Lerebours E, Hecketsweiler P, et al. Controlled trial of bowel rest in the treatment of severe acute colitis. Gut. 1986;27(5):481–5.
106. Fung J, Moore J. The complete guide to fasting: heal your body through intermittent, alternate-day, and extended fasting. Las Vegas: Victory Belt Publishing; 2016.
107. Longo VD, Mattson MP. Fasting: molecular mechanisms and clinical applications. Cell Metab. 2014;19(2):181–92.
108. Mendelsohn AR, Larrick JW. Prolonged fasting/refeeding promotes hematopoietic stem cell regeneration and rejuvenation. Rejuvenation Res. 2014;17(4):385–9.
109. Lee C, Raffaghello L, Brandhorst S, Safdie FM, Bianchi G, Martin-Montalvo A, et al. Fasting cycles retard growth of tumors and sensitize a range of cancer cell types to chemotherapy. Sci Transl Med. 2012;4(124):124ra27.
110. Safdie FM, Dorff T, Quinn D, Fontana L, Wei M, Lee C, et al. Fasting and cancer treatment in humans: a case series report. Aging. 2009;1(12):988–1007.
111. Mehanna HM, Moledina J, Travis J. Refeeding syndrome: what it is, and how to prevent and treat it. BMJ. 2008;336(7659):1495–8.
112. Lemmane D. Carbohydrate restriction in cancer therapy. Lecture presented at: Low Carb Breck 2017; Colorado; 2017.

Dietary Supplements: Understanding the Complexity of Use and Applications to Health

Eric R. Secor

43.1 Trends in Dietary Supplement Use – 756

43.2 A Growing Marketplace – 756

43.3 Nomenclature and Formulations – 756

43.4 Dietary Supplement Labels – 758

43.5 Nutrition Facts Panel Selection and Dose – 758

43.6 The Role of Government Regulations – 759

43.7 Supplement Selection and Education – 760

43.8 Quality Control and Safety – 760

43.9 Engaging Manufacturers – 761

43.10 Clinical Relevance and Controversies – 762

43.11 Summary – 764

References – 764

© Springer Nature Switzerland AG 2020
D. Noland et al. (eds.), *Integrative and Functional Medical Nutrition Therapy*,
https://doi.org/10.1007/978-3-030-30730-1_43

43.1 Trends in Dietary Supplement Use

Complementary and integrative healthcare includes a selection of modalities, products, and dietary approaches that usually begin outside the biomedical or modern healthcare system. This may include chiropractic, naturopathic medicine, functional medicine, massage therapy, acupuncture and traditional Oriental medicine, nutritional medicine, and mind-body therapies [1, 2]. When evaluating the use of integrative healthcare in the population, published reports consistently place the frequency from 32% to 62% [3, 4]. In an analysis of the 2012 National Health Interview Survey, it was determined that 59 million Americans had at least 1 payment dedicated to complementary health with $30.2 billion in total out-of-pocket expenditures, which included $14.7 billion for visits to practitioners and $12.8 billion for purchases of natural products [5]. The total out-of-pocket expenditures for purchases of natural supplements increased significantly as family income increased and the total spent ($12.8 billion) represented approximately 24% of out-of-pocket expenditures ($54.1 billion) for prescription drugs.

From a sample of 89,000 Americans, it was determined that nonvitamin, nonmineral dietary supplements remained the most prevalent complementary health approach in use when examining trends from 2002, 2007, and 2012 [3]. Investigators determined that participants reported taking at least one of the following supplements within the prior 30 days: fish oil, glucosamine or chondroitin, probiotics or prebiotics, melatonin, Coenzyme Q–10 (CoQ10), *Echinacea*, cranberry (pills or capsules), garlic supplements, ginseng, *Ginkgo biloba*, green tea pills (not brewed tea), or EGCG pills [4], "combination" herb pills, MSM (methylsulfonylmethane), milk thistle (*Silybum*), Saw palmetto, and Valerian [3, 6]. Supplement use was also found to be associated with female gender, older age, more education, any physical activity, and being of normal weight or slightly underweight [4]. The top-selling herbal supplements in 2016 included Ashwagandha (*Withania somnifera*), cranberry, and turmeric (*Curcuma longa*) [7], which are summarized in Table 43.1. Ashwagandha sales were found to increase approximately 55% from 2015 to 2016, and sales of turmeric alone topped $47 million. In individual populations, supplement use was found to be common in cardiac patients [8], cancer patients [9], hospitalized patients [10, 11], the military [12], those with sickle cell disease [13], and among athletes [14, 15]. Although supplement use continues to be widespread, there remains considerable confusion about how best to navigate the marketplace, wade through the myriad of products and their formulations, and arrive at the most appropriate product choice for an individual's healthcare needs [16–18].

43.2 A Growing Marketplace

Analysts have been assessing the ebb and flow of the dietary supplements mentioned above as well as the broader nutraceutical marketplace for decades. According to PricewaterhouseCoopers Global (Global PwC), nutraceutical sales were estimated at $142 billion in 2011 and were expected to grow to $180 billion into 2017 [19]. BCC Research projected that the market would grow to $199 billion in 2016 while adding growth of functional beverages and functional foods upward of $71.5 and $64.6 billion, respectively [20]. According to a recent report by Zion Market Research [21], the global, nonvitamin, nonmineral dietary supplement market was valued at $133 billion in 2016, increasing to well over $220 billion (US) by 2022. These analyses suggest dietary supplements are a very large and expanding market. The use of the Internet for patient-directed purchasing and expanding markets in Eastern Europe, India, and China [22] will ensure steady growth in all of these sectors for decades to come.

43.3 Nomenclature and Formulations

One of the first legally acceptable definitions of dietary supplements was put forth in the federal Dietary Supplement Health and Education Act (DSHEA) of 1994. The historical underpinnings of DSHEA are rooted in the Food Supplement Amendment or the Proxmire Amendment, also known as "the Vitamin Bill," S.2801, introduced in the 93rd Congress between 1973 and 1974 [23]. The Proxmire Amendment was proposed to set forth the potency of ingredients of vitamin and mineral products that are not inherently dangerous. When introduced in 1973, its language included, "shall not limit the potency, number, combination, amount, or variety of any synthetic or natural vitamin, mineral, or other nutritional substance, or ingredient of any food for special dietary uses if the amount recommended to be consumed does not ordinarily render it injurious to health." Two decades later, DSHEA amended the

Table 43.1 Change in sales of top selling herbal supplements in 2016

Ingredient	Latin binomial	Total sales	Change from 2015
Ashwagandha	*Withania somnifera*	$8,732,489	55%
Cranberry	*Vaccinium macrocarpon*	$7,513,172	36%
Turmeric	*Curcuma longa*	$47,654,008	32%
Horsetail	*Equisetum* spp.	$5,334,706	16%
Dandelion	*Taraxacum officinale*	$2,520,049	15%
Mushrooms	Order *Agaricales*	$4,527,372	14%
Cherry fruit	*Prunus* spp.	$3,507,680	13%
Kava	*Piper methysticum*	$3,232,327	10%
Ginger	*Zingiber officinale*	$2,454,767	9%

Federal Food, Drug, and Cosmetic Act [23–25] and defined a dietary supplement as:

> …a product other than tobacco, intended to supplement the diet, that contains a vitamin, mineral, herb or botanical, dietary substance, or a concentrate, metabolite, constituent, extract, or combination of the above ingredients; that is intended for ingestion, is not represented as food or as a sole item of a meal or diet, and is labeled as a dietary supplement; that includes an article approved as a new drug, certified as an antibiotic, or licensed as a biologic and that was, prior to such approval, certification or licensure, marketed as a dietary supplement or food, unless the conditions of use and dosages are found to be unlawful; and excludes such articles which were not so marketed prior to approval unless found to be lawful. Deems a dietary supplement to be a food.

This definition lays the groundwork for the regulatory aspects of DSHEA, which will be discussed below, but it does little to provide practical education to healthcare providers, practitioners, and patients. For example, the common term "dietary supplement" may refer to a large range of ingredients including vitamins, minerals, botanical and/or herbal products (concentrates, metabolites, constituents, and extracts), fish oils, amino acids, glucosamine, creatine, and essential fatty acids [6]. These products can be found sold and marketed as single ingredients or in a multitude of combinations [26], as well as added as ingredients to medical foods [27], functional foods [28], and/or beverages. In contrast to dietary supplements, the Food and Drug Administration (FDA) distinguishes prepared tube feedings as medical food in the use and dietary management of a specific disease or health-related condition that causes distinctive nutritional requirements different from healthy people and "formulated to be consumed or administered *enterally* under the supervision of a physician [28]." However, "functional" foods and beverages do not have a formal FDA designation and are commonly defined based on ingredients, formulations, and label or health claims.

In addition to taking into account the broad definition of dietary supplements put forth under DSHEA, when selecting or advising patients on the most appropriate choice for them, it is critical to understand the common forms by which dietary supplements are delivered and find their way into foods, beverages, and the multitude of products in the marketplace [25, 29]. An overview of some of the most common forms that product ingredients are available is provided in ▶ Box 43.1. From bulk herbs (which can be purchased by the pound) to essential oils (by the ounce) to common lozenges, tablets, and topicals (creams, gels, salves, poultices, clay, and lotion), most individual products limit themselves to one "form," such as in a Calendula cream, a cranberry (fruit) or turmeric (root /rhizome) extract [25, 30–34]. Other manufacturers may combine numerous forms into complex combination products with 15 to 25 or more individual components, not including binders, fillers, and additives.

Box 43.1 Common Forms and Delivery of Dietary Supplements
- *Bulk herbs* – Minimally processed herbal material (sort, cut, and dry)
- *Capsules* – Encapsulating a defined amount of liquid or powder into a gelatin capsule
- *Essential oil* – Distilled aromatic oils, applied orally, via inhalation, or topically within a fixed oil
- *Extracts* – Process of concentrating an herb with a specific solvent
- *Extract, decoction* – Raw herbal material (stems, roots, leaves) boiled in water to concentrate
- *Extract, fluid* – Herbal mixtures (alcohol, glycerine, vinegar, water) concentrated via evaporation
- *Extract, solid* – Fluid extracts concentrated via evaporation to form a solid (soft or crystalline)
- *Extract, tea* – Steeping fine herbal material in water to concentrate water-soluble material
- *Extract, tincture* – Soaking herb in alcohol and water to concentrate alcohol-soluble components
- *Glandulars* – Animal gland extracts (thyroid, adrenal)
- *Homeopathics* – Ultradilution of a variety of materials (animal, vegetable, or minerals)
- *Lozenges* – Candied, gummy, tableted, sublingual
- *Spa therapy* – Baths, colonics, steam, wraps, sauna
- *Suppositories* – Material combined and delivered in a vehicle (glycerine)
- *Tablets/pill* – Compressing a defined amount of active material under force
- *Topicals* – Balm, creams, gels, slaves, poultice, clay, and lotion

---

These combinations may include vitamins; minerals; glandulars; fatty acids; probiotics; bulk botanicals parts (fruit, leaf, or root); botanicals in the form of dried, crystallized solid extracts; and essential oils [4, 30, 35–37].

In addition to the delivery form and/or route, it is critical to consider that different forms may be easier to absorb or deliver a higher dose as compared to another. For example, although not set in stone, a good rule of thumb to follow is that as the form of (a botanical) product becomes more concentrated, the strength and dose may increase. So the bulk herb may be less potent when compared to its solid extract, i.e., potency of bulk herb < tea < decoction < tincture < fluid extract < solid extract [38]. In addition, scientific advances in the formulation of plant extracts have led to changes in how healthcare practitioners approach drug delivery. These new systems, which include nanoparticles, nanocapsules, liposomes, phytosomes, nanoemulsions, and microspheres, are transforming product solubility, bioavailability, and stability, which ultimately will lead to enhanced biological activity and novel clinical applications [39]. Therefore, the forms, dosage, and standardization information should be clearly presented on all dietary supplement bottles and delivery devices to help guide the consumer, patient, and healthcare team. Caution should be exercised as the same product formulated differently may impart a variety of biological effects. For example, consider green tea and its L-theanine (γ-glutamylethylamide) extract, which is now widely used as a mild anxiolytic as opposed to a low-dose caffeine stimulant [40].

43.4 Dietary Supplement Labels

The Department of Health and Human Services' Office of Inspector General (OIG) outlined several key aspects of dietary supplement labels that are required to ensure that consumers make the best possible choices to increase safety and lower risk when selecting products [41]. In the report "Adverse Event Reporting for Dietary Supplements, an Inadequate Safety Valve," the OIG presents a template composed of two main elements: [1] label content (ingredients, intended use, safety information, directions for use, and product information) and [2] label presentation (standardized format, distinct product features, readability, balance, and constructive use of space). Ingredients should be listed fully and clearly and reflect the accurate amount per serving and per bottle as much as possible. Intended use would distinguish "support to health" from health claims, which require scientific validation. Information on safety should include potential interactions or contraindications, such as whether it is safe in pregnancy. Directions for use may include suggested dosage, if known. Product information would identify the manufacturer, lot, and possibly gross and net amounts of the ingredients listed. The label presentation template provides details on consistency, similar types of information in a similar order, design, and format, made easy to understand by a wide range of consumers and provides information in a balanced manner (benefits, risks, claims, and facts) while using product packaging to provide additional sources of information beyond the nutrition facts panel.

43.5 Nutrition Facts Panel Selection and Dose

In addition to being able to identify the list of key components on dietary supplement labels, a more pressing issue is selecting and distinguishing a product based on the individual ingredient(s), their forms or parts used, as well as additional additives such as dyes, binders, and fillers. In ◘ Table 43.2, examples of some of the most common forms of vitamin, mineral, and botanical ingredients are highlighted. Vitamin A can be found as acetate, beta carotene, mixed carotenoids, palmitate [42], and vitamin B12, in either the cyanocobalamin or methylcobalamin form [43]. Calcium carbonate is said to be cheaper and harder to absorb than citrate, but a clinical determination must be made when choosing citrate over malate or pantothenate [44]. When choosing iron for a patient with gastrointestinal upset [45], would the best choice be ferric (citrate, sulfate) or ferrous (aspartate, fumarate, sulfate, gluconate) conjugates, and are these best delivered in liquid, tablet, or fortified in food [46]?

To many, interpreting botanicals within dietary supplements is even more overwhelming when compared to vitamin and mineral ingredients [6, 36, 47]. Once the botanical therapeutic is chosen based on the health indication, the genus and species should be verified, and then a determination must be made to validate that the dietary supplement contains the correct part and therapeutic strength of the extract [36, 37, 48]. For example, with *Camellia sinensis* or green tea, identifying that material as a leaf extract is very straightforward, since tea is derived from or is the leaf product. Next is the therapeutic determination of a dose and delivery provided by a wide range (10–50%) of total polyphenols or the individual green tea polyphenols, such as catechin, epicatechin gallate, epigallocatechin, gallic acid, and/or epicatechin [49]. Both the ranges (%)

◘ Table 43.2 Common forms of dietary supplement ingredients

Vitamin/mineral ingredient	Common forms
Vitamin A	Acetate, beta carotene, mixed carotenoids, palmitate
Vitamin B12	Methylcobalamin, cyanocobalamin
Calcium	Carbonate, citrate, dibasic calcium phosphate, malate, pantothenate
Vitamin D	Cholecalciferol-D3, ergocalciferol-D2
Vitamin E	Tocopherols – alpha, beta, delta and/or gamma
Folate	Folic acid, L-5-MTHF- Metafolin®
Iron	Ferric (citrate, sulfate), ferrous (aspartate, fumarate, sulfate, gluconate)
Magnesium	Glycinate, malate, oxide, stearate, sulfate
Selenium	Sodium selenite, selenomethionine
Zinc	Citrate, gluconate, oxide, sulfate
Botanical ingredient	**Form, part, and standardization**
Camellia sinensis	Green tea: leaf extract – 30–50%, polyphenols, EGCG, picatechin, epicatechin gallate, epigallocatechin, catechin, gallic acid
Echinacea purpurea	Herbal: stem, leaf, and/or flower extract, 3–4% phenolics
Curcuma longa	Turmeric: rhizome or root extract. 70–95% curcuminoids
Allium sativum	Garlic: bulb and/or oil extract. % Alliin, allicin, diallyl sulfide, or ajoene
Ginkgo biloba	Ginkgo: leaf extract (1:4–50:1). 6–24% heterosides, terpene lactones
Hypericum perforatum	Hypericum: aerial/herb. 0.3% hypericins
Polygonum cuspidatum	Resveratrol: root extract. 5–50% trans resveratrol
Rhodiola rosea	Rhodiola: root extract. 0–30% rosavins (rosarin, rosavin, rosin)
Serenoa repens	Saw palmetto: fruit extract. 45% fatty acids
Tanacetum parthenium	Feverfew: aerial, leaf extract. 0.4% to 0.7% parthenolide

and content of polyphenols may vary widely. With *Allium sativum*, or garlic, the consumer must distinguish between the garlic bulb and oil, if it was extracted, and what percentage of common sulfur-containing compounds (alliin, allicin, diallyl sulfide, or ajoene) is represented in the product [50]. For all botanical ingredients, high-quality products should clearly distinguish the part (aerial, bulb, fruit, leaf, herb, rhizome, or root), as well as if there is a particular marker compound to which the botanical ingredient has been standardized. Most should be clearly defined, such as *Boswellia serrata* gum extract (standardized to alpha and beta boswellic acids), *Curcuma longa* rhizome extract (standardized to total curcuminoids), *Crataegus oxyacantha* leaf and flower extract (standardized to total flavonoids), *Coleus forskohlii* root extract (standardized to forskolin), and *Tanacetum parthenium* feverfew leaf extract (standardized parthenolide). It is also important to consider that marker compounds [51] may be different from the active compounds, if they are known, and the product should be dosed accordingly.

Also as part of the manufacturing process, additives are commonly used to preserve, bind, absorb moisture, add color, smooth or distribute flavor, and provide a coating and moisture barrier. See Table 43.3. When selecting a product, the educated consumer should question whether additives are necessary, cause harm, or should be avoided if sensitive or intolerant [52]. For example, gelatin, gellan gum, and microcrystalline cellulose are naturally derived additives, whereas butylated hydroxytoluene or BHT [53], polyethylene glycol, polyvinyl alcohol, and FD&C colorants are not [54]. Consider if the effectiveness and quality of the dietary supplement are altered when individual additives are removed or remain in the product. Additionally, as the dietary supplement market continues to expand, manufacturers are constantly adding new products and formulations. This requires that healthcare providers and patients continually educate themselves and be aware of the benefits and risks of the individual ingredients as well as therapeutic goal of the combined product.

Table 43.3 Common additives found in dietary supplements

Common additives	Function	Avoid or consider
Butylated hydroxytoluene (BHT)	Preservative – prevents oil oxidation	Avoid
Corn Starch, modified	Binder/filler – absorbs moisture	Avoid
FD&C Blue 1, Yellow 6, Red 40	Colorant	Avoid
Gelatin	Collagen containing gelatin capsule	Consider
Gellan gum	Stabilizer-bacterial fermentation	Consider
Lactose	Milk sugars – thickener, filler	Consider; avoid if intolerant
Magnesium stearate	Excipient, binder	Avoid
Maltodextrin	Thickener, filler – prevents crystallization	Consider
Microcrystalline cellulose	Bulking agent and filler	Consider
Parabens	Preservative and anti-microbial agent	Avoid
Polyethylene glycol	Flavor distribution (GRAS)	Avoid
Polyvinyl alcohol	Coating agent- moisture barrier	Avoid
Silicon dioxide	Anti-caking agent	Avoid >2% by product weight
Talc	Absorbs moisture, debulking agent	Avoid

43.6 The Role of Government Regulations

Since DSHEA, government agency oversight of dietary supplements and their regulation is complex and still evolving. The Congress of the United States granted authority to the FDA and FTC to provide oversight of manufacturing and health claims on food labels as well as false advertising [29, 55, 56]. The FDA regulates foods, drugs, and cosmetics in interstate commerce, and new drugs are not allowed to be introduced into interstate commerce until they are proven safe and effective for their intended use. Foods have different regulatory requirements and commonly bypass safety and efficacy trials. Currently dietary supplements and botanical new drugs in the United States are entitled to a waiver when conducting preclinical and pharmacology/toxicology studies. As with new pharmaceutical drugs in development, genotoxicity testing is required. Due to the lack of metabolic safety data, the FDA and NIH are considering mandating that genotoxicity testing be carried out on botanicals and herbal supplements in initial clinical trials under an IND application as an added layer of regulation and mechanistic testing.

As products become more and more concentrated (pineapple vs bromelain or green tea vs L-theanine) and their biological activity enhanced through specialty delivery mechanisms, the historical divide between drugs and foods and drugs become blurred. DSHEA allows manufacturers to distribute dietary supplements without having to prove safety and efficacy, so long as the manufacturers use good production practices and make no specific claims linking the supplements to treating a specific disease state [55–57]. State laws also play a small role in the regulation use and distribution of supplements through scope of practice, professional licensure, and malpractice coverage of prescribing healthcare providers. Regarding licensure, each state has enacted unique medical licensing that prohibits the unlicensed practice of medicine and thereby criminalizes activity by unlicensed providers who offer healthcare services to patients. Similarly, scope of practice defines how these nutritional interventions may be applied. Discussions of liability have arisen when providers prescribe biologically active dietary supplements, which have drug-like activity (with potential side effects) but are regulated as food. The definitions of liability are no different in integrative medicine practice as compared general medicine; however, its application for use as a tool to provide legal protection and fill the regulatory gap produced by DSHEA raises novel questions that require consideration.

The Food and Drug Administration's Center for Food Safety and Applied Nutrition Adverse Event Reporting System, or CAERS [29], is one of the only mechanisms in which surveillance is carried out and reports on safety are generated. According to Schmitz, the CAERS reports were relatively small in number, had incomplete validation, and had inconsistencies within and are currently inadequate. He concluded that the concept and application of *Nutravigilance* or "the science and activities relating to the detection, assessment, understanding, and prevention of adverse effects related to the use of a food, dietary supplement, or medical food" require the entire industry to be proactive, systematic, and risk-based and must apply a scientific approach to all product safety. Only then can the public have full confidence that the products they choose are safe.

patients. By asking straightforward commonsense questions within the patient intake or via discussion, providers can engage patients, gain their trust, and learn about their home health habits and the degree to which they may be self-prescribing appropriate or inappropriate dietary supplements.

In investigating the attitudes and beliefs providers and patients have toward supplements, Tarn and colleagues [59] discovered that there were several topics that were identified as important to all parties. Summarized in ▶ Box 43.3, all supplements should be discussed and the potential for supplement-supplement or supplement-drug interactions outlined, and providers need to advise and provide an evidenced-based opinion about the supplements their patients take. Patients have numerous concerns about dose ranges, what is therapeutically optimal, and what, if any, risk is there in overdosing. Physicians and the healthcare team ought to provide additional options or alternatives to prescription medications and discuss the accuracy of information acquired from outside sources. Lastly, there is an increasing interest for consumers to understand more details about such supplement characteristics as their composition, if they should be natural or synthetic, and their purity and quality.

43.7 Supplement Selection and Education

In addition to suggestions from family and friends, patients who use dietary supplements are largely dependent on advice from online websites and manufacturer's label claims, both of which tend to diminish useful information on potential prescription-supplement interaction, which can be observed in the clinic [17, 58]. As outlined in ▶ Box 43.2, ask patients open-ended questions without judgment: Why are you taking this? Who suggested you take it? How did you arrive at the product, dose, and form? Do you have clear expectations of health outcomes to be evaluated, and do you understand interactions or potential side effects? It is critical that patients discuss their intended or current use with their entire healthcare team and practitioners must be proactive in querying

Box 43.3   Common Supplement-Related Topics Important to Providers and Patients
— All supplements taken by a patient should be mentioned during medical visits
— Supplement-drug interactions and potential effects or adverse events should be discussed
— Importance of providers advising or providing an opinion about supplements
— Potential benefits and risks of overdosing on supplements should be discussed
— Patients believed that doctors should talk about alternatives to prescription medications
— Providers discuss accuracy of information (friends, iInternet, television, other providers)
— Supplement characteristics, composition, purity, quality, natural or synthetic

43.8 Quality Control and Safety

One of the most concerning aspects of dietary supplements is fully understanding quality control and being able to guarantee product safety [32, 60–62]. There is an urgency to acquire reputable information assuring both that the individual is selecting the best product for their health interest and that manufacturers provide premium product formulated with the highest quality, safety, and stability. As an overview of this process, consider the life cycle of a dietary supplement or an individual ingredient as it goes into making single or combination products (◘ Fig. 43.1). In this example, we will use *Ananas comosus*, the common pineapple. During every step of the process, the source product and extracted material (bromelain, the enzymatically active pineapple

Box 43.2   Questions to Ask When Taking Dietary Supplements
1. Why am I taking this?
2. Who suggested I take this?
3. What are the health or medical indications for use?
4. What dose and duration are appropriate?
5. What is the optimal form to take?
6. What other ingredients contribute or detract from effectiveness?
7. Are there any safety concerns or interactions?
8. Are laboratory tests required?
9. Do I have clear expectations of outcomes and need to reevaluate?
10. Do I understand signs of any potential side effects?

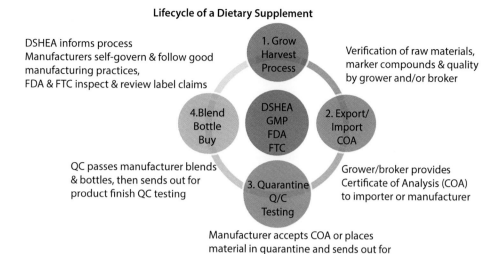

■ Fig. 43.1 Life cycle of a dietary supplement

extract) require verification [63, 64]. Beginning with step #1, the pineapple is grown and harvested and bromelain processed. Pineapple is verified through organoleptic techniques (inspection via sense organs such as sight, taste, odor color, texture) to determine maturity and ripeness. To obtain bromelain, the primary raw material, mature stems are then processed via maceration, centrifugation, extraction, and drying. The marker compounds (bromelain) are verified via high-performance liquid chromatography (HPLC) and Fourier-transform infrared spectroscopy (FTIR) after which product activity or potency (strength of enzymatic activity) is assessed. In step #2, this information, as well as basic quality control testing (for heavy metals and microbial profile), is summarized in a certificate of analysis (COA) by grower and/or broker, and the product is available for import to the manufacturer. Upon acquiring the raw material, step #3, the dietary supplement manufacturer decides to accept the COA and may begin product formulation, or the material is placed into quarantine and independent quality control (QC) testing begun. This is one of the key differentiators of what determines the quality of finished products on the shelf. Since many of the raw materials, such as bromelain, are supplied by the same companies, it is the verification and validation of the COA and by testing each lot that makes the difference. Once the product passes this secondary QC testing, the manufacturer can, in step #4, blend, formulate, bottle, and send out the finished product for one final round of authentication testing. Those companies that pay for testing of each lot of every ingredient that they use are generally very willing to share this information. Lastly, this life cycle of a dietary supplement, albeit complex, is similar for almost all ingredients described under DSHEA. The manufacturers largely self-govern and are required to follow good manufacturing practices. The FDA can inspect facilities and hold manufacturers accountable for documenting and maintaining strict oversight of the life cycle, whereas the Federal Trade Commission reviews label claims and issues surrounding advertising.

One of the best series in understanding the complex process of dietary manufacturing and quality control was published in "Integrative Medicine: A Clinician's Journal" from 2007 to 2009 [31, 65–68]. In this brief series, the issues of quality control and quality assurance are clarified. "Simply put, quality control is conducting *testing* and performing the procedures to ensure that established quality specifications for raw material and finished products are met. Whereas quality assurance is the *strict regulation* of all of the *processes* used to create the final product." It is very important to consider that a manufacturer may have a robust quality assurance program while actually conducting very little quality controls testing. A dietary supplement manufacturer can have established raw material, and finished product specifications as part of a GMP certification process have assurance processes in place, but some or all its products may not meet standards of acceptable quality control. Such products can be low or high potency, contaminated, and adulterated (contain unwanted ingredients such as pharmaceutical drugs) and have poor stability and lose potency over the duration of their shelf life. Consider that the FDA cGMPs do not mandate that manufacturers establish high-quality specifications; they just must meet specifications, which may unfortunately turn out to be low quality.

43.9 Engaging Manufacturers

One of the best ways to ensure that dietary supplements are of highest quality is to ask the salesperson to differentiate between products or directly ask the manufacturer to answer your questions. As practitioners, patients, or providers who use dietary supplements, the goal is to ensure that products exceed all regulatory requirements and are clinically effective and safe for use. It is difficult to improve health and succeed without reputable, high-quality materials and services. Therefore, we want to work with outstanding dietary supplement manufacturers and their products that meet and exceed national standards set forth by DSHEA, FDA and USP, and GMP and published in the Federal Register. It is good to recognize suppliers who provide legitimate proof through documentation that their products consistently meet label claims, are authentic, and lack contamination or adulteration and assure raw material and finished product authenticity and stability. When trying to ensure that the supplements you are taking are safe, here are ten questions to consider asking dietary supplement manufacturers (▶ Box 43.4). This list is a summary from Dr. Liva's quality control and quality assurance series, which provides a more comprehensive and detailed list in the Supplier Certification Questionnaire, which he created and is accessible online [65, 67, 68].

> **Box 43.4 Ensuring Supplement Quality via Manufacturer-Directed Questions**
> 1. Does your quality control department have the authority to approve and reject procedures, specifications, test methods and results, raw ingredients and components, finished ingredients, packaging materials, and labels?
> 2. Do you have a cGMP system? Which do you follow: Ffood cGMPs or FDA cGMPs for dietary supplements?
> 3. Have you been independently certified for cGMP compliance via USP, NSF, and/or NPA?
> 4. Do you use an in-house lab or contract lab? How are they audited?
> 5. When conducting testing of raw materials, are they checked for microbiology profile (bacteria, yeast, and mold), identity (authenticate material or botanical genus and species), potency (if potency claim exists), heavy metals (lead, mercury, cadmium, arsenic), chemical solvent residue, aflatoxins, and herbicide and pesticide residue?
> 6. Do you accept a certificate of analysis (COA) in lieu of independent testing of raw materials?
> 7. Are your finished products tested for label claim potency prior to release for sale?
> 8. Do you put expiration dates or a use-by date on your products?
> 9. Do you conduct label claim stability testing, verifying product through its expiration date?

Reputable manufacturers should be able to provide written examples of #2, #3, #5, #8, #9, and #10 (▶ Box 43.4) above by providing the testing information for two or three lots of a random product you select from their catalog. The manufacturers may need assurance that this is for patient safety and educational purposes only and, if reputable, should respond quickly. It has been the experience of the author that if companies readily conduct these tests, PDFs can be emailed within 2–24 hours. If companies choose to exclude product details that may reveal intellectual property on a research or patent basis, then select another basic product from their catalog, such as a multivitamin, B complex, or a 1–5 botanical combination, to determine the degree to which the manufacturer is adhering to or exceeding quality control testing standards.

43.10 Clinical Relevance and Controversies

The health outcomes and clinical benefits attributed to dietary supplements remain highly controversial and debated topics. It is important to note that the field is regulated (see ▶ Box 43.5). In terms of the general benefit to society and individual populations, there are numerous studies which suggest that dietary supplements may improve overall health and specific aspects of quality of life. A report by the US Centers for Disease Control and Prevention suggests that three in ten people experience selected nutritional deficiencies [69]. Blumberg et al. reviewed nutrient intake data in more than 10,000 adults recorded in the National Health and Nutrition Examination Surveys and determined that, as compared to food alone, the added use of dietary supplements significantly increased an average of 15 out of 18 nutrients assessed and reduced nutrient inadequacy rates [69]. Elia and colleagues examined the cost-effectiveness of using nutritional supplements (energy-/protein-enriched functional foods, vitamins, and/or minerals) in community and home care environments [70]. Their group conducted a systematic review of high-quality publications and found that both short-term and long-term use produced significant cost savings. In addition to cost benefits, the authors noted clinically relevant benefits such as improved quality of life, lower infections, and postoperative complications.

> **Box 43.5 Government, Regulatory, and Association Resources**
> - *National Institutes of Health Office of Dietary Supplements:* ▶ https://ods.od.nih.gov/HealthInformation/makingdecisions.sec.aspx NIH ODS provides a variety of information for consumers, researchers, and professionals including dietary supplements fact sheets, consumer safety updates, and a dictionary of terms used throughout healthcare and supplement industry
> - *The National Center for Complementary and Integrative Health (NCCIH):* ▶ https://nccih.nih.gov/ NIH center focused on scientific research into diverse medical and healthcare systems, practices, and products
> - *The National Cancer Institute Office of Cancer Complementary and Alternative Medicine (OCCAM):* ▶ https://cam.cancer.gov/ NCI's office focused on research in CAM relating to cancer prevention, diagnosis, treatment, and symptom management
> - *The US Food and Drug Administration:* ▶ https://www.fda.gov/Food/DietarySupplements/ Information for consumers and industry on dietary supplement products, ingredients, recalls, and adverse event reporting
> - *The Herbal Medicine Institute and the American Botanical Council:* ▶ http://abc.herbalgram.org/site/PageServer Council provides reliable information on botanicals and natural products and advocates for consumers and industry.
> - *Consumer Healthcare Products Association (CHPA):* ▶ https://www.chpa.org/ CHPA advocates for consumers and industry and provides educational tools to ensure safe use of dietary supplements and OTC medicines
> - *Dietary Supplement Label Database:* ▶ http://dsld.nlm.nih.gov/dsld/ Searchable database of ingredients, products, and ~2500 manufacturers

There is also growing clinical evidence for the use of select categories of supplements in well-defined clinical populations. For example, the effect of dietary prebiotics on gut microflora and metabolism in human subjects was reviewed by Kellow and Reid [71, 72]. In 26 trials with more than 830 subjects, prebiotics were found to significantly reduce postprandial glucose and insulin, whereas effectiveness against lipids, inflammatory markers, and immune function was inconsistent. Green tea and its polyphenols have been determined to have numerous beneficial health effects [49]. Tea and its extracts act as antioxidants, reduce oxidative stress, provide defense against inflammation in cardiovascular disease, and modulate key metabolic pathways such as NF-kB and cyclooxygenase. Mechanistic studies have implicated a role for green tea in inflammatory bowel disease, apoptosis, neurodegeneration, fatty liver disease, cognitive function, and metabolic weight loss.

Turmeric, or *Curcuma longa*, has a long history of therapeutic use as a natural botanical in the Indian healing system of Ayurveda. In modern times, turmeric and its concentrated extracts are being investigated for its ability to act as an anti-inflammatory agent and improve dermatologic diseases [73], radiation-induced dermatitis [74], and arthritis [75]. Due to a lack of comprehensive and definitive research on efficacy, some investigators are calling for the public to abstain from use curcumin until both the benefits and potential side effects are thoroughly documented [76]. In addition to turmeric, one of the most commonly used and top selling herbs (from ◘ Table 43.1) is *Withania somnifera*, or Ashwagandha, also known as Indian ginseng. The therapeutic use of Ashwagandha spans psychiatric disorders to fertility. Modern research is beginning to validate its plant extracts to support patients in obsessive-compulsive disorder [77], fertility [78], sexual function [79], and anxiety [80].

When reviewing the strength of the literature and drawing clinical conclusions on individual supplements or categories, consider the robustness of the research design. ► Box 43.6 lists some common peer-reviewed sources for information. Is the research conducted in cell culture systems, in acceptable preclinical or translational animal models, with human subjects as uncontrolled observation study, or in an experimental research design (the most rigorous) such as a prospective randomized controlled trial [81]? Before making final judgments on clinical application, also take into consideration if the dietary supplement product categories are generally considered safe and supportive to the clinical area of concern.

In addition to the numerous potentially beneficial dietary supplements listed above, there remain several controversial categories that should be avoided. The majority of complaints to both FTC and FDA come from the dietary supplement specialty areas of sexual enhancement [82–88], weight loss [89–93], and sports/fitness [94–98]. Complaints in these categories range from false advertising, overaggressive sales tactics, and unsubstantiated health claims to product adulteration, contamination with prescription medications, "filler" dietary supplements, unnecessary window dressing, or poor-quality and untested product. Healthcare practitioners can check for regular FDA online alerts on fraudulent, adulterated, or contaminated supplements at ► fda.gov. Examples of products which have received warnings include:

> **Box 43.6   Peer-Reviewed Journals**
> - *Journal of Natural Products*: ► http://pubs.acs.org/journal/jnprdf
> - *Journal of Dietary Supplements*: ► http://www.tandfonline.com/toc/ijds20/current
> - *The American Journal of Clinical Nutrition*: ► http://ajcn.nutrition.org/
> - *Journal of Nutrition*: ► http://jn.nutrition.org/
> - *Advances in Nutrition*: ► http://advances.nutrition.org/
> - *Journal of the American College of Nutrition*: ► http://www.americancollegenutrition.org/content/the-journal
> - *Journal of Medicinal Food*: ► http://www.liebertpub.com/overview/journal-of-medicinal-food/38/
> - *The Chemistry of Natural Compounds*: ► https://link.springer.com/journal/10600
> - *Natural Product Reports*: ► http://www.rsc.org/journals-books-databases/about-journals/npr/
> - *Natural Product Research*: ► http://www.tandfonline.com/toc/gnpl20/current
> - *Journal of Nutrition and Diet Supplements*: ► http://www.scienceinquest.com/journal-of-nutrition-and-diet-supplements/journal-home.php
> - *Evidence-Based Complementary and Alternative Medicine*: ► https://www.hindawi.com/journals/ecam/
> - *BMC Complementary and Alternative Medicine*: ► https://bmccomplementalternmed.biomedcentral.com/
> - *Journal of Pharmacognosy and Natural Products*: ► https://www.omicsonline.org/pharmacognosy-natural-products.php

1. Sexual enhancement: Tyrannosaurus (contained Viagra), Rhino 69 (contained sildenafil), Triple Green Capsules (contained Viagra), Big Penis Male Sexual Stimulant (contained Viagra), Black Mamba (contained Viagra), Kingdom Honey for Him (contained Cialis), African Viagra (contained Viagra), Duramaxxx (contained Viagra), My Steel Woody (contained Viagra), and Own the Knight 1750 (contained tadalafil and dapoxetine).
2. Weight loss: Asia Slim (contained sibutramine, diazepam, and bisacodyl), Queen Slimming Soft Gel (contained fluoxetine and sibutramine), Physic Candy (contained sibutramine), Skinny Bee Diet (contained desmethylsibutramine), ABX Weight Loss (contained sibutramine), Ultimate Lean (contained sibutramine), Dream Body Extreme Gold (contained sibutramine, fluoxetine, and sildenafil), Ultimate Body Tox (contained sibutramine), and Fruta Planta Life or Garcinia Cambogia Premium (contained sibutramine).
3. Sports/fitness: the ► bodybuilding.com company alone conducted a nationwide recall of more than 50 supplements because of possible adulteration by prescription and nonprescription steroids marketed as "Superdrol," "Madol," "Tren," "Androstenedione," and/or "Turinabol." Other individual products to avoid include Msten Extreme Mass Builder, Penta Built, Cannibal Pro Cutting Stack, Mass Destruction, Mayhem and Wolverine Xtreme, or Beast Stack.

If the pitches and marketing, such as a 110% money back guarantee, or the product itself seems too good to be true, then it probably is [48, 59]. It is important to use common sense by avoiding these categories of dietary supplements altogether and focusing on lifestyle modification such as healthy diet, exercise, stress reduction, and maintaining proper sleep-work balance.

Clinically it is critical to address the patient's interest in these products and bring questions back to the entire healthcare team. An appropriate care plan may include laboratory testing that could rule in or out the need for taking specific nutrients, as well as aid in product dosage, duration of use, and need for retesting. It is now common to test for many vitamins and minerals, such as B6, B12, iron, zinc, magnesium, calcium, and specialty supplements that may be related to biologically active biomarkers (homocysteine, folate, CoQ-10, essential fatty acids), as well as for steroid hormone cascades (DHEA, pregnenolone, testosterone, and cortisol). Also, with increasing age and prescription drug use, the potential for interactions increases. Evaluating the potential mechanisms of action of supplements and their pathways of detoxification, such as inhibition or induction of Cytochrome P450 enzymes (CYP450), may be critical, especially for those on life-sustaining medications which act through similar pathways. For supplements to be the most clinically relevant and beneficial, they should be used with good clinical judgment in the context of the best available information.

43.11 Summary

For consumers and patients who commonly use integrative medicine modalities, the overall goal of adding dietary supplements into a treatment plan is to maintain health, prevent future health issues, lower the risk of requiring heroic interventions, and reduce reliance on prescription medication. The added goals of the treatment team are to help patients navigate the complex and fast-moving regulatory terrain of the supplement industry and to help choose the most appropriate products in the proper form at the right dosage. At the same time, product safety must be assured while limiting any potential side effects or drug interactions and validating that the treatment course will be effective or efficacious when possible. When selecting dietary supplements, consider that there are numerous issues relating to nomenclature. Understand the unique and varied forms, dosage, potential combinations, and additives. Always read supplement labels carefully, and clarify any questions on formulations or unfamiliar delivery systems such as phytosomes and nanoemulsions. The nutrition facts label should be clear and easy to read and provide information about ingredients, dosage, serving size, and any known contraindications. The healthcare team must openly discuss, in a non-threatening and non-judgmental manner, all health-related information on use of dietary supplements. If needed, both consumers and practitioners can actively engage and request clarifying and quality control testing information that covers the entire life cycle of the supplement from manufacturers within the nutraceutical marketplace. Product safety depends on numerous factors which include growing conditions, herbicides and pesticides used in cultivation, formulation, additives, importation, the manufacturing processes and the use of multistage testing procedures, as well as industry manufacturing and government regulation. Those manufacturers, who are reputable and of high quality, who utilize good manufacturing practices, and who perform independent quality control testing, should make testing results available (barring any intellectual property issues) for each lot that they manufacture. Many will provide this information directly to the healthcare practitioner, which is critical to truly ensure product quality. In order to avoid unwanted exposure to adulterated and potentially contaminated dietary supplements, avoid the "terrible three" categories which include sexual enhancement, weight loss, and sports and fitness. Also avoid products that advertise by way of 2:30 a.m. infomercials. Acquiring a more comprehensive understanding of the dietary supplement landscape will help to ensure safer use, improve the clinical experience for those participating in healthcare, and maximize opportunities for health benefits.

References

1. Cohen MH. Complementary and integrative medical therapies, the FDA, and the NIH: definitions and regulation. Dermatol Ther. 2003;16:77–84.
2. IOM report. Complementary and alternative medicine in the United States. National Academies Publication. Washington, D.C. 2005. http://www.nationalacademies.org/hmd/Reports/2005/Complementary-and-Alternative-Medicine-in-the-United-States.aspx.
3. Clarke TC, Black LI, Stussman BJ, Barnes PM, Nahin RL. Trends in the use of complementary health approaches among adults: United States, 2002–2012. Natl Health Stat Report. 2015;79:1–16.
4. Radimer K, Bindewald B, Hughes J, Ervin B, Picciano MF. Dietary supplement use by US adults: data from the National Health and nutrition examination survey, 1999–2000. Am J Epidemiol. 2004;160(4):339–49.
5. Nahin RL. Expenditures on complementary health approaches: United States, 2012. Natl Health Stat Report. 2016;(95):1–11.
6. Egan B, Hodgkins C, Shepherd R, Timotijevic L, Raats M. An overview of consumer attitudes and beliefs about plant food supplements. Food Funct. 2011;2(12):747–52. https://doi.org/10.1039/c1fo10109a7.
7. Smith T, Kawa K, Eckl V, Morton C, Stredney R. Herbal supplement sales in US increase 7.7% in 2016. HerbalGram. 2017;(115):56–65.
8. Karny-Rahkovich O, Blatt A, Elbaz-Greener GA, Ziv-Baran T, Golik A, Berkovitch M. Dietary supplement consumption among cardiac patients admitted to internal medicine and cardiac wards. Cardiol J. 2015;22(5):510–8.
9. Tuna S, Dizdar O, Calis M. The prevalence of usage of herbal medicines among cancer patients. J BUON. 2013;18(4):1048–51.
10. Levy I, Attias S, Ben Arye E, Goldstein L, Schiff E. Interactions between dietary supplements in hospitalized patients. Intern Emerg Med. 2016;11(7):917–27. https://doi.org/10.1007/s11739-015-1385-3.
11. Qadour E, Ben-Arye E, Goldstein L, Attias S, Schiff E. [Dietary supplements use during hospitalization] English abstract. Harefuah. 2015;154(1):39–42, 67. http://europepmc.org/abstract/MED/25796674.
12. Gonsalves CDRS, Stavinoha MAJT, Usaf SP, Hite CL, Costa CJ, Dilly CCLG, et al. Dietary supplements in the department of defense: possible solutions to optimizing force readiness. Mil Med. 2012;177(12):1464–71.

13. Busari AA, Mufutau MA. High prevalence of complementary and alternative medicine use among patients with sickle cell disease in a tertiary hospital in Lagos, South West, Nigeria. BMC Complement Altern Med. 2017;17(1):1–8.
14. Knapik JJ, Steelman RA, Hoedebecke SS, Austin KG, Farina EK, Lieberman HR. Prevalence of dietary supplement use by athletes: systematic review and meta-analysis. Sport Med. 2016;46(1):103–23.
15. Rawson ES, Miles MP, Larson-Meyer DE. Dietary supplements for health, adaptation, and recovery in athletes. Int J Sport Nutr Exerc Metab. 2018;28:188–99. https://doi.org/10.1123/ijsnem.2017-0340.
16. Owens C, Baergen R, Puckett D. Online sources of herbal product information. Am J Med. 2014;127(2):109–15. http://www.sciencedirect.com/science/article/pii/S0002934313008401.
17. Raynor DK, Dickinson R, Knapp P, Long AF, Nicolson DJ. Buyer beware? Does the information provided with herbal products available over the counter enable safe use? BMC Med. 2011;9(1):94. http://www.biomedcentral.com/1741-7015/9/94.
18. Temple NJ. The Marketing of Dietary Supplements in North America : the emperor is (almost) naked. J Altern Complement Med. 2010;16(7):803–6. https://doi.org/10.1089/acm.2009.0176.
19. PricewaterhouseCooper-worlds-newsletter-foods-final 2012. https://www.pwc.com/gx/en/retail-consumer/pdf/rc-worlds-newsletter-foods-final.pdf.
20. Research B. Nutraceuticals: global markets. Vol. FOD013F, A BBC Research Food & Beverage Report. 2017.
21. Global Dietary Supplements Market will reach USD 220.3 Billion in 2022. [Internet] Zion Market Research. Sarasota; 2017. https://www.zionmarketresearch.com/news/dietary-supplements-market.
22. PricewaterhouseCoopers. Food as pharma. [Internet] 2012. https://www.pwc.com/gx/en/retail-consumer/pdf/rc-worlds-newsletter-foods-final.pdf.
23. Jiang T. Re-thinking the dietary supplement laws and regulations 14 years after the dietary supplement health and education act implementation. Int J Food Sci Nutr. 2009;60(4):293–301.
24. Wollschlaeger B. The dietary supplement and health education act and supplements. Diet Nutr Suppl. 2003;22:387–91.
25. Mcguffin M. Dietary supplements stakeholder forum. In: US pharmacopeial convention. [internet] 2016. Last accessed 8 Nov 2018. http://www.usp.org/sites/default/files/usp/document/get-involved/stakeholder-forums/briefing-material-2017-06-07_1.pdf.
26. Nicoletti M, Toniolo C. Analysis of multi-ingredient food supplements by fingerprint HPTLC approach. J Chem Chem Eng. 2015;9:239–44. https://doi.org/10.17265/1934-7375/2015.04.001.
27. Lewis CA, Jackson MC. Understanding medical foods under FDA regulations. In: Bagchi D, editor. Nutraceutical and functional food regulations in the United States and around the world. 2nd ed. Academic Press a division of Elsevier: Washington, D.C.; 2014. p. 169–82.
28. Ross S. Functional foods: the food and drug administration perspective 1–3. Am J Clin Nutr. 2000;71(suppl):1735S–8S.
29. Schmitz SM, Lopez HL, Mackay D. Nutravigilance: principles and practices to enhance adverse event reporting in the dietary supplement and natural products industry. Int J Food Sci Nutr. 2014;65(2):129–34. https://doi.org/10.3109/09637486.2013.836743.
30. Guideline for assigning titles to USP Herbal Medicines Compendium monographs. Herbal Medicines Compendium, US Pharmacopeial Convention; 2014. p. 1–5. https://hmc.usp.org/sites/default/files/documents/HMC/GN_Resources/HMC%20Nomenclature%20Guidelines%20v.%201.0.pdf.
31. Liva R. Facing the problem of dietary-supplement heavy-metal contamination: how to take responsible action. Integr Med. 2007;6(3):36–8.
32. Bell A. The importance of dietary supplement verification. [internet 2018] Last accessed 8 Nov 2018. http://www.usp.org/news/importance-of-dietary-supplement-verification.
33. Ehrenpreis ED, Kulkarni P, Burke C. FDA-related matters Committee of the American College of gastroenterology. What gastroenterologists should know about the Gray market, herbal remedies, and compounded pharmaceuticals and their regulation by the Food and Drug Administration. Am J Gastroenterol. 2013;108(5):642–6. https://doi.org/10.1038/ajg.2012.348.34.
34. Bucchini L, Rodarte A, Restani P. The PlantLIBRA project: how we intend to innovate the science of botanicals. Food Funct. 2011;2(12):769. http://xlink.rsc.org/?DOI=c1fo10150a.
35. Nisly NL, Gryzlak BM, Zimmerman MB, Wallace RB. Dietary supplement polypharmacy: an unrecognized public health problem. Evid Based Complement Alternat Med. 2010;7(1):107–13.
36. Silano V, Coppens P, Larrañaga-Guetaria A, Minghetti P, Roth-Ehrang R. Regulations applicable to plant food supplements and related products in the European Union. Food Funct. 2011;2(12):710. http://xlink.rsc.org/?DOI=c1fo10105f.
37. Franz C, Chizzola R, Novak J, Sponza S. Botanical species being used for manufacturing plant food supplements (PFS) and related products in the EU member states and selected third countries. Food Funct. 2011;2(12):720. http://xlink.rsc.org/?DOI=c1fo10130g.
38. Wren R. Potter's new encyclopedia of botanical drugs & preparations. Harper Col: Harper & Row; 1972. 400 p.
39. Ajazuddin SS. Applications of novel drug delivery system for herbal formulations. Fitoterapia. 2010;81(7):680–9. https://doi.org/10.1016/j.fitote.2010.05.001.
40. Mu W, Zhang T, Jiang B. An overview of biological production of L-theanine. Biotechnol Adv. 2015;33(3–4):335–42.
41. Department of Health and Human Services OIG. Adverse event reporting for dietary supplements: an inadequate safety valve [Internet]. 2001. https://oig.hhs.gov/oei/reports/oei-01-00-00180.pdf.
42. Albertini B, Di Sabatino M, Calogerà G, Passerini N, Rodriguez L. Encapsulation of vitamin a palmitate for animal supplementation: formulation, manufacturing and stability implications. J Microencapsul. 2010;27(2):150–61.
43. Thakkar K, Billa G. Treatment of vitamin B12 deficiency–Methylcobalamine? Cyancobalamine? Hydroxocobalamin?—clearing the confusion. Eur J Clin Nutr. 2015;69(1):1–2. https://doi.org/10.1038/ejcn.2014.165.
44. Argiratos V, Samman S. The effect of calcium carbonate and calcium citrate on the absorption of zinc in healthy female subjects. Eur J Clin Nutr. 1994;48(3):198–204. http://search.ebscohost.com/login.aspx?direct=true&db=mnh&AN=8194505&site=ehost-live&scope=site.
45. Yip R. The challenge of improving iron nutrition: limitations and potentials of major intervention approaches. Eur J Clin Nutr. 1997;51:S16–24.
46. Walczyk T, Tuntipopipat S, Zeder C, Sirichakwal P, Wasantwisut E, Hurrell RF. Iron absorption by human subjects from different iron fortification compounds added to Thai fish sauce. Eur J Clin Nutr. 2005;59(5):668–74.
47. van den Berg SJPL, Serra-Majem L, Coppens P, Rietjens IMCM. Safety assessment of plant food supplements (PFS). Food Funct. 2011;2(12):760. http://xlink.rsc.org/?DOI=c1fo10067j.
48. Krochmal R, Hardy M, Bowerman S, Lu Q-Y, Wang H-J, Elashoff RM, et al. Phytochemical assays of commercial botanical dietary supplements. Evid Based Complement Alternat Med. 2004;1(3):305–13.
49. Oz HS. Chronic inflammatory diseases and green tea polyphenols. Nutrients. 2017;9(6):1–14.
50. Londhe VP, Gavasane AT, Nipate SS, Bandawane DD, Chaudhari PD. Review role of garlic (Allium Sativum) in various diseases: an overview. J Pharm Res Opin. 2011;4:129–34.
51. Shi Z-Q, Song D-F, Li R-Q, Yang H, Qi L-W, Xin G-Z, et al. Identification of effective combinatorial markers for quality standardization of herbal medicines. J Chromatogr A. 2014;1345:78–85. https://doi.org/10.1016/j.chroma.2014.04.015.
52. Szucs V, Banati D. [trends in the utilization of food additives]. English abstract. Orv Hetil. 2013;154(46):1813–9.
53. Li YO, Lam J, Diosady LL, Jankowski S. Antioxidant system for the preservation of vitamin a in ultra rice. Food Nutr Bull. 2009;30(1):82–9.
54. Evaluation of certain food additives and contaminants. Eightieth report of the Joint FAO/WHO Expert Committee on Food Additives.

55. Hoadley JE, Craig Rowlands J. FDA perspectives on food label claims in the United States. In: Nutraceutical and functional food regulations in the United States and Around the World. 2nd ed. Elsevier Inc.; 2014. p. 121–40.
56. Ellwood KC, Trumbo PR, Kavanaugh CJ. How the US food and drug administration evaluates the scientific evidence for health claims. Nutr Rev. 2010;68(2):114–21.
57. Barrett S. How the dietary supplement health and education act of 1994 weakened the FDA [Internet] Quackwatch 2015. https://www.quackwatch.org/02ConsumerProtection/dshea.html.
58. Barrett S. Why Consumers Need More Protection Against Claims. Int J Toxicol. 2003;22:391–2.
59. Tarn DM, Guzmán JR, Good JS, Wenger NS, Coulter ID, Paterniti DA. Provider and patient expectations for dietary supplement discussions. J Gen Intern Med. 2014;29(9):1242–9. https://doi.org/10.1007/s11606-014-2899-5.
60. Soni MG, Thurmond TS, Miller ER, Spriggs T, Bendich A, Omaye ST. Safety of vitamins and minerals: controversies and perspective. Toxicol Sci. 2017;118(2):348–55.
61. Gad SC, Gad SE. Are dietary supplements safe as currently regulated? The great debate. Int J Toxicol. 2003;22:381–5.
62. Navarro VJ, Khan I, Björnsson E, Seeff LB, Serrano J, Hoofnagle JH. Liver injury from herbal and dietary supplements. Hepatology. 2017;65(1):363–73.
63. Secor ER, Szczepanek SM, Singh A, Guernsey L, Natarajan P, Rezaul K, et al. LC-MS/MS identification of a bromelain peptide biomarker from *Ananas comosus merr*. Evid Based Complement Alternat Med. 2012;2012:548486.
64. Secor ER, Singh A, Guernsey LA, McNamara JT, Zhan L, Maulik N, et al. Bromelain treatment reduces CD25 expression on activated CD4+ T cells in vitro. Int Immunopharmacol. 2009;9(3):340–6.
65. Liva R. Which road to superior quality? Integr Med. 2007;6(4):18–9.
66. Liva R. The difference between quality control and quality assurance: using an example of pomegranate extract. Integr Med. 2009;8(3):48–50.
67. Liva R. When it comes to quality assurance, Assume Nothing. 2007;6(1):30–33.
68. Liva R. Seeking high-quality products: whose definition should we believe? (part I). Integr Med. 2008;7(5):46–8.
69. Blumberg JB, Frei B, Fulgoni III VL, Weaver CM, Zeisel SH. Contribution of dietary supplements to nutritional adequacy in race/ethnic population subgroups in the United States. Nutrients. 2017;9(12). pii: E1295. https://doi.org/10.3390/nu9121295. http://www.ncbi.nlm.nih.gov/pubmed/29182574.
70. Elia M, Normand C, Laviano A, Norman K. A systematic review of the cost and cost effectiveness of using standard oral nutritional supplements in community and care home settings. Clin Nutr. 2016;35(1):125–37. https://doi.org/10.1016/j.clnu.2015.07.012.
71. Kellow NJ, Coughlan MT, Savige GS, Reid CM. Effect of dietary prebiotic supplementation on advanced glycation, insulin resistance and inflammatory biomarkers in adults with pre-diabetes: a study protocol for a double-blind placebo-controlled randomised crossover clinical trial. BMC Endocr Disord. 2014;14:55.
72. Kellow NJ, Coughlan MT, Reid CM. Metabolic benefits of dietary prebiotics in human subjects: a systematic review of randomised controlled trials. Br J Nutr. 2014;111(7):1147–61.
73. Vaughn AR, Branum A, Sivamani RK. Effects of turmeric (*Curcuma longa*) on skin health: a systematic review of the clinical evidence. Phytother Res. 2016 Aug;30(8):1243–64.
74. Palatty PL, Azmidah A, Rao S, Jayachander D, Thilakchand KR, Rai MP, et al. Topical application of a sandalwood oil and turmeric based cream prevents radiodermatitis in head and neck cancer patients undergoing external beam radiotherapy: a pilot study. Br J Radiol. 2014;87(1038):20130490.
75. Daily JW, Yang M, Park S. Efficacy of turmeric extracts and Curcumin for alleviating the symptoms of joint arthritis: a systematic review and meta-analysis of randomized clinical trials. J Med Food. 2016;19(8):717–29.
76. Unlu A, Nayir E, Dogukan Kalenderoglu M, Kirca O, Ozdogan M. Curcumin (Turmeric) and cancer. J BUON. 2016;21(5):1050–60.
77. Jahanbakhsh SP, Manteghi AA, Emami SA, Mahyari S, Gholampour B, Mohammadpour AH, et al. Evaluation of the efficacy of *Withania somnifera* (Ashwagandha) root extract in patients with obsessive-compulsive disorder: a randomized double-blind placebo-controlled trial. Complement Ther Med. 2016;27:25–9.
78. Sengupta P, Agarwal A, Pogrebetskaya M, Roychoudhury S, Durairajanayagam D, Henkel R. Role of *Withania somnifera* (Ashwagandha) in the management of male infertility. Reprod Biomed Online. 2018;36(3):311–26. https://doi.org/10.1016/j.rbmo.2017.11.007.
79. Dongre S, Langade D, Bhattacharyya S. Efficacy and safety of Ashwagandha (*Withania somnifera*) root extract in improving sexual function in women: a pilot study. Biomed Res Int. 2015;2015:284154.
80. Pratte MA, Nanavati KB, Young V, Morley CP. An alternative treatment for anxiety: a systematic review of human trial results reported for the Ayurvedic herb ashwagandha (*Withania somnifera*). J Altern Complement Med. 2014;20(12):901–8.
81. Hartung DM, Touchette D. Overview of clinical research design. Am J Health Syst Pharm. 2009;66(4):398–408.
82. Zou P, Hou P, SS-Y O, Ge X, Bloodworth BC, Low M-Y, et al. Identification of benzamidenafil, a new class of phosphodiesterase-5 inhibitor, as an adulterant in a dietary supplement. J Pharm Biomed Anal. 2008;47(2):255–9.
83. Reepmeyer JC, Woodruff JT, d'Avignon DA. Structure elucidation of a novel analogue of sildenafil detected as an adulterant in an herbal dietary supplement. J Pharm Biomed Anal. 2007;43(5):1615–21.
84. Poplawska M, Blazewicz A, Zolek P, Fijalek Z. Determination of flibanserin and tadalafil in supplements for women sexual desire enhancement using high-performance liquid chromatography with tandem mass spectrometer, diode array detector and charged aerosol detector. J Pharm Biomed Anal. 2014;94:45–53.
85. Pages G, Gerdova A, Williamson D, Gilard V, Martino R, Malet-Martino M. Evaluation of a benchtop cryogen-free low-field (1) H NMR spectrometer for the analysis of sexual enhancement and weight loss dietary supplements adulterated with pharmaceutical substances. Anal Chem. 2014;86(23):11897–904.
86. Low M-Y, Li L, Ge X, Kee C-L, Koh H-L. Isolation and structural elucidation of flibanserin as an adulterant in a health supplement used for female sexual performance enhancement. J Pharm Biomed Anal. 2012;57:104–8.
87. Cohen PA, Venhuis BJ. Adulterated sexual enhancement supplements: more than mojo. JAMA Intern Med. 2013;173(13):1169–70.
88. Balayssac S, Gilard V, Zedde C, Martino R, Malet-Martino M. Analysis of herbal dietary supplements for sexual performance enhancement: first characterization of propoxyphenyl-thiohydroxyhomosildenafil and identification of sildenafil, thiosildenafil, phentolamine and tetrahydropalmatine as adulterants. J Pharm Biomed Anal. 2012;63:135–50.
89. Zeng Y, Xu Y, Kee C-L, Low M-Y, Ge X. Analysis of 40 weight loss compounds adulterated in health supplements by liquid chromatography quadrupole linear ion trap mass spectrometry. Drug Test Anal. 2016;8(3–4):351–6.
90. Pawar RS, Grundel E. Overview of regulation of dietary supplements in the USA and issues of adulteration with phenethylamines (PEAs). Drug Test Anal. 2017 Mar;9(3):500–17.
91. Kim JY, Park HJ, Kim JW, Lee JH, Heo S, Yoon C-Y, et al. Development and validation of UPLC and LC-MS/MS methods for the simultaneous determination of anti-obesity drugs in foods and dietary supplements. Arch Pharm Res. 2016;39(1):103–14.
92. Kim HJ, Lee JH, Park HJ, Cho S-H, Cho S, Kim WS. Monitoring of 29 weight loss compounds in foods and dietary supplements by LC-MS/MS. Food Addit Contam Part A Chem Anal Control Expo Risk Assess. 2014;31(5):777–83.

93. Hachem R, Assemat G, Martins N, Balayssac S, Gilard V, Martino R, et al. Proton NMR for detection, identification and quantification of adulterants in 160 herbal food supplements marketed for weight loss. J Pharm Biomed Anal. 2016;124:34–47.
94. Gabriels G, Lambert M, Smith P, Wiesner L, Hiss D. Melamine contamination in nutritional supplements – is it an alarm bell for the general consumer, athletes, and "weekend warriors"? Nutr J. 2015;14:69.
95. Eichner A, Tygart T. Adulterated dietary supplements threaten the health and sporting career of up-and-coming young athletes. Drug Test Anal. 2016;8(3–4):304–6.
96. Cox HD, Eichner D. Detection of human insulin-like growth factor-1 in deer antler velvet supplements. Rapid Commun Mass Spectrom. 2013;27(19):2170–8.
97. Cho S-H, Park HJ, Lee JH, Do J-A, Heo S, Jo JH, et al. Determination of anabolic-androgenic steroid adulterants in counterfeit drugs by UHPLC-MS/MS. J Pharm Biomed Anal. 2015;111:138–46.
98. Attipoe S, Cohen PA, Eichner A, Deuster PA. Variability of stimulant levels in nine sports supplements over a 9-month period. Int J Sport Nutr Exerc Metab. 2016;26(5):413–20.

Clinical Approaches to Monitoring and Evaluation of the Chronic Disease Client

Cynthia Bartok and Kelly Morrow

44.1   Monitoring and Evaluation as a Step in the Nutrition Care Process – 771
44.1.1   Definitions of Monitoring and Evaluation – 771
44.1.2   Purpose of Monitoring and Evaluation – 772
44.1.3   Relationship to Other NCP Steps – 772
44.1.4   Process of Monitoring and Evaluating Outcomes – 773
44.1.5   Critical Thinking Aspects of Monitoring and Evaluation – 774

44.2   Monitoring and Evaluation Data – 774
44.2.1   Nutrition Care Outcomes and Indicators – 774
44.2.2   Selection of Nutrition Care Outcomes and Indicators – 775
44.2.3   Monitoring and Evaluation of Integrative and Functional Nutrition Data – 775

44.3   Practical Application of Monitoring and Evaluation Principles – 776
44.3.1   Developing and Charting the Monitoring and Evaluation Plan – 776
44.3.2   Developing and Charting Professional Goals – 776
44.3.3   Developing and Charting the Follow-Up Plan – 777
44.3.4   Developing and Charting the After-Visit Summary Goals – 777
44.3.5   Alignment and Integration of the Nutrition Diagnosis, Intervention, Professional Goal, and After-Visit Summary in an Outpatient Clinic Example – 778
44.3.6   Alignment and Integration of the Nutrition Diagnosis, Intervention, Professional Goal, and After-Visit Summary in a Nutrition Support Example – 779

© Springer Nature Switzerland AG 2020
D. Noland et al. (eds.), *Integrative and Functional Medical Nutrition Therapy*,
https://doi.org/10.1007/978-3-030-30730-1_44

44.4 Conclusion – 779
44.4.1 Assessment – 779
44.4.2 Diagnosis/PES Statement – 780
44.4.3 Intervention – 780
44.4.4 Professional Goals – 780
44.4.5 Follow-Up Plan – 780
44.4.6 After-Visit Summary – 780
44.4.7 Assessment – 781
44.4.8 Diagnosis/PES Statements – 781
44.4.9 Intervention – 781

44.5 Interventions – 781
44.5.1 Professional Goals – 781
44.5.2 Follow-Up Plan – 782
44.5.3 After-Visit Summary – 782

References – 782

Clinical Approaches to Monitoring and Evaluation of the Chronic Disease Client

Learning Objectives

By the end of this chapter, the reader will be able to:

- Describe the role of monitoring and evaluation from a functional medicine perspective in providing high-quality care to clients
- Explain how nutrition professionals use nutrition care indicators to monitor and evaluate the progress of clients
- Apply monitoring and evaluation principles including professional goal setting in a case example

44.1 Monitoring and Evaluation as a Step in the Nutrition Care Process

44.1.1 Definitions of Monitoring and Evaluation

Monitoring and evaluation is considered the fourth and final step of the four-step Nutrition Care Process (NCP) approach (◘ Fig. 44.1). This stage of the NCP is unique in that it often spans multiple visits. At the initial visit, monitoring and

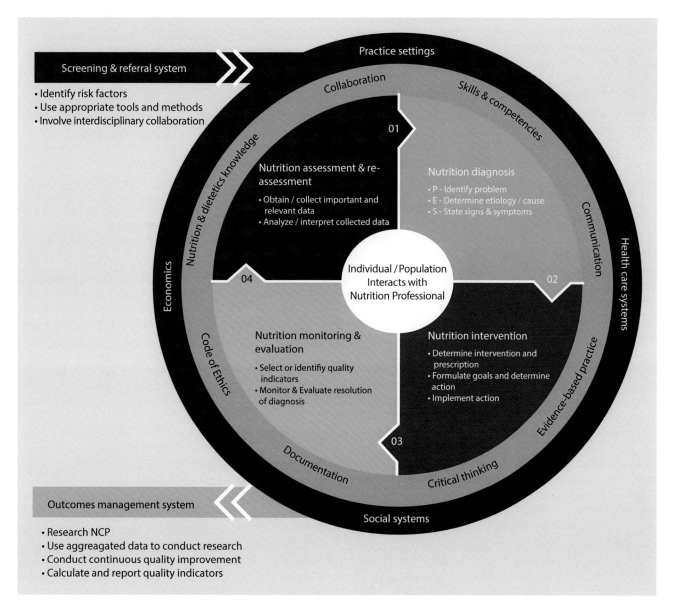

◘ Fig. 44.1 The Nutrition Care Process (NCP) and Model. (Reprinted from Swan et al. [11]. With permission from Elsevier)

evaluation activities often include working with the client to assess whether the intervention delivered that day has had an impact on client outcomes, as well as setting a plan for follow-up care. The plan for follow-up may include decisions about when the client will return to the clinic (outpatient setting) or be seen again by the clinician (inpatient setting). In addition, the clinician often will develop (or "stage") a plan for interventions that will occur at subsequent follow-up visits and discuss that plan with the client.

At the follow-up visit, the clinician continues to engage in monitoring and evaluation activities. The clinician reassesses data points from the initial visit and gathers new information to determine whether the intervention was successful. If it was not, the clinician explores the reasons why.

Nutrition monitoring is defined as a "preplanned review and measurement" of key health and nutrition-related data points that can provide information about the nutritional status of the client [1]. Collection of monitoring data can begin at the initial visit. For interventions that are largely focused on educating the patient about a new diet that needs to be adopted, the clinician can assess the client's understanding of key aspects of the diet and its application throughout the session. Many counseling interventions are aimed at promoting self-efficacy of the client and/or increasing motivation to make changes. The clinician also can assess the client's self-efficacy or motivation for making a behavior change during the initial visit. In both these examples, by reassessing key variables during the initial session, the clinician not only gains immediate feedback about the effectiveness of the intervention (nutrition monitoring) but also increases the likelihood that the client will be successful in implementing the desired behavioral changes in the interim period until follow-up time. In all cases, the clinician also will assess key data points at the follow-up visit to determine if the patient continues to understand key concepts about the new therapeutic diet, continues to have self-efficacy and motivation for change, has implemented new health behaviors, and has experienced any potential impact of these changes on health and nutritional status.

After collection of this key data at the initial and follow-up visits, the clinician engages in nutrition evaluation, which includes a "systematic comparison" of this collected data to baseline (pre-intervention) data or desired outcomes of the intervention [1]. As with nutrition monitoring, it is possible to begin nutrition evaluation at the first visit to evaluate the immediate impact and effectiveness of education and counseling interventions with the client. However, since the ultimate goal is to impact client knowledge, motivation, self-efficacy, and behavior change in the long-term, nutrition evaluation conducted at follow-up visits is the highest priority feedback. Data collected at follow-up visits help the clinician to optimally evaluate the effectiveness of the intervention on the client's nutritional status and an opportunity to reflect on whether a different approach would be more effective for the client's subsequent interventions.

44.1.2 Purpose of Monitoring and Evaluation

While clinicians have always been interested in providing high-impact, effective interventions that result in improved health of their patients, the current healthcare environment is highly focused on demonstrating the cost-effectiveness of prevention and treatment efforts for chronic disease [2]. By engaging in monitoring and evaluation activities, the clinician is committing to ensuring that the client is receiving the most effective care possible. The clinician and client both receive feedback about progress at key points (initial visit, follow-up visit) and can respond to that feedback by revisiting interventions or engaging in different interventions that may be more effective. In addition, when used in combination with NCP-driven terminology and a focus on nutrition-specific outcomes, clinicians can contribute to an internationally unified approach to provision of nutrition care and measurement of outcomes of care. This type of data can be analyzed and published to document optimal types of interventions for a variety of clinical conditions as well as document the value of nutrition care with the chronic disease patient.

An integrative and functional nutrition (IFN) approach can be compatible with the standardized approach recommended by the NCP. There is substantial overlap between the data gathered during assessment, monitoring, and evaluation in both approaches. When approaches are not completely aligned, the NCP terminology is general enough to be adapted to functional assessment parameters, including digestion, metabolism, inflammation, oxidative stress, lifestyle influences, and the nutrition-focused physical exam. In 2011, the Academy of Nutrition and Dietetics published the Standards of Practice and Standards of Professional Performance for Registered Dietitians in Integrative and Functional Medicine [3]. This document provides specific guidance on how to merge the two approaches, including the use of the Integrative and Functional Medical Nutrition Therapy (IFMNT) Radial (◘ Fig. 44.2).

44.1.3 Relationship to Other NCP Steps

Monitoring and evaluation is the fourth and final step of the NCP. However, in practical application, outcomes-based clinical care means that monitoring and evaluation is considered at every stage of the NCP. During assessment, potential key outcome variables are identified and measured to serve as a baseline value from which to mark improvement (see ► Chap. 40). For the nutrition diagnosis, the defining characteristics section of the nutrition diagnosis terminology reference sheets of the Electronic Nutrition Care Process Terminology (eNCPT) specifically lists the assessment data points that are most indicative of the nutrition problem. As such, these data points not only are included in the signs/symptoms portion of the PES (problem-etiology-signs and symptoms) statement to provide direction for the monitoring statement but also serve as optimal data points for moni-

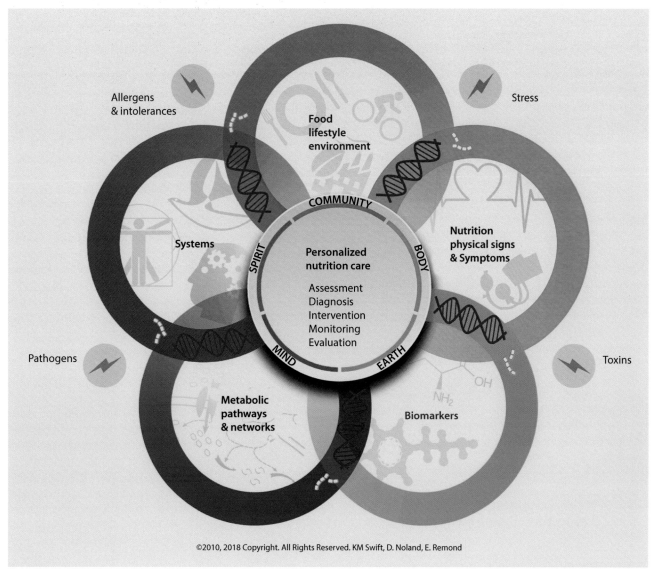

■ Fig. 44.2 The Radial: Integrative and Functional Medical Nutrition Therapy. (Courtesy of Kathie Madonna Swift, Diana Noland, Elizabeth Redmond)

toring the resolution of the nutrition problem. The intervention is specifically designed to address the problem stated in the PES statement, as well as result in improvement in the outcomes listed in the signs/symptoms portion. During the monitoring and evaluation stage, the clinician will reassess these outcome variables (monitoring) and evaluate them against predetermined standards to determine whether the client's health is improving (evaluation).

44.1.4 Process of Monitoring and Evaluating Outcomes

As described earlier, monitoring and evaluation activities can begin even at the initial visit. The clinician may reassess key variables such as nutrition knowledge, ability to implement a dietary change (skills), or motivation, self-efficacy, and/or readiness for behavior change. In addition, the clinician and patient collectively discuss a plan for following up, such as when the next visit will occur, when the patient is likely to see results of their behavior changes, when they will need a change of approach, and how long they may need to follow a specific therapeutic diet. Clinicians need to be aware that a patient may not ever return or return on a different time frame than recommended. Clearly communicating about the follow-up plan will help the patient receive maximum benefit and limit the potential harms from an intervention by helping them understand time limits on interventions such as temporary dietary changes or therapeutic supplement use.

At subsequent visits, the clinician often will start the appointment by asking specific and detailed questions that assess the patient's:

— Understanding of the specific intervention delivered at the previous visit
— Success with behavior changes discussed at the previous visit

- Ability to fully implement the intervention
- Barriers to fully implementing the intervention
- Experience of positive and negative outcomes from the intervention

The clinician uses this data to determine whether the nutrition intervention delivered at the previous visit has resulted in behavior change and improvement in nutritional and health status. If the reassessment reveals that the nutrition problem named in the PES statement still exists and is still a priority, the clinician will continue to focus that visit on exploration of the nature, cause, and significance of that problem. As described above, this could include a more in-depth investigation of IFN indicators that may be contributing to the patient's lack of improvement and additional interventions aimed at addressing IFN contributions to the nutrition problem.

The clinician may identify health and behavioral problems during reassessment that are outside of the clinician's expertise or scope of practice. The clinician needs to be vigilant and attentive to situations when a patient would benefit from a referral to other care providers for integrative medical treatment. If the nutrition problem identified at the previous visit (PES statement) has resolved or become a lower priority, the clinician may spend time during the appointment completing a full reassessment of the patient. The clinician determines if new data are necessary to identify the nature, cause, and significance of other potential high-priority nutrition problems.

44.1.5 Critical Thinking Aspects of Monitoring and Evaluation

Critical thinking skills are needed to successfully use monitoring and evaluation concepts, including an understanding of the interconnected nature of body systems and influences on health. As described previously, the clinician considers outcomes-based care at all stages of the Nutrition Care Process to increase the likelihood of the patient experiencing an improvement in nutritional and health status from the intervention. This begins with the selection of optimal outcomes at the outset and assessing those variables during initial data-gathering activities.

A related decision point is choosing the appropriate reference or standard by which to evaluate the collected data. For example, nutrition clinicians often use the Dietary Reference Intakes (e.g., RDA) to evaluate the adequacy of intake, despite the fact that this reference standard was not intended to be used to describe the needs of individuals with ongoing disease processes or those with unique nutritional needs [4]. Using body mass index (BMI) to evaluate the weight status and adiposity of adults is one of the most controversial areas in the practice of anthropometric assessment. Alternative approaches to evade misuse of available reference standards include comparing the client with a previous/baseline value to assess change or comparing the client to an individualized recommendation stated in the nutrition prescription (see ▶ Chap. 42). If clinicians use standards or criteria such as the Dietary Reference Intakes, prescriptive diets (e.g., Mediterranean diet, elimination diets, or low-FODMAP diets), or reference data (e.g., BMI classifications, laboratory values), they need to be thoroughly versed in the limitations of those standards and the populations in which they have been validated. If using functional laboratory data, it is important to ensure the lab is CLIA (Clinical Laboratory Improvement Amendments) certified.

The clinician uses critical thinking skills to piece together the puzzle of why the patient may not have experienced as much improvement as expected. A monitoring and evaluation appointment lends itself to a deeper discovery of the interconnected relationships of functional core areas when trying to solve a complex health challenge. The clinician thoroughly explains in the chart any shortfalls from the expected outcomes and the factors that have affected progress. This explanation provides documentation that is helpful to both the clinician (for understanding the patient's needs over time) and other healthcare professionals who participate in the patient's care.

While nutrition clinicians often have long-term working relationships with patients with chronic diseases, there are circumstances when the patient may be discharged from care. The most obvious example is the patient who has had complete resolution of significant nutrition problems. With chronic disease patients, this situation typically occurs after a prolonged phase of "weaning off" of visits with the healthcare provider during which time the clinician and client have successfully navigated behavior change principles such as lapse and relapse prevention. Other patients with chronic diseases may occasionally go through periods of time when they would prefer to take a "vacation" from self-care. While this may be alarming for the nutrition clinician, it is important to express empathy with the client's feelings, acknowledge the goal to support the client through any situation, and indicate an ongoing interest in providing nutrition care whenever the patient is interested in returning to care. The nutrition care provider documents this situation, communicates it to other key providers in the patient's medical home, and provides the patient with a referral to any other care providers that may be helpful during this time (e.g., mental health specialist).

44.2 Monitoring and Evaluation Data

44.2.1 Nutrition Care Outcomes and Indicators

Nutrition care outcomes are measurable data points that may change with implementation of the nutrition intervention or provision of care by the nutrition professional. *Nutrition care indicators* are a subset of nutrition care outcomes that are considered the most responsive and most sensitive data points for demonstrating improvement in or resolution of a nutrition diagnosis [5].

The Nutrition Assessment and Monitoring and Evaluation Terminology lists provide examples of nutrition care outcomes and indicators in the established domains of assessment data [6]:
- Food/nutrition-related history
- Biochemical data, medical tests, and procedures
- Anthropometric measurements
- Nutrition-focused physical findings

Even though the terminology is largely shared between assessment and monitoring/evaluation steps in the NCP, there are two distinct differences. First, not all data points that are included in assessment are considered to be nutrition care outcomes or indicators. Often these are data points that may provide feedback to the care provider about the nature and severity of the disease process, not necessarily a specific nutrition problem. Examples include such lab and test values as resting metabolic rate and toxicology reports or such physical exam data as body temperature or heart rate. Second, client history data (i.e., demographics and social history) are unlikely to be impacted by nutrition interventions and therefore are not optimal nutrition outcomes or indicators [7].

Nutrition care outcomes and indicators are distinctly different from other types of data discussed below (disease outcomes, healthcare outcomes, patient outcomes). Nutrition care outcomes and indicators are those data points that the nutrition practitioner can directly impact [7]. Nutrition care outcomes and indicators are important to select and use within the medical chart as they demonstrate the ability of nutrition clinicians and nutrition interventions to provide measurable and documentable benefits to patients and healthcare systems. Over time, this will lead to the ability to demonstrate the cost-effectiveness and efficacy of nutrition care in improving patient health and well-being, a desperately needed evidence base for the nutrition profession.

As nutrition care outcomes and indicators change (improve), the client is likely to experience improvements in other measurable healthcare outcomes such as:
- Health and disease status: changes in the severity, duration, and risk level of a disease process
- Cost outcomes: changes in the cost of providing care such as medications, medical interventions and tests, and length of stay
- Patient outcomes: changes in factors that impact the patient, such as quality of life and functional status [7]

Broadly trained medical providers such as naturopathic doctors, medical doctors, physician's assistants, and nurse practitioners may feel more comfortable tracking these healthcare outcomes. However, since they are likely to occur only after successful implementation of nutrition and physical activity changes over longer periods of time, we encourage all healthcare providers to track nutrition care outcomes and indicators when providing nutrition care.

44.2.2 Selection of Nutrition Care Outcomes and Indicators

Nutrition care outcomes and indicators can be gathered from the health record or referral letters, from patient interviews, and through direct measurements made by the clinician. Most clinicians face an overwhelming volume of data to sort through when completing an assessment and reassessment. Skilled clinicians become adept at identifying and focusing on the data points that are the highest priority for monitoring. Clinicians determine which information is the highest priority by considering the patient's medical diagnoses and their severity, the nutrition diagnoses that have been charted, the nutrition interventions delivered, and the desired outcomes of nutrition care (see [8] as an example). In addition, practical considerations, such as the patient's health insurance coverage, the practice setting, and time limits on appointments, may impact the selection of nutrition care outcomes and indicators. For example, in a hospital or integrative clinical setting, the clinician may have easier access to laboratory values and medical tests, making these readily available and potentially valid indicators of the outcomes of nutrition care. In an outpatient setting, where the client has the freedom to self-select food and beverage intake, food and nutrient intake data tend to be more important outcomes of care.

As clinicians monitor data, it is important to ensure the data are valid and evaluated against appropriate standards or criteria (see discussion above). As clinicians review data from the medical chart and interview the patient, they need to consider the possibility that the data are not valid. Patients may intentionally or unintentionally offer erroneous information. Medical errors can and do occur, including errors in test procedure results and laboratory data. Furthermore, as clinicians collect their own data and add it to the chart, they should always use techniques on which they are well-trained; make measurements using high-quality, calibrated equipment; and follow an established standardized measurement protocol. As the nutrition professional is likely to contribute unique, detailed data on the patient's food and nutrition intake to the chart, it is of utmost importance that this data be collected using as reliable and valid techniques as possible (see ▶ Chap. 9).

44.2.3 Monitoring and Evaluation of Integrative and Functional Nutrition Data

Listening to the patient's story is a foundation of an IFN assessment. When a patient returns for a follow-up visit, the process of monitoring and evaluation allows for a deepening of the relationship between provider and client and a more thorough examination of the patient's personal and medical history. As the clinician gains the patient's trust, the patient's story may unfold, and its inherent complexities may become more apparent. This allows the clinician to identify the affected IFN core areas of health, draw connections between them, and continue to build a comprehensive treatment plan. An intake form that gathers details from each functional core area is very helpful.

Often a patient will have a chief complaint, such as hypertension, that needs to be addressed at the first visit. Using an IFN assessment tool such as *NIBLETS* (nutrition, inflammation, biochemical individuality, lifestyle, energy and metabolism, toxic load, and stress), the clinician may identify that the patient also has several inflammatory conditions (arthritis, elevated antinuclear antibody test), poor sleep, a high level of psychological stress, and exposure to environmental toxins that potentially increase blood pressure. While initial interventions may center around direct dietary interventions for hypertension, such as increasing fruit and vegetable consumption to increase potassium or decreasing ultra-processed foods to reduce sodium, subsequent visits will allow for greater education and discussion around the potential interconnected nature of inflammation, stress, sleep, and toxins on blood pressure. Because most health conditions have multifactorial causes, a clinician can skillfully use multiple visits to expand the treatment plan to assess a wide scope of influences in a stepwise fashion that is not overwhelming to the patient.

When using the Nutrition Care Process, IFN assessments may lead to nutrition diagnoses such as food- and nutrition-related knowledge deficit, undesirable food choices, predicted suboptimal intake, predicted excessive intake, and altered nutrition-related laboratory values. In most cases, the information that is provided in the etiology and signs and symptoms of the diagnosis or PES statement provides direction for the monitoring and evaluation phase.

44.3 Practical Application of Monitoring and Evaluation Principles

44.3.1 Developing and Charting the Monitoring and Evaluation Plan

Bastyr Center for Natural Health (BCNH) is one of Bastyr University's two nonprofit teaching clinics. Located in a Seattle neighborhood, BCNH is the largest natural health clinic in the Pacific Northwest, offering as many as 35,000 patient-visits a year [9]. The clinic serves as a training site for many of the university's health professions students, including Didactic Program in Dietetics students (students training to become registered dietitian nutritionists) as well as coordinated nutrition-counseling psychology students. University educators provide information and training about the NCP and our institutional charting standards through the use of case studies in clinical nutrition courses during the first year of the program. In the second year of the program, students attend two rotations at BCNH, applying the NCP to patients and charting those encounters through an electronic charting system that reflects NCP steps and terminology. The Nutrition Department at BCNH set up the monitoring and evaluation process to accomplish the following:

− Focus the nutrition clinician's appointment on outcome-based care
− Set up professional benchmarks that measure the level of success of nutrition care
− Communicate to the healthcare team and billing which medical diagnoses or problems the nutrition clinician is collaboratively treating
− Plan and deliver complex nutrition interventions over multiple visits
− Collaborate with patients in developing goals surrounding behavior change and follow-up treatment

44.3.2 Developing and Charting Professional Goals

For each intervention, the nutrition clinician writes at least one professional goal. The professional goal serves as an official record of the desired nutrition care outcome that results from the intervention delivered. Stating the desired nutrition care outcomes encourages the nutrition clinicians to develop realistic, achievable, high-impact goals that provide a measure of the success of nutrition care. In addition, student clinicians are encouraged to consider potential professional goals throughout all stages of the NCP, so they continually remain focused on the desired goal of the nutrition intervention (see ▸ Sect. 44.1.3).

At BCNH, nutrition clinicians include three required elements when charting professional goals:

1. The professional goal must state the primary health concern or medical diagnosis that is being collaboratively treated through nutrition care. This communicates to the healthcare team, clinic administration, and insurance companies the value nutrition care can add when managing patient illnesses and diseases.
2. The goal must include a comparative standard or reference value that follows the client through the NCP (see ▸ Sect. 44.1.5). This "criteria" value flows through the chart's assessment, diagnosis, nutrition prescription, and intervention sections to maintain cohesion in approach and thinking around managing the nutrition problem. This topic is covered in greater detail later in this chapter.
3. The goal should include *SMART goal* elements:
   − *Specific*: The goal includes the comparative standard whenever possible. For saturated fat, "no more than 7 percent of calories" is more specific than "reduced saturated fat."
   − *Measurable*: The goal is written so that when the patient is reassessed, the nutrition clinician should be able to say "yes" or "no" about whether the client has achieved the goal.

- *Accountable:* The goal states who is responsible for achieving it. Usually this is the client, but some goals, such as referrals (coordination of care), require the clinician to act.
- *Reachable/realistic:* The goal needs to be realistic for the client or nutrition clinician. Behavior change goals for patients often include interim steps toward achieving the optimal outcomes as well as an understanding that clinicians don't expect perfection in behavior change (e.g., "5 out of 7 days per week").
- *Timely:* The goal states a specific time frame for goal achievement, such as the end of today's visit, at the next visit, or some point in the future. Be practical; don't wait a year to see if your client has lost 30 pounds.

See ▶ Box 44.1 for examples of professional goals for a variety of settings, nutrition problems, and desired outcomes for interventions. See ▶ Box 44.2 for a self-check tool based on BCNH's quality control standards for charting professional goals.

Box 44.1
- To improve blood lipid levels, at 1 month follow-up, a 24-hour recall will show that the patient's reported consumption of saturated fat is <7% of kcal.
- To reduce cancer risk, food logs gathered at the 2-week follow-up will show the client has consumed at least three cups/day of vegetables on at least 5 of 7 days per week.
- To improve coordination of care for iron deficiency anemia, the RDN will notify patient's physician about food-drug interactions with iron supplement within 48 hours.
- To manage hypertension, at the end of today's visit, patient will express a willingness to modify diet toward DASH diet standards.

Box 44.2
Each professional goal within the chart:
- Forms a single sentence that is grammatically correct
- Includes a health outcome (when possible)
- Includes a comparative standard or criteria (when possible)
- Includes all five SMART form elements: specific, measurable, accountable, realistic, and timely
- Is a high-impact outcome that is (1) within the provider's scope of practice and (2) likely to result from the nutrition intervention
- Supports a logical flow with the related PES, nutrition prescription, intervention, and after-visit summary

44.3.3 Developing and Charting the Follow-Up Plan

As clinicians complete Step 2 of the NCP, the nutrition diagnosis (see ◘ Fig. 44.1), a key critical thinking activity is to prioritize the patient's nutrition diagnosis. At integrative clinics, nutrition clinicians typically select and intervene on only the top one or two nutrition diagnoses at each visit.

However, the clinicians chart a list of additional nutrition problems, topics for education, and/or progressive steps in a complex nutrition intervention that should occur at subsequent visits in the "follow-up" section of the chart. When the patient returns and is reassessed during monitoring and evaluation activities, this follow-up plan may still be relevant, and its presence in the chart saves the clinician time. If the reassessment reveals new problems or indicates the need for reprioritization of problems, the follow-up plan would be adapted to reflect the most current assessment information.

44.3.4 Developing and Charting the After-Visit Summary Goals

The after-visit summary is a written document provided to clients at their visit that outlines specific instructions about how to implement the nutrition intervention to improve their health. For example, this document may include specific behavior changes, such as increasing consumption of particular therapeutic foods, engaging in additional self-education related to their health concerns or therapeutic diet, monitoring food patterns, journaling about concerns and feelings related to food and eating, and/or seeking out the help of other healthcare providers. Whenever possible, nutrition clinicians work with patients to write positive goals ("Do this") rather than negative goals ("Don't do this").

At BCNH, nutrition clinicians include three required elements when charting the after-visit summary:

- The goals focus on concrete, realistic, behaviors that the client has collaboratively developed with the clinician.
- The goals, if completed, directly lead to successful accomplishment of the nutrition clinician's professional goals.
- The goals are written in language that is accessible to the client – both in terms of general literacy, health literacy, and nutrition literacy.

See ▶ Box 44.3 for examples of after-visit summary goals for a variety of settings, nutrition problems, and desired outcomes for interventions. See ▶ Box 44.4 for a self-check tool based on BCNH's quality control standards for charting after-visit summary goals.

Box 44.3
Thank you for coming in today, [insert patient name]. Here are the goals we discussed for today's visit on 8/10/2017:
- Go shopping on Sundays and purchase at least 14 pieces of fruit.
- Wash and dry fruit when you get home, and put ten pieces in a bag to take to work, and place the remainder in a bowl on the kitchen counter.
- Your goal is to consume two pieces of fresh fruit per day either while at work and/or at home at least 5 days each week.
- Check off on a calendar the days you met this goal, and bring the calendar to the next appointment for feedback to the clinician.
- Note any barriers that prevented you from achieving this goal on the calendar.

> **Box 44.4**
> My after-visit summary:
> — Is about concrete, realistic, behavioral changes the client will do, using positive language instead of "don't do this" goals
> — Is in simple, nontechnical language and simple steps that improve patient compliance
> — Includes important and effective behavior changes that will lead to achievement of the nutrition clinician's professional goal
> — Supports a logical flow with the related PES, nutrition prescription, intervention, and professional goal

44.3.5 Alignment and Integration of the Nutrition Diagnosis, Intervention, Professional Goal, and After-Visit Summary in an Outpatient Clinic Example

Case Study 44.1 provides an example showing how a nutrition clinician uses outcomes-based care to align and integrate all stages of the NCP. During the assessment phase of the intervention, the clinician selects optimal criteria for evaluating the patient's data after considering the patient's situation, disease state, and the clinician's setting, resources, and skills. In this case, the clinician has selected the Dietary Approaches to Stop Hypertension (DASH) Food Plan and the DASH food and nutrient targets as an optimal disease-specific set of standards for assessing the patient's baseline food and nutrient intake [10]. The clinician assesses both food and nutrient intake using a 24-hour recall against DASH standards and then documents that suboptimal intake of both fruits and vegetables is leading to a suboptimal intake of potassium, a key therapeutic nutrient needed to regulate blood pressure. The clinician also charts the etiology of these suboptimal nutrition intake patterns, to support the PES statement charted and to place the first nutrition intervention in context.

> **Case Study 44.1**
> Situation: Patient would benefit from optimizing potassium intake to manage hypertension; goal is foods-based solution focusing on increasing intake of fruit and vegetables.
> *Excerpts from chart sections below show alignment and integration across all NCP steps and chart note sections.*

In the diagnosis phase of the NCP, the clinician could opt to write a PES statement noting the discrepancy between actual food intake patterns and the prescribed 2000 kcal DASH diet food recommendations (nutrition prescription) using an "undesirable food choices" diagnosis (NB-1.7). This is an optimal approach in a setting where the clinician prefers food-based approaches and/or lacks access to dietary analysis software or the time to engage in this activity. In the example provided, the clinician has elected to approach this situation as a nutrient-based problem (inadequate mineral intake), works in a setting that has structural resources to measure potassium intake (dietary assessment software, time to complete analysis), and intends to monitor and evaluate this nutrition care indicator at follow-up points.

The two interventions that the nutrition clinician delivers build the patient's knowledge of the therapeutic diet, address known barriers to implementation, and increase the patient's self-efficacy and motivation for making the necessary dietary changes. Note that the nutrition clinician has delivered education and behavior change instruction around food choices, rather than nutrients, even though the clinician will be monitoring nutrient intake "behind the scenes." The intervention includes instruction in patient recordkeeping behaviors that will provide the nutrition clinician with helpful feedback about the interventions, including how often the patient has successfully followed through on grocery shopping, fruit intake, and vegetable intake goals, as well as barriers encountered when implementing the interventions.

For monitoring and evaluation, the clinician can intermittently assess patient understanding of key aspects of the nutrition interventions at the initial visit, such as:
— Does the patient understand why more fruits, vegetables, and potassium intake are therapeutic for treating hypertension?
— Has the patient resolved the discrepancy between saving money when grocery shopping and desiring a non-pharmaceutical resolution to their hypertension?
— Can the patient verbally state (with support of nutrition education handouts) the amount of fruits and vegetables that equal to one DASH serving?
— Can the patient verbally state the ultimate goals for fruit and vegetable intake and the interim goals set for this first intervention?
— How likely would the patient rate (on a scale of 1–10) the accomplishment of key behaviors listed on their after-visit summary sheet?

Upon the patient's return to the clinic in 3 weeks, the nutrition clinician will first review the patient's log to evaluate whether the patient has successfully completed grocery shopping to support the dietary change and discuss any barriers encountered. Without successful achievement of this goal, it is unlikely food and nutrient patterns have changed and time may need to be diverted to other types of data collection and interventions at that visit. If a quick review of grocery shopping and food intake behaviors shows relative success, the clinician could invest the time in collecting and analyzing a 24-hour recall to evaluate potassium intake and determine if the professional goal was achieved. Based on reviews of all this data, the nutrition clinician is in a good position to determine whether items on the follow-up list (other DASH food groups, physical activity) could be implemented at that visit or if the client needs to continue to work on fruit and vegetable intake. Additionally, at the follow-up visit, the clinician can address functional core areas that may contribute to hypertension, including psychological health (stress level), lack of sleep, potential toxic exposures, inflammation, and oxidative stress, if not addressed during the first visit.

44.3.6 Alignment and Integration of the Nutrition Diagnosis, Intervention, Professional Goal, and After-Visit Summary in a Nutrition Support Example

Case Study 44.2 provides an example showing how a nutrition clinician uses outcomes-based care to align and integrate all stages of the NCP for a patient that is enterally fed. In the assessment phase of the NCP, the clinician gathers key data regarding the patient's significant weight loss over prior 5 years and the reasons that the patient's enteral prescription has failed to meet his needs (limited involvement of an RDN in care). Since there are few established quantitative criteria for intake of anti-inflammatory nutrients in patients experiencing gastrointestinal inflammatory disorders, the clinician qualitatively assesses the diet as lacking in key dietary components.

> **Case Study 44.2**
>
> Situation: Patient on nutrition support (enteral feeding) for the past 5 years would benefit from optimizing fatty acid and phytonutrient intake to manage risks of long-term enteral feeding and severe inflammatory state; initial goals include weight stabilization and improved intake of anti-inflammatory fatty acids and phytonutrients.
>
> *Excerpts from chart sections show alignment and integration across all NCP steps and chart note sections.*

In the diagnosis phase of the NCP, the clinician identifies the highest concern as the patient's significant weight loss and likely malnutrition from 5 years of underfeeding. Secondary concerns include optimizing anti-inflammatory fatty acids and phytochemicals. The clinician writes three separate PES statements to describe the nature and etiology of these nutrition problems.

All of these concerns are addressed today in the two interventions conducted by the RDN. The RDN provides education to the patient and his wife for how to modify the formula to increase the calorie content and provide anti-inflammatory fatty acids and plant phytochemicals. After the visit, the RDN will communicate with the patient's doctor to request both short- and long-term changes in the nutritional composition of the patient's diet to better meet his nutritional needs.

For monitoring and evaluation, the clinician has already assessed the patient's understanding of how to prepare the new formula safely at today's visit and has instructed the patient in the use of a log to gather key outcome data for the next visit. Upon the patient's return to the clinic in 2 weeks, the nutrition clinician will first review the patient's log to evaluate whether the patient has prepared and consumed the new formula and then to review the impact of the intervention on patient quality of life and indicators of bowel health (bowel movement frequency). The clinician also will measure the patient's weight to determine if additional adjustments to calorie intake are warranted. Based on the review of all this data, discussions with the patient's doctor, and review of any prior laboratory tests, the nutrition clinician will be able to determine whether there should be any additional changes to the nutrition prescription at the follow-up visit.

44.4 Conclusion

Monitoring and evaluation is the fourth and final step of the Nutrition Care Process (NCP). However, in the current environment of outcomes-based care, nutrition clinicians are encouraged to think about the value of monitoring and evaluation across all stages of the NCP. This would include collection of critical baseline data during nutrition assessment, use of this data in the problem-etiology-signs and symptoms (PES) statement written as a nutrition diagnosis, and intervention activities directly intended to impact or address the nutrition problem and its key signs and symptoms. Utilization of an integrative and functional approach will allow for a deeper exploration of contributors of poor health using the NIBLETS mnemonic (nutrition, inflammation, biochemical individuality, lifestyle, energy and metabolism, toxic load, and stress). Monitoring and evaluation can occur during and immediately after the delivery of interventions designed to impact nutrition attitudes, beliefs, values, knowledge, and practical skills. At subsequent visits, the clinician can anticipate progress on food intake or physical activity patterns that may eventually lead to changes in anthropometry, body biochemistry or functioning, physical exam parameters, quality of life, disease process indicators, and medical costs. Well-developed systems for charting nutrition care can enhance the clinician's critical thinking skills in providing effective, outcomes-based nutrition care that enhances the quality of life for patients.

44.4.1 Assessment

Typical intake patterns do not match DASH goals for patient's estimated calorie needs. Fruit intake (2 servings/d) is suboptimal (50% of goal 4 servings/d). Vegetable intake (3 servings/d) is suboptimal (75% of goal 4 servings/d). Estimated intake of potassium is 2000 mg (42% of goal 4700 mg/d). Patient perceives eating more fruits and vegetables will be costly and time-consuming. Patient motivated to find non-pharmaceutical method to treat hypertension. IFN assessment reveals presence of elevated ANA, arthritis, poor sleep, occupational toxin exposure (carbon dioxide), and high levels of psychological stress.

44.4.2 Diagnosis/PES Statement

Inadequate mineral intake (Potassium, NI-5.10.1.5) related to perception that consumption of fruits and vegetables is expensive and time-consuming as evidenced by estimated intake of potassium of 2000 mg/d (42% of DASH goal 4700 mg/d).

44.4.3 Intervention

Nutrition prescription 2000 kcal DASH diet food plan

- **Interventions**

Nutrition Counseling – Strategies – Problem Solving (C-2.4): Opened a conversation about patient's concerns about the cost of therapeutic foods (vegetables and fruit) and desire for management of hypertension without medication. Supported client in generating a list of potential strategies to lower the cost of purchasing potassium-rich produce – patient offered shopping weekly specials at store, purchasing produce in season, and selecting frozen fruits and vegetables. After discussing pros and cons of various approaches and specific weekly amount of produce the patient would need to purchase and consume to meet intake goals, the patient decided that purchasing weekly specials and frozen produce would be the most cost-effective way to support achieving an intake approaching four servings per day of fruit and four servings per day of vegetables. Provided the patient with a list of high-potassium fruits and vegetables to guide choices in the grocery store. Patient agreed to shop once per week for produce.

Nutrition Education – Content – Priority Modifications (E-1.2): Reviewed the DASH Eating Plan handout with patient, focusing on fruit and vegetable groups, especially those with high naturally occurring nitrates (beets, carrots, arugula) to boost nitric oxide concentration for vasodilation. Reinforced previous intervention's content regarding the serving size for different types of fruits and vegetables and choices that were high in potassium. Collaboratively generated a goal for daily intake of three servings per day of both fruits and vegetables as an interim step toward the ultimate goal of four servings per day of each food group. Patient agreed to log the number of servings of fruit and vegetable consumed each day on a simple food log and make notes about any barriers encountered at follow-up in 3 weeks.

44.4.4 Professional Goals

To manage hypertension, a 24-hour recall at the 3-week follow-up visit will show that the patient is consuming 3600 mg potassium (77% of nutrition prescription) through foods.

44.4.5 Follow-Up Plan

3 weeks; potential topics: new food groups in DASH diet pattern, brainstorm substitutions for ultra processed foods to increase whole foods intake, physical activity of 150 minutes/week brisk walking, increasing anti-inflammatory compounds in the diet, managing environmental exposures including supporting organs of elimination, exploring the need for referrals for improving sleep and reducing stress.

44.4.6 After-Visit Summary

- Thank you for coming in today, [insert patient name]. Here are the goals we discussed for today's visit on 8/1/2017:
  - Go shopping on Sundays, and purchase enough fruit and vegetables to support your goal of three servings per day of fruit and three servings per day of vegetables with an emphasis on high nitrate options as discussed.
  - Fruit: 3 servings per day = 21 servings per weekly shopping trip. This could look like:
    - 21 pieces of medium-sized fruit
    - 2–3 pounds of frozen fruit
    - A mix of fresh and frozen fruit such as 10 pieces of fresh fruit and 1 pound frozen fruit
  - Vegetables: 3 servings per day = 21 servings per weekly shopping trip. This could look like:
    - 10 cups of chopped, mixed fresh vegetables
    - 2–3 pounds of frozen vegetables
    - A mix of fresh and frozen fruit such as 5 cups fresh vegetables and 1 pound frozen vegetables
  - Consult your list of high-potassium fruits and vegetables to aid in selecting high-potassium options whenever possible.
  - Check off on a calendar the Sundays that you met this shopping goal, and bring the calendar to your next appointment
  - Starting today, aim to consume a minimum of 3 servings of fruit and 3 servings of vegetables each day:
    - One serving fruit = 1 medium fruit, ½ cup frozen fruit
    - One serving vegetables = ½ cup raw, frozen, or cooked vegetables, 1 cup raw leafy vegetables
  - Try a breakfast fruit and vegetable smoothie, salads at lunch and dinner, and fruit for snacks to help you reach this goal.
  - Check off on your simple food log when you consume a serving of fruit or vegetables to track the number of days you achieve the target of three servings each of fruits and vegetables. Bring the log back with you to the next appointment.

- Return to clinic in 3 weeks for follow-up so we can review the plan and talk about other ways you can decrease blood pressure by increasing anti-inflammatory compounds in your diet, improving your sleep, reducing stress, and managing environmental exposures.

44.4.7 Assessment

Client Hx: A 56-year-old male with history of type 2 diabetes, iron-deficiency anemia, GERD and esophageal lesions requiring G-tube feeding, unintended weight loss/cachexia, COPD with asthma, chronic disease of teeth. *Food/nutrition:* Current prescribed intake is 8 × 8 oz cans Glucerna 1.5 enteral formula/day (1896 mL × 1.5 kcal/mL = 2844 kcal/d), which is 89% of estimated kcal needs. PUFA intake (n–3) is 3% of kcal (9.5 g). Enteral nutrition originally initiated during hospital visit; no referral to dietitian. Patient and wife seeking dietitian consult independently. RDN confirmed G-tube is greater than the 14 French diameter to allow for homemade tube feedings.
Medications include:
- Ferrous Sulfate Iron 325 mg (60 mg elemental iron) BID, monitored quarterly
- Omega-3 fish oil: 684 mg 1 cap QD
- Triamcinolone acetonide 0.5% topical cream
- Glipizide 5 mg BID before meals
- Pantoprazole SOD DR 40 mg 1 tab QD
- Metformin HCL 500 mg 1 tab BID before meals
- Tamsulosin HCL 0.4 mg 1 cap ½ hour following AM meal QD
- Proscar 5 mg 1 tab QD
- Enalapril maleate 2.5 mg QD

Anthropometric: height, 74.0 in./187.9 cm; current weight, 154 lb/70.0 kg (87% of IBW); previous weight 250 lb/113.6 kg 5 years ago (38% weight loss in 5 years). *Biochemical:* Hgb 15.2 g/dL (Low end WNL); HCT 43% (Low end WNL); HbA1C 6.8% (WNL for DM); FBS 92 mg/dL (WNL for DM); CRP 13.6 mg/L (H). *Nutrition-focused physical exam:* complaints of fatigue and dyspnea. Reports constipation (BM Q 3–5 days); improved BM frequency with two cups of coffee per day (BM Q 1–2 days).

44.4.8 Diagnosis/PES Statements

Unintended weight loss (NC-3.2) due to mismatch between enteral formula prescription and patient nutritional needs as evidenced by 38% weight loss over the past 5 years.

Intake of types of fats inconsistent with needs (NI-5.5.3, omega-3 fatty acids) due to long-term use of enteral formula with suboptimal fatty acid profile for inflammatory conditions as evidenced by inadequate omega-3 fatty acid intake compared to nutrition prescription.

Inadequate bioactive substance intake (NI-4.1) due to long-term use of enteral diet as evidenced by low estimated intake of plant phytonutrients and presence of inflammation (elevated CRP).

44.4.9 Intervention

Nutrition prescription:
- REE/kcals: 40–45 kcal/kg × 70 kg = 2800–3150 kcal/d
- Protein (g/kg): 2 g/kg = 140 g
- Fluids (ml/kg): 40 mL/kg = 3240 mL

Other: 6 × 325 ml Kate Farms (brand) 1.5 Whole Foods Peptide Formula (3000 kcals/144g/protein/38mg Fe). Recommended additional nutrients added to the formula via a blender to meet needs for calories, iron, essential fatty acids, antioxidants and anti-inflammatory nutrients: 2 opened capsules multistrain probiotic (10–50 billion CFU/capsule), 500 mg buffered vitamin C powder, 1 avocado, substitute fish oil with 2500 mg of an Omega 3-6-9 supplement. Additionally, provide 2 cups coffee/d via tube for bowel motility. Flush tube with water after each feeding. Twice a day, add 40 mg liquid ferrous bisglycinate to 240 mL water and flush into tube.

44.5 Interventions

Bioactive substance management – Other (essential fatty acids, ND-3.3.9): instructed patient on therapeutic additions to current enteral nutrition feeding plan. Patient will mix 5 × 325 mL containers of Kate Farms 1.5 formula with the ingredients listed in nutrition prescription and divide into two portions (breakfast and dinner), plus two cups of coffee per day for bowel function. Provided detailed instructions on how to make formula and answered patient questions. Patient and wife express willingness to prepare new formula and knowledge of how to safely prepare it. Patient and wife agreed to complete provided log to record preparation and consumption of new formula. Will also record any concerns or positive/negative outcomes including bowel movements. Follow-up visit in 2 weeks.

Coordination of nutrition care by a nutrition professional – Collaboration and Referral of Nutrition Care – Collaboration with other providers (RC-1.4): RDN to contact PCP to request diet prescription change to support restoration of nutritional status and stabilization of weight. RDN also will request labs from prior 2 years and request PCP obtain Spectracell Lab Micronutrient Analysis or Genova Nutreval to assist with evaluation of patient's nutritional status.

44.5.1 Professional Goals

To manage weight loss and inflammation, food logs collected at the 2-week follow-up will show consumption of formula preparation that meets the nutrition prescription on at least 6 of 7 days per week.

To coordinate care, the RDN will follow up with patient's PCP within 48 hours of visit to discuss recommendations for diet prescription change and request laboratory tests.

44.5.2 Follow-Up Plan

2 weeks; assess weight change and bowel habits, review logs with patient/wife to determine if new formula was well-tolerated, review labs (if obtained) with patient, discuss any change in enteral formula (if approved). Assess for GERD and blood glucose levels in response to diet changes. Consider switching iron to a bisglycinate iron, which is more absorbable.

44.5.3 After-Visit Summary

- Thank you for coming in today, [insert patient name]. Here are the goals we discussed for today's visit on 8/1/2017:
  - Follow food safety procedures for formula preparation and storage:
    - Wash hands with soap and water prior to preparing foods.
    - Thoroughly clean blender in hot soapy water after each use.
    - Prepare formula in a clean workspace using clean utensils and a clean can opener.
    - Wash the lids of all cans to remove dust and debris.
    - Thoroughly clean avocado with a scrub brush under running water.
    - Refrigerate all prepared formula that is not immediately consumed.
    - Discard any unused formula within 48 hours.
  - Using your blender, mix 5 × 325 mL containers of Kate Farms 1.5 enteral formula with the following additional ingredients:
    - 1 small avocado
    - 500 mg buffered vitamin C powder
    - 2500 mg Omega 3-6-9 (liquid or squeezed from a capsule)
    - Contents of two opened capsules probiotic 10–50 billion per CFU/capsule (discard capsule)
  - Divide mixed formula into 4–5 portions; consume the modified enteral formula for evenly spaced meals and a snacks (for example ~500 mL at 8 am, 11 am, 2 pm, 5 pm and 8 pm).
  - Have two cups of coffee per day to help with bowel function
  - 30 minutes before breakfast and again before bed, add 40 mg liquid ferrous bisglycinate to 90 mL water and infuse into tube, then flush with 90 mL water
  - Add at least 1.5 liters of additional fluid to the tube in divided doses each day as a flush before and after feedings (90 mL each) or when feeling thirsty.
  - Please use the log to record your observations:
    - Record your preparation and consumption of the new formula.
    - Record any problems or concerns with preparing or consuming new formula.
    - Record positive/negative outcomes.
    - Record bowel movements.
  - Schedule and attend a follow-up visit in 2 weeks.

References

1. Academy of Nutrition and Dietetics. Nutrition care process and nutrition monitoring and evaluation. In: Nutrition Terminology Reference Manual (eNCPT): Dietetics Language for Nutrition Care [Internet]. Chicago: Academy of Nutrition and Dietetics; 2017 [cited 2017 Aug 15]. Available from: https://ncpt.webauthor.com/pubs/idnt-en/page-067.
2. Academy of Nutrition and Dietetics. Position of the Academy of Nutrition and Dietetics: the role of nutrition in health promotion and chronic disease prevention. J Acad Nutr Diet. 2013;113:972–9.
3. Ford D, Raj S, Batheja RK, DeBusk R, Grotto D, Noland D, et al. American dietetic association: standards of practice and standards of professional performance for registered dietitians (competent, proficient, and expert) in integrative and functional medicine. J Am Diet Assoc. 2011;111(6):902–913.e23.
4. Academy of Nutrition and Dietetics. Practice paper of the American Dietetic Association: Using the dietary reference intakes. J Am Diet Assoc. 2011;111:762–70.
5. Academy of Nutrition and Dietetics. Nutrition care indicators. In: Nutrition Terminology Reference Manual (eNCPT): Dietetics Language for Nutrition Care [Internet]. Chicago: Academy of Nutrition and Dietetics; 2017 [cited 2017 Aug 15]. Available from: https://ncpt.webauthor.com/pubs/idnt-en/page-070.
6. Academy of Nutrition and Dietetics. Nutrition assessment domains. In: Nutrition Terminology Reference Manual (eNCPT): Dietetics Language for Nutrition Care [Internet]. Chicago: Academy of Nutrition and Dietetics; 2017 [cited 2017 Aug 15]. Available from: https://ncpt.webauthor.com/pubs/idnt-en/page-033.
7. Academy of Nutrition and Dietetics. Nutrition care outcomes. In: Nutrition Terminology Reference Manual (eNCPT): Dietetics Language for Nutrition Care [Internet]. Chicago: Academy of Nutrition and Dietetics; 2017 [cited 2017 Aug 15]. Available from: https://ncpt.webauthor.com/pubs/idnt-en/page-068.
8. Gallagher Allred CR, Voss AC, Finn SC, McCamish MA. Malnutrition and clinical outcomes: the case for medical nutrition therapy. J Am Diet Assoc. 1996;96(4):361–9.
9. Bastyr Center for Natural Health (Internet). Seattle: Bastyr University; c2017 [cited 2017 August 15]. Available from: http://bastyrcenter.org/about-our-clinic.
10. U.S. Department of Health and Human Services, National Institutes of Health, National Heart, Lung, and Blood Institute. Your guide to lowering your blood pressure with DASH. Bethesda: National Institutes of Health; 2006.
11. Swan WI, Vivanti A, Hakel-Smith NA. Nutrition care process and model update: toward realizing people-centered care and outcomes management. J Acad Nutr Diet. 2017;117(12):2003–14.

Ayurvedic Approach in Chronic Disease Management

Sangeeta Shrivastava, Pushpa Soundararajan and Anjula Agrawal

45.1 Origin and History of Ayurveda – 784

45.2 Advancement of Ayurveda in the USA – 784

45.3 Ayurvedic Education in India, the USA, and Beyond – 784

45.4 Introduction to Ayurvedic Approach of Treatment – 785

45.5 Descriptions of the Doshas – 785
45.5.1 Three Supporting Pillars of Health (Trayopasthambas) – 786

45.6 Overview of Ayurvedic Nutrition and Principles – 786
45.6.1 The Six Tastes – 786
45.6.2 Basic Ayurvedic Principles – 786

45.7 Ayurvedic Concept of Food Is Medicine – 787
45.7.1 Ayurvedic Lifestyle – 790
45.7.2 Dinacharya: Daily Routine – 790
45.7.3 Ritucharya: Seasonal Routine – 790
45.7.4 Sadvritta: Ethical Regimen – 790
45.7.5 Panchakarma, or Ayurvedic Detox Treatment – 792

45.8 Disease Management in Ayurveda – 792
45.8.1 Obesity per Ayurveda – 792
45.8.2 Causes of Obesity – 792
45.8.3 Diabetes and Other Related Disorders – 794
45.8.4 Ayurvedic Dietary and Lifestyle Recommendations for Obesity and Metabolic Disorders – 794
45.8.5 Cancer and Ayurveda – 796
45.8.6 Integrating Ayurveda in Modern Practice – 796
45.8.7 Comparison of Ayurvedic and Modern Approaches to Nutrition – 797

References – 797

© Springer Nature Switzerland AG 2020
D. Noland et al. (eds.), *Integrative and Functional Medical Nutrition Therapy*,
https://doi.org/10.1007/978-3-030-30730-1_45

> **Definitions**
>
> **Ayurveda** - Science and wisdom of life
> **Samhitas** - Textbooks of Ayurveda
> **Agni** - Digestive fire
> **Prakruti** - Type of body constitution
> **Vikruti** - Doshic imbalance
> **Dosha** - Energy that circulates in the body around bodily tissues
> **Dhatu** - Body tissues
> **Mala** - Metabolic wastes (urine, sweat, feces)
> **Srotas** - Elaborate network of channels carrying nutrients in the body and wastes out of the body
> **Vata** - Wind energy
> **Pitta** - Heat energy
> **Kapha** - Building energy
> **Dinacharya** - Daily routine and practices
> **Ritucharya** - Seasonal routines and practices
> **Brahmacharya** - Self-control
> **Panchakarma** - Five Ayurvedic detoxifying treatments
> **Ama** - Toxic substances
> **Rasayana** - Rejuvenating herbs
> **Samprapti** - Pathogenesis of the disease
> **Roga** - Disease
> **Rasa** - Taste

45.1 Origin and History of Ayurveda

Ayurveda is defined as the science of life. The name comes from Sanskrit, where ayur means life and veda means science or knowledge. Dating back about 5000 years, it is one of the world's oldest and most complete medical systems, originating from India. It was derived from the Vedas [1], texts composed of mantras that embody the very laws and energies of the universe, which are not merely inanimate forces but powered by consciousness itself. There are four Vedas: Rig, Yajur, Sama, and Atharva. Although all Vedas contain references to Ayurvedic concepts, the Atharva Veda contains the most. Therefore Ayurveda is considered an Upaveda, or a branch, of Atharva Veda [2, 3].

About 1500 B.C., Acharya Charaka, representing the Atreya School of physicians, compiled the medical material scattered in the Vedas into the Charaka Samhita. This first main textbook outlines the fundamental principles of Ayurvedic medicine, including treatment modalities, diet, and herbs for diseases. Sushruta Samhita, representing the Dhanvantari School of surgeons, was later compiled by Acharya Sushruta, regarded as the father of surgery. It contains description of surgical instruments and procedures, including plastic surgery. It also contains description of vital points, or Marmas, equivalent to acupuncture meridians in Chinese medicine. In about the sixth century, Vagbhata compiled Ashtanga Hridaya, which mentions Buddhist deities, teachers, and Tibetan medicine. These three ancient books, written in Sanskrit, are known as the *Great Trilogy or Brihat Trayi*. They have gained respect in the last 2000 years and are considered the main and oldest texts on Ayurvedic medicine [4, 5].

It is clear that Ayurveda is not some folk medicine that was thrown together, but a systematic medical practice that has served to restore health and wellness at the physical and mental level. Yoga, already popular in the West, originated with Ayurveda as part of the greater wisdom of Vedic science. Yoga as a therapy for treating disease, whether physical or mental, is part of Ayurveda, and it is traditionally used according to Ayurvedic diagnosis and recommendations [2].

In the olden days, Ayurveda was learned in the Gurukul tradition of education, which involved a long period of intense study under accomplished masters. The student stayed with the teacher in the Gurukul residential school for years to study and learn. Ayurvedic medicine was originally an oral tradition, taught and passed directly from teacher to apprentice. The modern system of Ayurvedic education evolved in the early 1950s, requiring the Ayurvedic physician to complete at least a 5-year postgraduate degree program [6].

It should be remembered that Ayurveda is not just the science of health. Healthful living, prevention of disease, and personal and social hygiene all come under its wings and not merely the cure of diseases. The fundamentals on which the Ayurvedic system is based are essentially true for all times and do not change from age to age. These are based on human factors and intrinsic causes. Being a natural and holistic life science, it can fit to any region, person, occupation, day, and age. In the West, Ayurveda is recognized as a complementary and alternative health system by the National Institutes of Health, and it is blossoming in various educational institutions [3, 7].

45.2 Advancement of Ayurveda in the USA

Interest in Ayurveda in the USA began in the early 1970s because of Maharishi Mahesh Yogi's Organization of Transcendental Meditation. Interest continued to grow as Indian physicians came to the USA in the 1980s and spread the awareness about Ayurveda. In the late 1980s, Dr. Deepak Chopra wrote *Perfect Health*, a book that helped open the door to India's ancient science for many in the West. Among those were Dr. David Frawley, of the American Institute of Vedic Studies, and Dr. Robert Svoboda, who studied and completed India's BAMS program. In 1995, two of Dr. Frawley's students founded the first two schools of Ayurveda in the USA [8].

45.3 Ayurvedic Education in India, the USA, and Beyond

Ayurvedic medicine is a well-established educational system. There are around 247 BAMS colleges in India imparting Ayurvedic knowledge and wisdom and preparing Ayurvedic doctors for the future [9]. Students still study from traditional books and *Samhitas* [10]. In the 1970s and 1980s, *Ayurveda* started spreading in the West, as many Indians started migrating to the USA and Europe. Today numerous colleges and institutes offer *Ayurveda* courses in the USA, Europe, South America, and Australia. In the USA, there are two main professional associations catering to Ayurvedic professionals and those interested in *Ayurveda*. The *National Ayurvedic Medicine Association (NAMA,* ▶ ayurvedanama. org) is a professional organization representing Ayurvedic professionals in the USA. The *Association of Ayurvedic Professionals of North America (AAPNA,* ▶ aapna.org), founded in 2007, is another professional organization in the USA whose focus is to create a community of Ayurvedic professionals with the common goal of growing the presence of Ayurveda in integrative healthcare. Both organizations have lists on their websites of colleges/institutes for anyone wishing to pursue Ayurvedic education in the USA. The *Council for Ayurvedic Counseling (CAC)* is an independent group of trained, highly experienced, academically qualified and professional Ayurvedic clinicians and teachers who are deeply committed to the field of Ayurvedic education in the USA (▶ cayurvedac.org). Several experts in the field of Ayurvedic education and licensure indicated in interviews that there is precise work going on to establish licensure in Ayurveda (MD Jayarajan, Ayurvedic approach in chronic disease management, personal communication, 21 Mar 2017; A Raut, Ayurvedic approach in chronic disease management, personal communication, 24 Mar 2017; V Jain, Ayurvedic approach in chronic disease management, personal communication, 4 Apr 2017). In Europe, Ayurveda is growing at a fast pace and being recognized as a detailed traditional college of holistic medical system. The nonprofit *Rosenberg European Academy of Ayurveda (REAA,* ▶ ayurveda-academy. org) was founded by Kerstin and Mark Rosenberg in 1993 [11]. In South America, Ayurveda has been growing rapidly in the last 20 years, and courses are being offered in major medical schools in Argentina. *Fundación de Salud Ayurveda Prema, Argentina,* has been working to advance Ayurveda throughout Latin America. The organization is collaborating with Indian universities and has been accredited as a collaborating center for teaching and research in the field of Ayurveda in Argentina [12]. Based on the data from India, Europe, South America, and the USA, it is evident that Ayurveda is growing and becoming a prominent part of today's medical science. Integrative and functional medical nutrition therapies have a great potential to incorporate Ayurveda in practice.

45.4 Introduction to Ayurvedic Approach of Treatment

Ayurveda uses an individualized approach. It is based on the principle that everybody has a unique constitution or body type or humor, commonly referred to as dosha. The three main body types are Vata, Pitta, and Kapha, each defined by the composition of its dominant elements and their respective qualities, or gunas. There are 20 such qualities, described as pairs of opposites: heavy-light, dull-sharp, cold-hot, wet-dry, smooth-rough, dense-liquid, soft-hard, firm-mobile, gross-subtle, and sticky-clear. Every individual has a unique constitution in which one or two of the doshas can be dominant. In rare cases, all three doshas can be equally balanced; this is called Sama Prakruti. An examination of the individual based on their prakruti and subsequent imbalance, or vikruti, will shed light on the treatment modality for each individual.

45.5 Descriptions of the Doshas

Vata (air and space): It is responsible for any movement in the body and cooling energy. A Vata-dominant person will have a small frame and small eyes; a tall, thin, or lanky appearance; prominent bones and cracking joints; and wavy, rough, thick hair, to name a few. They are usually quick in movements, walking fast and talking fast, and constantly changing their ways and interests. They are enthusiastic and creative. Their digestion is usually irregular with bouts of hunger and not feeling hungry at times. They are prone to constipation, bloating, and joint pains. Their response to stress is usually anxiety, nervousness, or fear. Their memory is short-lived, and although they grasp things quickly, they also forget easily. They have very disturbed sleep and wake up easily.

Pitta (water and fire): It is responsible for heating energy, metabolism, and vision. A Pitta-dominant person will have a moderate frame, angular face, colored eyes, thinning or early graying hair, sharp and penetrating eyes, and pink nails. Their personality is one of passion and intensity, given the fire element. They are driven to do many things that they are passionate about and are very focused in their efforts. They are competitive, aggressive, controlling, very organized, and intelligent, and they assimilate knowledge very well. They have leadership qualities, but when stressed they tend to get irritable and angry. They have a strong appetite and can digest anything when in balance. Nightly sleep is moderate, about 7–8 hours, restful but not deep.

Kapha (earth and water): It is responsible for stability, binding, water balance, lubrication, healing properties, and cooling energy. A Kapha-dominant person usually has a large frame; solid muscles; fat; short bones; a flatter nose; large, beautiful eyes; a gentle look; dark-brown thick hair; a soft, oily clear complexion; a round face; and perfect teeth. They are very slow in movement and learning, but retention is excellent due to strong memory. Some other characteristics: being grounded, inertia, great stamina, a loving nature, compassion, and affection. They are miserly in spending money and hoard things because they can't let go. If imbalanced, they can become possessive, jealous, and depressed. Metabolism is very slow, so they tend to be overweight. They can skip meals, have cravings for sweet taste, find it hard to lose weight, and don't drink much water. Sleep is usually very deep, 9–10 hours, and they have difficulty waking up in the morning. When stressed, they become quiet and withdrawn (► Box 45.1).

Ayurveda teaches that what exists in the macrocosm is also present in the microcosm. We see this correlation in the natural world around us. In the earth's ecosystem, Vata is expressed in the motion of the wind and in water currents. Pitta is present as sunshine and fire. Kapha brings us the solid structure of the earth, rocks, and water (V Jain, Ayurvedic approach in chronic disease management, personal communication, 4 Apr 2017). Therefore, understanding nature and our relationship with it will help guide us intuitively in providing the right diet and lifestyle for healing and wellness.

> **Box 45.1**
> *Ayurvedic approach to health, or Swastha: Definition of health:*
>
> » *sama dosha sama agnischa sama dhatu mala kriyaaha | Prasanna atma indriya manaha swastha iti abhidheeyate || Sushruta Samhita 15:38* [13]
>
> One who is established in the self, who has balanced doshas, balanced agni (digestive fire), properly formed dhatus (tissues), proper elimination of malas (wastes), well-functioning bodily processes, and whose mind, soul and senses are full of bliss, is called a healthy person [23]

45.5.1 Three Supporting Pillars of Health (Trayopasthambas)

There are three pillars for an individual's utmost health and well-being, as per Ayurveda:
1. Ahara *(diet):* This encompasses anything which we ingest to nourish our body and mind, which includes food, water, breath, emotions, and information through sense organs [14].
2. Nidra *(sleep):* Includes proper sleep habits, including duration and timing of sleep.
3. Brahmacharya *(self-control):* A self-controlled life, including sexual activity, is essential to attain physical, mental, and spiritual balance [4, 15].
   Overindulgence in any of the above pillars could potentially lead to disease (► Fig. 45.1).

45.6 Overview of Ayurvedic Nutrition and Principles

Ahara, or food, is essential for Prana, or life force. Ahara Rasa, the essence of the food we eat, provides the building blocks to create new tissues or the seven dhatus in the body. A great deal of emphasis is placed on nutrition, digestion, and metabolism, which correlate with the sense of taste and digestive fire or Agni.

Modern nutrition classifies food into food groups like fruits and vegetables, dairy, meats and beans, starches, and fats and oils. Ayurveda, however, classifies foods based on their Rasa (six tastes: sweet, sour, salty, bitter, pungent, and astringent), their Virya (potency, heating or cooling), and their Vipaka (post-digestive effect, sweet or pungent). Food recommendations in Ayurveda are based on body types and the tastes that are most suitable for pacifying their respective doshas or imbalances.

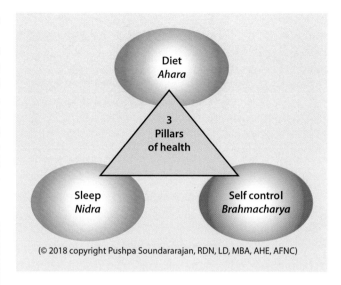

■ **Fig. 45.1** Pillars of Health. (Based on data from Refs. [15, 16])

45.6.1 The Six Tastes

Nature has packaged all possible food sources into six tastes. Before the discovery of proteins, carbohydrates, fats, minerals, vitamins, and trace elements, Ayurveda taught that we need all six tastes present to make a balanced meal. The six tastes are sweet, sour, salty, pungent, bitter, and astringent [17]. Each taste is a combination of two dominant elements, and each has a different effect on the doshas. Diet is recommended based on the type of doshic imbalance, incorporating the appropriate taste to be included in each meal. ■ Figure 45.2 lists some examples of foods from the six tastes and their impact on the doshas denoted by a "+"for aggravating the dosha or a "−" for a pacifying effect on the dosha. V, P, and K stand for Vata, Pitta and Kapha respectively.

45.6.2 Basic Ayurvedic Principles

A. *Agni,* the digestive fire, is a major principle in Ayurveda. Digestive fire is needed to digest anything we bring into our body. If the Agni is high, even a suboptimal intake can be fully digested, absorbed, and assimilated in the body to create energy. However, if the Agni is low, then even the best-quality foods do not provide the right nutritive value as the food is neither properly digested nor properly assimilated. The strength of appetite reflects the underlying state of our digestive fire (V Jain, Ayurvedic approach in chronic disease management, personal communication, 4 Apr 2017). When Agni is weakened or disturbed, food is not properly digested. The undigested, unabsorbed food particles start to build up in the GI tract

SIX Tastes of Ayurvedic Nutrition			
Rasa Dosha +/-	Virya (elements)	Vipaka	Food sources
Sweet Madura V - P - K+	Cooling (Earth + Water)	Sweet	Most grains like wheat, rice, barley, corn, and most bread. • Most legumes, such as beans, lentils, and peas. • Milk and sweet milk products, such as cream, butter, and ghee. • Sweet fruits like dates, figs, grapes, pears, and man-goes. • Certain cooked vegetables, potatoes, sweet potatoes, carrots, and beets. • All cooking oils including ghee, olive oil, coconut oil, sesame oil, and butter. • Almonds, walnuts, pistachios, sunflower seeds, and nut milks. • Sugar in any form except honey
Sour Amala V - P + K+	Heating (Earth + Fire)	Sour	Sour fruits like lemon, lime, and sour oranges. • Sour milk products like yogurt, cheese, sour cream, and whey. Fermented substances like wine, vinegar, soy sauce, and sour cabbage.
Salty Lavana V - P + K+	Heating (Water + Fire)	Sweet	Any kind of salt. • Foods to which a large amount of salt is added. • Most sea vegetables and animals. Pungent is a Kapha-pacifying taste. A sampling of the heating-pungent foods includes:· Spices like chili, black pepper, mustard seeds, ginger, cumin, and garlic.· Certain vegetables, like radish and onion. A sampling of the cooling-pungent foods includes:· Spices, including coriander, fennel, basil, and dill.
Pungent Katu V+ P++ K-	Heating (Fire + Air)	Pungent	Spices like chili, black pepper, mustard seeds, ginger, cumin, and garlic. • Certain vegetables, like radish and onion. A sampling of the cooling-pungent foods includes: • Spices, including coriander, fennel, basil, and dill.
Astringent Kshaya V+ P- K-	Cooling (Air+ Earth)	Pungent	Legumes, beans, and lentils. • Walnuts, hazelnuts, cashews, and pumpkin seed. • Honey and black, green, and white teas. • Sprouts, lettuce, and other green leafy vegetables; rhubarb; and most raw vegetables.
Bitter Thikta V+ P- K-	Cooling (Air+ Space)	Pungent	Certain fruits, like olives and grapefruits. • Green, leafy vegetables like spinach, green cabbage, Brussels sprouts, and zucchini. • Eggplant, bitter gourd, chicory, chocolate, and coffee. • Certain spices, like turmeric and fenugreek.

Rasa - Taste
Virya - Potency; heating or cooling to the body
Vipaka - Post digestive effect; long term action on the body

Prabhava - Special action
Unique action of a substance after consumption. Eg. **Amla (gooseberry)** is sour and heating but sweet post digestive effect which is cooling.

(© 2018 copyright Pushpa Soundararajan, RDN, LD, MBA, AHE, AFNC)

Fig. 45.2 Six tastes in Ayurvedic nutrition. (Based on data from reference (V Jain, Ayurvedic approach in chronic disease management, personal communication, 4 Apr 2017) and Refs. [15, 18])

and other subtle sites in the body, turning into a toxic, sticky, foul-smelling substance called Ama [19]. These toxins are similar to cholesterol that forms plaque in the arteries and need to be removed by detoxification modalities described as Panchakarma in Ayurveda.

B. Another principle that is observed in Ayurveda is that *"like increases like,"* which means that a food that is dominant in water element will increase water in a Prakruti dominant in a water element like Kapha. The logical intervention would be to restrict foods with that property in the corresponding Prakruti.

45.7 Ayurvedic Concept of Food Is Medicine

Ayurveda does not classify any specific food as a "superfood," as all foods can be modified to fit the individual's needs, based on Prakruti/Vikruti or body type/imbalance. However, there are certain foods that are considered beneficial in many ways for all doshas, and these are supported by plentiful scientific evidence. Figure 45.3 illustrates eight of those foods, although this is in no way a comprehensive list.

Ghee, as described in Ayurveda, is not technically clarified butter, which is made by melting butter available in stores. It is a more complex process that involves extracting the cream from cultured yogurt, making butter from that, then heating it to clarify it, and removing the brown residue by filtering it. This ghee has micronutrients and is called medhya, which is nourishing to the brain. Studies revealed no blood lipids elevated with the consumption of ghee, despite the presence of saturated fat [29].

Uses: It is used as a medium or Anupana for giving herbs, called medicated ghee. Also:

– It is a healthy fat and has a high smoke point, so it's suitable for cooking and frying.
– It can be used as a spread on toast instead of butter.

Fig. 45.3 Food is medicine. (Based on data from Refs. [15, 20–28])

- It is used for frying seasonings and adding them to foods.
- It is commonly used to make nasya for nasal cleansing and for the eyes as well [26].
- Take one spoon of ghee mixed in warm water or warm milk at night to relieve constipation. It helps with oleation therapy [15, 26].

Turmeric, or Curcuma longa, has a bitter taste with heating potency. It contains the diverse nutraceutical curcumin, which has been shown to prevent many chronic diseases – including obesity, diabetes, certain types of cancers, and autoimmune disease – by fighting inflammation [15, 23, 26]. It is a blood purifier and antiseptic, and it is very useful for liver disorders.

It has multifold uses:

- *Cooking:* It is a common spice used in Indian cooking. It can be added to soups, lentils, curry, and rice noodles.
- *Golden milk:* Popular nowadays, this drink is made by boiling 1/4 tsp. of turmeric in a cup of organic milk. You can add raw sugar to it if hot or honey if lukewarm and drink it. A pinch of cinnamon or nutmeg can be added as well and taken in the morning or at night for a good night's sleep.
- Mix 1/4 tsp. of turmeric with lemon juice and honey in warm water and drink on an empty stomach in the morning as a daily detox and to enhance metabolism.
- *For inflammation:* Make a paste of dry roasted turmeric and apply on the area of a sprain.
- *For pimples or acne:* Make a paste of turmeric, sandalwood, and milk. Apply it;
- *For type 2 diabetes:* Mix a tsp. each of turmeric and amalaki powder in water and drink it (A Raut, Ayurvedic approach in chronic disease management, personal communication, 24 Mar 2017).

Tulsi, or holy basil, is bitter and pungent with a heating potency. It is a powerful adaptogenic herb referred to as "queen of herbs" and is revered as a sacred plant with great healing powers. Tulsi has immense benefits for people with respiratory problems, including asthma, allergies, cough, cold, and fever. It is an antibacterial and antiviral herb. The reviewed studies reinforce traditional uses and suggest Tulsi as an effective treatment for lifestyle-related chronic diseases, including diabetes, metabolic syndrome, and psychological stress [15, 22].

Uses and other health benefits per Ayurvedic literature:

- It can be made into a tea for respiratory benefits, either by using the leaves themselves or by steeping Tulsi powder in hot water for 5–10 minutes.

- The leaves can be eaten raw for their medicinal properties and as a mouth freshener.
- The leaves can be crushed and the juice extracted to be used for fever, cough, and cold.
- It is also a water purifier and is considered very auspicious in spiritual practices in India.
- A decoction of Tulsi with ginger juice and honey is good for sore throat.

Fenugreek is pungent in taste and has a post-digestive effect with a heating Virya. It is a native spice of India and South Europe. Fenugreek seeds and leaves are anticholesterolemic, anti-inflammatory, antitumor, carminative, expectorant, hypoglycemic, laxative, parasiticidal, and restorative. They serve as a uterine tonic and are useful for burning sensations. Fenugreek is used by breastfeeding mothers to increase milk production. Fenugreek and alpha-lipoic acid is associated with a blood glucose-lowering effect, according to strong scientific evidence [15, 27].

The seeds and leaves can be used in many ways:
- Soak 1 tsp. of the seeds in water and drink it in the morning on an empty stomach for blood sugar control for diabetics [15, 27].
- The seeds can also be sprouted and eaten as a curry.
- It is used in cooking as well, roasted in oil or dry roasted, or as a powder added to soups, gravies, lentils, mixed grains, etc.
- The leaves can be mixed into the dough to make unleavened bread called paratha.
- A few leaves can be added to lentils or soup as well.

Mint, or pudina, has a sweet and mildly pungent taste with a warming potency and pungent post-digestive effect. Plant-derived natural compounds have the potential to ameliorate the causes and symptoms of neuroinflammation because of their various antioxidant and anti-inflammatory activities, without completely muting the immune defenses. *Scutellaria* is a perennial plant in the mint family that has been used to treat diseases in Asia and Eastern Europe throughout history [21].

Here are some ways it can be used:
- It has antiemetic properties and is used for nausea in the form of a tea or juice in the dosage amount of 10–15 ml.
- It can be used as a mouth freshener.
- Pudina chutney can be made using the leaves.
- It is also added to soups, lentils, curries, etc.

Ginger, also called ardraka (fresh) or shunti (dry), is called Vishwa Bhaishajya, or universal medicine, in Ayurveda. It is a rhizome (root) and considered pungent in taste and heating in potency, but its post-digestive effect is sweet, which makes it very unique. It plays an excellent role in improving digestion by increasing Agni, as well as in respiratory problems and circulation. It is a natural blood thinner, so if one is on warfarin, caution should be exercised in consumption. Studies have shown that ginger has a potential preventive property against some chronic diseases, especially hypertension and CHD, as well as an ability to reduce the probability of illness [28].

Ginger has many uses as food and medicine:
- *Cooking:* It is used in cooking with soups, noodles, rice, curries, and stir-fries.
- Make into an herbal tea by boiling ginger for a few minutes in hot water or adding it to black tea while steeping. Honey can be added if it is warm, not hot.
- *Cough:* 1/4 tsp. each of ginger powder, turmeric, and black pepper mixed with 1 tsp. of honey can be taken at night before bed to help with cough. Or this mixture can be added to 1/4 cup of warm water and taken as a shot [15, 26].
- *Sore throat*: Same as above with ginger, turmeric, and honey.
- *Diarrhea*: Add a pinch of ginger to cooked rice and a tsp. of ghee (A Raut, Ayurvedic approach in chronic disease management, personal communication, 24 Mar 2017).

Black pepper, or Piper nigrum, is pungent and has a heating potency that aids in digestion. It is great for respiratory disorders, for deep cleansing of sinuses, and as a cough expectorant. Piperine, an alkaloid present in black pepper, has many pharmacological effects and health benefits, especially against chronic diseases, such as reduction of insulin-resistance, anti-inflammatory effects, and improvement of hepatic steatosis [20].

Besides uses in cooking, it is also useful in increasing the bioavailability of herbs:
- It is one of the three ingredients in trikatu, which helps with reducing Kapha and burns up Ama (toxins) and is good for obesity [15, 26].
- It makes other herbs in the preparation more bioavailable. For instance, turmeric is more available when taken with black pepper mixed in honey, for coughs and colds.
- It can be used in cooking in many recipes, in powder form, and is part of many cuisines of the world.

Honey, or madhu, is a sweet and viscous liquid produced by bees from the nectar of flowers. It is primarily sweet in taste with a secondary astringent taste which gives it the quality of lekhana, or "scraping out the cholesterol," which gives it its medicinal property. The health benefits of honey range from antioxidant, immunomodulatory, and anti-inflammatory activity to anticancer action, metabolic and cardiovascular benefits, prebiotic properties, human pathogen control, and antiviral activity [15, 26].

Uses and other benefits per Ayurvedic literature:
- It is a sweetener that can be used in different foods. Because of its astringent nature, it is the only sweetener suitable for diabetics, although moderation is key.
- It can be added to warm water and taken in the morning or with a few drops of lemon juice.
- It is used as a medium or anupana for delivery of many herbs.
- *Caution:* There are certain toxic effects with honey if taken with other foods:

Fig. 45.4 Healthy eating habits. (Based on data from Ref. [31])

Healthy Eating Habits

1. Choose foods as per your body type or Prakruti.
2. Choose foods according to the season.
3. Eat fresh quality foods. Avoid old, stale, processed, canned and microwaved foods.
4. Listen to your hunger signals and do not eat unless hungry.
5. It is advised to sit and eat, not stand and eat for better digestion.
6. Chew well, as suggested 32 times per mouthful. This enables mouth enzymes to work properly.
7. Eat at a moderate speed, do not gobble.
8. Fill one third stomach with food, one third with water and leave one third empty for better digestion.
9. Eat the amount of food you can hold in two cupped hands (Anjalis). Do not over eat as that can expand the stomach and produce toxins in the digestive tract.

Important Tips:

- During meals **DON'T DRINK ICE COLD WATER, JUICE OR DRINK,** sip little warm water between mouthfuls of foods.
- **Honey** should never be cooked as heating honey causes its helpful enzymes to integrate and make it gluey which are toxins called Ama that can clog the subtle channels or strotasmi.

(© 2018 copyright Pushpa Soundararajan, RDN, LD, MBA, AHE, AFNC)

- Honey and ghee should NOT be taken in equal amounts [15, 26].
- Honey should NOT be mixed with hot water and consumed because that leads to a toxic substance that is very sticky and very difficult to remove even with detoxification therapy [30].
- Honey should not be heated or cooked with, and any packaged foods made with honey should be avoided [26] (Fig. 45.4).

45.7.1 Ayurvedic Lifestyle

One of more than 82 natural systems of healing worldwide, Ayurveda emphasizes prevention and gives guidelines for how to make your system optimal to prevent any imbalance without falling sick.

According to Vedic system, there are four major goals in life, called Purusharthas:

» *Dharma artha kama mokshanam*
 Aarogyam mulam Uttamam. [26]
 Health is the foundation (mulam) to achieve the four goals of life, namely, dharma (right conduct), artha (wealth and prosperity), kama (desires), and moksha (liberation).

An Ayurvedic lifestyle prescription includes dinacharya, or daily routine; *ritucharya*, or seasonal routine; and sadvritta, or mode of conduct, to achieve these four goals of life.

45.7.2 Dinacharya: Daily Routine

Dina means "daily" and charya means "routine" or "regimen." This covers the mental, speech, and physical aspects of well-being that one should follow every day [15].

Figure 45.5 gives a brief summary of the daily routine that one is encouraged to follow.

45.7.3 Ritucharya: Seasonal Routine

Ritu means "season," and charya means "regimen." This routine deals with diet and lifestyle adjustments for the seasonal variations. All activities in the universe are governed by the two energy principles of heat and cold. The ancient Ayurvedic textbooks are tailored toward an Indian climate, which consists of three seasons: winter, summer, and rainy. However, in temperate parts of the world, where cold is a detrimental factor, these two periods follow equinoxes [2]. In the USA, we talk about four seasons. Figure 45.6 gives a brief summary of the four seasons and suggested recommendations.

45.7.4 Sadvritta: Ethical Regimen

The third aspect of lifestyle management is Sadvritta, which means "ethical regimen" [2] or code of conduct. A healthy mind is as important as a healthy body for wellness. Many dos and don'ts are outlined in ancient texts for the right conduct. Practicing them will give balance and peace to the mind.

Fig. 45.5 Dinacharya daily routine

Dinacharya - Prescribed Daily Routine

1. Wake up time: Ninety minutes before sunrise or at least before sunrise.
2. Pray before leaving the bed to start the day.
3. Evacuate bowel and bladder ideally within an hour of waking up.
4. Oral hygiene:
    a. Brush teeth with natural toothpaste or powder which is astringent in nature. (like neem)
    b. Tongue scraping with metal scrapper, preferrabley copper to remove toxins and increase Agni.
    c. Oil pulling or Gandosha: Swish your mouth with sesame oil or coconut oil till it changes color. This exercises gums and jaws and gives strength.
5. Face cleaning: Wash face with warm water for Vata and Kapha or cool water for Pitta. Wash eyes as well.
6. Vyayaama or Exercise: Yoga and Pranayama (breathing). Other forms of exercise can be done like treadmill, bike etc.
7. Abhyanga or Self Massage with sesame oil starting from the head and going down to the feet.
8. Shower or bathe. This can be done before oil massage too if you want to keep the oil in the body.
9. Drink 1–2 glasses of warm water in an empty stomach.
10. Breakfast should be light and eaten only if hungry. Just fruits can be eaten if not hungry.
11. Lunch should be the main meal of the day preferrably between 10 and 2 pm, around noon.
12. Take 100 steps, light walking and then rest (not sleep) for better digestion.
13. Dinner should be very light more like soups eaten between 6 and 7 pm preferrably.
14. Go to bed by 10 pm preferrably. Staying up late leads into Pitta time which makes you hungry.
15. Meditation and Pranayama before bed helps get a restful sleep.

(© 2018 copyright Pushpa Soundararajan, RDN, LD, MBA, AHE, AFNC)

Fig. 45.6 Ritucharya – seasonal routine. (Based on data from Ref. [2])

45.7.5 Panchakarma, or Ayurvedic Detox Treatment

Panchakarma literally means five actions. Those actions are vamana (induced vomiting), virechana (purgation), nasya (nasal drops with herbs), basti (medicated enema), and raktamokshana (bloodletting, usually using leeches). Although all of these are practiced in India, only the first four are done in the USA. There are several centers in the USA that perform this kind of treatment, and their durations range from 3 days to even a month, depending on need and affordability. The concept behind this unique therapy is to remove toxins (Ama) built up throughout the body from poor diet and lifestyle. The main procedure is called panchakarma. There is a preparatory stage prior to that called purvakarma, which lasts for a week or two as prescribed by the Ayurvedic physician. After the panchakarma, rejuvenating herbs or rasayana are prescribed to be taken, as the body is detoxing during that phase as well. Special restricted diet and lifestyle recommendations are also made, based on the client's needs for maintenance of good health. This whole procedure is administered with the supervision of an Ayurvedic physician who has had extensive training in the field. Specially trained massage therapists also play a vital role in this process [2, 26].

45.8 Disease Management in Ayurveda

The National Health and Nutrition Examination Survey indicates that from 2011 to 2014, the prevalence of obesity was just over 36% in adults and 17% in youth. Obesity-related conditions include heart disease, stroke, type 2 diabetes, and certain types of cancer, all of which are among the leading causes of preventable mortality [32].

Metabolic syndrome, a group of risk factors related to the disorders listed above, bears striking resemblance to the santarpanjanya vikaras (comprised of diseases due to overnutrition and defective tissue metabolism) described in classical Ayurvedic texts. Ayurveda focuses mainly on conservation of health rather than disease eradication. It presumes that improper dietary habits and deranged functions of different sets of Agni (Metabolic fire) give rise to the formation of Ama (reactive antigenic factor), which in turn causes disease [33].

In the management of lifestyle diseases, Ayurveda offers various regimens, including dinacharya (daily regimen), ritucharya (seasonal regimen), panchakarma (five detoxification and bio-purification therapies), and rasayana (rejuvenation) therapies. The sadvritta (ideal routines) and achara rasayana (code of conduct) are of utmost importance to maintain a healthy and happy psychological perspective. Including several of these modalities is very effective in controlling these disorders. Moreover, the application of organ-specific rasayana herbs, which is unique to Ayurveda, helps not only in healing but also in the rejuvenation process [34].

Ayurveda describes metabolism of food in a unique way, from ingestion of food called Ahara to the formation of tissues in the body. There are seven tissues or dhatus that are formed in succession every 5 days. These are rasa (plasma and lymph), rakta (blood, RBCs, and platelets), mamsa (muscle), meda (adipose tissue), asthi (bone), majja (nerves and bone marrow), and shukra (reproductive tissue). The food eaten mixes with the digestive enzymes, and the essence called Ahara rasa gets metabolized with the help of the different dhatu agnis, or digestive fires, to form the seven bodily tissues in the order given. These provide support and strength to the body. These, along with the organs, form an elaborate network of channels called Strotas, to carry the nutrients to the different parts of the body and wastes for elimination. There are 14 such srotasmi described in Ayurveda, similar to the systems in modern medicine. Diseases occur when proper flow is obstructed through these channels. Obesity, as an epidemic in today's society, is discussed first. Weight management can go a long way in controlling many of the metabolic diseases.

45.8.1 Obesity per Ayurveda

The Ayurvedic term for obesity, or overaccumulation of fat under the skin and around certain internal organs, is called Sthaulya. Another name for obesity in Ayurveda is medaroga, as the meda or fat is the main cause of this condition and roga means disease [35]. Charaka Samhita has given the description as large hips, excess bulk over the midsection, pendulous breasts, and too much flesh over the chest [26]. Obesity is divided into central obesity, where the belly is big; truncated obesity, where the breasts are big; and peripheral obesity, where the extremities are large. These conditions are the result of displacement of fat to different regions by Vata energy because of the person's food, lifestyle, and emotions. Therefore, if a person has deep-seated grief and sadness in the lungs, which are Vata emotions, they are carried lower down to lead to chubby thighs [19]. We can deduce from this that some Vata-pacifying techniques may be beneficial in treatment.

45.8.2 Causes of Obesity

In Ayurveda, medaroga is caused by aggravation of Kapha. Kapha is the humor or dosha that is dense, heavy, slow, sticky, wet, and cold in nature. It governs all structure and lubrication in the mind and body, apart from controlling weight and formation of all seven tissues (dhatus).

In a balanced state, Kapha gives nourishment to these tissues through micro-channels. However, when it is aggravated, it leads to production of toxins in the body. These toxins or Ama are heavy and dense in nature and accumulate in weaker channels of the body, causing blockage. In the case of an obese person, toxins accumulate in medovaha strotas (fat channels), thereby leading to an increase in the production of fat tissues, which causes an increase in weight. The other problem that ensues as a result is that these blockages prevent air or vata from moving freely through the channels, and they get redirected to the stomach. Here they further stimulate the digestive fire (jatharagni) to burn more, like a fan that flames a fire, and thus leads to constant hunger [36]. This makes the person eat more, and this turns into a vicious cycle.

The degree of obesity in a person is dependent on the accumulation of fatty globules or cells. As long as the accumulated fat is stored in the adipose tissue, it does no harm. But when it begins to enter into the cellular elements of the body, especially the muscles, it becomes a source of danger. Since the fat channels get blocked, the next dhatu, which is asthi or bone tissue, is not formed well, and this results in osteoporosis.

As per Ayurveda, common causes of obesity are heavy intake of sweet, cooling, and unctuous food; lack of physical exercise; sleep during the day; lack of mental exercise; and heredity.

Habitual drinking of cold drinks and eating fatty fried foods; eating dairy products such as cheese, yogurt, and ice cream; and consuming excess sugar and carbohydrates are all causative factors [19]. It is believed that "food is the food for the body; love is the food for the soul." When a person is not feeling love within, he or she tends to turn to food to fill that void, commonly referred to as emotional eating in Western society.

Ayurveda approaches these issues from a deeper understanding of digestion, which started the whole process in the first place. If the digestion is improved, then the fat channels will clear up, and then the food that is consumed will be used to nourish all the dhatus instead of being stuck at the meda dhatu level. Clearing the Ama and improving digestion are the major approaches to treatment. Since Vata may also be involved in transporting these fat molecules, a comprehensive treatment could include Vata-pacifying techniques like oil massage, shirodhara (oil dripping over the forehead), yoga, and meditation, to name a few [35] (◘ Fig. 45.7).

Case Study

Pushpa Soundararajan, RDN, LD, AHC, AFNC

A 60 year old Caucasian female with Diabetes, Chronic renal failure Stage 3 and Obesity Grade II (BMI38.2), Fibromyalgia, joint problems, chronic pain, exhausted majority of the time and Constipation came in for Nutrition Counseling.

Anthropometrics: Height: 5'8", Weight: 252 lbs.

Weight history showed she used to be 129 lbs for majority of her life. She claimed to have gained 100 lbs due to Medications Paxil and Prednisone. She does not do any regular exercise but just some routine household chores. She had trouble falling asleep usually woke up exhausted even after 8-10 hours. She said her response to stress was to eat.

She had tried Atkins diet 8 years ago, lost 30 lbs but gained it all back. She tried another commercial diet program 12 years ago but did not notice any change.

Diet Recall revealed eating breakfast and dinner and often skipped lunch. If she did eat lunch it included a sandwich or leftovers. Breakfast usually consisted of cold cereal with almond milk or bagel and cream cheese with iced tea. Dinner included a salad, some vegetables and some pasta with meat. She rarely ate after dinner but if she did she had a craving for sweets.

Ayurvedic Assessment:

It was very clear based on her symptoms she had a Vata imbalance given her choice of cold foods, tiredness and constipation. Her dietary habits and lifestyle contributed to the imbalance and having kidney failure she wanted help with restrictions as well.

Recommendations:

Reviewed renal restrictions from Modern nutrition perspective. Suggested switching to warm water and starting with 1 glass in the morning before breakfast to aid in cleansing and facilitating bowel movement. Recommended eating her main meal at lunch time with cooked foods including small portion of meat, kidney friendly vegetables and limited carbohydrate foods. Suggested dinner of soup or something light with more vegetables and less meat. Counseled on mindful eating, paying attention to hunger cues, avoiding distractions while eating and enjoying her meal and portion control by not going for seconds. Sipping hot water or herbal teas like lemon ginger was suggested for better digestion. Encouraged exercise of walking for at least 15 minutes per day and work up to 30 minutes gradually as medically possible.

Follow Up Visit:

Patient came back in 3 months. She was very happy, energetic and had lost 13 lbs. Her weight was 239 lbs, BMI 36.3 down from 38.2. She said she had made the switch to main meal at dinner, eating only cooked foods, warm water and paying attention to her hunger signals. She said she was very satisfied with lunch and didn't even need any dinner most days. She also had no cravings. She was very upbeat and said she had more energy and felt so good that she is motivated to continue. She did not incorporate exercise as suggested but yet with her Vata getting better she found improvement. Suggested to continue with the dietary regimen, suggested some kidney friendly Ayurvedic recipes with spices and digestive CCF (Cumin, Corriander, Fennel) tea. Encouraged walking even 10 minutes after meals for consistent control of BS levels.

(© 2018 copyright Pushpa Soundararajan, RDN, LD, MBA, AHE, AFNC)

◘ **Fig. 45.7** Case study 1

45.8.3 Diabetes and Other Related Disorders

Prameha is the term used for diabetes and other related syndromes. There are 20 types of diabetes identified by Ayurveda. It is a metabolic disorder of agni, the energy of transformation. When there is a dysfunction of the agnis, especially the kloma (pancreas) agni, the carbohydrate metabolism is disturbed [37]. Any of the doshas may be imbalanced, but the first step is always to rebalance the agni to regulate nutrient metabolism and water-electrolyte balance. Dietary and lifestyle changes can go a long way in doing that.

Following are some suggested guidelines that could help kindle the agni and aid in weight loss. Since generally Kapha dosha is disturbed in obesity and other metabolic disorders, these suggestions would be useful. If other doshas are disturbed, adjustments need to be made accordingly. An Ayurvedic practitioner will be better equipped to make that assessment (Fig. 45.8).

45.8.4 Ayurvedic Dietary and Lifestyle Recommendations for Obesity and Metabolic Disorders

1. Generally, a Kapha-pacifying diet is recommended.
2. One can use basmati rice, barley, green gram, red gram, and horse gram.
3. Honey is the only sweetener that is recommended while treating obesity, but it should be original honey, directly collected from honeycombs.
4. Drink warm water instead of refrigerated water.
5. For cooking purposes, use sesame oil or mustard oil with spices like turmeric, black pepper, ginger, and rock salt.
6. Use vegetables with astringent, bitter, and pungent tastes. If you are particular about your weight, it might be helpful to fast for 1 to 2 days a week, or to partially fast while drinking fruit juices, warm water, and honey.

Clinical Case study
Vaidya Ashlesha, Raut BAMS, MD (Ayurveda)

38 year old female, 63 inches tall and weight is 124 lbs with Narcolepsy with Cataplexy (Sleep Paralysis).

Symptoms: Extensive sleepiness during daytime, intermittent uncontrollable episode of falling asleep at workplace, frequent episodes of partial to total loss of muscle tone (Cataplexy). Sleep paralysis was experienced during day- time. She also used to suffer from post- menstrual severe headaches, which further used to trigger her Cataplexy. Symptoms were overwhelming and disturbing her daily life as described by patient. She was treated with many drug prescriptions along with antidepressants from her MD. Patient did not experience significant improvement and decided to try Ayurveda.

Ayurvedic Diagnosis:

Patient's Dosha Prakruti showed the Vata-pitta type. The area of Doshik- imbalance was of strong Vata imbalance covering Sadhak pitta. The dhatu imbalance involved Mamsa dhatu and Rasa dhatu. She also showed signs of AMA (Toxins in Ayurveda) and low Ojas (Vital energy).

Ayurvedic Treatment approach:

Ayurveda treatment included Diet, lifestyle and herbs recommendations. She was started with Amapachaka herbs (Herbs that digest ama- toxins), along with Vata pacifying treatment. Her diet was carefully checked to rule out Vata provoking foods. Also Ojas (Energy) increasing foods were recommended. In lifestyle changes she was given specific Pranayama (Breathing exercise) along with many Apana soothing yoga postures (wind releasing poses). She was encouraged to perform very specific daily abhyanga (Self oil massage). After a period of time she was given authentic Panchakarma treatment, which was followed by Ayurveda Rasayana therapy which included focus on Vata dosha, Rasa and Mamsa dhatu. Within 3 months significant improvement was noticed, postmenstrual headaches stopped, episodes of cataplexy reduced. After panchakarma treatment her energy level increased and with Rasayana treatment her quality of life improved. The whole treatment took 9 to 10 months but tremendous improvement noticed.

Fig. 45.8 Case study 2

7. Avoid eating substances prepared with refined wheat flour, including white bread, cakes, and pastries; and abstain from dairy products and sweets prepared from milk and sugar.
8. You should also avoid cold drinks, alcohol, and deep-fried foods, and follow a vegetarian diet. If you do eat meat, choose chicken or fish just once a week [38].
9. Follow an active pattern of life by increasing work and mental activity. You should avoid taking too much rest, sleeping during the day (particularly after meals), and bathing with cold water.
10. Vyayama (special exercise) is definitely helpful in burning some calories, but the most important factor is that it brings the body into shape and creates a general feeling of well-being. Walking is good for all doshas.
11. Whatever the ultimate cause of obesity in your case, the immediate cause is energy imbalance, and weight reduction can be achieved only by reducing energy intake or by increasing output or by a combination of the two.
12. Yoga recommendations: Sun Salutation or Surya Namaskar (SN) is a classic yoga exercise consisting of 12 asanas performed sequentially and synchronized with breathing [39]. It is considered a comprehensive and ideal practice for physical and spiritual well-being. SN is performed as a prayer to the Sun, ideally at sunrise and in open air, facing east. It has aerobic and dynamic components and has been shown to improve strength, body composition, and general body endurance. Performing 6–8 rounds will achieve the energy expenditure of light exercise intensity, and a 10-minute practice may improve cardiorespiratory fitness in unfit or sedentary individuals. Fish, camel, cobra, and cow or other chest-opening poses would also be helpful.
13. Pranayama, especially Bhastrika Pranayama (Breath of Fire), will increase the rate at which your body burns off fat.
14. Meditation is also a strong tool in the Ayurvedic lifestyle to cope with various diseases and improve general well-being (Fig. 45.9).

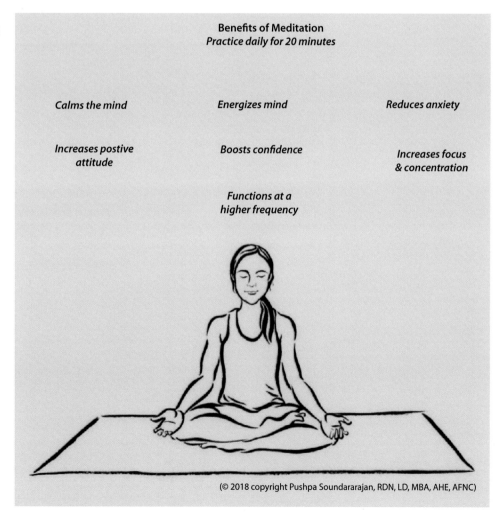

Fig. 45.9 Benefits of meditation: Ancient literature and current scientific evidence suggest that meditation can be a very powerful tool in managing chronic diseases [2, 15, 26, 40]. Potential benefits are listed in the figure

45.8.5 Cancer and Ayurveda

Arbuda is the name given to a malignant tumor or cancerous growth. Traditional Ayurvedic literature has a detailed description of various types of tumors, and the treatment is complicated and based on balancing three doshas, seven dhatus, and three malas (urine, waste, and sweat) [37]. It takes an experienced Ayurvedic doctor to administer treatment effectively (Fig. 45.10).

45.8.6 Integrating Ayurveda in Modern Practice

Integrative health is a system of health in which the intangible as well as the more material aspects of life are given equal importance. Our existence is physical as well as mental, emotional, energetic, and spiritual. Our emotions are intertwined with our physiology. Stress is implicated in 85% of chronic diseases. A healthcare system where therapies are directed to each of these aspects of our existence is called integrative health. This fundamental approach is crucial to life's wholeness. Treating just the physical body with drugs and surgery may be effective, but not enough to solve the problem of chronic diseases. No integrative health system is complete without a consciousness-based model of health and disease. This is what is found in the ancient system of Ayurveda. Integrating Ayurveda with advances in modern medicine will help provide a better healthcare regimen than any one of them practiced separately (MD Jayarajan, Ayurvedic approach in chronic disease management, personal communication, 21 Mar 2017; A Raut, Ayurvedic approach in chronic disease management, personal communication, 24 Mar 2017; V Jain, Ayurvedic approach in chronic disease management, personal communication, 4 Apr 2017).

The Ayurvedic approach consists of diet, lifestyle, and medicine in equal parts. So the integration of the Ayurvedic approach in medical nutrition therapy has great potential. Registered dietitians, nutrition professionals, and other healthcare practitioners can easily apply Ayurvedic principles in their practices. We spoke to several eminent Ayurvedic practitioners around the USA, and they all agreed that holistic and nutrition-focused healthcare practitioners can and should certainly incorporate Ayurvedic principles in their counseling and delivery of nutritional services. As Ayurveda focuses on an individualized approach, its use may provide a strong tool. Interviews were conducted with successful practitioners and eminent faculty at leading Ayurvedic institutions around the country (MD Jayarajan, Ayurvedic approach in chronic disease management, personal communication,

Fig. 45.10 Case study 3

> **Clinical Case Study**
> **Dr Vijay Jain MD**
>
> 78 years old male patient with Non-Hodgkins Lymphoma Stage 3.
>
> **Ayurvedic diagnosis:**
> Pitta/vata imbalance with Ama accumulation in the Rakta dhatu (Blood tissue).
>
> **Ayurvedic Approach of intervention:**
> Patient was put on special anti-inflammatory diet in addition to other treatments like lifestyle modification, meditation and pranayama for stress reduction. Patient also underwent a full Panchakarma treatment followed by rasayana therapy with Bhasmas and herbo-mineral compounds. He was in complete remission in six months.

Fig. 45.11 Comparison of Ayurvedic and modern nutrition

Criteria	Modern nutrition	Ayurvedic nutrition
Why do we eat?	Personal preference, habit, body image	To take in Prana to live
Nutrient element	Calories, carbohydrates, protein & fats	Panchamahabhutas or 5 elements (space, air, fire, water and earth)
Main focus for weight loss	Counting calories	Evaluating metabolism based on Agni or digestive fire
Importance give to	Caloric value	Individual Constitution (Prakruti) and imbalance (Vikruti)
Intervention based on	Nutrient deficiency	Balancing Prakruti and Vikruti
Summary	You are what you eat	You are what you digest

21 Mar 2017; A Raut, Ayurvedic approach in chronic disease management, personal communication, 24 Mar 2017; V Jain, Ayurvedic approach in chronic disease management, personal communication, 4 Apr 2017). They use an integrative Ayurvedic approach in their practice and teaching. Each of these experts emphasized the need to stabilize the disease state prior to beginning treatment. Agni is the hallmark of Ayurvedic nutrition (V Jain, Ayurvedic approach in chronic disease management, personal communication, 4 Apr 2017). Eating healthy foods does not serve you effectively if digestive fire (agni) and digestion, absorption, or assimilation are compromised. In conclusion, integrating Ayurveda with modern nutrition would be beneficial as a holistic treatment modality for patients. ◘ Figure 45.11 summarizes the differences in the two approaches, demonstrating the value in integrating the two in practice.

45.8.7 Comparison of Ayurvedic and Modern Approaches to Nutrition

> **Take-Home Message**
>
> If we healthcare practitioners were to imagine a form of medicine designed especially for twenty-first-century Americans, it might have the following traits:
> 1. *Preventive:* It would cultivate wellness before illness takes root, strengthening a person's resilience to disease.
> 2. *Low-intervention*: It would pursue health through diet, exercise, and lifestyle, using drugs and surgery as a last resort.
> 3. *Affordable*: It would be cost-effective by reducing surgeries and hospital procedures for chronic disease.
> 4. *Holistic*: It would see a person's body, mind, and spirit – and their environment – as deeply interconnected.
> 5. *Personal:* Practitioners would build relationships with patients, viewing them as unique individuals.
> In short, it would look a lot like Ayurvedic medicine [25], the traditional Indian health system growing robustly among Americans looking for wellness-focused healthcare providers like integrative healthcare practitioners (V Jain, Ayurvedic approach in chronic disease management, personal communication, 4 Apr 2017).

References

1. Medindia. History of Ayurveda. History of Medicine; c2017 [cited 2017 Apr 17]. Available from: medindia.net/ayurveda/.
2. Frawley D, Ranade S. Ayurveda: nature's medicine. Twin Lakes: Lotus Press; 2012. p. 53181.
3. National Institutes of Health, U.S. Department of Health and Human Services. Ayurvedic Medicine: In Depth; c2017 [cited 2017 Apr 7]. Available from: nccih.nih.gov/health/ayurveda/introduction.htm.
4. Ayurvedic Academy, Inc. Kerala Ayurveda academy course manual 101: Ayurvedic philosophy and anatomy; 2009. p. 6–7.
5. Narayanaswamy V. Origin and development of ayurveda (a brief history). Anc Sci Life. 1981;1(1):1–7.
6. Menon I. The Ashtavaidya physicians of Kerala: a tradition in transition. Ashtavaidya tradition| History and science of Indian systems of knowledge; c2017 [cited 2017 Oct]. Available from: ncbs.res.in/HistoryScienceSociety/content/ashtavaidya-tradition.
7. Nimivaggi FJ. Nutrition and diet in Ayurveda. Ayurveda: a comprehensive guide to traditional Indian medicine for the west. 2010th ed. Plymouth: Rowman and Littlefield; 2010. p. 171–72.
8. Halpern M. Status & development of Ayurveda in the United States. CA College of Ayurveda; c2017 [cited 2017 Apr]. Available from: ayurvedacollege.com/articles/drhalpern/Status_Development_Ayurveda_USA.
9. Ayurveda colleges in India. List of government and private Ayurveda colleges in India; c2015 [cited 2017 Apr 13]. Available from: medindia.net/education/ayurveda_colleges.asp
10. Patwardhan K, Sangeeta G, Singh G, Rathore HC. Ayurvedic education in India. How well are the graduates exposed to basic clinical skills? Evid Based Complement Alternat Med. 2011:197391. PMCID: PMC 3095267.
11. Rosenberg M. The European academy of Ayurveda: 20 years of Ayurvedic education in Germany. Anc Sci Life. 2012;32(1):63–5. c2017 [cited 2017 Apr 13].
12. Berra JL, Molho R. Ayurveda in Argentina and other Latin American countries. J Ayurveda Integr Med. 2010;1(3):225.
13. Acharya YT. Sushrut Samhita Sutrasthana Ch 15:38. Varansi: Chowkhamba Surabharti Prakashana; 1994.
14. Badge AB, Sawant RS, Pawar JJ, Ukhalkar VP, Qadri MJ. Trayopasthambas. Three supportive pillars of Ayurveda. J Biol Sci Opin. 2013;1(3):250–4.
15. Srikantha Murthy KR. Vagbhata's Astanga Hrdayam, vol. 1. 8th ed. Varanasi: Chowkhamba Krishnadas Academy; 2011.
16. National Health Portal. Triad of health/Ayurveda; c2016 [cited 2017 May 2]. Available from: www.nhp.gov.in.
17. Ambika Devi MA, Jain Vijay Unfolding happiness. Mythologem Press: Jensen Beach, FL; 2016.
18. Lad V. The complete book of Ayurvedic home remedies. New York: Three Rivers; 1999. p. 235–7.
19. Lad V. Textbook of Ayurveda: fundamental principles: volume 2. Albuquerque: The Ayurvedic Press; 2007. p. 257–8.
20. Derosa G, Maffioli P, Sahebkar A. Piperine and its role in chronic diseases. Adv Exp Med Biol. 2016;928:173–84.
21. Hussain F, Mittal S, Joshee N, Parajuli P. Application of bioactive compounds from Scutellaria in neurologic disorders. In: Advances in neurobiology. The benefits of natural products for neurodegenerative diseases; 2016. p. 79–94.
22. Jamshidi N, Cohen MM. The Clinical Efficacy and Safety of Tulsi in Humans: A Systematic Review of the Literature. Evid Based Complement Alternat Med. 2017; 2017: 9217567. https://doi.org/10.1155/2017/9217567.
23. Kunnumakkara AB, Bordoloi D, Padmavathi G, Manisha J, Roy NK, Prasad S, Aggarwal BB. Curcumin, the golden nutraceutical: multi-targeting for multiple chronic diseases. Br J Pharmacol. 2017;174(11):1325–48.
24. Mbikay M. Therapeutic potential of Moringa Oleifera leaves in chronic hyperglycemia and dyslipidemia: a review. Front Pharmacol. 2012;3:24.
25. Miguel MG, Antunes MD, Faleiro ML. Honey as a complementary medicine. Integr Med Insights. 2017;12:1178633717702869.
26. Sharma RK, Das B. Charaka Samhita, vol. 1. Reprint ed. Varanasi: Chowkhamba Sanskrit Series; 2008.
27. Smith JD, Valerie BC. Natural products for the management of type 2 diabetes mellitus and comorbid conditions. J Am Pharm Assoc. 2014;54:5.

28. Wang Y, Yu H, Zhang X, Feng Q, Guo X, Li S, et al. Evaluation of daily ginger consumption for the prevention of chronic diseases. Nutrition. 2017;36:79–84.
29. Sharma H, Zhang X, Dwivedi C. The effect of ghee (clarified butter) on serum lipid levels and microsomal lipid peroxidation. AYU. 2010;134:31–2.
30. Vagbhata. Astanga Hradayan Sutrasthana; c2017 [cited 2017 May 7]. Available from: ayur-veda.guru.pdf.
31. Ayurvedic Academy, Inc. Kerala Ayurveda Academy course manual 105. Swasthavritta. 2009. 11–17.
32. Prevalence of obesity among adults and youth: United States, 2011–2014; c2014 [cited 2017 May]. Available from: cdc.gov/nchs/data/databriefs/db219.pdf.
33. Jaspreet S, Pandey AK. Metabolic syndrome and its management through Ayurveda and yoga. IOSR-JDMS. 15(6):36–41. Ver. XI. c2016.
34. Chandola HM. Lifestyle disorders: Ayurveda with lots of potential for prevention. Āyurvedāloka. 2012;33(3):327.
35. Sanjeevan Ayurvedashram. Obesity treatment in Ayurveda, Ayurvedic medicine for weight loss; c2015 [cited 2017 Mar 20]. Available from: sanjivaniayurvedashram.com/obesity.php.
36. Chauhan Pratap. Jiva Ayurveda; c2017 [cited 2015 Mar 15]. Available from: jiva.com/diseases/obesity/.
37. Lad V. Ayurvedic perspective on selected pathologies. Albuquerque: The Ayurvedic Press; 2012. p. 59–72.
38. Chirumamilla, MM. Obesity and Ayurvedic treatment; c2017 [cited 2017 May]. Available from: muralimanohar.com/Articles,%20English/Diseases%20and%20Conditions/Obesity.htm.
39. Bhavanani AB, Udupa K, Madanmohan, Ravindra P. A comparative study of slow and fast suryanamaskar on physiological function. Int J Yoga. 2011;4(2):71–6.
40. Lad V. Textbook of Ayurveda: fundamental principles: volume 1. Albuquerque: The Ayurvedic Press; 2002. p. 135.

Suggested Websites

Bastyr University. Master of Science in Ayurveda, bastyr.edu/academics/ayurvedic-sciences/masters/masters-ayurvedic-sciences.
California College of Ayurveda. ayurvedacollege.com/programs/doctor-ayurvedic-medicine-school.
International University of Yoga and Ayurveda. iu-ya.org/.
John Douillard's Life Spa. lifespa.com.
Kerala Ayurveda Academy. ayurvedaacademy.com/academy/index.
Kripalu Center of Yoga and health. kripalu.org/.
National Ayurvedic Medical Association. ayurvedanama.org.
Resource website on Ayurveda, easyayurveda.com.
Southern California University Health Sciences. scuhs.edu/academics/ayurveda-certificate-programs/.
The Ayurvedic Institute. ayurveda.com.
Uses of ghee in cooking handout, pureindianfoods.com/how-to-use-ghee-a/256.htm.
VPK by Maharishi Ayurveda. mapi.com.

Cases & Grand Rounds

Contents

Chapter 46 **Cardiometabolic Syndrome – 801**
Anup K. Kanodia and Diana Noland

Chapter 47 **Revolutionary New Concepts in the Prevention and Treatment of Cardiovascular Disease – 823**
Mark C. Houston

Chapter 48 **Immune System Under Fire: The Rise of Food Immune Reaction and Autoimmunity – 843**
Aristo Vojdani, Elroy Vojdani, and Charlene Vojdani

Chapter 49 **Amyotrophic Lateral Sclerosis (ALS): The Application of Integrative and Functional Medical Nutrition Therapy (IFMNT) – 863**
Coco Newton

Chapter 50 **Gastroenterology – 913**
Jason Bosley-Smith

Chapter 51 **Respiratory – 927**
Julie L. Starkel, Christina Stapke, Abigail Stanley-O'Malley, and Diana Noland

Chapter 52 **The Skin, Selected Dermatologic Conditions, and Medical Nutrition Therapy – 969**
P. Michael Stone

Chapter 53 **Movement Issues with Chronically Ill or Chronic Pain Patients – 1003**
Judy Hensley, Julie Buttell, and Kristie Meyer

Cardiometabolic Syndrome

Anup K. Kanodia and Diana Noland

46.1 Introduction – 802

46.2 Pathophysiology: Overview – 804

46.3 Homeostasis and Allostasis – 804

46.4 Community and Social Impacts on Psychosocial Determinants of Health – 804

46.5 Inflammation – 804

46.6 Inflammatory Mediators – 806

46.7 Oxidative Stress – 807

46.8 Cardiometabolic Syndrome and Comorbidities – 807
46.8.1 Metabolic Disorder Comorbidities – 808
46.8.2 Psychiatric Disorders – 809
46.8.3 Autoimmune Disorders – 809
46.8.4 Rheumatic Diseases – 809

46.9 Dietary Considerations in Inflammation, Oxidative Stress, and Cardiometabolic Disorder – 809
46.9.1 The Microbiome – 810
46.9.2 Chronic Stress and Cardiometabolic Syndrome – 811
46.9.3 Hypercoagulability – 811
46.9.4 Insulin Resistance – 811
46.9.5 Sarcopenia/Sarcopenic Obesity – 812
46.9.6 Cardiometabolic Syndrome – 813
46.9.7 Adipocytes in Cardiometabolic Syndrome – 813

46.10 Anthropometric Assessment Methodologies – 815

46.11 Assessment of Physical Performance – 815
46.11.1 Biomarkers for Cardiometabolic Syndrome – 815
46.11.2 Diet – 816
46.11.3 Protocols of Treatment for Chronic Conditions – 817

46.12 Conclusion – 818

References – 818

© Springer Nature Switzerland AG 2020
D. Noland et al. (eds.), *Integrative and Functional Medical Nutrition Therapy*,
https://doi.org/10.1007/978-3-030-30730-1_46

46.1 Introduction

In 2010, the World Health Organization (WHO) announced that, for the first time, chronic noncommunicable diseases have overtaken infectious communicable diseases as the chief causes of mortality worldwide [1, 2]. Cardiovascular diseases (CVD), including coronary heart disease and stroke, are currently the leading cause of death in the world [3]. However, cancer deaths have recently reached an almost equal status to cardiovascular mortality [4]. The most common of these noncommunicable chronic diseases are cardiometabolic disease, cancer, neurological disease, autoimmune disease, metabolic syndrome, diabetes, developmental health conditions, and obesity, which are resulting in unsustainable impacts on global healthcare economies and needless suffering [5–8].

Further, the WHO reports that chronic (noncommunicable) diseases are responsible for 70% of all deaths worldwide, with cardiovascular disease as the primary underlying cause of death. Cardiovascular disease is followed by cancer, respiratory disease, and diabetes. These noncommunicable diseases are also characterized by modifiable risk factors that exist for these conditions [9].

These risk factors include:
- Tobacco use
- Excess sodium intake
- Alcohol use
- Lack of physical activity
- Unhealthy diet

The WHO also lists the metabolic risk factors intimately involved with the development of chronic disease pathophysiology:
- Increased blood pressure
- Obesity or overweight
- Central adiposity with sarcopenic obesity
- Hyperinsulinemia
- Hyperglycemia
- Hyperlipidemia

Characterization of Metabolic Syndrome
- Central adiposity
- Hypertension
- Systemic, chronic inflammation
- Compromised kidney function
- Type 2 diabetes
- Sarcopenia
- Neurodegeneration
- Fetal and infant chronic disease pre-programming
- Increased cancer risk
- Increased cardiovascular disease risk
- Sleep disturbance/apnea
- Increased autoimmune disease risk

These metabolic risk factors – modifiable to different extents with appropriate dietary, physical, and lifestyle interventions – are the same metabolic risk factors associated with the development of cardiometabolic syndrome [10–12] (Tables 46.1 and 46.2).

Table 46.1 Metabolic syndrome: 2017 American Heart Association

Risk factor	Defining level	Exceptions
Abdominal central obesity Waist circumference Men Women	Waist circumference Men > 102 cm (>40″) Women > 88 cm (>35″)	Height <60″ or >74″ Then measure height/weight ratio
Triglycerides (elevated)	Reference range(s) ≤9 years <75 mg/dL 10–19 years <90 mg/dL ≥20 years ≤150 mg/dL	If >150, check Carnitine status to rule out carnitine deficiency Check dietary intake of alcohol, sugar, and sugar-rich foods
HDL (low)	Male <1.03 mmol/l (<40 mg/dl) Female <1.29 mmol/l (<50 mg/dl) Specific treatment for this lipid abnormality	Check HDL/total cholesterol ratio
Hypertension	Systolic >130 Diastolic <85 Treatment of previously diagnosed hypertension	If BP <120/80, check mineral status of sodium and chloride, autoimmune, and thyroid status
Plasma fasting glucose (elevated)	Fasting reference 13–47 mg/dL Non-fasting reference 13–87 Fasting >126 mg/dL diagnosis diabetes	If glucose >5.6 mmol/l or 100 mg/dl, consider oral glucose tolerance test. Compare to HgA1C
Additional insulin resistance markers (not diagnostic for metabolic syndrome)		
HgbA1C (elevated)	HgbA1C normal 4.8–5.6% Hgb/A1C >5.7% prediabetes Hgb/A1C >6.5% diabetes	
Insulin, fasting (elevated)	Reference 2–12, 88–94 µIU/mL	If >10, check clinical waist circumference, body fat %, risk of hyperinsulinemia

Based on data from: Quest Diagnostics, ▶ www.questdiagnostics.com 2019; American Heart Association, ▶ www.heart.org; American College of Cardiology, ▶ www.acc.org [88]

This chapter will present the impact on healthcare of the underlying pathophysiology and nutritional relationships that lead to chronic disease and discuss the current science of chronic metabolic disease processes. These disorders are considered "lifestyle and dietary-related diseases [8, 89]," where the metabolic microenvironment promotes a phenotype characterized by insulin resistance, inflammation, and elevated visceral adipose tissue (VAT). The focus will be on evidence-based nutritional and lifestyle factors that can be individually assessed from where interventions are possible that enable management of the complex human metabolism.

The practitioner uses a functional systems assessment of the individual to identify metabolic priorities and develop therapies, including the application of nutritional medicine principles to improve health outcomes for the patient.

In addition to the metabolic risk factors listed above, chronic disease is becoming understood to include these related foundational risk factors: abdominal obesity (visceral adipose tissue), sarcopenia, insulin resistance, stress, nutrition, and lifestyle. As this understanding expands, there is increasing recognition of the need to address factors of any success quelling the epidemic of chronic disease. These factors are related to the development of most chronic diseases. In this section of the textbook, various metabolic systems are presented in more detail in the specific chapters. The current protocols of conventional healthcare are discussed, as well as integrative and functional medical nutrition therapy approaches that address the root causes of chronic diseases.

In addition, a growing recognition of diabetes and metabolic syndrome being a mediator of several chronic disease conditions like Alzheimer's, heart disease, osteoporosis, and cancer. They are sometimes referred to as "type 3 diabetes." Intervening with the dietary and lifestyle components of cardiometabolic syndrome may benefit all chronic diseases and can be used as a model therapeutic approach for all chronic diseases.

The societal burden of chronic metabolic disorders is expected to steadily increase. It is projected that 75% of the US population will be obese or overweight by 2030. There are currently 371 million people worldwide who are obese or overweight, and the number is predicted to reach 552 million by 2030. A new case of diabetes is diagnosed in the United States every 30 seconds; more than 1.9 million people are diagnosed each year [90, 91] (Fig. 46.1).

Table 46.2 Body composition measurements

Sarcopenic obesity	Loss of muscle		Independent of BMI
Body type % body fat	Females	Males	BMI
Essential fat % body fat	10–13%	2–5%	All – Independent of BMI
Athletes % body fat	14–20%	6–13%	All – Independent of BMI
Fitness % body fat	21–24%	14–17%	All – Independent of BMI
Acceptable % body fat	25–31%	18–25%	All – Independent of BMI
Obese	>31%		All – Independent of BMI

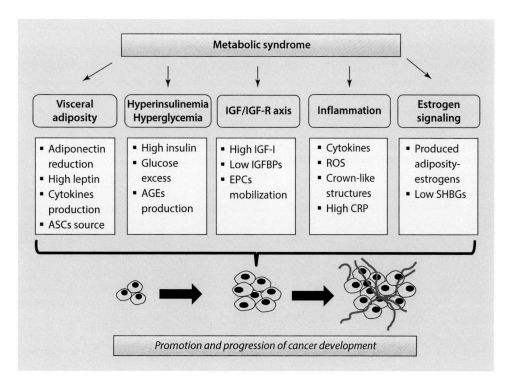

Fig. 46.1 Metabolic Syndrome and Cancer [92]. (Reprinted from Micucci et al. [92]. With permission from Creative Commons License 3.0: ▶ https://creativecommons.org/licenses/by/3.0/)

ICD-10 Data	
Medical Disorder	*ICD-10*
– Metabolic Syndrome	E88.81
– Coronary Heart Disease	I25
– Hypertension	I10/I16.9
– Hypercholesterolemia	E78.0
– Hypertriglyceridemia	E78.1
– Obesity	E66.9
– Hypo HDL cholesterol	E78.6

Chronic Diseases are Diseases of Lifestyle
- Nutrient insufficiency
- Low vegetable intake
- Low omega-3 fat
- Epigenomics
- Oxidative stress
- Poor sleep quality
- Environmental toxins
- Altered microbiome
- Community-socio-genomics

46.2 Pathophysiology: Overview

At the base of many chronic diseases, there are alterations in blood glucose and insulin metabolism, as well as a state of chronic inflammation. These metabolic changes begin to affect the sensitivity of cellular insulin receptors resulting in a growing insulin resistance. Unique to each individual are factors, including nutritional deficiencies and lifestyle choices, that lead to metabolic alterations and that are subject to many influences. Consideration of intra-systemic influences is of central importance when assessing for the etiology of an individual's condition. These influences begin to impact various systems and homeostatic mechanisms, eventually overcoming homeostatic and allostatic controls.

46.3 Homeostasis and Allostasis

Homeostasis may be defined as "the ongoing maintenance and defense of vital physiological variables such as blood pressure and blood sugar [93]". Feedback control, which may be positive or negative, is a core regulatory process that enables homeostasis.

Allostasis is a process that may be thought of as an anticipatory homeostatic mechanism that, in some ways, supplants the classical homeostatic concept of control of systems. Peters and McEwen [94] describe allostasis as an "active process by which living organisms adapt to potential threats to their survival and changes in their environment (often referred to as 'stressors') in order to maintain homeostasis and promote survival." This definition manifests a holistic approach because stressors can be biological, biochemical, and/or psychological, including the concepts of mind-body medicine. A further concept, allostatic load, describes the cumulative effects of physiological stress and disruptions, eventually resulting in a dysregulation of a system. The system whose dysregulation is often the first clinically noted is the cardiovascular system, though pre-existing perturbations can be seen in glucose and lipid control [95] (◘ Fig. 46.2).

46.4 Community and Social Impacts on Psychosocial Determinants of Health

Who your friend, family, or co-worker is can directly affect your health. These are new findings that bring important focus to consider when working with patients to bring into the conversation hearing the patient's story. And if a person does not have meaningful relationships, loneliness can be a primary mediator of illness and have a wide range of negative effects on all chronic diseases (◘ Fig. 46.3).

> These findings indicate that the influence of others' choices on risk-taking behavior is large, direct, cannot be explained by an economic utility model of risky decision-making, and goes against one's own better judgment [97].

46.5 Inflammation

Swelling, redness, pain, and heat classically characterize inflammation. Loss of function of the tissue or joint may also be involved. This characterization, while valid, primarily focuses on the inflammatory response surrounding a wound or superficial injury. Inflammation involves both the innate and acquired aspects of the immune response and includes a vascular component.

It is important to remember that inflammation is a normal, protective response to injury, but it is the loss of control and prolonged presence of the inflammatory response without resolution that is the underlying pathophysiologic mechanism of damage (see ◘ Fig. 46.4).

The inflammatory response involves the cellular and humoral aspects of the immune system and includes endothelial cells, circulating leukocytes and immune and mast cells, and macrophages and lymphocytes embedded within the connective tissue. The extracellular matrix, including glycoproteins, fibrous proteins, and proteoglycans, also plays an important role.

Endothelial cells constitute a selective barrier, controlling the extravasation of leukocytes by the regulation of cellular

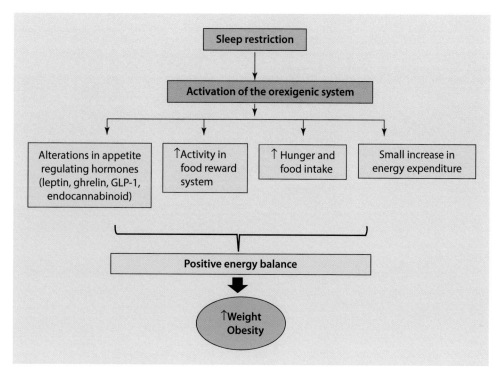

Fig. 46.2 Sleep influences on weight management or obesity. Putative pathways linking insufficient sleep and obesity risk. (Reprinted from Reutrakul and Van Cauter [96]. With permission from Elsevier.)

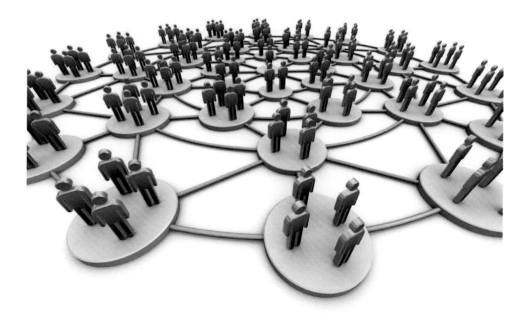

Fig. 46.3 Structure of social networks with nodes depicting relationships that can affect health diet and lifestyle. Social networks are groups of people performing a series of social professional and educational activities in the place of living and housing place of work or other locations. (Reprinted with permission from PIXTA)

adhesion molecules. The secretion of various cytokines and colony-stimulating factors regulates immune proliferation.

Neutrophils, macrophages, monocytes, and other immune cells produce vasoactive amines (e.g., histamine, serotonin) and pro-inflammatory cytokines including prostaglandins, platelet-activating factor, growth factors, interleukins (e.g., IL-6), TNF-α, and others. In addition to vasoactive amines and pro-inflammatory cytokines, the complement system, opsonization, phagocytosis, autophagy, and the intracellular release of lysozymes along with reactive oxygen and reactive

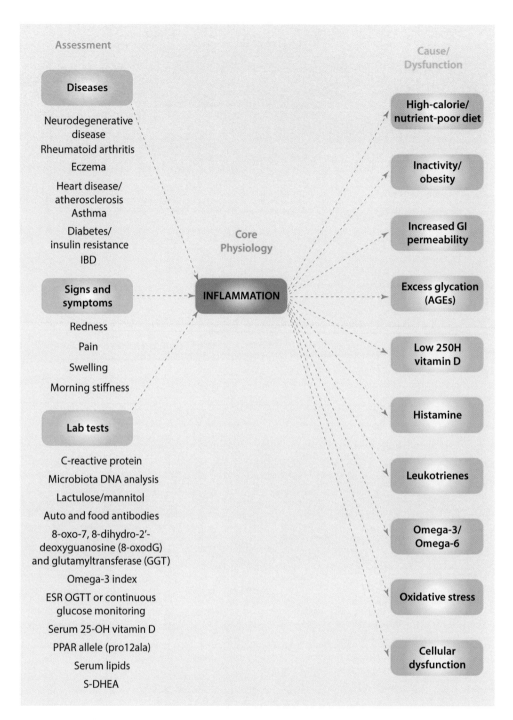

Fig. 46.4 Core physiology of inflammation. (Reprinted from Pizzorno and Katzinger [98]. With permission from Mind Publishing.)

nitrogen species (ROS/RNS) play a central role in the control of inflammation in response to infection with pathogenic organisms.

46.6 Inflammatory Mediators

Pro-inflammatory cytokines favor inflammation and include IL-1, TNF-α, IFN-γ, IL-6, IL-8, IL-18, and others. In metabolic syndrome and other disorders associated with adiposity and inflammation, adipocytes produce a variety of pro-inflammatory mediators (e.g., TNF-α, IL-1β, IL-6, IL-8), acute-phase response factors (e.g., C-reactive protein (CRP), ceruloplasmin, ferritin, fibrinogen, plasminogen activator inhibitor-1, haptoglobin, and serum amyloid A). High-sensitivity C-reactive protein (hs-CRP) is a very useful inflammation marker related to infection, trauma, and elevated visceral adipose tissue (VAT) [99]. In addition, hormones such as adiponectin, omentin, and ghrelin, GI-derived hormones (e.g., glucagon-like peptide-1 and other incretins), and possibly leptin are often elevated during inflammation. The following are cardiometabolic inflammatory markers:

- C-reactive protein with a half-life of 6–8 hours. It is reflective of bacterial infection, musculoskeletal stress/trauma, or high visceral adipose inflammatory cytokine-secreting tissue. It has more recently been considered a prognostic indicator of heart disease when elevated and often referred to as a "cardio CRP" [100, 101].
- Erythrocyte sedimentation rate (ESR): ESR can indicate inflammatory activity often associated with elevated fibrinogen, autoimmune responses, and viral infections. It remains high for an extended time after removal of the inflammatory trigger.
- TNF-α: In adipocytes, TNF-α is a paracrine mediator. TNF-α induces insulin resistance by inhibiting the insulin receptor substrate 1 signaling pathway and induces adipocyte apoptosis [102]. TNF-α is associated with increased body weight and increased leukocytes and triglycerides.
- IL-1β: IL-1 is a family of at least 11 cytokines. IL-1β is also known as leukocyte pyrogen and is produced primarily by macrophages. IL-1β increases body temperature and induces interferon production.
- IL-6: IL-6 is produced by adipose tissue and in skeletal tissue but has context-dependent pro- and anti-inflammatory actions; IL-6 is associated with increased BMI and the development of type 2 diabetes while inversely associated with HDL-C [103].
- IL-8: IL-8 is chemotactic for neutrophils, basophils, and T cells and is released from monocytes, macrophages, neutrophils, and cells of the intestine, kidney, placenta, and bone marrow. IL-8 has been found to be increased in inflammatory conditions, including cardiometabolic syndrome [104].
- Plasminogen activator inhibitor-1 (PAI-1): PAI-1 is an atherogenic adipokine, induced by TNF-α [105].
- Other mediators
  - Toll-like receptors: Toll-like receptors (TLRs) are pro-inflammatory immune receptors. Activation and expression of TLRs follows recognition of pathogen-associated molecular patterns (PAMPs) and endogenous, host-derived ligands. TLR2 and TLR4 are well-characterized inducers of insulin resistance and systemic inflammation [106].
  - Adiponectin: Adiponectin, involved in the regulation of glucose and fatty acid levels, is decreased in obese individuals as well as those with insulin resistance, type 2 diabetes, and coronary heart disease. Lowered levels of adiponectin are seen in cardiometabolic syndrome [107].
  - Leptin: Leptin is a hormone that inhibits hunger and is opposed by ghrelin, which increases hunger. Obesity appears to be related to leptin resistance. Leptin can directly induce IL-6 production and thus, indirectly, hepatic CRP production [108]. Leptin and adiponectin have opposing effects on inflammation: leptin upregulates pro-inflammatory cytokines such as TNF-α and IL-6, while adiponectin downregulates pro-inflammatory cytokines [106].
  - Omentin: Omentin is an insulin-sensitizing adipokine, with reduced expression in obesity, type 2 diabetes, and insulin resistance. Omentin is associated with anti-inflammatory, anti-atherogenic, and antidiabetic properties and appears to be decreased in cardiometabolic syndrome [109].

During inflammatory states, there are also "negative" acute-phase proteins that decrease (e.g., albumin, transferrin, etc.) in response to inflammation. This is especially important to consider with the blood level of albumin, as historically it was a biomarker of nutrition status but is now recognized as a marker of inflammation when low (Fig. 46.5).

46.7 Oxidative Stress

Oxidative stress can be described as the overaccumulation of oxidants versus antioxidants, leading to cellular and tissue damage [111]. Oxidative stress closely correlates with inflammation, and inflammation is closely correlated with oxidative stress [112, 113]. Both reactive oxygen species (ROS) and reactive nitrogen species (RNS) free radicals play a role in the development of oxidative stress. A variety of factors can induce oxidative stress, including inflammation, xenobiotics, poor nutrition (e.g., simple carbohydrates, saturated fats), tobacco use, stress (physical, biochemical, mental, and emotional), and lack of physical activity, among others.

In oxidative stress, the endogenous antioxidant defense systems, which include superoxide dismutase (SOD), catalase (CAT), and glutathione peroxidase (GSHPx), as well as natural antioxidants such as vitamins C and E, plant-derived polyphenols, carotenoids, and sacrificial proteins such as albumin, are overwhelmed, and levels of free radicals (ROS/RNS and others) increase, enhancing the potential for damage ranging from DNA mutations, lipids, proteins, cellular organelles, tissues, and organs.

Oxidative stress has been associated with a growing number of disorders, and the list is larger when inflammation is considered part of the underlying pathology. Oxidative stress with concomitant mitochondrial dysfunction is associated with obesity; insulin resistance; diabetes; cardiometabolic syndrome; neurodegenerative disorders such as Parkinson's disease, Huntington's disease, and Alzheimer's disease; cancer; and aging [114–121].

46.8 Cardiometabolic Syndrome and Comorbidities

In addition to the coexistence of major conditions in cardiometabolic syndrome, this disorder is also associated with psychiatric disorders, autoimmune disorders, rheumatic dis-

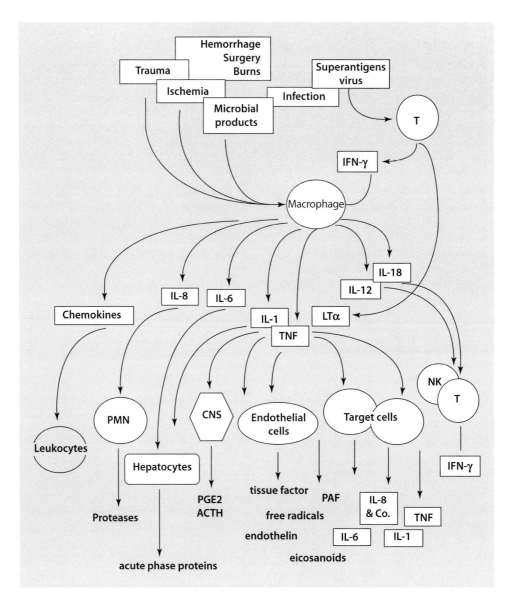

◘ Fig. 46.5 The pro-inflammatory cascade. (Reprinted from Cavaillon and Adib-Conquy [110]. With permission from Springer Nature.)

orders, and endocrine disorders. It is becoming clear that the underlying inflammatory and oxidative stress is at the root of these relationships.

46.8.1 Metabolic Disorder Comorbidities

- Metabolic/cardiometabolic syndrome
  - Diabetes
  - Obesity/sarcopenic obesity
  - Hyperuricemia
  - Hypertension
  - Dyslipidemia
  - Chronic kidney disease
  - Glucose dysregulation
  - Insulin resistance
  - Fatty liver, nonalcoholic fatty liver disease (NAFLD)
  - Acanthosis nigricans, peripheral neuropathy, retinopathy, and hirsutism (seen in patients with insulin resistance)
  - Xanthomas or xanthelasmas (seen in patients with severe dyslipidemia)
  - Erectile dysfunction
- Pain
- Fatigue
- Cardiovascular
  - Heart disease
  - Stroke; ischemic event
- Infection: can promote metabolic dysfunction (see
  - ► Chap. 20 – Infection)
  - Periodontitis
  - *Chlamydophila pneumoniae*; promotes increased rate of atherosclerotic plaque formation [122]

- Cancer
    - Solid tumor cancers
    - Hematological cancers
- Neurological
    - Neuropathies
    - Parkinson's disease
    - Alzheimer's disease
    - Seizure syndromes
    - Amyotrophic lateral sclerosis (ALS)
- Endocrine
    - Insulin resistance
    - Gonadal hormone dysfunction
    - Adrenal hormone dysfunction
- Mood and psychiatric disorders

There is a growing body of evidence indicating that Alzheimer's disease is, at its core, a metabolic disease that shares a number of features including insulin resistance, insulin-like growth factor resistance, increased oxidative stress, and local and systemic inflammation [123–128].

Alzheimer's disease is a neurodegenerative disease clinically characterized by the progressive loss of cognitive function and memory. Pathological characteristics include the presence of extracellular amyloid-β (Aβ) found as neuritic plaques and neurofibrillary tangles (NFT) made of abnormal and hyperphosphorylated tau protein. The brains of patients with Alzheimer's disease show increased levels of lipid peroxidation, oxidized proteins, RNA and DNA oxidation, carbohydrate modification, and posttranslational protein modifications, including nitrosylation and carbonylation [129, 130]. Insulin resistance has been shown to increase amyloid-β aggregates and neuritic plaque formation in addition to the decreased utilization of glucose in the CNS. Patients with type 2 diabetes are at an increased risk of developing Alzheimer's disease – the converse is also true [130]. Inflammatory mediators associated with increased adipose tissue including Ilβ, IL-6, and IL-18 [131] as well as TNF-α and adiponectin are believed to contribute to Alzheimer's disease.

46.8.2 Psychiatric Disorders

Patients with schizophrenia have an increased risk of cardiometabolic syndrome, with an overall rate of 32.5% (95% CI 30.1–35.0%) [132]. The highest rates were seen in patients who were prescribed clozapine (51.9%), while the lowest rates were associated with patients on no antipsychotic medications [112]. Patients with bipolar disorder are also at high risk for cardiometabolic syndrome, possibly related to polypharmaceutical treatment [133, 134]. Overall, a recent study found that the pooled prevalence of cardiometabolic syndrome in individuals with severe mental illness, including schizophrenia, bipolar disorder, and major depressive disorder, was 32.6% (95% CI: 30.8–34.4%; n = 52,678) [114]. Highest risk was also associated with the use of clozapine and olanzapine, although the use of clozapine carried a significantly higher risk. Lowest risk was associated with the use of aripiprazole [114]. Studies have indicated that post-traumatic stress disorder (PTSD) also increases the risk of cardiometabolic syndrome [135].

46.8.3 Autoimmune Disorders

Cardiometabolic syndrome is associated with autoimmune disorders such as psoriasis, psoriatic arthritis, rheumatoid arthritis, Hashimoto's thyroiditis, and Graves' disease [136, 137]. Several studies link psoriasis and psoriatic arthritis [138] with cardiometabolic syndrome, though the relationship is not clear. A recent meta-analysis indicated that psoriasis patients are at a greater risk of cardiometabolic syndrome, with a pooled odds ratio (OR) of 2.14 (95% CI 1.84–2.48) [139]. Chronic inflammation in psoriasis is believed to contribute to insulin resistance and an inflammatory cascade which ultimately may result in cardiometabolic syndrome [140].

Systemic lupus erythematosus (SLE) is also linked to an increased risk of cardiometabolic syndrome [13, 141, 142]. Cardiovascular events are a leading cause of mortality in SLE, and it is considered likely that both the inflammatory and autoimmune mechanisms may interact with a variety of factors, including genetic, environmental, and treatment-related factors, and that this interplay may trigger and/or exacerbate damage to the vascular wall [127, 128].

46.8.4 Rheumatic Diseases

Rheumatic diseases, including rheumatoid arthritis [14, 15], osteoarthritis, systemic lupus erythematosus, and ankylosing spondylitis, are associated with an increased risk of cardiometabolic syndrome. Cardiovascular disease in general is associated with nearly 50% increased mortality in those patients with rheumatic diseases [16]. The link between these rheumatic disorders has been related to increased adipose tissue, increased levels of adipokines, increased biomechanical loading (particularly in osteoarthritis), and an overall increased level of inflammatory markers [17]. A recent meta-analysis indicated that the association was somewhat criteria-dependent. If the National Cholesterol Education Program Adult Treatment Panel III (NCEP-ATP III) criteria were used to diagnose cardiometabolic syndrome, there was an association between rheumatoid arthritis and cardiometabolic syndrome (OR = 1.38, 95% CI: 1.04–1.83, P = 0.02), but this association was less evident when using the International Diabetes Federation (IDF) criteria [18].

46.9 Dietary Considerations in Inflammation, Oxidative Stress, and Cardiometabolic Disorder

The American Heart Association (AHA) recommends regular consumption of fruits and vegetables [19]. The AHA also recommends a diet high in whole grains, low-fat dairy products, skinless poultry, fish, nuts, legumes, and nontropical vegetable oils. The Academy of Nutrition and Dietetics (AND) similarly recommends a diet high in vegetables and fruit [20]. Part of the rationale for these recommendations includes evidence that a diet high in plant antioxidants can reduce mortality from cardiovascular disease [21, 22] and diabetes [23–25, 115]. Plant polyphenols and carotenoids, among other plant constituents, scavenge free radicals and may act indirectly by interacting with signaling cascades and cellular membranes.

46.9.1 The Microbiome

The microbiome may be defined as the sum total of all microorganisms residing in a host in a symbiotic and commensal relationship. In common usage, the microbiome often refers to the pattern of specific microorganisms, particularly bacteria and some fungi/yeasts, that reside in the human gut. These number over 10^{14} [26].

Methods such as high-throughput gene sequencing, proteomics, metagenomics, and metabolomics have indicated that the specific pattern(s) of microbial distribution play(s) a critical role in health and in the development of cardiometabolic diseases. This pattern, when disturbed, is termed "dysbiosis" and can have a significant and potentially causative impact on overall health. Dysbiosis is associated with cardiometabolic disease, obesity, insulin resistance, dyslipidemia, hypertension, and systemic inflammation [27–30].

Studies in mice fed a high-fat diet show the development of insulin resistance which can be correlated with the patterns of the microbiome [31]. In both genetically obese mice and mice fed a high-fat diet, these changes can be ameliorated by the use of antibiotics [32] with associated improved insulin sensitivity, decreased inflammatory biomarkers, and a reduction in weight gain, supporting the concept that the patterns or composition of the gut microbiome can significantly impact the metabolic phenotype and emphasizing the effect of diet – in this case a high-fat diet – on the microbiome. Obese mice undergoing the transfer of fecal bacteria derived from lean germ-free mice show a normalization of their metabolic phenotype, increased insulin sensitivity, and decreased inflammation and weight loss [33, 34].

Human studies have also found links between the composition of the microbiome with obesity, insulin resistance, dyslipidemia, adipose, and systemic inflammation [36–38] (see ◘ Fig. 46.6). The obese-type microbiome is characterized by a higher *Firmicutes/Bacteroidetes* ratio and higher levels of short-chain fatty acid excretion [39]. These patterns can be altered by dietary factors. An interventional small human study has shown that a high-fiber, strictly vegetarian diet could improve insulin resistance and dyslipidemia and induce weight loss in patients with type 2 diabetes and insulin resistance [40]. Larger five-center interventional studies of individuals at risk of cardiometabolic syndrome (both men and women between the ages of 30 and 65 years, based on BMI, plasma glucose, insulin, triglyceride, and HDL-C levels) showed that the type and quantity of dietary fat and

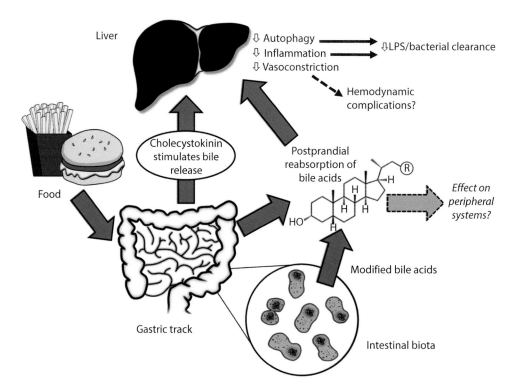

◘ Fig. 46.6 The role of bile acids with dietary intake and microbiome. (Reprinted from van Niekerk et al. [35]. With permission from Creative Commons License 4.0: ► https://creativecommons.org/licenses/by/4.0/.)

carbohydrate could alter the profile of fecal bacteria and alter short-chain fatty acid excretion levels [41].

The transfer of gut microorganisms (fecal transfer) is a possible therapeutic tool to treat insulin resistance, type 2 diabetes, cardiometabolic syndrome, and obesity [45]. In a recent study, patients with cardiometabolic syndrome who were overweight received either an allogeneic (from lean, healthy individuals) or an autologous transfer. These patients experienced improved hepatic insulin sensitivity and peripheral insulin sensitivity, particularly after allogeneic transfer [42, 43]. These results could be transitory and depend to a degree on baseline fecal microbial patterns [41, 44].

46.9.2 Chronic Stress and Cardiometabolic Syndrome

Chronic stress, whether physical, mental, emotional or spiritual, results in the hypersecretion of stress mediators, particularly cortisol, and autonomic imbalance. The presence of high levels of glucocorticoids supports the development of central adiposity, low levels of growth hormone, and hypogonadism [45]. Glucocorticoids also stimulate the activities of enzymes involved in fatty acid metabolism, induce preadipocyte differentiation, increase lipid oxidation, and induce the hepatic gluconeogenic pathway. Hypersecretion of glucocorticoids is associated with hypertension, dyslipidemia, visceral obesity, sarcopenia, and insulin hypersecretion and resistance [46, 47, 100]. A high-sugar and high-fat diet, particularly with comorbid chronic stress, has been described as a "more potent driver of visceral adiposity than diet alone, a process mediated by peripheral Neuropeptide Y (NPY)" [48]. NPY is expressed in specific neurons of the brainstem, hypothalamus, and limbic system. An important tract controlling the release of corticotropin-releasing hormone (CRH) is located between the hypothalamic arcuate nucleus (ARC) and the paraventricular nucleus of the hypothalamus (PVH), the major source of CRH. The biological effects of NPY depend on which of five receptors (Y1–Y5) it interacts and may be either anxiolytic or anxiogenic. NPY plays a role in food intake, energy homeostasis, circadian rhythm, and cognition [49]. NPY is increased in acute stress and visceral inflammation, while its expression in chronic stress is not entirely clear [50]. In addition, post-traumatic stress disorder (PTSD) is associated with lowered NPY, increased rates of obesity, and cardiometabolic disorder [51].

An often-overlooked chronic biological stress is the genetic hemochromatosis iron overload possibility for a patient with cardiometabolic syndrome. Iron overload over long term can damage the heart, liver, and kidney, even to the point of requiring organ transplantation. Newer research is associating iron overload with increased risk of Alzheimer's and cancer. For a patient with high hemoglobin/hematocrit and ferritin blood tests, it is prudent to test for the hemochromatosis gene (High Iron Fe (HFE)).

46.9.3 Hypercoagulability

The three factors considered most critical to thrombotic events are hypercoagulability, hemodynamic dysfunction, and endothelial dysfunction. Hypercoagulability may be hereditary or acquired. With acquired hypercoagulability, risk factors include age, immobilization or reduced physical activity (e.g., sedentary), inflammation, obesity, diabetes, pregnancy, hormone replacement therapy, and oral contraceptive use. Cardiometabolic syndrome is characterized by increased levels of clotting factors such as tissue factor, factor VII, and fibrinogen and by both increased plasminogen activator inhibitor-1 and decreased tissue plasminogen activator activity [52]. The cardiometabolic syndrome "presents a prothrombotic state as a result of endothelial dysfunction, the presence of a hypercoagulability state produced by an imbalance between coagulation factors and the proteins that regulate fibrinolysis and increased platelet reactivity" [105].

46.9.4 Insulin Resistance

Insulin resistance describes the lack of responsiveness of cells to the binding of insulin to its receptors. Insulin resistance can also be described as a prediabetic state and is intimately associated with obesity, glucose intolerance, metabolic syndrome, diabetes, cardiovascular disease, hyperlipidemia, as well as a variety of other conditions [53–56].

Insulin resistance is a heterogeneous and progressive clinical condition. It results in increased insulin secretion in a physiological attempt to maintain normal glucose levels. A number of mediators are involved in regulating insulin secretion – these include glucose; free fatty acids; appetite- and fat-regulating hormones including adiponectin, leptin, and ghrelin; GI-derived hormones (e.g., glucagon-like peptide-1 and other incretins); and the autonomic nervous system. Insulin regulation also is dependent on genetic, immunologic, and metabolic factors [57–59]. Dietary and nutritional influences are of obvious importance in the development and progression of insulin resistance.

Insulin resistance is primarily related to poor cell signaling at the insulin receptor on the surface of every human cell. Two things determine this insulin receptor sensitivity:
1. Structure: The structure of the cell bilayer membrane that houses the insulin receptor influences the ability of the insulin receptor's degree of sensitivity. The structure of the cell membrane is made up of 55–75% (depending on cell type) fatty acids and phospholipids. The fatty acids and phospholipids are primarily determined by those that come from diet and the milieu of hormone balance (especially insulin and cortisol), likely related to a person's inflammatory load, and that are most commonly determined by elevated visceral adipose tissue (VAT) that secretes inflammatory molecules.

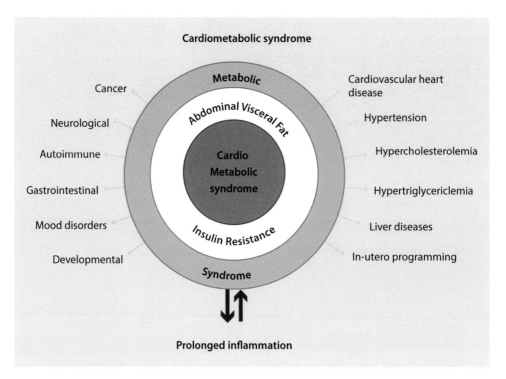

■ Fig. 46.7 Cardiometabolic syndrome and insulin resistance

2. Function: The dynamic biochemistry and energy flow within the individual's microenvironment by the presence of essential and conditionally essential nutrients and the energy flow largely dependent on a person's mineral status. If there is a significant toxin load (toxic metals, toxic chemicals, EMF exposure), there can be a negative interruption of nutrient flow to maintain healthy tissue. There are a plethora of environmental toxins individuals have been exposed to over the past 50–70 years that override the natural protective detoxification systems of a human and can be a source of metabolic perturbation exhibiting disease conditions like diabetes and metabolic syndrome [60].

Inflammation is a central component of insulin resistance [61, 62]. Both intrinsic and extrinsic factors inhibit insulin signaling. For example, serine phosphorylation of the insulin receptor substrates 1 and 2 by c-Jun N-terminal kinases (JNK) and inhibition of NF-κB (IKK) inhibits insulin signaling. In addition, both JNK and IKK stimulate the production of inflammatory cytokines, which in turn further activate JNK and IKK via autocrine and paracrine pathways. An individual's toxin load is now recognized as a contributor to the inflammation of insulin resistance [63–68]. Inflammation increases the risk of insulin resistance and metabolic syndrome as well as other metabolic conditions by three main pathways:
- Direct inhibition of insulin receptor substrates [69]
- Secretion of inflammatory cytokines [70]
- Recruitment of inflammatory leukocytes [71]

Other pathways involve excess intake of sugars, with a resultant hypertrophy in adipocytes, tissue inflammation, and white blood cell recruitment, increased insulin resistance, increased serum glucose levels, and tissue damage, with ensuing clinical consequences (■ Fig. 46.7).

46.9.5 Sarcopenia/Sarcopenic Obesity

Sarcopenia is defined as age-related loss of muscle [72]. Sarcopenia, along with cachexia, defined as the loss of weight due to underlying disease, is of great importance in the aging population but is often overlooked [112]. Sarcopenia is associated with a loss of muscle mass and muscle strength, while cachexia is associated with the loss of both muscle and fat mass. In sarcopenia, there is a mismatch between the decline in muscle strength which exceeds the decline in muscle mass.

The atrophy in sarcopenia is related especially to the atrophy of the fast, type II fibers. The number of motor units decrease, and fat is accumulated within the muscle tissues. In the geriatric population especially, inactivity may lead to increased muscle loss, and increased muscle loss may lead to greater inactivity.

Muscle mass represents the balance of protein anabolism and catabolism. Anabolic mechanisms depend in part on the activation of the serine/threonine kinase Akt, insulin-like growth factor-1 (IGF-1), insulin, testosterone, and branched-chain amino acids (BCAA). In addition, exercise and physical mobility can influence muscle mass. Catabolic mechanisms include the ubiquitin proteasome and caspase pathways as well as the myostatin pathway [73].

In aging, sarcopenia may be related to nutritional factors, genetic factors, autophagy, apoptosis, mitochondrial dysfunction, and possibly inflammation. With respect to nutrition, decreased food intake as well as decreased protein and energy intake can contribute to loss of muscle mass. Nutrients of concern include high-quality protein, vitamin D, vitamin E, and trace minerals such as selenium, carotenoids, and vitamin C [74]. As inflammation is an important underlying mechanism, consideration should also be given to intake of the omega-3 fatty acids.

Mortality risk, independent of obesity, is higher in older women with sarcopenia [75]. Approaches to sarcopenia should always also include a resistance training program to maintain or increase muscle strength. Resistance training is an effective approach, particularly in the elderly [76].

46.9.5.1 Sarcopenic Obesity

Sarcopenic obesity was originally defined as a muscle mass index that is less than 2 standard deviations below the reference gender and age-dependent reference value. An NHANES study defined sarcopenic obesity as the two lower quintiles of muscle mass and the two highest quintiles of fat mass [77]. This corresponds to:
- For men: Muscle mass < 9.12 kg/m^2 and fat mass > 37.16%
- For women: Muscle mass < 6.53 kg/m^2 and fat mass > 40.01%

There are no standard criteria for muscle strength, though grip strength is often tested because it is easier to measure than muscle mass.

Fat mass tends to increase with age, with a peak between 60 and 75 years, while muscle mass tends to decrease beginning in early middle age. This effect is correlated with reduced physical activity, reduced energy intake, and neurologic, hormonal, and environmental signals. As in the case of insulin resistance and obesity, inflammation plays a central role [78, 79].

Evidence from NHANES III also suggests that sarcopenia may be an early indicator of diabetes risk and was associated with increased insulin resistance in the non-obese and the obese. Sarcopenia was associated with dysglycemia in obese individuals, as well [80]. The InCHIANTI study, a prospective study of participants over the age of 65 in the Chianti region of Italy, found that sarcopenic obesity was associated with elevated levels of pro-inflammatory cytokines including IL-6, C-reactive protein, IL-1 receptor antagonist, and soluble IL-6 receptor ($P < 0.05$) [81].

As noted, sarcopenia is associated with increased mortality. Sarcopenic obesity is associated with higher total mortality as well as mortality associated with cardiovascular disease [82]. Thus, building and maintaining muscle mass over time is an important aspect of cardiometabolic health.

46.9.6 Cardiometabolic Syndrome

Metabolic syndrome can be defined as a cluster of abnormalities that increase the risk of cardiovascular disease, stroke, diabetes, nonalcoholic steatohepatitis, and obesity. Metabolic syndrome includes hypertension, hyperglycemia/dysglycemia, increased central obesity, hypercholesterolemia, and hypertriglyceridemia. Other signs and symptoms associated with metabolic syndrome include chest pain, shortness of breath, reduced HDL-C, and symptoms associated with dyslipidemia and insulin resistance.

Diagnostic criteria include three of the following [83]:
- Fasting glucose ≥100 mg/dL (or receiving drug therapy for hyperglycemia)
- Blood pressure ≥130/85 mm Hg (or receiving drug therapy for hypertension)
- Triglycerides ≥150 mg/dL (or receiving drug therapy for hypertriglyceridemia)
- HDL-C <40 mg/dL in men or <50 mg/dL in women (or receiving drug therapy for reduced HDL-C)
- Waist circumference USA ≥102 cm (40 in) in men or ≥88 cm (35 in) in women
- Waist circumference Asian ≥80 cm (31.5 in) in men or ≥70 cm (30 in) in women

Insulin resistance and adipose dysfunction are widely accepted as the underlying pathologic changes in metabolic syndrome [84, 85]. Risk factors include family history, poor nutrition, and low levels of physical activity.

46.9.7 Adipocytes in Cardiometabolic Syndrome

Distribution and dysfunction of the adipose tissue impacts the role adipose tissue plays in the metabolic syndrome: visceral adiposity correlates with inflammation, while subcutaneous tissue does not. There are significant ontogenetic differences between visceral and subcutaneous fat. For example, evidence suggests there are differences in genetic expression in Wilms tumor 1 (Wt1), a gene which codes a zinc finger DNA-binding protein required for development of the genitourinary system and mesothelial tissues. The differential expression of other genes including Wt1 may explain these differences [86]. Overall, it appears that expression of pro-inflammatory genes is significantly greater in visceral adipose tissue as compared to subcutaneous adipose tissue [87].

In addition to its relationship to metabolic syndrome [143], visceral adipose tissue (VAT) and the associated inflammation have been associated with atherosclerosis [144], heart disease [145], asthma [146], reflux esophagitis [147], GI disorders [148, 149], and others.

Adipocytes are the source of many biologically active substances known as adipokines. These include TNF-α [150], resistin [151], leptin [152], adiponectin [153], and plasminogen activator inhibitor-1 (PAI-1) [154]. These adipokines play a role in the pathogenesis of obesity, diabetes, and cardiometabolic syndrome [103]. For example, the increased production of PAI-1 increases the risk of thrombotic events, while elevated levels of TNF-α are associated with insulin resistance [103, 155–157]. Adiponectin, on the other hand, is associated with increased insulin sensitivity [158, 159].

In vascular diseases, oxidative stress and inflammation play significant roles in the pathophysiology of endothelial dysfunction, the proliferation of vascular smooth muscle cells, and foam cell formation. A recent review emphasized the "sequential and progressive" nature of atherogenesis and involves "cross talks at many levels" [160].

Clinical Presentation: Cardiometabolic Syndrome

George W. is a 57-year-old white male whose chief complaint at presentation is dissatisfaction with his weight. He has not seen a physician for approximately 2.5 years, seeing no need. His waist circumference is 103 cm. His height is 71 inches (180.3 cm) and his weight is 225 pounds (102.1 kg). BMI is 33.2 kg/m² with a waist circumference of 36.2 inches (91.95 cm). Patient has no history of tobacco use. Patient reports moderate alcohol use and no recreational drug use. He has in the past used cannabis to "take the edge off" but has not used cannabis in the last 5 years. Patient does not report the use of any medications and supplements or any history of chronic disease, but was in an MVA approximately 3 years ago where he sustained a concussion with an overnight stay in the local hospital. He reports a diet of fast food, high of red meat, processed foods, and few vegetables/fruits. He has a craving for salt- and sugar-rich foods. Patient is recently divorced with three children and has been self-employed as a criminal defense attorney for 22 years. He reports 5–6 hours of disrupted sleep per night.
Blood pressure: 135/85 mm Hg Heart rate: 88
Labs: Complete blood count (CBC): Hgb/Hct 16.5/32; MCV/MCH 101/31
- Thyroid stimulating hormone (TSH), T3/T4, normal
- Complete metabolic panel:
    - Fasting blood glucose (FBG): 110 mg/dL
    - Hemoglobin A_1c: 8.9
    - Lipid panel:
        - Total cholesterol: 238 mg/dL
        - HDL cholesterol: 35 mg/dL
        - LDL cholesterol: 180 mg/dL
        - Triglycerides: 452 mg/dL
        - ApoB: 145 mg/dL
        - LDL-P: 1325
        - hs-CRP: 10 mg/L
        - Fibrinogen: 398 mg/dL
    - Nutritional status markers:
        - Vitamin D 25OH: 15 ng/mL
        - Albumin: 3.6 g/dL
        - Retinol-binding protein (RBP): 42 mcg/mL
        - Transferrin: 165 mg/dL
        - Homocysteine: 126 mcmol/L
        - Vitamin B12: 157 pg/mL
        - Omega-3 index: 0.9% (range 2.5–11.8%)
    - Other:
        - Adiponectin: 0.78 mcg/mL
- Medical Symptoms Questionnaire/Signs and Symptoms: 92 (goal <50)

Assessment/diagnosis:
- Metabolic syndrome with high FBG and A_1c with potential comorbidities
- Hypertriglyceridemia, hypercholesterolemia (LDL cholesterol, LDL-P, HDL)
- hs-CRP (systemic inflammation)
- Fibrinogen
- Altered nutritional markers depleted
- Adiponectin low

Intervention:
- Structural assessment referral to a cranial-sacral trained practitioner related to MVA

Lifestyle
- Stress management education
- Exercise prescription
- Sleep (goal 7/8 hours per night)

Dietary recommendations: Low carbohydrate, healthy fats, and glycemic carbohydrates in a 40:40:30 ratio. Education on avoidance of processed foods, simple sugars, artificial sweeteners, charred meats, and high-heat processed vegetable oils.

Supplements:
- Multivitamin iron-free with bioactive forms of B complex
- B12 hydroxycobalamin/adenosylcobalamin 1000 mcg/day lozenge/AM
- Berberine 500 mg one capsule with each meal
- CoQ10 200 mg daily near food/meal
- Phosphatidylcholine/phosphatidylethanolamine formula 3000 mg daily (capsules or liquid)
- Magnesium threonate 200–400 mg daily
- Further recommendations after test results

Recommend further blood and urine tests to support nutritional assessment:
- Organic acids (urine)
- Lipoprotein particle lipid panel
- C-reactive protein (CRP)
- Ferritin (to rule out hemochromatosis; if ferritin is high, request HFE genomic testing)
- Methylmalonic acid (MMA) (functional B12 marker)
- After reviewing above tests, may recommend further testing of gut ecology, imaging.

Monitoring/evaluation:
- Review new test results
- Review referral consults
- Monitor anthropometrics, previously abnormal blood tests, signs, and symptoms.

Goals:
- Complete referral consults
- Weight loss BMI 25–27
- Blood labs
- $M < 30$
- Sleep 7–8 hours each night
- Consistent moderate exercise program
- Stress management routine in place

46.10 Anthropometric Assessment Methodologies

Anthropometry is the use of body measurements that can be used as part of a comprehensive nutrition assessment. Anthropometric measurements are influenced by a number of factors including age, nutritional status, genetics, the environment, social influences, lifestyle, and others, but they provide an inexpensive, noninvasive approach to obtaining important information regarding overall health status. Anthropometric assessments provide highly reliable information while avoiding the complexities of newer (and costlier) methodologies [161] (Table 46.3).

- Waist circumference
  - A waist circumference greater than 102 cm (40″) in men and 88 cm (35″) in women is associated with a higher risk of cardiometabolic syndrome, T2D, hypertension, and cardiovascular disease.
- Waist-to-hip ratio
  - The waist circumference is measured just below the ribs, and the hips are measured at the widest portion of the hips. For men, a normal waist/hip ratio is ≤0.90, while for women a normal waist/hip ratio is ≤0.80. Waist/hip ratios ≥1.0 are associated with an increased risk for cardiovascular disease, T2D, and cardiometabolic syndrome.
- Waist-to-height ratio
  - The waist-to-height ratio is a measure of body fat distribution. A healthy waist-to-height ratio ranges from 0.42 to 0.48. A waist-to-height ratio of 0.51 is equivalent to the risk associated with a BMI of 25, while a waist-to-height ratio of 0.57 is equivalent to the risk associated with a BMI of 30.
- Body mass index (BMI)
  - The BMI is weight (in kg) divided by height^2 (m). BMI is an estimate of body fat and can be used to screen for health risks in populations.
- Body fat percentage
  - The body fat percentage is the fat mass divided by the total body mass. The percentage of essential fat is 2–5% in men and 10–13% in women [162].
- Body cell mass (BCM)
  - BCM consists of the total mass of all the cellular elements including tissues of the muscles, organs, and bones, as well as extracellular and intracellular water. BCM decreases in wasting conditions such as AIDS, cancer, and other chronic diseases. The preferred method for the assessment of BCM is bioelectrical impedance analysis.
- Bioelectric impedance analysis (BIA)
  - Bioelectrical impedance is the resistance to electrical flow through body tissues and can be used to estimate body fat and body cell mass. BIA has been validated under a number of conditions and circumstances [163, 164].
- Imaging
  - Computerized tomography (CT) scan
  - Magnetic resonance imaging (MRI)
  - Dual-energy X-ray absorptiometry (DEXA, DXA) can be used to quantify total, subcutaneous, and visceral body fat.

Table 46.3 Body mass index height/weight

BMI	Weight status
Below 18.5	Underweight
18.5–24.9	Normal or healthy weight
25.0–29.9	Overweight
30.0–34.9	Obese: Class I
35.0–39.9	Obese: Class II
≥ 40.0	Obese: Class III

46.11 Assessment of Physical Performance

A number of different tests of physical performance and muscle strength may be used. These include gait speed, the 6-minute walk test, the stair climb power test, and the short physical performance battery (SPPB). The SPPB consists of three separate tests that assess balance, gait, and strength. The individual is tested for their ability to stand with their feet in tandem, semi-tandem, and parallel positions [165].

46.11.1 Biomarkers for Cardiometabolic Syndrome

46.11.1.1 Pro-Inflammatory State
The pro-inflammatory state may be assessed in several ways, but common measures are serum levels of hs-CRP, fibrinogen, IL-6, TNF-α, and leukocytosis.

46.11.1.2 Acute-Phase Reactants of Inflammation
- C-reactive protein high sensitivity
- Ferritin
- Albumin (negative)
- Ceruloplasmin (bound copper)
- Sedimentation rate (ESR)

46.11.1.3 Glucose/Insulin
- Fasting glucose
- Hemoglobin A_1c
- Fasting insulin
- Proinsulin
- Fasting C-peptide
- HOMA-IR

46.11.1.4 Dyslipidemia
- Triglycerides
- HDL
- Triglyceride HDL ratio
- VLDL and LDL
- Lipoprotein (a)
- Lipoprotein particle number and size

46.11.1.5 Visceral Adipose Tissue
- Men
  - Waist circumference (WC) USA >40 inches/102 cm
  - Asian BMI >21 kg/m^2
  - WC levels ≥90 cm men
  - Waist-to-hip ratio: >0.90
- Women
  - Waist circumference >35 inches/88 cm
  - Asian waist circumference ≥ 70 cm women
  - Waist-to-hip ratio: >0.85

46.11.1.6 Hypercoagulation
- Prothrombin time (PT)/international normalized ratio (INR)
- Fibrinogen
- Platelets
- Genomic factor V
- hs-CRP

46.11.1.7 Immune Compromised
- Complete blood count (CBC): white blood cells and differential
- Secretory IgA
- Natural killer cell count and activity
- Protein electrophoresis

46.11.1.8 Blood Labs
- CBC
- Differential: WBC, eosinophils, neutrophils, lymphocytes
- hs-CRP
- ESR/sed rate
- Lipid panel
- Liver enzymes: ALT, AST, Alk Phos, GGT
- Bilirubin, direct and total
- Ferritin
- Vitamin D: 25-OH vitamin D, 1,25 vitamin D
- Vitamin A: retinol
- Homocysteine
- Apo b/apo a1/Apob/apoa1
- Lp(a)

46.11.1.9 Further Testing
- Rheumatoid factor (RF)
- Cytokine panel: IL-8, IL-10, IL-6, IL-2
- Fibrinogen (platelets)
- Antitrypsin
- Organic acids
- Toxic metals
- Cleveland heart labs

46.11.1.10 Imaging: CT/Pet CT Scan/MRI/Ultrasound
- Liver – rule out NASH, cirrhosis, fatty liver
- Tumors
  - Benign: lipoma, fluid-filled cysts
  - Neoplastic: size, location, diagnosis
- Osteopenia/osteoporosis via DEXA bone density
- Joints: osteoarthritis
- Gastrointestinal
- Coronary calcium score
- Vo2max testing

46.11.1.11 Biopsy
Any tumor or lesion of GI inspection

46.11.1.12 Bioelectric Impedance Analysis
- Phase angle
- Capacitance
- Body cell mass
- Intracellular water (ICW)
- Total body water

46.11.1.13 Obesity (See Assessment Sarcopenia)
- Metabolically healthy obese
- Chronic disease obesity
- Increased disease associations with obesity

46.11.1.14 Lifestyle
- Sleep
- Exercise
- Relationships

46.11.2 Diet

Diets which are high in simple carbohydrates (including those high in high fructose corn syrup (HFCS), high ratios of omega-6/omega-3 essential fatty acids, high sodium foods, and food low in polyphenols and both soluble and insoluble fiber are associated with a higher risk of chronic disease [166]. Conversely, diets that are high in vegetables, fruits, lean meats, fish, legumes, nuts, and seeds, high in omega-3, high in fiber, and low in simple carbohydrates or processed foods tend to be protective against chronic disease [166–168].

46.11.2.1 Nutrient Insufficiencies, Deficiencies, or Excesses

Nutrient
- Vitamin D
- Vitamin A
- Vitamin K1 and K2
- Glutathione
- Essential fatty acids (w3 and w6)
- Beneficial MUFA and saturated
- Phospholipids (phosphatidylcholine, phosphatidylethanolamine)
- Cholesterol/sterols
- Methyl nutrients:
-     (B vitamins, trimethylglycine (TMG), phosphatidylcholine (PC)) (not lecithin)
- Magnesium (ensure adequate calcium food or supplement if needed)
- Zinc
- Selenium
- Vitamin C
- Vitamin E: gamma-tocopherol and tocotrienols
- ATP: CoQ10, d-ribose, l-carnitine
- Avoid damaged oils
- Excess calcium can be problematic

46.11.2.2 Dental/Oral Health
- Periodontal tissue disease (indicator for risk of folate, carotenoids, vitamin C deficits) [169]
- Tooth decay
- Amalgam dental material fillings
- Oral galvanization
- Composite dental material fillings
- Temporomandibular joint (TMJ) misalignment
- Tongue depapillation (indicator for poor B vitamin status) [170]

46.11.2.3 Gut Health
- Bowel movement frequency, color, and form
- Fermented food or probiotic intake
- Gastrointestinal symptoms
- Healthy toilet posture (squatting vs. sitting)
- *Lab testing*: Comprehensive Digestive Stool Analysis

46.11.2.4 Renal Health
- Kidney clearance
- Fatty acid composition of renal tubules
- Toxin load
- Electrolyte status

46.11.2.5 Gonadal Health
- Men: prostate, breast
- Women: breast, uterus, ovary, thyroid

46.11.2.6 Physical Activity
- *Health maintenance*
  - Intermittent intense exercise several times a day
  - Reduce sitting for long periods
  - Joyful movement
- *Sports* athletic *performance*
  - (See ▶ Chap. 35)

46.11.2.7 Sleep
- Associations with disease
- Sleep apnea
- Sleep hygiene

46.11.2.8 Happiness
- Relationships
- Community
- Mental health
- Blue zones

46.11.2.9 Loneliness
- Biggest lifestyle factor for decreased lifespan

46.11.2.10 Toxin Load
- Dioxin
- Glyphosate (roundup)
- Endocrine disruptors: BPA [171]

46.11.2.11 Physical Exam (See ▶ Chap. 39 Nutrition Physical Exam)
- Hair/scalp
- Ears
- Face
- Eyes
- Oral cavity: palate, tongue, tonsils, adenoids/throat
- Sinus
- Thyroid, neck, parotid
- Gag reflex, vagal nerve tone

46.11.2.12 Posture and Spinal Alignment
- Injury history: head, neck, birth trauma
- Posture habits
- Headache history

46.11.3 Protocols of Treatment for Chronic Conditions

46.11.3.1 Conventional Treatment
- Rx: statins, metformin, hypertensive meds, omega-3 fish oil, etc.
- Diet: optional
- Exercise: optional, 10,000 steps per day

46.11.3.2 Integrative and Functional Medicine Approaches

- Exercise
- Mediterranean diet/low-carb whole foods/ketogenic or modified ketogenic diet
- Glucose/insulin nutrient support
- Support for detoxification metabolism
- Stress management
- Sleep management
- Rx, if indicated

Healthy Human
- Healthy body fat %
  - Male: 8–16%
  - Female: 16–20%
- Muscle strength
- Healthy blood pressure
- Healthy protective inflammation
- Healthy kidney function
- Managed blood glucose
- Nervous system integrity
- Fetal and infant healthy pre-programming
- Minimal cancer risk
- Cardiovascular integrity
- Retain self vs. nonself
- Sexual vitality
- Healthy sleep

46.12 Conclusion

Metabolic disorders are abnormalities that are not isolated conditions but are metabolically closely linked to each other. The concept of systems biology describes the interrelationship of all body systems that need consideration when assessing an individual's health condition. For the past 100 years, there has been a rise in the incidence of chronic diseases that differ from acute diseases, primarily through the long-term development of disorders in complex ways. During early development of chronic diseases, the symptoms are often unrecognized while metabolic perturbations occur. Newer approaches are recognizing biomarkers of early signs of chronic diseases while they are more easily treated and prevented from developing into a serious disorder. The pathophysiology of chronic diseases is continuing to be better understood, with scientific advancements in biology, physiology, biochemistry, genomics, mind-body, and toxicology that are clarifying important differences from acute to chronic disease. This chapter presented evidence that the cardiometabolic syndrome is an underlying foundation for the development of all chronic diseases throughout the lifespan. Chronic diseases are largely due to influences from diet, lifestyle, genomics, and the environment. Chronic diseases are becoming epidemic globally. This chapter presented the principles and practices of identifying the components of cardiometabolic syndrome to influence medical and health practitioner's clinical practice. This knowledge has the potential to effect changes in clinical implementation to prevent and treat chronic diseases to restore wellness for all individuals and populations.

References

1. Beaglehole R, et al. Priority actions for the non-communicable disease crisis. Lancet. 2011;377(9775):1438–47.
2. Beaglehole R, et al. UN high-level meeting on non-communicable diseases: addressing four questions. Lancet. 2011;378(9789):449–55.
3. Santulli G. Epidemiology of cardiovascular disease in the 21st century: updated numbers and updated facts. J Cardiovasc Dis. 2013;1(1):1–2.
4. Patnaik JL, et al. Cardiovascular disease competes with breast cancer as the leading cause of death for older females diagnosed with breast cancer: a retrospective cohort study. Breast Cancer Res. 2011;13(3):R64.
5. Muka T, et al. The global impact of non-communicable diseases on healthcare spending and national income: a systematic review. Eur J Epidemiol. 2015;30(4):251–77.
6. Kankeu HT, et al. The financial burden from non-communicable diseases in low-and middle-income countries: a literature review. Health Res Policy Syst. 2013;11(1):31.
7. Di Cesare M, et al. Inequalities in non-communicable diseases and effective responses. Lancet. 2013;381(9866):585–97.
8. Bodai BI, et al. Lifestyle medicine: a brief review of its dramatic impact on health and survival. Perm J. 2018;22:17–025.
9. Kontis V, et al. Regional contributions of six preventable risk factors to achieving the 25 × 25 non-communicable disease mortality reduction target: a modelling study. Lancet Glob Health. 2015;3(12):e746–57.
10. Grundy SM, et al. Diagnosis and management of the metabolic syndrome: an American Heart Association/National Heart, Lung and Blood Institute Scientific Statement. Circulation. 2005;112(17):2735–52.
11. Tian YM, Ma N, Jia XJ, Lu Q. The "hyper-triglyceridemic waist phenotype" is a reliable marker for prediction of accumulation of abdominal visceral fat in Chinese adults. Eat Weight Disord. 2019:1–8.
12. Huang PL. A comprehensive definition for metabolic syndrome: NCEP-ATP. Dis Model Mech. 2009;2(5–6):231–7.
13. Mirabelli G, et al. One year in review 2015: systemic lupus erythematosus. Clin Exp Rheumatol. 2015;33(3):414–25.
14. Hallajzadeh J, et al. Metabolic syndrome and its components among rheumatoid arthritis patients: a comprehensive updated systematic review and meta-analysis. PloS One. 2017;12(3):e0170361.
15. Kerekes G, et al. Rheumatoid arthritis and metabolic syndrome. Nat Rev Rheumatol. 2014;10(11):691–6.
16. Nurmohamed MT. Cardiovascular risk in rheumatoid arthritis. Autoimmun Rev. 2009;8(8):663–7.
17. Abella V, et al. Adipokines, metabolic syndrome and rheumatic diseases. J Immunol Res. 2014;2014:343746.
18. Duan H, et al. Association of rheumatoid arthritis and the prevalence of metabolic syndrome: an update meta-analysis. Int J Clin Exp Med. 2016;9(9):17334–44.
19. American Heart Association. The American heart association diet and lifestyle recommendations. Retrieved from: http://www.heart.org/HEARTORG/HealthyLiving/HealthyEating/Nutrition/The-American-Heart-Associations-Diet-and-Lifestyle-Recommendations_UCM_305855_Article.jsp#. Accessed 12/2017.

20. Freeland-Graves JH, Nitzke S. Position of the academy of nutrition and dietetics: total diet approach to healthy eating. J Acad Nutr Diet. 2013;113(2):307–17.
21. Wang X, et al. Fruit and vegetable consumption and mortality from all causes, cardiovascular disease, and cancer: systematic review and dose-response meta-analysis of prospective cohort studies. BMJ. 2014;349:g4490.
22. Afshin A, et al. Consumption of nuts and legumes and risk of incident ischemic heart disease, stroke, and diabetes: a systematic review and meta-analysis. Am J Clin Nutr. 2014;100(1):278–88.
23. Bahadoran Z, Mirmiran P, Azizi F. Dietary polyphenols as potential nutraceuticals in management of diabetes: a review. J Diabetes Metab Disord. 2013;12(1):43.
24. KJC C, de ARS O, Marreiro D. Antioxidant role of zinc in diabetes mellitus. World J Diabetes. 2015;6(2):333.
25. Anhê FF, et al. A polyphenol-rich cranberry extract protects from diet-induced obesity, insulin resistance and intestinal inflammation in association with increased Akkermansia spp. population in the gut microbiota of mice. Gut. 2015;64(6):872–83.
26. Fouhy F, Ross RP, Fitzgerald GF, Stanton C, Cotter PD. Composition of the early intestinal microbiota: knowledge, knowledge gaps and the use of high-throughput sequencing to address these gaps. Gut Microbes. 2012;3:203–20.
27. Aroor AR, et al. Maladaptive immune and inflammatory pathways lead to cardiovascular insulin resistance. Metabolism. 2013;62(11):1543–52.
28. Sáez-Lara MJ, et al. Effects of probiotics and synbiotics on obesity, insulin resistance syndrome, type 2 diabetes and non-alcoholic fatty liver disease: a review of human clinical trials. Int J Mol Sci. 2016;17(6):928.
29. Kallio KA, et al. Endotoxemia, nutrition, and cardiometabolic disorders. Acta Diabetol. 2015;52(2):395–404.
30. Lee MS. ED 05-2 interaction of gut dysbiosis and innate immune dysfunction in the development of metabolic syndrome. J Hypertens. 2016;34:e187.
31. Cani PD, et al. Metabolic endotoxemia initiates obesity and insulin resistance. Diabetes. 2007;56:1761–72.
32. Carvalho BM, et al. Modulation of gut microbiota by antibiotics improves insulin signalling in high-fat fed mice. Diabetologia. 2012;55:2823–34.
33. Turnbaugh PJ, Bäckhed F, Fulton L, Gordon JI. Diet-induced obesity is linked to marked but reversible alterations in the mouse distal gut microbiome. Cell Host Microbe. 2008;3:213–23.
34. Ridaura VK, et al. Gut microbiota from twins discordant for obesity modulate metabolism in mice. Science. 2013;341:1241214.
35. van Niekerk G, Davis T, de Villiers W, Engelbrecht AM. The role of bile acids in nutritional support. Crit Care. 2018;22(1):231. https://doi.org/10.1186/s13054-018-2160-4.
36. Karlsson FH, et al. Gut metagenome in European women with normal, impaired and diabetic glucose control. Nature. 2013;498:99–103.
37. Zhang X, et al. Human gut microbiota changes reveal the progression of glucose intolerance. PLoS One. 2013;8:e71108.
38. Le Chatelier E, et al. Richness of human gut microbiome correlates with metabolic markers. Nature. 2013;500:541–6.
39. Fava F, et al. The type and quantity of dietary fat and carbohydrate alter faecal microbiome and short-chain fatty acid excretion in a metabolic syndrome 'at-risk' population. Int J Obes. 2013;37(2):216–23.
40. Kim MS, Hwang SS, Park EJ, Bae JW. Strict vegetarian diet improves the risk factors associated with metabolic diseases by modulating gut microbiota and reducing intestinal inflammation. Environ Microbiol Rep. 2013;5:765–75.
41. Aron-Wisnewsky J, Clément K. The gut microbiome, diet, and links to cardiometabolic and chronic disorders. Nat Rev Nephrol. 2016;12(3):169–81.
42. Vrieze A, Van Nood E, Holleman F. Transfer of intestinal microbiota from lean donors increases insulin sensitivity in individuals with metabolic syndrome (vol 143, p 913, 2012). Gastroenterology. 2013;144(1):250.
43. Kootte RS, et al. Improvement of insulin sensitivity after lean donor feces in metabolic syndrome is driven by baseline intestinal microbiota composition. Cell Metab. 2017;26(4):611–9.
44. Khan MT, Nieuwdorp M, Bäckhed F. Microbial modulation of insulin sensitivity. Cell Metab. 2014;20(5):753–60.
45. Charmandari E, Tsigos C, Chrousos G. Endocrinology of the stress response. Annu Rev Physiol. 2005;67:259–84.
46. Chrousos GP. Stress and disorders of the stress system. Nat Rev Endocrinol. 2009;5(7):374–81.
47. Wulsin L, Herman J, Thayer JF. Stress, autonomic imbalance, and the prediction of metabolic risk: a model and a proposal for research. Neurosci Biobehav Rev. 2018;86:12–20.
48. Aschbacher K, et al. Chronic stress increases vulnerability to diet-related abdominal fat, oxidative stress, and metabolic risk. Psychoneuroendocrinology. 2014;46:14–22.
49. Heilig M. The NPY system in stress, anxiety and depression. Neuropeptides. 2004;38(4):213–24.
50. Reichmann F, Holzer P. Neuropeptide Y: a stressful review. Neuropeptides. 2016;55:99–109.
51. Farr OM, et al. Stress-and PTSD-associated obesity and metabolic dysfunction: a growing problem requiring further research and novel treatments. Metab Clin Exp. 2014;63(12):1463.
52. Nieuwdorp M, Stroes ES, Meijers JC, Büller H. Hypercoagulability in the metabolic syndrome. Curr Opin Pharmacol. 2005;5(2):155–9.
53. Tangvarasittichai S. Oxidative stress, insulin resistance, dyslipidemia and type 2 diabetes mellitus. World J Diabetes. 2015;6(3):456.
54. Tchernof A, Després JP. Pathophysiology of human visceral obesity: an update. Physiol Rev. 2013;93(1):359–404.
55. Shulman GI. Ectopic fat in insulin resistance, dyslipidemia, and cardiometabolic disease. N Engl J Med. 2014;371(12):1131–41.
56. DeMarco VG, Aroor AR, Sowers JR. The pathophysiology of hypertension in patients with obesity. Nat Rev Endocrinol. 2014;10(6):364–76.
57. Rorsman P, Braun M. Regulation of insulin secretion in human pancreatic islets. Annu Rev Physiol. 2013;75:155–79.
58. Rutter GA, et al. Pancreatic β-cell identity, glucose sensing and the control of insulin secretion. Biochem J. 2015;466(2):203–18.
59. Molina J, et al. Control of insulin secretion by cholinergic signaling in the human pancreatic islet. Diabetes. 2014;63(8):2714–26.
60. Lauretta R, Sansone A, Sansone M, Romanelli F, Appetecchia M. Endocrine disrupting chemicals: effects on endocrine glands. Front Endocrinol. 2019;10:178.
61. Odegaard JI, Chawla A. Pleiotropic actions of insulin resistance and inflammation in metabolic homeostasis. Science. 2013;339(6116):172–7.
62. McArdle MA, et al. Mechanisms of obesity-induced inflammation and insulin resistance: insights into the emerging role of nutritional strategies. Front Endocrinol. 2013;4:52.
63. Schwabl P, et al. Assessment of microplastic concentrations in human stool – preliminary results of a prospective study, presented at UEG week 2018 Vienna, October 24, 2018.
64. European Chemicals Agency. Microplastics. 2018. Available at: https://echa.europa.eu/hot-topics/microplastics. Accessed 21 Aug 2018.
65. Hohenblum P, Liebmann B, Liedermann M. Plastic and microplastic in the environment. Vienna: Environment Agency Austria; 2015. Available at: http://www.umweltbundesamt.at/fileadmin/site/publikationen/REP0551.pdf.
66. Powell JJ, Thoree V, Pele LC. Dietary microparticles and their impact on tolerance and immune responsiveness of the gastrointestinal tract. Br J Nutr. 2007;98(Suppl 1):S59–63.
67. Geyer R, Jambeck JR, Law KL. Production, use, and fate of all plastics ever made. Sci Adv. 2017;3(7):e1700782.
68. Romeo T, Pietro B, Peda C, Consoli P, Andaloro F, Fossi MC. First evidence of presence of plastic debris in stomach of large pelagic fish in the Mediterranean Sea. Mar Pollut Bull. 2015;95(1):358–61.

69. Boucher J, Kleinridders A, Kahn CR. Insulin receptor signaling in normal and insulin-resistant states. Cold Spring Harb Perspect Biol. 2014;6(1):a009191.
70. Kahn SE, Hull RL, Utzschneider KM. Mechanisms linking obesity to insulin resistance and type 2 diabetes. Nature. 2006;444(7121):840–6.
71. Lee BC, Lee J. Cellular and molecular players in adipose tissue inflammation in the development of obesity-induced insulin resistance. Biochim Biophys Acta. 2014;1842(3):446–62.
72. Ali S, Garcia JM. Sarcopenia, cachexia and aging: diagnosis, mechanisms and therapeutic options-a mini-review. Gerontology. 2014;60(4):294–305.
73. White TA, LeBrasseur NK. Myostatin and sarcopenia: opportunities and challenges-a mini-review. Gerontology. 2014;60(4):289–93.
74. Robinson S, Cooper C, Sayer AA. Nutrition and sarcopenia: a review of the evidence and implications for preventive strategies. In: Clinical nutrition and aging: sarcopenia and muscle metabolism, vol. 1; 2016.
75. Batsis JA, et al. Sarcopenia, sarcopenic obesity and mortality in older adults: results from the National Health and Nutrition Examination Survey III. Eur J Clin Nutr. 2014;68(9):1001–7.
76. Liu CJ, Latham NK. Progressive resistance strength training for improving physical function in older adults. Cochrane Database Syst Rev. 2009;3:CD002759.
77. Davison KK, Ford E, Cogswell M, Dietz W. Percentage of body fat and body mass index are associated with mobility limitations in people aged 70 and older from NHANES III. J Am Geriatr Soc. 2002;50:1802–9.
78. Stenholm S, et al. Sarcopenic obesity-definition, etiology and consequences. Curr Opin Clin Nutr Metab Care. 2008;11(6):693.
79. Roubenoff R. Sarcopenic obesity: the confluence of two epidemics. Obesity. 2004;12(6):887–8.
80. Srikanthan P, Hevener AL, Karlamangla AS. Sarcopenia exacerbates obesity-associated insulin resistance and dysglycemia: findings from the National Health and Nutrition Examination Survey III. PLoS One. 2010;5(5):e10805.
81. Schrager MA, et al. Sarcopenic obesity and inflammation in the InCHIANTI study. J Appl Physiol. 2007;102(3):919–25.
82. Wannamethee SG, Atkins JL. Conference on 'nutrition and age-related muscle loss, sarcopenia and cachexia' symposium 4: sarcopenia and cachexia and social, clinical and public health dimensions. Muscle loss and obesity: the health implications of sarcopenia and sarcopenic obesity. Proc Nutr Soc. 2015;74:405–12.
83. Grundy SM, et al. Diagnosis and management of the metabolic syndrome: an American Heart Association/National Heart, Lung and Blood Institute Scientific Statement. Circulation. 2005;112(17):2735–52.
84. Guo S. Insulin signaling, resistance, and metabolic syndrome: insights from mouse models into disease mechanisms. J Endocrinol. 2014;220(2):T1–T23.
85. Ruderman NB, et al. AMPK, insulin resistance, and the metabolic syndrome. J Clin Invest. 2013;123(7):2764.
86. Chau YY, et al. Visceral and subcutaneous fat have different origins and evidence supports a mesothelial source. Nat Cell Biol. 2014;16(4):367–75.
87. Spoto B, et al. Pro-and anti-inflammatory cytokine gene expression in subcutaneous and visceral fat in severe obesity. Nutr Metab Cardiovasc Dis. 2014;24(10):1137–43.
88. IDF Consensus Statement: https://www.idf.org/e-library/consensus-statements/60-idfconsensus-worldwide-definitionof-the-metabolic-syndrome.html. Accessed 2 Jan 2019.
89. Ruiz-Núñez B, et al. Lifestyle and nutritional imbalances associated with Western diseases: causes and consequences of chronic systemic low-grade inflammation in an evolutionary context. J Nutr Biochem. 2013;24:1183–201.
90. Flegal KM, et al. Trends in obesity among adults in the United States, 2005 to 2014. JAMA. 2016;315(21):2284–91.
91. Guariguata L, et al. Global estimates of diabetes prevalence for 2013 and projections for 2035. Diabetes Res Clin Pract. 2014;103(2):137–49.
92. Micucci C, Valli D, Matacchione G, Catalano A. Current perspectives between metabolic syndrome and cancer. Oncotarget. 2016;7(25):38959–72. https://doi.org/10.18632/oncotarget.8341.
93. Ramsay DS, Woods SC. Clarifying the roles of homeostasis and allostasis in physiological regulation. Psychol Rev. 2014;121(2):225–47.
94. Peters A, McEwen BS. Introduction for the allostatic load special issue. Physiol Behav. 2012;106(1):1–4.
95. Reutrakul S, Van Cauter E. Interactions between sleep, circadian function, and glucose metabolism: implications for risk and severity of diabetes. Ann N Y Acad Sci. 2014;1311:151–73. https://doi.org/10.1111/nyas.12355.
96. Reutrakul S, Van Cauter E. Sleep influences on obesity, insulin resistance, and risk of type 2 diabetes. Metabolism. 2018;84:56–66.
97. Helfinstein SM, et al. If all your friends jumped off a bridge: the effect of others' actions on engagement in and recommendation of risky behaviors. J Exp Psychol. 2015;144(1):12–7.
98. Pizzorno J, Katzinger J. Clinical pathophysiology a functional perspective. Mind Publishing; 2012
99. Landry A, Docherty P, Ouellette S, Cartier LJ. Causes and outcomes of markedly elevated C-reactive protein levels. Can Fam Physician. 2017;63(6):e316–e23.
100. Pearson TA, Mensah GA, Alexander RW, Anderson JL, Cannon RO, Criqui M, Fadl YY, Fortmann SP, Hong Y, Myers GL, Rifai N, Smith SC, Taubert K, Tracy RP, Vinicor F. AHA/CDC scientific statement: markers of inflammation and cardiovascular disease. Circulation. 2003;107:499.
101. C-reactive protein test to screen for heart disease: Why do we need another test? Harvard Health Publishing Harvard Medical School. https://www.health.harvard.edu/heart-health/c-reactive-protein-test-to-screen-for-heart-disease. Accessed 17 Jul 2018.
102. Kaur J. A comprehensive review on metabolic syndrome. Cardiol Res Pract. 2014;2014:943162.
103. Hunter CA, Jones SA. IL-6 as a keystone cytokine in health and disease. Nat Immunol. 2015;16(5):448–57.
104. Bremer AA, Jialal I. Adipose tissue dysfunction in nascent metabolic syndrome. J Obes. 2013;2013:393192.
105. Ahn J, Lee H, Kim S, Ha T. Resveratrol inhibits TNF-α-induced changes of adipokines in 3T3-L1 adipocytes. Biochem Biophys Res Commun. 2007;364(4):972–7.
106. Jialal I, Kaur H, Devaraj S. Toll-like receptor status in obesity and metabolic syndrome: a translational perspective. J Clin Endo Met. 2014;99(1):39–48.
107. Fuentes E, et al. Mechanisms of chronic state of inflammation as mediators that Link obese adipose tissue and metabolic syndrome. Mediat Inflamm. 2013;2013:136584.
108. López-Jaramillo P, et al. The role of leptin/adiponectin ratio in metabolic syndrome and diabetes. Horm Mol Biol Clin Invest. 2014;18(1):37–45.
109. Jialal I, et al. Increased chemerin and decreased omentin-1 in both adipose tissue and plasma in nascent metabolic syndrome. J Clin Endocrinol Metab. 2013;98(3):E514–7.
110. Cavaillon JM, Adib-Conquy M. The pro-inflammatory cytokine cascade. In: Marshall JC, Cohen J, editors. Immune response in the critically ill. Update in intensive care medicine, vol. 31. Berlin, Heidelberg: Springer; 2002. p. 37–66.
111. Sies H. Oxidative stress: oxidants and antioxidant. Exp Physiol. 1997;82(2):291–5.
112. Siti HN, Kamisah Y, Kamsiah J. The role of oxidative stress, antioxidants and vascular inflammation in cardiovascular disease (a review). Vasc Pharmacol. 2015;71:40–56.
113. Chaudhari N, et al. A molecular web: endoplasmic reticulum stress, inflammation, and oxidative stress. Front Cell Neurosci. 2014;8:213.

114. Tjalkens RB, Streifel KM, Moreno JA. Neuroinflammation and oxidative stress in models of Parkinson's disease and protein-Misfolding disorders. In: Oxidative stress and redox signalling in Parkinson's disease; 2017. p. 184–209.
115. Zhang H, Tsao R. Dietary polyphenols, oxidative stress and antioxidant and anti-inflammatory effects. Curr Opin Food Sci. 2016;8:33–42.
116. Furukawa S, et al. Increased oxidative stress in obesity and its impact on metabolic syndrome. J Clin Invest. 2017;114(12):1752–61.
117. Gross MD, Sanchez OA, Jacobs DR. Abstract P029: relationships among indicators of oxidative stress and antioxidants: the coronary artery risk development in young adults (CARDIA) study. Circulation. 2016;133:AP029.
118. Ramalingam L, et al. The renin angiotensin system, oxidative stress and mitochondrial function in obesity and insulin resistance. Biochim Biophys Acta. 2017;1863(5):1106–14.
119. Pahwa R, Adams-Huet B, Jialal I. The effect of increasing body mass index on cardio-metabolic risk and biomarkers of oxidative stress and inflammation in nascent metabolic syndrome. J Diabetes Complicat. 2017;31(5):810–3.
120. Hecht F, et al. The role of oxidative stress on breast cancer development and therapy. Tumor Biol. 2016;37(4):4281–91.
121. Odegaard AO, et al. Oxidative stress, inflammation, endothelial dysfunction and incidence of type 2 diabetes. Cardiovasc Diabetol. 2016;15(1):51.
122. Górska EB, Galoch E, Jankiewicz U, Kowalczyk P. *Chlamydophila pneumoniae* as a cause of respiratory disease. Borgis - Medycyna Rodzinna. 2013:99–105.
123. Ríos JA, et al. Is Alzheimer's disease related to metabolic syndrome? A Wnt signaling conundrum. Prog Neurobiol. 2014;121:125–46.
124. Luque-Contreras D, et al. Oxidative stress and metabolic syndrome: cause or consequence of Alzheimer's disease? Oxidative Med Cell Longev. 2014;2014:497802.
125. Yuyama K, Mitsutake S, Igarashi Y. Pathological roles of ceramide and its metabolites in metabolic syndrome and Alzheimer's disease. Biochim Biophys Acta. 2014;1841(5):793–8.
126. Kim B, Feldman EL. Insulin resistance as a key link for the increased risk of cognitive impairment in the metabolic syndrome. Exp Mol Med. 2015;47(3):e149.
127. Yang H, et al. Association between the characteristics of metabolic syndrome and Alzheimer's disease. Metab Brain Dis. 2013;28(4):597–604.
128. Suzanne M, Tong M. Brain metabolic dysfunction at the core of Alzheimer's disease. Biochem Pharmacol. 2014;88(4):548–59.
129. Drake J, Link CD, Butterfield DA. Oxidative stress precedes fibrillar deposition of Alzheimer's disease amyloid β-peptide (1-42) in a transgenic Caenorhabditis elegans model. Neurobiol Aging. 2003;24(3):415–20.
130. Ferreira ST, et al. Inflammation, defective insulin signaling, and neuronal dysfunction in Alzheimer's disease. Alzheimers Dement. 2014;10(1):S76–83.
131. Ojala JO, Sutinen EM. The role of Interleukin-18, oxidative stress and metabolic syndrome in Alzheimer's disease. J Clin Med. 2017;6(5):55.
132. Mitchell AJ, et al. Prevalence of metabolic syndrome and metabolic abnormalities in schizophrenia and related disorders—a systematic review and meta-analysis. Schizophr Bull. 2013;39(3):306–18.
133. Weinstock LM, et al. Medication burden in bipolar disorder: a chart review of patients at psychiatric hospital admission. Psychiatry Res. 2014;216(1):24–30.
134. Vancampfort D, et al. Risk of metabolic syndrome and its components in people with schizophrenia and related psychotic disorders, bipolar disorder and major depressive disorder: a systematic review and meta analysis. World Psychiatry. 2015;14(3):339–47.
135. Rosenbaum S, et al. Physical activity in the treatment of post-traumatic stress disorder: a systematic review and meta-analysis. Psychiatry Res. 2015;230(2):130–6.
136. Tienhoven-Wind LJN, Dullaart RPF. Low–normal thyroid function and the pathogenesis of common cardio-metabolic disorders. Eur J Clin Investig. 2015;45(5):494–503.
137. Harada PHN, et al. Impact of subclinical hypothyroidism on cardiometabolic biomarkers in women. J Endocr Soc. 2017;1(2):113–23.
138. Puig L, et al. Psoriasis beyond the skin: a review of the literature on cardiometabolic and psychological co-morbidities of psoriasis. Eur J Dermatol. 2014;24(3):305–11.
139. Singh S, Young P, Armstrong AW. An update on psoriasis and metabolic syndrome: a meta-analysis of observational studies. PLoS One. 2017;12(7):e0181039.
140. Boehncke WH, Boehncke S, Tobin AM, Kirby B. The 'psoriatic march': a concept of how severe psoriasis may drive cardiovascular comorbidity. Exp Dermatol. 2011;20(4):303–7.
141. MDP E, et al. The relation between, metabolic syndrome and quality of life in patients with Systemic Lupus Erythematosus. PloS One. 2017;12(11):e0187645.
142. Ammirati E, et al. Cardiometabolic and immune factors associated with increased common carotid artery intima-media thickness and cardiovascular disease in patients with systemic lupus erythematosus. Nutr Metab Cardiovasc Dis. 2014;24(7):751–9.
143. Esser N, et al. Inflammation as a link between obesity, metabolic syndrome and type 2 diabetes. Diabetes Res Clin Pract. 2014;105(2):141–50.
144. Alexopoulos N, Katritsis D, Raggi P. Visceral adipose tissue as a source of inflammation and promoter of atherosclerosis. Atherosclerosis. 2014;233(1):104–12.
145. Mathieu P, Boulanger MC, Després JP. Ectopic visceral fat: a clinical and molecular perspective on the cardiometabolic risk. Rev Endocr Metab Disord. 2014;15(4):289–98.
146. Ahangari F, et al. Chitinase 3-like-1 regulates both visceral fat accumulation and asthma-like Th2 inflammation. Am J Respir Crit Care Med. 2015;191(7):746–57.
147. Nam S, et al. The effect of abdominal visceral fat, circulating inflammatory cytokines, and leptin levels on reflux esophagitis. J Neurogastroenterol Motil. 2015;21(2):247.
148. Yi L, et al. Influence of exclusive enteral nutrition therapy on visceral fat in patients with Crohn's disease. Inflamm Bowel Dis. 2014;20(9):1568–74.
149. Kredel LI, Siegmund B. Adipose-tissue and intestinal inflammation–visceral obesity and creeping fat. Front Immunol. 2014;5:462.
150. Uysal KT, Wiesbrock SM, Marino MW, Hotamisligil GS. Protection from obesity-induced insulin resistance in mice lacking TNF-alpha function. Nature. 1997;389:610–4.
151. Hsu BG, et al. High serum resistin levels are associated with peripheral artery disease in the hypertensive patients. BMC Cardiovasc Disord. 2017;17(1):80.
152. Meek TH, Morton GJ. The role of leptin in diabetes: metabolic effects. Diabetologia. 2016;59(5):928–32.
153. Gorgui J, et al. Circulating adiponectin levels in relation to carotid atherosclerotic plaque presence, ischemic stroke risk, and mortality: a systematic review and meta-analyses. Metabolism. 2017;69:51–66.
154. Yarmolinsky J, et al. Plasminogen activator inhibitor-1 and type 2 diabetes: a systematic review and meta-analysis of observational studies. Sci Rep. 2016;6:17714.
155. Peng S, et al. A long-acting PAI-1 inhibitor reduces thrombus formation. Thromb Haemost. 2017;117(07):1338–47.
156. Engstler AJ, et al. Plasminogen activator Inhibitor-1 is regulated through dietary fat intake and heritability: studies in twins. Twin Res Hum Genet. 2017;20(4):338–48.
157. Tofler GH, et al. Plasminogen activator inhibitor and the risk of cardiovascular disease: the Framingham heart study. Thromb Res. 2016;140:30–5.

158. Han SJ, et al. Low plasma adiponectin concentrations predict increases in visceral adiposity and insulin resistance. J Clin Endocrinol Metab. 2017;102:4626.
159. Arslanian S, et al. Adiponectin, insulin sensitivity, β-cell function, and racial/ethnic disparity in treatment failure rates in TODAY. Diabetes Care. 2017;40(1):85–93.
160. Varghese JF, Patel R, Yadav UCS. Novel insights in the metabolic syndrome-induced oxidative stress and inflammation-mediated atherosclerosis. Curr Cardiol Rev. 2018;14(1):4–14.
161. Sanchez-Garcia S, et al. Anthropometric measures and nutritional status in a healthy elderly population. BMC Public Health. 2007;7:2.
162. Myint PK, et al. Body fat percentage, body mass index and waist-to-hip ratio as predictors of mortality and cardiovascular disease. Heart. 2014;100:1613–9.
163. Macfarlane DJ, et al. Agreement between bioelectrical impedance and dual energy X-ray absorptiometry in assessing fat, lean and bone mass changes in adults after a lifestyle intervention. J Sports Sci. 2016;34(12):1176–81.
164. Mialich MS, Sicchieri JMF, Jordao AJ. Analysis of body composition: a critical review of the use of bioelectrical impedance analysis. Int J Clin Nutr. 2014;2(1):1–10.
165. Marzetti E, et al. Sarcopenia: an overview. Aging Clin Exp Res. 2017;29(1):11–7.
166. Shivappa N, et al. Associations between dietary inflammatory index and inflammatory markers in the Asklepios study. Br J Nutr. 2015;113(4):665–71.
167. Casas R, Sacanella E, Estruch R. The immune protective effect of the Mediterranean diet against chronic low-grade inflammatory diseases. Endocr Metab Immune Disord Drug Targets. 2014;14(4):245–54.
168. Kaulmann A, Bohn T. Carotenoids, inflammation, and oxidative stress—implications of cellular signaling pathways and relation to chronic disease prevention. Nutr Res. 2014;34(11):907–29.
169. Kossioni AE. The association of poor oral health parameters with malnutrition in older adults: a review considering the potential implications for cognitive impairment. Nutrients. 2018;10:1709. https://doi.org/10.3390/nu10111709.
170. WebMD. WebMD Symptom checker. Mouth cracks. Retrieved from: https://symptomchecker.webmd.com/single-symptom?symptom=cracks-at-corner-of-mouth&symid=321.
171. Bertoli S, et al. Human bisphenol A exposure and the "diabesity phenotype". Dose Response. 2015;13(3):1559325815599173.

Resources

Das UN. Metabolic syndrome pathophysiology: the role of essential fatty acids and their metabolites. Wiley Blackwell, USA. 1 Jan 2010.

Das UN. Molecular basis of health and disease springer, NY, USA. 1 Jan 2011.

Escott-Stump S. Nutrition & diagnosis-related care. 8th Ed. Wolters Kluwer; 2015.

Food Science and Human Wellness.

International Movement and Nutrition Society.

Revolutionary New Concepts in the Prevention and Treatment of Cardiovascular Disease

Mark C. Houston

47.1 Introduction – 824

47.2 Revolutionizing the Treatment of Cardiovascular Disease – 825
47.2.1 The Endothelium, Endothelial Function, and Endothelial Dysfunction – 825
47.2.2 The Pathophysiology of Vascular Disease [2, 4] – 827
47.2.3 Interrupting the Finite Pathways – 828

47.3 Hypertension – 828

47.4 Dyslipidemia – 830

47.5 Insulin Resistance, Metabolic Syndrome, Dysglycemia, and Diabetes Mellitus – 831

47.6 Nutritional Interventions – 832
47.6.1 Mediterranean Diet (TMD: Traditional Mediterranean Diet) [22–30] – 832
47.6.2 DASH Diets (DASH 1 and 2) – 833
47.6.3 Portfolio Diet [34, 35] – 833

47.7 Conclusions – 833

References – 840

© Springer Nature Switzerland AG 2020
D. Noland et al. (eds.), *Integrative and Functional Medical Nutrition Therapy*,
https://doi.org/10.1007/978-3-030-30730-1_47

Learning Objectives

Hypertension
1. Understand the pathogenesis of hypertension and the role of vascular biology, the three finite vascular responses of inflammation, oxidative stress, and vascular immune function, plasma renin activity (PRA), and aldosterone to select optimal integrative antihypertensive therapy to reduce blood pressure and decrease cardiovascular disease.
2. Review diagnostic testing for hypertension: 24-hour ambulatory blood pressure monitor (ABM), endothelial function, and arterial compliance testing.
3. Be able to prioritize and personalize the most important nutrition, nutraceutical supplements, and lifestyle treatments for hypertension.
4. Apply micronutrient testing (MNT) in the treatment of hypertension.
5. Review treatment with nutrition, lifestyle, and nutraceutical supplements.

Dyslipidemia
- Review the underlying causes and mechanisms of dyslipidemia and dyslipidemia-induced cardiovascular disease.
- Understand and apply advanced lipid testing (lipid particle number and size and HDL functionality and quality) for the diagnosis and treatment of dyslipidemia in clinical practice.
- Prioritize effective methods to promote vascular repair, reduce vascular damage, and define the interrelationships of the cardiovascular system, gastrointestinal tract, and microbiome.
- Evaluate nutrition and nutritional supplements to treat dyslipidemia.

Coronary Heart Disease
1. Review the top 5 coronary heart disease risk factors, the details of the correct analysis of each, and the top 25 modifiable key coronary heart disease risk factors.
2. Review and understand coronary heart disease and congestive heart failure clinical presentation, diagnosis, and treatment.
3. Prioritize which laboratory and noninvasive cardiovascular tests should be evaluated in patients in the primary care setting and how to interpret and treat.
4. Review integrative medical treatments with nutrition, nutritional supplements, and drugs for cardiovascular disease.

47.1 Introduction

Cardiovascular medicine needs a complete functional and metabolic reevaluation related to diagnosis, prevention, and integrative treatments. We have reached a limit in our ability to appropriately treat cardiovascular disease (CVD) [1]. Current treatments are not effective in reducing vascular inflammation. CVD remains the leading cause of morbidity and mortality in the United States [2]. Statistics show that we spend approximately $320 billion a year in both direct and indirect costs treating CVD alone [2] and over 2200 US citizens die from stroke or MI each day [2–5]. CHD includes angina, MI, ischemic heart disease, and ischemic cardiomyopathy with both systolic (low ejection fraction) and diastolic congestive heart failure (CHF) with a normal ejection fraction with a stiff and noncompliant left ventricle. The most common cause of CHF in the United States is ischemic heart disease. About 80 percent of CVD (myocardial infarction, angina, coronary heart disease, and congestive heart failure) can be prevented by:

- *Optimal nutrition*
- *Optimal exercise (both aerobic and resistance training)*
- *Optimal weight and body fat (total body fat, visceral fat, and epicardial fat)*
- *Avoiding all tobacco products*

The traditional evaluation, prevention, and treatment strategies for the top five cardiovascular risk factors – hypertension, diabetes mellitus, dyslipidemia, obesity, and smoking – have resulted in what is now referred to as a "CHD gap" [4]. Approximately 50% of patients will have CHD or MI despite having "normal" levels of these risk factors as currently defined in the medical literature [2, 5]. Traditional medicine continues to maintain a cholesterol-centric approach to the management of CHD but does not address the basic etiologies of CHD such as inflammation, oxidative stress, and immune vascular dysfunction. However, there are important details within each of these top five risk factors that are not being measured by physicians and are thus ignored in the prevention and treatment of CHD [2]. In fact, there are 25 top modifiable risk factors (▶ Box 47.1) and at least 395 other CHD risk factors that physicians' either aren't aware of or aren't using appropriate techniques to identify and treat them. Thus, it is imperative that we now begin to examine other methods to prevent and treat CVD [2].

Box 47.1 Top 25 modifiable CHD risk factors
- Hypertension (24-hour ABM)
- Dyslipidemia (advanced lipid analysis)
- Hyperglycemia, metabolic syndrome, insulin resistance, and diabetes mellitus
- Obesity
- Smoking
- Hyperuricemia
- Renal disease
- Elevated fibrinogen
- Elevated serum iron
- Trans fatty acids and refined carbohydrates
- Low dietary omega-3 fatty acids
- Low dietary potassium and magnesium with high sodium
- Inflammation: Increased hsCRP, MPO, interleukins
- Increased oxidative stress and decreased defense
- Increased immune dysfunction
- Lack of sleep
- Lack of exercise
- Stress, anxiety, and depression

- Homocysteinemia
- Subclinical hypothyroidism
- Hormonal imbalances in both genders
- Chronic clinical or subclinical infections
- Micronutrient deficiencies: numerous ones such as low vitamin D and K, etc.
- Heavy metals
- Environmental pollutants

47.2 Revolutionizing the Treatment of Cardiovascular Disease

The blood vessel has three finite responses to a large number of internal and external insults [2]. Those responses are inflammation, oxidative stress, and vascular immune dysfunction. Tracking backwards from those three finite responses brings us to the genesis of CVD, with the goal of starting effective treatments to resolve the downstream mediators and abnormalities.

Cell membrane physiology and cell membrane dysfunction are keys to this treatment strategy. This membrane barrier separates the outside and the inside of every cell, such as the endothelium, enterocyte, the blood brain barrier or any other membrane, and determines all of the signaling mechanisms that occur from the external to the internal milieu [2].

Any cell membrane insult such as high blood pressure, LDL cholesterol, oxLDL, glucose, microbes, toxins, heavy metals, or homocysteine results in a reaction diffusion wave ("tsunami effect") throughout the cell membrane that disrupts the receptors and signaling mechanisms that induce membrane damage and dysfunction [6, 7]. One small insult becomes a heightened response (metabolic memory) to create further cell damage [6, 7]. The blood vessel is really an innocent bystander in a correct but often dysregulated vascular response to these infinite insults.

In the acute setting, any vascular insult results in a correct defensive response by the endothelium. The vascular immune dysfunction, oxidative stress, or inflammatory responses are usually short-lived, appropriate, and regulated [2]. However, chronic insults result in a chronic exaggerated and dysregulated vascular dysfunction with preclinical, then clinical, CVD, due to maladaptation of various systems such as the renin-angiotensin-aldosterone (RAAS) system, sympathetic nervous system (SNS), and others [2].

Most diseases are arbitrarily defined with a specific abnormal level. Hypertension is defined as an arterial blood pressure greater than 140/90 mmHg, dyslipidemia as an LDL-cholesterol level over 100 mg/dL, and glucose intolerance as a fasting glucose over 99 mg/d [2]. However, it is very clear that there exists a continuum of CHD risk starting at lower levels of BP, LDL cholesterol and glucose, as well as for most other CHD risk factors [2]. For example, the blood pressure risk for CVD actually starts at 110/70 mmHg, and the risk for LDL-cholesterol causing reduction in nitric oxide in the endothelium starts at 60 mg/dL, and fasting glucose risk starts at 75 mg/dL [2]. There is a progressive continuum of risk with all of the CVD risk factors and mediators that affect the blood vessel, leading initially to functional abnormalities (endothelial dysfunction), then to structural abnormalities of the vascular (loss of arterial compliance with arterial stiffness, increased pulse wave velocity, increased augmentation index) and cardiac muscle (diastolic dysfunction, left ventricular hypertrophy, reduced ejection fraction), and later to preclinical and clinical CVD, such as MI, CHD, angina, and CHF.

Finally, it is important to understand the concept of "translational vascular medicine." Do the risk factors that are measured actually translate into vascular disease and illness? Vice versa, does the absence of those risk factors actually define vascular health? At this time, we often do not use appropriate or early noninvasive testing of functional and structural markers of endothelial dysfunction or vascular smooth muscle and cardiac dysfunction to identify the vascular effects of CHD risk factors, or the presence of vascular disease. Instead, we are relying only upon risk factors or some risk factor scoring system, such as Framingham, COSEHC (Consortium of Southeastern Hypertension Centers) or Rasmussen CHD scoring. We assume that if patients have risk factors, they also have vascular disease. If they don't, they may have vascular health. It is important to measure sensitive indicators of endothelial dysfunction and vascular structural disease that are induced by the insults. Early detection with aggressive treatment will reduce CVD.

47.2.1 The Endothelium, Endothelial Function, and Endothelial Dysfunction

The endothelium is a very thin lining of vascular cells, which forms an interface between the circulating blood in the lumen and the vascular smooth muscle [2, 4, 8] (◘ Fig. 47.1).

When the endothelium is working correctly (endothelial function) with adequate amounts of nitric oxide, all the blood elements and the vascular smooth muscle remain normal. However, when endothelial dysfunction occurs, the results are inflammation, oxidative stress, immune dysfunction, abnormal growth, vasoconstriction, increased permeability, thrombosis, and ultimately CVD [2, 4, 8, 10].

◘ Figure 47.2 illustrates LDL-cholesterol's role in atherosclerotic plaque formation [12]. Once inside the vessel wall, LDL-cholesterol becomes susceptible to oxidation, glycosylation, acetylation or modifications by free radicals and glucose, etc. [12]. Oxidized-LDL and glycated LDL are toxic to the vessel wall. The modified LDL is consumed by scavenger receptors (SR-A and CD-36) on macrophages to form foam cells. Foam cells are not able to process the oxidized-LDL or modified LDL and continue to accumulate oxidized and modified-LDL. It forms a plaque, which may rupture and cause acute coronary thrombosis. This is the progression that needs to be interrupted. There are actually 45 or more different steps in this process that are important in the treatment of dyslipidemia-induced vascular disease [12].

Vascular disease is a balance of vascular injury (angiotensin II and endothelin) versus vascular repair with endothelial

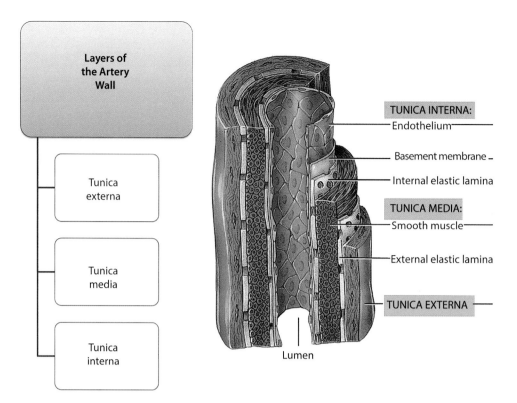

Fig. 47.1 The blood vessel showing the endothelial lining with connective tissue (the intima), vascular smooth muscle (media), and adventitia. Nitric oxide (NO) is released from the endothelial lining. (Reprinted from Tortora and Derrickson [9]. With permission from John Wiley & Sons.)

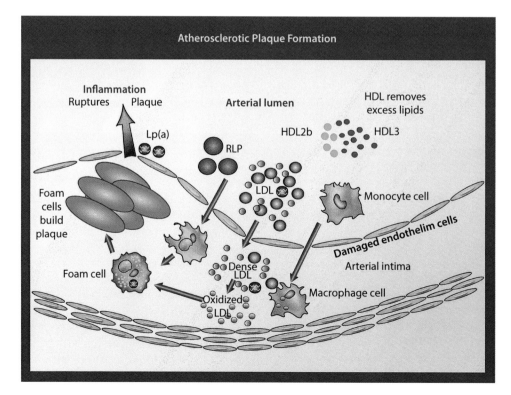

Fig. 47.2 A simplified illustration of atherosclerotic plaque formation in the subendothelial layer of the artery. (Reprinted with permission from Houston [11].)

progenitor cells (EPCs) produced in the bone marrow [2, 4]. The infinite insults result in preconditioned and heightened "metabolic memory" responses that trigger the three finite downstream responses that have a bidirectional communication involving endothelial dysfunction, vascular smooth muscle, and cardiac dysfunction [4, 6]. Once endothelial dysfunction has developed, a smaller insult occurring at a later date can result in a heightened response that induces more vascular damage [4, 6]. The concept of metabolic memory was demonstrated by Youssef-Elabd et al., who found that short-term exposure of adipose cells to uncontrolled levels of saturated fatty acids and glucose leads to a long-term inflammatory insult within adipocytes [6].

47.2.2 The Pathophysiology of Vascular Disease [2, 4]

What are the causes of vascular disease? The major causes are:
- Oxidative stress – reactive oxygen species (ROS) and reactive nitrogen species (RNS) are increased in the arteries and kidneys and with a decreased oxidative defense
- Inflammation – increased in the vasculature and kidneys: increased high-sensitivity C-reactive protein (hsCRP), leukocytosis, increased neutrophils and decreased lymphocytes, and increased renin–angiotensin–aldosterone system (RAAS) in the kidney
- Autoimmune dysfunction – of the arteries and kidneys: increased white blood count (WBC) and involvement of CD4+ (T-helper cells) and CD 8+ (cytotoxic T-cells)

These problems result in abnormal vascular biology with endothelial dysfunction and vascular smooth muscle hypertrophy and dysfunction. Of course, genetics, genomics, and epigenetics also play a role in the pathophysiology of vascular disease [3].

Figure 47.3 offers insight into the infinite insults that bombard the endothelium. These insults are divided into two major categories: biomechanical (blood pressure, pulse pressure, shear stress, and oscillatory pressure within the arterial system – most plaques form at the bifurcation of arteries) and biochemical (e.g., nutritional factors, metabolic, microbes, sterile antigens, and environmental toxins).

Endothelial cells express various receptors that determine the interaction between the insults and the downstream

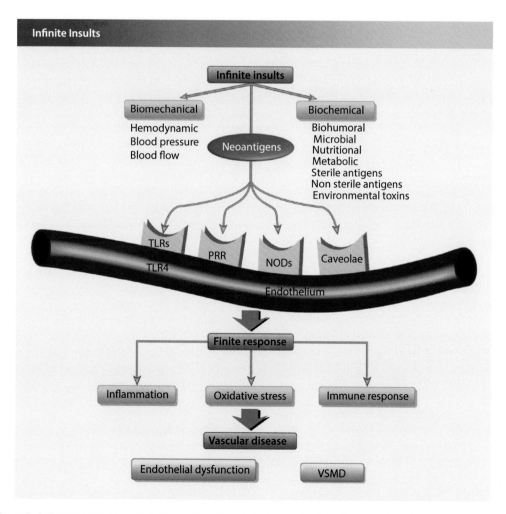

Fig. 47.3 The endothelium is subject to an infinite number of insults but can only elicit a finite number of responses to those insults which include inflammation, oxidative stress, and vascular immune dysfunction. (Reprinted with permission from Houston [11].)

mediators. These include pattern recognition receptors (PRRs), toll-like receptors (TLRs), NOD-like receptors (NLRs), and caveolae [13]. The TLRs and NLRs are membrane receptors that react to external insults with appropriate intracellular signaling that usually induces inflammation, oxidative stress, and immune dysfunction within the cell. The caveolae are membrane lipid microdomains that contain eNOS that determines nitric oxide levels and alterations in blood pressure, inflammation, dyslipidemia, oxidative stress, immune dysfunction, CHD, CVD, and atherosclerosis. The various risk factors and risk mediators attach to one of the receptors in the membrane and then set off a cascade of the three finite responses (inflammation, oxidative stress, and immune dysfunction), which leads to endothelial dysfunction and ultimately CVD [13].

47.2.3 Interrupting the Finite Pathways

The key to the successful prevention and treatment of CVD is both recognition of the risk factors and identification of treatments that will interrupt the pathways that connect the risk factors to these receptors. The TLR 1, 2, and 4 are the most common type of the PRR-TLRs related to the vascular membrane and endothelial dysfunction. The NLRs, NOD 1 and NOD 2, are also a type of PRRs that involve the vascular membrane. The scientifically proven nutraceuticals and dietary factors that accomplish this are listed below [14]:

- Curcurmin (turmeric): TLR 4, NOD 1 (NLR), and NOD 2 (NLR) (these are all PRR) [14]
- Cinnamaldehyde (cinnamon): TLR 4 [14]
- Sulforaphane (broccoli): TLR 4 [14]
- Resveratrol (nutritional supplement, red wine, grapes): TLR 1 [14]
- Epigallocatechin gallate (green tea): TLR 1 [14]
- Luteolin (celery, green pepper, rosemary, carrots, oregano, oranges, olives): TLR 1 [14]
- Quercetin (tea, apples, onion, tomatoes, capers): TLR 1 [14]
- Chrysin: TLR 1 [14]
- Omega-3 fatty acids: Interrupt caveolae lipid microdomains TLRs and NODs, decrease inflammation and hsCRP, lower BP, decrease LDL P, increase LDL and HDL size, improve glycation parameters, decrease immune vascular dysfunction, decrease CHD plaque formation, and improve CHD and CHF symptoms and outcomes [14].

The goal is to use a systematic (dynamic systems biology), functional and metabolic medicine approach to establish cardiovascular ecology, balance, and allostasis (achieve stability through change) and minimize chronic internal and external cardiovascular stressors, mediators, and risk factors that insult the blood vessel. Clinical interventions are needed to reduce the allostatic load, prevent, regulate, and treat the "abnormal" downstream finite responses.

The polygenetic codes for CVD identify 30 separate loci associated with MI and CHD, but most of those loci have nothing to do with the top five cardiovascular risk factors [3]. The majority of those loci deal directly with inflammatory pathways. Evaluation and treatment of only the top five risk factors and how they interact with our genome will never reduce CVD, and the CHD gap will persist.

Atherosclerosis, endothelial dysfunction, and vascular disease are postprandial phenomena [15]. Ingestion of sodium chloride, refined carbohydrates, and some foods containing trans fats may trigger glucotoxicity, triglyceride toxicity, vascular endotoxemia, inflammation, oxidative stress, and immune dysfunction [6, 15]. The data on saturated fats and endothelial dysfunction and cardiovascular disease remain controversial [6, 15]. Furthermore, these responses may be perpetuated long after the original insult with a heightened continued inflammatory response (metabolic memory) [6]. Fortunately, studies have shown that eating a diet that is rich in low-glycemic foods, monounsaturated fats, omega-3 fatty acids, polyphenols, and antioxidants can help to prevent postprandial endothelial dysfunction [12]. Early evidence of CVD in the form of fatty streaks has been documented in children in the first and second decades of life but may actually begin in utero (Fig. 47.4) [2]. The vascular disease is subclinical for 10–30 years or more prior to any cardiovascular event [2, 4, 8]. Patients with CVD may present with substernal chest pain with radiation to the neck, shoulders, arms, or back. Often, dyspnea with exertion may be the only symptom. New onset or severe fatigue, edema, dizziness, or syncope may be present. Women tend to have more atypical symptoms than men. Endothelial dysfunction is the earliest functional abnormality, followed by changes in arterial compliance, stiffness, and elasticity. It is important to begin using technologies that allow earlier identification of cardiovascular dysfunction before any structural changes have occurred.

Figure 47.5 illustrates the vessel changes that occur as CHD progresses. On the left is a fairly normal artery. In the middle, the CHD has progressed from minimal to moderate with the subendothelial layer becoming thick, but the lumen is still the same size. This extraluminal plaque and inflammation could be seen with computed tomography angiogram (CTA) but missed by conventional coronary arteriogram. In the image on the right in Fig. 47.5, there is extensive extraluminal and intraluminal disease.

Lack of the proper type of imaging, ignoring the majority of the other 395 or so CHD risk factors, and not properly evaluating the top five risk factors are some of the reasons for the persistence of the CHD gap [2].

47.3 Hypertension

The majority of patients with hypertension do not have any clinical symptoms until the blood pressure is very high or they have some clinical event such as myocardial infarction, TIA symptoms, stroke, congestive heart failure, or renal failure. If symptoms do occur, they could include chest pain,

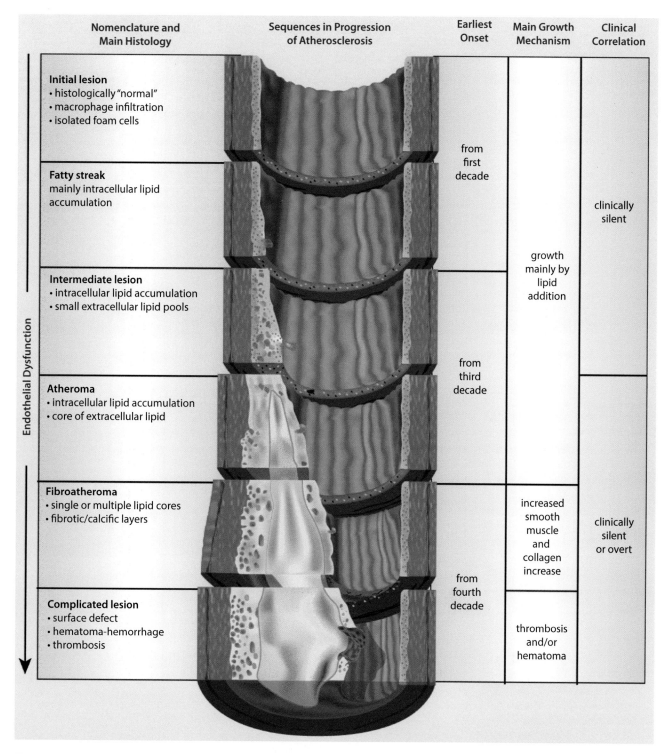

Fig. 47.4 Progression of atherosclerosis from the earliest initial lesion with endothelial dysfunction through the subintimal fatty streak, atheroma, fibroatheroma and final the complex lesion with stable and unstable plaque within the arterial lumen. (Reprinted with permission from Houston [11].)

dyspnea, headache, blurred vision, edema, or dizziness. Only a 24-hour ABM (ambulatory blood pressure monitor) can identify specific BP risks for CVD, such as nocturnal BP, dipping, nondipping, BP surges, BP load, and BP variability. Nondipping is defined as having less than a 10% reduction in BP at night. Nocturnal BP is the primary determinant of CVD, related to BP measurements. The BP load is the number of BP readings over 140/90 mmHg in 24 hours. The normal BP load is less than 15% of the total BP readings. BP surges that are high and rapid between 3 AM and 9 AM, as well as labile or variable BP, will increase CVD [8]. Excessive dipping is associated with an increased risk of ischemic

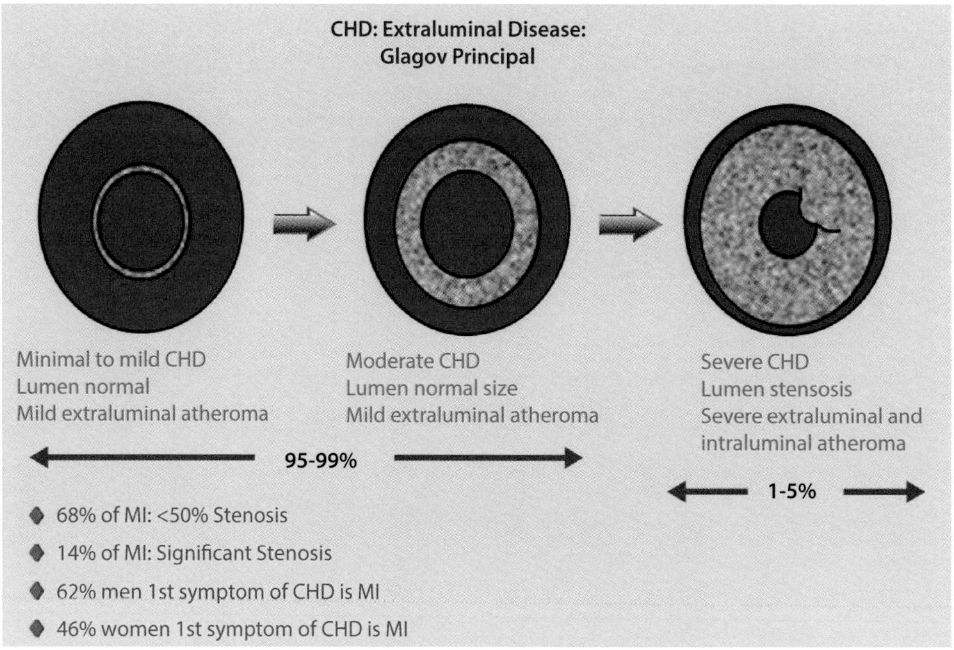

CHD: Extraluminal Disease: Glagov Principal

Minimal to mild CHD
Lumen normal
Mild extraluminal atheroma

Moderate CHD
Lumen normal size
Mild extraluminal atheroma

Severe CHD
Lumen stenosis
Severe extraluminal and intraluminal atheroma

← 95–99% →   ← 1–5% →

- 68% of MI: <50% Stenosis
- 14% of MI: Significant Stenosis
- 62% men 1st symptom of CHD is MI
- 46% women 1st symptom of CHD is MI

Fig. 47.5 Illustration of the vessel changes that occur as coronary heart disease progresses from an extra-lumenal disease at the beginning to the extra-lumenal and lumenal stenosis. (Reprinted with permission from Houston [11].)

stroke and reverse dipping is associated with an increased risk of intracerebral hemorrhage (ICH). Nocturnal blood pressure is more clinically important than day blood pressure (27/15 mmHg difference is optimal) [8]. Furthermore, morning blood pressure surges (level and rapidity) increase the risk of ischemic stroke, MI, and left ventricular hypertrophy [8]. Hypertension is not a disease; it is a marker for vascular dysfunction. Therefore, it is crucial that it is correctly identified. The following points should always be considered when evaluating blood pressure [8]:

- Normal blood pressure is 120/80 mmHg, but there is a continuum of risk for CVD starting at 110/70 mmHg.
- Each increase of 20/10 mmHg doubles cardiovascular risk.
- Before age 50, the diastolic blood pressure predicts risk best.
- After age 50, the systolic blood pressure predicts risk best.
- 24-hour ambulatory blood pressure monitoring is more accurate than office-based blood pressure measurements and should be the standard of care for defining hypertension and CVD risk.
- Mercury sphygmomanometers are the most accurate. Electronic machines with arm cuffs are good. Do not use wrist or finger monitors.
- Blood pressure load: percent over 140/90 mmHg should be less than 15%.

47.4 Dyslipidemia

Dyslipidemia is asymptomatic until a cardiovascular event, such as angina or MI, occurs. It is one of the top five cardiovascular risk factors, but proper measurement using advanced lipid profiles is rarely ordered to verify risk and optimal treatment [12, 16, 17]. An advanced lipid profile will measure:

- LDL-C total.
- LDL-P particle number (drives CHD risk).
- LDL size (dense type B versus large type A has a greater CHD risk).
- Modified LDL (oxidized, glycated, glyco-oxidized, and acetylated) increase CHD risk.
- Antibodies to oxLDL and modified LDL predict increase in CHD risk.
- Apolipoprotein (APO) B, if elevated, increases CHD risk.
- APO B antibodies and immune complexes increase CHD risk.
- Lp(a) is one of the hidden risk factors for CHD and MI.
- HDL-C total reduces risk for CHD.
- HDL-P particle number reduces CHD risk.
- HDL size (large 2b versus small type 3) lowers CHD risk.
- Dysfunctional HDL reduces reverse cholesterol transport and increases CHD risk.
- Pro-inflammatory and pro-atherogenic HDL increases CHD risk.
- Myeloperoxidase (MPO) and dysfunctional APO A increase oxidative stress, inflammation, and CHD risk as well as plaque rupture and hypertension.
- Low APO A increases CHD risk.
- Low paraoxonase (PON)-1 and PON-2 in HDL increases CHD risk.
- Increased APO-CIII in LDL or HDL increases CHD risk.
- Serum-free fatty acids, if elevated, increase insulin resistance, glucose, and CHD.

- VLDL (especially large VLDL) and triglyceride (TG) total increase thrombosis and CHD risk.
- Large VLDL increases CHD and risk for thrombosis.
- VLDL-P particle number increases CHD risk.
- Remnant particles increase CHD risk.

The primary driving cardiovascular risk related to LDL-cholesterol is the LDL-particle number (LDL-P and apolipoprotein B particles) [12]. HDL-P (particle number) is most protective, with larger HDL type 2b being a second important protective mechanism [12]. Larger number and size of HDL are more efficient at reverse cholesterol transport, and more protective to the vascular system in numerous other ways. It is also important to analyze dysfunctional HDL [12, 16, 17]. Patients who have a HDL of 85 mg/dL or more often have dysfunctional HDL that is not protective against CHD and MI [16, 17]. VLDL, triglycerides, and remnant particles are very atherogenic and thrombogenic [12].

47.5 Insulin Resistance, Metabolic Syndrome, Dysglycemia, and Diabetes Mellitus

A fasting blood sugar (FBS) of greater than 75 mg increases CHD by 1% per increase of 1 mg/dL and induces endothelial dysfunction [2]. If a patient has a FBS of 100 mg (often considered a normal level), the risk of CHD is increased by 25% [2]. A 2-hour glucose tolerance test (GTT) over 110 mg increases CHD by 2% per 1 mg/dL increase in glucose [2]. The current definition of an abnormal 2-hour GTT is >140 mg. If a patient's result is 140 mg, which again is currently classed as "normal," CHD and MI are increased by 60%. Hyperinsulinemia is also an independent risk factor for CHD [2]. Insulin resistance creates inflammation, reduces nitric oxide levels, and causes endothelial dysfunction and vascular disease through the mitogen-activated protein kinase (MAPK) pathway, which is atherogenic and induces hypertension. In contrast, the phosphatidylinositol 3-kinase (PI3K) pathway is anti-inflammatory, antihypertensive, and anti-atherogenic [2]. It is important to measure all glycation parameters including fasting glucose, 2-hour GTT, insulin levels, C-peptide, glycated serum protein, adiponectin and proinsulin, depending on the clinical setting [18].

Obesity with increased levels of inflammatory and oxidative stress-related adipokines contributes to CHD. Measurement of not just body weight but BMI and body composition with visceral fat total and regional body fat, epicardial fat (with chest CT scan), and lean body mass will help predict CHD risk [18].

The top 25 other CHD risk factors are listed below in ▶ Box 47.1. Some of the most neglected and important CHD risk factors to evaluate include gender-specific hormones, thyroid function, toxins, homocysteine, and vitamin D. In the other sections that follow, I will present algorithms, protocols, and treatments for hypertension, dyslipidemia, endothelial dysfunction, and coronary heart disease. Fortunately, there are a number of advanced biomarkers and noninvasive tests to determine vascular pathology before it actually starts [2] (▶ Boxes 47.2 and 47.3).

A discussion of these techniques is beyond the scope of this chapter. However, I would urge the reader to investigate these technologies, particularly EndoPAT, which is very accurate at assessing endothelial function and diagnosing endothelial dysfunction, computerized arterial pulse wave analysis (CAPWA) for endothelial function and arterial compliance, carotid intimal medial thickness (IMT) and plaque, magnetic cardiography (MCG) and cardiac CT angiograms (CTA) and CAC for cardiac calcium scoring [2, 8, 18–21] (▶ Box 47.3). The EndoPAT is the most cost-effective and accurate noninvasive test to identify early endothelial dysfunction and predict future CVD and CHD. This test, along with 24-hour BP, advanced lipid testing and glycation measurements are the best initial ways to evaluate the CV patient and are very cost-effective.

Diagnosis, prevention, and treatment of cardiovascular disease, coronary heart disease, hypertension, dyslipidemia, and diabetes mellitus can now be refined by adding selective cardiovascular genetics with the CHD risk factors, biomarkers, and noninvasive and invasive cardiovascular tests mentioned above (▶ Box 47.4).

Box 47.2 Noninvasive vascular testing for cardiovascular disease

- EndoPAT (endothelial dysfunction, augmentation index, and heart rate variability)
- CAPWA (computerized arterial pulse wave analysis)
- Mobile 0 graph for central and brachial BP, AI, PWV
- 24-hour ABM
- HRV: Heart rate variability and HRRT (heart rate recovery time)
- EKG: Electrocardiogram
- Chest X-ray
- TMT (treadmill test)
- CPET: Cardiopulmonary exercise test
- MCG (magnetocardiography)
- Carotid IMT/duplex
- CT angiogram (CTA)
- Sleep studies for obstructive sleep apnea (OSA)
- CAC (coronary artery calcium scoring)
- Cardiac MRI (CMRI)
- ECHO: Rest and exercise
- ABI: (ankle brachial index): Resting and exercise
- Retinal scan and OPA (ocular pulse amplitude)
- Cardiac PET, SPECT
- FDG-PET/CT: Vascular inflammation/plaque/biologic activity
- PET/CT/F-NaF for coronary plaque/inflammation/morphology
- PET/MRI for coronary plaque morphology and inflammation
- IVUS: Intravascular ultrasound

Box 47.3 Other comprehensive lab testing for cardiovascular disease
- Advanced lipid testing
- 24-hour ABM
- PRA and aldosterone
- Dysglycemia labs: Adiponectin, FBS, 2 hr. GTT, insulin, proinsulin, C-peptide, HBA1C, and glycated serum protein (GSP)
- BIA with total fat, regional fat, and LMM
- CT scan of chest for epicardial fat
- Markers for inflammation, oxidative stress, and immune function
- Micronutrient tests
- Thrombosis markers
- Renal function markers: CrCl (creatinine clearance), MAU (microalbuminuria), cystatin C, and SDMA (symmetric dimethyl arginine)
- Toxicology, heavy metals, POPs screen, organocide
- Omega-3 index
- Telomere test
- CV genetic SNP testing with Apo E genotypes with cardia-X
- CV genetic expression testing: Corus
- Gluten testing
- CBC with differential
- Urinalysis
- CMP 12
- APO B and APO AI and AII
- Free T4, T3, TSH, RT3, thyroid antibodies
- Magnesium
- Iron, TIBC, and ferritin
- Fibrinogen
- hsCRP
- Homocysteine
- Uric acid
- Microalbuminuria
- GGTP (gamma glutamyl transpeptidase) and hepatic profile
- Myeloperoxidase (MPO)
- Plasma viscosity
- Vitamin D3
- Hormone profile: Free testosterone, SHBG, estradiol, estriol, progesterone, DHEA, and DHEAS

Box 47.4 Cardiovascular genetics
1. 9p21 (GG/CC): CHD, MI, ASCVD, DM, IR AAA, thrombosis, plaque rupture, inflammation, and intracranial aneurysms
2. 6p21.4: CHD, MI, DVT)
3. 4q25: Atrial fibrillation, long QT, and PR intervals
4. ACE I/D (DD allele): HBP, LVH, CRF, MAU, nephro angiogenesis, carotid IMT, MI, and CHD.
5. COMT: Catecholamines, CHD, MI, HBP, ASA, and vitamin E responses.
6. 1q25 (GLUL): CHD in DM, enterocytes, and ED
7. 7 APO E: Dyslipidemia, CHD, MI, nitric oxide, statin response.
8. MTHFR: Methylation (1298C and 677 T): hypertension, CHD, MI, CVA, thrombosis, homocysteine, ED.
9. CYP 1A2: Caffeine, HBP, MI, aortic stiffness, PWV, AI, tachycardia, arrhythmias, vascular inflammation, catecholamines.
10. Corin: Hypertension, CHF, volume overload, sodium sensitive, CVD, CRF, preeclampsia), ANP, and BNP.
11. CYP 11 B2 (TT allele): HBP, aldosterone and response to spironolactone.
12. Glutathione peroxidase: CHD, MI, hypertension, LVH, CHF, glutathione, ALA 6 alleles, selenium.
13. ADR B2: HBP, PRA, inflammation and DASH diet with ACEI, ARB, or DRI.
14. APO A1: Lipids, HDL, CHD, MI, obesity.
15. APO A2: Lipids, HDL, CHD, MI, obesity.
16. APC C 3: Dyslipidemia, CHD, MI, dysfunctional HDL, inflammation, DM.
17. CYP 4 A11: Hypertension, ENaC and sodium, volume overload, CHD, and amiloride.
18. CYP 4F 2: Hypertension, ENaC and sodium, volume overload, CHD and amiloride.
19. AGTR1 (ATR1AA): HBP, ARBs, and potassium..
20. NOS 3: Nitric oxide, hypertension, MI, CHDCVA, thrombosis, ED, oxidative stress, inflammation.
21. SCARB1: Lipids, dysfunctional HDL with high HDL, CHD, MI.

47.6 Nutritional Interventions

47.6.1 Mediterranean Diet (TMD: Traditional Mediterranean Diet) [22–30]

In the 4.8-year primary prevention (PREDIMED), the rate of major cardiovascular events from MI, CVA, or total CV deaths was reduced by 30% with nuts and 30% with extra virgin olive oil (EVOO). The reduction in CVA was 39% ($p < 0.003$) with a 33% reduction from EVOO and a 46% reduction from nuts. The reduction in MI was 23% ($p = 0.25$) with a 20% reduction from EVOO and a 26% reduction from nuts. Total CV deaths were reduced by 17% ($p = 0.8$) [22–25]. New onset type 2 diabetes mellitus was decreased by 40% with EVOO and 18% with mixed nuts [22–25]. This reduction was associated with decreases in hsCRP and IL-6.

The high content of nitrate (NO3), at an average of 400 mg per day, is converted to nitrite (NO2), which eventually forms nitric oxide. Also, the increased amount of omega-3 FA, good omega-6 FA, and polyphenols (such as quercetin, resveratrol, and catechins, in grapes and wine) provide many of the beneficial outcomes in CHD.

Secondary prevention post-MI in the Lyon Diet Heart Study demonstrated significant reductions in all events, including cardiac death, nonfatal MI, unstable angina, CVA, CHF, and hospitalization at 4 years using the Mediterranean-style diet supplemented with alpha-linolenic acid, compared to a prudent western diet. Compared to the control, the Mediterranean-style diet demonstrated a 76% lower risk of cardiac death and nonfatal MI during the study period. Olive oil was associated with a decreased risk of overall mortality and an important reduction in CVD mortality in a large Mediterranean cohort of 40,622 subjects. For each increase in olive oil by 10 g, there was a 13% decrease in CV mortality. In the highest quartile of olive oil intake, there was a 44% decrease in CV mortality. One of the mechanisms by which the TMD, particularly if supplemented with virgin olive oil at 50 g per day, can exert CV health benefits is through changes in the transcriptomic response of genes related to cardiovas-

cular risk that include genes for atherosclerosis, inflammation, oxidative stress vascular immune dysfunction, type 2 diabetes mellitus, and hypertension. This includes genes for ADR-B2 (adrenergic beta 2 receptor), IL7R (interleukin 7 receptor), IFN gamma (interferon), MCP1 (monocyte chemotactic protein), TNFα (tumor necrosis factor alpha), IL-6 (interleukin 6), and hsCRP (high-sensitivity C reactive protein). In summary, the TMD has been shown to have the following effects:
- Lowers blood pressure
- Improves serum lipids: lowers TC, LDL, TG, increases HDL, lowers oxLDL and Lp(a), improves LDL size, and lower LDL P to less atherogenic profile
- Improves T2 DM and dysglycemia
- Improves oxidative defense and reduces oxidative stress: F-2 isoprostanes and 8OHDG
- Reduces inflammation: lowers hsCRP, IL 6, sVCAM, sICAM
- Reduces thrombosis and factor VII after meals
- Improves BNP
- Increases nitrates/nitrites
- Improves membrane fluidity
- Reduces MI, CHD, and CVA
- Reduces homocysteine

47.6.2 DASH Diets (DASH 1 and 2)

The DASH (Dietary Approaches to Stop Hypertension) diets reduce BP and CHD. Both DASH 1 and DASH 2 emphasize fruits, vegetables, whole grains, beans, fiber, low-fat dairy products, poultry, fish, seeds, and nuts but limit red meat, sweets, and sugar-containing beverages, while increasing the intake of potassium, magnesium, and calcium – and including variable restriction in dietary sodium [31, 32]. Both DASH diets reduce blood pressure within 4 weeks by about 10/5 mmHg or more, which is at least as effective as one antihypertensive medication. In the Nurses' Health Study, adherence to the DASH dietary pattern was associated with a lower risk of CHD by 14% in those with the highest adherence to the diet [33].

47.6.3 Portfolio Diet [34, 35]

The first Portfolio diet study consisted of 34 patients with dyslipidemia treated with three diets for 1 month, each in random order [34]:
- Control Diet: very low saturated fat
- Diet 2: control diet, plus lovastatin 20 mg
- Diet 3 (Portfolio): plant sterols, soy foods, almonds, viscous fibers, okra, eggplant

The results showed:
- Control Diet: reduced LDL 8.5%
- Diet 2: reduced LDL 33.3% and TG 11%
- Portfolio Diet: reduced LDL 29.6% and TG 9.3%

The second Portfolio diet study [35] was performed over a 6-month period with 351 subjects [34]. It included a low saturated fat diet as control, which reduced LDL-C by 3%. The dietary portfolio with variable counseling showed an LDL reduction of 13.1%. The dietary portfolio with two clinic visits resulted in an LDL decrease of 13.1%. The dietary portfolio with seven clinic visits had an LDL reduction of 13.8%.

47.7 Conclusions

The top five cardiovascular risk factors, as they are currently defined, are not an adequate explanation for CHD. In order to close the CHD gap, the top 5 risk factors must be better defined while assessing the top 25 modifiable risk factors and the other 395 risk factors and mediators. Early detection and aggressive prevention and treatment of cardiovascular disease are needed to identify functional CV disease and early CVD risk biomarkers before any cardiovascular structural changes occur or clinical disease becomes manifest. Improved utilization of new laboratory techniques are required, such as the advanced lipid profiles, 24-hour BP monitoring, and specific tests to identify inflammation, such as hsCRP; oxidative stress such as oxLDL and myeloperoxidase; and immune vascular dysfunction such as thyroid antibodies and other immune markers. In addition, vascular translational medicine will need to be evaluated with new imaging technologies, such as EndoPAT, CAPWA, carotid IMT, MCG, and CT angiogram.

In order to truly revolutionize the treatment of CVD, new therapies will need to involve management of the pathophysiologic risk factors, mediators and their downstream effects, as well as the finite vascular responses. This will be achievable by using a combination of targeted personalized and precision treatments with genomics, proteomics, metabolomics, nutrition, nutraceutical supplements, vitamins, minerals, antioxidants, anti-inflammatory agents, anti-immunological agents, and pharmacologic agents. Future studies must begin to measure all of the pertinent risk factors that have been reviewed here to correlate their direct relationship with CHD. Only by addressing all of these factors will we be able to decrease or halt subsequent vascular damage.

Case Presentations

Hypertension

Clinical History, Physical Exam, and Laboratory
- A 55-year-old black female with new onset hypertension
- BP 148/94 mmHg in office, averaged with BP readings x 3
- 24-hour ambulatory BP monitor: Mean BP 146/92 mmHg, BP load 52%, AM surges, nondipper
- FH negative for hypertension and CHD. Nonsmoker, no ETOH, no caffeine
- PE normal. Normal weight and body fat
- Lab normal except for microalbuminuria (MAU) of 66, FBS 104 mg/dL, and TG 298 mg/dL
- EKG mild left ventricular hypertrophy
- PRA 0.20 ng/ml/hr. (low)
- Plasma aldosterone: 10
- ARR (Aldo/renin ratio): 50
- Micronutrient deficiencies: GLA (gamma linoleic acid), vitamin D (25 ng/ml), zinc, magnesium, and lipoic acid

Concepts of Diagnosis and Treatment
- Patient has low renin hypertension (LRH).
- Endothelial dysfunction by EndoPAT at 1.44. Normal is >1.67.
- Her genetic profile (SNPs) for hypertension is negative.
- Use nutraceuticals and/or drugs that work in LRH.
- DASH 2 diet with 50 g of carbohydrates.
- Combined aerobic and resistance exercise 6 days per week at 60 minutes per session.
- GLA 500 mg twice per day.
- Vitamin D3 at 4000 IU per day.
- Zinc 50 mg per day.
- R-Lipoic acid 100 mg per day with biotin 2 mg per day.
- Magnesium chelates at 500 mg twice per day.
- Omega-3 fatty acids at 1.5 g bid (DHA and EPA).
- Replete nutrient deficiencies: GLA (gamma linoleic acid), vitamin D, zinc, magnesium, and lipoic acid.
- Start high-dose treatment with nutritional supplements that improve LRH: GLA (gamma linoleic acid), vitamin D, zinc, magnesium, and lipoic acid or others.
- Treat the insulin resistance and hypertriglyceridemia.
- Improve MAU.

Results
- 6 weeks: BP 126/84 mmHg.
- 3 months: BP 118/78 mmHg, 24-hour ABM is normal with mean BP 116/76 mm Hg, BP load <5%, no AM surges, normal dipping pattern.
- FBS 88 mg/dL.
- TG 110 mg/dL.
- MAU 34.
- Vitamin D3 62 ng/dl.
- MNT normal for nutrient deficiencies.
- EndoPAT is increased to 1.96.
- LVH on EKG unchanged.

Dyslipidemia

Clinical History, Physical Exam, and Laboratory
- A 38-year-old white male in for physical exam.
- Family history is positive for CHD early in life (age below 50 years).
- Normal weight, nonsmoker, excellent diet, and exercise program.
- History and PE are normal.
- All labs are normal, including a routine lipid profile. Note the normal TC, LDL, HDL, and TG.
- Expanded lipid profile done below (Fig. 47.6a, b):
- Lp(a) is 120 (normal is <30).
- hsCRP is 1.5 (borderline).
- Treatment: Should this patient be treated?

Treatment and Results at 2 Months
- Niacin B3 1000 mg bid
- NAC 1000 mg bid
- Omega-3 FA 4 g per day
- Vitamin C buffered at 5000 mg per day
- Lysine 1000 mg per day
- Proline 500 mg per day
- Red yeast rice at 800 mg hs
- Berberine 500 mg HS
- Pomegranate seeds ¼ cup bid
- LDL: 70–55 mg%
- LDL(p): 999–710
- LDL size: 20.3–22
- HDL: 41–50 mg%
- HDL 2b: 16–38 mg%
- TG: 70–55 mg%
- Lp(a) from 120 to 70

Coronary Heart Disease

Clinical Presentation, History, Physical Exam, and Laboratory
- A 58-year-old white male attorney.
- CC: fatigue for 8 years and mild DOE with steps and inclines only for 6 months. No chest pain. Mild memory issues for 5 years. Frequent URIs for 6 years. Erectile dysfunction for 4 years.
- PMH: negative except for GERD for 5 years, history of EBV and hepatitis A.
- Nonsmoker, 4 cups of coffee per day, one glass of red wine per day.
- Stressful job.
- Married with four children.
- Good diet. Eats lots of chicken and rice.
- Exercises 4 days per week for 1 hour without symptoms.
- Medications: nexium 40 mg per day for 5 years and Cialis 5 mg per day.
- No allergies.
- FH positive for hypertension, dyslipidemia, DM, CHD, and CVA.
- BP 142/88 mmHg HR 78.
- 6 ft. 1 inch Weight 220 pounds.
- WC 40 inches.
- Body Fat 25%.
- HEENT: mild glossitis.
- Cardiac: no murmur, rubs, or gallops.
- No edema.
- Skin: dry without rash.
- Otherwise normal exam.
- CHEM 12 normal except FBS of 102 mg/dL (high).
- CBC: PCV 38 (low) with MCV 104 (high).
- MAU 340 mg.(high), Cystatin C elevated at 1.7 (1.04).
- Homocysteine 12 (high).
- Fibrinogen 490 (high).

- hsCRP 3.0 (high).
- Ferritin: 162 (high).
- 2 h GGT 182 mg/dL (abnormal).
- HBAIC 6.2 (high).
- MNT: low in B 12, chromium, pantothenic acid, Co AQ 10, copper, vitamin C, and vitamin D.
- B 12 level 143 (low).
- Vitamin D 15 ng/ml (low).
- TFTs normal except for TSH of 3.4 (high)/positive TPO antibodies.
- Coenzyme Q 10 1.5 ug/ml (low).
- LPP advanced lipid profile: LPP 891, dense LDL, low HDL 2b.
- Genotype: heterozygote hemochromatosis, 2 SNPs for MTHFR (677 and 1298), Apo E 4/4, 9pq21, heterozygote SOD and GSH, homozygote CYP A12.
- ADMA elevated.
- Adiponectin low.
- MPO high at 730.
- Increased 8-0hdG and MDA with low PAO (plasma antioxidant capacity).
- GGTP 82 (elevated).
- OxLDL 99 (elevated).
- Hormones normal: free testosterone and DHEAS.
- Toxin screen: positive for arsenic.
- Gluten sensitivity: negative.
- Positive IGG for EBV and hepatitis A.
- TMT shows ventricular quadrigeminy, anterior and lateral ST depression of 1–2 mm.
- EndoPAT: endothelial dysfunction at 1.44, augmentation index is increased (stiff arteries with increased pulse wave velocity), and HRV is abnormal.
- CWPWA: C1 6 (low), C2 5 (low).
- Carotid duplex: increased IMT with less than 50% plaque bilateral.
- ECHO: normal EF, mild PI.
- MCG: abnormal/positive.
- Retinal scan: Grade 1 KW changes.
- 24-hour ABM: BP average 134/86 mm Hg, nondipper, BP load 35% over 140/90 mm Hg.
- Left main had 50–60% stenosis.
- Left anterior descending: multiple stenosis at 35%, 60%, and 95% high-grade obstruction.
- Left circumflex had 50% stenosis.
- Successfully stented the LAD 95% obstruction.

Treatment
- Mediterranean diet, stopped all caffeine, increased dark green leafy vegetables to eight servings per day with beets.
- Cardiac rehab and then ABCT exercise interval aerobics and resistance exercise.
- Stress management and relaxation therapy and HRV breathing training.
- Crestor 10 mg HS, berberine HCL 500 mg bid, niacin 500 mg hs with food and with quercetin 500 mg.
- Metoprolol 25 mg bid.
- ASA 81 mg per day.
- Quinapril 80 mg HS and melatonin 3 mg hs.
- Armour thyroid 60 mg per day.
- B-12 lozenges sublingual 1000 mcg/day.
- Vitamin D3 10,000 IU per day for 1 month and then 4000 IU per day.
- Chromium 800 ug per day.
- Coenzyme Q 10 200 mg per day.
- B complex with methyl folate 5000 ug per day.
- Vitamin C 500 mg bid with copper 1 mg/zinc 50 mg per day.
- R lipoic acid 100 mg, NAC 1000 mg, and whey protein 40 g for glutathione.
- Oral chelation for arsenic.
- Omega-3 FA 5000 mg, vitamin K 2 MK 7200 mcg, Kyolic garlic 600 mg bid, curcumin 1000 mg bid, magnesium chelates 500 mg bid, and Epsom salts bath.
- Phlebotomy for hemochromatosis.
- Probiotic 50 billion cfu per day with glutamine 1000 mg bid. Stopped PPI.

6-Month Results
- All symptoms resolved
- Weight 195 lbs., WC 34 inches, 20%
- BP 122/78 mm Hg, dipper, HR 62 b/m
- All labs normal
- LPP 608, LDL type A, HDL 2b and HDL P normal
- MNT normal
- EndoPAT normal 2.66 and CAPWA normal
- TMT normal

Algorithms, Protocols, and Treatments

Hypertension
- Get appropriate labs with PRA and aldosterone to determine renin status.
  - High PRA: High renin hypertension. Start Natural ACEI, ARB, DRI, or BB
  - Low PRA: Low renin hypertension. Start Natural diuretic or CCB
- Use combinations of supplements to achieve BP goal using office readings and 24-hour ABM.
- Stop all drugs that induce hypertension.
- Exclude all secondary forms of hypertension, especially OSA (obstructive sleep apnea).
- Start DASH 2 diet with high potassium and magnesium and low sodium intake.
- Start weight loss program.
- Start supervised exercise program.
- Stop alcohol.
- Stop all tobacco.
- Stop caffeine, if slow metabolizer.
- Sleep 8 hours per night.
- Employ relaxation techniques.
- Reassess BP at 4 weeks and adjust supplements and protocols.
- BP goal is based on 24-hour ABM.

Treatment
Nutraceutical Supplements for Hypertension
Many of the natural compounds in food, certain nutraceutical supplements, vitamins, antioxidants, or minerals function in a similar fashion to a specific class of antihypertensive drugs. Although the potency of these natural compounds may be less and it may take longer to work than the antihypertensive drug, when used in combination with other nutrients and nutraceutical supplements, the antihypertensive effect is magnified.

Natural Diuretics
- Vitamin B-6 (Pyridoxine): 200 mg per day
- Taurine: 2–3 g twice per day

- Celery: one to two sticks per day
- GLA: 1000 t0 2000 mg per day
- Vitamin C (Ascorbic Acid): 500 mg per day
- Potassium: 5–10 g per day
- High Gamma/Delta tocopherols and tocotrienols: 250 mg per day
- Magnesium 500 mg chelated form twice per day
- Protein: high-quality organic protein at 1.5 g per kg per day
- Fiber 50 g of mixed fiber per day
- Coenzyme Q-10: 100–200 mg per day
- L-Carnitine: 2–6 g per day
- Hawthorn berry extract standardized: 2 per day

Beta-Blockers (BB)
- Hawthorn

Central Alpha Agonists (CAA)
- Taurine: 2–3 g twice per day
- Potassium: 5–10 g per day
- Zinc 50 mg per day
- Na^+ restriction to 2000 mg per day
- Fiber: fiber 50 g of mixed fiber per day
- Protein: high-quality organic protein at 1.5 g per kg per day
- Vitamin C: 500 mg per day
- Vitamin B-6: 200 mg per day
- Coenzyme Q-10: 100–200 mg per day
- Celery: one to two sticks per day
- GLA/DGLA: 1500–3000 mg per day
- Aged Kyolic garlic: 600 mg twice per day

Direct Vasodilators
- Omega-3 FA: 4–5 g per day
- MUFA (omega-9 FA): 50 g per day of EVOO
- Potassium: 5–10 g per day
- Magnesium 500 mg chelated form twice per day
- Soy: fermented soy
- Fiber 50 g of mixed fiber per day
- Aged Kyolic garlic: 600 mg twice per day
- Flavonoids
- Vitamin C: 500 mg per day
- Vitamin E: Gamma/delta tocopherol: 200–500 mg per day
- Coenzyme Q-10: 100–200 mg per day
- L-Arginine: 5–10 g per day as food or supplement
- Taurine: 2–3 g twice per day
- Celery: one to two sticks per day

Natural Calcium Channel Blockers (CCB)
- Alpha Lipoic Acid (ALA): 100–200 mg per day
- Magnesium: 500 mg chelated twice per day
- Vitamin B-6 (Pyridoxine) 200 mg per day
- Vitamin C 500 mg per day
- Vitamin E: high gamma/delta E with alpha tocopherol at 200–500 mg per day
- (↑ cytosolic Mg^{++} with ↓ Ca^{++}), also diuretic
- N-Acetyl cysteine (NAC) 500–1000 mg twice per day
- Hawthorn standardized extract
- Celery one to two sticks per day
- Omega-3 fatty acids (EPA + DHA): 4–5 g per day
- Aged Kyolic garlic: 600 mg twice per day
- Taurine: 2–3 g twice per day

Natural Angiotensin-Converting Enzyme Inhibitors (ACEI)
- Aged Kyolic garlic: 600 mg twice per day
- Wakame seaweed: 3–4 g per day
- Tuna protein/muscle
- Sardine protein/muscle
- Hawthorn berry
- Bonito Fish (dried): 1500 mg per day
- Pycnogenol: 200 mg per day
- Hydrolyzed whey protein: 40 g per day
- Sour milk and milk peptides
- Gelatin
- Sake
- Omega-3 FA: 4–5 g per day
- Chicken egg yolks
- Zein
- Dried salted fish and fish sauce
- Zinc: 50 mg per day
- Melatonin: 3–6 mg at night
- Pomegranate: ½ to one cup per day
- Quercetin: 500–1000 mg twice per day
- Berberine: 500 mg twice per day

Natural Angiotensin Receptor Blockers (ARB)
- Potassium: 5–10 g per day
- Taurine: 2–3 g twice per day
- Resveratrol: 250 mg trans resveratrol per day
- Fiber: 50 g mixed fiber per day
- Aged Kyolic garlic: 600 mg twice per day
- Vitamin C: 500 mg per day
- Vitamin B-6: (Pyridoxine) 200 mg per day
- Co Enzyme Q-10: 100 t0 200 mg per day
- Celery: one to two sticks per day
- Gamma-linolenic acid (GLA and DGLA): 2–3 g per day

Dyslipidemia
- Identify underlying causes of dyslipidemia, remove, and treat them. These include poor micronutrient and macronutrient intake, toxins, and infections as well as other causes, such as hypothyroidism, obesity, smoking, and lack of exercise.
- Evaluate global CV risk with risk a scoring system.
- Evaluate for concomitant CHD risk factors.
- Measure noninvasive vascular testing such as EmdoPAT, CAPWA, and coronary artery calcium score and repeat testing in 6 months.
- Evaluate also carotid artery duplex, ECHO, EKG, exercise treadmill test, ABI at rest and exercise.
- Use advanced lipid testing to measure all particle sizes and particle numbers and Lp(a) and repeat testing in 2 months.
- LDL-P drives CHD risk.
- Reverse cholesterol transport, HDL-P, and HDL functionality drives reduction in CHD risk.
- HDL is often dysfunctional with inflammation and oxidative stress.
- HDL levels over 85 mg/dL are often dysfunctional.
- Measuring MPO and hsCRP help define dysfunctional HDL.
- Use scientifically proven nutraceuticals with the 45 different mechanisms of action to reduce dyslipidemia-induced vascular disease. Think/treat beyond just numbers. Look at mechanisms.

General Treatment Considerations
1. PREDIMED diet.
2. High-quality organic protein, especially fish.
3. Reduce refined carbohydrates to 50–75 g per day or less.
4. Eight servings of vegetables per day.
5. Four servings of berries or other low glycemic fruits per day.
6. Omega-3 fatty acids.
7. Monounsaturated fatty acids such as olive oil, olives, and mixed nuts. Consume 40 g of EVOO (extra virgin olive oil) per day.
8. Reduce saturated fats.
9. Eliminate trans fats.
10. Mixed aerobic and resistance training.
11. Ideal body weight and composition (BF for women <22% and less than 16% for men). Reduce visceral fat.

Nutritional Supplement Treatment
Supplements and Dose
- Red yeast rice: 2400–4800 mg at night with food.
- Plant sterols: 2.5 g per day.
- Berberine: 500 mg per day to twice per day.
- Niacin (nicotinic acid B3): 500–3000 mg per day as tolerated pretreated with quercetin, apples, ASA. Take with food and avoid alcohol. Never interrupt therapy.
- Omega-3 fatty acids with EPA/DHA at 3/2 ratio: 4 g / day with GLA at 50% of total EPA and GLA and gamma/delta tocopherol.
- Gamma delta tocotrienols: 200 mg hs.
- Aged garlic-Kyolic standardized: 600 mg twice per day.
- Sesame: 40 g per day.
- Pantethine: 450 mg bid.
- MUFA: 40–50 g per day (EVOO 4 tablespoons/day).
- Lycopene: 20 mg per day.
- Luteolin: 10 per day.
- Astaxanthin: 15 mg per day.
- Trans-resveratrol: 250 mg per day.
- NAC: 500 mg twice per day.
- Carnosine: 500 mg twice per day.
- Citrus bergamot: 1000 mg per day.
- Quercetin: 500 mg twice per day.
- Probiotics standardized: 15–50 billion organisms bid.
- Curcumin: 500–1000 mg twice per day.
- EGCG: 500–1000 mg bid or 60–100 ounces of green tea/day.
- Pomegranate: ½ to one cup of seeds/day or 6 ounces of juice per day.

Endothelial Dysfunction
Concepts
- The nitrate-nitrite NO pathway in which endogenous nitrate (NO3-) undergoes reduction to nitrite (NO2-) and then to NO in various tissues, including blood with the production of bioactive NO that is important for ED and CVD.
- Vegetables are the primary source of nitrates (80–85%).
- This is the alternate pathway to the arginine NO path.
- Nitrate-to-nitrite reduction is carried out by commensal bacteria present on dorsal surface of tongue and GI tract and requires healthy microbiome and gastric acid.
- Increased with low oxygen.
- Decreased with mouthwash and GI disease and use of PPIs and H2 blockers that reduce gastric acid.
- Conversion to NO by deoxyhemoglobin, xanthine oxidoreductase, deoxymyoglobin, protons, vitamin C, and polyphenols.

Protocol
1. EndoPAT for endothelial function, augmentation index, and heart rate variability (HRV).
2. CAPWA: computerized arterial pulse wave analysis for arterial compliance.
3. ADMA (asymmetric dimethyl arginine): competitive inhibitor of nitric oxide. One of the best markers for nitric oxide bioavailability.
4. SDMA (symmetric dimethyl arginine): competes with arginine for cell uptake.
5. Nitro-species in erythrocytes: most sensitive marker of nitric oxide.
6. BH4 (tetrahydrobiopterin): primary cofactor for eNOS to make nitric oxide. BH4 is cofactor for conversion of phenylalanine and tyrosine to DOPA, dopamine, NE, and EPI. Low CNS NT lead to hypertension and CVD. BH4 is cofactor for conversion of phenylalanine to tyrosine and thyroid hormones. Hypothyroidism leads to hypertension and CVD. BH4 is cofactor for conversion of L-tryptophan to 5HTP and if deficient increases kynurenate and quinolinate which stimulate glutamate NMDA receptor to increase CNS inflammation (and neurodegeneration) which increases BP and CVD risk.
7. BH2: oxidized form that provides no protection from CHD and is not active as eNOS cofactor.
8. BH2/BH4 ratio.
9. DDAH (dimethyarginine dimethylaminohydrolase): breaks down ADMA and increases NO.
10. Citrulline: increase production of nitric oxide with arginine.
11. Arginine: precursor for nitric oxide via eNOS.
12. Arginase 1: breaks down arginine and reduces NO.
13. Nitrates and nitrites: alternate pathway via food and oral microbes to produce nitric oxide.
14. Folate: cofactor for eNOS to make nitric oxide.
15. MTHFR.
16. Vitamin B 12: cofactor for methylation and improves endothelial function. Lowers homocysteine with other B vitamins.
17. Homocysteine.
18. Neopterin (macrophages, CMI, CVD, and autoimmune).
19. Biopterin.
20. Neopterin/biopterin ratio: normal is 0.5–6.0 (marker for CVD, vascular disease, inflammation, and autoimmune.disease).

Treatment
1. If EndoPAT is low or ADMA and SDAM are high, start treatments below first:
    - Beetroot extract with arginine, citrulline, and hawthorn
    - PREDIMED diet
    - VasculoSirt
    - Methyl folate

2. Repeat ADMA, SDMA, and EndoPAT in 4–6 months. If still abnormal, add the following to the initial treatment: Vitamin C 1000 mg bid, omega-3 fatty acids at 5 g per day (EFA-Sirt Supreme 5 BID), resveratrol 250 mg per day) and Kyolic Aged garlic extract (AGE): 600 mg twice per day and Co Q 10 200 mg per day.

Coronary Heart Disease
1. EndoPAT
2. EKG
3. Magnetic resonance EKG
4. chest x-ray
5. Echo
6. Computerized arterial pulse wave analysis
7. Treadmill exercise test
8. Exercise echo as indicated
9. Nuclear medicine scans as indicated
10. Coronary artery calcium score
11. Computerized CT angiogram as indicated
12. MRI and MRI of heart as indicated
13. 24-hour Holter monitor as indicated

Treatment
- DASH diet
- Mediterranean diet with EVOO
- Nut consumption: 20% reduction death with 7 servings per wk
- Vitamin D3 4000IU per day
- Vitamin C 500 mg per day
- Beetroot extract with arginine, citrulline, and hawthorn
- Dietary nitrate at 0.1 mmol/kg of body weight per day (high intake of F and V) reduces DBP 3.5 mm Hg
- Effect is potentiated by Vitamin C and polyphenols
- 500 mg beetroot juice with 45 mmol/L of 2.79 g/L of inorganic nitrate lowers BP 10.4/8.1 mm Hg, inhibits platelet aggregation by 20% and increased FMD 30%
- Lycopene 20 mg per day
- Omega-3 fatty acids 5 g per day EFA-Sirt Supreme
- Polyphenols, flavonoids, and flavonoid-rich foods. Best data with flavones and flavonols
- Resveratrol: 250 mg trans form per day
- Grape seed extract: 500 mg twice per day
- EGCG: 500–1000 mg twice per day
- Co Enzyme Q 10: 100 mg twice per day
- Cacao and dark chocolate: 30 g per day
- Tea and catechins
- Curcumin: 1000–2000 mg twice per day
- Berry anthocyanins and pomegranate seeds
- Orange juice and hesperidin
- Wine polyphenols
- Rhodiola extract: 200–500 mg/day
- Kyolic Aged Garlic extract (AGE): 600 mg twice per day
- VasculoSirt

Congestive Heart Failure
3. EndoPAT
4. EKG
5. Magnetic resonance EKG
6. Chest x-ray
7. Echo
8. Computerized arterial pulse wave analysis
9. Treadmill exercise test
10. Exercise echo as indicated
11. Nuclear medicine scans as indicated
12. Coronary artery calcium score
13. Computerized CT angiogram as indicated
14. MRI and MRI of heart as indicated
15. 24 hour Holter monitor as indicated
16. NT-proBNP and BNP: BNP and CVD Risk
17. Galectin-3

Treatment
1. PREDIMED diet.
2. High-quality organic protein, especially fish.
3. Reduce refined carbohydrates to 50–75 g per day or less.
4. 8 servings of vegetables per day.
5. 4 servings of berries or low-glycemic fruits per day.
6. Reduce fruit juices.
7. Increase fiber to 50 g per day.
8. Omega-3 fatty acids.
9. Monounsaturated fatty acids, such as olive oil, olives and mixed nuts.
10. Reduce saturated fats.
11. Eliminate trans fats.
12. Mixed aerobic and resistance training.
13. Ideal body weight and composition (BF for women less than 22% and less than 16% for men). Reduce visceral fat. Increase lean muscle mass.
14. Sleep 8 hours per night.

Specific Nutritional Supplements
1. Co Enzyme Q 10: 200–300 mg bid
2. D-Ribose: 5 g tid
3. Taurine: 3 g bid
4. Carnitine: 3 g bid
5. Magnesium: 500 mg bid of chelated magnesium
6. Vitamin K2 MK 7: 200–400 micrograms per day
7. Aged Garlic Kyolic: 600 mg bid
8. Omega-3 fatty acids: EFA-SIRT SUPREME
9. B vitamins mixed and thiamine: 200 mg per day

Revolutionary New Concepts in the Prevention and Treatment of Cardiovascular Disease

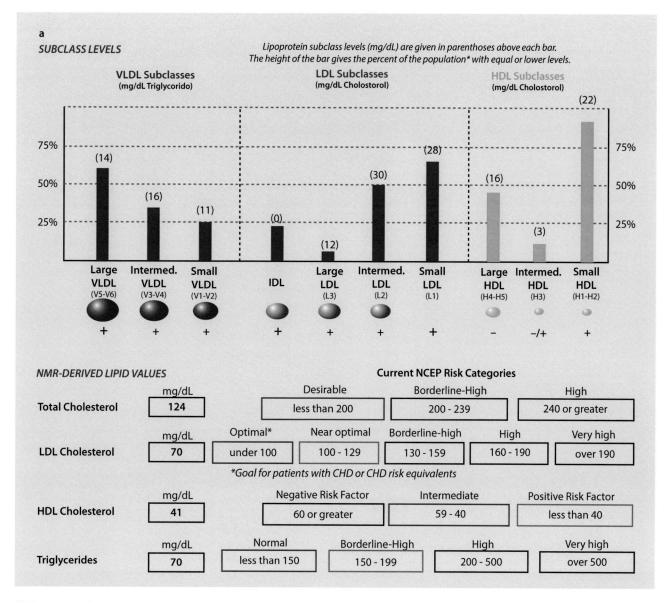

Fig. 47.6 a This is the VAP (vertical analytic profile) advanced lipid test indicating the various particle sizes of HDL, VLDL, and LDL. b This is the VAP (vertical analytic profile) advanced lipid test giving the actual LDLP number as well as VLDL and HDL

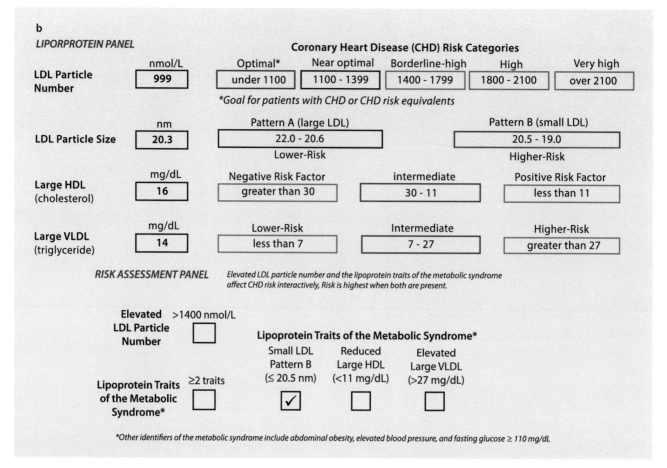

Fig. 47.6 (continued)

References

1. Yusuf S, Hawken S, Ounpuu S, Dans T, Avezum A, Lanas F, McQueen M, Budaj A, Pais P, Varigos J, Lisheng L, INTERHEART Study Investigators. Effect of potentially modifiable risk factors associated with myocardial infarction in 52 countries (the INTERHEART study): case-control study. Lancet. 2004;364(9438):937–52.
2. Houston MC. What Your Doctor May Not Tell You About Heart Disease. The revolutionary book that reveals the truth behind coronary illnesses and how you can fight them. New York: Grand Central Life and Style, Hachette Book Group; 2012.
3. O'Donnell CJ, Nabel EG. Genomics of cardiovascular disease. N Engl J Med. 2011;365(22):2098–109.
4. Houston MC. Nutrition and nutraceutical supplements in the treatment of hypertension. Expert Rev Cardiovasc Ther. 2010;8:821–33.
5. ACCORD Study Group, Gerstein HC, Miller ME, Genuth S, Ismail-Beigi F, Buse JB, Goff DC Jr, Probstfield JL, Cushman WC, Ginsberg HN, Bigger JT, Grimm RH Jr, Byington RP, Rosenberg YD, Friedewald WT. Long-term effects of intensive glucose lowering on cardiovascular outcomes. N Engl J Med. 2011;364(9):818–28.
6. Youssef-Elabd EM, McGee KC, Tripathi G, et al. Acute and chronic saturated fatty acid treatment as a key instigator of the TLR-mediated inflammatory response in human adipose tissue, in vitro. J Nutr Biochem. 2012;23:39–50.
7. El Khatib N, Génieys S, Kazmierczak B, Volpert V. Mathematical modelling of atherosclerosis as an inflammatory disease. Philos Transact A Math Phys Eng Sci. 2009;367(1908):4877–86.
8. Houston MC. Handbook of hypertension. Oxford, UK: Wiley–Blackwell; 2009.
9. Tortora GJ, Derrickson BH, editors. Principles of anatomy and physiology, 13th ed. Wiley: New Jersey; 2011.
10. Della Rocca DG, Pepine CJ. Endothelium as a predictor of adverse outcomes. Clin Cardiol. 2010;33(12):730–2.
11. Houston MC. New concepts in cardiovascular disease. J Restor Med. 2013;2(1):30–44.
12. Houston M. The role of nutraceutical supplements in the treatment of dyslipidemia. J Clin Hypertens (Greenwich). 2012;14(2):121–32.
13. Lundberg AM, Yan ZQ. Innate immune recognition receptors and damage-associated molecular patterns in plaque inflammation. Curr Opin Lipidol. 2011;22(5):343–9.
14. Zhao L, Lee JY, Hwang DH. Inhibition of pattern recognition receptor-mediated inflammation by bioactive phytochemicals. Nutr Rev. 2011;69(6):310–20.
15. Mah E, Bruno RS. Postprandial hyperglycemia on vascular endothelial function: mechanisms and consequences. Nutr Res. 2012;32(10):727–40.
16. Fazio S, Linton MF. High-density lipoprotein therapeutics and cardiovascular prevention. J Clin Lipidol. 2010;4(5):411–9.
17. van der Steeg WA, Holme I, Boekholdt SM, Larsen ML, Lindahl C, Stroes ES, Tikkanen MJ, Wareham NJ, Faergeman O, Olsson AG, Pedersen TR, Khaw KT, Kastelein JJ. High-density lipoprotein cholesterol, high-density lipoprotein particle size, and apolipoprotein A-I: significance for cardiovascular risk: the IDEAL and EPIC-Norfolk studies. J Am Coll Cardiol. 2008;51(6):634–42.
18. Houston M. What your doctor may not tell you about hypertension. The revolutionary nutrition and lifestyle program to help fight high blood pressure, Wellness central. New York: Hachette Book Group; 2003.

19. Bonetti PO, Pumper GM, Higano ST, Holmes DR Jr, Kuvin JT, Lerman AJ. Noninvasive identification of patients with early coronary atherosclerosis by assessment of digital reactive hyperemia. Am Coll Cardiol. 2004;44(11):2137–41.
20. Rozanski A, Gransar H, Shaw LJ, Kim J, Miranda-Peats L, Wong ND, Rana JS, Orakzai R, Hayes SW, Friedman JD, Thomson LE, Polk D, Min J, Budoff MJ, Berman DS. Impact of coronary artery calcium scanning on coronary risk factors and downstream testing the EISNER (early identification of subclinical atherosclerosis by noninvasive imaging research) prospective randomized trial. J Am Coll Cardiol. 2011;57(15):1622–32.
21. Kandori A, Ogata K, Miyashita T, Takaki H, Kanzaki H, Hashimoto S, Shimizu W, Kamakura S, Watanabe S, Aonuma K. Subtraction magnetocardiogram for detecting coronary heart disease. Ann Noninvasive Electrocardiol. 2010;15(4):360–8.
22. Sofi F, Abbate R, Gensini GF, Casini A. Accruing evidence on benefits of adherence to the Mediterranean diet on health: an updated systematic review and meta-analysis. Am J Clin Nutr. 2010;92(5):1189–96.
23. Estruch R, Ros E, Salas-Salvadó J, Covas MI, Corella D, Arós F, Gómez-Gracia E, Ruiz-Gutiérrez V, Fiol M, Lapetra J, Lamuela-Raventos RM, Serra-Majem L, Pintó X, Basora J, Muñoz MA, Sorlí JV, Martínez JA, Martínez-González MA, PREDIMED Study Investigators. Primary prevention of cardiovascular disease with a Mediterranean diet. N Engl J Med. 2013;368(14):1279–90.
24. Nadtochiy SM, Redman EK. Mediterranean diet and cardioprotection: the role of nitrite, polyunsaturated fatty acids, and polyphenols. Nutrition. 2011;27(7–8):733–44.
25. Salas-Salvadó J, Bulló M, Estruch R, Ros E, Covas MI, Ibarrola-Jurado N, Corella D, Arós F, Gómez-Gracia E, Ruiz-Gutiérrez V, Romaguera D, Lapetra J, Lamuela-Raventós RM, Serra-Majem L, Pintó X, Basora J, Muñoz MA, Sorlí JV, Martínez-González MA. Prevention of diabetes with Mediterranean diets: a subgroup analysis of a randomized trial. Ann Intern Med. 2014;160(1):1–10.
26. de Lorgeril M, Salen P, Martin JL, Monjaud I, Delaye J, Mamelle N. Mediterranean diet, traditional risk factors, and the rate of cardiovascular complications after myocardial infarction: final report of the Lyon Diet Heart Study. Circulation. 1999;99(6):779–85.
27. Buckland G, Mayén AL, Agudo A, Travier N, Navarro C, Huerta JM, Chirlaque MD, Barricarte A, Ardanaz E, Moreno-Iribas C, Marin P, Quirós JR, Redondo ML, Amiano P, Dorronsoro M, Arriola L, Molina E, Sanchez MJ, Gonzalez CA. Olive oil intake and mortality within the Spanish population (EPIC-Spain). Am J Clin Nutr. 2012;96(1):142–9.
28. Castañer O, Corella D, Covas MI, Sorlí JV, Subirana I, Flores-Mateo G, Nonell L, Bulló M, de la Torre R, Portolés O, Fitó M. PREDIMED study investigators. In vivo transcriptomic profile after a Mediterranean diet in high-cardiovascular risk patients: a randomized controlled trial. Am J Clin Nutr. 2013;98(3):845–53.
29. Konstantinidou V, Covas MI, Sola R, Fitó M. Up-to date knowledge on the in vivo transcriptomic effect of the Mediterranean diet in humans. Mol Nutr Food Res. 2013;57(5):772–83.
30. Corella D, Ordovás JM. How does the Mediterranean diet promote cardiovascular health? Current progress toward molecular mechanisms: gene-diet interactions at the genomic, transcriptomic, and epigenomic levels provide novel insights into new mechanisms. BioEssays. 2014;36(5):526–37.
31. Appel LJ, Moore TJ, Obarzanek E, Vollmer WM, Svetkey LP, Sacks FM, Bray GA, Vogt TM, Cutler JA, Windhauser MM, Lin PH, Karanja N. A clinical trial of the effects of dietary patterns on blood pressure. DASH collaborative research group. N Engl J Med. 1997;336(16):1117–24.
32. Sacks FM, Svetkey LP, Vollmer WM, Appel LJ, Bray GA, Harsha D, Obarzanek E, Conlin PR, Miller ER 3rd, Simons-Morton DG, Karanja N, Lin PH. DASH-sodium collaborative research group effects on blood pressure of reduced dietary sodium and the dietary approaches to stop hypertension (DASH) diet. DASH-sodium collaborative research group. N Engl J Med. 2001;344(1):3–10.
33. Fung TT Chiuve SE, McCullough ML, Rexrode KM, Logroscino G, Hu FB. Adherence to a DASH-style diet and risk of coronary heart disease and stroke in women. Arch Intern Med. 2008;168(7):713–20.
34. Jenkins DJ, Kendall CW, Marchie A, Faulkner DA, Wong JM, de Souza R, Emam A, Parker TL, Vidgen E, Trautwein EA, Lapsley KG, Josse RG, Leiter LA, Singer W, Connelly PW. Direct comparison of a dietary portfolio of cholesterol-lowering foods with a statin in hypercholesterolemic participants. Am J Clin Nutr. 2005;81(2):380–7.
35. Jenkins DJ, Jones PJ, Lamarche B, Kendall CW, Faulkner D, Cermakova L, Gigleux I, Ramprasath V, de Souza R, Ireland C, Patel D, Srichaikul K, Abdulnour S, Bashyam B, Collier C, Hoshizaki S, Josse RG, Leiter LA, Connelly PW, Frohlich J. Effect of a dietary portfolio of cholesterol-lowering foods given at 2 levels of intensity of dietary advice on serum lipids in hyperlipidemia: a randomized controlled trial. JAMA. 2011;306(8):831–9.

Immune System Under Fire: The Rise of Food Immune Reaction and Autoimmunity

Aristo Vojdani, Elroy Vojdani, and Charlene Vojdani

48.1   Introduction – 844

48.2   The Hygiene Hypothesis – 844

48.3   Alteration of Human Gut Microbiome by Western Diet – 844

48.4   Xenobiotic Residues in Food Commodities – 845

48.5   Genetically Modified Foods – 847

48.6   Food Hybridization and Change in Amino Acid Sequences of Proteins – 848

48.7   Industrial Food Additives – 849
48.7.1   Salt – 849
48.7.2   Sugars – 849
48.7.3   Emulsifiers or Surfactants – 849
48.7.4   Organic Solvents – 850
48.7.5   Nanoparticles – 850
48.7.6   Microbial Transglutaminase (Meat Glue, Food Glue) – 851
48.7.7   Food Colorants – 851
48.7.8   Gums – 852
48.7.9   Mechanisms of Food Additive-Induced Inflammation and Autoimmunities – 853

48.8   Mimicry or Cross-Reactivity: The Mechanism Responsible for Immune Reactivity and Autoimmunities – 854
48.8.1   Cross-Reactivity Between Wheat and Human Tissue – 854
48.8.2   Cross-Reactivity of Dairy with Human Tissue – 855
48.8.3   Cross-Reactivity of Aquaporin with Human Tissue – 856
48.8.4   Cross-Reactivity of Pectin with Human Tissue – 857
48.8.5   Cross-Reactivity of Food Antigens with Diabetes and Thyroid Antigens – 858

48.9   Conclusions: A Word to the Practitioners – 859

References – 860

© Springer Nature Switzerland AG 2020
D. Noland et al. (eds.), *Integrative and Functional Medical Nutrition Therapy*,
https://doi.org/10.1007/978-3-030-30730-1_48

Learning Objectives
- Alteration of the gut microbiome by the modern Western lifestyle is responsible for the rise of food immune reactivities and autoimmunities.
- Exposure to xenobiotics, genetically modified foods, food hybridization, and industrial food additives contributes not only to microbiome alteration but also to the surge of food immune reactivities and autoimmunities.
- Molecular mimicry or cross-reactivity between different foods and human tissue is an additional mechanism for the rise of autoimmunities.
- Detoxification and implementation of dietary protocols based on food-targeted tissue antigens are a strategy to reduce autoimmune reactivities.

48.1 Introduction

In our companion chapter (▶ Chap. 19), we established that the breakdown or failure of immune tolerance due to an unfavorable lifestyle can result in immune reaction to dietary components. Here we will explain different mechanisms that result not only in food immune reaction but possibly also in autoimmunities. For the sake of discussion, I am asking the following questions:
- Why are allergies and food immune reactivities on the rise?
- Why are autoimmune diseases reaching epidemic proportions?

The answers to these interrelated questions are discussed below.

48.2 The Hygiene Hypothesis

A 28-year-old idea claims that the modern lifestyle of Western societies, the widespread use of antibiotics, a zeal for cleanliness, and a lower number of children per family have purged the growth dynamics of gut microbiota and shifted the development of the immune system from Th1 to Th2 imbalances, setting the stage for the possible development of allergy and asthma.

The hygiene hypothesis states that the rise in allergy and asthma that has been observed in affluent countries since the Second World War is caused by reduced "infectious pressure" from the Western lifestyle environment. The mechanism behind this association has been linked to an imbalance in the immune system, favoring pathogenic Th2 immunity in the absence of counterbalancing Th1 immunity, natural killer T cells, or regulatory T cells that are often induced by infections [1, 2]. A recent investigation [3] provided evidence that environmental protective factors can also influence the threshold for allergen recognition by suppressing the activation of epithelial cells (ECs) and dendritic cells (DCs) via induction of the ubiquitin-modifying enzyme called A20. This regulatory mechanism is also seen in the gut, where colonizing microbiota induce the expression of A20 shortly after birth, thus dampening overt inflammatory reaction to the commensals. It is well known that growing up on a farm and being exposed to the farm microbiome or their antigens protects many individuals from allergy, hay fever, and asthma. It seems that chronic exposure to low doses of bacterial antigens activates regulatory T cells, suppressing Th2 immunity to house dust mites and preventing allergy and asthma. Thus, the farming environment protects from allergy by modifying the communication between epithelial cells and DCs through A20 induction [3].

48.3 Alteration of Human Gut Microbiome by Western Diet

Many years ago when I took my clinical microbiology course and started growing different bacteria in the lab, I asked the question why we needed so many different media and why we couldn't have just one medium on which all bacteria could grow. The answer to this question was very simple: because each bacterium thrives on different nutrients. For example, some bacteria grow on simple media such as nutrient agar; others, such as *Enterobacter*, love to grow on agar and sugar and others on agar and potato, agar and rice, agar and soy, agar and beef extracts, or agar and blood. Some fastidious bacteria will not grow without the addition of heart or even brain tissue to the basic agar medium. The best way to explain this is that we are what we eat, and that is why diet alters the human microbiome and why there is such a huge difference in the microbial communities of different populations [4, 5]. There is growing concern that the current prevailing lifestyle trends of modern civilization, most notably the high-fat/high-sugar "Western diet," have negatively altered the genetic composition and metabolic activity of the resident microorganisms in the human gut. Such diet-induced changes to the gut microbiomes are now suspected of contributing to growing epidemics of chronic illness in the developed world, including obesity, inflammatory bowel disease (IBD), and other autoimmunities [6–8]. It has been shown [9] that the short-term consumption of diets composed entirely of animal or plant products alters the microbial community structure and overwhelms interindividual differences in microbial gene expression. The animal-based diet increased the abundance of bile-tolerant microorganisms (*Alistipes*, *Bilophila*, and *Bacteroides*) and decreased the levels of *Firmicutes* that metabolize dietary plant polysaccharides (*Roseburia*, *Eubacterium rectale*, and *Ruminococcus bromii*). The differences between herbivores and carnivores were mirrored in microbial activity, reflecting trade-offs between carbohydrate and protein fermentation. Foodborne microbes from both diets transiently colonized the gut, including bacteria, fungi, and even viruses. Finally, increases in the abundance and activity of *Bilophila wadsworthia* on the animal-based diet

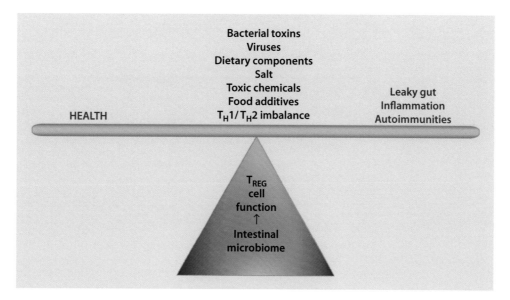

● Fig. 48.1 The host-microbiota homeostatic balance

support a link between dietary fat, bile acids, and the outgrowth of microorganisms capable of triggering gut barrier disruption inflammation and autoimmunities, including IBD and celiac disease (CD) [9, 10].

This is because intestinal dysbiosis is a hallmark of several immune disorders, including IBD and CD. Increased levels of pathobionts may activate the pro-inflammatory pathways that trigger a breakdown in food tolerance and promote the onset of disease. This, in turn, further drives the gut microbial community toward a state of dysbiosis, resulting in a vicious circle that disrupts the host-microbiota homeostatic balance [10] (see ● Fig. 48.1).

Environmental triggers such as viral infections have been proposed to have a role in the pathological role of CD. In support of a relationship between environmental triggers, CD, and non-celiac gluten sensitivity (NCGS), reovirus infection has been shown to trigger inflammatory responses to dietary antigens and to be involved in the development of celiac disease [11]. This is because the reovirus, a so-called avirulent pathogen, can disrupt intestinal immune homeostasis at inductive and effector sites of oral tolerance by suppressing peripheral regulatory T cell (pTreg) conversion and promoting Th1 immunity to dietary antigens. Initiation of Th1 immunity to dietary antigens was dependent on interferon regulatory factor 1 and dissociated from suppression of pTreg conversion, which was mediated by type 1 interferon. The finding of this study supports a role for infection in addition to dietary protein in triggering the development of autoimmune disease [11]. According to the hygiene hypothesis, the increasing incidence of autoimmune diseases in Western countries may be explained by changes in early microbial exposure, leading to altered immune maturation. This is mainly due to microbiome lipopolysaccharide immunogenicity, which may contribute to autoimmunity in humans [12]. For example, the *Bacteroides* species in the microbiota of children from countries with high susceptibility to autoimmunity produce a type of lipopolysaccharide (LPS) with immunoinhibitory properties. These properties may preclude early immune education and contribute to the development of type 1 diabetes and other autoimmune disorders. Therefore, early colonization by microbiota with a capacity of changing the Th1/Th2 balance may be responsible for a lack of immune fitness in many children and adults in Western societies.

48.4 Xenobiotic Residues in Food Commodities

Unfortunately, during the past several decades, we have been presented with the incorrect notion that the chemical insecticides applied to the plants, fruits, and vegetables we eat will not penetrate them but will be washed away. Similarly it is widely believed that when chemicals in our air, food, and water get into our body, they will be out within 24 hours. However, in reality, chemicals originating from the environment or applied to different products form a stable bond with food proteins and other macromolecules. Similarly, some chemicals, after entering the human body, directly or indirectly manage to bind to albumin, hemoglobin, and other tissue proteins. This process – the binding of low-molecular-weight chemicals to proteins or chemical reactivity to proteins – is called haptenation. Haptens such as pesticides must complex with proteins and form neo-antigens in order to be recognized by the immune system. It is during this immune recognition of chemicals bound to human tissue that the immune system will attack the tissue-bound chemicals, setting the stage for autoimmunities [13–16].

One of the best examples of a chemical binding to human body proteins such as albumin or hemoglobin is aflatoxin (see ● Fig. 48.2). Aflatoxins, especially aflatoxin B1 (AFB_1), are potent carcinogens produced by strains of fungi such as

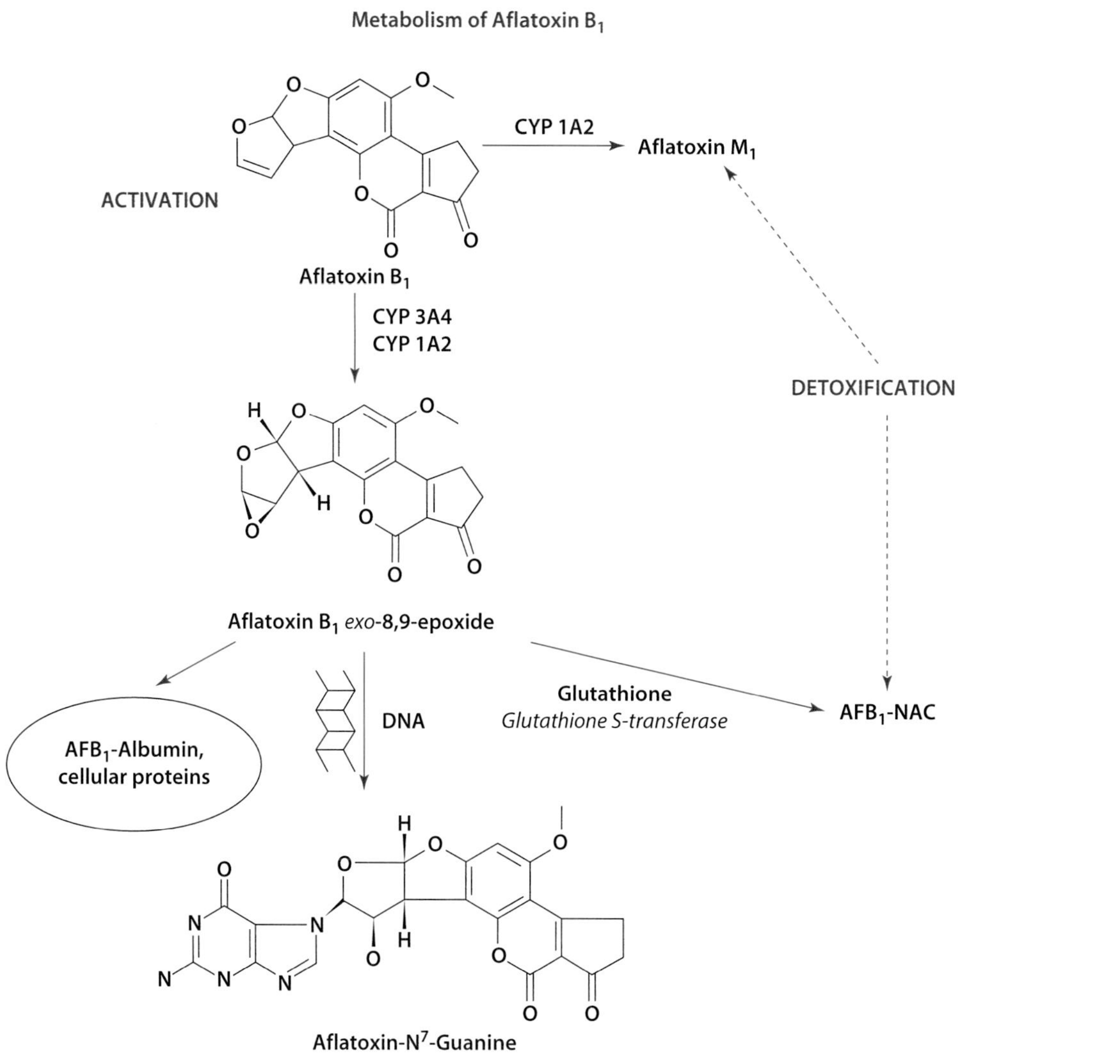

Metabolism of Aflatoxin B_1

□ **Fig. 48.2** Neo-antigen formation between aflatoxin, proteins, and DNA results in antibody production against DNA and proteins, setting the stage for autoimmunities

Aspergillus flavus. This organism grows on important food groups such as peanuts, other ground nuts, maize, and other oilseeds, and the toxin contaminates these foods. Consumption of these aflatoxin-contaminated foods results in high levels of AFB_1, proteins, and DNA adducts, which are the mechanisms responsible for the induction of immune dysfunction and cancer. In many countries there is not much data available regarding exposure to aflatoxin, but in Ghana, when AFB_1-albumin adducts were measured in 61 participants, 100% of them had very high levels [17].

Aflatoxin can form neo-antigens with cellular proteins and DNA; immune response to this combination in certain individuals may result in antibody production against it. In our own study from 2015 [16], we aimed to examine the percentage of blood samples from 400 healthy donors in which chemical agents mounted immune challenges and produced antibodies against HSA-bound chemicals. In fact, we found that at 2 SD above the mean of the participants, 17% produced IgM and 10% produced IgG against aflatoxin-HSA. Furthermore, when we measured these antibodies against 11 other chemicals bound to HSA or neo-antigens – such as formaldehyde-HSA, isocyanate-HSA, trimellitic anhydride-HSA, dinitrophenyl-HSA, bisphenol-A-HSA, tetrabromobisphenol-A-HSA, tetrachloroethylene-HSA, mercury-HSA, mixed heavy metals-HSA, parabens-HSA, and pyrethroid-HSA – in the same 400 subjects, elevation in IgG ranged from 8% to 22%, while IgM ranged from 13% to 18%. We concluded that this protein adduct formation could be one of the mechanisms by which environmental chemicals induce autoimmune reactivity in a significant percentage of the population [16]. Therefore, these types of studies prove the concept that chemicals are not in and out; some percentage of chemicals or their metabolites bind to human tissue antigens and set the stage for autoimmune attack and even cancer. Therefore, a xenobiotic-bound residue is a residue which is associated with one or more classes of endogenous macromolecules. It cannot be disassociated from the natural macromolecule using exhaustive extraction or digestion without significantly changing the nature of either the exocon or the associated endogenous macromolecules

[13]. This change in the endogenous macromolecules by a toxic chemical seems to be responsible for the rise of food immune reaction, allergies and autoimmunities. Therefore, it is the opinion of the senior author of this chapter that the mere detection of chemicals in urine or blood has no pathological significance; the human body has an enormous body burden of chemicals as an ongoing state, and detected chemical levels in blood would indicate exposure to the chemicals, while detection in urine would indicate their excretion or elimination from the body. It is only when the same chemicals or metabolites form neo-antigens with human proteins that they have the capacity to induce autoimmunities. Consequently, measurements of antibodies against chemicals bound to human albumin is the proper indication of the body burden of chemicals. The senior author has developed tests for these chemical antibodies that are offered as a panel from an independent laboratory.

48.5 Genetically Modified Foods

Genetically modified foods (GM foods) are foods that had changes introduced into their DNA using genetic engineering methods. Put another way, they are foods into whose organism genome more genes were introduced in order to confer them with new traits. Genetically modified in crops include corn, soybeans, canola, cotton, fruits, vegetables, and many more. Crops are modified to be resistant to pathogens and herbicides and for better nutrient content or even to increase their shelf life. For instance, *Bacillus thuringiensis* (Bt) is a Gram-positive soil-dwelling bacterium known to have insecticidal properties, used as a biological pesticide. Bt genes are now used in the genetic engineering of plants so that the crop itself will produce pesticides, eliminating the need for pesticide spraying. According to the USDA, the number of fields that used genetically engineered organisms exceeded 17,000 by September 2013 [18]. Without getting into the politics of GM foods, as an immunologist I firmly believe that any changes in DNA molecules also cause changes in RNA and protein structures, and it is for this exact reason that the changes in protein structures of food proteins may cause failure of the tolerance mechanism to the new antigenic materials, resulting in food immune reaction, allergies and autoimmunities (see ◘ Fig. 48.3).

Although combining genetically modified plants with pesticide-associated GM foods allows the protection of desirable crops from pests and weeds by making them insect-resistant or by reducing the desire for nutrients [19], there is a huge concern about the direct threat of genes and residues derived from herbicide-resistant GM crops such as glyphosate and its metabolites. Glyphosate [N-(phosphonomethyl) glycine] is a nonselective herbicide used for the control of a variety of weeds on non-crop land as well as in a great variety of crops. An enzyme named 5-enolpyruvylshikimate-3-phosphate synthase required for the synthesis of many plant metabolites, including some amino acids, is the major target of glyphosate action. Glyphosate binds to this plant enzyme,

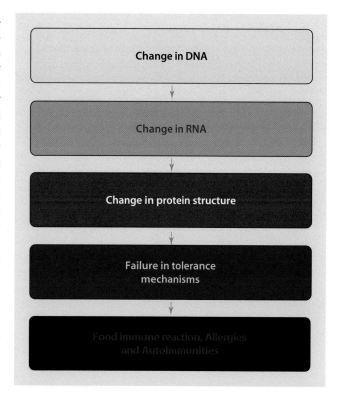

◘ Fig. 48.3 How change in DNA can lead to food immune reactivities and autoimmunities

interferes with its metabolism, and eventually causes cell death. Glufosinate [ammonium DL-homoalanin-4-yl(methyl)phosphinate] is known to be a broad spectrum contact herbicide that is also used to dry out or desiccate crops before harvest. Glufosinate inhibits the activity of the enzyme glutamine synthetase, which is necessary for the production of amino acid glutamine or Q, therefore leading to a decrease in the level of glutamine and cell death [20, 21]. Truncated Cry1Ab gene encoding Cry1Ab toxin is an insecticide produced by the previously mentioned soil bacterium and biological pesticide *Bacillus thuringiensis*, which has been genetically transferred into the maize genome in order to make it more insect-resistant, especially against Lepidoptera infestation [22]. Upon the application of Cry1Ab, it is solubilized by the gut proteases of insect larvae, and then it binds to specific receptors on the brush border of epithelial cells, leading to disruption of the epithelial barrier and larval death [23, 24]. The mechanisms of these chemicals or toxins – especially the binding of glyphosates and glufosinate to plant enzymes, neo-antigen formation, and their interference with amino acid synthesis or the binding of Cry1Ab toxins to the brush border proteins of epithelial cells – raise the following serious questions for immunologists:

- Is there a relationship between the use of these xenobiotics in our food, changes in the human microbiome, and many gut-associated disorders in children and in adults?
- Is there any correlation between the rise of CD, NCGS, Crohn's disease, ulcerative colitis, small intestinal bacterial overgrowth (SIBO), IBS, and the increased use

of genetically modified foods and their associated toxins?
- What is the relationship between GMO-associated xenobiotics in humans and the epidemic of metabolic syndrome and obesity that affects up to 30% of the world population?
- How about the autoimmune epidemic that currently affects 10% of the world population?
- And finally, is there any relationship between the use and consumption of GMO-associated xenobiotics by humans, cardiovascular disease, neurodegenerative disorders, and cancer?

Only these types of questions can pave the way for more research and hopefully the eventual removal of all xenobiotics from our environment, particularly from our foods, since the basis of better health is prevention. There is an urgent need to find an economical and effective way to grow crops and produce food without glyphosate, glufosinate, Cry1Ab-toxin, and other xenobiotics. In search of answers to these valid questions, I came across a very fascinating review article by Anthony Samsel and Stephanie Seneff published in the *Journal of Biological Physics and Chemistry 2016* [24]. In this article I found the answer to the above questions by learning about the mechanism by which these chemicals can replace amino acids such as glycine and glutamine and form new proteins, both in plants and in human tissue. The study showed that many enzymes in plants and in the human body are targets of structural modification. Therefore, this xenobiotic-induced modification of enzymatic structure is the additional mechanism by which our immune system attacks food or tissue proteins that it normally would tolerate or ignore. This may explain the rise of food immune reactions and autoimmunities in epidemic proportions in the industrialized world [24]. In support of this concept, a summary of this 36-page article is provided here:

» Epidemiological studies have revealed a strong correlation between the increasing incidence in the United States of a large number of chronic diseases and the increased use of glyphosate herbicide on corn, soy, and wheat crops. Glyphosate, acting as a glycine analogue, may be mistakenly incorporated into peptides during protein synthesis. A deep search of the research literature has revealed a number of protein classes that depend on conserved glycine residues for proper function. Glycine, the smallest amino acid, has unique properties that support flexibility and the ability to anchor to the plasma membrane or the cytoskeleton. Glyphosate substitution for conserved glycines can easily explain a link with diabetes, obesity, asthma, chronic obstructive pulmonary disease (COPD), pulmonary edema, adrenal insufficiency, hypothyroidism, Alzheimer's disease, amyotrophic lateral sclerosis (ALS), Parkinson's disease, prion diseases, lupus, mitochondrial disease, non-Hodgkin's lymphoma, neural tube defects, infertility, hypertension, glaucoma, osteoporosis, fatty liver disease and kidney failure. The correlation data together with the direct biological evidence make a compelling case for glyphosate action as a glycine analogue to account for much of glyphosate's toxicity. Glufosinate, an analogue of glutamate, likely exhibits an analogous toxicity mechanism [24].

48.6 Food Hybridization and Change in Amino Acid Sequences of Proteins

Due to the extensive hybridization of wheat and a possible change in the amino acid sequences of wheat proteins, we use wheat as the classical example of protein modification and oral tolerance breakdown, leading to an increase in the prevalence of an autoimmune disease in the gut called celiac disease (CD).

The wheat that is universally consumed today is far different genetically from the simple grain man used to eat. Like most modern food plants, it has been modified to increase productivity, yield, nutrient content, and longevity. With all this genetic tampering, it is little wonder that the immune systems of some individuals are sometimes unable to recognize wheat as a harmless and supposedly beneficial food staple. Gluten proteins from wheat can induce CD in genetically susceptible individuals. Specific gluten peptides can be presented by antigen-presenting cells to gluten-sensitive T-cell lymphocytes, leading to CD. A significant increase in the prevalence of CD has been observed during the last few decades. This may be partly attributed to an increase in awareness and to improved diagnostic techniques, but increased wheat and gluten consumption is also considered a major cause. One should also consider the possibility that the genetic modifications made to today's wheat may have made it intolerable to some segment of the population.

To analyze whether wheat breeding contributed to the increase of the prevalence of CD, Van den Broeck et al. [25] have compared the genetic diversity of gluten proteins for the presence of two CD epitopes (Glia-9 and Glia-20) in 36 modern European wheat varieties and in 50 landraces representing the wheat varieties grown up to around a century ago. Looking back over the last five decades, several trends are apparent in wheat consumption: an increase in CD-related T-cell stimulatory epitopes in wheat, an increase in the use of gluten in food processing, and an increase in the consumption of processed foods. Given the relation between the incidence of CD and exposure to cereals, it cannot be ruled out that an increased content of T-cell stimulatory epitopes has also contributed to the increased prevalence. Considering the epitope impact of CD patients of the major immunodominant Glia-α9 epitope, it is concluded from these data that in general the toxicity of modern wheat varieties has increased. It stands to reason that reduction of T-cell stimulatory epitopes in wheat may directly contribute to increasing the quality of life of many individuals. Further the application of advanced breeding technologies, including resynthesizing of hexaploids and specific gene silencing, will additionally be helpful [25]. This increase in the toxicity of modern wheat varieties and its impact on alpha-gliadin epitope may be one reason for oral tolerance failure to these new peptides and the increased prevalence of food immune reaction and autoimmunity, such as celiac disease and other associated autoimmunities. Wheat hybridization and its effect on the amino acid sequences of alpha-gliadin is only an example of changes in the protein structure. These changes may apply to other hybridized products whose AA sequences may have changed for better or worse.

48.7 Industrial Food Additives

The role of environmental triggers in the development of autoimmune disease (AD) is becoming clearer year after year [26, 27]. Epidemiological data provide strong evidence of a steady rise in AD throughout Westernized societies over the last three decades. Multiple sclerosis, type 1 diabetes, inflammatory bowel diseases (mainly Crohn's disease), systemic lupus erythematosus, primary biliary cirrhosis, myasthenia gravis, autoimmune thyroiditis, hepatitis, rheumatic diseases, bullous pemphigoid, and celiac disease are several examples [28–33].

For example, in the USA, type 1 diabetes increased 3–4% per annum, and undiagnosed CD mortality increased by fourfold [34, 35]. When autoimmune diseases were grouped as a major disease class, the highest net increase percentage per year was noted in the neurological autoimmunities (14%), with gastrointestinal autoimmunities at 6%, endocrinological autoimmunities at 5%, and rheumatological autoimmunities at 3.5%. This rise in the incidence of AD was associated with the percentage increase per year of industrial food additive usage as described in an elegant review article [36]. In that review it was hypothesized that commonly used industrial food additives abrogate human epithelial barrier function, thus increasing intestinal permeability through the open tight junction, resulting in the entry of foreign immunogenic antigens and activation of the autoimmune cascade [36]. The following paragraphs will deal with eight nutritional ingredients being increasingly added during industrial food processing that find their way to market shelves. These food additives have the capacity to affect the gut microbiome, inducing leaky gut, changing the integrity of the immune system, and impacting overall human health.

48.7.1 Salt

Salt is considered a silent killer. Excessive consumption is associated with hypertension, strokes, left ventricular hypertrophy, renal diseases, obesity, renal stones, and stomach cancer. Overconsumption of salt is very real, crossing multiple populations and demographics, ages, gender, and countries. Unfortunately, sodium intake around the world is five- to tenfold in excess of physiological need, and its indirect use is dominated by the sodium added to manufactured foods.

Cereals and baked goods are the single largest contributors to dietary sodium intake in UK and US adults. In Japan and China, the largest source remains salt added at home and in soy sauce. The salt content in processed foods can be more than 100 times higher than that in similar homemade meals [37]. However, while everyone feels that they already know about the harmful effects of salt and are apparently willing to risk their health for added flavor, they might think twice if they actually knew about the impact that excessive consumption of salt has on their immune system. Excess intake of salt can affect the innate immune system, in particular macrophage function, and affects the differentiation of naïve $CD4^+$ T cells. Under normal conditions, the naïve T cells have a significant plasticity that, depending on the presence of different cytokines, enables them to differentiate into the various types of T helper cells. High concentrations of salt – in addition to change in osmolarity, the influence of IL-23 receptor signaling, and the activation of various enzymes – drive the expression of Th17-associated cytokines and the formation of the pathogenic Th17 phenotype. This immune pathway plays a pivotal role in autoimmune disease [26]. Such increased salt concentrations were shown to drive neuropathy in a mouse model of multiple sclerosis by the induction of pathogenic Th17 cells [38] (see Fig. 48.4).

48.7.2 Sugars

The increase in the availability of sweeteners, especially sugar, during the past four decades has contributed significantly to the epidemic of obesity not only in the USA and central Europe but also in countries that used to adhere to the Mediterranean diet [39].

Glucose is a known absorption enhancer. A major portion of intestinal glucose absorption occurs through tight junctions (TJ), not by saturable transcellular active transport. This requires increased junctional permeability, a very high intraluminal glucose concentration, and a sufficient osmotic gradient to promote volume flow [40]. Glucose increases permeability and produces changes in the distribution of the main protein of the TJ in the human cell line Caco-2, indicating intercellular leakage. This effect of sucrose on increased intestinal permeability to macromolecules has been shown in patients with Crohn's disease and with a higher intake of sugars and refined carbohydrates [41]. Therefore, we should not be surprised that the overconsumption of sugar is partially responsible for changes in the microbiome, inflammation, and autoimmunities.

48.7.3 Emulsifiers or Surfactants

Emulsifiers, also known as surfactants, are a group of substances that stabilize an emulsion; they do this by concentrating at the interface between oil and water and reducing the surface of interfacial tension. Emulsifiers are considered to be the fastest-growing segment of the food

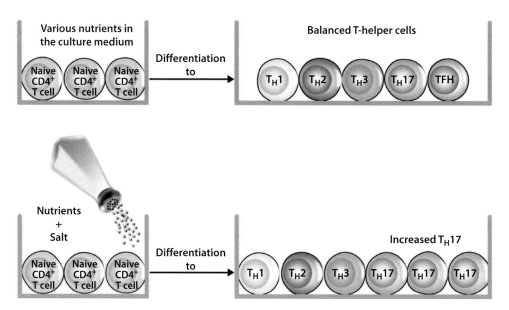

◘ Fig. 48.4 High concentrations of salt affect the differentiation of naïve T cells toward pathogenic Th17 cells

additive industry. They are widely used in butter, margarine, ice cream, liquors, meat, gums, coffee, beverages, dairy, confectionery, baked goods, fat, oil, sauces, chocolate, and many other products. The emulsifier market is largely driven by diglycerides and derivatives, lecithin, and stearoyl lactylates. Other emulsifiers include polyglycerol esters, polyglycerol polyricinoleate, polysorbate, and sucrose esters [42, 43]. Numerous synthetic surfactant food additives have been shown to increase intestinal permeability through paracellular and/or transcellular mechanisms; some of these additives were also shown to inhibit P-glycoprotein, which is known as a membranous transport protein. For example, sucrose monoester fatty acids, which are major and potent surfactants used in the food industry, including in infant formula, induce actin disbandment and structural separation of TJ or enhanced permeability to large macromolecules. Surface-active compounds, like oleic and docosahexaenoic acids, compromise the integrity of the intestinal epithelium and enhance the paracellular absorption of unwanted molecules. Moreover, self-microemulsifying systems used to improve drug delivery via a microemulsion achieved by chemical rather than mechanical means were found to open the TJ and change the distribution of ZO-1 and actin and therefore play a significant role in the induction of autoimmunities. In Japan, a major player in the food industry's utilization of emulsifiers, a positive correlation was shown between the annual sales of emulsifiers for food and beverage production and an increased incidence of Crohn's disease [44, 45].

48.7.4 Organic Solvents

Organic solvents are a group of liquid chemicals that are capable of dissolving other substances. Gluten, for instance, is not water-soluble, but the addition of alcohol makes it completely soluble. Many proteins can likewise be dissolved in hexane, acetone, toluene, turpentine, benzene, or dimethyl sulfoxide. Organic solvents are considered toxic chemicals and are dangerous to human health and as such are labeled as poisons. They are used for solubilization, extraction of active ingredients, or removal of unwanted substances. For example, for the extraction of 60 million tons of soy oil and other edible oils, hexane extraction technology is used. This extensive use of solvents in the food industry was reported by one study as a serious risk factor for autoimmune diseases including multiple sclerosis, systemic sclerosis, vasculitis, and liver autoimmunity [46]. The author concluded that individuals with non-modifiable risk factors for autoimmune disease should avoid any exposure to organic solvents in order to prevent the risk of developing autoimmune disease.

48.7.5 Nanoparticles

Nanoparticles are very small molecules, those with dimensions between 1 and 100 nm. Due to their tiny size, such materials usually acquire unusual chemical and biological properties and functions; they are prepared by nanoprecipitation of poly(lactic-co-glycolic acid) as carrier material and surface-modified by methoxy poly(ethylene glycol) and chitosan. While nanomolecules offer solutions to technological problems that cannot be achieved by conventional systems, the future impact of these nanofood particles on human health is not yet clear. Many reports confirm that nanoparticulate systems with unique properties can increase the transport of poorly water-soluble compounds across the GI barrier by enhancing paracellular transport via opening of the TJ and enhancing permeability. For example, thiolated chitosan is used for enhancing gut permeability and mucoadhesivity and increasing the intestinal absorption of active agents. Permeation studies have shown that nanoparticles open the tight junction through ZO1 redistribution, F-actin reorganization, claudin-4 downregulation, and enhancement of the paracellular permeation of macromolecules [47–49].

Because of this, there are health concerns associated with increasing the oral bioavailability of bioactive components that exhibit deleterious effects when consumed at excessive levels. If one of these bioactive components normally has a very low bioavailability, its absorption by the human body can be increased substantially by encapsulating it within lipid nanoparticles; it could subsequently have unexpected toxic effects. Nanotechnology usage in the food sector has been hindered by concerns about the safety of the engineered nanoparticles, as well as ethical, policy, and regulatory issues [50]. Their safe use in food requires knowledge of their absorption, distribution, metabolism, excretion, and toxicological profiles. Even though the FDA is currently monitoring developments in this relatively new field with undeniably exciting possibilities, the truth is that the safety of extensive use of nanomaterials is still largely unknown.

48.7.6 Microbial Transglutaminase (Meat Glue, Food Glue)

Transglutaminases form a group of enzymes belonging to the class of transferases. Transglutaminases catalyze the formation of an isopeptide bond between two different proteins by cross-linking glutamine to lysine. In the human body, transglutaminase-2 is involved in celiac disease, transglutaminase-3 in dermatitis herpetiformis, and transglutaminase-6 in gluten ataxia [51].

Transglutaminase is also biosynthesized by microbes such as *Streptoverticillium* SP. This transglutaminase is named microbial transglutaminase (mTG), more commonly called "meat glue." mTG has become a very useful tool for modifying the functionality of many food proteins in food products [52]. The extensive use of mTG in the food industry, whether to glue two different pieces of meat or other food components together, became possible due to transgenic procedures and mass production using microbes [53, 54]. Currently this enzyme is applied to a variety of food products and occupies a significant segment of the food industry. Its purposes include improvement of the texture, appearance, hardness, and preservability of meat; increased hardness of fish products; improved quality and texture of milk and dairy products; decreased calories and improved texture and elasticity of sweet foods; and protein film stability and appearance and improved texture and volume in the bakery industry. The amount of mTG used for each kilogram of food is about 50–100 mg; this means that on the average every one of us is exposed to about 15 mg of mTG each day [36]. In patients with intestinal permeability to large molecules, this daily exposure to mTG is enough to be detected by the immune system and to result in the production of IgA, IgG, or IgM antibodies against it. These antibodies may cross-react with human transglutaminases that play a role in CD, NCGS, dermatitis herpetiformis, and gluten ataxia. Furthermore, the mTG antibodies may contribute to false-positive results in CD and NCGS. Indeed, in our laboratory we performed tests for IgG and IgA transglutaminase antibodies on many blood specimens; we detected antibodies against tTGs, but not against any components of wheat proteins. Therefore, this autoimmune reactivity against human tTGs could be due to extensive consumption of mTG by individuals with leaky gut syndrome. In this regard some observations about the role of mtG in increased intestinal permeability are presented:

- mTG may change the balance of the microbiome, and the subsequent release of toxins could increase intestinal permeability to macromolecules, including mTG.
- By mediating nonspecific linking of sulfur containing amino acids, cysteine, and methionine, mTG contributes to enhanced TJ permeability.
- The cross-linking of different proteins by the use of meat glue or food glue acts as a major modifier of the physicochemical structure, antigenicity, and other characteristics of food proteins, which may make them more resistant to digestive enzymes.
- mTG has the ability to catalyze lipidation of proteins, thus providing them with emulsifying activity [42]. Moreover, mTG induces cross-linking of various dietary proteins originating from casein, pork myofibrils, peanut, and fish, which has been shown to improve their emulsifying capacity.
- mTG is also used as biological glue, in cross-linking nanoparticles, and in delivery systems. The potential of the nanoparticles as the enhancers of intestinal permeability was discussed in an earlier section.
- Multiple mTG-linked proteins, in bread and other food products, have the capacity to break the oral tolerance mechanisms, resulting in possible CD, NCGS, and other food immune reactivities [55–59].
- mTG is used in a vast array of food products, from steaks to sushi to faux crab, to hold pieces of food together to achieve a more appealing appearance. In the case of steaks, miscellaneous pieces of meat are mixed together with meat glue and placed on fat, then vacuum-packed together, and left overnight. A cross-section cut from the resulting mass the next morning looks convincingly like a ribeye steak.

48.7.7 Food Colorants

The way food looks has a lot to do with how much we want to eat it. It should surprise no one, then, that food manufacturers and purveyors go to a great deal of trouble to make the food we buy from them look appetizing. Foods with bright colors and fresh-looking appearances appeal to our senses, but in most cases these are achieved with artificial food colorants. Food dyes are made from petroleum – hardly an appetizing idea. Synthetic food colorants are cheaper, more easily available, and last longer than natural colors, and they can achieve hues not possible naturally. But as the use of artificial food coloring has increased through the years, so has the reported incidence of many allergic and other immune-

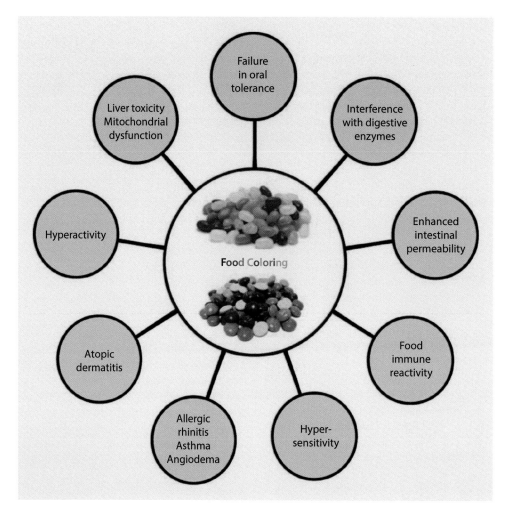

Fig. 48.5 Food colorants may make foods like Snow White's poison apple: beautifully delicious to look at but deathly toxic on the inside. Above are disorders that have been linked to food colorants

reactive disorders. In the past 50 years, the amount of artificial food colorants in food has increased by 500%. Alarmingly, there has also been an accompanying increase in child behavioral problems such as ADHD, ADD, and aggression. The molecules of food dyes are small and are difficult for the immune system to defend against. They can also bind to food or body proteins, and thus disguised are able to slip past the barriers and disrupt the immune system in various ways. They can activate the inflammatory cascade; they can induce intestinal permeability to large antigenic molecules; and they can lead to cross-reactivities, autoimmunities, and even neurobehavioral disorders (see ▶ Fig. 48.5). This is not as shocking as the legal amounts of artificial colorants actually allowed by the FDA in the foods, drugs, and cosmetics that we consume and use every day [60]. Even though the Color Additive Amendments of 1960 required that only color additives listed as suitable and safe could be used in foods and had no Generally Recognized as Safe (GRAS) or grandfather provision, in effect all color additives at the time were actually provisionally approved in order to avoid a ban of all colorants, which were then largely untested [61]. The consuming public unfortunately is still largely unaware of how widespread the use of food colorants is or the fact that their consumption carries the possibility of developing one of these devastating diseases:

- Failure in oral tolerance
- Interference with digestive enzymes
- Enhanced intestinal permeability
- Live toxicity
- Mitochondrial dysfunction
- Food immune reactivity
- Hypersensitivity
- Atopic dermatitis
- Allergic rhinitis, asthma, angioedema
- Neurobehavioral disorders
- Interference with neurotransmission
- Reproductive abnormalities

48.7.8 Gums

You shouldn't think of gums as just something kids chew or that unpleasant gunk that won't come off your shoe. Gums make up perhaps the most widely used and traded category of non-wood plants that are not consumed directly as foods, fodders, and medicines. In 1993 alone the world market for gums as food additives was about 10 billion US dollars. They have been used for thickening and stabilizing since ancient times and enjoy widespread use in everything from food to pharmaceuticals to printing and adhesives. There are different kinds of gums. Seed gums come from plant seeds and have versatile applications in paper, textiles, petroleum, drilling, food, pharmaceutics, cosmaceutics, explosives, and many more. Exudate gums come from plant sap and were used as far back as 5000 years ago as thickening and stabilizing agents. Marine gums come from seaweed and are used in a variety of foods such as "ready-mix" cakes and "instant" pie fillings. Microbial gums are sometimes referred to as "synthetic gums" since they are fermented from plants by bacteria and are the products of biosynthesis. Most gums are composed of complex and variable mixtures of oligosaccharides, polysaccharides, and glycoproteins with an extremely high-molecular-weight polysaccharide attached to a hydroxyproline-rich polypeptide backbone that accounts for about 2% of the molecular size. The result diagrammatically looks something like a bottlebrush (see ◘ Fig. 48.6). As opposed to food colorants, gum molecules are huge (200–2000 kDa), comprising from 4000 to 40,000 amino acids. Their molecular structure is similar to those of a number of common foods, so there is a possibility of cross-reactivity. Gums are generally recognized as safe by the FDA, but they do have a troublesome history of association with sensitive or allergic reactions [62]. To investigate the possible immunogenic effects of these supposedly safe gums on the general population, our own lab conducted a study using the sera of 288 nominally healthy individuals aged 18–65 years obtained from a commercial source [62]. The sera was screened for IgG and IgE antibodies against extracts of mastic gum, carrageenan, xanthan gum, guar gum, gum tragacanth, locust bean gum, and β-glucan, using indirect enzyme-linked immunosorbent assay (ELISA) testing. Of the 288 samples, 4.2–27% of the specimens showed a significant elevation in IgG antibodies against various gums. For the IgE antibody, 15.6–29.1% of the specimens showed an elevation against the various gums. A significant percentage of the specimens, 12.8%, simultaneously produced IgE antibodies against all seven tested extracts (see ◘ Table 48.1). The results showed that there may be apparently healthy individuals who may actually suffer from hidden food immune reactivities and sensitivities to gums. These findings indicate not only that is there a significant percentage of the healthy population exposed to various gum products but also that these individuals immunologically react against them and produce IgE-mediated and non-IgE-mediated immune reactivity.

48.7.9 Mechanisms of Food Additive-Induced Inflammation and Autoimmunities

The use of food colorants, gums, meat glue, food glue, organic solvents, glucose, salt, emulsifiers, and nanoparticles is extensively increasing from year to year. This widespread rise in the use of food additives may be for the improvement of food quality, as the manufacturers claim, but there is one thing to consider. All these food additives not only increase intestinal permeability and the transfer of macromolecules into the circulation but also contribute to neo-antigen formation, oral tolerance failure, and chemical and food immune reactivities. In fact, tight junction dysfunction, zonulin release, and the production of occludin/zonulin antibodies and other tight junction proteins have been described in autoimmune diseases [36, 63–66].

It is hypothesized that commonly used industrial food additives cause failure in the oral tolerance mechanism and abrogate human intestinal barrier function. This results in the entry of foreign immunogenic antigens, including neo-antigens, into the circulation, resulting in the production of antibodies against them and activation of the autoimmune cascade (see ◘ Fig. 48.7). While we are enhancing our knowledge about the mechanism responsible for the contribution of food additives to intestinal permeability and autoimmunities, exposure to food additives should be a part of the medical history taken from individuals with possible

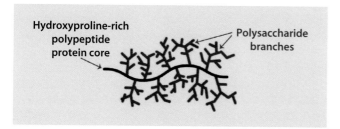

◘ Fig. 48.6 The bottlebrush structure of gums. Most gums are composed of complex and variable mixtures of oligosaccharides, polysaccharides, and glycoproteins with an extremely high molecular weight polysaccharide attached to a hydroxyproline-rich polypeptide backbone that accounts for about 2% of the molecular size

◘ Table 48.1	% Elevation in IgG and IgE antibodies against various gum antigens in 288 healthy subjects at 2 SD above the mean						
Antibody	Mastic gum	Carrageenan	Xanthan gum	Guar gum	Gum tragacanth	Locust bean gum	β-Glucan
IgG	42/288 (14.6)	78/288 (27)	30/288 (10.4)	12/288 (4.2)	15/288 (5.2)	33/288 (11.5)	18/288 (6.2)
IgE	84/288 (29.1)	57/288 (19.8)	48/288 (16.6)	51/288 (17.7)	48/288 (16.6)	45/288 (15.6)	66/288 (22.9)

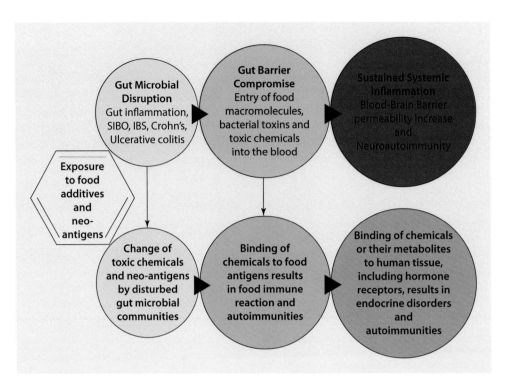

Fig. 48.7 Mechanism of food additive-induced gut inflammation, barrier disruption, food immune reaction, endocrine disorders, and autoimmunities

autoimmunities. This can facilitate the implementation of a program to decrease exposure to food additives and other chemicals in order to improve a patient's chance to recover from devastating and life-altering autoimmune diseases.

48.8 Mimicry or Cross-Reactivity: The Mechanism Responsible for Immune Reactivity and Autoimmunities

48.8.1 Cross-Reactivity Between Wheat and Human Tissue

As an immunologist, although I was familiar with and studied molecular mimicry or cross-reactivity between pathogens and human tissue [67], I have to admit that at one time I was not very knowledgeable about the cross-reactivity of food proteins with human tissue antigens. However, in 2002, I became familiar with an article in which investigators used commercially available IgA antigliadin antibody to show reactivity with human cerebellar and rat CNS tissue by indirect immunohistochemistry [68, 69]. They demonstrated that IgA antigliadin antibody reacted strongly with both rat and human cerebellar tissue. Furthermore, sera from 12 of 13 patients with gluten ataxia were able to strongly stain Purkinje cells. With this indirect evidence, it was suggested that patients with gluten ataxia have antibodies against Purkinje calls, and humoral immune responses are involved in the pathogenesis of gluten ataxia [69]. However, this cross-reaction between cerebellar tissue, Purkinje cells, and gliadin was done by using impure commercially available whole rabbit serum made against the gliadin molecule and not against its immunodominant peptides. My colleagues and I conducted our own study using affinity-purified antibodies made against the gliadin peptide and demonstrated its reactivity against the recombinant brain protein of Purkinje cells [70]. We showed that the molecular target for these antibodies was an epitope consisting of six amino acids (VPLLED) expressed predominantly in neuroectodermal tissue, as well as in α-gliadin peptide. We concluded that molecular mimicry between gliadin and cerebellar peptide may be a mechanism by which dietary peptides induce autoimmunity. Furthermore, we used highly specific monoclonal antibody made against α-gliadin peptide and reacted it against 180 food and 24 human tissue antigens [71].

We demonstrated a significant cross-reactivity between α-gliadin 33-mer and human tissues, including cytochrome p450, glutamic acid decarboxylase, collagen, ovary, thyroid peroxidase, asialoganglioside, myelin basic protein (MBP), cerebellar tissue, and synapsin [71] (see Fig. 48.8). With the degree of cross-reactivity that has been shown between gliadin and neuronal tissues, healthcare professionals can link gluten reactivity to autoimmune reactivities that target the nervous system.

A different study was able to establish that gliadin peptides carrying the QQQPFP epitope interact directly with actin or smooth muscle, which can alter the actin cytoskeleton and possibly contribute to autoimmune reactivity to actin [72]. Thus, untreated patients who continue to consume wheat will have circulating IgG and IgA antibodies to gliadin, which can cross-react with human tissue and lead to autoimmunity within and beyond the gut. Moving away from the

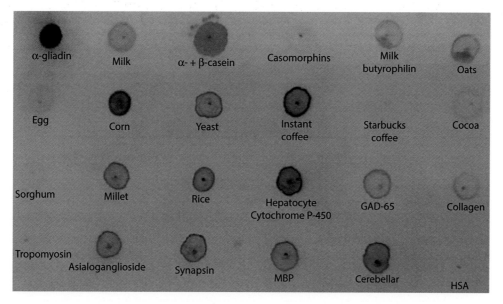

■ Fig. 48.8 Reaction of polyclonal anti-α-gliadin 33-mer to gliadin and various tissue and food antigens by dot blot

gluten family of proteins found in wheat, it is important to discuss yet another wheat trigger of autoimmunity, wheat germ agglutinin (WGA). According to Freed [73], WGA has the capacity to bind to multiple tissues in the body, from the stomach and colonic mucosa to myelin. WGA has an affinity for human tissues. It binds to the tissue, sometimes generating a neo-antigen. The immune system identifies the WGA bound to tissue, where it does not belong, and begins the process of eliminating the WGA. In the attempt to destroy the WGA, tissue is also broken away and enters the bloodstream. Tissue proteins in the circulation ignite production of antibodies against the tissue. Soon, autoantibodies begin attacking the tissue to which the WGA was bound. Untreated patients who continue to consume wheat will have circulating WGA, which can bind to human tissue and lead to autoimmunity within and beyond the gut.

48.8.2 Cross-Reactivity of Dairy with Human Tissue

Currently, public attention seems to have shifted focus from the old "star" of problematic foods, dairy, to the new media darling, gluten. However, dairy still tops the list of antigenic foods, especially for children. In studies previously reviewed by the senior author of this chapter, researchers found that early consumption of cow's milk may present a risk for the development of certain autoimmune diseases, such as Behçet's disease, CD, Crohn's disease, lupus, uveitis, and type 1 diabetes (see ■ Fig. 48.9). They also concluded that exposure to cow's milk can activate immune responses that lead to autoimmunity.

In the case of diabetes, the protein β-casein peptide in cow's milk has been shown to have antigenic similarity to islet cell autoantigen [74] (see ■ Fig. 48.10); thus, cow's milk may trigger diabetes in some individuals and different auto-

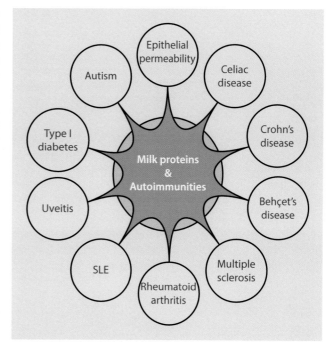

■ Fig. 48.9 Milk-related autoimmunities. Milk proteins have been linked to a variety of disorders that affect different tissues and organs. SLE = systemic lupus erythematosus

immunities in others. Milk has far-reaching effects. It has been shown that increased milk consumption correlates with a higher incidence of multiple sclerosis (MS). Furthermore, in a comparison of MS mortality rates to food consumption, a high correlation between milk consumption and MS mortality has been found [75–79].

The connection between milk and MS appears to be due to molecular mimicry or cross-reactivity. The basic protein antibody to human myelin showed a calcium-mediated interaction between the recombinant protein and caseins in milk

Fig. 48.10 Amino acid sequence similarity between casein and islet cells. A structural similarity exists between β-casein protein in cow's milk and islet cell antigen. That fact supports the link between milk reactivity and diabetes

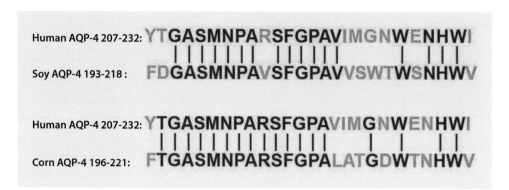

Fig. 48.11 Human AQP4 shares homology with plant aquaporin. Human aquaporin-4 (AQP4), found in astrocytic food processes within the brain and nervous system, has very significant similarities in amino acid sequence to four food aquaporins: soybean, corn, spinach, and tomato. Antibodies produced against these food aquaporins may result in autoimmune attacks on human AQP4, leading to neuroautoimmunity. Only the similarities with soy and corn are shown above

from transgenic (genetically modified) cows [80]; and milk butyrophilin, a milk-fat protein, has a structural similarity to myelin oligodendrocyte glycoprotein (MOG) [81, 82]. Guggenmos et al. [83] found that one-third of patients with MS had immune reactivity to both MOG and milk butyrophilin. For this reason we screened 400 so-called healthy subjects for the presence of antibodies against different wheat and milk components and neural antigens and found that approximately half of the sera with antibody elevation against gliadin reacted significantly with GAD-65 and cerebellar peptides. About half of the sera with elevated antibodies against α + β-casein and milk butyrophilin also showed antibody elevation against MBP and MOG. MBP and MOG are very commonly detected in the blood of patients with MS. Our inhibition studies showed that only two out of four of the samples with elevated cerebellar or MOG antibodies could be inhibited by gliadin or α + β-casein, confirming individual variation in epitope recognition.

From this study we concluded that in about 20% of the so-called healthy subjects, the oral tolerance mechanism was dysfunctional; about half of these individuals (about 10%) produced antibodies against MBP and MOG. Detection of high levels of antibodies against neural antigens may put these individuals at risk of developing neuroimmune disorders [84]. In addition to neural tissue antigens, casein from milk has been shown to be an essential co-factor in the induction of antibodies against C-terminal SmD1 peptide in systemic lupus erythematosus [85]. Based on these results, it is the recommendation of the senior author that food antibodies should be measured by a reliable methodology and that when they are detected in the blood, items such as gluten and dairy should be removed from the diet of individuals in the early stages of autoimmunity, as they could otherwise exacerbate or progress the autoimmune disorder. Reliable food panels can be obtained from only a few specialty labs offering this type of testing.

48.8.3 Cross-Reactivity of Aquaporin with Human Tissue

Neuromyelitis optica (NMO) is a neurologic disease greatly similar to MS and is characterized by simultaneous inflammation of the optic nerve and the spinal cord. Aquaporin-4 (AQP4) has been implicated as a triggering factor in the development of NMO. This water channel protein is expressed in the astrocytic foot processes that line the nervous system's side of the blood-brain barrier (BBB). AQP4 is also contained in some plants, including corn, soy, spinach, tomato, and even tobacco. Indeed, food-sourced AQP4 has amino acid sequences similar to human AQP4 (see Fig. 48.11), enough to cause significant cross-reactivity [86–89]. Highly stable in digestion, food aquaporins can survive as intact proteins. In cases of increased intestinal permeability, these intact aquaporins can infiltrate the body and become antigenic, and the immune reaction against them could ignite antibody production. If those cross-reactive antibodies penetrate the BBB, they could attack aquaporin in the brain astrocytes, resulting in NMO [89]. In addition to aquaporin antibody elevation in NMO, we reported a significant elevation in IgG, IgM, and IgA antibodies against human AQPS in patients with relapsing-remitting MS (RRMS) [90]. Because of this mechanism of cross-reactivity between soy, corn, spinach, tomato, and human AQP4, it is suggested that patients with neuroimmune disorders eliminate these food items from their diet if antibodies against these foods are detected in their blood.

Because these cross-reactive antibodies cross the barrier in susceptible individuals, the immune response could result in NMO and/or MS. Indeed, when the sera of NMO patients were applied to both AQP4 and various plant peptides, a significant reactivity was observed against both human and plant AQP4 [90]. These results further delineate the role of the environment in NMO etiology. These naturally expressed proteins should be exploited in therapeutic interventions, such as the sublingual low-dose introduction of dietary antigens [91] and the development of guidelines for dietary modifications in NMO and other neuroimmune disorders. ◻ Figure 48.12 shows the association between various dietary proteins, peptides, and environmental AQP4 and the development of NMO and other neuroautoimmune disorders including MS.

48.8.4 Cross-Reactivity of Pectin with Human Tissue

Researchers have demonstrated that carbohydrate antigen-specific autoantibodies may play a role in rheumatoid arthritis. Using pectin, they found arthritogenic potential for α1, 4-polygalacturonic acid (PGA) moiety in many pectin-containing foods such as apples, quince, oranges, grapefruits, and berries. The researchers concluded that "the PGA cross-reactive moiety represents a major autoantigen in the joints and can be targeted by autoantibodies capable of triggering arthritogenic responses in vivo" [92]. A variety of food proteins can trigger autoimmunity. Whether by causing molecular mimicry or cross-reactivity, or by binding to tissues, the end result is autoimmune reactivity to human tissues. Clearly, dietary protocols based on targeted tissues can be implemented as a strategy to reduce tissue autoimmune reactivity and even to prevent the onset of disease when a patient has a known genetic susceptibility for autoimmunity.

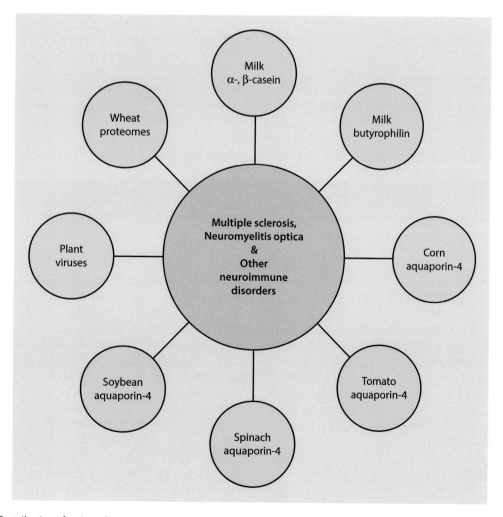

◻ Fig. 48.12 Contribution of various dietary proteins/peptides/environmental AQP4 in the development of neuroautoimmune disorders

48.8.5 Cross-Reactivity of Food Antigens with Diabetes and Thyroid Antigens

In search of additional cross-reactivity between human tissue target antigens such as diabetes and thyroid, we purchased a series of monoclonal and affinity-purified polyclonal antibodies and reacted them with 204 different food antigens [93, 94].

For diabetes we measured the reactivity of antibodies to insulin, insulin receptor alpha (IR-A), insulin receptor beta, zinc transporter 8 (ZnT8), tyrosine phosphatase or islet antigen 2 (IA2), and glutamic acid decarboxylase (GAD) 65 and 67 against 204 dietary proteins that are commonly consumed. These included unmodified (raw) and modified (cooked and roasted) foods, herbs, spices, food gums, brewed beverages, and additives. We identified strong to moderate immunological reactivity with antibodies against insulin receptor alpha (IR-A), ZnT8, IA2, GAD-65, and GAD-67 with several dietary proteins. For IR-A the strongest reactions were with milk butyrophilin (5+), potato (3+), amaranth (3+), quinoa (2+), and potato (1+); for GAD-65 the most reactive were buckwheat (3+), amaranth (3+), rice (3+), corn (3+), yeast (3+), potato (2+), quinoa (2+), and oats (2+) (◘ Table 48.2).

With IA2 antibody, the strongest reactions were with seaweed (5+), guar gum (5+), pea lectin (3+), cooked white and brown rice (3+), and fish (2+).

With ZnT8 antibody, the strongest reactions were with seaweed (5+), cooked lentil (5+), pea protein (5+), guar gum (4+), wheat (4+), cooked pea (4+), oleosin (3+), roasted peanut (3+), and fish (3+).

The results of our study identified immune reactivity between antibodies to insulin, insulin receptors, islet cell antigens, and food antigens. Potential tissue antibody binding with various food antigens or food antibody binding to specific pancreatic sites can lead to the possibility that some dietary proteins may play an antigenic role with autoimmune diabetes.

Even though many of the individual proteins in these groups may be considered safe or of low glycemic index, the consumption of these foods by a sensitive or predisposed individual may trigger immune reactions or autoimmunities.

Regarding the thyroid antigens, we identified immune reactivity between dietary proteins and target sites on the thyroid axis that included thyroid hormones, thyroid receptors, enzymes, transport proteins, thyroid-stimulating hormone (TSH) receptor, 5′deiodinase, thyroid peroxidase, thyroglobulin, thyroxine-binding globulin, thyroxine, and triiodothyronine against the same 204 purified dietary proteins commonly consumed in cooked and raw forms. Specific antigen-antibody immune reactivity was identified with several purified food proteins with triiodothyronine, thyroxine, thyroglobulin, and 5′deiodinase.

We selected sixteen foods to see if the same immune reactivity with a specific food – for instance, latex hevein, elicited by a monoclonal antibody made against T3 – would occur with a different monoclonal antibody. The data clearly shows that each monoclonal antibody presents different patterns of immune reactivity to the food antigens (see ◘ Table 48.3). In fact, TBG had no reactions to any food at all.

Our laboratory study found dietary proteins that share amino acid sequence homology with thyroid tissue antigen and have the potential to play a role in immune reactivity with thyroid target sites. The result was a significant list of many food proteins that demonstrated immune reactivity with thyroid antigens. This potential antigen-antibody against specific thyroid axis sites reacting with food antigens can lead to the possibility that some dietary proteins may play an antigenic role in autoimmune thyroid disease.

The results of our research provide a list of susceptible dietary proteins that may immunologically impact thyroid interactions and warrant further study. They also provide a first step in narrowing down a list of specific dietary proteins that, due to protein cross-reactivity, may potentially have an impact on autoimmune thyroid disease and diabetes [93, 94].

◘ **Table 48.2** Example of 17 selected foods and their degree of reactivity with antibodies made against diabetes target antigens

	Anti-insulin receptor-α	Anti-GAD-65	Anti-islet antigen 2	Anti-ZnT8
Buckwheat	+	+++	–	–
Potato	+++	++	–	–
Amaranth	+++	+++	–	–
Quinoa	++	++	–	–
Rice	–	+++	+++	–
Corn	–	+++	–	–
Oats	–	++	–	–
Yeast	–	+++	–	–
Milk butyrophilin	+++++	–	–	
Seaweed	–	–	+++++	+++++
Guar gum	–	–	+++++	++++
Wheat	–	–	–	++++
Pea lectin	–	–	+++	+
Pea protein	–	–	–	+++++
Lentil	–	–	–	+++++
Peanut	–	–	–	+++
Fish	–	–	++	+++

Table 48.3 Example of 16 selected foods and their degrees of reactivity with monoclonal antibodies against Tg, T3, T4, and TPO

	Tg	T₃	T₄	TPO		Tg	T₃	T₄	TPO
Latex hevein	+++	+++	−	−	Cashew, vicilin	−	+	+++	−
Mushroom	−	−	−	++	Coffee protein	−	+++	−	−
Kamut	−	+++	−	−	Brazil nut	−	+	+++	−
Soy sauce	−	+	+++	−	Almond	−	−	++	−
Gelatin	−	++	+++	−	Wheat germ agglutinin	−	−	−	+++
Halibut	−	−	−	+++	Pea lectin	−	−	−	+++
Scallops	−	+++	+++	−	Lentil lectin	−	−	−	++
Cashew, roasted	−	++	+++	−	Pineapple bromeliad	−	−	−	++

The list of reactive and cross-reactive foods presented in this section is just the tip of an iceberg that should be explored further so as to shed light on an entire world of autoimmunities hidden below.

48.9 Conclusions: A Word to the Practitioners

The food and drink that we consume daily are constantly being affected and transformed by the food and beverage industries with ever-new advancements in food processing technology. Some of the reasons may be desirable, such as improving taste, nutritional value, or shelf life. Others are external or cosmetic sleight of hand aimed at our pockets or designed to persuade us to choose one brand over others. Whatever the purpose, the result of all this food tampering and manipulation is neo-linked neo-antigens, transformed molecules, and delivery systems, representing an intestinal mucosal load with altered physiochemical and immunogenic properties that challenge both mucosal and central immunity. The many layers of the immune system are masterpieces of signaling networks, molecules, and mediators. Beneath the mucosal layer lies the intestinal barrier. A breach in the intestinal barrier brings out the agents of the innate immune system and the patrolling immune cells of central tolerance. The authors have written that fact into each section of this chapter and, because it is so important to understand the concept, will repeat it one more time: When a failure in one of the layers occurs, it can cause a cascade of inflammation, opening protective barriers and exposing the body and the brain to environmental antigens that trigger autoimmune reactivity, tissue degradation, and disease. Normally the brain and the body's immune system work together in harmony, guiding each other, supporting each other, giving and taking in a healthy symbiotic relationship. In ◘ Fig. 48.13, Lymphia (representing lymphocytes) loves

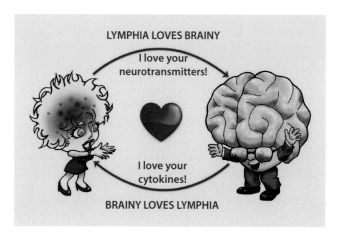

◘ Fig. 48.13 Lymphia loves Brainy and vice versa. Yes, but will their mutually beneficial relationship be able to survive the assault and interference of the dreaded environmental triggers?

Brainy, and Brainy says, "Right back at you." However, when all these environmental triggers, all these infections and reactive foods and toxic chemicals, come down on poor Lymphia's and Brainy's heads like a sea of troubles, as Shakespeare said, how can this poor couple manage to fight the tide of this onslaught and still maintain a healthy, working relationship?

This is what can happen when the assault of environmental triggers comes in a massive horde that overwhelms the guardians of our immune system. Thus, intestinal and blood barrier permeability is increased through damage done to the tight junctions, resulting in the entry of foreign immunogenic antigens such as food, infections, and toxic chemicals into the circulation, where antibody production against the invaders activates the autoimmune cascade (see ◘ Fig. 48.14). Because autoantibody development predates clinical autoimmune diagnosis by several years [75], the measurement of antibodies against food antigens, toxic chemicals, infections, tight junction and BBB proteins, and various tissue antigens

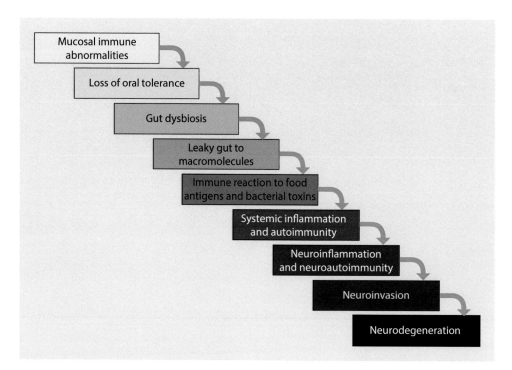

Fig. 48.14 Dysfunction from the gut to brain. The figure shows the downward, step-by-step process that takes an individual from loss of mucosal immune function to neurodegeneration. When the process is left unaddressed throughout the pathogenesis, the result is loss of neurological function

can help practitioners to modify or remove the triggers. The senior author of this chapter was one of the first functional immunologists to develop such tests. Panels are available for intestinal permeability, gluten reactivity, cross-reactivity and sensitivity, and chemical immune reactivity; tests for diabetes, neurological, joint, and other tissue autoantibodies; and multiple food immune reactivity panels. Tests such as these enable clinicians to help patients avoid years of suffering by detecting these predictive antibodies while there is still time. Simply by detecting first, removing the triggers, and then healing the intestinal barrier and optimizing the mucosal and innate immune system, we can reverse many disorders. If mucosal or intestinal barrier dysfunction is detected and addressed early, the onset of disease can be prevented.

References

1. Eder W, Ege MJ, von Mutius E. The asthma epidemic. N Engl J Med. 2006;355(21):2226–35.
2. Pearce N, Asher I, Billo N, Bissell K, Ellwood P, El Sony A, et al. Asthma in the global NCD agenda: a neglected epidemic. Lancet Respir Med. 2013;1:96–8.
3. Schuijs MJ, et al. Farm dust and endotoxin protect against allergy through A20 induction in lung epithelial cells. Science. 2015;349(6252):1106–10.
4. Wu GD, et al. Linking long-term dietary patterns with gut microbial enterotypes. Science. 2011;334(6052):105–8.
5. Muegge BD, et al. Diet drives convergence in gut microbiome functions across mammalian phylogeny and within humans. Science. 2011;332(6032):970–4.
6. Ley RE, Turnbaugh PJ, Klein S, Gordon JI. Microbial ecology: human gut microbes associated with obesity. Nature. 2006;444:1022–3.
7. Walker AW, et al. Dominant and diet-responsive groups of bacteria within the human colonic microbiota. ISME J. 2011;5(2):220–30.
8. Turnbaugh PJ, et al. The effect of diet on the human gut microbiome: a metagenomic analysis in humanized gnotobiotic mice. Sci Transl Med. 2009;1(6):6ra14.
9. David LA, et al. Diet rapidly and reproducibly alters the human gut microbiome. Nature. 2014;505(7484):559–63. https://doi.org/10.1038/nature12820.
10. Sanz Y. Microbiome and gluten. Ann Nutr Metab. 2015;67(suppl 2):27–42.
11. Bouziat R, et al. Reovirus infection triggers inflammatory responses to dietary antigens and development of celiac disease. Science. 2017;356(6333):44–50.
12. Vatanen T, et al. Variation in microbiome LPS immunogenicity contributes to autoimmunity in humans. Cell. 2016;165(4):842–53.
13. Skidmore M, et al. Bound xenobiotic residues in food commodities of plant and animal origin. Pest Manag Sci. 58:313. https://doi.org/10.1002/ps.442.
14. Hajeb P, et al. Toxic elements in food: occurrence, binding, and reduction approaches. Compr Rev Food Sci Food Saf. 13:457–72. https://doi.org/10.1111/1541-4337.12068.
15. Chipinda I, Hettick JM, Siegel PD. Haptenation: chemical reactivity and protein binding. J Allergy (Cairo). 2011;2011:839682. https://doi.org/10.1155/2011/839682.
16. Vojdani A, Kharrazian D, Mukherjee PS. Elevated levels of antibodies against xenobiotics in a subgroup of healthy subjects. J Appl Toxicol. 2015;35(4):383–97.
17. Jiang Y, et al. Aflatoxin B1 albumin adduct levels and cellular immune status in Ghanaians. Int Immunol. 2005 Jun;17(6):807–14.
18. Fernandez-Cornejo J, Wechsler SJ, Livingston M, Mitchell L. Genetically engineered crops in the United States. *Economic Research Report No.* (ERR-162) 60 pp, February 2014.

19. Aris A, Leblanc S. Maternal and fetal exposure to pesticides associated to genetically modified foods in Eastern Townships of Quebec, Canada. Reprod Toxicol. 2011;31(4):528–33.
20. Magaña-Gómez JA, de la Barca AM. Risk assessment of genetically modified crops for nutrition and health. Nutr Rev. 2009;67(1):1–16. https://doi.org/10.1111/j.1753-4887.2008.00130.x.
21. Watanabe S. Rapid analysis of glufosinate by improving the bulletin method and its application to soybean and corn. Shokuhin Eiseigaku Zasshi. 2002;43(3):169–72.
22. Estruch JJ, Warren GW, Mullins MA, Nye GJ, Craig JA, Koziel MG. Vip3A, a novel Bacillus thuringiensis vegetative insecticidal protein with a wide spectrum of activities against lepidopteran insects. Proc Natl Acad Sci U S A. 1996;93(11):5389–94.
23. Aranda E, Sanchez J, Peferoen M, Güereca L, Bravo A. Interactions of Bacillus thuringiensis crystal proteins with the midgut epithelial cells of Spodoptera frugiperda (Lepidoptera: Noctuidae). J Invertebr Pathol. 1996;68(3):203–12.
24. Samsel A, Seneff S. Glyphosate pathways to modern diseases V: amino acid analogue of glycine in diverse proteins. J Biol Phys Chem. 2016;16:9–46. https://doi.org/10.4024/03SA16A.jbpc.16.01.
25. van den Broeck HC, et al. Presence of celiac disease epitopes in modern and old hexaploid wheat varieties: wheat breeding may have contributed to increased prevalence of celiac disease. Theor Appl Genet. 2010 Nov;121(8):1527–39.
26. Vojdani A. A potential link between environmental triggers and autoimmunity. Autoimmune Diseases. 2014:437231, 18 pages. https://doi.org/10.1155/2014/437231.
27. Parks CG, Miller FW, Pollard KM, Selmi C, Germolec D, Joyce K, Rose NR, et al. Expert panel workshop consensus statement on the role of the environment in the development of autoimmune disease. Int J Mol Sci. 2014;15(8):14269–97.
28. Selmi C. The worldwide gradient of autoimmune conditions. Autoimmun Rev. 2010;9(5):A247–50. https://doi.org/10.1016/j.autrev.2010.02.004.
29. Okada H, Kuhn C, Feillet H, Bach JF. The 'hygiene hypothesis' for autoimmune and allergic diseases: an update. Clin Exp Immunol. 2010;160(1):1–9. https://doi.org/10.1111/j.1365-2249.2010.04139.x.
30. Rees F, Doherty M, Grainge M, Davenport G, Lanyon P, Zhang W. The incidence and prevalence of systemic lupus erythematosus in the UK, 1999-2012. Ann Rheum Dis. 2016;75(1):136–41. https://doi.org/10.1136/annrheumdis-2014-206334.
31. Molodecky NA, Soon IS, Rabi DM, Ghali WA, Ferris M, Chernoff G, et al. Increasing incidence and prevalence of the inflammatory bowel diseases with time, based on systematic review. Gastroenterology. 2012;142(1):46–54.e42; quiz e30. https://doi.org/10.1053/j.gastro.2011.10.001.
32. Lebwohl B, Ludvigsson JF, Green PH. The unfolding story of celiac disease risk factors. Clin Gastroenterol Hepatol. 2014;12(4):632–5. https://doi.org/10.1016/j.cgh.2013.10.031.
33. Widdifield J, Paterson JM, Bernatsky S, Tu K, Tomlinson G, Kuriya B, et al. The epidemiology of rheumatoid arthritis in Ontario, Canada. Arthritis Rheumatol. 2014;66(4):786–93. https://doi.org/10.1002/art.38306.
34. Patterson CC, Gyürüs E, Rosenbauer J, Cinek O, Neu A, Schober E, et al. Trends in childhood type 1 diabetes incidence in Europe during 1989-2008: evidence of non-uniformity over time in rates of increase. Diabetologia. 2012;55(8):2142–7. https://doi.org/10.1007/s00125-012-2571-8.
35. Rubio-Tapia A, Kyle RA, Kaplan EL, Johnson DR, Page W, Erdtmann F, et al. Increased prevalence and mortality in undiagnosed celiac disease. Gastroenterology. 2009;137(1):88–93. https://doi.org/10.1053/j.gastro.2009.03.059.
36. Lerner A, Matthias T. Changes in intestinal tight junction permeability associated with industrial food additives explain the rising incidence of autoimmune disease. Autoimmun Rev. 2015;14(6):479–89. https://doi.org/10.1016/j.autrev.2015.01.009.
37. Brown IJ, Tzoulaki I, Candeias V, Elliott P. Salt intakes around the world: implications for public health. Int J Epidemiol. 2009 Jun;38(3):791–813. https://doi.org/10.1093/ije/dyp139.
38. Kleinewietfeld M, Manzel A, Titze J, Kvakan H, Yosef N, Linker RA, et al. Sodium chloride drives autoimmune disease by the induction of pathogenic Th17 cells. Nature. 2013;496:518–22. https://doi.org/10.1038/nature11868.
39. Kearney J. Food consumption trends and drivers. Philos Trans R Soc Lond B Biol Sci. 2010;365(1554):2793–807. https://doi.org/10.1098/rstb.2010.0149.
40. Ballard ST, Hunter JH, Taylor AE. Regulation of tight-junction permeability during nutrient absorption across the intestinal epithelium. Annu Rev Nutr. 1995;15:35–55.
41. Mahmud N, Weir DG. The urban diet and Crohn's disease: is there a relationship? Eur J Gastroenterol Hepatol. 2001;13(2):93–5.
42. Kralova I, Sjoblom J. Surfactants used in food industry: a review. J Dispers Sci Technol. 2009;30(9):1363–83.
43. Kamba EA, Itodo AU, Ogah E. Utilization of different emulsifying agents in the preparation and stabilization of emulsions. Int J Mater Chem. 2013;3(4):69–74.
44. Traunmüller F. Etiology of Crohn's disease: do certain food additives cause intestinal inflammation by molecular mimicry of mycobacterial lipids? Med Hypotheses. 2005;65(5):859–64.
45. Sha XY, Fang XL. Effect of self-microemulsifying system on cell tight junctions. Yao Xue Xue Bao. 2006;41(1):30–5.
46. Barragán-Martínez C, Speck-Hernández CA, Montoya-Ortiz G, Mantilla RD, Anaya JM, Rojas-Villarraga A. Organic solvents as risk factor for autoimmune diseases: a systematic review and meta-analysis. PLoS One. 2012;7(12):e51506. https://doi.org/10.1371/journal.pone.0051506.
47. Saremi S, Dinarvand R, Kebriaeezadeh A, Ostad SN, Atyabi F. Enhanced oral delivery of docetaxel using thiolated chitosan nanoparticles: preparation, in vitro and in vivo studies. Biomed Res Int. 2013;2013:150478. https://doi.org/10.1155/2013/150478.
48. Hsu LW, Lee PL, Chen CT, Mi FL, Juang JH, Hwang SM, Ho YC, Sung HW. Elucidating the signaling mechanism of an epithelial tight-junction opening induced by chitosan. Biomaterials. 2012;33(26):6254–63. https://doi.org/10.1016/j.biomaterials.2012.05.013.
49. Yu SH, Tang DW, Hsieh HY, Wu WS, Lin BX, Chuang EY, Sung HW, Mi FL. Nanoparticle-induced tight-junction opening for the transport of an anti-angiogenic sulfated polysaccharide across Caco-2 cell monolayers. Acta Biomater. 2013;9(7):7449–59. https://doi.org/10.1016/j.actbio.2013.04.009.
50. Borel T, Sabliov CM. Nanodelivery of bioactive components for food applications: types of delivery systems, properties, and their effect on ADME profiles and toxicity of nanoparticles. Annu Rev Food Sci Technol. 2014;5:197–213. https://doi.org/10.1146/annurev-food-030713-092354.
51. Malandain H. Transglutaminases: a meeting point for wheat allergy, celiac disease, and food safety. Eur Ann Allergy Clin Immunol. 2005;37(10):397–403.
52. Kieliszek M, Misiewicz A. Microbial transglutaminase and its application in the food industry. A review. Folia Microbiol (Praha). 2014;59(3):241–50. https://doi.org/10.1007/s12223-013-0287-x.
53. Yokoyama K, Nio N, Kikuchi Y. Properties and applications of microbial transglutaminase. Appl Microbiol Biotechnol. 2004;64(4):447–54. Epub 2004 Jan 22.
54. Miguel ASM, Martins-Meyer TS, Figueiredo EVDC, Lobo BWP, et al. Chapter 14: Enzymes in bakery: current and future trends. In: Muzzalupo I, editor. Food industry; 2013. p. 287–321. https://doi.org/10.5772/53168.
55. Li N, Neu J. Glutamine deprivation alters intestinal tight junctions via a PI3-K/Akt mediated pathway in Caco-2 cells1,2. J Nutr. 2009;139(4):710–4.
56. Mullin JM, Skrovanek SM, Valenzano MC. Modification of tight junction structure and permeability by nutritional means. Ann N Y Acad Sci. 2009;1165:99–112. https://doi.org/10.1111/j.1749-6632.2009.04028.x.

57. Kaufmann A, Köppel R, Widmer M. Determination of microbial transglutaminase in meat and meat products. Food Addit Contam Part A Chem Anal Control Expo Risk Assess. 2012;29(9):1364–73. https://doi.org/10.1080/19440049.2012.691557. Epub 2012 Jun 29.
58. Lerner A, Matthias T. Hypothesis: increased consumption of food industry bacterial transglutaminase explains the surge in celiac disease incidence. Clin Chem Lab Med. 2015;53(11):eA115–6.
59. Das S, Chaudhury A. Recent advances in lipid nanoparticle formulations with solid matrix for oral drug delivery. AAPS PharmSciTech. 2011;12(1):62–76. https://doi.org/10.1208/s12249-010-9563-0.
60. Vojdani A, Vojdani C. Immune reactivity to food coloring. Altern Ther Health Med. 2015;21(Suppl 1):52–62.
61. Roberts MT. Food law in the United States. New York, NY: Cambridge University Press; 2016.
62. Vojdani A, Vojdani C. Immune reactivities against gums. Altern Ther Health Med. 2015;21(Suppl 1):64–72.
63. Ulluwishewa D, Anderson RC, McNabb WC, Moughan PJ, Wells JM, Roy NC. Regulation of tight junction permeability by intestinal bacteria and dietary components. J Nutr. 2011;141(5):769–76. https://doi.org/10.3945/jn.110.135657.
64. Selmi C, Tsuneyama K. Nutrition, geoepidemiology, and autoimmunity. Autoimmun Rev. 2010;9(5):A267–70. https://doi.org/10.1016/j.autrev.2009.12.001.
65. Sapone A, de Magistris L, Pietzak M, Clemente MG, Tripathi A, Cucca F, Lampis R, Kryszak D, Cartenì M, Generoso M, Iafusco D, Prisco F, Laghi F, Riegler G, Carratu R, Counts D, Fasano A. Zonulin upregulation is associated with increased gut permeability in subjects with type 1 diabetes and their relatives. Diabetes. 2006;55:1443–9. [PMID: 16644703. https://doi.org/10.2337/db05-1593.
66. Vojdani A, Vojdani E, Kharrazian D. Fluctuation of zonulin levels in blood versus stability of antibodies. World J Gastroenterol. 2017;23(31):5669–79.
67. Vojdani A, Rahimian P, Kalhor H. Immunological cross-reactivity between Candida albicans and human tissue. J Clin Lab Immunol. 1996;48:1–15.
68. Hadjivassiliou M, Grünewald RA, Lawden M, Davies-Jones GA, Powell T, Smith CM. Headache and CNS white matter abnormalities associated with gluten sensitivity. Neurology. 2001;56(3):385–8.
69. Hadjivassiliou M, Boscolo S, Davies-Jones GA, Grünewald RA, Not T, Sanders DS, Simpson JE, Tongiorgi E, Williamson CA, Woodroofe NM. The humoral response in the pathogenesis of gluten ataxia. Neurology. 2002;58(8):1221–6.
70. Vojdani A, O'Bryan T, Green JA, Mccandless J, Woeller KN, Vojdani E, Nourian AA, Cooper EL. Immune response to dietary proteins, gliadin and cerebellar peptides in children with autism. Nutr Neurosci. 2004;7(3):151–61.
71. Vojdani A, Tarash I. Cross-reaction between gliadin and different food and tissue antigens. Food Nutr Sci. 2013;4:20–32.
72. Reinke Y, Behrendt M, Schmidt S, Zimmer KP, Naim HY. Impairment of protein trafficking by direct interaction of gliadin peptides with actin. Exp Cell Res. 2011;317(15):2124–35. https://doi.org/10.1016/j.yexcr.2011.05.022.
73. Freed DLJ. Dietary lectins and disease. In: Brostoff J, Challacombe SJ, editors. Food allergy and intolerance. 2nd ed. London: Saunders; 2002.
74. Cavallo MG, Fava D, Monetini L, Barone F, Pozzilli P. Cell-mediated immune response to beta casein in recent-onset insulin-dependent diabetes: implications for disease pathogenesis. Lancet. 1996;348(9032):926–8.
75. Kristjánsson G, Venge P, Hällgren R. Mucosal reactivity to cow's milk protein in coeliac disease. Clin Exp Immunol. 2007;147(3):449–55.
76. Triolo G, Accardo-Palumbo A, Dieli F, et al. Humoral and cell mediated immune response to cow's milk proteins in Behçet's disease. Ann Rheum Dis. 2002;61(5):459–62.
77. Triolo G, Triolo G, Accardo-Palumbo A, Carbone MC, Giardina E, La Rocca G. Behçet's disease and coeliac disease. Lancet. 1995;346(8988):1495.
78. Kolb H, et al. Cow's milk and type I diabetes: the gut immune system deserves attention. Immunol Today. 1999;20(3):108–10.
79. Vaarala O, Klemetti P, Savilahti E, Reijonen H, Ilonen J, Akerblom HK. Cellular immune response to cow's milk beta-lactoglobulin in patients with newly diagnosed IDDM. Diabetes. 1996;45(2):178–82.
80. Al-Ghobashy MA, Cucheval A, Williams MA, Laible G, Harding DR. Probing the interaction between recombinant human myelin basic protein and caseins using surface plasmon resonance and diffusing wave spectroscopy. J Mol Recognit. 2010;23(1):84–92. https://doi.org/10.1002/jmr.991.
81. Gardinier MV, Amiguet P, Linington C, Matthieu JM. Myelin/oligodendrocyte glycoprotein is a unique member of the immunoglobulin superfamily. J Neurosci Res. 1992;33(1):177–87.
82. Pham-Dinh D, Mattei MG, Nussbaum JL, et al. Myelin/oligodendrocyte glycoprotein is a member of a subset of the immunoglobulin superfamily encoded within the major histocompatibility complex. Proc Natl Acad Sci U S A. 1993;90(17):7990–4.
83. Guggenmos J, Schubart AS, Ogg S, et al. Antibody cross-reactivity between myelin oligodendrocyte glycoprotein and the milk protein butyrophilin in multiple sclerosis. J Immunol. 2004;172(1):661–8.
84. Vojdani A, Kharrazian D, Mukherjee PS. The prevalence of antibodies against wheat and milk proteins in blood donors and their contribution to neuroautoimmune reactivities. Nutrients. 2014;6:15–36. https://doi.org/10.3390/nu6010015.
85. Riemekasten G, Marell J, Hentschel C, et al. Casein is an essential cofactor in autoantibody reactivity directed against the C-terminal SmD1 peptide AA 83-119 in systemic lupus erythematosus. Immunobiology. 2002;206(5):537–45.
86. Jarius S, Wildemann B. AQP4 antibodies in neuromyelitis optica: diagnostic and pathogenetic relevance. Nat Rev Neurol. 2010;6(7):383–92. https://doi.org/10.1038/nrneurol.2010.72.
87. Kim SH, Kim W, Li XF, Jung IJ, Kim HJ. Clinical spectrum of CNS aquaporin-4 autoimmunity. Neurology. 2012;78(15):1179–85. https://doi.org/10.1212/WNL.0b013e31824f8069.
88. Bradl M, Lassmann DH. Anti-aquaporin-4 antibodies in neuromyelitis optica: how to prove their pathogenetic relevance? Int MS J. 2008;15(3):75–8.
89. Vaishnav RA, Liu R, Chapman J, et al. Aquaporin 4 molecular mimicry and implications for neuromyelitis optica. J Neuroimmunol. 2013;260(1–2):92–8. https://doi.org/10.1016/j.jneuroim.2013.04.015. Epub 2013 May 9.
90. Vojdani A, Mukherjee PS, Berookhim J, Kharrazian D. Detection of antibodies against human and plant aquaporins in patients with multiple sclerosis. Autoimmune Diseases. 2015:905208, 10 pages. https://doi.org/10.1155/2015/905208.
91. Smith DW, Nagler-Anderson C. Preventing intolerance: the induction of nonresponsiveness to dietary and microbial antigens in the intestinal mucosa. J Immunol. 2005;174(7):3851–7.
92. Dai H, Dong HL, Gong FY, Sun SL, Liu XY, Li ZG, Xiong SD, Gao XM. Disease association and arthritogenic potential of circulating antibodies against the α1,4-polygalacturonic acid moiety. J Immunol. 2014;192(10):4533–40. https://doi.org/10.4049/jimmunol.1303351.
93. Kharrazian D, Herbert M, Vojdani A. Detection of islet cell immune reactivity with low glycemic index foods – Is this a concern for type 1 diabetes? J Diabetes Res. 2017;2017:4124967.
94. Kharrazian D, Herbert M, Vojdani A. Immunological reactivity using monoclonal and polclonal antibodies of autoimmune thyroid target sites with dietary proteins. J Thyroid Res, Vol. 2017, Article ID 4354723. https://doi.org/10.1155/2017/4354723.

Amyotrophic Lateral Sclerosis (ALS): The Application of Integrative and Functional Medical Nutrition Therapy (IFMNT)

Coco Newton

49.1 Introduction – 866

49.2 pALS Advance Nutrition Research – 866

49.3 Hope, Nutrition, and ALS – 866

49.4 What Is Amyotrophic Lateral Sclerosis (ALS)? – 867

49.5 Altered Metabolism and Hypermetabolism – 867

49.6 Blood-Brain Barrier – 868

49.7 ALS Defined as Metaphor for Nourishment – 870

49.8 Current Limitations in Clinical Dietetics and Nutrition Care – 870

49.9 Conventional Dietary and Nutritional Protocols – 871

49.10 Complementary and Alternative Medicine – 871

49.11 Integrative and Functional Medical Nutrition Therapy – 871

49.12 Weight as a Risk Factor – 871

49.13 Calorie Requirements and Nutrition Assessment – 872

49.14 Macronutrients: Carbohydrates, Protein, and Fats – What Do Dietary Studies Tell Us? – 872

49.15 Sugar – 873
49.15.1 Sugar: Neurotransmitter Imbalances (Dopamine and Serotonin) – 873
49.15.2 Sugar: Inflammation and Autoimmunity – 873
49.15.3 Sugar: Hormonal Imbalances (Glucose and Insulin) – 874
49.15.4 Sugar: Brain-Derived Neurotrophic Factor (BDNF) – 874

© Springer Nature Switzerland AG 2020
D. Noland et al. (eds.), *Integrative and Functional Medical Nutrition Therapy*,
https://doi.org/10.1007/978-3-030-30730-1_49

49.16	Gluten	– 874
49.17	Protein	– 875
49.18	Dairy and Soy as Protein Sources in Oral and Tube Feedings – 876	
49.18.1	Dairy – 876	
49.18.2	Soy – 876	
49.19	Amino Acids – 876	
49.20	Fats – 878	
49.21	Optimal Sources and Amounts of Dietary Fat – 879	
49.21.1	Omega-3 PUFAs (ALA, EPA, DHA) – 879	
49.21.2	Omega-6 PUFAs (LA, AA) – 880	
49.21.3	Medium-Chain Triglycerides – 880	
49.21.4	Saturated Fatty Acids (SFAs) – 880	
49.21.5	Cholesterol – 881	
49.21.6	Phospholipids (PLs) – 882	
49.22	Ketogenic Diet (KD) and Medium-Chain Triglycerides (MCTs) – 882	
49.23	The Deanna Protocol (DP) – 882	
49.24	Commercial Enteral Formulas – 883	
49.25	Healthy Enteral Feedings – 884	
49.26	Foods Containing Phytonutrients – 884	
49.27	Environmental Toxins – 884	
49.28	Detoxification of Heavy Metals – 886	
49.29	Natural Chelators – 886	
49.30	Essential Metals – 887	
49.30.1	Iron – 887	
49.30.2	Copper – 887	
49.31	BMAA – 887	
49.32	Gastrointestinal System – 887	
49.33	Microbiome – 888	
49.34	Probiotics – 889	
49.35	Butyrate – 889	

49.36	Small Intestinal Bacterial Overgrowth (SIBO) – 889
49.37	Laboratory Biomarkers – 890
49.38	Metabolomics – 891
49.39	Urinary Organic Acid Analysis (OAA) – 891
49.40	Impaired Glycolysis, TCA Cycle, and Oxidative Phosphorylation – 891
49.41	Methylation and ALS – 892
49.42	Additional Considerations for Laboratory Testing – 892
49.42.1	Erythrocyte Fatty Acid Analysis – 892
49.42.2	Liver Function and ALS – 892
49.42.3	Thyroid and ALS – 892
49.42.4	Adrenal and ALS – 892
49.42.5	Autoantibodies, Antibodies, and ALS – 892
49.42.6	Infectious Diseases and ALS – 892
49.43	Supplements – 893
49.43.1	Reductionist Versus Systems Biology Approach – 893
49.44	Conclusion – 899
	References – 899

49.1 Introduction

A diagnosis of ALS follows the descriptors of most diseases: cause(s), diagnostic criteria, symptoms, progression, management, treatments, and prognosis. The dilemma with ALS is that we don't know the cause(s), we have no effective treatments, but we do know the prognosis. Every ALS patient receiving a diagnosis hears the same script that the disease is usually fatal within 2–5 years, and there are no drugs or diets to avert the progressive deterioration. Less than 1% of patients with ALS will have significant improvement in function after 1 year, only 10% of patients will live past 10 years and 5% past 20 years. As stated on their website, "ALS is 100% fatal" [1], and the diagnosis of ALS is frequently described as a "death sentence" [2].

Nutrition counseling for persons with ALS (pALS) lags far behind other common neurodegenerative diseases such as Parkinson's disease, Alzheimer's disease, and Multiple Sclerosis, all conditions with a much better prognosis and life expectancy. One can easily surmise why most healthcare practitioners haven't had a role in working with pALS. They are dissuaded and intimidated by the futility of treatment and fatality of the diagnosis. Primary care physicians consign management of pALS to *neurologists who distances pALS* from other healthcare practitioners. The Ice Bucket Challenge in the summer of 2014 brought ALS out of relative obscurity and into the national spotlight [3], infusing funds into scientific research and empowering pALS to network across the globe for support. An increasing number of pALS and their caregivers explore the Internet, use social media, and seek a range of resources, healthcare professionals, and healers beyond the purview of the neurologist.

A holistic approach to ALS recognizes the multifactorial, metabolic, genetic, epigenetic, inflammatory, immune, environmental, excitotoxic, gastrointestinal, mitochondrial, substrate altered, nutritionally challenged, emotional, and spiritual crisis confronting the care of the patient. ALS presents the Integrative and Functional Medical Nutrition Therapy (IFMNT) practitioner the responsibility of leveraging metabolic pathways through diet and nutritional supplements to hopefully delay progression, improve symptoms, or even reverse disease.

49.2 pALS Advance Nutrition Research

Some pALS engage in their own (*n* = 1) case-controlled studies by researching and designing their own protocols, sharing experiences with other pALS, and hiring selected practitioners to support them. pALS and their loved ones are the stimulus driving nutritional practices and research for ALS. They refuse the death sentence and, out of necessity, become their own students, researchers, patients, and physicians. This phenomenon is seen in pALS like Dr. Craig Oster [4], Eric Edney [5], ALS Winners [6], pioneering advocates [7], blogs [8], and *The Deanna Protocol: Hope for ALS and Other Neurological Conditions* [9].

There are patients who have accomplished brief reversals [10] and others who have healed significantly [11]. The study of ALS reversals by neurologist Dr. Richard Bedlack at Duke University has included demographics, disease characteristics, treatments, and comorbidities to try to understand the mechanisms of disease resistance [12]. It is not false hope. It is hope that pALS draw from selected preclinical and clinical scientific research that they interpret and translate into their own self-styled treatment. Consider that 17 years is the typical lag time for translational research [13], and pALS know that only 5–10% of them might still be alive at that point. pALS are not weighing "alternative" treatments over conventional medical treatments. There is nothing to dispute, as there are no available medical treatments to halt or reverse ALS.

As IFMNT practitioners, we must serve as "carrier molecules" for a higher standard of dietary/nutritional care for pALS into our own practices and ultimately into the conventional healthcare system. We must advance education, mentorship, and training among our colleagues. IFMNT practitioners uphold these principles of practice: First, do no harm. Second, apply best available practice-based and evidence-based research to inform clinical decision-making. Third, respect the values and desires of every pALS. Fourth, practice with hope.

49.3 Hope, Nutrition, and ALS

What about the willingness to adopt a strict nutrient-dense diet? An online survey was conducted to study whether ALS patients would follow such a diet if it were to increase longevity. The survey found that dietary self-care choices appear to be related to hope issues [4]. Dr. Oster's work was referenced again in a study on illness trajectories in pALS as determined by illness perception, trust in medical care, self-construction, and the distribution of dependency [14].

Dr. Craig Oster, the lead author of the nutrient-dense diet survey, was diagnosed with ALS in 1994, yet he persevered to obtain his PhD in Psychology from the University of Michigan in 1996. By late 2008, he entered hospice care as he had lost over 45 pounds of muscle, refused a feeding tube, and needed a mechanical ventilator to get through the night [15]. In tremendous agony and deep prayer, he was released from hospice in 2009 and then came "back to the future" to continue his healing journey [16]. He regained weight, muscle, and stamina and reversed some of his ALS symptoms, as of 2017. He never used commercial oral formulas, tube feedings, or any medications for ALS. He follows an individualized diet plan with nutritional supplements, massage, spiritual practice, psychotherapy, yoga, advanced technologies for exercise and movement like pacer gait trainers, neurofeedback, and LED light therapy, among other therapeutic modalities [17].

pALS and their loved ones are the driving force behind research in nutritional interventions. They become their own students, researchers, patients, and physicians. Out of their discoveries, they are able to advise the research community about promising therapies.

49.4 What Is Amyotrophic Lateral Sclerosis (ALS)?

ALS, also named motor neuron disease or Lou Gehrig's disease, is a progressive and fatal neurodegenerative disease. ALS usually strikes between the ages of 40–70, with an average age of 55 years at diagnosis. Males comprise 60% of ALS cases and Caucasians make up 93% of the people in those cases. Today, more than 6000 Americans are diagnosed each year, and about 20,000 Americans have ALS at any given time. ALS afflicts an estimated 350,000 people around the world [18]. According to the World Health Organization, the World Bank, and Harvard School of Public Health, neurodegenerative diseases will become the 8th leading cause of death by mid-century [19], with a projected 69% global increase in ALS cases from 2015 to 2040 [20].

ALS was first identified in 1869 by French neurologist Jean-Martin Charcot. The official recognition of the disease began in 1939 when Lou Gehrig, one of the most famous baseball players of all time, was diagnosed with ALS at 36 years of age and died 2 years later. Readers of Mitch Albom's New York Times bestseller in 2000, "Tuesdays with Morrie," might recall that ALS drained his old professor's life in 1 year over the 14 visits the author made to his home [21]. Famous theoretical physicist Stephen Hawking lived with ALS for more than 50 years before his death in 2018 [22]. ALS is notoriously conspicuous in professional sports and the military, presumably from traumatic brain injuries (TBI) in athletes and both TBI and toxic exposures in the military [23].

ALS is a disease of motor neurons, the neurologic cells that initiate and control the movement of muscles. Motor neurons that originate in the brain are characterized as upper, and those originating in the spinal cord are characterized as lower. ALS affects both upper and lower motor neurons, although each person with ALS has varying amounts of upper and lower motor neuron involvement. Bulbar ALS, also called progressive bulbar palsy, prominently affects the muscles involved in speech, swallowing, and tongue movements. Upper motor neuron disease causes stiffness or spasticity. Lower motor neuron disease causes weakness, loss of muscle (atrophy), and muscle twitching (fasciculations). ALS may begin with abnormalities of upper or lower motor neurons.

As motor neurons gradually die, the brain is unable to initiate and control voluntary muscle movement. ALS does not affect the heart muscle or the smooth muscle of the digestive tract, bladder, and other internal organs. An exception is a secondary complication that involves the gradual weakening of the muscles of the lungs, leading to atelectasis or collapse [24]. A patient's cognition, sexual function, and eyesight, along with the senses of smell, taste, hearing, sight, and touch, remain intact.

Most ALS cases are sporadic (SALS), but 5–10% of the cases are familial (FALS) [1, 25]. However, the boundary between familial and sporadic ALS is somewhat blurred since up to 95% of "apparently" sporadic ALS cases have involvement of the same ALS-associated genes (SOD1, TARDP, FUS, C9orf72) as seen in familial ALS [26].

Copper-zinc superoxide dismutase (Cu-Zn SOD1) reduces superoxide (O2-) to yield hydrogen peroxide and oxygen (H2O2 and O2). H2O2 is less reactive than O2- but still highly reactive. O2 reacts with nitric oxide to yield peroxynitrites. In the case of SOD1 ALS, the mutation allows it to bind with copper but not with zinc, which causes the conversion of H2O2 back to O2- in a reversed enzymatic step. This gain-of-function mutation results in misfolded proteins that aggregate and become toxic [27]. Only 2% of all ALS patients have the SOD1 ALS mutation, while 20% of the familial cases have been found to carry this mutation [27–28]. Rotunno and colleagues [28] support the controversial hypothesis that misfolded forms of wild-type SOD1 contribute to sporadic ALS pathogenesis, whereas others have reported the opposite to be true [29]. Saccon and colleagues [30] bring together historical and experimental findings to suggest that SOD1 gain and loss of function could complement each other in both FALS and SALS. Cu-ZnSOD1 is differentiated from manganese superoxide dismutase (MnSOD2), which has no genetic association with ALS [31]. It is interesting to note that higher concentrations of Mn exists in spinal cords of ALS patients and is known to inhibit neuronal transmission [32].

There are five medical classifications of ALS as seen in Fig. 49.1 [1, 25, 33].

Riluzole, the only approved drug since 1995, may extend life by 3–5 months [34], and Radicava, just approved in 2017, may slow progression of ALS by 33% [35]. Riluzole decreases glutamate excitotoxicity and Radicava reduces oxidative stress, but outcomes are limited. Although the scientific research on ALS describes multisystem metabolic pathways, pharmaceutical research only approaches a pathway at a time. Important neurodegenerative pathways such as detoxification, inflammation, immunity, oxidative stress, nutritional deficiencies, gut-brain, mitochondrial, energy homeostasis, and others lack successful therapeutic applications. The majority of research has been done on the human mutant transgenic SOD-1 G93A mouse model developed in 1994 [36]; however, the metabolic and nutritional underpinnings of FALS and SALS share many common multifactorial pathways.

ALS can no longer be viewed as a single disease entity guided by reductionist single target pathways for medications. ALS involves biological heterogeneity and is a multisystem disorder [37]. It is also a complex clinical syndrome with a rapidly evolving understanding of biomarkers and potential therapeutic treatments [38]. Approaching an ALS patient from an individualized and multisystem metabolic framework aptly places ALS in the IFMNT model of therapeutic assessment and interventions.

49.5 Altered Metabolism and Hypermetabolism

Hypermetabolism in ALS patients is an early and persistent phenomenon with studies confirming approximately 50% of pALS are hypermetabolic and 80% of them remain hypermetabolic over time [39]. Hypermetabolism is largely a result

- **Classical ALS** - a progressive neurological disease characterized by a deterioration of upper and lower motor neurons (nerve cells). This type of ALS affects more than two-thirds of those with the disease.
- **Primary Lateral Sclerosis (PLS)** - a progressive neurological disease in which the upper motor neurons (nerve cells) deteriorate. If the lower motor neurons are not affected within two years, the disease usually remains a pure upper motor neuron disease. This is the rarest form of ALS.
- **Progressive Bulbar Palsy (PBP)** - a condition that starts with difficulties in speaking, chewing and swallowing due to lower motor neuron (nerve cell) deterioration. This disorder affects about 25% of those with ALS.
- **Progressive Muscular Atrophy (PMA)** - a progressive neurological disease in which the lower motor neurons (nerve cells) deteriorate. If the upper motor neurons are unaffected within two years, the disease usually remains a pure lower motor neuron disease.
- **Familial** - a progressive neurological disease that affects more than one member of the same family. This type of ALS accounts for a very small number of people with ALS in the United States (between five and ten percent).

Fig. 49.1 How is ALS classified? (Based on data from Johns Hopkins Medicine [33])

of mitochondrial dysfunction being fueled by oxidative stress in motor neurons and skeletal muscle with broader systemic effects [39]. pALS have altered expression of metabolic proteins and adipokines [40], which profoundly affect appetite and satiety. A decrease in ghrelin (appetite) with no change in leptin (satiety) levels causes reduced food intake due to low appetite along with increased satiety. Impaired insulin action is thought to contribute to reduced glucose tolerance and reduced ability to use glucose as energy [41]. Increased inflammatory cytokines (IL-6, IL-8) may contribute to insulin resistance, and increased TNF-alpha induces lipolysis, causing the breakdown of fat stores, weight loss, and lower BMI [41].

The brain has the highest energy requirements of all organs in the body, accounting for 2% of the total body mass and consuming 20% of the body's oxygen [41]. Within the brain, energy consumption is predominantly demanded by the neurons and less by the astrocytes. This results in a vicious cycle of bioenergetic deficits that exacerbate the pathogenesis of ALS as depicted in Fig. 49.2 [41]. The utilization of glucose by neurons and skeletal muscle is impaired in ALS, seemingly driven by insulin resistance and glucose intolerance. As a result, the increased reliance on fat as an energy substrate depletes endogenous energy storage, promotes weight loss, and perpetuates higher metabolic demands.

Decreased production of adenosine triphosphate (ATP) or decreased glucose metabolism in neurons and decreased glucose metabolism in the skeletal muscle may contribute to the hyperexcitability and selective degeneration of upper and lower motor neurons and muscle pathology/denervation in ALS, respectively [41]. Insulin resistance and glucose intolerance may underpin an inability to efficiently use glucose as an energy substrate. Overall, an inability to use glucose in the periphery, in neurons and in skeletal muscle, will result in an increased dependence on the use of fat as an energy substrate to offset energy deficit. With escalating metabolic pressure, the rapid depletion of endogenous energy stores will result in a catastrophic failure to meet increased metabolic demand. Thus, a vicious cycle of bioenergetic deficit may underpin or exacerbate disease pathogenesis in ALS [41].

49.6 Blood-Brain Barrier

The blood-brain barrier (BBB) is the main focus of nutritional research and potential therapeutic targets for ALS. The central nervous system is tightly sealed from the changeable milieu of blood by the BBB, the blood-cerebrospinal fluid barrier (BCSFB), and the blood-spinal cord barrier (BSCB) [42]. However, neurodegeneration creates vulnerability of the BBB, which disrupts the barrier function [43]. A short overview of the BBB is presented to better appreciate how macronutrients, micronutrients, and other nutritional interventions impact brain health, as the BBB will be frequently referred to throughout this chapter.

Picture the BBB as an organized structure of framing, trusses, insulation, ducts, locks, leaks, and weathering. Also, picture the BBB as an infrastructure of gatekeepers, guards, patrol officers, defenders, escorts, and conservators. Endogenously produced cholesterol and peripherally derived dietary fats comprise the main structure and function of the cell membrane, especially arachidonic acid (AA) and docosahexaenoic acid (DHA) [44]. Other fats that cross the BBB are oleic acid, palmitic acid, and alpha-linolenic acid [45]. It cannot be overstated how crucial lipids are to the health of cell membranes to provide structure, fluidity, transmission of electrical signals, movement of membrane proteins, stabilization of synapses, and protection [44].

The BBB provides protection from invading microbial pathogens, antigens, antibodies, inflammatory cytokines, and toxins and controls the microenvironment of the tissues by tightly regulating the movement of molecules and ions between the cellular spaces. The basic process of BBB break-

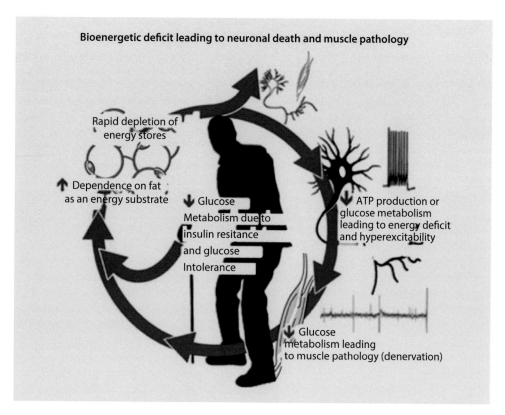

Fig. 49.2 Bioenergetic deficits. (Reprinted from Ngo and Steyn [41]. With permission from Creative Commons License 4.0: ▶ https://creativecommons.org/licenses/by/4.0/)

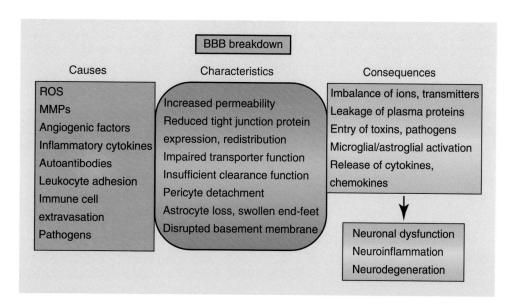

Fig. 49.3 BBB breakdown leading to neuronal dysfunction, neuronal inflammation, and neurodegeneration. (Reprinted from Obermeier et al. [43]. With permission from Springer Nature)

down leading to neuronal dysfunction, neuronal inflammation, and neurodegeneration is depicted in Fig. 49.3 [43].

The brain barrier function has many similarities to the gut barrier function. Daneman and colleagues [46] describe how the gut immune barrier (GIB) protects the blood from entry of pathogens and toxins ingested in food and how the BBB protects the CNS from pathogens and toxins derived from blood. The GIB and BBB share several structural and functional features such as basement membranes, endothelial cells, tight junctions with their structural proteins, claudins, occludin, and junctional adhesion molecule [46]. Impaired permeability of gut and brain barriers share similar descriptive terminology in the medical and popular literature such as "leaky" and "leakage" or "leaky gut" and "leaky brain" [47].

The integrity of the gut and brain barriers is vulnerable to bacterial lipopolysaccharides, inflammatory cytokines, antigens, antibodies, and toxins causing permeability [48]. A molecule called microRNA-155 is responsible for cleaving epithelial cells to create microscopic gaps that allow toxins and other molecules access to the brain [49], hence the term leaky BBB. Further supportive evidence describes various mechanisms of BBB structural and molecular disruption occurring in CNS pathologies [50].

From an IFMNT perspective, the goal is to minimize allergens, toxins, microbes, and other inflammatory molecules from harming the BBB while optimizing nutrients that protect the BBB. The idiom "where the rubber meets the road" has many analogies to "where the brain meets the membrane." IFMNT is judiciously positioned to support this crucial interface at the BBB.

49.7 ALS Defined as Metaphor for Nourishment

As found on the ALSA website [51], the meaning of amyotrophic lateral sclerosis is derived from Greek roots: "A" means no, "myo" refers to muscle, and "trophic" means nourishment, so amyotrophic means "no muscle nourishment," referring to the loss of nerve cells that normally signal to muscle cells. This trophic definition of nourishment includes the provision of food and nutrients while encompassing sustenance and maintenance of function. When a muscle has no nourishment, it atrophies or wastes away and loses function. "Lateral" means "to the side" which refers to the location of damaged portions of these nerve cells. As the spinal cord degenerates, it leads to "sclerosis," which is scarring or hardening in the region. The word "diet" derives from the Greek word "diata," which means a way of life. The IFMNT practitioner recognizes that diet and nutritional management becomes a way of life for ALS patients who want to extend their life span and quality of life on the physical, emotional, and spiritual levels.

49.8 Current Limitations in Clinical Dietetics and Nutrition Care

As discussed above, it is not surprising that a dietitian, nutritionist, many physicians, and other healthcare providers may never experience directly caring for an ALS patient throughout their career. Patients with abnormal neurological symptoms are referred to and managed by neurologists; thus dietetic management typically focuses on high-calorie diets, chewing and swallowing issues, and tube feeding management under the auspices of the neurologist. Dietary goals are to support feeding and provide enough calories to prevent weight loss. Dietitians lack sufficient education and training in neurodegenerative diseases, and standardized protocols lack personalized application. Most dietitians are not trained in the therapeutic application of nutritional supplements. As reported in Today's Dietitian [52], "Dietitians play an important role in the care of ALS patients, yet there is little evidence on how we should be caring for these patients." This perspective is reiterated in the following statement: "Training and support in ALS/MND nutrition should be made available as part of post-dietetic registration."

Rio and colleagues [53] report further dietetic research is required to stimulate nutritional care. Fast-forward 10 years when data were analyzed from the data extraction tool (DET) provided by the Academy of Nutrition and Dietetics (AND). "This systematic review shows that there is a lack of high-quality evidence regarding the efficacy of any dietary interventions for promoting survival in ALS or slowing disease progression; therefore, more research is necessary related to effects of nutrition interventions" [54]. This conclusion leaves doctors, dietitians, other clinicians, families, and pALS feeling hopeless and under the mistaken impression that there is no specific nutrition information or intervention for pALS.

One review [55] offers the long-standing conventional recommendations for feeding support and adequate nutrition; however, another review has made an effort to improve the suggestions for high-calorie and easy-to-chew recipes [56]. Prior to these suggestions, the ALS Association (ALSA) website only offered "Maintaining Good Nutrition with ALS" [57] that included many detrimental foods made from processed and toxic ingredients (sugar, monosodium glutamate, preservatives, artificial flavors and dyes, GMO ingredients, etc.). There was no regard for the nutritional value of food beyond calories, texture, and convenience. This limited perspective on ALS and nutrition was established in 1998 by a physician expert on the American Academy of Neurology (Committee to Study and Develop Guidelines for the Role of Nutrition in ALS), who stated that "In ALS, it's not so much what you eat, but how much you eat" [58].

Advances in scientific research are showing that there are many metabolic targets for nutritional intervention in ALS. As practitioners, we must evaluate the practice-based and evidence-based research for insight into best practices and take a leadership role in providing individualized therapeutic plans for our patients. The validation of an individualized multisystem IFMNT approach to ALS is alluded to in this article [59], where the authors hypothesized that there may be subgroups of patients, eventually defined by a specific underlying etiology or clinical presentation, which selectively respond to a particular regimen, supporting the concept of a personalized nutritional approach.

Many of our patients and their loved ones scour the scientific literature on diet and nutrition and seek out integrative and functional practitioners. They are not willing to passively accept their death sentence diagnosis or be limited by antiquated dietetic recommendations. Drugs are not likely to effectively treat or cure this disease in the near future, as discussed in "ALS Clinical Trials Review: 20 Years of Failure. Are We Any Closer to Registering a New Treatment?" [60]. For pALS and their caregivers, while waiting for new drugs, the most hopeful and empowering therapeutic intervention is diet and nutrition.

Patients will experiment on themselves and design protocols, and some will successfully stall, reduce, or reverse symptoms of ALS. As IFMNT practitioners, we do not myopically fight ALS but holistically address metabolic imbalances throughout the body. We are obligated to support our patients on this healing journey by combining our hope with the best practice-based and evidence-based research and therapeutic plans in tandem with our patient's goals and value systems. The purpose of this chapter is to provide translation of scientific research into possible therapeutic applications for ALS.

49.9 Conventional Dietary and Nutritional Protocols

From the conventional perspective, ALS is deemed a fatal disease of progressive deterioration. The goal of food, oral formulas, or tube feeding is to maximize caloric intake to offset weight loss and malnutrition. There is scant research on nutritional status and ALS disease progression based on the macronutrient composition of calories and even less research on the micronutrient composition. The consumer and professional resources provided by the ALS Association (ALSA), Motor Neurone Disease Association (MNDA), American Academy of Neurology (AAN), and Academy of Nutrition and Dietetics (AND) are standardized and formulaic. They don't guide the practitioner to establish therapeutic goals much beyond feeding and swallowing assessments, estimating energy requirements, and anthropometrics. A 2009 Report of the Quality Standards Subcommittee of the AAN provides a nutrition management algorithm based on the assessment of choking, dysphagia, aspiration, weight loss, and dehydration to determine food consistency, oral liquid supplements, and alternative routes for delivering calories and nutrients [61]. This nutrition management algorithm is what the majority of ALS healthcare professionals still use to determine dietary and nutritional care of their patients.

49.10 Complementary and Alternative Medicine

A majority of pALS and/or their families search beyond the knowledge and practice-based limitations of conventional healthcare professionals and their corresponding professional organizations to find complementary and alternative (CAM) practitioners and resources to hopefully achieve symptom reduction and increased longevity. In an excellent review of CAM therapies, the prevalence of supplement users is even greater in the ALS community (75%) as compared to the American public (50%) [62]. Another study of 121 pALS [63] reports that 52% of those ALS patients who took supplements reported greater vitality, physical and social functioning, and an overall improved quality of life in the early and mid-stages of the disease.

The relationship between pALS and their conventional healthcare providers may deteriorate if CAM therapies are glibly dismissed as quackery, unscientific, useless, or, at worst, dangerous. Many patients have reported mockery and intimidation when discussing CAM therapies with their doctors or conventional dietitians, and these patients either seek new healthcare providers or simply decide to keep their CAM usage private. A relationship of trust and shared decision-making, along with the appreciation that patients often make independent choices for their health, will foster a supportive alliance between patient and healthcare provider. The patient might pursue a variety of CAM therapies and therapeutic technologies such as hyperbaric oxygen, neurofeedback, LED light therapy, sonic vibration therapy, infrared sauna, ultraviolet light therapy, and others. It is not expected that the primary care physician or dietetic-nutrition professional be knowledgeable in all CAM and technological therapies; however, a collaborative approach that honors the patient's desire to explore options will support a positive provider-patient relationship.

49.11 Integrative and Functional Medical Nutrition Therapy

Integrative and Functional Medical Nutrition Therapy (IFMNT) model of care emphasizes a more patient-centered approach and less of a diagnosis-centered approach in all aspects of assessment and treatment. The IFMNT practitioner searches to discover underlying causes and mechanisms that result in the signs and symptoms of a disease. Regarding ALS, the IFMNT practitioner evaluates the following factors as they contribute to neuroinflammation and neurodegeneration: chewing and swallowing, food allergies and intolerances, alterations in substrate utilization, genomics, inflammation, oxidative stress, immune function, hormonal balance, gastrointestinal function and microbiota, mitochondrial function, and detoxification.

The IFMNT approach supports optimizing function by addressing the previously mentioned contributing factors that drive progression of ALS. This must be further scrutinized based on the patients' pre-diagnostic history, looking at antecedents, triggers, and mediators. For example, an antecedent might be a family history of neurological disorders, a trigger could be a head injury or toxic exposure, and mediators could be common toxins, stress, dietary, and other factors. Each patient must be assessed based on evaluating biochemical individuality, physical function, and psychosocial variables in collaboration with the patient's goals and expectations. It is essential that the family and caregivers be on board with the patient's nutrition and health goals to uphold a patient-centered approach.

49.12 Weight as a Risk Factor

Malnutrition is an independent prognostic factor for survival in ALS with a 7.7-fold increase in risk of death and is estimated to develop in up to one-half of patients [59]. It is very clear that weight loss and malnutrition are risk factors for disease progression. It was concluded that increased body fat prior to diagnosis is associated with a decreased risk of ALS mortality [64]. However, it is not known if maintaining weight (neutral balance) is more or less protective than gaining weight, regardless of start weight. A clinical trial [65] was commenced in 2008 to examine this question through the US National Institutes of Health titled "Quantitative Measurement for Nutritional Substrate Utilization in Patients with Amyotrophic Lateral Sclerosis (ALS)" but was terminated in 2015 due to lack of funding and support staff.

Factors that contribute to weight loss include impaired neuronal mechanisms, dysphagia, gastric discomfort and delayed emptying, fatigue and prolonged mealtimes, social isolation, upper extremity weakness, and impaired release of gut hormones that regulate appetite [66].

49.13 Calorie Requirements and Nutrition Assessment

The absence of specific methods for nutritional assessment of ALS patients shows the need to integrate clinical history, anthropometric measurement, dietary history, and laboratory tests for classification of nutritional status. Beyond this basic-level assessment, the IFMNT approach should consider how inflammation, hormonal status, oxidative stress, environmental factors, gastrointestinal function, food allergies, infections, and emotional health influence nutritional status.

There is a general agreement that energy expenditure is increased 10–20% in the majority of pALS, with estimated daily energy needs of 35 kcal/kg [67]. Coupled with these increased energy requirements is the finding that energy intakes are below recommended dietary allowances in at least 70% of ALS patients [68]. Several studies have found that standard equations used to calculate energy expenditure were not valid for pALS and indirect calorimetry has been suggested as an optimal measurement [69]. Yet indirect calorimetry is not practical in the outpatient or home setting. Another study [70] reported that clinicians should choose prediction equations that incorporate sex and age as predictor variables, such as the Harris Benedict and Mifflin-St Jeor equations that compare well with indirect calorimetry measurements. Multiple groups have reported an association between nutritional status determined by body mass index (BMI) and survival [41].

Loss of muscle is a more accurate prognostic factor in ALS than BMI [71]. Regression analysis demonstrated progressive decreases in body fat, lean body mass, muscle power, and nitrogen balance with an increase in resting energy expenditure as death approached [68]. The authors concluded that ALS patients require augmentation of energy intake rather than the consumption of high-protein nutritional supplements.

Bioelectrical impedance analysis (BIA) is a simple technique that is valid for use in ALS patients, both for single exam measurement and for longitudinal monitoring, with the use of an adapted equation and frequency of 50 KHz [72]. A more recent study [70] found that fat-free mass (FFM) by BIA was the strongest determinant of metabolic cart resting energy expenditure (mREE), along with sex and age variables. Another study [73] analyzed the relationship between phase angle (PA) obtained by BIA and found that patients with a PA <2.5 degrees had significantly poorer survival rates than patients with a PA >2.5 degrees. The study concluded that in ALS patients, PA is greatly decreased. It is related to the nutritional and cellular status of patients, and it is also an independent prognostic factor of survival.

While the nutritional status of pALS has been shown to be associated with mortality, fewer studies have evaluated the association of nutritional status and disease severity. The ALS Functional Rating Scale-Revised (ALSFRS-R) is a validated rating instrument for monitoring the progression of ALS in areas of speech, salivation, swallowing, handwriting, cutting food, dressing, hygiene, turning in bed, walking, climbing stairs, dyspnea, orthopnea, and respiratory insufficiency relative to years since onset of symptoms [74]. Park and colleagues [75] determined that nutritional status assessed by BMI and the geriatric nutritional risk index (GNRI) was negatively associated with disease severity using the ALSFRS-R and that intake of nutrients decreases with disease progression in ALS patients.

An ongoing challenge is determining the ideal balance between underfeeding and overfeeding. Underfeeding can cause diaphragm impairment, while overfeeding can increase the ventilatory load [69]. A review [76] of the therapeutic uses of the ketogenic diet (KD) reported benefits for patients with respiratory insufficiency, since the KD decreases carbon dioxide body stores and increases the efficiency of oxygen utilization. This is of particular importance in ALS patients with compromised respiratory function. Below there will be more discussion of the benefits of the KD.

49.14 Macronutrients: Carbohydrates, Protein, and Fats – What Do Dietary Studies Tell Us?

Caamaño [77] eloquently states, "nutrition is a non-genetic factor that has a transcendental influence on the prevention and treatment of neurodegenerative diseases." Yet, there is no consensus among researchers and clinicians on the best method of determining caloric requirements. A literature search reveals contradicting research on the optimal macronutrient composition of calories and scant attention on micronutrient and other nutritional requirements. Dietary studies on pALS are very difficult to obtain due to the logistics of reliable reporting and monitoring of food intake, so the majority of dietary studies rely on tube feedings for accurate measurements of quantity and composition of calories. In one recent study, investigators [78] compared three enteral diets (isocaloric, high-calorie/high-carbohydrate, and high-calorie/high-fat) and concluded that the high-calorie/high-carbohydrate diet (54% carbohydrate, 30% fat, 17% protein) resulted in less weight loss and slowed progression of ALS. The authors reported that the higher-fat diet was poorly tolerated due to gastrointestinal side effects and surmised that the same concerns would exist for a ketogenic diet (KD). This study has been widely quoted in the literature and continues to promote the customary administration of high-calorie/high-carbohydrate diets that include highly processed ingredients. Another study [79] found the opposite results based on a self-administered food frequency questionnaire. Their findings suggest that diets consisting of high intakes of carbohydrate with low intakes of fat could increase the risk of ALS. A prospective interventional study [80] showed that a high-caloric food supplement with high fat is suitable to establish body weight compared to a high-carbohydrate formula. A mouse study found that a high-fat jelly diet reduced bioenergetic stress, attenuated excess AMPK activation, and extended survival of TDP-43-associated ALS, compared to a low-fat jelly diet or standard pellets [81].

The customary standard of practice dietary guidelines for pALS emphasizes high-calorie and high-carbohydrate foods and liquid oral or tube feeding formulas containing highly refined and processed ingredients. There is currently no scientific evidence-based research to support sugar and other processed ingredients as optimal sources of calories in treatment of ALS. Physicians and dietitians follow dietary and nutritional standards of practice to prevent weight loss and negative nitrogen balance and manage dysphagia through diet, enteral or parenteral feedings. The standard of care does not include a nutritional assessment that includes vitamin, mineral, antioxidant, genomic, or digestive analysis and how these factors impact nutritional status and health parameters. The IFMNT practitioner recognizes that nutritional status impacts mitochondrial function, oxidative stress, inflammation, excitotoxicity, immunity, and other systemic parameters of ALS.

49.15 Sugar

The accumulation of sugar in neurons may explain the origin of several neurodegenerative diseases [82]. Studies in mice indicate that changes in glucose and glycogen levels occurred in peripheral tissues before any overt signs of disease and that glycogen accumulation in spinal cord tissue is a common feature of ALS [83]. Approximately 30% of pALS will show signs of frontotemporal degeneration (FTD) [84], and FTD is linked to excessively high cravings and consumption of sugar [85]. Glucose intolerance, insulin resistance, and hyperlipidemia have all been reported in ALS, as previously noted. A high-sugar diet may add to an already dysregulated metabolism in pALS. Sugar and refined carbohydrates are ubiquitous in the typical diets of pALS, yet they do not report specific sugar cravings or increased appetite.

There might be one caveat to the case against sugar found in a very interesting study on pure maple syrup in a *C. elegans* model of ALS, demonstrating neuroprotective and antioxidant properties [86]. The concentration of phenolic compounds in maple syrup (catechol, gallic acid, 3,4-dihydroxybenzaldehyde, syringaldehyde) may protect against neurodegeneration. It was shown that the combination of sugars with polyphenols in the pure maple syrup was more powerful than larger doses tested individually. The sugars, antioxidants, and phenols may have a broader efficacy when combined [86].

This discussion around the impact of sugar for pALS will focus on direct and indirect factors that impact neurodegeneration and declining health: (a) neurotransmitter imbalances, (b) inflammation and autoimmunity, (c) hormonal imbalances, and (d) brain-derived neurotrophic factor (BDNF).

49.15.1 Sugar: Neurotransmitter Imbalances (Dopamine and Serotonin)

Dopamine is a key intercommunication regulator of the immune and nervous systems [87]. ALS patients show abnormalities in dopamine metabolism [88], and antibodies from ALS patients inhibit dopamine release [89]. Sugar stimulates the release of dopamine in the nucleus accumbens, an area associated with reward and pleasure, yet sugar can lead to tolerance with a decrease in reward [90].

It has been found that platelet serotonin levels are significantly decreased in ALS and that platelet serotonin level is a strong predictor of survival independent of age [91]. Low serotonin levels usually intensify sugar cravings, because sugar produces insulin which helps tryptophan (precursor to serotonin) enter the brain [92].

Since signs of serotonin deficiency include anxiety, panic, insomnia, and depression, could the typical ALS diet be, by default, promoting desensitization to serotonin and dopamine? Could dysregulated neurotransmitter status in ALS with blunted responses to dopamine (pleasure) and lowered levels of serotonin (happiness) be exacerbated by the typical Western diet? With these concerns raised, the IFMNT practitioner should recommend a diet for pALS without sugar and refined carbohydrates to avoid aggravating neurotransmitter imbalances.

49.15.2 Sugar: Inflammation and Autoimmunity

Dietary patterns associated with the Western diet include excess refined starches, sugar, and saturated and trans-fatty acids, as well as low levels of natural antioxidants and fiber from fruits, vegetables, nuts, whole grains, and omega-3 fatty acids. These dietary patterns may cause an activation of the innate immune system, most likely by an excessive production of pro-inflammatory cytokines associated with a reduced production of anti-inflammatory cytokines [93]. The Western diet is also associated with inflammatory autoimmune diseases, including neurodegeneration [94]. Evidence is presented that several preexisting autoimmune disorders are associated with a small increased risk of ALS [95] and that autoimmunity in the form of SALS potentially underlies pathogenic mechanisms [96]. IFMNT practitioners should recommend a diet for pALS without sugar and refined carbohydrates to avoid autoimmunity, systemic inflammation, and neuroinflammation.

49.15.3 Sugar: Hormonal Imbalances (Glucose and Insulin)

The brain is highly dependent on glucose for fuel. The majority of glucose uptake by the brain is independent of insulin, with only 10–20% of uptake insulin-dependent. Insulin resistance will inhibit this glucose-stimulated uptake into the brain and may decrease neuronal energy availability [97]. Glucose intolerance and insulin resistance have been reported in a significant percentage of pALS, suggesting defective carbohydrate metabolism [98]. Subsequent studies have confirmed multiple pathways linking diabetes and ALS [99]. However, conflicting data associate diabetes with both increased risk for ALS [100] and reduced risk for ALS. Reduced risk was hypothesized to be through a compensating neuroprotective role that was also associated with reduced cardiovascular risk factors [101]. But it is certainly not recommended to promote diabetes or cardiovascular risk factors in patients with ALS by adopting unhealthy dietary patterns, since the downside is chronic systemic inflammation caused by excessive and inappropriate innate immune system activity.

49.15.4 Sugar: Brain-Derived Neurotrophic Factor (BDNF)

Brain-derived neurotrophic factor (BDNF) is a neurotrophic hormone that plays an important role in survival, maintenance, and growth of many types of neurons promoting function and plasticity of synapses. BDNF is expressed mostly in the hippocampus, hypothalamus, and cerebral cortex. High-sugar and high-fat diets and obesity reduce BDNF, leading to impaired memory and learning [102]. BDNF serum levels were significantly lower in ALS patients who had lower ALSFRS-R scores. BDNF serum levels might be a biomarker and possibly contribute in part to ALS phenotypic variability [103]. A high-sugar and refined carbohydrate diet results in neurodegeneration by lowering BDNF, leading to another compelling reason to eliminate sugar and simple carbohydrates for pALS.

49.16 Gluten

It is not surprising that the subject of gluten is somewhat controversial in ALS, as it is in the general medical profession. It is known that celiac disease can mimic ALS [104]. There were significantly more cases than expected of ALS associated with a prior diagnosis of celiac disease, raising the possibility of shared genetic or environmental risk factors [95]. White matter lesions suggestive of ALS have also been attributed to celiac disease [105].

Gluten sensitivity may result in neurologic manifestations. One proposed mechanism is through stimulation and elevation of transglutaminase-6 antibodies (TG6), a neurological-specific marker, in the serum of patients with ALS [106]. Non-celiac gluten sensitivity refers to patients with gluten sensitivity in the absence of enteropathy but demonstrating neurological manifestations similar to patients with celiac disease [107]. It appears that TG6 mediates this response with or without HLA-DQ2 or HLA-DQ8 (Hadjivassiliou, 2016). Other studies report no association between gluten sensitivity, celiac disease, and ALS [108, 109].

Wheat should be avoided regardless of gluten content since residues of the organophosphate herbicide, glyphosate, have been found in wheat-containing products. Glyphosate is the active ingredient in Roundup™, which has been researched in isolation from the more toxic product actually used [110]. Glyphosate disrupts sulfate transport from the gut to the liver and may lead over time to severe sulfate deficiency throughout all the tissues, including the brain [111]. Sulfate deficiency in the brain has been associated with ALS. Glyphosate interacts synergistically with such factors as insufficient sun exposure, nutrient deficiencies of sulfur and zinc, and synergistic exposure to other xenobiotics whose detoxification is impaired by glyphosate [112].

A comprehensive article detailed the role of glyphosate as an amino acid analogue of glycine; the authors state that almost all cases of sporadic ALS share a common neuropathology characterized by transactive response DNA-binding protein 43 (TDP-43) protein inclusions [112]. Multiple missense mutations have been implicated in association with ALS regarding TDP-43 encoding of the C-terminal glycine-rich domain. Therefore it is likely glyphosate acts as an analogue to glycine, thus replacing glycine and exerting toxicity to increase the risk of ALS. ◘ Figure 49.4 shows the proposed downstream consequences of glyphosate exposure leading to the possible development of sporadic ALS [19].

Crop desiccation with glyphosate just prior to harvest may increase residues in foods prepared with wheat [113]. Glyphosate was declared a probable carcinogen by the International Agency for Research on Cancer for the World Health Organization (WHO). The WHO report and the associated researchers have suffered an intense backlash from Monsanto, which has accused them of "junk science" [114]. Glyphosate exposure is associated with oxidative damage and neurotoxicity in the hippocampus [115] and potentially disrupts the microbiome [116]. The IFMNT practitioner does not defend known toxins as part of anyone's diet, especially in neurodegenerative and high-risk patients.

Recommended alternative flours include organic quinoa, amaranth, buckwheat, legume-based flours, nut-based flours, millet, tapioca, coconut flour, oats (GF), flax meal, and other non-gluten products. Consuming organic potato, rice, and corn is allowed. Carbohydrate recommendations as part of

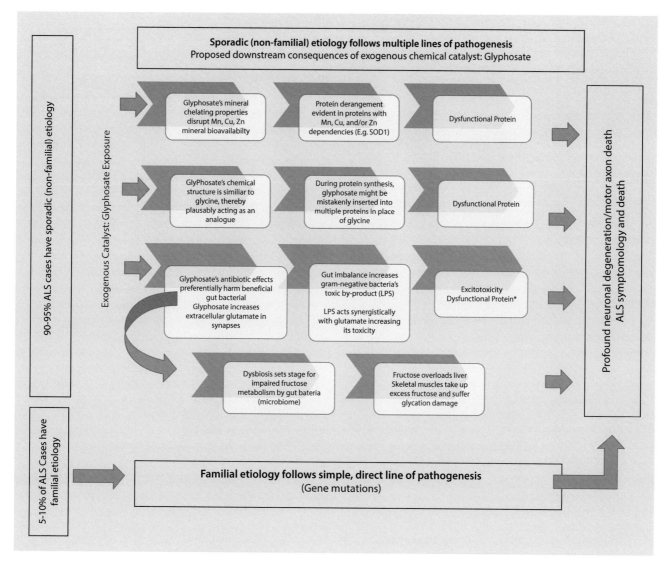

☐ **Fig. 49.4** Downstream consequences of glyphosate exposure. ∗ ALS hallmark sign of increased TDP-43 mislocation and aggregation. (Reprinted from Seneff et al. [19]. Retrieved from ▶ https:// bangmosnowdotcom.files.wordpress.com/2016/08/glyphosate-als.pdf. With permission from Creative Commons License 4.0: ▶ https:// creativecommons.org/licenses/by/4.0/)

total calories will be discussed below. In summary, there is no evidence-based research to support the common inclusion of sugars, refined carbohydrates, or wheat in the diets of pALS. On the contrary, because of potential neurotoxic effects, metabolic dysregulation, gluten sensitivity, and glyphosate exposure, diets excluding sugars, refined carbohydrates, and gluten/gliadin would be the better option.

49.17 Protein

There is no consensus among researchers and healthcare professionals on protein requirements for pALS. ALS is a catabolic process of destructive metabolism that costs the body energy as molecules break down. The anabolic process cannot stay ahead of the net catabolic loss. Piquet and colleagues report in the absence of available data on protein requirements in ALS an intake ranging between 1 and 1.5 g/kg/day seems reasonable. This was with a caveat that there is a risk of deficiency when intake is less than 1 g/kg/day and that an increase to 1.5 g/kg/day, considered as harmless, could be useful in the event of hypercatabolism [67]. The calculation of protein requirements was based on the recommendations of the Brazilian Society for Food and Nutrition suggesting 1.0 g protein/kg/day with intakes ranging 1–1.2 g/kg/day. The authors go on to state food intake with a source of high biological value protein should be highlighted and comprise roughly 70% of the total protein intake per day [117].

It seems paradoxical that protein recommendations in terms of the amount, forms, and sources of protein cannot effectively address the critical reduction of skeletal protein muscle mass and function seen in ALS. As previously noted, the common approach of providing adequate calories to spare muscle mass is considered more effective than focusing on specific protein requirements.

49.18 Dairy and Soy as Protein Sources in Oral and Tube Feedings

Dairy or soy comprises up to approximately 50% of protein intake in the typical ALS patient on oral/enteral commercial formulas [118]. There is no scientific evidence that either of these protein sources are optimal for pALS, yet by default these protein sources are standard of practice for ALS.

49.18.1 Dairy

There is evidence that milk-derived foods appear to have a negative effect on ALS disease symptoms according to a study on 302 patients in the ALS Multicenter Cohort Study of Oxidative Stress, although the reasons are not understood (Anderson, 2016). Another study found that a diet enriched in casein may induce astrocyte activation through exaggerated blood-brain barrier (BBB) permeability [119].

Amen explains that casein is an excitotoxin that can lead to brain inflammation [120]. Blaylock further explains that sodium caseinate, calcium caseinate, whey protein, whey protein concentrate, and whey protein isolate are ingredients with hidden free glutamic acid excitotoxins [121].

An FDA report highlights findings that commercial milk contains permitted antibiotics, while some samples contain illegal antibiotics [122]. Antibiotic-induced neurotoxicity may affect at-risk patients causing neuropathy, neuromuscular blockade, nonspecific encephalopathy, and seizures [123]. Other contaminants in dairy products include exogenous estrogens [124], and detergents, sanitizers, disinfectants, and melamine, all of which pose a threat to human health [125].

Adding to the evidence presented above, the organophosphate herbicide glyphosate is the most widely used GMO herbicide in dairy cattle feed. In the first ever testing of glyphosate in feeding tube liquid, glyphosate may have untoward effects in ALS and several other neurodegenerative pathologies [112]. The FDA has suspended testing for glyphosate residues in food, including milk, pending further validation of testing methods [126]. ALS patients are prone to mucus plugs that occur when bronchial secretions accumulate and obstruct airflow [127]. If the cough is weak and the plug cannot be cleared, the situation can be deadly. Many pALS report that milk exacerbates phlegm; however, the scientific literature does not support the connection between milk and mucus or phlegm production. The IFMNT practitioner would assess milk sensitivity or allergy on an individualized basis.

There is no evidence-based research to support dairy products for pALS. In fact, there is evidence-based research to advise against dairy products in the diet for pALS [128]. This report showed unexplained negative effects of milk-derived products when included in ALS diets. Some substitutions for dairy protein to consider are vegetarian-derived proteins from pea, quinoa, almond, and hemp. An IFMNT perspective must consider the potential risk of dairy products due to associations with allergies, intolerances, excitotoxins, hormones, antibiotics, glyphosate, and other contaminants as we educate and nutritionally support the ALS patient.

49.18.2 Soy

pALS are frequently ingesting soy-derived products in processed foods and oral/tube feedings, especially as a protein source. Soy products are highly allergenic and genetically modified and contain glutamic acid which may potentiate excitotoxicity. Based on USDA survey data, the acres planted of herbicide-tolerant GMO soy has risen from 17% in 1997 to 94% in 2014 [129]. There is no evidence-based research to support the use of soy protein for pALS. However, inclusion of organic tofu and fermented soy (tempeh, miso, natto) might offer a healthy alternative for pALS without soy allergies.

Oral/tube feeding supplements frequently contain soy protein isolate, which has processed free glutamic acid [130]. Glutamate is a primary CNS excitatory neurotransmitter. In 1968, monosodium glutamate (MSG) was identified as the cause of "Chinese restaurant syndrome," also known as "MSG symptom complex," following numerous complaints made to the FDA of dizziness, headaches, flushing, chest pain, bronchospasms, numbness, and other symptoms [131]. There is an ongoing controversy about the impact of glutamic acid and MSG on human health. Many scientific articles report that neither dietary MSG as a food additive, free glutamic acid in processed foods, free glutamic acid in natural foods, or bound glutamic acid in whole protein foods affect brain function or neurotransmission [132]. However, MSG and glutamic acid produced for use in supplements and processed foods can cause neurodegenerative diseases in animals and humans, especially with long-term intake [133].

An excellent mechanistic review describes the luminal (blood facing) and abluminal (brain facing) membranes and their role in glutamate transport. The review describes how glutamate traverses the luminal membrane but cannot pass the abluminal membrane to enter the brain in any appreciable quantities except in the circumventricular organs [134]. This phrase "any appreciable quantities" acknowledges some vagueness that exists around the BBB. Could genetic alterations in glutamate metabolism [135], in addition to BBB susceptibility in neurodegeneration, predispose pALS to toxicity from dietary glutamates?

In summary, there is no evidence-based research to support the use of dietary soy-derived proteins for ALS. In fact, due to the high allergic potential, GMO contamination, and the presence of glutamates in soy products, it is advised that the IFMNT practitioner exclude processed soy for pALS.

49.19 Amino Acids

Studies have examined the role of individual amino acids and ALS to determine their role in muscle (maintenance, growth, and breakdown) and neurotransmitter balance. Several studies agree that ALS shows imbalances in plasma amino acids compared to controls and imbalances in excitatory versus inhibitory neurotransmitters. The ability to personalize amino acid requirements is not easily determined by blood, urine, or CSF biomarkers of amino acids.

It is known that plasma amino acid composition of pALS differs from the general population. For example, there are significantly lower percentages of plasma tyrosine, valine, methionine, leucine, and isoleucine, as well as significantly higher percentages of glutamine and serine in ALS than in controls [136]. The clinical state significantly influenced the percentage of plasma phenylalanine and alanine [136]. Evidence of glutamate imbalances in ALS is associated with CNS alterations, and it was found that ALS patients had statistically significant elevations in plasma levels of aspartate, glutamate, and glycine [137]. Another study [138] found that higher levels of glutamate were correlated with early onset (<55 years.) compared to later onset (>74 years.) of ALS. Since plasma amino acids might show biological correlates of the age of onset in ALS, this could potentially provide insight into possibly aberrant biochemical pathways that might unlock key pathological pathways.

The biosynthesis of amino acids in the brain is regulated by the concentration of amino acids in plasma, although the levels of CSF and plasma amino acids are not directly correlated [139]. Essential amino acids cannot be synthesized by the brain and must be supplied from diet and protein breakdown. Other amino acids, including taurine, have various rates of movements into the brain, since many are synthesized endogenously which restricts peripheral entry into the brain [140].

Table 49.1 shows particular amino acids that have been the subject of interest in ALS: Branched Chain Amino Acids (BCAA), L-threonine, L-arginine, glutamine, serine, and taurine.

Table 49.1 Amino acids

	Proposed action	Evidence for or against	References
Branched-chain amino acids (BCAA), leucine, isoleucine, valine	Stimulate muscle growth and recovery after exercise; transported into the brain using the same pathway as serotonin and may compete for access to a limited number of transporters	BCAAs did not show clinical benefit and might worsen pulmonary symptoms; showed lack of benefit in improving survival time, muscle strength, or disability; altered immune profile of microglia as possible mechanism by which BCAAs might turn into toxicants and facilitate neurodegeneration; proposed decreases serotonin (*Not* recommended)	[91, 144–146, 147–149]
L-Threonine	Conversion of L-threonine to glycine, inhibitory amino acid blocking excitatory effect of aspartate	Conflicting evidence for improved vocal ability, swallowing, energy, and reduced drooling and spasticity; possible worsened pulmonary function (possibly ineffective)	[150–152]
L-Arginine	In vitro and in vivo demonstration of neuroprotective effects and prolonged life span; enters the Krebs cycle and may provide energy and promote muscle protein synthesis and strength	(No evidence for or against use)	[9, 142, 153]
L-Glutamine	In neurons, mitochondrial metabolism of exogenous glutamine is mainly responsible for the net synthesis glutamate; necessary for the synthesis of glutathione	Limited research; no clear guidelines; concerns for glutamate synthesis (caution)	[141, 154–155]
L-Serine	Low levels are associated with demyelination and neurological symptoms; endogenously produced by astrocytes and abundant in diet	In vitro and in vivo studies to be neuroprotective for ALS. In Phase I study, L-serine in doses up to 15 g twice daily appears to be safe in patients with ALS; possible dose-related slowing of disease progression; currently in Phase II trial (cautious recommended use in amounts used in Phase I study)	[156–158]
Taurine	Neuroprotective, prevents glutamate toxicity in cultured neurons, modulates intracellular calcium excess	May be a good candidate for therapeutic trials in ALS	[467]

Dietary protein should be from organic sources whenever available. The best animal protein sources are 100% pasture-raised beef and chicken, wild meats, wild-caught Alaskan salmon, herring, sardines, and eggs. Other good quality proteins are from nuts, nut butters, seeds, hemp, quinoa, organic soy, and lentils.

49.20 Fats

Lipids are a lifeline for the proper structure and function of cell membrane bilayers and as lipid rafts in subdomains of the plasma and mitochondrial membranes [159]. Abnormal lipid metabolism plays a central role in neurodegeneration. ◘ Figure 49.5 shows the various lipid systems in ALS, other CNS disorders, and brain injuries [160]. Every neurodegenerative disease has its own signature abnormalities in lipid metabolism [160]; therefore, a clinician cannot generalize a lipid-based treatment program to effectively address all neurodegenerative diseases.

◘ Figure 49.6 lists some interesting brain and fat facts to "keep in mind" [161–172].

◘ Figure 49.7 describes the important ways that lipids protect brain cell membranes [44].

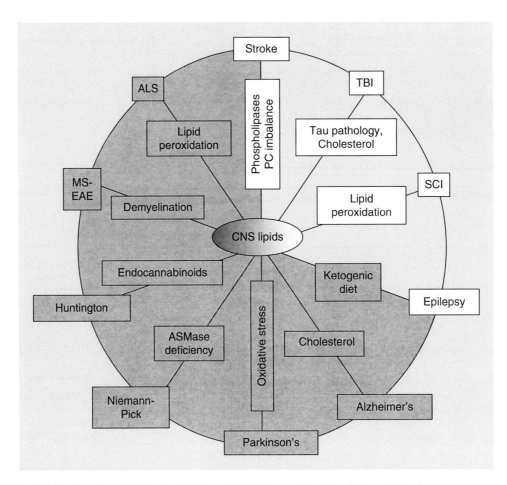

◘ Fig. 49.5 Lipid wheel. (Reprinted from Adibhatla and Hatcher [160]. With permission from Springer Nature)

- 2% of total body weight
- 20% of total caloric expenditure
- 60% fat & fattiest organ
- 25% of total body cholesterol
- 70% of CNS cholesterol resides in myelin
- >95% of cholesterol synthesized in brain
- <5% of plasma cholesterol crosses BBB
- Brain contains 1:1 ratio AA:DHA
- DHA, AA, phosphatidylcholine and phosphatidylserine cross BBB
- BBB integrity is compromised in ALS
- Brain prefers glucose and ketone bodies for energy

◘ Fig. 49.6 Interesting brain and fat facts

- SUPPORT cell membrane structure and function
- SUPPORT lipid raft substructure
- SUPPORT energy homeostasis
- SUPPORT cell signaling
- SUPPORT synaptic function
- SUPPORT myelin sheath
- REDUCE protein aggregation and folding
- REDUCE neuroinflammation
- REDUCE oxidative stress
- REDUCE excitotoxicity
- REDUCE toxin exposure

◘ Fig. 49.7 Important ways lipids protect brain cell membranes. (Based on data from Ref. [44])

49.21 Optimal Sources and Amounts of Dietary Fat

Most preclinical and clinical dietary research on lipid metabolism and ALS have not been translated into therapeutic interventions. Dietary fatty acids that freely cross the blood-brain barrier and contribute to the composition of brain fats are arachidonic acid, DHA, phosphatidylcholine, and phosphatidylserine [173], yet common practice does not manipulate these fats for therapeutic benefit. Dietary studies typically limit the role of fat as a macronutrient percentage of total calories, and fewer yet have examined the types and sources of fatty acids. The conventional standard of practice in ALS dietary treatment does not specify the balance, source, or composition of dietary fat via oral or tube feedings. We will discuss the potential benefits of the ketogenic diet (KD) for pALS and other chronic disorders here and elsewhere in this text.

A consistently ignored variable in most dietary studies is an analysis and description of the quality of the fat sources. We should be asking if lipid sources from organic cold-pressed oils behave differently in the body compared with industrially derived lipids that are oxidized, bleached, deodorized, solvent extracted, and genetically modified. How does the quality of the industrial-made lipid products influence immune and inflammatory responses, and how might these factors skew published literature outcomes regarding the anti-inflammatory and inflammatory balance? In other words, we need to understand the quantitative versus qualitative impact of lipids derived from various sources.

49.21.1 Omega-3 PUFAs (ALA, EPA, DHA)

Important omega-3 polyunsaturated fatty acids (PUFAs) include alpha-linolenic acid (ALA), eicosapentaenoic acid (EPA), and docosahexaenoic acid (DHA). The circulating levels of DHA and EPA available for uptake by the brain represent a combination of those derived from the diet and those biosynthesized in the liver from alpha-linolenic acid (ALA) [176]. Recent research highlights the critical role of omega-3 fatty acids in preserving the integrity of the BBB to protect the brain from blood-borne bacteria, toxins, and other pathogens [174].

Consumption of foods high in omega-3 PUFAs (alpha-linolenic and fish oils) may help prevent or delay the onset of ALS. Inflammatory processes that destroy motor neurons in the spinal cord, brain stem, and cortex during ALS could be reduced by DHA administration due to its potent anti-inflammatory properties [175]. It is known that DHA increases total glutathione levels in microglia cells, thus enhancing their antioxidant capabilities. In parallel, DHA causes a reduction of pro-inflammatory cytokines, such as IL-6 or TNF-α [175]. A prospective study evaluated five large cohort studies on a pooled analysis of 995 individuals with ALS, finding that overall total omega-3 intake was associated with a 34% reduced risk for ALS in the multivariable model comparing the highest with the lowest quintile [176]. The men in this study had a median dietary omega-3 intake of 1.40–1.85 g/day, and women had a mean dietary omega-3 intake of 1.14–1.43 g/day [176].

DHA does more than just protect the brain from neuroinflammation. It can also promote resolution of neuroinflammation. Neuroprotectin-D1 and Resolvin-D1 are specialized pro-resolving lipid mediators (SPMs) derived from DHA and dependent on sufficient DHA intake, conferring protection and resolution of inflammation [177]. Omega-3 PUFA may reduce neuroinflammation through prevention of microglial activation by pro-inflammatory cytokines and the production of anti-inflammatory and anti-oxidant Neuroprotectin-D1 [178]. The study findings on Resolvin-D1 may offer a new approach to attenuating the inflammation in ALS.

In the face of dietary (n-3) PUFA deprivation, homeostatic mechanisms slow brain DHA metabolism while increasing AA metabolism. These changes may exacerbate neuroinflammation and excitotoxicity [179]. Conversely, in the face of dietary (n-6) PUFA deprivation, homeostatic mechanisms increase brain DHA metabolism and their anti-inflammatory metabolites that likely promote neuroprotection [180]. The above examples illustrate how the body coordinates availability of dietary fatty acid intake with upregulation and downregulation of fatty acid metabolism in a dynamic balancing process within the brain.

In our enthusiasm about the therapeutic role of omega-3 PUFAs, it is important to caution against excessive or imbalanced dosages. All omega-3 PUFAs can be potentially oxidized by various environmental factors such as presence of microbes, UV light, heat, metals, and other factors [181]. One study found that EPA accelerated ALS progression in a mouse model of ALS, suggesting that prolonged high exposure of EPA should be contraindicated for ALS patients [182]. That same study did not evaluate DHA. The study used ethyl esters (70% EPA, 10% DHA), whereas eating fish generally provides higher levels of DHA [183].

The numerous publications supporting the neuroprotective effects of omega-3 PUFAs (especially DHA) warrant supplementation for most pALS. Keeping fish oil doses between 2 and 4 g/day might be therapeutic for pALS. pALS should be provided with the following resources on fish to avoid and fish that is safe [184–185]. A vegetarian or vegan diet is not recommended for pALS; however, healthy vegan choices of omega-3 PUFAs are derived from foods and oils containing alpha-linolenic acid (ALA) such as walnuts, flax seeds, pumpkin, chia seeds, hemp seeds, natto, cashews, Brazil nuts, Brussels sprouts, kale, spinach, watercress, mustard oil, hemp oil, and walnut oil [186]. There is only a 2–5% conversion rate of ALA to DHA [187], so the best direct vegan DHA source comes from marine-sourced algae and algal oil [188]. It is very challenging to achieve optimal omega-3 PUFA balance of EPA:DHA in vegan and vegetarian diets, especially considering the unique requirements for pALS. For more personalized recommendations on the use of fish oil, the IFMNT practitioner can assess erythrocyte fatty acid status through laboratory testing for corresponding brain and erythrocyte levels of fatty acids (AA, EPA, DHA) and the total balance of cell membrane fatty acids.

49.21.2 Omega-6 PUFAs (LA, AA)

There is an ongoing international debate among the scientific and medical community regarding the inflammatory role of linoleic acid (LA). The intake of LA has increased 20-fold over the twentieth century to our current intake at approximately 6% of total calories [189]. The Academy of Nutrition and Dietetics (AND) advises LA dietary intake at 3–10% of total calories and concludes no risk of inflammation based on its analysis of 15 clinical trials [190–191]. Opposing scientific literature indicates that omega-6 PUFAs promote inflammation and augment many disease processes, whereas omega-3 PUFAs seem to counter these adverse effects [192]. The deleterious effects of LA-rich diets promote oxidized fatty acids, causing alterations in the gut microbiome that have potential to foster systemic inflammation [193]. The plethora of contradicting research on the anti- vs. pro-inflammatory roles of LA may reflect unaccounted for variables such as (a) substitution of LA for saturated, omega-3 or trans-fatty acids, (b) total ratio of omega-6/omega-3 dietary PUFAs, or (c) variations in desaturase and elongase enzyme functions in individuals, and genomic variability.

Most dietary studies in ALS patients use commercial oral and enteral formulas that contain processed omega-6 PUFAs containing mostly LA from soy, corn, and canola oil, with lesser amounts of omega-3 PUFAs from fish oil. Although they are called vegetable oils, they are actually derived from tough seeds and legumes. These types of lipids are unstable and susceptible to oxidation. They contain residues of pesticides and solvents and are genetically modified, which alter the microbiome and promote inflammation [194].

From an IFMNT perspective, pALS should optimize the quality of their dietary fat intake by avoiding oxidized, bleached, deodorized, solvent extracted, or genetically modified sources of fatty acids. pALS should choose organic, unrefined, cold-processed, and expeller-pressed oils, preferably in dark glass bottles. Assuming adequate intake of omega-6 fats to prevent deficiency (1–2% of calories), pALS should focus on healthy omega-3 fats (e.g., salmon, sardines, mackerel, herring, chia seeds, flax seeds, walnuts, hemp seeds), omega-7 fats (macadamia nut oil, sea buckthorn oil, avocado oil), omega-9 fats (e.g., olives, olive oil, grapeseed oil, avocado, avocado oil, almonds, macadamia nuts, eggs), and saturated fats (e.g., coconut, grass-fed butter, ghee, and grass-fed animal proteins). pALS can reduce inflammation by optimizing ratios of omega-6 PUFA to omega-3 PUFA from the typical Western diet of 25:1 toward optimal levels of 3–1:1 [195–196].

49.21.3 Medium-Chain Triglycerides

Medium-chain triglycerides are defined as triglycerides that contain two or three fatty acids on the glycerol backbone that are medium chained: C6, C8, C10, and C12 from palm or coconut oil [197]. Coconut oil is preferred because palm oil isn't considered socially or environmentally sustainable [198], and there is questionable reliability even when palm oil is verified through certification by the Roundtable on Sustainable Palm Oil [199].

Coconut oil, a medium-chain lipid, contains more than 50% lauric acid (C12) and equal parts caproic acid (C6), caprylic acid (C8), and capric acid (C10) [200]. Caprylic acid (C8) converts to ketones the most efficiently and may be easiest on digestion; thus it is preferable to use fractionated coconut oil as a combination of caproic and caprylic acids or caprylic acid alone [201]. Caprylic acid has been specifically studied in ALS mice showing reduced progression of weakness, protection of spinal cord motor neurons, and a lower mortality rate, which may translate to a high impact on quality of life for ALS patients [202]. A clinical trial was designed to determine safety and tolerability of caprylic acid in ALS patients, with the goal for subsequent studies to determine if caprylic acid slows or stops ALS disease progression [203]. A consequence of solely using caprylic acid is that it might decrease glycolysis and the levels of TCA intermediates in the brain [204].

MCTs do not require lipase for digestion or bile acids for emulsification [205]. They are rapidly absorbed into the portal circulation, rather than being transported through chylomicrons into the lymphatic system. They convert into ketones (beta-hydroxybutyrate, acetoacetate, acetone) in the liver, yet are never metabolized by the liver. These ketone bodies provide energy for the peripheral body and cross the BBB to provide energy to the brain [205].

Triheptanoin is synthesized from heptanoic acid (C7) as an alternative MCT fuel source, having the unique role of sparing glycolysis and refilling the TCA cycle (anaplerosis) by providing pyruvate for carboxylation into oxaloacetate, a key intermediate. Triheptanoin slows motor neuron loss, muscle wasting, and the onset of motor symptoms in ALS mice, which is likely mediated by reducing oxidative stress and preserving mitochondrial function [204]. ▶ Clinicaltrials.gov lists 28 trials for triheptanoin in various mitochondrial diseases, Huntington's disease, ALS, and other conditions [206].

49.21.4 Saturated Fatty Acids (SFAs)

The Western diet (high in saturated fat, trans fats, and omega-6 fatty acids) causes damage to the brain, including oxidative stress, insulin resistance, inflammation, and changes to vascularization and BBB integrity [207]. Saturated fatty acids (SFAs) are derived from (a) animal foods that contain cholesterol, (b) processed foods that also contain trans fats, and (c) vegetarian sources (coconut oil, palm oil). In this analysis of dietary fat substitutions, replacing the equivalent calorie value of SFAs with PUFAs and monounsaturated fatty acids (MUFAs) was associated with a 27% combined reduction in mortality from neurodegenerative disease. This effect was not recognized when SFAs were substituted with carbohydrates [208].

Blood cell palmitoleic acid (MUFA16:1) to palmitic acid (SFA16:0) is an independent prognostic factor for ALS, with higher ratios extending life for 10 months independent of BMI [209]. The highest content of dietary palmitoleic acid is found in macadamia nut oil, sea buckthorn oil, and lesser amounts in avocado [197]. Palmitoleic acid has beneficial effects of lowering hsCRP and TNF-alpha and improving glucose and insulin regulation [210].

The balance of fatty acid intake is emphasized for optimal brain health, and though current dietary guidelines suggest SFA intake should comprise less than 10% total calories daily [211], individual assessment of fatty acid status is best for personalizing requirements during altered metabolic states such as ALS.

49.21.5 Cholesterol

The brain synthesizes its own cholesterol. About 25% of total body cholesterol resides in the brain, and the majority (70%) of brain cholesterol resides in myelin that surrounds axons and facilitates the transmission of electrical signals [212]. Other crucial functions of brain cholesterol are to build and maintain membranes, modulate membrane fluidity, and support synapse development, formation and function, dendrite differentiation, axonal elongation, and long-term potentiation [213]. Cholesterol has a remarkable dual effect on membranes by making them stiffer but retaining the fluidity required for membrane function, similar to an "antifreeze agent" [214]. The maintenance of cholesterol homeostasis is crucial for the metabolic and structural benefits in neuronal and glial plasma membranes and in the composition of myelin. The impact of modifying free cholesterol and cholesteryl ester levels in the brain to decrease neurodegeneration is an active area of scientific and pharmacologic research [215].

Cholesterol balance is dysregulated in several neurodegenerative diseases, yet the brain disease sites and mechanisms of cholesterol metabolism are variable, thereby lacking a common therapeutic approach [216]. A recent report measured the serum and CSF in 20 ALS patients to compare levels of cholesterol and cholesterol metabolites [217]. As neurons die in ALS, cholesterol is released into the CNS and becomes toxic to surrounding neuronal cells, while saturated metabolic pathways are insufficient at removal of excess cholesterol [217].

Demyelination, or the loss of "insulation" of oligodendrocytes, damages the ability and efficiency of nerve transmission. Demyelination is a primary symptom of Multiple Sclerosis (MS), and the brain shows disturbed lipid metabolism with serum lipid profiles in the normal range. In MS mouse models, increasing dietary cholesterol actually supports remyelination and reduced axonal injury [218]. Demyelination is considered a secondary symptom of ALS, which occurs after neuronal death has begun [219]. However, researchers at Johns Hopkins University found that significant changes also occur in oligodendrocytes prior to neuronal death; thus demyelination is both a cause and result of damage to motor neurons. It is important to appreciate that motor neurons are dependent on healthy oligodendrocytes and that oligodendrocytes are dependent on sufficient availability of cholesterol [220]. Therefore, adequate cholesterol in the diet and serum is important for healthy oligodendrocytes.

The generalized message throughout the literature is that the brain is highly dependent on dietary fatty acids but is quantitatively less dependent on dietary cholesterol since the brain (neurons and astrocytes) synthesizes cholesterol endogenously and is independent of the peripheral cholesterol levels [213]. However, we might question this rigid assumption of the blood-brain barrier (BBB), as it has been found through radiolabeled dietary cholesterol feedings that <5% does cross the BBB to comprise brain cholesterol [166]. Exogenous cholesterol enters the CNS through an impaired BBB, resulting in enhanced repair and amelioration of the neurological phenotype in two distinct models of remyelination [218]. The data suggest that dietary cholesterol directly facilitates repair of myelin by modulating the profile of growth factor expression and promoting oligodendrocyte precursor cell (OPC) proliferation [218]. These findings emphasize the safety of dietary cholesterol and its nutritional importance in both prevention and remediation of remyelination.

Does dyslipidemia improve or worsen ALS? One study found that dyslipidemia (elevated LDL:HDL) was possibly protective for ALS patients, associated with an extra 12 months of survival [221]. Another study found that dyslipidemia was associated with an extra 14 months of survival and that a diet rich in lipids and calories should be considered [222]. A subsequent study found that hypolipidemia was associated with ALS even when the diet was intact [223]. A review on 11 antecedent diseases in a large-scale human ALS study found that the prevalence of hyperlipidemia and all diseases was less in ALS compared to the age and gender ratio-matched control conditions [224]. Another study found that there were no significant differences between mean total cholesterol, LDL, LDL:HDL, or triglycerides between ALS patients and controls; however, ALS patients with higher triglycerides (>127.5 mg/dl) had increased life expectancy by 5.8 months [225]. Over the past decade, studies have found varying results and associations between serum lipid measurements, morbidity, mortality, and ALS, resulting in no evidence-based consensus regarding lipid measurements as biomarkers for ALS.

Weighing the risk/benefit of dietary cholesterol in the presence or absence of elevated lipid markers, the ALS patient will likely benefit from more dietary cholesterol. Eggs from organic pasture-raised hens would be the healthiest and easiest way to raise serum cholesterol in pALS.

49.21.6 Phospholipids (PLs)

Neurodegenerative diseases incur extensive oxidative damage to mitochondrial cell membranes. Mitochondrial dysfunction, a hallmark of ALS, is highly influenced by derangements in lipid structures and metabolism, especially involving glycerophospholipids such as phosphatidylcholine (PC), phosphatidylethanolamine (PE), phosphatidylserine (PS), phosphatidylinositol (PI,) phosphatidic acid (PA), and phosphatidylglycerol (PG) [159]. PC accounts for approximately 50% of the phospholipids that make up cell membranes [141]. These phospholipids are essential components of the inner and outer mitochondrial cell membranes, governing membrane integrity, fluidity, fission, fusion, and cell-to-cell signaling [226].

Phospholipids are fundamental building blocks of all cell membranes, forming stable bilayers, with their polar head groups exposed to water and their hydrophobic tails buried in the interior of the membrane [227]. The fluid-mosaic model of membrane structure describes the structure of the plasma membrane as a mosaic of components, which includes the phospholipids, cholesterol, proteins, and carbohydrates and provides the structure to give the membrane a fluid character [228–229].

Beneficial effects of dietary PLs have been known since the early 1900s as they relate to different illnesses and symptoms [230]. Dietary PLs are found in egg yolk, soy, milk, and marine sources, and the average diet contains 2–8 grams [230]. Phosphatidylcholine is sometimes used interchangeably with lecithin, although the two are different. Choline is a component of PC, which is also a component of lecithin, but lecithin is also a mixture of several lipids and phospholipids [231]. Various phospholipid products are used to replace damaged lipids in cellular membranes and organelles and are commercially available. These include intravenous preparations available pharmaceutically and over-the-counter oral preparations.

Veterans of the 1991 Gulf War had an "outbreak" of ALS for 10 years presumably due to exposure to toxins [232]. Phospholipid profiling of plasma from Gulf War illness showed dysfunction within DHA and AA-containing phospholipids and inflammatory imbalances [233]. There exist numerous peer-reviewed publications on membrane lipid replacement for cancer and other fatigue-causing illnesses; however, no specific studies on ALS. A Phase I clinical trial that was underway at the time of this writing might advance our understanding of dietary phospholipids in ALS: "The Combination of Phospholipids and Medical Herbs for the Treatment of Patients with Amyotrophic Lateral Sclerosis (ALS): A Pilot Study" [234].

Most clinical studies have used 1.5–3 g daily of dietary phospholipids or 0.5–2 g daily of IV administration [229]. Commercially available dietary supplements contain approximately 1.5 g per dose, and therapeutic doses are upward of 4 g daily without any side effects [235]. It is more expensive and invasive to receive IV phospholipids, and less accessible due to travel for appointments and a paucity of physicians competent to administer this therapy. Dietary phospholipids are affordable and convenient and can be given orally or in tube feedings. To date, peer-reviewed scientific studies on phospholipid therapy for mitochondrial membrane support have focused on dietary but not IV phospholipids. Many IFMNT physicians use intravenous or dietary phospholipids as a therapeutic intervention for pALS and other neurodegenerative disorders. This is a good example of applying evidence-based research to treat mitochondrial damage, a hallmark of ALS, for a potentially therapeutic and personalized intervention.

49.22 Ketogenic Diet (KD) and Medium-Chain Triglycerides (MCTs)

The diagnosis of ALS is concomitant with a race against malnutrition mediated by metabolic derangements in aerobic and anaerobic metabolism, substrate utilization, impaired mobility, chewing and swallowing, frequent gastrointestinal disturbances, inflammation, breathing abnormalities, sleep disturbance, and mental and emotional factors. The ketogenic diet (KD) optimizes energetic efficiency and beneficially modulates various metabolic pathways to help protect against neurodegeneration. Despite the relative lack of clinical data, there is an emerging scientific literature supporting the broad use of the KD (and its variants) against a variety of neurological conditions. Regarding the KD, the authors ask, "How can a simple dietary change lead to improvement in disorders with such a huge span of pathophysiological mechanisms?" [236].

The KD is the intentional or therapeutic provision of ketones for energy as part of a calorie-sufficient diet consisting primarily of fatty acids, limited protein, and very low carbohydrate intake. The KD and the following variants may be neuroprotective as seen in ◘ Fig. 49.8 [237]. For further information regarding the KD, please see ▶ Chap. 22.

In conclusion, there is very little clinical research on the KD in ALS patients, so ultimately the therapeutic understanding and application of the KD diet for ALS will require large randomized, placebo-controlled clinical trials [238]. An unknown consideration is the timing of implementation of the KD relative to the advancement of ALS disease progression [236]. From the IFMNT perspective, using each of our ALS patients as their own case controls, we can individually determine if the KD or a variation of the KD is helping to improve signs and symptoms. For pALS to follow the KD, it is most practical to aim somewhere between an MCT or low glycemic index meal plan.

49.23 The Deanna Protocol (DP)

The Deanna Protocol (DP) was developed by Dr. Vincent Tedone to treat his daughter with ALS [239]. Winning the Fight, Inc. helps pALS implement the DP and is a collaborative effort of Dr. Tedone, researchers, and advisors. The primary ingredients in the DP are a TCA cycle intermediate

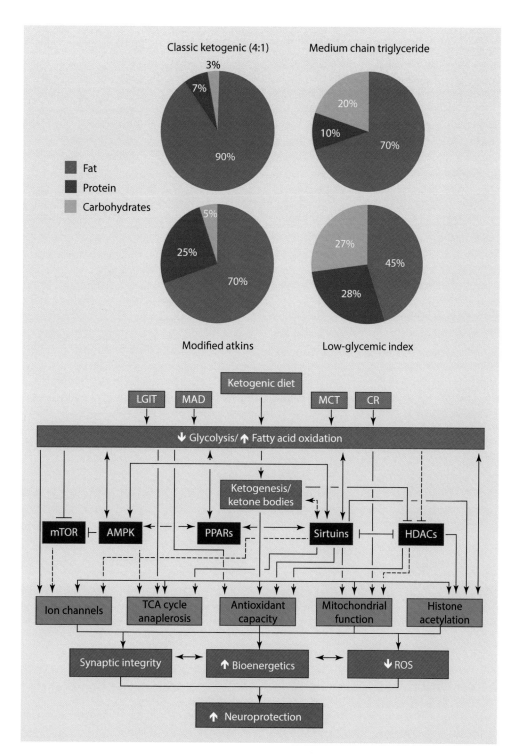

Fig. 49.8 The KD and the following variants may be neuroprotective: low glycemic index treatment (LGIT), modified Atkins diet (MAD), medium-chain triglycerides (MCT), and calorie restriction (CR) [241]. (Reprinted from Gano et al. [237]. With permission from the American Society for Biochemistry and Molecular Biology)

(arginine alpha-ketoglutarate (AAKG)), the ketogenic diet with 10% MCT oil, gamma-aminobutyric acid (GABA), and ubiquinol. There were positive results in a preclinical study [240]. The true effects of the DP in pALS are worth studying in a Phase I clinical trial; however, this will be very difficult, because most patients are taking other dietary supplements and vitamins [241].

49.24 Commercial Enteral Formulas

The enteral formulas [242] used in the comparison study of isocaloric, high-calorie/high-carbohydrate, and high-calorie/high-fat formulas [78] contain a combination of highly processed, genetically modified, and artificial ingredients, such as soy, milk, corn, sugar, corn syrup solids, maltodextrin, artificial flavors, and glutamates. The fats in the high-calorie, high-carbohydrate formula are from corn, canola, and medium-chain triglyceride (MCT) oils. The fats in the high-calorie, high-fat formula are from omega-3 fatty acids (anchovy, cod, menhaden, pollock, sardine, and tuna) and omega-6 (borage oil), omega-9 (canola oil), and MCT oil. The fish sources typically contain contaminants [243] from GMO ingredients, pesticides such as DDT and dieldrin, heavy metals such as mercury, and PCBs. Canola oil is refined, bleached, deodorized, genetically modified, and processed with hexane, an industrial solvent [244]. Borage oil may contain hepatotoxic and carcinogenic pyrrolizidine alkaloid constituents [245]. Furthermore, this study on hypercaloric enteral nutrition in patients with ALS [78] didn't consider the classification of fatty acids; the ratio of omega-3-6-9 fatty acids; the specific composition of MCT oil, such as amount of C6, C8, C10, or C12 fatty acids; the quality of the fatty acids in terms of oxidation or contamination; or the ketogenic diet.

Other commercially available widely used enteral formulas contain refined carbohydrates, sugars, damaged and toxic fats, processed free glutamates, and allergenic proteins. They are chemically adulterated and contain genetically modified ingredients. They are not tailored to the unique metabolic demands of pALS and may be contributing to worsening inflammation and metabolic derangements. Dietary studies on tube feedings only evaluate macronutrients while disregarding the immune and inflammation-modulating variables of the total formula. The Academy of Nutrition and Dietetics has concluded the lack of any efficacious diets for ALS [54]. Could this conclusion result from a generic approach to nutrition intervention that uses oral/enteral formulas with imbalanced substrates and toxic ingredients?

49.25 Healthy Enteral Feedings

There are convenient whole food-based oral/enteral products. These are unique formulations, but all share the same commitment to nutritional quality.

Homemade oral/enteral formulas are another favorable option. Steve Smith, an ex-NFL player, was on commercial tube feedings for 6 years before changing to a whole foods homemade tube feeding. Enough formula is made to last for 2 weeks, frozen, and then used as needed [246]. From the author's communication with his wife, Chie Smith, while Steve was on commercial feedings, he was diabetic and prescribed metformin twice daily, constipated, bloated and had fingernail fungus. After he switched to whole foods homemade feedings, his diabetes was reversed and all other symptoms cleared up. Chie Smith granted permission to include this information in this reading.

49.26 Foods Containing Phytonutrients

A recent *JAMA Network* report described a cohort of 302 ALS patients studied for 5 years and found there was higher function at the time of diagnosis when antioxidants, carotenes, fruits, and vegetables were in the diet [247]. Many phytonutrients can cross the BBB to increase the endogenous antioxidant capacity of the brain to protect against oxidative stress in neurodegenerative disorders [248]. Because ALS is a multifactorial disease, bioactive compounds like flavonoids that target more than one aspect are needed to fight this devastating disease [249] (Table 49.2).

49.27 Environmental Toxins

There is a broad scientific consensus that sporadic ALS is caused by gene-environment interactions. In fact, given that only about 10% of ALS diagnosis has a genetic basis, gene-environment interactions may account for the remaining sporadic cases of ALS. Unfortunately, there's been relatively little attention paid to environmental toxins and lifestyle factors that may trigger the cascade of motor neuron degeneration leading to ALS [270]. Ultimately, the relationship between genetic background and xenobiotics may be the avenue to pursue in order to clarify the influence of environmental factors in ALS [271]. Accumulating data indicate that environmental exposure may influence the development of the ALS phenotype, suggesting that specific populations of neurons may be more sensitive to specific toxins; however, specific exposures have not yet been linked to precipitating ALS [272].

Several occupations are linked to an increased risk of ALS such as welders, electrical workers, athletes from head injuries and chemicals used on athletic fields, carpenters, cockpit workers, construction workers, farm workers, hairdressers, house painters, laboratory technicians, leather workers, machinists, medical service workers, military personnel, power production plant workers, tobacco workers, precision metal workers, rubber workers, veterinarians, nurses, funeral directors, truck drivers, flight attendants, firefighters, and miners [23, 273–277].

The following environmental exposures have been identified as potential risk factors for ALS, such as electromagnetic fields, heavy metals, pesticides, beta-methylamino-L-alanine and cyanobacteria, head trauma, metabolic diseases, cancer, viral infections, formaldehyde, diesel exhaust, polychlorinated biphenyls (PCB), polycyclic aromatic hydrocarbons (PAH) and dioxin, selenium, zinc, copper, smoking, and aromatic solvents in residential ambient air [23, 270, 273–274, 278–279]. Many studies have shown elevated glutamate levels in different brain areas due to electromagnetic radiation, aluminum, cyanide, aspartame, and MSG. In addition to glutamate neurotoxicity, glutamate has been found to mediate the hazardous effects of some environmental pollutants [280].

US statewide mortality for motor neuron diseases are significantly correlated with the percentage of the state's population that uses well water, a known reservoir for *Legionella* infections that may cause some cases of ALS. For example, Vermont has the highest age-adjusted motor neuron disease

Table 49.2 Phytonutrients

Phytonutrient	Proposed activity	Reference
Pomegranate juice	Contains many polyphenols; inhibits excitotoxicity and calcium influx in human neurons	[251]
Avocados	Contains glutathione, phytosterols, carotenoids, flavonoids, omega-3 and oleic fatty acids; protects against oxidative stress, mitochondrial dysfunction, and neurodegeneration	[252]
Walnut, date, and fig extracts	Protects against quinolinic acid (QUIN) and 1-methyl-4-Phenyl-1,2,3,6-tetrahydropyridine (MPTP)-induced excitotoxicity in human neurons	[253]
Cruciferous vegetables	Contains sulforaphane, activates Nrf2 and antioxidant response element (ARE), protects against oxidative stress, regulates several genes (Keap1/Nrf2) that protect against neurodegeneration, and penetrates the BBB to deliver neuroprotective effects in the CNS	[254]
Blueberries, grapes	Contains resveratrol, improves mitochondrial bioenergetics, improves autophagy in ALS, activates survival pathways (sirtuin-1, AMP-kinase), delays ALS disease onset, and extends survival	[255]
Asparagus	Contains sarsasapogenin, increases proteins called neurotrophic factors, and mimics their effects in the CNS, and orphan designation of sarsasapogenin was granted in the United States for the treatment of ALS	[256]
Green tea	Contains epigallocatechin gallate (EGCG) – protects motor neurons from stress, significantly prolonged symptom onset, and increased life span; neutralizes ferric iron to form redox-inactive iron in neuronal cells	[250, 257]
Cherries	Contains anthocyanins – protect neurons from oxidative stress, especially tart cherries	[258]
Pecans	Delayed ALS progression, increased survival of motor neurons	[259]
Saffron	Contains crocin, protects neurons from demyelination, and suppresses inflammation	[260]
Cocoa	Contains polyphenols (catechin and epicatechin), downregulates excitotoxicity, protects neurons from neurotoxins, reduces neuroinflammation, promotes neuronal survival and synaptic plasticity	[261]
Artichoke, parsley, celery, peppers, olive oil, rosemary, lemons, peppermint, sage, thyme	Contains luteolin and apigenin, prevents neurotoxic inflammatory mediators, reduces stimulation of microglial cells, neutralizes free radicals, promotes nerve and muscle function, reduces microglial CD40 expression	[262–263]
Lemon peel essential oil	Contains limonene and other flavonoids, inhibits lipid peroxidation by inhibiting quinolinic acid-induced malondialdehyde production, scavenges free radicals, and chelates ferrous iron	[264]
Turmeric	Contains curcumin, inhibits cytotoxicity, modulates microglial gene expression and migration, inhibits lipopolysaccharide neurotoxicity, ameliorates cellular damage by heavy metals, attenuates reactive oxygen species COX-2 expression in ALS	[265–266]
Coffee	Contains chlorogenic acid, protects degenerating motor neurons, reduces upper and lower motor neuron excitability, increases antioxidant enzyme capacity; not associated with increased ALS risk in humans	[267–268]
Cannabis sativa	Contains cannabinoids, provides antioxidant, anti-inflammatory, and neuroprotective effects in the CNS, and humans studies indicate cannabinoids may ameliorate spasticity, appetite loss, depression, pain, and drooling	[269]

mortality rate and historically has had a high prevalence of legionellosis [281]. Bisphenol A (BPA) and phthalates pose a theoretical risk of ALS, and 6–10% of ALS is apparently caused by any of the following toxins: DDT, lead, mercury, and organophosphate pesticides [110].

Many heavy metals are potential risk factors for ALS. Lead, manganese, iron, selenium, copper, aluminum, arsenic, cadmium, cobalt, zinc, uranium, and vanadium have been found in higher levels in the CSF of ALS patients when compared with healthy controls [23]. Single metal evalua-

Table 49.3 Heavy metals

Heavy metal	Adverse effects	References
Mercury	Mercury toxicity listed in differential diagnosis of ALS; vapor from mercury amalgams absorbed by motor neurons leads to increased oxidative stress and enhanced glutamate toxicity Accidental mercury exposure may be associated with ALS Retrospective study reported a statistically significant association between an increased number of amalgam fillings and the risk of motor neuron diseases pALS should have mercury amalgam fillings removed under the care of a biological dentist accredited through the International Academy of Oral Medicine and Toxicology Exposure to mercury from amalgams removed incorrectly may hasten onset or worsen ALS Methylmercury from fish may hasten the onset of ALS and worsened outcomes for pALS eating fish high in methylmercury	[87, 270, 284–289]
Aluminum	Aluminum adjuvants in vaccines are neurotoxic and change DNA expression Animal models after injections of aluminum adjuvants are associated with neurological deficits and lead to an ALS phenotype Link in aluminum miners and development of ALS Avoid aluminum in food products such as baking powder and processed foods, food additives with the word aluminum, and aluminum cookware Aluminum can be found in toothpastes, antiperspirants, dental amalgams, nasal sprays, antacids, and popular prescription and OTC medicines	[290–293]
Cadmium	Impairs the BBB Reduces brain Cu-Zn SOD Enhances excitotoxicity of glutamate Suggested as a cause-effect relationship in a worker at a battery factory diagnosed with ALS Elevated levels are found in cigarette smoke, electronics, seafood and organ meats, and plant foods grown in contaminated soils Cocoa powder generally contains four times amount Cadmium and lead have been found in 45 out of 70 popular chocolate products	[294–297]

tions may underestimate their relevance for health risks compared to the additive synergistic effects of various metal interactions, and more studies are needed to investigate the role of environmental factors on the development of ALS [282] (Table 49.3).

A study involving 156 ALS patients and 128 healthy controls evaluated for occupational and residential exposure by obtaining surveys of exposure and blood levels of 122 persistent neurotoxic organic pollutants [298]. Seven pollutants were significantly associated with ALS, including three organochlorine pesticides (OCPs), two polychlorinated biphenyls (PCBs), and two brominated flame retardants (BFRs). Although there was weak concordance of survey data and blood measurements, there were strong associations of specific environmental toxins and ALS, which may represent modifiable risk factors [298]. A large population case-control study found that long-term exposure to traffic-related air pollution increases susceptibility to ALS [299].

Although the literature overwhelmingly implicates environmental toxins with risk of ALS, a contradictory CDC study found that "white-collar" jobs are linked to a 55–67% increase in ALS deaths, compared to a 31% decrease in ALS deaths in occupations more likely exposed to environmental toxins [300]. These statistics are confounding and counterintuitive yet implore us to continue examining the interrelationships of toxins, both separately and cumulative, toward the risk of ALS.

It is important to remember that the IFMNT approach does not require unequivocal proof that specific toxins cause ALS, as neurotoxins and all toxins are harmful for pALS. Most pALS that we see in the clinic are being cared for at home. It is our responsibility to help our patients remove existing unintended exposures in their homes while reducing their toxic load through responsible detoxification support and elimination.

49.28 Detoxification of Heavy Metals

The use of chelating agents to treat acute metal poisoning is now well established; however, adaptation of chelation therapy to neurodegenerative diseases is directed toward brain metal redistribution rather than brain metal scavenging and removal [301]. This review found that chelation therapy has not been reported to help ALS patients and may cause harm in some cases [62]. Chelation therapy should be undertaken by a professional trained in the use of chelating agents in neurodegenerative disorders, both over-the-counter and pharmaceutical.

49.29 Natural Chelators

Some examples of natural chelators are curcuminoids and catechins (green tea, berries, cocoa, onions) which permeate the BBB and combat neurotoxicity [250]. Modified citrus pectin as well as dietary fibers from grains and fruit may interrupt enterohepatic circulation of metals, with findings of reduced levels of mercury in the brain and blood [302]. Sulfur-containing foods, such as allium from garlic, and brassicas, such as broccoli, support glutathione [302].

Beyond the effort to chelate specific heavy metals, we must try to support the liver's detoxification pathways for pALS. Diet should focus on optimizing the body's metabolic Phase I and Phase II detoxification pathways and metallothionein, through a variety of foods and food-based nutrients known to contain bioactive compounds with demonstrated or potential clinical impact on detoxification systems [303]. The authors report clinical evidence of effects from cruciferous vegetables, allium vegetables, apiaceous vegetables, grapefruit, resveratrol, fish oil, quercetin, daidzein, and lycopene [303].

49.30 Essential Metals

49.30.1 Iron

Iron dyshomeostasis is linked to ALS with brain iron accumulation in the motor cortex, and the SOD-1 mutation has been shown to cause damage via iron-induced oxidative stress [304]. Also, there appears to be a link between mutations in the hemochromatosis gene (HFE; principally the H63D polymorphism) and SALS [305]. Iron accumulation in the motor cortex corresponds to the hand knob region, presumably causing small hand muscle weakness seen in these patients [23]. Higher concentrations of serum ferritin correlate to poor prognosis in ALS patients [305].

If a pALS has elevated ferritin, then it is recommended to avoid high-iron foods [306]. Curcumin, green tea, and blueberries provide direct binding and complex formation with toxic metals such as iron, with enhanced free radical scavenging effects protecting the neurons from toxicity and inflammation [301].

49.30.2 Copper

Copper and iron are described as "peas in a pod" since they are both highly reactive redox metals, where a deficiency or excess of either one can damage mature neurons and neuronal activities [307]. Both excessive [308] and deficient [309] copper have been associated with ALS. Copper is a regulator of the contents of iron in the CNS [310]. The balance of metals is regulated by the integrity of the BBB and Blood-CSF Barrier (BCB), which are damaged in neurodegenerative diseases and from environmental interferences [307].

A patient-driven ALS free copper pilot study showed that free copper was elevated in ALS, postulating that low ceruloplasmin was connected to initiation of the disease process; thus high free copper would correlate with neuronal damage [311]. Free copper initiates free radicals which is potentially harmful for brain circuits, whereas ceruloplasmin scavenges free radicals and participates in many important enzyme systems [310].

Dr. George Brewer, an expert in copper metabolism, describes two forms of copper: Copper 1 (Cu+) cuprous which is organic and found in nature and Copper 2 (Cu++) cupric which is inorganic and found in supplements [312]. Cu++ is also found in drinking water exposed to copper pipes (Oskarsson, 2011). To lower overall copper load and especially free copper, it is advised to reduce intake of red meat, monitor copper in one's water supply, and avoid copper in supplements. Also, N-acetylcysteine has been proposed as a copper chelator in ALS treatment [270].

In the potential case of copper insufficiency or deficiency, dietary copper in the organic Copper 1 cuprous (Cu+) form can be found in foods such as sesame seeds, cashews, soybeans, sunflower seeds, shitake mushrooms, tempeh, garbanzo beans, lentils, lima beans, and walnuts [313].

49.31 BMAA

Beta-methylamino-L-alanine (BMAA) is a cyanobacterial neurotoxin that has been found on autopsy of SALS patients, suggesting a gene-environment interaction triggering ALS [314]. It was first identified 50 years ago in Guam where one in three adults died of ALS. In short, they determined that the cyanobacteria biomagnified from cycad tree seeds eaten by fruit bats that subsequently were eaten by natives [315]. Cyanobacteria are some of the oldest organisms on the planet and can occur wherever there is moisture. Blooms are fed largely by nutrients in agricultural and urban runoff, and fish and shellfish from contaminated waters may be another way that people ingest BMAA [316]. The body may mistake BMAA for the amino acid L-serine, which is mistakenly inserted into proteins that become misfolded [316]. Chronic dietary exposure to BMAA is a cause of neurodegenerative illness; therefore a clinical trial will determine if L-serine is a safe and efficacious treatment to reduce disease progression in ALS [317].

49.32 Gastrointestinal System

ALS and the gastrointestinal (GI) system have some mutual patterns of pathological expression, and the care of ALS patients should include evaluation and treatment of the GI system. The bidirectional gut-brain axis consists of the entirety of the intestinal microbiota, enteral nervous system (ENS), parasympathetic and sympathetic nervous systems, CNS, neuroendocrine connections, humoral pathways, cytokines, neuropeptides, and signaling molecules [318]. ENS deficits are reported to accompany an increasing number of CNS disorders, from neurodevelopmental to neurodegenerative, and dysfunctional GI manifestations might occur even before CNS symptoms become evident [319]. Neuronal connections and the immune system might provide conduits that allow diseases acquired in the gut to spread to the brain, such as ALS, a disorder with both gastrointestinal and neurological consequences [319]. The gut is both a pathological marker of neurodegeneration and a site for studying the pathology of neurodegeneration [320] (◘ Table 49.4).

Table 49.4 Gut-brain pathologies in neurodegeneration

Interactions		References
Misfolded proteins	SOD1 and TDP43 occur in both the ENS and CNS reflecting common neurodegenerative mechanisms	[320]
GI phenotype of ALS G93A mice	Show abnormal tight junctions, increased mucosal permeability, decreased Paneth cell expression of the antibacterial peptide defensin-5	[319]
Microbiome abnormality of ALS G93A mice	Show abnormal intestinal microbiome and decreased butyrate-producing bacteria associated with increased inflammatory cytokines (IL-17) and intestinal permeability before ALS onset and during ALS progression; cyanobacteria from the GI tract may produce elevated levels of ß-N-methylamino-L-alanine (BMAA) in the brain, an excitatory neurotoxin which depletes glutathione and induces protein misfolding	[321, 323–324]
Lipopolysaccharides from gut-associated microbial translocation	LPS are elevated in the plasma of sporadic ALS patients; LPS are inversely correlated with anti-inflammatory monocyte IL-10 expression in sporadic ALS patients which may be associated with disease progression	[325]
Neurotoxins	Produced in the gut, such as clostridial species; might cause sporadic ALS	[322]
Xenobiotics	Can disturb microbial balance; become more toxic to the brain if microbial demethylation is suppressed or the absence of necessary GI microbiota reduces fecal excretion; mercury is a neurotoxin that may increase the risk of ALS	[287, 326–327]

49.33 Microbiome

Gut microbiota influence neurodegenerative diseases through such pathways as the vagal nerve, cytokines, neurotransmitters, bacterial metabolites, and the HPA axis [328]. Figure 49.9 depicts how changes in the gut microbiota impact neuropathology and neurodegenerative diseases through neurological, endocrine, and immune pathways [318].

It is rare for individual researchers to conduct both animal and human ALS studies of the microbiome. The collaboration between preclinical and clinical researchers strengthens the quest for scientific understanding of gut-brain pathologies and potential therapeutic treatments of dysbiosis in pALS, which may affect onset and progression of ALS. The University of Illinois at Chicago, in collaboration with Wellness and Integrative Neurology in Westchester, Illinois, published the first comprehensive examination of the gut microbiome in pALS [329]. They compared five ALS patients with a large database of healthy cohorts to evaluate bacterial diversity and abundance, infection, short-chain fatty acids, inflammation, and immunology through stool testing [329]. The most notable results included low microbial diversity in all ALS patients, low *Firmicutes/Bacteroidetes* ratio and elevated gut inflammatory markers (calprotectin, secretory IgA, and/or eosinophilic protein X) in most patients, celiac disease in one patient, and candida infection in another patient [329]. All patients had preexisting GI symptoms prior to their ALS diagnosis. The authors encouraged further research using comprehensive stool analysis to evaluate microbiome and other gut health indices in ALS, in addition to studying how other inflammatory markers commonly reported in ALS patients (serum LPS and CSF cytokines IL-17 and IL-23) might correlate to gut inflammation in pALS [329].

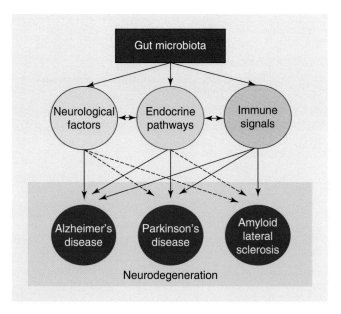

Fig. 49.9 Gut microbiome [322]. (Reprinted from Westfall et al. [318]. With permission from Springer Nature)

There are other active investigations into the microbiome of ALS patients. The Microbiome Assessment in People with ALS (MAP ALS) at the time of this writing is seeking 100 pALS and 100 healthy volunteers for a one-time stool collection for microbiome assessment [330–331]. In contradiction to the findings that the microbiome could be involved in ALS, a study of 25 ALS patients and 32 controls found that microbial diversity is not altered in ALS and argues against a primary role of the gut microbiota in it [332].

Given that several ALS G93 mouse studies indicate an associated role of the gut microbiota and inflammatory markers with onset and progression of ALS [329] and that pALS often report gut symptoms before and after diagnosis of ALS, it is timely and vital for continued research in the gut-brain relationship in ALS. It may be important to study the microbiome as it relates to gut-associated inflammatory and immune markers, short-chain fatty acids, and intestinal permeability to better understand the gut-brain connection in ALS.

> **Box 49.1 Suggestions to Assess and Support the Gut-Brain Connection in pALS**
> - *Obtain comprehensive stool testing* to assess microbiome diversity, beneficial, pathogenic, potentially pathogenic and commensal bacteria, yeast, viruses and parasites, digestive enzymes, inflammatory and immune markers, malabsorption, intestinal permeability, short-chain fatty acids and other factors.
> - *Individualize probiotic recommendations* based on results of stool testing.
> - *Individualize antimicrobial recommendations* based on specimen sensitivity testing.
> - *Evaluate for small intestinal bacterial overgrowth (SIBO)* through breath tests to detect hydrogen and methane gas-producing bacteria.
> - *Remove allergenic foods and xenobiotics* to reduce gut and systemic immune activation resulting in neuroinflammation.
> - *Add digestive enzymes and HCL* if required.
> - *Add supplemental nutrients and herbs* to heal mucosa, reduce gas and bloating, and decrease abdominal pain.
> - *Eat organic, low glycemic index, or ketogenic diet with high levels of phytonutrients* to heal the gut and protect the brain.

49.34 Probiotics

One study found that archaebiotics, which are probiotics derived from prokaryotic single-cell bacteria, could reduce mutant SOD1 misfolded proteins and cellular toxicity in ALS cell cultures [333]. Since then, research in animals has evaluated specific probiotic strains in neuroprotection, MS, Alzheimer's disease, Parkinson's disease, and neurodegenerative disorders in general, though less specifically in ALS [318]. Positive effects, such as improved antioxidant capacity, reduced inflammation, reduced microglial activation, increased GABA, downregulating genes involved in neurodegeneration and reducing neuronal apoptosis, are associated with specific probiotic strains [318], which hopefully will lead to researching specific effects of probiotic strains that might be therapeutic in ALS.

49.35 Butyrate

Diet influences intestinal microflora composition, which has an important role in the fermentation of dietary fibers leading to the production of short-chain fatty acids (SCFAs) such as butyrate [334]. Butyrate is primarily used as an energy source for colonocytes, and some butyrate exits the portal vein to the liver, but it is the remaining butyrate at the distal end of the colon that enters the systemic circulation and reaches the BBB [335]. Butyrate exerts multiple beneficial effects at intestinal and extraintestinal levels, including neuroprotection, related at least in part to the epigenetic regulation of gene expression [334]. Phenylbutyrate (PB) acts as a prodrug to easily penetrate the BBB [336].

In ALS mice with abnormal microbiomes, damaged junctions, and leaky guts, supplementation with butyrate restored the microbiome, improved neuromuscular function, delayed onset of ALS symptoms, and extended life span compared to untreated ALS control mice [321]. Additionally, sodium butyrate was associated with protection of the endothelial cells of the BBB through increased expression of tight junctions [337]. This role of butyrate working to protect both the gut and brain endothelial barriers nicely illustrates a congruent linkage between the gut and the brain.

A Phase II clinical study of sodium phenylbutyrate (NAPB) in ALS significantly increased histone acetylation, and the majority of patients tolerated very high doses, yet it was found that the lowest dose of 9 grams daily was therapeutically beneficial [338]. Another Phase II clinical trial is underway at the time of this writing and will evaluate a combination compound of PB and tauroursodeoxycholic acid (TUDCA) for the treatment of ALS [339]. TUDCA is a bile acid that may have cytoprotective and anti-apoptotic action with potential neuroprotective activity, having shown significantly higher ALSFRS-R scores in pALS who received TUDCA in a proof of principle study [340]. PB and TUDCA combination therapy is designed to reduce neuronal death through blockade of key cellular death pathways originating in the mitochondria and endoplasmic reticulum while measuring the rate of decline from the ALSFRS-R and monitoring biomarkers of neuronal death and neuroinflammation [339].

The following foods will promote butyrate production: whole grains; fiber-rich fruits and vegetables such as green beans, legumes, leafy greens, apples, kiwi, and oranges; milk fats (butter, ghee); and prebiotics such as raw jicama, underripe bananas, Jerusalem artichokes, and raw dandelion greens [341]. Over-the-counter products are available in time-release formulations that maximize absorption and have better taste and smell compared to the common mineral salt butyrate products using sodium, potassium, calcium, or magnesium.

49.36 Small Intestinal Bacterial Overgrowth (SIBO)

The involuntary muscles such as those that control the GI tract are not directly affected in ALS. However, prolonged immobility can lead to indirect GI issues in ALS [342]. Gastrointestinal motor dysfunction or motility disorders, such as delayed gastric emptying and delayed colonic transit times, can occur in ALS, even if patients do not complain of GI symptoms [343], and motility disorders are a risk factor for SIBO [344].

Clinical manifestations of SIBO seriously impact nutritional status via weight loss; steatorrhea; deficiency of fat-soluble vitamins, vitamin B12, and iron; excess folate; low total protein and albumin; and decreased xylose absorption when evaluating carbohydrate malabsorption [344]. Further discussion of SIBO can be found in ▶ Chaps. 23 and 24.

49.37 Laboratory Biomarkers

Laboratory biomarkers have been a missing piece in the diagnosis and monitoring of ALS. At present, the diagnosis of ALS is based on a differential clinical diagnosis and may include electromyography, nerve conduction studies, muscle biopsy, MRI, and genetic testing [283]. The key monitoring measures are the ALSFRS-R, clinical examination, and spirometry [25], with outcomes determined by survival rates [345]. From standard laboratory measurements to in-depth metabolomics, biomarkers have become the focus of intense research in the field of ALS, with the hope for therapeutic development [346].

Few studies have evaluated deficiency or overload of essential elements in ALS using blood, urine, and hair. A study of 34 ALS patients and cohorts evaluated calcium, zinc, magnesium, iron, copper, magnesium, and selenium to determine if different concentrations were related to a higher frequency of ALS. Levels of Ca and Cu in blood and Se and Zn in hair were significantly higher in pALS than controls, while urinary excretion of Mg and Se was significantly lower in pALS [347]. The authors suggest that multiple metals involved in multiple mechanisms have a role in ALS degeneration and might serve as good biomarkers [347] (◘ Table 49.5).

◘ **Table 49.5** ALS serum biomarkers that increase risk

Laboratory assessment	Elevated levels	Lowered levels	References
Inflammation and immunity	Copper Sedimentation rate Fibrinogen hs/wr C-reactive protein TNF-alpha IL-2, IL-6, IL-8, IL-10, IL-18 Interferon-gamma CD4+ T lymphocytes CD16+ leukocytes Natural killer cell T lymphs Neutrophils/monocytes C16:1/C16:0 ratio Arachidonic acid	Albumin CD14+ monocytes CD8+ T lymphocytes IL-2	[209, 345, 354–361, 376]
Excitotoxicity and neurodegeneration	Ammonia Quinolinate Glutamate	Kynurenate Glutamine Vitamin D	[143, 348, 360]
Glycolysis and oxidative phosphorylation	Pyruvate Carnitine Phosphate		[349]
Lipids and liver function	LDL/HDL Triglycerides SGOT, SGPT		[222, 225, 352]
Collagen degradation		Hydroxyproline Type 1V collagen	[348]
Oxidative stress	Ferritin Homocysteine 8-OH-2'-Deoxyguanosine 4-OH-2,3-Nonenal (HNE) GSSG/GSH Prostaglandin (PgE2)	Folate Vitamin E Uric acid GSH GPx Transferrin	[345, 350, 359, 362]
Detoxification		7-Methylxanthine	[359]
Muscle mass	Creatine	Creatinine Creatine kinase	[348, 363]

Pathologically elevated and lowered serum biomarkers reflect ALS disease and progression. Many values may change from early to late progression of ALS. Some biomarkers appear paradoxical, such as the suggested protective effect of raised LDL/HDL and triglycerides; however, in this table elevated lipids are listed as detrimental biomarkers of ALS.

The following discussion will cover metabolomics, blood biomarkers, and the potential relevance of urinary organic acid (OAA) laboratory testing to extend understanding the metabolic and nutritional imbalances in ALS. It should be noted that the CSF represents a very valuable source of biomarkers, which contain proteins and metabolites at relatively higher concentrations than other fluids [345], but the invasive nature of CSF assessment makes it largely unavailable for IFMNT practitioners.

49.38 Metabolomics

Metabolomics is the study of small molecules in biological systems such as cells, tissues, blood, urine, and CSF. Metabolomic biomarkers reflect the overall metabolic status of an individual as the result of highly complex metabolic exchanges between diverse biological compartments, including organs, biofluids, and microbial symbionts [348]. Applied to a biofluid such as plasma, metabolomics could provide an understanding of biochemical changes in ALS and identify a panel of biomarkers for diagnosis and disease progression [349]. At present, we lack consensus and independent validation of metabolomics biomarkers for ALS [350]. Appropriate application of multimodal approaches, international collaboration, presymptomatic studies, and biomarker integration into future therapeutic trials are among the essential priorities going forward [364].

49.39 Urinary Organic Acid Analysis (OAA)

Does the urinary organic acid analysis (OAA) have clinical relevance for the IFMNT practitioner when caring for pALS? Urine is not subject to homeostatic mechanisms as much as the blood and CSF in neurodegenerative diseases; however, every neurodegenerative disease carries its own metabolic signature which is a potential biomarker [365]. There are limited studies using urine from pALS, since biomarker research has primarily focused on blood and CSF. A recent review on ALS-associated urinary biomarkers could only positively validate collagen degradation biomarkers and 8-OH2'dG [366]. The OAA reflects metabolic pathways that can't move forward due to inadequate nutrients, toxins, genomic influences, or disease states, causing buildup of pathway intermediates to overflow from the blood into the urine. It is challenging to discern OAA markers of nutritional imbalances from inherent ALS metabolic defects; however, a better understanding of ALS metabolism will help the clinician determine nutritional requirements. The urinary OAA analysis is helpful to evaluate requirements for Coenzyme Q10, B vitamins, and other nutrients in energy metabolism and assessment of metabolic contributions of environmental toxins, methylation, dysbiosis, oxidative stress, and other parameters. Interpretation and therapeutic intervention must be carefully considered in light of the signature metabolic abnormalities of ALS.

> **Box 49.2 The Above-Described Defects in Energy Metabolism May Be Reflected in the Urinary OAA**
> - *Pyruvate* may be increased during impaired metabolism to acetyl-CoA.
> - *Lactate* may be increased since more pyruvate goes to lactic acid, and there is failure of the ATP-dependent muscle-neuronal-lactate shuttle to regulate the flow of lactate from muscle to neurons and vice versa. This leads to lactate toxicity and degeneration of nerve endings at the neuromuscular junction [367].
> - *Citrate* may be lowered during impaired metabolism of oxaloacetate convergence with acetyl-CoA to form citrate.
> - *Beta-hydroxybutyrate* may be increased during ketosis.
> - *Alpha-ketoglutarate* may be increased during fatty acid metabolism and ketosis or may be decreased due to both impaired glycolysis, fatty acid oxidation, and deficient glutamate transamination.
> - *Adipate and suberate* may be increased during impaired fatty acid oxidation and carnitine deficiency. The demand for carnitine is increased to shuttle long-chain fatty acids from the cytosol to mitochondria for beta-oxidation [349].

> **Box 49.3 Impaired Urinary Neurotransmitters and Excitotoxicity**
> - *Quinolinic acid* may be increased
> - *Kynurenic/quinolinic ratio* may be decreased, reflecting excitotoxicity and neurodegeneration. The kynurenine pathway, the major biochemical pathway for tryptophan metabolism, is dysregulated in ALS. Accumulation of the neuroactive and cytotoxic intermediate, quinolinic acid, contributes to the pathology of ALS [143].
> - *Ammonia* may be increased. Neurotoxic ammonia accumulates in ALS during impaired glycolysis, along with reduced ammonia removal and hepatic urea cycle dysfunction [353], and during excessive protein intake.
> - *Glutamate* may be increased [368].
> - *Proline* may be increased due to weakened collagen tissues, especially noticeable in the skin. The increased urinary excretion of collagen metabolites is highly associated with ALS [348].

49.40 Impaired Glycolysis, TCA Cycle, and Oxidative Phosphorylation

A combination of defective energy metabolism, decreased glucose use in the cortex and spinal cord, reduced expression of TCA cycle intermediates in the brain and spinal cord, damaged neuronal mitochondria, and mitochondrial Electron Transport Chain (ETC) dysfunction suggest a bioenergetic limitation of ATP generation [41]. This is due to reduced phosphofructokinase, pyruvate dehydrogenase [241], citrate synthase, succinate dehydrogenase [369, 370], and impaired oxidative phosphorylation ETC complexes I, II, III, and IV [371]. These defective metabolic pathways shift energy utilization through lipolysis, beta-oxidation of fatty acids, and ketosis [44, 372, 373].

49.41 Methylation and ALS

It is necessary to understand the causal relationship between DNA methylation and ALS and how environmental factors trigger the DNA modifications. For example, hypomethylation of promoter regions of SOD1 and VEGF causes no transcriptional silencing in ALS, whereas hypermethylation of the promoter regions of EAAT2 and ATXN2 causes functional loss of excitatory amino acid transporters and pathogenic CAG expansions [351]. MTHFR C677T and A1298C polymorphisms are reported as genetic risk factors for SALS women [374]. Homocysteine exerts multiple neurotoxic mechanisms relevant to ALS pathogenesis [375], and elevated plasma homocysteine is a biomarker of ALS [376]. The pathway from homocysteine to S-adenosylmethionine (SAMe) may be blocked by deficiencies in folate, B6, and B12 [351]; therefore the IFMNT practitioner should assess vitamin status and supplement accordingly.

49.42 Additional Considerations for Laboratory Testing

49.42.1 Erythrocyte Fatty Acid Analysis

The IFMNT practitioner might obtain the ALS patient's fatty acid status through an essential fatty acid analysis based on red cell biopsy to help guide therapeutic interventions. On ▶ ClinicalTrials.gov, a study was commenced in 2015 titled "Evaluating the Potential of the Diet as Disease Modifier in Amyotrophic Lateral Sclerosis" to study dietary intake and erythrocyte fatty acid composition as it relates to disease progress or survival [377]. The study aims to develop beneficial interventions and therapy options for ALS patients based on the understanding that PUFAs, especially DHA, are an important structural components in neuronal membranes, support neurogenesis and neuroprotection, and exert potent anti-inflammatory effects in the brain [377].

The erythrocyte fatty acid analysis could inform the practitioner of existing omega-6 to omega-3 ratios as well as levels of other parent fatty acids that contribute to these ratios. The omega-6 to omega-3 ratio of the average Western diet is 20:1, whereas the diet of our paleolithic ancestors contained a 1:1 ratio, the same level that is found in balanced neuronal lipids. This ratio (between 2:1 and 1:1) should be the target for human nutrition [378]. Also, the increased ratio of palmitoleic acid (C16:1) to palmitic acid (C16:0) increases ALS risk associated with misfolded proteins [209]. The IFMNT practitioner can assess erythrocytes fatty acids and adjust diet or supplements to optimize fatty acid ratios for pALS.

49.42.2 Liver Function and ALS

A high incidence of mild liver dysfunction and ultrastructural changes in hepatocytes has been found in ALS patients [379]. Elevated lipids and hepatic steatosis are frequent and unique phenomena of ALS [352]. These abnormalities are concomitant with elevated homocysteine, ammonia levels, and a faulty urea cycle [353]. Although it has been suggested that impaired liver function serves a protective effect, we should caution against inadvertent clinical dismissal of this phenomenon. The typical ALS diet contains very high levels of refined carbohydrates and sugars, especially in the form of fructose, which may contribute to impaired liver function, along with inherent lipid metabolism abnormalities in ALS. A low glycemic load or ketogenic diet would be advised to help protect the liver.

49.42.3 Thyroid and ALS

Repeated studies have shown that thyroid function in ALS patients does not differ from controls, and clinical parameters of the disease do not influence the measured hormones [380]. However, two large retrospective studies observed that approximately one-fifth of pALS reported a personal or family history of thyroid disease or had biochemical evidence of thyroid disease [95]. Each patient should be assessed with a complete thyroid panel, including autoantibodies to determine thyroid involvement.

49.42.4 Adrenal and ALS

ALS is associated with significant distress linked to changes in hypothalamic-pituitary-adrenal axis activity, possibly due to a loss of cortisol circadian rhythmicity. Increased morning plasma cortisol levels have been associated with spinal-onset ALS and in those with intermediate to rapidly progressing disease [381]. The researchers determined that the changes were too small to make it a sensitive biochemical biomarker [381]. Alternatively, researchers have found a blunted cortisol awakening response in pALS that correlated with disease and depression severity [382]. The adrenal status of each patient should be assessed through dried urine or salivary hormone testing measuring different time points during the day for cortisol levels.

49.42.5 Autoantibodies, Antibodies, and ALS

ALS lacks the prominent infiltrates of blood-derived mononuclear cells that characterize primary autoimmune diseases [383]. A panel of highly expressed serum IgG antibodies served as very sensitive and specific biomarkers to differentiate pALS from controls and monitor disease severity [386]. It is important to learn where, in the course of ALS disease progression, inflammation and immune responses are conferring protection or damage [384].

49.42.6 Infectious Diseases and ALS

Infections may play a role in the etiology or constitute a risk factor for ALS. A main cause of death in ALS is bronchial pneumonia and pneumonia [385]. The diagnosis of Lyme disease can mimic other neurological disorders such as ALS [386]. Some pALS who have Lyme with coinfections can improve with antibiotics and nutritional and herbal support (such as the Cowden Protocol), especially when Lyme, Babesia, and mycoplasma are present [387]. Hepatitis B virus antibodies [388] and hepatitis C virus antibodies [389] have been suggested to be involved in the pathogenesis of ALS. ALS is highly associated with human endogenous retroviral (HERV) sequences, notably HERV-K compared to controls [390]. There has been increasing evidence for disseminated fungal infection in CSF and brain tissue from pALS [391], and 11 ALS patients revealed numerous varieties of intracellular and intranuclear yeast in neural tissue [392]. The common parasite, *Toxoplasma gondii* (toxoplasmosis), disrupts glutamate function, potentially leading to ALS in predisposed individuals [393]. It has been hypothesized that ALS can be caused by motor neuron toxins from a gut clostridial species [322]. Infections may be antecedents, triggers, or mediators of ALS; thus IFMNT practitioners must assess the potential role of viral, bacterial, fungal, and parasitic infections in pALS.

49.43 Supplements

49.43.1 Reductionist Versus Systems Biology Approach

The ALS research on specific vitamins, minerals, metabolic intermediates, cofactors, neurotransmitters, and other nutraceuticals are presented in ◘ Table 49.6. It is beyond the scope of this chapter to include all supplements that have been referenced in research, clinical practice, or public popularity. It is not surprising that no single nutraceutical has been proven efficacious for ALS treatment in humans. With the exception of the preclinical Deanna Protocol study [142] and the clini-

◘ **Table 49.6** Supplements

Supplement/ diet	Research status	Rationale	Use in pALS	References
Deanna protocol	Preclinical	The metabolic therapy with DP supplementation delays disease progression and extends survival in ALS mouse model. The premise of the DP is to provide targeted metabolic therapies that address mitochondrial deficits and excitotoxicity, to increase survival and quality of life	Anecdotal reports (26 testimonials) from ALS patients describe symptomatic improvement in motor control following DP, and the ingredients (arginine-alpha-ketoglutarate, GABA, CoQ10, MCT oil) have a good safety profile without negative side effects being reported	[9, 142]
Arginine-alpha-ketoglutarate (AAKG) and alpha-ketoglutarate (AKG)	Preclinical	A study showed that AAKG and AKG, key ingredients of the Deanna Protocol (DP), are TCA cycle intermediates that stall ALS progression, improve motor performance and survival time, conceivably through protection of mitochondria in muscle fibers, motor neurons, and astroglial cells. Alpha-ketoglutarate can refill deficient C4 carbon TCA cycle intermediates (anaplerosis), which thereby will improve oxidative phosphorylation of any fuel in addition to providing alternative fuel.	Since AAKG and AKG have a good safety profile and are available OTC, pALS can consider this supplement individually or as part of the DP, recognizing that dosages may vary per individual. AAKG at 16–18 g in divided doses is required to have an effect, beginning at a lower dose to offset possible bloating and diarrhea	[142, 241]
Glutamate oxaloacetate-transaminase (GOT) and oxaloacetate (OAA)	Preclinical	OAA is a TCA cycle intermediate and glutamate scavenger in blood, reduces CSF glutamate by concentration gradient, and requires very high-dose injections to achieve this effect. OAA is pro-mitochondrial, yet researchers need to address the unknown brain and blood levels of OAA that cause this effect before clinical trials can be planned. The combined treatment of GOT and OAA provides significant neuroprotection through glutamate scavenging without needing to cross the BBB and may have high clinical significance for the future treatment of chronic neurodegenerative disease	A safe source of GOT and OAA for human consumption is not yet available for pALS; however, OAA is available for human use, and the authors cautiously advise prescribing OAA for ALS patients, until clinical trials show it is safe and effective	[9, 396, 397]

(continued)

Table 49.6 (continued)

Supplement/diet	Research status	Rationale	Use in pALS	References
Acetyl-L-carnitine (ALC)	Clinical	ALC increases intracellular levels of carnitine and serves as a major transporter of fatty acids into the mitochondria and animal studies found that ALC counteracts motor neuron death. This study compared Riluzole alone to riluzole plus ALC, finding that patients on the combination treatment had significantly better results on the ALSFRS-R and forced vital capacity (FVC), and there were no differences in adverse events between the groups. There have not been any trials using ALC alone. The acetyl group allows it to cross the BBB which does not occur with L-carnitine	The study used ALC powder which was safe at 1000 mg TID, and 4000–8000 mg in divided doses is recommended on the ALS Worldwide website	[398–400]
Gamma-aminobutyric acid (GABA)	None	GABA is the major inhibitory neurotransmitter, and supplements that block glutamate release, enhance reuptake, or protect cells against the damaging effects of excessive glutamate may aid in the medicinal management of ALS. There is conflicting evidence with regard to GABA's BBB permeability and if GABA exerts any biologically relevant effect, even though there are hundreds of consumers who report a calming effect of GABA	OTC 250 mg BID of a sustained-release GABA. GABA is likely safe. Liposomal drug delivery in neurodegenerative diseases are potentially more effective at crossing the BBB and possibly a liposomal form of GABA would be more effective	[400–403]
Coenzyme Q10 (CoQ10) and MitoQ	Clinical (CoQ10) and preclinical (MitoQ)	CoQ10 is essential for ETC oxidative phosphorylation, an antioxidant, and is anti-inflammatory, apoptotic, and neuroprotective. An ALS Phase II clinical trial of CoQ10 as ubiquinone (1800 mg daily or 2700 mg daily) versus placebo found insufficient evidence to support a pPhase III trial. It should be noted that ubiquinone (oxidized state) formulations are significantly less bioavailable than ubiquinol (reduced state) formulations due to poor intestinal absorption. MitoQ differs structurally from CoQ10 by the addition of a fat soluble, positively charged molecule, facilitating ubiquinone to pass through the BBB and negatively charged mitochondrial membrane for conversion to the active form, ubiquinol. Preclinical ALS research on MitoQ demonstrated reduced oxidative stress, improved generation of ATP, preservation of the neuromuscular junction, and significantly extended life span, yet there are no human studies	CoQ10 has not shown efficacy in human ALS; however, daily doses can be as high as 3000 mg for pALS. Only one preclinical ALS study has shown that MitoQ readily crossed the BBB, entering deep inside mitochondria to quench free radicals and reduce oxidative stress, but human studies are lacking. CoQ10 and MitoQ are safe, well tolerated, and available OTC. MitoQ shows potential for pALS. Clinical trials of MitoQ in other conditions have used 40 mg and 80 mg doses of MitoQ	[404–409]

◼ Table 49.6 (continued)

Supplement/diet	Research status	Rationale	Use in pALS	References
Glutathione (GSH)	Clinical	ALS patients have lower brain GSH concentrations which suggest ongoing oxidative stress and compromised endogenous antioxidant defenses, thus promoting neuronal injury and functional decline in ALS. Clinically, GSH levels and the activities of glutathione reductase (GR) are reduced in ALS patients which correlate with disease progression. In addition to antioxidant defense, GSH is critical for detoxification of environmental toxins, which pose added risk to pALS. ALS patients received reduced GSH IM at 600 mg daily for 12 weeks, and results found no significant effect of GSH in modifying ALS progression. The evidence for GSH is mostly indirect in ALS, but hypothetically very promising.	Although there are no clinical trials proving the efficacy of GSH for ALS, there is potential therapeutic utility based on known metabolic roles of GSH in neurodegeneration. The best delivery system of GSH is IV 3000 mg once weekly, according to what is currently being used by neurologists and wellness clinics, and oral liposomal or sustained-release GSH is another option. Approximately 430 mg daily oral GSH would equal 3000 mg weekly, considering variations in GI absorption and mitochondrial availability. Although it is commonly assumed that supplemental GSH is degraded in the gut by gamma-glutamyl transpeptidase (GGT), a substantial amount of evidence shows that it is bioavailable and absorbed into the plasma. GSH has been extensively studied as a feasible and promising approach for targeted drug delivery across the BBB, especially effective with GSH attached to liposomes. A logical but unproven inference is that OTC liposomal GSH formulations would enhance the oral delivery of GSH	[410–418]
N-acetylcysteine (NAC)	Clinical	NAC boosts GSH production by providing cysteine, the rate-limiting amino acid in GSH production. Other mechanisms of NAC include modulation of oxidative stress, mitochondrial dysfunction, detoxification, neuroinflammation, and dysregulation of glutamate. Preclinical studies of NAC on ALS mice showed significantly delayed impairment and extended life span. The ability of NAC to cross the BBB is still disputed in the literature. A systematic review of NAC found that the largest study failed to show any benefit of subcutaneous 50 mg/kg/day of NAC in survival or disease progression when compared to pALS getting the placebo, and 3 other studies showed either no benefit or mixed results	NAC has not been proven efficacious in ALS yet is possibly supportive. 2–2.4 g oral NAC total daily dosing was safe and well tolerated, except mild GI symptoms were reported in some pALS	[419–421]
Alpha-lipoic acid (ALA)	Preclinical	ALA is an antioxidant and cofactor for several mitochondrial enzymes, and an animal study showed a significant delay in onset of ALS, increased survival, and attenuated weight loss compared to controls	ALA is well tolerated and safe and can be given orally or IV. Dosages range from 300 mg to 1200 mg daily in divided doses	[422–423]
Creatine monohydrate	Clinical	Creatine has demonstrated beneficial effects in AD, PD, and HD. Creatine is phosphorylated by ATP to the high-energy phosphocreatine which prevents energy depletion and stabilizes mitochondrial membranes by binding to phospholipids. Many preclinical studies were promising, though several clinical trials failed to show significant beneficial effects in ALS progression or survival. There are multiple anecdotal reports of beneficial effects and some positive results from a phase II trial	Creatine monohydrate might be helpful	[241]

(continued)

Table 49.6 (continued)

Supplement/diet	Research status	Rationale	Use in pALS	References
Vitamin D	Clinical	Vitamin D has genomic and non-genomic effects on ALS: cell signaling, glutamate, matrix metalloproteinases, mitogen-activated protein kinase (MAPK) pathways, prostaglandins, oxidative stress, and other mechanisms. Severe vitamin D deficiency in pALS reduces life expectancy and accelerates the rate of decline by 4X, serving as a reliable prognostic factor of ALS. Low vitamin D levels were associated with the clinical severity of ALS. A study found that ALS patients receiving 2000 IU daily had a slower decline in ALSFRS-R at 3, 6, and 9 months after treatment initiation compared to untreated ALS patients even after data were adjusted for age and baseline vitamin D levels. Low vitamin D levels were associated with lower gross motor scores, but low vitamin D didn't predict disease progression over 12 months. Another study found a negative effect of higher vitamin D levels on prognosis in pALS. A recent study found that pALS receiving supplemental cholecalciferol 100,000 IU weekly did not result in any clinical differences or improve prognosis compared to untreated patients. It has been verified that vitamin D receptor gene polymorphisms are linked to ALS	Although studies on vitamin D are conflicting, and most don't consider genomic variables or diet, pALS might benefit with vitamin D as part of the total nutritional approach. Since pALS have reduced or no mobility and spend less time in the sun, the need for vitamin D and other synergistic nutrients (calcium, magnesium, vitamin K2, boron, etc.) is important for bone health too. A minimum of vitamin D3 2000–3000 IU daily is recommended on the ALS Worldwide website	[360–361, 424–429]
Vitamin E	Clinical	Vitamin E (alpha-tocopherol) protects cell membranes from lipid peroxidation, by decreasing pro-inflammatory cytokines such as IL-1ß and TNF-alpha. The year after Lou Gehrig was diagnosed with ALS in 1939, a preliminary report on vitamin E was published in the *Journal of the American Medical Association*, claiming that early treatment of ALS animals with alpha-tocopherol yielded "spectacular" results with the "possibility of not only arresting the process but of reversing it.". Numerous preclinical and clinical studies since then, and a pooled analysis of 5 prospective cohort studies, found that vitamin E supplementation was associated with lower ALS rates and that higher serum alpha-tocopherol was associated with lower risk of ALS	Although vitamin E is promising as an antioxidant, we still lack clinical trials that demonstrate benefits for survival outcomes, functional status, or quality of life. Only alpha-tocopherol has been used in human ALS studies, and 500 mg (1100 IU) twice daily is suggested on the ALS Worldwide website. It should be noted that alpha-tocopherol combined with mixed tocopherols (alpha, gamma, delta) protects cells from lipid peroxidation, more than alpha-tocopherol alone	[38, 61, 362, 430–433]
Vitamin B12	Clinical	Cell culture studies have shown that methylcobalamin can ameliorate oxidative stress, glutamate toxicity, and apoptosis that might be useful for ALS. High doses of B12 can inhibit DNA methylation which is known to be increased in the spinal cords of pALS and is potentially responsible for harmful transcriptional dysregulation. B12 lowers plasma homocysteine levels which are characteristically elevated in pALS and can be toxic to motor neurons in culture. B12 is a cofactor in the conversion of methylmalonyl-CoA (MMA) to succinyl-CoA, and elevated MMA is an indication of impaired myelin synthesis that affects neuronal function. Researchers from Japan presented results of a phase III clinical trial where early administration of ultrahigh-dose methylcobalamin (50 mg IM twice weekly) could reduce progression and increase survival by a median of 600 days compared with placebo; however, the latest new drug application for B12 as a treatment for ALS was withdrawn	B12 status should be assessed in each patient with a serum B12 and MMA. B12 IM shots allow for better bioavailability than oral, but oral may be satisfactory. Genomic variations might determine if methyl, hydroxyl, or adenosyl forms of B12 are preferred	[434–437]

Supplement/ diet	Research status	Rationale	Use in pALS	References
Niacin (B3) as nicotinamide riboside (NR)	None	A protein, SARM1, underlies axonal degeneration both through intrinsic removal of damaged axons and in neurodegenerative diseases such as ALS through rapid breakdown and depletion of NAD+. This causes a massive loss of energy supply within axons. This detrimental effect was reversed by supplementing with nicotinamide riboside (NR), a precursor of NAD, which maintained the energy status in axons. NR is a naturally occurring form of vitamin B3 present in trace amounts in some foods, and studies in AD models indicate bioavailability and protection to brain. Several clinical trials have assessed safety, tolerability, and efficacy of NR to raise NAD levels, yet there haven't been any clinical trials completed on neurodegeneration	NAD depletion is implicated in the pathogenesis of ALS, and supplying NR as a precursor to NAD has been shown to increase NAD levels in humans. Dosages of 250 mg to 1 g will be used in a clinical trial studying brain function and cognition with an estimated completion date of December 2018	[438–444]
Magnesium (Mg)	None	Mg is an essential mineral and cofactor for over 300 biochemical reactions, including energy metabolism, detoxification, nucleic acid and protein synthesis, ion transport, muscle function, bone health, and cell signaling. Inhibition of vital enzymatic activities in the CNS due to Mg deficiency was suggested as a causative factor in CNS degeneration, and patients with ALS had significantly lower levels of Mg in the CNS compared to controls. A synthetically developed compound, magnesium-L-threonate (MgT), shows enhanced brain bioavailability in preclinical studies. According to 5 large prospective cohort studies, the relation between Mg intake from diet and supplements was not associated with either ALS risk or protection, although Mg deficiency is a possible risk factor for ALS in individuals who are genetically vulnerable or exposed to heavy metals. This study did not evaluate Mg status through serum, RBC, or CNS Mg levels	It is suggested to obtain an RBC Mg to determine Mg status, but this does not necessarily reflect CNS Mg. There are many forms of Mg available orally and transdermally. MgT is the most bioavailable form for the brain. Products include Mg chelates, MgT, bath salts, lotions, sprays OTC	[445–448]
Hydrogen (H2)	None	Molecular hydrogen may be neuroprotective against ALS, possibly through abating oxidative and nitrosative stress and preserving mitochondrial. H2 has been studied in numerous diseases with multiple therapeutic targets, including a clinical trial in PD, showing significant symptomatic differences between the treatment and placebo groups. H2 is safe at high concentrations, modulates gene expression, penetrates biological membranes including the BBB, and is easy to administer dissolved in water	Although there are no clinical studies of H2 and ALS, it is safe, provides superior antioxidant bioavailability, and has shown beneficial effects in PD. Numerous studies suggest that H2 may be useful for prevention, treatment, and mitigation of neurological disorders	[449–452]
Lunasin	Clinical	Lunasin is a peptide found in soy and some cereal grains that can reduce oxidative stress and inflammation and alter histone acetylation and gene expression. There are no preclinical studies using lunasin in ALS; however mice studies indicate that lunasin crosses the BBB. The success story of an individual pALS using lunasin showed significant improvement in the ALSFRS. A clinical trial including 50 pALS, "ALS Reversals – Lunasin Regimen" was recently completed with results showing that Lunasin doesn't slow progression	Lunasin is non-GMO and considered safe	[394, 453]

(continued)

Table 49.6 (continued)

Supplement/diet	Research status	Rationale	Use in pALS	References
Multivitamin mineral			It is recommended that pALS take a multivitamin mineral supplement free of copper and iron and with bioavailable forms of B vitamins and other nutrients	
Other nutraceuticals		*Bee venom* has a neuroprotective effect against motor neuron cell death and suppresses neuroinflammation-induced disease progression in symptomatic ALS mice *Berberine* was able to reverse TDP-43 aggregates and neuropathology through autophagy clearance in mice *Vitamin K2* occurs predominantly in all brain regions as menaquinone-4 (MK-4), influencing cell growth, survival, and apoptosis. Vitamin K also participates in the synthesis of sphingolipids for brain cell membrane structure, proliferation, differentiation, senescence, and cell-cell interactions. Vitamin K2 may protect against oxidative stress and inflammation *Vitamin C* is highly concentrated in the brain and is dysregulated in neurodegenerative disorders. A damaged BBB may lead to diffusion of ascorbic acid from the CNS and impair the brain's ability to maintain a high CSF/plasma ascorbic acid ratio. Vit C reduces abnormal proteins in human astrocyte glial cells; participates in antioxidant defense, biosynthesis of collagen, carnitine, tyrosine, peptide hormones, and myelin; supports neurotransmission; and protects against neurotoxicity and excitotoxicity. Vit C might paradoxically act as a prooxidant to produce hydroxyl radicals in the CSF of SALS patients while potentially reducing SOD activity. 5 pooled prospective clinical studies showed that neither a high Vit C diet nor long-term Vit C supplementation affected risk of ALS. Assessing Vit C status in pALS is challenging, as plasma levels only reflect recent dietary intake. Leukocyte levels more closely reflect tissue stores, but do not accurately reflect muscle and brain tissue stores. It would seem prudent to make sure that pALS are replete in Vit C and take enough supplemental Vit C to saturate tissue levels, approximately 200–400 mg daily *Vitamin A* supplementation is not therapeutically understood in ALS. Motor neurons in ALS patients have altered retinoid signaling genes and receptors which promote oxidative-induced cell death *Inosine* raises endogenous urate, a promising therapeutic target for antioxidant and neuroprotection, with expected completion of a phase II trial in 2020 *Lithium* may have neuroprotective properties as a pilot trial reported remarkable improvements; then subsequent clinical trials have failed to replicate the positive results. Trace elements: zinc and selenium in blood were inversely correlated with ALS in patients with worse function, whereas copper was positively associated with ALS. Over exposure to selenium in the inorganic forms has been hypothesized as a risk factor for ALS [462]. In another study, selenium showed a protective role through antioxidant mechanism		[60, 377, 454–463]

cal Lunasin Regimen study [394], other studies evaluate a single nutrient at a time, similar to the reductionist pharmaceutical model of studying a single compound at a time. The Deanna Protocol used a more holistic approach, which studied four supplements and a controlled diet to evaluate the synergistic effects in ALS mice [142]. It is necessary to recognize the importance of an integrated research approach to ALS by emphasizing the multifactorial pathological pathways of ALS and the need for developing therapeutic strategies based on a systems biology approach [395].

49.44 Conclusion

The role of nutrition science with translation into ALS treatment is slowly evolving. As stated 30 years ago, "Nutritional etiology of amyotrophic lateral sclerosis (ALS), and other motor neuron disorders, cannot be ignored even though many trials of nutritional therapy have not, thus far, yielded any benefit to the sufferers. Possibly the nutrients available, or the dosages used, have not been adequate to the circumstances; perhaps an as-yet-unknown nutrient is deficient, poorly absorbed or under-utilized" [464]. Similar hurdles are expressed in recent testimony from Dr. D'Agostino in *The Deanna Protocol*: "Ordinarily, there is no significant depth of exposure to nutritional biochemistry and metabolomics in a general medical school curriculum. So, it is not surprising that doctors don't understand metabolomics, unless they choose to specialize in this area. I don't think that neurologists are ready to accept that a patient with ALS can manage this disease by diet. So, there is going to be a natural resistance to any claims that metabolomics delays the progression of ALS" [465].

pALS experiment with diet and nutrition in their fight against ALS, sometimes with profound beneficial outcomes. As IFMNT practitioners, we must obligate ourselves to champion nutrition research and practice to its fullest therapeutic potential for pALS. In "Non-Familial ALS: A tangled web," the authors state "when insights from genetics, cell biology, and epidemiology can be brought together, this disease will give up its secrets" [466]. Nutrition underlies genetic expression, cell biology, and epidemiology, and nutrition may be one of the most important keys to unlocking the secrets of ALS.

References

1. ALS Association. About ALS. 2013 [Internet]. Last accessed 23 July 2018, from ALS Association: http://www.alsa.org/about-als/2013-aam/.
2. Smaldino C. The Huffington Post. Last accessed 9 June 2018, from The Huffington Post. 2015 [internet]. http://www.huffingtonpost.com/carol-smaldino/dying-of-als-or-not_b_6511248.html.
3. ALS Association. ALS ice bucket challenge – FAQ. 2017 [internet]. Last accessed 9 June 2018 from ALS Association: http://www.alsa.org/about-us/ice-bucket-challenge-faq.html?.
4. Oster CP, Pagnini F. Resentment, hate, and hope in amyotrophic lateral sclerosis. Front Psychol. 2012;3:530. https://doi.org/10.3389/fpsyg.2012.00530.
5. Edney E. Eric is winning!! Revised 9th edition ed. USA: Xlibris Corporation; 2006.
6. ALS Winners http://www.alswinners.com/ ALS winners – the road to recovery: Mr. Kim Cherry. [internet]. Last accessed 5/24/2018.
7. Health Advocates. 2014. Healing ALS. Retrieved from YouTube: https://www.youtube.com/watch?v=quEikibrswo. Last accessed 9 June 2018.
8. Wot D. Amyotrophic lateral sclerosis. 2017 [internet]. Last accessed 9 June 2018 from making sense of my world: http://makingsenseofmyworld.blogspot.com.
9. Tedone T. The Deanna protocol®: Hope for ALS and other neurological conditions. Henderson: Paradise/Inspire a Paradise Publishing Company; 2015.
10. Bedlack RS, Vaughn T, Wicks P, Heywood J, Sinani E, Selsov R, Macklin EA, Schoenfeld D, Cudkowicz M, Sherman A. How common are ALS plateaus and reversals? Neurology. 2016;86(9):808–12. https://doi.org/10.1212/WNL.0000000000002251.
11. Healed/Healing PALS. 2017 [internet]. Last accessed 9 June 2018 from https://healingals.org.
12. Harrison D, Mehta P, van Es M, Stommel E, Drory V, Nefussy B, van den Berg L, Crayle J, Bedlack R. ALS reversals: demographics, disease characteristics, treatments, and co-morbidities. Amyotroph Lateral Scler Frontotemporal Degener. 2018; https://doi.org/10.1080/21678421.2018.1457059.
13. Morris ZW. The answer is 17 years, what is the question: understanding time lags in translational research. J R Soc Med. 2011;104(12):510–20.
14. Cipoletta SG. Illness trajectories in patients with amyotrophic lateral sclerosis: how illness progression is related to life narratives and interpersonal relationships. J Clin Nurs. 2017;26(23–24):5033–43. https://doi.org/10.1111/jocn.14003.
15. Oster C. Craig oster part 1: grateful to show up. 2009 [Internet]. Retrieved from YouTube: https://www.youtube.com/watch?v=YnKI6hPe2ds. Last accessed 9 June 2018.
16. Oster C. Craig oster part 2: grateful to show up 15 years of ALS. 2009 [Internet]. Retrieved from YouTube: https://www.youtube.com/watch?v=YnKI6hPe2ds. Last accessed 9 June 2018.
17. The Healers. 2017 [internet]. Last accessed 9 June 2018. http://healingwithdrcraig.com.
18. US News and World Report. Researchers say they've found common cause of all types of ALS. 2011 [internet]. http://health.usnews.com/health-news/managing-your-healthcare/research/articles/2011/08/21/researchers-say-theyve-found-common-cause-of-all-types-of-als.
19. Seneff S, Morley WA, Hadden MJ, Michener MC. Does glyphosate acting as a glycine analogue contribute to ALS? J Bioinform Proteomics. 2016;2(3):1–21; Review.
20. Arthur KC, Calvo A, Price TR, Geiger JT, Chio A, Traynor BJ. Projected increase in amyotrophic lateral sclerosis from 2015 to 2040. Nat Commun. 2016;7:12408. https://doi.org/10.1038/ncomms12408.
21. Wikipedia. Tuesdays with Morrie. 2017 [internet]. Retrieved from Wikipedia: http://en.wikipedia.org/wiki/Tuesdays_with_Morrie.
22. Wikipedia. Stephen Hawking. 2018 [internet]. Retrieved from Wikipedia: http://en.wikipedia.org/wiki/Stephen_Hawking.
23. Ingre C, Roos PM, Piehl F, Kamel F, Fang F. Risk factors for amyotrophic lateral sclerosis. Clin Epidemiol. 2015;7:181–93. https://doi.org/10.2147/CLEP.S37505. eCollection 2015.
24. Sivak E. What everyone with ALS should know about breathing. 2000 [internet] from MDA/ALS Newsmagazine: http://alsn.mda.org/article/what-everyone-als-should-know-about-breathing.
25. ALS Therapy Development Institute. What is ALS? October 11, 2017, [internet] from ALS Therapy Development Institute: https://www.als.net/what-is-als/.
26. Blitterswijk MV, Vlam L, van Es MA, van der Pol WL, Hennekam EAM, Dooijes D, Schelhaas HJ, van der Kooi AJ, de Visser M, Veldink JH, van der Berg LH. Genetic overlap between apparently sporadic motor neuron diseases. PLOSOne. 2012;7(11):e48983. https://doi.org/10.1371/journal.pone.0048983.

27. CureFFI.org. How do SOD1 mutations cause ALS? 2015 [internet]. Last accessed 9 June 2018 from Cureffi.org: http://www.cureffi.org/2015/04/30/how-do-sod1-mutations-cause-als/.
28. Rotunno MB, Bosco DA. An emerging role for misfolded wild-type SOD1 in sporadic ALS pathogenesis. Front Cell Neurosci. 2013;7:253. https://doi.org/10.3389/fncel.2013.00253.
29. Da Cruz SB, Bui A, Saberi S, Lee SK, Stauffer J, McAlonis-Downes M, Schulte D, Pizzo DP, Parone PA, Cleveland DW, Ravits J. Misfolded SOD1 is not a primary component of sporadic ALS. Acta Neuropathol. 2017;134(1):97–111. https://doi.org/10.1007/s00401-017-1688-8.
30. Saccon RA, Bunton-Stasyshyn RK, Fisher EM, Fratta P. Is SOD1 loss of function involved in amyotrophic lateral sclerosis? Brain. 2013;136(8):2342–58.
31. Tomkins JB, Banner SJ, McDermott CJ, Shaw PJ. Mutation screening of manganese superoxide dismutase in amyotrophic lateral sclerosis. Neuroreport. 2001;12(11):2319–22.
32. Miyata S, Nakamura S, Nagata H, Kameyama M. Increased manganese level in spinal cords of amyotrophic lateral sclerosis determined by radiochemical neutron activation analysis. J Neurol Sci. 1983;61(2):283–93.
33. Johns Hopkins Neurology and Neurosurgery Centers/ALS Clinic. How is ALS classified? 2017 [Internet]. Last accessed 12 July 2018 from: https://www.hopkinsmedicine.org/neurology_neurosurgery/centers_clinics/als/conditions/als_amyotrophic_lateral_sclerosis.html.
34. Rilutek (riluzole) ALS Worldwide. 2015 [internet]. http://alsworldwide.org/research-and-trials/article/rilutek-riluzole. Last accessed 9 June 2018.
35. ALS IV therapy Radicava launched in U.S. 2017 [internet]. Last accessed 9 June 2018. http://www.pharmalive.com/first-fda-approved-treatment-for-als-in-22-years-now-available-in-u-s/.
36. Acevedo-Arozena A, Kalmar B, Essa S, Ricketts T, Joyce P, Kent R, Rowe C, Parker A, Gray A, Hafezparast M, Thorpe JR, Greensmith L, Fisher EM. A comprehensive assessment of the SOD1G93A low-copy transgenic mouse, which models human amyotrophic lateral sclerosis. Dis Model Mech. 2011;4(5):686–770.
37. Strong MJ. Progress in clinical neurosciences: the evidence for ALS as a multisystem disorder of limited phenotypic expression. Canad J Neurol Sci. 2001;28(4):283–98. https://doi.org/10.1017/S0317167100001505.
38. Rosenfeld J, Strong MJ. Challenges in the understanding and treatment of amyotrophic lateral sclerosis/motor neuron disease. Neurotherapeutics. 2015;12(2):317–25. https://doi.org/10.1007/s13311-014-0332-8.
39. Bouteloup C, Desport JC, Clavelou P, Guy N, Derumeaux-Burel H, Ferrier A, Couratier P. Hypermetabolism in ALS patients: an early and persistent phenomenon. J Neurol. 2009;256(8):1236–42. https://doi.org/10.1007/s00415-009-5100-z.
40. Ngo ST, Steyn FJ, Huang L, Mantovani S, Pfluger CM, Woodruff TM, O'Sullivan JD, Henderson RD, McCombe PA. Altered expression of metabolic proteins and adipokines in patients with amyotrophic lateral sclerosis. J Neurol Sci. 2015;357(1–2):22–7. https://doi.org/10.1016/j.jns.2015.06.053.
41. Ngo ST, Steyn FJ. The interplay between metabolic homeostasis and neurodegeneration: insights into the neurometabolic nature of amyotrophic lateral sclerosis. Cell Regen (Lond). 2015;4(1):5. https://doi.org/10.1186/s13619-015-0019-6.
42. Engelhardt B, Sorokin L. The blood-brain and the blood-cerebrospinal fluid barriers: function and dysfunction. Semin Immunopathol. 2009;31(4):497–511. https://doi.org/10.1007/s00281-009-0177-0.
43. Obermeier B, Daneman R, Ransohoff RM. Development, maintenance and disruption of the blood-brain barrier. Nat Med. 2013;19(12):1584–96. https://doi.org/10.1038/nm.3407.
44. Schmitt F, Hussain G, Dupuis L, Loeffler JP, Henriques A. A plural role for lipids in motor neuron diseases: energy, signaling and structure. Front Cell Neurosci. 2014;8:25. https://doi.org/10.3389/fncel.2014.00025.
45. NutrientsReview.com. The transport of nutrients across the cell membranes. 2016 [internet]. Retrieved 18 Aug 2017, from NutrientsReview.com: http://www.nutrientsreview.com/articles/transport-nutrients-blood-brain-barrier-placenta.html.
46. Daneman R, Rescigno M. The gut immune barrier and the blood-brain barrier: are they so different? Immunity. 2009;31(5):722–35. https://doi.org/10.1016/j.immuni.2009.09.012.
47. van de Haar HJ, Burgmans S, Jansen JF, van Osch MJ, van Buchem MA, Muller M, Hofman PA, Verhey FR, Backes WH. Blood-brain barrier leakage in patients with early Alzheimer disease. Radiology. 2016;281(2):527–35.
48. FxMedicine. Leaky gut, leaky brain: the role of zonulin. 2015 [internet]. Retrieved from FxMedicine: https://www.fxmedicine.com.au/content/leaky-gut-leaky-brain-role-zonulin.
49. Lopez-Ramirez MA, Wu D, Pryce G, Simpson JE, Reijerkerk A, King-Robson J, Kay O, de Vries HE, Hirst MC, Sharrack B, Baker D, Male DK, Michael GJ, Romero IA. MicroRNA-155 negatively affects blood-brain barrier function during neuroinflammation. FASEB J. 2014;28(6):2551–65. https://doi.org/10.1096/fj.13-248880.
50. Stamatovic SM, Keep RF, Andjelkovic AV. Brain endothelial cell-cell junctions: how to "open" the blood brain barrier. Curr Neuropharmacol. 2008;6(3):179–92.
51. ALS Association. Who gets ALS? Retrieved 11 Feb 2017 [internet]. ALS Association: http://www.alsa.org/about-als/facts-you-should-know.html.
52. Commare C. Caring for patients with ALS: implications for dietitians. 2007 [internet]. Retrieved Feb 2017, from Today's Dietitian: http://www.todaysdietitian.com/newarchives/tdoct2007pg84.shtml.
53. Rio A, Cawadias E. Nutritional advice and treatment by dietitians to patients with amyotrophic lateral sclerosis/motor neurone disease: a survey of current practice in England, Wales, Northern Ireland and Canada. J Hum Nutr Diet. 2007;20(1):3–13.
54. Kellog J, Bottman L, Arra EJ, Selkirk SM, Kozlowski F. Nutrition management methods effective in increasing weight, survival time and functional status in ALS patients: a systematic review. Amyotroph Lateral Scler Frontotemporal Degener. 2018;19(1–2):7–11. https://doi.org/10.1080/21678421.2017.1360355.
55. Knoche C. Maintaining adequate nutrition: a continuing challenge in ALS. 2016 [internet]. Retrieved 6 Feb 2017, from ALS Association: http://www.alsa.org/als-care/living-with-als/maintaining-adequate-nutrition-2015.html.
56. Interns S. FYI high calorie and easy to chew recipes. 2012 [internet]. Retrieved 6 Feb 2017, from ALS Association: http://www.alsa.org/als-care/resources/publications-videos/factsheets/recipes.html.
57. Tanenbaum B. Maintaining good nutrition with ALS. 1999 [internet]. Retrieved 6 Feb 2017, from ALS Association: http://www.alsa.org/assets/pdfs/brochures/nutrition.pdf.
58. Wahl M. Keep calories coming, expert says. 2006 [internet]. Retrieved 6 Feb 2017, from MDA/ALS News Magazine: http://alsn.mda.org/article/keep-calories-coming-expert-says.
59. Sanaie S, Mahmoodpoor A. High caloric diet for ALS patients: high fat, high carbohydrate or high protein. Adv Biosci Clin Med. 2015;3:1–3.
60. Petrov D, Mansfield C, Moussy AM, Hermine O. ALS clinical trials review: 20 years of failure. Are we any closer to registering a new treatment? Retrieved from Frontiers in Aging. Front Aging Neurosci. 2017;9:68. https://doi.org/10.3389/fnagi.2017.00068.
61. Miller RG, Jackson CE, Kasarskis EJ, England JD, Forshew D, Johnston W, Kalra S, Katz JS, Mitsumoto H, Rosenfeld J, Shoesmith C, Strong MJ, Woolley SC. Quality Standards Subcommittee of the American Academy of Neurology. Practice parameter update:

62. Bedlack R, Joyce N, Carter GT, Pagonoi S, Karam C. Complementary and alternative therapies in ALS. Neurol Clin. 2015;33(4):909–36.
63. Körner S, Hendricks M, Kollewe K, Zapf A, Dengler R, Silani V, Petri S. Weight loss, dysphagia and supplement intake in patients with amyotrophic lateral sclerosis (ALS): impact on quality of life and therapeutic options. BMC Neurol. 2013;13:84. https://doi.org/10.1186/1471-2377-13-84.
64. Gallo V, Wark PA, Jenab M, Pearce N, Brayne C, Vermeulen R, Andersen PM, et al. Prediagnostic body fat and risk of death from amyotrophic lateral sclerosis: The EPIC cohort. Neurology. 2013;80(9):829–38. https://doi.org/10.1212/WNL.0b013e3182840689.
65. Drexel University College of Medicine. Quantitative Measurement of Nutritional Substrate Utilization in Patients with Amyotrophic Lateral Sclerosis (ALS). 2014 [internet]. Retrieved 5 Feb 2017, from National Institutes of Health: https://clinicaltrials.gov/show/NCT00714220.
66. Ngo ST, Henderson RD, McCombe PA, Steyn FJ. Exploring targets and therapies for amyotrophic lateral sclerosis: current insights into dietary interventions. Degener Neurol Neuromusc Dis. 2017;2017(7):95–108.
67. Piquet MA. Nutritional approach for patients with amyotrophic lateral sclerosis. Rev Neurol (Paris). 2006;162 Spec No 2:4S177–87. Abstract English Article French.
68. Kasarskis EJ, Berryman S, Vanderleest JG, Schneider AR, McClain CJ. Nutritional status of patients with amyotrophic lateral sclerosis: relation to the proximity of death. Am J Clin Nutr. 1996;63(1):130–7.
69. Sherman MS, Pillai A, Jackson A, Heiman-Patterson T. Standard equations are not accurate in assessing resting energy expenditure in patients with amyotrophic lateral sclerosis. JPEN J Parenter Enteral Nutr. 2004;28(6):442–6.
70. Ellis AC, Rosenfeld F. Which equation best predicts energy expenditure in amyotrophic lateral sclerosis? J Am Diet Assoc 2011;111(1):1680–1687.
71. Hughes S. Serum creatinine and albumin predict ALS survival. 2014 [internet]. Last accessed 9 June 2018: http://www.medscape.com/viewarticle/828681.
72. Desport JC, Preux PM, Bouteloup-Demange C, Clavelou P, Beaufrère B, Bonnet C, Couratier PP. Validation of bioelectrical impedance analysis in patients with amyotrophic lateral sclerosis. Am J Clin Nutr. 2003;77(5):1179–85.
73. Desport JC, Marin B, Funalot B, Preux PM, Couratier P. Phase angle is a prognostic factor for survival in amyotrophic lateral sclerosis. Amyotroph Lateral Scler. 2008;9(5):273–80. https://doi.org/10.1080/17482960801925039.
74. The ALS Care Program. The ALS care program. 2005 [internet]. Last accessed 11 June 2018, from ALS Functional Rating Scale: https://www.outcomes-umassmed.org/ALS/.
75. Park Y, Park J, Kim Y, Baek H, Kim SH. Association between nutritional status and disease severity using the amyotrophic lateral sclerosis (ALS) functional rating scale in ALS patients. Nutrition. 2015;31(11–12):1362–7. https://doi.org/10.1016/j.nut.2015.05.025.
76. Paoli A, Rubini A, Volek JS, Grimaldi KA. Beyond weight loss: a review of the therapeutic uses of very-low-carbohydrate (ketogenic) diets. Eur J Clin Nutr. 2013;67(8):789–96. https://doi.org/10.1038/ejcn.2013.116.
77. Caamaño D, de la Garza A, Beltrán-Ayala P, Chisaguano AM. Nutrition and neurodegenerative diseases: the role of carbohydrates and gluten. Int J Nutr Sci. 2016;1(2):1007.
78. Wills AM, Hubbard J, Macklin EA, Glass J, Tandan R, Simpson EP, Brooks B, Gelinas D, Mitsumoto H, Mozaffar T, Hanes GP, Ladha SS, Heiman-Patterson T, Katz J, Lou JS, Mahoney K, Grasso D, Lawson R, Yu H, Cudkowicz M, MDA Clinical Research Network. Hypercaloric enteral nutrition in patients with amyotrophic lateral sclerosis: a randomised, double-blind, placebo-controlled phase 2 trial. Lancet. 2014;383(9934):2065–72. https://doi.org/10.1016/S0140-6736(14)60222-1.
79. Okamoto K, Kihira T, Kondo T, Kobashi G, Washio M, Sasaki S, Yokoyama T, Miyake Y, Sakamoto N, Inaba Y, Nagai M. Nutritional status and risk of amyotrophic lateral sclerosis in Japan. Amyotroph Lateral Scler. 2007;5:300–4.
80. Dorst J, Cypionka J, Ludolph AC. High-caloric food supplements in the treatment of amyotrophic lateral sclerosis: a prospective interventional study. Amyotroph Lateral Scler Frontotemporal Degener. 2013;(7–8):533–6. https://doi.org/10.3109/21678421.2013.823999.
81. Coughlan KS, Halang L, Woods I, Prehn JHM. A high-fat jelly diet restores bioenergetic balance and extends lifespan in the presence of motor dysfunction and lumbar spinal cord motor neuron loss in TDP-43A315T mutant C57BL6/J mice. Dis Model Mech. 2016;9(9):1029–37. https://doi.org/10.1242/dmm.024786.
82. Institute for Research in Biomedicine. Accumulation Of Sugar In Neurons May Explain Origin Of Several Neurodegenerative Diseases. 2007 [internet]. Last accessed 12 June 2018. Retrieved August 16, 2017, from ScienceDaily: https://www.sciencedaily.com/releases/2007/10/071021142327.htm.
83. Dodge JC, Treleaven CM, Fidler JA, Tamsett TJ, Bao C, Searles M, Taksir TV, Misra K, Sidman RL, Cheng SH, Shihabuddin LS. Metabolic signatures of amyotrophic lateral sclerosis reveal insights into disease pathogenesis. Proc Natl Acad Sci U S A. 2013;110(26):10812–7. https://doi.org/10.1073/pnas.1308421110.
84. AFTD. FTD with motor neuron disease (FTD/MND). 2017 [internet]. Last accessed 12 June 2018. http://www.theaftd.org/understandingftd/disorders/ftdal.
85. Hughes S. Frontotemporal dementia linked to abnormal eating. 2016 [internet]. Last accessed 12 June 2018: http://www.medscape.com/viewarticle/858757#vp_3.
86. Chu W. Maple syrup polyphenols may protect against neurodegenerative effects: Study. 2016 [internet]. Last accessed 12 June 2018: http://www.nutraingredients.com/Research/Maple-syrup-polyphenols-may-protect-against-neurodegenerative-effects-Study.
87. Armon C. 2017. Amyotrophic lateral sclerosis. 2017 [internet]. Last accessed 12 June 2018: http://emedicine.medscape.com/article/1170097-overview.
88. Eisen A, Weber M. Amyotrophic lateral sclerosis: a synthesis of research and clinical practice. Cambridge, NY: Cambridge University Press; 2001. https://doi.org/10.1002/mus.1042.
89. Offen D, Halevi S, Orion D, Mosberg R, Stern-Goldberg H, Melamed E, Atlas D. Antibodies from ALS patients inhibit dopamine release mediated by L-type calcium channels. Neurology. 1998;51(4):1100–3.
90. Greenberg M. why our brains love sugar – and why our bodies don't. 2013 [internet]. Last accessed 12 June 2018: https://www.psychologytoday.com/blog/the-mindful-self-express/201302/why-our-brains-love-sugar-and-why-our-bodies-dont.
91. Dupuis L, Spreux-Varoquaux O, Bensimon G, et al. Platelet serotonin level predicts survival in amyotrophic lateral sclerosis. Egles C, ed. PLoS ONE. 2010;5(10):e13346. https://doi.org/10.1371/journal.pone.0013346.
92. Wurtman RJ, Wurtman JJ. Brain serotonin, carbohydrate-craving, obesity and depression. Obes Res. 1995;3(Suppl 4):477S–80S.
93. Giugliano D, Ceriello A, Esposito K. The effects of diet on inflammation: emphasis on the metabolic syndrome. J Am Coll Cardiol. 2006;48(4):677–85.
94. Manzel A, Muller DN, Hafler DA, Erdman SE, Linker RA, Kleinwietfeld M. Role of "Western diet" in inflammatory autoimmune diseases. Curr Allergy Asthma Rep. 2014;14(1):404. https://doi.org/10.1007/s11882-013-0404-6.
95. Turner MR, Goldacre R, Ramagopalan S, Talbot K, Goldacre MJ. Autoimmune disease preceding amyotrophic lateral sclerosis. Neurology. 2013;81(14):1222–5. https://doi.org/10.1212/WNL.0b013e3182a6cc13.

96. Pagani MR, Gonzalez LE, Uchitel OD. Autoimmunity in amyotrophic lateral sclerosis: past and present. Neurol Res Int. 2011;2011:497080: 11 pages. https://doi.org/10.1155/2011/497080.
97. Scheibye-Knudsen M. Nourishing the aging brain. 2015 [internet]. The Scientist. Last accessed 12 June 2018: http://www.the-scientist.com/?articles.view/articleNo/42273/title/Nourishing-the-Aging-Brain/.
98. Reyes ET, Perurena OH, Festoff BW, Jorgensen R, Moore WV. Insulin resistance in amyotrophic lateral sclerosis. J Neurol Sci. 1984;63(3):317–24.
99. Logroscino G. Motor neuron disease: are diabetes and amyotrophic lateral sclerosis related? Nat Rev Neurol. 2015;11(9):488–90. https://doi.org/10.1038/nrneurol.2015.145.
100. Sun Y, Lu CJ, Chen RC, Hou WH, Li CY. Risk of amyotrophic lateral sclerosis in patients with diabetes: a nationwide population-based cohort study. J Epidemiol. 2015;25(6):445–51. https://doi.org/10.2188/jea.JE20140176.
101. Hollinger SK, Okosun IS, Mitchell CS. Antecedent disease and amyotrophic lateral sclerosis: what is protecting whom? Front Neurol. 2016;7:47. https://doi.org/10.3389/fneur.2016.00047.
102. Beilharz JE, Maniam J, Morris MJ. Diet-induced cognitive deficits: the role of fat and sugar. Potential Mechan Nutr Intervent Nutr. 2015;7(8):6719–38. https://doi.org/10.3390/nu7085307.
103. Tremolizzo L, Pellegrini A, Conti E, Arosio A, Gerardi F, Lunetta C, Magni P, Appollonio I, Ferrarese C. BDNF serum levels with respect to multidimensional assessment in amyotrophic lateral sclerosis. Neurodegener Dis. 2016;16(3–4):192–8. https://doi.org/10.1159/000441916.
104. Turner MR, Chohan G, Quaghebeur G, Greenhall RC, Hadjivassiliou M, Talbot K. A case of celiac disease mimicking amyotrophic lateral sclerosis. Nat Clin Pract Neurol. 2007;3(10):581–4.
105. Brown KJ, Jewells V, Herfarth H, Castillo M. White matter lesions suggestive of amyotrophic lateral sclerosis attributed to celiac disease. AJNR Am J Neuroradiol. 2010;5:880–1. https://doi.org/10.3174/ajnr.A1826.
106. Gadoth A, Nefussy B, Bleiberg M, Klein T, Artman I, Drory VE. Transglutaminase 6 antibodies in the serum of patients with amyotrophic lateral sclerosis. JAMA Neurol. 2015;72(6):676–81. https://doi.org/10.1001/jamaneurol.2015.48.
107. Hadjivassiliou M, Rao DG, Grìnewald RA, et al. Neurological dysfunction in coeliac disease and non-coeliac gluten sensitivity. Am J Gastroenterol. 2016;111(4):561–7. https://doi.org/10.1038/ajg.2015.434.
108. Visser AE, Pazoki R, Pulit SL, van Rheenen W, Raaphorst J, van der Kooi AJ, Ricaño-Ponce I, Wijmenga C, Otten HG, Veldink JH, van den Berg LH. No association between gluten sensitivity and amyotrophic lateral sclerosis. J Neurol. 2017;264(4):694–700. https://doi.org/10.1007/s00415-017-8400-8.
109. Ludvigsson JF, Mariosa D, Lebwohl B, Fang F. No association between biopsy-verified celiac disease and subsequent amyotrophic lateral sclerosis – A population-based cohort study. Eur J Neurol: Off J Eur Fed Neurol Soc. 2014;21(7):976–82. https://doi.org/10.1111/ene.12419.
110. Pizzorno J. The toxin solution. HarperOne: New York; 2017.
111. Baden-Mayer A. Monsanto's Roundup. Enough to make you sick. 2015 [internet]. Last accessed 12 June 2018: https://www.organicconsumers.org/news/monsantos-roundup-enough-make-you-sick.
112. Samsel A, Seneff S. Glyphosate's suppression of cytochrome P450 enzymes and amino acid biosynthesis by the gut microbiome: pathways to modern diseases. Entropy. 2013;15(4):1416–63. https://doi.org/10.3390/e15041416.
113. Samsel A, Seneff S. Glyphosate, pathways to modern diseases II: celiac sprue and gluten intolerance. Interdiscip Toxicol. 2013;6(4):159–84. https://doi.org/10.2478/intox-2013-0026.
114. Gillam C. IARC scientists defend glyphosate cancer link; Surprised by industry assault. 2016 [internet]. Last accessed 12 June 2018: http://www.huffingtonpost.com/carey-gillam/iarc-scientists-defend-gl_b_12720306.html.
115. Cattani D, de Liz Oliveira Cavalli VL, Heinz Rieg CE, Domingues JT, Dal-Cim T, Tasca CI, Mena Barreto Silva FR, Zamoner A. Mechanisms underlying the neurotoxicity induced by glyphosate-based herbicide in immature rat hippocampus: involvement of glutamate excitotoxicity. Toxicology. 2014;320:34–45. https://doi.org/10.1016/j.tox.2014.03.001.
116. Myers JP, Antoniou MN, Blumberg B, Carroll L, Colborn T, et al. Concerns over use of glyphosate-based herbicides and risks associated with exposures: a consensus statement. Environ Health. 2016;15:19. https://doi.org/10.1186/s12940-016-0117-0; https://ehjournal.biomedcentral.com/articles/10.1186/s12940-016-0117-0.
117. Santos Salvioni CC. Nutritional care in motor neurone disease/amyotrophic lateral sclerosis. Arquivos de Neuro-Psiquiatria. 2014;72(2):157–63. English version. https://doi.org/10.1590/0004-282X20130185.
118. Abbott. Adult. 2017 [internet]. Last accessed 12 June 2018: https://abbottnutrition.com/adult.
119. Snelson M, Mamo JCL, Lam V, Giles C, Takechi R. Differential effects of high-protein diets derived from soy and casein on blood–brain barrier integrity in wild-type mice. Front Nutr. 2017;4:35. https://doi.org/10.3389/fnut.2017.00035.
120. Amen DG, Amen T. The brain warrior's way. New York: Penguin Random House; 2016.
121. Blaylock R. Excitotoxins: the Taste that kills. Albuquerque: Health Press NA Inc.; 1997.
122. Feeney N. Illegal antibiotics could be in your milk, FDA finds. 2015 [internet]. Last accessed 12 June 2018 from Time Health: http://time.com/3738069/fda-dairy-farmers-antibiotics-milk/.
123. Grill MF, Maganti RK. Neurotoxic effects associated with antibiotic use: management considerations. Br J Clin Pharmacol. 2011;72(3):381–93. https://doi.org/10.1111/j.1365-2125.2011.03991.x.
124. Maruyama K, Oshima T, Ohyama K. Exposure to exogenous estrogen through intake of commercial milk produced from pregnant cows. Pediatr Int. 2010;52(1):33–8. https://doi.org/10.1111/j.1442-200X.2009.02890.x.
125. Fischer WJ, Schilter B, Tritscher A, Stadler R. Contaminants of milk and dairy products: contamination resulting from farm and dairy practices. Encyclopedia of Dairy Sciences. 2015:887–97. https://doi.org/10.1016/B978-0-12-374407-4.00104-7.
126. Gillam C. FDA suspends testing for glyphosate residues in food. 2016 [internet]. Last accessed 12 June 2018: http://www.huffingtonpost.com/carey-gillam/fda-suspends-glyphosate-r_b_12913458.html.
127. Madsen A. Managing mucus plugs. 2009 [internet]. Last accessed 12 June 2018 from MDA/ALS Newsmagazine: http://alsn.mda.org/article/managing-mucus-plugs.
128. Anderson P. Dietary factors linked to better – or worse – function in ALS. 2016 [internet]. Last accessed 12 June 2018: http://www.medscape.com/viewarticle/871750#vp_2.
129. USDA. Recent trends in GE adoption. 2017 [internet]. Last accessed 13 June 2018 from USDA Economic Research Service: https://www.ers.usda.gov/data-products/adoption-of-genetically-engineered-crops-in-the-us/recent-trends-in-ge-adoption.aspx.
130. Truth in Labeling. Names of ingredients that contain processed free glutamic acid (MSG). 2014 [internet]. Last accessed 13 June 2018 from Truth in Labeling: http://www.truthinlabeling.org/hiddensources.html.
131. U.S. Department of Health and Human Services. FDA and monosodium glutamate. Food and Drug Administration, FDA Backgrounder. 1995 [internet]. http://vm.cfsan.fda.gov/~lrd/msg.html.
132. Forbes. Alarmism About Monosodium Glutamate (MSG) in your diet may be ill-informed. 2017 [internet]. Last accessed 13 June 2018 from Forbes: https://www.forbes.com/sites/

133. Sharma V, Deshmukh R. Ajimomoto (MSG): a fifth taste or a bio bomb. Ejpmr. 2015;2(2):381–400. http://www.ejpmr.com/admin/assets/article_issue/1425900864.pdf.
134. Hawkins RA, Vina JR. How glutamate is managed by the blood–brain barrier. Biology. 2016;5(4):37. https://doi.org/10.3390/biology5040037. https://www.ncbi.nlm.nih.gov/pmc/articles/PMC5192417/.
135. Riva N, Clarelli F, Domi T, et al. Unraveling gene expression profiles in peripheral motor nerve from amyotrophic lateral sclerosis patients: insights into pathogenesis. Sci Rep. 2016;6:39297. https://doi.org/10.1038/srep39297.
136. Ilżecka J, Stelmasiak Z, Solski J, Wawrzycki S, Szpetnar M. Plasma amino acids percentages in amyotrophic lateral sclerosis patients. Neurol Sci. 2003;24(4):293–5.
137. Iwasaki Y, Ikeda K, Kinoshita M. Plasma amino acid levels in patients with amyotrophic lateral sclerosis. J Neurol Sci. 1992;107(2):219–22.
138. Cecchi M, Messina P, Airoldi L, Pupillo E, Bandettini di Poggio M, Calvo A, et al. Plasma amino acids patterns and age of onset of amyotrophic lateral sclerosis. Amyotroph Lateral Scler Frontotemp Degeneration. 2014;15(5–6):371–5.
139. Rajagopal S. Modulatory effects of dietary amino acids on neurodegenerative diseases. In: Essa MM, Abar M, Guillemin G, editors. The benefits of natural products for neurodegenerative diseases, vol. 12. Springer: Switzerland; 2016. p. 401–14.
140. Laterra J, Keep R, Betz LA, Goldstein GW. Chapter 32. Blood—brain—cerebrospinal fluid barriers. In: Siegel G, Agranoff BW, Albers RW, Fisher SK, Uhler MD, editors. Basic neurochemistry, Molecular, cellular and medical aspects. 6th ed. Philadelphia, PA: Lippincott-Raven; 1999.
141. Supplements in Review. Phosphatidylcholine as a nootropic. 2016 [internet]. Last accessed 22 June 2018. Supplements in review: http://supplementsinreview.com/nootropic/phosphatidylcholine-nootropic/.
142. Ari C, Poff AM, Held HE, Landon CS, Goldhagen CR, et al. Metabolic therapy with Deanna protocol Supplementation Delays disease progression and extends survival in amyotrophic lateral sclerosis (ALS) mouse model. PLoS One. 2014;9(7):e103526. https://doi.org/10.1371/journal.pone.0103526.
143. Lee JM, Tan V, Lovejoy D, Braidy N, Rowe DB, Brew BJ, Guillemin GJ. Involvement of quinolinic acid in the neuropathogenesis of amyotrophic lateral sclerosis. Neuropharmacology. 2017;112:346–64.
144. Tandan R, Bromberg MB, Forshew D, Fries TJ, Badger GJ, Carpenter J, et al. A controlled trial of amino acid therapy in amyotrophic lateral sclerosis: I. clinical, functional, and maximum isometric torque data. Neurology. 1996;47(5):1220–6.
145. Parton M, Mitsumoto H, Leigh PN. Amino acids for amyotrophic lateral sclerosis/motor neuron disease. Cochrane Database of Syst Rev. 2003;(4):3457.
146. Parton M, Mitsumoto H, Leigh PN. Withdrawn: amino acids for amyotrophic lateral sclerosis / motor neuron disease. Cochrane Database Syst Rev. 2008;2:CD003457. https://doi.org/10.1002/14651858.CD003457.pub2.
147. Simone R, Vissicchio F, Mingarelli C, De Nuccio C, Visentin S, Ajmone-Cat MA, et al. Branched-chain amino acids influence the immune properties of microglial cells and their responsiveness to pro-inflammatory signals. Mol Basis Dis. 2013;1832(5):650–9. https://doi.org/10.1016/j.bbadis.2013.02.001.
148. Manuel M, Heckman CJ. Stronger is not always better: could a bodybuilding dietary supplement lead to ALS? Exp Neurol. 2011;228(1):5–8. https://doi.org/10.1016/j.expneurol.2010.12.007.
149. Busch S. Side effects of too many amino acids. 2017 [Internet]. Last accessed 11 July 2018 from Livestrong: http://www.livestrong.com/article/500895-side-effects-of-too-many-amino-acids/.
150. Patten BM, Klein LM. L threonine and the modification of ALS. Neurology. 1988;38(3 Suppl 1):354–5.
151. Blin O, Pouget J, Aubrespy G, Guelton C, Crevat A, Serratrice G. A double-blind placebo-controlled trial of L-threonine in amyotrophic lateral sclerosis. J Neurol. 1992;239:79–81.
152. Natural Medicines Comprehensive Database. Threonine. 2017 [Internet]. Last accessed 11 July 2018 from Natural Medicines Comprehensive Database: http://naturaldatabase.therapeuticresearch.com/nd/Search.aspx?cs=&s=ND&pt=100&id=1083&ds=&name=L-Threonine+(THREONINE)&searchid=64010163.
153. Lee J, Ryu H, Kowall NW. Motor neuronal protection by L-arginine prolongs survival of mutant Sod1 (G93A) Als mice. Biochem Biophys Res Commun. 2009;384(4):524–9. https://doi.org/10.1016/j.bbrc.2009.05.015.
154. D'Alessandro G, Calcagno E, Tartari S, Rizzardini M, Invernizzi RW, Cantoni L. Glutamate and glutathione interplay in a motor neuronal model of amyotrophic lateral sclerosis reveals altered energy metabolism. Neurobiol Dis. 2011;43(2):346–55. https://doi.org/10.1016/j.nbd.2011.04.003.
155. Dharmananda S. Amino acid supplements I: Glutamine. 2017 [Internet]. Last accessed 11 July 2018 from Institute for Traditional Medicine: http://www.itmonline.org/arts/glutamine.htm.
156. Levine TD, Miller RG, Bradley WG, Moore DH, Saperstein DS, Flynn LE, et al. Phase I clinical trial of safety of L-serine for ALS patients. Amyotroph Lateral Scler Frontotemporal Degener. 2017;18(21–2):107–11. https://doi.org/10.1080/21678421.2016.1221971.
157. Bradley W, Moore D, Miller R, Saperstein D, Forshew D. Trials of L-serine in ALS. Neurology. 2017;88(16 Suppl):P3.128.
158. Business Wire. New studies aimed at L-serine as management of Lou Gehrig's disease symptoms. 2016 [Internet]. Last accessed 11 July 2018 from Business Wire: http://www.businesswire.com/news/home/20160429005860/en/Studies-Aimed-L-Serine-Management-Lou-Gehrigs-Disease.
159. Aufschnaiter A, Kohler V, Diessl J, Peselj C, Carmona-Gutierrez D, Keller W, Büttner S. Mitochondrial lipids in neurodegeneration. Cell Tissue Res. 2017;367(1):125–40. https://doi.org/10.1007/s00441-016-2463-1.
160. Adibhatla RM, Hatcher JF. Altered lipid metabolism in brain injury and disorders. Subcell Biochem. 2008;49:241–68. https://doi.org/10.1007/978-1-4020-8831-5_9.
161. Raichle ME, Gusnard DA. Appraising the brain's energy budget. PNAS. 2002;99(16):10237–9.
162. Albhan D. 72 amazing human brain facts (based on the latest science). LLC: Be Brain Fit & Blue Sage; 2018. Retrieved from: https://bebrainfit.com/human-brain-facts/.
163. Björkhem I, Meaney S. Brain cholesterol: long secret life behind a barrier. Arterioscler Thromb Vasc Biol. 2004;24:806–15.
164. Saher G, Brügger B, Lappe-Siefke C, et al. High cholesterol level is essential for myelin membrane growth. Nat Neurosci. 2005;8(4):468–75.
165. Petrov AM, Kasimov MR, Zefirov AL. Brain cholesterol metabolism and its defects: linkage to neurodegenerative diseases and synaptic dysfunction. Acta Nat. 2016;8(1):58–73.
166. Orth M, Bellosta S. Cholesterol: its regulation and role in central nervous system disorders. 2012;2012:292598.
167. Bradbury J. Docosahexaenoic acid (DHA): an ancient nutrient for the modern human brain. Nutrients. 2011;3(5):529–54.
168. Farooqui AA. Transport, synthesis, and incorporation of n–3 and n–6 fatty acids in brain glycerophospholipids. In: Beneficial effects of fish oil on human brain. New York: Springer; 2009. p. 47–78.
169. Kim H-Y, Huang BX, Spector AA. Phosphatidylserine in the brain: metabolism and function. Prog Lipid Res 2014; 0: 1–18.
170. Wurtman RJ, Cansev M, Ulus IH. Choline and its products acetylcholine and phosphatidylcholine. In: Lajtha A, Tettamanti G, Goracci G, editors. Handbook of neurochemistry and molecular neurobiology. Boston: Springer; 2009. p. 443–501.

171. Garbuzova-Davis S, Sanberg PR. Blood-CNS barrier impairment in ALS patients versus an animal model. Front Cell Neurosci. 2014;8:21.
172. Bélanger M, Allaman I, Magistretti PJ. Brain energy metabolism: focus on astrocyte-neuron metabolic cooperation. Cell Metab. 2011;14(6):724–38.
173. Schmidt MA. Brain-building nutrition: how dietary fats and oils affect mental, physical, and emotional intelligence. Berkeley: Frog, Ltd.; 2006.
174. Harvard Medical School. "Unlocking the barrier: Surprising role of omega-3 fatty acids in keeping the blood-brain barrier closed." 2017 [internet]. ScienceDaily. www.sciencedaily.com/releases/2017/05/170505085009.htm. Accessed 16 June 2018.
175. Zárate R, El Jaber-Vazdekis N, Tejera N, Pérez JA, Rodríguez C. Significance of long chain polyunsaturated fatty acids in human health. Clin Transl Med. 2017;6(1):25. https://doi.org/10.1186/s40169-017-0153-6.
176. Fitzgerald KC, O'Reilly ÉJ, Falcone GJ, McCullough ML, Park Y, Kolonel LN, Ascherio A. Dietary ω-3 polyunsaturated fatty acid intake and risk for amyotrophic lateral sclerosis. JAMA Neurol. 2014;71(9):1102–10. https://doi.org/10.1001/jamaneurol.2014.1214.
177. Bannenberg G, Serhan CN. Specialized pro-resolving lipid mediators in the inflammatory response: an update. Biochim Biophys Acta. 2010;1801(12):1260–73. https://doi.org/10.1016/j.bbalip.2010.08.002.
178. Bazan NG, Molina MF, Gordon WC. Docosahexaenoic acid signalolipidomics in nutrition: significance in aging, neuroinflammation, macular degeneration, Alzheimer's, and other neurodegenerative diseases. Annu Rev Nutr. 2011;31:321–51. https://doi.org/10.1146/annurev.nutr.012809.104635.
179. Rapoport SI. Arachidonic acid and the brain. J Nutri. 2008;138(12):2515–20. https://doi.org/10.1093/jn/138.12.2515.
180. Igarashi M, Kim HW, Chang L, Ma K, Rapoport SI. Dietary N-6 polyunsaturated fatty acid deprivation increases docosahexaenoic acid metabolism in rat brain. J Neurochem. 2012;120(6):985–97.
181. Grant R, Guest J. Role of omega-3 PUFAs in neurobiological health. Essa MM, Mohammed A, Guillemin G. The benefits of natural products for neurodegenerative diseases. 1st edition. Series 12. Switzerland: Springer International. 2016. (112, pp. 247–274)
182. Yip PK, Pizzasegola C, Gladman S, Biggio ML, Marino M, Jayasinghe M, Ullah F, Dyall SC, Malaspina A, Bendotti C, Michael-Titus A. The omega-3 fatty acid eicosapentaenoic acid accelerates disease progression in a model of amyotrophic lateral sclerosis. PLoS One. 2013;8(4):e61626. https://doi.org/10.1371/journal.pone.0061626.
183. Coila B. Fish containing highest levels of EPA & DHA. 2015 [internet]. Last accessed 18 June 2018 from Livestrong: http://www.livestrong.com/article/309847-fish-containing-highest-levels-of-epa-dha/.
184. Dr. Axe. 17 fish you should never eat + safer seafood options. 2017 [internet]. Last accessed 18 June 18, 2018 from Dr. Axe food is medicine: https://draxe.com/fish-you-should-never-eat/.
185. Monterey Bay Aquarium Seafood Watch. Consumer guides. Retrieved from Monterey Bay Aquarium Seafood Watch. 2017 [internet]. Last accessed 18 June 2018: http://www.seafoodwatch.org/seafood-recommendations/consumer-guides.
186. Dr. Axe. 15 omega-3 foods your body needs now. 2017 [internet]. Last accessed 18 June 2018: https://draxe.com/omega-3-foods/.
187. Kresser C. Why vegetarians and vegans should supplement with DHA. 2016 [internet]. Last accessed 18 June 2018: https://chriskresser.com/why-vegetarians-and-vegans-should-supplement-with-dha/.
188. Winwood R. Recent developments in the commercial production of DHA and EPA rich oils from micro-algae. OCL. 2013;20(6):D604. https://doi.org/10.1051/ocl/2013030. Last accessed 18 June 2018: https://www.ocl-journal.org/articles/ocl/pdf/2013/06/ocl130011.pdf.
189. Jandacek RJ. Linoleic acid: a nutritional quandary. Healthcare. 2017;5(2):25. https://doi.org/10.3390/healthcare5020025.
190. Vannice G, Rasmussen H. Position of the academy of nutrition and dietetics: dietary fatty acids for healthy adults. J Acad Nutr Diet. 2014;114(1):136–53. https://doi.org/10.1016/j.jand.2013.11.001.
191. Johnson GH, Fritsche K. Effect of dietary linoleic acid on markers of inflammation in healthy persons: a systematic review of randomized controlled trials. J Acad Nutr Diet. 2012;112(7):1029–41, 1041.e1–15. https://doi.org/10.1016/j.jand.2012.03.029.
192. Lawrence GD. Dietary fats and health: dietary recommendations in the context of scientific evidence. Adv Nutr. 2013;4(3):294–302. https://doi.org/10.3945/an.113.003657.
193. Fritsche KL. The science of fatty acids and inflammation. Adv Nutr. 2015;6(3):293S–301S. https://doi.org/10.3945/an.114.006940.
194. Lipman F. 4 reasons to avoid cooking with vegetable oils. 2016 [internet]. Last accessed 18 June 2018 from The Be Well Blog: https://www.bewell.com/blog/4-reasons-to-avoid-vegetable-oil/.
195. GB Health Watch. Omega-3: Omega-6 balance. 2017 [internet]. Last accessed from GB Health Watch: http://www.gbhealthwatch.com/Science-Omega3-Omega6.php.
196. Wahls T. The Wahls protocol. Avery: New York City; 2014.
197. Wikipedia. Omega-7 fatty acid. 2017 [internet]. Last accessed 18 June 2018: https://en.wikipedia.org/wiki/Omega-7_fatty_acid.
198. Wikipedia. Social and environmental impact of palm oil. 2017 [internet]. Last accessed 18 June 2018: https://en.wikipedia.org/wiki/Social_and_environmental_impact_of_palm_oil.
199. RSPO: Roundtable on Sustainable Palm Oil 2017 [internet]. Last retrieved 18 June 2018. Roundtable on sustainable palm oil: http://www.rspo.org/certification.
200. Wikipedia. Medium-chain triglyceride. 2017 [internet]. Retrieved from Wikipedia: https://en.wikipedia.org/wiki/Medium-chain_triglyceride.
201. Mercola J. Fat for fuel: Hay House US, Inc; 2017.
202. Zhao W, Varghese M, Vempati P, Dzhun A, Cheng A, Wang J, Lange D, Bilski A, Faravelli I, Pasinetti GM. Caprylic triglyceride as a novel therapeutic approach to effectively improve the performance and attenuate the symptoms due to the motor neuron loss in ALS disease. PLoS One. 2012;7(11):e49191. https://doi.org/10.1371/journal.pone.0049191.
203. Cornell University. Safety of caprylic triglycerides in ALS: a pilot study. 2016 [internet]. Retrieved from clinicaltrials.gov: https://clinicaltrials.gov/ct2/show/NCT02716662.
204. Tefera TW, Wong Y, Barkl-Luke ME, Ngo ST, Thomas NK, McDonald TS, Borges K. Triheptanoin protects motor neurons and delays the onset of motor symptoms in a mouse model of amyotrophic lateral sclerosis. PLoS One. 2016;11(8):e0161816. https://doi.org/10.1371/journal.pone.0161816.
205. Designs for Health. MCTs & ALS. 2015 [internet]. Last accessed 18 June 2018 from Designs for Health Research & Education Blog: http://blog.designsforhealth.com/blog/mcts-als-0.
206. Clinicaltrials.gov. ALS and triheptanoin ClinicalTrials.gov Identifier: NCT03506425. Last accessed 18 June 2018 from Clinicaltrials.gov: https://clinicaltrials.gov/ct2/show/NCT03506425?term=triheptanoin&cond=ALS&rank=1
207. Freeman LR, Haley-Zitlin V, Rosenberger DS, Granholm AC. Damaging effects of a high-fat diet to the brain and cognition: a review of proposed mechanisms. Nutr Neurosci. 2014;17(6):241–51. https://doi.org/10.1179/1476830513Y.0000000092.
208. Wang DD, Li Y, Chiuve SE, Stampfer MJ, Manson JE, Rimm EB, Willett WC, Hu FB. Association of specific dietary fats with total and cause-specific mortality. JAMA Intern Med. 2016;176(8):1134–45. https://doi.org/10.1001/jamainternmed.2016.2417.
209. Henriques A, Blasco H, Fleury MC, Corcia P, Echaniz-Laguna A, Robelin L, Rudolf G, Lequeu T, Bergaentzle M, Gachet C, Pradat PF, Marchioni E, Andres CR, Tranchant C, Gonzalez De Aguilar JL, Loeffler JP. Blood cell palmitoleate-palmitate ratio is an independent prognostic factor for amyotrophic lateral sclerosis.

210. Stockton C. Omega-7 protects against metabolic syndrome. 2014 [internet]. Retrieved from LifeExtension: http://www.lifeextension.com/Magazine/2014/4/Omega-7-Protects-Against-Metabolic-Syndrome/Page-01.
211. USDA Dietary Guidelines 2015–2020. Shifts needed to align with healthy eating patterns. 2017 [internet]. Dietary Guidelines 2015–2020: https://health.gov/dietaryguidelines/2015/guidelines/chapter-2/a-closer-look-at-current-intakes-and-recommended-shifts/#figure-2-9.
212. Harvard Health Publications. Cholesterol, the mind, and the brain. 2007 [internet]. Last accessed 22 June 2018: https://www.health.harvard.edu/newsletter_article/cholesterol-the-mind-and-the-brain.
213. Zang J, Qiang L. Cholesterol metabolism and homeostasis in the brain. Protein Cell. 2015;6(4):254–64. https://doi.org/10.1007/s13238-014-0131-3.
214. Mouritsen OG, Bagatolli LA. Life- as a matter of fat: lipids in a membrane biophysics perspective. Switzerland: Springer; 2016.
215. Anchisi L, Dessi S, Pani A, Mandas A. Cholesterol homeostasis: a key to prevent or slow down neurodegeneration. Front Physiol. 2013 [internet];. https://doi.org/10.3389/fphys.2012.00486
216. Vance JE. Dysregulation of cholesterol balance in the brain: contribution to neurodegenerative diseases. Dis Model Mech. 2012;5(6):746–55. https://doi.org/10.1242/dmm.010124.
217. Abdel-Khalik J, Yutuc E, Crick PJ, Gustafsson JA, et al. Defective cholesterol metabolism in amyotrophic lateral sclerosis. J Lipid Res. 2017;58:267–78. https://doi.org/10.1194/jlr.P071639.
218. Berghoff SA, Gerndt N, Winchenbach J, Stumpf SK, Hosang L, Odoardi F, Ruhwedel T, Böhler C, Barrette B, Stassart R, Liebetanz D, Dibaj P, Möbius W, Edgar JM, Saher G. Dietary cholesterol promotes repair of demyelinated lesions in the adult brain. Nat Commun. 2017;8:14241. https://doi.org/10.1038/ncomms14241.
219. Weatherspoon D. Multiple sclerosis vs. ALS: similarities and differences. 2018 [internet]. Last accessed 22 June 2018 Healthline: http://www.healthline.com/health/multiple-sclerosis/ms-vs-als#overview1.
220. Johns Hopkins Medicine. Researchers Discover New Clues about how Amyotrophic Lateral Sclerosis (ALS) Develops. 2013 [internet]. https://www.hopkinsmedicine.org/news/media/releases/researchers_discover_new_clues_about_how_amyotrophic_lateral_sclerosis_als_develops.
221. Dupuis L, Corcia P, Fergani A, Gonzalez De Aguilar JL, Bonnefont-Rousselot D, et al. Dyslipidemia is a protective factor in amyotrophic lateral sclerosis. Neurology. 2008;70(13):70324–7. https://doi.org/10.1212/01.wnl.0000285080.70324.27.
222. Dorst J, Kühnlein P, Hendrich C, Kassubek J, Sperfeld AD, Ludolph AC. Patients with elevated triglyceride and cholesterol serum levels have a prolonged survival in amyotrophic lateral sclerosis. J Neurol. 2011;258(4):613–7. https://doi.org/10.1007/s00415-010-5805-z.
223. Yang JW, Kim SM, Kim HJ, Kim JE, Park KS, et al. Hypolipidemia in patients with amyotrophic lateral sclerosis: a possible gender difference? J Clin Neurol. 2013;9(2):125–9. https://doi.org/10.3988/jcn.2013.9.2.125.
224. Mitchell CS, Hollinger SK, Goswami SD, Polak MA, Lee RH, Glass JD. Antecedent disease is less prevalent in amyotrophic lateral sclerosis. Neurodegener Dis. 2015;15:109–13.. https://doi.org/10.1159/000369812
225. Huang R, Guo X, Chen X, Zheng Z, Wei Q, Cao B, Zeng Y, Shang H. The serum lipid profiles of amyotrophic lateral sclerosis patients: A study from south-west China and a meta-analysis. Amyotroph Lateral Scler Frontotemporal Degener. 2015;16(5–6):359–65. https://doi.org/10.3109/21678421.2015.1047454. Epub 2015 Jun 29. https://onlinelibrary.wiley.com/doi/full/10.1111/ncn3.143.
226. Meija EM, Hatch GM. Mitochondrial phospholipids: role in mitochondrial function. J Bioenerg Biomembr. 2016;48(2):99–112. https://doi.org/10.1007/s10863-015-9601-4.
227. Cooper GM. The cell. In: A molecular approach. 2nd ed. Boston: Sinauer Associates; 2000.
228. Nicolson GL. The fluid—mosaic model of membrane structure: still relevant to understanding the structure, function and dynamics of biological membranes after more than 40 years. Biochim Biophys Acta. 2014;1838(6):1451–66. https://doi.org/10.1016/j.bbamem.2013.10.019.
229. Nicolson GL, Ash ME. Lipid replacement therapy: a natural medicine approach to replacing damaged lipids in cellular membranes and organelles and restoring function. Biochim Biophys Acta. 2014;1838(6):1657–79. https://doi.org/10.1016/j.bbamem.2013.11.010.
230. Küllenberg D, Taylor LA, Schneider M, Massing U. Health effects of dietary phospholipids. Lipids Health Dis. 2012;11:3. https://doi.org/10.1186/1476-511X-11-3.
231. WebMD. Phosphatidylcholine. 2017 [Internet]. Last accessed 7/10/2018: http://www.webmd.com/vitamins-supplements/ingredientmono-501-phosphatidylcholine.aspx?activeingredientid=501&activeingredientname=phosphatidylcholine.
232. White RF, Steele L, O'Callaghan JP, Sullivan K, et al. Recent research on Gulf War illness and other health problems in veterans of the 1991 Gulf War: effects of toxicant exposures during deployment. Cortex. 2016;74:449–75.
233. Emmerich T, Zakirova Z, Klimas N, Sullivan K, et al. Phospholipid profiling of plasma from GW veterans and rodent models to identify potential biomarkers of Gulf War illness. PLoS One. 2017;12(4):e0176634. https://doi.org/10.1371/journal.pone.0176634.
234. ClinicalTrials.gov. Food supplement for the treatment of patients with Amyotrophic Lateral Sclerosis (ALS-PHL). 2017 [Internet] Retrieved from U.S. National Institutes of Health ClinicalTrials.gov: https://clinicaltrials.gov/ct2/show/NCT02588807.
235. Nicolson GL, Rosenblatt S, Ferreira de Mattos G, Settineri R, Breeding PC, Ellithorpe RR, Ash ME. Clinical uses of membrane lipid replacement supplements in restoring membrane function and reducing fatigue in chronic diseases and cancer. Discoveries. 2016;4(1):e54. https://doi.org/10.15190/d.2016.1.
236. Stafstrom CE, Rho JM. The ketogenic diet as a treatment paradigm for diverse neurological disorders. Front Pharmacol. 2012; 3(59). doi: https://doi.org/10.3389/fphar.2012.00059.
237. Gano LB, Patel M, Rho JM. Ketogenic diets, mitochondria, and neurological diseases. J Lipid Res. 2014;55(11):2211–28. https://doi.org/10.1194/jlr.R048975.
238. Paganoni S, Willis AM. High-fat and ketogenic diets in amyotrophic lateral sclerosis. J Child Neurol. 2013;28(8):989–92. https://doi.org/10.1177/0883073813488669.
239. Winning the Fight. The Deanna protocol plan. 2016 [Internet]. Last accessed 10 July 2018 from Winning the Fight: https://www.winningthefight.org.
240. Csilla A, Poff AM, Held HE, Landon CS, Goldhagen CR, et al. Metabolic therapy with Deanna protocol supplementation delays disease progression and extends survival in amyotrophic lateral sclerosis (ALS) mouse model. PLoS One. 2014;9(7):e103526. https://doi.org/10.1371/journal.pone.0103526.
241. Tefera TW, Borges K. Metabolic dysfunctions in amyotrophic lateral sclerosis pathogenesis and potential metabolic treatments. Front Neurosci. 2017;10:611.. https://doi.org/10.3389/fnins.2016.00611
242. Abbott. Adult jevity. 2016 [Internet]. Last accessed 10 July 2018 from Abbott Nutrition: https://abbottnutrition.com/search/result?q=jevity.
243. Environmental Defense Fund. Common questions about contaminants in seafood. 2017 [Internet]. Last accessed 10 July 2018 from EDF Seafood Selector: http://seafood.edf.org/common-questions-about-contaminants-seafood.

244. Crosby G. Ask the expert: concerns about canola oil. 2017 [Internet]. Last accessed 10 July 2018 from Harvard: Nutrition Source: https://www.hsph.harvard.edu/nutritionsource/2015/04/13/ask-the-expert-concerns-about-canola-oil/.
245. Vacilotto G, Favretto D, Seraglia R, Pagiotti R, Traldi P, Mattoli L. A rapid and highly specific method to evaluate the presence of pyrrolizidine alkaloids in Borago officinalis seed oil. J Mass Spectrom. 2013;48(10):1078–82. https://doi.org/10.1002/jms.3251.
246. Steve Smith, former NFL player with ALS fighting for full settlement. 2017 [Internet]. Dallas Morning News Interview. Last accessed 10 July 2018 from https://www.youtube.com/watch?v=iq_aROVgv5s.
247. Nieves JW, Gennings C, Factor-Litvak P, Hupf J, Singleton J, Sharf V, et al. Association between dietary intake and function in amyotrophic lateral sclerosis. JAMA Neurol. 2016;73(12):1425–32. https://doi.org/10.1001/jamaneurol.2016.3401.
248. Joshi G, Johnson JA. The Nrf2-ARE pathway: a valuable therapeutic target for the treatment of neurodegenerative diseases. Recent Pat CNS Drug Discov. 2012;7(3):218–29. PMID:22742419.
249. Solanki I, Parihar P, Mansuri ML, Parihar MS. Flavonoid-based therapies in the early Management of Neurodegenerative Diseases. Adv Nutr. 2015;6(1):64–72. https://doi.org/10.3945/an.114.007500.
250. Mitra J, Vasquez V, Hegde PM, Boldogh I, Mitra S, et al. Revisiting metal toxicity in neurodegenerative diseases and stroke: therapeutic potential. Neurol Res Ther. 2014;1(2):107. https://www.ncbi.nlm.nih.gov/pmc/articles/PMC4337781/.
251. Braidy N, Selvaraju S, Essa MM, Vaishnav R, Al-Adawi S, et al. Neuroprotective effects of a variety of pomegranate juice extracts against MPTP-induced cytotoxicity and oxidative stress in human primary neurons. Oxidative Med Cell Long. 2013:685909. 12 pages. https://doi.org/10.1155/2013/685909.
252. Ameer K. Avocado as a major dietary source of antioxidants and its preventive role in neurodegeneration. In: Ezza MM, editor. The benefits of natural products for neurodegenerative diseases. Springer: Switzerland; 2016.
253. Selvaraju S, Essa M, Braidy N, Al-Adawi S, Al-Asmi A, et al. Antioxidant and anti-excitotoxic effects of date, fig and walnut extracts in human neurons. J Alzheimer Assc. 2013;9(4):P801. https://doi.org/10.1016/j.jalz.2013.05.1654.
254. Tarozzi A, Angeloni C, Malaguti M, Morroni F, Hrelia S, Hrelia P. Sulforaphane as a Potential Protective Phytochemical against Neurodegenerative Diseases. Oxidative Med Cell Longevity. 2013:415078:10 pages. https://doi.org/10.1155/2013/415078.
255. Topcuoglu MA. Neuronutrition: an emerging concept. In: Arsava E, editor. Nutrition in neurologic disorders: a practical guide. Springer: Cham; 2017. p. 188.
256. European Medicine Agency. Public summary of opinion on orphan designation: Sarsasapogenin for the treatment of amyotrophic lateral sclerosis. 2012 [Internet]. Last accessed 11 July 2018 from Committee for Orphan Medicinal Products: http://www.ema.europa.eu/docs/en_GB/document_library/Orphan_designation/2009/10/WC500005789.pdf.
257. Koh SH, Lee SM, Kim HY, Lee KY, Lee YJ, Kim HT, Kim J, et al. The effect of epigallocatechin gallate on suppressing disease progression of ALS model mice. Neurosci Lett. 2006;395(2):103–7.
258. Kim DO, Heo HJ, Kim YJ, Yang HS, Lee CY. Sweet and sour cherry phenolics and their protective effects on neuronal cells. J Agric Food Chem. 2005;53(26):9921–7.
259. Suchy J, Sangmook L, Ahmed A, Shea TB. Dietary supplementation with pecans delays motor neuron pathology in transgenic mice expressing human superoxide dismutase-1. Curr Topics Nutraceut Res. 2010;8(1):45–49.
260. University of Alberta Faculty of Medicine & Dentistry. Medical researchers make important research link between active ingredient in saffron and MS. 2011 [Internet]. Last accessed 11 July 2018 from EurekAlert: https://www.eurekalert.org/pub_releases/2011-11/uoaf-mrm110311.php.
261. Nehlig A. The neuroprotective effects of cocoa flavanol and its influence on cognitive performance. Br J Clin Pharmacol. 2013;75(3):716–27. https://doi.org/10.1111/j.1365-2125.2012.04378.x.
262. Natural Health 365. The amazing power of bioflavonoids. 2012 [Internet]. Accessed from Natural Health 365: https://www.naturalhealth365.com/bioflavonoids.html/.
263. Rezai-Zadeh K, Ehrhart J, Bai Y, Sanberg PR, Bickford P, Tan J, Shyte RD. Apigenin and luteolin modulate microglial activation via inhibition of STAT1-induced CD40 expression. J Neuroinflammation. 2008;5:41. https://doi.org/10.1186/1742-2094-5-41.
264. Oboh G, Olasehinde TA, Ademosun AO. Essential oil from lemon peels inhibit key enzymes linked to neurodegenerative conditions and pro-oxidant induced lipid peroxidation. J Oleo Sci. 2014;63(4):373–81. https://doi.org/10.5650/jos.ess13166.
265. Karlstetter M, Lippe E, Walczak Y, Moehle C, Aslanidis A, Mirza M, Langmann T. Curcumin is a potent modulator of microglial gene expression and migration. J Neuroinflammation. 2011;8:125. https://doi.org/10.1186/1742-2094-8-125.
266. Kim DS, Kim JY, Han Y. Curcuminoids in neurodegenerative diseases. Recent Pat CNS Drug Discov. 2012;7(3):184–204.
267. Beghi E, Pupillo E, Messina P, Giussani G, Chiò A, Zoccolella S, Moglia C, Corbo M, Logroscino G, EURALS Group. Coffee and amyotrophic lateral sclerosis: a possible preventive role. Am J Epidemiol. 2011;174(9):1002–8. https://doi.org/10.1093/aje/kwr229.
268. Fondell E, O'Reilly EJ, Fitzgerald KC, Falcone GJ, Kolonel LN, Park Y, Gapstur SM, Ascherio A. Intakes of caffeine, coffee and tea and risk of amyotrophic lateral sclerosis: results from five cohort studies. Amyotroph Lateral Scler Frontotemporal Degener. 2015;16(0):366–71. https://doi.org/10.3109/21678421.2015.1020813.
269. Giacoppo S, Mazzon E. Can cannabinoids be a potential therapeutic tool in amyotrophic lateral sclerosis? Neural Regenerat Res. 2016;11(12):1896–9. https://doi.org/10.4103/1673-5374.197125.
270. Trojsi F, Monsurro MR, Tedeschi G. Exposure to environmental toxicants and pathogenesis of amyotrophic lateral sclerosis: state of the art and research perspectives. Int J Mol Sci. 2013;14(8):15286–311. https://doi.org/10.3390/ijms140815286.
271. Bozzoni V, Pansarasa O, Diamanti L, Nosari G, Cereda C, Ceroni M. Amyotrophic lateral sclerosis and environmental factors. Funct Neurol. 2016;31(1):7–19. https://doi.org/10.11138/FNeur/2016.31.1.007.
272. Cannon JR, Greenamyre JT. The role of environmental exposures in neurodegeneration and neurodegenerative diseases. Toxicol Sci. 2011;124(2):225–50. https://doi.org/10.1093/toxsci/kfr239.
273. Roberts AL, Johnson NJ, Cudkowicz ME, Eum KD, Weisskopf MG. Job-related formaldehyde exposure and ALS mortality in the USA. J Neurol Neurosurg Psych. 2015;87(7) https://doi.org/10.1136/jnnp-2015-310750.
274. Pamphlett R, Rikard-Bell A. Different occupations associated with amyotrophic lateral sclerosis: is diesel exhaust the link? PLoS One. 2013;8(11):e80993. https://doi.org/10.1371/journal.pone.0080993
275. Aviation Travel Writer. Mortality from ALS neurodegenerative diseases in flight attendants. 2016 [Internet]. Last accessed 10 July 2018 from: https://aviationtravelwriter.wordpress.com/2016/10/15/mortality-from-als-neurodegenerative-diseases-in-a-flight-attendants/.
276. Vanacore N, Cocco P, Fadda D, Dosemeci M. Job strain, hypoxia and risk of amyotrophic lateral sclerosis: results from a death certificate study. Amyotrophic Lateral Sclerosis Amyotroph Lateral Scler. 2010;11(5):430–4. https://doi.org/10.3109/17482961003605796.
277. Henderson W. ALS rate in miners exposed to McIntyre powder causing concern. 2017 [Internet]. Last accessed 10 July 2018 from ALS News Today: https://alsnewstoday.com/2017/05/04/als-rate-miners-exposed-mcintyre-powder-causing-concern/.

278. Ash PEA, Stanford EA, Al Abdulatif A, Ramirez-Cardenas A, Ballance HI, Boudeau S, et al. Dioxins and related environmental contaminants increase TDP-43 levels. Mol Neurodegener. 2017;12(1):35. https://doi.org/10.1186/s13024-017-0177-9.
279. Malek AM, Barchowsky A, Bowser R, Heiman-Patterson T, Lacomis D, Rana S, Youk A, Talbott EO. Exposure to hazardous air pollutants and the risk of amyotrophic lateral sclerosis. Environ Pollut. 2015;197:181–6. https://doi.org/10.1016/j.envpol.2014.12.010.
280. Ezza HAS, Khadrawyb YA. Glutamate excitotoxicity and neurodegeneration. J Mol Genet Med. 2014;08:141. https://doi.org/10.4172/1747-0862.1000141.
281. Schwartz GG, Klug MG. Motor neuron disease mortality rates in U.S. states are associated with well water use. Amyotroph Lateral Scler Frontotemporal Degener. 2016;17(7-8):528–34. PMID: 27324739.
282. Stewart J. Higher levels of trace metals detected in the blood of some ALS patients. 2017 [internet]. Last accessed 10 July 2018 from ALS News Today: https://alsnewstoday.com/2017/03/31/trace-metals-blood-als-patients/.
283. Henderson W. 4 tests that help diagnose ALS. 2017 [Internet]. Last accessed 11 July 2018 from ALS News Today: https://alsnewstoday.com/2017/06/15/ways-determine-als-diagnosis/.
284. Mutter J. Is dental amalgam safe for humans? The opinion of the scientific committee of the European Commission. J Occup Med Toxicol. 2011;6:2. https://doi.org/10.1186/1745-6673-6-2.
285. IAOMT. International Academy of Oral Medicine & Toxicology Position Statement against Dental Mercury Amalgam Fillings. 2013 [Internet]. Last accessed 12 July 2018 from International Academy of Oral Medicine & Toxicology: https://iaomt.org/wp-content/uploads/IAOMT-Position-Paper-Dental-Mercury-Amalgam-Full.pdf.
286. Sienko DG, Davis JP, Taylor JA, Brooks BR. Amyotrophic lateral sclerosis. A case-control study following detection of a cluster in a small Wisconsin community. Arch Neurol. 1990;47(1):38–41.
287. American Academy of Neurology. Mercury in fish, seafood may be linked to higher risk of ALS. 2017 [Internet]. Last accessed 12 July 2018 from ScienceDaily: https://www.sciencedaily.com/releases/2017/02/170223092345.htm.
288. Patrick L. Mercury toxicity and antioxidants: part 1: role of glutathione and alpha-lipoic acid in the treatment of mercury toxicity. Altern Med Rev. 2002;7(6):456–71. PMID: 12495372.
289. Environmental Working Group. EWG'S consumer guide to seafood. 2014 [Internet]. Last accessed 12 July 2018 from Environmental Working Group: http://www.ewg.org/research/ewgs-good-seafood-guide/#.WcbM7a3MwQ9.
290. Shaw CA, Li D, Tomljenovic L. Are there negative CNS impacts of aluminum adjuvants used in vaccines and immunotherapy? Immunotherapy. 2014;6(10):1055–71. https://doi.org/10.2217/imt.14.81.
291. Shaw CA, Tomljenovic L. Aluminum in the central nervous system (CNS): toxicity in humans and animals, vaccine adjuvants, and autoimmunity. Immunol Res. 2013;56(2–3):304–16. https://doi.org/10.1007/s12026-013-8403-1.
292. Chandler S. What foods contain harmful aluminum? 2017 [Internet]. Last accessed 12 July 2018 from Livestrong: http://www.livestrong.com/article/540321-what-foods-contain-harmful-aluminum/.
293. Hupston F. Avoid aluminum – locate the unexpected sources of aluminum in products. 2011 [Internet]. Last accessed 12 July 2018 from Natural News: http://www.naturalnews.com/033431_aluminum_personal_care_products.html.
294. Bar-Sela S, Reingold S, Richter ED. Amyotrophic lateral sclerosis in a battery-factory worker exposed to cadmium. Int J Occup Environ Health. 2001;7(2):109–12. https://doi.org/10.1179/107735201800339470.
295. Gregor M. How to reduce your dietary cadmium absorption. 2015 [Internet]. Last accessed 12 July 2018 from Nutrition Facts. org: https://nutritionfacts.org/2015/10/15/how-to-reduce-your-dietary-cadmium-absorption/.
296. ConsumerLab.com. Why is there so much cadmium, a toxin, in cocoa powders but not in dark chocolate? 2017 [Internet]. Last accessed 12 July 2018 from ConsumerLab.com: https://www.consumerlab.com/answers/why-is-there-so-much-cadmium-in-cocoa-powders-but-not-in-dark-chocolate/cadmium_in_dark_chocolate/.
297. As You Sow. Toxic chocolate. 2017 [Internet]. Last accessed 12 July 2018 from As You Sow: http://www.asyousow.org/our-work/environmental-health/toxic-enforcement/lead-and-cadmium-in-food/.
298. Su FC, Goutman SA, Chernyak S, Mukherjee B, Callaghan BC, Batterman S, Feldman EL. The role of environmental toxins on ALS: a case-control study of occupational risk factors. JAMA Neurol. 2016;73(7):803–11. https://doi.org/10.1001/jamaneurol.2016.0594.
299. Seelen M, Toro Compos RA, Veldnik JH, Visser AE, Hoek G, Brunkreef B, et al. Long-term air pollution exposure and amyotrophic lateral sclerosis in Netherlands: a population-based case–control study. Environ Health Perspect. 2017;097023:1–7. https://doi.org/10.1289/EHP1115.
300. News from the Centers for Disease Control and Prevention. Professional jobs linked with ALS and Parkinson disease deaths. JAMA. 2017;318(8):691. https://doi.org/10.1001/jama.2017.10459.
301. Muralidhar LH, Bharathi P, Suram A, Venugopal C, Jagannathan R, et al. Challenges associated with metal chelation therapy in Alzheimer's disease. J Alzheimers Dis. 2009;17(3):457–68.
302. Sears ME. Chelation: harnessing and enhancing heavy metal detoxification—a review. Sci World J. 2013;2013:219840. https://doi.org/10.1155/2013/219840.
303. Hodges RE, Minich DM. Modulation of metabolic detoxification pathways using foods and food-derived components: a scientific review with clinical application. J Nutr Metabol. 2015;ID 760689:23 pages. https://doi.org/10.1155/2015/760689.
304. Chen P, Miah MR, Aschner M. Metals and neurodegeneration. F1000Res. 2016;5:F1000 Faculty Rev-366. https://doi.org/10.12688/f1000research.7431.1.
305. Lovejoy DB, Guillemin GJ. The potential for transition metal-mediated neurodegeneration in amyotrophic lateral sclerosis. Front Aging Neurosci. 2014;6:173. https://doi.org/10.3389/fnagi.2014.00173.
306. The George Mateljan Foundation. Iron. 2017 [Internet]. Last accessed 10 July 2018 from The World's Healthiest Foods: http://www.whfoods.com/genpage.php?tname=nutrient&dbid=70.
307. Zheng W, Monnot AD. Regulation of brain Iron and copper homeostasis by brain barrier systems: implication in neurodegenerative diseases. Pharmacol Ther. 2012;133(2):177–88. https://doi.org/10.1016/j.pharmthera.2011.10.006.
308. Weihl CC, Lopate G. Motor neuron disease associated with copper deficiency. Muscle Nerve. 2006;34(6):789–93.
309. Desai V, Kaler SG. Role of copper in human neurological disorders. Am J Clin Nutr. 2008;88(3):855S–8S.
310. Manto M. Abnormal copper homeostasis: mechanisms and roles in neurodegeneration. Toxics. 2014;2(2):327–45.. https://doi.org/10.3390/toxics2020327
311. Hall JC, Mattila R. Elevated level of non-ceruloplasmin bound copper found in patients with amyotrophic lateral sclerosis. 2017 [internet]. Last accessed 11 July 2018 from Studies on ALS: http://alsanesthetics.org/copperstudy/pilotreport/freecopperstudy_report.html.
312. Brewer G. Personalizing copper-2 ingestion and copper levels to avoid Alzheimer's Disease (AD). 2016 [Internet]. Presentation American College of Nutrition. San Diego. http://www.naturalhealthresearch.org/wp-content/uploads/2016/12/Personalizing-Copper-2-Ingestion-and-Copper-Levels-to-Avoid.pdf.
313. Byer S. Zinc and copper. 2015 [Internet]. Last accessed 11 July 2018 from ALS Worldwide: http://alsworldwide.org/care-and-support/article/zinc-and-copper.

314. Pablo J, Banack SA, Cox PA, Johnson TE, Papapetropoulos S, Bradley WG, et al. Cyanobacterial neurotoxin BMAA in ALS and Alzheimer's disease. Acta Neurologica. 2009; published online.; https://doi.org/10.1111/j.1600-0404.2008.01150.x.
315. Greger M. What is the cause of ALS? 2017 [Internet]. Last accessed 11 July 2017 from NutritionFacts.org: https://nutritionfacts.org/2017/02/16/what-is-the-cause-of-als/.
316. Konkel L. Closing in on ALS? Link between lethal disease and algae explored. 2014 [Internet]. Last accessed 11 July 2018 from Environmental Health News: https://truthout.org/articles/closing-in-on-als-link-between-lethal-disease-and-algae-explored/.
317. Cox PA, Davis DA, Mash DC, Metcalf JS, Banack SA. Dietary exposure to an environmental toxin triggers neurofibrillary tangles and amyloid deposits in the brain. Proc R Soc B. 2015;283:20152397. https://doi.org/10.1098/rspb.2015.2397. http://rspb.royalsocietypublishing.org/content/283/1823/20152397.
318. Westfall S, Lomis N, Kahouli I, Dia SY, Singh SP, Prakash S. Microbiome, probiotics and neurodegenerative diseases: deciphering the gut brain axis. Cell Mol Life Sci. 2017;74(20):3769–87. https://doi.org/10.1007/s00018-017-2550-9.
319. Rao M, Gershon MD. The bowel and beyond: the enteric nervous system in neurological disorders. Nat Rev Gastroenterol Hepatol. 2016;13(9):517–28. https://doi.org/10.1038/nrgastro.2016.107.
320. Natale G, Pasquali L, Paparelli A, Fornai F. Parallel manifestations of neuropathologies in the enteric and central nervous systems. Neurogastroenterol Molti. 2011;23(12):1056–65. https://doi.org/10.1111/j.1365-2982.2011.01794.x.
321. Zhang YG, Wu S, Yi Y, Jin D, Zhou J, Sun J. Target intestinal microbiota to alleviate disease progression in amyotrophic lateral sclerosis. Clin Ther. 2017;39(2):322–36. https://doi.org/10.1016/j.clinthera.2016.12.014.
322. Longstreth WT Jr, Meschke JS, Davidson SK, Smoot LM, Smoot JC, Koepsell TD. Hypothesis: a motor neuron toxin produced by a clostridial species residing in gut causes ALS. Med Hypotheses. 2005;64(6):1153–6.
323. Wu S, Yi J, Zhang YG, Zhou J, Sun J. Leaky intestine and impaired microbiome in an amyotrophic lateral sclerosis mouse model. Physiol Rep. 2015;3(4):e12356. https://doi.org/10.14814/phy2.12356.
324. Ghasisas S, Maher J, Kanthasamy A. Gut microbiome in health and disease: linking the microbiome-gut-brain axis and environmental factors in the pathogenesis of systemic and neurodegenerative diseases. Pharmacol Ther. 2016;158:52–62. https://doi.org/10.1016/j.pharmthera.2015.11.012.
325. Zhang R, Miller RG, Gascon R, Champion S, Katz J, Lancero M, et al. Circulating endotoxin and systemic immune activation in sporadic Amyotrophic Lateral Sclerosis (sALS). J Neuroimmunol. 2009;206(1–2):121–4. https://doi.org/10.1016/j.jneuroim.2008.09.017.
326. Koppel N, Rekdal VM, Balskus EP. Chemical transformation of xenobiotics by the human gut microbiota. Science. 2017;356(6344):eaag2770. https://doi.org/10.1126/science.aag2770.
327. Claus SP, Guillou H, Ellero-Simatos S. The gut microbiota: a major player in the toxicity of environmental pollutants? NPJ Biofilms Microbiomes. 2016;2:16003. https://doi.org/10.1038/npjbiofilms.2016.3. https://www.nature.com/articles/npjbiofilms20163.
328. Tillisch K. The effects of gut microbiota on CNS function in humans. Gut Microbes. 2014;5(3):404–10. https://doi.org/10.4161/gmic.29232.
329. Rowin J, Xia Y, Jung B, Sun J. Gut inflammation and dysbiosis in human motor neuron disease. Physiol Rep. 2017;5(18):e13443. https://doi.org/10.14814/phy2.13443.
330. Neurological Clinical Research Institute at Massachusetts General Hospital. Microbiome Assessment in People with ALS (MAP ALS). 2017 [Internet]. Last accessed 11 July 2018 from PARTNERS Healthcare: https://clinicaltrials.partners.org/study/10168.
331. Pflumm M. Scientists go for the gut in ALS. 2017 [Internet]. Last accessed 11 July 2018 from The ALS Research Forum: http://www.alsresearchforum.org/scientists-go-for-the-gut-in-als/.
332. Brenner D, Hiergeist A, Adis C, Mayer B, Gessner A, Ludolph AC, Weishaupt. The fecal microbiome of ALS patients. Neurobiol Aging. 2018;61:132–7. https://doi.org/10.1016/j.neurobiolaging.2017.09.023.
333. Yamada SI, Niwa JI, Ishigaki S, Takahashi M, Ito T, Sone J, Doyu M, Sobue G. Archaeal proteasomes effectively degrade aggregation-prone proteins and reduce cellular toxicities in mammalian cells. J Biol Chem. 2006;281:23842 51. https://doi.org/10.1074/jbc.M601274200.
334. Canani RB, Di Costanzo M, Leone L. The epigenetic effects of butyrate: potential therapeutic implications for clinical practice. Clin Epigenetics. 2012;4:4. https://doi.org/10.1186/1868-7083-4-4.
335. Bourassa MW, Alim I, Bultman SJ, Ratan RR. Butyrate, neuroepigenetics and the gut microbiome: can a high fiber diet improve brain health? Neurosci Lett. 2016;625:56–63. https://doi.org/10.1016/j.neulet.2016.02.009.
336. Butchbach MER. Butyrate-based neuroprotectants as therapeutics for amyotrophic lateral sclerosis. 2010 [Internet]. Last accessed 11 July 2018 from ALS Association: http://web.alsa.org/site/PageServer?pagename=ResearchArchive_080410#Butyrate.
337. Al-Asmakh M, Hedin L. Microbiota and the control of blood-tissue barriers. Tissue Barriers. 2015;3:e1039691. https://doi.org/10.1080/21688370.2015.1039691.
338. Cudkowicz ME, Andres PL, Macdonald SA, Bedlack RS, Choudry R, et al. Phase 2 study of sodium phenylbutyrate in ALS. Amyotroph Lateral Scler. 2009;10(2):99–106. https://doi.org/10.1080/17482960802320487.
339. ClinicalTrials.gov. Evaluation of the Safety, Tolerability, Efficacy and Activity of AMX0035, a Fixed Combination of Phenylbutyrate (PB) and Tauroursodeoxycholic Acid (TUDCA), for the Treatment of ALS. 2017 [Internet]. Last accessed 11 July 2018 from https://clinicaltrials.gov/ct2/show/NCT03127514.
340. Elia AE, Lalli S, Monsurro MR, Sagnelli A, Taiello AC, Reggiori B, et al. Tauroursodeoxycholic acid in the treatment of patients with amyotrophic lateral sclerosis. Eur J Neurol. 2016;23(1):45–52. https://doi.org/10.1111/ene.12664.
341. Axe J. What is butyric acid? 6 butyric acid benefits you need to know about. 2017 [Internet]. Last accessed 11 July 2018 from Dr. Axe Food is Medicine: https://draxe.com/butyric-acid/.
342. Muscular Dystrophy Association. ALS signs and symptoms. 2017 [Internet]. Last accessed 11 July 2018 from MDA: https://www.mda.org/disease/amyotrophic-lateral-sclerosis/signs-and-symptoms.
343. Toepfer M, Folwaczny C, Klauser A, Riepl RL, Muller-Felber W, Pongratz D. Gastrointestinal dysfunction in amyotrophic lateral sclerosis. Amyotroph Lateral Scler Other Motor Neuron Disord. 1999;1(1):15–9.
344. Dukowicz AC, Lacy BE, Levine GM. Small intestinal bacterial overgrowth. Gastroenterol Hepatol. 2007;3(2):112–22.
345. Robelin L, De Aguilar LG. Blood biomarkers for amyotrophic lateral sclerosis: myth or reality? BioMed Res Int. 2014:525097:pages 11. https://doi.org/10.1155/2014/525097.
346. Benatar M, Boylan K, Jeromin A, Rutkove SB, Berry J, Atassi N, Bruijn L. ALS biomarkers for therapy development: state of the field and future directions. Muscle Nerve. 2016;53(2):169–82. https://doi.org/10.1002/mus.24979.
347. Forte G, Bocca B, Oggiano R, Clemente S, Asara Y, Sotgiu MA, Farace C, et al. Essential trace elements in amyotrophic lateral sclerosis (ALS): results in a population of a risk area of Italy. Neurol Sci. 2017;38(9):1609–15. https://doi.org/10.1007/s10072-017-3018-2.
348. Kumar A, Ghosh D, Singh RL. Amyotrophic lateral sclerosis and metabolomics: clinical implication and therapeutic

348. approach. J Biomarkers. 2013;2013:538765:15 pages. https://doi.org/10.1155/2013/538765.
349. Lawton KA, Cudkowicz ME, Brown MV, Alexander D, Caffrey R, Wulf JE, et al. Biochemical alterations associated with ALS. J Amyotroph Lat Scler. 2012;13(1):110–8. https://doi.org/10.3109/17482968.2011.619197.
350. Blasco H, Patin F, Madji Hounoum B, Gordon PH, Vourc'h P, et al. Metabolomics in amyotrophic lateral sclerosis: how far can it take us? Eur J Neurol. 2016;23(3):447–54. https://doi.org/10.1111/ene.12956.
351. Lu H, Liu X, Deng Y, Qing H. DNA methylation, a hand behind neurodegenerative diseases. Front Aging Neurosci. 2013;5:85. https://doi.org/10.3389/fnagi.2013.00085.
352. Nodera H, Takamatsu N, Muguruma N, Ukimoto K, Nishio S, Oda M, et al. Frequent hepatic steatosis in amyotrophic lateral sclerosis: implication for systemic involvement. Neurol Clin Neurosci. 2014;3(2):58–62. https://doi.org/10.1111/ncn3.143.
353. Parekh B. A(a)LS: ammonia-induced amyotrophic lateral sclerosis. F1000Res. 2015;4:119. https://doi.org/10.12688/f1000research.6364.1.
354. Italiani P, Carlesi C, Giungato P, Puxeddu I, Borroni B, Bossu P, et al. Evaluating the levels of interleukin-1 family cytokines in sporadic amyotrophic lateral sclerosis. J Neuroinflammation. 2014;11:94. https://doi.org/10.1186/1742-2094-11-94.
355. Guo J, Yang X, Gao L, Zang D. Evaluating the levels of CSF and serum factors in ALS. Brain Behav. 2017;7(3):e00637.
356. Murdock BJ, Bender DE, Kashlan SR, Figueroa-Romero C, Backus C, Callaghan BC, Goutman SA, Feldman EL. Increased ratio of circulating neutrophils to monocytes in amyotrophic lateral sclerosis. Neurol Neuroimmunol Neuroinflamm. 2016;3(4):e242. https://doi.org/10.1212/NXI.0000000000000242.
357. Chiò A, Calvo A, Bovio G, Canosa A, Bertuzzo D, Galmozzi F, Cugnasco P, et al. Amyotrophic lateral sclerosis outcome measures and the role of albumin and creatinine: a population-based study. JAMA Neurol. 2014;71(9):1134–42. https://doi.org/10.1001/jamaneurol.2014.1129.
358. Keizman D, Rogowski O, Berliner S, Ish-Shalom M, Maimon N, Nefussy B, Artamonov I, Drory VE. Low-grade systemic inflammation in patients with amyotrophic lateral sclerosis. Acta Neurol Scand. 2009;119(6):383–9. https://doi.org/10.1111/j.1600-0404.2008.01112.x.
359. Lawton KA, Brown MV, Alexander D, Li Z, Wulff JE, Lawson R, Jaffa M, et al. Plasma metabolomic biomarker panel to distinguish patients with amyotrophic lateral sclerosis from disease mimics. Amyotroph Lateral Scler Frontotemporal Degener. 2014;15(5–6):362–70. https://doi.org/10.3109/21678421.2014.908311.
360. Cortese R, D'Errico E, Introna A, Schirosi G, Scarafino A, Distaso E, Nazzaro P, et al. Vitamin D levels in serum of amyotrophic lateral sclerosis patients. Neurol 2015;84(14S):P2.069.
361. Libonati L, Onesti E, Gori MC, Ceccanti M, Cambieri C, Fabbri A, Frasca V, Inghilleri M. Vitamin D in amyotrophic lateral sclerosis. Funct Neurol. 2017;32(1):35–40.
362. Freedman DM, Kuncl RW, Weinstein SJ, Malila N, Virtamo J, Albanes D. Vitamin E serum levels and controlled supplementation and risk of amyotrophic lateral sclerosis. Amyotroph Lateral Scler Frontotemporal Degener. 2013;14(4):246–51. https://doi.org/10.3109/21678421.2012.745570.
363. Tai H, Cui L, Guan Y, Liu M, Li X, Shen D, Li D, et al. Correlation of creatine kinase levels with clinical features and survival in amyotrophic lateral sclerosis. Front Neurol. 2017;8:322. https://doi.org/10.3389/fneur.2017.00322.
364. Turner MR. The role of immune and inflammatory mechanisms in ALS. Muscle Nerve. 2015;51(1):14–8.
365. An M, Gao Y. Urinary biomarkers of brain diseases. Genomics Proteomics Bioinformatics. 2015;13(6):345–54. https://doi.org/10.1016/j.gpb.2015.08.005.
366. Vu LT, Bowser R. Fluid-based biomarkers for amyotrophic lateral sclerosis. Neurotherapeutics. 2017;14(1):119–34. https://doi.org/10.1007/s13311-016-0503-x.
367. Vadakkadath M, Atwood CS. Lactate dyscrasia: a novel explanation for amyotrophic lateral sclerosis. Neurobiol Aging. 2012;33(3):569–81. https://doi.org/10.1016/j.neurobiolaging.2010.04.012.
368. Lewerenz J, Maher P. Chronic glutamate toxicity in neurodegenerative diseases—What is the evidence? Front Neurosci. 2015;9:469. https://doi.org/10.3389/fnins.2015.00469.
369. Jiang Z, Wang W, Perry G, Zhu X, Wang X. Mitochondrial dynamic abnormalities in amyotrophic lateral sclerosis. Transl Neurodegen. 2015;4:14. https://doi.org/10.1186/s40035-015-0037-x.
370. Wilkins HM, Morris JK. New therapeutics to modulate mitochondrial function in neurodegenerative disorders. Curr Pharm Des. 2017;23(5):731–52. https://doi.org/10.2174/1381612822666161230144517.
371. Smith EF, Shaw PJ. De Voss KJ. Neuroscience Letters: The role of mitochondria in amyotrophic lateral sclerosis; 2017.; ; 17 pages. https://doi.org/10.1016/j.neulet.2017.06.052.
372. Paoli A, Bianco A, Damiani E, Bosco G. Ketogenic diet in neuromuscular and neurodegenerative diseases. Biomed Res Int. 2014;2014:474296. https://doi.org/10.1155/2014/474296.
373. Loeffler JP. Metabolomic biomarkers for amyotrophic lateral sclerosis (ALS) in patients and animal models of ALS: 5th International Conference and Exhibition on Metabolomics. Metabolomics: Open Access. 2016;6(2Suppl) https://doi.org/10.4172/2153-0769.C1.031.
374. Sazci A, Ozel MD, Emel E, Idrisoglu HA. Gender-specific association of methylenetetrahydrofolate reductase gene polymorphisms with sporadic amyotrophic lateral sclerosis. Genet Test Mol Biomarkers. 2012;16(7):716–21. https://doi.org/10.1089/gtmb.2011.0313.
375. Zoccolella S, Bendotti C, Beghi E, Logroscino G. Homocysteine levels and amyotrophic lateral sclerosis: a possible link. J Amyotrophic Lateral Sclerosis. 2010;11(1–2):140–7. https://doi.org/10.3109/17482960902919360.
376. Valentino F, Bivona G, Butera D, Paladino P, Fazzari M, Piccoli T, Ciaccio M, La Bella V. Elevated cerebrospinal fluid and plasma homocysteine levels in ALS. Eur J Neurol. 2010;17(1):84–9. https://doi.org/10.1111/j.1468-1331.2009.02752.x.
377. Jena University Hospital Clinical Trials.gov. Evaluating the potential of the diet as disease modifier in amyotrophic lateral sclerosis. 2017 [Internet]. Last accessed 11 July 2018 from: https://clinicaltrials.gov/ct2/show/NCT02572479.
378. Simopoulos AP. The omega-6/omega-3 fatty acid ratio: health implications. OCL. 2010;17(5):267–75. https://doi.org/10.1051/ocl.2010.0325.
379. Nakano Y, Hirayami K, Terao K. Hepatic ultrastructural changes and liver dysfunction in amyotrophic lateral sclerosis. Arch Neurol. 1987;44(1):103–6.
380. Ilzecka J, Stelmasiak Z. Thyroid function in patients with amyotrophic lateral sclerosis. Ann Univ Mariae Curie-Sklodowska Med. 2003;58(1):343–7.
381. Spataro R, Volanti P, Vitale F, Meli F, Colletti T, Di Natale A, La Bella V. Plasma cortisol level in amyotrophic lateral sclerosis. J Neurol Sci. 2015;358(1–2):282–6. https://doi.org/10.1016/j.jns.2015.09.011.
382. Roozendaal B, Kim S, Wolf OT, Kim MS, Sung KK, Lee S. The cortisol awakening response in amyotrophic lateral sclerosis is blunted and correlates with clinical status and depressive mood. Psychoneuroendocrinology. 2012;37(1):20–6. https://doi.org/10.1016/j.psyneuen.2011.04.013.
383. McCombe PH. The role of immune and inflammatory mechanisms in ALS. Curr Mol Med. 2011;11(3):246–54.
384. May C, Nordhoff E, Casjens S, Turewicz M, Eisenacher M, Gold R, et al. Highly immunoreactive IgG antibodies directed against a

385. Nicolson GL, Haier J. Role of chronic bacterial and viral infections in neurodegenerative, neurobehavioral, psychiatric, autoimmune and fatiguing illnesses: part 1. Brit J Med Practitioners. 2009;2(4):20–8.
386. Burakgazi AZ. Lyme disease –induced polyradiculopathy mimicking amyotrophic lateral sclerosis. Int J Neurosci. 2014;124(11):859–62. https://doi.org/10.3109/00207454.2013.879582.
387. Horowitz RI. Why can't I get better? Solving the mystery of Lyme and chronic disease. New York: St. Martin's Press; 2013.
388. Hino H, Kusuhara T, Kaji M, Shoji H, Oizumi K. Significance of hepatitis B virus antibody in motor neuron disease. Rinshō shinkeigaku (Clinical Neurology). 1995;35(4):341–3. Abstract in English/article in Japanese.
389. Akhvlediani T, Kvirkvelia N, Shakarishvili R, Tsertsvadze T. ALS-like syndrome in the patient with chronic hepatitis C. Georgian Med News. 2009;172-173:70–2. Abstract in English/article in Russian.
390. Alfhahad T, Nath A. Retroviruses and amyotrophic lateral sclerosis. Antivir Res. 2013;99(2):180–7. https://doi.org/10.1016/j.antiviral.2013.05.006.
391. Alonso R, Pisa D, Marina AI, Morato E, Rábano A, Rodal I, Carrasco L. Evidence for fungal infection in cerebrospinal fluid and brain tissue from patients with amyotrophic lateral sclerosis. Int J Biol Sci. 2015;11(5):546–58. https://doi.org/10.7150/ijbs.11084.
392. Alonso R, Pisa D, Fernández-Fernández AM, Rábano A, Carrasco L. Fungal infection in neural tissue of patients with amyotrophic lateral sclerosis. Neurobiol Dis. 2017;108:249–60. https://doi.org/10.1016/j.nbd.2017.09.001.
393. Clément ND, Frias ES, Szu JI, Vieira PA, Hubbard JA, Lovelace J, et al. GLT-1-dependent disruption of CNS glutamate homeostasis and neuronal function by the protozoan parasite toxoplasma gondii. PLoS Pathog. 2016;12(6):e1005643. https://doi.org/10.1371/journal.ppat.1005643.
394. Bedlack R.. ClinicalTrials.gov. ALS reversals – Lunasin Regimen. 2017 [Internet]. Last accessed 11 July 2018 from ClinicalTrials.gov: https://clinicaltrials.gov/ct2/show/study/NCT02709330.
395. Wood LB, Winslow AR, Strasser SD. Systems biology of neurodegenerative diseases. Integr Biol (Camb). 2015;7(7):758–75. https://doi.org/10.1039/c5ib00031a.
396. Wilkins HM, Harris JL, Carl SM, E L, Lu J, Eva Selfridge J, Roy N, et al. Oxaloacetate activates brain mitochondrial biogenesis, enhances the insulin pathway, reduces inflammation and stimulates neurogenesis. Hum Mol Genet. 2014;23(24):6528–41. https://doi.org/10.1093/hmg/ddu371.
397. Ruban A, Malina KC, Cooper I, Graubardt N, Babakin L, Jona G, Teichberg VI. Combined treatment of an amyotrophic lateral sclerosis rat model with recombinant GOT1 and oxaloacetic acid: A novel neuroprotective treatment. Neurodegener Dis. 2015;15(4):233–42. https://doi.org/10.1159/000382034.
398. Beghi E, Pupillo E, Bonito V, Buzzi P, Caponnetto C, Chiò A, Corbo M, et al. Randomized double-blind placebo-controlled trial of acetyl-L-carnitine for ALS. Amyotroph Lateral Scler Frontotemporal Degener. 2013;14(5–6):397–405. https://doi.org/10.3109/21678421.2013.764568.
399. Veronese N, Stubbs B, Solmi M, Ajnakina O, Carvalho AF, Maggi S. Acetyl-l-carnitine supplementation and the treatment for depressive symptoms: a systematic review and meta-analysis. Psychosom Med. 2018;80(2):154–9. https://doi.org/10.1097/PSY.0000000000000537.
400. Rosenfeld J. Supplements and vitamins. 2015 [Internet]. Last accessed 12 July 2018 from ALS Worldwide: http://alsworldwide.org/care-and-support/article/supplements-and-vitamins.
401. Boonstra E, de Kleijn R, Colzato LS, Alkemade A, Forstmann BU, Nieuwenhuis S. Neurotransmitters as food supplements: the effects of GABA on brain and behavior. Front Psychol. 2015;6:1520. https://doi.org/10.3389/fpsyg.2015.01520.
402. Natural Medicines. Gamma-aminobutyric acid. 2017 [Internet]. Natural Medicines: https://naturalmedicines.therapeuticresearch.com/databases/food,-herbs-supplements/professional.aspx?productid=464#scientificName.
403. Vieira DB, Gamarra LF. Getting into the brain: liposome-based strategies for effective drug delivery across the blood–brain barrier. Int J Nanomedicine. 2016;11:5381–414.
404. Mancuso M, Orsucci D, Calsolaro V, Choub A, Siciliano G. Coenzyme Q10 and neurological diseases. Pharmaceuticals (Basel). 2009;2(3):134–49.
405. Kaufmann P, Thompson JL, Levy G, Buchsbaum R, Shefner J, Krivickas LS, Katz J, et al. Phase II trial of CoQ10 for ALS finds insufficient evidence to justify phase III. Ann Neurol. 2009;66(2):235–44. https://doi.org/10.1002/ana.21743.
406. Failla ML, Chitchumroonchokchait C, Aoki F. Increased bioavailability of ubiquinol compared to that of ubiquinone is due to more efficient micellarization during digestion and greater GSH-dependent uptake and basolateral secretion by Caco-2 cells. J Agric Food Chem. 2014;62(29):7174–82. https://doi.org/10.1021/jf5017829.
407. Moore C. MitoQ proprietary antioxidant selected for study by National Institute of Aging's Interventions Testing Program. 2015 [Internet]. Last accessed 12 July 2018 from Alzheimer's News Today: https://alzheimersnewstoday.com/2015/12/17/mitoq-proprietary-antioxidant-selected-study-national-institute-agings-interventions-testing-program/.
408. Miquel E, Cassina A, Martínez-Palma L, Souza JM, Bolatto C, Rodríguez-Bottero S, Logan A, et al. Neuroprotective effects of the mitochondria-targeted antioxidant MitoQ in a model of inherited amyotrophic lateral sclerosis. Free Radic Biol Med. 2014;70:204–13. https://doi.org/10.1016/j.freeradbiomed.2014.02.019.
409. Smith RA, Murphy MP. Animal and human studies with the mitochondria-targeted antioxidant MitoQ. Ann N Y Acad Sci. 2010;1201:96–103. https://doi.org/10.1111/j.1749-6632.2010.05627.x.
410. Choi IY, Lee P, Statland J, McVey A, Dimackie M, Brooks W, Barohn R. Reduction in cerebral antioxidant, glutathione (GSH), in patients with ALS: a preliminary study (P6.105). Neurol. 2015;84(14S):P6.105.
411. Weiduschat N, Mao X, Hupf J, Armstrong N, Kang G, Lange DJ, Mitsumoto H, Shungu DC. Motor cortex glutathione deficit in ALS measured in vivo with the J-editing technique. Neurosci Lett. 2014;570:102–7. https://doi.org/10.1016/j.neulet.2014.04.020.
412. Aoyama K, Nakaki T. Impaired glutathione synthesis in neurodegeneration. International J Mol Sci. 2013;14(10):21021–44. https://doi.org/10.3390/ijms141021021.
413. Pizzorno JE, Katzinger JJ. Glutathione: physiological and clinical relevance. J Restorative Med. 2012;1(1):24–37.
414. Chiò A, Cucatto A, Terreni AA, Schiffer D. Reduced glutathione in amyotrophic lateral sclerosis: an open, crossover, randomized trial. Ital J Neurol Sci. 1998;19(6):363–6.
415. Muyderman H, Chen T. Mitochondrial dysfunction in amyotrophic lateral sclerosis – a valid pharmacological target? Br J Pharmacol. 2014;17(8):2191–205. https://doi.org/10.1111/bph.12476.
416. Natural Medicine Journal. The Health Dividend of Glutathione. 2011 3(2) [Internet]. Last accessed 12 July 2018 from Natural Medicine Journal: https://www.naturalmedicinejournal.com/journal/2011-02/health-dividend-glutathione.
417. Englert C, Trützschler AK, Raasch M, Bus T, Borchers P, Mosig AS, Traeger A, et al. Crossing the blood-brain barrier: glutathione-conjugated poly(ethylene imine) for gene delivery. J Control Release. 2016;241:1–14. https://doi.org/10.1016/j.jconrel.2016.08.039.
418. Scudellari M. Penetrating the brain. 2013. [Internet] Last accessed 12 July 2018 from TheScientist: http://www.the-scientist.com/?articles.view/articleNo/37957/title/Penetrating-the-Brain/
419. Shahripour RB, Harrigan MR, Alexandrov AV. N-acetylcysteine (NAC) in neurological disorders: mechanisms of action and thera-

peutic opportunities. Brain Behav. 2014;4(2):108–22. https://doi.org/10.1002/brb3.208.
420. Zoccolella S, Santamato A, Lamberti P. Current and emerging treatments for amyotrophic lateral sclerosis. Neuropsychiatr Dis Treat. 2009;5:577–95.
421. Deepmala, Slattery J, Kumar N, Delhey L, Berk M, Dean O, Spielholz C, Frye R. Clinical trials of N-acetylcysteine in psychiatry and neurology: a systematic review. Neurosci Biobehav Rev. 2015;55:294–321. https://doi.org/10.1016/j.neubiorev.2015.04.015.
422. Rosenfeld J, Ellis A. Nutrition and dietary supplements in motor neuron disease. Phys Med Rehabil Clin N Am. 2008;19(3):573. https://doi.org/10.1016/j.pmr.2008.03.001.
423. Natural Medicines. Alpha-lipoic acid. 2017 [Internet] from Natural Medicines: https://naturalmedicines.therapeuticresearch.com/databases/food,-herbs-supplements/professional.aspx?productid=767#scientificName.
424. Khanh vinh quốc Lương, Lan Thi Hoàng Nguyễn. Roles of vitamin D in amyotrophic lateral sclerosis: possible genetic and cellular signaling mechanisms. Molecular Brain. 2013;6:16. https://doi.org/10.1186/1756-6606-6-16.
425. Camu W, Tremblier B, Plassot C, Alphandery S, Salsac C, Pageot N, Juntas-Morales R, et al. Vitamin D confers protection to motoneurons and is a prognostic factor of amyotrophic lateral sclerosis. Neurobiol Aging. 2014;35(5):1198–205. https://doi.org/10.1016/j.neurobiolaging.2013.11.005.
426. Neurology Reviews. What role does vitamin D play in neurologic diseases? 2012 [Internet]. Last accessed 12 July 2018 from Neurology Reviews: http://www.mdedge.com/neurologyreviews/article/73564/alzheimers-cognition/what-role-does-vitamin-d-play-neurologic.
427. Paganoni S, Macklin EA, Karam C, Yu H, Gonterman F, Fetterman KA, Cudkowicz M, et al. Vitamin D levels are associated with gross motor function in amyotrophic lateral sclerosis. Muscle Nerve. 2017;56(4):726–31. https://doi.org/10.1002/mus.25555.
428. Blasco H, Hounoum BM, Dufour-Rainfray D, Patin F, Maillot F, Beltran S, Gordon PH, et al. Vitamin D is not a protective factor in ALS. CNS Neurosci Therap. 2015 [Internet]; https://doi.org/10.1111/cns.12423.
429. Török N, Török R, Klivényi P, Engelhardt J, Vécsei L. Investigation of vitamin D receptor polymorphisms in amyotrophic lateral sclerosis. Acta Neurol Scand. 2015;133(4):302–8. https://doi.org/10.1111/ane.12463.
430. Dadhania VP, Trivedi PP, Vikram A, Tripathi DN. Nutraceuticals against neurodegeneration: a mechanistic insight. Curr Neuropharmacol. 2016;14(6):627–40. https://doi.org/10.2174/1570159X14666160104142223.
431. Wechsler I, RECOVERY IN. Amyotrophic lateral sclerosis treated with tocopherols (vitamin E): preliminary report. JAMA. 1940;114(11):948–50.
432. Wang H, O'Reilly ÉJ, Weisskopf MG, Logroscino G, McCullough ML, Schatzkin A, Kolonel LN, Ascherio A. Vitamin E intake and risk of amyotrophic lateral sclerosis: a pooled analysis of data from 5 prospective cohort studies. Am J Epidemiol. 2011;173(6):595–602. https://doi.org/10.1093/aje/kwq416.
433. Liu M, Wallin R, Wallmon A, Saldeen T. Mixed tocopherols have a stronger inhibitory effect on lipid peroxidation than alpha-tocopherol alone. J Cardiovasc Pharmacol. 2002;39(5):714–21.
434. The ALSUntangled Group. ALSUntangled No. 30: Methylcobalamin. Amyotroph Lateral Scler Frontotemporal Degener. 2015;16(7-8):536–9. https://doi.org/10.3109/21678421.2015.1070574.
435. Tsiminis G, Schartner EP, Brooks JL, Hutchinson MR. Measuring and tracking vitamin B12: a review of current methods with a focus on optical spectroscopy. Appl Spectrosc Rev. 2017;52(5):439–55. https://doi.org/10.1080/05704928.2016.1229325.
436. Kaji R, Kuzuhara S, Iwasaki Y, Okamoto K, Nakagawa M, Imai T, Takase T, et al. Ultra-high dose methylcobalamin (E0302) prolongs survival of ALS: randomized double-blind, phase 3 clinical trial (ClinicalTrials.gov NCT00444613) P7.060. Neurology. 2015;84(14Supp):P7.060.
437. Azevedo M. Eisai withdraws application for ultra-high dose of mecobalamin, a B12 form, as ALS treatment. 2016. Last accessed 12 July 2018 from ALS News Today: https://alsnewstoday.com/2016/03/23/eisai-withdraws-new-drugs-application-for-mecobalamin-ultra-high-dose-preparation-as-treatment-for-amyotrophic-lateral-sclerosis/.
438. Gerdts J, Brace EJ, Sasaki Y, DiAntonio A, Milbrandt J. SARM1 activation triggers axon degeneration locally via NAD+ destruction. Science. 2015;348(6233):453–7. https://doi.org/10.1126/science.1258366.
439. Inacio P. Mechanism behind nicotinamide riboside could be translated into future ALS, neurodegenerative disease therapy. 2015 [Internet]. Last accessed 12 July 2018 from Mitochondrial Disease News: https://mitochondrialdiseasenews.com/2015/04/30/mechanism-behind-nicotinamide-riboside-translated-future-als-neurodegerative-disease-therapy/.
440. Conze DB, Crespo-Barreto J, Kruger CL. Safety assessment of nicotinamide riboside, a form of vitamin B3. Hum Exp Toxicol. 2016; pii: 0960327115626254.
441. Chi Y, Sauve AA. Nicotinamide riboside, a trace nutrient in foods, is a vitamin B3 with effects on energy metabolism and neuroprotection. Curr Opin Clin Nutr Metab Care. 2013;16(6):657–61. https://doi.org/10.1097/MCO.0b013e32836510c0.
442. Heilbronn LK. Clinical trials corner. Nutr Healthy Aging. 2017;4(2):193–4. https://doi.org/10.3233/NHA-170001.
443. ClinicalTrials.gov. Nicotinamide riboside and mitochondrial biogenesis. https://clinicaltrials.gov/ct2/show/NCT03432871.
444. Clinical Trials.gov. The Effects of Nicotinamide Adenine Dinucleotide (NAD) on Brain Function and Cognition (NAD). ClinicalTrials.gov: https://clinicaltrials.gov/ct2/show/NCT02942888
445. Linus Pauling Institute. Magnesium. 2017 [Internet]. Last accessed 12 July 2018 from Linus Pauling Institute Oregon State University: http://lpi.oregonstate.edu/mic/minerals/magnesium.
446. Yasui M. Magnesium-related neurological disorders. In: Yasui M, Verity MA, editors. Mineral and metal neurotoxicology. New York: CRC Press; 1996. p. 219–26.
447. Slutsky I, Abumaria N, Wu LJ, Huang C, Zhang L, Li B, Zhao X, Govindarajan A, et al. Enhancement of learning and memory by elevating brain magnesium. Neuron. 2010;65(2):165–77. https://doi.org/10.1016/j.neuron.2009.12.026.
448. Fondell E, O'Reilly EJ, Fitzgerald KC, Falcone GJ, McCullough ML, Park Y, Kolonel LN, Ascherio A. Magnesium intake and risk of amyotrophic lateral sclerosis: results from five large cohort studies. Amyotroph Lateral Scler Frontotemporal Degener. 2013;14(5-6):356–61. https://doi.org/10.3109/21678421.2013.803577.
449. Zhang Y, Li H, Yang C, Fan DF, Guo DZ, Hu HJ, Meng XE, Pan SY. Treatment with hydrogen-rich saline Delays disease progression in a mouse model of amyotrophic lateral sclerosis. Neurochem Res. 2016;41(4):770–8. https://doi.org/10.1007/s11064-015-1750-7.
450. Ohta S. Molecular hydrogen as a preventive and therapeutic medical gas: initiation, development and potential of hydrogen medicine. Pharmacol Ther. 2014;144(1):1–11. https://doi.org/10.1016/j.pharmthera.2014.04.006.
451. Nicolson GL, Ferreira de Mattos G, Settineri R, Costa C, Ellithorpe R, Rosenblatt S, La Valle J, et al. Clinical effects of hydrogen administration: from animal and human diseases to exercise medicine. IJCM [Internet]. 2016;7(1) https://doi.org/10.4236/ijcm.2016.71005.
452. Iketani M, Ohsawa I. Molecular hydrogen as a neuroprotective agent. Current Curr Neuropharmacol. 2017;15(2):324–31. https://doi.org/10.2174/1570159X14666160607205417.
453. The ALSUntangled Group. ALSUntangled No. 26: Lunasin. Amyotrophic Lateral Sclerosis and Frontotemporal Degeneration. 2014;15(7-8): 622–626, DOI: https://doi.org/10.3109/21678421.2014.959297.

454. Yang EJ, Jiang JH, Lee SM, Yang SC, Hwang HS, Lee MS, Choi SM. Bee venom attenuates neuroinflammatory events and extends survival in amyotrophic lateral sclerosis models. J Neuroinflamm. 2010;7(69). https://doi.org/10.1186/1742-2094-7-69
455. Chang CF, Lee YC, Lee KH, Lin HC, Chen CL, Shen CKJ, Huang CC. Therapeutic effect of berberine on TDP-43-related pathogenesis in FTLD and ALS. J Biomed Sci. 2016;23:72. https://doi.org/10.1186/s12929-016-0290-z.
456. Ferland G. Vitamin K, an emerging nutrient in brain function. Biofactors. 2012;38(2):151–7. https://doi.org/10.1002/biof.1004.
457. Covarrubias-Pinto A, Acuña AI, Beltrán FA, Torres-Díaz L, Castro MA. Old things new view: ascorbic acid protects the brain in neurodegenerative disorders. Int J Mol Sci. 2015;16(12):28194–217. https://doi.org/10.3390/ijms161226095.
458. Linus Pauling Institute. Vitamin C. 2017 [Internet]. Last accessed 12 July 2018 from Oregon State University: http://lpi.oregonstate.edu/mic/vitamins/vitamin-C.
459. Martin A, Joseph JA, Cuervo AM. Stimulatory effect of vitamin C on autophagy in glial cells. J Neurochem. 2002;82(3):538–49. PMID: 12153478.
460. Kocot J, Luchowska-Kocot D, Kiełczykowska M, Musik I, Kurzepa J. Does vitamin C influence neurodegenerative diseases and psychiatric disorders? Nutrients. 2017;9(7):659. https://doi.org/10.3390/nu9070659.
461. Kolarcik CL, Bowser R. Retinoid signaling alterations in amyotrophic lateral sclerosis. Am J Neurodegener Dis. 2012;1(2):130–45.
462. Peters TL, Beard JD, Umbach DM, Allen K, Keller J, Mariosa D, Sandler DP, et al. Blood levels of trace metals and amyotrophic lateral sclerosis. Neurotoxicology. 2016;54:119–26. https://doi.org/10.1016/j.neuro.2016.03.022.
463. Oggiano R, Solinas G, Forte G, Bocca B, Farace C, Pisano A, Sotgiu MA, et al. Trace elements in ALS patients and their relationships with clinical severity. Chemosphere. 2018;197:457–66. https://doi.org/10.1016/j.chemosphere.2018.01.076.
464. Norris F. Nutritional supplements in amyotrophic lateral sclerosis. In: Cosi VK, editor. Amyotrophic lateral sclerosis. 1st ed. London: Springer; 1987. p. 183–9.
465. D'Agostino D. Dr. D'Agostino's testimony. In: Tedone VT-G, editor. The Deanna protocol. Tampa: paradies/inspire,llc; 2015. p. 130.
466. Brown C. Non-familial ALS: a tangled web. Nature. 2017;550(7676):S109–11. https://doi.org/10.1038/550S109a.
467. https://www.ncbi.nlm.nih.gov/pubmed/28849508.

Gastroenterology

Jason Bosley-Smith

50.1 Gastrointestinal Disorders: Clinical Features and Prevalence – 914
50.1.1 Introduction – 914
50.1.2 Digestion and Absorption of Nutrients – 914
50.1.3 Regulation of Motility – 914
50.1.4 Food-Mediated Disease – 915
50.1.5 Structural Abnormalities – 915
50.1.6 Prevalence of Gastrointestinal Disease and Global Impact on Health – 915
50.1.7 Organic Gastrointestinal Disease – 916
50.1.8 Motility Disorders – 916
50.1.9 Functional Disorders – 916
50.1.10 Prevalence of Irritable Bowel Syndrome – 916

50.2 Pathophysiology Review and Conventional Approaches to Irritable Bowel Syndrome – 917
50.2.1 Irritable Bowel Syndrome Pathophysiology – 917
50.2.2 Altered Motility – 919
50.2.3 Visceral Sensation – 919
50.2.4 Brain-Gut Interaction – 919

50.3 Integrative and Functional Medical Approach Beyond Conventional Protocols – 920
50.3.1 Integrative and Functional Medicine Approach – 920
50.3.2 Dietary Intervention – 920
50.3.3 Herbal and Nutritional Supplement Interventions – 921

50.4 Nutritional and Lifestyle Clinical Application Principles – 921

50.5 Grand Rounds Case Presentation – 922
50.5.1 Assessment – 922

References – 925

© Springer Nature Switzerland AG 2020
D. Noland et al. (eds.), *Integrative and Functional Medical Nutrition Therapy*,
https://doi.org/10.1007/978-3-030-30730-1_50

50.1 Gastrointestinal Disorders: Clinical Features and Prevalence

50.1.1 Introduction

The role of the integrative nutritionist in maintaining and optimizing gastrointestinal health is paramount, especially given the expanding body of scientific research elucidating the influence of digestive function on systemic health. Nutritional status is associated with risk for digestive disease, while, in return, gastrointestinal conditions affect food tolerability and nutrient requirement.

Diet is inextricably linked to health outcomes and optimal digestive function. Epidemiological data demonstrate the association between inadequate dietary intake, low in fruit and vegetable consumption, and increased risk of stomach and colorectal cancer [1]. Concurrently, the presence of existing gastrointestinal dysfunction impacts nutrient demand and the potential for frank deficiency or subclinical insufficiency. This direct influence on nutrient uptake may ultimately manifest in myriad symptoms and potential sequelae in the GI patient. For example, in Crohn's disease patients, where surgical resections, fibrotic tissue, villous atrophy, excess fluid loss from diarrhea, and anorexia may occur, both macro- and micronutrient deficiencies are prevalent. (◘ Table 50.1).

The digestion and absorption of nutrients, regulation of motility, potential impact of food-mediated disease, and structural abnormalities that influence dietary intake are all critical considerations for the integrative nutritionist supporting patients with a diagnosed digestive disorder, as well as those exhibiting symptoms of gastrointestinal disturbance.

50.1.2 Digestion and Absorption of Nutrients

Due to the direct interface between consumed nutrients and gastrointestinal structures, clinical nutritionists serve as essential care providers for patients experiencing digestive symptoms and chronic GI conditions. The transport, breakdown, and absorption of macromolecules rely on appropriate physiological conditions. Many digestive diseases result in impaired digestion and absorption of nutrients, creating a cascading effect on other systems of the body and overall function. Overt nutrient deficiencies and subclinical insufficiencies are the resulting outcome of this impairment and are prevalent in gastrointestinal patients [4]. Addressing and mitigating the effects of maldigestion and malabsorption is vital for optimizing patient outcomes. Without such intervention, patients may develop sequelae that increase the complexity of their condition and require additional treatments and providers. As such, integrative nutritionists serve as provider for the GI patient and a linchpin to prevent downstream pathophysiology.

50.1.3 Regulation of Motility

Regulation of motility is an essential consideration influencing dietary consumption and the resulting prospect for GI symptoms. Transit time, while variable in the individual, generally falls within assessable parameters absent any pathophysiological factors. Under appropriate physiological conditions, motility through the intestinal tract adheres to a pattern as outlined in ◘ Table 50.2.

◘ **Table 50.1** Common nutrient deficiencies associated with Crohn's disease

Nutritional deficiency	Pathological basis	Sequelae
Calories	Anorexia, bile acid malabsorption	Cachexia, poor growth (pediatric)
Protein	Enteropathy, impaired gastric acid and pancreatic enzyme secretion, proinflammatory cytokine activity, increased energy expenditure	Cachexia, poor growth (pediatric)
Vitamin D, calcium	Long-term and systematic steroid use, malabsorption	Osteopenia/osteoporosis, increased risk for GI cancer (vitamin D)
Vitamin K	Modulation of gastrointestinal microflora	Poor coagulation
Zinc	Intestinal losses, malabsorption	Poor wound healing
B12	Ileal resection	Anemia, hyperhomocysteinemia
Folate	Medication	Anemia, hyperhomocysteinemia
Magnesium	Chronic diarrhea	Arrhythmia, muscle spasm, anxiety, seizure
Iron	Blood loss, exudative enteropathy	Anemia
Selenium	Malabsorption	Oxidative stress, poor thyroid function

Based on data from Massironi et al. [2] and Wędrychowicz et al. [3]

Table 50.2 Gastrointestinal transit times under normal physiological conditions

Gastric emptying	2–4 hours
Small bowel transit	4–6 hours
Colonic transit	24–72 hours

Based on data from Maurer [5, 6]

Motility disorders involve alteration in the transit of nutrients, as well as secretions into the digestive tract. Transit time impacts nutrient delivery and can manifest in a variety of symptoms:
- Abdominal distension
- Severe abdominal colicky pain
- Visceral hypersensitivity
- Constipation
- Diarrhea
- Gastroesophageal reflux disease
- Intractable, recurrent vomiting

Source (► http://emedicine.medscape.com/article/179937-overview).

Examples of relevant dietary modifications in such cases may include considerations around the intake of fiber, fermentable carbohydrates, lipid load, alcohol, caffeine, and electrolyte balance.

50.1.4 Food-Mediated Disease

In addition to optimizing nutrient absorption, integrative nutritionists help identify and manage food-mediated disease in the digestive patient. A growing body of evidence supports the role of food antigens as mediators of chronic disease in susceptible patients [7–9]. Consequently, clinicians may seek to modify dietary intake to remove food compounds that irritate the GI tract, exacerbate symptoms, and cause reactivity in the patient experiencing immune imbalance. Lack of development in oral tolerance during early childhood, mediated by factors such as neonatal feeding, method of delivery, and inoculation of commensal gut bacteria, may determine eventual immune regulation and susceptibility to food-mediated disease [10]. This predisposition may carry over to adulthood and manifest in the appearance of symptoms when specific triggers result in perturbation of immune function. Antibiotic intervention, pathogenic infection, biopsychosocial stress, or a combination of these factors may serve as initiators of food-mediated dysfunction and symptom presentation. The proximity and activity of gut immunological structures, namely, the gut-associated lymphoid tissue (GALT), under modulation by the local commensal microbial population, play a crucial role and may, in part, dictate dietary modification.

50.1.5 Structural Abnormalities

Structural and mechanical abnormalities in the digestive patient are additional considerations for the integrative nutritionist. Each structure along the alimentary canal and in accessory digestive organs must work in concert to transform foodstuffs into a food bolus, chyme, and waste matter for eventual elimination. Structure dictates function; therefore, abnormalities in the tissues, glands, and cells integral to digestion can derail optimal gastrointestinal function and lead to symptoms. A comprehensive understanding of the complexity of GI anatomy and physiology is fundamental to developing appropriate treatment plans.

50.1.6 Prevalence of Gastrointestinal Disease and Global Impact on Health

In the USA alone, digestive disease accounts for nearly $142 billion in healthcare costs annually and affects upward of 60–70 million individuals [11]. Patients suffering from GI diseases encounter not only the physical burden of their symptoms but significant impairment in productivity, social interaction, and quality of life.

The data located in ◘ Fig. 50.1 highlight the significance of digestive disease in the USA and support the need for clinical dietetic intervention in this population:

Gastrointestinal disease can be divided into three primary clinical domains:
1. Organic or structural disease
2. Motility disorders
3. Functional disorders

- Prevalence: 60 to 70 million people affected by all digestive diseases
- Ambulatory care visits: 48.3 million (2010)
- Hospitalizations: 21.7 million (2010)
- Mortality: 245,921 deaths (2009)
- Diagnostic and therapeutic inpatient procedures: 5.4 million—12 percent of all inpatient procedures (2007)
- Ambulatory surgical procedures: 20.4 million—20 percent of all "write-in" surgical procedures (2010)
- Costs: $141.8 billion (2004)
- $97.8 billion, direct medical costs (2004)
- $44 billion, indirect costs—for example, disability and mortality (2004)
- Primary diagnosis at office visits: 36.6 million (2010)
- Primary diagnosis at emergency department visits: 7.9 million (2010)
- Primary diagnosis at outpatient department visits: 3.8 million (2010)

◘ Fig. 50.1 Digestive disease statistics. (Based on data from The National Institute of Diabetes and Digestive and Kidney Diseases (NIDDK) [11].)

50.1.7 Organic Gastrointestinal Disease

Organic gastrointestinal disease is classified based on histology and organ morphology [12]. Pathology in gastrointestinal tissue with observable damage to its form and structure marks this category of disease. Diagnosis of disease presence and measurement of its severity are identified via analysis of the affected tissues. Examples of organic gastrointestinal diseases include inflammatory bowel disorder (IBD) and esophagitis. In the presence of organic gastrointestinal disease, dietary restrictions may be necessary to alleviate irritation to the affected tissues. In specific scenarios, such as IBD, damage to absorptive sites warrants enhanced nutritional intake and may require supplementation where appropriate.

50.1.8 Motility Disorders

Motility disorders are defined in terms of organ function – specifically altered propulsion and movement of contents along the gastrointestinal tract. Abnormal visceral muscle activity results in dysregulated peristalsis or segmentation of the digestive tract, altering appropriate transit time of gastrointestinal contents. Assessment of motility disorders is conducted via the use of physiological testing, where transit time can be measured. Small bowel manometry, gastric emptying study, and whole-gut transit scintigraphy are diagnostic procedures available to analyze motility in the presenting patient. Gastroparesis and colonic inertia are two conditions classified within the clinical domain of motility disorders. Based upon the precise mechanism and location of motility disturbance, modification of dietary intake may be prudent to alleviate symptoms. For example, in the case of gastroparesis, which involves delayed gastric emptying and slowed movement of food from the stomach to the small intestine, intervention may require an initial reliance on a liquid diet, graduation to a small particle diet, and eventual progression to small frequent meals that are lower in fat and fiber content [13].

50.1.9 Functional Disorders

Whereas organic and motility disorders are diagnosed by clinical procedure, the patient's subjective illness experience distinguishes functional GI disorders (FGIDs). FGIDs are the most common GI disorders among the general population [14]. Estimates vary, but approximately 1 in 4 people or more in the USA have one of these disorders [15]. In the evaluation of functional GI disorders, the clinician relies on the patient-reported symptoms for classification. According to Drossman et al. "A symptom is a noticeable experiential change in the body or its parts that is reported by the patient as being different from normal and may or may not be interpreted as meaningful" [12]. This interpretation is a critical clinical distinction, as many patients become accustomed to familiar patterns of digestion and elimination, skewing their perspective and subjective experience. The task of the astute clinical nutritionist is to differentiate between what a patient deems as their "normal" state and ideal physiological function. This distinction often requires examination of health history and the use of the functional medicine timeline to identify initiating events where physiological function may have first shifted (◘ Fig. 50.2).

A functional GI disorder is characterized as a syndrome based on a constellation of symptoms and is diagnosed by Rome criteria. Rome IV defines FGIDs as follows:

» Functional GI disorders are disorders of gut–brain interaction. It is a group of disorders classified by GI symptoms related to any combination of the following: motility disturbance, visceral hypersensitivity, altered mucosal and immune function, altered gut microbiota, and altered central nervous system (CNS) processing. [12]

This definition highlights the variety of mechanisms at play within FGIDs, including both physiological components and biopsychosocial elements. In addition, the array of symptoms relates to the primary considerations for nutritionists discussed earlier (digestion and absorption of nutrients, regulation of motility, potential impact of food-mediated disease, and structural abnormalities).

Due to its prevalence, varied etiological factors, pathophysiological considerations, and influence on nutritional intake, this Grand Rounds in Gastroenterology will examine the case of a patient presenting with one of the most common, and often complex, functional gastrointestinal disorders: irritable bowel syndrome.

50.1.10 Prevalence of Irritable Bowel Syndrome

Irritable bowel syndrome affects between 7% and 21% of the population globally, and approximately 30% of people experiencing their IBS symptoms will consult physicians for medical intervention [16, 17]. Patients who seek medical attention do not necessarily have significantly different or more severe abdominal symptoms, but often report high levels of anxiety and greater impact of their symptoms on overall quality of life. Patients often do not pursue physician intervention at the onset of IBS symptoms and may wait months, if not years, before soliciting medical care. This hesitation to seek care may stem in part from the variable nature of IBS symptoms, which may wax and wane over time. Additionally, patient knowledge gap surrounding appropriate patterns of digestion and elimination may result in patients mistaking or attributing IBS symptoms to more general experiences of GI upset. Initial symptoms may at first be attributed to variation in food intake or acquisition of a transient stomach virus, and not necessarily be associated with an actual pathological condition. Finally, patients may forego medical intervention during initial prodrome with IBS, opting to manage the symptoms through self-medicating efforts using over-the-counter digestive aids or dietary modifica-

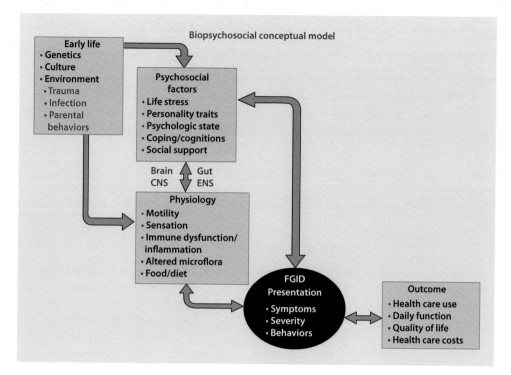

Fig. 50.2 Biopsychosocial conceptual model. A biopsychosocial conceptualization of the pathogenesis, clinical experience, and effects of functional GI disorders. There is a relationship between early life factors that can influence the psychosocial milieu of the individual, their physiological functioning, as well as their mutual interaction (brain-gut axis). These factors influence the clinical presentation of the disorder and the clinical outcome. Modified from Rome III. (Reprinted from Drossman [12]. With permission from Elsevier.)

tions. Inevitably when these measures fall short of remedying the situation, and symptoms increase in frequency and severity, impacting quality of life, individuals pursue medical care.

50.2 Pathophysiology Review and Conventional Approaches to Irritable Bowel Syndrome

50.2.1 Irritable Bowel Syndrome Pathophysiology

As a syndrome, IBS involves a constellation of signs, symptoms, and deranged physiology along the GI tract. Although symptoms can vary among patients, hallmarks of IBS include the following:

- Abdominal cramping
- Abdominal distension, bloating
- Change in bowel habits
- Exaggerated gastrocolonic response
- Visceral hypersensitivity and pain

Stratification and categorization of patients are based on predominant stool pattern (◘ Fig. 50.3):

- IBS with predominant constipation (IBS-C)
- IBS with predominant diarrhea (IBS-D)
- IBS with mixed bowel habits (IBS-M)
- IBS unclassified (IBS-U)

Rome IV diagnostic criteria for irritable bowel syndrome:

Recurrent abdominal pain[1] at least 1 day per week for the past 3 months associated with two or more of the following (◘ Fig. 50.4):

1. Associated with defecation
2. Associated with a change in frequency of stool
3. Associated with a change in form (appearance) of stool

In pathophysiology research and clinical trials, a pain frequency of at least two days a week during screening evaluation is recommended for subject eligibility.

The diagnosis of IBS should be made based on the following four key features: clinical history, physical examination, minimal laboratory tests, and, when clinically indicated, colonoscopy, or other appropriate tests.

Evaluation of symptom severity and quality of life for the presenting patient can be ascertained through the use of validated assessment tools, such as the Irritable Bowel Syndrome Severity Scoring System (IBS-SSS) [19] and the Irritable Bowel Syndrome Quality of Life (IBS-QOL) questionnaire [20]. IBS-SSS primarily evaluates the intensity of IBS symptoms during a 10-day period: abdominal pain, distension, stool frequency and consistency, and interference with life in general. The IBS-SSS calculates the sum of these five items; each scored on a visual analog scale from 0 to 100 [19]. The IBS-QOL questionnaire is a 34-item instrument designed from a needs-based conceptual paradigm, and scoring is

---

1 Criterion fulfilled for the last 3 months with symptom onset at least 6 months prior to diagnosis.

> **Diagnostic criteria for IBS subtypes**
>
> Predominant bowel habits are based on stool form on days with at least one abnormal bowel movement.
>
> IBS with predominant constipation: More than one-fourth (25%) of bowel movements with Bristol stool form types 1 or 2 and less than one-fourth (25%) of bowel movements with Bristol stool form types 6 or 7. Alternative for epidemiology or clinical practice: Patient reports that abnormal bowel movements are usually constipation (like type I or 2 in the picture of Bristol Stool Form Scale (BSFS), see Figure 2A).
>
> IBS with predominant diarrhea (IBS-D): more than one-fourth (25%) of bowel movements with Bristol stool form types 6 or 7 and less than one-fourth (25%) of bowel movements with Bristol stool form types 1 or 2. Alternative for epidemiology or clinical practice: Patient reports that abnormal bowel movements are usually diarrhea (like type 6 or 7 in the picture of BSFS, see Figure 2A).
>
> IBS with mixed bowel habits (IBS-M): more than one-fourth (25%) of bowel movements with Bristol stool form types I or 2 and more than one-fourth (25%) of bowel movements with Bristol stool form types 6 or 7. Alternative for epidemiology or clinical practice: Patient reports that abnormal bowel movements are usually both constipation and diarrhea (more than one-fourth of all the abnormal bowel movements were constipation and more than one-fourth were diarrhea, using picture of BSFS, see Figure 2A).
>
> IBS unclassified (IBS-U): Patients who meet diagnostic criteria for IBS but whose bowel habits cannot be accurately categorized into 1 of the 3 groups above should be categorized as having IBS unclassified.

Fig. 50.3 Diagnostic criteria for IBS subtypes. (Reprinted from Lacy et al. [18]. With permission from Elsevier.)

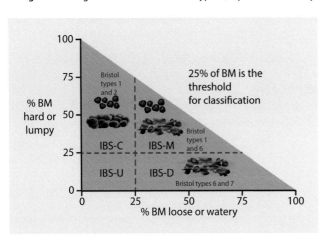

Fig. 50.4 IBS subtyping based upon Bristol Stool Form Scale (BSFS). IBS subtypes should be established according to stool consistency, using the BSFS. Accuracy in IBS subtyping is more readily achieved when patients have at least four days of abnormal bowel habits per month. Bowel habit subtypes should be based on BSFS for days with irregular bowel habits. (Reprinted from Lacy et al. [18]. With permission from Elsevier.)

composed of a total score and eight subscale scores that have been validated for a general IBS patient population [20].

The pathophysiology associated with IBS is multifactorial and often variable – presenting a challenge in both assessment and intervention. Identifying the relevant underlying and potentially unique factors in the presenting patient is necessary to determine the most appropriate course(s) of treatment. Deducing the critical drivers unique to an individual's presentation is particularly significant for the integrative nutritionist as the heterogenic nature of irritable bowel syndrome means that no single physiological anomaly accounts for symptoms in all patients.

Historically, the pathogenesis of IBS has focused on abnormalities in motility, visceral sensation, brain-gut interaction, and psychosocial distress [16]. More recently, altered gut immune activation, intestinal permeability, and intestinal and colonic microbiome have been identified in some IBS patients [21, 22]. Several other etiological factors may drive the IBS phenotype adding to the complexity of the condition:
- Celiac disease
- Non-celiac gluten sensitivity
- Food intolerance (e.g. lactose, fructose)
- Small intestinal bacterial overgrowth (SIBO)

Reportedly 20–23.3% of treated CD patients fulfill the symptom-based Rome criteria for IBS [23]. This may be due to an incomplete resolution of the inflammatory response, even upon intervention with a gluten-free diet. This chronic, low-grade inflammatory state may mediate elevated immunological status and visceral hypersensitivity that results in IBS symptoms.

The debate over the role of gluten and wheat-based products among those who present with gastrointestinal symptoms but test negative for celiac disease continues in the clinic setting. Non-celiac gluten sensitivity (NCGS) is often a self-administered diagnosis by the patient, who, after adopting a gluten-free diet, experiences full or partial symptom resolution. However, not all patients respond as such, highlighting the possibility that other dietary factors may be at play in patients experiencing the gastrointestinal and extragastrointestinal symptoms common to both NCGS and IBS. Aside from gluten, other constituents found in wheat-based foods, such as amylase-trypsin inhibitors and fructans, could contribute to IBS symptoms [24]. A low FODMAP (fermentable oligosaccharides, disaccharides, and monosaccharides, and polyols) dietary intervention has demonstrated efficacy in IBS, supporting the associated intolerance and malabsorption of these short-chain carbohydrates as a causative factor in patients [25].

In addition to digestive difficulty with gluten and wheat derivatives, intolerance to other substrates has been implicated in IBS symptomology. Both lactose and fructose intolerance are cited as potential drivers in IBS pathophysiology, falling under the broader umbrella of FODMAPs as disaccharides and monosaccharides, respectively [26]. Researchers posit that intolerance develops by one of three potential pathways: food hypersensitivity (immune-mediated), food chemicals (bioactive molecules), and luminal distension [27].

Various studies evaluating the frequency of SIBO among patients with IBS as compared with controls have been conducted, elucidating the prevalence of small intestinal bacterial overgrowth among patients with irritable bowel syndrome. The majority of case-control studies reveal that SIBO was more common among IBS than controls, suggesting a significant association between SIBO and IBS.

Additional meta-analyses further support an association between SIBO pathogenicity in IBS clinical presentation. In a meta-analysis by Ford et al., of the 12 studies including 1921 patients with IBS, pooled prevalence of a positive lactulose-hydrogen breath test (LHBT) and glucose-hydrogen breath test (GHBT) was 54% [28]. In another meta-analysis that included 11 studies, breath testing was found to be abnormal among patients with IBS over that of controls [29].

Characteristically, SIBO disrupts normal gastrointestinal physiology, resulting in pathophysiological features present in IBS, including increased intestinal permeability, immune activation, carbohydrate malabsorption, visceral hypersensitivity, and GI dysmotility.

50.2.2 Altered Motility

Debate exists in the scientific community as to whether abnormal motility is a cause or a resulting effect of other pathogenic factors in irritable bowel syndrome. Mechanistically, peripheral abnormalities in enteric nervous signaling may lead to altered small bowel or colonic motility or both. The resulting motor dysfunction contributes to some symptoms of IBS, such as abdominal pain, defecatory urgency, and postprandial bowel movements. In addition to irregular motility patterns during the fed state, abnormal periods of migrating motility complex (MMC) contractions during the fasting/interdigestive phase have been observed in IBS patients [30]. Researchers have noted that the MMC of IBS-C and IBS-D patients can be changed, as compared with that in healthy people, and that abnormal stage III contractions in the jejunum may be a dominant change in IBS gastroenteric motility [30]. These irregular contractions result not only in impaired movement of matter through the GI tract but also in the accompanying sensations of cramping and visceral pain symptomatology in IBS patients.

50.2.3 Visceral Sensation

Visceral hypersensitivity of the peripheral and central nervous system is a principal etiological driver of symptoms in irritable bowel syndrome. Alteration of neurological signaling results in persistent visceral hyperalgesia and allodynia – elevated nociceptive sensation in response to normal stimuli. Changes in contents of the alimentary canal vary the pressure and volume within tissues, resulting in mechanical stimuli. Under normal physiological conditions, these changes would occur without any perception of pain. IBS patients experience enhanced perception of mechanical triggers from distensions and contractions of GI structures. The severity of visceral pain symptoms may be mediated, in part, by intake of certain dietary substrates as well as biopsychosocial stress. While the underlying cause of visceral hypersensitivity has yet to be fully elucidated, several factors, including microbial infections, intestinal dysbiosis, psychological factors, inflammation and immunological factors, brain-gut communication, diet, as well as genes, appear to be involved in the manifestation elevated pain perception [31].

50.2.4 Brain-Gut Interaction

Neurogastroenterology is a burgeoning area of clinical investigation, with clear associations elucidated between GI disease and behavioral disorders. The study of neurogastroenterology examines the physiological features of the brain-gut axis and clinical treatment targets for functional GI disorders. The connectedness between gastrointestinal health and neurological function is particularly relevant. At the foundation of this relationship is the enteric nervous system – replete with vast innervated tissues. The enteric nervous system and central nervous system communicate along vagal and autonomic pathways to modulate many gastrointestinal (GI) functions.

Stemming between the brain and these tissues runs the vagus nerve, a conduit of signaling that can serve to enhance or disrupt physiological function. Exogenous stress can activate the sympathetic nervous system and hypothalamic-pituitary-adrenal (HPA) axis, resulting in a release of glucocorticoids and catecholamines that have a downstream effect on the enteric nerves and digestive function.

In addition to these signaling molecules from primary neurological and endocrine tissues, several crucial neurotransmitters are synthesized locally within the GI tract, including serotonin (5-HT) – 90% of which is manufactured in the gut. In the GI tract, serotonin activates as many as 14 different 5-HT receptor subtypes [32] located on enterocytes [33], enteric neurons [34], and immune cells [35]. Serotonin influences small intestinal motility, a feature illustrated by the fact that the administration of serotonin agonists and antagonists alters intestinal function. Neuromodulators, such as selective serotonin reuptake inhibitors, are regularly administered in IBS patients to both modulate peristalsis and treat common comorbidities in

mood and behavior. In clinical trials, patients with IBS have exhibited significantly higher levels of anxiety and depression than healthy controls [36]. In addition, common clinical, psychosocial manifestations in patients presenting with IBS include the following:
1. IBS patients are more likely to have anxiety, depression, and somatization disorders.
2. They are more likely to have been physically or sexually abused in childhood.
3. Both factors can influence care-seeking.
4. Psychiatric disturbance and abuse history can influence the severity of symptoms and level of disability [37].

Expanding on the concept of the gut-brain connection, and overlapping with the clinical significance of gut commensal microbes in IBS, is the construct of the microbiota-gut-brain axis. Preclinical data have exhibited that changes in the microbiome can influence aspects of behavior and brain function [38]. Some of this influence may be a result of specific bacterial species that promote 5-HT biosynthesis from colonic enterochromaffin cells (ECs), which supply 5-HT to the mucosa and gastrointestinal lumen [39]. Serotonin synthesis may act locally to influence motility as well as distally to modulate pain perception and mood (◘ Fig. 50.5).

50.3 Integrative and Functional Medical Approach Beyond Conventional Protocols

Standard treatment of care for irritable bowel syndrome includes a variety of potential interventions, based mainly on the wide range of clinical manifestations and variable nature of etiological factors. Customizing treatment ensures greater effectiveness in addressing the symptoms in a presenting patient and their specific subtype.

Pharmacological interventions include medications to treat diarrhea, constipation, cramping/spasm, dysmotility, pain, and aspects of psychological distress. ◘ Table 50.3 lists examples of pharmaceuticals used in the treatment of IBS.

50.3.1 Integrative and Functional Medicine Approach

Integrative and functional medicine practitioners may rely on some initial pharmacological interventions in cases where severity of symptoms and long-standing pathophysiology warrant some degree of symptom relief for the patient, but ultimately seek to resolve the causative factors in the IBS patient. During treatment, integrative providers often initiate combination protocols utilizing dietary and nutritional/herbal supplement interventions that support appropriate mucosal barrier function, eradicate any infectious pathogens, ameliorate inflammatory activity, modulate immune function, and rebalance gastrointestinal microbial composition.

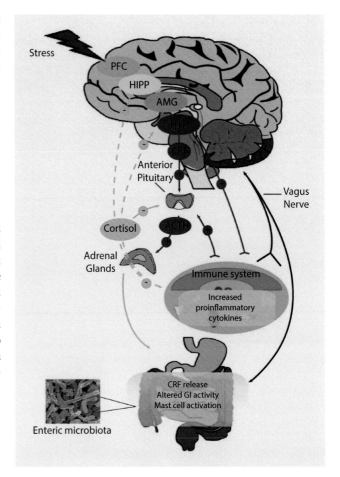

◘ **Fig. 50.5** Microbiome-gut-brain axis. The central nervous system (CNS) and enteric nervous system (ENS) communicate along vagal and autonomic pathways to modulate many gastrointestinal (GI) functions. The enteric microbiota influence the development and function of the ENS and immune system which affects CNS function. The hypothalamic-pituitary-adrenal (HPA) axis forms a key component of brain-gut signaling, responding to stress or heightened immune activity. Mood and various cognitive processes can mediate top-down bottom/bottom-up signaling. The HPA axis can be activated in response to environmental stress or by elevated systemic proinflammatory cytokines. Cortisol released from the adrenal glands feeds back to the pituitary, hypothalamus (HYP), amygdala (AMG), hippocampus (HIPP), and prefrontal cortex (PFC) to shut off the HPA axis. Cortisol released from the adrenals has a predominantly anti-inflammatory role on the systemic and GI immune system. In response to stress, GI activity can be altered, and corticotropin releasing factor (CRF) increased. Stress can increase systemic proinflammatory cytokines which can act at the pituitary to activate the HPA axis and can signal to the central nervous system via the vagus nerve, which also transmits changes due to mast cell activation in the GI tract. (Reprinted from Kennedy et al. [40]. With permission from Baishideng Publishing Group Inc.)

50.3.2 Dietary Intervention

From a dietary perspective, a low FODMAP diet has proven to be most efficacious among potential nutritional therapies and, in many parts of the world, is now considered a frontline therapy for IBS. FODMAPs (fermentable oligosaccharides, disaccharides, monosaccharides, and polyols) are categorized as short-chain carbohydrates that are either slowly absorbed

Table 50.3 Pharmaceutical applications utilized in the treatment of irritable bowel syndrome

Medication category	Treatment target
Antidiarrheal agents	Diarrhea
Laxatives	Constipation
Bulking agents (psyllium)	Constipation
Antispasmodic	Pain, cramping
Anticholinergics	Pain, cramping
Tricyclic antidepressants	Pain, visceral hypersensitivity, comorbid psychological symptoms
Selective serotonin reuptake inhibitors	Pain, visceral hypersensitivity, comorbid psychological symptoms

or incompletely digested in the small intestine. Evidence has accumulated over several years, supporting the effectiveness of a low FODMAP diet in patients with IBS. Historically, individual short-chain carbohydrates were identified as inducing or exacerbating abdominal symptoms. Eventually, these various substrates were combined in trials where their elimination as part of the composite low FODMAP protocol produced the most significant symptom resolution. A review of research shows that up to 86% of patients with IBS find improvement in overall gastrointestinal symptoms as well as such individual symptoms as abdominal pain, bloating, constipation, diarrhea, abdominal distension, and flatulence following the diet [25]. According to Nanayakkara et al., FODMAP restriction reduces the osmotic load and gas production in the distal small bowel and the proximal colon, providing symptomatic relief in patients with IBS [25]. Clinical trials testing the efficacy of low FODMAP dietary intervention demonstrate two vital for successful outcomes: (1) patient education and (2) dietitian-led intervention [41].

Despite substantial evidence of its efficacy in IBS patients, concerns over long-term implications exist, primarily related to the diet's effects on the gut microbiota and synthesis of anti-inflammatory short-chain fatty acids (butyrate, propionate, acetate) – both of which are influenced by the consumption of fermentable carbohydrates. The role of the integrative nutritionist applying a low FODMAP diet with IBS patients is to help address other pathophysiological underpinnings and carefully reintroduce these beneficial carbohydrates to the patient based on tolerability and optimization of overall digestive function.

50.3.3 Herbal and Nutritional Supplement Interventions

In the treatment of irritable bowel syndrome, integrative and functional medical providers often leverage various novel compounds to alleviate symptoms and address underlying pathophysiology. Compounds that display significant evidence in the scientific literature of ameliorating some of the physiological abnormalities associated with IBS include the amino acid l-glutamine, serotonin precursor 5-hydroxytryptophan, and a variety of herbal antimicrobials that have shown efficacy in the eradication of pathogenic and overabundant microbial species. Researchers at Johns Hopkins conducted a comparative trial placing a multi-ingredient herbal compound against standard course of treatment in IBS mediated by small intestinal bowel overgrowth (SIBO) and reported that the herbal antimicrobials were of similar efficacy to rifaximin for the resolution of SIBO, and were also as effective as triple antibiotic therapy for SIBO rescue therapy for rifaximin non-responders [42].

50.4 Nutritional and Lifestyle Clinical Application Principles

Whether driven by immunological imbalance or impaired digestive function, it is evident that IBS patients benefit from dietary modifications. The challenge for the integrative nutritionist is to strategically implement the appropriate evidence-based intervention, assess its efficacy in their individual patient, and progress the patient through phases of dietary adjustment as symptoms resolve and underlying pathophysiology is attenuated. A few dietary considerations to try for patients experiencing irritable bowel syndrome based on the influence of substrates and the presence of impaired digestive capacity:

- Avoidance of large meals – spread food intake throughout day
- Reduction or elimination/challenge of lactose
- Reduction or elimination/challenge of sorbitol, mannitol, xylitol
- Reduction or elimination/challenge of fructose especially high-concentration fruits, HFCS, honey
- Reduction in raffinose intake
- Maintenance of food journal citing intake and response (GI symptoms, bowel movements, etc.)
- Introduction to food preparation techniques that aid digestibility, e.g. soaking, and inclusion of fermented foods upon exclusion of bacterial overgrowth

Integrative nutritionists should also consider patient stress levels in the context of patient treatment. Essential in this component of care is the recognition that stress should be defined in the broader context of the biopsychosocial model, where stress includes not only mental-emotional duress but also endogenous and exogenous exposures.

Endogenous influences may arise from metabolic dysfunction – the by-products of which can increase oxidative stress and inflammatory cytokine activity that may contribute to IBS pathophysiology. Endogenous stress on the system may also arise from qualitative and quantitative gut microbial divergence, as well as endotoxins released from bacterial cells upon cell degradation.

Myriad exogenous factors impart stress on the body, and collectively these factors constitute part of the exposome. Per the Centers for Disease Control and Prevention, the exposome has been defined as "the measure of all the exposures of an individual in a lifetime and how those exposures relate to health [43]". These exposures begin before birth and include insults from environmental sources. In the case of the patient with IBS, influential factors may consist of peri-/postnatal exposure to antibiotics, microbial interaction during delivery method, food introduction, trauma/early adverse life events, and pathogenic infection. Post-infectious IBS, which develops in 4–32% of patients with bacterial gastroenteritis [44], appears to be a nonspecific response to infection caused by a variety of enteric pathogens and has been documented after illness due to *Campylobacter* species, *Salmonella* species, diarrheagenic strains of *Escherichia coli*, and *Shigella* species [45].

Psychological stress, either in conjunction with the appearance of symptoms or in subsequent years after initial diagnosis, is associated with patient-reported increases in symptom severity. Comorbidity of IBS and psychological distress is common, and prevalence of psychiatric disorders alongside IBS is evident. Research examining the psychosocial determinants of IBS has been explored, with studies reporting a significant increase in stress score just before progression from IBS non-patient to IBS patient [46]. Major life traumas (e.g. disruption of a close relationship, a marital separation, a family member leaving home, or termination of romantic relationship) were frequently reported 38 weeks prior to onset of IBS symptoms [46]. Techniques to mitigate mental-emotional stress such as cognitive-behavioral therapy and meditation are a vital component of care for IBS patients.

50.5 Grand Rounds Case Presentation

50.5.1 Assessment

50.5.1.1 Patient Profile: History and Subjective Information

A 53-year-old female engaged in care, presenting with significant abdominal bloating, distension, belching, and irregular bowel movements. The patient tended toward constipation, but would periodically (approximately every 7–10 days) have a large bowel movement she deemed as a "massive evacuation" with considerable fecal mass. Using the Bristol Stool Chart, the patient identified bowel movement as type 5/6 in consistency – oriented toward loose stool. Between these periodic eliminations, patient reported that bowel movements may not occur for 3–5 days at a time. Her chief complaint was bloating and abdominal distension, followed by the variability and discomfort associated with her bowel habits. IBS-SSS questionnaire was administered to gauge symptom severity and as a metric for reassessment and protocol efficacy. Patient IBS-SSS score was 360 out of 500 – designating her as a "severe case" per survey criteria.

Patient provided subjective history stating symptoms had initiated 24 years ago concurrent with the birth of her first daughter. Predominant symptoms at the time were severe abdominal cramping and diarrhea. After four years of self-management, patient sought medical care and was prescribed selective serotonin reuptake inhibitor (sertraline Hcl), along with meditation techniques to treat depression and manage the stress that seemed to precipitate an increase in gastrointestinal symptoms. The prescription and meditation seemed to alleviate symptoms for a period, but eventually, symptoms returned, and new symptoms of bloating and belching appeared.

50.5.1.2 Pertinent Medication History

Patient-reported medication list is included in ◘ Table 50.4.

50.5.1.3 Medical Assessment and Treatment

Patient had sought medical intervention with a gastroenterologist for the first time two years prior to initial nutrition consultation. The treating physician ran standard blood chemistries, including 25(OH)D, comprehensive metabolic panel, complete blood count, and lipid panel. Lipid panel indicated potential familial hypercholesterolemia due to total cholesterol at 299 mg/dL and LDL-C of 211.00 mg/dL. Familial hypercholesterolemia is suspected when fasting LDL cholesterol is above 189 mg/dL or non-HDL cholesterol is above 219 mg/dL. Collection of family history indicating high cholesterol and heart disease in first-degree relatives can support presence of FH, which was confirmed by patient. Status of serum 25(OH)D was deemed insufficient at 21 ng/mL. All other blood chemistries analyzed were within reference range. The patient's gastroenterologist also conducted a gastric emptying study, which returned unremarkable, with a conclusion of normal gastric emptying and absence of gastroesophageal reflux. Gastroenterologist provided a diagnosis of irritable bowel syndrome.

To investigate the underlying etiology of the patient's IBS symptoms, specifically bloating and distension, gastroenterologist conducted hydrogen breath test to assess for potential presence of small intestinal bacterial overgrowth (SIBO).

Clinical assessment involved hydrogen breath test with administration of a lactulose solution and testing of hydrogen levels measured in ppm at 15-minute intervals for 120 minutes. Test result indicated a fivefold increase above basal hydrogen breath level at 60 minutes and early increase (within 90 minutes) greater than 20 ppm (22 ppm recorded at 60-minute mark) – both considered definitive for a positive test result. Technician's interpretation, supported by review of the physician, was positive indication for small intestinal bacterial overgrowth.

Upon confirmation of SIBO, patient's gastroenterologist prescribed 2-week course of Xifaxan (rifaximin) 550 mg. Patient reported resolution of symptoms for approximately three months, upon which time all symptoms returned with the same severity and profile. Patient revisited her physician who readministered Xifaxan prescription. Again, the patient experienced alleviation of symptoms for a short time, only to have the bloating, distension, and vacillating constipation and diarrhea return.

Table 50.4 Patient medication list

Current medications (over-the-counter and prescription)

Name	Dosage	Frequency	Length of time	Reason for taking
Enablex (darifenacin)	7.5 mg	Nightly	4 years	Frequent urination sensation
Crestor (rosuvastatin calcium)	10 mg	Daily	10 years	High cholesterol
Zoloft (sertraline HCl)	50 mg	Daily	5 years	Depression
Align probiotic (*B. infantis* 35624™)	1 capsule (1 billion CFU)	Daily	10 years	IBS
Dulcolax (bisacodyl)	300 mg	Daily	5 years	IBS
Lo/Ovarol-28 (ethinyl estradiol and norgestrel)		28 days/month	35 years	Heavy menstruation

After a third course of Xifaxan, physician advised against additional administration of the antibiotic therapy due to potential complications. Patient continued to see mild improvement immediately following course of Xifaxan, only to have symptoms return, upon which time she sought out clinical nutrition intervention and adjunct therapeutic options.

50.5.1.4 Diagnosis

Collection of health history data illustrated several additional points of interest identified via use of the IFM framework of antecedents, triggers, and mediators (Table 50.5).

Antecedents

Early, and particularly repeated, childhood antibiotic treatment influences the development of the gut microbiota both quantitatively and qualitatively, leading to modulation of immune function and long-term health implications [47, 48]. Patient also reported a high incidence of first-degree relatives with alcohol/substance abuse history and her challenges with eating disorders. At the time patient sought my services, she was enrolled in and actively attending Food Addicts Anonymous (FAA).

Triggers

Psychological stress is acutely associated with irritable bowel syndrome. Stress serves as a trigger for IBS symptoms by activating the HPA axis and resulting release of cortisol, epinephrine, and norepinephrine, all of which may influence gastrointestinal activity.

Declining ovarian hormones (i.e., estrogen and progesterone) are associated with an increase in GI symptoms [49]. The patient reported discernable changes in IBS symptom severity during two distinct life events: (1) initiation of menarche and (2) onset of menopause. Both stages are associated with variation in ovarian hormone and specifically with fluctuation of estrogen and progesterone – with decreased physiological levels experienced during menses and transition to menopause.

Table 50.5 Relevant IBS patient antecedents, triggers, and mediators

Antecedents	1. Multiple courses of antibiotic intervention in childhood 2. Potential genetic predisposition to psychosocial substance use/abuse as indicated by patient-reported family history of alcohol dependency and own experience with eating disorder
Triggers	1. Biopsychosocial stress 2. Menopause
Mediators	1. Psychological stress: caregiver for terminally ill mother with cancer 2. Foods – patient-reported suspicion of certain foods but unable to identify precise food triggers

Mediators

Around the time of the increase in IBS symptoms, the patient had assumed a more active role as the caregiver for her mother, who was terminally ill with metastatic cancer. This psychic trauma coincided with a significant intensification of her IBS symptoms, owing to the increased psychological stress.

Although the patient had made some degree of dietary changes over time to self-manage her symptoms, she was inconsistent with these nutritional modifications and was also unclear as to which foods seemed to mediate increases in her bloating, pain, gas, and bowel irregularities. The patient had prepared a 3-day food log for evaluation and intake, which reflected a standard, generalized diet with the following inadequacies:

- Inadequate fiber intake
- Inappropriate intake of fats (inadequate essential fatty acids relative to total fat intake)
- Inadequate protein intake
- Inadequate mineral intake (iron, zinc, calcium, potassium, magnesium)
- Inadequate vitamin intake (A, D, E, K, folate, B12)

After a thorough review of biomarkers, health history, food journal, and patient-reported symptoms, the following PES statement was developed:

PES Statement: Disordered eating pattern related to altered GI function as evidenced by 3-day food intake survey and patient-reported symptoms of excess bloating, gas, stool frequency and type (Bristol type 5/6), and positive test result for presence of small intestinal bacterial overgrowth per clinical parameters.

50.5.1.5 Intervention

The therapeutic nutritional intervention consisted of a structured low FODMAP diet with accompanying education and resources for practical implementation. As discussed earlier, the low FODMAP dietary intervention has proven to be among the most efficacious therapies for patients with IBS. Based upon this patient's clinical presentation, the elimination of fermentable fibers was clinically relevant owing to the presence of bacterial overgrowth in the small intestine. Guidelines for dietary compliance were provided, with a specific focus on consumption of bioactive substances lacking from the existing diet, specifically, strategies to enhance protein status, bioflavonoids, omega-3 fatty acids, and aforementioned micronutrients. Recommendations were also provided to ease the digestibility of foods through avoidance of raw vegetables and cooking of relevant foods.

From a behavioral perspective, the patient was encouraged to continue attendance at Food Addicts Anonymous meetings. Patient attributed these meetings and support system as integral in providing insight into her eating habits.

50.5.1.6 Dietary Supplement Protocol

To rectify existing vitamin D insufficiency, as evidenced by serum 25(OH)D, patient began supplementation of vitamin D3 (cholecalciferol) at 2000 IU daily. As 2000 IU (50 mcg) per day is generally required to increase vitamin D blood levels by increments of 20 ng/ml (50 nmol/L) over the course of several months [50, 51], selection of this dosage would provide clinical utility and replete the patient to sufficient status. Supplementation in this scenario was warranted owing to the patient's lab chemistries, inadequate intake, potential GI malabsorption issues, and lack of daily UVB exposure resulting in insufficient cutaneous production.

In addition to vitamin D3 supplementation, a targeted supplement protocol was developed to address the patient's refractory SIBO.

Various herbal compounds exhibit antimicrobial properties and have been used in traditional medicine for centuries. Caprylic acid has displayed the ability to eradicate resistant gram-positive, gram-negative, and fungal biofilms [52]. Berberine sulfate has also been shown to decrease bacterial adherence to mucosal or epithelial surfaces and be an effective antimicrobial and has exhibited antisecretory activity in vivo [53]. A search of the US National Library of Medicine National Institutes of Health (▶ PubMed.gov) database results in over 2600 publications exploring the antimicrobial activity of essential oils, including oil of oregano (*Origanum vulgare*) [54]. The antibacterial effects of oil of oregano are associated with the presence of their phenolic components, carvacrol and thymol, which have shown potent bactericidal effects against an impressive array of bacterial strains:

- *Staphylococcus aureus*
- *Escherichia coli*
- *Bacillus cereus*
- *Pseudomonas aeruginosa*
- *Klebsiella pneumoniae*
- *Acinetobacter baumannii*
- *Aeromonas sobria*
- *Enterococcus faecalis*
- *Salmonella Typhimurium*
- *Serratia marcescens*
- *Yersinia enterocolitica*
- *Shigella flexneri*

An in vivo trial employing a compounded herbal formula exhibited efficacy in the resolution of bacterial overgrowth in the small intestine and remission of associated IBS symptoms [42].

As repeated courses of Xifaxan were unable to resolve entrenched dysbiotic gut ecology, alternatives were leveraged to remediate the bacterial overgrowth and gastrointestinal symptoms. Patient was provided with recommendations to incorporate 1 capsule qd of oil of oregano (36 mg carvacrol and thymol; 60–75% carvacrol oregano oil) and 1 capsule qd of a proprietary blend of botanical extracts (GI Microb-X™ from Designs For Health™) that includes the following:

- Tribulus extract (*Tribulus terrestris*)(aerial) [standardized to contain 40% saponins] 200 mg
- Magnesium caprylate (yielding 120 mg caprylic acid; 10 mg magnesium) 150 mg
- Berberine sulfate (*Berberis aristata*)(root) 100 mg
- Grapefruit extract (Citrus paradisi)(seed) 100 mg
- Barberry extract (*Berberis vulgaris*)(bark) [standardized to contain 6% berberine] 50 mg
- Bearberry Extract (*Arctostaphylos uva-ursi*)(leaf) [standardized to contain 20% arbutin] 50 mg
- Black walnut powder (*Juglans nigra*)(hull) 50 mg
- Artemisinin (from sweet wormwood) (*Artemisia annua*) (herb) 15 mg

50.5.1.7 Monitoring

Throughout the entire intervention period, the patient maintained detailed food journals noting dietary intake and symptoms, with attention given to the timing of GI symptoms relative to meal consumption and psychological stress. The patient reported this as being an invaluable tool as it allowed her to become more mindful with her eating habits and more keenly aware of her body response and mental state.

Monitoring of IBS symptoms was continuous through use of the iPhone application mySymptoms Food and Symptoms Tracker (© 2016 SkyGazer Labs Ltd), which provides recording, tracking, and analysis of dietary intake, along with various metrics around physical symptoms of digestion, stress, and energy levels.

At eight weeks of the dietary and supplement intervention, patient reported near remission of IBS symptoms. Bloating, gas, abdominal pain, and distension had completely resolved, belching was considerably decreased, and bowel function had improved with the patient experiencing near-daily bowel movement of Bristol type 3/4.

50.5.1.8 Evaluation

The protocol was continued for another four weeks, at which time the patient returned to her gastroenterologist for a follow-up hydrogen breath test. Diagnostic report exhibited hydrogen never exceeding 3 ppm. Technician and physician confirmed negative readings and the absence of SIBO.

Nutritionally, the patient continued exclusion diet of high FODMAP foods, with gradual reintroduction and periodic inclusion of certain vegetables and fruits to tolerance. Abstention from wheat-/gluten-based foods and cow's milk products continued as patient recognized a marked improvement when those substrates were avoided.

IBS-SSS was administered again, with severity score decreasing to mild severity at 125. From a psychological perspective, patient reported a significant improvement to mood and quality of life. Patient also expressed satisfaction for the resolution of many of her symptoms and gratitude for the nutritional and lifestyle techniques that conferred her with a sense of self-efficacy in managing her IBS.

References

1. Bradbury KE, Appleby PN, Key TJ. Fruit, vegetable, and fiber intake in relation to cancer risk: findings from the European prospective investigation into Cancer and nutrition (EPIC). Am J Clin Nutr. 2014;100(Suppl 1):394S–8S.
2. Massironi S, Rossi RE, Cavalcoli FA, Della Valle S, Fraquelli M, Conte D. Nutritional deficiencies in inflammatory bowel disease: therapeutic approaches. Clin Nutr. 2013;32(6):904–10.
3. Wędrychowicz A, Zając A, Tomasik P. Advances in nutritional therapy in inflammatory bowel diseases: review. World J Gastroenterol. 2016;22(3):1045–66.
4. Hogenauer C, Hammer H. Maldigestion and malabsorption. In: Feldman M, Friedman L, Brandt L, editors. Sleisenger and Fordtran's gastrointestinal and liver disease. 9th ed. Philadelphia: Saunders; 2010. p. 1735–68.
5. Maurer AH. Gastrointestinal motility, part 1: esophageal transit and gastric emptying. J Nucl Med. 2015;56(8):1229–38.
6. Maurer AH. Gastrointestinal motility, part 2: small-bowel and colon transit. J Nucl Med. 2015;56(9):1395–400.
7. Nowak-Wegrzyn A, et al. Food allergy and the gut. Nat Rev Gastroenterol Hepatol. 2017;14(4):241–57.
8. Krigel A, Lebwohl B. Non-celiac gluten sensitivity. Adv Nutr. 2016;7(6):1105–10.
9. Nowak-Węgrzyn A, et al. Non-IgE-mediated gastrointestinal food allergy. J Allergy Clin Immunol. 2015;135(5):1114–24.
10. Miniello LV, et al. Gut microbiota biomodulators, when the stork comes by the scalpel. Clin Chim Acta. 2015;451(Part A):88–96.
11. The National Institute of Diabetes and Digestive and Kidney Diseases (NIDDK). Digestive diseases statistics for the United States. Bethesda: National Institutes of Health; 2014. NIH Publication No. 13–3873.
12. Drossman DA, et al. Rome IV, the functional gastrointestinal disorders. Gastroenterology. 2016;150:1262–79.
13. Olausson EA, Störsrud S, Grundin H, et al. A small particle size diet reduces upper gastrointestinal symptoms in patients with diabetic gastroparesis: a randomized controlled trial. Am J Gastroenterol. 2014;109:375–85.
14. International Foundation for Functional Gastrointestinal Disorders. 2016. Functional GI disorders. Retrieved from http://www.iffgd.org/functional-gi-disorders.html.
15. Talley NJ. Functional gastrointestinal disorders as a public health problem. Neurogastroenterol Motil. 2008;20(Suppl 1):121–9.
16. Chey WD, Kurlander J, Eswaran S. Irritable bowel syndrome a clinical review. JAMA. 2015;313(9):949–58.
17. Canavan C, West J, Card T. The epidemiology of irritable bowel syndrome. Clin Epidemiol. 2014;6:71–80.
18. Lacy B, Mearin F, Chang L, Chey W, Lembo A, Simren M, Spiller R. Bowel disorders. Gastroenterology. 2016;150:1393–407.
19. Francis CY, et al. The irritable bowel severity scoring system: a simple method of monitoring irritable bowel syndrome and its progress. Aliment Pharmacol Ther. 1997;11(2):395–402.
20. Patrick DL, Drossman DA, Frederick IO, DiCesare J, Puder KL. Quality of life in persons with irritable bowel syndrome: development of a new measure. Dig Dis Sci. 1998;43:400–11.
21. Simrén M, Barbara G, Flint HJ, et al. Intestinal microbiota in functional bowel disorders: a Rome foundation report. Gut. 2013;62(1):159–76.
22. Dupont HL. Review article: evidence for the role of gut microbiota in irritable bowel syndrome and its potential influence on therapeutic targets. Aliment Pharmacol Ther. 2014;39(10):1033–42.
23. Hauser W, Musial F, Caspary WF, Stein J, Stallmach A. Predictors of irritable bowel-type symptoms and healthcare-seeking behavior among adults with celiac disease. Psychosom Med. 2007;69:370–6.
24. Catassi C, et al. Non-celiac gluten sensitivity: the new frontier of gluten-related disorders. Nutrients. 2013;5(10):3839–53.
25. Nanayakkara WS, Skidmore PM, O'Brien L, Wilkinson TJ, Gearry RB. Efficacy of the low FODMAP diet for treating irritable bowel syndrome: the evidence to date. Clin Exp Gastroenterol. 2016;9:131–42.
26. Portincasa P, et al. Irritable bowel syndrome and diet. Gastroenterol Rep. 2017;5(1):11–9.
27. Gibson PR. Food intolerance in functional bowel disorders. J Gastroenterol Hepatol. 2011;26(Suppl 3):128–31.
28. Ford AC, Spiegel BM, Talley NJ, Moayyedi P. Small intestinal bacterial overgrowth in irritable bowel syndrome: systematic review and meta-analysis. Clin Gastroenterol Hepatol. 2009;7:1279–86.
29. Shah ED, Basseri RJ, Chong K, Pimentel M. Abnormal breath testing in IBS: a meta-analysis. Dig Dis Sci. 2010;55:2441–9.
30. Wang SH, Dong L, Luo JY, Li L, Zhu YL, Wang XQ, Zou BC, Gong J. A research of migrating motor complex in patients with irritable bowel syndrome. Zhonghua Nei Ke Za Zhi. 2009;48(2):106–10.
31. Farzaei MH, Bahramsoltani R, Abdollahi M, Rahimi R. The role of visceral hypersensitivity in irritable bowel syndrome: pharmacological targets and novel treatments. J Neurogastroenterol Motil. 2016;22(4):558–74.
32. Gershon MD, Tack J. The serotonin signaling system: from basic understanding to drug development for functional GI disorders. Gastroenterology. 2007;132:397–414.
33. Hoffman JM, Tyler K, MacEachern SJ, Balemba OB, Johnson AC, Brooks EM, Zhao H, Swain GM, Moses PL, Galligan JJ, et al. Activation of colonic mucosal 5-HT(4) receptors accelerates propulsive motility and inhibits visceral hypersensitivity. Gastroenterology. 2012;142:844–54.
34. Mawe GM, Hoffman JM. Serotonin signaling in the gut – functions, dysfunctions and therapeutic targets. Nat Rev Gastroenterol Hepatol. 2013;10:473–86.
35. Baganz NL, Blakely RD. A dialogue between the immune system and brain, spoken in the language of serotonin. ACS Chem Neurosci. 2013;4:48–63.
36. Fond G, Loundou A, Hamdani N, et al. Anxiety and depression comorbidities in irritable bowel syndrome (IBS): a systematic review and meta-analysis. Eur Arch Psychiatry Clin Neurosci. 2014;264:651.

37. Olden KW. Diagnosis of irritable bowel syndrome. Gastroenterology. 2002;122(6):1701–14.
38. Moloney RD, Desbonnet L, Clarke G, Dinan TG, Cryan JF. The microbiome: stress, health and disease. Mamm Genome. 2014;25: 49–74.
39. Yano JM, et al. Indigenous bacteria from the gut microbiota regulate host serotonin biosynthesis. Cell. 2015;161(2):264–76.
40. Kennedy PJ, Cryan JF, Dinan TG, Clarke G. Irritable bowel syndrome: a microbiome-gut-brain axis disorder? World J Gastroenterol. 2014;20(39):14105–25.
41. Gibson PR. The evidence base for efficacy of the low FODMAP diet in irritable bowel syndrome: is it ready for prime time as a first-line therapy? J Gastroenterol Hepatol. 2017;32:32–5.
42. Chedid V, Dhalla S, Clarke JO, Roland BC, Dunbar KB, Koh J, Mullin GE. Herbal therapy is equivalent to rifaximin for the treatment of small intestinal bacterial overgrowth. Glob Adv Health Med. 2014;3(3):16–24.
43. Centers for Disease Control and Prevention: National Institute for Occupational Safety and Health, Division of Applied Research and Technology. 2014. Exposome and exposomics. Retrieved from https://www.cdc.gov/niosh/topics/exposome/.
44. Spiller RC, Campbell E. Post-infectious irritable bowel syndrome. Curr Opin Gastroenterol. 2006;22:13–7.
45. Ericsson CD, et al. Postinfectious irritable bowel syndrome. Clin Infect Dis. 2008;46(4):594–9.
46. Surdea-Blaga T, Băban A, Dumitrascu DL. Psychosocial determinants of irritable bowel syndrome. World J Gastroenterol. 2012;18:616–26.
47. Munyaka PM, et al. External influence of early childhood establishment of gut microbiota and subsequent health implications. Front Pediatr. 2014;2:109.
48. Korpela K, et al. Intestinal microbiome is related to lifetime antibiotic use in Finnish pre-school children. Nat Commun. 2016; 7:10410.
49. Heitkemper MM, Chang L. Do fluctuations in ovarian hormones affect gastrointestinal symptoms in women with irritable bowel syndrome? Gend Med. 2009;6(Suppl 2):152–67. https://doi.org/10.1016/j.genm.2009.03.004.
50. Shab-Bidar S, et al. Serum 25(OH)D response to vitamin D3 supplementation: a meta-regression analysis. Nutrition. 2014;30(9): 975–85.
51. Talwar SA, et al. Dose response to vitamin D supplementation among postmenopausal African American women. Am J Clin Nutr. 2007;86(6):1657–62.
52. Rosenblatt J, et al. Caprylic acid and glyceryl trinitrate combination for eradication of biofilm. Antimicrob Agents Chemother. 2015;59(3):1786–8.
53. Chen C, et al. Effects of berberine in the gastrointestinal tract — a review of actions and therapeutic implications. Am J Chin Med. 2014;42(05):1053–70.
54. Sakkas H, Papadopoulou C. Antimicrobial activity of basil, oregano, and thyme essential oils. J Microbiol Biotechnol. 2017;27(3): 429–38.

Respiratory

Julie L. Starkel, Christina Stapke, Abigail Stanley-O'Malley, and Diana Noland

51.1 Impact of Nutrition and Lifestyle on Respiratory Health and Disease – 929
51.1.1 Overview – 929

51.2 Anatomy and Physiology – 929
51.2.1 General Anatomy – 929
51.2.2 Cellular Physiology: Membrane Structure Determines Function – Very Important Barrier – 929
51.2.3 Gas Exchange – 931
51.2.4 Oral Health Connection – 932
51.2.5 Microbiome – 932
51.2.6 Autophagy – 933
51.2.7 Category of Respiratory Diseases – 933

51.3 Micronutrients Important for Lung Function: Food-Based and Supplemental – 933
51.3.1 Iron – 933
51.3.2 The B Vitamins – 935
51.3.3 Acid/Alkaline – 936
51.3.4 Vitamin A – 936
51.3.5 Vitamin D – 936
51.3.6 Vitamin C – 937
51.3.7 Vitamin E – 937
51.3.8 Phytonutrients – 937
51.3.9 Minerals – 938
51.3.10 Alpha Lipoic Acid – 938
51.3.11 NAC – 938
51.3.12 Micronutrient Testing – 938

51.4 Macronutrients Important for Lung Function: Food-Based and Supplemental – 938
51.4.1 Fats and Fatty Acids – 938
51.4.2 Protein – 940
51.4.3 Carbohydrates – 940

51.5 Endogenous Essential Respiratory Metabolites – 940
51.5.1 Glutathione – 940
51.5.2 CoQ10 – 942

© Springer Nature Switzerland AG 2020
D. Noland et al. (eds.), *Integrative and Functional Medical Nutrition Therapy*,
https://doi.org/10.1007/978-3-030-30730-1_51

51.6 Anti nutrients and Inhibitors of Lung Physiology – 942
51.6.1 Toxic Metals – 942
51.6.2 Air Pollutants – 943
51.6.3 Chemicals – 944

51.7 Stress – 944
51.7.1 Stress Overview – 944
51.7.2 Biological Stress – 944

51.8 Disease States – 944
51.8.1 Asthma – 944
51.8.2 Chronic Infection and Respiratory Health – 955
51.8.3 Alpha-1 Antitrypsin Deficiency (A1AT Deficiency) – 956
51.8.4 Pulmonary Fibrosis – 957

51.9 Conclusion – 962

References – 962

51.1 Impact of Nutrition and Lifestyle on Respiratory Health and Disease

51.1.1 Overview

Lung disease is far more prevalent worldwide than commonly thought. In fact, death from chronic lung disease is increasing, and as of 2017, chronic obstructive pulmonary disease (COPD) has become the third leading cause of death in the United States in the past decade, disproportionately affecting the elderly [1]. Another lung disease, asthma, affects 1 in 13, or about 25 million Americans, according to the Centers for Disease Control and Prevention and the National Center for Health Statistics [2]. This is 7.6% of adults, more women than men, and 8.4% of children. Asthma is the leading chronic disease in children [3]. This disease has been increasing since the early 1980s in all age, sex, and racial groups. In Europe, lung disease represents 15% of all deaths – the fourth leading cause. According to the World Health Organization (WHO), in 2008, 9.5 million people died from acute or chronic lung disease, representing one sixth of the global total deaths [4].

Worldwide, four respiratory disease categories appear in the top ten leading causes of death in 2010 [5]. Specifically, COPD was the third leading cause of death, followed by lower respiratory infections as the fourth, lung cancer as the fifth, and tuberculosis as the tenth [4]. The major risk factor is smoking, leading to 50% of all lung disease-related deaths in Europe, where smoking is more prevalent (28% prevalence) than in the United States (15% prevalence) by nearly twofold [6, 7]. Lung cancer, particularly non-small-cell lung cancer (NSCLC) subtype, is the leading cause of cancer-related death worldwide [8]. Added together, lung disease rivals the position for the top cause of death.

Throughout the life cycle, diet and lifestyle are important modifiable risk factors in the development, progression, and management of obstructive lung diseases, such as asthma and COPD [9], as well as restrictive lung diseases such as pulmonary fibrosis and sarcoidosis. Inflammation, in particular, seems to be the leading contributor toward the progression of lung diseases. As with many diseases, maintaining a healthy lifestyle, including sufficient sleep, low stress, regular exercise, a whole foods diet rich in phytonutrients from plants (fruits and vegetables), and potential anti-inflammatory supplements, is beneficial in supporting the body during these difficult diseases.

> Inflammation, in particular, seems to be the leading contributor toward the progression of lung diseases.

High inflammatory foods should be avoided, such as fried foods and foods disproportionately high in carbohydrates, sugar, alcohol, and excessive protein. A healthier suggestion would be a diet with more than half of all food consumed as vegetables, about one third as protein, and the remainder (one sixth) as other foods, such as fruits, dairy, grains, or starches. Some dietary supplements may also be recommended for their anti-inflammatory benefits, which will be discussed later in the chapter.

As human life expectancy increases, we can expect to see more chronic disease. The World Health Organization estimates that by 2030, chronic lung disease will account for 20% (one fifth) of all deaths [10], up from one sixth in 2008. Despite these growing numbers, relatively little human nutrition research exists for respiratory health, compared to other, less prevalent, diseases. Investigators in the areas of aging and lung biology suggest some hope, using genetics and animal models, as well as epidemiological research, to further the general medical approach to lung disease.

51.2 Anatomy and Physiology

51.2.1 General Anatomy

The pulmonary system is composed of the upper and lower respiratory tracts. Air flows in through the nose or mouth, past the frontal and maxillary sinuses, down the pharynx (throat), past the larynx (voice box), and then down the trachea. This makes up the upper respiratory tract. Once past the trachea, the air divides into the left and right bronchi, which supply the left and right lungs, each divided into five sections called lobes. The bronchi then divide into smaller bronchioles, at the end of which are air sacs called the alveoli. This makes up the lower respiratory tract [11] (◘ Fig. 51.1).

The diaphragm is the central muscle that is used for breathing. The intercostal muscles, located between the ribs, and the abdominal muscles are helpful for breathing out when the breath becomes labored, such as during exercise. The neck muscles and the muscles in the collarbone area help with breathing when the other muscles are compromised or impaired. In some neurological diseases, such as amyotrophic lateral sclerosis (ALS) [see ▶ Chap. 51, Newton], nerve damage from the brain to breathing muscles can result in impaired movement of these muscles and thus impaired breathing.

In certain cases, such as in lung cancer when a lobectomy, removal of part of the lung, is required, there is an expected decrease in short- and long-term pulmonary function and oxygenation. However, respiratory muscle strength may be preserved [13]. In a pneumonectomy, removal of the entire lung, dramatic changes in thoracic anatomy take place, such as elevation of the hemidiaphragm, hyperinflation of the remaining lung, and influx of fluid into the postpneumonectomy space [14, 15].

51.2.2 Cellular Physiology: Membrane Structure Determines Function – Very Important Barrier

There are phagocytic macrophages on the cellular surface of the alveoli, Type I epithelial cells and Type II epithelial cells. Phagocytic macrophages destroy inhaled bacteria and serve

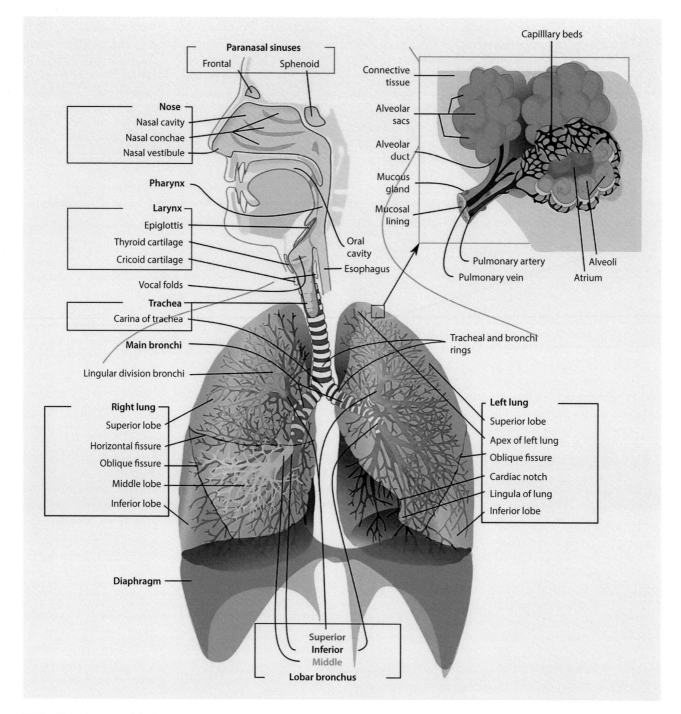

Fig. 51.1 Anatomy of the human respiratory system [12]. (Reprinted with permission from: ► https://en.wikipedia.org/wiki/Respiratory_system#/media/File:Respiratory_system_complete_en.svg)

an important role in suppressing or activating the immune response to antigens and pathogens, similar to dendritic cells discussed below. Macrophage function has been shown to be inhibited by cigarette smoke [16]. Alveolar macrophages also secrete enzymes, arachidonic acid metabolites, growth factors, immune response components, cytokines, and lymphocytes [17].

Type I cells are responsible for maintaining the structure of the alveolar wall, whereas Type II cells and Clara cells are responsible for the production of pulmonary surfactant (composed of 85–90% lipid and 10–15% protein as lecithin and myelin), which is essential for lung function. The surfactant reduces surface tension, facilitating easier stretching and collapsing of alveoli during respiration [18]. Diseases associated with inadequate surfactant production are acute/adult respiratory distress syndrome (ARDS) and infant respiratory distress syndrome (IRDS) [19]. IRDS is seen in premature babies born prior to 32 weeks of gestation due to immature development of pulmonary surfactant, which only begins to develop around the 20th week of gestation [18].

Dipalmitoylphosphatidylcholine, phosphatidylglycerol, and cholesterol compose the lipid portion of the surfactant, where apoproteins and proteins found in blood plasma compose the protein portion [18, 20]. The importance of cholesterol is minimized in today's medical community. Those with higher levels of cholesterol tend to have more in their fatty cell membranes which resist pathogenesis at a cellular level. Low cholesterol predicts a greater risk of dying from gastrointestinal, neoplastic, or respiratory diseases. It occupies 30–40% of our cell membranes, enhances the mechanical strength of the membrane, and reduces permeability [21]. It suppresses main-phase transition of the lipid bilayer [22].

Collagen, a fibrous protein, along with elastin and proteoglycans, is a fundamental component of the connective tissue that composes the lungs, and collagen is present in the blood vessels, bronchi, and alveolar interstitium [23]. Connective tissue in the lung is key for the passive diffusion of oxygen and carbon dioxide that characterizes alveolar-capillary gas [18]. Collagen homeostasis is vital to maintaining respiratory function, where collagen production and degradation are balanced. Dysregulated collagen homeostasis that favors collagen production over degradation can lead to pulmonary fibrosis and compromised lung function [24]. Some key nutrients to consider for collagen synthesis and cross-linking to maintain connective tissue integrity are vitamin C, vitamin B6, iron, copper, zinc [25], riboflavin, thiamin, and pantothenic acid [11].

The airways of the respiratory system (with the exception of parts of the nose and mouth) have cilia, special hairs coated with mucus that trap pathogens and other particles that enter with the air that is inhaled. Cilia are responsible for triggering this mucus upward toward the pharynx where these particles or bacteria can be coughed out or swallowed. Mucus present in the lungs can also trap inhaled particles such as viruses, bacteria, and smoke particulates [11, 12].

Along the lining of the respiratory tract, there are several types of cells that are involved in immune response, such as secretory cells (i.e., goblet cells and Clara cells) and mast cells. Ciliated epithelium and mucus secreted by glands present on airways, goblet cells, and the secretory products of Clara cells serve an important mechanism for lung protection. However, excessive goblet cells or hypertrophy of mucous glands may result in increased viscosity of mucus seen in pathologies like bronchitis [16]. Ciliary function is also impaired by cigarette smoke [16].

Dendritic cells are also found in the airway lining from the trachea to the alveoli. Immature dendritic cells phagocytize bacteria or other antigens, where they then mature and travel to lymphoid tissues to communicate with the immune system. This delivery of antigens can promote tolerance of the antigen by releasing anti-inflammatory cytokines. Conversely, this delivery can also trigger the opposite response if the antigen is recognized as a pathogen, where T lymphocytes are activated and inflammatory cytokines are released [16].

One potential cause of infections in the upper respiratory tract or bronchial tubes, such as bronchitis, or deep in the lungs, such as pneumonia, is when cilia become damaged and do not trap inhaled germs and particles as effectively. In diseases such as cystic fibrosis, thick mucus secretions can accumulate in the airways and lungs, making it hard to clear and thus increasing risk for infection. In asthma, specific inhaled particles can trigger a reaction causing the airways to narrow, restricting breathing [12].

Surface enzymes and factors can also be found in the lining of the airways that compose the majority of the innate immune system of the respiratory tract. These include:

- Lysozymes: found in leukocytes with bactericidal properties
- Lactoferrin: a bacteriostatic agent (inhibits bacterial reproduction) synthesized by lymphocytes and glandular mucosal cells
- Alpha-1 antitrypsin: an antiprotease to protect lung tissue from excessive enzymatic activity
- Interferon: an antiviral substance that may be produced by lymphocytes and macrophages
- Complement: participates as a cofactor in antigen-antibody reactions [16]

51.2.3 Gas Exchange

Gas exchange takes place in the alveoli so oxygen can enter the body to support metabolic function and the carbon dioxide product from these functions can be removed. This is accomplished through millions of capillaries in the alveoli. These capillaries in the alveoli then connect to arteries and veins that move blood throughout the body. The pulmonary artery supplies carbon dioxide-rich blood to these capillaries within the alveoli to remove carbon dioxide, and the oxygen-rich blood then gets delivered to the heart through the pulmonary vein. The lungs also serve the vital function of maintaining acid-base balance through changes in minute ventilation. These changes affect the pH of the blood by either retaining or excreting carbon dioxide [11].

Poor physiologic management of CO_2 and bicarbonate can lead to the conditions of respiratory acidosis and respiratory alkalosis. Respiratory acidosis is characterized by higher blood concentrations of CO_2 and H^+, caused by hypoventilation or decreased rate of breathing. Hypoventilation can have acute or chronic etiologies, resulting from COPD, interstitial lung diseases, respiratory muscle fatigue (i.e., extended asthma attack), or mechanical abnormalities (i.e., deformities).

Respiratory alkalosis is characterized by lower blood concentrations of CO_2 and H^+ due to hyperventilation, or increased rate of breathing. Possible causes of hyperventilation can also be chronic or acute, such as pneumonia and fever, increased stress and anxiety, liver disease, stroke or meningitis, pregnancy, overuse of aspirin and/or caffeine, excessive mechanical ventilation, or increases in altitude [18].

A pulse oximeter tool can be used to measure the percentage of oxygenated hemoglobin in an individual's blood to determine their overall respiratory status. Typically, oxygen

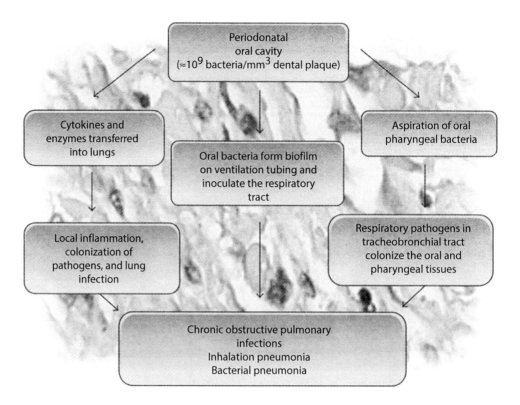

Fig. 51.2 Possible role of periodontal infection in respiratory disease. (Reprinted from Nagpal et al. [27]. With permission from Creative Commons)

saturations of 92% or less are indicative of central hypoxia [26]. Pulse oximetry is especially useful for assessing individuals with asthma and COPD [26].

51.2.4 Oral Health Connection

Oral health must also be considered as a contributing factor to respiratory health [27]. For example, in patients affected with periodontal disease, 1 mm of dental plaque could contain around 10^9 of bacteria. One potential mechanism of this connection is aspiration of bacteria from the oropharynx into the upper or lower respiratory tracts, leading to their adherence to the alveolar and bronchial lining, potentially colonizing respiratory ducts and causing respiratory infections. In addition, cytokines and enzymes associated with inflammation of periodontal tissues can be transferred into the lungs, potentially triggering or exacerbating lung infections [27] (Fig. 51.2).

A systematic review done in 2013 examined oral health in the elderly and its association with risk of aspiration pneumonia. This review suggested that maintaining oral health, such as brushing after each meal, cleaning dentures once per day, and professional oral healthcare, potentially reduced the amount of potential respiratory pathogens that resulted in lower incidence of aspiration pneumonia [28]. Several other systematic reviews have found that adequate oral hygiene plays an important role in preventing pneumonia, particularly in clinical settings where there is increased risk for hospital-acquired pneumonia (HAP) and ventilator-associated pneumonia (VAP), as well as in older populations [29]. In addition, associations have been made between COPD and the risk of periodontitis, although systematic reviews have established that these associations are preliminary and further studies are needed [29].

Another important consideration in respiratory health is orofacial development and structure. Anatomical obstructions at the level of the nose and pharynx, such as those caused by allergic rhinitis and hypertrophy of the tonsils, pose an increased risk for obstructive sleep apnea syndrome and respiratory infections due to lack of airflow through the upper respiratory system [30].

51.2.5 Microbiome

It has been established that the lung has a microbiome of its own that may have a large impact on health and disease [31]. The fungal microbiome, or mycobiome, may also have a significant impact on respiratory health, although more research is needed to determine definitive associations [31]. Dysbiosis may occur in the lungs with a bacterial infection. A few specific bacterial strains have been studied, and one, in particular, *Pseudomonas aeruginosa*, seems to grow in inflammatory conditions. It then seems to encode inflammatory components causing further inflammation. Anti-inflammatory nutrients could help stop the cycle, and vitamin D use has some research supporting this. Recurrent bacterial respiratory infections may damage lungs and lead to worse outcomes in future lung disease [32].

An increased interest in research of the relationship of the airway and gut microbiome is indicating potentially positive results regarding the use of probiotics in pediatric populations that may aid in asthma prevention and intervention

[33, 34]. The gut-lung axis has also been established, where the microbiomes of the lung and gut have been immunologically linked and are thought to have an impact on respiratory disease [35, 36].

51.2.6 Autophagy

The autophagy mechanism within our microenvironment provides a constant "cleanup" system to recycle cell debris from microscopic biowaste generated by dynamic cellular biochemistry [37]. Enzymes such as neutrophil elastase function like garbage disposals recycling waste molecules. Alpha-1 antitrypsin is a thermostat-like control factor that signals the proteolytic enzymes to stop and protect healthy tissue from being affected. Antiproteases in the lung, such as alpha-1 antitrypsin, are required to prevent the overactivity of neutrophil elastase to prevent the degradation of healthy lung tissue. Those with the genetic mutations of A1AT deficiency are at disadvantage, and subsequent lung tissue damage can occur promoting lung diseases like COPD, asthma, bronchitis, and emphysema.

Key components of lung structure are elastin and collagen, which provide support for the bronchioles and clusters of alveoli (acini). The key enzyme present in these cells is neutrophil elastase, which is responsible for the destruction of respiratory bacteria. Protease and antiprotease imbalance in the lung resulting in emphysema can be caused by alpha-1 antitrypsin deficiency and nicotine in cigarette smoke or polluted inhalant exposure [18]. IFMNT approaches to the A1AT-deficient patient assess for nutrient insufficiencies for some of the important connective tissue, collagen, and elastin system key nutrients: vitamin C, vitamin D, biotin, balanced fatty acids, and gut microbiome. When insufficiencies or deficiencies are identified, appropriate food and dietary supplementation interventions can be recommended. It should be noted that if an individual is identified with A1AT deficiency genotype, the status of liver health should also be assessed, as A1AT pathophysiology can express in liver cirrhosis.

More recent studies of respiratory disease [38] have revealed the relationship with bacterial or viral infections exacerbating the individual's genotype eliciting expression of the associated diseases. One of the most recognized inherited conditions of altered autophagy mechanisms is alpha-1 antitrypsin deficiency, with 80–100 genetic variants affecting severity of lung expression.

Low levels of circulating A1AT allow potentially harmful enzymes like neutrophil elastase to remain in the lungs unchecked. Low levels of A1AT, and the consequent proliferation of neutrophil elastase, leave lung tissue vulnerable to destruction, resulting in a decline in lung function.

51.2.7 Category of Respiratory Diseases

There are several categories of lung disease and many diseases within those categories (● Table 51.1).

51.3 Micronutrients Important for Lung Function: Food-Based and Supplemental

Some micronutrients and phytonutrients have important antioxidant and methyl-donating properties important for the lungs and therefore have great role in a nutritional approach to lung health.

51.3.1 Iron

Iron's interaction with the lungs is essential. It carries oxygen from the lungs to the peripheral parts of the body, as well as carbon dioxide back to the lungs to be exhaled. However, too little or too much iron can pose a problem for the lungs. Before iron administration, it is important to rule out hemochromatosis, or iron overload, for an individual.

Iron-deficiency anemia often presents in many chronic diseases including those of the lung, such as COPD, lung cancer, and IPF [57]. Increased mortality, decreased quality of life, increased hospital admissions, and cost of treatment have been reported for those with chronic disease and low iron [58]. Anemia of chronic disease (ACD) is usually at the root of this. ACD is often the result of inflammation. Inflammatory proteins, including IL-6, stimulate the production of hepcidin in the liver, which inhibits absorption and increases storage of iron resulting in a functional iron deficiency. Typical iron markers, such as transferrin saturation, total iron binding capacity (TIBC), and ferritin, are also affected by inflammation and are less useful markers in chronic disease. Soluble transferrin receptor (sTFR) seems to be a lesser known marker that is less affected by inflammation [59].

Because of the difficulty with iron absorption, intravenous iron is often used to replete deficiencies. As iron is a pro-oxidant, researchers studied any negative repercussions. There does not seem to be any increased oxidative stress with intravenous iron, but glutathione, the body's endogenous super antioxidant, does seem to decrease, likely in response to the pro-oxidative activity of iron. In a recent study, administration with vitamin E was seen to eliminate these negative effects [57].

Excessive iron can also be problematic for lung health for those with the genetic mutation for hemochromatosis (HFE). Disorders of iron overload are increasingly being recognized as risk factors for most of the chronic diseases like cardiovascular, Alzheimer's, and cancer [60]. High iron can catalyze the formation of highly reactive hydroxyl radicals, oxidative stress, and programmed cell death. In the instance of lung cancer and other cancers affecting the lungs, tumors sequester iron for their own growth, usually leaving the patient with iron-deficiency anemia. In fact, 90% of cancer patients undergoing chemotherapy are iron deficient. Inflammation also plays a role in iron homeostasis. The pro-inflammatory cytokines cascade down to affect the proteins that regulate

Table 51.1 Table of lung diseases

Category	Disease	Definition	Source
Obstructive lung disease	Asthma	Chronic inflammatory lung disease, triggered by either an IGE allergic reaction or nonallergic factors and results in reversible airway obstruction and inflammation of the airway	[11]
	Chronic obstructive pulmonary disease (COPD)	Disease that restricts airflow through either inflammation of the lining of the bronchial tubes or destruction of alveoli. Increased risk of emphysema if genetic variant of alpha-1 antitrypsin deficiency and smoking or exposed to high levels of air pollution	[11]
	Bronchiectasis	A disorder of the airways that leads to airway dilation and destruction, chronic sputum production, and a tendency toward recurrent infection	[39]
	Bronchiolitis	Airway injury that can be caused by infections, irritants, toxic fumes, drug exposures, pneumonitis (typically viral), organ transplants, connective tissue disorders, vasculitis, or other insults	[40]
	Dyspnea	Shortness of breath or difficulty breathing	[11]
	Emphysema	Thinning and destruction of the alveoli, resulting in decreased oxygen transfer into the bloodstream and shortness of breath. Increased risk of emphysema if genetic variant of alpha-1 antitrypsin deficiency and smoking or exposed to high levels of air pollution	[11]
	Alpha-1 antitrypsin deficiency	A deficiency of A1AT, a protein produced in the liver that protects the lungs from excessive neutrophil elastase, an autophagic enzyme. A1AT may also accumulate in liver and cause liver disease	[55]
	Obstructive sleep apnea syndrome (OSAS)	A sleep disorder characterized by repetitive upper airway obstructions despite respiratory effort causing sleep fragmentation caused by repetitive arousals	[56]
Restrictive pathophysiology-parenchymal disease	Idiopathic pulmonary fibrosis (IPF)	Disease in which tissue deep in the lungs becomes thick and stiff, or scarred, over time. The formation of scar tissue is called fibrosis	[41]
	Asbestosis	Fibrotic lung disease resulting from extensive inhalation of asbestos fibers	[42]
	Desquamative interstitial pneumonitis (DIP)	Form of idiopathic interstitial pneumonia that is more common in cigarette smokers but may be seen in nonsmokers, in patients with underlying connective tissue diseases or those exposed to inorganic dust/particles	[43]
	Sarcoidosis	Immune-mediated systemic disorder that is characterized by granuloma formation of the lung parenchyma and the skin	[44]
Restrictive pathophysiology-neuromuscular weakness	Amyotrophic lateral sclerosis (ALS)	Progressive neurological disease that affects the motor neurons of the nervous system	[11]
	Guillain-Barre syndrome	Progressive immune system attack on the peripheral nerves, usually following an infectious illness such as a respiratory infection. May eventually cause respiratory distress syndrome	[11]
Restrictive pathophysiology-chest wall/pleural disease	Kyphoscoliosis	Kyphoscoliosis: a deformity of the thoracic cage that results in restriction of the lungs and impairs pulmonary function	[45]
	Ankylosing spondylitis	Autoimmune inflammatory disorder characterized by inflammation of the axial skeleton and peripheral joints	[46]
	Chronic pleural effusions	Chronic accumulation of fluid between the two outer membranes surrounding the lungs	[11]
Pulmonary vascular disease	Pulmonary embolism	Blood clot that typically originates from thrombi in the deep venous system of the legs and travels to the lungs	[47]
	Pulmonary arterial hypertension (PAH)	Progressive disorder of primary pulmonary arterial vasculopathy characterized by a mean pulmonary arterial pressure >25 mm Hg at rest (>30 mmHg during exercise)	[48]

Table 51.1 (continued)

Category	Disease	Definition	Source
Malignancy	Adenocarcinoma	About 40% of lung cancers are adenocarcinomas. These cancers start in early versions of the cells that would normally secrete substances such as mucus	[49]
	Squamous cell (epidermoid) carcinoma	About 25–30% of all lung cancers. These start in early versions of squamous cells, which are flat cells that line the inside of the airways in the lungs. Often linked to a history of smoking and tend to be found in the central part of the lungs, near the bronchus	[50]
	Large cell (undifferentiated) carcinoma	About 10–15% of lung cancers. It can appear in any part of the lung and tends to grow and spread quickly. A subtype of large cell carcinoma, known as large cell neuroendocrine carcinoma, is a fast-growing cancer that is very similar to small-cell lung cancer	[51]
	Small-cell lung cancer (SCLC)	About 10–15% of lung cancers are SCLC. Typically start in the cells lining the bronchi and parts of the lung such as the bronchioles or alveoli	[52]
Infectious diseases	Pneumonia	Inflammation of the lungs, usually caused by bacteria, viruses, or fungi	[11]
	Bronchitis	Inflammation and eventual scarring of the lining of the bronchial tubes accompanied by restricted airflow, excessive mucus production, and persistent cough	[11]
	Tracheitis	Bacterial infection that can develop in the trachea	[53]
	Infant respiratory distress syndrome	Also known as hyaline membrane disease (HMD) or respiratory distress syndrome, this condition affects the alveolar ducts and terminal bronchioles in which the hyaline membrane is a fibrinous material composed of blood and cellular debris, caused by the absence of proper surfactant production due to an immature or poorly developed lung	[54]
	Upper respiratory infection (URI)	Acute infections involving the nose, sinuses, pharynx, larynx, trachea, and bronchi, referred to as the common cold	[11]
	Bronchopulmonary dysplasia (BPD)	Chronic lung disorder which may affect infants who have been exposed to high levels of oxygen therapy and ventilator support	[11]
Other	Cystic fibrosis	Disease characterized by abnormally thick mucus secretions from the epithelial surfaces of many organ systems, including the respiratory tract, the gastrointestinal tract, the liver, the genitourinary system, and the sweat glands	[11]
	Acute lung injury	Clinical and radiographic changes in lung function associated with critical illness (acute respiratory distress syndrome is most severe form)	[11]

iron homeostasis [61]. Iron can also impair cytokine secretion, which can leave those with an iron overload much more susceptible to infection, increasing the morbidity and mortality of infectious diseases, including those of the lung [59].

Oxidative stress may contribute to injury of lung tissue, causing further fibrosing in those lung diseases with that characteristic. Allele variants in the genes associated with iron homeostasis (C282Y, S65C, and H63D HFE) are significantly more common in those with idiopathic pulmonary fibrosis (IPF) than those without IPF (40.4% IPF patients vs 22.4% non-IPF) and are associated with higher iron-dependent oxygen radical generation [62].

Iron is implicated in lung pathology. Monitoring iron status and using supplements or diet to aid the body in increasing or decreasing the iron load are imperative for the nutritionist working with lung disease patients. Choosing a good non-constipating form of iron is important, such as iron glycinate.

51.3.2 The B Vitamins

The B vitamins are also important to monitor for lung health. Vitamin B6 and its bioactive form, P-5-P, are typically known to protect DNA from mutation or damage [63]. However, there is mixed evidence on its role for lung cancer. Some research has shown that it is helpful for lung cancer patients as it is important for apoptosis when using chemotherapy, because it sensitizes cancer cells to apoptosis [63]. However, research in 2017 showed that adult male smokers taking greater than 20 mg vitamin B6/day for long periods tended to have a greater risk for lung cancer. Many variables, including genetic variants, form of B6, and the status of other co-nutrients may be at play [64]. Other studies showed that men in the top quintile of vitamin B6 serum concentration had about one half the risk of lung cancer, and specifically, vitamin B6 and folate were inversely associated with risk of lung cancer [65].

Because of disagreement in research, particularly with smokers or former smokers, using food first for B vitamins may be a prudent way forward. Good sources of vitamin B6 are fish, chickpeas, chicken, potatoes, turkey, bananas, ground beef, and winter squash.

Pyridoxal kinase (PDXK) is the enzyme that converts pyridoxine and other vitamin B6 precursors to its bioactive form of P-5-P. Dysfunction of this enzyme is a good prognostic for lung cancer and other lung diseases. *MTHFR 1298AA* genotype is associated with a higher risk of lung cancer in women but not in men. The *MTHFR 677TT* genotype was associated with a significantly decreased risk of lung cancer in women but not in men. In contrast, the *MTHFR C677T* and *A1298C* polymorphisms interacted with smoking status in men but not in women [66]. Methylation gene testing is imperative to understand the patient's status.

Some studies suggest that a higher intake of riboflavin (vitamin B2) may protect against lung cancer in smokers [67]. Folate deficiency was also associated with asthma and attacks of shortness of breath [8].

51.3.3 Acid/Alkaline

Correcting acidosis may preserve muscle mass in diseases where wasting is an issue, such as COPD or IPF. For those receiving chemotherapy, a higher pH (more alkaline status) is helpful for muscle mass protection. High alkaline diets contain more fruits and vegetables, and those supply more magnesium, which is needed to activate vitamin D. As discussed below, vitamin D is extremely helpful for lung health.

Sleep quality involves maintaining adequate 7–8 hours with good sleep hygiene (see ▶ Chap. 34). Good REM cycling, feeling refreshed upon awakening, and other characteristics of good sleep play significant roles in maintaining healthy acid-base balance.

Dietary intake of the minerals magnesium, potassium, sodium, chloride, and calcium promotes the balance of acid-base microenvironment. After exposure and tissue retention of toxic minerals and metals, these substances can contribute to perturbations in the acid-base metabolic milieu.

Some conditions reduce oxygen intake and should be addressed. One of the most common oxygen-impairing conditions is sleep apnea, altered sleep with random halting of breathing during sleep that is often accompanied by snoring. Other limiting conditions are respiratory diseases like COPD, A1AT deficiency, asthma, cystic fibrosis, etc.

51.3.4 Vitamin A

Vitamin A is an important antioxidant and a general umbrella term for several fat-soluble retinoids, including retinol, retinal, and retinyl esters. There are also other substances that are provitamin A carotenoids or precursors to vitamin A. Two forms are found in foods, the preformed forms of retinol or retinyl esters, which are found in dairy, fish, caviar, and meats (especially liver), and the provitamin A carotenoids, including the most important and common provitamin A carotenoid, beta-carotene, as well as others including alpha-carotenes and cryptoxanthin, which are found in plant-based foods. Our bodies must convert these two forms within our cells to retinal and retinoic acid, the active forms of vitamin A in the body. New studies of the gene, β-carotene 15,15′-monooxygenase (BCMO1), which is responsible for the enzymatic conversion of β-carotene to vitamin A, are revealing that individuals with heterozygous or homozygous BCMO1 SNPs have 30–60% less efficient conversion than those with normal gene function (see ▶ Chap. 17) [68]. Other carotenoids found in food, such as lycopene, lutein, and zeaxanthin, are not converted to vitamin A but have other antioxidant benefits in the body. Most vitamin A is stored in the liver as retinyl esters, and deficiency is not visible until these stores are nearly depleted.

Vitamin A's role as an antioxidant helps the lungs in several ways, including maintaining alveolar epithelium cells and preventing development of respiratory tract infections. Most of the developed world's population does not have a risk of deficiency due to sufficient vitamin A intake. However, most people with cystic fibrosis have pancreatic insufficiency, which reduces the ability to absorb fat and therefore the fat-soluble vitamins A, D, E, and K. According to a study in 2002, between 15% and 40% of people with cystic fibrosis had a vitamin D deficiency, also a fat-soluble vitamin. With the addition of pancreatic replacement treatments, better nutrition, and vitamin A supplementation, deficiency has become rare. However, improved vitamin A status has not been thoroughly studied as of 2018, and therefore it is largely unknown if an improved vitamin A status has any effect on cystic fibrosis [69].

Vitamin A deficiency has been shown to be associated with emphysema in rats. Smoke exposure significantly decreases vitamin A concentration in lung tissue, significantly more in those with COPD [70].

Retinoic acid seems to play a beneficial role in the treatment of IPF. A review showed that in all studies, retinoic acid decreased fibrosing, the formation of collagen, and reduced the expression of alpha-smooth muscle actin (alpha-SMA), all hallmarks of IPF [71].

It is important to not take large doses of vitamin A if one is in a malnourished state as it can cause toxicity and should be monitored with blood testing of vitamin A retinol. Nourish the body with all foods and all nutrients slowly.

The non-provitamin A carotenoids have also shown some benefit. Lycopene, found in high amounts in guavas, watermelon, tomatoes, papaya, grapefruit, sweet red peppers, asparagus, purple cabbage, mangos, and carrots, slowed forced expiratory volume (FEV) decline in former smokers [70].

51.3.5 Vitamin D

Vitamin D's importance with lung health cannot be understated. Vitamin D deficiency, or even insufficiency, is linked to accelerated decline in lung function, increased inflammation, and reduced immunity in chronic lung diseases. Vitamin D has a role in the regulation of inflammation, immunity, cellular proliferation, senescence, differentiation, and apoptosis. Sufficient vitamin D levels are correlated with better asthma control, better immune response related to respiratory infections, and reduced severity of exacerbations with COPD and asthma when exposed to inflammation-causing pathogenic activity [72].

Vitamin D is obtained through sunlight on the skin (without sunscreen) and very few dietary sources. Therefore, supplementation is generally recommended. Higher vitamin D levels are shown to be protective in many lung disease states. Sufficient levels improve treatment response with medications and reduce asthma severity [68]. With infectious diseases of the lung, higher vitamin D concentrations are shown to have a protective action [6]. Vitamin D has a protective effect on lungs of smokers, and higher levels of vitamin D inhibit the pro-fibrotic phenotype of lung fibroblasts and epithelial cells. Current data suggest an inverse association between serum vitamin D and lung cancer risk, and vitamin D deficiency at 16–20 weeks' gestation is associated with impaired lung function and asthma at 6 years of age [73].

Lower levels of vitamin D are associated with an increased risk for respiratory infections, cystic fibrosis, chronic obstructive pulmonary disease, and interstitial lung disease [74].

51.3.6 Vitamin C

Vitamin C is an important antioxidant that helps decrease oxidative damage in the body, including in lung tissue. It is also essential for lipid metabolism. It is present in the airway surface liquid and creates an interface between the epithelial cells and the external environment. Vitamin C is a cofactor in collagen synthesis, which can aid in repair of bronchial and alveolar tissue when damaged. It also provides beneficial control of lipid peroxidation of cellular membranes, including those surrounding as well as those within intracellular organelles. Vitamin C has some of the best lung protective capabilities, according to current research.

Vitamin C may also diminish oxidative attack on non-lipid nuclear material and is an antioxidant component of plasma and extracellular fluids surrounding the lungs. It is an antioxidant that not only fights oxidative stress but also reduces oxidized vitamin E and glutathione, allowing them to become active as antioxidants again. Vitamin C is anti-inflammatory and is helpful in all inflammatory states of the lung, even allergies.

There are many ways in which vitamin C, along with its antioxidant partners, glutathione, vitamin E, vitamin A, and plant-based phytonutrients, affects lung health. It is well established that increased levels of vitamin C in the diet improve health outcomes for smokers and their offspring, as smoking depletes vitamin C [75, 76]. Vitamin C is also helpful in fighting infectious diseases such as respiratory infections and pneumonia, COPD regardless of smoking status, asthma, and lung cancer [77]. Specifically, in certain lung cancers, vitamin C, along with other nutrients such as lysine, proline, epigallocatechin gallate, and zinc, can inhibit the proliferation of certain carcinoma lines and induce apoptosis, as well as inhibit lung cancer metastasis [78]. Even in lung transplants, vitamin C is helpful against oxidative stress by reducing glutathione and lowering lipid peroxidation, along with vitamins A and E [79, 80].

The literature suggests these benefits can be achieved at 500–3000 mg/day. Check iron status before administering vitamin C supplementation as vitamin C doubles iron absorption from foods.

51.3.7 Vitamin E

Vitamin E's primary role is as an antioxidant, breaking free radical chain damage and preventing peroxidation of lipid molecules. This vitamin also is promising with regard to beneficial effects on lung function preservation. Oxidative stress and inflammation are key features in many lung diseases; therefore nutrients with antioxidant capacity can be useful. A few studies suggest that alpha-tocopherol found in sunflower and olive oils has a beneficial effect on FEV (forced expiratory volume), whereas gamma-tocopherol found in canola, soybean, and corn oils has a negative effect on FEV [81]. However, from these authors' perspective, this is likely due to the source and type of the oils, which can be inflammatory, rather than the form of vitamin E. For example, a recent study showed that gamma-tocopherol was protective in allergic asthma [82]. In addition, sufficient levels of vitamin E, in the alpha-tocopherol form, were found to reduce susceptibility of the elderly to acquiring pneumonia. Some of the positive effects of vitamin E are synergistic with vitamin C [83].

51.3.8 Phytonutrients

Phytonutrients have been found to have two effects with respect to lung disease: one is a symptom-improving pattern, and the other is a rate-reducing pattern [84]. Idiopathic pulmonary fibrosis (IPF) is largely characterized by reduced antioxidant and increased inflammatory action. Recent literature is showing the ability of certain flavonoids, in particular quercetin, to reduce inflammation and act as a strong antioxidant countering the pro-oxidant environment of IPF. Quercetin is recognized as the most potent ROS scavenger. Taken together with glutathione, the impact is even greater, and it seems to help improve the antioxidant and inflammatory status more for those with IPF than non-diseased controls [85].

Curcumin has been shown to slow or limit fibrosing in murine studies related to lung, liver, or kidney fibrosing [86–89]. It has also been shown to attenuate metastatic melanoma in the lungs when delivered in a nanoparticle [90]. The potential for curcumin is interesting and hopeful.

Fisetin and fenugreek have also been studied as useful phytonutrients that help combat inflammation in lungs

[91, 92]. Fisetin is found in apples, strawberries, persimmons, cucumbers, and onions, among many other fruits and vegetables. Fenugreek is a plant used frequently in South and Central Asian cooking, where both the seeds and leaves are used. There are now supplements available for both of these phytonutrients. This is a reminder to eat a primarily plant-based diet when combating inflammation and to broaden our palates to include healthy foods and ingredients from other cultures than our own.

Lastly, the powerful antioxidant cannabidiol (CBD), from the cannabis and closely related hemp plants, is a powerful shield against oxidative stress, prevalent in lung disease [93].

51.3.9 Minerals

The research is not robust regarding lung function and minerals, and most has been done with regard to cystic fibrosis where bone density is associated with general nutritional status, including minerals. There have also been many studies trying to determine a correlation between mineral status and COPD, where, again, the research shows that mineral status is not predictive but overall nutrient status may fall if not monitored. In contrast, one study in Japan showed an inverse association between dietary calcium and the risk for COPD [94]. In an NIH-AARP Diet and Health Study, magnesium, iron, selenium, zinc, and copper intakes, both dietary and supplemental, were studied with respect to lung cancer. Mineral *supplementation* did not affect lung cancer risk, yet *dietary* intake of calcium, along with vitamin D, and iron reduced the risk, and dietary intake of magnesium increased risk [95]. Boron has been shown to be protective against lung cancer, along with other nutrients, at levels of 3 mg/day [96].

There is some research showing that selenium is helpful, particularly for smokers, for improved FEV. Higher magnesium status is correlated to better FEV but is not yet seen as an association. This may be due to magnesium's role as the vitamin D activator. There have been a few studies showing increased copper levels are related to decreased FEV. Some recent research has also shown that dietary zinc and iron are associated with reduced lung cancer, but the same was not seen with calcium, copper, magnesium, or selenium [97]. Low mineral bone density is prevalent at a higher rate among cystic fibrosis patients, and therefore supplementation with vitamin D, vitamin K2, magnesium, calcium, and the trace minerals can be helpful [98].

51.3.10 Alpha Lipoic Acid

Alpha-lipoic acid (ALA) is a powerful antioxidant endogenously produced in the human body from foods such as yeast, organ meats, spinach, broccoli, and potatoes and is both water- and fat-soluble. ALA, along with N-acetyl cysteine (NAC), glycine, and vitamin C, is an important precursor to glutathione, which is a powerful endogenous antioxidant and the primary antioxidant in the lungs. ALA has been shown to be anti-inflammatory in lung tissue in those with acute lung injury, and the proposed action is via inhibition of the NF-kappaB signaling pathway [99].

ALA has also been shown to downregulate some cancer-promoting actions prevalent in lung cancer, likely by this same pathway [100]. It also may alleviate nicotine-induced lung oxidative stress [101].

51.3.11 NAC

N-acetyl cysteine (NAC), another precursor to glutathione, is a powerful antioxidant on its own as well. In relation to the lungs, NAC helps the clearance of mucus in the lungs by pulmonary cilia. This has been shown to be effective at 400–600 mg/day in divided doses [102]. There is significant research on NAC and lung health, showing improvement with nearly all lung issues, including nearly 40 studies showing improvement for bronchitis [103], infectious diseases by reducing the bacterial count [104], smokers, and people with asthma and COPD, through both its antioxidant effects and by reducing the viscosity of sputum and mucus. At an oral dose of 1800 mg/day, the mean glutathione concentration in lung tissue increased by 49% on one study [105]. There are additional studies showing improvement for those with COPD, asthma, cystic fibrosis, pulmonary fibrosis, and symptoms related to allergies or other infections. The dose that has been studied and has been shown to be most useful is 600 mg twice daily and more effective if nebulized [106, 107]. Both ALA and NAC supplementation should be accompanied by vitamin B6 and the complex of B vitamins to prevent an elevation in liver enzymes (◘ Fig. 51.3).

51.3.12 Micronutrient Testing

There are several specialty labs that conduct micronutrient analysis and functional testing, such as Genova Diagnostics and SpectraCell. These tests can be useful for evaluating levels of individual nutrients as they function in the body, rather than just in serum, which is not an accurate indicator of tissue or functional status.

51.4 Macronutrients Important for Lung Function: Food-Based and Supplemental

51.4.1 Fats and Fatty Acids

Patients suffering from COPD, interstitial lung disease, and other diseases tend to have muscle and weight loss related to respiratory acidosis, and increasing weight and muscle mass helps with quality of life. Respiratory acidosis occurs with CO_2 buildup where the lungs are no longer able to effectively exchange O_2 and CO_2. Nutritional supplementation should attempt to reduce metabolic CO_2 production. Fat

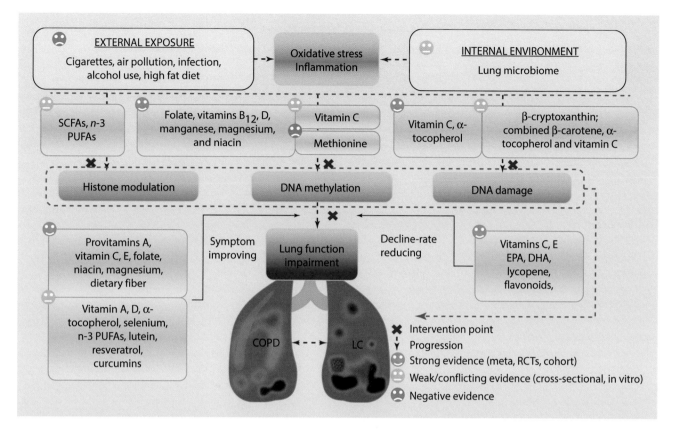

Fig. 51.3 Micronutrients and phytochemicals in the pathogenesis of chronic obstructive pulmonary disease (COPD) and lung cancer (LC). Summary of potential protective micronutrients and phytochemicals in the pathogenesis of chronic obstructive pulmonary disease (COPD) and lung cancer (LC). External and internal factors lead to oxidative stress and inflammation and thus initiate COPD and LC pathogenesis. Interventions against external exposure are not satisfying, while targeting the lung microbiome is promising. Recent studies have revealed strong evidence on protective nutrients in DNA methylation and damage, but studies on histone modulation are limited to animal or cell experiments. Epidemiological studies about micronutrients and phytochemicals in later key event lung function impairment are abundant and have identified different intervention patterns, including symptom improvement and a decline rate-reducing pattern. PUFA, polyunsaturated fatty acid; EPA, eicosapentaenoic acid; DHA, docosahexaenoic acid; SCFA, short-chain fatty acid; RCT, randomized controlled trials [108, 109]. (Reprinted from Zhai et al. [70]. with permission from Creative Commons License 4.0: ▶ https://creativecommons.org/licenses/by/4.0/)

metabolism produces less CO_2 than carbohydrate metabolism, so emphasizing a higher fat, lower carbohydrate diet can be helpful [110].

In general, a high intake of omega-6 fatty acids is associated with poorer forced expiratory volume (FEV) in patients with lung disease because of their pro-inflammatory nature. However, a complete fatty acid panel or a red blood cell membrane fatty acid test would reveal more details about the status of an individual's omega-6 pathway. Certain omega-6s and the work of their corresponding metabolizing enzymes such as elongase and delta-5 or delta-6-desaturase may allow healthful omega-6s (linoleic (LA), gamma-linolenic (GLA), lipoxins [111], prostaglandin 1 series metabolites) to flow down an anti-inflammatory pathway instead. Important cofactors for this pathway are vitamin B2, vitamin B3, vitamin B5, vitamin B6, biotin, vitamin C, zinc, and magnesium.

Lipid metabolism dysregulation is understood to be part of the pathogenesis of idiopathic pulmonary fibrosis. In IPF, free fatty acids play a role in the proliferation of fibroblasts. Certain fats, in particular palmitic acid, oleic acid, and linoleic acid, are elevated in the lungs of those with IPF, whereas stearic acid is low. Stearic acid is found in meat, poultry, fish, grain products, and milk and milk products. The palmitic, oleic, and linoleic acids enhance the TGF-ß1-induced expression of α-smooth muscle actin (SMA) and collagen type 1 in MRC-5 cells, which can lead to fibrosis. Stearic acid inhibits the levels of these fibrosing cells. Stearic acid also improves the thrombogenic and atherogenic risk factor profiles [112].

In one study on patients with COPD, omega-3 fatty acids were found to reduce inflammation in bacterial infections of the lungs without suppressing the ability to clear the bacteria. Those taking EPA, DHA, ALA, and GLA had improved exercise capacity and had lower risk of developing COPD [113].

Although results have been mixed over the years possibly due to doses used in studies, a recent 2018 prospective study showed that PUFAs (omega-3s) from fish help prevent lung cancer and can be part of treatment during lung cancer. In general, the strongest evidence for improved lung function and slowing decline is with the EPA and DHA forms of omega-3 fatty acids [114]. Because of toxicity issues in fish, increasing quality supplements vs fish intake may be more prudent.

51.4.2 Protein

Protein is essential for all lung conditions, and lack of it can result in poorer pulmonary function, decreased exercise capacity, and increased risk exacerbations. Since many lung diseases have oxidative stress as a characteristic, it can cause protein carbonylation which may negatively affect DNA expression and lipid membranes. Nutritional supplementation with added protein and healthy carbohydrates can increase body weight and muscle strength and improve quality of life. Those with COPD, interstitial lung diseases, and others that affect oxygen absorption and CO_2 exhalation have greater levels of hypoxia and sometimes respiratory acidosis, which exacerbates the loss of muscle through oxidative stress and inflammation.

Supplementation of free essential amino acids versus complete proteins has been shown to help prevent muscle wasting among COPD patients. Muscle-building exercise is often prescribed for those with COPD and interstitial lung diseases [115]. Supplemental L-carnitine at 2–6 g/day for 1–2 weeks increased the capacity of COPD patients to rehabilitate and build muscle and helped inspiratory muscle strength.

51.4.3 Carbohydrates

Carbohydrates should be monitored for sufficient but not excessive levels. More CO_2 is produced with the utilization of carbs versus fats for energy. Therefore, with gas exchange being an issue with most lung disorders, a slightly higher fat and lower carbohydrate diet may be indicated. It is worth mentioning fiber for a moment, as it is mostly delivered in carbohydrate-rich foods. There is evidence that consuming whole fruits and vegetables higher in dietary fiber is associated with reduced severity of asthma and COPD [116]. A diet that derives its carbohydrates from vegetables and fruits rather than from processed carbohydrates such as grains, breads, pasta, or added sugars will deliver fewer carbohydrate grams.

51.5 Endogenous Essential Respiratory Metabolites

51.5.1 Glutathione

Glutathione (GSH), a tripeptide composed of cysteine, glutamine, and glycine and produced from methionine, is in every cell in the body. It is the most powerful and abundant endogenous antioxidant in the airway epithelial lining and is responsible for detoxification of electrophilic compounds, the scavenging of free radicals, and modulation of cellular processes such as DNA synthesis and repair, differentiation, apoptosis, and immune function [117]. It is also a heavy metal chelator. It is more effective than some other antioxidants because it is intracellular and extracellular. In isolated type II alveolar epithelial cells, extracellular glutathione inhibits hyperoxia-induced injury, inhibits pro-inflammatory cytokine release, and promotes cell growth. It is obviously very important to maintaining lung function as this is the inflammatory process that begins lung cell or tissue damage, as mentioned above. The highest levels of glutathione concentrations in the body are in the lungs, liver, and brain. GSH depletion leads to activation of NF-kB (pro-inflammatory signaling) and increased pro-inflammatory gene transcription and cytokine release from histone deacetylase suppression in epithelial cells. Total and reduced GSH concentrations are much lower in people with ARDS, pulmonary fibrosis, and hypersensitivity pneumonitis than observed in healthy adults. Alterations in alveolar and lung GSH metabolism are widely recognized as a central feature of many inflammatory lung diseases such as idiopathic pulmonary fibrosis, acute respiratory distress syndrome, cystic fibrosis, and asthma [118].

We make glutathione in the body with cysteine and methionine, and it is difficult to take exogenously because digestion can destroy it. The precursors of cysteine (essential), glutamine, and glycine and cofactors (vitamin C, vitamin E, vitamins B1, B2, B6, and B12, folate (B9), minerals selenium, magnesium, and zinc, and alpha-lipoic acid, see below) are therefore recommended so that the body can produce it on its own. The two enzymes necessary to produce it, gamma-glutamylcysteine synthetase and glutathione synthetase, must also be functioning well. We also recycle glutathione if the precursors and cofactors are available. Cysteine is usually the most rate-limiting precursor, and many people supplement with N-acetylcysteine to provide the body with this nutrient. Although glutathione is produced in every cell of the body, the greatest production is in the liver, so focusing on liver health is important to maintain good glutathione production. Production declines with age and with lung disease, as well as other conditions.

There are very few foods containing glutathione; they are raw or very rare meat, especially liver, unpasteurized milk and other unpasteurized dairy products, and freshly picked fruits and vegetables, such as avocado and asparagus. However, as mentioned earlier, it may be destroyed during digestion. Glutathione contains sulfur molecules, which may be why foods high in sulfur help to boost its natural production in the body. These foods include:
- Cruciferous vegetables, such as broccoli, cauliflower, Brussels sprouts, and bok choy
- Allium vegetables, such as garlic and onions
- Eggs
- Nuts
- Legumes
- Lean protein, such as fish and chicken

Other foods and herbs that help to naturally boost glutathione levels include:
- Milk thistle (a liver-regenerating herb)
- Flaxseed
- Guso seaweed
- Whey

Glutathione is also negatively affected by insomnia. Getting enough rest on a regular basis can help increase levels.

Addressing a drop in glutathione for lung health involves maintaining good levels of the precursors and cofactors mentioned above. A good way to bring in the less abundant amino acid cysteine is to take N-acetylcysteine (NAC). Doses of 400–600 mg were more effective than placebo in reducing symptoms [117]. Supplemental selenium can also help with glutathione production. Glutathione supplementation has also become more effective. There are several forms, from capsules to topical liposomal, which have shown good absorption.

Inhaled GSH has good research for use in cystic fibrosis (CF), chronic otitis media with effusion (OME), HIV seropositive individuals, idiopathic pulmonary fibrosis (IPF), and chronic rhinitis. It is not recommended for asthma due to significant side effects, and additional evidence is needed to determine if use with emphysema is recommended although theoretically it should be useful. It is also not recommended to use inhaled GSH during cancer chemotherapy treatment as it may interfere with the medication's actions. The mechanism of action of inhaled glutathione is limited to the upper airways and lungs and does not seem to affect serum levels. Before considering inhaled GSH treatment, the patient should undergo urine sulfite sensitivity testing using a readily available special test strip called "EM-Quant 10013 Sulfite Test." If positive, inhaled GSH should not be used as bronchoconstriction may occur.

The recommended dose is 600–5000 mg per day, depending on response, and whether inhaled GSH is considered safe. Efficacy should be tested using a baseline pulmonary function test and a follow-up test after a prescribed time later [119] (◘ Fig. 51.4, ► Box 51.1). There are also serum tests for glutathione levels.

> **Box 51.1 Glutathione Cofactors**
> These cofactors are vitamin C; vitamin E; vitamins B1, B2, B6, and B12; folate (B9); minerals selenium, magnesium, and zinc; and alpha-lipoic acid.
> **What do the glutathione cofactors do that makes them so important?**
> - Direct cysteine toward glutathione production and increase cellular uptake of cysteine
> - Help form the glutathione molecule out of the three precursor amino acids
> - Help recycle glutathione from its oxidized GSSG form back to its reduced (active) GSH form
> - Help maintain glutathione levels and keep the GSSG-GSH ratio balanced
> - Recycle each other, improving overall antioxidant activity
> - Stimulate the activity of the whole glutathione enzymatic system

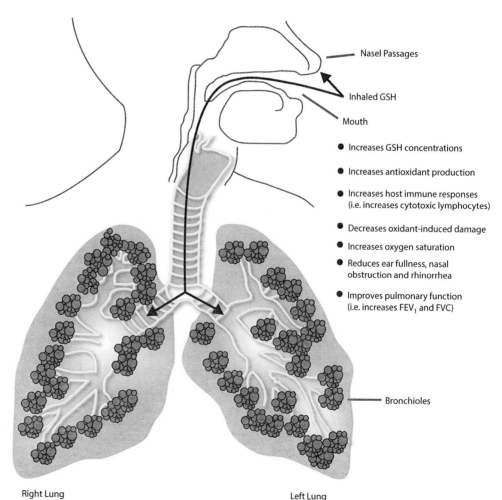

◘ Fig. 51.4 Inhaled GSH's mechanism of action. GSH, reduced glutathione; FEV1, forced expiratory volume in 1 s; FVC, forced vital capacity [119]. (Reprinted from Prousky [35]. With permission from Creative Commons License 3.0: ► https://creativecommons.org/licenses/by/3.0/)

Vitamin C – As an antioxidant, it assists glutathione in this function and has been shown scientifically to raise glutathione levels short term; it is recycled by glutathione from its oxidized state back to its active state, thus strengthening antioxidant defenses; vitamin C also recycles vitamin E and alpha-lipoic acid

Vitamin E – As an antioxidant it also assists glutathione in eliminating free radicals much like vitamin C; it is also required for the proper functioning of glutathione enzymes; it recycles vitamin C and alpha-lipoic acid

B vitamins – Vitamins B1 and B2 maintain glutathione and its enzymes in their active forms; vitamin B2 participates in the formation of a glutathione molecule; vitamin B6 influences glutathione synthesis indirectly as it is important for the proper functioning of amino acids including GSH precursors; vitamin B6 increases the amount of magnesium (a vital cofactor) that can enter cells; folate (B9) pushes cysteine toward glutathione production rather than homocysteine production; folate and vitamin B12 work together in amino acid metabolism and protein synthesis. You can read more vitamin B12 deficiency and its effect on immune health at ▶ http://www.immunehealthscience.com/vitamin-b12-deficiency.html

Selenium – Part of the enzyme glutathione peroxidase (GPx). Glutathione peroxidases, also known as selenoproteins, are a family of antioxidant enzymes that speed up the reaction between glutathione and free radicals

Magnesium – Required for the proper functioning of the enzyme gamma-glutamyl transpeptidase (GGT) involved in the synthesis of glutathione

Zinc – Zinc deficiency reduces glutathione levels, especially in red blood cells. However, zinc levels above normal have pro-oxidant properties and reduce glutathione too

Alpha-lipoic acid – An antioxidant produced by the body; it has been scientifically proven to enhance and maintain glutathione levels by stimulating enzymes involved in the synthesis of glutathione; it also helps increase the cellular uptake of cysteine, the crucial building block of glutathione; in addition, alpha-lipoic acid recycles vitamins C and E

Based on data from Ref. [124]

51.5.2 CoQ10

Co10 is a fat-soluble compound produced endogenously and also available through food and supplementation. It is required in the production of ATP, is a powerful antioxidant, and therefore is helpful against oxidative stress, an important issue in lung disease. CoQ10 achieves its strong effects through a set of different mechanisms. It influences genes through its epigenetic effect to reduce inflammation, helps with the immune system, and even reduces aging by reducing systemic oxidative stress and mitochondrial aging [120].

Lungs are the most susceptible organ to oxidant damage because they interact directly with oxygen. Therefore, it makes sense that antioxidants, and those that especially affect the lungs, are helpful in tissue and lung cell preservation [121].

CoQ10 levels are significantly lower in those with COPD and asthma with insignificant amounts of research on the levels of CoQ10 with other lung issues. It has been shown that supplementing patients with CoQ10 resulted in measurable benefits. In one study, patients with COPD using steroids to reduce inflammation were able to reduce their steroid dosage when using CoQ10 [122]. In another study, benefits were shown for COPD patients during exercise, measuring performance, tissue oxygenation, and heart rate at a low dosage of 90 mg/day [123]. The levels of CoQ10 in the blood have been shown to indicate the degree of systemic oxidative stress, which implies it could be used as a marker to assess COPD [121]. Several studies confirm the beneficial role of CoQ10 in decreasing oxidative stress, cardiovascular risk, and modulating inflammation during aging. Dosage levels of 1200 mg/day of CoQ10 have been shown to be therapeutic. However, in the reduced, more absorbable form, ubiquinol, 400 mg/day, was shown to be as effective.

51.6 Anti nutrients and Inhibitors of Lung Physiology

There is a wide range of toxins and anti-nutrients that can significantly impact the respiratory system. This can occur through acute or chronic exposure to these agents.

The Earth's air is the source of oxygen, and the lungs provide access to that oxygen to support life. The human need for oxygen is precarious because humans can only survive for about 6 minutes without the precious gas. From about 1760 to sometime between 1820 and 1840 in Europe and the United States, the ramp-up of new industrial revolution manufacturing processes opened a new era of increasing chemical and heavy metal atmospheric contamination. These pollutants can enter the body through breathing the polluted air. The more concentrated atmospheric pollutant densities cluster around areas of dense population. The dirty air provides a serious direct threat to those with respiratory diseases. An integrative and functional approach to assessing an individual with respiratory disease needs to include consideration of potential environmental contributors to the etiology of a condition. ◘ Table 51.2 lists environmental pollutants that are known to promote lung pathology.

A 2016 study published in the *Canadian Respiratory Journal* examined exhaled fractional nitric oxide (FeNO) – an indicator of inflammation in the lungs – in school children at three different schools located three different distances from a large steel mill [127]. Steel processing is known to be a source of ambient iron, nickel, lead, copper, vanadium, and zinc. The study found statistically significant differences in FeNO between the two closer schools compared to the farthest school from the mill, indicating potential increased lung inflammation caused by heavy metals and/or air pollutants [127].

51.6.1 Toxic Metals

Although acute metal toxicity is possible, chronic, low-grade exposure is more common and may contribute to respiratory complications and disease. An individual's ability to

Table 51.2 Environmental pollutants related to promoting respiratory disease

"Silent hazards" ingested or inhaled hazardous substances undetected until disease and death results and investigation is mounted to identify the cause [125]	
Natural sources – weather, geology, and pathogen exposure	
Dust	All respiratory stress
Volcanogenic air pollution	All respiratory stress
Mold/mycotoxins	Pneumoconiosis, COPD, all respiratory stress
Infection: viral, bacterial	Pneumococcal pneumonia Viral pneumonia
Anthropogenic – caused by human activity	
Smoking/tobacco	Lung cancer
Dental: mercury amalgam	Acute mercury inhalation poisoning
Dental: fluorosis/fluorine vapor	Pulmonary fluorosis [126]
Asbestos (construction materials/dust particles)	Lung cancer, mesothelioma
Nuclear radiation accidents/job exposure	Pulmonary inflammation, scarring, cancer
Coal mining	Pneumoconiosis, black lung disease
Coal combustion/mercury vapor	Acute mercury inhalation poisoning

eliminate these metals via detoxification in conjunction with gastrointestinal health and other factors can serve as important factors in whether or not these metals accumulate in the body.

51.6.1.1 Arsenic

Chronic arsenic exposure may be linked to respiratory complications [128]. Chronic arsenic ingestion via contaminated drinking water may be connected to respiratory symptoms such as chronic cough, shortness of breath, blood in sputum, and abnormal breath sounds [129]. Arsenic can also be ingested through foods such as rice and rice products, shellfish, and seaweeds, which have been shown to have high levels of inorganic arsenic (more toxic than organic arsenic found in fish) [120]. However, ingested inorganic arsenic is typically biotransformed and excreted in the urine [130]. That said, altered biotransformation has been observed depending on an individual's age, gender, nutritional status, and genetic polymorphisms responsible for the biotransformation of inorganic arsenic [130]. Chronic inhalation versus ingestion may result in irritation of the throat and respiratory tract [131]. Individuals most affected by arsenic exposure are children, nursing children, and infants of exposed pregnant mothers [132].

51.6.1.2 Cadmium

Acute inhalation of cadmium may lead to dyspnea and coughing [133]. Long-term exposure to cadmium has been reported to contribute to emphysema, dyspnea, and inflammation of the nose, pharynx, and larynx [123]. Individuals most affected by cadmium toxicity are those with occupations with cadmium exposure, such as those who work in certain types of factories, women, due to higher intestinal absorption because of low iron stores, and residents of Asia due to high intake of rice grown in contaminated soil [134].

The 2013 US National Health and Nutrition Examination Survey (NHANES) demonstrated an association between obstructive lung disease and serum lead and cadmium concentrations in the blood, where cadmium was shown to partially mediate the association between smoking and obstructive lung disease [135]. In the 2015 Korean NHANES, obstructive lung function was found to be associated with higher serum blood levels of cadmium and lead as well [136].

The specific mechanism of heavy metal burden and its effects on respiratory health must be further investigated. Although testing and treatment of heavy metal burden have its limitations, it is worth considering as heavy metal accumulation can wreak havoc on the body. An example of heavy metal testing that can be used in practice is urine provocation testing with a chelating agent, such as FDA-approved DMSA. Eliminating heavy metals from the body can be potentially harmful and requires careful monitoring and guidance by an experienced healthcare professional.

51.6.2 Air Pollutants

Air pollutants that are used as indicators of air quality are carbon monoxide, lead, nitrogen dioxide, ozone, particles, and sulfur dioxide [137]. Air pollution has been shown to have adverse effects on human health [138]. A 2017 systematic review and meta-analysis done in China showed an association between respiratory disease and ambient nitrogen dioxide, which is increased through fuel combustion, industrial production, and fuel exhaust [129]. Diesel exhaust particles in particular have been associated with an increase in cytokines such as IL-2, IL-6, and IgE in nasal mucosa [139]. Nitrogen dioxide in particular can potentially contribute to respiratory disease as it is a free radical that is highly reactive and poorly water-soluble and can be deposited in the lungs when inhaled [138]. In another study performed in England, air concentration of nitrogen dioxide was significantly associated with respiratory hospital admissions [140].

Other pollutants, such as fine particulate matter and ozone, have been shown to significantly affect respiratory function in COPD patients [141]. Increased ozone exposure has also been associated with increased airway inflammation and respiratory symptoms along with decreased respiratory function in children [142].

Optimization of nutrition and antioxidant status is essential to combating the potential health effects of air pollutants.

Several studies have shown that nutrients such as vitamin C, vitamin E, vitamin D, omega-3 PUFA, and B vitamins have demonstrated a protective effect against the damage done by particulate matter [143]. It would be reasonable to assume having adequate stores and ability to utilize these nutrients may protect against other insults to the respiratory system as discussed in this section through their anti-inflammatory properties.

51.6.3 Chemicals

Acute and chronic exposure to certain chemicals can also pose a risk to respiratory health. Obtaining a full occupational and social history when assessing individuals is important in order to identify any potential exposure to chemicals.

One of the most well-known and common toxic chemical exposures that affects respiratory health is cigarette smoke. Smoking cigarettes has been identified as a main cause of COPD [144]. Increased oxidative stress from inhaling cigarette smoke appears to activate the NF-KB inflammatory pathway, increasing the production of pro-inflammatory cytokines such as interleukin (IL)-1, IL-6, and IL-8 and tumor necrosis factor-α (TNF-α) [144]. It also appears to reduce anti-inflammatory cytokines such as IL-10 [145].

Electronic cigarettes, or E-cigs, have been increasing in popularity in recent years and are marketed as a better alternative to tobacco cigarettes. However, recent evidence suggests that the vapor and associated chemicals produced by E-cigs may be harmful to the respiratory system, although further research is needed to determine the mechanism [146, 147].

Exposure to metalworking fluid aerosols has been associated with asthma, hypersensitivity pneumonitis, impaired lung function, allergic alveolitis, and sinusitis [148]. A 2015 review also identified an association between occupational exposure to pesticides and increased risk of asthma and chronic bronchitis [149].

There are many chemicals that are toxic when inhaled. For example, inhalation of chlorine is toxic to the lungs, where low doses can cause airway injury and high doses can cause both airway and alveolar injury [150]. These injuries can manifest as dyspnea, hypoxemia, pulmonary edema, and pneumonitis [150]. High doses of carbon dioxide, such as that released from dry ice, can also induce respiratory failure.

51.7 Stress

51.7.1 Stress Overview

Stress may also play a role in respiratory health and the body's ability to combat insults imposed on the respiratory system.

From a physiological standpoint, it is worth noting that acute stress via activation of the sympathetic nervous system increases ventilation through the production of glucocorticoids [139]. Repeated acute stress may also affect growth and repair mechanisms [139].

Chronic biological stress in the form of infections can also be inflammatory and negatively affect the immune system and may affect an individual's susceptibility to respiratory complications. See the *Chronic Infections and Respiratory Health section on page #* below for further information on this association.

However, appropriate amounts of physical stress, such as in the form of exercise, can be beneficial to respiratory health. Some research has indicated a benefit of aerobic exercise to respiratory muscle strength in cystic fibrosis patients [151].

51.7.2 Biological Stress

Chronic stress can be defined as recurrent acute stress or inability to moderate acute stress responses [139]. This can be in the form of physical or emotional stress. Chronic stress and negative emotions such as depression, anxiety, and anger may be linked to endocrine and immune processes [152].

Immunoglobulin E (IgE) and cytokine production, as well as respiratory inflammation, are markers that characterize the asthma response and have been shown to respond to stress in some capacity [139]. It has been hypothesized that increased stress may increase susceptibility to air pollution given its effects on the inflammatory response [139].

Another connection between emotions and respiratory health is acknowledged in East Asian medicine, noting the association between the lungs and feelings of sadness, grief, and anxiety [153] (◘ Table 51.3).

51.8 Disease States

51.8.1 Asthma

51.8.1.1 Background

Asthma is a chronic inflammatory lung disease, triggered by either an IgE allergic reaction or nonallergic factors, and results in reversible airway obstruction and inflammation of the airway [11]. It is characterized by recurrent episodes of wheezing, breathlessness, coughing, and chest tightness [11]. Severe asthma or asthma that is chronic or poorly controlled may lead to airway and lung remodeling that involves deposition of fibrotic tissue which leads to constriction of the bronchi [18].

Although the exact mechanisms have not yet been identified, compromised nutritional status, such as deficiencies in selenium, zinc, and vitamins A, C, D, and E, has been connected to asthma [155]. The pathophysiology of asthma, nutrition considerations, genotypic characteristics, and lifestyle influences will be discussed in this section.

There are numerous potential triggers to the development and/or exacerbation of asthma which can be summarized in ► Box 51.2.

The various causes of asthma have led to the classification of several different subtypes and endotypes of asthma in hopes of choosing more targeted treatments.

Table 51.3 Anti-nutrients of the lung and potential mechanisms

Anti-nutrient	Mechanism/hypothesized mechanisms
Toxic metals (e.g., arsenic, cadmium)	Inhaled cadmium (Cd) is deposited in the alveoli where it is then absorbed into the bloodstream 　Cd is transported to erythrocytes or bound to albumin, where it is then taken up by the liver to form a complex with metallothionein (MT) 　Cd interferes with the absorption of zinc and competes for the same enzyme binding sites 　Enzymatic activity of zinc-dependent enzymes reduces 　Preferential binding of Cd to MT can cause zinc deficiency Altered biotransformation and excretion of ingested arsenic via contaminated water are linked to respiratory complications Chronic inhalation of arsenic may result in irritation of respiratory tract
Air pollutants	Diesel exhaust particles in particular have been associated with increase in cytokines such as IL-2, IL-6, and IgE in nasal mucosa [139] Nitrogen dioxide is a free radical that is highly reactive and poorly water-soluble and can be deposited in the lungs when inhaled [102] Rising pollen and mold counts [154] Increasing ozone [154]
Chemicals	Increased oxidative stress from inhaling cigarette smoke may activate the NF-KB inflammatory pathway, increasing the production of pro-inflammatory cytokines such as interleukin (IL)-1, IL-6, IL-8, tumor necrosis factor-α(TNF-α) Cigarette smoke may reduce anti-inflammatory cytokines such as IL-10 [145]
Stress	Repeated acute stress may also affect growth and repair mechanisms [139] IgE and cytokine production, as well as respiratory inflammation, are markers that characterize the asthma response and have been shown to respond to stress in some capacity [139] Increased stress may increase susceptibility to air pollution given its effects on the inflammatory response [139]

Box 51.2 Suspected Triggers and/or Risk Factors for the Development or Exacerbation of Asthma (1) [18]; (2) [25]; (3) [156]

(1)
- Cigarette smoke
- Air pollution
- Grass, mold, plants
- Pet dander
- Cockroach droppings
- Cold temperatures
- Viral infections (i.e., respiratory syncytial virus (RSV))
- Stress
- Physical exertion
- Decreased exposure to dirt
- Frequent use of antibiotics in early childhood and adolesence
- Overuse of aspirin and acetaminophen in early childhood and adolesence

(2)
- Food allergy
- Monosodium glutamate (MSG)
- Sulfites
- Reactive hypoglycemia
- Sodium chloride
- Trans fatty acids
- Obesity

(3)
- Maternal obesity during pregnancy
- Low vitamin D

51.8.1.2 Pathophysiology

The pathophysiology of asthma is complex and not fully understood, due in part to its heterogeneous nature, which necessitates its organization into individual phenotypes and endotypes. This organization is important to be able to utilize targeted treatments by identifying the root causes of the symptoms. However, more research is needed to more clearly identify the specific pathological mechanisms of each phenotype and particular treatment responses [156].

Two of the most common asthma phenotypes are allergic and nonallergic asthma [147]; allergic is characterized by increased Th2 immunity (Th2 high) and nonallergic defined by varying mechanisms depending on the trigger (Th2 low) [157] (see also ▶ Chap. 19).

Allergic asthma involves the ingestion of typically harmless environmental triggers (listed in ◻ Table 51.3) by antigen-presenting cells in the bronchi, which interact with immature helper T cells that, in turn, trigger an unwarranted allergic response [18]. This reaction occurs from repeated exposure to a trigger and is referred to as the type 1 hypersensitivity response [18]. This increased Th2 immunity upregulates eosinophilic inflammation, tissue damage, airway hyperresponsiveness, and bronchoconstriction [113]. Mast cell activation disorders, which is characterized by diseases and conditions related to mast cell mediators and the activation of mast cells, must also be considered when addressing allergic asthma [158].

In contrast, nonallergic asthma can be caused by other factors such as anxiety, exercise, stress, dry air, cold air, viruses, hyperventilation, smoke, or other irritants [11].

Table 51.4 A few of the proposed phenotypes and endotypes and their characteristics

Proposed phenotype or endotype	Clinical findings	Biomarkers	Epidemiology	Proposed mechanisms/ genetics	Medications
Phenotypes					
Allergic asthma	Allergic rhinitis Allergen-associated symptoms	Positive skin prick tests (SPT) High IgE High FeNO	Childhood onset History of eczema	Th2 dominant Th2 pathway single-nucleotide polymorphisms	Less responsive to inhaled corticosteroids IgE antagonists (omalizumab) are typically more effective [159]
Nonallergic asthma [156]	Asthma not associated with allergic	May be neutrophilic, eosinophilic, or contain only a few inflammatory cells	Additional research needed	Additional research needed	Less responsive to inhaled corticosteroids
Severe late-onset hypereosinophilic asthma (see box below on eosinophilic asthma)	Severe exacerbations, late-onset disease	Blood and sputum eosinophils	20% of severe asthmatics	Nonatopic Genetics unknown	Oral corticosteroids, IL-5 antibody therapy potential treatment
Endotypes					
Allergic bronchopulmonary mycosis (ABPM)	Severe Mucus production	Blood eosinophils High IgE High FeNO	Adult onset Long duration Poor prognosis	Colonization of airways Human leukocyte (HLA) and rare cystic fibrosis (CF) variants	Oral corticosteroids (not inhaled) and oral antifungal agents can be effective [159]
Cross-country skiing-induced asthma (CCSA) [107]	Exposure to dry, cold air provoking wheezing Airway remodeling, thickening of basement membrane	Increased lymphocytes, macrophages, neutrophils Seldom eosinophils	Induced at very cold temperatures during strenuous exercise	Nonatopic Unknown	Usually not responsive to glucocorticoids [159]

Based on data from Ref. [157]

Individuals suffering from nonallergic asthma will tend to be less responsive to Th2-targeted treatments due to a differing immune response at play [157].

Some of the additional proposed phenotypes are eosinophilic, exacerbation-prone, exercise-induced, fixed obstruction/airflow limitation, poorly steroid-responsive, and adult-onset obesity-related [159]. Several of the proposed endotypes are summarized in Table 51.4.

> The American Partnership for Eosinophilic Disorders defines eosinophilic asthma as a type of asthma characterized by especially high levels of eosinophils, more commonly developed later in adulthood, although may occur in some children [160]. Many with eosinophilic asthma do not have underlying allergies or history of allergic conditions such as eczema, food allergy, and hay fever, which are thought to be seen more in people with allergic asthma [160]. In contrast to allergic asthma, the cause of eosinophilic asthma is still unknown.

Histamine intolerance must also be considered in assessing the root cause of asthma. Ingesting histamine-rich foods and beverages such as bananas, grapes, strawberries, citrus fruits, tomatoes, nuts, chocolate, pineapples, fish, spinach, fermented foods, and beverages [161] has been shown to provoke a histamine response that may result in asthma exacerbations, among many other potential signs and symptoms [162].

Disruptions in redox, or oxidation/reduction, reactions in addition to hindered antioxidant defense have been found to be a risk factor for asthma severity and development [163]. The levels of glutathione, one of the lung's most predominant antioxidants in both reduced and unreduced forms, are thought to be important for lung homeostasis and tied to asthma [163]. More research is needed to determine the exact differences in the pathophysiologies of the various subtypes of asthma in order to develop more targeted treatments.

51.8.1.3 Key Nutrient Cofactors for Respiration and How to Modulate Toward Optimum

Minerals such as zinc, selenium, copper, and manganese may serve as cofactors to major enzymes with antioxidant activity in the lung, such as superoxide dismutase, catalase, and glutathione peroxidase [164]. Asthma has been associated with decreased activity of these enzymes [165].

Low selenium intake has been associated with multiple chronic diseases including asthma [163]. Selenium serves as a cofactor to glutathione peroxidase, an enzyme with antioxidant activity in the lung that is responsible for maintaining GSH/GSSG redox balance [163].

Imbalance between oxidants and antioxidants seems to serve an important role in asthma. Levels of nonenzymatic antioxidants glutathione, ascorbic acid, alpha-tocopherol, lycopene, and beta-carotene, in addition to antioxidant enzymes superoxide dismutase (SOD) and glutathione peroxidase, were significantly lower in asthmatic children compared to healthy controls [165]. The amino acids glycine and glutamine, which are important in glutathione synthesis, were also found to be significantly lower in children with asthma [165].

DHA has also been found to be abundant in airway mucosa, where it is decreased in individuals with asthma and cystic fibrosis [166].

Magnesium is known to elicit the relaxation of bronchial smooth muscle, decrease responsiveness to histamine, have an anti-inflammatory effect, and decrease the susceptibility of animals to developing anaphylactic reactions [25]. It is estimated that two-thirds of the population in the Western world is not consuming the recommended daily allowance of magnesium [167]. Magnesium can be used intravenously as an effective treatment of acute asthma attacks. One double-blind controlled trial that used 1.2 g of magnesium sulfate when patients did not respond to treatment with beta-agonists found decreased likelihood of hospitalization and improved lung function [168]. Magnesium sulfate as an adjunct therapy with bronchodilators and steroids has also been shown to have a benefit in children with moderate to severe asthma [168]. Although the exact mechanism is not yet known, magnesium is thought to increase glutathione concentrations in the lung [169].

More research is needed to determine additional associations between specific nutrients and asthma. However, optimization of the nutrients discussed in this section has the potential to reduce the severity and/or progression of asthma (◘ Fig. 51.5).

51.8.1.4 Key Genotypic Characteristics of Topic and Nutritional Influence

Asthma has a strong genetic component, with more than 100 genes associated with it in varying degrees across many populations [18]. More recent potential genetic associations include Filaggrin, which encodes for the epithelial barrier; ORMDL3, which encodes transmembrane protein; beta-2 adrenergic receptor gene, expressed throughout smooth muscle and epithelial cells of the lung; and interleukin-4 receptor gene, which has a variant associated with elevated IgE [171].

51.8.1.5 Key Dietary and Food Patterns to Promote Disease or Wellness in Asthma Nutrition Status

A 2011 systematic review and meta-analysis showed that deficiencies in selenium, zinc, vitamins A, C, D, and E, and low fruit and vegetable intake could be associated with the development of asthma [155]. Although this data is tenuous due to lack of randomized controlled trials, it does give some indication of the relationship between nutrition status and dietary patterns with respect to asthma development. More research needs to be done to isolate the impact of these nutrients and dietary patterns on asthma prevention and development.

Dietary Patterns

A 2015 review conducted by Berthon and Wood noted the protective effects of the Mediterranean diet for allergic respiratory diseases as evidenced by epidemiological studies. This diet emphasizes minimally processed plant foods in the form of fruit, vegetables, cereals, beans, breads, nuts, seeds, and olive oil and low to moderate intake of dairy, poultry, fish, and wine, as well as low intake of red meat [172]. This association was the strongest in children, where the Mediterranean diet had a protective effect on atopy, wheezing, and asthma symptoms [172]. However, there is less data available to support this pattern in adults.

The same review noted an association between the "Western" diet, which emphasizes refined grains, red and cured meats, French fries, sweets and desserts, and high-fat dairy products and increased risk of asthma in children [172]. A meta-analysis and systematic review done in 2014 showed a reduction of risk in childhood wheezing with high fruit and vegetable intake and also showed negative association between fruit and vegetable intake and asthma risk in adults and children [173].

Food Allergy

In contrast, food allergy has been especially linked with allergic asthma in children [161]. A study examining food allergy in asthmatic children identified higher serum levels of IgE in asthmatic children compared to healthy controls, where all asthmatic children in the study were also identified as having a positive skin prick test (SPT) to various food allergens [174].

A study done on 322 children under the age of 1 diagnosed with asthma, with or without allergic rhinitis, was

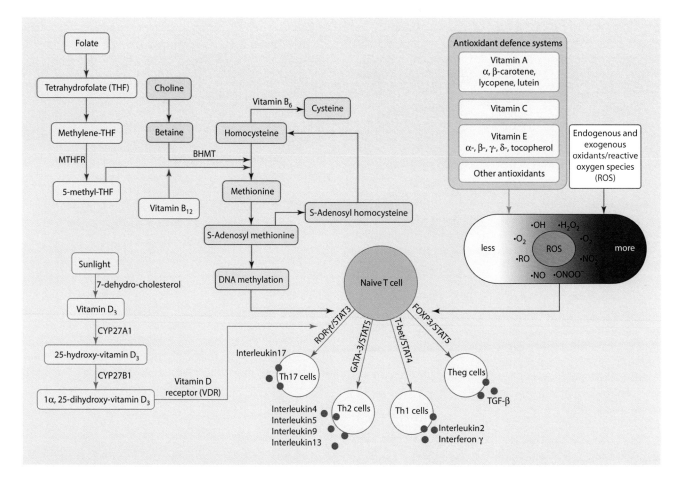

Fig. 51.5 Potential mechanisms of methyl donors and vitamins. Potential mechanisms of action of methyl donors and vitamins a, c, e, and d on Th1 and Th2 immune responses. BHMT, betaine-homocysteine methyl-transferase; RORγt, retinoic acid-related orphan receptor γt; GATA-3, GATA binding protein 3; T-bet, T-box transcription factor; FOXP3, forkhead box P3; STAT, signal transducer and activator of transcription . (Reprinted from Han et al. [170]. With permission from Elsevier)

placed on a meat-based formula of carrots, beef, broccoli, and apricots for 6 weeks. It was found that 61% had nearly complete resolution of symptoms [25]. This same study also found that the most common food triggers were milk, egg, chocolate, soy, legumes, and grains [25].

While food allergy as a cause of asthma is more common in children, hidden food allergy has been reported to be the root cause of asthma in around 40% of adults [25]. Improvement in respiratory symptoms was also seen in a small study of adults given an antigen-free elemental diet in a hospital setting [25].

Removal of food triggers has also been linked to improvement in exercise-induced asthma [25].

Identifying food allergies can be a complicated process because many of the testing methodologies such as skin prick tests (SPTs) and blood tests can yield false-positive results for up to 50–60% of cases, according to the Food Allergy Research & Education Organization [175]. A food elimination diet and/or oral food challenge can be a powerful tool in determining food allergy specific to asthma symptoms, where a dietitian or nutritionist in conjunction with physician and/or allergist can serve an important role through this process to support the individual.

51.8.1.6 Mechanisms and Relevance to Asthma as a Chronic Disease

Oxidative stress may play a key role in the development of asthma, which can also be true for the development of chronic diseases such as cardiovascular disease, diabetes, and cancer [117].

It has been shown that obesity may be a risk factor for people with and without allergy and may worsen pre-existing asthma [159]. Individuals with asthma are twice as likely to have gastroesophageal reflux disease (GERD) than people who do not have asthma, especially those resistant to treatment [159].

Celiac disease and asthma have also been linked. An Italian cohort study was done that showed a significant association between treated asthma and celiac disease, where antibiotic exposure in the first year of life was controlled for and not found to contribute to this association [176]. It has also been found that individuals with celiac disease following a gluten-free diet experienced improvement in asthma symptoms [25].

Key Toxin-Related Influences

It is well-known that toxic exposure to particulate matter, airborne pollutants, or cigarette smoke can trigger asthma symptoms [165]. More specifically, a dose-dependent

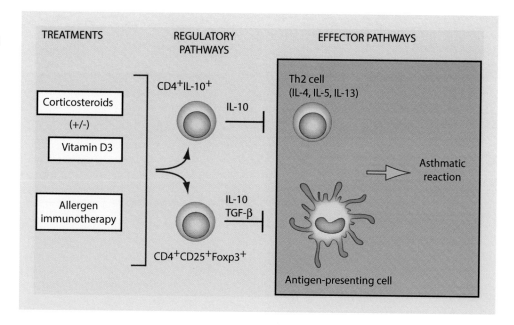

Fig. 51.6 Effect of asthma treatments on regulatory pathways. (Reprinted from Lloyd and Hawrylowicz [177]. With permission from Elsevier)

relationship between cigarette smoke exposure and rates of asthma has been shown [165]. Traffic density and asthma exacerbations have also been clearly demonstrated [165].

Certain medications may also serve as triggers to asthma. Aspirin-exacerbated respiratory disease (AERD) is considered another asthma subtype caused by nonsteroidal anti-inflammatory drugs (NSAIDS) and is characterized by asthma, chronic rhinosinusitis, and acute respiratory reactions [159]. In addition, overuse of antibiotics in childhood has been linked to asthma [18], indicating a connection between the microbiome and asthma development.

Allergic bronchopulmonary mycosis (ABPM) noted in Table 51.4 is caused by a hypersensitivity reaction to fungal colonization of the airways [159]. This is typically caused by the fungus *Aspergillus fumigatus*. Without treatment, this may lead to fixed airflow obstruction and bronchiectasis [159].

51.8.1.7 Key Lifestyle Influences

The progression of asthma is complex and multifaceted, from preconception through childhood and adulthood. Research suggests that early life events are largely predictive for regulatory mechanisms within the pulmonary immune system [177]. For example, prenatal exposure to a farming environment, one rich in microbial compounds, is thought to influence innate immune patterning in the mother which may affect the development of the neonatal immune system [177]. This influence in immune patterning can be seen through higher expression of Toll-like receptors 2 and 4 and CD14 on peripheral blood cells, which implies possible desensitization to allergens in children [178]. T regulatory cells, which serve an important role in immune regulation and are thought to play an important role in asthma by suppressing the Th2 inflammatory response to harmless air particles, have been shown to be impaired in the cord blood of neonates at hereditary risk for allergy [179].

In the 2017 study performed by Singh et al. looking at serum IgE and cutaneous sensitivity to food allergens in asthmatic children here was a negative correlation of total IgE and duration of breastfeeding, indicating a connection between breastfeeding and the immune response [174].

Additionally, reduced maternal intake of vitamins D and E and zinc during pregnancy has been associated with increased asthma symptoms in children [180, 181]. Vitamin D has been associated with the maintenance and/or development of the T regulatory cells stated earlier in mice; however more research is needed to determine a definitive association in humans [177].

A clinical trial performed on non-smoking asthmatic patients showed higher vitamin D levels were associated with greater lung function; furthermore, supplementation with vitamin D showed improved treatment response to glucocorticoids [182]. Vitamin D may also directly increase the anti-inflammatory cytokine, interleukin (IL)-10 and also enhance steroid-induced IL-10 production (see Fig. 51.6) [177]. More research is needed to determine the exact mechanism of vitamin D in asthma and respiratory disease.

51.8.1.8 Conventional Assessments

Beta-agonists, combined with corticosteroids, serve as the primary conventional therapy [183]. Typically, a short-acting beta-agonist will first be prescribed to manage symptoms as needed, where low-dose inhaled corticosteroids may also be prescribed [156]. If symptoms persist, it is recommended to evaluate problems such as adherence to use, inhaler technique, or persistent allergen exposure and comorbidities [156]. Once these are ruled out, the step-up treatment is a combination of an inhaled corticosteroid with a long-acting beta-agonist [156]. A summary of other conventional treatments and their mechanisms can be found in Table 51.5 below.

Unfortunately, conventional methods for the treatment of asthma may have harmful side effects. For example, the use of

Table 51.5 Conventional medications used in the treatment of asthma

Medication type	Medication name	Mechanism
Short-acting β2-agonists (inhaled)	Albuterol Levalbuterol Pirbuterol	Counteract the inhibitory effect on the beta-2 adrenergic receptor resulting in dilation of bronchial passages and relaxation of bronchial smooth muscle; lasts for 4–6 hours
Long-acting β2-agonists (inhaled)	Bambuterol Formoterol Salmeterol	Same as short acting, except effect lasts for about 12 hours

Corticosteroids (oral and inhaled)	Inhaled:
	  Budesonide
	  Flunisolide
	  Fluticasone propionate
	  Mometasone
	Oral:
	  Dexamethasone
	  Hydrocortisone
	  Methylprednisolone
	  Prednisolone
		Bind to glucocorticoid receptor which leads to expression of anti-inflammatory proteins, some of which block expression of pro-inflammatory modulators

Anticholinergics/muscarinic antagonists	Ipratropium bromide	Blocks acetylcholine, which leads to dilation of bronchial airways and relaxation of bronchial smooth muscle
Leukotriene antagonists	Montelukast Zafirlukast	Block binding of leukotrienes to receptors on bronchial cells
Oral methylxanthines	Theophylline Oxtriphylline	Methyl xanthine found in tea Used less commonly due to side effects Relaxes airways due to inhibition of phosphodiesterases; acts as a functional antagonist in airway smooth muscle [171]

Based on data from Ref. [18]

systemic glucocorticoids may lead to immunosuppression, cataracts, and osteoporosis, where long-acting beta-agonists have the potential of increasing asthma exacerbation risk and death [25]. Beta-agonist desensitization is thought to be one of the reasons for increasing asthma exacerbation risk and death [184].

51.8.1.9 Integrative and Functional Medical Nutrition Therapy (IFMNT) Assessment of Asthma

Related to several subtypes of asthma and their differing pathophysiologies, it is important to first determine the subtype before deciding on treatment. For example, in an individual with allergic asthma, this could be a potentially simple fix once the allergen that exacerbates symptoms is identified. A more conventional approach may involve starting the individual on an inhaled corticosteroid or an IgE antagonist (i.e., omalizumab) [159], rather than identifying the root cause of the patient's symptoms. While medications may be warranted until the trigger is identified, finding the underlying causes may not be common practice in many conventional settings.

In contrast, the IFMNT assessment takes a much deeper dive into identifying triggers and any nutrient insufficiencies, inflammation or immune dysregulation, biochemical individuality, lifestyle, energy dysfunction, toxic load, sleep, and stress issues are taken into account. With this information, the practitioner can make more targeted dietary, lifestyle, and supplement recommendations to obtain sustained resolution of symptoms by treating the root cause (Table 51.6, Fig. 51.7, ▶ Box 51.3).

51.8.1.10 IFMNT Case Study: Asthma

Client Overview

A 26-year-old female presented with a complaint of reactive airway disease, which was diagnosed as asthma and had been prescribed inhalers. She reported that she felt like she had difficulty breathing most of her life, especially when exercising. However, her condition was not severe enough to seek help until she was 25 years of age. She reported a lot of stress during this time related to applying for a postgraduate training position. She also reported 1 year prior to diagnosis developing new allergic symptoms.

Her past medical history was significant for conditions related to airways, including chronic sinus infections, strep throat, bronchitis, and recurrent pneumonia.

She could not remember the last time she felt well but assumed it was sometime as a young child. Her nutrition and health goals were to breathe better and to not have to rely on inhalers. The following data was collected on her initial visit.

Table 51.6 Summary of an integrative and functional medical nutrition therapy assessment

Nutrient insufficiencies	**Intake-digestion-utilization (IDU):** Adequacy of nutrient-dense foods to begin to assess nutritional status Organic or nonorganic to assess toxic load and nutrient intake Food preparation and processing to assess nutrient content and identify potential contaminants (e.g., plastic endocrine disruptors) Assess food sensitivities or intolerances to identify potential triggers Microbiome status: assess comprehensive digestive stool analysis for microbiology and fermented food intake; history of antibiotics or microbiota agonists (medications, toxins, stress, etc.) Toxin intake via plastics or inhalation and skin absorption which may affect immune response Assess flavonoids intake as they are antioxidant and anti-inflammatory compounds with mast cell inhibitory action; adequacy may reduce airway reactivity Consider celiac disease and gluten intake as potential inflammatory antigens **Mineral** Assess and restore zinc, selenium, magnesium, manganese, iron, and iodine status to normal reference. Caution to not supplement or intake of food sources higher than reference **Antioxidants** Assess and restore antioxidant balance; vitamins A, C, D, and E and glutathione Assess quercetin intake (leafy vegetables, broccoli, red onions, peppers, apples, grapes, black and green tea, red wine) as it may act as mast-cell stabilizing agent inhibiting release of histamine, TNF-alpha release, formation of prostaglandin D2, reducing interleukin production Consider supplementation of quercetin if quercetin intake is low [185] **Protein status** Assess and restore to support connective tissue and immune status Ensure adequate glutamine and glycine intake **Oils/lipid/fatty acids** Assess fatty acid balance as DHA important in lung tissue integrity Assess adequate serum cholesterol and fat intake to support lipid bilayer important for cellular function in lung (epithelial cells, surfactant production, etc.) **Methylation** Assess methylation status and detoxification capacity of toxins related to asthma exacerbation; important assessment biomarkers suggested: MCV/MCH, homocysteine, methylmalonic acid, RBC Folate, genomic methylation SNPs
Inflammation/immune dysregulation	Assess asthma biomarkers to help identify root cause (see Fig. 51.7, and Quote Box: What is Periostin?) Eosinophils Exhaled nitric oxide (FeNO) Periostin IgE: Total IgE, IgE specific foods, and chemicals Diamine oxidase (DAO) Assess Th2 immunity: Th1 and Th2 Cytokine Blood Test Panel [187]
Biochemical individuality	Signs and symptoms Assess when the individual experiences wheezing, breathlessness to identify the cause, and when did the initial symptom occur? Phenotype Consider the various asthma phenotypes Associated genes: Filaggrin: codes for epithelial barrier ORMDL3: encodes transmembrane protein Beta-2 adrenergic receptor gene: expressed through smooth muscle and epithelial cells of lung IL-4 receptor gene: variant associated with elevated IgE IL-10 gene promoter SNPs A1AT (alpha-1 antitrypsin) gene Environmental history Prenatal exposure to allergens with influence on immune patterning Developmental BPA exposure [188]
Lifestyle	Activity Assess whether asthma is exercise-induced Community Evaluate hobbies, occupation, household environment, and potential exposures to allergens or asthmatic triggers as listed in Fig. 51.3
Energy dysfunction	Assess overweight or obesity, including inflammatory visceral adiposity

(continued)

Table 51.6 (continued)	
Toxic load	Evaluate exposure to fungus to identify allergic bronchopulmonary mycosis Assess individual's medication history, considering short- and long-term use of conventional treatments Evaluate exposure to particulate matter, airborne pollutants, cigarette smoke, or toxic metals such as cadmium and arsenic
Sleep and stress	Assess sleep adequacy (7–9 hours with 5-hour REM sleep) and quality (good sleep hygiene with little light/sound/EMF disturbance) to support detoxification of toxins that may worsen respiratory status and aid in repair of damaged lung tissue

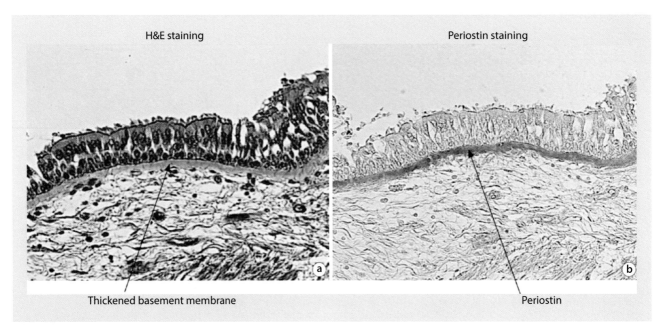

Fig. 51.7 Functions of periostin in inflammation. Functions of periostin in inflammation. Expression of periostin in asthmatic patients. The histochemical localization of periostin in asthmatic patients is depicted. The left and right panels show bronchial tissues from an asthmatic patient in H&E staining **a** and immunostaining **b** of periostin. Periostin in the right panel is stained brown and is localized in the thickened basement membrane in asthmatic patients. (Reprinted from Izuhara et al. [186]. With permission from The Korean Academy of Asthma, Allergy and Clinical Immunology)

Box 51.3 What Is the Inflammatory Influence of Periostin on Tissues and Organs?
Periostin plays a role in the pathogenesis of allergic diseases, including asthma, as it is associated as a downstream molecule of the cytokine, IL-13. Periostin is used as a biomarker for type 2 immunity and can be used to determine the potential effectiveness of medications used to treat asthma, such as anti-IgE antibodies and anti-IL-13 antibodies. Asthmatic patients with high serum periostin tend to be aspirin intolerant, eosinophilic, late asthma onset, and have a high nitric oxide fraction. High periostin can also indicate a reduced response to inhaled corticosteroids [186, 224].

Symptoms/Complaints at First Visit

Extreme exhaustion, depression, ADD, anxiety (accompanied by panic attacks), constipation, pain in legs, neuropathy in feet (numbness and tingling), rapid heartbeat, and a very severe rash on feet known as chilblains.

Past Medical History

- UTIs – recurrent as a child.
- Poor immune function (frequent infections).
- Antibiotic use (very frequent from childhood into adulthood).
- Sinus infections, strep throat, and bronchitis – she had recurrent sinus infections and strep throat about once a year every year and often this would lead to bronchitis, she could not remember if she had these issues before middle school.
- Depression, anxiety, ADD.
- Acne.
- Peptic ulcers.
- Yeast infections – multiple throughout college.
- Eczema.
- Two recent episodes of pneumonia the last episode resulted in her asthma diagnosis.
- Asthma.

- Several medical consults (cardio, rheumatologist, etc.). Most recent, rheumatologist, with suspicion of lupus. Diagnosis was not lupus; no recommendations for further care made.
- Past Surgical History
- Wisdom teeth removed
- Deviated septum surgery

Pharmaceutical Use
- Albuterol inhaler for acute asthma attacks taken prn
- Inhaled corticosteroid inhaler for long-term use; taken once daily
- Yaz birth control for acne and to prevent pregnancy; taken once daily

Timeline
Birth
- Vaginal birth, breastfed for 6 months

Childhood
- Hx of cyst on face, for which she had to undergo several treatments to remove (unsure of what the treatment was or what type of cyst)
- Parents divorced at 2 years old
- Frequent UTIs

Adolescence
- 8th grade: severe case of strep throat, undiagnosed for several weeks, led to being immobile for almost 2 weeks
- 8th–12th grade: was often sick (strep, sinus infections, bronchitis); described it as being constantly sick from fall through winter every year; also developed eating disorder during this time; had severe menstrual cramps (induced vomiting) accompanied by acne, which led to being put on birth control at age 17 as a precursor to Accutane (never prescribed); chronic constipation starting during this time

University
- Freshman – sophomore year: eating disorder was most severe during this time.
- First semester of freshman year: developed digestion issues, after eating certain foods (especially Mexican or salsas), stomach would become distended, experienced pain, and often would result in vomiting. Pain so severe during finals week she was admitted to ER with no diagnosis. CT scan revealed possible peptic ulcers.
- Junior–senior year: depression, anxiety, and inability to focus were most severe during this time which resulted in missing a lot of class and struggling as a student; suffered multiple panic attacks; gained a lot of weight (from 120 to 180 lb); end of senior year became engaged to be married – moved to Dallas, TX.

Young Adulthood
- Lived in Dallas for 6 months, continued to experience depression and anxiety and weight gain, and moved back to home state
- Initially started running (~2 miles a day) and experiencing inability to breathe, diagnosed with pneumonia, prescribed inhaler to help with running; other symptoms: eczema around the eyes and neck (after running outside), pain in calves, numbness and poor circulation in feet (pulse not detected by several health professionals), and development of chilblain rash (very painful, itching, lasts about 3–4 weeks from development to resolution); increased running – ran a half-marathon. Visited PCP and several specialists for help with chilblain rash with no resolution or diagnosis; lost a lot of weight (from 170 to 140 lb).
- Ongoing increased depression, anxiety, and inability to focus; PCP Rx Cymbalta (depression and anxiety); Cymbalta discontinued after ~2 months (did not tolerate side effects), continued psychological therapy for several months; chilblain rash continued. Stopped running long distances. Gained weight back (from 140 to 170 lb); subsequently saw blog for integrative RD and followed suggestion to eliminate gluten and focusing on whole foods diet.
- Chilblains and eczema began to resolve while following integrative RD recommendations of gluten-free diet with some improvements. However, difficulty breathing got worse, and diagnosis of asthma was made with fast-acting inhaler used for exercise; as time progressed, breathing continued to worsen, led to daily inhaler use. Weight at this time is still at around 170 lb.

Social
- Stress: very high (graduate student and completing internship)
- Sleep: variable, sometimes <6 hours and sometimes >9 hours
- Physical activity: some yoga and HIIT running
- Exposures: pet dander and Teflon cookware

Nutrition (In General)
- History of chronic dieting and disordered eating
- No known food allergies but suspected gluten sensitivity
- Consult with IFMNT nutritionist, diet whole foods, completely gluten-free diet. Minimal dairy whole fat (mostly from Greek yogurt, butter, and cheese)

Fats and Oils Questionnaire: Initial Survey
- Omega 9-MUFAs: High
- Omega-6: High (mainly from commercial meats and eggs)
- Omega-3: Low
- Saturated: High (mainly from dairy sources)
- Damaged: High (trans, hydrogenated, deep fried, charred meats)

Nutritional and Lifestyle Data Collected

Clinical

Anthropometrics
- Height: 5′7″
- Weight: 170 lbs
- BMI: 26.6
- Waist circumference: 34″
- Waist/height ratio: 0.51
- Blood pressure: 122/74
- Body composition: not completed at initial intake

Nutrition Physical Exam
- Acne present along the jawline and on the neck
- Appeared overweight and bloated
- Appeared tired and had trouble recalling certain details
- Tongue, nails, etc. not checked at initial intake visit

Genotypic Risks
- IL-13 c112T (+/−)
- HLA-DQA2 (+/+), HLA-DQA1 (+/−)
- VDR (+/+)
- Several (+/+) for phase II related genes
- MTHFR C677T (+/−) and several other (+/+) (+/−) for genes related to methylation and methionine/homocysteine pathways

Biochemical

Blood Lab Results
- MCV 101 (H) (79.3–98.6)/MCH: 32.0 (H) [27.0–31.0]
- Na: 137 (lower end of normal) [133–143]
- K: 3.8 (lower end of normal) [3.5–5.0]
- Bicarbonate: 23 (L) [24–32]
- Glucose: 85 (higher end of normal) [60–97]
- Creatinine: 0.54 (L) [0.70–1.40]
- ALT: 29 (L) [30–65]
- HDL: 74 WNL

Structural
- Hx eczema – breakdown of the skin
- Gut barrier likely compromised evidenced by Hx of ulcers, poor immune function, yeast infections, cyst, constipation, etc.
- Nail structure very good
- Hair: reported frequent shedding, but hair structure appeared healthy

Signs/Symptoms/Medical Symptom Questionnaires (MSQ)
- MSQ total score of 87 [REFERENCE > 50 significant imbalances]
- High MSQ categories: lungs, skin, and weight

Nutrition Assessment: NIBLETS

Intake
- High dairy diet (consumed dairy products at most meals and snacks), consumed three smaller meals with three snacks in between
- Meals and snacks balanced with protein, fat, and carbs, with carbs coming from fruits and vegetables and fat mainly from full fat cheese, Greek yogurt, and butter
- Mostly nonorganic produce and commercially raised meats

Digestion, Assimilation, and Elimination
- Hx of peptic ulcers and chronic constipation (BM ~1–2 times a month)
- BMs currently at about 2 × per week on encounter

Utilization, Cellular, and Molecular (MAPDOM)
- Hx of likely gluten sensitivity.
- Presented symptoms of possible dairy sensitivity (bloating, acne, asthma).
- Evidence for compromised intestinal barrier.
- Minerals: infrequent BMs could indicate low fiber or low mineral status (Mg); when BMs do occur, they are hard and dry (low Mg); severe menstrual cramps (low Mg); labs showed low K and Na, on Yaz birth control (low zinc and low B vitamins).
- Antioxidants: consumed adequate fruits and vegetables each day.
- Protein: has some evidence of poor/slowed wound healing as evidenced by sore on leg that has not completely healed after a year; cuts that take months to heal.
- D and fat-soluble A, E, and K vitamins – Hx of poor immune function (low D), VDR +/+ (low D and possibly A).
- Oils/fatty acids: high omega-6/omega-3 ratio, higher intake of damaged fats, very low intake of omega-3.
- Methylation: symptoms of depression, anxiety, ADD combined with MTHFR C677T snp and on Yaz (low B6 and folate).

Inflammation
- Eicosanoid fatty acids status – suspect issues with PGE1 series pathway to control inflammation due to following signs and symptoms: allergies, autoimmune condition (asthma), peptic ulcers, eczema, and severe menstrual cramps
- Immune function – suspect gut dysbiosis due to following S&S: poor immune function, yeast infections, Hx frequent antibiotic use, cyst, and constipation

Body Composition
- Genetic makeup that indicated prone to gluten and dairy sensitivity, low vitamin D status, and impairment in methylation

Lifestyle
- Low exercise, high stress, and food sensitivities

Energy
- Fatigue

Toxin Load
- Known genetic SNPs in phase 2 detoxification.

- Dairy could be considered a toxin contributing to overall toxic load on body.

Sleep and Stress
- Either sleeps too much or not enough
- Reported high stress due to completing dietetic internship and master's degree at the same time, also Hx of long-term stress with parents' divorce at young age and stress related to relationships

Nutrition Diagnosis
- Suspect dairy sensitivity related to impaired GI structure and genetic susceptibility evidenced by autoimmune condition (asthma), Hx of strep, and HLA-DQ2 & 1 SNPS
- Altered GI function related to gut microbe dysbiosis evidenced by poor immune function, constipation, Hx of yeast infections, cyst, and frequent antibiotic use

Plan
1. Trial of dairy-free diet for 3 months. Then add back sources of dairy (separately) to see if symptoms return. Log any symptoms experienced during the reintroduction of dairy in a food journal.
2. Supplements:
   - Vit D + K2 (5000 IU + 90 mcg) daily
   - Natural Calm Mg (daily, morning and night)
   - Fish oil (2000 mg daily)
   - Broad spectrum probiotic + fermented foods
   - BioActive B complex (includes 50 mg P5P B6 and 800 mcg 5 THF)
   - 260 mg GLA evening primrose oil and zinc
3. Aim to eat three larger meals a day, allowing space in between of ~ 5 hours; increase omega-3 intake by adding in small fatty fish, such as sardines or anchovies, once per week and taking fish oil; decrease omega-6 intake, switch from conventionally raised meats to organic, pasture-raised; and replace fat in diet from dairy with coconut sources, more nuts, and avocados.

Follow-Up
- Patient presented ~6 months after the initial visit (September 2015). Her breathing had improved immensely. She was able to stop taking her Albuterol inhaler before exercise, recently stopped daily inhaler.
- After dairy-free diet for 3 months, reintroduced dairy (cheese, butter, yogurt). Asthmatic symptoms returned about 2–3 days after the addition of each. Noticed the more dairy consumed, the worse her symptoms became.
- At time of appointment, diet whole foods, gluten-free, and dairy-free. Weight loss 10 lb within the first month of going dairy-free, continued to lose some weight. When reintroduced dairy symptoms of bloating and increase in weight, which resolved returning to dairy-free diet.
- BMs are regular now at ~ 2 × *daily*.

Anthropometrics
- Height: 5′7″
- Weight: 155 lb
- BMI: 24.2
- Waist circumference: 32 inches
- Waist/height ratio: 0.477

Medical Symptom Questionnaire (MSQ)
- Highest categories: lungs and eyes (had recent accidental exposure to a little bit of dairy)

Fats and Oils
- Omega-9: HIGH, healthy
- Omega-6: Lowered and increased GLA
- Omega-3: Normal, healthy
- Saturated: Normal, balanced healthy sources
- Damaged: Reduced 45%

Outcome
This patient case followed some common patterns in the development of chronic disease and the comorbidities that are common, especially autoimmune conditions like asthma. The first is the genetic susceptibility of the individual; several SNPs are prone to dairy sensitivity. Second, significant evidence for gut dysbiosis, promoted compromised gut barrier, can contribute to the development of dairy sensitivity. Third is the exposure to dairy protein antigen. Diet history evidenced trigger for asthmatic condition.

Additionally, inflammation, immune dysfunction, and methylation issues present. Signs and symptoms significant for decrease in PGE1 series anti-inflammatory pathways. Low dietary omega-3s potential contributor to asthma. Immune dysfunction evidenced by extensive history of infection-antibiotic use. Genomic SNP MTHFR C667T gene, which indicated a greater need for folate. The use of Yaz birth control and symptoms of depression, anxiety, and ADD known further to deplete B6 and folate.

The diet and supplements recommended targeted control of inflammation, restore gut ecology, promote proper methylation, and replete nutrient insufficiencies. Results from 6-month follow-up showed successful outcome in helping improve breathing and wean her off of inhalers.

This case is an example of the IFMNT approach able to address the complexity of the whole patient story and bring the metabolic priorities into a manageable intervention program for the individual.

51.8.2 Chronic Infection and Respiratory Health

51.8.2.1 Upper Respiratory Tract Infection
One study found that the composition of the nasopharyngeal microbiota in children was linked to the frequency of upper respiratory tract infections and acute sinusitis [189]. A study that intranasally inoculated mice with *Lactobacillus*

fermentum reduced the amount of *S. pneumoniae* in the respiratory tract and increased the number of macrophages in the lung and lymphocytes in the trachea [189]. These findings may indicate a benefit of manipulating the upper respiratory tract microbiota with orally or nasally administered probiotics in the prevention and/or treatment of upper respiratory tract infections.

51.8.2.2 Fungal, Viral, and Bacterial Infections

Allergic bronchopulmonary mycosis (ABPM) is caused by a hypersensitivity reaction to fungal colonization of the airways. This is typically caused by the fungus *Aspergillus fumigatus*. Without treatment this may lead to fixed airflow obstruction and bronchiectasis [159].

Guillain-Barre syndrome (GBS) is a rare neurological disorder in which the body's immune system attacks the peripheral nervous system, known as the network of nerves located outside of the brain and spinal cord [190]. It is often preceded by a bacterial or viral infection. There are several potential mechanisms in which these infections trigger GBS. If an individual contracts a *Campylobacter jejuni* bacterial infection, antibodies made to fight this infection can attack axons in motor nerves, which can potentially cause paralysis and respiratory failure [190]. *Campylobacter* can be ingested via contaminated food or other exposures [190].

Pérez-Guzmán 2005 states that hypocholesterolemia is common among tuberculosis patients and suggests that cholesterol should be used as a complementary measure in antitubercular treatment [8].

51.8.3 Alpha-1 Antitrypsin Deficiency (A1AT Deficiency)

Alpha-1 antitrypsin (A1AT) deficiency is an underrecognized disease in the United States, with around documented 100,000 people suffering from it, according to the Alpha-1 Foundation. This deficiency is inherited through autosomal codominant transmission, meaning affected individuals have inherited an abnormal AAT gene from each parent [191]. Individuals with this deficient allele present with AAT levels at less than 35% to low-end normal levels [191].

However, it is also possible for individuals with a variant of this allele to be asymptomatic given different environmental conditions or lifestyle factors, such as refraining from smoking to reduce lung disease development risk [191] (▶ Box 51.4).

51.8.3.1 Lungs

A1AT deficiency most often manifests in the lungs as chronic obstructive pulmonary disease (COPD) (i.e., emphysema or bronchiectasis or "genetic COPD"). A1AT deficiency is often undiagnosed because people with genetic COPD experience the same symptoms as people with COPD, such as [191]:
- Shortness of breath
- Wheezing

Box 51.4 Genomic Variants of Alpha-1 Antitrypsin Deficiency
Normal genotype M M
- Most common abnormal genes are called S and Z
- Abnormal variant combinations:
  - ZZ (highest risk)
  - SZ (lower risk increasing if smoker, inhalant pollutants)
  - MZ (lower risk of carrying an A1AT gene variant; considered "carriers")
- Alpha-1 is the most commonly known genetic risk factor for emphysema
- Up to 3% of all people diagnosed with COPD may have undetected Alpha-1
- Alpha-1 can also lead to liver disease. The most serious liver diseases are cirrhosis and liver cancer
- The World Health Organization (WHO), American Thoracic Society (ATS), and the European Respiratory Society (ERS) recommend that everyone with COPD be tested for Alpha-1

Alpha-1 is a progressive disease that benefits from early detection. It can cause serious lung diseases, such as COPD and emphysema when undiagnosed. In some cases, Alpha-1 can also cause liver disease [225]
Symptoms related to the lung [225]:
- Shortness of breath
- Wheezing
- Chronic bronchitis, which is cough and sputum (phlegm) production that lasts for a long time
- Recurring chest colds
- Less exercise tolerance
- Year-round allergies
- Bronchiectasis

- Recurring chest colds
- Low exercise tolerance
- Year-round allergies

The only way you will know for sure if you have genetic COPD due to alpha-1 is to get tested.

51.8.3.2 Liver

A1AT deficiency can manifest in the liver as cirrhosis.

Symptoms Related to the Liver
- Unexplained liver disease or elevated liver enzymes
- Eyes and skin turning yellow (jaundice)
- Swelling of the abdomen (ascites) or legs
- Vomiting blood (from enlarged veins in the esophagus or stomach)

51.8.3.3 Skin

A1AT expresses sometimes in the skin as panniculitis [191]. Panniculitis typically appears as raised red spots on the skin, which may break down and give off an oily discharge. While panniculitis spots (called nodules) may appear anywhere on the body, the most common places are the thighs, buttocks, and areas subject to injury or pressure.

51.8.3.4 A1AT Biochemical Mechanisms

The alpha-1 antitrypsin (A1AT) protein protects the body, especially fragile lung tissues, from the damaging effects of a powerful enzyme called neutrophil elastase that is released from white blood cells. In A1AT deficiency, a genetic mutation reduces levels of the protective protein in the bloodstream. A1AT deficiency can lead to chronic obstructive pulmonary disease (COPD), specifically emphysema, and liver disease. Smoking, which can inhibit what little A1AT protein an affected person does have, increases the risk of lung disease.

Alpha-1 antitrypsin deficiency is completely determined by mutations in a single gene. The severity of symptoms is mostly a function of which mutations a person has and how many copies. However, smoking can greatly increase the risk of lung disease due to AAT mutations.

23andMe reports data only for the PI∗M, PI∗S, and PI∗Z versions of the gene that encodes AAT. Keep in mind that it is possible to have another mutation that causes this condition that is not included in this report [192]. A1AT deficiency is a genetic disorder that reduces circulating levels of a protein that protects the lungs by trapping A1AT in the liver, where the protein is produced, and prevents A1AT from entering circulation. A1AT deficiency can lead to chronic obstructive pulmonary disease (COPD), specifically emphysema, and liver disease.

When a disease-causing mutation is fairly common, as the PI∗S and PI∗Z mutations are in Europeans, it suggests that the mutation actually conferred an evolutionary advantage at one time. Some researchers have suggested that several thousand years ago when the PI∗Z and PI∗S mutations first arose, these versions of the gene for A1AT gave people a survival advantage by creating an environment in their lungs that helped fight off infections. The scientists theorize that the antimicrobial benefits of the AAT mutations outweighed the cost of an increased risk of COPD and liver disease in the era before antibiotics were available [193].

51.8.3.5 Liver Cirrhosis

In contrast to lung disease, manifestation of liver disease related to A1AT can be referred to as a "toxic gain of function," due to accumulation of mutant A1AT protein rather than protease deficiency within the liver [144].

51.8.4 Pulmonary Fibrosis

51.8.4.1 Overview

When taken together, fibrotic lung diseases are the leading cause of mortality worldwide. Under the umbrella of interstitial lung disease (ILD), pulmonary fibrosis (PF) is the most common. Any ILD that involves scarring of the lungs falls in the pulmonary fibrosis category. Pulmonary fibrosis is the scarring of lungs, which destroys tissue over time, making it impossible to transfer oxygen from inhaled air into the bloodstream. There are more than 200 different diseases under the pulmonary fibrosis umbrella. Because PF is often misdiagnosed or goes undiagnosed, there is not an accurate count of those with these diseases. However, it is estimated that as many as 1 in 200 adults over 60, or 200,000 people in the United States, are affected [184]. There are more than 50,000 deaths from IPF every year in the United States. More people die each year from idiopathic pulmonary fibrosis than from breast cancer [194].

There are other forms of interstitial lung disease including the newly identified pleuroparenchymal fibroelastosis, cryptogenic organizing pneumonia (COP), desquamative interstitial pneumonitis, nonspecific interstitial pneumonitis, hypersensitivity pneumonitis, acute interstitial pneumonitis, interstitial pneumonia, sarcoidosis, and asbestosis [195].

Symptoms include cough and dyspnea, restrictive pulmonary function tests with impaired gas exchange, and progressive lung scarring. The disease progresses with an initiation of inflammation. Fibrosing starts with the action of transforming growth factor-β (TGF-β)-dependent differentiation of fibroblasts to myofibroblasts, which then express α-SMA (smooth muscle actin) [196].

After the TGF-β-dependent differentiation of fibroblasts to myofibroblasts, which express α-SMA, there is sustained, excessive deposition of collagen by the myofibroblasts in the lung interstitium leading to the progressive lung damage in patients with PF [185]. Research published in 2011 supported the idea that dysfunctional type II AECs (alveolar epithelial cells) facilitate lung fibrosis through increased susceptibility to injury, leading to excessive and dysregulated remodeling [197]. The disease seems to progress in steps, and inflammation is not typically present continuously, except during certain periodic episodes of deterioration (◘ Fig. 51.8).

There are five main categories of PF causes: drug-induced, radiation-induced, environmental, autoimmune, and occupational. Of these five, four have identifiable causes. Some of the autoimmune diseases that can lead to PF are rheumatoid arthritis, scleroderma, Sjogren's syndrome, polymyositis, dermatomyositis, and antisynthetase syndrome.

Idiopathic pulmonary fibrosis (ILP) is defined as PF with an unknown cause, including a genetic cause for some families [see ◘ Fig. 51.9]. The symptoms of ILP are a dry, hacking cough, shortness of breath, fatigue, chest discomfort, loss of appetite, and unexplained weight loss, all caused by the fibrosing of the lungs.

Diagnosis can be difficult, and PF is often misdiagnosed as COPD or other more common lung diseases. In addition, in the recent past, path to a true diagnosis was invasive. Since damage to the lungs, even through a diagnostic biopsy, can trigger further lung damage or a period of fibrosis, many physicians or patients are cautious with a biopsy approach to diagnosis. Since the current treatments are limited, one must evaluate whether defining the exact form of PF is necessary for treatment and follow-up. Difficulty breathing, crackling sounds while breathing, and low oxygen levels are the first indicators. Clubbed fingernails may also be a symptom.

High-resolution CT scans are performed, which can show scarring. The pulmonologist will ask many questions

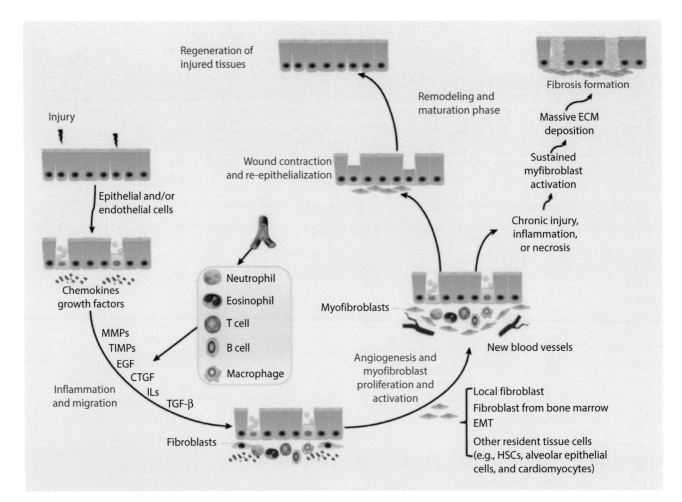

■ **Fig. 51.8** The cellular and molecular mechanisms of fibrosis in multiple organs. The cellular and molecular mechanisms of fibrosis in multiple organs. Once an injury occurs in an organ, epithelial and/or endothelial cells are impaired, which results in the release of chemokines and growth factors, including IL-13 and TGF-b1. Macrophages and monocytes are recruited and activated, both of which further release cytokines and chemokines and further induce fibroblast activation. Activated fibroblasts transform into a-SMA-expressing myofibroblasts and migrate into the wound along the fibrin lattice. ECM is excessively accumulated, and some parenchymal cells (hepatic stellate cells in the liver, tubular epithelial cells in the kidney, alveolar epithelial cells in the lung, or cardiomyocytes in the heart) are further differentiated into myofibroblasts or fibroblasts by the stimulation of cytokines and chemokines, especially for TGF-b1. After the inflammatory phase, two events occur. One is the regeneration of injured tissues followed by wound contraction and reepithelialization. In contrast, once chronic injury, inflammation, and necrosis occur, myofibroblasts are perpetually activated, and excessive ECM is deposited, finally resulting in fibrosis formation. CTGF, connective tissue growth factor; ECM, extracellular matrix; EGF, epidermal growth factor; EMT, epithelial-mesenchymal transition; HSC, hepatic stellate cell; IL, interleukin; MMP, matrix metalloproteinase; TGF, transforming growth factor; TIMP, tissue inhibitors of metalloproteinase. (Reprinted from Chen et al. [198]. With permission from Elsevier)

and order more blood tests to try to distinguish between the 200 forms of PF. The future is pointing to molecular endotyping as a more accurate way to diagnose. Molecular endotyping includes genetic, metabolic, transcriptional, and environmental factors to help determine the pathophysiology [199].

Genetic research has been progressing for a couple decades with illuminating results. There are more than a dozen genetic variants that have been associated with this family of diseases. Researchers now believe at least 20% of idiopathic pulmonary fibrosis (IPF) patients with multiple family members suffering from IPF have some common familial genetic variants, which may allow researchers to eventually drop the term idiopathic and further define various forms or categories, with differing progression or outcome. The name given to this version of interstitial pneumonias is familial interstitial pneumonia (FIP) [200] [see ■ Fig. 51.10].

Currently two categories of genetic focus have been defined: those genes related to telomere biology (shorter telomeres) and those related to surfactant protein processing. The genes related to shorter telomeres are *TERT*, *TERC*, *hTR*, *DKC1*, and *RTEL1*. More mutations have been found in the *TERT* gene, which encodes the protein component of telomerase, than any other gene. Further research may allow targeted therapies to affect the genetic expression associated with the development of IPF [201, 202]. A common variant within the promoter of the *MUC5B* gene is the most replicated single-nucleotide polymorphism related to familial and sporadic forms of IPF as well as early radiographic findings of IPF [203] (■ Figs. 51.9 and 51.10).

Fig. 51.9 Summary of common genetic variants linked to IPF. (Reproduced with permission of the © ERS 2019: European Respiratory Journal 33(1):99–106. ▶ https://doi.org/10.1183/09031936.00091607. Published 31 December 2008)

Locus	Gene	SNP	IPF risk	IPF survival
2q14	IL1RN	rs408392	Yes	
		rs419598	Yes	
		rs2637988	Yes	
3q26	hTR	rs6793295	Yes	
4q13	IL8	rs4073	Yes	
		rs2227307	Yes	
4q22	FAM13A	rs2609255	Yes	
4q35	TLR3	rs3775291		Harmful
5p15	TERT	rs2736100	Yes	
6p21	CDKN1A	rs2395655	Yes	Harmful
6p21	HLA-DRB1		Yes	
6p24	DSP	rs2076295	Yes	
7q22	Intergenic	rs47274443	Yes	
10q24	OBFC1	rs11191865	Yes	
11p15	MUC5B	rs35705950	Yes	Protective
	MUC2	rs7934606	Yes	
	TOLLIP	rs111521887	Yes	
	TOLLIP	rs5743894	Yes	
	TOLLIP	rs2743890	Yes	Protective
13q34	ATP11A	rs1278769	Yes	
14q21	MDGA2	rs7144383	Yes	
15q14-15	Intergenic	rs2034650	Yes	
17q13	TP53	rs12951053	No	Harmful
	TP53	rs12602273	No	Harmful
17q21	MAPT	rs19819997	Yes	
17q21	SPPL2C	rs17690703	Yes	
19q13	DPP9	rs12610495	Yes	
19q13	TGFB1	rs1800470	No	Harmful

Gene	Reported % of FIP
TERT	8-15%
RTEL1	5%
hTR	<1%
DKC1	<1%
TINF2	<1%
SFTPC	2-25%
SFTPA2	<1%
ABCA3	<1%
Unknown	75-85%

Fig. 51.10 Rare genetic variants linked to familial interstitial pneumonia (FIP). (Reproduced with permission of the © ERS 2019: European Respiratory Journal. 33(1):99–106. ▶ https://doi.org/10.1183/09031936.00091607. Published 31 December 2008)

51.8.4.2 Conventional Treatment

Conventional treatment is typically palliative. The American Thoracic Society recognizes that supplemental oxygen and transplantation are the only suggested treatments for IPF. Supplemental oxygen is prescribed, and the need for oxygen increases over the progression of the disease. Keeping the oxygen saturation level over 90% (normal is in the upper 90s) is ideal and is how healthcare providers determine the level of supplemental oxygen to be used. Cardiovascular exercise, in this case called pulmonary rehabilitation, is recommended to maintain as much use of the lungs as possible.

Infrequently, nutrition and counseling are recommended and are placed into the category of symptom management. Nutrition can have a significant role in the management of this disease, but little implementation exists in some of the proposed protocols.

There are currently two medications available in the United States with minor impact on the disease progression: nintedanib (commonly called Ofev) and pirfenidone (Esbriet). Histopathological quantification showed similar amounts of dense collagen fibrosis, fibroblast foci, and alveolar macrophages in untreated or pirfenidone- or nintedanib-treated IPF patients [204]. Both have significant side effects, including fatigue and GI issues, and patients may have to evaluate their quality of life versus length of life. Other anti-inflammatories or immune-suppressing medications used are corticosteroids, mycophenolate mofetil/mycophenolic acid (CellCept®), or azathioprine (Imuran®). Immune-suppressing drugs may be harmful for those with short telomeres, and researchers are exploring this potentially contradictory recommendation [205].

Lung transplantation is a final effort. About 1,000 lung transplants in the United States go to those with PF, which is half of all transplants. With the prevalence of this disease closer to 200,000, this is a small fraction of those with the disease. Some of those with the transplant go on to live productive lives, while others develop PF again, in the transplanted lungs. Overall, there is a shorter life expectancy in those with PF, because of telomere shortening. Bone marrow or immune response abnormalities have been found in some IPF cases before and after lung transplantation, which increases the associated morbidity.

51.8.4.3 Integrative and Functional Nutrition Medical Therapy

As stated above, inflammation occurs at the beginning and throughout the progression of all fibrosing diseases, including those of the lungs. Therefore, reducing inflammation is one wise strategy to slow fibrosing. There are several nutrients that can help slow or reverse the inflammation involved in the fibrosing process. The following two-part diagram shows where in the fibrosing pathogenesis each phytonutrient acts [198] (◘ Fig. 51.11).

A few of those compounds are discussed in more detail here.

Curcumin, the active constituent in the common spice turmeric, has been shown to reduce fibrotic activity in several studies. In mice, curcumin inhibited collagen secretion of IPF fibroblasts. It affects the signaling of TGF-β, in a dose-specific manner, resulting in reduced expression of α-SMA, which is responsible for inappropriate fibrosing. This was shown in vitro and in vivo in mice, with intraperitoneal, but not oral, administration. At the time of the study, oral ingestion of curcumin was not adequately absorbed into plasma, and there was greater than ten times plasma concentration of curcumin following an intraperitoneal injection [88]. However, some new oral products on the market are showing greater absorption.

The results of this study suggest more research into curcumin, including improved delivery into patients. For example, some delivery options may include nebulized curcumin directly into the lungs, binding it to highly absorbable agents for oral use or liposome-encapsulated curcumin suitable for intravenous use (already shown to be effective in an animal model).

According to manufacturers of curcumin products, some are more readily absorbed than others. One study on fibrosing suggested that a dose of around 2200 mg curcumin split into three doses taken with meals including pepper (bioperine) achieved doses that were sufficient to exert the desired therapeutic effect.

Research into using quercetin also has some promising results in slowing the progression of IPF. Quercetin reversed lung fibrosing in mice and reversed the disease progression normally caused by typical pulmonary senescence markers [206].

It is worth mentioning that N-acetylcysteine (NAC), a long-used therapeutic agent for breaking down mucus in the lungs, has not been found to be effective in those with IPF. In fact, due to its acidic nature, it has even been shown to be harmful when used in the inhaled form [207].

Several of the drugs being developed have a natural product as a model or foundation. Until a drug or gene therapy is developed that stops or reverses this disease, it may make sense for the patient to focus on anti-inflammation and reducing myofibroblast activation, the extracellular matrix (ECM) accumulation, and the epithelial-mesenchymal transition (EMT) process. The phytochemicals listed in ◘ Fig. 51.11 would be good ones to investigate.

51.8.4.4 Genetics and Telomeres

With the recent identification of genes associated with ILD, a call for gene-related therapies both related to telomere lengthening and connective tissue disease has been initiated, and this type of therapy, as with any disease, could be personalized [208]. One recent study looked at various biomarker values as a more precise way of diagnosing. The biomarker molecules were classified according to their involvement into alveolar epithelial cell injury, fibroproliferation, and matrix remodeling as well as immune regulation. Furthermore, genetic variants of TOLLIP, MUC-5B, and other genes associated with a differential response to treatment and with the development and/or the prognosis of IPF were identified. Research into personalized medicine for treatment is starting [209].

Although controversial, because of the lack of research on interpretation of the results, telomere length testing is available directly to consumers and through healthcare

---

◘ **Fig. 51.11** Antifibrosis therapy. The molecular mechanisms and therapeutic targets of natural products against fibrosis. **a** TGF-b exerts a profibrotic effect through Smad-dependent [Target (1)] and Smad-independent pathways [Target (2)]. In the Smad-dependent pathway, TGF-b1 directly phosphorylates and activates the downstream mediator Smad2 and Smad3 through TGF-b receptor I, and then Smad2 and Smad3 bind Smad4, which forms a complex that moves into the nucleus and initiates gene transcription. Smad7, transcribed by Smad3, is a negative regulator of TGF-b/Smad signaling, and the imbalance between Smad3 and Smad7 contributes to fibrosis. PI3K, ERK, and p38 MAPK are downstream mediators of the Smad-independent TGF-b pathway. PPARg [Target (3)] could inhibit TGF-b to reduce fibrosis, while CTGF [Target (4)], a matricellular protein, contributes to wound healing and virtually all fibrotic pathology. Additionally, Gas6 contributes to fibrosis through the TAM receptor, which further activates the PI3K/Akt pathway. Similarly, LPA triggers fibrosis through the LPA1 receptor [Target (5)] that stimulates b-catenin to induce fibrogenesis. The activation of the hedgehog pathway [Target (6)] induces the transcriptional activity of Gli to express target genes, which have an important role in interstitial fibrosis, undergoing myofibroblast transformation and proliferation. IL pathway [Target (7)] stimulates NF-kB [Target (8)] to activate TGF-b to induce fibrogenesis, while Nrf2 [Target (9)] antagonizes NF-kB activity to protect against fibrosis. **b** The chemical structures of isolated compounds and their therapeutic targets are presented. CTGF, connective tissue growth factor; IL, interleukin; LRP, low-density lipoprotein receptor-related protein; RI, transforming growth factor-b receptor I; RII, transforming growth factor-b receptor II; SARA, Smad anchor for receptor activation; STAT, signal transducer and activator of transcription; TCF, T-cell factor; TGF, transforming growth factor. (Reprinted from Chen et al. [198]. With permission from Elsevier)

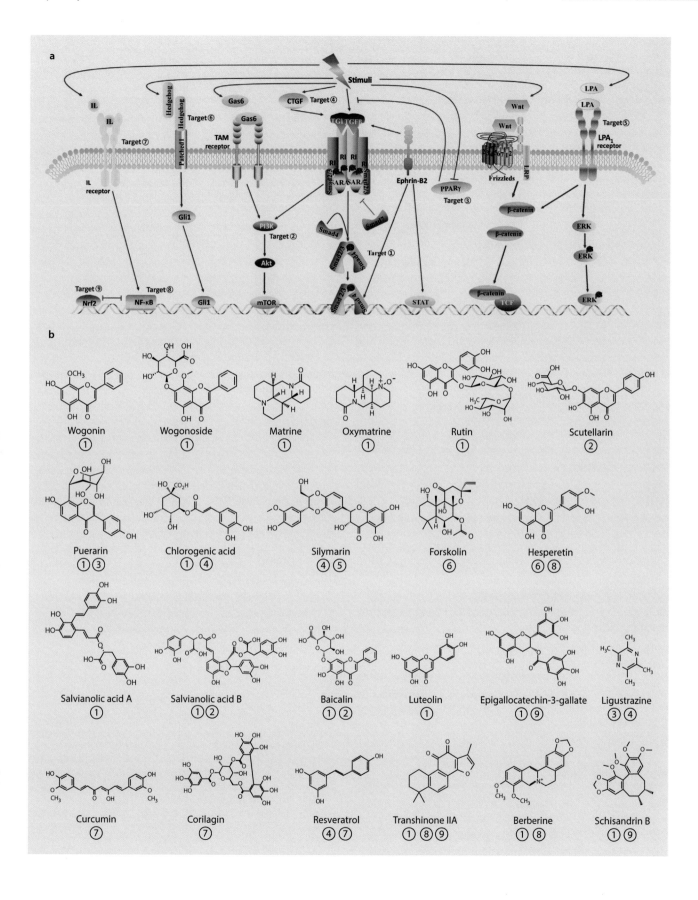

practitioners. There are a few different methods: quantitative polymerase chain reaction, or qPCR, which has a 20% variability rate, and flow cytometry and fluorescent in situ hybridization, or Flow-FISH, which has a 5% variability rate. Most research labs use Flow-FISH for research.

Telomere length is a hot topic in research, the antiaging industry, and with popular health blogs. Shorter-than-average telomeres have also been linked to heart disease and heart failure [163, 210, 211], cancer [212], diabetes [213], and osteoporosis [214]. Research has shown ways to slow telomere shortening. Some include reducing stress, meditation, practicing loving kindness (a technique encouraging compassion) [215], reducing exposure to air pollution and toxins [216], cardiovascular exercise [217], and a healthy fat and high vegetable diet [218, 219]. One study showed that 45 minutes of cardiovascular exercise three times per week resulted in longer telomeres representing 10 years of biological age, similar to those of marathon runners, compared to those who didn't exercise much or at all [220]. Intermittent fasting, which reduces oxidative stress and keeps weight in check, has exploded in the scientific literature as a way to increase longevity and slow telomere shortening [221, 222].

Nicotinamide adenine dinucleotide (NAD+) supplements may also help maintain telomere length by activating sirtuins, the antiaging enzymes; PARPs, which are involved in DNA repair; and CD38, which plays a role in insulin production. Another supplement, cycloastragenol, derived from the herb astragalus, has also been shown to activate telomerase in mice. An ingredient called TA-65 has been derived and is used in supplements [223].

Overall, a healthy lifestyle and diet seem to delay the shortening of telomeres. With relation to PF, the gene mutations involved in telomere shortening may or may not be influenced by the above interventions. More research is needed for this.

Pulmonary fibrosis is a devastating disease with no management or a known cure. The integrative and functional medicine nutritionist can help her/his patient by managing weight, encouraging a healthy diet full of anti-inflammatory foods and encouraging a healthy lifestyle with exercise and stress reduction. There is some promising research into natural supplement use to target the different areas of progression within the disease process and some ongoing drug and gene therapy development to follow.

51.9 Conclusion

The prevalence of lung disease in the United States and worldwide is growing and will continue to grow rapidly with the deterioration of Earth's atmosphere, which is caused by pollutants such as industrial and construction toxins and volcanic and wildfire particulates. Poor maternal, childhood, and adult nutrition from micronutrient-poor diets resulting in nutrient insufficiencies, not necessarily nutrient *deficiencies*, is also contributing to increased lung disease diagnoses or poorer results during treatment [226, 227]. Lifestyle choices and habits also play a role in the development of many of the lung diseases in today's world, such as smoking or vaping, which uses chemicals that are poorly studied to date. Other lung diseases have their roots in genetics.

Some key processes drive many lung diseases, with the inflammatory process being the most important, according to current literature. Nutrition can be of great help with inflammation, using a diet rich in whole foods providing micronutrients and phytonutrients. Understanding genetics is also key to unraveling the causes and potential future treatments for many lung diseases. Those patients with both genetic and environmental determinants, such as in those who smoke and have genes associated with COPD, are at the greatest risk [228].

Despite the prevalence of lung disease, there is a general lack of nutrition knowledge among practitioners, including familiarity with the research about the use of nutrition for prevention, slowing disease progression, or as a treatment of lung disease. Historically, nutrition has been used in a supportive role, primarily monitoring macronutrients to prevent weight loss, muscle atrophy, and acid/alkaline balance. Although this is extremely important, more attention needs to be directed toward emphasizing micronutrients and phytonutrients. Research is strong regarding the benefits of vitamins, minerals, and pre- and probiotics, and indeed, some integrative and functional practitioners are using vitamin and mineral nutritional therapy in oral, intramuscular, and intravenous applications, when allowed, in practice. A newer area of research is around nutraceuticals, including targeted vitamins, minerals, and plant-derived constituents concentrated to therapeutic doses. Some exciting research around the use of curcumin and quercetin, for example, has been shown to dampen inflammation to the point of disrupting the disease process (see above). The expanding knowledge of the microbiome is identifying the importance of the lung and airway microbiome in respiratory health.

More research, and indeed more education for nutritionists around the existing research, is needed to fully understand the best opportunities for the use of nutrition in the treatment or prevention of lung disease.

References

1. Budinger G, Kohanski R, Gan W, et al. The intersection of aging biology and the pathobiology of lung diseases: a Joint NHLBI/NIA Workshop. J Gerontol A Biol Sci Med Sci. 2017;72(11):1492–500.
2. Chronic Respiratory Disease | Gateway to Health Communication | CDC [Internet]. Cdc.gov. 2019 [cited 10 June 2019]. Available from: https://www.cdc.gov/healthcommunication/toolstemplates/entertainmented/tips/ChronicRespiratoryDisease.html.
3. Asthma | Healthy Schools | CDC [Internet]. Cdc.gov. 2019 [cited 10 June 2019]. Available from: https://www.cdc.gov/healthyschools/asthma/.
4. World Health Organization Global Status Report on Noncommunicable Diseases. 2014. Accessed at: https://apps.who.int/iris/bitstream/handle/10665/148114/9789241564854_eng.pdf.
5. Nichols H, Timothy J, Legg C. The top 10 leading causes of death in the United States [Internet]. Medical News Today. 2019 [cited 10 June 2019]. Available from: https://www.medicalnewstoday.com/articles/282929.php.

6. Berry D, Hesketh K, Power C, Hyppönen E. Vitamin D status has a linear association with seasonal infections and lung function in British adults. Br J Nutr. 2011;106(9):1433–40.
7. Gibson J, Loddenkemper R, Lundback B, Sibille Y. Respiratory health and disease in Europe: the new European Lung White Book. Eur Respir J. 2013;42:559–63. https://doi.org/10.1183/09031936.00105513.
8. Pérez-Ramírez C, Cañadas-Garre M, Alnatsha A, Villar E, Delgado J, Calleja-Hernández M, Faus-Dáder M. Impact of DNA repair, folate and glutathione gene polymorphisms on risk of non-small cell lung cancer. Pathol Res Pract. 2017. pii: S0344-0338(17)30895-6. https://doi.org/10.1016/j.prp.2017.11.015.
9. Berthon B, Wood L. Nutrition and respiratory health—Feature review. Nutrients. 2015;7(3):1618–43.
10. Lopez AD, Mathers CD. Measuring the global burden of disease and epidemiological transitions: 2002–2030. Ann Trop Med Parasitol. 2006;100(5–6):481–99.
11. Nelms M, Sucher K, Lacey K, Roth S. Nutrition therapy & pathophysiology. 2013;53:650. https://doi.org/10.1017/CBO9781107415324.004
12. Respiratory system [Internet]. En.wikipedia.org. 2019 [cited 10 June 2019]. Available from: https://en.wikipedia.org/wiki/Respiratory_system#/media/File:Respiratory_system_complete_en.svg.
13. Brocki B, Westerdahl E, Langer D, Souza DSR, Andreasen J. Decrease in pulmonary function and oxygenation after lung resection. ERJ Open Res. 2018;4(1):00055-2017.
14. Kopec S, Irwin R. Sequelae and complications of pneumonectomy. UpToDate.com:2018. Available online at: https://www.uptodate.com/contents/sequelae-and-complications-of-pneumonectomy.
15. Ganti S, Milton R, Anikin V, Thorpe A. Surgery for tumor recurrence in a pneumonectomy space. Eur J Cardiothorac Surg. 2006;30(4):683–5. https://doi.org/10.1016/j.ejcts.2006.06.021.
16. Levitzky M. Chapter 10. Function and structure of the respiratory system. In: Levitzky M, editor. Pulmonary physiology, 8e. New York: McGraw-Hill; 2013.
17. Levitzky M. Chapter 1. Function and structure of the respiratory system. In: Levitzky M, editor. Pulmonary physiology, 8e. New York: McGraw-Hill; 2013.
18. Huang H. Chapter 17. The respiratory system. In: Janson L, Tischler M, editors. The big picture: medical biochemistry. New York: McGraw-Hill; 2012.
19. Levitzky M. Chapter 2. Function and structure of the respiratory system. In: Levitzky MG, editor. Pulmonary physiology, 8e. New York: McGraw-Hill; 2013.
20. Halling K, Ramstedt B, Nyström J, Slotte J, Nyholm T. Cholesterol interactions with fluid-phase phospholipids: effect on the lateral organization of the bilayer. Biophys J. 2008;95(8):3861–71. https://doi.org/10.1529/biophysj.108.133744. Epub 2008 Jul 18.
21. Pathak P, London E. The effect of membrane lipid composition on the formation of lipid ultrananodomains. Biophys J. 2015;109: 1630–8.
22. McMullen T, McElhaney R. Physical studies of cholesterol-phospholipid interactions. Curr Opin Colloid Interface Sci. 1996;1(1):83–90.
23. Bradley K, McConnell S, Crystal R. Lung collagen composition and synthesis. Characterization and changes with age. J Biol Chem. 1974;249(9):2674–83.
24. McKleroy W, Lee T, Atabai K. Always cleave up your mess: targeting collagen degradation to treat tissue fibrosis. Am J Physiol Lung Cell Mol Physiol. 2013;304(11):L709–21. https://doi.org/10.1152/ajplung.00418.2012. Epub 2013 Apr 5.
25. Gaby AR. Nutritional medicine. 2nd ed. Concord: Fritz Perlberg Publishing; 2017. pp 84, 104, 156, 274, 225, 258, 646.
26. Ingram G, Munro N. The use (or otherwise) of pulse in general practice. Br J Gen Pract. 2005;55(516):501–2.
27. Nagpal R, Yamashiro Y, Izumi Y. The two-way association of periodontal infection with systemic disorders: an overview. Mediators Inflamm. 2015;2015:793898. https://doi.org/10.1155/2015/793898. Epub 2015 Aug 3.
28. van der Maarel-Wierink C, Vanobbergen J, Bronkhorst E, Schols J, de Baat C. Oral health care and aspiration pneumonia in frail older people: a systematic literature review. Gerodontology. 2013;30(1):3–9. https://doi.org/10.1111/j.1741-2358.2012.00637.x. Epub 2012 Mar 6.
29. Linden GJ, Lyons A, Scannapieco FA. Periodontal systemic associations: review of the evidence. J Clin Periodontol. 2013;40(Suppl 14):S8–19. https://doi.org/10.1111/jcpe.12064.
30. Grippaudo C, Paolantonio EG, Antonini G, Saulle R, La Torre G, Deli R. Association between oral habits, mouth breathing and malocclusion. Acta Otorhinolaryngol Ital. 2016;36(5):386–94. https://doi.org/10.14639/0392-100X-770.
31. Cui L, Morris A, Huang L, Beck J, Twigg L 3rd, von Mutius E, Ghedin E. The microbiome and the lung. Ann Am Thorac Soc. 2014;11 Suppl 4:S227–32. https://doi.org/10.1513/AnnalsATS.201402-052PL.
32. Martineau AR, Jolliffe DA, Hooper RL, Greenberg L, Aloia JF, Bergman P, Dubnov-Raz G, Esposito S, Ganmaa D, Ginde AA, Goodall EC. Vitamin D supplementation to prevent acute respiratory tract infections: systematic review and meta-analysis of individual participant data. BMJ. 2017;356:i6583.
33. Lin J, Zhang Y, He C, Dai J. Probiotics supplementation in children with asthma: a systematic review and meta-analysis. Paediatr Child Health. 2018;54(9):953–61. https://doi.org/10.1111/jpc.14126. Epub 2018 Jul 27.
34. Mennini M, Dahdah L, Artesani MC, Fiocchi A, Martelli A. Probiotics in asthma and allergy prevention. Front Pediatr. 2017;5:165. Published 2017 Jul 31. https://doi.org/10.3389/fped.2017.00165.
35. He Y, Wen Q, Yao F, Xu D, Huang Y, Wang J. Gut-lung axis: the microbial contributions and clinical implications. Crit Rev Microbiol. 2017;43(1):81–95. https://doi.org/10.1080/1040841X.2016.1176988. Epub 2016 Oct 26.
36. Marsland B, Trompette A, Gollwitzer E. The gut-lung axis in respiratory disease. Ann Am Thorac Soc. 2015;12(Suppl 2):S150–6. https://doi.org/10.1513/AnnalsATS.201503-133AW.
37. Liu JN, Suh DH, Trinh H, Chwae YJ, Park HS, Yoo S. The role of autophagy in allergic inflammation: a new target for severe asthma. Exp Mol Med. 2016;48:e243. https://doi.org/10.1038/emm.2016.38.
38. Pragman AA, Kim HB, Reilly CS, Wendt C, Isaacson RE. The lung microbiome in moderate and severe chronic obstructive pulmonary disease. PLoS One. 2012;7(10):e47305. https://doi.org/10.1371/journal.pone.0047305. https://www.alpha1.org/what-is-alpha1. Accessed 30 Nov 2018.
39. Learn About Bronchiectasis [Internet]. American Lung Association. 2019 [cited 10 June 2019]. Available from: https://www.lung.org/lung-health-and-diseases/lung-disease-lookup/bronchiectasis/learn-about-bronchiectasis.html.
40. Learn About Bronchiolitis [Internet]. American Lung Association. 2019 [cited 10 June 2019]. Available from: https://www.lung.org/lung-health-and-diseases/lung-disease-lookup/bronchiolitis/learn-about-bronchiolitis.html.
41. Idiopathic Pulmonary Fibrosis | National Heart, Lung, and Blood Institute (NHLBI) [Internet]. Nhlbi.nih.gov. 2019 [cited 10 June 2019]. Available from: https://www.nhlbi.nih.gov/health/health-topics/topics/idiopathic-pulmonary-fibrosis.
42. Asbestosis [Internet]. American Lung Association. 2019 [cited 10 June 2019]. Available from: https://www.lung.org/lung-health-and-diseases/lung-disease-lookup/asbestosis/.
43. Tazelaar HD, Wright JL, Churg A. Desquamative interstitial pneumonia. Histopathology. 2011;58(4):509–16. https://doi.org/10.1111/j.1365-2559.2010.03649.x. Epub 2010 Sep 21.
44. Sarcoidosis [Internet]. American Lung Association. 2019 [cited 10 June 2019]. Available from: https://www.lung.org/lung-health-and-diseases/lung-disease-lookup/sarcoidosis/.
45. Chokroverty S. Kyphoscoliosis [Internet]. Kyphoscoliosis – an overview. ScienceDirect Topics [cited 15 May 2019] Available from: https://www.sciencedirect.com/topics/veterinary-science-and-veterinary-medicine/kyphoscoliosis.

46. Kim Y, et al. Diagnosis and treatment of inflammatory joint disease. Hip Pelvis. 2017;29(4):211–22. https://doi.org/10.5371/hp.2017.29.4.211.
47. Pulmonary Embolism (PE): Practice Essentials, Background, Anatomy [Internet]. Emedicine.medscape.com. 2019 [cited 10 June 2019]. Available from: https://emedicine.medscape.com/article/300901-overview.
48. American Lung Association (ALA). Pulmonary Arterial Hypertension (PAH). Retrieved from: https://www.lung.org/lung-health-and-diseases/lung-disease-lookup/pulmonary-arterial-hypertension/.
49. What Is Non-Small Cell Lung Cancer? [Internet]. Cancer.org. 2019 [cited 10 June 2019]. Available from: https://www.cancer.org/cancer/non-small-cell-lung-cancer/about/what-is-non-small-cell-lung-cancer.html.
50. Lung Carcinoid Tumor [Internet]. Cancer.org. 2019 [cited 10 June 2019]. Available from: https://www.cancer.org/cancer/lung-carcinoid-tumor.html.
51. What Is Small Cell Lung Cancer? [Internet]. Cancer.org. 2019 [cited 10 June 2019]. Available from: https://www.cancer.org/cancer/small-cell-lung-cancer/about/what-is-small-cell-lung-cancer.html.
52. Non-Small Cell Lung Cancer [Internet]. Cancer.org. 2019 [cited 10 June 2019]. Available from: https://www.cancer.org/cancer/non-small-cell-lung-cancer.html.
53. Burton LV, et al. Bacterial tracheitis. [Updated 17 Dec 2018]. In: StatPearls [Internet]. Treasure Island (FL): StatPearls Publishing; 2019. Available from: https://www.ncbi.nlm.nih.gov/books/NBK470240/.
54. Smith J. [Internet]. Samples.jbpub.com. 2019 [cited 10 June 2019]. Available from: http://samples.jbpub.com/9780763755461/58370_CH02_Harrison.pdf.
55. Alpha-1 Foundation | Alpha1.org [Internet]. Alpha1.org. 2019 [cited 10 June 2019]. Available from: https://www.alpha1.org.
56. Obstructive sleep apnea (OSA) in adults. [serial online]. n.d. Available from: Dynamed Plus, Ipswich, MA. Accessed 30 Aug 2018.
57. Vasquez A, Logomarsino J. Anemia in chronic obstructive pulmonary disease and the potential role of iron deficiency. COPD. 2016;13(1):100–9.
58. Ghio A. Disruption of iron homeostasis and lung disease. Biochim Biophys Acta. 2009;1790(7):731–9.
59. Gozzelino R, Arosio P. Iron homeostasis in health and disease. Int J Mol Sci. 2016;17 https://doi.org/10.3390/ijms17010130.
60. Steinbicker A, Muckenthaler M. Out of balance – systemic iron homeostasis in iron-related disorders. Nutrients. 2013;5:3034–61. https://doi.org/10.3390/nu5083034.
61. Gozzelino R, Arosio P. Iron homeostasis in health and disease. Inflammation and Neurodegeneration Laboratory, Chronic Diseases Research Center (CEDOC), Nova Medical School (NMS)/Faculdade de Ciências Médicas, University of Lisbon, Lisbon 1150-082, Portugal. Department of Molecular and Translational Medicine (DMMT), University of Brescia, Brescia 25123, Italy. Received: 13 December 2015 / Revised: 4 January 2016 / Accepted: 12 January 2016 / Published: 20 January 2016.
62. Sangiuolo F, Puxeddu E, et al. HFE gene variants and iron-induced oxygen radical generation in idiopathic pulmonary fibrosis. Eur Respir J. 2015;45(2):483–90. https://doi.org/10.1183/09031936.00104814.
63. Galluzzi L, Vitale I, et al. Prognostic Impact of Vitamin B6 in Lung Cancer. Cell Rep. 2012;2(2):257–69.
64. Galluzzi L, Vacchelli E, et al. Effects of vitamin B6 metabolism on oncogenesis, tumor progression and therapeutic responses. Oncogene. 2013;32:4995–5004.
65. Hartman T, Woodson K, et al. Association of the B-vitamins pyridoxal 5′-phosphate (B6), B12, and folate with lung cancer risk in older men. Am J Epidemiol. 2001;153(7):688–94. https://doi.org/10.1093/aje/153.7.688.
66. Shi Q, Zhang Z, et al. Sex differences in risk of lung cancer associated with methylene-tetrahydrofolate reductase polymorphisms. Cancer Epidemiol Biomarkers Prev. 2005;14(6):1477–84. https://doi.org/10.1158/1055-9965.EPI-04-0905.
67. Bassett JK, Hodge AM, English DR, Baglietto L, Hopper JL, Giles GG, Severi G. Dietary intake of B vitamins and methionine and risk of lung cancer. Eur J Clin Nutr. 2012;66(2):182.
68. Leung W, Hessel S, Méplan C, Flint J, Oberhauser V, Tournaire J, Hesketh E, von Lintig J, Lietz G. Two common single nucleotide polymorphisms in the gene encoding β-carotene 15,15′-monoxygenase alter β-carotene metabolism in female volunteers. FASEB J. 2008;23:1041. https://doi.org/10.1096/fj.08-121962.
69. Borowitz D, Baker RD, Stallings V. Consensus report on nutrition for pediatric patients with cystic fibrosis. J Pediatr Gastroenterol Nutr. 2002;35:246–59.
70. Zhai T, Shizhen L, et al. Potential micronutrients and phytochemicals against the pathogenesis of chronic obstructive pulmonary disease and lung cancer. Nutrients. 2018;10(7):813.
71. Zhou T, Drummen G, et al. The controversial role of retinoic acid in fibrotic diseases: analysis of involved signaling pathways. Int J Mol Sci. 2013;14:226–43. https://doi.org/10.3390/ijms14010226.
72. Sundar I, Rahman I. Vitamin D and susceptibility of chronic lung diseases: role of epigenetics. Front Pharmacol. 2011;2:50. https://www.frontiersin.org/article/10.3389/fphar.2011.00050.
73. Zosky G, Hart P, et al. Vitamin D deficiency at 16 to 20 weeks' gestation is associated with impaired lung function and asthma at 6 years of age. Ann Am Thorac Soc. 2014;11(4). https://doi.org/10.1513/AnnalsATS.201312-423OC.
74. Pfeffer P, Hawrylowicz C. Vitamin D and lung disease. Thorax. 2012;67:1018–20.
75. Park H, Byun M, Kim H, et al. Dietary vitamin C intake protects against COPD: the Korea National Health and Nutrition Examination Survey in 2012. Int J Chron Obstruct Pulmon Dis. 2016;11:2721. https://doi.org/10.2147/COPD.S119448.
76. Shin JY, Shim JY, Lee DC, Lee HR. Smokers with adequate vitamin C intake show a preferable pulmonary function test. J Am Coll Nutr. 2015;34:385. https://doi.org/10.1080/07315724.2014.926152.
77. Chen Y, Luo G, Yuan J, Wang Y, Yang X, Wang X, et al. Vitamin C mitigates oxidative stress and tumor necrosis factor-alpha in severe community-acquired pneumonia and LPS-induced macrophages. Mediators Inflamm. 2014;2014:426740. https://doi.org/10.1155/2014/426740.
78. Ibrahim A, Borai I, Ali M, Ghanem H, Hegazi A, Mousa A. Inhibitory effect of a mixture containing vitamin C, lysine, proline, epigallocatechin gallate, zinc and alpha-1-antitrypsin on lung carcinogenesis induced by benzo(a) pyrene in mice. J Res Med Sci. 2013;18(5):427–34.
79. Madill J, Aghdassi E, Arendt BM, et al. Thoracic transplantation: oxidative stress and nutritional intakes in lung patients with bronchiolitis Obliterans syndrome. Transplant Proc. 2009;41:3838–44. https://doi.org/10.1016/j.transproceed.2009.04.012.
80. Zhang Y, Ma C, et al. Dual role of vitamin C utilization in NO2-induced oxidative stress in lung tissues of mice. Bull Environ Contam Toxicol. 2010;84(6):662–6. https://doi.org/10.1007/s00128-010-0021-1.
81. Cook-Mills JM, Abdala-Valencia H, et al. Two faces of vitamin E in the lung. Am J Respir Crit Care Med. 2013;188(3):279–84. https://doi.org/10.1164/rccm.201303-0503ED.
82. Peh HY, Ho WE, et al. Vitamin E isoform γ-tocotrienol downregulates house dust mite–induced asthma. J Immunol. 2015;195:437. https://doi.org/10.4049/jimmunol.1500362.
83. Hanson C, Lyden E, Furtado J, et al. Serum tocopherol levels and vitamin E intake are associated with lung function in the normative aging study. Clin Nutr. 2015;35(1):169–74.
84. Leng S, Picchi MA, Tesfaigzi Y, Wu G, James Gauderman W, Xu F, Gilliland FD, Belinsky SA. Dietary nutrients associated with preservation of lung function in Hispanic and non-Hispanic white smokers from New Mexico. Int J Chron Obstruct Pulmon Dis. 2017;12:3171–81.
85. Veith C, Drent M, et al. The disturbed redox-balance in pulmonary fibrosis is modulated by the plant flavonoid quercetin. Toxicol Appl Pharmacol. 2017;336(1):40–8. https://doi.org/10.1016/j.taap.2017.10.001.
86. Daishun L, Ling G, Guichuan H, et al. Curcumin inhibits transforming growth factor β induced differentiation of mouse lung fibroblasts to myofibroblasts. Front Pharmacol. 2016;7 [serial online]. 2016; Available from: Directory of Open Access Journals, Ipswich, MA. Accessed 28 Mar 2017.

87. Cho YJ, Yi CO, Jeon BT, et al. Curcumin attenuates radiation-induced inflammation and fibrosis in rat lungs. Korean J Physiol Pharmacol [serial online]. 2013;17(4):267.. Available from: Publisher Provided Full Text Searching File, Ipswich, MA. Accessed 4 Apr 2017.
88. Annisa R, Setiyono A, Juniantito V. Curcumin effect on Bleomycin-induced pulmonary fibrosis in Mus musculus. Jurnal Ilmu Ternak Dan Veteriner. 2015;20(2):148–57.. [serial online]. 2015;(2):148. Available from: Directory of Open Access Journals, Ipswich, MA. Accessed 28 Mar 2017.
89. Zhang D, Huang C, Brömme D, et al. Antifibrotic effects of curcumin are associated with overexpression of cathepsins K and L in bleomycin treated mice and human fibroblasts. Respir Res [serial online]. 2011;12(1):154. Available from: CINAHL Complete, Ipswich, MA. Accessed 4 Apr 2017.
90. Loch-Neckel G, Santos-Bubniak L, et al. Orally administered chitosan-coated polycaprolactone nanoparticles containing curcumin attenuate metastatic melanoma in the lungs. J Pharm Sci. 2015;104:3524–34. https://doi.org/10.1002/jps.24548.
91. Kandhare A, Bodhankar S, et al. Effect of glycosides based standardized fenugreek seed extract in bleomycin-induced pulmonary fibrosis in rats: decisive role of Bax, Nrf2, NF-κB, Muc5ac, TNF-α and IL-1β. Chem Biol Interact. 2015;237(25):151–65.
92. Srabani K, Silpak B, Ena RB. Evaluating the ameliorative potential of plant flavonoids and their nanocomposites in bleomycin induced idiopathic pulmonary fibrosis. Biomed Res Ther. 2016;3(7):707–22.. [serial online]. 2016;(7):707. Available from: Directory of open Access Journals, Ipswich, MA. Accessed 28 Mar 2017.
93. Hampson AJ, et al. Cannabidiol and (-)Delta9-tetrahydrocannabinol are neuroprotective antioxidants. Proc Natl Acad Sci U S A. 1998;95(14):8268–73. https://doi.org/10.1073/pnas.95.14.8268.
94. Hirayama F, Lee AH, Oura A, Mori M, Hiramatsu N, Taniguchi H. Dietary intake of six minerals in relation to the risk of chronic obstructive pulmonary disease. Asia Pac J Clin Nutr. 2010;4:572.
95. Mahabi S, Forman M, Dong Y, Park Y, Hollenbeck A, Schatzkin A. Mineral intake and lung cancer risk in the NIH-American Association of Retired Persons Diet and Health Study. Cancer Epidemiol Biomarkers Prev. 2010;19(8):1976–83. https://doi.org/10.1158/1055-9965.EPI-10-0067.
96. Pizzorno L. Nothing boring about boron. Integr Med (Encinitas). 2015;14(4):35–48.
97. Kraja B, Ruiter R, Lahousse L, Keyser C, Stricker B, Kiefte-de Jong J, et al. Dietary mineral intake and lung cancer risk: the Rotterdam study. Eur J Nutr. 2017;56(4):1637–46.
98. Bravo MP, Balboa P, Torrejón C, Bozzo R, Boza ML, Contreras I, et al. Bone mineral density, lung function, vitamin D and body composition in children and adolescents with cystic fibrosis: a multicenter study. Nutr Hosp. 2018;35(4):789–95.
99. Shoji Y, Takeuchi H, Fukuda K, Fukunaga K, Nakamura R, Takahashi T, et al. The alpha-lipoic acid derivative DHLHZn: a new therapeutic agent for acute lung injury in vivo. Inflamm Res. 2017;66(9):803–11.
100. Phiboonchaiyanan PP, Chanvorachote P. Suppression of a cancer stem-like phenotype mediated by alpha-lipoic acid in human lung cancer cells through down-regulation of β-catenin and Oct-4. Cell Oncol (Dordr). 2017;40(5):497–510.
101. Ateyya H, Nader MA, Attia GM, El-Sherbeeny NA. Influence of alpha-lipoic acid on nicotine-induced lung and liver damage in experimental rats. Can J Physiol Pharmacol. 2017;95(5):492.
102. Todisco T, Polidori R, Rossi F, et al. Effect of N-acetylcysteine in subjects with slow pulmonary mucociliary clearance. Eur J Respir Dis. 1985;66(Suppl 139):136–41.
103. Stey C, Steurer J, Bachmann S, Medici TC, Tramer MR. The effect of oral N-acetylcysteine in chronic bronchitis: a quantitative systematic review. Eur Respir J. 2000;16(2):253–62.
104. Riise GC, Larsson S, Larsson P, Jeansson S, Andersson BA. The intrabronchial microbial flora in chronic bronchitis patients: a target for N-acetylcysteine therapy? Eur Respir J. 1994;7(1):94–101.
105. Meyer A, Buhl R, Magnussen H. The effect of oral N-acetylcysteine on lung glutathione levels in idiopathic pulmonary fibrosis. Eur Respir J. 1994;7(3):431–6.
106. DeFlora S, Grassi C, Carati L. Attenuation of influenza-like symptomatology and improvement of cell-mediated immunity with long-term N-acetylcysteine treatment. Eur Respir J. 1997;10(7):1535–41.
107. Tirouvanziam R, Conrad CK, Bottiglieri T, Herzenberg LA, Moss RB, Herzenberg LA. High-dose oral N-acetylcysteine, a glutathione prodrug, modulates inflammation in cystic fibrosis. Proc Natl Acad Sci U S A. 2006;103(12):4628–33.
108. National Institutes of Health. How the lungs work. https://www.nhlbi.nih.gov/health/health-topics/topics/hlw/system. Accessed 22 Feb 2017.
109. Zhai T, Li S, Hu W, Li D, Leng S. Potential micronutrients and phytochemicals against the pathogenesis of chronic obstructive pulmonary disease and lung cancer. Nutrients. 2018;10(7):813.
110. McKeever T, Lewis S, Cassano P, et al. The relation between dietary intake of individual fatty acids, FEV1 and respiratory disease in Dutch adults. Thorax. 2008;63:208–14.
111. Chandrasekharan JA, Sharma-Walia N. Lipoxins: nature's way to resolve inflammation. J Inflamm Res. 2015;8:181–92. Published 2015 Sep 30. https://doi.org/10.2147/JIR.S90380.
112. Kim H-S, Lee K, et al. The role of free fatty acids in idiopathic pulmonary fibrosis. Eur Respir J. 2017;50(Suppl 61):OA4635. https://doi.org/10.1183/1393003.congress-2017.OA4635.
113. Batlle J, Sauleda J, Balcells E, Gómez F, Méndez M, Rodriguez E, et al. Association between Ω3 and Ω6 fatty acid intakes and serum inflammatory markers in COPD. J Nutr Bio. 2012;23(7):817–21. https://doi.org/10.1016/j.jnutbio.2011.04.005.
114. Luu HN, Cai H, et al. A prospective study of dietary polyunsaturated fatty acids intake and lung cancer risk. Int J Cancer. 2018;143(9):2225–37. https://doi.org/10.1002/ijc.31608. Epub 2018 Aug 7.
115. Jonker R, Deutz N, et al. Effectiveness of essential amino acid supplementation in stimulating whole body net protein anabolism is comparable between COPD patients and healthy older adults. Metabolism. 2017;69:120–9.
116. Dreher ML. Whole fruits and fruit fiber emerging health effects. Nutrients. 2018. [Internet]. [cited 2018 Dec 10];10(12):1833.
117. Fitzpatrick AM, Jones DP, Brown LAS. Glutathione redox control of asthma: from molecular mechanisms to therapeutic opportunities. Antioxid Redox Signal. 2012;17(2):375–408. https://doi.org/10.1089/ars.2011.4198.
118. Rahman I, Swarska E, Henry M, Stolk J, MacNee W. Is there any relationship between plasma antioxidant capacity and lung function in smokers and in patients with chronic obstructive pulmonary disease? Thorax. 2000;55(3):189–93.
119. Prousky J. The treatment of pulmonary diseases and respiratory-related conditions with inhaled (nebulized or aerosolized) glutathione. Evid Based Complement Alternat Med. 2008;5(1):27–35.
120. Santos-Gonzalez M, Gomez Diaz C, Navas P, Villalba JM. Modifications of plasma proteome in long-lived rats fed on a coenzyme Q10-supplemented diet. Exp Gerontol. 2007;42(8):798–806.
121. Wada H, Hagiwara S, Saitoh E, et al. Increased oxidative stress in patients with chronic obstructive pulmonary disease (COPD) as measured by redox status of plasma coenzyme Q10. Pathophysiology. 2006;13(1):29–33.
122. Gvozdjakova A, Kucharska J, Bartkovjakova M, Gazdikova K, Gazdik FE. Coenzyme Q10 supplementation reduces corticosteroids dosage in patients with bronchial asthma. Biofactors. 2005;25(1–4):235–40.
123. Fujimoto S, Kurihara N, Hirata K, Takeda T. Effects of coenzymeQ 10 administration on pulmonary function and exercise performance in patients with chronic lung diseases. Clin Investig. 1993;71(8):S162–6.
124. Adapted from: Glutathione cofactors. Available at: http://www.immunehealthscience.com/glutathione-cofactors.html.

125. Skinner HC, Berger AR, editors. Geology and health: closing the gap. Oxford University Press; 2003.
126. Ameeramja J, Perumal E. Pulmonary fluorosis: a review. Environ Sci Pollut Res. 2017;24:22119. https://doi.org/10.1007/s11356-017-9951-z.
127. Acat M, Aydemir Y, Çetinkaya E, et al. Comparison of the fractional exhaled nitric oxide levels in adolescents at three schools located three different distances from a large steel mill. Can Respir J. [serial online]. 2017;2017:6231309. Available from: MEDLINE Complete, Ipswich, MA. Accessed 22 Aug 2018.
128. Sanchez T, Perzanowski M, Graziano JH. Inorganic arsenic and respiratory health, from early life exposure to sex-specific effects: a systematic review. Environ Res. 2016;147:537. https://doi.org/10.1016/j.envres.2016.02.009.
129. Naujokas M, Anderson B, Ahsan H, et al. The broad scope of health effects from chronic arsenic exposure: update on a worldwide public health problem. Environ Health Perspect. 2013;121(3):295–302; Guha Mazumder DN. Chronic arsenic toxicity & human health. Indian J Med Res. 2008;128(4):436–47.
130. Cubadda F, Jackson BP, Cottingham KL, Van Horne YO, Kurzius-Spencer M. Human exposure to dietary inorganic arsenic and other arsenic species: state of knowledge, gaps and uncertainties. Sci Total Environ. 2017;579:1228. https://doi.org/10.1016/j.scitotenv.2016.11.108.
131. Toxic Substances Portal – Arsenic. Centers for Disease Control and Prevention. https://www.atsdr.cdc.gov/MMG/MMG.asp?id=1424&tid=3. Published October 21, 2014. Accessed 19 Sept 2018.
132. DynaMed Plus [Internet]. Ipswich (MA): EBSCO Information Services. 1995 - . Record No. T114977, Chronic arsenic poisoning; [updated 2018 Nov 30]. Available from https://www.dynamed.com/topics/dmp~AN~T114977. Accessed 6 Sept 2018.
133. Usuda K, Kono K, Ohnishi K, et al. Toxicological aspects of cadmium and occupational health activities to prevent workplace exposure in Japan: a narrative review. Toxicol Ind Health. 2011;27(3):225–33.
134. Järup L, Akesson A. Current status of cadmium as an environmental health problem. Toxicol Appl Pharmacol. 2009;238(3):201–8.
135. Rokadia H, Agarwal S. Serum heavy metals and obstructive lung disease: results from the National Health and Nutrition Examination Survey. Chest. 2013;143:388. https://doi.org/10.1378/chest.12-0595.
136. Jung J, Leem A, Kim S, et al. Relationship between blood levels of heavy metals and lung function based on the Korean National Health and Nutrition Examination Survey IV–V. Int J Chron Obstruct Pulmon Dis. 2015;1559. https://doi.org/10.2147/copd.s86182.
137. Department of the Environment and Energy [Internet]. Department of the Environment and Energy. 2019 [cited 10 June 2019]. Available from: http://www.environment.gov.au/protection/air-quality/air-pollutants.
138. Sun J, Barnes A, He D, Wang M, Wang J. Systematic review and meta-analysis of the association between ambient nitrogen dioxide and respiratory disease in China. Int J Environ Res Public Health. 2017;14(6):646. https://doi.org/10.3390/ijerph14060646.
139. Clougherty J, Kubzansky L. A framework for examining social stress and susceptibility to air pollution in respiratory health. Environ Health Perspect. 2009;117:1351. https://doi.org/10.1289/ehp.0900612.
140. Pannullo F, Lee D, Neal L, et al. Quantifying the impact of current and future concentrations of air pollutants on respiratory disease risk in England. Environ Health. 2017;16:29. https://doi.org/10.1186/s12940-017-0237-1.
141. Kariisa M, Foraker R, Pennell M, et al. Short- and long-term effects of ambient ozone and fine particulate matter on the respiratory health of chronic obstructive pulmonary disease subjects. Arch Environ Occup Health. 2015;70:56. https://doi.org/10.1080/19338244.2014.932753.
142. Karakatsani A, Samoli E, Rodopoulou S, et al. Weekly personal ozone exposure and respiratory health in a panel of Greek school children. Environ Health Perspect. 2017;125:077017. https://doi.org/10.1289/EHP635.
143. Péter S, Holguin F, Wood L, Clougherty J, Raederstorff D, Antal M, Weber P, Eggersdorfer M. Nutritional solutions to reduce risks of negative health impacts of air pollution. Nutrients. 2015;7(12):10398–416.
144. Rom O, Avezov K, Aizenbud D, Reznick AZ. Cigarette smoking and inflammation revisited. Respir Physiol Neurobiol. 2013;187:5. https://doi.org/10.1016/j.resp.2013.01.013.
145. Arnson Y, Shoenfeld Y, Amital H. Effects of tobacco smoke on immunity, inflammation and autoimmunity. J Autoimmun. 2012;34:258–65.
146. Heydari G, Ahmady A, Chamyani F, Masjedi M, Fadaizadeh L. Electronic cigarette, effective or harmful for quitting smoking and respiratory health: a quantitative review papers. Lung India. 2017;34:25. https://doi.org/10.4103/0970-2113.197119.
147. Lødrup Carlsen K, Skjerven H, Carlsen K-H. Review: the toxicity of E-cigarettes and children's respiratory health. Paediatr Respir Rev. 2018;28:63. https://doi.org/10.1016/j.prrv.2018.01.002.
148. Liaw SH, Hashim Z, Lye MS, Zainuddin H. Respiratory health and allergies from chemical exposures among machining industry workers in Selangor, Malaysia. Iran J Public Health. 2014;43(3):94–102.
149. Ye M, Beach J, Martin JW, Senthilselvan A. Occupational pesticide exposures and respiratory health. Int J Environ Res Public Health. 2013;10:6442. https://doi.org/10.3390/ijerph10126442.
150. Hoyle GW, Svendsen ER. Persistent effects of chlorine inhalation on respiratory health. Ann N Y Acad Sci. 2016;1378:33. https://doi.org/10.1111/nyas.13139.
151. Dassios T, Katelari A, Doudounakis S, Mantagos S, Dimitriou G. Respiratory muscle function in patients with cystic fibrosis. Pediatr Pulmonol. 2013;48:865. https://doi.org/10.1002/ppul.22709.
152. Kiecolt-Glaser J, McGuire L, Robles T, Glaser R. Emotions, morbidity, and mortality: new perspectives from psychoneuroimmunology. Annu Rev Psychol. 2002;53:83–107.
153. Lee Y, Ryu Y, Jung WM, Kim J, Lee T, Chae Y. Understanding mind-body interaction from the perspective of east Asian medicine. Evid Based Complement Alternat Med. 2017;2017:1. https://doi.org/10.1155/2017/7618419.
154. Ayres JG, Forsberg B, Annesi-Maesano I, Dey R, Ebi KL, Helms PJ, Medina-Ramon M, Windt M, Forastiere F. Climate change and respiratory disease: European Respiratory Society position statement. Eur Respir J. 2009;34(2):295–302.
155. Nurmatov U, Devereux G, Sheikh A. Nutrients and foods for the primary prevention of asthma and allergy: systematic review and meta-analysis. J Allergy Clin Immunol. 2011;127(3):724–733.e30. https://doi.org/10.1016/j.jaci.2010.11.001.
156. Global Initiative For Asthma (GINA). Global strategy for asthma management and prevention. Glob Initiat Asthma. 2017. http://ginasthma.org/2017-gina-report-global-strat. https://doi.org/10.1183/09031936.00138707
157. Skloot GS. Tests for assessing asthma. Allergy Clin Immunol. 2015;16:77–81.
158. Broesby-Olsen S, Carter M, Kjaer HF, Mortz CG, Møller MB, Kristensen TK, Bindslev-Jensen C, Agertoft L. Pediatric expression of mast cell activation disorders. Immunol Allergy Clin North Am. 2018;38(3):365–77.
159. Corren J. Asthma phenotypes and endotypes: an evolving paradigm for classification. Discov Med. 2013;15(83):243–9.
160. E-Asthma is Eosinophilic Asthma. n.d. Retrieved 13 Jan 2018, from https://www.easthma.com/#how-is-e-asthma.
161. Son JH, Chung BY, Kim HO, Park CW. A histamine-free diet is helpful for treatment of adult patients with chronic spontaneous urticaria. Ann Dermatol. 2018;30(2):164–72.
162. Manzotti G, Breda D, Di Gioacchino M, Burastero SE. Serum diamine oxidase activity in patients with histamine intolerance. Int J Immunopathol Pharmacol. 2016;29(1):105–11.
163. Fitzpatrick AL, Kronmal RA, Gardner JP, Psaty BM, Jenny NS, Tracy RP, Walston J, Kimura M, Aviv A. Leukocyte telomere length and

cardiovascular disease in the cardiovascular health study. Am J Epidemiol. 2006;165(1):14–21.
164. Quinlan T, Spivack S, Mossman BT. Regulation of antioxidant enzymes in lung after oxidant injury. Environ Health Perspect. 1994;102(Suppl 2):79–87.
165. Sahiner UM, Birben E, Erzurum S, Sackesen C, Kalayci O. Oxidative stress in asthma. World Allergy Organ J. 2011;4(10):151–8. https://doi.org/10.1097/WOX.0b013e318232389e.
166. Levy BD, Serhan CN. Resolution of acute inflammation in the lung. Annu Rev Physiol. 2014;76(1):467–92. https://doi.org/10.1146/annurev-physiol-021113-170408.
167. Schwalfenberg G. The alkaline diet: is there evidence that an alkaline pH diet benefits health? J Environ Public Health. 2012;2012:727630, 7 pages. https://doi.org/10.1155/2012/727630.
168. Gröber U, Schmidt J, Kisters K. Magnesium in prevention and therapy. Nutrients. 2015;7(9):8199–226. https://doi.org/10.3390/nu7095388.
169. Bede O, Nagy D, Surányi A, Horváth I, Szlávik M, Gyurkovits K. Effects of magnesium supplementation on the glutathione redox system in atopic asthmatic children. Inflamm Res. 2008;57(6):279–86. https://doi.org/10.1007/s00011-007-7077-3.
170. Han YY, Blatter J, Brehm J, Forno E, Augusto A, Celedón J. Diet and asthma: vitamins and methyl donors. Respir Med. 2013;1(10): 813–22.
171. Barnes PJ. Drugs for asthma. Br J Pharmacol. 2006;147(Suppl 1). https://doi.org/10.1038/sj.bjp.0706437.
172. Wood LG, Shivappa N, Berthon BS, Gibson PG, Hebert JR. Dietary inflammatory index is related to asthma risk, lung function and systemic inflammation in asthma. Clin Exp Allergy. 2015;45(1):177–83.
173. Seyedrezazadeh E, Moghaddam MP, Ansarin K, Vafa MR, Sharma S, Kolahdooz F. Fruit and vegetable intake and risk of wheezing and asthma: a systematic review and meta-analysis. Nutr Rev. 2014;72:411–28.
174. Singh M, Agarwal A, Chatterjee B, Chauhan A, Das RR, Paul N. Correlation of cutaneous sensitivity and cytokine response in children with asthma. Lung India: Off Organ Indian Chest Soc. 2017;34(6):506–10. https://doi.org/10.4103/lungindia.lungindia_357_16.
175. Blood Tests. n.d. Retrieved 14 Jan 2018, from https://www.foodallergy.org/life-food-allergies/food-allergy-101/diagnosis-testing/blood-tests.
176. Canova C, Pitter G, Ludvigsson JF, Romor P, Zanier L, Zanotti R, Simonato L. Coeliac disease and asthma association in children: the role of antibiotic consumption. Eur Respir J. 2015;ERJ-01857
177. Lloyd CM, Hawrylowicz CM. Regulatory T cells in asthma. Immunity. 2009;31(3):438–49. https://doi.org/10.1016/j.immuni.2009.08.007.
178. Ege M, Bieli C, Frei R, van Strien R, Riedler J, Ublagger E, Schram-Bijkerk D, Brunekreef B, van Hage M, Scheynius A, et al. Prenatal farm exposure is related to the expression of receptors of the innate immunity and to atopic sensitization in school-age children. J Allergy Clin Immunol. 2006;117:817–23.
179. Ly NP, et al. Characterization of regulatory T cells in urban newborns. Clin Mol Allergy. 2009;7:8. https://doi.org/10.1186/1476-7961-7-8.
180. Litonjua AA, Weiss ST. Is vitamin D deficiency to blame for the asthma epidemic? J Allergy Clin Immunol. 2007;120:1031–5.
181. Willers S, Wijga A, Brunekreef B, Kerkhof M, Gerritsen J, Hoekstra M, de Jongste JC, Smit H. Maternal food consumption during pregnancy and the longitudinal development of childhood asthma. Am J Respir Crit Care Med. 2008;178(2):124–31.
182. Sutherland R, Goleva E, Jackson L, Stevens A, Leung D. Vitamin D levels, lung function, and steroid response in adult asthma. Am J Respir Crit Care Med. 2010;181(7):699–704. https://doi.org/10.1164/rccm.200911-1710OC.
183. Billington CK, Ojo OO, Penn RB, Ito S. cAMP regulation of airway smooth muscle function. Pulm Pharmacol Ther. 2013;26(1):112–20. https://doi.org/10.1016/j.pupt.2012.05.007.
184. Townsend E, Emala C Sr. Quercetin acutely relaxes airway smooth muscle and potentiates β-agonist-induced relaxation via dual phosphodiesterase inhibition of PLCβ and PDE4. Am J Physiol Lung Cell Mol Physiol. 2013;305(5):L396–403.
185. Bielory L. UpToDate [Internet]. Uptodate.com. 2019 [cited 10 June 2019]. Available from: https://www.uptodate.com/contents/complementary-and-alternative-therapies-for-allergic-rhinitis-and-conjunctivitis.
186. Izuhara K, Ohta S, Ono J. Using periostin as a biomarker in the treatment of asthma. Allergy Asthma Immunol Res. 2016;8(6): 491–8.
187. Biomarkers in asthma | AAAAI [Internet]. The American Academy of Allergy, Asthma & Immunology. 2019 [cited 10 June 2019]. Available from: https://www.aaaai.org/global/latest-research-summaries/Current-JACI-Research/biomarker-asthma.
188. Xu J, Huang G, Guo T. Developmental bisphenol A exposure modulates immune-related diseases. Toxics. 2016;4(4):23.
189. Santee CA, Nagalingam NA, Faruqi AA, DeMuri GP, Gern JE, Wald ER, Lynch SV. Nasopharyngeal microbiota composition of children is related to the frequency of upper respiratory infection and acute sinusitis. Microbiome. 2016;4(1):34.
190. Guillain-Barré Syndrome Fact Sheet. National Institute of Neurological Disorders and Stroke. https://www.ninds.nih.gov/Disorders/Patient-Caregiver-Education/Fact-Sheets/Guillain-Barré-Syndrome-Fact-Sheet. Accessed 21 Oct 2018.
191. Stoller JK. Clinical manifestations, diagnosis, and natural history of alpha-1 antitrypsin deficiency. In: Post TW, Barnes PJ, Hollingsworth H (editors). Uptodate. 2016. Available from http://www.uptodate.com/contents/clinical-manifestations-diagnosis-and-natural-history-of-alpha-1-antitrypsin-deficiency.
192. 23andMe. Let's talk about Alpha-1 Antitrypsin Deficiency. Retrieved from: https://www.23andme.com/topics/health-predispositions/alpha-1-antitrypsin-deficiency/.
193. Silverman EK, Sandhaus RA. Alpha1-antitrypsin deficiency. N Engl J Med. 2009;360:2749–57.
194. What is Pulmonary Fibrosis | Pulmonary Fibrosis Foundation [Internet]. Pulmonaryfibrosis.org. 2019 [cited 10 June 2019]. Available from: https://www.pulmonaryfibrosis.org/life-with-pf/about-pf.
195. Home Page – IPF Foundation [Internet]. IPF Foundation. 2019 [cited 10 June 2019]. Available from: https://ipffoundation.org/support-ipf-reasearch-donate-now.
196. PFF Clinical Trial Finder [Internet]. A clinical evaluation of Baofeikang Granule in combined pulmonary fibrosis and emphysema treatment - Pulmonary Fibrosis Foundation. [cited 11 Nov 2018]. Available from: https://pulmonaryfibrosis-org.clinicaltrialconnect.com/trials/NCT02805699.
197. Lawson W, Cheng D, Degryse A, Tanjore H, Polosukhin V, Xu X, et al. Endoplasmic reticulum stress enhances fibrotic remodeling in the lungs. Proc Natl Acad Sci. 2011;108(26):10562–7. https://doi.org/10.1073/pnas.1107559108.
198. Chen DQ, Feng YL, Cao G, Zhao YY. Natural products as a source for antifibrosis therapy. Trends Pharmacol Sci. 2018;39(11):937–52.
199. King C, Aryal S, Nathan S. Idiopathic pulmonary fibrosis: phenotypes and comorbidities. In: Meyer K, Nathan S, editors. Idiopathic pulmonary fibrosis. Respiratory medicine. Cham. First Online15 December 2018.: Humana Press; 2019. https://doi.org/10.1007/978-3-319-99975-3_11.
200. Kropski JA, Blackwell TS, Loyd JE. The genetic basis of idiopathic pulmonary fibrosis. Eur Respir J. 2015;45(6):1717–27.
201. Armanios M, et al. Telomerase mutations in families with idiopathic pulmonary fibrosis. (March 29, 2007). N Engl J Med. 2007;356:1317–26. https://doi.org/10.1056/NEJMoa066157.
202. Armanios M. Telomerase and idiopathic pulmonary fibrosis. Mutat Res. Fundamental and Molecular Mechanisms of Mutagenesis. 2012;730(1):52–8.
203. Adams TN, Garcia CK. Genetics of pulmonary fibrosis. In: Idiopathic pulmonary fibrosis. Cham: Humana Press; 2019. p. 183–206.

204. Zhang Y, Jones KD, Achtar-Zadeh N, Green G, Kukreja J, Xu B, Wolters PJ. Histopathological and molecular analysis of idiopathic pulmonary fibrosis lungs from patients treated with pirfenidone or nintedanib. Histopathology. 2019;74:341–9. https://doi.org/10.1111/his.13745.
205. Molina-Molina M. Telomere shortening is behind the harm of immunosuppressive therapy in idiopathic pulmonary fibrosis. Am J Respir Crit Care Med. 2019;200(3):274–5.
206. Hohmann MS, Habiel DM, Coelho AL, Verri WA Jr, Hogaboam CM. Quercetin enhances ligand-induced apoptosis in senescent idiopathic pulmonary fibrosis fibroblasts and reduces lung fibrosis in vivo. Am J Respir Cell Mol Biol. 2019;60(1):28–40.
207. Conrad C. Application of N-Acetylcysteine in pulmonary disorders. In: The therapeutic use of N-Acetylcysteine (NAC) in medicine. Singapore: Adis; 2019. p. 255–76.
208. Adegunsoye A, Vij R, Noth I. Integrating genomics into management of fibrotic interstitial lung disease. Chest. 2019;155(5):1026–40.
209. Inchingolo R, Varone F, Sgalla G, Richeldi L. Existing and emerging biomarkers for disease progression in idiopathic pulmonary fibrosis. Expert Rev Respir Med. 2019;13(1):39–51.
210. Zee RY, Michaud SE, Germer S, Ridker PM. Association of shorter mean telomere length with risk of incident myocardial infarction: a prospective, nested case–control approach. Clin Chim Acta. 2009;403(1-2):139–41.
211. Van Der Harst P, van der Steege G, de Boer RA, Voors AA, Hall AS, Mulder MJ, van Gilst WH, van Veldhuisen DJ, MERIT-HF Study Group. Telomere length of circulating leukocytes is decreased in patients with chronic heart failure. J Am Coll Cardiol. 2007;49(13):1459–64.
212. McGrath M, Wong JY, Michaud D, Hunter DJ, De Vivo I. Telomere length, cigarette smoking, and bladder cancer risk in men and women. Cancer Epidemiol Biomarkers Prev. 2007;16(4):815–9.
213. Sampson MJ, Winterbone MS, Hughes JC, Dozio N, Hughes DA. Monocyte telomere shortening and oxidative DNA damage in type 2 diabetes. Diabetes Care. 2006;29(2):283–9.
214. Valdes AM, Richards JB, Gardner JP, Swaminathan R, Kimura M, Xiaobin L, Aviv A, Spector TD. Telomere length in leukocytes correlates with bone mineral density and is shorter in women with osteoporosis. Osteoporos Int. 2007;18(9):1203–10.
215. Hoge EA, Chen MM, Orr E, Metcalf CA, Fischer LE, Pollack MH, DeVivo I, Simon NM. Loving-Kindness Meditation practice associated with longer telomeres in women. Brain Behav Immun. 2013;32:159–63.
216. Hoxha M, Dioni L, Bonzini M, Pesatori AC, Fustinoni S, Cavallo D, Carugno M, Albetti B, Marinelli B, Schwartz J, Bertazzi PA. Association between leukocyte telomere shortening and exposure to traffic pollution: a cross-sectional study on traffic officers and indoor office workers. Environ Health. 2009;8(1):41.
217. Werner C, Fürster T, Widmann T, Pöss J, Roggia C, Hanhoun M, Scharhag J, Büchner N, Meyer T, Kindermann W, Haendeler J. Physical exercise prevents cellular senescence in circulating leukocytes and in the vessel wall. Circulation. 2009;120(24):2438–47.
218. Farzaneh-Far R, Lin J, Epel ES, Harris WS, Blackburn EH, Whooley MA. Association of marine omega-3 fatty acid levels with telomeric aging in patients with coronary heart disease. JAMA. 2010;303(3):250–7.
219. Shen J, Gammon MD, Terry MB, Wang Q, Bradshaw P, Teitelbaum SL, Neugut AI, Santella RM. Telomere length, oxidative damage, antioxidants and breast cancer risk. Int J Cancer. 2009;124(7):1637–43.
220. Arsenis NC, You T, Ogawa EF, Tinsley GM, Zuo L. Physical activity and telomere length: impact of aging and potential mechanisms of action. Oncotarget. 2017;8(27):45008.
221. Hanjani NA, Vafa M. Protein restriction, epigenetic diet, intermittent fasting as new approaches for preventing age-associated diseases. Int J Prev Med. 2018;9:58.
222. Schilling R. Fasting mimicking diet is very relevant for health and longevity [Internet]. Medical Articles by Dr. Ray. 2019 [cited 10 June 2019]. Available from: http://www.askdrray.com/fasting-mimicking-diet-is-very-relevant-for-health-and-longevity.
223. Ip FC, Ng YP, An HJ, Dai Y, Pang HH, Hu YQ, Chin AC, Harley CB, Wong YH, Ip NY. Cycloastragenol is a potent telomerase activator in neuronal cells: implications for depression management. Neurosignals. 2014;22(1):52–63.
224. Izuhara K, Nunomura S, Nanri Y, Ono J, Takai M, Kawaguchi A. Periostin: an emerging biomarker for allergic diseases. Allergy. 2019;00:1–13. https://doi.org/10.1111/all.13814.
225. Zemaira® | Alpha-1 Antitrypsin Deficiency Diagnosis [Internet]. Zemaira.com. 2019 [cited 10 June 2019]. Available from: https://www.zemaira.com/alpha-1/blood-test-diagnosis.
226. Pathak SK, Jaiswal AK. Assessment of quality of life in children suffered from asthma. Assessment. 2018;4(10):1–3.
227. Robinson CB, Dunn DC. Identifying maternal conditions affecting altered embryologic development. Neonatal Advanced Practice Nursing: A Case-Based Learning Approach. 2016;25:25.
228. Pouwels SD, Hesse L, Faiz A, Lubbers J, Bodha PK, ten Hacken NH, van Oosterhout AJ, Nawijn MC, Heijink IH. Susceptibility for cigarette smoke-induced DAMP release and DAMP-induced inflammation in COPD. Am J Physiol Lung Cell Mol Physiol. 2016;311(5):L881–92.

The Skin, Selected Dermatologic Conditions, and Medical Nutrition Therapy

P. Michael Stone

52.1 Introduction – 970

52.2 Overview of Skin Function and Composition – 970
52.2.1 Epidermis – 970
52.2.2 Dermis – 970
52.2.3 Skin Immunity – 971

52.3 Nutrition Deficiency and the Skin – 971

52.4 Skin Conditions and Medical Nutrition Therapy – 973

52.5 Macro- and Micronutrient Deficiencies Influencing Skin Health – 974
52.5.1 Protein – 974
52.5.2 Essential Fatty Acid – 974
52.5.3 Fat-Soluble Vitamins – 983
52.5.4 Water-Soluble Vitamins – 983
52.5.5 Minerals – 986
52.5.6 Nutrition Excesses and Dermatologic Conditions – 988

52.6 Dietary Interventions – 988
52.6.1 Mediterranean Diet – 988
52.6.2 Modified Elimination Diet – 989
52.6.3 Modified Elimination Diet in Autoimmunity, IgE Reactions, IgG Hypersensitivity, and Other Immune Reactions – 990
52.6.4 Dermatologic Findings in Obesity – 992
52.6.5 Therapeutic Uses of Probiotics and Fermented Food in Skin Disorders – 992

52.7 Autoimmune or Hypersensitivity Reactions to Food Proteins – 993
52.7.1 Autoimmunity-Triggered Dietary Antigens – 993
52.7.2 Dermatologic Findings in Inflammatory Bowel Disease – 993
52.7.3 Food Bioactives in the Treatment of Conditions – 994
52.7.4 Conclusion – 994

References – 996

© Springer Nature Switzerland AG 2020
D. Noland et al. (eds.), *Integrative and Functional Medical Nutrition Therapy*,
https://doi.org/10.1007/978-3-030-30730-1_52

52.1 Introduction

Skin conditions in the United States are a common concern, accounting for nearly 110 million patient visits in a recent survey [1]. Complementary and alternative medicines were used for eight million of those visits, with recommendations for fish oil or omega-3 fatty acids, lutein, glucosamine, and Reishi mushrooms, to name a few [1]. Included in these recommendations are medical nutrition interventions and diet suggestions, individual nutrients, and a growing number of bioactive dietary components and plant extracts. These interventions are beginning to be recommended by dermatologists [2]. The skin has unique physiology, endocrine, and immune responses. Understanding this uniqueness helps direct the clinician to successful therapeutic interventions.

52.2 Overview of Skin Function and Composition

The skin is complex, highly specialized and serves multiple functions. It provides barrier function to prevent internal water loss and protects the body from external environmental agents [3, 4]. The adult human skin covers approximately a 2-square-meter area with appendages of hair and nails [4]. The skin has two main layers: dermis and epidermis. Embryologically, ectoderm, neuroectoderm, and mesoderm contribute to constituents of the skin. From the ectoderm arise the epidermis, hair, sebaceous glands, sweat glands, and nails. The neuroectoderm contributes to the melanocytes, nerves, and neuroreceptors. The mesoderm gives rise to collagen, reticulin, elastic fibers, blood vessels, muscle, and fat.

52.2.1 Epidermis

The epidermis has microscopically distinct layers. The stratum basale abuts and communicates with the dermis, followed by the stratum spinosum, stratum granulosum, and stratum lucidum. The final layer, the stratum corneum, interfaces with the external environment and is an external stratified, non-vascularized, epithelium of between 75 and 150 micrometers and can be up to 600 micrometers thick on the palms of the hand and soles of the feet. This outermost layer of the epidermis, stratum corneum, consists of keratinocytes which are dead and flattened remnants imbedded in a lipid matrix, connected by desmosomes. This unit is difficult to penetrate both by internal-to-external water loss and external-to-internal transfer of environmental materials. The lipids of the stratum corneum form a lamellar matrix, which has a unique composition that provides important properties to this organ [5, 6]. The stratum corneum contains equimolar quantities of ceramides, cholesterol, and fatty acids. Maintaining this ratio is imperative for normal epidermal barrier homeostasis, as well as for normal lamellar organization [5]. The ceramides, cholesterol, and fatty acids comprise 50%, 25%, and 15%, respectively, of the total lipid mass [5–7]. Both essential and nonessential fatty acids play roles in normal and proper skin function.

Plant- and animal-derived essential fatty acids (EFA), such as eicosapentaenoic acid (C20:5, n-3) and docosahexaenoic acid (C22:6, n-3), are transported to the epidermis from the capillaries (please see ▶ Chap. 10 for Essential Fatty Acid Biology and ▶ Chap. 12 for Structure and Function). There is also evidence that some fatty acids used in the formation of the epidermis are intrinsically made. The keratinocytes transport EFAs, such as arachidonic acid and linoleic acid, with higher specificity than for nonessential fatty acids like oleic acid (C18:1,n9) [6, 8]. EFAs are integral for normal stratum corneum function in that they help form three acylceramide species (linoleic acid esterified to the omega hydroxyl group of very long chain fatty acids –C>28–34). The ceramides are key to the barrier function and contribute to minimal transepidermal water loss [5, 9]. When considering the cofactors for the elongases and desaturases which allow intrinsic conversion of these fats, it makes intuitive sense that when the mineral and vitamin cofactors are deficient, some skin conditions worsen.

In skin conditions where there is alteration of the keratinocyte fatty acid composition, such as in atopic dermatitis and psoriasis, modulating the levels with EFA intake improves barrier defects, proliferative epidermal changes, and inflammatory processes [10–13]. On the other hand, when oleic acid has a higher predominance in the diet or is substituted for essential fatty acids, certain skin conditions worsen because of changes in proliferation and inflammation [4].

The integrity of the epidermis is dependent on the pH of the skin [14]. An acidic pH at the skin surface helps modulate the microbial barrier function and inflammation, along with the rate and degree of desquamation. Free fatty acids are a major contributor to the maintenance of the pH [15, 16]. This helps explain why environmental and lifestyle habits, such as excessive water exposure, solvents, and hygiene products, change the pH and can adversely alter the integrity and the turnover of the skin.

52.2.2 Dermis

The dermis is located beneath the epidermis and varies in thickness, depending on location. While it is usually less than 2 mm thick, it may be as thick as 4 mm in the adult back and as thin as 0.3 mm at the eyelid. The dermis houses many of the skin's "business centers" and is a rich repository of various tissue types, including the vascular system, lymphatics, and most nerves [4, 17]. Within the dermis, there are three types of connective tissue, with the most abundant being collagen, followed by elastic tissue, and reticular fibers. Also found are the Meissner's and Vater-Pacini corpuscles, which collect and interpret touch and pressure. Unmyelinated nerve endings in the papillary dermis capture and transmit the sensations of pain, itch, and temperature. Sebaceous glands and adrenergic fibers that innervate the blood vessels, hair erector muscles, and apocrine glands are located in this important layer.

52.2.3 Skin Immunity

The immune reactions in the skin are both innate and acquired, and all Gell-Coombs reactions types 1 through 4 can be identified (please see ▶ Chap. 19). Histiocytes are macrophages that wander the skin landscape. Histiocytes accumulate hemosiderin, melanin, and the debris of inflammation. Mast cells are located around blood vessels and release histamine, heparin, and more than 100 different chemotactic and adaptogenic molecules [4]. Immune surveillance and maintenance of barrier integrity for the skin is provided by the memory T cells with the imprinting signals from the resident dendritic cells.

Chemokine expression helps drive and modulate skin immunity. Chemokines and their receptors are important for signaling migration of various cell types into areas of inflammation. One example is expression of chemokine receptor 8, CCR8, in the skin resident memory T cells resulting from skin-specific factors. Of interest, this is not seen in the mesentery or intestinal mucosal layer, which suggests this is unique to the keratinocyte layer in the skin [18]. The factors leading to chemokine expression are soluble and independent of vitamins A and D, although these vitamins play a significant role in the control of T-cell homing to both the small intestine and skin tissue [18]. In brief, vitamin A and vitamin A metabolites were shown to play a crucial role in the induction of the gut-homing chemokine receptors CCR9 and $\alpha 4\beta 7$ and a feature of local $CD103^+$ dendritic cells [18]. In contrast, the situation for skin-homing T cells is more complicated. The vitamin D_3 metabolite $1\alpha,25$-dihydroxyvitamin D_3 was shown to be necessary to induce chemokine CCR10 in human T cells [18]. Furthermore, vitamins A and D, along with calcium, are important for cathelicidin and kallikrein expression. Cathelicidin is critical for innate immunity of the skin, providing antimicrobial action and immunomodulation. Cathelicidin is enzymatically controlled by the family of kallikreins, and expression is important for control of keratinocytes' immune protection [19]. These variables may further control the functions of antimicrobial peptides in the skin, which can modulate the severity of psoriasis and rosacea. Because of nutrient insufficiencies, excesses, or other bioactive immune-modulating compounds, many common skin conditions may be affected.

The epithelium and the underlying connective tissue of the dermis together form a pliable, strong, communication-rich organ system. All the functions discussed above, along with development and replacement of these tissues, are dependent on adequate nutrition from conception to the last breath.

52.3 Nutrition Deficiency and the Skin

The classic nutritional deficiencies all have skin signs and symptoms. This makes functional sense as the macronutrients consisting of protein, fats (cholesterol, essential fatty acids, ceramides, sphingomyelins), and carbohydrates (simple and complex) form the structural architecture and the micronutrients, consisting of minerals, vitamins, and phytonutrients, are the cofactors and modulators of physiologic mechanisms and healthy tissue turnover. When there is inadequacy of macronutrients, micronutrients, phytonutrients, and/or bioactives to sustain and modulate genetic, epigenetic, and physiologic conditions, the signs and symptoms of inadequacy develop. See ◘ Table 52.1. With growing understanding of genetic single nucleotide polymorphisms (SNPs) and epigenetic influences changing the expression of genes and altering tissue requirements, it is no wonder that there are so many classic nutritional deficiency signs and symptoms seen with variable occurrence. See ◘ Table 52.2. The increasing recognition of SNPs for each enzyme system modulated by nutrient cofactors and phytonutrient adaptogens is increasing the complexity of our understanding.

An exciting new frontier in health modulation and treatment is the understanding of microRNA (miRNA). The miRNA from food alters and blocks the activity of endogenous miRNA associated with disease processes [24–26]. The trans-kingdom transfer of miRNA in each bite of our food is epigenetically altering our individual health and disease processes [24–26]. This understanding gives insight to the variable penetrance of different diet and nutritional treatments when applied to inflammatory conditions of the skin. For example, in psoriasis, atopic dermatitis, and other autoimmune diseases, there are increasing or decreasing levels of miRNA with variable regulation, which seems to drive the immune balance [24]. Also, miRNAs of curcumin and resveratrol impact the degree of disease by modulating immune function [27]. This understanding has brought about an emerging frontier for the use of plants, roots, herbs, and spices, with their associated miRNA, to modulate the immune response.

When considering the results of a skin-related physical exam, a single finding may be contributed by one or a multitude of nutrient insufficiencies. A good example is cheilitis—dry, scaly, and fissured lips. See ◘ Table 52.2. Cheilitis can be caused by a number of nutritional deficiencies, including protein, zinc, or B vitamins. In contrast, a single food may trigger varied immune reactions that are manifested in many skin findings. Two examples are the gluten and milk proteins, whose proteomes trigger autoimmune reactions manifested throughout the body and skin. See ◘ Table 52.3. Just as chronic disease is associated with many underlying conditions, skin diseases are associated with a variety of intersecting causes.

Table 52.1 Protein, fat, carbohydrate, minerals, vitamins, phytonutrient deficiencies, and presenting signs

Nutrient	Signs and symptoms
Protein	
Marasmus: wasting and stunting less than 60% body weight for age- protein fat and carbohydrate-deficient diet	Infants <1 year, failure to thrive Dry, thin, loose, wrinkled skin Hair loss; fine brittle hair, alopecia Fissuring and impaired growth of nails Loss of subcutaneous fat and muscle mass Loss of buccal fat (monkey facies) Alopecia Angular cheilitis
Kwashiorkor: body weight less than 60–80% predicted weight with a protein- or fat-deficient diet	Children between 6 months and 5 years Failure to thrive, edema Generalized dermatitis, "flaking enamel paint" dermatosis, "cracked pavement" dermatosis Increased pigmentation on arm and legs Hair color changes (dark brown hair becomes a rusty red, light-colored hair becomes gray-white) flag sign, alopecia
Marasmic kwashiorkor-stunted growth	Combination of the above symptoms with edema of extremities, face
Protein calorie malnutrition	Skin: generalized dermatitis, "flaking enamel paint" dermatosis or "cracked pavement" dermatosis, edema, increased pigmentation on arms and legs Hair: dark brown hair becomes a rusty red, light-colored hair becomes blonde (flag sign)
Fats	
Essential fatty acids	Skin: xerosis, scaly, diffuse erythema, poor wound healing, traumatic purpura Hair: Hyper or hypopigmentation of hair, alopecia Nails: brittle nails
Minerals	
Copper (genetic—Menkes disease)	Skin: hypopigmentation, follicular hyperkeratosis, inelastic depigmented skin at nape of neck, axillae, trunk Hair: kinky or steel wool hair: short, sparse, lusterless, tangled, depigmented
Copper deficiency—acquired	Skin: poor wound healing Hair: hair loss, alopecia
Iron	Skin: generalized pruritus Hair: lusterless, dry, focally narrow and split hair shafts, heterochromia of black hair, hair loss (alopecia) or thinning Nails: fragile, longitudinally ridged, lamellate brittle nails with thinning, flattening of nail plates, koilonychias Mouth: aphthous stomatitis, angular stomatitis, glossodynia, atrophied tongue papillae
Selenium	Skin: loss of hair and nails, leukonychia

Table 52.1 (continued)

Nutrient	Signs and symptoms
Zinc deficiency (genetic—defect in intestinal zinc transporter (ZIP4)	Skin: eczematous and erosive dermatitis, preferentially localized to periorificial and acral areas, alopecia, superinfection with *Candida albicans* and *Staphylococcus aureus*
Zinc deficiency—acquired	Skin: psoriasiform dermatitis involving the hands and feet and occasionally the knees; dry, scaly, eczematous skin around buttocks, enlarged fingers with paronychia, bright erythema of the terminal phalanges, poor wound healing, warts
Vitamins	
Vitamin A	Skin: xerosis, skin fissuring (dermatomalacia), phrynoderma, perifollicular hyperkeratosis Ocular: impaired dark adaptation, exophthalmos, corneal xerosis, ulceration, keratomalacia, corneal perforation, blindness Mouth: xerostomia, hypotonia, hypogeusia
Vitamin D	Skin: pale, increased inflammation. Tendency for thicker and increased basal cell, squamous cell, and melanoma cancers compared to those with normal vitamin D levels Pain: hyperesthetic pain response in skin lesions, increased pain sensitivity
Vitamin E	No skin-associated deficiency signs
Vitamin K	Bruising, perifollicular hemorrhages
Thiamin	Skin: edema of face, sacrum; decreased sweating or hyperhidrosis; atrophy of the skin with distal extremity hair loss
Riboflavin	Acute: Skin: erythema, epidermal necrolysis, mucositis Chronic: Skin: seborrheic dermatitis of nasolabial folds, nostrils, nasal bridge, forehead, cheeks, posterior auricular areas, flexural areas of limbs and genitalia Mouth: angular stomatitis, cheilosis with erythema, xerosis, fissuring, glossitis, magenta tongue
Niacin (nicotinic acid)	Pellagra Skin: painful pruritic dermatitis of sun-exposed skin—scaly, dry, atrophic, in intertriginous areas, dorsum of the hands (gauntlet); dorsum of feet (gaiter) macerations and abrasions may occur; Casal's necklace in the neckline exposed to sun leads to dermatitis with erythema and hyperpigmentation; malar suborbital pigmentation (butterfly distribution), scrotal dermatitis, erythema, and hyperpigmentation Mouth: angular stomatitis, cheilitis, glossitis

Table 52.1 (continued)

Nutrient	Signs and symptoms
Folate (B9)	Skin: perirectal ulcerations, perineal seborrheic dermatitis, diffuse brown hyperpigmentation concentrated in the palmar creases and flexures Mouth: glossitis with atrophy of the filiform papillae, angular cheilitis, mucosal ulceration
Cobalamin (B12)	Skin: cutaneous hyperpigmentation—diffuse, symmetric, or scattered macules with greatest concentration on hands, nails, face, palmar creases, flexural regions, and pressure points; vitiligo or hypopigmentation Hair: hair depigmentation—localized or diffuse Mouth: glossitis—atrophic red painful with atrophy of the filiform papillae in early states may be linear glossitis, angular cheilitis
Biotin and multiple decarboxylase deficiency	Skin: erythematous, crusting, scaly dermatitis around eyes, nose, mouth and periorificial areas, alopecia, glossitis, conjunctivitis
Vitamin C	Skin: follicular keratotic plugging, perifollicular purpura, lower-extremity edema with ecchymosis, poor wound healing, dehiscence Hair: corkscrew hairs, swan-neck hairs Mouth: mucosal swelling, ecchymosis and bleeding of gingiva, hemorrhagic gingivitis, necrosis, loss of teeth

Based on data from Refs. [20, 21]

Table 52.2 Classic clinical signs of potential nutrient deficiency

Classical signs	Consider deficiency
Skin	
Xerosis	Essential fatty acids, ceramides
Petechia (pinhead-sized hemorrhages)	Vitamin A, vitamin C
Pigmentation increased in sun-exposed areas	Niacin, B12
Follicular hyperkeratosis (goose flesh)	Vitamin A, possibly essential fatty acids, B vitamins, vitamin C
"Flaky-paint" dermatosis, subcutaneous fat loss, fine wrinkling	Protein
Subcutaneous fat loss, fine wrinkling	Protein-energy
Decreasing tissue turgor	Water, protein (hypoalbuminemia)
Edema	Protein, thiamin (wet beriberi)
Purpura	Vitamin C, vitamin K
Perifollicular hemorrhage	Vitamin C
Pallor	Folacin, iron, vitamin B12, copper, biotin
Seborrheic dermatitis	Essential fatty acid, pyridoxine, zinc, biotin
Thickening of the skin	Essential fatty acid
Hair	
Easily pluckable	Protein, biotin
Sparse	Protein
Straight, dull	Protein
Coiled, corkscrew-like	Vitamin A, vitamin C
Skin of the mouth, lip, oral structures	
Angular fissures, scars, stomatitis	B complex, iron, protein, riboflavin
Cheilosis (cheilitis)	Protein, riboflavin, niacin, pyridoxine, biotin, folate, zinc
Ageusia, dysgeusia	Zinc
Tongue	
Magenta tongue	Riboflavin
Fiery, raw	Niacin
Glossitis	Pyridoxine, folacin, iron, vitamin, B12
Pale	Iron, vitamin B12
Atrophic papillae	Riboflavin, niacin, iron

Based on data from Refs. [21–23]

52.4 Skin Conditions and Medical Nutrition Therapy

Many skin conditions are seen in the primary care setting and are amenable to nutrition interventions [29]. Skin conditions reviewed below have been improved by therapeutic dietary changes and oral or topical administration of nutrition therapies. See Table 52.4. The quality of evidence from data ranges from Grades 1A to 2C. In the clinic setting, the opportunity to meld medical nutrition therapy with the pharmaceutical model is increasing as clinicians are becoming more adept at uncovering the root causes of conditions. The treatments for chronic conditions are provided with the

Table 52.3 Wheat and milk proteome association in autoimmune diseases with skin manifestations

Food protein	Autoimmunities	Signs
Wheat proteomes	Multiple sclerosis Autoimmune thyroiditis Addison's disease Alopecia Type 1 DM Dermatitis herpetiformis Arthritis osteoarthritis Lupus erythematosus Autism Osteopenia-osteoporosis Autoimmune myocarditis	Xerosis, pruritus Distal temperature changes, xerosis, pruritus Pigment changes Hair thinning, loss Vascular or temperature change, ulcerations Bullous rash Tissue warmth Macular rash, hair loss Pruritus, rashes, compulsive itching, biting Tenderness over areas of osteopenia/malacia Peripheral vascular constriction
Milk proteins	Epithelial permeability Celiac disease Crohn's disease Behcet's disease Multiple sclerosis Systemic lupus erythematosus Uveitis Type 1 diabetes Autism	Increased contact, environmental dermatitis Mouth mucosal lesions Mouth mucosal lesions Mouth, peripheral ulcerations Xerosis Macular rash, hair loss, ulcerations Photophobia Vascular or temperature changes, ulcerations Pruritus, rashes, compulsive itching, biting

Based on data from Ref. [28]

are useful tools for addressing obesity-associated skin changes, autoimmune or hypersensitivity reactions to food proteins, and other selected dermatologic conditions.

52.5 Macro- and Micronutrient Deficiencies Influencing Skin Health

The skin, hair, and nail disorders have long been associated with identified nutrition deficiencies. It is well understood that the quality and quantity of food nutrients vary. Food nutrient concentrations have decreased over the last 80 years in response to soil changes, processing, and storage practices [158, 159]. It is vital to recognize the relationship of the quality of dietary intake and nutrient deficiencies in association with dermatologic findings on physical exam. Selected references expand the role of nutrients in skin health, disease, and associated findings in insufficiency, or deficiency [2, 40, 156, 157]. An abbreviated nutrient review follows.

52.5.1 Protein

Protein deficiency is evident in many clinical settings. Causes range from inadequate intake, increased tissue requirements due to turnover, altered genetics, increased need for repair, or loss through increased excretion. The essential amino acids and conditionally essential amino acids form the peptide and protein structures which allow structural collagen and keratins to form. Collagen represents 70% of skin protein and is composed of three repeated amino acids: glycine, hydroxyproline derived from proline, and hydroxylysine derived from lysine [160]. Deficiencies of total protein, glycine, and lysine lead to altered skin integrity. In association, a diet deficient in lysine is associated with hair loss [160]. Elastin, the protein that allows skin to stretch and bounce back, is composed of leucine. Diets inadequate in protein, sulfur-rich amino acids, glycoproteins, and transport proteins promote the loss of tissue integrity. Protein deficiency and insufficiency physical exam findings are summarized in Tables 52.1 and 52.2.

52.5.2 Essential Fatty Acid

Essential fatty acid (EFA) deficiency is seen in malnutrition but also in the imbalanced fat consumption patterns in the standard American diet (SAD). SAD can be deficient in essential fatty acids or have a high omega-6 to omega-3 ratio. The predominant EFA deficiencies include linoleic, linolenic, and arachidonic acid, and plasma levels are often diminished with palmitoleic and oleic acids increased. The presence of mead acid (5,8,11-eicosatrienoic acid in the plasma) is also a marker of essential fatty acid deficiency.

Eczematous and pruritic skin changes are associated with increased cell turnover, especially the periorificial areas.

increased understanding of how individual genetics, the epigenetic influences, time course of the illness, and the power of modifiable lifestyle choices, including nutrition, sleep, exercise, thoughts and emotions, and personal relationships, all converge to impact skin conditions.

Understanding of nutrient or bioactive function is important to be able to predict the signs and symptoms of insufficiency or excesses related to skin structure, immunity, and protective barrier. Understanding nutrient and bioactive function helps direct therapeutic interventions. Whole foods in the Mediterranean diet have been used successfully to treat most of the conditions listed in Table 52.4. The frank nutrient deficiencies (pellagra, scurvy, acrodermatitis enteropathica) are specifically treated with single nutrients. Excellent extensive reviews are available to provide more in-depth information for interested readers [2, 40, 156, 157].

Macro and micronutrient deficiencies impact skin health. The application of Mediterranean food plans, the modified elimination diet, and the use of probiotic supplementation

Table 52.4 Skin conditions amenable to nutrition interventions: practical summary

Condition	Practical summary of the condition	Medical nutrition therapy
Acanthosis Nigricans Type 1 hereditary benign Type 2 benign Type 3 pseudo-AN Type 4 drug induced Type 5 malignant ICD10: L83	*Epidermal-metabolic disorder* 1: No associated endocrine disorder— can present at birth 2: Endocrine disorder associated insulin resistance, insulin-resistant type II diabetes mellitus, hyperandrogenic states, acromegaly, Cushing's disease, hypogonadal syndromes with insulin resistance, Addison's disease, and hypothyroidism (DeWitt) 3: Pseudo-acanthosis nigricans—associated with obesity, patients with darker pigmentation, metabolic syndrome, obesity producing insulin resistance 4: Drug induced—nicotinic acid in high doses, stilbestrol in young males, glucocorticoid therapy, diethylstilbestrol-oral contraceptives, growth hormone therapy 5: Malignant—paraneoplastic usually adenocarcinoma of the gastrointestinal or genitourinary tract, less commonly bronchial carcinoma and lymphoma [30–32]	1. Diet: low-glycemic index diet to decrease basal insulin levels [33] 2. Nutraceutical Supplementation 2a. EFA/DHA—fish oil 10–20 g 6 months [34] Improvement in insulin resistance and HOMA-IR score, IGF-1, and the acanthosis nigricans [35] 2b. Topical vitamin A (tretinoin 0.1% gel) applied to localized areas for 2 weeks [36, 37] 2c. Topical vitamin D (calcipotriol-calcipotriene 0.005% cream) twice a day for 3 months led to improvement and resolution [38]

Acne rosacea
ICD10:L71
	Disorders of the sebaceous and apocrine glands
	Chronic inflammatory acneiform disorder of the facial pilosebaceous units
	Increased reactivity of capillaries leading to flushing and telangiectasias
	Rubbery thickening of the nose, cheeks, forehead, and chin skin due to sebaceous hyperplasia edema and fibrosis
	Occurs in 10% of fair-skinned people
	Age of onset: 30–50 years with peak incidence between 40 and 50 years
	Females predominate
	Rhinophyma predominates in males
	Celtic person's skin phototypes 1 and 2 but also southern Mediterranean's, less frequently in skin types V and VI
	Stages: Plewig and Kligman's classifications
	   Stage 1: persistent erythema with telangiectases
	   Stage 2: persistent erythema, telangiectases, papules, tiny pustules
	   State 3: persistent deep erythema, dense telangiectases, papules, pustules, nodules, rarely persistent solid edema of the central face
	Special lesions:
	   Rhinophyma—enlarged nose
	   Metophyma—enlarged cushion-like swelling of the forehead
	   Blepharophyma–swelling of the eyelids
	   Otophyma—cauliflower-like swelling of the ear lobes
	   Gnathophyma— swelling of the chin
	Bacterial culture: *Staph aureus*
	Scapings: *Demodex folliculorum*
	Treatment: topical or systemic antibiotics, oral isotretinoin for severe disease, ivermectin for demodex, or erythema, and telangiectasias beta-blocker carvedilol 6.5 mg BID to reduce erythema and telangiectasia [39]
		1. Diet: Consider dietary triggers, alcohol, coffee, tea, spice, dairy rich in branch chain amino acids and triggers of mtorc1. Diets low in essential fatty acids [40–42]
		2. Lifestyle: emotional stress
		3a. Associated conditions treated with Diet
		3b. Consider Crohn's, celiac disease, Ulcerative colitis, SIBO, IBS (all > 1.34–1.46 RR) [42] or *H. pylori*
		4. Nutraceutical supplementation [43–49]
		4a. Zinc sulfate—23 mg TID × 3 mo decreased compared to placebo
		   Zinc sulfate—50 mg BID × 3 mo no difference
		   Zinc 60–75 mg/d for 3-mo improvement in skin compared to controls
		4b. Pancreatin (8–10 × USP) 350–500 mg before meals—daily dose led to improvement in skin compared to controls [47]
		4c. Brewer's yeast 2 g TID. Hansen CBS 5926 inhibits bacterial growth 60% [50]
		4d. B complex 100 mg/d avoid higher doses of niacin [51]
		4e. Probiotics
		   E. coli Nissle 1917 25 Billion CFU 4/d × 1 month. Marked improvement with the addition of the probiotic [52]
		   Eradication of *H. Pylori*. Improved rosacea [53]
		Topical:
		4f. Azelaic acid—15–20% cream topical BID for 15 weeks. More improvement then with metronidazole [54, 55]

(continued)

Table 52.4 (continued)

Condition	Practical summary of the condition	Medical nutrition therapy

Acne vulgaris ICD10:L70.0	*Disorders of the sebaceous and the apocrine glands*
	Occurs in 85% of young people in the last half of the twentieth century. Rare in the nineteenth century
	Acne vulgaris affects more than 80% of teenagers and persists beyond the age of 25 years in 3% of men and 12% of women [56]
	Typical lesions of acne include comedones, inflammatory papules, and pustules. Nodules and cysts occur in more severe acne and can cause scarring and psychological distress
	Most frequently in the adolescent appearing on the face, trunk, and rarely the buttocks
	Manifests as comedones, papulopustules, nodules, or cysts
	Can result in pitted, depressed, hypertrophic scars
	More severe in males than females
	Lower incidence in Asians and Africans
	Pathogenesis and triggers: follicular keratinization, androgens, *Propionibacterium* acnes.
	Bacterial lipase converts lipids to fatty acids and produces proinflammatory mediators (IL-1 and TNF alpha) leading to the inflammatory response
	Contributors: acnegenic mineral oils
	Drugs: Lithium, hydantoin, isoniazid, glucocorticoids, oral contraceptives, iodides, bromides, testosterone androgens
	Emotional stress, bowel dysbiosis [57]
		[48, 49, 58–79]
		1. Diet:
		1a. Modified elimination diet: Milk, low-glycemic index, low trans fatty acids.
		  Diet composition: 45% protein, 35% Carbohydrate, 20% fat. This increases the saturated/monounsaturated ratio which is associated with decreasing the lesion counts and increasing the sebum flow [64]
		  Milk consumption: Nurses' health Study II. Severe acne 22% higher in highest milk drinkers vs lowest milk drinkers [48, 49, 62, 79]
		  Changes in the insulin levels and the associated lesions are seen within 7 days of a low-glycemic index and more binding of the insulin-like growth factors
		1b. Avoid high carbohydrate diet (10% protein, 70% carbohydrate, and 20% fat) encourages high insulin response and promotes elevated IGF-1 promoting acne
		  Insulin-like growth factor (IGF) may play a role in acne [66, 67]. Milk consumption has been associated with increased levels of serum IGF [68]. IGF is also increased by ingestion of high glycemic loads and could potentially link diet and acne. A 12-week randomized trial that compared low- and high-glycemic load diets in 43 male patients with acne found a greater reduction in lesion counts with the low-glycemic load diet [63]
		1c. Remove skim milk > 1% > 2% since they contain endogenous hormones and increase acne levels
		2. Nutraceutical supplementation
		2a. Zinc—inhibits Toll-like receptor 2 expression. Most trials supplementation of zinc 30–150 mg (elemental) positive effect
		  Zinc gluconate 30 mg/d—66 patients. Decreased inflammatory scores compared to control [48]
		  Zinc sulfate (68 mg elemental) twice a day for 12 weeks. 56% improvement, decrease number papules, infiltrates, and cysts [49]
		  Data on favorable effects of dietary factors such as zinc, omega-3 fatty acids, antioxidants, vitamin A, and dietary fiber on acne vulgaris are limited [58, 59]
		2b. Selenium: (uncontrolled trial) 29 patients. 400 mcg/d of selenium 20 iu vitamin E for 6 weeks with improvement [69]
		2c. Vitamin A—vitamin A levels are lower in the skin and plasma of acne victims compared to controls. High-dose vitamin A has been used 100,000 iu/d for 3 months with improvements
		  300,000–500,000 iu/d of water miscible vitamin A for 3–5 months (136 patients) found half the patients improved >75% and another 40% improved >50%. Partial relapse 3–6 weeks following the discontinuation. Side effects: mild hair loss and dryness of the skin and mucous membranes. Mild elevation in triglycerides [71, 72]
		2d. Pyridoxine—uncontrolled study 50 mg/d of pyridoxine 1 week prior to menses and continuing during menstruation for three cycles, 72% of patients with premenstrual acne flares [70]
		2e. Topical niacinamide: 4% was as effective as topical 1% clindamycin in patients with inflammatory acne. 82% improvement compared to 69% improvement in the clindamycin group Applied twice a day for 8 weeks [73–75]
		2f. Topical retinoids
		  Considerations of isotretinoin
		  Side effects:
		  Increase homocysteine levels (doubling after 45 days of 0.5 mg/kg/d) [76]
		  Decrease serum 5-MTHF levels by 13.5–17.7% within 28 days [77]
		  Carnitine-responsive hypertransaminasemia, myalgia, muscular weakness, and hypotension on 0.5 mg/kg/d isotretinoin responded to oral L carnitine dose (weight-dependent)
		  25 kg 2.5 g carnitine/day
		  50 kg 5.0 g carnitine/day
		  75 kg 7.5 g carnitine/day
		  100 kg 10.0 g carnitine/day with decreased liver enzymes and symptoms within 45 days [78]

Table 52.4 (continued)

Condition	Practical summary of the condition	Medical nutrition therapy
Acrodermatitis enteropathica ICD10:E60	*Nutritional disease* Genetic: autosomal recessive. Occurs in infants bottle-fed with bovine milk or in breastfed infants following weaning Approximately 10–40% of dietary zinc is absorbed in the small bowel; absorption is inhibited by the presence of phytates and fiber in the diet that bind to zinc, as well as dietary iron and cadmium [80] Acquired zinc deficiency in older individuals due to dietary deficiency, failure of intestinal absorption of zinc (malabsorption, pancreatic insufficiency, alcoholism, prolonged parenteral nutrition, gastric bypass surgeries—Roux-en-Y—biliopancreatic diversion) Skin: patches of plaques of dry, scaly, sharply marginated and brightly red, eczematous dermatitis gradually evolving into vesiculobullous, pustular, erosive, and crusted lesions Occur periorificial (mouth, anus) then on the scalp, hands, feet, flexural regions trunk. Fingertips glistening erythematous, with fissures and secondary paronychia, perleche secondary infected with *Candida albicans* or *Staph aureus* Diffuse alopecia, graying of hair, nail ridging, and loss of nails Red glossy tongue, superficial aphthous like erosions, secondary oral candidiasis Diagnosis Lab: low serum/plasma zinc, reduced zinc excretion Acquired zinc deficiency (cystic fibrosis, anorexia nervosa, celiac disease, alcoholic cirrhosis, Roux-en-Y, chronic diarrhea). Similar dosing Evaluate copper levels since long-term zinc supplementation can lead to zinc deficiency [40] Aberrant fatty acid metabolism, with some unknown impairment in the elongation-desaturation system of fatty acid biosynthesis. In particular, low levels of serum arachidonic acid have been reported [81, 82]	Zinc Therapy: Dietary or IV 3 × the requirement for age, daily for days to weeks. Fix root cause of the zinc deficiency 1a. Zinc therapy: 1–2 mg/kg of body weight per day. Higher doses in the presence of intestinal infections or other diarrheal states [83, 84] Higher replacement doses (approximately 3 mg/kg/day of elemental zinc) are used for acrodermatitis enteropathica, to overcome the defect in intestinal zinc absorption 1b. Adjusting for diarrhea—WHO Supplementation reduces the severity and duration of acute diarrhea in children from populations in which zinc deficiency is common. For infants and children with acute diarrhea in developing countries, the supplements are given at a dose of 20 mg/day for children or 10 mg/day for infants younger than 6 months old, for 10–14 days [80, 85–87] Recovery: 1–2 weeks diarrhea ceases, irritability, and depression can improve within 24–48 hours

(continued)

☐ Table 52.4 (continued)

Condition	Practical summary of the condition	Medical nutrition therapy

Atopic dermatitis ICD10:L20	*Eczema-dermatitis*
	Acute, chronic, chronic relapsing skin disorder. Common in infancy, peaks 15–20% in early childhood
	Dry skin and pruritus, increased inflammation and lichenification with rubbing to further the itch-scratch cycle
	Diagnosis based on symptoms
	Family history, allergic rhinitis, asthma all related. 35% of infants develop asthma in later life
	Associated with skin barrier dysfunction
	IgE reactivity
	Genetic basis influenced by environmental factors. Alterations in immunologic responses to T cells, antigen processing, inflammatory cytokine release, allergen sensitivity, infection, miRNA-associated [88]
		1b. Anti-inflammatory diet
		   High maternal intake of margarine and vegetable oil during pregnancy through the first 2 years of life was associated with increased eczema in the offspring. A high maternal fish intake is inversely associated with eczema during the first 2 years in the offspring [89]
		1b. Modified elimination diet
		   IgE directed with lowering of the IgE causes stabilization and partial regression of skin manifestations [90]
		2. Probiotics
		2a. *Lactobacillus rhamnosus* 10–50 cfu/d decreased risk of eczema vs control ($n = 474$ infants) DBCP with mothers taking probiotics for 35 weeks to 6 months lactating postpartum and infant taking through 2 years of age. No reduced atopy [91]
		2b. *L. reuteri*—100 million CFU for 36 weeks gestation to delivery. Less IgE associated eczema during the 2nd year 8% compared to 20% placebo. Skin prick test less common 14% vs 31% placebo [92]
		2c. *L. rhamnosus* GG and *L. reuteri*
		3. Zinc 10–40 mg/day
		4. Riboflavin—if scrotal dermatitis is present and persistent [93]
		5. Essential Fatts
		   Fish oil 2–5 g/d- DHA 5.4 g/day ($n = 21$) for 8 weeks showed clinical improvement in the SCORAD (severity scoring of atopic dermatitis) compared with saturated fat control [94]
		6. Vitamin D oral 1000–5000 iu/d
		   Topical
		   Zinc (if determined to be low <10 umol/L) has more recurrent infections than atopics with higher zinc levels. They also have higher copper levels, lower ferritin than controls [95, 96]

Dermatitis herpetiformis ICD10:L13.0	*Genetic, acquired bullous disease* IgA deposition in the skin. Antibodies to transglutaminases are major autoantigens. IgA activates complement via alternative pathway. Associated with HLA-B8, DR, DQ Distribution: scalp, extensor surface arms, back, upper gluteal area, thighs Remove wheat, gliadin Consider pharmacologic interventions: (diaminodiphenyl sulfone, sulfapyridine) (Ronaghy) [97] Associated with celiac disease, rheumatoid arthritis, hypothyroidism and Sjogren's syndrome. IgA autoantibodies toward an enzyme type 3 epidermal transglutaminase [97]	1. Modified elimination diet Gluten-free 2. Inositol—600–1800 mg/day ($n = 5$) 4/5 patients with marked improvement taking the supplement 4–12 months [98] 3. Niacin 50–200 mg ($N = 12$) 50% suppressed disease, 10/12 improvement [99]
Follicular hyperkeratosis Keratosis pilaris ICD10:E50.8	Disorder of follicular keratinization. Keratotic follicular papules and variable perifollicular erythema. Manifests in childhood or adolescence with spiny keratotic papules on the extensor aspects of the back of upper arms and distal extremities. May include phrynoderma caused by linoleic acid insufficiency, vitamin A deficiency, vitamin E, or general undernutrition, including B vitamin insufficiency [1, 100, 101]	1. Essential fatty acids 1a. Flaxseed oil 21 cc/d or linoleic acid 5 cc/d ($n = 61$) 100% lesion disappearance after 16–20 weeks with flax and 12 weeks with the linoleic acid [102, 103] 2. Vitamin A deficiency (Olson) Normalization to skin of children with follicular hyperkeratosis at >100,000 iu/d for 2–4 months (range 100,000–300,000 iu/day) [104, 105] 3. B vitamin and vitamin A injections (unlisted levels) over 10–14 days with >80% resolution [106, 107] 4. Vitamin C insufficiency associated with development of hyperkeratosis [108, 109]

The Skin, Selected Dermatologic Conditions, and Medical Nutrition Therapy

Table 52.4 (continued)

Condition	Practical summary of the condition	Medical nutrition therapy
Keratosis		
Actinic ICD10:L57.0	Summary [110] Precancerous lesions on a spectrum from photodamaged skin to squamous cell cancer Strongest predictor of development of non-melanoma skin cancer and melanoma Increased susceptibility with cumulative UV radiation exposure, immunosuppression, prior skin cancers Risk of progression of actinic keratosis to squamous cell cancer ranges from less than 1 to 20% Risks: older age, male, fair skin that burns, blond or red hair, light-colored eyes, immunosuppression Common non-nutrition treatments: 5 fluorouracil solution or cream, imiquimod cream, diclofenac gel, cryotherapy, curettage with or without electrosurgery, shave excision	1. Diet: high fat intake increases the incidence of ultraviolet, light-induced skin cancers, reducing dietary fat intake decreased the incidence of actinic keratosis. If 40% of the calories are fat, decreasing to 20% of calories being fat decreased the incidence within 2 years of increasing number of additional actinic keratosis [111] 2. Niacinamide 500–1000 mg a day in divided dose decreased the number of new actinic keratosis and induced regression of some or all of actinic keratosis when compared to placebo groups [112–115] 3. Retinoids 3a. Oral: Vitamin A 100,000 iu/d of vitamin A for 19.8 months led to improvement in 90% of actinic keratosis and complete disappearance in 26% of the cases [116]. No side effects of the vitamin A were mentioned in the study 25,000 iu vitamin A for 3.8 years compared to placebo. 26% decrease in new squamous cell carcinoma, nonsignificant 6% increase in basal cell carcinoma in 2297 patients with at least 10 actinic keratosis (double-blind study) [117, 118] 3b. Topical retinoids have been studied for the treatment of AKs. In a 9-month randomized trial of 90 patients with multiple AKs, adapalene gel (0.1 or 0.3%) applied daily as field therapy for 4 weeks and twice daily thereafter significantly but modestly decreased the number of AKs compared with placebo and also appeared to improve the appearance of photodamaged skin [119]. In contrast, a randomized trial including 1131 patients found that long-term use of topical tretinoin 0.1% cream was ineffective in reducing the number of AKs [120]
Arsenical ICD10:L85.8	Summary [110] Precancerous lesions associated with chronic arsenic. Increased risk of becoming squamous cell cancers Highest risk for SCC (squamous cell carcinoma) with MTHFR TT polymorphism and low folate intake [121] Exposure to arsenic results from medicinal, occupational, and water source or other environmental exposures The risk of skin lesion due to arsenic exposure is modifiable by nutritional factors, such as folate and selenium status, lifestyle factors, including cigarette smoking and body mass index, and genetic polymorphisms in genes related to arsenic metabolism [122] Keratotic, yellow papules overlying pressure points on the hands and feet Latency period can up to 40 years Normally cryosurgery or curettage. 5-FU not as effective as when treating actinic keratosis	1. Check for sources of arsenic- water, diet, living conditions Remove arsenic source 2. Testing consider pre- and post-provocation testing for arsenic load determination 3. Improve methylation factors 4. Significant associations between low intakes of various nutrients (retinol, calcium, fiber, folate, iron, riboflavin, thiamin, vitamins A, C, and E) and keratotic skin lesion incidence in people exposed to environmental arsenic [122] Greater intakes of methionine, cysteine, protein, and vitamins such as thiamin and niacin increased arsenic secretion [123] Curcumin 500 mg BID × 3 months. Significantly less DNA, antioxidant damage, increased clearance (Biswas). Studies up to 10 g/day [124] Increased skin lesions in the undernourished underweight, lower BMI patients (decreased clearance) 1.65×
Seborrheic keratosis Dermatosis papulosa nigra ICD10:L82.1	Most common benign epithelial tumors. They are hereditary, small, to barely elevated with a stuck-on appearance Onset rarely before 30 years and more extensive in males Tan to black macules. Distribution on the face, trunk, upper extremities, submammary intertriginous. Darker pigmented in darker-skinned people	1. Increased association with VDR Taq1 [125] Consideration of increasing 25 OH vitamin D [126] With 30% response when placed topically [127, 128]

(continued)

◻ Table 52.4 (continued)

Condition	Practical summary of the condition	Medical nutrition therapy
Pemphigus ICD10: L10	Serious acute or chronic, bullous autoimmune disease of the skin, and mucous membranes due to acantholysis Pemphigus vulgaris—flaccid blisters of the skin and erosions on mucous membranes Pemphigus foliaceus—scaly and crusted skin lesions IgG autoantibodies to desmogleins, transmembrane desmosomal adhesion molecules There can be environmental triggers Standard of care: modified elimination diet. Immunosuppressive agents Skin lesions: flabby vesicles, easy ruptures and weeping, scattered, discrete, localized or generalized. Erosions bleed easily. Crusts on the scalp Sites of predilection: scalp, face, chest, axilla, groin, umbilicus., mucous membranes- nose, pharynx, larynx, vagina Drug-induced pemphigus: captopril, D-penicillamine Brazilian pemphigus: black fly, *Simulium nigramanum*, triggered in the south-central part of Brazil	1. Modified elimination diet 1a. Both environmental and dietary triggers in acantholysis. The dietary factors implicated include allelic compounds in the genus *Allium* (garlic, onion, leek) 1b. The thiolic compounds can be acantholytic [129]. There is a high association of pemphigus in communities with high intake of mustard, red and black pepper, coriander, cumin seeds (India), or in communities with high tannins in the drinking water (Amazon basin) [129] 1c. Antigliadin antibodies were found in serum of the majority of patients with bullous pemphigoid raising the possibility of gluten sensitivity playing a role in pemphigoid [28, 40, 130] 2. Niacinamide or Niacinamide with tetracycline treatments. Niacinamide: case reports Niacinamide 500 mg TID [131] Niacinamide + tetracycline (N = 18) Niacinamide 500 mg TID + tetracycline 500 mg QID [132]
Pellagra ICD10; E52	Diet deficient in niacin or tryptophan or both. Consumption of a maize-based diet is often implicated Characteristic features: dermatitis in sun-exposed areas, diarrhea, cognitive changes leading to dementia and finally death Symmetric itching, painful erythema on the dorsa of the hands, neck, and face Vesicles and bullae may erupt and break, crusting occurs and lesions become scaly. Later skin becomes indurated, lichenified, rough, covered with dark scales and crusts Cracks and fissures and sharp demarcation with normal skin Distribution: dorsa of the hands and fingers, band like around the neck (Casal's necklace), dorsa of the feet up to malleoli and sparing the heel, and butterfly region of the face [133]	1. Niacinamide orally 100–300 mg and the other B vitamins 2. Pellagra Treatment Adult Oral: 50–100 mg 3–4 times daily maximum: 500 mg daily [134–136] Some experts prefer niacinamide for treatment due to more favorable side effect profile [81, 82, 137] Pediatric 0–5 months: 2 mg daily 6–11 months: 3 mg daily Oral: 50–100 mg 3 times daily Some experts prefer niacinamide (nicotinamide) for treatment due to more favorable side effect profile [81, 82, 137] 3. Pellagra with comorbidities 3a. Diabetes: Niacin (nicotinic acid) may increase fasting blood glucose, although clinical data suggest increases are generally modest (<5%) [138] Use niacin with caution in patients with diabetes. Monitor glucose; adjustment of diet and/or hypoglycemic therapy may be necessary. Niacin should not be used if patient experiences persistent hyperglycemia during therapy [139] 3b. Gout: May be associated with hyperuricemia. Use with caution in patients predisposed to gout. Niacin should not be used if patient experiences acute gout during therapy [139] 3c. Hepatic impairment: Use with caution in patients with a past history of hepatic impairment; monitor liver function tests. Contraindicated with active liver disease or unexplained persistent transaminase elevation. Niacin should not be used if hepatic transaminase elevations >2–3 times upper limit of normal occur during therapy [139]

Table 52.4 (continued)

Condition	Practical summary of the condition	Medical nutrition therapy

Psoriasis vulgaris ICD10: L40.0	1.5–2% of the population in western countries
	Chronic disorder—polygenic predisposition and triggering environmental factors including bacterial infection, trauma, drugs, alcohol, stress, weight gain
	Typical lesions chronic, recurring, scaly papules, plaques. Pustular eruptions and erythroderma occur
	There can be localized or generalized plaques, pustular psoriasis, palmoplantar pustulosis (L40.1)
	Peak incidence 22.5 years, children 8 years, older adults 55 years. Early onset predicts more severe long-lasting disease
	No gender preference
	Race incidence: low incidence in west Africans, Japanese, Inuits, absence in north and native south American
	Genetics: one parent—offspring 8%. Both parents 41%
	HLA-B13,B27,B57, Cw6. PSORS1 consistent susceptibility locus
	Triggers: physical trauma. Acute strep infections, stress flares psoriasis in 40% adults and higher in children
	Drugs: glucocorticoids, oral lithium, antimalarial drugs, interferon, beta-adrenergics, alcohol
	Pathogenesis: altered cell kinetics—shortening of cell cycle resulting in 28 times the normal production of epidermal cells and CD8+ T cells and overwhelming T-cell populations in lesions. Psoriasis is a T-cell-driven disease in the Th1 response of cytokine spectrum
	Acute guttate: disseminated, generalized mainly on the trunk
	Chronic stable: single lesions or lesions on the elbows, knees, sacral gluteal region, scalp, palm/soles
	Nails: fingernails and toenails involved 25%, especially with concomitant arthritis. Pitting, subungual hyperkeratosis, onycholysis, yellowish-brown spots under the nail plate, oil spot
	Skin: thickening of the epidermis (acanthosis)
	Arthritis (L40.5) seronegative spondyloarthropathies. Asymmetric peripheral joint involvement of the upper extremities, distal interphalangeal joints, with dactylitis-sausage fingers
	Enthesitis: inflammation of ligament insertion into the bone
	Bone erosions. Often associated with psoriatic nails
	MHC class 1 antigens, RA is MHC II antigens
	Incidence 5–8% rare before age 20
	10% of individuals without any visible psoriasis—but a family history
	Increase BMI, sarcopenic obesity
	Alcohol, smoking
	Low Mediterranean diet score
		1. Modified elimination diet
		1a. Remove refined sugar and alcohol
		1b. Gluten-free diet
		   Observational studies: heavy alcohol consumption is associated with increased severity of psoriasis, in males but not females [69, 140, 141]
		1c. Food allergy: gluten intolerance increased serum IgA and IgG antigliadin antibody positive improved psoriasis and antigliadin antibody titers with a gluten-free diet
		   Improvement on a vegan diet [142]
		   Treatments for Psoriasis Presentations
		A. *Plaque psoriasis*
		1. Whey Protein
		   Some clinical evidence shows that taking a specific whey protein extract product (Dermylex, Advitech, Inc.) 5 grams daily for 8 weeks can decrease the severity of psoriasis symptoms when compared with placebo
		2. Omega-3 fatty acids: 3.6–14 g/d
		   There is some evidence that administering fish oil intravenously can decrease the severity of symptoms in patients with acute, extended guttate psoriasis and chronic plaque psoriasis. Fish oil seems to be superior to omega-6 fatty acids for this use. However, fish oil taken orally doesn't seem to have a significant effect on psoriasis despite promising preliminary research
		   But when administered in combination with suberythemal ultraviolet B (UVB) therapy, oral fish oil therapy seems to decrease the total body surface area of psoriasis compared to placebo
		   Preliminary evidence suggests drinking saffron tea daily, along with a specific diet rich in fruits and vegetables, may reduce the severity of psoriasis
		3. Topical treatment
		   Vitamin D analogues
		   Applying vitamin D in the form of calcitriol or other vitamin D analogues such as calcipotriene, maxacalcitol, and paricalcitol topically effectively treats plaque-type psoriasis in some patients, including those with chronic plaque psoriasis [143]
		B. *Psoriatic arthritis* [144, 145]
		1. Zinc
		   Marked improvement in arthritis, biomarkers, pain scores after 12–24 weeks of 125–600 mg of zinc sulfate in divided doses. Copper levels significantly dropped
		2. Inositol—6 g/day for lithium-induced psoriasis
		3. Vitamin D: 1,25 dihydroxycholecalciferol [146]. 85 patients who received oral calcitriol, 88.0% had some improvement in their disease; 26.5, 36.2, and 25.3% had complete, moderate, and slight improvement in their disease, respectively. The mean baseline psoriasis area severity index score (PASI) of 18.4 ± 1.0 was reduced to 9.7 ± 0.8 and 7.8 ± 1.3 after 6 and 24 months on oral calcitriol therapy

(continued)

Table 52.4 (continued)

Condition	Practical summary of the condition	Medical nutrition therapy
Scurvy ICD10: E54	Acute or chronic disease caused by deficiency of ascorbic acid Associated symptoms: lassitude, weakness, arthralgia, myalgia Skin lesions petechial, follicular hyperkeratosis, perifollicular hemorrhage especially of the legs Hair de-pigmented, buried in perifollicular hyperkeratotic papules Extensive ecchymosis Splinter hemorrhages Gingivitis, purple, spongy, gums that bleed easily. Loosening teeth Hemarthrosis, epiphyseal separation, scorbutic rosary Vitamin C has a half-life of 10–20 days. However, signs of deficiency generally develop after 1–3 months of inadequate vitamin C intake [147] Labs: normocytic, normochromic anemia Functional test: Hess test	1. Vitamin C. Supplementation with 1 g/day of oral vitamin C for 2 weeks is the usual treatment [147–149] Vitamin C 100–200 mg/day for a longer period [149]
Urticaria ICD10: L50.9	Diet: Food allergy—IgE—intervention. Modified elimination diet after identification of IgE reacting foods Removal of salicylates, benzoates, histamines from the diet if the removal of foods triggering igE is not therapeutic Common urticarial evokers: dairy, chocolate, eggs, fish, yeast, and wheat Removal of food additives: tartrazine (FD&C yellow #5), sodium benzoate, sunset yellow, amaranth, patent blue, indigotin, choline yellow, yellow orange, butylated hydroxyanisole (BHA), butylated hydroxytoluene (BHT), parabens, gum tragacanth and sulfites. Fluoride from toothpaste, municipal water supplies, and vitamin preparations Penicillin in consumed cow's milk Nutraceutical supplementation	[150] 1. Modified elimination diet Modified elimination diet after identification of IgE reacting foods Removal of salicylates, benzoates, histamines from the diet if the removal of foods triggering igE is not therapeutic Common Urticarial evokers: dairy, chocolate, eggs, fish, yeast, and wheat. 2. Removal of food additives: tartrazine (D and C yellow #5), sodium benzoate, sunset yellow, amaranth, patent blue, indigotin, choline yellow, yellow orange, butylated hydroxyanisole (BHA), butylated hydroxytoluene (BHT), parabens, gum tragacanth and sulfites. 3. Fluoride from toothpaste, municipal water supplies, and vitamin preparations 4. Association with antibiotics in the food supply. Pennicillin consumed in cow's milk is associated with urticaria.
Xerosis ICD10:L85.2	Further genetic evaluation: consider FADS polymorphisms FADS1 (FEN1 rs174537G > T), FADS2 (rs174575, rs2727270), and FADS3 (rs1000778). Over time rs174537T had lower arachidonic, AA/linoleic acid and higher IL-6 levels than rs174537GG, and higher oxidized LDL, lower EPA, and oxidative stress markers. The higher the arachidonic acid, the greater the urinary F2 isoprostanes, oxidized LDL and IL6. Serum and urine markers of inflammation. $N = 122$. No intervention studies. FADS SNP may play a role in incidence of xerosis	1. Nutraceutical supplementation [151–154] 1a. Flax seed oil 2.2 grams orally decreased sensitivity, total evaporative water loss, skin roughness and scaling, while smoothness and hydration increased. Water loss decreased by 10% 2.2 g fatty acids with flaxseed (alpha-linolenic acid/alpha-linoleic acid) and borage oil (alpha-linolenic acid/gamma-linolenic acid) had similar effects with the flaxseed oil group with less TEWL by 12 weeks 1b. Topical: Twice a day mineral oil or coconut oil with significant improvement in total evaporative water loss and skin pH [155]

With the increasing temporal length of EFA deficiency, there is gradual skin depigmentation and alopecia (telogen effluvium). In younger patients, there is delayed healing, capillary fragility, and a gradual diminishment of hepatic, renal, and neurologic function. Deficiencies or imbalances in free unsaturated fatty acids are also involved in psoriasis and acne. Deficiency of linoleic acid and 18-carbon polyunsaturated fatty acid results in scaly skin disorder and excessive epidermal water loss [10, 11]. In psoriasis, omega-6 arachidonic acid ratio to omega-3 EFA are up to 22 times higher than in non-affected skin [10, 11].

In other conditions, omega-6 arachidonic acid to omega-3 ratios are elevated with an inflammatory presentation most often characterized as atopic dermatitis. Atopic dermatitis is a complex chronic inflammatory disease with altered lipid composition in the stratum corneum. This altered composition leads to decreased levels of filaggrin and excessive or lessened amounts of proteases leading to respective thinner or thicker stratum corneum. With deficiency, there is associated increased water loss and poor barrier function.

When considering atopic dermatitis, you have to consider essential fatty acid metabolism and the role of skin ceramides in skin changes [161, 162]. There appears to be a hydrocarbon chain length deficiency in the ceramides and alterations in the free fatty acid fraction of the stratum corneum lipids in patients with atopic dermatitis. Also, ceramide species greater than 50 carbon atoms in length are expressed at a much lower level than in normal skin, which correlates with higher total evaporated water loss [161]. In addition, there is altered expression of delta 5 and delta 6 desaturases, which result in deficiencies in downstream metabolites of linoleic acid [161]. Gamma-linolenic acid, safflower oil as one type, is useful in mild atopic conditions with improved function.

52.5.3 Fat-Soluble Vitamins

Vitamin A deficiency leads to ichthyosis-like skin changes with generalized fine scaling and a thickening of the stratum corneum (phrynoderma is an alternative name) [163]. The pronounced hyperkeratosis of the follicular openings is associated with effluvium and fragility of hair. There is associated impaired night vision, inability for rhodopsin reconstitution in response to bright light. The conjunctival epithelium becomes metaplastic and thickened with keratoconjunctivitis sicca (Bitot's macules) which progresses to keratomalacia with scarring and blindness [22, 163]. The constellation of vitamin A deficiency includes increased epidermal neoplasms, extracutaneous growth failure, mental retardation, asthma, and immune dysfunction. Plasma retinol levels are diagnostic.

Beta-carotene is converted to vitamin A in the intestine by the enzyme beta-carotene-15,15' monooxygenase (BCMO1) to support vision, reproduction, immune function, and cell differentiation in many tissues, including the skin. Beta-carotene monooxygenase single nucleotide polymorphisms (SNPs), BCMO1 and BCMO2, are associated with decreased availability in conversion of beta-carotene to retinol. There is also considerable variability in the activity of BCMO1. There are five single nucleotide polymorphisms (SNPs) involved in the enzymatic conversion of beta-carotene to retinol [164]. Up to 71% of the people have at least a 50% reduction in their ability to convert beta-carotene to retinol and therefore may require supportive therapy [88].

Vitamin D, as a vitamin and a hormone, has a cascade of bioactive molecules that emanate from the skin [165]. Vitamin D deficiency has only recently been associated with skin signs or symptoms, including increased inflammation, higher associations with basal, squamous cell, and melanoma cancers [166], and increased pain and sensitivity or hyperesthetic pain response to lesions of the skin [167].

Vitamin D inhibits the activation and induces the generation of regulatory T cells. It promotes tolerance and inhibits immunity after stimulation with specific antigens. It promotes the migration of T cells from the vessels of the dermis to the epidermis. Vitamin D suppresses the expression of Class 2 MHC molecules and stimulates the production of IL-10. In the keratinocyte, it is a major regulator of cathelicidins and defensins as noted above [167]. Adequacy of vitamin D is imperative for normal cell immunity [167]. Therapeutically, it is used to treat morphea and psoriasis [146, 167]. With greater understanding of vitamin D single nucleotide polymorphisms, it is shown that the Taq1 SNP is associated with increased solar keratosis, BSM SNP with increased squamous cell cancers, and the Fok-1 Snps with increased risk of malignant melanoma, while the BSML SNP is associated with a decreased risk of malignant melanoma [121, 126, 168, 169].

52.5.4 Water-Soluble Vitamins

Thiamin is involved in carbohydrate metabolism, and a deficiency of thiamine gives rise to the classic deficiency disorder, beriberi. Moreover, thiamin deficiency is seen in patients who have had gastric bypass surgery, GI disease, alcoholism, pregnancy and lactation, diabetes mellitus, and a diet restricted to polished rice. In fact, up to 29% of patients seeking bariatric surgery for obesity are deficient at presentation [170]. Many of these patients have acanthosis nigricans, stasis dermatitis, and insulin resistance with xerosis. The obese patient has a greater incidence of thiamin insufficiency, especially when consuming a higher-carbohydrate diet [170]. It becomes a cycle of thiamine inadequacy and energy insufficiency, especially when there is a glucose-induced downregulation of thiamine transporters in the kidney. The thiamin transporters in the kidney are found in the proximal tubular epithelium (THTR-1, THTR-2) and are downregulated by 37%, which produces thiamine insufficiency in diabetes [171]. The mucocutaneous changes including edema and glossitis with glossodynia. There is associated peripheral neuropathy, confabulation (Korsakoff syndrome), and encephalopathy (Wernicke's encephalopathy). Low excretion levels of urinary aneurin are diagnostic.

As an organ, the skin uses a lot of energy. Conditions of increased turnover, such as psoriasis and atopic dermatitis, only increase its energy use. Thiamin is critical in energy production. Supplementing with benfotiamine (fat-soluble thiamin) has been associated with blocking three pathways which perpetuate tissue damage in type 2 diabetes [172]. Benfotiamine treatment leads to an increased intracellular thiamine diphosphate level, which is a critical cofactor for transketolase. Transketolase directs the advanced glycation end products (AGE) and lipoxidation end products substrates to the pentose phosphate pathway. Therefore, benfotiamine is important for reducing tissue AGE and has been shown to decrease hemoglobin A1c levels [173].

Riboflavin has impaired absorption in many settings. Diets poor in riboflavin, primary deficiencies, and diminished intestinal transport are not the only cause of inadequate riboflavin. Endocrine abnormalities (aldosterone and thyroid insufficiency), specific medications used in treatments of acne (tetracycline antibiotics) and depression (tricyclic antidepressants), other medications (galactoflavin, phenothiazines), and alcohol abuse interfere with the utilization of riboflavin [174]. Riboflavin deficiency can lead to seborrheic dermatitis scaling on the face (nasolabial folds), behind the ears, in the eyebrows, and on the scalp as well as in the genitocrural regions. There is often cheilitis, perleche, pallor, and atrophy of the tongue. Ophthalmologic blepharitis, conjunctivitis, and corneal vascularization are often seen. Riboflavin is a cofactor of flavin mononucleotide (FMN) and flavin adenine dinucleotide (FAD), which is involved in redox reactions. Laboratory testing in riboflavin deficiency include evaluating normochromic anemia and decreased RBC glutathione reductase activity. To correct deficiency, riboflavin can be dosed at 3–10 mg by mouth or 2 mg IV daily and is adequate for reversal of deficiency [174].

Niacin (nicotinic acid) and niacinamide (nicotinamide) are converted to nicotinamide adenine dinucleotide (NAD+), which has many metabolic roles. It is an essential cofactor for ATP production. When added to therapeutic regimens, niacinamide is a cofactor in glycosylceramide and sphingomyelin synthesis and increases the biosynthesis of ceramides and stratum corneum lipids to improve the epidermal permeability barrier. Altered barrier function from inadequate niacinamide allows the "outside to get in" and plays a significant role in the pathogenesis of atopic dermatitis and acne [175].

Niacin aids in the treatment of acne, atopic dermatitis, and psoriasis [175]. It is also therapeutic in autoimmune blistering conditions, such as pemphigus [112, 113], and in pellagra dermatitis clinically seen in undernourished populations where food access is challenged. Nutrient absorption is diminished, or there are increased requirements in many patient groups. There is a greater metabolic requirement in the chronically ill, alcoholic, those with psychiatric disorders, or those with increased exposure to environmental toxins [176]. In the metabolic disorder, Hartnup disease, or pellagra-like dermatosis, photosensitive rashes are seen. Photosensitive skin eruptions are seen in frank niacin or niacinamide deficiency.

It is felt that patients with skin cancers have a greater susceptibility to the immunosuppressive effects of sunlight [112, 113, 177]. Niacin acts as a glycolytic blockade induced by UV radiation and has been shown to be a skin-cancer chemopreventive in a randomized double-blind placebo-controlled phase 3 trial [113]. In 12 months, the rate of new non-melanoma skin cancers was 23% lower than the placebo group. Similarly, new squamous cell carcinomas, basal cell cancers, and actinic keratosis were, respectively, 30%, 20%, and 13% lower than controls.

Because of the role of niacin in so many metabolic processes, there is often clinical association between neurologic and dermatologic conditions that can be treated with niacin or niacinamide. For example, although dermatologic findings in autistic spectrum children are multifactorial, niacinamide is a key nutrient in improving mitochondrial dysfunction with associated beneficial improvements in skin disorders in this population [178]. Autism is also associated with inflammation and increased TNF alpha in lymphocytes. When niacinamide and other B vitamins are used therapeutically for mitochondrial resuscitation, inflammation and increased TNF alpha are positively impacted [178]. Niacinamide improvement in rash-associated mast cell activation of autistic children is also well documented [179]. Therefore, it is not surprising that niacin treatment, while benefiting many of the neurologic conditions, also has a positive impact in skin conditions. Tobin cleverly states "the skin is our brain on the outside" [4]. Niacinamide and niacin are safe at higher doses [180].

Pyridoxine is a cofactor for the enzymes, transaminases, synthetases, and hydroxylases important for amino acid metabolism [22, 181]. Pyridoxine also plays a role as the cofactor for enzymes involved in the essential fatty acids, linoleic and arachidonic acid metabolism. Deficiency of pyridoxine is associated with seborrheic dermatitis-like skin changes in the periorificial distribution affecting the eyes, nose, mouth, and the cheilitis of the lips and the glossitis of the tongue. Knowing of the cofactor roles in amino acid and essential fatty acid metabolism, this is no surprise. Checking pyridoxine levels helps in the diagnosis when deficiency is suspected [22].

Pantothenic acid is a component of coenzyme A, which is involved in the Krebs cycle, fatty acid oxidation and synthesis, and synthesis of heme, cholesterol, and acetylcholine. It also plays a role in amino acid catabolism [182]. Deficiency is associated with hair loss, cheilitis, abnormal gait, ulcerations of the colon, fatty infiltrations of the liver, and altered mineral metabolism. Painful burning sensation in the feet is also reported to improve with pantothenic acid supplementation [182]. Glossitis and cheilosis that were unresponsive to other B vitamin supplementations are reported to respond to calcium pantothenate [183]. Pantothenic acid has been used in the treatment of acne and lupus when deficiency is identified [184, 185].

Biotin (vitamin H) is a cofactor in carboxylases, including biotinidase and holocarboxylase [22]. Urinary organic acid accumulation is the consequence of multiple carboxylase deficiency resulting from failure to recycle biotin. Biotin deficiency may be genetic or acquired with accompanying diagnosis by hyperammonemia and organic aciduria. Biotin is beneficial in the treatment of exfoliative dermatitis of acral skin, cheilitis, and periorificial dermatitis. In deficient newborns, erythroderma and alopecia are identified. Enteritis may also be associated with biotin deficiency states. Extracutaneous symptoms include metabolic acidosis, developmental delay, hearing loss, paresthesias, seizures, and conjunctivitis. Also known to occur in biotin deficiency states is impaired cellular immunity with a predisposition for infections, including *Candida*.

Folate deficiency is associated with mucocutaneous changes similar to B12 deficiency with hyperpigmentation of skin, glossitis, cheilitis, and mucosal erosions [22]. Diagnosis of deficiency may be made with either serum folate levels or formiminoglutamic acid (FIGLU), which is an intermediate metabolite in L-histidine catabolism in the conversion of L-histidine to L-glutamic acid. It may be an indicator of vitamin B12-folic acid deficiency, or liver disease. Folate by itself is vital to the skin, supporting many biochemical processes that impact the maintenance and function of healthy skin [186]. Medical conditions negatively impacting small intestine folate hydrolase essential for normal nutrient balance include alcoholism, celiac disease, and Crohn's disease [186] (◘ Fig. 52.1).

Folates are photosensitive compounds. They have a role in regulating melanin production through the synthesis of guanosine 5′-triphosphate, a necessary cofactor in the synthesis of tyrosinase inhibitor tetrahydrobiopterin [188]. An increased number of cancers have been linked to folate deficiencies, including those of the colon, breast, pancreas, stomach, cervix, bronchus, blood, and skin [186]. The mechanism of cancer formation is thought to be related to DNA synthesis and repair for which adequate folate is essential. Atrophy of oropharyngeal mucous membrane is an early marker of

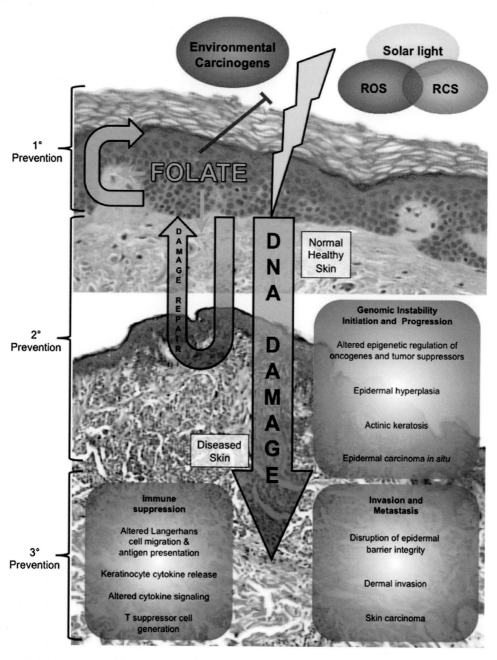

◘ Fig. 52.1 Proposed roles for optimum folate nutrition in the maintenance of skin health and the prevention of skin cancer. (Adapted from Neyestani [187]. With permission from Springer Nature)

folate deficiency [189]. Supplementation is beneficial and generally safe.

B12 deficiency can lead to atrophic glossitis, angular cheilitis, mucositis, and symmetric acral (dorsal fingers and toes) and flexural hyperpigmentation. Poliosis, vitiligo, and alopecia areata may also occur. The differential for these symptoms includes Addison's disease [22]. Many commonly used medications decrease the absorption and bioavailability or increase the requirements of B12. All of the noted signs of B12 insufficiency are more frequent when there are single nucleotide polymorphisms, increasing the baseline requirement of B12 for the patient. Frequent antibiotic use, small intestinal bacterial overgrowth, and dysbiosis should be considered if dermatologic signs of B12 inadequacy are clinically present [20, 22].

Vitamin C deficiency results from general undernutrition, alcoholism, mental illness, or chronic modified elimination diet of foods containing vitamin C. Follicular hyperkeratosis of the extensor surfaces of extremities and perifollicular hemorrhages are frequently identified. Often petechia over mechanical pressure points or under the blood pressure cuff following evaluation is a clue. Urinary tract bleeding or microhematuria without infection can be a subtle sign. Subperiosteal hemorrhages are painful and seen in the affected. Impaired wound healing, bleeding gums, gingivitis, gingival hypertrophy, and loose teeth are frequently seen clinical indicators. The accompanying fatigue, myalgias, and muscle weakness are associated with mitochondrial dysfunction. Other symptoms can include arthralgias, diarrhea, and anemia [29, 190]. Long-standing vitamin C deficiency is associated with diffuse edema, oliguria, anemia, dyspnea, and even neuropathy (see Table 52.4—Scurvy—for details).

Clinically, supplementation is key while supportive dietary intake is added and repletion of deficiency is achieved. Vitamin C is transported into the cell, whether as ascorbic acid through sodium cotransporter or as dehydroascorbic acid via the glucose transporters GLUT1, GLUT2, and GLUT4. There are two sodium-dependent transporters SVCT1 and SVCT2 that move ascorbic acid into epithelial cells [190–193]. If a person has SVCT1 with SVCT2 SNPs, they may have an increased requirement of 200 mg/d vitamin C [191]. Women with an SVCT2 SNP (SLC23A2) have an increased incidence of preterm labor [193].

52.5.5 Minerals

Zinc is an essential component in more than 300 metalloenzymes and more than 2000 transcription factors that regulate lipid, protein, nucleic acid metabolism, and gene transcription [43, 44, 194]. Zinc deficiency is known to result in dermatitis, alopecia, and diarrhea, however only 20% with zinc deficiency present with all three. Deficiency may be hereditary or acquired. Intestinal absorption is dependent on pancreatic function, zinc-binding factors, and albumin transport. About 99% of zinc is intracellular, but zinc storage is notoriously poor. Within a month of inadequate zinc intake, deficiency (zinc < 70 mg/dl) is seen [43]. Zinc deficiency is common in conditions of epidermal barrier dysfunction, activation of the innate immune system, altered cutaneous microflora, and follicular fusiform bacterial overgrowth. Long-standing zinc deficiency is associated with diarrhea, poor wound healing, growth retardation, anorexia, anemia, hypogonadism, altered mental status, and infertility. Whether the cause of deficiency is related to poor intestinal absorption, increased excretion, increased tissue catabolism, or altered transport of zinc between tissues, organs, or physiologic compartments, zinc adequacy can be affected by many conditions (See Table 52.5).

Perioral dermatitis, perleche, migrates to involve the entire face, scalp, acral sites, and diaper area [194]. The oral mucosa shows ulcerations and glossitis. Palmar erythema with annular or collarette-like scaling is present. Alopecia is associated with the dermatitis and hair shedding, also known as telogen effluvium. Telogen effluvium without the skin changes is not as frequently associated with zinc deficiency. Zinc is also associated with seborrheic dermatitis. The constellation of cutaneous infections, weeping dermatitis, delayed wound healing, secondary infections, alopecia, nail defects paronychia, *Candida* dermatitis, blepharitis, conjunctivitis, and onychodystrophy T cell impaired immunodeficiency are related to zinc deficiency. Besides acquired or genetic zinc deficiency, periorificial dermatitis can be caused by reduced protein levels, essential fatty acid deficiency, riboflavin, pyridoxine, and biotin deficiencies but also glucagonoma syndrome, a very rare tumor of the pancreas [195]. Associated non-cutaneous findings include photophobia and night blindness.

Dietary zinc is thought to activate nuclear factor kappa B (NF-kB), expression of proinflammatory cytokines (IL-1b), tumor necrosis factor alpha, and neutrophil infiltration critical for early wound healing. Many foods can block NF-kB-mediated activation of inflammatory cytokines which may play a role in the inoculation of cutaneous immunity [43].

In a recent review, zinc was used therapeutically in dermatologic conditions of infection (leishmaniasis, warts), inflammatory dermatoses (acne vulgaris, rosacea), pigmentary disorders (melasma), and neoplasias (basal cell cancer) [43]. Table 52.5 is a useful starting point for therapeutic considerations.

Iron deficiency is associated with skin pallor, dry scaly skin, perleche, glossitis, and dull-shaggy hair. If the iron deficiency is long-standing, there is telogen effluvium and nail changes, including koilonychias [196]. The laboratory will show microcytic hypochromic anemia, low ferritin and serum iron with changes in the protein-bound transport transferrin, and low intracellular storage ferritin. Subconjunctival pallor continues to be the most predictive sign that anemia is present with a likelihood ratio of over 16. If present, the underlying cause of the physical exam finding and the anemia needs then to be pursued.

Table 52.5 Topical, oral, or injected zinc in therapy for skin conditions

Selected conditions	Dose/vehicle
Infection	
Cutaneous leishmaniasis	Oral: 2.5, 5, 10 mg/kg/day for 45 days of zinc sulfate effective and safe treatment 83–96% cure rates Intralesional: clinical cure with 2% intralesional zinc sulfate comparable to meglumine antimoniate
Warts	Oral: 10 mg/kg/day oral zinc sulfate for 2 months 87% clearance at 2 months Topical: 5–10% zinc sulfate lotion, 20% zinc oxide paste apply three times a day with 50% clearance
Herpes genitalis	Topical: 1–4% zinc sulfate for 3 months was effective in treating and preventing recurrences of herpes type 2. More studies showed 4% solution most effective
Dermatophytosis (ringworm)	Topical: 20% zinc undecylenate powder twice daily for 4 weeks showed marked clinical improvement in tinea pedis- negative tinea culture in 80% by 4 weeks
Bromhidrosis (usually from *Corynebacterium* sp.)	Topical: 15% zinc sulfate solution manages bromhidrosis and foot malodor apply once daily for 2 weeks, then three times a week for 2 weeks. 70% cure rate
Pityriasis versicolor	Topical 15% zinc sulfate solution applied once daily for 3 weeks effective in treatment
Hidradenitis suppurativa	Oral: zinc gluconate 90 mg a day showed clinical improvement
Inflammatory	
Acne vulgaris	Topical: 5% zinc sulfate in mild-to-moderate acne, treating daily Oral: Zinc sulfate and zinc gluconate useful in moderate and severe acne- equal efficacy to antibiotics minocycline and oxytetracycline
Rosacea	Oral: Zinc sulfate 100 mg three times a day effective after 3 months of therapy. But the results in other studies are mixed
Psoriasis	Topical 0.25% zinc pyrithione applied twice a day to psoriatic plaques
Psoriatic arthritis	Oral: zinc sulfate 125–200 mg BID
Pigmentary	
Melasma	Topical: 10% zinc sulfate solution applied twice a month for 3 months reduces the MASI (melasma area severity index) score
Vitiligo	Oral: zinc sulfate shows moderate efficacy when given as an adjuvant with topical steroids
Neoplasia	
Basal cell cancer	Intralesional 2% zinc gluconate efficacious in treating basal cell carcinoma
Xeroderma pigmentosum	Topical: 20% zinc sulfate solution twice daily cleared solar keratosis and small malignancies
Androgenic alopecia	Topical: zinc pyrithione 1% solution considerable hair grows, though less effective than 5% minoxidil solution
Alopecia areata	Oral: Zinc sulfate 5 mg/kg/day divided dose 62% patients with hair growth

Based on data from Ref. [43]

Copper in the serum is 80–90% associated with ceruloplasmin and albumin [22]. Heritable copper deficiency results in Menkes syndrome, kinky hair syndrome, or steely hair syndrome and is due to X-chromosomal recessive mutation in copper (2+)-transporting ATPase (ATP7A). Features include saggy and hypopigmented skin; follicular hyperkeratosis; sparse, hypopigmented, and brittle hair (pili torti, monilethrix); and occasionally formation of trichorrhexis nodosa, nodes along the hair shaft that contribute to hair breakage. Frequently the patients lack eyebrows and eyelashes. On examination, the skin has tortuous elongated arteries due to immature elastin fibers. There are extracutaneous and neurodegenerative changes speaking to the scope of copper's role in normal metabolic function. Characteristically the affected newborns are normal at birth and over the first few months, followed by gradual hypotonia, seizures, and failure to thrive [22].

Copper is a critical component of lysyl oxidase, a copper amine oxidase that initiates covalent cross-linkage formation in elastin and collagen and is important in melanin synthesis. When deficient, there is faulty connective tissue formation. Lack of copper also impairs the function of the enzymes of catecholamine synthesis and metabolism. The biomarker changes include an increase in the dihydroxyphenylacetic acid to dihydroxyphenylglycol ratio. Copper supplementation should be undertaken with the need to balance with zinc and serum levels should be followed for both.

Selenium deficiency leads to effluvium and a whitish discoloration of nails. Cardiomyopathy, muscle pain, and weakness are seen in extreme cases [22]. Selenium is essential for glutathione peroxidase. A low activity of the enzyme and low plasma selenium are diagnostic of selenium deficiency. There is a relative narrow therapeutic intervention for selenium, and an intake of greater than 600 mcg/day can lead to toxicity.

52.5.6 Nutrition Excesses and Dermatologic Conditions

Hartnup disease and phenylketonuria are genetic amino acid metabolic conditions that respond to niacinamide and restriction of phenylalanine, respectively. The excess amino acids associated with inborn and acquired conditions have been identified and presented in Table 52.6. The more common mineral and vitamin excesses seen in the primary care clinic and on the hospital wards are outlined in Tables 52.7 and 52.8.

Table 52.6 Amino acids associated with inborn and acquired conditions

Amino acids	Skin
Hartnup disease	Autosomal recessive disorder of intestinal and renal amino acid transport. Sunburn-like photosensitive eruption similar to pellagra, sometimes blistering, onset less than 13 years of age. Post inflammatory hypopigmentation. Intermittent ataxia, nystagmus, tremor. Treat with a high-protein diet, oral nicotinamide supplementation Nicotinamide is photoprotective (up to 500 mg twice a day)
Phenylketonuria (PKU)	Autosomal recessive abnormality. Lack of downstream metabolic product tyrosine with the accumulation of phenylalanine. Paucity of tyrosine results in diffuse hypopigmentation of the skin and hair (blond-haired and blue-eyed). Eczematous and early-onset atopic dermatitis, scleroderma-like skin lesions. Phenylalanine accumulation is toxic to the brain, mental retardation, developmental delay, microcephaly, seizures, behavioral psychiatric problems. Urinary phenylalanine increased. Restriction of phenylalanine in diet
Tyrosinemia	Rare disorder of tyrosine metabolism. Tyrosinemia type II (Richner-Hanhart syndrome, oculocutaneous tyrosinemia), photophobia, corneal ulcerations onset during the first year of life), painful focal palmoplantar keratoderma occur in childhood and may be delayed until adolescence. Restrict diet in tyrosine and phenylalanine will clear the keratitis and keratoderma and improve the cognitive impairment
Argininosuccinic aciduria	Fragile hair (trichorrhexis nodosa); neurologic symptoms in the first days of life
Alkaptonuria	Body cannot process amino acids phenylalanine and tyrosine; autosomal recessive defect. Grayish blue discolorations of the ears, axilla, perioral area (ochronosis), arthritis, darkening of the urine
Eosinophilia-myalgia syndrome	Scleroderma-like disorder, woody induration of distal extremities, in individuals consuming large quantities of L tryptophan

Based on data from Ref. [22]

Table 52.7 Skin findings of excess mineral/element consumption, absorption, or decreased excretion

Mineral/element	Cutaneous and systemic findings
Zinc	Vomiting, nausea, upper GI hemorrhage, vertigo, neutropenia; mucosal irritation of the gastric mucosa secondary high intake of zinc-associated copper deficiency (plum); cutaneous infections, weeping dermatitis, delayed wound healing, secondary infections, alopecia, nail defects paronychia, *Candida* dermatitis, blepharitis, conjunctivitis
Iron	Primary hemochromatosis: deposited in the liver, heart, and skin. Diffuse bronze color or hyperpigmentation of the skin in sun-exposed areas. Melanogenesis, ichthyosis-like scaling, alopecia, koilonychia. Diabetes, cardiomyopathy, liver cirrhosis. Cutaneous palmar erythema, telangiectasia
Lead/mercury	Blue-gray discolorations of the skin and gums
Silver/argyrosis	Gray and bluing of skin; grayish color with predilection of sun-exposed areas, sclera, mucous membranes, fingernails (toenails not affected) silver deposits on dark-field microscopy
Aluminum	Porphyria-like bullous changes
Arsenic	Bowen's disease, basal cell carcinoma, Mees' lines, lung cancer, vomiting, diarrhea, hepatic and renal damage, peripheral neuropathy
Chrysiasis	Deposition of gold in the skin. Diffuse bluish gray. The mucous membranes are not affected, unlike in argyrosis. Maculopapular vesiculobullous and urticarial eruptions occasionally with erythema multiforme-like rash

Based on data from Ref. [22]

52.6 Dietary Interventions

52.6.1 Mediterranean Diet

The Mediterranean diet focuses on reduced intake of meats and processed sugar-containing carbohydrates and incorporating moderate proportions of poultry, eggs, cheese, and yogurt on a weekly basis. The diet emphasizes eating fish or seafood at least twice a week. On a daily basis with each meal, there is the inclusion of vegetables, fruits, whole grains, olive oil, beans, nuts, legumes, seeds, herbs, and spices. Adults may include low-to-moderate red wine consumption and plenty of water daily. There is an emphasis on physical activity and enjoying meals with others [198].

Of note, there is variability in the food consumption patterns seen between the 16 Mediterranean countries. For example, Italy has an average of 30% total lipid intake, while Greece has 40% lipid intake, with the predominant fats for both coun-

Table 52.8 Hypervitaminosis skin findings

Vitamin	Skin and symptomatic findings
A	Pruritus, generalized scaling, dry mucous membranes, alopecia (telogen effluvium), cephalea, nausea, hyperostosis Increased transaminases and lipids Hypercarotenemia- Orange- yellow skin tint with sparing of the sclera, eyelids, ears, and axillary folds. The skin color of the palms and soles depends on the thickness of the epithelium. Common in children. Also seen in vegetarians, those with renal disease, diabetes mellitus, thyroid disease where there is either primary or secondary decreased ability to convert beta-carotene to vitamin A. Hypervitaminosis A is seen in Inuit populations
C	Vitamin C interferes with vitamin B12 metabolism with eventual symptoms of vitamin B12 deficiency especially the darkening of skin
D	Hypercalcemia, vomiting, diarrhea, anorexia, cephalea, calcinosis cutis
E	Gastrointestinal upset, cephalea, icterus in premature neonates

Based on data from Refs. [22, 108, 197]

tries being monounsaturated fats. The bioactive molecules of the Mediterranean diet include high levels of vitamin C, vitamin E, beta-carotene, selenium, flavonoids, and phenolic compounds. When considering all of the skin conditions aided by increasing essential fatty acids, bioactive components in vegetables, herbs, and spices, an initial consideration for dietary intervention should be a trial of the Mediterranean diet adjusted for the condition. For example, in patients with metabolic syndrome who demonstrate an acanthosis nigricans, psoriasis, or xerosis, shifting to a Mediterranean diet improves the markers of metabolic syndrome [199].

The Mediterranean diet is high in multiple B vitamins, most notably high levels of folate. In addition, there is a greater anti-inflammatory/photooxidative fatty acid profile with higher omega-3 and omega-9 poly- and monounsaturated fatty acids. This results in lower intake of proinflammatory omega-6 polyunsaturated fatty acids and lower omega-6/omega-9 and omega 6/omega-3 fatty acid ratios. There are lower exposure levels of prooxidants and carcinogens from iron that is heme derived from red meat, and there are reduced polycyclic aromatic hydrocarbons produced with high-temperature grilling. Instead there is a preference for gently grilled and steamed fish, low fat dairy and poultry, high levels of complex carbohydrates from grains and legumes, and small to moderate amounts of alcohol, primarily from antioxidant-rich wines [198].

The Mediterranean diet is higher in vitamin E than the average American diet, because of the inclusion of high vitamin E foods. Vitamin E decreases prostaglandin E2 production, and, as a result, T-cell proliferation and function may be enhanced. Vitamin E is a common plant-based, fat-soluble antioxidant and is found in combination with water-soluble antioxidants like vitamin C; mineral antioxidants including zinc, selenium, and others; and other fat-soluble antioxidants such as alpha-lipoic acid [198]. These compounds join the group of phytonutrients in the carotenoid family that are key to modulating immune-heightened inflammation. More than 600 different carotenoids exist in nature, and, through the Mediterranean diet, it is common to consume 40 types of carotenoids, especially when ingesting vegetables. The human GI tract can efficiently absorb 12 carotenoids, including alpha- and beta-carotene, lutein, zeaxanthin, beta-cryptoxanthin and lycopene [198].

The Mediterranean diet is recommended for many of the conditions in Table 52.4 as a primary intervention and medical nutrition therapy. Further modifying gluten and dairy intake acts to modulate treatments of acne, psoriasis, and atopic dermatitis. Rich helpings of bioactives found in the Mediterranean diet, implemented over months to years, aid in the overall balance of the essential fatty acids, carotenoids, vitamin A, and minerals including zinc, magnesium, calcium, and chromium. Targeted additions of specific essential fats, minerals, vitamins, and phytonutrient spices when indicated are also useful (Table 52.4).

By identifying the role of food, spices, herbs, and foodborne bioactive compounds provided by the Mediterranean diet, there has been rapid expansion of the understanding of mechanisms with refinement of applications for health promotion and disease intervention. For example, curcumin, berberine, and resveratrol are on a long list of nutraceuticals that, when incorporated into the treatment regimen, help shift balance from disease to health [27]. With increased research of these and similar dietary compounds influencing microRNAs, associated mechanisms of aging, cancer development, tissue restoration, and stem cell activity are coming to light. The epigenetic diet, which incorporates foods high in curcumin, resveratrol, genistein, garlic, selenium, and folate, is able to modulate methylation, modify histones, and promote overall health [200]. What is old is new again, and the incorporation of the right food, in the right setting, for the desired effect will become more mainstream. Let food be thy medicine, and medicine thy food will be realized in the Western and American health system.

52.6.2 Modified Elimination Diet

The modified elimination diet has been used by practitioners since the days of early Egyptian, Chinese, and Greek healers. There are skin conditions in which dietary intake is causative and where removal of dietary components is therapeutic (Table 52.4). Identifying foods, spices, or herbs associated with development of specific skin conditions can be difficult. In the clinical setting, clinicians trying to distinguish food allergy and immune food sensitivity by diet diary are often left to random elimination trials, since there is no perfect test for food allergy and sensitivity [201]. However, the use of modified elimination diet without prior testing, or in response to laboratory immune globulin testing, or skin testing is well-established.

Modified elimination diet, with empiric removal of the six leading food-associated triggers for eosinophilic mucosal response without laboratory testing, is established as a therapeutic intervention [202]. Even removal of one family of foods can remove the trigger for the uncomfortable symptom of scar pruritus [203]. Changing to an oligoantigenic and histamine-free diet has been used to treat chronic idiopathic urticaria successfully and shows decrease in associated histamine levels [204]. In treating eczema and atopic dermatitis with elimination diets, 51% showed a positive response with subsequent food challenges used for verification of reaction. Food allergy prevalence was 70.8% (85/120) in the elevated eosinophil group and 34.7% (50/144) in the normal blood eosinophil group [205]. The response to an elimination diet may be fairly rapid. In one open-label trial of a modified elimination diet during which milk and milk products, nuts and nut-containing foods, egg and egg-containing foods, seafish and prawns, eggplant, and soybeans were removed, there was a significant lowering of all parameters of disease activity after 3 weeks ($P < 0.001$) for SCORAD, surface area, and severity of itching [206].

52.6.3 Modified Elimination Diet in Autoimmunity, IgE Reactions, IgG Hypersensitivity, and Other Immune Reactions

Our understanding of the immune system has evolved, and various bioactive compounds in food have been identified as contributory or causal in all four Gell-Coombs responses, which lead to type 1 allergic or type 2, 3, or 4 hypersensitivity immunologic skin reactions (please see ▶ Chaps. 19 and 48). It is true that nutrient insufficiency or excess can cause alterations in the immune reactions that have been previously discussed.

The laboratory evaluation using IgE serum testing for individual foodborne antibody reactions can be variable. The skin testing to determine wheal-and-flare reaction can have both false positives and negatives and only identify Gell-Coombs type I reactions. The traditional skin prick tests have marked variability if there is not a wheal-and-flare reaction less than 8 mm in diameter. These skin prick tests are influenced by many over-the-counter antihistamines, anti-inflammatory, and prescription medications including corticosteroids and leukotriene inhibitors leading to false-negative readings. Type 3 reactions, modulated mainly by IgG immunoglobulins, do not classically have the same wheal-and-flare reactions. Type 4 hypersensitivity testing is frequently determined by patch testing.

There are a variety of autoimmune responses and hypersensitivity reactions to dietary antigens, which respond to modification of the diet and removal of the known triggering antigens. The modified elimination diet has been used therapeutically to diminish immune mediated skin disorders by downregulating the antibody or T-cell response burden. Examples of conditions responsive to modified elimination diet include dermatitis herpetiformis, pemphigus, atopic dermatitis, contact dermatitis, contact urticaria, protein contact dermatitis, and photoallergic contact dermatitis [129].

In the two most common bullous disorders of the skin that are associated with diet-triggered autoimmune diseases, dermatitis herpetiformis and pemphigus, and the modified elimination diet is an essential part of the therapeutic intervention [129]. The foods that trigger dermatitis herpetiformis and pemphigus are listed in ◘ Tables 52.9 and 52.10 [129]. Foods recommended in the modified elimination diet of dermatitis herpetiformis and pemphigus are listed in ◘ Tables 52.9 and 52.10 [129]. Atopic dermatitis has responded to a modified elimination diet by removing multiple nut varieties and other foods as shown in ◘ Table 52.11.

Certain foods coming in contact with the mucosa of the mouth, esophagus, and intestine can cause degranulation of mucosal mast cells leading to urticarial reactions, just as the same foods can cause degranulation of dermal mast cells resulting in contact urticaria of the skin. ◘ Table 52.12 lists

◘ **Table 52.9** Foods recommended for elimination diet

Condition	Avoidance
Dermatitis herpetiformis	Wheat, barley, buckwheat, rye, oat Cross reactivity of alpha gliadin and different foods Gliadin, casein, casomorphin, milk chocolate, oats, brewer's yeast, instant coffee, millet, corn, rice (Vojdani Tarish)

Based on data from Ref. 129]

◘ **Table 52.10** Foods recommended for elimination diet

Pemphigus	Fruit: monica Vegetable: garlic, leek, onion, radish, red pepper Spice/herb: black pepper, coriander, cumin seeds, horseradish, mustard, red chilies, tea Fat: evening primrose oil (gamma-linolenic acid), monounsaturated (olive oil), short-chain fatty acids (butter, ghee)

Based on data from Ref. [129]

◘ **Table 52.11** Foods recommended for elimination diet

Condition	Avoidance
Atopic dermatitis	Almonds, carob, hazelnuts, peanuts, cow's milk, crustacean, eggs, maize, wheat

Based on data from Ref. [129]

Table 52.12 Foods associated with contact urticaria of the mucosal membranes and skin

Food category	Food
Fruit	Apple, apricot, banana, coffee, fig, grapefruit, kiwi, lemon, lime, mango, melon, orange, peach, pear, pineapple, plum, pomegranate, strawberry, watermelon
Nut	Almond, brazil, cashew, hazelnut, peanut, sunflower seeds
Vegetable	Artichoke, arugula, asparagus, cabbage, carrot, cauliflower, celery, chives, corn, cucumber, dill, endive, garlic, green pepper, lettuce, maize, mushroom, onion, parsley, parsnip, potato, rutabaga, seaweed, shallots, spinach, tomato, watercress, winged bean
Legume	Beans, soy, tofu
Grains	Amaranth, buckwheat, chicory rice, wheat, wheat bran
Spice/herb	Caraway seed, coriander, curry, dill, fennel, thyme
Animal protein	Beef, egg, fish, lamb, liver, milk, pork, shellfish, turkey, venison

Based on data from Ref. [129]

Table 52.13 Foods associated with contact dermatitis

Food category	Food
Fruit	Ginkgo fruit, lemon, lime, mango, orange
Nut/seed	Cashew nut oil
Vegetable	Artichoke, asparagus, broccoli, brussel sprouts, cabbage, carrots, cauliflower, celery, corn, cucumber, kale, lettuce, leek, mushroom, onion, parsnips, parsley, potato, radish, tomato, turnip
Spice/herb	Anise, basil, bay (laurel) leaf, capsicum, caraway oil, cardamom, cassia, chamomile tea, chicory, cinnamon, cloves, dill, horseradish, garlic, ginger, Jamaican pepper, mace, mustard, nutmeg, paprika, peppermint, rosemary, spearmint, turmeric, vanilla

Based on data from Ref. [129]

Table 52.14 Foods associated with the protein contact dermatitis

Food category	Food
Fruit	Banana, fig, kiwi fruit, lemon, pineapple
Nut	Almond, hazelnut, peanuts
Vegetable	Bean, carrot, cauliflower, celery, cress, cucumber, eggplant, endive, garlic, mushroom, onion, parsley, parsnip, potato, tomato
Legume	Peanuts
Grains	Barley, rye, wheat
Spice/herb	Caraway, curry, dill, garlic, paprika,
Animal product	Blood (pig, cow), gut (pig), mesenteric fat (pig) skin (turkey, chicken)
Animal protein	Cheese, egg yolk, fish, liver (calf, chicken), meat (cow, pig, horse, lamb)

Based on data from Ref. [129]

foods documented to cause contact urticaria of the mucosal membranes and skin.

Contact dermatitis, diagnosed with patch testing, is seen most commonly with metals, including nickel, tin, and mercury fillings. Occasionally IgG reactions involving metals found in orthodontia materials such as titanium have also been documented. Common causes of contact dermatitis induced by foods are listed in Table 52.13.

Protein components of different foods can stimulate immune-mediated delayed-type hypersensitivity. The combination of the immediate type I and delayed Type IV allergic responses have been identified in butchers and slaughterhouse workers. There are four principle groups responsible for the protein contact dermatitis, outlined in Table 52.14, including fruit, nuts, vegetables, legumes, grains, spices, and animal proteins [207]. The foods trigger is a Type IV Gell-Coombs reaction. It is an IgG-associated response, in which antibodies are directed against desmogleins known to be cell-cell adhesion molecules [129].

Ultraviolet A radiation from sunlight can interact with proteins that have accumulated in the skin either following consumption or after direct skin contact. Food exposed to ultraviolet A radiation has the potential to trigger photoallergic contact dermatitis after coming in contact with the skin. Hapten formation between the skin protein and the activated antigen has to occur to incite the delayed hypersensitivity reaction [129].

There are other compounds in foods that are biologically active and known to elicit photosensitive skin reactions. Psoralens are responsible for some frequent reactions and are found in the umbelliferae foods, such as carrot, celery, fennel, parsley, and the parsnip family [129]. Psoralens are compounds that bind with DNA and potentially inhibit DNA synthesis and cell division, especially when used with therapeutic phototherapy. Rutaceae, found in lemon, lime, bitters, bergamot, orange, grapefruit, and fig families, also have photoallergic contact dermatitis history [129]. Garlic exposure has also been linked to such reactions [129]. Avoidance of these foods, once the trigger has been identified, is the therapeutic preventative. Foods associated with skin phenomenon of contact photoallergy can be eliminated when identified to ease the dermatitis (Table 52.15).

Table 52.15 Foods associated with photoallergic contact dermatitis

Food category	Food
Fruit	Fig, grapefruit, lemon, orange
Vegetable	Carrot, celery, fennel, parsley, parsnip
Spice/herb	Bergamot

Based on data from Ref. [129]

52.6.4 Dermatologic Findings in Obesity

No discussion is complete without focusing on dietary interventions for obesity and herein as it focuses on skin changes [30]. Skin changes are common in the obese (>30.0 BMI and >23% body fat). One notable change is pseudo-acanthosis nigricans, an indicator of insulin resistance and metabolic syndrome. In acanthosis nigricans, it is important to determine the root cause. The clinician is aided by determining when the condition began. If developed in early life from childhood through puberty, it may be characterized as hereditary type I, which is benign. With weight gain and increased insulin resistance, the disorder may be either benign type 2 or pseudo-acanthosis nigricans type 3. The most common forms of acanthosis nigricans are types 2 and 3, which are associated with systemic metabolic conditions, insulin resistance, diabetes, and obesity. The majority of times (80%) acanthosis nigricans occurs idiopathically or in benign conditions such as endocrinopathies and/or heritable diseases [30].

If acanthosis nigricans develops following treatment with glucocorticoids, steroids, or niacin, it is considered drug-induced type 4. However, if it occurs de novo without any of the other of the hallmarks of types 1 through 4, the consideration of the rarest form, malignant type 5, must be entertained [208]. In malignant acanthosis nigricans, 80% occur in individuals over 40 years of age and 90% of the tumors are intra-abdominal adenocarcinoma, of which 60% are gastric [30, 31].

Any condition which results in excess insulin binding to keratinocytes and skin fibroblast IGF-1 receptors promotes acanthosis nigricans [32]. These IGF-1 receptors are upregulated in proliferative conditions. Epidermal and fibroblast growth factor receptors are both members of the tyrosine kinase receptor family, which are upregulated in the basal membranes of acanthosis-affected tissue. Epiliths or skin tags are often present in the hyperpigmented field of acanthosis nigricans. Hyperinsulinemia is more important than hyperandrogenism in the development of acanthosis nigricans [32]. This disorder improves with management of insulin resistance and weight loss as long as there are no other underlying conditions. See ◘ Table 52.4.

Intertriginous dermatitis is an inflammatory condition at skin folds, caused by heat, moisture, friction, infections, and lack of air circulation. Intertriginous eczema may be secondary to overgrowth of bacteria, erythrasma, dermatophyte, and yeast infections. Skin folds of obese patients are subject to increased friction and result in hyperpigmented areas, most commonly the inner thighs and submammary regions and often have associated skin tags (acrochordon). Other skin findings include hyperhidrosis, striae distensae, stasis dermatitis, and leg ulcers.

There is increasing association of exogenous toxicity with metabolic syndrome and obesity, and these toxins or xenobiotics are referred to as obesogens. Environmental obesogens and their effects are becoming well-known and widely characterized [176]. Since metabolism and excretion of obesogens are dependent on vitamin, mineral, and amino acid cofactors, nutritional status of the affected is critical in removing xenobiotics from fat, muscle, and bone stores. Skin changes have been associated with obesogens. For example, characteristic rashes are found on the hands and feet from arsenic toxicity known as arsenic keratosis. Rashes on the hands may also be seen in exposures to DDT, dioxins, organochlorine, and organophosphate pesticides. Careful exposure history needs to be taken and toxicity considered in the differential diagnosis of obesity [176].

52.6.5 Therapeutic Uses of Probiotics and Fermented Food in Skin Disorders

The skin microbiome is unique, as are the biofilms and resident bacteria in all organ systems. The skin microflora is altered in response to the external and internal environment. Skin microflora is influenced by topical pH and helps determine the barrier function of the skin. The immune response to the resident microflora is determined by the balance in the bacteria. The addition of certain species modulates skin allergic reactions, atopic dermatitis, and even skin sensitivity and pain [129, 209].

Certain probiotic strains have been shown to be efficacious in quenching skin allergic reactions. For example, prenatal administration of Lactobacillus rhamnosus GG has been shown to reduce atopic dermatitis in at-risk infants. The addition of the probiotic is felt to increase Th1 immune response by influencing interferon-gamma production [129, 210, 211].

A diet supplemented with *Lactobacillus johnsonii* effects the cutaneous immune system by modulating the production of serum interleukin 10. Interleukin 10 production maintains the Langerhans cells, cells important for skin immunity, and increases protective cell density when exposed to UV light. Specifically, UV radiation's deleterious effect on skin defenses is prevented by *L. johnsonii* [129, 212]. In a randomized double-blind placebo-controlled trial using 10^{10} CFU of each *Lactobacillus paracasei* and *Bifidobacterium lactis* for 8 weeks, skin sensitivity was decreased. When compared to controls, participants receiving the active probiotic mix were found to have a 50% decrease in skin sensitivity and cutaneous neurosensitivity by day 57 [129].

52.7 Autoimmune or Hypersensitivity Reactions to Food Proteins

52.7.1 Autoimmunity-Triggered Dietary Antigens

This section expands the discussion above. When homology or similarity of dietary proteins compared to the human protein structure is greater than 65%, there is a greater incidence of human protein cross-reactivity to food components [28]. For both pemphigus and dermatitis herpetiformis, there is an autoimmune association with these bullous conditions. In dermatitis herpetiformis, IgA autoantibodies are formed against the gluten protein, type 3 epidermal transglutaminase, which is homologically similar to tissue transglutaminase. The homology is greatly responsible for the cross-antibody reaction in patients with gluten/gliadin enteropathy. With greater understanding of the multitude of proteins homologous to gluten/gliadin, it becomes clear that dietary elimination of gluten/gliadin is therapeutically necessary to treat dermatitis herpetiformis. See Table 52.9.

There is a triad required to develop a food-related autoimmune condition. Foods contain proteomes, proteins expressed by the particular food. The triad needs, first, increased mucosal permeability and translocation of the food proteomes across the GI tract; second, an individual's genetic predisposition for inflammation or autoimmunity; and, third, the above two coupled with an environmental trigger. With the understanding of the triad, proteomes of different foods have been identified and associated with different autoimmune conditions. For example, there is an association of wheat and milk proteomes with autoimmunity, many of which have manifestations ranging from increased xerosis, pigment changes, hair loss, vascular changes, or bullous rashes. When any of the skin manifestations are identified with autoimmune conditions, a consideration of the role of wheat or milk proteomes can be determined by a modified elimination diet. See Table 52.3.

There are many acnegenic components in the Western diet (Bowe 2010; Melnik). These are outlined in Table 52.16. Encouraging a Paleolithic diet as nutrition therapy for acne has been shown to be of benefit. This is detailed in Table 52.17. The dysbiotic gastrointestinal tract bacteria combined with brain-diet interactions are key in worsening acne vulgaris [58, 59, 213]. Improving gut-associated microflora balance and dietary intake are known to improve acne vulgaris.

Table 52.16 Acnegenic components of the Western diet

Nutrients	Metabolic and nutrigenomic effects	Diet sources
Hyperglycemic carbohydrates	Postprandial hyperinsulinemia; insulin-mediated hepatic IGF-1 synthesis; reduction of IGFBP3; increased bioavailability of free circulating IGF-1; reduction of SHBG; increased bioavailability of free circulating testosterone; reduced nuclear activity of FoxO1; increased expression of sebocyte SReBP-1c; activation of mTORC1; glucose-mediated microRNA-21 expression	Sugar Sweets Soft drinks Pizza Pasta Wheat bread Wheat Rolls Cornflakes
Milk and dairy products	Postprandial hyperinsulinemia; increased levels of circulating IGF-1; leucine-mediated activation of mTORC1; glutamine-mediated activation of mTORC1; palmitate-mediated activation of mTORC1; milk-/dairy-related microRNA-21-mediated proliferation and inflammation	Whole and skim milk Pasteurized fresh milk Yogurt Ice cream Whey and casein supplements Cheese
Saturated fats	Palmitate-mediated activation of mTORC1; palmitate-driven inflammasome activation	Butter Cream
Trans fats	Possible mTORC1 activation; proinflammatory signaling	Fast food French fries

Based on data from Ref. [213]
Abbreviations: *IGF-1* insulin-like growth factor 1, *IGFBP3* IGF-binding protein 3, *SHBG* sex hormone-binding globulin, *FoxO1* forkhead box O1, *SReBP-1c* sterol response element-binding protein 1c, *mTORC1* mechanistic target of rapamycin complex

52.7.2 Dermatologic Findings in Inflammatory Bowel Disease

When the patient has a chronic medical condition, nutritional requirements can be altered. As highlighted earlier with zinc, influences affecting zinc absorption, transport, and catabolism can lead to altered requirements and are eventually associated with deficiency [22, 156]. See Table 52.18. In autoimmune-related inflammatory bowel disease (IBD), chronic malabsorption of nutrients increases the incidence of nutrient deficiency-associated dysfunction (Table 52.19) [214] and can manifest in skin lesions (Table 52.20) [244]. Nutrient malabsorption in IBD may lead to skin findings as outlined below. When considering medical nutrition therapy, the diagnosis influences the consideration for repletion of specific nutrients. Common nutrient deficiencies are seen in association with IBD and

Table 52.17 Paleolithic-type diet for acne

Nutrients	Metabolic effects	Sources
Carbohydrates with low-glycemic index	Reduced insulin signaling; reduction of free IGF-1; increase of IGFBP3 and sex hormone binding globulin; increase in nuclear FoxO1; reduction of SReBP-1c; attenuation of mTORC1	Salads Vegetables
ω-3-Fatty acids (docosahexaenoic acid and eicosapentaenoic acid)	Inhibition of mTORC1 Inhibition of SReBP-1c; reduction of proinflammatory eicosanoids (LTB4, PGe2); inhibition of NLRP3 inflammasome activation	Sea fish Omega-3 fatty acid containing oils
Plant products and spices	Inhibition of mTORC1; activation of nuclear FoxO1; inhibition of *P. acnes* biofilm	Green tea (EGCG) Berries (resveratrol) Curcumin

Abbreviations: *IGF-1* insulin-like growth factor 1, *IGFBP-3* IGF binding protein 3, *SHBG* sex hormone binding globulin, *FoxO1* forkhead box O1, *SReBP-1c* sterol response element-binding protein 1c, *mTORC1* mechanistic target of rapamycin complex 1, *LTB4* leukotriene B4, *PGe2* prostaglandin e2, *EGCG* epigallocatechin 3-gallate

Table 52.18 Medical conditions that affect absorption, excretion, requirement, or transport of zinc

Area of abnormality	Condition associated
Poor intestinal absorption of zinc	Malabsorption due to rapid transit, pancreatic hypofunction Chronic liver and pancreatic disease Other gastrointestinal disease Alcoholism Unbalanced diet (e.g., only high fiber) Parenteral nutrition lacking zinc supplementation
Increased zinc excretion	Liver cirrhosis Renal disease Diabetes mellitus Dialysis
Increased catabolism	Cancer Chronic recurrent infections AIDS Burns, trauma
Decreased serum albumin binding	Nephrotic syndrome Liver cirrhosis

Based on data from Ref. [22]

can guide practitioners in the choice of therapy. In addition, nutrient insufficiencies can worsen the symptoms of IBD if not corrected.

52.7.3 Food Bioactives in the Treatment of Conditions

Topical application of food bioactives is being used in dermatologic treatments [245]. The advantage of topical application over oral intake is the direct application of the bioactive substance to the skin lesions and avoids GI tract absorption and assimilation. For example, resveratrol is used topically in acne vulgaris and psoriasis. When resveratrol-carboxymethylcellulose-based gel is applied topically, there is a greater than 50% reduction in symptoms after 60 days compared with 6% in the controls [246]. In psoriasis, where there is a marked increase in plaque-associated TNF alpha, 1% resveratrol ointment topically applied is associated with 50% rapid 2-week decrease in inflammatory symptoms. Following the treatment period, there is 80% significant improvement and 20% acceptable improvement compared to no improvement in controls [245]. The use of specific amino acids also has been effective. Taurine, for example, has been used in the treatment of acne, psoriasis, and atopic dermatitis [245]. Many spices and foods block NF kappa B, a main modulator of TNF alpha, including turmeric, red pepper, cloves, ginger, cumin, anise, fennel, basil, rosemary, garlic, and pomegranate [247]. Many of these spices are in chutneys, and other traditionally prepared foods and upon scientific evaluation are found to be therapeutic.

52.7.4 Conclusion

Medical nutrition therapy for the skin conditions, as reviewed above, focuses on improving deficiency, removing excess, modulating immune responses and autoimmune reactions, and eliminating hypersensitivity reactions to food protein triggers. The uses of the Mediterranean diet or the modified food elimination diet are powerful modulators of the immune response and promote health and healing of common skin conditions. Skin healing occurs by altering and improving the availability of nutrients and bioactive molecules. The patient with a wide range of dermatologic conditions can be helped by the targeted and precise application of diet, supplements, and/or topical nutrients and bioactives. Therapeutic interventions should be considered for patients whose chronic conditions may seem remote to dermatology yet may manifest in dermatologic changes such as those with inflammatory bowel disease. By leveraging medical nutrition therapy integrated with the knowledge of the underlying condition and coupled with dietary and exposure history, many patients will have marked improvement in their dermatologic condition from nutrient intervention without the need for pharmaceuticals.

Table 52.19 Nutrient deficiencies in inflammatory bowel disease

Niacin	Low plasma concentrations of niacin are common among patients with Crohn's disease (CD), but clinically apparent disease is rare. In one study of adults with CD in remission, low plasma concentrations of niacin were found in 77%. Pellagra (the clinical manifestations of niacin deficiency) has been described in adult patients with CD [214–217]
Folate	Folate deficiency was observed in 20–60% of adults with inflammatory bowel disease (IBD) in older series but is uncommon in studies from the last decade [214, 218, 219]
B12	Laboratory evidence of vitamin B12 deficiency has been reported in about 20% of adult and pediatric patients with CD [214, 220, 221]
Vitamin A	Deficiency has been described in IBD. In one study of adult patients with CD, 5% had clinically significant vitamin A deficiency characterized by impaired dark adaptation [214, 222]
Vitamin D	Insufficiency is common in patients with IBD. In one study, 25% of adults with CD had deficient serum 25-hydroxyvitamin D (25OHD, calcidiol) concentrations (<10 ng/mL). Studies in children report that 6–36% of children with IBD have deficient 25 OH D concentrations (<15 ng/mL). Several factors probably contribute to the variability in reported vitamin D concentrations, including genetic and environmental factors such as diet and sun exposure, as well as factors related to the IBD itself with malabsorption and anorexia. In addition, a study in children suggests that IBD is associated with abnormal vitamin D metabolism, reflected by the absence of a secondary elevation of parathyroid hormone (PTH) concentrations in the subset of children with 25OHD <30, and impaired conversion of 25OHD to 1,25-hydroxyvitamin D. These effects may be mediated by inflammatory cytokines, which suppress PTH and renal 1-α hydroxylase activity. One study suggests that only about 10% of patients with IBD malabsorb vitamin D [214, 223–228]
Iron	Literature reviews suggest that 35–90% of adults with IBD are iron-deficient. Iron deficiency is probably the primary cause of the anemia that affects 16% of outpatients and up to 70% of inpatients. Iron deficiency has a significant negative impact on quality of life and can lead to developmental and cognitive abnormalities in children and adolescents. A study in adults with IBD and anemia demonstrated that oral iron supplementation increased hemoglobin concentration and improved quality of life without changing disease activity [214, 229–231]
Zinc	As many as 65% of patients with CD have decreased serum zinc concentrations. However, serum zinc levels vary with albumin and correlate poorly with total body zinc stores; clinically significant zinc deficiency is probably much less common. A deficient state is often reflected by a decreased serum alkaline phosphatase concentration since alkaline phosphatase is a zinc metalloenzyme [214, 215, 232–234]
Calcium	Approximately 13% of adults with CD malabsorb calcium. This can occur due to the binding of calcium to undigested fats in the intestinal lumen, loss of the ileum leading to vitamin D deficiency, and possibly as a result of genetic factors and the effects of inflammatory cytokines. The recommended intake of elemental calcium for children and adults with IBD is the same as for the general population. 1–3 years—700 mg daily 4–8 years—1000 mg daily 9–18 years—1300 mg daily Men and premenopausal women—1000 mg daily Postmenopausal women and men older than 70 years—1200 mg daily [214, 233, 235]
Selenium	Concentrations in whole blood and glutathione peroxidase activity are often somewhat lower in adults or children with IBD as compared with healthy individuals, but biochemical or clinical evidence of deficiency is rare. Deficient levels of selenium are more common among patients with CD who have undergone small bowel resection of >200 cm and in those receiving enteral nutrition exclusively [214, 236–238]
Copper	Clinical manifestations of copper deficiency include abnormally formed hair, depigmentation of the skin, and microcytic anemia. The neurologic manifestations include ataxia, neuropathy, and cognitive deficits that can mimic vitamin B12 deficiency. High flow fistulas or resections can lead to copper deficiency especially in the setting of aggressive zinc supplementation [214, 239, 240]
Magnesium	Deficiency in IBD may result from decreased oral intake, malabsorption, increased intestinal losses, or low concentrations of magnesium in a formula used for enteral nutrition. Magnesium deficiency may contribute to osteopenia [214, 241–243]

Table 52.20 Secondary nutrition-related skin lesions associated with inflammatory bowel disease

Secondary cutaneous lesions	Symptoms	Crohn's disease (CD)	Ulcerative colitis (UC)	Causation
Zinc deficiency acrodermatitis enteropathica	Erythematous patches, plaques, progress to crusted vesicles, bullae, or pustules Location: mouth, anus, limbs, fingers scalp	40%	No difference from control	Reduced mucosa available for absorption and chronic diarrhea
Iron deficiency with anemia	Koilonychias, angular cheilitis, pale skin	39%	81%	Malabsorption of iron and chronic intestinal bleeding
Essential Fatty Acid deficiency	Xeroderma, dry skin, unspecified eczema	CD > UC	UC=CD	Malabsorption

Based on data from Ref. [244]

References

1. Landis ET, Scott A, Davis SA, Feldman SR, Taylor S. The complementary and alternative medicine use in dermatology in the United States. J Altern Complement Med. 2014;20(5):392–8.
2. Watson RR, Zibadi S, editors. Bioactive dietary factors and plant extracts in dermatology. Nutrition and health. New York: Springer Science+Business Media, Humana Press; 2013.
3. Proksch E, Brandner JM, Jensen JM. The skin: an indispensable barrier. Exp Dermatol. 2008;17:1063–72.
4. Tobin DJ. Biochemistry of human skin—our brain on the outside. Chem Soc Rev. 2006;35(1):52–67.
5. Khnykin D, Miner JH, Jahnsen F. Role of fatty acid transporters in epidermis: implications for health and disease. Dermatoendocrinol. 2011;3(2):53–61.
6. Feingold KR. Thematic review series: skin lipids. The role of epidermal lipids in cutaneous permeability barrier homeostasis. J Lipid Res. 2007;48:2531–46.
7. Man MQ, Feingold KR, Thornfeldt CR, Elias PM. Optimization of physiological lipid mixtures for barrier repair. J Invest Dermatol. 1996;106:1096–101.
8. Schurer NY, Stremmel W, Grundmann JU, Schliep V, Kleinert H, Bass NM, Williams ML. Evidence for a novel keratinocyte fatty acid uptake mechanism with preference for linoleic acid: comparison of oleic and linoleic acid uptake by cultured human keratinocytes, fibroblasts and a human hepatoma cell line. Biochim Biophys Acta. 1994;1211:51–60.
9. Uchida Y, Holleran WM. Omega-O-acylceramide, a lipid essential for mammalian survival. J Dermatol Sci. 2008;51:77–87.
10. Ziboh VA. The significance of polyunsaturated fatty acids in cutaneous biology. Lipids. 1996;31:S249–53.
11. Ziboh VA, Miller CC, Cho Y. Metabolism of polyunsaturated fatty acids by skin epidermal enzymes: generation of antiinflammatory and antiproliferative metabolites. Am J Clin Nutr. 2000;71:361–6.
12. McCusker MM, Grant-Kels JM. Healing fats of the skin: the structural and immunologic roles of the omega-6 and omega-3 fatty acids. Clin Dermatol. 2010;28:440–51.
13. Sala-Vila A, Miles EA, Calder PC. Fatty acid composition abnormalities in atopic disease: evidence explored and role in the disease process examined. Clin Exp Allergy. 2008;38:1432–50.
14. Hachem JP, Crumrine D, Fluhr J, Brown BE, Feingold KR, Elias PM. pH directly regulates epidermal permeability barrier homeostasis and stratum corneum integrity/cohesion. J Invest Dermatol. 2003;121:345–53.
15. Bouwsta JA, Gooris GS, Dubbelaar FE, Ponec M. Phase behaviour of skin barrier model membranes at pH 7.4. Cell Mol Biol (Noisy-le-Grand). 2000;46:979–92.
16. Fluhr JW, Kao J, Jain M, Ahn SK, Feingold KR, Elias PM. Generation of free fatty acids from phospholipids regulates stratum corneum acidification and integrity. J Invest Dermatol. 2001;117:44–51.
17. Thomas P. Habif. In: Clinical dermatology. 5th ed. Philadelphia: Mosby Elsevier; 2010.
18. McCully M, Ladell K, Hakobyan S, Mansel RE, Price DA, Moser B. Epidermis instructs skin homing receptor expression in human T cells. Blood. 2012;120(23):4591–8.
19. Morizane D, Yamasaki K, Kabigting ED, Gallo RL. Kallikrein expression and cathelicidin processing are independently controlled in keratinocytes by calcium, vitamin D3, and retinoic acid. J Invest Dermatol. 2010;130(5):1297–306.
20. Stone PM, Boham E. Functional nutrition evaluation: the skin. Federal Way Washington: The Institute for Functional Medicine; 2016. www.FxMed.org.
21. Stone PM. Functional nutrition evaluation: the 8 step mouth exam. Federal Way Washington: The Institute for Functional Medicine; 2014. www.FxMed.org.
22. Schmuth M, Fritsch PO. Cutaneous changes in nutritional diseases. In: Krutmann J, Humbert P, editors. Nutrition for healthy skin; strategies for clinical and cosmetic practice. Berlin Heidelberg: Springer-Verlag; 2011. p. 3–14.
23. Heymsfield SB, Williams PJ. Nutritional assessment by clinical and biochemical methods. In: Modern nutrition in health and disease. 7th ed. Philadelphia: Lea and Febiger; 1988.
24. Sonkoly E. Pivarcsi: advances in microRNAs: implications for immunity and inflammatory diseases. J Cell Mol Med. 2009;13(1):24–38.
25. Botchkareva N. MicroRNA/mRNA regulatory networks in the control of skin development and regeneration. Cell Cycle. 2012;11(3):468–74.
26. Notay M, Foolad N, Vaughn AR, Sivamani RK. Probiotics, prebiotics, and synbiotics for the treatment and prevention of adult dermatological diseases. Am J Clin Dermatol. 2017;18:721–32. https://doi.org/10.1007/s40257-017-0300-2.
27. McCubrey JA, Lertpiriyapong K, Steelman LS, Abrams SL, Yang LV, Murata RM, Rosalen PL, Scalisi A, Neri LM, Cocco L, Ratti S, Martelli AM, Laidler P, Dulińska-Litewka D, Rakus D, Gizak A, Lombardi P, Nicoletti F, Candido S, Libra M, Montalto G, Cervello M. Effects of resveratrol, curcumin, berberine and other nutraceuticals on aging, cancer development, cancer stem cells and microRNAs. Aging (Albany NY). 2017;9(6):1477–536.

28. Vojdani A, Tarach I. Cross-reaction between gliadin and different food and tissue antigens. Food Nutr Sci. 2013;4:20–32.
29. Wolff K, Johnson RA, Saavedra AP. Fitzpatrick's color atlas and synopsis of clinical dermatology. 7th ed. New York: McGraw-Hill Companies; 2013.
30. DeWitt CA, Buescher LS, Stone SP. Cutaneous manifestations of internal malignant disease: cutaneous paraneoplastic syndromes. Chapter 153: 1880–1900. In: Goldsmith LA, Katz SI, Gilchrest BA, Paller AS, Leffell DJ, Wolff K, editors. Fitzpatrick's: dermatology in general medicine. 8th ed. New York: McGraw Hill; 2012.
31. Chung VQ, et al. Clinical and pathologic findings of paraneoplastic dermatosis. J Am Acad Dermatol. 2006;54:745.
32. Kalus AA, Chien AJ, Olerud JE. Diabetes mellitus and other endocrine diseases. Chapter 151: 1840–1869. In: Goldsmith LA, Katz SI, Gilchrest BA, Paller AS, Leffell DJ, Wolff K, editors. Fitzpatrick's: dermatology in general medicine. 8th ed. New York: McGraw-Hill; 2012.
33. Schwartz RA. Acanthosis nigricans. J Am Acad Dermatol. 1994;31:1.
34. Sherertz EF. Improved acanthosis nigricans with lipodystrophic diabetes during dietary fish oil supplementation. Arch Dermatol. 1988;124:1094–6.
35. Malisiewicz B, Boehncke S, Lang V, et al. Epidermal insulin resistance as a therapeutic target in acanthosis nigricans? Acta Derm Venereol. 2014;94:607.
36. Berger BJ, Gross PR. Another use for tretinoin-pseudoacanthosis nigricans. Arch Dermatol. 1973;108:133.
37. Darmstadt GL, Yokel BK, Horn TD. Treatment of acanthosis nigricans with tretinoin. Arch Dermatol. 1991;127:1139.
38. Gregoriou S, Anyfandakis V, Kontoleon P, et al. Acanthosis nigricans associated with primary hypogonadism: successful treatment with topical calcipotriol. J Dermatolog Treat. 2008;19:373.
39. Pelle M. Rasacea. Chapter 81. In: Goldsmith LA, Katz SI, Gilchrest BA, Paller AS, Leffell DJ, Wolff K, editors. Fitzpatrick's dermatology in general medicine. 8th ed. New York: McGraw-Hill; 2012. p. 918–25.
40. Gaby A. Nutritional medicine. 2nd ed. Concord: Fritz Perlberg Publishing; 2017.
41. Gravina A, Federico A, Ruocco E, Lo Schiavo A, Masarone M, Tuccillo C, Peccerillo F, Miranda A, Romano L, de Sio C, de Sio I, Persico M, Ruocco V, Riegler G, Loguercio C, Romano M. Helicobacter pylori infection but not small intestinal bacterial overgrowth may play a pathogenic role in rosacea. United European Gastroenterol J. 2015;3(1):17–24.
42. Egeberg A, Weinstock LB, Thyssen EP, Gislason GH, Thyssen JP. Rosacea and gastrointestinal disorders: a population-based cohort study. Br J Dermatol. 2017;176(1):100–6. https://doi.org/10.1111/bjd.14930. Epub 2016 Oct 31.
43. Gupta M, Mahajan VK, Mehta KS, Chauhan PS. Zinc therapy in dermatology: a review. Dermatol Res Pract. 2014;2014:709152.
44. Ogawa Y, Kinoshita M, Shimada S, Kawamura T. Zinc and skin disorders. Nutrients. 2018;10:199. https://doi.org/10.3390/nu10020199.
45. Sharquie KE, Najim RA, Al-Salaman HD. Oral zinc sulfate in the treatment of rosacea: a double-blind placebo-controlled study. Int J Dermatol. 2006;45:857–61.
46. Bamford JTM, Gessert CE, Haller IV, et al. Randomized, double-blind trial of 220 mg zinc sulfate twice daily in treatment of rosacea. Int J Dermatol. 2012;51:459–62.
47. Barba A, Rosa B, Angelini G, et al. Pancreatic exocrine function in rosacea. Dermatologica. 1982;165:601–6.
48. Dreno B, Amblard P, Agache P, et al. Low doses of zinc gluconate for inflammatory acne. Acta Derm Venereol. 1989;69:541–3.
49. Verma KC, Sainin AS, Dhamija SK. Oral zinc sulfate therapy in acne vulgaris: a double-blind trial. Acta Derm Venereol. 1980;60:337–40.
50. Blumenthal M, Busse WR, Goldberg A, et al. The complete German Commission E monographs. In: Therapeutic guide to herbal medicines. Austin: The American Botanical Council; 1998.
51. Tulipan L. Acne rosacea; a vitamin B complex deficiency. Arch Derm Syphilol. 1947;56(5):589–91. https://doi.org/10.1001/archderm.1947.01520110035005.
52. Manzhalii E, Hornuss D, Stremmel W. Intestinal-borne dermatoses significantly improved by oral application of Escherichia coli Nissle 1917. World J Gastroenterol. 2016;22(23):5415–21. https://doi.org/10.3748/wjg.v22.i23.5415.
53. Iijima K, Sekine H, Koike T, Imatani A, Ohara S, Shimosegawa T. Long-term effect of Helicobacter pylori eradication on the reversibility of acid secretion in profound hypochlorhydria. Aliment Pharmacol Ther. 2004;19:1181–8. https://doi.org/10.1111/j.1365-2036.2004.01948.x.
54. Nazzaro-Porro M. Azelaic acid. J Am Acad Dermatol. 1987;17:1033–41.
55. Nguyen QH, Bui TP. Azelaic acid. Pharmacokinetic and pharmacodynamics properties and its therapeutic role in hyperpigmentary disorders and acne. Int J Dermatol. 1995;34:75–84.
56. Purdy S, de Berker D. Acne vulgaris. BMJ Clin Evid. 2011;2011. pii: 1714. 2015;6:147–53.
57. Zaenglein AL, Graber EM, Thiboutot DM. Acne vulgaris and acneiform eruptions. Chapter 80. In: Goldsmith LA, Katz SI, Gilchrest BA, Paller AS, Leffell DJ, Wolff K, editors. Fitzpatrick's dermatology in general medicine. 8th ed. New York: McGraw-Hill; 2012. p. 897–917.
58. Bowe WP, Joshi SS, Shalita AR. Diet and acne. J Am Acad Dermatol. 2010;63:124.
59. Bowe WP, Logan AC. Acne vulgaris, probiotics and the gut-brain-skin axis—back to the future? Gut Pathog. 2011;3:1–3.
60. Kappas A, Anderson K, Conney A, et al. Nutrition-endocrine interactions: induction of reciprocal changes in the delta 4-5 alpha-reduction of testosterone and the cytochrome P-450-dependent oxidation of estradiol by dietary macronutrients in man. Proc Natl Acad Sci U S A. 1983;80:7646–9.
61. Berra B, Rizzo AM. Glycemic index, glycemic load: new evidence for a link with acne. J Am Coll Nutr. 2009;28 Suppl:450S–4S.
62. Adebamowo CA, Spiegelman D, Berkey CS, et al. Milk consumption and acne in teenaged boys. J Am Acad Dermatol. 2008;58(5):787–93.
63. Smith RN, Mann NJ, Braue A, et al. The effect of a high-protein, low glycemic-load diet versus a conventional, high glycemic-load diet on biochemical parameters associated with acne vulgaris: a randomized, investigator-masked, controlled trial. J Am Acad Dermatol. 2007;57:247.
64. Smith RN, Braue A, Varigos GA, et al. The effect of a low-glycemic diet on acne vulgaris and the fatty acid composition of skin surface triglycerides. J Dermatol Sci. 2008;50(1):41–52.
65. Jarrousse V, Castex-Rizzi N, Khammari A, et al. Zinc salts inhibit in vitro Toll-like receptor 2 surface expression by keratinocytes. Eur J Dermatol. 2007;17(6):492–6.
66. Thiboutot D. Acne: hormonal concepts and therapy. Clin Dermatol. 2004;22:419.
67. Cappel M, Mauger D, Thiboutot D. Correlation between serum levels of insulin-like growth factor 1, dehydroepiandrosterone sulfate, and dihydrotestosterone and acne lesion counts in adult women. Arch Dermatol. 2005;141:333.
68. Holmes MD, Pollak MN, Willett WC, Hankinson SE. Dietary correlates of plasma insulin-like growth factor I and insulin-like growth factor binding protein 3 concentrations. Cancer Epidemiol Biomarkers Prev. 2002;11:852.
69. Michaelsson G, Edqvist LE. Erythrocyte glutathione peroxidase activity in acne vulgaris and effect of selenium and vitamin E treatment. Acta Derm Venereol. 1984;64:9–14.
70. Snider B, Dieteman DF. Pyridoxine therapy for premenstrual acne flare. Arch Dermatol. 1974;110:130–1.
71. Kligman AM, Mills OH Jr, Leyden JJ, et al. Oral vitamin A in acne vulgaris. Int J Dermatol. 1981;20:278–85.
72. Kligman AM, Leyden JJ, Mills O. Oral Vitamin A (Retinol) in Acne Vulgaris. In: Orfanos CE, Braun-Falco O, Farber EM, Grupper C, Polano MK, Schuppli R. (eds) Retinoids. Berlin, Heidelberg: Springer; 1981.

73. Shalita AR, Smith JG, Parish LC, et al. Topical nicotinamide compared with clindamycin gel in the treatment of inflammatory acne vulgaris. Int J Dermatol. 1995;34:434–7.
74. Khodaeiani E, Foulad RF, Amirnia M, et al. Topical 4% nicotinamide vs 1% clindamycin in moderate inflammatory acne vulgaris. Int J Dermatol. 2013;52:999–1004.
75. Shahmoradi Z, Iraji F, Siadat AH, Ghorbaini A. Comparison of topical 5% nicotinamide gel versus 2% clindamycin gel in the treatment of the mild–moderate acne vulgaris. A double-blinded randomized clinical trial. J Res Med Sci. 2013;18:115–7.
76. Schulpis KH, Karikas GA, Georgala S, et al. Elevated plasma homocysteine levels in patients on isotretinoin therapy for cystic acne. Int J Dermatol. 2001;40(1):33–6.
77. Chanson A, Cardinault N, Rock E, et al. Decreased plasma folate concentration in young and elderly healthy subjects after a short-term supplementation with isotretinoin. J Eur Acad Dermatol Venereol. 2008;22(1):94–100.
78. Georgala S, Schulpis KH, Georgala C, et al. L-carnitine supplementation in patients with cystic acne on isotretinoin therapy. J Eur Acad Dermatol Venereol. 1999;13(2):205–9.
79. Ayres S, Mihan R. Acne vulgaris: therapy directed at pathophysiologic defects. Cutis. 1981;28:41–2.
80. Lönnerdal B. Dietary factors influencing zinc absorption. J Nutr. 2000;130:1378S.
81. Jen M, Yan AC. Syndromes associated with nutritional deficiency and excess. Clin Dermatol. 2010;28(6):669–85.
82. Jen M, Yan AC. Sin in nutritional, metabolic, and heritable disease: cutaneous changes in nutritional disease. Chapter 130. In: Goldsmith LA, Katz SI, Gilchrest BA, Paller AS, Leffel DJ, Wolff K, editors. Fitzpatrick's dermatology in general medicine. 8th ed. New York: McGraw-Hill; 2012. p. 1499–525.
83. Moynahan EJ. Acrodermatitis enteropathica: a lethal inherited human zinc-deficiency disorder. Lancet. 1974;2:399–400.
84. Gaby A. Acrodermatitis enteropathica. Chapter 170. In: Gaby A, editor. Nutritional medicine. 2nd ed. Concord: Fritz Perlberg Publishing; 2017. p. 712–3.
85. WHO/CAH Diarrhoea treatment guidelines including new recommendations for the use of ORS and zinc supplementation for clinic-based healthcare workers. UNICEF, MOST, USAID, Geneva, 2005. http://www.who.int/child-adolescent-health/New_Publications/CHILD_HEALTH/WHO_FCH_CAH_03.7.pdf.
86. Bhutta ZA, Bird SM, Black RE, et al. Therapeutic effects of oral zinc in acute and persistent diarrhea in children in developing countries: pooled analysis of randomized controlled trials. Am J Clin Nutr. 2000;72:1516.
87. Lukacik M, Thomas RL, Aranda JV. A meta-analysis of the effects of oral zinc in the treatment of acute and persistent diarrhea. Pediatrics. 2008;121:326.
88. Leung WC, Hessel S, Méplan C, Flint J, Oberhauser V, Tourniaire F, Hesketh JE, von Lintig J, Lietz G. Two common single nucleotide polymorphisms in the gene encoding beta-carotene 15,15′-monooxygenase alter beta-carotene metabolism in female volunteers. FASEB J. 2009;23(4):1041–53.
89. Sausenthaler S, Koletzko S, Schaaf B, et al. Maternal diet during pregnancy in relation to eczema and allergic sensitization in the offspring at 2 y of age. Am J Clin Nutr. 2007;85(2):530–7.
90. Samsonov MA, Kalinina AA. Effect of diet therapy on the serum immunoglobulin level in chronic eczema. Vopr Pitan. 1982;2:12–5.
91. Wickens K, Black PN, Stanley TV, et al. A differential effect of 2 probiotics in the prevention of eczema and atopy: a double-blind randomized, placebo-controlled trial. J Allergy Clin Immunol. 2008;122(4):788–94.
92. Abrahamsson TR, Jakobsson T, Bottcher M, et al. Probiotics in prevention of IgE-associated eczema: a double blind, randomized, placebo-controlled trial. J Allergy Clin Immunol. 2007;119(5):1174–80.
93. Frankland AW. Deficiency scrotal dermatitis in P.O.W.s. Br Med J. 1948;1(4560):1023–6.
94. Koch C, Dolle S, Metzger M, et al. Docosahexaenoic acid (DHA) supplementation in atopic eczema: a randomized double-blind, controlled trial. Br J Dermatol. 2008;158(4):786–92.
95. David TJ, Wells FE, Sharpe TC, et al. Low serum zinc in children with atopic eczema. Br J Dermatol. 1984;111(5):597–601.
96. David TJ, Wells FE, Sharp TC, et al. Serum levels of trace metals in children with atopic eczema. Br J Dermatol. 1990;122(4):485–9.
97. Ronaghy A, Katz SI, Hall RP III. Dermatitis herpetiformis. Chapter 61. In: Goldsmith LA, Katz SI, Gilchrest BA, Paller AS, Leffel DJ, Wolff K, editors. Fitzpatrick's dermatology in general medicine. 8th ed. New York: McGraw-Hill; 2012. p. 642–9.
98. Welsh AL, Ede M. Inositol hexanicotinate for improved nicotinic acid therapy. Preliminary report. Int Rec Med. 1961;174:9–15.
99. Johnson HH Jr, Binkley GW. Nicotinic acid therapy of dermatitis herpetiformis. J Invest Dermatol. 1950;14:233–7.
100. Maronn M, Allen DM, Esterly NB. Phrynoderma: a manifestation of vitamin A deficiency?...The rest of the story. Pediatr Dermatol. 2005;22(1):60–3.
101. Chia MW, Tay Y, Liu TT. Phrynoderma: a forgotten entity in a developed country. Singapore Med J. 2008;49(6):e160.
102. Bagchi K, Halder K, Chowdhury SR. The etiology of phrynoderma. Am J Clin Nutr. 1959;7:251–8.
103. Ghafoorunissa, Vidyasagar R, Krishnaswamy K. Phrynoderma: is it an EFA deficiency disease? Eur J Clin Nutr. 1988;42(1):29–39.
104. Lehman E, Rapaport HG. Cutaneous manifestations of vitamin A deficiency in children. JAMA. 1940;114:386–93.
105. Shrank AB. Phrynoderma Br. Medizinhist J. 1966;1:29–30.
106. Ragunatha S, Kumar VJ, Murugesh SB. A clinical study of 125 patients with phrynoderma. Indian J Dermatol. 2011;56(4):389–92. https://doi.org/10.4103/0019-5154.84760.
107. Ragunatha S, Jagannath Kumar V, Murugesh SB, Ramesh M, Narendra G, Kapoor M. Therapeutic response of vitamin A, vitamin B complex, essential fatty acids (EFA) and vitamin E in the treatment of phrynoderma: a randomized controlled study. J Clin Diagn Res. 2014;8(1):116–8.
108. Barthelemy H, Couvet B, Cambazard F. Skin and mucosal manifestations in vitamin deficiency. J Am Acad Dermatol. 1986;15:1263–74.
109. Loewenthal LJA. A new cutaneous manifestation in the syndrome of vitamin A deficiency. Arch Derm Syphilol. 1933;28(5):700–8.
110. Duncan KO, Geisse JK, Leffell DJ. Epithelial precancerous lesions. Chapter 113. In: Goldsmith LA, Katz SI, Gilchrest BA, Paller AS, Leffel DJ, Wolff K, editors. Fitzpatrick's dermatology in general medicine. 8th ed. New York: McGraw-Hill; 2012. p. 1261–82.
111. Black HS, Herd JA, Goldberg LH, et al. Effect of a low-fat diet on the incidence of actinic keratosis. N Engl J Med. 1994;330:1272–5.
112. Chen AC, Damian DL. Nicotinamide and the skin. Australas J Dermatol. 2014;55:169–75.
113. Chen AC, Martin AJ, Choy B, et al. A phase 3 randomized trial of nicotinamide for skin-cancer chemoprevention. N Engl J Med. 2015;373:1618–26.
114. Drago F, Ciccarese G, Parodi A. Nicotinamide for skin cancer chemoprevention. N Engl J Med. 2016;374:789–90.
115. Surjana D, Halliday GM, Martin AJ, et al. Oral nicotinamide reduces actinic keratosis in phase II double-blinded randomized controlled trials. J Invest Dermatol. 2012;132:1497–500.
116. Dublin WB, Hazen BM. Relation of keratosis seborrheica and keratosis senilis to vitamin A deficiency. Arch Dermatol Syph. 1948;57:178–83.
117. Gaby A. Actinic keratosis. Chapter 171. In: Gaby A, editor. Nutritional medicine. 2nd ed. Concord: Fritz Perlberg Publishing; 2017. p. 713–5.
118. Moon TE, Levine N, Carmet B, et al. Effect of retinol in preventing squamous cell skin cancer in moderate-risk subjects: a random-

119. Kang S, Goldfarb MT, Weiss JS, et al. Assessment of adapalene gel for the treatment of actinic keratoses and lentigines: a randomized trial. J Am Acad Dermatol. 2003;49:83.
120. Weinstock MA, Bingham SF, VATTC Trial Group. High-dose topical tretinoin for reducing multiplicity of actinic keratoses. J Invest Dermatol. 2010;130(Suppl 1):S63.
121. Han J, Colditz GA, Hunter DJ. Polymorphisms in the MTHFR and VDR genes and skin cancer risk. Carcinogenesis. 2007;28:390–7.
122. Chen Y, Parvez F, Gamble M, Islam T, Ahmed A, Argos M, Graziano JH, Ahsan H. Arsenic exposure at low-to-moderate levels and skin lesions, arsenic metabolism, neurological functions, and biomarkers for respiratory and cardiovascular diseases: review of recent findings from the Health Effects of Arsenic Longitudinal Study (HEALS) in Bangladesh. Toxicol Appl Pharmacol. 2009;239(2):184–92.
123. Mekonian S, Argos M, Chen Y, et al. Intakes of several nutrients are associated with incidence of arsenic-related keratotic skin lesions in Bangladesh. J Nutr. 2012;142:2126–34.
124. Biswas J, Sinha D, Mukherjee S, Roy S, Siddiqi M, Roy M. Curcumin protects DNA damage in a chronically arsenic-exposed population of West Bengal. Hum Exp Toxicol. 2010;29:513–24.
125. Carless MA, Kraska T, Lintell N, Neale RE, Green AC, Griffiths LR. Polymorphisms of the VDR gene are associated with presence of solar keratoses on the skin. Br J Dermatol. 2008;159:804–10.
126. Lu'o'ng K, Nguyen LT. The roles of vitamin D in seborrhoeic keratosis: possible genetic and cellular signalling mechanisms. Int J Cosmet Sci. 2013;35(6):525–31. https://doi.org/10.1111/ics.12080.
127. Mitsuhashi Y. New aspects on vitamin D3 ointment; treatment of senile warts with topical application of active forms of vitamin D. Clin Calcium. 2004;14:141–4.
128. Mitsuhashi Y, Kawaguchi M, Hozumi Y, Kondo S. Topical vitamin D3 is effective in treating senile warts possibly by inducing apoptosis. J Dermatol. 2005;32:420–3.
129. Kumar Y, Bhatia A. Immune-mediated disorders of the skin: role of dietary factors and plant extracts? Chapter 2. In: Watson RR, Zibadi S, editors. Bioactive dietary factors and plant extracts in dermatology. New York: Humana Press; 2013. p. 13–25.
130. Kieffer M, Barnetson RS. Increased gliadin antibodies in dermatitis herpetiformis and pemphigoid. Br J Dermatol. 1983;108(6):673–8.
131. Honl BA, Elston DM. Autoimmune bullous eruption localized to a breast reconstruction site: response to niacinamide. Cutis. 1998;62:85–6.
132. Fivenson DP, Breneman DL, Rosen GB, et al. Nicotinamide and tetracycline therapy of bullous pemphigoid. Arch Dermatol. 1994;130:753–8.
133. Belenky P, Bogan KL, Brenner C. NAD+ metabolism in health and disease. Trends Biochem Sci. 2006;32(1):12–9.
134. Prousky JE. Pellagra may be a rare secondary complication of anorexia nervosa: a systematic review of the literature. Altern Med Rev. 2003;8(2):180–5.
135. Delgado-Sanchez L, Godkar D, Niranjan S. Pellagra: rekindling of an old flame. Am J Ther. 2008;15(2):173–5.
136. Oldham MA, Ivkovic A. Pellagrous encephalopathy presenting as alcohol withdrawal delirium: a case series and literature review. Addict Sci Clin Pract. 2012;7:12.
137. Hegyi J, Schwartz RA, Hegyi V. Pellagra: dermatitis, dementia, and diarrhea. Int J Dermatol. 2004;43(1):1–5.
138. Guyton JR, Bays HE. Safety considerations with niacin therapy. Am J Cardiol. 2007;99(6A):22C–31C.
139. Stone NJ, Robinson J, Lichtenstein AH, et al. 2013 ACC/AHA guideline on the treatment of blood cholesterol to reduce atherosclerotic cardiovascular risk in adults: a report of the American College of Cardiology/American Heart Association task force on practice guidelines. Circulation. 2014;63:2889–934.
140. Gaby A. Psoriasis. Chapter 192. In: Gaby A, editor. Nutritional medicine. 2nd ed. Concord: Fritz Perlberg Publishing; 2017. p. 762–71.
141. Ricketts JR, Rothe MJ, Grant-Kels JM. Nutrition and psoriasis. Clin Dermatol. 2010;28(6):615–26.
142. Lithell H, Bruce A, Gustafsson IB, Höglund NJ, Karlström B, Ljunghall K, Sjölin K, Venge P, Werner I, Vessby B. A fasting and vegetarian diet treatment trial on chronic inflammatory disorders. Acta Derm Venereol. 1983;63(5):397–403.
143. Osier E, Wang AS, Tollefson MM, et al. Pediatric psoriasis comorbidity screening guidelines. JAMA Dermatol. 2017;153(7):698–704.
144. Frigo A, Tambalo C, Bambara LM, Biasi D, Marrella M, Milanino R, Moretti U, Velo G, De Sandre G. Zinc sulfate in the treatment of psoriatic arthritis. Recenti Prog Med. 1989;80:577–81.
145. Clemmensen OJ, Siggaard-Andersen J, Worm AM, Stahl D, Frost F, Bloch I. Psoriatic arthritis treated with oral zinc sulphate. Br J Dermatol. 1980;103(4):411–5.
146. Perez A, Raab R, Chen TC, Turner A, Holick MF. Safety and efficacy of oral calcitriol (1,25 dihydroxyvitamin D) for the treatment of psoriasis. Br J Dermatol. 1996;134:1070–8.
147. Alganatish JT, Algahtani F, Alsewairi WM, Al-kenaizan S. Childhood scurvy: an unusual cause of refusal to walk in a child. Pediatr Rheumatol. 2015;13:23.
148. Fain O. Musculoskeletal manifestations of scurvy. Joint Bone Spine. 2005;72(2):124–8.
149. Weinstein M, Babyn P, Zlotkin S. An orange a day keeps the doctor away: scurvy in the year 2000. Pediatrics. 2001;108:E55.
150. Gaby A. Urticaria. Chapter 199. In: Gaby A, editor. Nutritional medicine. 2nd ed. Concord: Fritz Perlberg Publishing; 2017. p. 779–83.
151. De Spirt S, Stahl W, Tronnier H, Sies H, Bejot M, Maurette JM, Heinrich U. Intervention with flaxseed and borage oil supplements modulates skin condition in women. Br J Nutr. 2009;101(3):440–5.
152. Hong SH, Kwak JH, Paik JK, Chae JS, Lee JH. Association of polymorphisms in FADS gene with age-related changes in serum phospholipid polyunsaturated fatty acids and oxidative stress markers in middle-aged nonobese men. Clin Interv Aging. 2013;8:585–96.
153. Neukam KH, Hambidge KM. Zinc therapy of acrodermatitis enteropathica. N Engl J Med. 1975;292:879–82.
154. Neukam K, De Spirt S, Stahl W, Bejot M, Maurette JM, Tronnier H, Heinrich U. Supplementation of flaxseed oil diminishes skin sensitivity and improves skin barrier function and condition. Skin Pharmacol Physiol. 2011;24(2):67–74. https://doi.org/10.1159/000321442.
155. Agero AL, Verallo-Rowell VM. A randomized double-blind controlled trial comparing extra virgin coconut oil with mineral oil as a moisturizer for mild to moderate xerosis. Dermatitis. 2004;15(3):109–16.
156. Krutmann J, Humbers P, editors. Nutrition for health skin. Strategies for clinical cosmetic practice. New York: Springer; 2011.
157. Pappas A, editor. Nutrition and skin: lessons for anti-aging, beauty and healthy skin. New York: Springer; 2011.
158. Davis D, Epp M, Riordan H. Changes in USDA food composition data for 43 garden crops, 1950 to 1999. J Am Coll Nutr. 2004;23(6):669–82.
159. Mayer AM. Historical changes in the mineral content of fruits and vegetables. Br Food J. 1997;99(6):207–11.
160. Rushto DH. Nutritional factors and hair loss. Clin Exp Dermatol. 2002;27(5):396–404.
161. Rawlings AV. Essential fatty acids and atopic dermatitis. Chapter 11. In: Pappas A, editor. Nutrition and skin: lessons for anti-aging, beauty and healthy skin. New York: Springer; 2011. p. 159–75.

162. Schafer L, Kragballe K. Abnormal epidermal lipid metabolism in patients with atopic dermatitis. J Invest Dermatol. 1991;96:10–5.
163. Elewa RM, Zoubouli CC. Vitamin A and the skin. In: Pappas A, editor. Nutrition and skin. Lessons for anti-aging, beauty and healthy skin. New York: Springer; 2011. p. 7–24.
164. Lobo GP, Amengual J, Baus D, Shivdasani RA. Genetics and diet regulate vitamin A production via the homeobox transcription factor ISX. J Biol Chem. 2013;288(13):9017–27.
165. Bikle DD. Vitamin D metabolism and function in the skin. Mol Cell Endocrinol. 2011;347(1–2):80–9.
166. Kamradt J, Rafi L, Mitschele T, Meineke V, Gartner BC, Tilgen W, Holick MF, Reichrath J. Analysis of the Vitamin D system in cutaneous malignancies. Recent Results Cancer Res. 2003;164: 259–69.
167. Wadhwa B, Relhan V, Goel K, Kochhar AM, Garg VK. Vitamin D and skin diseases: a review. Indian J Dermatol Venereol Leprol. 2015;81(4):344–55.
168. Hutchinson PE, Osborne JE, Lear JT, Bowers PW, Morris PN, Jones PW, et al. Vitamin receptor polymorphisms are associated with altered prognosis in patients with malignant melanoma. Clin Cancer Res. 2000;6:498–504.
169. Trémezaygues L, Reichrath J. From the bench to emerging new clinical concepts. Our present understanding of the importance of the vitamin D endocrine system (VDES) for skin cancer. Dermatoendocrinol. 2011;3:11–7. https://doi.org/10.4161/derm.3.1.14875.
170. Kerns JC, Arundel C, Chawla LS. Thiamin deficiency in people with obesity. Adv Nutr. 2015;6:147–53.
171. Larkin JR, Zhang F, Godfrey L, Motostvov G, Zehnder D, Rabbani N, Thornalley PJ. Glucose-induced down regulation of thiamine transporters in the kidney proximal tubular epithelium produces thiamine insufficiency in diabetes. PLoS One. 2012;7(12): e53175.
172. Hammes HP, Du X, Edelstein D, Taguchi T, Matsumura T, Ju Q, Lin J, Bierhaus A, Nawroth P, Hannak D, Neumaier M, Bergfeld R, Giardino I, Brownlee M. Benfotiamine blocks three major pathways of hyperglycemic damage and prevents experimental diabetic retinopathy. Nat Med. 2003;9(3):294–9.
173. Lin J, Alt A, Liersch J, Bretzel RG, Brownlee M. Benfotiamine inhibits intracellular formation of advanced glycation end products in vivo. Diabetes. 2000;49(Suppl 1):(A143) 583.
174. Pinto JT, Zempleni J. Riboflavin. Adv Nutr. 2016;7:973–5.
175. Tanno O, Ota Y, Kitamura N, Katsube T, Inoue S. Nicotinamide increases biosynthesis of ceramides as well as other stratum corneum lipids to improve the epidermal permeability barrier. Br J Dermatol. 2000;143:524–31.
176. Pizzorno J. Symptoms by toxins. In: The toxin solution. New York, NY: Harper Collins; 2017. p. 236–7.
177. Yiasemides E, Sivapirabu G, Halliday GM, Park J, Damian DL. Oral nicotinamide protects against ultraviolet radiation induced immunosuppression in humans. Carcinogenesis. 2009;30:101–5.
178. Rossignol DA, Frye RE. Evidence linking oxidative stress, mitochondrial dysfunction, and inflammation in the brain of individuals with autism. Front Physiol. 2014;5(150):1–15.
179. Theoharides TC, Angelidou A, Alysandratos K-D, Zhang B, Asadi S, Francis K, Toniato E, Kalogeromitros D. Mast cell activation and autism. Biochim Biophys Acta. 2012;1822:34–41.
180. Knip M, Douek IF, Moore WP, et al. Safety of high-dose nicotinamide: a review. Diabetologia. 2000;43:1337–45.
181. Klatt KC, Caudill MA. Folate, choline, vitamin B12, vitamin B6 Ch. 22. In: Stipanuk MH, Caudill MA, editors. Biochemical, physiological and molecular aspects of human nutrition. 4th ed. St. Louis: Elsevier; 2019. p. 614–58.
182. Trumbo PR. Pantothenic acid. In: Shils ME, Shike M, Ross AC, et al., editors. Modern nutrition in health and disease. 10th ed. Baltimore: Lippincott Williams and Wilkins; 2006. p. 462–9.
183. Field H Jr, Green ME, Wilkinson CW Jr. Glossitis and cheilosis healed following the use of calcium pantothenate. Am J Dig Dis. 1945;12:246–50.
184. Leung LH. Pantothenic acid deficiency as the pathogenesis of acne vulgaris. Med Hypotheses. 1995;44:490–2.
185. Goldman L. Intensive panthenol therapy of lupus erythematosus. J Invest Dermatol. 1950;15:291–3.
186. Bermudez Y, Cordova K, Williams JD. Folate nutrition in skin health and skin cancer prevention. In: Watson RR, Zibadi S, editors. Bioactive dietary factors and plant extracts in dermatology, nutrition and health. New York: Springer Science+Business Media; 2013. p. 229–56.
187. Neyestani TR, Vitamin D. Skin cancer: meet sunshine halfway. In: Watson R, Zibadi S, editors. Bioactive dietary factors and plant extracts in dermatology. nutrition and health. Totowa: Humana Press; 2013. p. 257–68.
188. Schallreuter KU, et al. Regulation of melanogenesis-controversies and new concepts. Exp Dermatol. 2008;17(5):395–404.
189. Bjorkegren K, Svardsudd K. Reported symptoms and clinical findings in relation to serum cobalamin, folate, methylmalonic acid and total homocysteine among elderly Swedes: a population-based study. J Intern Med. 2003;254(4):343–52.
190. Saini R, Badole SL, Zanwar AA. Vitamin C (L-Ascorbic Acid): antioxidant involved in skin care. Ch. 6. In: Watson RR, Zibadi S, editors. Bioactive dietary factors and plant extracts in dermatology. New York: Humana Press; 2013. p. 61–6.
191. Kohlmeier M. Nutrigenetics: applying the science of personal nutrition. Chapter 4. In: How nutrients are affected by genetics. Waltham: Elsevier; 2013. p. 167.
192. Eck P, Erichsen HC, Taylor JG, Yeager M, Hughes AL, Levine M, et al. Comparison of the genomic structure and variation in the two human sodium-dependent vitamin C transporters, SLC23A1 and SLC23A2. Hum Genet. 2004;115(4):285–94.
193. Savini I, Rossi A, Pierro C, Avigliano L, Catani MV. SVCT1 and SVCT2: key proteins for vitamin C uptake. Amino Acids. 2008;34(3): 347–55.
194. Centilli M, Legacy M, Legacy M, Seiger E. Acrodermatitis enteropathica. Consultant. 2014;54(7):543–5.
195. Gehrig KA, Dinulos JG. Acrodermatitis due to nutritional deficiency. Curr Opin Pediatr. 2010;22(1):107–12.
196. Winkler P. Minerals and the skin. Chapter 7. In: Pappas A, editor. Nutrition and skin. Lessons for anti-aging, beauty and healthy skin. New York: Springer; 2011. p. 91–109.
197. Bleasel NR, Stapleton KM, Lee MS, Sullivan J. Vitamin A deficiency phrynoderma: due to malabsorption and inadequate diet. J Am Acad Dermatol. 1999;41:322–4.
198. Primavesi L, Piantanida M, Pravettoni V. Mediterranean diet and skin health. In: Watson R, Zibadi S, editors. Bioactive dietary factors and plant extracts in dermatology. New York: Humana Press; 2013.
199. Kastorini CM, Milionis HJ, Esposito K, Giugliano D, Goudevenos JA, Panagiotakos DB. The effect of Mediterranean diet on metabolic syndrome and its components: a meta-analysis of 50 studies and 534,906 individuals. J Am Coll Cardiol. 2011;57(11):1299–313.
200. Hardy TM, Tollefsbol TO. Epigenetic diet: impact on the epigenome and cancer. Epigenomics. 2011;3(4):503–18.
201. Chafen JJ, et al. Diagnosing and managing common food allergies: a systematic review. JAMA. 2010;303(18):1848–56.
202. Lucendo AJ, Arias A, Gonzalez-Cervera J, Yague-Compadra JL, Guagnozzi D, Angueira T, Jimenez-Contreras S, Gonzalez-Castillo S, Rodriguez-Domingez B, DeRezende LC, Tenias JM. Empiric 6-food elimination diet induced and maintained prolonged remission in patients with adult eosinophilic esophagitis: a prospective study on the food cause of the disease. J Allergy Clin Immunol. 2013;131(3):797–804.
203. Alonso PE, Rioja LF. Solanidine and tomatidine trigger scar pruritus. Burns. 2016;42(3):535–40.
204. Guida B, De Martino CD, De Martino SD, Tritto G, Patella V, Trio R, D'Agostino C, Pecoraro P, D'Agostino L. Histamine plasma levels and elimination diet in chronic idiopathic urticaria. Eur J Clin Nutr. 2000;54(2):155–8.

205. Noh G, Jin H, Lee J, Noh J, Lee WM, Lee S. Eosinophilia as a predictor of food allergy in atopic dermatitis. Allergy Asthma Proc. 2010;31(2):e18–24.
206. Dhar S, Malakar R, Banerjee R, Chakraborty S, Chakraborty J, Mukherjee S. An uncontrolled open pilot study to assess the role of dietary eliminations in reducing the severity of atopic dermatitis in infants and children. Indian J Dermatol. 2009;54(2):183–5.
207. Janssens V, Morren M, Dooms-Goossens A, et al. Protein contact dermatitis: myth or reality? Br J Dermatol. 1995;132:1–6.
208. Sander I. Acanthosis nigricans. www.uptodate.com. Wolters Kluwer; 2017. Accessed 5-1-2017.
209. Ozdemir O, Zanwar AA. The role of probiotics in atopic dermatitis and skin allergy reactions: prevention and therapy. Chapter 4. In: Watson RR, Zibadi S, editors. Bioactive dietary factors and plant extracts in dermatology. Nutrition and health. New York: Springer science+Business Media; 2013. p. 493–529.
210. Pohjavuori E, Vilanen M, Korpela R, Kuitunen M, Sarnesto A, Vaarala O, Savilahti E. Induction of inflammation as a possible mechanism of probiotic effect in atopic eczema-dermatitis syndrome. J Allergy Clin Immunol. 2005;115:1254–9.
211. Viljanen M, Pohjavuori E, Haahtela T, Korpela R, Kuitunen M, Sarnesto A, Vaarala O, Savilahti E. Induction of inflammation as a possible mechanism of probiotic effect in atopic eczema-dermatitis syndrome. J Allergy Clin Immunol. 2005;115:1254–9.
212. Gueniche A, Benyacoub J, Buetler T, Smola H, Blum S. Supplementation with oral probiotic bacteria maintains cutaneous immune homeostasis after UV exposure. Eur J Dermatol. 2006;16:511–7.
213. Melnik BC. Linking diet to acne metabolomics, inflammation and comedogenesis: an update. Clin Cosmet Investig Dermatol. 2015;8:371–88.
214. Teitelbaum JE. Nutrient deficiencies and inflammatory bowel disease. UptoDate. Accessed 4 Aug 2017.
215. Filippi J, Al-Jaouni R, Wiroth JB, et al. Nutritional deficiencies in patients with Crohn's disease in remission. Inflamm Bowel Dis. 2006;12:185.
216. Abu-Qurshin R, Naschitz JE, Zuckermann E, et al. Crohn's disease associated with pellagra and increased excretion of 5-hydroxyindoleacetic acid. Am J Med Sci. 1997;313:111.
217. Pollack S, Enat R, Haim S, et al. Pellagra as the presenting manifestation of Crohn's disease. Gastroenterology. 1982;82:948.
218. Kulnigg S, Gasche C. Systematic review: managing anaemia in Crohn's disease. Aliment Pharmacol Ther. 2006;24:1507.
219. Vagianos K, Bector S, McConnell J, Bernstein CN. Nutrition assessment of patients with inflammatory bowel disease. JPEN J Parenter Enteral Nutr. 2007;31:311.
220. Seidman E, LeLeiko N, Ament M, et al. Nutritional issues in pediatric inflammatory bowel disease. J Pediatr Gastroenterol Nutr. 1991;12:424.
221. Headstrom PD, Rulyak SJ, Lee SD. Prevalence of and risk factors for vitamin B(12) deficiency in patients with Crohn's disease. Inflamm Bowel Dis. 2008;14:217.
222. Main AN, Mills PR, Russell RI, et al. Vitamin A deficiency in Crohn's disease. Gut. 1983;24:1169.
223. Driscoll RH Jr, Meredith SC, Sitrin M, Rosenberg IH. Vitamin D deficiency and bone disease in patients with Crohn's disease. Gastroenterology. 1982;83:1252.
224. Pappa HM, Grand RJ, Gordon CM. Report on the vitamin D status of adult and pediatric patients with inflammatory bowel disease and its significance for bone health and disease. Inflamm Bowel Dis. 2006;12:1162.
225. Pappa HM, Langereis EJ, Grand RJ, Gordon CM. Prevalence and risk factors for hypovitaminosis D in young patients with inflammatory bowel disease. J Pediatr Gastroenterol Nutr. 2011;53:361.
226. Sentongo TA, Semaeo EJ, Stettler N, et al. Vitamin D status in children, adolescents, and young adults with Crohn disease. Am J Clin Nutr. 2002;76:1077.
227. Prosnitz AR, Leonard MB, Shults J, et al. Changes in vitamin D and parathyroid hormone metabolism in incident pediatric Crohn's disease. Inflamm Bowel Dis. 2013;19:45.
228. Vogelsang H, Schöfl R, Tillinger W, et al. 25-hydroxyvitamin D absorption in patients with Crohn's disease and with pancreatic insufficiency. Wien Klin Wochenschr. 1997;109:678.
229. Gisbert JP, Gomollón F. Common misconceptions in the diagnosis and management of anemia in inflammatory bowel disease. Am J Gastroenterol. 2008;103:1299.
230. Cronin CC, Shanahan F. Anemia in patients with chronic inflammatory bowel disease. Am J Gastroenterol. 2001;96:2296.
231. Wells CW, Lewis S, Barton JR, Corbett S. Effects of changes in hemoglobin level on quality of life and cognitive function in inflammatory bowel disease patients. Inflamm Bowel Dis. 2006;12:123.
232. Alkhouri RH, Hashmi H, Baker RD, et al. Vitamin and mineral status in patients with inflammatory bowel disease. J Pediatr Gastroenterol Nutr. 2013;56:89.
233. McClain C, Soutor C, Zieve L. Zinc deficiency: a complication of Crohn's disease. Gastroenterology. 1980;78:272.
234. Sturniolo GC, Di Leo V, Ferronato A, et al. Zinc supplementation tightens "leaky gut" in Crohn's disease. Inflamm Bowel Dis. 2001;7:94.
235. Institute of Medicine, Food and Nutrition Board. Dietary reference intakes for calcium and vitamin D. Washington DC: National Academy Press; 2011. https://www.nap.edu/search/?term=13050.
236. Rannem T, Ladefoged K, Hylander E, et al. Selenium status in patients with Crohn's disease. Am J Clin Nutr. 1992;56:933.
237. Kuroki F, Matsumoto T, Iida M. Selenium is depleted in Crohn's disease on enteral nutrition. Dig Dis. 2003;21:266.
238. Johtatsu T, Andoh A, Kurihara M, et al. Serum concentrations of trace elements in patients with Crohn's disease receiving enteral nutrition. J Clin Biochem Nutr. 2007;41:197.
239. Goldschmid S, Graham M. Trace element deficiencies in inflammatory bowel disease. Gastroenterol Clin North Am. 1989;18:579.
240. Spiegel JE, Willenbucher RF. Rapid development of severe copper deficiency in a patient with Crohn's disease receiving parenteral nutrition. JPEN J Parenter Enteral Nutr. 1999;23:169.
241. Galland L. Magnesium and inflammatory bowel disease. Magnesium. 1988;7:78.
242. Habtezion A, Silverberg MS, Parkes R, et al. Risk factors for low bone density in Crohn's disease. Inflamm Bowel Dis. 2002;8:87.
243. Park RH, Galloway A, Shenkin A, et al. Magnesium deficiency in patients on home enteral nutrition. Clin Nutr. 1990;9:147.
244. Huang BL, Chandra S, Shih DQ. Skin manifestations of inflammatory bowel disease. Front Physiol. 2012;3:13.
245. Das DK, Vasanthi H. Resveratrol in dermal health. Chapter 18. In: Watson RR, Zibadi S, editors. Bioactive dietary factors and plant extracts in dermatology. Nutrition and health. New York: Springer Science+Business Media; 2013. p. 177–87.
246. Fabbrocini G, De Vita V, Monfrecola A, De Padova MP, Brazzini B, Teixeira F, Chu A. Percutaneous collagen induction: an effective and safe treatment for post-acne scarring in different skin phototypes. J Dermatolog Treat. 2014;25(2):147–52. https://doi.org/10.3109/09546634.2012.742949.
247. Aggarwal BB, Shishodia S. Suppression of nuclear factor-Kappa B activation pathway by spice-derived phytochemicals: reasoning for seasoning. Ann N Y Acad Sci. 2004;1040:434–41.

Key Textbook Resources for Further Reading

Gaby A, editor. Nutritional medicine. 2nd ed. Concord: Fritz Perlberg Publishing; 2017.
Goldsmith LA, Katz SI, Gilchrest BA, Paller AS, Leffel DJ, Wolff K, editors. Fitzpatrick's dermatology in general medicine. 8th ed. New York: McGraw Hill; 2012.

Habif TP, editor. Clinical dermatology. 5th ed. Edinburgh: Mosby Elsevier; 2010.

Krutmann J, Humbers P, editors. Nutrition for healthy skin. Strategies for clinical cosmetic practice. New York: Springer; 2011.

Pappas A, editor. Nutrition and skin: lessons for anti-aging, beauty and healthy skin. New York: Springer; 2011.

Watson RR, Zibadi S, editors. Bioactive dietary factors and plant extracts in dermatology. Nutrition and health. New York: Springer Science+Business Media, Humana Press; 2013.

Wolff K, Johnson RA, Saavedra AP, editors. Fitzpatrick's color atlas and synopsis of clinical dermatology 7th edition. New York: McGraw-Hill Companies; 2013.

Internet Resources

dermnet.com.
fxmed.org/n-sight.
NaturalMedicineComprehansiveDatabase.com.

Movement Issues with Chronically Ill or Chronic Pain Patients

Judy Hensley, Julie Buttell, and Kristie Meyer

53.1 Introduction – 1004

53.2 Tissue Healing – 1004

53.3 Chronic Staircase of Change – 1005

53.4 Metabolic Effects of Activity – 1007

53.5 Effects of Bed Rest Resulting in Chronic Dysfunction – 1008

53.6 Rehabilitation of Chronic Dysfunction – 1010

53.7 Conclusion – 1011

References – 1011

© Springer Nature Switzerland AG 2020
D. Noland et al. (eds.), *Integrative and Functional Medical Nutrition Therapy*,
https://doi.org/10.1007/978-3-030-30730-1_53

53.1 Introduction

As a practitioner seeing chronically ill patients, have you been frustrated by the person who lands in your clinic after being ill for more than 6 months and having already visited multiple practitioners before you? This patient has not been able to recieve an effective treatment plan helping their recovery. If you are nodding your head, you are not alone! Many patients experience some level of pain, illness, or dysfunction without finding out the source of what is wrong with them or being given a solution to help them recover or reverse their condition. Counter to current medical wisdom, there are multiple factors that an individual can control in preventing and reversing many health conditions. Most ailments stem from poor physical activity and can lead to many chronic musculoskeletal and neuromuscular impairments. According to the Centers for Disease Control and Prevention, such chronic diseases and conditions as heart disease, stroke, cancer, type 2 diabetes, obesity, and arthritis rank among the most common, costly, and preventable of all health problems. As of 2012, about half of all American adults, 117 million people, had one or more chronic health conditions. One in four adults had at least two. In this chapter, we will discuss the many physiological changes that may occur in chronic illness and/or chronic pain due to inactivity [1]. There is little guidance on proper rehabilitation of this population in a safe and effective manner. We will discuss normal and abnormal tissue healing, the effects of bed rest on the body, metabolic changes associated with inactivity, and rehabilitation considerations. Besides treatment, movement alone may play a role in prevention of chronic dysfunctions. Further research needs to be conducted on effective treatment approaches and protocols for proper rehabilitation of a patient with chronic dysfunction.

53.2 Tissue Healing

Tissue healing can be divided into three phases: inflammation, repair, and remodeling [2]. Phase one, *inflammation*, is the body's first response and is essential for appropriate healing after an injury. New blood vessel formation and collagen production are critical steps that occur during this first stage and are important for healing. A healthy environment for new tissue regeneration and formation is essential for preventing prolonged inflammation, as this could disrupt blood vessel and collagen production. For the body to repair, it is vital to prevent tissue disruption and prolonged inflammation during the acute inflammatory stage. The main goal for practitioners during an inflammatory state is to protect the injured area while continuing to maintain full function in the unaffected areas.

Phase two, *repair*, can take up to 8 weeks if the proper amount of restorative stress is applied [2] or longer if this stress is inadequate. Restorative stress is defined as maintaining a careful balance in which disruption of newly formed collagen fibers is avoided, while low load stresses are gradually introduced to allow increased collagen synthesis and prevent loss of joint motion. Early gentle motion allows for alignment of collagen fibers and promotes tissue mobility. The treatment goal for this phase is to prevent excessive muscle atrophy and joint deterioration. To protect weak collagen fibers, caution should be used when adding load or resistance. Too little activity can have negative effects, causing new fibers to misalign forming adhesions. As a result, it can prevent a full range of motion.

Phase three, *remodeling*, occurs when damaged tissue is replaced with new collagen fibers. After these fibers are laid down, the body can begin to remodel and strengthen the new tissue and allow for gradual return to full activity. Closed chain [2] exercises, such as squatting, stair climbing, transferring from seated to standing positions, and balancing on one leg, have been found to best replicate these common movements. In these exercises, the distal part of the joint segment is fixed to the ground. This type of exercise is said to enhance collagen remodeling, increasing joint stability and neuromuscular control.

Healing cannot take place when inflammation, repair, and remodeling do not occur. With acute illnesses or injuries, the body can usually heal itself within 6–8 weeks. In addition, as nutrition becomes depleted on a macro- and micronutrient level, healing becomes more difficult [3]. If pain or dysfunction persists greater than 6 months, it is labeled as a chronic condition. Chronic pain can be episodic or continuous and neuropathic in nature, leading to myofascial and soft tissue adaptations, as well as biopsychosocial changes. Neuropathic pain results when nerve fibers become injured, damaged, or dysfunctional and send faulty signals to other areas and systems.

Although the pain or dysfunction may have started from disruption of one tissue, such as a muscle tear, pain begins to spread to surrounding tissues and multiple areas of the body. It takes less stimulation to trigger firing from the nerves around the structure [4]. This creates changes in sensation, muscle control, and proprioception. Muscles and fascia react, becoming shortened or lengthened, pulling the body's frame into positions that increase stress on its structures.

As the functional limitation progresses, it is common for individuals to develop emotional and psychological dysfunctions coinciding with the changes in their bodies [5]. Time is no longer a factor as the healing trajectory is disrupted; more areas in the body are involved as the response progresses from localized to widespread. As we will discuss, chronic musculoskeletal pain issues affect the heart, gut, brain, immunity, and genetics [6]. Patients experiencing functional limitations may have impairment in ambulation, require wheelchairs, or even become bedridden. For the sake of this chapter, both chronic illness and chronic pain will be discussed as a chronic dysfunction, unless differentiation is specified.

53.3 Chronic Staircase of Change

Many changes occur to the body with onset of chronic dysfunction. Postural changes are one of the first characteristics seen in any injury or illness. The figure below demonstrates an ideal static posture in the sagittal and frontal plane (◘ Fig. 53.1).

Any position outside that is represented in ◘ Fig. 53.1 can create excessive stress on the body. Postural asymmetries may develop due to prolonged positioning, repetitive tasks, hand dominance, or uneven loading in the body. Always sitting on one hip, crossing your legs, or sitting in a slouched position may cause these postural asymmetries to occur. Posture imbalances may also develop with repetitive motions or activities such as carrying a child on one hip.

Postural asymmetries may be asymptomatic but could foreshadow further disease or injury in a weakened muscular or spinal system. When muscles or the spinal system are subjected to stress or disease, changes in posture or alignment cause decreased flexibility and range of motion in the body. This leads to restricted lung volume, decreased chest wall and rib mobility, sluggish gut motility, and altered diaphragmatic function, making these systems work less efficiently. The results of these changes can lead to decreased function and premature morbidity.

When postural asymmetries begin to develop, muscle imbalances occur. Movement and function require a balance of muscle length and strength between opposing muscles surrounding a joint. Muscles, along with the fascia around it, may become weakened, become lengthened, or develop adhesions, creating muscle imbalances. Fascia is a thin sheath of connective tissue, primarily collagen, which attaches, stabilizes, encloses, and separates muscles and other internal organs. Fascial sheaths surrounding muscle fibers can cross multiple joints and different regions of the body. Stecco et al. [7] found with dissection of the pectoralis muscle, its fascial attachments are connected to both arms across the sternum, the cervical spine, the contralateral abdominals, and the ipsilateral trunk muscles. Fascial restrictions and adhesions are frequently found in the pectoralis muscles, latissimus muscles, hip flexors, and thoracolumbar and iliotibial band fascial sheaths.

> Healing cannot take place when inflammation, repair, and remodeling do not occur. With acute illnesses or injuries, the body can usually heal itself within 6 to 8 weeks. In addition, as nutrition becomes depleted on a macro- and micronutrient level, healing becomes more difficult. If pain or dysfunction persists greater than 6 months, it is labeled as a chronic condition. Chronic pain can be episodic or continuous and neuropathic in nature, leading to myofascial and soft tissue adaptations, as well as biopsychosocial changes. Neuropathic pain results when nerve fibers become injured, damaged, or dysfunctional and send faulty signals to other areas and systems.

Progression of muscle imbalances leads to the development of musculoskeletal issues. The postural muscles [8], including the deep cervical flexors, middle and lower trapezius, transverse abdominis, gluteus medius and maximus, and multifidi (the stabilizing muscles that extend and rotate the spine), become lengthened, weak, and atrophied. These muscles are meant to provide stabilization for the body and fire aerobically and continuously. During an injury, surgery, or chronic pain, this process becomes altered. The global muscles [8] become dominant and use more energy to sustain posture, causing more fatigue. The most common global muscles include the erector spinae, hip flexors, pectoralis major and minor, rectus abdominis, and latissimus dorsi. The goal is to re-establish dominance of the postural muscles and establish length and flexibility of the global muscles.

For the body to function optimally, muscles do not work individually, they work in synergistic patterns and help create specific movement strategies. When muscle imbalances are present, altered firing patterns are seen. Greig et al. [9] studied 24 elderly women with osteoporosis, half of whom had a vertebral fracture. The women stood and quickly raised their arms in response to a light that indicated the direction of movement. The authors found that the women with vertebral fractures tended to fire both global and postural muscles simultaneously, creating an inefficient muscle pattern, leading to muscle imbalances. Those that demonstrated these muscle imbalances were shown to use poor movement strategies to keep from falling. The other group could perform the task without stiffening, due to improved balance reactions.

The body was designed to move in the upright position, against gravity, and interact with gravitational forces that

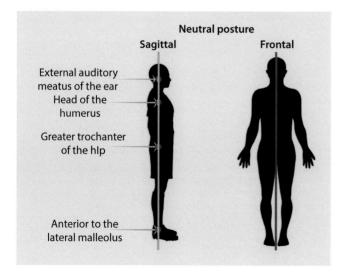

◘ Fig. 53.1 Neutral posture

benefit the entire body, from the bones and muscles to the circulatory system. Wasting of muscles, specifically postural muscles, in the neck, back, abdomen, and legs, will occur after a few days spent in a horizontal position such as with bed rest or hospitalization (see section below). This process can happen faster or slower, depending on the individual's fitness level and lifestyle. One week of complete bed rest can lead to a decrease of 20–30% in muscle strength and cause structural changes to the muscles, bones, nerves, and blood vessels [10]. This will directly affect an individual's muscle tone, balance, coordination, and walking. One becomes more prone to falling and incurring fractures after spending several days in bed, and, as a result, it is important to start an exercise program as soon as possible. Deep vein thrombosis, swelling, and muscle weakness become a risk with prolonged bed rest. Transfers become increasingly difficult due to changes in blood pressure, muscle atrophy, and decreased tissue oxygenation.

Poor posture and muscle imbalances often cause movement patterns to become inefficient. These muscle inefficiencies lead to balance deficits. Any movement of the body or external force creates a change in the body's ability to balance. Balance is the ability of the body to right itself when tipped in any direction. The act of walking requires a significant amount of balance. While transitioning from left to right, one must be able to fully support their body on one leg and propel themselves through space. When balance is compromised, an assistive device may be used to avoid limping or compensation and prevent falling.

Muto et al. [11] found postural and balance changes in 67 women reporting chronic pain and diagnosed fibromyalgia. When compared with the control group, those with fibromyalgia and chronic pain demonstrated increased sway and decreased balance regardless of visual input and surface type. Similarly, when Radebold et al. [12] studied 16 adults with chronic low back pain in an unstable sitting position, subjects

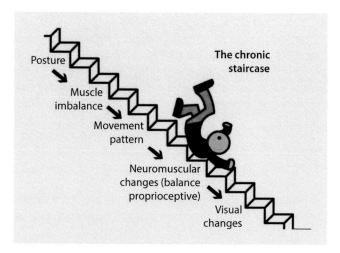

Fig. 53.2 Chronic staircase. (Courtesy of Judy Hensley, PT, DPT, MHS, OCS, MTC)

demonstrated delayed reaction time, decreased balance reactions, and required more visual feedback to weight-shift and balance compared to healthy matched controls. When balance reactions have been compromised, there is less spatial awareness and a need for more conscious effort to maintain balance. See Fig. 53.2.

As balance changes, one side of the body becomes more dominant, and visual perceptual field changes occur [13]. The eye muscles on the dominant side shift the brain's perception of midline as larger on the non-affected or now-dominant side. The ear on the affected side may bend toward the shoulder to allow the dominant eyes to take control. The eye muscles in this position give the brain input that the head is neutral. In this way, one-sided neglect is possible over time in an orthopedic dysfunction without brain trauma. Those that are inactive the majority of time, whether bedridden, wheelchair-bound, or simply deconditioned, improper positioning may also create midline shifts. (See Case ▶ 53.1.)

Case 53.1

TB, a 70-year-old male physical therapy patient, was seen for right shoulder pain from a chronic rotator cuff injury. He has a history of right total knee and hip arthroplasties. TB stands with minimal weight-bearing on his right leg and holds his right shoulder lower than his left. His shoulders are positioned behind his hips, and his shoulder and scapula are tipped forward. Atrophy appears in his right gluteals, as well as weakness in his scapular muscles and tightness in his hip flexors, pectorals, and latissimus dorsi bilaterally. Spinal stabilizers, including the multifidi and the transverse abdominis, are weak and delayed in their firing response. He leans to the right while attempting to stand on his right foot. Findings include changes in his posture, muscle balance, movement patterns, stabilization, balance, and body perception. Proprioception, his body's awareness of where it is in space, is decreased on the right, and he believes his weight is equally distributed. His perception of midline has moved more to the right, allowing his brain to perceive his left side as larger. His left eye is more dominant as it sees the right side of the body. This demonstrates a case of right-sided neglect. His recent complaint is pain reaching overhead with his right arm. A physical therapy treatment began with posture education, manual therapy to improve shoulder joint and soft tissue mobility, spinal and scapular stabilization, balance training, and spatial awareness exercises. He became more aware of a neutral posture but was unable to raise his arm standing. With the appropriate training, he became more aware of his right side and was able to control reaching overhead with the appropriate balance reactions and body mechanics.

53.4 Metabolic Effects of Activity

Physical activity is integral to our general health. In years past, increasing heart rate and exertion were the goals of most exercise scientists in promotion of heart health, weight loss, mental acuity, and disease prevention. Many bedridden and chronically ill individuals are unable to perform such rigorous exercise. There was uncertainty whether low-impact, low-intensity, or short-duration movements were worthwhile, but they have been found to be highly effective on a wide range of systems including cardiovascular, cardiopulmonary, musculoskeletal, gastrointestinal, neural, and endocrine [14]. These findings are relevant for many individuals, including those who are bedridden, recovering from surgery, suffering an illness, aging, and physically challenged.

Lipoprotein lipase (LPL) and myokines are two key factors highlighted in research regarding the effectiveness of gentle movement [6]. LPL is a molecule that is produced during muscle contraction. It plays a central role in how the body processes fats. Levels of LPL rapidly decrease after 4 hours of inactivity and are abolished completely after 18 hours of inactivity. Pederson et al. [6] found that this decrease in LPL leads to the deposit of fats being stored in our adipose tissue, rather than being metabolized by muscle for energy use. Even after moderate exercise, such as a long run in the morning, inactivity for 4 hours afterward has major effects on how we metabolize fat. Inactivity for most of the day, for example, sitting at a desk, is now considered a risk factor for metabolic disease and disorders involving poor lipid metabolism. This research reinforces the need for short, frequent periods of movement at least every 4 hours, no matter if long periods of intense exercise are part of one's daily routine [15].

Myokines are proteins/cytokines that are produced, expressed, and released by muscle fibers upon contraction with movement and exercise. They play an important role in cross talk between skeletal muscle and other organs including the adipose tissue, liver, pancreas, and gut [16]. As the body moves, muscles release myokines that then break down fat tissue to be used for energy instead of being stored around organs as fat. Myokines help regulate inflammation and prevent insulin resistance and metabolic syndrome.

The most studied myokine is interleukin-6 (IL-6). It has an important part in the transition from acute to chronic inflammation, acting as both a pro-inflammatory cytokine and anti-inflammatory myokine. With physical activity, there is an acute increase in IL-6 resulting in an increase in fatty acid oxidation, followed by an increase in systemic lipolysis. IL-6's main function is fatty acid metabolism within the muscle tissue [16].

Inflammation is a common problem for patients in the intensive care unit, frequently associated with serious or prolonged critical illnesses. (Please refer to ▶ Chaps. 42 and 48 for anti-inflammatory diets.) Winkelman et al. [17] examined the relationship between type and duration of physical activity and serum levels of IL-6 and IL-10. A study compared ten critically ill patients who were mechanically ventilated for an average of 10 days. An average of 14.7 minutes of passive physical activity, typically done with multiple bed turns associated with hygiene, increased the anti-inflammatory action of IL-6, which is associated with decreased mortality rate. The authors concluded cytokine balance may be improved by low levels of activity during prolonged critical illnesses.

Brain-derived neurotrophic factor (BDNF) is a myokine that is found in the brain and plays a key role in the regulation, survival, growth, and maintenance of neurons. Physical exercise has been shown to increase BDNF production three times its regular rate [18]. This contributes to increased cognitive function and postexercise-related neurogenesis. Those with type 2 diabetes, regardless of BMI, have decreased circulating levels of BDNF. Lack of physical activity appears to be the main contributing factor in development of type 2 diabetes [18].

The control of inflammation is key in prevention of the disease process in the human body. In the nonobese population, skeletal muscle represents the largest organ of the body and is an active endocrine organ. There is cross talk between the endocrine and the immune system. Inflammatory cytokines are proteins produced during inflammation that activate the hypothalamic-pituitary-adrenal (HPA) axis. When the HPA axis continues to be activated, this leads to diseases that are triggered and maintained by chronic inflammation [6]. These diseases include but are not limited to type 2 diabetes, cardiovascular disease (CVD), colon cancer, breast cancer, dementia, and depression. These have been classified in literature by Pederson [6] as "diseasome of physical inactivity," leading to abdominal adiposity reflected in an accumulation of visceral fat mass. This accumulation is associated with systemic inflammation. Chronic inflammation plays a role in the pathogenesis of insulin resistance, atherosclerosis, neurodegeneration, and tumor growth. Exercise and movement have an anti-inflammatory effect by decreasing visceral fat mass and increasing LPL and other myokine expressions [6] (◘ Fig. 53.3).

Tumor necrosis factor alpha (TNFα) is an inflammatory cytokine that is stimulated with sympathetic nervous system stimulation and is a key player in the pathobiology of inflammatory disorders including rheumatoid arthritis (RA), irritable bowel disorder (IBS), and Crohn's disease. There is evidence to point toward TNFα playing a key role in metabolic syndrome. Those with type 2 diabetes have high protein expression of TNFα in skeletal muscle and plasma. In vitro studies by Pederson [6] demonstrate that TNFα has a direct inhibitory effect on insulin. Infusing TNFα in healthy humans creates insulin resistance in skeletal muscle. Starkie et al. [19] showed that TNFα levels were blunted after 3 hours

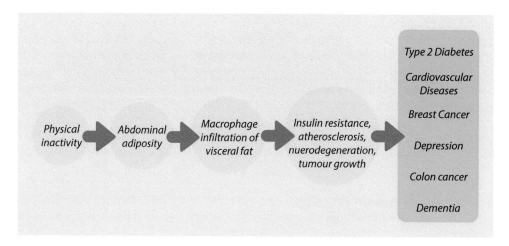

Fig. 53.3 Diseasome of inactivity. (Courtesy of Julie Buttell, PT, MPT, COMT)

of ergometer cycling. This indicates that exercise can reverse the TNF effects of insulin resistance.

If physical activity is crucial in controlling inflammation, one must ask what role mental state plays? Recent research is providing evidence that the mind and body work together to establish homeostasis. Moliere stated, "The mind has a great influence over the body, and maladies often have their origin there" [20]. Mental stress is a common thing in our lives, yet chronic unremitting stress is deleterious to our health.

Stress stimulates our sympathetic nervous system (SNS), which is responsible for the upregulation of the neuronal and hormonal response known as fight or flight. The stress response starts with the release of acetylcholine, activating the secretion of epinephrine from the adrenal gland. Blood vessels dilate in the skeletal muscle and constrict in the gastrointestinal organs. The adrenal glands produce cortisol, an anti-inflammatory response, decreasing our immune response and the body's ability to absorb calcium. Low-grade production of cortisol, due to constant stress, decreases immune function, playing a role in developing autoimmune disease, as well as disrupting the brain's gray and white matter connectivity [20]. This conductivity is vital for the four lobes of the brain to communicate. Decreases in communication lead to mental illness, memory loss, and behavioral disorders [20].

The vagus nerve is responsible for regulating the parasympathetic nervous system and controlling the body's immune response. It does this by activating the HPA axis, which gives information from the body to the brain in an afferent pathway and stimulates a response from the brain to the body in an efferent pathway through the cholinergic anti-inflammatory pathway (CAP) [20]. Immune receptors are located on the vagus nerve. When diseased cells proliferate, an increase in inflammatory cytokines stimulates our immune system to respond. Vagus nerve stimulation increases serotonin and norepinephrine in the presence of inflammatory cytokines and acts as an anti-inflammatory pathway to release corticosteroids by the adrenal glands. The efferent response inhibits inflammation by suppressing inflammatory cytokine production after the release of acetylcholine, which is a neurotransmitter that delivers messages between neural structures and helps with muscle contractions.

The vagus nerve plays a key role in the neuroendocrine-immune axis to maintain homeostasis, but is only activated when the SNS is turned off. Turning off the SNS is a difficult task, as it is stimulated by both high-grade stressors as well as low-grade stressors that require one to remain alert and focused. This prevents the transfer of information to the brain from the parasympathetic nervous system via the vagus nerve [20].

From this research, we know any activity to increase the parasympathetic nervous system will decrease stress and systemic inflammation while improving our immunity [20]. Slipping into the parasympathetic nervous system must be a mindful effort, as we are not frequently in stress-free situations. Meditation, mindfulness, and diaphragmatic breathing show promising research on diminishing our stress response, therefore decreasing risk for initiation and maintenance of the disease process. The diaphragm is innervated by the vagus nerve. When breathing diaphragmatically, air is taken into the low belly, letting the abdomen rise fully before the chest. The vagus nerve is stimulated and sends messages to the brain that the system is safe and relaxed. This allows the vagus nerve to have its anti-inflammatory effect through its afferents activating the HPA axis. Impaired HPA axis due to stress will decrease the communication between the brain and the vagus nerve, creating a risk for such human inflammatory diseases and disorders as rheumatoid arthritis, irritable bowel syndrome, multiple sclerosis, asthma, and dermatitis, and others [20].

53.5 Effects of Bed Rest Resulting in Chronic Dysfunction

Bed rest adds another dimension to this sequela. Some individuals are on bed rest for only a short period after surgery or injury, while others become bedfast for extended periods. The risk of developing osteoporosis and osteopenia is increased in conditions with lack of mechanical challenge to the skeletal system, as seen with immobility. When there is not a sufficient amount of load or strain on the bone, the trabeculae lose its strength, and bone density is lost. The bone loss of osteoporosis reflects the disrupted balance of formation and resorption activities of remodeling, where removal of bone driven by osteoclasts exceeds the ability of osteoblasts to replace it. To protect or recover bone quantity and quality in osteoporotic patients, pharmaceutical agents and strategic exercise programs have been used [21].

Changes in spinal morphology and atrophy with bed rest are being researched in spaceflight. Belavy et al. [21] kept nine male subjects on bed rest with their heads downwardly tilted for 60 days. Axial magnetic resonance imaging of the lumbar spine showed increased disc volume, increased spinal length, change in lumbar spine curvature, and decrease of the cross-sectional area (CSA) of the multifidi and increased CSA of the psoas. This decrease was greatest at L4–L5 in the multifidus, one of the most frequent sites for spinal instability and fusion. Stabilizing the spine at this level is imperative for prevention of lumbar pain. Lumbar instability leads to decreased function of the transverse abdominis and gluteal muscles. These are postural muscles that are responsible for pelvic, hip, and spine stability. When weakness develops in these muscles, the psoas becomes overactive, pulling the lumbar spine into an extended unstable position. Five of the nine subjects reported low back pain during the study or upon ambulating after bed rest. Posterior disc height increased while multifidi CSA decreased. Having periods of inactivity due to bed rest causes changes including pain, postural asymmetries, balance dysfunction, muscle imbalances, and delayed anticipatory response due to decreased spinal stabilization.

Sarabon and Rosker [22] studied 16 healthy older males (59.6 years +/− 3.4) who were kept on bed rest for 14 days to determine trunk muscle stabilization. The subjects were instructed to raise their arms overhead as quickly as possible when a light was observed in their visual field, testing anticipatory responses. The researchers determined the subject's response to being pushed off balance and that was labeled a reflex response. To initiate the reflex response, pressure was placed at the subject's elbow randomly and quickly released. Reactive reflex responses in the trunk were measured with EMG, testing the abdominal obliques and trunk extensor muscle activity. All muscles demonstrated a delayed anticipatory and reflex response following bed rest that persisted for more than 14 days. This study suggests delayed reflexive responses in the elderly persist after 2 weeks of bed rest.

In another study of 19 elderly individuals (66 +/− 1 year) kept on bed rest for 10 days, Coker et al. [23] found body weight, muscle strength, speed of stair climbing, speed of sit to stand transfer, and walking speed significantly reduced. Findings also included decreased VO_2 max and lean muscle mass, as well as an increase in body fat. The authors theorized that the infiltration of lipids into skeletal muscle may play a role in precipitating or affecting the decline of strength and/or functional capacity.

In a study by Fardo et al. [24], 28 healthy female students were confined to a bed for 90 minutes and compared to people in a control group who were sitting for the same period of time. The two groups were compared for pain responses with electrical stimulation to determine which area of the brain was most affected. The horizontal bed rest group demonstrated an inhibition in the central prefrontal portion of the brain making them more sensitive to pain. This region is involved with cognitive, affective, and motor aspects of pain processing. Although this was not conducted with chronic pain patients, it may indicate individuals on bed rest have less inhibition for pain control.

Chronic low back pain, along with any ongoing pain or illness, has been demonstrated to show changes in diaphragmatic breathing. The diaphragm precedes any movement of the body by lowering and subsequently establishing abdominal pressure which helps stabilize the lumbar spine. The lower ribs must be stabilized by the abdominals and can only expand to the sides. Vostatek et al. [25] used magnetic resonance imaging to study the breathing patterns of 35 individuals, half of which had chronic low back pain. The group with chronic low back pain was found to have diaphragmatic changes including increased respiration rate, increased anterior expansion, decreased total excursion, expansion more horizontal than vertical, and a higher resting position of the diaphragm in the thorax. The group with pain suffered from insufficient abdominal strength which resulted in poorly stabilized lower ribs. Abdominal weakness is associated with decreases in proper muscle balance, spinal stabilization, pelvic floor, and respiratory control. Smith et al. [26] found disorders of urinary incontinence and respiration were stronger predictors of frequent back pain than physical activity and obesity. Furthermore, the researchers found patients with only low back pain are likely to develop urinary or respiratory dysfunctions within 2–3 years of onset of back pain.

Chronic neck pain has been associated with decreased pulmonary function. Dimitriadis et al. [27] compared 45 individuals with chronic neck pain with 45 matched controls. The authors found the pain group had significantly reduced vital capacity (the volume of air expressed from the lungs), forced vital capacity, expiratory reserve volume (the additional amount of air that has been left in the lungs after expiration), and maximum voluntary ventilation (the volume of gas that has been inhaled 15 seconds when a person breathes as quickly and deeply as possible). The reduction in lung volumes resembled restrictive pulmonary dysfunctions, more than obstructive. Muscle imbalances and spinal instability, also found in the pain group, included weak neck flexors/extensors and decreased range of motion. The diaphragm is an integral part of the core. It may be helpful to visualize the core as a cylindrical container with the top of the container being the vocal cords, the bottom being the pelvic floor, and the sides being the internal obliques, the quadratus lumborum, and the diaphragm. This cylindrical container provides pressure against external forces, providing stability for the trunk.

Along with physical and respiratory changes, mental changes often occur in individuals with chronic dysfunction. Social networks have a great influence on our physical health. Christakis and Fowler [28] in 2007 found a 171% increased likelihood of becoming obese if a friend became obese. It was only 41% more likely someone would become obese if a sibling became obese. This gives reason to think mindset in community, and physical activities within communities, are essential. A Harvard study showed body position can influence hormone levels. Assuming a powerful posture opening shoulders, with a straight spine, increased testosterone 20% and decreased cortisol levels 25% [29]. This gives feedback that physical activity does not need to be intense to be effective. Many options from chair yoga to Tai Chi can be effective for the chronically ill individual [30].

Genetic changes have been noted in individuals with physical inactivity. DNA sequencing does not change, but how the cells read the gene can be influenced by disease state, environment/lifestyle, and age of the individual, also known as epigenetics. Bed rest studies have shown us physical inactivity is one of the main contributing factors in metabolic disturbances. "Transcriptional analysis of human skeletal muscle after 9 days of bed rest revealed changes in the expression of more than 4500 genes, and the downregulation of a total of 54% of genes involved in the oxidative phosphorylation pathway" [16]. This supports the theory suggesting physical inactivity is one of the main factors in the start of many chronic diseases. When we are inactive, fat is no longer used as an energy source but stored around our organs. Visceral fat has been shown to initiate the onset of almost all inflammatory diseases [6].

The above evidence poses the question of the significance of myokine and LPL deficiency in the inactive population and the role for anti-inflammatory diets coupled with movement. The imperative for intervention, including passive range of motion, submaximal movement, and balance training among those on bed rest or who are physically inactive, can no longer be ignored.

53.6 Rehabilitation of Chronic Dysfunction

Returning to exercise after being bedfast, chronically ill, or a period of immobility should happen progressively to allow the body to adapt to a new routine. As described above, being bedridden can lead to deconditioning, negatively affecting health. For this reason, it is important to discuss appropriate exercise prescription to allow the chronically ill and bedridden patient to regain strength, range of motion, and stability required for daily tasks.

A referral to physical therapy can be of great benefit to those with chronic dysfunction. Manual therapy, including joint and soft tissue mobilization, can decrease pain, adhesions, and inflammation and increase mobility, allowing the patient to move into normal postures. The effects of sympathetic nervous system (SNS) overactivation must also be decreased, awakening the parasympathetic nervous system to allow the vagus nerve to give input to the brain regarding the bodies' immune state and stabilize the blood pressure. This can be addressed through diaphragmatic breathing, mindfulness training, relaxation, and physical exercise. Massage may be needed for myofascial issues including muscle and soft tissue holding. Acupuncture or dry-needling techniques may be beneficial in calming the nervous system [31]. Inflammation must be addressed by a nutritionist to improve diet and resolve gut issues. These are all provided in conjunction with physical therapy, and the result is facilitation of improved muscle balance, spinal stabilization, postural awareness, and functional movement.

Introducing diaphragmatic breathing and meditation benefits the individual with chronic dysfunction, creating metabolic and physical changes. Breathing and mindfulness training will allow the nervous system to calm, causing anti-inflammatory effects allowing for pain reduction and aiding in healing. Lung capacity, thoracic extension, resting posture, and muscle tone can all be improved by learning appropriate breathing patterns. There are many programs and apps available to assist in this training, such as Headspace™, Muse™, Calm™, and others.

Proper positioning, correcting muscle imbalances, and education in correct body mechanics contribute to decreasing stress. Shahidi et al. [31] studied 60 asymptomatic workers performing a computer task that became progressively more difficult and stressful. As stress increased, the workers developed an increase in their forward head posture with activation of the upper trapezius, resulting in pulling the shoulders upward. If this response occurs in pain-free workers, it is likely that those with chronic pain will have an exaggerated response.

Various types of vibration therapy have been used with patients with chronic dysfunctions. As it is known that bed rest and inactivity lead to osteoporosis, degenerative disc disease, and chronic pain, vibration therapy may have a role to play in mitigating the damage. To increase bone density, whole-body vibration therapy may be administered passively in supine, seated, and standing positions. In a study by Holguin et al. [32], healthy individuals were kept on bed rest for 90 days. Decreased disc swelling and a return of disc volume were seen in the group who underwent 10 minutes per day of whole-body vibration after 7 days of being taken off bed rest. Compared with the control group who did not receive vibration therapy, those receiving the therapy retained disc convexity throughout the lumbar spine and reduced the incidence of back pain by 46%. There has also been some potential for whole-body vibration to serve as a surrogate for exercise for those that are unable to participate in physical activity and as an adjunct to therapy. Evidence demonstrates multiple significant advantages using vibration therapy as an intervention [10].

Another type of vibration therapy can be used locally to address pain using a TENS unit. Lundberg et al. [33] enrolled 731 patients, 596 with chronic pain, where there was placement of an electrode or a vibrator near or at the area of pain. Treatment was localized to the antagonist muscle or a trigger point near the painful area for 30 to 45 minutes, and pain was found to be reduced in 82% of the patients. Even patients with moderate to severe pain typically had at least 50% reduction in complaints. Vibration is thought to have a greater effect on the central nervous system to dampen pain.

Assessment and treatment of pain generators is most important to help calm the SNS and allow for appropriate relaxation during treatment. Limitations in soft tissue and joint mobility should be addressed. Active range of motion and passive range of motion are among the safest form of exercise that can be performed during bed rest and periods of low activity. Passive range of motion improves joint mobility, improves synovial fluid, and prevents joint contractures, allowing the individual to maintain range of motion until it can be performed actively. Active range of motion is used to help the individual gain control of their musculoskeletal system and progressively expose the body to gravity regaining muscle strength, flexibility, coordination, and agility. Benefits of returning to an active lifestyle can be experienced almost immediately when beginning an exercise program.

Submaximal isometric exercises are added in the early phases of the exercise prescription to ensure deconditioning does not worsen. Submaximal isometric exercises occur when one can contract a muscle without movement of the joint. These exercises help maintain neuromuscular function and strength with low intensities that will not disrupt newly formed collagen fibers. Strengthening exercises should be added to the program once flexibility has been achieved and isometric control is established. These exercises begin with only body weight and then progress to external loads, such as light weights. Muscle strengthening assists in reversing the negative effects of prolonged bed rest and chronic inactivity, preventing muscle atrophy and improving circulation.

Focus on even distribution of weight is essential, easing into simple movements as walking, squatting, and reaching. Proper movement control and utilization of correct movement patterns is also necessary. Progression moves into endurance tasks, allowing for return to activities of daily living. Balance training, motor control, and proprioception need to be addressed to restore efficient movement strategies, prevent injury, and allow appropriate energy expenditure during daily tasks. This can be achieved through closed chain exercises designed to enhance neuromuscular control and replicate common movements performed in activity of daily living. Visual and spatial awareness, balance reactions, and postural control need to be addressed so one can return to movement without risk of falling or reinjury. Attention to core and trunk stabilization exercises is imperative for improvement of posture and balance. One must focus on proper alignment to ensure the body can achieve proper motor patterning/control. These tasks listed above vary from patient to patient. Range of motion and strength gains may be seen in the first 4-6 weeks of rehabilitation, but neuromuscular control, balance training, and postural awareness may take much longer for the patient to master. Addressing all the above issues ranges from months to years of physical therapy.

When developing an exercise program for the bedridden or chronically ill, careful consideration needs to be taken to ensure the exercise prescription is appropriate. Inappropriate exercise can be detrimental to the healing process, creating unnecessary discomfort. Prior level of function, current state of physical/mental and metabolic status, balance, proprioception, and neuromuscular control need to be considered in development of the best exercise program for the individual.

Caution is also raised to guard against those who recommend over-training. Overall, exercise can help decrease the chance of developing heart disease, increase bone density, and improve immunity. Exercise is beneficial, but caution needs to be taken to not overdo. Heavy long-term exercise causes harm to the body and suppresses the immune system [14]. The innate immune system responds negatively to chronic stress of intensive exercise. Natural killer cell activity is enhanced, destroying abnormal cells and viruses in the body while neutrophil function (the immune response to bacteria) is suppressed. In athletes, however, investigators have had little success in linking long-term intense exercise with higher incidence of infection and illness.

Many components of the immune system exhibit change after prolonged heavy exertion. An altered immunity may last between 3 and 72 hours, giving viruses and bacteria an advantage increasing the risk of infection. Studies on the influence of moderate exercise training on cellular protection and immune function have shown that near-daily brisk walking reduces the number of sick days by half. Compared to the inactive population, no data suggests moderate exercise training is linked with improved immunity [14].

53.7 Conclusion

In this chapter, we have discussed the stages of normal tissue healing and what abnormal healing looks like and described the many changes that occur due to pain or inactivity. Many physiological and metabolic changes are associated with inactivity and bed rest. This chapter aims to help the reader understand the physiological changes that occur due to pain and inactivity, assisting in improving treatment strategies for individuals with chronic dysfunction. Further research needs to be conducted on effective treatment approaches and protocols for proper rehabilitation of the chronically ill patient. Education of the public should continue to focus on prevention of these chronic conditions.

References

1. Ward BW, Schiller JS, Goodman RA. Multiple chronic conditions among US adults: a 2012 update. Prev Chronic Dis. 2014;11:E62.
2. Baechle TR, Earle Roger W. Essentials of strength training and conditioning. 3rd ed. Champaign, IL: Human Kinetics; 2008. p. 529–37.
3. Goodman CC, Snyder TEK. Differential diagnosis in physical therapy. 3rd ed. Philadelphia, PA: W.B. Saunders; 2000. p. 17.
4. Xie RG, Chu WG, Hu SJ, Luo C. Characterization of differential types of excitability in large somatosensory neurons and its plastic

changes in pathological pain states. Int J Mol Sci. https://doi.org/10.3390/ijms19010161. Pubmed:PMID:29303989.
5. Minami M. Neuronal mechanisms underlying pain-induced negative emotions. Brain Nerve. 2012;64(11):1241–7.
6. Pederson BK. The diseasome of physical inactivity – and the role of myokines in muscle-fat cross talk. J Physiol. 2009;587(Pt 23):5559–68.
7. Stecco A, Masiero S, Macchi V, Stecco C, Porzionato A, DeCaro R. The pectoral fascia: anatomical and histological study. J Bodyw Mov Ther. 2009;13(3):255–61.
8. Bandy WD, Sanders B. Therapeutic exercise. Baltimore, MD: Lippincott, Williams and Wilkins; 2001. p. 264–7.
9. Greig AM, Briggs AM, Bennell KL, Hodges PW. Trunk muscle activity is modified in osteoporotic vertebral fracture and thoracic kyphosis with potential consequences for vertebral health. PLoS One:2014;9(10). https://doi.org/10.1371/journal.pone.0109515. PubMedCentral:PMID: 25285908;PMCIC:PMC4186857.
10. Boeselt T, Nell C, Kehr K, Holland A, Dresel M, Greulich T, Tackenberg B, Kenn K, Boeder J, Klapdor B, Kirschbaum A, Vogelmeier C, Alter P, Koczulla R. Whole-body vibration therapy in intensive care patients: a feasibility and safety study. J Rehabil Med. https://doi.org/10.2340/16501977-2052. PubMed:PMID:26805786.
11. Muto LHA, Sauer JF, Yuan SLK, Sousa A, Mango APC, Marques AP. Postural control and balance self-efficacy in women with fibromyalgia: are there differences? Euro J Phys Rehab Med. 2015;51:149–54.
12. Radebold A, Cholewicki J, Polzhofer G, Green H. Impaired postural control of the lumbar spine is associated with delayed muscle response times in patients with chronic idiopathic low back pain. Ovid. 2001;26(7):724–30.
13. Puentedura E, Flynn T. Combining manual therapy with pain neuroscience education in the treatment of chronic low back pain: a narrative review of literature. Physiother Theory Pract. 2016;32(5):408–14.
14. Burton C, Clerckx B, Robbeets C, Ferdinande P, Langer D, Troosters T, Hermans G, Decramer M, Gosselink R. Early exercise in critically ill patients enhances short-term functional recovery. Crit Care Med. 2009;37(9):2499–505. https://doi.org/10.1097/CCM.ob013e3181a38937. PubMed:PMID:19623052.
15. Bey L, Hamilton MT. Suppression of skeletal muscle lipoprotein lipase activity during physical inactivity: a molecular reason to maintain daily low-intensity activity. J Physiol. 2013;551(Pt 2):673–82.
16. Eckardt K, Gorgens SW, Raschke S, Jurgen E. Myokines in insulin resistance and type 2 diabetes. Diabetologia. 2014;57(6):1087–99.
17. Winkelman C, Higgins PA, Chen YJ, Levine A. Cytokines in chronically ill patients after activity and rest. Biol Res Nurs. 2007;8(4):261–71. https://doi.org/10.1177/1099800406298168. PubMed:PMID:17456587.
18. Hamilton MT, Hamilton DG, Zderic TW. Role of low energy expenditure and sitting in obesity, metabolic syndrome, type 2 diabetes, and cardiovascular disease. Diabetes. 2007;56(11):2655–67.
19. Starkie R, Ostrowski SR, Jauffred S, Febbraio M, Pedersen BK. Exercise and IL-6 infusion inhibit endotoxin-induced TNFα production in humans. FASEB. 2003;17(8):884–6.
20. Bonaz B, Picq C, Sinniger V, Mayol JF. Vagus nerve stimulation: from epilepsy to the cholinergic anti-inflammatory pathway. Neurogastroenterol. Motil. 2013;25(3):208–21.
21. Belavy et DL, Armbrecht G, Richardson CA, Felsenberg D, Hides JA. Muscle atrophy and changes in spinal morphology. Spine J. 2011;36(2):137–45.
22. Sarabon N, Rosker J. Effects of fourteen-day bed rest on trunk stabilizing in aging adults. Biomed Res Int. 2015. https://doi.org/10.1155/2015/309386. PubMedCentral:PMID:26601104;PMCID:PMC4637013.
23. Coker RH, Hays NP, Williams RH, Wolfe RR, Evans WJ. Bed rest promotes reductions in walking speed, functional parameters, and aerobic fitness in older, healthy adults. J Gerontol A Biol Sci Med Sci. 2015;70(1):91–6.
24. Fardo F, Spironelli C, Angrilli A. Horizontal body position reduces cortical pain-related processing: evidence from late ERPs. PLoS One. 2013;8(13). https://doi.org/10.1371/journal.pone.0081964. PubMedCentral:PMID:24278467;PMCID:PMC3835670.
25. Vostatek P, Novak D, Rychnovsky T, Rychnovsky S. Diaphragm postural function analysis using magnetic resonance imaging. PLoS One. 2013;8(3). https://doi.org/10.1371/journal.pone.006724. PubMedCentral:PMCID:PMC3597716.
26. Smith MD, Russell A, Hodges PW. Disorders of breathing and continence have a stronger association with back pain than obesity and physical activity. Aust J Physiol. 2006;52:11–6.
27. Dimitriadis Z, Kapreli E, Strimpakos N, Oldham J. Pulmonary function of patients with chronic neck pain: a spirometry study. Respir Care. 2014;59(4):543–9.
28. Cristakis NA, Fowler JH. The spread of obesity in a large social network over 32 years. N Engl J Med. 2007;357(4):370–9.
29. Carney D, Cuddy A, Yap A. Power posing: brief nonverbal displays affect neuroendocrine levels and risk tolerance. Psychol Sci. 2010;21(10):1363–8. https://doi.org/10.1177/0956797610383437. [Epub ahead of print].
30. Bonakdar R. Integrative pain management. Med Clin N Am. 2017;101(5):987–1004.
31. Shahidi B, Haight A, Maluf K. Differential effects of mental concentration and acute psychological stress on cervical muscle activity and posture. J Electromyogr Kinesiol. 2013;23(5):1082–9.
32. Holuguin N, Muir J, Rubin C, Judex S. Short applications of very low-magnitude vibrations attenuate expansion of the intervertebral disc during extended bed rest. Spine J. 2009;9(6):470–7.
33. Lundberg TC. Vibratory stimulation for the alleviation of chronic pain. Acta Physiol Scand Suppl. 1983;523:1–51.

Practitioner Practice Resources

Contents

Chapter 54 Systems Biology Resources – 1015
Jeanne A. Drisko, Diana Noland, and Leigh Wagner

Chapter 55 Initial Nutrition Assessment Checklist – 1019
Leigh Wagner, Diana Noland, and Jeanne A. Drisko

Chapter 56 Nutritional Diagnosis Resources – 1043
Leigh Wagner, Diana Noland, and Jeanne A. Drisko

Chapter 57 Specialized Diets – 1045
Leigh Wagner, Diana Noland, and Jeanne A. Drisko

Chapter 58 Motivational Interviewing – 1051
Leigh Wagner, Diana Noland, and Jeanne A. Drisko

Chapter 59 Authorization for the Release of Information – 1055
Leigh Wagner, Diana Noland, and Jeanne A. Drisko

Chapter 60 Patient Handouts – 1057
Leigh Wagner, Diana Noland, and Jeanne A. Drisko

Systems Biology Resources

Jeanne A. Drisko, Diana Noland, and Leigh Wagner

Systems Biology
National Institutes of Health, Intramural Research
Systems Biology as Defined by NIH
▶ https://irp.nih.gov/catalyst/v19i6/systems-biology-as-defined-by-nih

Institute of Systems Biology: biomedical research organization (nonprofit)

ISB is a holistic approach that encompasses all metabolic systems and their networks to bring a usable understanding of the complexity of the biological systems of whole of living organisms, where they are perturbed, and a basis for developing interventions that restore structure and function. There is a peer collaboration among healthcare and technology disciplines. Integrative medicine, and functional medicine, has evolved on the foundation of systems biology. It is important for integrative and functional healthcare practitioners to have a robust appreciation of the concept of systems biology to build their paradigm of integrative and functional medicine (◘ Figs. 54.1 and 54.2).

▶ https://systemsbiology.org/

The Meaning of Systems Biology (Cell)
▶ https://www.cell.com/action/showPdf?pii=S0092-8674%2805%2900447-2

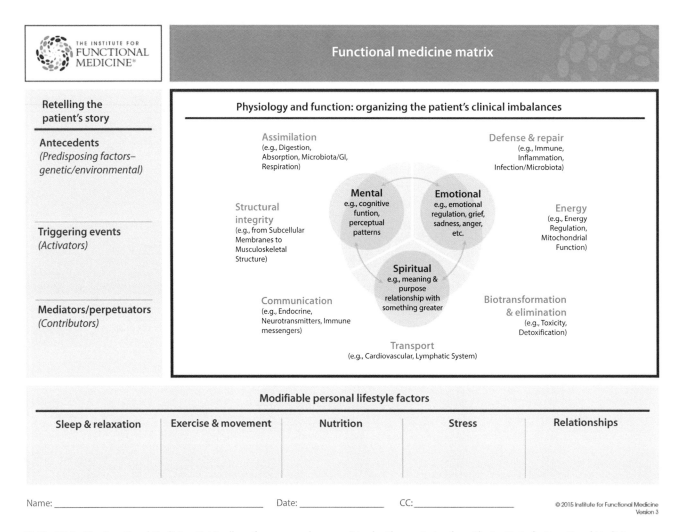

◘ Fig. 54.1 The Functional Medicine Matrix allows for a comprehensive evaluation of the patient's life history leading to chronic disease and provides a roadmap for interventions leading back to good health. (Used with permission from The Institute for Functional Medicine ©2015)

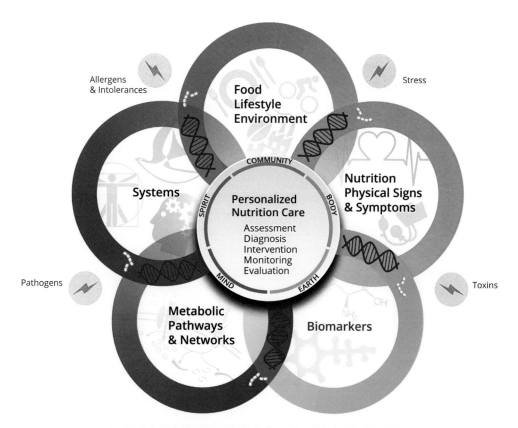

© 2010, 2018 Copyright. All Rights Reserved. KM Swift, D. Noland, E. Redmond

Fig. 54.2 The Radial provides a detailed interconnected algorithm to deliver personalized nutrition care in an integrative and functional medicine therapy approach. (Courtesy of Kathie Madonna Swift, Diana Noland, Elizabeth Redmond)

Initial Nutrition Assessment Checklist

Leigh Wagner, Diana Noland, and Jeanne A. Drisko

55.1 **Initial Nutrition Assessment Process Background** – 1020
55.1.1 Chart Review and Preparation – 1020
55.1.2 Before Appointment – 1020
55.1.3 Nutrition Assessment Process – 1020
55.1.4 Chronic Disease Pathophysiology – 1022

55.2 **Glossary** – 1022

55.3 **Gastrointestinal Assessment (Fig. 55.6, Table 55.1)** – 1038
55.3.1 Testing for Low Stomach Acidity (Hypochlorhydria) – 1038
55.3.2 Common Symptoms of Hypochlorhydria – 1038
55.3.3 Diseases Associated with Hypochlorhydria – 1038
55.3.4 Self-Testing and Treatment for Low HCL/Hypochlorhydria – 1038

References – 1041

© Springer Nature Switzerland AG 2020
D. Noland et al. (eds.), *Integrative and Functional Medical Nutrition Therapy*,
https://doi.org/10.1007/978-3-030-30730-1_55

55.1 Initial Nutrition Assessment Process Background

The initial nutrition assessment is based on the Nutrition Care Process established by the Academy of Nutrition and Dietetics (the Academy) and is summarized as follows:
- Assessment
- Diagnosis
- Intervention
- Monitoring
- Evaluation (◘ Fig. 55.1)

55.1.1 Chart Review and Preparation

The RDN reviews the paper and/or electronic medical record for information about the client's health history, nutrition history, anthropometrics, and any other information that could be helpful to begin to evaluate the person's nutritional status.

Papers or documents that can be helpful for nutrition consultations:
- Nutrition intake forms submitted prior to nutrition consultation.
- While reviewing the chart, you can fill in the "blank nutrition assessment and consultation summary form."
- Blank "progress note" to write notes during appointment (lined paper with the client's name, medical record number (MRN), and date of birth (DOB)).
- Charge sheet (varies by practice).
- Nutritional supplement list and company contact information packet (varies by practice).
- Updated list of supplements.

55.1.2 Before Appointment

Prior to the actual consultation/appointment, the patient service representative (PSR) will need to take care of technical information, paperwork, and other items:
- PSR: Confirm that patient has filled out all necessary forms and take photo (if approved and appropriate). If not, prepare forms that need to be completed and explain each one in detail, if necessary. Chart should be prepared ahead of time.
    - Notice of privacy practices (PSR)
    - HIPAA authorization (PSR)
    - Diet prescription or specific referral form (if needed)
    - Integrative nutrition intake form (self- or physician-referred)
    - Diet diary (1- to 3-day history)
    - Fats and oils questionnaire
    - Medical symptoms questionnaire
- Photograph (taken by PSR): Take a photograph of the client to be placed in the chart. This will be used for ease of recognition and better recall of client story and team approach to care.
- Dietitian (RDN): Introduce yourself, welcome, and ask how the person prefers to be addressed. Lead client to private, quiet consultation room. Invite the client to sit in designated chair in counseling area. If a family member or friend of the client wants to join the consultation and is 18 years or older, ask client for permission.
- When necessary, ask the client whether a dietetic intern can observe the appointment.

55.1.3 Nutrition Assessment Process

55.1.3.1 Background and Explanation of Assessment

- Explain to client, when necessary, the uniqueness of dietetics and integrative medicine by providing the handout "What is DIM?".
- Ask client to state his or her most important goals to achieve during the session. Be sure to address these goals during the session and in the visit summary.
- Briefly explain the nutrition assessment process, when necessary:
    - In-office assessments: Bioelectric impedance analysis (BIA), weight, waist circumference, nutrition physical exam, and others.
    - Review of nutrition intake forms, medical symptoms questionnaire (MSQ), fats and oils questionnaires, and any others.
    - Ask client to sign MSQ.
    - Review diet diary. If not completed, take a 24-hour dietary recall.
    - Develop intervention of a nutrition action plan and supplement plan.

55.1.3.2 In-Office Assessments and Measurements

In-office assessments (not all of these tests are completed at every visit on every client):
- Nutrition physical exam
- Anthropometrics: height, weight, and waist circumference
- Bioelectric impedance analysis (BIA): measurements taken on a massage or acupuncture table or other surface client can easily and comfortably lay supine
- 02 Sat (respiratory issues)
- Handgrip strength measurement
- Blood pressure and pulse
- Frame size assessment

Other assessments/measurements:
- Other circumferences (pediatric, head; sports, bicep; muscle injury, atrophy of specific muscle(s); amputation)

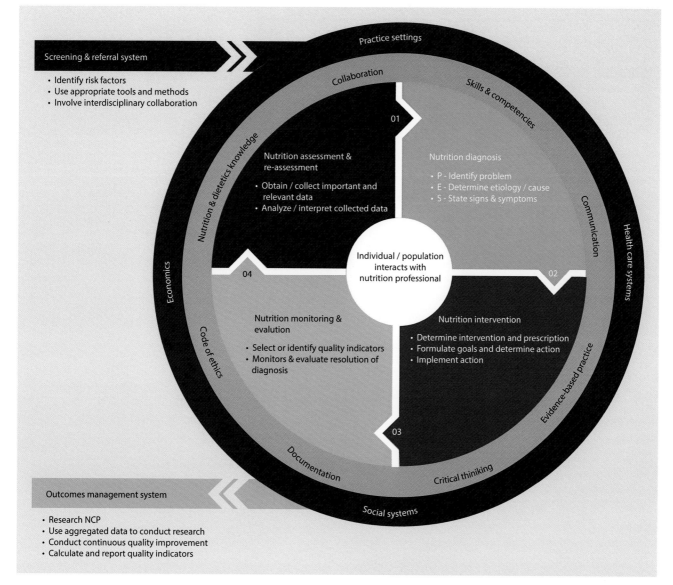

Fig. 55.1 The Nutrition Care Process highlights the importance of systematically combining elements of assessment, diagnosis, intervention, monitoring, and evaluation (ADIME) in the long-term management of patients. (Reprinted from Swan et al. [13]. With permission from Elsevier.)

- Zinc tally
- Iodine patch test
- Urinary pH
- Urinary ascorbic acid

55.1.3.3 Review Forms with Client

- Nutrition intake forms (initial comprehensive or brief follow-up form)
  - Fill in blanks and ask questions in areas needing clarification or expansion.
  - Ask additional questions, if necessary.
- Diet diary
- Review important findings in lab work/note labs needing to be monitored/retested and timeline.
- Review BIA results and other measurements.

55.1.3.4 Develop Priorities and Create Nutrition Action Plan

- Review most important nutrition/lifestyle changes to accomplish goals within anticipated time and client adherence to plan.
- Review supplement recommendations for understanding, tolerance, and manageability.
- Discuss important lifestyle changes (exercise, sleep, stress management, food preparation, procurement).
- Discuss whether further laboratory testing is recommended.
- Set measurable goals for next appointment.
- Write and provide a copy to the client of the nutrition action plan for the client to have record and reference to his/her nutritional goals and actions.

55.1.3.5 Review Other Recommendations

- If recommended, explain any further nutritional laboratory testing recommended by you, the physician, or advanced practice registered nurse (APRN).
- Schedule follow-up, "touch base," and phone call within a week for 10–15 minutes to answer questions and monitor progress of nutrition action plan.
- Plan time for follow-up visit: 1–6 months, depending on the complexity of the case and nutrition action plan.

55.1.3.6 Documentation and Follow-Up

- After appointment, organize the chart, notes and laboratory results.
- Ensure action items are followed-up and completed (email articles, "to-dos," recipes, laboratory requisition, testing kits, etc.).
- Document (write, type, dictate) the nutrition note and finalize by signing and submitting to the referring doctor for review and/or signature.
- Write the nutrition and supplement plan, if necessary.
- What should be done prior to the next visit (labs, diet diary, etc.)?
- Document treatment goal(s) and plan for client.
- Based on nutrition assessment, develop nutrition diagnostic statements and intervention (plan) by following the Nutrition Care Process (NCP) form for Nutrition Diagnostic Statements, PES:
  - Problem
  - Etiology
  - Signs/symptoms (◘ Figs. 55.2, 55.3, and 55.4)

Recommended Websites:

- Tooth-Organ Relationships - ▶ http://www.secretofthieves.com/tooth-chart/tooth-meridian-14.cfm
- Dams, Inc. - ▶ www.dams.cc
- The International Academy of Oral Medicine and Toxicology - ▶ www.iaomt.org
- Information on fluoride - ▶ www.fluoridealert.org
- Information on mercury-free dentistry and fluoride - ▶ www.bioprobe.com

55.1.4 Chronic Disease Pathophysiology

55.1.4.1 Nutrient Insufficiencies
1. IDU/MAPDOM
2. Nutritional triage theory
3. Long latency nutritional insufficiencies

55.1.4.2 Inflammation/Immune Dysregulation
1. Nutritional influences on inflammation: eicosanoids and bioflavonoids
2. Nutritional influences on immune regulation: nutrient control and cytokine metabolism

55.1.4.3 Biochemical Individuality
1. Signs and symptoms
2. Phenotype
3. Family history
4. Genomics/genotype/methylation
5. Environmental history

55.1.4.4 Lifestyle
1. Activity
2. Sleep
3. Community (beliefs/self-care)

55.1.4.5 Energy Dysfunction
1. Mitochondrial ATP production (glycolysis, gluconeogenesis, ketosis)
2. Dietary intake and output
3. Body composition

55.1.4.6 Toxic Load
1. Metabolic influences of toxic substances (heavy toxic metals, petrochemicals, etc.)
2. Metabolic influences of toxic stress factors (lack of sleep, emotional, etc.)
3. Toxic load = total toxic influence on an individual's metabolism
4. Toxic epigenetic influence on gene expression

55.1.4.7 Sleep and Stress
1. Sleep quality reflects state of circadian rhythm affecting all systems positive or negative.
2. Stress perturbs immune and hormonal and circadian rhythm and impacts nutrient status.

55.2 Glossary

Bioflavonoid A plant pigment responsible for the colors found in fruits and vegetables that has antioxidant properties beneficial to humans. Bioflavonoids include flavonols (quercetin), favanones (hesperetin), isoflavones (genistein), and anthocyanidin (cyanidin) [1].

Body composition An individual's body percentage of fat, bone, muscle, and intra/extracellular hydration. Chronic disease expression of altered body fat distribution is an important consideration in body composition and usually measured as waist circumference, waist-to-hip ratio, and waist-to-height ratio.

Cytokine metabolism Cytokines are four families of mostly lymphoid cells: interleukins (ILs), tumor necrosis factors (TNFs), interferons (IFNs), colony stimulating factors (CSFs). Cytokines are produced in response to immunological challenges from nutritional deficiency, infection,

Initial Nutrition Assessment Checklist

General Information　　　　　　　　　　　　　　　　　　　　　　　　　　　　　　Date:

Name		Preferred Name:	
Date of Birth	Age:	Gender: ☐ M ☐ F	Height __'__" Weight ____
Genetic Background	☐ African American ☐ Native American ☐ Mediterranean	☐ Hispanic ☐ Caucasian ☐ Northern European	☐ Asian ☐ Other *(please note)*
Family Status	Marital Status: M S	Do you have children: Y N	Children Ages:
ABO Blood Type	*(circle one)* O A B AB	Have you ever had a blood transfusion?	Y N
Address			
Home Phone		Cell Phone:	
Work Phone		Occupation:	
Fax			
Email			
Best Way to Reach?			
Primary Physician	Name:		
	City:	Phone:	
Secondary Physician	Name:		
	City:	Phone:	
Referred by			

Complaints/Concerns

What do you hope to achieve in your visit?

If you had a magic wand and could erase three problems, what would they be?
(list your three main <u>health</u> concerns)

1												
2												
3												

Fig. 55.2 Nutrition new patient intake form

If you had a magic wand and could erase three problems, what would they be? (list your three main <u>nutrition</u> concerns)										
1										
2										
3										
When was the last time you felt well?										
Did something trigger your change in health?										
The biggest challenge(s) to reaching my nutrition goals is/are?										
In the past, I have tried the following techniques, diets, behaviors, etc. to reach my nutrition goals…										
What makes you feel better?										
What makes you feel worse?										
What is the lowest body weight that you have been comfortably able to maintain for at least 2 years in your adult life, since around age 30?										
Please note any past or current injuries:										

Fig. 55.2 (continued)

Initial Nutrition Assessment Checklist

Readiness Assessment

Rate on a scale of 5 (very willing) to 1 (not willing)
In order to improve your health, how willing are you to:

Significantly modify your diet	☐ 5 ☐ 4 ☐ 3 ☐ 2 ☐ 1
Take several nutritional supplements each day	☐ 5 ☐ 4 ☐ 3 ☐ 2 ☐ 1
Keep a record of everything you eat each day	☐ 5 ☐ 4 ☐ 3 ☐ 2 ☐ 1
Modify your lifestyle (e.g., work demands, sleep habits, exercise)	☐ 5 ☐ 4 ☐ 3 ☐ 2 ☐ 1
Practice a relaxation technique	☐ 5 ☐ 4 ☐ 3 ☐ 2 ☐ 1
Engage in regular exercise/physical activity	☐ 5 ☐ 4 ☐ 3 ☐ 2 ☐ 1
Have periodic lab tests to assess your progress	☐ 5 ☐ 4 ☐ 3 ☐ 2 ☐ 1

How much ongoing support and contact (e.g., telephone, e-mail) from the nutritionist would be helpful to you as you implement your personal health program?

Allergy Information

Please list <u>food</u> allergies	
Please list <u>non-food</u> allergies	
What type of allergic symptoms do you experience?	
Notes:	

Fig. 55.2 (continued)

Medical History

	Height:	Weight:	Waist:

Please check those health conditions that your doctor has diagnosed (provide the date of onset)

Gastrointestinal	Inflammatory/Autoimmune
☐ Irritable Bowel Syndrome	☐ Chronic Fatigue Syndrome
☐ Inflammatory Bowel Disease	☐ Rheumatoid Arthritis
☐ Crohn's Disease	☐ Lupus SLE
☐ Ulcerative Colitis	☐ Poor Immune Function *(frequent infections)*
☐ Gastric or Peptic Ulcer Disease	☐ Severe Infectious Disease
☐ GERD (reflux/heartburn)	☐ Herpes-Genital
☐ Celiac Disease	☐ Multiple Chemical Sensitivities
☐ Hepatitis C or Liver Disease	☐ Gout
☐ Other Digestive:	☐ Other:

Cardiovascular	Metabolic/Endocrine
☐ Heart Disease (heart attack)	☐ Diabetes ☐ Type 1 or ☐ Type 2
☐ Stroke	☐ Metabolic Syndrome (insulin resistance)
☐ Elevated Cholesterol	☐ Hypoglycemia
☐ Irregular heart rate – Pacemaker	☐ Hypothyroidism (low thyroid)
☐ High Blood Pressure	☐ Hyperthyroidism (overactive thyroid)
☐ Mitral Valve Prolapse/heart murmur	☐ Polycystic Ovarian Syndrome (PCOS)
☐ Other Heart & Vascular:	☐ Genetic Disorder: _____
	☐ Other:

Respiratory		Musculoskeletal/Pain	
☐ Asthma	☐ Bronchitis	☐ Osteoarthritis	☐ Fibromyalgia
☐ Chronic Sinusitis	☐ Emphysema	☐ Chronic Pain	☐ Migraines
☐ Pneumonia	☐ Tuberculosis	☐ Other:	
☐ Sleep Apnea	☐ Other:		

Notes:

Fig. 55.2 (continued)

Medical History (continued)

Please note any past or current injuries:

Neurological/Mood		Cancer
☐ Depression	☐ Bipolar Disorder	☐ Cancer (*please describe type and treatment*)
☐ Anxiety	☐ ADD/ADHD	
☐ Autism	☐ Multiple Sclerosis	
☐ Seizures	☐ Other:	

Other (use separate sheet if necessary)		
☐ Kidney stones	☐ Anemia	Please any other diseases or health conditions
☐ Eczema	☐ Urinary (UTIs)	
☐ Psoriasis	☐ Frequent Yeast	Have you ever had genetic testing? ☐ Y ☐ N
☐ Acne	☐ Other:	If yes, please note type and results.

Medications (Please list all prescribed medications you are taking and note reason.)

Name:	Reason:
Name:	Reason:
Name:	Reason:
Name:	Reason:
Name:	Reason:
Name:	Reason:
Name:	Reason:
Name:	Reason:

| Have you had prolonged or regular use of Tylenol? ☐ Y ☐ N |
| Have you had prolonged or regular use of acid-blocking drugs (Tagamet, Zantac, Prilosec, etc.)? ☐ Y ☐ N |
| Frequent antibiotics >3 times per year? ☐ Y ☐ N   Long term antibiotics? ☐ Y ☐ N |

Fig. 55.2 (continued)

Environmental Information

Do you have known adverse food reactions or sensitivities? ☐ Y ☐ N	If yes, please describe symptoms.
Are you exposed regularly to any of the following? *(check all that apply)*	What is your occupation?

☐ Cigarette smoke	☐ Perfumes	Please note any regular exposure to harmful chemicals/substances.
☐ Auto exhaust/fumes	☐ Paint fumes	
☐ Dry-cleaned clothes	☐ Mold	
☐ Nail polish/hair dyes	☐ Pesticides	Please not any past exposure to harmful chemicals/substances.
☐ Heavy metals	☐ Fertilizers	
☐ Teflon Cookware	☐ Pet dander	
☐ Aluminum Cookware	☐ Chemicals	

Do you use any recreational drugs? If so, please note.	

Notes:

Lifestyle Information

Do you engage in moderate cardiovascular physical activity at least 3 days a week, for a minimum of 20 minutes duration? (brisk walking, jogging, hiking, cardio exercise classes, cycling, stair-climbing, etc.)

☐ Y ☐ N

Activity	Type/Intensity (low-moderate-high)	# Days/Week	Duration (minutes)
Stretching/Yoga			
Cardio/Aerobics			
Strength Training			
Sports or Leisure			

Rate your level of motivation for including exercise in your life? ☐ Low ☐ Med ☐ High

Note any problems that limit your physical activity.

Fig. 55.2 (continued)

Lifestyle Information (continued)

Do you smoke? ☐ Y ☐ N	How many years?
Packs per day?	2nd hand smoke exposure? ☐ Y ☐ N
Excess stress in your life? ☐ Y ☐ N	Easily handle stress? ☐ Y ☐ N
Daily Stressors: *Rate on a scale of 1 (low) to 10 (high)* ☐ Work____ ☐ Family____ ☐ Social____ ☐ Finances____ ☐ Health____ ☐ Other:____	
Do you feel your life has meaning and purpose? ☐ Y ☐ N ☐ unsure	Do you believe stress is presently reducing the quality of your life? ☐ Y ☐ N
Average number of hours you sleep per night <u>during the week</u>?	Average number of hours you sleep per night on <u>weekends</u>?
Trouble falling asleep? ☐ Y ☐ N	Rested upon waking? ☐ Y ☐ N
Do you wake up during the night? ☐ Y ☐ N If yes, how many times?	
Note the approximate times you generally wake during the night.	
How would you rate the overall quality of your sleep? *low quality* 1 2 3 4 5 *high quality*	

Family History

Please note any family history of the following diseases: heart disease, cancer, stroke, high blood pressure, overweight, lung disease, kidney disease, diabetes, cancer, mental illness or addiction.

Family Member:	Health Condition:
Family Member:	Health Condition:
Family Member:	Health Condition:
Family Member:	Health Condition:
Family Member:	Health Condition:
Family Member:	Health Condition:
Family Member:	Health Condition:
Family Member:	Health Condition:

Genetic Disorders Known:

Fig. 55.2 (continued)

Surgeries/Hospitalizations

Please list any surgeries or hospitalizations (include dates and your ages if known).

Dental History					
Do you have any silver/mercury amalgam fillings?		☐ Y ☐ N If **Y**, how many?			
Do you have any	☐ Gold fillings	☐ Root canals	☐ Implants	☐ Bridges	☐ Crowns
Do you have any	☐ Tooth pain	☐ Bleeding gums	☐ Gingivitis	☐ Chewing problems	
Do you visit a dentist regularly (twice per year)?		☐ Y ☐ N			
Have you ever had an infection in your jawbone?		☐ Y ☐ N			
TMJ : ☐ grinding teeth	☐ jaw clicking	☐ braces? If yes, what age ____		☐ surgery	☐ jaw pain
Teeth: ☐ extraction? How many? _____		☐ Which teeth are missing? (# or name)_____			

Ingestion: Nutrition History

Have you ever had a nutrition consultation? ☐ Y   ☐ N

Have you made any changes in your eating habits because of your health? ☐ Y   ☐ N

Please describe.

Do you currently follow a special diet or nutritional program? ☐ Y   ☐ N

Check all that apply.

☐ Low fat	☐ Low Carb	☐ High protein	☐ Low sodium
☐ No Gluten	☐ Vegetarian	☐ Vegan	☐ Diabetic
☐ No Dairy	☐ No Wheat	☐ Weight Loss	☐ Other _____

How often do you weigh yourself?

Fig. 55.2 (continued)

Initial Nutrition Assessment Checklist

Have you had any recent history of weight loss or weight gain? If so, please describe.

How many meals per day do you eat?	How many snacks?	
Do you avoid any particular foods? *If yes, describe.*		
If you could only eat a few foods a week, what would they be?		
How many meals do you eat out per week?	☐ 0–1 ☐ 1–3 ☐ 3–5 ☐ more than 5 per week	

Check all the factors that apply to your current lifestyle and eating habits:

☐ Fast eater	☐ Family member have different tastes
☐ Erratic eating patterns	☐ Love to Eat
☐ Eating too much	☐ Eat because I have to
☐ Late night eating	☐ Have a negative relationship to food
☐ Dislike healthy food	☐ Struggle with eating issues
☐ Time constraints	☐ Emotional eater (stress, bored, etc.)
☐ Travel frequently	☐ Confused about food/nutrition
☐ Do not plan meals or menus	☐ Frequently eat fast foods
☐ Rely on convenience items	☐ Poor snack choices

Current Eating Habits

Mark the meals you eat regularly: ☐ Breakfast   ☐ Lunch   ☐ Dinner   ☐ Snacks

Where do you obtain your food from: ☐ home prepared from whole foods ____%   ☐ organic ___%
☐ home prepared convenience food ____%   ☐ eat out ____%

Mark how many times you eat or drink the following items <u>per week</u>:

___ Soda (regular)	___ Fast food	___ Dried fruit	___ Crackers
___ Soda (diet)	___ Candy	___ Canned fruit	___ Pasta
___ Alcohol	___ Ice cream	___ Fresh Fruit	___ Brown rice
___ Hot tea	___ Pudding	___ Jelly/jam	___ White rice

Fig. 55.2 (continued)

___ Cold tea
___ Coffee (regular)
___ Coffee (decaf.)
___ Sugar in coffee
___ Coffee drinks
___ Sweetened drinks
___ Sparkling water
___ Purified water
___ Tap water
___ Fruit juice
___ Lemonade
___ Milk (cow)
___ Milk (goat)
___ Soy Milk
___ Rice Milk
___ Nut Milk
___ Herbal teas

___ Refined sugars
___ Tuna fish
___ Swordfish
___ Sushi/sashimi
___ Salmon/other fish
___ Lunch meats
___ Bacon
___ Hot dogs
___ Whole eggs
___ Red meat
___ Poultry
___ Tofu
___ Tempeh/Miso

Sweeteners:
___ Equal/Nutrasweet (Aspartame)
___ Splenda (sucralose)
___ Saccharin
___ Stevia/Xylitol

___ Sweets (cookies)
___ Green Salads
___ Raw veggies
What kind?

___ Cooked veggies
What kind?

___ Potatoes
___ Yams/Sweet Potatoes
___ Popcorn
___ Cereals
___ Oatmeal
___ Bagels/pretzels
___ White bread
___ Sprouted Br.
___ Wheat Bread

___ Corn tortillas
___ Flour tortillas
___ Potato Chips
___ Tortilla Chips
___ Pizza
___ Yogurt (plain)
___ Yogurt (sweet)
___ Prepared meals (Lean cuisine, etc.)
___ Microwave meals/soups
___ Restaurant meals (healthy)
___ Restaurant meals (unhealthy)
___ Airplane meals
___ Legumes (beans, lentils)

Notes:

■ **Fig. 55.2** (continued)

Ingestion: Nutrition History (continued)

What are the top three dietary changes do you think would make the most difference in your overall health?	1. 2. 3.
How committed are you to making dietary changes in order to improve your health?	*not committed* 1 2 3 4 5 *very committed*

Please list all <u>nutritional supplements</u> you currently take daily. Please include brand names and amounts as well as any herbs/botanical products.

_____ | _____
_____ | _____
_____ | _____
_____ | _____
_____ | _____
_____ | _____

Do you drink alcohol? ☐ Y   ☐ N   If yes, how many drinks per week?

Do you drink coffee or other caffeinated beverages? ☐ Y   ☐ N   If yes, # daily?

Do you use artificial sweeteners? ☐ Y   ☐ N   If yes, which ones?

Digestion:

Do you feel like belching or are you bloated after eating? ☐ Y   ☐ N

Do you have (or had) any eating disorders? ☐ Y   ☐ N   If yes, please describe.

Bowel Movements: How often? _____   Color? _____   Consistency? _____

Your Birth: ☐ Natural/vaginal   ☐ C-Section | Were you breastfed as an infant (if known)?   ☐ Y   ☐ N

Please note anything additional about your nutrition/eating habits.

Fig. 55.2 (continued)

Fats and Oils

Please indicate how many times PER WEEK you eat the following fats/oils.

OMEGA 9 *(stabilizer)* ~50% of daily fat calories Oleic Fatty Acid	___ Almond Oil ___ Almonds/Cashews ___ Almond butter ___ Avocados ___ Peanuts ___ Peanut butter (natural/soft)	___ Olives ___ Olive Oil ___ Sesame Seeds/Tahini ___ Hummus (tahini oil) ___ Macadamia Nuts ___ Pine Nuts
OMEGA 6 *(controllers)* Essential Fatty Acid Family ~30% of daily fat calories LA ☐ GLA ☐ DGLA ☐ AA	___ Eggs (whole), organic (AA) ___ Meats (commercial) (AA) ___ Meats (grass-fed, org) (AA) ___ Brazil nuts (raw) ___ Pecan (raw) ___ Hazelnuts/Filberts (raw) ___ Hemp Seeds	___ Evening Primrose (GLA) ___ Black Currant Oil (GLA) ___ Borage Oil (GLA) ___ Hemp Oil ___ Grapeseed Oil ___ Sunflower Seeds (raw) ___ Pumpkin seeds (raw)
OMEGA 3 *(fluidity/communicators)* Essential Fatty Acid Family ~10% of daily fat calories ALA ☐ EPA ☐ DHA	___ Fish Oil capsule: ↑DHA ___ Fish Oil capsule: ↑EPA ___ Fish (salmon/fin-fish) ___ Fish (shellfish) ___ Flax seeds/meal	___ Flax Oil ___ UDO's DHA Oil ___ Algae ___ Greens Powder w/algae ___ Chia seeds
BENEFICIAL SATURATED *(structure)* ~10% of daily fat calories Short Chain/Medium-chain Triglycerides	___ Coconut Oil ___ Butter, organic ___ Ghee (clarified butter) ___ Dairy, raw & organic	___ Meats, grass-fed ___ Wild game ___ Poultry, organic ___ Eggs, whole organic
DAMAGED FATS/OILS (promoting stress to cells & tissues) Should be <5% (try to avoid) Trans Fats Acrylamides Odd-Chain Fatty Acids VLCFA/damaged	___ Margarine ___ Reg. vegetable oils (corn, sunflower, canola) ___ Mayonnaise (Commercial) ___ Hydrogenated Oil (as an ingredient) ___ "Imitation" cheeses ___ Tempura	___ Doughnuts (fried) ___ Deep-fried foods ___ Chips fried in oil ___ Reg. Salad dressing ___ Peanut Butter (JIF, etc) ___ Roasted nuts/seeds ___ Non-dairy products

Fig. 55.2 (continued)

Nutrition Physical Exam

Review the following and check, circle, and make notes as you proceed through the nutrition assessment.

General Appearance:	Face/Neck:	Skin:
☐ Age (looks older/younger?)	☐ Eyes (clear, red, spots, discharge, dry, tearing?)	☐ Easy bruising
☐ Body type (apple/pear?)	☐ Eyes (bulging, swollen)	☐ Rashes/hives
☐ Central Adiposity	☐ Eyelid (drooping over eye)	☐ Bumps on back of arms?
☐ Fatness (excessive/diminished)	☐ Eyelid/inside (pallor, very red)	☐ Excessive Tan/Pale color?
☐ Neat?	☐ Wrinkles (excessive)	☐ Dry/Flaky/Oily?
☐ Sick?	☐ Skin color/pigmentation	☐ Scars
☐ Tremor/Tic	☐ Cheeks rosacea/redness	☐ Tatoos
☐ Vitality/energy	☐ Neck/armpit/legs (dark skin) acanthosis nigricans (AN)	☐ Warts/ abnormal spots
	☐ Skin (acne, freckles)	☐ Piticulae (deep red spots)
	☐ Eyebrows (hairless ends?)	☐ Front thighs (pustules) dermatitis herpetiforme
	☐ Ears (crease in lobe)	
Handshake:	☐ Nasal function (stuffy, clear)	**Nails:**
☐ Temperature (hot, cold, moist)	☐ Thyroid (goiter?)	☐ Spots, ridges, pale nail beds, white lines
☐ Strength/Tremor		☐ Strong curve?
		☐ Fungus (fingernails, toes?)
Head/Hair/Scalp:	**Mouth:**	**Structure:**
☐ Head shape	☐ Lips (cracking, shape)	☐ Neck/posture
☐ Hair (thin/thick/color/dyes)	☐ Lips (angular cracks)	☐ Gait
☐ Hair loss (head/body)	☐ Tongue (fissures, decreased taste, smooth?)	☐ Muscles/tone
☐ Flaky scalp?	☐ Breath odor (halitosis, acetone, garlic, musty, ETOH)	☐ Joints/connective tissue
☐ Scalp: sores/scabs	☐ Bleeding gums	☐ Physical disability
Hydration:		
☐ Skin, tongue		
☐ BIA: ICW/ECW/TBW		
☐ Ankle/wrist signs of edema/dehydration - "pitting"		
☐ Constipation/diarrhea		

Notes:

◨ **Fig. 55.2** (continued)

Nutrition Assessment Summary Form					
Reason for Nutrition Appointment or follow-up, "Chief Complaint" or relevant Medical Diagnoses					

Any Foods Withdrawn

Age	Wt	Ht	Waist Circ:	Phase <	% IBW

Nutrition Physical Exam, Dental History (I, D, U)

Subjective (Including Past Medical History (PMH) and History of Present Illness (HPI)):

Goal(s) of Appointment:

MSQ Summary (may use top ~3 categories)	Medications/Supplements
Digestion (BM frequency/form, symptoms) (D)	Labs: Abnormal or Meaningful "Normal" Values
Diet (I, D, U)	Lifestyle: Stress, Environmental Exposure, Physical Activity (I, D, U)

Assessment of Nutrition Status // Progress and Monitoring/Evaluation of previous plan

Plan // Goals

I: Ingestion   D: Digestion   U: Utilization

Fig. 55.3 Nutrition assessment summary form

genotype, toxic influences, stress, and/or physical injury. Chronic inflammatory cytokine production contributes to all chronic disease.

Eicosanoid Arachidonic acid (AA) and eicosapentaenoic acid (EPA) serve as precursors for the synthesis of eicosanoids (prostaglandins, prostacyclins, leukotrienes, and thromboxanes) which have an anti-inflammatory effect.

Energy dysfunction Mitochondrial energy metabolism, dietary intake and output, effect on body composition, physical activity, and influence of stress on blood glucose/insulin/hormonal energy relationships.

Epigenetics Environmental influences that control gene activity within a cell without changing the gene structure [2]. Each gene has the capability of about 40,000 different expressions depending on the environmental messaging. Epigenetic effects can be passed on from generation to generation. The discovery of epigenetics has revealed a nongenetic mode of inheritance that influences our health. The most vulnerable periods in the lifespan for epigenetic changes are in utero/pregnancy, early childhood, teenage years, menopause (female), and andropause (male).

Gluconeogenesis During low-carbohydrate intake or starvation, protein is used as de novo synthesis of glucose [3]. Gluconeogenesis is about 60% efficient in ATP production.

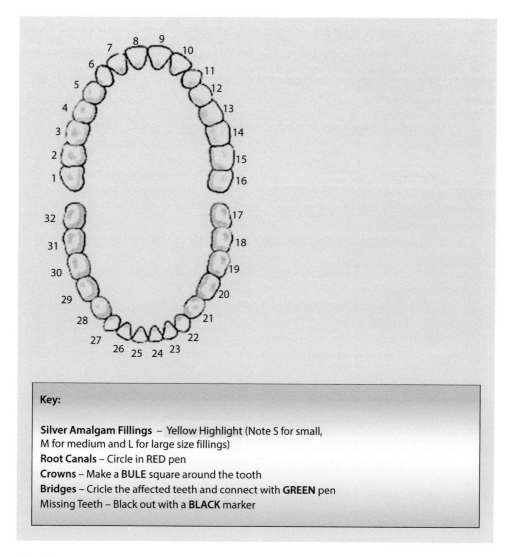

Fig. 55.4 Dental health impacts the entire physiologic function from neurologic to gastrointestinal and multiple organ systems. It is a critical part of the nutritional physical examination

Glycolysis Anaerobic pathway to release glucose without the presence of oxygen; lactic acid is the end product [4]. Glycolysis is about 30% efficient in ATP production.

I-D-U Dietetic and integrative medicine nutritional assessment heuristic of an individual's nutrition status including intake, digestion and utilization of nutrients.

Ketosis To maintain metabolic homeostasis, ketone metabolism is the most energetically efficient metabolic fuel to synthesize glucose using fatty acids and ketone bodies under periods of prolonged food restriction or low-carb/high-fat dietary regimen.

Lack of sleep Underlying deregulation of neurotransmitters and hormones that help flip the on-off switch between waking and sleep. Lack of sleep quality and duration disturbs the 24-hour circadian rhythm and management of wake-time energy. Lack of sleep associations are allergies, central adiposity, diabetes, hypertension, heart disease, and other physical ailments [5].

Lifestyle and community Most chronic disease often has a social basis as well as a biological basis. An individual's social network can influence their health and alter gene expression [6]. An individual is more likely to be healthy if they are a part of a community that supports a healthy diet, lifestyle, and emotional environment.

Long latency nutritional insufficiencies The presumption has been that if the intake of the nutrient is sufficient to prevent the expression of the index disease (i.e., iodine and goiter, vitamin D and rickets, etc.), then the nutrient intake is sufficient for the entire organism. But a nutrient deficiency could have long-term consequences and contribute to chronic disease, either by a similar mechanism to that inducing the index disease or by an entirely different mechanism [7].

MAPDOM Within the "U" of the I-D-U assessment heuristic, the MAPDOM acronym represents the six most important nutritional categories determining an individual's health. MAPDOM stands for: minerals, antioxidants, proteins, D-fat-soluble vitamins (D, E, A, K), oils/fats, and methylation nutrition.

Methylation Denotes the addition of a methyl group to a substrate for the substation of an atom or group by a methyl group (-CH3). Can be involved in modification of heavy metals, regulation of gene expression, and regulation of protein function and RNA metabolism [8]. Primary nutrients involved as cofactors in methylation metabolism are B6, B12, folate, B2, zinc, and magnesium.

Nutritional triage theory When the availability of a micronutrient is inadequate, metabolic resources will ensure that functions needed for short-term survival are protected at the expense of functions whose lack of a micronutrient only has long-term consequences [9].

Petrochemicals Derived from petroleum or natural gas (ethane, propane, butane, and other hydrocarbons), petrochemicals are used to manufacture artificial products including plastic, medicine, medical devices, cosmetics, furniture, cars, and computers [10]. Humans can ingest phytochemicals or it can be absorbed through the skin. When they accumulate in tissues and organs, they are responsible for brain, nerve, and liver damage, birth defects, cancer, asthma, hormonal disorders, and allergies.

Toxic heavy metals Metals and metal compounds that have a negative effect on human health by building up in biological system [11]. They include arsenic, beryllium, cadmium, hexavalent chromium, lead, and mercury. Other toxic metals may include essential metals if imbalanced (zinc and copper), medicinal metals (platinum and bismuth), metals in emerging technology (indium and uranium), toxic metalloids (arsenic and antimony) and certain nonmetallic elements if imbalanced (selenium and iodine) [12] (◘ Fig. 55.5).

55.3 Gastrointestinal Assessment (◘ Fig. 55.6, ◘ Table 55.1)

55.3.1 Testing for Low Stomach Acidity (Hypochlorhydria)

A normal stomach acid level creates a pH of 1.5–2.5. As we age, the parietal cells in the stomach lining produce less hydrochloric acid (HCl) or "stomach acid." In fact, half of people over the age of 60 have low stomach acid. Low stomach acid can also be caused by long-term use of acid-blocking medications (i.e., Pepcid®, Prilosec®, etc.), which increases stomach pH to 3.5 or higher. This change in pH inhibits pepsin, which is necessary for protein digestion. Interestingly, symptoms of low acidity often mimic those of high acidity.

Stomach acid is necessary for absorption of minerals (magnesium, calcium, iron, zinc, and copper) and vitamin B12. Also, stomach acid provides our first defense against food poisoning, *H. pylori*, and parasitic and other infections. Without adequate acid, we leave ourselves open to decreased immune resistance. Overgrowth of bacteria in the intestinal tract occurs in 20 percent of people aged 60 to 80 and in 40 percent of people over 80.

The following may cause hypochlorhydria: pernicious anemia, chronic *H. pylori* infection, long-term use of proton pump inhibitors (like Prilosec®), autoimmune gastritis, and mucolipidosis type IV.

55.3.2 Common Symptoms of Hypochlorhydria

Bloating, belching, burning, or flatulence immediately after meals	Indigestion, diarrhea, or constipation
A sense of fullness after eating	Nausea after taking supplements
Dilated blood vessels in cheeks/nose	Itching around the rectum
Acne	Undigested food in stool
Iron deficiency	Chronic candida infections
Chronic intestinal parasites or abnormal flora	

55.3.3 Diseases Associated with Hypochlorhydria

Addison's disease	Diabetes	Hypothyroidism
Asthma	Eczema	Lupus erythematosus
Celiac disease	Gallbladder disease	Myasthenia gravis
Chronic autoimmune disorders	Graves disease	Osteoporosis
Chronic hives	Hepatitis	Pernicious anemia
Dermatitis herpetiformis (herpes)	Hyperthyroidism	

55.3.4 Self-Testing and Treatment for Low HCL/Hypochlorhydria

1. Begin by taking one 350–750 mg capsule of betaine HCL with a protein-containing meal. A normal response in a healthy person would be discomfort – basically heartburn. If you do not feel a burning sensation, begin taking two capsules with each protein-containing meal. Meals should last a minimum of 20 minutes and a maximum of 60 minutes. Remember to chew every bite well (your stomach does not have teeth).
2. If there are no reactions after 2 days, increase the number of capsules with each meal to three.

Initial Nutrition Assessment Checklist

Anthropometrics: Height ____ Weight ____ BMI ____ Body Fat % ____ BP ____/____ Pulse ____ Waist Circum _____ Waist Circum/Height Ratio: _____					
Assessment	**History/Physical**	**Labs/Procedures**	**Nutrition Diagnosis**	**Nutrition Intervention**
Intake - food, liquids, organic vs. non-, food preparation, toxin intake, processing, packaging, air/skin intake (toxin)				
Food Nutrients Meal Timing & Mastication Toxin Intake Fluids				
Digestion - Adeq HCL, digestive enzymes, bioavailability, meds, emotional stress, toxins, genetics or acquired deficits				
Protein/Carb Digestion Digestion of Fat Enzyme deficiency Endotoxins / Pathogens Microbiome GI Structure				
Utilization - MAPDOM -Intestinal competence, food allergies, malabsorption, meds, toxins, microbial overgrowth				
Cellular & Molecular metabolism Hydration/fluid status Nutrient absorption				
MAPDOM - Nutrient Requirements -affected by Disease states, stress, environment, age, drugs, genetic differences				

Mineral Status
  Fiber
  BM
frequency, consistency				

Antioxidants Water Soluble Vitamin C / Phytonutrients				
Protein Status Wound healing Connective tissue Glucose transport Immune status				
D / Vitamin D, A, K Contraindications Vitamin E				
Oils/Lipid/Fatty Acids EFA/Omega-9/ Beneficial Saturated Fats Inflammation status				
Methylation - Clinical symptoms - genomics or Fam Hx - biomarkers - r/o Mood Disorder/Cancer				
Sum total of previous factors influence assessment -Nutritional Status				

Fig. 55.5 Nutrition status assessment I-D-U worksheet

Subjective	Objective	GI tract
Medical symptoms questionnaire GI symptoms: constipation, diarrhea, stomach pain, etc. Other themes in systems: headaches, sinuses/nasal, frequent illness, emotions... **Diet history** Themes of diet (repetitive intake of foods + symptoms) Food symptoms Genomic risk **Patient report of food sensitivities** **Nutrition-focused physical** Nails, hair, skin (rashes) Abdominal exam: RUQ, LUQ, RLQ, LLQ Oral cavity exam **Water balance (BIA, edema)**	**Conventional labs** Small bowel biopsy IgE testing: anaphylaxis IgG: primarily systemic symptoms IgA: primarily GI symptoms Micronutrients: suspected malabsorption, general screening (Fe, B12, Vitamin D) Hydrogen breath Mannitol/lactulose **Genetic testing functional labs** CDSA Neurotransmitters: emotional indication, insomnia Organic acids Stomach acid test	**Mouth** Saliva Enzymes (amylase, lipase) Mastication quality/ time Dental health **Esophagus** Eosinophilic esophagitis Swallow difficulties GERD or sensation of discomfort **Stomach** pH (too high/low) **Biliary - Bile acids pancreas** Elastase, amylase, etc. **Small intestine** Flora (beneficial, imbalanced, dysbiosis) Surgeries/disease (IBD/IBS) **Large intestine** Short-chain fatty acids Surgeries/disease (IBD/IBS) β-glucuronidase Inflammation (lactoferrin, calprotectin)

Fig. 55.6 Gastrointestinal assessment. GI assessment is a fundamental component of integrative and nutritional nutrition therapy and should be evaluated at every visit. Documentation at each visit is critical to assess progress

3. Continue increasing the number of capsules every 2 days, using up to eight capsules with each meal if necessary. These dosages may seem large, but a normally functioning stomach manufactures considerably more. You'll know you've taken too much if you experience tingling, heartburn, diarrhea, or any type of discomfort including a feeling of unease, digestive discomfort, neck ache, backache, headache, or any new odd symptoms. If you experience tingling, burning, or any symptom that is uncomfortable, you can neutralize the acid with 1 teaspoon of baking soda in water or milk.
4. When you reach a state of tingling, burning, or any other type of discomfort, cut back by one capsule per meal. If the discomfort continues, discontinue the HCL and consult with your healthcare professional.
5. Once you have established a dose (not to exceed eight capsules without consulting with your healthcare provider), continue this dose.
6. With smaller meals, you may require less HCL, so you may reduce the amount of capsules taken.
7. Individuals with very moderate HCL deficiency generally show rapid improvement in symptoms and have early signs of intolerance to the acid. This typically indicates a return to normal acid secretion.
8. Individuals with low HCL/pepsin typically do not experience such quick improvement, so to maximize the absorption and benefits of the nutrients you take, it is important to be consistent and continue your HCL supplementation.

Precautions: Administration of HCL/pepsin is contraindicated in peptic ulcer disease. HCL can irritate sensitive tissue and can be corrosive to teeth; therefore, capsules should not be emptied into food or dissolved in beverages.

When you have adequate HCL, you will have good absorption of all your nutrients and can then watch the rapid regeneration of health in every brain/body system.

Table 55.1 Fats and oils survey

Fats and oils survey	Name	Date
Please indicate how many times per week you eat the following fats/oils		
Omega-9 (*stabilizer*) ~50% of daily fat calories Oleic fatty acid	___Almond oil ___Almonds/cashews ___Almond butter ___Avocados ___Peanuts ___Peanut butter (natural/soft)	___Olives ___Olive oil ___Sesame seeds/tahini ___Hummus (tahini oil) ___Macadamia nuts ___Pine nuts
Omega-6 (*controllers*) Essential fatty acid family ~30% of daily fat calories LA>GLA>DGLA>AA	___Eggs (whole), organic (AA) ___Meats (commercial) (AA) ___Meats (grass-fed, org) (AA) ___Brazil nuts (raw) ___Pecan (raw) ___Hazelnuts/filberts (raw) ___Hemp seeds	___Evening primrose (GLA) ___Black currant oil (GLA) ___Borage oil (GLA) ___Hemp oil ___Grapeseed oil ___Sunflower seeds (raw) ___Pumpkin seeds (raw)
Omega-3 (*fluidity/communicators*) Essential fatty acid family ~10% of daily fat calories ALA>EPA>DHA	___Fish oil capsule: ↑DHA ___Fish oil capsule: ↑EPA ___Fish (salmon/finfish) ___Fish (shellfish) ___Flax seeds/meal	___Flax oil ___UDO's DHA oil ___Algae ___Greens powder w/ algae ___Chia seeds
Beneficial saturated (*structure*) ~10% of daily fat calories Short-chain/medium-chain triglycerides	___Coconut oil ___Butter, organic ___Ghee (clarified butter) ___Dairy (raw and organic?)	___Meats ___Wild game ___Poultry (organic) ___Eggs, whole organic
Damaged fats/oils (promoting stress to cells and tissues) *Should be <5% (try to avoid)* Trans fats Acrylamides Odd-chain fatty acids VLCFA/damaged	___Margarine ___Reg. vegetable oils (corn, sunflower, canola) ___Mayonnaise(commercial) ___Hydrogenated oil (as an ingredient) ___"Imitation" cheeses ___Tempura	___Mustard seed/pate ___Deep-fried foods/ ___Chips fried in oil ___Reg. salad dressing ___Peanut butter (JIF, etc.) ___Roasted nuts/seeds ___Nondairy products

Notes:

References

1. Lucock M. Molecular nutrition and genomics: nutrition and the ascent of humankind, vol. 84. Hoboken: Wiley; 2007.
2. Francis RC. Epi-genetics: the ultimate mystery of inheritance. New York: W.W. Norton & Company, Inc.; 2011.
3. Gallagher ML. Macronutrients: carbohydrates. In: Krause's food and the nutrition care process. 13th ed. St. Louis: Elsevier Saunders; 2012. p. 51.
4. Dorfman L. Nutrition in exercise and sports performance. In: Krause's food and the nutrition care process. 13th ed. St. Louis: Elsevier Saunders; 2012. p. 508.
5. Lombardo GT, Ehrlich H. Sleep to save your life. New York: HarperCollins; 2005. p. 1–5.
6. Hyman M. Can social networks cure disease? Part I. Retrieved from: http://drhyman.com/blog/2012/04/20/can-social-networks-cure-disease-part-i/. Feb 2013.
7. Heaney RP. Long-latency deficiency disease: insights from calcium and vitamin D. Am J Clin Nutr. 2003;78:912.
8. Wikipedia. Methylation. Retrieved from: http://en.wikipedia.org/wiki/Methylation. June 2013.
9. McCann JC, Ames BN. Vitamin K, an example of triage theory: is micronutrient inadequacy linked to diseases of aging? Am J Clin Nutr. 2009;90:889.
10. American Fuel & Petrochemical Manufacturers (AFPM). Petrochemicals. Retrieved from: http://www.afpm.org/petrochemicals/. Aug 2013.
11. Occupational Safety & Health Administration, United States Department of Labor. Toxic metals. Retrieved from: https://www.osha.gov/SLTC/metalsheavy/. Aug 2013.
12. Costa LG. Toxic effects of pesticides. In: Casarett & Doull's essentials of toxicology. New York: McGraw-Hill; 2010. p. 324.
13. Swan WI, Vivanti A, Hakel-Smith NA. Nutrition care process and model update: toward realizing people-centered care and outcomes management. J Acad Nutr Diet. 2017;117(12):2003–14.

Nutritional Diagnosis Resources

Leigh Wagner, Diana Noland, and Jeanne A. Drisko

56.1 Abbreviations – 1044

Nutritional diagnosis (ND) as a terminology beyond ICD CM codes related to malnutrition that emerged in 2000 as defined by the Medicare Medical Nutrition Therapy (MNT) legislation, in coordination with the American Dietetic Association (Academy of Nutrition and Dietetics).

ND assists the nutrition practitioner in defining nutrition-related metabolic injuries to guide therapy, intervention, and counseling for the purpose of disease management. MNT involves in-depth individualized nutrition assessment and a duration and frequency of care using the Nutrition Care Process to manage diseases as outlined by Medicare law.

56.1 Abbreviations

CPT codes: Current Procedural Terminology codesprocedure codes that describe the healthcare professional services

ICD: International Classification of Disease

MNT Codes 97802, 97803, and 97804 (CPT codes that RNDs use).

Because codes change over time and across insurance programs, it is important with the current codes related to nutrition professionals are up to date:

Medicare MNT
- ▶ cms.gov/medicare/coding/icd10/2017-icd-10-cm-and-gems.html
- ▶ eatrightpro.org/payment/coding-and-billing/diagnosis-and-procedure-codes/diagnosis-codes-for-medicare-mnt

Malnutrition
- ▶ eatrightpro.org/-/media/eatrightpro-files/practice/patient-care/medical-nutrition-therapy/mnt/medicare_malnutrition_drg_for_inpatient_care.pdf

Academy of Nutrition and Dietetics
- ▶ eatright.org

Hospital Reimbursement for Malnutrition Expanded Effective October 1, 2012
- ▶ eatrightpro.org/payment/coding-and-billing/diagnosis-and-procedure-codes/malnutrition-codes-characteristics-and-sentinel-markers

ICD-10 Coding and Billing
- ▶ eatrightpro.org/payment/coding-and-billing/diagnosis-and-procedure-codes/icd-10-cm

Disclaimer   Any codes used should be verified with a coding professional when billing Medicare, Medicaid, and insurance carriers.

Specialized Diets

Leigh Wagner, Diana Noland, and Jeanne A. Drisko

57.1 How to Help Your Friends and Family Support Your Lifestyle Changes When You Follow a Specialized Diet (Table 57.1) – 1046
57.1.1 Explain What You Are Doing and Why – 1049
57.1.2 Nutrition Approaches to Wellness Is "The Road Less Traveled" – 1049
57.1.3 Make Light of It – 1049
57.1.4 Be Confident – 1049
57.1.5 Plan Ahead: Social Setting and Restaurants – 1049
57.1.6 My Loved Ones Don't Want to Eat What I Need to Eat – 1049

57.2 How to Embrace Your Decision to Follow a Specialized Diet – 1049

57.1 How to Help Your Friends and Family Support Your Lifestyle Changes When You Follow a Specialized Diet (◻ Table 57.1)

Table 57.1 Specialized diets for GI healing: Allowed foods and forbidden foods (*italics = none*)

	Comprehensive elimination diet	Gluten-free/casein-free	Specific carbohydrate diet	Gut and psychology syndrome diet	Anti fungal diet	FODMAP diet	Restoration diet
Protein	ALL unprocessed meats: chicken, turkey, duck, goose, quail, ostrich, fish, shellfish, lamb, venison, rabbit, eggs. Wild game	ALL unprocessed meats	ALL unprocessed meats: beef, pork, chicken, turkey, duck, goose, quail, ostrich, fish, shellfish, lamb, venison, rabbit, eggs. Processed meats that do not have any SCD forbidden ingredients	Eggs, fresh (if tolerated) Fresh meats (not preserved), fish, shellfish Broths with every meal Canned fish in oil or water only	ALL unprocessed meats: beef, pork, chicken, turkey, duck, goose, quail, ostrich, fish, shellfish, lamb, venison, rabbit, eggs. Tofu, tempeh, texturized vegetable protein	All unprocessed meats Eggs	All unprocessed meats in small amounts: pureed, well-cooked, stews, soups
Dairy products and dairy alternatives	*NONE* Dairy alternatives are allowed: coconut, hemp, rice	*NONE* Dairy alternatives are allowed: nut, coconut, hemp, rice, soy	All natural cheeses *except for ricotta, mozzarella, cottage cheese, cream cheese, feta, processed cheeses and spreads.* Homemade yogurt cultured 24 hours	All natural cheeses Yogurt – homemade	Eggs, plain yogurt (cow, sheep, goat) with live cultures, organic soy milk, soy cheese, coconut milk, unaged goat cheese	Lactose-free dairy products: milk, cottage cheese, rice milk, almond milk, hemp milk	Goat milk or sheep milk kefir. Dairy alternatives as coconut kefir
Fats and oils	Sunflower, olive, flax, ghee, coconut, avocado, nut oils	ALL	Avocados, olive oil, coconut oil, corn oil, avocado oil, etc.	Butter, ghee, coconut, avocado oil, olive oil	ALL	ALL	Ghee, coconut, olive, Sam Queen's restorative ghee
Nuts and seeds	Coconut, pine nuts, chia seeds, flaxseeds, almonds, Brazil nuts, walnuts, chestnuts, filberts, pecans, nut flours, and meals	ALL that are non-processed with dairy or gluten	Almonds, Brazil nuts, walnuts, chestnuts, filberts, pecans, nut flours and meals	Almonds, avocado, Brazil nuts, coconut, filberts, walnuts, chestnuts, pecans, nut flours and meals, peanuts, nut butters	ALL raw. Can roast at home or cook them	*Nuts and seeds in moderation* *Nut butters in moderation*, Psyllium	*Nut butters in tiny amounts*

Table 57.1 (continued)

	Comprehensive elimination diet	Gluten-free/casein-free	Specific carbohydrate diet	Gut and psychology syndrome diet	Anti-fungal diet	FODMAP diet	Restoration diet
Non-starchy vegetables	ALL	ALL	Most: fresh, frozen, raw or cooked. Asparagus, broccoli, cauliflower, artichokes, beets, Brussels sprouts, cabbage carrots, celery, cucumbers, eggplant, zucchini, summer squash, rhubarb, peppers, garlic, lettuce, spinach, mushrooms (unless candidiasis), onions, turnips, watercress. NO canned vegetables	Most: fresh, mostly cooked, some raw	ALL	Alfalfa, avocado, bamboo shoots, bean shoot, beets, bok choy, broccoli, chili peppers, carrots, celery, chive, corn, cucumber, eggplant, fennel, kohlrabi, lettuce, olive, parsnip, mushroom, snow peas, spinach, squash, water chestnut, watercress	Well-cooked
Starchy vegetables	ALL except corn	ALL	*NONE: potatoes, yams*	Beets, winter squash *NONE: potatoes, yams*	*NONE: Exclude corn, yams, potatoes*	peas, potato, sweet potato, taro, turnip, pumpkin,	Well-cooked
Legumes	ALL	ALL	Dried navy beans, lentils, peas, split peas, unroasted cashews, peanuts in shell, natural peanut butter, lima beans, string beans	*Lima beans, peas (dried split, fresh green) These are consumed in later stages of the diet only, best sprouted*	Small amounts, not more than 1 cup cooked per day	Sweet peas, peanuts, peanut butter	Dahl
Fruits	ALL	ALL	ALL. Juices with no additives	ALL, fresh and dried	*Restricted: only whole/fresh or frozen in protein smoothie*	RESTRICTED QUANTITY:1/2 cup serving/no more often than every 2 hours Berries, citrus fruits, cantaloupe, banana, jackfruit, kiwi, grapes, passion fruit, pineapple, rhubarb, guava, pawpaw, lychee	Cooked, smoothies
Grains	Quinoa, millet, amaranth, teff[a], oat[a], tapioca, rice, sorghum	Quinoa, millet, amaranth, teff[a], oat[a], tapioca, rice, sorghum	*NONE*	*NONE*	*NONE*	Barley, oats, quinoa millet, teff[a], oat[a], tapioca, rice, sorghum, seitan, amaranth, buckwheat, arrowroot, sago, oat bran, barley bran NO: WHEAT/RYE	Rice congee

(continued)

Table 57.1 (continued)

	Comprehensive elimination diet	Gluten-free/casein-free	Specific carbohydrate diet	Gut and psychology syndrome diet	Anti-fungal diet	FODMAP diet	Restoration diet
Herbs and spices	All pure spices, fresh or dried	All pure spices, fresh or dried	All pure spices, fresh or dried	All pure spices, fresh or dried	Fresh only	All pure spices, fresh or dried *No onion* *Minor amts of garlic tolerated*	Not at first, then add: turmeric, ginger, cumin, coriander, and other spices
Beverages	Water, broths. Uncaffeinated herbal teas, seltzer, mineral water. Diluted juices, vegetable juices	ALL without dairy or gluten	Water, tea, weak, freshly made water. Broths	Water, tea, weak, freshly made water. Broths	Water, herbal tea	Tea, herbal teas, herbal infusions, hot water, coconut water. Coffee: <2 cups daily. Chicory/roasted	Broths. Water. Herbal teas. Seltzer, mineral water, diluted juices, diluted vegetable juices
Sweeteners	Use sparingly: brown rice syrup, agave nectar, honey, stevia, fruit sweetener, blackstrap molasses	ALL	Honey if tolerated. Saccharine	Honey	Stevia	Maple syrup, rice syrup, Treacle, golden syrup, glucose syrup, NutraSweet, sucralose, aspartame, stevia, saccharine	Use sparingly
Miscellaneous	Broths. Medical foods (non-dairy, soy, or gluten-containing). Fermented and cultured foods. Vinegar (not white vinegar)	Broths. Medical foods (non-dairy, soy, or gluten-containing). Fermented and cultured foods. Vinegar	Broths. Gelatin. Pickles (without additives)	Soups. Stews. Cellulose in supplements. Gin, scotch occasionally. Pickles (without additives). Tea, weak, freshly made. Vinegar. Wine (dry)	Lemon and lime and vitamin C crystals as replacements for vinegar. Herbal tea. *Tequila and mead in small amounts*	Jam, marmalade, vegemite, marmite. Alcohol: clear refined spirits such as gin and vodka in moderation	Medical Foods. Broths. Herbal Infusions. Coconut kefir. Coconut water

[a]Certified gluten-free
Comprehensive elimination diet: IFM Tool Kit
Specific carbohydrate diet: ▶ http://www.breakingtheviciouscycle.info/
Gut and psychology syndrome diet: ▶ http://gapsdiet.com/The_Diet.html
Restoration diet: *Digestive Wellness*, 4th ed.
Anti-fungal diet: IFM Tool Kit
Yeast questionnaire: ▶ http://cassia.org/candida.htm
FODMAP diet: ▶ http://www.fodmapsdiet.com

57.1.1 Explain What You Are Doing and Why

- As you begin your nutrition and lifestyle changes, you may receive questions or even criticism from family and friends. Sometimes all that is needed to gain their support is to have an honest talk and explain *why* you're making these changes.
- Ask for support. Be positive about your desire to change, and let your friends/family know that encouraging words help you a lot.
- Tell them the positives you are experiencing. "I am avoiding _____ because it makes me feel better, I have more energy, my digestion is better, and I no longer struggle with difficult symptoms."
- If friends and family are interested, provide them with resources that you've gotten from your healthcare team to help your friends and family understand more. They may want to join you in your health quest!

57.1.2 Nutrition Approaches to Wellness Is "The Road Less Traveled"

- By choosing to approach your health and wellness with food and nutrition, you are using "food as medicine" and not a pill to control your symptoms. Be realistic and know that this takes time. These changes will not happen overnight, so be patient and work closely with your nutritionist to continue to steadily heal and achieve your best health and self!

57.1.3 Make Light of It

- If you feel there is unwanted attention on you, make light of the situation. If someone asks, "Why are you not eating _____?", casually respond with "Oh, I'm making some nutrition changes for my health." Then ask a question about them.
- Tell your friends and family that you have a boot camp/dictator nutritionist who demands you make these changes, and you have no other choice. *She's/he's scary!* ☺ We're always happy to be the scapegoat.

57.1.4 Be Confident

- These health changes are ultimately for *YOU*! Friends and family may give you hard time or pressure you to stray from your plan. Regardless, you are the one who will see and feel benefits, so it's important to remind yourself that you're doing it for you! If necessary, talk to them about it. They may not even realize what they are doing.

57.1.5 Plan Ahead: Social Setting and Restaurants

- For social events: bring a dish you know you can eat, eat before you go, or pack a snack to take with you.
- Make suggestions for a restaurant that you trust will have selections for you.
- Gain the support of servers by conveying that you want to *partner* with them to develop your meal as opposed to *telling* them what you need. Ask their opinion. Be positive and cooperative to make it an enjoyable experience for everyone. (If you're unsure whether they take your request seriously, tell them you have a food allergy, which could result in a medical emergency if exposed to a particular food.)
- At restaurants, it doesn't hurt to smile at your server and say, "I promise I'll tip, can I get your help modifying my meal to fit my current health needs?".

57.1.6 My Loved Ones Don't Want to Eat What I Need to Eat

- Explain to them that these changes are about using food to heal your body. Compromise by making meals that they can substitute in items they prefer versus casseroles or mixed dishes.
- If you must, make special meals for you and freeze the rest for future meals while others eat their meal of choice.
- Over time, your friends/family may begin to notice the positive impact of a healthy diet by seeing the physical, mental, and emotional changes *in you* and be more open to trying new foods.

57.2 How to Embrace Your Decision to Follow a Specialized Diet

- **I can't do this. It's too hard** *"The dangerous thing about excuses is that if we recite them enough times, we actually come to believe they are true"*
- Accepting responsibility for your life is a must. Once you take the responsibility, real change is within your reach. Ask yourself this question: Is it harder to make these changes and feel well or to not make changes and continue to struggle with my symptoms?
- Tip: Begin each day with positive self-talk. Post your favorite inspirational or motivational quotes in the house, pray, practice deep breathing, listen to calming music, exercise, or enjoy the company of others.

- **I can't see a difference** *"You don't have to see the whole staircase, just take the first step"*
  - You are a work in progress, which means you get there a little at a time and not all at once. Your symptoms didn't develop in a week and will probably not go away in a week.
  - Tip: Consider journaling your symptoms or experiences so you have a self-reported history to reflect back on. You may be improving more than you realize.

- **I can't change my lifestyle** *"If you want something different, you have to do something different"*
  - Slowly but surely, you'll learn the little things you can do to make these changes easier. Over time, you'll realize what use to seem like a huge inconvenience is now a routine part of your day.
  - Tip: Identify your less helpful habits. Instead of focusing on breaking them, focus on replacing them with new positive habits.

- **I can't cook** *"Keep calm and cook on"*
  - Mistakes are an important part of life. Keep experimenting and you will slowly begin to learn the ways of the kitchen. We provide many recipes and resources that are very basic and can get you started.
  - Tip: Stick with what you are comfortable with and try a new recipe once a week.
  - Take advantage of cooking classes and YouTube videos on "how-to" basics in the kitchen.

- **I can't do this with my work/school schedule** *"You are confined only by the walls you build yourself"*
  - You will likely be confronted with barriers at some point. Not every day will go according to plan and that's okay. Strive to make the best decision when faced with challenges and use our resources/handouts to provide strategies for various situations.
  - Tip: Let leftovers be your best friend. Cook extra portions at the evening meals or on the weekend to use for the week.

- **I can't see the point of this anymore** *"Believe you can and you're halfway there"*
  - You're in the driver seat! We strive to help you in any way we can, but you are the one who makes the final decision. Discover the reason *why* you want to make these changes and focus on that why instead of focusing on the challenges.
  - Tip: Find your drive. Find your inspiration. Find your why.

- **I can't afford it** *"Don't judge each day by the harvest you reap, but by the seeds you plant"*
  - Remind yourself that this is about working toward wellness. If you think healthy is expensive, you haven't priced illness lately.
  - Tip: Use the coupons at the front of the store, shop for what is in season, and plan ahead to avoid spoilage.

Motivational Interviewing

Leigh Wagner, Diana Noland, and Jeanne A. Drisko

58.1 The Practitioner (Healthcare Provider) – 1053

58.2 Observing the Client – 1053

© Springer Nature Switzerland AG 2020
D. Noland et al. (eds.), *Integrative and Functional Medical Nutrition Therapy*,
https://doi.org/10.1007/978-3-030-30730-1_58

Motivational interviewing (MI) is a collaborative person-centered form of guiding to elicit and strengthen a person's motivation for change. It is a counseling technique that is client-centered and goal-oriented. Integrative and functional medicine nutrition practitioners and other practitioners use MI mainly in the clinical setting when working with behavior change one-on-one with clients. The following document summarizes some of the key principles of MI. When observing clinical nutrition patients, look for MI in action, and questions below will help identify whether the practitioner is using MI (Fig. 58.1).

Working with clients with chronic diseases means often practitioners are working toward behavior change. In MI, *ambivalence* is a foundational challenge. In MI, ambivalence refers to both the desire to change and the desire to maintain the status quo (Fig. 58.2).

Refer below for reflections as a practitioner when using MI with patients. Start by asking these questions (Fig. 58.3):
1. Who is making the argument for change? Patient? Practitioner? Parent? Child? Spouse? Other?
2. Is the encounter patient-centered?
3. Is the encounter goal-directed?

Fig. 58.1 Four principles of motivational interviewing

> **Expressing empathy:** Reflective listening
> **Developing discrepancies:** How do client behaviors fit with who the client wants to be?
> **Rolling with resistance:** ↑ resistance, ↓ likely change; validate
> **Supporting self-efficacy:** Bulid confidence and hope

Fig. 58.2 Motivational interviewing spirit

> **Collaborative:** Client and practitioner working togther to set goals & formulate solutions
> **Evocation:** The practitioner inspires or elicits motivation from the client
> **Respect:** The practitioner respects the client's choices, resources, ablity to follow-through to change, and decision *not* to change, if that is a result of the session.

Fig. 58.3 Skills of motivational interviewing

> **Open questions:** To elicit evocative spirit and avoid yes/no questions
> **Affirmations:** Affirm the client's strengths and efforts. Sincerity is important.
> **Reflections:** Reflections of what the client wants, needs or feels
> **Summary:** Summarize what the patient believes. Repeat a summary of a compilation of what the patient needs or wants in terms of behavior change or non-change.

58.1 The Practitioner (Healthcare Provider)

Does the practitioner:
- Use open questions?
- Practice reflective listening?
- Show the "righting reflex?"
- Fall into the "expert trap?"

58.2 Observing the Client

Does the patient demonstrate ambivalence (the desire to change with the desire to remain the same)?
Do you notice the client using "change talk?"
Change talk:
1. *Desire for change*: What are the patient's desires, preferences, and wishes for change?
2. *Ability to change*: If the patient were to change, what strategies would he or she use to achieve change? How would the patient change?
3. *Reasons for change*: Most predictive of an actual change made by patient/client, e.g., get in shape, reduce risk of disease, for kids, etc.
4. *Need for change*: Visceral/emotional reasons for change. Values-based.

Do you notice the client using:
1. "Commitment language?"
   - Client's statements of commitment to actual change
2. "Sustain talk?"
   - Clients' own arguments or reasons against change or desire for status quo, inability for change

Authorization for the Release of Information

Leigh Wagner, Diana Noland, and Jeanne A. Drisko

I, _____, hereby authorize the use or disclosure of my health information from the listed healthcare practitioner as described below to the requesting practitioner.

Patient Information

Name _____ Date of Birth _____

Address _____

City _____ State _____ Zip Code _____

Phone _____ Social Security Number _____

Healthcare Practitioner

Healthcare Practitioner Name _____

Address _____

City _____ State _____ Zip Code _____

Phone _____ Fax Number _____

I authorize for **[practitioner name]** _____ to release and/or disclose the medical information as indicated below to the healthcare provider, entity, or person I have indicated above.

DURATION: This authorization shall become effective immediately and shall remain in effect until _____ (date) or for 1 year from the date of signature if no date entered.

REVOCATION: This authorization may be revoked in writing by the undersigned at any time prior to the release of information from the disclosing party. Written revocation will not affect any action taken in reliance on this authorization before the written revocation was received.

INITIAL and check the box for which types of information are to be released and/or disclosed:

_____ General Medical Information from _____ to _____ (dates)

_____ Laboratory Tests (serum, urine) from _____ to _____ (dates)

_____ Information regarding specific diagnosis or treatment from _____ to _____

_____ Other _____

Requesting Practitioner Information

[insert practitioner name here] _____

[insert practitioner address, phone and fax here] _____

Patient Name (printed): _____

_____ Date: _____

Signature of Patient

ALL PATIENT INFORMATION IS HANDLED UNDER THE HIPPA PRIVACY ACT
CONFIDENTIAL/HIPPA Approved Form

Patient Handouts

Leigh Wagner, Diana Noland, and Jeanne A. Drisko

60.1 Diet Diary Instructions – 1059
60.1.1 Date and Time – 1060
60.1.2 Food or Beverage Consumed – 1060
60.1.3 How Much – 1060
60.1.4 Where – 1060
60.1.5 Activity While Eating – 1060
60.1.6 Mood – 1060
60.1.7 Symptoms – 1060

60.2 Basic Rules to Remember – 1060
60.2.1 Write Everything Down – 1060
60.2.2 Do It Now – 1060
60.2.3 Be Specific – 1060
60.2.4 Estimate Amounts – 1060

60.3 Gluten-Free Dairy-Free Guide – 1060
60.3.1 General Guidelines: What Can I Eat and Drink? – 1061
60.3.2 Cleansing Beverages – 1062
60.3.3 Healthy Carbohydrates – 1062
60.3.4 Clean Protein – 1062
60.3.5 Balanced Fats and Oils – 1062
60.3.6 Preparation – 1062
60.3.7 Breakfast Ideas – 1063
60.3.8 Lunch/Dinner Ideas – 1065
60.3.9 Lunch/Dinner Ideas – 1065
60.3.10 Snacks – 1065

60.4 Food Sensitivity Resources – 1067
60.4.1 Organizations – 1067
60.4.2 Web Resources – 1067
60.4.3 Book Resources – 1068
60.4.4 The Plate: Meal Guide – 1068

60.5 Protein Guide – 1069
60.5.1 Why Are Proteins Important? (Table 60.4) – 1069

© Springer Nature Switzerland AG 2020
D. Noland et al. (eds.), *Integrative and Functional Medical Nutrition Therapy*,
https://doi.org/10.1007/978-3-030-30730-1_60

60.6 Carbohydrate Guide – 1070
60.6.1 Why Are Carbohydrates Important? – 1070

60.7 Fat/Oil Guide – 1071
60.7.1 Why Are Fats and Oils Important? (Table 60.6) – 1071

60.1 Diet Diary Instructions

The information you record in your food diary will help you and your health care provider identify patterns in your diet that may correlate to your health condition. This information can help them to design an eating program to meet your special needs (◻ Fig. 60.1).

These instructions will help you get the most out of your food diary. Generally, food diaries are meant to be used for a whole week, but studies have shown that keeping track of

Name:					
Date/Time	Location/ Activity	Food or Beverage Consumed	Amount (cup, etc)	Mood	Symptoms
Was this a typical day for you? Yes No If no, why not:					

Reviewed by _____   Date/Time _____

◻ **Fig. 60.1** Diet diary

what you eat for even 1 day can help you make changes in your diet.

60.1.1 Date and Time

Write the date and time of day you ate the food.

60.1.2 Food or Beverage Consumed

In this column, write down the type of food you ate or drank. Be as specific as you can. Don't forget to write down "extras," such as butter, oils, salad dressing, mayonnaise, sour cream, sugar, and ketchup. Please include brand names when possible or indicate if an item was homemade.

60.1.3 How Much

In this space indicate the amount of the particular food item you ate. Give your best estimate of the size (2″ x 1″ x 1″), the volume (1/2 cup), the weight (2 ounces), and/or the number of items (12) of that type of food.

60.1.4 Where

Write what room or part of the house you were in when you ate. If you ate in a restaurant, fast-food chain, your desk, or your car, write that location down.

60.1.5 Activity While Eating

In this column, list any activities you were doing while you were eating (e.g., working on the computer, driving, watching TV, sitting at the dinner table).

60.1.6 Mood

How were you feeling while you were eating (e.g., sad, happy, rushed, stressed, bored)?

60.1.7 Symptoms

In this column, make a note of any symptoms (good or bad) you experience throughout the day to help tune into how certain foods make you feel. *Include* bowel/urine habits such as formed stool, loose stool, hard stool, scant urination, or frequent urination; feelings of discomfort such as gas, bloating, heartburn, headaches, brain fog, low energy, or sinus congestion; and feelings of wellness such as increased energy, mental clarity, and/or relief of previous symptoms. Try to correlate the entries as closely as possible with the times listed to the left on the diet diary form.

60.1.7.1 Hints

- Do not change your eating habits while you are keeping your food diary unless your doctor or dietitian has given you specific instructions to do so.
- Tell the truth. Your doctor or dietitian can help only if you record what you really eat.
- Record what you eat on all days your doctor or dietitian recommends.
- Be sure to bring the completed forms back with you to your next appointment.

60.2 Basic Rules to Remember

60.2.1 Write Everything Down

Keep your form with you all day, and do your best to write down everything you eat or drink. A piece of candy, a handful of pretzels, a can of soda pop, or a small donut may not seem like much at the time, but over a week these foods may add up!

60.2.2 Do It Now

Don't depend on your memory at the end of the day. Record your eating as you go.

60.2.3 Be Specific

Make sure you include "extras," such as gravy on your meat or cheese on your vegetables. Do not generalize. For example, record French fries as French fries, not as potatoes.

60.2.4 Estimate Amounts

If you had a piece of cake, estimate the size (2″ × 1″ × 2″) or the weight (3 ounces). If you had a vegetable, record how much you ate (1/4 cup). When eating meat, remember that a 3-ounce cooked portion is about the size of a deck of cards.

If you have any questions, contact your doctor or dietitian (◘ Fig. 60.2).

60.3 Gluten-Free Dairy-Free Guide

I have recommended that you avoid gluten and/or dairy to determine whether one or the other may be affecting your health. Withdrawing these foods is often recommended to:
1. Alleviate any potential stress or inflammation in your body
2. Determine which foods/food groups may be affecting your health

By withdrawing gluten and/or dairy, you may alleviate stress on your whole body and immune system. This dietary plan

Date	Time	Food or Drink: What kind	How much	Where	Activity	Mood	Symptoms
1/3/08	8am	Quaker instant oatmeal made with water	1 cup	Car	Driving to work	Rushed/anxious	Loose bowel at 7am

Fig. 60.2 Sample food diary entry

will help minimize symptoms, promote vitamin and mineral absorption, and generally calm down your immune system and overall decrease inflammation.

Did you know that an estimated 70% of your immune system is associated with your digestive tract? When we consistently consume foods that we are unable to process, our immune system is constantly on high alert. This stress makes our body vulnerable to other illnesses or diseases: depression, anxiety, diabetes, cancer, cardiovascular disease, and others.

With that said, it is important to remain 100% committed to this plan. Keep in mind that you have been eating these foods for 10, 20, 30, and more years. It takes a lot of time (months and *years*) for your body to recover from these offending foods. Since it takes time, patience, and commitment, I suggest you set aside a 5–6-week period that gives you the freedom to prepare or plan meals and/or menus. Depending on the time you have available and your comfort in the kitchen, you may want to prepare meals each night of the week or cook large batches on weekends to eat throughout the week.

Common foods that all clients are asked to eliminate and/or minimize during the trial are caffeine (for certain individuals, 2 cups of green tea or organic coffee daily may be beneficial), alcohol, and added sugars. Plus, the two proteins that are difficult for many bodies to process are gluten and casein. Gluten is a protein formed when water mixes with certain grains (wheat, rye, barley, etc.) and their flours. Casein is a major protein in dairy foods.

For the next _____ **weeks/months**, you will remove gluten and casein as well from your diet. Consuming these "offending foods" can be associated with symptoms including:

- Joint/muscle pain, aches, or twitching
- Fatigue/low energy
- Digestive upset: bloating, gas, diarrhea, constipation, reflux, abdominal pain, etc.
- Headaches/migraines (may be brief or prolonged)
- Sore throat, stuffy nose, runny nose, itchy nose or eyes

> **Symptoms** associated with the food "reintroduction" or "challenge" period may not be the same symptoms experienced before the food withdrawal trial. For example, before the diet, a person's symptom may be chronic sinus pain. Once the food is reintroduced/challenged, their main symptom may be abdominal pain or diarrhea. This does not mean that the food group being challenged was not causing the sinus pain. It means the body and immune system react differently when the offending agent is removed and then reintroduced.

- Skin rash or redness
- Lower back pain
- Anemia or low iron levels
- Vitamin/mineral imbalances
- Brain "fog," lack of focus and/or concentration
- Sleepiness, insomnia, fatigue, apathy
- Mood disturbances: irritability, depression, anxiety
- Excitability (feeling hyper or "buzzed")
- Inflammation of any kind

60.3.1 General Guidelines: What Can I Eat and Drink?

Food withdrawal recommendations are based on an individual assessment which evaluates specific symptoms, lab values, and health recommendations. Food withdrawal and specific nutrition recommendations are necessary to allow the GI tract to heal while providing optimal levels of nutrients to support metabolism, reduce inflammation, and promote healing. The types and amounts of carbohydrates, protein, and fats recommended for each individual will vary based on the individual's GI tract ability to digest and absorb nutrients. For impaired GI tracts, elemental forms of food may be used (simple sugars, amino acids, oils), while a more functional GI tract will allow consumption of whole foods (fruits, vegetables, meats). Instead of focusing on those foods you *can't* eat, we will focus on those you *can enjoy*! The following is a summary of whole foods, supplemental support, and beverages that you will enjoy during your trial:

60.3.2 Cleansing Beverages

- *Water, mineral water, and green and herbal tea:* Keep a *glass* or *stainless steel* water bottle or other container to keep you supplied with water and clean liquids throughout the day. Water helps deliver nutrients to your cells, and subsequently water helps remove waste and toxins from the body. In other words, drinking water and green or herbal teas helps you take out the body's trash!
- *Organic coffee*: If you tolerate coffee and caffeine, 1–2 cups daily (8 ounces).

60.3.3 Healthy Carbohydrates

- *Guidelines:* Organic, when possible (dirty dozen/clean 15), varied colors, and types.
- *Avoid/minimize:* High-fructose corn syrup, processed/packaged foods, simple sugars, and drinks.
- *Low-starch vegetables:* Should make up 50% of your plate at each meal.
- *Starchy vegetables and fruits:* Sweet potatoes, potatoes, corn, fresh or frozen (not canned) fruit.
- *Beans/legumes:* All dried beans/green/waxed beans.
- Soy may or may not be recommended or tolerated by all people. Thus, ask your dietitian or physician whether or not soy should be consumed.
- Peanuts should be organic.
- *Whole grains:* Quinoa, rice, amaranth, millet, Teff, buckwheat, and others.
- *Nut, seed milks:* Almond, hazelnut, coconut, soy (if tolerated), oat (if tolerated), rice, or other nondairy milks.
- *Supplemental carbohydrates/fiber:* Ground flaxseed, psyllium, modified citrus pectin, oat bran.

60.3.4 Clean Protein

- *Guidelines:* Organic meats, BPA-free canned beans, raw nuts and seeds.
- *Avoid:* High-temperature cooking or grilling, charred meats, improperly stored meats, processed meats, roasted nuts, processed nut butters.
- *Poultry, meat, seafood:* Free-range/organic (if possible) poultry, wild game (bison/buffalo, deer/venison, lamb, etc.), wild Alaskan salmon, protein powders.
- *Eggs:* Pasture-raised, omega-3, and organic eggs are preferable choices.
- *Beans/legumes, lentils:* Plant-based protein.
- *Nuts and seeds:* Plant-based protein.
- *Supplemental protein:* Whey protein powder, amino acid complex, other protein powders.

60.3.5 Balanced Fats and Oils

- *Guidelines:* Organic meats, cold pressed oils, temperature-resistant cooking oils, organic butters.
- *Avoid:* Damaged fats and oils, trans fats (packaged foods), processed butters, fast food.
- *Nuts and seeds:* All peanuts should be organic. Pistachios may or may not be recommended.
- *Fats and oils:* Olive oil, avocado, organic pasture butter, coconut oil. See Fats and Oils Handout, which lists options: Healthy *Omega 9, 6, 3, and beneficial saturated fats.*
- *Supplemental fats and oils:* On an individual basis, MCT oil, fish oil, evening primrose, borage oil, black currant oil.

60.3.6 Preparation

- **Week 1**

1. *Clean the cupboards:* When possible, clear your cupboards and refrigerator of foods that are not "approved" on your particular withdrawal trial. Read labels, and look for added sugars, high-fructose corn syrup, and any of the "red flag" words we discussed.
2. *Shop for foods and stock your pantry:* Make sure that your refrigerator and cupboards are prepared with "approved" items so that you're always able to make a safe choice.
3. *Supplements:* Order any supplements or protein powders recommended by your doctor or dietitian.
4. *Caffeine and your genes:* Individuals with SNP's CYP1A1, CYP1A2, and CYP1B1 should avoid caffeine or consume no caffeine after noon.
5. *Wean off caffeine, if necessary:*
   - Begin on a weekend so you're able to take naps, as needed.
   - For the first 3 days, cut down to 1/2 normal amount of coffee, soda, black tea, or other caffeinated beverages.
   - For the next 4 days, drink 1 cup caffeinated green tea steeped in boiling water for 5 minutes.
   - You may consider taking 1000–2000 mg buffered vitamin C powder (Emergen-C is one brand that would be ok, even though it has added sugar, it would still help you adjust to decreased caffeine intake).
   - Drink a minimum of 6–8 glasses filtered water daily.
6. *Diet diary:* You will need to keep a diet diary for the 3 days prior to reintroduction or "challenge" period and throughout the reintroduction process. This process is time-consuming, but it is very important. You want to make your efforts worth it by noting any symptom changes. This helps determine next steps for continued healing (◘ Tables 60.1 and 60.2).

Table 60.1 Diet snapshot

Yes: These items are gluten–/dairy-free

Naturally gluten- and dairy-free foods

Vegetables and fruits

Beans (lentils, navy, kidney, black beans, garbanzo beans/chickpeas, etc.)

Nuts/seeds (almonds, walnuts, pecans, cashews, sunflower seeds, etc.)

Meat, poultry, seafood, eggs

Gluten-free grains and gluten alternatives (not recommended if grain-free)

Amaranth	GF oats (may not be recommended for celiac disease)	Rice (brown, wild, white)	Tapioca
Arrowroot		Rice bran	Teff
Buckwheat		Sago	
Flax	Millet	Sorghum	
Flours made from nuts, beans, and seeds	Organic, non-GMO corn	Soy	
	Quinoa		

Dairy-free alternatives

Almond milk[a]	Coconut milk yogurt	Hazelnut milk[a]	Rice milk
Cocoa butter	Daiya (tapioca-based cheese shreds)	Hemp seed milk[a]	*So Delicious* brand
Coconut butter and oil		100% fruit sorbet	Yogurts, desserts
Coconut milk[a]	Ghee (if guaranteed casein-free)	Imagine brand soups	Milks
Coconut milk creamer		Oat milk[a]	

For those who avoid dairy, kosher "pareve" products are considered milk-free under kosher dietary law; however, they may contain a very small amount of milk protein. Individuals who have a dairy allergy will want to look closer at the ingredients list for confirmation of safety before consuming the product

Make sure to read labels. If there is insufficient information on the label to make an informed decision, avoid the food. You can also call food companies to gather manufacturing information

[a]Preferably unsweetened version

Condiments: Any condiments made from apple cider vinegar are safe; Heinz and Organicville brand condiments are gluten-free.

Deli meats: Applegate Farms, Boar's Head, and Hormel Naturals are gluten-free.

Cosmetics (shampoos and others): Desert Essence and Burt's Bees are gluten-free

Medications: Celiac Central website (▶ celiaccentral.org/resources)
- ▶ http://www.glutenfreedrugs.com/
- Ingredients in medications that may indicate gluten: Wheat, modified starch (source not specified), pregelatinized starch (source not specified), pregelatinized modified starch (source not specified), dextrates (source not specified), dextrimaltose (when barley malt is used), caramel coloring (when barley malt is used), dextrin (source not specified, but usually by corn or potato)

60.3.7 Breakfast Ideas

- *Eggs and greens:* Two omega-3 or organic eggs with 2 cups sautéed greens with onions in vegetable or chicken broth. Drizzle with olive oil after sautéed. Add poached or sliced hard-boiled egg over the greens and season with sea salt, pepper, or other spices as desired.
- *Warm apple cinnamon quinoa:* 1/2 cup cooked quinoa with diced apples, 1/2 teaspoon cinnamon, and 1/4 cup almonds/walnuts or drizzle of natural peanut butter or almond butter.
- *Nutty cereal:* 1 1/2 cups gluten-free cereal with 1 cup unsweetened soy milk or other unsweetened milk alternative, 1/4 cup nuts or 2 tablespoons nut butter (peanut, almond, etc.), 1 tablespoon ground flaxseed.
- *Protein smoothie:* 1 cup unsweetened milk alternative with 1 serving hemp seed protein, RAW protein, or egg protein powder (Jay Robb brand is a staple at Whole

Table 60.2 Possible gluten-containing food items

Caution: These items contain gluten or dairy and must be avoided

Gluten-containing food items		Dairy- or casein-containing food items	
Barley	Matzo flour/meal	Butter	Recaldent®
Barley malt/extract	Pastas (wheat-based)	Cheeses (most, except	Rennet casein
Bran	Panko	some soy brands)	Sherbet
Bread crumbs	Rye	Cream	Soup bases
Bulgur	Seitan	Creamed soups and	Sour cream
Cereal extract	Semolina	vegetables	Tagatose
Club wheat	Spelt	Curds	Whey
Couscous	Sprouted wheat	Custards	White or milk chocolate
Durum	Triticale	Diacetyl	Yogurt
Einkorn	Vital wheat gluten	Ghee	
Emmer	Udon	Half-and-half	
Farina	Wheat	Ice cream	
Faro	Wheat bran	Ice milk	
Graham flour	Wheat germ/germ oil	Lactose, lactulose,	
Hydrolyzed wheat	Wheat grass	lactoferrin,	
protein	Wheat protein isolate	lactalbumin	
Kamut	Wheat starch	Milk	
Malt vinegar/flavoring	Whole wheat berries	Puddings	

Maybe: These food items require further inspection. They may or may not contain gluten or dairy

Items that may contain gluten		Foods that may contain casein	
Ales, beers, lagers	Marinades/thickeners	Artificial flavorings	Tuna fish
Breading mixes	Pasta	Bacterial cultures	[a]Many nondairy foods
Brown rice syrup	Roux	Caramel candies	contain casein proteins
Brown sugar	Sauces	Cosmetics, medicines	Avoid foods that
Caramel coloring	Soup bases and broths	Dairy-free cheese (most	contain any ingredient
Coating mixes	Stuffing	brands)	with casein or
Communion wafers	Self-basting poultry	Dairy-free may contain	caseinate
Condiments (mustard, ketchup, BBQ sauce, etc.)	Soy sauce (wheat-free tamari is gluten-free)	casein[a] Ghee	
	Medications	Hot dogs	
Croutons	Modified food/veg.	Lactic acid	
Candy	starch	Lunch meats	
Deli/luncheon meats	Over-the-counter meds	Margarine	
Herbal supplements	Prescription meds	Nisin	
Imitation bacon	Surim	Nougat	
Imitation seafood	Vitamin and mineral	Sausage	
Lotions, body care	supplements	Semisweet chocolate	

- Foods), 1 cup frozen or fresh berries, 1 cup fresh spinach or other green or 2–3 stems from leafy greens, 1 tablespoon olive oil or Balance 3-6-9 oil or flaxseed oil (depending on those recommended by your healthcare practitioner); however, you can always default to olive oil, 1/4 avocado, or 1 tablespoon ground flax or flaxseed oil.
- *Breakfast roll-up:* 3.5 ounces leftover fish or chicken wrapped in a gluten-free rice or Teff tortilla filled with 2 tablespoons guacamole and 3 tablespoons salsa. Enjoy with 1 cup fresh fruit such as cubed cantaloupe.
- *Rice cakes or crackers with hummus and a side of chicken sausage or nut butter:* Two rice cakes topped with hummus and 3 ounces leftover meat or gluten-free turkey or chicken sausage (1 link Applegate Farms chicken sausage) or spread with 2 tablespoons almond or organic peanut butter. Serve with a whole grapefruit or 1 cup fresh berries or mixed fruit.
- *Salad with herbs:* 2 cups mixed salad greens (romaine, leaf lettuce, spinach, cabbage) and herbs: (basil, cilantro, parsley, mint), top with your favorite chopped vegetables (tomatoes, onions, bell peppers, asparagus, shredded beets, sprouts, broccoli, cauliflower). Add 1/4 cup quinoa or brown rice and 1/2 beans, and sprinkle with raw nuts and 1 tablespoon dried fruit (raisins, cranberries, diced figs, etc.). Dress with a citrus or balsamic vinaigrette or vinegar and extra virgin olive oil. Optional: Top with 2–3 ounces leftover chicken, canned or fresh cooked salmon, or canned chunk-light tuna in water.

60.3.8 Lunch/Dinner Ideas

- *Hummus and veggies with gluten-free crackers:* 1/4 cup hummus (chickpeas/garbanzo bean garlic dip) or baba ghanouj (same as hummus only with eggplant instead of chickpeas). Dip the fresh vegetables and/or gluten-free crackers. Choose a variety of brightly colored and flavored vegetables: bell peppers, carrots, celery, snap peas, mini-sweet peppers, cucumber, broccoli, cauliflower, zucchini, etc.
- *Lentil tacos with mixed greens:* Sauté onions and garlic and add cooked lentils. Season with Mexican seasoning, lime juice, and a pinch of salt, and serve with a brown rice tortilla or corn tortilla (make sure the corn tortilla is 100% gluten-free – no added wheat).
- *Gluten-free pasta with veggie-boosted sauce and side salad:* Sauté 1/2 to one whole onion (chopped), 1/4 cup zucchini, 2 cloves garlic with 1–2 tablespoons olive oil, a pinch of salt, and Italian seasoning with 3 ounces of chicken. Add 1/2 cup Classico brand (or other brand with no sugar added) marinara sauce and 1/4 cup white, black, or garbanzo beans. Serve over 3/4 cup brown rice, gluten-free pasta. Serve with a side salad.
- *Salmon salad four ways:* Mix canned wild Alaskan salmon (e.g., Bear & Wolf or Kirkland brands from Costco) or chunk light tuna in water mixed with 1–2 tablespoons light mayonnaise or vegenaise, 2 teaspoons lemon or lime juice, 1/4 cup diced grapes or apples, and 1 tablespoon walnuts; season with pepper to taste.
    - Serve over a bed of greens and mixed vegetables.
    - Dip or top gluten-free crackers.
    - Dip with vegetables: carrots, broccoli, cucumber, etc.

60.3.9 Lunch/Dinner Ideas

- Turkey spinach meatloaf with mashed cauliflower
- Gluten- and dairy-free fajita/burrito bowls
- Asian chicken salad
- Italian Tuscan vegetable soup
- Grilled citrus trout crunchy Mediterranean slaw
- Whole roast chicken with sweet apple walnut kale
- Black bean cakes with mango salsa
- Quinoa cabbage soup
- Curry chicken breast with tomato and coconut milk served over brown rice or quinoa
- Chicken salad wraps/roll-ups
- Quick lemon and garlic quinoa salad and spiced collard greens
- Spaghetti squash
- Quick sautéed chicken with sautéed vegetables with cashews
- Ginger peanut or almond chicken with sweet potatoes and snap peas
- Fish tacos with citrus slaw
- Salmon salad
- Quick turkey pasta sauce
- Cumin-crusted salmon and cauliflower rice

60.3.10 Snacks

- Hummus and veggies or gluten-free crackers/chips (Beanitos – bean chips)
- Trail mix
- Hard-boiled eggs
- Apple or pear with almonds/nut butter
- Cinnamon mixed nuts
- Roasted chickpeas
- Quick and easy kale chips
- Chocolate avocado pudding
- Whipped sweet potatoes (◘ Table 60.3)

Table 60.3 Gluten- and dairy-free shopping list

Vegetables

Artichoke	Carrots	Kohlrabi	Pumpkin	Swiss chard
Asparagus	Cauliflower	Leeks	Radishes	Tomatoes
Avocado	Celery	Mushrooms	Romaine lettuce	Turnips
Beets/beet greens	Collard greens	Mustard greens	Rutabaga	Winter squash
Bell peppers	Cucumber	Okra	Salad greens	Yams
Bok choy	Eggplant	Olives	Snap peas	Yellow squash
Broccoli	Fennel	Onions	Snow peas	Zucchini
Brussels sprouts	Green beans	Parsnips	Spinach	
Cabbage	Kale	Potatoes	Sweet potatoes	

Herbs and spices

Fresh or dried herbs and spices

Italian (blends of basil, oregano, marjoram, parsley, thyme, sage, and rosemary)

Fruits

Apples	Currants	Lemons	Peaches	Raspberries
Apricots	Dates	Limes	Pears	Strawberries
Bananas	Dried fruits	Mango	Pineapple	Tangerines
Blackberries	Figs	Melons	Plums	Watermelon
Blueberries	Grapes	Nectarines	Pluots	
Cherries	Grapefruit	Oranges	Pomegranate	
Cranberries	Kiwi	Papaya	Prunes	

Dairy alternatives

Unsweetened almond milk, coconut milk, hemp seed milk, hazelnut milk, etc.

Daiya tapioca-based cheese shreds

Meat, fish, poultry, protein

Fresh or frozen poultry (turkey, chicken, etc.)

Wild game (deer (venison), elk, duck, buffalo, etc.)

Grass-fed beef

Wild Alaskan salmon or other wild Pacific seafood

Eggs (preferably pasture raised, organic)

Clean deli meats: Boar's head, Applegate farms deli meats

Beverages

Water	Vegetable juice, up to 8–16 oz. daily
Green, white, and herbal teas (chamomile, peppermint, ginger, tulsi)	Mineral water (still or sparkling) Organic coffee 1–2 cups daily (8 oz. cups)
Coconut water	

Fats and oils

Nuts, nut butters, and nut flours (e.g., almond butter, cashew butter, pumpkin butter)

Organic peanuts and peanut butters

Extra virgin olive oil

Table 60.3 (continued)

Vegetables	
Unrefined coconut oil	
Organic, pasture butter Grapeseed oil, flaxseed oil, etc.	
Frozen foods	
Variety of frozen vegetables, fruits (preferably organic)	
Snacks	
Terra brand vegetable chips (sweet potato, taro, yuca, batata, parsnip, and ruby taro)	
Baby carrots, celery sticks or other veggies	
Hummus made with olive oil	
Rice crackers	
Mary's gone crackers brand gluten-free crackers	
Fresh fruit	
Nuts and trail mixes (without yogurt/chocolate-covered pieces)	
Cereals, grains, and beans/legumes	
Amaranth	Brown or white rice
Beans/legumes: Lentils, peanuts, soybeans, black, navy, garbanzo (aka chickpeas), black-eyed peas, great northern, kidney, etc.	Cream of rice Brown rice tortillas Rice crackers
Buckwheat	Rice cakes
Quinoa	La tortilla brand Teff tortillas
Quinoa flakes	Corn
100% buckwheat soba noodles	Corn tortillas
Buckwheat kasha	Polenta
Cream of buckwheat	Millet
Wild rice (not the same as rice)	Tapioca
Enjoy life brand cereals (nutty flax and Rice cereals, GF granolas)	Sorghum Flaxseed
Condiments	
Condiments made with apple cider vinegar	
Vinegars, oils	

60.4 Food Sensitivity Resources

60.4.1 Organizations

- Food Allergy Research and Education ▶ http://www.foodallergy.org
- Celiac Disease Foundation ▶ www.celiac.org
- Celiac Sprue Association ▶ www.csaceliacs.org
- National Foundation for Celiac Awareness ▶ www.celiaccentral.org
- Gluten Intolerance Group ▶ www.gfco.org

60.4.2 Web Resources

- Elana's Pantry: Gluten-Free Recipes ▶ www.elanaspantry.com
- Salted Plains: Gluten + Dairy-Free Recipes ▶ www.saltedplains.com
- Cookie + Kate: Gluten/Dairy-Free Variations ▶ www.cookieandkate.com
- Detoxinista: Gluten + Dairy-Free Recipes ▶ www.detoxinista.com
- Nourishing Meals: Gluten + Dairy-Free Recipes ▶ www.nourishingmeals.com

- Raw Food Recipes and Desserts: Gluten-Free ▶ www.rawmazing.com
- Gluten-Free Living ▶ www.glutenfreeliving.com
- Grocery: Whole Foods Market – Download List of Gluten-Free Products ▶ www.wholefoodsmarket.com
- Restaurants: Urbanspoon website – Link to Gluten-Friendly Restaurants ▶ www.urbanspoon.com

60.4.3 Book Resources

- These books are full of great ideas and menus. However, the information related to health conditions may or may not be appropriate for all patients as we individualize a plan specific to each patient.
- Take caution when using cured meats or meats with nitrates/nitrites. Opt for organic grass-fed meat if possible.
- Be sure to note other ingredients you may not currently eat (i.e., dairy, soy, grains, eggs, etc.)
- *Well Fed* by Melissa Joulwan
- *Wheat Belly* by William Davis
- *Eating on the Wild Side* by Jo Robinson
- *Practical Paleo* by Diane Sanfilippo
- *Nom Nom Paleo* by Michelle Tam and Henry Fong
- *The Ultimate Food Allergy Cookbook and Survival Guide* by Nicolette Dumke
- *The Gluten-Free Gourmet Cooks Fast and Healthy* by Bette Hagman
- *The Gluten-Free Gourmet, Living Well without Wheat* by Bette Hagman
- *Gluten-Free Diet: A Comprehensive Resource Guide* by Shelley Case
- *The Gluten Connection* by Shari Lieberman

60.4.4 The Plate: Meal Guide

This meal guide will help you build a meal. The ultimate goal is for you to meet your nutritional and health needs with real, *whole* foods. If your body is accustomed to eating more convenience, processed, or packaged foods, with time, your body will begin to crave these whole foods your body *needs to live*. Eventually, you will be able to listen to your body's cravings and signals of hunger or fullness to decide when and how much to eat.

Please refer to handouts below for P (protein), C (carbohydrate), and F (fats and oils). If you need to avoid dairy foods (casein), please eat two servings daily of the calcium-rich dairy-free foods (see ▫ Fig. 60.3).

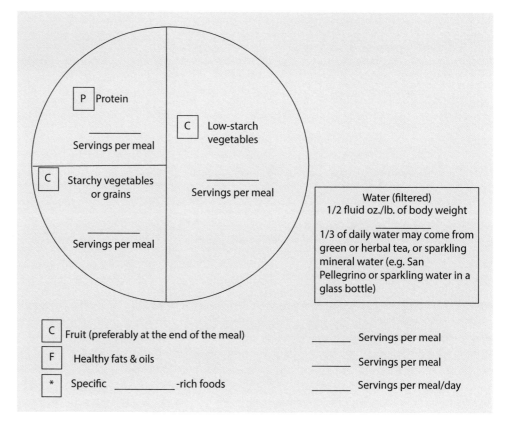

▫ Fig. 60.3 The Plate guide is the first page of a 4 page guide to helps patients build a healthy and balanced meal that includes protein, complex carbohydrates and healthy fats and oils. There is an astrisk section to include any specific nutrient-dense foods that the patient should try to include in their regular diet. This can be personalized for each patient by filling in the blanks for their specific needs

60.5 Protein Guide

60.5.1 Why Are Proteins Important? (Table 60.4)

Table 60.4 Protein guide

Proteins make up healthy cell membranes	Proteins are a major part of muscle tissue
Proteins provide energy	Proteins supply essential amino acids our body cannot make
Proteins support immune defense	Some hormones/chemical messengers are proteins
Bile salts contain proteins needed for nutrient absorption and detoxification	Enzymes are proteins required proper metabolism

include protein with every meal and start your meals with two bites of protein

Plant-based protein	Example	#Servings
Grains and beans Canned, dry, frozen Organic when possible Look for BPA-free cans	GF grains: Quinoa, rice, amaranth, Teff, buckwheat, millet Legumes/beans: Kidney, black, lima, navy, pinto, white, lentils, green peas, black-eyed peas, split peas, etc. Soy beans and soy products: Soy beans, tofu, tempeh, soy milk Vegetable protein powders	
Seeds and seed butters Raw, unsalted, preferable sealed, and refrigerated	Seeds: Pumpkin seeds, sesame seeds, sunflower seeds, chia seeds, flaxseeds Seed butters: Sesame tahini	
Nuts and nut butters Raw, unsalted, preferable sealed, and refrigerated	Nuts: Almonds, cashews, hazelnuts, peanuts, pecans, pistachios, walnuts, macadamia nuts Nut butters: Peanut, almond, and cashew butters Excluding peanut butter, nuts are not a primary protein source	

Table 60.4 (continued)

Clean animal protein	Example	#Servings
Lean red meat Organic, free range/grass-fed	Beef: Medallions, filet mignon, steaks – Flank, skirt, top round Pork: Tenderloin Lamb and game meats: Lamb shoulder and shoulder steak, bison	
Poultry Lean, organic, free range	Lean cuts include the breast and dark meat Chicken, duck, goose, Turkey	
Fish/seafood Fresh, cold water	Wild-caught, northern Pacific fish and seafood Sardines and anchovies Scallops, mussels, oysters	
Egg white Organic, free-range Omega-3-rich eggs	Chicken eggs, duck eggs purchase locally where possible	
Organic dairy Grass-fed, GMO-, pesticide-, hormone-, and antibiotic-free	Milk, cheese, yogurt, cream, butter, cottage cheese, cream cheese, sour cream, ice cream Purchase locally where possible	

Type of protein	Serving size estimate	Protein grams
Grains	1/2 cup	6 grams
Beans and lentils	3/4 —1 cup	7 grams
Whole egg/egg white	2 each	12 grams/ 8 grams
Meat, poultry, seafood	3–6 ounces	7 grams/ ounce
Seeds/peanut butter	3/4 cup/2 tbsp.	8 grams
Protein powders	Variable	10–15 grams per ~2 tablespoons

60.6 Carbohydrate Guide

60.6.1 Why Are Carbohydrates Important?

- Complex carbohydrates (carbs) are the body's primary energy source and preferred source for the brain and nervous system.
- Complex carbs provide essential vitamins and minerals needed for good health.
- Complex carbs contain many phytonutrients ("Phyto" = plant, nutrients) known to help our bodies remove toxins, fight infection, and prevent disease.
- Complex carbs provide fiber and bulk to our diet, promoting satiety and healthy digestion.
- Complex carbs help remove toxins, slow absorption, and promote colon health.
- Choose mostly complex carbs: Low-starchy vegetables (raw, fermented, steamed, sautéed, boiled, pureed) (◘ Table 60.5).

Table 60.5 Carbohydrate guide

Carbohydrate category	Example	#Servings
Low-starchy vegetables		
Organic, fresh, local – When possible Fresh, raw, cooked, frozen (when cooking add oil/fats just before serving)	Artichoke, asparagus, baby corn, bamboo shoots, bean sprouts, beets, broccoli, Brussels sprouts, cabbage, carrot, cauliflower, celery[a], cucumber, greens[a] (collard, kale, mustard, turnip, kohlrabi, lettuce[a]), leeks, mushrooms, okra, onions, pea pods, rhubarb, radish, rutabaga, sauerkraut, spinach[a], summer and spaghetti squash, sugar snap peas, Swiss chard, turnips, water chestnuts Nightshades: Tomatoes, sweet and hot peppers[a], eggplant, tomatillos, pepinos, pimentos, paprika, cayenne peppers	
Starchy vegetables		
Organic, fresh, local – When possible Fresh, raw, cooked, frozen These can fill 1/4 of your plate	Beans/lentils: Garbanzo beans, kidney beans, black beans, lentils, lima beans, navy beans, pinto beans, white beans Peas: Black-eyed peas, green peas, split peas, chickpeas, parsnips, plantains, sweet potatoes, taro, yams, pumpkin, squash Nightshades: Potatoes[a]	
Starchy grains		
Organic, non-GMO These can fill 1/4 of your plate	Rice: Wild, brown, long grain, basmati, arborio Quinoa, amaranth, Teff, buckwheat, millet, corn, oats **Avoid, if gluten-free:** Wheat, barley, rye, spelt	
Fruit		
Organic, fresh, local Eat with meals Fresh, dried, frozen	Grapefruit, apples[a], apricots, blueberries[a], pears, plums, strawberries[a], oranges, nectarines[a], peaches[a], pineapple, grapes[a], bananas, kiwi, papaya, figs, raisins, cantaloupe, watermelon	
Carbs and sweeteners to avoid	High-fructose corn syrup Artificial sweeteners: Splenda, equal, sweet'N low (sucralose, aspartame, saccharin, acesulfame, etc.) Avoid processed, packaged carbohydrates Limit 100% fruit juices to 3–4 ounces and only with meals Avoid sugar-sweetened drinks	
Type of carbohydrate		**Serving size estimate**
Low-starchy vegetables, spaghetti squash		1 cup or 2 cups lettuces
Starchy vegetables and grains		1/2 cup or 1 small potato
Fruit (apple, orange, melon, berries)		1 small or 1/2 cup

[a]Organic when possible

Patient Handouts

60.7 Fat/Oil Guide

60.7.1 Why Are Fats and Oils Important? (Table 60.6)

Table 60.6 Fats and oils guide

Fats provide energy	Fats make up a large portion of our brain and nervous system	
Fats build healthy cell membranes	Dietary fats supply essential fatty acids our body cannot make	
Fats make hormones	Fats make bile salts for nutrient absorption and detoxification	
Fats help absorb fat soluble vitamins	Fats insulate, protect, and cushion our organs	
Fat or oil category	**Example**	**#Servings**
Omega-9s Oleic fatty acids *Stabilizers* ~50% daily fat calories	Raw nuts and seeds: Almonds, cashews, peanuts, sesame seeds Walnuts, macadamia nuts, pine nuts Peanut butter (natural, oil on top), almond butter Tahini (sesame seed butter), olives Almond oil, olive oil, avocado, hummus	
Omega-6s Linoleic acid *Controllers* ~30% daily fat calories	Eggs (whole, organic) Meats (commercial, grass-fed) Raw nuts and seeds: Brazil nuts, pecans, hazelnuts, filberts, walnuts, hemp seeds, sunflower seeds, pumpkin seeds Oils: Evening primrose, black currant, borage, hemp, grapeseed	
Omega-3s Alpha-linolenic acid *Fluidity/communicators* ~10% daily fat calories	Fish oil (high DHA or EPA), 3-6-9 balanced or DHA oils Fish (salmon/fin-fish), fish (shellfish) Flaxseeds (ground/meal), chia seeds Flax oil Algae	
Beneficial saturated fats SCT and MCT *Structure* ~10% daily fat calories	Butter (organic, pasture), ghee (clarified butter – Indian cuisine) Dairy (preferably organic, raw) Meats (preferably grass-fed) wild game, poultry (preferably organic) Coconut oil and butter, MCT oil (medium-chain triglyceride) Eggs (whole, organic)	
Fats to avoid Trans fats and damaged LCFA *Stressors* <5% daily fat calories	Most packaged foods and fast foods Margarine, vegetable oils (corn, sunflower, canola), mayonnaise (commercial), hydrogenated oil (as an ingredient), "imitation" cheeses tempura, doughnuts (fried), deep-fried foods, chips, regular salad dressing, peanut butter (Jif, Skippy, etc.), roasted nuts/seeds, dairy substitutes (not including almond, coconut milk, etc.) Avoid acrylamides formed with high-temperature cooking – browning during grilling, baking, frying, or deep-frying will produce acrylamide	Zero (less than 5%)
Type of fat or oil		**Serving size estimate**
Hummus		1/3 cup
Nuts, seeds, ground flaxseeds, olives		1/4 cup
Nut butters (peanut butter, almond butter, cashew, coconut butter, etc.)		2 tablespoons
Oils (olive oil, almond oil, avocado oil, etc.)		1 tablespoon
Butter or ghee (clarified butter – Indian cuisine)		1 teaspoon
Eggs		2 each
Fatty ocean fish, meat, poultry		3–4 ounces
Avocado		1/2 avocado

Supplementary Information

Glossary – 1074

Index – 1077

© Springer Nature Switzerland AG 2020
D. Noland et al. (eds.), *Integrative and Functional Medical Nutrition Therapy*,
https://doi.org/10.1007/978-3-030-30730-1

Glossary

AMI Any mental illness – a mental, behavioral, or emotional disorder (excluding developmental and substance use disorders)

Behavioral health nutrition An umbrella term incorporating conditions with altered, mental, and intellectual states that may be influenced by nutrients or nutritional status.

BDNF gene The BDNF gene provides instructions for making a protein found in the brain and spinal cord called brain-derived neurotrophic factor. This protein promotes the survival of nerve cells (neurons) by playing a role in the growth, maturation (differentiation), and maintenance of these cells. In the brain, the BDNF protein is active at the connections between nerve cells (synapses), where cell-to-cell communication occurs. The synapses can change and adapt over time in response to experience, a characteristic called synaptic plasticity. The BDNF protein helps regulate synaptic plasticity, which is important for learning and memory. The BDNF protein is found in regions of the brain that control eating, drinking, and body weight; the protein likely contributes to the management of these functions [1].

Bioflavonoid A plant pigment responsible for the colors found in fruits and vegetables that has antioxidant properties beneficial to humans. Bioflavonoids include flavonols (quercetin), favanones (hesperetin), isoflavones (genistein), and anthocyanidin (cyanidin) [2].

Body composition An individual's body percentage of fat, bone, muscle, and intra/extracellular hydration. Chronic disease expression of altered body fat distribution is an important consideration in body composition and usually measured as waist circumference, waist-to-hip ratio, and waist-to-height ratio.

Chromatin The matter of chromosomes, includes DNA and all the factors physically associated with it: small chemical groups covalently attached to DNA (e.g., methyl groups), bound histone and nonhistone proteins, and associated RNA molecules [3].

Chromosomes Chromosomes are thread-like molecules that carry hereditary information for everything from height to eye color. They are made of protein and one molecule of DNA, which contains an organism's genetic instructions, passed down from parents [4].

CNS Central nervous system

Cytokine metabolism Cytokines are four families of mostly lymphoid cells: interleukins (ILs), tumor necrosis factors (TNFs), interferons (IFNs), colony stimulating factors (CSFs). Cytokines are produced in response to immunological challenges from nutritional deficiency, infection, genotype, toxic influences, stress, and/or physical injury. Chronic inflammatory cytokine production contributes to all chronic disease.

DHA Docosahexaenoic acid – 20 C; N-3 fatty acid; plays a role in membrane structure. DHA is present in large amounts in neuron membrane phospholipids, a structural component of the human brain, cerebral cortex, skin, sperm, testicles, and retina.

Eicosanoid Arachidonic acid (AA) and eicosapentaenoic acid (EPA) serve as precursors for the synthesis of eicosanoids (prostaglandins, prostacyclins, leukotrienes, and thromboxanes) which have an anti-inflammatory effect.

Energy dysfunction Mitochondrial energy metabolism, dietary intake and output, effect on body composition, physical activity, and influence of stress on blood glucose/insulin/hormonal energy relationships.

EPA Eicosapentaenoic acid – 22 C; N-3 fatty acid; plays a role in membrane function; Can be synthesized from alpha linolenic acid (ALA), although not efficiently (<5%)

Epigenetics Epigenetics is the study of heritable changes in gene expression (active versus inactive genes) that does not involve changes to the underlying DNA sequence – a change in phenotype without a change in genotype – which, in turn, affects how cells read the genes. Epigenetics affects how genes are read by cells and subsequently how they produce proteins [5]. Epigenetics is the reason why a skin cell looks different from a brain cell or a muscle cell. All three cells contain the same DNA, but their genes are expressed differently (turned "on" or "off"), which creates the different cell types [4].

Environmental influences that control gene activity within a cell without changing the gene structure [6]. Each gene has the capability of about 40,000 different expressions depending on the environmental messaging. Epigenetic effects can be passed on from generation to generation. The discovery of epigenetics has revealed a nongenetic mode of inheritance that influences our health. The most vulnerable periods in the lifespan for epigenetic changes are in utero/pregnancy, early childhood, teenage years, menopause (female), and andropause (male).

Epigenetics control genes Certain circumstances in life can cause genes to be silenced or expressed over time. In other words, they can be turned off (becoming dormant) or turned on (becoming active) [5]. The control of gene expression is the result of the *epigenetic marks* on the genetic material of that cell and termed the epigenetic signature of that cell type. The genome does not change; it is the expression that changes [7].

Epigenetic inheritance Occurs when phenotypic variations that do not stem from variations in DNA base sequences are transmitted to subsequent generations of cells or organisms [3]. The transmittance of information from one generation of an organism to the next (e.g., parent-child transmittance) that affects the traits of offspring without alteration of the primary structure of DNA (i.e., the sequence of nucleotides) [8].

Epigenome The total of all the epigenetic markings in a cell type [7]. The *epigenome* is a multitude of chemical compounds that can tell the *genome* what to do [9]. The epigenome consists of chemical compounds that modify, or mark, the genome in a way that tells it what to do, where to do it, and when to do it. Different cells have different epigenetic marks. These epigenetic marks, which are not part of the DNA itself, can be passed on from cell to cell as cells divide and from one generation to the next [9].

Eugenics The study of or belief in the possibility of improving the qualities of the human species or a human population, especially by such means as discouraging reproduction by persons having genetic defects or presumed to have inheritable undesirable traits (negative eugenics) or encouraging reproduction by persons presumed to have inheritable desirable traits (positive eugenics) [10].

GALT Gut-associated lymphatic tissue. Bacterial colonization of the gut is essential for proper development of the GALT. Polysaccharides produced by bacteria activate T-cell-dependent immune response, specifically CD4 T cells, B cells, immunoglobulins, balance of TH1, TH2, TH17 cells, etc.

Germline The cellular lineage of a sexually reproducing organism from which eggs and sperm are derived; the genetic material contained in this cellular lineage which can be passed to the next generation [11].

Glossary

Gluconeogenesis During low-carbohydrate intake or starvation, protein is used as de novo synthesis of glucose [12]. Gluconeogenesis is about 60% efficient in ATP production.

Glycolysis Anaerobic pathway to release glucose without the presence of oxygen; lactic acid is the end product [13]. Glycolysis is about 30% efficient in ATP production.

Gut-immune system It is comprised of GALT, MALT, TLR, microbes, PANTH cells, and organs of the digestive system which can be triggered by environmental factors such as stress, drugs, infection, LPS, etc.

Histones Proteins that DNA wraps around. (Without histones, DNA would be too long to fit inside cells.) If histones squeeze DNA tightly, the DNA cannot be "read" by the cell. Modifications that relax the histones can make the DNA accessible to proteins that "read" genes [4].

HUFA Highly unsaturated fatty acids – include N-3 and N-6 fatty acids, EPA and DHA

I-D-U Dietetic and integrative medicine nutritional assessment heuristic of an individual's nutrition status including intake, digestion and utilization of nutrients.

Ketosis To maintain metabolic homeostasis, ketone metabolism is the most energetically efficient metabolic fuel to synthesize glucose using fatty acids and ketone bodies under periods of prolonged food restriction or low-carb/high-fat dietary regimen.

Lack of sleep Underlying deregulation of neurotransmitters and hormones that help flip the on-off switch between waking and sleep. Lack of sleep quality and duration disturbs the 24-hour circadian rhythm and management of wake-time energy. Lack of sleep associations are allergies, central adiposity, diabetes, hypertension, heart disease, and other physical ailments [14].

Lifestyle and community Most chronic disease often has a social basis as well as a biological basis. An individual's social network can influence their health and alter gene expression [15]. An individual is more likely to be healthy if they are a part of a community that supports a healthy diet, lifestyle, and emotional environment.

Lifestyle medicine An approach to decreasing disease risk and illness burden by utilizing lifestyle interventions. The practice of medicine that extends beyond primary prevention, to modifying existing risk factors that cause disease, and rehabilitation from the disease and prevention of recurrence [16].

Long latency nutritional insufficiencies The presumption has been that if the intake of the nutrient is sufficient to prevent the expression of the index disease (i.e., iodine and goiter, vitamin D and rickets, etc.), then the nutrient intake is sufficient for the entire organism. But a nutrient deficiency could have long-term consequences and contribute to chronic disease, either by a similar mechanism to that inducing the index disease or by an entirely different mechanism [17].

MALT The mucosal-associated lymphoid tissue in the mucosal lining of the digestive lumen.

MAPDOM Within the "U" of the I-D-U assessment heuristic, the MAPDOM acronym represents the six most important nutritional categories determining an individual's health. MAPDOM stands for: minerals, antioxidants, proteins, D-fat-soluble vitamins (D, E, A, K), oils/fats, and methylation nutrition.

Methylation The addition of a methyl (CH_3) group from S-adenosylmethionine to a cytosine nucleotide or lysine or arginine residue [5]. The addition of a methyl group, or a "chemical cap," to part of the DNA molecule, which prevents certain genes from being expressed [4].

DNA hypomethylation refers to the loss of the methyl group in the 5-methyl cytosine nucleotide. Methylation is a natural modification of DNA and mainly affects the cytosine base (C) when it is followed by a guanosine (G) in mammals (methylation) [19].

Hypermethylation is an increase in the epigenetic methylation of cytosine and adenosine residues in DNA [20].

Methylation Denotes the addition of a methyl group to a substrate for the substation of an atom or group by a methyl group (-CH3). Can be involved in modification of heavy metals, regulation of gene expression, and regulation of protein function and RNA metabolism [18]. Primary nutrients involved as cofactors in methylation metabolism are B6, B12, folate, B2, zinc, and magnesium.

MET minute A metabolic equivalent of a task. It is a physiological measure expressing the energy cost of physical activities. One MET is the energy equivalent expended by an individual while seated at rest. The MET equivalent is the energy expended compared to rest, so MET values indicate the intensity. An activity with a MET value of 5 means five times the energy is being expended than would be at rest. MET minutes is the time engaged in an activity with consideration to the number of METs [21].

Microbiome The collection of microbes in our digestive system which includes bacteria, fungi, viruses, and parasites.

Nutraceutical *Derived* from *nutrition* and *pharmaceutical*, broadly a food or part of a food that provides medical or health benefits, including disease treatment and prevention.

Nutraceuticals range from specific nutrients to dietary supplements, herbal products, and processed foods and include beta-carotene, fish oil, garlic, green tea, oat bran, olive oil, and various herbs. Sometimes called functional foods. Also, "standardized pharmaceutical-grade nutrients, known as nutraceuticals"

Nutritional triage theory When the availability of a micronutrient is inadequate, metabolic resources will ensure that functions needed for short-term survival are protected at the expense of functions whose lack of a micronutrient only has long-term consequences [22].

Petrochemicals Derived from petroleum or natural gas (ethane, propane, butane, and other hydrocarbons), petrochemicals are used to manufacture artificial products including plastic, medicine, medical devices, cosmetics, furniture, cars, and computers [23]. Humans can ingest phytochemicals or it can be absorbed through the skin. When they accumulate in tissues and organs, they are responsible for brain, nerve, and liver damage, birth defects, cancer, asthma, hormonal disorders, and allergies.

Plasticity The adaptability of an organism to changes in its environment or differences between its various habitats [24].

Prebiotic A substrate that is selectively utilized by host microorganisms conferring a health benefit.

Probiotic A live microorganism that, when administered in adequate amounts, confers a health benefit on the host.

Secretory IgA Protective antibody which comprises 80% of the gut antibodies and is primarily in the gut lumen. It is the earliest of the antibodies to respond to an immune trigger.

SMI *Serious mental illness* – is defined as a mental disorder with serious functional impairment which substantially interferes with or limits one or more major life activities.

Suboptimal The negative effects of levels of consumption that lie above the minimum, but under the optimal level of consumption for these vitamins.

Telomeres Caps at the end of each strand of DNA that protect chromosomes, like the plastic tips at the end of shoelaces. Telomeres determine how cells age. Without the caps, telomeres become frayed until they can no longer do their job, DNA strands become damaged, and cells can't do their job [25].

TLR Toll-like receptor.

Toxic heavy metals Metals and metal compounds that have a negative effect on human health by building up in biological system [26]. They include arsenic, beryllium, cadmium, hexavalent chromium, lead, and mercury. Other toxic metals may include essential metals if imbalanced (zinc and copper), medicinal metals (platinum and bismuth), metals in emerging technology (indium and uranium), toxic metalloids (arsenic and antimony) and certain nonmetallic elements if imbalanced (selenium and iodine) [27].

Transgenerational epigenetic inheritance is the transmittance of information (epigenetic marks such as DNA methylation, histone modifications) from one generation of an organism to the next (e.g., parent-child transmittance) that affects the traits of offspring without alteration of the primary structure of DNA (i.e., the sequence of nucleotides) [28].

References

1. NIH US Library of Medicine. BDNF gene brain derived neurotrophic factor. Available at: https://ghr.nlm.nih.gov/gene/BDNF. Accessed 14 Dec 2016.
2. Lucock M. Molecular nutrition and genomics: nutrition and the ascent of humankind, vol. 84. Hoboken: Wiley; 2007.
3. Jablonka E, Raz G. Transgenerational epigenetic inheritance: prevalence, mechanisms and implications for the study of heredity and evolution. Q Rev Biol. 2009;84(2):131–76.
4. LiveScience. Epigenetics: definitions & examples. Available at: http://www.livescience.com/37703-epigenetics.html. Accessed 10 Dec 2016.
5. What is epigenetics? Epigenetics: Fundamentals. Available at: http://www.whatisepigenetics.com/fundamentals/. Accessed 10 Dec 2016.
6. Francis RC. Epi-genetics: the ultimate mystery of inheritance. New York: W.W. Norton & Company, Inc.; 2011.
7. Mahan KL, Raymond JL, editors. Krause's food and the nutrition care process. 14th ed. St. Louis: Elsevier; 2017.
8. Transgenerational Epigenetic Inheritance. Google Definition Search: Available at: https://www.google.com/search?q=transgenerational+epigenetic+inheritance&ie=utf-8&oe=utf-8. Accessed 13 Dec 2016.
9. National Human Genome Research Institute. Epigenome. Available at: https://www.genome.gov/27532724/epigenomics-fact-sheet/. Accessed 10 Dec 2016.
10. Dictionary.com: Eugenics. Available at: http://www.dictionary.com/browse/eugenics. Accessed 10 Dec 2016.
11. Merriam-Webster: Germ Line. Available at: https://www.merriam-webster.com/dictionary/germ%20line. Accessed 10 Dec 2016.
12. Gallagher ML. Macronutrients: carbohydrates. In: Krause's food and the nutrition care process. 13th ed. St. Louis: Elsevier Saunders; 2012. p. 51.
13. Dorfman L. Nutrition in exercise and sports performance. In: Krause's food and the nutrition care process. 13th ed. St. Louis: Elsevier Saunders; 2012. p. 508.
14. Lombardo GT, Ehrlich H. Sleep to save your life. New York: HarperCollins; 2005. p. 1–5.
15. Hyman M. Can social networks cure disease? Part I. Retrieved from: http://drhyman.com/blog/2012/04/20/can-social-networks-cure-disease-part-i/. Feb 2013.
16. Egger GJ, Binns AF, Rossner SR. The emergence of "lifestyle medicine" as a structured approach for management of chronic disease. Med J Aust. 2009;190(3):143–5.
17. Heaney RP. Long-latency deficiency disease: insights from calcium and vitamin D. Am J Clin Nutr. 2003;78:912.
18. Wikipedia. Methylation. Retrieved from: http://en.wikipedia.org/wiki/Methylation. June 2013.
19. Hypomethylation. Google Definition Search. Available at: https://www.google.com/search?q=hypomethylation+definition&ie=utf-8&oe=utf-8. Accessed 14 Dec 2016.
20. Wiktionary. Hypermethylation. Available at: https://en.wiktionary.org/wiki/hypermethylation. Accessed 14 Dec 2016.
21. The Cooper Institute. MET minutes: a simple common value to track exercise progress. Available at: http://www.cooperinstitute.org/2012/04/met-minutes-a-simple-common-value-to-track-exercise-progress. Accessed 30 Dec 2016.
22. McCann JC, Ames BN. Vitamin K, an example of triage theory: is micronutrient inadequacy linked to diseases of aging? Am J Clin Nutr. 2009;90:889.
23. American Fuel & Petrochemical Manufacturers (AFPM). Petrochemicals. Retrieved from: http://www.afpm.org/petrochemicals/. Aug 2013.
24. Plasticity. Google Definition Search. Available at: https://www.google.com/search?q=plasticity+definition&ie=utf-8&oe=utf-8. Accessed 10 Dec 2016.
25. Telomere Activation Sciences. What is a Telomere? Available at: https://www.tasciences.com/what-is-a-telomere/. Accessed 13 Dec 2016.
26. Occupational Safety & Health Administration, United States Department of Labor. Toxic metals. Retrieved from: https://www.osha.gov/SLTC/metalsheavy/. Aug 2013.
27. Costa LG. Toxic effects of pesticides. In: Casarett & Doull's essentials of toxicology. New York: McGraw-Hill; 2010. p. 324.
28. Transgenerational. Google Definition Search. Available at: https://www.google.com/search?q=transgenerational&ie=utf-8&oe=utf-8. Accessed 13 Dec 2016.

Index

A

Aarogyam mulam Uttamam 790
A1AT biochemical mechanisms 957
Abnormal vaginal flora (AVF) 546
Abnormal vaginal microbiota (AVM) 546
Academy of Nutrition and Dietetics (AND) 810, 870
Acceptable Macronutrient Distribution Range (AMDR) 179, 338, 355
Accreditation Council for Education in Nutrition and Dietetics (ACEND) 622
Acetaminophen 462
Acetoacetate 343
Acetone 343
Acetyl-11-keto-β-boswellic acid (AKBA) 460
Acetyl-L-carnitine (ALCAR) 476
Achara rasayana 792
Acid/alkaline balance 936
Acidosis 357, 358
Acoustic nerve (CN8) examination 660
Acquired immune deficiency syndrome (AIDS) 141
Acquired immunity 7
Acrodynia 653
Activation energy 93
Active Hexose Correlated Compound (AHCC) 423
Active range of motion 1011
Activities of Daily Living Questionnaire 701
Acute/adult respiratory distress syndrome (ARDS) 930
Acute illness/communicable disease 7
Acute immune response 125
Acute phase reactants 307
Adaptive immune system
- central tolerance 297, 298
- diagnostic features 295
- IgE and allergic reactions 294, 295
- macrophages and dendritic cells 292
- oral tolerance
  - allergies and autoimmunities 295
  - child at birth 295
  - control ingested antigens 295
  - dietary proteins 296
  - food immune reactivities 296, 297
  - IgA and IgM 295, 296
  - induction/disturbance 296
- peripheral tolerance 298, 299
- primary immune response 293
- secondary immune response 293, 294
Adaptogens 525
Adenosine diphosphate (ADP) 263
Adenosine triphosphate (ATP) production 598
Adenosylcobalamin 259
Adiponectin 520, 521
Adjuvant analgesics 462
Adrenal dysfunction
- adrenal bank account 524
- adrenal fatigue 525
- adrenal glands 524
- adrenal steroid hormones 524
- causes 525
- salivary cortisol test 525
- stages 524, 525
- three-legged stool approach
  - adaptogens 525
  - hormonal supplementation 526
  - nutraceuticals 525
  - nutrition and lifestyle 525
Adrenal fatigue 525
Adrenal glands 524
Adult Intake Questionnaire 700
Adult Treatment Panel III 519
Advanced Sleep Phase Syndrome (ASPS) 581
Adverse Childhood Events (ACE) 699
Aerobic exercise 598
Aerobic vaginitis 546
Age-related macular degeneration (AMD) 180, 181
Agni 786
Agouti viable yellow (Avy) gene 237
Ahara (diet) 786, 792
Air displacement plethysmography (ADP) method 331, 685
Air pollutants 943
Alcoholic ketoacidosis (AKA) 358
Alginate 383
Allergic asthma 945, 946
Allergic bronchopulmonary mycosis (ABPM) 946, 949, 956
Allodynia 449
Allostasis 737
Aloe 383
Aloe vera gel 387
Alpha lipoic acid (ALA) 181, 417, 453
Alpha-1 antitrypsin (A1AT) deficiency 933
- A1AT biochemical mechanisms 957
- liver 956
- lungs 956
- skin 956
Alpha-linolenic acid (ALA) 158
Alpha-lipoic acid (ALA) 476, 938
ALS Functional Rating Scale-Revised (ALSFRS-R) 872
Alternative Healthy Eating Index 126
Aluminum 496
Alveolar macrophages 930
Alzheimer's disease (AD) 166–167, 240, 317, 347, 399, 400, 809
Ama 787
American Diabetes foundation 455
American Heart Association (AHA) 809
American Psychiatric Association Committee on Research on Psychiatric Treatments 478
Amiloride binding protein 1 (ABP1/AOC1) 258
Amino acids 479, 877
Amyloid precursor protein (APP) 240
Amyotrophic lateral sclerosis (ALS) 330, 929
- adrenal 892
- age of onset 867
- altered metabolism and hypermetabolism 867–868
- amino acids 877
- antibodies 892
- autoantibodies 892
- bioelectrical impedance analysis 872
- biological heterogeneity 867
- blood-brain barrier 868–870
- bulbar ALS 867
- butyrate 889
- carbohydrate 872, 873
- cause of death 867
- cholesterol 881
- classifications 867, 868
- clinical dietetics and nutrition care 870, 871
- commercial enteral formulas 883, 884
- complementary and alternative (CAM) medicine 871
- conventional dietary 871
- dairy 876
- Deanna Protocol 882
- demyelination 881
- diagnosis 866, 867
- dietary fat 879–882
- dietary self-care choices 866
- DNA methylation 892
- energy expenditure 872
- environmental toxins 884
- erythrocyte fatty acid analysis 892
- essential metals
  - beta-methylamino-L-alanine 887
  - copper 887
  - iron dyshomeostasis 887
- familial 867
- fats 872, 873, 878
- gastrointestinal (GI) system 887
- gluten 874
- gut-brain interactions 888
- healthy enteral feedings 884
- heavy metals 885, 886
- human mutant transgenic SOD-1 G93A mouse model 867
- IFMNT practitioners 866, 870
- impaired glycolysis 891
- indirect calorimetry measurements 872
- infections 892, 893
- integrative and functional medical nutrition therapy 871
- ketogenic diet 872, 882
- laboratory biomarkers 890, 891
- limitations 870
- liver dysfunction 892
- loss of muscle 872
- medium-chain triglycerides 880, 882
- metabolomics 891
- microbiome 888
- microRNA-155 870
- multisystems disorder 867
- multisystems metabolic pathways 867
- natural chelators 886
- neurodegenerative pathways 867
- nourishment 870
- nutrient-dense diet survey 866
- nutrition management algorithm 871
- nutritional assessment 872
- nutritional interventions 866
- omega-3 PUFAs 879

Amyotrophic lateral sclerosis (ALS) (cont.)
- omega 6 PUFAs 880
- oxidative phosphorylation 891
- PALS advance nutrition research 866
- phospholipids 882
- phytonutrients 884, 885
- probiotics 889
- protein 872, 875
- reductionist versus systems biology approach 893, 899
- Riluzole 867
- risk factor 871, 872
- saturated fatty acids 880
- small intestinal bacterial overgrowth 889–890
- soy 876
- sporadic 867
- sugar
  - accumulation of 873
  - dopamine 873
  - hormonal imbalances (glucose and insulin) 874
  - inflammation and autoimmunity 873
  - neuroprotective and antioxidant properties 873
  - serotonin deficiency 873
- supplements 893–898
- TCA cycle 891
- thyroid function 892
- upper and lower motor neurons 867
- urinary organic acid analysis 891
Analgesics 462
Androgens 524
Anemia of chronic disease (ACD) 933
Angiogenesis 440
Angiotensin converting enzyme (ACE) 238
Angiotensin-converting enzyme (ACE) inhibitors 208
Angiotensin II receptor blockers (ARBs) 208
Angiotensinogen gene (AGT) 238
Anorexia nervosa 750
Anthropometrics 680–691
Anti-endomysium-IgA antibodies (EMA) 244
Anti-fungal diet 1046
Antigen-presenting cells (APCs) 288
Anti-inflammatory activities 141
Anti-inflammatory agents 728
Anti-inflammatory diet (AID) 410, 454
Antimicrobial resistance (AMR) 306
Antioxidants 179
Anupana 787
Anxiety 242
Anxious ADD 506
Any mental illness (AMI) 474
- See Also Psychiatric disorders
Apgar Score 512
Apolipoprotein A2 (APOA2) gene 238
Apolipoprotein E (APOε) 238
Apoptosis 178
Approximate Macronutrient Distribution Ranges (AMDR) 49
Arachidonic acid (AA) 115, 116, 142, 393, 544, 729, 734, 1036
Arachidonic/di-homo-gamma-linolenic acid ratio (AA/DGLA ratio) 160
Arbuda 796
Arizona Center for Integrative Medicine 625
Arm muscle area (AMA) 684
Aromatherapy 589

Aromatic L-amino acid decarboxylase (AADC) 244, 277
Arsenic 541, 943
Artemisinin 440
Arthritis 448, 454, 456, 458, 460
Aryl hydrocarbon receptor (AhR) 126, 299, 542
Ascorbic acid 299
Ascorbyl palmitate vitamin C 504
Ashwagandha 756, 763
Asperger syndrome 474
Aspirin-exacerbated respiratory disease (AERD) 949
Assessment-based interventions 120
Assisted reproductive technologies (ART) 534
Association of Ayurvedic Professionals of North America (AAPNA) 785
Asterixis 381
Asthma
- allergic asthma 945, 946
- allergic bronchopulmonary mycosis 946, 949
- aspirin-exacerbated respiratory disease 949
- chronic disease 929
- cross country skiing-induced asthma 946
- definition 944
- dietary patterns 947
- food allergy 947, 948
- genotypic characteristics 947
- histamine intolerance 946
- IFMNT assessment 950–952
- IFMNT case study 950–955
  - adolescence 953
  - anthropometrics 955
  - biochemical 954
  - birth 953
  - body composition 954
  - childhood 953
  - clinical 954
  - college 953
  - digestion, assimilation and elimination 954
  - energy 954
  - fats and oils questionnaire 953
  - follow-up 955
  - genotypic risks 954
  - inflammation 954
  - lifestyle 954
  - nutrient cofactors 947
  - nutrition 953
  - nutrition Dx 955
  - nutrition physical exam 954
  - outcome 955
  - past medical history 952–953
  - past surgical history 953
  - pharmaceutical use 953
  - plan 955
  - signs/symptoms/medical symptom questionnaire (MSQ) 954
  - sleep and stress 955
  - social 953
  - structural 954
  - symptoms/complaints at first visit 952
  - toxin Load 954–955
  - utilization, cellular & molecular 954
  - young adulthood 953
- lifestyle influences 949
- nonallergic asthma 945, 946
- oxidative stress 948
- pathophysiology 944, 945
- primary conventional therapy 949

- severe late-onset hypereosinophilic asthma 946
- suspected triggers/risk factors 944, 945
- symptoms 944
- toxic exposure 948
- traffic density and asthma exacerbations 949
Atkins diet 354
Atopic dermatitis 990
ATSDR 2013 Substance Priority List 69
Attention deficit hyperactivity disorder (ADHD) 144
- anxious 506
- classic 505
- comprehensive fatty acids test 505
- comprehensive stool analysis 505
- definition 505
- dietary and lifestyle strategies 506, 507
- gene testing 505
- gluten/casein peptides test 505
- IgE/G food allergy test 505
- inattentive 505
- lab testing 505
- limbic 506
- metals hair test or toxin testing 505
- organic acid testing 505
- over-focused 505
- ring of fire 506
- temporal lobe 506
- treatment 505
Autism 400
Autistic spectrum disorder (ASD) 474
- allergy 501
- associated conditions 498
- autoimmune testing 501
- biomedical assessment for 500
- birth history 504
- blood chemistry 500
- DHFR 502
- diabetes mellitus 499
- diet history 504
- DSM-V criteria 498
- environmental factors 499
- family history 501, 504
- folate and methylation genes 502
- FUT2 502
- gastrointestinal symptoms 499, 500
- genes affecting detoxification 502
- genetic screening 500, 501
- gluten/casein peptides test 504
- HLA genes 502
- hormones 501
- immune function and inflammation markers 500
- incidence 498
- MAO B 504
- maternal immune activation 499
- MCM6 502
- mitochondrial dysfunction 501
- mitochondrial issues in 500
- MTHFR 502
- neonatal intensive care unit 499
- neurotransmitter genes 502
- niacin and CoQ10 504
- nutritional support 503
- oxidative stress 501
- past medical history 504
- physical examination 504
- polymorphisms in 501
- prenatal factors 499

A–B

Autistic spectrum disorder (ASD) (cont.)
- prenatal history 504
- pyroluria test 501
- RFC polymorphisms 502
- severity assessment 498
- thyroid hormones 499
- toxins 501
- transcobalamin II gene 502
- transsulfuration pathway genes 502
- treatment 504
- urine for organic acid testing 500
- urine for peptide testing 500

Autoimmune microenvironment (AIM) 441
Autoimmune Paleo (AIP) diet 426, 427
Autoimmunity 167
Autoimmunity-triggered dietary antigens 993, 994
Autoinflammatory diseases, development 292
Autointoxication 380
Autonomic dysfunction 523
Autonomic nervous system (ANS) 738
Autophagy mechanism 933
Ayurveda 744
Ayurvedic approach
- Agni 786
- Ahara Rasa 786
- for cancer 796
- detox treatment 792
- for diabetes 794
- dietary and lifestyle recommendations 794–796
- Dinacharya 790
- disease management 792
- dosha (See Dosha)
- food is medicine
  - black pepper/piper nigrum 789
  - fenugreek 789
  - ghee 787
  - ginger/shunti 789
  - honey/madhu 789, 790
  - mint/pudina 789
  - tulsi 788, 789
  - turmeric/curcuma longa 788
- in India 784–785
- lifestyle 790
- like increases like principle 787
- in modern practice 796–797
- nutrition and principles 786–787
- for obesity 792, 793
- origin and history 784
- Panchakarma 792
- Ritucharya 790
- Sadvritta 790
- six tastes 786, 787
- Trayopasthambas 786
- in USA 784–785

B

Bacillus subtilis 393
Bacterial vaginosis (BV) 546
Baicalin 440
Barrett's esophagus 423
Barriers
- biological structure 185
- evaluation/assessment 185
- pathophysiology 185
- prevention/treatment 186

Basal metabolic rate analyzers 330
Baseline nutrient status 101
Basic fibroblast growth factors (bFGFs) 438
Basti 792
Bastyr (Robert's) Formula B 387
Bastyr Center for Natural Health (BCNH) 776, 777
Beau's lines 652, 687
Beck Depression Inventory 485
Behavioral competencies 76, 78
Behavioral evaluation report 79
Behavioral profiling tools 79
Behavioral styles 75
Bentonite clay 504
Beta 2 adrenergic receptor ADRB2 463
Beta-agonists 949
Beta carotene 983
Beta-hydroxybutyrate (βHB) 343
Betaine hydrochloride 383
Beta-methylamino-L-alanine (BMAA) 887
Bhastrika Pranayama 795
BIA instruments 158
Bifidobacterium sp. 230, 397, 402, 462
- B. breve 393
- B. infantis 398
- B. lactis 545
- B. longum 393, 397, 545
Bile 228, 229
Bile canaliculus 228
Biochemical individuality 101, 327
Biochemical individuality and genetic/epigenetic influences on chronic disease 127
Biochemical mechanism
- activation energy 93
- active site 93
- B-complex and metabolic briefing 94, 95
- B-complex and metabolic correction 95
- cofactors 93
- enzyme-catalyzed reactions 93
- gene mutation 93
- Law of Mass Action 94
- physiological impairments 94
- Principle of Le Chatelier 94
- substrate 93
Bioelectrical impedance analysis (BIA) 128, 327, 328, 685, 872
- body capacitance 328
- body cell mass 328
- total body water 329
Bioenergetic deficits 868
Biofeedback 615
Bioflavonoid 1022
Bio-identical hormone replacement 528
Bioimpedance spectroscopy (BIS) 327
Biological heterogeneity 867
Bioperine (black pepper) 460
BioPsychoSocial (BPS) model 615
Biopsychosocial conceptual model 916, 917
Biotin (vitamin H) 984
Biotinidase deficiency 508
Biotransformation 226–228
- dietary composition and bioactive constituents 194
- factors 194
- mechanisms 193
- multiple chemical sensitivity 194
- phases 193
Bipolar disorder (BD) 396, 475, 476
Bisphenol A (BPA) 420, 421

Black tea 539
Blood-brain barrier (BBB) 182, 183, 399, 868–870
Blood cell palmitoleic acid (MUFA16:1) 881
Blood-cerebrospinal fluid barrier (BCSFB) 868
Blood-spinal cord barrier (BSCB) 868
Blue nails 654
Body cell mass (BCM) 328
Body composition 1022
- ACSM age-adjusted body fat classifications 333
- ACSM and ACE body fat classifications 332
- analysis 128, 683
- basal metabolic rate analyzers 330
- bioelectrical impedance analysis
  - body capacitance 328
  - phase angle 328
  - total body water 329
- body mass index 325
- body weight 325
- extra-, intracellular and total body water 329
- fat and fat-free compartments 324
- field methods 324
- girth measurements
  - bioelectrical impedance analysis 327, 328
  - bioimpedance spectroscopy 327
  - body fat percentage 326
  - skinfold measurement 327
  - waist circumference 326
  - waist-to-height ratio 326
- height 325
- laboratory methods
  - air displacement plethysmography method 331
  - computed tomography 332
  - dual-energy X-ray absorptiometry 331
  - hydrostatic weighing 331
  - magnetic resonance imaging 332
  - ultrasound technology 332
- weight 325
Body Ecology Diet 503
Body mass index (BMI) 325
Body's inherent detoxification pathways 208
Boswellia tree 460
Boswellin 387
Brahmacharya (self-control) 786
Brain
- biological structure 182
- blood-brain barrier 182
- cerebral spinal fluid 183
- evaluation/assessment 184
- glymphatic circulatory system 183
- pathophysiology 183, 184
- treatment 184, 185
Brain cells 182
Brain-derived neurotrophic factor (BDNF) 396, 398–400, 477, 874, 1007
Brain maturation delay theory 505
Branched chain amino acids (BCAA) 65, 66
Breastfeeding infant nutrition 34
Bredesen Protocol™ 348
Bristol Stool Form Scale (BSFS) 918
Brittle nails 653
Brown/gray nails 654
Brown–Vialetto–Van Laere Syndrome 2 (BVVLS2) 479
Bulbar amyotrophic lateral sclerosis 867
Butterbur 461
Butyrate 889

B vitamin deficiencies 459
B vitamins 104–107, 935, 936
B12 deficiency 443, 986

C

CACNA1C 474
Cadmium 486, 943
Caffeine 460
Caffeine intoxication 477
Calcidiol 260
Calcitriol 260
Calorie restriction 350
Cancer-related pain 451
Cancer stem cells (CSC) 413
Cannabinoid receptor type 1 (CB1-R) 535
Cannabinoids 461
Cannabis 461
Capillary refill test 655
Carbohydrate diet 746
Carbohydrates 177
Carcinoembryonic antigen (CEA) 348
Carcinogenesis-associated fibroblasts (CAFs) 438
Cardiac diet 745
Cardiometabolic syndrome 601, 602
– acute phase reactants 815
  – bioelectric impedance analysis 816
  – biopsy 816
  – blood labs 816
  – CT/PET CT scan/MRI/ultrasound 816
  – dental/oral health 817
  – diets 816
  – dyslipidemia 816
  – glucose/insulin 816
  – gonadal health 817
  – gut health 817
  – happiness 817
  – hypercoagulation 816
  – immune compromised 816
  – lifestyle 816
  – loneliness 817
  – nutrient insufficiencies, deficiencies/excesses 817
  – obesity 816
  – physical activity 817
  – physical examination 817
  – posture and spinal alignment 817
  – renal health 817
  – sleep 817
  – toxin load 817
  – visceral adipose tissue 816
– adipocytes 813, 814
– allostasis 804, 805
– American Heart Association 802
– anthropometry 815
– APOa2 genotypes 238
– apolipoprotein E 238
– and autoimmune disorders 809
– body composition measurements 803
– body mass index height/weight 815
– characterization 238, 802
– and chronic stress 811
– and comorbidities 809
– conventional treatment 817
– definition 813
– diagnostic criteria 813

– dietary considerations 809, 810
– eNOS/NOS3 239
– evidence-based nutritional and lifestyle factors 803
– fasting 239
– FTO 239, 240
– hemochromatosis 240
– homeostasis 804
– hypercoagulability 811
– hypertension 238, 239
– inflammation 804, 806
– inflammatory markers 807
– inflammatory mediators 809
– insulin resistance 811, 812
– integrative and functional medicine approaches 818
– iron overload 811
– metabolic disorder comorbidities 808
– metabolic risk factors 802, 803
– microbiome 810, 811
– oxidative stress 807
– pathophysiology 804
– PCOS 240
– PON1 239
– proinflammatory state 815
– and psychiatric disorders 809
– psychosocial determinants of health 804, 805
– and rheumatic diseases 809
– sarcopenia 812, 813
– short physical performance battery (SPPB) 815
– type 2 diabetes 239
Cardiovascular disease (CVD)
– cardiovascular genetics 831, 832
– causes 827
– cell membrane dysfunction 825
– cell membrane physiology 825
– CHD risk factors 831
– clinical interventions 828
– comprehensive lab testing 832
– computed tomography angiogram (CTA) 828
– congestive heart failure (CHF) 824
– coronary heart disease 824
– DASH 1 and 2 diets 833
– dyslipidemia 824, 830, 831
– EndoPAT 831
– endothelial cells 827
– endothelial dysfunction 825, 827
– endothelium 825
– finite responses 825
– hyperinsulinemia 831
– hypertension 824, 825, 828, 830
– LDL-cholesterol 825, 826
– Mediterranean diet 832–833
– metabolic memory 827
– mitogen-activated protein kinase (MAPK) pathway 831
– noninvasive vascular testing 831
– nutraceuticals and dietary factors 828
– obesity 831
– oxidative stress/inflammatory responses 825
– phosphatidylinositol 3-kinase (PI3K) pathway 831
– polygenetic codes 828
– Portfolio diet 833
– post-prandial phenomena 828
– prevention 824
– risk factors 824–825

– symptom 828
– translational vascular medicine 825
– treatment 824
– tsunami effect 825
– vascular immune dysfunction 825
– vessel changes 828
Cardiovascular/cognitive decline 716
Carnitine 503
Carotenoids 936
Carrageenan 386
Casomorphin peptides 504
Casomorphins 384
Catecholamines 275, 276
Catechol-O-methyltransferase (COMT) 109, 241, 275, 276, 463, 502
Cathelicidin 971
CBD hemp oil 461
CD8+ T cell response 438
CDC research statistics 32
Celiac disease (CD) 110, 244, 454
Celiac gene 394
Cell membrane damage 177
Cell membrane permeability 158
Cellular hydration 157
Cellular interventions 716
Cellular pathology 7
Centers for Disease Control and Prevention 451
Central hepatic venule 223
Central nervous system (CNS) 395, 399, 449, 477, 482
Central sensitization (CS) 449
Ceramides 970
Certified Nutrition Specialist (CNS) 625
Ceruloplasmin 113
Cervical cancer 444
Chamomile 385
CHD8 gene 501
Cheilitis 971
Chelation therapy 886
Chemical Abstracts Service (CAS) registry system 192
Childhood disintegrative disorder 474
Childhood wellness 36
Chinese restaurant syndrome 876
Chlamydia trachomatis 535, 546
Cholesterol 158, 176, 478, 479, 881
Cholesteryl ester transfer protein (CETP) 264
Chondrocytes 441
Chondrosarcoma 440
Chronic disease
– environmental toxins 570–571
– food insecurity 570
– history 564
– legislative policy
  – arsenic exposure 571
  – asbestos 571
  – lead 571
– primary origin
  – early life experiences 565
  – environmental factors 564
  – epigenetic 564, 565
  – genetic predisposition 564
  – preconception and intrauterine 565
  – social disadvantage 564
– secondary origin
  – alcohol 566
  – dietary patterns 568

Index

- physical activity 568
- sleep 566
- smoking 566
- social determinants 567, 568
- stress 565
- tertiary origins 571–574
- types of 564
- years lived with disability (YLD) 564

Chronic dysfunction
- balance reactions 1006
- bedrest 1008–1010
- dominant and visual perceptual field changes 1006
- fascial restrictions and adhesions 1005
- fascial sheaths 1005
- global muscles 1005
- individual's fitness level and lifestyle 1006
- muscle imbalances 1005
- postural asymmetries 1005
- postural changes 1005
- posture imbalances 1005
- rehabilitation 1010, 1011

Chronic idiopathic constipation 102
Chronic inflammation 1007
Chronic iodine ingestion 104
Chronic myelomonocytic leukemia (CMML) 420
Chronic noncommunicable disease 8
Chronic pain
- anti-inflammatory diet 454
- Boswellia tree 460
- butterbur 461
- B vitamin deficiencies 459
- cancer-related pain 451
- cannabinoids 461
- central sensitization 449
- definition 448
- <u>depression</u> 451
- elimination diet 454
- epigenetics 463, 464
- essential nutrients 459
- feverfew 461
- food sensitivities 453
- functional labs to assess nutritional status 454
- genetics 463
- ginger (*Zingiber officinale*) 460
- gut microbiome 452, 453
- hydration 460
- inflammation 452
- language of pain 464, 465
- lifestyle 464
- magnesium 458, 459
- medical treatment
  - non-opioid/adjuvant analgesics 462
  - <u>non-opioids</u> 462
  - opioids 462, 463
- mediterranean diet 454, 455
- neuropathic pain 449
- nociceptive pain 449
- nutrients and supplementation in 456
- nutrition and 451
- nutritional assessment 453, 454
- obesity and 452
- omega-3 essential fatty acids 456, 458
- omega 6 fatty acids 458
- oral facial pain 453
- oxalate-containing foods 456
- paleo diet 455
- prevalence 448

- psychological pain 449
- psychosocial mediators 449, 450
- <u>sleep</u> 451
- therapy 448
- turmeric (*Curcuma longa*) 460
- vegan, vegetarian diet 455, 456
- vitamin D/vitamin A 458
- willow bark 461

Chronic/noncommunicable disease 7
Chronic psychological stress 395
Chronobiology 579
Chvostek's sign 657
Ciliary function 931
Circadian rhythm 165
- bright light exposure 579
- central and peripheral circadian clocks 580
- chrononutrition 579
- chronotypes and personalized nutrition 590
- circadian rhythm sleep disorders (CRSDs) (*See* Circadian rhythm sleep disorders (CRSDs))
- definition 579
- functional genomics 579
- gold-standard measures 579
- lifestyle influences 589, 590
- melatonin therapy 579
- metabolism and nutrition status 579
- pathophysiology-mechanism
  - cancer 586, 587
  - cardiovascular issues 587
  - diabetes 587
  - endocrine system 588
  - immune system 588
  - metabolism and obesity 586
  - renal function 588
  - reproductive system 588
  - sleep hygiene 589
- risk factors 579
- sleep-wake cycle 581
- suprachiasmatic nucleus (SCN) 579
- toxin-related influences 589
- zeitgebers 579

Circadian rhythm shifts 131
Circadian rhythm sleep disorders (CRSDs)
- advanced sleep phase syndrome (ASPS) 581
- circadian clock genes 585, 586
- delayed sleep phase syndrome (DSPS) 581
- irregular sleep-wake pattern 581
- non-24 hour sleep pattern 581
- sleep apnea 581
- treatment
  - circadian clock system 583
  - feeding/fasting cycle 583, 584
  - light, melatonin, and vitamin B12 581, 582
  - metabolism and clock coupling 583
  - nutrition 582
  - time-of-day restricted feeding 584, 585

cis and *trans* isomerism 136
Classic attention deficit hyperactivity disorder 505
Clay play 504
Cleveland Clinic's Center for Integrative & Lifestyle Medicine 622
Clinical medicine 7
Clopidogrel 264
Clostridium difficile infection (CDI) 392, 397, 401, 402
Clostridium perfringens 392
Clubbing 686

Cobalamin 259
Coconut oil 509
Codeine 462
Coenzyme Q10 (CoQ10) 128, 179, 476, 942
Coffee 460
Cognitive behavioral therapy 615, 616
COLEVA project 706
Colitis 385
Collaborative and participatory approaches 14
Collagen 438, 439, 442
Colonic inertia 916
Colonic motility 131
Comfrey 387
Commitment language 1053
Common lymphoid progenitor (CLP) 289
Communication skills 78
Community of wellness 36, 37
Complementary and alternative (CAM) medicine 871
Complementary feeding 42
Complete blood count (CBC) 443
Complex active transporter (pump) system 193
Comprehensive elimination diet 1046
Compressed eating window (CEW) 751
Congenital urinary disorder 77
Congestive heart failure (CHF) 838
- treatment 838
Conjugated linoleic acid (CLA) 116, 117, 137, 147, 387
Conscientious behavior (C) 75
Consortium of Southeastern Hypertension Centers (COSEHC) 825
Consumer-centric healthcare delivery system 5
Contact dermatitis 991
Contact urticaria 991
Continuous positive airway pressure (CPAP) 716
Conventional assessments 949–950
Cooking oils 154
Copper 483, 887
Copper-to-zinc ratio, in cancer 114
Copper-zinc superoxide dismutase (Cu-Zn SOD1) 867
Cordyceps 543
Coronary heart disease (CHD) 838
- clinical presentation 834, 835
- physical examination and laboratory 834, 835
- results 835
- treatment 835, 838
Cortical thymic epithelial cell (cTEC) 298
Corticosterone 398
Cortisol 539
Cortisol-releasing hormone (CRH) 399
Council for Ayurvedic Counseling (CAC) 785
Covalently-linked carbon atoms bearing hydrogen atoms 136
C-reactive protein (CRP) 452, 454, 723
C-reactive protein-high sensitivity (CRP-hs) 314, 315
Creatine monohydrate (CM) 476
Crohn's disease 258
Cross country skiing-induced asthma(CCSA) 946
Crunch test 606
C-terminal glycine-rich domain 874
Curcumin 387, 440–442, 460, 937
Current conventional care system models 5
Current education and treatment systems 8
Current healthcare paradigm 81
Current United States health system "dysfunctional" 4

CXC motif cytokines (CXCL12) 439, 440
Cyanocobalamin 259
Cyclooxygenase 456
Cyclooxygenase 2 (COX-2) 439, 545
Cystathionine beta synthase (CBS) 271, 502
Cystathionine beta synthase (CBS) gene 261
Cytochrome C oxidase (COX) 263
Cytochrome enzymes 263
Cytochrome P450 enzymes (CYP 450) 193, 228
Cytochrome P450 genes 502
Cytokine metabolism 1022
Cytokine production 944

D

Dairy 876
Dandelion root extract (DRE) 420
Data extraction tool (DET) 870
D-beta-hydroxybutyrate (D-βHB) 343
Deanna Protocol (DP) 882
Deglycyrrhizinated licorice (DLC) 383
Delayed sleep phase syndrome (DSPS) 581
Delirium 381
Delta-5-desaturase (D5D) 159, 160
Delta-9-tetrahydrocannabinol (THC) 461, 507
Demyelination 881
Denaturing gradient gel electrophoresis (DGGE) 258
Dendritic cells (DCs) 288, 931
Dental assessment
– biochemical individuality 1022
– energy dysfunction 1022
– lifestyle 1022
– nutrient Insufficiencies 1022
– sleep and stress 1022
– toxic load 1022
– websites 1022
Depression 243, 244, 475
– dopamine deficiency 243, 244
– enzyme deficiency 244
– serotonin deficiency 243
Depression Anxiety Stress Scale (DASS) 706
Dermatitis herpetiformis 990
Dermis 970, 971
Desaturase enzymes delta-6-desaturase (D6D) 159, 160
Desaturation process 139
Desmodium styracifolium 536
Detoxification 410
– alcoholic binge 206
– definition 206
– exhaled breath 206
– kidney's filtering system 206
– phases 206
– primary approach 206, 207
– rich blood flow 206
– secondary approach
  – kidney 208
  – liver 207, 208
– tertiary approach
  – alcohol 208
  – artificial sweeteners 208
  – body's inherent detoxification pathways 208
  – corn sweeteners 208
  – dairy consumption 209
  – fiber 208

– fruit 208
– gluten 209
– healthy fats 209
– stevia 208
– sugar 208, 209
– supplementation use 209–210
– vegetables 208
– wheat 209
Detractors 453
Detroit Water and Sewer Department (DWSD) 731
Dexamethasone 538
Diabetes mellitus type 2 519
Diabetes Prevention Program Research Group 2002 study 522
Diabetic ketoacidosis 748
Diagnostic and Statistical Manual of Mental Disorders, 5th Edition (DSM-V) 240
Diamine oxidase (DAO) 258, 259
Diaphragmatic breathing 1008, 1009
Dietary and lifestyle influences
– antioxidants 179
– coenzyme Q10 (CoQ10) 179
– L-carnitine 180
– lipoic acid 179
– mitochondria
  – cardiolipin 178
  – evaluation/assessment 179
  – inner and the outer mitochondrial membranes 178
  – mitochondrial dysfunction 178
  – oxidative phosphorylation 178
  – powerhouse of cell 178
  – prevention/treatment 179
– mitochondrial biogenesis 180
– pathophysiology 177
– phosphotidylethanolamine 180
– reactive oxygen species 177
Dietary Approaches to Stop Hypertension (DASH) diet 717, 778, 833
Dietary fat
– history 153–154
– human history 153–154
Dietary fiber 102
Dietary flavonoids 440
Dietary intervention 920, 921
Dietary Reference Intakes (DRIs) 338
Dietary Supplement Health and Education Act of 1994 (DSHEA) 734, 756, 757
Dietary supplements
– Academy of Nutrition and Dietetics 734
– additives 759
– and pathways 764
– botanical therapeutics 758, 759
– dietary prebiotics 763
– dietary supplement labels 758
– DMAA 734
– drug-nutrient interactions 735
– DSHEA 734
– FDA 734
– filler dietary supplements 763
– genetic factors 735
– GMPs 734
– Government regulations 759, 760
– Government, regulatory and association resources 762
– green tea 763

– health outcomes and clinical benefits 762
– maintenance 735
– manufacturers 761, 762
– need for nutrients 735
– nomenclature and formulations 756, 757
– nutraceutical marketplace 756
– nutritional gaps 735
– nutritional supplements 762
– patient-directed purchasing and expanding markets 756
– peer reviewed sources 763
– post-market surveillance 734
– quality control and safety 760, 761
– rationale 734
– risk reduction 734
– roles and responsibilities 735
– selection and education 760
– sexual enhancement 763
– sports /fitness 763
– top-selling herbal supplements 756
– turmeric/Curcuma longa 763
– US Pharmacopeia and NSF International 734
– vitamin, mineral and botanical ingredients 758
– weight loss 763
Diet diary instructions
– activities 1060
– amount of food 1060
– date and time 1060
– food/beverage consumed 1060
– location 1060
– mood 1060
– symptoms 1060
Dietetic and integrative medicine nutritional assessment 1037
Dietitians in Integrative and Functional Medicine (DIFM) 625
Diet ratio 354, 355
Diet supplementation
– antioxidants 353
– digestive aids 353
– gut health 353
– herbs and botanicals 353
– minerals 353
– vitamins 353
Digestive disease 915
Digestive enzymes 385
Diglycerides 137
Dihomo-gamma-linolenic acid (DGLA) 458
Dihydrofolate reductase (DHFR) 271, 499, 502
Dimethylamylamine (DMAA) 734
Dinacharya 790–792
Dioxins 542
Diphtheroids 544
DISC paradigm 75
D-lactate 271
DNA methylation 892
DNA methyltransferases (DNMTs) 237
Docosahexaenoic acid (DHA) 393, 474–476, 478
Dominant behavior (D) 75
Dominant payer system 4
Dopamine 397, 873
Dopamine deficiency 243
Dopamine-hydroxylase (DBH) activity 483
Doshas
– Kapha 785
– Pitta 785
– Sama Prakruti 785

Index

- Vata 785
Drug-nutrient interactions
- albumin 214
- altered body composition 214
- food and nutrient kinetics 214–218
- hypoalbuminemia 214
- nutrition professional 215–218
- nutrition status 215
Dual-energy X-ray absorptiometry (DEXA) 128, 331, 685
Dutch Hunger Winter Study 237, 238
Dutch Winter Famine study 565
Dysbiosis 392–394, 397, 503, 932
- amino acid metabolism 258
- celiac disease 244
- Crohn's disease 258, 259
- FUT2 polymorphisms 258
- histamine degradation 258
- microbiota 244, 258
- MUC2 258
- NCGS 244
Dysbiotic gastrointestinal tract bacteria 993
Dysfunctional medical system 8
Dysfunctional system 6
Dyslipidemia
- causes 836
- clinical history 834
- noninvasive vascular testing 836
- physical examination and laboratory 834
- supplements and dose 837
- treatment 834, 837
Dysregulated collagen homeostasis 931
Dysthymia 477

E

Educating practitioners 74
Effective communication 79
E4 Method 80, 81
Eicosanoid cascade 158
- alpha-linolenic acid 158
- 20-carbon eicosanoids 158
- direct source 159
- fatty acid desaturation 159–160
- fatty acid elongation 159
- IFMNT practitioner 158
- linoleic acid 158
- omega-6 and omega-3 fatty acids 158
- specialized pro-resolving mediators 158
Eicosanoids 141–143
Eicosapentaenoic acid (EPA) 474–476, 478, 1036
Electrical impedance myography (EIM) 330
Electromagnetic frequencies (EMF) 129, 416
Electron transport chain (ETC) 537
Electronic cigarettes (E-cigs) 944
Electronic health record (EHR) technology 4
Electronic Nutrition Care Process Terminology (eNCPT) 772
Eleutherococcus senticosus 525
Elimination diet 454
Elimination/exclusion diets
- adverse food reactions 745
- autoimmune diseases 745
- food intolerances 745
- food sensitivities 745
- implementation 745
- low FODMAP diet 745

- non-celiac gluten sensitivity 745
- Paleo Diet 745
- risk factors 746
Emodi 440
Emotional freedom technique (EFT) 616
Endocannabinoid system (ECS) 535
Endocrine active chemicals (EACs) 518
Endocrine-disrupting chemicals (EDC) 198, 416, 570
Endocrine disruptors 518, 522, 529
Endogenous essential respiratory metabolites
- CoQ10 (Coenzyme Q10) 942
- glutathione 940–942
- micronutrient analysis and functional testing 938
Endogenous stress 921
Endogenous toxins 129
Endometriosis 534, 535
EndoPAT 831, 837
Endothelial dysfunction
- nitrate-nitrite NO pathway 837
- nitrate-to-nitrite reduction 837
- protocol 837
- treatment 837, 838
Endothelial nitric oxide synthase (eNOS) 239, 598
Endotoxins 392
Endurance/slow twitch 180
Energy 128, 129
Energy dysfunction 1036
Enteric-coated peppermint oil 385
Enteric endocrine cells 231
Enteric nervous system (ENS) 230, 394, 395, 403, 453
Enterohepatic circulation (EHC)
- bile 228, 229
- bile ducts and gallbladder 223
- clinical nutrition 222
- conventional Western medicine 222
- cytochrome P450 enzymes (CYP 450) 228
- definition 222
- enteric endocrine cells 231
- enteric nervous system 230
- gut-associated lymphoid tissue 231
- gut microbiome 222
- hepatic sinusoids 223, 225
- hepatocyte 223
- hepatocyte cell membrane 224
- large intestine/colon 223, 231
- liver 223
- liver function 225, 228
- liver/hepatic lobule 223
- metabolic zonation/organization 224–227
- metabolism of xenobiotics 222
- microbiome/metabolome 230
- portal tracts 222, 224
- portal vein 222, 223
- recirculation and conservation 222
- single nucleotide polymorphism 228
- small and large intestines 229
- small intestine 223
- terminal ileum 223, 231
- xenobiotic
  - biotransformation 226–228
  - CDC report and scan 227
  - classes 228
  - definition 226
  - environmental chemical exposures with fetal development 227

- harmful xenobiotic exposures 227
Enteropathic arthropathies 442
Environmental chemical exposures 227
Environmental/exogenous toxins 129
Environmental pollutants 942, 943
Environmental toxins 884
Enzymatic detoxification processes 209
Epidermis 970
Epigallocatechin gallate (EGCG) 439–440, 535
Epigenetic diet 989
Epigenetics 463, 464, 497–499, 513, 549, 1036
- agouti gene 237
- cancer
  - CBS 261
  - glutathione 261, 262
  - NAT1 and NAT2 262
  - PTEN function 262
  - TNF-α 262
  - transsulfuration 262
- cardiometabolic disease
  - APOa2 genotypes 238
  - apolipoprotein E 238
  - characteristics 238
  - eNOS/NOS3 239
  - fasting 239
  - FTO 239, 240
  - hemochromatosis 240
  - hypertension 238, 239
  - PCOS 240
  - PON1 239
  - type 2 diabetes 239
- definition 236
- DNA methylation 237
- Dutch Hunger Winter Study 237, 238
- histone modification 237
- IBD and dysbiosis
  - celiac disease 244
  - Crohn's disease 258, 259
  - FUT2 polymorphisms 258
  - histamine degradation 258
  - microbiota 244
  - MUC2 258
  - NCGS 244
- maternal diet, smoking status, mental state and social environment 236
- metabolomics and microbiome analyses 236
- microbiota 258
- mitochondrial function 262, 263
- non-coding RNAs 237
- nutrients
  - adenosylcobalamin 259
  - cyanocobalamin 259
  - GIF deficiency 260
  - hydroxocobalamin 259
  - methylcobalamin 259
  - TCN1 259, 260
  - VDR 260, 261
  - vitamin B12, vitamin D and autoimmune disease 259
- pharmacogenomics
  - acetylation 263
  - CETP 264
  - cytochrome enzymes 263
  - glucuronidation 263
  - glutathione 263
  - mercaptopurine 264
  - warfarin genetics 264

- psychiatric and neurodegenerative disease
  - ALZ 240, 241
  - anxiety 242
  - COMT 241
  - depression 243, 244
  - DSM-V 240
  - GAD 241
  - GAD enzyme 243
  - glutamate 242, 243
  - magnesium deficiency 241, 242
  - methylation 241
  - panic disorder 242
  - PNMT 241
  - s-adenosyl-homocysteine 241
- SNPs 245–257
- tags 236

Epithelial cells (ECs) 290
Epsom salt baths 503
Equilibrium Energy Intake (EEI) 49, 50
Erectile dysfunction (ED) 278
Erythrocyte fatty acid analysis 892
Erythrocyte fatty acid profiles 118
Escherichia coli 535
Essential fatty acid (EFA) deficiency 118, 974
Essential fatty acids 478
Essential metals
- beta-methylamino-L-alanine 887
- copper 887
- iron dyshomeostasis 887

Essential nutrients 459
Essential oils (EO) 425, 426
Estimated glomerular filtration rate (eGFR) 208
Estrogen 538
Estrogen-DNA adducts 529
Estrogen metabolism 518, 520, 522, 528, 529
Estrogen receptor beta (ERbeta) activation 529
Estrovera 529
Evidence-based medicine (EBM) 4, 9, 13, 14
Exercise 522, 546, 547
Exercise training (ET) 596
- adenosine triphosphate (ATP) production 598
- aerobic conditioning training 603
- for cancer 610
- cardiometabolic syndrome 601, 602
- daily movement 596
- for dementia 608–609
- dietary supplements 596
- excessive sitting 597
- fitness assessments 606
- healthcare savings 596
- IGF-1 markers 606
- integrative lifestyle therapies 606
- lifestyle habits 596
- lifestyle intervention programs 606
- lymphatic system 599
- moderate aerobic exercise training 605
- movement and detoxification 596–597
- for multiple sclerosis (MS) 610
- for NAFLD 607, 609, 610
- nitric oxide (NO) production 598
- omega-6 to omega-3 ratio 596
- oxygen, nutrition and movement 597, 598
- for Parkinson's patients 610
- physical therapy 607
- physical therapy and targeted training 605
- planned exercise training 604
- public health recommendations 603
- pyramid 603
- for rheumatoid arthritis (RA) 610
- for sarcopenia 608–610
- for Schizophrenia 610
- for spinal cord injuries 610
- TETR 606, 607
- traditional exercise training 603
- 12-12 rule recommendation 605
- types 602
- walking 603

Exogenous stress 919
Eye 180

F

Faces Pain Scale 462
Facial Nerve (CN7) examination 657
Familial interstitial pneumonia (FIP) 958
Fascial sheaths 1005
Fasting 455, 456
Fasting glucose 518, 519, 521
Fasting plasma glucose (FGP) 400
Fat-free parenteral nutrition (PN) 143
Fat mass and obesity gene (FTO) 239
Fats 878
Fat-soluble vitamins 109, 983
Fatty acids 164
- analysis 118
- elongation 159

Fatty liver/hepatic steatosis 207
Fear of movement (FOM) 464
Fecal microbiota for transplantation (FMT) 317, 402
Federal Food, Drug, and Cosmetic Act 757
Fenugreek 938
Fermentable fiber products 103
Fermentable oligo-, di-, monosaccharides and polyols (FODMAP) diet 455
Fermentable oligosaccharides, disaccharides, monosaccharides and polyols (FODMAP) diet 101, 384, 717
Fermented foods 402
Fetal neurodevelopment 103
Fetal nutrition status 34
Fetal origins of adult-onset diseases (FOAD) hypothesis 22
Feverfew 461
Fiber 102, 103, 208
Fiber aids 102
Fibroids 535
Fibromyalgia 449, 451–454, 456–460
Fibrosarcoma cells 440
Field methods 324
Fine particulate matter 943
Firmicutes 544
Fisetin 937, 938
Fish oil 454, 457
FITTPRO model 605
Flavin adenine dinucleotide (FAD) 275
Flax seed 529
5-Fluorouracil (5-FU) 424
FODMAP diet 1046
Folate (vitamin B9) 108, 459
- conversion system 271
- deficiency 128, 165, 316, 985
- supplementation 548

Folate receptor (FOLR) 502
Folate receptor 1 (FOLR1) gene 508
Folate-sensitive fragile genetic sites 475
Folinic acid 271
Follicle-stimulating hormone (FSH) 537
Food
- antibacterial and antiviral activity 722
- antioxidant activity 721
- blood platelet aggregation 722
- cholesterol metabolism 722
- culture 717
- detoxification enzymes 721
- *vs.* dietary nutrient 719, 720
- dysbiosis 732
- factors 717
- fatty acids
  - biological activities 727
  - health and well-being 727
  - impact 727
  - MUFA 728
  - n-3 PUFAs 729, 730
  - omega-6 PUFAs 728, 729
  - saturated fatty acids 727, 728
  - trans fats 730
  - types 727
- fiber intake 723
  - cancer 724
  - cardiovascular 723
  - diabetes 724
  - endogenous hormones 723
  - gastrointestinal tract 723, 724
  - immune system 724
  - obesity 723, 724
  - soluble fiber 723
  - water-insoluble organic compounds 723
  - water-soluble organic compounds 723
- food triggers, food allergens, and intolerances 724–727
- gut microbiota 731, 732
- immune system 721
- intestinal microbiota 732
- low-fat animal products 722, 723
- meal timings and intervals 720, 721
- nutrition 717
- nutrition intervention 733, 734
- nutritionist/dietitian 717
- personalized and precision medicine 717
- prebiotics 732, 733
- probiotics 732
- processed foods 717, 718
- recommendations 717
- social relations 717
- steroid hormone concentrations 722
- therapeutic diets 717
- water intake
  - autoimmune thyroiditis 731
  - DWSD 731
  - filtering options 731
  - Flint River 731
  - functional medicine assessment 730
  - functional physiology 730
  - health implications 731
  - hemodynamics and functional reserve 730
  - hydration 730, 731
  - intercellular water binds 730
  - long-term health consequences 731
  - SDWA 731
  - social and economic conditions 731
  - water discoloration 731

Index

- whole foods, consumption
  - epidemiological studies 718
  - fibers 719
  - Nrf pathway 718, 719
  - organosulfur compounds 719
  - phytochemicals 718
  - polyphenols 719
  - terpenoids 719
- Food addicts anonymous (FAA) 923
- Food allergy 947, 948
- Food and Drug Administration (FDA) 734
- Food aversion 546
- Food bioactives 994
- Food immune reaction and autoimmunity
  - cross-reactivity
    - of aquaporin with human tissue 856, 857
    - between wheat and human tissue 854, 855
    - of dairy with human tissue 855, 856
    - of food antigens with diabetes and thyroid antigens 858, 859
    - of pectin with human tissue 857
  - emulsifiers/surfactants 849, 850
  - food colorants 851–853
  - genetically modified foods (GM foods) 847, 848
  - gluten peptides 848
  - gums 852, 853
  - hygiene hypothesis 844
  - industrial food additives 849, 853, 854
  - nanoparticles 850, 851
  - oral tolerance breakdown 848
  - organic solvents 850
  - protein modification 848
  - salt 849, 850
  - sugars 849
  - transglutaminases 851
  - Western diet 844, 845
  - wheat breeding 848
  - xenobiotic residues 845–847
- Food-insufficient adolescents 477
- Food-mediated disease 915
- Food sensitivities 453
- Food sensitivity resources 1067–1069
- Foreign direct investments (FDI) 20
- Formiminoglutamate (FIGLU) 272, 273
- Free plasma copper 483
- Frontotemporal degeneration (FTD) 873
- Fructo-oligosaccharides (FOS) 385
- Fucosyltransferase 2 (FUT2) 258, 271, 502
- Full-body (hand-to-foot) multi-frequency BIA 327
- Functional GI disorders (FGIDs) 916
- Functional Medicine Matrix (IFM Matrix) 614, 1016
- Functional medicine model 596
- Functional Nutrition Evaluation 691
- Furancarbonylglycine 504

G

- Galen's program 6
- Gamma linolenic acid (GLA) 458
- Gamma-aminobutyric acid (GABA) 242, 398, 400, 449
- Gamma-linolenic acid (GLA) 116
- Gas exchange 931
- Gas liquid chromatography (GLC) 478
- GAS6 protein 483
- Gastric cancers 423

Gastroesophageal reflux disease (GERD)
- alginate 383
- aloe 383
- deglycyrrhizinated licorice 383
- diet for 382
- hypochlorhydria 383
- limonene 383
- melatonin 383
- mucilaginous substances 383
- probiotics 383
- proton pump inhibitor 382
- stomach acid 383
- STW 5 (Iberogast) 383

Gastrointestinal (GI) tract 723, 887
- on brain 381
- on endocrine effects 381
- on immunity 380, 381
- mechanistic model 380
- on metabolism 381, 382
- microbiome 380, 381
- microbiota 380

Gastrointestinal assessment 1038–1040

Gastrointestinal disorders
- assessment
  - diagnosis 923–924
  - dietary supplement protocol 924
  - evaluation 925
  - medical assessment and treatment 922–923
  - monitoring 924–925
  - patient profile 922
  - pertinent medication history 922, 923
  - PES statement 924
  - therapeutic nutritional intervention 924
- dietary intervention 920, 921
- digestive disease 915
- endogenous influences 921
- food-mediated disease 915
- functional GI disorders 916
- Global Impact on Health 915–916
- herbal and nutritional supplement interventions 921
- impaired digestive capacity 921
- integrative and functional medicine practitioners 920
- irritable bowel syndrome
  - altered motility 919
  - brain-gut interaction 919, 920
  - diagnostic criteria 917, 918
  - IBS-QOL questionnaire 917
  - IBS-SSS 917
  - non-celiac gluten sensitivity 918
  - pathogenesis 918
  - pathogenicity 919
  - pathophysiology 918
  - patient stratification and categorization 917
  - prevalence 916
  - quality of life 917
  - symptomology 919
  - symptoms 917
  - visceral hypersensitivity 919
- motility disorders 916
- myriad exogenous factors 922
- nutritional and lifestyle clinical application principles 921–922
- organic gastrointestinal disease 916
- pharmaceutical applications 920, 921
- prevalence 914, 915

- primary clinical domains 915
- regulation of motility 914, 915
- structural and mechanical abnormalities 915
- transit time 914
- transit times 915

Gastrointestinal dysbiosis 306

Gastrointestinal inflammatory bowel disease
- anxiety and depression 386
- comfrey 387
- diet 386
- fish oil 387
- food additives 386
- functional medicine strategies advantages 387
- hygiene hypothesis 386
- iron 387
- nutraceuticals 387
- oral butyrate supplements 386
- paleo diet 386
- probiotics 386, 387
- specific carbohydrate diet 386
- vitamin D 387

Gastrointestinal reflux (GERD) 462
Gastroparesis 916
Gate control theory 448
Gender benders 381
Generalized anxiety disorder (GAD) 241
Genetic polymorphisms 193
Genetic susceptibility 200
Genetically modified foods (GM foods) 847, 848
Genetically modified organisms (GMOs) 244
Genetics 463
Genistein 539
Geographic information systems (GIS) 20
Geometric framework for nutrition (GFN)
- balanced foods 44
- complementary feeding 44
- geometry of 43
- intake target 44
- model selection 43, 44
- nutritional rails 44
- nutritionally balanced 44
- nutritionally imbalanced 44
- protein prioritization 44, 45
- rules of compromise 44, 45

Geometric Framework for Nutrition (GFN) 43
Geriatric nutritional risk index (GNRI) 872
Germ theory 7
Ghee 787
GIF deficiency 260
Ginger (*Zingiber officinale*) 210, 460

Girth measurements
- bioelectrical impedance analysis 327, 328
- bioimpedance spectroscopy 327
- body fat percentage 326
- skinfold measurement 327
- waist circumference 326
- waist-to-hip ratio 326

Gliadorphins 384
Global phenomenon 19, 20, 82
Global positioning systems (GPS) 20
Global Strategy on Diet 24
Global technological connectivity vehicle 82
Globally Harmonised System of classification and labelling of chemicals (GHS) 192
Glucocorticoids 524, 811
Gluconeogenesis 1036
Glucosamine sulfate 187

Glucose 479
Glucose-dependent insulinotropic polypeptide (GIP) 724
Glucose-hydrogen breath test (GHBT) 919
Glucose-6-phosphate dehydrogenase (G6PD) deficiency 339
Glucose transporter 1 (GLUT1) 340
Glucose transporter 2 (GLUT2) 340
Glucose transporter 3 (GLUT3) 340
Glucose transporter 4 (GLUT4) 340
Glucose transporter 5 (GLUT5) 340
Glutamate 242, 243, 503, 506, 510, 512
Glutamatergic NMDA receptors 398, 400
Glutamic acid decarboxylase 1 (GAD1) 242
Glutathione (GSH) 210, 261, 263, 504, 940–942
Glutathione S-Transferase Mu 1 (GSTM1) 261
Glutathione S-transferase theta 1 deletion (GSTT1) 261
Glutathione S-transferases (GSTs) 261, 422, 502
Gluten 874
Gluten-free dairy-free guide
– balanced fats and oils 1062
– breakfast ideas 1063–1065
– cleansing beverages 1062
– general guidelines 1061–1062
– healthy carbohydrates 1062
– lunch/dinner ideas 1065
– preparation 1062–1064
– shopping list 1066–1067
– snacks 1065–1067
– symptoms 1061
Gluten-free diet 500, 503, 1046
Gluteomorphin peptides 504
Glycerophospholipids (GPLs) 117
Glycolysis 340, 537, 1037
Glycyrrhetinic acid 383
Glyphosate (GLY) 244, 506, 544
Glyphosate exposure 874
Goiter 103
Gonadotropin-releasing hormone (GnRH) 537
Grandmaternal nutrition status 34
Grape seed 510
Green Revolution 19
Green tea 763
Group A Streptococcal (GAS) infection 511
GSTM1 gene 534
Guillain-Barre syndrome (GBS) 956
Gunas 785
Gut microbiome 222
Gut and Psychology Syndrome (GAPS) diet 503, 717, 1046
Gut-associated immune system (GALT)
– biofilm layer 373
– diet 372
– dietary and lifestyle recommendations
  – acidity with diluted vinegar 369
  – betaine HCl 369
  – eating habits 370
  – Swedish bitters 370
  – umeboshi plum 369
– dietary supplementation recommendations
  – *Bovine colostrum* 370
  – medium chain triglycerides (MCT oil) 370
  – protease enzymes 370
  – *Saccharomyces boulardii* 370
  – transfer factor 370
  – Vitamin A 370
– digestive process
  – absorption 368
  – assimilation 368
  – digestion 368
  – eating 368
  – elimination 369
  – motility 368
  – secretion 368
– digestive tract 368
– functional laboratory testing 369–370
– gastric acids 369
– gut microbiome 368
– gut microbiota 372
– hydrochloric acid (HCl) 369
– and MALT 370
– Matzinger's Danger model 368
– microbiome 372–374
– nutrition 373–375
– prebiotics 375
– probiotics 375, 376
– rest and relaxation 370
– 5R protocol 373
– small intestine 371
– stomach acid level 369
– supplements 372
– therapeutic foods 375
Gut-associated lymphoid tissue (GALT) system 231, 372, 915
Gut-brain interactions 888
Gut immune barrier (GIB) 869
Gut microbiome 8, 102, 442, 452, 453
Gut microbiota
– Alzheimer's disease 399, 400
– autism 400
– BDNF 398, 399, 401
– bipolar disorder 396
– blood brain barrier 401
– dysbiosis 392, 393
– enteric nervous system 394
– fecal microbiota transplantation 402
– food sources 402, 403
  – antidepressant foods 403
  – fatty acids 403
  – fermented foods 402
  – LPS 402
  – traditional fermented whole foods 402
– gut brain interactions 394, 395
– inflammaging 401
– MGB axis 394
– microbiota composition 401
– mood disorders 395–397
– neuroimmune interactions 394
– neurological disorders 399
– non-steroidal anti-inflammatory drugs 401
– Parkinson's disease 400
– probiotics 393, 394
– psychobiotics 397, 398
– schizophrenia 396
Gut motility 131
Gut response/gut decision 394
Gycerol-containing lipids 137
Gymphatic circulatory system (GCS) 183

H

Half-and-half nails 653, 688
Hangnails 655
Hapalonychia 654
Hard skills 74
Harmful xenobiotic exposures 227
Hartnup disease 988
Hashimoto's thyroiditis 441
Healing Skills in Medical Practice 700
Health action plan 25
Healthcare dollar expenditure 5
Healthcare management model 96
Healthcare practitioner 1056
Healthcare systems 4
Healthcare workforce 6
Healthful food-based patterns 10
Healthy enteral feedings 884
Healthy gluten-free, casein-free, soy-free (HGSCF) diet 503
Heart rate variability (HRV) 738
HeartMath 706, 707
Heavy metals 886
Helicobacter pylori (H. pylori) 535
Hemochromatosis (HFE) 240
Hemoglobin A1c (HbA1c) 519, 521
Hemp 461
Hepatic sinusoids 223
Hepatocyte 223
Hepatocyte cell membrane 224
Herbal and nutritional supplement interventions 921
Herbal therapies 410
Herceptin 2 (HER2) protein 264
Hif1a 441
High-density lipoprotein (HDL) cholesterol 141
High-intensity interval training (HIIT) 605, 703, 737
High sensitivity C-reactive protein (hs-CRP) 163
HIPPA PRIVACY ACT 1056
Histamine 385
Histamine methyltransferase (HNMT) 258
Histiocytes 971
History of medicine 6–8
Homeostatic Model Assessment (HOMA) levels 475
Homeostatic model of assessment for insulin resistance (HOMA-IR) 398, 521
Homocysteine
– BHMT 274
– CBS 274, 275
– cystathionine 275
– MTR and MTRR 273, 274
– transsulfuration pathway 273
Homovanillate (HVA) 241
Honokiol 440
Hormonal health
– adrenal glands (*See* Adrenal dysfunction)
– insulin (*See* Insulin resistance)
– sex steroid hormones (*See* Sex steroid hormones)
Hormone replacement therapy 528
Hospital-acquired pneumonia (HAP) 932
Human chorionic gonadotropin (hCG) 537
Human leukocyte antigen (HLA) 244, 317, 502
Human mutant transgenic SOD-1 G93A mouse model 867
Human papillomavirus 535
Hydration 460
Hydrochloric acid 385
Hydrogen sulfide 261
Hydrogenated oils 119
Hydrostatic weighing 331

Index

Hydroxocobalamin 259
Hydroxyhemopyrrolin-2-one 501
5-hydroxyindoleacetic acid (5-HIAA) 243
5-hydroxy-tryptophan (5-OH-TRP) 276
Hyperalgesia 449
Hyperbaric oxygen therapy (HBOT) 428
Hyperglycemia 479
Hyperhomocysteinemia 479
Hyperinsulinemia 126, 519
Hypermagnesemia 485
Hyperparathyroidism 111
Hypertension 238, 239, 600
- angiotensin-converting enzyme inhibitors (ACEI) 836
- angiotensin receptor blockers (ARB) 836
- beta-blockers (BB) 836
- calcium channel blockers (CCB) 836
- central alpha agonists (CAA) 836
- clinical history 834
- diagnosis and treatment 834
- direct vasodilators 836
- high renin hypertension 835
- low renin hypertension 835
- natural diuretics 835, 836
- nutraceutical supplements 835
- physical examination and laboratory 834
- results 834
Hypervitaminosis skin findings 989
Hypnotherapy 616
Hypochlorhydria 383, 1038
- causes 1038
- disease association 1038
- low stomach acidity 1038
- self-testing and treatment 1038–1040
- symptoms 1038
Hypocholesterolemia 956
Hyponatremia 477
Hypothalamic-pituitary-adrenal (HPA) axis 1007
Hypothyroidism 95
Hypoxia inducible factor 1 alpha (Hif1á) degradation 439

I

Iatrogenic deaths 92
Iberogast 368
Idiopathic pulmonary fibrosis (IPF) 937, 957
IFM Adult Intake Questionnaire 705
IFM Exercise History and Activities of Daily Living questionnaires 700
IFM Matrix tool 697
IL1-β activation disorders 292
Illegitimate nonscientific approaches 7
Immune dysregulation 125
Immune system 1011
- adaptive immune system
  - central tolerance 297, 298
  - diagnostic features 295
  - IgE and allergic reactions 294, 295
  - macrophages and dendritic cells 292
  - oral tolerance 295, 296
  - peripheral tolerance 298, 299
  - primary immune response 293
  - secondary immune response 293, 294
- dietary intervention
  - AhR 299
  - probiotics 300
  - tryptophan 299, 300

- vitamin A 299
- vitamin B 299
- vitamin C 299
- Vitamin D/1,25-(OH)$_2$D$_3$ 299
- innate immunity
  - basophils 289, 290
  - dendritic cells 288
  - diagnostic features 292
  - eosinophils 290
  - epithelial cells 290
  - first responders 288
  - food components 288
  - ILCs 289
  - macrophages 288, 289
  - mast cells 289
  - monocytes 288
  - neutrophils/neutrocytes 289, 290
  - NK cells 289
  - paneth cells 290, 292
  - physiologic, anatomic, and cellular components 288
- mucosal immune system
  - diagnostic features 287, 288
  - IgA 286, 287
  - intestinal permeability-large macromolecules 287
  - pathogens and antigens 286
  - role of 286
  - SIgA 286
  - protective layers 286
Immunoglobulin E (IgE) 944
Immunoglobulin G (IgG) 62
Immunomodulatory activities 141
Immunosenescence 401
Impaired digestive capacity 921
Impaired glucose tolerance 518, 519
In vitro fertilization (IVF) 534
Inattentive ADD 505
Indole-3-carbinol (I3C) 722
Inducible nitric oxide synthase (iNOS) 278
Industrial Revolution 192
Infant respiratory distress syndrome (IRDS) 930
Infection
- AMR 306
- autoimmune 317
- causes of 311
- CRP-hs 314, 315
- defense and repair systems 305, 306
- diagnosis
  - history 310
  - inflammation 309
  - intravenous (IV) technology 310
  - NSAIDs 309
  - severe sepsis 310
  - steroids 309
  - vitamin C requirements 309
- digestion, absorption, microbiota/GI, respiration 313
- endogenous structure and function 315
- folate metabolism 315
- gastrointestinal dysbiosis 306
- heart disease/cardiovascular association 316
- homocysteine catabolism 316
- immunonutrition assessment investigation tools 310, 311
- inflammatory response 308
- laboratory data 311, 314
- lipids 315

- malnutrition 305, 307, 311
- methionine metabolism 316
- methyl nutrients 315
- minerals 316
- neurology 317
- oncology 316
- phytonutrients 316
- respiratory 317
- sleep 316
- stress 306–308, 316
- tuberculosis 308, 309
- vaccination
  - adverse effects 317
  - aluminum 318
  - antibiotics and antimicrobials 317
  - FMT 317
  - mercury 318
  - principle of 317
  - safety 318
- vitamin A retinol 314
- vitamin D, A, E 315
- vitamin D 25-hydroxy 314
Infectious disease 892, 893
Inflammaging 401
Inflammasomes 291
Inflammation 165, 452, 1007
- Alternative Healthy Eating Index 126
- aryl hydrocarbon receptors 126
- chronic diseases 125
- complete nutritional assessment 125
- control 155
- diet 126
- host defense and survival 125
- hyperinsulinemia 126
- immune dysregulation 125
- initiation and resolution 125
- integrative assessment 126, 127
- Mediterranean or Paleolithic pattern of dieting 126
- micro- and macronutrient deficiencies 125
- oxidative stress 125
- pro-inflammatory adipokines 126
- undernutrition 125
- visceral adiposity 126
- western lifestyle behaviors 126
Inflammatory autoimmune diseases 873
Inflammatory bowel disease (IBD) 258, 402, 442, 993–996
Inflammatory load assessment 163
Inflammatory prostaglandins 116
Inflammatory response 339
Inflammatory tumor-promoting cytokines as interleukin 6 (IL-6) 439
Influential behavior (I) 75
Information and communication technologies (ICTs) 20
Initial nutrition assessment checklist
- background 1020–1022
- before appointment 1020
- chart review and preparation 1020
- clinical background and explanation 1020
- dental assessment
  - biochemical individuality 1022
  - energy dysfunction 1022
  - lifestyle 1022
  - nutrient Insufficiencies 1022
  - sleep and stress 1022
  - websites 1022

- dental health impacts 1037
- develop priorities, create nutrition action plan 1021
- documentation and follow-up 1022
- fats and oils survey 1041
- gastrointestinal assessment 1038–1040
- hypochlorhydria
  - causes 1038
  - disease association 1038
  - low stomach acidity 1038
  - self-testing and treatment 1038–1040
  - symptoms 1038
- I-D-U worksheet 1039
- in-office assessments and measurements 1020
- review forms with client 1021
- review other recommendations 1022
- summary form 1036

Innate immune system 873, 1011
- basophils 289, 290
- dendritic cells 288
- diagnostic features 292
- eosinophils 290
- first responders 288
- food components 288
- ILCs 289
- macrophages 288, 289
- mast cells 289
- monocytes 288
- neutrophils/leucocytes 289, 290
- NK cells 289
- paneth cells 290
  - cellular component 291
  - exogenous factors 291
  - inflammasomes 291
  - inflammatory and autoimmune diseases 292
  - neuroinflammation 292
  - physiological and pathological conditions 291
  - small intestine 290
  - tissue homeostasis 291
- physiologic, anatomic, and cellular components 288

Innate lymphoid cells (ILCs) 289, 299
Inner mitochondrial membrane (IMM) 178
Insecticide resistance 192
Insoluble fiber 102
Insomnia 448, 451, 458–461
Institute for Functional Medicine (IFM) Matrix 624, 634–636
Insulin-like growth factor (IGF) 427
Insulin resistance 339
- adiponectin 521
- assessment protocol 520
- cardiovascular disease 518
- children, in diabetes 518
- detoxification 522
- diabetes mellitus type 2 519
- dietary management 522
- exercise 522
- HbA1c 521
- HOMA-IR 521
- impaired glucose tolerance 518, 519
- insulin production 518
- laboratory evaluation 520
- laboratory testing 520
- metabolic syndrome vs. 519
- micronutrient recommendations 523
- patient risk identification 519
- prediabetes 519
- proinsulin 521
- sleep 523
- staging the patient 520
- stress and autonomic dysfunction 523
- treatment protocol 522

Insurance payer market 4
Integrated standing, movement and exercise training (ISMET) 602, 605, 606
Integrative and functional approaches to therapeutic diets (IFMNT)
- advantages 744
- assessment tools 744
- cardiac diet 745
- elimination/exclusion diets
  - adverse food reactions 745
  - autoimmune diseases 745
  - carbohydrate diet 746
  - food intolerances 745
  - food sensitivities 745
  - implementation 745
  - low FODMAP diet 745
  - non-celiac gluten sensitivity 745
  - Paleo diet 745
  - risk factors 746
- IF (See Intermittent fasting (IF))
- LCHF (See Low carbohydrate high fat diet (LCHF))
- low histamine diet (See Low histamine diet)
- nutrition care process (NCP) 744

Integrative and functional medical nutrition therapy (IFMNT) 58, 101, 622, 634, 639, 716, 772, 866, 871, 960
- assessment tools and measures 623
- asthma 950
- biomarkers 629
- conventional tests 629–630
- core imbalances 630
- description 622
- diet therapy 622
- dietitians and certified nutritionists 622
- DIFM 625
- education and training 622
- evidence-based approach 624
- functional medicine 624
- functional tests 630
- integrative and functional approaches 624
- lifestyle 628
- medical nutrition therapy (MNT) 622
- metabolic function 623
- metabolic pathways 630
- mind-body interventions 624
- mind-body-spirit connection 623
- nutrition-based therapies 622
- nutrition care process 627
- nutrition-focused physical exam 628, 629
- organizing bodies 622
- patient-centered care 622
- precipitating factors 630
- radial 628, 630
- SOP 627, 628
- SOPP 627, 628
- therapeutic relationship 622
- voluntary actions 623

Integrative and functional medicine practitioners 920
Integrative and functional nutrition 101

Integrative Health and Wellbeing program 622
Integrative medicine 622
Integrative/naturopathic practitioner 103
Integrative therapy
- AIP diet 426, 427
- algae 423, 424
- anti-cancer potential 424
- artemisia 425
- bee products 423
- bio-individual protocol 428, 429
- cancer 409
  - alkylating agents 412
  - anti-angiogenesis drugs 413
  - antimetabolites 412
  - antitumor antibiotics 412
  - apoptosis 413
  - chemotherapy 411–413
  - emotional aspect of 415, 416
  - hallmark traits 411
  - metastases 411
  - monoclonal antibodies 413
  - optimization and protection 414
  - perspectives 410
  - plant alkaloids 412
  - proton therapy 413, 414
  - radiation therapy 413
  - statistics 409
  - surgical resection 414
  - synergistic therapy 414, 415
  - target pathways 413
  - topoisomerase inhibitors 412
  - treatment options 411
  - types 409
  - vs. terrain 414
- carotenoids 425
- cruciferous vegetables 421, 422
- dandelion 420
- detoxification 416
  - ALA 417
  - amalgam fillings 416
  - antigenic triggers 416
  - EDCs 416
  - EMF exposure 416
  - high glycemic-load foods 416
  - intravenous nutrition 416, 417
  - LDN 417
  - microbiome 416
  - mold exposure 416
  - wellness 416
- dietary interventions 426
- essential oils 425, 426
- exercise 430
- fear and empowerment 430, 431
- fermented probiotic-rich foods 425
- fiber diet 424
- and functional nutrition 417, 418
- ginger 419, 420
- honokiol 421
- hyperthermia 429
- iodine sufficiency 423
- ketogenic diet 427, 428
- limonene 421
- liquid oral/enteral nutrition 418
- massage therapy 430
- MCP 424
- melatonin-rich foods 421
- methyl-rich foods 420, 421
- mindfulness meditation 429

- movement 430
- mushrooms 423
- polyphenols 425
- probiotics 425
- SCFAs 424
- seasonally balanced eating 426
- seaweed 424
- seeds 422, 423
- sepsis 417
- tea therapy 418
- touch therapy 430
- turmeric 418, 419
- vegetarian diet 427

Interconnected algorithm 1017
Interdisciplinary medical team 79
Interleukin-6 (IL-6) 1007
Intermittent fasting (IF)
- advantages 750, 751
- anorexia nervosa 750
- breaking a fast 751
- compressed eating window (CEW) 751
- electrolyte imbalances 751
- 5/2 fast 751
- fasting models 751, 752
- in fasting patients 751
- heartburn 751
- managing feelings of hunger 751
- readiness and appropriate stage 751
- sugar-lowering medication 751

International Agency for Research on Cancer (IARC) 195, 197–198
International Association for the Study of Pain (IASP) 448
International Diabetes Federation (IDF) criteria 809
Interprofessional collaborative practice 11
Interprofessional teams 11
Interstitial lung disease 957
Interventions, monitoring and evaluation 739
Intestinal barrier function 231
Intestinal firewall 231
Intestinal permeability 209
Intraepithelial lymphocyte (IEL) cells 299
Investigational New Drug (IND) application 402
Iodine 103, 104, 528
Iron 933, 935
Iron-deficiency anemia 933
Iron dyshomeostasis 887
Iron homeostasis 933
Irregular sleep-wake pattern 581
Irritable bowel disease (IBD)
- amino acid metabolism 258
- celiac disease 244
- Crohn's disease 258, 259
- FUT2 polymorphisms 258
- histamine degradation 258
- microbiota 244, 258
- MUC2 258
- NCGS 244

Irritable bowel syndrome (IBS) 396, 402, 449
- altered motility 919
- botanical agents 385
- brain-gut interaction 919, 920
- chamomile 385
- diagnostic criteria 917, 918
- enteric-coated peppermint oil 385
- fiber 384
- FODMAP diet 384

- food allergy testing 384
- gluten elimination 384
- histamine 385
- holistic treatment 384
- hydrochloric acid 385
- IBS-C/IBS-D 383
- IBS-QOL questionnaire 917
- IBS-SSS 917
- magnesium citrate 385
- natural antimicrobials 385
- non-celiac gluten sensitivity 918
- pathogenesis 918
- pathogenicity 919
- pathophysiology 918
- patient stratification and categorization 917
- prebiotics 385
- prevalence 916
- probiotics 385
- SIBO 384
- skin testing and IgE RAST testing 384
- symptom severity 917
- symptomology 919
- symptoms 384, 917
- visceral hypersensitivity 919

Irritable Bowel Syndrome Quality of Life Questionnaire (IBS-QOL) 917
Irritable Bowel Syndrome Severity Scoring System (IBS-SSS) 917
Isotretinoin 483

J

Jatharagni 793
Joint pain 460, 461
Journal of the American Geriatric Society 608

K

Kaolin 385
Kapha 785, 792
Karolinska Institute study 599
Kayser-Fleischer ring 483
KetoDietCalculator (KDC) 355
Ketogenic diet 119, 509, 872, 873, 879, 882
- breakdown products 508
- challenge 509
- dietitian 508
Ketones 509
Ketosis 1037
Kidney's filtering system 206
Kinky hair syndrome 987
Koilonychia 652, 687
Korsakoff's syndrome 479
Krebs citric acid cycle (CAC) 428
Kryptopyrrole quantitative urine 501

L

Laboratory profiles 118
Lack of sleep 1037
Lactic acid 440
Lactobacillus sp. 402, 462
- *L. acidophilus* 545
- *L. helveticus* 393, 397
- *L. plantarum* 398
- *L. reuteri* 545
- *L. rhamnosus* 393, 398

- *L. salivarius* 545
- *L. sporogenes* 398

Lactulose-hydrogen breath test (LHBT) 919
Lambert-Eaton Myasthenic Syndrome 398
Large intestine/colon 231
Laser interstitial thermal therapy (LITT) 429
Law of Mass Action 94
L-carnitine 180
LD50 540
Leaky gut 371, 392, 393, 396, 397
Leaky gut syndrome 371
Leptin 130
Leukotrienes (LT) 142
Leydig cell tumor (LT) 536
Lifestyle
- assimilation 737
- and behavioral risks 11
- biotransformation and elimination 737
- communication 737
- and community 1037
- defense and repair 737
- energy 737
- exercise and movement 736
- factors 736
  - alcohol 128
  - cigarette smoking 128
  - folate deficiency 128
  - physical activity 128
  - sleep and stress 127
  - substance abuse 128
- interventions 716
- relaxation response 738
- sleep 736
- squared breathing 738
- stress and resilience 737, 738
- structural integrity 737
- transport 737

Light exposure 130
Limbic ADD 506
Limonene 383
Lindsay's nails 653, 688
Linoleic acid (LA) 158, 729
Lipid bilayer 174, 175
Lipid metabolism dysregulation 939
Lipid therapy
- cell membrane permeability and integrity 163
- defense and repair 163
- dental 163
- fatty acid/nutrition status 160
- laboratory principles 163
- structural integrity 163
- structural-spinal alignment 163

Lipidomics
- after world war II 154
- application 154
- body composition and function
  - cell membrane structure 156
  - cellular hydration 157
  - cellular, organelle, and nuclear bilayer membranes 156
  - eicosanoid cascade 158
  - membrane barriers 157, 158
- chronic diseases 154
- clinical imbalances 155–156
- complex neurological condition 169
- cooking oils 154

- defense and repair assessment
  - gut microbiome 163
  - vitamin A (Retinol) 163
  - vitamin D 25-hydroxy 163
- dietary fat
  - history 153–154
  - human history 153–154
- fat/oil-rich foods 154
- fatty acids 164
- heart disease / cardiovascular association 166
- IFMNT 155
- inflammation control 155
- inflammatory load assessment 163
- integrative and functional medicine 153
- lifestyle factors
  - movement 165
  - sleep 165
  - stress 165
- lipid therapy
  - cell membrane permeability and integrity 163
  - defense and repair 163
  - dental 163
  - fatty acid/nutrition status 160
  - laboratory principles 163
  - structural integrity 163
  - structural-spinal alignment 163
- lipids 163
- maintaining lipid balance and homeostasis 154
- membrane structure 155
- metabolic stressors 160
- minerals
  - B12 165
  - folate deficiency 165
  - magnesium 164
  - methyl nutrients 165
  - niacin 165
  - vitamin A 165
  - vitamin C 165
  - vitamin D and A receptors 165
  - zinc 164, 165
- neurological conditions
  - Alzheimer's disease 166–167
  - developmental plasticity 167
  - mitochondrial dysfunction 166
- oncology 166
- phospholipids 164
- phytonutrients 165
- respiratory 167
- simple– childhood asthma 168
- sterols 164

Lipids 163
Lipoic acid 179
Lipophilic xenobiotics 193
Lipopolysaccharide binding protein (LBP) 396
Lipopolysaccharide endotoxin (LPS) 392, 399, 402, 403
Lipopolysaccharides (LPS) 381, 732
Lipoprotein lipase (LPL) 1007
Lipoxygenase 456
Lithium 483
Lithium salts 476
Live blood cell analysis 443
Liver acinus 225
Liver cirrhosis 957
Liver dysfunction 892
L-methylfolate/folinic acid 459

Long latency nutritional insufficiencies 1037
Longitudinal melanonychia 655
Lou Gehrig's disease, see Amyotrophic lateral sclerosis (ALS)
Low- and middle-income countries (LMICs) 20
Low carbohydrate high fat diet (LCHF)
- avoidance 750
- definition 749
- in cancer therapy 749
- macronutrient percentages 749

Low-density lipoprotein (LDL) cholesterol 141
Low dose naltrexone (LDN) 417
Lower motor neuron disease 867
Low-fat vegan diet 119
Low FODMAP diet 745
Low glycemic index treatment (LGIT) 508
Low histamine diet
- definition 746
- histamine degradation 747
- histamine load 747
- histamine overload 748
- in human function 747
- intolerance 747
- mast cell activation disorder (MCAD) 747
- monoamines 747
- polyamines 747

Low magnesium status 111
Low-molecular weight citrus pectin (LCP) 424
L-tyrosine 505
Lumbrokinase 536
Lung physiology 942
Lunula 653
Luteinizing hormone (LH) 537, 538
Lymphatic system 599

M

Macrophage activation syndrome 292
Macrophage function 930
Macrophage inflammatory protein 2-alpha (MIP2-alpha) 439
Magnesium 164, 485, 947
Magnesium citrate 510
Magnesium deficiency 241, 242, 458, 459
Magnesium glycinate 276
Magnesium L-Threonate 510
Magnolia officinalis 506
Major Depressive Disorder (MDD) 475
Major depressive disorder (MMD) 243, 244, 475
Major histocompatibility complex (MHC) proteins 244
Malnutrition Universal Screening Tool (MUST) 686
Malondialdehyde (MDA) 400
Mammalian target of rapamycin (mTOR) 426, 427
Manganese toxicity 485
MAO B + (an X-linked gene) 504
Marijuana 507
Mast cell activation disorder (MCAD) 747
Maternal nutrition status 34
Matrix metalloproteinases (MMPs) 440, 441
Matzinger's Danger model 368
Mauve factor 421
McGill Pain Questionnaire 462
Meat-sweet diet 717
Medaroga 792
Medhya 787

Medical education post-Flexner report 11
Medical education system 8
Medical information 1056
Medical marijuana 461
Medical nutrition therapy (MNT) 622, 639, 973–982
- biochemical pathways 64
- biomarkers 61
- cardiovascular diseases 67
- DNA strands 59
- endocrine system 67
- fatty acid breakdown 64–66
- food 59
- food allergy 70
- food intolerance 70
- gastrointestinal tract 67
- immune system-defense and repair 67
- laboratory testing 61
  - assessment 63
  - detoxification 63, 64
  - digestion and absorption 63
  - direct testing 61
  - energy metabolism 61
  - evaluation 61
  - IgE reactions 62
  - IgG 62
  - immune system's reaction 61
  - indirect testing 61
  - inflammation 61
  - integrative and functional clinicians 61
  - intestinal permeability 63
  - IOM 62, 63
  - metabolomics 61
  - nutrigenomics 62, 63
  - specimen selection 61
- lifestyle and environmental factors 59, 60
- NPE 60
- overview 58
- pathogens 69, 70
- personalized nutrition care 58–60
- potential triggers 68, 69
- radial core 59
- respiratory system 67, 68
- signs and symptoms 60, 61
- stress 69
- systems biology 66, 67
- toxins and toxicants 69

Medical symptoms questionnaire (MSQ) 60, 614
Medieval aspirin, see Feverfew
Mediterranean diet 119, 454, 455, 975–982, 988, 989
Medium chain triglycerides (MCT) 137, 138, 164, 880, 882
Medium-chain triglyceride (MCT) diet 508–510
Medovaha strotas 792
Mee's lines 654, 689
Melancholia 380
Melatonin 383, 387, 476, 507, 510, 538
Membrane barriers 157, 158
Membrane fluidity 175
Membrane permeability 156, 158
Membrane rafts 153
Membrane structure
- carbohydrates 177
- cholesterol 176
- function 174
- lipid bilayer 174, 175
- membrane fluidity 175

- organelle 175
- phosphatidylcholine 175
- phosphatidylethanolamine 176
- phosphatidylserine 176
- phospholipids 174
- proteins 176
- stability 175

Mendelian genetics 240
Menkes syndrome 987
Menopause 518, 528
Mental/behavioral/neurological health
- amino acids 479–481
- cholesterol 478, 479
- essential fatty acids 478
- glucose 479
- minerals
  - copper 483
  - lithium 483
  - magnesium 485
  - manganese 485
  - selenium 485
  - zinc deficiency 485, 486
- psychotropic drugs
  - antidepressants medications 486
  - phenytoin 487
  - protein pump inhibitors 486
  - riboflavin 487
  - weight gain 486
- thiamin deficiency 479
- toxic minerals
  - cadmium 486
  - methyl mercury 486
- vitamin A 483
- vitamin B1 (thiamin) 479
- vitamin B2 (riboflavin) 479
- vitamin B3 (niacin) 479
- vitamin B6 (pyridoxine) 479, 482
- vitamin D 483
- vitamin K 483

Mental-emotional stress 922
Mental stress 1008
Mercaptopurine 264
Mercury 486
Metabolic and nutritional balance
- B vitamins 104–107
- fat-soluble vitamins 109
- fatty acids and phospholipids
  - conjugated linoleic acid 116, 117
  - erythrocyte fatty acid profiles 118
  - gamma-linolenic acid 116
  - glycerophospholipids 117
  - hydrogenated oils 119
  - increasing omega-3 fatty acid 118
  - increasing omega-6 fatty acid 118
  - omega-6 to omega-3 fatty acid 114–116
  - short-chain fatty acids 117, 118
- fiber 102, 103
- folate (vitamin B9) 108
- iodine 103, 104
- microbiome 101–102
- minerals
  - calcium and magnesium 110–112
  - copper 112–114
  - copper-to-zinc ratio 114
  - sodium-to-potassium (Na–K) ratio on hypertension 112
  - zinc 112–114
- niacin (vitamin B3) 107

- pantothenic acid 107
- riboflavin (vitamin B2) 107
- thiamine (vitamin B1) 107
- tocopherols 110
- tocotrienols 110
- vitamin A 109
- vitamin B12 108, 109
- vitamin B6 107, 108
- vitamin D 109
- vitamin K 109

Metabolic cart resting energy expenditure (mREE) 872
Metabolic correction (MC) therapy
- adverse side effects 92
- biochemical mechanism
  - activation energy 93
  - active site 93
  - B-complex and metabolic briefing 94, 95
  - B-complex and metabolic correction 95
  - cofactors 93
  - enzyme-catalyzed reactions 93
  - gene mutation 93
  - *Law of Mass Action* 94
  - physiological impairments 94
  - *Principle of Le Chatelier* 94
  - substrate 93
- conceivable genetic defects 93
- disease states 90–91
- functional biochemical/physiological concept 90
- healthcare management model 96
- hidden hunger (or occult hunger) 91
- iatrogenic death 92
- medication-induced nutient depletion 92
- metabolic optimization therapy 91
- minerals 91
- nutrient-dense foods 92
- nutrient dependency 90
- nutrient imbalances 93
- nutrient requirements 93
- nutritional demands 93
- nutritional insufficiency 90
- nutritional triage 92
- principles
  - biochemical individuality 96
  - detect nutrient deficiencies 96
  - environmental pollution 96
  - inspire active role-taking responsibility 96
  - metabolic correctors 95
  - monitor and update metabolic correction over time 96
  - nutrient-related disorders 96
  - nutrigenomics 96
  - pharmacogenomics 96
  - recommended daily allowances for diseased individuals 96
- vitamins 91

Metabolic optimization therapy 91
Metabolic stressors 160
Metabolic syndrome (MetS) 160, 519
Metabolomics 230, 891
Metformin 522
Methionine metabolism 316
Methionine synthase reductase (MTRR) 259, 502
5-Methylenetetrahydrofolate reductase (MTHFR) 547
Methyl mercury 486
Methyl nutrients 165, 315

Methylation 497, 501, 502, 504, 512, 1038
- catecholamines 275, 276
- CBS 271
- D-lactate 271
- epigenetics 270, 271
- homocysteine 273
  - BHMT 274
  - CBS 274, 275
  - cystathionine 275
  - MTR and MTRR 273, 274
  - transsulfuration pathway 273
- homozygous mutation 270
- iron 276–278
- Mendelian inheritance 270
- NIH 273
- nitric oxide 277, 278
- nutritional genomics 270
- one-carbon metabolism 271–274
- pharmaceuticals 273
- Punnett square 270
- serotonin 276
- SNPs 270, 271, 279–282
- stress 273
- tetrahydrobiopterin 275

Methylcobalamin 259, 272
Methylenetetrahydrofolate reductase (MTHFR) 259, 271, 273, 476, 497, 499, 502, 504, 505
Methylmalonic acid (MMA) 61, 64, 273
Microbiome 101–102, 230, 888
Microbiota-gut-brain (MGB) axis 394, 399, 401
Micrococcus 544
Microdomains 153
Microencapsulation technology 174
Microenvironment of chronic diseases
- arthritic conditions 441, 442
- autoimmune disease 440, 441
- cancer 441
- cellular components
  - CAFs 438, 439
  - TAMs 439
- dark-field illumination 442
- inflammatory influences 440
- influence, phase-contrast illumination 442
- live blood cell analysis 443
- in malignancies 438
- molecular components
  - NFκβ 439
  - p53 439
- nutritional influence
  - AIM 441
  - angiogenesis 440
  - arthritic conditions 442
  - extracellular matrix 439, 440
- pH 443
- phase-contrast illumination 442
- TME components 438
- white blood cell morphology 443

MicroRNA (miRNA) 971
MicroRNA-155 molecule 399, 870
Microscopic colitis 385
1-mile walk test 606
Mind-body techniques 241, 614
- biofeedback 615
- goal of 614
- neuro emotional technique (NET) 614, 615
- Neuro Emotional Technique (NET) 615
- neurofeedback 615

Mild cognitive impairment 347
Mindfulness meditation 429
Mindfulness-based stress reduction (MBSR) 738
Mineralocorticoids 524
Minerals 938, 947
- B12 165
- calcium and magnesium 110–112
- copper 112–114, 987
- copper-to-zinc ratio 114
- folate deficiency 165
- iron deficiency 986
- magnesium 164
- methyl nutrients 165
- niacin 165
- selenium 987
- sodium-to-potassium (Na–K) ratio on hypertension 112
- vitamin A 165
- vitamin C 165
- vitamin D and A receptors 165
- zinc 112–114, 164, 165, 986
Mini Nutritional Assessment (MNA) 686
Minichromosome maintenance complex 6 (MCM6) 502
Mini-mental state examination (MMSE) scores 400, 479, 483
Minnesota Coronary Experiment (MCE) 728
3-minute step test 606
Mitochondria 537
- apoptosis 178
- cardiolipin 178
- evaluation/assessment 179
- inner and outer mitochondrial membranes 178
- mitochondrial dysfunction 178
- oxidative phosphorylation 178
- powerhouse of cell 178
- prevention/treatment 179
Mitochondrial biogenesis 180
Mitochondrial dysfunction 166, 178
Mitochondrial function 262, 263
Mitochondrial health 340
Mitochondrial issues, autism 500
Mitochondrial metabolic disease 347
Mitochondrial modulators (MM) 476
Modifiable lifestyle factors
- ACE 699
- assessment questionnaire tools
  - benefits for clients 699
  - benefits for practitioners 698, 699
- change and transformation 710, 711
- exercise and movement 697
- exercise/movement assessment
  - Activities of Daily Living Questionnaire 701
  - Adult Intake Questionnaire 700
  - breath and mindfulness techniques 701
  - dance 701
  - FITT questions 700, 701
  - health benefits 700
  - HIIT 703
  - at home 701, 702
  - IFM Exercise History and Activities of Daily Living questionnaires 700
  - metaphorical level 701
  - Pickleball 701
  - Silver Sneakers 701
  - walk and talk therapy 701
  - at work 702, 703
- GOTOIT 698
- Healing Skills in Medical Practice 700
- IFM Adult Questionnaire 711
- physiological impacts 697
- questionnaire-induced insights 699
- relationships and beliefs
  - back story 707
  - lifestyle changes 707, 708
  - open-ended questions 710
  - social networks and health outcomes 708, 709
  - spirituality and religion 709, 710
- sleep and restoration habits 697
- sleep and restorative activities
  - laughter 704
  - leading-edge science 704
  - meditation and mindfulness-based stress reduction techniques 704
  - mortality 703
  - questions for restorative activities 704
  - screening 703
  - sleep deprivation 703
  - sleep questionnaire 703, 704
- smart goal-setting 711
- storytelling and healing 699
- stress and resilience
  - abuse and neglect 705, 706
  - anxiety and depression 706
  - emotional awareness 707
  - external coping strategies 705
  - external sources of stress 704
  - functional medicine evaluations 705
  - IFM Adult Intake Questionnaire 705
  - internal locus of control 707
  - optimism 707
  - positive attitude 707
  - QEEG neurofeedback harnesses 707
  - questions 705
  - resources 707
  - Self-Care Assessment Questionnaire 705
  - sense of humor 707
  - social support 707
  - stress level assessment questions 706
  - toolkit 707
  - twelve-step programs 707
- stress levels and perception 697
- therapeutic relationship 697, 698
Modified Atkins diet (MAD) 508–509
Modified citrus pectin (MCP) 424
Modified elimination diet
- atopic dermatitis 990
- contact dermatitis 991
- contact urticaria 991
- dermatitis herpetiformis 990
- food-associated triggers 990
- hypersensitivity reactions 990
- IgE serum testing 990
- immune system 990
- pemphigus 990
- photoallergic contact dermatitis 991, 992
- protein contact dermatitis 991
- removal of dietary components 989
Monoamine oxidase A (MAO-A) 243, 276, 512
Monoamine oxidase B (MAO-B) 275
Monoamine oxidase inhibitors (MAOIs) 486
Monoculture farming 206
Monoglycerides 137
Monosodium glutamate (MSG) 243, 497, 509

Monounsaturated fatty acid (MUFA) 139, 728, 881
Monroe Institute 707
Monsanto's Roundup herbicide 506
Mood disorder spectrum 474
Mood disorders 395–397
Morningness-Eveningness Questionnaire 590
Morphine 462
Motility disorders 915, 916
Motivational interviewing (MI) 727
- ambivalence 1052
- change talk 1053
- client observation 1053
- open questioning 1053
- practitioner 1053
- principles 1052
- skills 1052
- spirit 1052
Motor Neuron Disease, see Amyotrophic lateral sclerosis (ALS)
MSG Symptom Complex, see Chinese restaurant syndrome
Mu agonists 462
Mucilaginous substances 383
Mucin 2 (MUC2) 258
Mucins 102
Mucosal-associated lymphoid tissue (MALT) 370
Mucosal immune system
- diagnostic features 287, 288
- IgA 286, 287
- intestinal permeability-large macromolecules 287
- pathogens and antigens 286
- role of 286
- SIgA 286
Muehrcke's lines 654, 688
Multidisciplinary Pain Research 449
Multifactorial behavior modification approach 24
Multiple chemical sensitivity 194
Multiple sclerosis (MS) 399, 441
Multisystems disorder 867
Multisystems metabolic pathways 867
Muscle imbalances 1005
Muscle quality 330
Muscular dystrophy 330
Musculoskeletal system
- glucosamine sulfate 187
- omega-3 fatty acids 187
- osteoarthritis
  - definition 186
  - evaluation/assessment 187
  - prevention/treatment 187
- osteoporosis and ostopenia
  - definition 187
  - evaluation/assessment 188
  - pathophysiology 187
  - treatment/prevention 188
- sarcopenia 186
- strontium 187
- vitamin C 187
Mycoplasma genitalium 535
Myokines 1007

N

N-acetyl cysteine (NAC) 476, 938
N-acetyl glucosamine 387

M–N

N-acetyltransferase 1 and 2 (NAT1 and NAT2) 262, 263
Nail pitting 687
Nasya 792
National Ayurvedic Medicine Association (NAMA) 785
National Cholesterol Education Program (NCEP) 519
National Health and Nutrition Evaluation Survey (NHANES) 519, 570, 792
National Health Interview Survey 756
National Institute of Health (NIH) 273, 398
National Sleep Foundation 589
Nattokinase 536
Natural chelators 886
Natural killer cells (NK cells) 289
Naturally occurring patient tribes 81, 82
Nature-deficit disorder 507
Nature's aspirin, *see* Willow bark
Neural tube defects (NTDs) 548
Neuro emotional technique (NET) 614, 615
Neurodegenerative disease 184, 347
– ALZ 240, 241
– anxiety 242
– COMT 241
– depression 243, 244
– DSM-V 240
– GAD 241
– GAD enzyme 243
– glutamate 242, 243
– magnesium deficiency 241, 242
– methylation 241
– panic disorder 242
– PNMT 241
– s-adenosyl-homocysteine 241
Neurodegenerative pathways 867
Neurodevelopmental disorders in children
– ASD (*See* Autistic spectrum disorder (ASD))
– C-section births 495
– electrolyte status 495
– family history and genetics 497
– immune system 495
– lifestyle factors 497, 498
– microbiome colonization 495, 496
– skeleton system 495
– toxins and toxicants 495, 496
Neurofeedback 615
Neurogastroenterology 919
Neurological disorders 399
Neuronal nitric oxide synthase (nNOS) 278
Neurons, *see* Brain cells
Neuropathic pain 449
Neuroprotective and antioxidant properties 873
Neurotransmitters 479–481
Neutral posture 1005
NHANES study 813
Niacin (vitamin B3) 107, 984
Niacin deficiency 479
Nicotinamide adenine dinucleotide (NAD+) supplements 962
Nidra (sleep) 786
Nightshade vegetables 456
Nightshades 456
Nitric oxide (NO) 277
Nitric oxide synthase (NOS) 277, 278
Nitroprusside 360
N-methyl-D-aspartate (NMDA) receptors 243, 396, 398, 400

Nociception 449, 456
Nociceptive pain 449
Nonalcoholic fatty liver disease (NAFLD) 207, 208, 607, 609
Nonalcoholic steatohepatitis (NASH) 208
Non-allergic asthma 945, 946
Non-celiac gluten sensitivity (NCGS) 244, 874, 918
Non-communicable disease (NCD) 317
Non-*H. pylori* related gastritis 260
Non-24 Hour Sleep Pattern 581
Nonionizing radiation (NIER) 543, 544
Non-opioids 462
Non-steroidal anti-inflammatories (NSAIDs) 143, 192, 309, 451, 453, 459, 462
Non-synthetic hormones 529
Nuclear factor Kappa Beta (NF-κβ) expression 439–442
Numerical Rating Scale 462
Nutraceuticals 525
Nutrient deficiencies/insufficiencies
– dietary intake 124
– lab testing 125
– medical history/current condition/diagnosis 124
Nutrient-dense diet survey 866
Nutrient dependency 90
Nutrient therapy 130
Nutrient-dense foods 92
Nutrients
– and diet 451
– ATP production 600, 601
– energy and health 600
– oxygen transport 600
Nutrient-specific appetites 42
Nutritional assessment 453, 454
Nutritional counseling 1044
Nutritional diagnosis (ND) resources 1044
Nutritional ecology
– anthropology 41
– appetite interactions 47, 50–52
– appetites interaction 42, 43
– complementary feeding 45
– compositions 45
– dietary macronutrient balance 48, 49
– energy balance 49, 50
– energy intakes 49, 50
– GFN (*See* Geometric framework for nutrition)
– human health 41
– human nutrition 40
– macronutrient regulation 46
– mixture hierarchy 47, 48
– multiple appetites 42
– non-human animals 42
– NPE intake target
– nutritional geometry 47
– nutritional rails 45, 46
– petite interactions 51
– PLH 47
– prioritization pattern 46
– protein intake 50, 51
– protein leverage 46, 47
– protein prioritization rule 47
– protocols 45
– randomized control trials 46
– RMT 47, 48
Nutritional insufficiency 90
Nutritional interventions 716, 1044

Nutritional ketosis 343, 748
Nutritional neuropathies 456
Nutritional supplements 838
Nutritional therapy 167, 1044
Nutritional triage theory 1038
Nutritional wellness 36
Nutrition care process (NCP) 58, 214, 474, 627, 638, 744, 1021
– Integrative and Functional Medical Nutrition Therapy (IFMNT) Radial 772
– monitoring and evaluation 772
  – cost-effectiveness 772
  – critical thinking skills 774
  – developing and charting 776–778
  – education and behavior change instruction 778
  – Electronic Nutrition Care Process Terminology (eNCPT) 772
  – follow-up plan 773
  – follow-up visit 778
  – functional assessment parameters 772
  – hypertension treatment 779–781
  – in-depth investigation 774
  – integrative and functional nutrition (IFN) approach 772, 775, 776
  – nutrient-based problem 778
  – nutrition interventions 778
  – nutrition knowledge 773
  – in nutrition support example 779
  – outcomes-based clinical care 772, 778
  – patient assessment 773
  – PES statement 778
  – preplanned review and measurement 772
  – systematic comparison 772
  – weight loss and inflammation management 781, 782
– nutrition care outcomes and indicators
  – client history data 775
  – cost outcomes 775
  – health and disease status 775
  – lab and test values 775
  – medical chart 775
  – patient outcomes 775
  – selection of 775
Nutrition education
– barriers 10
– clinical problems 8
– complex biochemical mammalian environment 9
– dietary research and interventions 8
– diverse pathways 8
– economic burdens 8
– evidence-based approaches 10
– evidence-based medicine 9
– gut microbiome 8
– healthful food-based dietary recommendations 10
– healthful food-based patterns 10
– historical emphasis 9
– homeostatics 9
– poor diet quality 10
– poor nutrition 9
– radical reform 9
– risk factor 8
Nutrition-focused physical exam (NFPE)
– abdomen 672
– accessibility 691
– anthropometrics 680–691

- blood pressure 646
- body composition analysis 683
- chest
  - auscultation 671
  - cardiac exam 670
  - peak expiratory flow 671
  - pulmonary exam 670
  - pulmonary inspection 671
  - respiratory rate and rhythm 671
- components 639
- comprehensive nutritional assessment 638
- cost 692
- diagnostic accuracy 691
- ears 660
- examination tools and techniques 639
- eyes
  - conjunctiva 658
  - eyelids 658
  - iris 659
  - lens and cornea 659
  - lesions 658
  - optic (CN2) and oculomotor (CN3) examination 659
  - pupils 658
  - vision changes 660
- face
  - acne rosacea 657
  - affect 657
  - asymmetry 657
  - brown, patchy pigmentation of the skin 657
  - Chvostek's sign 657
  - dilated capillaries on cheeks and nose 657
  - facial nerve (CN7) examination 657
  - flushing 657
  - hair growth 657
  - loss of lateral 1/3 of eyebrow 657
  - myxedema 657
  - seborrheic dermatitis 657
  - tics/abnormal movements 657
  - trigeminal nerve examination (CN5) examination 657
  - vertical creases 657
  - yellow skin 657
- goal of 639
- hair
  - baldness 656
  - corkscrew hair 656
  - dandruff 656
  - dry/brittle 656
  - flag sign 656
  - hair loss 656
  - Menkes steely hair 656
  - premature graying 656
  - sparse/thin hair 656
- head-to-toe exam 639
- hospital-based internships 638
- integrative nutrition 638
- internet-based survey 638
- jaw
  - chewing and swallowing 669
  - hypoglossal (CN12) examination 668
  - movement 668
- medical nutrition therapy (MNT) 639
- mouth
  - bitter strips 666
  - breath odor 666
  - buccal mucosa 664
  - dental materials 665
  - fruity/ammoniacal odor 666
  - glossopharyngeal nerve (CN9) and vagus (CN10) examination 664
  - gums 664
  - intraoral inflammation 666
  - lips 663
  - lymph glands (neck) 667
  - palate 663
  - parotid gland 667
  - pharynx 664
  - taste 665
  - teeth 665
- musculoskeletal 674
- nails
  - acrodynia 653
  - anatomy 651
  - Beau's lines 652
  - blue nails 654
  - brown/gray nails 654
  - capillary refill test 655
  - clubbing 651
  - examination 651
  - half-and-half nails 653
  - hangnails 655
  - Hapalonychia 654
  - Koilonychia 652
  - longitudinal melanonychia 655
  - lunula 653
  - Mees' lines 654
  - Muehrcke's lines 654
  - onycholysis 652
  - onychomadesis 653
  - onychophagia 655
  - onychorrhexis 653
  - onychotillomania 655
  - pallor 654
  - paronychia 655
  - pitting 652
  - punctate leukonychia 655
  - ragged cuticles 655
  - red nails 654
  - ridging of nails 653
  - splinter hemorrhage 655
  - telangiectasia 655
  - Terry's nails 653
  - white nails 654
  - Yellow nail 654
- NCP model 638
- neck 669
- neurological 676
- nose
  - external nose 660
  - nasal canal/Intranasal 661
  - smell 662
- nutrition screening questionnaires 686
- nutritional assessment 639
- ongoing care and prognosis 691
- oxygen saturation 647
- patient assessments 638
- patient contact 691
- patient safety 692
- patient-provider relationship 639, 641–679
- pedagogic value 691
- proponents 638
- pulse 645
- pulse pressure 647
- respiratory rate 646
- scalp 656
- sinuses 662
- skin
  - color 648
  - edema 650
  - hydration/turgor 650
  - lesions 648
  - moisture 648
  - pruritus 651
  - rashes 650
  - temperature 648
  - texture 649
  - vascularity 651
  - wounds and ulcers 649
- subjective global assessment (SGA) 645
- system-based examination 639
- temperature 645
- thyroid 670
- tongue
  - buds/papillae 667
  - coat 667
  - color 667
  - Lesions 668
  - ridges/furrow 667
  - swelling 667
- US medical schools reports 638

Nutrition, inflammation, biochemical individuality, lifestyle, energy and metabolism, toxic load, and stress (NIBLETS) 776
Nutrition new patient intake form 1023
Nutrition physical exam (NPE) 60
Nutrition policy 24
Nutrition professional 58
Nutrition screening questionnaires 686
Nutrition transition (NT)
- behavior change stage 24
- definition 18
- degenerative disease and behaviour stage 19
- developmental programming and metabolic adaptation 22, 23
- dietary and physical activity patterns 21
- dietary convergence 21
- double burden of disease 21, 22
- famine and receding famine stage 18, 19
- global epidemiological data 21
- global phenomenon 19, 20
- Global Strategy on Diet 24
- Health action plan 24, 25
- healthy diet and physical activity lifestyle 24
- healthy eating and physical activity 24
- hunter-gathering stage 19
- inflammation 23, 24
- intra- and inter-country heterogeneity 21
- multifactorial behavior modification approach 24
- nutrition transition spectrum 21
- obesity rates 21
- pathophysiological consequences 23
- personalized lifestyle medicine 24
- physical activity 24
- populations' lifestyles 20–21
- principles 25
- specialized assessment techniques 24
- strong governance 24
- transition phenomenon 21
- undernutrition and overnutrition 22
- Westernized dietary pattern 21

O

Obesity 344, 992
Obstructive sleep apnea syndrome (OSAS) 716
Omega 3 alpha linolenic acid (ALA) 164
Omega-3 essential fatty acids 456–458
Omega-6 essential fatty acids (EFAs) 452
Omega-3 fatty acids 187, 440, 453, 461, 939
Omega 6 fatty acids 458, 939
Omega-3 index 145
Omega 6 linoleic Acid (LA) 164
Omega 9 monounsaturated fatty acids (MUFAs) 164
Omega-3 polyunsaturated fatty acids 879
Omega-6 (n-6) polyunsaturated fatty acids (PUFAs) 728, 729
Oncology 166
ONE Research Foundation (ONE) 615
One-carbon metabolism 271–274
Onycholysis 652
Onycholysis nail bed 687
Onychomadesis 653
Onychophagia 655
Onychorrhexis 653
Onychotillomania 655, 690
Opioids 462, 463
Optic (CN2) and oculomotor (CN3) examination 659
Oral cannabis 461
Oral facial pain (OFP) 453
Oral health connection 932
Oral/tube feeding supplements 876
Organ structure and function
– dry form 181
– eye 180
– reactive oxygen species
  – lutein 181
  – omega-3 fatty acids 181
  – risk factors 181
  – taurine deficiency 181
  – zeaxanthin 181
– wet form 181
Organ system interventions 716
Organic gastrointestinal disease 916
Orphan receptors 195
Osmotic laxative (lactulose) 381
Osteoarthritis (OA) 449
– definition 186
– evaluation/assessment 187
– prevention/treatment 187
Ostopenia
– definition 187
– evaluation/assessment 188
– pathophysiology 187
– treatment/prevention 188
Oteoporosis
– definition 187
– evaluation/assessment 188
– pathophysiology 187
– treatment/prevention 188
Overactive maternal immune system 544
Over-focused ADD 505
Oxalates 456, 500, 503
Oxidative stress 125, 935, 948
Oxytocin (OT) 537, 538
Ozone 943

P

Pain Project 456
Paleo Autoimmune Protocol (AIP) 455
Paleo diet 455, 745
Pallor 654
Palmitic acid 136
Palmitoleic acid 881
PALS advance nutrition research 866
Panax ginseng 525
Panchakarma 787, 792
Pancreatic elastase 1 (PE1) 63
Panic disorder 242
Panniculitis 956
Pantothenic acid 107, 525, 984
Paraoxonase 1 (PON1) 239
Parathyroid hormone test 111
Parkinson's disease (PD) 400
Paronychia 655, 689
Paternal nutrition status 34
Pathogen-associated molecular patterns (PAMPs) 291
Pathophysiological consequences 23
Patient healing process 74
Patient information 1056
Patient self-care record-keeping and instructions
– carbohydrate guide 1070–1071
– diet diary 1059–1060
– fat/oil guide 1071
– food sensitivity resources 1067–1069
– gluten-free dairy-free guide
  – balanced fats and oils 1062
  – breakfast ideas 1063–1065
  – clean protein 1062
  – cleansing beverages 1062
  – general guidelines 1061–1062
  – healthy carbohydrates 1062
  – lunch/dinner ideas 1065
  – preparation 1062–1064
  – symptoms 1061
– protein guide 1069–1070
– sample food diary entry 1061
Patient service representative (PSR) 1020
Patient story
– comprehensive patient story 634
– evidence-based healthcare 635
– food sensitivities 636
– IFM matrix 634–636
– integrative and functional medicine 634
– nutritional lab testing 636
– patient interview 634
– PEECE 635
– practitioner's interviewing skill 634
– relationship-centered care 635
– therapeutic approach 635
Patient therapy 74
Patient-centered care 14
Patient-centered diet and lifestyle approach 12
Pattern of malabsorption 111
Pattern-recognition skills 717
Peak expiratory flow rate (PEFR) technique 690
Pectin 385
Pediatric autoimmune acute-onset neuropsychiatric syndrome (PANS) 510
– factors contributing to exacerbations 511
– NIMH guideline for 511
– treatment 511, 512

Pediatric autoimmune neuropsychiatric disorders associated with Streptococcal infections (PANDAS)
– birth and previous medical history 512
– CBS genes 512
– COMT genes 512
– diagnosis 510
– laboratory testing 511
– MAO-A 512
– MTHFD1 enzyme 512
– NIMH guidelines 510
– SNP 512
– symptoms by NIMH 511
– treatment 511, 512
Pellagra 479
Pemphigus 990
Perimenopause 518, 528
Periodontal disease 535
Periportal hepatocytes 225
Peristalsis 368
Peroxisome proliferator-activated receptor (PPAR) 716
Peroxisome proliferator-activated receptor gamma (PPARγ) 116, 117
Personalized medicine 12
Pervasive Developmental Disorder Not Otherwise Specified 474
Petasites hybridus 461
Petrochemicals 1038
Peyer's patches 231, 393
Phagocytic macrophages 929
Pharmacogenomics
– acetylation 263
– CETP 264
– cytochrome enzymes 263
– glucuronidation 263
– glutathione 263
– mercaptopurine 264
– warfarin genetics 264
Phase angle (PA) 129, 158, 328
Phase I biotransformation (detoxification) 228
Phase II biotransformation enzyme pathways 228
Phellodendron amurense 506
Phenol sulfotransferase (PST) 503
Phenylalanine hydroxylase (PAH) 275, 277
Phenylethanolamine N-methyltransferase (PNMT) 241, 273
Phenylketonuria 988
Phenytoin 487
Phosphatase and tensin homolog (PTEN) 262
Phosphatidylcholine (PC) 175, 387
Phosphatidylethanolamine (PE) 175, 176
Phosphatidylserine (PS) 117, 175, 176
Phosphodiesterase 5 (PDE5) inhibitor 278
Phospholipids (PL) 117, 164, 174, 882
Phosphotidylethanolamine 180
Photoallergic contact dermatitis 991, 992
Phthalates (PHT) 542, 543
Physical activity 128, 1007
Physical exercise 1007
Physical inactivity 596
Phytonutrient status 165
Phytonutrients 165, 884, 885, 937
Phytonutrition 410
Pitta 785
Plant- and animal-derived essential fatty acids (EFA) 970

Plasma triene/tetraene ratio 144
Platelet-derived growth factor (PDGF) 438, 722
Platelets 443
Poikilocytosis 443
Political power 8
Polychlorinated aromatic hydrocarbons (PAHs) 542
Polychlorinated biphenyls (PCBs) 542
Polycyclic aromatic hydrocarbons (PAHs) 129
Polycystic ovary syndrome (PCOS) 240, 345, 537
Polymorphisms, autism 501
Polyunsaturated fatty acids (PUFA) 177, 442, 456, 475
- biosynthesis 139–140
- cholesterol 141
- deficiency 143, 144
- eicosanoids 141–143
- food sources 146
- functions 140–143
- gene expression 141
- immune response 141
- infection 141
- requirements 144
- skin 141
- *trans*-fatty acids 146, 147
- W_3 fatty acids 143–145
Portfolio diet 833
Positive prognosis, Empathy, Empowerment, Connection, and Education (PEECE) 635
Postmortem dissection 7
Postprandial hyperglycemia 479
Postprandial insulin 521, 523
Post-traumatic stress disorder (PTSD) 448, 811
Postural asymmetries 1005
Practitioner-patient relationship 77, 78
Prameha 794
Prediabetes 519
Pregnane X receptor (PXR) 193
Presenilin 1(PSEN1) 240
Presenilin 2 (PSEN2) 240
Prevotella copri 442
Prilosec 383
Principle of Le Chatelier 94
Proanthocyanidins 440–442
Probiotics 383, 393, 400, 889, 992, 993
Procalcitonin (PCT) 310
Progesterone 528, 529, 535
Progestins 529
Progressive bulbar palsy, *see* Bulbar ALS
Pro-inflammatory adipokines 126
Proinflammatory cytokines 452
Proinsulin 520
Prolactin 536–538
Proprotein convertase subtilisin/kexin type 9 (PCSK9) 264
Prostacyclins (PGI) 142
Prostaglandin 1 series (PG1) metabolites 116
Prostaglandins (PG) 142
Protein 176, 875, 940
Protein contact dermatitis 991
Protein deficiency 974
Protein kinase C (PKC) 440
Protein leverage hypothesis (PLH) 47
Protein pump inhibitors 486
Proton pump inhibitors (PPIs) 462
Pseudoaldosteronism 383
Psychiatric disorders 809
- ALZ 240, 241
- anxiety 242
- bipolar disorder 475, 476
- brain-derived neurotropic factor 477
- caffeine intoxication 477
- Celiac disease 477
- COMT 241, 242
- depression 243, 244, 475
- drugs
  - antidepressant medications 486
  - nutrition 487
  - phenytoin 487
  - protein pump inhibitors 486
  - riboflavin 487
  - weight gain 486
- DSM-V 240
- food insecurity 477
- GAD 241
- GAD enzyme 243
- glutamate 242, 243
- hyponatremia / water intoxication 477
- metabolic syndrome 477
- methylation 241
- panic disorder 242
- PNMT 241
- s-adenosyl-homocysteine 241
- schizophrenia 476, 477
- serious mental illness and genetics 474, 475
- spectrum of 474
- spectums of 474
Psychobiotics 397, 398
Psychological pain 449
Psychological stress 922
Psychometric assessment tools 79
Psychometric behavioral assessments 79, 80
Psychoneuroimmunology 129, 614
Psychosis spectrum syndrome, *see* Schizophrenia
Psychosocial mediators 449, 450
Pteroharps 443
Pulmonary fibrosis
- bone marrow or immune response abnormalities 960
- causes 957
- conventional treatment 959
- curcumin 960
- genetic focus 958
- integrative and functional medicine nutrition therapy 960
- lengthening telomeres or reducing shortening 960–962
- lung transplants 959
- medications 959
- molecular endotyping 958
- mortality 957
- N-acetylcysteine 960
- quercetin 960
- symptoms 957
Pulse oximetry 932
Punctate leukonychia 655, 689
Punnett square 270
Purusharthas 790
Purvakarma 792
Pushup test 606
Pyridoxal kinase (PDXK) 936
Pyridoxal phosphate (PLP) 107
Pyridoxal-5-phosphate (P5P) 242
Pyridoxine 984
Pyridoxine deficiency 479, 482
Pyridoxine-dependent epilepsy 508
Pyridoxine-dependent seizures 508
Pyroluria 501
Pyrrole disorder 421, 501
Pyrroloquinoline quinone (PQQ) 180

Q

Quadruple Aim 708
Quercetin 440, 442, 539, 937
Quetelet index 325, 682

R

Raft-forming agent 383
Ragged cuticles 655
Raktamokshana 792
RAND Corporation study 4
Rapamycin 458
Rasa 786
Rasayana 792
RBC fatty acid profiles 118
Reactive medical approach 12
Reactive nitrogen species (RNS) 536
Reactive oxygen species (ROS) 177, 536
- alpha lipoic acid 181
- lutein 181
- omega-3 fatty acids 181
- risk factors 181
- taurine deficiency 181
- zeaxanthin 181
Recombination activating gene (RAG) 289
Recommended daily allowance (RDA) 111, 306, 735
Recommended dietary allowances (RDAs) 338, 456
Red blood cell (RBC) fatty acid analysis 158
Red nails 654
Redesign professional health education 11
Reduced folate carrier (RFC) 502
Referrals 739
Registered dietitian 58
Renal disorders 7
Reproduction
- acquired genetic issues 549
- energy production 537
  - aerobic metabolism 537
  - mitochondria 536, 537
  - oxygen 537
  - PCOS 537
- energy utilization 537
- exercise 546, 547
- folate supplementation 548
- folic acid, risk factors 548, 549
- genetics 547
- hormonal factors
  - anti-inflammatory medication 538
  - cortisol 539
  - fallopian tube 538
  - FSH 538
  - LH 538
  - melatonin 538
  - natural aromatase inhibitors 539
  - oxytocin 538
  - sex hormones 537
- information layer 547
- microenvironment
  - ART techniques 534

- CB1-R ligands 535
- CYP1A1 gene 534
- emotional factors 535
- endometriosis 534, 535
- etiological and pathophysiological factors 535
- fibroids 535
- GSTM1 gene 534
- infertility etiologies 535
- MTHFR 547, 548
- neural tube defects 548
- nutrient factors
  - carbohydrates 539
  - folic acid 540
  - iron and copper 539
  - magnesium 540
  - trans fatty acid 539
  - vitamin C (AA) 540
  - vitamin D 540
  - Zinc 539
- nutrients 534
- reproductive immunology
  - abnormal vaginal microbiota 546
  - aversions 546
  - chlamydia trachomatis (CT) infection 545
  - COX-2 inhibitors 545
  - immune cells 544
  - microbiota 545
  - mycoplasma species 546
  - prophylactic antibiotics 546
  - protozoa 546
  - pTregs 544
  - Treg activity 545
  - vaginal microbiota 546
- reproductive toxicology
  - administration route 540
  - aluminum 541, 542
  - arsenic 541
  - cadmium 541
  - carbamates 542
  - dioxins 542
  - EPA 540
  - in food 541
  - glyphosate 544
  - lead 541
  - magnitude of toxic substance 540
  - mercury 541
  - NIER 543
  - organophosphates 542
  - phthalates 542, 543
  - polychlorinated aromatic hydrocarbons 542
  - polychlorinated biphenyls 542
- structural layer 536
- Requesting practitioner information 1056
- Respiratory alkalosis 931
- Respiratory disease
  - air pollutants 943
  - alpha-1 antitrypsin (A1AT) deficiency 933
    - A1AT biochemical mechanisms 957
    - liver 956
    - liver cirrhosis 957
    - lungs 956
    - skin 956
  - alveolar macrophages 930
  - anatomy 929
  - anti-nutrients 945
  - asthma
    - adolescence 953
    - allergic asthma 945, 946
    - allergic bronchopulmonary mycosis 946, 949
    - anthropometrics 955
    - aspirin-exacerbated respiratory disease 949
    - biochemical 954
    - birth 953
    - body composition 954
    - childhood 953
    - clinical 954
    - college 953
    - complaints at first visit 952
    - cross country skiing-induced asthma 946
    - definition 944
    - dietary patterns 947
    - digestion, assimilation and elimination 954
    - energy 954
    - fats and oils questionnaire 953
    - follow-up 955
    - food allergy 947, 948
    - genotypic characteristics 947
    - genotypic risks 954
    - histamine intolerance 946
    - IFMNT assessment 950–952
    - IFMNT case study 950–955
    - inflammation 954
    - lifestyle 954
    - lifestyle influences 949
    - nonallergic asthma 945, 946
    - nutrient cofactors 947
    - nutrition 953
    - nutrition Dx 955
    - nutrition physical exam 954
    - outcome 955
    - oxidative stress 948
    - past medical history 952–953
    - past surgical history 953
    - pathophysiology 944, 945
    - pharmaceutical use 953
    - plan 955
    - primary conventional therapy 949
    - severe late-onset hypereosinophilic asthma 946
    - signs/symptoms/medical symptom questionnaire (MSQ) 954
    - sleep and stress 955
    - social 953
    - structural 954
    - suspected triggers/risk factors 944, 945
    - symptoms 944, 952
    - toxic exposure 948
    - toxin Load 954–955
    - traffic density and asthma exacerbations 949
    - utilization, cellular & molecular 954
    - young adulthood 953
  - autophagy mechanism 933
  - categories 933–935
  - causes of death 929
  - chemicals 944
  - chronic infection and respiratory health
    - fungal, viral, and bacterial infections 956
    - upper respiratory tract infection 955–956
  - ciliary function 931
  - collagen 931
  - connective tissue 931
  - dendritic cells 931
  - dietary supplements 929
  - dysregulated collagen homeostasis 931
  - endogenous essential respiratory metabolites
    - CoQ10 (Coenzyme Q10) 942
    - glutathione 940–942
    - micronutrient analysis and functional testing 938
  - environmental pollutants 942, 943
  - excessive goblet cells/hypertrophy 931
  - fungal microbiome/mycobiome 932
  - gas exchange 931
  - human life expectancy 929
  - lung physiology 942
  - macronutrients
    - carbohydrates 940
    - fats and fatty acids 939
    - fats and fatty Acids 938–939
    - protein 940
  - mast cells 931
  - nutritional approach
    - acid/alkaline balance 936
    - alpha-lipoic acid (ALA) 938
    - B vitamins 935, 936
    - carotenoids 936
    - iron 933, 935
    - minerals 938
    - N-acetyl cysteine 938
    - phytonutrients 937
    - vitamin A 936
    - vitamin C 937
    - vitamin D 936, 937
    - vitamin E 937
  - oral health connection 932
  - phagocytic macrophages 929
  - pulmonary fibrosis
    - bone marrow or immune response abnormalities 960
    - causes 957
    - conventional treatment 959
    - curcumin 960
    - genetic focus 958
    - integrative and functional medicine nutrition therapy 960
    - lengthening telomeres or reducing shortening 960–962
    - lung transplants 959
    - medications 959
    - molecular endotyping 958
    - mortality 957
    - N-acetylcysteine 960
    - quercetin 960
    - symptoms 957
  - pulse oximeter tool 931
  - risk factors 929
  - secretory cells 931
  - stress
    - acute stress 944
    - chronic biological stress 944
    - respiratory health and body's ability 944
  - surface enzymes and factors 931
  - toxic metal
    - arsenic 943
    - cadmium 943
  - type I cells 930
  - type II and clara cells 930
- Restoration diet 1046
- Restrictive respiratory diseases 68
- Resveratrol 440
- Retinoic acid (RA) 299

Retinoic acid receptor alpha (RARA) 483
Retinol 440
Retroconversion 140
Rett syndrome 508
Reynolds/aldrich lines 654, 689
Rheum rhaponticum (ERr 731) 529
Rheumatoid arthritis (RA) 145, 441, 448, 449
Rhodiola rosea 507, 525
Riboflavin deficiency 479
Riboflavin (vitamin B2) 107, 459, 984
Rich blood flow 206
Ridging of nails 688
Right-angled mixture triangle (RMT) 47, 48
Ring of Fire ADD 506
Ritucharya 790–792
Rosenberg European Academy of Ayurveda (REAA) 785
Rouleaux 443

S

S-adenosyl-L-methionine (SAM) 241, 271, 272, 316, 476, 547
s-adenosyl-methyltransferase (SAM) 241
Sadvritta 790, 792
Safe Drinking Water Act (SDWA) 731
Salicin 461
Salivary cortisol test 525
Salutogenesis 37
Salutogenic paradigm 37
Sama Prakruti 785
Sample food diary entry 1061
Santarpanjanya vikaras 792
Sarcopenia 186, 812, 813
Sargassum fulvellum 536
Saturated fatty acids 164
– conjugated linoleic acid 147
– long-chain C13-C21 and very-long-chain fatty acids C22 138–139
– medium chain triglycerides 137, 138
– monounsaturated fatty acids 139
– polyunsaturated fatty acids
  – biosynthesis 139–140
  – cholesterol 141
  – deficiency 143, 144
  – eicosanoids 141–143
  – food sources 146
  – functions 140–143
  – gene expression 141
  – immune response 141
  – infection 141
  – requirements 144
  – skin 141
  – *trans*-fatty acids 146, 147
  – W$_3$ fatty acids 143–145
– short-chain saturated fatty acids (SCFA) 137, 138
Saturated fatty acids (SFAs) 164, 880
Schamroth's sign 686
Schisandra chinensis 525
Schistocytosis 443
Schizophrenia 396, 476–477
Scientific approaches 6
Scientific medicine 6
Scottish enlightenment 7
Scutellaria baicalensis 440
Second brain 394, 395

Secretory immunoglobulin A (SIgA) 286, 287, 370
Seizures disorders
– biotinidase deficiency 508
– EEG 508
– ketogenic diet
  – breakdown products 508
  – calcium and vitamin D 510
  – carnitine 510
  – check labs 509
  – constipation 510
  – contraindications 509
  – DHA and omega-3 oils 510
  – dietitian 508
  – grape seed 510
  – guidelines 510
  – melatonin 510
  – multivitamins 509
  – nutrient depletions 510
  – oral citrates 510
  – selenium 510
– low sodium 508
– MAD 508, 509
– MCT diet 509
– pyridoxine-dependent seizures 508
– vitamin B6 508
Selective serotonin reuptake inhibitors (SSRIs) 243, 475
Selenium 485, 528, 947
Selenium deficiency 987
Selenium-dependent glutathione peroxidase (Se-GPX) 485
Self-Care Assessment Questionnaire 705
Self-management methods 10
Sepsis 417
Serious mental illness and genetics 474
Serotonin 397
Serotonin deficiency 243, 873
Serotonin seeping 400
Serotonin transporter gene (*SLC6A4*) 463
Severe cutaneous adverse drug reactions (SCARs) 264
Severe late-onset hypereosinophilic asthma 946
Sex hormone binding globulin (SHBG) 529
Sex steroid hormones
– assessment 528
– DHEA 529
– estrogen-DNA adducts 529
– Estrovera 529
– hormone replacement therapy 528
– menstruating women 528
– SHBG 529
– synthetic hormones 529
– VMS 528
Shared medical appointments (SMAs) 708
Short-chain fatty acids (SCFA) 117, 118, 136, 164, 386, 424
Short-chain saturated fatty acids (SCFA) 137, 138
Short physical performance battery (SPPB) 815
Signal transducer and activator of transcription 3 (STAT3) 439
Silver Sneakers 701
Silymarin 440
Single nucleotide polymorphism (SNP) 228
Single nucleotide polymorphisms (SNP) 63, 199, 200, 228, 270, 279–282, 463, 497, 502, 504, 505, 508, 512
Situational communication 78
Skin

– autoimmunity-triggered dietary antigens 993, 994
– biological structure 181
– dermis 970, 971
– epidermis 970
– excess mineral/element consumption, absorption, or decreased excretion 988
– folds 128
– food bioactives 994
– hypervitaminosis skin findings 989
– immune reactions 971
– inborn and acquired conditions 988
– inflammation 182
– inflammatory bowel disease 993–996
– *Lactobacillus johnsonii* effects 992
– macro- and micronutrient deficiencies
  – essential fatty acid deficiency 974
  – fat-soluble vitamins 983
  – minerals 986–987
  – protein deficiency 974
  – water soluble vitamins 983–986
– medical nutrition therapies 973–982
– Mediterranean diet 975–982, 988, 989
– microflora 992
– modified elimination diet
  – atopic dermatitis 990
  – contact dermatitis 991
  – contact urticaria 991
  – dermatitis herpetiformis 990
  – food-associated triggers 990
  – hypersensitivity reactions 990
  – IgE serum testing 990
  – immune system 990
  – pemphigus 990
  – photoallergic contact dermatitis 991, 992
  – protein contact dermatitis 991
  – removal of dietary components 989
– nutritional deficiencies
  – signs and symptoms 972–973
  – signs ansd symptoms 971
  – wheat and milk proteome association 971, 974
– obesity 992
– pathophysiology 182
– phytonutrient phenol foods 182
– probiotics 992, 993
– skin health status 182
– treatment 182
Skinfold measurement 327
Skulpt® device 330
Sleep 130, 131, 165, 523
Sleep apnea 581
Sleep complaints 451
Sleep hygiene 589
Small intestinal bacterial overgrowth (SIBO) 381, 384, 452, 889–890
Small intestinal bowel overgrowth (SIBO) 921
SMART goal 776
Smartphone apps 330
Social media 82
Social networks 1010
Sodium-to-potassium (Na–K) ratio 112
Soft skills 74
Solanaceae family 456
Soluble dietary fiber 102
Soluble fiber 102
Soluble transferrin receptor (sTFR) 933
Soluble viscous fibers 102
Soy 454, 876

Specialized assessment techniques 24
Specialized diets 1046–1050
Specialized pro-resolving lipid mediators (SPMs) 23, 879
Specific carbohydrate diet (SCD) 386, 503, 717, 1046
Specific kinase response modulators (SKRMs) 522
Splinter hemorrhage 655, 689
Spontaneous generation theory 7
Spoon nails 652
Sprint intensity training (SIT) 605
Standard American diet (SAD) 568, 717
Standards of Process (SOP) 627, 628
Standards of Professional Performance (SOPP) 627, 628
Starvation 750
STAT3 signaling 444
Steady behavior (S) 75
Steely hair syndrome 987
Steroids 512
Sterols 164
Sthaulya 792
Stomach cancer 423
Strain gauge scales 325
Stratum corneum 970
Strengthening exercises 1011
Stress 129, 130, 165, 464
– and autonomic dysfunction 523
– acute stress 944
– chronic biological stress 944
– respiratory health and body's ability 944
Strontium 187
Strotas 792
STW 5 (Iberogast) 383
Subcellular/gene expression interventions 717
Subcellular/mitochondrial interventions 716, 717
Subclinical dietary magnesium deficiency 111
Subjective global assessment (SGA) 645
Subjective Units of Distress Scale (SUDS) 616
Submaximal isometric exercises 1011
Sugar
– accumulation of 873
– brain-derived neurotrophic factor 874
– dopamine 873
– hormonal imbalances (glucose and insulin) 874
– inflammation and autoimmunity 873
– neuroprotective and antioxidant properties 873
– serotonin deficiency 873
Sulfate oxidase (SUOX) 275
Sulfotransferase 2A1 (SULT2A1) 240, 263
Sun Salutation/Surya Namaskar (SN) 795
Superoxide dismutase (SOD) 261, 502
Suprachiasmatic nucleus (SCN) 720, 736
Supra-physiological oxidative stress 194
Sustainable health system 13
Sympathetic nervous system (SNS) 1008, 1010
Syndrome X 519
Synthetic fertilizers 19
Synthetic hormones 529
Systemic lupus erythematosus (SLE) 809
Systems biology 153
Systems biology resources 1016

T

Targeted supplementation 410
Taurine deficiency 181

Tea polyphenols 539
Teaching behavioral competencies 74
Telangiectasia 655
Temporal lobe ADD 506
Temporomandibular syndrome (TMJD) 449
Terminal ileum 231
Terry's nails 653, 688
Testosterone 528, 529
Tetrahydrobiopterin (BH4) 275
Tetrahydrofolate (THF) 271
The Goddess of Neuroscience 614
The Healing Encounter 635
The Mother of Psychoneuroimmunology 614
Therapeutic diets
– conventional nutritional approaches 744
– definition 744
– IFMNT (See Integrative and functional approaches to therapeutic diets (IFMNT))
– traditional Chinese medicine (TCM) 744
– traditional/historical perspective 744
Therapeutic ET for rehabilitation (TETR) 606, 607
Therapeutic ketogenic diet
– absolute contraindications 348, 349
– acetoacetate 343
– acetone 343
– amino acids 342
– applications for 344
– Atkins diet 354
– baseline laboratory evaluation 348
– βHB 343
– cancer 345, 346
– carbohydrate restriction 341
– D-βHB 343
– diabetes 344, 345
– diet macros 355
– diet ratio 354, 355
– diet supplementation
    – antioxidants 353
    – digestive aids 353
    – gut health 353
    – herbs and botanicals 353
    – minerals 353
    – vitamins 353
– DRIs 338
– energy equivalent 338
– epigenetic influences 339
– eucaloric ketogenic diet 344
– fatty acids 342
– feast/famine cycling 354
– genetic variants 339
– gluconeogenic activity 341
– glucose 340
– history 343, 344
– hormones insulin and glucagon 338
– inflammatory response 339
– insulin 340
– insulin resistance 339, 344
– macronutrient calculations 349
    – alcohols 352
    – calorie restriction 350
    – carbohydrate 350
    – dairy 352
    – fats 350, 351
    – fats and oils 353
    – food trackers and meal planners 351, 352
    – fruits 352
    – grains 352
    – nuts and seeds 352

– protein target 349
– proteins 352
– sugars 352
– vegetables 352
– mitochondria 342
– mitochondrial health 340
– neurodegenerative diseases 347
– obesity 344
– pathologically high levels 343
– PCOS 345
– physiologically normal levels 343
– relative contraindications 349
– risk/benefit analysis 348
– short-term vs. long-term maintenance 355, 356
– transition
    – acidosis 357, 358
    – breath analyzers 360
    – constipation 358
    – dizzy, lightheaded/shaky 358
    – exercise tolerance/physical performance 358
    – fasting 356
    – heart rate/rhythm changes, palpitations 358
    – hunger and cravings 357
    – hypoglycemia 357
    – Keto Flu 357
    – Keto rash 358
    – make or break period 359
    – rigorous ketogenic diet 356
    – risk of kidney stones and gout 358, 359
    – slow transition 356, 357
    – testing blood glucose 359
    – testing blood ketones 360
    – testing glucose and ketones 359
    – urine testing 360
– triglyceride ester 341–342
– troubleshooting
    – boredom/dissatisfaction with food choices 362
    – carbohydrate content of medications 361
    – dropping out of ketosis 361
    – food craving 361
    – glucose spikes 361
    – high blood glucose levels 362
    – hunger and craving 360
    – nausea/vomiting 361
    – ongoing flu-like symptoms 360
    – steroid medications 361
    – unsustainable/rapid weight loss 361
– weight loss 339
Thiamin deficiency 91, 479
Thiamine diphosphate 479
Thiamine (vitamin B1) 107, 983
Thrombomodulin (TM) 310
Thromboxanes (TX) 142
Thyroid function 892
– TRH 526
Thyroid hormone (TH) 483
– assessment 527
– consequences of low thyroid hormone 526
– inhibit conversion of T4 to T3 526, 527
– synthesis 103
– treatment 527, 528
– TRH disruptors 526
– TSH 526
Thyroid-releasing hormone (TRH) 526

Thyroid-stimulating hormone (TSH) 526
Time-restricted feeding (TRF) 119
Tissue healing
– chronic pain 1004
– functional limitations 1004
– inflammation 1004
– neuropathic pain 1004
– remodeling 1004
– repair 1004
Titanium dioxide 386
Tocopherols (vitamin E) 110, 719
Tocotrienols 110, 719
Top selling herbal supplements 756
Total body water 329
Toxic heavy metals 1038
Toxic load
– endogenous toxins 129
– environmental/exogenous toxins 129
Toxic metal
– arsenic 943
– cadmium 943
Toxoplasma gondii 396, 546
Trachyonychia 654
Traditional Chinese medicine (TCM) 415
Traditional fermented whole foods 402
Traditional healthcare model 5
Transactive response DNA binding Protein 43 (TDP-43) protein inclusions 874
Transcobalamin I (TCN1) 259, 260
Transcobalamin II gene 502
Transcription Factor 7 Like 2 (TCF7L2) locus 239
Transcriptional analysis 1010
Trans-fatty acids 146, 147
Transforming growth factor beta (TGFβ) 439
Transglutaminase-6 antibodies (TG6) 874
Transition phenomenon 21
Transsulfuration pathway genes 502
Treg cells 441
Triage hypothesis 92
Tricarboxylic acid cycle (TCA) 428
Trigeminal nerve examination (CN5) examination 657
Triglycerides 137
Triiodothyronine (T3) 95
Triphala 103
Triple Aim 708
Tryptophan 299, 300
Tryptophan hydroxylase (TPH2) 243, 483
Tsunami effect 825
Tuberculosis (TB) infection 308
Tumor necrosis factor alpha (TNFα) 262, 452, 1007
Tumor-associated macrophages (TAMs) 439, 440
Turmeric (*Curcuma longa*) 460
Type 2 diabetes (DM2) 239
Tyrosine 275

U

Ulcerative colitis (UC) 385
United States Department of Agriculture (USDA) 569
Unmetabolized folic acid (UMFA) 549
Unsaturated fatty acid palmitoleic acid 457
Unsaturated fatty acids 137, 177
Urinary excretion 206
Urinary organic acid analysis (OAA) 891
Urine for organic acid testing 500
Urine peptide testing 500

V

Vaccination
– antibiotics and antimicrobials 317
– FMT 317
Vagus nerve 393, 397, 1008
Vamana 792
Vanilmandelate (VMA) 241
Vascular adhesion protein 1 (VAP1) 442
Vascular endothelial growth factor (VEGF) 438–441, 444
Vasomotor symptoms (VMS) 528
Vata 785
Ventilator-associated pneumonia (VAP) 932
Verbal Rating Scale (VRS) 462
Vibration therapy 1010, 1011
Vipaka 786
Virechana 792
Virta Health system 12
Virya 786
Visceral hypersensitivity 919
Vishwa Bhaishajya 789
Visual Analogue Scale (VAS) 462
Vitamin A 109, 314, 483, 936
Vitamin A deficiency 983
Vitamin B1 (thiamin) 479
Vitamin B12 108, 109
Vitamin B12 deficiency 259
Vitamin B2 (riboflavin) 479
Vitamin B6 (pyridoxine) 107, 108, 479, 482, 486
Vitamin B6 (pyridoxine) deficiency 508
Vitamin B6 deficiency 299
Vitamin BIII 756
Vitamin C 187, 937
Vitamin C deficiency 261, 986
Vitamin D 109, 440, 441, 483, 936, 937, 983
Vitamin D 25-hydroxy 314
Vitamin D receptors (VDR) 260, 261
Vitamin D-resistant rickets 111
Vitamin E 937
Vitamin K 109, 483
Vitamin K antagonists (VKA) 109
Vitamin K1 deficiency 109
Vitamin K2 458
Vitex agnus-castus (VAC) 538
Volatile organic compounds (VOCs) 207
Voltage-gated Ca2+ channels (VGCCs) 496
Vyayama 795

W

Waist circumference (WC) 326, 682
Waist-to-Height Ratio (WHR) 326
Waist-to-height ratio (WHtR) 683
Waist-to-hip ratio 683
Waist-to-hip ratio (WHR) 326
Wall-mounted measuring devices 325
Warburg effect 345
Warburg Effect 428, 749
Warfarin 483
Water intoxication 477
Water-soluble vitamins 91
– B12 deficiency 986
– biotin (vitamin H) 984
– folate deficiency 985
– niacin 984
– pantothenic acid 984
– pyridoxine 984
– riboflavin 984
– thiamin 983
– vitamin C deficiency 986
Weight gain 486
Wellness
– biomarkers 32, 33
– CDC research statistics 32
– centenarians/supercentenarians 100+ years 36
– community 36, 37
– definition 32
– health and vitality 32
– infant (birth -6 to 24 months)
  – childhood (3–12 years) 34
  – teenage (13–19 years) 34–35
  – toddler (12–36 months) 34
– integrative and functional medicine 33
– markers 36
– middle-age adult (35–54 years) 35
– nutrients 36
– nutritional status assessment 33
– obesity 33
– old-old seniors 75–99 year 35–36
– seniors 54–74 years 35
– in utero 34
– young adult 20–34 years 35–36
Wernicke-Korsakoff encephalopathy (WKE) 479
Wernicke-Korsakoff's syndrome 128
Western diet 844–845, 873, 880
Western lifestyle behaviors 126
Westernized dietary pattern 21
Western-style diet 717
W₃ fatty acid deficiency 144
W₃ fatty acids 143–145
While opiates stem 462
White blood cell morphology 443
White fish 387
White nails 654
Whole-body interventions 716
Whole-person medical evaluation 1016
Willow bark 461
Wilson's disease 113, 483
Withania somnifera 525
W numbering system 136
Women's Health Initiative study 529
Women's Healthy Eating and Lifestyle (WHEL) 749
Women's Interventional Nutrition Study (WINS) 749
World Cancer Research Fund International's ongoing NOURISHING framework 24
World Health Organization (WHO) 486, 723

X

Xanthurenate 242
Xenobiotic exposures 227
Xenobiotic nuclear receptors 194
Xenobiotics 721
– adverse effects 192
– biotransformation 226–228
  – dietary composition and bioactive constituents 194
  – factors 194

- mechanisms 193
- multiple chemical sensitivity 194
- phases 193
- CDC report and scan 227
- chronic toxicity 192
- classes 228
- clinical considerations
  - diet and lifestyle 200
  - endocrine disrupting chemicals 198
  - genetic susceptibility 200
  - potential exposure 198
  - risks determination 198
- clinical strategies
  - body's detoxification capacity 201
  - reducing or avoiding exposure 200–201
- definition 192, 226
- environmental chemical exposures with fetal development 227
- exposure 192, 193
- harmful xenobiotic exposures 227
- health hazard 192
- insecticide resistance 192
- organs and body systems 192
- pathophysiology
  - antibiotics and oral pharmaceuticals 198
  - chronic diseases 195
  - excitotoxins and neurotoxins 198
  - implications 198
  - inflammation 195
  - interference with critical biotransformation 194
  - long-term prescriptions and polypharmacy 198
  - organs, tissues and organelles 198
  - orphan receptors 195
  - proven/inferred associations 195
  - soft-touch decisions 195
  - steroid receptors 195
- supra-physiological oxidative stress 194
- xenobiotic nuclear receptors 194
Xenobiotic-sensing receptors 193

Y

Yellow nail 654
Yield-enhancing methods 92
Yoga recommendations 795

Z

Zantac 383
Zearalenone 541
Zeaxanthin 181
Zellweger syndrome 144
Zinc 164, 496, 528
Zinc deficiency 440, 485, 486
Zonulin 385